AINSWORTH & BISBY'S

DICTIONARY OF THE FUNGI

CAB INTERNATIONAL is an intergovernmental organization providing services worldwide to agriculture, forestry, human health and the management of natural resources.

The information services maintain a computerized database containing over 2.7 million abstracts on agricultural and related research with 150,000 records added each year. This information is disseminated in 47 abstract journals, and also on CD-ROM and online. Other services include supporting development and training projects, and publishing a wide range of academic titles.

The four scientific institutes are centres of excellence for research and identification of organisms of agricultural and economic importance: they provide annual identifications of over 30,000 insect and microorganism specimens to scientists worldwide, and conduct international biological control projects.

International Mycological Institute
An Institute of CAB INTERNATIONAL

The Institute, founded in 1920, and employing over 70 staff:
- Is the largest mycological centre in the world, and is housed in a complex specially designed to support it requirements;
- Carries out research on a wide range of systematic and applied problems involving fungi (including lichens, mushrooms, and yeasts), plant and soil bacteria, and also on the preservation of fungi, the biochemical, physiological and molecular characterization of strains, biodiversity inventorying and monitoring, bioindication and on crop protection, environmental and industrial mycology, food spoilage, public health, biodeterioration and biodegradation;
- Provides an authoritative identification service, especially for microfungi of economic and environmental importance (other than certain human and animal pathogens), and for plant pathogenic bacteria and spoilage yeasts;
- Undertakes extensive computerization and indexing of bibliographic work, including the preparation of 7 serial publications;
- Provides advice and undertakes a wide range of project and culture work on crop protection, environmental, food spoilage and industrial topics;
- Offers training in pure and applied aspects of mycology and bacteriology, both at the Institute and overseas.

The Institute's dried reference collection numbers in excess of 370,000 specimens representing about 32,000 different species, and the genetic resource collection holds more than 19,000 living isolates represent about 4,500 species maintained by a variety of the most modern methods. The library receives about 600 current journals, and has extensive book and reprint holdings reflecting the Institute's interests.

International Mycological Institute   *Tel:* +44 (0) 1784 470111
Bakeham Lane   *Fax:* +44 (0) 1784 470909
Egham   *E-mail:* imi@cabi.org
Surrey TW20 9TY   *Telex:* 9312102252
UK

For full details of the services provided, please contact the Institute Director.

AINSWORTH & BISBY'S

DICTIONARY OF
THE FUNGI

by

D.L. Hawksworth, P.M. Kirk, B.C. Sutton
and D.N. Pegler

with the assistance of

G.C. Ainsworth, G.L. Benny, P.F. Cannon, V. Demoulin, M.J. Dick,
W.C. Elsik, R.W. Lichtwardt, J.E. Longcore, J.E.M. Mordue,
F. Pando, I.I. Tavares, M.A.J. Williams

and others

Eighth Edition

prepared by the INTERNATIONAL MYCOLOGICAL INSTITUTE

CAB INTERNATIONAL

First Edition 1943
Second Edition 1945
Third Edition 1950
Fourth Edition 1954
Fifth Edition 1961
Sixth Edition 1971
Seventh Edition 1883
Eighth Edition 1995
Reprinted 1996

CAB INTERNATIONAL
Wallingford
Oxon OX10 8DE
UK

Tel: +44 (0)1491 832111
Fax: +44 (0)1491 833508
E-mail: cabi@cabi.org

CAB INTERNATIONAL
198 Madison Avenue
New York, NY 10016-4341
USA

Tel: +1 212 726 6490
Fax: +1 212 686 7993
E-mail: cabi-nao@cabi.org

A catalogue record for this book is available from the British Library, London, UK
A catalogue record for this book is available from the Library of Congress, New York

ISBN 0 85198 885 7

Printed and bound in the UK at the University Press, Cambridge

Dedicated to
GEOFFREY C. AINSWORTH
on the occasion of his 90th birthday

This eighth edition of the *Dictionary* is dedicated to Dr Geoffrey Ainsworth as a token of respect for his immense contribution to mycology worldwide. He has already published an account of the origins of the *Dictionary* in the Preface to the seventh edition: leafing through the mycological literature and filling in index cards while 'fire-watching' at Kew in World War II, the adoption of Basic English, a special dispensation to use paper, and the first copy sent to Winston Churchill in September 1943.

He appreciated the value of synthetic reference works early in his career, producing *The Plant Diseases of Great Britain* in 1937. The five volume *The Fungi: An Advanced Treatise* (co-edited with A.S. Sussman and F.K. Sparrow) of 1965-73 provided an unparalleled in-depth reference work, including the most complete keys to genera to have been compiled since 1931. His intimate and wide-ranging interests in the history of the subject, are evident in the scholarly triumvirate, *Introduction to the History of Mycology* (1976), *Introduction to the History of Plant Pathology* (1981), and *Introduction to the History of Medical and Veterinary Mycology* (1986).

His desire to strengthen mycology worldwide led to his proposal to hold the First International Mycological Congress (IMC1) in Exeter in 1971, the Organizing Committee for which he chaired. The International Mycological Association (IMA) was founded at IMC1 following on his suggestion; he was made an Honorary President of the IMA in 1977.

Other honours he has received include a DSc from the University of Exeter in 1978, the Linnean Medal in 1980, and honorary membership of several mycological societies, including the American (1965), British (1965), and Indian (1984). IMI's new laboratory block at Egham was named after him in 1992 in recognition of his work in developing the Institute, especially during his time as Assistant Editor (1957-61), Assistant Director (1961-64) and Director (1964-68).

His industry and scholarship in the service of mycology have provided an example to all who know him. Those of us associated with this edition of the *Dictionary*, join in sending him and Frances our best wishes on his ninetieth anniversary.

Preface

This *Dictionary*, now in its 52nd year, aims to provide all those who work with fungi a way into our accumulated knowledge on them. We regard fungi as all organisms studied by mycologists, including lichens, mushrooms, slime moulds, water moulds and yeasts. The inclusion of names of lichens (more correctly lichen-forming fungi) in the sixth edition occasioned some surprise; we trust that in the 1990s it can be assumed that wherever we use 'fungi', we are automatically including those that form lichens.

In preparing and revising entries for this eighth edition we are fortunate in not only being able to build on the pioneering compilation of G.C. Ainsworth and G.R. Bisby (q.v), but also to have had the benefit of a wide range of contributors (p. viii) and other help from numerous mycologists. At the same time, we have personally revised many of the entries, so that all errors found are our responsibility.

This edition is the first to accept that fungi have to be dispersed through three kingdoms of eukaryotes: *Chromista*, *Fungi*, and *Protozoa*. It also abandons the deuteromycetes as a formal systematic category and endeavours to incorporate them in the system of ascomycetes and basidiomycetes; this is the only defensible treatment as these fungi can be assigned to the different phyla on ultrastructural, molecular and biochemical grounds. Our decision to use 'mitosporic fungi' has been extensively debated at international workshops (Portland, Oregon 1993; Paris 1994). The system adopted for the ascomycetes here is that developed in collaboration with O.E. Eriksson (University of Umeå), with whom D.L.H. has been privileged to cooperate in this endeavour since 1985. The system of higher categories of chromistan fungi used is deliberately more conservative than proposed by M.W. Dick (*Mycol. Pap.*, in press) as we feel that classifications are hypotheses that need a period of testing before they can be commended for general use.

We also wish to record our particular gratitude to D.N. Pegler (Royal Botanical Gardens, Kew) for joining us to take primary responsibility for the system of and updating entries on *Basidiomycetes*, and further for compiling the key to accepted families in that group.

This eighth edition also has several other new features: (a) the inclusion of a key to the accepted families for the first time since the fifth edition in 1961; (b) a synopsis of the genera referred to particular orders and families (also the first time since the fifth edition); (c) notes on major mycological collections and cross references to their acronyms; (d) the revision of most general entries by specialists; (e) the separate listing of many prefixed terms; (f) the inclusion of entries for all accepted families and some frequently used synonyms; (g) the inclusion of illustrations adjacent to the appropriate entries; and (h) a larger page size to help its stay-open factor.

The updates for this edition were greatly aided by the computerization of the previous edition arranged in 1982 by J. Gilmore (Director-General of CAB INTERNATIONAL from 1994) using BASIS software. The database was then adapted to run under SMARTWARE II on the personal computer network at IMI so that we could update entries on screen. The database is now one of the suite of major systematic mycological databases available in IMI.

Having been intimately involved in the compilation and proof-stage revisions, we are acutely aware of imperfections and improvements that we would have liked to have made. As remarked recently of another major compilatory work, our aspiration is that it will at least prove to be 'a marvellously imperfect work needed by all'.

Do send us your corrections and comment so that the ninth edition will be less imperfect and of even more value to mycologists of all disciplines world-wide.

D.L. Hawksworth
P.M. Kirk
B.C. Sutton

13 October 1995
IMI, Egham

Contributors

The following kindly provided new material or revised entries on the subjects indicated:

G.C. Ainsworth   (ex IMI), biographical notes
D. Allsopp   (IMI), biodeterioration
E.J.M. Arnolds   (Landbouwuniversiteit Wageningen), conservation
G.L. Benny   (University of Florida), *Zygomycetes*
P.D. Bridge   (IMI), electrophoresis, molecular biology, neural networks, numerical taxonomy
C.K. Campbell   (Public Health Laboratory Service), medical and veterinary mycology, mycetism
P.F. Cannon   (IMI), ascomycete order and family diagnoses, ascomycete key, biodiversity
S.T. Chang   (Hong Kong University), mushroom cultivation
K. Clay   (Indiana University), endophytes
J.A. Cooper   (IMI), Internet
R.R. Davenport   (IMI Consultant), brewing, yeasts
J.C. David   (IMI), cladistics, Latin, terms
V. Demoulin   (Liege), *Gasteromycetes*
M.W. Dick   (University of Reading), chromistan fungi, chromistan key
P. Döbbeler   (Institut für Systematische Botanik, München), bryophilous fungi
W.C. Elsik   (Mycostrat), fossil fungi
H.C. Evans   (International Institute of Biological Control), biological control, entomogenous fungi, insects and fungi, mycopesticides, predacious fungi
J.C. Frisvad   (Technical University of Denmark), chemotaxonomy
W. Gams   (CBS), soil fungi
G.F. Godwin   (IMI), microscopy
G.S. Hall   (IMI), societies and organizations
D.J. Hill   (University of Bath), dyeing
M.A. Holderness   (IMI), seed-borne fungi
J. Kelley   (IMI), biodegradation, bioremediation, biotransformation, patent protection
P. Jeffries   (University of Kent), fungicolous fungi
D.H. Jennings   (University of Liverpool), nutrition, physiology
D. Jones   (Macauly Land Use Research Institute), rock weathering
E.B.G. Jones   (University of Portsmouth), marine fungi
S.C. Jong   (ATCC), fermentated food and drinks
J.A. Kolkowski   (IMI), media, mites, safety
Z. Kozakiewicz   (IMI), food spoilage, mycotoxicosis, scanning electron microscopy, wine making
J. Lacey   (Rothamsted Experimental Station), air spora, allergy, spore discharge and dispersal
R.W. Lichtwardt   (University of Kansas), *Trichomycetes*
D. Langley   (Glaxo Research), antibiotics
D.J. Lodge   (USDA Forest Service), decomposition
J.E. Longcore   (University of Maine), *Chytridiomycota*
M.A. Mayo   (Scottish Crop Research Institute), viruses
D.W. Minter   (IMI), *Rhytismatales*
R.T Moore   (University of Ulster), ultrastructure
J.E.M. Mordue   (IMI), rusts and smuts
F. Pando   (Madrid), *Myxomycetes*
R.R.M. Paterson   (IMI), metabolic products, mycotoxins
D.J. Read   (University of Sheffield), mycorrhizas

A.D.M. Rayner   (University of Bath), population biology
P.J. Roberts   (K), resupinate basidiomycetes
G.S. Saddler   (IMI), actinomycetes, bacteria
M.R.D. Seaward   (University of Bradford), conservation
D. Smith   (IMI), genetic resource collections, preservation, safety
I.I. Tavares   (University of California), *Laboulbeniales*
T.N. Taylor   (Ohio State University), fossil fungi
A.J.P. Trinci   (University of Manchester), growth, tropism
J.M. Waller   (IMI), fungicides, plant pathogenic fungi
J. Webster   (University of Exeter), aquatic fungi
M.A.J. Williams   (ex IMI), terms
V. Winchester   (University of Oxford), lichenometry

Acknowledgements

The following kindly provided information on corrections or omissions in the seventh edition: T. Ahti (University of Helsinki), G.W.R. Arnold (University of Jena), V. Barkham (ex IMI), E.R. Boa (IMI), J.C. Dianese (Federal University of Brasilia), W. Gams (CBS), H. Hagedorn (Biologische Bundesanstalt), P. Holliday (ex IMI), R.P. Korf (Cornell University), R. Lowen (New York Botanical Garden), E. McKenzie (Landcare, NZ), R.T. Moore (University of Ulster), S. Redhead (Agriculture Canada), A.Y. Rossman (USDA Research Service), C.V. Subramanian (National Taiwan University).

For helpful comments on early drafts of the key to the *Ascomycota* the following are gratefully acknowledged: A. Aptroot (CBS), M.E. Barr (Sidney, Canada), J.C. David (IMI), H. Dissing (University of Copenhagen), O.E. Eriksson (University of Umeå), W. Gams (CBS), A. Henrici (London), R.P. Korf (Cornell University), T. Læssøe (Copenhagen), N. Lundqvist (University of Stockholm), A. Sivanesan (IMI), B.M. Spooner (K), M.T. Smith (CBS).

E. Punithalingam and A.E. Ansell (both of IMI) assisted in the bibliographic checking of citations; the IMI Librarians (L. Ragab, E. Wheater) located copies of elusive references, and M.S. Rainbow (IMI) keyboarded many of the corrections, revisions and new entries. I. Smith (CABI Information Technology Services) prepared the camera-ready copy from output supplied.

Longer Entries

General

Authors
Cladistics
Classification
Coevolution
Collection and Preservation
Colour
Common names
Conidia
Genetic resource collections
Fungicides
Genetics
Growth
History

Hyphal analysis
Kingdoms of fungi
Latin
Lichens
Macromycetes
Media
Metabolic products
Methods
Mitosporic fungi
Nomenclature
Numbers of fungi
Nutrition
Phylogeny

Physiology
Reference collections
Safety
Serology
Sex
Societies and organizations
Spore
States of fungi
Sterilization
Symbiosis
Viruses
Yeasts

Ecology and Distribution

Acid rain
Air pollution
Air spora
Aquatic fungi
Biodiversity
Bioindication
Bryophilous fungi
Coprophilous fungi
Conservation
Ecology

Endophytes
Entomogenous fungi
Fossil fungi
Fungicolous fungi
Geographical distribution
Growth rates
Insects and fungi
Lichenicolous fungi
Lichenometry
Marine fungi

Mycorrhiza
Phytosociology
Plant pathogenic fungi
Population biology
Predacious fungi
Rock weathering
Sand dune fungi
Seed-borne fungi
Soil fungi
Spore discharge and dispersal

Applied Mycology

Allergy
Antibiotics
Biodeterioration
Biological control
Bioremediation
Brewing
Biotechnology
Dyeing
Edible fungi

Fermented foods
Food & beverage mycology
Food spoilage
Fungicides
Industrial mycology
Medical & veterinary mycology
Mushroom culture
Mycetism

Mycotoxicoses
Phytotoxic mycotoxins
Plant pathogenic fungi
Seed-borne fungi
Starters
Truffles
Wine making
Wood-attacking fungi

KINGDOMS OF FUNGI is the starting point for all systematic entries; see also CHROMISTA, FUNGI, PROTOZOA.

User's Guide

To extract the maximum amount of information from this *Dictionary* with the minimum of effort it is necessary to understand the scope of the compilation and certain conventions used to compress the data recorded and minimize repetition; the full citation of any publication or the statement of any fact is rarely given more than once, except where that greatly facilitates accessibility.

Content. The longest series of entries are those of the generic names (both accepted names and synonyms) complied to the end of *Index of Fungi* **6** (10) July 1995. Every accepted generic name is referred to a higher group (family, order, class, or phylum) and brief descriptions are given of these higher taxa. The systematic entries are supplemented by a glossary of terms, English common names, and the names of important fungal antibiotics, toxins, etc. In addition, there are entries on general mycological topics, ecology and distribution, applied mycology, and historical and biographical notes on some well known mycologists and major reference collections. See opposite (p. x) for information on larger entries.

Suprageneric names. Author citations are given only for accepted names to the rank of family as only these have been comprehensively catalogued to date (see Literature, Catalogues of names).

Generic names. Every generic name is followed by the name (abbreviated according to Kirk & Ansell, 1992; see Author) of the author(s) who first proposed the genus and the year of publication. The place of publication of a generic name can be found from the catalogues of names listed under Literature. For example, '*Curvularia* Boedijn (1933)' is found via the cumulative index to *Petrak's Lists* to have been complied in *List 7* where the reference is given as 'Bull. Jard. Bot. Buitenzorg, 3 sér., XIII, Livr. **1**: 123'; that for the earlier '*Coniothyriella* Speg. (1889)' can be found from Saccardo's *Sylloge fungorum*, Clements & Shear's *Genera of Fungi*, or the *Index nominum genericorum (plantarum)*, provided that it was validly published (see Nomenclature). The available Catalogues of names are listed under 'Literature'.

The list of generic names is as complete as possible. Some dates and authorities differ from those that may be found in the literature, many of which have been checked in the original for this edition, some names omitted from previous compilations are included, as are some which are not validly published (included as nevertheless present in the mycological literature). Where genera are not properly typified, p.p. (*pro parte*) has been used when species cited in the protologue are currently dispersed among more than one accepted genus.

For generic names consigned to synonymy, the authority for the disposition is usually given. For each accepted genus estimates are given for the number of its species and its geographical distribution. Where possible these data are based on recent revisions or the personal knowledge of specialists, but in the majority of cases they have not been updated in the absence of such authorities. In the case of larger genera particularly, we have not revised species numbers upwards even though many may have been described since the last edition, in the absence of modern treatments (see Numbers of fungi). This policy is adopted as critical reassesments in such genera usually result in reductions in species numbers.

The distributions given are approximate, especially for genera not critically revised in recent years, and should be regarded as indicative rather than comprehensive. Whenever possible users should verify the facts for themselves and draw their own conclusions.

Coding. The coding used for Mitosporic fungi is explained under that entry.

Systematic arrangement. The synopsis of generic names by order and families following the keys can be used in collaboration with the keys to attempt the identification to genus of an unknown fungus.

References. For the full citation of a reference given by author's name and date only see Literature, or the literature list for the group, e.g. 'Buller, **1-6**' under 'Spore discharge and dispersal' is found expanded under Literature as 'A. H. R. Buller, *Researches on fungi*, vols. **1-6**, 1909-34'; Eriksson (1981) under *Dothideales* is found expanded under *Ascomycota*, the next higher level.

Abbreviations. See p. 1.

Dictionary of the Fungi

a- (an-) (prefix), not having; not; as in acaudate, anaerobe, aniso-.

AAA pathway, alpha- aminoadipic acid pathway for lysine synthesis (cf. DAP pathway).

Aaosphaeria Aptroot (1995), Dothideales (inc. sed.). 1, Colombia, C. Afr.

ab- (prefix), position away from.

Abacina Norman (1853) ≡ Diplotomma (Physc.).

Abaphospora Kirschst. (1939) = Massarina (Lophiostom.) fide Bose (1961).

abaxial (of a basidiospore), the side away from the long axis of the basidium (Corner, 1948); cf. adaxial.

Abbreviations and signs frequently used in this work are:
adj(ective)
Afr(ica)
Am(erica)
Ann(ales) Myc(ologici)
Auct(ores), authors; used esp. as the authority of a name to indicate frequent (and usually incorrect) usage
Austr(alasia)
bibl(iography)
biogr(aphy)
B(ulletin Trimestriel de la) S(ociété) M(ycologique de) F(rance)
C(anadian) J(ournal of) B(otany)
C(entral)
(International) Code (of Botanical Nomenclature)
c(irca), approximately
c(on)f(er), compare; make a comparison with
cosmop(olitan), probably in almost all countries
D(ematiaceous) H(yphomycetes) (1971)
E(ast)
Ed(itor)
Ed(itor)s
ed(itio)n
et al(ia), and others
e(xempli) g(ratia), for example
em(ended by)
esp(ecially)
Eur(ope)
Fam(ily, -ilies)
fide, used for 'on the authority of'
Fig(ure)
f(orm) cat(egory)
gen(us, -era)
Hemisph(ere)
hypog(eous)
I(ndex) N(ominum) G(enericorum)
Isl(and, -s)
L(ichen-forming)
Lit(erature)
Mediterr(anean region)
M(ore) D(ematiaceous) H(yphomycetes) (1976)
Mycol(ogia)
Mycol(ogical) Pap(ers)
M(ycological) R(esearch)
n(oun)
N(orth)
nom(en) cons(ervandum), nom(en) rej(iciendum); see Nomenclature
Obit(uary)
obsol(ete), no longer in use
p(atho)v(ar)
Philipp(ine Islands)
pl(ural)
portr(ait)
pos(itio)n
p(ro) p(arte), in part
q(uod) v(ide), which see
R(eview of) A(pplied) M(ycology)
R(eview of) P(lant) P(athology)
S(ystema) A(scomycetum)
s(ensu) l(ato), in the broad sense; widely
s(ensu) str(icto), in the strict sense; narrowly
S(outh)
sp(ecies), spp. (pl.)
syn(onym, -s) (q.v.)
T(axonomic) L(iterature) (edition)-2
T(ransactions of the) B(ritish) M(ycological) S(ociety)
temp(erate parts)
trop(ics), -(ical)
v(erb)
W(est)
widespr(ead), in a number of countries
O, I, II, III, see *Uredinales*
=, is heterotypic (taxonomic, facultative) a synonym of
≡, is homotypic (nomenclatural, obligate) a synonym of
(), sign for 'is the cause of'; e.g. *Ascochyta pinodella* (foot rot of pea)
±, more or less
µm, micron,
:, in references precedes page number; in author citations, see Nomenclature.

See also Mitosporic fungi for abbreviations for conidiomatal types (1-9), spore groups (A1, B1, etc.), and conidiogenous events (1-44).
Most abbreviations of names of periodicals, except for those noted above, are taken from the *World List of Scientific Periodicals*, 1952 and 1965-67.
If the abbreviation of a Family given with a synonym is not clear the cross-reference gives it in full.
And see Authors' names.

Abeliella Mägd. (1937), Fossil fungi (mycel.). 2 (Cretaceous, Oligocene), Eur.

aberrant an organism that deviates in one or more ways from the norm.

Abgliophragma R.Y. Roy & Gujarati (1966), Mitosporic fungi, 1.C1.10. 1, India. ? = Wiesneriomyces (Mitosp. fungi) fide Pirozynski (*Mycol. Pap.* **129**, 1972).

abhymenial, opposite the spore-producing surface.

abjection, the separating of a spore from a sporophore or sterigma by an act of the fungus.

abjunction, the cutting off of a spore from a hypha by a septum.

Abkultur, see Normkultur.

aboospore, a parthenogenetic oospore.

Abortiporus Murrill (1904), Coriolaceae. 1, USA. See Ryvarden (*Synopsis Fung.* **5**: 104, 1991).

abraded (of lichen thalli), having the surface worn; eroded.

Abropelta B. Sutton (1986), Mitosporic fungi, 5.C1.15. 1, India.

Abrothallomyces Cif. & Tomas. (1953) = Dactylospora (Dactylospor.).

Abrothallus De Not. (1845), Ascomycota (inc. sed.). *c.* 12 (on lichens), widespr. See Bellèmere *et al.* (*Cryptogamie, Mycol.* **7**: 47, 1986; ultrastr.). Anamorph *Vouauxiomyces*.

abrupt, as if cut off transversely; truncate.

abscission, separating by disappearance of a joining layer or wall, as of conidia from a conidiogenous cell.

Absconditella Vězda (1965), Stictidaceae (L). *c.* 8, Eur.,
N. Am. See Vězda & Vivant (*Folia geobot. phytotax.*
10: 205, 1975; key 5 spp.).

Absidia Tiegh. (1878), Mucoraceae. *c.* 20 (esp. in soil),
widespr. See Ellis & Hesseltine (*Sabouraudia* **5**: 59,
1966), Hesseltine & Ellis (*Mycol.* **56**: 568, 1964,
cylindrical-spored spp.; **57**: 234, 1965, globose-
spored spp.; **58**: 761, 1966, ovoid-spored spp.), Not-
tebrock *et al.* (*Sabouraudia* **12**: 64, 1974), Zycha *et al.*
(1969), Váňová (*Česká Myk.* **37**: 151, 1983), Schipper
(*Persoonia* **14**: 133, 1990; key), Hesseltine *et al.*
(*Mycol.* **82**: 523, 1990; key), Burmester *et al.* (*Curr.
Genetics* **17**: 155, 1990; transformations), Wöstemeyer *et al.* (*Curr. Genetics* **17**: 163, 1990; somatic
hybrids), Kayser & Wöstemeyer (*Curr. Genetics* **19**:
279, 1991; karyotype), Ginman & Young (*Microbios*
66: 39, 1991; ultrastr.), Wöstemeyer & Burmester
(*Microbiol. Res.* **149**: 407, 1994; rDNA).

Absidiaceae, see *Mucoraceae.*

absorb, to obtain food by taking up water and dissolved
substances across a membrane. Cf. ingest.

Abstoma G. Cunn. (1926), Lycoperdaceae. 2, Asia,
Australasia, N. & S. Am. See Wright & Suarez
(*Cryptog. Bot.* **1**: 372, 1990; key).

abstriction, abjunction and then abscission, esp. by constriction.

Abyssomyces Kohlm. (1970), Ascomycota (inc. sed.). 1
(marine), S. Atlantic.

Acallomyces Thaxt. (1902), Laboulbeniaceae. 3, N.
Am., Jamaica, Philipp., Japan. See Tavares (*Mycol.*
65: 929, 1973).

Acalyptospora Desm. (1848) , nom. dub.; based on
gland-like hairs.

acantha, a sharp pointed process; a spine.

Acantharia Theiss. & Syd. (1918), Venturiaceae. 5,
widespr. See Hsieh *et al.* (*MR* **99**: 917, 1995; key),
Bose & Müller (*Indian Phytopath.* **18**: 340, 1965),
Sivanesan (*TBMS* **82**: 507, 1984; anamorphs).
Anamorphs *Fusicladium, Stigmina* s.l.

Acanthobasidium Oberw. (1966), Aleurodiscaceae. 3,
Eur.

Acanthocystis (Fayod) Kühner (1926) = Hohenbuehelia
(Tricholomat.) fide Singer (1975).

acanthocyte, spiny cell produced on a short branch from
the vegetative mycelium of *Stropharia* spp. (Farr,
Mycotaxon **11**: 241, 1980).

Acanthoderma Syd. & P. Syd. (1917), Mitosporic
fungi, ?.C1.?. 1, Philipp.

Acanthodochium Samuels, J.D. Rogers & Nagas.
(1987), Anamorphic Xylariaceae, 3.A1.10. Teleomorph *Collodiscula.* 1, Japan.

Acanthographina Walt. Watson (1929) ≡
Acanthothecis (Graphid.).

Acanthographis (Vain.) Walt. Watson (1929) =
Acanthothecis (Graphid.).

acanthohyphidium, see hyphidium.

Acanthomyces Thaxt. (1892) [non *Akanthomyces*
Lebert (1858)] ≡ Rhachomyces (Laboulben.).

Acanthonitschkea Speg. (1908), Nitschkiaceae. 4 (2 on
lichens), widespr. See Nannfeldt (*Svensk bot. Tidskr.*
69: 49, 1975).

Acanthophiobolus Berl. (1893), Dothideales (inc. sed.).
1, widespr. See Walker (*Mycotaxon* **11**: 1, 1980).

Acanthophysellum Parmasto (1967), Aleurodiscaceae.
2, Eur. = Aleurodiscus (Aleurodisc.) fide Eriksson &
Ryvarden (1973).

acanthophysis, see hyphidium.

Acanthophysium (Pilát) G. Cunn. (1963), Aleurodiscaceae. 26, cosmop.

Acanthorhynchus Shear (1907) = Physalospora
(Hyponectr.) fide Barr (*Mycol.* **68**: 611, 1976).

Acanthorus Bat. & Cavalc. (1967), Mitosporic fungi,
5.A1.?. 1, Brazil.

Acanthosphaeria Kirschst. (1939), Trichosphaeriaceae.
2, Eur. See Petrak (*Ann. Myc.* **38**: 198, 1940).

Acanthostigma De Not. (1863) ? = Nematostoma
(Pseudoperispor.) fide v. Arx & Müller (1975).

Acanthostigmella Höhn. (1905), ? Tubeufiaceae. 5,
Eur., N. & S. Am. See Barr (*Mycotaxon* **6**: 17, 1977;
key), Untereiner (*MR* **99**: 897, 1995; posn).

Acanthostigmella Rick (1933) = Acanthostigma
(Pseudoperispor.).

Acanthostigmina Höhn. (1909) = Tubeufia (Tubeuf.)
fide v. Arx & Müller (1975).

Acanthostoma Theiss. (1912) = Phaeodimeriella
(Pseudoperispor.) fide Müller & v. Arx (1975).

Acanthotheca Clem. & Shear (1931) [non DC. (1838),
Compositae] ≡ Acanthotheciella (Sordariales, inc.
sed.).

Acanthotheciella Höhn. (1911), Sordariales (inc. sed.).
3, S. Am., Asia. See Barr (*Mycotaxon* **39**: 43, 1990;
posn.), Nag Raj (*CJB* **55**: 1518, 1977). Anamorph
Ypsilonia.

Acanthotheciopsis Zahlbr. (1923) = Acanthothecis
(Graphid.).

Acanthothecis Clem. (1909), Graphidaceae (L). 4, S.
Am.

Acanthothecium Speg. (1889) ? = Ypsilonia (Mitosp.
fungi) fide Sutton (1977).

Acanthothecium Vain. (1890) ≡ Acanthothecis
(Graphid.).

Acanthothecomyces Cif. & Tomas. (1953) ≡
Acanthothecis (Graphid.).

Acarella Syd. (1927), Mitosporic fungi, 5.A1.?. 1, C.
Am.

Acarellina Bat. & H. Maia (1960), Mitosporic fungi,
5.A1.?. 1, Brazil.

Acariniola T. Majewski & J. Wiśn. (1978) = Pyxidiophora (Pyxidiophor.) fide Lundqvist (1980).

Acarocybe Syd. (1937), Mitosporic fungi, 2.A2.28. 2,
Afr. See Ellis (*Mycol. Pap.* **76**, 1960; key).

Acarocybella M.B. Ellis (1960), Mitosporic fungi,
1.C2.28. 1, trop. Afr., Trinidad.

Acarocybellina Subram. (1992), Mitosporic fungi,
1.C2.26. 1, trop.

Acaropeltis Petr. (1937), Mitosporic fungi, 5.A1.?. 1, C.
Am.

Acarospora A. Massal. (1852), Acarosporaceae (L). *c.*
100, cosmop. See Castello & Nimis (*Lichenologist*
26: 283, 1994; Antarct.), Clauzade & Roux (*Bull.
Mus. Hist. nat. Marseille* **41**, 1981; key 69 Eur. spp.),
Magnusson (*K. svenska VetenskAkad. Handl.* ser. 3 **7**
(4), 1929; *Ann. Crypt. Exot.* **6**: 13, 1933; *Rabenh.
Krypt.-Fl.* **9**(5,1): 104, 1935; *Göteborgs K. Vetensk.-
o. VitterhSamh. Handl.* ser. 6B, **6** (17), 1956), Weber
(*Lichenologist* **4**: 16, 1968; sect. *Xanthothallia*),
Golubkova & Shapiro (*Nov. Sist. niz. Rast.* **13**: 150,
1976; sect. *Trochia*). See also *Pleopsidium.*

   Lit.: Golubkova (*Lishaĭniki semeĭstva
Acarosporaceae Zahlbr. v. SSSR*, 1988; keys 8 gen.,
91 spp.).

Acarosporaceae Zahlbr. (1906), Lecanorales (L). 12
gen. (+ 21 syn.), 173 spp. Thallus usually crustose,
often squamulose; ascomata usually deeply
immersed, rarely almost perithecial, with a varied
margin; paraphyses simple, immersed in gel; asci usually with a well-developed apical dome, usually I-
though sometimes with an outer layer of I+ gel,
polysporous; ascospores small, hyaline, usually
aseptate. Lichenized with green algae.

Acarosporina Sherwood (1977), Stictidaceae. 4, N. &
S. Am., Asia. See Johnston (*Mycotaxon* **24**: 359,
1985; anamorph).

Acarosporium Bubák & Vleugel ex Bubák (1911),
Anamorphic Sclerotiniaceae, 8.B1-2.39. Teleomorph
Pycnopeziza. 2, Eur., N. Am.

Acarosporomyces Cif. & Tomas. (1953) = Pleospidium (Lecanor.).

Acarothallium Syd. (1937) = Wentiomyces (Pseudoperispor.) fide Müller & v. Arx (1962).

acaryallagic, see caryallagic.

acaudate, not having a tail.

Acaulium Sopp (1912) ≡ Scopulariopsis (Mitosp. fungi) fide Raper & Thom (*Manual of the Penicillia*, 1949).

Acaulopage Drechsler (1935), Zoopagaceae. 27, Kenya, N. Am, U.K. See Drechsler (*Mycol.* **27**: 185, 1935; **28**: 363, 1936; **30**: 137, 1938; **31**: 128, 1939; **33**: 248, 1941; **34**: 274, 1942; **37**: 1, 1945; **38**: 120, 1946; **38**: 253, 1947; **40**: 85, 1948; **47**: 364, 1955; **51**: 747, 1959; *Am. J. Bot.* **49**: 1089, 1962), Juniper (*TBMS* **36**: 356, 1953), Jones & Peach (*TBMS* **42**: 95, 1959), Saikawa & Morikawa (*CJB* **63**: 1386, 1985).

Acaulospora Gerd. & Trappe (1974), Acaulosporaceae. 33, widespr. See Mosse (*Arch. Microbiol.* **70**: 167, 1970, **74**: 120, 146, 1970; life cycle, ultrastr.), Schenck *et al.* (*Mycol.* **76**: 685, 1984; key), Berch (*Mycotaxon* **23**: 409, 1985; emend.), Morton (*Mycol.* **78**: 787, 1986; effect of mountants & fixatives on spores), Sieverding & Toro (*Angew. Bot.* **61**: 217, 1987), Sieverding (*Angew. Bot.* **62**: 373, 1988), Błaszkowski (*Karstenia* **27**: 32, 1987, *Mycol.* **82**: 794, 1990, *Mycorrhiza* **4**: 173, 1994), Maia & Kimborough (*MR* **97**: 1183, 1993; spore wall ultrastr.), Ingleby *et al.* (*Mycotaxon* **50**: 99, 1994), Yao *et al.* (*Kew Bull.* **50**: 349, 1995).

Acaulosporaceae J.B. Morton & Benny (1990), Glomales. 2 gen., 37 spp.

accumbent, resting against anything.

acellular, not divided into cells, e.g. a myxomycete plasmodium.

Acephalis Badura & Badurowa (1964) = Syncephalis (Piptocephalid.) fide Skirgiełło & Zadara (*Sydowia Beih.* **8**: 366, 1979).

acephalous, not having a head.

Acerbia (Sacc.) Sacc. & P. Syd. (1899) ? = Rosenscheldia (Dothideales) fide Eriksson & Yue (*SA* **13**: 129, 1994).

Acerbiella Sacc. (1905), Ascomycota (inc. sed.). 2 or 3, S. Am., Java.

acerose, needle-like and stiff; like a pine needle (Fig. 37.3).

acervate, massed up; heaped; growth in heaps or groups.

Acerviclypeatus Hanlin (1990), Anamorphic Phyllachoraceae, 8.E1.?. Teleomorph *Ophiodothella*. 1, USA.

Acervulopsora Thirum. (1945) = Maravalia (Chacon.) fide Cummins & Hiratsuka (1983).

acervulus (pl. **-i**; adj. **-lar**), a ± saucer-shaped conidioma (embedded in host tissue) in which the hymenium of conidiogenous cells develops on the floor of the cavity from a pseudoparenchymatous stroma beneath an integument of host tissue which ruptures at maturity; acervular conidioma (Fig. 10 O).

Acervus Kanouse (1938), Otideaceae. 2, N. & S. Am., Asia, Afr. See Pfister (*Occ. Pap. Farlow Herb. Crypt. Bot.* **8**: 1, 1974; key), Pant (*TBMS* **71**: 326, 1978), Pfister & Bessette (*Mycotaxon* **22**: 435, 1985).

Acetabula (Fr.) Fuckel (1870) = Helvella (Helvell.) fide Dissing (1966).

Acetabularia (Berk.) Massee (1893) [non J.V. Lamour. (1812), *Algae*] ≡ Cyphellopus (Agaric.) fide Singer (1951).

acetabuliform, saucer-like in form.

Achaetobotrys Bat. & Cif. (1963), Antennulariellaceae. 2, trop. See Hughes (1976).

Achaetomiaceae, see *Chaetomiaceae*.

Achaetomiella Arx (1970) = Chaetomium (Chaetom.) fide Udagawa (*Trans. mycol. Soc. Japan* **21**: 34, 1980), Cannon (*TBMS* **87**: 50, 1986).

Achaetomium J.N. Rai, J.P. Tewari & Mukerji (1964), Chaetomiaceae. 3, Asia, Afr. See Cannon (*TBMS* **87**: 50, 1986; key), Kauser *et al.* (*Biologia* Lahore **34**: 257, 1988; Pakistan spp.).

Acharius (Erik; 1757-1819). Country doctor at Vadstena, Sweden, a pupil of Linnaeus (q.v.) and correspondent of Fries (q.v.). Laid scientific basis for the study and classification of lichens. Responsible for the terms thallus, podetium, apothecium, perithecium, soredium, cyphella as applied in lichen morphology. Described many new species, esp. from Europe. Main collections in **H**, other material in **BM**, **UPS**, **LINN** (Smith Herb.). Main works include the *Methodus qua omnes detectos Lichenes*, 1803; *Lichenographia universalis*, 1810; and *Synopsis methodica Lichenum*, 1814. See Galloway (*Bull. Br. Mus. nat. Hist., Bot.* **18**: 149, 1988; influence on Br. lichenology, specs in BM), González Bueno & Rico (*Acta Bot. Malast.* **16**: 141, 1991; impact on Spanish lichenology), Grummann (1974: 469), Vitikainen (Intro., *Lich. univ.* [reprint], 1976), Stafleu & Cowan (*TL-2* **1**: 4, 1976), Stafleu & Mennega (*TL-2, Suppl.* **1**: 14, 1992), Tibell (*Ann. bot. fenn.* **24**: 257, 1987; *Caliciales*).

Achitonium Kunze (1819) = Pactilia (Mitosp. fungi).

Achlya Nees (1823), Saprolegniaceae. 35, N. temp. See Johnson (*The genus Achlya*, 1956).

Achlyella Lagerh. (1890), ? Chytridiales (inc. sed.). 1, Eur.

Achlyites Mesch. (1902), Fossil fungi. 1 (Silurian, Tertiary), Atlantic.

Achlyogeton Schenk (1859), Chytridiales (inc. sed.). 1, N. temp. See Dick (*in press*).

Achlyogetonaceae, see *Chytridiales* (inc. sed.).

Achlyopsis De Wild. (1896) nom. dub. (? Pythiales).

Achorella Theiss. & Syd. (1915), ? Dothideales (inc. sed.). 4, widespr.

Achorion Remak (1845) = Trichophyton (Mitosp. fungi).

Achorodothis Syd. (1926), Mycosphaerellaceae. 1, Costa Rica.

Achoropeltis Syd. (1929), Mitosporic fungi, 5.A1.?. 1, Costa Rica.

achroic (**achromatic**, **achrous**), having no colour or pigment; see Colour.

Achroomyces Bonord. (1851) = Platygloea (Platygl.) fide Donk (*Persoonia* **4**: 145, 208, 1966).

Achrotelium Syd. (1928), Chaconiaceae. 4, Philipp., USA, India.

Acia P. Karst. (1879) [non Schreb. (1791), *Rosaceae*] ≡ Mycoacia (Merul.).

acicular, slender and pointed; needle-shaped (Fig. 37.33).

Aciculariella Arnaud (1954), Mitosporic fungi, 1.E2.?. 2, Eur.

Aciculoconidium D.S. King & S.C. Jong (1976), Mitosporic fungi, 1.A1.3. 1, USA.

Aciculosporium I. Miyake (1908), Clavicipitaceae. 1, Japan. See Kao & Leu (*Plant Prot. Bull. Taiwan* **18**: 276, 1976).

acid rain, the wet acidic deposition of air pollutants, can affect lichen-forming and other fungi. Lichens with cyanobacterial partners are most at risk and have declined dramatically in some parts of Europe (Farmer *et al.*, *in* Bates & Farmer, 1992: 284); nitrogenase activity may be affected (Fritz-Sheridan, *Lichenologist* **17**: 27, 1985). Reductions in many mycorrhizal fungi in Europe have been correlated with acid rain, though it is not often clear whether this is a cause of or a result from damage seen in the trees. The decline in fruiting of *Cantharellus cibarius* has been especially noticeable (Jansen & van Doben, *Ambio* **16**: 211, 1987; Derbsch & Schmitt, *Atlas der Pilze des Saarlandes* **2**, 1987). *Russula mustelina* fruiting has been singled out as a valuable early indicator of acid rain problems in European forests (Felher, *Agric. Ecosyst. Envir.* **28**: 115, 1990).

Lit.: Arnolds (*in* Hawksworth (Ed.), *Frontiers in mycology*: 243, 1991), Bates & Farmer (Eds) (*Bryophytes and lichens in a changing environment*, 1992), Pegler *et al.* (Eds) (*Fungi of Europe*, 1993), Richardson (*Pollution monitoring with lichens*, 1992).
See Air pollution, Bioindication.

acid-fast (of bacteria), keeping carbol fuchsin stain after the addition of 25 per cent sulphuric acid (H_2SO_4).

acidiphilous (**acidophilous, acidophilic**), growing on or in conditions of low hydrogen ion concentration (q.v.); e.g. *Scytalidium acidophilum* with an optimum pH for growth of 3, with good growth even at pH 1 (Miller *et al.*, *Internat. Biodet.* **20**: 27, 1984); also used of lichens on peaty soils or bark of a pH below 5.

Aciella (P. Karst.) P. Karst. (1899) [non Tiegh. (1894), *Loranthaceae*] = Asterodon (Asterostrom.) fide Donk (1956).

Aciesia Bat. (1961), Mitosporic fungi (L), 8.A2.?. 1, Brazil.

Acinophora Raf. (1808) nom. dub. (Tulostomatales, inc. sed.).

Acinula Fr. (1822), Mitosporic fungi, 9.-.-. 1, Eur.

Acitheca Currah (1985), Gymnoascaceae. 1, N. Am. = Gymnoascus (Gymnoasc.) fide v. Arx (*Persoonia* **13**: 173, 1986).

Ackermannia Pat. (1902) = Sclerocystis (Glom.) fide Zycha *et al.* (1969).

Acladium Link (1809), Anamorphic Thelephoraceae, 1.A1.3. Teleomorph *Botryobasidium*. 4, widespr.

Acleistia Bayl. Ell. (1917), Anamorphic Leotiaceae, 7.A1.15. 1, Eur. Teleomorph *Ombrophila*.

Acleistomyces Bat. (1961), Mitosporic fungi L, 8.A1.?. 2 (L), Brazil.

Acmosporium Corda (1839) = Aspergillus (Mitosp. fungi) fide Hughes (1958).

Acoliomyces Cif. & Tomas. (1953) = Thelomma (Calic.).

Acolium (Ach.) Gray (1821) = Cyphelium (Calic.).

Acolium Trevis. (1862) ≡ Pseudacolium (Calic.).

Acompsomyces Thaxt. (1901), Laboulbeniaceae. 6, Eur., N. & S. Am. See Benjamin (*Mem. N.Y. Bot. Gdn* **49**: 20, 1989; key, ontogeny).

Acontiopsis Negru (1961) = Cylindrocladium (Mitosp. fungi).

Acontium Morgan (1902), Mitosporic fungi, 1.A1.?. 2 or 3, N. Am.

acquired immunity, see immune.

acquired resistance, see resistance.

Acrasaceae, *see Acrasiaceae.*

Acrasiales, Acrasiomycota. 4 fams., 4 gen. (+ 3 syn.), 12 spp. The only order of *Acrasiomycota* (q.v.) comprising fams:
(1) **Acrasiaceae**.
(2) **Copromyxaceae**.
(3) **Guttulinopsidaceae**.
(4) **Fonticulaceae**.

acrasin, a chemotactically active substance which controls the streaming together of the myxamoebae of *Dictyostelium discoideum* (Bonner, *J. exp. Zool.* **110**: 259, 1949) and other *Acrasales.*

Acrasiomycetes, see *Acrasiomycota.*

Acrasiomycota (Acrasea, Acrasiomycetes, Acrasiales, Sorophorae), Protozoa; the acrasid cellular slime moulds; acrasids. 1 ord., 3 fam., 4 gen. (+ 3 syn.), 12 spp. Trophic phase amoeboid, pseudopodia lobose; aggregating without streaming; nuclei with a compact centrally placed nucleolus; sporocarp sessile, independent and dividing when vegetative, some with simple supportive stalks; multispored, in chains or

delimited sori; flagellate cells usually absent; sexual reproduction unknown. On dung and isolated from a wide range of decaying plant materials and macromycetes, and also soil. Ord.:
Acrasiales.

Raper (1973) included the dictyostelids (see *Dictyosteliomycota*) and protostelids (see *Myxomycota*) in the class *Acrasiomycetes*, later excluding the protostelids as a separate class (Raper, 1984). Molecular data show these three groups not to be closely allied (see Phylogeny, *Protozoa*) so they are treated in separate phyla here.

Lit.: Olive (1975), Raper (*in* Ainsworth *et al.* (Eds), *The fungi* **4B**: 9, 1973; keys gen.; *The dictyostelids*, 1984).

Acrasis Tiegh. (1880), Acrasiaceae. 2 (on beer yeast), widespr.

Acremoniella Sacc. (1886), Mitosporic fungi, 1.A2.1. 5, widespr. See Groves & Skolko (*Can. J. Res.* **24**: 74, 1946). Nom. illegit. = Harzia (Mitosp. fungi) fide Holubová-Jechová (1974).

Acremonites Pia (1927), Fossil fungi. 1 (Oligocene), Eur.

Acremoniula G. Arnaud (1954) ≡ Acremoniula (Mitosp. fungi).

Acremoniula G. Arnaud ex Cif. (1962), Mitosporic fungi, 1.A2.1. 1 (on *Schiffnerula* and *Meliola*), widespr., trop. See Deighton (*Mycol. Pap.* **118**, 1969).

Acremonium Link (1809), Mitosporic fungi, 1.A1.15. *c.* 105 (sometimes as endophytes, then important in *Gramineae*), cosmop. Teleomorphs reported in many ascomycete groups, probably polyphyletic (Kendrick & DiCosmo, *in* Kendrick (Ed.), *The whole fungus*, 1979). See Gams (*Cephalosporium-artige Schimmelpilze*, 1971, keys; *TBMS* **64**: 389, 1975), Lowen (*Mycotaxon* **53**: 81, 1995; key 9 spp. sect. *Lichenoidea*), Samuels (*N.Z. Jl Bot.* **14**: 231, 1976; teleomorphs), Chesson *et al.* (*TBMS* **70**: 345, 1978; electrophoresis), Bałazy (*Bull. Soc. Annls Sci. Lett. Poznań* D **14**: 101, 1973; entomogenous spp.), White & Morgan-Jones (*Mycotaxon* **30**: 87, 1987; keys to sect. *Albo-lanosum*), Peberdy (Ed.) (*Biotech. Handb.* **1**, 1988; taxonomy and morphology), Pitson *et al.* (*MR* **95**: 352, 1991; β-glucanase production), Christensen *et al.* (*MR* **95**: 918, 1991; variation in endophytic spp.), Walz (*Bibl. Mycol.* **147**: 1, 1992; molecular analysis of *A. chrysogenum*), Quisenberry & Joost (Eds) (*Proc. Int. Symp. Acremonium/grass Interactions*, 1990), Leuchtmann (*MR* **98**: 25, 1994; isozyme relationships of *Festuca* endophytes), Okada *et al.* (*Trans. Mycol. Soc. Jap.* **34**: 171, 1993; *A. alcalophilum*, an alkalophilic cellulolytic sp.), Christensen *et al.* (*MR* **97**: 1083, 1993; taxonomy of fescue endophytes).

Acriasaceae Poche (1913), Acrasiales. 1 gen., 2 spp. Sorocarp cells differentiated into spores and stalk cells.

acro- (combining form), at the end; apical; terminal.

acroauxic (of conidiophores), growth in length restricted to the apical region.

Acrocalymma Alcorn & J.A.G. Irwin (1987), Anamorphic Lophiostomataceae, 4.A1.15. Teleomorph *Massarina*. 1 (on *Medicago*), Australia. See Shoemaker *et al.* (*CJB* **69**: 569, 1991; teleomorph).

acrochroic, see Colour.

Acrocladium Petr. (1949) [non Mitt. (1869), *Musci*] = Periconiella (Mitosp. fungi) fide v. Arx (*Persoonia* **11**: 39, 1981).

Acroconidiella J.C. Lindq. & Alippi (1964), Mitosporic fungi, 1.C2.26. 1, widespr.

Acroconidiellina M.B. Ellis (1971), Mitosporic fungi, 1.B2.26. 4, trop.

Acrocordia A. Massal. (1854), Monoblastiaceae (L). *c.* 30, mainly N. temp. See Coppins & James (*Lichenologist* **10**: 179, 1978; UK spp.).

Acrocordiaceae, see *Monoblastiaceae*.

Acrocordiella O.E. Erikss. (1982), Pyrenulaceae. 1 (on *Ribes*), Sweden. = Requienella (Pyrenul.) fide Boise (*Mycol.* **78**: 37, 1986). See Eriksson & Hawksworth (*SA* **5**: 115, 1986).

Acrocordiomyces Cif. & Tomas. (1953) = Acrocordia (Monoblast.).

Acrocordiopsis Borse & K.D. Hyde (1989), Melanommataceae. 1, India.

Acrocorelia R. Doll (1982) nom. nud. (? Dothideales (L), inc. sed.).

Acrocylindrium Bonord. (1851), Mitosporic fungi, 1.A1.?. 3, Eur. See Gams & Hawksworth (*Kavaka* **3**: 60, 1976). ? = Sarocladium (Mitosp. fungi) fide Gams (*in litt.*).

Acrocystis Ellis & Halst. ex Halst. (1890) nom. dub. (Plasmodiophorales).

Acrodesmis Syd. (1926) = Periconiella (Mitosp. fungi) fide Ellis (1967).

Acrodictyopsis P.M. Kirk (1983), Mitosporic fungi, 1.D2.1. 1, U.K.

Acrodictys M.B. Ellis (1961), Mitosporic fungi, 1.D2.1/19. 26, widespr. See Ellis (*DH, MDH*).

Acrodontium de Hoog (1972), Mitosporic fungi, 1.A1.1. 9, widespr.

Acrogenospora M.B. Ellis (1971), Mitosporic fungi, 1.A2.19. 4, widespr. See Hughes (*N.Z. Jl Bot.* **16**: 312, 1978).

Acrogenotheca Cif. & Bat. (1963), Dothideales (inc. sed.). 2, trop. See Hughes (*N.Z. Jl Bot.* **5**: 504, 1967; 1976). Anamorph *Hiospira*.

acrogenous, development at the apex.

Acrogynomyces Thaxt. (1931), Laboulbeniaceae. 6, Afr.

acronema, extension of flagellum tip containing the two central microtubules but none of the nine peripheral elements.

acropetal, (1) describes chains of conidia in which the youngest is at the apex, basifugal; cf. basipetal; (2) a pattern of apical growth.

Acrophialophora Edward (1961), Mitosporic fungi, 1.A1.15. 3, Asia, Eur., Tahiti. See Samson & Mahmood (*Acta bot. neerl.* **19**: 804, 1970; key).

Acrophragmis Kiffer & Reisinger (1970), Mitosporic fungi, 1.C2.19. 1, Congo, Australia.

Acrophytum Sacc. (1883) = Cordyceps (Clavicipit.).

acropleurogenous, formed at the end and on the sides.

Acrorixis Trevis. (1860) = Thelenella (Thelenell.) p.p., Diploschistes (Thelotremat.) p.p., and ? Anthracothecium (Pyrenul.) p.p.

Acroscyphus Lév. (1846), Caliciaceae (L). 1, C. & S. Am., Asia.

Acrospeira Berk. & Broome (1857), Mitosporic fungi, 1.D2.1. 4, N. temp. See Wiltshire (*TBMS* **21**: 211, 1938).

Acrospermaceae Fuckel (1870), Ascomycota (inc. sed.). 2 gen. (+ 3 syn.), 11 spp. Stromata ± superficial, hyaline to brown, pulvinate or stipitate, composed of gelatinous pseudoparenchymatous tissue, often containing only a single ascoma; ascomata perithecial, thin-walled, the ostiole large, not periphysate; interascal tissue composed of narrow paraphyses; asci cylindrical, elongated, ? with separable wall layers, not fissitunicate, usually with a capitate apical thickening with a narrow pore; ascospores hyaline, elongate, multiseptate, not fragmenting, without a sheath.

Family of doubtful affinity; included in *Pyrenulales* by Eriksson (1982) and *Xylariales* by Barr (1994).

Acrospermoides J.H. Mill. & G.E. Thomps. (1940), Lasiosphaeriaceae. 1, USA. See Barr (*Mycotaxon* **39**: 43, 1990; posn).

Acrospermum Tode (1790), Acrospermaceae. 10, widespr. See Tonolo (*R.C. 1st Sup. Sanit.* **20**: 842, 1957), Eriksson (*Ark. Bot.* II, **6**: 381, 1967), Sherwood (*Mycotaxon* **5**: 39, 1977; posn), Webster (*TBMS* **39**: 361, 1956; conidia).

Acrosphaeria Corda (1842) = Xylaria (Xylar.) fide Læssøe (*SA* **13**: 43, 1994).

Acrospira Mont. (1857), Mitosporic fungi, 1.?.?. 1, Eur.

acrospore, an apical spore.

Acrosporella Riedl & Ershad (1977) = Cladosporium (Mitosp. fungi) fide Sutton (*in litt.*).

Acrosporium Nees (1816) nom. rej. = Oidium (Mitosp. fungi).

Acrosporium Bonord. (1851), Mitosporic fungi, 1.A1.?. 1, Germany.

acrosporogenous (of conidial maturation), cells delimited and maturing in sequence from base to apex as the tip of the conidium expands (Luttrell, 1963).

Acrostalagmus Corda (1838) = Verticillium (Mitosp. fungi) fide Hughes (1958).

Acrostaphylus G. Arnaud ex Subram. (1956), Mitosporic fungi, 1.A2.6. 8, widespr. = Nodulisporium (Mitosp. fungi) fide Jong & Rogers (1972).

Acrostaurus Deighton & Piroz. (1972), Mitosporic fungi, 1.G2.19. 1 (on ascomycete), trop.

Acrostroma Seifert (1987), Anamorphic Batistiaceae, 2.A1.15. Teleomorph *Batistia*. 1, Venezuela.

Acrotamnium Nees (1816) nom. dub. (? 'basidiomycetes') fide Donk (*Taxon* **11**: 103, 1962).

Acrotellomyces Cif. & Tomas. (1953) ≡ Acrotellum (Verrucar.).

Acrotellum Tomas. & Cif. (1952) = Thelidium (Verrucar.).

Acrothamnium, see *Acrotamnium*.

Acrotheca Fuckel (1860) = Pleurophragmium (Mitosp. fungi) fide Ellis (*MDH*).

Acrotheciella Koord. (1907), Mitosporic fungi, 3.C2.?24. 1, Java.

Acrothecium (Corda) Preuss (1851), Mitosporic fungi, 1.C1.?. *c.* 15, widespr.

acroton, a spinule in lichens bearing side branches.

actidione, trade name for cycloheximide (q.v.).

Actidium Fr. (1815), Mytilinidiaceae. 9, Eur., N. Am. See Zogg (*Ber. Schweiz. bot. Ges.* **70**: 195, 1960; key).

Actigea Raf. (1814) = Scleroderma (Sclerodermat.).

Actigena Raf. (1814) ≡ Actigea (Sclerodermat.).

actin, and **mycosin** are proteins associated with contraction and relaxation of muscle; also present in several lower eukaryotic organisms and responsible for the periodic reversal of protoplasmic streaming in the plasmodium of myxomycetes.

Actiniceps Berk. & Broome (1876), Pterulaceae. 5, trop. See Boedijn (*Persoonia* **1**: 11, 1959).

Actiniopsis Starbäck (1899) = Trichothelium (Trichothel.) fide Santesson (1952).

Actinobotrys H. Hoffm. (1856) = Bremia (Peronospor.) fide Saccardo (1888).

Actinocephalum Saito (1905) = Cunninghamella (Cunninghamell.) fide Hesseltine (1955).

Actinochaete Ferro (1907) nom. conf. (Mitosp. fungi) = Aspergillus (Mitosp. fungi) p.p. and Septobasidium (Septobasid.) p.p. fide Ellis (*in litt.*).

Actinocladium Ehrenb. (1819), Mitosporic fungi, 1.G2.1. 1, widespr.

Actinocymbe Höhn. (1911), Chaetothyriaceae. 1 or 2, trop.

Actinodendron G.F. Orr & Kuehn (1963) = Oncocladium (Mitosp. fungi) fide Hughes (*CJB* **46**: 939, 1968).

Actinodermium Nees (1816) ≡ Sterbeeckia (Sclerodermat.).

Actinodochium Syd. (1927), Mitosporic fungi, 3.A1.3. 2, C. Am., India.

Actinodothidopsis F. Stevens (1925) = Venturia (Ventur.) fide Müller & v. Arx (1962).

Actinodothis Syd. & P. Syd. (1914) = Amazonia (Meliol.) fide Hansford (1961).

Actinoglyphis Mont. (1856) = Sarcographa (Graphid.).

Actinogyra Schol. (1934) = Umbilicaria (Umbilicar.).

actinogyrose (actinogyr) (of apothecia), disc gyrose and having no proper margin.

actinolichen, a lichen-like association between a green alga and an actinomycete (e.g. Chlorella and Streptomyces sp.; Lazo & Klein, Mycol. **57**: 804, 1965) occurring in nature and also in mixed laboratory cultures. See Kalakoutskii et al. (Actinomycetes, n.s. **1**(2): 27, 1990; lab. expts, bibliogr.).

Actinomadura H. Lechev. & M.P. Lechev. (1968), Actinomycetes, q.v.

Actinomma Sacc. (1884) = Atichia (Mitosp. fungi).

Actinomortierella Chalab. (1968) = Mortierella (Mortierell.) fide Gams (Nova Hedw. **18**: 30, 1969).

Actinomucor Schostak. (1898), Mucoraceae. 2, widespr. See Benjamin & Hesseltine (Mycol. **49**: 240, 1957), Jong & Yuan (Mycotaxon **23**: 261, 1985).

Actinomyce Meyen (1827) nom. dub. (? Fungi).

Actinomyces Harz (1877), Actinomycetes, q.v.

Actinomycetes ('Ray Fungi'). A group of morphologically diverse but usually filamentous gram positive bacteria which have sometimes been classified as Mitosporic fungi and which are still frequently referred to a mycologist for identification, esp. those pathogenic for humans to medical mycologists. Actinomycetes are typically saprobes (esp. in soil) but a few are pathogenic for humans, animals, and plants; some (esp. Streptomyces) are important sources of antibiotics (see amphotericin, cycloheximide, nystatin, streptomycin); some form lichen-like associations with green algae (see actinolichen). Some of the more common or recently described genera of Actinomycetes, which were omitted from the previous edition of this Dictionary, are included in the present edition for reference only.

　Lit.: The literature on actinomycetes is extensive. Generic names are listed by Skerman et al. (Approved lists of bacterial names, Amended edn, 1989). See Williams et al. (Eds) (Bergey's manual of systematic bacteriology **4**, The actinomycetes, 1989), Balows et al. (The procaryotes, 2nd edn, 1992), Goodfellow et al. (Eds) (Biology of the actinomycetes, 1984), Ortiz-Ortiz et al. (Eds) (Biological, biochemical, and biomedical aspects of actinomycetes, 1984), Goodfellow et al. (Eds) (Actinomycetes in biotechnology), Goodfellow & Williams (Ann. Rev. Microbiol., **37**: 189, 1983).

Actinomycites D. Ellis (1916), Fossil fungi (Actinomycetes). 1 (Jurassic), UK.

Actinomycodium K.M. Zalessky (1915), Fossil fungi (? Mitosp. fungi or Actinomycetes). 1 (Permo-Carboniferous), former USSR.

Actinomyxa Syd. & P. Syd. (1917), Microthyriaceae. 1, Australia.

Actinonema Pers. (1822) nom. dub. (Mitosp. fungi). See Sutton (Mycol. Pap. **141**, 1977), where interpreted as sterile mycelium, but often used for Marssonina rosae (teleomorph Diplocarpon rosae) (black spot of rose).

Actinonema Fr. (1849) = Spilocaea (Mitosp. fungi) fide Sutton (Mycol. Pap. **141**, 1977).

Actinonemella Höhn. (1916) = Asteroma (Mitosp. fungi) fide Sutton (Mycol. Pap. **141**, 1977).

Actinopelte Stizenb. (1861) = Solorinella (Solorinell.).

Actinopelte Sacc. (1913) ≡ Tubakia (Mitosp. fungi).

Actinopeltella Doidge (1924) = Actinopeltis (Microthyr.) fide v. Arx & Müller (1975).

Actinopeltis Höhn. (1907), Microthyriaceae. 10, Afr., S. Am., Eur. See Ellis (TBMS **68**: 145, 1977), Spooner & Kirk (MR **94**: 223, 1990).

Actinophora Merr. (1943) ≡ Acinophora (Tulostomatales, inc. sed.).

Actinoplaca Müll. Arg. (1891), Gomphillaceae (L). 2, trop. See Vĕzda & Poelt (Folia geobot. phytotax. **22**: 180, 1987).

Actinoplacomyces Cif. & Tomas. (1953) ≡ Actinoplaca (Gomphill.).

Actinoplanes Couch (1950), Actinomycetes, q.v.

Actinopolyspora Gochn., K.G. Johnson & Kushner (1975), Actinomycetes, q.v.

Actinoscypha P. Karst. (1888) = Micropeziza (Dermat.) fide Nannfeldt (1976).

Actinosoma Syd. (1930) ? = Actinopeltis (Microthyr.). See Eriksson & Hawksworth (SA **9**: 6, 1991; status), Spooner & Kirk (MR **94**: 223, 1990).

Actinospira Corda (1854) ≡ Myxotrichum (Myxotrich.).

Actinospora Ingold (1952), Anamorphic Otideaceae, 1.G1.23. Teleomorph Miladina. 1 (in water), UK. See Descals (TBMS **67**: 208, 1976), Descals & Webster (TBMS **70**: 466, 1978; teleomorph).

Actinostilbe Petch (1925) = Sarcopodium (Mitosp. fungi) fide Sutton (TBMS **76**: 97, 1981). See Samuels & Seifert (in Sugiyama (Ed.) (Pleomorphic fungi: 29, 1987).

Actinostroma Klotzsch (1843) = Cymatoderma (Podoscyph.) fide Donk (1957).

Actinosynnema T. Haseg., H. Lechev. & M.P. Lechev. (1978), Actinomycetes, q.v.

Actinoteichus Cavalc. & Poroca (1971), Mitosporic fungi, 5.A1.?. 4 (L), Brazil.

Actinotexis Arx (1960), Mitosporic fungi, 5.E1.?. 1, Brazil.

Actinothecium Ces. (1854), Mitosporic fungi, 5.A1.?. 5, widespr.

Actinothecium Flot. (1855) = Verrucaria (Verrucar.).

Actinothyrella Edward, Kr.P. Singh, S.C. Tripathi, M.K. Sinha & Ranade (1974) nom. dub. (Mitosp. fungi) fide Sutton (Mycol. Pap. **141**, 1977).

Actinothyrium Kunze (1832), Mitosporic fungi, 5.E1.?. 10, widespr.

Actinotrichum Wallr. [not traced] nom. nud. (Mitosp. fungi) fide Sutton (Mycol. Pap. **141**, 1977).

Actonia C.W. Dodge (1935) nom. dub. (Fungi, inc. sed.) fide Batra (1978).

Actycus Raf. (1815) nom. dub. (Fungi, inc. sed.).

aculeate, having narrow spines (Fig. 29.3).

aculeolate, having spine-like processes.

acuminate, gradually narrowing to a point.

Acumispora Matsush. (1980), Mitosporic fungi, 1.C1/2.1. 2, Taiwan.

Acurtis Fr. (1849) nom. dub.; a sterile form of Entoloma (Entolomat.) fide Donk (Taxon **11**: 76, 1962).

acute, (1) pointed (Fig. 37.41); (2) less than a right angle.

Acutocapillitium P. Ponce de León (1976), ? Lycoperdaceae. 2, trop. Am. ? = Glyptoderma (Lycoperd.).

Acytosteliaceae Raper ex Raper & Quinlan (1958) (Acytosteliidae), Dictyosteliales. 1 gen., 4 spp. Sporocarp stalk slender and narrow.

Acytostelium Raper (1956), Acytosteliaceae. 4, widespr.

Adamson's fringe, the downward growing hyphae of a dermatophyte in the region above the bulb of a hair.

adapted race (Magnus), see physiologic race.

adaxial (of a basidiospore), the side next to the long axis of the basidium, usually that with the apiculus (Corner, 1948); cf. abaxial.

Adea Petr. (1928) = Seiridium (Mitosp. fungi) fide Nag Raj & Kendrick (1986).

Adella Petr. (1936) = Wojnowicia (Mitosp. fungi) fide Sutton (1975).

Adelococcaceae Rambold (1993), Verrucariales. 2 gen., 8 spp. Thallus absent; ascomata perithecial, not clypeate, ± globose; peridium dark brown esp. at the apex, composed of pseudoparenchymatous cells; hymenium gelatinous, blueing in iodine; interascal tissue of narrow anastomosing paraphysis-like hyphae and apical paraphyses, the ostiole periphysate; asci ± clavate, thick-walled. Biotrophic or necrotrophic on lichens, widespr.

Adelococcus Theiss. & Syd. (1918), Adelococcaceae. 2 (on lichens), Eur. See Matzer & Hafellner (*Bibl. Lich.* **37**, 1990).

Adelodiscus Syd. (1931), Leotiales (inc. sed.). 1, Philipp.

Adelolecia Hertel & Hafellner (1984), Bacidiaceae (L). 2, Eur., N. Am. See Hertel & Rambold (*Bibl. Lich.* **57**: 211, 1995).

Adelomyces Thaxt. (1931) = Phaulomyces (Laboulben.) fide Tavares (1985).

Adelomycetes, see Mitosporic fungi (Langeron, *Précis de Mycologie*, edn 1, 1945).

Adelopus Theiss. (1918) = Phaeocryptopus (Ventur.) fide v. Arx & Müller (1975).

adelphogamy, pseudomictic copulation of mother and daughter cells, as in some yeasts (Gäumann & Dodge, 1928: 13).

adenose, having glands; gland-like.

Aderkomyces Bat. (1961), Mitosporic fungi (L), 8.A1.15. 1, Brazil.

Adermatis Clem. (1909) = Lecania (Bacid.).

adherance (of fungicides), the ability of a fungicide (or other crop protectant) to stick to a surface. Cf. retention.

adhesive disc, see holdfast.

adhesorium, the organ developed from a resting zoospore of *Plasmodiophora* for attachment to, and penetration of, the host (Aist & Williams, *CJB* **49**: 2023, 1971).

Adhogamina Subram. & Lodha (1964) = Gilmaniella (Mitosp. fungi) fide Barron (1968).

adiaspiromycosis, pulmonary infection in animals (particularly soil-burrowing rodents) and rarely humans by *Emmonsia* spp., esp. *E. parva* (syn. *Haplosporangium parvum*) and *E. crescens* (Jellison, *Adiaspiromycosis* (syn. *Haplomycosis*), 1969); haplomycosis. Cf. adiaspore.

adiaspore, a large spherical chlamydospore produced in the lungs of animals by the enlargement of an inhaled conidium of *Emmonsia* spp.; cf. adiaspiromycosis. *Chrysosporium pruinosum* produces similar spores in culture (Carmichael, *CJB* **40**: 1167, 1962).

adjunct (in brewing), any legally permitted substance lacking nutritional properties added to the fermentation.

Adlerocystis Feldm.-Muhs. & Havivi (1963), Zygomycetes (inc. sed.). 2 (symbiotic of *Ornithodorus*), Israel, ? USA. See Feldman-Muhsam & Havivi (*Parasitology* **53**: 183, 1963).

adnate (of lamellae or tubes), joined to the stipe; if lamellae, proximal end not notched (cf. sinuate); sometimes restricted to lamellae widely joined to the stipe (Fig. 19C) (cf. adnexed); (of pellicle, scales, etc.), tightly fixed to the surface.

adnexed (of lamellae), narrowly joined to the stipe (Fig. 19B) (cf. adnate); an ambiguous term.

Adomia S. Schatz (1985), Lasiosphaeriaceae. 1 (marine, on *Avicennia*), Egypt, Australia.

adpressed, see appressed.

adspersed, of wide distribution; scattered.

aduncate, bent; hooked; crooked.

Adustomyces Jülich (1979), Hyphodermataceae. 1, Eur., Afr.

adventitious septum, see septum.

adventive branching (of fruticose lichens), branching not of the normal pattern; e.g. regenerate branches produced after damage to the original branches in *Cladonia*.

Aecidiconium Vuill. (1892), Uredinales (inc. sed.). 1 (on *Pinus*), France.

Aecidiella Ellis & Kelsey (1897) = Pucciniosira (Pucciniosir.) fide Arthur (*N. Am. Fl.* **7**: 126, 1907).

Aecidiolum Unger (1832), Anamorphic Uredinales. A name for (0) only.

aecidiospore, see *Uredinales*.

Aecidites Debey & Ettingsh. (1859), Fossil fungi. 4 (Cretaceous, Tertiary), Eur.

aecidium, see *Uredinales*.

Aecidium Pers. (1796), Anamorphic Uredinales. 600 (on angiosperms), widespr. A name for (0) and (I) only, used esp. in areas where there is little knowledge of the life-histories of rusts. A number may be 'duplicate' names; some may be species of *Endophyllum* (q.v.). As with other anamorphic fungi, an *Aecidium* name is sometimes used even when there is a named teleomorphic (telial) state.

aeciospore, see *Uredinales*.

aeciotelium, see *Uredinales*.

aecium, see *Uredinales*.

Aeciure Buriticá & J.F. Hennen (1994), Phakopsoraceae. 1 (on *Croton*), Brazil.

Aedycia Raf. (1808) nom. rej. = Mutinus (Phall.).

Aegerita Pers. (1801), Anamorphic Hyphodermataceae, 3.A1.1. Teleomorph *Bulbillomyces*. 1, Eur. See Hennebert (*Persoonia* **7** 191, 1973), Julich (*Persoonia* **8** 59, 1974).

Aegeritella Bałazy & J. Wiśn. (1974), Mitosporic fungi, 3.A1.1. 2 (on ants), Poland, Brazil.

Aegeritina Jülich (1984), Anamorphic Hyphodermataceae, 3.?.?. Teleomorph *Subulicystidium*. 1, Eur.

Aegeritopsis Höhn. (1903) = Aegerita (Mitosp. fungi) fide Clements & Shear (1931).

Aenigmatomyces R.F. Castañeda & W.B. Kendr. (1993), Mitosporic fungi, 1.A1.1. 1 (on *Pythium*), Canada.

aequi-hymeniiferous (of hymenial development in agarics), having basidia which mature and shed their spores evenly over the surface of each lamella; the non-*Coprinus* type (Buller, *Researches* **2**: 19, 1922). cf. inequi-hymeniiferous.

aero-aquatic fungi, fungi that grow under water but produce spores in the air above (van Beverwijk, *TBMS* **34**: 280, 1951). See Aquatic fungi.

aerobe, an organism needing free oxygen for growth; cf. anaerobe.

aerobiological pathway, the process (comprising the source, liberation, dispersion, deposition, and impact on another living organism) by which air-borne microorganisms are dispersed (Edwards, *Aerobiology*, 1979).

aerogenic, describes an organism that produces detectable gas during the breakdown of carbohydrate.

aerole (of lichens), a scale-like area on the thallus delimited by cracks or depressions.

Aerophyton Eschw. (1824) nom. dub. (Mitosp. fungi).

Aeruginospora Höhn. (1908) = Camarophyllus (Hygrophor.) fide Singer (1975); = Hygrophorus (Hygrophor.).

Aessosporon Van der Walt (1970), Sporidiobolaceae. 2, Netherlands. Anamorphs *Bullera*, *Sporobolomyces*.

Aethaliopsis Zopf (1885) = Fuligo (Physar.).

aethalium (of myxomycetes), a sessile fruit-body made by a massing of all or a part of the plasmodium.

Aethalium Link (1809) = Fuligo (Physar.).

aetiology, the science of the causes of disease; etiology (Amer.).

Aetnensis Lloyd (1910) nom. nud.

aflatoxins, a series of toxic polybutole metabolites (mycotoxins) esp. of *Aspergillus flavus* strains when growing on groundnuts, cereals, etc.; most well known mycotoxin; most developed countries have statutory limits; gene probes available; the cause of **aflatoxicosis** in poultry and cattle and carcinogenic for rats and humans.

  Lit.: Hesseltine *et al.* (*Bact. Rev.* **30**: 795, 1966), *Aflatoxin bibliography, 1960-67*, 1968), Goldblatt (Ed.) (*Aflatoxin: scientific background, control and implications*, 1969), Racovitza (*J. gen. Microbiol.* **57**: 379, 1969; aflatoxin toxic to the mite *Glyciphagus domesticus*), Heathcote & Hibbert (*Aflatoxins: chemical and biological aspects*, 1978), Eaton & Groopman (*The toxicology of aflatoxins*, 1994), Flannigan (Ed.) (*Internat. Biodet.* **22** (Suppl.), 1986; in cereals and stored products), Wylie & Morehouse (Eds) (*Mycotoxic fungi, mycotoxins, mycotoxicoses* **1-3**, 1977-8), Mycotoxicoses.

African histoplasmosis, infection of humans or animals by *Histoplasma capsulatum* var. *duboisii*.

Afroboletus Pegler & T.W.K. Young (1981), Strobilomycetaceae. 2, trop. Afr.

agamic (agamous), asexual.

agar (agar-agar), a substance from certain red algae (*Gelidium* (Japan, USA), *Gracilaria* (USA), *Gigartina* (UK), *Pterocladia* (NZ), etc.) used to make culture media into gels which few microorganisms can liquefy. See Chapman (*Seaweeds and their uses*, 1950), Newton (*Seaweed utilization*, 1951), Humm (*Econ. Bot.* **1**: 317, 1947); a possible substitute using granulated tapioca or tapioca pearls (*Manihot esculenta*, cassava) has been proposed for use where agar is unavailable or prohibitively priced (Nene & Sheila, *Indian J. mycol. Pl. Path.* **24**: 159, 1994). Cf. gelatin, Media.

agaric, (1) one of the *Agaricales*; **fly -**, *Amanita muscaria*; **honey -**, *Armillaria mellea*; (2) (in early medicine, obsol.), species of *Fomes* or *Polyporus*; **female, white**, or **purging -** (agaricum), *F. officinalis*; **male -**, *Phellinus igniarius* (*F. igniarius*).

Agaricaceae Chevall. (1826), Agaricales. 42 gen. (+ 31 syn.), 591 spp. Basidioma with velar structures; spore colour varies but never rusty- or cinnamon-brown.

Agaricales, Basidiomycetes. 15 fam., 297 gen. (+ 253 syn.), *c.* 6,000 spp. Mushrooms and Toadstools, Gill Fungi, Agarics; terrestrial or lignicolous, saprobic, mycorrhizal (see Mycorrhiza), rarely pathogenic (e.g. *Armillaria* and *Marasmiellus* spp. on seed plants); edible (see Edible fungi, Mushroom culture), hallucinogenic, and poisonous (see Mycetism); cosmop.

  The mycelium, which is frequently seen in leaf mould and decaying wood, may be perennial (frequently as Fairy rings, q.v.); sometimes there are rhizomorphs (*Armillaria*) or sclerotia (*Collybia tuberosa*, *Agrocybe tuberosa*). The characteristic macroscopic basidioma (or basidiocarp) (Fig. 4) is fleshy, generally stipitate, and has a pileus bearing hymenium-covered lamellae on the underside. The young basidioma may be covered by a universal veil, which becomes broken down by the growth of the stipe and pileus, but part may remain as a volva at the base of the stipe and as fragments on the upper surface of the mature pileus. The developing hymenium may be covered by a partial veil which later becomes a cortina (as in *Cortinarius*) or an annulus around the mature stipe. There may be cystidia of various kinds, setae, or hyphidia among the basidia producing the 1-celled hyaline or coloured ballistospores, typically in fours.

  Classification. Fries in his *Systema* (1821-32) put almost all the agarics in the genus *Agaricus*, his 'tribes' (subgenera) being the common genera of today. Later he made genera of some of his 'tribes' but others, esp. Karsten, Kummer and Quélet, made most of the changes. Fries based his genera on macroscopic characters of the basidiocarp and spore colour and his system has been widely used and replaced with reluctance because it has the advantage of allowing both genera and species to be identified on field characters. Recent microscopic studies of basidiocarp structure and development have shown many of Fries's groupings to be unnatural and new families and genera have been proposed.

  In previous editions of this *Dictionary* the *Agaricales* was treated very broadly to include the polypores, *Cantharellales*, etc. The families now accepted approximate to those of Singer (1986), although genera of the *Boletales* (q.v.), *Cortinariales* (q.v.) and *Russulales* (q.v.) are excluded. Fams:

(1) **Agaricaceae**.
(2) **Amanitaceae**.
(3) **Bolbitiaceae**.
(4) **Coprinaceae**.
(5) **Entolomataceae** (syn. *Rhodophyllaceae*).
(6) **Galeropsidaceae**.
(7) **Hydnangiaceae**.
(8) **Hygrophoraceae**.
(9) **Pluteaceae**.
(10) **Podaxaceae**.
(11) **Richoniellaceae**.
(12) **Secotiaceae**.
(13) **Strophariaceae**.
(14) **Torrendiaceae**.
(15) **Tricholomataceae**.

  Lit.: See also under *Basidiomycetes*, Macromycetes.

  General: Josserand (*La description des champignons supérieurs*, 1952 (revised, 1983); technique, glossary), Reijnders (*Les problèmes du développement des carpophores des Agaricales et de quelques groupes voisins*, 1963), Douwes & v. Arx (*Acta Bot. Neerl.* **14**: 197, 1965; morphogenesis and structure), Bonnet (*BSMF* **75**: 215, 1959; chromatographic analysis), Disbery & Watling (*Mycopathologia* **32**: 81, 1967; histological and histochemical techniques), Heim (*TBMS* **30**: 161, 1948), Savile (*CJB* **33**: 60, 1955; phylogeny), Pegler & Young (*Beih. Nova Hedw.* **35**: 1971; spore morphology), Largent *et al.* (*How to identify mushrooms to genus* I-IV, 1977), Kühner (*Les Hyménomycètes agaricoïdes, études générales et classification*, 1980 (*Bull. Soc. Linn. Lyon, numéro spéciale*); classification lines). Moore *et al.* (*Developmental biology in the Higher Fungi* [agarics], 1985).

  Taxonomic: Lebedeva (*Opredelitel' shlyapochnykh gribov (Agaricales)*, 1949 [Russ.]; key), Pilát (*Agaricalium europaeorum clavis dichotomica*, 1951 [Czech]), Singer (*The Agaricales in modern taxonomy*, edn 4, 1986), Singer (*Keys for the determination of Agaricales*, 1962 [reprinted from edn 2 of preceding]), Singer (keys *A-Cantharellula*, *Sydowia* **30**: 192, 1978; *Catathelasma-Clitopilus*).

  Regional: see also under Macromycetes: **Africa**, Pegler (*Persoonia* **4**: 73, 1966; tropical); **—, East**, Pegler (*Kew Bull., addit. ser.* **6**, 1977). **Kenya**, Pegler & Rayner (*Kew Bull.* **23**: 347, 1969). **Morocco**, G. Malençon & R. Bertault (*Flore des champignons* *supérieurs du Maroc* **I**, 1970; **II**, 1975). **America, North**, Kauffman (*Agaricaceae of Michigan*, 2 vol., 1918 [reprint 1964]); —, **Cuba**, Pegler (*A revision of the Agarics of Cuba*, 1988); —, **Venezuela**, Dennis (*Kew Bull., addit. ser.* **3**, 1970); —, **Lesser Antilles**, Pegler (*Agaric Flora of the Lesser Antilles*, 1983); —, **South**, Singer & Digilio (*Lilloa* **25**, 1952), Singer (*Mycoflora Australis*, 1969); **Chile**, Garrido (*Index Agaricalium Chilensium*, 1985). **Asia, India**, Natarajan & Ramau (*S. Indian Agaricales*, 1983); **Sri Lanka**, Pegler, (*Agaric Flora of Sri Lanka*, 1986). **Australasia: Australia**, Pegler (*Austr. J. Bot.* **3**: 323, 1965; examination types 117 spp.). **New Zealand**, Stevenson (*Kew Bull.* **16**: 65, 227, 373, 1962-3; **19**, 1964), Horak (*N.Z. Jl Bot.* **9**: 403, 463, 1971). **Papua New Guinea**, Horak & Kobayasi (*Trans. mycol. Soc. Japan* **19**: 103, 1978). **Japan**, Ito (*Mycological flora of Japan* **II**, 1959). **Europe**, Fries (*Hymenomycetes Europaei*, 1874 [reprint 1963]), Kühner & Romagnesi (*Flore analytique des champignons supérièurs*, 1953; Compléments (9 papers), *Bibl. Mycol.* **56**, 1977), Moser (*Die Röhrlinge und Blätterpilze* [Gams, *Kl. KryptogFl.* **IIb**(2), 1978]). **British Isles**, Dennis *et al.* (*TBMS* **43**, Suppl., 1960; check list), Dennis (*Fungi of the Hebrides*, 1986); Henderson *et al.* (*British fungus flora, Part 1*, 1969; keys fam. gen., glossary, colour chart), Hora (*Reading Nat.* **15**, 1950; key 400 spp.). **France**, Gillet (*Hyménomycètes de France*, 1874-78), Quélet (*Flore mycologique de la France et des pays limitrophes*, 1888; *Champignons du Jura et des Vosges*, 1872-1902). **Germany**, Jahn (*Pilze rundum*, 1979, reprint). **Greenland**, Lange (*Meddr. Grønl.* **14** (7), 1955). **Italy**, Balletto (*Flora micologica analitica, funghi superiori*, 1972). **Portugal**, Da Camara (1956-58; **I**).

agaricic acid, a hydroxylated tribasic acid from *Fomes officinalis*; used to control tubercular night sweats (Milner, *Med. Klin.* **62**: 1443, 1967).

agaricicolous, living on agarics.

Agaricites Mesch. (1891), Fossil fungi. 4 (Tertiary, Quaternary), Eur.

Agarico-carnis Paulet (1793) = Fistulina (Fistulin.) fide Donk (1974).

Agaricochaete Eichelb. (1906), Tricholomataceae. 3, Afr. See Pegler (1977).

Agaricodochium X.J. Liu (1981), Mitosporic fungi, 3.A1.15. 1, China.

Agarico-igniarium Paulet (1793) = Fomes (Coriol.).

Agaricon Tourn. ex Adans. (1763) = Agaricum (Coriol.).

Agarico-pulpa Paulet (1793) = Agaricum (Coriol.) fide Donk (1974).

Agaricostilbaceae Oberw. & R. Bauer (1989), Agaricostilbales. 1 gen. (+ 1 syn.), 2 spp. Basidioma stilboid; spores statismosporic; multiple branching.

Agaricostilbales, Basidiomycetes. 1 fam., 1 gen. (+ 1 syn.), 2 spp. Fam.:
Agaricostilbaceae.
Lit.: Oberwinkler & Bauer (*Sydowia* **41**: 224, 1989).

Agaricostilbum J.E. Wright (1970), Agaricostilbaceae. 2, Argentina, India. See Wright *et al.* (*Mycol.* **73**: 880, 1981), Bauer *et al.* (*Syst. Appl. Microbiol.* **15**: 259, 1992; ultrastr.), Brady *et al.* (*TBMS* **83**: 540, 1984; nomencl.).

Agarico-suber Paulet (1793) = Daedalea (Coriol.) fide Donk (1974).

Agaricum Paulet (1812) ≡ Agaricon (Coriol.).

Agaricum P. Micheli ex Maratti (1822), Coriolaceae. 1, Eur. See Donk (*Proc. K. ned. Akad. Wet.* C **74**: 125, 1971).

Agaricus L. (1753), Agaricaceae. 200, cosmop. The genus was at one time used for almost all agarics.

A. campestris is the common or field mushroom; *A. bisporus* (= *A. brunnescens* fide Malloch *et al.*, *Mycol.* **68**: 912, 1976), the cultivated mushroom (see Mushroom cultivation). See Möller (*Friesia* **4**: 1-60, 135-200, 1950-52; Danish species, as *Psalliota*), Pilát (*Acta mus. nat. Prag.* **7**, 1951), Essette (*Atlas mycologiques I. Les Psalliotes*, 1964; keys 58 spp.), Cappelli (*Fungi Europaei 1. Agaricus*, 1984; keys 70 spp.), Wasser *et al.* (*Acta bot. Acad. Sci. Hung.* **22**: 249, 1976; infragen. taxa), Heinemann (*Sydowia* **30**: 6, 1978, keys 191 spp.; *Bull. Jard. bot. natn. belg.* **50**: 3, 1980, Malaysia, Indonesia); Freeman (*Mycotaxon* **8**: 1, 1979, N. Am. types; *Mycotaxon* **8**: 50, 1979, keys 40 S.E. USA spp.), Cappelli (*The genus Agaricus*, 1983 [*Fungi Europaei* **2**]).

Agaricus Raf. (1830) ? = Amanita (Amanit.).

Agaricus Murrill (1905) = Daedalea (Coriol.).

agaritine, an amino acid from *Agaricus bisporus*.

Agarwalia D.P. Tiwari & P.D. Agrawal (1974), Mitosporic fungi, 1.A2.3. 1 (from soil), India.

Agarwalomyces R.K. Verma & Kamal (1987), Mitosporic fungi, 2.A2.3. 1, India.

agglutinate, fixed together as if with glue.

agglutinin, see antigen.

aggregate, (1) (in taxonomy; 'agg.' or 'aggr.'), see species; (2) (in descriptions), near together, crowded.

aggregate plasmodium, see plasmodium.

Aglaocephalum Weston (1933) nom. nud. = Pulchromyces (Mitosp. fungi) fide Pfister *et al.* (*Mycotaxon* **1**: 137, 1974).

Aglaopisma De Not. ex Bagl. (1856) = Caloplaca (Teloschist.).

Aglaospora De Not. (1844) = Massaria (Massar.) fide Eriksson (*SA* **5**: 116, 1986).

Aglaothecium Groenh. (1962) nom. rej. prop. = Gyalidea (Solorinell.) fide Hafellner (*Beih. Nova Hedw.* **79**: 241, 1984).

Agmocybe Earle (1909) = Inocybe (Cortinar.) fide Kauffman (1918).

Agonimia Zahlbr. (1909), Verrucariaceae (L). 2 or 3, Eur. See Coppins & James (*Lichenologist* **10**: 179, 1978).

Agonimiella H. Harada (1993), Verrucariaceae (L). 1, Japan, Taiwan.

Agonium Oerst. (1844) nom. dub. (? Fungi or Cyanobacteria).

Agonomycetales (Mycelia sterilia), see Mitosporic fungi. 28 gen. (+ 30 syn.), 200 spp. True conidia absent, but non-dehiscent propagules (allocysts, bromatia, bulbils, chlamydospores, sclerotia etc.) produced in some genera. *Agonomycetes* may be states of basidiomycetes, ascomycetes or other mitosporic fungi. *Rhizoctonia* and *Sclerotium* include important plant pathogens.
Lit.: Watling (*in* Kendrick (Ed.), *The whole fungus* **2**: 453, 1979; states of basidiomycetes), v. Arx (*Genera of fungi sporulating in pure culture*, 1981; keys gen.), Domsch *et al.* (*Compendium of soil fungi*, 1980; identification, refs.).

Agostaea (Sacc.) Theiss. & Syd. (1915) = Anhellia (Myriang.) fide v. Arx (1963).

Agrestia J.W. Thomson (1961) = Aspicilia (Hymenel.) fide Weber (*Aquilo, Bot.* **6**: 43, 1967).

agroclavine, a clavine alkaloid (an intermediate in the biosynthesis of ergoline alkaloids) which is a major alkaloidal constituent of *Claviceps fusiformis* sclerotia. Cf. ergot.

Agrocybe Fayod (1889), Bolbitiaceae. 48, cosmop. See Singer (1978), Romagnesi (*BSMF* **78**: 337, 1963), Watling (*Fl. Illustr. Champ. Afr. centr.* **3**: 64, 1974, key 8 Afr. spp.; *British fungus flora* **3**: 1982, key 17 Br. spp.), Bon (*Bull. Féd. Mycol. Dauphiné-Savoie*

76: 32, 1980; keys), Singer (*Sydowia* **30**: 194, 1977; key world spp.).

Agrogaster D.A. Reid (1986), Bolbitiaceae. 1, NZ.

Agyriaceae Corda (1838), Lecanorales (± L). 5 gen. (+ 9 syn.), 12 spp. Thallus usually absent, sometimes with weakly developed immersed stromatic tissue; ascomata apothecial, sometimes elongated, often domed, without a well-developed outer wall; hymenium gelatinous, often blueing in iodine; interascal tissue of branched paraphyses with a well-developed epithecial layer; asci with a strongly thickened apical cap but sometimes without an ocular chamber, not blueing in iodine but rarely with an I+ outer gelatinous layer, usually polysporous; ascospores small, hyaline, aseptate. Anamorphs pycnidial. Saprobic on bark and wood, esp. on conifers.

Agyriales, see *Lecanorales*.

Agyriella Sacc. (1884), Mitosporic fungi, 3.A1-2.15. 1, Eur. See Ellis (*DH*).

Agyriella Ellis & Everh. (1897) ≡ Agyriopsis (Stictid.).

Agyriellopsis Höhn. (1903), Mitosporic fungi, 8,A1.15. 2, Eur.

Agyrina (Sacc.) Clem. (1909) = Steinia (Aphanopsid.) fide Nannfeldt (*Nova Acta R. Soc. Scient. upsal.* 4, **8**(2), 1932).

Agyriopsis Sacc. & Syd. (1899) = Schizoxylon (Stictid.) fide Sherwood (1977).

Agyrium Fr. (1822), Agyriaceae. 2, Eur.

Agyrona Höhn. (1909) = Molleriella (Elsin.) fide v. Arx (1963).

Agyronella Höhn. (1909) = Schizothyrium (Schizothyr.) fide v. Arx & Müller (1975).

Agyrophora (Nyl.) Nyl. (1896) = Umbilicaria (Umbilicar.).

Ahlesia Fuckel (1870) = Thelocarpon (Acarospor.) fide Poelt & Hafellner (1975).

Ahmadia Syd. (1939), Mitosporic fungi, 6.C1.15. 1, Pakistan.

Ahmadinula Petr. (1953) = Truncatella (Mitosp. fungi) fide Sutton (1977).

Ahtia M.J. Lai (1980) ≡ Cetrariopsis (Parmel.).

Ahtiana Goward (1986), Parmeliaceae (L). 1, N. Am.

AIDS. Acquired immunity deficiency syndrome. See Bossche *et al.* (Eds) (*Mycoses in AIDS patents*, 1989; infections by fungi in AIDS patients). See Medical and Veterinary mycology, *Pneumocystis*.

Aigialus Kohlm. & S. Schatz (1986), Massariaceae. 5 (marine, on *Rhizophora*), C. & N. Am., India. See Barr (*N. Am. Fl.* ser. 2, **13**, 1990; posn), Boise (*TBMS* **83**: 424, 1987; key 4 spp.), Hawksworth (*SA* **6**: 338, 1987; status).

Ainsworthia Bat. & Cif. (1962) [non Boiss. (1844), *Umbelliferae*] = Phaeosaccardinula (Chaetothyr.) fide v. Arx & Müller (1975).

Aipospila Trevis. (1857) = Lecania (Bacid.) fide Hafellner (1984).

Air pollution. Affects many foliicolous and stem fungi, and also lichen-forming species on all substrata. The algae or cyanobacteria in lichens are particularly sensitive to pollutants such as sulphur dioxide which disrupt membranes leading to chlorophyll breakdown. Lichens are arguably the most sensitive organisms to sulphur dioxide known, some being affected at mean levels of about 30 µg m^{-3}. Nylander (q.v.) suggested lichens could be used to monitor air quality in 1866 and there is now a vast literature on this subject. Fluorides are also highly toxic to lichens but particulate deposits (e.g. smoke), heavy metals, and photochemical smog components have less effect. Differential sensitivity due to physiological, structural, and chemical characters enables zones to estimate pollution levels to be constructed (Hawksworth & Rose, *Nature* **227**: 145, 1970; Gilbert, *New Phytol.*

69: 629, 1970); recolonization in response to falling sulphur dioxide levels can be dramatic (Hawksworth & McManus, *Bot. J. Linn. Soc.* **100**: 99, 1989; London); statistical and computer assisted approaches are increasingly used (e.g. Nimis *et al.*, *Stud. Geobot.* **11**, 1991).

Erysiphales and *Uredinales* are amongst the other most sensitive fungi; *Diplocarpon rosae* (Saunders, *Ann. appl. Biol.* **58**: 103, 1966) and *Rhytisma acerinum* (Bevan & Greenhalgh, *Environ. Pollut.* **10**: 271, 1976) can also be used as pollution monitors.

Leaf-dwelling yeasts (*Sporobolomyces, Tilletiopsis*) can be cultured and the density of sporing has been found to be directly related to acidic air pollution (Dowding, *in* Richardson, *Biological indicators of pollution*: 137, 1987). The amount of metal and radionuclides taken up by lichens can be used to map the extent of affected areas; this was especially so in relation to the 1986 Chernobyl disaster (Steinne *et al.*, *J. Environ. Radioact.* **21**: 65, 1993). Increases in lead contents from traffic, and falls since the introduction of unleaded fuel, are documented by Lawrey (*Bryologist* **96**: 339, 1993).

Lit.: Bates & Farmer (Eds) (*Bryophytes and lichens in a changing environment*, 1992), Ferry *et al.* (Eds) (*Air pollution and lichens*, 1973; incl. reviews effects on all plants and fungi), Hawksworth & Rose (*Lichens as pollution monitors*, 1976), Henderson (*Lichenologist* 1974-; twice-yearly bibl.), Nash & Wirth (Eds) (*Lichens, bryophytes and air quality*, [*Bibl. Lich.* **30**], 1988), Nieboer *et al.* (*in* Mansfield, 1976: 61; review sulphur dioxide toxicity), Richardson (*Bot. J. Linn. Soc.* **96**: 31, 1988; *Pollution monitoring with lichens*, 1992). See also Acid rain, Bioindication, Ecology, Index of Atmospheric Purity, lichen desert.

Air spora. Airborne particles of microbial, plant and animal origin are collectively referred to as the air spora or bioaerosol. Fungal spores are important components of the air spora. Knowledge of their occurrence in air has been revolutionized by use of continuously operating volumetric samplers (Hirst, *Ann. appl. Biol.* **39**: 257, 1952) out of doors and a realization of the importance of the sampling and collection efficiencies of different trapping methods in determining what is caught. The Hirst and subsequent Burkard traps have revealed the importance in the air spora of ascospores and basidiospores that were previously underestimated by using exposed horizontal sticky slides and open Petri dishes. Indoors, fungal spores are often abundant when stored products are handled but their sampling and enumeration require different methods from those used out of doors because of their smaller size and greater concentrations (see Cox & Wathes, *Bioaerosols handbook*, 1994).

Out of doors, fungal spores are almost always present in the air but their numbers and types depend on time of day, weather, season, geographical location and the nearness of large local spore sources. Total spore concentrations may range from fewer than 200 to 2 million m^{-3}. Spores of different species exhibit characteristic circadian periodicities in their occurrence in the air spora because they are liberated at different times of day determined by the method of their liberation (see Spore discharge and dispersal). Spores with active mechanisms requiring water are usually most numerous in the air at night, following dew formation, or rain; those dependent on drying are most numerous in the early morning as the sun dries foliage; and those released through mechanical disturbance occur during the middle of the day, when temperatures are highest and wind speeds, turbulence and convection are greatest. However, some discomycetes release their spores after sunrise, the species with large apothecia later than those with smaller, perhaps because drying is required to increase pressure on the asci. *Cladosporium* is the most numerous daytime spore type throughout most of the world although, in some seasons it may be exceeded by *Alternaria* in warm dry climates or by *Curvularia* or *Drechslera* in humid climates. At night time, ascospores, basidiospores and the ballistospores of *Sporobolomyces* and related 'mirror' yeasts become most numerous. Rain initially causes an increase in spore concentrations through 'tap and puff' (Hirst & Stedman, *J. gen. Microbiol.* **33**: 335, 1963), then washes spores from the air, and, afterwards, stimulates the release of ascospores. Large seasonal differences in spore concentrations occur in temperate regions, with few airborne spores in winter. In tropical regions, spores may be numerous all the year round although some types may be particularly favoured by wet or dry seasons. Growing crops form large sources of spores, especially of phytopathogenic fungi, whose occurrence may be correlated with crop growing seasons (see Lacey, *in* Cole & Kendrick (Eds), *Biology of conidial fungi*: 373, 1981). Sometimes, fungi pathogenic to humans can become airborne in dust in desert areas (e.g., *Coccidioides immitis*) or when deposits of guano beneath bird roosts are disturbed (*Histoplasma capsulatum*) (see also Medical mycology).

Indoors, numbers and types of airborne spores are determined by their source and, with stored products, the conditions in which they have been stored, the degree of disturbance of the substrate and the position and amount of ventilation. Concentrations of fungal spores may exceed 100 million m^{-3} air when mouldy hay and grain are handled, with *Aspergillus* and *Penicillium* spp. predominant. *Aspergillus fumigatus*, an opportunistic pathogen and frequent cause of asthma and mycotic abortion in cattle, may also be abundant. Concentrations of oyster mushroom (*Pleurotus ostreatus*) basidiospores may reach 27 million m^{-3} in growing sheds while up to 14 million m^{-3} *Penicillium* spores can be released when mouldy cork is handled. These concentrations may cause occupational allergies (see Allergy).

Lit.: Gregory (*Microbiology of the atmosphere*, 2nd edn, 1973), Edmonds (*Aerobiology, the ecological systems approach*, 1979), Dimmick & Akers (Eds) (*An introduction to experimental aerobiology*, 1969).

Aithaloderma P. Syd. (1913), ? Capnodiaceae. 4, trop. See Hughes (1976). Anamorphs *Asbolisia, Leptoxyphium, Microxyphium.*

Aithalomyces Woron. (1926) = Euantennaria (Euantennar.) fide Hughes (1972).

Aivenia Svrček (1977), Dermateaceae. 1, former Czechoslovakia.

Ajellomyces McDonough & A.L. Lewis (1968), Onygenaceae. 2, mainly trop. *A. dermatitidis* (anamorph *Blastomyces zymonema* (syn. *B. dermatitidis*); see blastomycosis), *A. capsulata* (anamorph *Histoplasma capsulatum*; see histoplasmosis). See Eriksson & Hawksworth (*SA* **11**: 58, 1992; nomencl.), *Emmonsiella.*

Ajrekarella Kamat & Kalani (1964), Mitosporic fungi, 8.A1.19. 1, India. See Sutton (*Mycopath. mycol. appl.* **33**: 76, 1967; redescr.).

Akanthomyces Lebert (1858), Mitosporic fungi, 3.A1.?. 4 (on insects), widespr. See Mains (*Mycol.* **42**: 566, 1950), Samson & Evans (*Acta bot. Neerl.* **23**: 28, 1974).

Akaropeltella M.L. Farr (1972) = Stomiopeltis (Micropeltid.) fide v. Arx & Müller (1975).

Akaropeltis Bat. & J.L. Bezerra (1961) [non *Acaropeltis* Petr. (1937)] ≡ Akaropeltella (Micropeltid.).

Akaropeltopsis Bat. & Peres (1966) ? = Stomiopeltis (Micropeltid.) fide v. Arx & Müller (1975).

akaryote (of *Plasmodiophoraceae*), the stage in the nuclear cycle before meiosis in which no or little chromatin is seen in the nucleus.

Akenomyces G. Arnaud (1954) nom. inval. (Mitosp. fungi). = Akenomyces G. Arnaud ex D. Hornby.

Akenomyces G. Arnaud ex D. Hornby (1984), Mitosporic fungi, 9.-.-. 1 (with clamp connexions), Eur.

akinete, (1) a non-motile reproductive structure; (2) a resting cell.

Akrophyton Lebert (1858) = Cordyceps (Clavicipit.) fide Tulasne & Tulasne (*Sel. fung. carp.* **3**: 4, 1865).

Alacrinella Manier & Ormières ex Manier (1968), Eccrinaceae. 2 (in *Isopoda*), France, USA. See Hibbets (*Syesis* **11**: 213, 1978; morphology, development), Lichtwardt (1986; key).

alate, winged.

Alatosessilispora K. Ando & Tubaki (1984), Mitosporic fungi, 1.G1.1. 1, Japan.

Alatospora Ingold (1942), Mitosporic fungi, 1.G1.15. 5 (aquatic), UK, USA. See Marvanová & Descals (*Bot. J. linn. Soc.* **91**: 1, 1985; key).

Albatrellopsis Teixeira (1993), Scutigeraceae. 1, Sweden, Portugal.

Albatrellus Gray (1821), Scutigeraceae. 12, N. temp. See Nuss (*Hoppea* **39**: 127, 1980; position). = Scutiger (Scutiger.) fide Donk (1960);

Albertiniella Kirschst. (1936), Pseudeurotiaceae. 1 (on *Ganoderma*), Eur., Japan. See Lundqvist (*Svensk bot. Tidskr.* **86**: 261, 1992). Anamorph *Phialemonium*-like.

Albigo Ehrh. ex Steud. (1824) ? = Sphaerotheca (Erysiph.).

Albocrustum Lloyd (1925) = Biscogniauxia (Xylar.) fide Læssøe (*SA* **13**: 43, 1994).

Alboffia Speg. (1899) = Corynelia (Corynel.) fide Fitzpatrick (*Mycol.* **12**: 239, 1920).

Alboffiella Speg. (1898) = Itajahya (Phall.).

Alboleptonia Largent & R.G. Benedict (1970), Entolomataceae. 4, mainly N. Am. See Largent & Benedict (*Mycol.* **62**: 440, 1970; key).

Albomyces I. Miyake (1908) = Aciculosporium (Clavicipit.).

Albophoma Tak. Kobay., Masuma, Omura & K. Watan. (1994), Mitosporic fungi, 4.A1.19. 1 (from soil), Japan.

Albosynnema E.F. Morris (1967), Mitosporic fungi, 2.C2.1. 2, Panama, Windward Isl.

Albotricha Raitv. (1970) = Lachnum (Hyaloscyph.).

Albuginaceae J. Schröt. (1893), Peronosporales. 1 gen. (+ 1 syn.), 30 spp.

Albugo (Pers.) Roussel ex Gray (1821), Albuginaceae. 30, cosmop. The 'White Blisters' or 'White Rusts'; *A. candida* (white blister of crucifers). See Biga (*Sydowia* **9**: 339, 1955; key), Pound & Williams (*Phytopath.* **53**: 1146, 1963; physiologic races *A. candida*).

Aldona Racib. (1900), Parmulariaceae. 3 (esp. on *Pterocarpus*), Indonesia, Dominican Republic, India. See Müller & Patil (*TBMS* **60**: 117, 1973; key).

Aldonata Sivan. & A.R.P. Sinha (1989), Parmulariaceae. 1, India.

Aldridgea Massee (1892) nom. dub. fide Donk (*Taxon* **6**: 18, 1957).

Aldridgiella, see *Aldrigiella.*

Aldrigiella Rick (1934) ? nom. dub. See Donk (*Taxon* **6**: 18, 1957).

ale, see beer.

Alectoria Ach. (1810), Alectoriaceae (L). 8, montaneboreal and bipolar. See Brodo & Hawksworth (*Op. bot. Soc. bot. Lund* **42**, 1977; key).

Alectoria Link (1833) = Usnea (Parmel.).
Alectoriaceae (Hue) Tomas. (1949), Lecanorales (L). 3 gen. (+ 5 syn.), 41 spp. Thallus fruticose, erect or pendant, the branches terete, smooth, hollow; ascomata apothecial, with a well-developed lecanorine margine, often absent; interascal tissue of narrow anastomosing pseudoparaphysis-like hyphae; asci elongate or clavate, strongly I+, with a well-developed apical cap, usually with fewer than 8 ascospores; ascospores large, thick-walled, brown, aseptate to muriform, usually with a distinct perispore. Pseudocyphellae often present.
  Lit.: Kärnefelt & Thell (*Pl. Syst. Evol.* **180**: 181, 1992).
Alectoriomyces Cif. & Tomas. (1953) = Alectoria (Alector.).
Alectoriopsis Elenkin (1929) = Ramalina (Ramalin.) fide Eriksson & Hawksworth (*SA* **6**: 112, 1987).
Alectorolophoides Batt. ex Earle (1909) = Cantharellus (Cantharell.).
alepidote, having no scales or scurf; smooth.
aleukia disease (alimentary toxic aleukia; ATA), see trichothecenes.
Aleuria Fuckel (1870), Otideaceae. 10, N. temp. See Wakefield (*TBMS* **23**: 281, 1939), Kaushal (*Mycol.* **68**: 1021, 1976; Indian spp.), Rifai (1968; Australian spp.), Moravec (*Česká Myk.* **26**: 74, 1972), Häffner (*Rheinl.-Pfälz. Pilzj.* **3**: 6, 1993).
Aleuria (Fr.) Gillet (1879) ≡ Peziza (Peziz.).
Aleuriaceae, see *Pezizaceae*.
Aleuriella P. Karst. (1871) = Mollisia (Dermat.) fide Saccardo (*Syll. Fung.* **8**: 1889).
Aleurina Massee (1898), Otideaceae. 10, widespr. See Zhuang & Korf (*Mycotaxon* **26**: 361, 1986; key).
Aleurina (Sacc.) Sacc. & P. Syd. (1902) = Peziza (Peziz.) fide Eckblad (1968).
aleuriospore (obsol.), formerly used for a thick-walled and pigmented but sometimes thin-walled and hyaline conidium developed from the blown-out end of a conidiogenous cell or hyphal branch from which it secedes with difficulty, as in *Aleurisma*, *Mycogone*, *Microsporum*; 'chlamydospore' sensu Hughes (1953); gangliospore. Since introduced by Vuillemin (1911), aleuriospore has been used in various senses, see Mason (1933, 1937) and Barron (1968), and finally rejected as a confused term (Kendrick, *Taxonomy of Fungi imperfecti*, 1971).
Aleurisma Link (1809) = Trichoderma (Mitosp. fungi) fide Hughes (1958); sensu Vuillemin (1911) = Chrysosporium (Mitosp. fungi) fide Carmichael (1962).
Aleurobotrys Boidin (1986), Aleurodiscaceae. 1, cosmop.
Aleurocorticium P.A. Lemke (1964) = Dendrothele (Cortic.) fide Lemke (1965).
Aleurocystidiellum P.A. Lemke (1964), Aleurodiscaceae. 2, widespr.
Aleurocystis Lloyd ex G. Cunn. (1956), Aleurodiscaceae. 2, widespr.
Aleurodiscaceae Jülich (1982), Stereales. 8 gen. (+ 4 syn.), 60 spp. Resupinate; monomitic or dimitic; dendrohyphidia often present; spores large, amyloid.
Aleurodiscus Rabenh. ex J. Schröt. (1888) nom. cons., Aleurodiscaceae. *c.* 30, cosmop. See Lemke (*CJB* **42**: 213, 1964; key 26 amyloid-spored spp.); 2 spp. are sometimes pathogenic to *Abies* and *Quercus*. See *Cyphella*.
Aleurodomyces Buchner (1912), ? Mitosporic fungi. 1 (on *Insecta*), ? Eur.
Aleurophora O. Magalh. (1916) = Aleurisma (Mitosp. fungi) fide Dodge (*Medical mycology*, 1935).

Aleurosporia Grigoraki (1924) = Trichophyton (Mitosp. fungi) fide Dodge (*Medical mycology*, 1935).
Alexopoulos (Constantine John; 1907-86). American university teacher (Michigan, Iowa, Texas). Wrote books on general mycology (see Literature) and myxomycetes (q.v.) which became standard texts. See Brodie (*Mycol.* **79**: 163, 1986), Blackwell (*TBMS* **90**: 153, 1988; portr.), Grummann (1974: 201), Stafleu & Mennega (*TL-2, Suppl.* **1**: 67, 1992).
algae (fungi as parasites and mutualists of), see Kohlmeyer (*Veröff. Inst. Meersforsch. Bremerh., Suppl.* **5**: 339, 1974), Kohlmeyer & Kohlmeyer (*Marine mycology*, 1979), Lichens, mycophycobiosis, photobiont.
algal-layer (of lichen thalli), the photobiont-containing layer (usually between the upper cortex and the medulla) of the thallus (Fig. 40).
algicolous, living on algae; **- fungi** see van Donk & Brumsz (*in* Reisser (Ed.), *Algae and symbiosis*: 567, 1992; review), algae.
Algincola Velen. (1939), ? Leotiales (inc. sed.). 1, former Czechoslovakia.
Algonquinia R.F. Castañeda & W.B. Kendr. (1991), Mitosporic fungi, 3.A1.12. 1, Canada.
Algorichtera Kuntze (1891) ≡ Scorias (Capnod.).
aliform, wing-like in form.
Alina Racib. (1909) [non Adans. (1763) nom. rej., *Menispermaceae*], Parodiopsidaceae. 1, Java. Anamorph *Septoidium*.
Alinocarpon Vain. (1928) = Thelocarpon (Acarospor.).
aliquot part, a portion that is contained an exact number of times in the whole; not the equivalent of 'sample' in which the concepts of both uniformity and representation are implicit (Emmons, *Bact. News* 1960: 17).
alkaphilic, used or organisms growing well at high pH values; e.g. *Fusarium* sp. at pH 10 (Hiura & Tanimura, *in* Horrikoshi & Grant (Eds), *Superbugs: microorganisms in extreme environments*: 287, 1991).
allantoid (esp. of spores), slightly curved with rounded ends; sausage-like in form (Fig. 37.8).
Allantomyces M.C. Williams & Lichtw. (1993), Legeriomycetaceae. 1 (in *Ephemeroptera*), Australia.
Allantonectella, see *Allonectella*.
Allantonectria Earle (1901) = Nectria (Hypocr.) fide Rossman *et al.* (*Mycol.* **85**: 685, 1993).
Allantoparmelia (Vain.) Essl. (1978), Parmeliaceae (L). 1, arctic.
Allantophoma Kleb. (1933) nom. inval. (Mitosp. fungi) fide Sutton (*Mycol. Pap.* **141**, 1977).
Allantophomopsis Petr. (1925), Mitosporic fungi, 8.A1.15. 2, widespr. See Carris (*CJB* **68**: 2283, 1990; generic revision).
Allantoporthe Petr. (1921), Melanconidaceae. *c.* 2, Eur., N. Am. See Barr (1978).
Allantosphaeriaceae, see *Diatrypaceae*.
Allantospora Wakker (1895) = Cylindrocarpon (Mitosp. fungi) fide Booth (1966).
Allantozythia Höhn. (1923) = Phlyctema (Mitosp. fungi). See Petrak (*Ann. Myc.* **27**: 370, 1929), Sutton (1977).
Allantozythiella Danilova (1951) = Endothiella (Mitosp. fungi) fide Sutton (1977).
Allantula Corner (1952), Pterulaceae. 1, Brazil.
Allarthonia (Nyl.) Zahlbr. (1903) = Arthonia (Arthon.).
Allarthoniomyces E.A. Thomas (1939) nom. inval. = Arthonia (Arthon.).
Allarthotheliomyces Cif. & Tomas. (1953) ≡ Allarthothelium (Arthon.).
Allarthothelium (Vain.) Zahlbr. (1908) = Arthonia (Arthon.).
Allelochaeta Petr. (1955) = Seimatosporium (Mitosp. fungi) fide Sutton (1977).

Alleppeysporonites Ramanujam & Rao (1979), Fossil fungi (*f. cat.*; Grallomyces). 1 (Miocene), India.

Allergy. An acquired, specific, altered capacity to react. It is acquired by exposure to allergenic particles; the sensitivity acquired from a single exposure is specific to one or to closely related species, although multiple exposures may result in multiple sensitivities; and subsequent re-exposure results in an altered capacity to react or allergic reaction. The form of that reaction depends on the nature of the allergenic particle, for instance, its size and chemical characteristics, the immunological reactivity of the subject and the circumstances of exposure. The two forms of allergy of most concern in this context are an immediate reaction, characterized by rhinitis and hay fever-like symptoms and a late reaction, characterized by alveolitis or pneumonitis. Fungal spores have been implicated as causative agents of both types of allergic reaction. Rhinitis and asthma are caused by normal everyday exposure to airborne allergens in subjects who are constitutionally predisposed (atopic) and who produce specific IgE antibodies against the allergen. Symptoms occur within a few minutes of exposure and may be provoked by 10^4 spores/m^{-3} air, or fewer, typically of fungi with spores larger than 10 µm. The spores may be components of the normal air spora, including *Cladosporium, Alternaria* and *Didymella*, or they may be associated with work environments, for instance cereal rusts and smuts and *Verticillium lecanii* when harvesting, *Agaricus bisporus* and *Boletus edulis* spores when preparing mushroom soup, and *Aspergillus flavus* and *A. awamori* from surface fermentations. Asthma may also be associated with exposure to fungal enzymes during their production. Allergic alveolitis occurs in non-atopic subjects after intense exposures to spores, typically 10^6-10^{10} spores/m^{-3}. At least 10^8 spores/m^{-3} may be required for sensitization but species differ in their antigenicity. Symptoms occur about 4 h after exposure and persist for 24-36 h if there is no further exposure. They include influenza-like symptoms, feverishness, chills, a dry cough, breathlessness and weight loss. With repeated exposure, breathlessness becomes increasingly severe and eventually permanent lung damage may occur with fibrosis and the increased load on the heart may lead to death. Specific IgG antibodies develop and may be an aid to diagnosis although implication of a fungus in the disease may require further tests. The disease is typically occupational and associated with poorly stored agricultural products. The classic form is farmer's lung, usually caused by thermophilic actinomycetes but sometimes by fungi, including *Aspergillus flavus, A. versicolor* and *Eurotium rubrum* (syn. *Aspergillus umbrosus*). Other forms of allergic alveolitis include cheese-washer's lung (*Penicillium casei*), malt-worker's lung (*Aspergillus clavatus, A. fumigatus*), suberosis (*Penicillium frequentans*), maple-bark stripper's lung (*Cryptostroma corticale*), sawmill worker's lung (*Rhizopus rhizopodiformis, Penicillium* spp., *Aspergillus fumigatus, Trichoderma viride*), sequoiosis (*Aureobasidium pullulans, Graphium* spp.), mushroom picker's lung (*Pleurotus ostreatus, Pholiota nameko, Aspergillus fumigatus, Doratomyces stemonitis*, and allergic alveolitis from citric acid fermentations (*Aspergillus fumigatus, A. niger, Penicillium* spp.). Mouldy lichens have also been reported to cause allergic alveolitis.

Allergic skin reactions may be caused by spores of the *Arthrinium arundinis* state of *Apiospora montagnei* in workers cutting the canes of *Arundo donax* in France, by contact with lichens in wood-cutters and people using lichens in decorations (Richardson, *in* Galun (Ed.), *CRC Handbook of lichenology* **3**: 98, 1988; review), and secondary to dermatophyte infections (see mycid).

For further information, see Pepys (*Hypersensitivity diseases of the lungs due to fungi and organic dusts*, 1969), Wilken-Jensen & Gravesen (*Atlas of moulds in Europe causing respiratory allergy*, 1984), Lacey (*in* Hawksworth (Ed.), *Frontiers in mycology*: 157, 1991), Lacey & Crook (*Ann. occup. Hyg.* **32**: 515, 1988), Lacey & Dutkiewicz (*J. Aerosol Sci.*, 1994).

Allescheria R. Hartig (1899) ≡ Hartigiella (Mitosp. fungi) fide Vuillemin (*Ann. Myc.* **3**: 341, 1905).

Allescheria Sacc. & P. Syd. (1899) = Monascus (Monasc.) fide Malloch (*Mycol.* **62**: 727, 1970).

Allescheriella Henn. (1897), Mitosporic fungi, 1.A2.1. 1, widespr. See Hughes (*Mycol. Pap.* **41**, 1951), Petrak (*Sydowia* **23**: 265, 1970).

Allescherina Berl. (1902) = Cryptovalsa (Ascomycota, inc. sed.) fide Clements & Shear (1931).

Allewia E.G. Simmons (1990) = Lewia (Pleospor.) fide Eriksson & Hawksworth (*SA* **9**: 2, 1991).

alliaceous, having a taste or smell of onions or garlic; cepaceous.

alliance, see phytosociology.

Alliospora Pim (1883) ? = Aspergillus (Mitosp. fungi). See Bisby (*TBMS* **27**: 101, 1944).

Allocetraria Kurok. & M.J. Lai (1991), Parmeliaceae (L). 3, China, India, Nepal, Taiwan. See Kärnefelt *et al.* (*Acta bot. fenn.* **150**: 79, 1994).

allochronic, occurring at different time periods, e.g. contemporary and fossil specimens.

allochrous (allochroous), changing from one colour to another.

allochthonous, transported to the place where found; not indigenous; cf. autochthonous.

Allochytridium Salkin (1970), Endochytriaceae. 2, N. Am.

allocyst, a chamydospore-like structure in *Flammula gummosa* (Kühner, 1946).

Allodium Nyl. (1896) = Chaenotheca (Coniocyb.).

Allodus Arthur (1906) = Puccinia (Puccin.) fide Arthur (1934).

Allographa Chevall. (1824) ? = Graphina (Graphid.).

Allomyces E.J. Butler (1911), Blastocladiaceae. 6 (in soil), widespr., esp. trop. See Emerson (*Lloydia* **4**: 77, 1941; life cycle, taxonomy), Teter (*Mycol.* **36**: 194, 1944; sexuality), Emerson & Wilson (*Mycol.*, **46**: 393, 1954; cytogenetics and cytotaxonomy), Taylor *et al.* (*Nature* **367**: 601, 1994; fossil from Devonian).

Allonecte Syd. (1939), Tubeufiaceae. 1, Ecuador. See Rossman (*Mycotaxon* **8**: 485, 1979; 1987).

Allonectella Petr. (1950), Ascomycota (inc. sed.). 1, Ecuador.

Allonema Syd. (1934), Mitosporic fungi, 1.A1.?. 1, Eur. Sartory & Meyer (*Ann. Myc.* **33**: 101, 1935).

Alloneottiosporina Nag Raj (1993), Mitosporic fungi, 4.C1.19. 2, USA, Australia.

allopatric, occurring in different geographical regions. Cf. sympatric.

Allophoron Nádv. (1942), Caliciales (inc. sed., L). 1, Colombia.

Allophylaria (P. Karst.) P. Karst. (1870), Leotiaceae. *c.* 5, Eur. See Arendholz (*Mycotaxon* **36**: 283, 1989; nomencl.), Carpenter (*Mem. N.Y. bot. Gdn* **33**: 17, 1981).

Allopuccinia H.S. Jacks. (1931) = Sorataea (Uropyxid.) fide Cummins & Hiratsuka (1983).

Allosoma Syd. (1926), Dothideales (inc. sed.). 1, C. Am. Anamorph *Periconiella*. See *Acrodesmis*.

Allosphaerium Link (1826) = Rhizoctonia (Mitosp. fungi) fide Saccardo (1901).

Allotelium Syd. (1939) = Diorchidium (Ravenel.) fide Cummins & Hiratsuka (1983).

Allothyriella Bat., Cif. & Nascim. (1959), Mitosporic fungi, 5.C2.?. 2, C. Am., Afr.

Allothyrina Bat. & J.L. Bezerra (1964), Mitosporic fungi, 5.C1.?. 1, Brazil.

Allothyriopsis Bat., Cif. & H. Maia (1959), Mitosporic fungi, 5.C2.?. 1, Ghana.

Allothyrium Syd. (1939), Asterinaceae. 1, Ecuador.

Almbornia Essl. (1981), Parmeliaceae (L). 2, S. Afr. See Brusse (*Mycotaxon* **40**: 265, 1991).

Almeidaea Cif. & Bat. (1962) [non Post & Kuntze (1903), *Rutaceae*] = Chaetothyrium (Chaetothyr.) fide v. Arx & Müller (1975).

Alnicola Kühner (1926) = Naucoria (Cortinar.) fide Reid (1984).

Alocospora J.C. Krug (1990), Xylariaceae. 1, Eur.

Aloysiella Mattir. & Sacc. (1908) = Antennularia (Ventur.) fide v. Arx & Müller (1975).

Alpakesa Subram. & K. Ramakr. (1954), Mitosporic fungi, 1.A1.1. 3, India. See Morgan-Jones *et al.* (*CJB* **50**: 877, 1972).

alpha-spore (**A-spore**, α-**spore**), a fertile, fusoid to oblong, biguttulate spore of an anamorph of the *Valsaceae* (*Phomopsis*). Cf. beta-spore.

Alphitomorpha Wallr. (1819) = Erysiphe (Erysiph.) fide Fries (*Syst. mycol.* **3**: 234, 1829).

Alphitomyces Reissek (1856) = Paecilomyces (Mitosp. fungi) fide Samson (1974).

alpine mycology, see Polar and alpine mycology.

Alpova C.W. Dodge (1931), Melanogastraceae. 13, N. temp. See Trappe (*Beih. Nova Hedw.* **51**: 279, 1975).

Alternaria Nees (1816), Anamorphic Pleosporaceae, 1.D2.26. Teleomorph *Lewia*. 50, widespr. *A. brassicae* (leaf spot of crucifers), *A. cucumerina* (cucurbit leaf spot; Jackson & Weber, *Mycol.* **51**: 401, 1959), *A. longipes*, and others, on tobacco, *A. solani* (early blight of potato) which produces the highly phytotoxic antibiotic alternaric acid (q.v.). Some species may cause onychomycosis (Wadhwani & Srivastava (*Mycopath.* **92**: 149, 1985). A number are common cosmop. saprobes.

Lit.: Neergaard (*Danish species of Alternaria and Stemphylium*, 1945), Joly (*BSMF* **75**: 149, 1959; speciation), Joly (*Le genre Alternaria*, 1964; 22 spp. + 94 uncertain or rejected), Simmons (*Mycol.* **59**: 73, 1967; typification), Simmons (*CMI Descriptions*, set 25, 1970; 10 plant pathogenic spp.), Rao (*Nova Hedw.* **17**: 219, 1969; 57 Indian spp.), Ellis (*DH* and *MDH*; ill. and descr. 44 spp.), Simmons (*Mycol.* **61**: 1, 1969; *Sydowia* **38**: 284, 1986; teleomorphs), Irwin *et al.* (*Austr. Jl Bot.* **34**: 281, 1986; Australian spp.), Tal & Robeson (*Z. Naturforsch.* **41**: 1032, 1986; production of pyrenocine), Simmons (*Mycotaxon* **13**: 16, 1981; **14**: 17, 44, 1982; **25**: 195, 203 and 287 [teleomorphs], 1986; **37**: 79, 1990, **46**: 171, 1993, and **48**: 91, 1993, **50**: 219 and 409; themes and variations), Simmons & Roberts (*Mycotaxon* **48**: 109, 1993, themes and variations), Klimešová & Prášil (*Novit. Bot. Univ. Carol.* **5**: 7, 1989; morphological variation in *A. alternata*), Tohyama & Tsuda (*Trans. mycol. Soc. Jap.* **31**: 501, 1990; spp. on *Cruciferae*), Samson & Frisvad (*Proc. Jap. Assoc. Mycotoxic.* **32**: 3, 1990; species concepts and mycotoxins), Müller (*Zentralbl. Mikrobiol.* **147**: 207, 1992; mycotoxins), Chelkowski & Visconti (eds) (*Alternaria: biology, plant diseases and metabolites*, 1992), Simmons in Chelkowski & Visconti (eds) (*Alternaria: biology, plant diseases and metabolites*, 1992), Rotem (*The Genus Alternaria*, 1994; biology, epidemiology, pathogenicity).

alternaric acid, a metabolite produced by *Alternaria solani* which inhibits spore germination in some fungi and causes wilting and necrosis in higher plants.

alternate host, one or other of the two unlike hosts of an heteroecious rust. See *Teliomycetes*.

alternation of generations, the succession of gametophyte and sporophyte or sexual and asexual phases in a life cycle: **homologous** when the two generations are like in form; **antithetic** if unlike, when the gametophyte is named the **protophyte** and the sporophyte the **antiphyte** (Celakovsky).

Althornia E.B.G. Jones & Alderman (1972), Thraustochytriaceae. 1, UK.

alutaceous, the colour of buff leather.

alveola, (1) a small surface cavity or hollow; (2) a pore of a polypore (obsol.).

Alveolaria Lagerh. (1892), Pucciniosiraceae. 2 (on *Cordia*), trop. Am.

alveolate, marked with ± 6-sided (honey-comb-like) hollows; faveolate.

Alveolinus Raf. (1815) nom. dub. (Fungi, inc. sed.). No spp. included.

Alveomyces Bubák (1914) = Uromyces (Puccin.) fide Nattrass (*First list Cyprus fungi*, 1937).

Alveophoma Alcalde (1952), Mitosporic fungi, 4.A1.10. 1, Spain. See Sutton (*TBMS* **47**: 497, 1964).

Alwisia Berk. & Broome (1873) = Tubifera (Lycogal.).

Alysia Cavalc. & A.A. Silva (1972) = Vouauxiella (Mitosp. fungi) fide Sutton (1980).

Alysidiopsis B. Sutton (1973), Mitosporic fungi, 1.A/B2.3. 1, Canada.

Alysidium Kunze (1816), Anamorphic Botryobasidiaceae, 1.A2.3. Teleomorph *Botryobasidium*. 2, Eur. See Ellis (*DH*).

Alysisporium Peyronel (1922) = Phragmotrichum (Mitosp. fungi) fide Sutton & Pirozynski (1966).

Alysphaeria Turpin (1827) ? Fungi (inc. sed., L).

Alytosporium Link (1824) nom. dub. fide Donk (*Taxon* **12**: 156, 1963). See also Stalpers (*Rev. Mycol.* **39**: 99, 1975).

Alyxoria Gray (1821) = Opegrapha (Roccell.).

AM, arbuscular mycorrhiza; see Mycorrhiza.

amadou, the context of *Fomes fomentarius* or *Phellinus igniarius* after the addition of saltpetre ($NaNO_3$); tinder; touchwood; punk.

Amallocystis Fage (1936) = Thalassomyces (Algae) fide Kane (1964).

Amallospora Penz. (1897), Mitosporic fungi, 3.G1.?. 1, Java.

Amandinea M. Choisy (1950), Physciaceae (L). 4, N. Hemisph., Antarctic.

Amanita Dill. ex Boehm. (1760) ≡ Agaricus (Agaric.).

Amanita Adans. (1763) nom. dub. (Agaricales, inc. sed.) fide Donk (1962).

Amanita Pers. (1797), Amanitaceae. 200, cosmop. See Gilbert (*Le Genre Amanita*, 1918 and Suppl.), Bresadola (*Icon. Mycol.* **27**, 1941), Bas (*Persoonia* **5**: 285, 1969; gen. account, keys 93 spp. Sect. *Lepidella*), Beeli (*Fl. Icon. Champ. Congo* **1**, 1935), Corner & Bas (*Persoonia* **2**: 241, 1962; Singapore, Malaya), Jenkins (*A taxonomic and nomenclatural study of the genus Amanita sect. Amanita for North America*, 1977), Courtillot & Staron (*Rev. Path. comp. med. exp.* **809-810**, 1971; poisonous spp., review), Reid (*Austr. J. Bot., Suppl.* **8**, 1979; Australia), Reid & Eicker (*MR* **95**: 80, 1991; key S. Afr. spp.), Garrido & Bresinsky (*Bot. Jahrb. Syst.* **107**: 521, 1985; key S. Am. spp.). See also Mycetism. (Fig. 4A-E).

amanita factor B, see pantherine; **- - C**, see ibotenic acid.

Amanitaceae R. Heim ex Pouzar (1983), Agaricales. 5 gen. (+ 23 syn.), 234 spp. Hymenophoral trama bilateral, divergent; spores white or cream.

Amanitaria J.-E. Gilbert (1940) = Amanita (Amanit.) fide Singer (1975).

Amanitella Earle (1909) = Amanita (Amanit.) fide Singer (1975).

Amanitella Maire (1913) = Limacella (Amanit.) fide Singer (1975).

amanitin, see amatoxins.

Amanitina J.-E. Gilbert (1940) = Amanita (Amanit.) fide Singer (1975).

Amanitopsis Roze (1876) nom. cons. = Amanita (Amanit.) fide Singer (1975).

Amarenographium O.E. Erikss. (1982), Mitosporic fungi, 8.D2.15. 1, Eur. See Nag Raj (*CJB* **67**: 3169, 1989; redescription).

Amarenomyces O.E. Erikss. (1981), ? Botryosphaeriaceae. 1, Eur.

Amastigis Clem. & Shear (1931) ≡ Amastigosporium (Mitosp. fungi).

Amastigomycètes, fungi lacking a motile phase; cf. *Mastigomycètes.*

Amastigomycota, the zygo-, asco-, and basidiomycetes (Whittaker, 1969).

Amastigosporium Bond.-Mont. (1921) = Mastigosporium (Mitosp. fungi) fide Hughes (*Mycol. Pap.* **36**, 1951).

amatoxins, cyclic octopeptides (including α- and β-amanitin, amanin, and the non-toxic amanillin) toxic to humans from *Amanita phalloides*, etc. See Wieland (*Science* **159**: 951, 1968), Wieland (*Peptides of poisonous Amanita mushrooms*, 1986). Cf. phallotoxins.

Amaurascopsis Guarro, Gené & De Vroey (1992), Ascodesmidaceae. 2, Burundi, Honduras.

Amauroascus J. Schröt. (1893), ? Onygenaceae. 7, widespr. See v. Arx (*Persoonia* **6**: 374, 1971), Currah (*in* Hawksworth (Ed.), *Ascomycete systematics*: 370, 1994).

Amaurochaete Rostaf. (1873), Stemonitidaceae. 5, cosmop.

Amauroderma Murrill (1905), Ganodermataceae. 20, trop. See Furtado (*Mem. N.Y. bot. Gdn* **34**: 1-109, 1980), Ryvarden & Johansen (1980, key 11 Afr. spp.), Corner (*Beih. Nova Hedw.* **75**: 45, 1983; keys S. Am. & Malaysian spp.).

Amauroderma (Pat.) Torrend (1920) = Amauroderma Murrill (Ganodermat.).

Amaurodon J. Schröt. (1888), Corticiaceae. 1, Eur.

Amaurohydnum Jülich (1978), Stereaceae. 1, Australia.

Amauromyces Jülich (1978), Stereaceae. 2, Australia, Réunion.

Amazonia Theiss. (1913) [non L. f. (1782) nom. cons., *Verbenaceae*], Meliolaceae. 10, Am., Asia, Afr. See Stevens (*Ann. Myc.* **25**: 405, 1927).

Amazoniella Bat. & H. Maia (1960) = Amazonia (Meliol.) fide Hughes (*in litt.*).

Amazonomyces Bat. & Cavalc. (1964) = Eremothecella (Arthon.) fide Sérusiaux (*SA* **11**: 39, 1992).

Amazonotheca Bat. & H. Maia (1959), Schizothyriaceae. 1, Philipp.

amber, for reports of fungi on arthropods in amber, including *Entomophthora* sp. on *c.* 25 million year old winged termite from Oligocene-Miocene (Dominican Republic), see Fossil fungi. See Poinar & Thomas (*Mycol.* **74**: 332, 1982) for lichens. See also Fossil fungi.

ambimobile, systemic fungicides which can move upward in the xylem or downward in the phloem.

ambiregnal (of organisms), ones that can be classified in more than one kingdom according to different systematic viewpoints; esp. of those which can potentially be treated under different *Codes.* See Nomenclature, Corliss (*BioSystems* **28**: 1, 1993), Patterson & Larsen (*Regnum veg.* **123**: 197, 1991).

Ambivina Katz (1974), Corticiaceae. 1, USA.

Amblyosporiopsis Fairm. (1922) = Oedocephalum (Mitosp. fungi) fide Clements & Shear (1931).

Amblyosporium Fresen. (1863), Mitosporic fungi, 1.A1.40. 3, Eur. *A. botrytis* (on agarics, esp. *Lactarius*). See Nicot & Durand (*BSMF* **81**: 623, 1966), Pirozynski (*CJB* **47**: 325, 1969).

Ambrodiscus S.E. Carp. (1988), Leotiales (inc. sed.). 1, USA.

ambrosia fungi, fungi, e.g. yeasts, *Ambrosiozyma*, *Ascoidea* and *Dipodascus* spp., etc., that grow in tunnels of ambrosia beetles (wood-boring *Scolytidae*) and serve as food for larvae and adults; many are specific for the particular insects (Batra, *Trans. Kansas Acad. Sci.* **66**: 213, 1963; *Mycol.* **59**: 981, 1968; key gen.); **- gall,** see gall.

Ambrosiaemyces Trotter (1934), Fungi (inc. sed.). 1 (on wood damaged by ambrosia beetles), Sri Lanka.

Ambrosiella Brader (1964), Mitosporic fungi, 2.A1.1/38. 9, widespr. See Batra (*Mycol.* **59**: 986, 1968; key).

Ambrosiozyma Van der Walt (1972), ? Endomycetaceae. 4 (2 fide Barnett *et al.*, 1990), widespr. See Goto & Takami (*J. gen. appl. Microbiol.* **32**: 271, 1986), Kreger van Rij (1984), v. Arx (1977), ambrosia fungi.

Ameghiniella Speg. (1888), Leotiaceae. 2, N. & S. Am. See Zhuang (*Mycotaxon* **31**: 261, 1988; key), Gamundí (*MR* **95**: 1131, 1991).

amend, the act and result of making an alteration, not necessarily to correct a fault or error. Cf. emend.

Amepiospora Locq. & Sal.-Cheb. (1980), Fossil fungi. 5, Cameroon.

Ameris Arthur (1906) = Phragmidium (Phragmid.) fide Arthur (1934).

Amerobotryum Subram. & Natarajan (1976) = Agaricostilbum (Agaricostilb.) fide Subramanian & Natarajan (*Mycol.* **69**: 1224, 1977).

Amerodiscosiella M.L. Farr (1961), Mitosporic fungi, 5.A1.15. 1, Cambodia, Brazil. See Sutton (*TBMS* **60**: 525, 1973), Nag Raj (*CJB* **53**: 2435, 1975), Farr (*Taxon* **26**: 580, 1977; typification).

Amerodiscosiellina Bat. & Cavalc. (1966), Mitosporic fungi, 5.A1.?. 1, Brazil.

Amerodothis Theiss. & Syd. (1915) = Botryosphaeria (Botryosphaer.) fide v. Arx & Müller (1954).

Ameromassaria Hara (1918), Ascomycota (inc. sed.). 1, Japan.

Ameropeltomyces Bat. & H. Maia (1967), Mitosporic fungi (L), 5.A1.?. 1, Brazil.

amerospore, a 1-celled (i.e. non-septate) spore with a length/width ratio < 15:1 (cf. scolecospore); if elongated, axis single and not curved through more than 180° (cf. helicospore); any protuberances < ¹/₄ spore body length (cf. staurospore). See Mitosporic fungi.

Amerosporiella Höhn. (1916) nom. illegit. Mitosporic fungi, 1.A1/2.?. 1, Eur.

Amerosporina (Petr.) Petr. (1965) = Amerosporium (Mitosp. fungi) fide Sutton (1980).

Amerosporiopsis Petr. (1941), Mitosporic fungi, 4.A1.15. 1, Iran. See Sutton (1980).

Amerosporis Clem. & Shear (1931) ≡ Amerosporiella (Mitosp. fungi).

Amerosporium Speg. (1882), Mitosporic fungi, 8.A2.15. 2, widespr. See Sutton (1980).

Amerostege Theiss. (1916) ? Fungi (inc. sed.) fide v. Arx & Müller (*Beitr. Krypt. fl. Schw.* **11**, 1954).

Amethicium Hjortstam (1983), Corticiaceae. 1, Tanzania.

ametoecious, see autoecious (q.v.; de Bary).

Amicodisca Svrček (1987), Hyaloscyphaceae. 1, Eur.

Amidella J.-E. Gilbert (1940) = Amanita (Amanit.) fide Singer (1975).

amixis, see heterothallism.

ammonia fungi, a chemoecological group in which reproductive structures develop after the addition of ammonia, urea, etc. or alkalis to the soil (Sagara, *Contrib. biol. Lab. Kyoto Univ.* **24**: 205, 1975).

Amoebidiaceae J.L. Licht. (1916), Amoebidiales. 2 gen., 10 spp. Thalli unbranched, non-septate with a basal, secreted holdfast; holocarpic, producing either rigid or amoeboid sporangiospores; released amoebae encyst and produce infestive cytospores; sexual reproduction unknown. Ectocommensals (*Amoebidium*) or endocommensals (*Paramoebidium*) of freshwater *Arthropoda*.

Amoebidiales, Trichomycetes. 1 fam., 2 gen., 10 spp. Fam.:
Amoebidiaceae.
  Lit.: Manier (*Ann. Sci. nat., Bot.* sér. 12, **10**: 565, 1969; taxonomy), Trotter & Whisler (*CJB* **43**: 869, 1965; wall structure), Whisler (*J. Protozool*. **13**: 183, 1966; *Devel. Biol.* **17**: 562, 1968; culture and development), Moss (*in* Batra (Ed.), *Insect-fungus symbiosis*: 175, 1979), Lichtwardt (1986).

Amoebidium Cienk. (1861), Amoebidiaceae. 5 (on *Crustacea* and *Insecta*), widespr. See Whisler (*Am. J. Bot.* **49**: 193, 1962; life cycle, culture & nutrition; *Devel. Biol.* **17**: 562, 1968; development), Whisler & Fuller, *Mycol.* **60**: 1068, 1968; ultrastr.), Trotter & Whisler (*CJB* **43**: 869, 1965; wall), Lichtwardt (1986; taxonomy, biology), Lichtwardt & Williams (*Mycol.* **84**: 376, 1992). Cf. *Paramoebidium*.

Amoeboaphelidium Scherff. (1925) , Protozoa, see monads.

Amoebochytrium Zopf (1884), Cladochytriaceae. 1, Eur.

amoeboid, not having a cell wall and changing in form, like an amoeba.

amoeboid cell (of *Ameobidiales*), uninucleate cells, formed by protoplasmic cleavage within the fungal thallus, which lack a rigid wall and when released usually encyst, the cysts, in time, producing cystospores.

Amoebomyces Bat. & H. Maia (1965), Mitosporic fungi (L), 5.C1.?. 1, Brazil.

Amoebophilus P.A. Dang. (1910), Cochlonemataceae. 2, Eur., N. Am. See Drechsler (*Mycol.* **27**: 33, 1935, **51**: 787, 1959), Barron (*CJB* **61**: 3091, 1983).

Amoebosporus Ivimey Cook (1933), ? Chytridiales (inc. sed.). 2, W. Indies.

Amogaster Castellano (1995), ? Rhizopogonaceae. 1, USA.

Amorphomyces Thaxt. (1893), Laboulbeniaceae. 11, widespr.

Amorphotheca Parbery (1969), Ascomycota (inc. sed.). 1, widespr. *A. resinae* (anamorph *Hormoconis resinae*; kerosene fungus, q.v.). See Sheridan *et al.* (*Tuatara* **19**: 130, 1972), Redhead & Malloch (1977), Domsch *et al.* (1980).

Amorphothecaceae Parbery (1969), Ascomycota (inc. sed.). 1 gen., 1 sp. Stromata absent; ascomata cleistothecial, dark brown to black, ± globose; peridium amorphous, sometimes incorporating a few hyphae, often variable in thickness, evanescent; interascal tissue absent; asci irregularly arranged, ± globose, very thin-walled, evanescent; ascospores usually pale brown, aseptate, thin-walled, without germ pores, without a sheath. Anamorph *Hormoconis*. In soil, or associated with hydrocarbons, widespr.

Amparoina Singer (1958), Tricholomataceae. 2, S. Am. See Singer (*Rev. Mycol.* **40**: 57, 1976).

Ampelomyces Ces. ex Schltdl. (1852), Mitosporic fungi, 4.A2.15. 1 (on *Erysiphales*), widespr.

amphi- (prefix), the two (sorts, sides).

Amphiblistrum Corda (1837) = Oidium (Mitosp. fungi) fide Linder (*Lloydia* **5**: 165, 1942).

Amphichaeta McAlpine (1904) = Seimatosporium (Mitosp. fungi) fide Shoemaker (1964).

Amphichaete Kleb. (1914) ≡ Amphichaetella (Mitosp. fungi).

Amphichaetella Höhn. (1916), Mitosporic fungi, 3.A1.?. 1, Eur. See Morgan-Jones (*CJB* **51**: 1431, 1973).

Amphichorda Fr. (1825) = Isaria (Mitosp. fungi) fide Fries (*Syst. mycol.* **3**, 1832).

Amphiciliella Höhn. (1919) nom. dub. fide Sutton (*Mycol. Pap.* **141**, 1977).

Amphiconium Nees (1816) nom. dub.; based on algae fide Fries (*Syst. mycol.* **3** (index): 51, 1832).

Amphicypellus Ingold (1944), Chytridiaceae. 1, Eur., Am. = Chytriomyces (Chytrid.) fide Dogma (*Kalikasan* **5**: 136, 1976).

Amphicytostroma Petr. (1921), Anamorphic Valsaceae, 8.A1.15. Teleomorph *Amphiporthe*. 1, Eur.

Amphididymella Petr. (1928) = Acrocordia (Monoblast.) fide Yue & Eriksson (*Mycotaxon* **24**: 293, 1985).

Amphidium Nyl. (1891) [non Schimp. (1856), *Musci*] = Epiphloea (Hepp.) fide Gyelnik (1940).

Amphiernia Grüss (1926) = Sporobolomyces (Mitosp. fungi) fide Derx (*Ann. Myc.* **28**: 1, 1930).

amphigenous, making growth all round or on two sides.

amphigynous (of *Pythiaceae*), having an antheridium through which the oogonial incept grows (Fig. 28F).

Amphiloma Körb. (1855) ≡ Gasparrinia (Teloschist.).

Amphiloma Nyl. (1855) = Lepraria (Mitosp. fungi).

Amphilomopsis Jatta (1905) = Chrysothrix (Chrysothric.).

amphimixis, copulation of two cells and nuclei which are not near relations, e.g. egg and sperm; cf. apomixis, automixis and pseudomixis.

Amphimyces Thaxt. (1931), Laboulbeniaceae. 1, W. Afr.

Amphinectria Speg. (1924) ? Fungi (inc. sed., L). See Rossman (*Mycol. Pap.* **157**, 1987).

Amphinema P. Karst. (1892), Atheliaceae. 4, widespr.

Amphinomium Nyl. (1888) = Pannaria (Pannar.) fide Galloway & Jørgensen (*Lichenologist* **19**: 345, 1987).

Amphiporthe Petr. (1971), Valsaceae. 3, Eur., N. Am. See Barr (*Mycol. Mem.* **7**, 1978). Anamorph *Amphicytostroma*.

Amphischizonia Mont. (1856) nom. inval. = Cryptodictyon (Lecid.) fide Santesson (*Symb. bot. upsal.* **12**(1), 1952).

Amphisphaerella (Sacc.) Kirschst. (1934), Amphisphaeriaceae. 3, Eur. See Eriksson (*Svensk bot. Tidskr.* **60**: 315, 1966).

Amphisphaerellula Gucevič (1952), Ascomycota (inc. sed.). 1, former USSR.

Amphisphaeria Ces. & De Not. (1863) nom. cons., Amphisphaeriaceae. 80, widespr.

Amphisphaeriaceae G. Winter (1885), Xylariales. 36 gen. (+ 22 syn.), 150 spp. Stromata crustose, often clypeate, rarely absent; ascomata perithecial, immersed or erumpent, ± globose, the ostiole papillate, periphysate; interascal tissue of numerous thin-walled paraphyses; asci cylindrical, persistent, with a small, usually I+ apical ring; ascospores hyaline to brown, elongate, usually transversely septate, often with germ pores. Anamorphs coelomycetous, usually acervular. Saprobic or necrotrophic in leaves, stems etc., widespr.
  Lit.: Samuels *et al.* (*Mycotaxon* **28**: 473, 1987; anamorphs).

Amphisphaerina Höhn. (1919), Ascomycota (inc. sed.). 3, Eur., N. Am.

amphispore, a second, special type of urediniospore; see Uredinales.

Amphisporium Link (1815) = Didymium (Didym.).

amphithallism, see homothallism.

amphithecium, the thalline margin of an apothecium (L).

Amphitiarospora Agnihothr. (1963) = Dinemasporium (Mitosp. fungi) fide Sutton (1977).

amphitrichous (amphitrichiate), having one flagellum at each pole.

Amphitrichum T. Nees (1818) nom. dub. fide Hughes (*CJB* **36**: 727, 1958).

Amphobotrys Hennebert (1973), Anamorphic Sclerotiniaceae, 1.A1/2.7. Teleomorph *Botryotinia*. 1, USA.

Amphopsis (Nyl.) Hue (1892) = Pyrenopsis (Lichin.).

Amphoridium A. Massal. (1852) = Verrucaria (Verrucar.).

Amphoridium Servít (1955), see Amphoridium A. Massal. Cited in *ING*, but an incorrect entry.

Amphoroblastia Servít (1953) = Polyblastia p.p., Thelidium p.p., Verrucaria p.p. (Verrucar.).

Amphoromorpha Thaxt. (1914) = Basidiobolus (Entomophth.) fide Blackwell & Malloch (*Mycol.* **81**: 735, 1989).

Amphoropsis Speg. (1918) ? = Pyxidiophora (Pyxidiophor.) fide Blackwell (*Mycol.* **86**: 1, 1994). = Basidiobolus (Entomophthor.) fide Blackwell & Malloch (*Mycol.* **81**: 735, 1989).

Amphoropycnium Bat. (1963), Mitosporic fungi, 4.A1.15. 2, Brazil, Philipp.

Amphorula Grove (1922) = Chaetoconis (Mitosp. fungi) fide Sutton (1968). See also Petrak (*Sydowia* **13**: 180, 1959).

Amphorulopsis Petr. (1959), Ascomycota (inc. sed.). 1, former Yugoslavia.

amphotericin A and B, polyene antibiotics from actinomycetes (*Streptomyces* spp.); antifungal; - **B** (fungizone) is used in the therapy of systemic mycoses of humans.

Amplariella J.-E. Gilbert (1940) = Amanita (Amanit.) fide Singer (1975).

amplectant, covering; embracing.

ampliate, made greater; enlarged.

ampoule effect, Corner's (*New Phytol.* **47**: 48, 1948) term for the normal working of a basidium which is compared to an ampoule from which the contents are discharged into the basidiospores by the enlargement of a basal vesicle.

ampoule hypha, see hypha.

ampulla, (1) the swollen tip of a conidiogenous cell which produces synchronous blastic conidia (as in *Gonatobotryum*); (2) a conidiophore which develops a number of short branches or discrete conidiogenous cells (as in *Aspergillus*).

Ampullaria A.L. Sm. (1903) [non Couch (1963), *Actinomycetes*] = Melanospora (Ceratostomat.) fide Cannon & Hawksworth (1982).

Ampullariella Couch (1964), *Actinomycetes*, q.v.

Ampullifera Deighton (1960), Mitosporic fungi, 1.A1.4. 6 (on foliicolous lichens), Afr., Brazil. See Hawksworth (*Bull. Br. Mus. nat. hist., Bot.* **6**: 183, 1979).

Ampulliferella Bat. & Cavalc. (1964) = Ampullifera (Mitosp. fungi) fide Hawksworth (1979).

Ampulliferina B. Sutton (1969), Mitosporic fungi, 1.B2.38. 1, Canada.

Ampulliferinites Kalgutkar & Sigler (1995), Fossil fungi (Mitosp. fungi). 1 (Eocene), Canada.

Ampulliferopsis Bat. & Cavalc. (1964) = Ampullifera (Mitosp. fungi) fide Hawksworth (1979).

ampulliform, flask-like in form (Fig. 37.30).

Ampullina Quél. (1875) = Leptosphaeria (Leptosphaer.) fide v. Arx & Müller (1975).

Amygdalaria Norman (1852), Porpidiaceae (L). 3, temp. See Inoue (*J. Hattori bot. Lab.* **56**: 321, 1984; key), Brodo & Hertel (*Herzogia* **7**: 493, 1987; key 8

spp.), Esnault & Roux (*An. Jard. bot. Madrid* **44**: 211, 1987). ? = Porpidia (Porpid.) fide Purvis *et al.* (*Lichen flora of Great Britain and Ireland*, 1992).

Amylaria Corner (1955), Amylariaceae. 1, Bhutan.

Amylariaceae Corner (1970), Bondarzewiales. 1 gen., 1 sp. Basidiomata clavarioid; spores amyloid, ridged.

Amylascus Trappe (1971), Pezizaceae. 2 (hypogeous), Australasia. See Trappe (*TBMS* **65**: 496, 1975; key).

Amylirosa Speg. (1920) nom. dub. (Dothideales, inc. sed.) fide v. Arx & Müller (1975).

Amylis Speg. (1922), Ascomycota (inc. sed.). 1, S. Am.

Amylo process (Amylomyces process), a method for the commercial production of alcohol by the saccharification of starchy materials by *Amylomyces rouxii* or *Rhizopus* spp. See Erb & Hildebrandt (*Industr. engin. Chem.* **38**: 792, 1946), Hesseltine (*Mycologist* **5**: 166, 1991).

Amyloathelia Hjortstam & Ryvarden (1979), Amylocorticiaceae. 3, Eur., S. Am.

Amylobasidium Ginns (1988), Hyphodermataceae. 1, USA.

Amylocarpus Curr. (1859), Ascomycota (inc. sed.). 1, Eur. See Currah (*Mycotaxon* **24**: 1, 1985).

Amylocorticiaceae Jülich (1982), Stereales. 6 gen. (+ 1 syn.), 11 spp. Basidioma resupinate; dendrohyphidia present; spores small, amyloid.

Amylocorticium Pouzar (1959), Amylocorticiaceae. 10, widespr.

Amylocystis Bondartsev & Singer (1944), Coriolaceae. 1, Eur.

Amylodontia M.I. Nikol. (1967), Hydnaceae. 1, former USSR.

Amyloflagellula Singer (1966), Tricholomataceae. 2, trop. Am., Asia.

Amylohyphus Ryvarden (1978), Stereaceae. 1, Rwanda.

amyloid (of asci, spores, etc.), stained blue by iodine (see Iodine, Stains); cf. dextrinoid. See Dodd & McCracken (*Mycol.* **64**: 1341, 1972; nature of fungal starch), amylomycan, lichenan.

amylomycan, a name proposed for the I+ blue or red compounds associated with asci (Common, *Mycotaxon* **41**: 67, 1991).

Amylomyces Calmette (1892), Mucoraceae. 1, Asia. See Ellis *et al.* (*Mycol.* **68**: 131, 1976).

Amylonotus Ryvarden (1975), Auriscalpiaceae. 3, trop.

Amylophagus Scherff. (1925), Monad (q.v.).

Amyloporia Singer (1944), Coriolaceae. 5, widespr. fide Pegler (*in litt.*). = Poria (Coriol.) fide Lowe (1966).

Amyloporiella A. David & Tortič (1984), Coriolaceae. 5, Eur. N. Am. See David & Tortič (*TBMS* **83**: 659, 1984; key).

Amylora Rambold (1994), Rimulariaceae (L). 1, Austria, Italy, Switzerland.

Amylosporomyces S.S. Rattan (1977), Gloeocystidiellaceae. 3, widespr.

Amylosporus Ryvarden (1973), Auriscalpiaceae. 2, trop.

Amylostereum Boidin (1958), Stereaceae. 5, widespr.

Amylotrogus Roze (1896) nom. dub. (Myxomycetes).

an-, see a-.

anaerobe, an organism able to grow without free oxygen. An **obligate -** grows only without free oxygen; a **facultative -** grows with or without free oxygen. See Zehnder (Ed.) (*Biology of anaerobic micro-organisms*, 1988).

Anaerobic fungi, fungi able to grow only in the absence of oxygen, were unknown until 1975 when motile cells in the rumen of sheep were found to be zoospores of an obligately anaerobic fungus, *Neocallimastix frontalis* (Orpin, *J. gen. microbiol.* **91**: 249, 1975). These fungi are now recognized as participating in the digestive processes of both foregut- (ruminants; e.g. cattle, deer, kangaroo, reindeer, sheep) and hindgut-fermenting herbivores (e.g. elephant, horse, rhinoceros, zebra). The genera so far identified include also *Anaeromyces*, *Caecomyces*, *Piromyces* and *Orpinomyces*. See Theodorou *et al.* (*in* Carroll & Wicklow, *The fungal community*, edn 2 : 43, 1992), Trinci *et al.* (*MR* **98**: 129, 1994; review, bibliogr.).

Anaeromyces Breton, Bernalier, Dusser, Fonty, B. Gaillard & Guillot (1990), Neocallimastigaceae. 2, France, Australia.

analogous, showing a resemblance in form, structure, or function which is not considered to be evidence of evolutionary relatedness; cf. homologous.

anamorph, (1) (of shapes), a deformed figure appearing in proportion when correctly viewed; (2) (of fungi), see States of fungi.

Anamylopsora Timdal (1991), ? Trapeliaceae (L). 1, Alaska, ? Eur., Asia.

anaphylaxis, manifestation of a change (immediate hypersensitivity) in a living animal from the uniting of an antibody with its antigen which may result in the death of the animal; cf. allergy.

anaphysis, a thread-like conidiophore persisting in apothecia of *Ephebe*.

Anaphysmene Bubák (1906), Mitosporic fungi, 6.B1.19. 2, Eur., Guatemala. See Sutton (*TBMS* **59**: 285, 1972), Sutton & Hodges (*Mycol.* **82**: 313, 1990).

Anaptychia Körb. (1848), Physciaceae (L). *c.* 10, widespr. See Kurokawa (*Beih. Nova Hedw.* **6**, 1962), Poelt (*Nova Hedw.* **9**: 21, 1965), Kurokawa (*J. Hattori bot. Lab.* **37**: 563, 1973), Swinscow & Krog (*Lichenologist* **8**: 103, 1976; Africa), Kashiwadani *et al.* (*Bull. Natn. Sci. Mus.*, B, **16**: 147, 1990; 23 spp., Peru).

Anaptychiomyces E.A. Thomas (1939) nom. inval. ≡ Anaptychia (Physc.).

Anapyrenium Müll. Arg. (1880) nom. conf. = Buellia (Physc.) p.p. fide Eriksson (1981).

Anarhyma M.H. Pei & Z.W. Yuan (1986), Mitosporic fungi, 8.D2.1. 1, China.

Anariste Syd. (1927), Asterinaceae. 1, C. Am.

Anastomaria Raf. (1820), Boletales (inc. sed.). 1, Eur.

anastomosing, joining irregularly to give a vein-like network.

anastomosis (pl. **anastomoses**), the fusion between branches of the same or different hyphae (or other structures) to make a network.

Anastrophella E. Horak & Desjardin (1994), Tricholomataceae. 2, NZ, Hawaii.

Anatexis Syd. (1928) = Englerula (Englerul.) fide Müller & v. Arx (1962).

Anatolinites Elsik, V.S. Ediger & Bati (1990), Fossil fungi. 1 (Eocene-Holocene), China, Turkey, USA.

Anatomy, see Lohwag (*Anatomie der Asco-und Basidiomyceten*, 1941 [reprint, 1965]; for details of English translation see *Mycol.* **39**: 249, 1947), Talbot (*Bothalia* **6**: 1, 1951; lower *Basidiomycetes*), Reijnders (*Les problèmes du développement des carpophores des Agaricales et de quelques groupes voisins*, 1963). See also Langeron (1952), hyphal analysis, Lichens, tissue types.

Anavirga B. Sutton (1975), Anamorphic Vibrisseaceae, 1.G2.1/10. Teleomorph *Vibrissea*. 2, Eur. See Hamad & Webster (*Sydowia* **40**: 60, 1988).

anbury, see club root.

Ancistroporella G. Thor (1995), Roccellaceae (L). 1, Australia.

Ancistrospora G. Thor (1991) [non C.A. Menéndez & Azcuy (1972), fossil sporae-dispersae] ≡ Ancistroporella (Roccell.).

Anconomyces Cavalc. & A.A. Silva (1972), Mitosporic fungi (L), 8.E1.?. 1, Brazil.

Ancoraspora Mig. Rodr. (1982), Mitosporic fungi, 1.C2.1. 1, Cuba.

Ancylistaceae J. Schröt. (1893), Entomophthorales. 3 gen. (+ 2 syn.), 32 spp. Sporophores without subsporangial vesicle; nucleus difficult to observe during mitosis, nucleolus prominent.
  Lit.: Humber (1989; emend.).

Ancylistales, see *Entomophthorales*; Ubrizsy & Vörös (1973).

Ancylistes Pfitzer (1872), Ancylistaceae. 3 (on *Closterium*) N. temp. See Berdan (*Mycol.* **30**: 396, 1938), Sparrow (*Aquatic Phycomycetes*: 1065, 1960; key), Tucker (1981; key).

Ancylospora Sawada (1944) = Pseudocercospora (Mitosp. fungi) fide Deighton (1976).

Ancyrophorus Raunk. (1888) = Enerthenema (Stemonitid.).

Andohaheloa Manier (1955) = Enterobryus (Eccrin.) fide Manier & Lichtwardt (*Ann. Sci. nat. Bot. sér.* 12, **9**: 519, 1968).

Andreaea Palm & Jochems (1923) [non Hedw. (1801), *Musci*] ≡ Andreaeana (Mitosp. fungi).

Andreaeana Palm & Jochems (1924), Mitosporic fungi, 1.A1.?. 1, Sumatra. = Acremonium (Mitosp. fungi) fide Gams (*in litt.*).

Andreanszkya Tóth (1968) = Podospora (Lasiosphaer.) fide Lundqvist (1972).

androgynous, having the antheridium and its oogonium on one hypha; in de Bary's original sense (*Bot. Zeit.* **46**: 597, 1888) covers hypogynous, etc. (see Fig. 28A-D). Cf. monoclinous.

androphore, a branch forming antheridia, as in *Pyronema*.

Androsaceus (Pers.) Pat. (1887) = Marasmius (Tricholomat.) fide Saccardo (1887).

Anekabeeja Udaiyan & V.S. Hosag. (1992), Microascaceae. 1, India. See Korf (*Mycotaxon* **54**: 413, 1995; nomencl.). ? = Pseudeurotium (Pseudeurot.) fide Eriksson & Hawksworth (*SA* **12**: 24, 1993).

Anellaria P. Karst. (1879), Strophariaceae. 2, widespr. = Panaeolus (Strophar.) fide Dennis *et al.* (1960).

Anema Nyl. (1885) nom. cons., Lichinaceae (L). 7, cosmop. See Moreno & Egea (*Acta Bot. Barcin.* **91**: 1, 1992; key).

Anematidium Gronchi (1931), Mitosporic fungi, 1.C2.38. 1, Eur. = Zasmidium (Mitosp. fungi) fide Ciferri & Montemartini (*Atti Ist. bot. Univ. Critt. Pavia*, ser. 5, **17**: 274, 1959).

anemophilous (of spores), taken about by air currents.

aneuploid, having a chromosome number which is not a multiple of the haploid set.

Angatia Syd. (1914), Saccardiaceae. 4 or 5, trop.

Angelina Fr. (1849), Dermateaceae. 1, N. Am. See Durand (*J. Mycol.* **8**: 108, 1906).

angio- (of a sporocarp), closed at least till the spores are mature. Cf. endo-, gymno-, hemi-angiocarpous, and cleistocarp.

angiocarpous (of a basidiome), hymenial surface at first exposed but later covered by an incurving pileus margin and/or excrescences from the stipe (Singer, 1975: 26); also used in a parallel way for *Ascomycota*.

Angiococcus Jahn (1924) nom. dub. (? Fungi). See Peterson & McDonald (*Mycol.* **58**: 962, 1967).

Angiophaeum Sacc. (1898) ≡ Phaeangium (Otid.).

Angiopoma Lév. (1841) nom. rej. = Drechslera (Mitosp. fungi) fide Sutton (*Mycotaxon* **3**: 377, 1976).

Angiopomopsis Höhn. (1912), Mitosporic fungi, 4.C2.19. 1, Java. See Sutton (*Česká Myk.* **29**: 97, 1975).

Angiopsora Mains (1934) = Phakopsora (Phakopsor.) fide Ono *et al.* (*MR* **96**: 825, 1992).

Angioridium Grev. (1827) = Physarum (Physar.).

Angiosorus Thirum. & M.J. O'Brien (1974) = Thecaphora (Ustilagin.) fide Mordue (*Mycopathol.* **103**: 177, 1988).

Angiotheca Syd. (1939) = Dictyonella (Saccard.) fide v. Arx (1963).

-angium (-ange, suffix), a structure having no opening; a cavity.

ang-kak (red rice), an Oriental food colouring obtained by growing *Monascus purpureus* on polished rice; see Fermented food and drinks.

Anguillospora Ingold (1942), Mitosporic fungi, 1.C1.2. 4 (aquatic), widespr. See Petersen (*Mycol.* **54**: 117, 1962; key), Jooste & van der Merwe (*S. Afr. J. Bot.* **56**: 319, 1990; ultrastr.).

anguilluliform, worm- or eel-like in form.

angular septum, see septum.

Angulimaya Subram. & Lodha (1964), Mitosporic fungi, 1.A1.19. 1 (coprophilous), India.

Angulospora Sv. Nilsson (1962), Mitosporic fungi, 1.E1.2. 1 (aquatic), Venezuela.

Angusia G.F. Laundon (1964) = Maravalia (Chacon.) fide Ono (*Mycol.* **76**: 892, 1984).

angustate, narrowed.

anheliophilous, preferring diffuse light. Cf. heliophilous.

Anhellia Racib. (1900), ? Myriangiaceae. 7, trop. See v. Arx (*Persoonia* **2**: 421, 1963).

Animal mycophagists. Rabbits and squirrels (Hastings & Mottram, *TBMS* **5**: 364, 1916; Buller, *TBMS* **6**: 355, 1920; *Researches* **2**: 195, 1922); slugs (Elliott, *TBMS* **8**: 34, 1922); snails (*Polygyra thyroides*) (Wolf & Wolf, *Bull. Torrey bot. Cl.* **66**: 1, 1939); as seed dispersal mutualists (Pirozynski & Malloch, *in* Pirozynski & Hawksworth (Eds), 1988: 227). See also ambrosia fungi, Coevolution, Insects and fungi, reindeer lichen, termite fungi.

Aniptodera Shearer & M.A. Mill. (1977), Halosphaeriaceae. 9 (aquatic and marine), widespr. See Shearer (*Mycol.* **81**: 139, 1989), Volkmann-Kohlmeyer & Kohlmeyer (*Bot. Mar.* **37**: 109, 1994; table chars 9 spp.).

aniso- (prefix), unequal.

Anisochora Theiss. & Syd. (1915) = Apiosphaeria (Phyllachor.).

Anisochytridiales, see *Hyphochytriales*; Karling (1977).

anisogamy, the copulation of gametes of unlike form or physiology, i.e. of **-gametes**; heterogamy; cf. isogamy.

Anisogramma Theiss. & Syd. (1917), Valsaceae. 5, Eur., N. Am.

anisokont, having flagella of unequal length; heterokont.

Anisolpidiaceae Karling (1943), Anisolpidiales. 1 gen., 3 (or 6) spp. See Dick (*in press*; key).

Anisolpidiales, Oomycota. 1 fam., 1 gen., 3 spp. Fam.: Anisolpidiaceae.

Anisolpidium Karling (1943), Anisolpidiaceae. 3 or 6 (in marine algae), USA, Eur.

Anisomeridium (Müll. Arg.) M. Choisy (1928) nom. cons., ? Monoblastiaceae (L). *c.* 15, temp.

Anisomyces Theiss. & Syd. (1914), Valsaceae. 1, trop. Am.

Anisomyces Pilát (1940) nom. nud. = Osmoporus (Polypor.) fide Donk (1960).

Anisomycopsis I. Hino & Katum. (1964), Diaporthales (inc. sed.). 1, Japan.

Anisomyxa Němec (1913) = Ligniera (Plasmodiophor.).

anisospory, having spores of more than one kind.

Anisostomula Höhn. (1919) = Hyponectria (Hyponectr.) fide Barr (*Mycol.* **68**: 611, 1976).

anisotomic dichotomic branching, branching where one dichotomy becomes stouter and forms a main stem so that the other branch of the dichotomy appears to be lateral, as in *Alectoria ochroleuca* ; cf. isotomic dichotomic branching.

Anixia Fr. (1819) nom. dub. (? 'gasteromycetes') fide Demoulin (*in. litt.*).

Anixia H. Hoffm. (1862) = Orbicula (Otid.) fide Hughes (1951).

Anixiella Saito & Minoura ex Cain (1961) = Gelasinospora (Sordar.) fide v. Arx (1973).

Anixiopsis E.C. Hansen (1897) = Aphanoascus (Onygen.) fide Cano & Guarro (1990). See de Vries (*Mykosen* **12**: 111, 1969), Guého & de Vroey (*CJB* **64**: 2207, 1986; SEM ascospores).

Ankistrocladium Perrott (1960) = Casaresia (Mitosp. fungi) fide Ellis (*DH*).

Ankultur, see Normkultur.

Annajenkinsia Thirum. & Naras. (1955) = Puttemansia (Tubeuf.) fide Pirozynski (*Kew Bull.* **31**: 595, 1977).

Annella S.K. Srivast. (1976), Fossil fungi. 2 (Jurassic), UK.

annellate (of asci), ones with a thickened apical pore (e.g. *Leotiales*); see ascus; **annellations**; see annellidic.

annellidic (of conidiogenesis), holoblastic conidiogenesis in which the conidiogenous cell (**annellide,** annellophore) by repeated enteroblastic percurrent proliferation produces a basipetal sequence of conidia (**annelloconidia,** annellospores) leaving the distal end marked by transverse bands (**annellations**). See Conidial nomenclature.

Annellodentimyces Matsush. (1985), Mitosporic fungi, 1.C2.19. 1, Japan.

Annellodochium Deighton (1969), Mitosporic fungi, 4.B2.19. 1 (on *Diatrype*), Sierra Leone.

Annellolacinia B. Sutton (1964), Mitosporic fungi, 6.A2.19. 2, trop. See Frölich *et al.* (*MR* **97**: 1433, 1993).

Annellophora S. Hughes (1952), Mitosporic fungi, 1.C2.19. 4, trop. See Ellis (*Mycol. Pap.* **70**: 1958; key).

annellophore, see annellidic.

Annellophorella Subram. (1962), Mitosporic fungi, 1.D2.19. 1, S. Afr.

Annellophragmia Subram. (1963), Mitosporic fungi, 2.C2.19. 1, India.

annular, ring-like; ring-like arrangement.

Annularia Raf. (1815) nom. dub. (Fungi, inc. sed.). No spp. included.

Annularia (Schulzer) Gillet (1876) [non Sternb. (1825), fossil *Pteridophyta*] = Chamaeota (Plut.).

Annularius Roussel (1806) = Coprinus (Coprin.).

Annulatascus K.D. Hyde (1992), Lasiosphaeriaceae. 2 (marine), Australia.

annulus, (1) (of basidiomata), a ring-like partial veil, or part of it, round the stipe after expansion of the pileus (Fig. 4C); hymenial veil; apical veil; ring; an - near the top of the stipe is **superior** (an **armilla,** fide Gäumann & Dodge, 1928: 453), one lower down, **inferior;** (2) (in *Papulospora*), the ring of cells around a bulbil; (3) (of asci), the apical ring; anneau apicale; (4) (in *Alternaria*), thickening in apices of conidiogenous cells, fide Campbell (*Arch. Mikrobiol.* **69**: 60, 1970).

anoderm, having no skin.

Anodotrichum (Corda) Rabenh. (1844) = Blastotrichum (Mitosp. fungi) fide Saccardo (1886).

Anomalemma Sivan. (1983), ? Melanommataceae. 1, Eur. Anamorph *Exosporiella*.

Anomalographis Kalb (1992), Graphidaceae (L). 1, Madeira.

Anomomorpha Nyl. ex Hue (1891) = Graphis (Graphid.).

Anomomyces Höhn. (1928) nom. dub. fide Sutton (*Mycol. Pap.* **138**, 1975).

Anomoporia Pouzar (1966) = Poria (Coriol.) fide Donk (1974).

Anomothallus F. Stevens (1925) Fungi (inc. sed.). See Petrak (*Sydowia* **5**: 328, 1951).

Anopeltis Bat. & Peres (1960), ? Capnodiaceae. 1, Venezuela.

Anopodium N. Lundq. (1964), Lasiosphaeriaceae. 2, Sweden. ? = Podospora (Lasiosphaer.) fide Mirza & Cain (1969).

Ansatospora Newhall (1944) nom. inval. = Mycocentrospora (Mitosp. fungi) fide Deighton (1972).

Anserina Velen. (1934) [non Dumort. (1827, *Chenopodiaceae*] = Ascobolus (Ascobol.) fide Eckblad (1968).

antabuse, tetraethylthiuramdisulphate (disulfiram); after ingestion reacts with alcohol to give unpleasant symptoms; used in the treatment of chronic alcoholism; see coprine.

antagonism, a general name for associations of organisms damaging to one or more of the associates (cf. antibiosis, symbiosis). Though parasitism is an example of antagonism, the term is used esp. for the effects of toxic metabolic products (see Staling substances) or of undetermined causes on fungi and bacteria in competition. Much experimental work has been done on the antagonism between bacteria, bacteria and fungi, and fungi; and esp. on the competition between microorganisms in the soil; for example, on the effect of saprobic soil fungi on pathogenic species, e.g. *Trichoderma viride* on *Rhizoctonia*, *Pythium*, and other damping-off fungi.
   Lit.: Waksman (*Soil Sci.* **43**: 51, 1937; *Bact. Rev.* **5**: 231, 1941); Porter & Carter, and Weindling (*Bot. Rev.* **4**: 165, 475, 1938) give long reference lists, and Hawksworth (*in* Cole & Kendrick, *Biology of conidial fungi* **1**: 171, 1981) more recent ones; Moreau & Moreau (*BSMF* **72**: 250, 1956) (types of association and antagonism). Cf. antibiotic substances.

antarctic mycology, see Polar and alpine mycology.

Antarctomia D.C. Linds. (1975) = Placynthium (Placynth.) fide Henssen (*Lichenologist* **13**: 307, 1981).

Antenaglium F.C. Albuq. (1969), Mitosporic fungi, 1.A1.15. 1, Brazil. = Gliocephalotrichum (Mitosp. fungi). See Carmichael *et al.* (1980).

Antennaria Link (1809) [non Gaertn. (1791), *Compositae*] = Antennularia (Ventur.).

Antennariella Bat. & Cif. (1963), Mitosporic fungi, 4.A1.?. *c.* 5, trop. See Hughes (*Mycol.* **68**: 693, 1976), Sutton (*Mycol. Pap.* **141**, 1977).

Antennataria Rchb. (1841) ≡ Antennularia (Ventur.).

Antennatula Fr. ex F. Strauss (1850), Anamorphic Euantennariaceae, 1.C2.1. Teleomorph *Euantennaria.* 8, widespr. See Hughes (*N.Z. Jl Bot.* **12**: 299, 1974).

Antennella Theiss. & Syd. (1918) = Scorias (Capnod.) fide v. Arx & Müller (1975).

Antennellina J.M. Mend. (1925) ? = Scorias (Capnod.) fide v. Arx & Müller (1975).

Antennellopsis J.M. Mend. (1930) = Phragmocapnias (Capnod.) fide Reynolds (1978).

Antennina Fr. (1849) ≡ Antennularia (Ventur.).

Antennopsis R. Heim (1952), Mitosporic fungi, 1.C2.1. 1 (on termites), France. See *Gloeohaustoriales.*

Antennospora Meyers (1957), Halosphaeriaceae. 1 (marine), Eur., N. Am. See Jones *et al.* (*Bot. Mar.* **27**: 129, 1984).

Antennula, see *Antennatula.*

Antennularia Rchb. (1828), Venturiaceae. *c.* 30, widespr. See Müller & v. Arx, 1962). Nom. dub. fide Hughes (*N.Z. Jl Bot.* **8**: 156, 1970). See also *Protoventuria.*

Antennulariella Woron. (1915), Antennulariellaceae. 1, Corsica. See Hughes (1976). = Wentiomyces (Dothid.) fide v. Arx & Müller (1975).

Antennulariellaceae Woron. (1925), Dothideales. 2 gen. (+ 2 syn.), 3 spp. Superficial mycelium irregular, dark, smooth- or rough-walled, adpressed or erect; ascomata small, perithecial, ± globose, stalked or sessile, opening by a small poorly-defined lysigenous pore, sometimes with hyphal appendages; interascal tissue absent; asci small, ovoid, fissitunicate, I-; ascospores hyaline to brown, usually 1-septate, sheath lacking. Anamorphs coelomycetous and hyphomycetous, if the latter then conidia elongate, multiseptate. Saprobic, usually epiphytic, cosmop.
   Lit.: Hughes (*Mycol.* **68**: 693, 1976; gen. names, anamorphs), Reynolds (*Mycotaxon* **27**: 377, 1986; status).

anterior, (1) at or in the direction of the front; (2) (of lamellae), the end at the edge of the pileus.

Anthasthoopa Subram. & K. Ramakr. (1956) = Coniella (Mitosp. fungi) fide Sutton (*CJB* **47**: 603, 1969).

antheridiol, a sex hormone (sterol) of *Achlya bisexualis* which induces antheridial formation in male strains of *Achlya* (McMorris & Barksdale, *Nature* **215**: 320, 1967; Barksdale, *Science* **166**: 831, 1969).

antheridium (pl. **-a, antherid**), the male gametangium, either formed from a haplophase thallus, or in which meiosis occurs after delimitation.

antherozoid, a motile male cell; a sperm.

Anthina Fr. (1832), Mitosporic fungi, -.-.-. 5, temp. *A. citri* and *A. brunnea* ('leaf felt' in *Citrus*).

Anthomyces Dietel (1899), Raveneliaceae. 1 (on *Leguminosae*), S. Am.

Anthomyces Grüss (1918) = Metschnikowia (Metschnikow.) fide v. Arx *et al.* (1977).

Anthomycetella Syd. & P. Syd. (1916), Raveneliaceae. 1 (on *Canarium*), Philipp.

Anthopeziza Wettst. (1885) = Microstoma (Sarcoscyph.) fide Eckblad (1968).

Anthopsis Fil. March., A. Fontana & Luppi Mosca (1977), Mitosporic fungi, 1.A2.15. 2, Eur. & Japan. See Bonfante-Fasolo & Marchisio (*Allionia* **23**: 13, 1970; ultrastr. phialide).

Anthoseptobasidium Rick (1943), Auriculariaceae. 2, S. Am.

Anthostoma Nitschke (1867) = Cryptosphaeria (Diatryp.) fide Læssøe & Spooner (*Kew Bull.* **49**: 1, 1994). See also Eriksson (*Svensk bot. Tidskr.* **60**: 315, 1966), Rappaz (*Mycol. Helv.* **5**: 21, 1992).

Anthostomaria (Sacc.) Theiss. & Syd. (1918), Ascomycota (inc. sed.). 1 (on lichens, *Umbilicaria*), former USSR.

Anthostomella Sacc. (1875), Xylariaceae. 50, widespr. See Eriksson (*Svensk bot. Tidskr.* **60**: 315, 1966), Francis (*Mycol. Pap.* **139**, 1975; key 30 Eur. spp.), Rappaz (*Mycol. Helv.* **7**: 99, 1995; on hardwoods, Eur., N. Am.).

Anthostomellina L.A. Kantsch. (1928), Ascomycota (inc. sed.). 1, former USSR.

anthracnose, a plant disease having characteristic limited lesions, necrosis, and hypoplasia, generally caused by one of the acervular coelomycetes. See Jenkins (*Phytopathology* **23**: 389, 1933); **spot -**, a disease caused by *Elsinoë* or its anamorph *Sphaceloma* (Jenkins; see *RAM* **26**: 255, 1947).

Anthracobia Boud. (1885), Otideaceae. *c.* 15, N. temp. See Hohmeyer & Schnacketz (*Beitr. Kennt. Pilze Mitteleur.* **3**: 427, 1987; key 9 spp.), Svrĉek (*Acta Mus. Nat. Prague* **4B**, 6 (Bot. 1): 75, 1948), Delattre-Durand & Parguey-Leduc (*BSMF* **95**: 355, 1979; ontogeny).

anthracobiontic, obligately inhabiting burnt areas; **anthracophilous**, sporulation favoured by burnt areas (*see* Pyrophilous fungi); **anthracophobic**, sporulation suppressed or checked on burnt areas; **anthracoxe-nous**, incidence and growth not affected by burnt areas (Moser, 1949).

Anthracocystis Bref. (1912) = Ustilago (Ustilagin.) fide Dietel (1928).

Anthracoderma Speg. (1888), Mitosporic fungi, 8.A1.?. 3, S. Am. See Petrak & Sydow (*Ann. Myc.* **33**: 188, 1935).

Anthracoidea Bref. (1895), Ustilaginaceae. *c. c.* 60 (on *Cyperaceae*), widespr., temp. See Kukkonen (*Ann. Bot. zool.-bot. fenn. Vanamo* **34** (3), 1963; *Ann. Bot. fenn.* **1**: 161, 1964; keys), Braun & Hirsch (*Feddes Repert.* **89**: 43, 1978; keys), Kukkonen (*TBMS* **47**: 273, 1964; spore germination; *Ann. Bot. fenn.* **1**: 257, 1964; homothallism), Nannfeldt (*Symb. bot. upsal.* **22** (3), 1979; 34 Nordic spp.), Vánky (1987), Salo & Sen (*CJB* **71**: 1406, 1993; isoenzyme analysis), Ingold (*MR* **92**: 245, 1989; spore germination, posn).

Anthracomyces Renault (1898), Fossil fungi (mycel.). 2 (Carboniferous), France.

Anthracophlous Mattir. ex Lloyd (1913) = Rhizopogon (Rhizopogon.).

Anthracophyllum Ces. (1879), Tricholomataceae. 8, trop. See Pegler & Young (*MR* **93**: 352, 1989; key).

Anthracostroma Petr. (1954), Dothideales. 1, Australia.

Anthracothecium Hampe ex A. Massal. (1860), Pyrenulaceae (L). *c.* 70, mainly trop. See Harris (*Mem. N.Y. bot. Gdn* **49**: 74, 1989; key 5 N. Am. spp.), Malme (*Ark. Bot.* **22A** (11), 1929), Johnson (*Ann. Mo. bot. Gdn* **27**: 1, 1940), Singh (*Feddes Repert.* **93**: 67, 1982, *Geophytology* **14**: 69, 1984; **15**: 98, 1985; Singh & Raychaudhury, *New Bot.* **9**: 32, 1983; India).

Anthracothecomyces Cif. & Tomas. (1953) = Pyrenula (Pyrenul.) fide Harris (1989).

Anthropomorphus Seger (1745) nom. inval. = Geastrum (Geastr.). Used by Lloyd but see Donk (*Reinwardtia* **1**: 205, 1951).

anthropophilic (of dermatophytes, etc.), preferentially pathogenic for man. Cf. zoophilic.

Anthurus Kalchbr. & MacOwan (1880) = Clathrus (Clathr.) fide Dring (1980).

anti- (in combination), against.

antiamoebin, an antibiotic from *Emericellopsis poonensis*, *E. synnematicola*, and '*Cephalosporium*' *pimprinum*; anti-protozoa and helminths (*Hindustan Antibiot. Bull.* **11**: 27, 1968).

antibiosis, antagonism (q.v.) between two organisms resulting in one overcoming the other.

antibiotic, (1) (adj.) damaging to life; esp. of substances produced by microorganisms which are damaging to other microorganisms; (2) (n.) any antibiotic substance, esp. one used as a therapeutant, cf. toxin. See Waksman (*Mycol.* **39**: 565, 1947) for a discussion on the use of this term. **- substances** are produced by fungi (esp. *Penicillium* and *Aspergillus*, actinomycetes (esp. *Streptomyces*; see amphotericin, blasticidin, cycloheximide, streptomycin), and other microorganisms.

Lit.: Grayon (Ed.) (*Antibiotics, chemotherapeutics and antibacterial agents for disease control*, 1982), Chadwick & Whelan (Eds) (*Secondary metabolites: their function and evolution*, 1992), Demain et al. (Eds) (*Novel microbial products for medicine and agriculture*, 1989), Jong et al. (Eds) (*ATCC names of industrial fungi*, 1994),

Many fungi when grown under appropriate conditions produce antibiotics; see the reviews by Brian (*Bot. Rev.* **17**: 357, 1951) and Broadbent (*PANS* **B 14**: 120, 1968). For some important or interesting antibiotics from fungi see: antiamoebin, alternaric acid, calvacin, cephalosporins, dendrochin, flammulin,

fumigillin, fumigatin, fusidic acid, gliotoxin, griseofulvin, helenin, lepiochlorin, patulin, penatin, penicillic acid, penicillin, phomin, poricin, proliferin, sparassol, statolin, trichomycin, trichothecin, trypacidin, ustilagic acids, variecolin, viridin, wortmannin.

Some lichen products (q.v.) are antibiotics. In general they are most effective against gram-positive bacteria. Usnic acid is used commercially ('Usno', 'Binan', 'Usniplant') and strongly inhibits *Mycobacterium*. Sodium usnate is effective against tomato canker (*Corynebacterium michiganense*) and several lichen acids are active against *Trichosporon*. Usnic acid inhibits *Neurospora crassa* and this and lichen extracts inhibit wood-rotting fungi (Henningsson & Lundström, *Mater. Organ.* **5**: 19, 1970). Hale (*Biology of lichens*, 1967; edn 2, 1974; review), Virtanen et al. (*Suomen Kem.* **B27-B30**, 1954-7; many papers on 'Usno'), Vartia (in Ahmadjian & Hale (Eds), *The lichens*: 547, 1974; review), Lowe & Elander (*Mycol.* **75**: 361, 1983; antibiotic industry in USA).

antibody, see antigen.

anticlinal, perpendicular to the surface; cf. periclinal.

antigen, a substance which when introduced into the tissues of a living animal induces the development in the blood serum (see **-serum**) of another substance (see Drouhet et al. (Eds), *Fungal antigens*, 1988). (the **-body**) with which it reacts specifically; antibodies may be classified according to whether they cause lysis (**lysins**), agglutination (**agglutinins**), or precipitation (**precipitins**) of the antigen; see anaphylaxis, complement-fixation, ELISA, Serology.

Antilyssa Haller ex M. Choisy (1929) = Peltigera (Peltiger.).

Antimanoa Syd. (1930), Ascomycota (inc. sed.). 1, S. Am.

Antimanopsis Petr. (1948) = Monostichella (Mitosp. fungi) fide v. Arx (1957).

antimetabolite, a substance which resembles in chemical structure some naturally occurring compound essential in a living process and which specifically antagonizes the biological action of such an essential compound. See Woolley (*Science, NY* **129**: 615, 1959; review).

Antinoa Velen. (1934) ? = Pezizella (Leot.) fide Lizon (*Mycotaxon* **45**: 1, 1992).

antiphyte, see alternation of generations.

Antipodium Piroz. (1974), Anamorphic Hypocreaceae, 1.C1.15. Teleomorph *Ophionectria*. 1, Jamaica.

antiserum, blood serum (the fluid fraction of coagulated blood) containing antibodies to one or more antigens (q.v.).

antithetic, see alternation of generations.

Antlea P.A. Dang. (1890) nom. dub (? Fungi or Protozoa).

Antonigeppia Kuntze (1898) = Lycogala (Myxom.).

Antrocarpon A. Massal. (1856) = Ocellularia (Thelotremat.) p.p. and Thelotrema (Thelotremat.) p.p. fide Hale (1981).

Antrocarpum G. Mey. (1825) ≡ Thelotrema (Thelotremat.).

Antrodia P. Karst. (1879), Coriolaceae. 8, Eur., N. Am. See Donk (*Persoonia* **4**: 339, 1966), Niemelä & Ryvarden (*TBMS* **65**: 427, 1975; typification), Lombara (*Mycol.* **82**: 185, 1990; culture).

Antrodiella Ryvarden & I. Johans. (1980), Coriolaceae. 1, USA.

Antromyces Fresen. (1850), Mitosporic fungi, 2.A1.3/39. 2, Eur., S. Am.

Antromycopsis Pat. & Trab. (1897), Anamorphic Lentinaceae, 2.A1.41. Teleomorph *Pleurotus*. 3, widespr. See Moore (*CJB* **55**: 1251, 1977; *TBMS* **82**: 377, 1984), Pollack & Miller (*Mem. N.Y. bot. Gdn* **28**: 174,

1976; teleomorph), Stalpers *et al.* (*CJB* **69**: 6, 1991; generic revision, key).

antrorse, directed upwards or forwards.

Anulohypha Cif. (1962), Mitosporic fungi, 1.-.-. 1, Dominican Republic.

Anulomyces Bydgosz (1932) nom. dub. (Fungi, inc. sed.).

Anulosporium Sherb. (1933) = Dactylaria (Mitosp. fungi) fide Drechsler (*Mycol.* **26**: 135, 1934).

Anungitea B. Sutton (1973), Mitosporic fungi, 1.B2.3/9. 7, widespr.

Anungitopsis R.F. Castañeda & W.B. Kendr. (1990), Mitosporic fungi, 1.C2.?28. 4, Cuba, Japan.

Anzia Stizenb. (1861) nom. cons., Parmeliaceae (L). 28, cosmop. See Kurokawa & Jinzenji (*Bull. natn. Sci. Mus., Tokyo* **8**: 369, 1965), Culberson (*Brittonia* **13**: 381, 1961), Yoshimura & Elix (*J. Hattori bot. Lab.* **74**: 287, 1993).

Anzia Garov. (1868) ≡ Lichenothelia (Lichenothel.).

Anziella Gyeln. (1940) = Placynthium (Placynth.).

Anzina Scheid. (1982), Trapeliaceae (L). 1, Eur. See Scheidegger (*Nova Hedw.* **41**: 191, 1985).

Aorate Syd. (1929) = Titaea (Mitosp. fungi) fide Boedijn (*Sydowia* **5**: 211, 1951).

Aoria Cif. (1962), Mitosporic fungi, 8.A1.10. 1, Dominican Republic.

apandrous, forming oospores when no antheridia are present.

Aparaphysaria Speg. (1922), Otideaceae. 2, India, Tierra del Fuego. See Korf (1973).

Apatelomyces Thaxt. (1931), Laboulbeniaceae. 1, W. Afr.

Apatomyces Thaxt. (1931), Laboulbeniaceae. 1, Philipp.

Apatoplaca Poelt & Hafellner (1980), Teloschistaceae (L). 1, N. Am.

Aphanandromyces W. Rossi (1982), Laboulbeniaceae. 1, Italy, Poland. See Rossi (*SA* **6**: 114, 1987; posn).

Aphanistis Sorokīn (1889), ? Chytridiales (inc. sed.). 1 or 2, former USSR.

Aphanoascus Zukal (1890), Onygenaceae. 12, widespr. See Cano & Guarro (*MR* **94**: 455, 1990; key). Anamorphs *Chrysosporium, Malbranchea.*

Aphanobasidium Jülich (1979), Xenasmataceae. 10, widespr.

Aphanocladium W. Gams (1971), Mitosporic fungi, 1.A1.15. 4, widespr.

Aphanodictyon Huneycutt ex M.W. Dick (1971), Leptolegniellaceae. 1, USA.

Aphanofalx B. Sutton (1986), Mitosporic fungi, 8.A1.1. 2, Zambia, Pakistan. See Sutton & Abbas (*TBMS* **87**: 640, 1987).

Aphanomyces de Bary (1860), Saprolegniaceae. 25 (aquatic), widespr. *A. euteiches* (pea root rot); *A. astaci* (pathogenic for European crayfish, *Astacus*); *A. cochlioides* (sugar beet black rot). See Scott (*Tech. Bull. Va agric. Exp. Stn* **151**, 1961; keys).

Aphanomycopsis Scherff. (1925), Leptolegniellaceae. 8, Eur., N. Am. See Dick (*in press*; key).

Aphanopeltis Syd. (1927), Asterinaceae. 7, trop. Am., Indonesia.

aphanoplasmodium, see plasmodium.

Aphanopsidaceae Printzen & Rambold (1995), Ascomycota (inc. sed., L). 2 gen. (+ 2 syn.), 2 spp. Thallus crustose; ascomata apothecial, ± flat, without a well-developed wall; interascal tissue of rarely branched thin-walled paraphyses, the apices not swollen; asci thick-walled, with a well-developed apical dome, ocular chamber poorly-developed, with a well-developed I+ plug or tube and an outer I+ gelatinized layer; ascospores hyaline, aseptate, without a sheath. Anamorphs pycnidial. Lichenized with green algae, on soil etc., Eur.

Aphanopsis Nyl. ex P. Syd. (1887), Aphanopsidaceae (L). 1, Eur. See Coppins & James (*Lichenologist* **16**: 241, 1984), Printzen & Rambold (*Lichenologist* **27**: 91, 1995).

Aphanostigme Syd. (1926), ? Pseudoperisporiaceae. *c.* 10, widespr. See Hansford (1946), Müller (*Sydowia* **18**: 86, 1965). = Nematostoma (Pseudoperisporiopsid.) fide Rossman (*Mycol. Pap.* **157**, 1987).

Apharia Bonord. (1864), Ascomycota (inc. sed.). 1, Eur.

Aphelaria Corner (1950), Aphelariaceae. 15, widespr.

Aphelariaceae Corner (1970), Cantharellales. 3 gen., 19 spp. Basidioma ramarioid; small, slender, terrestrial; hymenophore unilateral.

Aphelariopsis Jülich (1982), Platygloeaceae. 2, Sarawak, S. Am.

Aphelidium Zopf (1885) nom. dub. (Protozoa or fungi in algal cells).

Aphidomyces Brain (1923), Saccharomycetaceae. 5 (in *Insecta*), widespr.

Aphotistus Humb. (1793) = Rhizomorpha (Mitosp. fungi) fide Mussat (*Syll. Fung.* **15**, 1901); nom. dub. fide Donk (*Taxon* **11**: 79, 1962).

Aphragmia Trevis. (1880) [non Nees (1836), *Acanthaceae*] = Ionaspis (Hymenel.).

Aphyllophorales. Order proposed by Rea (after Patouillard) for basidiomycetes having macroscopic basidiocarps in which the hymenophore is flattened (*Thelephoraceae*), club-like (*Clavariaceae*), toothlike (*Hydnaceae*) or has the hymenium lining tubes (*Polyporaceae*) or sometimes on lamellae, the poroid or lamellate hymenophores being tough and not fleshy as in the *Agaricales*. Traditionally the order has had a core of 4 fam. (as indicated above) based on hymenophore shape but recent detailed microscopic studies of basidiocarp structure has shown these groupings to be unnatural. Keys to 550 spp. in culture are given by Stalpers (*Stud. mycol.* **16**, 1978).

Much of the literature of the order is based on the traditional family groupings and as one fam. may exhibit several different types of hymenophore (e.g. *Gomphaceae* has effuse, clavarioid, hydnoid, and cantharelloid hymenophores) reference to the lit. is complicated because information about the gen. of any one fam. may occur in apparently unrelated monographs. Specific references are given under many fam. and gen.

Aphyllotus Singer (1974), Tricholomataceae. 1, Colombia.

Aphysa Theiss. & Syd. (1917) = Coleroa (Ventur.) fide Müller & v. Arx (1962).

Aphysiostroma Barrasa, A.T. Martínez & G. Moreno (1986), Hypocreaceae. 1 (coprophilous), Spain. Anamorph *Verticillium*.

apical, at the end (or **apex**); **- granule**, a deeply staining granule at the hyphal apex, esp. in *Basidiomycetes*; the 'Spitzenkorper' of Brunswik (1924); **- veil**, see annulus; **- wall building**, see wall building.

apiculate, having an apiculus.

apiculus (of a spore), a short projection at one end; a projection by which it was fixed to the sterigma (Josserand); apicule; hilar appendage.

apileate, having no pileus; resupinate.

Apinisia La Touche (1968), Onygenaceae. 2 or 3, Eur., Australia. See Guarro *et al.* (*Mycotaxon* **42**: 193, 1991).

Apiocamarops Samuels & J.D. Rogers (1987), Boliniaceae. 2, S. Am.

Apiocarpella Syd. & P. Syd. (1919), Mitosporic fungi, 4.B1.1. 8, widespr. See Melnik (*Nov. Sist. niz. Rast.* **13**: 93, 1976), Vanev & Sofia (*Fitologiya* **29**: 39, 1985; key). = Ascochyta (Mitosp. fungi) fide Punithalingam (1979).

Apioclypea K.D. Hyde (1994), ? Hyponectriaceae. 1, Papua New Guinea.

Apiocrea Syd. & P. Syd. (1921), Hypocreaceae. 5, mainly temp. Anamorph *Sepedonium.* = Hypomyces (Hypocr.) fide Rogerson & Samuels (1989).

Apiodiscus Petr. (1940), ? Rhytismatales (inc. sed.). 1, Iran.

Apiodothina Petr. & Cif. (1932) = Coccoidea (Coccoid.) fide Müller & v. Arx (1962).

Apiognomonia Höhn. (1917), Valsaceae. 8, Eur., N. Am. *A. erythrostoma* (cherry leaf scorch), *A. quercina* (oak anthracnose). See v. Arx (*Ant. v. Leeuwenhoek* **17**: 259, 1951), Barr (*Mycol. Mem.* **7**, 1978; key), Monod (*Sydowia* **37**: 222, 1984).

Apioplagiostoma M.E. Barr (1978), Valsaceae. 3, Eur., N. Am.

Apioporthe Höhn. (1917) = Anisogramma (Vals.) fide Müller & v. Arx (1973).

Apioporthella Petr. (1929), Valsaceae. 1, Eur.

Apiorhynchostoma Petr. (1923), ? Clypeosphaeriaceae. 1, Eur.

Apiosordaria Arx & W. Gams (1967), Lasiosphaeriaceae. 5, widespr. See Krug *et al.* (*Mycotaxon* **17**: 553, 1983), Guarro & Cano (*TBMS* **91**: 587, 1988), Mouchacca & Gams (*Mycotaxon* **48**: 415, 1993; anamorphs). Anamorph *Cladorrhinum.*

Apiosphaeria Höhn. (1909), Phyllachoraceae. 2 or 3, widespr. Anamorph *Oswaldina.*

Apiospora Sacc. (1875), ? Lasiosphaeriaceae. 3, Eur. See Barr (*Mycotaxon* **39**: 43, 1990; posn), Cannon (*in* Hawksworth (Ed.), *Ascomycete systematics*: 377, 1994; posn), Müller (*Boln Argent. Bot.* **28**: 201, 1992; key). Anamorphs *Arthrinium, Cordella.*

Apiosporella Speg. (1910) ≡ Apiocarpella (Mitosp. fungi).

Apiosporella Speg. (1912) = Aplosporidium (Mitosp. fungi).

Apiosporella Höhn. ex Theiss. (1917) = Pseudomassaria (Hyponectr.) fide Barr (1976).

Apiosporina Höhn. (1910), Venturiaceae. 2, N. Am. *A. collinsii* (witches' broom of *Amelanchier*), *A. morbosa* (syn. *Dibotryon morbosum*) (black knot of *Prunus*).

Apiosporina Petr. (1925) ≡ Pseudapiospora (Hyponectr.).

Apiosporium Kunze (1817) = Sclerotium (Mitosp. fungi) fide Höhnel (*Sber. Akad. Wiss. Wien* **118**: 1159, 1909).

Apiosporopsis (Traverso) Mariani (1911), ? Melanconidaceae. 1, Eur. See Reid & Dowsett (*CJB* **68**: 2398, 1990).

apiosporous (of two-celled spores), where one cell is markedly smaller then the other.

Apiothecium Lar.N. Vassiljeva (1987) = Cryptodiaporthe (Vals.) fide Eriksson & Hawksworth (*SA* **7**: 60, 1989).

Apiothyrium Petr. (1947), Hyponectriaceae. 1, Finland.

Apiotrabutia Petr. (1929) = Munkiella (Polystomell.) fide Müller & v. Arx (1962).

Apiotrichum Stautz (1931) = Candida (Mitosp. fungi) fide Diddens & Lodder (*Die anaskosporogenen Hefen* **2**, 1942).

Apiotypa Petr. (1925) Ascomycota (inc. sed.). 1, Philipp. = Didymosphaeria (Didymosphaer.) fide Clements & Shear (1931).

Aplacodina Ruhland (1900) = Pseudomassaria (Hyponectr.) fide Barr (1976).

Aplanes de Bary (1888), Saprolegniaceae. 3, Eur., N. Am. ? = Achlya (Saprolegn.) fide Johnson (1956).

aplanetism, the condition of having non-motile spores in place of zoospores.

Aplanochytrium Bahnweg & Sparrow (1972), Thraustochytriaceae. 1 (marine), Kerguelen Isls.

aplanogamete, a non-motile gamete.

Aplanopsis Höhnk (1952), Saprolegniaceae. 2, Eur.

aplanospore, (1) a naked, amoeboid or non-amoeboid mobile cell; (2) a sporangiospore.

Aplectosoma Drechsler (1951), Cochlonemataceae. 1, USA. See Drechsler (*Mycol.* **43**: 173, 1951).

aplerotic, of an oospore which occupies < 60% of the oogonial volume (Shahzad *et al., Bot. J. Linn. Soc.* **108**: 143, 1992). See Fig. 28B.

Aplopsora Mains (1921), Chaconiaceae. 2, N. Am., Russia.

Aplosporella Speg. (1880), Mitosporic fungi, 8.A2.1. 44, widespr. esp. trop. See Petrak (*Sydowia* **6**: 336, 1952), Tilak & Ramchandra Rao (*Mycopath. mycol. appl.* **24**: 362, 1964), Ramchandra Rao (*Mycopath. mycol. appl.* **28**: 45, 68, 1966; Indian spp.), Pandey (*Perspectives in mycological research* II: 77, 1990; review), Pande & Rao (*Nova Hedw.* **60**: 79, 1995; key to 44 spp.).

Aplosporidium Speg. (1912) = Asteromella (Mitosp. fungi) fide Sutton (1977).

Aplotomma A. Massal. ex Beltr. (1858) ? = Buellia (Physc.).

Apoa Syd. (1931) = Pachypatella (Parmular.) fide v. Arx & Müller (1975).

apobasidiomycete, a gasteromycete having apobasidia.

apobasidium, see basidium.

Apocoryneum B. Sutton (1975) = Massariothea (Mitosp. fungi) fide Sutton (1977).

apocyte, multinucleate cell in which the multinucleate condition is accidental, transitory or secondary. See coenocyte.

Apocytospora Höhn. (1924) = Plectophomella (Mitosp. fungi) fide Petrak (*Ann. Myc.* **27**: 368, 1929).

Apodachlya Pringsh. (1883), Leptomitales (inc. sed.). 3, widespr. See Jacobs (*Ant. v. Leeuwenh.* **48**: 389, 1982; key).

Apodachlyella Indoh (1939), Apodachlyellaceae. 1, N. Am., Eur., Japan.

Apodachlyellaceae M.W. Dick (1986), Leptomitales. 2 gen., 2 spp. *Lit.*: Dick (*in press*).

apodial, having no stalk; sessile.

Apodospora Cain & J.H. Mirza (1970), Lasiosphaeriaceae. 4 (coprophilous), N. Am., Eur. See Barr (*Mycotaxon* **39**: 43, 1990; posn), Lundqvist (1972).

Apodothina Petr. (1970), Ascomycota (inc. sed.). 1, N. Am.

Apodus Malloch & Cain (1971), Sordariaceae. 1 (coprophilous), N. Am.

Apodya Cornu (1872) = Leptomitus (Leptomit.) fide Sparrow (1960).

apogamy, the apomictic development of diploid cells.

Apogloeum Petr. (1954), Mitosporic fungi, 8.A1.?. 1, Tasmania.

Apomelasmia Grove (1937), Anamorphic Diaporthaceae, 8.A1.15. Teleomorph ? *Aporhytisma.* 1, Eur.

Apomella Syd. (1937) = Botryosphaeria (Botryosphaer.) fide Sutton (*in litt.*).

apomixis (adj. **apomictic**), the development of sexual cells into spores, etc., without being fertilized. Cf. amphimixis, automixis, and pseudomixis.

Aponectria (Sacc.) Sacc. (1883) = Nectria (Hypocr.) fide Petch (*TBMS* **21**: 243, 1938).

apophysis, a swelling or a swollen filament, e.g. at the end of a sporangiophore below the sporangium in Mucorales (cf. columella) or on the stem of some species of *Geastrum*; (in basidiomycetes), the swelling at the tip of a sterigma from which the basidiospore develops and which becomes the hilar appendage (q.v.).

Apophysomyces P.C. Misra (1979), Mucoraceae. 1, India. See Misra *et al.* (*Mycotaxon* **8**: 377, 1979) Ellis

& Ajello (*Mycol.* **74**: 144, 1982), Lakshmi *et al.* (*J. Clin. Microbiol.* **31**: 1368, 1993; zygomycosis).

apoplasmodial (of *Acrasiales*), having non-fusion of the myxamoebae.

apoplastic, movement of substances via the cell walls, not entering the living cell; cf. symplastic.

Aporella Syd. (1939) [non Podp. (1916), *Musci*] ≡ Aporellula (Coelom.) fide Sutton (1976).

Aporellula B. Sutton (1986), Mitosporic fungi, 8.B/C1.15. 1, Ecuador.

Aporhytisma Höhn. (1917) = Diaporthopsis (Diaporth.) fide v. Arx & Müller (1954). See also Petrak (*Sydowia* **24**: 249, 1971).

Aporia Duby (1862) ? = Lophodermium (Rhytismat.).

Aporomyces Thaxt. (1931), Laboulbeniaceae. 8 (on *Limnichideae* and *Strophylinidae*), trop., Eur., S. Am., Asia. See Benjamin (*Aliso* **12**: 335, 1989; key).

Aporophallus A. Møller (1895), Phallaceae. 1, Brazil.

Aporothielavia Malloch & Cain (1973), Pseudeurotiaceae. 1, UK.

Aporpiaceae Bondartsev & Bondartseva (1960), Tremellales. 2 gen. (+ 1 syn.), 6 spp. Context coriaceous, dimitic; hymenophore tubulate.

Aporpium Bondartsev & Singer (1944) = Elmerina (Aporp.) fide Reid (1992).

Aposphaeria Berk. (1860) nom. rej. See Sutton (*Mycol. Pap.* **141**, 1977).

Aposphaeria Sacc. (1880) nom. cons., Anamorphic Melanommataceae, 4.A1.15. Teleomorph *Melanomma*. 100, widespr. See Chesters (*TBMS* **22**: 116, 1938).

Aposphaeriella Died. (1912) = Zignoella (Lasiosphaer.) fide Höhnel (*Sber. Akad. Wiss. Wien* **126**: 283, 1917).

Aposphaeriopsis Died. (1913) = Cephalotheca (Cephalothec.) fide Chesters (*TBMS* **19**: 261, 1935).

Aposporella Thaxt. (1920), Mitosporic fungi, 1.A2.38. 1 (on *Insecta*), Afr.

apospory, direct incorporation in a spore of an oogonial or antheridial diploid nucleus with cytoplasm uninfluenced by any meiosis at the time of spore wall formation (Dick, 1972).

Apostemidium P. Karst. (1871) ≡ Apostemium (Vibriss.).

Apostemium (P. Karst.) P. Karst. (1870) = Vibrissea (Vibriss.) fide Sánchez & Korf (1968). See Graddon (*TBMS* **48**: 639, 1965; key).

Apostrasseria Nag Raj (1983), Anamorphic Phacidiaceae, 8.A1.15. Teleomorph *Phacidium*. 4, NZ, N. Am. ? = Allantophomopsis (Mitosp. fungi) fide Sutton (*in litt.*).

Apotemnoum Corda (1833) = Clasterosporium (Mitosp. fungi) fide Saccardo (*Syll. Fung.* **4**: 382, 1886).

apothecium (pl. **apothecia**), a cup- or saucer-like ascoma in which the hymenium is exposed at maturity, sessile or stipitate, the stipes sometimes lichenized (podetium; q.v.). See the following for terminology of anatomical structures of apothecia: Degelius (*Sym. bot. upsal.* **13** (2), 1954; tabulation of terms), Korf (*Sci. Rep. Yokohama nat. Univ.* II **7**: 7, 1958; *in* Ainsworth *et al.* (Eds), *The Fungi* **4A**: 249, 1973), Letrouit-Galinou (*Bryologist* **71**: 297, 1969), Maas Geesteranus (*Blumea* **6**: 41, 1947), Sheard (*Lichenologist* **3**: 328, 1967).

Apoxona Donk (1969) = Hexagonia (Coriol.) fide Bondartsev & Singer.

Appelia (Sacc.) Trotter (1931) = Trichoconis (Mitosp. fungi).

appendage, a process (outgrowth) of any sort. For coelomycete conidial appendage terminology see Nag Raj (*Coelomycetous anamorphs*, 1993).

Appendichordella R.G. Johnson, E.B.G. Jones & S.T. Moss (1987), Halosphaeriaceae. 1 (marine), Eur., N. Am.

Appendicularia Peck (1885) [non DC. (1828), *Melastomataceae*] ≡ Appendiculina (Laboulben.).

appendiculate, (1) (of an agaric basidioma), having the edge of the expanded pileus fringed with tooth-like remains of the veil, as in *Psathyrella candolleana*; (2) (of a spore), having one or more setulae.

Appendiculella Höhn. (1919), Meliolaceae. 250, trop. See Hughes (*Mycol. Pap.* **166**, 1993).

Appendiculina Berl. (1889) = Stigmatomyces (Laboulben.) fide Thaxter (*Proc. Am. Acad. Arts Sci.* **25**: 8, 1890).

Appendispora K.D. Hyde (1994), Dothideales (inc. sed.). 1, Brunei.

applanate, flattened.

apple canker, disease caused by *Nectria galligena*.

appressed (**adpressed**), closely flattened down.

appressorium, a swelling on a germ-tube or hypha, esp. for attachment in an early stage of infection, as in certain *Uredinales* and in *Colletotrichum*; the '.... expression of the genotype during the final phase of germination', whether or not morphologically differentiated from vegetative hyphae, as long as the structure adheres to and penetrates the host (Emmett & Parbery, *Ann. Rev. Phytopath.* **13**: 146, 1975); the term hyphopodium (q.v.) is probably best treated as a synonym.

Apra J.F. Hennen & F.O. Freire (1979), Raveneliaceae. 1 (on *Mimosa*), Brazil.

apud, in; sometimes used to indicate a name published by one author in the work of another; cf. ex.

Apus Gray (1821) ≡ Schizophyllum (Schizophyll.).

Apyrenium Fr. (1849) nom. dub. (Anamorphic Hypocreales). See Donk (*Taxon* **7**: 164, 1958).

Aquadiscula Shearer & J.L. Crane (1985), Leotiaceae. 1 (aquatic), USA.

Aqualinderella R. Emers. & W. Weston (1967), Rhipidiaceae. 1 (anaerobic), Costa Rica. See Held (*Mycol.* **62**: 339, 1970; nutrition), Emerson (*Mycol.* **62**: 359, 1970; oogonium production).

Aquamortierella Embree & Indoh (1967), Mortierellaceae. 1, NZ, Japan. See Embree & Indoh (*Bull. Torrey bot. Club* **94**: 464, 1967), Indoh (*Trans. Mycol. Soc. Japan* **8**: 28, 1967).

Aquascypha D.A. Reid (1965), Podoscyphaceae. 1, C. & S. Am.

Aquathanatephorus C.C. Tu & Kimbr. (1978), Ceratobasidiaceae. 1 (aquatic), Panama.

Aquatic fungi. Living in water, esp. freshwater in contrast to Marine fungi (q.v.). The chief zoosporic fungi of freshwater are *Chytridiomycota* and *Oomycota*, esp. *Chytridiales* and *Saprolegniales*: Sparrow (*Aquatic phycomycetes*, 1943 [edn 2, 1960]; *Mycol.* **50**: 797, 1959, phylogeny), Emerson (*Mycol.* **50**: 589, 1959; culture), Fuller & Jaworski (*Zoosporic fungi in teaching and research*, 1987). A number are parasites of freshwater plankton: Canter & Lund (*Ann. Bot., Lond.*, n.s. **14-15**, 1950-51; *New Phytol.* **47**: 238, 1948; *TBMS* **36**: 13, **37**: 111, 1953-54), Paterson (*Mycol.* **50**: 85, 483, 1958), Cook (*Am. J. Bot.* **50**: 580, 1943, on desmids).

'Hyphomycetes' (frequently having branched or sigmoid spores; Ingold, *Mycol.* **58**: 43, 1966) of freshwater have received much attention (Ingold, *TBMS* **25**: 339, 1942) and *c.* 100 anamorph gen. and 300 spp. (fide Descals *et al.*, pers. comm.) are recorded (Ingold, *Am. J. Bot.* **66**: 218, 1979; *An illustrated guide to aquatic and water-borne hyphomycetes* [*Publs Freshwater biol. Assn* 30], 1975, keys, illustr.; *Biol. J. Linn. Soc.* **7**: 1, 1975, convergent evolution), Nilsson (*Symb. bot. upsal.* **18** (2), 1964), Webster &

Descals (*in* Cole & Kendrick, *Biology of conidial fungi* 1: 295, 1981). **Ecology**: Bärlocher (Ed.) (*The ecology of aquatic hyphomycetes*, 1992). **Teleomorphs**: Webster (*in* Bärlocher, *The ecology of aquatic hyphomycetes*: 99, 1992). **Regional surveys**: Peterson (*Mycol.* **54**: 117, **55**: 18, 570, 1962-63; **N. Am.**; gen. key); Zhu & Yu (*Acta Mycol. Sin.* **11**: 43, 1992; **China**), Marvanová & Marvan (*Česká Myk.* **23**: 135, 1969; **Cuba**), Dixon (*TBMS* **42**: 174, 1959; **Ghana**), Ranzoni (*Mycol.* **71**: 786, 1979; **Hawaii**), Johnson (*J. Elisha Mitch. sci. Soc.* **84**: 179, 1968; **Iceland**), Hudson & Ingold (*TBMS* **43**: 469, 1960; **Jamaica**), Tubaki (*Bull. Nat. Sci. Mus. Tokyo* **41**: 149, 1957; **Japan**), Nawawi (*Malayan Nature Journal* **39**: 75, 1985; **Malaysia**), Ingold (*TBMS* **39**: 108, **42**: 479, 1956-59; **Nigeria**), Le'John (*TBMS* **48**: 261, 1965; **Sierra Leone**), Ingold (*TBMS* **41**: 109; **Uganda**, **Zimbabwe**), Dudka ([*Aquatic hyphomycetes of the Ukraine*], 1974; **Ukraine**), Nilsson (*Svensk bot. Tidskr.* **56**: 351, 1962; **Venezuela**), Brathen (*Nord. Jl Bot.* **4**: 375, 1984; **Norway**); Aimer & Segedin (*N.Z. Jl Bot.* **23**: 273, 1985; **New Zealand**). See also aeroaquatic fungi.

Over 200 ascomycetes have also been recorded from freshwater habitats (Shearer, *Nova Hedw.* **56**: 1, 1993) and the tropics are now proving extremely rich in novel ascomycete genera (e.g. Hyde, *MR* **98**: 719, 1994).

Some saxicolous lichens, mainly of the Lichinaceae and the gen. *Dermatocarpon, Hymenelia, Placynthium, Polyblastia, Staurothele, Verrucaria* (q.v.), occur in freshwater; some may be always submerged (e.g. *Collema fluviatile, Hydrothria venosa*). They can form zones on river and lake margins related to the frequency of submersion (Rosentreter, *Northwest Sci.* **58**: 108, 1984; Santesson, *Medd. Lunds Univ. Limnol. Inst.* **1**, 1939, Sweden; Scott, *Lichenologist* **3**: 368, 1967, Zimbabwe), and can be used in the determination of river channel capacity (Gregory, *Earth Surface Processes* **1**: 273, 1976; Australia); a 'lichen-line' on trees can also indicate highwater levels (Hale, *Bryologist* **87**: 261, 1984).

Diverse fungi are found in polluted water and sewage: Cooke (*Sydowia, Beih.* **1**: 136, 1957, list; *A laboratory guide to fungi in polluted waters, sewage and sewage treatment systems*, 1963; *Our mouldy earth*, 1970 [reprints and summarizes his studies in this field]).

Review: Gareth-Jones (Ed.) (*Recent advances in aquatic mycology*, 1976).

Arachnion Schwein. (1822), Lycoperdaceae. 6, warm areas. See Demoulin (*Nova Hedw.* **21**: 641, 1972).

Arachniopsis Long (1917) [non Spruce (1882), *Hepaticae*], ? Lycoperdaceae. 1, N. Am.

Arachniotus J. Schröt. (1893), Gymnoascaceae. 1, Poland. See Orr *et al.* (*Mycol.* **69**: 126, 1977), Currah, 1985). = Gymnoascus (Gymnoasc.) fide v. Arx (*Persoonia* **13**: 173, 1986).

Arachnocrea Z. Moravec (1956), Hypocreaceae. 1, Eur.

arachnoid, covered with, or formed of, delicate hairs or fibres; araneose.

Arachnomyces Massee & E.S. Salmon (1902), Ascomycota (inc. sed.). 6, Eur., Am. See Currah (*Mycotaxon* **24**: 1, 1985), Malloch & Cain (*CJB* **48**: 839, 1970).

Arachnopeziza Fuckel (1870), ? Hyaloscyphaceae. 12, N. temp. See Huhtinen (*Mycotaxon* **30**: 9, 1987).

Arachnopezizella Kirschst. (1938) = Arachnopeziza (Hyaloscyph.) fide Korf (1951).

Arachnophora Hennebert (1963), Mitosporic fungi, 1.C2.1. 2, Belgium, Canada. See Hughes (*N.Z. Jl Bot.* **17**: 139, 1979; descr.).

Arachnoscypha Boud. (1885) = Arachnopeziza (Hyaloscyph.) fide Korf (1951). See Svrček (*Česká Myk.* **41**: 193, 1987).

Arachnotheca Arx (1971), Onygenaceae. 3, widespr. See Currah (1985). Anamorphs *Chrysosporium, Malbranchea*.

Arachnula Cienk. (1876), Hydromyxales, q.v.

Araeocoryne Corner (1950), Clavariadelphaceae. 1, Malaysia.

Araispora Thaxt. (1896), Rhipidiaceae. 4, Am., Eur.

Araneomyces Höhn. (1909), Anamorphic Tubeufiaceae, 1.G1.1. Teleomorph *Paranectriella*. 1 (mycoparasitic), Brazil. See Sutton (*TBMS* **83**: 399, 1984).

Araneosa Long (1941), Secotiaceae. 1, USA. ? = Endoptychum (Podax.).

araneose (**araneous**), see arachnoid.

Arberia Nieuwl. (1916) [non C.D. White (1908), fossil ? *Pteridophyta*] ≡ Asteridium (Meliol.).

Arborella Zebrowski (1936), ? Fossil fungi (? Chytrid.). 2 (Cambrian to ? Recent), Australia.

arboricolous, growing on trees.

Arborispora K. Ando (1986), Mitosporic fungi, 1.G1.1/10. 3 (aquatic), Japan.

arbuscle (**arbuscule**), see mycorrhiza.

Arbuscula Bat. & Peres (1965) [non H.A. Crum, Steere & L.E. Anderson (1964), *Musci*] ≡ Neoarbuscula (Mitosp. fungi).

Arbusculidium B. Sutton (1982) [non J. Deunff (1968), fossil *Acritarcha*] ≡ Neoarbuscula (Mitosp. fungi).

Arbusculina Marvanová & Descals (1987), Mitosporic fungi, 1.G1.19. 2 (aquatic), UK, USA.

Arbusculites Paradkar (1976), Fossil fungi. 1 (Cretaceous), India.

Arcangelia Sacc. (1890) = Didymella (Dothideales) fide v. Arx & Müller (1975).

Arcangeliella Cavara (1900), Elasmomycetaceae. 5, widespr. See Thiers (*Sydowia* **37**: 296, 1984; key spp. W. USA).

Archaea (**archaebacteria**), an heterogeneous group of prokaryotic organisms belonging to the Domain Archaea. See bacteria.

archaeascus, see ascus.

Archagaricon Hancock & Atthey (1869), Fossil fungi (mycel.). 5 (Carboniferous), UK.

Archecribraria Locq. (1983), Fossil fungi. 2, Sahara.

Archemycota. Name in the rank of phylum including the groups treated in this *Dictionary* as Chytridiomycota and Zygomycota (incl. Trichomycetes); see Cavalier-Smith (*in* Rayner *et al.* (Eds), *Evolutionary biology of fungi*: 339, 1987; *in* Osawa & Honjo (Eds), *Evolution of life*: 271, 1991).

Archeomycelites Bystrov (1959), Fossil fungi (mycel.). 1 (Devonian), former USSR.

Archeplax Locq. (1985), Fossil fungi. 1, Sahara.

Archeterobasidium Koeniguer & Locq. (1980), Fossil fungi (basidiomycetes). 1 (Miocene), Libya.

Archiascomycetes. Class of *Ascomycota* provisionally proposed by Nishida & Sugiyama (*Mycoscience* **35**: 361, 1994) for *Pneumocystis, Protomyces, Saitoella, Schizosaccharomyces* and *Taphrina* based on 18S rRNA sequences; considered by the authors to perhaps not be monophyletic but to have originated before *Euascomycetes* and *Hemiascomycetes*. *Neolecta* can also be placed here (Eriksson, *SA* **14**: 61, 1995).

archicarp (of ascomycetes), the cell, hypha, or coil which later becomes the ascoma or part of it.

Archilegnia Apinis (1935) nom. conf. (Saprolegn.).

Archilichens, lichens in which the algae are bright green (obsol.).

Archimycetes (obsol.). Name used rarely for *Plasmodi-ophoromycota* and *Chytridiomycota*. *Myxochytridiales*.

Architrypethelium Aptroot (1991), Trypetheliaceae (L). 2, trop.

archontosome, an electron-dense body occurring near nuclei at all stages from crozier formation to the development of young ascospores in *Xylaria polymorpha*. See Beckett & Crawford (*J. gen. Microbiol.* **63**: 269, 1970).

arctic mycology, see Polar and alpine mycology.

Arcticomyces Savile (1959) = Exobasidium (Exobasid.) fide Donk (*Persoonia* **4**: 287, 1966).

Arctocetraria Kärnefelt & Thell (1993), Parmeliaceae (L). 2, Eur.

Arctoheppia Lynge (1938) ≡ Theligyna (Lichin.) fide Jørgensen & Henssen (*Taxon* **39**: 343, 1990).

Arctomia Th. Fr. (1860), Arctomiaceae (L). 2, Eur., N. Am. See Henssen (*Svensk bot. Tidskr.* **63**: 126, 1969).

Arctomiaceae Th. Fr. (1860), Lecanorales (L). 2 gen., 3 spp. Thallus crustose or fruticose, gelatinized, with rhizoids; ascomata apothecial, the outer wall often not well-developed, sometimes becoming compound; interascal tissue of paraphyses or anastomosing pseudoparaphysis-like hyphae; asci cylindrical, with a well-developed apical cap but no ocular chamber, 8-spored, with an I+ mucous outer layer; ascospores hyaline, elongate, transversely septate, often with attenuated apices. Lichenized with cyanobacteria. Anamorph pycnidial. Associated with bryophytes, esp. arctic and subarctic.

Arctoparmelia Hale (1986) Parmeliaceae. 5, widespr. ? = Parmelia (Parmel.)

Arctopeltis Poelt (1983), Lecanoraceae (L). 1, Canada, Greenland, Novaja Zemlya, Spitzbergen.

Arctosporidium Thor (1930) nom. dub. (? Fungi, inc. sed.).

Arcuadendron Sigler & J.W. Carmich. (1976), Mitosporic fungi, 1.A1.40. 2, India, former Yugoslavia.

arcuate, arc-like.

Arcyodes O.F. Cook (1902), Trichiaceae. 2, N. Am., Eur.

Arcyrella (Rostaf.) Racib. (1884) = Arcyria (Myxom.).

Arcyria Hill ex F.H. Wigg. (1780), Arcyriaceae. 30, cosmop. See Robbrecht (*Bull. Jard. bot. Natn. Belg.* **44**: 303, 1974; key 13 spp.).

Arcyriaceae Rostaf. ex Cooke (1877), Trichiales. 2 gen. (+ 6 syn.), 31 spp.

Arcyriatella Hochg. & Gottsb. (1989), Trichiaceae. 1, Brazil.

ardella, a small spot-like apothecium, as in the lichen *Arthonia*.

Ardhachandra Subram. & Sudha (1978) = Rhinocladiella (Mitosp. fungi) fide Onofri & Castagnola. But see Chen & Tzean (*MR* **99**: 364, 1995; key to 4 spp.), Keates & Carris (*Crypt. Bot.* **4**: 336, 1994; from *Vaccinium macrocarpon*).

ardosiaceous (ardesiaceous), slate-coloured.

Aregma Fr. (1815) = Phragmidium (Phragmid.).

Arenaea Penz. & Sacc. (1901) = Lachnum (Hyaloscyph.).

Arenariomyces Höhnk (1954), Halosphaeriaceae. 4 (marine), widespr. See Jones *et al.* (*Bot. J. Linn. Soc.* **87**: 193, 1983), Kohlmeyer & Volkmann-Kohlmeyer (*MR* **92**: 413, 1989; key).

Arenicola Velen. (1947), Agaricaceae. 1, Eur.

Areolinia Kalchbr. (1884) = Phellorinia (Phellorin.).

areolate, having division by cracks into small areas.

Areolospora S.C. Jong & E.E. Davis (1974), Xylariaceae. 1, Afr., Asia, S. Am. See Hawksworth (*Norw. J. Bot.* **27**: 97, 1980), Krug *et al.* (*Mycol.* **86**: 581, 1994).

arescent, becoming crustose on drying.

Argomyces Arthur (1912) ≡ Argotelium (Puccin.).

Argomycetella Syd. (1922) = Maravalia (Chacon.) fide Mains (*Bull. Torrey bot. Cl.* **66**: 173, 1939).

Argopericonia B. Sutton & Pascoe (1987), Mitosporic fungi, 1.A1.6/10. 1, Australia.

Argopsis Th. Fr. (1857), ? Brigantiaeaceae (L). 3, subantarctic. See Lamb (*J. Hattori bot. Lab.* **38**: 447, 1974).

Argotelium Arthur (1906) = Puccinia (Puccin.) fide Arthur (*Am. J. Bot.* **5**: 485, 1918).

Argylium Wallr. (1833) = Melanogaster (Melanogastr.).

Argynna Morgan (1895), Argynnaceae. 1, N. Am. See Shearer & Crane (*TBMS* **75**: 193, 1980).

Argynnaceae Shearer & J.L. Crane (1980), Dothideales. 2 gen., 2 spp. Ascomata initially formed from clusters of pseudoparenchymatous cells, sessile, cleistothecial, the peridium parenchymatous or thick and fragmenting into well-defined plates; interascal tissue composed of branched paraphyses or pseudoparaphyses; asci thin- or thick-walled, if the latter then apparently fissitunicate but not discharging actively; ascospores brown, 1-septate, the septa sometimes thickened, inequilateral, smooth. Saprobic on wood, sometimes aquatic. The two genera currently included may not be closely related.
    Lit.: Hawksworth (*SA* **6**: 153, 1987).

arid, dry.

Ariefia Jacz. (1922), Meliolaceae. 1, former USSR.

Ariella J.-E. Gilbert (1941) = Amanita (Amanit.) fide Singer (1975).

Aristadiplodia Shirai (1919) nom. dub. fide Sutton (*Mycol. Pap.* **141**, 1977).

Aristastoma Tehon (1933), Mitosporic fungi, 4.C1.1. 5, USA, Afr. *A. oeconomicum* (zonate leaf spot of cowpea, *Vigna*). See Sutton (*Mycol. Pap.* **97**, 1964; key)

Arkoola J. Walker & Stovold (1986), Venturiaceae. 1, Australia.

Armata W. Yamam. (1958), ? Micropeltidaceae. 1, Japan.

Armatella Theiss. & Syd. (1915), Meliolaceae. 3, trop. See v. Arx (*Fungus* **28**: 1, 1958).

armilla, see annulus.

Armillaria (Fr.) Staude (1857), Tricholomataceae. 40, cosmop. *A. mellea* (the honey fungus), the cause of serious root rots in trees and other plants, is common and variable and has 'shoe-string' rhizomorphs. See Raabe (*Hilgardia* **33**: 25, 1962; host list), Watling *et al.* (*TBMS* **78**: 271, 1982; nomencl., typification, speciation), Morquer *et al.* (*BSMF* **83**: 124, 1967; rhizomorph formation), Anderson & Ullrich (*Mycol.* **71**: 402, 1979; biol. spp. in N. Am), Shaw & Aikle (Eds) (*Armillaria root disease*, 1991), Singer (1962, only 2 spp., temp.; see *Armillariella*), Guillaumin *et al.* (*Eur. J. For. Path.* **23**: 321, 1993; distr. Eur. spp.).

Armillariella (P. Karst.) P. Karst. (1881) = Armillaria (Tricholomat.) fide Dennis *et al.* (*TBMS* **37**: 33, 1954). 26 spp. fide Singer (1978). See also Singer (*Flora neotrop.* **3**: 7, 1970).

armillate, edged; fringed; frilled.

Arnaudia Bat. (1960) = Acantharia (Ventur.) fide Müller & v. Arx (1962).

Arnaudiella Petr. (1927), Microthyriaceae. 3 or 4, Eur., Turkey.

Arnaudina Trotter (1931), Mitosporic fungi, 3.C2.?. 1, Brazil. See Carmichael *et al.* (1980).

Arnaudovia Valkanov (1963) = Polyphagus (Chytrid.) fide Karling (1977).

Arniella Jeng & J.C. Krug (1977) [non Lindb. (1887), *Hepaticae*], Lasiosphaeriaceae. 2, Am.

Arnium Nitschke ex G. Winter (1873), Lasiosphaeriaceae. 11, widespr. See Krug & Cain (*CJB* **50**: 367, 1972; key).

Arnoldia A. Massal. (1856) [non Cass. (1824), *Compositae*] = Lempholemma (Lichin.).

Arnoldia D.J. Gray & Morgan-Jones (1980) ≡ Arnoldiomyces (Mitosp. fungi).

Arnoldiella R.F. Castañeda (1984) [non V.V. Mill. (1928), *Chlorophyta*], Mitosporic fungi, 1.F2.10. 1, Cuba.

Arnoldiomyces Morgan-Jones (1980), Mitosporic fungi, 1.C1.10. 1, USA.

Arongylium, see *Strongylium*.

Aropsiclus Kohlm. & Volkm.-Kohlm. (1994), Ascomycota (inc. sed.). 1, USA.

Arpinia Berthet (1974), Otideaceae. 5, Eur. See Hohmeyer (*Mycol. Helv.* **3**: 221, 1988; key), Häffner (*Mitt. Arbeitsg. Pilzk. Niederrhein* **7**: 132, 1989; key).

arrect, stiffly upright.

Arrhenia Fr. (1849), Tricholomataceae. 12, temp. See Redhead (*CJB* **62**: 865, 1984).

Arrhenosphaera Stejskal (1974), Ascosphaeraceae. 1, Venezuela.

Arrhytidia Berk. & M.A. Curtis (1849) = Dacrymyces (Dacrymycet.) fide Donk (*Persoonia* **4**: 269, 1966).

Arscyria, see *Arcyria*.

arsenic detection, see *Scopulariopsis*.

Artallendea Bat. & H. Maia (1960) = Armatella (Meliol.) fide Katumoto (*Bull. Fac. Agr. Yamag. Univ.* **13**: 291, 1962).

Artheliopsis Vain. (1896) = Echinoplaca (Gomphill.).

Arthonaria Fr. (1825) ? = Enterographa (Roccell.).

Arthonia Ach. (1806) nom. cons., Arthoniaceae (±L). c. 500, cosmop. See Coppins (*Lichenologist* **21**: 195, 1989; Br. Isl.), Grube *et al.* (*Lichenologist* **27**: 25, 1995; key 9 spp. on lichens), Lücking (*Lichenologist* **27**: 127, 1995; Costa Rica), Redinger (*Rabenh. Krypt.-Fl.* **9**: 2 (1), 1936), Santesson (*Symb. bot. upsal.* **12** (1), 1952; foliicolous spp.).

Arthoniaceae Rchb. (1841), Arthoniales (± L) . 7 gen. (+ 49 syn.), 620 spp. Thallus usually absent or poorly developed; ascomata with rudimentary walls, often elongated and branched; hymenium reddish or brownish. Mainly lichen-forming; anamorph coelomycetous where known.

  Lit.: Lücking *Lichenologist* **27**: 127, 1995; 25 foliicolous spp. Costa Rica), Redinger (*Rabenh. Krypt.-Fl.* **9**(2,1), 1936-38; Eur.), Santesson (1952; foliicolous spp.).

Arthoniactis (Vain.) Clem. (1909) = Lecanactis (Roccell.).

Arthoniales, Ascomycota (±L). 4 fam., 61 gen. (+ 123 syn.), 1201 spp. Thallus varied, crustose but sometimes very poorly developed or absent; ascomata usually apothecial but sometimes with a poroid opening, often elongated and branched, ascomatal wall poorly to well-developed. Interascal tissue composed of branched paraphysoids in a gel matrix. Asci thick-walled, ± fissitunicate, usually with a large apical dome, blueing in iodine; ascospores simple or septate, sometimes becoming brown and ornamented, without sheath. Anamorph pycnidial. Forming crustose lichens with green photobionts (esp. trentepohlioid), lichenicolous or saprobes; on a wide range of substrata, incl. many trop. foliicolous and corticolous spp. Fams:

(1) **Arthoniaceae**.

(2) **Chrysothricaceae**.

(3) **Roccellaceae** (syn. *Opegraphaceae*).

  Lit.: Letrouit-Galinou *et al.* (*Bull. Soc. linn. Provence* **45**: 389, 1994; ultrastr. asci), Renobales & Barreno (*Anales jard. bot. Madrid* **46**: 263, 1989; asci), Tehler (*CJB* **68**: 2458, 1990, cladistics; *Crypt. Bot.* **5**: 82, 1995, molec. & morph. cladistics).

Arthoniomyces E.A. Thomas ex Cif. & Tomas. (1953) ≡ Arthonia (Arthon.).

Arthoniopsis Müll. Arg. (1890) = Arthonia (Arthon.) fide Santesson (1952).

Arthopyrenia A. Massal. (1852), Arthopyreniaceae (±L). c. 100, cosmop. See Coppins (*Lichenologist* **20**: 305, 1988; Br. Is.), Foucard (*Graphis Scripta* **4**: 49, 1992; key 13 spp. on bark, Sweden), Gams (*Taxon* **41**: 99, 1992; nomencl.), Vainio (*Acta Soc. Fauna Flora fenn.* **49** (2), 1921), Swinscow (*Lichenologist* **3**: 55, 1965), Keissler (*Rabenh. Krypt. -Fl.* **9**, 1(2), 1938), Harris (*Mich. Bot.* **12**: 1, 1973), Riedl (*Sydowia* **29**: 115, 1977; *A. punctiformis*-group).

Arthopyreniaceae Walt. Watson (1929), Dothideales (±L). 7 gen. (+ 30 syn.), c. 195 spp. Thallus absent or poorly developed; ascomata with a single perithecial locule, black, spherical or flattened, composed of pseudoparenchymatous cells, with a broad ostiole; interascal tissue of narrow cellular pseudoparaphyses, sometimes immersed in gel, sometimes evanescent; asci clavate or elongate, fissitunicate, without a clear apical structure; ascospores hyaline or brown, sometimes transversely septate and/or weakly ornamented, thin-walled. Anamorph (where known) coelomycetous, probably spermatial. Saprobic, especially in smooth bark, or lichenized.

  Lit.: Harris (1973).

Arthopyreniella J. Steiner (1911) = Mycoglaena (Dothideales) fide Harris (*in litt.*).

Arthopyreniomyces Cif. & Tomas. (1953) ≡ Pyrenyllium (Ascomycota; inc. sed.). See Aguirre-Hudson (*Bull. Br. Mus. nat. Hist., Bot.* **21**: 85, 1991).

Arthotheliomyces Cif. & Tomas. (1953) = Arthothelium (Arthoniales).

Arthotheliopsidomyces Cif. & Tomas. (1953) ≡ Arthotheliopsis (Gomphill.).

Arthotheliopsis Vain. (1896) = Echinoplaca (Gomphill.).

Arthothelium A. Massal. (1852), ? Arthoniales (±L). c. 80, cosmop. See Santesson (*Symb. bot. upsal.* **12** (1), 1952), Coppins & James (*Lichenologist* **11**: 27, 1979; key Br. spp.), Tehler (*CJB* **68**: 2451, 1990; posn).

arthric (of conidiogenesis), thallic conidiogenesis characterized by the conversion of a pre-existing, determinate hyphal element into a conidium (**arthroconidium**, thallic-arthroconidium, arthrospore), as in *Geotrichum*. See arthrocatenate.

Arthriniites Babajan & Tasl. (1977), Fossil fungi. 1 (Tertiary), former USSR.

Arthrinium Kunze (1817), Anamorphic Lasiosphaeriaceae, 1.A2.37. Teleomorph *Apiospora*. 18, temp., widespr. See Ellis (*Mycol. Pap.* **103**, 1965; key), Minter (*Proc. Ind. Acad. Sci.* (Pl. Sci.) **94**: 281, 1985; relationships with other mitosporic fungi), Rai (*Mycoses* **32**: 472, 1989; *A. phaeospermum* var. *indicum* in humans).

arthro- (prefix), jointed.

Arthroascus Arx (1972), Saccharomycopsidaceae. 2, Eur. See Smith *et al.* (*Ant. v. Leeuwenhoek* **58**: 249, 1990).

Arthrobotryella Sibilia (1928), Mitosporic fungi, 1.B2.6. 1, Eur. See Hughes (*N.Z. Jl Bot.* **16**: 326, 1978). ? = Cordana (Mitosp. fungi).

Arthrobotryomyces Bat. & J.L. Bezerra (1961), Mitosporic fungi, 2.C1.?. 1 (L), Brazil.

Arthrobotrys Corda (1839), Mitosporic fungi, 2.B1.6. 28, widespr. See Haard (*Mycol.* **60**: 1140, 1969; key), Jarowaja (*Acta Mycol.* **6**: 337, 1970; key), van Oorschot (*Stud. Mycol.* **26**: 61, 1985; key), Werthmann-Cliemas & Lysek (*TBMS* **87**: 656, 1987; synnema formation). See also Predacious fungi.

Arthrobotryum Ces. (1854), Mitosporic fungi, 2.C2.19. 1, Eur. See Hughes (*Naturalist* (Hull) 1951: 171, 1951).

Arthrobotryum O. Rostr. (1916) ≡ Gonyella (Mitosp. fungi).

arthrocatenate (of thalloconidia), formed in chains by the simultaneous or random fragmentation of a hypha.

Arthrocladia Golovin (1956) [non Duby (1830), *Algae*] ≡ Arthrocladiella (Erysiph.).

Arthrocladiella Vassilkov (1960), Erysiphaceae. 1, Eur., Asia, N. Am., NZ. Anamorph *Oidium*.

Arthrocladium Papendorf (1969), Mitosporic fungi, 1.C2.1/10. 1 (from soil), S. Afr.

arthroconidium, see arthric.

Arthrocristula Sigler, M.T. Dunn & J.W. Carmich. (1982), Mitosporic fungi, 1.A2.40. 1, USA.

Arthroderma Curr. (1860), Arthrodermataceae. 23, widespr. See Currah (1985), Kawasaki *et al.* (*Mycopath.* **118**: 95, 1992), Padhye & Carmichael (*CJB* **49**: 1525, 1971; key 13 spp.), Takashio *et al.* (*Mycol.* **77**: 166, 1985; ontogeny), Weitzman *et al.* (*Mycotaxon* **15**: 505, 1986). Anamorphs *Trichophyton*, *Chrysosporium*.

Arthrodermataceae Locq. ex Currah (1985), Onygenales. 2 gen., 24 spp. Ascomata not stipitate; peridium hyphal, usually with complex appendages; ascospores ± oblate, smooth. Anamorphs *Chrysosporium*, *Microsporum* or *Trichophyton*. Keratinophilic, on hair, skin etc., sometimes pathogenic, cosmop.

Arthrodochium R.F. Castañeda & W.B. Kendr. (1990), Mitosporic fungi, 3.A1.38. 1 (with clamp connexions), Cuba.

Arthrographis G. Cochet ex Sigler & J.W. Carmich. (1976), Mitosporic fungi, 1.A1.38. 4, cosmop. See Sigler & Carmichael (*Mycotaxon* **18**: 495, 1983), Sigler *et al.* (*Can. J. Microbiol.* **36**: 77, 1990; n.sp.), Ayer & Nozawa (*Can. J. Microbiol.* **36** : 83, 1990; inhibitory metabolite), Uchida *et al.* (*Trans. mycol. Soc. Jap.* **34**: 275, 1993; *A. cuboidea*).

Arthrographium Ces. (1854) = Arthrobotryum Ces. (Mitosp. fungi) fide Mussat (1901).

Arthromitus Leidy (1849) nom. dub.; bacteria occurring as trichomes in intestines of millipedes, cockroaches and toads; formerly incorrectly placed in *Trichomycetes*.

Arthroon Renault (1894), ? Fossil fungi. 1 (Carboniferous), France.

Arthropsis Sigler, M.T. Dunn & J.W. Carmich. (1982), Mitosporic fungi, 1.A1/2.38. 1, Peru.

Arthropycnis Constant. (1992), Anamorphic Trichosphaeriaceae. 4.A2.39. Teleomorph *Rhynchostoma*. 1, N. Eur., NZ.

Arthrorhaphidaceae Poelt & Hafellner (1976), Patellariales (±L). 1 gen. (+ 4 syn.), 7 spp. Thallus crustose; ascomata often strongly cupulate; peridium composed of hyphae with swollen walls; ascospores hyaline, elongated, multiseptate. Anamorphs not known; lichenized with green algae or parasitic on other lichens, sometimes eventually free-living; widespr. in temp. regions.

Arthrorhaphis Th. Fr. (1860) nom. cons. prop., Arthrorhaphidaceae (±L). 7, cosmop., temp., alpine. See Obermayer (*Nova Hedw.* **58**: 275, 1994; key 5 Eur. spp.), Galloway & Bartlett (*N.Z. Jl Bot.* **24**: 393, 1986; NZ spp.), Hafellner & Obermayer (*Crypt., bryol. lich.* **16**: 177, 1995; key fungi on).

Arthrorhynchus Kolen. (1857), Laboulbeniaceae. 6, Eur., Afr., Asia. See Benjamin (1971), Blackwell (*Mycol.* **72**: 159, 1980; morphology).

Arthrospora Th. Fr. (1861) ≡ Arthrosporum (Catill.).

arthrospore, (1) see arthric; (2) a specialized uninucleate cell functioning as a spore and derived from the disarticulation of cells of a formerly vegetative branch (*Asellariales*).

Arthrosporella Singer (1970), Tricholomataceae. 1, S. Am.

Arthrosporia Grigoraki (1925) = Trichophyton (Mitosp. fungi).

Arthrosporium Sacc. (1880), Mitosporic fungi, 2.C1.10. 2, Italy, N. Am. See Wang (*Mycol.* **64**: 1175, 1972).

Arthrosporum A. Massal. (1853), Catillariaceae (L). 1, Eur. See Timdal (*Opera Bot.* **110**, 1991).

arthrosterigma (of lichens), a septate conidiophore (spermatiophore) (obsol.).

Arthur (Joseph Charles; 1850-1942). Botanist, Agricultural Experiment Station, Geneva, N.Y., 1884-7; Purdue University, 1887-1915. Noted for his work on the *Uredinales* and the 'Arthur Herbarium' at Purdue is one of the most important collections of rusts (75,000 specimens; Baxter & Kern, *Proc. Indiana Acad. Sci.* **71**: 228, 1962). Chief writings: *Uredinales, N. Am. Flora* **7**, 1907-27; *Plant rusts*, 1929; *Manual of the rusts in United States and Canada*, 1934. See Mains (*Mycol.* **34**: 601, 1942), Kern (*Phytopath.* **32**: 833, 1942), Cummins (*Ann. Rev. Phytopath.* **16**: 19, 1978), Urban (*Česká Myk.* **25**: 185, 1971), Stafleu & Cowan, (*TL-2* **1**: 70, 1976) Stafleu & Mennega (*TL-2, Suppl.* **1**: 173, 1992).

Arthurella Zebrowski (1936), ? Fossil fungi (Chytrid.). 1 (Cambrian to ? Recent), Australia.

Arthuria H.S. Jacks. (1931), Phakopsoraceae. 3 (on *Euphorbiaceae*), Am., India.

Arthuriomyces Cummins & Y. Hirats. (1983), Phragmidiaceae. 1 (on *Rubus*), N. Am.

Articularia Höhn. (1909), Mitosporic fungi, 2.A1.?. 1, N. Am. See Charles (*Mycol.* **27**: 74, 1935).

Articulariella Höhn. (1909) = Microstoma (Sarcoscyph.) fide v. Arx (1970).

articulated, jointed.

Articulis Clem. & Shear (1931) ≡ Articulariella (Mitosp. fungi).

Articulophora C.J.K. Wang & B. Sutton (1982), Mitosporic fungi, 1.D2.11. 1, USA.

Articulospora Ingold (1942), Mitosporic fungi, 1.G.10. 4 (aquatic), N. temp. See Petersen (*Mycol.* **54**: 143, 1962; key).

Artocreas Berk. & Broome (1875) nom. ambig.

Artolenzites Falck (1909) = Trametes (Coriol.) fide Donk (1974).

Artomyces Jülich (1982), Hericiaceae. 8, widespr.

Artotrogus Mont. (1845) = Pythium (Pyth.) fide Middleton (1943).

Artymenium Berk. ex E. Fisch. (1933) nom. inval. = Secotium (Secot.) s.l.

Arualis Katz (1980), Mitosporic fungi, 1.1/2A.1. 1 (with clamp connexions), USA.

Arundinella L. Léger & Duboscq (1905) [non Raddi (1823), *Gramineae*] = Arundinula (Eccrin.).

Arundinula L. Léger & Duboscq (1906), Eccrinaceae. 6 (in *Decapoda*), widespr. See Lichtwardt (*Mycol.* **54**: 440, 1962; morphology), Hibbets (*Syesis* **11**: 213, 1978; development), Lichtwardt (1986; key), Van Dover & Lichtwardt (*Biol. Bull.* **171**: 461, 1986).

Arwidssonia B. Erikss. (1974), Hyponectriaceae. 2, Eur.

Arx (Joseph Adolf, von; 1922-88). Swiss by birth, and with Emil Müller were both students of E. Gäuman at Zurich. In 1949 Arx was appointed to the Willie Commelin Scholten Phytopathological Laboratories, Baarn, Netherlands, and in 1963 succeeded Dr A.L. van Beverwijk as Director of the **CBS**. Arx's thesis was on *Mycosphaerella*, a genus in which he maintained a life-long interest, and in 1957 in his monograph on *Gloeosporium* he synonymised numerous generic names. He is also noted for his review of the fungal kingdom, 1970 (expanded in 1974 and 1987) and his standing as a general mycologist. He published in collaboration with E. Müller for many years. *Obit.*: van der Aa *et al.* (*Stud. Mycol.* **31**, 1989; portr., bibl.), Arnold (*Boletus* **13**: 24, 1989; portr.).

Arxiella Papendorf (1967), Mitosporic fungi, 1.B1.9. 2, S. Afr., NZ, India.

Arxiomyces P.F. Cannon & D. Hawksw. (1983), Ceratostomataceae. 2, Eur., Japan. See Horie *et al.* (*Mycotaxon* **25**: 229, 1986).

Arxiozyma Van der Walt & Yarrow (1984), Saccharomycetaceae. 1, S. Afr.

Arxula Van der Walt, M.T. Sm. & Y. Yamada (1989), Mitosporic fungi, 1.A1.38. 2 (anamorphic yeasts), widespr. See Yamada & Nogawa (*J.Gen. Appl. Microbiol.* **36**: 425, 1990; molecular phylogeny), Kunze & Kunze (*Microbiol. Eur.* **2**: 24, 1994; comparative morphology), Kunze & Kunze (*Ant. v. Leeuwenhoek* **65**: 29, 1994; DNA fingerprinting of *A. adeninivorans*).

Asahina (Yasuhiko; 1881-1975). Japanese natural product chemist and later lichenologist; Professor Univ. Tokyo; a founder of lichen chemotaxonomy who introduced the use of PD (q.v.) in 1934 and of microcrystal tests in 1936-40 (*J. Jap. Bot.* **12-16**); author of *Lichens of Japan* (3 vol., 1950-56), (with Shibata) *Chemistry of lichen substances* (1954 [reprint 1971]), *Atlas of Japanese Cladoniae* (1971). See Culberson & Culberson (*Bryologist* **79**: 258, 1976; portr.), Grummann (1979: 585), Kurokawa (*Lichenologist* **8**: 93, 1976; portr.), *Dr. Yasuhiko Asahina's lichenological bibliography* (1980; 281 titles), Stafleu & Cowan (*TL-2* **1**: 72, 1976), Stafleu & Mennega (*TL-2, Suppl.* **1**: 184, 1992).

Asahinea W.L. Culb. & C.F. Culb. (1965), Parmeliaceae (L). 2, circumpolar. See Gao (*Nordic J. Bot.* **11**: 483, 1991).

Asaphomyces Thaxt. (1931), Laboulbeniaceae. 2 or 3, N. & S. Am., Eur.

Asbolisia Speg. (1918) nom. dub. fide Sutton (*Mycol. Pap.* **141**, 1977).

Asbolisia Bat. & Cif. (1963), Anamorphic Capnodiaceae, 4.A1.?. Teleomorph *Aithaloderma*. 6, C. & S. Am., Tonga.

Asbolisiomyces Bat. & H. Maia (1961), Mitosporic fungi (L), 4.A1.?. 1, Brazil.

Ascagilis K.D. Hyde (1992), Dothideales (inc. sed.). 1 (aquatic), Australia.

ascending (**ascendent**) (of an annulus), having the free edge above attached, cf. descending; (of conidiophores), curving up, cf. erect; (of lamellae), on a cone-like or an unexpanded pileus.

Aschersonia Endl. (1842) nom. rej. ≡ Laschia (Polypor.), = Junghuhnia (Polypor.) fide Ryvarden (1972).

Aschersonia Mont. (1848) nom. cons., Anamorphic Clavicipitaceae, 8.A/B1.15. Teleomorph *Hypocrella*. 20 (on whiteflies (*Aleyrodidae*) and scale insects (*Coccidae*)), warmer areas. See Mains (*Lloydia* **22**: 215, 1960), Petch (*Ann. R. bot. Gdn Peraden.* **7**: 167, 1921).

Aschersoniopsis Henn. (1902) = Munkia (Mitosp. fungi) fide Höhnel (1917).

Aschion Wallr. (1833) ≡ Tuber (Tuber.).

Aschizotrichum Rieuf (1962) = Wiesneriomyces (Mitosp. fungi) fide Carmichael *et al.* (1980).

Ascidiophora Rchb. [not traced] ? = Mucor (Mucor.) fide Mussat (1901).

Ascidium Tode (1782) nom. dub.; based on insect eggs fide Fries (*Syst. mycol.* **3**, Index: 52, 1832).

Ascidium Fée (1824) nom. rej. ≡ Ocellularia (Thelotremat.) fide Hale (1981).

ascigerous, having asci.

ascigerous centrum, the special tissue which produces the asci and hamathecium.

Ascluella DiCosmo, Nag Raj & W.B. Kendr. (1983), Dermateaceae. 1, India.

asco- (prefix), pertaining to an ascus.

Ascoblastomycetes, see *Blastomycota*.

Ascobolaceae Boud. ex Sacc. (1884), Pezizales. 9 gen. (+ 10 syn.), 106 spp. Ascomata apothecial, rarely cleistothecial, usually pulvinate, fleshy, brightly coloured, not setose; asci broad, operculate, protruding above the hymenium when mature, rarely with an I+ wall; ascospores biseriately arranged, thick-walled, usually ornamented, often with purple/brown epispore, sometimes with a sheath, sometimes dispersed as a compound unit; mostly coprophilous, esp. temp.
Lit.: van Brummelen (*Persoonia, Suppl.* **1**, 1967; monogr.).

Ascobolus Pers. (1791), Ascobolaceae. *c.* 50 (mainly coprophilous), widespr. See van Brummelen (1967; key), Kempken (*Bibl. Mycol.* **128**, 1989; extrachromosomal DNA), Paulsen & Dissing (*Bot. Tidsskr.* **74**: 67, 1979; key 20 spp.), Kaushal & Thind (*J. Indian bot. Soc.* **62**: 16, 1983; W. Himalayas, key 12 spp.), Parrettini (*Boln. Gr. Micol. Bresad.* **28**: 140, 1985; col. pl.), Wells (*Univ. Calif. Publs Bot.* **62**, 1972; ontogeny), Prokhorov (*Mikol. i Fitopatol.* **28**: 17, 1994; key to Russian spp.).

Ascocalathium Eidam ex J. Schröt. (1893), ? Otideaceae. 1, Eur.

Ascocalvatia Malloch & Cain (1971), Onygenaceae. 1 (coprophilous), Canada.

Ascocalyx Naumov (1926), Leotiaceae. 6, widespr. See Groves (*CJB* **46**: 1273, 1968; key), Schlapfer-Bernard (*Sydowia* **22**: 1, 1969), Müller & Dorworth (*Sydowia* **36**: 193, 1983; key, 6 spp.), Petrini *et al.* (*CJB* **67**: 2805, 1989). Anamorph *Bothrodiscus*.

ascocarp, see ascoma.

Ascocephalophora Matsush. (1995), ? Endomycetaceae. 1, Japan. Anamorphs *Fusidium*, *Trichosporiella*.

Ascochyta Lib. (1830), Anamorphic Dothideales, Mycosphaerellaceae and Leptosphaeriaceae, 4.B1.15. Teleomorphs *Didymella*, *Mycosphaerella*, *Leptosphaeria*. *c.* 350, cosmop. *A. fabae* (on *Vicia*), *A. phaseolorum* (on *Phaseolus*), *A. pisi* (on pea), *A. rabiei* (on chickpea), and others. See Chaube & Mishra in Singh *et al.* (Eds) (*Plant Diseases of International Importance: Diseases of Cereals and Pulses*, 1992), Melnik (*Key to the species of Ascochyta* (Leningrad), 1977; key 328 spp.; many redispositions), Punithalingam (*Mycol. Pap.* **142**, 1979; 79 graminicolous spp.; *Mycol. Pap.* **159**, 1988; 63 monocotyledonous spp.), Buchanan (*Mycol. Pap.* **156**, 1987; reappraisal of *Ascochyta* and *Ascochytella*), El Shanawani *et al.* (*Minufiya Jl Agr. Res.* **8**: 39, 1984; TEM of conidiogenesis), Upadhyay *et al.* (*CJB* **69**: 797, 1991; *A. cypericola* and leaf blight of *Cyperus*), Morgan-Jones *et al.* (*Mycotaxon* **42**: 53, 1991; *A. andropogonivora*), Djiwanti, Kobayashi & Oniki (*Mycoscience* **35**: 161, 1994; *A. rhei* on rhubarb).

Ascochytella Tassi (1902) = Ascochyta (Mitosp. fungi) fide Buchanan (1987).

Ascochytites Babajan & Tasl. (1973), Fossil fungi. 1 (Tertiary), former USSR.

Ascochytites Barlinge & Paradkar (1982), Fossil fungi (Mitosp. fungi). 1 (Cretaceous), India.

Ascochytopsis Henn. (1905), Mitosporic fungi, 7.E1.15. 4 (on *Leguminosae*), widespr. See Sutton (1980).

Ascochytula (Potebnia) Died. (1912) = Ascochyta (Mitosp. fungi) fide Buchanan (1987).

Ascochytulina Petr. (1922), Mitosporic fungi, 4.B2.15. 1, Eur.

Ascoclavulina Y. Otani (1974), Leotiaceae. 1, Japan. Anamorph *Gliomastix*.

Ascocoma H.J. Swart (1987), ? Phacidiaceae. 1, Australia.

ascoconidiophore, the phialide bearing an ascoconidium in *Ascoconidium* (Seaver, *Mycol.* **34**: 412, 1942).

ascoconidium, a conidium formed directly from an ascospore, esp. when still within the ascus (e.g. *Clausenomyces*).

Ascoconidium Seaver (1942), Anamorphic Leotiaceae, 3.C1.15. Teleomorph *Sageria*. 2, N. temp. See Nag Raj & Kendrick (*A monograph of Chalara and allied genera*, 1975), Funk (*CJB* **44**: 39, 1966).

Ascocorticiaceae J. Schröt. (1893), Leotiales. 3 gen., 5 spp. Stromata absent; ascomata effuse, forming an indefinite palisade, white or pinkish; peridium ± absent; interascal tissue not well defined, composed of irregular undifferentiated hyphae interspersed with the asci; asci with an I- apical ring; ascospores small, hyaline, aseptate. Anamorphs unknown. Saprobic, esp. on conifer bark, temp.

Ascocorticiellum Jülich & B. de Vries (1982), Ascomycota (inc. sed.). 1, Eur.

Ascocorticium Bref. (1891), Ascocorticiaceae. 3, temp. See Cooke (*Ohio J. Sci.* **68**: 161, 1968), Eriksson *et al.* (*Göteborgs Svampkl. Årsskr.* **1981**: 1, 1981).

Ascocoryne J.W. Groves & D.E. Wilson (1967), Leotiaceae. *c.* 6, Eur., N. Am. See Roll-Hansen & Roll-Hansen (*Norw. J. Bot.* **26**: 193, 1979), Christiansen (*Friesia* **7**: 75, 1963; anamorph, Danish spp.), Verkley (*MR* **99**: 187, 1995; asci). Anamorph *Coryne*.

Ascocorynium S. Ito & S. Imai ex S. Imai (1934) = Neolecta (Neolect.) fide Korf (*Phytol.* **21**: 201, 1971).

Ascocratera Kohlm. (1986) = Anisomeridium (Monoblast.). See Harris (*in* Aptroot, *Bibl. Mycol.* **44**, 1991).

Ascocybe D.E. Wells (1954) = Cephaloascus (Cephaloasc.) fide v. Arx (1972).

Ascodesmidaceae J. Schröt. (1893), Pezizales. 2 gen. (+ 1 syn.), 12 spp. Ascomata ± globose, formed from coiled antheridial and ascogonial branches; peridium absent; interascal tissue poorly developed; asci saccate, with a very large operculum; ascospores pigmented, with ornamentation derived directly from the secondary wall layer. Anamorphs unknown. From soil or coprophilous etc., widespr.
   Lit.: van Brummelen (1972, *Persoonia* **14**: 1, 1989; ascus ultrastr.). Von Arx (*Persoonia* **13**: 273, 1987) suggested that the *Amauroascaceae* belongs with this fam.

Ascodesmis Tiegh. (1876), Ascodesmidaceae. 6, widespr. See v. Brummelen (*Persoonia* **11**: 377, 1981), Obrist (*CJB* **39**: 943, 1961; key), Delattre-Durand & Janex-Favre (*BSMF* **95**: 49, 1979; ontogeny), Patil & Ghadge (*Indian Phytopath.* **40**: 30, 1987; 5 spp.).

Ascodesmisites Trivedi, Chaturv. & C.L. Verma (1973), Fossil fungi. 1 (Eocene), Malaysia. See Korf (*Mycotaxon* **6**: 193, 1977).

Ascodichaena Butin (1977), Ascodichaenaceae. 2, Chile, Eur., N. Am. See Hawksworth (*Taxon* **32**: 212, 1983; nomencl.). Anamorph *Polymorphum*.

Ascodichaenaceae D. Hawksw. & Sherwood (1982), Rhytismatales. 3 gen. (+ 4 syn.), 26 spp. Ascomata apothecial, erumpent in clusters, elongate, opening by a longitudinal split, the wall tissue stromatal in origin, composed of vertically oriented pseudoparenchymatous tissue, black; interascal tissue of simple paraphyses; asci clavate or saccate, thin-walled, with a wide apical ring; ascospores ellipsoidal, hyaline aseptate, without a sheath. Anamorphs coelomycetous, disseminative. Saprobic on bark, N. temp.
   Lit.: Hawksworth & Sherwood (*Mycotaxon* **16**: 262, 1982).

ascogenous (ascogenic), ascus-producing or -supporting.

ascogonium, the cell or group of cells in Ascomycotina fertilized by a sexual act.

Ascographa Velen. (1934), ? Leotiales (inc. sed.). 1, former Czechoslovakia.

Ascohansfordiellopsis D. Hawksw. (1979) = Koordersiella (Dothideales) fide Hawksworth (1989).

Ascohymeniales. *Ascomycota* having asci (and paraphyses) developing as a hymenium and not in a pre-formed stroma, as in *Pyrenomycetes* and *Discomycetes* (Nannfeldt, 1932); *Hymenoascomycetes*. Cf. *Ascoloculares*.

Ascoidales, see *Saccharomycetales*.

Ascoidea Bref. (1891), Endomycetaceae. 4, widespr. See Batra & Francke-Grossman (*Mycol.* **56**: 632, 1964; key), v. Arx & Müller (*Sydowia* **37**: 6, 1984).

Ascoideaceae, see *Endomycetaceae*.

Ascoideales, see *Saccharomycetales*.

Ascolanthanus Cailleux (1967) = Pyxidiophora (Pyxidiophor.) fide Lundqvist (1980).

Ascolectus Samuels & Rogerson (1990), Saccardiaceae. 1, Brazil.

Ascoloculares. *Ascomycota* having asci (and paraphyses) developing in cavities in a pre-formed stroma, as in *Loculoascomycetes* (Nannfeldt, 1932). Cf. *Ascohymeniales*.

ascoma (pl. **ascomata**), an ascus-containing structure, ascocarp.

Ascomyces Mont. & Desm. (1848) = Taphrina (Taphrin.) etc. fide Mussat (*Syll. Fung.* **15**: 51, 1901). Sometimes used for *Ginanniella* (Tillet.) anamorphs.

ascomycete, one of the *Ascomycota*.

Ascomycetella Peck (1881) = Cookella (Cookell.).

Ascomycetella Sacc. (1886) ≡ Myriangiopsis (Dothideales).

Ascomycetes, see *Ascomycota*.

Ascomycota (Ascolichenes, Ascomycetes, Ascomycotina, Ascomycoptera, Thecamycetes). Fungi (±L); sac fungi; ascomycetes. 46 ord., 264 fam., 3,266 gen. (+ 3,173 syn.), 32,267 spp.; saprobes, parasites (esp. of plants), or lichen-forming; cosmop. The largest group of *Fungi*, for which the ascus (q.v.) is the diagnostic character. The presence of lamellate hyphal walls with a thin electron-dense outer layer and a relatively thick electron-transparent inner layer also appears diagnostic; this enables Mitosporic fungi to be recognized as ascomycetes even in the absence of asci (v. Arx, 1987). In the past they have often been grouped on how the asci were arranged (*Hemiascomycetes*, *Plectomycetes*, *Pyrenomycetes*, *Discomycetes*, *Laboulbeniomycetes*, *Loculoascomycetes*; q.v.); in recent decades the development of the ascomata, and especially the structure and method of discharge of the asci, have been given increasing importance. The size of the group makes it difficult to embrace the enormous range of structures in the group, and to determine which features should be stressed in the recognition of higher categories. A particular problem with early classifications was that lichen-forming fungi (q.v.), almost half of all ascomycetes, had often been classified separately. A simple intercalation of them into a hierarchial system based only on variation in non-lichenized fungi was not to be expected.

There is considerable variation between the systems of higher categories proposed for *Ascomycota* (Table 1), and in the last (1983) edition of this *Dictionary* no categories above order were accepted; this practice is also adopted in this edition, following the *Outline of the ascomycetes - 1993* (Eriksson & Hawksworth, 1993). This approach more accurately reflects what we can have confidence in (i.e. well-circumscribed and increasingly monophyletic orders related

TABLE 1. Selected classification proposed for the *Ascomycota* 1951-1993

Luttrell (1951)	Ainsworth *et al.* (1973)	Barr (1976)
SYNASCOMYCETES	HEMIASCOMYCETES	HEMIASCOMYCETES
Protomycetales	Endomycetales	Protomycetales
	Protomycetales	Taphrinales
HEMIASCOMYCETES	Taphrinales	Spermophthorales
Dipodascales		Cephaloascales
Endomycetales	PLECTOMYCETES	Dipodascales
Taphrinales	Eurotiales	Ascoideales
EUASCOMYCETES	PYRENOMYCETES	EUASCOMYCETES
Bitunicatae	Erysiphales	*Plectoascomycetidae*
Myriangiales	Meliolales	Eurotiales
Dothideales	Coronophorales	*Laboulbeniomycetidae*
Pseudosphaeriales	Sphaeriales	Spathulosporales
Hysteriales		Laboulbeniales
Trichothyriales	DISCOMYCETES	*Parenchemycetidae*
Unitunicatae	Mediolariales	Erysiphales
Plectomycetes	Cyttariales	Meliolales
Aspergillales	Tuberales	Diaporthales
Microascales	Pezizales	Sordariales
Onygenales	Phacidiales	Coronophorales
Pyrenomycetes	Ostropales	*Anoteromycetidae*
Xylariales	Helotiales	Ostropales*
Hypocreales		Gyalectales*
Diaporthales	LABOULBENIOMYCETES	Clavicipitales
Erysiphales	Laboulbeniales	Hypocreales
Coronophorales		Chaetomiales
Coryneliales	LOCULOASCOMYCETES	*Elaphomycetidae*
Laboulbeniomycetes	Myriangiales	Mediolariales
Laboulbeniales	Dothideales	Cyttariales
Discomycetes	Pleosporales	Coryneliales
Pezizales	Hysteriales*	Pezizales
Tuberales	Hemisphaeriales	Tuberales
Cyttariales		Helotiales*
Ostropales		Lecanorales*
Helotiales		Caliciales*
Lecanorales*		Rhytismatales
Caliciales*		Phyllachorales*
		Xylariales
		LOCULOASCOMYCETES
		Loculoplectascomycetidae
		Myriangiales*
		Loculoparenchemycetidae
		Asterinales
		Dothideales*
		Loculoanoterromycetidae
		Chaetothyriales
		Verrucariales*
		Loculoedaphomycetidae
		Hysteriales*
		Pleosporales*
		Melanommatales*

* Orders including at least some lichen-forming fungi (as circumscribed by the author indicated).

TABLE 1. (Cont.)

Dictionary (1983)	Barr (1987, 1990)[1]	Eriksson & Hawksworth (1993)[2]
Arthoniales*	MONOCARYOMYCOTERA	Arthoniales*
Ascosphaerales	ENDOMYCOTA	Caliciales*
Caliciales*	ENDOMYCETES	Calosphaeriales
Clavicipitales		Coryneliales
Coryneliales	DICARYOMYCOTERA	Cyttariales
Cyttariales	TAPHRINOMYCOTA	Diaporthales
Diaporthales	TAPHRINOMYCETES	Diatrypales
Diatrypales		Dothideales*
Dothideales*	ASCOMYCOTA	Elaphomycetales
Elaphomycetales	ASCOPHAEROMYCETES	Erysiphales
Endomycetales	LABOULBENOMYCETES	Eurotiales
Erysiphales	HYMENOASCOMYCETES	Glaziellales
Eurotiales	*Plastoascomycetidae*	Gyalectales*
Graphidales*	Onygenales	Halosphaeriales
Gyalectales*	Eurotiales	Hypocreales
Gymnoascales	*Parenchymatomycetidae*	Laboulbeniales
Helotiales*	Erysiphales	Lahmiales
Hypocreales	Meliolales	Lecanorales*
Laboulbeniales	Coryneliales	Leotiales
Lecanidiales*	Spathulosporales	Lichinales
Lecanorales*	Diaporthales	Medeolariales
Microascales	Halosphaeriales	Meliolales
Opegraphales*	Sordariales	Microascales
Ophiostomatales	Microascales	Neolectales
Ostropales*	*Anoteromycetidae*	Onygenales
Peltigerales*	Hypocreales	Ophiostomatales
Pertusariales*	*Edaphomycetidae*	Ostropales*
Pezizales	Clavicipitales	Patellariales*
Polystigmatales	Xylariales	Peltigerales*
Pyrenulales*	Calosphaeriales	Pertusariales*
Rhytismatales		Pezizales
Sordariales	LOCULOASCOMYCETES	Phyllachorales
Sphathulosporales	*Loculoplectoascomycetidae*	Protomycetales
Sphaeriales*	Myriangiales	Pyrenulales*
Taphrinales	Arthoniales*	Rhytismatales
Teloschistales*	*Loculoparenchymatomycetidae*	Saccharomycetales
Verrucariales*	Asterinales	Schizosaccharomycetales
	Capnodiales	Sordariales
	Dothideales*	Sphathulosporales
	Loculoanteromycetidae	Taphrinales
	Chaetothyriales	Teloschistales*
	Loculoedaphomycetidae	Triblidiales
	Opegraphiales*	Trichosphaeriales
	Patellariales	Verrucariales*
	Pleosporales	Xylariales
	Melanommatales	

*Orders include at least some lichen-forming fungi (as circumscribed by the authors indicated).
[1]Exclusively lichenized and non-pyrenocarpous orders not detailed; see Barr (1976).
[2]This system is adopted in this edition of the *Dictionary* with the delection of *Glaziellales*, and the addition
  of two recently recognized orders, *Pneumocystidales* and *Trichotheliales*. 29 fams not assigned to any order.

to their biology and ecology; see Dick & Hawksworth, *Bot. J. Linn. Soc.* **91**: 175, 1985) in contrast to that which is as yet controversial or speculative. A fresh *Outline* is now published every four years, based on inputs from numerous specialists; preparation of that for 1993 involved a workshop of 140 participants (Hawksworth, 1994). Proposals for changes at the generic level and above are detailed in twice-yearly issues of *Systema Ascomycetum* (1982 on).

46 orders are accepted here, compared with 39 in the last *Dictionary*; this is due to an increasing awareness of the fundamental levels of difference of certain families from other members of the order in which they were placed, but 5 have been merged with other accepted orders as more data have become available. The accepted orders are listed below with indications of the current placements of ones not now used that were accepted in the last edition of the *Dictionary* in brackets. However, note that 29 accepted families are not referred to any accepted order, and over 500 genera could not be assigned with confidence to any family. Diagnostic characters are given in the separate order entries. * = orders incl. some or only of lichen-forming spp.

(1) **Arthoniales** (incl. *Opegraphales*)*.
(2) **Caliciales***.
(3) **Calosphaeriales**.
(4) **Coryneliales**.
(5) **Cyttariales**.
(6) **Diaporthales**.
(7) **Diatrypales**.
(8) **Dothideales***.
(9) **Elaphomycetales**.
(10) **Erysiphales**.
(11) **Eurotiales** (incl. *Ascosphaerales*).
(12) **Gyalectales***.
(13) **Halosphaeriales**.
(14) **Hypocreales** (incl. *Clavicipitales*).
(15) **Laboulbeniales**.
(16) **Lahmiales**.
(17) **Lecanorales***.
(18) **Leotiales** (syn. *Helotiales*).
(19) **Lichinales***.
(20) **Medeolariales**.
(21) **Meliolales**.
(22) **Microascales**.
(23) **Neolectales**.
(24) **Onygenales** (syn. *Gymnoascales*).
(25) **Ophiostomatales**.
(26) **Ostropales** (incl. *Graphidales*)*.
(27) **Patellariales** (syn. *Lecanidiales*)*.
(28) **Peltigerales***.
(29) **Pertusariales***.
(30) **Pezizales** (incl. *Glaziellales*).
(31) **Phyllachorales** (syn. *Polystigmatales*).
(32) **Pneumocystidales**.
(33) **Protomycetales**.
(34) **Pyrenulales***.
(35) **Rhytismatales**.
(36) **Saccharomycetales** (syn. *Endomycetales*).
(37) **Schizosaccharomycetales**.
(38) **Sordariales**.
(39) **Spathulosporales**.
(40) **Taphrinales**.
(41) **Teloschistales***.
(42) **Triblidiales**.
(43) **Trichosphaeriales**.
(44) **Trichotheliales**.
(45) **Verrucariales***.
(46) **Xylariales** (syn. *Sphaeriales*).

Molecular data are adding a new dimension to understanding of the relationships between the different ascomycete orders. In general there is good correspondance with molecular phylogenies, and some groupings are becoming clearer. It is now evident that there is a basal group of ascomycetes including *Pneumocystidales*, *Protomycetales*, *Schizosaccharomycetales*, and *Taphrinales* ('Archiascomycetes'; q.v.); that the *Saccharomycetales* is quite separate from that group; that *Eurotiales*, *Lecanorales*, *Leotiales*, and *Pezizales*, are separate from *Arthoniales*, *Dothideales*, *Ophiostomatales* and *Sordariales* (Eriksson *et al.*, *SA* **11**: 119, 1993; Gargas *et al.*, *Science* **268**: 1492, 1995; Gargas & Taylor, *Exp. Mycol.* **19**: 7, 1995). Berbee & Taylor (*Molec. Biol. Evol.* **9**: 278, 1992) argued that the class names *Plectomycetes* and *Pyrenomycetes* should be reinstated, but this has not been widely accepted as the circumscriptions differ from those based only on ascomatal type (cf. Hawksworth, 1994). Almost half the orders and many families have yet to have any members in them sequenced, including some which have been the subject of phylogenetic speculations (e.g. *Lichinales*, *Peltigerales*, *Verrucariales*; cfr. Eriksson, 1991; Hawksworth, *J. Hattori bot. Lab.* **52**: 323, 1982). It will become clearer what higher categories will be most appropriate to use when more groups have been sequenced, but in the interim for general purposes it remains pragmatic to use only ordinal names.

Lit.: **General**: v. Arx (*in* Kendrick (Ed.), *The whole fungus* **1**: 201, 1979, classif., anamorphs; *Genera of fungi sporulating in pure culture*, edn 3, 1981; keys gen., lit.; *Plant Pathogenic Fungi*, 1987), v. Arx & Müller (*Beitr. Krypt.-Fl. Schweiz* **11** (1), 1954; gen. amerospored pyrenom.; *Stud. mycol.* **9**, 1975; bitunicate gen., keys), Barr (*Mem. N.Y. bot. Gdn* **28** (1): 1, 1976, classif.; *Prodromus to class Loculoascomycetes*, 1987, keys gen.; *Mycotaxon* **39**: 43, 1990, keys pyren. gen.; *in* Parker, 1982, **1**: 201), Benny & Kimbrough (*Mycotaxon* **12**: 1, 1980; plectomycete gen.), Clements & Shear (1931), Eriksson (*Opera Bot.* **60**, 1981; bitunicate fams), Eriksson & Hawksworth (*SA* **12**: 51, 1993; outline classif., orders, fam., gen.), Gäumann (*Die Pilze*, edn 2, 1964), Hafellner (*in* Galun, *CRC Handbook of lichenology* **3**: 41, 1988; lichenized gps), Hanlin (*Illustrated genera of ascomycetes*, 1990), Hansford (*Mycol. Pap.* **15**, 1946; foliicolous spp.), Hawksworth (*Proc. Indian Natn Acad. Sci., Pl. Sci.* **94**: 319, 1985; development classif. systems; (Ed.), *Ascomycete systematics: problems and perspectives in the nineties* [NATO ASI Ser. A **269**], 1994), Henssen & Jahns (*Lichenes*, ['1974'] 1973), Kohlmeyer & Kohlmeyer (*Marine mycology*, 1979; *Bot. Mar.* **34**: 1, 1991, keys 255 spp.), Korf (*The Fungi* **4A**: 249, 1973; keys discomycete gen.), Luttrell (*Univ. Miss. Stud.* **24** (3), 1951 [reprint 1969]), Müller & v. Arx (*Beitr. Krypt.-Fl. Schweiz* **11** (2), 1962, didymospored gen.; *The Fungi* **4A**: 87, 1973; keys pyrenomycete gen.), Munk (*Dansk bot. Arkiv.* **15** (2), 1953; system), Nannfeldt (*Nova Acta Reg. Soc. Sci. upsal.*, iii, **8** (2), 1932; inoperculate discom.), Poelt (*in* Ahmadjian & Hale, *The lichens*: 599, ['1973'] 1974; fams), Reynolds (Ed.) (*Ascomycete systematics*, 1981; ascus, centrum types), Santesson (*Symb. bot. upsal.* **12** (1), 1952; foliicolous L), Sivanesan (*The bitunicate ascomycetes and their anamorphs*, 1985; keys), Wehmeyer (*The pyrenomycetous fungi*, 1975), Zahlbruckner (*Nat. Pflanzenfam.* **8**: 61, 1926; L gen.), *Systema ascomycetum* (1982 on; twice-yearly, incl. *Outline* classif., papers, Notes [2023 to Aug. 1995]).

Regional: **British Isles**, Cannon *et al.* (*The British Ascomycotina: an annotated checklist*, 1985; 5100 spp.), Dennis (*British Ascomycetes*, 1968; edn 2, 1978), Ellis & Ellis (*Microfungi on land plants*, 1985; *Microfungi on miscellaneous substrates*, 1988). **Brazil**, Da Silva & Minter (*Mycol. Pap.* **169**, 1995; Batista & co-workers collns). **Denmark**, Munk (*Dansk bot. Arkiv* **17** (1), 1957). **Germany**, Schmidt & Schimdt (*Ascomyceten im Bild* **1-**, 1990 on). **Hungary**, Bánhegyi *et al.* (*Magyarország* **1-3**, 1985-87; keys). **North America**, Ellis & Everhart (*North American Pyrenomycetes*, 1892). **Pakistan**: Ahmad (*Monogr. Biol. Soc. Pakistan* **7-8**, 1978; keys). **Romania**, Sandu-Ville (*Ciuperci Pyrenomycetes - Sphaeriales din România*, 1971). **Spain**, López (*Aportacion al conocimento de los ascomicetes (Ascomycotina) de Cataluña*, **1**, 1987). **Sweden**, Eriksson (*The non-lichenized pyrenomycetes of Sweden*, 1992; 1524 spp.). **Switzerland**, Breitenbach & Kränzlin (*Pilze der Schweiz* **1**, 1981). **U.S.A.**, Farr *et al.* (*Fungi on plants and plant products in the United States*, 1989; checklist). **Venezuela**, Dennis (*Fungus flora of Venezuela and adjacent countries*, 1970).

See also under *Discomycetes*, Geographical distribution, Lichens, *Loculoascomycetes*, Plant pathogenic fungi, and Yeasts.

Ascomycotina, see *Ascomycota*.

ascoparaphysis see paraphysis.

Ascophanella Faurel & Schotter (1965), Ascobolaceae. 2, Eur. = Thecotheus (Peziz.) fide Korf (1973).

Ascophanopsis Faurel & Schotter (1965) = Thecotheus (Ascobol.) fide Krug & Khan (1987).

Ascophanus Boud. (1869), Ascobolaceae. 20, temp. See Kimbrough (*CJB* **44**: 697, 1966), Pouzar & Svrček (*Česká Myk.* **26**: 25, 1972; typification), Moravec (*Česká Myk.* **25**: 150, 1971). See also *Thelebolus*, *Coprotus*.

Ascophora Tode (1790) nom. rej. = Mucor (Mucor.) p.p. and Rhizopus (Mucor.) p.p. fide Hesseltine (1955).

ascophore, (1) an ascus-producing hypha, esp. the stalk-like hyphae supporting asci in *Cephaloascus*; (2) apothecium (obsol.).

ascophyte, hypothetical autotrophic ancestor of the *Ascomycota* (Cain, 1972), see Phylogeny, cf. basidiophyte.

ascoplasm, epiplasm (q.v.).

Ascopolyporus A. Møller (1901), Clavicipitaceae. 3, C. & S. Am. See Heim (*BSMF* **69**: 417, 1954), Doi *et al.* (*Bull. natn Sci. Mus. Tokyo* **3**: 22, 1977).

Ascoporia Samuels & A.I. Romero (1993), Patellariaceae. 1, Pará.

Ascorhiza Lecht.-Trnka (1931), Ascomycota (inc. sed.). 1, Eur. Referred to *Ascosphaerales* by Benny & Kimbrough (1980).

Ascorhizoctonia Chin S. Yang & Korf (1985), Anamorphic Otideaceae, 9.-.-. Teleomorph *Tricharina*. 7, cool temperate.

Ascoscleroderma Clémencet (1932) = Elaphomyces (Elaphomycet.) fide Trappe (1979).

Ascosorus Henn. & Ruhland (1900), Ascomycota (inc. sed.). 1, N. Am.

Ascosparassis Kobayasi (1960), Otideaceae. 1, Japan, Asia, Venezuela. See Korf (*Lloydia* **26**: 23, 1963), Pfister & Halling (*Mycotaxon* **35**: 283, 1989).

Ascospermum Schulzer (1863) nom. dub.; based on sterile mycelium.

Ascosphaera L.S. Olive & Spiltoir (1955), Ascosphaeraceae. 11, N. temp. (esp. Eur.). *A. apis* on larvae of honey bees causing chalk brood. See Spiltoir (*Am. J. Bot.* **42**: 501, 1955), Hitchcock & Christensen (*Mycol.* **64**: 1193, 1972), Bisset (*CJB* **66**: 2541, 1988; key), Rose *et al.* (*Mycotaxon* **19**: 41, 1984; key 7 spp., N. Am.), Skou (*Mycotaxon* **31**: 173, 1988; 7 spp. nov.

Japan). See also Skou (1972, 1984), MacManus & Youssof (*Mycol.* **76**: 830, 1984).

Ascosphaeraceae Olive & Spiltoir (1955), Eurotiales. 3 gen. (+ 1 syn.), 13 spp. Ascomata acellular, consisting of a brown hollow non-ostiolate cyst-like organ, lacking appendages; interascal tissue absent; asci thinwalled, evanescent at a very early stage (wall ? never formed); ascospores simple, hyaline, smooth, lacking pores, compacted into spore 'balls'; anamorphs arthroconidial; associated with pollen in honeycombs.

Lit.: Benny & Kimbrough (1980; key gen., position), Berbee & Taylor (*BioSystems* **28**: 117, 1992), Gäumann (1964), Kowalska (*Polskie Arch. Wet.* **24**: 7, 1984; biochem. syst.), Skou (*Friesia* **10**: 1, 1972, monogr.; *Mycotaxon* **15**: 487, 1982, emended concept); *Austr. J. Bot.* **32**: 225, 1984, spp.; *Mycotaxon* **31**: 191, 1988; rank).

Ascosphaerales, see *Eurotiales*.

Ascosphaeromycetes, see Eurotiales; used by Skou (*Mycotaxon* **31**: 191, 1988) to accommodate the single order (and family) *Ascosphaerales* (*Ascosphaeraceae*).

Ascospora Fr. (1825) = Mycosphaerella (Mycosphaerell.).

Ascospora Mont. (1849) nom. conf. (Mitosp. fungi) fide Sutton (1977).

ascospore, a spore produced in an ascus by 'free cell formation'; the ascospore wall is multilayered, it consists of an outer **perispore**, an intermediary layer, the proper wall (**epispore**) and sometimes an internal **endospore**; major differences in which layers are thickened, folded or pigmented can give rise to considerable variation even in a single family (e.g. *Lasiosphaeriaceae*); see Bellemère (*in* Hawksworth (Ed.), *Ascomycete systematics*: 111, 1994), basidiospore, spore wall.

Ascosporium Berk. (1860) = Taphrina (Taphrin.) fide Saccardo (*Syll. Fung.* **8**: 817, 1889).

ascostome, a pore in the apex of an ascus (obsol.).

Ascostratum Syd. & P. Syd. (1912), ? Dothideales (inc. sed.). 1, S. Afr.

ascostroma, a stroma in or on which asci are produced, usually restricted to groups with ascolocular ontogeny.

Ascostroma Bonord. (1851) = Kretzschmaria (Xylar.) fide Læssøe (*SA* **13**: 43, 1994).

Ascotaiwania Sivan. & H.S. Chang (1992), Amphisphaeriaceae. 1, Taiwan.

Ascotremella Seaver (1930), Leotiaceae. 1, N. Am., Eur. See Seaver (*Mycol.* **22**: 51, 1930), Gamundí & Dennis (*Darwiniana* **15**: 14, 1969; status).

Ascotremellopsis Teng & S.H. Ou ex S.H. Ou (1936) = Myriodiscus (Leotiales) fide Liu & Guo (*Acta Mycol. Sin.., Suppl.* **1**: 97, 1988).

Ascotricha Berk. (1838), Xylariaceae. 10, widespr. See Hawksworth (*Mycol. Pap.* **126**, 1971; key), Horie *et al.* (*TMS Japan* **34**: 123, 1993), Læssøe (*SA* **13**: 43, 1994; posn), Udagawa *et al.* (*Mycotaxon* **52**: 215, 1994). Anamorph *Dicyma*.

Ascotrichella Valldos. & Guarro (1988), ? Coniochaetaceae. 1, Chile. See Læssøe (*SA* **13**: 43, 1994; posn). Anamorph *Humicola*-like.

Ascoverticillata Kamat, Subhedar & V.G. Rao (1979) = Crocicreas (Leot.) fide Eriksson (*SA* **5**: 119, 1986).

Ascoxyta Lib. (1830) nom. dub. (Ascomycota, inc. sed.). ≡ Pseudovalsa (Melanconid.) fide Mussat (*Syll. Fung.* **15**: 52, 1901), but see Holm (*Taxon* **24**: 475, 1975).

Ascozonus (Renny) E.C. Hansen (1876), Ascobolaceae. c. 3, N. temp. See Kimbrough (*CJB* **44**: 693, 1966).

ascus (pl. **asci**), term introduced by Nees (*Syst. Pilze*: 164, 1817) for the typically sac-like cell (first figured in *Pertusaria* by Micheli in 1729; q.v.) characteristic of *Ascomycota* (q.v.), in which (after karyogamy and meiosis) ascospores (generally 8) are produced by 'free cell formation' (Fig. 11). Asci vary considerably in structure, and work in the last two decades has shown previous separation into only 2-3 categories (e.g. **bitunicate, prototunicate, unitunicate**) to be an over simplification. Sherwood (1981) illustrated 9 main types distinguishable by light microscopy (reproduced on p. 36 of edn 7 of this *Dictionary*): prototunivate, bitunicate, astropalean, annellate, hypodermataceous, pseudoperculate, operculate, lecanoralean, and verrucariod). Eriksson (1981) distinguished 7 types of dehiscence in bitunicate asci with an ectotunica and distinct endotunica (see p. 37 of edn 7). These classifications mask a much wider range of variation; Bellemère (1994) recognized 3 predehiscence types and 11 dehiscence categories (Fig. 1). The details of the asci are stressed in ascomycete systematics, esp. in lichen-forming orders where reactions with iodine are emphasized (q.v.) (Hafellner, 1984).

Bitunicate asci with two functional wall layers; those splitting at discharge (**fissitunicate**; 'jack-in-the-box') had been correlated with an ascolocular ontogeny by Luttrell (1951). Reynolds (1989) critically examined this paradigm and found the term to be applied to different ascus types and that an exclusive link to ascostromatic fungi could not be upheld; he also introduced the term **extenditunicate** for asci which extend without any splitting of the wall layers (Reynolds, *Cryptog. Mycol.* **10**: 305, 1989).

Much variation depends on the modifications in the various wall layers, especially the thickness of the walls and the *c* and *d* layers, and the details of apical differentiation (Bellemère, 1994) (Fig. 2). Caution is needed in comparing ascus staining reactions (see iodine) and structures in the absence of ultrastructural data. For terms used to describe the various structures see Fig. 2.

Also encountered are - **crown** (annular thickenings in *Phyllachora*), and - **plug** (thickening in the apex through which the spores are forcibly discharged).

Lit.: Bellemère (*Ann. Sci. nat., Bot.* **12**: 429, 1971; *Rev. Mycol.* **41**: 233, 1977, bitunicate discom.; *in* Hawksworth, 1994: 111, review), Bellemère & Letrouit-Galinou (*Bibl. Lich.* **25**: 137, 1987; ultrastr.), van Brummelen (*Persoonia* **10**: 113, 1978; operculate), Chadefaud (*Rev. Mycol.* **7**: 57, 1942; **9**: 3, 1944; apical apparatus), Chadefaud *et al.* (*Mem. Soc. bot. Fr.* **79**, 1968; lichen asci), Eriksson (*Opera bot.* **60**, 1981; bitunicate types), Griffiths (*TBMS* **60**: 261, 1973; unitunicate pyrenom.), Hafellner (*Beih Nova Hedw.* **79**: 24, 1984), Hawksworth (*J. Hattori bot. Lab.* **52**: 323, 1982; evolution types; (Ed.), *Ascomycete systematics: problems and perspectives in the nineties*, 1994), Holm (*Symb. bot. upsal.* **30**(3): 21, 1995; history of term), Honneger (*Lichenologist* **10**: 47, 1978; lecanoralean, *Peltigera*; **12**: 157, 1980; *Rhizocarpon*; **14**: 205, 1982, *Pertusaria*; **15**: 57, 1983, *Baeomyces, Cladonia, Leotia* etc.; *J. Hattori bot. Lab.* **52**: 417, 1982, review lecanoralean types), Janex-Favre (*Revue bryol. Lichen.* **37**: 421, 1971; lich. pyrenom.), Letrouit-Galinou (*Bryologist* **76**: 30, 1973; archaeasceous), Parguey-Leduc (*Ann. Sci. nat., Bot.* XII, **7**: 33, 1966; ascoloc.; *Rev. Mycol.* **41**: 281, 1977; pyrenom.), Parguey-Leduc & Janex-Favre (*Cryptogamie Mycol.* **5**: 171, 1984; ultrastr. 'unitunicate' types), Reynolds (Ed.) (*Ascomycete systematics: the Luttrellian concept*, 1981; *Bot. Rev.* **55**: 1, 1989; bitunicate paradigm), Sherwood (*Bot. J. Linn. Soc.* **82**: 15, 1981; main types), Ziegenspeck (*Bot. Arch. Koenigsberg* **13**: 341, 1926).

ascus plug, thickening in the apex through which the spores are forcibly discharged.

Aseimotrichum Corda (1831) nom. dub (? Fungi).

Asellaria R. Poiss. (1932), Asellariaceae. 6 (in *Isopoda*), widespr. See Lichtwardt (*Mycol.* **65**: 1, 1973; morphology, 1986; revision, key), Manier (*C.R. Hebd. Séanc. Acad. Sci. Paris* **276**: 3429, 1973; ultrastr.),

Asellariaceae Manier ex Manier & Lichtw. (1968), Asellariales. 2 gen. (+ 1 syn.), 10 spp. Thalli branched, septate with a polymorphic basal cell secreting, or functioning as, a holdfast; asexual reproduction by uninucleate arthrospores; sexual reproduction unknown. Endocommensals of aquatic and terrestrial *Crustacea* and *Insecta*.

Asellariales, Trichomycetes. 1 fam., 2 gen. (+ 1 syn.), 10 spp. Fam.:
Asellariaceae.
Lit.: Manier (*Ann. Sci. nat., Bot.*, Sér. 12, **10**: 565, 1969; taxonomy), Scheer (*Z. binnenfischerei DDR* **19**: 369, 1972; taxonomy), Lichtwardt & Manier (*Mycotaxon* **7**: 441, 1978; taxonomy), Moss & Young (*Mycol.* **70**: 944, 1978; phylogeny), Moss (*in* Batra (Ed.), *Insect-fungus symbiosis*: 175, 1979), Lichtwardt (1986; taxonomy, key).

aseptate, having no cross walls.

aseptic, free from damaging microorganisms.

Aseroë Labill. (1800), Clathraceae. 2, trop.

Aserophallus Mont. & Lepr. (1845) = Clathrus (Clathr.) fide Dring (1980).

asexual, without sex organs or sex spores; vegetative.

Ashbia Cif. & Gonz. Frag. (1928) ≡ Ashbya (Metschnikow.).

Ashbya Guillierm. (1928), Metschnikowiaceae. 1, Afr., C. & N. Am. *A. gossypii* (stigmatomycosis of cotton bolls). See Batra (*USDA Tech. Bull.* **1469**, 1973).

Ashtaangam Subram. (1992), Mitosporic fungi, 1.G2.1. 1, Malaysia.

Asiosordariopsis, see *Lasiosordariopsis*.

Asirosiphon Nyl. (1873) = Spilonema (Coccocarp.) fide Henssen (1963).

Asordaria Arx, Guarro & Aa (1987) = Sordaria (Sordar.) fide Eriksson & Hawksworth (*SA* **7**: 61, 1988).

asperate, rough with projections or points.

Aspergillaceae, see *Trichocomaceae*.

Aspergillales, see *Eurotiales*.

aspergilliform (of a sporulating structure), resembling that of an *Aspergillus* conidiophore.

aspergillin, (1) a black, water-insoluble pigment of *Aspergillus niger* spores (Linossier, 1891); (2) various antibiotics produced by *Aspergillus* spp. See Tobie (*Nature* **158**: 709, 1946).

Aspergillites Trivedi & C.L. Verma (1969), Fossil fungi. 1 (Tertiary), Malaysia.

Aspergilloides Dierckx (1901) = Penicillium (Mitosp. fungi) fide Raper & Thom (*A manual of the penicillia*, 1949).

aspergilloma, a 'fungus ball' composed principally of hyphae of *Aspergillus*, found in a pre-existing cavity (esp. in an upper lobe of the lung) or a bronchus, which usually has a relatively benign or asymptomatic effect; cf. aspergillosis.

Aspergillopsis Speg. (1910) = Aspergillus (Mitosp. fungi).

Aspergillopsis Sopp (1912) nom. dub. fide Raper & Thom (*A manual of the penicillia*, 1949).

aspergillosis, any disease in humans or animals caused by *Aspergillus* (esp. *A. fumigatus*); esp. common in birds (see Chute *et al.* (1971), Ainsworth & Austwick (1973) under Medical and veterinary mycology). See Austwick, (*in* Raper & Fennell, Eds, *The genus Aspergillus*, 1965), Bossche *et al.* (Eds) (*Aspergillus and aspergillosis*, 1988).

Aspergillus Link (1809), Anamorphic Trichocomaceae, 2.A1/2.32. Teleomorphs *Eurotium, Neosartorya*,

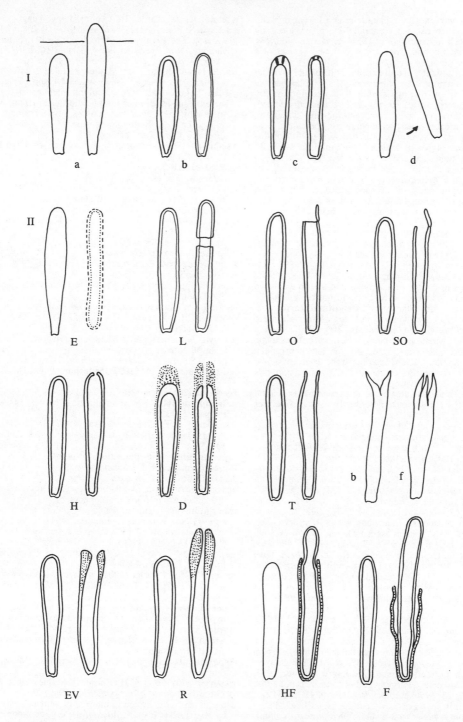

Fig. 1. I. Predehiscence stage of asci. a = protruding ascus; b = ascus wall becoming thinner; c = change in apical structure; d = ascus liberation. II. Dehiscence stage of asci; evanescent ascus (E); rupture of lateral wall (L); subapical rupture (O, operculate, and SO, suboperculate dehiscence); rupture by apical wall without extrusion (H, pore-like dehiscence); D, *Dactylospora*-type; T, *Teloschistes*-type = extenditunicate (b = bivalve, f = fissurate variants); rupture with extrusion (EV, eversion; R, rostrate; HF, hemifissitunicate; F, fissitunicate). After Bellemére, *in* Hawksworth (Ed.) (*Ascomycete Systematics*: 111, 1994).

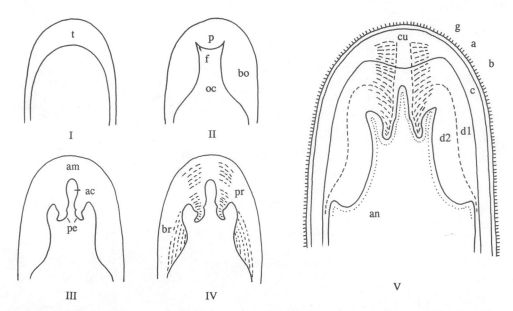

Fig. 2. I-V. Ascus apex components. ac = axial canal; am = axial mass; bo = bourrelet; br = ring in bourrelet; f = furrow; oc = ocular chamber; p = plug; pe = pendant; pr = rings in the plug and pendant; t = tholus; V, ascus apex structure. a = a layer; an = apical nasse; b = b layer; c = c layer; cu = cushion; d1 and d2 = sublayers of the d layer. After Bellemére (*in* Hawksworth (Ed.), *Ascomycete Systematics*: 111, 1994).

Emericella. 185 (fide Pitt & Samson, *Regnum Veg.* **128**: 13 1993), cosmop. See Fig. 3. *A. niger* ('smut' of fig and date; black mould of cotton bolls, fruits, vegetables, etc.).

Lit.: **General**: Thom & Church (*The aspergilli*, 1926), Thom & Raper (*A manual of the aspergilli*, 1945), Benjamin (*Mycol.* **47**: 669, 1955), Raper & Fennell (*The genus Aspergillus*, 1965), Al-Musallam (*Revision of the black Aspergillus species* (Utrecht), 1980; keys 15 spp.), Smith & Pateman (Eds) (*Genetics and physiology of Aspergillus*, 1977), Christensen (*Mycol.* **74**: 210, 226, 1982; synoptic keys *A. ochraceus* and *A. nidulans* groups), Gams *et al.*, *in* Samson & Pitt (Eds), *Advances in Penicillium and Aspergillus systematics*; 337, 1985), Samson & Pitt (Eds) (*Modern concepts in Penicillium and Aspergillus classification*, 1989), Samson (*Stud. Mycol.* **18**, 1979; compilation spp. described since 1965), Kozakiewicz (*TBMS* **70**: 175, 1978; SEM of conidia), Locci (*Rivta Patol. Veget.* ser. 4, **8** (Suppl.), 1972; SEM of teleomorphs), Kozakiewicz (*Mycol. Pap.* **161**: 1, 1989; spp. on stored products), Tzean *et al.* (*Mycol. Monogr.* **1**: 1, 1990; *Aspergillus* and related teleomorphs from Taiwan), Bilai & Koval (*Aspergilly*: 1, 1988; spp. from former USSR), Bennett (*Biology of industrial microorganisms*: 359, 1985; taxonomy and biology), Klich & Pitt (*TBMS* **91**: 99, 1988; differentiation of *A. flavus* from *A. parasiticus*), Klich & Mullaney (*Mycol.* **81**: 159, 1989; differentiation of *A. parasiticus* from *A. sojae*), Bezjak (*Mycoses* **32**: 187, 1989; abnormal conidial structures), Weidenbörner *et al.* (*J. Phytopath.* **126**:1, 1989; preparation for SEM), Pitt (*J. Appl. Bact.* **67** (Suppl. Symp. Ser. 18): 375, 1989; recent developments in systematics), Bennett & Papa (*Adv. Pl. Path.* **6**: 263, 1988; aflatoxigenic spp.), Samson & Frisvad (*Proc. Jap. Assoc. Mycotoxic.* **32**: 3, 1990; species concepts and mycotoxins), Fragner (*Česká Myk.* **45**: 113, 1991; spp. from humans and animals), Chang *et al.* (*J.Gen. Appl. Microbiol.* **37**: 289, 1991; phylogeny), Bossche *et al.* (Ed.) (*Aspergillus and aspergillosis*, 1988), Kozakiewicz *et al.* (*Taxon* **41**: 109, 1992; spp. nom. cons. prop.), Klich (*Mycol.* **85**: 100, 1993; sect. *Versicolores*), Tiedt (*MR* **97**: 1459, 1993; ultrastr. conidiogenesis in *A. niger*), Smith (ed.) (*Aspergillus [Biotechnology Handbooks* 7], 1994).

Molecular Biology, Genetics & Biochemistry: Kurtzman *et al.* (*Mycol.* **78**: 955, 1986; DNA relatedness), Sekhon *et al.* (*Diagn. immunol.* **4**: 112, 1986; exoantigen grouping), Bojovic-Cvetic & Vujicic (*TBMS* **91**:619, 1988; polysaccharide cytochemistry), Moody & Tyler (*Appl. Env. Microbiol.* **56**: 2441, 2453, 1990; restriction enzyme analysis of mtDNA, and DNA RLFPs of *A. flavus* group), Clutterbuck (*Fungal Genetics Newsl.* **37**: 80, 1990; bibliography of *A. nidulans*), Hull *et al.* (*Molecular Microbiol.* **3**: 553, 1989; L-proline catabolism gene cluster in *A. nidulans*), Wirsel *et al.* (*Molecular Microbiol.* **3**: 3, 1989; amylase genes in *A. oryzae*), Novak & Kohn (*Exp. Mycol.* **14**: 339, 1990; developmental proteins), Kálmán *et al.* (*Can. J. Microbiol.* **37**: 391, 1991; interspecific hybrids in *A. nidulans* group by isoenzyme analysis), Meyer *et al.* (*Curr. Gen.* **19**: 239, 1991; DNA fingerprinting and differentiation of strains), Gomi *et al.* (*J. Gen. Appl. microbiol.* **35**: 225, 1989; differentiation of koji moulds by AGE), Wicklow & Kurtzman in Natori *et al.* (Eds) (*Mycotoxins and Phycotoxins 1988*: 1, 1989; nucleic acid relatedness), Sugiyama *et al.* (*Jap. J. Med. Myc.* **32**: 39, 1991; chemotaxonomy), Kirimura *et al.* (*FEMS Microbiol. Lett.* **90**: 235, 1992; mapping mitochondrial DNA in *A. niger*), Keller *et al.* (*Curr. Gen.* **21**: 371, 1992; karyotypes in sect. *Flavi*), Bennett & Klich (Eds) (*Aspergillus: biology and industrial applications*, 1992), Powell *et al.* (Eds) (*The genus Aspergillus* (FEMS Symposium no. 69), 1994; taxonomy, genetics and industrial applications). See aflatoxins, aspergillin, aspergillosis, Industrial Mycology.

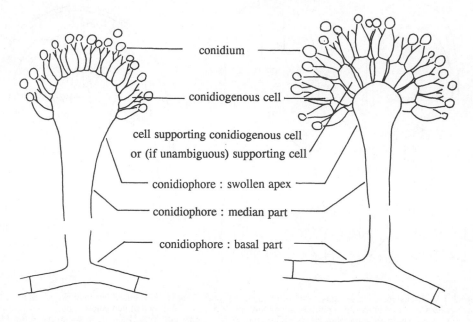

conidium

conidiogenous cell

cell supporting conidiogenous cell
or (if unambiguous) supporting cell

conidiophore : swollen apex

conidiophore : median part

conidiophore : basal part

Fig. 3. Terms recommended to describe the different conidiogenous structures in *Aspergillus*. See Minter *et al.*, *in* Samson & Pitt (Eds) (*Advances in Penicillium and Aspergilllus Systematics*: 71, 1985).

Asperisporium Maubl. (1913), Mitosporic fungi, 3.B2.10. 4, Am. *A. caricae* (*Carica papaya* leaf spot). See Sutton & Hodges (*Mycol.* **82**: 313, 1990).

Asperopilum Spooner (1987), Hyaloscyphaceae. 1 (on *Juncus*), Australasia.

Asperotrichum, see *Asporothrichum*.

asperulate, delicately asperate.

Aspicilia A. Massal. (1852) nom. cons., Hymeneliaceae (L). *c.* 100, widespr. See also Clauzade & Roux (*Bull. Soc. bot. Cent. Ouest* n.s., **15**: 127, 1984; Eur., gen. concept.), Laundon & Hawksworth (*Taxon* **37**: 478, 1988; nomencl.).

Aspiciliella M. Choisy (1932), Hymeneliaceae (L). 1, Eur., Asia.

aspicilioid (of lecanorine apothecia), more or less immersed in the thallus, at least when young.

Aspiciliomyces Cif. & Tomas. (1953) ≡ Pachyospora (Hymenel.).

Aspiciliopsis (Müll. Arg.) M. Choisy (1929) = Placopsis (Trapel.).

Aspidelia Stirt. (1900) = Parmelia (Parmel.) s. str. fide Culberson (*Bryologist* **69**: 113, 1966), Hale (1987).

Aspidella J.-E. Gilbert (1940) = Amanita (Amanit.) fide Singer (1975).

Aspidopyrenis Clem. & Shear (1931) ≡ Aspidopyrenium (Aspidothel.).

Aspidopyrenium Vain. (1890) = Aspidothelium (Aspidothel.).

Aspidothea Syd. (1927) = Inocyclus (Parmular.) fide Müller & v. Arx (1962).

Aspidotheliaceae Räsänen ex J.C. David & D. Hawksw. (1991), Ascomycota (inc. sed.). 2 gen. (+ 6 syn.), 8 spp. Thallus crustose; ascomata solitary, perithecial, ± globose, sometimes clypeate, the wall sometimes ornamented, the ostiole periphysate; interascal tissue of sometimes branched and anastomosing paraphyses; asci clavate, persistent, with a multilayered wall, ? fissitunicate, the apex thickened, not blueing in iodine, with a small ocular chamber; ascospores hyaline, transversely septate or muriform, with a mucous sheath. Anamorphs pycnidial. Lichenized with green algae, trop.

Lit.: Aptroot & Sipman (*Lichenologist* **25**: 121, 1993).

Aspidotheliomyces Cif. & Tomas. (1953) ≡ Aspidothelium (Aspidothel.).

Aspidothelium Vain. (1890), Aspidotheliaceae (L). 4, trop. See Santesson (1952).

Aspilaima Bat. & H. Maia (1961), Mitosporic fungi, 5.A2.?. 1, Brazil.

A-spore, see alpha-spore

asporogenic (**asporogenous**), not forming spores.

asporogenic yeasts, see Yeasts.

Asporomyces Chaborski (1918) = Torulopsis (Mitosp. fungi). See Mrak *et al.* (*Mycol.* **34**: 139, 1942).

Asporothrichum Link (1809) nom. dub. (Mitosp. fungi); based on mycelium fide Fries (*Syst. mycol.* **3**, index: 1832). = Sporotrichum (Mitosp. fungi) fide Streinz (*Nom. fung.*, 1862).

Asproinocybe R. Heim (1970), Tricholomataceae. 3, trop. Afr. See Heinemann & Thoen (*Fl. Illustr. Champ. Afr. centr.* **5**: 102, 1977).

Aspropaxillus Kühner & Maire (1934) = Leucopaxillus (Tricholomat.) fide Singer (1975).

assimilative, (1) taking in; (2) (of hyphae) having to do with the growth phase before reproduction; non-reproductive; vegetative.

Assoa Urries (1944) [non Cothen. (1790), *Byttneriaceae*], Ascomycota (inc. sed.). 1, Spain.

association, see phytosociology.

astatocoenocytic (of nuclear behaviour in basidiomycetes), haplont mycelium cells coenocytic, diplont binucleate but coenocytic and without clamps when aeration insufficient, basidioma binucleate; in contrast to **holocoenocytic** (haplont and diplont coenocytic, only developing basidium binucleate), **heterocytic**

(haplont regularly coenocytic), and the **normal** condition when the haplont is uninucleate, the diplont binucleate (Boidin, *in* Petersen (Ed.), *Evolution in the higher basidiomycetes*: 129, 1971).

Astelechia Cif. (1962), Mitosporic fungi, 1.C2.?. 2, Dominican Republic.

Asterella (Sacc.) Sacc. (1891) [non P. Beauv. (1805), *Hepaticae*] = Asterina (Asterin.) fide Müller & v. Arx (1962).

Asterella Hara (1936) = Astrosphaeriella (Melanommat.) fide Hawksworth (1981).

Astereptonema, see *Astreptonema*. Cited in *ING*, in error.

Asteridiella McAlpine (1897), Meliolaceae. 250, trop. See Hansford (*Sydowia* **10**: 41, 1956), Hughes (*Mycol. Pap.* **166**, 1993).

Asteridiellina Seaver & Toro (1926) = Actinopeltis (Microthyr.) fide v. Arx & Müller (1975).

Asteridium (Sacc.) Speg. ex Sacc. (1891) = Meliola (Meliol.) fide Höhnel (*Sber. Akad. Wiss. Wien* I, **109**: 414, 1910).

Asterina Lév. (1845), Asterinaceae. *c*. 200 (on leaves), mainly trop. See Doidge (1942), Hansford (1946), Reynolds (*Crypt., Mycol.* **8**: 251, 1987; asci), Theissen (*Abh. zool. -bot. Ges. Wein* **7** (3), 1913). Anamorph *Asterostomella*.

Asterinaceae Hansf. (1946), Dothideales. 37 gen. (+ 37 syn.), 389 spp. Mycelium superficial, often hyphopodiate; ascomata round or elongate, strongly flattened, either opening by radiating or longitudinal splits, or by a lysigenous pore; peridium usually composed of radiating isodiametric cells; hymenium often gelatinous, I+ blue; interascal tissue absent; asci ovoid to cylindrical, thick-walled, fissitunicate; ascospores brown, transversely septate. Anamorph hyphomycetous or coelomycetous. Biotrophic on leaves, esp. trop.
    Lit.: Doidge (*Bothalia* **4**: 273, 1942; S. Afr.).

Asterinales, see *Dothideales*.

Asterinella Theiss. (1912), Microthyriaceae. *c*. 3, warmer areas.

Asterinema Bat. & Gayão (1953), Microthyriaceae. 1, Brazil. See Farr (*Mycol.* **75**: 1036, 1983).

Asterinites Krassilov (1967), Fossil fungi. 2 (Cretaceous), former USSR.

Asterinites Doub. & D. Pons (1973), Fossil fungi. 2 (Paleocene), Colombia.

Asterinopeltis Bat. & H. Maia (1958) = Platypeltella (Microthyr.) fide v. Arx & Müller (1975).

Asterinotheca Bat. & H. Maia (1958) ? = Asterina (Asterin.). See Müller & v. Arx (1962).

Asterinothyriella Bat. & Cif. (1959), Mitosporic fungi, 5.C1.?. 1, Uganda.

Asterinothyrium Bat., Cif. & H. Maia (1959), Mitosporic fungi, 5.A1.?. 1, S. Afr.

Asterinula Ellis & Everh. (1889) = Leptothyrella (Mitosp. fungi) fide Saccardo (1892).

Asterisca G. Mey. (1825) = Sarcographa (Graphid.).

Asteristion Leight. (1870) = Thelotrema (Thelotremat.) fide Hale (1981).

Asteristium Clem. (1909) ≡ Asteristion (Thelotremat.).

Asteritea Bat. & R. Garnier (1961), ? Microthyriaceae. 1, Brazil.

Asterobolus Redhead & P.W. Perrin (1972) = Valdensia (Mitosp. fungi) fide Redhead & Perrin (*CJB* **50**: 2083, 1972).

Asterocalyx Höhn. (1912) [non Ettingsh. (1888), fossil *Dioscoreaceae*], Sclerotiniaceae. 2, Asia, Australia, Eur., C. & N. Am. See Dumont & Carpenter (*Mycol.* **70**: 68, 1978), Spooner (*Bibl. Mycol.* **116**, 1987).

Asterochaete (Pat.) Bondartsev & Singer (1941) [non Nees (1834), *Cyperaceae*] ≡ Echinochaete (Polypor.).

Asteroconium Syd. & P. Syd. (1903), Mitosporic fungi, 6.G1.1/10. 2, C. Am., India, China. See Sutton (1980).

Asterocyphella W.B. Cooke (1961), Cyphellaceae. 3, Argentina, S. Afr., Papua New Guinea. See Cooke (1961: 118; key).

Asterocystis De Wild. (1893) = Olpidium (Olpid.) fide Sampson (1939).

Asterodon Pat. (1894), Asterostromataceae. 1, N. temp. See Corner (*TBMS* **31**: 234, 1948), Rick (*Ann. Myc.* **38**: 56, 1940).

Asterodothis Theiss. (1912), Asterinaceae. 1, Afr. Anamorph *Asterostromina*.

Asterogastraceae, see *Elasmomycetaceae*.

asteroid body, a stellate cell of *Sporothrix schenckii* (more rarely *Aspergillus* or other pathogens) in animal tissues resulting from an antigen-antibody complex precipitate deposited on the cell wall (Lurie & Snell, *Sabouraudia* **7**: 64, 1969).

Asteroides Puntoni & Léon (1940) [non Mill. (1754), *Compositae*], nom. dub. (Fungi, inc. sed.).

Asterolibertia G. Arnaud (1918), Asterinaceae. *c*. 15, warmer areas.

Asteroma DC. (1815), Mitosporic fungi, 6.A1.15. 14, esp. N. temp. See Sutton (1980).

Asteromassaria Höhn. (1917), Pleomassariaceae. 9, Eur., N. Am. See Barr (1982; *Mycotaxon* **49**: 129, 1993, key 8 N. Am. spp.), Sivanesan (*TBMS* **91**: 317, 1988; key 9 spp.).

Asteromella Pass. & Thüm. (1880), Anamorphic Mycosphaerellaceae, 4.A1.15. Teleomorph *Mycosphaerella*. 140, widespr. See Batista *et al.* (*Saccardoa* **1**: 17, 1960), Sutton (1980).

Asteromellopsis H.E. Hess & E. Müll. (1951), Anamorphic Polystomellaceae, 8.A1.15. Teleomorph *Dothidella*. 1, Switzerland.

Asteromidium Speg. (1888), Mitosporic fungi, 6.C1.10. 2, Brazil. See Petrak & Sydow (*Ann. Myc.* **34**: 14, 1936), Ferreira & Muchovej (*Mycotaxon* **30**: 97, 1987; addit. spp.).

Asteromyces Moreau & M. Moreau ex Hennebert (1962), Mitosporic fungi, 1.A2.11/14. 1, France.

Asteromyxa Theiss. & Syd. (1918) ≡ Dimeriella (Parodiopsid.).

Asteronaevia Petr. (1929) = Diplonaevia (Dermat.) fide Hein (1983).

Asteronectrioidea Cant. (1949), Mitosporic fungi, 8.B1.15. 1, Afr.

Asteronema Trevis. (1845) nom. dub. (? Fungi).

Asteronia (Sacc.) Henn. (1895), Microthyriaceae. 2, Brazil. See Sutton (*Mycol. Pap.* **141**, 1977).

Asteropeltis Henn. (1904) = Trichothelium (Trichothel.).

Asterophlyctis H.E. Petersen (1903), Chytridiales (inc. sed.). 1, widespr. = Diplophlyctis (Chytrid.) fide Dogma (*Nova Hedw.* **25**: 121, 1974).

Asterophoma D. Hawksw. (1981), Anamorphic Mycocaliciaceae, 4.A1.15. 1 (on *Calicium*), Afr., N. Am., Eur. See Tibell (*CJB* **69**: 2427, 1991; ultrastr.). Teleomorph *Chaenothecopsis*.

Asterophora Ditmar (1809), Anamorphic Tricholomataceae, 2.A1.41. Teleomorph *Nyctalis*. 1, temp. See Lundqvist (*Bot. Notiser* **133**: 121, 1980), Kollerz & Jahrmann (*Ant. v. Leeuwenhoek* **51**: 255, 1985; life cycle, physiology).

asterophysis, see seta.

Asteroporomyces Cif. & Tomas. (1953) = Asteroporum (Pyrenulales).

Asteroporum Müll. Arg. (1884), Pyrenulales (inc. sed., ±L). 3, trop.

Asteropsis Gonz. Frag. (1917) [non Less. (1832), *Compositae*] Mitosporic fungi, 4.A2.?. 1, Spain.

Asteroscutula Petr. (1948), Mitosporic fungi, 5.A2.?. 1, Ecuador.

asteroseta, (1) see cystidium; (2) see seta.

Asterosphaeria (Höhn.) Syd. (1913) = Astrosphaeriella (Melanommat.) fide Hawksworth (1981).

Asterosporales, see *Russulales*.

Asterosporium Kunze (1819), Mitosporic fungi, 6.G2.1. 4, temp. See Murvanishvili & Dekanoidze (*Mikol. Fitopatol.* **26**: 27, 1992; key).

Asterostomella Speg. (1886), Anamorphic Asterinaceae, 5.A2.?. Teleomorph *Asterina*. 21, trop. See Batista & Ciferri (*Mycopathol.* **11**: 44, 1959).

Asterostomidium Lindau (1900) ≡ Asteromidium (Mitosp. fungi).

Asterostomopora Bat. & H. Maia (1960), Mitosporic fungi, 5.A2.?. 1, Jamaica.

Asterostomopsis Bat., Cif. & H. Maia (1959), Mitosporic fungi, 3.A2.?. 1, Ghana.

Asterostomula Theiss. (1916), Mitosporic fungi, 5.A2.?. 4, trop.

Asterostomulina Bat., J.L. Bezerra & H. Maia (1964), Mitosporic fungi, 5.A1.?. 1, Brazil.

Asterostroma Massee (1889), Asterostromataceae. 7, widespr.

Asterostromataceae (Donk) Pouzar (1983), Hymenochaetales. 2 gen. (+ 3 syn.), 21 spp. Basidioma resupinate; monomitic; asterosetae present.

Asterostromella Höhn. & Litsch. (1907) = Vararia (Lachnoclad.) fide Burt (1922).

Asterostromina Bat. & A.F. Vital (1957), Anamorphic Asterinaceae, 5.A1.?. Teleomorph *Asterodothis*. 1, S. Afr.

Asterotexis Arx (1958), Asterinaceae. 1, trop. Am.

Asterotheca I. Hino (1938) [non C. Presl (1846), fossil *Pteridophyta*] ≡ Astrotheca (Melanommat.).

Asterothecium Wallr. (1836) ≡ Stephanoma (Mitosp. fungi).

Asterothelium, see *Astrothelium*.

Asterothrix Kütz. (1843) [non Cass. (1827), *Compositae*] ≡ Asteronema (? Fungi).

Asterothyriaceae, see *Thelotremataceae*.

Asterothyriomyces Cif. & Tomas. (1953) = Asterothyrium (Thelotremat.).

Asterothyrites Cookson (1947), Fossil fungi. 4 (Tertiary), Australia, USA, India. = Phragmothyrites (Fossil fungi) fide Selkirk.

Asterothyrium Müll. Arg. (1890), Thelotremataceae (L). 8, trop. See Santesson (*Symb. bot. upsal.* **12**(1), 1952), Vězda & Poelt (*Phyton, Horn* **30**: 47, 1990; posn).

Asterothyrium Henn. (1904) ≡ Septothyrella (Mitosp. fungi).

Asterotrema Müll. Arg. (1884) ? = Arthonia (Arthon.) fide Aptroot (*SA* **12**: 25, 1993).

Asterotrichum Bonord. (1851) = Asterophora (Mitosp. fungi) fide Saccardo (1886).

Asterotus Singer (1943) = Resupinatus (Tricholomat.) fide Singer (1975).

Asterula (Sacc.) Sacc. (1891) = Venturia (Ventur.) fide v. Arx & Müller (1975).

Astiothyrium Bat. (1964) = Eudimeriolum (Pseudoperispor.) fide v. Arx & Müller (1975).

Astoma Gray (1821) = Sclerotium (Mitosp. fungi) fide Rabenhorst (*Deutsch. Krypt. Fl.* **1**, 1844).

astomate (astomous), lacking an ostiole.

Astomella Thirum. (1947), Ascomycota (inc sed.). 1, India.

Astrabomyces Bat. (1961), Mitosporic fungi (L), 1.C2.?. 1, Brazil.

Astraeaceae Zeller ex Jülich (1982), Sclerodermatales. 1 gen. (+ 1 syn.), 2 spp. Gasterocarp geasteroid, splitting stellately; peridium stratified.

Astraeus Morgan (1889), Astraeaceae. 2, widespr. *A. hygrometricus*, a mycorrhizal earth-star common in dry places.

Astragoxyphium Bat., Nascim. & Cif. (1963) = Leptoxyphium (Mitosp. fungi) fide Hughes (1976).

Astreptonema Hauptfl. (1895), Eccrinaceae. 5 (in *Amphipoda*), Eur., USA. See Moss (*TBMS* **65**: 115, 1975; ultrastr.), Hibbets (*Syesis* **11**: 213, 1978; development), Lichtwardt (1986; key).

Astrocitum Raf. (1806) ≡ Astrycum (Geastr.).

Astrocystis Berk. & Broome (1873), Xylariaceae. 15 (esp. on bamboo), trop. See Læssøe & Spooner (*Kew Bull.* **49**: 1, 1994; key). Anamorph *Acanthodochium*.

Astrodochium Ellis & Everh. (1897), Mitosporic fungi, 3.A2.?. 1, N. Am.

astrogastraceous fungi, gasteroid members of the *Russulales*. See also *Hymenogastrales*, *Podaxales*.

Astronatelia Bat. & H. Maia (1962), Mitosporic fungi, 5.A1.?. 1, USA.

Astroplaca Bagl. (1858) = Placolecis (Lecan.) fide Hafellner (1984).

Astrosphaeriella Syd. & P. Syd. (1913), ? Melanommataceae. 12, Afr., Asia, S. Am. See Hawksworth (*Bot. J. Linn. Soc.* **82**: 35, 1981), Hawksworth & Boise (*Sydowia* **38**: 114, 1986; key 10 spp.).

Astrosporina J. Schröt. (1889) = Inocybe (Cortinar.) fide Kauffman (1918). See Horak (*Persoonia* **10**: 157, 1979; key to 30 spp. from Indomalaya, Australasia).

Astrotheca I. Hino (1938) = Astrosphaeriella (Melanommat.) fide Hawksworth (1981).

Astrotheliaceae, see *Trypetheliaceae*.

Astrothelium Eschw. (1824), Trypetheliaceae (L). *c.* 40, trop. See Harris (*Acta Amazon.* Suppl. **14**: 55, 1986; key 13 spp. Brazil).

Astrycum Raf. (1809) ? = Geastrum (Geastr.).

Astylospora Fayod (1889) = Psathyrella (Coprin.) fide Singer (1975).

asymmetric (of spores), having one side flattened or concave.

Asyregraamspora Locq. & Sal.-Cheb. (1980), Fossil fungi. 1, Cameroon.

ATBI (All-Taxon Biodiversity Inventory), a record of the total diversity of living organisms present in one area. See Cannon (*Inoculum* **46**(4): 1, 1995), Inventorying.

ATCC. American Type Culture Collection (Rockville, Md, USA); a not-for-profit service collection founded in 1925; see *American Type Culture Collection profile* (1992).

Ateleothylax M. Ota & Langeron (1923), Mitosporic fungi, 1.?.?. 1, Sweden.

Atelocauda Arthur & Cummins (1933), Pileolariaceae. 5 (on *Leguminosae*), C. Am., Austr., Hawaii.

Atelosaccharomyces Beurm. & Gougerot (1909) = Cryptococcus (Mitosp. fungi) fide v. Arx *et al.* (*Stud. Mycol.* **14**, 1977).

Atestia Trevis. (1861) = Oropogon (Alector.).

Athecaria Nyl. (1897) ? = Aspicilia (Hymenel.). See Santesson (*ING* **1**: 155, 1979; typific.).

Athelia Pers. (1822), Atheliaceae. *c.* 30, widespr. See Jülich (1972), Donk (*Fungus* **27**: 12, 1957), Christiansen (*Dansk bot. Arkiv* **19**: 137, 1960), Jülich (*Persoonia* **10**: 149, 1978; key 'lichenized' spp.). *A. arachnoidea* is an important pathogen of lichens, esp. *Lecanora conizaeoides*, also epiphytic green algae (causing brownish white lesions in its colonies; see Arvidsson, *Svensk bot. Tidskr.* **72**: 285, 1979). See also *Sclerotium*.

Atheliaceae Jülich (1982), Stereales. 17 gen. (+ 4 syn.), 65 spp. Basidioma pellicular; monomitic; no cystidia, spores inamyloid.

Athelicium K.H. Larss. & Hjortstam (1986), Atheliaceae. 1, Eur.

Athelidium Oberw. (1966), Atheliaceae. 1, Eur.

Athelium Nyl. (1886) = Thelocarpon (Acarospor.).

Atheloderma Parmasto (1968), Hyphodermataceae. 2, Eur, Asia.

Athelopsis Oberw. ex Parmasto (1968), Atheliaceae. 8, widespr. See Hjortstam (*Mycotaxon* **42**: 149, 1991).

Athrismidium Trevis. (1860) = Tomasellia (Arthopyren.).

Atichia Flot. (1850), Anamorphic Seuratiaceae, 2/7.G1/2.1. Teleomorph *Seuratia*. 6, Java, Dominican Republic, Eur. See Meeker (*CJB* **53**: 2483, 1975).

Atichiopsis R. Wagner (1900) = Seuratia (Seurat.) fide Meeker (1975).

Atkinsiella Vishniac (1958), Haliphthoraceae. 2 (on *Crustacea*), Eur., N. Am. See Sparrow & Gotelli (*Mycol.* **61**: 199, 1969), Martin (*Am. J. Bot.* **64**: 760, 1977; posn), Kitancharoen *et al.* (*Mycoscience* **35**: 265, 1994; disease of abalone), Dick (*in press*).

Atkinson (George Francis; 1854-1918). Professor of Botany, Cornell University, 1896-1918. Chief writings: *Mushrooms edible and poisonous*, 1901; *Phylogeny and relationships in the ascomycetes* (*An. Mo. bot. Gdn* **2**: 315, 1915), and other papers on *Agaricaceae*, phylogeny, and plant diseases. See Farlow *et al.* (*Am. J. Bot.* **6**: 301, 1919), Stafleu & Cowan (*TL-2* **1**: 78, 1976), Stafleu & Mennega (*TL-2, Suppl.* **1**: 200, 1992).

Atkinsonella Diehl (1950), Clavicipitaceae. 2, N. temp. See Leutchmann & Clay (*Mycol.* **81**: 692, 1989), Morgan-Jones & White (*Mycotaxon* **35**: 455, 1989). Anamorphs *Ephelis*, *Sphacelia*.

Atkinsonia Lloyd (1916) [non F. Muell. (1865), *Loranthaceae*] = Sebacina (Exid.) fide Donk (*Persoonia* **4**: 305, 1966).

atlantic, confined to the Atlantic seaboard. For classification of different types of atlantic distribution in Europe see Ratcliffe (*New Phytol.* **67**: 365, 1968).

Atmospheric pollution, see Air pollution.

atomate, having a powdered surface.

Atopospora Petr. (1925), Venturiaceae. 2, N. temp. See Barr (*Sydowia* **41**: 25, 1989).

Atractiella Sacc. (1886), Hoehnelomycetaceae. 5, widespr. See Donk (*Persoonia* **4**: 209, 1966).

Atractiellales, Basidiomycetes. 5 fam., 8 gen. (+ 6 syn.), 13 spp. Fams:
(1) **Atractogloeaceae**.
(2) **Chionosphaeraceae**.
(3) **Hoehnelomycetaceae**.
(4) **Pachnocybaceae**.
(5) **Phleogenaceae**.
    Lit.: Oberwinkler & Bauer (*Sydowia* **41**: 224, 1989).

Atractilina Dearn. & Barthol. (1924), Mitosporic fungi, 1/2.C1/2.10. 2 (on leaf ascomycetes), trop. See Deighton & Pirozynski (*Mycol. Pap.* **128**, 1972).

Atractina Höhn. (1904) = Sterigmatobotrys (Mitosp. fungi) fide Hughes (1958).

Atractium Link (1809), Mitosporic fungi, 2.C1.?. 5, widespr.

Atractobasidium G.W. Martin (1935) = Patouillardina (Exid.) fide Rogers (*Mycol.* **28**: 398, 1936).

Atractobolus Tode ex P. Beauv. (1805), Ascomycota (inc. sed.). 1, Eur. See Spooner (*Bibl. Mycol.* **116**: 1987).

Atractodorus Klotzsch (1832) nom. dub. (Fungi, inc. sed.).

Atractogloea Oberw. & Bandoni (1982), Atractogloeaceae. 1, USA.

Atractogloeaceae Oberw. & R. Bauer (1989), Atractiellales. 1 gen., 1 spp. Basidioma pulvinate, gelatinous; hyphae with simple septal pores, clamp-connexions present; spores statismosporic.

Atrichophyton Castell. & Chalm. (1919) = Chrysosporium (Mitosp. fungi) fide Carmichael (*in* Kendrick & Carmichael, 1973).

Atricordyceps Samuels (1983), Clavicipitaceae. 1 (on *Insecta*), NZ. Anamorph *Harposporium*.

Atrocybe Velen. (1947), ? Leotiales (inc. sed.). 1, former Czechoslovakia.

Atropellis Zeller & Goodd. (1930), Dermateaceae. 4, N. Am. See Reid & Funk (*Mycol.* **58**: 428, 1966; key). *A. pinicola* (pine canker).

Atroporus Ryvarden (1973), Polyporaceae. 2, trop.

Atroseptaphiale Matsush. (1995), Mitosporic fungi, 1.E1.15. 1, Peru.

Atrotorquata Kohlm. & Volkm.-Kohlm. (1993), Amphisphaeriaceae. 1, USA.

Attamyces Kreisel (1972), Anamorphic Agaricaceae, 9.-.-. Teleomorph *Leucoagaricus*. 1 (in ants nests), Cuba. See Singer (*Nova Hedw.* **26**: 435, 1975).

attenuate, (1) narrowed; (2) (of a pathogen), having lowered pathogenicity or virulence.

Atylospora, see *Astylospora*.

atypical, not normal.

Auerswaldia Rabenh. (1857) = Melanospora (Ceratostomat.).

Auerswaldia Sacc. (1883), ? Dothideaceae. 30, widespr. = Bagnisiella (Dothid.) fide v. Arx & Müller (1975).

Auerswaldiella Theiss. & Syd. (1914), Botyrosphaeriaceae. 4, trop. See Eriksson & Hawksworth (*SA* **14**: 45, 1995; posn), Sivanesan & Hsieh (*MR* **93**: 340, 1989; key).

Auerswaldiopsis Henn. (1904) ? = Patouillardiella (Mitosp. fungi) fide Höhnel (*Sber. Akad. Wiss. Wien* **119**: 432, 1910).

Aulacographa Leight. (1854) = Graphis (Graphid.).

Aulacostroma Syd. & P. Syd. (1914), Parmulariaceae. 3, Philipp., Hawaii, India. See Luttrell & Muthappa (*Mycol.* **66**: 563, 1974).

Aulaxina Fée (1825), Gomphillaceae (L). 9, trop. See Santesson (*Symb. bot. upsal.* **12** (1), 1952), Vězda & Poelt (*Folia geobot. phytotax.* **22**: 179, 1987).

Aulaxinomyces Cif. & Tomas. (1953) = Opegrapha (Roccell.).

auleate (of gasteromycete basidiomata), a closed basidioma in which pleated plates of trama project into the glebal cavity from top and sides. See Dring (1973); after Kreisel (1969).

Aulographaceae Luttr. (1973), Dothideales. 2 gen., 31 spp. Mycelium superficial, brown, without hyphopodia; ascomata strongly flattened, elongate, opening by an irregular split, composed of epidermoid cells, merging at the edge with the vegetative mycelium by irregularly branched hyphae; interascal tissue absent; asci numerous, clavate to broadly cylindrical; ascospores transversely septate, hyaline. Anamorph unknown. Saprobic, usually on dead plant material.
    Lit.: Batista (*Publs Inst. micol. Recife* **56**, 1959).

Aulographella Höhn. (1917) = Morenoina (Asterin.) fide Müller & v. Arx (1962).

Aulographina Arx & E. Müll. (1960), ? Asterinaceae. 3, widespr. See Wall & Keane (*TBMS* **82**: 257, 1984).

Aulographopsis Petr. (1938) nom. nud. (Mitosp. fungi).

Aulographum Lib. (1834), Aulographaceae. 30, widespr.

Aulospora Speg. (1909), Ascomycota (inc. sed.). 1, Argentina. See Eriksson & Hawksworth (*SA* **5**: 120, 1986).

Aurantiosacculus Dyko & B. Sutton (1979), Mitosporic fungi, 8.E1.1. 1 (on *Eucalyptus*), Australia.

Aurantiporellus Murrill (1905) = Pycnoporellus (Hymenochaet.) fide Pegler (1973).

Aurantiporus Murrill (1905) = Tyromyces (Coriol.) fide Donk (1974).

Aureobasidium Viala & G. Boyer (1891), Mitosporic fungi, 2.A2.16. 7, cosmop. *A. pullulans*, a variable sp. with many syn. See Cooke (*Mycopath.* **17**: 1, 1962), Joly (*BSMF* **81**: 402, 1965; 149 refs), Pugh & Buckley

(*TBMS* **57**: 227, 1971; endophytic in trees), Herma-nides-Nijhof (*Stud. Mycol.* **15**: 141, 1977; 14 spp., distinction from *Hormonema* and *Sarcinomyces*), Park (*TBMS* **78**: 385, 1982; Y-M dimorphism), Yoshikawa & Yokoyama (*Ann. Phytop. Soc. Jap.* **53**: 606, 1987; *A. microstictum* on *Hemerocallis*), Elinov *et al.* (*Mikol. Fitopatol.* **23**: 425, 1989; physiology and biochemistry), Mokrousov & Bulast (*Genetika* **28**: 31, 1992; DNA unhybridizable UP-PCR patterns in *A. pullulans*).

Aureobasis Clem. & Shear (1931) ≡ Aureobasidium (Mitosp. fungi).

Aureoboletus Pouzar (1957) = Pulveroboletus (Bolet.) fide Singer (1975).

Aureomyces Ruokola & Salonen (1970) = Cephaloas-cus (Cephaloasc.) fide v. Arx *et al.* (1972).

Auricula Battarra ex Kuntze (1891) [non Castrac. (1873) nom. cons., *Algae*] ≡ Auricularia (Auricar.).

Auricula Lloyd (1922) = Punctularia (Cortic.).

Auricularia Bull. ex Juss. (1789), Auriculariaceae. *c.* 15, widespr. The edible *A. polytricha* is cultured on poles of *Quercus* in China; *A. auricula-judae*, Jew's ear fungus, is sometimes parasitic, esp. on elder (*Sambucus*). See Lowy (*Mycol.* **44**: 656, 1952; **43**: 351, 1951; key), Donk (*Taxon* **7**: 168, 1958; *Persoonia* **4**: 154, 209, 1966; nomencl.), McLaughlin (*Am. J. Bot.* **67**: 1225, 1980; metabasidium ultrastr.).

Auriculariaceae Fr. (1838), Auriculariales. 5 gen. (+ 10 syn.), 16 spp. Basidium cylindrical, transversely sep-tate; basidioma gelatinous; spores cylindrical, ballis-tosporic.

Auriculariales, Basidiomycetes. 1 fam., 5 gen. (+ 10 syn.), 16 spp. Basidiocarps hemiangiocarpous and sessile; metabasidium cylindrical and horizontally septate, 1-4 cells each bearing a sterigma and basidio-spore; hyphae with septal dolipores. Fam.: **Auriculariaceae**.

    Lit.: Donk (1951-63) VIII; (1966: 208), Bandoni (*Trans. mycol. Soc. Japan* **25**: 521, 1984).

Auriculariella (Sacc.) Clem. (1909) = Auricularia (Auricular.) fide Donk (*Persoonia* **4**: 158, 1966).

Auriculariopsis Maire (1902), Meruliaceae. 2, widespr. See Donk (*Persoonia* **1**: 76, 1959).

Auriculora Kalb (1988), Lecanorales (inc. sed., L). 1, S. Am. See Henssen & Titze (*Bot. Acta* **101**: 131, 1990).

Auriculoscypha D.A. Reid & Manim. (1985), Platygloeaceae. 1, India.

Aurificaria D.A. Reid (1963), Hymenochaetaceae. 3, Afr., Asia, S. Am. See Reid (*Kew Bull.* **17**: 278, 1963).

Auriporia Ryvarden (1973), Coriolaceae. 3, USA, for-mer USSR. See Parmasto (*Mycotaxon* **11**: 173, 1980; key).

Auriscalpiaceae Maas Geest. (1963), Hericiales. 4 gen. (+ 1 syn.), 13 spp. Basidioma tough, coriaceous; hymenophore spinose.

    Lit.: Donk (1964: 245). See also *Lit.* under *Hydnaceae*.

Auriscalpium Gray (1821), Auriscalpiaceae. 5, widespr. See Maas Geesteranus (*Persoonia* **9**: 493, 1978; key).

Aurophora Rifai (1968), Sarcoscyphaceae. 1, Cuba, Madagascar, Australia.

Australohydnum Jülich (1978), Stereaceae. 1, Austra-lia.

Australoporus P.K. Buchanan & Ryvarden (1988), Coriolaceae. 1, Australia.

Austroblastenia Sipman (1983), ? Megalosporaceae (L). 2, Australasia.

Austroboletus (Corner) Wolfe (1980), Strobilo-mycetaceae. 13, widespr. See Wolfe (*Bibl. Mycol.* **69**, 1980).

Austroclitocybe Raithelh. (1972), Tricholomataceae. 1, Argentina.

Austrogaster Singer (1962), Paxillaceae. 1, Argentina.

Austrogautieria E.L. Stewart & Trappe (1985), Gau-tieriaceae. 6, Australia. See Stewart & Trappe (*Mycol.* **77**: 674, 1985; key).

Austrolecia Hertel (1984), Catillariaceae (L). 1, Antarct.

Austrolentinus Ryvarden (1991), Lentinaceae. 1, Aus-tralia, Solomon Isl.

Austroomphaliaster Garrido (1988), Hygrophoraceae. 1, Chile.

Austropeltum Henssen, Döring & Kantvilas (1992), Stereocaulaceae (L). 1, Australasia.

Austropezia Spooner (1987), Hyaloscyphaceae. 1, NZ.

Austrosmittium Lichtw. & M.C. Williams (1990), Legeriomycetaceae. 4, Australia, NZ. See Lichtwardt & Williams (*Mycol.* **84**: 384, 1992), Williams & Lichtwardt (*CJB* **68**: 1045, 1990).

autecology, ecological studies on a single species and its relationship to the biological and physiochemical aspects of its environment.

aut-eu-form, an autoecious rust having all the spore stages.

authentic (of specimens, cultures, etc.), identified by the author of the name of the taxon to which they are referred.

author citations, see Nomenclature.

Authors' names. It is customary to abbreviate many authors' names when cited as authorities for the scien-tific names of taxa in order to provide a clue as to where the name was published. There is frequently much variation and ambiguity in the abbreviations used by different writers and uniformity in usage is desirable. Kirk & Ansell (*Authors of fungal names*, *Index of Fungi Supplement*, 1992) provide a list of over 9,000 authors of scientific names of fungi with recommended standard forms of their names, includ-ing abbreviations. The 'standard form' for an author is the surname, or an abbreviation of it, or rarely a con-traction of it, with or without initials or other distin-guishing appendages. Among the more important criteria used in determining a standard form are: (1) names are in Roman characters; (2) every standard form must be unique to one person; (3) the same surname (i.e. identical spelling) must always be given in the same form, unless it is part of a compound name, and different surnames must not be given the same form; (4) all abbreviations and contractions are terminated by a full-stop but the full-stop does not make a standard form different from the same spelling without a full-stop; (5) the standard forms recom-mended in *TL-2* are retained in the majority of cases, one of a few exceptions being conflict with particu-larly well established abbreviations used elsewhere; (6) names are never abbreviated before a consonant; (7) names are usually not abbreviated unless more than two letters are eliminated and replaced by a full-stop.

    The above cited list forms part of a larger compila-tion (Brummitt & Powell (Eds), *Authors of plant names*, 1992) covering names of authors of all taxa whose nomenclature is governed by the *Code*.

    The following list of deceased authors for which there are **biographical notices** in this *Dictionary* pro-vides a representative series of examples of author abbreviations.

Ach(arius, E. 1757-1819)
Alexop(oulos, C.J. 1907-1986)
Arthur (J.C. 1850-1942)
Arx (J.A. von 1922-1988)
Asahina (Y. 1881-1975)
G.F. Atk(inson, 1854-1918)

Bat(ista, A.C. 1916-1967)
Berk(eley, M.J. 1803-1889)
Berl(ese, A.N. 1864-1903)
E.A. Bessey (1877-1957)
Bisby (G.R. 1889-1958)
Bolton (J. 1750-1799)
Bondartsev (A.S. 1877-1968)
Boud(ier, J.L.É. 1828-1920)
Bourdot (H. 1861-1937)
Bref(eld, J.O. 1839-1925)
Bres(àdola, G. 1847-1929)
W. Br(own, 1888-1975)
Buller (A.H.R. 1874-1944)
Bull(iard, J.B.F. 1752-1793)
E.J. Butler (1874-1943)

Cif(erri, R. 1897-1964)
Cooke (M.C. 1825-1914)
Corda (A.K.J. 1809-1849)
Costantin (J.N. 1857-1936)
G.(H.) Cunn(ingham 1892-1962)
M.A. Curtis (1808-1872)

P.(C.)A. Dangeard (1862-1947)
Dearn(ess, J. 1852-1954)
de Bary (H.A. 1831-1888)
De Not(aris, G. 1805-1877)
Desm(azières, J.B.H.J. 1786-1862)
Dietel (P. 1860-1947)
Dill(enius, J.J. 1684-1747)
Doidge (E.M. 1887-1965)
Donk (M.A. 1908-1972)

Ellis (J.B. 1829-1905)
Erikss(on, J. 1848-1931)

Farl(ow, W.G. 1844-1919)
Fée (A.L.A. 1789-1874)
E. Fisch(er, 1861-1939)
Fitzp(atrick, H.M. 1886-1949)
Fr(ies, E.M. 1794-1878)
Th.(M.) Fr(ies, 1832-1913)
Fuckel (K.W.G.L. 1821-1876)

Gäum(ann, E.A. 1893-1963)
Grev(ille, R.K. 1794-1866)
Grove (W.B. 1848-1938)
J.W. Groves (1906-1970)
Gruby (D. 1810-1898)
Guillierm(ond, M.A.A. 1876-1945)
Gyeln(ik, V.K. 1906-1945)

Hale (M.E. 1928-1990)
E.C. Hansen (1842-1909)
(H.J.A.)R. Hartig (1839-1901)
R. Heim (1900-1979)
Henn(ings, P.C. 1841-1908)
Höhn(el, F.X.R. von 1852-1920)

Jacz(ewski, A.L.A. 1863-1932)

P.(A.) Karsten (1834-1917)
Kauffman (C.H. 1869-1931)
Kniep (K.J.H. 1881-1930)
Körb(er, G.W. 1817-1885)
J.G. Kühn (1825-1910)
Kusano (S. 1874-1962)

J.E. Lange (1864-1941)
Langeron (M.C.P. 1874-1950)
Lév(eillé, J.-H. 1796-1870)
Lindau (G. 1866-1923)

Linds(ay, W.L. 1829-1880)
Link (J.H.F. 1767-1851)
L(innaeus, C. 1709-1778)
Liro (J.I. 1872-1943)
Lister (A. 1830-1908)
G. Lister (1860-1949)
Lloyd (C.G. 1859-1926)

McAl(pine, D. 1849-1932)
(A.)H. Magn(usson 1885-1964)
Maire (R.C.J.E. 1878-1949)
G.W. Martin (1886-1971)
E.W. Mason (1890-1975)
Massal(ongo, A.B. 1824-1862)
Massee (G.E. 1850-1917)
P.(A.) Micheli (1679-1737)
Mont(agne, J.P.F.C. 1784-1866)
Müll(er) Arg(oviensis, J. 1828-1896)
Mundk(ur, B.B. 1896-1952)
Murrill (W.A. 1869-1957)

Nannf(eldt, J.A. 1904-1985)
Nees (von Esenbeck, C.G.D. 1776-1858)
T.(F.L.) Nees (von Esenbeck 1787-1837)
Niessl (von Meyendorf, G. 1839-1919)
Nyl(ander, W. 1822-1899)

Pasteur (L. 1822-1895)
Pat(ouillard, N.T. 1854-1926)
Peck (C.H. 1833-1917)
Pers(oon, C.H. 1761-1836)
Petch (T. 1870-1948)
Petr(ak, F. 1886-1973)
Pilát (A. 1903-1974)
Poelt (J. 1924-1995)

Quélet (L. 1832-1899)

Rabenh(orst, G.L. 1806-1881)
Racib(orski, M. 1863-1917)
Ramsb(ottom, J. 1885-1974)
Rostaf(iński, J.T. 1850-1928)
Rostrup (E. 1831-1907)

Sabour(aud, R. 1864-1938)
Sacc(ardo, P.A. 1845-1920)
Săvul(escu, T. 1889-1963)
Schwein(itz, L.D. von 1780-1830)
Schwend(ener, S. 1829-1919)
Seaver (F.J. 1877-1970)
Shear (C.L. 1865-1956)
Singer (R. 1906-1994)
A.H. Sm(ith 1904-1986)
A.L. Sm(ith 1854-1937)
E.F. Sm(ith 1854-1927)
W.G. Sm(ith 1935-1917)
Sorauer (P.C.M. 1839-1916)
Sowerby (J. 1757-1822)
Sparrow (F.K. 1903-1977)
Speg(azzini, C.L. 1858-1926)
Stakman (E.C. 1885-1979)
P. Syd(ow 1851-1925)
Syd(ow, H. 1879-1946)

Thaxter (R. 1858-1932)
Theiss(en, F. 1877-1919)
Thom (C. 1872-1956)
Tode (H.J. 1733-1797)
Trevis(an, V. 1818-1887)
Tuck(erman, E. 1817-1886)
Tul(asne, L.R. 1815-1885)
C. Tul(asne 1816-84)

Unger (F. 1800-1870)

Viégas (A.R. 1906-1986)
Vain(io, E.A. 1853-1929)
Vuill(emin, P. 1861-1932)

Wakefield (E.M. 1886-1976)
Westerd(ijk, J. 1883-1961)
Weston (W.H. 1890-1978)
Whetzel (H.H. 1877-1944)
(H.)G. Winter (1848-1887)
Wormald (H. 1879-1953)
Woronin (M.S. 1838-1903)

Zahlbr(uckner, A. 1860-1938)
Zopf (W. 1846-1909).

For information on particular authors see also *Lit.* cited under History of mycology, Literature (Bibliographies), Medical and veterinary mycology, Reference Collections. Currently active mycologists are listed in society membership lists and regional compilations (e.g. Anon, *Revta Iberoamer. Micol.* **10**: ix, 1993 [Latin Am.]; Buyck & Hennebert, *Directory of African mycology*, 1993; and Holmgren & Holmgren, *Plant specialist index* [*Regnum Veg.* **124**], 1992: 11-36. See also Mycologists Online available via e-mail on huh.harvard.edu (Biodiversity and Biological Collections Gopher).

auto- (prefix), self-inducing, -producing, etc.

autobasidium, see basidium.

autochthonous, (1) indigenous; cf. allochthonous; (2) (of soil organisms), continuously active, as opposed to **zymogenous** organisms which become active when a suitable substrate becomes available (Winogradsky, 1924); cf. exochthonous (Park, 1957).

autodeliquescent (of lamellae and pileus of *Coprinus*), becoming liquid by **-digestion**.

autoecious, completing the life cycle on one host (esp. of rusts; cf. heteroecious); ametoecious (de Bary).

autogamy, the fusion of nuclei in pairs within the female organ, without cell fusion having taken place.

Autoicomyces Thaxt. (1908), Ceratomycetaceae. 27, widespr.

autolysis, self digestion of a cell or tissue by endogenous enzymes.

automictic sexual reproduction, karyogamy between daughter nuclei of different meioses in the same gametangium (Dick, 1972).

automixis, self-fertilization by the fusion of two closely related sexual cells or nuclei; cf. amphi-, apo-, pseudomixis.

Autophagomyces Thaxt. (1912), Laboulbeniaceae. 18, widespr.

autotroph (adj. **autotrophic**) (of a living organism), one not using organic compounds as primary sources of energy, i.e. using energy from light or inorganic reactions as do green plants, lichen-forming fungi, and the photosynthetic iron and sulphur bacteria. See Fry & Peel (Eds) (*Autotrophic micro-organisms*, 1954), Lees (*Biochemistry of autotrophic bacteria*, 1955); cf. heterotrophic.

auxanogram, the differential growth of a yeast in Petri dishes prepared by the auxanographic method of Beijerinck (as modified by Lodder, *Die anaskosporogenen Hefen*, 1934, and Langeron, 1952: 430) for determining the carbon and nitrogen requirements of the organism. See also Lodder & van Rij (1952), Pontecorvo (*J. gen. Microbiol.* **3**: 122, 1949; auxanographic techniques in biochemical genetics).

Auxarthron G.F. Orr & Kuehn (1963), Onygenaceae. 8, widespr. See Orr (1977), Samson (*Acta Bot. neerl.* **21**: 517, 1972). Anamorph *Chrysosporium.*

auxiliary zoospore, first-formed zoospore, formed and flagellate within the sporangium, in a species with dimorphic zoospores (Dick, 1973); flagellar insertion apical or sub-apical.

auxotroph, a biochemical mutant which will only grow on the minimal medium (q.v.) after the addition of one or more specific substances.

avenacein, see enniatin.

avenacin, a fungus inhibitor from oats (*Avena*) (Turner, *Nature* **186**: 325, 1960).

aversion, the inhibition of growth at the adjacent edges of colonies of microorganisms, esp. in a culture of one species. Cf. antagonism; barrage.

Avettaea Petr. & Syd. (1927), Mitosporic fungi, 4.A2.15. 3, Australia, Philipp., Pakistan.

Awasthia Essl. (1978), Physciaceae (L). 1, India.

Awasthiella Kr.P. Singh (1980), Verrucariaceae (L). 1, India.

axenic (of cultures), consisting of one organism; uncontaminated; a pure culture. Cf. gnotobiotic.

axeny, inhospitality; 'passive' as opposed to 'active' resistance of a plant to a pathogen (Gäumann, 1946).

axial canal (- **mass**), see ascus.

axoneme, the main core of a flagellum composed of 2 central microtubules surrounded by 9 double microtubules.

Aylographum, see *Aulographum.*

Azbukinia Lar.N. Vassiljeva (1989), Ascomycota (inc. sed.). 1, former USSR. See Eriksson & Hawksworth (*SA* **8**: 99, 1990).

Azosma Corda (1831) = Cladosporium (Mitosp. fungi) fide Fries (*Syst. mycol.* **3** (Index): 55, 1832).

azotodesmic nitrogen-fixing (Pike & Carroll, *in* Alexopoulos & Mims, *Introductory mycology*, edn 3, 1980).

Azureothecium Matsush. (1989), Pseudeurotiaceae. 1, Australia.

Azygites Fr. (1832) = Syzygites (Mucor.) fide Hesseltine (1955).

azygospore, a parthenogenetic zygospore; characteristic of some *Mucorales*. See Benjamin (*Aliso* **5**: 235, 1963; list).

Azygozygum Chesters (1933) = Mortierella (Mortierell.) fide Plaats-Niterink *et al.* (*Persoonia* **9**: 85, 1976).

Azymocandida E.K. Novák & Zsolt (1961) = Candida (Mitosp. fungi) fide Lodder (1970).

Azymohansenula E.K. Novák & Zsolt (1961) = Pichia (Endomycet.).

Azymomyces E.K. Novák & Zsolt (1961) = Torulaspora (Saccharomycet.) p.p. and Saccharomyces (Saccharomycet.) p.p. See Lodder (1970), Batra (1978).

Azymoprocandida E.K. Novák & Zsolt (1961) = Candida (Mitosp. fungi) fide Lodder (1970).

B, Botanischer Garten und Botanisches Museum Berlin-Dahlem (Berlin, Germany); founded 1815; from 1995 part of the Free University of Berlin; see Kohlmeyer (*Willdenowia* **3**: 63, 1962).

Babjevia Van der Walt & M.T. Sm. (1995), Saccharomycetaceae. 1, Eur.

baccate, soft throughout like a berry.

baccatin, a wilt toxin from *Gibberella baccata* (Gäumann *et al.*, *Phytopath. Z.* **36**: 114, 1959); antibacterial.

Bachmannia Zschacke (1934) [non Pax (1897), *Capparaceae*] = Verrucaria (Verrucar.) fide Swinscow (1968).

Bachmanniomyces D. Hawksw. (1981), Mitosporic fungi, 4.A1.19. 1 (on lichen, *Cladonia*), Eur., N. Am.

Bacidia De Not. (1846), Bacidiaceae (±L). *c.* 200, cosmop. See Awasthi & Mathur (*Proc. Indian Acad. Sci.*

(Pl. Sci.) **97**: 481, 1987; key 8 spp. India), Sérusiaux (*Nordic J. Bot.* **13**: 447, 1993; diffs. 9 segr. gen.), 1952), Stizenberger (*Nova Acta Acad. Leop.-Carolin.* **34** (2), 1867), Vainio (*Acta Soc. Fauna Fl. fenn.* **53** (1), 1922), Vězda (*Čas. slezsk. Mus. Opave* ser. A, **10**: 103, 1961; *Folia geobot. phytotax.* **15**: 75, 1980, key foliicolous spp.).

Bacidiaceae Walt. Watson (1929), Lecanorales (±L). 23 gen. (+ 24 syn.), 394 spp. Thallus crustose; ascomata apothecial, pale to black, without a well-developed thalline margin, ± superficial on the thallus; interascal tissue of paraphyses, usually branched at the apex, sometimes anastomosing, often with a well-developed epithecium; asci with a well-developed apical cap, strongly blueing in iodine apart from the apical cushion, ocular chamber small, often with an outer I+ mucous layer; ascospores ellipsoidal to elongated, usually septate, without a sheath. Lichenized with green algae, occ. lichenicolous, widespr.
Lit.: See under *Bacidia*.

Bacidiactis M. Choisy (1931) ? = Lecanactis (Roccell.).

Bacidina Vězda (1991), Lecanoraceae (L). 11, widespr. = Bacidia (Bacid.) fide Santesson (*Lichens and lichenicolous fungi of Sweden and Norway*, 1993).

Bacidiomyces Cif. & Tomas. (1953) ≡ Bacidia (Bacid.).

Bacidiopsis Bagl. (1861) = Pachyphiale (Gyalect.).

Bacidiospora Kalb (1988), Bacidiaceae (L). 1, S. Am.

bacillar, (**bacilliform**), rod-like in form (Fig. 37.4*a*).

Bacillaria Mont. (1840) [non J.F. Gmel. (1791), *Angiospermae*] ≡ Camillea (Xylar.).

Bacillina Nyl. (1897) = Toninia (Catillar.).

Bacillispora Sv. Nilsson (1962), Mitosporic fungi, 1.A1.1. 1 (aquatic), Sweden. = Cylindrocarpon (Mitosp. fungi) fide Descals & Marvanová (*TBMS* **89**: 501, 1987).

Bacillopeltis Bat. (1957), Mitosporic fungi, 5.A1.?. 1, Brazil.

Bacillopsis Petsch. (1908) nom. dub. (? Saccharomycetales).

Backusella Hesselt. & J.J. Ellis (1969), Thamnidiaceae. 3, widespr. See Benny & Benjamin (*Aliso* **8**: 301, 1975; key), Stalpers & Schipper (*Persoonia* **11**: 39, 1980).

Backusia Thirum., M.D. Whitehead & P.N. Mathur (1965) = Monascus (Monasc.) fide Cole & Kendrick (*CJB* **46**: 987, 1968).

Bacteria. Heterogeneous group of (usually) unicellular prokaryotic organisms. Used in the strictest sense to describe all prokaryotes belonging to the Domain *Bacteria* (formerly *Eubacteria*), which excludes the *Archaea* (formerly *Archaebacteria*). See Embley *et al.* (*Phil. Trans. R. Soc. Lond.* B **345**: 21-33, 1994). Some bacteria are pathogenic for plants, a few are pathogenic for fungi: *Bacillus polymyxa* (bacterial pit) of cultivated mushroom; *Pseudomonas agarici* (drippy gill) of cultivated mushroom (see Geels *et al.*, *Jl Phytopathol.* **140**: 249, 1994); *Pseudomonas tolaasi* (brown blotch) of cultivated mushroom (see Cole & Skellerup, *TBMS* **87**: 314, 1986); *Pantoea agglomerans* pv. *uredovora* (syn. *Erwinia uredovora*) a parasite of rust uredinia. Bacteria, in particular actinomycetes, *Bacillus* spp. and *Pseudomonas* spp. produce a variety of biologically active molecules (antibiotics, enzymes etc.) and have been used as biological control agents against fungal plant diseases.

bacteriostatic, of a substance, or a concentration of a bactericide, which will not let growth of bacteria take place but which is not bactericidal.

bactivory, bacteria-feeding; known in fungi only amongst certain marine *Thraustochytriales* (Raghukumar, *Mar. Biol.* **113**: 165, 1992).

bactobiont, see photobiont.

Bactrexcipula Höhn. (1918) ? = Rhizothyrium (Mitosp. fungi) fide Petrak (*Sydowia* **15**: 185, 1962).

Bactridiopsis Henn. (1904) = Coccospora (Mitosp. fungi) fide Damon & Downing (*Mycol.* **46**: 209, 1954).

Bactridiopsis Gonz. Frag. & Cif. (1927) = Phillipsiella (Phillips.) fide Rossman *et al.* (*Sydowia* **46**: 66, 1994).

Bactridium Kunze (1817), Mitosporic fungi, 3.C1.1. 5, widespr. See Hughes (*N.Z. Jl Bot.* **4**: 522, 1966).

Bactriexta Preuss (1852) nom. dub (? Protozoa).

Bactroboletus Clem. (1909) ≡ Filoboletus (Tricholomat.).

Bactrodesmiastrum Hol.-Jech. (1984), Mitosporic fungi, 1.C2.1. 1, former Czechoslovakia. ? = Janetia (Mitosp. fungi).

Bactrodesmiella M.B. Ellis (1959), Mitosporic fungi, 3.C2.19/40. 1, UK.

Bactrodesmiites Babajan & Tasl. (1977), Fossil fungi. 1 (Tertiary), former USSR.

Bactrodesmium Cooke (1883), Anamorphic Dothideales, 3.C2.1. Teleomorph *Stuartella*. 20, widespr. See Ellis (*Mycol. Pap.* **72**, 1959; key), Hughes & White (*Fungi Canad.* 251-261, 1983).

Bactropycnis Höhn. (1920) = Coleophoma (Mitosp. fungi) fide Sutton (1977).

Bactrosphaeria Penz. & Sacc. (1897), Ascomycota (inc. sed.). 1, Java.

Bactrospora A. Massal. (1852), Roccellaceae (L). 20, mainly temp. See Egea & Torrente (*Lichenologist* **25**: 211, 1993; key).

Bactrosporaceae, see *Roccellaceae*.

Bactrosporomyces Cif. & Tomas. (1953) ≡ Bactrospora (Roccell.).

baculate, (1) (**baculiform**) (of spores), rod-shaped; (2) (of surface ornamentation), rod-shaped (Fig. 29.8).

Baculospora Zukal (1887), Ascomycota (inc. sed.). 1, Eur. See Lundqvist (*SA* **7**: 62, 1988).

Badarisama Kunwar, J.B. Manandhar & J.B. Sinclair (1986), Mitosporic fungi, 1.D2.1. 1 (bulbil-forming), USA.

Badhamia Berk. (1853), Physaraceae. 29, cosmop. See Sekhon (*J. Indian bot. Soc.* **58**: 56, 1979; key 8 Indian spp.).

Badhamiopsis T.E. Brooks & H.W. Keller (1976) = Badhamia (Physar.) fide Pando (*in litt.*).

Badimia Vězda (1986) nom. cons. prop., Ectolechiaceae (L). 11 (on leaves), trop. See Lücking *et al.* (*Bot. Acta* **107**: 393, 1994), Lücking & Vězda (*Taxon* **44**: 227, 1995).

Badimiella Malcolm & Vězda (1994), Ectolechiaceae (L). 1, NZ.

Baeoderma Vain. (1922) = Sphaerophorus (Sphaerophor.).

Baeodromus Arthur (1905), Pucciniosiraceae. 7 (5 on *Senecio*), Am., Russian Far East; 0, III. See Buriticá (*Rev. Acad. Colomb. Cienc.* **18**(69): 131, 1991).

Baeomyces Pers. (1794), Baeomycetaceae (L). *c.* 8, cosmop. See Honegger (*Lichenologist* **15**: 57, 1983; asci), Jahns (*Herzogia* **2**: 133, 1971; ontogeny), Thomson (*Bryologist* **70**: 285, 1967), Sérusiaux (*Taxon* **32**: 646, 1983; gen. nomencl.), Kumar (*Geophytology* **15**: 159, 1985; key 4 spp. India), Gierl & Kalb (*Herzogia* **9**: 593, 1993; concept). See also *Dibaeis*, *Phyllobaeis*.

Baeomycetaceae Dumort. (1829), Leotiales (L). 2 gen. (+ 7 syn.), 12 spp. Thallus varied, usually crustose or squamulose; ascomata sessile or shortly stipitate, sometimes clustered, formed on specialized generally non-lichenized thalline branches, flat or convex, pink or brown, the wall of interwoven hyphae; interascal tissue of simple or sparingly branched paraphyses, often swollen at the apices; hymenial gel I+ or I-; asci with an I+ or I- apical pore; ascospores hyaline, simple or transversely septate. Anamorph pycnidial. Lichenized with green algae, widespr.

Baeomycomyces E.A. Thomas ex Cif. & Tomas. (1953) = Baeomyces (Baeomycet.).

Baeopodium Trevis. (1857) = Gomphillus (Gomphill.).

Baeospora Singer (1938), Tricholomataceae. 5, N. temp., trop.

Baeostratoporus Bondartsev & Singer (1944) = Flaviporus (Polypor.) fide Donk (1960).

Baeumleria Petr. & Syd. (1927) ? = Coniella (Mitosp. fungi) fide Clements & Shear (1931), but see Sutton (1977).

Bagcheea E. Müll. & R. Menon (1954), Valsaceae. 1, India.

Baggea Auersw. (1866), ? Patellariaceae. 1, Eur.

Bagliettoa A. Massal. (1853) = Verrucaria (Verrucar.), but maintained by Poelt & Vězda (1981) and Santesson (*Lichens and lichenicolous fungi of Sweden and Norway*, 1993).

Bagnisiella Speg. (1880), ? Dothideaceae. 13, S. Am., India. See Patil & Patil (*Indian J. Mycol. Pl. Path.* **13**: 169, 1985).

Bagnisimitrula S. Imai (1942), ? Geoglossaceae. 1, Japan.

Bagnisiopsis Theiss. & Syd. (1915) = Coccodiella (Phyllachor.) fide Katumoto (1968).

Bahianora Kalb (1984), Lecideaceae (L). 1, Brazil.

Bahuchashaka Subram. (1978), Mitosporic fungi, 1.C2.1. 1, Japan.

Bahugada K.A. Reddy & V.G. Rao (1984), Mitosporic fungi, 3.D2.10. 1, India. ? = Monodictys (Mitosp. fungi) fide Sutton (*in litt.*).

Bahukalasa Subram. & Chandrash. (1979), Mitosporic fungi, 1.A1.10/12. 1, India.

Bahupaathra Subram. & Lodha (1964) = Cladorrhinum (Mitosp. fungi) fide Mouchacca & Gams (*Mycotaxon* **48**: 415, 1993).

Bahusaganda Subram. (1971) nom. inval. Mitosporic fungi, 1.C2.3. 1, India.

Bahusakala Subram. (1958), Mitosporic fungi, 1.C2.39/40. 1, Sri Lanka, India. See Sigler & Carmichael (*Mycotaxon* **4**: 349, 1976).

Bahusandhika Subram. (1956), Mitosporic fungi, 1.C2.3. 2, India.

Bahusutrabeeja Subram. & Bhat (1977), Mitosporic fungi, 1.A1.15. 1, India.

Bainieria Arnaud (1952) nom. inval. Mitosporic fungi, 1.A1.19. 1, Eur.

Bakanae disease, of rice (*Gibberella fujikuroi*); see gibberellin.

Bakeromyces Syd. & P. Syd. (1917) = Trichosphaeria (Trichosphaer.) fide Höhnel (*Ann. Myc.* **16**: 77, 1918).

Bakerophoma Died. (1916) nom. dub. fide Sutton (*Mycol. Pap.* **141**, 1977).

Balaniopsis P.M. Kirk (1985), Mitosporic fungi, 1.A2.40. 1, Afr.

Balanium Wallr. (1833), Mitosporic fungi, 1.B2.8. 1, Eur. See Hughes (*CJB* **39**: 1505, 1961).

balanoid, acorn-shaped.

Balansia Speg. (1885), Clavicipitaceae. 20, warmer areas. See Diehl (*USDA Agric. Monogr.* **4**, 1950). See *Ephelis*.

Balansiella Henn. (1904) = Claviceps (Clavicipit.) fide Rogerson (1970).

Balansina G. Arnaud (1918) = Dothidasteromella (Asterin.) fide Müller & v. Arx (1962).

Balansiopsis Höhn. (1910), Clavicipitaceae. 3, Am. See Diehl (*USDA Agric. Monogr.* **4**, 1950).

Balazucia R.K. Benj. (1968), Laboulbeniaceae. 2, Mexico, Japan. See Terada (*Trans. mycol. Soc. Jap.* **21**: 193, 1980).

Balladyna Racib. (1900), Parodiopsidaceae. *c.* 15 (on *Rubiaceae*), trop. See Müller & v. Arx (1962), Eboh & Cain (*CJB* **51**: 61, 1973).

Balladynastrum Hansf. (1941) = Balladynopsis (Parodiopsid.) fide Sivanesan (1981).

Balladynella Theiss. & Syd. (1918) = Dysrhynchis (Parodiopsid.) fide Müller & v. Arx (1962).

Balladynocallia Bat. (1965), Parodiopsidaceae. 3, Afr., S. Am. See Sivanesan (*Mycol. Pap.* **146**, 1981; key).

Balladynopsis Theiss. & Syd. (1918), Parodiopsidaceae. 12, trop. See Sivanesan (*Mycol. Pap.* **146**, 1981; monogr., key). = Balladyna (Dothid.) fide v. Arx & Müller (*Sydowia* **37**: 6, 1984). Anamorph *Tretospora*.

ballistospore, a forcibly abjected basidiospore. See Nyland (*Mycol.* **41**: 688, 1950); - **discharge,** see Buller (*Researches* **2-6**, 1922-34), Olive (*Science, N.Y.* **146**: 524, 1964), Ingold (*Friesia* **9**: 66, 1969; review). See also Buller's drop, *Itersonilia*.

Ballistosporomyces Nakase, G. Okada & Sugiy. (1989), Mitosporic fungi, 1.A1.21. 2 (ballistosporic yeasts), Japan.

Ballocephala Drechsler (1951), Meristacraceae. 3 (on tardigrades), N. Am., Scotland, Japan. See Richardson (*TBMS* **55**: 307, 1970), Pohlad & Bernard (1978), Saikawa & Oyama (*Trans. mycol. Soc. Japan* **33**: 305, 1992; EM of infection of tardigrades), Saikawa & Sakuramata (*Trans. mycol. Soc. Japan* **33**: 237, 1992; zygospores), Tucker (1981; key).

Balsamia Vittad. (1831), Balsamiaceae. 6, N. Am., Eur., N. Afr. See Hawker (*Phil. Trans. R. Soc. Lond.* **237**: 429, 1954; key Eur. spp.), Gilkey (*N. Am. Fl.* **2** (1), 1954; key N. Am. spp.), Pegler *et al.* (*British truffles*, 1993).

Balsamiaceae E. Fisch. (1897), Pezizales. 3 gen. (+ 3 syn.), 13 spp. Ascomata large, cleistothecial, ± globose, solid, with pockets of fertile tissue separated by sterile undifferentiated cells; asci cylindrical to globose, the walls not blueing in iodine, persistent, indehiscent, randomly produced or in distinct hymenial areas; ascospores ± ellipsoidal, hyaline or brown, smooth, without a sheath. Anamorphs unknown. Hypogeous, N. temp.
   Lit.: Donadini (*BSMF* **102**: 373, 1986; status), Trappe (1979).

Balzania Speg. (1899), Hypocreaceae. 1, S. Am.

Banhegyia Zeller & Tóth (1960), Ascomycota (inc. sed.). 1, Eur. See Kohlmeyer & Kohlmeyer (*Icones fung. maris*, 1968).

Bankera Coker & Beers ex Pouzar (1955), Bankeraceae. 7, widespr.

Bankeraceae Donk (1961), Thelephorales. 2 gen., 17 spp. Basidioma pileate, stipitate; hymenophore spinose; spores white.
   Lit.: Donk (1964: 246); see also under *Hydnaceae*.

Banksiamyces G.W. Beaton (1982), ? Leotiaceae. 4 (on *Banksia* cones), Australia. ? = Encoelia (Leot.) fide Zhuang (*Mycotaxon* **32**: 97, 1988).

Bapalmuia Sérus. (1993), Lecanorales (inc. sed., L). 4, trop.

Barbariella Middelh. (1949) = Asaphomyces (Laboulben.) fide Benjamin (1955).

Barbarosporina Ķirulis (1942), Mitosporic fungi, 6.A1.?. 1 (on *Rhytisma*), former USSR. ? Anamorph of *Rhytisma*.

barbate, having one or more groups of hairs; bearded.

Barbeyella Meyl. (1914), Clastodermataceae. 1, widespr. See Curtis (*Mycol.* **60**: 708, 1968).

Barclayella Dietel (1890) = Chrysomyxa (Coleospor.) fide Dietel (1928).

Bargellinia Borzí (1888) = Wallemia (Mitosp. fungi) fide v. Arx (1981).

Barklayella Sacc. (1892) [non *Barclayella* Dietel (1890)] = Polynema (Mitosp. fungi) fide Sutton (1977); = Neobarclaya (Mitosp. fungi) fide Nag Raj (*CJB* **56**: 706, 1978).

Barlaea Sacc. (1889) [non Rchb. (1876), *Orchidaceae*] ≡ Barlaeina (Otid.).

Barlaeina Sacc. & P. Syd. (1899) = Lamprospora (Otid.). See Eckblad (1968).

barm, (1) the froth on the surface of fermenting malt liquids; (2) baker's yeast.

Barnettella D. Rao & P. Rag. Rao (1964), Mitosporic fungi, 8.D2.43. 3, India. See Satyanarayana & Rao (*Mycopath. Mycol. appl.* **45**: 267, 1971), Rao & Rao (*Indian Phytopath.* **26**: 233, 1973; key).

Barnettia Bat. & J.L. Bezerra (1962) = Microcallis (Chaetothyr.) fide v. Arx & Müller (1975).

barrage, the space between two mycelia which have an aversion for one another (Vandendries & Brodie, 1933). Cf. zone lines.

Barria Z.Q. Yuan (1994), ? Phaeosphaeriaceae. 1, China.

Barrmaelia Rappaz (1995), Xylariaceae. 6, Eur., N. Am. Anamorph *Libertella*.

Barssia Gilkey (1925), Balsamiaceae. 2, N. Am., Eur., Japan. See Ławrynowicz & Skirgiełło (*Acta mycol.* **20**: 277, 1986).

Bartalinia Tassi (1900), Mitosporic fungi, 1.C1.19. 1, Italy. See Morgan-Jones *et al.* (*CJB* **50**: 877, 1972), Roux & Van Warmelo (*MR* **94**: 109, 1990; conidioma ontogeny.

Bartaliniopsis S.S. Singh (1972) = Doliomyces (Mitosp. fungi) fide Sutton (1977).

Bartheletia Arnaud (1954) nom. inval. Mitosporic fungi, 3.?.?. 1, France.

Bartlettiella D.J. Galloway & P.M. Jørg. (1990), Lecanorales (inc. sed., L). 1, NZ.

Barubria Vězda (1986), Ectolechiaceae (L). 1 (on leaves), Afr.

Barya Fuckel (1864) [non Klotzsch (1854), *Begoniaceae*] ≡ Neobarya (Clavicipit.).

Baryeidamia H. Karst. (1888) = Papulaspora (Mitosp. fungi) fide Saccardo (*Syll. Fung.* **9**: 339, 1891).

Baryella Rauschert (1988) ≡ Neobarya (Clavicipit.).

basal body, (1) (of *Blastocladiaceae*), the part of the thallus fixed to the substratum by rhizoids at the lower end (Indoh, 1940); (2), see blepharoplast.

basal frill (of a spore), the apical part of a conidiogenous cell, or basal part of a cell which is carried away with the detached conidium following rhexolytic secession (Sutton, *CJB* **45**: 1251, 1967). Cf. marginal frill.

Basauxia Subram. (1992), Mitosporic fungi, 1.C2.1/21. 1, Malaysia.

basauxic (of conidiophores), elongating by a basal growing point (Hughes, *CJB* **31**: 650, 1953). See meristem blastospore.

base ratio, see Molecular biology.

Basiascella Bubák (1914) ? = Piggottia (Mitosp. fungi) fide Sutton (1977).

Basiascum Cavara (1888) = Spilocaea (Mitosp. fungi) fide Hughes (1958).

Basididyma Cif. (1962), Mitosporic fungi, 1.A1.10. 1, Dominican Republic.

Basidiella Cooke (1878) = Aspergillus (Mitosp. fungi) fide Subramanian (*Hyphomycetes*, 1971). See Seifert (*TBMS* **85**: 123, 1985).

Basidioblastomycetes. Class used by Moore (*Bot. Mar.* **23**: 361, 1980; key 13 gen.) for *Malasseziales* and *Sporobolomycetales*, and placed by him in *Blastomycota*.

Basidiobolaceae Engl. & E. Gilg (1924), Entomophthorales. 1 gen. (+ 1 syn.), 4 spp. Sporophores with subsporangial vesicle; all cells uninucleate.

Lit.: Humber (1995).

Basidiobolus Eidam (1886), Basidiobolaceae. 4, cosmop. *B. ranarum*, a widespr. saprobe; *B. meristosporus* (subcutaneous phycomycosis in humans). See Drechsler (*Mycol.* **48**: 655, 1956), Greer & Friedman (*Sabouraudia* **4**: 231, 1966), Srinivasan & Thirumalachar (*Mycopath. Mycol. Appl.* **33**: 56, 1967), Coremans-Pelseneer (*Acta Zool. Path.* **60**: 1, 1974; biology), Benjamin (*Aliso* **5**: 223, 1962; key), Cochrane *et al.* (*Mycol.* **81**: 504, 1989; isozymes), Nelson *et al.* (*Experimental Mycol.* **14**: 197, 1990; ribosomal DNA), Dykstra (*Mycol.* **86**: 494, 1994; spore formation), Dykstra & Bradley-Kerr (*Mycol.* **86**: 336, 1994; ultrastr.).

Basidiobotrys Höhn. (1909) = Xylocladium (Mitosp. fungi) fide Jong & Rogers (1972).

basidiocarp, see basidioma.

Basidiodendron Rick (1938), Exidiaceae. 15, cosmop. See Wells (*Mycol.* **51**: 541, 1960; key), Oberwinkler (*Ber. bayer bot. Ges.* **36**: 42, 1963; Bavarian spp.), Luck-Allen (*CJB* **41**: 1033, 1963; key N. temp. spp.), Wells & Raitviir (*Mycol.* **67**: 909, 1975).

basidiograph, the straight-line graph obtained by plotting the ratio of the length (*l*) to the width (*w*) against the length of the basidia of a species of agaric (Corner, 1947: 214). Cf. sporograph.

basidiole, a basidium-like hymenial element that lacks sterigmata because it is either young or permanently sterile; best restricted to immature basidia fide Singer (1962).

Basidiolum Cienk. (1861) nom. dub. (? Mitosp. fungi).

basidioma (pl. -ata), a basidium-producing organ; basidiome (Donk, *Taxon* **18**: 666, 1969); basidiocarp; carpophore; fruit-body; hymenophore; sporophore. See Fig. 4.

basidiomycete, one of the *Basidiomycota*.

Basidiomycetes, Basidiomycota. 32 ord., 140 fam., 473 gen. (+ 373 syn.), 13,857 spp. By accepting the distinction made by Talbot (1968) between 'primary' and 'adventitious' septa two subclasses may be distinguished:

Phragmobasidiomycetidae (approx. syn. *Heterobasidiomycetes*) in which the metabasidium is divided by primary septa, usually cruciate or horizontal. Orders:

(1) **Agaricostilbales**.
(2) **Atractiellales**.
(3) **Auriculariales**.
(4) **Heterogastridiales**.
(5) **Tremellales**.

Lit.: Martin (*Univ. Ia Stud. nat. Hist.* **19** (3), 1952; N. Am. spp.; *Mycol.* **37**: 527, 1945, *Tremellales* s.l. classification).

Holobasidiomycetidae (approx. syn. *Homobasidiomycetes*) in which the metabasidium is not divided by primary septa but may sometimes become adventitiously septate. Orders:

(6) **Agaricales**.
(7) **Boletales**.
(8) **Bondarzewiales**.
(9) **Cantharellales**.
(10) **Ceratobasidiales**.
(11) **Cortinariales**.
(12) **Dacrymycetales**.
(13) **Fistulinales**.
(14) **Ganodermatales**.
(15) **Gautieriales**.
(16) **Gomphales**.
(17) **Hericiales**.
(18) **Hymenochaetales**.
(19) **Hymenogastrales**.
(20) **Lachnocladiales**.
(21) **Lycoperdales**.

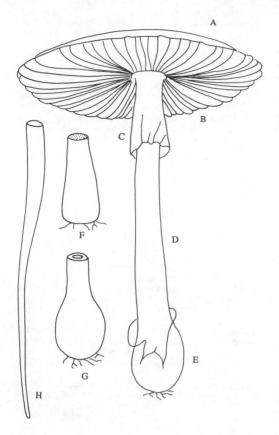

Fig. 4. Basidioma (sporophore) of *Amanita phalloides*. A, pileus (cap); B, lamellae (gills); C, annulus (ring); D, stipe (Stalk); E, volva. Other types of stipe: F, stuffed, with base truncate; G, hollow, with base bulbous; H, solid with base radicate.

(22) **Melanogastrales**.
(23) **Nidulariales**.
(24) **Phallales**.
(25) **Poriales**.
(26) **Russulales**.
(27) **Schizophyllales**.
(28) **Sclerodermatales**.
(29) **Stereales**.
(30) **Thelephorales**.
(31) **Tulasnellales**.
(32) **Tulostomatales**.

Lit. (see also under Macromycetes): **General**: Donk (1951-63), Generic names proposed for Hymenomycetes, I ('Cyphellaceae'), II (Hymenolichenes), III ('Clavariaceae'), IV (Boletaceae), *Reinwardtia* **1**: 199, **2**: 435, **3**: 275, 1951-58, V ('Hydnaceae'), *Taxon* **5**: 69, 95, 1956, VI (Brachybasidiaceae, Cryptobasidiaceae, Exobasidiaceae), *Reinwardtia* **4**: 113, 1956, VII ('Thelephoraceae'), VIII (Auriculariaceae, Septobasidiaceae, Tremellaceae, Dacrymycetaceae), *Taxon* **6**: 17, 68, 106, **7**: 164, 193, 236, 1957-58, IX ('Meruliaceae', *Cantharellus*), *Fungus* **28**: 7, 1958, X ('Polyporaceae'), *Persoonia* **1**: 173, 1960 (additions and corrections, **2**: 201, 1962): XI (Agaricaceae); *Beih. Nova Hedw.* **5**, 1962, XII (Deuteromycetes), XIII (additions and corrections); *Taxon* **11**: 75, **12**: 113, 1962-63. [I-IX, XII, XIII,

reprinted as 1 vol., 1966; X reprinted, 1968. In this valuable series of papers many taxonomic points are also discussed.] Donk (1954-62) Notes on resupinate hymenomycetes: I (*Pellicularia*), *Reinwardtia* **2**: 425, 1954; II (Tulasnelloid fungi), **3**: 363, 1956; III, IV, V, *Fungus* **26**: 3, **27**: 1, **28**: 16, 1956-58; VI, *Persoonia* **2**: 217, 1962. Rea (1922), Bourdot & Galzin (1927), Killerman (1928), Eriksson (*Symb. bot. upsal.* **16**(1): 1-172, 1958; N. Sweden), Donk (1954-62; *Reinwardtia* **2**: 425, 1954; **3**: 363, 1956; *Fungus* **26**: 3, **27**: 1, **28**: 16, 1956-58; *Persoonia* **2**: 217, 1962; resupinates), Donk (*Persoonia* **3**: 199, 1964; conspectus of families), Shaffer (*in* Parker, 1982, **1**: 248), Stephanova-Kartavenko ([Aphyllophorous fungi of the Urals], 1967; gen. keys), Parmasto (*The Lachnocladiaceae of the Soviet Union with a key to boreal species*, 1970 [*Scripta mycol.* **2**]), Pegler (*The polypores*, 1973 [*Bull. BMS Suppl.*]; keys world gen., Br. spp.), Strid (*Aphyllophorales of N. Central Scandinavia*, 1975 [*Wahlenbergia* **1**]), Domański (*Mala Flora Grzylów* **1**, *Aphyllophorales*, 1975), Rattan (1977), Stalpers (1978). Clémençon (Ed.) (*The species concept in Hymenomycetes*, 1977). Donk (1966), *Persoonia* **4**: 145, 1966; **8**: 33, 1974; checklists of European heterobasidiomycetes, annotations, ref., index. Lowy, *Taxon* **17**: 118, 1968; (heterobasidiomycete taxonomy); Talbot, *Taxon* **17**: 620, 1968. Kühner (*TBMS* **68**: 1, 1977; nuclear behaviour, review), Moser (*Röhrlinge und Blätterpilze*, 1978), Jülich (*Bibl. Mycol.* **85**, 1992), Jülich & Stalpers (*The resupinate non-poroid Aphyllophorales of the temperate Northern hemisphere*, 1980), Kühner (*Les Hyménomycetètes agaricoïdes (Agaricales, Tricholomatales, Pluteales, Russulales)*, 1980), Parmasto (*Windhalia* **16**: 3, 1986), Corner (*Ad Polyporaceas* **1-7** [*Beih. Nova Hedw.*], 1983-1991), Moser & Jülich (*Farbatlas der Basidiomyceten* **1-12**, 1994).

Regional: **America, North**, Shaffer (*Keys to genera of higher fungi*, edn 2, 1968; mostly hymenomycetes), **South**, Singer (*Beih. Nova Hedw.* **29**, 1969; Agaricales, Aphyllophorales, Gasteromycetes). **Europe**, Donk (1966); **Great Britain**, Rea (*British Basidiomycetae*, 1922; Suppl. *TBMS* **12**: 205, **17**: 35, 1927-32, incl. gasteromycetes), Reid & Austwick (*Glasgow Nat.* **18**: 255, 1963; annot. list of Scottish basidiomycetes, incl. gasteromycetes, excl. rusts and smuts). **France**, Bourdot & Galzin (*Hyménomycetes de France, Hetérobasidiés, Homobasidiés gymnocarpes*, 1927). **Portugal**, Da Camara (*Catalogus systematicus fungorum omnia Lusitaniae*. I, *Basidiomycetes*. Pars 1, *Hymeniales*, 1956; Pars 2, *Gasterales, Phalloidales, Tremelloidales, Uredinales et Ustilaginales*, 1958). **former USSR**, Raitviir [Key to Heterobasidiomycetidae of the USSR, 1967].

Basidiomycota (Basidiomycotina, basidiomycetes). Fungi. 3 class., 41 ord., 165 fam., 1,428 gen. (+ 1,237 syn.), 22,244 spp., cosmop. The diagnostic character of this phylum is the presence of a basidium (q.v.) bearing basidiospores. A typical basidium is aseptate and has four 1-celled haploid basidiospores (ballisto- or statismospores) dispersed by air currents but the basidium may be transversely or longitudinally septate and the number of spores (which may be statismospores) are occasionally fewer or more than four. Other diagnostic characters are clamp connexions (Fig. 9), dolipore septa (Fig. 13), and a double-layered wall, lamellate and electron-opaque in electron microscopy. *Basidiomycota* are typically mycelial but some are yeasts (or have a yeast-like state). Such yeasts may be distinguished from ascomycetous yeasts by the morphology of the bud scars, giving a red colour with diazonium blue B, being urease +, having a high GC percentage (see base ratio), and other ultrastructural characters (see Moore, *Bot. Mar.* **23**: 361, 1980; v. Arx, 1987) some of which may apply to basidiomycetes in general; molecular sequence data can also clearly place other fungi in the phylum.

The typical life cycle (on which there are variations) involves the germination of the basidiospore to give a septate primary (haploid) mycelium which may produce 'oidia' but anamorphic states (except for yeasts, rusts, and smuts) have been neglected (see Kendrick & Watling, *in* Kendrick (Ed.), *The whole fungus* **2**: 473, 1979). Later by diploidization, the homo- or heterothallic primary mycelium becomes a secondary (dikaryotic) mycelium which frequently has clamp connexions. There is nuclear fusion in the young basidium and meiosis before basidiospore development. The mycelium may be perennial in soil or wood and may form 'fairy rings', sclerotia, rhizomorphs or mycorrhizas. The basidiomata are typically macroscopic and take a variety of forms.

While there is widespread agreement on the use of certain families within the *Basidiomycota* there is less uniformity in grouping the families into orders and esp. supraordinal taxa because of the lack of good differential characters. During recent years the acceptance of supraordinal taxa by specialists has been based on microscopical basidial characters. Talbot (*Taxon* **17**: 620, 1968; 1971), after separating the *Uredinales* and *Ustilaginales* as the *Teliomycetes* grouped the remaining orders as the *Phragmobasidiomycetes* and *Holobasidiomycetes* and distinguished 'gasteromycetes' within the last. Donk (*Proc. K. nederl. Akad. Wet.* C **75**: 365; **76**: 109, 1972-73) separated the *Ustilaginales* as the *Hemibasidiomycetes* and classified the remaining forms as the *Heterobasidiomycetes* (which included the *Uredinales*) and the *Homobasidiomycetes*. Shaffer (*Mycol.* **67**: 1, 1975) advocated no supraordinal groupings of the 14 orders (59 fams.) of *Basidiomycetes* which he recognized.

The analysis underlying the subdivisions used here is pragmatic. By using the presence or absence of a macroscopic basidioma, the life-form and life-style (including host specialization) and by retaining the traditional 'hymenomycetes' and 'gasteromycetes' as informal not monophyletic categories the resulting subdivisions correspond to the special interests of many of those who consult this *Dictionary* and also with the divisions of much of the literature dealing with these fungi. The classes used here are:
(1) **Basidiomycetes.**
(2) **Teliomycetes.**
(3) **Ustomycetes.**

Teleomorphic or anamorphic yeasts (see Prillinger *et al.*, *Sydowia* **43**: 170, 1991; Prillinger, *Stud. Mycol.* **30**: 33, 1987; Laaser, *Bibl. Mycol.* **130**, 1989) with basidiomycete affinities could also be included here as **Blastobasidiomycetes** for teleomorphs (included in the *Ustilaginomycetes* (*Sporidiales*), *Tremellales* and *Septobasidiales* in this *Dictionary*) and as **Basidioblastomycetes** for anamorphs (included as Mitosporic fungi in the *Dictionary*).

Khan & Kimbrough (*Mycotaxon* **15**: 103, 1982) proposed the adoption of four classes based mainly on septal ultrastructure: *Teliomycetes*, *Hemibasidiomycetes*, *Phragmobasidiomycetes* and *Holobasidiomycetes*.

Lit.: see under Classes, Fungi, Literature.

Basidiomycotina, see *Basidiomycota*.

Basidiophora Roze & Cornu (1869), Peronosporaceae. 3, N. temp, S. Am. See Barreto & Dick (*Bot. J. Linn. Soc.* **107**: 313, 1991; key).

basidiophytes, Cain's (1972) term for hypothetical autotrophic ancestors of basidiomycetes; see Phylogeny. Cf. ascophyte.

Basidioradulum Nobles (1967), Hyphodermataceae. 1, Eur., N. Am. See Nobles (*Mycol.* **59**: 192, 1967).

basidiospore, a propagative cell (typically a ballistospore but in gasteromycetes a statismospore) containing one or two haploid nuclei produced, after meiosis, on a basidium (Fig. 6). The colour, form and ornamentation of the basidiospore are fundamental to basidiomycete classification and an essential part of any specific description. Greater use of electron microscopy has revealed an increasing complexity of the wall layers or teguments (for a comparison of the terminologies applied to these layers see Fig. 5). Also see Spore wall.

Lit.: Perreau (*Ann. Sci. nat., Bot.* sér. 12, **8**: 639, 1967; homobasidiomycetes), Clemençon (*Z. Pilzkde* **36**: 113, 1970; wall ultrastr.), Pegler & Young (*Beih. Nova Hedw..* **35**, 1971; morphology in *Agaricales*), Kühner (*Persoonia* **7**: 217, 1973), Locquin (*Bull. Soc. bot. Fr., Coll. Palyn.*: 135, 1975).

Basidiosporites Elsik (1968), Fossil fungi (*f. cat.*). 1 (Paleocene), USA.

basidium (pl. **-ia**), (1) the cell or organ, diagnostic for basidiomycetes, from which, after karyogamy and meiosis, basidiospores (generally 4) are produced externally each on an extension (sterigma, q.v.) of its wall (Fig. 6); (2) a conidiophore or phialide (obsol.).

The confused terminology applied to basidia (sense 1) and their parts has been traced by Clémençon (*Z. Mykol.* **54**: 3, 1988) and is analyzed by Talbot (*TBMS* **61**: 497, 1973) whose recommended usage (basically that of Donk) and synonymy is adopted in the series of definitions which follow (see Fig. 7): **pro-**, the morphological part or developmental stage of the basidium in which karyogamy occurs; primary basidial cell; probasidial cyst; **hypo-** (Martin) p.p.; teliospore of Uredinales. **meta-**, (1) the morphological part or developmental stage in which meiosis occurs; hypo- (Martin) p.p.; **epi-** (Martin) p.p.; promycelium of Uredinales. (When the whole meta- includes proremnants the distal and functional part may be distinguished as a **pario-** (Talbot, 1973).) (2) See protobelow. **holo-**, a basidium (e.g. of *Agaricus*) in which the meta- is not divided by primary septa (see septum) but may become adventitiously septate (see septum) (Talbot, *Taxon* **17**: 625, 1968). A holo- may be a **sticho-**, cylindrical, with nuclear spindles longitudinal and at different levels, or a **chiasto-**, clavate, with nuclear spindles across the basidium and at the same level (see Fig. 6A,B). **phragmo-**, a basidium in which the meta- is divided by primary septa, usually cruciate (e.g. *Tremella*) or transverse (e.g. *Auricularia*) (Talbot, 1968).

Among other terms applied to basidia are: **apo-**, one with non-apiculate spores borne symmetrically on the sterigmata and not forcibly discharged (Rogers, *Mycol.* **39**: 558, 1947); **auto-**, one with spores borne asymmetrically and forcibly discharged; **endo-**, one developing within the basidioma, as in gasteromycetes; **epi-**, Martin's term for protosterigma, see sterigma; **hetero-**, a basidium of the *Heterobasidiomycetes*, usually a phragmo-; **homo-**, a basidium of the *Homobasidiomycetes*, usually a holo-; **hypo-**, = pro- (Donk), meta- (Martin); (of *Septobasidium*) = basidium (Martin); **pleuro-**, one relatively broad at the base and with bifurcated spreading 'roots', as in *Pleurobasidium* (Donk); **proto-**, a primitive basidium; the opposite of meta- in the sense of changed or degenerate basidium; **repeto-**, see Chadefaud (*Rev. mycol.* **39**: 173, 1975); **sclero-**, the thick-walled, encysted, gemma-like pro- of the *Uredinales*

	PERREAU (1967)	BESSON (1972)	CLEMENÇON (1970)		CLEMENÇON (1973)		KELLER (1974)
BASIDIAL REMNANTS	ECTOSPORIUM			**EPITUNICA**	SPOROTHECIUM	**MYXOSPORIUM**	SPOROTHECIUM
	PERISPORIUM	MYXOSPORIUM	SPOROTHECIUM		MUCOSTRATUM		MYXOLEMMA
	EXOSPORIUM	EPITUNICA	TECTUM		PODOSTRATUM		PODOSTRATUM
		MEDIOSTRATUM					
SPORE PROPER	EPISPORIUM	SCLEROSPORIUM	TUNICA	**CORIOTUNICA**	B2	**EUSPORIUM**	TUNICA
					B1		CORIOTUNICA
	ENDOSPORIUM	ENDOSPORIUM	CORIUM		A2		CORIUM
					A1		ENDOCORIUM

Fig. 5. Basidiospore wall-layer terminology, based on transmission electron microscopy (TEM) sections.

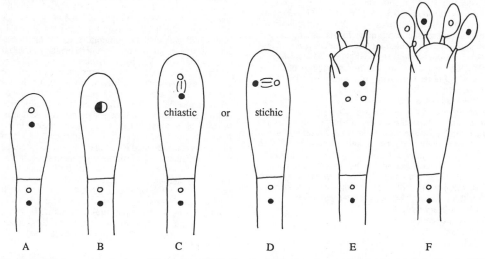

Fig. 6. Basidiospore development (diagrammatic). A-E meiosis (C, chiastic, and D, stichic). E, diploid probasidium; F, basidium (metabasidium) with four basidiospores on sterigmata.

(teliospore) and the *Auriculariales* (Janchen, 1923). See also Wells & Wells (*Basidium and basidiocarp evolution, cytology, function and development* 1982).

Basidopus Earle (1909) = Mycena (Tricholomat.) fide Singer (1975).

Basifimbria Subram. & Lodha (1968), Mitosporic fungi, 1.A1.13. 1 (coprophilous), India. = Dicyma (Mitosp. fungi) fide v. Arx (1982).

basifugal, development from the base up, acropetal.

Basilocula Bubák (1914) ? = Ceuthospora (Mitosp. fungi) fide Sutton (1977).

basionym (basinym, basonym) (in nomenclature, q.v.), the name- or epithet- bringing synonym on which a new transfer or new combination is based. Donk (*Bull. Jard. bot. Buitenz.*, sér. 3, **18**: 274, 1949) uses **isonym** for a name derived from a basionym. Cf. synisonym, synonym, toponym.

basipetal, describes a chain of conidia in which new spores are formed at the base, the oldest at the apex, cf. acropetal.

Basipetospora G.T. Cole & W.B. Kendr. (1968), Anamorphic Monascaceae, 1.A1.36. Teleomorph *Monascus*. 1, widespr.

Basipilus Subram. (1961) = Seimatosporium (Mitosp. fungi) fide Sutton (*Mycol. Pap.* **88**, 1963), Shoemaker (*CJB* **42**: 411, 1964).

Basisporium Molliard (1902) = Nigrospora (Mitosp. fungi) fide Mason (*Mycol. Pap.* **3**: 60, 1933).

Basitorula Arnaud (1954) = Gliomastix (Mitosp. fungi) fide Dickinson (1968).

basket fungi, *Clathrus* spp.

basocatenate (of conidia), formed in chains with the youngest conidium at the basal or proximal end of the chain.

Basramyces Abdullah, Abdulk. & Goos (1989), Mitosporic fungi, 1.F2.1. 1, Germany, Iraq.

Bassi (Agostino; 1773-1856). Italian lawyer turned farmer who by elucidating the etiology of the muscardine disease of silkworms (*Beauveria bassiana*) was the first to prove by experiment the pathogenicity of a fungus for an animal (*Del mal del segno*, 1835-6 [for Engl. transl. of part I, 1835, see *Phytopath. Classics* **10**, portr., 1958]. Collected works: *Opere do Agostino Bassi*, 1925 (Pavia); bicentenary tributes: Verona (*Agostino Bassi nel 200 anno dalla nascita*, portr., bibl., 1973), Porter (*Bact. Rev.* **37**: 284, portr., 1973).

Bastien treatment, treatment for amanitin poisoning, involving: (1) twice-daily injection of 1g vitamin C, (2) 2 capsules of nifurazide, three times a day, (3) 2 tablets of dihydrostreptomycin, three times a day, (4) penicillin therapy, (5) maintenance of fluid and electrolyte balance.

Batarrea, see *Battarrea*.

Batcheloromyces Marasas, P.S. van Wyk & Knox-Dav. (1975) = Stigmina (Mitosp. fungi) fide Sutton & Pascoe (1989).

Bathelium Ach. (1803) = Trypethelium (Trypethel.).

Bathelium Trevis. (1861) ≡ Laurera (Trypethel.).

Bathyascus Kohlm. (1977), Halosphaeriaceae. 1 (marine), USA.

Bathystomum Füisting (1868) ? = Massaria (Massar.) fide Saccardo [not traced].

Batista (Augusto Chaves; 1916-67). Professor of phytopathology, Escola Superior de Agricultura, and director, Instituto de Micologia, Universidade do Recife [now Univ. Federal de Pernambuco], Brazil, was author of more than 600 papers and monographs, mostly accounts and revisions of tropical fungi from Brazil, and with various coworkers. The majority of these publications appeared as a numbered series, *Publicações do Instituto de Micologia, Universidade do Recife* (1954 on), in a range of journals, etc., and in a long series of offset printed pamphlets published by I.M.U.R. (*Publ.* **301** lists 1-300; *Publ.* **674**, 301-673). Visited IMI 1951 (Batista, *Bol. S.A.I.C.* **22**: 29, 1953). Collections in Recife (**URM**); few in **IMI**. *Obit.*: Carmeiro (*Mycol.* **60**: 1137, portr., 1969), Singer (*Sydowia* **22**: 343, 1969). See also Aguilar (*Acta Amazonica* **18**: 39, 1988; types in INPA), Da Silva & Minter (*Mycol. Pap.* **169**, 1995; records 3,340 spp., host/substrate/state index, bibliogr.); Grummann (1974: 771).

Batistaella Cif. (1962) = Phaeosaccardinula (Chaetothyr.) fide v. Arx & Müller (1975).

Batistamnus J.L. Bezerra & Cavalc. (1967) Fungi (inc. sed.). See v. Arx & Müller (*Stud. Mycol.* **9**, 1975).

Batistia Cif. (1958), Batistiaceae. 1, Brazil. See Samuels & Rodrigues (*Mycol.* **81**: 52, 1989). Anamorph *Acrostroma*.

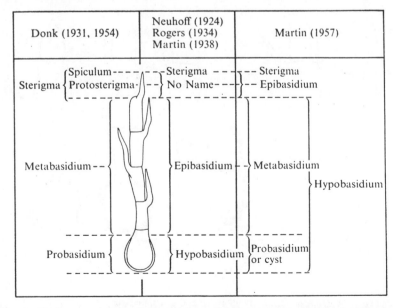

Donk (1931, 1954)		Neuhoff (1924) Rogers (1934) Martin (1938)	Martin (1957)
Sterigma	Spiculum / Protosterigma	Sterigma / No Name	Sterigma / Epibasidium
Metabasidium		Epibasidium	Metabasidium / Hypobasidium
Probasidium		Hypobasidium	Probasidium or cyst

Fig. 7. Basidium terminology, to compare the terminology of different authors, illustrated with reference to the *Septobasidium*-type (Talbot, 1973). Note that in this extreme case the metabasidium of Donk coincides with that of Martin.

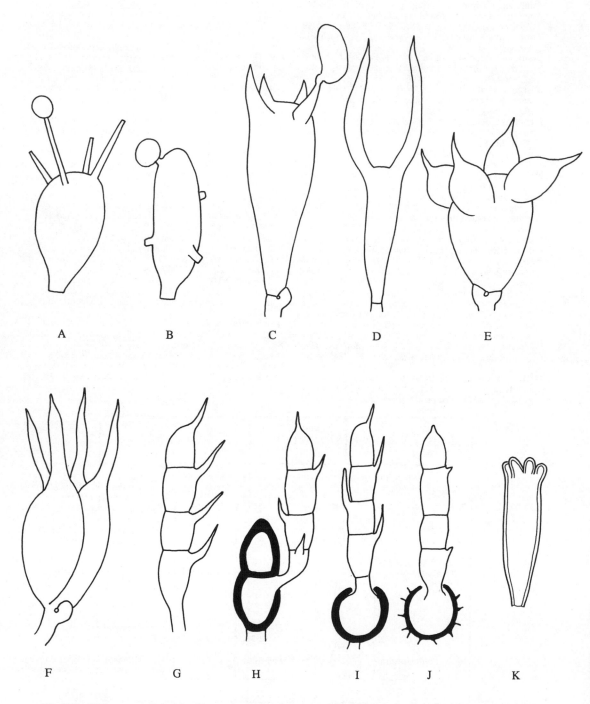

Fig. 8. Basidial types. A-E, holobasidial (A-B, apobasidial; C-E, autobasidial). A, *Lycoperdales*; B, *Tulostomatales*; C, *Agaricales*; D, *Dacrymycetales*; E, *Tulasnellales*. F-K, phragmobasidial (F-G, *Basidiomycetes*; H-I, *Teliomycetes*; J-K, *Ustomycetes*). F, *Tremellales*; G, *Auriculariales*; H, *Uredinales*; I, *Septobasidiales*; J, *Ustilaginales*; K, *Cryptobasidiales*.

Batistiaceae Samuels & K.F. Rodrigues (1989), Sordariales. 1 gen., 1 sp. Ascomata non-ostiolate, black, thick-walled, cephalothecoid, with a long transversely ridged stalk; interascal tissue absent; asci minute, deliquescent; ascospores translucent brown, simple, without germ pores; anamorph hyphomycetous, synnematal; on rotten wood.

Batistina Peres (1961), Mitosporic fungi, 5.B1.?. 1, Brazil.

Batistinula Arx (1960), Asterinaceae. 1, Brazil. Anamorph *Triposporium*.

Batistopsora Dianese, R.B. Medeiros, L.T.P. Santos, Furlan., M. Sanchez & A.C. Dianese (1993), Phakopsoraceae. 1, Brazil.

Batistospora J.L. Bezerra & M.M.P. Herrera (1964), Ascomycota (inc. sed.). 1, Brazil. See v. Arx & Müller (1975).

Batkoa Humber (1989), Entomophthoraceae. 4 (pathogens of *Homoptera*, *Hemiptera* and other *Insecta*), widespr.

Batschiella Kirschst. (1938) = Clypeoporthella (Vals.) fide Barr (1978).

Battareopsis, see *Battarreopsis*.

Battarraea, see *Battarrea*

Battarraeastrum Heim & Herrera (1960) ≡ Battarreoides (Battar.).

Battarraeoides, see *Battarreoides*.

Battarrea Pers. (1801), Battarreaceae. 3, esp. warm parts. The volvate stipe to 30 cm. See Maublanc & Malençon (*BSMF* **46**: 43, 1930), Rea (*Mycol.* **34**: 563, 1942), Long (*Mycol.* **35**: 546, 1943).

Battarreaceae Corda (1842), Tulostomatales. 2 gen (+ 3 syn.), 4 spp. Gleba first lacunose (compact in typical *Tulostomatales*), at maturity with characteristic elaters of hymenial origin. Warm dry areas.

Battarreoides T. Herrera (1953), Battarreaceae. 1, N. Am. deserts.

Battarreopsis Henn. (1902) = Dictyocephalos (Phellorin.) fide Long & Plunkett (*Mycol.* **32**: 696, 1940).

Battarrina (Sacc.) Clem. & Shear (1931), Hypocreaceae. 1, Eur.

Bauch test. A macroscopic test for determining whether monosporidial lines of smuts are compatible (the mixed culture has white aerial mycelium) or incompatible (aerial mycelium absent) (Bauch, *Biol. Zbl.* **42**: 9, 1922).

Bauhinus R.T. Moore (1992), Ustilaginaceae. 6 (on dicots), widespr.

Baumanniella Henn. (1897) = Physalacria (Physalacr.) fide Corner (1950).

Baumiella Henn. (1903) = Leptosphaeria (Leptosphaer.) fide v. Arx & Müller (1975).

Bayrhofferia Trevis. (1857) = Lecania (Bacid.) p.p., Catillaria (Catillar.) p.p. and Ramonia (Gyalect.) p.p.

Bdellospora Drechsler (1935), Zoopagaceae. 1, N. Am. See Drechsler (*Mycol.* **27**: 25, 1935).

beaded (of a lamella), having a line of small drops of liquid on the edge.

beak (of ascoma or conidioma), an elongated neck through which the spores are discharged. See rostrum.

beard moss, species of *Alectoria*, *Bryoria*, *Ramalina* and *Usnea*.

Beauveria Vuill. (1912), Mitosporic fungi, 1.A1.10. 7 (on *Insecta*), widespr.; sometimes used in biocontrol (q.v.). See de Hoog (*Stud. Mycol.* **1**, 1972), v. Arx (*Mycotaxon* **25**: 153, 1986), McLeod (*CJB* **32**: 818, 1954), Riba *et al.* (*Fundamental and applied aspects of invertebr. pathology*: 205, 1986; isoesterase variability), Mugnai *et al.* (*MR* **92**: 199, 1989; chemotaxonomic re-evaluation), Bridge *et al.* (*Mycopath.* **111**: 85, 1990; chemotaxonomy of *B. bassiana*), Rakotonirainy *et al.* (*Jl Invert. Path.* **57**: 17, 1991; rRNA sequence comparison with *Tolypocladium*),

Shimizu & Aizawa (*Jl Invert. Path.* **52**: 348, 1988; serological classification of *B. bassiana*), Kosir *et al.* (*Can. J. Microbiol.* **37**: 534, 1991; RFLPs of virulent and avirulent *B. bassiana*), St Leger *et al.* (*MR* **96**: 1007, 1992; genetic variation), Pfeifer & Khachatourians (*Jl. invert. Pathol.* **61**: 231, 1993; electrophoretic karyotyping).

Beauveriphora Matsush. (1975), Mitosporic fungi, 1.A1.10. 1, Japan.

Beccaria Massee (1892) [non Müll. Hal. (1872), *Musci*] ≡ Beccariella (Podoscyph.) fide Donk (1957).

Beccariella Ces. (1879) = Cymatoderma (Podoscyph.) fide Donk (1957).

Beccopycnidium F. Stevens (1930), Mitosporic fungi, 4.E1.?. 1, S. Am.

Beckhausia Hampe ex Körb. (1865) nom. inval. = Tomasellia (Arthopyren.).

beef-steak fungus (or **liver fungus**), basidioma of the edible *Fistulina hepatica*; cf. brown oak.

Beejadwaya Subram. (1978), Mitosporic fungi, 1.A2.3. 1, Japan.

Beejasamuha Subram. & Chandrash. (1977), Mitosporic fungi, 3.A1.10. 1, India. = Beauvaria (Mitosp. fungi) fide Gams (*in litt.*).

Beeli formulae. Numerical designations for coding the ascospore, perithecial, setae, and hyphopodial characteristics of *Meliola* spp. together with their sizes devised by Beeli (*Bull. Jard. bot. Brux.* **8**: 89, 1920) and modified by Stevens (*Ann. Myc.* **25**: 405, 1927) Hansford (*Sydowia* **2**, 1961), and Farr (*Mycopath.* **43**: 161, 1971); e.g. *Asteridiella westermannii* 3101.5340.

Beelia F. Stevens & R.W. Ryan (1925), Elsinoaceae. 1, Hawaii. See Petrak (*Sydowia* **7**: 321, 1953).

Beenakia D.A. Reid (1956), Beenakiaceae. 5, trop. See Nuñez & Ryvarden (*Sydowia* **46**: 321, 1994; key).

Beenakiaceae Jülich (1982), Boletales. 1 gen., 5 spp. Hymenophore smooth to spinose; spores hyaline or yellowish, verrucose.

beer, an alcoholic drink obtained by the fermentation of wort. The two main series of beers are the **ales** (produced by top-yeast (*Saccharomyces cerevisiae*) fermentation) and the **lagers** (produced by bottom-yeast (*S. carlsbergensis*) fermentation). See also Brewing, cider, porter, wine, yeast.

behind (of lamellae), the end nearest the stipe.

Belaina Bat. & Peres (1961) = Polynema (Mitosp. fungi) fide Punithalingam (*Nova Hedw.* **49**: 297, 1989).

Belainopsis Bat. & H. Maia (1965) nom. dub. fide Punithalingam (*Nova Hedw.* **49**: 297, 1989).

Belemnospora P.M. Kirk (1981), Mitosporic fungi, 1.A/B2.19. 4, widespr. See Sutton *et al.* (*MR* **92**: 354, 1989; key).

Belizeana Kohlm. & Volkm.-Kohlm. (1987), Dothideales (inc. sed.). 1 (marine), Belize.

Bellemerea Hafellner & Cl. Roux (1984), Porpidiaceae (L). 6, Eur.

Bellulicauda B. Sutton (1967), Mitosporic fungi, 8.A1.19. 1, Afr.

Belonia Körb. (1856), Gyalectales (inc. sed., L). 12, Eur., N. Am., NZ. See Hafellner & Kalb (*Bibl. Lich.* **57**: 161, 1995; posn), Henssen (*in* Brown *et al.* (Eds), *Lichenology: progress and problems*: 107, 1976; ontogeny), Jørgensen *et al.* (*Lichenologist* **15**: 45, 1983, Eur.; *SA* **5** 121, 1986, posn), Vězda (*Prirodov. Čas. slezsky* **20**: 241, 1959).

Belonidium Mont. & Durieu (1846) = Lachnum (Hyaloscyph.) fide Raitviir (*Eesti NSV Tead. Akad. Toim., Biol.* **36**: 313, 1987).

Beloniella Th. Fr. (1877) = Belonia (Gyalectales, inc. sed.).

Beloniella (Sacc.) Boud. (1885) ≡ Odontura (Odontotremat.).

Beloniella Rehm (1892) = Calloria (Dermat.)

Beloniomyces Cif. & Tomas. (1953) ≡ Belonia (Gyalectales, inc. sed.).

Belonioscypha Rehm (1892) = Cyathicula (Leot.) fide Baral & Krieglsteiner (*Beih. Z. Mykol.* **6**, 1985).

Belonioscyphella Höhn. (1918), Leotiaceae. 1, Eur. See Döbbeler (*Ber. bayer bot. Ges.* **57**: 153, 1986). ? = Gloeopeziza (Leot.) fide Carpenter (*Mem. N.Y. bot. Gdn* **33**, 1981).

Belonium Sacc. (1884) = Pyrenopeziza (Dermat.) fide Baral (*SA* **13**: 113, 1994).

Belonopeziza Höhn. (1917) ? = Calloria (Dermat.).

Belonopsis (Sacc.) Rehm (1891), Dermateaceae. 4, Eur. See Aebi (*Nova Hedw.* **23**: 49, 1972), Nannfeldt (*Sydowia* **38**: 194, 1986).

Belonopsis (Sacc.) Sacc. & P. Syd. (1902), Dermateaceae. 4, Eur. See Nannfeldt (*Sydowia* **38**: 194, 1986).

Belospora Clem. (1909) ≡ Hymenoscyphus (Leot.).

Beltraminia Trevis. (1857) = Dimelaena (Physc.).

Beltrania Penz. (1882), Mitosporic fungi, 1.A2.10. 4, widespr. See Hughes (*Mycol. Pap.* **47**, 1951), Pirozynski (*Mycol. Pap.* **90**, 1963; key).

Beltraniella Subram. (1952), Anamorphic Amphisphaeriaceae, 1.A2.10. Teleomorph Leiosphaerella. 1, India. See Pirozynski (*Mycol. Pap.* **90**, 1963).

Beltraniopsis Bat. & J.L. Bezerra (1960), Mitosporic fungi, 1.A2.10. 2, S. Am. See Pirozynski (*Mycol. Pap.* **90**, 1963).

Benedekiella Negru & Verona (1964) = Physalospora (Hyponectr.) fide Müller & v. Arx (1973).

Benekea Bat. & J.L. Bezerra (1960) ? = Geastrumia (Mitosp. fungi) fide Sutton (1977).

Benguetia Syd. & P. Syd. (1917), ? Leotiales (inc. sed.). 1, Philipp.

Beniowskia Racib. (1900), Mitosporic fungi, 1.A2.1. 1, trop. See Mason (*Mycol. Pap.* **2**: 26, 1928).

Benjaminella I.I. Tav. (1981), Laboulbeniales (inc. sed.). 4, N. & S. Am. See Eriksson & Hawksworth (*SA* **5**: 122, 1986; nomencl.).

Benjaminia S. Ahmad (1967) [non Benjamina Vell. (1835), *Rutaceae*], Mitosporic fungi, 4.D2.?. 1, Pakistan.

Benjaminia Pidopl. & Milko (1971) [non *Benjamina* Vell. (1835), *Rutaceae*] ≡ Benjaminiella (Mycotyph.).

Benjaminiella Arx (1981), Mycotyphaceae. 3, Canary Isls, India, USA. See Cole *et al.* (*Can. J. Microbiol.* **26**: 35, 1980; dimorphism & wall chemistry), Forst & Pillinger (*Z. Mykol.* **54**: 139, 1988; dimorphism), Benny *et al.* (*Mycotaxon* **22**: 119, 1985), Kirk (*Mycotaxon* **35**: 121, 1989).

Benjpalia Subram. & Bhat (1989), Mitosporic fungi, 1.B2.10/12/13. 1, India.

benomyl ('Benlate'), the first systemic fungicide: one of the benzimidazoles, has very low toxicity to plants and animals, and controls ascomycetous fungal plant pathogens (incl. their anamorphs) by interfering with spindle formation during nuclear division.

Bensingtonia Ingold (1986), Mitosporic fungi, 1.A1.1/10. 7, widespr. See Nakase & Boekhout (*J. Gen. Appl. Microbiol.* **34**: 433, 1988; emend.), Nakase *et al.* (*J. Gen. Appl. Microbiol.* **35**: 53, 1989).

Berengeria Trevis. (1853) = Rinodina p.p. and Dimelaena p.p. (Physc.).

Berggrenia Cooke (1879), Ascomycota (inc. sed.). 1 or 2, NZ.

Bergorea Nieuwl. (1916) ≡ Robergea (Stictid.).

Berkelella (Sacc.) Sacc. (1891), Herpotrichiellaceae. 2, Asia.

Berkeley (Miles Joseph; 1803-89). A great British mycologist, was at Cambridge University, took Holy Orders and went to Margate, later going to Northamptonshire and living at King's Cliffe till 1868 and then at Sibbertoft.

Berkeley's interests in natural history were very wide; though his chief works are on fungi (including lichens) he gave his attention to molluscs, mosses, algae, and gardening. His first printed mycological work was the important *British Fungi* (for which he sent out exsiccati) in J.E. Smith's *English Flora* (**5** (2), 1836) [= Hooker's *British Flora*, **2** (2)]. This he supplemented (with Broome) by *Notices of British Fungi*, 1837-85 (index in *TBMS* **17**: 308, 1933; reprint, 1967). In other papers he gave accounts of fungi from the then British Colonies, America (esp. collections of M.A. Curtis, q.v.), and other countries, and names to about 6,000 new species. He was in touch with Fries, Montagne, and other mycologists, and from 1848 did much work with C.E. Broome. Berkeley's early work was done with great care but later, because of the number of microfungi sent to him, there were more errors. In 1879 his collection was put with that **K**.

Of Berkeley's 400 papers on mycology, some were short notes, such as those in the *Gdnrs' Chron.*, in which, from 1845, a number of his writings were printed (e.g. the group on '*Vegetable Pathology*', 1854-7; see *Phytopath. Classics* **8**, with biogr., portr.) and some were important books, e.g. *Introduction to cryptogamic botany* (1857), *Outlines of British Fungology* (1860). He was said to have been 'the last of the old race of mycologists'. See Ainsworth (*Mycologist* **21**: 126, 1987; portr.), Grummann (1974: 368), Stafleu & Cowan (*TL-2* **1**: 192, 1976), Stafleu & Mennega (*TL-2, Suppl.* **2**: 98, 1993).

Berkeleyna Kuntze (1898) ≡ Cephalotrichum (Mitosp. fungi).

Berkleasmium Zobel (1854), Mitosporic fungi, 3.D2.1. 12, Eur., N. Am., India. See Moore (*Mycol.* **51**: 734, 1961; key), Ellis (*DH*, *MDH*).

Berlese (Augusto Napoleone; 1864-1903). Professor of plant pathology, School of Agriculture, Milan; responsible for parts of Saccardo's *Sylloge* and for starting (with his brother, Antonio Berlese, the zoologist) *Rivista di Patologia vegetale*, 1892. Chief writings: *Fungi moricolae*, 1885-9; *Icones fungorum ad usum Sylloges Saccardianae accomodatae*, 1890-1905; papers on *Pleospora* (1888), *Lophiostomataceae* (1890), *Dematophora* and *Rosellinia* (1892), *Cladosporium* and *Dematium* (1895), and many more. See Antonio Berlese (*Riv. pat. Veg.* **10**: 347, 1904), Cavara (*Ann. Myc.*. **1**: 178, 1903), Grummann (1974: 513), Stafleu & Cowan (*TL-2* **1**: 197, 1976), Stafleu & Mennega (*TL-2, Suppl.* **2**: 101, 1993).

Berlesiella Sacc. (1888) = Capronia (Herpotrichiell.) fide Müller *et al.* (1987).

Berteromyces Cif. (1954) = Cercosporidium (Mitosp. fungi) fide Deighton (1967).

Bertia De Not. (1844), Nitschkiaceae. 7, widespr. See Corlett & Krug (*CJB* **62**: 2561, 1985).

Bertiella (Sacc.) Sacc. & P. Syd. (1899) = Massarina (Lophiostom.) fide Eriksson & Yue (*Mycotaxon* **27**: 247, 1986).

Bertiella Kirschst. (1906) ≡ Kirschsteinia (Nitschk.).

Bertossia Cif. & Tomas. (1953) = Mycoglaena (Dothideales).

Bertramia Mesnil & Caullery (1897), Chytridiomycota (inc. sed.). 6, widespr. See Weiser & McCauley (*Z. Parasitkde* **43**: 299, 1974).

Bertrandia R. Heim (1936), Hygrophoraceae. 1, Madagascar, Philipp.

Bertrandiella R. Heim (1959) = Lactocollybia (Tricholomat.) fide Singer (1975).

Bessey (Ernst Athearn; 1877-1957). Son of Charles E. Bessey (1845-1915); Professor of botany and mycology, Michigan Agricultural College, East Lansing, 1910-46; author of the first American textbook of mycology, 1935, 1950 (see Literature). *Lit.*: Barnett (*Mycol.* **50**: 1, portr., bibl., 1958), Grummann (1974: 204), Stafleu & Cowan (*TL-2* **1**: 219, 1976), Stafleu & Mennega (*TL-2, Suppl.* **2**: 141, 1993).

beta-spore (**B-spore**, β-**spore**), a fertile, usually hamate, spore of an anamorph of the *Valsaceae* (*Phomopsis*). Cf. alpha-spore.

Bettsia Skou (1972), Ascosphaeraceae. 1 (in pollen on honeycomb), Eur. See Skou (*Friesia* **11**: 62, 1975). Anamorph *Chrysosporium*-like.

Betulina Velen. (1947), Hyaloscyphaceae. 2, Eur. See Graddon (*TBMS* **63**: 477, 1974).

Beverwykella Tubaki (1975), Mitosporic fungi, 1.D1.1. 1 (aquatic), Japan.

Bhargavaella Sarj. Singh & K.S. Srivast. (1980), Mitosporic fungi, 1.D2.3. 1, India.

bi- (prefix), twice; having two; two-.

biallelic (of an incompatibility system), having 2 alleles per locus; cf. multiallelic.

Biannularia Beck (1922) = Catathelasma (Tricholomat.) fide Singer (1975).

Biatora Ach. (1810) = Stenhammarella (Lecan.) fide Hertel (*Beih. Nova Hedw.* **24**, 1967).

Biatora Fr. (1817), Lecanoraceae (L). *c.* 10, temp. See Hafellner (*Herzogia* **8**: 53, 1989).

Biatorella De Not. (1846), Biatorellaceae (L). *c.* 30, cosmop. See Magnusson (*Rabenh. Krypt.-Fl.* **9**, 5(1): 15, 1935; *Annls Crypt. Exot.* **7**: 115, 1935).

Biatorellaceae Choisy ex Hafellner & Casares (1992), Lecanorales (L). 1 gen. (+ 3 syn.), 30 spp. Thallus crustose, often poorly developed; ascomata apothecia, developing superficially, pale yellowish to brown, without a well-developed wall, usually slightly domed; hymenium blueing in iodine; interascal tissue of paraphyses, branched near the apex and forming an epithecium; asci cylindrical to clavate, with a well-developed usually weakly I+ apical dome which has a darker staining basal layer, no ocular chamber, and a outer gelatinous layer which blues strongly in iodine, multispored; ascospores small, hyaline, aseptate. Anamorph not known. Lichenized with green algae, N. temperate.
   Lit.: Hafellner & Casares-Porcel (*Nova Hedw.* **55**: 309, 1992).

Biatorellina Henn. (1903) ? = Tryblidiopsis (Rhytismat.).

Biatorellopsis C.W. Dodge (1965) = Pleopsidium (Lecanor.) fide Castello & Nimis (*Lichenologist* **26**: 283, 1994).

Biatoridina Schczedr. (1964) = Epithyrium (Mitosp. fungi) fide Sutton (1977).

Biatoridium J. Lahm ex Körb. (1860), Lecanorales (inc. sed., L). 2, temp. See Hafellner (*Acta bot. fenn.* **150**: 39, 1994).

Biatorina A. Massal. (1852) nom. rej. prop. = Catinaria (Bacid.) fide Jørgensen & Santesson (*Taxon* **42**; 881, 1993).

biatorine (of apothecia), of the lecideine type, pale or more or less coloured and soft in consistency.

Biatorinella Dechatres & Werner (1974) = Fuscidea (Fuscid.).

Biatorinopsis Müll. Arg. (1881) = Dimerella (Gyalect.).

Biatoropsis Räsänen (1934), Platygloeales (inc. sed.). 1 (causing galls, 'carpoids', on *Usnea*), cosmop. See Diederich & Christiansen (*Lichenologist* **26**: 47, 1994). Anamorph *Hormomyces*-like.

Biatriospora K.D. Hyde & Borse (1986), Dothideales (inc. sed.). 1 (on mangrove wood), Seychelles.

Bibanasiella R.F. Castañeda & W.B. Kendr. (1991), Mitosporic fungi, 3.G1.12. 1, Cuba.

Bibbya J.H. Willis (1956) = Toninia (Catillar.) fide Santesson (*Muelleria* **1**: 91, 1959).

bibulous (of the surface of a pileus), able to take up water.

bicampanulate, like two bells arranged mouth to mouth (Fig. 37.20).

Biciliospora Petr. (1952) = Nitschkia (Nitschk.). See v. Arx (*Sydowia* **34**: 13, 1981).

Biciliosporina Subram. & Sekar (1993), Coronophoraceae. 1, India.

Bicilium H.E. Petersen (1910) = Olpidiopsis (Olpidiopsid.) fide Sparrow (1960).

biconic (of spores), like two cones attached base to base (Fig. 37.27).

Biconiosporella Schaumann (1972), Lasiosphaeriaceae. 1 (marine), Eur.

Bicricium Sorokīn (1889) nom. conf. See Dick (*in press*).

Bicrouania Kohlm. & Volkm.-Kohlm. (1990), Melanommataceae. 1 (marine), France.

Bidenticula Deighton (1972) = Fusarium (Mitosp. fungi) fide Booth & Sutton (*in litt.*).

Bidonia Adans. (1763) = Hydnum (Hydn.).

bifarious, in two lines or series; distichous.

bifid, having a crack or division near the middle; forked.

Bifidocarpus Cano, Guarro & R.F. Castañeda (1994), ? Onygenaceae. 1, Cuba.

Biflagellospora Matsush. (1975), Mitosporic fungi, 1.A1.1/19. 1, Japan.

Biflagellosporella Matsush. (1993), Mitosporic fungi, 1.G1.1. 1, Peru.

Biflua J. Koch & E.B.G. Jones (1989), Ascomycota (inc. sed.). 1, Denmark.

Bifrontia Norman (1872), Dothideales (inc. sed.). 2, Norway. = Naetrocymbe (Coccodin.) fide Keissler (*Rabenh. Krypt.-Fl.* **9**, 1(1), 1933).

Bifusella Höhn. (1917), Rhytismataceae. 5, N. Am., India, Okinawa. *B. faullii* (needle-cast in *Abies*), *B. linearis* (on pine).

Bifusepta Darker (1963), Rhytismataceae. 1, N. Am.

Biharia Thirum. & Mishra (1953) = Stenella (Mitosp. fungi) fide Ellis (*DH*).

bilabiate, (1) two-lipped; (2) (of asci), ones in which the ectotunica splits in a lip-like manner to expose the endotunica (e.g. *Pertusaria*).

bilaminate, two-layered.

Bilboque Viégas (1960), Mitosporic fungi, 1.B2.?. 1, Brazil.

Bilgramia Panwar, Purohit & Chouhan (1974), Mitosporic fungi, 1.D2.3. 1, India. See Carmichael *et al.* (1980).

Bilimbia De Not. (1846) [non Rchb. (1837), *Averrhoaceae*] ≡ Mycobilimbia (Porpid.).

Bilimbiospora Auersw. (1861) nom. rej. = Leptosphaeria (Leptosphaer.) fide Holm (*Taxon* **24**: 475, 1975).

Bimeris Petr. (1949), Mitosporic fungi, 4.B1.?. 1, Ecuador.

Bimuria D. Hawksw., Chea & Sheridan (1979), ? Melanommataceae. 1 (from soil), NZ. See Hawksworth (*SA* **6**: 238, 1987; posn).

binate, in two parts.

binding hyphae, see hyphal analysis.

binucleate-phase, the dikaryo-phase.

bio- (prefix), pertaining to life.

biocide, a substance which kills living organisms. Cf. biostat.

Bioconiosporium Bat. & J.L. Bezerra (1964), Mitosporic fungi, 1.D2.1. 3, N. & S. Am., Afr.

biocontrol, see Biological control.

Bioconversion. The conversion of one material, usually a waste, into a product of increased value (e.g. of lignocellulosic residues for ethanol production); see Saddler (Ed.) (*Bioconversion of forest and agricultural plant residues*, 1993), Biodegradation, Biodeterioration, Biotechnology.

Biodegradation. A term sometimes used synonymously with Biodeterioration (q.v.). However, it is more correctly employed to describe the beneficial breakdown of materials, e.g. the removal of and/or utilization of wastes. It has been defined as 'the harnessing, by man, of the decay abilities of organisms to render a waste material more useful or acceptable' (Allsopp & Seal, *Introduction to biodeterioration*, 1986), and as such covers the use of microorganisms in solid or liquid state fermentations to improve digestibilities of, for example, lignocellulosic wastes to ruminants or to produce single cell proteins from wastes. The concept of biodegradability has come to be used more colloquially in recent years to describe detergents, plastics etc. which will break down when disposed of in the environment. There is as yet no universally accepted definition of the term **biodegradable**, particularly when applied to plastics. There are outstanding questions regarding whether the term implies complete mineralization and hence removal of the waste from the environment or whether degradation to a small non-toxic molecule will suffice. Should there be time limits set within which this must happen? A number of bodies are working to produce a satisfactory definition, e.g. International Biodeterioration Research Group (IBRG), American Society for the Testing of Materials (ASTM).

See Bioconversion, Biodegradation, Bioremediation.

Biodeterioration. Any undesirable change in the properties of a material caused by the vital activities of organisms (Hueck, 1965). Fungi play an important part in biodeterioration. Some representative examples of fungal damage are: to **animal feeding stuffs,** Snow *et al.* (*Ann. appl. Biol.* **31**: 102, 1944); **building materials,** Batista *et al.* (*Atas Inst. Mycol. Recife* **5**: 311, 1967; fungi, incl. lichens), Martin & Johnson (*Biodet. Abstr.* **6**: 101, 1992; bibl. control lichens), May *et al.* (*Biodet. Abstr.* **7**: 109, 1993; review), Richardson (*The vanishing lichens*, 1975), Singh (*Building mycology*, 1994), see also monuments (below), weathering; **stone,** May *et al.* (*Biodet. abstr.* **7**: 109, 1993); **electrical equipment,** Wasserbauer (*Internat. Biodet. Bull.* **3**: 1, 1967); **food,** see Food spoilage; **fuel,** see kerosene fungus; **glass and optical equipment,** Ohtsuki (*Bot. Mag. Tokyo* **75**: 221, 1962), Nagamuttu (*Internat. Biodet. Bull.* **3**: 25, 1967); **grain,** Christensen & Kauffman (*Grain storage. The role of fungi in quality and loss,* 1969); **gunpowder,** Lacey (*TBMS* **74**: 195, 1980); **leather,** Musgrave (*Ann. appl. Biol.* **34**, 1947), Gordon (*in* O'Flaherty *et al.* (Eds) (*The chemistry and technology of leather,* **4**, 1965); **meat,** Jensen (*Microbiology of meats,* 1942); **monuments,** Nimis *et al.* (*Licheni i conservazione dei monumenti,* 1992), see also building materials, weathering; **paint,** Eveleigh (*Ann. appl. Biol.* **49**: 403, 1961; *TBMS* **44**: 573, 1961, esp. *Phoma violacea*), 7th Paint Research Institute Symposium (*J. Coatings Technology* **50**: 35, 1978); **paper,** Sée (*Les maladies du papier piqué,* 1919), Wang (*Tech. Publ. Sta. Univ. Coll. Forestry Syracuse* **87**, 1965; fungi of pulp and paper; cellulolytic fungi); **polyurethane,** Pathirana & Seal (*Internat. Biodet.* **20**: 163, 229, 1984); **structural timber,** see Wood- attacking fungi; **textiles,** Thaysen & Bunker (*The microbiology of cellulose, hemicellulose, pectins and gums,* 1927), Prindle (*Microbiology of textile fibres, Textile Rev.* 1933-6), Morris (*J. Text.*

Inst. **18**: T99, 1927), Galloway (*J. Text. Inst.* **26**: T123, 1935); **tobacco,** Papavassiliou *et al.* (*Mycopath.* **44**: 117, 1971; cigarettes); **wood,** **archaeological,** Blanchette (*Biodet. Abstr.* **9**: 113, 1995). See also Seaward & Brightman (*in* Seaward, *Lichen ecology*: 253, 1977; lichens on man-made substrata, etc.).

The Biodeterioration Society covers all aspects of the biodeterioration of materials and bidegradation of wastes and holds regular meetings and international symposia, with published proceedings.

Lit.: Allsopp & Seal (*Introduction to biodeterioration,* 1986; Japanese edn, 1991), Hueck (*Material und Organismen* **1**: 5, 1965), Singh (Ed.) (*Building mycology,* 1994), *Biodeterioration Abstracts* (1987-; quarterly), *International Biodeterioration* (1984-; formerly *International Biodeterioration Bulletin,* 1965-), Biodegradation, Bioremediation.

Biodiversity (Biological diversity). The variety and value of Life on Earth from the genetic through the organismal to the ecological levels. The UN Convention on Biological Diversity agreed in Rio de Janeiro in 1992 and became effective at the end of 1993; this had been ratified by 128 governments by Sept. 1995. The Convention has focused the subject in political circles; many countries now have or are developing programmes to survey and value their natural biotic resources and investigate how they can best be conserved and sustainably used. Initiatives such as the IUBS/UNESCO/SCOPE DIVERSITAS programme, the GEF/UNEP Global Biodiversity Assessment and the Species 2000 and Systematics Agenda 2000 projects are prominent elements of the current research and development efforts.

Fungal diversity has received particular attention, due to the large numbers of species (see Numbers of fungi), the small proportion which are satisfactorily known (Cannon & Hawksworth, *Adv. Pl. Path.* **11**: 277, 1995), and their importance in ecosystem functioning (Christensen, *Mycol.* **81**: 1, 1989). Inventories (q.v.) are one of the principal ways of assessing biodiversity, and there is considerable interest in obtaining new products from natural ecosystems (see Bioprospecting).

Lit.: Allsopp *et al.* (Eds.) (*Microbial diversity and ecosystem function,* 1995), Galloway (*Biodiv. Conserv.* **1**: 312, 1992; lichens), Groombridge (Ed.) (*Global biodiversity,* 1992), Hawksworth (*MR* **95**: 641, 1991), Hawksworth (Ed.) (*Biodiversity: measurement and estimation,* 1995), Heywood (Ed.) (*Global biodiversity assessment,* 1995), Isaac *et al.* (Eds) (*Aspects of tropical mycology,* 1993), Oberwinkler (*Biodiv. Conserv.* **1**: 293, 1992), Schulze & Mooney (Eds) (*Biodiversity and ecosystem function,* 1994), Solbrig *et al.* (Eds) (*Biodiversity and global change,* edn 2, 1994).

biogenous, living on another living organism; parasitic.

Biographical notices, for list see Authors' names.

Bioindication. The use of organism(s) (**bioindicators**) expressing particular symptoms or responses to indicate changes in some environmental influence; lichens are used as bioindicators of air pollution (q.v.), ammonium eutrophication (Brown, *in* Bates & Farmer, 1992: 259), dating surfaces (see Lichenometry), ecological continuity (see RIEC), fire (Wolseley *et al., Global Ecol. Biogeogr. Lett.* **4**: 116, 1995), heavy metals (Gordon *et al., J. Trop. Ecol.* **11**: 1, 1995), radionuclides (see air pollution), and water levels (see Aquatic fungi). Other fungi and lichens are sensitive to acid rain (q.v.).

Lit.: Bates & Farmer (Eds) (*Bryophytes and lichens in a changing environment,* 1992), Boddy *et al.* (Eds) (*Fungi and ecological disturbance* [*Proc. R. Soc.*

Edinb. B **94**], 1988), Burton (*Biological monitoring of environmental contaminants: plants*, 1986), Ellenberg *et al.* (*Biological monitoring: signals from the environment*, 1991), Hawksworth (*in* Swaminathan & Jana, *Biodiversity: implications for global food security*: 184, 1992), Jeffrey & Madden (Eds) (*Bioindicators and environmental management*, 1991), Richardson (*Pollution monitoring with lichens*, 1992).

biologic form (Marshall Ward) or **race** (Klebahn), see physiologic race.

Biological control (Biocontrol). The use of one or more organisms (agents) to maintain another organism (pest) at a level at which it is no longer a problem. Fungal pathogens, parasites and antagonists are being exploited to control a range of agricultural pests, including arthropods, nematodes, weeds and crop diseases (Burge, *Fungi in biological control systems*, 1988; Association of Applied Biologists, *The exploitation of microorganisms in applied biology*, 1990). Two distinct approaches can be adopted: classical, involving the release of a coevolved fungal pathogen into an exotic environment where the target pest is an alien or non-indigenous species; inundative, through the application of a mass-produced, typically necrotrophic fungus as a mycopesticide (q.v.). Entomopathogenic fungi in genera such as *Beauveria*, *Metarhizium*, *Paecilomyces* and *Verticillium*, are being used for inundative biological control (see mycopesticides), whilst *Entomophthora radicans* has been classically introduced into Australia for control of Lucerne aphid (Milner *et al.*, *J. Aust. Ent. Soc.* **21**: 113, 1982).

Nematophagous fungi include the nematode trapping fungi (*Arthrobotrys*, *Dactylella*, *Geniculifera*, *Monacrosporium*), endoparasites (*Hirsutella*, *Catenaria*, *Meria*, *Nematoctonus*, *Nematophthora*) and highly specific egg parasites (*Dactylella oviparasitica*, *Paecilomyces lilacinus*, *Verticillium chlamydosporium*) (Stirling, *in* Burge, 1988). Immediate prospects for the exploitation of nematophagous fungi as biological control agents are uncertain since problems of formulation and application still have to be overcome (Stirling, *Biological control of plant parasitic nematodes*, 1992; Mankau, *J. Nematology* **12**: 244, 1980).

Fungal pathogens for the biological control of weeds have been extensively investigated (Charudattan & Walker (Eds), *Biological control of weeds with plant pathogens*, 1982; TeBeest (Ed.), *Microbial control of weeds*, 1991), using both the classical approach with obligate pathogens such as rusts and smuts, and the inundative approach, with necrotrophic pathogens in the genera *Colletotrichum* and *Phytophthora* (see mycopesticides). Successful control of skeleton weed has been achieved in Australia following release of the European rust *Puccinia chondrillina* (Hasan, *Ann. appl. Biol.* **99**: 119, 1981), and of mistflower in Hawaii with the white smut *Entyloma ageratinae* from the Caribbean (Trujillo, *Proc. VI Int. Symp. Biol. Control Weeds* 25, 1985; Barreto & Evans, *TBMS* **91**: 81, 1988), and of blackberry in Chile with the imported rust *Phragmidium violaceum* (Oehrens, *FAO Pl. Prot. Bull.* **25**: 26, 1977).

Mycoparasites and antagonistic fungi, particularly in the genera *Gliocladium*, *Sphaerellopsis*, *Trichoderma* and *Verticillium*, are currently being evaluated for biological control of crop diseases (Cool & Baker, *The nature and practice of biological control of plant pathogens* 1984; Hornby (Ed.), *Biological control of soil-borne plant pathogens*, 1990) (see mycopesticides for more details); bacteria and protozoa as well as other fungi have been tried against

Phytophthora cinnamomi (Scott, *Adv. Pl. Path.* **11**: 131, 1995).

bioluminescence, see Luminescent fungi.

biomass, the quantity (vol., wt, etc.) of organisms (or living material) in a particular environment (e.g. fungi in soil); sometimes extended to the quantity of organic matter in a material (e.g. domestic refuse). See Boucher & Stone (*in* Carroll & Wicklow, *The fungal community*, edn 2 : 538, 1992; in lichens on trees), Frankland (*Soil Biol. Biochem.* **7**: 339, 1975; **10**: 323, 1978; estimation live biomass), Newell (*in* Carroll & Wicklow, 1992: 521; in litter), Ritz *et al.* (Eds) (*Beyond the biomass*, 1994).

biomass support particles (ESPs), large open structures made from knitted stainless steel and crushed into spheres or reticulated polyurethane foam cut into cubes used for immobilization (q.v.) of fungal cells.

Biomyxa Leidy (1875), *Hydromyxales*, q.v.

Bionectria Speg. (1919), Hypocreaceae. 1, Indonesia. See Rossman *et al.* (*Mycol.* **85**: 685, 1993).

BioNET-INTERNATIONAL (BI), established in June 1993, aims to provide a permanent mechanism for mobilizing, pooling and strengthening the world's wealth of biosystematic resources, and is a mechanism for facilitating their effective deployment. Initially focused on bacteria, fungi, insects and nematodes, and supported by a Technical Secretariat funded by CABI (located at IMI), BI is promoting the development of technical co-operation networks (Locally Organized Operational Partnerships, LOOPs) in many parts of the world; e.g. CARINET (1993; Caribbean), and EUROLOOP (1994; Europe). Others are currently under development in southern (SAFRONET) and eastern (EAFRINET) Africa, Asia (ASEANET) and the Pacific (PACINET).

Lit.: Jones (*in* Ainsworth & Hawksworth, *Biodiversity in the Caribbean*: 15, 1992), Jones (*in* Hawksworth, *Identification and characterization of pest organisms*: 8, 1994).

biont, a living organism; commonly used as a suffix to a word indicating the nature or position of the biont; see Symbiosis.

biophagous, see biogenous.

biophilous, see biogenous.

Biophomopsis Petr. (1931), Mitosporic fungi, 8.A1.?. 1, W. Indies.

Bioporthe Petr. (1929) ? = Plagiostigme (Melanconid.) fide Müller & v. Arx (1973), but see Barr (*Mycol. Mem.* **7**, 1978).

Bioprospecting. The action of surveying natural ecosystems for economically valuable biotic products. For fungi, such products might include novel edible fungi, valuable enzymes for biotechnology companies, metabolites for pharmaceutical investigation, or new biological control agents. The extent to which the provisions of the Convention on Biological Diversity (1992) will impinge on bioprospecting will depend on national legislation and regulation systems relating to indigenous intellectual property rights (yet to be developed by most countries); for aspects of the Convention relating to microbial groups see Kelley (*in* Allsopp *et al.* (Eds), *Microbial diversity and ecosystem function*: 415, 1995), Sands (*in* Kirsop & Hawksworth (Eds), *The biodiversity of microorganisms and the role of microbial resource centres*: 9, 1994), Reid *et al.* (Eds) (*Biodiversity prospecting*, 1993). See Patent protection.

Bioremediation. The use of microorganisms to remove, reduce or ameliorate pollution or potentially polluting materials from the environment. This may be brought about by adding suitable nutrients to the substrate or by seeding with selected strains or mixtures of micro-organisms; they may be naturally occurring or genetically manipulated. Bacteria have tended to be the favoured organisms in the past, however fungi (particularly white rot fungi) have been used for the degradation of lignocellulosic wastes and more recently for xenobiotics. See Alexander (*Biodegradation and bioremediation*, 1994), Baker & Herson (Eds) (*Bioremediation*, 1994), Kerr (*Handbook of bioremediation*, 1994), Lamar *et al.* (*in* Leather (Ed.), *Frontiers in industrial mycology*: 127, 1992), Scheremaker *et al.* (*in* Betts (Eds), *Biodegradation: natural and synthetic materials*: 157, 1992).

bios, a mixture of aneurin (thiamin, vitamin B1), 'biotin', and other substances in yeasts which, on addition to culture media, gives a better growth of yeast (Wildiers, 1901; see Bonner, *Bot. Rev.* **3**: 616, 1937).

Bioscypha Syd. (1927), Leotiaceae. 1, C. Am. See Carpenter (*Mem. N.Y. bot. Gdn* **33**, 1981).

biostat, a substance which causes living organisms to stop growing. Cf. biocide

Biostictis Petr. (1950), Stictidaceae. 4, Indonesia, N. & S. Am. See Sherwood (*Occ. Pap. Farlow Herb.* **15**: 105, 1980; key).

biosystematics, (1) biological systematics (see systematics); (2) (in botany), experimental taxonomy, including genetical, cytological and ecological aspects. The first usage has a wide currency amongst zoologists and is the preferred term for use in a general context.

biotechnology, (1) (in microbiology), 'all lines of work by which products are produced from raw materials with the aid of living organisms' (Ereky, 1919); see Bud (*Nature* **337**: 10, 1989), Coombs (*Macmillan dictionary of biotechnology*, 1986), Hui & Khachatourians (Eds) (*Food biotechnology: microorganisms*, 1995), Wainwright (*An introduction to fungal biotechnology*, 1992); (2) technology concerned with machines in relation to human needs (obsol.).

See Genetic engineering, Industrial mycology, Molecular biology.

Biotransformation. Also known as biological or microbial transformation, or more generally bioconversion; the use of microorganisms to modify organic compounds to produce industrially, medically or environmentally important products. These are usually enzymatic reactions where the substrate may be metabolized or co-metabolized. The most useful reactions are quoted as oxidations, reductions, hydrolysis, condensation, isometisation, formation of new c-c bonds and introduction of hetero functions (Crueger & Crueger, 1990). Many transformations have been described but few are used industrially (e.g. transformations of antibiotics, steroids and sterols). *Rhizopus stolonifer* has been used to produce 11α-hydroxyprogesterone from progesterone. In several processes fungal spores are used directly to catalyze biotransformations. See Crueger & Crueger (*Microbial transformations in biotechnology: a textbook of industrial microbiology*, 1990), O'Sullivan (*in* Fogarty & Kelley (Eds), *Microbial enzymes and biotechnology*, edn 2: 295, 1990).

biotroph (adj. **-trophic**), an obligate parasite (cf. necrotroph, saprotroph), growing on another organism, in intimate association with its cytoplasm.

Biotyle Syd. (1929) = Pseudomeliola (Hypocr.).

biotype, (1) (Scheibe) = physiologic race; (2) one individual; a group of individuals having a like genetic make up (Christensen & Rodenhiser, *Bot. Rev.* **6**: 389, 1940; Waterhouse & Watson, *Proc. Linn. Soc. NSW* **66**: 269, 1941).

bipartite, having division into two.

bipolar, (1) (of spore), at the two ends (poles); (2) (of an incompatibility system), having 1 locus; unifactorial; cf. tetrapolar; (3) occurring in both Arctic and Antarctic regions;

Bipolaris Shoemaker (1959), Anamorphic Pleosporaceae, 1.C2.26. Teleomorph *Cochliobolus*. 52, widespr. See Alcorn (*Mycotaxon* **13**: 339, 1981), Luttrell (*Rev. Mycol.* **41**: 271, 1977), Sivanesan (*Mycol. Pap.* **158**, 1987; keys), Muchovej *et al.* (*Fitopat. Bras.* **13**: 211, 1988; keys), Alcorn (*Mycotaxon* **17**: 1, 1983; generic concepts), Alcorn (*Ann. Rev. Pl. Path.* **26**: 37, 1988; generic taxonomy), Hanau *et al.* (*Exp. Mycol.* **13**: 337, 1989; conidiogenous cell development), Alcorn (*Mycotaxon* **39**: 361, 1990; additions to genus), Alcorn (*Mycotaxon* **41**: 329, 1991; n. combs and syns), Khasanov (*Opredelitel' Gribov-Vozbul. 'Gelmintosporiozov' Rasten. iz Rodov Bipolaris, Drechslera i Exserohilum*, 1992).

Biporisporites Ke & Shi (1978), Fossil fungi. 1 (Tertiary), China.

birch canker, Siberian chaga fungus, sterile basidiomata of *Inonotus obliquus*; **- fungus**, *Piptoporus betulinus*.

bird's nest fungi, the *Nidulariales*.

Bireticulasporis Potonié & Sah (1960), Fossil fungi. 1 (Miocene), India.

Birsiomyces F. Schaarschm. (1966), Fossil fungi (Ascomycota). 1 (Triassic), Switzerland.

Bisby (Guy Richard; 1889-1958). Professor of plant pathology, Manitoba Agricultural College, Winnipeg, 1920-36; mycologist, Commonwealth Mycological Institute, Kew, 1937-54. See Gregory (*TBMS* **42**: 129, 1959; portr.), Johnson (*Phytopath.* **49**: 323, 1959; portr.), Stafleu & Cowan (*TL-2* **1**: 219, 1976), Stafleu & Mennega (*TL-2, Suppl.* **2**: 174, 1993).

Bisbyella Boedijn (1951) ≡ Agyriopsis (Stictid.).

Bisbyopeltis Bat. & A.F. Vital (1957), Mitosporic fungi, 5.E1.?. 1, USA.

Biscladinomyces Cif. & Tomas. (1953) = Cladonia (Cladon.).

Biscogniauxia Kuntze (1891), Xylariaceae. 25, widespr. See Callan & Rogers (*CJB* **64**: 842, 1986; anamorphs), González & Rogers (*Mycotaxon* **47**: 229, 1993; key 13 spp. Mexico), Granmo *et al.* (*Opera Bot.* **100**: 59, 1989; Nordic country keys), Petrini & Müller (*Mycol. Helv.* **1**: 501, 1986; key 5 spp. Eur.), Pouzar (*Česká Myk.* **40**: 1, 1986; Eur. spp.), Whalley *et al.* (*MR* **94**: 237, 1990). Anamorphs *Nodulisporium, Periconiella*.

biseriate (**biserial**), in two series.

Biseucladinomyces Cif. & Tomas. (1953) ≡ Cladonia (Cladon.).

Bispora Corda (1837), Mitosporic fungi, 1.B2.4. 5, temp. See Sutton (*CJB* **47**: 609, 1969), Wang (*Mem. N.Y. bot. Gdn.* **49**: 20, 1989; pleomorphism).

Bispora Fuckel (1870) ≡ Bisporella (Leot.).

Bisporella Sacc. (1884), Leotiaceae. c. 12, widespr. See Korf & Carpenter (*Mycotaxon* **1**: 57, 1974), Korf & Bujakiewicz (*Agarica* **6**: 302, 1985). Anamorphs *Bloxamia* (Johnston, *Mycotaxon* **31**: 345, 1988), *Eustilbum* (Seifert & Carpenter, *CJB* **65**: 1262, 1987).

Bisporomyces J.F.H. Beyma (1940) = Chloridium (Mitosp. fungi) fide Hughes (1958).

Bisporostilbella Brandsb. & E.F. Morris (1971), Mitosporic fungi, 2.B2.?. 1, USA.

Bisseomyces R.F. Castañeda (1985), Mitosporic fungi, 1.D2.10. 1, Cuba.

Bitancourtia Thirum. & Jenkins (1953) = Elsinoë (Elsin.) fide v. Arx & Müller (1975).

Bitrimonospora Sivan., Talde & Tilak (1974) = Monosporascus (Sordariales) fide v. Arx (*Kavaka* **3**: 33, 1976).

bitunicate, (1) having two walls; (2) (of asci), with two functional layers, that may or may not rupture or extend at discharge; see ascus.

Bitunicostilbe M. Morelet (1971), Mitosporic fungi, 2.C2.?. 2, USA.

Bitzea Mains (1939) = Chaconia (Chacon.) fide Thirumalachar & Cummins (*Mycol.* **41**: 523, 1949).

biuncinate, two-hooked.

Bivallum P.R. Johnst. (1991), Rhytismataceae. 6, Australasia, Chile.

bivalvate (1) (of spores), lens-shaped and having a hyaline rim, as in *Arthrinium*; (2) (of asci), see ascus.

Biverpa (Fr.) Boud. (1907) ? = Helvella (Helvell.). See Eckblad (1968).

biverticillate (of a penicillus), having branching at two levels, i.e. having metulae bearing phialides.

Bivonella (Sacc.) Sacc. (1891) = Thyridium (Thyrid.) fide Cannon (*SA* **8**: 78, 1989).

Bizozzeria, see *Bizzozeria*.

Bizozzeriella Speg. (1888), Mitosporic fungi, 3.A1. ?. 1, S. Am. See Donk (*Persoonia* **1**: 189, 1960; nomencl.).

Bizzozeria Sacc. & Berl. (1885), Lasiosphaeriaceae. 1, Italy. ? = Lasiosphaeria (Lasiosphaer.), but see Lundqvist (1972).

Bizzozeria Speg. (1889) ≡ Thaxteria (Lasiosphaer.).

Bjerkandera P. Karst. (1879), Coriolaceae. 2, N. temp.

black blotch, of clovers (*Cymadothea trifolii*): - **crottle**, see crottle; - **dot** of potato (*Colletotrichum coccodes*); - **jelly fungus**, basidioma of edible *Auricularia* spp.; - **knot** of plum and cherry (*Apiosporina morbosa*); - **leg** of beet (*Phoma betae*, *Pythium*, etc.); of pelargonium (*Pythium* spp.); - **line**, see zone lines; - **mildews**, *Meliolales*; - **piedra**, infection of hair shafts by *Piedraia hortae*; - **pustule** of *Ribes* (*Plowrightia ribesia*); - **root rot** of tobacco and other plants (*Thielaviopsis basicola*); of grapes (*Vitis*) (*Guignardia bidwellii*); - **scurf** of potato (*Thanatephorus cucumeris*, syn. *Corticium solani*); - **slime** of hyacinth (*Sclerotinia bulborum*); - **spot** of apple, see scab, apple; of rose (*Diplocarpon rosae*); - **stem rust** of cereals (*Puccinia graminis*); - **tip** of banana (*Musa*) (*Deightoniella torulosa*); - **tree lichen** (*Bryoria fremontii*) (Turner, *Econ. Bot.* **31**: 461, 1977); - **yeasts**, see yeasts.

blackfellows' bread (or native bread), the sclerotium (*Mylitta australis*) of the Australian *Polyporus mylittae*. See McAlpine (*J. Dep. Agric. Vict.* **2**, 1904), Willis (*Muelleria* **1**: 203, 1967; bibliogr.), Macfarlane *et al.* (*TBMS* **71**: 359, 1978; structure). There are similar sclerotia in India ('little mans' bread') and China.

Blakeslea Thaxt. (1914), Choanephoraceae. 2, esp. trop. See Thaxter (*Bot. Gaz.* **58**: 353, 1914), Mehrotra & Baijal (*J. Elisha Mitch. Sci. Soc.* **84**: 207, 1968), Kirk (1984; key), Zheng & Chen (*Acta Mycol. Sinica*, Suppl. **1**: 40, 1986).

Blarneya D. Hawksw., Coppins & P. James (1980), Mitosporic fungi (L), 1.B1.38. 1, W. Eur.

Blasdalea Sacc. & P. Syd. (1902), Vizellaceae. 1, Brazil. See Petrak (*Sydowia* **7**: 343, 1953).

Blastacervulus H.J. Swart (1988), Mitosporic fungi, 6.A2.4. 1, Australia.

Blastenia A. Massal. (1852) = Caloplaca (Teloschist.).

Blasteniomyces Cif. & Tomas. (1953) ≡ Protoblastenia (Psor.).

Blasteniospora Trevis. (1853) = Xanthoria (Teloschist.).

blasteniospore, a polarilocular (q.v.) ascospore.

blastic (of conidiogenesis), one of the two basic sorts of conidiogenesis (cf. thallic), characterized by a marked enlargement of a recognizable conidial initial *before* the initial is delimited by a septum. The conidium is differentiated from *part* of a cell (Kendrick, 1971:

255); **entero-**, when the inner wall (see tretic) or neither wall (see phialidic) of the blastic conidiogenous cell contributes to the formation of the **conidium** (blastic conidium) (cf. holoblastic); **holo-**, when both outer and inner walls of the blastic conidiogenous cell contribute to the formation of the conidium (cf. enteroblastic and see annellidic); **mono-**, when a conidiogenous cell has only one conidiogenous locus; **poly-**, when a conidiogenous cell has several conidiogenous loci.

Blasticomyces I.I. Tav. (1985), Laboulbeniaceae. 2, Asia.

blastidium, a lichen propagule produced by the budding of thalli in a yeast-like manner (Poelt, *Flora, Jena* **169**: 23, 1980). Fig. 22E.

Blastobasidiomycetes, see *Basidiomycota, Ustomycetes*.

Blastobotrys Klopotek (1967), Mitosporic fungi, 1.A1.6. 6, widespr. See Marvanová (*TBMS* **66**: 217, 1976), de Hoog *et al.* (*Ant. v. Leeuwenhoek* **51**: 79, 1985; keys), de Hoog (*Pleomorphic fungi*: 221, 1987; developmental cycle).

Blastocapnias Cif. & Bat. (1963) = Aithaloderma (Capnod.) fide v. Arx & Müller (1975).

Blastocatena Subram. & Bhat (1989), Mitosporic fungi, 2.C2.4. 1, India.

blastocatenate (of blastoconidia), formed in chains with the youngest at the apical or distal end of the chain.

Blastocladia Reinsch (1877), Blastocladiaceae. *c.* 13 (saprobes in water), widespr. See Emerson & Cantino (*Am. J. Bot.* **35**: 157, 1948), Das Gupta & John (*Indian Phytopath.* **41**: 521, 1988).

Blastocladiaceae H.E. Petersen (1909), Blastocladiales. 5 gen. (+ 4 syn.), 35 spp. Thallus mono- or polycentric, non-colonial, usually with a ± prominent basal part bearing one or more reproductive structures; saprobic.

Blastocladiales, Chytridiomycetes. 5 fam., 13 gen. (+ 7 syn.), 126 spp. Thallus mono- or polycentric; zoospores lacking lipid globules; freshwater or terricolous, saprobic or parasitic (*Coelomomyces* on mosquito larvae); cosmop. Fams:

(1) **Blastocladiaceae**.
(2) **Catenariaceae**.
(3) **Coelomomycetaceae**.
(4) **Physodermataceae**.
(5) **Sorochytriaceae**.

Lit.: Sparrow (1960: 605; 1973), Fitzpatrick (1930: 130), Emerson & Robertson (*Am. J. Bot.* **61**: 303, 1974), Karling (1977, 1978), Dewel & Dewel (*CJB* **68**: 1968, 1977), Lange & Olson (*TBMS* **74**: 449, 1980), Olson (*Opera Bot.* **73**: 1, 1984; key fams), Dewel (*CJB* **63**: 1525, 1985).

Blastocladiella V.D. Matthews (1937), Blastocladiaceae. 13, N. & C. Am., Eur. See Sparrow (1960: 660; key), Couch & Whiffen (*Am. J. Bot.* **29**: 582, 1942), Cantino (*in* Meynell & Gooder (Eds), *Microbial reaction to environment*: 243, 1961; morphogenesis).

Blastocladiopsis Sparrow (1950), Blastocladiaceae. 2, USA, Cuba.

blastoconidium, a blastic (q.v.) conidium.

Blastoconium Cif. (1931), Mitosporic fungi, 1.D2.?. 2, trop.

Blastocystis Alexeev (1911), Zygomycetes (inc. sed.). 1 (in alimentary tract of humans and animals), widespr. See Lee (*TBMS* **54**: 313, 1970; ultrastr.), Lavier (*Ann. Paras. hum. comp.* **27**: 339, 1952; synonymy).

Blastodendrion (M. Ota) Cif. & Redaelli (1925) = Candida (Mitosp. fungi). See Zobel (*Arch. Hyg. Berlin* **130**: 205, 1943). Teleomorph *Pichia*.

Blastoderma B. Fisch. & Brebeck (1894) nom. ambig. (Mitosp. fungi).

Blastodesmia A. Massal. (1852), ? Pyrenulales (inc. sed., L). 1 (on *Fraxinus*), Eur. See Keissler (*Rabenh. Krypt.-Fl.* **9**, 1 (2): 384, 1937), Aptroot (*SA* **12**: 25, 1993; posn).

Blastodictys M.B. Ellis (1976), Mitosporic fungi, 1.D2.10. 1, Uganda.

Blastomyces Costantin & Rolland (1888) ? = Histoplasma (Mitosp. fungi); = Chrysosporium (Mitosp. fungi) fide v. Oorschot (1980). See Watts *et al.* (*Am. Jl Clin. Path.* **93**: 575, 1990; giant forms of *B. dermatitidis*), Geber *et al.* (*J. Gen. Microbiol.* **138**: 395, 1992; rRNA sequence and phylogeny in *B. dermatitidis*). See *Ajellomyces*, blastomycosis.

Blastomyces Gilchrist & W.R. Stokes (1898) = Zymonema (Mitosp. fungi) fide Dodge (*Medical Mycology*, 1935).

Blastomycetes. Class often used for anamorphic yeasts (q.v.) and then divided into two Orders, (1) *Cryptococcales* (reproduction by budding, ballistospores absent; ascomycetous affinities, but see under Order), and (2) *Sporobolomycetales* (reproduction by budding and ballistospores; basidiomycetous affinities). This simplistic scheme is no longer tenable (see Kendrick, *The fifth kingdom*, edn 2, 1992). Included in Mitosporic fungi in this edition of the *Dictionary*.

blastomycin, an antigen made from *Blastomyces dermatitidis*, esp. for skin testing; **- S**, an antifungal antibiotic from *Streptomyces griseochromogenes* (Fukunage *et al.*, *Bull. agric. chem. Soc. Japan* **19**: 181, 1955) used against rice blast (*Pyricularia oryzae*).

Blastomycoides Castell. (1928) ≡ Zymonema (Mitosp. fungi).

blastomycosis, (1) a disease in humans caused by *Blastomyces dermatitidis* (teleomorph *Ajellomyces dermatitidis*; see Al-Doory & Di Salvo, *Blastomycosis*, 1992); **N. American -**; Gilchrist's disease; (2) any mycotic disease in humans having budding cells in the parasitized tissues; **cheloidal -**, see lobomycosis; **European -**, see cryptococcosis; **S. American -**, see *Paracoccidioides*.

Blastomycota. Proposed for *Ascoblastomycetes* and *Basidioblastomycetes*, anamorphic yeasts with ascomycetous and basidiomycetous affinities respectively (Moore, *Bot. Mar.* **23**: 361, 1980).

Blastophoma Kleb. (1933) ? = Sclerophoma (Mitosp. fungi) fide Sutton (1977).

Blastophorella Boedijn (1937), Mitosporic fungi, 2.B2.6. 1, Java, Sumatra.

Blastophorum Matsush. (1971), Mitosporic fungi, 1.C1.10/18. 1, Papua New Guinea.

Blastophragma Subram. (1993), Mitosporic fungi. ?.?.?. 2, S.E. Asia.

Blastoschizomyces Salkin, M.A. Gordon, Sams. & Rieder (1982), Mitosporic fungi, 1.A1.19. 1 (from sputum), USA. See Polachek *et al.* (*J. Clin. Microbiol.* **30**: 2318, 1992; tax. review).

Blastospora Dietel (1908), Uredinales (inc. sed.). 3 (on *Smilax, Betula*), Japan. See Mains (*Am. J. Bot.* **25**: 677, 1938), Kaneko & Hiratsuka (*Mycol.* **73**: 577, 1981).

blastospore, a spore formed by marked enlargement of a recognizable conidium initial before the initial is delimited by a septum. The conidium differentiates from part of the cell. See Kendrick (Ed.) (*Taxonomy of fungi imperfecti*, 1971).

Blastosporidium M. Hartmann (1912) ? = Coccidioides (Mitosp. fungi) fide Sutton (*in litt.*).

Blastotrichum Corda (1838), Mitosporic fungi, 1.C1.?. 5, esp. Eur. Nom. dub. fide Gams & Hoozemans (*Persoonia* **6**: 99, 1970).

Blastulidiopsis Sigot (1931) = Blastulidium (Leptomitales, inc. sed.) fide Manier (1976).

Blastulidium Pérez (1903), Leptomitales (inc. sed.). 1, France. See Manier (*Protistologia* **12**: 225, 1976), Dick (*in press*; key).

blematogen (**blematogen layer**), the undifferentiated tissue which becomes the universal veil in agarics (Atkinson, *Am. J. Bot.* **1**: 3, 1914).

Blennorella Kirschst. (1944) ? = Colletotrichum (Mitosp. fungi) fide Sutton (1977).

Blennoria Moug. & Fr. (1825), Mitosporic fungi, 8.A1.15. 1, Eur. See Sutton (*Taxon* **21**: 319, 1972).

Blennoriopsis Petr. (1920), Mitosporic fungi, 1.A1.?. 1, Eur.

Blennothallia Trevis. (1853) = Collema (Collemat.).

Blepharia (Pers.) Ainsw. & Bisby (1943) nom. inval. = Dematium (Mitosp. fungi) fide Mussat (*Syll. Fung.* **15**: 62, 1901).

blepharoplast (of zoospores), the basal body or granule (**kinetosome**) from which arise the longitudinal fibres constituting the axoneme of a flagellum; joined to the nucleus by a **rhizoplast.**

Blepharospora Petri (1917) = Phytophthora (Pyth.) fide Buisman (1927).

Bleptosporium Steyaert (1961), Mitosporic fungi, 4.C2.19. 1, Argentina. See Sutton (*Mycol. Pap.* **88**, 1963).

blewits, (blewitt, blue leg, bluette), basidiomata of the edible *Lepista saeva* (syn. *Tricholoma personatum*); **wood -**, *L. nuda* (syn. *T. nudum*).

blight, a common name for a number of different diseases of plants (and for insect attack), esp. when leaf damage is sudden and serious; **potato -**, **late -** (*Phytophthora infestans*); **early -** (*Alternaria solani*).

blister rust (of 5-needled pines), Cronartium ribicola.

Blistum B. Sutton (1973) = Polycephalomyces (Mitosp. fungi) fide Seifert.

Blitridium De Not. (1863) = Triblidium (Triblid.) fide Nannfeldt (*Nova Acta R. Soc. Scient. upsal.* 4 **8**(2), 1932).

Blodgettia Harv. (1858), Mitosporic fungi, 1.-.-. 2 (mycophycobionts; marine), widespr. See Kohlmeyer & Kohlmeyer (*Marine mycology*, 1979), Hawksworth (*Notes R. bot. Gdn Edinb.* **44**: 549, 1987).

Blodgettia E.P. Wright (1881) ≡ Blodgettia Harvey (Mitosp. fungi).

Blodgettiomyces Feldmann (1939) ≡ Blodgettia (Mitosp. fungi) fide Hawksworth (1987).

Blogiascospora Shoemaker, E. Müll. & Morgan-Jones (1966), Amphisphaeriaceae. 1, Eur. Anamorph *Seiridium*.

Bloxamia Berk. & Broome (1854), Anamorphic Leotiaceae, 3.A2.22. Teleomorph *Bisporella*. 4, Eur., India, S. Am. See Nag Raj & Kendrick (*A monograph of Chalara and allied genera*, 1975), Arambarri *et al.* (*Mycotaxon* **43**: 327, 1992; key).

blue cheeses, ripened and flavoured by *Penicillium roqueforti* (Mitosp. fungi), e.g. Roquefort, Stilton, Gorgonzola, Danish Blue etc.

blue stain, blue-grey colouration of wood caused by the growth of brown fungal hyphae in the surface layers.

Blumenavia A. Møller (1895), Clathraceae. 2, S. Am., Afr.

Blumeria Golovin ex Speer (1975), Erysiphaceae. 1, widespr. *B. graminis* (cereal and grass mildew) fide Braun (*Beih. Nova Hedw.* **89**, 1987). Anamorph *Oidium*.

Blumeriella Arx (1961), Dermateaceae. 5 (on *Prunus* and *Kerria*), N. Am., Eur. *B. jaapii* (anamorph *Phloeosporella padi*) (cherry leaf spot; Jakobsen & Jørgensen, *Tidsskr. Plant.* **90**: 161, 1986). See v. Arx (*Phytopath. Z.* **42**: 161, 1961), Williamson & Bernard (*CJB* **66**: 2048, 1988). Anamorph *Phloeosporella*.

blusher, the, basidioma of the edible *Amanita rubescens*.

Blytridium, see *Blitridium*.

Blyttiomyces A.F. Bartsch (1939), Chytridiaceae. 9, widespr. See Dogma & Sparrow (*Mycol.* **61**: 1149, 1970).

BM. The Natural History Museum (London, UK); founded 1753; known by official name The British Museum (Natural History) up to 1989; governed by a Board of Trustees and funded through the Office of Arts and Libraries; most non-lichenized fungi were transferred to **K** in 1969 (Brenan & Ross, *Lichenologist* **4**: 157, 1970); see Anon. (*The history of the collections contained in the Natural History Department of the British Museum*, **1**, 1904), Stearn (*The Natural History Museum at South Kensington*, 1981).

boathook hair, characteristic terminally bifid hair, produced intracellularly in principal-form zoospore of *Saprolegnia*, and which ornaments the cyst formed by this zoospore; principal-form cyst or secondary cyst (see Beakes, 1983).

Bodinia M. Ota & Langeron (1923) = Trichophyton (Mitosp. fungi) fide Sutton (*in litt.*).

Boedijnopeziza S. Ito & S. Imai (1937) = Cookeina (Sarcoscyph.) fide Denison (*Mycol.* **59**: 306, 1967) and Pfister (1974).

Boehmia Raddi (1807) ? = Leptoglossum (Tricholomat.) fide Donk (*Taxon* **6**: 22, 1957), = Arrhenia (Tricholomat.).

Boeomycomyces, see *Baeomycomyces*.

Boerlagella Penz. & Sacc. (1897) [non Cogn. (1891), *Sapotaceae*] ≡ Boerlagiomyces (Tubeuf.).

Boerlagellopsis C. Ramesh (1988), nom. nud. (? Dothideales). See Eriksson & Hawksworth (*SA* **8**: 62, 1989).

Boerlagiomyces Butzin (1977), Tubeufiaceae. 2, trop. See Rossman (1987). = Thaxteriellopsis (Tubeuf.) fide Sivanesan (*in litt.*).

Bogoriella Zahlbr. (1928), ? Verrucariaceae (L). 1, Java.

Bogoriellomyces Cif. & Tomas. (1954) ≡ Bogoriella (Verrucar.).

Bohleria Trevis. (1860) = Placidiopsis (Verrucar.).

Boidinia Stalpers & Hjortstam (1982), Gloeocystidiellaceae. 4, widespr.

Bojamyces Longcore (1989), Legeriomycetaceae. 1, USA.

Bolacotricha Berk. & Broome (1851) = Chaetomium (Chaetom.).

Bolbitiaceae Singer (1948), Agaricales. 10 gen. (+ 11 syn.), 154 spp. Pileipellis epithelial; spores ochraceous to rusty brown, with a germ pore.
    Lit.: van Waveren (*Coolia* **15**: 101, 1972; keys gen.), Watling & Gregory (*Census catalogue of world members of the Bolbitiaceae*, 1981).

Bolbitius Fr. (1838), Bolbitiaceae. 19, widespr. See Singer (*Sydowia* **30**: 216, 1977; keys world spp.), Watling (*British fungus flora* **3**, 1982; key 6 Br. spp.), Watling & Gregory (1981).

Boletaceae Chevall. (1926), Boletales. 13 gen (+ 23 syn.), 243 spp. Basidioma stout, with ornamented stipe; hymenophore tubulate, sinuato-adnexed; spores smooth or ridged.

Boletales, Basidiomycetes. 11 fam., 70 gen. (+ 68 syn.), 727 spp. The 'boletes'; terrestrial; usually saprobic, occasionally fungicolous, or mycorrhizal; most are edible, a few poisonous; cosmop. The poroid basidiocarp differs from that of the poroid 'aphyllophorales' by having a soft, fleshy cortex similar to the *Agaricales* and in the tubes being easily separated from the pileus, which usually has a central stipe.
    Major differences of opinion exist over the generic and family limits, although Singer (1975) is the most widely accepted classification with four families,

Boletaceae, Gomphidiaceae, Paxillaceae and Strobilomycetaceae retained in the *Agaricales*. The order was first proposed by Gilbert (1931) to include all epigeal, putrescent basidiocarps with tubulate hymenophores and their allies. Kreisel (1969) included agaricoid, boletoid and gasteroid genera in the order, reflecting the parallel evolutionary trends found in the *Russulales* and *Agaricales*. Fams:

(1) **Beenakiaceae**.
(2) **Boletaceae**.
(3) **Chamonixiaceae**.
(4) **Coniophoraceae**.
(5) **Gomphidiaceae**.
(6) **Gyrodontaceae**.
(7) **Hygrophoropsidaceae**.
(8) **Paxillaceae**.
(9) **Rhizopogonaceae**.
(10) **Strobilomycetaceae**.
(11) **Xerocomaceae**.

    Lit.: **General**: Snell (*Mycol.* **33**: 415, 1941; gen. characters), Donk (1951-63; **IV**; gen. list), Singer (1975), Arpin & Kühner (*Bull. mens. Soc. Linn. Lyon* **46**: 81, 181, 1977; history of classification), Baroni (*Mycol.* **70**: 1064, 1978; chemical tests), Corner (*Gdns Bull., Singapore* **26**: 159, 1972; spore formation), Pegler & Young (*TBMS* **76**: 103, 1981; outline classification and spore structure).

    Regional: [see also under Macromycetes]. **Africa**: Heinemann (*Bull. Jard. bot. Etat Brux.* **21**: 223, 1951; **30**: 21, 1960; **34**: 425, 1964; Katanga, Zaire, Uganda). **America, North**: Snell & Dick (*The Boleti of Northeastern North America*, 1970), Smith & Thiers (*The Boletes of Michigan,* 1971). **Belgium**: Heinemann (*Naturalistes belges* **42**: 333, 1961). **Europe, Central**: Singer (*Die Röhrlinge*, Teil 2, *Die Boletoideae und Strobilomyceteae*,1967 [*Die Pilze Mitteleuropas*, **6**]). **British Isles**: Pearson (*Naturalist Hull*, 1946; key), Watling (*British Fungus Flora* **2**, 1970). **France**: Gilbert (*Les bolets*, 1931), Blum (*Les bolets*, 1963; key 66 spp.; suppl. and revisions, *BSMF* **80**: 297, 1964; **81**: 478, 1965; **84**: 309, 577, 1969; *Revue mycol.* **34**: 249, 1970), Leclair & Essette (*Les bolets*, 1968; col. pl.). **Malaysia**: Corner (*Boletus in Malaysia*, 1972). **New Zealand**: McNabb (*N.Z. Jl Bot.* **6**: 137, 1968). **Nova Scotia**: Grund & Harrison (*Nova Scotia boletes*, 1976). **Poland**: Skirgiello (*Grzyby (Fungi), Podstawczaki (Basidiomycetes), Barowikwe (Boletales)*, 1960 [Engl. transl. see *Mycol.* **68**: 1136]). **USA**: Coker & Beers (*The Boletaceae of North Carolina*, 1943), Singer (*The Boletineae of Florida*, 1945-47 [reprinted 1970 from *Farlowia*, **2** and *Am. midl. Nat.* **37**]), Thiers (*California mushrooms: a field guide to the boletes*, 1975), Mycogeography in the South Pacific Region (*Austr. J. Bot.* Suppl. Ser. No. 10, 1983), Singer, Araujo & Ivory (*Beih. Nova Hedw.* **77**, 1983).

bolete, one of the *Boletales*.

Boletellites P. Briot, Lar.-Coll. & Locq. (1983), Fossil fungi. 1, Australia.

Boletellus Murrill (1909), Xerocomaceae. 38, trop. & sub. trop. See Pantidou (*CJB* **39**: 1149, 1961; culture), Singer (*Sydowia* **30**: 221, 1978; key), Singer, Garcia & Gomez (*Beih. Nova Hedw.* **105**: 3, 1992; key C. Am. spp.).

Boletinellus Murrill (1909) = Gyrodon (Gyrodont.) fide Singer (1975).

boletinoid (of hymenophores), having a structure intermediate between pores and gills.

Boletinus Kalchbr. (1867), Gyrodontaceae. 3 (with conifers), N. temp. = Suillus (Bolet.) fide Smith & Thiers (1971). See Singer (*Sydowia* **30**: 227, 1977; key).

Boletium Clem. (1909) ≡ Volvoboletus (Amanit.).

Boletochaete Singer (1944), Boletaceae. 1, Singapore.
Boletogaster Lohwag (1926) = Boletellus (Xerocom.) fide Singer (1945).
Boletolichen Juss. (1789) = Helvella (Helvell.) fide Mussat (*Syll. Fung.* **15**, 1901).
Boletopsis Fayod (1889), Thelephoraceae. 2, Eur.
Boletopsis Henn. (1898) ≡ Suillus (Bolet.) fide Singer (*Lilloa* **22**: 654, 1951).
Boletus L. (1753) = Phellinus (Hymenochaet.) fide Donk (*Persoonia* **1**: 173, 1960).
Boletus Tourn. ex Adans. (1763) = Morchella (Morchell.) fide Donk (*Reinwardtia* **3**: 275, 1955).
Boletus Fr. (1821) nom. cons., Boletaceae. 130 (ectomycorrhizal), cosmop. Some are edible (see cep). See Singer (*Sydowia* **30**: 227, 1978; key). Sensu Karsten (1881) = Suillus (Bolet.) fide Smith & Thiers (1964).
Bolinia (Nitschke) Sacc. (1882) = Camarops (Bolin.) fide Nannfeldt (1972).
Boliniaceae Rick (1931), Sordariales. 3 gen. (+ 10 syn.), 21 spp. Stromata immersed to erumpent, crustose or pulvinate, sometimes absent, usually soft-textured, composed of thin-walled hyphal tissue; ascomata perithecial, long-necked, sometimes vertically elongate, the ostiole periphysate; interascal tissue of narrow paraphyses, sometimes thin-walled and evanescent; asci cylindrical, persistent, thin-walled, with a small, usually I- apical ring; ascospores brown, sometimes transversely septate, sometimes with germ pores. Anamorphs not known. Saprobic in wood and bark, widespr.
  Lit.: Andersson *et al.* (*SA* **14**: 1, 1995; posn), Lundqvist (*Symb. bot. upsal.* **20**(1), 1972).
Bolosphaera Syd. & P. Syd. (1917) = Phaeostigme (Pseudoperispor.). See Hansford (*Mycol. Pap.* **15**, 1946).
Bolton (James; 1750-99). An amateur mycologist (probably in the weaving trade), author of *An history of fungusses, growing about Halifax* [Yorks., England], (4 parts) 1788-91 (in German [by Willdenow] as *Geschichte der merkwurdigsten Pilze*, 1795-1820). See Sartory & Maire (*Interpretation des planches de J. Bolton An history of fungusses, vols. 1 & 2*, n.d., Paris), Petersen (*Mycotaxon* **5**: 498, 1977; index), Shear (*TBMS* **17**: 302, 1933), Laplanche, (*Dictionnaire iconographique des champignons superieurs (Hymenomycetes)*, 1894; gives Friesian names for Figures by Bolton, Bulliard, Paulet, Persoon, Sowerby, and others), Watling & Seaward (*Arch. nat. Hist.* **10**: 89, 1981; biogr. data, bibliogr.), Grummann (1974: 257), Stafleu & Cowan (*TL-2* **1**: 264, 1976), Stafleu & Mennega (*TL-2, Suppl.* **2**: 296, 1993).
Bombardia (Fr.) P. Karst. (1873), Lasiosphaeriaceae. 1, Eur., N. Am.
Bombardiastrum Pat. (1893), Ascomycota (inc. sed.). 1, S. Am.
Bombardiella Höhn. (1909), Sordariales (inc. sed.). 1, Java. See Eriksson & Hawksworth (*SA* **6**: 110, 1987).
Bombardioidea C. Moreau ex N. Lundq. (1972), Lasiosphaeriaceae. 4, widespr. See Krug & Scott (*CJB* **72**: 1302, 1994; key).
Bombyliospora De Not. (1852) = Megalospora (Megalospor.) fide Hafellner & Bellemère (1982).
Bombyliosporomyces Cif. & Tomas. (1953) ≡ Bombyliospora (Megalospor.).
bombysine, like silk.
Bommerella Marchal (1885), Chaetomiaceae. 1, Eur., Japan, N. Am.
Bomplandiella Speg. (1886), Mitosporic fungi, 3.A2.?. 1, S. Am.
Bonanseja Sacc. (1906), ? Rhytismatales (inc. sed.). 1, Mexico.
Bonaria Bat. (1959), Micropeltidaceae. 4, trop.

Bondartsev (Apollinaris Semenovich; 1877-1968). Russian mycologist and plant pathologist; head of the Phytopathological Dept., Leningrad Botanical Garden, 1913-31, where he collaborated with Singer (q.v.). Chief writings: *The Polyporaceae of the European USSR*, 1953 (Engl. transl. 1971); [Key to dry rot fungi], 1956 [Russ.]. See *Mikol. i Fitopatol.* **3**: 550, 1969; portr., bibl.
Bondarzewia Singer (1940), Bondarzewiaceae. 3, widespr. See Corner (*Beih. Nova Hedw.* **78**: 205, 1984; key S. E. Asia spp.).
Bondarzewiaceae Kotl. & Pouzar (1957), Bondarzewiales. 1 gen., 3 spp. Pilei flabellate; hymenophore tubulate; spores amyloid, ridged. See Donk (1964: 247).
Bondarzewiales, Basidiomycetes. 3 fam., 3 gen., 5 spp. Fams:
  (1) **Amylariaceae**.
  (2) **Bondarzewiaceae**.
  (3) **Hybogasteraceae**.
Bondiella Piroz. (1972), Mesnieraceae. 1 (on *Palmae*), Tanzania.
Bonia Pat. (1892) [non *Bonia* Balansa (1890), *Gramineae*], Tremellales (inc. sed.). 2, Eur.
Boninogaster Kobayasi (1937), ? Hysterangiaceae. 1, Bonin Isl.
Boninohydnum S. Ito & S. Imai (1940) = Gyrodontium (Coniophor.) fide Maas Geesteranus (1964).
Bonordenia Schulzer (1866) = Hypomyces (Hypocr.) fide Rogerson (1970).
Bonordeniella Penz. & Sacc. (1901) = Coniosporium (Mitosp. fungi) fide Ellis (*DH*).
booted, see peronate.
Boothiella Lodhi & Mirza (1962), Chaetomiaceae. 1 (from soil), Asia. See Udagawa & Furuya (*Trans. mycol. Soc. Japan* **18**: 302, 1977).
boot-lace fungus, the honey agaric, *Armillaria mellea*.
Bordea Maire (1916) = Autophagomyces (Laboulben.) fide Thaxter (1931).
Bordeaux mixture. A spray first used by Millardet in 1883-85 against vine (*Vitis*) mildew (*Plasmopara*), and still in general use for controlling numbers of plant diseases. A common mixture is the '4-4-50': copper sulphate 1.8 kg, quick lime 1.8 kg (or hydrated lime 2.7 kg), water 227 l. When making small amounts the copper sulphate is put in some of the water, the lime in the rest, and the two liquids then mixed. See Ainsworth (*Introduction to the history of plant pathology*, 1981).
Boreoplaca Timdal (1994), Lecanoraceae (L). 1, Siberia.
Boreostereum Parmasto (1968), Stereaceae. 2, N. temp.
Borinquenia F. Stevens (1917) nom. dub. (Fungi, inc. sed.). See Rossman (*Mycol. Pap.* **157**, 1987).
Bornetina L. Mangin & Viala (1903) = Diacanthodes (Coriol.) fide Donk (1962).
Borrera Ach. (1810) nom. rej. [non *Borreria* G. Mey. (1818) nom. cons., *Rubiaceae*] = Teloschistes (Teloschist.) fide Kurokawa (*Beih. Nova Hedw.* **6**, 1962).
Bostrichonema Ces. (1867), Mitosporic fungi, 1.B1.?. 3, Eur., N. Am.
Bostrychia Fr. (1821) = Cytospora (Mitosp. fungi) fide Fries (*Syst. mycol.* **2**, 1823).
Bostrychonema, see *Bostrichonema*.
Botanamphora Nograsek & Scheuer (1990) = Trematosphaeria (Melanommat.) fide Nograsek & Scheuer (*SA* **12**: 31, 1993).
Bothrodiscus Shear (1907), Anamorphic Leotiaceae, 8.C1.10. Teleomorph *Ascocalyx*. 3, N. Am., former USSR. See Groves (*CJB* **46**: 1273, 1968).
bothrosome (**sagenogen**, **sagenogenetosome**), an invaginated organelle at the cell surface which connects the plasma membrane to the network membranes in *Labyrinthulomycota* (see Porter, 1990).

Botryandromyces I.I. Tav. & T. Majewski (1976), Laboulbeniaceae. 2, Eur., Afr., N. & S. Am.

Botrydiella Badura (1963) = Staphylotrichum (Mitosp. fungi) fide Ellis (*DH*).

Botrydina Bréb. ex Menegh. (1844) nom. rej. prop. = Omphalina (Tricholomat.). Combination of fungus with *Coccomyxa* (*Algae*). See Gams (*Öst. bot. Z.* **109**: 376, 1962), Poelt & Jülich (*Herzogia* **1**: 331, 1969), Oberwinkler (*Deutsch. bot Ges.* N.F. **4**: 139, 1970).

Botrydiplis Clem. & Shear (1931) ≡ Botryodiplodia (Mitosp. fungi).

Botryella Syd. & P. Syd. (1916) = Sphaerellopsis (Mitosp. fungi) fide Sutton (1977).

botryo-aleuriospore, one of an apical cluster of aleuriospores developed basipetally from the conidiogenous cells.

Botryoascus Arx (1972), Endomycetaceae. 1, S. Afr., India.

Botryobasidiaceae (Parmasto) Jülich (1982), Stereales. 5 gen. (+ 2 syn.), 21 spp. Basidioma resupinate, smooth; monomitic, hyphae with right-angled branching; spores inamyloid.

Botryobasidium Donk (1931), Botryobasidiaceae. 48, widespr. See Donk (*Fungus* **28**: 26, 1958), Eriksson & Hjortstam (*Friesia* **9**: 14, 1969; key Swedish spp.), Eriksson & Ryvarden (*Corticiaceae of northern Europe* **2**: 145, 1973; key 11 Eur. spp.), Langer (*Bibl. Mycol.* **158**, 1994; world key).

Botryobasidium Rick (1959) [non Donk (1931)], Thelephoraceae. 3, Brazil.

botryoblastospore, clusters of conidia borne on the swollen apex (ampulla) of a conidiogenous cell, arising synchronously or asynchronously, either singly or in chains.

Botryobolus Arnaud (1952) = Ballocephala (Meristacr.) fide Humber (*in litt.*).

Botryochaete Corda (1854) ≡ Phleogena (Phleogen.) fide Donk (*Persoonia* **4**: 160, 1966).

Botryochaete Rick (1959), Phleogenaceae. 3, Brazil.

Botryochora Torrend (1914), ? Dothioraceae (? L). 1, Afr. See v. Arx & Müller (1975).

Botryocladium Preuss (1851) = Nematogonum (Mitosp. fungi) fide Saccardo (*Syll. Fung.* **4**: 170, 1886).

Botryoconis Syd. & P. Syd. (1906), Cryptobasidiaceae. 3, Brazil, Japan. See Donk (*Reinwardtia* **4**: 114, 1956).

Botryocrea Petr. (1949) = Fusarium (Mitosp. fungi) fide Samuels *et al.* (*Mycol. Pap.* **164**, 1991).

Botryoderma Papendorf & H.P. Upadhyay (1969), Mitosporic fungi, 1.A1.14. 2, S. Afr., Brazil.

Botryodiplis Clem. & Shear (1931) ≡ Botryodiplodia (Mitosp. fungi).

Botryodiplodia (Sacc.) Sacc. (1884), Mitosporic fungi, 8.B2.?. 17, cosmop. See Zambettakis (*BSMF* **70**: 219, 1954), Stevens (*Mycol.* **33**: 69, 1941). For *B. theobromae* see Lasiodiplodia.

Botryodiplodina Dias & Sousa da Câmara (1954), Mitosporic fungi, 8.B2.?. 1, Portugal.

Botryodontia (Hjortstam & Ryvarden) Hjortstam (1987), Botryobasidiaceae. 2, trop.

Botryogene Syd. & P. Syd. (1917) nom. dub. fide Sutton (*Mycol. Pap.* **141**, 1977).

Botryohypochnus Donk (1931) = Botryobasidium (Botryobasid.) fide Langer (1994).

Botryohypoxylon Samuels & J.D. Rogers (1986), Dothideales (inc. sed.). 1, Brazil. Anamorph *Iledon*.

Botryola Bat. & J.L. Bezerra (1964), Nitschkiaceae. 1, Brazil.

Botryomonilia Goos & Piroz. (1975), Mitosporic fungi, 1.A1.38. 1, Panama.

Botryomyces Greco (1916), Mitosporic fungi, 1.?.?. 1 (from man), Argentina.

Botryomyces de Hoog & C. Rubio (1982) ≡ Ybotryomyces (Mitosp. fungi).

Botryonipha Preuss (1852) nom. rej. prop. Sterile mycelium fide Hughes (*Friesia* **9**: 61, 1969).

Botryophialophora Linder (1944) = Myrioconium (Mitosp. fungi) fide v. Arx (*Genera of fungi sporulating in pure culture*, 1970).

Botryophoma (P. Karst.) Höhn. (1916) = Sclerodothiorella (Mitosp. fungi) fide Sutton (1977).

Botryorhiza Whetzel & Olive (1917), Chaconiaceae. 1 (on *Hippocrateaceae*), Puerto Rico, Brazil.

botryose, racemose; grouped like grapes.

Botryosphaeria Ces. & De Not. (1863), Botryosphaeriaceae. *c.* 12, widespr. *B. ribis* (die-back of currants; *Dothiorella* rot of citrus fruit) and other spp.; probably plurivorous. See v. Arx & Müller (1954), Barr (*Contr. Univ. Mich. Herb.* **9**: 523, 1972), Tilak & Gaikwad (*Botanique, Nagpur* **5**: 113, 1974). Anamorphs *Botryodiplodia, Dothiorella, Lasiodiplodia*.

Botryosphaeriaceae Theiss. & P. Syd. (1918), Dothideales. 5 gen. (+ 16 syn.), *c.* 28 spp. Stromata varied in development, ± immersed in or erumpent from plant material; ascomata perithecial, often strongly aggregated, with a well-developed ostiole, the peridium composed of large pseudoparenchymatous cells; interascal tissue of pseudoparaphyses; asci widely clavate, fissitunicate, without well-developed apical structures; ascospores hyaline to pale yellow, usually aseptate. Anamorphs coelomycetous, esp. *Diplodia*-like. Biotrophic, necrotrophic or saprobic, esp. on woody plants.

Lit.: v. Arx & Müller (1954), Barr (1987), Sivanesan (1984).

Botryosphaerostroma Petr. (1921) = Botryodiplodia (Mitosp. fungi) fide Sutton (1977).

Botryosphaerostroma Petr. & Syd. (1926) ≡ Sphaeropsis (Mitosp. fungi) fide Sutton (1977).

Botryosporium Corda (1831), Mitosporic fungi, 1.A1.6. 4, Eur., Am., Sumatra. See Mason (*Mycol. Pap.* **2**: 27, 1928), Vincent & Blackwell (*Taxon* **36**: 158, 1987; nomencl.), Zhang & Kendrick (*Acta Mycol. Sin.* **9**: 31, 1990; key 4 spp.).

Botryosporium Schwein. (1832) ? = Dictyosporium (Mitosp. fungi) fide Sutton (*in litt.*).

Botryostroma Höhn. (1911), Venturiaceae. 2, trop. Am.

Botryothecium Syd. (1937) = Rosenscheldiella (Ventur.) fide Hansford (1946).

Botryotinia Whetzel (1945), Sclerotiniaceae. *c.* 18, widespr. See Hennebert & Groves (*CJB* **41**: 341, 1963), Jarvis (*Can. Dept. Agric. Monogr.* **15**, 1977; taxonomy, physiology, pathogenicity). Anamorph *Botrytis*.

Botryotrichum Sacc. & Marchal (1885), Anamorphic Chaetomiaceae, 1.A2.?. Teleomorphs *Chaetomium, Farrowia. c.* 3, widespr. See Kushwaha & Agrawal (*Trans. mycol. Soc. Japan* **17**: 18, 1976), Hawksworth (*Persoonia* **8**: 167, 1975).

Botryoxylon Cif. (1962) = Conoplea (Mitosp. fungi) fide Hughes (1958).

Botryozyma Shann & M.T. Sm. (1992), Mitosporic fungi, 1.A1.1. 1, Italy.

Botrypes Preuss (1852) = Ciliciopodium (Mitosp. fungi) fide Saccardo (*Syll. Fung.* **4**: 577, 1886), nom. dub. fide Seifert (*TBMS* **88**: 123, 1985).

Botrysphaeris Clem. & Shear (1931) = Sphaeropsis (Mitosp. fungi) fide Sutton (*in litt.*).

Botrytis P. Micheli ex Pers. (1794), Anamorphic Sclerotiniaceae, 1.A1.6. Teleomorph *Botryotinia*. 50, cosmop. *B. cinerea* (the common grey mould, frequently parasitic); *B. allii* and other spp. (neck rot of onions); *B. paeoniae* (paeony blight); *B. tulipae* (tulip fire). See Coley-Smith *et al.* (*The biology of Botrytis,*

1981), Hennebert (*Persoonia* **7**: 185, 1973; segregates), Jarvis (*Can. Dept. Agric. Monogr.* **15**, 1977), Harrison (*Pl. Path.* **37**: 168, 1987; review of spp. on beans), Shirare *et al.* (*Phytopath.* **79**: 728, 1989; nuclei and mitotic chromosomes in *Botrytis* spp.), Lu & Wu (*Acta Mycol. Sin.* **10**: 27, 1991; *Botryotinia fabae* teleomorph of *Botrytis fabae*, chocolate spot), Verhoeff *et al.* (*Recent advances in Botrytis research*, 1992); Van der Vlugt-Bergmans *et al.* (*MR* **97**: 1193, 1993; genetic variation and DNA polymorphisms in *B. cinerea*), van Kan, Goverse & van der Vlugt-Bergmans (*Neth. J. Pl. Path.* **99** (Suppl. 3): 119, 1993; electrophoretic karyotypes in *B. cinerea*).

Botrytites Mesch. (1892), Fossil fungi. 1 (Oligocene), Eur.

Botrytoides M. Moore & F.P. Almeida (1937) ≡ Rhinocladiella (Mitosp. fungi) fide Schol-Schwarz (*Ant. v. Leeuwenhoek* **34**: 140, 1968).

Bottaria A. Massal. (1856) = Mycoporum (Mycopor.) fide Harris (*Evansia* **4**: 28, 1987).

Bottariomyces Cif. & Tomas. (1953) = Pyrenula (Pyrenul.) fide Harris (1984).

botuliform, cylindrical with rounded ends; sausage-like in form; see allantoid.

Boubovia Svrček (1977), Otideaceae. 1, Eur.

Boudier (Jean Louis Émile; 1828-1920). A pharmacist of Montmorency, France. Chief writings: *Histoire et classification des discomycètes d'Europe*, 1907 [reprint 1968]; *Icones mycologicae ou iconographie des champignons de France, principalement discomycètes*, 1905-10 [reprint 5 vols, 1981-1986]. Boudier's microscopic measurements are usually *c*. 10% too high (van Brummelen, *Persoonia* **5**: 233, 1969). See Mangin (*BSMF* **36**: 181, 1920), Lamy (*BSMF* **139**, 1984; correspondence), Grummann (1974: 268), Stafleu & Cowan (*TL-2* **1**: 290, 1976).

Boudiera Cooke (1877), Pezizaceae. 10, Eur., N. Am. See Hirsch (*Wissen. Z. Fried.-Schiller-Univ.* **32**: 1013, 1983).

Boudiera Lázaro Ibiza (1916) = Phellinus (Hymenochaet.) fide Donk (1974).

Boudierella Sacc. (1895), ? Otideaceae. 1, Eur. See Kimbrough & Korf (*Am. J. Bot.* **54**: 9, 1967).

Boudierella Costantin (1897) ≡ Delacroixia (Ancylist.).

bouillon, meat broth used as a culture medium.

Bourdot (Hubert; 1861-1937). From 1898 until his death, curé of Saint-Priest-en-Murat, Allier, France. Noted for his studies on the *Hymenomycetes*, esp. *Thelephoraceae* and resupinate fungi; see *Hymenomycètes de France* (with A. Galzin), 1928 [reprint 1969]. Collections in **PC**. *Obit.*: Gilbert (*BSMF* **55**: 137, 1939; portr., bibl.), Grummann (1974: 268), Stafleu & Cowan (*TL-2* **1**: 294, 1976), Stafleu & Mennega (*TL-2, Suppl.* **2**: 391, 1992).

Bourdotia (Bres.) Trotter (1925), Exidiaceae. 1, widespr.

bourrelet, see ascus.

Bouvetiella Øvstedal (1986), ? Hymeneliaceae (L). 1, Antarct. See Timdal (*in* Hawksworth (Ed.), *Ascomycete systematics*: 381, 1994; posn).

Bovetia Onofri & Persiani (1982), Mitosporic fungi, 2.A2.15/16. 1, Ivory Coast. = Sarophorum (Mitosp. fungi) fide Samson & Seifert (Eds) (*Advances in Penicillium and Aspergillus systematics*, 1985).

Bovicornua J. Koch & E.B.G. Jones (1993), Halosphaeriaceae. 1 (marine), Eur.

Bovilla Sacc. (1883) = Cercophora (Lasiosphaer.) fide Lundqvist (1972).

Bovista Pers. (1794), Lycoperdaceae. 46, widespr. See Kreisel (*Beih. Nova Hedw.* **25**, 1967; monogr.).

Bovistaria (Fr.) P. Karst. (1889) ≡ Langermannia (Lycoperd.).

Bovistella Morgan (1892), Lycoperdaceae. 4, N. temp. See Kreisel & Calonge (*Mycotaxon* **48**: 13, 1993).

Bovistina Long & Stouffer (1941) = Disciseda (Lycoperd.) fide Ponce (*Fieldiana, Bot.* **38**: 23, 1976).

Bovistoides Lloyd (1919) = Disciseda (Lycoperd.) fide Ponce (*Fieldiana, Bot.* **38**: 23, 1976).

Boydia A.L. Sm. (1919) = Vialaea (Vialaeac.) fide Redlin (*Sydowia* **41**: 296, 1989).

BPI. US National Fungus Collection (Beltsville, Md, USA); founded 1869, part of the United States Department of Agriculture's (USDA) Agricultural Research Service (ARS): see Cross *et al.* (*Systematic collections of the Agricultural Research Service*, 1977).

BR. National Botanic Garden of Belgium (Meise, Belgium); founded 1870; supported by the Ministry of Agriculture.

Brachiosphaera Nawawi (1976), Mitosporic fungi, 1.G1.19. 2 (aquatic), Malaysia, Jamaica.

brachy- (prefix), short.

Brachyascus Syd. (1917) ≡ Microdiscus (Leotiales).

Brachybasidiaceae Gäum. (1926), Exobasidiales. 3 gen., 3 spp. Erumpent from host stroma; probasidium persistent; metabasidium cylindrical, basisterigmate; spores ballistosporic.
    Lit.: Cunningham *et al.* (*Mycol.* **68**: 642, 1976; key), Donk (1951-63, VI; *Taxon* **3**: 243, 1964), Gäumann (1964: 358).

Brachybasidiales, see *Exobasidiales*.

Brachybasidium Gäum. (1922), Brachybasidiaceae. 1 (on *Palmae*), Java.

Brachycarphium Berk. (1849), Fossil fungi (Mitosp. fungi). 1 (Oligocene), Baltic.

Brachycladites Mesch. (1892) ≡ Brachycarphium (Fossil fungi).

Brachycladium Corda (1838) = Dendryphion (Mitosp. fungi) fide Saccardo (*Syll. Fung.* **4**: 489, 1886).

Brachycladium Berk. (1848) ≡ Brachycarphium (Fossil fungi).

Brachyconidiella R.F. Castañeda & W.B. Kendr. (1990), Mitosporic fungi, 2/3.G2.1. 1, Cuba.

brachycyclic, see *Uredinales*.

Brachydesmiella G. Arnaud ex S. Hughes (1961), Mitosporic fungi, 1.C2.26. 3, France. See Nicot (*BSMF* **86**: 705, 1971).

Brachydesmium (Sacc.) Costantin (1888) = Clasterosporium (Mitosp. fungi) fide Saccardo (*Syll. Fung.* **4**: 386, 1886).

brachyform, see *Uredinales*.

Brachyhelicoon Arnaud (1952), Mitosporic fungi, 1.F1.?. 1, Eur.

brachymeiosis (obsol.), a third division, once claimed to occur in the ascus.

Brachymyces G.L. Barron (1980), Helicocephalidaceae. 1 (from soil), Canada.

Brachysporiella Bat. (1952), Mitosporic fungi, 1.C2.1. 3, widespr. See Ellis (*Mycol. Pap.* **72**, 1959; key).

Brachysporiellina Subram. & Bhat (1989), Mitosporic fungi, 1.C2.10. 1, India.

Brachysporisporites R.T. Lange & P.H. Sm. (1971), Fossil fungi. 2 (Eocene), Australia, Canada.

Brachysporium Sacc. (1886), Mitosporic fungi, 1.C8.11. 9, cosmop. See Ellis (*DH*; key).

Branchiomyces Plehn (1912), ? Saprolegniales (inc. sed.). 2 (on fish), Eur., Asia, USA, Ukraine. See Peduzzi (*Mem. Ist. ital. Idrobiol.* **30**: 81, 1973), Neish & Hughes (*Diseases of Fish* **6**: 50, 1980).

brand, a leaf disease caused by a microscopic fungus, esp. a rust or smut (sometimes named the **-fungus**) (obsol.); **- spore**, urediniospore; smut spore.

brandy, see spirits.

Brasiliomyces Viégas (1944), Erysiphaceae. 3, N. & S. Am., Pacific, S. Afr. See Braun (*Beih. Nova Hedw.*

89, 1987; key), Hanlin & Tortolero (*Mycol.* **76**: 439, 1984; key), Zheng (*Mycotaxon* **19**: 281, 1984).

Brassia A. Massal. (1860) [non R. Br. (1813), *Orchidaceae*] = Thelotrema (Thelotremat.) fide Hale (1981).

Braunia Rick (1934) [non Bruch & Schimp. (1846), *Musci*] = Brauniella (Plut.).

Brauniella Rick ex Singer (1955), Pluteaceae. 1, S. Am. See Singer (*Proc. K. Nederl. Akad. Wet.* C **66**: 115, 1963).

Brauniellula A.H. Sm. & Singer (1959), Gomphidiaceae. 3, N. Am.

breathing pore, raised aperture in the upper cortex of *Parmelia exasperata* from the medulla to the exterior.

Brefeld (Julius Oscar; 1839-1925). Son of a Westphalian pharmacist. In 1876 he became Professor at the Forestry Academy, Eberswalde, then Professor of Botany at Münster, and later at Breslau. Responsible for more than 40 papers on mycology and plant diseases in addition to the great series *Botanische Untersuchungen über Schimmelpilze* (later *Untersuchungen aus dem Gesammtgebeit der Mykologie*), 15 parts, 1872-1912, in which he gave, with the help of beautiful figures, detailed accounts of his investigations. For some 25 years he had the use of only one eye and from about 1914 was blind.

Brefeld is noted for his development of pure culture methods in connexion with work on the life-histories and growth of fungi. From the first he saw how necessary it was for culture media and apparatus to be sterile and 10 years before the time of Koch he was using gelatin to make solid media. Working in turn with different groups of fungi, his writings are still a mine of details on spore germination, and the growth and development of fungi. See (*Nature* **116**: 369, 1925), Sopp (*Norske Vidensk-Akad. Årbok* **1925**: 83, 1926), Grummann (1974: 16), Stafleu & Cowan (*TL-2* **1**: 314, 1976), Stafleu & Mennega (*TL-2*, *Suppl.* **3**: 47, 1995).

Brefeldia Rostaf. (1873), Stemonitidaceae. 1, N. temp.

Brefeldiella Speg. (1889), Brefeldiellaceae. 2, S. Am., Australasia. See Müller & v. Arx (1962).

Brefeldiellaceae (Theiss.) E. Müll. & Arx (1962), Dothideales. 1 gen., 2 spp. Stromata strongly flattened, skin-like, often irregularly shaped, composed of irregular cells with pores in their walls; ascomata flattened, thin-walled, circular, opening by a lysigenous pore, not periphysate; interascal tissue of ? pseudoparaphyses; asci ± clavate, short-stalked, with a thick apical cap and a narrow ocular chamber, ? fissitunicate; ascospores hyaline, 1-septate, thinwalled, without a sheath. Anamorphs unknown. Biotrophic on leaves, neotropics.

Brefeldiellites Dilcher (1965), Fossil fungi. 2 (Cretaceous, Eocene), Argentina, USA.

Brefeldiopycnis Petr. & Cif. (1932), Mitosporic fungi. ?.A1.?. 1, W. Indies.

Bremia Regel (1843), Peronosporaceae. 2, N. temp. *B. lactucae* (lettuce downy mildew). See Ling & Tai (*TBMS* **28**: 16, 1945), Crute & Dixon (*in* Spencer (Ed.), 1981).

Bremiella G.W. Wilson (1914), Peronosporaceae. 3, Eur., N. Am. See Constantinescu (*TBMS* **72**: 510, 1979).

Brencklea Petr. (1923) = Scolecosporiella (Mitosp. fungi) fide Sutton (1968).

Brenesiella Syd. (1929), Ascomycota (inc. sed.). 1, C. Am.

Bresàdola (Giacopo; 1847-1929). An amateur mycologist of Trento, Italy. Chief writings: *Fungi Tridentini novi et nondum delineati, descripti et iconibus illustrati*, 1881-92; *Iconographia mycologica* [mostly *Agaricales*], 1927-60 (28 vols.) [reprint 1981-2].

Main collection in Stockholm (**S**). See Bresadola (*Icon. mycol.* **26**: v, 1933; portr.), Zalin & Lazzari (*Carteggio Bresadola-Saccardo*, 1987; corresp.), Grummann (1974: 431), Stafleu & Mannega (*TL-2*, *Suppl.* **3**: 66, 1995).

Bresadolella Höhn. (1903) = Neorehmia (Trichosphaer.) fide Rogerson (*Mycol.* **62**: 865, 1970).

Bresadolia Speg. (1883) = Polyporus (Polypor.) fide Donk (1974).

Bresadolina Brinkmann (1909), Stereaceae. 1, Eur.

Bresadolina Rick (1928), Ascomycota (inc. sed.). 1, Brazil.

Bretonia Bertrand & Hovel. (1892), Fossil fungi (mycel.). 1 (Permian), Eur.

Brettanomyces N.H. Claussen ex Custers (1940), Anamorphic Saccharomycetaceae. Teleomorph *Dekkera*. 9 (fide Kreger-van Rij, 1984), widespr. = Dekkera (Mitosp. fungi) fide Barnett *et al.* (1990).

Brevicellicium K.H. Larss. & Hjortstam (1978), Sistotremataceae. 7, widespr.

brevicollate, short-necked.

Brevilegnia Coker & Couch (1927), Saprolegniaceae. 13, widespr. See Johnson (*Mycol.* **69**: 287, 1977; 6 Scandinavian spp.).

Brevilegniella M.W. Dick (1961), Leptolegniellaceae. 1 (from soil), USA, UK.

Brewing. The process of beer making. Classical brewing comprises a number of stages: (1) **malting**, when water soaked barley grain is allowed to germinate and endogenous enzymes attack the starch and certain proteins of the grain; (2) grinding the malted grain to form **grist**; (3) **mashing**, mixing the grist with water when there is further enzyme action after which (4) the resulting liquid, **wort**, is boiled with female hop (*Humulus lupulus*) flowers and on cooling (5) fermented by *S. cerevisiae* or another yeast. See Briggs *et al.* (*Malting and brewing science*, 2 vols, 1971; edn 2, 1981), Corran (*A history of brewing*, 1975), Helbert (*in* Reed, *Prescott & Dunn's industrial microbiology*, edn 4, : 403, 1982), Polloch (Ed.) (*Brewing science* 2 vols, 1979, 1981), Priest & Campbell (Eds) (*Brewing microbiology*, 1987). See beer, Wine making, yeast.

Briania D.R. Reynolds (1989), Anamorphic Meliolinaceae, 1.A1.15. Teleomorph *Meliolina*. 1, Hawaii.

Briardia Sacc. (1885) = Duebenia (Dermat.) fide Nannfeldt (*Svensk bot. Tidskr.* **23**: 316, 1929).

Briarea Corda (1831) = Aspergillus (Mitosp. fungi) fide Hughes (1958).

Bricookea M.E. Barr (1982), Phaeosphaeriaceae. 1 (on *Juncus*), Eur., N. Am. See Shoemaker & Babcock (*Stud. Mycol.* **31**: 165, 1989).

bridging hypha, a branch hypha joining two other hyphae (Buller, *Researches* **4**: 152, 1931); **- species** or **- host**, a plant by which a specialized parasite went, in Marshall Ward's opinion (*Ann. Bot.* **15**: 560, 1902; *Ann. Myc.* **1**: 132, 1903; but see Bean *et al.*, *Ann. Bot.* N.S. **18**: 129, 1954), from a susceptible to a resistant host.

Brigantiaea Trevis. (1853), Brigantiaeaceae (L). 5, widespr. See Hafellner & Bellemère (*Nova Hedw.* **35**: 237, 1982, **38**: 169, 1983; conidia from ascospores).

Brigantiaeaceae Hafellner & Bellem. (1982), Lecanorales (L). 2 gen. (+ 2 syn.), 8 spp. Thallus crustose or lobed; ascomata apothecial, with well-developed margins; interascal tissue of long narrow anastomosing paraphyses, often with a well-developed epithecium; asci thick-walled, with a well-developed I+ apical cap, and I+ inner and outer ascus walls, ocular chamber hardly developed, 1- to 2-spored; ascospores thin-walled, hyaline, muriform, sometimes germinating to form conidia. Lichenized with green algae.

Brigantiella (Sacc.) Sacc. (1905) ? = Lophiostoma (Lophiostomat.).

Briosia Cavara (1888), Mitosporic fungi, 2.A2.38. 2, widespr. See Sutton (*Mycol. Pap.* **132**, 1973), Sigler & Carmichael (*Mycotaxon* **4**: 349, 1976).

Brochospora Kirschst. (1944) = Sporormia (Sprorm.) fide v. Arx & Müller (1975).

Brodoa Goward (1987), Parmeliaceae (L). 3, N. Hemisph. (arctic-alpine).

bromatia, the rounded swellings at the ends of hyphae of ant fungi (see Insects and fungi) which are used by the ants as food.

Bromicolla Eichw. (1843) = Sclerotium (Mitosp. fungi) fide Mussat (*Syll. Fung.* **15**, 1901), but also considered an alga (*Chlorophyceae* or *Palmellaceae*).

bronchomycosis, see mycosis.

Brooksia Hansf. (1956), Dothideales (inc. sed.). 1, trop. See Deighton & Pirozynski (*Mycol. Pap.* **105**, 1966). Anamorphs *Hiospira, Overeemia*.

broom cells (of agarics), cells bearing apical appendages to give a broom-like appearance on pileus or edge of lamella, as in *Marasmius rotula*; cellules en brosse (Singer, 1962: 62).

Broomeia Berk. (1844), Broomeiaceae. 2, Afr., S. Am.

Broomeiaceae Zeller (1948), Lycoperdales. 1 gen., 2 spp. Basidioma with well-defined peristomes (*Geastrum*-like) embedded in a common stroma.

Broomella Sacc. (1883), ? Amphisphaeriaceae. 6, Eur., Asia, N. Am. See Müller (*in* Yuan & Zhao, *Sydowia* **44**: 90, 1992; key). Anamorphs *Pestalotia* s.l., *Truncatella*.

Broomeola Kuntze (1891), Mitosporic fungi. 3.B1.?. 1, UK.

broth, a liquid nutrient culture medium, esp. one containing meat extract.

Brown (William; 1888-1975). Worked at the Imperial College of Science & Technology, Univ. London, 1912-53 (Professor of botany, 1938-53). Noted for researches on the physiology of parasitism and as a teacher. See Garrett (*Biogr. Mem. Fellows roy. Soc.* **21**: 155, 1975; portr., bibl.), Hawker (*TBMS* **65**: 343, 1975; portr.).

brown oak, oak wood stained by *Fistulina hepatica* (Cartwright, *TBMS* **21**: 68, 1937).

brown rot fungi, species of *Monilinia* causing fruit rots and other damage to fruit trees. See Wormald (*Tech. Bull. Ministr. Agric.* **3**, 1954), Byrde & Willetts (*The brown rot fungi of fruit*, 1977).

Brunaudia (Sacc.) Kuntze (1898) ? = Rhytidhysteron (Patellar.).

Brunchorstia Erikss. (1891), Anamorphic Leotiaceae, 8.C1.15. 1, Norway. Teleomorph *Gremmeniella*. *B. pinea* (teleomorph *Gremmeniella abietina*). See Punithalingam & Gibson (*CMI Descriptions* no. 369, 1973).

Bruneaudia, see *Brunaudia*.

Brunneospora Guarro & Punsola (1987), Onygenaceae. 1, Spain. Anamorph *Chrysosporium*.

Brunnipila Baral (1985) = Lachnum (Hyaloscyph.) fide Raitviir (*Eesti NSV Tead. Akad. Toim., Biol.* **36**: 313, 1987).

Brycekendrickia Nag Raj (1973), Mitosporic fungi, 8.A1.15. 1, India.

Bryocaulon Kärnefelt (1986), Parmeliaceae (L). 3, N. temp.

Bryochiton Döbbeler & Poelt (1978), ? Pseudoperisporiaceae. 4 (on *Musci*), Eur.

Bryochysium Link (1833) = Rhizoctonia (Mitosp. fungi) fide Rabenhorst (*Deutsch. Krypt. Fl.* **2** (index): 15, 1853).

Bryocladium Kunze (1830) = Pisomyxa (Eurotiales) fide Saccardo (*Syll. Fung.* **9**: 374, 1891).

Bryodiscus B. Hein, E. Müll. & Poelt (1971), Odontotremataceae. 3 (on *Musci*), Eur. See Döbbeler & Poelt (*Svensk bot. Tidskr.* **68**: 369, 1974; key).

Bryoglossum Redhead (1977), Geoglossaceae. 1, Eur., N. Am.

Bryomyces Döbbeler (1978) [non Miq. (1839), *Musci*], ? Pseudoperisporiaceae. 9 (on *Musci*), widespr.

Bryonora Poelt (1983), Lecanoraceae (L). 11, Eur., Greenland, Mongolia, Nepal. See Poelt & Obermayer (*Nova Hedw.* **53**: 1, 1991; key), Holtan-Hartwig (*Mycotaxon* **40**: 295, 1991).

Bryopelta Döbbeler & Poelt (1978), Dothideales (inc. sed.). 1 (on *Hepaticae*), Norway, Sweden.

Bryophagus Nitschke ex Arnold (1862), ? Gyalectaceae (L). 1, Eur. See Hawksworth et al. (*Lichenologist* **12**: 18, 1980; nomencl.).

Bryophilous fungi, fungi growing on *Bryophyta* (mosses and liverworts). A wide range of fungi are restricted to bryophyte colonies, for example, *Cyphellostereum laeve* (esp. on moss setae), *Eocronartium muscicola*, *Galerina hypnorum* growing on the gametophytes of mosses and hepatics have been recorded. They represent an array of taxa of quite different systematic position (e.g. *Dothideales, Hypocreales, Leotiales, Ostropales, Pezizales*), mode of nutrition and host selection. Several genera are unknown elsewhere (e.g. *Bryodiscus, Bryoscyphus, Bryosphaeria, Epibryon, Hypobryon, Octospora, Octosporella*). In other cases mainly non-bryophilous genera contain obligate parasites of mosses and hepatics, for example *Acrospermum adeanum, Dactylospora heimerlii, Muellerella frullaniae, Nectria egens*.

The bryophilous habit has evoked surprising adaptations, e.g. the formation of tiny, frequently gelatinous ascomata which generally prefer those parts of a plant which prevent too rapid loss of moisture and allow at the same time effective spore discharge (leaf axils, border of the ventral leaf side, or perforation of the leaves in hepatics, interspaces of the photosynthetic leaf lamellae in *Polytrichales*). Some species are necrotrophic (*Belonioscypha hypnorum, Bryostroma necans, Lizonia emperigonia, Nectria muscivora*) with intracellular hyphae, causing necrotic lesions. Necrotic rings in moss cushions in polar regions can be conspicuous (Longton, *Bull. Br. Antarct. Surv.* **32**: 41, 1973). Lesions in liverwort colonies are known, too (Hawksworth, *Fld Stud.* **4**: 391, 1976).

Most species represent biotrophic parasites which do not cause severe damage to their hosts. Hyphae of these species grow over the cell walls or within or between them. Endophytic VA mycorrhizal-like fungi regularly occur inside hepatics; most are sterile and part of a coevolved mutualism (Boullard, 1988; see mycothalli). *Octospora* infects the underground rhizoids of acrocarpic mosses with large appressoria and intracellular haustoria, sometimes causing conspicuous rhizoid galls. *Lizonia* is specific to antheridial cups of *Polytrichum*. Systematically different species colonizing the spaces betwen the leaf lamellae of *Polytrichales*, a phylogenetically ancient and stable microhabitat, offer striking examples of convergent evolution. Even heavy infections with hundreds of fruitbodies in a single leaf of *Dawsonia superba* do not induce visible symptoms. Many species have a restricted host range and are specific to a certain host species or group of hosts. The presence of bryophilous fungi in colonies of mosses or hepatics seems to be a very frequent, universal phenomenon. Some hosts apparently never occur without their parasites. These fungi are almost totally neglected despite their number and frequency. *Nectria phycophora* has algae in

special packets in the wall. *Vezdaea aestivalis* is ± lichenized but intimately associated also with moss leaves. Some lichens can overgrow and kill mosses (Faegri, *Lichenologist* **12**: 248, 1980).
Lit.: Barrio *et al.* (*Boln. Soc. micol. Castellana* **9**: 73, 1985; macromycetes, Spain), Döbbeler (*Mitt. bot. StSamml., Münch.* **14**: 1, 1978, pyrenocarps; *Nova Hedw.* **31**: 817, 1980, parasitic *Pezizales*; *Mitt. bot. StSamml., Münch.* **17**: 393, 1981, monogr., key 21 spp. on *Dawsonia*); *Arctic Alp. Mycol.* **2**: 87, 1987, ascomycetes on *Polytrichum sexangulare*), Duckett *et al.* (*New Phytol.* **118**: 233, 1991; ultrastr.), Felix (*Bot. Helv.* **98**: 239, 1988; review), Poelt (*Sydowia* **38**: 241, 1986; bryophilous lichens), Pocock & Duckett (*Bryol. Times* **31**: 2, 1985; review), Racovitza (*Mém. Mus. natn Hist. Nat., Paris*, n.s. B, **10**, 1959; monogr. including Mitosporic fungi and sporophytes as substrata).

Bryophytomyces Cif. (1953), Anamorphic Leotiaceae. Teleomorph *Hymenoscyphus*. 1, Eur., Russia. See Bauch (*Ber. dt. bot. Ges.* **56**: 73, 1938), Redhead & Spicer (*Mycol.* **73**: 940, 1981).

Bryopogon Link (1833) ≡ Alectoria (Alector.).

Bryopogon Th. Fr. (1860) = Bryoria (Parmel.) fide Brodo & Hawksworth (1977).

Bryorella Döbbeler (1978), Dothideales (inc. sed.). 9 (on *Musci*), widespr.

Bryoria Brodo & D. Hawksw. (1977), Parmeliaceae (L). 45, widespr., mainly boreal and cool temp. See Brodo & Hawksworth (*Op. bot. Soc. bot. Lund* **42**, 1977; key 27 N. Am. spp.).

Bryoscyphus Spooner (1984), Leotiaceae. 4 (on *Bryophyta*), Eur.

Bryosphaeria Döbbeler (1978), Dothideales (inc. sed.). 8 (on *Musci*), Eur.

Bryostigma Poelt & Döbbeler (1979) = Arthonia (Arthon.) fide Coppins (*Lichenologist* **21**: 195, 1989).

Bryostroma Döbbeler (1978), Dothideales (inc. sed.). 6 (on *Musci*), Eur.

B-spore, see beta-spore. Cf. alpha spore.

Bubacia Velen. (1922) nom. nud. (Thelephor.). See Donk (*Taxon* **6**: 17, 1957).

Bubakia Arthur (1906) = Phakopsora (Phakopsor.) fide Cummins & Hiratsuka (1983).

Buchholtzia Lohwag (1924) = Macowanites (Elasmomycet.) fide Singer & Smith (*Mem. Torrey bot. Cl.* **21**, 1960).

Buchwaldoboletus Pilát (1969) = Pulveroboletus (Bolet.) fide Singer (1961).

buck-eye rot, a disease of tomato fruits (*Phytophthora nicotianae* var. *parasitica*).

buckle, see clamp-connexion.

budding, a process of multiplication in 1-celled fungi or in spores, in which there is a development of a new cell from a small outgrowth; cf. fission.

Buellia De Not. (1846) nom. cons., Physciaceae (±L). *c.* 400, cosmop. See Hafellner (*Beih. Nova Hedw.* **62**, 1979; gen. concept), Schauer (*Mitt. bot. StSamml., Münch.* **5**: 609, 1965), Scheidegger (*Lichenologist* **25**: 315, 1993; key 36 spp. on rock, Eur., *Bot. Chron.* **10**: 211, 1991; distrib. Medit. spp.), Scheidegger & Ruef (*Nova Hedw.* **47**: 433, 1988; xanthone spp. Eur.), Sheard (*Lichenologist* **2**: 225, 1964; UK, *Bryologist* **72**: 220, 1969), Lamb (*Br. Antarct. Surv. Sci. Rep.* **61**, 1968).

Buelliastrum Zahlbr. (1930), Lecanorales (inc. sed., L). 2, China.

Buelliella Fink (1935), Dothideales (inc. sed.). 3 (on lichens), Eur., C. & N. Am. See Hafellner (*Beih. Nova Hedw.* **62**, 1979; key).

Buelliomyces E.A. Thomas ex Cif. & Tomas. (1953) ≡ Diploicia (Physc.).

Buelliopsis A. Schneid. (1897) = Buellia (Physc.).

Buergenerula Syd. (1936), Magnaporthaceae. 2, Eur., N. Am. See Barr (*Mycol.* **68**: 611, 1976). Anamorph *Pyricularia*-like.

Buglossoporus Kotl. & Pouzar (1966), Coriolaceae. 26, cosmop. See Corner (*Beih. Nova Hedw.* **78**: 1984).

Buglossus Wahlenb. (1826) = Fistulina (Fistulin.).

bulbil, a discrete, compact, multicellular, thalloidic propagule initiated in one of several ways but always homogeneous throughout development, with all cells acropetally produced and expanding more or less synchronously to many times (e.g. 4-10×) the diameter of the colourless, thin-walled hyphae from which they arise; pseudoparenchymatous at least at maturity, and lacking internal differentiation (found in certain basidiomycetes such as *Burgoa* and *Minimedusa*, distinguished from sclerotia; see Weresub & LeClair, *CJB* **49**: 2203, 1971); lichenized in *Multiclavula vernalis* (Poelt & Obermayer, *Herzogia* **8**: 289, 1990).

bulbillate (of a stipe), having a small or not clearly marked bulb at the base.

Bulbillomyces Jülich (1974), Hyphodermataceae. 1, Eur. Anamorph *Aegerita*.

bulbillosis, the condition, in *Agaricales*, in which basidiome sporulation is suppressed and the basidial function taken on by bulbils, as in *Rhacophyllus* (Singer, 1962: 27).

Bulbithecium Udagawa & T. Muroi (1990), Pseudeurotiaceae. 1, Peru. Anamorph *Acremonium*.

Bulbomicrosphaera A.Q. Wang (1987), Erysiphaceae. 1 (on *Magnolia*), China.

Bulbomollisia Graddon (1984), Dermateaceae. 3, UK.

Bulbopodium Earle (1909) = Cortinarius (Cortinar.) fide Singer (1975).

Bulborrhizina Kurok. (1994), Parmeliaceae (L). 1, Mozambique. See Eriksson & Hawksworth (*SA* **13**: 188, 1995).

Bulbothamnidium J. Klein (1870) = Helicostylum (Thamnid.) fide Upadhyay (1973).

Bulbothrix Hale (1974) Parmeliaceae. 32, mainly trop. See Hale (*Smithson. Contr. bot.* **32**, 1976; monogr.). ? = Parmelia (Parmel.)

Bulbouncinula R.Y. Zheng & G.Q. Chen (1979), Erysiphaceae. 1, China.

bulbous, (1) bulb-like; (2) (of a stipe), having a swelling at the base (Fig. 4G).

Bulgaria Fr. (1822), Leotiaceae. 1, N. temp. *B. inquinans*, common saprobe on bark of hardwoods after felling, but sometimes parasitic. See Gamundí & Arambarri (*Revta Fac. Agron. Univ. Nac. La Plata* **59**: 17, 1983), Pišut (*Annot. zool. bot. Slovenské narod. Múz.* **172**, 1986), Verkley (*Persoonia* **15**: 3, 1992). Anamorph *Endomelanconium*.

Bulgariastrum Syd. & P. Syd. (1913) ? = Dermea (Dermat.) fide Korf (*in* Ainsworth *et al.*, *The Fungi* **4A**: 249, 1973).

Bulgariella P. Karst. (1885), Leotiaceae. 1 or 2, Eur., S. Am.

Bulgariopsis Henn. (1902), Leotiaceae. 1 or 2, Brazil.

Bulla Battarra ex Earle (1909) = Agrocybe (Bolbit.) fide Singer (1975).

Bullardia, see *Bulliardia*.

Bullaria DC. (1805) = Puccinia (Puccin.) fide Arthur (1934).

Bullaserpens Bat., J.L. Bezerra & Cavalc. (1965), Mitosporic fungi, 5.E1.?. 1, Brazil.

bullate, (1) having bubble- or blister-like swellings; (2) (of a pileus), having a rounded projection at the centre (Fig. 19K).

Bullatina Vězda & Poelt (1987), Gomphillaceae (L). 1, trop.

Buller (Arthur Henry Reginald; 1874-1944). Professor of botany, Manitoba Univ., 1904-36. A most versatile mycologist, noted esp. for his researches on spore discharge and sexuality in the larger fungi. Chief writings: see Literature. His library is at the Research Station, Agriculture Canada, Winnipeg (Dowding, *Mycol.* **50**: 794, 1958; Oliver, *Catalogue of the Buller memorial library*, 1965). See Ainsworth (*Mycologist* **2**: 83, 1988; portr.), Bisby (*Nature* **154**: 173, 1944), Brodie & Lowe (*Science* **100**: 305, 1944), Brooks (*Obit. Not. roy. Soc.* **5**: 51, 1945; portr.), Stafleu & Cowan (*TL-2* **1**: 401, 1976), Stafleu & Mennega (*TL-2, Suppl.* **3**: 217, 1995).

Buller phenomenon. The dikaryotization, in basidiomycetes and ascomycetes, of a homokaryon by a dikaryon (Quintanilha's (1933) term for Buller's discovery; see Buller, 1941); 'di-mon' matings (Papazian, *Bot. Gaz.* **112**: 143, 1950).

Bullera Derx (1930), Anamorphic Tremellaceae, 1.A1.1/12/15. Teleomorph *Bulleromyces*. 13, N. Am., Eur. See Barnett *et al.* (1990; 15 spp.), Stadelmann (*Ant. v. Leeuwenhoek* **41**: 575, 1975; generic emend.), Ingold & Young (*TBMS* **76**: 165, 1981; spore discharge), Nakase *et al.* (*J. Gen. Appl. Microbiol.* **36**: 33, 1990; Japan), Boekhout (*Stud. Mycol.* **33**: 1, 1991; revision).

Bulleromyces Boekhout & A. Fonseca (1991), Tremellaceae. 1, N. temp. Anamorph *Bullera*.

Buller's drop. A drop of liquid on the hilar appendix of a ballistospore (q.v.) developed immediately before discharge, studied by Buller (*Researches* **2-6**, 1922-34); the drop contains mannitol and hexoses in concentrations enabling water to be taken in from a saturated atmosphere; discharge is effected by the Buller's drop expanding and contacting another expanding drop on tha adaxial spore surface (Webster *et al.*, *MR* **99**: 833, 1995).

Bulliard (Jean Baptiste François ('Pierre'); 1752-93). For an account of the life of this eminent French naturalist, the author of *Histoire des champignons de la France*, 1791-92 [index: Petersen, *Mycotaxon* **6**: 127, 1977], with its important coloured plates, see Gilbert (*BSMF* **68**: 1, 1952; portr.), who also gives a detailed bibliographical catalogue of Bulliard's publications and manuscripts and of writings about Bulliard. See Bolton (q.v.), Grummann (1974: 269), Stafleu & Cowan (*TL-2* **1**: 402, 1976).

Bulliardella (Sacc.) Paoli (1905) = Actidium (Mytilinid.). See Lohman (*Pap. Mich. Acad. Sci.* **23**: 155, 1938).

Bulliardia Jungh. (1830) nom. rej. = Melanogaster (Melanogastr.).

Bulliardia Lázaro Ibiza (1916) [non *Bulliarda* DC. (1801), *Crassulaceae*] ≡ Cerrena (Coriol.) fide Donk (1974).

Bunodea A. Massal. (1855) = Pyrenula (Pyrenul.).

Bunodophoron A. Massal. (1861), Sphaerophoraceae (L). 20, mainly S. temp. See Wedin (*Pl. Syst. Evol.* **187**: 213, 1993; 1994, key).

bunt, a wheat disease (*Tilletia caries* and *T. laevis*, syn. *T. foetida*); stinking smut; **dwarf -,** *T. controversa*. See also *Ustilaginales*.

Burcardia Schmidel (1797) [non *Burchardia* Schreb. (1789), *Liliaceae*] ≡ Bulgaria (Leot.).

Burenia M.S. Reddy & C.L. Kramer (1975), Protomycetaceae. 2, Eur., former USSR.

Burgoa Goid. (1938), Anamorphic Coriolaceae, 1.D1.1. Teleomorph *Cerrena*. 1 (propagules bulbils), Italy. See Weresub & LeClair (*CJB* **49**: 2203, 1971).

Burgundy mixture, made as Bordeaux mixture (q.v.) but with sodium carbonate (Na_2CO_3) in place of lime (copper sulphate 1.8 kg, sodium carbonate 2.3 kg, water 227 l).

Burrillia Setch. (1891), Tilletiaceae. 6 (on aquatic plants), C. Am., India. See Thirumalachar (*Mycol.* **34**: 602, 1947), Vánky (*Sydowia* **34**: 167, 1981).

bursiculate, bag-like.

bursiform, bag-like.

Bursulla Sorokīn (1876), Echinosteliopsidaceae. 1 (coprophilous), former USSR. See Olive (1975: 111).

Buscalionia Sambo (1940), ? Trypetheliaceae (L). 1, Brazil.

Busseëlla Henn. (1902) = Cephaleuros (Algae) fide Höhnel (*Sber. Akad. Wiss. Wien* **120**: 411, 1911).

Butler (Edwin John; 1874-1943). Imperial Mycologist (India), 1905-19; founding Director of the Imperial Mycological Institute, Kew, 1920-35. Chief writings: ... the genus *Pythium* ... (*Mem. Dep. Agric. India, Bot. Ser.* **5**, **1**, 1907), *Fungi and disease in plants*, 1918; (with Sydow) *Fungi Indiae orientales* (*Ann. Myc.*, 1906-16); (with Jones) *Plant pathology*, 1950. See (*Obit. Not. roy. Soc.* **4**: 455, 1943; portr., bibl.), Güssow (*Phytopath.* **34**: 149, 1944; portr.), Johnston (*Rev. Trop. Pl. Path.* **7**: 1, 1993), Kulkarni (*Pl. Path. Newsletter* **1**: 5, 1983), Stafleu & Mennega (*TL-2, Suppl.* **3**: 285, 1995).

Butlerelfia Weresub & Illman (1980), Atheliaceae. 1, Canada.

Butleria Sacc. (1914), Elsinoaceae. 1, India. See Petrak & Sydow (*Ann. Myc.* **27**: 87, 1929).

button, a young mushroom (esp. *Agaricus bisporus*) before the pileus has expanded.

Byliana Dippen. (1930) = Palawaniella (Parmular.) fide Müller & v. Arx (1962).

Byrrha Bat., F. Monnier & J.S. Silveira (1959) = Pichia (Saccharomycet.) fide Batra (1978).

Byrsalis Neck. ex Kremp. (1869) = Peltigera (Peltiger.) fide Santesson (*Taxon* **3**: 236, 1954).

Byrsomyces Cavalc. (1972), Ascomycota (inc. sed., L). 1, Brazil.

Byssiplaca A. Massal. (1860) = Lecanora (Lecanor.).

byssisede, see byssoid.

Byssitheca Bonord. (1864) ≡ Rosellinia (Xylar.).

Byssoascus Arx (1971), Myxotrichaceae. 1, Canada.

Byssocallis Syd. (1927), Tubeufiaceae. 2 (on *Meliola*), Costa Rica, S. Afr. See Rossman (*Mycotaxon* **8**: 485, 1979; 1987, key).

Byssocaulon Mont. (1835), Chrysotrichaceae (L). 6, trop.

Byssochlamys Westling (1909), Trichocomaceae. 4, widespr. *B. fulva* may damage tinned fruits (Hull, *Ann. appl. Biol.* **26**: 800, 1939). See Stolk & Samson (*Persoonia* **6**: 341, 1971), Subramanian & Rajendran (*Kavaka* **5**: 83, 1978; ontogeny). Anamorph *Paecilomyces*.

Byssocladiella Gaillon (1833) ≡ Byssocladium (Mitosp. fungi).

Byssocladium Link (1815) nom. dub. (Mitosp. fungi). See Donk (*Taxon* **11**: 81, 1962).

Byssocorticium Bondartsev & Singer (1944), Atheliaceae. 7, Eur. See Jülich (1972).

Byssocristella M.P. Christ. & J.E.B. Larsen (1970), Corticiaceae. 1, Eur.

Byssocystis Riess (1853) = Ampelomyces (Mitosp. fungi) fide Rogers (1959).

Byssogene Syd. (1922), Dothideales (inc. sed.). 1, Indonesia.

byssoid, cotton-like; made up of delicate threads; floccose. **- lichens** , see Egea *et al.* (*Lichenologist* **27**: 351, 1995; in *Arthoniales*), Hafellner & Vězda (*Nova Hedw.* **55**: 183, 1992; key 17 gen. thalli).

Byssolecania Vain. (1921), Pilocarpaceae (L). 2, Afr., C. & S. Am. See Santesson (*Symb. bot. upsal.* **12** (1), 1952; *SA* **10**: 137, 1991).

Byssoloma Trevis. (1853), Pilocarpaceae (L). 15, cosmop. See Kalb & Vězda (*Nova Hedw.* **51**: 435, 1990; key 11 spp. neotropics), Vězda (*Folia geobot. phytotax.* **22**: 71, 1987; key 10 Afr. spp.).

Byssolophis Clem. (1931), Cucurbitariaceae. 1, widespr. See Barr (*Mycotaxon* **45**: 191, 1992; posn), Müller & v. Arx (1962), Holm (*Windahlia* **16**: 49, 1986).

Byssomerulius Parmasto (1967), Meruliaceae. 9, widespr. See Parmasto (1968; emend.)

Byssonectria P. Karst. (1881) = Inermisia (Otid.). See Pfister (*Mycol.* **85**: 952, 1993; key 4 spp. N. Am.).

Byssoonygena Guarro, Punsola & Cano (1987), Onygenaceae. 1, Spain. Anamorph *Malbranchea*.

Byssopeltis Bat., J.L. Bezerra & T.T. Barros (1970), Microthyriaceae. 1, Brazil.

Byssophoropsis (Vain.) Tehler (1993), = Sagenidiopsis (Arthoniales, inc. sed.) fide Egea *et al.* (1995).

Byssophragmia M. Choisy (1931) = Megalospora (Megalospor.) fide Sipman (1986).

Byssophytum Mont. (1848), Ascomycota (inc. sed., L, sterile). 2, Java, Tahiti. See Groenhart (*Nederl. Kruid. Arch.* **46**: 774, 1936).

Byssoporia M.J. Larsen & Zak (1978), Atheliaceae. 1, USA.

Byssopsora A. Massal. (1861) = Bacidia (Bacid.).

Byssosphaeria Cooke (1879), Melanommataceae. *c.* 10, widespr. See Barr (*Mycotaxon* **20**: 1, 1984; key, Barr (*N. Am. Fl.* II **13**, 1990; N. Am.). = Herpotrichia (Melanommat.) fide v. Arx & Müller (*Sydowia* **37**: 6, 1984).

Byssostilbe Petch (1912), Clavicipitaceae. 3, trop.

Byssotheciella Petr. (1923), Ascomycota (inc. sed.). 1 or 2, Eur., S. Am.

Byssothecium Fuckel (1861), Dacampiaceae. 2, temp. See Boise (*Mycol.* **75**: 666, 1983), Semeniuk (*Mycol.* **75**: 744, 1983; on alfalfa). Anamorph *Chaetophoma*-like.

Byssus L. (1753) nom. rej. prop. ≡ Trentepohlia (*Algae*). Formerly used for some filamentous lichenized and other fungus mycelia.

C, (1) see Metabolic products. (2) Botanical Museum and Herbarium (Copenhagen, Denmark); founded 1759; part of the University of Copenhagen.

CABI. CAB INTERNATIONAL (formerly Commonwealth Agricultural Bureaux; founded 1929), an intergovernmental organization established by treaty lodged with the UN, and of which IMI (q.v.) is an Institute; see Scrivenor (*CAB - the first 50 years*, 1980).

Cacahualia Mercado & R.F. Castañeda (1984) = Arachnophora (Mitosp. fungi) fide Sutton (*in litt.*).

Caccobius Kimbr. (1967), Thelebolaceae. 1 (coprophilous), Canada.

Cacosphaeria Speg. (1888) = Kacosphaeria (Calosphaer.).

Cacumisporium Preuss (1851), Mitosporic fungi, 1.C2.19. 5, Eur., N. Am. See Goos (*Mycol.* **61**: 52, 1969), Sutton (*Mycol. Pap.* **132**, 1973), Kirk (*Mycotaxon* **43**: 231, 1992).

cadavericole, an organism living on corpses.

Cadophora Lagerb. & Melin (1927) = Phialophora (Mitosp. fungi) fide Conant (*Mycol.* **29**: 597, 1937).

caducous (of spores, etc.), falling off readily, deciduous.

Cadyexinis Stach (1957), Fossil fungi. 3 (Miocene, Carboniferous), Taiwan, Germany.

Caecomyces J.J. Gold (1988), Neocallimastigaceae. 2, UK, Canada. Wubah *et al.* (*Mycol.* **83**: 303, 1991; morphology).

Caenomyces E.W. Berry (1916), ? Fossil fungi (leaf spots). 7 (Tertiary), Brazil, USA.

Caenothyrium Theiss. & Syd. (1918) ? = Actinopeltis (Microthyr.). See Spooner & Kirk (*MR* **94**: 223, 1990).

caeoma (pl. **caeomata**), an aecium as in *Caeoma*, i.e. without peridial cells and with or without paraphyses.

Caeoma Link (1809), Anamorphic Uredinales. 15, widespr. A name for spp. having (0) and (I) only; aecium without peridial cells and with or without paraphyses.

caeomatoid (of aecia), resembling caeomata; sometimes, incorrectly, 'caeomoid'.

Caeomurus Gray (1821) nom. rej. = Uromyces (Puccin.).

Caerulicium Jülich (1982), Atheliaceae. 2, N. Am. Asia.

Caesar's mushroom, basidioma of the edible *Amanita caesarea*.

caespitose (cespitose), in groups or tufts like grass; cf. gregarious.

caespitulus (pl. **caespituli**), a tuft of spores.

Caesposus Nuesch (1937), Tricholomataceae. 1, Eur. See Cooke (1953; typific.).

Cainea S. Hughes (1951) nom. nud. = Apiosordaria (Lasiosphaer.).

Cainia Arx & E. Müll. (1955), Amphisphaeriaceae. 2, Eur., N. Am. See Krug (*Sydowia* **30**: 122, 1978), Parguey-Leduc (*Rev. Mycol.* **28**: 200, 1963; asci, affinities).

Cainiaceae, see Amphisphaeriaceae.

Cainiella E. Müll. (1957), Sordariaceae. 2, Canada, Eur. See Barr (*Mycotaxon* **39**: 43, 1990; posn).

Calathaspis I.M. Lamb & W.A. Weber (1972), Cladoniaceae (L). 1, Papua New Guinea.

Calathella D.A. Reid (1964) [non Florin (1929), *Algae*], Tricholomataceae. 2, Eur., N. Am.

Calathinus Quél. (1886) [non Rafin. (1836), *Amaryllidaceae*] = Pleurotellus (Crepidot.) fide Singer (1975).

Calathiscus Mont. (1841) ? = Lysurus (Clathr.) fide Dring (1980).

Calbovista Morse (1935), Lycoperdaceae. 1, USA.

calcarate, having a projection or spur.

calcareous, containing lime.

Calcarispora Marvanová & Marvan (1963), Mitosporic fungi, 1.E1.1. 1 (aquatic), former Czechoslovakia.

Calcarisporiella de Hoog (1974), Mitosporic fungi, 1.A1.10. 1, UK.

Calcarisporium Preuss (1851), Mitosporic fungi, 1.A1.10. 2 (on agarics, etc.), widespr. See Barnett (*Mycol.* **50**: 497, 1958), Barnett & Lilly (*Bull. W. Va. Exp. Stn* **420T**, 1958), Nicot (*BSMF* **84**: 85, 1968).

calceiform, shoe-like in form.

Calceispora Matsush. (1975), Mitosporic fungi, 1.A1.15. 2, Japan, Malawi. See Sutton (*Mycol. Pap.* **167**: 11, 1993).

calceolate, see calceiform.

Calceomyces Udagawa & S. Ueda (1988), Xylariaceae. 1, Japan.

calcicolous, an organism (**calcicole**) growing on substrates rich in calcium; esp. of spp. on limestone or chalky rocks or soils.

Caldariomyces Woron. (1926) = Leptoxyphium (Mitosp. fungi) fide Hughes (1976).

Caldesia Trevis. (1869) [non Parl. (1860), *Alismataceae*] = Arthonia (Arthon.).

Caldesia Rehm (1889) ≡ Holmiella (Patellar.).

Caldesiella Sacc. (1877) = Tomentella (Thelephor.) fide Larsen (*Taxon* **16**: 510, 1967). See also Nikolajeva (*Mikol. i Fitopatol.* **2**: 198, 1968).

Calenia Müll. Arg. (1890), Gomphillaceae (L). 12, trop. See Santesson (*Symb. bot. upsal.* **12**(1), 1952), Vězda & Poelt (*Folia geobot. phytotax.* **22**: 179, 1987).

Caleniomyces Cif. & Tomas. (1953) ≡ Calenia (Gomphill.).

Caleniopsis Vězda & Poelt (1987), Gomphillaceae (L). 1, trop.

Caleutypa Petr. (1934), Ascomycota (inc. sed.). 1, Eur.

Caliciaceae Chevall. (1826), Caliciales (±L). 9 gen. (+ 28 syn.), 51 spp. Thallus crustose where present; ascomata stalked; asci cylindrical, formed from croziers, thin-walled, evanescent; ascospores 0- to 1-septate, dark brown, with ornamentation formed from rupturing of the outer wall layers, forming a dry mazaedial mass. Almost all lichen-forming, rarely saprobic or lichenicolous.

Caliciales, Ascomycota (±L). 7 fam., 31 gen. (+ 61 syn.), 202 spp. Thallus crustose, often poorly-developed or absent; ascomata apothecial, often long-stalked; interascal tissue absent; asci ± cylindrical, thin-walled but rarely with a thickened apex, evanescent in most cases; ascospores usually septate and ornamented, accumulating in a mazaedial mass above the ascus layer. Most lichen-forming, some lichenicolous (often on other *Caliciales*), fungicolous, or saprobic on wood or bark; optimally developed in austral and boreal forests. Several genera are not certainly referred to families. Fams:
(1) **Caliciaceae** (syn. *Cypheliaceae, Tholurnaceae*).
(2) **Calycidiaceae**.
(3) **Coniocybaceae**.
(4) **Microcaliciaceae**.
(5) **Mycocaliciaceae**.
(6) **Sphaerophoraceae**.
(7) **Sphinctrinaceae**.
Lit.: Hutchinson (*Mycol.* **79**: 786, 1987; spp. on polypores), Middelborg & Mattsson (*Sommerfeltia* **5**, 1987; Norway), Nádvornik (*Studia bot. čsl.* **5**: 6, 1942), Pant & Awasthi (*Biovigyanam* **15**: 3, 1989; key 31 spp. India & Nepal), Puntillo (*Webbia* **43**: 145, 1989; key 59 spp. Italy), Santesson (*Ark. Bot.* **30A** (14), 1943; S. Am.), Sato (*Miscnea bryol. lich.* **7**: 39, 1975; gen. concepts), Schmidt (*Mitt. St. Inst. allg. Bot. Hamburg* **13**: 111, 1970), Tibell (*Symb. bot. upsal.* **21** (2), 1975; boreal N. Am., *Lichenologist* **13**: 161, 1981; Afr., **14**: 219, 1982; Costa Rica, *Beih. Nova Hedw.* **79**: 597, 1985; system, *Symb. bot. upsal.* **27**(2), 1987; keys 18 gen., 78 spp. Australia, *Bot. J. Linn. Soc.* **116**: 159, 1994; distrib. 162 spp., dispersal patterns), Tobolewski (*Pr. Kom. biol. Poznán* **24** (5), 1966; Poland), Vainio (*Acta Soc. Fauna Fl. fenn.* **57**(1), 1927; Finland).

Calicidium, see *Calycidium*.

Caliciella Vain. (1927) = Calicium (Calic.).

Caliciomyces E.A. Thomas ex Cif. & Tomas. (1953) = Calicium (Calic.).

Caliciopsis Peck (1880), Coryneliaceae. 25, widespr. See Benny *et al.* (*Bot. Gaz.* **146**: 437, 1985; key).

Calicium Pers. (1794), Caliciaceae (±L). 25, cosmop. See Keissler (1937), Nádvornik (1942), Tibell (1975; *Svensk bot. Tidskr.* **71**: 239, 1977; 1984, gen. concept), Tibell & Kalb (*Nova Hedw.* **55**: 11, 1992; 9 trop. Am. spp.), Tobolewski (1966).

Calidia Stirt. (1876) = Byssoloma (Pilocarp.) fide James (*Lichenologist* **5**: 175, 1971).

Calidion Syd. & P. Syd. (1919), Anamorphic Uredinales. 1 (on *Lindsaea*), Brazil; II only.

Californiomyces U. Braun (1981) = Brasiliomyces (Erysiph.) fide Zheng (1984).

Calkinsia Nieuwl. (1916) ≡ Pterygium (Placynth.).

Callebaea Bat. (1962), ? Capnodiaceae. 1, Uganda.

Callimastix Weissenb. (1912) , *Protista*, q.v.

Callimothallus Dilcher ex Janson. & Hills (1977), Fossil fungi (Microthyr.). 7 (Cretaceous, Tertiary), Australia, Colombia, India, USA, but see Hansen (*Grana* **19**: 67, 1980).

Calliospora Arthur (1905) = Uropyxis (Uropyxid.) fide Arthur (1934).

Callistodermatium Singer (1981), Tricholomataceae. 1, Brazil.

Callistospora Petr. (1955), Mitosporic fungi, 1.C2.19. 1, Australia. See Nag Raj (*CJB* **67**: 3169, 1989; redescr.).

Callistosporium Singer (1944), Tricholomataceae. 4, widespr. See Singer (*Flora neotrop.* **3**: 47, 1970; *Sydowia* **30**: 261, 1978, key).

Callolechia Kremp. (1869) ≡ Collolechia (Placynth.).

Callopis (Müll. Arg.) Gyeln. (1933) = Phyllopsora (Bacid.).

Callopisma De Not. (1847) nom. rej. ≡ Caloplaca (Teloschist.).

Calloria Fr. (1836), Dermateaceae. 3, Eur., N. Am. See Hein (*Willdenowia, Beih.* **9**, 1976).

Calloriella Höhn. (1918), Dermateaceae. 1, Eur.

Callorina Korf (1971) ≡ Calloria (Dermat.) fide Hein (1976).

Calloriopsis Syd. & P. Syd. (1917), Leotiaceae. 1 (on *Meliolaceae*), trop. See Pfister (*Mycotaxon* **4**: 340, 1976).

Callorites Fiore (1932), Fossil fungi. 1 (Eocene), Italy.

callose, hard or thick and sometimes rough.

Callosisperma Preuss (1855) nom. dub. fide Sutton (*Mycol. Pap.* **141**, 1977).

callosities (of fungi), wall thickenings associated with the penetration of fungicolous parasites (Swart, *TBMS* **64**: 511, 1975). See papillae.

Calocera (Fr.) Fr. (1828), Dacrymycetaceae. 11, widespr. See McNabb (*N.Z. Jl Bot.* **3**: 31, 1965; key), Kennedy (*CJB* **50**: 413, 1972; basidioma devel.), Reid (*TBMS* **62**: 437, 1974; key Br. spp.), Mathiesen (*Svampe* **25**: 35, 1992; key Danish spp.).

Caloceras Fr. ex Wallr. (1833) = Calocera (Dacrymycet.) fide Kennedy (*Mycol.* **50**: 884, 1958).

Calochaetis Syd. (1935) = Wentiomyces (Pseudoperispor.) fide Müller & v. Arx (1962).

Calocladia Lév. (1851) [non Grev. (1836), *Rhodophyta*] ≡ Microsphaera (Erysiph.).

Calocline Syd. (1939), Mitosporic fungi, 8.A1.15. 1, Ecuador.

Calocybe Kühner ex Donk (1962), Tricholomataceae. 24, widespr. *C. gambosum*, St. George's mushroom. See Singer (*Sydowia* **30**: 264, 1978; key).

Caloderma Petri (1900) = Scleroderma (Sclerodermat.) fide Guzmán (*Darwiniana* **16**: 233, 1970).

Calodon P. Karst. (1881) = Hydnellum (Thelephor.) fide Donk (1956).

Calogloeum Syd. (1924) = Fusamen (Mitosp. fungi) fide v. Arx (1957).

Calolepis Syd. (1925) = Pycnoderma (Cookell.) fide v. Arx (1963).

Calomyxa Nieuwl. (1916), Dianemataceae. 2, widespr.

Calonectria De Not. (1867), Hypocreaceae. *c.* 7, cosmop. See Rossman (*Mycotaxon* **8**: 321, 1979), Hansford (1946), Pirozynski (*Kew Bull.* **31**: 595, 1977), Rossman (*Mycotaxon* **8**: 485, 1979, excl. names; *Mycol. Pap.* **150**, 1983), Subramanian & Bhat (*Cryptogamie, Mycol.* **4**: 269, 1983). Anamorph *Cylindrocladium*.

Calonema Morgan (1893), Trichiaceae. 3, Eur., N. Am., ? Japan.

Calopactis Syd. & P. Syd. (1912) = Endothiella (Mitosp. fungi) fide Sutton (1977).

Calopadia Vězda (1986), Ectolechiaceae (L). 6 (on leaves), trop.

Calopeltis Syd. (1925) = Cyclotheca (Mycrothyr.) fide Müller & v. Arx (1962).

Calopeziza Syd. & P. Syd. (1913) = Dictyonella (Saccard.) fide v. Arx (1962).

Caloplaca Th. Fr. (1860) nom. cons., Teloschistaceae (L). *c.* 450, cosmop. See Alon & Galun (*Israel J. Bot.* **20**: 273, 1971; Israel), Arup (*Bryologist* **97**: 377, 1994; key 5 N. Am. maritime spp.), Egea (*Collect. bot.* **15**: 173, 1984; key 59 spp., Spain), Hafellner &

Poelt (*J. Hattori bot. Lab.* **46**: 1, 1979; key 17 polarilocular spp.), Hansen *et al.* (*Meddel. Grønl., Biosci.* **25**, 1987; key 43 spp. Greenland), Kärnefelt (*Monogr. Syst. Bot., Missouri Bot. Gdn* **25**: 439, 1988; S. Africa), Magnusson (*Bot. Notiser* 1944: 63, 1944, *Göteb. K. Vetensk.-o VittSamh. Handl.* ser. B **3**(1), 1944; *Bot. Notiser* 1950: 369, 1950), Malme (*Ark. Bot.* **20A** (9), 1926), Nimis (*Not. Soc. lich. Ital.* **5**: 9, 1992; key 10 spp. Italy); Nordin (*Caloplaca, sect. Gasparrinia i Nordeuropa*, 1972), Poelt (*Mitt. bot. StSamml., Münch.* **2**: 11, 1951), Ryan (*Bryologist* **92**: 513, 1989; monogr. sect. *Endochloris*), Poelt & Pelleter (*Pl. Syst. Evol.* **148**: 51, 1984; key, 10 fructescent spp.), Søchting (*Opera Bot.* **100**: 241, 1989; key 13 spp.), Wade (*Lichenologist* **3**: 1, 1965; UK), Wetmore (*Mycol.* **86**: 813, 1994; key 17 spp. dark apothecia N. & C. Am.), Wunder (*Bibl. Lich.* **3**, 1974; dark apothecia).

Caloplacaceae, see *Teloschistaceae*.

Caloplacomyces E.A. Thomas (1939) nom. inval. = Caloplaca (Teloschist.).

Caloplacopsis (Zahlbr.) de Lesd. (1932) = Candelariella (Candelar.).

Caloporia P. Karst. (1893) ≡ Caloporus (Steccherin.).

Caloporus P. Karst. (1881) = Merulius (Merul.) fide Donk (1963).

Caloporus Quél. (1886) = Polyporus (Polypor.) fide Donk (1960); = Albatrellus (Polypor.) fide Pegler (1973); ≡ Ovinus (Polypor.).

Calopposis Lloyd (1925) = Calocera (Dacrymycet.) fide McNabb (1965).

Caloscypha Boud. (1885), Otideaceae. 1, Eur. Anamorph *Geniculodendron*.

Calospeira G. Arnaud (1949) nom. inval. (? Myxomycota).

Calosphaeria Tul. & C. Tul. (1863), Calosphaeriaceae. 25, widespr. See Barr (1985).

Calosphaeriaceae Munk (1957), Calosphaeriales. 7 gen. (+ 7 syn.), 41 spp. Stroma varied in development, not clearly 2-layered; ascomata often aggregated, the ostioles usually converging; asci with an I- apical ring. Anamorphs hyphomycetous, *Acremonium*-like where known.

Calosphaeriales, Ascomycota. 2 fam., 8 gen. (+ 7 syn.), 42 spp. Stromatic tissues almost absent to well-developed, usually pseudostromatic; ascomata perithecial, immersed, often clustered, with separate or convergent ostioles; interascal tissue composed of a few elongate paraphyses; asci formed in fascicles or spicate clusters, croziers sometimes absent, sessile or long-stalked, sometimes polysporous; usually with an inconspicuous I- apical ring; ascospores hyaline or pale brown, ellipsoidal or allantoid, thin-walled. Saprobes on bark or wood, mainly temp. Fams:
(1) **Calosphaeriaceae**.
(2) **Graphostromataceae**.
Lit.: Barr (*Mycol.* **77**: 549, 1985; *Mycotaxon* **39**: 43, 1990), Rogers (*in* Hawksworth (Ed.), *Ascomycete systematics*: 321, 1994).

Calosphaeriopsis Petr. (1941), Ascomycota (inc. sed.). 1, Eur.

Calospora Nitschke ex Niessl (1875) = Macrodiaporthe (Melancon.) fide Barr (1978). See Holm (*Taxon* **24**: 475, 1978).

Calospora Sacc. (1883) = Prosthecium (Melanconid.) fide Barr (1978).

Calosporella J. Schröt. (1897) = Prosthecium (Melanconid.) fide Barr (1978).

Calostilbe Sacc. & P. Syd. (1902) = Nectria (Hypocr.) fide Rogerson (1970), Samuels (*CJB* **51**: 1275, 1973).

Calostilbella Höhn. (1919), Anamorphic Hypocreaceae, 2.C2.1. Teleomorph *Nectria*. 1, W. Afr., W. Indies.

See Mason (*Mycol. Pap.* **2**: 29, 1925), Hewings & Crane (*Mycotaxon* **20**: 245, 1984).

Calostoma Desv. (1809), Calostomataceae. 15, warm temp., trop., E. Am., Asia. See Boedijn (*Bull. Jard. bot. Buitenz.*, ser. 3 **16**: 64, 1938; Indonesia, key), Liu (*J. Shansi Univ., nat. sci. ed.* **1**: 109, 1979; world key [Chinese]).

Calostomataceae E. Fisch. (1900), Tulostomatales. 1 gen. (+ 3 syn.), 15 spp. Gleba compact with pleurosporous basidia in a cartilaginous endoperidium, usually seated on a stipe of agglomerated cartilaginous cords; exoperidium complex.

Calothricopsis, see *Calotrichopsis*.

Calothyriella Höhn. (1917) = Microthyrium (Microthyr.) fide Müller & v. Arx (1962).

Calothyriolum Speg. (1919) = Asterina (Asterin.) fide Müller & v. Arx (1962).

Calothyriopeltis F. Stevens & R.W. Ryan (1925) Fungi (inc. sed.). See Petrak (*Sydowia* **5**: 169, 1951).

Calothyriopsis Höhn. (1919), Microthyriaceae. 2, trop. Am.

Calothyris Clem. & Shear (1931) ≡ Calothyriopeltis (Fungi, inc. sed.).

Calothyrium Theiss. (1912) = Asterinella (Microthyr.) fide v. Arx & Müller (1975).

Calotrichopsis Vain. (1890), Lichinaceae (L). 3, S. Am. See Henssen (*Symb. bot. upsal.* **18**(1), 1963).

calvacin, a non-diffusible mucoprotein antibiotic from *Langermannia gigantea*; active against mouse, rat, and hamster tumours (Beneke, *Mycol.* **55**: 257, 1963).

Calvarula Zeller (1939), Protophallaceae. 1, Florida.

Calvatia Fr. (1849) nom. cons., Lycoperdaceae. 35, widespr. See Zeller & Smith (*Lloydia* **27**: 148, 1964; N. Am. spp., keys), Lange (*Blyttia* **51**: 141, 1993; infr. generic taxa), Kreisel (*Nova Hedw.* **48**: 241, 1989; key to segr. *Handkea*, not accepted here), Calonge & Martin (*Bol. Soc. Micol. Madrid* **14**: 181, 1990; gen. limits), Demoulin (*Mycotaxon* **46**: 77, 1993).

Calvatiella C.H. Chow (1936) = Bovistella (Lycoperd.) fide Kreisel & Calonge (*Mycotaxon* **48**: 13, 1993).

Calvatiopsis Hollós (1929), Lycoperdaceae. 1, Eur. Perhaps a 'monstrosity'.

calvescent, becoming bare or bald.

Calvocephalis Bainier (1882) = Syncephalis (Piptocephalid.) fide Benjamin (1959).

calvous, naked, bare.

Calycella (Fr.) Boud. (1885) ≡ Calycina (Leot.).

Calycella Quél. (1886) = Bisporella (Leot.) fide Korf & Carpenter (1974).

Calycella (Sacc.) Sacc. (1899), Leotiaceae. 1, Eur. See Korf & Carpenter (*Mycotaxon* **1**: 52, 1974).

Calycellina Höhn. (1918), Hyaloscyphaceae. c. 26, widespr. See Baral (*Beitr. Kenntnis Pilze Mitteleur.* **5**: 209, 1989; 4-spored spp.), Lowen & Dumont (*Mycol.* **76**: 1003, 1984; key).

Calycellinopsis W.Y. Zhuang (1990), Dermateaceae. 1, China.

Calycidiaceae Elenkin (1929), Caliciales (L). 1 gen. (+ 2 syn.), 1 sp. Thallus foliose; ascomata sessile, marginal; asci thin-walled, evanescent; ascospores spherical, aseptate, brown, forming a dry mazaedial mass.

Calycidiomyces Cif. & Tomas. (1953) ≡ Calycidium (Calycid.).

Calycidium Stirt. (1877), Calycidiaceae (L). 1 (on *Nothofagus*), NZ. See Sato (*Miscnea bryol. lichen., Nichinan* **4**: 150, 1968), Tibell (1984).

calyciform, cup-like.

Calycina Nees ex Gray (1821), Leotiaceae. c. 40, widespr. See Baral (*SA* **13**: 113, 1993).

Calycium DC. (1805) ≡ Calicium (Calic.).

calycular, cup-like.

Calyculosphaeria Fitzp. (1923) = Nitschkia (Nitschk.) fide Nannfeldt (1971). See Gaikwad (*Sydowia* **26**: 290, 1974).

calyculus, a cup- or calyx-like structure at the base of the sporangium in myxomycetes.

Calyptella Quél. (1886), Tricholomataceae. 11, cosmop. See Singer (*Sydowia* **30**: 270, 1978, key).

Calyptellopsis Svrček (1986), Hyaloscyphaceae. 1, former Czechoslovakia.

Calyptospora J.G. Kühn (1869) = Pucciniastrum (Pucciniastr.) fide Cummins & Hiratsuka (1983), but used by Hiratsuka *et al.* (1992).

calyptra, a cap or hood.

Calyptra Theiss. & Syd. (1918), Dothideales (inc. sed.). 2, Am.

Calyptralegnia Coker (1927), Saprolegniaceae. 2, N. Am., Eur.

Calyptromyces H. Karst. (1849) = Mucor (Mucor.) fide Hesseltine (1955). See Sumstine (*Mycol.* **2**: 125, 1910).

Calyptronectria Speg. (1909), Hypocreaceae. 1 or 2, S. Am.

Calyssosporium Corda (1831) nom. dub. (? Myxomycetes).

Camaroglobulus Speer (1986), Anamorphic Mytilinidiaceae, 8.A2.1. Teleomorph *Mytilinidion*. 1, Brazil.

Camarographium Bubák (1916), Mitosporic fungi, 8.D2.15. 3, Eur.

Camarophyllopsis Herink (1958), Hygrophoraceae. 5, N. temp. See Printz & Læssøe (*Svampe* **14**: 83, 1986; key).

Camarophyllus (Fr.) P. Kumm. (1871) = Hygrophorus (Hygrophor.) fide Saccardo (1887). See Singer (*Sydowia* **30**: 271, 1978; key 50 spp.).

Camarops P. Karst. (1873), Boliniaceae. 14, widespr. See Eriksson & Hawksworth (*SA* **7**: 64, 1988; posn), Nannfeldt (*Svensk bot. Tidskr.* **66**: 335, 1972), Pouzar (*Česká Myk.* **40**: 218, 1987; 4 sp. Czech.), Vassilyeva (*Mikol. i Fitopatl.* **22**: 388, 1988).

Camaropycnis E.K. Cash (1945), Mitosporic fungi, 8.A1.15. 1, USA.

Camarosporellum Tassi (1902), Mitosporic fungi, 4.D2.1. 3, Italy, Sweden, USA. See Sutton & Pollack (*Mycopath. Mycol. appl.* **52**: 331, 1974), van Warmelo & Sutton (*Mycol. Pap.* **145**, 1981).

Camarosporium Schulzer (1870), Anamorphic Cucurbitariaceae, 8.D2.1/19. Teleomorph *Cucurbitaria, Leptosphaeria, Pleospora.* 100, widespr. (esp. temp.).

Camarosporula Petr. (1954), Anamorphic Dothideales, 8.D2.1. Teleomorph *Anthracostroma*. 1, Australia.

Camarosporulum Tassi (1902), Mitosporic fungi, 4.D2.?. 1, widespr.

Camarotella Theiss. & Syd. (1915) = Coccodiella (Phyllachor.) fide Müller & v. Arx (1973).

Camillea Fr. (1849), Xylariaceae. 29, trop. See González & Rogers (*Mycotaxon* **47**: 229, 1993; key 14 spp., Mexico), Læssøe *et al.* (*MR* **93**: 121, 1989; key 27 spp.), Rogers *et al.* (*Mycol.* **83**: 274, 1991), Silveira & Rogers (*Acta Amazonica* Suppl. **15**: 7, 1987; Brazil). Anamorph *Xylocladium*.

Campanella Henn. (1895), Tricholomataceae. 15, mostly trop. See Singer (*Nova Hedw.* **26**: 847, 1976; key).

Campanularius Roussel (1806) ? = Panaeolus (Strophar.).

campanulate, bell-like in form (Fig. 37.25).

Campbellia Cooke & Massee (1890) [non Wight (1849), *Orobanchaceae*] ≡ Rodwaya (Gyrodont.).

campestroid, agarics having a pileus with a diam. : stipe ratio of 1 or >1. See Freeman (*Mycotaxon* **8**: 1, 1979). Cf. placomycetoid.

Campoa Speg. (1921), Parmulariaceae. 2, S. Am., Philipp.

Camposporidium Nawawi & Kuthub. (1988), Mitosporic fungi, 1.C2.19. 1 (aquatic), Malaysia.

Camposporium Harkn. (1884), Mitosporic fungi, 1.C2.10. 14, widespr. See Hughes (*Mycol. Pap.* **36**, 1951), Peek & Solheim (*Mycol.* **50**: 844, 1959), Watanabe (*Trans. mycol. Soc. Japan* **34**: 71, 1993; key 14 spp.).

Campsotrichum Ehrenb. (1819) = Myxotrichum (Myxotrich.) fide Hughes (1968).

Camptobasidium Marvanová & Suberkr. (1990), Platygloeaceae. 1, USA. Anamorph *Crucella*.

Camptomeris Syd. (1927), Mitosporic fungi, 3.C2.10. 8, trop. See Hughes (*Mycol. Pap.* **49**, 1952), Bessey (*Mycol.* **45**: 364, 1953; key).

Camptomyces Thaxt. (1894), Laboulbeniaceae. 8, Eur., Am., Asia.

Camptosphaeria Fuckel (1870), Lasiosphaeriaceae. 4, Eur., Asia, N. Am. See Krug & Jeng (*Sydowia* **29**: 71, 1977; key).

Camptosporium Link (1818) nom. dub. (Mitosp. fungi) fide Kirk (*in litt.*). = Menispora fide Hughes (*CJB* **36**: 744, 1958).

Camptoum Link (1824) ≡ Arthrinium (Mitosp. fungi) fide Hughes (1958).

Campylacia A. Massal. ex Beltr. (1858) = Leptorhaphis (Arthopyren.).

campylidium, helmet-shaped conidiomata occurring in various, mainly foliicolous, tropical lichenized genera (e.g. *Badimia, Loflammia, Sporopodium*); the name *Pyrenotrichum* (Mitosp. fungi) has been applied to many of these conidiomata. See Sérusiaux (*Lichenologist* **18**: 1, 1986).

Campylobasidium Lagerh. ex F. Ludw. (1892) nom. rej. = Septobasidium (Septobasid.).

Campylospora Ranzoni (1953), Mitosporic fungi, 1.G1.1. 2 (aquatic), N. temp.

Campylostylus Genev. (1873) nom. dub. (? Fungi).

Campylothecium Ces. (1846) = Cordyceps (Clavicipit.) fide Tulasne & Tulasne (*Sel. Fung. Carp.* **3**: 18, 1865).

Campylothelium Müll. Arg. (1883), Trypetheliaceae (L). 9, trop. See Tucker & Harris (*Bryologist* **83**: 1, 1980).

Canadian tuckohoe, see stone-fungus.

canal, sometimes applied to the pore connecting the two cells of a polarilocular spore.

canaliculate, having longitudinal grooves (Fig. 29.16).

Canalisporium Nawawi & Kuthub. (1989), Mitosporic fungi, 3.D2.1. 3, widespr.

Canariomyces Arx (1984), Microascaceae. 1, Canary Isls. See v. Arx *et al.* (*Beih. Nova Hedw.* **94**, 1988; posn).

Cancellaria Brongn. (1825) = Roestelia (Anamorphic Uredinales); aecial states of *Gymnosporangium*.

cancellate, reticulate; like a network, as the basidioma of *Clathrus*.

Cancellidium Tubaki (1975), Mitosporic fungi, 1.D2.1. 1 (aero-aquatic, conidia hollow), Japan.

Canceromyces Niessen [not traced] nom. dub. based on a mould from a cancer.

Candelabrella Rifai & R.C. Cooke (1966), Mitosporic fungi, 1.B/C1.10. 6 (nematode trapping), widespr. See Cooke (*TBMS* **53**: 475, 1969). = Arthrobotrys (Mitosp. fungi) fide Schenck *et al.* (*CJB* **55**: 977, 1977).

Candelabrochaete Boidin (1970), Botryobasidiaceae. 7, widespr.

Candelabrum Beverw. (1951), Mitosporic fungi, 1.G1.1. 1 (aero-aquatic), Eur.

Candelaria A. Massal. (1852), Candelariaceae (L). 10, cosmop. See Hillmann (*Rabenh. Krypt.-Fl.* **9**, 5(3): 19, 1936), Poelt (*Phyton. Horn* **16**: 189, 1974).

Candelariaceae Hakul. (1954), Lecanorales (L). 5 gen. (+ 7 syn.), 61 spp. Thallus crustose, fruticose or foliose, yellow-green to orange; ascomata apothecial, with or without a clear margin, usually orange or yellow, the ascomatal wall formed from closely septate twisted hyphae; interascal tissue of simple paraphyses; asci with a thick non-amyloid cap above a blue-staining region with a central less strongly stained cushion, ocular chamber ± absent, often multispored; ascospores small, hyaline, usually aseptate. Conidiomata pycnidial. Lichenized with green algae, usually on nitrogen-rich substrata, polar and cold temperate.

Candelariella Müll. Arg. (1894), Candelariaceae (L). *c.* 45, cosmop. See Castello & Nimis (*Acta bot. fenn.* **150**: 5, 1994; Antarctic spp.), Gilbert *et al.* (*Lichenologist* **13**: 249, 1981; citrine spp.), Hakulinen (*Annls bot. Soc. zool.-bot. fenn. Vanamo* **27** (3), 1954; monogr.), Laundon (*Lichenologist* **4**: 297, 1970; UK), Poelt & Reddi (*Ergebn. ForschUnternehmens Nepal Himalaya* **6**: 1, 1969), Harris & Buck (*Mich. Bot.* **17**: 155, 1978; N. Am.).

Candelariellomyces E.A. Thomas ex Cif. & Tomas. (1953) ≡ Candelariella (Candelar.).

Candelariellopsis Werner (1936) = Candelariella (Candelar.).

Candelariopsis (Sambo) Szatala (1959) nom. inval. ? = Caloplaca (Teloschist.).

Candelina Poelt (1974), Candelariaceae (L). 3, C., N. & S. Am., Afr.

Candelospora Rea & Hawley (1912) = Cylindrocladium (Mitosp. fungi) fide Boedijn (*in litt.*).

Candelosynnema K.D. Hyde & Seifert (1992), Mitosporic fungi, 2.C1.10. 1, Australia.

candicidin, an antibiotic from the actinomycete *Streptomyces griseus*; antibacterial and antifungal (esp. against *Candida albicans*); Lechevalier *et al.* (*Mycol.* **45**: 155, 1953), Kligman & Lewis (*Proc. Soc. exper. Biol. Med.* **82**: 399, 1953).

Candida Berkhout (1923) nom. cons., Anamorphic Saccharomycetales, 1.A1.?. 165 (fide Barnett *et al.*, 1990), pseudomycelium or mycelium present, cosmop. *C. albicans* (candidiasis, q.v.) and other spp. are pathogenic for humans and animals; *C. utilis*, food yeast.

Lit.: Systematic: Lodder (1970: 900; key), Skinner (*Bact. Rev.* **11**: 227, 1947), Ramirez (*Microbiol. Exp.* **27**: 15, 1974; species compilation), Davis (*Mycopath.* **96**: 171, 1986; description), Weijman *et al.* (*Ant. v. Leeuwenhoek* **54**: 545, 1988; redefinition of genus), Kamiyama *et al.* (*J. Med. Vet. Myc.* **27**: 229, 1989; Adansonian taxonomy of spp.), Merson-Davies & Odds (*J. Gen. Microb.* **135**: 3143, 1989; morphology index), Odds *et al.* (*Jl Gen. Microbiol.* **136**: 761, 1990; no basis for separation from *Torulopsis*), Viljoen & Kock (*Syst. Appl. Microbiol.* **12**: 183, 1989; history of delimitation), Yarrow & Meyer (*Int. J. Syst. Bact.* **28**: 611, 1978; generic emend.), Lacher & Lehmann (*Ann. Clin. Lab. Sci.* **21**: 94, 1991; numerical taxonomy).

Medical: Mendling (*Vulvo-vaginal Candidosis*, 1988), Shepherd (*Oral Candidosis* (Ed. Samaranayake & MacFarlane):10, 1990; biology), Hendriks *et al.* (*J. Gen. microbiol.* **137**: 1223, 1991; phylogeny of medical spp.), Barns (*J. Bact.* **173**: 2250, 1991; evolutionary relationships among pathogenic spp.), Jensen, Hau, Aalbaek & Schonheyder (*Mycoses* **33**: 519, 1990; crossed immunoelectrophoresis to differentiate *C. albicans*), Samaranayake & MacFarlane (Ed.) (*Oral Candidosis*, 1990), Samaranayake & Yaacob (*Oral Candidosis*: 124, 1990; classification of oral candidosis).

Non-Medical: Skinner & Fletcher (*Bact. Rev.* **24**: 397, 1960; non-medical aspects), Shepherd *et al.* (*Ann. Rev. Microbiol.* **39**: 579, 1985; biology, genetics, pathogenicity).

Techniques: Gunasekaran & Hughes (*Mycol.* **72**: 505, 1980; gas-liquid chromatography), Gabriel-Bruneau & Guinet (*Int. J. Syst. Bact.* **34**: 227, 1984; antigenic relatedness), Bruneau *et al.* (*System. Appl. Microbiol.* **6**: 210, 1984; antigenic specificity of enzymes), Srivastava *et al.* (*Microb. Ecol.* **11**: 71, 1985; differentiation of biotypes), Belov & Kamanev (*Mikrobiol.* **55**: 473, 1986; wall components and morphology), Magee *et al.* (*J. Bact.* **169**: 1639, 1987; RFLP in rDNA), Lehmann *et al.* (*TBMS* **88**: 199, 1987; killer fungi characterize species and biotypes), Paulovicova & Sandula (*Biol. (Bratisl.)* **41**: 759, 1986; immunoanalysis of wall mannans), Farid *et al.* (*Assiut Vet. Med. J.* **16**: 355, 1986; identification by fatty acid content), Montrocher & Claisse (*Cell & Mol. Biol.* **33**: 313, 1987; spectrophotometric analysis), Weijman *et al.* (*Ant. v. Leeuwenhoek* **54**: 535, 1988; carbohydrate patterns), Viljoen *et al.* (*Syst. Appl. Microb.* **10**: 116, 1988; long chain fatty acids), Hamijima *et al.* (*Microbiol. Immunol.* **32**: 1013, 1988; DNA analysis of *C. tropicalis*), Kamiyama *et al.* (*Mycopath.* **107**: 3, 1989; DNA homology between strains), Su & Meyer (*Yeast* (special issue) **5**: S355, 1989; restriction endonuclease analysis of DNA), Montrocher *et al.* (*Yeast* (special issue) **5**: S385, 1989; biochemical analysis), Rustchenko-Bulgac *et al.* (*J. Bact.* **172**: 1276, 1990; genetic variation in *C. albicans*), Odds (*Bull. Soc. Fr. Myc. Med.* **19**: 5, 1990; molecular biology), Boiron (*Bull. Soc. Fr. Myc. Med.* **19**: 13, 1990; electrophoretic karyotypes), Hendriks *et al.* (*Syst. Appl. Microbiol.* **12**: 223, 1989; nucleotide sequence of *C.albicans*), Scherer & Magee (*Microbiol. Rev.* **54**: 226, 1990; genetics of *C. albicans*), Bezjak *et al.* (*Jl Med. Vet. Mycol.* **28**: 267, 1990; cell constituent polyamines), Smit *et al.* (*Jl Med. Vet. Mycol.* **28**: 303, 1990; QBASIC programme for whole cell protein profiles), Sadhu *et al.* (*J. Bact.* **173**: 842, 1991; telomeric and dispersed repeat sequences in strain identification), Su & Meyer (*Int. Jl Syst. Bact.* **41**: 6, 1991; characterization of mitochondrial DNA), Calderone & Braun (*Microbiol. Rev.* **55**: 1, 1991; adherence and receptor relationships on *C. albicans*), Kurtzman (*Ant. v. Leeuwenhoek* **57**: 215, 1990; diversity in *C.shehatae*), Marumo & Aoki (*Jl Clin. Microbiol.* **28**: 1509, 1990; cellular fatty acids), Kirsch *et al.* (Eds) (*The genetics of Candida*, 1990), Iwaguchi *et al.* (*J. Gen. Microbiol.* **136**: 2433, 1990; karyotypes).

candidiasis, a cosmop. disease of humans (including thrush, mouget, etc.) and animals caused by species of *Candida* (syn. *Monilia* auct.), esp. *C. albicans*; moniliasis; candidosis. See Winner & Hurley (*Candida albicans*, 1964; (Eds), *Symposium on Candida infections*, 1966), Odds (*Candida and candidosis*, edn. 2, 1988), Turnbay *et al.* (Eds) (*Candida and candidamycosis*, 1991).

candidosis, see candidiasis.

candle-snuff fungus, stromata of *Xylaria hypoxylon*.

canescent, becoming hoary or grey.

caninoid venation, see veins.

canker, a plant disease in which there is sharply-limited necrosis of the cortical tissue, e.g. **apple canker** (*Nectria galligena*).

CANL. Lichen Herbarium, Canadian Museum of Nature (Ottawa, Canada); founded 1882; a government corporation.

Cannanorosporonites Ramanujam & Rao (1979), Fossil fungi (*f. cat.*). 1 (Miocene), India.
Canomaculina Elix & Hale (1987) Parmeliaceae. 6, S. Am., S. Afr. ? = Parmelia (Parmel.).
Canoparmelia Elix & Hale (1986) Parmeliaceae. *c.* 33, N. & S. Am., Afr. ? = Parmelia (Parmel.)
Canteria Karling (1971), ? Endochytriaceae. 1, UK.
Canteriomyces Sparrow (1960), Hyphochytriaceae. 1 (on freshwater algae), Eur., N. Am.
Cantharellaceae J. Schröt. (1888), Cantharellales. 4 gen. (+ 4 syn.), 76 spp. Basidioma, pileate, agaricoid; hymenophore with radial folds and interveining; thickening hymenium.
Cantharellales, Basidiomycetes. 12 fam., 52 gen. (+ 50 syn.), 677 spp. Basidioma either funnel-shaped or tubular or stalked and pileate, monomitic, the hymenophore smooth, wrinkled, or folded to form thick gill-like structures; spores smooth, hyaline, non-amyloid; terrestrial, humicolous. Fams:
(1) **Aphelariaceae**.
(2) **Cantharellaceae**.
(3) **Clavariaceae**.
(4) **Clavariadelphaceae**.
(5) **Clavulinaceae**.
(6) **Craterellaceae**.
(7) **Hydnaceae**.
(8) **Physalacriaceae**.
(9) **Pterulaceae**.
(10) **Scutigeraceae**.
(11) **Sparassidaceae**.
(12) **Typhulaceae**.
Lit.: Donk (1964: 247), Corner (*A monograph of the cantharelloid fungi* [*Ann. Bot. Mem.* 2], 1966; *New Phytol.* 67: 219, 1968; *Nova Hedw.* 27: 325, 1976), Bigelow (*Mycol.* 70: 707, 1978; New England spp.).
Cantharellopsis (Weinm.) Kuyper (1986), Tricholomataceae. 1, Eur.
Cantharellula Singer (1936), Tricholomataceae. 2, temp.
Cantharellus Fr. (1821), Cantharellaceae. 70, widespr. *C. cibarius*, the edible chanterelle. See Smith & Morse (*Mycol.* 39: 497, 1947), Corner (*Sydowia Beih.* 1, 1957), Heinemann (*Fl. Icon. Champ. Congo* 8: 154, 1959, keys 17 spp. Congo).
Cantharocybe H.E. Bigelow & A.H. Sm. (1973), Tricholomataceae. 1, temp.
Cantharomyces Thaxt. (1890), Laboulbeniaceae. 25, widespr.
Cantharosphaeria Thaxt. (1920) = Eriosphaeria (Trichosphaer.) fide Müller & v. Arx (1962).
cap, see pileus.
capillaceous, see capilliform.
Capillaria Pers. (1822) [non Roussel (1806), *Algae*], Mitosporic fungi, 1.?.?. 4, temp.
Capillaria Velen. (1947) [non Roussel (1806), *Algae*] = Lycoperdon (Lycoperd.) fide Pegler (*in litt.*).
Capillataspora K.D. Hyde (1989), Dothideales (inc. sed.). 1, Brunei.
capilliconidium, a secondary conidium produced on a long capillary tube in *Entomophthorales*.
capilliform, hair-like; thread-like; capillaceous.
Capillipes R. Sant. (1956), Leotiaceae. 1, Lapland.
capillitium (of myxomycetes and gasteromycetes), a mass of sterile, thread-like elements, tubes or fibres among the spores.
Capillus Granata (1908) = Enterobryus (Eccrin.) fide Manier & Lichtwardt (*Ann. Sci. nat., Bot., sér.* 12, 9: 519, 1968).
capitate, having a well-formed head (Figs 29.9, 37.18).
capitate-fastigiate (of macrolichens), having a thallus cortex of erect, parallel hyphae terminated by swollen and pigmented apical cells.
capitellum, a little head.

Capitoclavaria Lloyd (1922) ? = Clavaria (Clavar.).
Capitorostrum Bat. (1957), Mitosporic fungi, 4.A1.15. 2, Australia, Papua New Guinea. See Hyde & Philemon (*Mycotaxon* 42: 95, 1991).
Capitotricha (Raitv.) Baral (1985) = Lachnum (Hyaloscyph.) fide Raitviir (*Eesti NSV Tead. Akad. Toim., Biol.* 36: 313, 1987).
Capitularia Flörke (1807) = Cladonia (Cladon.).
Capitularia Rabenh. (1851) = Uromyces (Puccin.) fide Dietel (1928).
capitulum, a stalked globose apical apothecium, as in the *Caliciales*. Cf. mazaedium.
Capnia Vent. (1799) = Dermatocarpon (Verrucar.) p.p. and Umbilicaria (Umbilicar.) p.p.
Capniomyces S.W. Peterson & Lichtw. (1983), Legeriomycetaceae. 1, USA. See Lichtwardt (1986).
Capnites Theiss. (1916) [non (DC.) Dumort. (1827), *Papaveraceae*] = Phaeosaccardinula (Chaetothyr.).
Capnobatista Cif. & F.B. Leal ex Bat. & Cif. (1963) = Trichomerium (Capnod.).
Capnobotryella Sugiy. (1987), Mitosporic fungi, 1.B2.1/12. 2, Japan. See Titze & de Hoog (*Ant. v. Leeuwenhoek* 58: 265, 1990; *C. renispora* on roof tile).
Capnobotrys S. Hughes (1970), Anamorphic Metacapnodiaceae, 1.C2.10. Teleomorph *Metacapnodium*. 1, Austria.
Capnocheirides J.L. Crane & S. Hughes (1982), Mitosporic fungi, 1.A/B/C2.38. 1, Eur.
Capnociferria Bat. (1963) = Antennulariella (Antennulariell.) fide Hughes (1976).
Capnocrinum Bat. & Cif. (1963) = Antennulariella (Antennulariell.) fide Hughes (1976).
Capnocybe S. Hughes (1966), Anamorphic Metacapnodiaceae, 1.C2.10. Teleomorphs *Ophiocapnocoma*, *Limacinia*. 3, NZ, USA.
Capnodaria (Sacc.) Theiss. & Syd. (1918), Capnodiaceae. 1, Eur.
Capnodendron S. Hughes (1976), Anamorphic Antennulariaceae, 1.C2.3. Teleomorph *Antennulariella*. 1, Brazil.
Capnodenia (Sacc.) Theiss. & Syd. (1917) = Capnodium (Capnod.) fide v. Arx & Müller (1975).
Capnodiaceae (Sacc.) Höhn. ex Theiss. (1916), Dothideales. 13 gen. (+ 19 syn.), 57 spp. Mycelium superficial, well-developed, dark, composed of ± cylindrical hyphae with mucous coating; ascomata small, sometimes vertically elongated, thin-walled, covered in a mucous layer, sometimes setose, usually with a clearly-defined ostiole; interascal tissue absent; asci saccate, fissitunicate; ascospores brown, septate, sometimes muriform. Anamorphs pycnidial, elongate, sometimes stipitate. Saprobic, usually on insect exudates on leaves and branches.
Lit.: Batista & Ciferri (*Saccardoa* 2, 1963), Hughes (*Mycol.* 68: 693, 1976; gen. names, anamorphs), Reynolds (*Taxon* 20: 759, 1971, hyphal morph.; *Nova Hedw.* 26: 179, 1975, growth forms; *Mycotaxon* 8: 417, 1979, stalked taxa; *Mycotaxon* 27: 377, 1988, cladistics; *in* Sugiyama (Ed.), *Pleomorphic fungi*: 157, 1987).
Capnodiales, see *Dothideales*; sooty moulds.
Capnodiastrum Speg. (1886), Mitosporic fungi, 4.B2.1. 5, esp. S. Am.
Capnodiella (Sacc.) Sacc. (1905) ≡ Sorica (Corynel.).
Capnodina (Sacc.) Sacc. (1926) = Antennulariella (Antennulariell.).
Capnodinula Speg. (1918) = Wentiomyces (Pseudoperispor.) fide v. Arx & Müller (1975).
Capnodinula Bat. & Cif. (1963), Dothideales (inc. sed.). 1, Australia.
Capnodiopsis Henn. (1902) = Molleriella (Elsin.) fide v. Arx (1963).

Capnodium Mont. (1848), Capnodiaceae. *c.* 15, widespr. See Reynolds (*Bull. Torrey bot. Cl.* **97**: 253, 1970).

Capnogoniella Bat. & Cif. (1963) nom. conf. fide Sutton (*Mycol. Pap.* **141**, 1977).

Capnogonium Bat. & Peres (1961) = Brooksia (Dothideales) fide Deighton & Pirozynski (1966).

Capnokyma S. Hughes (1975), Mitosporic fungi, 1.C2.1. 1, NZ.

Capnophaeum Speg. (1918), ? Capnodiaceae. 2 or 3, Asia.

Capnophialophora S. Hughes (1966), Mitosporic fungi, 1.A1.15. 1, NZ.

Capnosporium S. Hughes (1976), Anamorphic Metacapnodiaceae, 1.B/C2.28. Teleomorph *Metacapnodium*. 1, NZ.

Capnostysanus Speg. (1918) = Stysanus (Mitosp. fungi) fide Clements & Shear (1931).

Cappellettia Tomas. & Cif. (1952) = Gyalidea (Solorinell.).

Caprettia Bat. & H. Maia (1965), ? Dothideales (inc. sed., L). 1, Brazil.

Capricola Velen. (1947), ? Leotiales (inc. sed.). 1, Eur.

Capronia Sacc. (1883), Herpotrichiellaceae. 27, Eur., N. Am. See Müller *et al.* (*TBMS* **63**: 71, 1987; key), Barr (*Mycotaxon* **49**: 419, 1991; key N. Am. spp.), Untereiner *et al.* (1995; molec. taxonomy).

Caproniella Berl. (1896), Ascomycota (inc. sed.). 2, Eur. See Holm (*Taxon* **24**: 475, 1975).

Caproniella Berl. (1899) = Capronia (Herpotrichiell.) fide Müller *et al.* (1987).

Capsicumyces Gamundí, Aramb. & Giaiotti (1979), Mitosporic fungi, 1.A1.11. 1, Argentina.

capsidiol, a phytoalexin (q.v.) from spur pepper (*Capsicum frutescens*).

Capsulasclerotes Malan (1959), Fossil fungi (*f. cat.*). 1 (Permian), former Czechoslovakia.

capsule, a hyaline gelatinous sheath surrounding the cell of certain yeasts and bacteria.

Capsulotheca Kamyschko (1960), ? Trichocomaceae. 1, former USSR. See Benny & Kimbrough (1980).

Carbomyces Gilkey (1954), Carbomycetaceae. 3, USA. See Trappe (*TBMS* **57**: 85, 1971), Gilkey (*N. Am. Fl.* **2**(1), 1954; key).

Carbomycetaceae Trappe (1971), Pezizales. 1 gen., 3 spp. Ascomata large, cleistothecial, ± globose, smooth; peridium of intertwining hyphae with strongly inflated cells; interior solid, with fertile pockets surrounded by sterile veins; asci ± globose, randomly arranged, the wall not blueing in iodine, ? usually evanescent at maturity; ascospores ellipsoidal, hyaline or pale brown, ornamented. Anamorphs unknown. Hypogeous but frequently emergent, in deserts, USA.
Lit.: Trappe (1979).

carbonaceous, dark-coloured and readily broken; charcoal- or cinder-like.

Carbonea (Hertel) Hertel (1983), Lecanoraceae (±L). 5 (mainly on lichens), widespr.

carbonicolous, living on burnt ground; pyrophilous (q.v.).

Carbosphaerella I. Schmidt (1969), Halosphaeriaceae. 2 (marine), Baltic Sea. See Johnson *et al.* (*Bot. Mar.* **27**: 557, 1984).

Carcerina Fr. (1849) = Diderma (Didym.), etc.

Carcinomyces Oberw. & Bandoni (1982) = Syzygospora (Syzygospor.) fide Ginns (1986).

Carestiella Bres. (1897), Stictidaceae. 1, Eur. See Sherwood (*Mycotaxon* **5**: 1, 1977).

Caribaeomyces Cif. (1962), Microthyriaceae. 1, Dominican Republic.

carinate, keeled; boat-like.

Carinispora K.D. Hyde (1992), Phaeosphaeriaceae. 1 (on *Nypa*), Brunei.

cariose, decayed.

carioso-cancellate, becoming latticed by decay.

Caripia Kuntze (1898), Podoscyphaceae. 1, trop. Am. See Singer (1962: 792), Corner (1950).

Carlia Rabenh. (1857) nom. dub. (Fungi, inc. sed.) fide Wakefield (*TBMS* **23**: 215, 1939).

Carlosia Samp. (1923) = Thelomma (Calic.).

Carlosia G. Arnaud (1954) = Isthmospora (Mitosp. fungi) fide Kendrick & Carmichael (1973).

Carmichaelia N.D. Sharma (1980) [non R. Br. (1825), *Leguminosae*], Mitosporic fungi, 1.A2.10. 1, India.

carminophilic (of basidia), becoming densely granular (= siderophilous (or carminophilous) granulation) after treatment with aceto-carmine stain.

Carneopezizella Svrček (1987), Leotiaceae. 1, former Czechoslovakia.

Carnia Bat. (1960), Pezizaceae. 1, Brazil.

carnose (**carnous**), fleshy.

Carnostroma Lloyd (1919) = Xylaria (Xylar.) fide v. Arx & Müller (1954).

Carnoya Dewèvre (1893) = Mortierella (Mortierell.) fide Hesseltine (1955).

carnulose, somewhat fleshy.

Caromyxa Mont. (1856) nom. inval. = Mutinus (Phall.).

carotene, a mixture of pigments, chiefly the carotenoid β-carotene, found in various fungi, e.g. *Phycomyces blakesleeanus* (Lilly *et al.*, *Bull. W. Va. agric. Exp. Stn* **441T**, 1960), *Choanephora cucurbitarum* (Chu & Lilly, *Mycol.* **52**: 80, 1961); **carotenoids**, a large group of related polyene compounds, mostly with C_{40}, yellow, red or more rarely colourless. Many have been given trivial names, e.g. torularhodin, neurosporene. See Hesseltine (*Tech. Bull. USDA* **1245**, 1961; *Mucorales*), Arpin (*Bull. mens. Soc. linn. Lyon* **38**, 1968; *BSMF* **84**: 427, 1969; discomycetes), Shibata *et al.* (*List of fungal products*, 1964; 24 refs.), Valadon (*TBMS* **67**: 1, 1976; taxonomic value of carotenoids). See Metabolic products.

Carothecis Clem. (1931) ? = Cephalotheca (Cephalothec.).

Carouxella Manier, Rioux & Whisler (1965), Harpellaceae. 1 (in *Diptera*), France. See Lichtwardt (1986).

Carpenteles Langeron (1922) = Eupenicillium (Trichocom.) fide Stolk & Scott (1967).

Carpenterella Tehon & H.A. Harris (1941) [non Mun.-Chalm. ex L. Morellet & J. Morellet (1922), fossil *Algae*], Synchytriaceae. 2, USA, India.

Carpobolus P. Micheli ex Paulet (1808) [non Schwein. (1822), *Hepaticae*] ≡ Sphaerobolus (Sphaerobol.).

carpogenous, living on fruit.

carpogonium (generally of algae, sometimes of fungi, e.g. *Erysiphaceae*), the female sex organ.

carpoid, see *Biatoropsis*.

Carpomycetes, fungi having sporocarps; esp. ascomycetes and basidiomycetes.

carpophore, (1) stalk of the sporocarp; (2) sometimes (esp. in France) = basidioma.

carpophoroid, a sterile carpophore-like body, in agarics, of unknown function (Singer, 1962: 22).

Carpophoromyces Thaxt. (1931), Laboulbeniaceae. 1, Sri Lanka.

Carpozyma L. Engel (1872) = Hanseniaspora (Saccharomycod.) fide Lodder (1970).

carrier, an organism harbouring a parasite without itself showing disease (Anon., *TBMS* **33**: 154, 1950).

Carrionia Bric.-Irag. (1938) = Rhinocladiella (Mitosp. fungi) fide Schol-Schwarz (*Ant. v. Leeuwenhoek* **34**: 119, 1968).

cartilaginous, firm and tough but readily bent.

cartilaginous layer, sometimes applied to the sterome in *Cladonia* and the chondroid axis (q.v.) in *Usnea*.

Cartilosoma Kotl. & Pouzar (1958) = Antrodia (Coriol.) fide Donk (1974). Accepted by Pegler (1973).

caryallagic (of reproduction), having nuclear change; **acaryallagic**, not having nuclear change, as in clone development (Link, *Bot. Gaz.* **88**: 1, 1929).

caryo-, see karyo.

Caryospora De Not. (1855), Zopfiaceae. 5, widespr. See Barr (*Mycotaxon* **9**: 17, 1979), Hawksworth (*TBMS* **79**: 69, 1982).

Caryosporella Kohlm. (1985), ? Melanommataceae. 1 (on *Rhizophora*), Belize.

Casaresia Gonz. Frag. (1920), Mitosporic fungi, 1.G2.1. 1 (aquatic), N. temp.

Cashiella Petr. (1951), Dermateaceae. 2, N. Am.

cassideous, helmet-shaped.

Castagnella G. Arnaud (1914) = Rhagadostoma (Nitschk.) fide Müller & v. Arx (1973).

Castanoporus Ryvarden (1991), Meruliaceae. 1, Japan.

Castellania C.W. Dodge (1935) = Candida (Mitosp. fungi) fide Diddens & Lodder (*Die anaskosporogenen Hefen* **2**, 1942).

Castoreum Cooke & Massee (1887), Mesophelliaceae. 2, Australia. See Cunningham (*Proc. Linn. Soc. N.S.W.* **57**: 313, 1932), Beaton & Weste (*TBMS* **82**: 665, 1984).

Catabotrydaceae Petr. ex M.E. Barr (1990), Sordariales. 1 gen. (+ 1 syn.), 1 sp. Ascomata perithecial, immersed in a soft pulvinate stroma, globose, necked; interascal tissue composed of thin-walled paraphyses; asci cylindrical, with a non-amyloid ring; ascospores ellipsoidal, aseptate, hyaline to yellowish. Saprobic on palms.

The family might be more closely allied to the *Phyllachorales*.

Lit.: Barr (*Mycotaxon* **39**: 43, 1990).

Catabotrys Theiss. & Syd. (1915), Catabotrydaceae. 1 (on *Palmae*), trop. See Seaver & Waterston (*Mycol.* **38**: 180, 1946), Petrak (*Sydowia* **8**: 287, 1954).

Catacauma Theiss. & Syd. (1914) = Phyllachora (Phyllachor.) fide Petrak (*Ann. Myc.* **22**: 1, 1924). See Cannon (*Mycol. Pap.* **165**, 1991).

Catacaumella Theiss. & Syd. (1915) = Vestergrenia (Dothid.) fide v. Arx & Müller (1954).

Catachyon (Ehrenb. ex Fr.) Fr. (1832) = Podaxis (Podax.).

catahymenium, see hymenium.

cataphyses, pseudoparaphyses (Groenhart, *Persoonia* **4**: 11, 1965); see hamathecium.

Catapyrenium Flot. (1850), Verrucariaceae (L). 40, mainly temp. See Breuss & Hansen (*Pl. Syst. Evol.* **159**: 59, 1988; key 7 spp. Greenland), Breuss (*Stapfia* **23**: 1, 1990, key 27 spp. Europe; *Pl. Syst. Evol.* **185**: 17, 1993, key 13 spp. S. Am.; *Nova Hedw.* **58**: 229, 1994, key 12 N. Afr. spp.), Breuss & McCune (*Bryologist* **97**: 365, 1994; key 18 spp. N. Am.), Harada (*Nat. Hist. Res.* **2**: 113, 1993; gen. concept).

Catarraphia A. Massal. (1860), Arthoniales (inc. sed., L). 1, Malesiana-Pacific. See Egea & Torrente (*Crypt., bryol. lich.* **14**: 329, 1993).

Catarrhospora Brusse (1994), Porpidiaceae (L). 2, S. Afr.

cata-species, see Uredinales.

Catastoma Morgan (1892) = Disciseda (Lycoperd.).

catathecium, a flattened ascoma, having the wall more or less radial in structure, and with a basal plate, e.g. *Trichothyrina*; cf. thyriothecium.

Catathelasma Lovejoy (1910), Tricholomataceae. 5, Eur., N. Am.. See Singer (*Sydowia* **31**: 193, 1979; key).

Catatrama Franco-Mol. (1991), Tricholomataceae. 1, Costa Rica.

Catenaria Sorokīn (1889) [non Roussel (1806), *Algae*], Catenariaceae. c. 10, temp. See Birchfield (*Mycopath.* **13**: 331, 1960), Olson & Reichle (*TBMS* **70**: 423, 1978; meiosis), Dick (*in press*).

Catenariaceae (Sparrow) Couch (1945), Blastocladiales. 3 gen. (+ 1 syn.), 12 spp. Thallus polycentric, rhizomycelial with catenulate swellings separated by sterile isthmi; saprobic or parasitic.

catenate (**catenulate**), in chains or end-to-end series. See arthro-, baso-, blastocatenate.

Catenella Bat. & Peres (1963) [non Grev. (1830), *Algae*], Mitosporic fungi, 8.A1.?. 1, Italy.

Catenochytridium Berdan (1939), Endochytriaceae. 5, N. Am., Japan.

Catenomyces A.M. Hanson (1944), Catenariaceae. 1, USA.

Catenomycopsis Tibell & Constant. (1991), Anamorphic Mycocaliciaceae, 1.A1.3. Teleomorph *Chaenothecopsis*. 1, S. Am., Australasia.

Catenophlyctis Karling (1965), Catenariaceae. 2, widespr.

Catenophora Luttr. (1940), Mitosporic fungi, 6.A1.1. 3, USA. See Nag Raj & Kendrick (*CJB* **66**: 898, 1988; key).

Catenophoropsis Nag Raj & W.B. Kendr. (1988), Mitosporic fungi, 6.A1.10. 1, Australia.

catenophysis, a persistent chain of utricular, thin-walled cells formed by the vertical separation of the pseudoparenchyma in the centrum of certain ascomycetes, e.g. some *Halosphaeriaceae* (see Kohlmeyer & Kohlmeyer, *Mycol.* **63**: 857, 1971).

Catenospegazzinia Subram. (1991), Mitosporic fungi, 1.D2.37/5. 2, Australia.

Catenosubulispora Matsush. (1971), Mitosporic fungi, 1.E1.3. 1, Guadalcanal.

Catenularia Grove (1886), Anamorphic Lasiosphaeriaceae, 1.A2.17. Teleomorph *Chaetosphaeria*. 4, temp. See Hughes (*N.Z. Jl Bot.* **3**: 136, 1965).

Catenulaster Bat. & C.A.A. Costa (1959), Mitosporic fungi, 5.A1.?. 1, Brazil.

catenulate, see catenate.

catenuliform, chain-like.

Catenulopsora Mundk. (1943) = Cerotelium (Phakopsor.) fide Cummins & Hiratsuka (1983), Laundon (*Mycotaxon* **3**: 133, 1975).

Catenuloxyphium Bat., Nascim. & Cif. (1963) nom. dub. fide Sutton (*Mycol. Pap.* **141**, 1977). See also Hughes (*Mycol.* **68**: 693, 1976).

caterpillar fungi, see vegetable caterpillars.

Catharinia (Sacc.) Sacc. (1895) = Julella (Thelenell.) fide Clements & Shear (1931).

Cathisinia Stirt. (1888) = Sarcogyne (Acarospor.).

Catilla Pat. (1915), Cyphellaceae. 1, Eur.

Catillaria A. Massal. (1852), Catillariaceae (L). c. 150, cosmop. See Vainio (*Acta Soc. Fauna Flora fenn.* **53**(1), 1922), Pant & Awasthi (*Proc. Indian Acad. Sci. Pl. Sci.* **99**: 369, 1989; key 10 Indian spp.), Lamb (*Rhodora* **56**: 105, 1956), Kilias (*Herzogia* **5**: 209, 1981; monogr. Eur. saxic. spp.), Vězda (*Folia geobot. phytotax.* **15**: 75, 1980; key foliicolous spp.).

Catillariaceae Hafellner (1984), Lecanorales (±L). 8 gen. (+ 19 syn.), 209 spp. Thallus crustose, sometimes poorly developed; ascomata sessile, dark, small, with a poorly-developed proper and no thalline margin; interascal tissue of sparsely branched paraphyses, often pigmented at the apices; asci with a I+ apical cap, ocular chamber absent or poorly developed, with an outer I+ gelatinous layer, 8- or 16-spored; ascospores hyaline, aseptate or transversely septate, without a sheath. Anamorphs pycnidial, usually inconspicuous. Lichenized with green algae, a few lichenicolous, widespr.

Catillariomyces E.A. Thomas ex Cif. & Tomas. (1953) ≡ Catillaria (Catillar.).

Catillariopsis (Stein) M. Choisy (1950) = Rhizocarpon (Rhizocarp.).

Catinaria Vain. (1922) nom. cons. prop., Bacidiaceae (L). 2, temp. See Poelt & Vězda (1981).

Catinella Boud. (1907), Dermateaceae. 1 or 2, Eur., N. Am., Sri Lanka. See Durand (*Bull. Torrey bot. Cl.* **49**: 15, 1922).

Catinella Kirschst. (1924) = Unguicularia (Hyaloscyph.) fide Raschle (*Sydowia* **29**: 170, 1977).

Catinopeltis Bat. & C.A.A. Costa (1957), Mitosporic fungi, 5.C1.?. 1, Brazil.

Catinula Lév. (1848), Mitosporic fungi, ?4.?.?. 10, widespr.

Catocarpon Hellb. (1884) ≡ Catocarpus (Rhizocarp.).

Catocarpus (Körb.) Arnold (1871) = Rhizocarpon (Rhizocarp.).

Catolechia Flot. (1850), ? Rhizocarpaceae (L). 1 (montane), Eur. See Hafellner (*Nova Hedw.* **30**: 673, 1978).

Catopyrenium Flot. ex Körb. (1855) ≡ Catapyrenium (Verrucar.).

Catosphaeropsis Tehon (1939) = Sphaeropsis (Mitosp. fungi) fide Sutton (1980).

catothecium, see catathecium.

Cattanea Garov. (1875) = Dictyosporium (Mitosp. fungi) fide Damon (*Lloydia* **15**: 118, 1952).

Catulus Malloch & Rogerson (1978), Dothideales (inc. sed.). 1 (on *Seuratia*), Canada.

cat's ear, basidiome of *Clitopilus passeckerianus*, an invader of mushroom beds.

cauda, tail; tail-like appendage.

caudate, having a tail

Caudella Syd. & P. Syd. (1916), Microthyriaceae. 2, Am. See Hansford (*Mycol. Pap.* **15**, 1946).

Caudellopeltis Bat. & H. Maia (1960) = Maublancia (Microthyr.) fide Müller & v. Arx (1962).

Caudomyces Lichtw., Kobayasi & Indoh (1988), Legeriomycetaceae. 1, Japan. See Lichtwardt *et al.* (*Trans. Mycol. Soc. Japan* **28**: 376, 1987).

Caudophoma B.V. Patil & Thirum. (1968) = Phyllosticta (Mitosp. fungi) fide van der Aa (*Stud. Mycol.* **5**, 1973).

Caudospora Starbäck (1889), Diaporthales (inc. sed.). 1 (on *Quercus*), Eur., N. Am. See Rogers (*Mycotaxon* **21**: 475, 1984). = Hercospora (Diaporth.) fide Barr (1978). Anamorph *Phomopsis*-like.

Caudosporella Höhn. (1914) = Harknessia (Mitosp. fungi) fide Sutton (*Mycol. Pap.* **123**, 1971).

caulescent, having a stem; becoming stemmed.

caulicolous, living on herbaceous stems; **caulicole**, a fungus which does this.

Caulocarpa Gilkey (1947) = Sarcosphaera (Peziz.) fide Trappe (*Mycotaxon* **2**: 109, 1975).

Caulochora Petr. (1940) = Glomerella (Phyllachor.) fide v. Arx & Müller (1954).

Caulochytriaceae Subram. (1974), Spizellomycetales. 1 gen., 2 spp.
Lit.: Olive (*Am. J. Bot.* **67**: 568, 1980).

Caulochytrium Voos & L.S. Olive (1968), Caulochytriaceae. 2 (on *Cladosporium* and *Gloeosporium* conidia), USA. See Voos (*Am. J. Bot.* **56**: 898, 1969).

caulocystidium, see cystidium.

Caulogaster Corda (1831) nom. dub.; based on insect eggs fide Tulasne & Tulasne (*Selecta fungorum carpologa* **1**: 125, 1861).

Cauloglossum Grev. (1823) = Podaxis (Podax.).

cauloplane, the stem surface.

Caulorhiza Lennox (1979), Tricholomataceae. 1, USA.

Caumadothis Petr. (1971) = Botryosphaeria (Botryosphaer.) fide v. Arx & Müller (1975).

Causalis Theiss. (1918) = Coccodiella (Phyllachor.) fide Müller & v. Arx (1973).

Cautina Maas Geest. (1967), Coriolaceae. 1, Chile.

Cavaraella Speg. (1923), Rhytismatales (inc. sed.). 1, Cuba.

Cave fungi. There are possibly no fungi peculiar to caves. Lagarde (*Arch. Zool. exp. gen.* **53**: 277, 1913; **56**: 279, and *Notes et Revue*: 129, 1917; **60**: 593, 1922) recorded *c.* 50 taxa (including most groups of fungi) from caves in France, Spain and N. Africa. Went (*Science* **166**: 385, 1969) and Hasselbring *et al.* (*Mycol.* **67**: 171, 1975) describe fungi associated with stalactites. Polypores and other wood-destroying fungi are common in mines where the basidiomata are frequently abnormal (Fassatiová, *Česká Myk.* **24**: 162, 1970; mines of Příbram, former Czechoslovakia).

cavernose, having hollows or cavities.

cavernula (pl. **-ae**), cavity; esp. the cavities in the lower cortex of *Cavernularia*.

Cavernularia Degel. (1937), Parmeliaceae (L). 2, Eur., N. Am. See Ahti & Henssen (*Bryologist* **68**: 85, 1965).

Cavimalum Yoshim. Doi, Dargan & K.S. Thind (1977), Clavicipitaceae. 2, Asia.

Cavosteliaceae L.S. Olive (1964) (Cavosteliidae), Protosteliales. 4 gen., 7 spp. Flagellate motile stage present, sporocarps single.
Lit.: Spiegel (1990).

Cavostelium L.S. Olive (1965), Cavosteliaceae. 2, trop. See Olive & Stoianovitch (*Am. J. Bot.* **56**: 979, 1969).

Cazia Trappe (1989), Helvellaceae. 1 (hypogeous), USA.

CBS. Centraalbureau voor Schimmelcultures (Baarn and Delft, Netherlands); founded 1904; an Institute of the Royal Netherlands Academy of Arts and Science; see van Beverwijk (*Ant. v. Leeuwenh.* **25**: 1, 1959), de Hoog (Ed.) (*Centraalbureau voor Schimmelcultures: 75 years culture collection*, 1979).

cecidium, a gall (q.v.); caused by an animal (**zoo-**), esp. an insect; caused by a fungus (**myco-**).

Cecidonia Triebel & Rambold (1988), Lecideaceae. 1 (on lichens), Eur.

Cejpia Velen. (1934), Dermateaceae. 1, former Czechoslovakia.

Cejpomyces Svrček & Pouzar (1970) = Thanatephorus (Ceratobasid.) fide Langer (1994).

Celidiopsis A. Massal. (1856) = Arthonia (Arthon.).

Celidium Tul. (1852) = Arthonia (Arthon.) fide Santesson (*Symb. bot. upsal.* **12**(1): 69, 1952).

cell, a unit of cytoplasm containing one or more nuclei, limited by a membrane (the **cell membrane**), and, in fungi, usually enclosed by a wall (see cell wall chemistry). Adjacent fungal cells do not necessarily contain individual protoplasts (q.v.) because cytoplasm and nuclei may pass from one cell to another.

Cell wall chemistry. See Wessels (*Int. Rev. Cytol.* **104**: 37, 1986; *New Phytologist* **123**: 397, 1993), Bartnicki Garcia (*in* Rayner *et al.* (Eds), *Evolutionary biology of the fungi*: 389, 1987), Gooday (*in* Gow & Gadd, *The growing fungus*: 41, 1995). Polymers found in fungal walls are given in Table 2. The wall (e.g. zygospore wall of *Mucorales*) may also contain melanin (a dark coloured pigment which protects spores from UV light and microbial lysis) and sporopollenin (the most resistant biopolymer known).

cellar fungus, *Coniophora puteana* or *Rhinocladiella ellisii* (sterile mycelium *Zasmidium cellare*; syn. *Racodium cellare*; Hawksworth & Riedl, *Taxon* **26**: 208, 1977).

cellular slime moulds, Dictyosteliomycota (q.v.).

Cellularia Bull. (1788) = Coriolus (Coriol.) fide Donk (1974).

Cellulasclerotes Stach & Pickh. (1957), Fossil fungi (*f. cat.*). 6 (Permo-Carboniferous), Eur.

Table 2. Fungal cell wall polymers.

Group	Alkali insoluable	Alkali soluable
Oomycota	Cellulose, (1→3)-β / (1→6)β-glucan *	(1→3)-β / (1→6)β-glucan
Chytridiomycota	Chitin, glucan	Glucan
Zygomycota	Chitosan, Polyglucronic acid	Glucoronomanno-protein polyphosphate
Ascomycota	Chitin, (1→3)-β / (1→6)β-glucan	(1→3)α-glucan, (Galacto)-manno-protein
Basidiomycota	Chitin, (1→3)-β / (1→6)β-glucan	(1→3)α-glucan

* Chitin and chitin synthase have recently been identified in *Saprolegnia monoica* (Bulone *et al.*, *Exp. Mycol.* **16**: 8, 1992).

cellulin, a chitan-glucan complex which occurs as granules in the cells and plugs (**- plugs**) at hyphal constrictions in *Leptomitales* (see Lee *et al.*, *Mycol.* **68**: 87, 1976).

cellulolysis adequacy index, an estimate, derived by dividing the rate of cellulolysis by the mycelial growth rate on an agar plate, as to whether the rate of cellulose decomposition by a fungus is adequate to supply its needs for saprobic survival (Garrett, *TBMS* **49**: 59, 1966; Deacon, *TBMS* **72**: 469, 1979).

cellulolytic fungi, fungi able to utilize cellulose-containing materials (incl. plant cellulose, paper, cloth, etc.), e.g. *Chaetomiaceae* (Chahal & Wang, *Mycol.* **70**: 160, 1978). Cellophane or filter paper is often used in the culture of these fungi. See Biodeterioration.

Cellulosporium Peck (1879) nom. dub. (Mitosp. fungi) fide Sutton (1977).

Cellypha Donk (1959), Tricholomataceae. 1, Eur., N. Am. See Reid (*Persoonia* **3**: 131, 1964).

Celothelium A. Massal. (1860), Pyrenulales (inc. sed., L). 8, subtrop. & trop. See Aguirre-Hudson (*Bull. Br. Mus. nat. Hist., Bot.* **21**: 85, 1991).

Celtidia J.D. Janse (1897), Zopfiaceae. 1, Java. See Hawksworth (*CJB* **57**: 91, 1979).

Celyphus Batten (1973), ? Fossil fungi. 1 (Late Jurassic-Early Cretaceous), UK, USA.

CEM, see Societies and organizations.

Cenangella Sacc. (1884) = Dermatea (Dermat.) fide Nannfeldt (1932).

Cenangiella Lambotte (1887) = Scleroderris (Leot.).

Cenangina Höhn. (1909) = Cenangium (Leot.).

Cenangiomyces Dyko & B. Sutton (1979), Mitosporic fungi, 7.A1.41. 1 (with clamp connexions), UK.

Cenangiopsis Rehm (1912), Leotiaceae. 1, Eur.

Cenangiopsis Velen. (1947), ? Leotiales (inc. sed.). 1, former Czechoslovakia.

Cenangites Mesch. (1892), Fossil fungi. 1 (Tertiary), Eur.

Cenangium Fr. (1818), Leotiaceae. 25, widespr. *C. ferruginosum* (syn. *C. abietis*) (die-back of pines). See Verkley (*MR* **99**: 187, 1995; asci).

ceno-, see coeno-.

Cenococcum Moug. & Fr. (1825), Mitosporic fungi, 9.-.-. 1, Eur., N. Am. *C. geophilum* is mycorrhizal (*RAM* **5**: 316). See LoBuglio *et al.* (*CJB* **69**: 2331, 1992; variation in rDNA), Massicotte *et al.* (*CJB* **70**: 125, 1992; morphology).

Cenomyce Ach. (1810) ≡ Cenomyces (Cladon.).

Cenomyces Ach. (1809) = Cladonia (Cladon.).

Cenozosia A. Massal. (1854) = Ramalina (Ramalin.).

Centonites Peppers (1964), Fossil fungi. 1 (Carboniferous), USA.

central body (of ascomycetes), the cell structure (central apparatus) from which astral rays emanate and initiate a cleavage of the cytoplasm (Harper, 1905). See Lindegren *et al.* (*Can. J. Genet. Cytol.* **7**: 37, 1965).

centric (**central**) (of a stipe), at the centre of the pileus; (of oogonium of *Saprolegniaceae*), having one or two layers of fat droplets surrounding the central cytoplasm in contrast to **sub-**, having the cytoplasm surrounded by one layer of droplets on one side, and by two or three layers on the other and **ex-**, having one large drop or a lunate row of droplets on one side (Coker, 1923).

Centridium Chevall. (1826) = Gymnosporangium (Puccin.). Aecial states fide Dietel (1928).

centrifugal, from the centre outwards.

centriole, short-cylindrical or barrel-shaped cell organelle 25 µm long × 4 µm diam. (kinetosome lacking an axoneme).

centripetal, towards the centre.

Centrospora Neerg. (1942) [non Trevis. (1845), *Algae*] ≡ Mycocentrospora (Mitosp. fungi).

centrum, the structures within an ascoma, i.e. the asci and interascal tissue (the hamathecium); for types of centrum organization see Reynolds (Ed.) (*Ascomycete systematics*, 1981), and under Ascomycota.

cep, basidioma of the edible *Boletus edulis*; **fungi suilli** of Pliny and other classical writers.

cepaceous, see alliaceous.

Cephalacladium, see *Cephalocladium*.

Cephaleuros Kunze (1832) , Algae (*Trentepohliaceae*) parasitic on vascular plants. *C. virescens* (syn. *Mycoidea parasitica*) is common on tea (**red-rust**) and other trop. plants; photobiont of *Strigula* (q.v.) and other foliicolous lichens. See Printz (*Nytt Mag. Naturvid.* **80**: 137, 1940; keys), Santesson (*Symb. bot. upsal.* **12**(1), 1952).

Cephaliophora Thaxt. (1903), Mitosporic fungi, 1.C1.6. 4, trop. See Barron *et al.* (*CJB* **68**: 685, 1990; 2 n.spp.).

Cephaloascaceae L.R. Batra (1973), Saccharomycetales. 2 gen., 2 spp. Vegetative cells ± ellipsoidal, with multilateral budding; asci formed at the apex of an erect hyaline or brown hyphal seta-like stalk which is diploid (at least in *Cephaloascus*); either ellipsoidal, formed in branched chains and evanescent or elongate and solitary, releasing spores from the

apex; ascospores hyaline, with an eccentric thickening. On coniferous wood, etc., or on other fungi, at least sometimes insect-associated, N. temp.
Lit.: v. Arx & van der Walt (*Stud. Mycol.* **30**: 167, 1987), Hausner *et al.* (*Mycol.* **84**: 870, 1992; mol. phylogeny).

Cephaloascales, see *Saccharomycetales*.

Cephaloascus Hanawa (1920), Cephaloascaceae. 1, Japan, Canada. See Schippers-Lammertse & Heyting (*Ant. v. Leeuwenhoek* **28**: 5, 1962), v. Arx *et al.* (1972), Udagawa *et al.* (*Trans. mycol. Soc. Japan* **18**: 399, 1977), Cannon & Minter (*Trans. mycol. Soc. Japan* **33**: 51, 1992).

Cephalocladium Rchb. (1828) = Botrytis (Mitosp. fungi) fide Mussat (*Syll. Fung.* **15**: 82, 1901).

Cephalodiplosporium Kamyschko (1961), Mitosporic fungi, 1.B1/2.?. 1, former USSR. = Fusarium (Mitosp. fungi) fide Gams (*in litt.*).

cephalodium (pl. -ia), a delimited region within (**internal** -), or a warty, squamulose, or fruticose structure on the surface of, a lichen thallus containing a photobiont different from that characteristic of the rest of the thallus. Generally cephalodia contain cyanobacteria (e.g. *Nostoc*) whilst the rest of the thallus contains a green alga (e.g. *Trebouxia*). *Nostoc* cephalodia fix atmospheric nitrogen (see Millbank & Kershaw, *New Phytol.* **68**: 721, 1969). Known in about 400 lichenized spp. in diverse orders; see cyanotrophy, gall, Lichens, paracephalodium, photomorph, phycotype.

Cephalodochium Bonord. (1851), Mitosporic fungi, 3.A1.?. 1, Eur.

Cephaloedium Kunze (1828) = Exosporium (Mitosp. fungi) fide Mussat (*Syll. Fung.* **15**, 1901), nom. nud. fide Hughes (*CJB* **36**: 744, 1958).

Cephalomyces Bainier (1907) = Cephaliophora (Mitosp. fungi) fide Ellis (*DH*).

Cephalophorum Nees [not traced] nom. dub. (Mitosp. fungi). See Benjamin (*Taxon* **17**: 524, 1968).

Cephalophysis (Hertel) H. Kilias (1985), Teloschistaceae (L). 1, Eur.

Cephaloscypha Agerer (1975), Cyphellaceae. 1, UK.

cephalosporin, one of a series of antibacterial antibiotics from *Acremonium* spp., esp. *A. chrysogenum* (cephalosporin C) and *A. salmosynnematum*); see Roberts (*Mycol.* **44**: 292, 1952), Grosklags & Swift (*Mycol.* **49**: 305, 1957), Kavanagh *et al.* (*Mycol.* **50**: 370, 1958), Sassiver & Lewis (*Adv. appl. Microbiol.* **13**: 163, 1970; structure), Abraham & Newton (*Adv. Chemotherapy* **2**: 23, 1969); cephalosporin N = penicillin N.

Cephalosporiopsis Peyronel (1915), Mitosporic fungi, 1.B1.15. 2, Eur., Afr. ? = Acremonium or Fusarium (Mitosp. fungi) fide Carmichael *et al.* (1980).

Cephalosporium Corda (1839) = Acremonium (Mitosp. fungi) fide Gams (1967, 1971).

Cephalotelium Syd. (1921) = Ravenelia (Ravenel.) fide Dietel (1928).

Cephalotheca Fuckel (1871), Cephalothecaceae. 4, widespr. See Booth (*Mycol. Pap.* **83**, 1961; key), Fennell (*in* Ainsworth *et al.* (Eds), *The Fungi* **4A**: 45, 1973).

Cephalothecaceae Höhn. (1917), Ascomycota (inc. sed.). 1 gen. (+ 6 syn.), 4 spp. Stromata absent; ascomata cleistothecial, small, thick-walled, covered with dense yellow hyphae when young, fragmenting into preformed polygonal plates; interascal tissue absent; asci irregularly disposed, ± globose, thin-walled, evanescent; ascospores brown, ellipsoidal, aseptate, smooth. Anamorph hyphomycetous. Saprobic on wood, N. temp.
Probably allied to the *Pseudeurotiaceae* (q.v.).

Lit.: Chesters (*TBMS* **19**: 261, 1935; life history), Fennell (1973).

Cephalothecium Corda (1838) = Trichothecium (Mitosp. fungi) fide Hughes (1958).

cephalothecoid, fragmenting along a series of predefined suture lines, as in ascomata of *Cephalotheca* (q.v.).

Cephalothricum, see *Cephalotrichum*.

Cephalotrichum Link (1809), Mitosporic fungi, 2.A2.19. 5, widespr. See Hughes (1958), Morton & Smith (*Mycol. Pap.* **86**, 1963).

Cephalotrichum Berk. ex Sacc. (1886) ≡ Trichocephalum (Mitosp. fungi).

Ceracea Cragin (1885), Mitosporic fungi, 3.A1.?. 1, USA. See Martin (*Mycol.* **41**: 77, 1949), Donk (*Persoonia* **4**: 334, 1966).

Ceraceohydnum Jülich (1978) = Mycoaciella (Steccherin.) fide Hjortstam *et al.* (1990).

Ceraceomerulius (Parmasto) J. Erikss. & Ryvarden (1973) = Byssomerulius (Merul.) fide Hjortstam & Larsson (*Windahlia* **21**, 1994).

Ceraceomyces Jülich (1972), Meruliaceae. 10, widespr.

Ceraceopsora Kakish., T. Sato & S. Sato (1984), Chaconiaceae. 1, Japan; 0, I on *Anemone*; II, III on *Elaeagnus*.

Ceraceosorus B.K. Bakshi (1976), Brachybasidiceae. 1, India.

ceraceous (**cereous**), wax-like.

Ceraiomyces Thaxt. (1901) = Laboulbenia (Laboulben.) fide Thaxter (1915).

Ceramoclasteropsis Bat. & Cavalc. (1962), ? Capnodiaceae. 1, Brazil.

Ceramothyrium Bat. & H. Maia (1957), Chaetothyriaceae. 19, trop. See Hughes (1976). Anamorph *Stanhughesia*.

Cerania Gray (1821) = Thamnolia (Lecanorales, inc. sed.).

ceranoid, having horn-like branches.

Cerapora, see *Ceriporia*.

Ceraporia Donk (1933) ≡ Ceriporia (Coriol.).

Ceraporus Bondartsev & Singer (1941) nom. nud. ≡ Ceriporia (Coriol.).

Cerasterias Reinsch (1867) nom. dub. (? Algae).

Cerastoma Quél. (1875), Ascomycota (inc. sed.). 1, Eur.

Cerastomis Clem. (1931) = Endoxyla (Bolin.).

Ceratelium Arthur (1906) ≡ Cerotelium (Phakopsor.).

Ceratella Pat. (1887) [non Hook. f. (1844), *Compositae*] ≡ Ceratellopsis (Clavariadelph.).

Ceratella (Quél.) Bigeard & H. Guill. (1913), Pterulaceae. 4, Eur.

Ceratellopsis Konrad & Maubl. (1937), Clavariadelphaceae. 3, Eur. See Berthier (*Bull. Soc. linn. Lyon*, num. spec.: 187, 1976).

Ceratiomyxa J. Schröt. (1889), Ceratiomyxaceae. 3, cosmop.

Ceratiomyxaceae J. Schröt. (1889) (Ceratiomyxidae), Protosteliales. 1 gen. (+ 3 syn.), 3 spp. Sporocarps arising on a common slime, spores borne singly at tips of fine stalks, spore germination resulting in 8 haploid flagellate swarm spores.
Lit.: Spiegel (1990).

Ceratiomyxella L.S. Olive & Stoian. (1971), Cavosteliaceae. 1, Brazil.

Ceratiomyxomycetes, see *Protosteliomycetes*.

Ceratiopsis De Wild. (1896) = Ceratiomyxa (Ceratiomyx.) fide Saccardo (1899).

Ceratitium Rabenh. (1851) = Gymnosporangium (Puccin.).

Ceratium Alb. & Schwein. (1805) [non Schrank (1793), *Algae*] ≡ Ceratiomyxa (Ceratiomyx.).

Ceratobasidiaceae G.W. Martin (1948), Ceratobasidi-
ales. 12 gen. (+ 6 syn.), 40 spp. Pellicular;
homobasidious; ballistosporic.
Ceratobasidiales, Basidiomycetes. 1 fam., 12 gen. (+ 6
syn.), 40 spp. Fam.:
Ceratobasidiaceae.
Ceratobasidium D.P. Rogers (1935), Cer-
atobasidiaceae. 16, widespr. See Currah & Zelmer
(Rep. Tottori mycol. Inst. 30: 43, 1992), Talbot (Per-
soonia 3: 382, 1965). Anamorph Ceratorhiza.
Ceratocarpia Rolland (1896), Dothideales (inc. sed.). 2,
Eur., Afr. See Benny & Kimbrough (Mycotaxon 12: 1,
1980).
Ceratochaete Syd. & P. Syd. (1917) = Dysrhynchis
(Parodiopsid.) fide Müller & v. Arx (1962).
Ceratochaetopsis F. Stevens & Weedon (1927) =
Chaetothyrina (Micropeltid.) fide v. Arx & Müller
(1975).
Ceratocladia Schwend. (1860) = Alectoria (Alector.).
Ceratocladium Corda (1839), Mitosporic fungi,
1.A1.?6. 1, Eur. See Hughes (Mycol. Pap. 47, 1951).
Ceratocladium Pat. (1898) ≡ Xylocladium (Mitosp.
fungi).
Ceratocoma Buriticá & J.F. Hennen (1991), Puccini-
osiraceae. 2 (on Gompholobium, Xylopia), Australia,
Afr.
Ceratocystiopsis H.P. Upadhyay & W.B. Kendr. (1975)
= Ophiostoma (Ophiostomat.) fide Hausner et al. (MR
97: 625, 1993), Wingfield (in Wingfield et al., 1993:
21). See also van Wijk & Wingfield (CJB 71: 1212,
1993; ultrastr.), Wolfaardt et al. (S. Afr. Jl Bot. 58:
277, 1992; synoptic key; database).
Ceratocystis Ellis & Halst. (1890), ? Microascales (inc.
sed.). 14, mainly temp. See Seifert et al. (in Wingfield
et al. (Eds), Ceratocystis and Ophiostoma, 1993),
Grylls & Seifert (in Wingfield et al. (Eds), 1993: 261;
key), Hausner et al. (CJB 71: 52, 1993; posn), Hunt
(Lloydia 19, 1956; monogr., keys), Upadhyay (A monograph of Ceratocystis and Ceratocystiopsis, 1981;
monogr., keys), Kowalski & Butin (J. Phytopath. 124:
236, 1989; 6 spp. on Quercus), Wolfaardt et al. (S.
Afr. Jl Bot. 58: 277, 1992; synoptic key, database).
Anamorph Chalara.
Ceratogaster Corda (1841) = Elaphomyces (Elaphomycet.) fide Mussat (Syll. Fung. 15: 83, 1901).
Ceratomyces Thaxt. (1892), Ceratomycetaceae. 19, N.,
C. & S. Am.
Ceratomycetaceae S. Colla (1934), Laboulbeniales. 12
gen., 84 spp. Stroma (thallus) present; ascomata
formed directly from successive intercalary cells of
the primary thallus; outer wall layer of ascoma composed of many short usually ± equal cells; asci 4-
spored; ascospores with a submedian septum; usually
monoecious.
Ceratonema Pers. (1822) nom. dub.; based on mycelium, etc. fide Mussat (Syll. Fung. 15: 83, 1901).
Ceratophacidium J. Reid & Piroz. (1966), Rhytismataceae. 1, USA.
Ceratophoma Höhn. (1917), Mitosporic fungi, ?8.A1.?.
1, Eur.
Ceratophora Humb. (1793) = Gloeophyllum (Coriol.)
fide Donk (1974).
Ceratophorum Sacc. (1880), Mitosporic fungi, 1.C2.1.
5, widespr. See Hughes (lupin leaf spot). See Hughes
(Mycol. Pap. 36, 1951), Ellis (Mycol. Pap. 70, 1958).
Ceratophyllum M. Choisy (1951) [non L. (1753), Cer-
atophyllaceae] = Hypogymnia (Parmel.).
Ceratopodium Corda (1837) = Graphium (Mitosp.
fungi) fide Saccardo (Syll. Fung. 4: 609, 1886).
Ceratoporthe Petr. (1925), Melanconidaceae. 1 (on
Sarothamnus), Eur.

Ceratopycnidium Maubl. (1907), Mitosporic fungi,
4.A-C1.10. 2, Afr., S. Am. See Bertoni & Cabral (MR
95: 1014, 1991; amend. of genus, n.sp.).
Ceratopycnis Höhn. (1915), Mitosporic fungi, 4.C2.15.
1, Eur. See Morgan-Jones (Mycotaxon 2: 167, 1975).
Ceratopycnium Clem. & Shear (1931) ≡ Cer-
atopycnidium (Mitosp. fungi).
Ceratorhiza R.T. Moore (1987), Anamorphic Cer-
atobasidiaceae. Teleomorph Ceratobasidium. 5,
widespr. See Currah & Zelmer (Rep. Tottori mycol.
Inst. 30: 43, 1992), Moore (Ant. v. Leeuwenhoek 55:
393, 1989).
Ceratosebacina P. Roberts (1993), Exidiaceae. 1, Eur.
Ceratosperma Speg. (1918) = Nematostoma
(Pseudoperispor.) fide v. Arx & Müller (1975).
Ceratospermopsis Bat. (1951), Meliolaceae. 2, Brazil.
Ceratospermum P. Micheli (1729) nom. dub. (? Asco-
mycota, inc. sed.).
Ceratosphaeria Niessl (1876), ? Lasiosphaeriaceae. 8,
widespr. See Tsuda & Ueyama (Trans. mycol. Soc.
Japan 18: 413, 1977), Barr (1978).
Ceratosporella Höhn. (1923), Mitosporic fungi, 1.G2.1.
5, widespr. See Hughes (TBMS 35: 243, 1952),
Kuthubutheen & Nawawi (MR 95: 159, 1991; key).
Ceratosporium Schwein. (1832), Mitosporic fungi,
1.G2.1. 4, widespr. See Hughes (Mycol. Pap. 39,
1951; N.Z. Jl Bot. 2: 305, 1965).
Ceratostoma Fr. (1818) nom. rej. = Melanospora (Ceratostomat.) fide Cannon & Hawksworth (Taxon 32:
476, 1983).
Ceratostoma Sacc. (1876) ? = Arxiomyces (Ceratostomat.).
Ceratostomataceae G. Winter (1885), Sordariales. 12
gen. (+ 25 syn.), 53 spp. Ascomata translucent, yellow
to pale brown, ostiolate or not, often long-necked and
with smooth ostiolar setae; interascal tissue absent;
asci clavate, without apical apparatus, deliquescing;
ascospores usually large, brown, usually 2-pored,
non-septate, smooth or strongly ornamented, sheath
absent. Often fungicolous.
  Lit.: Cannon & Hawksworth (Bot. J. Linn. Soc. 84:
115, 1982; key gen.), Doguet (Botaniste 39: 1, 1955;
monogr.).
Ceratostomella Sacc. (1878), Clypeosphaeriaceae. c.
10, mainly temp. See Barr (Mycotaxon 46: 45, 1993;
posn), Untereiner (Mycol. 85: 294, 1993).
Ceratostomina Hansf. (1946) = Rhynchomeliola
(Trichosphaer.) fide Müller & v. Arx (1973).
Ceraunium Wallr. (1833) ≡ Elaphomyces (Elapho-
mycet.).
Cercidospora Körb. (1865), Dothideales (inc. sed.). 9
(on lichens), Eur. See Eriksson & Yue (SA 11: 166,
1993), Hafellner (Herzogia 7: 353, 1987; key).
Cercocladospora G.P. Agarwal & S.M. Singh (1974) =
Pseudocercospora (Mitosp. fungi) fide Deighton
(Mycol. Pap. 140, 1976).
Cercodeuterospora Curzi (1932) = Mycovellosiella
(Mitosp. fungi) fide Deighton (Mycol. Pap. 137,
1974).
Cercophora Fuckel (1870), Lasiosphaeriaceae. 20,
widespr. See Lundqvist (1972; coprophilous spp.),
Hilber & Hilber (Z. Mykol. 45: 209, 1979; on wood),
Ueda (Mycoscience 35: 287, 1994; anamorph
Chrysosporium-like), Cladorrhinum.
Cercoseptoria Petr. (1925) = Pseudocercospora
(Mitosp. fungi) fide Deighton (TBMS 88: 365, 1987).
Cercosperma G. Arnaud ex B. Sutton & Hodges
(1982), Mitosporic fungi, 1.C2.10. 1, Brazil.
Cercosphaerella Kleb. (1918) = Mycosphaerella
(Mycosphaerell.) fide v. Arx & Müller (1975).
Cercospora Fresen. (1863) [Apr.] nom. rej. prop. =
Pseudocercospora (Mitosp. fungi) fide Sutton & Pons
(Taxon 112: 643, 1991).

Cercospora Fresen. (1863) [Aug.] nom. cons. prop. Anamorphic Mycosphaerellaceae, 1.E1.31. *c.* 1270, widespr. parasites. Leaf spot of banana (*C. musae*, see *Mycosphaerella*), beet (*C. beticola*), celery (*C. apii*), groundnut (*Arachis*) (*C. arachidicola*, teleomorph *Mycosphaerella arachidis*), tobacco (*C. nicotianae*). See Chupp (*Monograph of the fungus genus Cercospora*, 1954; arranged by host fam.), Sutton & Pons (*Taxon* **116**: 643, 1991; conservation).
  Lit.: **Regional: Brazil**, Hino & Tokeshi (*Tech. Bull. TARC* **11**, 1978), Viégas (*Bol. Soc. Brasil. Agron.* **8**(1), 1945). **India**, Vasudeva (*Indian Cercosporae*, 1963; key 260 spp.), Govindu & Thirumalachar (*Sydowia* **18**: 18, 1965). **Indonesia**, Boedijn (*Nova Hedw.* **3**: 411, 1961; 90 spp.). **Japan**, Katsuki (*Trans. mycol. Soc. Japan, extra issue* I, 1965). **Cuba**, Arnold (*Feddes Rep.* **100**: 639, 1989). **Taiwan**, Hsieh & Goh (*TMS Rep. China* **4**: 9, 1989; *Cercospora and similar fungi from Taiwan*: 1, 1990).
  See also Fajola (*Nova Hedw.* **29**: 912, 922, 1983; cultural behaviour), Thaung (*Mycotaxon* **19**: 425, 1984; generic criteria), Pons *et al.* (*TBMS* **85**: 405, 1985; ultrastructure of *C. beticola*), Sivanesan (*TBMS* **85**: 397, 1985; *C. sesami* and related spp.), Assante *et al.* (*Riv. Pat. Veg.* ser. IV, **22**: 41, 1986; secondary metabolites), Pollack (*Mycol. Mem.* **12**: 1987; annotated compilation of species names), Pons & Sutton (*Mycol. Pap.* **160**: 1, 1988; spp. on *Dioscorea*), Sutton & Hodges (*Mycol.* **82**: 313, 1990; *Cercospora*-like spp. on conifers), Nelson & Campbell (*Pl. Dis.* **74**: 874, 1990; host range and culture characters of *C. zebrina*), Stewart & Pfleger (*J. Minn. Acad. Sci.* **53**: 34, 1989; synoptic key gen.), Braun (*Crypt. Bot.* **3**: 235, 1993; subgen. *Hyalocercospora*).
Cercosporella Sacc. (1880), Mitosporic fungi, 2.E1.10. 80, widespr. *C. pastinacae* (parsnip (*Pastinaca*) leaf spot). See Deighton (*Mycol. Pap.* **133**, 1973).
Cercosporidium Earle (1901), Mitosporic fungi, 2.C2.10. 17, widespr. See Deighton (*Mycol. Pap.* **112**, 1967; key).
Cercosporina Speg. (1911) = Cercospora (Mitosp. fungi) fide Sutton & Pons (*Mycotaxon* **12**: 201, 1980; redisp. original spp.).
Cercosporiopsis Miura (1928) nom. illegit. ≡ Pseudocercospora (Mitosp. fungi). See Carmichael *et al.* (1980).
Cercosporites E.S. Salmon (1903), Fossil fungi. 1 (Miocene), Italy.
Cercosporites Stopes (1913), Fossil fungi. 1 (Cretaceous), former Czechoslovakia.
Cercosporula Arnaud (1954) nom. inval. Mitosporic fungi, 1.E1.15. 2, France, Cuba.
Cercostigmina U. Braun (1993), Mitosporic fungi, 3.C2.19. 7, widespr.
Cerebella Ces. (1851) = Epicoccum (Mitosp. fungi) fide Schol-Schwarz (*TBMS* **42**: 149, 1959), but see Ellis (*DH*).
cerebriform, brain-like; convoluted.
Cereicium Locq. (1979), Cortinariaceae. 1, Eur.
Cereolus (Körb.) Boistel (1903) = Stereocaulon (Stereocaul.).
Cericium Hjortstam (1995), Corticiaceae. 1, S. Am.
Cerillum Clem. (1931) ≡ Colletomanginia (Xylar.).
Cerinomyces G.W. Martin (1949), Cerinomycetaceae. 6, temp. See McNabb (*N.Z. Jl Bot.* **2**: 403, 1964; key), Donk (*Persoonia* **4**: 267, 1966).
Cerinomycetaceae Jülich (1982), Dacrymycetales. 1 gen., 6 spp. Basidioma resupinate, membranous-ceraceous.
Cerinosterus R.T. Moore (1987), Mitosporic fungi, 1.A1.1. 2, widesp.
Ceriomyces Corda (1837) nom. dub. = Ptychogaster (Mitosp. fungi) fide Patouillard (1900).

Ceriomyces Murrill (1909) = Boletus (Bolet.) fide Singer (1945).
Cerion Massee (1901), Rhytismataceae. 2, Australia, trop. Am. See Sherwood (*Occ. Pap. Farlow Herb.* **15**, 1980).
Ceriophora Höhn. (1919) = Ceriospora (Hyponectr.) fide Clements & Shear (1931).
Cerioporus Quél. (1886) = Polyporus (Polypor.) fide Saccardo (1887).
Ceriospora Niessl (1876), ? Hyponectriaceae. 3, temp. See Hyde (*Sydowia* **45**: 204, 1993).
Ceriosporella Berl. (1894) = Lophiostoma (Lophiostomat.) fide v. Arx & Müller (1975). See Hyde (*Sydowia* **45**: 204, 1993).
Ceriosporella A.R. Caval. (1966) ≡ Marinospora (Halosphaer.).
Ceriosporopsis Linder (1944), Halosphaeriaceae. 7 (marine), widespr. See Jones & Zainal (*Mycotaxon* **32**: 237, 1988; key), Yusoff *et al.* (*CJB* **72**: 1550, 1994; ultrastr.).
Ceriporia Donk (1933), Coriolaceae. 7, widespr. See Pegler (1973). = Poria (Coriol.) fide Lowe (1966).
Ceriporiopsis Domański (1963) = Poria (Coriol.) fide Lowe (1966).
cernuous, hanging down; drooping; nodding.
Cerocorticium Henn. (1900), Hyphodermataceae. 1, Eur.
Cerodothis Muthappa (1969), Dothideales (inc. sed.). 1, India.
Cerophora Raf. (1808) = Lycoperdon (Lycoperd.) fide Donk (*Taxon* **5**: 73, 1956).
Ceropsora B.K. Bakshi & Suj. Singh (1960), Chaconiaceae. 1 (on *Picea*), India.
Cerotelium Arthur (1906), Phakopsoraceae. 20, widespr. (warmer areas). *C. fici* (fig rust). See Ono *et al.* (*MR* **96**: 844, 1992; key). See also *Phakopsora*.
Cerradoa J.F. Hennen & Y. Ono (1978), Uredinales (inc. sed.). 1 (on *Palmae*), Brazil.
Cerrena Gray (1821), Coriolaceae. 2, widespr. See Pegler (1973), Westhuizen (*CJB* **41**: 1487, 1963; structure).
Cerrenella Murrill (1905) = Inonotus (Hymenochaet.) fide Pegler (1964).
Cesatia Rabenh. (1850) [non *Cesatia* Endl. (1838), *Umbelliferae*] = Trullula (Mitosp. fungi) fide Saccardo (1884).
Cesatiella Sacc. (1878), Ascomycota (inc. sed.). 3, Eur., W. Indies. See Rossman (*in press*; posn).
cespitose, see caespitose.
Cestodella Tuzet, Manier & Jolivet (1957) = Enterobryus (Eccrin.) fide Manier & Lichtwardt (*Ann. Sci. Nat., Bot.*, sér. 12, **9**: 519, 1968).
Ceteraria Ach. (1809) ≡ Cetraria (Parmel.).
Cetraria Ach. (1803) nom. cons., Parmeliaceae (L). 15, temp. to arctic. See Kärnefelt *et al.* (*Bryol.* **96**: 394, 1993), Awasthi (*Bull. bot. Surv. India* **24**: 1, 1983; key, India), Jahns & Schuster (*Beitr. Biol. Pflanzen* **55**: 427, 1981; morphogenesis), Kärnefelt (*Opera bot.* **46**, 1979; monogr. brown fruticose spp.), Kärnefelt *et al.* (*Pl. Syst. Evol.* **183**: 113, 1992; cladistics).
Cetrariastrum Sipman (1980), Parmeliaceae (L). 3, mainly trop. See Sipman (*Mycotaxon* **26**: 235, 1986).
Cetrariella Kärnefelt & Thell (1993), Parmeliaceae (L). 2, N. Hemisph.
Cetrariomyces E.A. Thomas (1939) nom. inval. ≡ Cetraria (Parmel.).
Cetrariopsis Kurok. (1980), Parmeliaceae (L). 1, Himalayas, Japan, Taiwan. See Kärnefelt *et al.* (*Pl. Syst. Evol.* **183**: 113, 1992), Randlane *et al.* (*Cryptogamie, bryol. lich.* **16**: 35, 1995; status).
Cetrelia W.L. Culb. & C.F. Culb. (1968), Parmeliaceae (L). 15, cosmop. (mainly Asia). See Culberson & Culberson (*Contr. US Natn. Herb.* **34**: 449, 1968).

Cetreliopsis M.J. Lai (1980) = Cetrelia (Parmel.) fide Lumbsch (*SA* **7**: 105, 1988).

Ceuthocarpon P. Karst. (1873) = Linospora (Vals.).

Ceuthodiplospora Died. (1912), Mitosporic fungi, 4.B1.15. 1, former Czechoslovakia.

Ceuthosira Petr. (1924), Mitosporic fungi, ?4.A1.?. 1, Eur.

Ceuthospora Fr. (1825) nom. rej. = Pyrenophora (Pleospor.).

Ceuthospora Grev. (1826) nom. cons., Mitosporic fungi, 8.A1.15. 100, cosmop. See Sutton (1980), Ando *et al.* (*Ann. Phytopath. Soc. Jap.* **55**: 391, 1989; brown zonate leaf blight of tea by *C. lauri*), DiCosmo *et al.* (*Mycotaxon* **21**: 1, 1984).

Ceuthosporella Petr. & Syd. (1923), Mitosporic fungi, 8.A1.?. 1, Japan.

Ceuthosporella Höhn. (1929) ≡ Helhonia (Mitosp. fungi).

CF (CFL), see Nomenclature, Societies and organizations.

Chaconia Juel (1897), Chaconiaceae. 6, trop. See Ono & Hennen (*Trans. Mycol. Soc. Japan* **24**: 369, 1983).

Chaconiaceae Cummins & Y. Hirats. (1983), Uredinales. 11 gen. (+ 7 syn.), 73 spp. Pycnia discoid, with bounding structures, subcuticular or subepidermal; aeciospores either catenulate (*Aecidium*) or pedicellate (*Uraecium*); teliospores sessile or pedicellate, unicellular, free, usually pale.

Chadefaudia Feldm.-Maz. (1957), Halosphaeriaceae (?L). 6 (marine), widespr. See Kohlmeyer & Kohlmeyer, 1979), Stegena & Kemperman (*Bot. Marina* **27**: 443, 1984; S. Afr.).

Chadefaudiella Faurel & Schotter (1966), Chaedefaudiellaceae. 2 (coprophilous), Afr.

Chadefaudiellaceae Faurel & Schotter ex Benny & Kimbr. (1980), Microascales. 2 gen., 3 spp. Ascomata perithecial, the upper part formed from anastomosing setae; interascal tissue of undifferentiated hyphae; ascospores striate. Anamorphs unknown. Coprophilous in deserts, Afr.

Chadefaudiomyces Kamat, V.G. Rao, A.S. Patil & Ullasa (1974), Valsaceae. 1, India. Anamorph *Lasmenia*.

Chaenocarpus Rebent. (1804) = Thamnomyces (Xylar.). See Dennis (*Revta Biol.* **1**: 175, 1958).

Chaenocarpus Spreng. (1831), Xylariaceae. 1, N. temp. See Læssøe (*SA* **13**: 43, 1994).

Chaenocarpus Lév. (1843) nom. dub. (Xylar.).

Chaenotheca (Th. Fr.) Th. Fr. (1860), Coniocybaceae (L). 20, mainly temp. See Honegger (*Sydowia* **38**: 146, 1986; asci), Middelborg & Mattsson (*Sommerfeltia* **5**, 1987; Norway), Tibell (*Symb. bot. upsal.* **23** (1), 1980, key 14 N. Hemisph. spp.; *Nova Hedw.* **79**: 597, 1984; gen. concept; *Nordic J. Bot.* **13**: 441, 1993, anamorphs).

Chaenotheciella Räsänen (1943) ? = Chaenothecopsis (Mycocalic.).

Chaenothecomyces Cif. & Tomas. (1953) ≡ Chaenotheca (Coniocyb.).

Chaenothecopsis Vain. (1927), Mycocaliciaceae. 25 (on lichens, other fungi or wood), Eur. See Schmidt (1970), Hawksworth (*TBMS* **74**: 650, 1980), Tibell (1984; *Symb. bot. upsal.* **27**(1), 1987, status, Austral. spp.; *CJB* **69**: 2427, 1991, anamorphs), Tibell & Constantinescu (*MR* **95**: 556, 1991; anamorph), Tibell & Ryman (*Nova Hedw.* **60**: 199, 1995; key 8 spp.). Anamorph *Asterophoma*, *Catenomycopsis*, *Phialophora*-like.

Chaetalysis Peyronel (1922) = Acarosporium (Mitosp. fungi) fide Sutton (1977).

Chaetantromycopsis H.P. Upadhyay, Cavalc. & A.A. Silva (1986), Mitosporic fungi, 2.A2.15. 1, Brazil.

Chaetapiospora Petr. (1947) = Pseudomassaria (Hyponectr.) fide Barr (1976).

Chaetarthriomyces Thaxt. (1931), Laboulbeniaceae. 3, Asia, Eur., W. Afr. See Scheloske (*Parasitol. Schr. Reihe* **19**: 97, 1969).

Chaetasbolisia Speg. (1918), Mitosporic fungi, 4.A1.15. 7, widespr.

Chaetaspis Syd. & P. Syd. (1917) = Rhagadolobium (Parmular.) fide Müller & v. Arx (1962).

Chaetasterina Bubák (1909) = Chaetothyrium (Chaetothyr.).

Chaetendophragmia Matsush. (1971), Mitosporic fungi, 1.C2.19. 3, Papua New Guinea, Afr.

Chaetendophragmiopsis B. Sutton & Hodges (1978), Mitosporic fungi, 1.C2.19. 2, Brazil, Papua New Guinea.

Chaethomites, see *Chaetomites*.

Chaetoamphisphaeria Hara (1918), Ascomycota (inc. sed.). 1, ? Japan.

Chaetobasidiella Höhn. (1918) nom. dub. fide Sutton (*Mycol. Pap.* **141**, 1977).

Chaetobasis Clem. & Shear (1931) ≡ Chaetobasidiella (Mitosp. fungi).

Chaetoblastophorum Morgan-Jones (1977), Mitosporic fungi, 1.A1.1. 1, USA.

Chaetobotrys Clem. (1931) ≡ Kusanobotrys (Dothideales).

Chaetocalathus Singer (1942), Tricholomataceae. 14, cosmop, predom. trop. See Singer (*Lilloa* **8**: 441, 1942, monogr; *Flora neotrop.* **17**: 53, 1976, key 6 neotrop. spp.).

Chaetocarpus P. Karst. (1889) [non Thwaites (1849) nom. cons., *Euphorbiaceae*] = Columnocystis (Ster.) fide Donk (1964).

Chaetoceratostoma Turconi & Maffei (1912) = Scopinella (Ceratostomat.) fide Hawksworth (*TBMS* **64**: 447, 1975).

Chaetoceris Clem. & Shear (1931) ≡ Chaetoceratostoma (Ceratostomat.).

Chaetochalara B. Sutton & Piroz. (1965), Mitosporic fungi, 1.A/B1.22. 6, Eur., Afr., N. Am. See Nag Raj & Kendrick (*A monograph of Chalara and allied genera*, 1975). = Chalara (Mitosp. fungi) fide Kirk (1984).

Chaetocladiaceae A. Fisch. (1892), Mucorales. 2 gen., 7 spp., widespr. N. Hemisph. Sporangiophores terminating in sterile spines; pedicellate sporangiola unispored, formed on fertile vesicles; zygospores rough-walled, suspensors opposed.
  Lit.: Benny & Benjamin (*Mycol.* **85**: 670, 1993; emend., key).

Chaetocladium Fresen. (1863), Chaetocladiaceae. 2 (facultative parasites of *Mucorales*), widespr. See Benny & Benjamin (*Aliso* **8**: 391, 1976).

Chaetoconidium Zukal (1887), Mitosporic fungi, 1.?.?. 1, Eur.

Chaetoconis Clem. (1909), Mitosporic fungi, 4.C1.15. 2, N. temp. See Sutton (*CJB* **46**: 183, 1968), Nag Raj (*Coelomycetous anamorphs with appendage-bearing conidia*: 188, 1993).

Chaetocrea Syd. (1927) nom. dub. (Fungi, inc. sed.). See Rossman (*Mycol. Pap.* **157**, 1987).

Chaetocypha Corda (1829) Fungi (inc. sed.). See Donk (*Reinwardtia* **1**: 208, 1951).

Chaetocytostroma Petr. (1920), Mitosporic fungi, 8.A1.?. 1, Eur.

Chaetoderma Parm. (1968), Stereaceae. 1, Eur.

Chaetodermella Rauschert (1988), Stereaceae. 1, Eur.

Chaetodimerina Hansf. (1946) = Rizalia (Trichosphaer.) fide Pirozynski (1977).

Chaetodiplis Clem. (1931), Mitosporic fungi, 4.A2.?. 1, Eur.

Chaetodiplodia P. Karst. (1884), Mitosporic fungi, 4.B2.15. 9, esp. trop. See Zambettakis (*BSMF* **70**: 219, 1954).

Chaetodiplodina Speg. (1910), Mitosporic fungi, 4.A2.1. 2, S. Am. See Petrak & Sydow (*Ann. Myc.* **33**: 181, 1935).

Chaetodiscula Bubák & Kabát (1910) = Hymenopsis (Mitosp. fungi) fide Sutton (1977).

Chaetodochis Clem. (1909) ? = Chaetostroma (Mitosp. fungi).

Chaetodochium Höhn. (1932) = Volutella (Mitosp. fungi) fide Tulloch (*Mycol. Pap.* **130**, 1972).

Chaetolentomita Maubl. (1915) = Chaetosphaeria (Lasiosphaer.) fide Müller & v. Arx (1962).

Chaetomastia (Sacc.) Berl. (1891), Ascomycota (inc. sed.). 7, widespr. See Barr (*Mycotaxon* **34**: 507, 1989; key N. Am. spp.), Eriksson & Hawksworth (*SA* **8**: 64, 1989; posn).

Chaetomelanops Petr. (1948) = Pyrenostigme (Dothideales) fide v. Arx & Müller (1975).

Chaetomelasmia Danilova (1951) = Diachorella (Mitosp. fungi) fide Sutton (1977).

Chaetomeliola (Cif.) Bat., H. Maia & M.L. Farr (1962) = Meliola (Meliol.) fide Hughes (*in litt.*).

Chaetomella Fuckel (1870), Mitosporic fungi, 4.A1.15. 4 (esp. in soil), widespr. See Sutton & Sarbhoy (*TBMS* **66**: 297, 1976).

Chaetomeris Clem. (1931) ≡ Treubiomyces (Chaetothyr.).

Chaetomiaceae G. Winter (1885), Sordariales. 11 gen. (+ 6 syn.), 127 spp. Ascomata pale or dark, thin-walled, often with complex ornamented hairs, ostiolate or not; interascal tissue absent; asci lacking apical structures, deliquescing when mature; ascospores brown, often discharged as a cirrhus, usually 1-pored, non-septate, smooth, sheath absent. Anamorph usually absent. On cellulose-rich substratra.
    Lit.: See under *Chaetomium* and v. Arx *et al.* (1988).

Chaetomiales, see *Sordariales*.

Chaetomidium (Zopf) Sacc. (1882), Chaetomiaceae. 7 + (mainly from soil or coprophilous), widespr. See v. Arx (*Stud. Mycol.* **8**, 1975; *Beih. Nova Hedw.* **94**, 1988; keys).

Chaetomiopsis Mustafa & Abdul-Wahid (1990), Chaetomiaceae. 1, Egypt.

Chaetomiotricha Peyronel (1914) = Chaetomium (Chaetom.).

Chaetomites Pampal. (1902), Fossil fungi. 1 (Tertiary), Italy.

Chaetomium Kunze (1817), Chaetomiaceae. 81 (coprophilous, on straw, wet paper, cloth, cotton fibres (esp. *C. ·globosum*), many cellulolytic, some mycotoxic), cosmop. See v. Arx *et al.* (*Beih. Nova Hedw.* **84**, 1986; keys), Guarro & Figueras (*Crypt. Bot.* **1**: 97, 1989; ontogeny), Hawksworth & Wells (*Mycol. Pap.* **134**, 1973; ornamentation terminal hairs), Hubálek (*Česká Myk.* **28**: 65, 1974; dispersal by birds), Millner (*Biologia* Lahore **21**: 39, 1975; Pakistan spp., keys), Millner (*Mycol.* **69**: 492, 1977; growth responses), Millner *et al.* (*Mycol.* **69**: 720, 1977; ascospores SEM), Udagawa (*in* Kurato & Ueno (Eds), *Toxigenic fungi*: 139, 1984; toxins), Wicklow (*TBMS* **72**: 107, 1979; ecology and hair types).

Chaetomonodorus Bat. & H. Maia (1961), Mitosporic fungi (L), 4.C2,?. 1, Brazil.

Chaetomyces Thaxt. (1893), Laboulbeniaceae. 1, N. & S. Am.

Chaetonaemosphaera Schwarzman (1968) = Ceratocystis (Microascales) fide Sutton (*Mycol. Pap.* **141**, 1977).

Chaetonaevia Arx (1951), Dermateaceae. 1, Eur.

Chaetopatella I. Hino & Katum. (1958) = Pseudolachnea (Mitosp. fungi) fide Sutton (1977).

Chaetopeltaster Katum. (1975), Mitosporic fungi, 5.A1.?. 1, Japan.

Chaetopeltiopsis Hara (1913) nom. dub. (Mitosp. fungi).

Chaetopeltis Sacc. (1898) [non Berthold (1878), *Algae*] ≡ Tassia (Mitosp. fungi).

Chaetopeltopsis Theiss. (1913) = Chaetothyrina (Micropeltid.) fide v. Arx & Müller (1975).

Chaetophiophoma Speg. (1910), Mitosporic fungi, 4.E1.1. 1, S. Am.

Chaetophoma Cooke (1878), Mitosporic fungi, 4.A1.15. 30, widespr. See Hughes (*Mycol.* **68**: 693, 1976).

Chaetophomella Speg. (1918) nom. conf. fide Sutton (*Mycol. Pap.* **141**, 1977).

Chaetophorites Pratje (1922), ? Fossil fungi (mycel.). 1 (Silurian, Jurassic, Tertiary), Australia, Eur., N. Am.

Chaetoplaca Syd. & P. Syd. (1917), ? Schizothyriaceae. 1, Philipp.

Chaetoplea (Sacc.) Clem. (1931), Leptosphaeriaceae. 17, mainly N. temp. See Barr (*Mem. N.Y. bot. Gdn* **62**: 1, 1990; key), Eriksson & Hawksworth (*SA* **14**: 48, 1995; posn). = Leptosphaeria (Leptosphaer.) fide Crivelli (*Ueber die heterogene Ascomycetengattung Pleospora Rabh.*, 1983).

Chaetoporellus Bondartsev & Singer (1944) = Hyphodontia (Hyphodermat.) fide Langer (*Bibl. Mycol.* **158**, 1994).

Chaetoporus P. Karst. (1890), Coriolaceae. 17, widespr. See Pegler (1971). = Poria (Polypor.) fide Killermann (1916), = Junghuhnia (Polypor.) fide Ryvarden (1972) and Donk (1974).

Chaetopotius Bat. (1951) = Aithaloderma (Capnod.) fide v. Arx & Müller (1975).

Chaetopreussia Locq.-Lin. (1977), Sporormiaceae. 1, Sahara.

Chaetopsella Höhn. (1930) ≡ Chaetopsis (Mitosp. fungi) fide Hughes (1951).

Chaetopsina Rambelli (1956), Anamorphic Hypocreaceae, 1.A1.15. Teleomorph *Nectria*. 7, widespr. See Kirk & Sutton (*TBMS* **85**: 709, 1985; key), Rambelli (*Mic. Ital.* **16**: 7, 1987; bibliogr.), Onofri & Zucconi (*Mycotaxon* **41**: 451, 1991; SEM *C. fulva*), Samuels (*Mycotaxon* **22**: 13, 1985; teleomorphs).

Chaetopsis Grev. (1825), Mitosporic fungi, 1.B2.15. 6, widespr. See DiCosmo *et al.* (*Mycol.* **75**: 949, 1983; related genera).

Chaetopyrena Pass. (1881), Mitosporic fungi, ?.?.?. 1 or 2, Eur.

Chaetopyrena Sacc. (1883) ≡ Chaetopyrenis (Lophiostom.).

Chaetopyrenis Clem. & Shear (1931) = Keissleriella (Lophiostom.).

Chaetosaccardinula Bat. (1962) = Strigopodia (Euantennar.) fide v. Arx & Müller (1975).

Chaetosartorya Subram. (1972), Trichocomaceae. 3, Costa Rica. See Subramanian & Rajendran (*Revue Mycol.* **43**: 193, 1979; ontogeny). Anamorph *Aspergillus*.

Chaetosclerophoma Petr. (1924), Mitosporic fungi, 8.A1.?. 1, Eur.

Chaetoscorias W. Yamam. (1955) = Phragmocapnias (Capnod.) fide Hughes (1976).

Chaetoscutula E. Müll. (1959), Dothideales (inc. sed.). 1, France, Scotland.

Chaetoscypha Syd. (1924) ? = Lachnum (Hyaloscyph.). 1, N.Z.

Chaetoseptoria Tehon (1937), Mitosporic fungi, 4.E1.?. 1, N. Am. See Sutton (*Mycol. Pap.* **97**, 1964).

Chaetosira Clem. (1931) ≡ Wiesneriomyces (Mitosp. fungi).

Chaetospermella Naumov (1929) = Chaetospermum (Mitosp. fungi) fide Sutton (1977).

Chaetospermella Chardón & Toro (1934) ≡ Spermochaetella (Mitosp. fungi).

Chaetospermopsis Katum. & Y. Harada (1979), Mitosporic fungi, 3.A1.15. 1, Japan.

Chaetospermum Sacc. (1892), Mitosporic fungi, 8.A1.1. 3, widespr. See Sutton (1980).

Chaetosphaerella E. Müll. & C. Booth (1972), Lasiosphaeriaceae. 2, Eur. Anamorph *Cladotrichum.*

Chaetosphaeria Tul. & C. Tul. (1863), Lasiosphaeriaceae. 20, widespr. See Barr (*Mycotaxon* **39**: 43, 1990; posn), Booth (*Mycol. Pap.* **68**, 1957), Barr & Crane (*CJB* **57**: 835, 1979; anamorph), Gams & Holubová-Jechová (*Stud. Mycol.* **13**, 1976). Anamorphs *Catenularia, Chloridium, Custingophora, Menispora.*

Chaetosphaerites Félix (1894), Fossil fungi. 3 (Carboniferous, Tertiary), Eur., India, N. Am.

Chaetosphaeronema Moesz (1915), Mitosporic fungi, 4.B1.15. 3, widespr.

Chaetosphaeropsis Curzi & Barbaini (1927) = Coniothyriopsis (Mitosp. fungi.) fide Clements & Shear (1931), but see Sutton (1977).

Chaetosphaerulina I. Hino (1938) = Herpotrichia (Lophiostom.) fide Rossman (*Mycol. Pap.* **157**, 1987).

Chaetospora Faurel & Schotter (1965) [non R. Br. (1918), *Cyperaceae*] = Neochaetospora (Mitosp. fungi) fide Sutton & Sankaran (1991).

Chaetosporium Corda (1833) nom. dub. fide Saccardo (*Syll. Fung.* **4**: 761, 1886).

Chaetosticta Petr. & Syd. (1925), Mitosporic fungi, 4.C1.15. 1, N. Am. See Crane (*CJB* **49**: 31, 1971).

Chaetostigme Syd. & P. Syd. (1917) = Wentiomyces (Pseudoperispor.) fide Müller & v. Arx (1962).

Chaetostigmella Syd. & P. Syd. (1917) = Phaeodimeriella (Pseudoperispor.) fide Müller & v. Arx (1962).

Chaetostroma Corda (1829), Mitosporic fungi, 3.?.?. 10, widespr. Nom. conf., nom. dub. fide Tulloch (*Mycol. Pap.* **130**, 1972), Holubová-Jechová (*Sydowia* **46**: 238, 1994).

Chaetostromella P. Karst. (1895) nom. conf. fide Petrak (*Sydowia* **7**: 299, 1953).

Chaetostylum Tiegh. & G. Le Monn. (1873) = Helicostylum (Thamnid.) fide Lythgoe (1958) and Upadhyay (1973).

Chaetotheca Zukal (1890), Trichocomaceae. 1, Eur.

Chaetothyriaceae Hansf. ex M.E. Barr (1979), Dothideales. 8 gen. (+ 27 syn.), 75 spp. Mycelium largely superficial, with narrow cylindrical brown hyphae, sometimes with setose appendages; ascomata often formed beneath a subiculum, spherical or flattened, often collapsing when dry, the apex ± papillate, with a periphysate ostiole; peridium thin-walled; hymenium usually I+; interascal tissue of short apical periphysoids; asci saccate, fissitunicate; ascospores hyaline or pale, transversely septate or muriform. Anamorphs hyphomycetous. Epiphytic or biotrophic on leaves; mostly trop.
Lit.: Batista & Ciferri (*Sydowia, Beih.* **3**, 1962).

Chaetothyriales, see *Dothideales.*

Chaetothyrina Theiss. (1913), Micropeltidaceae. 6, widespr. (trop.). *C. musarum* (sooty blotch of banana). See Bitancourt (*Arq. Inst. biol., S. Paulo* **7**: 5, 1936).

Chaetothyriolum Speg. (1919) nom. dub. fide Santesson (*Symb. bot. upsal.* **12** (1), 1952).

Chaetothyriopsis F. Stevens & Dorman (1927) ? = Actinopeltis (Microthyr.). See Eriksson & Hawksworth (*SA* **9**: 6, 1991; status), Spooner & Kirk (*MR* **94**: 223, 1990).

Chaetothyrium Speg. (1888), Chaetothyriaceae. 25, esp. trop. See Hansford (1946) and Hughes (1976). Anamorph *Merismella.*

Chaetotrichum Rabenh. (1844) ≡ Chaetosporium (nom. dub.).

Chaetotrichum Syd. (1927) ≡ Annellophora (Mitosp. fungi).

Chaetotyphula Corner (1950), Clavariadelphaceae. 2, S.E. Asia.

Chaetozythia P. Karst. (1888) nom. dub.; based on a mite fide v. Höhnel (*Öst. bot. Z.* **63**: 238, 1913).

Chaetyllis Clem. (1931) ≡ Raciborskiomyces (Pseudoperispor.).

chaga fungus, see birch canker.

Chailletia Fuckel (1863) [non DC. (1811), *Chailletiaceae*] ? = Truncatella (Mitosp. fungi) fide Sutton (*in litt.*).

Chailletia P. Karst. (1871) [non DC. (1811), *Chailletiaceae*] ≡ Karstenia (Rhytismatales).

Chailletia Jacz. (1913), Valsaceae. 1, Switzerland.

Chainia Thirum. (1955), *Actinomycetes*, q.v.

Chainoderma Massee (1890) = Podaxis (Podax.) fide Cunningham (*Gast. Aust. N.Z.*: 196, 1944).

Chalara (Corda) Rabenh. (1844), Mitosporic fungi, 1.A/B1.22. 71, cosmop. *C. quercina* (oak wilt). See Nag Raj & Kendrick (*A monograph of Chalara and allied genera*, 1975), Kile & Walker (*Austr. J. Bot.* **35**: 1, 1987; *C. australis*, pathogenicity), Nag Raj & Kendrick (*in* Wingfield *et al.* (Eds), *Ceratocystis and Ophiostoma*, 1993; anamorphs ophiostomatoid fungi), *Endoconidiophora.*

Chalarodendron C.J.K. Wang & B. Sutton (1984), Mitosporic fungi, 2.A1.22. 1, USA.

Chalarodes McKenzie (1991), Mitosporic fungi, 1.A1.15. 2, S.W. Pacific Isl.

chalaroplectenchyma, see plectenchyma.

Chalaropsis Peyronel (1916) = Chalara (Mitosp. fungi) fide Nag Raj & Kendrick (1975).

Chalazion Dissing & Sivertsen (1975), Thelebolaceae. 2, Eur.

Chalciporus J. Bataille (1908), Strobilomycetaceae. 6, widespr. See Cooke (1953), Singer (*Sydowia* **31**: 196, 1979; key).

Chalcosphaeria Höhn. (1917) = Plagiostoma (Vals.) fide Müller & v. Arx (1973).

Chalymmota P. Karst. (1879) ≡ Panaeolus (Strophar.) fide Singer & Smith (1946).

Chamaeascus L. Holm, K. Holm & M.E. Barr (1993), Hyponectriaceae. 1, Svalbard.

Chamaeceras Rebent. ex Kuntze (1898) = Marasmius (Tricholomat.) fide Singer (1975).

Chamaemyces Earle (1909), Agaricaceae. 1, Eur.

Chamaenema Kütz. (1833) nom. dub. (? Fungi).

Chamaeota (W.G. Sm.) Earle (1909), Pluteaceae. 2, Eur., Am. See Singer (*Sydowia* **31**: 197, 1978; key).

Chamonixia Rolland (1899), Chamonixiaceae. 4, Eur., N. Am. See Smith & Singer (*Brittonia* **11**: 205, 1959).

Chamonixiaceae Jülich (1982), Boletales. 1 gen., 4 spp. Basidioma. Gasterocarpic, hypogeous; peridium cyanescent; spores golden brown, fusoid, longitudinally ridged.

chantarelle, basidioma of the edible *Cantharellus cibarius.*

Chanterel Adans. (1763) = Cantharellus (Cantharell.).

Chantransiopsis Thaxt. (1914), Mitosporic fungi, 1.A1.1. 3 (entomogenous), Java, Eur.

Chaos L. (1753) nom. dub.; included protozoa believed to be derived from fungus spores.

Chapeckia M.E. Barr (1978), Melanconidaceae. 2, N. Am.

Chapsa A. Massal. (1860) = Ocellularia (Thelotremat.).

Characonidia Bat. & Cavalc. (1965), Mitosporic fungi, 5.E1.?. 1, Brazil.

Charcotia Hue (1915) = Arthonia (Arthon.) fide Hawksworth (*SA* **10**: 127, 1991).

Chardonia Cif. (1930), Mitosporic fungi, 2.B1.?. 1, S. Am.

Chardoniella F. Kern (1939), Pucciniosiraceae. 4 (on *Compositae*), S. Am.; 0, III. See Buriticá & Hennen (*Fl. Neotrop. Monogr.* **24**: 39, 1980).

Charomyces Seifert (1987), Mitosporic fungi, 1A2.40. 2, Hawaii, Cuba.

Charonectria Sacc. (1880) = Nectriella (Hypocr.) fide Seaver (*Mycol.* **1**: 45, 1909).

Charrinia Viala & Ravaz (1894) nom. dub. (Fungi, inc. sed.). See Müller (*SA* **6**: 121, 1987).

chartaceous, paper-like.

Chaudhuria Zahlbr. (1932) = Heterodermia (Physc.).

Chaunopycnis W. Gams (1979), Mitosporic fungi, 3/8.A1.15. 1 (from soil), Sweden.

Cheese, Fungi are used in the manufacture of many cheeses, some of which have characteristic spp., e.g. camembert (*Penicillium camembertii*), roqueforte (*P. roquefortii*). See Babel (*Econ. Bot.* **7**: 27, 1953), Marth (*in* Reed, *Prescott & Dunn's Industrial microbiology*, edn 4: 65, 1982), Thom (*Bull. Bur. Anim. Ind. USDA* **82**, 1906); blue cheeses.

Cheilaria Lib. (1830), Mitosporic fungi, 8.C1.15. 1, widespr. See Sutton (1980).

Cheilariopsis Petr. (1959) ≡ Apomelasmia (Mitosp. fungi).

cheilocystidium, see cystidium.

Cheilodonta Boud. (1885) = Orbilia (Orbil.) fide Baral (1994).

Cheilophlebium Opiz & Gintl (1856) nom. dub. (? Agaric.). See Donk (1962: 50).

Cheilymenia Boud. (1885), Otideaceae. c. 40, Eur., N. Am. See Denison (*Mycol.* **56**: 718, 1964; key), Moravec (*Česká Myk.* **22**: 32, 1968, *Česká Myk.* **38**: 146, 1984, *Mycotaxon* **36**: 169, 1989, *Czech. Mycol.* **97**: 7, 1993), Wu & Kimbrough (*Bot. Gaz.* **152**: 421, 1992; ascospores).

Cheimonophyllum Singer (1955), Tricholomataceae. 2, widespr.

Cheiroconium Höhn. (1910) = Sirothecium (Mitosp. fungi) fide Hughes (1958).

cheiroid, see chiroid.

Cheiromoniliophora Tzean & J.L. Chen (1990), Mitosporic fungi, 1.D2.10. 1, Taiwan.

Cheiromycella Höhn. (1910), Mitosporic fungi, 3.G1.?. 2, Eur. and Japan. See Sutton (*Proc. Ind. Acad. Sci.* (Pl. Sci.) **94**: 229, 1985; disposition of names).

Cheiromyceopsis Mercado & J. Mena (1988), Mitosporic fungi, 3.G2.1. 1, Cuba.

Cheiromyces Berk. & M.A. Curtis (1857), Mitosporic fungi, 3.G2.1. 2, India, USA. See Sutton (*Proc. Ind. Acad. Sci.* (Pl. Sci.) **94**: 229, 1985; disposition of names).

Cheiromycina B. Sutton (1986), Mitosporic fungi (L), 3.G2.1. 1, Sweden.

Cheiropodium Syd. & P. Syd. (1915) = Clasterosporium (Mitosp. fungi) fide Ellis (*DH*).

Cheiropolyschema Matsush. (1980), Mitosporic fungi, 1.G2.28. 1, Taiwan.

Cheirospora Moug. & Fr. (1825), Mitosporic fungi, 6.G2.1. 1, widespr.

Chelisporium Speg. (1910) = Sirothecium (Mitosp. fungi). See Sutton (*Proc. Ind. Acad. Sci.* (Pl. Sci.) **94**: 229, 1985).

chemical race, a group of chemically differentiated individuals or populations and not of any particular taxonomic rank (i.e. a chemical race may be a species, var., or chemotype); **- strain**, used as an infraspecific taxonomic rank in lichenized taxa distinguished only by chemical characters (Lamb, *Nature* **168**: 38, 1951).

chemosyndrome, a biogenetically meaningful set of major and minor natural metabolic products produced by a species (Culberson & Culberson, *Syst. Bot.* **1**: 325, 1977).

chemotaxis, a taxis (q.v.) in response to a chemical stimulus. Gooday (*in* Carlile (Ed.), *Primitive sensory and communication systems*, 155, 1975; chemotaxis in fungi), Bonner (*Mycol.* **69**: 443, 1977; in cellular slime moulds).

Chemotaxonomy (biochemical systematics), taxonomy utilizing chemical characteristics, in a broad sense including both primary and secondary metabolites and physiological/chemical tests.

  Lit.: **Fungi**: Vogel (*Am. Naturalist* **48**: 435, 1964; evolutionary implications of lysine pathways), Murray (*J. gen. Microbiol.* **52**: 213, 1968; biochemical tests, pathogenic fungi), Tyrrell (*Bot. Rev.* **35**: 305, 1969; review), Benedict (*Adv. appl. Microbiol.* **13**: 1, 1970; basidiomycetes), Meixner (*Chemische Farbreaktionen von Pilzen*, 1975; asco- and basidiomycetes), Gómez-Miranda *et al.* (*Exp. Mycol.* **12**: 258, 1988; cell wall composition), Rast & Pfyffer (*Bot. J. Linn. Soc.* **99**: 1989; polyols), Klich & Mullaney (*in* Arora *et al.*, *Handbook of Applied Mycology* **4**: 35, 1992; DNA methods), Micales *et al.* (*loc. cit.* p. 57, isozyme analysis), Frisvad (*Chemom. Intell. Lab. Syst.* **14**: 253, 1992; secondary metabolites, chemometrics), Frisvad, *in* Hawksworth (Ed.) (*Identification and characterization of pest organisms*: 303, 1994), Botha & Kock (*Int. J. Food Microbiol.* **19**: 39, 1993; fatty acid profiles, yeasts).

  Lichens: Hawksworth (*in* Brown *et al.* (Eds), 1976: 139; review, bibliogr., use), Culberson & Culberson (*Syst. Bot.* **1**: 325, 1976; chemosyndromes), Rogers (*Bot. J. Linn. Soc.* **101**: 229, 1989; chemistry and species concept); Culberson & Elix (*Meth. Plant Biochem.* **1**: 509, 1993; review), Feige & Lumbsch (Eds) (*Bibl. Lich.* **53**, 1993; review). See also Metabolic products, Phylogeny.

chemotropism, see tropism.

chemotype, a group of chemically differentiated individuals of a species of unknown or of no taxonomic significance.

Chermomyces Brain (1923) Ascomycota (inc. sed.). See Batra (*Taxonomy of fungi* **1**: 187, 1978; applied to yeast-like cells in insect mycetomas).

Cheshunt compound. Copper sulphate (CuSO$_4$) 2 parts, ammonium carbonate ((NH$_4$)$_2$CO$_3$) 11 parts; used (at least 24 hr after mixing) at the rate of 56.7 g per 4.5 l of water as a soil disinfectant against damping-off.

chestnut blight, *Cryphonectria parasitica* (syn. *Endothia parasitica*). See Anagnostakis (*Mycol.* **79**: 23, 1987).

Chevalieria G. Arnaud (1920) [non Gaudich. ex Beer (1857), *Bromeliaceae*] ≡ Chevalieropsis (Parodiopsid.).

Chevalieropsis G. Arnaud (1923), Parodiopsidaceae. 1, Afr. See Petrak (*Sydowia* **5**: 346, 1951). Anamorph *Septoidium*-like.

Chiajaea (Sacc.) Höhn. (1920), Hypocreaceae. 2, Eur.

chiastobasidium, see basidium.

Chiastospora Riess (1852), Mitosporic fungi, 6.G1.?. 1, Eur. See Sutton (*Mycol. Pap.* **141**, 1977).

Chiastosporum Dughi (1956) nom. inval. = Collema (Collemat.).

Chikaneea B. Sutton (1973), Mitosporic fungi, 1.C1.10. 1, Canada.

Chilemyces Speg. (1910) = Dimerina (Pseudoperispor.) fide Petrak & Sydow (*Ann. Myc.* **32**: 1, 1934).

Chiliospora A. Massal. (1860) = Biatoridium (Lecanorales, inc. sed.) fide Hafellner (1994).

Chiloella Syd. (1928) = Glomerella (Phyllachor.) fide von Arx & Müller (1954).

Chilonectria Sacc. (1878) = Nectria (Hypocr.). See Winter (*Rabenh. Krypt.-Fl.* **1** (2): 114, 1885).

chimeroid (of lichen thalli), see Lichens, Phycotype.

Chinese cheese, see sufu.

Chinese mushroom, see straw mushroom.

Chinese rice, see starters.

Chiodecton Ach. (1814), Roccellaceae (L). 16, mainly trop. See Thor (*Opera Bot.* **103**, 1990; monogr., key).

Chiodectonomyces Cif. & Tomas. (1953) ≡ Enterographa (Roccell.).

Chiographa Leight. (1854) = Phaeographis (Graphid.).

Chionaster Wille (1903) nom. dub.; based on an alga having no chlorophyll.

Chionomyces Deighton & Piroz. (1972), Mitosporic fungi, 1.C1.10. 3 (on *Meliolaceae*), trop.

chionophilous, see nitrophilous.

Chionosphaera D.E. Cox (1976), Chionosphaeraceae. 1, N. Am.

Chionosphaeraceae Oberw. & Bandoni (1982), Atractiellales. 3 gen., 3 spp. Basidioma stilboid; hyphae with simple septal pores, clamp-connexions absent; spores hyaline, thick-walled, rugose.
  Lit.: Oberwinkler & Bauer (*Sydowia* **41**: 224, 1989).

Chionyphe Thienem. (1839) = Mucor (Mucor.) fide Hesseltine (1955).

chiroid (**cheiroid**), shaped like a hand with the fingers together and not divergent, e.g. the conidia of *Cheiromyces*; cf. digitate, palmate.

Chithramia Nag Raj (1988), Mitosporic fungi, 8.B2.1. 1, India.

chitinoclastic, chitin decomposing.

Chitinonectria M. Morelet (1968) = Nectria (Hypocr.) fide Hawksworth (*in litt.*).

Chitinozoa, fossil chitinous organisms of uncertain affinity found in the Upper Precambian to the Devonian.

Chitonia (Fr.) P. Karst. (1879) [non D. Don (1823), *Melastomataceae*] ≡ Clarkeinda (Agaric.).

Chitoniella Henn. (1898) ≡ Clarkeinda (Agaric.) fide Singer (1951).

Chitonis Clem. (1909) ≡ Chitonia (Agaric.).

Chitonomyces Peyr. (1873), Laboulbeniaceae. 96, widespr. See Sugiyama (*Trans. mycol. Soc. Japan* **18**: 155, 1977; Japanese spp.), Sugiyama & Nagasawa (*Trans. mycol. Soc. Japan* **26**: 3, 1985; Borneo).

Chitonospora E. Bommer, M. Rousseau & Sacc. (1891), Amphisphaeriaceae. 1, Eur. See Eriksson (*Svensk bot. Tidskr.* **60**: 320, 1966).

chitosan, a partially deacetylated form of chitin characteristic of zygomycetes; **chitases**, enzymes (EC 3.2.1.99) from fungi and bacteria able to hydrolyse chitosan (Monaghan *et al.*, *Nature* (*New Biol.*) **245**: 78, 1973).

chitosome, a small spheroidal structure (40-70 nm diam.), containing chitin synthetase zymogen, found in many fungi (Bracker *et al.*, *Proc. Nat. Acad. Sci. USA* **73**: 4570, 1976).

Chlamydoabsidia Hesselt. & J.J. Ellis (1966), Mucoraceae. 2, India, USA. See Hesseltine & Ellis (*Mycol.* **58**: 761, 1966), Behera & Mukerji (*Norw. Jl Bot.* **21**: 1, 1974).

Chlamydoaleurospora Grigoraki (1924) = Trichophyton (Mitosp. fungi).

chlamydocyst, a two-walled resting zoosporangium of *Blastocladiaceae* within a hypha.

Chlamydomucor Bref. (1889) = Mucor (Mucor.) fide Hesseltine (1955). See Ellis *et al.* (*Mycol.* **68**: 131, 1976).

Chlamydomyces Bainier (1907), Mitosporic fungi, 1.B2.1. 2, widespr. See Mason (*Mycol. Pap.* **2**: 37, 1928), Claudia (*Mycotaxon* **27**: 255, 1986).

Chlamydomyxa W. Archer (1875) = Labyrinthula (Labyrinthul.) fide Olive (1975: 215).

Chlamydomyzium M.W. Dick (1995), Myzocytiopsidaceae. 4 (parasitic in *Aschelminthes*), widespr.

Chlamydopsis Hol.-Jech. & R.F. Castañeda (1986), Mitosporic fungi, 1.D/G2.1/3. 1, Cuba.

Chlamydopus Speg. (1898), Tulostomataceae. 1, widespr. in deserts. See Long & Stouffer (*Mycol.* **38**: 619, 1947).

Chlamydorubra K.B. Deshp. & K.S. Deshp. (1966), Mitosporic fungi, 1.A2.?. 1, India.

chlamydospore, an asexual 1-celled spore (primarily for perennation, not dissemination) originating endogenously and singly within part of a pre-existing cell, by the contraction of the protoplast and possessing an inner secondary and often thickened hyaline or brown wall, usually impregnated with hydrophobic material. Originally proposed by de Bary in 1859 for *Asterophora* anamorphs. See Griffiths (*Nova Hedw.* **25**: 503, 1974; origin, structure, function), Hughes (*in* Arai (Ed.), *Filamentous microorganisms*: 1, 1985; definition, occurrence).

Chlamydosporites Paradkar (1975), Fossil fungi (? Ustilagin.). 1 (Cretaceous), India.

Chlamydosporium Peyronel (1913) = Phoma (Mitosp. fungi) fide Mouchacca & Sutton (*Crypt., Mycol.* **12**: 251, 1991).

Chlamydotomus Trevis. (1879) nom. dub. (? Algae).

Chlamydozyma Wick. (1964) = Metschnikowia (Metschnikow.) fide Pitt & Miller (*Mycol.* **60**: 663, 1968).

Chlorangium Link (1849) nom. rej. = Aspicilia (Hymenel.).

Chlorangium Rabenh. (1857) nom. dub. (Lecanorales, inc. sed.).

Chlorea Nyl. (1855) = Letharia (Parmel.).

Chlorencoelia J.R. Dixon (1975), Leotiaceae. 2, N. temp.

Chloridiella Arnaud (1953) nom. nud. = Idriella (Mitosp. fungi) fide Nicot & Charpentié (*BSMF* **87**: 621, 1972).

Chloridium Link (1809), Anamorphic Lasiosphaeriaceae, 1.A1.15/18. Teleomorph *Chaetosphaeria*. 12, widespr. See Gams & Holubová-Jechová (*Stud. Mycol.* **18**, 1976).

Chlorocaulum Clem. (1909) = Stereocaulon (Stereocaul.).

Chlorociboria Seaver ex C.S. Ramamurthi, Korf & L.R. Batra (1958) typ. cons., Leotiaceae. 4, widespr. See Dixon (*Mycotaxon* **1**: 193, 1975; key), Korf (*Mycol.* **51**: 298, 1951; status). See Tunbridge Ware.

Chlorocyphella Speg. (1909) = Pyrenotrichum (Mitosp. fungi) fide Santesson (*Symb. bot. upsal.* **12** (1), 1952).

Chlorodictyon J. Agardh (1870) = Ramalina (Ramalin.).

Chlorodothis Clem. (1909) = Tomasellia (Arthopyren.) fide Harris (*in litt.*).

Chlorolepiota Sathe & S.D. Deshp. (1979), Agaricaceae. 1, India.

Chloroneuron Murrill (1911) = Gomphus (Gomph.) fide Petersen (1972).

Chloropeltigera (Gyeln.) Gyeln. (1934) nom. inval. = Peltigera (Peltiger.).

Chloropeltis Clem. (1909) = Peltigera (Peltiger.).

chlorophycophilous (of fungi), lichenized with a green photobiont (Pike & Carroll, *in* Alexopoulos & Mims, *Introductory mycology*, edn 3, 1980).

Chlorophyllum Massee (1898), Agaricaceae. 7, widespr. See Heinemann (*Bull. Jard. bot. nat. Belg.* **38**: 195, 1968; *Rev. Mycol.* **31**: 317, 1966; *C. molybdites*, poisonous), Sundberg (*Madroño* **21**: 15, 1971).

Chlorophyllum Murrill (1910) = Gomphus (Gomph.) fide Petersen (1972).

chloroplast endoplasmic reticulum (CER), a layer of ribosome studded membrane surrounding the plastid.

Chloroscypha Seaver (1931), Leotiaceae. *c.* 12 (on conifers), N. temp. See Gremmen (*Nova Hedw.* **2**: 547, 1963; Netherlands spp.).

Chlorosperma Murrill (1922) = Melanophyllum (Agaric.) fide Singer (1951).

Chlorospleniella P. Karst. (1885), ? Leotiales (inc. sed.). 1, Brazil. See Groves & Wilson (*Taxon* **16**: 39, 1967).

Chlorosplenium Fr. (1849), Dermateaceae. 4, widespr. See Dixon (*Mycotaxon* **1**: 65, 1974; key), *Chlorociboria.*

Chlorospora Speg. (1891) ? = Peronospora (Peronospor.) fide Fitzpatrick (1930).

Chlorospora Massee (1898) ≡ Melanophyllum (Agaric.).

Chlorovibrissea L.M. Kohn (1989), Vibrisseaceae. 4, Australia, NZ. See Korf (*Mycosystema* **3**: 19, 1991; posn).

Chmelia Svob.-Pol. (1966), Mitosporic fungi, 1.A2.?. 1 (on humans), former Czechoslovakia. Nom. dub. fide de Hoog & Hermanides-Nijhof (*Stud. Mycol.* **18**, 1976).

Chnoopsora Dietel (1906) = Melampsora (Melampsor.) fide Cummins & Hiratsuka (1983).

Choanatiara DiCosmo (1984), Mitosporic fungi, 8.A1.15. 2, India, Canada.

Choanephora Curr. (1873), Choanephoraceae. 2, widespr., esp. trop. *C. cucurbitarum* (blossom blight and fruit rot of cucurbits and other plants). See Thaxter (*Rhodora* **5**: 97, 1903), Barnett & Lilly (*Phytopath.* **40**: 80, 1950; *Mycol.* **47**: 26, 1955; culture), Hesseltine & Benjamin (*Mycol.* **49**: 723, 1957), Kirk (1984; key), Higham & Cole (*CJB* **60**: 2313, 1982; sporangiolum ultrastr.).

Choanephoraceae J. Schröt. (1894), Mucorales. 3 gen. (+ 3 syn.), 5 spp. Sporangia and sporangiola borne on separate and distinct sporangiophores; zygospores striate, borne on tongs-like suspensors.
Lit.: Kirk (*Mycol. Pap.* **152**: 1, 1984; revision, keys).

Choanephorella Vuill. (1904) = Choanephora (Choanephor.) fide Zycha *et al.* (1969).

Choanephoroidea I. Miyake & S. Ito (1935) ? = Choanephora (Choanephor.) fide Kirk & Waterhouse (*in litt.*).

chocolate spot, a disease of *Vicia* and other legumes (*Botrytis cinerea* and *B. fabae*); see Wilson (*Ann. appl. Biol.* **24**: 258, 1937), Harrison (*Pl. Path.* **37**: 168, 1988).

Choeromyces Tul. & C. Tul. (1862) = Choiromyces (Helvell.) fide Fischer (1938).

Choiromyces Vittad. (1831), Helvellaceae. 5 (hypogeous), widespr. See Zhang & Minter (*MR* **92**: 91, 1989; cytology).

choke, a disease of grasses (*Epichloë typhina*).

Chondrioderma Rostaf. (1873) = Diderma (Didym.), etc.

Chondroderris Maire (1937), Leotiales (inc. sed.). 1, Spain.

Chondrogaster Maire (1925), Hymenogastraceae. 1, Afr.

chondroid axis, the cartilaginous axis occupying the central portion of the medulla in *Usnea.*

Chondroplea Kleb. (1933) = Discosporium (Mitosp. fungi) fide Sutton (1977).

Chondropodiella Höhn. (1917), Anamorphic Leotiaceae, 8.A1.?. Teleomorph *Godronia.* 1, USA.

Chondropodiola Petr. & Cif. (1932) nom. dub. fide Sutton (*Mycol. Pap.* **141**, 1977).

Chondropodium Höhn. (1916) = Corniculariella (Mitosp. fungi) fide DiCosmo (*CJB* **56**: 1665, 1978).

Chondropsis Nyl. (1863) nom. cons., Parmeliaceae (L). 2, Australasia. See Rogers (*New Phytol.* **70**: 1069, 1971), Elix & Child (*Brunonia* **9**: 113, 1986), Lumbsch & Kothe (*Lichenologist* **20**: 25, 1988; anatomy etc.).

Chondrospora A. Massal. (1860) = Anzia (Parmel.).

Chondrostereum Pouzar (1959), Meruliaceae. 4, widespr. *C. purpureum* (silver leaf of plum, etc.).

Chondrostroma Syd. (1940) [non Gürich (1906), fossil ? *Algae*] = Phacidiopycnis (Mitosp. fungi) fide Rupprecht (*Sydowia* **13**: 10, 1959).

Chordecystia Foster (1979), Fossil fungi.

Chordostylum Tode (1790) nom. dub. (Thamnid.).

Choreospora Constant. & R. Sant. (1987), Mitosporic fungi, 3.G1.15. 1 (L), Australia.

Chorioactis Kupfer ex Eckblad (1968), Sarcosomataceae. 1, USA. See Bellemere *et al.* (*Nova Hedw.* **58**: 49, 1994; ultrastr.).

Choriphyllum Velen. (1922) ≡ Phaeolus (Coriol.) fide Donk (1975).

Chorostate (Nitschke ex Sacc.) Traverso (1906) = Diaporthe (Vals.) fide Barr (*Mycol. Mem.* **7**, 1978).

Chorostella (Sacc.) Clem. & Shear (1931) = Cryptodiaporthe (Vals.) fide Barr (*Mycol. Mem.* **7**, 1978).

Christiansenia Hauerslev (1969) = Syzygospora (Syzygospor.) fide Ginns (1986).

Christiaster Kuntze (1891) ≡ Gonatobotryum (Mitosp. fungi).

Chromatium Link (1824) = Dematium (Mitosp. fungi) fide Mussat (*Syll. Fung.* **15**: 86, 1901).

Chromatochlamys Trevis. (1860), ? Thelenellaceae (L). 17, widespr. See Mayrhofer (*Bibl. lich.* **26**, 1987; monogr., key).

chromatography, physico-chemical diffusion technique for the identification of chemical products; **gas -**, in a gaseous phase; **high-pressure liquid -** (h.p.l.c.), in solvent under pressure; **paper -**, on filter paper; **thin-layer -** (t.l.c.) in thin silicate layers on glass, aluminium or plastic plates. Much used in lichenology (see Chemotaxonomy, Metabolic products).
Lit.: **General:** Harborne (*Phytochemical methods*, 1973), Smith (*Chromatographic and electrophoretic techniques* **1**, 1960).
Lichens: Culberson (*J. Chromatogr.* **72**: 113, 1972; **97**: 107, 1974; t.l.c.), Culberson & Ammann (*Herzogia* **5**: 1, 1979; t.l.c.), Culberson *et al.* (*Bryologist* **84**: 16, 1981; t.l.c. β-orcinol depsidones), Culberson & Johnson (*J. Chromatogr.* **128**: 253, 1976; two dimensional t.l.c.), Culberson & Kristinsson (*J. Chromatogr.* **46**: 85, 1970; t.l.c.), Elix *et al.* (*Mycotaxon* **31**: 89, 1988; computer progr. lichen products), Feige *et al.* (*J. Chromatogr.* **646**: 417, 1993; h.p.l.c. lichen products), Mietzsch *et al.* (*Mycotaxon* **47**: 475, 1993; computer progr. lichen products), Walker & James (*Bull. Br. lichen Soc.* **46**: 13, 1980; t.l.c.), Wilkins & James (*Lichenologist* **11**: 271, 1979; two dimensional t.l.c. triterpenoids).

Chromelosporium Corda (1833) = Ostracoderma (Mitosp. fungi) fide Hughes (1958), but segregated from *Botrytis* by Hennebert (1973). *C. fulvum* (peat mould). See Hughes & Bisalputra (*CJB* **48**: 361, 1970; ultrastructure of conidiogenesis), Hennebert & Korf (*Mycol.* **67**: 214, 1975).

CHROMISTA (incl. Heterokonta, Pseudofungi, Pseudomycotina, straminopiles). Kingdom of *Eukaryota.* Predominantly unicellular, filamentous, or colonial primarily phototrophic organisms; cell walls not of chitin and β-glucan (often cellulosic); chloroplasts (when present) located in the lumen of a generally rough endoplasmic reticulum, lack starch and phycobilisomes and have a two-membraned envelope inside a periplastid membrance, chlorophylls when present *a* and *c*; mitochonidia generally with tubular cristae; Golgi bodies and peroxisomes always present; flagella when present, with at least one with rigid, tubular, and usually tripartite flagellar hairs or mastigonemes (except haptophytes); mostly free-living; many microscopic in size (with major exceptions, e.g. brown algae).

Formerly often included in *Protoctista* (syn. *Protista*; e.g. Margulis *et al.*, 1990), but that kingdom is now generally split into *Chromista* and *Protozoa*. Includes 10 phyla (Corliss, 1994) or 3 phyla and 8 sub- and infraphyla (Cavalier-Smith, 1993); encompasses a wide range of golden and brown algae, diatoms, chrysophytes and cryptomonads (but not chlorophyte and red algae; in *Plantae*). Dick (in press), however, separates the kingdom *Straminipila* (q.v.) from the *Chromista* s.str.

The 'fungal' phyla are interpreted as having lost chloroplasts secondarily, and are part of subkingdom *Heterokonta*. The classification within the kingdom is unsettled; 2 of the phyla Corliss (1994) accepted comprise organisms studied by mycologists: *Labyrinthomorpha* (incl. *Labyrinthulea* and *Thraustochytriaceae*) and *Pseudofungi* (incl. *Oomycetes* and *Hyphochytriomycetes*); Barr (1992) included the same fungi but used different ranks.

Various arrangements and ranks have been proposed for the main groupings of 'fungi' within the *Chromista*; selected ones are compared in Table 4 (Fungi, q.v.). Three fungal phyla belonging to the *Chromista* are accepted in this edition of the *Dictionary*:

(1) **Hyphochytriomycota**.
(2) **Labyrinthulomycota**.
(3) **Oomycota**.

Lit.: Barr (*Mycol.* **84**: 1, 1992), Cantino (*Ann. Rev. Microbiol.* **13**: 103, 1959), Cavalier-Smith (*Prog. Phycol. Res.* **4**: 309, 1986; *in* Rayner *et al.* (Eds), *Evolutionary biology of the fungi*: 339, 1987; *in* Osawa & Honjo (Eds), *Evolution of life*: 271, 1991; *Microbiol. Rev.* **57**: 953, 1993), Corliss (*Acta Protozool.* **33**: 1, 1994), Buczacki (Ed.) (*Zoosporic plant pathogens*, 1983), Cordá-Olmedo & Lipson (Eds) (*Phycomycetes*, 1987), Dick (*Mycol. Pap.*: *in press*), Fitzpatrick (1930), Fuller & Jaworski (Eds) (*Zoosporic fungi in teaching and research*, 1986), Margulis *et al.* (Eds) (*Handbook of Protoctista*, 1990) Sparrow (*Aquatic Phycomycetes*, edn 2, 1960; *in* Ainsworth *et al.* (Eds), *The Fungi* **4B**: 61, 1973; keys classes, orders, etc.).

See Kingdoms of fungi, Phylogeny, Protozoa, Straminipila.

chromo- (combining form), colour.
Chromocleista Yaguchi & Udagawa (1993), Trichocomaceae. 2, Japan. Anamorph *Geosmithia*, *Paecilomyces*.
Chromocrea Seaver (1910) ≡ Creopus (Hypocr.).
Chromocreopsis Seaver (1910) = Thuemenella (Xylar.) fide Corlett (*Mycol.* **77**: 272, 1985).
Chromocyphella De Toni & Levi (1888), Crepidotaceae. 2, widespr.
Chromocytospora Speg. (1910) = Phomopsis (Mitosp. fungi) fide Petrak (*Sydowia* **5**: 335, 1951). See Sutton (*Mycol. Pap.* **141**, 1977).
chromogen, a stain-producing organism (Erlich, 1941).
chromogenesis, colour production.
chromogenic (chromogenous), colour-producing.
chromomycosis (chromoblastomycosis), a skin disease in humans caused by species of *Phialophora*; dermatitis verrucosa; see Al-Doory (*Chromomycosis*, 1972).
chromophilous, deeply staining.

chromosome, see Chromosome maps, Chromosome numbers.
Chromosome maps. The first chromosome map for a fungus was of the sex chromosome of *Neurospora crassa* (Lindegren, *J. Genet.* **32**: 243, 1936), the first for a set of fungal chromosomes was for the 8 chromosomes of *Aspergillus nidulans* (Kafer, *Adv. Genet.* **9**: 105, 1958; for revised maps see Dorn, *Genetics* **56**: 619, 1967). See also Lindegren *et al.* (*Nature* **183**: 800, 1959; *Saccharomyces cerevisiae*). See also Molecular biology.
Chromosome numbers. In fungi, are generally low; $n=4$ is possibly the basic number. 2- and 3-ploid series are frequent among *Chromista*, and higher polyploids can occur. *Allomyces* has 8-50+ chromosomes (Emerson & Wilson, *Mycol.* **46**: 393, 1954); *Puccinia graminis* $n=3$ (fide McGinnis, *CJB* **31**: 522, 1953); *Dermatocarpon weberi* $n=6$ or 8; *Porpidia crustulata* $n=2$; *Agaricales*, $n=2-12$ (Ueda, *Trans. mycol. Soc. Japan* **10**: 23, 1969); see karyotype.

Rather few fungi have so far been studied in this respect, although pulse gel electrophoresis will accelerate this process (see Cytology). See Altman & Dittmer (*Biological data book*, 1964), Burges (*in* Lousley, *Species studies in the British flora*: 76, 1955), Ornduff (*Index to plant chromosome numbers* [*Regnum veg.* **50**], 1968), Rogers (*Evolution* **27**: 153, 1973; polyploidy in fungi).

See also pulsed field gel electrophoresis.
Chromosporium Corda (1829), Mitosporic fungi, 1.A1.15. 20, widespr.
Chromostylium Giard (1889) ? = Metarhizium (Mitosp. fungi) fide Tulloch (*TBMS* **66**: 409, 1976).
Chromotorula F.C. Harrison (1927) = Rhodotorula (Mitosp. fungi) fide Lodder (*Die anaskosporogenen Hefen* **1**, 1934).
Chroocybe Räsänen (1943) nom. inval. ? = Coniocybe (Coniocyb.).
Chroodiscus (Müll. Arg.) Müll. Arg. (1890), Thelotremataceae (L). 4, trop. See Lumbsch & Vězda (*Nova Hedw.* **50**: 245, 1990).
Chroogomphus (Singer) O.K. Mill. (1964), Gomphidiaceae. 13, N. temp. See Singer & Kultem (*Česká Myk.* **30**: 81, 1976).
Chrooicia Trevis. (1861) = Pyrenula (Pyrenul.).
Chroolepus C. Agardh (1824) = Cystocoleus (Mitosp. fungi).
Chroostroma Corda (1837) = Pactilia (Mitosp. fungi) fide Saccardo (*Syll. Fung.* **4**: 673, 1886).
Chrysachne Cif. (1938), Mitosporic fungi, 3.A1.?. 1, W. Indies.
Chrysalidopsis Steyaert (1961), Mitosporic fungi, 6.C2.?. 1, Chile.
Chryseidea Onofri (1981), Mitosporic fungi, 2.A1.6/10. 1, Ivory Coast. ? = Phaeoisaria (Mitosp. fungi) fide Sutton (*in litt.*).
Chrysella Syd. (1926), Pucciniaceae. 1, C. Am.; 0, III.
chryseous, golden yellow.
Chrysobostrychodes G. Kost (1985), Tricholomataceae. 1, N. Am.
Chrysocelis Lagerh. & Dietel (1913), Chaconiaceae. 6, widespr., esp. tropics. See Ono & Hennen (*Trans. Mycol. Soc. Japan* **24**: 369, 1983).
Chrysoconia McCabe & G.A. Escobar (1979), Coniophoraceae. 1, Réunion.
Chrysocyclus Syd. (1925), Pucciniaceae. 3, trop. Am. See Davidson (*Mycol.* **24**: 221, 1932).
chrysocystidium, see cystidium.
Chrysoderma Boidin & Gilles (1991), Corticiaceae. 1, Réunion.
Chrysogloeum Petr. (1959), Anamorphic Vizellaceae, 6.A1.?. Teleomorph *Blasdalea*. 1, Peru.

Chrysogluten Briosi & Farneti (1904) = *Nectria* (Hypocr.) fide Rogerson (1970).

chrysogonidia, photobiont cells of *Trentepohlia* (obsol.).

Chrysomma Acloque (1893) = *Caloplaca* (Teloschist.).

Chrysomphalina Clémençon (1982), Tricholomataceae. 2, Eur.

Chrysomyces Theiss. & Syd. (1917) = *Perisporiopsis* (Parodiopsid.) fide Müller & v. Arx (1962).

Chrysomyxa Unger (1840), Coleosporiaceae. 20, N. Hemisph. 0, I on *Picea*, II (I^{II}, with catenulate urediniospores), III on Dicots (= *Melampsoropsis*); or microcyclic, III on *Picea*. See Savile (*Can. J. Res.* **C28**: 318, 1950).

Chrysonilia Arx (1981), Anamorphic Sordariaceae, 1.A1.38. Teleomorph *Neurospora*. 3, widespr. *C. sitophila* (syn. *Monilia sitophila*), red bread mould.

Chrysophlyctis Schilb. (1896) = *Synchytrium* (Synchytr.) fide Fitzpatrick (1930).

Chrysopsora Lagerh. (1892), Pucciniaceae. 1 (on *Gynoxis*), trop. Am.; 0, III.

Chrysopsora (Vain.) M. Choisy (1951) = *Psora* (Psor.) fide Timdal (1984).

Chrysosplenium Allesch. (1898) [non L. (1753), *Saxifragaceae*] ≡ *Chlorosplenium* (Dermat.) fide Dixon (1974).

Chrysospora, see *Chrysopsora* (Vain.) M. Choisy.

Chrysosporium Corda (1833), Anamorphic Onygenaceae and Arthrodermataceae, 1.A1.1. Teleomorphs various. *c.* 22, widespr. See Carmichael (*CJB* **40**: 1137, 1962), Sigler *et al.* (*Mycotaxon* **10**: 133, 1979), Oorschot (*Stud. Mycol.* **20**, 1980; key 22 spp.), Chabasse (*Bull. Soc. Fr. mycol. Med.* **17**: 373, 1988; key French spp.), Boekhout *et al.* (*Stud. Mycol.* **31**: 29, 1989; septal ultrastr.), Skou (*Mycotaxon* **43**: 237, 1992; xerophilic spp., ? related to *Ascosphaeraceae*), Takizawa *et al.* (*Mycoscience* **35**: 327, 1995; ubiquinone system).

Chrysothallus Velen. (1934) [non K.I. Mey. (1930), *Algae*], Hyaloscyphaceae. 8, former Czechoslovakia. See Korf (*Lloydia* **14**: 134, 1951).

Chrysothricaceae Zahlbr. (1905), Arthoniales (L). 2 gen. (+ 4 syn.), 10 spp. Thallus granular, without a cortex, usually poorly developed; ascomata apothecial, ± round, with rudimentary walls, the hymenium usually yellowish. Lichenized with green algae.

Chrysothrix Mont. (1852) nom. cons., Chrysothrichaceae (L). 4, widespr. See Laundon (*Lichenologist* **13**: 101, 1981; key).

Chuppia Deighton (1965), Mitosporic fungi, 1.D2.1. 1, Venezuela.

Chytrella Kirschst. (1941) = *Unguicularia* (Hyaloscyph.) fide Raschle (1977).

chytrid, one of the *Chytridiales*. See Canter-Lund & Lund, *Freshwater Algae* (1995) for general account.

Chytridhaema Moniez (1887), ? Olpidiaceae. 1, France.

Chytridiaceae Nowak. (1878), Chytridiales. 34 gen. (+ 12 syn.), 322 spp. Thallus eucarpic, monocentric; sporangium endogenous to the zoospore cyst, operculate or inoperculate; rhizoids endo- or interbiotic; resting spore endogenous or exogenous.

Chytridiales, Chytridiomycetes. 5 fam., 77 gen. (+ 25 syn.), 566 spp. Thallus monocentric or polycentric, endogenous or exogenous; zoospores mostly with one conspicuous lipid globule; rhizoids tapering to fine (<0.5μm) tips; generally aquatic but also in soil; saprobic or parasitic on algae, microfauna, other fungi, pollen, plant debris, chitin and keratin, rarely terrestrial on higher plants (*Synchytrium*); cosmop.

The *Spizellomycetales* were removed from *Chytridiales sensu* Sparrow and *Chytridiales* emended using ultrastructural characters of zoospore (Barr,

1980): microtubules extending from one side of kinetosome in a parallel array; non-flagellate centriole parallel and connected to the kinetosome; rumposome usually present on surface of lipid globule; ribosomes usually in mass, which is more or less enclosed by endoplasmic reticulum; mitochondria usually associated with the microbody-lipid globule complex (MLC). Families in order are not phylogenetic. Sparrow (1960) divided the order into an operculate and inoperculate series; Karling (1977) used Whiffen's (1944) system based on thallus development types, a system followed by Barr which is used herein. Fams:
(1) **Chytridiaceae**.
(2) **Cladochytriaceae**.
(3) **Endochytriaceae**.
(4) **Harpochytriaceae**.
(5) **Synchytriaceae**.

The placement of many genera within the revised *Chytridiales* is still uncertain.

Lit.: Sparrow (1960), Karling (*Chytridiomycetarum iconographia*, 1977; 175 pl.), Barr (*CJB* **58**: 2380, 1980; 1990), Barr & Désauliniers (*CJB* **66**: 869, 1988; precise configuration of organelles in zoospore), Powell & Roychoudhury (*CJB* **70**: 750, 1992; redefinition of chytridialean types of MLC), Longcore (*CJB* **71**: 414, 1993; new type of MLC, *Mycol.* **87**: 25, 1995; discussion of taxonomic characters).

Chytridioides Tregnoboff (1913), ? Chytridiomycetes (inc. sed.). 1, Eur. See Karling (1977: 37).

Chytridiomycetes, see *Chytridiomycota*.

Chytridiomycota (Chytridiomycetes, Rumpomycetes). Fungi. 1 class, 5 ord., 18 fam., 112 gen. (+ 41 syn.), 793 spp. Thallus coenocytic, holocarpic or eucarpic, monocentric, polycentric, or mycelial; cell walls chitinous (at least in hyphal stages); mitochondial cristae flat; zoospores posteriorly (whiplash) monoflagellate or more rarely polyflagellate, lacking mastigonemes or scales, with unique flagellar 'root' systems and sometimes rumposomes (q.v.). Aquatic saprobes or parasites growing on decaying and living organic material (incl. nematodes, insects, plants, other chytrids and fungi), in freshwater and soils; a few are marine, some are obligate anaerobes in guts of herbivores.

The presence of flagellate zoospores has led to these fungi sometimes being placed with the fungal phyla of the *Chromista* rather than in *Fungi* (Kendrick, 1992; Kreissel, 1988). Retained in the *Fungi* (Barr, 1992; Bruns *et al.*, 1991; Corliss, 1994) on ultrastructural (lack of mastigonemes), cell wall chemistry (chitinous) and molecular evidence (see Phylogeny).

Four orders are recognized here; gen. formerly placed in *Harpochytriales* are now distributed between *Chytridiales* and *Monoblepharidales*:
(1) **Blastocladiales**.
(2) **Chytridiales**.
(3) **Monoblepharidales**.
(4) **Neocallimastigales**.
(5) **Spizellomycetales**.

Lit.: Barr (*in* Margulis *et al.* (Eds), *Handbook of Protoctista*: 454, 1990; *Mycol.* **84**: 1, 1992), Bruns *et al.* (*Ann. Rev. Ecol. Syst.* **22**: 525, 1991), Corliss (1994), Karling (*Sydowia Beih.* **6**, 1966; Indian spp., keys; *Chytridiomycetarum iconographia*, 1977), Li *et al.* (*CJB* **71**: 393, 1993; cladistics), Paterson (*in* Parker (Ed.), 1982, **1**: 173), Powell (*Mycol.* **85**: 1, 1993; roles in environment), Sparrow (1960; *in* Ainsworth *et al.*, *The Fungi* **4B**: 85, 1973).

Chytridiopsis W.G. Schneid. (1884), ? Chytridiomycetes (inc. sed.). 2, Eur., S. Am. See Jirovec (*Arch. Protistenk.* **94**: 84, 1940).

Chytridium A. Braun (1851), Chytridiaceae. 34, temp. See Karling (1977; in algae).

Chytriomyces Karling (1945), Chytridiaceae. 33, widespr. See Karling (1977), Johnson (*J. Elisha Mitchell sci. Soc.* **87**: 200, 1971; Icelandic spp.).

Cibalocoryne Hazsl. (1881) = Geoglossum (Geogloss.) fide Saccardo (*Syll. Fung.* **8**, 1889).

Cibdelia Juel (1925), ? Olpidiaceae. 1, Eur.

Ciboria Fuckel (1870), Sclerotiniaceae. *c.* 15, temp. See Schumacher (*Norw. J. Bot.* **25**: 145, 1978; Norwegian amenticolous spp.), Spooner (*Bibl. Mycol.* **116**, 1987; 3 spp. Australasia).

Ciboriella Seaver (1951) = Hymenoscyphus (Leot.) fide Schumacher (*Mycotaxon* **38**: 233, 1990).

Ciborinia Whetzel (1945), Sclerotiniaceae. 9, widespr. See Batra (*Am. J. Bot.* **47**: 819, 1960; key).

Ciboriopsis Dennis (1962) = Ciboria (Sclerotin.). See Dumont (*Mycol.* **68**: 233, 1976).

Cicadocola Brain (1923), Mitosporic fungi, 1.A1.4. 1 (in *Insecta*), S. Afr.

Cicadomyces Šulc (1911), ? Saccharomycetaceae. 10 (in *Insecta*), widespr.

cicatricose, with longitudinal ridges (Fig. 29.15).

cicatrized (of conidiogenous cells and conidia), having thickened scars.

Cicinnobella Henn. (1904) = Perisporina (Parodiopsid.) fide Höhnel (*Sber. Akad. wiss. Wien* **120**: 379, 1911), but used for *Dimerium* anamorphs.

Cicinobolus Ehrenb. (1853) = Ampelomyces (Mitosp. fungi) fide Rogers (*Mycol.* **51**: 96, 1959), but see Donk (*Taxon* **15**: 149, 1966).

Cidaris Fr. (1849), ? Helvellaceae. 1, N. Am. See van Brummelen (*in* Hawksworth (Ed.), *Ascomycete systematics*: 399, 1994; posn).

cider, an alcoholic drink obtained by the fermentation of apple juice by various yeasts. Cf. beer.

Cienkowskia Rostaf. (1873) [non Regel & Rach (1859), *Boraginaceae*], Physaraceae. 1, widespr. ? = Willkommlangea (Physar.).

Ciferri (Raffaele; 1897-1964). A most versatile Italian mycologist (Professor of Botany, University of Pavia, 1942-64) who made notable contributions to tropical mycology, *Ustilaginales* (q.v.), plant pathogenic fungi (q.v.), medical mycology (q.v.), and mycological bibliography in more than a thousand publications. See Baldacci (*Mycol.* **57**: 198-201, 1965; portr.), Tomaselli (*Atti Ist. bot. Univ. Pavia, ser.* 5, **21** (Suppl.), 1964; bibl.), Grummann (1974: 516), Stafleu & Cowan (*TL-2* **1**: 503, 1976).

Ciferria Gonz. Frag. (1925), Mitosporic fungi, ?.1E.?. 1, W. Indies.

Ciferriella Petr. (1930), Mitosporic fungi, 6.C2.19. 1, W. Indies.

Ciferrina Petr. (1932), Mitosporic fungi, 8.E1.?. 1, W. Indies.

Ciferriolichen Tomas. (1952) = Arthopyrenia (Arthopyren.).

Ciferriomyces Petr. (1932) ? = Pyrenostigme (Dothideales) fide Hansford (1946).

Ciferriopeltis Bat. & H. Maia (1965), Mitosporic fungi, 5.A2.?. 1, Dominican Republic.

Ciferriotheca Bat. & I.H. Lima (1959) = Metathyriella (Schizothyr.) fide Luttrell (*in* Ainsworth *et al.*, *The Fungi* **4A**: 213, 1973).

Ciferrioxyphium Bat. & H. Maia (1963), Anamorphic Capnodiaceae, 2.C1.15. Teleomorph *Aithaloderma*. 2, trop.

Ciferriusia Bat. (1962) = Yatesula (Chaetothyr.) fide v. Arx & Müller (1975).

Ciglides Chevall. (1826) = Gymnosporangium (Puccin.) fide Sydow (1912).

Ciliaria Quél. (1885) [non Stackh. (1809), *Algae*] ≡ Scutellinia (Otid.) fide Clements & Shear (1931).

ciliate, edged with hairs.

Ciliatula Velen. (1922) = Pezoloma (Leot.) fide Korf (*Phytologia* **21**: 201, 1971).

ciliatulate, thinly ciliate.

Cilicia Fr. (1825), Basidiomycetes (inc. sed., L). 1, Eur. fide Donk (*Reinwardtia* **2**: 435, 1954). See Laundon (*Lichenologist* **13**: 101, 1981).

Ciliciocarpus Corda (1831) = Gautieria (Gautier.) fide Pilát (*Fl. ČSR* **B1**: 211, 1958).

Ciliciopodium Corda (1831), Anamorphic Hypocreaceae, 5.A1.?. Teleomorph *Nectria*. 5, widespr. See Booth (*Mycol. Pap.* **73**, 1959).

Ciliciopus Clem. & Shear (1931) ≡ Ciliciopodium (Mitosp. fungi).

Ciliella Sacc. & P. Syd. (1902), Leotiales (inc. sed.). 1, Brazil.

Ciliochora Höhn. (1919), Mitosporic fungi, 8.A1.19. 2, India, Indonesia, Philipp. See Nag Raj (1993).

Ciliochorella Syd. (1935), Mitosporic fungi, 8.B2.15. 1, trop. See Subramanian & Ramakrishnan (*TBMS* **39**: 314, 1956).

Ciliofusa Clem. & Shear (1931) ≡ Ciliofusarium (Mitosp. fungi).

Ciliofusarium Rostr. (1892) = Menispora (Mitosp. fungi) fide Hughes (1958).

Ciliofusospora Bat. & J.L. Bezerra (1963), Ascomycota (inc. sed.). 1, Brazil.

Ciliolarina Svrček (1977), Hyaloscyphaceae. 6, N. Am., N. Eur. See Huhtinen (*Bibl. Mycol.* **150**: 93, 1993; key).

Ciliomyces Höhn. (1906) = Paranectria (Hypocr.) fide Hawksworth & Pirozynski (1977).

Ciliomyces I. Foissner & W. Foissner (1986), ? Lagenaceae. 1, Austria. See Dick (*in press*).

Ciliophora Petr. (1929), Mitosporic fungi, 8.A1.19. 1, C. Am.

Ciliophorella Petr. (1940), Mitosporic fungi, 4.A1.?. 1, Austria.

Cilioplea Munk (1953), Lophiostomataceae. 4, Eur., N. & S. Am. See Barr (*SA* **5**: 125, 1986; posn), Crivelli (*Ueber die heterogene Ascomycetengattung Pleospora*, 1983; key).

Ciliosculum Kirschst. (1941), ? Hyaloscyphaceae. 1, Germany.

Ciliosira Syd. (1942) = Acarosporium (Mitosp. fungi) fide Petrak (*Sydowia* **14**: 347, 1960).

Ciliospora Zimm. (1902) = Chaetospermum (Mitosp. fungi) fide Sutton (1977).

Ciliosporella Petr. (1927), Mitosporic fungi, 4.A1.15. 1, Eur.

cilium (pl. **cilia**), (1) an appendage of animal cells, e.g. protozoa; sometimes used for the flagellum (q.v.) of a zoospore; (2) a hair-like out-growth, e.g. from the edge of an apothecium or lichen thallus.

-cillin (suffix), for penicillins; derivatives of carboxy-6-amino-penicillanic acid (*WHO Chron.* **17**: 400, 1963).

cincinnate (**cincinnal**), rolled round; curled.

Cinereomyces Jülich (1982), Coriolaceae. 1, USA, Sweden.

cingulate, edged all round.

Cintractia Cornu (1883), Ustilaginaceae. *c.* 10, widespr., trop. & subtrop. The sorus of agglutinated ustilospores is characterized by a sterile stroma; until recently the name has been widely used for *Anthracoidea* (q.v.). See Ingold (*MR* **99**: 140, 1995; spore germination).

Cintractiella Boedijn (1937), Ustilaginaceae. 1 (induces hypertrophy in *Hypolytrum*), E. Indies.

Cintractiomyxa Golovin (1952) = Anthracoidea (Ustilagin.) fide Vánky (1987).

Cionium Link (1809) = Didymium (Didym.).

Cionothrix Arthur (1907), Pucciniosiraceae. 10, S. Am., Australia; 0, III. See Buriticá (*Rev. Acad. Colomb. Cienc.* **18**(69): 131, 1991).

circadian, pertaining to a day, e.g. a 24-hr rhythm. Cf. diel, diurnal.

Circinaria Link (1809) nom. rej. = Aspicilia (Hymenel.) fide Laundon & Hawksworth (1988).

Circinaria Fée (1825) = Coccocarpia (Coccocarp.).

Circinaria Bonord. (1851) = Valsa (Vals.).

Circinaria M. Choisy (1929) = Lecanora (Lecanor.).

Circinastrum Clem. (1909) ≡ Weinmannodora (Mitosp. fungi). See Sutton (1977).

circinate, twisted round; coiled (Fig. 37.37).

Circinella Tiegh. & G. Le Monn. (1873), Mucoraceae. 7, widespr. See Hesseltine & Fennell (*Mycol.* **47**: 193, 1955; key).

Circinoconis Boedijn (1942), Mitosporic fungi, 1.F2.1. 1, E. Indies.

Circinomucor Arx (1982) = Mucor (Mucor.) fide Kirk (*in litt.*).

Circinostoma Gray (1821) ? = Valsa (Vals.).

Circinotrichum Nees (1816), Mitosporic fungi, 1.A1.6. 3, widespr. See Pirozynski (*Mycol. Pap.* **84**, 1962; key).

Circinumbella Tiegh. & G. Le Monn. (1872) = Circinella (Mucor.) fide Hesseltine (1955).

Circulocolumella S. Ito & S. Imai (1957), Hysterangiaceae. 1, Bonin Isl.

circum- (prefix), all round; round about.

circumcinct, having a band around the middle.

circumscissile, opening or cracking along a circle.

cirrate (cirrose), rolled round (curled) or becoming so.

Cirrenalia Meyers & R.T. Moore (1960), Mitosporic fungi, 1.F2.1. 10 (marine and terrestrial), temp. See Goos (*Proc. Ind. Acad. Sci.* (Pl. Sci.) **94**: 245, 1985), Kohlmeyer (*Ber. dt. bot. Ges.* **79**: 27, 1966).

Cirrholus Mart. (1821) nom. dub. (Fungi, inc. sed.).

Cirrhomyces Höhn. (1903) = Chloridium (Mitosp. fungi) fide Hughes (1958).

Cirrosporium S. Hughes (1980), Mitosporic fungi, 8.C2.23. 1, NZ.

cirrus (cirrhus), a curl-like tuft; a tendril-like mass or 'spore horn' of forced-out spores.

Cirsosia G. Arnaud (1918), Asteriaceae. 6, trop. See Batista & Maia (*Rev. brasil. Biol.* **2**: 115, 1960).

Cirsosiella G. Arnaud (1918) = Cirsosia (Asterin.) fide Müller & v. Arx (1962).

Cirsosina Bat. & J.L. Bezerra (1960), ? Microthyriaceae. 2 (on *Rhododendron*), Eur., N. Borneo.

Cirsosiopsis Butin & Speer (1979), Microthyriaceae. 1 (on *Araucaria*), Brazil.

Cissococcomyces Bat. (1923), Mitosporic fungi, 1.?.?. 1 (in *Insecta*), S. Afr.

Cistella Quél. (1886) nom. cons., Hyaloscyphaceae. c. 20, temp. See Dennis (*Mycol. Pap.* **32**, 1949), Matheis (*Friesia* **11**: 85, 1976), Raitviir (*Scripta Mycol.* **1**, 1970; keys).

Cistellina Raitv. (1978), Hyaloscyphaceae. 2, Eur., former USSR.

cisternal ring, a ring-like arrangement of the endoplasmic reticulum which appears to bud and give rise to vesicles.

Citeromyces Santa María (1957), Saccharomycetaceae. 1, Spain, Hawaii.

citreoviridin, a polyene toxin of *Penicillium citreoviride*; the cause of cardiac beri-beri in humans.

citrinin, a toxic yellow pigment of *Penicillium citrinum*, *P. viridicatum*, etc.; the cause of nephrotoxicosis in pigs.

Citromyces Wehmer (1893) = Penicillium (Mitosp. fungi) fide Thom (*The Penicillia*, 1930).

Civisubramaniania Vittal & Dorai (1986), Mitosporic fungi, 1.A2.1. 1, India.

CK, see Metabolic products.

Cladaria Ritgen (1831) = Ramaria (Ramar.) fide Corner (1953).

Cladaspergillus Ritgen (1831) = Aspergillus (Mitosp. fungi) fide Mussat (*Syll. Fung.* **15**: 89, 1901).

clade, a monophyletic group of any magnitude (first used by Huxley, *Nature* **180**: 454, 1957); the recognition and hierarchical arrangement of such taxa constitutes the practice of cladistics (q.v.).

Cladia Nyl. (1870), Cladoniaceae (L). 9, E. Afr., S. Hemisph. See Ahti (*Regnum veg.* **128**: 58, 1993; names in use), Filson (*J. Hattori bot. Lab.* **49**: 1, 1981; monogr., key).

Cladidium Hafellner (1984), Lecanoraceae (L). 2, W. USA. See Ryan (*Mycotaxon* **34**: 697, 1989).

Cladina (Nyl.) Nyl. (1866) = Cladonia (Cladon.) fide Rouss & Ahti (*Lichenologist* **21**: 29, 1989). Cladoniaceae (L), 43 (fide Ahti, *Regnum veg.* **128**: 58, 1993; names in use), widespr. See Ahti (*Ann. Soc. zool.-bot. fenn. Vanamo* **32**(1), 1961, *Beih. Nova Hedw.* **79**: 25, 1984), Burgaz & Ahti (1994), Huovinen & Ahti (*Ann. bot. fenn.* **23**: 93, 1986; chemistry 30 spp.), Wei *et al.* (*Acta mycol. sin.* **5**: 240, 1986; 9 spp. China).

Cladinomyces Cif. & Tomas. (1953) = Cladonia (Cladon.).

Cladistics (Phylogenetic reconstruction). A method of systematics which aims to reconstruct the genealogical descent of organisms by means of objective and repeatable analysis, and from this pattern to propose a falsifiable hypothesis of natural classification or phylogeny. It is based upon three fundamental assumptions namely: that taxa are united into natural groups on the basis of shared derived characters (**synapomorphies**); that all groups recognized must descend from a single ancestor (i.e. they are monophyletic) and that the most **parsimonious** pattern (that is the one requiring the least number of steps to resolve the relationships of the taxa) is the one most likely to be correct. Groups defined using shared primitive characters (**symplesiomorphies**) are likely to be paraphyletic and thus not include all the descendants of an ancestor. Groups delimited that do not share a common ancestor are polyphyletic and often result from convergent (non-homologous) characters. The product of cladistic analysis is a branching diagram (**cladogram**) which shows the pattern of relationships between the organisms based on the characters used.

Many algorithms exist for deriving cladograms (Forey *et al.*, *Cladistics: a practical course in systematics*, 1992; Wiley, *Phylogenetics: the theory and practice of phylogenetic systematics*, 1981; Linder, *S. Afr. J. Bot.* **54**: 208, 1988, review for botanists; Scotland *et al.* (Eds), *Models in phylogeny reconstruction*, 1994). Most cladistic analysis is carried out using computers and a range of packages are available, the most common being PAUP, HENNIG 86 & PHYLIP (Platnik, *Cladistics* **3**: 121, 1987). This methodology has had an impact on many areas of systematics, particularly in biogeography (Wiley, *Syst. Bot.* **5**: 194, 1980); host parasite relationships and coevolution (Wheeler & Blackwell (Eds), *Fungus-insect relationships*, 1984; Humphries *et al.*, in Stone & Hawksworth, *Coevolution and systematics*: 55, 1986; *Nothofagus* and its parasites), as well as more recently in the field of molecular systematics (Williams in Forey *et al.*, 1992: 88, 102). Its effect on mycology has also been relatively recent but is becoming highly significant and, when combined with molecular data, has contributed to understanding of the relationships of higher taxa, see Tehler (*Cladistics* **4**: 227, 1988, *Eumycota*; *CJB* **68**: 2458, 1990, euascomycetes), Hawksworth (Ed.) (*Ascomycete systematics*, 1994; molecular systematics) as well as at lower ranks, e.g. Reynolds (*Mycotaxon* **27**: 377, 1986; *Capnodiaceae*), Schumacher (*Op. bot.* **101**, 1990; *Scutellinia* - phenetics and cladistics), Tehler (*Op. bot.* **70**, 1983; *Dirina* and *Roccellina*) and Kuyper (*Persoonia*, suppl. **3**, 1986; *Inocybe* subgen. *Inosperma*). For the use of molecular systematics in mycology see Li & Heath (*CJB* **70**: 1738, 1992; rRNA sequences in chytrids) and Gargas & Taylor (*Exper. mycol.* **19**: 7, 1995; discomycetes).

Cladobotryum Nees (1816), Anamorphic Hypocreaceae, 1.C1.15. Teleomorph *Hypomyces*. 8, widespr. See Gams & Hoozemans (*Persoonia* **6**: 96, 1970; key), de Hoog (*Persoonia* **10**: 33, 1978; key 11 spp.).

Cladobyssus Ritgen (1831) = Hypha (Mitosp. fungi) fide Mussat (*Syll. Fung.* **15**, 1901).

Cladochaete Sacc. (1912) [non DC. (1838), *Compositae*] = Chaetomium (Chaetom.) fide Petrak & Sydow (*Reprium nov. Spec. Regni veg., Beih.* **42**: 487, 1927).

Cladochytriaceae J. Schröt. (1892), Chytridiales. 11 gen. (+ 2 syn.), 40 spp. Thallus eucarpic, polycentric, comprising a rhizomycelium and reproductive bodies; sporangia operculate or inoperculate.

Cladochytrium Nowak. (1877), Cladochytriaceae. 11, widespr. See Sparrow (1960: 462; key).

Cladoconidium Bandoni & Tubaki (1985), Mitosporic fungi, 1.C1.10. 1, Japan.

Cladodendron Lázaro Ibiza (1916) ≡ Grifola (Coriol.) fide Donk (1974).

Cladoderris Pers. ex Berk. (1842) = Cymatoderma (Podoscyph.) fide Donk (1957).

Cladodium (Tuck.) Gyeln. (1934) [non Brid. (1826), *Musci*] ≡ Cladidium (Lecanor.).

cladogram, see Cladistics.

Cladographium Peyronel (1918), Mitosporic fungi, 2.A2.?. 1, Eur.

Cladolegnia Johannes (1955) = Saprolegnia (Saprolegn.) fide Dick (1973).

Cladomeris Quél. (1886) = Grifola (Coriol.) fide Pegler (1971).

Cladona Adans. (1763) nom. rej. = Cladonia (Cladon.).

Cladonia P. Browne (1756) nom. cons., Cladoniaceae (L). 402, cosmop. See Ahti (*Regnum veg.* **128**: 58, 1993; names in use), Archer & Bartlett (*N.Z. Jl Bot.* **24**: 581, 1986; key 44 NZ spp.), Asahina (*Lichens of Japan* **1**, 1950), Ahti (*Annls bot. fenn.* **10**: 163, 1973, *Unciales*; *Annls bot. fenn.* **15**: 7, 1978, Eur.; *Annls bot. fenn.* **17**: 195, 1980, *gracilis*-group; *Lichenologist* **12**: 125, 1980; *J. Hattori bot. Lab.* **52**: 331, 1982, evol. trends; *Lichenologist* **14**: 105, 1982, morphology), Bird & Marsh (*CJB* **50**: 915, 1972; Alberta), Burgaz & Ahti (1994), Hammer (*Bryologist* **96**: 299, 1993, sect. *Perviae* N. Am.; **98**: 1, 1995; 57 NW USA spp.), *Mycol.* **87**: 46, 1995, ontogeny), Honegger (*Lichenologist* **15**: 57, 1983; asci), Jahns (*Untersuchungen zur Entwicklungsgeschichte der Cladoniaceen*, 1970), Jahns & Beltman (*Lichenologist* **5**: 349, 1973; ontogeny), Culberson & Kristinsson (*Bryologist* **72**: 431, 1970), Sandstede (*Rabenh. Krypt.-Fl.* **9**(4,2), 1931; Eur.), Thomson (*The lichen genus Cladonia in North America*, 1968), Upreti (*Feddes Repert.* **98**: 469, 1987; key 62 spp. India & Nepal), Vainio (*Acta Soc. Fauna Flora fenn.* **4**, 1887; **10**, 1894; **14**, 1897; monogr. [reprint 1978]). See also *Cladina*.

Cladoniaceae Zenker (1827), Lecanorales (L). 13 gen. (+ 24 syn.), 435 spp. Thallus composed of a usually well-developed persistent foliose to squamulose primary thallus, and a ± vertical usually branched solid or hollow secondary thallus; ascomata apothecial, without a properly differentiated margin, usually brightly coloured; interascal tissue of sparsely branched paraphyses; asci with an I+ apical cap and an I+ gelatinous outer layer; ascospores usually aseptate; cephalodia absent. Lichenized with green algae, widspr. esp. boreal and austral.

Lit.: Ahti (*Lichenologist* **14**: 105, 1982; interpretation of thallus, *Regnum veg.* **128**: 58, 1993; gen., spp. names in use), Burgaz & Ahti (*Nova Hedw.* **59**: 399, 1994), Jahns (*Nova Hedw.* **20**: 1, 1970; ontogeny), Steenroos *et al.* (*Fl. Cript. Tierra del Fuego* **13**(7), 1992; keys 52 spp.).

Cladoniomyces E.A. Thomas ex Cif. & Tomas. (1953) ≡ Pyxidium (Cladon.).

Cladoniopsis Zahlbr. (1941) = Baeomyces (Baeomycet.).

Cladophialophora Borelli (1980), Mitosporic fungi, 1.2B4+1A.15. 4, widespr. See Honbo *et al.* (*Sabouraudia* **22**: 209, 1984; relationship to *Cladosporium carrionii*), Braun & Feiler (*Microbiol. Res.* **150**: 81, 1995; key, n. sp. and comb.).

Cladophytum Leidy (1849) nom. dub. (? Fungi).

Cladoporus (Pers.) Chevall. (1826) = Laetiporus (Polyp.) fide Donk (1960; based on an abnormal polypore).

Cladopsis Nyl. (1885) = Pyrenopsis (Lichin.) fide Henssen (*Lichenologist* **21**: 101, 1989).

Cladopycnidium H. Magn. (1940) = Lecidea (Lecid.) fide Hertel (*Ergebn. Forsch Unternehmens Nepal Himalaya* **6**: 145, 1977).

Cladorrhinum Sacc. & Marchal (1885), Anamorphic Lasiosphaeriaceae, 1.A1.15. Teleomorphs *Apiosordaria*, *Cercophora*. 5, widespr. See Domsch *et al.* (*Compendium of soil fungi*, 1980), Mouchacca & Gams (*Mycotaxon* **48**: 415, 1993; key).

Cladosarum E. Yuill & J.L. Yuill (1938) ? = Aspergillus (Mitosp. fungi) fide Raper & Fennell (1965).

Cladosphaera Dumort. (1822), ? Ascomycota (inc. sed.). 1, Eur.

Cladosphaeria Nitschke ex Jacz. (1894) = Cryptosphaeria (Diatryp.). See Petrak (*Sydowia* **15**: 186, 1962).

Cladosporiella Deighton (1965), Mitosporic fungi, 1.C2.3. 1, Malaysia.

Cladosporites Félix (1894), Fossil fungi. 4 (Tertiary), Eur., USA.

Cladosporium Link (1816), Anamorphic Mycosphaerellaceae, 1.C2.3. Teleomorphs *Mycosphaerella*, *Venturia*. 50, cosmop. *C. herbarum* (see McKemy & Morgan-Jones (*Mycotaxon* **42**: 307, 1991) is ubiquitous, generally saprobic, and its teleomorph is *Mycosphaerella tassiana* (v. Arx, *Sydowia* **4**: 320, 1950); *C. carpophilum* (peach scab), *C. cucumerinum* (cucumber gummosis). See de Vries (*Contribution to the knowledge of the genus Cladosporium Link ex Fr.*, 1952 [reprint 1967]), Ellis (*DH, MDH*), Yamamoto (*Sci. Rep. Hyogo Univ. Agr.* **4**: 1, 1959; spp. from Japan), Latgé *et al.* (*J. Microbiol.* **34**: 1325, 1988; wall ultrastr. *C. cladosporioides*), Kwon-Chung *et al.* (*J. Med. Vet. Mycol.* **27**: 413, 1989; taxonomy of *C. trichoides*), Yegres & Richard-Yegres (*in* Miyaji (Ed.), *Current problems of opportunistic fungal infections*: 12, 1989; epidemiology of *C. carrionii*), Morgan-Jones & McKemy (*Mycotaxon* **39**: 185, 1990; taxonomy of *C. uredinicola*), McKemy & Morgan-Jones (*Mycotaxon* **41**: 135, 1991; taxonomy of *C. chlorocephalum*), McKemy & Morgan-Jones (*Mycotaxon* **41**: 397, 1991; taxonomy of *C. oxysporum*), Morgan-Jones & McKemy (*Mycotaxon* **43**: 9, 1992; taxonomy of *C. vignae*), McKemy & Morgan-Jones (*Mycotaxon* **43**: 163, 1992; taxonomy of *C. cucumerinum*), Kawasaki *et al.*

(*Mycopathol.* **124**: 149, 1993; molecular epidemiology of *C. carrionii*). See also cellar fungus, *Fulvia*, *Hormoconis*.

Cladosporothyrium Katum. (1984) = Zelopelta (Mitosp. fungi) fide Sutton (*in litt.*).

Cladosterigma Pat. (1892), Mitosporic fungi, 2.C1.?. 1, Ecuador. See Petch (*TBMS* **8**: 212, 1923), Seifert (*TBMS* **85**: 123, 1985).

Cladotrichum Corda (1831) = Oedemium (Mitosp. fungi) fide Hughes (1958).

clamp-connexion (-connection, clamp, clamp-cell) (of basidiomycetes), a hyphal outgrowth which, at cell division, makes a connection between the resulting two cells by fusion with the lower; buckle; nodose septum; by-pass hypha (Fig. 9).

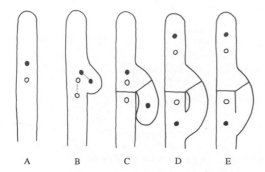

A   B   C   D   E

Fig. 9. Clamp connexion formation (diagrammatic). A, dikaryotic hyphal apex; simultaneous nuclear division and formation of a backwardly directed lateral branch into which one of the daughter nuclei passes; C, formation of two cross-walls cutting off an apical cell which contains two compatible nuclei, and a lateral branch with a single nucleus; D, fusion of lateral branch with subapical cell which then becomes dikaryotic; E, final stage.

Clarkeinda Kuntze (1891), Agaricaceae. 1, trop. Asia. See Pegler (*J. Linn. Soc., Bot.* **91**: 245, 1985).

Classification. The asigning of objects to defined categories; taxonomy. The application of scientific names to the categories into which fungi may be placed and the relative order of those categories is governed by the International Code of Botanical Nomenclature (see Nomenclature). The rank of **species** is basic (Art. 2), but as yet there is no universally applicable definition of species (q.v.). There is fairly general agreement on the ranks to be used for the main infraspecific ranks (although taxa based on pathogenic, physiological, or biochemical characters present difficulties), and on ranks above species from genus to order. Above order, and particularly above class there is more diversity in usage, not least because the rules of priority do not apply above family. In the example in Table 3, group-name endings in heavy type are those authorized for fungi in the *Code* (Arts 16-19). The relative order of the various ranks must not be altered (Art. 5) or the names become invalid (Art. 33.5; an exception is made in Art. 33.6 for wrong use of 'tribe' within a genus by Fries in the *Systema mycologicum*, 1821-32). Additional ranks may be intercalated or added, 'provided confusion or error is not thereby introduced' (Art. 4.3).

There is particular instability in the use of ranks above order, and different authors prefer to use or avoid particular categories. 'Division' is peculiar to botany, but from 1993 'phylum', familiar to zoologists, is authorized as an alternative that can be used in its place and is used by mycologists (see Kingdoms of fungi). Above the rank of kingdom, additional categories justified by evidence from molecular phylogeny (q.v.) are becoming necessary, 'domain' now being the most widely used.

Different suffixes are available for the terminations of higher taxonomic categories according to whether the Botanical or Zoological Code of nomenclature is used, and further under the Botanical Code whether the names are considered to be of fungi or algae (i.e. '-*phyta*' not '-*mycota*' as the phylum termination; '-*phyceae*' not '-*mycetes*' for class names; for consistency in this *Dictionary*, and to clearly indicate where phyla are ones traditionally studied by mycologists, we have retained the fungal termination throughout. This also has the practical advantage of keeping phylum names the same regardless of the kingdom in which they are placed.

It is important to note that all available ranks are not used for particular fungi, and there is an increasing tendency for mycologists not to use ranks between the principal ones, or below subspecies.

The rank of special form (q.v.; Art. 4) is available to mycologists wishing to separate morphologically identical fungi by host reactions. Special forms are not regulated by the Code, nor are notations for physiologic races (q.v.) designated by numbers by agreement between interested mycologists.

Any characters judged to represent significant discontinuities, whether biological, morphological, ultrastructural, or molecular, can be used in classifications. Particular emphasis is placed on reproductive structures, ultrastructure, and molecular evidence at levels above family.

Practice as regards the typography of scientific names has varied in different nations. The Codes all rule that species names should be differentiated, and *italic* is usually employed, but practice at the family level and above has been inconsistent. The current Botanical Code (Greuter *et al.*, *Regnum Veg.* **131**, 1994) uses *italic* for all scientific names regardless of rank (e.g. *Fungi, Ascomycota, Lecanorales, Russulaceae*); while this is not the subject of an Article, many leading journals have already followed this lead. The practice is expected to become increasingly widespread and has been adopted by IMI.

See Chemotaxonomy, Cladistics, Kingdoms of fungi, Nomenclature, Phylogeny, Species.

Clasterisporium, see *Clasterosporium*.

Clasteropycnis Bat. & Cavalc. (1963), Mitosporic fungi, 4.A2.?. 1, Brazil.

Clasterosphaeria Sivan. (1984), Magnaporthaceae. 1, Malaysia. Anamorph *Clasterosporium*.

Clasterosporites Pia (1927), Fossil fungi. 1 (Eocene), Eur.

Clasterosporium Schwein. (1832), Anamorphic Magnaporthaceae, 1.C2.1/19. Teleomorph *Clasterosphaeria*. 7, widespr. See Ellis (*Mycol. Pap.* **70**, 1958; key), Sutton *et al.* (*Pl. Pathol.* **43**: 1066, 1994; *C. flexum* on *Chamaecyparis*).

Clastoderma A. Blytt (1880), Clastodermataceae. 5, widespr.

Clastodermataceae Alexop. & T.E. Brooks (1971), Echinosteliales. 2 gen. (+ 2 syn.), 6 spp.

Clastostelium L.S. Olive & Stoian. (1977), Protosteliaceae. 1, Guam.

Clathraceae Chevall. (1826), Phallales. 11 gen. (+ 21 syn.), 38 spp. Gasterocarp multipileate, with stellate or latticed receptacle; gleba internal.
  Lit. Dring (*Kew Bull.* **35**: 1, 1980; monogr.).

clathrate (clathroid), like a network; latticed.

Clathrella E. Fisch. (1898) = Clathrus (Clathr.) fide Dring (1980).

Table 3. Principal, secondary and some other ranks in the nomenclatural hierarchy (botanical).

Domain .... *Eukaryota*
  Kingdom .... *Fungi*
    Subkingdom .... †
      Phylum (Division) .... *Basidiomycota*
        Subphylum (Subdivision) .... † *-mycotina*
          Class .... *Teliomycetes*
            Subclass .... † *-mycetidae*
              Order .... *Uredinales*
                Suborder .... † *-ineae*
                  Family .... *Pucciniaceae*
                    Subfamily .... † *-oideae*
                      Tribe .... *Puccinieae*
                        Subtribe .... † *-inae*
                          Genus .... *Puccinia*
                            Subgenus .... *Puccinia*
                              Section .... (*Hetero-Puccinia*)
                                Subsection .... †
                                Series .... †
                                  Subseries .... †
                                  Species .... *Puccinia graminis*
                                    Subspecies .... *Puccinia graminis* subsp. *graminis*
                                  Variety .... *P. graminis* var. *stackmanii*
                                  Subvariety .... †
                                  Form .... †
                                Subform .... †
                                Special form ...... §*Puccinia graminis* f.sp. *avenae*
                                Physiologic Race ...... *P. graminis* f.sp. *avenae* Race 1
                                Individual .... †

† Not necessary for this example.
§ See text.

Clathridium Berl. (1897) nom. illegit. ≡ Clethridium (Amphisphaer.). See Holm (*Taxon* **24**: 475, 1975). = Discostroma (Amphisphaer.) fide Brockman (1976).

Clathrina Müll. Arg. (1883) = Cladia (Cladon.).

Clathrococcum Höhn. (1911) ? = Epicoccum (Mitosp. fungi).

Clathroconium Samson & H.C. Evans (1982), Mitosporic fungi, 1.F2.10. 1 (on *Arachnida*), Ghana.

Clathrodastrum, see *Clathroidastrum*.

Clathrogaster Petri (1900), Hysterangiaceae. 2, Borneo.

Clathroidastrum P. Micheli ex Adans. (1763) ≡ Stemonitis (Stemonitid.).

Clathroidastrum Kuntze (1891) ≡ Stemonitis (Stemonitid.).

Clathroides Micheli (1729) = Arcyria (Arcyr.).

Clathroporina Müll. Arg. (1882), Trichotheliaceae (L). 4, trop. Am., Afr., Australasia (fide McCarthy, *Lichenologist* **27**: 321, 1995). = Porina (Trichothel.) fide McCarthy (*loc. cit.*).

Clathroporinopsis M. Choisy (1929) ? = Topelia (Stictid.) fide Hafellner & Kalb (*Bibl. Lich.* **57**: 161, 1995). = Clathroporina (Trichothel.) fide McCarthy (1995).

Clathroptychium Rostaf. (1875) = Dictydiaethalium (Lycogal.).

Clathrosorus Ferd. & Winge (1920) = Spongospora (Plasmodiophor.).

Clathrosphaera Zalewski (1888) nom. conf. fide van Beverwijk (*TBMS* **34**: 280, 1951). See *Clathrosphaerina*.

Clathrosphaerina Beverw. (1951), Mitosporic fungi, 1.G2.1. 1 (conidia hollow, net-like), Eur.

Clathrospora Rabenh. (1857), Diademaceae. 10, Eur., India, N. Am. See Shoemaker & Babcock (*CJB* **70**: 1617, 1992).

Clathrosporium Nawawi & Kuthub. (1987), Mitosporic fungi, 1.D2.1. 1 (aero-aquatic, conidia hollow, net-like), Malaysia.

Clathrotrichum Pat. (1921) ? = Beniowskia (Mitosp. fungi) fide Mason (*Mycol. Pap.* **2**: 27, 1928).

Clathrus P. Micheli ex L. (1801), Clathraceae. 16, warm temp., trop. See Burk (*Mycotaxon* **8**: 463, 1979), Dring (1980).

Clatroidastron, see *Clathroidastrum*.

Claudopus Gillet (1876), Entolomataceae. 20, N. temp. = Rhodophyllus (Agaric.) fide Singer (1951).

Claurouxia D. Hawksw. (1988), ? Candelariaceae (L). 1, Eur.

Clausaria Nyl. (1861) = Pertusaria (Pertusar.).

Claussenomyces Kirschst. (1923), Leotiaceae. c. 12, widespr. See Iturriaga (*Mycotaxon* **42**: 327, 1991), Verkley (*MR* **99**: 187, 1995; asci).

Claustria Fr. (1849) = Physarum (Physar.).

Claustula K.M. Curtis (1926), Claustulaceae. 1, NZ.

Claustulaceae G. Cunn. (1931), Phallales. 1 gen., 1 sp. Gasterocarp unipileate; gleba within unopened peridium.

Clautriavia (Pat.) Lloyd (1909) = Phallus (Phall.).

Clauzadea Hafellner & Bellem. (1984), Porpidiaceae (L). 2, Eur.

Clauzadeana Cl. Roux (1984), Lecanoraceae (L). 1, N. temp.

clava, a club-like fruiting structure, e.g. of *Cordyceps*.

clavacin, see patulin.

Clavaria Vaill. ex L. (1753) ? = Clavariadelphus (Clavariadelph.).

Clavaria Fr. (1821) nom. cons., Clavariaceae. 25, widespr. See Corner (1950).

Clavariaceae Chevall. (1826), Cantharellales. 8 gen. (+ 8 syn.), 125 spp.; Club or Coral Fungi. Basidioma clavarioid, simple or branched; monomitic; spores smooth or echinulate; saprobes, some *Clavulinopsis* species may be lichenized; terrestrial or lignicolous.
Lit.: Corner (1950), Donk (1964: 250), Coker (*The Clavarias of the United States and Canada*, 1923 [reprint, 1973]), Corner (*A monograph of Clavaria and allied genera*, 1950 [*Ann. Bot. Mem.* **1**]; addenda in *Ann. Bot., Lond.* N.S. **16, 17**; *TBMS* **35**: 285, 1952), Donk (1951-63, III; gen. names), Pilát (*Acta Mus. nat. Prague* **14B**: 129, 1958; European spp.), Thind (*The Clavariaceae of India*, 1961; 15 gen., 92 spp.; keys), Henry (*Ann. Carnegie Mus.* **39**: 125, 1967; keys 30 spp. W. Penn.), Petersen (*TBMS* **50**: 641, 1967; fam. interrelationships; *Sydowia* **21**: 105, 1968; *Friesia* **9**: 369, 1971; type studies of *Clavaria* and *Clavulinopsis*), Maas Geesteranus (*De fungi van Nederland Die Clavaroide fungi. Auriscalpiaceae, Clavariaceae, Clavulinaceae, Gomphaceae*, 1976 [*Wetensch. Med. KNVV* **113**]), Pilát (*Sborn. národ. muz. Praze* **27B**: 113, 1971; former Czechoslovakia), Berthier (*Monographie des Typhula Fr., Pistillaria Fr. et genres voisins*, 1976), Petersen (*The clavarioid fungi of New Zealand*, 1988).

Clavariachaeta Lloyd (1922) nom. nud. = Clavariachaete (Hymenochaet.).

Clavariachaete Corner (1950), Hymenochaetaceae. 2, trop. Am.

Clavariadelphaceae Corner (1970), Cantharellales. 4 gen. (+ 2 syn.), 21 spp. Basidioma clavarioid, robust; monomitic; spores smooth.

Clavariadelphus Donk (1933), Clavariadelphaceae. c. 15, temp. See Wells & Kempton (*Mich. Bot.* **7**: 35, 1968; key N. Am. spp.), Methven (*Mycotaxon* **34**: 153, 1989; N. Am. spp.).

Clavariana Nawawi (1976), Mitosporic fungi, 1.G1.10. 1 (aquatic), Malaysia.

Clavariella P. Karst. (1881) = Ramaria (Ramar.) fide Corner (1950).

Clavariopsis De Wild. (1895), Mitosporic fungi, 1.G1.1. 2, widespr. See Petersen (*Mycol.* **55**: 21, 1963; key).

Clavariopsis Holterm. (1898) ≡ Holtermannia (Tremell.).

Clavascina Beneš (1961), Fossil fungi (*f. cat.*). 1 (Carboniferous), former Czechoslovakia.

clavate, (1) club-like; narrowing in the direction of the base; (2) (of stipes of agarics), narrowing to the apex (Fig. 37.16). Cf. obclavate.

clavate, see clava.

clavatin, see patulin.

Clavatisporella K.D. Hyde (1995), Diaporthales (inc. sed.). 1, Irian Jaya.

Clavatospora Sv. Nilsson ex Marvanová & Sv. Nilsson (1971), Mitosporic fungi, 1.G1.1. 3 (aquatic), widespr.

Claviceps Tul. (1853), Clavicipitaceae. 36 (mainly on *Gramineae*, rarely *Cyperaceae*), cosmop. See Brady (*Lloydia* **25**: 1, 1962; hosts), Düvell (*Bibl. Mycol.* **126**, 1989; linear plasmid), Loveless (*TBMS* **47**: 205, 1964; identification from honey dew), Tanda (*J. Agric. Sci., Tokyo* **5**: 85, 1981; vars.), Taber (*in* Demain & Solomon (Eds), *Biology of industrial micro-organisms*: 449, 1985; review biology). See Ergot, Poisonous fungi.

Clavicipitaceae (Lindau) O.E. Erikss. (1982), Hypocreales. 27 gen. (+ 32 syn.), 237 spp. Ascomata superficial or immersed in a usually brightly coloured stroma, perithecial, fleshy; asci cylindrical, elongate, with a prominent apical cap with a narrow pore; ascospores filiform, multiseptate, usually fragmenting. Anamorphs *Acremonium, Aschersonia, Ephelis, Hirsutella, Paecilomyces, Sphacelia*, etc. Biotrophic on *Insecta*, other fungi or *Gramineae* (then often endophytic); mainly trop.
Lit.: Doguet (*BSMF* **76**: 171, 1960; status), Jones & Clay (*CJB* **65**: 1027, 1987; ascus ontogeny), Koval' (*Klavitsiptalnyie Grib SSSR*, 1984), Morgan-Jones *et al.* (*Mycotaxon* **43**: 401, 1992; on *Balansieae*), Rykard *et al.* (*Mycol.* **76**: 1095, 1984; conidiomata), Schardl *et al.* (*Pl. Syst. Evol.* **178**: 27, 1991; coevolution of endophytes and teleomorphs with hosts).

Clavicipitales, see *Hypocreales*.

Clavicorona Doty (1947), Clavicoronaceae. 11, widespr. See Dodd (*Mycol.* **64**: 746, 1972; key).

Clavicoronaceae Corner (1970), Hericiales. 1 gen., 11 spp. Basidioma clavarioid, with pyxidate branching.

Clavidisculum Kirschst. (1938), Hyaloscyphaceae. 1, Switzerland. See Svrček (*Česká Myk.* **41**: 193, 1987).

claviformin, see patulin.

clavine alkaloids, a group of ergoline alkaloids occurring in *Claviceps* sclerotia; also synthesized by *Aspergillus fumigatus* and *Penicillium chermesinum*. See agroclavine, ergoline alkaloids, ergot.

Clavispora Rodr. Mir. (1979), ? Metschnikowiaceae. 2, Israel. See Hendriks *et al.* (*Syst. Appl. Microbiol.* **15**: 98, 1992), Kurtzman (*in* Hawksworth (Ed.), *Ascomycete systematics*: 363, 1994), Lachance *et al.* (*Int. J. Syst. Bact.* **36**: 524, 1986). Anamorph *Candida*.

Clavocephalis Bainier (1882) = Syncephalis (Piptocephalid.) fide Kirk (*in litt.*).

Clavochytridium Couch & H.T. Cox (1939) = Blastocladiella (Blastoclad.) fide Couch & Whiffen (1942).

Clavogaster Henn. (1896), Secotiaceae. 1, NZ.

Clavomphalia E. Horak (1987), Tricholomataceae. 1, China

Clavularia P. Karst. (1882) = Cornucopiella (Mitosp. fungi) fide Seifert (*TBMS* **85**: 123, 1985).

clavulate, somewhat club-like.

Clavulicium Boidin (1957), Clavulinaceae. 3, widespr.

Clavulina J. Schröt. (1888), Clavulinaceae. 32, widespr. See Corner (1950), Petersen (*Mycol.* **59**: 39, 1967).

Clavulinaceae (Donk) Donk (1970), Cantharellales. 3 gen. (+ 1 syn.), 36 spp. Basidiomata clavarioid; basidia with two strongly curved sterigmata.
Lit.: Donk (1964: 253), Coker (*The Clavarias of the United States and Canada*, 1923 [reprint, 1973]), Corner (*A monograph of Clavaria and allied genera*, 1950 [*Ann. Bot. Mem.* **1**]; addenda in *Ann. Bot., Lond.* N.S. **16, 17**; *TBMS* **35**: 285, 1952), Pilát (*Acta Mus. nat. Prague* **14B**: 129, 1958; European spp.), Thind (*The Clavariaceae of India*, 1961; 15 gen., 92 spp.; keys), Henry (*Ann. Carnegie Mus.* **39**: 125, 1967; keys 30 spp. W. Penn.), Petersen (*TBMS* **50**: 641, 1967; fam. interrelationships), Maas Geesteranus (*De Fungi van Nederland Die Clavaroide Fungi. Auriscalpiaceae, Clavariaceae, Clavulinaceae, Gomphaceae* [*Wetensch. Med. KNVV* **113**], 1976).

Clavulinopsis Overeem (1923), Clavariaceae. 60, cosmop. See Petersen (*Mycol. Mem.* **2**, 1968; key 10 N.

Am. spp.). = Clavaria (Clavar.) fide Petersen (*Mycol.* **70**: 660, 1978).

clavus, the sclerotium of ergot (obsol.).

Cleidiomyces, see *Kleidiomyces*.

Cleistobombardia J.H. Mirza (1968) = Tripterosporella (Lasiosphaer.).

cleistocarp, see cleistothecium.

Cleistocystis Sousa da Câmara (1931), Mitosporic fungi, 4.A1.?. 1, Portugal.

Cleistoiodophanus J.L. Bezerra & Kimbr. (1976), Ascobolaceae. 1, USA.

Cleistonium Speer (1986), Anamorphic Hysteriaceae, 8.A1.?. Teleomorph *Glonium*. 1, Brazil.

Cleistophoma Petr. & Syd. (1927), Mitosporic fungi, 4.A1.?. 1, N. Am., Eur. See Sutton (*Mycol. Pap.* **141**, 1977).

Cleistosoma Harkn. (1884) = Emericella (Trichocom.) fide Peek & Solheim (*Mycol.* **50**: 844, 1959).

Cleistosphaera Syd. & P. Syd. (1916), Parodiopsidaceae. 1, S. Am.

Cleistotheca Zukal (1893) = Pleospora (Pleospor.) fide v. Arx & Müller (1975).

cleistothecium, a closed fruitbody having no predefined opening, e.g. an ascoma of *Thielavia*.

Cleistothecopsis F. Stevens & E.Y. True (1919) = Pleospora (Pleospor.) fide Clements & Shear (1931).

Cleistothelebolus Malloch & Cain (1971), Eoterfeziaceae. 1 (coprophilous), Canada. See van Brummelen (*in* Hawksworth (Ed.), *Ascomycete systematics*: 400, 1994; posn).

Clelandia Trappe (1979) [non J.M. Black (1932), *Violaceae*] ≡ Mycoclelandia (Peziz.).

Clematomyces Thaxt. (1900), Laboulbeniaceae. 4 or 5, Asia, W. Afr.

Cleptomyces Arthur (1918), Pucciniaceae. 2, S. Am. = Stereostratum (Puccin.) fide Thirumalachar (*Mycol.* **52**: 688, 1961).

Clethria, see *Cletria*.

Clethridium (Sacc.) Sacc. (1895), Amphisphaeriaceae. 3, Brazil, Eur. See Brockmann (*Sydowia* **28**: 275, 1976), Holm (*Taxon* **24**: 475, 1975).

Cletria P. Browne (1756) = Clathrus (Clathr.).

Clibanites (P. Karst.) P. Karst. (1871), ? Ascomycota (inc. sed.). 1, Finland.

Climacocystis Kotl. & Pouzar (1958), Coriolaceae. 1, N. temp.

Climacodon P. Karst. (1881), Hydnaceae. 6, widespr.

Clinoconidium Pat. (1898), Cryptobasidiaceae. 2, C. & S. Am. See Malençon (*BSMF* **69**: 94, 1953).

Clinotrichum Cooke (1871) = Acladium (Mitosp. fungi) fide Kendrick & Carmichael (1973).

Clintamra Cordas & Durán (1977), Tilletiaceae. 1 (on *Nolina*), N. Am.

Clinterium Fr. (1849) = Topospora (Mitosp. fungi) fide Groves (1965).

Clintoniella (Sacc.) Rehm (1900) = Hypomyces (Hypocr.) fide Rogerson (1970).

Cliostomum Fr. (1825), Bacidiaceae (L). 4, Eur., N. Am. See Gowan (*Mycologia* **82**: 766, 1990; key 4 N. Am. spp.).

Clisosporium Fr. (1821) nom. rej. = Coniothyrium (Mitosp. fungi).

Clistosoma Clem. & Shear (1931) ≡ Cleistosoma (Trichocom.).

Clithramia Nag Raj (1988), Mitosporic fungi, 8.B2.1. 1 (on *Oryza*), India.

Clithris (Fr.) P. Karst. (1870) = Cenangium (Leot.) fide Minter (*in litt.*).

Clitocybe (Fr.) Staude (1857), Tricholomataceae. 250, cosmop. *C. augeana* (frequently but incorrectly as *C. dealbata*), a mushroom-bed contaminant. See Bigelow (*Beih. Nova Hedw.* **72**, 1982; **81**, 1985; N. Am. spp.), Harmaja (*Karstenia* **10**: 5, 1969; key 43 spp.

Finland; *Karstenia* **15**: 13, 1976), Singer (*Sydowia* **31**: 199, 1979; key 202 spp.).

Clitocybula (Singer) Singer ex Métrod (1952), Tricholomataceae. 10, widespr. temp. See Bigelow (*Mycol.* **65**: 1101, 1973; N. Am. spp.), Singer (*Sydowia* **31**: 233, 1979; key).

Clitopilina Arnaud (1952) nom. nud. (Thelephor.).

Clitopilopsis Maire (1937) = Rhodocybe (Entolomat.) fide Singer (1946).

Clitopilus (Fr. ex Rabenh.) P. Kumm. (1871), Entolomataceae. 22, N. temp. *C. passeckerianus*, a mushroom-bed contaminant. See Singer (*Sydowia* **31**: 235, 1979; key), Ware (*Gdnrs' Chron.* **97**: 325, 1935). See cat's ear.

Clohesyomyces K.D. Hyde (1993), Mitosporic fungi, 4.A1.1. 1 (aquatic), Australia.

Clonophoromyces Thaxt. (1931), Laboulbeniaceae. 1, W. Indies.

Clonostachyopsis Höhn. (1907) = Clonostachys (Mitosp. fungi) fide Clements & Shear (1931).

Clonostachys Corda (1839), Mitosporic fungi, 1.A1.15. 7, esp. Eur. See Hawksworth & Punithalingam (*TBMS* **64**: 89, 1975).

Closteroaleurosporia Grigoraki (1924) ≡ Microsporum (Mitosp. fungi).

closterospore (in dermatology), a multinucleate phragmospore, as in *Trichophyton* (obsol.).

Closterosporia Grigoraki (1924) = Microsporum (Mitosp. fungi).

Closterosporium Sacc. (1883), Anamorphic Melanommataceae, 6.C2.?. Teleomorph *Melanomma*. 1, Eur.

club fungi, the *Clavariaceae*.

club root, a disease of crucifers (*Plasmodiophora brassicae*), see *Plasmodiophora*.

cluster-cup, see aecium.

clypeate, having a clypeus.

Clypeispora Ramaley (1991), Anamorphic Mycosphaerellaceae, 4.A1.1. Teleomorph *Mycosphaerella*. 1, USA.

Clypeocarpus Kirschst. (1941) = Mazzantia (Vals.) fide v. Arx & Müller (1954).

Clypeoceriospora Sousa da Câmara (1946), Ascomycota (inc. sed.). 1, Eur.

Clypeochorella Petr. (1923), Mitosporic fungi, ?8.A1.?. 1, Eur.

Clypeococcum D. Hawksw. (1977), Dacampiaceae. 6 (on lichens), Eur., N. Am.

Clypeodiplodina F. Stevens (1927) = Ascochytulina (Mitosp. fungi) fide Clements & Shear (1931).

Clypeolaria Tratt. (1825) nom. dub. See Rudolphi (*Linnaea* **4**: 395, 1829; det. as *Insecta*).

Clypeolella Höhn. (1910) = Schiffnerula (Englerul.) fide Hughes (*CJB* **62**: 2213, 1984; *in* Sugiyama (Ed.), *Pleomorphic fungi*: 103, 1987).

Clypeolina Theiss. (1918), Micropeltidaceae. 1, Brazil. = Stomiopeltis fide Müller & v. Arx (1963).

Clypeolina Speg. (1923) ≡ Clypeolopsis (Micropeltid.).

Clypeolinopsis Bat. (1959) ? = Stomiopeltis (Micropeltid.). 4, trop. See Müller & v. Arx (1963).

Clypeolopsis F. Stevens & Manter (1925) ? = Stomiopeltis (Micropeltid.) fide Müller & v. Arx (1962).

Clypeolum Speg. (1881), Ascomycota (inc. sed.). 8, trop. See Batista (1959). = Porina (Trichothel.) fide Müller & v. Arx (1962).

Clypeomyces Kirschst. (1935) Fungi (inc. sed.). See Petrak (*Sydowia* **1**: 61, 1947).

Clypeopatella Petr. (1942), Mitosporic fungi, 8.A1.?. 1, Rhodes.

Clypeophialophora Bat. & Peres (1962), Mitosporic fungi, 8.A1.?. 1, Brazil.

Clypeophysalospora H.J. Swart (1981), Amphisphaeriaceae. 1, Australia.

Clypeoporthe Höhn. (1919), Valsaceae. 2, Java, Vietnam.

Clypeoporthella Petr. (1924), Valsaceae. 2, Eur., N. Am. See Barr (*Mycol. Mem.* **7**, 1978).

Clypeopycnis Petr. (1925), Mitosporic fungi, 4.A1.15. 1, N. Am.

Clypeopyrenis Aptroot (1991), Pyrenulaceae (?L). 1, Costa Rica.

Clypeorhynchus Kirschst. (1936) = Diaporthe (Vals.) fide v. Arx & Müller (1954).

Clypeoseptoria F. Stevens & P.A. Young (1925), Mitosporic fungi, 8.E1.?. 3, Hawaii, Brazil. See Ciferri & Batista (*Atti Ist. bot. Univ. Lab. Crittog. Pavia* **14**: 53, 1957).

Clypeosphaeria Fuckel (1870), Clypeosphaeriaceae. 30, widespr. See Barr (*SA* **8**: 1, 1989).

Clypeosphaeriaceae G. Winter (1886), Xylariales. 9 gen. (+ 2 syn.), 49 spp. Stromatal tissues poorly developed, clypeate; ascomata perithecial, immersed or erumpent, sometimes aggregated, usually thick-walled, the ostiolar region thick-walled, widely papillate, periphysate; interascal tissue of numerous thin-walled paraphyses; asci cylindrical, persistent, with a usually I- apical ring; ascospores hyaline or brown, transversely septate, sometimes with a germ pore, sometimes with a mucous sheath. Anamorphs not known. Saprobic or necrotrophic in woody stems etc., widespr.

Clypeosphaerulina Sousa da Câmara (1939), Ascomycota (inc. sed.). 1, Portugal.

Clypeostagonospora Punith. (1981), Mitosporic fungi, 8.C1.1. 1, USA.

Clypeostigma Höhn. (1919) = Phyllachora (Phyllachor.) fide v. Arx & Müller (1954).

Clypeostroma Theiss. & Syd. (1914) nom. dub. (Fungi, inc. sed.). See v. Arx & Müller (1975).

Clypeothecium Petr. (1922) = Exarmidium (Hyponectr.) fide Barr & Boise (1985).

Clypeotrabutia Seaver & Chardón (1926) = Phyllachora (Phyllachor.) fide Cannon (*Mycol. Pap.* **163**, 1991).

Clypeum Massee (1896) = Parmularia (Parmular.) fide Batista & Vital (1960).

clypeus, a shield-like stromatic growth, with or without host tissue, over one or more ascomata or conidiomata.

Clypeus (Britzelm.) Fayod (1889) = Inocybe (Cortinar.) fide Mussat (1901).

CMA, see Media.

CMI, see IMI.

Cnazonaria Corda (1829) = Pistillaria (Typhul.) fide Corner (1950).

co- (prefix), together.

coacervate, massed (heaped), together.

coadnate, united, cohering, connate.

coalescent, joined together.

coarctate, crushed together, crowded, constricted.

Coccidiascus Chatton (1913), Metschnikowiaceae. 1 (in *Drosophila*), Eur., N. Am.

Coccidiodictyon Oberw. (1989), Septobasidiaceae. 1, Spain.

Coccidioides G.W. Stiles (1896), Mitosporic fungi, ?.A1.?. 1, widespr. (esp. Am.). *C. immitis* on humans and animals (**coccidiomycosis**, coccidioidal granuloma, San Joaquin Valley Fever). See Szaniszlo *et al.* (in Howard (Ed.), *Fungi pathogenic for man and animals* Part A, Biology, 1983; life cycle, dimorphism; Sigler & Carmichael (*Mycotaxon* **4**: 458, 1976; *Malbranchea* state), Fiese (*Coccidioidomycosis*, 1958), Al Doory (*Mycopathol.* **46**: 113, 1972; bibliogr.), Stevens (*Coccidioidomycosis. A text*, 1980). See also *Uncinocarpus*.

coccidioidin, an antigen prepared from *Coccidioides immitis*, esp. for skin testing. Cf. spherulin.

coccidiomycosis, see *Coccidioides*; mycosis.

Coccidomyces Buchner (1912), ? Saccharomycetaceae. 1 (in *Insecta*), Germany. See Stainhaus (*Insect microbiology*, 1946).

Coccidophthora Syd. & P. Syd. (1913) Fungi (inc. sed.). See Petch (*TBMS* **10**: 190, 1925).

Cocciscia Norman (1870), ? Dacampiaceae (? L). 1, Scandinavia.

Coccobolus Wallr. (1833) = Ceuthospora (Mitosp. fungi) fide Rabenhorst (*Dtschl. Krypt.-Fl.* **1**: 144, 1844), but see Sutton (1977).

Coccobotrys Boud. & Pat. (1900), Anamorphic Agaricaceae, 9.-.-. Teleomorph *Lepiota*. 1, Eur. See Donk (*Taxon* **11**: 82, 1962).

Coccocarpia Pers. (1827), Coccocarpiaceae (L). 21, cosmop. See Arvidsson (*Opera bot.* **67**, 1983; monogr., key), Awasthi (*Kavaka* **12**: 83, 1987; Indian spp.).

Coccocarpiaceae (Mont. ex Müll. Berol.) Henssen (1986), Lecanorales (L). 4 gen. (+ 5 syn.), 31 spp. Thallus varied but most often foliose, well-developed, loosely attached; ascomata apothecial, formed from the aggregation of short cells derived from adjacent thalline hyphae, flat to strongly convex, dark or concolorous with the thallus; hymenium not gelatinous; interascal tissue of sparsely branched paraphyses, sometimes pigmented at the apices or with a well-developed epithecium; asci with a well-developed apical cap, either wholly I+ or with a ring structure, usually with an I+ gelatinized outer layer; ascospores simple or transversely septate. Anamorphs pycnidial. Lichenized with cyanobacteria, esp. boreal and austral.

Coccochora Höhn. (1909), Dothideales (inc. sed.). 1, Asia. See Bose & Müller (*Indian Phytopath.* **17**: 3, 1964).

Coccochorella Höhn. (1910) ≡ Coccochora (Dothideales).

Coccochorina Hara ([pre-1927]) = Clypeosphaeria (Clypeosphaer.) fide Hara (*Byogaichu-hoten*, 1948).

Coccodiella Hara (1910), Phyllachoraceae. *c.* 15, trop.

Coccodiniaceae Höhn. ex O.E. Erikss. (1981), Dothideales. 3 gen. (+ 6 syn.), 13 spp. Mycelium at least partially superficial, with cylindrical brown hyphae; ascomata ± globose but frequently collapsing when dry, thin-walled, sometimes setose, with a periphysate ostiole; interascal tissue of short periphysoids; asci saccate, fissitunicate, with an I+ outer gelatinous layer; ascospores hyaline or brown, usually muriform. Anamorphs hyphomycetous, often poorly developed. Epiphytic or biotrophic on leaves and stems, widespr.

Coccodinium A. Massal. (1860), Coccodiniaceae. 1, Eur. See Eriksson (*Opera Bot.* **60**, 1981).

Coccodiscus Henn. (1904) = Coccoidea (Coccoid.) fide Müller & v. Arx (1962).

Coccodothella Theiss. & Syd. (1915) = Coccoidella (Coccoid.) fide Müller & v. Arx (1962).

Coccodothis Theiss. & Syd. (1914), Parmulariaceae. 2, Am. See Petrak (*Sydowia* **8**: 291, 1954).

Coccogloeum Petr. (1955), Mitosporic fungi, 6.A1.?. 1, Austria.

Coccoidea Henn. (1900), Coccoideaceae. 2, Japan, C. Am.

Coccoideaceae Henn. ex Sacc. & D. Sacc. (1905), Dothideales. 2 gen. (+ 4 syn.), 8 spp. Ascomata formed as locules in an erumpent pulvinate or peltate stroma with a short stalk, ± globose, the peridium spongy, two-layered, ostiole lysigenous, not periphysate; interascal tissue composed of cellular pseudoparaphyses, sometimes evanescent; asci ±

cylindrical, fissitunicate; ascospores relatively small, 1-septate, without sheath. Anamorph absent or acervular. Biotrophic on leaves.
Lit.: Sivanesan (*TBMS* **89**: 265, 1987).

Coccoidella Höhn. (1909), ? Coccoideaceae. 6, Am. See Müller & v. Arx (1962), Sivanesan (*TBMS* **89**: 265, 1987; posn). Anamorph *Colletogloeum*.

Coccoidiopsis Hara (1913) ≡ Coccodiella (Phyllachor.). See Cannon (*SA* **11**: 168, 1993), v. Arx & Müller (1973).

Coccomycella Höhn. (1917) = Coccomyces (Rhytismat.) fide Sherwood (1980).

Coccomyces De Not. (1847), Rhytismataceae. 74, widespr. See Sherwood (*Occ. Pap. Farlow Herb.* **15**, 1980; monogr., key), Spooner (*Kew Bull.* **45**: 451, 1990; key 11 spp., Sabah), Johnston (*N.Z. Jl Bot.* **24**: 89, 1986; key 19 spp., NZ).

Coccomycetella Höhn. (1917), Odontotremataceae. 1, Eur. See Sherwood-Pike (*Mycotaxon* **28**: 137, 1987).

Cocconia Sacc. (1889), Parmulariaceae. 6, trop. See Batista & Vital (*Atas Ist. micol. Recife* **1**: 167, 1960), Hansford (*Mycol. Pap.* **15**, 1946).

Cocconiopsis G. Arnaud (1918) = Cyclostomella (Parmular.) fide Müller & v. Arx (1962).

Coccopeziza Har. & P. Karst. (1890) = Arthonia (Arthon.) fide Sherwood (*Mycotaxon* **6**: 215, 1977).

Coccophacidium Rehm (1888) = Therrya (Rhytismat.) fide Nannfeldt (1937).

Coccophysium Link (1833) nom. dub. (? Fungi).

Coccopleum Ehrenb. (1818) = Sclerotium (Mitosp. fungi) fide Fries (*Syst. Mycol.* **2**: 256, 1823).

Coccospora Wallr. (1833) nom. dub. (Myxomycetes, inc. sed.) fide Carris (*Sydowia* **45**: 92, 1993).

Coccosporella P. Karst. (1893) = Mycogone (Mitosp. fungi) fide Hughes (1958).

Coccosporium Corda (1831) nom. dub. fide Hughes (*CJB* **36**: 752, 1958). = Corynespora (Mitosp. fungi) fide Holubová-Jechová (*Sydowia* **46**: 240, 1994).

Coccostroma Theiss. & Syd. (1914) = Coccodiella (Phyllachor.) fide Katumoto (1968).

Coccostromella Petr. (1968), ? Dothideaceae. 1, trop.

Coccostromopsis Plunkett (1924) = Coccodiella (Phyllachor.) fide Müller & v. Arx (1973).

Coccotrema Müll. Arg. (1888), Coccotremataceae (L). 8, Asia, Australia, N. & S. Am. See Brodo (*Bryologist* **76**: 260, 1973; N. Am. spp.).

Coccotremataceae Henssen ex J.C. David & D. Hawksw. (1991), Pertusariales (L). 2 gen. (+ 3 syn.), 9 spp. Thallus crustose or lobate; ascomata originating as a cavity within the primordium, opening by a pore, often perithecium-like; interascal tissue of apical and basal paraphyses, the basal paraphyses sometimes evanescent; hymenium non-amyloid; sometimes with cephalodia. Anamorphs pycnidial. Lichenized with green algae, scattered.
Lit.: Brodo (*Bryologist* **76**: 260, 1973), Henssen (*in* Brown *et al.* (Eds), *Lichenology: progress and problems*: 107, 1976; ontogeny).

Coccotrichum Link (1824) = Botrytis (Mitosp. fungi) fide Saccardo (*Syll. Fung.* **4**: 120, 1886) but used by Hughes (1958).

Coccularia Corda (1829) nom. dub.; rejected by Fries (*Summ. veg. scand.*: 522, 1849).

coccus, a spherical bacterium.

Cochlearia (Cooke) Lambotte (1888) [non L. (1753), *Cruciferae*] = Otidea (Otid.) fide Eckblad (1968).

cochleariform, spoon-like in form.

cochleate, shell-like in form; twisted like a shell.

cochliobolin, see ophiobolin.

Cochliobolus Drechsler (1934), Pleosporaceae. 11, widespr. Important plant pathogens include: *C. heterostrophus* (anamorph *Drechslera maydis*; leaf spot of maize (*Zea*)), Nelson (*Mycol.* **51**: 18, 24, 132,

1959; genetics), *C. miyabeanus* (anamorph *D. oryzae*; brown spot of rice), *C. sativus* (anamorph *Bipolaris sorokiniana*, spot blotch and root rot of temperate cereals), *C. stenospilus* (anamorph *D. stenospila*; brown spot of sugar-cane), *C. victoriae* (anamorph *D. victoriae*; foot rot of cereals); see victorin. See Luttrell & Rogerson (*Mycol.* **51**: 201, 1959), Ammon (*Phytopath. Z.* **47**: 244, 1963). Anamorphs *Bipolaris*, *Curvularia*.

Cochliomyces Speg. (1912), Euceratomycetaceae. 2, C. & S. Am.

Cochlonema Drechsler (1935), Cochlonemataceae. 18, N. Am, Kenya, UK. See Drechsler (*Mycol.* **27**: 185, 1935; **29**: 229, 1937; **31**: 128, 388, 1939; **33**: 248, 1941; **34**: 274, 1942; **37**: 1, 1945; **38**: 120, 1946; **43**: 161, 1951; **47**: 364, 1955; **51**: 787, 1959), Jones (*TBMS* **42**: 75, 1959; Kenya, **45**: 348, 1962; UK), Miura (*J. Jap. Bot.* **47**: 204, 1972; Japan), Dyal (1973; key), Saikawa & Sato (*Mycol.* **83**: 403, 1991; ultrastr.).

Cochlonemataceae Dudd. (1974), Zoopagales. 5 gen., 29 spp. Fertile hyphae only appearing outside the host; zygospores warty, borne on spirally twisted suspensors; ecto- or endoparasites.
Lit.: Dyal (*Sydowia* **27**: 293, 1973; keys parasites of nematodes and amoebae, bibliogr.).

Codinaea Maire (1937) = Dictyochaeta (Mitosp. fungi) fide Gamundí *et al.* (*Darwiniana* **21**: 96, 1977).

Codinaeopsis Morgan-Jones (1976), Mitosporic fungi, 1.A1.16. 1, N. Am., Japan.

Coeloanguillospora Dyko & B. Sutton (1978) = Filosporella (Mitosp. fungi) fide Dyko & Sutton (*Mycotaxon* **7**: 323, 1978).

Coelocaulon Link (1833), Parmeliaceae (L). 5, N. & S. temp. See Kärnefelt *et al.* (*Pl. Syst. Evol.* **183**: 113, 1992), Kärnefelt (*Op. Bot.* **86**: 1, 1986; key).

Coelographium (Sacc.) Gäum. (1920), Mitosporic fungi, 2.A2.?. 1, Eur.

Coelomomyces Keilin (1921), Coelomomycetaceae. 25 (in *Insecta*), widespr. See Couch & Bland (Eds) (*The genus Coelomomyces*, 1988), Couch (*J. Elisha Mitchell sci. Soc.* **61**: 124, 1945), Couch & Dodge (*J. Elisha Mitchell sci. Soc.* **63**: 69, 1947), Laird (*J. Elisha Mitchell sci. Soc.* **78**: 132, 1962), Sparrow (1960: 638; keys), Laird (*Can. J. Zool.* **37**: 781, 1959; key).

Coelomomycetaceae Couch ex Couch (1962), Blastocladiales. 1 gen. (+ 1 syn.), 25 spp. Sporophyte thallus in dipterans, unwalled, branched or lobed, converting into a mass of thick-walled, variously ornamented resting spores which function as zoosporangia; gametophyte thallus in microcrustaceans.
Lit.: Weiser & Žižka (*Česká Myk.* **28**: 227, 1975). See also *Lit.* under *Coelomomyces*.

Coelomorum Paulet (1793) = Helvella (Helvell.).

Coelomyces, see *Coelomomyces*.

Coelomycetes, Mitosporic fungi. 700 gen. (+ 775 syn.), *c.* 9000 spp., widespr., saprobic or parasitic on higher plants, fungi, lichens, vertebrates, also recovered from the widest range of ecological niches. The term 'coelomycetes' merely indicates that conidia are formed within a cavity lined by fungal or fungal/host tissue. The range of conidiogenous events is more limited than in hyphomycetes (q.v.), but solitary, 'phialidic', and 'annellidic' types predominate, and 'tretic' and 'basauxic' are absent. See Mitosporic fungi. For modern reassessments of coelomycete systematics see Sutton (*in* Ainsworth *et al.* (Eds), *The Fungi* **4A**: 513, 1973), Nag Raj (*in* Cole & Kendrick (Eds), *Biology of conidial fungi* **1**: 43, 1981), Nag Raj (*Coelomycetous anamorphs with appendage-bearing conidia*, 1993).

Lit.: Höhnel (*Mykol. Unters.* **1**: 301-369, 1923; classification), Grove (*British stem- and leaf- fungi* **1** & **2**, 1935, 1937), Petch (*TBMS* **26**: 53-70, 1943; British *Nectrioideae*), Sutton (*Coelomycetes* I - VII, *Mycol. Pap.* 1961-1981; VI, 1977, generic names proposed for *Coelomycetes*), Nag Raj *et al.* (*Icones generum coelomycetum* I-XIII, 1972-1982), Nag Raj *et al.* (*CJB* **49-67**; Genera coelomycetum I-XXVIII), Mathur (*The Coelomycetes of India*, 1979), Kendrick (*The whole fungus*, 1979), Michaelides *et al.* (*Icones generum coelomycetum* suppl.), Sutton (*The Coelomycetes*, 1980), Nag Raj (*in* Cole & Kendrick, *Biology of conidial fungi*: 43-84, 1981). See also under Mitosporic fungi.

This artificial class has been traditionally separated into three orders, Sphaeropsidales, characterized by pycnidial conidiomata, *Melanconiales* with acervular conidiomata, and *Pycnothyriales* with pycnothyrial conidiomata. An alternative system for Mitosporic fungi as a whole (now obsol.) was suggested by Sutton (1980) where differences in conidiogenesis were used for separation of taxa at the class, subclass and ordinal levels with conidiomatal structure used at the subordinal level. The traditional separation is:

(1) **Melanconiales** (fams. Melanconiaceae, Stilbosporaceae, Coryneaceae - see Sutton, 1973 for discussion for family names). Mycelium is within the host or substratum. Conidiogenous events are various. Conidiomata are subcuticular, epidermal, subepidermal, peridermal or subperidermal and the conidiogenous layer is formed within the substratum. Dehiscence is by rupture of the overlying tissues and conidial masses may be dry or slimy. Conidiomata become erumpent at maturity and grade into sporodochial conidiomata (tuberculariaceous hyphomycetes). In culture such fungi cannot be distinguished from many hyphomycetes.

Lit.: v. Arx (*A revision of the fungi classified as Gloeosporium*, 1970), Guba (*Monograph of Monochaetia and Pestalotia*, 1961), Vassiljevsky & Karakulin (*Fungi imperfecti parasitici*, Pt. II. Melanconiales, 1950).

(2) **Sphaeropsidales** (Phomales, Phyllostictales; fams. Sphaerioidaceae, Nectrioidaceae, Leptostromataceae, Excipulaceae, Discellaceae, Asbolisiaceae etc. - see Sutton, 1973 for discussion of family names). Mycelium may be immersed in the substratum or superficial. Conidia may be dry or slimy and conidiogenous events are various. Conidiomata are superficial, semi-immersed or immersed with the conidiogenous layer lining the walls of the locule(s). Diversity in conidiomatal structure is considerable in terms of tissue composition (see tissue types), number and arrangement of locules, relationship to the substrata and type of dehiscence (see Sutton, 1980; Nag Raj, 1981, for the range of variation).

Lit.: Allescher (*Rabenh. Krypt.-Fl.* **1** (6-7), 1900-03), Diedicke (*Krypt.-Fl. Mk Brandenb.* **9** (7), 1914), Petrak & Sydow (*Beih. Rep. spec. nov. regni veg.* **42**, 1927; *Macrophoma* and its segregates), Bender (*The Fungi Imperfecti: Order Sphaeropsidales*, 1933), Biga *et al.* (*Sydowia* **12**, 1959; *Coniothyrium*), Byzova *et al.* (*Flora Sporovyk rast. Kazak.* **5**, 1970), Punithalingam (*Studies on Sphaeropsidales in culture* I-III, *Mycol. Pap.* 1970-81).

(3) **Pycnothyriales** (fams. Pycnothyriaceae, Microthyriopsidaceae, Peltopycnidiaceae, Actinothyriaceae, Actinopeltaceae, Rhizothyriaceae, Peltasteraceae, see Katumoto, 1975, for discussion of family names). Mycelium may be immersed in the substratum or superficial; when superficial it may bear hyphopodia and/or setae. Conidia are produced

in several ways. Conidiomata are superficial or subcuticular, flattened, uni- or multi-locular, sometimes attached to the substratum by a central column of tissue or hypostroma, otherwise attached at the periphery; conidiogenous layer may be restricted to the upper or lower surface or occur on both; tissue structure of the pycnothyrium and nature of marginal cells are important generic criteria. Anamorphs of Dothideales.

Lit.: Naumov (*BSMF* **30**: 423-432, 1915), Arnaud (*Ann. Ecol. Nat. Agr. Montp.* **16**: 1-288, 1918), v. Höhnel (*Mykol. Unters.* **1**: 301-369, 1923), Tehon (*Trans.*

Coelomycidium Debais. (1919), Blastocladiales (inc. sed.). 1, Eur. See Loubès & Manier (*Prototistologica* **10**: 47, 1974), Weiser & Žižka (*Česká Myk.* **28**: 159 & 227, 1974).

Coelopogon Brusse & Kärnefelt (1991), Parmeliaceae (L). 2, S. Afr., S. Am., Antarctic.

Coelopus J. Bataille (1908) = Gyroporus (Gyrodont.) fide Singer (1946).

Coelorhopalon Overeem (1925) = Xylaria (Xylar.) fide Dennis (1956).

Coelosphaeria Sacc. (1873) ≡ Nitschkia (Nitschk.). See Holm (*Taxon* **24**: 275, 1975).

Coelosporidium Mesnil & Marchoux (1897) nom. dub. (Protozoa or Fungi).

Coelosporium Link (1824), Mitosporic fungi, 1.?.?. 1, Eur.

Coemansia Tiegh. & G. Le Monn. (1873), Kickxellaceae. 12, widespr. See Linder (*Farlowia* **1**: 49, 1943; key), Benjamin (*Aliso* **4**: 149, 1959), Chein (*Mycol.* **63**: 1046, 1971), Young (*Microbios* **63**: 187, 1990; ultrastr.).

Coemansiella Sacc. (1883) = Kickxella (Kickxell.) fide Linder (1943).

Coemurus, see *Coeomurus*. Cited in *ING* in error.

Coenicia Trevis. (1861) ? = Pyrenula (Pyrenul.) fide Aguirre-Hudson (*Bull. Br. Mus. nat. Hist.* Bot. **21**: 85, 1991).

coeno- (prefix), living together, e.g. multinucleate.

Coenocarpus Fr. (1825) ≡ Chaenocarpus (Xylar.).

coenocentrum (of oomycetes), a small deeply staining body at the centre of the multinucleate oosphere to which the egg-nucleus goes.

coenocyte (adj. **coenocytic**), a multinucleate mass of protoplasm; (adj., of fungi), non-cellular, in the sense of non-septate; Vuillemin (1912) used **coenocyte** for a cell usually multinucleate and **apocyte** for one temporarily or secondarily multinucleate; **coenocytium** originally used for a structure resulting from nuclear division not followed by cytoplasmic cleavage, in contrast to a **syncytium**, a multinucleate structure resulting from the fusion of several protoplasts.

coenogametes, multinucleate gametangia (q.v.) which, upon fusion, give a **coenozygote**.

Coenogoniaceae, see *Gyalectaceae*.

Coenogoniomycella Cif. & Tomas. (1954) = Coenogonium (Gyalect.).

Coenogoniomyces Cif. & Tomas. (1954) = Coenogonium (Gyalect.).

Coenogonium Ehrenb. (1820), Gyalectaceae (L). *c.* 15, cosmop. (mainly trop.). See Şantesson (1952), Uyenco (*Bryologist* **66**: 217, 1964; taxonomy; *Trans. Am. microsc. Soc.* **84**: 1, 1965, *Trentepohlia*), Wang Jang (*Taiwania* **17**: 40, 1972), Xavier Filho *et al.* (*Boln Soc. Bot.* II, **56**: 115, 1983; Brazil).

Coenogonium Clem. (1909) = Coenogonium (Gyalect.).

Coenoicia Trevis. (1861) = Melanotheca (Pyren.).

Coenomyces K.N. Deckenb. (1901), Cladochytriaceae. 1 (on marine cyanobacteria), Eur., N. Am.

Coenomycogonium Cif. & Tomas. (1954) = Coenogonium (Gyalect.).

Coenonia Tiegh. (1884), Dictyosteliaceae. 1, Eur.

Coenosphaeria Munk (1953) = Keissleriella (Lophiostom.) fide Bose (1961).

coenozygote, see coenogametes.

Coeomurus nom. rej. ≡ Uromyces (Puccin.).

Coevolution. Reciprocal evolutionary adaptations between disparate organisms leading to their interdependence. A process expected to be widespread in fungi as about two-thirds of the spp. form intimate relationships with other organisms as commensals, mutualists, or pathogens. Most studied in pathogenhost gene-for-gene (q.v.) 'arms-races'.

   Lit.: Cannon (*in* McKey & Sprent (Eds), *Advances in legume systematics* **5**, 1994; in *Phyllachoraceae and Leguminosae*), Hedberg (Ed.) (*Symb. bot. upsal.* **22**(4), 1979; parasites as taxonomists), Karatygin (*Koevolutseya gribov i rasteniǐ*, 1993), Kenneth & Palti (*Mycol.* **76**: 705, 1984; in *Compositae*), Pirozynski & Hawksworth (Eds) (*Coevolution of fungi with plants and animals*, 1988), Savile (*Bot. Rev.* **45**: 377, 1979), Stone & Hawksworth (Eds) (*Coevolution and systematics*, 1986), Thompson (*The coevolutionary process*, 1994).

   See also ambrosia fungi, Endophytes, Fungicolous fungi, galls, Insects and fungi, Lichens, Mycorrhiza, symbiosis.

Coilomyces Berk. & M.A. Curtis (1853) = Geastrum (Geastr.) fide Zeller (*Mycol.* **40**: 649, 1948).

Coinostelium Syd. (1939) = Prospodium (Uropyxid.) fide Thirumalachar & Kern (1955).

Cokeromyces Shanor (1950), Thamnidiaceae. 1, USA. See Shanor *et al.* (*Mycol.* **42**: 271, 1950), Benny & Benjamin (*Aliso* **8**: 391, 1976), Jeffries & Young (*Mycol.* **75**: 509, 1983; ultrastr.), Benny *et al.* (*Mycotaxon* **22**: 119, 1985), McGough *et al.* (*Clin. Microbiol. Newsl.* **12**: 113, 1980; human pathogenicity).

Colacogloea Oberw. & Bandoni (1991), Platygloeaceae. 1, N. temp.

Colemaniella Agnihothr. (1974), Mitosporic fungi, 1.C2.?. 1, India.

Colensoniella Hafellner (1979), Dothideales (inc. sed.). 1, NZ.

Coleocarpon Stubblef., T.N. Taylor, C.E. Mill. & G.T. Cole (1983), Fossil fungi. 1, USA.

Coleodictyospora Charles (1929), Mitosporic fungi, 1.D2.1. 1, W. Indies.

Coleodictys Clem. & Shear (1931) ≡ Coleodictyospora (Mitosp. fungi).

Coleoma Clem. (1909) ≡ Coleopuccinia (Uredinales, inc. sed.).

Coleomyces Moreau & M. Moreau (1937) = Cylindrocarpon (Mitosp. fungi) fide Booth (1966).

Coleonaema Höhn. (1924) = Coleophoma (Mitosp. fungi) fide Petrak & Sydow (*Beih. Rep. spec. nov. regni veg.* **42**: 465, 1927).

Coleophoma Höhn. (1907), Mitosporic fungi, 4.A1.15. 4, widespr. See Sutton (1980).

Coleopuccinia Pat. (1889), Uredinales (inc. sed.). 4, Japan, China; III on *Rosaceae*. See Thirumalachar & Whitehead (*Am. J. Bot.* **41**: 120, 1954). = Puccinia (Puccin.) fide Cummins & Hiratsuka (1983).

Coleopucciniella Hara ex Hirats. (1937), Uredinales (inc. sed.). 2, Asia, esp. Japan. See Laundon (*Mycotaxon* **3**: 134, 1975).

Coleoseptoria Petr. (1940), Mitosporic fungi, 8.E1.?. 1, Germany.

Coleosperma Ingold (1954), Dermateaceae. 1 (aquatic), UK.

Coleospora Cribb (1959) nom. dub. (? Fungi).

Coleosporiaceae Dietel (1900), Uredinales. 3 gen. (+ 5 syn.), 103 spp. Pycnia discoid, without bounding structures, subepidermal; aeciospores catenulate (*Peridermium*); teliospores sessile, in a single layer or pseudocatenulate to catenulate, unicellular, pale, commonly waxy or gelatinous, germination either internal or external.

Coleosporium Lév. (1847), Coleosporiaceae. 80, cosmop.; O, I on *Pinus* (Kern, *Mycol.* **20**: 60, 1928; key), II (I^{II}, with catenulate conidia), III on dicots (e.g. *Campanula* (*C. campanulae*), *Senecio* (*C. senecionis*), and *Solidago* and *Aster* (*C. solidaginis*), and sometimes monocots). See Laundon (*Mycotaxon* **3**: 154, 1975; typification), Kaneko (*Rep. Tottori mycol. Inst.* **19**: 1, 1981; keys 28 Jap. spp.).

Coleroa Rabenh. (1850), Venturiaceae. 8, widespr. See Barr (*Sydowia* **41**: 25, 1989; key 3 N. Am. spp.).

Colispora Marvanová (1988), Mitosporic fungi, 1.C1.21. 1 (aquatic), former Czechoslovakia.

collabent, falling in; collapsing.

Collacystis Kunze (1827) nom. dub. fide Sutton (*Mycol. Pap.* **141**, 1977).

collarette, a cup-shaped structure at the apex of a conidiogenous cell.

Collaria Nann.-Bremek. (1967) = Comatricha (Stemonitid.).

collariate, having a collar; see collarette.

collarium, the ring of tissue to which the proximal ends of remote lamellae are attached as in *Marasmius* spp., etc.

Collarium Link (1809) nom. dub. (? Fungi).

collateral host, see *Uredinales*.

Collecephalus J.A. Spencer (1972), Mitosporic fungi, 1.A1.6. 1, USA.

Collection and preservation. Carefully dried specimens of many fungi retain at least their microscopic structure for many years. Their preservation in dried reference collections as research and voucher material is therefore to be recommended. For most taxa, it is only practicable to retain fruit-bodies in dried collections, though mycelia can be preserved as living cultures. Most fungi with fleshy fruit bodies shrink considerably on drying, and such features as colour are frequently lost. Good field notes on size, shape, colour, texture etc. are thus important, and colour photographs are valuable. Freeze-drying (lyophilization) of fleshy fungi often gives satisfactory results (Kendrick, *Mycol.* **61**: 392, 1969), although the resulting specimens are fragile.

   The experienced worker when making a collection gives attention to (1) conservation: ensuring that the gathering is unlikely to affect the continuing presence of the fungus in that location; (2) quality: representative material must be gathered in a range of developmental stages including those with mature spores; (3) amount: where possible, enough material is collected for investigation by the collector and possibly others; and (4) field notes, ensuring that features which will deteriorate on drying are adequately described, and that as much information as possible is given about the location and timing of the collection, associated organisms (e.g. the species of plant on which the fungus occurs), the part of the organism (e.g. leaf) and details of the association (necrotrophic parasite etc.). Many lichens are very slow-growing, so particular care should be taken to ensure that populations are not unduly denuded by collectors. Field observations may suffice, and necessary chemical tests can sometimes be carried out in situ. It is possible to assess the condition of immersed and stromatic microfungi by cutting transversely across the fruit bodies and examining their contents; where spores are present the inner surface of the fruit-body is usually shiny, and the contents will swell if a drop of water is added. Collections are placed in packets or baskets as appropriate, ensuring that fragile specimens are protected from crushing, and that there is an unambiguous link between field notes and specimen. Plastic bags are not recommended due to the uncontrollable build-up of humidity levels. Where feasible, specimens should be collected along with their substratum, to minimize disturbance of their structures. Knives, folding saws or secateurs may be appropriate, and for rock-inhabiting lichens a geological hammer and masonry chisel will be needed. After as short a time as possible, and if necessary after microscopical examination, the material must be dried. Many leaf-fungi are best pressed between drying papers, although delicate hyphomycetes etc. may be damaged by such a process and should be air-dried. Specimens of microfungi on wood and bark, and most macrofungi, such as polypores and some gasteromycetes, should be dried in a current of warm dry air, ideally in a specially constructed dryer. Spore prints (on black paper if the spores are pale) of hymenomycetes are valuable, and can sometimes later be used for culturing.

Once dry, specimens are usually kept loose, to facilitate later observations. Microfungi should be protected by glassine inner bags in folded paper outer packets, which may be fixed onto herbarium sheets or kept in filing cabinets. Fragile specimens, especially myxomycetes and macrofungi, are normally pinned or glued onto small pieces of card and kept in small cardboard boxes. Nowadays, most collection data are stored in computerized databases, but packets must still be labelled with at least basic information in order to facilitate curation. Illustrations, photographs etc. must also be cross-referenced with the specimen. Semi-permanent slide preparations (see Mounting media) should always be kept if possible, either in protective cardboard boxes within the specimen packet or in a separate collection. This minimizes the need for future workers to deplete the specimen further, and is particularly important for type or authentic collections.

See also Genetic Resource Collections, Inventories, Reference collections.

Collema C.A. Browne (1756), ? Ascomycota (inc. sed., L). 1, Eur. See Laundon (*in litt.*).

Collema F.H. Wigg. (1780) nom. cons., Collemataceae (L). 77, cosmop. See Akhtar & Awasthi (*Biol. Mem.* **5**: 13, 1980; India), Degelius (*Symb. bot. upsal.* **13**(2), 1954, Europe, keys; **20**(2), 1974, extra-Europe, keys).

Collemataceae Zenker (1827), Lecanorales (L). 7 gen. (+ 34 syn.), 246 spp. Thallus usually foliose to fruticose, dark grey or green to black, gelatinized; ascomata sessile, usually strongly concave, the margin varied in form; interascal tissue of simple or branched paraphyses, immersed in an I+ gelatinous matrix; asci with a well-developed I+ apical cap; ascospores varied, hyaline. Anamorph pycnidial. Lichenized with cyanobacteria, widespr.
Lit.: Henssen (*Lichenologist* **3**: 29, 1965; simple-spored gen.).

Collematomyces E.A. Tomas ex Cif. & Tomas. (1953) ≡ Collema (Collemat.).

Collematospora Jeng & Cain (1976), Trichosphaeriaceae. 1 (coprophilous), Venezuela.

Collemis Clem. (1931) ≡ Collema (Collemat.).

Collemodes Fink (1918) = Collema (Collemat.).

Collemodiopsis (Vain.) de Lesd. (1910) = Collema (Collemat.).

Collemodium Nyl. (1878) = Leptogium (Collemat.).

Collemopsidiomyces Cif. & Tomas. (1953) ≡ Collemopsidium (Dothideales).

Collemopsidium Nyl. (1881), Dothideales (inc. sed., L). 3, Eur., N. Afr. See Henssen (*SA* **5**: 126, 1986), Gube & Hafellner (*Nova Hedw.* **51**: 283, 1990).

Collemopsis Nyl. ex Cromb. (1874) = Psorotichia (Lichin.) fide Ellis (*Lichenologist* **13**: 123, 1981).

Collemopsis Trevis. (1880), Lichinaceae (L). 1, Eur.

Colleptogium M. Choisy (1962) ≡ Leptogium (Collemat.).

Colletoconis de Hoog & Aa (1978), Mitosporic fungi, 6.A1/2.15. 1 (on aecia of *Uredinales*), Argentina.

Colletogloeum Petr. (1953), Mitosporic fungi, 6.C1/2,19. 9, widespr. See Sutton (1980), Sutton & Swart (*TBMS* **87**: 93, 1986).

Colletomanginia Har. & Pat. (1906) ? = Engleromyces (Xylar.) fide Lloyd (*Mycol. Writ.*, 1917). See also Rogers (*Mycol.* **73**: 28, 1981), Læssøe (*SA* **13**: 43, 1994).

Colletosporium Link (1824), Mitosporic fungi, ?.?.?. 2, Eur. See Hughes (1958).

Colletostroma Petr. (1953) = Colletotrichum (Mitosp. fungi) fide v. Arx (1957).

Colletotrichella Höhn. (1916) = Kabatia (Mitosp. fungi) fide Sutton (1977).

Colletotrichopsis Bubák (1904) ? = Colletotrichum (Mitosp. fungi) fide v. Arx (1957).

Colletotrichum Corda (1831), Anamorphic Phyllachoraceae, 6.A1.15. Teleomorph *Glomerella*. 39, widespr. *C. gloeosporioides* (600 syn.; teleomorph *Glomerella cingulata*) (anthracnose of citrus, banana, and many other plants), *C. kahawae* (coffee berry disease; Waller *et al.*, *MR* **97**: 989, 1993; Firman & Waller, *Phytopath. Pap.* **20**, 1977); *Phaseolus* (*C. lindemuthianum*, *CMI Descr.* **316**, 1971); cucumber (*C. orbiculare*, syn. *C. lagenarium*); *C. coccodes* (syn. *C. atramentarium*) (potato black dot and tomato root rot); *C. dematium* f. *circinans* (onion (*Allium*) smudge); *C. lini* (flax (*Linum*) seedling blight).
Lit.: Sutton (*in* Bailey & Jeger (Eds), *Colletotrichum: biology, pathology and control*, 1992; *Mycol. Pap.* **141**, 1977; generic synonymy), Park *et al.* (*Korean Jl Path.* **3**: 85, 1987; spp. separation by electrophoresis), Koch *et al.* (*Phytophyl.* **21**: 69, 1989; spp. on *Medicago* in S. Afr.), Baxter *et al.* (*S. Afr. Jl Bot.* **2**: 259, 1983; 11 S. Afr. spp.), Baxter *et al.* (*Phytophyl.* **17**: 15, 1985; lit. review), Wang & Li (*Acta mycol. Sin.* **6**: 211, 1987; key 14 spp. China), Panaccione *et al.* (*Mycol.* **81**: 876, 1989; conidial dimorphism in *C. graminicola*), Ali *et al.* (*Phytopath.* **79**: 1148, 1989; electrophoresis differentiates *C. graminicola*), Boland & Brochu (*Can. J. Pl. Path.* **11**: 303, 1989; cv. response to races of *C. destructivum*), Smith (*Pl. Dis.* **74**: 69, 1990; spp. from *Fragaria*), Jeffries *et al.* (*Pl. Path.* **39**: 343, 1990; biology and control of tropical fruit crop spp.), Bailey *et al.* (*MR* **94**: 810, 1990; species on *Vigna*), Braithwaite *et al.* (*Aust. Syst. Bot.* **3**: 733, 1990; ribosomal DNA as a taxonomic marker), Braithwaite *et al.* (*MR* **94**: 1129, 1990; RFLPs in *C. gloeosporioides*), Brooker *et al.* (*Phytopath.* **81**: 672, 1991; nitrate non-utilizing mutants and vegetative compatibility), Yang & Chuang (*Plant Prot. Bull.* **33**: 262, 1991; variation in *C. musae*), Bonde *et al.* (*Phytopath.* **81**: 1523, 1991; isozyme patterns in spp. on *Fragaria*), Walker *et al.* (*MR* **95**: 1175, 1991; spp. on *Xanthium*), Pain *et al.* (*Physiol. Mol. Pathol.* **40**: 111, 1992; monoclonal antibodies in), Bailey & Jeger (Eds) (*Colletotrichum: biology, pathology and control*, 1992), Gunnell & Gubler (*Mycol.* **84**: 157, 1992; spp. on strawberry), Lyanage *et al.* (*Phytopathol.* **83**: 113, 1993; curinase in *C. gloeosporioides*), Correll *et al.* (*Phytopath.* **83**: 1199, 1993; RAPD analysis of

C. orbiculare), Sreenivasaprasad *et al.* (*MR* **98**: 186, 1994; nucleotides and identification of *C. acutatum*), Freeman *et al.* (*Exp. Mycol.* **17**: 309, 1993; molecular genotyping), Sherriff *et al.* (*Exp. Mycol.* **18**: 121, 1994; ribosomal DNA analysis and species groups), Fabre *et al.* (*MR* **99**: 429, 1995; molecular markers and bean isolates), Freeman & Rodriguez (*MR* **99**: 501, 1995; spp. differentiation of strawberry isolates by PCR), Sherriff *et al.* (*MR* **99**: 475, 1995; rDNA sequence differentiates *C. graminicola* and *C. sublineolum*).

colliculose (**colliculous**), with rounded swellings; blistered.

Colligerites K.P. Jain & R.K. Kar (1979), Fossil fungi. 2 (Palaeocene), India.

Collodendrum Clem. (1909) ≡ *Tremellodendron* (Exid.).

Colloderma G. Lister (1910), Stemonitidaceae. 3, widespr.

Collodermataceae, see *Stemonitidaceae.*

Collodiscula I. Hino & Katum. (1955), Xylariaceae. 1, Japan. See Læssøe (*SA* **13**: 43, 1994), Læssøe & Spooner (*Kew Bull.* **49**: 1, 1994), Samuels *et al.* (*Mycotaxon* **28**: 453, 1987), Samuels (*SA* **6**: 121, 1987; status). Anamorph *Acanthodochium.*

Collodochium Höhn. (1902), Mitosporic fungi, 3.A1.?. 1, Eur.

Collolechia A. Massal. (1854) = *Placynthium* (Placynth.).

Collonaemella Höhn. (1915) = *Cornularia* (Mitosp. fungi) fide Clements & Shear (1931) but see Sutton (1977).

Collonema Grove (1886) nom. dub. (Ascomycota, inc. sed.); based on an effete ascoma fide Sutton (*Mycol. Pap.* **141**: 45, 1977).

Collopezis Clem. (1909) ≡ *Tjibodasia* (Platygl.).

Collopus Earle (1909) = *Mycena* (Tricholomat.) fide Singer (1951).

Collostroma Petr. (1947), Mitosporic fungi, 8.A1.?. 1, Austria.

collulum, the neck of a conidiogenous cell (Zaleski, 1927).

Collybia (Fr.) Staude (1857) nom. cons., Tricholomataceae. 70, cosmop. See Pegler (*Kew Bull., addit. ser.* **6**, 1977; key 10 spp. E. Afr.), Lennox (*Mycotaxon* **9**: 117, 1979; classification), Halling (*Mycol. Mem.* **8**, 1983; keys USA spp.).

Collybidium Earle (1909) ≡ *Flammulina* (Tricholomat.).

Collybiopsis (J. Schröt.) Earle (1909) nom. dub. fide Singer (1948).

Collyria Fr. (1849) nom. dub. fide Singer (1975).

colon (:) (in author citations), see Nomenclature.

Colonnaria Raf. (1808) = *Clathrus* (Clathr.) fide Dring (1980).

Colonomyces R.K. Benj. (1955), Euceratomycetaceae. 1, USA, Eur.

colony (of bacteria and yeasts), a mass of individuals, generally of one species, living together; (of mycelial fungi), a group of hyphae (frequently with spores) which, if from one spore or cell, may be one individual.

Colour. Names of colours should be given precision by reference to a colour standard, preferably the Munsell Color System as exemplified in the *Munsell book of color* (Cabinet edn, 1963). This work is expensive but the Munsell Color Co. issue a 'Mycologists' Color Kit' designed for individual ownership. The *Methuen handbook of colour* by Kornerup & Wanscher (an English translation of the Danish *Farver i Farver* by the same authors) is an inexpensive colour chart for which in edn 2 (1967 for both versions) the equivalent Munsell notation is given for each colour; Rayner's *A*

mycological colour chart, 1970, illustrates Dade's nomenclature (see below).

Since its publication in 1912 Ridgway's *Color standards and color nomenclature*, which covers 1,115 colours, has been much used by biologists but copies of this work must now be referred to with caution due to variations having occurred in some colours depending on the amount of use a copy has received and the care which has been given to its preservation. (Tables for Ridgway/Munsell conversions accompany Rayner's chart.)

Other general colour standards include Oberthur & Dauthenay, *Repertoire de couleurs*, 2 vols., 1905 (for Ridgway equivalents see Snell & Dick, *Glossary of mycology*, rev. edn. 1971); Klinksieck & Vallette, *Code de couleurs*, 1908; Maerz & Paul, *A Dictionary of color*, edn 2, 1950; and Wilson, *Horticultural colour chart*, 2 vols., 1939-42. Kelly & Judd (*The ISCC-NBS method of designating colors and a dictionary of color names*, 1955 [reprint 1963], National Bureau of Standards, Washington Circular 553) list the Inter-Society Color Council-NBS equivalents of 7,500 colour names including those of Ridgway and Maerz & Paul; a supplement provides standard colour patches (ISCC-NRSS, *Centroid color charts*, *Standard sample* No. 2106). Locquin, *Guide des couleurs naturelles* (*Obs. Disp. mycol.* **1** (2), 1975) has about 1,536 colour patches, Pfaff (*Z. Mykol.* **49**: 237, 1983; comparison colour atlases).

Saccardo (*Chromotaxia*, 1891; edn 3, 1912) gave the Latin, Italian, French, English and German names and synonyms of 50 colours of mycological interest, and Dade (*Mycol. Pap.* **6** (edn 2), 1949) related Saccardo's colours to the Ridgway standard, suggested additional Latin colour terms for mycological use, and gave an extended list of colour names. The colour terminology used by Fries for agarics was discussed by Wharton (*Grevillea* **13**: 25, 1884) whose paper has been reprinted in part by Stearn (*Botanical Latin*, edn 4, 1992) who also reprints Dade's chart and gives additional interesting historical information on colour terminology. Locquin, *Chromotaxis: Code mycologique et pedologique des couleurs*, 1958 (see *Mycol.* **50**: 447, 1958) related a number of colours to a series of colour transparencies. See Tocker *et al.* (*Taxon* **40**: 201, 1991; survey colour charts).

Colour names are omitted from this *Dictionary* although a few general colour terms such as dematiaceous, hyaline, inquinant, etc., and those given below have been compiled.

Corner (*Clavaria*, 1950) proposed the following series of terms for the more precise description of pigmentation of basidiomata and hyphae: **achroic**, without true pigmentation; **euchroic**, having true pigmentation as opposed to **epichroic**, discoloration due to injury; **hysterochroic**, slowly discoloured from base to apex in old age; euchroism may be **acrochroic**, coloured specially in the hyphal tips at the growing point, **metachroic**, changing colour through the appearance of a new pigment in maturer tissue, **ectochroic**, pigment on the outside of the hypha, **mesochroic**, pigment in the hyphal wall, or **endochroic**, pigment inside the cell. The last may be subdivided according to whether the pigment is diffused in the cytoplasm (**cytochroic**), in the cell vacuoles (**cystochroic**) or in oil drops (**lipochroic**). Cf. Pigments.

colour of the reverse, the colour of the underside of a fungus culture in a tube or Petri dish.

Colpoma Wallr. (1833), Rhytismataceae. 8, Chile, Eur., NZ, N. Am. See Twyman (*TBMS* **29**: 234, 1946). Anamorph *Conostroma.*

Colpomella Höhn. (1926) nom. dub. fide Sutton (*Mycol. Pap.* **141**, 1977).

Coltricia Gray (1821), Hymenochaetaceae. 13, widespr. See Pegler (1973).

Coltriciella Murrill (1904), Hymenochaetaceae. 2, subtrop., trop.

Coltriciopsis Teixeira (1993), Hymenochaetaceae. 2, S. Am.

columella, a sterile central axis within a mature fruitbody which may be uni- or multicellular, unbranched or branched, of fungal or host origin; (of gasteromycetes, after Cunningham) **axile -**, when an axis in the gleba; **dendroid -**, having lateral branches, as in *Gymnoglossum*; **percurrent -**, joining the peridium at the apex of gleba; **pseudo-**, embryonic tissue in the mature peridium of *Geastrum*; **simple -**, not branched, as in *Secotium*.

Columnocystis Pouzar (1959), = Veluticeps (Ster.) fide Hjortstam & Tellería (*Mycotaxon* **37**: 53, 1990).

Columnodomus Petr. (1941), Mitosporic fungi, 8.A1.?. 1, Eur.

Columnodontia Jülich (1979), Meruliaceae. 4, S.E. Asia, Australia.

Columnomyces R.K. Benj. (1955), Laboulbeniaceae. 1, USA.

Columnophora Bubák & Vleugel (1916), Mitosporic fungi, 1.A2.1. 1, Eur.

Columnosphaeria Munk (1953) = Guignardia (Mycosphaerell.) fide v. Arx & Müller (1972).

Columnothyrium Bubák (1916), Mitosporic fungi, 5.A1.?. 1, Eur.

Colus Cavalier & Séchier (1835), Clathraceae. 4, trop., warm temp.

Coma Nag Raj & W.B. Kendr. (1972), Anamorphic ? Phacidiaceae, 6.G2.19. Teleomorph *Ascocoma*. 1, Australia. See Sutton (*Nova Hedw.* **25**: 161, 1974), Swart (*TBMS* **87**: 603, 1987; teleomorph).

comate, having hairs; shaggy.

Comatospora Piroz. & Shoemaker (1971), Mitosporic fungi, 8.B1.19. 1, Canada.

Comatricha Preuss (1851), Stemonitidaceae. 38, cosmop.

Comatrichoides Hertel (1956) ≡ Comatricha (Stemonitid.).

Combea De Not. (1846), Roccellaceae (L). 1, S. Afr.

Combodia Fr. (1849) ? = Lasiodiplodia (Mitosp. fungi) fide Sutton (1977).

Comesella Speg. (1923), ? Dothideales (inc. sed.). 1 (on *Pteridophyta*), Cuba.

Comesia Sacc. (1884), ? Leotiales (inc. sed.). 2 or 3, Eur., N. Afr.

Cometella Schwein. (1835) ≡ Clasterosporium (Mitosp. fungi) fide Ellis (1971).

commensalism, a form of symbiosis (q.v.).

commissure, a closing join; a seam.

commixt, mixed with; intermingled.

Common names. In marked contrast to flowering plants few fungi (incl. lichens) have been given common names; only twenty-five or so fungi (e.g. fly agaric, blewits, morel, puff ball) have achieved recognition by the *Shorter Oxford Dictionary* and *Webster's Dictionary*. Historically, however, macrofungi have been recognized by common names from earliest times. Numerous attempts to coin common names for larger fungi (and occasionally for microfungi; see Cooke, *Handbook of British fungi*, 1871), frequently by more or less literal English versions of the scientific names, have had little appeal.

Vernacular names for diseases of plants, man, and animals caused by fungi are much more widely used in both technical and popular writing, and a number of the plant disease names are also applied to the pathogens, e.g. rust, smut, powdery mildew.

Lichens used by man for dyeing, or other purposes, or eaten by reindeer were also often given common names. See e.g. crottle, heather rags, lungwort, manna, old man's beard, orchil, rock tripe.

In scientific writing common names are frequently derived from the scientific names of genera or other taxa for use in both the singular or plural, e.g. aspergillus (aspergilli), ascomycete(s), fusarium (fusaria), etc., and this practice seems unobjectionable as it enables taxa to be distinguished from the individuals of which they are composed, e.g. phytophthoras = species of *Phytophthora*. Common names should be printed in Roman (not italic) type and decapitalized.

Lit.: Lundqvist & Persson (*Svenska Svampnamn*, 1987; 2584 Swedish names), Ulvinen *et al.* (*Karstenia* **29** (Suppl.), 1989; Finnish macromycetes).

See Classification, Dyeing, names, Nomenclature.

community, any phytosociological taxon.

Comocephalum Syd. (1939), Mitosporic fungi, 4.D2.?. 1, Ecuador.

Comocheila A. Massal. (1860) = Phlyctis (Phlyctid.).

Comoclathris Clem. (1909), Diademaceae. 21, Am., Eur. See Shoemaker & Babcock (*CJB* **70**: 1617, 1992; key). = Graphyllium (Diadem.) fide Barr (*SA* **12**: 27, 1993). Anamorph *Alternaria*.

comose, having hairs in groups or tufts.

compaginate, joined tightly together.

companion cell, contiguous donor gametangium of *Olpidiopsidales*.

compatible (of mating types, strains, etc.), able to be cross-mated; cross-fertile; **hemi-**, a homokaryon compatible with 1 of the 2 components of the dikaryon. See Sex.

complanate, flat; smooth.

complement-fixation test, a sensitive test by which antigen-antibody reaction may be detected and quantified. The test depends on the ability of antigens after reacting with their specific antibodies to 'fix' complement (a group of proteins normally present in freshly isolated serum) the presence of which is necessary for the lysis of red blood cells by haemolysin (red-cell-immune serum).

Completoria Lohde (1874), Completoriaceae. 1 (on prothalli of *Pteridophyta*), N. temp. See Atkinson (*Cornell Univ. Agric. Exp. Stn Bull.* **94**: 252, 1895), Tucker (1981).

Completoriaceae Humber (1989), Entomophthorales. 1 gen. 1 sp.

complex, sometimes used to designate a group of closely related species.

Complexipes C. Walker (1979), Mitosporic fungi, 1.A2.1. 1, USA, Korea. See Yang & Korf (*Mycotaxon* **23**: 457, 1985), Lee & Ka (*Korean J. Mycol.* **18**: 127, 1990; Korea).

complicate, bent upon itself.

compound, made up of a number of parts.

compressed (of a stipe), flattened transversely.

Compsocladium I.M. Lamb (1956), Bacidiaceae (L). 1, New Guinea.

Compsomyces Thaxt. (1894), Laboulbeniaceae. 6, widespr.

Compsosporiella Sankaran & B. Sutton (1991), Mitosporic fungi, 4.A1.1. 1, Sierra Leone.

Computing, has numerous applications in mycology, and pertinent papers are cited under appropriate topics. For coding of characters for use in microbial databases see Gams *et al.* (*J. gen. microbiol.* **134**: 1667, 1988) and Rogosa *et al.* (*Coding microbiological data for computers*, 1986). See also Authors, Genetic Resource Collections, Internet, Inventorying, Literature, Neural networks, Numerical taxonomy.

Concamerella W.L. Culb. & C.F. Culb. (1981), Parmeliaceae (L). 2, S. Am.

concatenate, in chains; catenulate.

concave (esp. of a pileus), hollowed out; basin-like.

concentric bodies, ultrastructures found in many lichenized fungi and also in some other fungi such as *Rhopographus*, *Sphaerotheca*, *Cercospora*, *Pseudopeziza*, *Sphaceloma*; 'elliptical bodies'. See Griffiths & Greenwood (*Arch. Microbiol.* **87**: 285, 1972), Pons *et al.* (*TBMS* **83**: 181, 1984), Beilharz (*TBMS* **84**: 79, 1985), Meyer (*Mycol.* **79**: 44, 1987; in granules in *Allomyces*).

conceptacle, any hollow structure producing spores or spermatia.

conchate (conchiform), like a bivalve shell.

Conchatium Velen. (1934) = Cyathicula (Leot.) fide Baral (*SA* **13**: 113, 1994).

Conchites Paulet (1791) = Auricularia (Auricular.) fide Donk (*Taxon* **7**: 174, 1958).

Conchomyces Overeem (1927) = Hohenbuehelia (Tricholomat.) fide Pegler (*in litt.*).

Conchyliastrum Zebrowski (1936), ? Fossil fungi (Chytrid.). 2 (Cambrian to Recent ?), Australia.

concolorous, of one colour.

concrescent, becoming joined.

concrete, joined by growth.

conditioning, the process by which fungi enzymically soften up substrata such as dead leaves before detritivorous animals can eat them.

conducting hypha, see hypha.

Condylospora Nawawi (1976), Mitosporic fungi, 1.E1.10. 3 (aquatic), Malaysia. See Nawawi & Kuthubutheen (*Mycotaxon* **23**: 329, 1988).

conferted, near together; crowded.

Conferticium Hallenb. (1980), Gloeocystidiellaceae. 3, N. temp.

Confertobasidium Jülich (1972) = Scytinostromella (Gloeocystidiell.) fide Ginns & Freeman (*Bibl. Myc.* **157**, 1994).

Confertopeltis Tehon (1933) nom. dub. fide Sutton (*Mycol. Pap.* **141**, 1977).

confervoid, composed of loose filaments or cells.

Confistulina Stalpers (1983), Anamorphic Fistulinaceae, 1.A1.1. Teleomorph *Fistulina*. 1, widespr.

confluent, (1) (of sori, etc.), coming together; running into one another; (2) (of the flesh of a stipe), continuous with the trama of the pileus.

congeneric, one of two or more taxa considered to belong to one genus; cf. synonym.

congested, very near together.

conglobate, (1) massed into a ball; (2) (of the bases of stipes), together making a fleshy mass.

conglutinate, glued together; esp. of paraphyses in *Lecanorales*.

Conia Vent. (1799) = Lepraria (Ascomycota, inc. sed.) fide Sutton (*in litt.*).

Coniangium Fr. (1821) = Arthonia (Arthon.).

Conicomyces R.C. Sinclair, Eicker & Morgan-Jones (1983), Mitosporic fungi, 2.E1.15. 2, S. Afr. See Illman & White (*CJB* **63**: 419, 1985; addit. sp.).

Conicosolen F. Schill. (1927) = Psorotheciopsis (Gomphill.).

Conida A. Massal. (1856) = Arthonia (Arthon.) fide Santesson (1952).

conidange, a small pycnidium (in a lichen thallus) having no stout wall (des Abbayes, 1951) (obsol.).

Conidella Elenkin (1901) ? = Arthonia (Arthon.).

Conidial nomenclature. The traditional approach to the nomenclature of the spores of Mitosporic fungi and anamorphs is that of Saccardo who differentiated 7 morphological types (amero-, dictyo-, didymo-, helico-, phragmo-, scoleco-, and staurospores), based on shape and septation, the terminology for which was further qualified according to whether the spores were pigmented (phaeo-amerospores, etc.) or not (hyalo-

amerospores, etc.); see Mitosporic fungi. Kendrick & Nag Raj (*in* Kendrick (Ed.), *The whole fungus* **1**: 43, 1979) discussed Saccardo's categories and offered precise definitions.

Since Hughes (*CJB* **31**: 577, 1953) drew attention to the systematic importance of conidiogenous events (q.v.), a nomenclature for conidia has evolved based on the methods by which the conidia develop, giving rise to terms such as annelloconidia, phialoconidia, tretoconidia etc. (cf.). These are not universally accepted, the preferred usage now being 'conidium' with qualifying adjectives or descriptors. Main contributions to conidial terminology have been by the first Kananaskis Workshop-Conference in 1969 (see Kendrick (Ed.), *Taxonomy of fungi imperfecti*, 1971), Ellis (*Dematiaceous hyphomycetes*, 1971) offered a series of definitions as did Kendrick & Carmichael (*in* Ainsworth *et al.* (Eds), *The Fungi* **4A**: 323, 1973), Cole (*CJB* **53**: 2983, 1975), Cole & Samson (*Patterns of development in conidial fungi*, 1979). The descriptive terminology in both conidial morphology and conidiogenous events was initiated by Minter *et al.* (*TBMS* **79**: 75, 1982; **80**: 39, 1983; **81**: 109, 1983), developed by Sutton (*in* Reynolds & Taylor (Eds), *The fungal holomorph*: 28, 1993) and Hennebert & Sutton, and Sutton & Hennebert (*in* Hawksworth (Ed.), *Ascomycete systematics*: 65, 77, 1994), and is used in this edition of the *Dictionary*. See Mitosporic fungi.

conidiangium, pycnidium (obsol.).

Conidiascus Holterm. (1898), ? Endomycetaceae. 1, Java.

Conidiobolus Bref. (1884), Ancylistaceae. 27, widespr. See King (*CJB* **54**: 45, 1285, 1976; key, **55**: 718, 1977), Remaudière & Keller (1980), Ben-Ze'ev & Kenneth (*Mycotaxon* **14**: 393, 1982; subgen. *Delacrouxia*), Latgé *et al.* (*CJB* **60**: 413, 1982; ultrastr., *Exp. Mycol.* **10**: 99, 1986; ultrastr.), Keller (1987), Huber (1989).

Conidiocarpus Woron. (1926), Mitosporic fungi, 4.A1.15. 2, widespr. See Hughes (*Mycol.* **68**: 693, 1976).

conidiogenesis, the process of conidium formation. Concepts of conidiogenesis have increasingly been used in mitosporic fungal systematics since Hughes (*CJB* **31**: 577, 1953) classified some hyphomycetes according to the different methods by which conidia develop from conidiophores and the ways in which conidiophores (and conidiogenous cells) grow before, during and after conidia are produced. The historical development of this approach has been covered by Kendrick (Ed.) (*Taxonomy of fungi imperfecti*, 1971), for hyphomycetes, by Sutton (*in* Ainsworth *et al.* (Eds), *The Fungi* **4A**: 513, 1973) for coelomycetes, and by Vobis (*Bibl. Lich.* **14**, 1980) for lichenized pycnidia. Cole & Samson (*Patterns of development in conidial fungi*, 1979) emphasized the contribution of ultrastructural data to developmental concepts, whilst Minter *et al.* (*TBMS* **79**: 75, 1982; **80**: 39, 1983; **81**: 109, 1983) reassessed optical and electron microscopy observations and demonstrated a continuum of developmental processes. See Mitosporic fungi, wall-building.

conidiogenous, producing conidia; - cell, any cell from or within which a conidium is directly produced; - locus, the place on a conidiogenous cell at which a conidium arises; Kendrick (1971: 258).

conidiole, a small conidium, esp. one on another; a secondary conidium, as in *Empusa*.

conidioma (pl. -ata), a specialized multi-hyphal, conidia-bearing structure (Kendrick & Nag Raj *in* Kendrick (Ed.), *The whole fungus* **1**: 51, 1979). See acervulus, pycnidium, sporodochium, synnema (all obsol. nouns, but used adjectivally, e.g. acervular conidioma). See Fig. 10. Cf. conidiophore.

conidiophore, a simple or branched hypha (a **fertile hypha**) bearing or consisting of conidiogenous cells from which conidia are produced; sometimes used when describing reduced structures for the conidiogenous cell.

Conidiosporomyces Vánky (1992), Tilletiaceae. 1 (on *Gramineae*), trop. Afr., Am., Asia.

Conidioxyphium Bat. & Cif. (1963) = Conidiocarpus (Mitosp. fungi) fide Hughes (*Mycol.* **68**: 693, 1976).

conidium, a specialized, non-motile (cf. zoospore), asexual spore, usually caducous, not developed by cytoplasmic cleavage (cf. sporangiospore) or free-cell formation (cf. ascospore); in certain *Oomycota* produced through the incomplete development of zoosporangia which fall off and germinate to produce a germination tube. See Sutton (*TBMS* **86**: 1, 1986; derivation etc.); - **initial**, a cell, or part of a cell from which a conidium develops; - **ontogeny**, conidiogenesis (q.v.).

Coniella Höhn. (1918), Anamorphic Melanconidaceae, 4.A2.15. Teleomorph *Schizoparme*. 9, widespr. See Sutton (1980; key).

Coniobotrys Pouzar (1958) ? = Jaapia (Coniophor.) fide Donk (1964).

Coniocarpon DC. (1805) = Arthonia (Arthon.).

Coniocephalum Brond. (1828) nom. dub. (Myxomycetes) fide Martin (*in litt.*).

Coniochaeta (Sacc.) Cooke (1887), Coniochaetaceae. 35, widespr. See Checa *et al.* (*Cryptogamie, Mycol.* **9**: 1, 1988; key 15 spp. Spain), Hawksworth (*in* Hawksworth (Ed.), *Ascomycete systematics*: 377, 1994; polyphyly), Hawksworth & Yip (*Austr. J. Bot.* **29**: 377, 1981; key 11 spp. in cult.), Mahoney & LaFavre (*Mycol.* **73**: 931, 1981; synopsis). Anamorphs *Nodulisporium*- or *Phialophora*-like.

Coniochaetaceae Malloch & Cain (1971), Sordariales. 5 gen. (+ 6 syn.), 43 spp. Ascomata usually perithecial, solitary or aggregated, sometimes on a poorly-developed subiculum; interascal tissue inconspicuous, of paraphyses; asci usually cylindrical, often with a small apical ring, sometimes I+ blue; ascospores aseptate, dark brown, with a germ-slit, sheath lacking. Coprophilous or on rotten wood; widespr.

Coniochaetidium Malloch & Cain (1971), Coniochaetaceae. 3, widespr. See v. Arx (*Stud. Mycol.* **8**: 25, 1975; key).

Coniochila A. Massal. (1860) ? = Ocellularia (Thelotremat.).

Coniocybaceae Rchb. (1837), Caliciales (L). 4 gen. (+ 13 syn.), 50 spp. Thallus crustose; ascomata stalked, the stalk composed of parallel hyphae; asci thin-walled, evanescent; ascospores usually spherical and aseptate, hyaline or pale brown, smooth or with inconspicuous ornamentation, forming a dry mazaedial mass.

Coniocybe Ach. (1816), Coniocybaceae (±L). 24, cosmop. (mainly temp.). See Honegger (*Lichenologist* **17**: 273, 1985; anamorph), Tibell (1975, *Svensk bot. Tidskr.* **72**: 171, 1978).

Coniocybomyces Cif. & Tomas. (1953) ≡ Coniocybe (Coniocyb.).

Coniocybopsis Vain. (1927) = Microcalicium (Microcalic.) fide Tibell (1978).

Coniodictyum Har. & Pat. (1909), Cryptobasidiaceae. 1, Afr. See Malençon (*BSMF* **69**: 77, 1953).

Conioloma Flörke (1815) = Arthonia (Arthon.).

Coniomela (Sacc.) Kirschst. (1934) = Coniochaeta (Coniochaet.) fide v. Arx & Müller (1954).

Coniomycetes (obsol.). Class for fungi with spores borne in a naked mass, including certain Mitosporic

fungi, *Uredinales* and *Ustilaginales* (Fries, *Syst. mycol.* **1**, 1821).

Coniophora DC. (1815), Coniophoraceae. 20, widespr. *C. puteana* (cellar fungus) is a cause of rot in wood. See Ginns (*CJB* **51**: 249, 1973; type studies, *Opera Bot.* **61**, 1982; world monogr.), Lentz (*Mycol.* **49**: 534, 1957; basidial development).

Coniophoraceae Ulbr. (1928), Boletales. 11 gen. (+ 8 syn.), 54 spp. Basidioma effuso-reflexed or resupinate; hymenophore smooth to merulioid; spores yellowish brown, smooth; lignicolous, humicolous.
    Lit.: Donk (1964: 254), Cooke (*Mycol.* **49**: 197, 1957; key), Hallenberg (*Lachnocladiaceae and Coniophoraceae of North Europe*, 1985), Besl *et al.* (*Zeit. Mykol.* **52**: 277, 1986; chemosyst.).

Coniophorafomes Rick (1934) nom. dub. (Coniophor.).

Coniophorella P. Karst. (1889) = Coniophora (Coniophor.) fide Ginns (1992).

Coniophoropsis Hjortstam & Ryvarden (1986), Coniophoraceae. 1, Argentina.

Coniophyllum Müll. Arg. (1892) = Calycidium (Calycid.).

Conioscypha Höhn. (1904), Mitosporic fungi, 1.A2.20. 3, N. Am., UK. See Shearer (*Mycol.* **65**: 128, 1973), Kirk (*TBMS* **82**: 177, 1984; n.sp.).

Coniosporiella Bat. (1966) = Schiffnerula (Englerul.) fide v. Arx & Müller (1975).

Coniosporiopsis Speg. (1918) nom. conf. fide Hughes (*CJB* **36**: 754, 1958).

Coniosporium Link (1809), Mitosporic fungi, 3.D2.23. 2 or 3, Eur.

Coniotheciella Speg. (1918) ? = Coniothecium (Mitosp. fungi).

Coniothecium Corda (1833) nom. dub. fide Hughes (*CJB* **36**: 727, 1958). See Hawksworth (*TBMS* **65**: 219, 1975; redisp. lichenicolous spp.).

Coniothele Norman (1868) = Trimmatothele (Verrucar.).

Coniothyriella Speg. (1889) nom. conf. fide Petrak & Sydow (*Beih. Rep. spec. nov. regn. veg.* **42**: 322, 1927).

Coniothyriella Speg. (1910) ≡ Coniothyrina (Mitosp. fungi).

Coniothyriites Babajan & Tasl. (1970), Fossil fungi. 1 (Tertiary), former USSR.

Coniothyrina Syd. (1912), Mitosporic fungi, 4.A2.15. 1 (on *Agave*), warm areas.

Coniothyrinula Petr. (1923) = Coniothyrium (Mitosp. fungi) fide Clements & Shear (1931).

Coniothyriopsiella Bender (1932) ? = Cyclothyrium (Mitosp. fungi) fide Sutton (1977).

Coniothyriopsis Speg. (1910) ? = Microsphaeropsis (Mitosp. fungi). See Sutton (1977).

Coniothyriopsis Petr. (1923) = Cyclothyrium (Mitosp. fungi) fide Sutton (1977).

Coniothyris Clem. (1909) ≡ Coniothyriella (Mitosp. fungi).

Coniothyrium Corda (1840) nom. cons., Anamorphic Leptosphaeriaceae, 4.A/B2.19. Teleomorph *Leptosphaeria*. 25, widespr. See Biga *et al.* (*Sydowia* **12**: 258, 1959), Sutton (*Mycol. Pap.* **123**, 1971; relationship to *Microsphaeropsis*). *C. wernsdorffiae* (brand canker of rose; Westcott, *Mem. Cornell agric. Exp. Stn* **153**, 1934). *C. minitans* (mycoparasite of sclerotia; Sesan, *Probl. Prot. Plant.* **17**: 29, 1989; Sesan & Crisan, *Stud. Cercet Biol.* **40**71, 1988; Sandys-Winsch *et al.*, *MR* **97**: 1175, 1993; world distrib.).

Conisphaeria Cooke (1879) nom. dub. (Fungi, inc. sed.). Based on various pyrenomycetes.

conjugate, joined; in twos; - **nuclei**, two nuclei in one cell which undergo division (- **division**) at the same time, as in basidiomycetes.

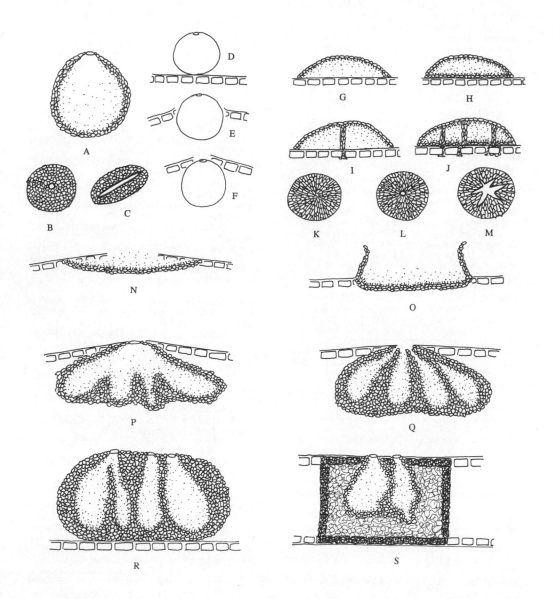

Fig. 10. Conidiomatal types. A-F, pycnidial; B, dehiscence by a central circular ostiole; C, dehiscence by a longitudinal ostiole (raphe); D, superficial; E, semi-immersed; F, immersed; G-M, pycnothyrial; G, with upper wall only; H, with upper and lower walls; I, with a central supporting column; J, multilocular with several supporting columns; K, dehiscence from the margin; L, dehiscence by a central ostiole; M, dehiscence by irregular fissures; N, acervular; O, cupulate; P-R, eustromatic; P, convoluted, immersed; Q, multilocular, immersed; R, multilocular, superficial; S, pseudostromatic.

conjugation, copulation (q.v.), esp. isogamic copulation; **- tube**, a tube between two copulating cells.

conk, the fruit-body (basidioma) of a wood-attacking fungus, esp. of a polypore.

connate, joined by growth.

connective, see disjunctor.

connective hyphae, hyphae of the connective tissue of the context (Fayod, 1889).

Connersia Malloch (1974), Pseudeurotiaceae. 1, UK.

connivent, (1) touching but not organically joined; (2) (of a pileus margin), touching the stipe.

Conocybe Fayod (1889) nom. cons., Bolbitiaceae. 70, widespr. See van Waveren (*Persoonia* **6**: 119, 1970), Watling (*Fl. Illustr. Champ. Afr. centr.* **3**: 57, 1974; key 13 Afr. spp.; *Revue mycol.* **40**: 31, 1976; *British Fungus Fl.* **3**, 1982; key 58 Br. spp.).

Conohypha Jülich (1975), Hyphodermataceae. 2, Eur., USA.

Conoplea Pers. (1801), Mitosporic fungi, 1.A2.10. 6, widespr. See Hughes (*CJB* **38**: 659, 1960; key).

Conostoma Bat. & J.L. Bezerra (1965), Mitosporic fungi, 1.A1.?. 1, Brazil.

Conostroma Moesz (1921), Anamorphic Rhytismataceae, 8.A1.10. Teleomorph *Colpoma*. 2, Eur., India.

Conotrema Tuck. (1848), Stictidaceae (L). 4, Eur., N. Am., N. Afr. See Gilenstam (*Ark. Bot.* ser. 2 **7**: 149, 1969), Sherwood (*Mycotaxon* **5**: 1, 1977).

Conotrematomyces Cif. & Tomas. (1953) ≡ Conotrema (Stictid.).

Conotremopsis Vězda (1977), Stictidaceae (L). 1, Tasmania.

Conservation (of fungi). There is an increasing need to pay attention to the conservation of rare and endangered species of fungi in view of the destruction of many habitats on a global scale. In addition, many groups of fungi, particularly lichenized forms, are sensitive to various kinds of environmental pollution. Conservation of fungi should therefore aim mainly at the conservation of the habitats and regulation of pollution. General introductions are provided by Arnolds and Richardson (*in* Hawksworth (Ed.), *Frontiers in mycology*: 187, 243, 1991). Hitherto conservation efforts have been directed mainly towards the visually more striking macromycetes and macrolichens. Basic activities include the collection of distribution and ecological data into regional and national databases, on the basis of which **Red Lists** can be drawn up, comprising condensed information on threatened species. National lists have been prepared for most European countries; in the case of fungi, these have been surveyed by Arnold & De Vries (*in* Pegler *et al.* (Eds), *Fungi of Europe: investigation, recording and conservation*: 211, 1993). On the basis of the Red Lists, it appears that some ecological groups of fungi are more vulnerable than others: of special concern are (1) wood-inhabiting fungi (esp. bracket spp.) on large logs in virgin and old-growth forests, endangered by intensive forestry (Høiland & Bendiksen, *in* Arnolds & Kreisel, *Conservation of fungi in Europe*: 51, 1993; Kotiranta & Niemelä, *Threatened polypores in Finland*, 1993); (2) fungi of peat bogs, marshland and boggy forests, due to reclamation and drainage (Winterhoff & Krieglsteiner, *Beig. Veröff. Natur. Landschaftspflege Bad. Württ.* **40**, 1984); (3) fungi of sand dunes, due to recreation, digging of sand and afforestation (Winterhoff & Krieglsteiner, 1984); (4) fungi of old, permanent pastures on soils poor in nutrients, due to increased fertilizer application, afforestation or lack of appropriate management (Arnolds, *Opera Bot.* **100**: 7, 1989; Nitare, *Svensk bot. Tidskr.* **82**: 341, 1988); (5) ectomycorrhizal fungi in forests on soils poor in nutrients, due to acidification and/or nitrogen accumulation from air pollution (Arnolds, *Agric. Ecosyst. Environm.* **35**: 209, 1991; Dighton & Jansen, *Environm. Pollut.* **73**: 179, 1991).

There is a wealth of evidence to show that lichens are also greatly affected by many of the above factors, but more particularly due to deforestation, agricultural practices and a wide variety of atmospheric pollutants (see Air pollution); a review of the principles and priorities of lichen conservation is provided by Seaward (*J. Hattori Bot. lab.* **52**: 401, 1982). While experiments and circumstantial evidence suggest that collection of mushrooms does not substantially contribute to their decline (Egli *et al.*, *Mycol. Helv.* **3**: 417, 1990), this is not the case with lichens, rarities having disappeared due to over-zealous collecting for economic reasons (e.g. as a component of curries, wreaths, decorations) or for reference collections and exsiccata. Maintenance of threatened species and species diversity of fungi in general may be dependent on management practices (Keizer, *in* Pegler *et al.* (Eds), 1993: 251).

The European countries cooperate in the European Council for Conservation of Fungi (ECCF) established by the 9th Congress of European Mycologists in Oslo, 1985. The ECCF is composed of national representatives of all countries. In addition, other persons active in fungal conservation can be placed on the mailing list by request. The ECCF organizes specialist meetings on fungal conservation in between the European Congresses and in short business meetings during these Congresses. It publishes a newsletter at irregular intervals (6 to date). A preliminary *Red List of macrolichens in the European Community* (Sérusiaux, 1989) has already been published and a *European Red List of threatened fungi* is in preparation (Ing, *in* Pegler *et al.* (Eds), 1993: 231). In 1990, the IUCN-The World Conservation Union confirmed the need for fungal conservation by the establishment of a Committee of Fungi to stimulate activities worldwide, including a *Fungi and Conservation Newsletter*; lichen conservation is similarly catered for by a subcommittee of the International Association for Lichenology (and via its *Newsletter*).

Consetiella Hol.-Jech. & Mercado (1982), Mitosporic fungi, 2.-.-. 1, Cuba.

consortium, a form of symbiosis (q.v.) in which two or more organisms live together in an interdependent way (obsol.).

conspecific, of two or more taxa considered to be one species; cf. synonym.

constipate, crowded together.

Constricta R. Heim & Mel.-Howell (1965), Agaricaceae. 1, Ivory Coast.

containment levels, see **Safety, Laboratory**.

contaminated, (1) bearing, or intermixed with, a pathogen, as spores on seeds, fungi in soil (c.f. infection and infested); (2) (of cultures), not pure.

context (of hymenomycetes), the hyphal mass between the superior surface and the subhymenium or the trama of basidiocarps.

contiguous, touching; joining.

contingent, touching.

continuous, (1) (of spores, hyphae, etc.), having no septa; (2) (of a stipe), one with the tissue of the pileus or peridium; (3) (of cultures), see culture.

control, to prevent or retard the development of a disease (Anon., *TBMS* **33**: 154, 1950).

convergence, describes two organisms with many characters in common but which are descended from widely separate origins. Difficult to distinguish from parallelism due to intermediate conditions.

convex (of a pileus), equally rounded; broadly obtuse; convexo-expanded, having the edge bent over; convexo-plane, convex when young, flat after expansion.

Cooke (Mordecai Cubitt; 1825-1914). Had, like his countryman Berkeley (q.v.), wide interests in natural history from which his mycological work was the chief development. He was responsible for a very great number of books, papers, and other writings, producing *Science Gossip* for some years and starting (in 1872) the cryptogamic journal *Grevillea*, most of which was written by him during the twenty years the journal was under his control. Among his important books are the *Handbook of British fungi* (1871) (supplemented by 1,300 numbers of *Fungi Britannici Exsiccati*), *Mycographia* (1875-78; pictures and accounts of discomycetes), *Illustrations of British fungi (Hymenomycetes)* (1881-91; with 1,200 coloured plates, see *TBMS* **20**: 33, 1935, for index), and the *Handbook of Australian fungi* (1892). In 1880 he took a position at the Royal Botanic Gardens, Kew, where, till 1892, he had charge of the Lower Cryptogams. He described a great number of new species, and new groupings, esp. of the polypores and pyrenomycetes (*Grevillea* 1884-90). His collection of some 46,000 specimens and 25,000 sheets of drawings, are now in the **K**. His last work, *Fungoid pests of cultivated plants*, came out in book form in 1906. See Ramsbottom (*TBMS* **5**: 169, 1915), English (*Mordecai Cubitt Cooke*, 1987), Stafleu & Cowan (*TL-2* **1**: 536, 1976).

Cookeina Kuntze (1891), Sarcoscyphaceae. 6, widespr., esp. trop. See Pfister & Kaushal (*Mycotaxon* **20**: 1, 1984; key).

Cookella Sacc. (1878), Cookellaceae. 1 (on *Articularia* and *Microstroma*), widespr.

Cookellaceae Höhn. ex Sacc. & Trotter (1913), Dothideales. 3 gen. (+ 11 syn.), 13 spp. Ascomata small, crustose or pulvinate, opening by irregular tears in the upper surface, consisting of thin-walled cells in a gelatinous matrix; interascal tissue absent; asci ± globose, scattered irregularly throughout the ascoma, fissitunicate; ascospores hyaline to brown, muriform. Anamorphs unknown. Fungicolous, on leaf-inhabiting spp.

Cooksonomyces H.J. Swart & D.A. Griffiths (1974), Mitosporic fungi, 6/8.C2.19. 1, Australia.

Coolia Huijsman (1943) = Squamanita (Agaric.) fide Bas (1965).

coolplate, a temperature-controlled plate on which cultures under a light source may be maintained without an undesirable rise in temperature. See Cooke (*FBPP News* **6**: 37, 1981).

Copelandia Bres. (1912), Coprinaceae. 3, warmer areas. See Boedijn (*Sydowia* **5**: 211, 1951), Weeks *et al.* (*J. Nat. Products* **42**: 469, 1979; key).

Copranophilus Speg. (1909) = Pyxidiophora (Pyxidiophor.) fide Lundqvist (1980).

Coprinaceae Gäum. (1926), Agaricales. 11 gen. (+ 22 syn.), 722 spp. Pileipellis epithelial; often with velar layers; spores fuliginous to black, with a germ pore.
Lit.: Watling & Gregory (*British Fungus Flora* **5**, 1987).

Coprinarius (Fr.) P. Kumm. (1871) nom. rej. = Panaeolus (Strophar.).

coprine, a disulfuran-like metabolite of the edible *Coprinus atramentarius* which gives a reaction in humans similar to that of antabuse (q.v.).

Coprinellus P. Karst. (1879) = Coprinus (Coprin.).

Coprinites Poinar & Singer (1990), Fossil fungi. 1, Dominican Republic.

Coprinopsis P. Karst. (1881) = Coprinus (Coprin.).

Coprinopsis Beeli (1929) ? = Oudemansiella (Tricholomat.) fide Singer (1975).

Coprinus Pers. (1797), Coprinaceae. The Ink-Caps. *c.* 100 (coprophilous, on wood, etc.), cosmop. Some are edible; *C. comatus*, the shaggy ink-cap or shaggymane, is a common mushroom in waste places. The lamellae and their cystidia 'deliquesce' by autodigestion after spore discharge giving 'ink' that may be used for writing. See Anderson (*The life history and genetics of Coprinus lagopus*, 1971), Seaver (*Mycol.* **27**: 83, 1935; lifting power), Buller (*Researches* **1-7**, 1909-50), Lange & Smith (*Mycol.* **45**: 747, 1953), Lange (*Dansk bot. Arkiv* **14** (6), 1952; species concept), Orton (*TBMS* **40**: 263, 1957), Kits van Waveren (*Persoonia* **5**: 131, 1968; 'stercorarius' group), Patrick (*Mycotaxon* **6**: 341, 1977; sectional ranks), Romagnesi (*BSMF* **92**: 189, 1976; *C. micaceus*-group), van der Bogart (*Mycotaxon* **4**: 233, 1976; Western N. Am. Sect. *Coprinus*; **8**: 243, 1979, Western N. Am. sect. *Seratuli*), Orton & Watling (*British Fungus Fl.* **3**, 1979; keys 92 spp.), Reijnders (*Persoonia* **10**: 384, 1979; devel. anatomy).

Coprinusella (Peck) Zerov (1979), Coprinaceae. 3, widespr.

Coprobia Boud. (1885) = Cheilymenia (Otid.) fide Moraveč (*Mycotaxon* **38**: 459, 1990).

Coprobolus Cain & Kimbr. (1970), Thelebolaceae. 1 (coprophilous), Canada.

coprogen, a growth factor in dung required by *Pilobolus* spp. (Hesseltine *et al.*, *Mycol.* **45**: 7, 1953; Pidacks *et al.*, *J. Am. Chem. Soc.* **75**: 6064, 1953).

Coprolepa Fuckel (1870) = Hypocopra (Xylar.) fide Cain (*Univ. Toronto Stud., Biol.* **38**, 1934).

coprome, a physically and chemically uniform unit (pellet) of faeces used in experimental studies of coprophilous fungi (Wood & Cooke, *TBMS* **83**: 337, 1984).

Copromyces N. Lundq. (1967), Sordariaceae. 1 (coprophilous), Sweden.

Copromyxa Zopf (1884), Copromyxaceae. 1, Eur., N. Am. See Nesom & Olive (*Mycol.* **64**: 1359, 1972), Spiegel & Olive (*Mycol.* **70**: 843, 1978), Raper (1984).

Copromyxaceae L.S. Olive & Stoian. (1975), Acrasales. 1 gen., 5 spp. Sorocarp cells alike; sorocarp arising directly from substratum.

Copromyxella Raper, Worley & Kurzynski (1978), Copromyxaceae. 4, USA, Costa Rica.

Coprophilous fungi. Fungi living on dung; fimicolous fungi. There are many coprophilous fungi, e.g. most *Acrasales*, some other *Myxomycota*, *Mucorales*, *Pezizales*, *Sordariales*, *Coprinaceae*, certain other *Basidiomycota*, and Mitosporic fungi. See Massee & Salmon (*Ann. Bot., Lond.* **15**: 313, **16**: 67, 1901-2), Fairman, (*The ascomycetous fungi of human excreta*, 1920), Cain (*Univ. Toronto Studies, biol. ser.* **38**, 1934 [reprinted 1968]; *Can. J. Res.* **C28**: 566; *CJB* **34**: 675; **35**: 255; **39**: 1633; **40**: 447, 1950-62; coprophilous ascomycetes I-VIII), Eliasson & Lundqvist (*Bot. Notiser* **132**: 551, 1979; coprophilous myxomycetes), Prokhorov (*Mikol. Fitopat.* **28**: 20; key 20 discom. gen.), Richardson & Watling (*Bull. BMS* **2**: 18, 1968; asco., basidio.; **3**: 86, 1969, phyco., key; edn 2, 1975), Webster (*TBMS* **54**: 161, 1970; review), Bell (*Dungi fungi*, 1983; NZ, illustr. guide), Dix & Webster (*Fungal ecology*, 1995; review), Wicklow (*in* Carroll & Wicklow, *The fungal community*, edn 2 : 715, 1992). See also ammonia fungi.

Coprotiella Jeng & J.C. Krug (1976), Thelebolaceae. 1, S. Am. See van Brummelen (*in* Hawksworth (Ed.), *Ascomycete systematics*: 400, 1994; posn).

Coprotinia Whetzel (1944) = Lambertella (Sclerotin.) fide Dumont (*Mycol.* **67**: 320, 1975).

Coprotrichum Bonord. (1851) = Sporendonema (Mitosp. fungi) fide Mason & Hughes (*in* Wood, *Nature* **179**: 328, 1957).

Coprotus Korf & Kimbr. (1967), Thelebolaceae. 18, widespr. See van Brummelen (*in* Hawksworth (Ed.), *Ascomycete Systematics*: 400, 1994; posn), Kimbrough *et al.* (*CJB* 50: 957, 1972; key), Thind *et al.* (*J. Ind. bot. Soc.* 57: 63, 1978; Indian spp.), Aas (*Nord. J. Bot.* 3: 253, 1983; Norway).

copulants, Kniep's name for copulating structures of like form.

copulation, the fusion of sexual elements; conjugation; gametangial -, the fusion of two sexual organs; heterogamic -, the fusion of gametes morphologically unlike; isogamic -, the fusion of gametes morphologically like; conjugation in the narrower sense; planogamic -, the fusion of motile gametes to give a motile zygote (a planozygote). Cf. merogamy.

Cora Fr. (1825) = Dictyonema (Merul.) fide Parmasto (1978).

Coraemyces Cif. & Tomas. (1954) ≡ Cora (Merul.).

coral fungi, basidiomata of *Clavariaceae*.

coral spot, a branch disease of shrubs and trees (*Nectria cinnabarina*).

Corallicola Volkm.-Kohlm. & Kohlm. (1992), Halosphaeriaceae. 1 (on coral), Belize.

Corallinopsis Lagarde (1917), Mitosporic fungi, 2.A2.?. 1 (on *Insecta*), France.

Coralliochytrium Domján (1937) = Scheffeliomyces (Chytrid.) fide Dick (*in press*).

Corallium G. Hahn (1883) = Ramaria (Ramar.) fide Donk (1954).

Corallochytrium Raghuk. (1987), ? Thraustochytriales (inc. sed.). 1, Arabian sea.

Corallocytostroma Y.N. Yu & Z.Y. Zhang (1980), Mitosporic fungi, 8.C1.?. 1, China.

Corallodendron Jungh. (1838) [non Mill. (1754), *Papilionaceae*] ≡ Corallomyces (Mitosp. fungi), but see Samson & Seifert (*in* Samson & Pitt (Eds), *Advances in Penicillium and Aspergillus systematics*: 397, 1985; represents a diatom].

Coralloderma D.A. Reid (1965), Podoscyphaceae. 1, Asia, Australia.

Corallofungus Kobayasi (1983), Hydnaceae. 1, Japan.

coralloid, much branched; like coral in form; esp. basidiomata of *Clavaria*. Cf. forate.

Coralloidea Roussel (1806) ≡ Ramaria (Ramar.).

Coralloides Wulfen (1776) nom. dub. (Fungi, inc. sed., L).

Coralloides Hoffm. (1789) nom. rej. ? = Stereocaulon (Stereocaul.).

Coralloides Tourn. ex Maratti (1822) nom. rej. prop. = Ramaria (Ramar.) fide Donk (*Persoonia* 8: 281, 1975).

Corallomorpha Opiz (1856) nom. dub. fide Sutton (*Mycol. Pap.* 141, 1977).

Corallomyces Fr. (1849), Mitosporic fungi, 2.A1.?. 2, Afr., Java.

Corallomyces Berk. & M.A. Curtis (1853) = Nectria (Hypocr.) fide Rogerson (1970).

Corallomycetella Henn. (1904) = Nectria (Hypocr.) fide Rogerson (1970).

corbiculae, protective structures forming a stroma around the telia of certain rusts (Kuhnholtz-Lordat, *Bull. mens. Acad. Sci. Lett. Montpellier* 71: 91, 1942; see *RAM* 26: 468, 1947); paraphyses; pseudoparaphyses.

Corbulopsora Cummins (1940), Pucciniaceae. 3 (on *Compositae*), Papua New Guinea, India.

Corda (August Karl Josef; 1809-49). A medical man. Keeper of the National Museum, Prague (which has Corda's collections, see Pilát, *Acta Mus. nat. Prague* 1B: 139, 1938). Chief writings: Pilze (in in *Sturm's Deutchlands Flora*, 1829-41); *Icones fungorum hucusque cognitorum* (6 parts, the last by Zobel, 1837-54); *Anleitung zum Studium der Mycologie*

(1842). See Šebek (*Česká Myk.* 38: 129, 1984; *Mykol. Sborník* 61: 113, 1984), Stafleu & Cowan (*TL-2* 1: 546, 1976).

Cordalia Gobi (1885) = Tuberculina (Mitosp. fungi) fide Saccardo (*Syll. Fung.* 4: 653, 1886).

Cordana Preuss (1851), Mitosporic fungi, 1.B2.6. 8, Eur., trop. *C. musae* (banana leaf spot). See Hughes (*CJB* 33: 259, 1955), de Hoog *et al.* (*Proc. K. Akad. Wet.* C 86 197, 1983), Priest (*MR* 94: 861, 1990; spp. on *Musa*, Australia).

Cordella Speg. (1886), Anamorphic Lasiosphaeriaceae, 1.A2.37. Teleomorph *Apiospora*. 4, S. Am. See Subramanian (*Proc. Indian Acad. Sci.* B, 55: 38, 1962).

Cordierites Mont. (1840), Leotiaceae. 3, Asia., Afr., S. Am. See Zhuang (*Mycotaxon* 31: 261, 1988; key).

Corditubera Henn. (1897), Melanogastraceae. 3, trop. Afr. See Demoulin & Dring (*Bull. Jard. bot. nat. Belg.* 45: 345, 1975), Malençon (*Crypt. Mycol.* 4: 1, 1983).

Cordycepioideus Stifler (1941), Hypocreaceae. 2 (on termites), Tanzania, Mexico. See Blackwell & Gilbertson (*Mycol.* 76: 763, 1984).

Cordyceps (Fr.) Link (1833) nom. cons., Clavicipitaceae. 100 (mostly on insects; caterpillar fungi, vegetable caterpillars); 5 (on *Elaphomyces*), cosmop. See Kobayasi & Shimizu (*Bull. natn Sci. Mus. Tokyo* 5: 69, 1960), Evans & Samson (*TBMS* 79: 431, 1982; 82: 127, 1984; on ants), Mains (*Mycol.* 50: 169, 1957; N. Am., key), Willis (*Muelleria* 1: 2, 1959; Australia, key), Petch (*TBMS* 10: 28, 1924; Sri Lanka), Moureau & Lacquemant (*Inst. Nat. col. Belge, Mém., coll.* 4°, 7 (5), 1949), Moureau (*Lejeunia, Mem.* 15, 1961; Congo), Kobayasi (*Bull. Biogeogr. Soc. Japan* 9: 271, 1939; Japan), Kobayasi & Shimizu (*Bull. natn Sci. Mus. Tokyo* B, 2: 133, 1976; New Guinea). Anamorph ? *Anthina*.

Cordylia Fr. (1818) [non Pers. (1807), *Leguminosae*] ≡ Cordyceps (Clavicipit.).

Cordyliceps Fr. (1832) ≡ Cordyceps (Clavicipit.).

Corella Vain. (1890) = Dictyonema (Merul.) fide Parmasto (1978).

Coremiales, see *Hyphomycetes*.

Coremiella Bubák & Willi Krieg. (1912), Mitosporic fungi, 2.A2.40. 1, widespr. See Ellis (*DH*).

Coremiopsis Sizova & Suprun (1957) = Paecilomyces (Mitosp. fungi) fide Samson (1974).

coremium, see synnema.

Coremium Link (1809) = Penicillium (Mitosp. fungi) fide Thom (*Bull. Bur. Anim. Ind. USDA* 118, 1910).

Corenohydnum Lloyd (1936) Fungi (inc. sed.).

Coreomyces Thaxt. (1902), Laboulbeniaceae. 20, widespr. See Majewski (*Acta mycol.* 9: 217, 1973; Polish spp.).

Coreomycetopsis Thaxt. (1920), ? Laboulbeniales (inc. sed.). 1 (on termites), W. Indies, USA. See Blackwell & Kimbrough (*Mycol.* 68: 541, 1976; structure and development).

Corethromyces Thaxt. (1892), Laboulbeniaceae. 74, widespr. See Tavares (1985).

Corethropsis Corda (1839), Mitosporic fungi, 2.A1.?. 3, Eur., Am.

Corethrostroma Kleb. (1933), Mitosporic fungi, 8.A1.?. 1, Eur.

coriacellate, somewhat coriaceous.

coriaceous, like leather in texture.

Corinophoros A. Massal. (1856) = Peccania (Lichin.).

Coriolaceae (Imazeki) Singer (1961), Poriales. 109 gen. (+ 91 syn.), 812 spp. Basidioma coriaceous; dimitic; hymenophore tubulate; spores hyaline.

Coriolellus Murrill (1905) = Antrodia (Coriol.) fide Donk (1974).

Coriolopsis Murrill (1905), Coriolaceae. 17, widespr. See Pegler (1971), Ryvarden & Johansen (1980; key 10 Afr. spp.).

Coriolus Quél. (1886), Coriolaceae. 20, widespr. See Singer (*Publ. Inst. Micol. Univ. Recife* **304**, 1961). = Trametes fide Domanski *et al.* (*Polyporaceae*, edn 2, 1973).

Coriscium Vain. (1890) nom. rej. prop. = Omphalina (Tricholomat.) lichenized squamules. See Gams (*Öst. bot. Z.* **109**: 376, 1962), Oberwinkler (*Deutsch. bot. Ges.* N.F. **4**: 139, 1970).

corium, see Spore wall and Fig. 5.

corkir, see cudbear.

Cormothecium A. Massal. (1854) = Rhizocarpon (Rhizocarp.).

Corneohydnum Lloyd (1924) nom. nud. (? basidiomycetes).

corneous, (1) horn-like in texture; (2) (of a substance), like horn.

Corneromyces Ginns (1976), Coniophoraceae. 1, Sabah.

Cornicularia (Schreb.) Hoffm. (1792), Parmeliaceae (L). 1, N. temp. See Kärnefelt (*Op. Bot.* **86**: 1, 1986), Kärnefelt *et al.* (*Pl. Syst. Evol.* **183**: 113, 1992).

Cornicularia Schaer. (1850) ≡ Alectoria (Alector.).

Cornicularia Bonord. (1851) = Clavulinopsis (Clavar.) fide Corner (1950).

Corniculariella P. Karst. (1884), Mitosporic fungi, 8.E1.15. 1, Sweden. See DiCosmo (*CJB* **56**: 1665, 1978), Illman (*Taxon* **34**: 512, 1985; typification).

corniform, shaped like a horn (Fig. 37.36).

Corniola Gray (1821) [non Adans. (1763), *Leguminosae*] = Arrhenia (Tricholomat.) fide Redhead (1984).

Cornucopiella Höhn. (1915), Mitosporic fungi, 4.A1.15. 1, Eur.

Cornuella Setch. (1891) [non Pierre (1891), *Sapotaceae*] ≡ Tracya (Tillet.).

Cornularia Sacc. (1884) = Corniculariella (Mitosp. fungi) fide DiCosmo (1978).

Cornumyces M.W. Dick (1995), Leptomitales (inc. sed.). 7, widespr.

Cornuntum Velen. (1947), ? Leotiales (inc. sed.). 1, Eur.

cornute, (1) horned; horn-like in form; (2) (of aecia), see roestelia.

Cornutispora Piroz. (1973), Mitosporic fungi, 4.G1.10. 5 (on lichens and *Rhytismatales*), Eur., N. Am., Australia. See Hawksworth (*TBMS* **67**: 151, 1976), Punithalingam (*in press*).

Cornutostilbe Seifert (1990), Mitosporic fungi, 2.A1.1. 1, Indonesia.

Cornuvia Rostaf. (1873), Arcyriaceae. 1, widespr.

Corollium Sopp (1912) ≡ Paecilomyces (Mitosp. fungi) fide Raper & Thom (*A manual of the Penicillia*, 1949).

Corollospora Werderm. (1922), Halosphaeriaceae. 13 (marine), widespr. See Jones *et al.* (*Bot. J. Linn. Soc.* **87**: 193, 1983), Nakagiri & Tokura (*Trans. mycol. Soc. Japan* **28**: 413, 1987, 1987; key 13 spp.), Kohlmeyer & Volkmann-Kohlmeyer (*TMBS* **88**: 181, 1987; key 6 spp.), Nakagiri (*Trans. mycol. Soc. Japan* **27**:197, 1986; anamorphs), Read *et al.* (*Bot. Mar.* **35**: 553, 1992; asci). Anamorphs *Clavariopsis*, *Sigmoidea*, *Varicospora*.

Coronasclerotes Stach & Pickh. (1957), Fossil fungi (*f. cat.*). 5 (Carboniferous), Germany.

coronate, crowned.

Coronella P. Crouan & H. Crouan (1867) = Kickxella (Kickxell.) fide Linder (1943).

Coronellaria P. Karst. (1870), Dermateaceae. 4, Eur.

Coronicium J. Erikss. & Ryvarden (1975), Xenasmataceae. 4, N. temp.

Coronium Bonord. (1864) nom. dub. (Mitosp. fungi). See Sutton (1977).

Coronopapilla Kohlm. & Volkm.-Kohlm. (1990), ? Zopfiaceae. 1 (marine), Belize. See Eriksson & Hawksworth (*SA* **8**: 7, 1991).

Coronophora Fuckel (1864), Nitschkiaceae. 4, Eur.

Coronophorales, see *Sordariales*.

Coronophorella Höhn. (1909) = Nitschkia (Nitschk.). See Müller & v. Arx (1973).

Coronoplectrum Brusse (1987), Parmeliaceae (L). 1, Namibia.

Coronospora M.B. Ellis (1971), Mitosporic fungi, 1.C2.10. 1, Myanmar.

Coronotelium Syd. (1921) = Puccinia (Puccin.) fide Dietel (1928).

correct (of names), see Nomenclature.

correlated species (of *Uredinales*), a species derived by reduction (of life cycle or morphology) from a parent heteroecious macrocyclic species/or the parent species itself.

Corrugaria Métrod (1949) = Mycena (Tricholomat.) fide Singer (1951).

corrugate, wrinkled.

cortex, a more or less thick outer covering; **epi-** (q.v.); **corticate**, having a cortex.

corticate, see cortex.

Corticiaceae Herter (1910), Stereales. 38 gen. (+ 10 syn.), 148 spp. Basidioma resupinate; monomitic; dendrohyphidia present; hymenophore smooth to tuberculate; spores smooth, inamyloid; lignicolous, herbicolous, or humicolous.
  Lit.: See *Stereales*.

Corticifraga D. Hawksw. & R. Sant. (1990), ? Lecanorales (inc. sed.). 2 (on lichens, *Peltigerales*), mainly temp.

Corticioides Lloyd (1908) = Tremella (Tremell.) fide Lloyd (1919).

Corticirama Pilát (1957), Corticiaceae. 2, Eur.

Corticium Pers. (1794), Corticiaceae. 150, cosmop. *C. solani* (anamorph *Rhizoctonia solani*), see *Thanatephorus*, and *C. salmonicolor* (pink disease of rubber, tea, and other tropical plants; see *Erythricium*). See Larsen & Gilbertson (*Norw. Jl Bot.* **24**: 99, 1977), Eriksson & Ryvarden (*Corticiaceae of North Europe* **4**: 759, 1976; key 5 Eur. spp.). Formerly used for many resupinate basidiomycetes; now restricted to the type and congeneric species.

Corticium Fr. (1835) = Phanerochaete (Cortic.) fide Donk (1964).

corticolous, living on bark; **corticole**, an organism which does this.

Corticomyces A.I. Romero & S.E. López (1989), Mitosporic fungi, 1.A1.2. 1 (with clamp-connexions), Argentina.

cortina (of agarics), a partial veil (or part of one), frequently web-like, covering the mature gills.

Cortinaria (Pers.) Gray (1821) nom. rej. ≡ Cortinarius (Cortinar.).

Cortinariaceae R. Heim ex Pouzar (1983), Cortinariales. 32 gen. (+ 36 syn.), 1150 spp. Pileipellis never epithelial; spores rusty to ferruginous, often ornamented.

Cortinariales, Basidiomycetes. 5 fam., 47 gen. (+ 48 syn.), c. 1360 spp. Fams:
  (1) **Cortinariaceae**.
  (2) **Crepidotaceae**.
  (3) **Cribbeaceae**.
  (4) **Hymenangiaceae**.

cortinarins, toxic fluorescent cyclic decapeptides produced by *Cortinarius* spp. (Laatsch & Matthies, *Mycol.* **83**: 492, 1991). See also mycetismus (Mycetism).

Cortinarius (Pers.) Gray (1821) nom. cons., Cortinariaceae. 400, esp. N. temp. See Garnier (*Docums mycol.* **3-8**, 1973-78; bibliogr.), Moser (*Sydowia* **5**:

488, 1951), Henry (*BSMF* **71**: 202; **73**: 18; **74**: 365; **79**: 277, 1956-63), Orton (*Naturalist, Hull* **1955** (Suppl.) [reprint *The Naturalist agaric keys*, 1978]), Bertaux (*Les Cortinares*, 1966), Moser (*Die Gattung Phlegmacium* [*Pilze Mitteleur.* **4**] 1960), Moser & Horak (*Beih. Nova Hedw.* **52**, 1975; S. Am.), Noculak (*Acta Mycol.* **15**: 183, 1979; atraquinones), Høiland (*Norw. J. Bot.* **27**: 101, 1980; 13 spp. subgen. *Leprocybe*; *Opera Bot.* **71**: 5, 1983; subgen. *Dermocybe*), Brandrud *et al.* (*Cortinarius: flora photographica*, 1990). Cortinarius poisoning, see Mycetism.

Cortinellus Roze (1876) = Tricholoma (Tricholomat.) fide Singer (1951).

Cortiniopsis J. Schröt. (1889) ≡ Lacrymaria (Coprin.).

Cortinomyces Bougher & Castellano (1993), Hymenogastraceae. 6, Australia.

Corymbomyces Appel & Strunk (1904) = Gliocladium (Mitosp. fungi) fide Kendrick & Carmichael (1973).

corymbose, arranged in clusters.

Corynascella Arx & Hodges (1975), Chaetomiaceae. 3, widespr. See v. Arx (*Beih. Nova Hedw.* **94**, 1988; key).

Corynascus Arx (1973), Ceratostomataceae. 4, widespr. See v. Arx (*Stud. Mycol.* **8**: 21, 1975; *Beih. Nova Hedw.* **94**, 1988, key). Anamorph *Myceliophthora*.

Coryne Nees (1816), Anamorphic Leotiaceae, 2.A1.15. Teleomorph *Ascocoryne*. 1, widespr., esp. temp. See Seifert (*Stud. Mycol.* **31**: 157, 1989).

Corynecystis Brusse (1985), Heppiaceae (L). 1, S. Afr.

Corynelia Ach. (1823), Coryneliaceae. 7 (on *Podocarpus*), mainly trop. and S. temp. See Benny *et al.* (*Bot. Gaz.* **146**: 238, 1985), Hawksworth (*SA* **14**: 48, 1995; nomencl.).

Coryneliaceae Sacc. ex Berl. & Voglino (1886), Coryneliales. 7 gen. (+ 7 syn.), 25 spp. Ascomata clustered, perithecial, black, thick-walled, opening by an ostiole or irregular split; interascal tissue absent; asci initially double-walled, the outer layer deliquescing early, becoming long-stalked and very thin-walled, without apical structures, evanescent; ascospores usually aseptate, dark, usually strongly ornamented, in a mazaedial mass above the ascus layer. Anamorph coelomycetous. Mostly biotrophic on leaves of *Podocarpaceae*, mainly trop. *Lit.*: see *Coryneliales*.

Coryneliales, Ascomycota. 1 fam., 7 gen. (+ 7 syn.), 31 spp. Fam.:
  Coryneliaceae.
  Lit.: Benny *et al.* (*Bot. Gaz.* **146**: 232, 238, 431, 437, 1985), Fitzpatrick (*Mycol.* **12**: 206, 1920; **34**: 464, 1942; keys, monogr.), Johnson & Minter (*MR* **92**: 422, 1989; asci), Müller & v. Arx (1973: key gen.).

Coryneliella Har. & P. Karst. (1890), Ascomycota (inc. sed.). 1, Mauritius. See Fitzpatrick (*Mycol.* **12**: 206, 1920).

Coryneliopsis Butin (1972), ? Coryneliaceae. 2 (with *Cyttaria*), Chile. See Benny *et al.* (*Bot. Gaz.* **146**: 437, 1985; key), Johnston & Minter (*MR* **92**: 422, 1989; asci).

Coryneliospora Fitzp. (1942), Coryneliaceae. 2, China, S. Afr., India. See Benny *et al.* (*Bot. Gaz.* **146**: 437, 1985; key).

Corynelites Babajan & Tasl. (1970), Fossil fungi. 1 (Tertiary), former USSR.

Corynella Boud. (1885) [non DC. (1825), *Leguminosae*] = Claussenomyces (Leot.) fide Korf & Abawi (*CJB* **49**: 1879, 1971).

Coryneopsis Grove (1933) = Seimatosporium (Mitosp. fungi) fide Sutton (1977).

Corynesphaera Dumort. (1822) nom. rej. ≡ Cordyceps (Clavicip.).

Corynespora Güssow (1906), Mitosporic fungi, 1.C2.25. 20, widespr. See Wei (*Mycol. Pap.* **34**,

1950), Ellis (*Mycol. Pap.* **65**, 1957; key), Ellis (*Mycol. Pap.* **76**, 1960).

Corynesporella Munjal & H.S. Gill (1961), Mitosporic fungi, 1.C2.?. 1, India.

Corynesporopsis P.M. Kirk (1981), Mitosporic fungi, 1.B/C2.29. 3, widespr. See Sutton (*Sydowia* **41**: 330, 1989).

Corynetes Hazsl. (1881) = Geoglossum (Geogloss.) fide Nitare (*Windahlia* **14**: 37, 1984). See also Spooner (*Bibl. Mycol.* **116**, 1987).

Coryneum Nees (1816), Anamorphic Melanconidaceae, 6.C2.19. Teleomorph *Pseudovalsa*. 21, mostly temp. See Sutton (*Mycol. Pap.* **138**, 1975; key), Sutton & Rizwi (*Nova Hedw.* **32**: 341, 1980).

Corynites Berk. & M.A. Curtis (1853) = Mutinus (Phall.).

Corynocladus Leidy (1850) nom. dub (? Fungi).

Corynodesmium Wallr. (1828) nom. nud. (Mitosp. fungi). See Hughes (1958).

Corynoides Gray (1821) = Calocera (Dacrymycet.) fide McNabb (1965).

Corynophoron Nyl. ex Müll. Arg. (1894) = Stereocaulon (Stereocaul.).

Corynophorus, see *Corinophoros*.

Coscinaria Ellis & Everh. (1886) = Oomyces (Acrosperm.) fide Rogerson (1970).

Coscinedia A. Massal. (1860) = Myriotrema (Thelotremat.) fide Hale (1981).

Coscinium Endl. [not traced] = Lamproderma (Stemonitid.) fide Mussat (*Syll. Fung.* **15**: 397, 1901).

Coscinocladium Kunze (1846), Ascomycota (inc. sed., L). 1, Eur.

coscinocystidium, a cystidium projecting as a pseudocystidium.

coscinoid, a pitted conducting element in *Linderomyces*.

Coscinopeltis Speg. (1909) = Munkiella (Polystomell.) fide Müller & v. Arx (1962).

Coscinospora Mirza (1963) nom. inval. ≡ Jugulospora (Lasiosphaer.).

Cosinopeltella Chardón (1930) = Vestergrenia (Dothid.) fide v. Arx & Müller (1954).

Cosmariospora Sacc. (1880), Mitosporic fungi, 3.B2.?. 1, Italy.

Cosmospora Rabenh. (1862) = Nectria (Hypocr.) fide Rogerson (1970).

Costanetoa Bat. & J.L. Bezerra (1963), Mitosporic fungi, 4.G1.?. 1, Brazil.

Costantin (Julien Noël; 1857-1936). Professor at the Natural History Museum, Paris. Chief writings: *Les Mucédinées simples. Histoire, classification, culture at rôle des champignons inférieurs dans les maladies des végétaux et des animaux*, 1888; *Atlas des champignons comestibles et vénéneux*, 1895; (with Dufour) *Nouvelle flore des champignons ... de France*, 1891 [and many later edns and reprints]. See Magrou (*BSMF* **53**: 245, 1937), Stafleu & Cowan (*TL-2* **1**: 555, 1976).

Costantinella Matr. (1892), Mitosporic fungi, 1.A2.10. 2, Eur., N. Am.

Costapeda Falck (1923) ≡ Helvella (Helvell.).

costate, veined or ribbed.

costiferous, see hypha.

cot death (Sudden Infant Death Sydrome; SIDS), hypothesis for involvement of *Scopulariopisis brevicaulis* through biodeteriogenic action on mattresses with subsequent release of toxic gasses was proposed by Richardson (*Lancet* **335**: 670, 1990) but not supported by later experiments (Kelley *et al.*, *Human Exp. Toxicol.* **11**: 347, 1992).

Cotylidia P. Karst. (1881), Podoscyphaceae. 7, widespr., esp. trop. See Reid (1965: 56; key).

cotyliform, plate- or wheel-like with an upturned edge.

Couchia W.W. Martin (1981), Saprolegniaceae. 1, USA.

Coulterella Zebrowski (1936) [non Vasey & Rose (1890), *Compostiae*] ? Fossil fungi (Chytrid.). 1 (Cambrian to ? Recent), Australia.

Courtoisia L. Marchand (1830) = Rinodina (Physc.) fide Hafellner (1984).

Coutinia J.V. Almeida & Sousa da Câmara (1903) = Botryosphaeria (Botryosphaer.) fide v. Arx & Müller (1975).

Coutourea Castagne (1845) nom. dub. fide Sutton (*Mycol. Pap.* **141**, 1977).

Cowlesia Nieuwl. (1916) ≡ Macropodia (Helvell.).

cramp balls, the ascomata of *Daldinia concentrica*.

Crandallia Ellis & Sacc. (1897), Anamorphic Rhytismataceae, 5.A1.15. Teleomorph *Duplicaria*. 1, USA.

Craneomyces Morgan-Jones, R.C. Sinclair & Eicker (1987), Mitosporic fungi, 1.A2.3. 1, S. Afr.

Craspedodidymum Hol.-Jech. (1972), Mitosporic fungi, 1.A2.15. 2, former Czechoslovakia, Afr.

Craspedon Fée (1825) = Strigula (Strigul.).

Crassoascus Checa, Barrasa & A.T. Martínez (1993), Clypeosphaeriaceae. 1, Spain. ? = Porosphaerellopsis (Lasiosphaer.).

Cratarellus, see *Craterellus*.

Craterella Pers. (1794) = Bresadolina (Ster.) fide Donk (1964).

Craterellaceae Herter (1910), Cantharellales. 1 gen. (+ 5 syn.), 20 spp. Basidioma infundibuliform, with hollow stipe; thickening hymenium; hymenophore smooth or wrinkled.

Craterellus Pers. (1825), Craterellaceae. 20, widespr. *C. cornucopioides* ('Horn of Plenty') is edible. See Corner (*Sydowia, Beih.* **1**: 266, 1957).

Crateriachea Rostaf. (1873) = Physarum (Physar.).

Crateridium Trevis. (1862) = Cyphelium (Calic.) fide Tibell, 1984).

crateriform, cup- or crater-like in form.

Craterium Trentep. (1797), Physaraceae. 10, cosmop.

Craterocolla Bref. (1888), Exidiaceae. 2, Eur. See Donk (*Persoonia* **4**: 164, 1966). Anamorph *Ditangium*.

Craterolechia A. Massal. (1860) ? = Arthonia (Arthon.) fide Zahlbruckner (1922).

Crateromyces Corda (1831) nom. dub.; ? based on insect eggs.

Crauatamyces Viégas (1944), Dothideales (inc. sed.). 1, Brazil.

Creangium Petr. (1950) = Saccardia (Saccard.) fide v. Arx (1963).

Crebrothecium Routien (1949), Metschnikowiaceae. 1, Sudan. See v. Arx *et al.* (*Stud. Mycol.* **14**, 1977), v. Arx (1981).

Cremasteria Meyers & R.T. Moore (1960), Mitosporic fungi, 1.A2.4. 1 (marine), USA.

Cremeogaster Mattir. (1924) = Leucophleps (Leucogastr.) fide Fogel (*CJB* **57**: 1718, 1979).

Crenasclerotes Stach & Pickh. (1957), Fossil fungi (*f. cat.*). 4 (Carboniferous), Germany.

crenate, having the edge toothed with rounded teeth (Fig. 37.43).

crenulate, delicately crenate (Fig. 37.44).

Creodiplodina Petr. (1957), Mitosporic fungi, 1.B1.?. 1, Australia.

Creographa A. Massal. (1860) = Phaeographina (Graphid.).

Creolophus P. Karst. (1879), Hericiaceae. 1, N. temp.

Creomelanops Höhn. (1920) = Botryosphaeria (Botryosph.) fide Eriksson & Yue (*SA* **6**: 241, 1987).

Creonecte Petr. (1949), Mitosporic fungi, 4.E1.?. 1 (on *Uredo*), S. Am.

Creonectria Seaver (1909) = Nectria (Hypocr.) fide Rogerson (1970).

Creopus Link (1833), Hypocreaceae. 9, N. temp., Java. See Boedijn (1964). = Hypocrea (Hypocr.) fide Rogerson (1970).

Creoseptoria Petr. (1937), Mitosporic fungi, 8.E1.?. 1, Caucasus.

creosote fungus, see kerosene fungus.

Creosphaeria Theiss. (1910), Xylariaceae. 2, N. & S. Am. See Ju *et al.* (*Mycotaxon* **48**: 219, 1993), Læssøe (*SA* **13**: 43, 1994). Anamorph *Libertella*-like.

Creothyriella Bat. & C.A.A. Costa (1957), Mitosporic fungi, 5.A1.?. 1, India.

Creothyrium Petr. (1925) = Cylindrocolla (Mitosp. fungi) fide Sutton (1977).

Crepidopus (Nees) Gray (1821) nom. rej. = Pleurotus (Polypor.).

Crepidotaceae (S. Imai) Singer (1951), Cortinariales. 12 gen. (+ 12 syn.), 207 spp. Basidioma often pleurotoid; spores pale cinnamon brown.

Crepidotus (Fr.) Staude (1857), Crepidotaceae. 150, cosmop. See Singer (*Lilloa* **13**: 59, 1947), Pilát (*TBMS* **33**: 215, 1950; key), Hesler & Smith (*North American species of Crepidotus*, 1965; 125 spp., keys), Pegler & Young (*Kew Bull.* **27**: 311, 1972; key 17 Br. spp.), Singer (*Beih. Nova Hedw.* **44**: 341, 1973; key 68 neotrop. spp.), Horak (*Nova Hedw.* **8**: 333, 1964; key 10 spp. S. Am.), Nordstein (*Synop. Fung.* **2**, 1990; key 8 Norway spp.).

Crepinula Kuntze (1891) ≡ Cephalotheca (Cephalothec.).

crescentic, see lunate.

Cresponea Egea & Torrente (1993), Roccellaceae (L). 11, Eur.

Cresporhaphis M.B. Aguirre (1991), ? Trichosphaeriaceae (±L). 6, temp. & mediteranean. See Barr (*Mycotaxon* **46**: 64, 1993).

Cribbea A.H. Sm. & D.A. Reid (1962), Cribbeaceae. 4, Australia, Argentina. See Smith & Reid (*Mycol.* **54**: 98, 1962).

Cribbeaceae Singer, J.E. Wright & E. Horak (1963), Cortinariales. 2 gen., 5 spp. Gasterocarp secotioid, with lamellate gleba; spores with echinulate exosporium plus myxosporium. Relationship to *Crepidotaceae* possible.
    Lit.: Smith (*in* Ainsworth *et al.* (Eds), *The Fungi* **4B**: 421, 1973), Heim (*in* Petersen (Ed.), *Evolution in the higher Basidiomycetes*: 505, 1973; keys gen.), Singer *et al.* (*Darwiniana* **12**: 598, 1963).

cribose (**cribriform**), having a network like a sieve.

Cribraria Schrad. ex J.F. Gmel. (1792) nom. dub. (Myxomycetes).

Cribraria Pers. (1794), Liceales (inc. sed.). 35, cosmop.

Cribrariaceae Corda (1838), Liceales. 2 gen. (+ 2 syn.), 37 spp.

Cribrites Lange (1978), Fossil fungi (Microthyr.). 1 (? Eocene), Australia.

Cribropeltis Tehon (1933), Mitosporic fungi, 5.A1.?. 1, USA.

Cricunopus P. Karst. (1881) = Suillus (Bolet.) fide Smith & Thiers (1971).

Criella (Sacc.) Henn. (1900), Rhytismataceae. 1, trop.

Crinigera I. Schmidt (1969), Ascomycota (inc. sed.). 1 (marine), Baltic, Pacific. See Koch & Jones (*CJB* **67**: 1183, 1989).

Crinipellis Pat. (1889), Tricholomataceae. 30, cosmop. *C. perniciosus* (witches' broom of cacao; *CMI Descr. Path. Fungi Bact.* **223**, 1970). See Singer (*Lilloa* **8**: 441, 1942; monograph, *Flora neotrop.* **17**: 9, 1976; key 14 neotrop. spp.).

Crinitospora B. Sutton & Alcorn (1985), Mitosporic fungi, 6.B1.15/19. 1, Australia.

Crinium Fr. (1819) ≡ Crinula (Mitosp. fungi).

Crinofera Nieuwl. (1916) ≡ Pilophora (Mucor.).

Crinula Fr. (1821), Anamorphic Leotiaceae, 2.A2.?. Teleomorph *Holwaya*. 1, Eur., N. Am.

Crinula Sacc. (1889) ≡ Holwaya (Leot.).

Criserosphaeria Speg. (1912), ? Leotiales (inc. sed.). 1, Argentina. See Petrak & Sydow (*Ann. Myc.* **33**: 157, 1935).

crispate, curled and twisted.

crista, tubular, pouch-like or shelf-like inwardly directed fold of the inner membrane of a mitochondrion; site of ATP production during aerobic metabolism.

Cristaspora Fort & Guarro (1984), Trichocomaceae. 1, Spain.

cristate, crested.

Cristella Pat. (1887) = Sebacina (Exid.) fide Rogers (see Weresub, *Taxon* **16**: 402, 1967; Donk, *Taxon* **17**: 278, 1968). = Trechispora (Sistotremat.) fide Liberta (*CJB* **51**: 1871, 1973).

Cristelloporia I. Johans. & Ryvarden (1979) = Trechispora (Sistotremat.) fide Larsson (*in litt.*).

Cristidium R. Sant. (1952) nom. nud. (Mitosp. fungi). 1 (*Gyalectidium*), trop.

Cristinia Parmasto (1968), Lindtneriaceae. 7, widespr. See Hjortstam & Grosse-Brauckmann (*Mycotaxon* **47**: 405, 1993; key).

Cristula Chenant. (1920), Mitosporic fungi, 1.G1.1. 1, Eur.

Cristularia (Sacc.) Costantin (1888) = Botrytis (Mitosp. fungi) fide Saccardo (*Syll. Fung.* **4**: 134, 1886).

Cristulariella Höhn. (1916), Anamorphic Sclerotiniaceae, 1.G1.1. Teleomorph *Grovesinia*. 2 (on *Acer* etc.), N. temp. See Redhead (*CJB* **53**: 700, 1975), Niedbalski *et al.* (*Mycol.* **71**: 722, 1979; devel. in *C. pyramidalis*).

Crocicreas Fr. (1849), Leotiaceae. 1, temp. (fide Baral & Kriegelsteiner (*Beih. Z. Mykol.* **6**, 1985), Carpenter (*Mem. N.Y. bot. Gdn* **33**, 1981).

Crocicreomyces Bat. & Peres (1964), Mitosporic fungi (L), 8.A1.?. 1, Amazon.

Crocodia Link (1833) nom. rej. = Pseudocyphellaria (Lobar.).

Crocynia (Ach.) A. Massal. (1860) nom. cons., Crocyniaceae (L). *c.* 5, trop. See Hue (*Bull. Soc. bot. Fr.* **71**: 311, 1924; s.l. monogr.).

Crocyniaceae M. Choisy ex Hafellner (1984), Lecanorales (L). 1 gen. (+ 1 syn.), *c.* 5 spp. Thallus crustose, often spongy; ascomata apothecial, sessile, pale, convex, with a poorly-developed margin; interascal tissue of simple paraphyses, the apices hardly swollen, in a gelatinous matrix; asci with a well-developed I+ apical cap, a darker-staining ring, a small ocular chamber and an outer I+ gelatinous layer; ascospores simple, hyaline, without a sheath. Lichenized with green algae, trop.

Crocysporium Corda (1837) = Aegerita (Mitosp. fungi) fide Saccardo (*Syll. Fung.* **4**: 662, 1886).

Cronartiaceae Dietel (1900), Uredinales. 2 gen., 24 spp. Pycnia intra- or subperidermal, without bounding structures, growth of indeterminate (often large) extent; aeciospores catenulate (*Peridermium*); uredinia with peridium, urediniospores shortly pedicellate; teliospores sessile, adhering to each other and forming erumpent, brownish, hair-like columns.

Cronartium Fr. (1815), Cronartiaceae. 20, widespr. 0, I on *Pinus*; II, III on Dicots. *C. ribicola* (blister rust of *Pinus strobus* and other 5-needle pines; II and III on *Ribes*); see Mielke (*Bull. Sch. Forest. Yale* **52**, 1943); *C. quercuum* (fusiform rust of pine; Bursdall & Snow, *Mycol.* **69**: 503, 1977; taxonomy). See also Peterson (*Rep. Tottori mycol. Inst.* **10**: 203, 1973), Hiratsuka & Powell (*Canad. Forest Serv. Tech. Rep.* **4**, 1976).

Crossopsora Syd. & P. Syd. (1919), Phakopsoraceae. 12, trop.

Crotalia Liro (1938), Anamorphic Ustilaginaceae. 1, Eur. Teleomorph *Anthracoidea*.

Crotone Theiss. & Syd. (1915), Venturiaceae. 1, S. Am.

Crotonocarpia Fuckel (1870) = Cucurbitaria (Cucurbitar.) fide v. Arx & Müller (1975).

crottle, Scottish term for many lichens (obsol.); often used collectively; **black -**, *Parmelia omphalodes*.

Crouania Fuckel (1870) [non J. Agardh (1842), *Algae*] = Lamprospora (Otid.) fide Eckblad (1968).

Crouaniella (Sacc.) Lambotte (1888) = Ascobolus (Ascobol.) fide van Brummelen (1967).

crowded (of gills), very close together; conferted.

crown rust (of oats), *Puccinia coronata*.

Crozalsiella Maire (1917), Ustilaginaceae. 2 (on *Gramineae*), Afr., S. Am. See Zambettakis (*Revue mycol.* **33**: 27, 1968). ? = Ustilago (Ustilagin.) fide Vánky (1987).

crozier, the hook of an ascogenous hypha before ascus-development; ascus crook (Fig. 11).

Crucella Marvanová & Suberkr. (1990), Mitosporic Auriculariaceae. Teleomorph *Camptobasidium*. 1, USA.

Crucellisporiopsis Nag Raj (1983), Mitosporic fungi, 7.G1.15. 2, Venezuela, NZ.

Crucellisporium M.L. Farr (1968), Mitosporic fungi, 6/7.G1.10. 2, USA, Tanzania. See Nag Raj & Kendrick (*CJB* **56**: 713, 1978), Punithalingam (*Nova Hedw.* **48**: 297, 1989; cytology).

cruciate, (1) in the form of a cross; (2) (of basidial septa), vertical and at right angles.

Crucibulum Tul. & C. Tul. (1844), Nidulariaceae. 3, temp.

cruciform (of nuclear division in *Plasmodiophora*), having the chromosomes in a ring around a dumbbell-shaped nucleolus.

cruciform division, see promitosis.

Crucispora E. Horak (1971), Agaricaceae. 1, NZ.

Crumenella P. Karst. (1890), Leotiaceae. 1 (on *Myrica*), Eur.

Crumenula De Not. (1864) = Godronia (Leot.) fide Groves (*CJB* **43**: 1195, 1965).

Crumenula Rehm (1889) ≡ Crumenulopsis (Leot.).

Crumenulopsis J.W. Groves (1969), Leotiaceae. 2, Eur.

crust, a general term for a hard surface layer, esp. of a sporocarp; crustose.

Crustoderma Parmasto (1968), Hyphodermataceae. 11, widespr. See Nakasone (*Mycol.* **76**: 40, 1984).

Crustodiplodina Punith. (1988), Mitosporic fungi, 8.B1.19. 1, UK.

Crustomollisia Svrček (1987), Leotiaceae. 1, Eur.

Crustomyces Jülich (1978), Corticiaceae. 5, widespr.

crustose (**crustaceous**), crust-like; used for lichens having a thallus stretching over and firmly fixed to the substratum by the whole of their lower surface; such thalli generally lack rhizinae and a lower cortex. (Fig. 21B).

Crustula Velen. (1934) = Mollisia (Dermat.) fide Baral (*SA* **13**: 113, 1994).

Cryocaligula Minter (1986), Anamorphic Rhytismataceae, 8.B1.10. Teleomorph *Ploioderma*. 1, C. Am.

cryopreservation, see Genetic Resource Collections.

Cryphonectria (Sacc.) Sacc. & D. Sacc. (1905), Valsaceae. 6, Asia, Eur., N. & S. Am. *C. parasitica* (chestnut (*Castanea*) blight or canker). See Barr (1978), Hodges (*Mycol.* **72**: 542, 1980; on *Eucalyptus*), Micales & Stipes (*Phytopath.* **77**: 650, 1987). Anamorph *Endothiella*.

crypta, a sleeve-like formation around a tree root (esp. evergreens) in tropics and subtropics developed by certain agarics (Singer, 1962: 20).

Cryptandromyces Thaxt. (1912), Laboulbeniaceae. 13 or 14, widespr. See Tavares (1985).

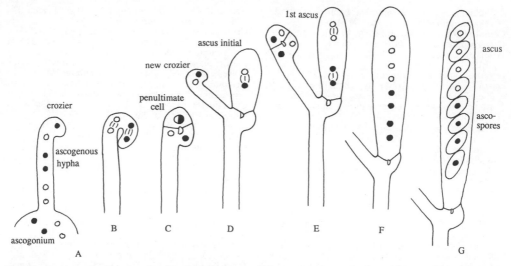

Fig. 11. Ascus and ascospore development (diagrammatic). A, ascogenous hypha with a crozier at the apex developing from an ascogonium; B, conjugate division of the two nuclei in the crozier; C, two septa cut off a binucleate penultimate cell, the nuclei in which fuse to form a diploid fusion nucleus, meanwhile the backwardly directed terminal cell fuses with the ascogenous hypha; D, penultimate cell enlarges to become the ascus within which the fusion nucleus begins to divide meiotically, and a new crozier develops from beneath the ascus and repeats the behaviour of the first; E, second division of meiosis occurs in the young ascus; F, mitotic division of the four haploid nuclei in the ascus; G, ascospores formed.

Cryptascoma Ananthap. (1988), Valsaceae. 1, India.

Cryptella Quél. (1875) = Robergea (Stictid.) fide Saccardo (*Syll. Fung.* **2**: 806, 1883).

Cryptendoxyla Malloch & Cain (1970), ? Pseudeurotiaceae. 1, Canada. See v. Arx & v. d. Walt (*Stud. Mycol.* **9**: 167, 1987; posn). Anamorph *Chalara*.

cryptic, inconspicuous or hidden.

Cryptica R. Hesse (1884) = Pachyphloeus (Terfez.) fide Fischer (1938).

Crypticola Humber, Frances & A.W. Sweeney (1989), Crypticolaceae. 2, Australia.

Crypticolaceae M.W. Dick (1995), ? Myzocytiopsidales. 1 gen., 2 spp.

Cryptoascus Petri (1909), Ascomycota (inc. sed.). 2, Eur. *C.oligosporus* on *Olea*, *C. graminis* on wheat.

Cryptobasidiaceae Malençon ex Donk (1956), Cryptobasidiales. 4 gen. (+ 2 syn.), 8 spp. On surface of hypertrophied host tissue. Holobasidia (ustidia) pallisadic, clavate with 2-8 sterigmata; spores thick-walled, septate, statismosporic.

Cryptobasidiales, Ustomycetes. 1 fam., 4 gen. (+ 2 syn.), 8 spp. Fam. **Cryptobasidiaceae**.

Cryptobasidium Lendn. (1921) = Botryoconis (Cryptobasid.) fide Malençon (*BSMF* **69**: 92, 1953).

Cryptoceuthospora Petr. (1921), Mitosporic fungi, 8.A1.?. 2, Eur.

Cryptochaete P. Karst. (1889) = Peniophora (Peniophor.) fide Donk (1964).

Cryptocline Petr. (1924), Mitosporic fungi, 6.A1.19. 14, N. temp. See Morgan-Jones (*CJB* **51**: 309, 1973), but see Sutton (1980).

Cryptococcales. Order used for *Blastomycetes* (q.v.) with ascomycetous teleomorphs, but *Cryptococcus* itself is now known to be polyphyletic and partly of basidiomycetous affinity. Included here in Mitosporic fungi.

cryptococcosis, a disease in humans and animals caused by *Cryptococcus neoformans* (teleomorph *Filobasidiella neoformans*); 'European blastomycosis'; torulosis. See Littman & Zimmerman

(*Cryptococcosis*, 1956), Al Dory (*Mycopathol.* **45**: 1, 1971; bibl.).

Cryptococcus Kütz. (1833) nom. rej. ≡ Cryptococcus Vuill.

Cryptococcus Vuill. (1901) [non Kütz., 1833] nom. cons. Anamorphic Filobasidiaceae, 1.A1.1/10. Teleomorphs *Filobasidiella*, *Filobasidium*. 37 (fide Barnett *et al.*, 1990), widespr. See Lodder (1970: 1092; key), Slodki *et al.* (*Can. J. Microbiol.* **12**: 489, 1966; extracell. polysaccharides, rel. with *Tremella*), Skinner (*Am. Midl. Nat.* **43**: 242, 1950), Lodder & van Rij (1952), Benham (*Bact. Rev.* **20**: 189, 1956), Weiss *et al.* (*Mykosen* **30**: 57, 1987; TEM of capsule), Mochizuki *et al.* (*J. Med. Vet. Mycol.* **25**: 223, 1987; ultrastructure of mitosis), Dufait *et al.* (*Mykosen* **30**: 483, 1987; varietal differentiation of *C. neoformans* by D-proline assimilation), Weijman *et al.* (*Ant. v. Leeuwenhoek* **54**: 545, 1988; emend. of genus), Almendros *et al.* (*MR* **94**: 211, 1990; pyrolysis-gas chromatography-mass spectrometry), Varma & Kwon-Chung (*J. Gen. Microbiol.* **135**: 3353, 1990; RFLPs in mtDNA *C. neoformans*), Marumo & Aoki (*Jl Clin. Microbiol.* **28**: 1509, 1990; cellular fatty acids), Martini (*Exp. Mycol.* **15**: 140, 1991; intraspecific discontinuity in *C. albidus* by nDNA/nDNA association). See also cryptococcosis; Fell *et al.* (*Taxon* **38**: 151, 1989).

Cryptocolax R.A. Scott (1956), Fossil fungi (Eurot.). 1 (Eocene), USA. See Pirozynski (1976).

Cryptocoryneopsis B. Sutton (1980), Mitosporic fungi, 1.G2.1. 1, Australia.

Cryptocoryneum Fuckel (1870), Mitosporic fungi, 8.G2.1. 4, temp.

Cryptocrea Petr. (1937) = Eudarluca (Phaeosphaer.) fide Müller & v. Arx (1962).

Cryptoderis Auersw. (1869) ≡ Pleuroceras (Vals.).

Cryptoderma Imazeki (1943), Hymenochaetaceae. 14, widespr. See Pegler (1973). = Phellinus (Hymenochaet.) fide Donk (1974).

Cryptodesma Leidy (1850) nom. dub. (? Fungi).

Cryptodiaporthe Petr. (1921), Valsaceae. 20, N. temp. See Wehmeyer (1933).

Cryptodictyon A. Massal. (1860), Lecideaceae (L). 2, Asia.

Cryptodidymosphaeria (Rehm) Höhn. (1917) = Didymosphaeria (Didymosphaer.) fide Aptroot (1995).

Cryptodiscus Corda (1838), Stictidaceae. 6, widespr. See Sherwood (*Mycotaxon* **5**: 1, 1977).

cryptoendolithic (of organisms, esp. lichens), surviving at low temperatures through modification of the thallus so that it can exist inside rock between rock crystals (Friedmann, *Science, N.Y.* **215**: 1045, 1982). Cf. endolithic.

Cryptogamia (obsol.). Division of the kingdom *Plantae* for the spore-producing plants, i.e. the *Thallophyta* (which traditionally included fungi), *Bryophyta* (mosses and liverworts), and *Pteridophyta* (ferns, etc.); the *Phanerogamia* (*Spermatophyta*) being the division for the flowering (or seed-producing) plants.

Cryptogene Syd. (1939) = Ascochytopsis (Mitosp. fungi) fide Sutton (1977).

Cryptogenella Syd. (1939) = Ascochytopsis (Mitosp. fungi) fide Sutton (1977).

Cryptohymenium Samuels & L.M. Kohn (1987), Dermateaceae. 1, NZ.

Cryptolechia A. Massal. (1853), Gyalectaceae (L). 8, mainly trop. See Hawksworth & Dibben (*Lichenologist* **14**: 98, 1982; nomencl.), Vězda (*Folia geobot. phytotax.* **4**: 443, 1969).

Cryptoleptosphaeria Petr. (1923), Hypocreaceae. 1 (on *Leptosphaeria*), Eur.

Cryptomela Sacc. (1884) = Cryptosporium (Mitosp. fungi) fide Höhnel (*Sber. Akad. Wiss. Wien* **125**: 76, 1916).

Cryptomphalina R. Heim (1966), Tricholomataceae. 1, Thailand.

Cryptomycella Höhn. (1925), Anamorphic Phacidiaceae, 8.A1.15. Teleomorph *Cryptomycina*. 1, Eur.

Cryptomyces Grev. (1825), Cryptomycetaceae. 1, N. temp. *C. maximus* (on *Salix*). See Alcock (*TBMS* **11**: 161, 1926).

Cryptomycetaceae Höhn. (1917), Rhytismatales. 4 gen. (+ 1 syn.), 5 spp. Stromata large, irregular, initially immersed, becoming erumpent, uniloculate, the covering layer sloughing off permanently; ascomata apothecial; interascal tissue of numerous narrow thin-walled paraphyses, swollen at the apices; asci elongate, persistent, thin-walled, the apex hardly differentiated, ? ascospores released though a small pore, without apical structures; ascospores elongate, hyaline, aseptate. Anamorphs not known. Saprobic or weakly parasitic on woody tissues, widespr.
  While the type genus at least belongs to *Rhytismatales* (Livsey & Minter, 1994), the other genera require more critical study.

Cryptomycina Höhn. (1917), Ascomycota (inc. sed.). 1 (? 3), N. temp. *C. pteridis* (common on bracken, *Pteridium aquilinum*). See Bache-Wiig (*Mycol.* **32**: 214, 1940). Anamorph *Cryptomycella*.

Cryptomycocolaceae Oberw. & R. Bauer (1956), Cryptomycocolales. 1 gen., 1 sp. Basidia in sori; metabasidium subulate; spores statismosporic; hyphal septa with Woronin-bodies; obligate parasites of ascomycetes.

Cryptomycocolacales, Ustomycetes. 1 fam., 1 gen., 1 sp. Fam. **Cryptomycocolacaceae**.

Cryptomycocolax Oberw. & R. Bauer (1990), Cryptomycocolacaceae. 1 (parasitic on an ascomycete), Costa Rica.

Cryptonectriella (Höhn.) Weese (1919) = Nectriella (Hypocr.) fide Rogerson (1970).

Cryptonectriopsis (Höhn.) Weese (1919) = Phomatospora (Ascomycota, inc. sed.) fide Barr (1978).

Cryptoniesslia Scheuer (1993), Niessliaceae. 1, U.K.

Cryptoparodia Petr. (1950) = Antennularia (Ventur.) fide Müller & v. Arx (1962).

Cryptopeltis Rehm (1906) = Porina (Trichothel.).

Cryptopeltosphaeria, see *Cryptoleptosphaeria*.

Cryptopezia Höhn. (1919), Leotiales (inc. sed.). 1, Samoa.

Cryptophaeella Höhn. (1917) ? = Microsphaeropsis (Mitosp. fungi) fide Sutton (1977).

Cryptophallus Peck (1897) = Phallus (Phall.) fide Lloyd (*Mycol. Writ.* **2**; *Mycol. Notes* **26**: 329, 1907).

Cryptophiale Piroz. (1968), Mitosporic fungi, 1.B1.15. 11, widespr. trop. See Sutton *et al.* (*MR* **92**: 354, 1989; key), McKenzie (*Mycotaxon* **49**: 307, 1993; NZ, New Caledonia spp.).

Cryptophialoidea Kuthub. & Nawawi (1987), Mitosporic fungi, 1.E1.15. 2, Malaysia.

Cryptoporus (Peck) Shear (1902), Coriolaceae. 1, N. Am., S.E. Asia.

Cryptopus Theiss. (1914) [non Lindl. (1824), *Orchidaceae*] ≡ Adelopus (Ventur.).

Cryptorhynchella Höhn. (1915) ? = Sphaerographium (Mitosp. fungi).

Cryptosordaria De Not. ex Sacc. (1891) nom. inval. = Anthostomella (Xylar.).

Cryptosphaerella Sacc. (1882) = Coronophora (Nitschk.) fide Müller & v. Arx (1973). ? = Endothia (Vals.) fide Walker (*in litt.*).

Cryptosphaeria Grev. (1822) nom. rej. = Diplodia (Mitosp. fungi) fide Bisby & Mason (*TBMS* **24**: 138, 1940).

Cryptosphaeria Ces. & De Not. (1863) nom. cons., Diatrypaceae. 4, mainly temp. See Rappaz (*Mycol. Helv.* **2**: 285, 1987; key).

Cryptosphaerina Lambotte & Fautrey ex Sacc. & P. Syd. (1902) = Cryptosphaeria (Diatryp.) fide Pirozynski (*in litt.*).

Cryptospora Tul. & C. Tul. (1863) [non Kar. & Kir. (1842), *Cruciferae*] = Winterella (Vals.). = Ophiovalsa (Vals.) fide Holm (*SA* **11**: 29, 1992).

Cryptosporella Sacc. (1877) = Winterella (Vals.) fide Reid & Booth (*CJB* **65**: 1320, 1987). = Wuestneia (Melanconid.) fide Ananthapadmanaban (1988).

Cryptosporellaceae, see *Melanconidaceae*.

Cryptosporina Höhn. (1905) = Botryosphaeria (Botryosphaer.) fide v. Arx & Müller (1954).

Cryptosporiopsis Bubák & Kabát (1912), Anamorphic Dermateaceae, 6/8.A1.15. Teleomorphs *Pezicula*, *Ocellaria*. c. 15, widespr., esp. temp. *C. malicorticis* (syn. *Gloeosporium perennans*) (perennial canker of apple). See Johnston & Fullerton (*N.Z. J. Exp. Agr.* **16**: 159, 1988; *C. citri* causing leaf spot of *Citrus*), Geñe *et al.* (*MR* **94**: 309, 1990; bud rot of *Corylus*), Dennis (*Kew Bull.* **29**: 157, 1974; key), Dugan, Grove & Rogers (*Mycol.* **85**: 551 and 565, 1993; morphology, pathogenicity, cytology etc. of *C. perennans* and *C. curvispora*).

Cryptosporium Kunze (1817), Mitosporic fungi, ?.?.?. 25, temp. See Sutton (*Mycol. Pap.* **141**, 1977).

Cryptostictella Grove (1912) = Discosia (Mitosp. fungi) fide Grove (1937) but see Petrak & Sydow (*Ann. Myc.* **23**: 209, 1925).

Cryptostictis Fuckel (1866) = Seimatosporium (Mitosp. fungi) fide Shoemaker (1964).

Cryptostroma P.H. Greg. & S. Waller (1952), Mitosporic fungi, 8.A2.19. 1, N. Am., Eur. *C. corticale* (sooty bark of *Acer*).

Cryptothamnium Wallr. (1842) = Chaenocarpus (Xylar.) fide Læssøe (*SA* **13**: 55, 1994).

Cryptothecia Stirt. (1876) nom. cons., Arthoniaceae (L). 18, trop. See Santesson (*Symb. bot. upsal.* **12** (1), 1952), Awasthi & Agarwal (*J. Indian bot. Soc.* **48**: 62, 1969), Lücking (*Lichenologist* **27**: 127, 1995; Costa Rica), Makhija & Patwardhan (*Biovigyanan* **11**: 1, 1985; key 24 spp.), Thor (*Bryologist* **94**: 278, 1991).

Cryptotheciaceae, see *Arthoniaceae.*

Cryptothecium Penz. & Sacc. (1897) [non Hübener (1851), fossil *Musci*], Hypocreaceae. 1, Java.

Cryptothele Th. Fr. (1866), Lichinaceae (L). 2, Eur., S. Afr. See Henssen & Büdel (*Beih. Nova Hedw.* **79**: 381, 1984).

Cryptotheliomyces Cif. & Tomas. (1953) ≡ Cryptothele (Lichin.).

Cryptothelium A. Massal. (1860) = Laurera (Trypethel.) fide Eriksson (*Opera Bot.* **60**, 1981). See also Harris (*Acta Amazon.* Suppl. **14**: 55, 1986; key 2 spp. Brazil), Eriksson & Hawksworth (*SA* **11**: 56, 1992; status).

Cryptovalsa Ces. & De Not. ex Fuckel (1870), Ascomycota (inc. sed.). 20, widespr.

Crystallocystidium (Rick) Rick (1940) nom. dub. (Ster.).

Ctenoderma Syd. & P. Syd. (1920) = Skierka (Uredinales, inc. sed.) fide Mains (*Mycol.* **31**: 175, 1939).

ctenoid, comb-like.

Ctenomyces Eidam (1880), Arthrodermataceae. 1, widespr. See Orr & Kuehn (*Mycopath.* **21**: 321, 1963), Apinis (1964), Currah (1985).

Ctenosporites Elsik & Janson. (1974), Fossil fungi (Mitosp. fungi). 2 (Eocene), Alaska, Australia, Pacific. See Lange & Smith (*N. Jb. Geol. Paläont. Mh.* **11**: 649, 1975).

Ctesium Pers. (1827) = Graphina p.p. and Phaeographina p.p. (Graphid.).

Cubamyces Murrill (1905) = Trametes (Coriol.) fide Overholts (1953).

Cubasina R.F. Castañeda (1986), Mitosporic fungi, 1.D2.1. 1, Cuba.

Cubonia Sacc. (1889), ? Ascobolaceae. 3, Eur. See Durand (*BSMF* **88**: 155, 1973; ontogeny), Eckblad (1968), Kimbrough & Korf (*Am. J. Bot.* **54**: 9, 1967), Pfister (*Mycol.* **76**: 843, 1984).

Cucujomyces Speg. (1917), Laboulbeniaceae. 12, S. Am., NZ, Eur., W. Afr.

Cucullaria Corda (1842) = Leotia (Leot.).

cucullate, hood- or cowl-like in form.

Cucullospora K.D. Hyde & E.B.G. Jones (1986) [non *Cucullispora* Scheuring (1970), fossil-sporae dispersae] ≡ Cucullosporella (Halosphaer.).

Cucullosporella K.D. Hyde & E.B.G. Jones (1990), Halosphaeriaceae. 1 (marine), Seychelles.

Cucurbidothis Petr. (1921) = Curreya (Cucurbitar.) fide v. Arx & Müller (1975).

Cucurbitaria Gray (1821), Cucurbitariaceae. *c.* 35 (on twigs), widespr. See Mirza (*Nova Hedw.* **16**: 161, 1968). Anamorphs *Camarosporium, Phoma.*

Cucurbitariaceae G. Winter (1885), Dothideales. 5 gen. (+ 9 syn.), 46 spp. Ascomata strongly aggegated, sometimes ± stromatic, erumpent, variable in shape, with a well-developed periphysate ostiole, thick-walled especially at the base; interascal tissue composed of cellular pseudoparaphyses; asci ± cylindrical, fissitunicate, sometimes with a distinct I- apical ring; ascospores pigmented, transversely septate or muriform. Anamorph pycnidial. Necrotrophic or saprobic on woody plants.

Cucurbitariaceites R.K. Kar, R.Y. Singh & Sah (1972), Fossil fungi (Dothid.). 2 (Tertiary), India.

Cucurbitariella Petr. (1917) = Coniochaeta (Coniochaet.) fide v. Arx & Müller (1954).

Cucurbitariopsis C. Massal. (1889) ? = Rhabdospora (Mitosp. fungi) fide Mussat (*Syll. Fung.* **10**, 1892) but see Sutton (1977).

Cucurbitariopsis Vassilkov (1960) ≡ Gemmamyces (Cucurbitar.).

Cucurbitopsis Bat. & Cif. (1957), Ascomycota (inc. sed.). 1, Portugal.

Cucurbitula Fuckel (1870) = Coniochaeta (Coniochaet.) fide Petrini (*Sydowia* **44**: 169, 1993).

cudbear (**corkir**), Scottish names for lichens used in making dye, esp. *Ochrolechia tartarea.* See Dyeing.

Cudonia Fr. (1849), ? Geoglossaceae. 8, temp. See Mains (*Am. J. Bot.* **27**: 322, 1940; *Mycol.* **48**: 694, 1956; N. Am. spp.), Nannfeldt (*Ark. Bot.* **30A**, 1942), Sharma & Rawla (*Bibl. Mycol.* **91**: 203, 1983; India).

Cudoniella Sacc. (1889), Leotiaceae. *c.* 15, widespr. See Dennis (*Persoonia* **3**: 72, 1964). = Hymenoscyphus (Leot.) fide Baral & Krieglsteiner (*Beih. Z. Mykol.* **6**, 1985).

Cudoniopsis Speg. (1925), Sclerotiniaceae. 1, S. Am.

Culcitalna Meyers & R.T. Moore (1960), Mitosporic fungi, 3.C2.1. 1 (marine), Newfoundland.

Culicicola Nieuwl. (1916) ≡ Lamia (Entomphthor.)

Culicidospora R.H. Petersen (1960), Mitosporic fungi, 1.G1.1. 2 (aquatic), N. Am. See Petersen (*Mycol.* **55**: 23, 1963).

Culicinomyces Couch, Romney & B. Rao (1974), Mitosporic fungi, 1.A1.15. 3 (in *Anopheles*), USA, Australia. See Sweeney *et al.* (*J. Inv. Path.* **42**: 224, 1983; ultrastructure), Inmann & Bland (*CJB* **61**: 2618, 1983; ultrastructure of conidiogenesis), Sigler *et al.* (*Mycol.* **79**: 493, 1987; on *Aedes kochi*).

culmicolous, living on stems, esp. those of grasses; **caulicolous**; **culmicole**, an organism which does this.

culmomarasmin, a wilt toxin of *Fusarium culmorum* (Gäumann *et al., Phytopath. Z.* **36**: 115, 1949).

cultivar, a variety in the horticultural (or agricultural) sense (see Art. 10, *International code of nomenclature for cultivated plants* [*Regnum veg.* **104**], 1980). Snyder *et al.* (*J. Madras Univ.* **27**: 185, 1957) used cultivar for infraspecific taxa of *Fusarium* but this term is only correctly used in mycology for trade varieties of cultivated mushrooms, etc.

cultivated mushroom basidioma of *A. bisporus* (syn. *A. brunnescens*; name in need of conservation; see Malloch, *Mycol.* **68**: 910, 1976, nomencl.);

culture, a growth of one organism or of a group of organisms for the purpose of experiment (esp. of microorganisms on laboratory media) or sometimes for trade (e.g. a **mushroom** -); **continuous** -, one in which the culture medium is simultaneously added and withdrawn (harvested) so that the volume remains constant; see Calcott (*Continuous culture of cells*, 2 vols, 1981); **enrichment** -, a culture which favours the growth of the desired organism in a mixed culture or population; **pure** -, a - of one sort of organism; - **medium**, see medium; **type** -, see type.

Culture collections, see Genetic resource collections.

Culture collections and maintenance. The chief **collections** of fungal cultures include: American Type Culture Collection, Washington, D.C., USA; Centraalbureau voor Schimmelcultures, Baarn, Netherlands (v. Beverwijk, *Ant. v. Leeuwenhoek* **25**: 1, 1959; historical account) [Yeasts: Laboratorium voor Microbiologie, Delft]; Institute for Fermentation, Osaka, Japan; International Mycological Institute, Egham [Yeasts; Food Research Inst., Norwich, Norfolk; Wood-destroying, Forest Products Res. Lab., Princes Risborough, Bucks; Medical, Mycological Reference Lab.]; Laboratoire de Cryptogamie, Muséum National d'Histoire Naturelle, Paris. See Takishima *et al.* (Eds.) (*Guide to World data center on microorganisms with a list of culture collections* *worldwide*, 1989), Staines *et al.* (Eds) (*World directory of collections of cultures of micro-organisms*, edn 3, 1986), *Directory of collections of microorganisms maintained in the United Kingdom* (CMI, 1978). See also Herbaria.

International collaboration is facilitated by meetings of the International Congress of Culture Collections and a World Federation of Culture Collections functions under the auspices of the IUBS.

Culture **maintenance** is traditionally on slopes of appropriate media stored at laboratory or low temperature but now cryopreservation (see below) when applicable is the method of choice. Lyophilization (freeze-drying) of the fungus spore is an ideal method to facilitate distribution. It is used, in particular, by service collections. Techniques for the preservation of living cultures include: *Aquatic fungi*: Goldie Smith (*J. Elisha Mitchell Sci. Soc.* **72**: 158, 1956), Fuller (*Lower fungi in the laboratory*, 1978). Cryopreservation by *frozen storage* at −20°C. Carmichael (*Mycol.* **48**: 378, 1956). Over *liquid nitrogen*: Hwang (*Mycol.* **60**: 612, (& Howells) 622, 1968), Smith, Tolerance to freezing and thawing, *in*, Jennings (Ed.) *Tolerance of fungi* pp. 145-171, New York, Marcel Dekker, Inc. By *lyophilization* (freeze drying): Raper & Alexander (*Mycol.* **37**: 499, 1945), Hesseltine *et al.* (*Mycol.* **52**: 762, 1961). Under *mineral oil*: Little & Gordon (*Mycol.* **59**: 733, 1967; refs.), Fennell (*Bot. Rev.* **26**: 1, 1960; gen. review). In anhydrous *silica gel*: Perkins (*Can. J. Microbiol.* **8**: 591, 1962), Gentles & Scott (*Sabouraudia* **17**: 415, 1979). In *soil*: Bakerspiegel (*Mycol.* **45**: 596, 1953). In *water*: Castellani (*Mycopath.* **20**: 1, 1963). See also Onions (*in* Booth (Ed.), *Methods in microbiology* **4**: 113, 1971), Smith & Onions (*TBMS* **81**: 535, 1983; comparison of mineral oil with other methods. *The preservation and maintainence of living fungi*, 1994 (IMI)), Smith (*TBMS* **79**: 415, 1982; 3,000 fungi preserved up to 13 years), *IMI Culture Collection Catalogue* 10th edn, 1992), Kirsop & Doyle (*Maintenance of microorganisms and cultured cells*, edn 2, 1991), Berry & Hennebert (*Mycol.* **83**: 605, 1992; freeze drying).

culture methods see Media.

Cumminsiella Arthur (1933), Pucciniaceae. 8, Eur., Am. See Baxter, *Mycol.* **49**: 864, 1957; key); autoecious on *Mahonia*.

Cumminsina Petr. (1955), Sphaerophragmiaceae. 1 (on *Grewia*), Angola.

cumulate, massed together; heaped up.

Cumulospora I. Schmidt (1985), Mitosporic fungi, 1.D2.?. 1 (marine), Baltic Sea.

cuneate, see cuneiform.

cuneiform (**cuneate**), thinner at one end than the other; wedge or axe-blade shaped (Fig. 37.23).

Cunningham (Gordon Herriot; 1892-1962). Pioneer New Zealand plant pathologist (*Plant protection by the aid of therapeutants*, 1935) and mycologist noted for his studies on the fungi of NZ, esp. *Ustilaginales* (1924), *Uredinales* (1931), *Gasteromycetes* (1944), *Thelephoraceae* (1963), *Polyporaceae* (1965), q.v. see Ramsbottom (*Biogr. Mem. roy. Soc.* **10**: 15, 1964; portr., bibl.), Stafleu & Cowan (*TL-2* **1**: 573, 1976).

Cunninghamella Matr. (1903), Cunninghamellaceae. 7, widespr. See Samson (*Proc. K. ned. Akad. Wet.* C **72**: 322, 1969; key), Hawker *et al.* (*J. gen. Microbiol.* **60**: 181, 1970; sporangiolum ultrastructure), Shipton & Lunn (*TBMS* **74**: 483, 1980; taxon. criteria, key), Baijal & Mehrotra (*Sydowia* **33**: 1, 1980; key), Lunn & Shipton (*TBMS* **81**: 303, 1983; key), Weitzman (*TBMS* **83**: 527, 1984), Zheng & Chen (*Mycosystema* **5**: 1, 1992), Dermoumi (*Mycoses* **36**: 293, 1993; zygomycoses).

Cunninghamellaceae Naumov ex R.K. Benj. (1959), Mucorales. 1 gen. (+ 3 syn.), 7 spp. Sporangia absent, sporangiola unispored; zygospores warty, borne on opposed suspensors.
Lit.: Benny *et al.* (*Mycologia* **84**: 639, 1992; emend.).

Cunninghamia Curr. (1873) [non R. Br. (1826), *Cycadaceae*] ≡ Choanephora (Choanephor.).

Cunninghammyces Stalpers (1985) = Xenasma (Xenasmat.) fide Hjortstam & Larsson (*Windahlia* **21**, 1994).

cup fungus, a discomycete (esp. *Leotiales* or *Pezizales*) ascoma. - **lichen**, a sp. of *Cladonia* having podetia expanded into goblet-like scyphi.

Cuphocybe R. Heim (1951), Cortinariaceae. 3, NZ. See Horak (*Beih. Nova Hedw.* **43**: 193, 1973; key).

Cuphophyllus (Donk) Bon (1985), Hygrophoraceae. 23, widespr.

Cupularia Link (1833) = Craterium (Myxom.).

cupulate, cup-like in form, as e.g. in conidiomata (Fig. 10 O).

Cupulomyces R.K. Benj. (1992), Laboulbeniaceae. 1, Grenada.

Curculiospora Arnaud (1954), Mitosporic fungi, 1.F2.1. 1, France.

curling factor, see griseofulvin.

Curreya Sacc. (1883), Cucurbitariaceae. 2 (on conifers), N. temp. See v. Arx & v. der Aa (*Sydowia* **36**: 1, 1983), Barr (*Mycotaxon* **29**: 501, 1987), Eriksson (*SA* **5**: 127, 1986). Anamorph *Coniothyrium*.

Curreyella Massee (1895) = Plicaria (Peziz.) fide Eckblad (1968).

Curreyella (Sacc.) Lindau (1897) ≡ Discostroma (Amphisphaer.).

Curtis (Moses Ashley; 1808-72). Minister of the Episcopal Church and student of American fungi. Lived in N. Carolina from 1835. Collaborated with Berkeley (q.v.) See Petersen, '*B. & C*': the mycological association of M.J. Berkeley and M.A. Curtis, [*Bibl. Mycol.* **72**, 1980]. Collections in FH. See Shear & Stevens (*Mycol.* **11**: 181, 1919), Snell & Dick (*Mycol.* **45**: 968, 1953), Petersen (*Mycotaxon* **9**: 459, 1979), Berkeley & Berkeley (*A Yankee botanist* ...., 1986), Stafleu & Cowan (*TL-2* **1**: 573, 1976).

Curucispora Matsush. (1981), Mitosporic fungi, 1.G1.1. 1, Ponape.

Curvidigitus Sawada (1943), Mitosporic fungi, 2.G.?. 1, Taiwan. See Pirozynski (*Mycol. Pap.* **129**, 1972).

Curvisporium, see *Curvusporium*.

Curvularia Boedijn (1933), Anamorphic Pleosporaceae, 2.C2.26. Teleomorph *Cochliobolus*. 33, widespr. *C. lunata* common on crop plants, esp. trop. (Lam-Quang-Bach, *Fiches Phytopath. trop.* **15**, 1964). See Kendrick & Cole (*CJB* **46**: 1279, 1968; spore devel.), Sivanesan (*Mycol. Pap.* **158**, 1987; key), Alcorn (*Mycotaxon* **39**: 361, 1990; additions to genus), Alcorn (*Mycotaxon* **41**: 329, 1991).

Curvulariopsis M.B. Ellis (1961), Mitosporic fungi, 2.C2.10. 1, Ecuador.

Curvusporium Corbetta (1963) = Curvularia (Mitosp. fungi) fide Sutton (*in litt.*).

cuspidate (e.g. of a pileus or cystidium), having a well-marked sharp outgrowth or point at the top.

Cuspidosporium Cif. (1955) = Exosporium (Mitosp. fungi) fide Ellis (1961).

Custingophora Stolk, Hennebert & Klopotek (1968), Anamorphic Lasiosphaeriaceae, 1.A2.15. Teleomorph *Chaetosphaeria*. 2, Eur., N. Am. See Barr & Crane (*CJB* **57**: 835, 1979).

Cuticularia Ducomet (1907), Mitosporic fungi, 1.-.-. 1, Eur.

cutis (**cuticle**) (of basidiomata), the outer layer consisting of compressed hyphae parallel to the surface; the upper and lower layers of the **cutis** are sometimes distinguished as **epi-** and **sub-**. See Shaffer (*Brittonia* **22**: 230, 1970; cuticular terminology in *Russula*), Singer (*Agaricales in Modern Taxonomy* edn 4: 69, 1986); also pellis.

Cutomyces Thüm. (1878) = Puccinia (Puccin.) fide Dietel (1928).

CYA, see Media.

cyanescent, becoming blue.

Cyanicium Locq. (1979), Cortinariaceae. 1, Eur.

Cyanisticta Gyeln. (1931) = Pseudocyphellaria (Lobar.).

Cyanobaeis Clem. (1909) = Baeomyces (Baeomycet.).

Cyanobasidium Jülich (1979) = Lindtneria (Lindtner.) fide Hjortstam (*Mycotaxon* **28**: 19, 1987).

cyanobiont, see photobiont.

Cyanocephalium Zukal (1893) ? = Thelocarpon (Acarospor.) fide Müller & v. Arx (1962).

Cyanochyta Höhn. (1915), Anamorphic Hypocreaceae, 4.B1.?. Teleomorph *Gibberella*. 1, Eur. See Sutton (1977).

Cyanoderma Höhn. (1919) [non Weber Bosse (1887), *Rhodophyceae*] ≡ Cyanodermella (Stictid.).

Cyanodermella O.E. Erikss. (1981), Stictidaceae. 2, Eur., N. Am. See Eriksson (*Ark. Bot.* II, **6**: 381, 1967; as *Cyanoderma*).

Cyanodiscus E. Müll. & M.L. Farr (1971), Saccardiaceae. 1 (or 2), trop. USA.

Cyanodontia Hjortstam (1987), Hyphodermataceae. 1, E. Afr.

Cyanohypha Jülich (1982) = Botryobasidium (Botryobasid.) fide Langer (*Bibl. Mycol.* **158**, 1994).

Cyanopatella Petr. (1949), Mitosporic fungi, 4.A1/2.?. 1, Iran.

cyanophilous (of spores, etc.), readily absorbing a blue stain such as cotton blue or gentian violet.

Cyanophomella Höhn. (1918), Anamorphic Hypocreaceae, 4.A/B1.?. Teleomorph *Gibberella*. 1, Eur.

cyanophycophilous (of fungi), ones lichenized with a cyanobacterium (Pike & Carroll, *in* Alexopoulos & Mims, *Introductory mycology*, edn 3, 1980); see photobiont.

Cyanoporina Groenh. (1951), ? Pyrenothricaceae (L). 1, Java.

Cyanopyrenia H. Harada (1995), Ascomycota (inc. sed., L). 1 (aquatic), Japan.

Cyanospora Heald & F.A. Wolf (1910) = Robergea (Stictid.) fide Nannfeldt (1932).

Cyanosporus McGinty (1909) = Polyporus (Polypor.) fide Stevenson & Cash (1936).

Cyanotheca Pascher (1914) nom. dub. (? Fungi).

cyanotrophic (of fungi, esp. lichen-forming spp.), obtaining nutrients (esp. nitrates fixed from the atmosphere) by forming regular connexions to free-living or ± lichenized cyanobacteria; see Poelt & Mayrhofer (*Pl. Syst. Evol.* **158**: 265, 1988); see also cephalodium.

Cyathela Raf. (1819) nom. dub. (Fungi, inc. sed.). No spp. included.

Cyathella Raf. (1815) nom. dub. (Fungi, inc. sed.). No spp. included.

Cyathia P. Browne (1756) = Cyathus (Nidular.).

Cyathicula De Not. (1863), Leotiaceae. *c.* 30, mainly N. temp. See Baral & Krieglsteiner (*Beih. Z. Mykol.* **6**, 1985), Carpenter (1981; as *Crocicreas*).

cyathiform, like a cup, a little wider at the top than at the bottom, and sometimes stalked.

Cyathipodia Boud. (1907) = Helvella (Helvell.) fide Dissing (1966).

Cyathisphaera Dumort. (1822) ≡ Cucurbitaria (Cucurbitar.) fide Nannfeldt (*Svensk bot. Tidskr.* **69**: 49, 1975).

Cyathodes P. Micheli ex Kuntze (1891) ≡ Cyathus (Nidular.).

Cyathus Haller (1768), Nidulariaceae. 44 (on soil, wood, etc.), cosmop. The Bird's Nest Fungi. See Brodie (*Bot. Notiser* **130**: 453, 1977; world key), Gomez & Pérez-Silva (*Rev. Mex Micol.* **4**: 161, 1988; key Mexican spp.).

Cybebe Tibell (1984) = Chaeonotheca (Coniocyb.) fide Middelborg & Mattson (*Sommerfeltia* **5**, 1987).

Cyclaneusma DiCosmo, Peredo & Minter (1983), Rhytismataceae. 2 (on *Pinus*), widespr.

Cycledium Wallr. (1833) ≡ Schizoxylon (Stictid.).

Cycledum, see *Cycledium*.

Cyclobium C. Agardh (1821) ≡ Clisosporium (Mitosp. fungi).

Cycloconium Castagne (1845) = Spilocaea (Mitosp. fungi) fide Hughes (1958).

Cyclocybe Velen. (1939) ? = Inocybe (Cortinar.).

Cyclocytospora Höhn. (1928) = Cytospora (Mitosp. fungi).

Cycloderma Klotzsch (1832) nom. dub. (Lycoperdales, inc. sed.) fide Lloyd (*Mycol. Writ.* **1**; *Mycol. Notes* **17**: 181, 1904); probably an unopened *Geastrum*.

Cyclodomella P.N. Mathur, V.V. Bhatt & Thirum. (1959) = Coniella (Mitosp. fungi) fide Petrak (*Sydowia* **14**: 352, 1960).

Cyclodomus Höhn. (1909), Mitosporic fungi, 4.A1.15. 1, USA.

Cyclodothis P. Syd. (1913) = Mycosphaerella (Mycosphaerell.) fide v. Arx & Müller (1975).

Cyclographa Vain. (1921) = Catarraphia (Arthon.) fide Egea & Torrente (1993).

Cyclographina D.D. Awasthi (1979), Graphidaceae (L). 9, widespr., trop.

cycloheximide (actidione), an antibiotic from *Streptomyces griseus* (*J. Am. Chem. Soc.* **69**: 174, 1947); antibacterial and antifungal. For use in isolating fungi pathogenic for humans, see Georg *et al.* (*J. Lab. Clin. Med.* **44**: 222, 1954). - **tolerance**, used to distinguish plant pathogenic fungi, e.g. *Ceratocystis* from *Ophiostoma* (Harrington, *Mycol.* **72**: 1123, 1981).

Cyclomarsonina Petr. (1965), Mitosporic fungi, 8.B1/2.?. 1, India.

Cyclomyces Kunze (1830), Hymenochaetaceae. 12, warmer parts.

Cyclomycetella Murrill (1904) = Cyclomyces (Hymenochaet.) fide Donk (1952).

Cyclopeltella Petr. (1953), Anamorphic Micropeltidaceae, 5.A1.?. Teleomorph *Cyclopeltis*. 1, Philipp.

Cyclopeltis Petr. (1953) [non Sm. (1846), *Aspidiaceae*], Micropeltidaceae. 1, Philipp.

Cyclophomopsis Höhn. (1920) = Phomopsis (Mitosp. fungi) fide Sutton (1977).

Cyclopleurotus Hasselt (1824) = Pleurotus (Lentin.) fide Singer (1986).

Cycloporellus Murrill (1907) ≡ Cyclomycetella (Hymenochaet.).

Cycloporus Murrill (1904), Hymenochaetaceae. 2, Eur. See Pegler (1971). = Coltricia (Hymenochaet.) fide Donk (1974), Ryvarden (*Khumbu Himal.* **6**: 380, 1977).

Cyclopus (Quél.) Barbier (1907) ≡ Agrocybe (Bolbit.).

Cycloschizella Höhn. (1919) = Cycloschizon (Parmular.) fide Müller & v. Arx (1962).

Cycloschizon Henn. (1902), Parmulariaceae. 11, trop.

cyclosis, cytoplasmic streaming; characteristic of eukaryotes.

cyclosporin (-**e**, **cyclosporin A**, Sandimmun), a ring-shaped polypeptide from *Tolypocladium inflatum* first reported by Dreyfuss *et al.* (*Eur. J. appl. Microbiol.* **3**: 125, 1976) which selectively inhibits the immune system in humans, especially affecting T cells; since 1983 approved for general use during kidney, heart, liver, pancreas, and bone marrow transplants, reducing organ rejection rates and increasing patient survival; see Winter (*in* Calhoun, *1986 Yearbook of science and the future*: 160, 1985; review). Also used for selective isolation of basidiomycetes.

Cyclostoma P. Crouan & H. Crouan (1867) ≡ Stictis (Stictid.).

Cyclostomella Pat. (1896), Parmulariaceae. 4, C. Am.

Cyclotheca Theiss. (1914) [non Moq. (1849), *Gyrostemonaceae*], Microthyriaceae. 8, trop.

Cyclothyrium Petr. (1923), Anamorphic Dothideales, 8.A2.15. Teleomorph *Thyridaria*. 1 (on *Juglans*), UK, India. See Sutton (1980).

Cylichnium Wallr. (1833) = Licea (Myxom.).

Cylicogone Emden & Veenb.-Rijks (1974) = Conioscypha (Mitosp. fungi) fide Sutton (*in litt.*).

Cylindrina Pat. (1886) ? = Stictis (Stictid.) fide Sherwood (1977).

Cylindrium Bonord. (1851) = Fusidium (Mitosp. fungi) fide Hughes (1958).

Cylindrobasidium Jülich (1974), Hyphodermataceae. 8, widespr.

Cylindrocarpon Wollenw. (1913), Anamorphic Hypocreaceae, 3.C1.15. Teleomorph *Nectria*. 27 (esp. in soil; sometimes pathogenic), widespr. See Booth (*Mycol. Pap.* **104**, 1966; key), Samuels (*N.Z. Jl Bot.* **16**: 73, 1978; *Nectria* teleomorphs), Brayford (*TBMS* **89**: 347, 1987; description of *C. bugnicourtii*), Brayford (*Mycopath.* **100**: 115, 1987; 10 descriptions and illustrations), Brayford & Samuels (*Mycol.* **85**: 612, 1993), Samuels & Brayford (*Sydowia* **45**: 55, 1993).

Cylindrocephalum Bonord. (1851) = Chalara (Mitosp. fungi) fide Hughes (1958).

Cylindrochytridium, see *Cylindrochytrium*.

Cylindrochytrium Karling (1941), Chytridiaceae. 1 or 2, USA.

Cylindrocladiella Boesew. (1982), Mitosporic fungi, 1.C1.15. 6, widespr. See Crous & Wingfield (*MR* **97**: 433, 1993; key).

Cylindrocladiopsis J.M. Yen (1979), Mitosporic fungi, 1.C1.10. 1, Malaysia.

Cylindrocladium Morgan (1892), Anamorphic Hypocreaceae, 1.C1.15. Teleomorph *Calonectria*. 19, widespr. See Alfenas (*Fitopat. bras.* **11**: 275, 1986; key), Stevens *et al.* (*Mycol.* **82**: 436, 1990; aminopepsidase specificity and identification), Peerally (*Mycotaxon* **40**: 323, 1991; key and review), Wormald (*TBMS* **27**: 71, 1944), Crous *et al.* (*S. Afr. For. Jl* **1576**: 69, 1991; spp. in S. Africa forest nurseries), Crous & Wingfield (*S. Afr. Jl Bot.* **58**: 397, 1992; states of *Calonectria*); El-Gholl *et al.* (*CJB* **71**: 466, 1993; *C.ovatum* n.sp. from *Eucalyptus*), Watanabe (*Mycol.* **86**: 151, 1994; 3 spp. on *Phellodendron*), Crous *et al.* (*Syst. appl. microbiol.* **16**: 266, 1993; techniques for characterization of 3-septate spp.).

Cylindrocolla Bonord. (1851), Anamorphic Dermateaceae, 3.A1.?. Teleomorph *Calloria*. 5, temp.

Cylindrodendrum Bonord. (1851), Mitosporic fungi, 1.A1.15. 3, Eur. See Petch (*TBMS* **27**: 81, 1944), Buffina & Hennebert (*Mycotaxon* **19**: 323, 1984; *C. album* synanamorph of *Cylindrocarpon*).

Cylindrodochium Bonord. (1851) ≡ Cylindrosporium (Mitosp. fungi).

Cylindrogloeum Petr. (1941), Mitosporic fungi, 6.A/B1.15. 1, Lapland.

Cylindronema Schulzer (1866) nom. dub. (Mitosp. fungi) fide Carmichael *et al.* (*Genera of Hyphomycetes*, 1980).

Cylindrophoma (Berl. & Voglino) Höhn. (1925), Mitosporic fungi, 4.A1.?. 1, Italy.

Cylindrophora Bonord. (1851) nom. dub. fide de Hoog (*Persoonia* **10**: 67, 1978).

Cylindrosporella Höhn. (1916) = Asteroma (Mitosp. fungi) fide Sutton (1977).

Cylindrosporium Grev. (1822), Anamorphic Dermateaceae, 6.A1.15. Teleomorph *Pyrenopeziza*. 1, N. temp. *C. concentricum*, light leaf spot of *Brassica*. See Rawlinson *et al.* (*TBMS* **71**: 425, 1978), Nag Raj & Kendrick (*CJB* **49**: 2119, 1971).

Cylindrosympodium W.B. Kendr. & R.F. Castañeda (1990), Mitosporic fungi, 1.C1.10. 6, widespr. ? = Solosympodiella (Mitosp. fungi).

Cylindrotaenium Thomé (1867) nom. dub. (? Fungi).

Cylindrotheca Bonord. (1864) [non Rabenh. (1859), *Algae*], Ascomycota (inc. sed.). 1, Eur.

Cylindrothyrium Maire (1907), Mitosporic fungi, 5.A1.?. 1, France.

Cylindrotrichum Bonord. (1851), Mitosporic fungi, 2.A1.16. 4, Eur. See Gams & Holubová-Jechová (*Stud. Mycol.* **13**, 1976; key). = Chaetopsis (Mitosp. fungi) fide DiCosmo *et al.* (*Mycol.* **75**: 949, 1983), but see Rambelli & Onofri (*TBMS* **88**: 393, 1987).

Cylindroxyphium Bat. & Cif. (1963), Mitosporic fungi, 4.A1.15. 1, USA.

Cylomyces Clem. (1931) ≡ Listeromyces (Mitosp. fungi).

Cymadothea F.A. Wolf (1935), Mycosphaerellaceae. 1, N. temp. *C. trifolii* (black blotch of clover). See Cannon (*Mycol. Pap.* **163**, 1991; nomencl.). Anamorph *Polythrincium*. = Mycosphaerella (Mycosphaerell.) fide v. Arx & Müller (1975).

Cymatella Pat. (1899), Tricholomataceae. 3, Antilles.

Cymatellopsis Parmasto (1985), Tricholomataceae. 1, E. Asia.

Cymatoderma Jungh. (1840), Podoscyphaceae. 9, widespr., trop. See Reid (1965: 95; key), Welden (*Mycol.* **52**: 856, 1962; Am. spp.).

Cymbanche Pfitzer (1869) ? = Ectrogella (Ectrogell.).

Cymbella Pat. (1886) [non C. Agardh (1830), *Bacillariophyta*] ≡ Chromocyphella (Crepidot.).

cymbiform, boat-shaped; navicular (Fig. 37.32).

Cymbothyrium Petr. (1947), Mitosporic fungi, 8.A2.1. 1, Eur.

Cynema Maas Geest. & E. Horak (1995), Tricholomataceae. 1, Papua New Guinea.

Cyniclomyces Van der Walt & D.B. Scott (1971), Saccharomycetaceae. 1, Eur.

Cynicus Raf. (1815) nom. dub. (Fungi, inc. sed.). No spp. included.

Cynophallus (Fr.) Corda (1842) nom. rej. ≡ Mutinus (Phall.).

Cypellomyces, see *Cypheliomyces*.

Cypeliaceae, see *Caliciaceae*.

Cypheliomyces E.A. Thomas ex Cif. & Tomas. (1953) = Cyphelium (Calic.) fide Tibell (1984).

Cypheliopsis (Zahlbr.) Vain. (1927) = Thelomma (Calic.) fide Tibell (1984).

Cyphelium Ach. (1815), Caliciaceae (L). *c.* 12, mainly N. & S. temp. See Tibell (1984; *Svensk bot. Tidskr.* **65**: 138, 1971; Eur.), Weber (*Bryologist* **70**: 197, 1967).

Cyphelium Chev. (1826) = Chaenotheca (Coniocyb.) fide Tibell (1984).

Cyphelium De Not. (1846), Coniocybaceae. 1, Eur.

cyphella (pl. -ae), a break in the lower (rarely upper) cortex of a lichen thallus which is roundish or ovate and in section appears as a cup-like structure lined with a layer of loosely connected, frequently globular, cells formed from the medulla, characteristic of *Sticta*; also used for pores open to the halliomedulla in *Oropogon*.

Cyphella Fr. (1822) nom. rej. = Aleurodiscus (Aleurodisc.) fide Donk (1951).

Cyphellaceae Lotsy (1907), Stereales. 16 gen., 73 spp. Basidioma cupulate; monomitic; no cystidia; spores often large. This family was used by Patouillard for polypores with disk or cup-shaped basidiomata and has been used for a diversity of fungi. Many of the genera have been distributed among other families of the *Stereales* and the *Agaricales* (see Donk, 1964: 290) but here most of these are retained in the '*Cyphellaceae*' because of the several recent treatments of the group s.l. Under a number of the genera an indication is, however, also given of the group to which it probably should be referred.

Lit.: Donk (1951-63) I; Cooke (*Mycol.* **49**: 680, 1957, *Porotheleaceae*; *Sydowia Beih.* **4**, 1961), Donk (*Persoonia* **1**: 25; **2**: 331, 1959-62), Reid (*Persoonia* **3**: 97, 1964; Michigan), Donk (*Acta bot. Neerl.* **15**: 95, 1966; reassessment), Cooke (*Sydowia Beih.* **4**, 1961), Pilát (*Ann. Myc.* **22**: 204, 1944; **23**: 144, 144, 1925).

Cyphellathelia Jülich (1972) = Pellidiscus (Crepidot.) fide Hjortstam (*Windahlia* **15**: 59, 1986).

Cyphellina Rick (1959) nom. dub. (Thelephor.) fide Donk (1964).

Cyphellocalathus Agerer (1981), Tricholomataceae. 1, Bolivia.

Cyphellomyces Speg. (1906) = Phellorinia (Phellorin.).

Cyphellophora G.A. de Vries (1962), Mitosporic fungi, 1.C2.15. 3 (on human skin), Eur. See Walz & de Hoog (*Ant. v. Leeuwenhoek* **53**: 143, 1987; n.sp. and cf. *Annellodentimyces*).

Cyphellopsis Donk (1931), Crepidotaceae. 2, Eur., Arctic. See Reid (*Kew Bull.* **15**: 265, 1961).

Cyphellopus Fayod (1889), Agaricales (inc. sed.). 1, UK. See Pearson (*TBMS* **20**: 521, 1935).

Cyphellopycnis Tehon & G.L. Stout (1929) = Phomopsis (Mitosp. fungi) fide Sutton (*TBMS* **47**: 497, 1964).

Cyphellostereum D.A. Reid (1965), Podoscyphaceae. 2, widespr. See Reid (1965; key).

Cyphidium Magnus (1875) = Olpidium (Olpid.) fide Sparrow (1960).

Cyphina Sacc. (1884) = Sarcopodium (Mitosp. fungi) fide Sutton (1977).

Cyphospilea Syd. (1926) = Coleroa (Ventur.) fide Müller & v. Arx (1962).

Cyptotrama Singer (1960), Tricholomataceae. 4, S. Am.

Cyrenella Goch. (1981), Mitosporic fungi, 3.A1.15. 1 (with clamp connexions), USA.

Cyrta Bat. & H. Maia (1961) [non Lour. (1790), *Styracaceae*], Mitosporic fungi (L), 8.C1.?. 1, Brazil.

Cyrtidium Vain. (1921), ? Dothideales (inc. sed.). 2, Eur.

Cyrtidula Minks (1876), Dothideales (inc. sed.). 1, Italy.

Cyrtocnon Link ex Rchb. (1828) nom. dub. (? Mitosp. fungi).

Cyrtographa Müll. Arg. (1894) = Minksia (Roccell.) fide Zahlbruckner (1926).

Cyrtopsis Vain. (1921), Dothideales (inc. sed.). 1, Eur.

cyst, (1) an encysted cell (? the product of meiosis after karyogamy), usually aggregated into a cystosorus which germinates to produce a zoospore (*Plasmodiophorales*); (2) an encysted zoospore (planospore) which becomes a gametangium (*Blastocladiales*); **macro-** (of *Myxomycota*), an encysted aggregate of myxoamoebae; the resting form of a young plasmodium; the alternative to the sorocarp in some cellular slime moulds (dictyostelids) (Nickerson & Raper, *Am. J. Bot.* **60**: 190, 1973); propagule, especially the walled structure of the encysted zoospore (*Peronosporales*); resting spores of chrysomonads etc.; **micro-** (of *Myxomycota*), an encysted myxamoeba or swarm spore; **spore-**, a cell, hollow organ or sac-like structure enclosing a mass of protoplasm containing spores as in the *Ascosphaeraceae* (Skou, 1982).

Cystangium Singer & A.H. Sm. (1960), Russulaceae. 3, Australia, S. Am.

cystesium, a cell which differentiates to adhere to a cystidium arising from the opposite hymenium (Horner & Moore, *TBMS* **88**: 488, 1987).

Cystidiella Malan (1943), Mitosporic fungi, 1.A1.1. 1 (with secondary conidia), Italy.

Cystidiodendron Rick (1943) nom. dub. (Hydn.) fide Donk (1964).

Cystidiodontia Hjortstam (1983), Corticiaceae. 1, Venezuela.

cystidiole (of hymenomycetes), a simple hymenial cell of about the same diameter as the basidia but remaining sterile and protruding beyond the hymenial surface.

Cystidiophorus Bondartsev & Ljub. (1963) nom. inval. (Polypor.).

cystidium (pl. **-ia**), a sterile body, frequently of distinctive shape, occurring at any surface of a basidioma, particularly the hymenium from which it frequently projects (Fig. 12). Cystidia have been classified and named according to their: (1) *origin*: **hymenial - (tramal -)**, originating from hymenial (tramal) hyphae; **pseudo-**, derived from a conducting element, filamentous to fusoid, oily contents, embedded not projecting; **coscino-**, see coscinoid; **skeleto-**, the apical part of a skeletal hypha (frequently ± inflated) projecting into or through the hymenium; false seta; **macro-**, arising deep in the trama in Lactario-Russulae; **hypho-**, hypha-like, derived from generative hyphae. (2) *position* (first by Buller, **2**): on the pileus surface (**pileo-, dermato-**, Fayod); at the edge (**cheilo-**), side (**pleuro-**), or within (**endo-**) a lamella; on the stipe (**caulo-**). (3) *form*: **lepto-**, smooth, thin-walled; **lampro-**, thick-walled, with or without encrustation (**setiform lampro-**, awl-shaped, wall pigmented; **asteroseta**, a radially branched lampro-; **microsclerid**, a versiform, endolampro-; **lyo-**, cylindrical to conical, very thick-walled, abruptly thin-walled at apex, not encrusted, colourless, as in *Tubulicrinis* (Donk, 1956); **monilioid gloeo-**, (torulose gloeo- (Bourdot & Galzin, 1928); moniliform paraphysis (Burt, 1918); pseudophysis; **schizo-** (Nikolayeve, 1956, 1961)), monilioid, frequently with a beaded apex (as in *Hericiaceae* and *Corticiaceae*). (4) *contents*: **gloeo-**, thin-walled, usually irregular, contents hyaline or yellowish and highly refractile; **chryso-**, like lepto- but with highly staining contents; **hypo-**, (Larsen & Burdsall, *Mem. N.Y. bot. Gdn* **28**: 123, 1976); **oleo-**, having an oily resinous exudate; **pseudo-**, see (1) above. See also hyphidium, seta. Reviews of cystidia include Romagnesi (*Rev. Mycol., Paris* **9** (suppl.): 4, 1944), Talbot (*Bothalia* **6**: 249, 1954), Lentz (*Bot. Rev.* **20**: 135, 1954), Smith (*in* Ainsworth & Sussman (Eds), *The fungi* **2**: 151, 1966), Price (*Nova Hedw.* **24**: 515, 1975; types in polypores).

Cystingophora Arthur (1907) = Ravenelia (Ravenel.) fide Arthur (1934).

Cystoagaricus Singer (1947), Agaricaceae. 2, subtrop. Am.

Cystobacter J. Schröt. (1886) = Polyangium (Myxobacteriales) fide Buchanan *et al.* (*Index Bergeyana*, 1966).

Cystobasidium (Lagerh.) Neuhoff (1924), Platygloeaceae. 3, Eur., S. Am. See Martin (*Mycol.* **31**: 507, 1939), Olive (*Mycol.* **44**: 564, 1952).

cystochroic, see colour.

Cystochytrium Ivimey Cook (1932), Hyphochytriaceae. 1 (on *Veronica* roots), Eur.

Cystocoleus Thwaites (1849), Mitosporic fungi, 1.-.-. 1 (L), Eur., Asia, N. Am.

Cystocybe Velen. (1921) ? = Cortinarius (Cortinar.) fide Singer (1975).

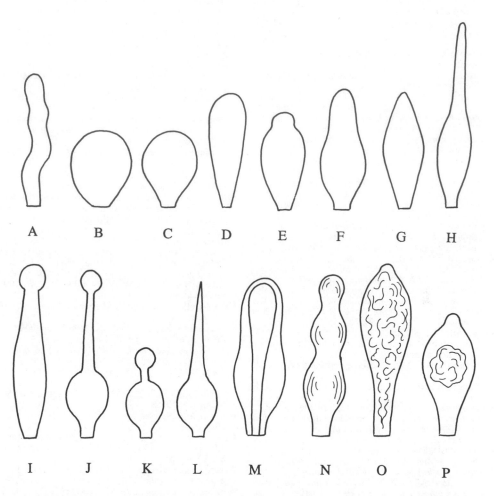

Fig. 12. Cystidia. A, hyphoid (*Collybia*); globose (*Agaricus*); C, pyriform (*Agaricus*); D, clavate (*Inocybe*); E, utriform (*Psathyrella*); F, lageniform (*Pholiota*); G, fusoid (*Psathyrella*); H, lanceolate (*Hypholoma*); I, capitate (*Hyphoderma*); J, tibiiform (*Galerina*); K, lecythiform (*Conocybe*); L, urticoid (*Naucoria*); M, metuloid (*Lentinus*); N, gloeocystidium (*Gloeocystidiellum*); O, macrocystidium (*Russula*); P, chrysocystidium (*Stropharia*). Not to scale.

Cystodendron Bubák (1914), Mitosporic fungi, 3.A1/2.15. 1, Eur.

Cystoderma Fayod (1889), Agaricaceae. 26, widespr. See Heinemann & Thoen (*BSMF* **89**: 5, 1973).

Cystodium Fée (1837) ? Ascomycota (inc. sed.). 1, trop.

Cystofilobasidium Oberw. & Bandoni (1983), Filobasidiaceae. 3 (some marine), widespr. See Oberwinkler *et al.* (*Syst. Appl. Microbiol.* **4**: 114, 1983).

Cystogomphus Singer (1942), Gomphidiaceae. 1, France.

Cystolepiota Singer (1952), Agaricaceae. 10, widespr. See Knudsen (*Bot. Tidsskr.* **73**: 124, 1978; key).

Cystolobis Clem. (1909) = Knightiella (Icmadophil.).

Cystomyces Syd. (1926), Raveneliaceae. 1 (on *Leguminosae*), C. Am.

Cystopage Drechsler (1941), Zoopagaceae. 7, N. Am., UK. See Drechsler (*Mycologia* **33**: 251, 1941, **37**: 1, 1945, **47**: 364, 1955, **49**: 387, 1957, **51**: 787, 1959), Dayal (1973; key).

Cystopezizella Svrček (1983) = Calycina (Leot.) fide Baral (*SA* **13**: 13, 1994).

Cystophora Rabenh. (1844) [non J. Agardh (1841), *Algae*] ≡ Voglinoana (Mitosp. fungi).

Cystopsora E.J. Butler (1910), Uredinales (inc. sed.). 1, India; 0, I, III. See Thirumalachar (*Bot. Gaz.* **107**: 74, 1945). = Zaghouania (Puccin.) fide Cummins & Hiratsuka (1983).

Cystopus Lév. (1847) = Albugo (Albugin.).

Cystosiphon Roze & Cornu (1869), Pythiaceae. 5, widspr. See Dick (1990, *in press*; key).

cystosorus (of *Chytridiales*), a group of united cysts or resting spores.

Cystospora J.E. Elliot (1916) nom. dub. (Plasmodioph.) fide Karling (*Plasmodiophorales*, 1968).

cystospore, (1) an encysted zoospore formed at the exit to the zoosporangium and germinating to produce a new zoospore (planospore) as in *Achlya* (*Oomycetes*), *Achlyogeton* and *Achlyella* (*Chytridiomycetes*); (2) (of

Amoebidiales) spores released from an encysted amoeboid cell.

Cystostereum Pouzar (1959), Corticiaceae. 6, widespr. See Hallenberg & Ryvarden (*Mycotaxon* **2**: 135, 1975).

Cystostiptoporus Dhanda & Ryvarden (1975), Coriolaceae. 1, India.

Cystotelium Syd. (1921) = Haploravenelia (Ravenel.) fide Dietel (1928).

Cystotheca Berk. & M.A. Curtis (1860), Erysiphaceae. 3, Asia, N. Am. See Braun (*Beih. Nova Hedw.* **89**, 1987; key), Katumoto (*Rep. Tottori mycol. Inst.* **10**: 437, 1973; key). Anamorph *Oidium*.

Cystothyrium Speg. (1888) nom. conf. fide Petrak & Sydow (*Ann. Myc.* **33**: 157, 1935).

Cystotricha Berk. & Broome (1850), Mitosporic fungi, 8.B1.1. 1, UK.

Cytidia Quél. (1888), Corticiaceae. 8, widespr., esp. N. Am. *C. salicina*, a common red fungus on *Salix*. See Donk (*Persoonia* **1**: 70, 1959), Cooke (*Mycol.* **43**: 196, 1951; key).

Cytidiella Pouzar (1954) = Auriculariopsis (Merul.) fide Stalpers (*Persoonia* **13**: 495, 1988).

Cytidium Morgan (1896) = Physarum (Physar.).

Cytispora Fr. (1823), Xylariaceae. 1, Eur.

cytochalasin, one of a series of related fungal metabolites (e.g. from *Helminthosporium*, *Metarhizium*, *Phoma*, *Xylariaceae*, *Zygosporium*) which inhibit cytokinesis so that multinucleate cells result (Carter, *Nature* **213**: 261, 1967; Turner (1971: 352); - **B** = phomin.

cytochroic, see colour.

Cytodiplospora Oudem. (1894) = Diplodina (Mitosp. fungi) fide Sutton (1977).

Cytodiscula Petr. (1931), Mitosporic fungi, 8.A1.?. 1, Madeira.

Cytogloeum Petr. (1925), Mitosporic fungi, 6.A1.15. 1, Eur.

Cytology. The study of cell contents, esp. the nucleus, is usually well covered in mycology textbooks. The nuclear membrane may (*Basidiobolus ranarum*) or may not (*Saccharomyces cerevisiae*) break down during mitosis. Pulse field gel electrophoresis (q.v.) has shown that fungal nuclei contain 3 (*Schizosaccharomyces pombe*) to *c.* 21 (*Ustilago hordei*) chromosomes (see Mills & McCluskey, *Molecular plant-microbe interactions* **3**: 351, 1990). Single or double stranded DNA or RNA mycoviruses and double stranded DNA or RNA plasmids may be present in fungal cytoplasm. DNA fingerprinting (Owen, *J. Med. Microbiol.* **30**: 898, 1989) and restriction fragment length polymorphisms (RFLP) (Kusters-van-Someren *et al.*, *Curr. Genet.* **19**: 21, 1991) have been used to type species and strains of fungi. Hyphal tips of extending hyphae contain a very high concentration of macrovesicles (100-250 nm diam.), microvesicles (40-70 nm diam.), and coated vesicles (vesicles surrounded by a basket-like lattice of fibrous protein called clathrin); the microvesicles may correspond to chitosomes (Bracker *et al.*, *Proc. Natl. Acad. Sci. USA* **73**: 4570, 1976); the concentration of vesicles at the hyphal tip can sometimes be observed as a phase dark body known as the Spitzenkörper. The fungal cytoskeleton contains microfilaments (4-9 nm diam.) formed by the polymerization of G-actin, microtubules (30-40 nm diam.) formed by the polymerization of α- and β-tubulin, and filasomes (aggregates of microfilaments coated with vesicles). The cytoplasm contains mitochondria, microbodies, hydrogenosomes (in anaerobic fungi), Golgi bodies (in *Chromista*), smooth and rough endoplasmic reticulum, 80S ribosomes, glycogen and lipid globules. Spherical Worinin bodies (e.g. in *Fusarium* spp.) or hexagonal crystals (*Neurospora crassa*) are associated with the septal pores of *Ascomycota*. See also Chromosome numbers.

cytolysis, breaking up or solution of the cell wall.

Cytomelanconis Naumov (1951), ? Melanconidaceae. 1, former USSR.

Cytonaema Höhn. (1914), Mitosporic fungi, 8.A1.15. 1, Austria. = Cytospora (Mitosp. fungi) fide Défago (*Phytopath. Z.* **14**: 103, 1944).

Cytophoma Höhn. (1914) = Cytospora (Mitosp. fungi.) fide Défago (*Phytopath. Z.* **14**: 103, 1944).

Cytophyllopsis R. Heim (1958) = Weraroa (Galeropsid.) fide Singer & Smith (*Bull. Torrey Bot. Cl.* **85**: 324, 1958).

Cytoplacosphaeria Petr. (1920), Mitosporic fungi, 8.C1.15. 1, Eur.

Cytoplea Bizz. & Sacc. (1885), Mitosporic fungi, 8.A2.15. 3, widespr. See Sutton (1980).

cytoskeleton, intracellular network of protein filaments that is insoluble in non-ionic detergents.

Cytosphaera Died. (1916), Mitosporic fungi, 8.A1.1. 1, Asia, Australia.

Cytospora Ehrenb. (1818), Anamorphic Valsaceae, 8.A1.15. Teleomorph *Valsa*. 100, cosmop. See Défago (*Phytopath. Z.* **14**: 103, 1944), Gvritishvili (*Fungi of the genus Cytospora Fr. in the USSR*, 1982), Gille (*Arch. Phytopath. Pflanzensch.* **26**: 237, 1990; spp. on *Prunus*).

Cytosporella Sacc. (1884), Mitosporic fungi, 8.A1.15. 30, temp.

Cytosporina Sacc. (1884) ≡ Dumortieria (Mitosp. fungi) fide Sutton (1977).

Cytosporites Babajan & Tasl. (1970), Fossil fungi. 1 (Tertiary), former USSR.

Cytosporium Sacc. (1884) ≡ Cellulosporium (Mitosp. fungi).

Cytosporopsis Höhn. (1918) = Cytospora (Mitosp. fungi) fide Sutton (1977).

Cytostaganis Clem. & Shear (1931) ≡ Cytostagonospora (Mitosp. fungi).

Cytostagonospora Bubák (1916), Mitosporic fungi, 4.E1.15. 2, Australia, Eur. See Sutton & Swart (*TBMS* **87**: 99, 1986; n. comb.).

Cytotriplospora Bayl. Ell. & Chance (1921) = Strasseria (Mitosp. fungi) fide Sutton (1977). See also Parmelee & Cauchon (*CJB* **57**: 1660, 1979).

Cyttaria Berk. (1842), Cyttariaceae. 10 (on *Nothofagus*), S. temp. Some edible. See Gamundí (*Darwiniana* **16**: 461, 1971; 7 spp.), Gamundí & de Lederkremer (*Ciencia Investig.* **43**: 4, 1989; evolution), Crisci *et al.* (*Cladistics* **4**: 279, 1988), Humphries *et al.* (in Stone & Hawksworth (Eds), *Coevolution and systematics*: 55, 1986; coevolution), Mengoni (*Boln Soc. Argent. Bot.* **24**: 393, 1986; asci), Minter *et al.* (*Mycologist* **1**: 7, 1987; col. illustr.), Gamundí (*Fl. cript. Tierra de Fuego* **10**(4), 1987; key 4 spp.), White (*TBMS* **37**: 431, 1954; ontogeny), Kobayasi (*Trans. mycol. Soc. Japan* **7**: 118, 1966, key; **19**: 473, 1978, history of use). Anamorphs, see Mengoni (*Boln Soc. Argent. Bot.* **26**: 7, 1989).

Cyttariaceae Speg. (1887), Cyttariales. 1 gen., 10 spp. Ascomata apothecial, formed within conspicuous brightly coloured fleshy compound stromata, ± spherical but with a wide opening; interascal tissue of numerous filiform paraphyses; asci developing synchronously or sequentially, cylindrical, with active discharge, the apex often with an I+ apical ring, opening irregularly; ascospores ± ellipsoidal, aseptate, pale grey, thin-walled, without a sheath. Anamorph pycnidial. Biotrophic on wood of *Nothofagus* spp., often gall-forming.

Lit.: see under *Cyttaria*.

Cyttariales, Ascomycota. 1 fam., 1 gen., 10 spp. Fam.: **Cyttariaceae**.
Lit.: Gamundí (*SA* **10**: 69, 1991; review), also see under *Cyttaria*. Close to *Leotiales* but usually kept separate.
Cyttariella Palm (1932), Anamorphic Cyttariaceae. ?.?.?. 1, S. Am. See Santesson (1945).
Cyttarophyllopsis R. Heim (1968), Galeropsidaceae. 1, India.
Cyttarophyllum (R. Heim) Singer (1936) = Galeropsis (Galeropsid.) fide Heim (*Rev. Mycol.* **15**: 3, 1950).
CZ, see Media.

Dabaryomyces Klöcker (1909) , see *Debaryomyces*.
Dacampia A. Massal. (1853), Dacampiaceae. 3 (on lichens), Eur. See Crivelli (*Ueber die heterogene Ascomycetengattungen Pleospora*, 1983), Eriksson (1981).
Dacampiaceae Körb. (1855), Dothideales. 13 gen. (+ 5 syn.), 51 spp. Ascomata globose to obpyriform, black, thick-walled, sometimes with hyphal appendages, with a well-developed periphysate ostiole; peridium thick, composed of three distinct layers; interascal tissue poorly to well-developed, composed of narrow cellular pseudoparaphyses, sometimes immersed in gel; asci ± cylindrical, fissitunicate, without clear apical structures; ascospores hyaline or brown, usually septate, sometimes muriform, sometimes with a gelatinous sheath. Anamorphs coelomycetous. Saprobic, necrotrophic or biotrophic, many lichenicolous.
Lit.: Barr (*Prodromus to class Loculoascomycetes*, 1987), Hawksworth & Diederich (*TBMS* **90**: 293, 1988).
Dacampiosphaeria D. Hawksw. (1980) = Pyrenidium (Dacamp.) fide Hawksworth (1983).
Dacrina Fr. (1825) nom. dub. (? Hydnaceae).
Dacrina Fr. (1832) = Strumella (Mitosp. fungi) fide Saccardo (*Syll. Fung.* **4**: 742, 1886).
Dacrydium Link (1809) nom. conf. (Mitosp. fungi) fide Sutton (*Mycol. Pap.* **144**, 1977).
Dacrymycella Bizz. (1885), Mitosporic fungi, 3.A1.13. 2, Eur., Java.
Dacrymyces Nees (1816), Dacrymycetaceae. 30, widespr. See McNabb (*N.Z. Jl Bot.* **11**: 461, 1973; key), Reid (*TBMS* **62**: 449, 1974; key Br. spp.), Mathiesen (*Svampe* **23**: 46, 1991; Danish spp.).
Dacrymycetaceae J. Schröt. (1888), Dactymycetales. 8 gen. (+ 8 syn), 63 spp. Basidioma pustulate to clavarioid; gelatinous.
Dacrymycetales, Basidiomycetes. 2 fam., 11 gen. (+ 8 syn.), 72 spp. Basidiomata gelatinous, with orange pigments, smooth or wrinkled; the furcate meta-basidium having 2 stout sterigmata each bearing a spore is characteristic; saprobes of dead wood. Fams:
(1) **Cerinomycetaceae**.
(2) **Dacrymycetaceae**.
Lit.: Donk (1951-63, VIII; *Proc. Kon. nederl. Akad. Wet.* C **67** (2), 1964; 1966: 264), Kennedy (*Mycol.* **50**: 874, 1959; gen. key), Reid (*TBMS* **62**: 433, 1974; Br. spp.; keys).
Dacryobasidium Jülich (1982) = Cristinia (Lindtner.) fide Hjortstam & Grosse-Brauckmann (*Mycotaxon* **47**: 405, 1993).
Dacryobolus Fr. (1849), Meruliaceae. 4, widespr. See Parmasto (1968: 98, taxonomy), Eriksson (*Symb. bot. upsal.* **16**(1): 115, 1958), Christiansen (*Dansk bot. Ark.* **19**: 244, 1960), Manjon *et al.* (*Anal. Jard. Bot. Madrid* **40**: 297, 1984; key Eur. spp.).
Dacryodochium P. Karst. (1896) = Graphiola (Graphiol.) fide Sutton (*in litt.*).
dacryoid, having one end rounded and the other more or less pointed; pear- or tear-like in form.

Dacryoma Samuels (1988), Anamorphic Hypocreaceae, 8.A/B1.15. Teleomorph *Nectria*. 1, Indonesia.
Dacryomitra Tul. & C. Tul. (1872) = Calocera (Dacrymycet.) fide McNabb (1965).
Dacryomyces, see *Dacrymyces*.
Dacryomycetopsis Rick (1958), Dacrymycetales (inc. sed.). 1, Brazil.
Dacryonaema Nannf. (1947), Dacrymycetaceae. 1, Eur.
Dacryopinax G.W. Martin (1948), Dacrymycetaceae. 10, esp. trop. McNabb (*N.Z. Jl Bot.* **3**: 59, 1965; key), Lowy (*Mycotaxon* **13**: 428, 1981; key trop. Am. spp.).
Dacryopsella Höhn. (1915) = Pistillina (Typhul.) fide Donk (1954).
Dacryopsis Massee (1891) = Ditiola (Dacrymycet.) fide Kennedy (*Mycol.* **56**: 298, 1964).
Dactuliochaeta G.L. Hartm. & J.B. Sinclair (1988), Mitosporic fungi, 4.A1.15. 1 (on *Glycine*), Afr. For *Pyrenochaeta* with *Dactuliophora* synanamorph.
Dactuliophora C.L. Leakey (1964), Mitosporic fungi, 3.D2.42. 4, mainly Afr. See Datnoff *et al.* (1988; as *Pyrenochaeta*).
Dactylaria Sacc. (1880), Mitosporic fungi, 1.B/C1.10. 55, widespr. See Bhatt & Kendrick (*CJB* **46**: 1253, 1968; type sp.), de Hoog & van Oorschot (*Proc. K. Ned. Akad. Wet.* C **86**: 55, 1983), de Hoog (*Stud. Mycol.* **26**: 1, 1985, key 41 spp.), de Hoog & v. Arx (*Kavaka* **1**: 55, 1973), Dixon & Salkin (*Jl Clin. Microbiol.* **24**: 12, 1986; medical spp.), Sawadogo & Cayrol (*Riv. Scient.*: 27, 1990), Castañeda Ruiz & Kendrick (*Univ. Waterloo Biol. Ser.* **35**: 1, 1991; Cuba).
Dactylariopsis Mekht. (1967) = Arthrobotrys (Mitosp. fungi) fide Schenck *et al.* (*CJB* **55**: 977, 1977).
Dactylella Grove (1884), Mitosporic fungi, 1.C1.10. 15 (mainly on nematodes), N. temp. See Subramanian (*J. Ind. bot. Soc.* **42**: 291, 1963), Zhang *et al.* (*Mycosystema* **7**: 111, 1995; review).
Dactylellina M. Morelet (1968), Mitosporic fungi, 1.C2.10. 1, USA.
Dactylifera Alcorn (1987), Mitosporic fungi, 1.G2.26. 1, Australia.
Dactylina Nyl. (1860), Parmeliaceae (L). c. 10, Eur., N. & S. Am., Asia. See Lynge (*Skr. Svalbard Ishavet* **59**, 1933), Follmann *et al.* (*Willdenowia* **5**: 7, 1968), Thomson & Bird (*CJB* **56**: 1602, 1978; N. Am. spp., key).
Dactylina G. Arnaud ex Subram. (1964) ≡ Lactydina (Mitosp. fungi).
Dactylium Nees (1816), Mitosporic fungi, 1.C1.10. 15, esp. temp. See Rifai (*Reinwardtia* **7**: 357, 1968), Carmichael *et al.* (1980; status).
Dactyloblastus Trevis. (1853) = Phlyctis (Phlyctid.).
dactyloid, finger-like.
Dactylomyces Sopp (1912), Trichocomaceae. 2 (thermophilic), Eur., N. Am. See Apinis (*TBMS* **50**: 576, 1967). = Thermoascus (Trichocom.) fide v. Arx (1981).
Dactyloporus Herzer (1893), Fossil fungi (Basidiomycetes). 1 (Carboniferous), N. Am., but see Seward (*Fossil plants* **1**: 211, 1898).
Dactylospora Körb. (1855), Dactylosporaceae. c. 30 (on lichens, hepatics or wood), widespr. See Hafellner (*Beih. Nova Hedw.* **62**, 1979), Bellemère & Hafellner (*Cryptogamie, mycol.* **3**: 71, 1982; asci), Döbbeler & Triebel (1985; on *Hepaticae*).
Dactylosporaceae Bellem. & Hafellner (1982), Ascomycota (inc. sed.). 2 gen. (+ 5 syn.), 31 spp. Stromata absent; ascomata apothecial, ± superficial, sometimes shortly stipitate, black, the wall composed of dark brown pseudoparenchymatous tissue, sometimes poorly developed; interascal tissue of sparsely branched narrow paraphyses, the apices swollen and pigmented; asci cylindric-clavate, persistent, thick-

walled esp. at the apex, not fissitunicate, with an outer I+ gelatinized layer, without a well-developed apical cap or ocular chamber; ascospores brown, septate, occ. ornamented. Anamorphs not known. Saprobic on wood, on liverworts, or commensalistic on lichens, widespr.
Lit.: Döbbeler & Triebel (*Bot. Jahrb. Syst.* **107**: 503, 1985).

Dactylosporangium Thiemann, Pagani & Beretta (1967), *Actinomycetes*, q.v.

Dactylosporina (Clémençon) Dörfelt (1985), Tricholomataceae. 2, S. Am.

Dactylosporites Paradkar (1976), Fossil fungi. 1 (Cretaceous), India.

Dactylosporium Harz (1872), Mitosporic fungi, 1.D2.10. 2, Eur., Cuba. See Hughes (*Naturalist*, Hull 1952: 63, 1952).

Dactylosporium Mekht. (1967) ≡ Dactylina (Mitosp. fungi).

dadih (**dadiah**), a fermented product of buffalo milk popular in Western Sumatra. See Gandjar *et al.* (*IMC3 Abstr.*: 452, 1983; microbiology).

Daedala Hazsl. (1887) = Hypodermina (Mitosp. fungi) fide Darker (1967).

Daedalea Pers. (1801), Coriolaceae. 6, cosmop. *D. quercina* (wood rot). See Corner (1987).

Daedaleites Mesch. (1892), Fossil fungi. 2 (Tertiary), Eur.

Daedaleopsis J. Schröt. (1888), Coriolaceae. 6, widespr. See Pegler (1971). = Phellinus (Hymenochaet.) fide Donk (1974).

Daedaloides Lázaro Ibiza (1916) = Cryptoderma (Hymenochaet.) fide Pegler (1971).

Daldinia Ces. & De Not. (1863) nom. cons., Xylariaceae. 13, cosmop. *D. concentrica* ('calico wood', 'cramp balls' on *Fraxinus*). See Child (*Ann. Mo. bot. Gdn* **19**: 429, 1932), Pérez-Silva (*Boln Soc. Mex. Micol.* **7**: 51, 1973; Mexican spp.), Petrini & Müller (*Mycol. Helv.* **1**: 501, 1986; key 5 spp. Eur.), Thind & Dargan (*Kavaka* **6**: 15, 1979; Indian spp.). = Hypoxylon (Xylar.) fide Læssøe (*SA* **13**: 43, 1994).

Daleomyces Setch. (1924) = Peziza (Peziz.) fide Korf (*Mycol.* **48**: 711, 1956).

Daloala Tuzet, Manier & Vog.-Zuber (1952) = Enterobryus (Eccrin.) fide Manier & Lichtwardt (*Ann. Sci. nat., Bot.* sér. 12 **9**: 519, 1968).

Damnosporium Corda (1842) = Bactridium (Mitosp. fungi) fide Saccardo (*Syll. Fung.* **4**: 691, 1886).

damping-off, a rotting of seedlings at soil level. In **pre-emergent** - the young plant is attacked at germination so that the seedling does not come up. Species of *Fusarium, Phytophthora, Pythium* and *Rhizoctonia* are common - **fungi.** Soil sterilization is frequently the best control measure (see Cheshunt compound).

Danaëa Caneva & Rambelli (1981) [non Sm. (1793), *Pteridophyta*] ≡ Kiliophora (Mitosp. fungi).

Dangeard (Pierre Clement Augustin; 1862-1947). A distinguished French mycologist, protozoologist, algologist, cytologist, and morphologist. He provided data for the classification of many fungi, esp. those with algal or protozoan affinities. *Obit.*: Moreau (*Rev. gén. Bot.* **57**: 193, 1950; portr.), Grummann (1974: 313), Stafleu & Cowan (*TL-2* **1**: 596, 1976).

Dangeardia Schröd. (1898), Chytridiaceae. 4, Eur., N. Am. See Batko (*Acta mycol.* **12**: 407, 1970).

Dangeardiana Valkanov ex Batko (1970), Chytridiaceae. 4 (on *Eudorina*), Bulgaria.

Dangeardiella Sacc. & P. Syd. (1899), Dothideales (inc. sed.). 2 (on *Pteridophyta*), Eur. See Obrist (*Phytopath. Z.* **35**: 379, 1959).

dangeardien (**dangeardium**), collective term for both asci and basidia; structures where diploid nuclei are formed, undergo meiosis and form haploid spores

(Moreau, *Botaniste* **34**: 315, 1949). See van der Walt & Johanssen (*Ant. v. Leeuwenhoek* **40**: 185, 1974; concept in yeast taxonomy).

Dangeardiomycetes, *Basidiomycetes, Protobasidiomycetes* and *Ascomycetes* excluding *Periascomycetes* (Moreau, 1953).

DAOM. Canadian National Mycological Herbarium (Ottawa, Canada); founded 1929; part of the Centre for Land and Biological Resources Research, Agriculture Canada; genetic resource collection **CCFC**; see Cody *et al.* (*Systematics in Agriculture Canada at Ottawa 1886-1986*, 1986).

DAP pathway, alpha, beta-diaminopimelic acid pathway for lysine synthesis (cf. AAA pathway).

Dapsilosporium Corda (1837) = Melanconium (Mitosp. fungi) fide Saccardo (1884). See also Sutton (1977).

DAR. Plant Pathology Branch Herbarium New South Wales (Rydalmere, NSW, Australia); founded 1890; collections vested in a Trust established by the NSW Parliament from 1983 and supported by the goverment.

Darbishirella Zahlbr. (1898) = Ingaderia (Roccell.) fide Feige & Lumbsch (*Mycotaxon* **48**: 381, 1993).

Darkera H.S. Whitney, J. Reid & Piroz. (1975), Rhytismatales (inc. sed.). 2, Eur., N. Am. Anamorph *Tiarosporella*.

Darluca Castagne (1851) = Sphaerellopsis (Mitosp. fungi) fide Sutton (1977).

Darlucella Höhn. (1919) = Sphaerellopsis (Mitosp. fungi) fide Sutton (1977).

Darlucis Clem. (1931) = Heteropatella (Mitosp. fungi) fide Sutton (1977).

Daruvedia Dennis (1988), Dothideales (inc. sed.). 1, U.K. See Barr (*Mycotaxon* **51**: 191, 1994; posn).

Darwiniella Speg. (1887) Fungi (inc. sed.). See Petrak & Sydow (*Ann. Myc.* **34**: 22, 1936).

Dasturella Mundk. & Khesw. (1943), Phakopsoraceae. 3, India. See Thirumalachar *et al.* (*Bot. Gaz.* **108**: 371, 1947).

Dasybolus Clem. & Shear (1931) ≡ Dasyobolus (Ascobol.).

Dasyobolus (Sacc.) Sacc. (1895) = Ascobolus (Ascobol.) fide v. Brummelen (1967).

Dasypezis Clem. (1909) = Lachnum (Hyaloscyph.) fide Nannfeldt (1932).

Dasyphthora Clem. (1909) = Nectria (Hypocr.).

Dasypyrena Speg. (1912) = Actinopeltis (Microthyr.) fide v. Arx & Müller (1975).

Dasyscypha Fuckel (1870) = Dasyscyphus (Hyaloscyph.) fide Holm (*TBMS* **67**: 333, 1976) but see Korf (*Mycotaxon* **5**: 515, 1977) and Holm (*Mycotaxon* **7**: 139, 1978).

Dasyscyphella Tranzschel (1898), Hyaloscyphaceae. 7, Asia, N. Am. See Baral (*Z. Mykol.* **59**: 3, 1993), Raitviir (*Eesti NSV Tead. Akad. Toim., Biol.* **26**: 33, 1977), Dennis (*Kew Bull.* **27**: 273, 1972).

Dasyscyphus Gray (1821) = Lachnum (Hyaloscyph.) fide Haines & Dumont (*Mycotaxon* **19**: 1, 1984). See Dennis (*Mycol. Pap.* **32**, 1949, Br. spp.; *Kew Bull.* **27**: 273, 1972, typification), Haines (*Mycotaxon* **11**: 189, 1980; trop. ferns), Luijt-Verheij (*Overzicht van Nederlande soorten van Dasyscyphus*, 1973), Singh (*Kavaka* **3**: 107, 1976; Indian spp.).

Dasysphaeria Speg. (1912), Ascomycota (inc. sed.). 1, S. Am.

Dasyspora Berk. & M.A. Curtis (1853), Uropyxidaceae. 1 (on *Annonaceae*), trop. Am. See Hennen & Figueredo (*Mycol.* **73**:350, 1981).

Dasysticta Speg. (1912), Mitosporic fungi, 4.A1.15. 1, S. Am.

Dasystictella Höhn. (1919) nom. dub. fide Sutton (*Mycol. Pap.* **141**, 1977).

Datronia Donk (1967), Coriolaceae. 2, Eur., USA. = Antrodia (Coriol.) sensu Murrill.

datum (pl. **data**), facts, figures, information and observations. Often used adjectivally, e.g. data bank, data matrix, database.

Davincia Penz. & Sacc. (1901) = Crocicreas (Leot.) fide Carpenter (*Mem. N.Y. bot. Gdn* **33**: 1, 1981).

Davinciella (Sacc. & D. Sacc.) Trotter (1928) ≡ Merodontis (Leotiales).

Davisiella Petr. (1924), Mitosporic fungi, 4.A1/D1.15. 1, USA.

Davisomycella Darker (1967), Rhytismataceae. 9, N. Am., Japan, NZ, Thailand.

Davisoniella H.J. Swart (1988), Mitosporic fungi, 8.A2.19. 1, Australia.

Dawsicola Döbbeler (1981), Leotiales (inc. sed.). 1 (on *Musci*), NZ.

Dawsomyces Döbbeler (1981), Dothideales (inc. sed.). 2 (on *Musci*), New Guinea, Australia.

Dawsophila Döbbeler (1981), Dothideales (inc. sed.). 2, New Guinea, NZ.

de Bary (Heinrich Anton; 1831-88). After a medical training and two years as a surgeon in Frankfurt-am-Main, the town of his birth, took up botany and in 1855 became a Professor at Freiburg; later he went to Halle, and from 1872 till his death was at Strasburg University. Though de Bary made important additions to general botany, it is his writings on mycology, such as *Die Brandpilze* (1853), *Die Mycetozoen* (1859), and the *Morphologie und Physiologie der Pilze* (1866; 2 edn, 1884; English edn, 1887), that take first place. His mycological interests were more biological and physiological than systematic; from his investigations on life histories, parasites and saprobes, myxomycetes (see Martin, *Proc. Iowa Acad. Sci.* **65**: 20, 1958), the nature of lichens, etc., he made new and important discoveries. Heteroecism in the *Uredinales* was made clear by his experiments and he gave accounts of development and of sex in a number of phycomycetes and ascomycetes. His teaching (numbers of his students became noted) and writings have had a very great effect on the later development of mycology, and it is true to say that he was 'the founder of modern mycology'. See Reess (*Ber. dtsch. bot. Ges.* **6**: viii, 1888), Smith (*Phytopath.* **1**: 2, 1911), Sparrow (*Mycol.* **70**: 222, 1978), Stafleu & Cowan (*TL-2* **1**: 135, 1976).

de Bary bubbles, air bubbles in ascospores; first described by de Bary (1884: 106). See Dodge (*Bull. Torrey bot. Cl.* **84**: 431, 1957).

de Notaris (Giuseppe; 1805-77). Trained as a medical man; Professor of Botany, University of Genoa, 1839-72; then to Rome. Chief writings: *Micromycetes italici novi vel minus cogniti*, 1839-55; *Sferiacei italici*, 1863; and other papers, esp. on lichenized and other and ascomycetes. Main collections in Rome (**RO**). See Graniti (*Acad. naz. Sci.* **40**, 1990; figures, illustr.; *Opinione, Rome* **34**, 1877), Stafleu & Cowan (*TL-2* **1**: 622, 1976).

Dearness (John; 1852-1954). Self-taught amateur mycologist, school inspector by profession, the first Canadian mycologist to receive international recognition. Collections in **DAOM** (Parmelee, *Mycol.* **70**: 509, 1978). *Obit.*: Tamblyn (*Mycol.* **47**: 909, portr., bibl., 1955), Ginns (*Mycotaxon* **26**: 47, 1986), Stafleu & Cowan, *TL-2* **1**: 605, 1976).

Dearnessia Bubák (1916), Mitosporic fungi, 4.C1.1. 1, Canada.

death cap, basidiomata of the highly poisonous *Amanita phalloides*.

Debarya Schulzer (1866) = Hypocrea (Hypocr.) fide Mussat (*Syll. Fung.* **19**: 552, 1910).

Debaryella Höhn. (1904), ? Trichosphaeriaceae. 2, Eur. See Barr (*Mycol. Mem.* **7**, 1978).

Debaryolipomyces Ramírez (1957) nom. nud. (Saccharomycet.) fide Batra (1978).

Debaryomyces Klöcker (1909) = Torulaspora (Saccharomycet.) fide van der Walt & Johanssen (1975).

Debaryomyces Lodder & Kreger ex Kreger (1984) nom. cons., Saccharomycetaceae. 10, widespr. See Barnett *et al.* (1990), Kreger van Rij (1984), Hendriks *et al.* (*Syst. Appl. Microbiol.* **15**: 98, 1992; posn), Yamada *et al.* (*J. gen. appl. Microbiol.* **38**; 623, 1992; gen. concept).

Debaryoscyphus Arendh. & R. Sharma (1986) = Hamatocanthoscypha (Hyaloscyph.) fide Huhtinen (1990).

Debaryozyma Van der Walt & Johannsen (1978) ≡ Debaryomyces (Saccharomycet.).

Decaisnella Fabre (1879), Massariaceae. 11, Eur., N. Am. See Barr (*Sydowia* **38**: 11, 1986, *SA* **5**: 127, 1986; status, *N. Am. Fl.* ser. 2 **13**, 1990; key, posn).

Decampia Mudd (1861) ≡ Dacampia (Dacamp.).

decay, the destruction of plant or animal material by fungi and other microorganisms.

Deccanodia Singhai (1974), Fossil fungi (Mitosp. fungi). 1 (Eocene), India.

deciduous (of spores, etc.), falling away at maturity; shed, either with (e.g. teliospores) or without (e.g. urediniospores) a fragment of the pedicel or sporophore; cf. persistent.

Deckenbachia Jacz. (1931) ≡ Coenomyces (Cladochyr.).

declinate, bent or curved down or forwards.

declivate (**declivous**), sloping.

decolourate, colourless.

Decomposition. In the strict sense, is the breakdown of organic materials through biological activity, although the physical processes of leaching and fragmentation are sometimes considered a part of decomposition. The products resulting from biological decomposition by fungi are energy for the decomposer, inorganic elements and compounds, and simple organic compounds such as CO_2 or alcohol which result from aerobic and anaeraobic respiration (fermentation), respectively. The release of mineral nutrients through decomposition is known as nutrient mineralization. When fungi decompose organic matter such as wood or straw in which the ratio of carbon to inorganic nutrients (especially nitrogen and phosphorus) is high, the nutrient mineralization phase is often preceded by a nutrient immobilization phase. During nutrient immobilization, the decomposers incorporate mineral nutrients from their organic substrate and sometimes from the surrounding environment into their biomass. See Carroll & Wicklow (Eds) (*The fungal community*, edn 2, 1992), Dix & Webster (*Fungal ecology*, 1995).

Deconica (W.G. Sm.) P. Karst. (1879) = Psilocybe (Strophar.) fide Singer (1951).

decorticate, having no cortex.

decumbent, resting on the substratum with the ends turned up.

decurrent (of lamellae), running down the stipe (Fig. 19E).

decurved (of the pileus edge), bent down.

decussate (of lichen thalli), having the surface divided and crossed by dark lines.

Dedalea Raf. (1815) = Daedalea (Coriol.).

dediploidization, see diplo-.

dediploidization, (in ascomycetes and basidiomycetes), the making of haploid cells (or hyphae) by a dikaryotic diploid mycelium or cell.

deer balls, (**Lycoperdon nuts** or **harts' truffles**), *Elaphomyces* ascomata.

Deflexula Corner (1950), Pterulaceae. 5, esp. trop.

Degelia Arv. & D.J. Galloway (1981), Pannariaceae (L). 9, temp. See Jørgensen & James (*Bibl. Lich.* **38**: 253, 1990; key).

dehiscence papilla, morphologically and ultrastructurally distinct protuberance on an undischarged sporangium that becomes converted into an exit tube.

dehiscent (**dehiscing**) (of asci or fruit-bodies), opening when mature, by pores or by becoming broken into parts.

Deichmannia Alstrup & D. Hawksw. (1990), Mitosporic fungi, 3.D2.1. 1 (on lichens), Greenland.

Deightonia Petr. (1947) = Vanderystiella (Mitosp. fungi) fide Petrak (*Sydowia* **5**: 328, 1951).

Deightoniella S. Hughes (1952), Mitosporic fungi, 1.B/C2.19. 8, widespr., esp. trop. *D. torulosa* (fruit spot ('speckle') of banana. See Meredith (*TBMS* **44**: 95, 265, 391, 487, 1961), Ellis (*DH*), Constantinescu (*Proc. K. Akad. Wet.* C **86**: 137, 1983), Ondřej (*Česká Myk.* **38**: 39, 1984; Czech. spp.).

Deigloria Agerer (1980), Physalacriaceae. 4, Colombia.

Dekkera Van der Walt (1964), Saccharomycetaceae. 5, widespr. See Molina *et al.*. (*Int. J. Syst. Bact.* **43**: 32, 1993; molec., key), Smith *et al.* (*Yeast* **6**: 299, 1990), Yamada *et al.* (*Biosci. Biotech. Biochem.* **58**: 1893, 1994; molec. syst.). Anamorph *Brettanomyces*.

Dekkeromyces Wick. & K.A. Burton (1956) nom. inval. = Kluyveromyces (Saccharomycet.) fide Batra (1978).

Dekkeromyces Santa María & C. Sánchez-Pinto (1970) = Kluyveromyces (Saccharomycet.) fide Batra (1978).

Delacourea Fabre (1879) = Lophiostoma (Lophiostomat.) fide Eriksson & Hawksworth (*SA* **6**: 123, 1987).

Delacroixia Sacc. & P. Syd. (1899) = Conidiobolus (Ancylist.) fide Tyrrell & McLeod (*J. invert. Path.* **20**: 11, 1972).

Delastreopsis Mattir. (1905) ≡ Lespiaultinia (Tuber.) fide Trappe (1975).

Delastria Tul. & C. Tul. (1843), ? Terfeziaceae. 1 (hypogeous), S. Eur., N. Afr. See Fischer (1938).

Delentaria Corner (1970), Ramariaceae. 1, Brazil.

Delicatula Fayod (1889), Tricholomataceae. 2, temp.

deliquescent, becoming liquid, e.g. after maturing.

Delisea Fée (1825) [non J.V. Lamour. (1819), *Algae*] ≡ Plectocarpon (Roccell.).

Delitescor Earle (1909) = Psilocybe (Strophar.) fide Singer (1951).

Delitschia Auersw. (1866), Sporormiaceae. 46, widespr. See Luck-Allen & Cain (*CJB* **53**: 1827, 1975; key), Barr (*N. Am. Fl.*, ser. 2, **13**, 1990; posn), Barrasa & Checa (*Rev. Iberoam. Micol.* **7**: 5, 1990; key 8 spp.), Eriksson & Hawksworth (*SA* **10**: 138, 1991), Parguey-Leduc (*BSMF* **94**: 409, 1978; ontogeny).

Delitschiella Sacc. (1905) = Delitschia (Spororm.) fide Müller & v. Arx (1962).

Delortia Pat. & Gaillard (1888), Mitosporic fungi, 3.F1.1. 1, trop. Am., Afr. See Pirozynski (*Mycol. Pap.* **129**, 1972).

Delphinella (Sacc.) Kuntze (1898), Dothioraceae. 6, widespr., esp. N. temp. See Barr (*Contr. Univ. Mich. Herb.* **9**: 523, 1972), Froidevaux (*Nova Hedw.* **23**: 679, 1973).

Delpinoella Sacc. (1899), Ascomycota (inc. sed.). 1, Afr.

Delpinoina Kuntze (1891), Ascodichaenaceae. 2, Eur. See Speer (*BSMF* **103**: 9, 1987). Anamorph *Macroallantina*.

Delpontia Penz. & Sacc. (1901), ? Stictidaceae. 1 (on *Pteridophyta*), Java. See Sherwood (*Mycotaxon* **5**: 1, 1977).

deltoid, triangular in shape.

Deltosperma W.Y. Zhuang (1988), Anamorphic Leotiaceae, 8.A2.15. Teleomorph *Unguiculariopsis*. 3, N. Am., Eur.

Dematiaceae. Mitosporic fungi (hyphomycetes) having dark-coloured hyphae and/or conidia (obsol.).

dematiaceous (of mycelium, spores, etc.), pigmented, more or less darkly. Cf. moniliaceous.

Dematioscypha Svrček (1977), Hyaloscyphaceae. 3, Eur., NZ. See Huhtinen (*Mycotaxon* **30**: 9, 1987; 3 spp.), Spooner (*Bibl. Mycol.* **116**, 1987). Anamorph *Haplographium*.

Dematium Pers. (1801) nom. conf. fide Hughes (*CJB* **36**: 727, 1958).

Dematoidium Stautz (1931) ? = Aureobasidium (Mitosp. fungi) fide Hermanides-Nijhof (*Stud. Mycol.* **15**: 143, 1977).

Dematophora R. Hartig (1883), Anamorphic Xylariaceae, 2.A2.11. Teleomorph *Rosellinia*. 1, widespr. *D. necatrix* (white root rot of apple and pear). See Watanabe (*Ann. Phytopath. Soc. Jap.* **58**: 65, 1992; sporulation in vitro).

-deme (suffix), a neutral term, always used with a prefix, and denoting any group of individuals within a taxon (q.v.; usually a species); first proposed by Gilmour & Gregor (*Nature* **144**: 333, 1939); occasionally used in mycology, e.g. agamodeme (predominantly apomictic), photosymbiodeme (of lichen thalli with different photobionts).

demicyclic, see *Uredinales*.

Demordium Link (1809), Myxomycetes (inc. sed.). 1, Eur.

Dencoeliopsis Korf (1971), Leotiaceae. 1, Eur., USA. = Rutstroemia (Leot.) fide Holm & Holm (*Symb. bot. upsal.* **21**(3): 6, 1977).

Dendrina Fr. (1825) nom. dub. fide Lindau (*Rabenh. Krypt. Fl.* **1** (8): 203, 1907).

Dendriscocaulon Nyl. (1888), Lobariaceae (L). 10, widespr. See James & Henssen (*in* Brown *et al.* (Eds), *Lichenology: progress and problems*: 27, 1976).

dendritic, irregularly branched; tree-like; dendroid.

Dendrochaete G. Cunn. (1965) = Echinochaete (Polypor.) fide Reid (1963).

dendrochin, an antifungal antibiotic from *Dendrodochium toxicum* toxic to farm animals (Bilai, *Antibiotic-producing microscopic fungi*: 139, 1963).

Dendrocladium (Pat.) Lloyd (1919) = Ramaria (Ramar.) fide Corner (1950).

Dendrocorticium M.J. Larsen & Gilb. (1974), Corticiaceae. 7, widespr. See Larsen & Gilbertson (*Norw. Jl Bot.* **24**: 99, 1977).

Dendrocyphella Petch (1922) = Cyphella (Aleurodisc.).

Dendrodochium Bonord. (1851), Mitosporic fungi, 3.A1.15. 20, esp. temp. See Tulloch (*Mycol. Pap.* **130**, 1972; status).

Dendrodomus Bubák (1915), Mitosporic fungi, 4.A1.15. 1, Eur.

Dendrodontia Hjortstam & Ryvarden (1980), Corticiaceae. 1, Afr.

Dendroecia Arthur (1906) = Haploravenelia (Ravenel.) fide Dietel (1928).

Dendrogaster Buchholz (1901) = Hymenogaster (Hymenogaster.) fide Smith (*Mycol.* **58**: 100, 1966), Fogel (*Mycol* **77**: 72, 1985).

Dendrographa Darb. (1895), Roccellaceae (L). 2, N. & C. Am.

Dendrographiella Agnihothr. (1972), Mitosporic fungi, 2.C2.28. 1 (on latex of *Hevea*), India.

Dendrographium Massee (1892), Mitosporic fungi, 2.C2.27. 2, S. Am., India.

dendrohyphidium, see hyphidium.

dendroid, tree-like in form; dendritic.

Dendroleptosphaeria Sousa da Câmara (1932) ? = Leptosphaeria (Leptosphaer.).

Dendromyceliates K.P. Jain & R.K. Kar (1979), Fossil fungi. 1 (Miocene), India. = Cryptophiale (Mitosp. fungi) fide Sutton (*in litt.*).

Dendromyces Libosch. (1810) = Battarrea (Battarr.).

Dendrophagus Murrill (1905) [non Toumey (1900), *Loranthaceae*] = Ganoderma (Ganodermat.) fide Donk (1974).

Dendrophoma Sacc. (1880) = Dinemasporium (Mitosp. fungi) fide Sutton (*TBMS* **48**: 611, 1965).

Dendrophora (Parmasto) Chamuris (1987), Peniophoraceae. 3, widespr.

Dendrophysellum Parmasto (1968), Aleurodiscaceae. 1, former USSR.

dendrophysis, see hyphidium.

Dendropleella Munk (1953) ? = Hendersonia (Coelom.) fide Barr & Holm (*Taxon* **33**: 109, 1984).

Dendropolyporus (Pouzar) Jülich (1982), Polyporaceae. 1, Eur.

Dendroseptoria Alcalde (1948), Mitosporic fungi, 4.G1.1. 1, Eur.

Dendrosphaera Pat. (1907), Trichocomaceae. 1, Indo-China, E. Indies. See Boedijn (*Bull. Jard. bot. Buitenz.* **13**: 472, 1935), Malloch (*in* Samson & Pitt (Eds), *Advances in Penicillium and Aspergillus systematics*: 365, 1985; posn).

Dendrospora Ingold (1943), Mitosporic fungi, 1.G1.1. 7 (aquatic), UK. See Descals & Webster (*TBMS* **74**: 135, 1980).

Dendrosporium Plakidas & Edgerton ex J.L. Crane (1972), Mitosporic fungi, 2.B1.10. 1, N. Am., India. See Crane (*TBMS* **58**: 423, 1972).

Dendrosporomyces Nawawi, J. Webster & R.A. Davey (1977), Mitosporic fungi, 1.G1.1. 1 (aquatic, with dolipore septa), Malaysia.

Dendrostilbe Dearn. (1924) nom. dub. (Mitosp. fungi). ? an error for *Dendrostilbella.*

Dendrostilbella Höhn. (1905), Mitosporic fungi, 3.A1.15. 5, Eur., Asia, N. Am.

Dendrothele Höhn. & Litsch. (1907), Corticiaceae. *c.* 30, cosmop. See Lemke (*CJB* **42**: 723, 1964; as *Aleurocorticium*).

Dendrotrichoscypha Svrček (1977) = Mollisina (Hyaloscyph.) fide Huhtinen (*SA* **11**: 170, 1993), Svrček (*SA* **11**: 170, 1993). See also Sharma (*Portug. Acta Biol.* **15**: 281, 1989; key).

Dendryphiella Bubák & Ranoj. (1914) = Dendryphion (Mitosp. fungi) fide Hughes (*CJB* **36**: 727, 1958) but used by Ellis (*DH*) for 2 spp. See Michaelis *et al.* (*Mycol.* **79**: 514, 1987; genetics and systematics), Mohamad *et al.* (*MR* **93**: 400, 1989; separation of spp. by ELISA).

Dendryphion Wallr. (1833), Anamorphic Pleosporaceae, 1.C/G2.26. Teleomorph *Pleospora.* 3, widespr.

Dendryphiopsis S. Hughes (1953), Anamorphic Pleomassariaceae, 1.C2.24. Teleomorph *Kirschsteiniothelia.* 3, widespr. See Hughes (*N.Z. Jl Bot.* **16**: 360, 1978).

Dendryphiosphaera Lunghini & Rambelli (1978), Mitosporic fungi, 1.C2.1. 2, Ivory Coast, India.

denigrate, blackened.

Dennisiella Bat. & Cif. (1962), Coccodiniaceae. 6, mainly trop. See Hughes (*Mycol.* **68**: 693, 1976). Anamorph *Microxyphium.*

Dennisiodiscus Svrček (1976), Dermateaceae. 9, Eur., Papua New Guinea. See Baral & Kriegelsteiner (*Beih. Z. Mykol.* **6**, 1985).

Dennisiomyces Singer (1955), Tricholomataceae. 2, S. Am.

Dennisiopsis Subram. & Chandrash. (1977), Thelebolaceae. 2, India. See van Brummelen (*in* Hawksworth (Ed.), *Ascomycete systematics*: 400, 1994; posn).

Dennisographium Rifai (1977), Mitosporic fungi, 2.A1.15. 1, Java.

dense body vesicle (DBV), cytoplasmic vesicle, associated with phosphorylated glucan metabolism, found in a variety of TEM morphological states ranging from a single (or several) electron-opaque core in an amorphous matrix to a highly structured, myelin-like arrangement of alternating electron-opaque and electron-translucent layers.

Densocarpa Gilkey (1954) = Stephensia (Otid.) fide Trappe (1975).

dentate, toothed (Fig. 37.45). Cf. denticulate.

denticle, a small tooth-like projection esp. one on which a spore is borne.

Denticularia Deighton (1972), Mitosporic fungi, 1.A2.3. 4, mainly trop. See de Hoog (*Persoonia* **10**: 51, 1978).

denticulate, having small teeth. Cf. dentate.

Dentinum Gray (1821), Hydnaceae. 2, widespr. See Hall & Stuntz (*Mycol.* **63**: 1113, 1971; key).

Dentipellis Donk (1962), Auriscalpiaceae. 3, widespr. See Ginns (*Windahlia* **16**: 35, 1986; key).

Dentipratulum Domański (1965), Hericiaceae. 1, Poland.

Dentocircinomyces R.F. Castañeda & W.B. Kendr. (1990), Mitosporic fungi, 1.A1.10. 1 (on *Eucalyptus*), Cuba.

Dentocorticium (Parmasto) M.J. Larsen & Gilb. (1974), Corticiaceae. 3, widespr. See Larsen & Gilbertson (*Norw. Jl Bot.* **24**: 99, 1977).

denuded, uncovered or glabrous by loss of scales, etc.

depauperate, poorly developed.

Depazea Fr. (1818) = Asteroma (Mitosp. fungi) fide Sutton (1977).

Depazites Geinitz (1855), Fossil fungi. 16 (Tertiary, Quaternary), Eur.

dependent, hanging down.

Dephilippia Rambelli (1959) = Circinotrichum (Mitosp. fungi) fide Pirozynski (1962).

deplanate, flat.

depressed (1) (of a pileus), having the middle lower than the edge (Fig. 19N); (2) (of lamellae), sinuate (q.v.).

depside, (depsidone), see Metabolic products.

Derexia Naras. (1970), Mitosporic fungi, 1.A1.1. 1 (on *Lecanium*), India, Java.

derived, of a character that has changed from the form in which it appeared in an ancestor.

derm (dermium) (of basidiomata), an outer layer in which the hyphae are perpendicular to the surface (Lowag, 1941); cf. cortex.

Dermapteromyces Thaxt. (1931), Laboulbeniaceae. 3, trop. Am.

Dermascia Tehon (1935) ? = Lophodermium (Rhytismat.) fide Darker (1967).

Dermatangium Velen. (1926) = Tremella (Tremell.) fide Donk (*Persoonia* **4**: 179, 1966).

Dermatea Fr. (1849) ≡ Dermea (Dermat.).

Dermateaceae Fr. (1849), Leotiales. 72 gen. (+ 46 syn.), 495 spp. Stroma (sclerotia) absent. Ascomata small, flat or concave, usually sessile, usually grey-brown or black, occasionally immersed in plant tissue, then sometimes with a specialized opening mechanism; usually with a well-defined margin which is often downy but rarely with distinct hairs; excipulum composed of ± brown, thin- or thick-walled, isodiametric cells; interascal tissue of simple paraphyses; asci usually with a well-developed I+ or I- ring; ascospores small, hyaline, septate or aseptate, often elongated.

Anamorphs pycnidial or various where known. Saprobic or parasitic on herbaceous and woody material, cosmop.

The limits of many genera need investigation.

Dermatella P. Karst. (1871) = Dermea (Dermat.) fide Nannfeldt (1932).

Dermateopsis Nannf. (1932), Dermateaceae. 1, N. Am., Eur.

Dermatina (Sacc.) Höhn. (1909) = Pezicula (Dermat.) fide Nannfeldt (1932).

Dermatina (Almq.) Zahlbr. (1922) = Arthonia (Arthon.) p.p. and Arthothelium (Arthon.) p.p. fide Harris (*Mich. Bot.* **12**: 3, 1973). See *Mycoporum*.

Dermatiscum Nyl. (1867), Physciaceae (L). 3, S. Afr., N. Am. See Brusse (*Mycotaxon* **25**: 161, 1986).

dermatitis verrucosa see chromomycosis.

Dermatocarpaceae, see *Verrucariaceae*.

Dermatocarpella H. Harada (1993), Verrucariaceae (L). 3, widespr. See Eriksson & Hawksworth (*SA* **12**: 27, 1993; status).

Dermatocarpon Eschw. (1824), Verrucariaceae (L). 12 (esp. damp rocks), widespr. See Awasthi & Upreti (*J. econ. tax. Bot.* **7**: 7, 1985; key 4 spp. India), Doppelbauer (*Nova Hedw.* **2**: 279, 1960; structure), Harada (*Nat. Hist. Res.* **2**: 113, 1993; gen. concept), Janex-Favre & Wagner (*BSMF* **102**: 161, 1986; pycnidia).

Dermatocarpon W. Mann (1825) = Endocarpon (Verrucar.).

dermatocyst (dermatocystidium), see cystidium (1).

Dermatodea Vent. (1799) = Nephroma (Nephromat.) p.p. and Peltigera (Peltiger.) p.p.

Dermatodothella Viégas (1944), Dothideales (inc. sed.). 1, Brazil.

Dermatodothis Racib. ex Theiss. & Syd. (1914), Dothideales (inc. sed.). 4, Asia, S. Am. See Müller (*Sydowia* **28**: 148, 1976; key).

Dermatoidium Stautz [not traced] nom. dub. (? Mitosp. fungi).

Dermatomeris Reinsch (1890) = Mastodia (Mastod.).

dermatomycosis, see mycosis.

Dermatophilus (Van Saceghem) M.A. Gordon (1964), *Actinomycetes*, q.v.

dermatophyte, a fungus parasitizing keratinized tissue (hair, skin, nails) of humans and animals and causing **dermatophytosis** (pl. **-es**) (ringworm, tinea). These fungi, which are typically Mitosporic fungi (hyphomycetes) with teleomorphs in the *Arthrodermataceae* (*Onygenales*), have frequently been treated as a special group the 'Dermatophytes' or Ringworm Fungi. Ringworm is cosmop. and dermatophytes, or non- or weakly pathogenic dermatophyte-like fungi, occur widely in soil and other keratin containing substrata such as bird's nests. There are 3 main gen. (*Epidermophyton, Microsporum, Trichophyton* distinguished by characteristic macroconidia), + *c.* 30 syn., *c.* 40 spp. (1,000 names).

    Lit.: Sabouraud (*Les Teignes*, 1910), Bruhns & Alexander (*in* Jadassohn, *Handbuch der Haut- und Geschlechtskrankheiten* **2**, 1930), Emmons (*Arch. Derm. Syph., Chicago* **30**: 337, 1934), Stockdale (*Biol. Rev.* **28**: 84, 1952; nutrition), Georg (*Animal ringworm in public health*, 1959 [US Dep. Health, Educ., & Welfare]), Gotz (*Die Pilzkrankheiten der Haut durch Dermatophyten*, 1962 [= *Jadassohn Handb., Ergänzungwerk* **4** (3)]), Dawson (*Rev. med. vet. Mycol.* **6**: 223, 1968; ringworm in animals), Hironaga (*Jap. J. med. mycol.* **24**: 283, 1983, anamorph-teleomorph connexions). See also Lewis *et al.* (1958), Conant *et al.* (1971) under Medical and veterinary mycology; griseofulvin; Wood's light.

dermatophytid, a pustular allergic eruption (**id-reaction**) of the skin at a distance from a primary infection by a dermatophyte.

Dermatosorus Sawada ex L. Ling (1949), Ustilaginaceae. 4 (on *Cyperaceae*), Asia, Austr. See Langdon (*TBMS* **68**: 447, 1977); Vánky (*TBMS* **89**: 60, 1987).

Dermea Fr. (1825), Dermateaceae. 22, esp. temp. See Groves (*Mycol.* **38**: 351, 1946), Funk (*CJB* **54**: 2852, 1976).

Derminus (Fr.) Staude (1857) nom. rej. prop. ≡ Pholiota (Strophar.) fide Donk (1949).

dermis (of lichens), the limiting layer of a thallus (obsol.).

Dermiscellum Hafellner, H. Mayrhofer & Poelt (1979), Physciaceae (L). 1, N. Am.

Dermocybe (Fr.) Wünsche (1877), Cortinariaceae. 15, widespr. See Moser (*Schweiz. Z. Pilzk.* **50**: 153, 1972; key), Gill (*Aust. Jl Chem.* **48**: 1, 1995; pigments in Australian spp.).

Dermocystidium Pérez (1908), Protozoa, see monads. 4, widespr. *D. marinum* on oysters (*Crassostrea virginica*) (see Ray & Chandler, *Exp. Parasitol.* **4**: 173, 1955); *D. cochliopodii* on *Cochliopodium bilimbosum* (see Valkanov, *Nova Hedw.* **12**: 393, 1967). See Dykova & Lom (*J. appl. Ichthyology* **8**: 180, 1992; evidence of fungal nature), Dick (*in press*).

Dermocystis Pérez (1907) ≡ Dermocystidium (Protozoa).

Dermodium Link (1809) nom. conf. (Myxom.).

Dermodium Rostaf. (1875) = Lycogala (Myxom.).

Dermoloma J.E. Lange ex Herink (1959), Tricholomataceae. 5, Eur., Am. See Svrček (*Česká Myk.* **20**: 256, 1966; key), Arnolds (*Persoonia* **14**: 519, 1992).

Dermomycoides Granata (1919), ? Chytridiales (inc. sed.). 2, Eur.

Dermosporidium Carini (1940) = Rhinosporidium (? Hyphochytriales) fide Azevedo (1958).

Dermosporium Link (1815) = Aegerita (Mitosp. fungi) fide Saccardo (*Syll. Fung.* **4**: 644, 1886).

Descalsia A. Roldán & Honrubia (1989), Mitosporic fungi, 1.G1.10. 1 (aquatic), Spain.

Descematia Nieuwl. (1916) ≡ Sphaerocephalum (Calic.).

descending (descendant) (of an annulus), having the free edge below the attached (cf. ascending).

Descolea Singer (1950), Bolbitiaceae. 8, S. Am., Australia. See Horak (*Persoonia* **6**: 231, 1971; key).

Descomyces Bougher & Castellano (1993), Hymenogastraceae. 3, Australasia (now widespr. having spread with *Eucalyptus*).

Describing, naming and publishing, see Hawksworth (*Mycologist's handbook*: 48, 1974), Classification, Nomenclature, Species, Systematics.

Desertella Mouch. (1979), Mitosporic fungi, 1.A1.10. 1 (from soil), Egypt.

Desetangsia Nieuwl. (1916) ≡ Sphaerotheca (Erysiph.).

Deshpandiella Kamat & Ullasa (1973), Phyllachoraceae. 1, India. See Hawksworth (*SA* **5**: 128, 1986). Anamorph *Mycohypallage*.

Deslandesia Bat. (1962) nom. inval. = Limacinula (Coccodin.) fide Reynolds (1971).

Desmaturus (Schltdl.) Kalchbr. (1880) = Lysurus (Clathr.) fide Dring (1980).

Desmazierella Lib. (1829), Sarcosomataceae. 2 (on conifer needles), Eur. Anamorph *Verticicladium*.

Desmazierella Crié (1878) nom. dub. (? Mitosp. fungi).

Desmazières (Jean Baptiste Henri Joseph; 1787-1862). A botanist of Lille, the 'possesseur d'une belle fortune'. Noted for *Plantes cryptogames du Nord de la France*, 1825-60, and his papers on these exsiccati in *Ann. Sci. nat., Paris* and *Mém. Soc. sci. Lille*. See

Anon. (*Bull. Soc. roy. bot. Belg.* **1**: 102, 1862), Grummann (1974: 274), Stafleu & Cowan (*TL-2* **1**: 630, 1976).

Desmazieria Mont. (1852) [non Dumort. (1822), *Poaceae*] ≡ Niebla (Ramalin.).

Desmella Syd. & P. Syd. (1919), Uredinales (inc. sed.). 8, S. Am. See Thirumalachar & Cummins (*Mycol.* **40**: 417, 1948).

Desmellopsis J.M. Yen (1969), Uredinales (inc. sed.). 1, Gabon. = Puccinia (Puccin.) fide Cummins & Hiratsuka (1983).

Desmidiospora Thaxt. (1891), Mitosporic fungi, 1.G2.1. 1, N. Am. See Clark & Prusso (*Mycol.* **78**: 865, 1986; on ants).

Desmopatella Höhn. (1924) = Phacidiella (Mitosp. fungi) fide Sutton (1977).

Desmotascus F. Stevens (1919) = Botryosphaeria (Botryosphaer.) fide v. Arx & Müller (1954).

Desmotelium Syd. (1937) = Chaconia (Chacon.) fide Laundon (*Mycotaxon* **3**: 132, 1975).

Desmotrichum Lév. (1843) = Gonatobotrys (Mitosp. fungi) fide Saccardo (*Syll. Fung.* **4**: 169, 1886).

destroying angel, the pure white agaric, *Amanita virosa*, ingestion of 1 mg of which can prove fatal; toxins are cyclic polypeptides, esp. amanitins.

Destuntzia Fogel & Trappe (1985), Hymenogastraceae. 5, N. Am. See Fogel & Trappe (*Mycol.* **77**: 732, 1985; key).

determinate, (1) clearly marked; definite; (2) (of conidiophores), growth ceasing with the production of terminal conidia.

detersile (of villosity), removable so that the surface becomes bare.

Detonia Sacc. (1889) = Plicaria (Peziz.) fide Korf (1960).

Detonina Kuntze (1891) ≡ Apiospora (Lasiosphaer.).

Detonisia Gonz. Frag. (1925) nom. dub. (? Oomycota).

detoxification, the conversion of a toxin (e.g. an inhibitory phytoalexin) to non-toxic (non-inhibitory) products.

deuteroconidium (of dermatophytes), a spore-like cell, the outcome of the division of a hemispore (protoconidium).

deuterogamy, the condition in which other processes replace fusion of gametes, as in some fungi, the macroalgae, and phanerogams; secondary pairing.

Deuterolichenes (obsol.). Introduced by Mameli-Calvino (*Nuovo G. bot. ital.* n.s. **27**: 379, 1930) for *Pyrenotrichum* (q.v.). Also later used for sterile leprose and filamentous lichens as well as lichenized Mitosporic fungi.

Deuteromycotina, see Mitosporic fungi, Kendrick (*Sydowia* **41**: 6, 1989; abandonment of term).

Deuterophoma Petri (1929) = Phoma (Mitosp. fungi) fide Kanciaveli & Ghikascvili (1948). See Sutton (1977).

devalidated (of names), names which would have been validly published under the *Code* except for the operation of Art. 13 (see Nomenclature). Such names may be 'revalidated' or 'taken up' by post-starting point authors. Formerly widely used in mycology prior to changes made in this Art. in 1983.

devil's cigar, *Urnula geaster* (see Seaver, *Mycol.* **29**: 60, 1937); - **snuffboxes**, puffballs.

Dexhowardia J.J. Taylor (1970), Mitosporic fungi, 1.A1.6. 1, USA.

Dexteria F. Stevens (1917) = Hyalosphaera (Dothideales) fide Rossman (*Mycol. Pap.* **157**, 1987).

Dextrinocystis Gilb. & M. Blackw. (1988), Corticiaceae. 1, USA.

Dextrinodontia Hjortstam & Ryvarden (1980), Corticiaceae. 1, Tanzania. = Trechispora (Sistotremat.) fide Larsson (*in litt.*).

dextrinoid (of spores, etc.), stained yellowish- or reddish-brown by Melzer's iodine (see Iodine); pseudoamyloid (Singer); cf. amyloid.

Dextrinosporium Bondartsev (1972), Coriolaceae. 1, Russia.

dhose, a Malaysian fermented food produced my means of yeasts and peas.

Diabole Arthur (1922), Raveneliaceae. 1, C. Am., W. Indies, Mex., Brazil; 0, III on *Mimosa*.

Diabolidium Berndt (1995), Pucciniaceae. 1, S. Am.

Diaboliumbilicus I. Hino & Katum. (1955), Ascomycota (inc. sed.). 1, Japan.

Diacanthodes Singer (1945), Coriolaceae. 1, widespr. Anamorph *Bornetina* (+ mealy bugs) causes phthiriosis (root disease) of coffee. See Fidalgo (*Rickia* **1**: 145, 1962).

Diachaeella Höhn. (1909) = Diachea (Stemonitid.).

Diachea Fr. (1825), Stemonitidaceae. 9, cosmop.

Diacheopsis Meyl. (1930), Stemonitidaceae. 8, widespr. See Kowalski (*Mycol.* **67**: 616, 1975; key).

Diachora Müll. Arg. (1893), Phyllachoraceae. 5 (on *Leguminosae*), temp. See Müller (*Trans. Bot. Soc. Edinb., Suppl.*: 69, 1986; key), Cannon (*Mycol. Pap.* **163**, 1991). Anamorph *Diachorella*.

Diachorella Höhn. (1918), Anamorphic Phyllachoraceae, 8.A/E1.15. Teleomorph *Diachora*. 3 (on *Leguminosae*), widespr. See Sutton (1980).

Diacrochordon Petr. (1955), Ascomycota (inc. sed.). 1, Germany.

Diadema Shoemaker & C.E. Babc. (1989), Diademaceae. 6 (esp. on *Gramineae*), Himalaya, W. & N. Am.

Diademaceae Shoemaker & C.E. Babc. (1992), Dothideales. 5 gen., 41 spp. Ascomata globose to ellipsoid-elongate, typically opening by a disc-like operculum but in some spp. by a lysigenous pore or slit; peridium composed of large thick-walled pseudoparenchyma; interascal tissue composed of pseudoparaphyses; asci cylindrical, fissitunicate; ascospores large, brown, muriform, usually radially asymmetrical. Saprobic in stems and leaves.
    Lit.: Barr (*SA* **12**: 27, 1993).

Diademosa Shoemaker & C.E. Babc. (1992), Diademaceae. 1, USA. = Graphyllium (Diadem.) fide Barr (1993).

Diademospora B.E. Söderstr. & Bååth (1979), Mitosporic fungi, 1.C1.1. 1 (from soil), Sweden.

diageotropism, tendency to horizontal growth in relation to the earth surface.

diagnosis, (1) an account, esp. the first (see Nomenclature), of the distinguishing characteristics of a taxonomic group; (2) determining a fungus or disease; **diagnostic**, characteristic; of use for identification.

Dialaceniopsis Bat. (1959), Mitosporic fungi, 4.B1.15. 1, Uganda.

Dialacenium Syd. (1930) = Rhytidenglerula (Englerul.) fide Müller & v. Arx (1962).

Dialhypocrea Speg. (1919), Hypocreaceae. 1, Brazil. = Neoskofitzia (Hypocr.) fide Weese [not traced].

Dialonectria (Sacc.) Cooke (1884) = Nectria (Hypocr.) fide Rogerson (1970).

Dialytes Nitschke (1867) nom. nud. = Diaporthe (Diaporth.) fide Wehmeyer (1933).

Diamphora Mart. (1821) nom. dub. (? Fungi).

Diandromyces Thaxt. (1918), Laboulbeniaceae. 1, Chile.

Dianema Rex (1891), Dianemataceae. 7, widespr. See Kowalski (*Mycol.* **59**: 1080, 1968). Cf. *Dianemina*.

Dianemataceae T. Macbr. (1899), Trichiales. 2 gen. (+ 3 syn.), 9 spp.

Dianemina A.R. Loebl. & Tappan (1961) ≡ Dianema (Dianemat.).

Diaphanium Fr. (1836), Mitosporic fungi, ?3.?A1.?. 2, Eur.

diaphanous, transparent or nearly so.

Diaphoromyces Thaxt. (1926), Laboulbeniaceae. 3, Sri Lanka, Am., Afr.

Diapleella Munk (1953) = Kalmusia (Dothideales) fide Eriksson (*SA* **9**: 8, 1990).

Diaporthaceae, see *Valsaceae*.

Diaporthales, Ascomycota. 2 fam., 98 gen. (+ 55 syn.), 425 spp. Ascomata perithecial, usually aggregated into a pseudostroma, usually long-necked; interascal tissue absent or of thin-walled unspecialized cells, deliquescing early; asci usually thick-walled but not fissitunicate, with a conspicuous I- apical ring; ascospores very varied. Anamorphs varied, coelomycetous. Saprobes and plant parasites, mainly on bark and wood. 13 gen. (+ 3 syn.), 60 spp., stated to belong here, but not referred to fams., may require transfer elsewhere. Fams:
(1) **Melanconidaceae**.
(2) **Valsaceae**.

Malloch (*in* Kendrick (Ed.), *The whole fungus* **1**: 153, 1979; *in* Reynolds (Ed.), *Ascomycete systematics*, 1981) used a much broader concept of this Order. *Lit.*: v. Arx & Müller (1954), Barr (*The Diaporthales in North America* [*Mycol. Mem.* **7**], 1978), Kobayashi (*Bull. Govt For. Exp. Stn Meguro* **226**, 1972; keys), Merezhko & Smyk (*Flora Gribov Ukrainȳ, Diaportal'nye Gribȳ*, 1991; Ukraine, keys), Müller & v. Arx (1962, 1973; keys gen.), Petrak (*Sydowia* **24**: 249 *et seq.*, 1971), Smyk *et al.* (*Ukr. bot. Zh.* **46**: 46, 1989; SEM), Wehmeyer (*The genus Diaporthe*, 1933; *Univ. Michigan Stud.* sci. ser. **14**, 1942).

Diaporthe Nitschke (1870), Valsaceae. 75, esp. N. temp. Some are parasites, at least in the *Phomopsis* state, e.g. *D. phaseolorum* var. *batatatis* (storage rot of sweet potato, *Ipomoea*), *D. citri* (*Phomopsis* stem-end rot of citrus fruits), *D. perniciosa* (a connexion with die-back of fruit trees), *D. phaseolorum* (pod blight of lima bean, *Phaseolus lunatus*), *D. woodii* (lupinosis of sheep). See Wehmeyer (*The genus Diaporthe*, 1933), Uecker (*Mem. N.Y. Bot. Gdn* **49**: 38, 1989; ontogeny). Anamorph *Phomopsis*.

Diaporthella Petr. (1924), Valsaceae. 2, Eur., N. Am. See Barr (*Mycol. Mem.* **7**, 1978).

diaporthin, a wilt toxin from *Endothia parasitica* (Boller *et al.*, *Helv. chim. Acta* **40**: 875, 1957); antibacterial.

Diaporthopsis Fabre (1883), Valsaceae. 8, Eur., N. Am., Philipp. Anamorph *Phomopsis*.

Diarimella B. Sutton (1980), Mitosporic fungi, 8.A1.15. 1, India.

Diarthonis Clem. (1909) = Arthonia (Arthon.).

Diasporangium Höhnk (1936), Pythiaceae. 1 (in soil), Eur., USA.

diaspore, (1) any unit of dissemination, e.g. a spore, fragment of mycelium, sclerotium (Sernander, 1927); (2) (of lichens), particularly applied to vegetative propagules. See hormocyst, isidium, soredium, etc. (Fig. 22).

Diathrypton Syd. (1922) = Schiffnerula (Englerul.) fide v. Arx & Müller (1975).

Diatractium Syd. & P. Syd. (1921), Phyllachoraceae. 2, C. & S. Am. See Cannon (*MR* **92**: 327, 1989), Barr (1978), Sydow (*Ann. Myc.* **33**: 85, 1935).

Diatrypaceae Nitschke (1869), Diatrypales. 9 gen. (+ 20 syn.), 204 spp. Ascomata perithecial, immersed in a well-developed eu- or pseudostroma, often long-necked, with sulcate ostioles. Interascal tissue composed of cylindrical paraphyses. Asci long-stalked, the apex ± truncate, with a small often I+ apical ring;

ascospores pale brown, allantoid. Anamorphs coelomycetous. Saprobes and parasites, esp. on branches, widespr.
Lit.: Rappaz (1987; keys to 8-spored spp.), Glawe (*Sydowia* **41**: 122, 1989; anamorphs).

Diatrypales, Ascomycota. 1 fam., 9 gen. (+ 20 syn.), 204 spp. The Order is perhaps closely related to the *Xylariales* (Rogers, *in* Hawksworth (Ed.), *Ascomycete systematics*: 321, 1994), but it is badly in need of a modern revision. Fam.:
Diatrypaceae.
Lit.: Carmarán & Romero (*Boln. Soc. Argent. Bot.* **28**: 139, 1992; gen. concepts), Eriksson & Hawksworth (*SA* **7**: 67, 1988; status), Höhnel (*Ann. Myc.* **16**: 127, 1918), Müller & v. Arx (1973), Munk (1957), Schrantz (*BSMF* **76**: 305, 1961; 27 gen., keys), Glawe & Rogers (*Mycotaxon* **20**: 401, 1984; Pacific NW, keys), Rappaz (*Mycol. Helv.* **2**: 285, 1987; monogr., keys), Vasil'eva (*Mikol. Fitopat.* **19**: 3, 1985; gen. relationships).

Diatrype Fr. (1849), Diatrypaceae. 56, widespr. See Rappaz (1987; key), Janex-Favre (*Revue mycol.* **42**: 265, 1978; spore formation), Glawe & Rogers (1984), Patil & Patil (*Indian J. Mycol. Pl. Path.* **13**: 134, 1985; 22 spp. India), Rappaz (*Mycotaxon* **30**: 209, 1987; *D. stigma* group), Dargan & Bhatia (*Nova Hedw.* **48**: 405, 1989; key 11 spp. Himalayas), Romero & Minter (*TBMS* **90**: 457, 1988; asci).

Diatrypella (Ces. & De Not.) De Not. (1863), Diatrypaceae. 30, widespr. See Croxall (*TBMS* **33**: 45, 1950), Glawe (*Mem. N.Y. bot. Gdn* **49**: 51, 1989; anamorph), Glawe & Rogers (1984).

Diatrypeopsis Speg. (1884) = Camillea (Xylar.) fide Læssøe *et al.* (1989).

Dibaeis Clem. (1909), Icmadophilaceae (L). 13, trop. See Gierl & Kalb (*Herzogia* **9**: 593, 1993).

Dibeloniella Nannf. (1932), Dermateaceae. 1, Eur. ? = Pyrenopeziza (Dermat.) fide Korf (*in* Ainsworth *et al.* (Eds), *The Fungi* **4A**: 249, 1973).

Dibelonis Clem. (1909) = Fabraea (Dermat.) fide Nannfeldt (1932).

Dibelonis Clem. & Shear (1931) = Leptotrochila (Dermat.). See Nannfeldt (1932).

Diblastia Trevis. (1857) = Xanthoria (Teloschist.).

Diblastospermella Speg. (1918) = Cicinnobella (Mitosp. fungi) fide Petrak (*Sydowia* **5**: 328, 1951), but see Sutton (1977).

Diblepharis Lagerh. (1900) = Monoblepharis (Monoblepharid.) fide Fitzpatrick (1930).

Dibotryon Theiss. & Syd. (1915), Venturiaceae. 1, N. Am. See Barr (*Prodromus to class Loculoascomycetes*, 1987; *Sydowia* **41**: 25, 1989).

Dicaeoma Gray (1821) = Puccinia (Puccin.) fide Dietel (1928).

Dicarpella Syd. & P. Syd. (1921) [non Bory (1824), *Rhodophyta*], Melanconidaceae. 3, N. Am. See Reid & Dowsett (*CJB* **68**: 398, 1990). Anamorph *Mastigosporella*.

Dicarphus Raf. (1808) nom. nud. ? = Hydnum (Hydn.).

Dicaryomycota. Division introduced by Schaffer (*Mycol.* **67**: 3, 1975) for ascomycetes and basidiomycetes; used by Kendrick (*The fifth kingdom*, edn 2, 1992).

dicaryon, see dikaryon.

Dicellaesporites Elsik (1968), Fossil fungi (f. cat.). 22 (Tertiary), China, Colombia, India, N. Am.

Dicellispora Sawada (1944), Mitosporic fungi, 1.A2.?3. 1, Taiwan. ? = Gonatobotryum (Mitosp. fungi) fide Carmichael *et al.* (1980).

Dicellomyces L.S. Olive (1945), Dacrymycetales (inc. sed.). 2, Eur., N. Am. See Parmasto (1968), Reid (*TBMS* **66**: 537, 1976). Referred to

Brachybasidiaceae by McNabb & Talbot (*in* Ainsworth *et al.* (Eds), *The Fungi* **4B**: 317, 1973).

Dicephalospora Spooner (1987), Sclerotiniaceae. 2, widespr.

Dichaena Fr. (1825) ≡ Polymorphum (Mitosp. fungi) fide Hawksworth & Punithalingam (1973).

Dichaenopsella Petr. (1952) ? = Polymorphum (Mitosp. fungi) fide Sutton (1977).

Dichaenopsis Paoli (1905) = Stagonospora (Mitosp. fungi) fide Butin (*Phytopath. Z.* **100**: 186, 1981).

Dichaetis Clem. (1931) ≡ Wentiomyces (Pseudoperispor.).

Dichantharellus Corner (1966), Lachnocladiaceae. 1, Malaysia.

Dicheirinia Arthur (1907), Raveneliaceae. 11 (on *Leguminosae*), trop. Am., Afr., New Caledonia. See Cummins (*Mycol.* **27**: 151, 1935, *Bull. Torrey bot. Cl.* **64**: 39, 1937).

Dichelostroma Bat. & Peres (1963), Mitosporic fungi, 5.A1.?. 1, USA.

Dichitonium Berk. & M.A. Curtis (1875) ? = Dendrodochium (Mitosp. fungi). See also Oichitonium (Mitosp. fungi).

Dichlaena Durieu & Mont. (1849), Trichocomaceae. 1, N. Afr.

Dichlamys Syd. & P. Syd. (1920) = Uromyces (Puccin.). See Thirumalachar (*Sydowia* **5**: 23, 1951).

Dichobotrys Hennebert (1973), Anamorphic Otidiaceae, 1.A1.7. Teleomorph *Trichophaea*. 4, widespr.

Dichodium Nyl. (1888) = Physma (Collemat.).

dichohyphidium, see hyphidium.

Dicholobodigitus G.P. White & Illman (1988), Mitosporic fungi, 1.C/G2.28. 1, Canada.

Dichomera Cooke (1878), Mitosporic fungi, 8.D2.1/19. 40, widespr.

Dichomitus D.A. Reid (1965), Polyporaceae. 2, Eur., N. Am.

Dichomyces Thaxt. (1893) = Peyritschiella (Laboulben.) fide Tavares (1985).

Dichonema Blume & T. Nees (1826) = Dictyonema (Merul.) fide Parmasto (1978).

dichophysis, see hyphidium.

Dichopleuropus D.A. Reid (1965), Lachnocladiaceae. 1, Malaysia.

Dichoporis Clem. (1909) = Strigula (Strigul.) fide Hafellner & Kalb (*Bibl. Lich.* **57**: 161, 1995).

Dichosporidium Pat. (1903), Roccellaceae (L). 6, trop. See Thor (*Opera Bot.* **103**, 1990).

Dichosporium Nees (1816) nom. dub. (Physaraceae).

Dichosporium Pat. (1899) ≡ Dichosporidium (Roccell.).

Dichostereaceae Jülich (1982), Lachnocladiales. 1 gen., 26 spp. Basidioma resupinate.

Dichostereum Pilát (1926), Dichostereaceae. 11, widespr. See Boidin & Lanquetin (*Mycotaxon* **6**: 277, 1977; *BSMF* **96**: 381, 1980; key).

Dichothrix Theiss. (1912) [non Zanardini ex Bornet & Flahault. (1886), *Cyanobacteria*] ≡ Schistodes (Parodiopsid.).

Dichotomella Sacc. (1914) = Nigrospora (Mitosp. fungi) fide Hughes (1958).

Dichotomocladium Benny & R.K. Benj. (1975), Chaetocladiaceae. 5, USA, India, Pakistan. See Benny & Benjamin (*Aliso* **8**: 338, 1975, *Mycol.* **85**: 660, 1993).

Dichotomomyces Saito ex D.B. Scott (1970), Trichocomaceae. 1, Japan. See Malloch (*SA* **6**: 124, 1987; posn). Anamorph *Polypaecilum*.

Dichotomophthora Mehrl. & Fitzp. ex P.N. Rao (1966), Mitosporic fungi, 1.C2.26. 2, trop. See de Hoog & van Oorschot (*Proc. K. Ned. Wet.* C **86**: 55,

1983), Hosoe *et al.* (*Phytochem.* **29**: 997, 1990; anthraquinone derivative from *D. lutea*).

Dichotomophthoropsis M.B. Ellis (1971), Mitosporic fungi, 1.F2.26. 2, India, USA.

dichotomous, branching, frequently successively, into two more or less equal arms.

Dicksonomyces Thirum., P.N. Rao & M.A. Salam (1957) nom. conf. (Peronosp.) fide Waterhouse (*Mycol.* **60**: 977, 1968).

Dicladium Ces. (1852) = Colletotrichum (Mitosp. fungi) fide Sutton (1966).

Diclasmia Trevis. (1869) = Sticta (Lobar.).

diclinous, having the oogonium and its antheridium on different hyphae (Fig. 28E). Cf. androgynous.

Diclonomyces Thaxt. (1931), Laboulbeniaceae. 3, W. Afr., Java, C. Am.

Dicoccum Corda (1829) nom. ambig. fide Hughes & Pirozynski (*CJB* **50**: 2521, 1972).

Dicollema Clem. (1909) = Collema (Collemat.).

Dicrandromyces Thaxt. (1931) = Tetrandromyces (Laboulben.) fide Tavares (1985).

Dicranidion Harkn. (1885), Mitosporic fungi, 3.G1.10. 1, Am. See Peek & Solheim (*Mycol.* **50**: 844, 1958).

Dicranocladium Sousa da Câmara (1931) nom. provis. (Mitosp. fungi).

Dicranophora J. Schröt. (1886), ? Thamnidiaceae. 1, (on decaying agarics), N. temp. See Vuillemin (*Ann. Myc.* **5**: 33, 1907), Dobbs (*TBMS* **21**: 167, 1938), Volgmayr & Krisai-Greilhuber (*MR, in press*).

Dicranophoraceae, see Thamnidiaceae.

Dicranotropis Breddin [not traced] nom. dub. ? based on fungi on an insect.

Dicteridium Raf. (1815) nom. dub. (Fungi, inc. sed.). No spp. included.

Dictydiaethalium Rostaf. (1873), Lycogalaceae. 2, cosmop.

dictydine granules, see plasmodic granules.

Dictydium Schrad. (1797) = Cribraria (Cribrar.) fide Pando (*in litt.*).

Dictyoarthrinium S. Hughes (1952), Mitosporic fungi, 1.D2.37. 2 or 3, trop.

Dictyoarthrinopsis Bat. & Cif. (1958), Mitosporic fungi, 1.D1.?. 1, Costa Rica.

Dictyoasterina Hansf. (1947), Microthyriaceae. 1, Afr.

Dictyobole G.F. Atk. & Long (1902) = Lysurus (Clathr.) fide Dring (1980).

Dictyocatenulata Finley & E.F. Morris (1967), Mitosporic fungi, 2.D1.23. 1, widespr. See Seifert *et al.* (*Mycol.* **79**: 459, 1987; generic redescr.).

Dictyocephala A.G. Medeiros (1962) = Pantospora (Mitosp. fungi) fide Deighton (*Mycol. Pap.* **140**, 1976).

Dictyocephalos Underw. ex V.S. White (1901), Phelloriniaceae. 1, USA, Afr., Japan. See Long & Plunkett (*Mycol.* **32**: 696, 1940).

Dictyochaeta Speg. (1923), Anamorphic Lasiosphaeriaceae, 1/2.A1/B1.16. Teleomorph *Chaetosphaeria*. 69, widespr. See Hughes & Kendrick (*N.Z. Jl Bot.* **6**: 331, 1968), Gamundí *et al.* (*Darwiniana* **21**: 95, 1977), Kuthubutheen & Nawawi (*MR* **95**: 1211, 1220, 1224, 1991; Malaysian spp., key 59 spp.).

Dictyochaetopsis Aramb. & Cabello (1990), Mitosporic fungi, 1.B/C1.16. 8, widespr.

dictyochlamydospore, a non-deciduous multicelled chlamydospore composed of an outer wall separable from the walls of the component cells which are rather easily separated from each other, as in some *Phoma* spp. formerly ascribed to *Peyronellaea* (Luedemann, *Mycol.* **51**: 778, 1961).

Dictyochora Theiss. & Syd. (1914) Fungi (inc. sed.).

Dictyochorella Theiss. & Syd. (1915) nom. dub. (Fungi, inc. sed.) fide v. Arx & Müller (1975).

Dictyochorina Chardón (1932) Fungi (inc. sed.). See Petrak (*Sydowia* **5**: 343, 1951).

Dictyocoprotus J.C. Krug & R.S. Khan (1991), Pyronemataceae. 1, Mexico, USA.

Dictyodesmium S. Hughes (1951), Mitosporic fungi, 3.D2.1. 2, USA, France.

Dictyodochium Sivan. (1984), Anamorphic Venturiaceae, 3.D2.1/10. Teleomorph *Gibbera*. 1, India.

Dictyodothis Theiss. & Syd. (1915), Dothideales (inc. sed.). 2, Am.

Dictyographa Müll. Arg. (1893) ? = Ingaderia (Roccell.).

Dictyographa Darb. (1897) = Darbishirella (Roccell.).

Dictyolus Quél. (1886) = Leptoglossum (Tricholomat.) fide Singer (1951), = Arrhenia (Tricholomat.).

Dictyomollisia Rehm (1909) = Uleomyces (Cookell.) fide v. Arx (1963).

Dictyomorpha Mullins (1961), Rozellopsidaceae. 2 (on aquatic 'phycomycetes'), N. Am., Eur. See Mullins (*Am. J. Bot.* **48**: 377, 1961), Mullins & Barksdale (*Mycol.* **57**: 352, 1965; parasitism), Dick (*in press*).

Dictyonella Höhn. (1909), Saccardiaceae. 5, trop. See v. Arx (*Persoonia* **2**: 421, 1963).

Dictyonema C. Agardh ex Kunth (1822), Meruliaceae (L). 5, mainly trop. See Parmasto (*Nova Hedw.* **29**: 99, 1978).

Dictyonema Reinsch (1875) nom. dub. (? Fungi, inc. sed.).

Dictyonematomyces Cif. & Tomas. (1954) = Dictyonema (Merul.).

Dictyonia Syd. (1904), Leotiaceae. 2, trop.

Dictyopanus Pat. (1900) = Panellus (Tricholomat.) fide Burdsall & Miller (1975).

Dictyopeltella Bat. & I.H. Lima (1959), Micropeltidaceae. 2, trop. See Batista (1959).

Dictyopeltis Theiss. (1913), Micropeltidaceae. 6, Java, C. Am., Afr. See Batista (1959), Müller & v. Arx (1962).

Dictyopeplos Kuhl & Hasselt (1824) = Phallus (Phall.).

Dictyophallus Corda (1842) = Phallus (Phall.).

Dictyophora Desv. (1809) = Phallus (Phall.) fide Dring (*Mycol. Pap.* **98**, 1964).

Dictyophrynella Bat. & Cavalc. (1964), Mitosporic fungi, 1.C2.?. 1, Brazil.

Dictyoploca Mont. ex Pat. (1890), Tricholomataceae. 2, S. Am. See Raithelhuber (*Metrodiana* **4**(3): 48, 1973). = Collybia (Tricholomat.) fide Singer (1951).

Dictyopolyschema M.B. Ellis (1976), Mitosporic fungi, 1.D2.24. 1, UK.

dictyoporospore, a deciduous, multicelled porospore the component cells of which are firmly united and not enclosed by an outer wall, as in *Alternaria* (Luedemann, *Mycol.* **51**: 778, 1961).

Dictyoporthe Petr. (1955), Melanconidaceae. 1, Pakistan.

Dictyopus Quél. (1886) = Boletus (Bolet.) fide Singer (1945).

Dictyorinis Clem. (1909) = Rinodina (Physc.).

dictyoseptate, having transverse and longitudinal cross walls (septa), like layers of cement between bricks; muriform.

dictyosomes, ± spherical vesicles associated with the edges of the membrane-bound sacs (cisternae) which constitute the golgi apparatus in *Oomycota* and other fungi as shown by electron microscopy.

Dictyosiropes M.B. Ellis (1976), Mitosporic fungi, 1.D2.10. 1, India.

dictyosporangium, a septate sporangium, as in *Dictyuchus*.

dictyospore, differs from an amerospore (q.v.) by being divided by intersecting septa in more than one plane; muriform spore. See Mitosporic fungi.

Dictyosporites Félix (1894), Fossil fungi. 2 (Paleocene), Asia, Australia, Eur.

Dictyosporium Corda (1836), Mitosporic fungi, 1.G2.1. 15, widespr. See Sutton (*Proc. Ind. Acad. Sci.* (Pl. Sci.) **94**: 229, 1985; disposition of names), Tzean & Chen (*MR* **92**: 497, 1989; 2 n. spp.).

Dictyosteliaceae Rostaf. ex Cooke (1877) (Dictyosteliidae), Dictyosteliales. 3 gen. (+ 1 syn.), 42 spp. Sporocarp stalk stout and filled with empty cells.

Dictyosteliales, Dictyosteliomycota. 2 fam., 4 gen. (+ 1 syn.), 46 spp. The only order of *Dictyosteliomycota* (q.v.) comprising Fams:
(1) **Actyosteliaceae**.
(2) **Dictyosteliaceae**.
   See Cavender (1990; key 15 spp. Ohio).

Dictyosteliomycetes, see *Dictyosteliomycota*.

Dictyosteliomycota (Dictyostelida, Dictyosteliomycetes, Dictyostelia, Dictyostelea, Dictyostelidae). Protozoa; cellular slime moulds; dictyostelids. 1 ord., 2 fam., 4 gen. (+ 1 syn.), 46 spp. Life-cycle with amoeboid bactivorous, multicellular sporocarp, spore producing, and aggregation phases; trophic phase amoeboid, pseudopodia mainly filose (sometimes also lobose), the pseudoplasmodium (slug) formed by aggregating streaming myxamoebas; nuclei with 2 or more peripheral nucleoli; sporocarps stalked, stalks (sorophores) branched or not, usually of vacuolate cells compacted in cellulosic tubes; spores dark and thick-walled; flagellate cells absent; some spp with sexual reproduction. Primarily soil organisms, also isolated from a wide variety of habitats, esp. ones with decaying plant and fungal materials; associated with bacteria on which they feed and on which they are cultured in the laboratory.

Easily separated from *Acrasomycota* in the production of well-developed and differentiated stalked sporocarps. Includes a single order:
Dictyosteliales.

Formerly sometimes united with the acrasid (see *Acrasomycota*) and(or) protostelid (see *Myxomycota*) slime moulds, but now known not to be closely allied to these groups (see Phylogeny). The development and life-cycle of *Dictyostelium* has been extensively investigated, and shown to need different mating types, and genes involved in cell differentiation have been identified (Loomis (Ed.), *The development of Dictyostelium discoideum*, 1982); of value also for study of microtubule formation (Roos & Camenzind, *Eur. J. Cell Biol.* **21**: 248, 1981).

Lit.: Bonner (*Researches on cellular slime moulds*, 1991; reprint 46 papers), Cavender (*in* Margulis *et al.* (Eds), *Handbook of Protoctista*: 88, 1990; bibliogr.), Olive (1975), Raper (*The dictyostelids*, 1984).

Dictyostelium Bref. (1870), Dictyosteliaceae. 35, widespr. See Raper (*The dictyostelids*, 1984), Olive (1975: 55; key), Raper (*Quart. Rev. Biol.* **26**: 169, 1951; culture), Lee (*Trans. mycol. Soc. Japan* **12**: 142, 1971; Japanese spp.), Loomis (*Dictyostelium discoideum: a developmental system*, 1975), Waddell (*Nature* **298**: 464, 1983; *D. caveatum* predacious on *Dictyostelium* spp.).

Dictyostomiopelta Viégas (1944), Micropeltidaceae. 1, Brazil.

Dictyothyriella Rehm (1914) = Micropeltis (Micropeltid.) fide Clements & Shear (1931).

Dictyothyriella Speg. (1924), Micropeltidaceae. 1, Cape Horn.

Dictyothyrina Theiss. (1913), Micropeltidaceae. 2, Am.

Dictyothyrium Theiss. (1912), Micropeltidaceae. c. 15, trop.

Dictyothyrium Grove (1932) ? = Mycoporum (Mycopor.). See Sutton (*Mycol. Pap.* **141**, 1977).

Dictyotopileos Dilcher (1965), Fossil fungi (Micropeltid.). 1 (Eocene), USA.

Dictyotremella Kobayasi (1971), Tremellaceae. 1, Papua New Guinea.

Dictyotrichiella Munk (1953) = Capronia (Herpotrichiell.) fide Müller *et al.* (*TBMS* **88**: 63, 1987).

Dictyuchus Leitg. (1868), Saprolegniaceae. 5, N. temp.

Dicyma Boulanger (1897), Anamorphic Xylariaceae, 2.A2.10. Teleomorph *Ascotricha*. 11, widespr. See v. Arx (*Proc. K. ned. Akad. Wet.* C **85**: 21, 1982), Hawksworth (*Mycol. Pap.* **126**, 1971).

Diderma Pers. (1794), Didymiaceae. 59, cosmop.

Didonia Velen. (1934), ? Leotiales (inc. sed.). 5, Eur. See Svrček (*Česká Myk.* **46**: 41, 1992; status).

Didothis Clem. (1931) ≡ Uleodothis (Ventur.).

Didymaria Corda (1842) = Ramularia (Mitosp. fungi) fide Hughes (1958).

Didymariopsis Speg. (1910) = Colletotrichum (Mitosp. fungi) fide Deighton (*TBMS* **59**: 185, 1972).

Didymascella Maire & Sacc. (1903), ? Hemiphacidiaceae. 3 (on conifers), Eur., N. Am. *D. thujina* damages *Thuja*. See Maire (*Bull. Soc. Hist. nat. Afr. N.* **18**: 117, 1927).

Didymascina Höhn. (1905) = Didymosphaeria (Didymosphaer.) fide Aptroot (1995).

Didymascus Sacc. (1896), ? Rhytismatales (inc. sed.). 1, Siberia. See Korf (*Mycol.* **54**: 24, 1962).

Didymaster Bat. & H. Maia (1967), Mitosporic fungi (L), 5.B1.?. 1, Brazil.

Didymella Sacc. (1880), Dothideales (inc. sed.). *c.* 75, widespr. *D. applanata* (raspberry spur blight), *D. citrullina* (on cucurbits), *D. lycopersici* (tomato stem and fruit rots). See Corbaz (*Phytopath. Z.* **28**: 375, 1956), Grube & Hafnellner (*Nova Hedw.* **51**: 283, 1990; redisp. lichenicolous spp.), Holm (*Taxon* **24**: 475, 1975; nomencl.). Anamorphs *Ascochyta*, *Phoma*.

Didymellina Höhn. (1918) = Mycosphaerella (Mycosphaerell.) fide v. Arx & Müller (1975).

Didymellopsis (Sacc.) Clem. & Shear (1931), Dothideales (inc. sed.). 4 (on lichens), Eur. See Grube & Hafellner (*Nova Hedw.* **51**: 283, 1990).

Didymiaceae Rostaf. ex Cooke (1877), Physarales. 6 gen. (+ 10 syn.), 124 spp.

Didymium Schrad. (1797), Didymiaceae. 54, cosmop.

Didymobotryopsis Henn. (1902) = Hirsutella (Mitosp. fungi) fide Samson & Evans (1991).

Didymobotrys Clem. & Shear (1931) ≡ Didymobotryopsis (Mitosp. fungi).

Didymobotryum Sacc. (1886), Mitosporic fungi, 2.B2.28. 5, N. Am., Asia.

Didymochaeta Sacc. & Ellis (1898) [non Steud. (1855), *Gramineae*], Mitosporic fungi, 4.B/C1.15. 1, N. Am.

Didymochaetina Bat. & J.L. Bezerra (1965), Mitosporic fungi, 4.B1.?. 1, Jamaica.

Didymochlamys Henn. (1897) [non Hook. (1872), *Rubiaceae*] ≡ Kuntzeomyces (Tillet.).

Didymochora Höhn. (1918), Anamorphic Mycosphaerellaceae, 8.A1.?. Teleomorph *Euryachora*. 1, Eur.

Didymocladium Sacc. (1886) = Cladobotryum (Mitosp. fungi) fide Hughes (1958).

Didymocoryne Sacc. & Trotter (1913), Leotiales (inc. sed.). 4, N. temp.

Didymocrater Mart. (1821) nom. dub (? Fungi, inc. sed.).

Didymocrea Kowalski (1965), Hypocreaceae. 1, India.

Didymocyrtidium Vain. (1921), Dothideales (inc. sed.). 2, Eur.

Didymocyrtis Vain. (1921), ? Dothideales (inc. sed.). 1 (on *Caloplaca*), Finland.

Didymolepta Munk (1953), Leptosphaeriaceae. 1, Eur. See Barr (*Mycotaxon* **43**: 371, 1992), Eriksson & Hawksworth (*SA* **7**: 68, 1988).

Didymopeltis Bat. & I.H. Lima (1959) = Schizothyrium (Schizothyr.) fide Müller & v. Arx (1962).

Didymopleella Munk (1953), Dothideales (inc. sed.). 1, Eur.

Didymoporisporonites Sheffy & Dilcher (1971), Fossil fungi (*f. cat.*). 10 (Eocene, Tertiary), China, USA.

Didymopsamma Petr. (1925) = Chaetosphaeria (Lasiosphaer.) fide Müller & Arx (1962).

Didymopsis Sacc. & Marchal (1885), Mitosporic fungi, 2.B1.1. 5, Eur., Am., Afr.

Didymopsora Dietel (1899), Pucciniosiraceae. 6, S. Am., Afr.; O, III. See Buriticá (*Rev. Acad. Colomb. Cienc.* **18**(69): 131, 1991).

Didymopsorella Thirum. (1950), Uropyxidaceae. 2, India, Afr.

Didymopycnomyces Cavalc. & A.A. Silva (1972), Mitosporic fungi (L), 4.B1.?. 1, Brazil.

Didymosamarospora T.W. Johnson & H.S. Gold (1957) nom. dub. (Fungi, inc. sed.) fide Kohlmeyer & Kohlmeyer (1979).

Didymosamarosporella, see *Didymosamarospora*.

Didymosira Clem. (1909) ≡ Pucciniosira (Pucciniosir.).

Didymosphaerella Cooke (1889) = Didymosphaeria (Didymosphaer.) fide Aptroot (1995).

Didymosphaeria Fuckel (1870) nom. cons. prop., Didymosphaeriaceae. 7, widespr. See Aptroot (*Stud. Mycol.* **37**, 1995, key; *Nova Hedw.* **60**: 325, 1995, excl. names), Kohlmeyer & Volkmann-Kohlmeyer (*MR* **94**: 685, 1990; redisp. marine spp.). Anamorph *Fuscladiella*-like or *Phoma*-like.

Didymosphaeriaceae Munk (1953), Dothideales. 2 gen. (+ 6 syn.), 8 spp. Ascomata perithecial, globose to flattened, immersed or erumpent, sometimes clypeate, with a clearly defined periphysate ostiole; peridium black, thick-walled esp. in the apical region, composed of closely interspersed hyphal cells; interascal tissue composed of trabeculate pseudoparaphyses in a gelatinous matrix; asci cylindrical, fissitunicate; ascospores brown, usually 1-septate. Anamorphs coelomycetous. Saprobic in woody or herbaceous material.

Lit.: Aptroot (*Stud. mycol.* **37**, 1995), Scheinpflug (*Ber. schweiz. bot. Ges.* **68**: 325, 1958).

Didymosphaerites Fiore (1932), Fossil fungi. 1 (Eocene), Italy.

didymospore, differs from an amerospore (q.v.) in having one transverse septum. See Mitosporic fungi.

Didymosporiella Traverso & Migliardi (1911) nom. dub. (Mitosp. fungi) fide Sutton (1977).

Didymosporina Höhn. (1916), Mitosporic fungi, 6.B2.19. 1, Eur.

Didymosporis Clem. & Shear (1931) ≡ Didymosporiella (Mitosp. fungi).

Didymosporium Nees (1816) nom. dub. (Mitosp. fungi) fide Sutton (1977).

Didymosporium Sacc. (1880), Mitosporic fungi, 6.A2.?. 1, Italy. See Sutton (1977).

Didymosporonites Sheffy & Dilcher (1971), Fossil fungi. 4, USA.

Didymostilbe Henn. (1902) ? = Kutilakesopsis (Mitosp. fungi) fide Carmichael & Kendrick (1973). See Seifert (*Stud. Mycol.* **27**: 130, 1985; 6 spp. Indonesia).

Didymostilbe Bres. & Sacc. (1902) nom. dub. fide Sydow & Sydow (*Ann. Myc.* **1**: 176, 1903).

Didymothozetia Rangel (1915), Mitosporic fungi, 3.B/C1.?. 1, Brazil.

Didymothyriella Bat. & I.H. Lima (1959) = Plochmopeltis (Schizothyr.) fide Müller & v. Arx (1962).

Didymotrichella Arnaud (1954), Mitosporic fungi, 1.B2.?27. 1, France.

Didymotrichia Berl. (1893) = Neopeckia (Dothideales) fide Barr (1984). = Herpotrichia (Lophiostom.) fide v. Arx & Müller (1975, 1984).

Didymotrichiella Munk (1953) = Capronia (Herpotrichiell.) fide Müller *et al.* (*TBMS* **88**: 63, 1987).

Didymotrichum Bonord. (1851) = Cladosporium (Mitosp. fungi) fide Hughes (1958).

Didymotrichum Höhn. (1914) = Dactylaria (Mitosp. fungi) fide Bhatt & Kendrick (1968).

Didymozoophaga Soprunov & Galiulina (1951) = Arthrobotrys (Mitosp. fungi).

Diedickea Syd. & P. Syd. (1913), Mitosporic fungi, 5.A1.?. 3, trop.

Diedickella Petr. (1922) = Stagonospora (Mitosp. fungi) fide Clements & Shear (1931).

Diehlia Petr. (1951) ? = Phaeangellina (Leot.) fide Korf (*in* Ainsworth *et al.* (Eds), *The Fungi* **4A**: 249, 1973).

Diehliomyces Gilkey (1955), Ascomycota (inc. sed.). 1, widespr. *D. microsporus*; false-truffle invading mushroom beds.

diel, a 24-hr periodicity. Cf. circadian, diurnal.

Dielsiella Henn. (1903) = Cycloschizon (Parmular.) fide Müller & v. Arx (1962).

Dietel (Paul; 1860-1947). A life-long German student of the *Uredinales*, he published from 1887 to 1943 more than 150 papers on rusts, including the systematic accounts in *Nat. Pflanzenfam.* (1898, 1928). His work was generally accepted as the most authoritative on rust classification. *Obit.*: Poeverlein (*Sydowia* **4**: 1, 1950; portr., bibl.). See also Stafleu & Cowan (*TL-2* **1**: 649, 1976).

Dietelia Henn. (1897), Pucciniosiraceae. 7 (on dicots), S. Am., Philipp.; 0, III. See Buriticá (*Rev. Acad. Colomb. Cienc.* **18**(69): 131, 1991).

Dievernia M. Choisy (1931) = Ramalina (Ramalin.).

differential hosts, the special species or cultivars of host plants the reactions of which are used for determining physiologic races.

diffluent, breaking up in water.

diffract (of a pileus surface), cracked into small areas; areolate.

Diffractella Guarro, P.F. Cannon & Aa (1991), Lasiosphaeriaceae. 1, Eur., Japan.

diffuse, widely or loosely spreading and having no distinct margin; **- wall building**, see wall building.

digitate, with deep radiating divisions, finger-like.

Digitatispora Doguet (1962), Amylocorticiaceae. 1 (marine), Eur., N. Am.

Digitellus Paulet (1791) ? = Lentinus (Lentin.) fide Donk (*Taxon* **11**: 82, 1962).

Digitodesmium P.M. Kirk (1981), Mitosporic fungi, 3.G2.1. 1, UK. See Sutton (*Proc. Acad. Ind. Acad. Sci.* (Pl. Sci.) **94**: 229, 1985).

Digitodochium Tubaki & Kubono (1989), Mitosporic fungi, 3.G1.1. 1, Japan.

Digitoramispora R.F. Castañeda & W.B. Kendr. (1990), Mitosporic fungi, 1.G2.19. 2, Cuba, Canada.

Digitosarcinella S. Hughes (1984), Mitosporic fungi, 1.G1.1. 1, Brazil.

Digitosporium Gremmen (1953), Anamorphic Leotiaceae, 8.G2.1. Teleomorph Crumenulopsis. 1, Finland.

Digitothyrea P.P. Moreno & Egea (1992), Lichinaceae (L). 3, Afr., C. Am.

Digraphis Clem. (1909) = Graphis (Graphid.).

Diheterospora Kamyschko (1952), Mitosporic fungi, 1.A1.15. 12, widespr. See Barron & Onions (*CJB* **44**: 861, 1966), Barron (*CJB* **63**: 211, 1984; 12 spp. from parasitized rotifers).

Dihyphis Locq. (1985), Fossil fungi. 1, Estonia.

dikaryon (adj. **dikaryotic**), a cell having two genetically distinct haploid nuclei.

dikaryoparaphysis, see hyphidium.

dikaryotization, the conversion of a homokaryon into a dikaryon typically by the fusion of 2 compatible homokaryons, but see Buller phenomenon; **illegitimate -**, the sporadic occurrence of a dikaryon in non-compatible di-mon matings.

dilacerate, torn asunder.

Dillenius (Johann Jacob; 1684-1747). First Sherardian Professor of botany, Oxford Univ. (1728). Paid special attention to fungi (incl. lichens), and mosses. Noted for *Historia muscorum*, 1742, which includes all the then known lichens, many of which are supported by specimens in **OXF**. See *Dict. Nat. Biogr.* (Compact edn) **1**: 541; Grummann (1974: 9), Jørgensen *et al.* (*Bot. J. Linn. Soc.* **115**: 261, 1994; typif. many lichen names), Petersen (*Mycotaxon* **5**: 415, 1977; mycological work), Stafleu & Cowan (*TL-2* **1**: 655, 1976).

Dilophia Sacc. (1883) [non Thomson (1853), *Cruciferae*] ≡ Lidophia (Dothideales).

Dilophospora Desm. (1840), Anamorphic Dothideales, 8.C1.15. Teleomorph *Lidophia*. 1, widespr. See Walker & Sutton (*TBMS* **62**: 231, 1974; teleomorph).

Dimargaris Tiegh. (1875), Dimargaritaceae. 7 (mycoparasites of *Mucorales*), N. Eur., N. Am., India. See Benjamin (*Aliso* **6**: 1, 1965; key), Mandelbrot & Erb (*Mycol.* **64**: 1124, 1972; hosts of *D. verticillata*), Saikawa (*J. Jap. Bot.* **52**: 200, 1977; septal ultrastr.), Jeffries & Young (*Ann. Bot.* **47**: 107, 1981; haustoria, *TBMS* **83**: 223, 1984; spore ultrastr.), Jeffries & Cuthbert (*Protoplasma* **121**: 129, 1984; ultrastr.), Kirk & Kirk (*TBMS* **82**: 551, 1984).

Dimargaritaceae R.K. Benj. (1959), Dimargaritales. 1 fam., 3 gen., 14 spp. Merosporangia bisporous, dry- or wet- spored, zygospores smooth or slightly ornamented; obligate mycoparasites of *Mucorales* (rarely *Chaetomium* spp.).
    For *Lit.* see *Dimargaritales*.

Dimargaritales, Zygomycetes. 1 fam., 4 gen., 15 spp. Fam.:
    Dimargaritaceae
    Lit.: Benjamin (*Aliso* **4**: 321, **5**: 273, **6**: 1, 1959-1965, *in* Kendrick (Ed.), *The whole fungus* **2**: 573, 1979).

Dimastigosporium Faurel & Schotter (1965), Mitosporic fungi, 7.A1.?. 1 (coprophilous), Sahara.

Dimaura Norman (1853) = Catolechia (Rhizocarp.) fide Hafellner (1978).

Dimelaena Norman (1853), Physciaceae (L). 3, widespr. See Leuckert *et al.* (*Nova Hedw.* **34**: 623, 1981; chemotypes), Sheard (*Bryologist* **80**: 100, 1977; palaeogeography, chemotaxonomy).

Dimera Fr. (1825) = Oedemium (Mitosp. fungi) fide Hughes (1958).

Dimerella Trevis. (1880), Gyalectaceae (L). *c.* 25, cosmop. See Lettau (*Beih. Feddes Repert.* **69** (2): 97, 1937), Santesson (1952), Vězda & Farkas (*Folia geobot. phytotax.* **23**: 187, 1988; key 12 spp. Afr.).

Dimeriaceae, see *Pseudoperisporiaceae*.

Dimeriella Speg. (1908), Parodiopsidaceae. 1, Brazil. See Barr (*Prodromus to class Loculoascomycetes*, 1987; posn), Farr (*Mycol.* **71**: 243, 1979).

Dimeriellina Chardón (1939) = Auerswaldiella (Dothid.) fide v. Arx & Müller (1954).

Dimeriellopsis F. Stevens (1927) = Nematostoma (Pseudoperispor.) fide Sivanesan (*SA* **6**: 201, 1987).

Dimerina Theiss. (1912), Pseudoperisporiaceae. *c.* 10 (on *Meliolaceae, Asterinaceae*), trop. See Hansford (1946). Anamorph Ectosticta.

Dimerinopsis Syd. & P. Syd. (1917) = Dimerina (Pseudoperispor.) fide v. Arx & Müller (1975).

Dimeriopsis F. Stevens (1917) = Dimerina (Pseudoperispor.) fide Müller & v. Arx (1962).

Dimerisma Clem. (1909) = Spheconisca (Verrucar.).

Dimerium Syd. & P. Syd. (1904), Parodiopsidaceae. 1, Chile. See Hughes (*Mycol. Pap.* **166**, 1993).

Dimerium (Sacc. & P. Syd.) Sacc. & D. Sacc. (1905) nom. dub. (Ascomycota, inc. sed.) fide Hughes (*Mycol. Pap.* **166**, 1993). See also *Phaeostigme*.

Dimeromyces Thaxt. (1896), Laboulbeniaceae. 100, widespr. See Santamaria (*Nova Hedw.* **58**: 177, 1994).

Dimerospora Th. Fr. (1860) = Lecania (Bacid.).

Dimerosporiella Speg. (1908), Ascomycota (inc. sed.). 1, Brazil. See Petrak & Sydow (*Ann. Myc.* **32**: 5, 1934).

Dimerosporiella Höhn. (1909) ≡ Dimerosporina (Parodiopsid.).

Dimerosporina Höhn. (1910) = Dysrhynchis (Parodiopsid.) fide Müller & v. Arx (1962).

Dimerosporiopsis Henn. (1901) = Antennularia (Ventur.) fide Müller & v. Arx (1962).

Dimerosporium Fuckel (1870) = Asterina (Asterin.) fide Müller & v. Arx (1962).

dimerous (of basidia), having a constriction between the probasidium and the metabasidium, as in *Brachybasidium*.

dimidiate, (1) shield-like; appearing to lack one half, or having one half very much smaller than the other; (2) (of a pileus), without a stalk and semi-circular; (3) (of lamellae), stretching only halfway to the stipe; (4) (of an ascomatal wall), having the outer wall covering only the top part.

dimitic, see hyphal analysis.

dimixis, see heterothallism.

di-mon, see Buller phenomenon.

dimorphic, having two forms; esp. of *Histoplasma*, *Sporothrix*, and other pathogens of humans and animals which have yeast and mycelial habits. See Romano (*in* Ainsworth & Sussman (Eds), *The Fungi* **2**: 181, 1966; review), Szaniszlo (Ed.) (*Fungal dimorphism* ...., 1985), San-Blas (*Handb. Appl. Mycol.* **2**. *Humans, animals and insects*: 459, 1991; molec. aspects).

Dimorphocystis Corner (1950) = Actiniceps (Pterul.) fide Boedijn (1959).

Dimorphomyces Thaxt. (1893), Laboulbeniaceae. 25, Eur., Am., Asia, W. Afr.

Dimorphospora Tubaki (1958), Mitosporic fungi, 1.A1.3+15. 1, widespr.

Dimorphotricha Spooner (1987), Hyaloscyphaceae. 1, Australia.

Dinemasporiella Speg. (1910) nom. dub. (Mitosp. fungi) fide Sutton (1977).

Dinemasporiella Bubák & Kabát (1912) = Pseudolachnea (Mitosp. fungi) fide Sutton (1977).

Dinemasporiopsis Bubák & Kabát (1914) = Pseudolachnea (Mitosp. fungi) fide Sutton (1977).

Dinemasporis Clem. & Shear (1931) ≡ Dinemasporiella (Mitosp. fungi).

Dinemasporium Lév. (1846), Anamorphic Ascomycota (inc. sed.), 7.A1.15. Teleomorph *Phomatospora*. 7, widespr. See Nag Raj (1993).

Dingleya Trappe (1979), Otideaceae. 7 (hypogeous), Australia. See Trappe *et al.* (*Austr. Syst. Bot.* **5**: 597, 1992).

dioecism (adj. **dioecious**), the condition in which the male and female sex structures are on different thalli, e.g. in certain *Laboulbeniales*; also reported in *Lecidea verruca* where the 'male' thalli are mostly smaller (Poelt, *Pl. Syst. Evol.* **135**: 81, 1980). cf. monoecism, heterothallism.

Dioicomyces Thaxt. (1901), Laboulbeniaceae. 29, widespr.

Diomedella Hertel (1984), Lecanoraceae (L). 2, NZ, Subantarctic Isl.

Dionysia Arnaud (1952) = Candelabrum (Mitosp. fungi) fide Bottomley (*TBMS* **37**: 234, 1954).

Diorchidiella Lindq. (1957), Raveneliaceae. 1 (on *Mimosa*), S. Am.

diorchidioid (of teliospores), 2-celled and with septum longitudinal.

Diorchidium Kalchbr. (1882), Raveneliaceae. *c.* 12, widespr.

Diorygma Eschw. (1824) = Graphina (Graphid.).

Diosporangium, see *Diasporangium*.

Dioszegia Zsolt (1957) = Cryptococcus (Mitosp. fungi) fide v. Arx *et al.* (*Stud. Mycol.* **14**, 1977).

Diphaeis Clem. (1909) = Rhizocarpon (Rhizocarp.).

Diphaeosticta Clem. (1909) = Pseudocyphellaria (Lobar.).

Diphanis Clem. (1909) = Rhizocarpon (Rhizocarp.).

Diphanosticta Clem. (1909) = Pseudocyphellaria (Lobar.) fide Galloway (1988).

Diphloeis Clem. (1909) = Toninia (Catillar.).

Diphragmium Boedijn (1960), Uredinales (inc. sed.). 1 (on *Leguminosae*), Java.

Diphratora Trevis. ex Jatta (1900) = Solenopsora (Bacid.).

Diphtherium Ehrenb. (1818) = Lycogala (Myxom.).

diphycophilous, fungi lichenized with both a green and a blue-green photobiont (Pike & Carroll, *in* Alexopoulos & Mims, *Introductory mycology*, edn 3, 1980).

Diphymyces I.I. Tav. (1985), Laboulbeniaceae. 5, Eur., S. Am., Asia, NZ.

Diplacella Syd. (1930), Valsaceae. 1, trop. Am.

Diplanes Leitg. (1868) = Saprolegnia (Saprolegn.) fide Coker (1923).

diplanetism (adj. **diplanetic**) (of zoospores of *Oomycota*), a sequence of two motile flagellate phases with an interspersed mobile aplanosporic phase in the zoosporic part of the life-history; the aplanosporic phase as a walled cyst; motile phases may be monomorphic or dimorphic.

diplo- (prefix), two; twice; double.

diplobiontic, see *Diplobionticae*.

Diplobionticae. Subclass introduced by Nannfeldt (1932) for *Ascomycota* in which the life cycle consists of two thalli (bionts), like the gametophyte and sporophyte of many algae, e.g. *Spermophthora* is diplobiontic, because of its two generations, free from each other; cf. *Haplobionticae*.

Diplocarpa Massee (1895), Hyaloscyphaceae. 1, Eur.

Diplocarpon F.A. Wolf (1912), Dermateaceae. 6, widespr. *D. rosae* (anamorph *Marssonina rosae*; black spot of rose), *D. earlianum* (anamorph *M. fragariae*; strawberry leaf scorch).

Diplocarponella Bat. (1957) = Stomiopeltis (Micropeltid.) fide Müller & v. Arx (1962).

Diploceras (Sacc.) Died. (1915) = Seimatosporium (Mitosp. fungi) fide Sutton (*TBMS* **64**: 483, 1975).

Diplochora Höhn. (1906) = Pseudophacidium (Ascodichaen.). See Petrak (*Sydowia* **5**: 193, 1951).

Diplochora P. Syd. (1913) ≡ Diplochorella (Mycosphaerell.).

Diplochorella P. Syd. (1913) = Microcyclus (Mycosphaerell.) fide v. Arx & Müller (1975).

Diplochorina Gutner (1933), Dothideales (inc. sed.). 1, former USSR.

Diplochytridium Karling (1971), Chytridiaceae. 21, Eur., Asia, Am.

Diplochytrium Tomaschek (1878) = Diplochytridium (Chytrid.) fide Karling (1977).

Diplocladiella G. Arnaud ex M.B. Ellis (1976), Mitosporic fungi, 1.G2.10. 3, Eur., Malaysia. See Nawawi (*Mycotaxon* **28**: 297, 1987).

Diplocladium Bonord. (1851) = Cladobotryum (Mitosp. fungi) fide Hughes (1958).

Diplococcium Grove (1885), Mitosporic fungi, 1.B2.28. 7, Eur. See Sinclair *et al.* (*TBMS* **85**: 736, 1985).

diploconidium, a binucleate conidium.

Diplocryptis Clem. (1909) = Cryptodiscus (Stictid.) fide Sherwood (1977).

Diplocystaceae Kreisel (1974), Sclerodermatales. 1 gen., 1 sp. Gasterocarpic with numerous globose gasterocarps with a small irregular opening joined on a common stroma.

Diplocystis Berk. & M.A. Curtis (1869), Diplocystaceae. 1, W. Indies. See Kreisel (*Feddes Repert.* **85**: 325, 1974).

Diploderma Link (1816) ? = Astraeus (Astr.).

Diplodia Fr. (1834), Anamorphic Botryosphaeriaceae, 4.B2.1. Teleomorph *Botryosphaeria*. 950, widespr. See Zambettakis (*BSMF* **70**: 219-350, 1954; sp. concept, 24 spp. accepted), Shear (*Mycol.* **25**: 274, 1933; teleomorphs). See also *Botryodiplodia*.

Diplodiella (P. Karst.) Sacc. (1884) nom. dub. fide Sutton (*Mycol. Pap.* **141**, 1977).

Diplodiella Petr. (1953) nom. illegit. (Mitosp. fungi) fide Sutton (*Mycol. Pap.* **141**, 1977).

Diplodina Westend. (1857), Anamorphic Valsaceae, 6.B/C1.15. Teleomorph *Cryptodiaporthe*. 3, widespr. See Sutton (1980), Wulf (*Nachricht. Deutsch. Pflanzenschutz.* **42**: 97, 1990; *D. acerina* endophytic and antagonistic to leaf-feeding insects).

Diplodinis Clem. (1931), Mitosporic fungi, 4.B1.?. 1, Scandinavia.

Diplodinula Tassi (1902), Mitosporic fungi, 4.B1.?. 68, widespr.

diplodioecious, see heterophytic.

Diplodiopsis Henn. (1904) = Parodiella (Parodiell.) fide Müller & v. Arx (1962).

diplodiosis, in cattle and sheep, a neuromuscular paretic syndrome caused by *Stenocarpella maydis* (syn. *Diplodia maydis*) infected maize in South Africa.

Diplodites Babajan & Tasl. (1970), Fossil fungi. 1 (Tertiary), former USSR.

Diplodothiorella Bubák (1916) = Sphaerellopsis (Mitosp. fungi) fide Sutton (1977).

Diplogelasinospora Cain (1961), Sordariaceae. 2, Asia, N. Am. See Udagawa & Horie (*J. Jap. Bot.* **47**: 297, 1972), Udagawa (*in* Subramanian (Ed.), *Taxonomy of fungi* **1**: 225, 1978; anamorph).

Diplogramma Müll. Arg. (1891), Graphidaceae (L). 1, Australia.

Diplogrammatomyces Cif. & Tomas. (1953) ≡ Diplogramma (Graphid.).

Diplographis Kremp. ex A. Massal. (1860) = Graphis (Graphid.).

diploheteroecious, see heterophytic.

Diploicia A. Massal. (1852), Physciaceae (L). *c.* 3, temp. See Elix *et al.* (*Mycotaxon* **33**: 457, 1988; chemistry).

diploid (1) (of a nucleus), having the 2n number of chromosomes; (2) (of a cell), having the 2n number of chromosomes in one (synkaryotic, 2n) or two (dikaryotic, n + n) nuclei; (3) (of a mycelium), made up of dikaryotic diploid cells.

Diploidium G. Arnaud (1923) = Septoidium (Mitosp. fungi) fide Ellis (*DH*).

diploidization, the process by which a haploid cell (or mycelium) becomes a diploid (dikaryotic) cell (mycelium) having conjugate nuclei (Buller, 1941); cf. heterokaryotise.

diplokaryon, see synkaryon.

Diplolabia A. Massal. (1854) = Graphis (Graphid.).

Diplolaeviopsis Giralt & D. Hawksw. (1991), Mitosporic fungi, 4.B1.19. 1 (on lichens), France, Spain.

Diplomitoporus Domański (1970), Coriolaceae. 2, Eur.

diplomitotic nuclear cycle, occurrence of two mitotic phases of different ploidy in the nuclear cycle (Dick, 1987); karyogamy - mitosis - meiosis - mitosis - karyogamy -.

diplomonoecious, see homophytic.

Diplomyces Thaxt. (1895), ? Laboulbeniaceae. 3, N. Am., Afr., Asia, Eur. See Rossi & Cesari Rossi (*Giorn. Bot. ital.* **112**: 63, 1978), Benjamin (*Aliso* **10**: 345, 1983).

Diplonaevia Sacc. (1889), Dermateaceae. 20, N. temp. See Hein (*Sydowia* **36**: 78, 1983; key), Nannfeldt (*Nord. Jl Bot.* **4**: 791, 1985; key 10 spp. *Juncaceae*).

Diplonema P. Karst. (1889) [non Kjellm. (1855), *Algae*] ≡ Amphinema (Athel.).

Diploneurospora K.P. Jain & R.C. Gupta (1970), Fossil fungi (Xylariales). 1 (Miocene), India.

diplont, the thallus of the diploid stage; the sporophyte.

Diploöspora Grove (1916), Mitosporic fungi, 2.B1.?. 2, Eur.

Diplopeltis Pass. (1889) [non Endl. (1837), *Sapindaceae*] ≡ Pycnoseynesia (Mitosp. fungi). See Sutton (1977).

Diplopeltopsis Henn. ex Höhn. (1911) = Asterothyrium (Thelotremat.).

diplophase, the part of a life-history in which the cells are diploid.

Diplophlyctis J. Schröt. (1892), Endochytriaceae. 5, widespr. See Sparrow (1960: 386; key), Dogma (*TBMS* **67**: 255, 1976; key chitinophilic spp.).

Diplophragmia Vain. (1934) = Lecidella (Lecanor.) fide Santesson (*The lichens and lichenicolous fungi of Sweden and Norway*, 1993).

Diplophrys J.S.F. Barker (1868), ? Labyrinthulomycota (inc. sed.). 2, N. temp. See Olive (1975: 227), Dick (*in press*).

Diplophysa J. Schröt. (1886) = Olpidiopsis (Olpidiopsid.) fide Sparrow (1960).

Diploplacis Clem. & Shear (1931) = Sphaerellopsis (Mitosp. fungi) fide Sutton (1977).

Diploplacosphaeria Petr. (1921) = Sphaerellopsis (Mitosp. fungi) fide Sutton (1977).

Diploplenodomopsis Petr. (1923) = Diplodina (Mitosp. fungi) fide Clements & Shear (1931).

Diploplenodomus Died. (1912), Mitosporic fungi, 8.B1.?. 2 or 3, Eur.

Diplopodomyces W. Rossi & Balazuc (1977), Laboulbeniaceae. 1, Eur.

Diplorhinotrichum Höhn. (1902) = Dactylaria (Mitosp. fungi) fide Bhatt & Kendrick (1968).

Diplorhynchus Arnaud (1952) [non Welw. ex Ficalho & Hiern (1881), *Phanerogamae*], nom. inval. Mitosporic fungi, 1.F1.1. 1, France.

Diploschistaceae, see Thelotremataceae.

Diploschistella Vain. (1926) nom. rej. = Gyalideopsis (Gomphill.) fide Lumbsch & Hawksworth (*Taxon* **36**: 764, 1987).

Diploschistes Norman (1853), Thelotremataceae (L). *c.* 35, cosmop. See Lettau (*Beih. Feddes Repert.* **69**, 1932-37), Lumbsch (*J. Hattori bot. Lab.* **66**, 133-196, 1989; key 14 holarctic spp., *Nova Hedw.* **56**: 227, 1993), Pant & Upreti (*Lichenologist* **25**: 33, 1993; key 14 spp. India & Nepal).

Diplosclerophoma Petr. (1923) = Diplodina (Mitosp. fungi). See Sutton (1977).

Diplosis Clem. (1909) = Toninia (Catillar.).

Diplosphaerella Grove (1912) = Delphinella (Dothior.) fide v. Arx & Müller (1975).

Diplosporis Clem. (1906) ≡ Geminispora (Phyllachor.).

Diplosporites Pia (1927), Fossil fungi. 1 (Oligocene), France.

Diplosporium Link (1824) = Oedemium (Mitosp. fungi) fide Hughes (1958).

Diplosporonema Höhn. (1917), Mitosporic fungi, 6.C1.10. 1, widespr.

diplospory, when a diploid nucleus is incorporated into cytoplasm influenced by adjacent meioses in the coenocytic gametangium, subsequently giving rise to oospores (in *Oomycetes*; Dick, *New Phytol.* **71**: 1151, 1972).

Diplostephanus Langeron (1922) = Emericella (Trichocom.) fide v. Arx (1981).

diplostichous, in two lines or groups.

Diplostoma, see *Diploderma*.

diplostromatic, see stroma.

diplosynoecious, see homophytic.

Diplotheca Starbäck (1893) [non Hochst. (1846), *Leguminosae*], Myriangiaceae. 1 (on *Cactaceae*), Am.

Diplotheca (Zahlbr.) Räsänen (1943) [non Hochst. (1846), *Leguminosae*] = Lecidea (Lecid.).

Diplothrix Vain. (1921) = Calloriopsis (Leot.) fide Santesson (*Svensk bot. Tidskr.* **45**: 300, 1951).

Diplotomma Flot. (1849), Physciaceae (±L). *c.* 15, widespr. See Singh & Awasthi (*Geophytology* **19**: 173, 1990; key 11 Indian spp.).

Diplotomma A. Massal. (1852) = Diploicia (Physc.).

Diplozythia Bubák (1904) nom. conf. (Mitosp. fungi) fide Sutton (1977).

Diplozythiella Died. (1916), Mitosporic fungi, 8.B1.15. 1, India.

Dipodascaceae Engl. & E. Gilg (1924), Saccharomycetales. 2 gen. (+ 3 syn.), 16 spp. Mycelium well-developed, lacking a polysaccharide sheath, the septa with clusters of minute pores, fragmenting to produce thallic conidia; asci formed by fusion of gametangia from adjacent cells or separate mycelia, usually elongate, ± persistent, 1- or multispored; ascospores released from the attenuated apex, usually ± ellipsoidal, rarely ornamented, usually with a mucous sheath, not blueing in iodine.

Dipodascales, see *Saccharomycetales*.

Dipodascopsis L.R. Batra & Millner (1978), Lipomycetaceae. 2, Eur., N. Am. See Batra (1978), v. Arx *et al.* (1977).

Dipodascus Lagerh. (1892), Dipodascaceae. 14, widespr. See de Hoog *et al.* (*Stud. Mycol.* **29**, 1986; key). Anamorph *Geotrichum*.

Dipodomyces Thaxt. (1931), Laboulbeniaceae. 2, W. Afr., Poland.

Diporicellaesporites Elsik (1968), Fossil fungi (*f. cat.*). 15 (Paleocene), China, India, USA.

Diporina Clem. (1909) = Strigula (Strigul.) fide Harris (*in* Hafellner & Kalb, *Bibl. Lich.* **57**: 161, 1995).

Diporisporites Hammen (1954), Fossil fungi (*f. cat.*). 20 (Cretaceous, Tertiary), China, Colombia, India, USA.

Diporopollis S. Dutta & Sah (1970), Fossil fungi (*f. cat.*). 1 (Eocene), India.

Diporotheca C.C. Gordon & C.G. Shaw (1961), Diporothecaceae. 1 (on roots), USA. See Mibey & Hawksworth (*SA* **14**: 25, 1995).

Diporothecaceae Mibey & D.Hawksw. (1995), Ascomycota (inc. sed.). 1 gen., 1 sp. Mycelium superficial, dark, not setose, with hyphopodia; stromata absent; ascomata perithecial, thin-walled, dark; interascal tissue absent, the ostiole periphysate; asci ± clavate, very thin-walled, without apical structures, evanescent, 8-spored; ascospores dark brown, 2-septate, the septa near the apices, with a separable perispore with poroid ends. Anamorph unknown. Biotrophic on roots.

Dipsacomyces R.K. Benj. (1961), Kickxellaceae. 1, Honduras. See Benjamin (*Aliso* **5**: 15, 1961).

Dipyrenis Clem. (1909) nom. dub. (Pyrenul.) fide Hawksworth (*Nova Hedw.* **43**: 1, 1986).

Dipyrgis Clem. (1909), Ascomycota (inc. sed.). 1, Australia. See Tibell (*Beih. Nova Hedw.* **79**: 597, 1984).

Dipyxis Cummins & J.W. Baxter (1967), Uropyxidaceae. 2, Mexico, Brazil.

direct (of fruit-body development), cell enlargement occurring at the same time as cell division; in **indirect**

development cell enlargement mainly occurs after the period of cell division (Corner, 1950).

Dirimosperma Preuss (1855) nom. dub. fide Sutton (*Mycol. Pap.* **141**, 1977).

Dirina Fr. (1825), Roccellaceae (L). 7, Eur., C., N. & S. Am., Afr. See Tehler (*Opera Bot.* **70**, 1983).

Dirinaria (Tuck.) Clem. (1909), Physciaceae (L). 26, trop. See Awasthi (*Bibl. Lich.* **2**, 1975; monogr.), Swinscow & Krog (*Norw. Jl Bot.* **25**: 157, 1978; E. Afr.).

Dirinastromyces Cif. & Tomas. (1953) ≡ Dirinastrum (Roccell.).

Dirinastrum Müll. Arg. (1893), Roccellaceae (L). 1, Australia.

Dirinella M. Choisy (1931) = Lecanora (Lecanor.).

Dirinopsis De Not. (1846) = Dirina (Roccell.) fide Tehler (1983).

Disaeta Bonar (1928) = Seimatosporium (Mitosp. fungi) fide Shoemaker (1964).

Disarticulatus G.F. Orr (1977) = Arachniotus (Gymnoasc.) fide v. Arx (*Persoonia* **9**: 393, 1977).

disc, (1) (**disk**) (of discomycetes), the round, plate-like or curved spore-producing part of the ascoma; (2) (of a pileus), the central part of the top surface.

Discales. Used by Le Gal (1953) for discomycetes.

Discaria (Sacc.) Sacc. (1889) [non Hook. (1830), *Rhamnaceae*] ≡ Plicaria (Peziz.).

Discella Berk. & Broome (1850) ? = Rhabdospora (Mitosp. fungi) fide Sutton (1977).

Dischloridium B. Sutton (1977), Mitosporic fungi, 1.A1.15. 4, widespr. See Holubová-Jechová (*Česká Myk.* **41**: 107, 1987).

disciform, round and flat.

Discina (Fr.) Fr. (1849) = Gyromitra (Helvell.) fide Kimbrough *et al.* (*CJB* **68**: 317, 1990). See Donadini (*Bull. Soc. Linn. Provence* **38**: 161, 1987; 17 spp.).

Discina Bonord. (1851) nom. dub. (Fungi, inc. sed.) fide Eckblad (*Nytt Mag. Bot.* **15**: 170, 1968).

Discinella Boud. (1885), Leotiaceae. 5, widespr. See Dennis (*Mycol. Pap.* **62**, 1956).

Discinella P. Karst. (1891), Pezizales (inc. sed.). 1, Eur.

Disciotis Boud. (1885), Morchellaceae. *c.* 5, Eur.

Disciseda Czern. (1845), Lycoperdaceae. 15, widespr. See Moravec (*Sydowia* **8**: 278, 1954).

Discoascina Beneš (1961), Fossil fungi (*f. cat.*). 1 (Carboniferous), former Czechoslovakia.

Discocainia J. Reid & A. Funk (1966), Rhytismataceae. 3, Eur., Greenland, N. Am. See Sherwood (*Occ. Pap. Farlow Herb.* **15**, 1980), Livsey & Minter (*CJB* **72**: 549, 1994).

discocarp, an ascoma in which the hymenium is uncovered when the asci and spores are mature; an apothecium.

Discocera A.L. Sm. & Ramsb. (1917) nom. rej. prop. ? = Trapelia (Trapel.). See Hawksworth & David (*Taxon* **38**: 493, 1989), Purvis *et al.* (*Lichen flora of Great Britain and Ireland*, 1992), Rambold & Triebel (*Notes R. bot. Gdn Edinb.* **46**: 375, 1990; status).

Discochora Höhn. (1918) = Guignardia (Mycosphaerell.)

Discocistella Svrček (1962) ≡ Cistella (Hyaloscyph.).

Discocolla Prill. & Delacr. (1894), Mitosporic fungi, 3.C1.?. 1 or 2, N. temp.

Discocurtisia Nannf. (1983), Dermateaceae. 1, N. Am.

Discocyphella Henn. (1900) nom. dub. fide Singer (1975). ? = Gloiocephala (Tricholomat.).

Discodiaporthe Petr. (1921) = Melanconis (Melanconid.) fide Wehmeyer (1941).

Discodothis Höhn. (1909) = Rhagadolobium (Parmular.) fide Müller & v. Arx (1962).

Discofusarium Petch (1921) = Fusarium (Mitosp. fungi) fide Wollenweber & Reinking (*Die Fusarien*, 1935).

Discogloeum Petr. (1923), Mitosporic fungi, 6.A1.15. 2, Eur. See v. Arx (*Bibl. Mycol.* **24**, 1970).

Discohainesia Nannf. (1932), Dermateaceae. 1, N. Am., Eur. See Shear & Dodge (*Mycol.* **13**: 135, 1921). Anamorphs *Hainesia*, *Pilidium.*

discoid, flat and circular; resembling a disk (Fig. 37.5*a*, *b*).

discolourous, of a different colour, as of the two surfaces of a foliose lichen thallus.

Discomycella Höhn. (1912), ? Leotiales (inc. sed.). 1, Java.

discomycete, one of the *Discomycetes.*

Discomycetella Sanwal (1953) = Inermisia (Otid.) fide Kimbrough (1970). See Pfister (*Am. J. Bot.* **60**: 355, 1973).

Discomycetes, cup fungi; class of *Ascomycota* formerly used for taxa with ascomata which are sessile, open, ± saucer- or cup-shaped apothecia, but these may be covered by a membrane at first or be permanently closed and hypogeous, or have the hymenium borne on stipitate convoluted structures; the apothecia are generally ascohymenial in ontogeny, with unitunicate asci (inoperculate or operculate). The *Leotiales*, *Elaphomycetales*, *Ostropales*, *Pezizales* and *Rhytismatales* are regularly included, but the name has also been applied to some *Dothideales*, *Patellariales* and *Lecanorales*. For usages of the name see Tables under *Ascomycota* (q.v.). The class is not accepted in modern classification but 'discomycetes' still has value as a colloquial descriptive term.

  Lit.: **General**: Boudier (1907, 1905-10; see Boudier), Nannfeldt (1932), Dennis (1978), Bellèmere (*BSMF* **83**: 393, 753, 1968; inoperc. ontogeny; *Revue mycol.* **41**: 233, 1977; ascus ultrastr.), Kimbrough (*Bot. Rev.* **36**: 91, 1970; classification; fam. key; *in* Parker, 1982, **1**: 232), Korf (1973), Kamaletdinova & Vassilyeva (*Cytology of discomycetes*, 1982; ultrastr.). See also Le Gal (1953), Seaver (1928) cited below, and under *Ascomycota*, ascus, Lichens and Macromycetes.

  Regional: America, North, Seaver (*North American cup-fungi. (Operculates)*, 1928; suppl. 1942; nomenclatural revision, Pfister, *Occ. Pap. Farlow Herb. Cryptog. Bot.* **17**, 1982); *North American cup-fungi. (Inoperculates)*, 1951). **Argentina**, Gamundí (*Darwiniana* **12**: 386, 1962, *Leotiales*; **13**: 568, 1964, *Pezizales*; keys). **Australasia**, Rifai (*Verh. K. ned. Akad. wet.* ser. 2 **57** (3), 1968; Operculates; Spooner (1987; *Leotiales*) **British Isles**, Ramsbottom & Balfour-Browne (*TBMS* **34**: 38, 1951; checklist), Dennis (1978), Cannon *et al.* (1981). **Canada**, Abbott & Currah (*The larger cup fungi and other ascomycetes of Alberta*, 1989; keys). **former Czechoslovakia**, Velenovský (*Monographia discomycetum Bohemiae*, 2 vols, 1934). **Europe**, see Boudier (1907, 1905-10). **France**, Gillet (*Champignons de France. Les Discomycetes*, 1879-83 [reprint 1979 (*Bull. Soc. bot., C.-o.*; num. spéc. 3)]). **Germany**, Baral & Krieglsteiner (*Z. Mykol., Beih.* **6**, 1985; inoperculates). **India**, Batra & Batra (*Kansas Univ. Sci. Bull.* **44**: 109, 1963; 185 spp. keys), Thind & Singh (*Res. Bull. Punjab Univ.* **22**: 51, 1971). **Japan**, Otani (*Trans. mycol. Soc. Japan* **31**: 117, 1990; keys gen.). **Madagascar**, Le Gal (*Les discomycètes de Madagascar* [*Prodr. Fl. mycol. Madagascar* 40], 1953). **Russia**, Naumov (*Flora gribov Leningradskovoblasti, Vypusk II. Diskomitsety*, 1964; keys). **Switzerland**, Breitenbach & Kränzlin (*Pilzäner Schweiz* **1**, 1984).

Discomycetoidea Matsush. (1993), Mitosporic fungi, 2.A2.22. 1, Ecuador.

Discomycopsella Henn. (1902) = Phyllachora (Phyllachor.) fide v. Arx & Müller (1954).

Discomycopsis Müll. Arg. (1893) ? = Euryachora (Mycosphaerell.). See Ferdinandsen & Rostrup (*Dansk. bot. Arkiv* **5** (20), 1928).

Discorehmia Kirschst. (1936), ? Dermateaceae. 6, Eur.

Discosia Lib. (1837), Mitosporic fungi, 8.C2.19. 11, esp. temp. See Nag Raj (1993), Subramanian & Reddy (*Kavaka* **2**: 57, 1974; types), Reddy (*in* Subramanian (Ed.), *Taxonomy of fungi*: 493, 1984), Nag Raj (*CJB* **69**: 1246, 1991; excluded spp.), Vanev (*Mycotaxon* **41**: 387, 1991; sections).

Discosiella Syd. & P. Syd. (1912) = Strigula (Strigul.) fide Eriksson & Hawksworth (*SA* **11**: 56, 1992). See also Subramanian & Reddy (*Proc. Ind. Acad. Sci.* B **75**: 111, 1972), Nag Raj (*CJB* **59**: 2519, 1981).

Discosiellina Subram. & K.R.C. Reddy (1972), Mitosporic fungi, 5.E1.1. 1, India. See Nag Raj (*CJB* **59**: 2531, 1981).

Discosiopsis Edward, Kr.P. Singh, S.C. Tripathi, M.K. Sinha & Ranade (1974) = Pestalotiopsis (Mitosp. fungi) fide Sutton (1977).

Discosiospora A.W. Ramaley (1989) = Discosia (Mitosp. fungi) fide Nag Raj (1993).

Discosphaera Dumort. (1822) nom. rej. = Hypoxylon (Xylar.).

Discosphaerina Höhn. (1917), Mycosphaerellaceae. 5, Eur., N. Am. See Barr (*Contr. Univ. Mich. Herb.* **9**: 523, 1972). = Guignardia (Mycosphaerell.) fide v. Arx & Müller (1975). Anamorph *Selenophoma.*

Discospora Arthur (1907) = Pileolaria (Pileolar.) fide Arthur (1934).

Discosporella Höhn. (1927) ≡ Conostroma (Mitosp. fungi). See Petrak (*Ann. Myc.* **27**: 371, 1929).

Discosporiella Petr. (1923) = Cryptosporiopsis (Mitosp. fungi) fide v. Arx (*Verh. K. ned. Akad. Wet. Amst.* **51**: 22, 1957).

Discosporina Höhn. (1927), Mitosporic fungi, 6.A1.15. 4, widespr.

Discosporiopsis Petr. (1921) = Phacidiopycnis (Mitosp. fungi) fide Sutton (1977).

Discosporium Sacc. & P. Syd. (1902), Mitosporic fungi, 3.A2.?. 1, Italy.

Discosporium Höhn. (1915), Mitosporic fungi, 8.A1.19. 3, widespr. See Sutton (1980).

Discostroma Clem. (1909), Amphisphaeriaceae. 10, widespr. See Brockman (*Sydowia* **28**: 275, 1976; key), Sivanesan (*TBMS* **81**: 325, 1983). Anamorph *Seimatosporium.*

Discostromella Petr. (1924) = Leptostromella (Mitosp. fungi) fide Clements & Shear (1931).

Discostromopsis H.J. Swart (1979) = Discostroma (Amphisphaer.) fide Sivanesan (*TBMS* **81**: 325, 1983).

Discotheciella Syd. & P. Syd. (1917), Mitosporic fungi, ?5.B1.?. 1, Philipp.

discothecium, an ascostroma resembling an apothecium but bearing cylindrical bitunicate asci and differing from a hysterothecium by the weathering away of the covering layer (Korf, *Mycol.* **54**: 25, 1962).

Discothecium Zopf (1897) = Endococcus (Dothideales) fide Hawksworth (1979).

Discothecium Syd. & P. Syd. (1916) ≡ Discotheciella (Mitosp. fungi).

Discoxylaria Lindq. & J.E. Wright (1964), Xylariaceae. 1, Argentina, Mexico. See Rogers *et al.* (*Mycol.* **87**: 41, 1995). Anamorph *Hypocreodendron.*

Discozythia Petr. (1922), Mitosporic fungi, ?.A1.?. 3, Eur.

discrete, (1) separate; not joining; (2) (of a conidiogenous cell), not subtended by a conidiophore; cf. integrated.

discrete body, a non-functional cleistothecial initial of a dermatophyte in culture; pseudocleistothecium.

Discula Sacc. (1884), Anamorphic Valsaceae, 6.A1.15. Teleomorph *Apiognomonia*. 12, Eur., N. Am. See v. Arx (*Verh. K. ned. Akad. Wet. Nat.* **51**: 32, 1957), Petrak (*Sydowia* **15**: 221, 1962; **24**: 270, 1971), Swart *et al.* (*Phytophyl.* **22**: 143, 1990; *D. platani* on plane trees in S. Afr.), Redlin (*Mycol.* **83**: 633, 1991; *Cornus* anthracnose), Haemmerli *et al.* (*Mol. Plant-Microbe Interact.* **5**: 479, 1992; differentiation by RAPDs), Toti *et al.* (*MR* **96**: 420, 1992; morphometry), McElreath *et al.* (*Curr. Microbiol.* **29**: 57, 1994; double-stranded RNA in *D. destructiva*).

Disculina Höhn. (1916), Mitosporic fungi, 8.A1.19. 1, Eur.

disinfectant, a substance for the destruction of pathogenic microorganisms.

disjunctor, a cell or projection, sometimes having a short existence, developing through the pores of septal lamellae of adjoining conidia in a chain (e.g. in *Monilinia*); a connective. See Batra (*Mycol.* **80**: 660, 1988).

disk, see disc.

Disperma Theiss. (1916) [non J.F. Gmel. (1792), *Rubiaceae*] ≡ Dicarpella (Melanconid.).

dispersal spore, a spore disseminated by wind, water, or other agent; diaspore.

Dispira Tiegh. (1875), Dimargaritaceae. 4 (mycoparasitic on *Mucorales* and *Chaetomium*), N. Eur., N. Am., S.E. Asia. See Benjamin (*Aliso* **5**: 248, 1963; key), Kurtzman (*Mycol.* **60**: 915, 1968; parasitism & culture), Misra & Lata (*Mycotaxon* **8**: 372, 1979).

dispore, one of the spores of a 2-spored basidium as opposed to a **tetraspore**, one of the spores of a 4-spored basidium (Corner, 1947). Cf. monospore.

Disporium Léman (1819), Myxomycetes (inc. sed.). 1, Eur.

Disporotrichum Stalpers (1984), Mitosporic fungi, 1.A1.10. 1, Netherlands.

dissepiment, a partition, e.g. that between the pores of a polypore.

Dissitimurus E.G. Simmons, McGinnis & Rinaldi (1987), Mitosporic fungi, 1.C2.3/10. 1 (from *Homo sapiens*), USA.

Dissoacremoniella Kiril. (1970), Mitosporic fungi, 1.A1.6. 1, former USSR.

dissociation, Leonian's name for mutation or saltation.

Dissoconium de Hoog, Oorschot & Hijwegen (1983), Mitosporic fungi, 1.A/B1.10. 2, Eur. See de Hoog & Takeo (*Ant. v. Leeuwenh.* **59**: 285, 1991; karyology).

Dissoderma (A.H. Sm. & Singer) Singer (1974), Agaricaceae. 1, Eur., Am.

Dissophora Thaxt. (1914), Mortierellaceae. 2, N. temp. See Gams & Carreiro (*Stud. Mycol.* **31**: 85, 1985; taxonomy), Carreiro & Koste (*CJB* **70**: 2177, 1982; growth temperature).

distal, situated away from either the centre of a body or the point of origin; terminal; cf. proximal.

Distichomyces Thaxt. (1905) = Rickia (Laboulben.) fide Thaxter (1912).

distichous, in two lines.

Distocercospora N. Pons & B. Sutton (1988), Mitosporic fungi, 1.C2.10. 1, widespr.

Distolomyces Thaxt. (1931), Laboulbeniaceae. 3, Eur., Asia.

Distopyrenis Aptroot (1991), Pyrenulaceae. 3, trop.

distoseptate (of septation), having the individual cells each surrounded by a sac-like wall distinct from the outer wall, as in *Drechslera* (Luttrell, *Mycol.* **55**: 672, 1963) (Fig. 35A-B). Cf. euseptate.

distribution, see Geographical distribution.

Ditangium P. Karst. (1867), Anamorphic Exidiaceae, ?.?.?. Teleomorph *Craterocolla*. 1, Eur. See Donk (*Persoonia* **4**: 165, 1966).

Dithelopsis Clem. (1909) = Thelopsis (Stictid.).

dithiocarbamates, organic fungicides; **dimethyl-** (DMDC): thiram, ferbam, ziram; **ethylene-bis-** (EBDC): maneb, mancozeb, zineb.

Dithozetia Clem. & Shear (1931) ≡ Didymothozetia (Mitosp. fungi).

Ditiola P. Browne (1756) ? = Schizophyllum (Schizophyll.) fide Donk (*Beih. Nova Hedw.* **6**: 89, 1962).

Ditiola Fr. (1822), Dacrymycetaceae. 2, widespr. See McNabb (*N.Z. Jl Bot.* **4**: 546, 1966).

Ditiola Schulzer (1860) ≡ Holwaya (Leot.).

Ditmaria Lühnem. (1809) nom. dub (? Fungi, inc. sed.).

Ditopella De Not. (1863), Melanconidaceae. 1 (on *Alnus*), Eur. See Barr (1978), Reid & Booth (*CJB* **45**: 1479, 1967).

Ditopellina J. Reid & C. Booth (1967), Valsaceae. 1, Portugal.

Ditopellopsis J. Reid & C. Booth (1967), Valsaceae. 3, N. Am. See Barr (*Mycol. Mem.* **7**, 1978, *Mycotaxon* **41**: 287, 1991), Monod (*Beih. Sydowia* **9**, 1993).

Ditremis Clem. (1909) nom. rej. prop. = Anisomeridium (Monoblast.).

Ditylis Clem. (1909) = Tylophoron (Caliciales) fide Tibell (1984).

diurnal, in daylight hours. Cf. circadian, diel.

divaricate, divergent at right angles.

diverticulum, a pocket-like side branch, as on mycelium of *Pythium*.

Divinia Cif. (1955), Mitosporic fungi, 1.C2.?. 1, Dominican Republic.

Dixidium R. Poiss. (1932) = Smittium (Legeriomycet.) fide Manier & Lichtwardt (*Ann. Sci. nat., Bot.* sér. 12 **9**: 519, 1968).

Dixomyces I.I. Tav. (1985), Laboulbeniaceae. 14, Asia, Afr., Am.

Dixophyllum Earle (1909) = Russula (Russul.) fide Singer (1951).

DNA, see Molecular biology.

Doassansia Cornu (1883), Tilletiaceae. *c.* 25 (on aquatic plants), widespr. Spore ball, immersed in host tissue, has sterile cortex. See Vánky (*Sydowia* **34**: 167, 1981).

Doassansiella Zambett. (1970) nom. inval. Anamorphic Tilletiaceae. Teleomorph *Doassansiopsis*. See Zambettakis (*Rev. Myc.* **35**: 164, 1970). ? = Paepalopsis (Mitosp. fungi).

Doassansiopsis (Setch.) Dietel (1897), Tilletiaceae. 8 (on aquatic plants), N. Am., Eur., Asia.

Dochmiopus Pat. (1887) = Crepidotus (Crepidot.) fide Singer & Smith (1946).

Dochmolopha Cooke (1878) = Seimatosporium (Mitosp. fungi) fide Shoemaker (1964).

Dodgea Malençon (1939) = Truncocolumella (Rhizopogon.) fide Smith & Singer (*Brittonia* **11**: 215, 1959).

Dodgella Zebrowski (1936), ? Fossil fungi (Chytrid.). 3 (Eocene to ? Recent), widespr.

dog lichen, *Peltigera* spp., e.g. *P. canina*; used in folk lore for treatment of bites of a rabid dog.

dog stinkhorn, basidioma of *Mutinus caninus*.

Doguetia Bat. & J.A. Lima (1960) = Trichasterina (Asterin.) fide Müller & v. Arx (1962).

Doidge (Ethel Mary; 1887-1965). Worked for the South African Public Service from 1908; principal plant pathologist, 1929-42. Made many contributions on S. African fungi (esp. rusts, *Bothalia* **2-4**) and phytopathology including the comprehensive list 'The South African fungi and lichens to the end of 1945' (*Bothalia* **5**, 1950). See Gunn (*Bothalia* **9**: 251, 1967; portr., bibl.), Grummann (1974: 396).

Dolabra C. Booth & W.P. Ting (1964), Dothideales (inc. sed.). 1, Malaysia.

dolabrate (dolabriform), hatchet-like in form (Fig. 37.24).

dolicho- (in Greek combinations), long.

Dolichoascus Thibaut & Ansel (1970), Saccharomycetales (inc. sed.). 1 (in *Insecta*), widespr. See Batra (1978; typification).

Dolichocarpus R. Sant. (1949), Roccellaceae (L). 1, Chile.

dolichospore, a long spore.

doliiform, barrel-like in form (Fig. 37.31).

Doliomyces Steyaert (1961), Mitosporic fungi, 4.C2.19. 2, India, S. Am. See Nag Raj & Kendrick (*CJB* **50**: 45, 1972).

dolipore septum, a septum of a dikaryotic basidiomycete hypha which flares out in the middle portion forming a barrel-shaped structure with open ends as shown by electron microscopy; see Markham (*MR* **98**: 1089, 1994; review), Moore (*in* Hawksworth (Ed.), *Identification and characterization of pest organisms*: 249, 1994; summary diagr., Fig. 13), Moore & McAlear (*Am. J. Bot.* **49**: 86, 1962); septal pore swelling; cf. parenthesome.

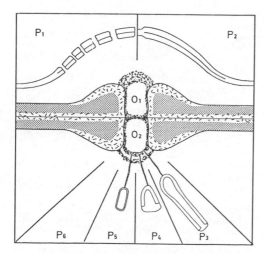

Fig. 13. Summary diagram of the several forms of the dolipore/parenthosome (d/p) septum in the Basidiomycota. The outer part of the mushroom-shaped occlusions can appear either as a granule (O_1) or a striated band (Osub(2")); the parenthosomes can be regularly perforate (P_1), imperforate (P_2), vesiculate (P_{3-5}), or absent (P_6). Particular taxa are generally associated with distinct d/p combinations: O_1 /P_1, subclass *Homobasidiomycetidae*; O_1/P_2, polypore tribes *Phellineae* and *Hirschioporae*, *Heterobasidiomycetidae* orders *Auriculariales*, *Dacrymycetales* and *Tulasnellales*, and suborder *Exidiineae* (note the central thin spot in which the intervening line in the double membrane morphology is absent); O_2/$P_{3,4,6}$, suborder *Tremellineae*; O_2/P_5, *Wallemia sebi*. Adapted from Moore, *in* Hawksworth (Ed.) (1994; 252).

dollar spot, a turf disease caused by *Sclerotinia homeocarpa*.

Domain (Empire, Superkingdom). A category in Classification (q.v.) above that of kingdom; all fungal phyla belong in kingdoms in the domain *Eukaryota*.

Domingoella Petr. & Cif. (1932), Mitosporic fungi, 2.A2.?. 2, trop. See Deighton & Pirozynski (*Mycol. Pap.* **128**, 1972).

Domin-scale. A 10-point scale used in ecological tables to indicate the approximate cover of a surface by different taxa: + = single individuals; 1 = a few individuals; 2 = sparsely distributed; 3 = frequent but cover less than 4%; 4 = cover 4-10%; 5 = cover 11-25%; 6 = cover 26-33%; 7 = cover 34-50%; 8 = cover 51-75%; 9 = 76-90%; 10 = cover 91-100%.

Donk (Marinus Anton; 1908-72). Mycologist at the herbarium, Botanic Gardens, Buitenzorg (Herb. Bogoriense), Java, 1934-40; head of the Mycological Dept, Rijksherbarium, Leiden, 1956-72. Taxonomist (esp. *Basidiomycetes*, q.v.), nomenclaturalist. See Maas Geesteranus (*Persoonia* **7**: 119, 1973; portr., bibl.), Singer (*Mycol.* **65**: 503, 1973; portr.), Stafleu & Cowan (*TL-2* **1**: 671, 1976).

Donkella Doty (1950) = Clavulinopsis (Clavar.) fide Donk. = Ramariopsis (Clavar.) fide Petersen (1978).

Donkia Pilát (1937) = Climacodon (Hydn.) fide Maas Geesteranus (1971).

Donkioporia Kotl. & Pouzar (1973), Coriolaceae. 1, Eur.

Dontuzia L.D. Gómez (1973), Ascomycota (inc. sed.). 1, trop.

Doratomyces Corda (1829), Mitosporic fungi, 2.A2.19. 8, widespr. See Morton & Smith (*Mycol. Pap.* **86**, 1963). = Cephalotrichum (Mitosp. fungi) fide v. Arx (*Genera of fungi sporulating in pure culture*, 3rd edn, 1981).

Doratospora J.M. Mend. (1930) = Rizalia (Trichosphaer.) fide Pirozynski (1977).

dorsal, back or upper surface; the surface facing away from the axis, cf. ventral; sometimes used for the upper surface of foliose lichens.

Dothichiza Lib. ex Roum. (1880), Anamorphic Dothioraceae, 8.A1.?. Teleomorph *Dothiora*. 15, cosmop. See Petrak (*Sydowia* **10**: 201, 1957), Sutton (1977).

Dothichloë G.F. Atk. (1894) = Balansia (Clavicipit.) fide Diehl (1950).

Dothiclypeolum Höhn. (1916) = Thyriopsis (Asterin.) fide Clements & Shear (1931).

Dothidasteris Clem. & Shear (1931) ≡ Dothidasteromella (Asterin.).

Dothidasteroma Höhn. (1909), Parmulariaceae. 2, Sri Lanka, Philipp.

Dothidasteromella Höhn. (1910), Asterinaceae. 3, Japan, S. Am., S. Afr.

Dothidea Fr. (1818), Dothideaceae. 20, temp.

Dothideaceae Chevall. (1826), Dothideales. 13 gen. (+ 19 syn.), 96 spp. Stromata multiloculate, immersed or erumpent, usually pulvinate or crustose; black, thick-walled, opening by an apical, usually lysigenous pore; interascal tissue lacking; asci saccate or clavate, fissitunicate; ascospores small, hyaline or brown, transversely septate. Biotrophic, necrotrophic or saprobic on plant tissue.

dothideaceous, having the asci in locules in a stroma, as in *Dothidea*.

Dothideales, Ascomycota (±L). 58 fam., 711 gen. (+ 718 syn.), 4,774 spp. Ascomata very varied, apothecial, perithecial or cleistothecial, at least nominally formed as lysigenous locules within stromatic tissue; hymenium sometimes gelatinous and blueing in iodine; interascal tissue frequently present, usually composed of branched or anastomosed paraphysoids or pseudoparaphyses, at least initially attached at both base and apex; asci usually cylindrical, thick-walled, often fissitunicate, rarely with apical structures; ascospores almost always septate, longitudinally asymmetrical, constricted at the primary septum but not always at the others, sometimes muriform; hyaline or brown, not often ornamented. Saprobes, parasites, lichen-forming, coprophilous, etc.

Includes *Asterinales*, *Capnodiales*, *Chaetothyriales*, *Dothiorales*, *Hysteriales*, *Melanommatales* p.p., *Myriangiales*, *Perisporiales*, *Pleosporales*, and *Pseudosphaeriales* as treated here (following Eriksson & Hawksworth, 1993).

The largest and most varied group of the *Ascomycota*, including most ascolocular ascomycetes with I- bitunicate asci. The classification of this group is unsettled and widely differing schemes have been proposed by Luttrell (1973), v. Arx & Müller (1975) and Barr (1979, 1987), as shown on pp. 000-000. v. Arx & Müller accepted 34 fams., and Eriksson (1982) 57 (+ 13 doubtfully here or in allied orders). For a detailed account of the type species of families see Eriksson (1981).

Not all fungi with functionally bitunicate (fissitunicate) asci are included here; see also under *Arthoniales*, *Erysiphales*, *Patellariales*, *Peltigerales*, *Pyrenulales* and *Verrucariales*.

The *Dothideales* will eventually require division into several orders but this is premature when the characters of so many genera are unclear. 180 genera (and 51 syns.) accepted in this *Dictionary* are not referred to a family. Barr (1987) accepted the *Myriangiales*, *Asterinales*, *Capnodiales*, *Dothideales*, *Chaetothyriales*, *Pleosporales* and *Melanommatales* as separate orders, and Eriksson (*SA* 6: 125, 1987) suggested that 8 suborders should be recognized. Fams:

(1) **Antennulariellaceae**.
(2) **Arthopyreniaceae** (syn. *Naetrocymbaceae*) (±L).
(3) **Argynnaceae**.
(4) **Asterinaceae**.
(5) **Aulographaceae**.
(6) **Botryosphaeriaceae**.
(7) **Brefeldiellaceae**.
(8) **Capnodiaceae**.
(9) **Chaetothyriaceae**.
(10) **Coccodiniaceae**.
(11) **Coccoideaceae**.
(12) **Cookellaceae**.
(13) **Cucurbitariaceae**.
(14) **Dacampiaceae**.
(15) **Diademaceae**.
(16) **Didymosphaeriaceae**.
(17) **Dimeriaceae** (syn. *Pseudoperisporiaceae*).
(18) **Dothideaceae**.
(19) **Dothioraceae**.
(20) **Elsinoaceae**.
(21) **Englerulaceae**.
(22) **Eremomycetaceae**.
(23) **Euantennariaceae**.
(24) **Fenestellaceae**.
(25) **Herpotrichiellaceae**.
(26) **Hypsostromataceae**.
(27) **Hysteriaceae**.
(28) **Lautosporaceae**.
(29) **Leptopeltidiaceae**.
(30) **Leptosphaeriaceae**.
(31) **Lichenotheliaceae**.
(32) **Lophiostomataceae** (syn. *Massarinaceae*).
(33) **Melanommataceae**.
(34) **Mesnieraceae**.
(35) **Metacapnodiaceae**.
(36) **Micropeltidaceae**.
(37) **Microtheliopsidaceae** (L).
(38) **Microthyriaceae** (syn. *Trichothyriaceae*).
(39) **Mycoporaceae**.
(40) **Mycosphaerellaceae**.
(41) **Myriangiaceae**.
(42) **Mytilinidiaceae** (syn. *Lophiaceae*).
(43) **Parmulariaceae**.
(44) **Parodiellaceae**.
(45) **Parodiopsidaceae**.
(46) **Phaeosphaeriaceae**.
(47) **Phaeotrichaceae**.
(48) **Piedraiaceae**.
(49) **Pleomassariaceae**.
(50) **Pleosporaceae** (syn. *Pseudosphaeriaceae*, *Pyrenophoraceae*).
(51) **Polystomellaceae** (syn. *Munkiellaceae*).
(52) **Pyrenothricaceae** (L).
(53) **Schizothyriaceae**.
(54) **Sporormiaceae**.
(55) **Tubeufiaceae**.
(56) **Venturiaceae**.
(57) **Vizellaceae**.
(58) **Zopfiaceae**.

Lit.: v. Arx & Müller (1954; *Stud. Mycol.* 9, 1975, keys fams, gen.), Barr (*Contr. Univ. Mich. Herb.* 9: 523, 1972, N. Am.; *Mycol.* 71: 935, 1979; *Prodromus to Class Loculoascomycetes*, 1987; orders, fams; in Parker, 1982, 1: 201), Benny & Kimbrough (1980; non-ostiolate gen.), Dennis (1978), Eriksson (1981, 1982), Eriksson & Hawksworth (1993), Hansford (1946; foliicolous spp.), Harris (*Mich. Bot.* 12: 3, 1973; L), Henssen & Jahns (1973 ['1974']; L), Janex (*Revue bryol. lichén.* 37: 421, 1971; ontogeny, L), Luttrell (1951, *Phytopathology* 55: 828, 1965, in Ainsworth *et al.* (Eds), *The Fungi* 4A: 135, 1973; keys, gen.), Müller & v. Arx (1962), Munk (1957), Parguey-Leduc (*Ann. Sci. nat.*, Bot. sér. 12, 7: 33, 1966; ontogeny), Pirozynski (*Kew Bull.* 31: 595, 1977; trop. fungicolous spp.), Poelt (*in* Ahmadjian & Hale (Eds), *The lichens*: 599, 1974 ['1973']; fams, L), Reynolds (Ed., 1981; centrum types, many chs), Theissen & Sydow (*Ann. Myc.* 13: 149, 1915; gen.), Walker (1980), also under *Ascomycota*, ascus, hamathecium, Lichens.

Dothideites Pat. (1893), Fossil fungi.

Dothidella Speg. (1880), Polystomellaceae. 3, trop. See Müller & v. Arx (1962), Wakefield (*TBMS* 24: 282, 1940).

Dothideodiplodia Murashk. (1927), Mitosporic fungi, 8.B/C2.19. 1, former USSR.

Dothideopsella Höhn. (1915) = Leptosphaeria (Leptosphaer.) fide v. Arx & Müller (1975).

Dothideovalsa Speg. (1909), Diatrypaceae. 3, C., S. & N. Am., W. Indies. See Rappaz (1987). = Eutypa (Diatryp.) fide Petrak (*Sydowia* 5: 169, 1951).

Dothidina Theiss. & Syd. (1915) = Coccodiella (Phyllachor.) fide Müller & v. Arx (1973).

Dothidites Mesch. (1892), Fossil fungi. 8 (Tertiary), Eur.

Dothidotthia Höhn. (1918), Botryosphaeriaceae. *c.* 10, temp. See Aptroot (*Nova Hedw.* 60: 325, 1995), Barr (*Mycotaxon* 34: 517, 1989; key 7 N. Am. spp.).

Dothiomyces Bat. & J.L. Bezerra (1961), Mitosporic fungi (L), 8.A1.15. 1, Brazil.

Dothiopeltis E. Müll. (1957), Leptopeltidaceae. 1, Switzerland. See Holm & Holm (*Bot. Notiser* 130: 115, 1977).

Dothiopsis P. Karst. (1884) nom. dub. (Mitosp. fungi) fide Sutton (1977).

Dothiora Fr. (1837) nom. rej. = Melanogramma (Fungi) fide Holm (*Taxon* 24: 475, 1975).

Dothiora Fr. (1849) nom. cons., Dothioraceae. 13, esp. temp. See Froidevaux (*Nova Hedw.* 23: 679, 1973; key). Anamorphs *Dothichiza*, *Hormonema*.

Dothioraceae Theiss. & P. Syd. (1917), Dothideales. 9 gen. (+ 16 syn.), 44 spp. Ascomata immersed to erumpent, globose to pulvinate, uniloculate, sometimes with a central sterile column, opening by a large irregular pore; peridium thick, of large-celled pseudoparenchymatous cells; interascal tissue absent but asci sometimes separated by stromatic tissue; asci clavate,

fissitunicate, sometimes polysporous; ascospores usually septate, sometimes muriform. Anamorphs coelomycetous where known. Biotrophic or necrotrophic, usually associated with woody plants.
Lit.: Froidevaux (*Nova Hedw.* **23**: 679, 1973).

Dothiorales, see *Dothideales*.

Dothiorella Sacc. (1880), Mitosporic fungi, ?.?.?. 50, widespr. See Petrak & Sydow (*Rep. Spec. nov. regni veg.* **42**: 214, 1927), Sutton (1977; typification).

Dothiorellina Bubák (1911) nom. dub. fide Sutton (*Mycol. Pap.* **141**, 1977).

Dothiorina Höhn. (1911), Mitosporic fungi, 8.A1.15. 3, Eur. See Riedl (*Sydowia* **29**: 146, 1977; key).

Dothioropsis Riedl (1974), Mitosporic fungi, 8.A1.?38. 1, Austria.

Dothisphaeropsis Höhn. (1919) = Microsphaeropsis (Mitosp. fungi) fide Sutton (1977).

Dothistroma Hulbary (1941), Anamorphic Mycosphaerellaceae, 8.E1.1/10. Teleomorph *Mycosphaerella*. 1, widespr. *D. septospora* causes defoliation of *Pinus*. See Gibson (*Ann. Rev. Phytopath.* **10**: 51, 1972), Evans (*Mycol. Pap.* **153**, 1984).

Dothithyrella Höhn. (1918) = Leptopeltis (Leptopeltid.) fide Holm & Holm (1977).

Dothivalsaria Petr. (1966), Massariaceae. 1, Eur. See Barr (*N. Am. Fl.* ser. 2, **13**, 1990).

Dothophaeis Clem. (1931) ≡ Englerodothis (Parmular.).

Dozya P. Karst. (1873) [non Sande Lac. (1866), *Musci*] ≡ Hypocreopsis (Hypocr.).

DR curve, a log-probit dosage-response curve (Horsfall, *Principles of fungicidal action*, 1956).

Drechmeria W. Gams & H.-B. Jansson (1985), Mitosporic fungi, 1.A1.15. 2 (from nematodes and protozoans), widespr.

Drechslera S. Ito (1930) nom. cons., Anamorphic Pleosporaceae, 1.C2.26. Teleomorph *Pyrenophora. c.* 23 (graminicolous), widespr. See Shoemaker (*CJB* **37**: 880, 1959; **40**: 809, 1962, key 14 spp.), Pandey & Gupta (*Acta Mycol.* **20**: 209, 1986; as a mycoparasite), Sivanesan (*Mycol. Pap.* **158**: 1, 1987; keys), Muchovej *et al.* (*Fitopat. Bras.* **13**: 211, 1988; keys), Strobel *et al.* (*Phytoparasitica* **16**: 145, 1988; phytotoxins), Alcorn (*Ann. Rev. Phytopath.* **26**: 27, 1988; generic taxonomy), Ondrej (*Česká Myk.* **43**: 45, 1989; key 20 Czech spp.), Khazanov (*Opredelitel' Gribov-Vozbul. 'Gel'mintosporiozov' Rast. iz Rodov Bipolaris, Drechslera i Exserohilum*, 1992).

Drechslerella Subram. (1964), Mitosporic fungi, 1.C1.1/10. 1, USA.

Drechsleromyces Subram. (1978), Mitosporic fungi, 1.E1.10. 1, USA.

Dremuspora Sal.-Cheb. & Locq. (1980), Fossil fungi. 1, Cameroon.

Drepanispora Berk. & M.A. Curtis (1875), Mitosporic fungi, 1.F2.1/10. 2, widespr. See Goos (*Mycol.* **81**: 356, 1989).

Drepanoconis J. Schröt. & Henn. (1896), Cryptobasidiaceae. 2, S. Am. See Linder (*Ann. Mo. bot. Gdn* **16**: 343, 1929).

Drepanomyces Thaxt. (1931), Ceratomycetaceae. 1, Sumatra, Sarawak.

Drepanopeziza (Kleb.) Höhn. (1917), Dermateaceae. 9, temp. See Rimpau (*Phytopath. Z.* **43**: 257, 1962; on *Ribes, Populus, Salix*; keys). Anamorphs *Gloeosporidiella, Marssonina*.

Drosella Maire (1935) = Chamaemyces (Agaric.) fide Singer (1951), = Lepiotella (Agaric.) fide Kühner & Maire (1934).

Drosophila Quél. (1886) = Psathyrella (Coprin.) fide Singer (1951) but used by Romagnesi (*BSMF* **91**: 137, 1975).

Drudeola Kuntze (1891), Mitosporic fungi, ?.?.?. 2, USA. See Sutton (1977).

Drummondia Bat. & H. Maia (1963) [non Hook. (1828), *Musci*] ≡ Mycousteria (Mitosp. fungi) fide Farr (*Mycol.* **78**: 280, 1986).

Drumopama Subram. (1957), Mitosporic fungi, 2.A2.10. 1, India.

Druparia Raf. (1808) nom. dub. (? 'gasteromycetes').

Drupasia Raf. (1809) ≡ Druparia (? 'gasteromycetes').

druse, a stellate cluster of large crystals in a lichen thallus.

dry rot fungus, (or **house fungus**), *Serpula lacrymans*; see Seehann & Hegarty (*Docs. Internat. Res. Group Wood Preserv.* IRG/WP/1337, 1988; bibliogr.), Watkinson (*Biodet. Abstr.* **8**: 161, 1994; review).

dry spore, a spore that becomes separated without slime from the cell producing it (Mason, 1937); cf. slime spore. See Xerosporae.

dryad's club, basidioma of *Clavaria pistillaris*.

dryad's saddle, basidioma of *Polyporus squamosus*.

Dryinosphaera Dumort. (1822) nom. dub. (Ascomycota, inc. sed.), used for diverse perithecioid fungi.

Dryodon Quél. ex P. Karst. (1881) = Hericium (Heric.) fide Donk (1956).

Dryophila Quél. (1886) = Pholiota (Strophar.) fide Patouillard (1900).

Dryophilum Schwein. (1834) nom. dub.; ? based on insect galls.

Dryosphaera J. Koch & E.B.G. Jones (1989), Ascomycota (inc. sed.). 2 (marine), trop. See Kohlmeyer & Volkmann-Kohlmeyer (*CJB* **71**: 992, 1993).

dual phenomenon (in Mitosporic fungi), the condition in which a fungus is made up of two culturally different elements or individuals (Hansen, *Mycol.* **30**: 442, 1938; Hansen & Snyder, *Am. J. Bot.* **30**: 419, 1943).

dual propagule (in a lichen), one comprising elements of both the fungal and the photosynthetic partner (e.g. isidium, soredium; see Lichens).

Dualomyces Matsush. (1987), Mitosporic fungi, 1.E1/2.10/11. 2, Taiwan.

Dubiocarpon S.A. Hutch. (1955), Fossil fungi (? Ascomycota). 4 (Carboniferous), Eur., USA. See Baxter (*Paleont. Contrib. Univ. Kansas* **77**, 1975), Pirozynski (1976).

Dubiomyces Lloyd (1921) = Ustilaginoidea (Mitosp. fungi) fide Diehl (*in litt.*). See Stevenson & Cash (*Lloyd, The new fungus names*, 1936).

Dubitatio Speg. (1882) = Passerinula (Hypocr.) fide v. Arx & Müller (1973).

Ducellieria Teiling (1957), Ducellieriaceae. 1 (aquatic), UK. See Kusel-Fetzmann & Novak (*Pl. Syst. Evol.* **138**: 199, 1981; position), Hesse *et al.* (*Pl. Syst. Evol.* **165**: 1, 1989; life cycle, ultrastr.), Dick (*in press*; key).

Ducellieriaceae M.W. Dick (1995), Leptomitales. 1 gen., 1 sp.

Ductifera Lloyd (1917), Exidiaceae. 10, widespr. See Wells (*Mycol.* **50**: 407, 1958; key).

Duddingtonia R.C. Cooke (1969) = Arthrobotrys (Mitosp. fungi) fide Schenk *et al.* (*CJB* **55**: 977, 1977), but see van Oorschot (*Stud. Mycol.* **26**: 61, 1985).

Duebenia Fr. (1849), Dermateaceae. 2, Eur. See Hein (*Willdenowia, Beih.* **9**, 1976; key).

Dufourea Ach. (1810) nom. rej. = Xanthoria (Teloschist.).

Dufoureomyces Cif. & Tomas. (1953) = Dactylina (Parmel.).

Dufouria Trevis. (1861) = Xanthoria (Teloschist.).

Dumontinia L.M. Kohn (1979), Sclerotiniaceae. 1 (on *Anemone* rhizomes), Eur.

Dumortiera Westend. (1857) [non Nees (1824), *Marchantiaceae*], Mitosporic fungi, 8.A1.?. 2, Belgium. See Sutton (1977).

Dumoulinia Stein (1883) = Megalospora (Megalospor.) fide Sipman (1983).

Duosporium K.S. Thind & Rawla (1961), Mitosporic fungi, 1.C2.26. 1, India, China.

duplex (of the context), in two layers, that adjacent to the lamellae or tubes being harder than the one over it.

Duplicaria Fuckel (1870), Rhytismataceae. 2, boreal and alpine. See Powell (*Mycol.* **65**: 1362, 1973). Anamorph *Crandallia.*

Duplicariella B. Erikss. (1970), Rhytismataceae. 1, Scandinavia.

Duportella Pat. (1915), Peniophoraceae. 12, widespr. See Cunningham (*Trans. r. Soc. N.Z.* **85**: 91, 1957), Hjortstam & Ryvarden (1990; key), Boidin *et al.* (*BSMF* **107**: 91, 1991; key trop. spp.).

Duradens Samuels & Rogerson (1990), Clypeosphaeriaceae. 1, Guyana.

Durandia Rehm (1913) [non Boeck. (1896), *Cyperaceae*] ≡ Durandiella (Dermat.).

Durandiella Seaver (1932), Dermateaceae. 9, Eur., N. Am. See Groves (*CJB* **32**: 116, 1954), Krieglsteiner (*Z. Mykol.* **44**: 277, 1978; key 5 spp. Europ.).

Durandiomyces Seaver (1928) = Peziza (Peziz.) fide Korf (*Mycol.* **48**: 711, 1956).

Durella Tul. & C. Tul. (1865) = Xylogramma (Leot.) fide Sherwood (*Mycotaxon* **5**: 1, 1977).

Durietzia Gyeln. (1935) = Ionaspis (Hymenel.) fide Petterson (*Bot. Not.*: 100, 1946).

Durispora K.D. Hyde (1994), Valsaceae. 1, Malaysia.

Durogaster Lloyd (1919) nom. dub. Based on *Helosis brasiliensis* (*Balanophoraceae*) fide Stevenson & Cash (*Bull. Lloyd Libr.* **35**: 178, 1936).

Durosaccum Lloyd (1924) ? = Pisolithus (Sclerodermat.).

Dussiella Pat. (1890), Clavicipitaceae. 2, Am. See White (*Am. J. Bot.* **80**: 1465, 1993).

Dutch elm disease, see elm disease.

duvet (of dermatophytes), a soft, thick layer of hyphae like brushed-up cloth.

Dwayaangam Subram. (1978), Mitosporic fungi, 1.G1.1. 3, N. Am., Malaysia. See Barron (*CJB* **69**: 1402, 1991).

Dwayabeeja Subram. (1958), Mitosporic fungi, 1.C/E2.1/10. 1, India.

Dwayaloma Subram. (1957), Mitosporic fungi, 3.B2.?. 1, India.

Dwayalomella Brisson, Piroz. & Pauzé (1975), Mitosporic fungi, 6/7.B1.15. 2, N. Am. See Carris (*Mycol.* **81**: 638, 1989).

Dwayamala Subram. (1956) = Dendryphiella (Mitosp. fungi) fide Kendrick & Carmichael (1973).

Dwibahubeeja N. Srivast., A.K. Srivast. & Kamal (1995), Mitosporic fungi, 1.G2.1. 1, India.

Dwibeeja Subram. (1992), Mitosporic fungi, ?.?.?. 1, Singapore.

Dwiroopa Subram. & Muthumary (1986), Mitosporic fungi, 8.A2.1. 1, India. ? = Lasiodiplodia (Mitosp. fungi), fide Sutton (*in litt.*).

Dwiroopella Subram. & Muthumary (1986), Mitosporic fungi, 8.B2.15/19. 1, India.

Dyadosporites Hammen (1954), Fossil fungi (*f. cat.*). 2 (Cretaceous/Paleocene, Tertiary), China, Colombia.

Dyadosporonites Elsik (1968), Fossil fungi (*f. cat.*). 11 (Cretaceous, Tertiary), India, N. Am.

Dycticia Raf. (1808) ? = Clathrus (Clathr.).

Dyctionella, see *Dictyonella.*

Dyctiostomiopelta, see *Dictyostomiopelta.*

Dyctyoblastus Kremp. (1869) = Phlyctis (Phlyctid.).

Dyeing. Some lichens were formerly used (esp. in N. Europe and N. America) as dyes of animal fibres (e.g. wool). **Orchil** (from *Roccella* spp.) is one of the best known giving a reddish purple as did **cudbear**

(*Ochrolechia tartarea*), both being direct dyes. These colours, which are fugitive to light, are a complex series of orcein derivatives formed from microaerophilic oxidation of orcinol-type secondary metabolites in the presence of ammonia. Other spp. such as **crottle** (*Parmelia saxatilis* and *P. omphalodes*) give yellowish buff to brown shades and are reactive dyes. These colours are light-fast and are produced by colourless lichen substances which have an aldehyde group such as the depsidone salazinic acid. The fruiting bodies of some macromycetes (e.g. agarics) have been less used and not historically, but they can give a range of pinks, blues, yellows, reds and browns (e.g. *Boletus, Cortinarius, Hydnellum, Hygrocybe* spp.) using mordants such as alum or iron. Lichens in particular should not be collected indiscriminately for dyeing purposes from natural or seminatural habitats as this is contrary to the conservation of biodiversity.

Lit.: Bolton (*Lichens for vegetable dyeing*, 1960; edn 2, 1982), Brough (*Syesis* **17**: 81, 1984; 250 tests on 42 Can. lichens), Cardon (*Guide des teintures naturelles*, 1990), Casselman (*Craft of the dyer: colour from plants and lichens of the Northeast*, 1980), Kok (*Lichenologist* **3**: 248, 1966; orchil dyes), Rice & Beebee (*Mushrooms for color*, edn 2, 1980), Richardson (*The vanishing lichens*, 1975), Solberg (*Acta Chem. Scand.* **10**: 1116, 1956; dyeing wool with lichens and lichen substances).

Dyonisia, see *Dionysia.*

Dyplolabia A. Massal. (1854) = Graphis (Graphid.).

Dyrithium M.E. Barr (1994), Amphisphaeriaceae. 1, N. Am., Puerto Rico. See Eriksson & Hawksworth (*SA* **13**: 191, 1995; nomencl.).

dysgonic (of dermatophytes), growing more slowly in culture, frequently with less aerial mycelium, than a normal, **eugonic**, strain. See Johnstone & La Touche (*TBMS* **39**: 442, 1956).

Dyslachnum Clem. (1909) = Lachnum (Hyaloschyph.). See Raitviir (1970).

Dyslecanis Clem. (1909) = Lecania (Bacid.).

Dysrhynchis Clem. (1909), Parodiopsidaceae. 7, trop.

Dysticta Clem. (1909) = Sticta (Lobar.).

Dystictina Clem. (1909) = Sticta (Lobar.).

dystrophic, inadequately nourished. Cf. eutrophic, oligotrophic.

e- (prefix) (**ex-**), from; out of; without; not having. See ex.

E, (1) the ratio of length to width (of spores, basidia, etc.); Corner (*New Phytol.* **46**: 195, 1947). Cf. Q; (2) Royal Botanic Garden, Edinburgh (UK); founded 1670; a direct grant institution of the Department of Agriculture and Fisheries for Scotland; see Fletcher & Brown (*The Royal Botanic Garden, Edinburgh, 1670-1970*, 1970), Hedge & Lamond (*Index of collectors in the Edinburgh herbarium*, 1970), Watling (*Trans. bot. Soc. Edinb.* **45**: 1, 1986).

Earlea Arthur (1906) = Phragmidium (Phragmid.) fide Arthur (1934).

Earliella Murrill (1905) = Trametes (Coriol.).

earth-balls, basidiomata of *Sclerodermatales.*

earth-stars, basidiomata of *Geastrum.*

earth-tongues, ascomata of *Geoglossum.*

Ebollia Minter & Caine (1980), Mitosporic fungi, 8.E1.1. 1, Chile.

Ecchyna Fr. ex Boud. (1885) = Phleogena (Phleogen.) fide Donk (*Persoonia* **4**: 160, 1966).

Eccilia (Fr.) P. Kumm. (1871), Entolomataceae. 35, esp. N. Am. = Rhodophyllus (Entolomat.) fide Singer (1951).

Eccrina Leidy (1852) = Enterobryus (Eccrin.) fide Manier & Lichtwardt (*Ann. Sci. nat., Bot.* sér. 12 **9**: 519, 1968).

Eccrinaceae L. Léger & Duboscq (1929), Eccrinales. 12 gen. (+ 14 syn.), 52 spp. Thalli unbranched, or branched only at the base, non-septate, sporangiospores of two basic types, formed singly in basipetal series of terminal sporangia; endocommensals of *Crustacea, Diplopoda, Insecta.*

Eccrinales, Trichomycetes. 3 fam., 16 gen. (+ 14 syn.), 62 spp. Thallus unbranched or branched only at the base, vegetatively coenocytic, with basal, secreted holdfast; asexual reproduction by endogenous sporangiospores produced in basipetal succession from the thallus tip, (a) multinucleate, thin-walled sporangiospores germinating within the digestive tract of the same host, (b) usually uninucleate, thick-walled sporangiospores which in aquatic forms are frequently appendaged and on release function as reinfestive spores; microthallus present in some genera; sexual reproduction known only in *Enteropogon*; endocommensals of aquatic and terrestrial *Crustacea, Diplopoda* and *Insecta*. Fams:
(1) **Eccrinaceae.**
(2) **Palavasciaceae.**
(3) **Parataeniellaceae.**
    Lit.: Lichtwardt (*Mycol.* **46**: 564, 1954, **52**: 410, 1960; taxonomy, 1986), Manier (1969, taxonomy), Hibbits (*Syesis* **11**: 213, 1978; development, sexual repr.), Moss (1979).

Eccrinella L. Léger & Duboscq (1933) = Astreptonema (Eccrin.) fide Manier (*Ann. Sci. nat., Bot.* sér. 12 **5**: 767, 1964).

eccrinid, one of the *Eccrinales* or *Amoebidiales*.

Eccrinidus Manier (1969), Eccrinaceae. 1 (in *Diplopoda*), France. See Manier & Grizel (*C.R. Acad. Sci. Paris* **274**: 1159, 1972; ultrastr.), Lichtwardt (1986; key).

Eccrinoides L. Léger & Duboscq (1929), Eccrinaceae. 4 (in *Isopoda* and *Diplopoda*), Eur. See Lichtwardt (1986).

Eccrinopsis L. Léger & Duboscq (1916) = Enterobryus (Eccrin.) fide Manier & Lichtwardt (*Ann. Sci. nat. Bot.* sér. 12 **9**: 519, 1968).

Echidnocymbium Brusse (1987), Bacidiaceae (L). 1, S. Afr.

Echidnodella Theiss. & Syd. (1918), Asterinaceae. 7, esp. trop.

Echidnodes Theiss. & Syd. (1918), Asterinaceae. *c.* 15, trop. ? = Lembosina (Asterin.).

echinate (dim. **echinulate**) (of spores, etc.), having sharply pointed spines (Fig. 29.2); spinose.

Echinella Massee (1895) = Pirottaea (Dermat.) fide Nannfeldt (1932).

Echinobotryum Corda (1831), Mitosporic fungi, 2.A2.10. 1, widespr. Synanamorph *Doratomyces*.

Echinocatena R. Campb. & B. Sutton (1977), Mitosporic fungi, 1.A2.3. 1, India.

Echinochaete D.A. Reid (1963), Polyporaceae. 4, trop. See Reid (*Mem. N.Y. bot. Gdn* **28**: 187, 1976), Ryvarden & Johansen (*Preliminary polypore flora of East Africa*: 315, 1980), Corner (*Beih. Nova Hedw.* **78**: 105, 1984).

Echinochondrium Samson & Aa (1975), Mitosporic fungi, 1.D2.1. 1 (bulbils with spines), Sri Lanka.

Echinoderma (Locq. ex Bon) Bon (1991), Tricholomataceae. 13, cosmop.

Echinodia Pat. (1919), Anamorphic Polyporaceae. 3.A1.?41. 1, Singapore.

Echinodontiaceae Donk (1961), Stereales. 1 gen. (+ 2 syn.), 6 spp. Basidioma resupinate to pileate; dimitic; hymenophore smooth to spinose; skeletohyphidia present; spores amyloid.

Lit.: Donk (1964: 263).

Echinodontium Ellis & Everh. (1900), Echinodontiaceae. 6, N. Am., Japan. *E. tinctorium* (Indian paint fungus, conifer wood rot). See Gross (*Mycopath.* **24**: 1, 1964; key), Thomas (*Publ. Dep. Agric. Can. For. Bio. Div.* **1041**, 1958).

Echinodothis G.F. Atk. (1894) ≡ Dussiella (Clavicipit.).

Echinomyces Rappaz (1987), Diatrypaceae. 2, Afr., Australia. ? = Fassia (Diatryp.) fide Læssøe (*SA* **13**: 43, 1994).

Echinophallus Henn. (1898), Phallaceae. 1, E. Indies. See Boedijn (*Bull. Jard. bot. Buitenz.*, sér. 3, **12**: 90, 1932); possibly based on *Phallus* 'egg' fide Demoulin (*in litt.*).

Echinoplaca Fée (1824), Gomphillaceae (L). 9, trop. See Santesson (*Symb. bot. upsal.* **12**(1), 1952), Vězda (*Folia geobot. phytotax.* **14**: 43, 1979), Vězda & Poelt (*Folia geobot. phytotax.* **22**: 179, 1987).

Echinopodospora B.M. Robison (1970) = Apiosordaria (Lasiosphaer.) fide Hawksworth (*in litt.*). See Jong & Davis (*Mycol.* **66**: 467, 1974).

Echinoporia Ryvarden (1980), Coriolaceae. 2, trop. = Hyphodontia (Hyphodermat.) fide Langer (*Bibl. Mycol.* **154**, 1994).

Echinospora Mirza (1963) nom. inval. ≡ Apiosordaria (Lasiosphaer.).

Echinosporangium Malloch (1967) [non Kylin (1956), *Rhodophyta*], Mortierellaceae. 1, Mexico, USA. See Malloch (*Mycol.* **59**: 326, 1967), Ranzoni (*Mycol.* **60**: 356, 1968).

Echinosporella Contu (1992), Tricholomataceae. 1, Eur.

Echinosporium Woron. (1913) = Petrakia (Mitosp. fungi) fide Petrak (*Sydowia* **20**: 186, 1968).

Echinosteliaceae Rostaf. ex Cooke (1877), Echinosteliales. 1 gen. (+ 3 syn.), 15 spp.

Echinosteliales, Myxomycota. 2 fam., 3 gen. (+ 5 syn.), 21 spp. Spore mass white, pale pink, or yellow, columella present, peridium fugaceous. Fams:
(1) **Clastodermataceae.**
(2) **Echinosteliaceae.**
    Lit.: Alexopoulos & Brooks (*Mycol.* **63**: 925, 1971).

Echinosteliopsidaceae L.S. Olive (1970), Echinosteliopsidales. 2 gen., 2 spp.

Echinosteliopsidales, Myxomycota. 1 fam., 2 gen., 2 spp. Differs from *Echinostelium* in having no flagellate stage and in having nuclei with many peripheral nucleoli instead of a single central one. Fam.:
Echinosteliopsidaceae.
    Lit.: See under *Myxomycota*.

Echinosteliopsis D.J. Reinh. & L.S. Olive (1967), Echinosteliopsidaceae. 1, widespr. See Reinhardt & Olive (*Mycol.* **58**: 967, 1967), Olive (1975: 110).

Echinostelium de Bary (1873), Echinosteliaceae. 15, cosmop. See Whitney (*Mycol.* **72**: 950, 1980; key). *E. lunatum* (*Mycol.* **63**: 1051, 1971), the smallest known myxomycete.

Echinothecium Zopf (1898), ? Capnodiaceae. 2 (on lichens), Eur., N. Am.

Echinotrema Park.-Rhodes (1955) = Trechispora (Sistotremat.) fide Larsson (*MR* **98**: 1153, 1994).

Echinula Graddon (1977), Hyaloscyphaceae. 1, UK.

echinulate, see echinate.

Echinus Haller (1768) ≡ Hydnum (Hydn.).

Echusias Hazsl. (1873) = Nitschkia (Nitschk.). See Höhnel (*Ann. Myc.* **17**: 130, 1919).

eclosion, an explosive series of movements which results in the release of a germinating inner spore from a rigid exosporium, as in *Hypoxylon fragiforme* (Chapela *et al.*, *CJB* **68**: 2571, 1990)

Ecology. Fungal ecology (**myco-**) is the entire field of ecological and coenological study in fungi. See Weir

(*Mycol.* **10**: 4, 1918; altitudinal range of forest fungi); Gilbert (*La mycologie sur le terrain*, 1928), Wilkins *et al.* (*Ann. appl. Biol.* **24**: 703, **25**: 472, **26**: 25, 1937-39; larger fungi), Cooke (*Ecology* **29**: 376, 1948, lit. fungus sociology, ecology; *Bot. Rev.* **24**: 241, 1958; *Mycopath. Mycol. appl.* **48** (1), 1972, IMC 1 papers), Morten Lange (*Dansk. bot. Arkiv* **13** (1), 1948; agarics of Maglemose), Ramsbottom (*Mushrooms and toadstools*, 1953), Ainsworth & Sussman (*The Fungi* 3, 1968; many chs), Harley (*J. Ecol.* **59**: 887, 1971; in ecosystems), Nuss (*Zur Oekologie der Porlinge*, 1975), Pugh (*TBMS* **74**: 1, 1980; strategies), Wicklow & Carroll (Eds) (*The fungal community*, 1981; wide ranging surveys, relationship to general ecological concepts; edn 2, 1992), Cooke & Rayner (*Ecology of saprotrophic fungi*, 1984), Watling (*TBMS* **90**: 1, 1988; macromycetes), Lynch & Hobbie (Eds) (*Microorganisms in action: concepts and applications in microbial ecology*, 1988), Christensen (*Mycol.* **81**: 1, 1989; review, 186 refs.), Grigorova & Norris (Eds) (*Techniques in microbial ecology*, 1990), Allsopp *et al.* (Eds) (*Microbial diversity and ecosystem function*, 1995), Dix & Webster (*Fungal ecology*, 1995).

Lichens are primary colonizers in plant succession, occurring in all pioneer terrestrial habitats. Single species may grow on a wide range of substrata or be restricted to particular types. Whole communities rather than individual species tend to be characteristic of particular substrata growing in a particular climatic region, though a few species are restricted to particular trees (e.g. *Lecanora populicola* on *Populus*), esp. those which are probably not really lichenized (e.g. *Arthopyrenia laburni* on *Laburnum, Stenocybe septata* on *Ilex*). On bark and rocks, pH, shade, nutrient enrichment, buffer and ion exchange capacities, air pollution, and disturbance affect the communities developed. Lichens tend to dominate habitats in which competition from other plants is minimal. Vertical zonation of communities occurs on freshwater and maritime rocks (see Aquatic fungi and Marine fungi) and trees (Kershaw, *Lichenologist* **2**: 263, 1964). See Barkman (*Phytosociology and ecology of cryptogamic epiphytes*, 1958), Brodo (*in* Ahmadjian & Hale, 1974: 401 ['1973']; substrate ecology), Seaward (Ed.) (*Lichen ecology*, 1977; comprehensive survey, glossary).

See also Air pollution, Aquatic fungi and lichens, Biodiversity, Bioindicators, Cave fungi, Coprophilous fungi, Entomogenous fungi, Endophyte, Fungicolous fungi, Insects and fungi, Lichenicolous fungi, Lichenometry, Growth rates, Mycorrhiza, Mycosociology, Plant pathogenic fungi, Phytosociology, RIEC, Sand dune fungi and lichens, Soil fungi, Wood-attacking fungi.

ecorticate, having no cortex.

ecotype, part of a population of a species showing morphological, chemical, or physiological characteristics which appear to be genetically determined and correlated with particular ecological conditions, but which are not considered of major taxonomic significance.

ectal, outer; outermost.

ectal excipulum (of ascomata), the outer layers including the subhymenium in a non-lichenized apothecium, sometimes multi-layered; see excipulum.

Ecteinomyces Thaxt. (1902), Laboulbeniaceae. 1, Am., Eur. See Benjamin (1971).

ecto- (prefix), outside.

ectoascus, the outer wall of a fissitunicate (q.v.) ascus, as in *Lecanidion*.

ectochroic, see colour.

Ectochytridium Scherff. (1925) ≡ Zygorhizidium (Chytrid.).

Ectographis Trevis. (1853) = Phaeographina p.p. and Graphina p.p. (Graphid.).

Ectolechia A. Massal. (1853) = Ocellularia (Thelotremat.) fide Hale (1981).

Ectolechia Trevis. (1853) = Sporopodium (Ectolech.).

Ectolechiaceae (Vain.) Zahlbr. (1905), Lecanorales (L). 11 gen. (+ 8 syn.), 95 spp. Thallus usually foliose or crustose; ascomata pale, convex, without a well-developed margin; interascal tissue of very narrow anastomosing pseudoparaphysis-like hyphae, variously pigmented, the apices thin-walled; asci with a I+ apical cap and a very well-developed ocular chamber, with an outer I+ gelatinous layer, mostly 1- or 2-spored; ascospores muriform, usually thin-walled; sometimes with cephalodia. Anamorphs pycnidial, often with campylidia. Lichenized with green algae, usually foliicolous, esp. trop.

The *Pilocarpaceae* is doubtfully distinct (Lücking *et al.*, *Bot. Acta* **107**: 393, 1994).

Lit.: Kalb & Vězda (*Folia geobot. phytotax.* **22**: 286, 1987; 13 spp. Brazil), Vězda (*Folia geobot. phytotax.* **21**: 199, 1986).

Ectomyces P. Tate (1927) = Termitaria (Mitosp. fungi) fide Tate (*Parasitology* **20**: 77, 1968).

ectomycorrhiza, see mycorrhiza.

ectoparasite, a parasite living on the outside of its host.

ectoplacodial, see stroma.

Ectosphaeria Speg. (1921) = Diatrype (Diatryp.) fide Petrak & Sydow (*Ann. Myc.* **32**: 23, 1934).

ectospore, (1) an exogenous spore; (2) a basidiospore (obsol.).

Ectosticta Speg. (1912) = Rhizosphaera (Mitosp. fungi) fide Clements & Shear (1931), but see Sutton (1977).

ectostroma, see stroma.

Ectostroma Fr. (1823) nom. dub. (Mitosp. fungi) fide Carmichael *et al.* (1980).

ectothecal (of ascomycetes), having the hymenium exposed.

Ectothrix, see Endothrix.

ectothrix, living on the surface of hair.

Ectotrichophyton Castell. & Chalm. (1919) = Trichophyton (Mitosp. fungi).

ectotroph (adj. **ectotrophic**), see mycorrhiza.

ectotropic, curving out.

ectotunica, the outer wall of a bitunicate ascus.

Ectrogella Zopf (1884), Ectrogellaceae. 8, N. temp. See Dick (*in press*; key).

Ectrogellaceae Cejp (1959), ? Myzocytiopsidales. 2 gen. (+ 1 syn.), 14 spp.

ED, effective dose; **ED$_{50}$**, the effective dose for a 50% (usually lethal) response (Horsfall, *Principles of fungicidal action*, 1956).

edaphic, pertaining to the soil.

Edible fungi. Macromycetes are an important food in some countries, particularly fleshy basidiomycetes (esp. species of *Agaricus, Coprinus, Lepista, Macrolepiota, Pleurotus, Termitomyces* and *Tricholoma* of the *Agaricales*, and *Lycoperdon, Boletus* and *Cantharellus*), and so are a number of ascomycetes, e.g. truffles (*Tuber*), terfas or kames (*Terfezia*) of N. Afr. and S. Spain, morels (*Morchella*), and *Cyttaria* of Tierra del Fuego. The larger fleshy ascomycetes (but not truffles) chiefly occur in the spring, while basidiomycetes are most frequent in the late summer and fall. The edible part is generally the basidioma but in *Polyporus mylittae* (blackfellows' bread of Australia), *Poria cocos* (Indian bread or tuckahoe of America; cf. stone-fungus), and some tropical species of *Lentinus* it is the sclerotium (which in *Lentinus* may weigh 10kg or more), while in Taiwan young *Ustilago esculenta* in *Zizania aquatica* (Canadian rice) is used and in Mexico *U. maydis* is commonly eaten. Yeast (*Saccharomyces cerevisiae*) after autolysis (under the trade names Marmite, Vegex, etc.) is a food for humans and dry yeast such as *Candida utilis* (q.v.) is coming into use, as is mycoprotein (q.v.). Button mushrooms (*Agaricus*), the padi-straw mushroom (*Volvariella volvacea*), the shii-take (*Lentinula edodes*) and the matsu-take (*Tricholoma matsutake*), and truffles are frequently cultured (see Mushroom culture). *Agrocybe aegerita* which, with fungi suilli (*Boletus edulis*), was among the fungi valued by the Romans, is sometimes obtained by watering old poplar (*Populus*) wood. Fungi also provide food for animals (see Animal mycophagists).

Though a very important food for reindeer and caribou (*Bryoria*, *Cladonia*, Iceland moss, q.v.), and also invertebrates, most lichens are not edible for humans (but see rock tripe, shaybah, manna) although not poisonous, except possibly those containing large amounts of vulpinic acid (e.g. *Letharia vulpina*, Wolf's moss, q.v.). See Llano (*Bot. Rev.* **10**: 1, 1944), Dahl (*Bot. Rev.* **20**: 463, 1954), Richardson (*The vanishing lichens*, 1975), Seaward (Ed.) (*Lichen ecology*, 1977).

Lit.: Christensen (*Common edible mushrooms*, 1943; USA), Krieger (*The mushroom handbook*, 1947 [reprint 1967]), Edible fungi of Britain (*Bull. Ministr. Agric. Lond.* **23**, edn 6, 1947), Rammeloo & Walleyn (*The edible fungi of Africa south of the Sahara*, 1993; 300 spp.), Herrera & Guzmán (*An. Inst. Biol. Univ. Mexico* **32**: 33, 1961; Mexico, incl. glossary common names), Fidalgo & Prance (*Mycol.* **68**: 201, 1976; Brazilian tribes), Guba (*Wild mushrooms - food and poison*, edn 2, 1977), Batra (*Biologia, Lahore* **29**: 293, 1983; Afghanistan, Pakistan, N. India), Purkayastha & Chandra (*Manual of Indian edible mushrooms*, 1985), Singer & Harris (*Mushrooms and truffles. Botany, cultivation and utilization*, 2nd edn, 1987), Huang ([*Edible fungi cyclopedia*], 1993; in Chinese).

See also Ethnomycology, Hallucinogenic fungi, Industrial mycology, Larger fungi, mycoprotein, Mushroom culture, soma.

Edmundmasonia Subram. (1958) = Brachysporiella (Mitosp. fungi) fide Kendrick & Carmichael (1973).

Edrudia W.P. Jord. (1980), Lecanoraceae (L). 1, USA. See Poelt & Hafellner (*Mitt. Bot. StSamml., Münch.* **16**: 503, 1980).

Edwardiella Henssen (1986), Lichinaceae (L). 1, Marion Isl.

Edythea H.S. Jacks. (1931), Uredinales (inc. sed.). 3 (on *Berberis*), S. Am. See Thirumalachar & Cummins (*Mycol.* **40**: 47, 1948).

Edyuillia Subram. (1972), Trichocomaceae. 1, widespr. See Subramanian & Rajendran (*Revue mycol.* **41**: 223, 1977; ontogeny). Anamorph *Aspergillus*.

Eeniella M.T. Sm., Bat. Vegte & Scheffers (1981), Mitosporic fungi, 1.A1.10/19. 1, Sweden. = Brettanomyces (Mitosp.) fide Berkhout *et al.* (*Int. J. Syst. Bact.* **44**: 781, 1994).

effete, (1) overmature, exhausted; (2) (of fruiting bodies), empty.

Effetia Bartoli, Maggi & Persiani (1984), ? Sordariaceae. 1, Ivory Coast. Anamorph *Virgariella*-like.

effigurate (of lichen thalli), obscurely lobed.

efflorescent, bursting out of.

effuse, stretched out flat, esp. as a film-like growth.

effused-reflexed (of *Basidiomycetes*), stretched out over the substratum but turned up at the edge to make a pileus.

Efibula Sheng H. Wu (1990), Meruliaceae. 5, Eur., Taiwan. = Phanerochaete (Merul.) fide Hjortstam & Larsson (*Windahlia* **21**, 1994).

Efibulobasidium K. Wells (1975), Exidiaceae. 2, widespr.

egg, (1) the female gamete; (2) (of phalloids, *Amanita*, etc.), the young basidioma before the volva is broken.

Eichleriella Bres. (1903), Exidiaceae. *c.* 10, widespr. See Wells & Raitviir (*Mycol.* **72**: 564, 1980).

Eidamella Matr. & Dassonv. (1901) = Myxotrichum (Myxotrich.) fide Orr *et al.* (*CJB* **41**: 1439, 1963).

Eidamia Lindau (1904) = Harzia (Mitosp. fungi) fide Carmichael *et al.* (1980).

Eiglera Hafellner (1984), Eigleraceae (L). 1, Eur., N. Am., esp. coastal.

Eigleraceae Hafellner (1984), Lecanorales (L). 1 gen., 1 sp. Thallus crustose, effuse, sometimes evanescent; ascomata apothecial, immersed, with a very poorly-developed margin; interascal tissue of narrow sparsely branched paraphyses; asci with a uniformly I+ apical cap and an I+ gelatinous outer layer, especially thick at the apex; ascospores hyaline, aseptate, without a sheath. Anamorph pycnidial. Lichenized with green algae.

Eiona Kohlm. (1968), Ascomycota (inc. sed.). 1 (marine), Denmark. See Kohlmeyer (*SA* **10**: 35, 1991; posn).

Ejectosporus S.W. Peterson, Lichtw. & M.C. Williams (1991), Lageriomycetaceae. 1 (in *Plecoptera*), USA.

Eklundia C.W. Dodge (1968) = Candelariella (Candelar.) fide Castello & Nimis (*Lichenologist* **26**: 283, 1994).

Ekmanomyces Petr. & Cif. (1932) = Dictyonella (Saccard.) fide v. Arx (1963).

Elachopeltella Bat. & Cavalc. (1964), Mitosporic fungi, 5.A1.?. 2, Brazil.

Elachopeltis Syd. (1927), Mitosporic fungi, 5.A1.?. 4, trop. Am.

Elachophyma Petr. (1931) = Molleriella (Elsin.) fide v. Arx (1963).

Eladia G. Sm. (1961), Mitosporic fungi, 1.A2.15. 1, widespr.

Elaeodema Syd. (1922), Mitosporic fungi, 1 or 6.A2.?. 2, China. See Sutton (*Mycol. Pap.* **141**, 1977).

Elaeomyces Kirchn. (1888) ? Ascomycota (inc. sed.).

Elaeomyxa Hagelst. (1942), Stemonitidaceae. 2, N. temp.

Elaphocephala Pouzar (1983), Sistotremataceae. 1, Eur.

Elaphomyces Nees (1820), Elaphomycetaceae. 20 (ectomycorrhizal, hypogeous), widespr. See Dodge (1929), Kobayashi (*Nagaoa* **7**: 35, 1960; key 14 Jap. spp.), Trappe (*Trans. mycol. Soc. Japan* **17**: 209, 1976; variability), Samuelson *et al.* (*Mycol.* **79**: 571, 1987; ultrastr. ascospore ornam.), Zhang (*MR* **95**: 973, 1991; key 7 spp. China), Zhang & Minter (*CJB* **67**: 909, 1989; key 5 spp. Canada), Pegler *et al.* (*British truffles*, 1993; key 5 Br. spp.).

Elaphomycetaceae Tul. ex Paol. (1889), Elaphomycetales. 1 gen. (+ 6 syn.), 20 spp. Ascomata large, cleistothecial, black, subterranean; peridium very thick, usually strongly ornamented; interascal tissue of copious unordered hyphae; asci often developing in locules, thin-walled, evanescent; ascospores globose, aseptate, strongly ornamented and pigmented. Anamorphs unknown. Ectomycorrhizal hypogeous fungi, remote from other ascomycetous truffles.
Lit.: see under *Elaphomycetales*.

Elaphomycetales, Ascomycota. 1 fam., 1 gen. (+ 6 syn.), 20 spp. Fam.:
Elaphomycetaceae:
Lit.: Dodge (*Ann. Myc.* **27**: 145, 1929), Ławrynowicz (*Fl. Polska, Grzyby* **18**, 1988, keys; *Acta mycol.* **25**: 3, 1989, maps Eur.), Trappe (*Mycotaxon* **9**: 297, 1979; posn).

Elasmomyces Cavara (1898), Elasmomycetaceae. 9, widespr. See Singer & Smith (*Mem. Torrey bot. Cl.* **21**: 56, 1960; key).

Elasmomycetaceae Locq. ex Pegler & T.W.K. Young (1979), Russulales. 6 gen. (+ 3 syn.), 86 spp. Gasterocarp stipitate or sessile; spores statismosporic.

Lit.: Singer & Smith (*Mem. Torrey Bot. Cl.* **21**, 1960), Pegler & Young (*TBMS* **72**: 353, 1973), Beaton *et al.* (*Kew Bull* **33**: 669, 1984; Austr. spp.), Smith (*Mycol.* **54**: 626, 1962).

elater, (1) a free capillitium-thread, e.g. in myxomycetes and *Farysia*; (2) a body with spiral or annular markings in the gleba of *Battarrea*.

Elateraecium Thirum., F. Kern & B.V. Patil (1966), Uredinales (inc. sed.). 2, India. See Laundon (*Mycotaxon* **3**: 133, 1975), Thirumalachar *et al.* (1975; teleomorphs).

Elateromyces Bubák (1912) ≡ Farysia (Ustilagin.).

Elattopycnis Bat. & Cavalc. (1964), Mitosporic fungi, 5.A1.?. 1, Brazil.

Elderia McLennan (1961) = Stephensia (Otid.) fide Trappe (1979).

Electron microscopy, see Ultrastructure.

electrophoresis, a method of chromatography (q.v.) in which charged particles (esp. DNA, RNA, proteins) are separated by their differential migration in an inert matrix by an electric field. Agarose provides a matrix of large pores particularly suitable for DNA and RNA electrophoresis, and polyacrylamide (**PAGE**) and starch provide matrices of smaller pore size more suitable for the separation of proteins. For DNA and RNA, migration is directly related to particle size, but for proteins migration may be due to both size and charge. The extraction methods depend on whether proteins or nucelic acids are being examined, and the bands are stained for study. Valuable for studying both total protein patterns and for particular enzymes (inter- and extracellular), especially variation in one enzyme between isolates; the variants are **isozymes** (**isoenzymes**) and about 140 can assayed (e.g. esterases, dehydrogenases, pectinases, phosphatases).

Automated and standardized units are available, which can operate with computer-linked gel-readers. The techniques have proved especially valuable in the characterization of closely related strains of plant pathogenic fungi.

See also **pulse gel -**.

Lit.: Gabriel & Gersten (*Analyt. Biochem.* **45**: 1741, 1992; staining), Loxdale & Den Hollander (Eds) (*Electrophoretic studies on agricultural pests*, 1989), Paterson & Bridge (*Biochemical techniques for filamentous fungi* [*IMI Techn. Handbk* **1**], 1994).

Elenkinella Woron. (1922) = Molleriella (Elsin.) fide v. Arx (1963).

Eleutherascus Arx (1971), Ascodesmidaceae. 4, Asia, Eur., S. Am. See Steffens & Jones (*CJB* **61**: 1599, 1983; asci).

Eleutheris Clem. & Shear (1931) ≡ Eleutheromycella (Mitosp. fungi).

Eleutheromycella Höhn. (1908), Mitosporic fungi, 4.A1.15. 1, Eur. See Seeler (*Farlowia* **1**: 119, 1943).

Eleutheromyces Fuckel (1870), Mitosporic fungi, 2.A1.15. 1 (on macromycetes), temp. See Seeler (*Farlowia* **1**: 119, 1943), Sigler (*Crypt. Bot.* **1**: 384, 1990; yeast-like synanamorph).

Eleutherosphaera Grove (1907) ? = Pyxidiophora (Pyxidiophor.). See Hawksworth & Webster (*TBMS* **68**: 329, 1977).

elf cups, ascomata of *Pezizales*.

Elfvingia P. Karst. (1889), Ganodermataceae. 6, widespr. See Pegler (1971).

Elfvingiella Murrill (1914) ≡ Fomes (Coriol.). = Ganoderma (Ganodermat.) fide Donk (1974).

Elina N.J. Artemczuk (1972), Thraustochytriaceae. 2, former USSR. But see Karling (*Mycol.* **71**: 829, 1979).

ELISA. (enzyme-linked immunosorbent assay). Currently a widely used method for detecting and quantifying antibodies (see antigen). Species-specific antigens in yeasts (Middelhoven & Notermans, *J. Gen. Appl. Bact.* **34**: 15, 1988).

Elletevera Deighton (1969), Mitosporic fungi, 1.C2.10. 1 (on *Phyllachora*), N. Am., Afr.

Ellimonia Syd. (1930) = Inocyclus (Parmular.) fide Müller & v. Arx (1962).

Elliottinia L.M. Kohn (1979), Sclerotiniaceae. 1, Eur.

ellipsoidal (of spores, etc.), elliptical in optical section (Fig. 37.1).

Ellis (Job Bicknell; 1829-1905). Pioneering N. American mycologist. See Kaye (*Mycotaxon* **26**: 29, 1986), Cash (*USDA Spec. Publ.* **2** (1-3); fungi named by J.B. Ellis), Rodrigues (*Mycotaxon* **34**: 577, 1989; index 202 types with amyloid rings), Grummann (1974: 187), Stafleu & Cowan (*TL-2* **1**: 742, 1976).

Ellisembia Subram. (1992), Mitosporic fungi, 1.C2.1/19. 12, widespr.

Ellisia Bat. & Peres (1965) [non L. (1763), *Hydrophyllaceae*], Mitosporic fungi, 1.C2.?. 1, Brazil. ? = Heteroconium (Mitosp. fungi) fide Carmichael *et al.* (1980).

Ellisiella Sacc. (1881) = Colletotrichum (Mitosp. fungi) fide Nag Raj (*CJB* **51**: 2463, 1973).

Ellisiella Bat. (1956), Mitosporic fungi, ?.?.?. 3, Portugal.

Ellisiellina Sousa da Câmara (1949) = Colletotrichum (Mitosp. fungi) fide Sutton (*in litt.*). See Sutton (*CJB* **44**: 887, 1966).

Ellisiodothis Theiss. (1914) ? = Muyocopron (Microthyr.) fide Sivanesan (*in litt.*).

Ellisiopsis Bat. (1956) = Beltraniella (Mitosp. fungi) fide Pirozynski & Patil (*CJB* **48**: 567, 1970).

Ellisomyces Benny & R.K. Benj. (1975), Thamnidiaceae. 1, USA. See Beakes *et al.* (*CJB* **62**: 2677, *TBMS* **83**: 593, 1984; thallospore formation and ultrastr.), Beakes & Campos-Takaki (*TBMS* **83**: 607, 1984; sporangiolum ultrastr.).

Ellobiocystis Coutière (1911) nom. dub. (? Fungi, inc. sed.).

Ellobiopsis Caullery (1910) nom. dub. (? Fungi, inc. sed.).

Ellula Nag Raj (1980), Mitosporic fungi, 8.A1.2. 1 (with clamp connexions), Brazil.

Ellurema Nag Raj & W.B. Kendr. (1986), ? Amphisphaeriaceae. 1, India. Anamorph *Hyalotiopsis*.

elm disease (**Dutch elm disease**), a vascular disease of elm (*Ulmus*) caused by *Ophiostoma ulmi* or the even more aggressive *O. novo-ulmi*; the last sp. may be related to one from Himalaya; anamorphs in *Pesotum*. See *IMI Descriptions* **361** (1973), *IMI Map* **36**, Holmes & Heybrock (*Phyt. classic* **13**, 1990; transl. early papers), Brasier (*Mycopath.* **115**: 155, 1991; *Nature* **372**: 227, 1994).

Elmeria Bres. (1912) [non Ridl. (1905), *Saxifragaceae*] ≡ Elmerina (Aporp.).

Elmerina Bres. (1912), Aporpiaceae. 6, cosmop., mostly trop. See Parmasto (*Nova Hedw.* **39**: 101, 1984; monogr.), Reid (*Persoonia* **14**: 465, 1992).

Elmerinula Syd. (1934), Dothideales (inc. sed.). 1, Philipp.

Elmerococcum Theiss. & Syd. (1915) = Plowrightia (Dothior.) fide v. Arx & Müller (1975).

Elosia Pers. (1822) = Alternaria (Mitosp. fungi) fide Hughes (1958).

Elpidophora Ehrenb. ex Link (1824) = Graphiola (Graphiol.) fide Mussat (*Syll. Fung.* **15**: 134, 1901).

Elsikisporonites P. Kumar (1990), Fossil fungi. 1, India.

Elsinoaceae Höhn. ex Sacc. & Trotter (1913), Dothideales. 10 gen. (+ 13 syn.), 66 spp. Ascomata immersed or erumpent, round or elongate, usually crustose, composed of pale gelatinous thin-walled hyphal or pseudo-parenchymatous cells, opening by unordered breakdown of the surface layers; specialized interascal tissue absent; asci arranged in individual locules, in a single layer or irregularly disposed, saccate to globose, fissitunicate; ascospores hyaline to brown, septate, sometimes muriform. Anamorphs acervular where known. Biotrophic or necrotrophic on plants, esp. trop.

Elsinoë Racib. (1900), Elsinoaceae. 40, esp. warmer parts. *E. fawcettii* (citrus scab), *E. canavaliae* (scab of *Canavalia*; *CMI Descr.* **313**, 1971), *E. phaseoli* (scab of Lima bean; *CMI Descr.* **314**, 1971), Gabel & Tiffany (*Mycol.* **79**: 737, 1987; development). Anamorph *Sphaceloma*.

Elvela, see *Helvella*.

Elytroderma Darker (1932), Rhytismataceae. 2, Eur., N. Am. *E. deformans* (witches' brooms on pine). See Minter & Fonseca (*Nova Hedw.* **37**: 181, 1983).

Elytrosporangium Morais, Bat. & Massa (1966) Fungi (inc. sed.).

emarginate, (1) (of lamellae), see sinuate; (2) (of apothecia), lacking a thalline exciple (excipulum thallinum) (lichens) or a raised proper exciple (excipulum proprium).

Embellisia E.G. Simmons (1971), Anamorphic Pleosporaceae, 2.C2.26. Teleomorph *Allewia*. 18, widespr. See Simmons (*Mycotaxon* **38**: 251, 1990; spp. and teleomorphs).

Emblemia Pers. (1827) ? = Phaeographina (Graphid.).

Emblemospora Jeng & J.C. Krug (1976), Lasiosphaeriaceae. 2, S. Am.

Embolidium Sacc. (1878) = Calicium (Calic.) fide Schmidt (1970).

Embolidium Bat. (1964) = Coniella (Mitosp. fungi) fide Sutton (1977).

Embolus Haller (1768) nom. dub. (Myxomycota).

Embolus Batsch (1783) [non Haller (1768)] ? = Chaenotheca (Coniocyb.).

emend, to correct an error. Cf. amend.

Emericella Berk. (1857), Trichocomaceae. 27, widespr. See Christensen & Raper (*TBMS* **71**: 177, 1978), Frisvad, *in* Samson & Pitt (Eds) (*Advances in Penicillium and Aspergillus systematics*: 437, 1985; chemotaxonomy). Anamorph *Aspergillus*.

Emericellopsis J.F.H. Beyma (1940), ? Trichocomaceae. 11, widespr. See Belyakova (*Mikol. i Fitopat.* **8**: 385, 1974; key), Davidson & Christensen (*TBMS* **57**: 385, 1971; key), Gams (1971: 21; keys), Malloch (*in* Hawksworth (Ed.), *Ascomycete systematics*: 374, 1994; posn), Wu & Kimbrough (*CJB* **68**: 1877, 1990; ultrastr.). Anamorph *Acremonium*.

Emilmuelleria Arx (1986), Chaetomiaceae. 1 (coprophilous), USA. See Barr (*Mycotaxon* **39**: 43, 1990; posn).

Emmonsia Cif. & Montemart. (1959) = Chrysosporium (Mitosp. fungi) fide Carmichael (1962), but accepted by van Oorschot (*Stud. Mycol.* **20**, 1980). See adiaspiromycosis, adiaspore.

Emmonsiella Kwon-Chung (1972) = Ajellomyces (Onygen.) fide McGinnis & Katz (*Mycotaxon* **8**: 157, 1979). See Fukushima *et al.* (*Mycopath.* **116**: 151, 1991).

Empusa Cohn (1855) [non Lindl. (1824), *Orchidaceae*] ≡ Entomophthora (Entomophor.).

Empusaceae, see *Entomophthoraceae*.

Enantioptera Descals (1983), Mitosporic fungi, 1.G1.10. 1 (from foam), UK.

Enantiothamnus Pinoy (1911), Mitosporic fungi, 1.?.?. 1 (on humans), Afr.

Enarthromyces Thaxt. (1896), Laboulbeniaceae. 1, Asia, Afr.

Encephalium Link (1816) = Tremella (Tremell.) fide Donk (*Persoonia* **4**: 178, 1966).

Encephalographa A. Massal. (1854), ? Hysteriaceae. 1 (on lichens), S. Eur. See Renobales & Aguirre (*SA* **8**: 87, 1990; nomencl., posn).

Encephalographomyces Cif. & Tomas. (1953) = Poeltinula (Rhizocarp.) fide Hawskworth (*SA* **10**: 36, 1991).

Enchnoa Fr. (1849), Nitschkiaceae. 5, widespr. See Barr (*SA* **13**: 192, 1995; posn).

Enchnosphaeria Fuckel (1870) = Herpotrichia (Lophiostom.) fide Barr (1984).

Enchylium Gray (1821) = Leptogium (Collemat.) p.p. and Collema (Collemat.) p.p.

Enchylium A. Massal. (1853) = Pterygiopsis (Lichin.).

Encliopyrenia Trevis. (1860) = Verrucaria (Verrucar.) p.p., Thelidium (Verrucar.) p.p. and Staurothele (Verrucar.) p.p.

Encoelia (Fr.) P. Karst. (1871) nom. cons., Leotiaceae. 6, N. temp. See Korf & Kohn (*Mem. N.Y. bot. Gdn* **28**: 109, 1976), Verkley (*MR* **99**: 187, 1995; asci).

Encoeliella Höhn. (1907) = Unguiculariopsis (Leot.) fide Zhuang (1988).

Encoeliopsis Nannf. (1932), Leotiaceae. 4, Eur., N. Am. See Groves (*CJB* **47**: 1319, 1969).

Endacinus Raf. (1814) = Pisolithus (Sclerodermat.) fide Saccardo (*Syll. Fung.* **7**: 148, 1888).

Endaematus Raf. (1814) nom. dub. (Fungi, inc. sed.).

Endematus Raf. (1815) ≡ Endaematus (Fungi, inc. sed.).

endemic, native to one country or geographical region.

Endemosarca L.S. Olive & Erdos (1971), Endemosarcaceae. 3, widespr. See Erdos (*Mycol.* **64**: 423, 1972; nuclear cycle), Olive (1975: 212).

Endemosarcaceae L.S. Olive & Erdos (1971), ? Plasmodiophorales. 1 gen., 3 spp. Osmotrophic at first; nuclear division not cruciate; zoospores biflagellate, mastigonemes absent. Endoparasites of ciliate protozoa.

endo- (prefix), inside.

endoascospores, spore-like cells produced within ascospores (see Morgan-Jones, *CJB* **51**: 493, 1972).

endoascus, the often extensible inner wall layers of a bitunicate (q.v.) ascus.

endobasidial (of a conidiophore in a lichenized pycnidium), having a secondary sporing branch (obsol.); cf. exobasidial.

endobasidium, see basidium.

Endobasidium Speschnew (1901), Exobasidiaceae. 1, Samarkand (Asia). See Donk (*Reinwardtia* **4**: 116, 1956; basidiomycete affinities doubtful).

endobiotic, making growth inside living organisms.

Endoblastidium Codreanu (1931), ? Blastocladiales (inc. sed.). 2 (in *Insecta*), Eur.

Endoblastoderma B. Fisch. & Brebeck (1894) = Candida (Mitosp. fungi) fide Lodder (*The yeasts*, 1970).

Endoblastomyces Odinzowa (1947) = Trichosporon (Mitosp. fungi) fide Camara-Sousa (*in* Lodder, 1970).

Endobotrya Berk. & M.A. Curtis (1874), Mitosporic fungi, 6.D2.1. 1, N. Am.

Endobotryella Höhn. (1909), Mitosporic fungi, 8.D2.1. 1, Eur.

Endocalyx Berk. & Broome (1876), Mitosporic fungi, 3.A2.10. 5, trop. See Montemartini Corte (*Atti Ist. bot. Univ. Pavia*, ser. 5, **20**: 260, 1963; key), Okada & Tubaki (*Mycol.* **76**: 300, 1984; cultural behaviour).

Endocarpidium Müll. Arg. (1862) = Placidiopsis (Verrucar.).

endocarpinoid (of lichenized perithecia), sunk into the tissues of the thallus, as in *Endocarpon* (obsol.).

Endocarpiscum Nyl. (1864) = Heppia (Hepp.).

Endocarpomyces E.A. Thomas ex Cif. & Tomas. (1953) ≡ Dermatocarpon (Verrucar.).

Endocarpon Hedw. (1789), Verrucariaceae (L). *c.* 35, cosmop. (mainly temp.). See Harada (*Nova Hedw.* **56**: 335, 1993; key 7 spp. Japan), McCarthy (*Lichenologist* **23**: 27, 1991; 10 spp. Australia), Singh & Upreti (*Candollea* **39**: 539, 1984; key 6 spp., India), Stocker-Wörgötter & Turk (*Pl. Syst. Evol.* **158**: 313, 1988; ultrastr.), Wagner (*CJB* **65**: 2441, 1987; ontogeny), Wagner & Letrouit-Galinou (*CJB* **66**: 2118, 1988; squamule development), Zschacke (*Rabenh. Krypt.-Fl.* **9**, 1(1): 645, 1933).

endocarpous (of gasteromycetes, etc.), having the mature hymenium covered over; angiocarpous.

Endocena Cromb. (1876), Acarosporaceae (L). 1 (sterile), S. Am.

Endochaetophora J.F. White & T.N. Taylor (1988), Fossil fungi (Ascomycota). 1 (Triassic), Antarct.

endochroic, see colour.

Endochytriaceae Sparrow ex D.J.S. Barr (1980), Chytridiales. 10 gen., 50 spp. Thallus eucarpic or holocarpic, monocentric; sporangium exogenous to the zoospore cyst, operculate or inoperculate; resting spore exogenous.

Endochytrium Sparrow (1933), Endochytriaceae. 6, temp. See Karling (*Mycol.* **33**: 356, 1941).

Endocladia Clem. & Shear (1931) ≡ Endoramularia (Mitosp. fungi).

Endococcus Nyl. (1855), Dothideales (inc. sed.). 12 (on lichens), mainly temp. See Hawksworth (*Bot. Notiser* **132**: 283, 1979).

Endocochlus Drechsler (1935), Cochlonemataceae. 4, N. Am, Kenya. See Drechsler (*Mycol.* **27**: 14, 1935, **28**: 363, 1936, **41**: 229, 1949), Dayal (1973; key).

Endocoenobium Ingold (1940), Endochytriaceae. 1, UK.

Endocoleroa Petr. (1969) = Venturia (Ventur.) fide Sivanesan (1977).

Endocolium Syd. (1937), Ascomycota (inc. sed.). 1, C. Am.

endocommensal, an organism living as a commensal (see symbiosis) inside another (e.g. *Trichomycetes* in the gut of *Insecta*).

Endoconia Raf. (1819) nom. dub. (? Basidiomycetes). See Murrill (*Index Rafinesq.*, 1949).

Endoconidiophora Münch (1907) = Ceratocystis (Microascales).

endoconidium, a conidium formed inside a hypha, e.g. as in *Thielaviopsis basicola*.

Endoconidium Prill. & Delacr. (1891), Anamorphic Sclerotiniaceae, 3.A1.15. Teleomorph *Gloeotinia*. 3, temp.

Endoconospora Gjaerum (1971), Mitosporic fungi, 3.A1.15. 1, Norway.

Endocoryneum Petr. (1922), Mitosporic fungi, 8.C2.15. 2, Austria, Italy. See Marras, Franceschini & Sutton, *in* Luisi *et al.*(Eds) (*Recent advances in oak decline*: 255, 1993).

Endocreas Samuels & Rogerson (1989), Hypocreaceae. 1, French Guiana.

Endocronartium Y. Hirats. (1969), Cronartiaceae. 4 (on *Pinus*), Eur., N. Am., Japan.

endocyanosis, the inclusion of cyanobacteria inside the cells of another organism; in fungi, see *Geosiphon*.

Endocycla Syd. (1927) = Schizothyrium (Schizothyr.) fide Müller & v. Arx (1962).

endocyclic, see *Uredinales*.

endocystidium, see cystidium.

endocystidium-form, see *Uredinales*.

Endodermophyton Castell. (1910) = Trichophyton (Mitosp. fungi) fide Sutton (*in litt.*).

Endodesmia Berk. & Broome (1871) [non Benth. (1862), *Clusiaceae*] ≡ Broomeola (Mitosp. fungi). 3.B1.?. 1, UK.

Endodesmidium Canter (1949), Synchytriaceae. 1 (on desmids), Eur.

Endodothella Theiss. & Syd. (1915) = Phyllachora (Phyllachor.).

Endodothiora Petr. (1929), Dothioraceae. 1 (on *Dothidea*), former USSR.

Endodromia Berk. (1840) ? = Echinostelium (Myxom.).

endoectothrix, making growth in and on a hair.

endogenous, (1) living inside; (2) immersed in the substratum; (3) undergoing development within.

Endogloea Höhn. (1915) = Phomopsis (Mitosp. fungi) fide Sutton (1977).

Endogonaceae Paol. (1889), Endogonales. 3 gen., 21 spp. Sporocarp mainly hypogeous; forming only zygospores; saprobes or form ectomycorrhiza; cosmop.
Lit.: Morton & Benny (1990; rev.), Yao *et al.* (1995).

Endogonales, Zygomycetes. 1 fam., 3 gen., 21 spp.
Endomycorrhizal taxa transferred to *Glomales* (q.v.) by Morton & Benny (1990). Fam.:
Endogonaceae.
Lit.: Morton & Benny (1990), Pegler *et al.* (*British Truffles. A revision of British hypogeous fungi*, 1993), Warcup (*MR* **94**: 173, 1990; Australia), Yao *et al.* (*Kew Bull.* **50**: 349, 1995; gen. names).

Endogone Link (1809), Endogonaceae. *c.* 20 (saprobic or forming ectomycorrhiza), widespr., esp. temp. See Gerdemann & Trappe (1974; key NW Am. spp.), Trappe & Gerdemann (*Mycol.* **71**: 206, 1979), Bonfante-Fasolo & Scannerini (*Mycopath.* **59**: 117, 1976; ultrastr.), Tandy (*Aust. J. Bot.* **23**: 849, 1975; Australia), Berch & Fortin (*Mycol.* **75**: 328, 1983; zygospore germination, *CJB* **61**: 899, 1983; axenic culture, *CJB* **62**: 170, 1984), Berch & Castellano (*Mycol.* **78**: 292, 1986; sporulation in culture), Gibson *et al.* (*Mycol.* **79**: 433, 1987; cytochemistry), Jabaji-Hare *et al.* (*Mycol.* **80**: 54, 1988; fatty acids), Dalpé (*CJB* **68**: 910, 1990; culture), Warcup (1990; ectomycorrhizal association), Yao *et al.* (1995).

Endogonella Höhn. (1913) = Glaziella (Glaziell.) fide Zycha *et al.* (*Mucorales*, 1969).

endogonidium, a gonidium having its development inside a receptacle or gonidangium (obsol.).

Endogonopsis R. Heim (1966), Sclerodermatales (inc. sed.). 1, S. Asia. See Heim (*Revue Mycol.* **31**: 150, 1966; **33**: 379, 1968; *C.R. Acad. Sci. Paris* D **268**: 1489, 1969). Possibly based on young *Astraeus* (Astr.).

Endohormidium Auersw. & Rabenh. (1869) = Corynelia (Corynel.) fide Arnaud [not traced].

endohypha, see hypha.

endokapylic, a thallus of a lichenicolous fungus in which no morphologically distinct lichenized structure is formed (Poelt & Vězda, 1984; Rambold & Triebel, *Bibl. Lich.* **48**, 1992).

Endolepiotula Singer (1963), Secotiaceae. 1, Argentina.

endolithic, in stone (Kobluk & Kahle, *Bull. Can. Pet. Geol.* **25**: 208, 1977; fungi, bibliogr.); cfr. epilithic. See also cryptoendolithic.

Endolpidium De Wild. (1894) = Olpidium (Olpid.) fide Sparrow (1960).

Endomelanconium Petr. (1940), Mitosporic fungi, 8.A2.1. 1, Eur.

Endomeliola S. Hughes & Piroz. (1994), Meliolaceae. 1, NZ.

Endomyces Reess (1870), Endomycetaceae. 1 (on *Armillaria*), temp. See v. Arx (*Ant. v. Leeuwenhoek* **43**: 33, 1977), Redhead & Malloch (1977).

Endomycetaceae J. Schröt. (1893), Saccharomycetales. 13 gen. (+ 1 syn.), 34 spp. Mycelium ranging from well-developed to ± absent, then vegetative cells proliferating by multilateral budding; asci formed singly or in short irregular chains, directly from vegetative cells, usually ± globose, usually evanescent, sometimes multispored and proliferating percurrently; ascospores usually with asymmetrical flanges or sheaths. Fermentation sometimes present, coenzyme system usually Q-8.

Suggested relationships with the *Ophiostomatales* now seem dubious (Weijman, *Ant. v. Leeuwenhoek* **42**: 315, 1976). The Fam. as currently circumscribed is almost certainly polyphyletic.

Lit.: Redhead & Malloch (*CJB* **55**: 1701, 1977).

Endomycetales, see *Saccharomycetales*.

Endomycetes (Endomycota). Class (or phylum) for sporogenous and asporogenous yeasts, *Taphrinales*, *Exobasidiales*, and *Ustilaginales* (v. Arx, *Pilzkunde*, 1967); emended to include *Trichomycetes*, etc. by Kreisel (1969), and emended to exclude *Exobasidiales*, *Taphrinales* and *Ustilaginales* by v. Arx (1981); supported by glyceraldehyde-3-phosphate dehydrogenase sequences (Smith, *Proc. natn. Acad. Sci., USA* **86**: 7063, 1989). See *Ustomycota*.

endomycobiont, a fungal biont in a symbiosis completely immersed in the tissues of the host (e.g. the fungal partner in a mycophycobiosis); an inhabitant (see symbiosis); also used of certain mycorrhizas; see also Endophyte.

Endomycodes Delitsch (1943), Saccharomycetales (inc. sed.). 2, widespr.

Endomycopsella Boedijn (1960), Saccharomycopsidaceae. 2 or 3, widespr. See Barnett *et al.* (1990).

Endomycopsis Dekker (1931) ≡ Guillermondella (Metschnikow.). See v. Arx & Yarrow (*Ant. v. Leeuwenhoek* **50**: 799, 1984).

endomycorrhiza, see mycorrhiza.

Endomycota, see *Endomycetes*.

Endonema Pascher (1929) ≡ Pascherinema (Mitosp. fungi).

Endonevrum Czern. (1845) = Mycenastrum (Mycenastr.).

Endonius Raf. (1819) nom. dub. (Fungi, inc. sed.). No spp. included.

endo-operculation (of sporangial dehiscence in chytrids), operculum forced off and carried away by the emerging sporogenous contents (cfr. exo-operculation).

endoparasite, a parasite living inside its host.

endoperidermal, within the periderm (Lambright & Tucker, *Bryologist* **83**: 170, 1980); endophloeodal + hypophloeodal.

endoperidium, the inner layer of the peridium.

Endoperplexa P. Roberts (1993), Exidiaceae. 4, Eur.

Endophallus M. Zang & R.H. Petersen (1989), Phallaceae. 1, China. ? a monstrosity from *Phallus*.

Endophis Norman (1853) = Leptorhaphis (Arthopyren.).

endophloeodic (endophloeodal, endophloeic) (of the thallus of a crustaceous lichen), almost entirely immersed in bark.

Endophragmia Duvernoy & Maire (1920), Mitosporic fungi, 2.C2.1. 6, Eur. See Ellis (*Mycol. Pap.* **72**, 1959; key) but rejected by Hughes (*N.Z. Jl Bot.* **17**: 139, 1979).

Endophragmiella B. Sutton (1973), Mitosporic fungi, 1.C2.1. 37, widespr. See Hughes (*N.Z. Jl Bot.* **17**: 139, 1979; key).

Endophragmiopsis M.B. Ellis (1966), Mitosporic fungi, 1.C2.1. 1, India. See Hughes (*CJB* **61**: 1727, 1984).

Endophyllachora Rehm (1913) = Phyllachora (Phyllachor.) fide Clements & Shear (1931).

Endophylloides Whetzel & Olive (1917), = Dietelia (Pucciniosir.) fide Buriticá (*Rev. Acad. Colomb. Cienc.* **18**(no. 69): 131, 1991).

endophyllous, living within (i.e., below the cuticle) leaves.

Endophyllum Lév. (1825), Pucciniaceae. 28, widespr. *E. sempervivi* (houseleek, *Sempervivum*, rust). The (telial) aeciospores (I^{II}) produce basidia on germination. Reduced forms, sometimes placed in *Puccinia* or *Uromyces*. See Buriticá (*Rev. Acad. Colomb. Cienc.* **18**(no. 69): 131, 1991).

Endophyte. Literally an organism that lives within a plant, analogous to an epiphyte. Endophyte is used in a variety of ways, giving rise to some semantic confusion and ambiguities. The term has been used to refer to bacteria and parasitic angiosperms (e.g. mistletoes) that exist completely or partially within plants. With respect to fungi, endophyte is used to refer to endomycorrhizal fungi in plant roots (Kinden & Brown, *Can. J. Microbiol.* **21**: 1930, 1975) and fungi occurring within gametophytes of nonflowering vascular plants (Peterson *et al.*, *CJB* **59**: 711, 1981). However, the most prevalent usage is in reference to asymptomatic fungi within aerial plant parts. Wilson (*Oikos* **68**: 379, 1993) defines them as fungi which invade the stems and leaves of plants but cause no symptoms of disease, but Wennström (*Oikos* **71**: 3, 1994) objects to its use in mycology. The distinction between latent pathogens that cause disease only under specific environmental circumstances or saprophytic fungi that have colonized living plant tissues is not always clear. Typically endophytic fungi are identified by plating healthy, surface-sterilized leaves on nutrient agar and observing the outgrowth of fungi (Carroll & Carroll, *CJB* **56**: 3034, 1978).

Endophyte is frequently used in reference to systemic clavicipitaceous fungi (*Balansia, Epichloë*) that occur intercellularly in the shoots of grasses (Clay, *Ann. Rev. Ecol. Syst.* **21**: 275, 1990; fig.). These fungi fruit on leaves or inflorescences and cause some degree of sterility of hosts. However, hosts are often vegetatively invigorated and exhibit increased resistance to herbivory due to the production of alkaloid toxins by the fungi. *Epichloë* hosts sometimes flower normally and the fungus is seed-transmitted (Sampson, *TBMS* **18**: 337, 1935). Many grasses are infected by asexual derivatives of *Epichloë* (*Acremonium* spp.) that do not sporulate and are completely seed-transmitted (White, *Pl. Dis.* **71**: 340, 1987). They appear to exist in a mutualistic symbiosis with their hosts. No agreement exists as to whether the term endophyte should be restricted only to completely asymptomatic and/or mutualistic fungi within aerial plant parts or to fungi that grow endophytically regardless of their symptoms or disease effects. Lewis (*in* Rayner *et al.* (Eds), *Evolutionary biology of the fungi*: 161, 1987) suggested the alternative term 'mycophylla' to refer to mutualistic fungi inhabiting aerial plant parts. Given its variety of usages, endophyte should be clearly defined when used.

Lit.: Carroll (*Ecology* **69**: 2, 1988), Clay (*Ecology* **69**: 10, 1988; *in* Pirozynski & Hawksworth (Eds), *Coevolution of fungi with plants and animals*: 79, 1988, grasses; *MR* **92**: 1, 1989, grasses; *in* Burdon & Leather (Eds), *Pests, pathogens and plant communities*: 11, 1990); Petrini (*in* Andrews & Hiano (Eds), *Microbial ecology of leaves*: 179, 1991), Wilson (*Oikos* **68**: 379, 1993). See Lolium endophyte.

Endoplacodium Petr. (1949), Mitosporic fungi, 8.A1.?. 1, Iran.

endopropagule, a propagule produced inside the body (medical mycology).

Endoptychum Czern. (1845), Secotiaceae. 4, dry areas. See Singer & Smith (*Brittonia* **10**: 216, 1958).

Endopyreniomyces E.A. Thomas (1939) nom. inval. ? ≡ Endocarpon (Verrucar.).

Endopyrenium Flot. (1855) = Catapyrenium (Verrucar.).

Endoraecium Hodges & D.E. Gardner (1984), Uredinales (inc. sed.). 2, Hawaii.

Endoramularia Petr. (1923), Mitosporic fungi, ?3 or 6.C1.?. 1, Eur.

endosaprophytism, the destruction of an alga by the fungus in a lichen (Elenkin).

endosclerotium, a sclerotium of endogenous origin.

Endoscypha Syd. (1924), Leotiales (inc. sed.). 1, NZ.

Endosphaerium D'Eliscu (1977), Pythiales (inc. sed.). 1 (on fish), USA.

Endospora Scherff. (1925) nom. dub. (? Chytridiomycetes).

endospore, (1) the inner wall of a spore (see ascospore, Spore wall); (2) an endogenous spore, e.g. a sporangiospore.

Endosporella Thaxt. (1920) = Pyxidiophora (Pyxid.) fide Blackwell (*Mycol.* **86**: 1, 1994).

Endosporostilbe Subram. (1958) = Bloxamia (Mitosp. fungi) fide Nag Raj & Kendrick (1975).

Endostelium L.S. Olive, W.E. Benn. & Deasey (1984), Protosteliaceae. 1, Papua New Guinea.

Endostigme Syd. (1923) = Venturia (Ventur.) fide v. Arx & Müller (1975).

Endostilbum Malençon (1964), Mitosporic fungi, 2.A1.15. 1, Eur., Morocco. See Korf & Candoussau (*BSMF* **90**: 209, 1974).

endosymbiont, an organism which lives in mutualistic symbiosis within the cells of another organism; the inhabitant.

Endothia Fr. (1849), Valsaceae. 10, widespr. See Micales & Stipes (*Phytopath.* **77**: 650, 1987), Roane (*in* Roane *et al.*, *Chestnut blight*: 28, 1986; key 11 spp.). See also *Cryphonectria*.

Endothiella Sacc. (1906), Anamorphic Valsaceae, 8.A1.15. Teleomorphs *Cryphonectria*, *Endothia*. 2, Eur., N. Am.

Endothlaspis Sorokīn (1890), Ustilaginales (inc. sed.). 2, Central Asia. = Sporisorium (Ustilagin.). See Langdon & Fullerton (*Mycotaxon* **6**: 421, 1978), Vánky (1987).

endothrix, living inside a hair.

Endothrix and **Ectothrix**, were Sabouraud's names for his two primary divisions of *Trichophyton*; *Ectothrix* covering the subgroups *Microides* and *Megaspore* (*Megalosporon*). All these group names have been wrongly used as though they were valid generic names.

Endotrabutia Chardón (1930) = Phyllachora (Phyllachor.). See Petrak (*Sydowia* **5**: 336, 1951).

Endotrichum Corda (1838) nom. dub. fide Sutton (*Mycol. Pap.* **141**, 1977).

endotrophic, see mycorrhiza.

Endotryblidium Petr. (1959), ? Leotiales (inc. sed.). 1, temp.

endotunica, endoascus (q.v.).

Endoxyla Fuckel (1871), Boliniaceae. 5, widespr. See Barr (1978, posn; *Mycotaxon* **46**: 45, 1993), Untereiner (*Mycol.* **85**: 294, 1993; key).

Endoxylina Romell (1892), Ascomycota (inc. sed.). 4, Eur., Am., Asia. See Mhaskar (*Botanique, Nagpur* **3**: 69, 1972).

endozoic, living inside an animal.

Endozythia Petr. (1959), Mitosporic fungi, 4.A1.?. 1 (in *Pleospora* perithecia), former Yugoslavia.

Enduria Norman (1885) ? Ascomycota (inc. sed.).

Endyllium Clem. (1931) ≡ Magnusiomyces (Dipodasc.).

Enerthenema Bowman (1830), Stemonitidaceae. 3, N. temp.

Enerthidium Syd. (1939), Mitosporic fungi, 6.A2.19. 1, Afr.

Engelhardtiella A. Funk (1973), Mitosporic fungi, 3.G1.15. 1 (on *Botryosphaeria*), N. Am.

Engizostoma Gray (1821) ≡ Valsa (Vals.).

Englera F. Stevens (1939) = Asterina (Asterin.) fide Müller & v. Arx (1962).

Englerodothis Theiss. & Syd. (1915), Parmulariaceae. 1, Afr.

Engleromyces Henn. (1900), Xylariaceae. 1, Afr. See Dennis (*Bull. Jard. bot. Brux.* **31**: 148, 1961).

Englerula Henn. (1904), Englerulaceae. 6, Afr., Am., Australasia. See Hansford (*Mycol. Pap.* **15**, 1946).

Englerulaceae Henn. (1904), Dothideales. 6 gen. (+ 10 syn.), 53 spp. Mycelium superficial, dark, usually hyphopodiate; ascomata superficial, sometimes stalked, globose to pulvinate; peridium thin-walled, deliquescing in the upper part to expose the asci; hymenium gelatinous; interascal tissue absent; asci ovoid to saccate, fissitunicate, with an I+ outer layer; ascospores brown, 1-septate, usually with a mucous sheath. Anamorphs hyphomycetous or coelomycetous, usually prominent. Biotrophic on leaves, esp. trop.

Lit.: Doidge (*Bothalia* **4**: 273, 1942; S. Afr.), Stevens & Ryan (*Ill. biol. monogr.* **17** (2), 1939).

Englerulaster Höhn. (1910) = Asterina (Asterin.) fide Müller & v. Arx (1962).

Englerulella Hansf. (1946) = Rhytidenglerula (Englerul.) fide Müller & v. Arx (1962).

Engyodontium de Hoog (1978), Mitosporic fungi, 1.A1.10. 6, widespr. See Gams *et al.* (*Persoonia* **12**: 135, 1984; key).

Enigma G. Arnaud (1949) nom. inval. (? Myxomycota).

enniatin A (lateratiin) and **B**, peptide antibiotics from *Fusarium orthoceras*; antibacterial (Gäumann *et al.*, *Experientia* **3**: 202, 325, 1947); avenacein, sambucinum.

enokitake (winter mushroom), the edible *Flammulina velutipes*, cultivated in Japan and Taiwan (Chang & Hayes, 1978).

enphytotic, a plant disease of which the damage is constant from year to year; cf. epiphytotic.

Ensaluta Zobel (1854) = Tuber (Tuber.) fide Fischer (1938).

ensate (**ensiform**), narrow and pointed; sword-like in form.

Enslinia Fr. (1835) [non *Enslenia* Raf. (1817), *Acanthaceae*] ≡ Porodiscus (Coriol.).

Entelexis Van der Walt & Johannsen (1973), ? Saccharomycetales (inc. sed.). 1, S. Afr. Anamorph *Torulopsis*.

Enteridiaceae, see *Lycogalaceae*.

Enteridium Ehrenb. (1819), Lycogalaceae. 7, cosmop., esp. N. temp. See Farr (*Taxon* **25**: 514, 1976)

enteroblastic, see blastic.

Enterobotryum Preuss (1853) nom. dub. (Fungi, inc. sed.) fide Lundqvist (1972).

Enterobrus, see *Enterobryus*.

Enterobryus Leidy (1850), Eccrinaceae. 26 (in *Coleoptera*, *Diplopoda* and *Decapoda*), widespr. See Lichtwardt (*Mycol.* **46**: 564, 1954; **49**: 463, 734, 1957; **50**: 550, 1958; **52**: 248, 410, 743, 1960), Manier, Gasc. & Bouix (*Biol. Gabon* **3-4**: 305, 1972; *Bull. Inst. Fund. Afrique Noire* **36**: 614, 1974), Tuzet & Manier (*Protistol.* **3**: 413, 1967; ultrastructure), Wright (*Proc.*

Helminthol. Soc. Wash. **46**: 213, 1979; ultrastr.), Lichtwardt (1986; key), Gorter (*Bothalia* **23**: 85, 1990).

Enterocarpus Locq.-Lin. (1977), Microascaceae. 2, Sahara.

Enterodictyon Müll. Arg. (1892), Roccellaceae (L). 2, E. Indies.

Enterodictyonomyces Cif. & Tomas. (1953) ≡ Enterodictyon (Roccell.).

Enterodictyum Clem. & Shear (1931) ≡ Enterodictyon (Roccell.).

Enterographa Fée (1824), Roccellaceae (L). *c.* 30, mainly trop. See Aptroot *et al.* (*Bibl. Lich.* **57**: 19, 1995; key 7 foliicolous spp.), Coppins & James (*Lichenologist* **11**: 27, 1979; Br. spp.), Redinger (*Feddes Repert.* **43**: 49, 1938), Santesson (*Symb. bot. upsal.* **12**(1), 1952), Torrente & Egea (1989).

Enteromyces Lichtw. (1961), Eccrinaceae. 1 (in *Decapoda*), Eur., Chile, USA, Japan. See McCloskey & Caldwell (*J. Elisha Mitchell Sci. Soc.* **81**: 114, 1965), Hibbets (*Syesis* **11**: 213, 1978).

Enteromyxa Ces. (1879) ? = Lycogalopsis (? Lycoperdales).

Enteropogon Hibbits (1978) [non Nees (1836), *Gramineae*], Eccrinaceae. 1 (in *Decapoda*), USA. See Lichtwardt (1986).

Enterostigma Müll. Arg. (1885) = Thelotrema (Thelotremat.) fide Salisbury (*Lichenologist* **5**: 319, 1972).

Enterostigmatomyces Cif. & Tomas. (1953) ≡ Enterostigma (Thelotremat.).

Enthallopycnidium F. Stevens (1925), Mitosporic fungi, 5.E1.?. 2, Hawaii, N. Am.

entheogen, a plant (or fungus) substance used by humans in prehistory associated with religous feelings; see ethnomycology, hallucinogenic fungi, soma.

entire (of edges of lamellae, etc.), not torn; having no teeth.

ento-, (prefix), inside.

Entoderma Hanula, Andreadis & M. Blackw. (1991), Mitosporic fungi, 6.A/B1.1. 1 (on *Insecta*), USA.

Entodesmium Riess (1854), Lophiostomataceae. 6 (on legumes), temp. See Barr (*Mycotaxon* **43**: 371, 1992; posn), Holm (*Symb. bot. upsal.* **14**(3), 1957; key), Shoemaker (*CJB* **62**: 2730, 1984; key 6 spp.).

Entoleuca Syd. (1922) ? = Rosellinia (Xylar.) fide Læssøe & Spooner (*Kew Bull.* **49**: 1, 1994), but see Læssøe (*SA* **13**: 43, 1994).

Entoloma (Fr.) P. Kumm. (1871), Entolomataceae. *c.* 400, mainly trop. See Horak (*Sydowia* **28**: 171, 1976; cuboid spored spp., **29**: 289, 1977, **30**: 40, 1978; 74 S. Am. spp., *Beih. Nova Hedw.* **65**, 1980; Indomalaya, Australasia), Nordeloos (*Persoonia* **10**: 207, 1979; key 11 spp. subgen. *Pouzaromyces*, **10**: 427, 1979; subgen. *Nolanea*, key Eur. spp., **11**: 121, 1980; subgen. *Entoloma*, key Eur. spp., **11**: 451, 1980; subgen. *Leptonia*, key Eur. spp.), Pegler & Young (*Beih. Sydowia* **8**, 1979), Manimohan *et al.* (*MR* **99**: 1083, 1995; key 21 Indian spp.). = Rhodophyllus (Entolomat.) fide Singer (1949).

Entolomataceae Kotl. & Pouzar (1972), Agaricales. 17 gen. (+ 12 syn.), 1282 spp. Basidiospores pink, faceted, angular or longitudinally ridged.

Entolomina Arnaud (1952), Thelephoraceae. 1, Eur.

entomo- (prefix), of *Insecta*.

Entomocorticium H.S. Whitney, Bandoni & Oberw. (1987), Corticiaceae. 1, Canada.

Entomocosma Speg. (1918) ? = Pyxidiophora (Pyxidioph.) fide Blackwell (*Mycol.* **86**: 1, 1994).

entomogenous, living in or on insects, esp. as pathogens.

Entomogenous fungi. Range from commensals or mutualists, through ectoparasites which do not seriously affect their arthropod hosts, to pathogens that

are lethal and include representatives of all the major groups of fungi. 750 spp. in 56 gen. known to be pathogens or parasites of arthropod pests. The terms entomopathogens, entomogenous mutualistic symbionts, entomogenous ectoparasites and entomogenous endoparasites have been proposed (Evans, *in* Pirozynski & Hawksworth (Eds), *Coevolution of fungi with plants and animals*: 149, 1988).

Septobasidium has symbiotic associations with scale insects, as has *Stereum sanguinolentum* with *Sirex* (Perkin, *Ann. appl. Biol.* **29**: 268, 1942), and a number of insects make use of fungi in their alimentary systems; e.g. Phaff *et al.* (*Ecology* **37**: 533, 1956; yeasts of *Drosophila*). See Brues (*Insect dietary*, 1946), Steinhaus (*Insect microbiology*, 1946; *Insect pathology*, edn 2, 1963), Shifrine & Phaff (*Mycol.* **48**: 41, 1956; yeasts and bark beetles), Madelin (*Endeavour* **19**: 181, 1960), Batra (Ed.) (*Insect-fungus symbiosis*, 1980).

Trichomycetes and *Laboulbeniales* have little effect on their hosts but the chytrid *Coelomomyces*, the oomycete *Lagenidium giganteum* (both mainly on mosquito larvae), and among zygomycetes the *Entomophthoraceae* (e.g. *Entomophthora, Erynia, Massospora, Neozygites, Zoophthora*) are important insect pathogens. Among ascomycetes *Ascosphaera, Cordyceps, Torrubiella,* and *Hypocrella* and its *Aschersonia* anamorph, frequently cause epizootics as do the hyphomycetes *Beauveria, Culicinomyces, Hirsutella, Metarhizium, Nomuraea, Paecilomyces,* and *Verticillium*. A number of these genera are being exploited as mycoinsecticides to control a range of arthropod pests (see mycopesticides).

Lit.: General: Tavares (*Laboulbeniales, Fungi, Ascomycetes*, 1985), Petch (*TBMS* **7-12**, 1921-27; *Studies* **16-27**, 1931-44, *Notes*), Steinhaus (1963), Müller-Kögler (*Pilzkrankheiten bei Insekten*, 1965), Madelin (*Ann. Rev. Entomol.* **11**: 423, 1966; *in* Ainsworth & Sussman (Eds), *The Fungi* **3**: 227, 1968), Brady (*Biocontrol News and Information* **2**: 281, 1981), McCoy (*J. Cell Biochem.* **13A** 156, 1989; review control of pests), Wheeler & Blackwell (*Fungus-insect relationships*, 1984; compr. review), Evans & Samson (*Mycologist* **1**: 152, 1987; fungi on spiders), Samson *et al.* (*Atlas of entomopathogenic fungi*, 1988), Kobayasi & Shimizu (*Iconography of vegetable wasps and plant worms*, 1983; taxonomy), Moore (*Biocontrol News and Information* **9**: 209, 1988; fungi on mealybugs), Evans & Prior (*in* Rosen (Ed.), *Armored scale insects*: 3, 1990; fungi on scales), Evans & Hywel-Jones (*in* Ben-Dov & Hodgson (Eds), *Soft scale insects*: 5, 1994; fungi on coccids), Hajek (*in* Lumsden & Vaughan (Eds), *Pest Management: biologically based technologies*: 54, 1993; prospects for insect control).

Regional: British Isles, Petch (*TBMS* **31**: 286, 1948), Leatherdale (*Entomophaga* **15**: 419, 1970; hosts). **Israel**, Kenneth *et al.* (*Israel J. agric. Res.* **21**: 63, 1971), Kenneth & Olmert (*Israel J. Entomol.* **10**: 105, 1975). **N. America**, Charles (*Insect Pest Surv. Bull. US* **21** (Suppl. 9), 1941). **Former USSR**, Koval (*Key to the entomogenous fungi of the USSR*, 1974), Khachatourians (*Handb. Appl. Mycol.: Humans, Animals & Insects* **2**: 613, 1991; physiology and genetics).

See ambrosia fungi, Coevolution, Insects and fungi, Laboulbeniales.

Entomopatella Petr. (1927) = Chaetospermum (Mitosp. fungi) fide Petrak (*Ann. Myc.* **32**: 447, 1934).

Entomopeziza Kleb. (1914) = Diplocarpon (Dermat.) fide Nannfeldt (1932).

Entomophaga Batko (1964), Entomophthoraceae. 9, widespr. *E. grylli* on locusts. See Remaudière & Keller (1980), Nolan (*Can J. Microbiol.* **33**: 808, 1987; protoplasts), Murrin & Nolan (*CJB* **65**: 169, 1987; ultrastr.), Keller (1987; key), Humber (*Mycotaxon* **21**: 265, 1984; 1989)

entomophilous (of fungi), having spores distributed by insects.

Entomophthora Fresen. (1856), Entomophthoraceae. 11 (on *Arthropoda*, *Insecta*), cosmop. *E. muscae* on house flies. See Waterhouse & Brady (*Bull. BMS* **16**: 113, 1982; keys), Gustafsson (*LantrHogsk Annlr* **31**: 103, 1965; taxonomy, 405, 1966; cultivation, physiology), Remaudière & Keller (1980; segregate genera), Eilenberg *et al.* (*J. Invert. Path.* **48**: 318, 1986; ultrastr.), Keller (1987; key), Humber (1989; redescript.).

Entomophthora Krenner (1961) nom. inval. = Entomophthora (Entomophthor.).

Entomophthoraceae Nowak. (1877), Entomophthorales. 11 gen. (+ 8 syn.), 131 spp. Sporophores without subsporangial vesicle; some cells multinucleate, nucleus visible during mitosis and interphase, nucleolus not prominent.
Lit.: Humber (1989).

Entomophthorales, Zygomycetes. 6 fam., 23 gen. (+ 12 syn.), 185 spp. Spores forcibly discharged; most are parasites of insects. Fams:
(1) **Ancylistaceae**.
(2) **Basidiobolaceae**.
(3) **Completoriaceae**.
(4) **Entomophthoraceae**.
(5) **Neozygitaceae**.
(6) **Meristacraceae**.
The order may be heterogenous from recent molecular data (Nagahama *et al.*, 1995) in which *Basidiobolus* clustered with some genera of *Chytridiales*, and *Conidiobolus*, *Entomophthora*, and *Zoopthora* with *Mucorales*.
Lit.: Pohlad & Bernard (*Mycol.* **70**: 130, 1978; key spp. on nemat. and tardigr.), Waterhouse (*in* Ainsworth *et al.* (Eds), *The Fungi* **4B**: 219, 1973; key gen.), Lakon (*Nova Hedw.* **5**: 7, 1963; gen. key), Remaudière & Keller (*Mycotaxon* **11**: 323, 1980), Tucker (*Mycotaxon* **13**: 481, 1981; key nonentomogen. spp.), Humber & Ramoska (*in* Samson *et al.*, *Fundamental and applied aspects of invertebrate pathology*: 190, 1986; life cycles), Latgé *et al.* (*in* Samson *et al.*, *Fundamental and applied aspects of invertebrate pathology*: 190, 1986; life cycles), Wolf (*Nova Hedw.* **46**: 121, 1988; parasitism), Humber (*Mycotaxon* **34**: 441, 1989; emend.), Mikawa (*Bull. natn Sci. Mus.* Tokyo B **15**: 49, 1989; Nepal), Papierok (*Ann. Entom. Fenn.* **55**: 63, 1989; Finland), Toriello *et al.* (*J. Inv. Path.* **53**: 358, 1989; immunological separation of gen.), Keller (*Sydowia* **40**: 122, 1987, **43**: 39. 1991; Switzerland), Bałazy (*Flora of Poland* **24**. *Entomophthorales*, 1993), Nagahama *et al.* (*Mycol.* **87**: 203, 1995; phylogeny by 18S RNA).

Entomospora Sacc. ex Jacz. (1926) = Taphrina (Taphrin.) fide Eriksson & Hawksworth (*SA* **7**: 70, 1988).

Entomosporium Lév. (1856), Anamorphic Dermateaceae, 6.D1.15. Teleomorph *Diplocarpon*. 1 (on *Rosaceae*), temp. See Sutton (1980), Muthumary (*Curr. Sci.* **57**: 195, 1987; conidiogenesis).

Entomyclium Wallr. (1833) = Dendryphion (Mitosp. fungi) fide Hughes (1958).

Entonaema A. Møller (1901), Xylariaceae. 5, widespr., mainly trop. See Rogers (*Mycol.* **73**: 28, 1981; key).

entoparasitic, parasitic inside the host.

Entopeltacites Selkirk (1972), Fossil fungi. 5 (Miocene), Australia.

Entopeltis Höhn. (1910) = Vizella (Vizell.) fide Hughes (1953), Swart (1971). See v. Arx & Müller (1975).

Entophlyctis A. Fisch. (1891), Endochytriaceae. *c.* 19, temp. See Barr (*CJB* **49**: 2215, 1971).

entoplacodial, see stroma.

Entorrhiza C.A. Weber (1884), Tilletiaceae. 5 (on *Juncaceae* and *Cyperaceae* roots), Eur., Algeria, NZ. See Fineran (*Nova Hedw.* **29**: 825, 1978; **30**: 1, 1979). = Melanotaenium (Tillet.) fide Thirumalachar & Whitehead (*Am. J. Bot.* **55**: 183, 1968).

Entosordaria (Sacc.) Höhn. (1920) = Clypeosphaeria (Clypeosphaer.) fide Barr (1989).

Entosordaria Speg. (1920) nom. nud. = Anthostomella (Xylar.).

Entosthelia (Wallr.) Hue (1915) = Dermatocarpon (Verrucar.).

entostroma, see stroma.

Entrophospora R.N. Ames & R.W. Schneid. (1979), Acaulosporaceae. 4, N. Am., NZ, Taiwan. See Ames & Schneider (*Mycotaxon* **8**: 347, 1979), Schenck *et al.* (*Mycol.* **76**: 685, 1984), Sieverding & Toro (*Mycotaxon* **28**: 209, 1987), Wu *et al.* (*Mycotaxon* **53**: 283, 1995), Yao *et al.* (1995).

Entyloma de Bary (1874), Tilletiaceae. *c.* 150, cosmop. Spores hyaline or pale, intercellular. The sori are generally in leaves (causing leaf spots) and there is frequently a mitosporic state; cf. *Entylomella* and *Cylindrosporium s.l. E. dahliae* (on *Dahlia*), *E. australe* (on *Physalis*), *E. ellisii* (on spinach, *Spinacia*).

Entylomella Höhn. (1924), Anamorphic Tilletiaceae. 45, widespr. Teleomorph *Entyloma*. See Ciferri (1959).

Eoagaricus L. Krieg. (1923) ≡ Physalacria (Physalacr.).

Eoaleurina Korf & W.Y. Zhuang (1986), Otideaceae. 1, C. & S. Am. See Korf (*SA* **6**: 127, 1987).

Eocronartium G.F. Atk. (1902), Platygloeaceae. 1 (on *Musci*), Eur., N. Am. See Stanley (*Trans. Am. micr. Soc.* **59**: 407, 1940), Khan & Kimbrough (*CJB* **58**: 642, 1980; ultrastr.).

Eoetvoesia Schulzer (1866) nom. dub. (? Mitosp. fungi).

Eolichen Zukal (1884) ? = Nectria (Hypocr.) fide Keissler (*Rabenh. Krypt.-Fl.* 9, 1(2), 1936-38).

Eolichenomyces Cif. & Tomas. (1954) ≡ Eolichen (Hypocr.).

Eomycenella G.F. Atk. (1902) = Mycena (Tricholomat.) fide Smith (1947).

Eomyces F. Ludw. (1894) nom. dub.; based on an achlorotic alga fide Batra (*in* Subramanian (Ed.), *Taxonomy of fungi* **1**: 187, 1978).

Eomycetopsis J.W. Schopf (1968), ? Fossil fungi.

Eopyrenula R.C. Harris (1973), Pleomassariaceae (L). 3, N. temp. See Aptroot (*Bibl. Mycol.* **44**, 1991; posn), Barr (1993).

Eosphaeria Höhn. (1917), Lasiosphaeriaceae. 1, Eur., N. Am. See Petrini *et al.* (*TBMS* **82**: 554, 1984). Anamorph *Phialophora*-like.

Eoterfezia G.F. Atk. (1902), Eoterfeziaceae. 1, N. Am.

Eoterfeziaceae G.F. Atk. (1902), Pezizales. 3 gen., 7 spp. Ascomata small, cleistothecial, ± globose; the peridium thin, membranous, either composed of coalescing hyphae or with a narrow layer of pseudoparenchymatous cells covered by a hyaline granular layer; interior sometimes divided into locules by radiating mycelial strands; asci ? formed from croziers, saccate, very thin-walled, evanescent; ascospores small, hyaline, aseptate. Anamorphs unknown. Fungicolous, N. and S. Am.
The affinities of this fam. are uncertain; possibly a placement near the *Amorphothecaceae* might be appropriate.

Epaphroconidia Calatayud & V. Atienza (1995), Mitosporic fungi, ?.?.?. 1, Spain.

epapillate, having no papillae.

Ephebe Fr. (1825), Lichinaceae (L). 12, cosmop. See Henssen (*Symb. bot. upsal.* **18** (1), 1963).

Ephebeia Nyl. (1875) = Ephebe (Lichin.) fide Henssen (1963).

Ephebella Itzigs. (1857) nom. dub. (Fungi, inc. sed.). Apparently based on an alga fide Currah (*SA* **5**: 130, 1986).

Ephebomyces Cif. & Tomas. (1953) = Ephebe (Lichin.).

Ephedracetes T.C. Huang (1981), Fossil fungi. 1 (Miocene), Taiwan.

Ephedrosphaera Dumort. (1822) nom. rej. = Nectria (Hypocr.).

Ephelidium Speg. (1920) nom. conf. fide Sutton (*Mycol. Pap.* **141**, 1977).

Ephelidium C.W. Dodge & E.D. Rudolph (1955), Mitosporic fungi, ?.?.?. 1 (L), Antarctic.

Ephelina Sacc. (1889) = Leptotrochila (Dermat.) fide Schüepp (1959).

Epheliopsis Henn. (1908) = Eutypa (Diatryp.) fide Petrak (*Sydowia* **5**: 169, 1951). See also Ciferri (*Atti Ist. bot. Univ. Pavia* **19**: 105, 1962).

Ephelis Fr. (1849), Anamorphic Clavicipitaceae, 3.E1.?. Teleomorph *Balansia*. 5 (mostly on grasses), esp. trop. See Govindu & Thirumalachar (*in* Subramanian (Ed.), *Taxonomy of fungi*: 328, 1984), Ullasa (*Mycol.* **61**: 572, 1969).

Ephemeroascus Emden (1973), Coniochaetaceae. 1 (from soil), Eur.

Ephemerocybe Fayod (1869) = Coprinus (Coprin.) fide Singer (1951).

epi- (prefix), upon.

epibasidium, see basidium.

Epibelonium E. Müll. (1963), Saccardiaceae. 1, France.

epibiotic, living on the surface of another organism.

Epibotrys Theiss. & Syd. (1915) = Gilletiella (Dothideales) fide v. Arx & Müller (1975).

Epibryon Döbbeler (1978), ? Pseudoperisporiaceae. 20 (on *Musci*), cosmop.

epibryophilous, growing over bryophytes.

Epichloë (Fr.) Tul. & C. Tul. (1865), Clavicipitaceae. 8, widespr. *E. typhina* (choke of grasses). See Doguet (*BSMF* **76**: 171, 1960; development), White (*Mycol.* **85**: 444, 1993; Br. spp.).

Epichloea Giard (1889) nom. dub. (Fungi, inc. sed.).

epichroic, see colour.

Epichysium Tode (1790) nom. dub; ? based on insect debris fide Fries (*Syst. mycol.* **3**: 293, 1832).

Epicladonia D. Hawksw. (1981), Mitosporic fungi, 4.A/B1.15. 3 (on lichens, esp. *Cladonia*), Eur.

Epiclinium Fr. (1849), Mitosporic fungi, 7.B.?. 2, Eur., Am.

Epicnaphus Singer (1960), Tricholomataceae. 1, S. Am. See Raithelhuber (*Metrodiana* **4** (3): 52, 1973).

Epicoccum Link (1815), Mitosporic fungi, 3.D2.1. 2, widespr. See Schol-Schwarz (*TBMS* **42**: 149, 1959). *E. nigrum*, *E. andropogonis* (syn. *Cerebella andropogonis*) on *Sphacelia* of *Claviceps* and frequently mistaken as *Ustilaginales*; see Langdon (*Phytopath.* **32**: 613, 1942; *Mycol. Pap.* **61**, 1955).

epicortex, a thin polysaccharide-like layer over the surface of the cellular upper cortex in thalli of some *Parmeliaceae* visible by SEM (Hawksworth, *in* Hale, *Smithson. Contr. bot.* **10**: 5, 1973) and which may have regular pores functioning in gas exchange (Hale, *Lichenologist* **13**: 1, 1981). See Hyvärinen (*Lichenologist* **24**: 267, 1992; environmental induction), Lumbsch & Kothe (*Mycotaxon* **43**: 277, 1992; *Coccocarpiaceae, Pannariaceae*).

Epicorticium Velen. (1926) = Phaeomarasmius (Cortinar.) fide Singer (1975).

Epicrea Petr. (1950), Hypocreaceae. 1, Ecuador.

epicutis, see cutis.

Epicymatia Fuckel (1870) = Stigmidium (Mycosphaerell.) fide Roux & Triebel (1994).

Epicyta Syd. (1926) = Aplosporella (Mitosp. fungi) fide Clements & Shear (1931).

epidemic, (1) (adj.), (of a disease of humans, but used of plants and animals), general and severe in a group for a time; (2) (n.), the disease itself; cf. epiphytotic, epizootic.

epidemiology, the study of disease incidence, distribution and control. See Plant pathogenic fungi.

Epidermella Tehon (1935) ? = Hypoderma (Rhytismat.).

Epidermomyces Loeffler (1983) ≡ Epidermophyton (Mitosp. fungi).

Epidermophyton Sabour. (1907) nom. cons., Mitosporic fungi, 1.C1.2. 2 (on humans), widespr. Macroconidia pyriform; microconidia absent. *E. floccosum* (syn. *E. inguinale* and *E. cruris*) on humans (glabrous skin) causing tinea cruris and tinea pedis.

epidermophytosis, see dermatophytosis (esp. tinea cruris and tinea pedis).

Epidochiopsis P. Karst. (1892), Mitosporic fungi, ?.A1.?. 1, Eur.

Epidochium Fr. (1849) = Tremella (Tremell.) fide Donk (*Taxon* **7**: 193, 1958).

Epidrolithus Raf. (1836) ? = Leptogium (Collemat.) fide Merrill (*Index Rafinesq.*, 1949).

epiflora, surface flora; sometimes applied (incorrectly) to the microbiota on seed surfaces; the epibiota.

epigeal (epigean, epigeic), on the earth.

epigeic (of lichens), not attached to any substrate but blowing about on the surface of the ground; see wandering lichens.

epigenous, growing on the surface.

Epiglia Boud. (1885) = Mniaecia (Leot.) fide Korf (*in* Ainsworth et al., The Fungi **4A**: 249, 1973).

Epigloea Zukal (1889), Epigloeaceae (?L). 10 (on *Algae*), Eur., Antarctic. See David (*SA* **6**: 217, 1987), Döbbeler (*Beih. Nova Hedw.* **79**: 203, 1984; key).

Epigloeaceae Zahlbr. (1903), Ascomycota (inc. sed., ? L). 1 gen (+ 2 syn.), 10 spp. Thallus inconspicuous or absent; ascomata immersed in a thin gelatinous algal film, perithecial, dark green to black, superficial, composed of thin-walled periclinally arranged hyphae immersed in a gelatinous matrix; interascal tissue of narrow rarely branched thin-walled paraphyses, the apices not swollen; asci cylindrical, persistent, at first thick-walled but becoming thin-walled, the wall blueing in iodine, without apical structures, releasing ascospores through a vertical split; ascospores hyaline, transversely septate, thin-walled, without a sheath. Anamorphs pycnidial. ? Lichenized or parasitic on algae, temp.

Lit.: David (*SA* **6**: 217, 1987).

Epigloeomyces Cif. & Tomas. (1957) ≡ Epigloea (Epigl.).

epigynous, having the antheridium above the oogonium on one hypha.

epihymenium, a thin layer of interwoven hyphae on the surface of the hymenium (Corner, 1950; epithecium).

epikapylic, a thallus of a lichenicolous fungus in which a morphologically distinct lichenized structure is formed (Poelt & Vězda, 1984).

Epilichen Clem. (1909), ? Rhizocarpaceae (L). 1 (on *Baeomyces*), N. temp. See Hafellner (*Beih. Nova Hedw.* **62**, 1979).

Epilithia Nyl. (1853), Anamorphic Asterothyriaceae. 1, Eur. See Seifert (*TBMS* **85**: 123, 1985).

epilithic, living on the surface of stones; cfr. endolithic.

epinecral layer, see necral layer.

Epinectria Syd. & P. Syd. (1917), Hypocreaceae. 1, Philipp.

Epinyctis Wallr. (1831) = Lepraria (Ascomycota).

Epipeltis Theiss. (1913) = Schizothyrium (Schizothyr.) fide v. Arx & Müller (1975).

Epiphegia G.H. Otth (1870) ? = Massarina (Lophiostom.) fide Saccardo (*Syll. Fung.* **11**: 332, 1895).

Epiphloea Trevis. (1880), Heppiaceae (L). 1, Eur. See Gyelnik (*Rabenh. Krypt.-Fl.* **9**, 2(2), 1940).

epiphloeodal, living upon bark.

Epiphora Nyl. (1876) [non Lindl. (1837), *Orchidaceae*] = Plectocarpon (Roccell.) fide Santesson (*The lichens and lichenicolous fungi of Sweden and Norway*, 1993).

epiphragm, the membrane over the young fruit-body in the *Nidulariaceae*.

epiphyllous, on the upper surface of a leaf; foliicolous.

Epiphyma Theiss. (1916) = Botryosphaeria (Botryosphaer.) fide v. Arx & Müller (1954).

epiphyte, a plant living on another, but not as a parasite.

epiphytic (adj.), frequently = corticolous.

epiphytotic, an epidemic among plants.

epiplasm (of an ascus), the cytoplasm not used up in the 'free cell formation' of ascospores.

Epiploca Kleb. (1918) = Epipolaeum (Pseudoperispor.) fide Müller & v. Arx (1962).

Epipolaeum Theiss. & Syd. (1918), Pseudoperisporiaceae. 17, N. & S. Am., Eur. See Farr (*Mycol.* **71**: 243, 1979), Shoemaker (*CJB* **43**: 631, 1965).

Episclerotium L.M. Kohn (1984), Leotiaceae. 2, Eur. See Malaval (*Docums mycol.* **19**: 9, 1989; ascus ultrastr.).

Episoma Syd. (1925) = Phaeostigme (Pseudoperispor.). See v. Arx & Müller (1975).

Episphaerella Petr. (1924), Pseudoperisporiaceae. 5, Am., Afr. See Müller & v. Arx (1962), Farr *et al.* (*CJB* **63**: 1983, 1985).

Episphaeria Donk (1962), Crepidotaceae. 1, Eur.

epispore, see ascospore, Spore wall.

episporium, see Spore wall.

Episporogoniella U. Braun (1994), Mitosporic fungi, 1.C2.1. 1 (on *Bryophyta*), Brazil.

Epistictum Trevis. (1869) = Dermatocarpon (Verrucar.).

Epistigme Syd. (1924), Mitosporic fungi, 4.A2.?. 1, S. Afr.

epistroma, see stroma.

Epitea Fr. (1832) = Phragmidium (Phragmid.) fide Dietel (1928).

epithalline, of a falsely thalline apothecial edge in lichenized fungi.

epithecium, tissue at the surface of an apothecium formed by the branching of the ends of the paraphyses above the asci; cf. epihymenium, pseudoepithecium.

Epithele (Pat.) Pat. (1900), Epitheliaceae. 13, widespr. See Boquiren (*Mycol.* **63**: 937, 1971; key).

Epitheliaceae Jülich (1982), Stereales. 4 gen., 14 spp. Basidioma resupinate, odontoid, with hyphal pegs; dimitic; spores inamyloid.

epithelium, see cutis.

Epithelopsis Jülich (1976), Epitheliaceae. 1, India, NZ.

epithet, (1) the second (specific) adjectival part of a Latinized binomial (the 'trivial' name of the zoologist); (2) the third or fourth (varietal, etc.) term.

Epithyrium (Sacc.) Trotter (1931), Anamorphic Agyriaceae, 8.A2.15. Teleomorph *Sarea*. 1 (on conifer resin), widespr. See Hawksworth & Sherwood (*CJB* **59**: 357, 1981).

epitunica, see exosporium; Spore wall.

epitype, see Nomenclature, type.

Epixyla Raf. (1806) nom. dub. (Fungi, inc. sed.).

epixylic (**epixylous**), living on wood; lignicolous.

Epixylon Füisting (1867) = Hypoxylon (Xylar.) fide Læssøe (*SA* **13**: 43, 1994).

epizoic, living on animals.

epizootic, an epidemic among animals.

epizootic lymphangitis, a disease of horses caused by *Histoplasma farciminosum*.

Epochniella Sacc. (1880) = Stemphylium (Mitosp. fungi) fide Lindau (*Rabenh. Krypt.-Fl.* **1** (9): 207, 1907).

Epochnium Link (1809) = Monilia (Mitosp. fungi) fide Hughes (*CJB* **36**: 727, 1958).

epruinose, having no pruina.

Epulorhiza R.T. Moore (1987), Anamorphic Tulasnellaceae. Teleomorph *Tulasnella*. 3, widespr. See Currah *et al.* (*CJB* **68**: 1171, 1990; mycorrhizal sp.), Currah & Zelmer (*Rep. Tottori mycol. Inst.* **30**: 43, 1992).

equal (of a stipe), having the same diameter throughout.

Erannium Bonord. (1860) = Coleosporium (Coleospor.) fide Saccardo (*Syll. Fung.* **18**: 774, 1906).

erasure phenomenon (in *Dictyostelium*), the loss by amoebae of the capacity to recapitulate when developing cultures are disaggregated and placed on a growth medium.

Erebonema A. Roem. (1845) nom. dub. (? Mitosp. fungi).

erect, upright; straight, not curved, up.

Eremascaceae Eng. & E. Gilg (1924), Eurotiales. 1 gen., 2 spp. Mycelium copious; ascomata absent; asci formed by anastomosis of equal cells on short coiled ascogenous hyphae, ± globose, thin-walled, evanescent. Ascospores hyaline, smooth, aseptate. Anamorph hyphomycetous, thallic, *Chrysosporium*-like. Saprobic, esp. on food.

Eremascus Eidam (1883), Eremascaceae. 2, Eur., India. See Berbee & Taylor (*Mol. Biol. Evol.* **9**: 278, 1992; posn), Harrold (*Ann. Bot., Lond.*, n.s. **14**: 127, 1950).

Eremastrella S. Vogel (1955), ? Psoraceae (L). 1, S. Afr., Australia. See Lumbsch & Kothe (*Nova Hedw.* **57**: 19, 1993; thalli), Pietschman (*Nova Hedw.* **51**: 521, 1990; posn.), Schneider (*Bibl. Lich.* **13**, 1980).

Eremodothis Arx (1976), Testudinaceae. 1, India, Japan. See Udagawa & Ueda (*J. Jap. Bot.* **56**: 289, 1981).

Eremomyces Malloch & Cain (1971), Eremomycetaceae. 2 (coprophilous), Afr., India, N. Am. See Malloch & Sigler (*CJB* **66**: 1929, 1988). Anamorphs *Arthrographis*, *Trichosporiella*.

Eremomycetaceae Malloch & Cain (1971), Dothideales. 2 gen. (+ 1 syn.), 3 spp. Ascomata cleistothecial, formed from solid pseudoparenchymatous initials, spherical, small, sometimes setose; peridium brown to black, pseudoparenchymatous or cephalothecoid; interascal tissue absent; asci irregularly disposed within the ascoma, clavate, thin-walled, not fissitunicate, evanescent; ascospores small, hyaline to brown, aseptate. Anamorph hyphomycetous, conidiogenesis thallic. From soil and coprophilous.
Lit.: Malloch & Sigler (*CJB* **66**: 1929, 1988).

Eremotheca Theiss. & Syd. (1917) = Schizothyrium (Schizothyr.) fide Müller & v. Arx (1962).

Eremothecella Syd. & P. Syd. (1917), Arthoniaceae (L). 4, trop. See Sérusiaux (*SA* **11**: 39, 1992).

Eremothecium Borzí (1888), Metschnikowiaceae. 2, widespr. See Batra (*USDA Tech. Bull.* **1469**, 1973; key), Rosing (*Mycol.* **79**: 157, 1987; ultrastr.).

ergoline alkaloids, from *Claviceps* sclerotia include both lysergic acid derivatives (esp. *C. purpurea*, *C. paspali*) and clavine alkaloids (esp. *C. fusiformis*, *C. gigantea*, *Sphacelia sorghi*). Cf. ergot.

ergometrine (D-lysergic acid propanolamide), an ergot alkaloid from *Claviceps purpurea* sclerotia (esp.

Spanish and Portuguese); used in medicine against migraine. Cf. ergotamine.

ergosterol, the commonest sterol of fungi (hence also in lichens) first isolated from *Claviceps purpurea* sclerotia; yeast ergosterol is converted to vitamin D_2 by ultraviolet radiation.

ergot, (1) the *Claviceps* disease of cereals and grasses, esp. ergot of rye (*Claviceps purpurea*); (2) an ergot fungus; (3) (in trade), the sclerotia of ergot fungi.

Infection (of ovaries) of cereals and grasses is by rain- or insect-borne conidia of the *Sphacelia* state which occur suspended in 'honey dew' exuded from the host florets in response to infection. Sclerotial development begins 2-3 weeks after infection. The sclerotia contain a range of alkaloids (of which the most important are lysergic acid derivatives; see Stoll & Hofmann, *in* Manske (Ed.) (*The alkaloids* **8**: 725, 1965); also ergometrine, ergotamine, ergotoxine, including the active principals not only of the poisonous properties of ergot but also of the therapeutic applications (e.g. in migraine, obstetrics, and as hallucinogens). The ergoline ring system is built up from *L*-triptophan and mevalonic acid. *N*-methyl group is derived from methionine.

Ergot of rye contaminating bread causes **ergotism** in humans which is of two main types: the gangrenous (St Anthony's Fire of the Middle Ages) and the spasmodic (see Barger, *Ergot and ergotism*, 1931; Bové, *The story of ergot*, 1970). Human exposure to low levels of ergolines still appears to be widespread and there have been severe outbreaks in France in 1951 (Fuller, *The day of St. Anthony's fire*, 1968), and more recently in Ethiopia and India (*WHO Environmental Health Criteria* 105, 1990).

Claviceps paspali causes paspalum staggers in cattle, sheep, and horses (Hopkirk, *NZ Jl Agric.* **53**: 103, 1936) and *C. fusiformis* on *Pennisetum typhoides* has been associated with agalactia in sows resulting in the death of new-borne piglets (Loveless, *TBMS* **50**: 15, 1967).

Ergot of rye was once the only source of the medicinal ergot alkaloids but since the early 1960s semisynthetic alkaloids have been prepared on a large scale from lysergic acid produced by *C. paspali* fermentations. Ergot alkaloids are also produced by other genera (Řeháček & Sajdl, *Ergot alkaloids*, 1990).

Ergotaetia E.J. Quekett (1841) = Sphacelia (Mitosp. fungi).

ergotamine, a cyclic tripeptide derivative of lysergic acid from *Claviceps purpurea* sclerotia; used in medicine against migraine. Cf. ergometrine.

ergotism, ergot poisoning. See ergot.

ergotoxine, a mixture of ergocornine, ergocristine, and ergokryptine; cyclic tripeptide derivatives of lysergic acid from *Claviceps purpurea* sclerotia.

Ericianella Brond. (1828) = Bactridium (Mitosp. fungi) fide Fries (*Summa veg. Scand.*, 1849).

Eriksson (Jakob; 1848-1931). Professor and Director, Department of Plant Physiology and Agricultural Botany, Academy of Agriculture, Stockholm, 1885-1913. He discovered of physiologic races in rusts (see *Ber. dtsch. bot. Ges.* **12**: 292, 1894; *Jb. wiss. Bot.* **29**: 499, 1896). Other writings: *Über die Mykoplasma-theorie ... (Biol. Zbl.* **30**: 618, 1910), *Die Pilzkrankheiten der landwirtschaftlichen Kulturgegewachse* (1926; in English, 1930 [first printed in Swedish, 1910 and English, 1912]); (with Hennings) *Die Getreideroste, ihre Geschichte und Natur, sowie Massregeln gegen dieselben* (1896, first printed in Swedish as *Medd. Kongl. Landtbruks Akad. Esper.* **38**, 1894). See Stafleu & Cowan (*TL-2* **1**: 798, 1976).

Erikssonia Penz. & Sacc. (1898), Phyllachoraceae. 3, trop. See Stevenson (*Mycol.* **35**: 629, 1943).

Erikssonopsis M. Morelet (1971), Leotiaceae. 1, Eur.

Erinacella, see *Ericianella*.

erinaceous, prickly like a hedgehog.

Erinaceus Dill. ex Maratti (1822) = Hydnum (Hydn.) fide Donk (*Persoonia* **8**: 279, 1975).

Erinella Quél. (1886) ≡ Lachnum (Hyaloscyph.).

Erinella Sacc. (1889) = Lachnum (Hyaloscyph.) fide Spooner (*Bibl. Mycol.* **116**, 1987).

Erinellina Seaver (1951) = Lachnum (Hyaloscyph.).

Erineum Pers. (1822), gall on an outgrowth caused by gall-mites (*Arachnida*; *Eriophydae*).

Eriocercospora Deighton (1969), Mitosporic fungi, 1.C2.10. 1 (on *Asterinaceae*), widespr., trop.

Eriocercosporella Rak. Kumar, A.N. Rai & Kamal (1994), Mitosporic fungi, 1,C2,?. 1, India.

Eriocladus Lév. (1846) = Lachnocladium (Lachnoclad.).

Eriocorys Quél. (1886) ≡ Strobilomyces (Strobilomycet.).

Erioderma Fée (1825), Pannariaceae (L). 16, cosmop., mainly trop.

Eriomene (Sacc.) Clem. & Shear (1931) = Menispora (Mitosp. fungi) fide Hughes (1958).

Eriomenella Peyronel (1918) = Menispora (Mitosp. fungi) fide Hughes (1958).

Eriomycopsis Speg. (1910), Mitosporic fungi, 1.C1.10. *c.* 10 (fungicolous), trop. See Deighton & Pirozynski (*Mycol. Pap.* **128**, 1972).

Erionema Penz. (1898), Physaraceae. 1, Asia.

Erionema Maire (1906) = Menispora (Mitosp. fungi) fide Hughes (1958).

Eriopezia (Sacc.) Rehm (1892), Hyaloscyphaceae. 2, Eur., NZ. See Korf (*Mycotaxon* **7**: 457, 1978).

Eriopeziza, see *Eriopezia*.

Erioscypha Kirschst. (1938) = Lachnum (Hyaloscyph.). See Korf (*Mycotaxon* **7**: 399, 1978).

Erioscyphella Kirschst. (1938) = Lachnum (Hyaloscyph.) fide Haines & Dumont (*Mycotaxon* **19**: 1, 1984). See Korf (*Mycotaxon* **7**: 399, 1978).

Eriosperma Raf. (1808) nom. dub. (? Lycoperdales).

Eriosphaera De Toni (1888) [non F. Dietr. (1817), *Compositae*] ≡ Lasiosphaera (Lycoperd.).

Eriosphaerella Höhn. (1906) = Eriosphaeria (Trichosphaer.) fide Müller & v. Arx (1962).

Eriosphaeria Sacc. (1875), Trichosphaeriaceae. 5, Asia, Eur. See Müller & v. Arx (1962).

Eriospora Berk. & Broome (1850), Mitosporic fungi, 4.E1.10. 5, Eur., S. Am. See Petrak (*Sydowia* **1**: 94, 1947).

Eriosporangium Bertero ex Ruschenb. (1831) = Puccinia (Puccin.) fide Jackson (*Mycol.* **24**: 62, 1932).

Eriosporella Höhn. (1916), Mitosporic fungi, 8.G1.15. 1, Eur.

Eriosporina Tognini (1894) ? = Sirothecium (Mitosp. fungi) fide Sutton (1977).

Eriosporopsis Petr. (1947), Mitosporic fungi, 4.E1.?. 1, Eur.

Eriothyrium Speg. (1888), Mitosporic fungi, 5.A1.?. 5, S. Am.

Erisiphites, see *Erysiphites*.

Erispora Pat. (1922), Hypocreaceae. 1, Philipp.

erogen, a substance controlling the induction and differentiation of sex organs; **erotactin**, a sperm attractant; **erotropin**, a substance inducing a chemotropic response in sex organs (Machlis, *Mycol.* **64**: 238, 1972). See hormones.

Eromitra Lév. (1846) ≡ Mitrophora (Morchell.).

erose (of a lamella, etc.), having delicate tooth-like projections from the edge.

Erostella (Sacc.) Sacc. (1906) = Pleurostoma (Calosphaer.) fide Barr *et al.* (*Mycotaxon* **48**: 529, 1993).

Erostrotheca G.H. Martin & Charles (1928) = Melanospora (Ceratostomat.) fide v. Arx & Müller (1954).

erratic (of lichen thalli), not fixed to the substratum and often blowing around, e.g. *Chondropsis semiviridis*, *Sphaerothallia esculenta* ('manna'); epigaeic; vagrant; wandering lichens.

erumpent, bursting through the surface of the substratum. Cf. perrumpent.

Erynia (Nowak. ex Batko) Remaud. & Hennebert (1980), Entomophthoraceae. 12 (on *Insecta*), widespr. See Remaudière & Hennebert (*Mycotaxon* **11**: 269, 1980), Humber (*Mycotaxon* **13**: 471, 1981, **15**: 167, 1982, 1989), Perry & Fleming (*Mycol.* **81**: 154, 1989; zygospore germination), Butt & Beckett (*Protoplasma* **120**: 61, 72, 1984; ultrastr. mitosis & spindle-pole body), Descals & Webster (*TBMS* **83**: 669, 1984; aquatic spp.), Li & Humber (*CJB* **62**: 653, 1984), Ben-Ze'ev (*Mycotaxon* **25**: 1, **27**: 263, 1986, **28**: 403, 1987), Kelley (1991, *Sydowia* **45**: 252, 1993).

Eryniopsis Humber (1984), Entomophthoraceae. 5, Eur., USA. See Keller & Eilenberg (*Sydowia* **45**: 264, 1993), Humber (*Mycotaxon* **21**: 257, 1984).

Eryporus Quél. (1886) = Boletinus (Gyrodont.) fide Murrill (1910).

Erysibe Theophr. ex Wallr. (1833) Anamorphic Uredinales. Mainly uredinial states.

Erysiphaceae Tul. & C. Tul. (1861), Erysiphales. 21 gen. (+ 24 syn.), 437 spp. Mycelium largely superficial, hyaline. Ascomata cleistothecial, globose, solitary or aggregated, becoming dark, usually with complex appendages; interascal tissue absent; asci broadly clavate, thin-walled, with two wall layers at the base but the inner layer absent towards the apex, dehiscence ? explosive; ascospores hyaline, ellipsoidal, aseptate, without a sheath. Anamorphs hyphomycetous, prominent, e.g. *Oidium*, *Ovulariopsis*. Biotrophic on leaves and stems, cosmop. *Lit.*: see under *Erysiphales*.

Erysiphales, Ascomycota. 1 fam., 21 gen. (+ 24 syn.), 437 spp. Fam.:
Erysiphaceae.
The powdery mildews. Included with the ascohymenial unitunicate *Ascomycota* in most systems (see Tables 1-2 under *Ascomycota*) but actually bitunicate (Nannfeldt, 1932) and perhaps best placed close to the *Dothideales* (cf. Eriksson, 1981, 1982). *Lit.*: **General**: Amano (*Host range and geographical distribution of the powdery mildew fungi*, 1987; compr. host & geogr. distr.), Boesewinkel (*Revue mycol.* **41**: 493, 1977; ident. by conidia, key), Braun (*Feddes Repert.* **88**: 655, 1978; taxonomy), Salmon (*Mem. Torrey bot. Cl.* **9**, 1900), Saenz *et al.* (*Mycol.* **86**:212, 1994; molec. syst. posn), Blumer (*Beitr. Krypt. Flora Schweiz* **7**(1), 1933; *Echte Mehltaupilze* (*Erysiphaceae*), 1967; keys; descriptions), Eliade (*Lucr. Grăd. Bot. Bucur.* 1990: 105, 1990; 147 spp. Romania), Yarwood (*Bot. Rev.* **23**: 235, 1957), Gelyuta (*Biol. Zh. Armen.* **41**: 351, 1988; fam. concepts), Junell (*TBMS* **45**: 539, 1965; nomenclature), Hirata (*Host range and geographical distribution of the powdery mildews*, 1968; *Sydowia* **25**: 100, 1972), Karis (*Eesti jahu kastelised* (*Erysiphaceae*), 1987; Estonia), Sałata (*Flora Polska. Grzyby* **15**, 1985; Poland), Spencer (Ed.) (*The powdery mildews*, 1978), Braun (*A monograph of the Erysiphales*, 1987 [*Beih. Nova Hedw.* **89**]; keys 435 sp.), Zheng (*Mycotaxon* **22**: 209, 1985; key).
Regional: **Canada**, Parmelee (*CJB* **55**: 1940, 1977; host index). **China**, Zheng & Yu (*Fl. Fung. Sin.* **1**, 1987; 214 spp.). **Europe**, see Blumer (1933, 1967 above), Braun (*The powdery mildews* (*Erysiphales*) *of Europe*, 1995). **Greece**, Pantidou (*Annls Inst.*

Phytopath. Benaki, n.s. **10**: 187, 1971). **India**, Sharma & Patel (*Mycol. Inform.* **4**, 1995; checklist and bibliog.). **Israel**, Chorin & Palti (*Israel J. agric. Res.* **12**: 153, 1963). **Italy**, Ciferri & Camera (*Quaderno Ist. bot. Univ. Pavia* **21**, 1962). **Japan**, Homma (*J. Fac. Agric., Hokkaido Imp. Univ. Sapporo* **38**: 183, 1937). **Kazakhstan**, Vasyagina *et al.* ([*Flora sporovikh rastenii Kazakhstana*] **3**, 1961). **New Zealand**, Hammett (*N.Z. Jl Bot.* **15**: 687, 1977). **Portugal**, de Varennes e Mendonca & de Sequeira (*Agron. lusit.* **24**: 87, 1963; **26**: 21, 1965; **33**: 151, 1972). **Romania**, Sandu-Ville (*Ciupercile Erysiphaceae din Romania*, 1967). **Russia**, Golovin ([*Powdery mildews parasitizing cultivated and useful plants*], 1960). **South Africa**, van Jaarsveld (*Phytophylactica* **16**: 155, 1984). **Spain**, Durrieu & Macé (*BSMF* **88**: 175, 1973). **Sweden**, Junell (*Symb. bot. upsal.* **19** (1), 1967). **UK**, Ing (*Mycologist* **4**: 46, 88, 125, 172, 1990, checklist; **5**: 24, 60, 1991, key).

Erysiphe R. Hedw. ex DC. (1805), Erysiphaceae. 102, widespr. *E. cichoracearum* (mildew of curcurbits and other plants), *E. polygoni* (more than 500 hosts, esp. *Leguminosae*; Stavely & Hanson, *Phytopathology* **56**: 309, 1966). See Braun, *Beih. Nova Hedw.* **89**, 1987; key), Martin & Gay (*CJB* **61**: 2472, 1983; conidiogenesis), Zeller (*Mycol.* **87**: 525, 1995; molec. syst.). Anamorph *Oidium*.

Erysiphella Peck (1876) = Erysiphe (Erysiph.) fide Salmon (1900).

Erysiphites Pampal. (1902), ? Fossil fungi. 1 (Miocene), Sicily. See Salmon (*J. Bot. Lond.* **41**: 127, 1903).

Erysiphites Mesch. (1902), Fossil fungi. 1 (Tertiary), former USSR.

Erysiphopsis Halst. (1899) = Erysiphe (Erysiph.) fide Salmon (1900).

Erysiphopsis Speg. (1910), Mitosporic fungi, ?.?.?. 1, S. Am. See Petrak & Sydow (*Ann. Myc.* **34**: 36, 1936).

Erythricium J. Erikss. & Hjortstam (1970), Hyphodermataceae. 2, widespr.

Erythrobasidium Hamam., Sugiy. & Komag. (1988), Mitosporic fungi, 1.A1.1/10. 1 (anamorphic yeast with clamp connexions occ. present), Japan. See Hamamoto *et al.* (*J. Gen. Appl. Microbiol.* **37**: 131, 1991; nomencl.).

Erythrocarpon Zukal (1885), Ceratostomataceae. 1, Eur.

Erythrocarpum Sacc. (1891) ≡ Erythrocarpon (Ceratostomat.).

Erythrodecton G. Thor (1991), Roccellaceae (L). 2, Indonesia, Sri Lanka, Australasia, Pacific, S. Am.

Erythrogloeum Petr. (1953), Mitosporic fungi, 8.A1.15. 1, Costa Rica. See Ferreira, Demuner & Rezende (*Fitopat. Bras.* **17**: 106, 1992; anthracnose disease of *Hymenaea* spp.).

Erythrogymnotheca Yaguchi & Udagawa (1994), Trichocomaceae. 1, Japan.

Erythromyces Hjortstam & Ryvarden (1990), Hymenochaetaceae. 1, Afr., S.E. Asia, Australasia.

Erythrosphaera Sorokīn (1871) = Cephalotheca (Cephalothec.) fide Hawksworth (*in litt.*).

Eschatogonia Trevis. (1853), ? Acarosporaceae (L). 1, Afr.

Escovopsis J.J. Muchovej & Della Lucia (1990), Mitosporic fungi, 1.A1.15. 2 (from attine ant nests), C., N. & S. Am., W. Indies. See Seifert *et al.* (*Mycol.* **87**: 407, 1995), Romero *et al.* (*Rev. Mex. Mic.* **3**: 231, 1987; as *Phialocladus*).

esculent, of use as a food; see Edible fungi.

Esdipatilia Phadke (1981), Mitosporic fungi, 2.C1.1. 1, India.

eseptate, see aseptate.

Esfandiaria Petr. (1955) [non Charif & Aellen (1955), *Chaenopodiaceae*] ≡ Esfandiariomyces (Ascomycota, inc. sed.).

Esfandiariomyces Ershad (1985), Ascomycota (inc. sed.). 1, Iran.

esorediate (esorediose), having no soredia.

Esslingeriana Hale & M.J. Lai (1980), Parmeliaceae (L). 1, USA. See Kärnefelt *et al.* (*Pl. Syst. Evol.* **183**: 113, 1992; status).

Etheirodon Banker (1902) ≡ Odontia (Steccherin.); sensu Fries, = Steccherinum (Steccherin.). See Banker (*Mycol.* **21**: 145, 1929).

Etheirophora Kohlm. & Volkm.-Kohlm. (1989), Halosphaeriaceae. 3 (marine), Atlantic, Pacific. See Kohlmeyer & Volkmann-Kohlmeyer (*MR* **92**: 416, 1989; key).

ethnomycology, mycology as a branch of ethnology. See Wasson & Wasson (*Mushrooms, Russia and history*, 2 vols. 1957), and the writings by Wasson and Heim on hallucinogenic fungi (q.v.); also Lowy (*Mycol.* **64**: 816, 1972, Maya codices; **66**: 188, 1974, *Amanita muscaria* and the thunderbolt legend; *Revista Interam. Rev.* **2**: 405, 1972, **5**: 110, 1975, **10**: 94, 1980, mushrooms and religion), Redlinger (Ed.) (*The sacred mushroom seeker*, 1990), soma.

etiology, see aetiology.

eu- (prefix), true; sometimes used, but wrongly for the subgenus or section including the type species of the generic name of which it is an infrageneric taxon.

Euacanthe Theiss. (1917) = Acanthonitschkea (Nitschk.) fide Nannfeldt (1975).

Euactinomyces Langeron (1922) nom. dub. (Fungi, inc. sed.).

Eualectoria (Th. Fr.) Gyeln. (1934) ≡ Alectoria (Alector.).

Euantennaria Speg. (1918), Euantennariaceae. 6, widespr. See Hughes (*N.Z. Jl Bot.* **12**: 299, 1974), Corlett *et al.* (*N.Z. Jl Bot.* **11**: 213, 1973).

Euantennariaceae S. Hughes & Corlett ex S. Hughes (1972), Dothideales. 4 gen. (+ 5 syn.), 15 spp. Mycelium superficial, dark, the hyphae cylindrical, forming a flattened mat but frequently with erect branches; ascomata ± spherical, small, superficial, with a small lysigenous pore, the peridium dark, with hyphal appendages; interascal tissue absent; asci saccate, fissitunicate; ascospores brown, transversely septate or muriform, sometimes attenuated at the apices. Anamorphs hyphomycetous. Epiphytic, widely distributed.
 Lit.: Batista & Ciferri (*Saccardoa* **2**, 1963), Hughes (*N.Z. Jl Bot.* **10**: 225, 1972, *Mycol.* **68**: 693, 1976; gen. names, anamorphs), Reynolds (*Taxon* **20**: 759, 1971; hyphal morph., *Nova Hedw.* **26**: 179, 1975; growth forms, *Mycotaxon* **8**: 417, 1979; stalked taxa, *Mycotaxon* **27**: 377, 1986; status).

Euascomycetes. Class for pyrenomycetes, discomycetes and laboulbeniomycetes; cfr. Hemiascomycetes, Loculoascomycetes.

Euascomycetidae, see *Ascomycota*.

Euaspergillus F. Ludw. (1892) = Aspergillus (Mitosp. fungi) fide Raper & Fennell (1965).

Eubelonis Clem. (1909) ? = Calycina (Leot.). See Arendholz (*Mycotaxon* **36**: 283, 1988).

Eubelonis Höhn. (1926), Leotiaceae. 2, Eur. See Arendholz (*Mycotaxon* **36**: 283, 1988).

Eucantharomyces Thaxt. (1895), Laboulbeniaceae. 21, widespr. See Santamaria (*MR* **98**: 1303, 1994; asci).

eucarpic, developing reproductive structures on limited portions of the thallus; residual nucleate protoplasm remaining and capable of further mitotic growth and regeneration.

Eucaryota (Eucarya), see *Eukaryota*.

Euceramia Bat. & Cif. (1962), Chaetothyriaceae. 1, Brazil. = Limacinia (Dothideales) fide v. Arx & Müller (1975).

Euceratomyces Thaxt. (1931), Euceratomycetaceae. 1, N. Am.

Euceratomycetaceae I.I. Tav. (1980), Laboulbeniales. 5 gen., 7 spp. Stroma (thallus) present; ascomata formed from successive cells of a lateral appendage of the primary thallus, the appendage extending beyond the base of the ascoma; outer wall cells of ascoma usually small, ± equal; asci 4-spored; ascospores with a submedian septum; usually monoecious.

euchroic, see colour.

Eucladoniomyces Cif. & Tomas. (1953) = Cladonia (Cladon.).

Eucollema (Cromb.) Horw. (1912) ≡ Collema (Collemat.).

Eucorethromyces Thaxt. (1900) = Corethromyces (Laboulben.) fide Thaxter (1931).

eucortex (of lichens), a cortex composed of well-differentiated tissue.

Eucyphelis Clem. (1909) = Chaenotheca (Coniocyb.).

Eudacnus Raf. ex Merr. (1943) ≡ Endacinus (Sclerodermat.).

Eudarluca Speg. (1908), ? Phaeosphaeriaceae. 3-4 (on rusts), widespr. See Eriksson (*Bot. Notiser* **119**: 33, 1966; biology, systematics), Kranz (*Nova Hedw.* **24**: 169, 1974; hosts). Anamorph *Sphaerellopsis*.

Eudimeriolum Speg. (1912), Pseudoperisporiaceae. *c.* 5, trop. See Hansford (*Mycol. Pap.* **15**, 1946), Farr (*Mycol.* **71**: 243, 1979).

Eudimeromyces Thaxt. (1918) = Dimeromyces (Laboulben.) fide Tavares (1985).

Euepixylon Füisting (1867), Xylariaceae. 1, Eur. See Læssøe (*SA* **13**: 43, 1994).

euform, see *Uredinales*.

eugonic, see dysgonic.

eugonidium, a bright green lichen photobiont (e.g. *Trebouxia*) (obsol.).

Euhaplomyces Thaxt. (1901), Laboulbeniaceae. 1, UK.

euhymenium, see hymenium.

Euhypoxylon Füisting (1867) ≡ Hypoxylon (Xylar.).

Eukaryota (Eukarya). The domain (empire, superkingdom) to which all **eukaryotes** belong; i.e. encompassing all organisms with one or more nuclei in their cells bounded by a nuclear membrane and with paired DNA-containing chromosomes (and also other complex organelles, e.g. Golgi bodies, mitochondria). The counterpart of the **Prokaryota (Prokarya, prokaryotes)** which is now generally divided into two separate kingdoms, *Archaea* (formerly *Archaebacteria*) and *Bacteria* (formerly *Eubacteria*; including also *Cyanobacteria*), and the viruses that lack the above structures. See Classification, Phylogeny.

eukaryote (adj. **eukaryotic**), one of the *Eukaryota* (q.v.); cf. prokaryote.

Eumela Syd. (1925), Pseudoperisporiaceae. 1, C. Am. See Hansford (*Mycol. Pap.* **15**, 1946).

Eumicrocyclus, see *Eumycrocyclus*.

Eumisgomyces Speg. (1912) = Laboulbenia (Laboulben.) fide Spegazzini (*An. Mus. nac. Hist. nat. B. Aires* **27**: 70, 1915), Tavares (1985).

Eumitria Stirt. (1881) = Usnea (Parmel.).

Eumonoicomyces Thaxt. (1901), Laboulbeniaceae. 2 or 3, Am., Pacific Isl., Afr.

eumorphic, well-formed.

Eumycetes (Eumycota), true fungi; see *Fungi*.

eumycetoma, see mycetoma.

Eumycota (Eumycophyta), see *Eumycetes*.

Eumycrocyclus Hara (1915) ≡ Coccoidella (Coccoid.).

Euoidium Y.S. Paul & J.N. Kapoor (1986) = Oidium (Mitosp. fungi).

Euopsis Nyl. (1875), Lichinaceae (L). 1, Eur. See Hafellner (*Beih. Nova Hedw.* **79**: 241, 1984).

Eupelte Syd. (1924), Asterinaceae. 3, trop. Anamorphs *Clasterosporium* or *Septoideum*-like.

Eupenicillium F. Ludw. (1892), Trichocomaceae. 43, widespr. See Berbee *et al.* (*Mycol.* **87**: 210, 1995; molec. data suggests affinity with *Aspergillus*), Pitt (*The genus Penicillium*, 1980; keys), Udagawa & Horie (*Ant. v. Leeuwenhoek* **39**: 313, 1973; ascospores), Stolk & Samson (*Stud. Mycol.* **23**, 1983). Anamorph *Penicillium*.

Eupezizella Höhn. (1926) = Hyaloscypha (Hyaloscyph.) fide Huhtinen (*Karstenia* **29**: 45, 1989).

Euphoriomyces Thaxt. (1931), Laboulbeniaceae. 13, Am., Asia, Eur. See Santamaria (*Revta Iberoamer. Micol.* **8**: 43, 1991; key).

Eupropolella Höhn. (1917), Dermateaceae. 7, Eur. See Défago (*Sydowia* **21**: 1, 1967).

Eupropolis De Not. (1863) = Phaeotrema (Thelotremat.) fide Sherwood (*Mycotaxon* **5**: 50, 1977).

Eupythium Nieuwl. (1916) ≡ Pythium (Pyth.).

Eurasina G.R.W. Arnold (1970), Mitosporic fungi, ?.?.?. 1 (on *Polyporus*), former USSR. = Helminthophora (Mitosp. fungi) fide de Hoog (1978).

Euricoa Bat. & H. Maia (1955) = Cylindrocarpon (Mitosp. fungi) fide Kendrick & Carmichael (1973).

European Council for Conservation of Fungi. see Conservation.

European mildew, see mildew.

Europhium A.K. Parker (1957) = Ophiostoma (Ophiostomat.) fide Benny & Kimbrough (1980). See also v. Arx (*SA* **5**: 310, 1986), Solheim (*Nordic Jl Bot.* **6**: 199, 1986). Anamorph *Verticicladiella*.

Eurotiaceae, see *Trichocomaceae*.

Eurotiales, Ascomycota. 4 fam., 52 gen. (+ 35 syn.), 232 spp. Ascomata small, cleistothecial, usually solitary, rarely absent; peridium usually thin, membranous, usually brightly coloured, varied in structure and rarely acellular and cyst-like; interascal tissue absent; asci clavate or saccate, thin-walled, evanescent, sometimes formed in chains; ascospores varied, small, aseptate, often ornamented and with equatorial thickening, without a sheath. Anamorphs prominent, many of industrial and medical importance (e.g. *Aspergillus*, *Penicillium*). Saprobic, mainly from soil or decaying plant materials (syn. *Aspergillales*, *Plectascales*).

Formerly used for ± all ascomycetes with cleistocarpic ascomata, many now placed within groups including ostiolate counterparts from which they have been derived (e.g. *Dothideales*, *Microascales*, *Sordariales*). Malloch (1981) places the *Trichocomaceae* in the *Hypocreales*. Von Arx (1987) accepted four families: *Eurotiaceae*, *Gymnoascaceae*, *Onygenaceae* (incl. *Trichocomaceae*) and *Amauroascaceae*, mainly on the basis of ascospore shape and ornamentation. 5 gen. (+ 1 syn.), 11 spp., included in *Eurotiales* here are not placed in any of the following Fams:

(1) **Ascosphaeraceae**.
(2) **Eremascaceae**.
(3) **Monascaceae**.
(4) **Trichocomaceae** (syn. *Aspergillaceae* auct., *Eurotiaceae*, *?Thermoascaceae*, 'Trichochomataceae').

Lit.: v. Arx (1981; *Persoonia* **13**: 273, 1987), Benny & Kimbrough (1980), Guarro *et al.* (*Mycotaxon* **42**: 193, 1991; key 8 spherical-spored gen.), Fennell (*in* Ainsworth *et al.* (Eds), *The Fungi* **4A**: 45, 1973; keys gen.), Malloch (*in* Reynolds (Ed.), *Ascomycete systematics*: 73, 1981).

Eurotiella Lindau (1900) ≡ Allescheria (Monasc.).

Eurotiopsis P. Karst. (1889) nom. dub. (? Mitosp. fungi).

Eurotiopsis Costantin ex Laborde (1897) = Monascus (Monasc.) fide Malloch (*Mycol.* **62**: 738, 1970).

Eurotites Mesch. (1892), Fossil fungi. 1 (Oligocene), Eur.

Eurotium Link (1809), Trichocomaceae. 19, cosmop. See Blaser (*Sydowia* **28**: 1, 1976; key), Kozakiewicz (*Mycol. Pap.* **161**, 1989), Pitt (*in* Pitt & Samson (Eds), *Advances in Penicillium and Aspergillus systematics*: 383, 1986; nomencl.). Anamorph *Aspergillus*.

Euryachora Fuckel (1870), Mycosphaerellaceae. c. 5, temp. See Obrist (*Phytopath. Z.* **35**: 382, 1959).

Euryancale Drechsler (1939), Cochlonemataceae. 4, Japan, N. Am. See Drechsler (*Mycol.* **31**: 410, 1939, **47**: 364, 1955, **51**: 787, 1959), Dayal (1973; key), Saikawa & Saito (*TBMS* **87**: 337, 1986; zygospores), Saikawa & Aoki (*Trans. Mycol. Soc. Japan* **32**: 509, 1991), Saikawa & Katsurashima (*Mycol.* **85**: 24, 1993; ultrastr.).

Eurychasma Magnus (1905), Eurychasmataceae. 1 (on *Ectocarpus*), Eur., Greenland. See Dick (*in press*; key).

Eurychasmataceae H.E. Petersen (1905), ? Myzocytiopsidales. 2 gen., 4 spp.

Eurychasmidium Sparrow (1936), Eurychasmataceae. 3 (on marine *Rhodophyceae*), N. Am., Eur. See Dick (*in press*; key).

Eurychasmopsis Canter & M.W. Dick (1994), Apodachlyellaceae. 1, UK. See Dick (*in press*; key).

Euryporus Quél. (1886) = Boletinus (Gyrodont.) fide Murrill.

Eurytheca De Seynes (1878), ? Myriangiaceae. 2, Eur., W. Indies.

euseptate (of conidial septation), having cells separated by multilayered walls of similar structure to lateral walls, as in *Pyricularia* (Luttrell, *Mycol.* **55**: 672, 1963) (Fig. 35C); cf. distoseptate.

Eusordaria Zopf (1883) ? = Sordaria (Sordar.). See Lundqvist (1972: 269).

Eustegia Fr. (1823) [non R. Br. (1810), *Asclepiadaceae*] ≡ Stegia (Fungi, inc. sed.).

Eustilbum Arnold (1885) nom. nud. = Dendrostilbella (Mitosp. fungi).

eustroma, see stroma.

Eusynaptomyces Thaxt. (1931), Ceratomycetaceae. 5, E. Afr., Eur., N. Am. See Scheloske (*Plant Syst. Evol.* **126**: 267, 1976).

euthecium, an ascoma (cleistothecium, perithecium, apothecium) of an euascomycete; cf. pseudothecium.

Euthrypton Theiss. (1916) = Seuratia (Suerat.) fide Meeker (1975).

euthyplectenchyma, see plectenchyma.

Euthythyrites Cookson (1947), Fossil fungi (Asterin.). 3 (Oligocene, Miocene), Australia, India.

Eutorula H. Will (1916) = Torulopsis (Mitosp. fungi).

Eutorulopsis Cif. (1925) ? = Torulopsis (Mitosp. fungi).

eutrophic, rich in nutrients; cf. dystrophic, oligotrophic.

eutrophication, nutrient enrichment, usually used when directly or indirectly caused by human influences.

Eutryblidiella (Rehm) Höhn. (1959) = Rhytidhysteron (Patellar.) fide Samuels & Müller (1979).

Eutypa Tul. & C. Tul. (1863), Diatrypaceae. 26, mainly temp. See Rappaz (*Mycol. Helv.* **2**: 285, 1987; key), Glawe & Rogers (1984).

Eutypella (Nitschke) Sacc. (1875) nom. cons., Diatrypaceae. 76, widespr. See Rappaz (*Mycol. Helv.* **2**: 285, 1987; key), Glawe & Rogers (1984).

eutypoid, having groups of perithecia in a stroma with the ostioles vertical and breaking through the surface individually. Cf. valsoid.

Eutypopsis P. Karst. (1878) = Endoxyla (Bolin.) fide Barr (1978).

Euzodiomyces Thaxt. (1900), Euceratomycetaceae. 2, Eur., Afr., USA, Japan.

evanescent, having a short existence; fugacious.

Evanidomus Caball. (1941), Mitosporic fungi, 8.A2.?. 1, Spain.

Everhartia Sacc. & Ellis (1882), Mitosporic fungi, 3.F1.?. 3, E. & N. Am. See Moore (*Mycol.* **47**: 90, 1955).

Evernia Ach. (1810), Parmeliaceae (L). *c.* 10, cosmop. See Ahlner (*Acta phytogeogr. suec.* **22**, 1948), Awasthi (*Bull. bot. Surv. India* **24**: 96, 1983; key 3 spp. India), Bird (*CJB* **52**: 2427, 1974; N. Am.), Golubkova & Shapiro (*Nov. Sist. niz. Rast.* **24**: 144, 1987; 9 spp. former USSR).

Everniastrum Hale ex Sipman (1986), Parmeliaceae (L). 29, mainly trop. See Eriksson & Hawksworth (*SA* **10**: 37, 1991; nomencl.), Jiang & Wei (*Lichenologist* **25**: 57, 1993; key 9 spp. China), Sipman (*Mycotaxon* **26**: 235, 1986; key).

Everniicola D. Hawksw. (1982), Mitosporic fungi, 4.B1.15. 1 (on lichens, esp. *Evernia* and *Nephroma*), Eur., N. Am. See Alstrup & Hawksworth (*Meddr Grønl., Biosci.* **31**, 1990).

Everniomyces Cif. & Tomas. (1953) ≡ Evernia (Parmel.).

Everniopsis Nyl. (1860), Parmeliaceae (L). 2, C. & S. Am.

Eversia J.L. Crane & Schokn. (1977), Mitosporic fungi, 1.D2.19. 1, widespr.

everted, turned inside out.

Evolution, see Coevolution, Fossil fungi, Kingdoms of fungi, Phylogeny.

Evulla Kavina (1939) ? = Neobulgaria (Leot.) fide Korf (*in* Ainsworth *et al.*, *The Fungi* **4A**: 249, 1973).

ex, (1) (in citations, e.g. G. Arnaud ex M.B. Ellis), from; first validly published by the second author(s), see Nomenclature; (2) (prefix), see e-.

ex situ (Lat.), **ex-situ** (Engl.) (of an organism), one taken from its natural habitat; used of living cultures isolated from nature and maintained in Genetic resource collections, and also of non-viable material held in Reference collections; cfr. *in situ*.

Exarmidium P. Karst. (1873), Hyponectriaceae. 5, Eur., N. Am., Philipp. See Barr & Boise (*Mycotaxon* **23**: 233, 1985).

exasperate, roughened with hard projecting points.

excavate, hollow out.

excentric (eccentric), (1) one sided; (2) (of a stipe), at one side or not in the centre of the pileus; cf. centric.

Excioconidium Plunkett (1925) = Chalara (Mitosp. fungi) fide Nag Raj & Kendrick (1975).

Excioconis Clem. & Shear (1931) ≡ Excioconidium (Mitosp. fungi).

exciple, see excipulum.

Excipula Fr. (1823) = Pyrenopeziza (Dermat.) fide Nannfeldt (1932). See also Sutton (*Mycol. Pap.* **141**, 1977)

Excipulaceae, see *Dermatiaceae*. Based on *Excipula*, a synonym of *Pyrenopeziza*, and not, therefore, applicable to Mitosporic fungi. See Sutton (*in* Ainsworth *et al.* (Eds), *The Fungi* **4A**: 553, 1973; *Coelomycetes*, 1980), Petch (*TBMS* **26**: 53, 1943).

Excipularia Sacc. (1884), Mitosporic fungi, 3.C2.10. 1, Eur., Russia. See Petrak (*Sydowia* **16**: 357, 1963), Spooner & Kirk (*TBMS* **78**: 247, 1982).

Excipulariopsis P.M. Kirk & Spooner (1982), Mitosporic fungi, 3.C2.1. 1, India.

Excipulella Höhn. (1915) = Heteropatella (Mitosp. fungi) fide Sutton (1977).

Excipulina Sacc. (1884) = Heteropatella (Mitosp. fungi) fide Sutton (1977).

Excipulites Göpp. (1836), Fossil fungi. 4 (Cretaceous, Tertiary), Eur.

excipulum (of ascomata), tissue or tissues containing the hymenium in an apothecium, or forming the walls of a perithecium; cf. **ectal -**, **medullary -**. **- proprium**, non-lichenized excipular tissue forming the margins of an apothecium of a lichenized fungus; **- thallinum**, lichenized excipular tissue of a lecanorine apothecium, external to an excipulum proprium (which may be much reduced), usually with a structure like that of the vegetative lichen thallus. See *Lit.* under apothecium, tissue types.

Exesisporites Elsik (1969), Fossil fungi (*f. cat.*). 1 (Pleistocene), USA.

exhabitant, see symbiosis.

Exidia Fr. (1822), Exidiaceae. *c.* 35, cosmop. See Lowy (*Fl. Neotrop.* **6**, 1970; *Nova Hedw.* **19**: 407, 1971; key), Donk (*Persoonia* **4**: 166, 1966; syns).

Exidiaceae R.T. Moore (1978), Tremellales. 32 gen. (+ 17 syn.), 110 spp. Basidioma resupinate to flabellate or clavarioid; haplophase mycelial; lignicolous or terrestrial.

Exidiopsis (Bref.) A. Møller (1895), Exidiaceae. *c.* 30, widespr. See Wells (*Mycol.* **53**: 317, 1962; key), Wells & Raitviir (*Mycol.* **69**: 987, 1977; former USSR), Roberts (*MR* **97**: 467, 1993; Br. spp.).

exigynous, having the antheridial stalk arising directly from the oogonial cell above the basal septum.

Exiliseptum R.C. Harris (1986), Trypetheliaceae (L). 1, Brazil.

Exilispora Tehon & E.Y. Daniels (1927) = Leptosphaeria (Leptosphaer.) fide v. Arx & Müller (1975).

exit tube, extension of the sporangium, produced prior to or during sporangial discharge, which enables sporangial contents to be released outside the host or substrate.

exo- (prefix), outside.

Exoascus Fuckel (1860) = Taphrina (Taphrin.) fide Mix (1949).

Exobasidiaceae J. Schröt. (1888), Exobasidiales. 6 gen. (+ 2 syn.), 61 spp. Stromatic, below host epidermis. Spores ballistosporic; blastoconidia common.

exobasidial, (1) having the basidia uncovered; (2) separated by a wall from the basidium; (3) (of a conidiophore in a lichenized pycnidium sporophore; obsol.), having no secondary branch (Steiner); cf. endobasidial.

Exobasidiales, Ustomycetes. 3 fam., 10 gen. (+ 2 syn.), 67 spp. Basidia forming a layer on leaf surfaces; gall-forming plant parasites, esp. of *Ericaceae* and *Commelinaceae*. Fams:
(1) **Brachybasidiaceae**.
(2) **Exobasidiaceae**.
(3) **Microstromataceae**.
Lit.: Donk (1951-63, VI: 1966: 280), Blanz (*Zeit. f. Myk.* **44**: 91, 1978; posn), Gäumann (1964: 358), Cunningham *et al.* (*Mycol.* **68**: 642, 1976; key).

Exobasidiellum Donk (1931), Exobasidiaceae. 1, Eur. See Donk (1966: 256), Bandoni & Johri (*CJB* **53**: 2561, 1975), McNabb & Talbot (*in* Ainsworth *et al.* (Eds), *The Fungi* **4B**, 1973).

Exobasidiopsis Karak. (1922) = Kabatiella (Mitosp. fungi).

Exobasidium Woronin (1867), Exobasidiaceae. 50 (on *Ericaceae*), widespr., esp. N. temp. *E. japonicum*, *Azalea* gall. See Savile (*CJB* **37**: 641, 1959; N. Am.), McNabb (*Trans. roy. Soc. N.Z., Bot.* **1**: 259, 1962; NZ), Nannfeldt (*Symb. bot. upsal.* **23**(2): 1, 1981; Eur.).

Exochalara W. Gams & Hol.-Jech. (1976), Mitosporic fungi, 1.A1.32. 2, widespr.

exochthonous (of soil organisms), invaders ill-adapted to live in soil (Park, 1957); cf autochthonous.

exogenization, an hypothetical process whereby endogenously formed spores become exogenously formed: a mechanism proposed to support the evolution of basidiomycetes from the ancestral ascomycetes (Clemençon, *Persoonia* **9**: 363, 1977).

exogenous, undergoing development outside.

Exogone Henn. (1908) = Agyrium (Agyr.) fide Höhnel (*Sber. Akad. Wiss. Wien*, I **120**: 8, 1911).

exolete (of perithecia, pycnidia, etc.), long over-mature; empty.

Exomassarinula Teng (1940) ? = Melchioria (Trichosphaer.) fide Müller & v. Arx (1962). See Petrak (*Sydowia* **13**: 23, 1959).

exomycology, mycology of outer space.

exo-operculation (of sporangial dehiscence in chytrids), the operculum is hinged to the rim of the pore; 'true operculation'; cf. endo-operculation.

exoperidium, the outer layer of the peridium.

Exophiala J.W. Carmich. (1966), Mitosporic fungi, 1.A1/B2.15. 7, widespr. See de Hoog (*Stud. Mycol.* **15**: 100, 1977), McGinnis & Ajello (*Mycol.* **66**: 518, 1974; 3 on fish), de Bievre *et al.* (*Bull. Soc. Fr. Myc. Med.* **16**: 345, 1987; physiological basis for taxonomy), de Hoog (*Pleomorphic fungi*: 221, 1987; developmental cycle), Pedersen & Langvad (*MR* **92**: 153, 1989; spp. on fish), Kawasaki *et al.* (*Mycopathol.* **110**: 107, 1990; mitochondrial DNA of spp.), Matsumoto *et al.* (*J. Med. Vet. Mycol.* **28**: 437, 1990; synanamorph for *E. dermatitidis* as *Wangiella*), de Hoog *et al.* (*Proc. 10th ISHAM Congr., Barcelona*: 168, 1988; taxonomy *E. jeanselmei* complex), de Hoog *et al.* (*Ant. v. Leeuwenh.* **65**: 143, 1994; pleoanamorphic cycle).

Exophoma Weedon (1926), Mitosporic fungi, ?.?.?. 1, N. Am. See Sutton (1977).

exopropagule, a propagule formed outside the body (medical mycology).

Exormatostoma Gray (1821) nom. dub. (? Diaporthales, inc. sed.).

exospore, see Spore wall.

Exosporella Höhn. (1912), Mitosporic fungi, 3.E1.?. 1, Java.

Exosporiella P. Karst. (1892), Mitosporic fungi, 3.C2.19. 1 (on *Corticium*), Eur.

Exosporina Oudem. (1904), Mitosporic fungi, 3.A2.?. 2 or 3, temp.

Exosporina G. Arnaud (1921) ≡ Arnaudina (Mitosp. fungi).

Exosporinella Bender (1932) ≡ Arnaudina (Mitosp. fungi).

exosporium, see Spore wall.

Exosporium Link (1809), Mitosporic fungi, 1/3.C2.26. 17, widespr. See Ellis (*DH, MDH*).

exotic, (1) (adj.), of another country; not indigenous; (2) (n), an - organism.

Exotrichum Syd. & P. Syd. (1914) ? = Crocicreas (Leot.) fide Höhnel (*Mykol. Unters.* **1**: 359, 1923).

expallant (of a pileus), becoming pale on drying.

expansin, see patulin.

expersate (of oospores in *Saprolegniaceae*), having one large refractive body surrounded by a homogeneous cytoplasm (Howard, *Mycol.* **63**: 684, 1971).

explanate, spread out.

explosive (of asci), see ascus.

Exserohilum K.J. Leonard & Suggs (1974), Anamorphic Pleosporaceae, 1.C2.26. Teleomorph *Setosphaeria*. 20, cosmop. See Sivanesan (*Mycol. Pap.* **158**: 1, 1987; keys), Muchovej *et al.* (*Fitop. Bras.* **13**: 211, 1988; keys), Alcorn (*Ann. Rev. Phytopath.* **26**: 37, 1988; generic taxonomy), Khazanov (*Opredelitel' Gribov-Vozbul. 'Gel'mintosp-oriozov' Rast. iz Rodov Bipolaris, Drechslera i Exserohilum*, 1992).

exserted, sticking out; protruding (e.g. a mature ascus of *Ascobolus*).

Exserticlava S. Hughes (1978), Mitosporic fungi, 1.C2.6. 2, widespr., esp. trop.

exsiccatus (adj.; Latin), dried or dry, e.g. fungus (-i) exsiccatus (-i), planta (-ae) exsiccata (-ae), specimen (specimina) exsiccatum (-a); **exsiccatum** (n.; pl. **-a**), a dried specimen; **exsiccata** (n.; pl. **-ae**; preferred abbreviation, Exs.), a set of dried specimens (fide Jackson, 1928, and Stearn, 1968). Exsiccatae distributed to major reference collections are generally cited in systematic works; those with printed descriptions issued before 1 January 1953 can be the places of valid publication of new taxa (Art. 30.3; see Nomenclature), and separately published labels are acceptable after that date (note that label data were not always cited in some major nomenclators as sources of names, e.g. by Saccardo, 1882-1972).

Lit.: General, Sayre (*Mem. N.Y. bot. Gdn* **19**: 1, 1969, general cryptogamic exsiccatae, lichens; **19**: 277, 1975; collectors), Stafleu & Cowen (*TL-2*, **1-7**, 1976-88). Fungi, Pfister (*Mycotaxon* **23**: 1, 1985; compr. catalogue), Stevenson (*Beih. Nova Hedw.* **36**, 1971; N. Am.). Lichens, Hawksworth (*in* Seaward, 1977: 498; issued 1969-76), Hawksworth & Ahti (*Lichenologist* **22**: 1, 1990; issued 1976-89), Hawksworth & Seaward (*Lichenology in the British Isles 1568-1975*, 1977; UK), Hertel (*Mitt. Bot. StSamml., München* **18**: 297, 1982; labels), Lynge (*Nyt. Mag. Naturvid.* **55-60**, 1915-22; **79**: 233, 1939), Sayre (1969).

extenditunicate, see ascus.

extramatrical, (1) living on or near the surface of the matrix or substratum; (2) VAM structures (mycelium, spores) developing outside roots of a phytobiont.

Extrawettsteinina M.E. Barr (1972), Pleosporaceae. 4, Eur., N. Am. See Barr (*Prodromus to class Loculoascomycetes*, 1987).

extrusome, membrane-bound structure derived from vesicle of the Golgi system and anchored to the cell membrane by proteinaceous particles; contents extruded in respose to stimuli.

fabiform, see reniform.

Fabospora Kudrjanzev (1960) ? = Guilliermondella (Metschnikow.) fide Batra (1978).

Fabraea Sacc. (1881) = Leptotrochila (Dermat.) fide Schüepp (*Phytopath. Z.* **36**, 1959).

Fabrella Kirschst. (1941), Hemiphacidiaceae. 1 (on *Tsuga*), N. Am. See Korf (*Mycol.* **54**: 12, 1962).

Fabreola Kuntze (1891) ≡ Urosporella (Amphisphaer.).

facial eczema, see sporidesmin.

facultative, (1) sometimes; not necessarily; not obligate (q.v.); (2) (of a parasite), having the power of living as a saprobe; able to be cultured on laboratory media; (3) - synonym, see synonym.

Faerberia Pouzar (1981), Lentinaceae. 2, Eur.

Fairmania Sacc. (1906) = Microascus (Microasc.) fide Malloch & Cain (*CJB* **49**: 859, 1971).

Fairmaniella Petr. & Syd. (1927), Mitosporic fungi, 6.A2.15. 1, widespr. See Sutton (*Mycol. Pap.* **123**, 1971).

fairy butter, basidiomata of *Tremella albida*.

Fairy rings. Fungus rings, which are generally of basidiomycetes (some 60 recorded species), are very frequent in grass and grassland, and not uncommon in woods. There are three chief types: (1) those in which the development of the sporocarps has no effect on the vegetation, e.g. *Chlorophyllum molybdites*, *Lepista sordida*, myxomycete rings; (2) those in which there is increased growth of the vegetation, e.g. *Calvatia* cyathiformis, Disciseda subterranea (Catastroma subterraneum), the basidiomata of which are at the outer edge of the ring, Lycoperdon gemmatum, Lepista personata; (3) those in which the vegetation is damaged, sometimes so badly as to have an effect on its value, e.g. Agaricus praerimosus (A. tabularis), Leucopaxillus giganteus, Marasmius oreades (in British Isles), Calocybe gambosa. Rings of the third type are frequently made up of outer and inner rings in which the growth of the vegetation is strong with a ring of dead or badly damaged vegetation between.

Rings are started from a mycelium, the growth of which is at all times on the outer edge because of the band of decaying mycelium and used-up soil within the ring of active hyphae. The mean growth of a ring of A. praerimosus is 12 cm in radius every year (0-30 cm in any one year); that of one of Calvatia cyathiformis about 24 cm. From this, the ages of rings of these two fungi in Colorado, 60 and more than 200 m diam., were thought to be 250 and 420 years; parts of A. praerimosus rings were possibly 600 years old.
  Lit.: Shantz & Piemeisel (J. Agric. Res. 11: 191, 1917), Bayliss Elliott (Ann. appl. Biol. 13: 277, 1926), Parker-Rhodes (TBMS 38: 59, 1955), Burnett & Evans (Nature 210: 1368, 1966), Stevenson & Thompson (J. theor. Biol. 58: 143, 1976; kinetics); Gregory (Bull. BMS 16: 161, 1982; 'free' and 'tethered' rings).

fairy-ring champignon, the edible Marasmius oreades.
falcarindiol, an antifungal compound produced by carrot roots (Garrod & Lewis, TBMS 72: 515, 1979).
falcate (falciform), curved like the blade of a scythe or sickle (Fig. 37.10).
Falciascina Beneš (1961), Fossil fungi (f. cat.). 1 (Carboniferous), former Czechoslovakia.
Falciformispora K.D. Hyde (1992), Pleosporaceae. 1 (marine), Mexico.
Falcipatella Gucevič (1952) = Heteropatella (Mitosp. fungi) fide Sutton (1977).
Falcipatellina Gucevič (1952) = Heteropatella (Mitosp. fungi) fide Sutton (1977).
falciphore, see falx.
Falcispora Bubák & Serebrian. (1912) = Selenophoma (Mitosp. fungi). See Petrak (Sydowia 5: 328, 1951).
Falcocladium Silveira, Alfenas, Crous & M.J. Wingf. (1994), Mitosporic fungi, 2/3.A/B1.15. 1, Brazil.
false membrane (of a smut), a tissue of sterile fungal cells limiting the sorus, as in Sphacelotheca.
false morel, see lorchel.
false truffle, see truffle.
falx, a 'fertile hypha' or conidiophore of Zygosporium, having the form of a bill-hook. Falces may be sessile or on special hyphae or falciphores (Mason, 1941).
Famintzinia Hazsl. (1877) = Ceratiomyxa (Ceratiomyx.) fide Martin & Alexopoulos (1969).
Fanniomyces T. Majewski (1972), Laboulbeniaceae. 2, Eur., N. Am. See Balazuc (Revue mycol. 43: 393, 1979).
farctate (of a stipe), having the centre softer than the outer layer; stuffed.
farinaceous (farinose), like meal in form or smell.
Farinaria Sowerby (1803) = Ustilago (Ustilagin.) etc. fide Fries (1829).
Farinodiscus Svrček (1987) = Proliferodiscus (Hyaloscyph.) fide Baral (SA 13: 113, 1994).
Farlow (William Gilson; 1844-1919), was a noted American mycologist. After a training under Asa Gray and de Bary he became Professor of Cryptogamic Botany at Harvard where he and B. Thaxter were responsible for training a number of men who were later important in the fields of mycology and plant pathology. His writings were on plant diseases, fungi (incl. lichens) and bibliography. The Farlow

Cryptogamic Herbarium (FH) and Library are now among the most extensive. See Clinton (Phytopath. 10: 1, 1920), Pfister (Bull. Boston mycol. Cl. 4: 6, 1975 [Farlow, Icones Farlowiana, 1927]), Grumman (1974: 188), Stafleu & Cowan (TL-2 1: 813, 1976).
Farlowia Sacc. (1883) [non J. Agardh (1876), Algae] ≡ Farlowiella (Hyster.).
Farlowiella Sacc. (1891), Hysteriaceae. 2, Eur., Tristan da Cunha. See Zogg (1962). Anamorph Acrogenospora.
Farnoldia Hertel (1983), Porpidiaceae (L). 4, widespr. See Pietschmann (Nova Hedw. 51: 521, 1990; posn).
Farringia Stafleu (1979) [as '1983'] nom. nud. (Fungi, inc. sed.).
Farriolla Norman (1885), Ascomycota (inc. sed.). 1, Norway. See Tibell (Beih. Nova Hedw. 79: 597, 1984).
Farriollomyces Cif. & Tomas. (1953) ≡ Farriolla (Ascomycota, inc. sed.).
Farrowia D. Hawksw. (1975), Chaetomiaceae. 3, widespr. = Chaetomium fide v. Arx et al. (1986). Anamorph Botryotrichum.
Farysia Racib. (1909), Ustilaginaceae. c. 8 (on Cyperaceae), widespr. Spore mass interspersed with elaterlike hyphae.
fasciate (fasciated), massed or joined side by side.
Fasciatispora K.D. Hyde (1991), Amphisphaeriaceae. 1 (on Nypa), Borneo.
fascicle, (1) (esp. of hyphae), a little group or bundle; (2) (of books or exsiccatae), one part, or collection of separate leaves, of a work issued in parts.
fasciculate, having growth in fascicles; - basidium, see basidium.
Fassia Dennis (1964), ? Diatrypaceae. 1, Congo. See Cannon (SA 5l: 130, 1986), Læssøe (SA 13: 43, 1994).
fastigiate, having parallel, massed, upright branches.
fastigiate cortex (of lichens), made up of parallel hyphae at right angles to the axis of the thallus; cf. fibrous cortex.
Fastigiella Benedix (1969) ≡ Neogyromitra (Helvell.).
fatiscent, cracked or falling apart.
fatty acids, a class of organic compounds, and generally the hydrophobic component of many microbial cellular membrane lipids. Quantitative and qualitative differences in the fatty acid content of microbial cells can be used in classification and identification; widely used in bacteria and also of value for some yeasts (Botha & Kock, Int. J. Food Microbiol. 19: 39, 1993), but yet to be fully explored with filamentous fungi.
Faurelina Locq.-Lin. (1975), ? Chadefaudiellaceae. 1 (coprophilous), Chad. See Parquey-Leduc & Locquin-Linard (Revue mycol. 40: 161, 1976; ontogeny).
Fauxylostoma McGinty (1923) nom. inval. (Basidiomycetes, inc. sed.).
Favaria Raf. (1815) nom. dub. (Fungi, inc. sed.). No spp. included.
faveolate (favose), honeycombed; alveolate.
Faveoletisporonites Ramanujam & Rao (1979), Fossil fungi (f. cat.). 1 (Miocene), India.
favic chandeliers, dichotomously branched, swollen, hyphal tips, growing submerged from the edge of the colony of Trichophyton schoenleinii.
Favillea Fr. (1848), Sclerodermataceae. 1, Australia. ? = Pisolithus (Sclerodermat.) or Scleroderma (Sclerodermat.).
favoid, like a honeycomb.
Favolaschia (Pat.) Pat. (1895), Tricholomataceae. 51, esp. trop. See Singer (Beih. Nova Hedw. 50, 1974).
Favolus P. Beauv. (1805), Polyporaceae. 2, esp. trop. See Donk (1960; nomencl.).
Favolus Fr. (1828) = Polyporus (Polypor.) fide Donk (1960).

Favomicrosporon Benedek (1967) Fungi (inc. sed.) fide Ajello (*Sabouraudia* **6**: 153, 1967).

Favostroma B. Sutton & E.M. Davison (1983), Mitosporic fungi, 4.A1.1. 1, Australia.

Favotrichophyton (Castell. & Chalm.) Neveu-Lem. (1921) = Trichophyton (Mitosp. fungi).

Favraea Sacc. (1881), Leotiales (inc. sed.). 1, Eur.

favus, a skin disease in humans (*Trichophyton schoenleinii*).

Fayodia Kühner (1930), Tricholomataceae. 10, N. temp. See Bigelow (*Mycotaxon* **9**: 38, 1979).

Fechtneria Velen. (1939) = Hymenogaster (Hymenogastr.) fide Svrček (in Pilát, *Flora ČSR* **B**, **1**: 143, 1958).

federation, see Phytosociology.

Fée (Antoine Laurent Apollinaire; 1789-1874) graduated at Strasburg, becoming Professor of botany there in 1832. Died exiled in Paris. Much important work on tropical cryptogams including lichens. Main collection at Rio de Janeiro, other material in **P** (C), **G**, and **BM**. Main works: *Nova Acta Acad. Caes. Leop.-Carol.* **18**, suppl. 1 (1841), *Essai sur les cryptogames des écorces* (1824-25; *Suppl.* 1837), *Méthode Lichénographique et genera* (1825). See Anon. (*J. Bot., Lond.* **12**: 223, 1874), Grummann (1974: 277), Stafleu & Cowan (*TL-2* **1**: 818, 1976).

Feigeana Mies, Lumbsch & Tehler (1995), Roccellaceae (L). 1, Socotra.

Felisbertia Viégas (1944), Dermateaceae. 1, S. Am.

Felixites Elsik ex Janson. & Hills (1990), Fossil fungi. 2 (Carboniferous), Eur., N. Am.

fellent, bitter like gall.

Fellhanera Vězda (1986), Pilocarpaceae (L). 19, trop. See Awasthi & Mathur (*Proc. Indian Acad. Sci.* (Pl. Sci.) **97**: 481, 1987; key 4 spp. India), Lücking *et al.* (*Bot. Acta* **107**: 393, 1994).

Fellneria Fuckel (1867) = Colletotrichum (Mitosp. fungi) fide Duke (*TBMS* **13**: 156, 1928).

Fellomyces Y. Yamada & I. Banno (1984), Mitosporic fungi, 1.A1.1. 4, widespr. See Yamada *et al.* (*J. Gen. appl. Microbiol.* **32**: 157, 1986; enzyme systems), Kurtzman (*Int. J. syst. Bact.* **40**: 56, 1990; DNA relatedness), Guého *et al.* (*Int. J. Syst. Bact.* **40**: 60, 1990; partial rRNA sequencies), Yamada *et al.* (*Agric. Biol. Chem.* **53**: 2993, 1989; phylogeny).

felt (of citrus), superficial saprobic fungi, such as *Septobasidium pseudopedicellatum*; **leaf** -, *Anthina citri* and other fungi; **root** -, *Helicobasidium mompa* (Japan).

Feltgeniomyces Dieder. (1990), Mitosporic fungi, 1.B1.15. 1 (on lichens), Luxembourg.

Femsjonia Fr. (1849), Dacrymycetaceae. 2, N. temp., Brazil. See McNabb (*N.Z. Jl Bot.* **3**: 223, 1965; key). = Ditiola (Dacrymycet.) fide Reid (*TBMS* **62**: 474, 1974).

Fenestella Tul. & C. Tul. (1863), Fenestellaceae. 2-3 (on twigs), temp. See Barr (*Rept. Kevo Subarctic Res. Stn* **11**: 12, 1974, *N. Am. Fl.* II **13**, 1990; N. Am.).

Fenestellaceae M.E. Barr (1979), Dothideales. 3 gen. (+ 2 syn.), 13 sp. Ascomata immersed or erumpent, often aggregated or in stromata with convergent ostioles, large, thick-walled, often turbinate, with a well-developed periphysate ostiole; peridium dark, consisting of small pseudoparenchymatous cells, often distinctly thicker at the base and/or the apex; interascal tissue trabeculate pseudoparaphyses in a gelatinous matrix; asci cylindrical, fissitunicate; ascospores large, complex, dark brown, transversely septate, muriform, constricted at the septa, sometimes fragmenting, often with a mucous sheath. Anamorphs not known. Saprobic or necrotrophic in woody tissue, widespr.

fenestrate, (1) having windows or openings; (2) (of spores), muriform.

Fennellia B.J. Wiley & E.G. Simmons (1973), Trichocomaceae. 3, widespr. Anamorph *Aspergillus*.

Fennellomyces Benny & R.K. Benj. (1975), Thamnidiaceae. 4, USA, Pakistan, India. See Benny & Benjamin (1975). Misra *et al.* (*Mycotaxon* **10**: 251, 1979), Mirza *et al.* (*Mucorales of Pakistan*, 1979).

Feracia Rolland (1905), Hypocreaceae. 1, Eur.

fermentation, chemical changes in organic substrates caused by enzymes, generally those of living microorganisms. See Fermented food and drinks.

Fermented food and drinks. Microorganisms, such as fungi (esp. yeasts) and bacteria, have been used in the preparation of fermented foods for many centuries, long before their function was recognized or understood. Every country in east Asia has indigenous fermented food, prepared on a scale that ranges from single households to large commercial operations. Often referred to as 'oriental fermentation', the process makes the starting material more digestible or palatable in terms of texture, flavour, aroma, pH and appearance, and furnishes essential nutrients in the form of vitamins, proteins, amino acids and calories.

Fermented foods are the result of the action of specific microbial enzymes. Any enzyme used in the food industry can be manufactured if pure cultures of the appropriate micro-organism are available. Much of our understanding of the microbiology and biochemistry of oriental fermentation stems from work with pure cultures isolated from naturally fermented food products. As a result, most indigenous fermented foods can now be made from raw materials inoculated with cultures which have been obtained through a long selection process. One microorganism or, more commonly, a combination of two or more microorganisms work together to produce the final product.

Various substrates are used in fermentations: the most popular are soybeans and rice. Others include milk, fish, corn, coconut, cassava and peanuts. All these foods are staples in the diets of the people who normally consume the fermented foods. A koji or starter is prepared for every fermentation. Portions of the starter culture are then used to inoculate a larger batch of the same or a different substrate. Enzymes and byproducts produced by the starter culture accelerate the rate of the next fermentation and provide a better growing environment for the succeeding organism(s). Cultures selected for starters are usually highly proteolytic, lipolytic, and/or amylolytic moulds and predominantly species of *Rhizopus*, *Aspergillus*, *Mucor*, *Actinomucor*, *Neurospora* or *Monascus*. The moulds break down macromolecules to produce amino acids, small fatty acids, vitamins and sugars, which ultimately add to the flavour and digestibility of the product. These smaller fermentation byproducts are then utilized by bacteria to produce organic acids that lower pH and provide a favourable environment for yeasts. The yeasts chosen to complete a fermentation differ according to the desired product. The relatively high sugar, salt or alcohol content of fermented food helps to prolong its shelf life in places where refrigeration is not common.

Asian and Oriental foods and drinks based on fungal fermentations (frequently by *Aspergillus*) include: ang-kak (*Monascus purpureus*), hama-natto, laochao, oncom merah, ontojam, sufu, tape, tempeh, and the Japanese sake (rice wine), shoyn (soy sauce), miso (soy cheese), scocho (a distilled spirit) and mizaume (a sugar syrup from rice). Fermented milk drinks involving fungi include kephir (Caucasus), kumiss (Russia), leben (Egypt), mazu (Armenia), yoghurt (Bulgaria); bread (bakers yeast), *S. cerevisiae*.

Lit.: Batra & Millner (*Mycol.* **66**: 942, 1974; fungi and fermentation in Asian foods and beverages),

Hessletine (*Mycologist* 5: 162, 1991; *Zygomycetes* in food fermentation), Bennett & Kuch (Eds) (*Aspergillus: biology and industrial applications*, 1992), Hesseltine & Wang (*Indigenous fermented food of nonwestern origin* [*Mycologia Memoir* 11], 1986), Nout (*in* Carroll & Wicklow, *The fungal community*, edn 2 : 817, 1992), Powell *et al.* (Eds) (*The genus Aspergillus from taxonomy and genetics to industrial application*, 1994), Samson (*in* Jones (Ed.), *Exploitation of microorganisms*: 321, 1993), Steinkraus (Ed.) (*Industrialization of indigenous fermented foods*, 1989), Yokotsuka (*in* Arora *et al.* (Eds), *Proteinaceous fermented foods and condiments prepared with koji molds.* [*Handb. Appl. Mycol.*: 329], 1991). See also Brewing, Mycoprotein, Wine-making.

Fermentotrichon Novák & Zsolt (1961) = Trichosporon (Mitosp. fungi) fide Lodder (1970).

Fernaldia Lynge (1937) [non Woodson (1932), *Apocynaceae*] = Theligyna (Lichin.) fide Jørgensen & Hensen (*Taxon* 39: 343, 1990).

Ferrarisia Sacc. (1919), Parmulariaceae. 4, trop.

fertile hypha, see conidiophore.

fertilization, the fusion of sex nuclei; - tube/hypha hypha developing from an antheridial gametic cell; passing through the antheridium wall and bridging the gap between non-contiguous gametangia to penetrate the oogonium.

FH. Farlow Reference Library and Herbarium of Cryptogamic Botany, Harvard University (Cambridge, Mass, USA); founded 1919; funded by endowment and the university.

Fibriciellum J. Erikss. & Ryvarden (1975) = Trechispora (Sistotremat.) fide Larsson (*in litt.*).

Fibricium J. Erikss. (1958), Steccherinaceae. 4, widespr. See Hayashi (*Bull. Govt For. Exp. Stn* 260, 1974; 4 spp.).

fibril, (1) a very small fibre; (2) (in *Usnea*), short simple branches perpendicular to the main branches.

fibrillar surface coat, fibrous component attached to the flagellar membrane and covering the entire surface of the flagellum.

Fibrillaria Sowerby (1803) nom. dub.; based on sterile hymenomycete mycelium fide Donk (*Taxon* 11: 84, 1962).

fibrillose, covered with silk-like fibres.

Fibrodontia Parmasto (1968) = Hyphodontia (Hyphodermat.) fide Langer (*Bibl. Mycol.* 154, 1994).

Fibroporia Parmasto (1968) = Poria (Coriol.) fide Donk (1974).

fibrous cortex (of lichens), made up of loosely woven distinct hyphae parallel with the long axis of the thallus; cf. fastigiate cortex.

Fibulobasidium Bandoni (1979), Sirobasidiaceae. 1, USA.

Fibulochlamys A.I. Romero & Cabral (1989), Mitosporic fungi, 1.A1.1. 1 (with clamp connexions), Argentina.

Fibulocoela Nag Raj (1978), Mitosporic fungi, 8.A1.1. 1 (with clamp connexions), India.

Fibulomyces Jülich (1972), Atheliaceae. 4, N. temp. = Leptosporomyces (Athel.) fide Hjortstam & Larsson (*Windahlia* 21, 1994).

Fibuloporia Bondartsev & Singer (1944) = Trechispora (Sistotremat.) fide Ryvarden (*Genera of polypores*, 1991).

Fibulosebacea K. Wells & Raitv. (1987), Exidiaceae. 1, Eur.

Fibulostilbum Seifert & Oberw. (1992), Chionosphaeraceae. 1, Brazil.

Fibulotaeniella Marvanová & Bärl. (1988), Mitosporic fungi, 1.E1.10. 1 (with clamp connexions, aquatic), Canada.

Fictoderma Preuss (1852) ? Protozoa.

field mushroom (**common mushroom**), basidioma of the edible *Agaricus campestris*.

filamentous, (1) thread-like; filamentose; (2) (of lichens), the photobiont forms a filament of cells which is surrounded by hyphae or cells of the mycobiont (e.g. *Cystocoleus*, *Racodium*, *Coenogonium*).

Filariomyces Shanor (1952), Laboulbeniaceae. 1, USA, Japan, Taiwan.

Filaspora Preuss (1855) nom. rej. = Rhabdospora (Mitosp. fungi).

filiform, thread-like (Fig. 37.2).

Filobasidiaceae L.S. Olive (1968), Tremellales. 5 gen., 13 spp. Basidioma absent; spores statismosporic, often catenulate; hyphae often with haustoria, septa with dolipores; yeast-like anamorph; mycoparasitic.

Filobasidiella Kwon-Chung (1976), Filobasidiaceae. 4 (or 1 fide Barnett *et al.*, 1990), widespr. Anamorph *Cryptococcus*.

Filobasidium L.S. Olive (1968), Filobasidiaceae. 4 (fide Barnett *et al.*, 1990), Eur., N. Am. See Kwon-Chung (*Int. J. Syst. Bact.* 27: 293, 1977). Anamorph *Cryptococcus*.

Filoboletus Henn. (1900), Tricholomataceae. 8, trop.

filoplasmodium, see plasmodium.

filopodium, a slender unbranched process (pseudopodium) from a plasmodium, as in *Schizoplasmodiopsis*. Cf. rhizopodium.

Filosporella Nawawi (1976), Mitosporic fungi, 1.E1.19. 2 (aquatic), widespr. See Dyko & Sutton (*Mycotaxon* 7: 323, 1978).

Fimaria Velen. (1934) = Pseudombrophila (Otid.) fide van Brummelen (*Libr. Bot.* 14, 1995).

fimbriate, edged; delicately toothed; fringed. Cf. fimbrillate.

fimbrillate, having a very small fringe. Cf. fimbriate.

Fimetaria Griffiths & Seaver (1910) ≡ Sordaria (Sordar.).

Fimetariella N. Lundq. (1964), Lasiosphaeriaceae. 1, Eur., Canada. See Barr (*Mycotaxon* 39: 43, 1990; posn).

fimicolous, living on animal droppings. Cf. coprophilous.

fine structure, see ultrastructure.

Finerania C.W. Dodge (1971) ? = Ramonia (Gyalect.) fide Galloway (*in litt.*).

finger-and-toe, see club root.

Finkia Vain. (1929), Lichinaceae (L). 1, C. Am.

Fioriella Sacc. (1905) = Diplodina (Mitosp. fungi) fide Sutton (1977).

fireplace fungi, fungi characteristic of burnt ground, etc. See pyrophilous fungi.

Fischer (Edward; 1861-1939). Professor at the University and Director of the Botanic Gardens, Berne, Switzerland, 1897-1933. Chief writings: *Die Uredineen der Schweiz* (1904); *Tuberaceae* (1897, 1938) and *Gasteromycetes* (1900, 1933) in Engler & Prantl, *Naturl. PflFam.* (1st and 2nd edn); *Tuberaceae* and *Hemiasci* in Rabenh. *Krypt.-Fl.* (1897); (with Gäumann) *Biologie der pflanzenbewohnenden parasitischen Pilze* (1929). See Rytz (*Ber. schweiz. bot. Ges.* 50: 793, 1940), Grummann (1974: 640), Stafleu & Cowan (*TL-2* 1: 834, 1976).

Fischerula Mattir. (1928), Helvellaceae. 2 (hypogeous), Italy, USA. See Trappe (*Mycol.* 67: 934, 1975; key).

fission, (1) becoming two by division of the complete organism; cf. budding; (2) (of conidial liberation), secession by the separation of a double septum; cf. fracture, lysis.

fissitunicate, see ascus.

Fissolimbus E. Horak (1979), Cortinariaceae. 1, Papua New Guinea.

Fissuricella Pore, D'Amatao & Ajello (1977), Mitosporic fungi, 1.D1.?. 1 (yeast-like), USA.

Fissurina Fée (1824) = Graphis (Graphid.).

fistular (**fistulose**), hollow, like a pipe.

Fistulariella Bowler & Rundel (1977) = Ramalina (Ramalin.) fide Krog & Østhagen (1980). See Bowler (*Mycotaxon* **29**: 345, 1987; 30 spp.).

Fistulina Bull. (1791), Fistulinaceae. 3, Eur., N. & S. Am. *F. hepatica*, the edible beef-steak fungus, is the cause of brown oak (q.v.).

Fistulinaceae Lotsy (1907), Fistulinales. 2 gen. (+ 3 syn.), 4 spp. Basidioma pileate; hymenophore of crowded, free tubes; spores hyaline; lignicolous. *Lit.*: Donk (1964: 263).

Fistulinales, Basidiomycetes. 1 fam., 2 gen. (+ 3 syn.), 4 spp. Fam.:
Fistulinaceae.

Fistulinella Henn. (1901), Strobilomycetaceae. 1, neotrop. See Singer (1946), Guzman (*Bol. Soc. Mex. Mic.* **8**: 53, 1974; taxonomy and geography), Pegler & Young (*TBMS* **76**: 103, 1981).

Fitzpatrick (Harry Morton; 1886-1949). Mycologist (Professor from 1913) at Cornell University from 1908 until his death. Best known for his book *The lower fungi. Phycomycetes* (1930). *Obit.*: Barrus (*Mycol.* **43**: 249, 1951; portr., bibl.). See Stafleu & Cowan (*TL-2* **1**: 842, 1976).

Fitzpatrickella Benny, Samuelson & Kimbr. (1985), Coryneliaceae. 1, Juan Fernández.

Fitzpatrickia Cif. (1928) ? = Acanthonitschkea (Nitschk.) fide Nannfeldt (1975).

Fixatives. Fixation is the process of killing cells while preserving their structure and organization. Chemical fixatives:

Formal-acet-alcohol.

Formalin	13.0 ml
Acetic acid (glacial)	5.0 ml
Ethyl alcohol (50 per cent)	200.0 ml

Flemming's weak solution.

Chromic acid (1 per cent)	25.0 ml
Acetic acid (1 per cent)	10.0 ml
Water	60.0 ml
With addition of Osmic acid (2 per cent) before use.	5.0 ml

Other fixatives:
Glutaraldehyde
Osmium tetroxide.

Flabellaria Pers. (1818) [non Cav. (1790), *Malpighiaceae*] ≡ Schizophyllum (Schizophyll.).

flabellate (**flabelliform**), like a fan; in the form of a half-circle.

Flabellimycena Redhead (1984), Tricholomataceae. 1, S. Am.

Flabellocladia Nawawi (1985), Mitosporic fungi, 1.G1.1. 2 (aquatic), Malaysia.

Flabellomyces Kobayasi (1982) = Coenogonium (Gyalect.) fide Kashiwadani (*Bull. natn Sci. Mus. Tokyo* B **9**: 159, 1983).

Flabellophora G. Cunn. (1965), Coriolaceae. 22, pantrop. See Corner (*Beih. Nova Hedw.* **86**: 18, 1987; key).

Flabellopilus Kotl. & Pouzar (1957) ≡ Meripilus (Coriol.) fide Donk (1974).

Flabellospora Alas. (1968), Mitosporic fungi, 1.G1.1. 3, trop.

flaccid, not stiff; limp; flabby.

Flagelloscypha Donk (1951), Tricholomataceae. 2, widespr. See Agerer (*Sydowia* **27**: 131, 1975; key).

Flagellosphaeria Aptroot (1995), Amphisphaeriaceae. 1, Portugal.

Flagellospora Ingold (1942), Mitosporic fungi, 1.E1.15. 3, widespr. See Petersen (*Mycol.* **55**: 570, 1963; key),

Jooste & van der Merwe (*S. Afr. Jl Bot.* **56**: 319, 1990; ultrastr. *F. penicillioides*).

flagellum (pl. **flagella**), cylindrical extension of a eukaryotic cell, bounded by a plasma membrane and containing an axoneme; two types can be distinguished by electron microscopy, the **whiplash** with a smooth continuous surface (as in *Chytridiomycota*) and the **tinsel**, characteristic of *Hyphochytriomycota*, with the surface covered with hair-like processes (**mastigonemes** or **flimmers**); **-ar apparatus**, complex consisting of one or more basal bodies which may bear flagella, may have microtubular and fibrous roots associated with their bases; **-ar fibrous roots**, roots composed of a bundle of filaments, frequently appearing cross-striated; **-ar hairs**, filamentous appendages usually arranged in one or more rows but not covering the entire surface of a flagellum; **-ar scales**, organic structures of discrete size and shape, often covering the whole surface of the flagellum external to the plasma membrane, usually assembled in the dictyosome. See Barr (*Mycol.* **84**: 1, 1992; terminology flagellar apparatus). See also axoneme, blepharoplast; and cf. cilium.

Flageoletia (Sacc.) Höhn. (1916) = Phomatospora (Ascomycota, inc. sed.) fide Reid & Booth (*CJB* **44**: 445, 1966).

Flahaultia Arnaud (1952), Anamorphic Exidiaceae, 1.A1.?. Teleomorph *Sebacina*. 1, Eur. See Watling & Kendrick (*Naturalist*, Hull **104**: 1, 1979).

Flakea O.E. Erikss. (1992), Ascomycota (inc. sed., L). 1, C. & S. Am., Afr., Australia, Pacific.

Flaminia Sacc. & P. Syd. (1902), Uredinales (inc. sed.). 1 (on *Xanthoxylon*), Brazil. See Sherwood (*Mycotaxon* **5**: 1, 1977).

Flammopsis Fayod (1889) = Pholiota (Strophar.) fide Singer (1951).

Flammula (Fr.) P. Kumm. (1871) [non (DC.) Fourr. (1868), *Ranunculaceae*] = Pholiota (Strophar.) fide Singer (1951).

Flammulaster Earle (1909) = Phaeomarasmius (Cortinar.) fide Singer (1956). See Watling (*Notes RBG Edin.* **28**: 65, 1967; key), Horak (*N.Z. Jl Bot.* **18**: 173, 1980; key 8 NZ spp.), Vellinga (*Persoonia* **13**: 1, 1986; key 9 Eur. spp.).

flammulin, an anti-tumour antibiotic from *Flammulina velutipes* (Watanabe *et al.*, *Bull. chem. Soc. Japan* **37**: 747, 1964).

Flammulina P. Karst. (1891), Tricholomataceae. 2, temp. See Singer (*Darwiniana* **13**: 180, 1964), Ingold (*TBMS* **75**: 107, 1980; *F. velutipes* mycelium, oidia, etc.), enokitake.

flask fungi, Ascomycota with perithecioid ascomata.

Flaviporellus Murrill (1905) = Flaviporus (Coriol.) fide Pegler (1973).

Flaviporus Murrill (1905), Coriolaceae. 1, widespr.

Flavocetraria Kärnefelt & Thell (1994), Parmeliaceae (L). 2, arctic-boreal, S. Am., Papua New Guinea.

Flavodon Ryvarden (1973) = Irpex (Steccherin.) fide Hjortstam & Larsson (*Windahlia* **21**, 1994).

Flavoparmelia Hale (1986) Parmeliaceae. *c.* 22, esp. temp. Australasia, S. Am. See Elix & Johnston (*Mycotaxon* **33**: 391, 1988). ? = Parmelia (Parmel.).

Flavophlebia (Parmasto) K.H. Larss. & Hjortstam (1977) = Radulomyces (Hyphodermat.) fide Jülich (*Persoonia* **10**: 325, 1979).

Flavopunctelia (Krog) Hale (1984) Parmeliaceae. 5, temp. & trop. ? = Parmelia (Parmel.)

Flavoscypha Harmaja (1974) = Otidea (Otid.) fide Dissing (*in litt.*).

Flegographa A. Massal. (1860) = Sarcographa (Graphid.).

Fleischeria Penz. & Sacc. (1901) = Hypocrella (Clavicipit.) fide Rogerson (1970).

Fleischhakia Auersw. (1869) = Preussia (Spororm.) fide Clements & Shear (1931).

Fleischhakia Rabenh. (1878) = Psilopezia (Otid.) fide Seaver (1928).

flesh, the trama, esp. of the pileus of an agaric or bolete.

fleshy (of sporocarp), soft, not cartilaginous- or wood-like.

flexuous (flexuose), wavy.

flexuous hyphae (of *Uredinales*), an unbranched or branched haploid hyphal projection from a pycnium, which may be diploidized by a pycniospore of opposite 'sex' (Craigie, see *Nature* **141**: 33, 1938). Cf. receptive body.

flimmergeissel, flagellum bearing two rows of tripartite tubular hairs. Cf straminipilous.

Floccaria Grev. (1827) = Penicillium (Mitosp. fungi) fide Fries (*Syst. mycol.* **3**: 409, 1832).

flocci, cotton like groups or tufts.

Floccomutinus Henn. (1895) = Mutinus (Phall.) fide Demoulin & Dring (*Bull. Jard. bot. nat. Belg.* **45**: 365, 1975).

floccose, cottony; byssoid.

Floccularia Pouzar (1957), Tricholomataceae. 1, N. temp.

flocculent (of a liquid-culture), having small masses of cells throughout or as a deposit.

Flocculina P.D. Orton (1960) = Phaeomarasmius (Cortinar.) fide Singer (1962).

flocculose, delicately cottony.

flor effect (of yeasts), formation of a pellicle; see yeasts.

flor yeasts, see yeast.

flora, (1) the plants of a particular geographical area or habitat; (2) a description, catalogue or list of all or some groups of plants in a particular area. Formerly applied to fungi and lichens (i.e. **fungus -**, **lichen -**), but as fungi are not plants the term mycobiota (q.v.) is preferred.

Flosculomyces B. Sutton (1978), Mitosporic fungi, 1.D2.1. 2, Japan, USA, Ivory Coast.

flowers of tan, the myxomycete *Fuligo septica*.

Fluminispora Ingold (1958) = Dimorphospora (Mitosp. fungi) fide Ingold (*Guide to aquatic hyphomycetes*, 1975).

fluorescent, giving out light when placed in ultraviolet (or other) radiation.

flush (of fungal growth), the sudden development of a large quantity of mycelium or a periodic surge of basidiomata emergence, esp. in mushroom cultures.

Fluviatispora K.D. Hyde (1994), Halosphaeriaceae. 2 (on *Livistona*), Papua New Guinea.

Fluviostroma Samuels & E. Müll. (1980), Trichosphaeriaceae. 1, C. & S. Am. Anamorph *Stromatostilbella*.

fly agaric (fly fungus, fly mushroom), basidioma of *Amanita muscaria* (see soma).

fly fungus (house fly fungus), *Entomophthora muscae*.

fly-speck fungi ascomata of *Microthyriaceae*, *Micropeltidaceae* and conidiomata of their anamorphs.

Foetidaria A. St.-Hil. (1835) = Lysurus (Clathr.) fide Dring (1980).

foliicolous, living on leaves; **- lichens**, see Santesson (*Symb. bot. upsal.* **12** (1), 1952; monogr.), Farkas & Sipman (*Trop. bryol.* **7**: 93, 1993; checklist 482 spp., bibliogr.), 81; bibliogr. 83 papers 1952-85), Lücking (*Beih. Nova Hedw.* **104**: 1, 1992; keys 228 spp. Costa Rica), Ferraro (*Bonplandia* **5**: 191, 1983; S. Am.), Santesson & Tibell (*Austrobaileya* **2**: 529, 1988; 66 spp. Australia), Sérusiaux (*Bot. J. Linn. Soc.* **100**: 87, 1989; review).

foliole, a small leaf-like excrescence on the surface of a foliose lichen.

Foliopollenites Sierotin (1961), ? Fossil fungi.

foliose, (1) leaf-like; (2) (of lichens), having a layered (stratose) thallus, usually with a lower cortex, and attached to the substratum either by rhizines or at the base, but not by the whole lower surface (e.g. *Parmelia*, *Peltigera*) (Fig. 21D).

Follmannia C.W. Dodge (1967) = Caloplaca (Teloschist.) fide Kárnfelt (*Crypt. Bot.* **1**: 147, 1989).

Fomes (Fr.) Fr. (1849) nom. cons. prop., Coriolaceae. 1, widespr. See Donk (*Persoonia* **8**: 278, 1975). See also *Heterobasidion*, *Phellinus*, Medical uses of fungi.

Fomesporites T.C. Huang (1981), Fossil fungi. 1 (Miocene), Taiwan.

Fominia Girz. (1927) = Colletotrichum (Mitosp. fungi) fide Sutton (1977).

Fomitella Murrill (1905) = Coriolopsis (Coriol.) fide Pegler (1971).

Fomites Locq. & Koeniguer (1982), Fossil fungi. 1, Libya.

Fomitiporella Murrill (1907) = Phellinus (Hymenochaet.) fide Pegler (1973).

Fomitiporia Murrill (1907) = Phellinus (Hymenochaet.) fide Pegler (1973).

Fomitopsis P. Karst. (1881), Coriolaceae. 18, widespr. See Pegler (1973).

Fonsecaea Negr. (1936) = Rhinocladiella (Mitosp. fungi) fide Schol-Schwarz (*Ant. v. Leeuwenhoek* **34**: 119, 1968).

Fontanospora Dyko (1978), Mitosporic fungi, 1.G1.1. 2 (aquatic), USA.

Fonticula Worley, Raper & M. Hohl (1979), Fonticulaceae. 1, USA. See Worley *et al.* (*Mycol.* **71**: 746, 1979).

Fonticulaceae Worley, Raper & M. Hohl (1979), Acrasales. 1 gen. (+ 1 syn.), 1 spp.

Food and beverage mycology. Filamentous fungi and yeasts are used commercially in the manufacture of a large variety of foods and drinks. Perhaps the most well known use of fungi is in cheese making, notably *Penicillium camembertii* in camembert cheese and *P. roquefortii* for the manufacture of blue-veined cheeses such as roquefort, stilton, and Danish blue. They can also be eaten as themselves (see Mushroom culture), and more recently as mycoprotein (q.v.). However, in S.E. Asia particularly, fungi have been used for centuries for the production of various food fermentations (see Fermented food and drinks). The most well-known of these products is soya sauce, produced by seeding soy beans with *Aspergillus oryzae*.

Yeasts are used for the manufacture of bread, beer and wine (see Brewing, Wine making). In most cases *Saccharomyces cerevisiae* is involved, causing a fermentation of sugar to alcohol (production of wine and beer) and carbon dioxide (in bread, the alcohol evaporates during the baking process). *S. carlsbergensis* is used in the manufacture of lager.

Lit.: Beuchat (Ed.) (*Food and beverage mycology*, edn 2, 1987), Pitt & Hocking (*Fungi and food spoilage*, 1985), Samson *et al.* (*Introduction to food-borne fungi*, edn 4, 1995), Hui & Khachatourians (Eds) (*Food biotechnology*, 1995). See also Food spoilage, Mycotoxicoses, Yeasts.

Food spoilage. Can be caused by many fungi which invade and decompose harvested and processed foods and beverages, and their production of toxic metabolites (see Mycotoxicoses). They are responsible for a loss of 5-10% in food production in developing countries. In most fresh, moist foods, fungi do not grow well due to competition from bacteria, in ones which have conditions such as lowered water activity (q.v.), pH or refrigerated temperatures, filamentous fungi and yeasts may proliferate. Fungi can contaminate dried foods such as nuts, spices, cereals, dried milk and meat, as well as salted fish, fruit, vegetables, meats, jams, confectionery and dairy products. Some fungi can resist pasteurization, e.g. *Byssochlamys*, *Talaromyces* and *Neosartorya* which can be found in canned and bottled fruits, vegetables and juices. Common food spoilers include *Absidia*, *Mucor*, *Rhizopus*, *Syncephalastrum*, *Paecilomyces*, *Byssochlamys*, *Aspergillus*, *Penicillium* (and their teleomorphs); the latter two genera dominate.

Yeasts are particularly preservative resistant and are found in beers, wines, cider, and soft drinks (*Saccharomyces cerevisiae*, *S. bailii*, *Brettanomyces intermedius*), pickles and sauces (*Candida krusei*, *Pichia membranaefaciens*).

Lit.: Beuchat (Ed.) (*Food and beverage mycology*, edn 2, 1987), Cannon (*in* Robinson (Ed.), *Developments in food microbiology* 3: 141, 1988), Fassatiová (*Moulds and filamentous fungi in technical microbiology*, 1986), *Food Science and Technology Abstracts* (1969 on; abstracts), Pitt & Hocking (*Fungi and food spoilage*, edn 2, 1985), Samson *et al.* (Eds) (*Introduction to food-borne fungi*, edn 4, 1995). See also Beer, Fermented foods and drinks, Food and beverage mycology, Mycotoxins, Wine making, Yeasts.

foot cell, (1) a basal cell supporting the conidiophore in *Aspergillus*; (2) the basal cell of the conidium in *Fusarium*. See Sutton (*TBMS* 86: 1, 1986; occurrence in Mitosporic fungi).

Foraminella S.L.F. Mey. (1982) ≡ Parmeliopsis (Parmel.).

forate (of 'gasteromycete' basidiomata), invagination of the primordial tissue resulting in a series of pits; the type of development generally known as 'coralloid' of which it is the opposite (Dring, 1973).

-form (suffix), shape.

form (pl. **-a**) [**f.**; **ff.**, pl.], the lowest formal taxonomic rank regulated by the Code (see Classification, Nomenclature); **- category**, see Fossil fungi; **- genus/- species**, one used for anamorphs (q.v.); **-a specialis** [**f. sp.**; **-ae speciales**, **f. spp.**, pl.], see special form.

fornicate, arched; (of *Geastrum*), having the fibrous and fleshy layers of the fruit body becoming arched over the cup-like mycelial layer.

Forssellia Zahlbr. (1906) = Pterygiopsis (Lichin.) fide Henssen (*Bryol.* 73: 617, 1970).

Fossil fungi (Fungi fossiles). The fossil record of Fungi extends back to the early Phanerozoic, and no doubt well into the Proterozoic (see Pirozynski, *Ann. Rev. Phytopath.* 14: 237, 1976). Some fossil fungi are assigned extant generic names (or sometimes names of higher taxa) modified by a suffix (usually *-ites*; e.g. *Pleosporites*, *Pleosporonites*), or are given new form-generic names. In addition, fossil fungi (typically dispersed spores) may be given terminological names denoting suprageneric form — **form categories** (*f. cat.*) within morphographic classifications developed by palynologists or coal petrologists.

There is little information about fossil lichens, but this is the result of an unfamiliarity with these organisms, and the fact that few have been searching for them as fossils. Because cyanobacteria are well represented in Precambrian biotas and most groups of fungi were highly diversified during the Paleozoic, it would appear that lichens were also components of the early terrestrial ecosystems. As many have a relatively durable thallus morphology and organization this argues that lichens should be preserved. There is some suggestion that certain Silurian and Devonian and even earlier enigmatic fossils may represent lichens (Retallack, *Palaebiol.* 20: 523, 1994; Stein, *Am. J. Bot.* 80: 93, 1993), but reports of Precambrian lichens appear to be based on abiotic features (see *Thuchomyces*). To date the best records are from Tertiary

sediments (e.g. Sherwood-Pike, *Lichenologist* 17: 114, 1985; Richardson & Green, *Lichenologist* 3: 89, 1965). There are reports of preservation amber for lichens (Mägdefrau, *Ber. dtsch. bot. Ges.* 70: 433, 1957, Baltic; Poinar *et al.*, *Science* 259: 222, 1993), and even gilled fungi (Hibbett *et al.*, *Nature* 377:487: 1995; Cretaceous *Marasmius*-like basidime).

Lit.: *General*: Meschinelli (*in* Saccardo, *Syll. Fung.* 10: 741, 1892; 11: 657, 1895) and in his *Fungorum fossilium hucusque cognitorum iconographia* (1902; edn 1, 1898) described numerous species of fossil fungi, as sporocarps, spores and symptoms like those caused by extant parasitic fungi. Graham (*J. Palynol.* 36: 60, 1962; spores), Taylor (*in*: *The Fossil Record* 2: 9, 1993; geologic ranges), Pia (*Arch. Hydrobiol.* 31: 264, 1937; microborings in calcareous substrates), Elsik *et al.* (*Annotated glossary of fungal polynomorphs*, AASP 11, 1983; terminology), Pirozynski (*Geosci. Canada* 16: 183, 1989; methods of study), Sherwood-Pike (*Biosystems* 25: 121, 1991; evolution), Stubblefield & Taylor (*New Phytol.* 108: 3, 1988; paleomycology). See amber.

Major groups: **Chytridiomycota**, Taylor *et al.* (*Amer. J. Bot.* 79: 1233, 1992; *Mycol.* 84: 901, 1992), **Zygomycetes**, Pirozynski & Dalpé (*Symbiosis* 7: 1, 1989), **Ascomycota**, Sherwood-Pike & Gray (*Lethaia*, 18: 1, 1985), Stubblefield & Taylor (*Am. J. Bot.* 70: 387, 1983), **Basidiomycota**, Dennis (*Mycologia* 62: 578, 1970).

Interactions: **Saprophytism**, Stubblefield *et al.* (*Am. J. Bot.* 72: 1765), Taylor & White (*Am. J. Bot.* 76: 389, 1989), **Mutualism**, Stubblefield *et al.* (*Am. J. Bot.* 74: 1904, 1987), **Parasitism**, Hass *et al.* (*Am. J. Bot.* 81: 29, 1994), Stidd & Cosentino (*Science* 190: 1092, 1975), Daghlian (*Palaeontology* 21: 171, 1978), **Animal Interactions**, White & Taylor (*Mycol.* 81: 643, 1989), Grahn (*Lethaia* 14: 135, 1981). **Geologic Interactions**, Wright (*Sedimentology* 33: 831, 1986).

Regional: **Africa**, Wolf (*Bull. Torrey bot. Cl.* 93: 104, 1966), **Antaractica**, Stubblefield & Taylor (*Bot. Gaz.* 147: 116, 1986), **Argentina**, Singer & Archangelsky (*Am. J. Bot.* 45: 194, 1958), **Australia**, Selkirk (*Proc. Linn. Soc. NSW* 100: 70, 1975), **Brazil**, Martill (*Merc. Geol.* 12: 1, 1989), **Canada**, Currah & Stockey (*Nature* 350: 698, 1991), **Germany**: Kretzschmar (*Facies* 7: 237, 1982), **India**, Kalgutkar (*Rev. Palaeobot. Palynol.* 77: 107, 1993), **Mexico**, Magallon-Puebla & Cevallos-Ferriz (*Am. J. Bot.* 80: 1162, 1993), **N. America**, Dilcher (*Palaeontographica* B116: 1, 1965), Wagner & Taylor (*Rev. Palaeobot. Palynol.* 37: 317, 1982), **Russia**, Krassilov (*Lethaia* 14: 235, 1981), **W. Indies**, Poinar & Singer (*Science* 248: 1099, 1990), **United Kingdom**, Taylor *et al.* (*Am. J. Bot.* 79: 1233, 1992, *Mycol.* 87: 560, 1995; arbuscular fungi), Smith (*Palaeontology* 23: 205, 1980), Kidston & Lang (*Trans. R. Soc. Edinb.* 52: 855, 1921).

Fouragea Trevis. (1880) = Opegrapha (Roccell.) fide Santesson (1952).

foveate, having small holes or cavities; pitted (Fig. 29.11).

Foveodiporites C.P. Varma & Rawat (1963), Fossil fungi (*f. cat.*). 1 (Tertiary), India.

foveolate, delicately pitted; dimpled, dim. of foveate.

Foveoletisporonites Ramanujam & K.P. Rao (1979), Fossil fungi. 1, India.

Foveostroma DiCosmo (1978), Mitosporic fungi, 8.C1.15. 6, widespr. See Hosagoudar & Balakrishnan (*J. Econ. Tax. Bot.* 15: 477, 1989).

Foxia Castell. (1908) = Exophiala (Mitosp. fungi) fide de Hoog (1977).

Fracchiaea Sacc. (1873) = Nitschkia (Nitschk.) fide Nannfeldt (1975).

Fractisporonites R.T. Clarke (1965), Fossil fungi (*f. cat.*). 3 (Cretaceous, Eocene), USA.

fracture (of conidial liberation), secession involving the rupture of the wall of an adjacent vegetative or degenerate cell at a point removed from the septum; cf. fission, lysis, rhexolytic.

fragmentation spores, conidia produced by hyphae breaking up into separate cells.

Fragosia Caball. (1928), Saccharomycetales (inc. sed.). 1, Eur.

Fragosoa Cif. (1926) ? = Hysterographium (Hyster.). See Clements & Shear (1931).

Fragosoella Petr. & Syd. (1927) ? = Scleropycnium (Mitosp. fungi) fide Clements & Shear (1931).

Fragosphaeria Shear (1923), Pseudeurotiaceae. 2, widespr. See Malloch & Cain (*CJB* **48**: 1815, 1970).

Frankia Brunch. (1886) [non auct.] = Frankiella (? Plasmodiophor.).

Frankia Brunch. (1886), *Actinomycetes*, q.v.

Frankiella Speschnew (1900) = Greeneria (Mitosp. fungi) fide Sutton (1977).

Frankiella Maire & A. Tison (1909) ? = Plasmodiophora (Plasmodiophor.) fide Hawker & Fraymouth (*J. gen. Microbiol.* **5**: 369, 1951).

Franzpetrakia Thirum. & Pavgi (1957), Ustilaginaceae. 2 (on *Gramineae*), India, China, Japan. See Guo et al. (*Mycosystema* **3**: 57, 1990)

Fraseria Bat. (1962) = Microcallis (Chaetothyr.) fide v. Arx & Müller (1975).

Fraseriella Cif. & A.M. Corte (1957), Anamorphic Monascaceae, 1.A/B.1. Teleomorph *Xeromyces*. 1, widespr.

Fraserula Syd. (1938) = Inocyclus (Parmular.) fide Müller & v. Arx (1962).

free (of lamellae or tubes), not joined to the stipe (Fig. 19A); cf. remote, seceding.

free cell formation, the process by which the 8 nuclei, each with some adjacent cytoplasm, are cut off by walls in the immature ascus to become ascospores.

freeze drying, see Genetic resource collections, Preservation.

Fremineavia Nieuwl. (1916), Melanconidaceae. 1, N. Am. See Wehmeyer (*Revision of Melanconis*, 1941).

Fresenia Fuckel (1866) [non DC. (1836), *Compositae*] ≡ Graphiothecium (Mitosp. fungi).

Freynella Kuntze (1891) ≡ Coccosporium (Mitosp. fungi).

friable, readily powdered.

Fries (Elias Magnus; 1794-1878). Has been said to be 'the Linnaeus of Mycology': names of Fungi Caeteri in his *Systema mycologicum* (3 vol. and the *Elenchus*, 1821-32) [reprint 1952] have protected status in nomenclature (q.v.) of fungi. Fries, like Linnaeus, came from Smaland, was at school in Wexiö, and at the University of Lund; he studied under Acharius (q.v.) and became in his turn Professor of botany at Uppsala, and was a great writer and taxonomist. An only child, Fries was first interested in flowering plants, following his father, but at the age of 12 his attention was taken by fungi. Within five years he had a knowledge of more than 300 species, to which he gave names, and on going to Lund he was able to see the works of Persoon and other mycologists and to get the current names for some of his specimens. After being given his degree for a thesis on flowering plants, he took a position in the University where he was stationed for twenty years. The *Systema*, though produced without the help of a compound microscope, is still one of the most important books on systematic mycology. It is specially important for hymenomycetes, but takes in all groups of fungi (e.g. there are more than 500 pyrenomycetes, in good order), and it put the taxonomy of most fungi far in front of that in any

earlier work. In addition, he distributed 450 numbers of his *Scleromyceti Sueciae* exsiccati [pyrenomycetes and coelomycetes; see Holm & Nannfeldt (*Friesia* **7**: 10, 1964) and Pfister (*Mycotaxon* **3**: 185, 1975) for annotated check list]. Among his other writings are the *Lichenographia europaea reformata* (1831), *Epicrisis systematis mycologici* (June 1838; edn 2 as *Hymenomycetes Europaei*, 1874 [reprint 1963]), *Summa vegetabilium Scandinaviae* (1846-49), *Observationes mycologicae*, 1815-18 (index, Petersen, *Mycotaxon* **17**: 87-147, 1983), *Monographia Hymenomycetum Sueciae* (1853 [reprint 1963]), together with the *Icones* made under his direction, and works on vascular plants. Fries was at Uppsala from 1834 and hard at work to the very end; then, as now, he was respected everywhere. See Lundström (*Trans. bot. Soc. Edinb.* **13**: 383, 1879), Dudley (*J. Mycol.* **2**: 91, 1886), Lloyd (*Mycological Writings* **1**: 161, 1904; **3**: 413, 1909), Krok (*Bibliotheca botanica Suecicana*, 1925; list 171 titles), T.M. & R. Fries (*Friesia* **5**: 135, 1955; Engl. transl. Fries' autobiogr.), Eriksson (*Elias Fries och den romantiska biologien*, 1962), Holm (*Symb. bot. upsal.* **30**(3): 21, 1995), Nannfeldt (*Symb. bot. upsal.* **22**(4): 24, 1979), Grummann (1974: 474), Stafleu & Cowan (*TL-2* **1**: 878, 1976), Strid (*Catalogue of fungus plates painted under the supervision of Elias Fries*, 1994).

Fries (Theodore Magnus: 1832-1913). Son of E.M. Fries who was docent of botany Univ. Uppsala 1837-62. Later docent Borgstromian Anjunct of botany and practical economy becoming Professor 1877. Edited correspondence of Linnaeus and wrote some 170 publications on many aspects of biology and geography. Probably responsible for accelerating acceptance of Schwenderian concept of lichens. Described many new species of lichens, especially from Europe. Collections in **UPS**. Main works included: *Lichenes arctoi* (1860), *Lichenographia Scandinavica* (1, 1871; 2, 1874). See Hemmendorff (*Svensk bot. Tidsk.* **8**: 109, 1914), Grummann (1974: 475), Stafleu & Cowan (*TL-2* **1**: 889, 1976).

Friesia Lázaro Ibiza (1916) [non Spreng. (1818), *Euphorbiaceae*] = Ganoderma (Ganodermat.) fide Donk (1974).

Friesites P. Karst. (1879) = Hericium (Heric.) fide Donk (1956).

Friesula Speg. (1880) nom. dub. (Thelephor.) fide Donk (1957).

Fritzea Stein (1879) ? = Psora (Psor.).

frog cheese, young puff-balls.

Frommea Arthur (1917) = Phragmidium (Phragmid.) fide Laundon (*Mycotaxon* **3**: 155, 1975).

Frommeëlla Cummins & Y. Hirats. (1983), Phragmidiaceae. 1 (on *Potentilla*), N. temp.

Frondicola K.D. Hyde (1992), Hyponectriaceae. 1 (on *Nypa*), Brunei. See Hyde (*Sydowia* **45**: 204, 1993; posn).

Frondispora K.D. Hyde (1993), Amphisphaeriaceae. 1, Eur.

Frostiella Murrill (1942) nom. nud. = Boletellus (Xerocom.) fide Singer (1945).

fructicolous, living on fruit.

fructification, see fruit-body.

fruit-body (fructification), a general term for spore-bearing organs in both macro- and microfungi. The more precise terms apothecium, ascoma (ascocarp), basidioma (basidiocarp), conidioma, perithecium, sporocarp, etc. are preferred usage.

fruticolous, living on shrubs.

fruticose, (1) shrub-like; (2) (of lichens), having an upright or hanging thallus of radiate structure (e.g. *Cladonia, Ramalina, Usnea*) (Fig. 21E).

fruticulose (of lichens), having a minutely shrubby habit (e.g. *Ephebe, Polychidium*).

Fuckel (Karl Wilhelm Gottlieb Leopold; 1821-76). Noted for his *Symbolae Mycologicae, Beiträge zur Kenntniss der Rheinischen Pilze*, 1870 (Supplements 1871, 1873, and 1875; publication dates, Rogers, *Mycol.* **46**: 533, 1954) for which he sent out exsiccati. The types for the new species he made in this work (and the rest of his collections) are in l'Herbier Boissier (**G**), with mass isotypes also distributed among the 2700 colls in his *Funghi rhenani exsiccati* (1863-74). See Stafleu & Cowan (*TL-2* **1**: 896, 1976).

Fuckelia Bonord. (1864), Anamorphic Leotiaceae, 8.B1.15. Teleomorph *Godronia*. 1, Eur.

Fuckelia (Nitschke ex Sacc.) Cooke (1869) = Euepixylon (Xylar.) fide Læssøe & Spooner (*Kew Bull.* **49**: 1, 1994).

Fuckelia Niessl (1875) nom. conf. fide Holm (*Taxon* **24**: 275, 1975).

Fuckelina Sacc. (1875) = Stachybotrys (Mitosp. fungi) fide Hughes (1958).

Fuckelina Kuntze (1891) ≡ Macropodia (Helvell.).

fugacious, see evanescent.

Fujimyces Minter & Caine (1980), Mitosporic fungi, 7.E1.1. 1 (on *Pinus*), UK.

fulcrum (of lichens), a conidiophore within a pycnidium (obsol.).

Fulgensia A. Massal. & De Not. (1853), Teloschistaceae (L). 8, Eur., Asia, N. Am. See Poelt (*Mitt. bot. StSamml., Münch.* **5**: 571, 1965), Klement (*Nova Hedw.* **11**: 495, 1965; key).

Fulgia Chevall. (1822) = Coniocybe (Coniocyb.) fide Trevisan (1862).

Fuligo Haller (1768), Physaraceae. 5, cosmop.

Fuligomyces Morgan-Jones & Kamal (1984), Mitosporic fungi, 1.D2.1. 1, India.

Fulminaria Gobi (1900) = Harpochytrium (Harpochytr.) fide Atkinson (1903).

Fulvia Cif. (1954), Mitosporic fungi, 1.C2.10. 1, widespr.

Fulvidula Romagn. (1936) nom. nud. = Gymnopilus (Cortinar.) fide Singer (1951).

Fulvifomes Murrill (1914) = Phellinus (Hymenochaet.) fide Pegler (1973).

fumagillin, an antibiotic (epoxide) from *Aspergillus fumigatus*; an amoebicide (McCowen *et al.*, *Science, N.Y.* **113**: 202, 1951).

Fumago Pers. (1822) nom. conf. fide Friend (*TBMS* **48**: 371, 1965).

Fumagopsis Speg. (1910) = Tridentaria (Mitosp. fungi) fide Kendrick & Carmichael (1973) but see v.d. Aa & v. Oorschot (*Persoonia* **12**: 415, 1985).

Fumagospora G. Arnaud (1911), Anamorphic Capnodiaceae, 4.D2.?. Teleomorph *Capnodium*. 1, Eur. See Hughes (*Mycol.* **68**: 693, 1976).

fumigatin, a benzoquinone antibiotic from *Aspergillus fumigatus*; antibacterial (Anslow & Raistrick, *Biochem. J.* **32**: 687, 1938).

fumitremorgin, a tremorgenic metabolite (indole derivative) of *Aspergillus fumigatus* (Yamazaki *et al.*, *Tetrahedron Lett.* **14**: 1241, 1975).

fumonisin, B1, B2 (B1, B2), toxic metabolites of *Fusarium moniliforme* (ear rot of maize). Infected maize eaten by horses causes equine leukocephalomalacia (Plattner *et al.*, *Mycol.* **82**: 698, 1990).

Funalia Pat. (1900), Coriolaceae. 5, pantrop. See Pegler (1973).

fungaemia (fungemia), fungi in the blood.

fungal, (1) (n.), a fungus (obsol.); (2) (adj.), see fungous.

FUNGI (Carpomycetes, Eumycota, Eumycophyta, Eumycetes, Fungales, Hysterophyta, Inophyta, Mycota, Mycetes, Mycetoideum, Mycetales, Mycetalia, Mycophyta, Mycophytes, Mycophycophytes). Kingdom of *Eukaryota*; the true fungi. 4 phyla, 103 ord., 484 fam., 4979 gen. (+ 4556 syn.), 56,360 spp. Eukaryotic organisms without plastids, nutrition absorptive (osmotrophic), never phagotrophic, lacking an amoeboid pseudopodial phase; cell walls containing chitin and β-glucans; mitochondria with flattened cristae and peroxiomes nearly always present; Golgi bodies or individual cisternae present; unicellular or filamentous and consisting of multicellular coenocytic haploid hyphae (homo- or heterokaryotic); mostly non-flagellate, flagella when present always lacking mastigonemes; reproducing sexually or asexually, the diploid phase generally short-lived, saprobic, mutualistic, or parasitic.

The organisms studied by mycologists, fungi, are mosly placed here, but there is now overwhelming evidence that some belong in the kingdoms *Chromista* and *Protozoa* (see Kingdoms of fungi, Phylogeny); this conclusion is fully accepted in this edition of the *Dictionary*. Some authors use *Eumycota* as the kingdom name which has the advantage of avoiding confusion with 'fungi' (e.g. Barr, 1992). The kingdom name *Fungi* is retained by Cavalier-Smith (1993) and Corliss (1994); as most fungi belong here, and as the name *Fungi* is immediately familiar to most students, we retain that here.

Various arrangements and ranks have been proposed for the main groupings of *Fungi*; selected ones are compared in Table 4. There is a general agreement that four major categories should be distinguished but the ranking allocated varies. The rank of phylum is adopted here as that is currently almost universally used (Barr, 1992; Bruns *et al.*, 1992; Corliss, 1994; Kendrick, 1992) and was already advocated by v. Arx (*Sydowia* **37**: 1, 1984); some authors nevertheless prefer lower ranks, e.g. classes (Nishida & Sugiyama, *Molec. Biol. Evol.* **10**: 431, 1993). The 4 phyla accepted here are (names in the seventh edition of the *Dictionary* in parentheses):

(1) **Ascomycota** (*Ascomycotina*).
(2) **Basidiomycota** (*Basidiomycotina*).
(3) **Chytridiomycota** (*Chytridiomycetes*).
(4) **Zygomycota** (*Zygomycotina*).

The *Deuteromycotina* is not accepted as a formal taxonomic category in this edition of the *Dictionary*; they are not a monophyletic unit, but are fungi which have either lost a sexual phase or which are anamorphs of other phyla (mainly *Ascomycota*; some *Basidiomycota*); with modern molecular or ultrastructural techniques such fungi can be assigned to existing classes; see under Mitosporic fungi. See Classification, Kingdoms of fungi, Literature, Phylogeny.

Fungi fossiles, see Fossil fungi.

Fungi Imperfecti, see Mitosporic fungi (and anamorphic fungi); fungi having no sexual state, presumptively mitotic.

fungi suilli, see cep.

fungicidal, able to kill fungus spores or mycelium.

fungicides, substances able to kill fungi, esp. if lethal at low concentration. The term is sometimes loosely applied to fungistatic and genestatic substances. **eradicant -**, those applied to a substratum in which a fungus is already present, or used in disease control after infection has been established; **protective -**, those used to protect an organism against infection by a fungal pathogen (*TBMS* **33**: 155, 1950); **systemic -**, substances which are fungicidal (or fungistatic) when taken up systemically by plants (see below). Heat, light, and other radiations have fungicidal properties, see sterilization.

Many chemical substances are fungicidal or inhibit fungal growth. Even pure water sometimes inhibits

TABLE 4. Proposed classification of higher categories including fungi[1]

Bessey (1950)	Kriesel (1969)	Ainsworth *et al.* (1973)	v. Arx (1981)	Dictionary (1983)	Kreisel (1988)
Mycetozoa Class Phycomyceteae Division **Carpomyceteae** Class Ascomyceteae 'The Pyrenomycetes' Class Basidiomyceteae Subclasses: Teliosporae Heterobasidiae Hymenomyceteae 'Gasteromycetes' The Imperfect Fungi Moniliales Sphaeropsidales Melanconiales	**PROTOBIONTA** [Myxomycota excluded from Fungi] **Eumycota** [Chytridiomycetes excluded from fungi] [Oomycetes included as a Class of Chrysophyta (Algae)] *Zygomycetes* *Endomycetes* *Ascomycetes* *Euascomycetidae* *Loculoascomycetidae* *Basidiomycetes* *Phragmobasidiomycetidae* *Hymenobasidiomycetidae* *Gasteromycetidae* Endomycetes imperfecti Ascomycetes imperfecti Basidiomycetes imperfecti	**FUNGI** Myxomycota Acrasiomycetes [Labyrinthulales] Myxomycetes Plasmodiophoromycetes **Eumycota** *Mastigomycotina* Chytridiomycetes Hyphochytriomycetes Oomycetes *Zygomycotina* Zygomycetes Trichomycetes *Ascomycotina* Hemiascomycetes Plectomycetes Discomycetes Pyrenomycetes Loculoascomycetes Laboulbeniomycetes *Basidiomycotina* Teliomycetes Hymenomycetes Gasteromycetes *Deuteromycotina* Blastomycetes Hyphomycetes Coelomycetes	**MYCOTA** Myxomycota Acrasiomycetes Plasmodiophoromycetes Labyrinthulomycetes Oomycota Oomycetes Hyphochytriomycetes Chytridiomycota Chytridiomycetes Eu-Mycota Zygomycetes Endomycetes Ustomycetes Ascomycetes Basidiomycetes Deuteromycetes	**FUNGI** **Myxomycota** Ceratiomyxomycetes Dictyosteliomycetes Acrasiomycetes Myxomycetes Plasmodiophoromycetes Labyrinthulomycetes **Eumycota** **Mastigomycotina** Chytridiomycetes Hyphochytriomycetes Oomycetes **Zygomycotina** Zygomycetes Trichomycetes **Ascomycotina** [No Classes recognized] **Basidiomycotina** Hymenomycetes Gasteromycetes Urediniomycetes Ustilaginomycetes **Deuteromycotina** Coelomycetes Hyphomycetes	**MYXOMYCOTA** Acrasiomycetes Ceratiomyxomycetes Myxomycetes Plasmodiophoromycetes Protosteliomycetes **LABRYINTHULOMYCOTA** Labyrinthulomycetes **OOMYCOTA** Hyphochytridiomycetes Oomycetes **CHYTRIDIOMYCOTA** Chytridiomycetes **EUMYCOTA** Ascomycetes Basidiomycetes Endomycetes Teliomycetes Trichomycetes Ustomycetes Zygomycetes [Deuteromycetes]

[1] In some cases only the principle ranks are included for simplicity.
[3] Listed as classes in an unnamed phylum for plasmodial slime moulds.
[2] Phyla including fungi only detailed here.
[4] All lichen-forming fungi together with their photobionts.

Cavalier-Smith (1991)	Kendrick (1992)	Barr (1992)	Margulis (1993)	Moore (1994)	Dictionary (1995)
PROTOZOA Mycetozoa **CHROMISTA** Heterokonta **FUNGI** **Archemycota** Chytridiomycetes Trichomycetes Zygomycetes **Ascomycota** **Basidiomycota**	**PROTOCTISTAN FUNGI** Myxostelida Dictyostelida Labyrinthulida Plasmodiophorida Chytridiomycota Hyphochytriomycota Oomycota **EUMYCOTAN FUNGI** Dikaryomycota Ascomycotina Basidiomycotina Zygomycota	**PROTOZOA** Myxomycota Plasmodiophoromycota **CHROMISTA** Heterokonta Pseudomycotina Oomycetes Hyphochytriomycetes Labyrinthista Labyrinthulea **EUMYCOTA** Ascomycota Basidiomycota Chytridiomycota Zygomycota	**PROTOCTISTA** *Acrasea*[2] Chytridiomycota Dictyostelida Hyphochytriomycota Labyrinthulomycota *Myxomycota*[3] Oomycota Plasmodiophoromycota *Protostelida*[3] **FUNGI** Ascomycota Basidiomycota Deuteromycota Mycophycophyta[4] Zygomycota	**FUNGI** *MASTIGOMYCETIA* **Oomycota** Saprolegniomycetes Peronosporomycetes **Hyphochytriomycota** **Chytridiomycota** *ZYGOMYCETIA* Zygomycota Trichomycota *ASCOMYCETIA* **Euascomycota** **Hemiascomycota** *BASIDIOMYCETIA* **Basidiomycota** *Homobasidiomycotina* Hymenomycetes *Heterobasidiomycotina* Heterobasidiomycetes Teliomycetes Ustomycota *DEUTEROMYCETIA* **Deuteromycota** Coelomycetes Hyphomycetes Agonomycetes **Blastomycota** Ascoblastomycetes Basidioblastomycetes	**PROTOZOA** Acrasiomycota **Dictyosteliomycota** **Myxomycota** Myxomycetes Protosteliomycetes **Plasmodiophoromycota** **CHROMISTA** **Hyphochytriomycota** **Labyrinthulomycota** **Oomycota** **FUNGI** Ascomycota **Basidiomycota** Basidiomycetes Teliomycetes Ustomycetes **Chytridiomycota** **Zygomycota** Trichomycetes Zygomycetes

spore germination (e.g. of *Sclerotinia fructicola*) while conversely very low concentrations of some fungicides may stimulate spore germination or even be a prerequisite for fungal growth by supplying an essential trace element (see Nutrition). Today hundreds of fungicides are available in thousands of formulations for specific uses. Sulphur (elemental or in combination) and inorganic compounds of copper and mercury were the first fungicides employed and these three elements, frequently in organic combination, are still ingredients of important fungicides. Some of the more important fungicides (with indications of their main uses) are:

Sulphur: elemental sulphur (as flowers of sulphur, colloidal sulphur, wettable sulphur, etc.), lime sulphur.

Copper: bordeaux and burgundy mixtures (q.v.), cheshunt compound (q.v.), copper sulphate, basic copper sulphates, chlorides and carbonate, copper oxides (seed treatment), organo-coppers (copper oleate and resinate) (plant protection).

Mercury: mercuric chloride (corrosive sublimate) (wood preservation, soil treatment), mercurous chloride (calomel) (seed and soil treatment). Organomercurials are not now used in agriculture because of non-target effects.

Other inorganic compounds: borax and boric acid (storage decay of fruit), calcium hypochlorite (bleaching powder) (surface sterilization of biological material), potassium iodide (sporotrichosis), potassium dichromate, sodium fluoride, zinc chloride, and arsenic compounds (wood preservation), triphenyl tin salts (crop protection).

Organic compounds: formaldehyde (formalin) (seed and soil disinfection), 5-fluorocytosine (candidiasis, cryptococcosis), guanidines (dodine), dithiocarbamates (thiram, mancozeb, propineb), carboximides (iprodione), imidayols (prochloraz, imazalil), chlorinated hydrocarbons (chlorathalonil, didoran), methyl bromide and cresylic acid (soil sterilization), propionic acid (food preservation), propylene oxide (sterilization of biological materials), salicylanide (Shirlan) (plant protection, textile spoilage); see also antibiotics, copper, mercury, sulphur, systemic fungicides above and below.

Antibiotics: amphotericin (systemic mycoses), kasugamycin (systemic; rice blast), cycloheximide (crop protection and differential isolation of fungi), nystatin (candidiasis).

Systemic fungicides: (against plant diseases) acylalanines (furaxyl, metalaxyl), benzimidazoles (benomyl, carbendazim, thiabendazole), oxathiins (carboxin), pyrimidine derivatives (dimethirimol, ethirimol), triazoles (triadimefon, propinconazole). See Marsh (Ed.) (*Systemic fungicides*, 1972 [edn 2, 1977]).

Lit.: Martin (*The scientific principles of plant protection* edn 6, 1973), *Guide to the chemicals used in crop protection* (edn 4, 1961; *Canada Dep. Agric., Publ. Res. Branch* **1093**), Horsfall (*Fungicides and their action*, 1945; *Principles of fungicidal action*, 1956), Torgeson (Ed.) (*Fungicides, an advanced treatise*, **1**, *Agricultural and industrial applications. Environmental interactions*, 1967; **2**, *Chemistry and physiology*, edn 2, 1983), IMI (*Plant pathologist's pocketbook*, edn 2, 1983: 161), Smith (Ed.) (*Fungicides in crop protection. 100 years of progress*. 2 vol., 1985 [BCPC Monogr. **31**], Nene & Thapliyal (*Fungicides in plant disease control*. edn 3, 1993), Tomlin (Ed.) (*The pesticide manual*, edn 10, 1994). For 'Definitions of fungicide terms' and 'Recommended methods' for testing fungicides, see *Phytopath*. (**33**: 624, 1943; **34**: 401, 1944); also Zehr (Ed.) (*Methods for*

evaluating plant fungicides, nematocides and bactericides, 1978).

Fungicolous fungi. Fungi growing on other fungi as parasites ('mycoparasites') commensals, or saprobes; fungicoles. Reliable estimates of the total number of fungi involved are not available; Hawksworth (1981) traced records of 1100 spp. of Mitosporic fungi alone on 2500 other spp. of fungi. The fungi may be 'necrotrophic' (destructive), or 'biotrophic' (forming balanced relationships). Destructive effects may be mediated at a distance, by contact and coiling via antibiotics or hyphal-wall degrading enzymes, or hyphal interference (contact necrotrophs) or by penetration of the hyphae of the host (invasive necrotrophs). Biotrophic mycoparasitism occurs externally via haustoria (haustorial biotrophs), or via specialized contact cells which may produce colacosomes (Bauer & Oberwinkler, *Bot. Acta* **104**: 53, 1991) and which achieve direct cytoplasmic continuity with the host through fine interhyphal channels (fusion biotroph), or internally by entry of the complete thallus of the mycoparasite into the host (intracellular biotrophs). For a description of these host-parasite interfaces see Jeffries & Young (*Interfungal parasitic relationships*, 1994).

Particularly destructive mycoparasites are *Gliocladium virens* and *Trichoderma* spp. (also of value in biocontrol; see esp. Papavizas (*Ann. Rev. Phytopathol.* **23**: 23, 1985); both have very wide host ranges (Domsch *et al.*, 1980). Others are restricted to particular host fams., gen. or spp. Examples (**host**: fungi) are:- **Myxomycota**: *Aphanocladium album, Polycephalomyces* spp., *Leucopenicillifer, Nectria myxomyceticola* anamorph *Verticillium rexianum*). See Ing (*Bull. BMS* **8**: 25, 1974; key), Rogerson & Stephenson (*Mycol.* **85**: 456, 1993). **Oomycota**: esp. other *Oomycota*: e.g. *Pythium oligandrum* on *Pythium ultimum* (Whipps & Lumsden, *Biocontrol Sci. Technol.* **1**: 75, 1991). **Chytridiomycota**: chytrids esp. other *Chytridiales*, e.g. *Allomyces* by *Catenaria allomycis* (Sykes & Porter, *Mycol.* **72**: 288, 1980; Powell, *Bot. Gaz.* **143**: 176, 1982), *Allomyces* and *Polyphagus* by *Rozella* and *Rozellopsis* (Held, *Bot. Rev.* **47**: 451, 1981; Powell, *Mycol.* **76**: 1039, 1984). **Zygomycota**: *Chaetocladium, Piptocephalis, Dimargaris* etc. On other *Mucorales* (Benjamin, *Aliso* **4**: 321, 1959; Jeffries, *Bot. J. Linn. Soc.* **91**: 135, 1985); *Stachybotrys chartarum* and *Anguillospora pseudolongissima* on spores of *Glomus* and *Gigaspora* (Paulitz & Menge, *Phytopath.* **76**: 351, 1986). **Ascomycota**: the zenith for fungicolous fungi. *Ampelomyces quisqualis* on numerous *Erysiphales*; *Nectria magnusiana* (anamorph *Fusarium epistromum*) on *Diatrypella* spp.; *Nematogonium ferrugineum* on *Nectria coccinea*; *Coniothyrium minitans* on *Botryotinia* and *Sclerotinia* sclerotia (Tribe, *TBMS* **40**: 489, 1957; Huang, *CJB* **55**: 289, 1977); *Cordyceps ophioglossoides* on *Elaphomyces*; *Tympanosporium parasiticum* on *Nectria cinnabarina*; trop. foliicolous spp. (esp. *Meliolaceae*) support a vast array of fungi (Hansford, *Mycol. Pap.* **15**, 1946; Deighton & Pirozynski, *Mycol. Pap.* **128**, 1972); Pirozynski, *Kew Bull.* **31**: 595, 1977; Hughes, *Mycol. Pap.* **166**, 1993). See also Lichenicolous fungi. **Basidiomycota**: *Eudarluca caricis* (anamorph *Sphaerellopsis filum*) on 226 spp. *Uredinales* (Kranz, *Nova Hedw.* **24**: 169, 1974); *Tuberculina persicina* on 26 spp. *Uredinales*; *Hypomyces* spp. (anamorphs *Apiocrea, Cladobotryum*) esp. common on decaying *Agaricales* and *Boletales* (Arnold, *Bibliogr. Mill. Univ. bibliotek Jena* **25**, 1976; bibliogr. host index; Rogerson & Samuels, *Mycol.* **86**: 839, 1994, on agarics); *Mycogone perniciosa* (wet bubble disease) on cult. mushrooms (*CMI Descr.* **499**, 1976); *Nyctalis* spp. on *Russula; Amblyosporium spongiosum* esp. on *Lactarius* (Nicot & Durand, *BSMF* **81**: 623, 1965); *Helminthosphaeria* spp. on *Clavariaceae; Xerocomus parasiticus* on *Scleroderma.* See Nicot (*BSMF* **31**: 393, 1967); key spp. on agarics, Tubaki (*Nagaoa* **5**: 11, 1955), de Hoog (*Persoonia* **10**: 33, 1978, keys), *Verticillium biguttatum* on *Rhizoctonia solani* (Boogert *et al., Soil Biol. Biochem.* **24**: 159, 1992) and mushroom culture. **Mitosporic fungi:** *Gonatobotrys simplex; Hansfordia pulvinata* on *Cercospora* etc.; *Pseudofusidium hansfordii* on *Mycovellosiella; Sphaerulomyces coralloides* on aquatic hyphomycetes; *Syspastospora parasitica* on *Beauveria, Hirsutella, Paecilomyces* and *Verticillium.*

Lit.: (*Ann. Rev. Phytopathol.* **28**: 59, 1990), Barnett & Binder (*Ann. Rev. Phytopath.* **11**: 273, 1973), Burge (*Fungi in biological control systems,* 1988), Buller (3), Cooke (*The biology of symbiotic fungi,* 1977), Domsch *et al.* (*Compendium of soil fungi,* 2 vols., 1980), Fletcher *et al.* (*Mushrooms- pest and disease control,* 1986), Hashioka (*Forsch. Geb. Pflanzenkrankh.* **8**: 179, 1973; *Rep. Tottori Mycol. Inst.* **10**: 473, 1973; SEM), Hawksworth (*in* Cole & Kendrick (Eds), *The biology of conidial fungi* **1**: 171, 1981, conidial spp.), Helfer (*Pilze auf Pilzffruchtkörpern. Libri Botanici* **1**: 1, 1991), Jeffries & Young (*Interfungal parasitic relationships,* 1994), Lumsden (*in* Carroll & Wicklow (Eds), *The fungal community,* 1992; review, examples), Madelin (*The Fungi* **3**: 253, 1968; review), Rudakov (*Mycol.* **70**: 150, 1978; physiol. groups), Seeler (*Farlowia* **1**: 119, 1943), Weindling (*Bot. Rev.* **4**: 475, 1938). See also Ecology, Lichenicolous fungi.

fungiform, mushroom-shaped.
fungistasis, (1) see mycostasis; (2) (**fungistatic**) (of a substance, or of a concentration of a fungicide), inhibiting fungus growth but not fungicidal. Cf. genestasis.
Fungites Hallier (1865), Fossil fungi (mycel.). 1 (Tertiary), Germany.
Fungites Casp. (1907), Fossil fungi (mycel.). 4 (Oligocene), Baltic.
fungivorous, using fungi as food; **fungivore,** an organism which does this.
fungizone, trade name for amphotericin B (q.v.).
fungi, higher, see higher fungi
fungi, lower, see lower fungi.
Fungodaster Haller ex Kuntze (1891) = Leotia (Leot.).
fungoid, fungus-like.
Fungoidaster Micheli (1729) nom. inval. = Craterellus (Craterell.) p.p.
Fungoides Tourn. (1719) nom. inval. (Ascomycota & Basidiomycota). Used for a wide range of fungi with cup-like fruits, including discomycetes and *Nidulariales.*
fungology, mycology (obsol.).
fungoma, fungus ball formation.
fungophobia, a horror or dread of fungi (Hay, *British Fungi*: 6, 1887) (obsol.).
fungous, of, or having to do with, fungi; fungal.
funguria, the presence of fungi, particularly yeasts, in urine.
fungus (pl. **fungi**), (Lat. **fungus,** orig. *sfungus,* cognate with *spongia* from Gk *sphongis,* a sponge), champignon (Fr.), ciupercă (Rumanian), cogumelo (Port.), fungo (Ital.), gljiva and guba (Croat.), goba (Slovene), gomba (Hung.), grÿbas (kremblys) (Lith.), grzyb (Pol.), hongo (Span.), houba (Czech.), huba (Slovak), 菌 (jun, chün) (Chin.), kavak (Hindi), kavaka (Sanskrit), këphurdhe (Albanian), 菌 類 (kin-rui) (Jap.), XXX (Gk), makunä (Kikuyu), Pilz (Germ.), poonjalam (Tamil), seen (Estonian), sēne (Latvian), sieni (Fin.), svamp (Scand.), zwam (Dutch), ГРИБ (Russ.), ГРЫБ; (Byeloruss.), ГЪЬИВА and ГУБА (Serb.), ГЪБА (Bulgar.); 582.28 (Universal Decimal Classification); MACR.003 (semantic code; Perry & Kent, *Tools for machine literature searching,* 1958), one of the *Fungi;* [in Medicine: an abnormal sponge-like growth].

Fungus Tourn. ex Adans. (1763) ≡ Agaricus (Agaric.).
Fungus Adans. ex Kuntze (1898) = Agaricus (Agaric.), etc.
fungus ball, see aspergilloma.
fungus gnats, the *Mycetophilidae* (*Insecta*); see Mushroom culture.
fungus root, see mycorrhiza.
fungus, ray, see ray fungi.
funicular, cord-like.
funicular cord, (funiculus), the cord of hyphae by which the peridioles in *Nidulariaceae* (e.g. *Cyathus*) are at first fixed to the inner wall of the peridium; see splash cup.
Funicularius K.K. Baker & Zaim (1979), Mitosporic fungi, 1.A1.10. 1 (on mosquito), USA.
funiculose (of hyphae), aggregated into rope-like strands, plectonematogenous (q.v.).
funoid, composed of rope-like strands or fibres; funicular (q.v.).
Furcaspora Bonar (1965), Mitosporic fungi, 8.G1.10. 1, USA.
furcate, forked.
Furcouncinula Z.X. Chen (1982) = Uncinula (Erysiph.) fide Braun (1987).
Furculomyces Lichtw. & M.C. Williams (1992), Harpellaceae. 2, Australia.
furfuraceous, covered with bran-like particles; scurfy.
Furia (Batko) Humber (1989), Entomophthoraceae. 12, widespr.
Fusamen (Sacc.) P. Karst. (1890), Mitosporic fungi, 6.A1.19+10. 5 (on *Salix*), Eur. See v. Arx (*Bibl. Mycol.* **24**, 1970), Sutton (1980).
fusaric acid (Fusarinsäure, Germ.), a pyridine-carboxylic acid from *Fusarium bulbigenum* var. *lycopersici, F. vasinfectum* and other *Hypocreaceae* able to induce wilt symptoms in tomato (Gäumann, *Phytopath.* **47**: 342, 1948, review; **48**: 670, 1958, mechanism of action). Cf. lycomarasmin.
Fusariella Sacc. (1884), Mitosporic fungi, 1.C2.15. 4, temp. See Hughes (*Mycol. Pap.* **28**, 1949).
Fusariellites Babajan & Tasl. (1977), Fossil fungi. 1 (Tertiary), former USSR.
Fusariopsis Horta (1919), Mitosporic fungi, 2.A1.?. 1 (on humans), France. See Dodge (1936: 860).
Fusarium Link (1809), Anamorphic Hypocreaceae, 4.C1.15. Teleomorphs *Gibberella, Nectria. c.* 50, widespr. saprobes and parasites. This important genus has fusoid, curved, septate macroconidia in slimy masses (sporodochia) on branched conidiophores. Smaller 0- or 1-septate microconidia and chlamydospores are common. A third conidial type, the mesoconidium, is recognized by Pascoe (*Mycotaxon* **37**: 121, 1990). The mycelium and spores are generally bright in colour. Important pathogens: *F. avenaceum* and *F. culmorum* (cereal foot rots), *F. bulbigenum* (*Narcissus* basal rot) and its var. *lycopersici* (tomato wilt), *F. caeruleum* (dry rot of potato tubers), *F. conglutinans* (cabbage yellows) and its var. *callistephi* (aster (*Callistephus*) wilt), *F. lini* (flax (*Linum*) wilt), *F. oxysporum* (potato wilt) and its var. *cubense* (Panama disease of banana), *F. vasinfectum* (see Kshirsagar & Patil (*Res. Pub. Mahatma Phule Krishi vidy.* **6**: 1, 1992) (cotton wilt), and others.

Lit.: **General:** Booth (*The genus Fusarium,* 1971; keys), Wollenweber & Reinking (*Die Fusarien,* 1935), Tousson & Nelson (*A pictorial guide to the* *identification of Fusarium species*, 1969), Booth (*Fusarium. Laboratory guide to the identification of the major species*, 1977), Gordon (*CJB* **32**: 576, 622; **34**: 833, 847; **37**: 257, 1954-59; list and host index; **38**: 643, 1960, trop. and temp. spp.), Gerlach (*in* Subramanian (Ed.), *Taxonomy of fungi* **1**: 115, 1978), Gerlach & Ershad (*Nova Hedw.* **20**: 725, 1970; Iran), Joffe (*Mycopath. Mycol. appl.* **53**: 201, 1974; 13 sect., 33 spp., 14 vars.), Nirenberg (*Mitt. biol. Bund. Ld-Forst.* **169**, 1976; sect. *Liseola*), Pattin (*Acta phytotox.* **6**, 1978; toxigenic spp.), Teetor-Barsch & Roberts (*Mycopath.* **84**: 3, 1983; entomogenous spp.), Joffe (*Fusarium species: their biology and toxicology*, 1986), Tiedt *et al.* (*TBMS* **87**: 237, 1986; TEM conidiogenesis), Van Wyk *et al.* (*TBMS* **88**: 347, 1987; TEM conidiogenesis), Neish (*CJB* **65**: 589, 1987; neotypification of *F. tricinctum*), Marasas *et al.* (*Bothalia* **17**: 97, 1987; S. Afr. bibliography 1945-1985), Moss & Smith (*Applied mycology of Fusarium*), Tiedt & Jooste (*TBMS* **90**: 531, 1988; ultrastructure of collarette formation in sect. *Liseola*), Van Wyk *et al.* (*TBMS* **91**: 611, 1988; delimitation of *F. crookwellense* conidia), Smith *et al.* (*J. Med. Vet. Mycol.* **27**: 83, 1989; *F. solani* in sharks), Brayford & Bridge (*Lett. Appl. Microbiol.* **9**: 9, 1989; differentiation between *F. oxysporum* and *F. solani* by growth and pigmentation), Brayford (*J. Appl. Bact. Symp. Suppl.*: 475, 1989; progress in *Fusarium* studies), Windels *et al.* (*Mycol.* **81**: 272, 1989; perithecial production by *F. graminearum*), Chelkowski (Ed.) (*Fusarium: mycotoxins, taxonomy and pathogenicity*, 1989), Van Wyk *et al.* (*MR* **95**: 284, 1991; conidial development in sect. *Sporotrichiella*), Chen (*Variation within Fusarium sect. Moniliforme*, 1991), Chelkowski (Ed.) (*Mycotoxin Research, 2nd European Seminar. Fusarium* **7**(I & II), 1991), Logrieco *et al.* (*Exp. Mycol.* **15**: 174, 1991; phylogeny of Sect. *Sporotrichiella*), Gams & Nirenberg (*Mycotaxon* **35**: 407, 1989; generic definition), Jeschke *et al.* (*Mycol.* **82**: 727, 1990; ecology of S. Afr. soil spp.), Zhang (*Acta mycol. Sin.* **10**: 85, 1991; taxonomic advances), Bai & Chen (*Acta Mycol. Sin.* **10**: 120, 1991; spp. on insects in China), Anon. (*Phytopath.* **81**: 1043, 1991; *Recent Advances in Fusarium Systematics*), Kwaśna *et al.* (*Flora Polska, Grzyby (Mycota)* **22** Fusarium, 1991), Madhosingh (*J. Phytopath.* **136**: 113, 1992; interspecific hybrids), Brayford (*Mycopath.* **118**: 39, 1992; descriptions of 10 f.spp.), Huang *et al.* (*Trans. Mycol. Soc. R.O.C.* **7**(1/2): 1, 1992; **7**(1/2): 23, 1992; **7**(3/4): 1, 1992; spp. in Taiwan).

Techniques: Ye & Guang (*Acta Phytopath. Sin.* **15**: 87, 1985; esterase enzymes), Rubidge (*TBMS* **8**: 463, 1986; plasmid DNA), Naiki (*Res. Bull. Fac. Agr. Gifu Univ.* **51**: 29, 1986; nuclear DNA of f.spp.), Wasfy *et al.* (*Mycopath.* **99**: 9, 1987; biochemical and physiological tests), Szécsi & Hornok (*Acta Phyt. Ent. hung.* **21**: 215, 1986; genetic distance by isoelectric focusing esterase profiles), Abd El-Halim (*Egypt. J. Microbiol.* **20**: 185, 1986; DNA base composition), Cristinzio *et al.* (*Pl. Path.* **37**: 120, 1988; agglutinating effect of lectins on 8 spp.), Ellis (*Mycol.* **80**: 734, 1988; DNA relatedness), Min (*Korean Jl Mycol.* **16**: chromosomes in *F. oxysporum* vars and f.spp.), Kuninaga & Yokosawa (*Ann. Phyt. soc. Jap.* **55**: 216, 1989; DNA-DNA reassociation kinetics and f.spp. in *F. oxysporum*), Thrane & Frisvad (*Mycotoxin Res.* (suppl.): 21, 1989; secondary metabolite species prolific profiles), Thrane, *in* Chelkowski (Ed.) (*Fusarium: mycotoxins, taxonomy and pathogenicity*: 199, 1989; secondary metabolites), Burmeister & Vesonder (*Appl. Env. Microbiol.* **56**: 3209, 1990; steroid sulphate metabolite from several spp.), Thrane (*Jl*

Microbiol. Methods **12**: 23, 1990; Sect. *Discolor* analysis of secondary metabolite production), Luz *et al.* (*Lett. Appl. Microbiol.* **11**: 141, 1990; identification of *F. vasinfectum* f.sp. *vasinfectum* by metabolites), Guadet *et al.* (*Mol. Biol. Evol.* **6**: 227, 1989; phylogeny determined by rRNA sequences), Jacobson & Gordon (*MR* **94**: 734, 1990; mDNA in populations in *F. oxysporum*), Kuninaga & Yokosawa (*Ann. Phytopath. Soc. Jap.* **57**: 9, 1991; DNA homology and *F. oxysporum*), Manicom *et al.* (*Phytophyl.* **22**: 233, 1990; identification of *F. oxysporum* by molecular methods), Marasas *et al.* (*Mycopath.* **113**: 191, 1991; toxicity and moniliformin production), Szécsi (*J. Phytopath.* **130**: 188, 1990; pectic enzyme zymograms), Thiel *et al.* (*Appl. Env. Microbiol.* **57**: 1089, 1991; fumonisin production), Miller *et al.* (*Mycol.* **83**: 121, 1991; trichothecene chemotypes), Min (*Korean J. Microbiol.* **27**: 342, 1989; chromosomes), Kim *et al.* (*Phytoparas.* **19**: 211, 1991; RFLPs in *F. oxysporum* f.sp. *niveum*), Rataj-Guranowska & Wolko (*J. Phytopath.* **132**: 287, 1991; cf. *F. oxysporum* and var. *redolens* by isozymes and serology), Rataj-Guranowska *et al.* (*J. Phytopath.* **132**: 294, 1991; heterokaryons between *F. oxysporum* and var. *redolens*), Nelson *et al.* (*Appl. Env. Microbiol.* **58**: 984, 1992; fumonisin B1 production), Lodolo *et al.* (*MR* **97**: 345, 1992; rapid molecular technique for identification).

Plant Pathology: Bosland & Williams (*CJB* **65**: 2067, 1987; pathotypes of *F. oxysporum* on Cruciferae), Tivoli (*Agronomie* **8**: 211, 1988; identification of spp. from potato in France), Wong (*Lett. Appl. Microbiol.* **6**: 51, 1988; differential medium for races of *F. oxysporum* f.sp. *cubense*), Baayen & Gams (*Neth. J. Pl. Path.* **94**: 273, 1988; sect. *Elegans* and carnation wilt), Bosland (*Adv. Pl. Path.* **6**: 281, 1988; *F. oxysporum*), Boyer & Charest (*Can. J. Pl. Path.* **11**: 14, 1989; lectins and differentiation of *F. oxysporum* f.sp. *radicis-lycopersici* from f.sp. *lycopersici*), Anwah & Lorbeer (*Mycol.* **81**: 278, 1989; variability in *F. oxysporum* f.sp. *apii* race 2), Ploetz (Ed.) (*Fusarium wilt of Banana*, 1990), Moore *et al.* (*Aust. J. Bot.* **39**: 161, 1991; characterization of *F. oxysporum* f.sp. *cubense* by volatiles), Ploetz (*CJB* **68**: 1357, 1990; variation in *F. oxysporum* f.sp. *cubense*), Ruppel (*Pl. Dis.* **75**: 486, 1991; variation in *F. oxysporum* from sugar beet), Crowhurst *et al.* (*Curr. Gen.* **20**: 391, 1991; differentiation of *F. solani* f.sp. *cucurbitae* races 1 and 2 by random amplification of polymorphic DNA), Ouellet & Seifert (*Phytopath.* **83**: 1003, 1993; RAPD and PCR amplification in *F. graminearum*), Nelson *et al.* (*Ann. Rev. Phytopathol.* **31**: 233, 1993; fumonisins and *Fusarium*), Sangalang *et al.* (*MR* **99**: 287, 1995; systematics of *F. avenaceum*).

Fuscidea V. Wirth & Vězda (1972), Fuscideaceae (L). *c.* 20, mainly temp. See Inuque (*Hikobia, Suppl.* **1**: 161, 177, 1981), Oberhollenzer & Wirth (*Beih. Nova Hedw.* **79**: 537, 1984; 9 spp., *Stuttg. Beitr. Naturk.* A **376**, 1985).

Fuscideaceae Hafellner (1984), ? Teloschistales (L). 6 gen. (+ 3 syn.), 37 spp. Thallus crustose, areolate; ascomata blackish, lacking anthraquinone pigments, without a well-developed thalline margin; interascal tissue of sparsely branched paraphyses, the apices often pigmented; asci with a well-developed outer I+ gelatinous layer and usually a conspicuous inner I+ layer surrounding the often poorly defined ocular chamber; ascospores usually aseptate, sometimes with attenuated apices. Lichenized with green algae, widespr.

Fuscoboletinus Pomerl. & A.H. Sm. (1962), Strobilomycetaceae. 6, mostly N. Am. = Suillus (Bolet.) fide

Singer *et al.* (*Mycol.* **55**: 362, 1963), but see Pomerleau & Smith (*Brittonia* **14**: 156, 1962).

Fuscocerrena Ryvarden (1982), Coriolaceae. 1, USA, C. & S. Am.

Fuscoderma (D.J. Galloway & P.M. Jørg.) P.M. Jørg. & D.J. Galloway (1989), Pannariaceae (L). 3, S. temp.

Fuscolachnum J.H. Haines (1989), Hyaloscyphaceae. 6 (on leaves), Eur., N. Am.

Fuscopannaria P.M. Jørg. (1993), Pannariaceae (L). 20, mainly temp. See Jørgensen (*Op. bot.* **45**, 1978; Eur. spp., as *Pannaria*).

Fuscophialis B. Sutton (1977), Mitosporic fungi, 1.C2.16. 1, trop.

Fuscoporella Murrill (1907) = Phellinus (Hymenochaet.) fide Pegler (1971).

Fuscoporia Murrill (1907) = Phellinus (Hymenochaet.) fide Donk (1974).

Fuscoscypha Svrček (1987), Hyaloscyphaceae. 1, Eur.

fuscous, dusky; too brown for a grey (Corner, 1958).

fuseau, a fusoid macroconidium of a dermatophyte (e.g. *Microsporum*); a spindle (obsol.).

Fusella Sacc. (1886), Mitosporic fungi, 1.A2.?. 3, Eur., Afr.

Fusellites Babajan & Tasl. (1977), Fossil fungi. 1 (Tertiary), former USSR.

Fusichalara S. Hughes & Nag Raj (1973), Mitosporic fungi, 1.C2.15. 3, NZ.

Fusicladiella Höhn. (1919), Mitosporic fungi, 1.B1.1. 4, trop. See Deighton (*Mycol. Pap.* **101**, 1965).

Fusicladiites Babajan & Tasl. (1973), Fossil fungi. 1 (Tertiary), former USSR.

Fusicladina Arnaud (1952) nom. inval. Mitosporic fungi, 1.B1.?. 1, Eur.

Fusicladiopsis Maire (1907) = Stemphylium (Mitosp. fungi) fide Hughes (1958).

Fusicladiopsis Karak. & Vassiljevsky (1937) ≡ Karakulinia (Mitosp. fungi).

Fusicladium Bonord. (1851), Anamorphic Venturiaceae, 3.B2.10. Teleomorphs *Acantharia, Venturia*. 40, widespr. See Hughes (*CJB* **31**: 560, 1953), Ondřej (*Česká Myk.* **25**: 165, 1971), Sutton & Pascoe (*Austr. Syst. Bot.* **1**: 79, 1988; on *Parahebe*).

fusicoccin, a tricarboxylic terpene from *Fusicoccum amygdali* which induces stomatal opening (Turner & Granati, *Nature* **223**: 1070, 1969) and promotes spore germination and cell elongation. See also Chain *et al.* (*Physiol. Pl. Path.* **1**: 495, 1971; toxicity).

Fusicoccum Corda (1829), Anamorphic Botryosphaeriaceae, 8.A1.1. Teleomorph *Botryosphaeria*. 50, widespr. *F. putrefaciens* (cranberry (*Vaccinium*) end rot). See Morgan-Jones & White (*Mycotaxon* **30**: 117, 1987; redescr. anamorph of *Botryosphaeria ribis*).

Fusicolla Bonord. (1851), Mitosporic fungi, 3.A1.?. 5, widespr.

Fusicytospora Gutner (1934) = Phomopsis (Mitosp. fungi) fide Petrak (*Sydowia* **1**: 61, 1947).

fusidic acid, an antibacterial antibiotic from *Fusidium coccineum* (Godfredsen, *Nature* **193**: 987, 1962); = ramycin (from *Micromucor ramanniana*).

Fusidites Mesch. (1902), Fossil fungi. 1 (Oligocene), Eur.

Fusidium Link (1809) nom. rej. = Cylindrocarpon (Mitosp. fungi) fide Booth (1966).

Fusidomus Grove (1929) = Fusarium (Mitosp. fungi) fide Sutton (*Mycol. Pap.* **141**, 1977).

fusiform, spindle-like; narrowing toward the ends (Fig. 37.1).

fusiform rust (of pine), *Cronartium quercuum* (Czabator, *U.S. For. Serv. Res. Pap.* **65**, 1971).

Fusiformisporites Rouse (1962), Fossil fungi (*f. cat.*). 8 (Tertiary), India, N. Am.

Fusispora Fayod (1889) = Lepiota (Agaric.) fide Singer & Smith (1946).

Fusisporella Speg. (1910), Mitosporic fungi, 3.B1.?. 1 or 2, Am., trop.

Fusisporium Link (1809) = Fusarium (Mitosp. fungi) fide Wollenweber & Reinking (*Die Fusarien*: 5, 1935). See also Hughes (1958).

Fusitheca Bonord. (1864) ? Fungi (inc. sed.).

fusoid, somewhat fusiform.

Fusoma Corda (1837) nom. dub. (Mitosp. fungi) fide Hughes (1958).

Fusticeps J. Webster & R.A. Davey (1980), Mitosporic fungi, 1.C2.1. 1 (aquatic), Malaysia.

fuzz-ball, see puff-ball.

fuzzy coat, the outer gelatinous coat of an ascus, esp. of one staining blue in iodine; see ascus.

G, Conservatoire et Jardin Botanique de la Ville de Genève (Geneva, Switzerland); founded 1817; funded by the city.

Gabarnaudia Samson & W. Gams (1974), Mitosporic fungi, 1.A1.15. 4, Eur.

Gabura Adans. (1763) = Collema (Collemat.).

Gaertneriomyces D.J.S. Barr (1980), Spizellomycetaceae. 2 (from soil), Eur.

Gaeumannia Petr. (1950) = Melchioria (Trichosphaer.) fide Müller & v. Arx (1973).

Gaeumanniella Petr. (1952), Ascomycota (inc. sed.). 1, Florida.

Gaeumannomyces Arx & D.L. Olivier (1952), Magnaporthaceae. *c.* 5, widespr. *G. graminis* (with 3 vars.; syn. *Ophiobolus graminis, O. oryzinus*; whiteheads and take-all (q.v.) of cereals). See Chambers (*Phytopath. Z.* **72**: 169, 1971), Walker (*TBMS* **58**: 427, 1972, *Mycotaxon* **11**: 1, 1980; taxonomy), Weste (*TBMS* **59**: 133, 1972; root infection), Simonsen (*Friesia* **9**: 361, 1971; anamorph). Anamorph *Phialophora*.

Gaillardiella Pat. (1895), Nitschkiaceae. 1, S. Am. See Petrak (*Sydowia* **4**: 158, 1950).

Galactinia (Cooke) Boud. (1885) = Peziza (Peziz.) fide Seaver (1928).

Galactomyces Redhead & Malloch (1977), Dipodascaceae. 2, widespr. See de Hoog *et al.* (*Stud. Mycol.* **29**, 1986; key).

Galactopus Earle (1909) = Mycena (Tricholomat.) fide Singer (1951).

galeate, hooded; hat- or helmet-shaped.

Galeoscypha Svrček & J. Moravec (1989), Otideaceae. 1, Eur.

Galera (Fr.) P. Kumm. (1871) [non Blume (1825), Orchidaceae] = Galerina (Cortinar.).

Galeraicta Preuss (1852) ? = Rabenhorstia (Mitosp. fungi) fide Saccardo (*Syll. Fung.* **3**, 1884), but see Sutton (1977).

Galerella Earle (1909), Bolbitiaceae. 4, widespr.

Galerina Earle (1909), Cortinariaceae. 200 (freq. bryophilous), cosmop. See Kühner (*Le genre Galera*, 1935), Smith & Singer (*A monograph on the genus Galerina Earle*, 1964), Pegler & Young (*Kew Bull.* **27**: 483, 1972; basidiospore form), Wells & Kempton (*Lloydia* **32**: 369, 1969; Alaskan spp.), Kühner (*BSMF* **88**: 41, 119, 1973; alpine spp.), Barkmann (*Coolia* **14**: 49, 1969; key 50 Dutch spp.).

Galeromycena Velen. (1947) = Macrocystidia (Tricholomat.) fide Pegler (*in litt.*).

Galeropsidaceae Singer (1962), Agaricales. 5 gen. (+ 3 syn.), 19 spp. Gasterocarp subagaricoid, stipitate; gleba sublamellate; spores clay brown to rusty brown, smooth, thick-walled, usually with germ pore.

Galeropsina Velen. (1947) = Psilocybe (Strophar.) fide Singer (1986).

Galeropsis Velen. (1930), Galeropsidaceae. 12 (xerophytic), widespr. See Singer (*Proc. K. Ned. Akad. Wet.* C **66**: 106, 1963; key).

Galerula P. Karst. (1879) ? = Galerina (Cortinar.) fide Singer & Smith (1946).

Galiella Nannf. & Korf (1957), Sarcosomataceae. 4, N. temp. See Korf (*Mycol.* **49**: 107, 1957). = Sarcosoma (Sarcosomat.) fide Boedijn (*Persoonia* **1**: 7, 1959).

gall, a predictable and consistent plant deformation that occurs in response to the stimulus of a foreign organism, often a swelling or outgrowth; a cecidium.

Fungi producing galls directly on plants include *Exobasidiales* (Ing, *in* Williams (Ed.), 1994: 67; key 35 Eur. sp.) and *Uredinales* (Preece & Hicks, *in* Williams (Ed.), 1994: 57; 87 Br. spp.). The gall midges (*Cecidomyiidae*) have mycophagous larvae implicated in the gall reaction (Harris, *in* Williams (Ed.), 1994: 201); some are only mycophagous and do not cause galls; **ambrosia -s** are caused by *Macrophoma* and anamorphs of *Botryosphaeria* (Bisset & Borkent, *in* Pirozynski & Hawksworth (Eds), 1988: 203).

Lichens have been interpreted as galls induced by an alga in a fungus (Moreau & Moreau, *BSMF* **34**: 84, 1918; see also cephalodium) but there are significant differences (Hawksworth & Honegger, *in* Williams (Ed.), 1994: 77; **-s on lichens**, may be caused by some lichenicolous fungi (e.g. *Biatoropsis*, *Plectocarpon*, *Polycoccum*, *Thamnogalla*), nematodes (Siddiqi & Hawksworth, *Lichenologist* **14**: 175, 1982), mites (Gerson & Seaward, *in* Seaward (Ed.), *Lichen ecology*: 69, 1977), or be of unknown origin (Grummann, *Bot. Jb.* **80**: 101, 1960).

See Mani (*Ecology of plant galls*, 1964), Williams (Ed.) (*Plant galls: organisms, interactions, populations*, 1994), Coevolution.

Gallacea Lloyd (1905) = Hysterangium (Hysterang.) fide Cunningham (*Gast. Austr. N.Z.*: 65, 1944).

Gallowaya Arthur (1906), Uredinales (inc. sed.). 2, N. Am., Siberia. *G. crowellii* (microcyclic *Coleosporium* having 0, III on *Pinus*). = Coleosporium (Puccin.) fide Cummins & Hiratsuka (1983).

Galoperdon F.H. Wigg. (1780) = Lycogala (Myxom.).

Galorrheus (Fr.) Fr. (1825) [non Haw. (1812), *Euphorbiaceae*] ≡ Lactarius (Russul.).

Galzinia Bourdot (1922), Sistotremataceae. 10, widespr. See Rogers (*Mycol.* **36**: 70, 1944), Olive (*Mycol.* **46**: 794, 1954), Boidin & Gilles (*BSMF* **110**: 185, 1994; key).

Galziniella Parmasto (1968) = Sistotrema (Sistotremat.) fide Eriksson *et al.* (*Corticiaceae of Northern Europe* **7**, 1984).

Gambleola Massee (1898), = Pucciniosira (Pucciniosir.) fide Buriticá (*Rev. Acad. Colomb. Cienc.* **18**)(69): 131, 1991).

gametangium (pl. **-ia**) (gametange), cell containing gametes or gametic nuclei; the gametangium may initially be diploid and is the site of meiosis, or haploid. See zygangium.

gamete, naked uninucleate haploid cell with the sole function of fusing with another gamete to produce a zygote; sometimes used for the sex-nuclei of coenogametes.

gametogenesis, the development of gametes.

gametophyte, a haploid or sexual plant; haplont or haplophase. Cf. sporophyte.

gametothallus, a thallus producing gametes; cf. sporothallus.

gamma particle, a DNA-containing cytoplasmic organelle in the zoospore of *Blastocladiella emersonii* (Myers & Cantino, *The gamma particle*, 1974; Barstow & Lovett, *Mycol.* **67**: 518, 1975).

Gamolpidium Vlădescu (1892), ? Chytridiales (inc. sed.). 2, Romania.

Gamonaemella Fairm. (1922) = Gamospora (Mitosp. fungi) fide Clements & Shear (1931) but see Sutton (1977).

Gamosphaera Dumort. (1822) ≡ Nemania (Xylar.).

Gamospora Sacc. (1885) ? = Wiesneriomyces (Mitosp. fungi) fide Sutton (*Mycol. Pap.* **141**, 1977).

Gamosporella Speg. (1888) nom. dub. fide Sutton (*Mycol. Pap.* **141**, 1977).

Gampsonema Nag Raj (1975), Mitosporic fungi, 8.C1.10. 1, widespr.

Gamsia M. Morelet (1969), Mitosporic fungi, 1.A2.19. 1, Eur.

Gamundia Raithelh. (1979), Tricholomataceae. 1, S. Am.

gangliform, having knots; knotted.

Gangliophora Subram. (1992), Mitosporic fungi, 1.C2.19. 1, Taiwan.

Gangliophragma Subram. (1979), Mitosporic fungi, 1.C1.10. 2, USA.

gangliospore, Subramanian's (*Curr. Sci.* **31**: 410, 1962) term for aleuriospore in the sense of 'holoblastic conidium'.

Gangliostilbe Subram. & Vittal (1976), Mitosporic fungi, 2.C2.1. 1, India.

Ganoderma P. Karst. (1881), Ganodermataceae. 50, esp. trop. *G. applanatum* (wood decay); *G. lucidum* (wood decay); *G. philippii* (syn. *G. pseudoferreum*; root rot of cacao, coffee, rubber, tea, etc.).

Ganodermataceae (Donk) Donk (1948), Ganodermatales. 5 gen. (+ 6 syn.), 81 spp. Basidioma sessile to stipitate, coriaceous with waxy crust; spores brown, with ornamented exosporium plus myxosporium; lignicolous, causing decay of standing timber.

Lit.: Donk (1964: 264), Steyaert (*Persoonia* **7**: 55, 1972; SE Asian spp., *Bull. Jard. bot. Natn. Belg.* **50**: 135, 1980; subgen. classif.).

Ganodermatales, Basidiomycetes. 2 fam., 6 gen. (+ 6 syn.), 83 spp. Fams:
(1) **Ganodermataceae**.
(2) **Haddowiaceae**.

Garethjonesia K.D. Hyde (1992), Lasiosphaeriaceae. 1 (aquatic), Australia.

gari, a fermentation product of manioc, bacteria and *Geotrichum candidum* in west Africa.

Garnaudia Borowska (1977), Mitosporic fungi, 1.B2.1. 2, Cuba, Poland.

Garovaglia Trevis. (1853) = Polychidium (Placynth.).

Garovaglina Trevis. (1880) ≡ Garovaglia (Placynth.).

Gasparrinia Tornab. (1848) nom. rej. [non *Gasparinia* Bertol. (1839), *Umbelliferae*] = Caloplaca (Teloschist.).

Gassicurtia Fée (1824) nom. rej. = Buellia (Physc.) fide Aptroot (*Taxon* **36**: 474, 1987).

Gasterella Zeller & L.B. Walker (1935), Gasterellaceae. 1, USA.

Gasterellaceae Zeller (1948), Hymenogastrales. 1 gen., 1 spp. Gasterocarp minute, lacking a stipe, hollow; spores dark brown, thick-walled, verrucose.

Gasterellopsis Routien (1940), Podaxaceae. 1, USA.

Gasteroagaricoides D.A. Reid (1986), Coprinaceae. 1, NZ.

gasteroconidium, see gasterospore.

Gasterohymeniales, see *Hymenogastrales* and *Podaxales*.

Gasteromycetes. Class of *Basidiomycota*, traditionally based on holobasidiomycetes which do not actively discharge their spores and as such can be applied to 1169 spp. distributed among 164 gen. (+ 171 syn.), 56 fam. and 14 ord. The polyphyletic nature of the group makes it inappropriate to maintain in a natural classification and 382 spp. are here placed in the mostly hymenomycetous orders *Agaricales*, *Boletales*,

Bondarzewiales, Cortinariales, Russulales and *Stereales*. When very close to hymenomycetous genera they have been placed in the same family but often they are in entirely gasteromycetous families (*Chamonixiaceae, Cribbeaceae, Elasmomycetaceae, Galeropsidaceae, Hybogasteraceae, Hydnangiaceae, Limnoperdaceae, Podaxaceae, Rhizopogonaceae, Richonelliaceae, Secotiaceae, Stephanosporaceae, Torrendiaceae*. There are, however, 8 orders encompassing 30 fam., 106 gen. (+ 147 syn.) and 787 spp. for which affinity is either unknown or not sufficiently evident to be disposed otherwise and which may be considered the 'true' gasteromycetes: *Gautieriales, Hymenogastrales, Lycoperdales, Melanogastrales, Nidulariales, Phallales, Sclerodermatales, Tulostomatales*.

The single-celled basidium (q.v.) is generally 4-spored but the basidiospores are not forcibly discharged and sterigmata may be absent. The basidia and basidiospores mature within the basidioma, which typically has a peridium (sometimes multi-layered) covering a ± fleshy mycelial tissue ('gleba') in which the basidia are borne, eventually lining hymenial cavities. The gleba is frequently crossed by a columella and/or veins. After the decay of the gleba, the columella and capillitium (q.v.) may facilitate spore dispersal.

Gasteromycetes are mainly terrestrial or hypogeous, some lignicolous or coprophilous, some are saprobes and some are mycorrhizal. They are cosmop., esp. in dry warm areas.

Lit.: **General**: Dring (*in* Ainsworth *et al.* (Eds), *The Fungi* **4B**: 421, 1973; keys orders, fams., gen.), Fischer (*Nat. Pflanzenfam.* edn 2, **7A**, 1933; unless otherwise stated, synonymies follow this work), Ponce de Léon (*in* Parker, 1982, **1**: 256), Zeller (*Mycol.* **40**: 639, 1948; **41**: 36, 1949; keys orders, fams., gen.), Miller & Miller (*Gasteromycetes. Morphological and developmental features*, 1988).

Regional: **Africa, South**, Bottomley (*Bothalia* **4**: 473, 1948; keys), **East**, Demoulin & Dring (*Bull. Jard. bot. Belg.* **45**: 339, 1975), **West**, Dring (*Mycol. Pap.* **98**, 1964; keys). **Canary Islands**, Beltrán Tejera & Wildpret (*Vieraea* **7**: 49, 1977). **China**, Liu Bo (*The Gasteromycetes of China* [*Beih. Nova Hedw.* **76**], 1984) **Congo**, Dissing & Lange (*Fl. Icon. Champ. Congo* **12-13**, 1963-64; keys). **Zaïre**, Dissing & Lange (*Bull. Jard. bot. Etat. Brux.* **32**: 325, 1962). **America, North**, Burk (*A bibliography of North American gasteromycetes. I Phallales*, 1980), Coker & Couch (*The gasteromycetes of the eastern United States and Canada*, 1928 [reprint 1968]), Smith (*Puffballs and their allies in Michigan*, 1951); **South**, Wright (*Holmbergia* **5**: 45, 1956; key), Fries (*Ark. Bot.* **8** (11), 1909). **Argentina**, Wright (*Lilloa* **21**: 191, 1949). **Brazil**, Homrich (*Rev. Mycol.* **34**: 3, 1969; Rio Grande do Sul). **Mexico**, Herrera (*An. Inst. Biol. Univ. Mex.* **35**: 9, 1965; keys), Guzmán & Herrera (*An. Inst. Biol. Univ. nal. autón. Mex., Bot.* **40**, 1969). **Uruguay**, Lohwag & Swoboda (*Rev. Sudam. Bot.* **7**: 1, 1942). **West Indies**, Dennis (*Kew Bull.* **1953**: 307), Reid (*Kew Bull.* **31**: 657, 1977). **Asia, Iran**, Safez (*Iran. J. Plant Pathol.* **22**: 25, 1986). **Israel**, Dring & Rayss (*Israel J. Bot.* **12**: 147, 1964). **Middle East**, Eckblad (*Nytt Mag. Bot.* **17**: 129, 1970; *Iranian J. Bot.* **1**: 65, 1976). **Mongolia**, Kreisel (*Feddes Rep.* **86**: 321, 1975), Dörfelt & Bumžaa (*Nova Hedw.* **43**: 87, 1986). **Nepal**, Kreisel (*Khumbu Himal.* **6**: 25, 1969; *Feddes Rep.* **87**: 83, 1976). **Pakistan**, Ahmad (*Publ. Dept. Bot. Univ. Punjab* **11**, 1952). **Surinam**, Fischer (*Ann. Myc.* **31**: 113, 1933). **Thailand**, Ellingsen (*Nordic Jl*

Bot. **2**: 283, 1982), Phanichapol (*Thai For. Bull.* **16**: 233, 1986). **Former USSR**, Sosin (*Opredelitel' Gasteromitsetov SSSR*, 1973), Järva (*Folia crypt. Eston.* **2**: 15, 1973; Estonia), Shvartsman & Filimonova (*Fl. Spor. Rast. Kazakhstana* **6**, 1970; Kazakhstan). **Australasia**, Cunningham (*The Gasteromycetes of Australia and New Zealand*, 1944 [reprint 1979]). **Europe**, **Belgium**, Demoulin (*Les gastéromycètes Belgique*, edn 2, 1975; *Bull. Jard. Bot. nat. Belg.* **38**: 1, 1968). **British Isles**, Palmer (*Nova Hedw.* **25**: 65, 1968; bibliogr.), Demoulin & Marriott (*Bull. BMS* **15**: 37, 1981; key), Pegler *et al.* (*Puffballs, Earthstars & Stinkhorns*, 1995). **Former Czechoslovakia**, Pilát (Ed.) (*Fl. CSR, B*, **1**, *Gasteromycetes*, 1958). **Germany**, Gross *et al.* (*Beih. Z. Mykol.* **2**, 1980). **Greenland**, Lange (*Meddr Grønl.* **147**(4), 1948). **Hungary**, Hollós (*Die Gasteromyceten Ungarns*, 1904). **Italy**, Petri (*Fl. Ital. Crypt.* **1**(5), 1909). **Netherlands**, Mass Geesteranus (*Coolia* **15**: 49, 1971). **Norway**, Eckblad (*Nytt Mag. Bot.* **4**: 19, 1955; *Astarte* **4**: 7, 1971, Finland). **Portugal**, Almeida (*Port. Acta Biol.* B **11**: 205, 1972). **Spain**, Calonge & Demoulin (*BSMF* **91**: 247, 1975). See also *Lit.* under Macromycetes.

gasterospore (**gasteroconidium**), a thick-walled, globose, chlamydospore of *Ganoderma*; probably apomictic (see Bose, *Mycol.* **25**: 432, 1933; *Sydowia, Beih.* **1**: 176, 1957).

Gastroboletus Lohwag (1926), Xerocomaceae. 10, widespr. See Thiers (*Mem. N.Y. bot. Gdn* **49**: 355, 1989).

Gastrocybe Watling (1968), Galeropsidaceae. 1, USA. Galeropsis (Galeropsid.) fide Moreno *et al.* (*Mycotaxon* **36**: 63, 1989).

Gastroleccinum Thiers (1989), Boletaceae. 1, N. Am.

Gastromyces R. Ludw. (1861), Fossil fungi (? Basidiomycetes). 1 (Carboniferous), former USSR.

Gastropila Homrich & J.E. Wright (1973), Lycoperdaceae. 4, Am. See Ponce de Leon (*Phytologia* **33**: 455, 1976). ? = Calvatia (Lycoperd.).

Gastrosporiaceae Pilát (1934), Hymenogastrales. 1 gen. (+ 1 syn.), 1 sp. Gasteroid; spores globose, warted; xerothermic areas of esp. Europe associated with *Gramineae*.

Gastrosporium Mattir. (1903), Gastrosporiaceae. 1, Eur., India, N. Am. See Pilát (*BSMF* **50**: 37, 1934), Monthoux & Röllin (*Candollea* **31**: 119, 1976).

Gastrosuillus Thiers (1989), Boletaceae. 4, N. Am. See Baura *et al.* (*Mycol.* **84**: 592, 1992; sp. as recent mutant from *Suillus grevillei*).

Gaubaea Petr. (1942), Mitosporic fungi, 8.A2.19. 1, Turkmenia.

Gäumann (Ernest Albert; 1893-1963). Professor of botany, Institut für Spezielle Botanik, Eidg. Technische Hochschule, Zürich, Switzerland, 1927-63; mycologist and plant pathologist. *Vergleichende Morphologie der Pilze* (1926), *Die Pilze* (1949, 1964), *Pflanzliche Infektionslehre* (1946, 1951), *Die Rostpilze Mitteleuropas* (1959). See Zobrist *et al.* (*Herrn Professor Dr. Ernst Gäumann zum siebzigsten Geburtstag...*, 1963). *Obits.*: Gardner & Kern (*Mycol.* **57**: 1, 1965; portr.), Landholt (*Verh. Schweiz. Naturforsch. Ges.* **1963**: 194; portr., bibl.). See Stafleu & Cowan (*TL-2* **1**: 903, 1976).

Gausapia Fr. (1825) nom. rej. = Septobasidium (Septobasid.).

Gauthieromyces Lichtw. (1983), Legeriomycetaceae. 1, France. See Lichtwardt (1986).

Gautieria Vittad. (1831) nom. cons., Gautieriaceae. 15, widespr. See Rauschert (*Hercynia* N.F. **12**: 217, 1975), Beaton *et al.* (*Kew Bull.* **40**: 193, 1985; key Austral. spp.).

Gautieriaceae Zeller (1948), Gauteriariales. 3 gen (+ 3 syn.), 22 spp. Gasterocarp hypogeous; gleba labyrinthoid, cavities of independent origin; spores yellowish-brown, ridged. Relationship to agarics or bolets probable but unclear.

Gautieriales, Basidiomycetes ('gasteromycetes'). 1 fam., 3 gen. (+ 3 syn.), 22 spp. Gasterocarp hypogeous, peridium often lacking at maturity; gleba with cavities of independent origin lined by an hymenium, gleba cartilaginous at maturity and traversed by a branched columella; spores large, ellipsoid, ribbed; saprobes or possibly mycorrhizal; widespr. (mostly temp. regions). Fam.:
Gautieriaceae.
   Lit.: see under *Gasteromycetes*. See also hypogeous fungi.

GC ratio; **GC content**; see base ratio, molecular biology.

Geaster P. Micheli ex Fr. (1829) ≡ Geastrum (Geastr.). See Demoulin (*Taxon* **36**: 498, 1984).

Geasterites Pia (1927), Fossil fungi. 1 (Miocene), USA. But see Tiffney (*TBMS* **76**: 493, 1981).

Geasteroides Long (1917), Geastraceae. 1, USA. See Long (*Mycol.* **37**: 601, 1945).

Geasteropsis Hollós (1903) = Geastrum (Geastr.) fide Ponce de Leon (*Fieldiana, Bot.* **31**: 311, 1968).

Geastraceae Corda (1842), Lycoperdales. 6 gen. (+ 11 syn.), 59 spp. Gasterocarp opening with stellate rays of peridium, exoperidium three-layered; capillitium unbranched; gleba powdery; spores small, warted.
   Lit.: Boiffard (*Docums. mycol.* **6**(24): 1, 1976; France), Ponce de Leon (*Fieldiana Bot.* **31**: 303, 1968; world monogr., broad spp. concept), Sunhede (*Geastraceae (Basidiomycotina)* [*Synopsis Fungorum* 1], 1989; monogr. esp. N. Eur.), Dörfelt (*Die Erdsteine Geastraceae und Asteraceae*, 1985; centr. Eur.).

Geastrum Pers. (1801), Geastraceae. 50, widespr. The earth stars. See Sunhede (*Synopsis Fung.* **1**, 1990).

Geastrumia Bat. (1960), Mitosporic fungi, 7.G1.1. 1, Dominican Republic, Afr. See Pirozynski (*Mycol.* **63**: 897, 1971).

Geisleria Nitschke (1861) ? = Strigula (Strigul.) fide Harris (*in litt.*). See Ernst (*Herzogia* **9**: 321, 1993).

Geisleriomyces Cif. & Tomas. (1953) ≡ Geisleria (Strigul.).

Geissodea Vent. (1799) = Cetraria (Parmel.), Parmelia (Parmel.) p.p., Physcia (Physc.) p.p., and Xanthoria (Teloschist.) p.p.

gel tissue, a mixture of gel and hyphae found in members of the *Leotiales* and *Tremellales*; the gel may arise either by direct secretion or by disintegration of hyphae: see Moore (*Mycol.* **57**: 114, 1965; *Am. J. Bot.* **52**: 389, 1965, ontogenesis; *Stain Technol.* **40**: 23, 1965, staining). Cf. gliatope.

Gelasinospora Dowding (1933), Sordariaceae. 20, widespr. See v. Arx (*Persoonia* **11**: 443, 1982; key), Cailleux (*BSMF* **87**: 536, 1972; key), Dowding & Bakerspigel (*Can. J. Microbiol.* **1**: 68, **4**: 295, **5**: 125, 1954-59; nuclear behaviour).

Gelasinosporites P. Briot, Lar.-Coll. & Locq. (1983), Fossil fungi. 1, Australia.

gelatin, product obtained by boiling collagen, soluble in water above *c.* 40°C. Gels of *c.* 4-12% used to test ability of some microorganisms to liquefy or hydrolyse gelatin.

Gelatina Raf. (1806) = Tremella (Tremell.). See Merrill (*Index Rafin.*: 68, 1949).

Gelatinaria Raf. (1815) nom. dub. (Fungi, inc. sed.). No spp. included.

Gelatinaria Flörke ex Wallr. (1831) nom. inval. ? = Nostoc (Algae).

Gelatinocrinis Matsush. (1995), Mitosporic fungi, 3.A1.15. 1, Japan.

Gelatinodiscaceae, see *Leotiaceae*.

Gelatinodiscus Kanouse & A.H. Sm. (1940), Leotiaceae. 1, USA. See Carpenter (*Mycotaxon* **3**: 209, 1976).

Gelatinopsis Rambold & Triebel (1990), Leotiaceae. 2 (on lichens and other fungi), Eur., N. Am.

Gelatinopycnis Dyko & B. Sutton (1979), Mitosporic fungi, 8.E1.15. 1, Germany.

Gelatinosporis Clem. & Shear (1931) ≡ Gelatinosporium (Mitosp. fungi).

Gelatinosporium Peck (1873), Mitosporic fungi, 8.B1.15. 1, USA. See DiCosmo (*CJB* **56**: 1665, 1978).

gelatinous, jelly-like; used for the hyphae of tissues which become partly dissolved and glutinous with moisture.

Gelatoporia Niemalä (1985), Coriolaceae. 1, N. temp.

Gelatosphaera Bat. & H. Maia (1959) = Rhizosphaera (Mitosp. fungi) fide Sutton (1977).

Geleenites Dijkstra (1949), Fossil fungi. 1 (Cretaceous), Canada, Netherlands. ? = Ascotricha (Xylar.); see Jansonius *et al.* (*Pollen et Spores* **23**: 557, 1981).

Gelimycetes. Class within the *Orthomycotina* (q.v.) including *Auriculariales* p.p., *Dacrymycetales*, *Tremellales* p.p. and *Tulasnellales* (Cavalier-Smith *in* Rayner *et al.* (Eds), *Evolutionary biology of the fungi*: 339, 1987).

Gelineostroma H.J. Swart (1988), Rhytismatales (inc. sed.). 2, Austr.

Gelona Adans. (1763) = Pleurotus (Lentin.) fide Donk (1962).

Gelopellaceae Zeller (1939), Phallales. 1 gen., 4 spp. Gasterocarp hypogeous, indehiscent, gelatinous layer continuous; columella unbranched; foetid odour.

Gelopellis Zeller (1939), Gelopellaceae. 4, S. Am., Japan. See Homrich (*Rev. mycol.* **34**: 6, 1969).

Geltingia Alstrup & D. Hawksw. (1990), Odontotremataceae. 3 (on lichens), mainly arctic.

Geminella J. Schröt. (1870) [non Turpin (1828), *Algae*] = Schroeteria (Mitosp. fungi).

Geminispora Pat. (1893), ? Phyllachoraceae. 2, Afr., S. Am. See Cannon (*Mycol. Pap.* **163**, 1991).

Geminoarcus K. Ando (1993), Mitosporic fungi, ?.?.?. 1, Japan.

gemma (pl. -ae), (1) an asexual propagule borne singly or in chains at the ends of hyphae, referred to in older literature as a chlamydospore (*Saprolegniaceae*); (2) another term for oidia in *Basidiomycetes* (Gäumann, *The Fungi*: 449, 1928), rejected for this usage by Kendrick & Watling (*in* Kendrick (Ed.), *The whole fungus* **2**: 477, 1979).

Gemmamyces Casagr. (1969) = Cucurbitaria (Cucurbitar.) fide Hawksworth (*in litt.*). Used by Yuan & Wang (*Mycotaxon* **53**: 371, 1995). See Petrak (*Sydowia* **23**: 265, 1970).

Gemmophora Schkorb. (1912) nom. dub. (Mitosp. fungi). Probably based on chlamydospores and sterile mycelium.

Gemmularia Raf. ex Steud. (1824), ? Pezizales (inc. sed.). 1, ? N.Am.

Genabea Tul. & C. Tul. (1844), Otideaceae. 5 (hypogeous), Eur., N. Am. See Gilkey (*N. Am. Fl.* **2**(1), 1954; key), Fischer (1938; key). = Genea (Otid.) fide Zhang (*MR* **95**: 986, 1991), Pegler *et al.* (1993; key Br. spp.).

gene probes, see Molecular biology: DNA fingerprinting.

Genea Vittad. (1831), Otideaceae. *c.* 24 (hypogeous), N. Am., Eur. See Fischer (1938), Gilkey (1954), Lazzara & Montecchi (*Revista Micol.* **34**: 44, 1991; 5 spp., Italy), Pfister (*Mycol.* **76**: 170, 1984; posn), Zhang (*MR* **95**: 986, 1991; concept, key 3 spp. China, *SA* **11**: 31, 1992; nuclei).

Geneaceae, see *Otideaceae*.

gene-for-gene (in host-parasite relationships), the correspondence for each gene determining resistance in the host of a specific and related gene determining virulence in the pathogen; first described by Flor in 1955 for *Melampsora lini* on flax (*Linum usitatissimum*), which has 29 resistance genes each of which the pathogen has avirulent counterparts (Lawrence *et al.*, *Phytopath.* **71**: 12, 1981). See Person (*CJB* **37**: 1101, 1959), Person *et al.* (*Nature* **194**: 561, 1962), Flor (*Ann. Rev. Phytopath.* **9**: 275, 1971), Parlevliet (*in* Pirozynski & Hawksworth (Eds), 1988: 19), Coevolution.

Geneosperma Rifai (1968) = Scutellinia (Otid.) fide Schumacher (1990). Accepted by Zhuang & Korf (*Acta mycol. Sin.* Suppl. **1**, 1986).

genera, see genus.

generative hyphae, see hyphal analysis.

genestasis, inhibition of sporulation; **genistat**, a substance preventing or reducing sporulation in fungi without materially affecting vegetative growth; 'antisporulator' (Horsfall, 1947); cf. fungistatic.

Genetic engineering. The insertion or removal of inheritable genetic material from an organism so that its properties are transformed; genes from fungi can also be inserted into plasmid genomes and expressed through *Escherichia coli* or *Saccharomyces cerevisiae*. See Bennett & Lasure (Eds) (*Gene manipulations in fungi*, 1985; *More gene manipulation in fungi*, 1991), Fincham (*Microbiol. Rev.* **53**: 148, 1989; review), Kinghorn & Turner (Eds) (*Applied molecular genetics of filamentous fungi*, 1992), Biotechnology.

Genetic resource collections. The genetic resource collections including fungal cultures are catalogued in Takishima *et al.* (Eds) (*Guide to World Data Center on Microorganisms with a list of culture collections worldwide*, 1989), and Sugawara *et al.* (Eds) (*World directory of collections of cultures of microorganisms*, edn 4, 1993). Together these collections maintained some 350,000 strains in 1993; those held in 1990 represented around 11,500 species (Hawksworth, *MR* **95**: 641, 1991). Collections are given standard acronyms or abbreviations by the World Data Center, and a selection of those for the larger collections are included in this edition of the *Dictionary*. See also Reference collections.

International collaboration is facilitated by meetings of the International Congress of Culture Collections held under the auspices of the World Federation for Culture Collections (WFCC; see Societies and organizations).

Culture **maintenance** is traditionally on slopes of appropriate media stored at laboratory or low temperature but now cryopreservation (see below) when applicable is the method of choice. Lyophilization (freeze-drying) of the fungus spore is an ideal method to facilitate distribution; this method is favoured by service collections. The properties of the fungus in culture may be unstable through loss of plasmids, spontaneous mutations or genetic recombination due to the presence of heterokaryons, the parasexual cycle or normal sexual events. Therefore conditions of storage should be selected that minimize the risk of such changes. Freezing and storage of fungi at ultra-low temperature such as in or above liquid nitrogen or in -140° to -150°C freezers appears to provide the ideal method. There are, however, several other methods that are used successfully. They range from continuous growth through methods that reduce rates of metabolism to the ideal situation where metabolism is halted, or to a level that for all practical purposes it can be treated as suspended. Microbial resource collections, large or small, set out to maintain organisms in a

pure, viable and stable condition to make them available for future use. The method selected may depend upon the requirements of the use of the organism. These vary according to the numbers and range of fungi to be preserved and the facilities available. The cost of materials and labour involved and the desired level of stability and longevity required is also taken into consideration. Preservation methods that allow growth or metabolism can only be used for short-term storage, such methods are subculture, storage under a layer of mineral oil and storage in sterile water. Drying and freeze-drying techniques can be used for long-term storage of fungi but not all will survive. Storage in soil or in or above silica gel produce dry conditions that can allow the desiccated fungus to survive for 8 and up to 20 years. Freeze-drying, the removal of water, dehydration of fungi under reduced pressure by the sublimation of ice is a method widely used but generally only allows the fungus spore or other robust structures, such as sclerotia, to survive. The methods for the above preservation technique have been published widely.

Lit.: **General**, Hawksworth & Kirsop (Eds) (*Living resources for biotechnology: filamentous fungi*, 1988), Kirsop & Kurtzman (Eds) (*Living resources for biotechnology: yeasts*, 1988), Kirsop & Doyle (Eds) (*Maintenance of micro-organisms and cultured cells: A manual of laboratory methods*, 1991), Smith & Onions (*The preservation and maintenance of living fungi*, edn 2, 1994).

Cryopreservation by frozen storage, Gulya *et al.* (*MR* **97**: 240, 1993), Holden & Smith (*MR* **96**: 473, 1992), Ito (*Inst. Fermen. Osaka Res. Commun.* **15**: 119, 1991), Morris *et al.* (*Cryobiology* **25**: 471, 1988; *J. Gen. Microbiol.* **134**: 2897, 1988), Pearson *et al.* (*Cryoletters* **11**: 205, 1990), Roquebert & Bury (*Wrld J. Myc. Res.* **9**: 651, 1992), Smith (*in* Chang *et al.* (Eds), *Genetics and breeding of edible mushrooms*, 1993; *in* Jennings (Ed.), *Tolerance of fungi*, 1993).

Cryopreservation over liquid nitrogen, Smith (*in* Jennings (Ed.), *Tolerance of fungi*: 145, 1993).

Lyophilization (freeze drying), Kolkowski & Smith (*in* Day & McLellan (Eds), *Methods Molec. Biol.* **38**: 49, 1995; protocols), Tan *et al.* (*Mycol.* **83**: 654, 1991; *Mycol.* **86**: 281, 1994).

Mineral oil, Little & Gordon (*Mycol.* **59**: 733, 1967; refs.), Fennell (*Bot. Rev.* **26**: 1, 1960; review).

Anhydrous silica gel, Perkins (*Can. J. Microbiol.* **8**: 591, 1962), Gentles & Scott (*Sabouraudia* **17**: 415, 1979).

Soil, Bakerspigel (*Mycol.* **45**: 596, 1953).

Water, Castellani (*Mycopath.* **20**: 1, 1963).

Comparisons of methods, Onions (*in* Booth, *Methods in microbiology* **4**: 113, 1971), Smith & Onions (*TBMS* **81**: 535, 1983; mineral oil), Smith & Onions (1994), Smith (*TBMS* **79**: 415, 1982; 3,000 fungi preserved up to 13 years), *IMI Culture Collection Catalogue* 10th edn, 1992), Berry & Hennebert (*Mycol.* **83**: 605, 1992; freeze drying).

See also Media, Mites, and Safety.

Genetics. In general, fungi appear to conform to the well-established patterns of genetical behaviour typical of other groups of organisms and species of *Saccharomyces, Neurospora*, and *Aspergillus* have been widely used in studies on formal genetics. They mostly exhibit haplo- and diplophases in their life-history, and thus undergo meiosis at some stage.

The use of monosporous cultures, and esp. of methods for the isolation of all the spores from one ascus or basidium, has made possible such detailed studies as the transmission of various mycelial characters and segregation of sex. Investigation of the stage of meiosis at which the allelomorphs segregate has been rendered possible in some ascomycetes (e.g. *Neurospora*) because the eventual products, the ascospores, exhibit a linear succession in the ascus, and the frequencies of genetical crossing-over for various characters have been deduced from such meiocyte analyses.

Incompatability mechanisms involving one (*Mucor*) or multiple genes (*Coprinus*) are known (see heterothallism). Mutants induced by X-rays, ultraviolet light, etc., have been shown to be deficient in their capacity to effect particular stages of protein synthesis (*Neurospora*) and fermentation (yeasts).

Lit.: Kniep (*Bibliogr. genet.*, 1929), Fincham, Day & Radford (*Fungal genetics*, edn 4, 1979), Esser & Kuenen (*Genetic der Pilze*, 1965), Sermonti (*Genetics of antibiotic-producing organisms*, 1969), Day (*Genetics of host-parasite interactions*, 1974), Burnett (*Mycogenetics*, 1975), Ullrich & Raper (*Taxon* **26**: 169, 1977; evol. genetic mechanisms, 80 refs), Koltin *et al.* (*Bact. Rev.* **36**: 156, 1972; evol. incomp. factors in higher fungi), Day & Jellis (Eds) (*Genetics and plant pathogens*, 1987), Sidhu (Ed.) (*Genetics of plant pathogenic fungi* [*Adv. Pl. Path.* **6**], 1988), Clutterbuck (*in* Gow & Gadd (Eds), *The growing fungus*: 239, 1995). See also Genetic engineering, Parasexual cycle, Sex, Variation.

Genicularia Rifai & R.C. Cooke (1966) [non Rouss. (1806), *Algae*] ≡ Geniculifera (Mitosp. fungi).

geniculate, bent like a knee.

Geniculifera Rifai (1975), Mitosporic fungi, 1.C1.10. 7, widespr. = Arthrobotrys (Mitosp. fungi) fide Schenck *et al.* (*CJB* **55**: 977, 1977), but see van Oorschot (*Stud. Mycol.* **26**: 61, 1985).

Geniculodendron G.A. Salt (1974), Anamorphic Otideaceae, 1.A1.10. Teleomorph *Caloscypha*. 1, Canada, UK. See Paden *et al.* (*CJB* **56**: 2375, 1978).

Geniculospora Sv. Nilsson ex Marvanová & Sv. Nilsson (1971), Mitosporic fungi, 1.G1.1. 2 (aquatic), widespr. See Nolan (*Mycol.* **64**: 1173, 1972). = Articulospora (Mitosp. fungi) fide Ingold (*Guide to aquatic hyphomycetes*, 1975).

Geniculosporium Chesters & Greenh. (1964), Anamorphic Xylariaceae, 1.A1.11. Teleomorph *Nemania*. 1, widespr.

Geniopila Marvanová & Descals (1985), Mitosporic fungi, 1.C1.15. 1 (aquatic), former Czechoslovakia, UK.

genistat, see genestasis.

Genistella L. Léger & M. Gauthier (1932) [non Ortega (1773), *Papilionaceae*] ≡ Legeriomyces (Legeriomycet.).

Genistellaceae, see *Legeriomycetaceae*.

Genistelloides S.W. Peterson, Lichtw. & B.W. Horn (1981), Legeriomycetaceae. 2 (in *Plecoptera*), USA. See Lichtwardt (1986), Williams & Lichtwardt (*Mycol.* **79**: 473, 1987).

Genistellospora Lichtw. (1972), Legeriomycetaceae. 1 (in *Diptera*), USA, Eur. See Lichtwardt (*Mycol.* **64**: 167, 1972), Moss & Lichtwardt (*CJB* **54**: 2346, 1976; **55**: 3099, 1977; ultrastr.).

genocentric, see reproductocentric.

genome, the total inheritable genetic material of an organims; a haploid set of chromosomes in eukaryotes. In *Saccharomyces cerevisiae* this is 12 Mgbases coding for some 6,000 genes; in *Neurospora crassa* the genome is much longer, 40 Mgbases; and *Aspergillus nidulans* has some 13,000 genes. See also Chromosome number.

genotype, the sum of the genetic potential of an organism; in some fungi only part of this is expressed at any given time. See holomorph, teleomorph, anamorph.

genus (pl. **genera**; adj. **generic**), (1) (in taxonomy), one of the principal ranks in the nomenclatural hierarchy (see Classification), the name of which forms the first part of a binomial species name (see species); (2) (more generally), a class of objects or concepts.

As in the case of the species (q.v.), there is no universally applicable criteria by which genera are distinguished, but in general the emphasis is now on there being several discontinuities in fundamental characters, especially concerning the nature of the reproductive structures. In the last century, however, features such as spore colour and septation were accorded a predominant role by some workers (see Mitosporic fungi).

Lit.: Clemençon (Ed.) (*Mycol. Helv.* **6**, 1993; esp. in macromycetes), Hale (*Beih. Nova Hedw.* **79**: 11, 1984; in lichens), Hawksworth (1974), Poelt (*in* Hawksworth (Ed.), *Frontiers in mycology*: 85, 1991; in lichens).

Geocoryne Korf (1978), Leotiaceae. 2, Canary Isl., India.

Geodina Denison (1965), Sarcoscyphaceae. 1, Costa Rica.

geofungi, soil fungi (Cooke, 1963).

Geoglossaceae Corda (1838), Leotiales. 16 gen. (+ 11 syn.), 73 spp. Stromata absent; ascomata clavate, spathulate or strongly stipitate, the hymenium usually not clearly separable from the stalk, usually dark; interascal tissue of paraphyses, often strongly pigmented, often with complex branching and swollen apices; the hymenium sometimes setose; asci with an I+ or I- pore; ascospores often elongated, brown, transversely septate. Anamorphs unknown. Saprobic on soil or rotten vegetation, esp. in grassland, widespr.

Lit.: Durand (*Ann. Myc.* **6**: 387, 1908), Imai (*J. Fac. Agric. Hokkaido Univ.* **45**: 155, 1941), Mains (*Mycol.* **46**: 586, 1954, brown-spored spp.; **47**: 846, 1955, hyaline-spored spp.; **48**: 694, 1956, N. Am. Cudonieae), Maas Geesteranus (*Persoonia* **4**: 47, 1965; keys Indian spp.), Benkert (*Gleditschia* **10**: 141, 1983; white-spored, Eur.), Verkley (*Persoonia* **15**: 405, 1994; asci).

Geoglossum Pers. (1794), Geoglossaceae. *c.* 20, temp. See Mains (*Mycol.* **46**: 586, 1954; N. Am.), Benkert (*Mykol. Mitt.* **20**: 47, 1976; Germany), Maas Geesteranus (*Persoonia* **3**: 89, 1964), Raitviir (*in* Parmasto, *Zhivaya priroda Dal'nego Vostoka*: 52, 1971; E. former USSR), Nannfeldt (1942; key, Eur. spp.), Olsen (*Agarica* **7**: 120, 1986; key 22 spp. Norway), Spooner (*Bibl. Mycol.* **116**, 1987; 9 spp. Australasia, nomencl.).

Geographical distribution. Knowledge of the geographical distribution of most fungi is inadequate. It is, however, possible to make a few generalizations. Some fungi are extremely widespread compared with plants, and this is especially true for some lichenized ascomycetes, myxomycetes, polypores, opportunistic moulds, and soil fungi. However, the majority of fungi are related to particular plant hosts and can be expected to have distributions within their ranges. *Uredinales* (Bisby, 1933) and *Erysiphales* (Hirata, 1966) geographical distribution has been stated to conform to Willis's 'age and area' theory which postulates that the commonest genera are the oldest, and that the older the genus the wider its distribution.

Some lichenized species have exceptionally broad distributions, and wide disjunctions are known; many follow particular forest zones or rock types, and some are even bipolar (Du Rietz, *Acta phytogeogr. succ.* **13**: 215, 1940; Lynge, *Naturen* **12**: 367, 1941). The wide distribution of some lichens has been related to patterns of continental drift (Sheard, *Bryologist* **80**: 100, 1977; Kärnefelt, *Bibl. Lich.* **38**: 291, 1990; Tehler,

Opera Bot. **70**, 1988) or glaciations (Brodo & Hawksworth, *Opera Bot.* **42**, 1977); isolated populations of various species existed by the Cretaceous (Kärnefelt, 1990). Small-spored genera are more likely to be distributed by long-range dispersal than large-spored ones in *Caliciales* (Tibell, *Bot. J. Linn. Soc.* **116**: 159, 1994).

Saprobic fungi have a wider distribution than parasitic ones; and many common air-borne and soil moulds are ubiquitous. The thesis that most fungi are everywhere and that the environment selects (Gams, 1993) is probably true only for non-host restricted species. The key factor determining fungal distribution is from the substratum, whether it be a particular host plant or animal or some material able to support their growth. Almost every ecological niche has a specialized mycobiota (see Ecology).

Because of the importance of the geographical distribution of plant pathogenic fungi for the design of regulations for disease control and the international movement of plant germ plasm the distribution of the more important plant pathogenic fungi is comparatively well known. The *IMI Distribution Maps of Plant Diseases* (1943 on) cover 663 spp. and are being constantly updated; new editions issued as required.

There has been a renewed interest in mapping distributions in the last two decades, facilitated by advances in computer techniques. However, most of these are at a national level, and concentrate on species that are most easily recorded (e.g. lichens). The European Mycological Congress has a project to map certain larger fungi for Europe. Kreisel has compiled a bibliography of published distribution maps for non-lichenized fungi (*Feddes Rep.* **82**: 589, 1971, hymeno- and gasteromycetes, 1930-69; **83**: 741, 1973, hemi- and phragmobasidiomycetes; **84**: 619, 1973, basid. suppl.; **85**: 161, 1974, mitosporic fungi and endomycetes, 1941-72; **86**: 329, 1975, phycom.; **87**: 109, 1976, suppl.), and Hawksworth & Ahti (*Lichenologist* **22**: 1, 1990) include publications with maps.

Distribution within many countries has, however, also been affected by acid rain, air pollution (q.v.), forest clearance, and accidental and deliberate disposal by humans.

Lit.: Diehl (*J. Wash. Acad. Sci.* **27**: 244, 1937), Bisby (*Am. J. Bot.* **20**: 246, 1933; *Bot. Rev.* **9**: 466, 1943), Gams (*in* Winterhoff (Ed.), *Fungi in vegetation science*: 183, 1992), Hirata (*Host range and geographical distributions of the powdery mildews*, 1966), Pirozynski (*in* Ainsworth & Sussman (Eds), *The Fungi*, **3**: 487, 1968), Pirozynski & Weresub (*in* Kendrick (Ed.), *The whole fungus* **1**: 93, 1979), Pirozynski & Walker (*Aust. J. Bot.* Suppl. **10**, 1983; Pacific).

Major regional studies are cited in this *Dictionary* under systematic entries and also Discomycetes, Inventories, Lichens, Literature, Macromycetes, Plant pathogenic fungi, and Pyrenomycetes. See also Biodiversity, Numbers of fungi, and particular substrata.

Geolegnia Coker (1925), Saprolegniaceae. 2, N. Am., Eur.

Geomorium Speg. (1922) = Helvella (Helvell.). See Gamundí (*Darwiniana* **11**: 418, 1957).

Geomyces Traaen (1914), Mitosporic fungi, 1.A2.1. 4, N. temp., Australia. See Sigler & Carmichael (*Mycotaxon* **4**: 376, 1976), v. Oorschot (*Stud. Mycol.* **20**, 1980), Hocking & Pitt (*Mycol.* **80**: 82, 1988).

Geopetalum Pat. (1887), Lentinaceae. 1, Eur.

Geopetalum Singer (1951) ≡ Geopetalum Pat. (Lentin.).

Geophila Quél. (1886) [non D. Don (1825), *Rubiaceae*] = Stropharia (Strophar.) fide Singer & Smith (1946).

geophilous, earth loving, e.g. of fungi having underground fruit bodies. Cf. terricolous.

Geopora Harkn. (1885), Otideaceae. 12, N. Hemisph. See Burdsall (*Mycol.* **60**: 504, 1968), Senn-Irlet (*Beitr. Kenntnis Pilze Mittel.* **5**: 191, 1989; key 8 spp.).

Geoporella Soehner (1951) = Hydnotrya (Helvell.) fide Trappe (1975).

Geopyxis (Pers.) Sacc. (1889), ? Otideaceae. 6, Eur., Australia, N. Am. See Garnoeidner *et al.* (*Z. Mykol.* **52**: 201, 1991), Kimbrough & Gibson (*CJB* **68**: 342, 1990; ultrastr.).

Georgefischeria Thirum. & Naras. (1963), Tilletiaceae. 2 (witches' broom of *Rivea*, *Argyreia*), India.

Geoscypha (Cooke) Lambotte (1888) = Peziza (Peziz.) fide Eckblad (1968).

Geosiphon F. Wettst. (1915), Zygomycota (inc. sed.). 1 (? arbuscular mycorrhizal), Eur. Sometimes considered lichenized as it includes cyanobacteria (*Nostoc*) in vesicles; the only known example of endocyanosis (q.v.) in fungi. See Mollenhauer (*in* Reisser (Ed.), *Algae and symbiosis*: 339, 1992), Mollenhauer *et al.* (*Bot. Acta* **107**: 36, 1994; posn), Schüsslet *et al.* (*Bot. Acta* **107**: 36, 1994; arbuscular mycorrhizal affinity).

Geosiphonomyces Cif. & Tomas. (1957) ≡ Geosiphon (Oomycota).

Geosmithia Pitt (1979), Anamorphic Trichocomaceae, 1.A1.?32/33. Teleomorph *Talaromyces*. 8, widespr.

Geotrichella Arnaud (1954), Mitosporic fungi, 1.A2.40. 1 (on paper), France. See Sigler & Carmichael (*Mycotaxon* **4**: 349, 1976; nom. dub.).

Geotrichites Stubblef., C.E. Mill., T.N. Taylor & G.T. Cole (1985), Fossil fungi. 1 (on arachnoid in amber), Dominican Republic.

Geotrichoides Langeron & Talice (1932) = Trichosporon (Mitosp. fungi) fide v. Arx *et al.* (*Stud. Mycol.* **14**: 30, 1977).

Geotrichopsis Tzean & Estey (1991), Mitosporic fungi, 1.A1.39. 1 (with dolipore septa), Canada.

geotrichosis, disease in humans or animals caused by *Geotrichum*.

Geotrichum Link (1809), Anamorphic Dipodascaceae, 2.A1.39. Teleomorphs *Dipodascus*, *Galactomyces*. 23, Eur., Am. *G. candidum* (often as *Oospora lactis*) in milk. See de Hoog *et al.* (*Stud. Mycol.* **29**: 1, 1986; key), Morenz (*Mykol. Schriftenreihe* **1**, 1963), Weijman (*Ant. v. Leeuwenhoek* **45**: 119, 1979; carbohydrates and taxonomy), Guého (*Ant. v. Leeuwenhoek* **45**: 199, 1979; base comp. and taxonomy), Olesen & Kier (*Nord. J. Bot.* **4**: 365, 1984; SEM and TEM structure), Guého *et al.* (*J. Clin. Microb.* **25**: 1191, 1987; DNA relatedness in *G. capitatum*), de Hoog & Amberger (*Ant. v. Leeuwenhoek* **58**: 101, 1990; protein patterns in *Geotrichum* and teleomorphs), Jensen *et al.* (*Mycoses* **33**: 519, 1991; crossed immunoelectrophoresis to differentiate *G. candidum*). See also gari.

geotropism, see tropisms.

Geotus Pilát & Svrček (1953) = Arrhenia (Tricholomat.) fide Redhead (1984).

Gerhardtia Bon (1994), Tricholomataceae. 2, Eur., N. Am.

Gerlachia W. Gams & E. Müll. (1980) = Microdochium (Mitosp. fungi) fide Samuels & Hallett (1983).

germ pore, a differentiated, frequently apical area, or hollow, in a spore wall (esp. in rusts) through which a **germ tube** (a germination hyphae) may come out; see Melendez-Howell (*Ann. Sci. nat. Bot.* sér. 12 **8**: 487, 1967; germ pore of basidiospores).

germ slit, a thin area of spore wall running the length of the spore.

germ tube, a germination hyphae which is formed by a germinating spore.

germicide, a substance causing destruction of microorganisms.

germination by repetition, producing secondary spores in place of germ tubes, as in *Heterobasidiomycetes* and *Sporobolomyces*.

Germslitospora Lodha (1978) = Coniochaetidium (Coniochaet.) fide v. Arx (1981).

Gerronema Singer (1951), Tricholomataceae. 30, widespr. See Singer (*Nova Hedw.* **7**: 53, 1964, keys; *Flora neotrop.* **3**: 24, 1970, key 20 neotrop. spp.).

Gerulajacta Preuss (1855) nom. dub. (Mitosp. fungi) fide Sutton (1977).

Gerwasia Racib. (1909), Phragmidiaceae. 3 (on *Rosaceae*), E. Indies, trop. Am., China. See Tai (*Farlowia* **3**: 95, 1947).

ghost fungus, *Pleurotus nidiformis*, an Australian luminous agaric. See Willis (*Muelleria* **1**: 213, 1967).

Giacominia Cif. & Tomas. (1953) = Arthopyrenia (Arthopyren.).

giant puff-ball, *Langermannia gigantea*; see record fungi.

giant stone-fungus, *Polyporus tumulosus*; the pseudosclerotium may exceed 1m³.

Gibbago E.G. Simmons (1986), Mitosporic fungi, 1.D2.26. 1, N. & S. Am.

gibber, gibbous (q.v.).

Gibbera Fr. (1825), Venturiaceae. *c.* 25, widespr. See Müller (*Sydowia* **8**: 60, 1954), Barr (*Sydowia* **41**: 25, 1989; key 4 N. Am. spp.), Eriksson (*Svensk bot. Tidskr.* **68**: 192, 1974; key 9 spp.), Sivanesan (*TBMS* **82**: 507, 1984; anamorphs).

Gibberella Sacc. (1877), Hypocreaceae. 10, widespr. *G. fujikuroi* (on cotton, maize, rice (Bakanae disease) and other crops in warm areas); *G. zeae* (frequently as *G. saubinetii*) (foot rot and ear blight (scab) of cereals). See gibberellin. Anamorph *Fusarium*.

gibberellin, a complex of hormone-like substances from *Gibberella fujikuroi* (anamorph *Fusarium moniliforme*) which causes overgrowth of higher plants, first recognized as the cause of Bakanae disease of rice. Gibberellin A$_1$, A$_2$, A$_3$ and other fractions have been distinguished including **gibberellic acid** which has a similar physiological action to gibberellin A$_1$. Gibberellin is manufactured commercially for use in horticulture. See Stodola (*Source book on gibberellin* 1828-1957, 1958; 632 abstracts), Knapp (Ed.) (*Eigenschaften und Wirkungen der Gibberelline*, 1962), MacMillan & Takahashi (*Nature* **217**: 170, 1968; allocation of trivial names), Jefferys (*Adv. appl. Microbiol.* **13**: 283, 1970).

Gibberellulina Sousa da Câmara (1950), Hypocreaceae. 1, Eur.

Gibberidea Fuckel (1870), Dothideales (inc. sed.). 1, Eur. See Holm (*Svensk bot. Tidskr.* **62**: 217, 1968). Anamorph *Pleurostomella*.

Gibberidea (Fr.) Kuntze (1898) ≡ Cucurbitaria (Cucurbitar.).

Gibberinula Kuntze (1898) ≡ Gibberidea (Dothideales).

gibbous (of a pileus), having a swelling or wide umbo, or having a convex top and a flat underside; gibber, gibbose.

Gibellia Sacc. (1885), ? Melanconidaceae. 1, Australia.

Gibellia Pass. (1886) ≡ Gibellina (Phyllachor.)

Gibellina Pass. (1886), Phyllachoraceae. 2, Asia, Eur. *G. cerealis* on wheat. See Glynne *et al.* (*TBMS* **84**: 653, 1985).

Gibellula Cavara (1894), Anamorphic Clavicipitaceae, 2.A1.15/16. Teleomorph *Torrubiella*. 8, widespr. See Petch (*Ann. Myc.* **30**: 386, 1932), Samson & Evans (*Mycol.* **84**: 300, 1992; on *Arachnida*).

Gibellulopsis Bat. & H. Maia (1959) = Verticillium (Mitosp. fungi) fide Kendrick & Carmichael (1973).

Gibsonia Massee (1909) ? = Melanospora (Ceratostomat.) fide Cannon & Hawksworth (1982).

Gigasperma E. Horak (1971), ? Hydnangiaceae. 1, Australasia. See Beaton & Malajczuk (*TBMS* **87**: 478, 1986), Castellano & Trappe (*Aust. Syst. Bot.* **5**: 613, 1992).

Gigaspora Gerd. & Trappe (1974), Gigasporaceae. 8, widespr. See Walker & Sanders (*Mycotaxon* **27**: 169, 1986; emend.), Spain *et al.* (*Mycotaxon* **34**: 667, 1989), Sward (*New Phytol.* **87**: 761, **88**: 661, 667, 1981; spore ultrastr.), Tommerup & Sivasithamparam (*MR* **94**: 897, 1991), Maia *et al.* (*Mycol.* **85**: 883, 1993, **86**: 343, 1994; spore wall ultrastr., germination), Yao *et al.* (1995).

Gigasporaceae J.B. Morton & Benny (1990), Glomales. 2 gen., 37 spp.

Gilbertella Hesselt. (1960), Gilbertellaceae. 1, trop. *G. persicaria* pathogenic to peach (*Amygdalus persicae*). See Hesseltine (*Bull. Torrey bot. Cl.* **87**: 21, 1960), O'Donnell *et al.* (*CJB* **55**: 662, 1977; zygospore ontogeny), Powell *et al.* (*CJB* **59**: 908, 1981; ultrastr. chlamydospore, *Protoplasma* **111**: 87, 1982; ultrastr. membrane), Whitney & Arnott (*Mycol.* **78**: 42, 1986, **80**: 707, 1988; calcium oxalate chrystals), Benny (*Mycol.* **83**: 150, 1991).

Gilbertellaceae Benny (1991), Mucorales. 1 gen., 1 sp. Sporangia columellate, sporangiospores appendaged; zygospore wall rough, suspensors opposed.

Gilbertia Donk (1934) nom. inval. = Amanita (Amanit.). See *Lepidella*.

Gilbertiella R. Heim (1965) [non Boutique (1951), *Annonaceae*] ≡ Gyrodon (Gyrodont.).

Gilbertina R. Heim (1966) = Gyrodon (Gyrodont.) fide Pegler (1981).

Gilchristia Redaelli & Cif. (1934) ≡ Zymonema (Mitosp. fungi).

gill (of an agaric), commonly used in English for lamella (q.v.) which is to be preferred as a more international term; **- fungi**, members of the *Agaricales*.

Gilletia Sacc. & Penz. (1882) = Basidiophora (Peronospor.). See Saccardo (*Syll. Fung.* **12**: 267, 1897).

Gilletia Torrend (1914) ≡ Telligia (Mitosp. fungi).

Gilletiella Sacc. & P. Syd. (1899), Dothideales (inc. sed.). 2 or 3 (on *Chusquea*), Am. See Eriksson (*SA* **7**: 72, 1988), Müller & v. Arx (1962). Anamorph *Ascochyta*-like.

Gillotia Sacc. & Trotter (1913), Mycosphaerellaceae. *c.* 3, trop. See v. Arx & Müller, 1975). Anamorph *Asteromella*-like.

Gilmania Bat. & Cif. (1962) = Chaetothyrium (Chaetothyr.) fide v. Arx & Müller (1975).

Gilmaniella G.L. Barron (1964), Mitosporic fungi, 1.A2.1/10. 5, widespr. Moustafa & Ezz-Eldin (*MR* **92**: 502, 1989; key).

gilvous, pale yellow.

gin, see spirits.

Ginanniella Cif. (1938), Tilletiaceae. 3 (on *Primulaceae*), Eur., Asia, N. Am. Usually included in *Urocystis*; see Vánky (1987), Nagler (1987).

ginger beer plant (Californian or American 'bees'), a mixture of a yeast (*Saccharomyces pyriformis*) and a bacterium (*Bacterium vermiforme*) used for fermenting a sugar solution to make a drink (Marshall Ward, *Phil. Trans. R. Soc., Lond.* **B 183**: 125, 1892). See Ramsbottom (*TBMS* **7**: 86, 1921); cf. tibi, tea fungus, teekwass.

Ginzbergerella Zahlbr. (1931) = Gyrocollema (Lichin.) fide Henssen (*SA* **5**: 131, 1986).

Giraffachitina Locq. (1985), Fossil fungi. 1, Estonia.

Girardia Gray (1821) = Bangia (Algae) fide Henssen (*Symb. bot. upsal.* **18**(1), 1963).

Giulia Tassi (1904), Mitosporic fungi, 4.A1.1. 1, Eur. See Pirozynski & Shoemaker (*CJB* **49**: 529, 1971).

Glabrocyphella W.B. Cooke (1961), Cyphellaceae. 12, widespr.

Glabrotheca Chardón (1939), Ascomycota (inc. sed.). 1, Venezuela.

glabrous, smooth; not hairy.

glaireous, slimy.

Glaphyriopsis B. Sutton & Pascoe (1987), Mitosporic fungi, 7.C1/2.15. 2, Australia.

Glaucinaria Fée ex A. Massal. (1860) = Graphina (Graphid.).

Glaucomaria M. Choisy (1929) ? = Lecanora (Lecanor.). See Hafellner (*Beih. Nova Hedw.* **79**: 241, 1984; status).

Glaucospora Rea (1922) = Melanophyllum (Agaric.) fide Singer (1951).

glaucous, having a bluish-grey waxy bloom.

Glaziella Berk. (1880), Glaziellaceae. 1, widespr., trop. See Gibson *et al.* (*Mycol.* **78**: 941, 1987).

Glaziellaceae J.L. Gibson (1986), Pezizales. 1 gen., 1 sp. Ascomata hollow, cleistothecial, lobed, with basal opening, orange to red; peridium thin and gelatinous; interascal tissue absent; asci embedded in ascomatal wall, clavate to subglobose, deliquescent, 1-spored; ascospores globose, large, smooth, orange. Anamorph unknown. ? Mycorrhizal, epigean. See Landvik & Eriksson (*SA* **13**: 13, 1984; posn).

Glaziellales, see *Pezizales*.

gleba, the sporing tissue in an angiocarpous sporocarp, esp. of gasteromycetes and hypogeous *Pezizales*; **glebal mass**, the projectile of *Sphaerobolus*.

glebula, a rounded process from a lichen thallus.

Glenospora Berk. & Desm. (1849) nom. rej. = Septobasidium (Septobasid.) but used in medical mycology for unrelated fungi (see Petch, *TBMS* **12**: 105, 1927).

Glenosporella Nann. (1931) = Chrysosporium (Mitosp. fungi) fide Carmichael (1962). = Geomyces (Mitosp. fungi) fide van Oorschot (1980).

Glenosporopsis O.M. Fonseca (1943), Mitosporic fungi, 1.A2.1/4. 1 (on humans), Brasil.

gleocystidium, see cystidium.

Gleophyllum, see *Gloeophyllum*.

gleoplerous hyphae (oil hyphae), hyphae with very long cells (or unicellular), with numerous oil drops in the plasma. See Jülich (*in* Gams (Ed.), *Kleine Kryptogamenflora* **II**(7), 1984).

gliatope, a site of heavy gel production (Moore, *Am. J. Bot.* **52**: 391, 1965). See gel tissue.

Glioannellodochium Matsush. (1989), Mitosporic fungi, 3.A1.19. 1, Australia.

Glioblastocladium Matsush. (1989), Mitosporic fungi, 1.A1.10. 1, Australia.

Gliobotrys Höhn. (1902) = Stachybotrys (Mitosp. fungi) fide Bisby (*TBMS* **26**: 133, 1943).

Gliocephalis Matr. (1899), Mitosporic fungi, 2.A1.15. 2, Asia, N. Am. See Hawksworth (*Bull. Br. Mus. nat. Hist., Bot.* **6**: 181, 1979).

Gliocephalotrichum J.J. Ellis & Hesselt. (1962), Anamorphic Hypocreaceae, 1.A1.15. Teleomorph *Leuconectria*. 4, widespr. See Wiley & Simmons (*Mycol.* **63**: 575, 1971; key).

Gliocladiopsis S.B. Saksena (1954) = Cylindrocladium (Mitosp. fungi) fide Barron (1968).

Gliocladium Corda (1840), Mitosporic fungi, 2.A1.15. 13, widespr. *G. roseum*, a mycoparasite. See Morquer (*BSMF* **79**: 137, 1963; key), Barnett & Lilly (*Mycol.* **54**: 72, 1962), Gullino (*Petria* **1**: 120, 1991; *IMC IV Trichoderma and Gliocladium Workshop*).

Gliocladochium Höhn. (1916) ≡ Periola (Mitosp. fungi).

Gliocoryne Maire (1909) = Pistillaria (Typhul.) fide Corner (1950).

Gliodendron Salonen & Ruokola (1969) = Sterigmatobotrys (Mitosp. fungi) fide Sutton (*Mycol. Pap.* **132**, 1973).

Gliomastix Guég. (1905), Mitosporic fungi, 1.A2.15. 10, widespr. See Dickinson (*Mycol. Pap.* **115**, 1968; key), Hammill (*Mycol.* **73**: 229, 1981; typification). = Acremonium (Mitosp. fungi) fide Gams (1971).

Gliophorus Herink (1958) = Hygrocybe (Hygrophor.) fide Singer (1962).

Gliophragma Subram. & Lodha (1964), Mitosporic fungi, 2.G1.10. 1, India.

Gliostroma Corda (1837) nom. dub. (? Fungi) fide Holubová-Jechová (*Sydowia* **46**: 242, 1994).

gliotoxin, an antibiotic from *Gliocladium virens* (Webster & Lomas, *TBMS* **47**: 535, 1964), *Aspergillus fumigatus, Penicillium cinerascens* (Brian & Hemming, *Ann. appl. Biol.* **32**: 214, 1945; *Biochem. J.* **41**: 570, 1947); antibacterial and antifungal (has been used as a seed dressing). Cf. viridin.

Gliotrichum Eschw. (1822) nom. dub. (? Fungi, inc. sed.).

Glischroderma Fuckel (1870), Mitosporic fungi, 8.A2.6. 1, Eur. See Hennebert (*Persoonia* **7**: 183, 1973).

Globaria Quél. (1873) ≡ Bovista (Lycoperd.) fide Demoulin (*Persoonia* **7**: 152, 1973).

Globifomes Murrill (1904), Coriolaceae. 1, N. Am.

Globipilea Beauseign. (1926) = Helvella (Helvell.) fide Nannfeldt (*Svensk bot. Tidskr.* **31**: 47, 1937).

Globoa Bat. & H. Maia (1962), Dothideales (inc. sed.). 1, Uganda.

Globoasclerotes Stach & Pickh. (1957), Fossil fungi. 1 (Carboniferous), Germany.

globoid (**globose, globular, globulose**), spherical or almost so (Fig. 37.1).

Globopilea Beauseign. (1926) = Helvella (Helvell.) fide Dissing (1966).

Globosasclerotes Stach & Pickh. (1957), Fossil fungi (*f. cat.*). 1 (Carboniferous, Tertiary), Eur. See Beneš (1969).

Globosomyces Jülich (1980), Stereales (inc. sed.). 1, Borneo.

Globosopyreno Lloyd (1923) nom. dub. (? Fungi, inc. sed.).

Globosphaeria D. Hawksw. (1990), Sordariales (inc. sed.). 1 (on lichens, esp. *Normandina*), Tasmania.

Globulicium Hjortstam (1973), Hyphodermataceae. 1, Eur.

Globuligera (Sacc.) Höhn. (1918), ? Patellariaceae. 1, N. Am.

Globulina Speg. (1888) [non Link (1820), *Chlorophyta*], Dothideales (inc. sed.). 1, Brazil. See Rossman (*Mycol. Pap.* **157**, 1987).

Globulina Velen. (1934) [non Link (1820), *Chlorophyta*] = Unguiculella (Hyaloscyph.) fide Svrček (*Česká Myk.* **41**: 193, 1987).

Globuliroseum Sullia & K.R. Khan (1983), Mitosporic fungi, 3.A1.1. 1, India.

glochidiate, covered with barbed bristles.

Gloeandromyces Thaxt. (1931), Laboulbeniaceae. 2, C. & S. Am.

Gloeoasterostroma Rick (1938) nom. conf. fide Rick (*Iheringia, Bot.* **4**: 116, 1959).

Gloeocalyx Massee (1901) = Plectania (Sarcosomat.) fide Korf (*Mycol.* **49**: 102, 1957).

Gloeocantharellus Singer (1945), Gomphaceae. 1, N. Am. See Petersen (*The genera Gomphus and Gloeocantharellus in North America*, 1972).

Gloeocercospora D.C. Bain & Edgerton ex Deighton (1971), Mitosporic fungi, 1.E1.1. 2, widespr. See Rawla (*TBMS* **60**: 283, 1973; comparison with *Ramulispora*).

Gloeocorticium Hjortstam & Ryvarden (1986), Corticiaceae. 1, Argentina.

Gloeocoryneum Weindlm. (1964), Mitosporic fungi, 6.C2.19. 2, USA. = Leptomelanconium (Mitosp. fungi) fide Morgan-Jones (*CJB* **49**: 1011, 1971).

Gloeocybe Earle (1909) = Lactarius (Russul.) fide Singer (1951).

Gloeocystidiellaceae (Parmasto) Jülich (1982), Hericiales. 12 gen. (+ 4 syn.), 57 spp. Basidioma resupinate to reflexed; hypochnoid to membranous; hymenophore smooth to tuberculate.
　　Lit. Ginns & Freeman (*Bibl. Mycol.* **157**, 1994; key N. Am. spp.).

Gloeocystidiellum Donk (1931), Gloeocystidiellaceae. 40, widespr. See Donk (*Fungus* **26**: 8, 1956), Eriksson & Ryvarden (*Corticiaceae of N. Europe* **3**: 405, 1975; key 10 Eur. spp.).

Gloeocystidiopsis Jülich (1982) = Gloeocystidiellum (Gloeocystidiell.) fide Hjortstam & Larsson (*Windahlia* **21**, 1994).

gloeocystidium, see cystidium.

Gloeocystidium P. Karst. (1889) = Dacryobolus (Gloeocystidiell.) fide Donk (1964).

Gloeodes Colby (1920), Mitosporic fungi, 5.A1.?. 1, widespr. G. pomigena (sooty blotch of apple and citrus).

Gloeodiscus Dennis (1961), ? Dothideales (inc. sed.). 1, NZ.

Gloeodontia Boidin (1966), Gloeocystidiellaceae. 4, Afr., N. & S. Am. See Burdsall & Lombard (*Mem. N.Y. bot. Gdn* **28**: 16, 1976).

Gloeoglossum E.J. Durand (1908) = Geoglossum (Geogloss.) fide Nannfeldt (*Ark. Bot.* **30A**: 4, 1942).

Gloeohaustoriales. Ordinal name proposed by Heim (*BSMF* **67**: 354, 1951) for *Antennopsis, Muiaria, Muiogone* and *Chantransiopsis*.

Gloeoheppia Gyeln. (1935), Gloeoheppiaceae (L). 4, desert areas. See Henssen (*Lichenologist* **27**: 261, 1995).

Gloeoheppiaceae Henssen (1995), Lichinales (L). 3 gen., 8 spp. Thallus squamulose to somewhat peltate, dark, attached by rhizoidal strands, homoiomerous, hyphae forming a reticulum surrounding cyanobacteria cells; apothecia laminal, immersed to adnate; interascal filaments sparsely branched or anastomosed, tips with enlarged cells; asci 8-16-spored; ascospores ellipsoid, 0-1-septate. Lichenized with cyanobacteria, desert areas.

Gloeohypochnicium (Parmasto) Hjortstam (1987) = Hypochnicium (Hyphodermat.) fide Hjortstam & Larsson (*Windahlia* **21**, 1994).

Gloeolecta Lettau (1937) = Bryophagus (Gyalect.) fide Hawksworth et al. (1980).

Gloeomucro R.H. Petersen (1980), Hydnaceae. 7, widespr.

Gloeopeniophora Höhn. & Litsch. (1907) = Peniophora (Peniophor.) fide Donk (1964) but used by Hayashi (*Bull. Govt. For. exp. Stn Meg.* **260**, 1974).

Gloeopeniophorella Rick (1934) nom. dub. (Basidiomycetes, inc. sed.).

Gloeopeziza Zukal (1891), Leotiaceae. 2 or 3 (on *Hepaticae*), Eur.

Gloeophyllum P. Karst. (1882), Coriolaceae. 8, widespr. See David & Fiasson (*Bull. mém. Soc. linn. Lyon* **46**: 304, 1977; chemotaxonomy), Corner (*Beih. Nova Hedw.* **86**: 61, 1987; key Malaysian spp.).

Gloeoporus Mont. (1842), Meruliaceae. 9, widespr. See David (*BSMF* **88**: 209, 1972).

Gloeopyrenia Zschacke (1937) = Protothelenella (Protothelenell.).

Gloeoradulum Rick (1959) nom. dub. (Basidiomycetes, inc. sed.).

Gloeosebacina Neuhoff (1924) = Stypella (Exid.) fide Donk (*Persoonia* **4**: 178, 1966).

Gloeosoma Bres. (1920) = Aleurodiscus (Aleurodis.) fide Donk (1964).

Gloeosporidiella Petr. (1921), Anamorphic Dermateaceae, 6.A1.15. Teleomorph *Drepanopeziza.* 11, widespr. See Rimpau (*Phytopath. Z.* **43**: 257, 1962).

Gloeosporidina Petr. (1921), Mitosporic fungi, 6.A1.15. 4, widespr. See Sutton & Pollack (*Mycol.* **65**: 1125, 1973), Kubono (*Trans. mycol. Soc. Jap.* **34**: 261, 1993).

Gloeosporidium Höhn. (1916) = Discula (Mitosp. fungi) fide v. Arx (*Bibl. Mycol.* **24**, 1970).

Gloeosporiella Cavara (1892), Mitosporic fungi, 3.G1.1. 1, Eur. See Sutton (*Nova Hedw.* **25**: 163, 1974).

Gloeosporina Höhn. (1916) = Asteroma (Mitosp. fungi) fide Sutton (1977).

Gloeosporiopsis Speg. (1910) = Colletotrichum (Mitosp. fungi). See Petrak & Sydow (*Ann. Myc.* **33**: 178, 1935).

Gloeosporium Desm. & Mont. (1849) nom. rej. prop. = Marssonina (Mitosp. fungi). Used for many very diverse fungi, v. Arx (1970) lists 735 names and refers 288 to the *Colletotrichum* anamorph of *Glomerella cingulata* and others to 48 other gen. For the plant pathogens G. album, see *Phlyctema*; G. concentricum, Cylindrosporium; G. musarum, Colletotrichum; G. perennans, Cryptosporiopsis.

Gloeostereum S. Ito & S. Imai (1933), Stereaceae. 1, Japan.

Gloeosynnema Seifert & G. Okada (1988), Mitosporic fungi, 2.A1.21. 1 (with clamp connexions), Indonesia, Japan.

Gloeotinia M. Wilson, Noble & E.G. Gray (1954), Sclerotiniaceae. 1, widespr. G. granigena (syn. G. temulenta) (Blind seed disease of grass seed). See Hardison (*Mycol.* **54**: 201, 1962; hosts), Wilson et al. (*TBMS* **37**: 29, 1954), Griffiths (*TBMS* **41**: 461, 1958; sexuality), Schumacher (*Mycotaxon* **8**: 125, 1975; nomencl.).

Gloeotrochila Petr. (1947) = Cryptocline (Mitosp. fungi) fide v. Arx (*Verh. K. Ned. Akad. Wet. Nat.* **51**: 24, 1957).

Gloeotromera Ervin (1956) = Ductifera (Exid.) fide Wells (1958).

Gloeotulasnella Höhn. & Litsch. (1908) = Tulasnella (Tulasnell.) fide Olive (*Mycol.* **49**: 668, 1957).

Gloiocephala Massee (1892), Tricholomataceae. 9, widespr., esp. trop. See Bas (*Persoonia* **2**: 77, 1961), Singer (*Flora Neotrop.* **17**: 284, 1976; monogr. neotrop. spp.).

Gloiodon P. Karst. (1879), Gloeocystidiellaceae. 1, Eur. See Maas Geesteranus (*Proc. Ned. Akad. Wet.* C **66**: 430, 1963).

Gloiosphaera Höhn. (1902), Mitosporic fungi, 1.A1.15. 2, Eur., N. Am. See Pollack & McKnight (*Mycol.* **64**: 415, 1972), Wang (*Mycol.* **63**: 890, 1971).

Gloiosporae. Mitosporic fungi (hyphomycetes) having slimy spores. Cf. *Xerosporae*. See Wakefield & Bisby (*TBMS* **25**: 50, 1941).

Gloiothele Bres. (1920), Gloeocystidiellaceae. 6, widespr. See Ginns & Freeman (*Bibl. Mycol.* **157**, 1994).

Glomaceae Piroz. & Dalpé (1989), Glomales. 2 gen. (+ 5 syn.), 86 spp. See Pirozynski & Dalpé (*Symbiosis* **7**: 1, 1989), Morton & Benny (1990; key).

Glomales, Zygomycetes. 3 fam., 6 gen. (+ 5 syn.), 160 spp. Endomycorrhizal on plants. Fams:
　(1) **Acaulosporaceae**.
　(2) **Gigasporaceae**.
　(3) **Glomaceae**.

Morton & Benny (*Mycotaxon* **37**: 471, 1990; rev., keys) recognized two suborders: *Glomineae* for the vesicular-arbuscular mycorrhiza (VAM) forming (1) and (2), and *Gigasporineae* for the arbuscular mycorrhiza forming (3).
Lit.: Thaxter (*Proc. Am. Acad. Arts & Sci.* **57**: 291, 1922; sporocarpic spp.), Gerdemann & Trappe (*Mycol. Mem.* **5**, 1974; rev.), Walker (*Mycotaxon* **18**: 443, 1983; spore morphology [murograph]), Powell & Bagyaraj (*VA Mycorrhiza*, 1984; review), Burdrett *et al.* (*CJB* **62**: 2128, 1984; arbuscule staining), Berch (*Front. Appl. Microbiol.* **2**: 161, 1986; review), Silvia & Huddell (*Symbiosis* **1**: 259, 1986; aeroponic culture), Bonfonte-Falso (*Symbiosis* **3**: 249, 1987; review), Stubblefield & Taylor (*New Phytol.* **108**: 3, 1988; palaeomycology), Jabaji-Hare (*Mycol.* **80**: 622, 1988; lipid & fatty acid profiles), Morton (*Mycotaxon* **32**: 267, 1988, checklist 126 spp. and classific., **37**: 493, 1990; evolution, *Mycol.* **82**: 192, 1990; phylogeny, *Mycorrhiza* **2**: 97, 1993; review), Pirozynski & Dalpé (*Symbiosis* **7**: 1, 1989; rev. & palaeomycology), Schenck & Peréz (*Manual for the identification of VA mycorrhizal fungi*, 3rd edn, 1990), Allen (*The evolution of mycorrhizae*, 1991; *Mycorrhizal functioning*, 1992), Bruns *et al.* (*Mol. Phylog. Evol.* **1**: 231, 1992; phylogeny), Millner & Kitt (*Mycorrhiza* **2**: 9, 1992; soil-less culture), Streussy (*Mycorrhiza* **1**: 113, 1992; phylogeny), Read *et al.* (*Mycorrhizas in ecosystems*, 1992), Morton *et al.* (*Mycotaxon* **48**: 491, 1993; INVAM culture collection), Maia *et al.* (*Mycol.* **85**: 323, 1993; fixation & embedding), Simon *et al.* (*Nature* **363**: 67, 1993; phylogeny), Walker & Trappe (*MR* **97**: 339; names & epithets), Sancholle & Dalpé (*Mycotaxon* **49**: 187, 1993; fatty acids), Hass *et al.* (*Am. J. Bot.* **81**: 29, 1994; palaeomycology), Giovannetti & Gianinazzi-Pearson (*MR* **98**: 705, 1994; review), Gianinazzi-Pearson *et al.* (*Mycol.* **86**: 478, 1994; wall chemistry & phylogeny), Morton & Bentivegna (*Plant & Soil* **159**: 47, 1994; taxonomy), Bentivegna & Morton (*in* Pfegler & Linderman (Eds), *Mycorrhizae and plant health*: 283, 1983; systematics), Robson *et al.* (*Management of mycorrhizas in agriculture, horticulture, and forestry*, 1994), Gianinazzi & Schüepp (Eds) (*Impact of arbuscular mycorrhizas on sustainable agriculture and natural ecosystems*, 1994; taxonomy, biodiversity, Eur. culture collection), Morton *et al.* (*in* Varma & Hock (Eds), *Mycorrhiza: structure, function, molecular biology and biochemistry*: 669, 1995), Yao *et al.* (*Kew Bull.* **50**: 349, 1995).

Glomerella Spauld. & H. Schrenk (1903), Phyllachoraceae. *c.* 5, widespr. *G. cingulata* on many hosts (*CMI Descript.* **315**, 1971); *G. gossypii* on cotton, *G. tucumanensis* on sugarcane. See Uecker (*Mycol.* **86**: 82, 1994; ontogeny). Anamorph *Colletotrichum*.

Glomerilla Norman (1869), ? Verrucariaceae (?L). 1, Norway.

Glomerula Bainier (1903) = Actinomucor (Mucor.) fide Hesseltine (1955).

Glomerularia H. Karst. (1849) = Gonatobotrys (Mitosp. fungi) fide Henderson (*Notes R. Bot. Gdn Edinb.* **23**: 497, 1961).

Glomerularia Peck (1879) ≡ Glomopsis (Mitosp. fungi).

Glomerulomyces A.I. Romero & S.E. López (1989), Mitosporic fungi, 1.A1.6. 1 (with clamp connexions), Argentina.

glomerulus (glomerule), a clump or cluster; frequently used for clusters of photobiont cells in lichens.

Glomites T.N. Taylor, W. Remy, Hass & Kerp (1995), Fossil fungi. 1 (Devonian), Scotland.

Glomopsis D.M. Hend. (1961), Anamorphic Platygloeaceae, 3.G1.1. Teleomorph *Herpobasidium*. 2, USA. See Donk (*Persoonia* **4**: 213, 1966).

Glomospora D.M. Hend. (1961) [non Butterw. & R.W. Williams (1958), fossil sporae dispersae], Mitosporic fungi, 3.D1.1. 1, UK.

Glomosporium Kochman (1939), ? Tilletiaceae. 2 (on *Chenopodiaceae, Amaranthaceae*), temp., Eur., Am., Austr. See Vánky (1987).

Glomus Tul. & C. Tul. (1845), Glomaceae. *c.* 85, widespr. See Gerdemann & Trappe (1974), Schenk & Peréz (1990), Berch & Fortin (*CJB* **61**: 2608, 1983, **62**: 170, 1984), Jabaji-Hare *et al.* (*Mycol.* **76**: 1024, 1984; lipid composition), Morton (*Mycol.* **77**: 192, 1985; morphological variation), Bonfante-Fasolo *et al.* (*Biol. Cell* **57**: 265, 1986; chitin in cell wall), Bonfante-Fasolo & Schubert (*CJB* **65**: 539, 1987; spore wall ultrastr.), Koske & Halvorsol (*Mycol.* **81**: 927, 1989), Koske & Gemma (*Mycol.* **81**: 935, 1989), Almeida & Schenk (*Mycol.* **82**: 703, 1990), Meier & Charvat (*Int. J. Pl. Sci.* **153**: 541, 1992; ultrastr. spore germ.), Chabot *et al.* (*Mycol.* **84**: 315, 1992; life cycle in root culture), Wu & Silvia (*Mycol.* **85**: 317, 1993; spore ontogeny), Cabello *et al.* (*Mycotaxon* **51**: 123, 1994), Gaspar *et al.* (*Mycotaxon* **51**: 129, 1994; lipid & fatty acid composition), Błaszkowski (*Mycorrhiza* **4**: 201, 1994), Miller & Jeffries (*MR* **98**: 307, 1994; spore wall ultrastr.), Maia & Kimborough (*Int. J. Pl. Sci.* **155**: 689, 1994; ultrastr.), Yao *et al.* (*Mycologist* **6**: 132, 1992; ultrastr., *Kew Bull.* **50**: 349, 1995).

Gloniella Sacc. (1883), Hysteriaceae. 6, temp.

Gloniopsis De Not. (1847), Hysteriaceae. 2 or 3, temp. See Zogg (1962).

Glonium Muhl. (1813), Hysteriaceae. *c.* 10, temp. See Speer (*BSMF* **102**: 101, 1986), *Delphinella*. Anamorph *Cleistonium*.

Glossifungites Lomnicki (1886), Fossil fungi. 1 (Cretaceous), Poland.

Glossodium Nyl. (1855) = Icmadophila (Icmadophil.) fide Rambold *et al.* (*Bibl. Lich.* **53**: 217, 1993).

glossoid cell, elongate (tongue-shaped) cell containing an elaborate extrusome (*Haptoglossa*).

Glotzia M. Gauthier ex Manier & Lichtw. (1969), Legeriomycetaceae. 4 (in *Ephemeroptera*), Eur., USA. See Lichtwardt (1986), Lichtwardt & Williams (*CJB* **68**: 1057, 1990), Williams & Lichtwardt (*CJB* **68**: 1045, 1990).

glucans, one of the main constituents of fungal walls. **R-glucans** are alkali insoluble; **S-glucans** are soluble.

Glukomyces Beij. [not traced] nom. dub. (? Fungi).

gluten, (1) a substance on the surface of some agarics, etc., which is sticky when wet; (2) spore mass in *Phallus*.

Glutinaster Earle (1909) = Tricholoma (Tricholomat.) fide Singer (1951).

Glutinisporidium Thor (1930) nom. dub. (? Fungi, inc. sed.).

Glutinium Fr. (1849), Mitosporic fungi. 3, Eur., N. Am.

Glutinoagger Sivan. & Watling (1980), Mitosporic fungi, 3.A1.10. 1 (with clamp connexions), Seychelles.

glutinous, sticky; made up of, or covered with, gluten.

glyceollin, a phytoalexin (q.v.) from soybean (*Glycine max*).

Glycydiderma Paulet (1808) ≡ Geastrum (Geastr.).

Glycyphila Mont. (1851), Mitosporic fungi. 2, Eur.

Glyphidium A. Massal. (1860) ? = Arthonia (Arthon.).

Glyphis Ach. (1814), Graphidaceae (L). *c.* 30, cosmop. (mainly trop.). See Pant (*Geophytology* **20**: 48, 1991; 2 spp. India).

Glyphium Nitschke ex F. Lehm. (1886), Mytilinidiaceae. 4 or 5, Eur., N. Am. See Goree (*CJB*

52: 1265, 1974), Sutton (*TBMS* **54**: 255, 1970). Anamorph *Peyronelia.*

Glypholecia Nyl. (1853), Acarosporaceae (L). 5, Eur., N. Am.

glypholecine, having particularly labyrinth-like lirella as in *Glypholecia.*

Glypholeciomyces Cif. & Tomas. (1953) ≡ Glypholecia (Acarospor.).

Glyphomyces Cif. & Tomas. (1953) ≡ Glyphis (Graphid.).

Glyphopeltis Brusse (1985), ? Lecideaceae (L). 1, Eur., S. Afr. See Timdal (*Mycotaxon* **31**: 101, 1988).

Glyptoderma R. Heim & Perr.-Bertr. (1971), Lycoperdaceae. 1, trop. Am.

Glyptospora Fayod (1889) = Psathyrella (Coprin.) fide Singer (1951).

Gnaphalomyces Opiz [not traced] = Lanosa (? Fungi) fide Streinz (*Nom. fung.* : 302, 1862).

Gnomonia Ces. & De Not. (1863), Valsaceae. 50, widespr. *G. platani* (plane (*Platanus*) scorch). See Barr (*Mycol. Mem.* **7**, 1978; key 35 spp.), Bolay (*Ber. schweiz. bot. Ges.* **81**: 398, 1972), Monod (1983).

Gnomoniaceae nom. cons. = Valsaceae (Diaporthales)

Gnomoniella Sacc. (1881), Valsaceae. 10, widespr. See Barr (*Mycol. Mem.* **7**, 1978), Monod (1983).

Gnomonina Höhn. (1917) ≡ Laestadia (Mycosphaerell.).

Gnomoniopsis Berl. (1892) = Gnomonia (Vals.) fide Bolay (1972).

Gnomoniopsis Stoneman (1898) ≡ Glomerella (Phyllachor.).

gnotobiotic (of cultures), ones in which all the living components are known. Cf. axenic.

Godal Adans. (1763) nom. dub. (? Mitosp. fungi).

Godfrinia Maire (1902) = Hygrocybe (Hygrophor.) fide Singer (1951).

Godronia Moug. & Lév. (1846), Leotiaceae. 26, N. temp. See Groves (*CJB* **43**: 1195, 1965; key). Anamorphs *Chondropodiella, Fuckelia, Topospora.*

Godroniella P. Karst. (1884) = Myxormia (Mitosp. fungi) fide Clements & Shear (1931). See Tulloch (*Mycol. Pap.* **130**, 1972).

Godroniopsis Diehl & E.K. Cash (1929), Leotiaceae. 2, N. Am. See Petrak (*Sydowia* **6**: 336, 1952).

Goidanichia Tomas. & Cif. (1952) = Staurothele (Verrucar.).

Goidanichia G. Arnaud (1954) ≡ Goidanichiella G. Arnaud ex G.L. Barron (Mitosp. fungi).

Goidanichiella G. Arnaud ex G.L. Barron (1968) ≡ Haplographium (Mitosp. fungi) fide Gams *et al.* (*Mycotaxon* **38**: 149, 1990).

Goidanichiella G.L. Barron ex W. Gams (1990), Mitosporic fungi, 1.A1.15. 1, widespr. See Gams *et al.* (*Mycotaxon* **38**: 149, 1990; discussion of *Goidanichia* and *Goidanichiella*).

Goidanichiomyces Cif. & Tomas. (1953) ≡ Goidanichia (Verruc.).

Golgi body. Dictyosome with large numbers of cisternae which may be visible using light microscopy after staining.

Golovinia Mekht. (1967), Mitosporic fungi, 1.C1.10. 6, widespr.

Golovinomyces (U. Braun) V.P. Gelyuta (1988) = Erysiphe (Erysiph.) fide Braun (1987).

Gomezina Chardón & Toro (1934) = Aphanostigme (Pseudoperispor.) fide v. Arx & Müller (1975).

Gomphaceae Donk (1961), Gomphales. 4 gen. (+ 6 syn.), *c.* 10 spp. Basidioma cantharelloid; spores yellowish brown, ornamented; terrestrial.
Lit.: Donk (1964: 267).

Gomphales, Basidiomycetes. 3 fam., 9 gen. (+ 13 syn.), 133 spp. Fams:
(1) **Gomphaceae**.

(2) **Lentariaceae**.
(3) **Ramariaceae**.
For *Lit.* see *Cantharellales.*

Gomphidiaceae Maire ex Jülich (1982), Boletales. 6 gen. (+ 2 syn.), 28 spp. Basidioma agaricoid; lamellate; spores fuscous, elongate.

Gomphidius Fr. (1836), Gomphidiaceae. 9, esp. N. temp. See Watling (*British fungus flora* **1**: 77, 1970; key), Miller (*Mycol.* **63**: 1129, 1971; key), Singer (*Taxon* **22**: 445, 1973; typific.).

Gomphillaceae Walt. Watson ex Hafellner (1984), Ostropales (L). 12 gen. (+ 17 syn.), 94 spp. Thallus crustose; ascomata apothecial, usually sessile, discoid, without a well-defined margin; interascal tissue of branched and anastomosing filiform paraphyses; asci cylindrical, with only a slight apical thickening, 1- to 8-spored; ascospores ellipsoidal to filiform, with transverse septa, sometimes muriform. Anamorphs coelomycetous, spine- or brush-like, or peltate. Lichenized with green algae, mostly foliicolous, esp. trop.
Lit.: Kalb & Vězda (*Bibl. Lich.* **29**, 1988; 43 spp. neotrop.), Vězda & Poelt (*Folia geobot. phytotax.* **22**: 179, 1987).

Gomphillus Nyl. (1855), Gomphillaceae (L). 3, Eur., N. & S. Am. See Kalb & Vězda (1988).

Gomphinaria Preuss (1851) = Acrotheca (Mitosp. fungi) fide Saccardo (*Syll. Fung.* **4**: 277, 1886).

Gomphogaster O.K. Mill. (1973), Gomphidiaceae. 1, USA.

Gomphora Fr. (1825) ≡ Gomphus (Gomph.) fide Petersen (1972).

Gomphos Kuntze (1891) ≡ Cortinarius (Cortinar.).

Gomphospora A. Massal. (1852) ? = Schismatomma (Roccell.).

Gomphus (Pers.) Pers. (1797), Gomphaceae. *c.* 7, temp. See Petersen (*The genera Gomphus and Gloeocantharellus in North America,* 1972).

Gomphus (Fr.) Weinm. (1826) = Gomphidius (Gomphid.).

Gonapodya A. Fisch. (1891), Gonapodyaceae. 2, temp. See Karling (1977).

Gonapodyaceae H.E. Petersen (1909), Monoblepharidales. 2 gen., 7 spp. Thallus a well-developed, differentiated hyphal system bearing numerous reproductive structures; zygote undergoing a period of motility before encystment, propelled by flagellum of male gamete.

Gonatobotrys Corda (1839), Anamorphic Ceratostomataceae, 1.A/B1.8. Teleomorph *Melanospora.* 2 (fungicolous), widespr. See Walker & Minter (*TBMS* **77**: 299, 1981; key), Whaley & Barnett (*Mycol.* **55**: 199, 1963; *G. simplex* on *Alternaria* and *Cladosporium*), Vakili (*MR* **93**: 67, 1989; teleomorph).

Gonatobotrytites Pia (1927), Fossil fungi. 1 (Eocene), Eur.

Gonatobotryum Sacc. (1880), Mitosporic fungi, 1.A1/2.5. 3 (fungicolous), widespr. See Walker & Minter (*TBMS* **77**: 299, 1981; key), Kendrick *et al.* (*CJB* **45**: 591, 1968; conidiogenesis).

Gonatophragmiella Rak. Kumar, A.N. Rai & Kamal (1994), Mitosporic fungi, 1.C2.?. 1, India.

Gonatophragmium Deighton (1969), Anamorphic Acrospermataceae, 1.C2.10. Teleomorph *Acrospermum.* 2, trop. See Takahashi & Teramine (*Ann. Phytopath. Soc. Jap.* **52**: 404, 1986; teleomorph).

Gonatopyricularia Z.D. Jiang & P.K. Chi (1989) ≡ Pyriculariopsis (Mitosp. fungi).

Gonatorhodis Clem. & Shear (1931) ≡ Gonatorrhodiella (Mitosp. fungi).

Gonatorrhodiella Thaxt. (1891) = Gonatobotryum (Mitosp. fungi) fide Walker & Minter (1981).

Gonatorrhodum Corda (1839), Mitosporic fungi, 1.A2.?3. 1 or 2, Eur.

Gonatosporium Corda (1839) = Arthrinium (Mitosp. fungi) fide Hughes (1958).

Gonatotrichum Corda (1842) ≡ Gonytrichum (Mitosp. fungi).

Gongromeriza Preuss (1851) = Chloridium (Mitosp. fungi) fide Hughes (*Friesia* **9**: 61, 1969).

Gongronella Ribaldi (1952), Mucoraceae. 2, widespr. See Hessletine & Ellis (*Mycol.* **53**: 406, 1961, **56**: 568, 1964; key), Upadhyay (*Nova Hedw.* **17**: 65, 1969; key).

Gongylia Körb. (1855) = Arthrorhaphis (Arthrorhaphid.) p.p. fide Hawksworth *et al.* (*Lichenologist* **12**: 13, 1980).

gongylidius (pl. **gongylidia**), a bulbous structure developed by fungi cultivated by termites.

Gongylocladium Wallr. (1833) ≡ Oedemium (Mitosp. fungi).

gonidial layer, photobiont layer in a lichen thallus (obsol.).

gonidimium, a hymenial alga (obsol.).

Gonidiomyces Vain. (1921), Ascomycota (inc. sed.). 1 ('mycosymbiosis'), Philipp. ? = Meliola (Meliol.) fide Sydow (*Ann. Myc.* **20**: 66, 1922); nom. dub. fide Santesson (*Svensk Bot. Tidskr.* **45**: 300, 1951).

gonidium, photobiont (obsol.).

gonimium, a cyanobacterial cell in a lichen thallus (obsol.).

Gonimochaete Drechsler (1946), Myzocytiopsidaceae. 4, widespr. See Dick (*in press*; key), Drechsler (*Bull. Torr. Bot. Cl.* **73**: 1, 1946), Newell *et al.* (*Bull. Mar. Sci.* **27**: 189, 1977), Barron (*CJB* **51**: 2451, 1973, *Mycol.* **77**: 17, 1985), Saikawa & Anazawa (*CJB* **63**: 2326, 1985; ultrastr.).

goniocyst (**goniocystula**), a group of algal cells derived from a single cell surrounded by a hyphal envelope forming a roundish structure which is not a soralium (e.g. the vegetative thallus '*Botrydina vulgaris*'; q.v.), (Fig. 22H). See Sérusiaux (*Lichenologist* **17**: 1, 1985).

goniocystangium, cup-like structure bearing goniocysts (q.v.) on foliicolous species of *Catillaria* (Vězda, 1980) and *Opegrapha* (Santesson, *Svensk Naturv.* 1968: 176, 1968; Sérusiaux, 1985).

Gononema Nyl. (1855) = Thermutis (Lichin.).

Goniopila Marvanová & Descals (1985), Mitosporic fungi, 1.A1.21. 1 (aquatic), USA, Eur.

Goniosporium Link (1824) = Arthrinium (Mitosp. fungi) fide Höhnel (*Mitt. bot. Lab. Techn. Hochsch. Wien* **2**: 9, 1925).

goniosporous, having angled spores.

Gonohymenia J. Steiner (1902) = Lichinella (Lichin.) fide Moreno & Egea (1992).

Gonohymeniomyces Cif. & Tomas. (1953) ≡ Gonohymenia (Lichin.).

Gonolecania Zahlbr. (1924) = Byssolecania (Pilocarp.) fide Santesson (*Svensk bot. Tidskr.* **43**: 547, 1949, *SA* **10**: 137, 1991).

gonoplasm (of *Peronosporales*), the protoplasm, at the centre of an antheridium, which later undergoes fusion with the oosphere.

gonosphere, a zoospore of the *Chytridiales* (obsol.).

Gonothecis Clem. (1909) = Gyalectidium (Gomphill.).

Gonothecium (Vain.) Clem. & Shear (1931) = Gyalectidium (Gomphill.) fide Santesson (1952).

gonotocont, the organ in which meiosis takes place.

Gonsala, see *Gonzala*.

Gonyella Syd. & P. Syd. (1919), Mitosporic fungi, 1.B2.?. 1, Eur.

Gonytrichella Emoto & Tubaki (1970), Mitosporic fungi, 1.A2.10. 1, Japan. = Dicyma (Mitosp. fungi) fide Arx (1982).

Gonytrichum Nees & T. Nees (1818), Mitosporic fungi, 2.A2.15. 4, widespr. See Gams & Holubová-Jechová (*Stud. Mycol.* **13**, 1976; key).

Gonzala Adans. ex Léman (1821) = Peziza (Peziz.) fide Streinz (*Nom. fung.*, 1862).

Goosiella Morgan-Jones, Kamal & R.K. Verma (1986), Mitosporic fungi, 1.F.2.10. 1, India.

Goosiomyces N.K. Rao & Manohar. (1989), Mitosporic fungi, 1.G2.12. 1, India.

Goossensia Heinem. (1958), Cantharellaceae. 1, Congo.

Goplana Racib. (1900), Chaconiaceae. 10, trop.; II, III. See Ono & Hennen (*Trans. Mycol. Soc. Japan* **24**: 369, 1983).

Gorgadesia Tav. (1964), Roccellaceae (L). 1, Cape Verde.

Gorgomyces M. Gönczöl & Révay (1985), Mitosporic fungi, 1.E1.1/10. 1 (nematophagous), Hungary.

Gorgoniceps (P. Karst.) P. Karst. (1871), Leotiaceae. 6, widespr. See Seaver (*Mycol.* **38**: 548, 1946), Korf (*Mycol.* **58**: 724, 1966).

Gorodkoviella Vassilkov (1969) = Pachyella (Peziz.) fide Dissing (*SA* **6**: 129, 1987).

gossamers, fine, floating mycelial nets produced by fungi on media lacking added carbon (Wainwright, *Mycologist* **1**: 182, 1987).

Goupilia Mérat (1834) ? = Scleroderma (Sclerodermat.).

Govindua Bat. & H. Maia (1960), Microthyriaceae. 1, India.

Graamspora Locq. & Sal.-Cheb. (1980), Fossil fungi. 1, Cameroon.

Gracea M.W. Dick (1995), Olpidiopsidales (inc. sed.). 2, UK. See Dick (*in press*; key).

Graddonia Dennis (1955), Dermateaceae. 1, Eur.

Graddonidiscus Raitv. & R. Galán (1992), Hyaloscyphaceae. 2, UK, Spain.

Grahamiella Spooner (1981), Leotiaceae. 1, Eur.

Grallomyces F. Stevens (1918), Mitosporic fungi, 2.G2.19. 1, trop. See Deighton & Pirozynski (*Mycol. Pap.* **105**, 1966).

Gram + or - (of bacteria), staining or not staining by Gram's method (see Methods).

Gramincola Velen. (1947), Agaricaceae. 1, Eur.

Graminella L. Léger & M. Gauthier ex Manier (1962), Legeriomycetaceae. 2 (in *Ephemeroptera*), France, USA. See Lichtwardt & Moss (*TBMS* **76**: 311, 1981), Lichtwardt (1986; key).

graminicolous, living on *Gramineae*.

Grammothele Berk. & M.A. Curtis (1868), Grammotheleaceae. 1, trop. See Ryvarden (*TBMS* **73**: 1, 1979).

Grammotheleaceae Jülich (1982), Poriales. 4 gen. (+ 1 syn.), 14 spp. Basidioma resupinate; dimitic; hymenophore irpicoid to poroid, bluish grey.

Grammothelopsis Jülich (1982), Coriolaceae. 1, Kenya.

Granatisporites Elsik & Janson. (1974), Fossil fungi (*f. cat.*). 4 (Paleocene), Canada. = Brachysporisporites (Fossil fungi) fide Kremp (*BMR Bull.* **192**: 76, 1978).

Grandigallia M.E. Barr, Hanlin, Cedeño, Parra & M. Rodr. (1987), Dothideales (inc. sed.). 1, Venezuela.

Grandinia Fr. (1838) nom. dub. fide Donk (*Taxon* **5**: 77, 1956) but see Jülich (*Int. J. Mycol. Lich.* **1**: 27, 1982).

Grandiniella P. Karst. (1895), Meruliaceae. 18, cosmop. See Burdsall (*Taxon* **26**: 327, 1977). = Phanerochaete (Cortic.) fide Eriksson *et al.* (*Taxon* **27**: 51, 1978).

Grandiniochaete Rick (1940) nom. dub. (Basidiomycetes). See Donk (*Taxon* **5**: 79, 1956).

Grandinioides Banker (1906) = Mycobonia (Polypor.) fide Donk (1957).

Granmamyces J. Mena & Mercado (1988) = Weufia (Mitosp. fungi). See Mercado (*AnaNet* **10**: 8, 1990).

granular (**granulate, granulose**) (of a surface), covered with very small particles. Fig. 29.4.

Granularia Roth (1791) = Nidularia (Nidular.) fide Palmer (*Taxon* **10**: 54, 1961).

Granularia Sowerby (1815) = Urocystis (Ustilagin.).

Granularia Sacc. & Ellis ex Sacc. (1882) Mitosporic fungi, 3.A1.?. 1, N. Am.

Granulina Velen. (1947) nom. dub. ('gasteromycetes').

Granulobasidium Jülich (1979) = Hypochnicium (Hyphodermat.) fide Hjortstam & Larsson (*Windahlia* **21**, 1994).

Granulocystis Hjortstam (1986), Meruliaceae. 1, N. temp.

Granulodiplodia Zambett. ex M. Morelet (1973) = Sphaeropsis (Mitosp. fungi) fide Sutton (1977).

granuloma, a nodule of firm tissue formed as a reaction to chronic irritation.

Granulomanus de Hoog & Samson (1978), Anamorphic Clavicipitaceae, 1.E1.15. Teleomorph *Torrubiella*. 1 (on *Arachnida*), UK. = Gibellula (Mitosp. fungi) fide Humber & Rombach (1987).

Granulopyrenis Aptroot (1991), Pyrenulaceae. 3, trop.

Graphidaceae Dumort. (1822), Ostropales (L). 15 gen. (+ 64 syn.), 998 spp. Ascomata elongated, curved or branched; marginal tissues well-developed, black, carbonized; ascospores mostly large, transversely and sometimes longitudinally septate, hyaline or brown. Anamorphs pycnidial. Lichenized with green algae (usually *Trentopohliaceae*), esp. on bark, mainly trop.
Lit.: Harris (*Some Florida lichens*, 1990; gen. concepts), Nakanishi (*J. Sci. Hiroshima Univ.* B (2), **11**: 51, 1966; Japan), Redinger (*Ark. Bot.* **26A**(1), **27A**(3), 1934-35; S. Am., *Rabenh. Krypt.-Fl.* **9**, 2(1), 1936-38, Eur.), Wirth & Hale (*Contr. US natn Herb.* **36**: 63, 1963; Mexico, *Smithson. Contr. bot.* **40**, 1978; Dominican Republic).

Graphidales, see *Ostropales*; Lumbsch (*SA* **6**: 129, 1987).

Graphidastra (Redinger) G. Thor (1990), Roccellaceae (L). 2, India, Indonesia, Australia, Pacific.

Graphidium Lindau (1909) ? = Paecilomyces (Mitosp. fungi) fide Kendrick & Carmichael (1973).

Graphidomyces E.A. Thomas ex Cif. & Tomas. (1953) = Graphis (Graphid.).

Graphidula Norman (1853) = Phaeographis (Graphid.).

Graphilbum H.P. Upadhyay & W.B. Kendr. (1975), Anamorphic ? Micrascales, 2.A1.19. Teleomorph *Ceratocystis*. 1, Alaska.

Graphina Müll. Arg. (1880), Graphidaceae (L). *c.* 270, cosmop. (mainly trop.). See Dharne & Raychoudhury (*Bull. Bot. Serv. India* **10**: 267, 1969), Nakanishi (1966), Wirth & Hale (1963), Redinger (*Ark. Bot.* **26A** (1), 1933).

Graphinella Zahlbr. (1923) = Spirographa (Odontrotrem.) fide Santesson (*The lichens and lichenicolous fungi of Sweden and Norway*, 1993).

Graphinomyces Cif. & Tomas. (1953) = Graphina (Graphid.).

Graphiocladiella H.P. Upadhyay (1981), Anamorphic ? Micrascales, 1/2.C1.10. Teleomorph *Ceratocystis*. 1, Canada. See Tsuneda & Hiratsuka (*CJB* **62**: 2618, 1984; SEM conidiogenesis).

Graphiola Poit. (1824), Graphiolaceae. *c.* 5, trop and glasshouses. *G. phoenicis* on *Phoenix*). See Cole (1983).

Graphiolaceae Clem. & Shear (1931), Graphiolales. 2 gen. (+ 3 syn.), 6 spp. Sori stromatic, black, simple or compound, cup-shaped, surrounding a fertile region of small, pale, basipetally produced spores; after meiosis the nuclei migrate into lateral whorls of propagules (readily detached) which germinate to form yeast-like cells (or hyphae) from which reinfection of the host (*Palmae*) is established. This small family has often been included in *Ustilaginales* but lacks the thick-walled probasidium (ustilospore) and the metabasidium morphology normal in that group; although it has been excluded from major recent compilations on smuts. Recent molecular data indicate that it is closely allied to them (Sugiyama *et al.*, *Proc. Int. Symp. Microbiol. Ecol.* **7**: in press).
Lit.: Cole (*Mycol.* **75**: 93, 1983), Oberwinkler *et al.* (*Pl. Syst. Evol.* **140**: 251, 1982), Blanz & Gottschalk (1986; affin. with *Exobasidium* and some graminicolous smuts from 5S ribosomal RNA data)

Graphiolales, Ustomycetes. 1 fam., 2 gen. (+ 3 syn.), 6 spp. Fam.:
Graphiolaceae.

Graphiolites Fritel (1910), Fossil fungi. 1 (Eocene), France.

Graphiopsis Trail (1889) = Cladosporium (Mitosp. fungi) fide Mason & Ellis (*Mycol. Pap.* **137**, 1953).

Graphiopsis Bainier (1907) = Phaeoisaria (Mitosp. fungi) fide Mason (1937).

Graphiothecium Fuckel (1870), Mitosporic fungi, 2.A2.?. 5, N. temp.

Graphis Adans. (1763), Graphidaceae (L). *c.* 300, cosmop. (mainly trop.). See Nakanishi (1966), Wirth & Hale (1963), Redinger (*Ark. Bot.* **27A** (3), 1935).

graphium (pl. -ia), the synnema of *Graphium*.

Graphium Corda (1837), Anamorphic Ophiostomataceae, 2.A1.19. Teleomorph *Ophiostoma*. *c.* 10, widespr. See Ellis (*DH*; key), Sutton & Laut (*Bi-mon. Res. Nts Can. For.* **26**: 25, 1970; conidiogenesis), Wingfield *et al.* (*MR* **95**: 1328, 1991; generic synonymy), Seifert & Okada (*in* Wingfield *et al.* (Eds), *Ceratocystis and Ophiostoma*: 27, 1993; anamorphs of *Ophiostoma*).

Graphostroma Piroz. (1974), Graphostromataceae. 1, N. Am. See Barr *et al.* (*Mycotaxon* **48**: 529, 1993; posn), Glawe & Rogers (*CJB* **64**: 1493, 1986). Anamorph *Nodulisporium*-like.

Graphostromataceae M.E. Barr, J.D. Rogers & Y.M. Ju (1993), Calosphaeriales. 1 gen., 1 sp. Stroma two-layered; ascomata short-necked with separate ostioles; asci with a faint I+ apical ring. Anamorph *Nodulisporium*-like.

Graphyllium Clem. (1901), Hysteriaceae. 3, N. Am. See Barr (*SA* **12**: 27, 1993), Shoemaker & Babcock (*CJB* **70**: 1617, 1992; posn).

Greeneria Scribn. & Viala (1887), Mitosporic fungi, 6.A2.15. 1, widespr. See Sutton (1980), Kao *et al.* (*Pl. Prot. Bull. Taiwan* **32**: 256, 1990; pathology of *G. uvicola*).

gregarious, in companies or groups but not joined together.

Greletia Donadini (1980) = Smardaea (Otid.) fide Korf & Zhuang (*Mycotaxon* **40**: 413, 1991). See Donadini (*Bull. Soc. linn. Provence* **36**: 139, 1985; posn), Pfister (*Sydowia* **38**: 235, 1986; key 2 N. Am. spp.).

Gremlia Nieuwl. (1916) ≡ Polycephalum (Fungi, inc. sed.).

Gremmenia Korf (1962) = Phacidium (Phacid.) fide Reid & Pirozynski (*Mycol.* **60**: 526, 1968).

Gremmeniella M. Morelet (1969), Leotiaceae. 1 (conifer canker), N. Hemisph. See Petrini *et al.* (*CJB* **67**: 2805, 1989). Anamorph *Brunchorstia*.

Greville (Robert Kaye; 1794-1866). An Edinburgh botanist; trained as a medical man. Chief writings: *Scottish cryptogamic flora* (1823-28), *Flora Edinensis* (1824). See Balfour (*Trans. bot. Soc. Edinb.* **8**: 463, 1866), Grummann (1974: 371), Stafleu & Cowan (*TL-2* **1**: 1009, 1976).

grex, (1) see plasmodium; (2) (of cultivated plants) progeny of an artificial cross between known parents.

Grifola Gray (1821), Coriolaceae. 1, temp.

Griggsia F. Stevens & Dalbey (1919), ? Dothideales (inc. sed.). 1, W. Indies.

Grilletia Renault & Bertrand (1885), Fossil fungi (Chytridiomycota). 1 (Carboniferous), France.

Grimmicola Döbbeler & Hertel (1983), Leotiaceae. 1 (on *Grimmia*), Marion Island.

Griphosphaerella Petr. (1927) = Monographella (Hyponectr.) fide Müller (*Rev. mycol.* **41**: 129, 1977).

Griphosphaeria Höhn. (1918) = Discostroma (Amphisphaer.) fide Brockmann (1976).

Griphosphaerioma Höhn. (1918), Amphisphaeriaceae. 1, N. Am. See Shoemaker (*CJB* **41**: 1419, 1963). Anamorph *Labridella*.

griseofulvin, ('fulvicin', 'grifulvin', 'grisactin'), a Cl-containing antibiotic from *Penicillium griseofulvum*, *P. nigricans*; described and named by Raistrick *et al.* in 1936, and independently as 'curling factor' by Brian *et al.* in 1946. See Brian (*Ann. Bot. Lond.* N.S. **13**: 59, 1949; *TBMS* **43**: 1, 1960; reviews); antifungal. Has been used as a systemic fungicide against plant pathogens (Rhodes *et al.*, *Ann. appl. Biol.* **45**: 215, 1957) and orally against dermatophyte infections in animals and humans (Gentles, *Nature* **182**: 476, 1958; Lauder & O'Sullivan, *Vet. Rec.* **70**: 949, 1958; Davies, *Antifungal chemotherapy*, 1980: 180).

Griseoporia Ginns (1984), Coriolaceae. 1, N. Am.

grisette, basidiomata of the edible *Amanita vaginata*.

grist, see Brewing.

Groenhiella J. Koch, E.B.G. Jones & S.T. Moss (1983), Nitschkiaceae. 1 (marine), Denmark.

Grosmannia Goid. (1936) = Ophiostoma (Ophiostomat.) fide de Hoog (1974).

Grove (William Bywater; 1848-1938). Headmaster of Birmingham High School for Boys, 1887-1900; Lecturer in Botany, Birmingham Municipal Technical School, 1905-27. Chief writings: *British rust fungi* (1913), *British stem- and leaf- fungi (coelomycetes)* (1935-7), Engl. transl. of Tulasne's *Carpologia* (1931). See Mason (*J. Bot., Lond.* **76**: 86, 1938), Buller (*North-west Naturalist* **13**: 30, 1938; portr.), Stafleu & Cowan (*TL-2* **1**: 1015, 1976).

Groveola Syd. (1921) = Uromyces (Puccin.) fide Dietel (1928).

Groveolopsis Boedijn (1951), Mitosporic fungi, 4.E1.19. 1, Java.

Groves (James Walton; 1906-1970), Canadian. Chief of Mycology Section, Canada Agriculture 1951-70. Published many papers on discomycetes (esp. *Dermateaceae*, *Sclerotiniaceae*) and seed-borne fungi (some with Skolko). *Obit.*: Shoemaker (*Mycol.* **63**: 1, 1971; portr.), Thomson (*Can. Fld Nat.* **86**: 177, 1972; portr.).

Grovesia Dennis (1960), Leotiaceae. 1, Venezuela.

Grovesiella M. Morelet (1969), Leotiaceae. 3, USA., Eur.

Grovesiella B. Erikss. (1970) ≡ Erikssonopsis (Leot.).

Grovesinia M.N. Cline, J.L. Crane & S.D. Cline (1983), Sclerotiniaceae. 1, N. temp. Anamorph *Cristulariella*.

growth form, see habit.

Growth rates. In fungi growth may be unrestricted or restricted.

(1) **Unrestricted growth** occurs when a fungus is grown in batch culture in a medium containing an excess of all nutrients. Under these conditions, growth is exponential and proceeds at the organism's maximum specific growth rate for the conditions (type of nutrients, temperature, pH etc.). Thus, $\frac{dM}{dt} = \mu_{max}M$ where M = fungal biomass, t = time and μ_{max} = maximum specific growth rate. The fastest μ_{max} recorded for a fungus in batch culture on a glucose-mineral salts medium is 0.61 h^{-1} (doubling time of 1.1 h) for *Geotrichum candidum*. During unrestricted growth, the total hyphal length and the number of tips of a mycelium increase at the same μ_{max}, and consequently the ratio (G, the hyphal growth unit) between these parameters is a constant (Trinci, *J. Gen. Microbiol.* **81**: 225, 1974). Thus, unrestricted growth of a mycelium involves the duplication of a physiological unit of growth (G) consisting of a hyphal tip and a certain length of hypha. G is species- and strain-specific (Trinci, *in* Jennings & Rayner (Eds), *Ecology and physiology of the fungal mycelium*: 23, 1984) varying from *c.* 48 μm (*Penicillium chrysogenum*) to *c.* 682 μm (*Fusarium vaucerium*).

(2) **Restricted growth** of a fungus occurs when not all nutrients (including O_2) are present in excess [e.g. in a chemostat culture (Righelato, *in* Smith & Berry (Eds), *The filamentous fungi* **1**: 79, 1975)] or when factors such as nutrient concentration, pH or mycelial morphology are altered sufficiently to affect μ_{max} [e.g. in submerged batch culture in which growth is decelerating because of nutrient depletion or in which mycelial pellets (spherical colonies) are formed (Trinci, *Arch. Mikrob.* **73**: 353, 1970)]. Batch growth of a fungus on a solid medium eventually results in the establishment of conditions (e.g. nutrient depletion, change in pH etc.) below the centre of the colony which are less favourable for growth than was initially the case. Consequently, mature colonies of fungi increase in radius at a linear rate (K_r). Thus, $R = R_0 + K_r(t_1 - t_0)$ where t_0 = time at onset of linear expansion of colony and R_0 = colony radius at t_0. For colonies expanding in radius at a linear rate, growth in a peripheral ring of biomass (w, the peripheral growth zone) occurs at approximately μ_{max} but growth proceeds at below μ_{max} elsewhere in the colony, often falling to zero or near zero at the colony centre, i.e. a colony is heterogeneous and growth of most of its biomass is restricted. Expansion in radius of a mature colony is described by the following equation (Trinci, *J. Gen. Microbiol.* **67**: 325, 1971) $K_r = w\mu$. It follows from this equation that K_r cannot be used to study the effect of an environmental variable on growth if the variable alters w.

Lichens are amongst the slowest growing organisms known; measured rates vary from 0.01-90 mm marginal growth p.a. (most in range 1.0-6.0 mm). Minutely crustose species are generally the slowest, and fruticose ones (e.g. *Ramalina menziesii*) the fastest. Some species have different growth rates in different geographical regions, and competition may limit growth of adjacent thalli (Pentecost, *Lichenologist* **12**: 135, 1980). Growth rates of some species are reported to be accelerated by chemicals (see Barashkova, *Problems of the North* **7**: 149, 1964). A few attempts at modelling lichen growth have been made (Hill, *Lichenologist* **13**: 265, 1981; Topham (*in* Seaward, *Lichen ecology*: 35, 1977; review).

For reviews see Gow & Gadd (Eds) (*The growing fungus*, 1995), Wessels (*in* Hawksworth (Ed.), *Frontiers in mycology*: 27, 1991; molec. aspect).

See also Ecology, Lichenometry, Longevity, Physiology.

Gruby (David; 1810-98). An Austrian by birth, studied medicine at Vienna and then took up research at the Hospital St Louis and the Foundling Asylum in Paris. During 1841-44 he published the first descriptions of the fungi causing favus (*Trichophyton schoenleinii*), thrush (*Candida albicans*), and ringworm of the beard (*T. mentagrophytes*) and of the scalp (*Microsporum audouinii*) in humans. From 1844 he practised medicine in Paris. See Zakon & Benedek (*Bull. Hist. Med.* **16**: 155, 1944; portr. and Engl. transl. five papers).

Grubyella M. Ota & Langeron (1923) = Trichophyton (Mitosp. fungi).

Guceviczia Glezer (1959) = Wojnowicia (Mitosp. fungi) fide Sutton (*Česká Myk.* **29**: 97, 1975).

Gudelia Henssen (1995), Gloeoheppiaceae (L). 1, Mexico.

Guedea Rambelli & Bartoli (1978), Mitosporic fungi, 1.C2.1. 2, Ivory Coast, NZ.

Gueguenia Bainier (1907) = Amblyosporium (Mitosp. fungi) fide Pirozynski (1969).

Guelichia Speg. (1886), Mitosporic fungi, 3.A1.?. 1, S. Am. See Petrak & Sydow (*Ann. Myc.* **34**: 38, 1936).

Guepinella Bagl. (1870) = Heppia (Hepp.).

Guepinia Fr. (1825) [non Bastard (1821), *Cruciferae*] = Tremiscus (Exid.) fide Donk (*Persoonia* **4**: 185, 1966).

Guepinia Hepp (1864) [non Bastard (1821), *Cruciferae*] = Heppia (Hepp.).

Guepiniopsis Pat. (1883), Dacrymycetaceae. 3 or 4, widespr. See Kennedy (*Mycol.* **50**: 874, 1959), McNabb (*N.Z. Jl Bot.* **3**: 159, 1965; monotypic).

Guignardia Viala & Ravaz (1892) nom. cons., Mycosphaerellaceae. *c.* 40, widespr. *G. aesculi* (leaf blotch of horse chestnut, *Aesculus*), *G. bidwellii* (black rot of grapes, *Vitis*), *G. camelliae* (copper blight of tea, *Camellia*), *G. vaccinii* (scald or blast of cranberry, *Vaccinium*). See Eriksson & Hawksworth (*SA* **5**: 132, 1986), Punithalingam (*Mycol. Pap.* **136**, 1974; culture), Petrak (*Sydowia* **11**: 435, 1958; status), van der Aa (*Stud. Mycol.* **5**, 1973; *Phyllosticta* anamorphs). Anamorphs *Phyllosticta, Leptodothiorella.*

Guignardiella Sacc. & P. Syd. (1902) ≡ Vestergrenia (Dothid.).

Guilliermond (Marie Antoine Alexandre; 1876-1945). Botanist, cytologist, mycologist. Worked at the University of Lyons and from 1921 in Paris where he became Professor of Botany at the Sorbonne. Membre de l'Institut, 1935. Noted as a mycologist for his studies on the cytology, sexuality, and phylogeny of yeasts. *Les levures* (1912; [English transl. Tanner, 1920]). Founder of *Revue de cytologie et de cytophysiologie vegetales* (1935 -). *The cytoplasm of the plant cell* (1941) gives special attention to fungi and has a long list of Guilliermond's publications. *Obit.*: Verona (*Mycopath.* **4**: 124, 1948), Heim (*Notice sur la vie et les travaux de Alexandre Guilliermond (1876-1945) déposée en la séance du 21 avril 1947.* Institut de France Acad. Sci.).

Guilliermondella Nadson & Krassiln. (1928), ? Metschnikowiaceae. 1, former USSR. See v. Arx & Yarrow (*Ant. v. Leeuwenhoek* **50**: 799, 1984).

Guilliermondia Boud. (1904), ? Sordariaceae. 1, Eur.

Guilliermondia Nadson & Konok. (1911) ≡ Nadsonia (Saccharomycod.).

Guizhounema X. Mu (1977), Fossil fungi. 1, China.

gummosis, a plant disease having secretion of 'gum' as a well-marked symptom; of cucumber (*Cladosporium cucumerinum*).

Gussonea Tornab. (1848) = Pleopsidium (Lecanor.) fide Hafellner (*Nova Hedw.* **56**: 281, 1993).

guttate, (1) having tear-like drops; (2) (of a pileus), marked as if by drops of liquid.

Guttularia W. Oberm. (1913) ? = Melanospora (Ceratostomat.).

guttulate (of spores), having one or more oil-like drops (**guttules**) inside.

Guttulina Cienk. (1874) [non Orb. (1839), *Invertebrata*], Guttulinaceae. 2, widespr.

Guttulinaceae Zopf ex Berl. (1888), Acrasales. 1 gen. (+ 2 syn.), 4 spp. Sorocarp cells alike; sorocarp arising from pseudoplasmodial plaques.

Guttulinopsidaceae, see *Guttulinaceae*.

Guttulinopsis Olive (1901), Guttulinaceae. 2, N. Am. See Raper *et al.* (*Mycol.* **69**: 1016, 1977).

Gutturomyces Rivolta (1884) ? = Aspergillus (Mitosp. fungi) fide Sutton (*in litt.*).

Gyalecta Ach. (1808), Gyalectaceae (L). 30, mainly temp. See Lettau (*Beih. Feddes Repert.* **69**(2), 1937), Vězda (*Sborn. Vys. Zemšd. Lesn. Brnŏ* **1**: 21, 1958; *Folia geobot. phytotax.* **4**: 443, 1969).

Gyalecta Eaton (1829) [non Ach. (1808)] = Diploschistes (Thelotremat.).

Gyalectaceae Stizenb. (1862), Gyalectales (L). 8 gen. (+ 18 syn.), 96 spp. Thallus filamentous or crustose; ascomata apothecial, cup-shaped, partly immersed in the thallus; interascal tissue of paraphyses, usually unbranched, with knob-like apices; asci cylindrical, thin-walled at the apex, the walls blueing in iodine, without a tholus or apical structure; ascospores usually hyaline, transversely septate or muriform. Anamorph pycnidial, not prominent. Mainly lichen-forming with green, esp. trentepohlioid photobionts; mainly trop. or in humid situations, some foliicolous. *Lit.*: Hansen *et al.* (*Herzogia* **7**: 367, 1987; Greenland), Lettau (*Beih. Repert. spec. nov. regni veg.* **69**, 1932-37), Santesson (1952), Vězda (*Folia geobot. phytotax.* **1**: 311, 1966; **4**: 443, 1969).

Gyalectales, Ascomycota (L). 1 fam., 9 gen. (+ 20 syn.), 108 spp. Fam.: **Gyalectaceae**. The *Orbiliaceae* (*Leotiales*) may be better placed as a second family in this Order. *Lit.*: see *Gyalectaceae.*

Gyalectella J. Lahm (1883) = Dimerella (Gyalect.).

Gyalectidium Müll. Arg. (1881), Gomphillaceae (L). 3, cosmop. (trop.). See Santesson (*Symb. bot. upsal.* **12** (1), 1952), Vězda (*Folia geobot. phytotax.* **14**: 43, 1979), Vězda & Poelt (*Folia geobot. phytotax.* **22**: 179, 1987).

Gyalectina Vězda (1970) ≡ Cryptolechia (Gyalect.).

Gyalectomyces E.A. Thomas ex Cif. & Tomas. (1953) = Gyalecta (Gyalect.).

Gyalidea Lettau ex Vězda (1966) nom. cons. prop., Solorinellaceae (L). 24, Antarctica, Eur., Afr., Asia, C., N. & S. Am., Australasia. See Vězda & Poelt (*Nova Hedw.* **53**: 99, 1991; key), Vězda (*Folia geobot. phytotax.* **1**: 311, 1966; 1979, key), Vězda & Poelt (*Phyton* **30**: 47, 1990; 21 spp.).

Gyalideopsis Vězda (1972) nom. cons., Gomphillaceae (L). 28, widespr. See Kalb & Vězda (*Nova Hedw.* **58**: 511, 1994; trop.), Vězda (*Folia geobot. phytotax.* **14**: 43, 1979; key).

Gyalolechia A. Massal. (1852) = Fulgensia (Teloschist.) fide Kärnefelt (*Crypt. Bot.* **1**: 147, 1989).

Gyelnik (Vilmos Köfaragó; 1906-45). Received PhD from Budapest Univ. 1929 and worked at the Budapest National Museum during the German occupation. Died in Austria as a result of a bomb attack. Published about 100 papers on lichens 1926-42 proposing new names mainly in *Alectoria, Nephroma, Parmelia,* and *Peltigera.* His work has been much criticized. Collections now in BP (? some missing); see Verseghy (*Typen-Verzeichnis der Flechtensammlung,* 1964; catalogue of types in BP). See Sjödin (*Acta Horti gothoburg.* **19**: 113, 1954; bibl., cat. new names), Hillmann (*Feddes Repert.* **46**: 132, 1939), Kušan (*Ann. Myc.* **32**: 57, 1934; criticism), Grummann (1974: 451), Stafleu & Cowan (*TL-2* **1**: 1027, 1976).

Gymnascella Peck (1884), Gymnoascaceae. 12, widespr. See Currah (1985).

Gymnoascaceae Baran. (1872), Onygenales. 9 gen. (+ 9 syn.), 21 spp. Ascomata not stipitate; peridium formed from intertwined hyphae, sometimes with complex thick-walled appendages; ascospores ± oblate, often with polar and/or equatorial thickenings, often minutely ornamented. Anamorphs absent, or arthric. Keratinophilic or cellulolytic, usually isol. ex soil, cosmop.

Included in the *Onygenaceae* by v. Arx (1981) and Malloch (1981).

Lit.: Currah (*Mycotaxon* **24**: 1, 1985; monogr.), Ghosh (*Proc. Indian Acad. Sci.* (Pl. Sci.) **94**: 197, 1986; physio-ecology, *Kavaka* **12**: 1, 1986; phylogeny).

Gymnoascales, see *Onygenales*.

Gymnoascoideus G.F. Orr, K. Roy & G.R. Ghosh (1977), Arthrodermataceae. 1, widespr. See Currah (*Mycotaxon* **24**: 1, 1985, *in* Hawksworth (Ed.), *Ascomycete systematics*: 291, 1994; posn).

Gymnoascopsis C. Moreau & M. Moreau (1959) ? = Ascosorus (Fungi, inc. sed.) p.p. and Myriangiella (Myriang.) p.p fide v. Arx & Müller (1975).

Gymnoascus Baran. (1872), Gymnoascaceae. 2, N. temp. See Orr (*Mycotaxon* **5**: 470, 1977; key), v. Arx (1981; *Persoonia* **13**: 173, 1986; key 14 spp. s.l.). Anamorphs *Chrysosporium* or *Malbranchea*-like.

gymnocarpous (of a sporocarp), open, with the hymenium appearing and developing to maturity exposed and not enclosed. Cf. angiocarpous.

Gymnocaulon P.A. Duvign. (1956) = Stereocaulon (Stereocaul.).

Gymnochilus Clem. (1896) [non Blume (1859), *Orchidaceae*] = Psathyrella (Coprin.) fide Singer (1951).

Gymnoconia Lagerh. (1894), Phragmidiaceae. 2, widespr.; 0, I, III. *G. nitens* (orange rust of *Rubus*). See Laundon (*Mycotaxon* **3**: 133, 1975).

Gymnocybe P. Karst. (1879) [non Fr. (1825) nom. rej., *Musci*], nom. dub. (Basidiomycetes) fide Singer & Smith (1946).

Gymnoderma Humb. (1793) nom. rej. (Thelephor.).

Gymnoderma Nyl. (1860) nom. cons., Cladoniaceae (L). 4, Asia, N. Am. See Ahti (1993), Yoshimura & Sharp (*Am. J. Bot.* **55**: 635, 1968).

Gymnodermatomyces Cif. & Tomas. (1953) ≡ Gymnoderma (Cladon.).

Gymnodiscus Zukal (1887) [non Less. (1831), *Compositae*] ≡ Zukalina (Ascobol.).

Gymnodochium Massee & E.S. Salmon (1902), Mitosporic fungi, 3.B1.?. 1, Eur.

Gymnoeurotium Malloch & Cain (1973) = Edyuilla (Trichocom.) fide Benny & Kimbrough (*Mycotaxon* **12**: 1, 1980).

Gymnogaster J.W. Cribb (1956), Podaxaceae. 1, Australia.

Gymnoglossum Massee (1891), Bolbitiaceae. 1, Australia. See Smith (*Mycol.* **58**: 122, 1966).

Gymnogomphus Fayod (1889) nom. dub. (Gomphid.) fide Singer & Smith (1946).

Gymnographa Müll. Arg. (1887) = Sarcographa (Graphid.).

Gymnographoidea Fink (1930), ? Arthoniaceae (L). 1, Puerto Rico.

Gymnographomyces Cif. & Tomas. (1953) ≡ Gymnographa (Graphid.).

Gymnographopsis C.W. Dodge (1967), ? Graphidaceae (L). 3, Chile.

Gymnohydnotrya B.C. Zhang & Minter (1989), Helvellaceae. 3, Australia.

Gymnomitrula S. Imai (1941) = Heyderia (Leot.) fide Maas Geesteranus (*Persoonia* **3**: 81, 1964).

Gymnomyces Massee & Rodway (1898), Elasmomycetaceae. 9, widespr. See Singer & Smith (*Mem. Torrey bot. Cl.* **21** (3), 1960; key).

Gymnomycota, see *Myxomycota*.

Gymnomyxa, see *Myxomycota*.

Gymnopaxillus E. Horak (1966), Xerocomaceae. 1, Chile.

Gymnopeltis F. Stevens (1924) = Lecideopsella (Schizothyr.) fide Eriksson & Hawksworth (*SA* **9**: 15, 1991).

Gymnopilus P. Karst. (1879), Cortinariaceae. *c.* 175, widespr. See Hesler (*Mycol. Mem.* **3**, 1969; 78 N. Am. spp., keys).

gymnoplast, see protoplast.

Gymnopuccinia K. Ramakr. (1951) = Didymopsorella (Uropyxid.) fide Cummins & Hiratsuka (1983).

Gymnopus (Pers.) Roussel (1806) = Mycena (Tricholomat.) fide Singer & Smith (1946).

Gymnopus (Quél.) Quél. ex Moug. & Ferry (1887) ≡ Rostkovites (Bolet.).

Gymnosphaera Tassi (1902) [non Blume (1828), *Cyatheaceae*] = Stagonospora (Mitosp. fungi) fide Saccardo & Saccardo (*Syll. Fung.* **18**: 361, 1906).

Gymnosporangium R. Hedw. ex DC. (1805), Pucciniaceae. 57, widespr; 0, I generally on *Rosaceae*; III on *Juniperaceae*. See Kern (*Revised taxonomic account of Gymnosporangium*, 1972; key, 400 refs.). *G. nootkatense* has II; *G. bermudianum* is autoecious; *G. juniperi-virginianae* (apple (*Malus*)-cedar (*Juniperus*) rust of USA); *G. clavipes* (quince (*Cydonia*) rust).

Gymnosporium Corda (1833) nom. dub. fide Hughes (*CJB* **36**: 727, 1958).

Gymnostellatospora Udagawa, Uchiy. & Kamiya (1993), Myxotrichaceae. 1, Japan.

Gymnotelium Syd. (1921) = Gymnosporangium (Puccin.) fide Kern (1972).

gymnothecium, an ascoma in which the peridium is a loose hyphal network, typical of *Gymnoascaceae*.

Gymnotrema Nyl. (1858) = Gyrostomum (Graphid.).

Gymnoxyphium Cif., Bat. & I.J. Araujo (1963), Mitosporic fungi, 8.B1.?. 2, USA, Brazil.

gynophore (of *Pyronemataceae*), the multinucleate female structure undergoing development.

Gyoerffyella Kol (1928), Mitosporic fungi, 1.F1.?1. 6 (aquatic), Eur., N. Am. See Marvanová *et al.* (*Persoonia* **5**: 29, 1967, key), Marvanová (*TBMS* **65**: 555, 1975), Dudka & Mel'nik (*Mikol. i Fitopatol.* **24**: 13, 1990; USSR spp.).

Gypsoplaca Timdal (1990), Gypsoplacaceae (L). 1, Greenland, N. Am., China.

Gypsoplacaceae Timdal (1990), Lecanorales (L). 1 gen., 1 sp. Thallus foliose, squamiform, with well-developed rhizoids; ascomata effuse, irregular, without a margin or wall, asci and interascal tissue developing directly from cortical hyphae; interascal tissue of branched and anastomosing paraphyses; asci with a well-developed apical cap and a stronger-staining ring around the apical cushion, usually with an ocular chamber; ascospores simple, hyaline, without a sheath. Anamorph pycnidial. Lichenized with green algae.

Gyraria Nees (1816) = Tremella (Tremell.) fide Donk (*Persoonia* **4**: 179, 1966).

gyrate (**gyrose**), curved to the back and to the front in turn; folded and wavy; convoluted like a brain; (of an apothecioid ascoma), concentrically folded, e.g. *Umbilicaria*.

Gyratylium Preuss (1855) ? = Sphaeropsis (Mitosp. fungi). See Sutton (1977).

Gyrocephalus Pers. (1824) nom. rej. = Gyromitra (Helvell.).

Gyrocephalus Bref. (1888) = Tremiscus (Exid.) fide Donk (1966).

Gyrocerus Corda (1837) ? = Circinotrichum (Mitosp. fungi).

Gyrocollema Vain. (1929), Lichinaceae (L). 1, C. Am.

Gyrocratera Henn. (1899) = Hydnotrya (Helvell.) fide Trappe (1975).

Gyrodon Opat. (1836), Gyrodontaceae. 8, widespr. = Uloporus (Gyrodont.) fide Donk, *Reinwardtia* **3**: 289, 1955).

Gyrodontaceae (Singer) Heinem. (1961), Boletales. 9 gen. (+ 11 syn.), 40 spp. Basidioma boletoid; spores brown, subglobose to ellipsoid.

Gyrodontium Pat. (1900), Coniophoraceae. 1, widespr. See Maas Geesteranus (*Persoonia* **3**: 187, 1964).

Gyrolophium Kunze ex Krombh. (1831) = Dictyonema (Merul.) fide Parmasto (1978).

Gyromitra Fr. (1849) nom. cons., Helvellaceae. *c.* 15, mainly N. temp. *G. esculenta* (the lorchel) has a heat-labile toxin, **gyromitrin** (q.v.), and is carcinogenic; see Azema (*Doc. Mycol.* **10**: 1, 1979). See also Gibson & Kimbrough (*CJB* **66**: 1743, 1988; ascosporogenesis, *MR* **95**: 421, 1991; septa), Kimborough (*MR* **95**: 421, 1991; septal structure). See also mycetism.

gyromitrin (N-formylhydrazone), heat-labile, carcinogenic, cellular toxin produced by *Gyromitra esculenta* (the lorchel; false morel); breaks down to monomethylhydrazine (MMH), which is also extremely toxic.

Gyromitrodes Vassilkov (1942) = Pseudorhizina (Helvell.) fide Pouzar (*Česká Mykol.* **15**: 42, 1961).

Gyromium Wahlenb. (1812) ≡ Umbilicaria (Umbilicar.).

Gyromyces Göpp. (1844), ? Fossil fungi. 1 (Carboniferous), Germany.

Gyrophana Pat. (1897) = Serpula (Coniophor.) fide Cooke (1957).

Gyrophanopsis Jülich (1979), Corticiaceae. 1, NZ.

Gyrophila Quél. (1886) = Tricholoma (Tricholomat.) fide Singer & Smith (1946).

Gyrophora Ach. (1803) ≡ Umbilicaria (Umbilicar.).

Gyrophora Pat. (1887) ≡ Gyrophana (Coniophor.).

Gyrophoromyces E.A. Thomas ex Cif. & Tomas. (1953) ≡ Umbilicaria (Umbilicar.).

Gyrophoropsis Elenkin & Savicz (1910) = Umbilicaria (Umbilicar.).

Gyrophragmium Mont. (1843), Podaxaceae. 3, widespr., warm temp. See Kreisel (*Feddes Repert.* **83**: 577, 1973).

Gyrophthorus Hafellner & Sancho (1990), Ascomycota (inc. sed.). 2 (on lichens), Eur.

Gyropodium E. Hitchc. (1825) = Calostoma (Calostomat.) fide Fischer (1938).

Gyroporus Quél. (1886), Gyrodontaceae. 9, widespr.

Gyrostomomyces Cif. & Tomas. (1953) ≡ Gyrostomum (Graphid.).

Gyrostomum Fr. (1825), Graphidaceae (L). 4, trop. See Hale (*Bull. Br. Mus. nat. Hist., Bot.* **8**: 227, 1981).

Gyrostroma Naumov (1914), Mitosporic fungi, 8.A1.?. 3, Am., former USSR.

Gyrothecium Nyl. (1855) = Sporostatia (Acarospor.).

Gyrothrix (Corda) Corda (1842), Mitosporic fungi, 1.A1.6. 15, widespr., mainly trop. See Cunningham (*Mycol.* **66**: 127, 1974; key).

Gyrothyrium Arx (1950) = Schizothyrium (Schizothyr.) fide Müller & v. Arx (1962).

Gyrotrichum Spreng. (1827) ≡ Circinotrichum (Mitosp. fungi).

H, Botanical Museum, University of Helsinki (Finland); founded *c.* 1750.

H bodies, pairs of sporidia of *Tilletia* fused in pairs while still attached to the promycelium.

Haasiella Kotl. & Pouzar (1966), Tricholomataceae. 2, Eur.

habitat, natural place of occurrence of an organism.

Habrostictis Fuckel (1870) = Orbilia (Orbil.) fide Baral (*SA* **13**: 113, 1994).

Habrostictis Clem. (1909) nom. dub (Mitosp. fungi). See Hein (*Willdenowia, Beih.* **9**, 1976).

Haddowia Steyaert (1972), Haddowiaceae. 2, pantrop.

Haddowiaceae Jülich (1982), Ganodermatales. 1 gen., 2 spp. Basidioma stipitate; spores ridged.

Hadotia Maire (1906) = Lophodermium (Rhytismat.) fide v. Arx & Müller (1975).

hadromycosis, term coined by Pethybridge (*Sci. Proc. Roy. Dublin Soc.* **15**: 63, 1916) for a plant disease in which the pathogen is confined to the xylem; tracheomycosis. 'Vascular wilt' is the preferred term when wilting is a symptom. Cf. wilt.

Hadronema Syd. & P. Syd. (1909), Mitosporic fungi, 1.B2.?. 1, Japan.

Hadrospora Boise (1989), Phaeosphaeriaceae. 2, Eur., N. Am.

Hadrosporium Syd. (1938), Mitosporic fungi, 3.C2.19. 1, Australia.

Hadrotrichum Fuckel (1865), Mitosporic fungi, 3.A2.10. 5, temp.

Haematomma A. Massal. (1852) typ. cons. prop., Haematommataceae (L). *c.* 10, cosmop. See Rogers (*Lichenologist* **14**: 115, 1982; Australia), Rogers & Bartlett (*Lichenologist* **18**: 247, 1986; NZ).

Haematommataceae Hafellner (1984), Lecanorales (L). 2 gen. (+ 2 syn.), 15 spp. Thallus crustose. Ascomata apothecial, sessile or immersed, with a thalline margin, the disc often red; hymenium blueing in iodine; interascal tissue of sparingly branched and anastomosing paraphyses; asci with a large I+ apical cap, sometimes with a very large conical ocular chamber, and an outer I+ gelatinized layer; ascospores elongate, multiseptate, hyaline, without a sheath. Anamorphs pycnidial. Lichenized with green algae, widspr.

Haematommatomyces Cif. & Tomas. (1953) ≡ Haematomma (Haematommat.).

Haematomyces Berk. & Broome (1875) Fungi (inc. sed.); based on resin.

Haematomyxa Sacc. (1884), Patellariaceae. 2, N. Am. See Pirozynski (*SA* **6**: 130, 1987).

Haematostereum Pouzar (1959) = Stereum (Ster.) fide Donk (1964).

haerangium, the sporulating organ of certain ascomycetes (e.g. *Fugascus*, *Ceratostomella*), classed by Falck as *Herangiomycetes*, in which the 8 ascospores developed from the **octophore** are contained by a membrane and surrounded by a circle of hairs (the **tentacle**) around the ostiole of the perithecium (Falck, 1947).

Hafellia Kalb, H. Mayrhofer & Scheid. (1986), Physciaceae (L). 5, Eur., Afr., N. Am. See Sheard (*Bryologist* **95**: 79, 1992; key 5 spp., N. Am.).

Hafellnera Houmeau & Cl. Roux (1984), Schaereriaceae (L). 1, Eur.

Hagenia Eschw. (1824) [non J.F. Gmel. (1791), *Angiospermae*] = Anaptychia (Physc.) fide Kurokawa (1962).

Haglundia Nannf. (1932), Dermateaceae. 5, Eur.

Hainesia Ellis & Sacc. (1884), Anamorphic Leotiaceae, 7.A1.15. Teleomorph *Pezizella*. 1, widespr. See Sutton & Gibson (*CMI Descriptions* no. **535**, 1977), Palm (*Mycol.* **83**: 787, 1992; synanamorph of *Pilidium*).

hair (in *Agaricales*), one of the hair-shaped epicuticular elements forming a pilose covering or down under a lens and not homologous with a cystidium, pseudoparaphysis, or seta, e.g. in *Lachnella*, *Crinipellis* (as restricted by Singer, 1962: 61).

Halbania Racib. (1889), Asterinaceae. 1, Java.

Halbaniella Theiss. (1917) = Actinopeltis (Microthyr.) fide v. Arx & Müller (1975).

Halbanina G. Arnaud (1918) = Cirsosia (Asterin.) fide v. Arx & Müller (1975).

Hale (Mason Ellsworth jr., 1928-90). American lichenologist born in Connecticut, influenced by A.W.

Evans (1868-1959) while at Yale University, received PhD from the University of Wisconsin in 1953 under J.W. Thomson, and learnt microchemical techniques from Asahina (q.v.); employed at the Smithsonian Institution from 1957 until his death. Made a unique contribution to the understanding of the *Parmeliaceae*, preparing monographs of numerous segregate genera and also later to ostropalean lichens. His keys and textbooks contributed to a revival of interest in N. American lichenology (*Lichen handbook*, 1957; *How to know the lichens*, edn 2, 1979; *The biology of lichens*, edn 3, 1983), and with V. Ahmadjian he edited *The lichens* (1974 ['1973']). Made some 80,000 collections from the arctic, antarctic, and esp. the tropics as well as the US; these are in the Smithsonian Institution (US). Also a collector and user of old printing machines.

Lit.: Ahmadjian (*Endocyt. Cell Res.* **7**: 1, 1990; portr.), Culberson (*Bryologist* **94**: 90, 1991; portr.), Grummann (1974: 189), Lawrey (*Lichenologist* **22**: 405, 1990; portr.), Sipman & Seaward (*Internat. Lich. Newsl.* **23**(2): 42, 1990).

Halecania M. Mayrhofer (1987), Catillariaceae (L). 6, Eur., N. Am.

Haleomyces D. Hawksw. & Essl. (1993), ? Verrucariaceae. 1 (on lichens, esp. *Oropogon*), neotrop.

Haligena Kohlm. (1961), Halosphaeriaceae. 1 (marine), widespr. See Johnson *et al.* (*CJB* **65**: 931, 1987).

haline, found near the sea shore (obsol.).

Haliphthoraceae Vishniac (1958), Salilagenidales. 2 gen., 5 spp. Holocarpic parasites of diatoms and marine invertebrates.

Lit.: Vishniac (*Mycol.* **50**: 75, 1958), Dick (*Pap. Mich. Acad. Sci.* **46**: 195, 1962).

Haliphthoros Vishniac (1958), Haliphthoraceae. 3 (on *Urosalpinx* eggs), USA. See Tharp & Bland (*CJB* **55**: 2936, 1977; biol. and host range), Dick (*in press*; key).

Halisaria Giard (1889) nom. dub (? Fungi).

hallucinogenic fungi (teonanácatl), basidiomata (of *Psilocybe* (*P. mexicana*, etc.; see Singer and Singer & Smith, *Mycol.* **50**: 239, 1958), *Stropharia*, *Panaeolus*, *Lycoperdon cruciatum*, *L. mixtecorum*, etc.) eaten by Mexican Indians during magical ceremonies to induce cerebral effects. A crystalline active principle has been isolated and named **psilocybin**.

Lit.: Heim, Wasson *et al.* (*Les champignons hallucinogènes du Mexique*, 1958; *Nouvelles investigations sur les champignons hallucinogènes*, 1967 [reprinted from *Archiv. Mus. Nat. d'Hist. Nat.* sér. 7 **6**, 9]). See ethnomycology, soma.

halmophagous (of ectotrophic mycorrhiza), having a mantle and a Hartig net (Burgeff, 1943).

Halobyssus Zukal (1893) = Monilia (Mitosp. fungi) fide Clements & Shear (1931).

Halocyphina Kohlm. & E. Kohlm. (1965), Cyphellaceae. 1 (marine), USA. See Ginns & Malloch (*Mycol.* **69**: 53, 1977).

Halographis Kohlm. & Volkm.-Kohlm. (1988), Roccellaceae (?L). 1 (marine), Australia, Belize. See Kohlmeyer & Volkmann-Kohlmeyer (*Crypt. Bot.* **2**: 367, 1992).

Haloguignardia Cribb & J.W. Cribb (1956), Phyllachoraceae. 5 (forming galls on marine *Fucales*), Australasia. See Kohlmeyer & Kohlmeyer (*Marine mycology*, 1979).

halonate, (1) (of a leaf-spot), having concentric rings; one of the 'frog-eye' type; (2) (of a spore), having a transparent coat around it.

Halonectria E.B.G. Jones (1965), Hypocreaceae. 1 (marine), UK.

Halonia Fr. (1849) [non Lindl. & Hutton (1833), *Pteridophyta*], Fungi (inc. sed.) fide v. Arx & Müller (*Beitr. Kryptfl. Schweiz.* **11**(1), 1954).

halophilic, tolerating salt; living in salt water; *Dendryphiella salina* is esp. well-studied and there appears to be no salt accumulation in its vacuoles (see Jennings, *in* Rodriguez-Valera (Ed.), *General and applied aspects of halophilic microorganisms*: 107, 1991; review, bibliogr.). See Marine fungi.

Halophiobolus Linder (1944) = Lulworthia (Halosphaer.) fide Kohlmeyer & Kohlmeyer (*Marine mycology*, 1979).

Halophytophthora H.H. Ho & S.C. Jong (1990), Pythiaceae. 11, widespr.

Halosarpheia Kohlm. & E. Kohlm. (1977), Halosphaeriaceae. 12 (marine or aquatic), Bermuda, N. & S. Am.

Halosphaeria Linder (1944), Halosphaeriaceae. 4 (marine), widespr. See Jones *et al.* (*Bot. Marina* **27**: 129-143, 1984), Kohlmeyer & Kohlmeyer (*Marine mycology*, 1979), Shearer & Crane (*Mycol.* **69**: 1218, 1978; anamorph). Anamorph *Periconia*-like, *Trichocladium*.

Halosphaeriaceae E. Müll. & Arx ex Kohlm. (1972), Halosphaeriales. 42 gen. (+ 6 syn.), 114 spp. Ascomata perithecial, usually solitary, immersed; peridium usually thin, membranous, hyaline or black; interascal tissue absent; asci clavate, thin-walled, evanescent, without apical structures; ascospores varied, hyaline, usually septate, occasionally ornamented, usually with complex mucous or cellular appendages. Anamorphs hyphomycetous, often absent. Marine, usually on submerged or intertidal wood.

Lit.: Farrant (*in* Moss (Ed.), *Biology of marine fungi*: 231, 1986; asci), Kohlmeyer (*CJB* **50**: 151, 1972; key 13 gen.), Kohlmeyer & Kohlmeyer (*Marine mycology*, 1979; monogr.), Jones *et al.* (*Bot. J. Linn. Soc.* **87**: 193, 1983; *Corollospora* s.l., *Bot. Marina* **27**: 129, 1984; *Halosphaeria* s.l.), Jones & Moss (*SA* **6**: 179, 1987; key 21 gen.), Nakagiri & Tubaki (*Bot. Mar.* **28**: 485, 1985; anamorphs).

Halosphaeriales, Ascomycota. 1 fam., 42 gen. (+ 6 syn.), 114 spp. Fam.:
    Halosphaeriaceae.
Lit.: Kohlmeyer & Volkmann-Kohlmeyer (*Bot. Marina* **34**: 1, 1991; key) and under fam.

Halosphaeriopsis T.W. Johnson (1958), Halosphaeriaceae. 1 (marine), widespr. See Jones *et al.* (*Bot. Marina* **27**: 129, 1984).

Halospora, see *Holospora*.

Halotthia Kohlm. (1963), ? Zopfiaceae. 1 (marine), Eur.

Halstedia F. Stevens (1920) = Phyllachora (Phyllachor.) fide Petrak (*Ann. Myc.* **32**: 317, 1934).

Halteromyces Shipton & Schipper (1975), Mucoraceae. 1, Australia. See Shipton & Schipper (*Ant. v. Leeuwenhoek* **41**: 337, 1975).

Halterophora Endl. (1836) nom. dub. (? Myxomycetes).

Halysiomyces E.G. Simmons (1981), Mitosporic fungi, 1.B/C1.4. 1, USA.

Halysium Corda (1837) nom. conf. fide Hughes (*CJB* **36**: 727, 1958).

hamanatto, an edible oriental product obtained by the fermentation of soybeans with *Aspergillus oryzae*; tao-cho (Malaysia); tao-si (Philippines).

hama-natto, see Fermented food and drinks.

Hamaspora Körn. (1877), Phragmidiaceae. 11 (on *Rubus*), Afr., Asia, Australasia. See Monoson (*Mycopath.* **37**: 263, 1969).

Hamasporella Höhn. (1912) = Hamaspora (Phragmid.) fide Dietel (1928).

hamate (**hamose, hamous**), hooked (Fig. 37.35); unciate. Cf. hamulate.

hamathecium (Eriksson, *Opera Bot.* **60**: 15, 1981), a neutral term for all kinds of hyphae or other tissues between asci, or projecting into the locule or ostiole of ascomata; usually of carpocentral origin; interascal tissues. Eriksson recognized seven categories (see Fig. 14A-F):

(A) **Interascal pseudoparenchyma**, carpocentral tissues unchanged or compressed between developing asci; e.g. *Wettsteinina*.

(B) **Paraphyses**, hyphae originating from the base of the cavity, usually unbranched and not anastomosed; e.g. *Pyrenula*, *Xylaria*.

(C) **Paraphysoids** (trabecular pseudoparaphyses; tinophyses), interascal or pre-ascal tissue stretching and coming to resemble pseudoparaphyses; often only remotely septate, anastomosing and very narrow (see Barr, *Mycol.* **71**: 935, 1979); e.g. *Patellaria*, *Melanomma*.

(D) **Pseudoparaphyses** (cellular pseudo-paraphyses; cataphyses), hyphae originating above the level of the asci and growing downwards between the developing asci, finally becoming attached to the base of the cavity and often also then free in the upper part; often regularly septate, branched and anastomosing and broader; e.g. *Pleospora*.

(E) **Periphysoids**, short hyphae originating above the level of the developing asci but not reaching the base of the cavity; e.g. *Nectria*, *Metacapnodium*.

(F) **Periphyses**, hyphae confined to the ostiolar canal; unbranched and not anastomosing; can occur in conjunction with (B), (D) or (E); e.g. *Gibberella*, *Pyrenula*.

(G) **Hamathecial tissue absent** (not figured), e.g. *Dothidea*.

Hamatocanthoscypha Svrček (1977), Hyaloscyphaceae. 7, temp. See Huhtinen (*Karstenia* **29**: 45, 1990; key).

Hamidia Chaudhuri (1942) nom. dub. (Saprolegn.).

Hamigera Stolk & Samson (1971) = Talaromyces (Trichocom.) fide Pitt (1980).

hamulate (hamulose), having little hooks. Cf. hamate.

Handkea Kreisel (1989) = Calvatia (Lycoperd.) fide Demoulin (*in litt.*).

hanging drop, see van Tieghem cell.

Hansen (Emil Christian; 1842-1909). Microbiologist and director of the physiological department of the Carlsberg Laboratories, Copenhagen, specialized in yeasts and coprophilous fungi. See Stafleu & Cowan (*TL-2* **2**: 46, 1979).

Hansenia P. Karst. (1879) [non Turcz. (1884), *Umbelliferae*] = Coriolus (Coriol.) fide Donk (1960).

Hansenia Zopf (1883) = Strattonia (Lasiosphaer.) fide Lundqvist (1972).

Hansenia Lindner (1905) = Hanseniaspora (Saccharomycod.) fide v. Arx (1981).

Hansenia Zikes (1911) = Hanseniaspora (Saccharomycod.) fide v. Arx (1981).

Hanseniaspora Zikes (1911), Saccharomycodaceae. 5 (fide Barnett *et al.*, 1990), widespr. See Meyer *et al.* (*Ant. v. Leeuwenhoek* **44**: 79, 1978). Anamorph *Kloeckera*.

Hansenula Syd. & P. Syd. (1919) = Pichia (Endomycet.) fide Kurtzman (*Ant. v. Leeuwenhoek* **50**: 209, 1984) but 30 spp. accepted in Kreger-van Rij (1984) and 3 spp. in Barnett *et al.* (1990).

Hansfordia S. Hughes (1951), Mitosporic fungi, 1.A2.11. 10, widespr. See Deighton (*TBMS* **59**: 531, 1972; synonymy). = Dicyma (Mitosp. fungi) fide v. Arx (1982).

Hansfordiella S. Hughes (1951), Mitosporic fungi, 1.C2.1. 5, Afr., Philipp. See Subramanian (*Proc. Ind. Acad. Sci.* B **45**: 282, 1957; key spp. on *Meliola* etc.).

Hansfordiellopsis Deighton (1960), Anamorphic Dothideales, 1.D2.1. Teleomorph *Koordersiella*. 5 (on lichens), Afr., S. Am. See Hawksworth (*Bull. Br. Mus. nat. Hist., Bot.* **6**: 181, 1979; key).

Hansfordina Bat. (1962) = Microcallis (Chaetothyr.) fide v. Arx & Müller (1975).

Hansfordiopeltis Bat. & C.A.A. Costa (1956), Mitosporic fungi, 5.A1.?. 2, Congo.

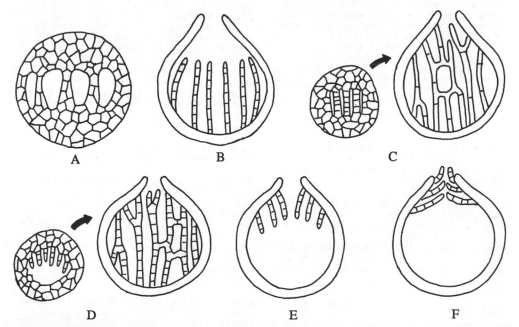

A   B   C

D   E   F

Fig. 14. Hamathecium terminology, following Eriksson (1981). See text for explanation.

Hansfordiopeltopsis M.L. Farr (1986), Mitosporic fungi, 5.G1.?. 1, Brazil.

Hansfordiopsis Bat. (1959), Micropeltidaceae. 1, Australia.

Hansfordiula E.F. Morris (1963) = Phaeoisaria (Mitosp. fungi) fide de Hoog & Papendorf (1976).

Hantzschia Auersw. (1862) [non Grunov (1877), *Bacillariophyceae*], nom. rej. = Phialocephala (Mitosp. fungi) fide Kendrick (1964).

Hapalocystis Auersw. ex Fuckel (1863), Melanconidaceae. 3, Eur., N. Am. See Barr (1978). Anamorph *Phoma*-like (Glawe, *Mycol.* **77**: 880, 1985).

Hapalopera Fott (1942) = Phlyctidium (Chytrid.) fide Sparrow (1960). But see Batko (1975).

Hapalophragmiopsis Thirum. (1950) = Hapalophragmium (Sphaerophragm.) fide Cummins & Hiratsuka (1983), Monoson (*Mycol.* **69**: 21, 1977), Lohsomboon *et al.* (*MR* **96**: 461, 1992).

Hapalophragmium Syd. & P. Syd. (1901), Sphaerophragmiaceae. 14 (on *Leguminosae*), trop. Afr., Asia. See Monoson (*Mycol.* **69**: 22, 1977; key), Lohsomboon *et al.* (*MR* **96**: 461, 1992).

Hapalopilus P. Karst. (1881), Coriolaceae. 5, widespr.

Hapalosphaeria Syd. (1908), Mitosporic fungi, 4.A1.15. 1 (in anthers), Eur.

Haplaria Link (1809) = Botrytis (Mitosp. fungi) fide Hennebert (*Persoonia* **7**: 188, 1973).

Haplariella Syd. & P. Syd. (1908), Mitosporic fungi, 4.A1.?. 1, S. Am.

Haplariella Sacc. (1931) = Haplariella Sydow (Mitosp. fungi) fide Carmichael *et al.* (1980).

Haplariopsis Oudem. (1903), Mitosporic fungi, 2.B1.?. 1, Eur.

Haplariopsis Henn. (1908) ≡ Haplariella (Mitosp. fungi).

haplo- (prefix), one only; single.

Haplobasidion Erikss. (1889), Mitosporic fungi, 1.A2.3. 3, Eur., Asia. See Ellis (*Mycol. Pap.* **67**, 1957; key).

haplobiontic, see Haplobionticae; cf. diplobiontic.

Haplobionticae. subclass of *Ascomycota* in which the life cycles consists of one thallus (biont); cf. Diplobionticae. Introduced by Nannfeldt (1932).

Haploblastia Trevis. (1860) = Strigula (Strigul.).

Haplocarpon M. Choisy (1936) = Porpidia (Porpid.).

Haplochalara Linder (1933) = Catenularia (Mitosp. fungi) fide Mason (*Mycol. Pap.* **5**: 121, 1941).

haploconidium (of *Tremellales*), a uninucleate conidium.

Haplocybe Clem. (1909) ? = Ombrophila (Leot.) fide Nannfeldt (*Nova Acta R. Soc. Scient. upsal.* 4, **8**(2), 1932).

Haplocystis Sorokīn (1874), ? Chytridiales (inc. sed.). 1, Italy.

Haplodina Zahlbr. (1930), Roccellaceae (L). 3, China.

haplodioicious, see heterothallic.

Haplodothella Werderm. (1923) = Vestergrenia (Dothid.) fide v. Arx & Müller (1954).

Haplodothis Höhn. (1911) = Mycosphaerella (Mycosphaerell.) fide v. Arx & Müller (1975).

haplogonidia (haplogonimia), gonidia (gonimia) in ones, not in groups (obsol.).

Haplographa Anzi (1860) = Lithographa (Rimular.).

Haplographites Félix (1894), Fossil fungi. 2 (Tertiary), Eur.

Haplographium Berk. & Broome (1859), Anamorphic Hyaloscyphaceae, 1.A1.10. Teleomorph *Hyaloscypha*. 15, esp. temp. See Zucconi & Pagano (*Mycotaxon* **46**: 11, 1993; gen. limits).

haploheteroecious, see heterothallic.

haploid, (1) (of a nucleus), having the *n* number of chromosomes; (2) (of a cell), having 1 haploid nucleus; (3) (of a mycelium), made up of haploid cells;

Haplolepis Syd. (1925), Mitosporic fungi, ?.?.?. 2, Eur. See Sutton (1977).

Haploloma Trevis. (1857), ? Lecanorales (inc. sed.). 1 (on lichens), Eur.

Haplomela Syd. (1925) = Melanconium (Mitosp. fungi) fide Clements & Shear (1931).

haplomitotic A nuclear cycle, occurrence of one mitotic haploid phase in the nuclear cycle (Dick, 1987); karyogamy - meiosis - mitosis - karyogamy -.

haplomitotic B nuclear cycle, occurrence of one mitotic diploid phase in the nuclear cycle (Dick, 1987); karogamy - mitosis - meiosis - karyogamy -.

haplomonoecious, see homothallic.

Haplomyces Thaxt. (1893), Laboulbeniaceae. 3, N. Am., Eur.

haplomycosis, see adiaspiromycosis.

haplont, the thallus of the haplophase; the gametophyte.

Haplopeltheca Bat., J.L. Bezerra & Cavalc. (1963), Micropeltidaceae. 1, Brazil.

Haplopeltis Theiss. (1914) = Muyocopron (Microthyr.) fide v. Arx & Müller (1975).

haplophase, the part of the life history in which the cells are haploid.

Haplophoma Riedl & Ershad (1977) = Phomopsis (Mitosp. fungi) fide Sutton (*in litt.*).

Haplophyse Theiss. (1916), ? Rhytismatales (inc. sed.). 1, Hawaii.

Haploporus Bondartsev & Singer (1944), Coriolaceae. 1, Eur. See Eriksson (*Symb. bot. upsal.* **16**(1): 160, 1958), Niemalä (*Ann. Bot. Fennici* **8**: 237, 1971).

Haploporus Bondartsev (1953) = Haploporus (Coriol.).

Haplopyrenula Müll. Arg. (1883) = Vizella (Vizell.) fide Santesson (*Svensk bot. Tidskr.* **43**: 547, 1949).

Haplopyrenulomyces Cif. & Tomas. (1953) ≡ Haplopyrenula (Vizell.).

Haplopyxis Syd. & P. Syd. (1920) = Uromyces (Puccin.) fide Baxter (1963).

Haploravenelia Syd. (1921), Raveneliaceae. 120, trop.; O, I, II, III on *Leguminosae*, etc. Not used by Arthur (1934) or Cummins (*Rust fungi on legumes and composites in North America*, 1978).

Haplospora Räsänen (1943) nom. inval. ≡ Haplopyrenula (Vizell.).

Haplosporangium Thaxt. (1914) = Mortierella (Mortierell.) fide Gams (1977).

Haplosporella, see *Aplosporella*.

Haplosporidium Caullery & Mesnil (1899) nom. dub. (? Fungi).

Haplosporidium Speg. (1912) [non Caullery & Mensil (1899)], Mitosporic fungi, 4.A1.?. 1, Argentina. ? = Chaetophoma (Mitosp. fungi) fide Trotter (*Syll. Fung.* **25**: 178, 1925), ? = Asteroma (Mitosp. fungi) fide Clements & Shear (*Gen. Fung.*: 357, 1931).

Haplosporium Mont. (1843) Fungi (inc. sed.) fide Petrak & Sydow (*Ann. Myc.* **27**: 114, 1929).

Haplostroma Syd. & P. Syd. (1916) = Coccodiella (Phyllachor.) fide Müller & v. Arx (1973).

haplostromatic, see stroma.

Haplostromella Höhn. (1917) nom. nud. = Strasseria (Mitosp. fungi) fide Sutton (1977).

haplosynoecious, see homothallic.

Haplotelium Syd. (1922) = Uromyces (Puccin.) fide Dietel (1928).

Haplotheciella Höhn. (1918) = Didymella (Dothideales) fide v. Arx & Müller (1975).

Haplothecium Theiss. & Syd. (1915) = Glomerella (Phyllachor.) fide v. Arx & Müller (1954).

Haplothelopsis Vain. (1921) = Thelopsis (Stictid.).

Haplotrichella Arnaud (1954) = Gliomastix (Mitosp. fungi) fide Carmichael *et al.* (1980).

Haplotrichum Link (1824) = Acladium (Mitosp. fungi) fide Hughes (1958). Used by Holubová-Jechová (*Česká Myk.* **30**: 3, 1976).

Haplotrichum Eschw. (1824) nom. dub. (Ascomycota, inc. sed.).

Haplovalsaria Höhn. (1919) = Didymosphaeria (Didymosphaer.) fide Aptroot (1995).

Hapsidascus Kohlm. & Volkm.-Kohlm. (1991), Ascomycota (inc. sed.). 1 (marine), Belize.

Hapsidomyces J.C. Krug & Jeng (1984), Pezizaceae. 1 (coprophilous), Venezuela.

Hapsidospora Malloch & Cain (1970), ? Pseudeurotiaceae. 1, Canada.

hapteron, (1) an aerial organ of attachment of some fruticose lichens (e.g. *Alectoria sarmentosa* subsp. *vexillifera*) formed by a secondary branch which becomes attached to the substratum; (2) attachment organ at base of a funicular cord in *Nidulariaceae*.

Haptocara Drechsler (1975), Mitosporic fungi, 1.C1.6. 1 (on nematodes), USA.

Haptoglossa Drechsler (1940), Ectrogellaceae. 6, N. Am., Eur. See Davidson & Barron (*CJB* **51**: 1317, 1973), Robb & Lee (*CJB* **64**: 1935, 1986; ultrastr. gun cell), Dick (*in press*; relationship with *Plasmodiophorales*).

haptonema, filamentous appendage (usually coiled) consisting of the plasma membrane, a sheath of endoplasmic reticulum, and a core of several microtubules anchored near the kinetosome.

Haptospora Barron (1991), Mitosporic fungi, 1.G1.15. 3 (endoparasitic in rotifers), W. Indies, NZ.

Haraea Sacc. & Syd. (1913), Meliolaceae. 1 or 2, Japan, W. Indies.

Haraella Hara & I. Hino (1955) [non Kudô (1930), *Orchidaceae*] ≡ Hinoa (Mitosp. fungi).

Harikrishnaella D.V. Singh & A.K. Sarbhoy (1972) = Chaetomella (Mitosp. fungi) fide Sutton & Sarbhoy (1978).

Hariotia P. Karst. (1889) [non Adans. (1763), *Cactaceae*] ≡ Delphinella (Dothid.).

Hariotula G. Arnaud (1917) = Cyclotheca (Microthyr.) fide Müller & v. Arx (1962).

Harknessia Cooke (1881), Anamorphic Melanconidaceae, 8.A2.19. Teleomorph *Cryptosporella*. 29, widespr. See Sutton (*Mycol. Pap.* **123**, 1971; key), Nag Raj & DiCosmo (*Bibl. Mycol.* **80**, 1981), Sutton & Pascoe (*MR* **92**: 431, 1989; development of ostiole), Nag Raj (1993), Crous *et al.* (*Mycol.* **85**: 108, 1993; S. Afr. spp.).

Harknessiella Sacc. (1889), Dothideales (inc. sed.). 1, N. Am.

Harmandiana de Lesd. (1914) nom. dub. (Ascomycota, inc. sed.).

Harmoniella Darling (1906), Mitosporic fungi, 1.A2.10. 1, Ukraine.

Harpagomyces Wilcz. (1911), Mitosporic fungi, 1.G2.1. 1, Eur.

Harpella L. Léger & Duboscq (1929), Harpellaceae. 2 (in *Diptera*), widespr. See Lichtwardt (*Mycol.* **59**: 482, 1967), Reichle & Lichtwardt (*Mycol.* **81**: 103, 1972; ultrastr.), Moss & Lichtwardt (*CJB* **58**: 1035, 1980; ultrastr., taxonomy), Lichtwardt (1986; key).

Harpellaceae L. Léger & Duboscq (1929), Harpellales. 5 gen., 24 spp. Thalli unbranched, producing trichospores; usually attached to the periotrophic membrane of aquatic *Diptera* larvae.

Harpellales, Trichomycetes. 2 fam., 26 gen. (+ 6 syn.), 103 spp. Thallus simple or branched, septate, with basal holdfast; asexual reproduction by lateral, monosporous, elongate sporangia (trichospores), which upon release exhibit one or more basally attached appendages; sexual reproduction by biconical zygospores; endocommensals or parasites of freshwater arthropods (normally *Insecta*). Fams:
  (1) **Harpellaceae**.
  (2) **Legeriomycetaceae**.
  Lit.: Lichtwardt (1986; fam. key, *Am. J. Bot.* **51**: 836, 1964; culture), Lichtwardt & Manier (*Mycotaxon* **7**: 441, 1978; taxonomy), Manier (1969; taxonomy), Moss & Lichtwardt (*CJB* **54**: 2346, 1976, **55**: 3099, 1977; ultrastr.), Moss & Young (*Mycol.* **70**: 944, 1978; phylogeny), Moss *et al.* (*Mycol.* **67**: 120, 1975; sexual reproduction), Moss (1979), Peterson & Lichtwardt (*TBMS* **88**: 189, 1987; antigenic variation).

Harpellomyces Lichtw. & S.T. Moss (1984), Harpellaceae. 1, Sweden, UK. See Lichtwardt (1986).

Harpezomyces Malloch & Cain (1973) ≡ Chaetosartorya (Trichocom.).

Harpidium Körb. (1855), Lichinaceae (L). 2, Eur., N. Am. See Henssen *et al.* (*Bot. Acta* **101**: 49, 1987; posn), Sancho & Crespo (*Lazaroa* **5**: 265, 1983).

Harpocephalum G.F. Atk. (1897) = Periconia (Mitosp. fungi) fide Ellis (*DM*, 1971).

Harpochytriaceae (Wille) R. Emers. & Whisler (1984), Chytridiales. 1 gen. (+ 2 syn.), 6 spp. Thallus eucarpic with a sub-basal holdfast, upper region forming zoospores; capable of repeated sporulation.
  Lit.: Barr (1990), Powell & Roychoudhury (1992).

Harpochytriales, see *Chytridiales*.

Harpochytrium Lagerh. (1890), Harpochytriaceae. 6, N. temp. See Jane (*J. Linn. Soc. Lond.* **53**: 28, 1946), Gaurilof *et al.* (*CJB* **58**: 2090, 1980; ultrastr.).

Harpographium Sacc. (1880), Mitosporic fungi, 2.A2.10. 10, widespr. See Morris (*Am. midl. Nat.* **68**: 319, 1962).

Harposporella Höhn. (1925) nom. dub. fide Sutton (*Mycol. Pap.* **141**, 1977).

Harposporium Lohde (1874), Mitosporic fungi, 1.E1.15. 6, temp. See Drechsler (*Mycol.* **38**: 1, 1946), Saikawa & Endo (*Trans. Mycol. Soc. Jap.* **27**: 341, 1986; TEM nematode infection), Barron & Szuarto (*CJB* **69**: 1284, 1991).

Harpostroma Höhn. (1928), Anamorphic Valsaceae, 8.E1.?. Teleomorph *Leptosillia*. 1, Eur.

Hartiella Massee (1910) = Calostilbella (Mitosp. fungi) fide Mason (1925).

Hartig (Heinrich Julius Adolph Robert; 1839-1901). Teacher at Forestry Academy Eberswalde, 1866-78, then Professor of Botany, Munich. Noted for his work on tree diseases. Chief writings: *Wichtige Krankheiten der Waldbäume* (1874 [Engl. transl. *Phytopath. Classics* **12**, 1975]), *Die Zersetzungserscheinungen des Holzes der Nadelholzbäume und der Eiche* (1878), *Lehrbuch der Baumkrankheiten* (1882) [in French, 1891; in English, 1894]; as *Lehrbuch der Pflanzenkrankheiten*, 1900. See Meinecke (*Phytopath.* **5**: 1, 1915; portr.), Merrill *et al.* (*Phytopath. Classics* **12**, 1975; portr., bibl.), Ostrofsky & Ostrofsky (*Rev. Trop. Pl. Path.* **7**: 237, 1993).

Hartig net, the intercellular hyphal network formed by an ectomycorrhizal fungus upon the surface of a root.

Hartigiella P. Syd. (1900) = Meria (Mitosp. fungi) fide Vuillemin (*Ann. Myc.* **3**: 340, 1905).

Harzia Costantin (1888), Mitosporic fungi, 1.A2.1. 2, widespr. See Holubová-Jechová (*Folia geobot. phytotax.* **9**: 315, 1974).

Harziella Kuntze (1891) ≡ Trichocladium (Mitosp. fungi).

Harziella Costantin & Matr. (1899), Mitosporic fungi, 1.A1.15. 1 (on *Lepista nuda*), Eur. See Fontana (*Allionia* **6**: 35, 1960).

Hasegawaea Y. Yamada & I. Banno (1987), Schizosaccharomycetaceae. 1, Japan.

Hassea Zahlbr. (1902), Dothideales (inc. sed., L). 1, USA.

Hasskarlinda Kuntze (1891) ≡ Corallodendron (Mitosp. fungi).

hastate, like a spear- or arrow-head in form.

Hastifera D. Hawksw. & Poelt (1986), Mitosporic fungi, 8.E1.1 (L). 1, Austria, Italy. See Hafellner (*Herzogia* **9**: 167, 1992; teleomorph ? *Micarea*).

haustorial cap, an electron-dense, cap-like mass at the end of a lobe of the haustorial apparatus of *Exobasidium camelliae* (Mims, *Mycol.* **74**: 188, 1982).

haustorium, a special hyphal branch, esp. one within a living cell of the host, for absorption of food (see Karling, *Am. J. Bot.* **19**: 41, 1932). Honegger (*New Phytol.* **103**: 785, 1986) distinguishes three main types of fungus-plant cell interactions (Fig. 15): (1) **wall-to-wall apposition** with no penetration; (2) **intracellular haustoria** where the fungus penetrates into the plant cell, with or without the formation of special sheath, neckband, or collar; (3) **intraparietal haustoria** where penetration is restricted to the wall layers (common in some groups of lichens).

Haustoria of *Phytophthora* (Blackwell, *TBMS* **36**: 138, 1953), of *Peronosporales* (Fraymouth, *TBMS* **39**: 79, 1956); in lichens (Honegger, 1986); evolution in rusts (Rajendren, *Bull. Torrey bot. Cl.* **99**: 84, 1972); ultrastructure (Beckett *et al.*, *Atlas of fungal ultrastructure*, 1974); cf. rhizoid. Used in addition, by de Bary, for organs of attachment (appressoria).

Hawksworthiana U. Braun (1988), Mitosporic fungi, 1.A/B1.10. 1 (on lichens), UK.

Haynaldia Schulzer (1866) [non Schur (1866), *Gramineae*] = Helicostylum (Thamnid.) fide Hesseltine (1955).

hazard groups, (of fungi), see Safety, Laboratory.

Hazlinszkya Körb. (1861) ? = Melaspilea (Melaspil.).

heart rot, decay of the inner wood of trees, caused by basidiomycetes.

heather rags, common name for *Hypogymnia physodes*.

Hebeloma (Fr.) P. Kumm. (1871), Cortinariaceae. 25, esp. N. temp. See Romagnesi (*BSMF* **81**: 321, 1965), Bruchet (*Bull. mens. Soc. linn. Lyon* **39** (Suppl. 6): 3, 1970; descr., keys), Bon (*BSMF* **86**: 79, 1970), Smith *et al.* (*The veiled species of Hebeloma in the Western United States*, 1983).

Hebelomatis Earle (1909) = Hebeloma (Cortinar.) fide Singer (1951).

Hebelomina Maire (1925), Cortinariaceae. 2, N. Afr., Eur. See Huijsman (*Persoonia* **9**: 485, 1978; distinction from *Hebeloma*).

Hectocerus Raf. (1806) nom. nud. See Merrill (*Index Rafinesq.*, 1949).

Heim (Roger; 1900-79). Held posts at the Muséum National d'Histoire Naturelle, Paris (1927-65) of which he became Director. Main interests: larger basidiomycetes (e.g. *Le genre Inocybe*, 1931), mycetism, termite fungi and ethnomycology (q.v.). See Dorst (*C.R. Acad. Sci., Paris* **290**: 120, 1980), Romagnesi (*BSMF* **96**: 117, 1980; portr.), Ainsworth (*TBMS* **76**: 177, 1981; portr.), Batra (*Mycol.* **72**: 1063, 1981; portr.), Grummann (1974: 326), Stafleu & Cowan (*TL-2* **2**: 137, 1979).

Heimatomyces Peyr. (1873) = Chitonomyces (Laboulben.) fide Thaxter (1896).

Heimerlia Höhn. (1903) nom. dub. (? Myxomycota) fide Whitney (*Mycol.* **72**: 950, 1980).

Heimiella Boedijn (1951) = Boletellus (Xerocom.) fide Singer (1946), Corner (1970), Pegler & Young (1981).

Heimiella Racov. (1959), Mitosporic fungi, 4.B1.?. 1, France, Romania.

Heimiodiplodia Zambett. (1955) = Botryodiplodia (Mitosp. fungi) fide Petrak (*Sydowia* **16**: 353, 1963).

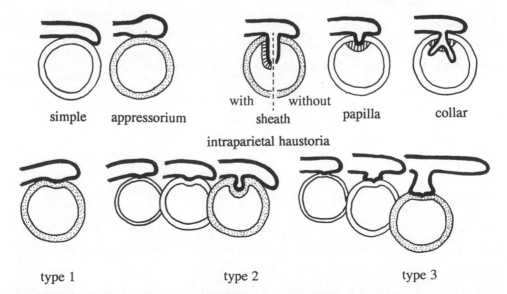

wall-to wall apposition

intracellular haustoria

simple   appressorium   with | without   papilla   collar

sheath

intraparietal haustoria

type 1   type 2   type 3

Fig. 15. Various types of fungus-plant cell interactions in parasitic and mutualistic symbioses. Whilst intracellular hustoria with a neckband are restricted to rusts and powdery mildews and interparietal haustoria of type 2 and 3 occur in certain groups of lichens, the others occur in different symbiotic systems (diagrammatic), after Honneger (1992).

Heimiodora Nicot (1960), Mitosporic fungi, 1.A1.10. 1 (from soil), Thailand. See Benjamin (*Aliso* **5**: 278, 1963; posn).

Heimiomyces Singer (1942), Tricholomataceae. 2, Eur., N. Am.

helenin (helenine), an antiviral antibiotic from *Penicillium funiculosum*; considered to be RNA of viral origin (Banks *et al.*, *Nature* **218**: 545, 1968).

Heleococcum C.A. Jørg. (1922), Hypocreaceae. 4, Denmark, Indonesia, Japan, Philip. See Udagawa (*Mycoscience* **36**: 37, 1995).

Helhonia B. Sutton (1980), Mitosporic fungi, 8.B1.15. 1, Eur.

Heliastrum Petr. (1931), Ascomycota (inc. sed.). 1, Philipp.

Helicascus Kohlm. (1969), Dothideales (inc. sed.). 1 (marine), Hawaii.

Helicia Dearn. & House (1925) [non Lour. (1790), *Proteaceae*] = Cylindrocolla (Mitosp. fungi) fide Sutton (1977).

Helicobasidium Pat. (1885), Platygloeaceae. 6, widespr. *H. brebissonii* (syn. *H. purpureum*) (anamorph *Rhizoctonia crocorum*) is the cause of violet root rot in a number of plants. See Donk (*Persoonia* **4**: 156, 1966; synonymy).

Helicobasis Clem. & Shear (1931) ≡ Helicobasidium (Platygl.).

Helicobolomyces Matzer (1995), Mitosporic Arthoniales, 1.F1.1. Teleomorph *Arthonia*. 1 (on lichens), Costa Rica.

Helicobolus Wallr. (1833) = Phloeospora (Mitosp. fungi) fide Sutton (1977).

Helicocephalidaceae Boedijn (1959), Zoopagales. 3 gen., 8 spp. Sporophores unbranched, rhizoidal; spores large, pigmented; zygospores unknown; obligate parasites of nematodes, rotifers and their eggs. *Lit.*: Boedijn (*Sydowia* **12**: 355, 1958).

Helicocephalum Thaxt. (1891), Helicocephalidaceae. 4 (obligate parasites of nematode eggs), widespr. See Thaxter (*Bot. Gaz.* **16**: 201, 1891), Drechsler (*Mycol.* **26**: 33, 1934, **35**: 134, 1943), Barron (*TBMS* **65**: 309, 1975), Watanabe & Koizumi (*Trans. Mycol. Soc. Japan* **17**: 1, 1976), Kitz & Embree (*Mycol.* **81**: 164, 1989).

Helicoceras Linder (1931), Mitosporic fungi, 2.C2.1. 4, N. temp. See Moore (*Mycol.* **47**: 90, 1955; key).

Helicocoryne Corda (1854) = Helicoma (Mitosp. fungi) fide Hughes (1958).

Helicodendron Peyronel (1918), Mitosporic fungi, 1.F2.3. 18, widespr. See Moore (*Mycol.* **47**: 90, 1955; key), Glen-Bott (*TBMS* **38**: 17, 1955), Abdullah (*TBMS* **81**: 638, 1983), Goos *et al.* (*TBMS* **84**: 423, 1985; keys), Abdullah (*Nova Hedw.* **44**: 339, 1987).

Helicodesmus Linder (1925) = Helicodendron (Mitosp. fungi) fide Linder (*Ann. Mo. bot. Gdn* **16**: 329, 1929).

Helicofilia Matsush. (1983), Mitosporic fungi, 1.F/G2.1. 1, India.

Helicogermslita Lodha & D. Hawksw. (1983), Xylariaceae. 5, Eur., India, Chile. See Læssøe & Spooner (*Kew Bull.* **49**: 1, 1994).

Helicogloea Pat. (1892), Platygloeaceae. 16, widespr. See Baker (*Mycol.* **38**: 630, 1946), Donk (*Persoonia* **4**: 157, 1966; accepts 3 spp.).

Helicogonium W.L. White (1942), ? Saccharomycetales (inc. sed.). 1 (on *Corticium*), Canada.

Helicogoosia Hol.-Jech. (1991), Mitosporic fungi, 1.F2.1. 1, Czech Republic.

Helicoma Corda (1837), Mitosporic fungi, 1.F1.10. 32, widespr. See Goos (*Mycol.* **78**: 744, 1986; key).

Helicomina L.S. Olive (1948) = Pseudocercospora (Mitosp. fungi) fide Deighton (1976).

Helicominites Barlinge & Paradkar (1982), Fossil fungi (Mitosp. fungi). 1 (Cretaceous), India.

Helicominopsis Deighton (1960), Mitosporic fungi, 1.F1.10. 1, Sierra Leone.

Helicomyces Link (1809), Mitosporic fungi, 1.F1.10. 11, widespr. See Goos (*Mycol.* **77**: 606, 1985; keys).

Helicoon Morgan (1892), Mitosporic fungi, 1.F1.10. 9, N. Am., Eur. See Moore (*Mycol.* **47**: 90, 1955; key), Goos (*TBMS* **87**: 115, 1986; key 8 spp.), van der Aa & Samson (*MR* **98**: 74, 1994; key amendment).

Helicoonites Kalgutkar & Sigler (1995), Fossil fungi (Mitosp. fungi). 1 (Eocene), Canada.

Helicopsis P. Karst. (1889) = Helicoma (Mitosp. fungi) fide Linder (*Ann. Mo. bot. Gdn* **16**: 302, 1929).

Helicorhoidion S. Hughes (1958), Mitosporic fungi, 1.F1/2.10. 3, widespr. See Sutton (*Sydowia* **41**: 336, 1989; relationships).

Helicosingula P.S. van Wyk, Marasas, Baard & Knox-Dav. (1985), Mitosporic fungi, 1.F2.1. 1, S. Afr.

Helicosporangium H. Karst. (1865) ? = Papulospora (Mitosp. fungi) fide Kendrick & Carmichael (1973).

helicospore, a non-septate or septate spore, with a single (usually elongated) axis curved through at least 180° but may describe one or more complete rotations, in two or three dimensions (cf. amerospore, scolecospore); any protuberances, other than setulae, $<^1/_4$ spore body length (cf. staurospore). See Goos (*Mycol.* **79**: 4, 1987; terminology). See also Mitosporic fungi.

Helicosporella Arnaud (1954) nom. inval. Mitosporic fungi, 1.F2.1. 1, Eur.

Helicosporiates Kalgutkar & Sigler (1995), Fossil fungi (Mitosp. fungi). 1 (Eocene), Canada.

Helicosporidium Keilin (1921), Fungi (inc. sed.). 2 (in *Insecta* and *Acari*), trop. See Avery & Undeen (*J. Invert. Path.* **49**: 246, 1987; effect on mosquitoes), Eriksson & Hawksworth (*SA* **8**: 68, 1989; posn), Lindegren & Hoffman (*J. Invert. Path.* **27**: 105, 1976; ? *Protozoa*), Wesier (*J. Protozool.* **17**: 436, 1970; ? *Ascomycota*).

Helicosporina Arnaud (1954) nom. inval. (Mitosp. fungi).

Helicosporina G. Arnaud ex Rambelli (1960), Mitosporic fungi, 1.F2.10. 1, Italy. ? = Troposporella (Mitosp. fungi) fide Sutton (*in litt.*).

Helicosporium Nees (1816), Mitosporic fungi, 1.F2.10. 16, widespr. See Goos (*Mycol.* **81**: 356, 1989; key).

Helicostilbe Höhn. (1902) Mitosporic fungi, 2.F2.10. 1, Austria.

Helicostilbe Linder (1929) = Trochophora (Mitosp. fungi) fide Moore (*Mycol.* **47**: 90, 1955).

Helicostylum Corda (1842), Thamnidiaceae. 2, Eur., India, former USSR, Japan. See Upadhyay (*Mycol.* **65**: 735, 1973; key), Benny (*Mycol.* **87**: 253, 1995; key).

Helicothyrium I. Hino & Katum. (1961), Mitosporic fungi, 5.F1.?. 1, Japan.

Helicotrichum Nees & T. Nees (1818) = Helicosporium (Mitosp. fungi) fide Hughes (1958).

Helicoubisia Lunghini & Rambelli (1979), Mitosporic fungi, 1.F2.10. 1, Ivory Coast.

Heliocephala V.G. Rao, K.A. Reddy & de Hoog (1984), Mitosporic fungi, 1.C2.1. 1, India.

Heliocybe Redhead & Ginns (1985), Polyporaceae. 1, N. temp.

Heliomyces Lév. (1844) = Marasmius (Tricholomat.) fide Singer (1951).

Heliomycopsis Arnaud (1949) nom. inval. (? Myxomycota).

heliophilous, preferring direct sunlight. Cf. anheliophilous.

heliozooid, amoeba-like, but having well-marked ray-like pseudopodia.

Heliscella Marvanová (1980), Mitosporic fungi, 1.G1.15. 2 (aquatic), widespr.

Heliscina Marvanová (1980), Mitosporic fungi, 1.B/G1.19. 2 (aquatic), Czech Republic.

Heliscus Sacc. (1880), Anamorphic Hypocreaceae, 3.C1.15. Teleomorph *Nectria*. 4 (aquatic), widespr. See Petersen (*Mycol.* **55**: 570, 1963; key).

Helminthascus Tranzschel (1898), Clavicipitaceae. 1 (on *Insecta*), former USSR.

Helminthocarpon Fée (1837), ? Graphidaceae (L). 2, S. Am. See Awasthi & Joshi (*Norw. J. Bot.* **29**: 165, 1979).

Helminthocarponomyces Cif. & Tomas. (1953) ≡ Helminthocarpon (Graphid.).

helminthoid, worm-like in form; vermiform.

Helminthopeltis Sousa da Câmara (1950), ? Microthyriaceae. 1, Eur.

Helminthophana Peyr. (1873) = Arthrorhynchus (Laboulben.) fide Thaxter (1901).

Helminthophora Bonord. (1851), Mitosporic fungi, 1.C1.15. 1 (fungicolous), temp. See de Hoog (*Persoonia* **10**: 55, 1978).

Helminthosphaeria Fuckel (1870), Sordariales (inc. sed.). 2 or 3 (on *Clavulina* and *Peniophora*), N. temp. See Kirschstein (*TBMS* **18**: 302, 1934), Parguey-Leduc (*BSMF* **77**: 15, 1961).

helminthosporal, a terpenoid mycotoxin from the *Drechslera* anamorph of *Cochliobolus sativus* toxic to wheat and barley.

Helminthosporiopsis Speg. (1880) nom. dub. fide Seifert (*Sydowia* **45**: 103, 1993).

Helminthosporites Pia (1927), Fossil fungi. 1 (Carboniferous), Eur.

Helminthosporites Chitaley & M.T. Sheikh (1971), Fossil fungi. 1 (Palaeocene), India.

Helminthosporium Link (1809) nom. cons., Mitosporic fungi, 1.C2.27. *c.* 20 (lignicolous), widespr. See Ellis (*Mycol. Pap.* **82**, 1961; key 10 spp.), Luttrell (*Mycol.* **55**: 643, 1963; spore development; terminology for characters). Cf. *Bipolaris, Drechslera, Exserohilum.*

helminthosporoside, a host-specific toxin produced by *Drechslera sacchari* in sugarcane (Strobel & Steiner, *Physiol. Pl. Pathol.* **2**: 129, 1972).

Helmisporium Link (1809) nom. rej. = Helminthosporium (Mitosp. fungi).

Helocarpon Th. Fr. (1860), Micareaceae (L). 1, arctic-alpine, Eur., N. Am. = Micarea (Micar.) fide Coppins (1983).

Helochora Sherwood (1979), ? Mesnieraceae. 1, Chile. See Cannon (*Stud. Mycol.* **31**: 49, 1989).

Helodiomyces F. Picard (1913), Ceratomycetaceae. 1, Eur., N. Afr.

Helolachnum Torrend (1910) = Lachnum (Hyaloscyph.) fide Nannfeldt (1932).

Helopodium Ach. ex Michx. (1803) = Cladonia (Cladon.).

Helostroma Pat. (1902) = Microstroma (Mitosp. fungi) fide Saccardo (*Michelia* **1**: 273, 1878).

Helote Hazsl. (1881) ≡ Microglossum (Geogloss.) fide Durand (*Ann. Myc.* **6**: 387, 1908).

Helotiaceae, see *Leotiaceae*.

Helotiales, see *Leotiales*.

Helotidium Sacc. (1884) = Allophylaria (Leot.) fide Nannfeldt (*Nova Acta R. Soc. Scient. upsal.* ser. 4 **8**(1), 1932).

Helotiella Sacc. (1884), Leotiales (inc. sed.). 1, Eur.

Helotiopsis Höhn. (1910) = Pithyella (Hyaloscyph.) fide Korf & Zhuang (*Mycotaxon* **29**: 1, 1987).

helotism, the physiologic relation of alga to fungus in a lichen (obsol.).

Helotium Tode (1790) nom. rej. = Omphalina (Tricholomat.) fide Donk (*Beih. Nova Hedw.* **5**: 122, 1962).

Helotium Pers. (1801) = Cudoniella (Leot.). See Dennis (*Persoonia* **3**: 29, 1964; redisp. spp.), Holm (*TBMS* **67**: 333, 1976; nomencl.).

Helvella L. (1753), Helvellaceae. *c.* 25 (saddle fungi, false morels), N. temp. See Abbott & Currah (*Mycotaxon* **33**: 229, 1988; key 16 spp. Alberta), Calonge & Arroya (*Mycotaxon* **39**: 203, 1990; key 21 spp. Spain), Dissing (*Dansk bot. Arkiv.* **25** (1), 1966; Gibson & Kimbrough (*CJB* **66**: 771, 1988; ascosporogenesis), Haffner (*Z. Mykol. Beih.* 7, 1987; key 43 spp., illustr.), Harmaja (*Karstenia* **14**: 102, 1974; gen. concept, **17**: 45, 1977, **19**: 33, 1979; cupulate spp.), Liu *et al.* (*Acta Mycol. Sin.* **4**: 208, 1985; key 16 Chinese spp.), Weber (*Beih. Nova Hedw.* **51**, 1975; W.N. Am.).

Helvellaceae Fr. (1823), Pezizales. 11 gen. (+ 36 syn.), 68 spp. Ascomata large, apothecial or cleistothecial; if apothecial varied in form, stalked, the hymenium cupulate or everted, either saddle-shaped or corrugated, pale to dark brown; if cleistothecial solid or chambered, often convoluted; asci cylindrical, persistent, operculate, not blueing in iodine (sometimes saccate and indehiscent in cleistothecial taxa); ascospores hyaline or brown, smooth, lacking appendages, tetranucleate.
   Lit.: Dissing (*Dansk Bot. Ark.* **25** (1), 1966), Kimbrough (*MR* **95**: 421, 1991).

Helvellella S. Imai (1932) = Pseudorhizina (Helvell.) fide Eckblad (1968).

hemi- (prefix), half; in part. Cf. semi-.

Hemialysidium Hol.-Jech. (1992) nom. inval. Mitosporic fungi, 1.?.?. 1, Eur.

hemiangiocarpous (of a sporocarp), opening before quite mature.

Hemiarcyria Rostaf. (1875) = Hemitrichia (Trich.).

Hemiasci (sensu Varitchak, *Botaniste* **25**: 370, 1933), *Dipodascus* + *Ascoidea*.

Hemiascomycetes. Ascomycota in which the asci are not produced in ascomata. In addition the thallus usually comprises poorly developed mycelium or is represented by separate cells; mainly included in *Saccharomycetales* here.

hemiascospore, ascospore of a hemiascus.

Hemiascosporium L.R. Batra (1973) = Eleutherascus (Ascodesmid.) fide v. Arx (1977) and Batra (1978).

hemiascus, the atypical multispored ascus of the *Hemiasci* (q.v.).

Hemibasidiomycetes = Uredinales and Ustilaginales, q.v.

Hemibeltrania Piroz. (1963), Mitosporic fungi, 2.A2.10. 2, Sierra Leone, Cuba.

Hemicarpenteles A.K. Sarbhoy & Elphick (1968), Trichocomaceae. 2, widespr. Anamorph *Aspergillus*.

hemicompatible, see compatible.

Hemicorynespora M.B. Ellis (1972), Mitosporic fungi, 1.A1/B2.25. 2, trop.

Hemicorynesporella Subram. (1992), Mitosporic fungi, 1.C2.24. 1, Solomon Isl.

Hemicybe P. Karst. (1879) = Lentinellus (Lentinell.) fide Singer (1951).

Hemicyphe Corda (1831) nom. dub (? Fungi, inc. sed.).

Hemidiscia Lázaro Ibiza (1916) = Tyromyces (Polypor.) fide Donk (1974).

Hemidothis Syd. & P. Syd. (1916), Mitosporic fungi, 8.E1.?. 1 or 2, trop. Am. See Petrak (*Ann. Myc.* **27**: 374, 1929), Sydow (*Ann. Myc.* **28**: 193, 1930).

hemifissitunicate, see ascus.

hemiform, see *Uredinales*.

Hemigaster Juel (1895), ? Coprinaceae. 1, Sweden. See Singer (*Agaricales*, edn 4: 845, 1986).

Hemiglossum Pat. (1890), ? Geoglossaceae. 2, China, Japan. See Imai (*J. Fac. Agric. Hokkaido Imp. Univ.*

45: 155, 1941), Zhuang (*Mycotaxon* **32**: 97, 1988; posn.).

Hemigrapha (Müll. Arg.) R. Sant. ex D. Hawksw. (1975), ? Hysteriaceae. 1 (on lichens), widespr.

Hemigyalecta, see *Semigyalecta*.

Hemileia Berk. & Broome (1869), Uredinales (inc. sed.). 35, esp. trop.; II, III only. *H. vastatrix* (coffee rust); see Fulton (Ed.) (*Coffee rust in the Americas*, 1984), Stevenson & Bean (*Spec. Publ.* **4**, *Div. Mycol. Dis. Survey, USDA*, 1953; annotated bibliogr.); *CMI Descr.* **1**. See Gopalkrishnan (*Mycol.* **43**: 271, 1951; morphology), Rajendren (*Mycol.* **59**: 918, 1967; telial urediniospores (II)^{III} in which meiosis occurs in *H. vastatrix*).

Hemileiopsis Racib. (1900) = Hemileia (Uredinales, inc. sed.) fide Dietel (1928).

Hemilichenes. Lichens of uncertain systematic position because sporocarps are unknown (obsol.); usually = *Phycolichenes, Deuterolichenes*.

Hemimycena Singer (1938), Tricholomataceae. 25, cosmop.

Hemimyriangium J. Reid & Piroz. (1966), Elsinoaceae. 1, Canada. = Molleriella (Elsin.) fide v. Arx & Müller (1975).

hemiparasite, a facultative parasite.

Hemiphacidiaceae Korf (1962), Leotiales. 5 gen. (+ 2 syn.), 12 spp. Stromata absent; ascomata immersed, becoming exposed by circumscissile or laciniate rupture of the overlying host tissues; peridium hardly developed; interascal tissue of simple paraphyses, sometimes swollen towards the apices; asci with an I+ or I- ring; ascospores usually septate, hyaline or brown, sometimes with a sheath. Anamorphs coelomycetous where known. Saprobic (endophytic) or parasitic, on conifers, temp.
    Lit.: Korf (*Mycol.* **54**: 481, 1962), Reid & Cain (*Mycol.* **54**: 481, 1962).

Hemiphacidium Korf (1962), Hemiphacidiaceae. 3, N. Am. See Reid & Cain (*Mycol.* **54**: 481, 1962).

Hemipholiota (Singer) Romagn. (1980), Strophariaceae. 4, Eur.

Hemisartorya J.N. Rai & H.J. Chowdhery (1976) = Neosartorya (Trichocom.) fide v. Arx (1981).

Hemiscyphe, see *Hemicyphe*.

Hemisphaeria Klotzsch (1843) nom. rej. ≡ Daldinia (Xylar.).

Hemisphaeriales, see *Dothideales*.

Hemisphaeropsis Petr. (1947), Mitosporic fungi, 5.B2.?. 1, ? USA.

Hemispora Vuill. (1906) = Wallemia (Mitosp. fungi) fide Carmichael *et al.* (1980).

hemispore, (esp. of dermatophytes), (1) a cell at the end of a filament, which later becomes by division deuteroconidia; protoconidium (after Vuillemin); (2) one of the two cells produced by a primary trans-septum in an ascospore (Eriksson, *Ark. Bot.* II **6**: 339, 1967), see septum.

Hemisynnema Subram. (1994), Mitosporic fungi. 2.C2.1. 1, Malaysia.

Hemithecium Trevis. (1853) = Graphina (Graphid.).

Hemitrichia Rostaf. (1873), Trichiaceae. 17, cosmop.

Hendersonia Berk. (1841) nom. rej. = Stagonospora (Mitosp. fungi). See Swart & Walker (*TBMS* **90**: 633, 1988; redisposition spp. on *Eucalyptus*).

Hendersonia Sacc. (1884) ≡ Hendersonia Berk. (Mitosp. fungi).

Hendersoniella Tassi (1900), Mitosporic fungi, 4.?.?. 1, Italy. See Sutton (1977).

Hendersoniella (Sacc.) Sacc. (1902) ≡ Hendersoniella Tassi (Mitosp. fungi). See Sutton (1977).

Hendersonina E.J. Butler (1913), Mitosporic fungi, 8.A/B2.15. 1, Asia.

Hendersoniopsis Höhn. (1918), Mitosporic fungi, 8.C2.15. 1, temp.

Hendersoniopsis Woron. (1922) = Stenocarpella (Mitosp. fungi) fide Sutton (1977).

Hendersonula Speg. (1880), Mitosporic fungi, 8.C2.19. 3 (on fungi), trop. See Sutton & Dyko (*MR* **93**: 466, 1989; revision, key).

Hendersonulina Tassi (1902), Mitosporic fungi, 4.C2.?. 43, widespr. See Sutton (1977).

Hendersonulina Petr. (1951) [non Tassi (1902)], Mitosporic fungi, 8.C2.?. 6, widespr.

Hendrickxia P.A. Duvign. (1942) = Everniopsis (Parmel.) fide Santesson (*Svensk Bot. Tidskr.* **43**: 547, 1949).

Henicospora P.M. Kirk & B. Sutton (1980), Mitosporic fungi, 1.C2.2. 3, widespr.

Hennebertia M. Morelet (1969), Mitosporic fungi, 1.A2.10 and A1.19. 1 (dimorphic), Eur. See Hennebert (*TBMS* **51**: 749, 1968).

Hennings (Paul Christoph; 1841-1908). At the Kiel Botanic Gardens (1860-4, 1874-9), then the Berlin Botanic Gardens (Kustos, 1891; Professor, 1902). Chief writings: papers on the fungi of tropical Africa, Amazon Basin (Batista *et al.*, Fungi Paraenses, *Publ. Inst. Micol. Univ. Recife* **506**, 1966, have listed Hennings' specimens in Emílio Goeldi Museum, Para), India, Japan, New Guinea, and other countries. See Lindau (*Hedwigia* **48**: 1, 1909; bibl.), Perkins (*Bot. Gaz.* **47**: 239, 1909), Grummann (1974: 18), Stafleu & Cowans (*TL-2* **2**: 157, 1979).

Henningsia A. Møller (1895), Coriolaceae. 3, neotrop. = Rigidoporus (Coriol.) fide Corner (*Mycologist* **9**: 127, 1995).

Henningsiella Rehm (1895), Schizothyriaceae. 2, S. Am.

Henningsina A. Møller (1901) = Phylacia (Xylar.) fide Müller & v. Arx (1973).

Henningsomyces Kuntze (1898), Schizophyllaceae. 30, widespr. See Agerer (*Persoonia* **7**: 389, 1973), Reid (*Persoonia* **3**: 118, 1964).

Henningsomyces Sacc. (1905) [non Kuntze (1898)] ≡ Dysrhynchis (Parodiopsid.).

Henrica de Lesd. (1921), Verrucariaceae (L). 1, Eur.

Henriquesia Pass. & Thüm. (1879) [non Spruce ex Benth. (1854), *Rubiaceae*] ≡ Delpinoina (Ascodichaen.).

Hepataria Raf. (1808) ? = Tremella (Tremell.). See Merrill (*Index Rafin.*: 69, 1949).

hepatic, (1) concerning the liver; (2) a liverwort.

hepaticolous, growing on liverworts (*Hepaticae*); see Bryophilous fungi.

Heppia Nägeli ex A. Massal. (1854), Heppiaceae (L). 6, widespr. (esp. desert or dry regions). See Henssen (*Acta bot. fenn.* **15**: 57, 1994; key), Egea (*Bibl. Lich.* **31**, 1989; W. Eur. & N. Afr.), Filson (*Muelleria* **6**: 495, 1988; Australia), Swinscow & Krog (*Norw. J. Bot.* **26**: 213, 1979; E. Afr.), Wetmore (*Ann. Mo. bot. Gdn* **57**: 158, 1971; N. Am.).

Heppiaceae Zahlbr. (1906), Lecanorales (L). 5 gen. (+ 10 syn.), 11 spp. Thallus foliose to fruticose, sometimes peltate, not gelatinized, with rather wide hyphae; ascomata apothecial, deeply immersed, sometimes initially cleistothecial; interascal tissue of thick-walled anastomosing paraphyses which are swollen at the apices; asci ± thin-walled, with an I+ apical cap; ascospores aseptate, hyaline, without a sheath. Anamorphs pycnidial. Lichenized with cyanobacteria, esp. trop., often in deserts.
    Lit.: Upreti & Büdel (*J. Hattori bot. Lab.* **68**: 269, 1990; India).

Heppiomyces Cif. & Tomas. (1953) ≡ Heppia (Hepp.).

Heppsora D.D. Awasthi & Kr.P. Singh (1977), Bacidiaceae (L). 1, India. See Poelt & Grube (*Nova Hedw.* **57**: 1, 1993; status).

Heptameria Rehm & Thüm. (1879), Dothideales (inc. sed.). 2, Eur. See Lucas & Sutton (*TBMS* **57**: 283, 1971).

Heptasporium Bref. (1908) = Sistotrema (Sistotremat.) fide Rogers (1944).

Heptaster Cif., Bat. & Nascim. (1956), Mitosporic fungi, 2.G2.?. 2, Brazil. ? = Tripospermum (Mitosp. fungi) fide Hughes (*Mycol.* **68**: 810, 1976).

Heraldoa Bat. (1959) = Lembosia (Asterin.) fide Müller & v. Arx (1962).

Herbampulla Scheuer (1993), ? Magnaporthaceae. 1, Austria.

herbarium (pl. -ia), (1) a collection of dried plants; (2) the place in which such a collection is stored. Often also used for dried Reference collections (q.v.) of fungi, especially when curated along with plant specimens.

herbarium beetle, *Stegobium paniceum*, may eat the spores of certain fungi, esp. *Lycoperdon* and smuts. See Gordon (*TBMS* **21**: 193, 1938), Bridson & Forman (*The herbarium handbook*, rev. edn, 1992; control), Pinniger (*Biodet. Abstr.* **5**: 125, 1991; control).

herbicolous, living on herbs.

Hercospora Fr. (1825), Melanconidaceae. 1, Eur. See Petrak (*Ann. Myc.* **36**: 44, 1938). Anamorph *Rabenhorstia*.

Herculea Fr. (1823) = Podaxis (Podax.).

Hericiaceae Donk (1964), Hericiales. 7 gen. (+ 10 syn.), 48 spp. Basidiocarp typically clavarioid and strongly branched, monomitic, gloeocystidia not darkening in sulphoaldehydes; spores colourless (spore print white); asperulate, amyloid.
Lit.: Donk (1964: 269).

Hericiales, Basidiomycetes. 5 fam., 25 gen. (+ 17 syn.), 137 spp. Fams:
(1) **Auriscalpiaceae**.
(2) **Clavicoronaceae**.
(3) **Gloeocystidiellaceae**.
(4) **Hericiaceae**.
(5) **Lentinellaceae**.

Hericium Pers. (1794), Hericiaceae. 7, N. temp.

Hericium Fr. (1825) ≡ Hericium Pers. (Heric.). See Hall & Stuntz (*Mycol.* **63**: 1103, 1971; subgen. key).

Hericius Juss. (1789) ≡ Hericium (Heric.).

Heringia Schwein. ex Berk. & Curtis (1853) [non (Kütz.) J. Agardh (1846), *Algae*], nom. nud. (? Fungi, inc. sed.).

Hermanniasporidium Thor (1930) nom. dub. (? Fungi, inc. sed.).

Hermatomyces Speg. (1910), Mitosporic fungi, 3.D2.1. 3, trop. See Hughes (*Mycol. Pap.* **50**: 100, 1953).

Herminia R. Hilber (1979) = Eosphaeria (Lasiosphaer.) fide Petrini *et al.* (*TBMS* **82**: 554, 1984). See Barr (*Mycotaxon* **39**: 43, 1990; status).

Herpobasidium Lind (1908), Platygloeaceae. 4, widespr. See Donk (*Persoonia* **4**: 158, 214, 1966), Oberwinkler & Bandoni (*TBMS* **83**: 639, 1984). Anamorph *Glomopsis*.

Herpocladiella J. Schröt. (1894) nom. dub. (? Mucor.) fide Hesseltine (1955).

Herpocladium J. Schröt. (1886) [non Mitt. (1873), *Hepaticae*] ≡ Herpocladiella (Mortierell.).

Herpomyces Thaxt. (1902), Herpomycetaceae. 25 (on cockroaches), widespr. See Tavares (*Mycol.* **57**: 104, 1965; development, *Mycotaxon* **11**: 485, 1980).

Herpomycetaceae (Thaxt.) I.I. Tav. (1981), Laboulbeniales. 1 gen., 25 spp. Stroma (thallus) present; with a usually 4-celled primary thallus and ascomata formed from secondary thalli; asci 8-spored;

ascospores with median septa; dioecious, on cockroaches.

Herposira Syd. (1938), Mitosporic fungi, 2/4.A2.?. 1, Australia.

Herpothallon Tobler (1937) = Cryptothecia (Arthon.) fide Thor (1991).

Herpothallonomyces Cif. & Tomas. (1954) ≡ Herpothallon (Arthon.).

Herpothrix Clem. (1909) ? = Chaetomastia (Ascomycota).

Herpotrichia Fuckel (1868), Lophiostomataceae. 17, widespr. *H. juniperi* (brown felt blight, snow moulds, of conifers in cold regions). See Sivanesan (*Mycol. Pap.* **127**, 1972; key), Barr (*Mycotaxon* **20**: 1, 1984; key 4 spp., gen. concept), v. Arx & Müller (*Sydowia* **37**: 6, 1984). Anamorph *Pyrenochaeta*, *Spilocaea*.

Herpotrichiella Petr. (1914) = Capronia (Herpotrichiell.) fide Müller *et al.* (1987).

Herpotrichiellaceae Munk (1953), Dothideales. 4 gen. (+ 6 syn.), 35 spp. Ascomata erumpent or superficial, small, ± globose, occasionally aggregated in small stromata, with a well-developed often periphysate ostiole, often setose; peridium thin, composed of compressed pseudoparenchymatous cells, varied in pigmentation; interascal tissue of short periphysoids; asci saccate to clavate, the inner wall layer often conspicuously thickened in the apical region, fissitunicate, sometimes polysporous; ascospores hyaline to greyish, septate, sometimes muriform. Anamorphs hyphomycetous, sometimes yeast-like (black yeasts, see yeasts). Saprobic on plants or other fungi, cosmop.
Lit.: Barr (*Rhodora* **78**: 53, 1976), Janex-Favre (*Cryptog., Mycol.* **9**: 133, 1988; ontogeny), Untereiner *et al.* (*MR* **99**: 897, 1995; molec. taxonomy).

Herpotrichiopsis Höhn. (1914) = Pyrenochaeta (Mitosp. fungi) fide Clements & Shear (1931).

Herpotrichum Fr. [not traced] = Protonema (? *Algae*) fide Fries (*Syst. orb. veg.*, 1825); based on mycelium fide Fries (*Summ. veg. Scand.*, 1849).

Herreromyces R.F. Castañeda & W.B. Kendr. (1991), Mitosporic fungi, 1.A1.15. 1, Cuba.

Herteliana P. James (1980), Bacidiaceae (L). 1, Eur.

Hertella Henssen (1985), Placynthiaceae (L). 2, S. Hemisph.

Hesperomyces Thaxt. (1891), Laboulbeniaceae. 5, Am., E. Afr., Eur., Philipp., Borneo.

Hesseltinella H.P. Upadhyay (1970), Radiomycetaceae. 1, Brazil. See Upadhyay (*Persoonia* **6**: 111, 1970), Stuart & Young (*TBMS* **89**: 392, 1987; morphology), Benny & Khan (1988), Benny & Samson (1989), Benny & Benjamin (1991).

Heterina Nyl. (1858) = Heppia (Hepp.).

hetero- (prefix), other; not normal; different.

Heteroacanthella Oberw. (1990), Ceratobasidiaceae. 2, Taiwan, USA.

Heterobasidiomycetes, see Donk (*Proc. K. Ned. Akad. Wet.* C **75**: 365, 376; **76**: 1, 109, 126, 1972-73; a 'reconnaissance'), Hymenomycetes.

Heterobasidion Bref. (1888), Coriolaceae. 2, temp. *H. annosum* (syn. *Fomes annosus*), serious root rot of conifers.

heterobasidium, see basidium.

Heterobasidium Massee (1889) nom. conf. (Basidiomycetes).

Heterobotrys Sacc. (1880) = Seuratia (Seurat.) fide Meeker (1975).

Heterocarpon Müll. Arg. (1885), Verrucariaceae. 1 (on lichen), U.S.A. See Harada (*SA* **10**: 1, 1991).

Heterocephalacria Berthier (1980), Tremellaceae. 1, USA. = Syzygospora (Syzygospor.) fide Ginns (1986).

Heterocephalum Thaxt. (1903), Mitosporic fungi, 2.A1.15. 1, trop. Onofri *et al.* (*TBMS* **87**: 551, 1987; SEM).

Heteroceras Sacc. (1915) ≡ Neoheteroceras (Mitosp. fungi) fide Nag Raj (1993).

Heterochaete Pat. (1892), Exidiaceae. *c.* 40, esp. trop. See Bodman (*Lloydia* **15**: 193, 1952).

Heterochaetella (Bourdot) Bourdot & Galzin (1928), Exidiaceae. 3, Eur., N. Am. See Luck-Allen (*CJB* **38**: 559, 1960; key).

Heterochlamys Pat. (1895) [non Turcz. (1843), *Euphorbiaceae*] ≡ Gilletiella (Dothideales).

Heteroconidium Sawada (1944), Mitosporic fungi, 2.C1 and A1.?. 1 (dimorphic), Taiwan.

Heteroconium Petr. (1949), Mitosporic fungi, 1.C2.3. 2, S. Am., S. Afr. See Ellis (*MDH*), Morgan-Jones (*Mycotaxon* **4**: 498, 1976).

Heterocyphelium Vain. (1927), Caliciales (inc. sed., L). 1, Australia, C. & S. Am., USA, W. Afr. See Tibell (1984, 1987).

heterocytic, see astatocoenocytic.

Heterodea Nyl. (1868), Heterodeaceae (L). 2, Australia. See Filson (*Lichenologist* **10**: 13, 1978; key).

Heterodeaceae Filson (1978), Lecanorales (L). 1 gen. (+ 1 syn.), 2 spp. Thallus foliose, irregular, sometimes becoming erect, the lower surface ecorticate, rhizoidal; ascomata apothecial, marginal, irregular, domed, without a well-developed margin; interascal tissue of unbranched paraphyses; asci with a well-developed I+ apical cap, lacking an ocular chamber; ascospores ellipsoidal, hyaline, aseptate, thin-walled, without a sheath. Anamorph pycnidial, superficial. Lichenized with green algae, Australia.

Heterodermia Trevis. (1868), Physciaceae (L). *c.* 80, mainly trop. See Kashiwadani *et al.* (*Bull. Natn Sci. Mus. Tokyo*, B, **16**: 147, 1990; Peru), Trass (*Folia Crypt. Estonica* **29**: 2, 1992; tab. key), Swinscow & Krog (*Lichenologist* **8**: 103, 1976; E. Afr. spp.).

Heterodictyon Rostaf. (1873) = Dictydium (Cribrar.).

Heterodoassansia Vánky (1993), Tilletiaceae. 4, widespr.

Heterodothis Syd. & P. Syd. (1914) = Strigula (Strigul.).

heteroecious (n., **heteroecism**), undergoing different parasitic stages on two unlike hosts, as in the *Uredinales* (q.v.).

heterogametes, gametes of different form.

heterogamy, the copulation of heterogametes; cf. isogamy.

Heterogastridiaceae Oberw. & R. Bauer (1990), Heterogastridiales. 1 gen., 1 sp. Basidioma pycnidial; basidium cylindrical, transversely septate; spores statismosporic, tetraradiate.

Heterogastridiales, Basidiomycetes. 1 fam., 1 gen., 1 sp. Fam.:
Heterogastridiaceae.

Heterogastridium Oberw. & R. Bauer (1990), Heterogastridiaceae. 1, N. temp. Anamorph Hyalopycnis.

Heterographa Fée (1824) = Polymorphum (Mitosp. fungi) fide Hawksworth & Punithalingam (1973).

heterokaryosis (adj. **heterokaryotic**), (1) (of cells), the condition of having two or more genetically different nuclei, sometimes as a result of anastomosis, cf. dikaryotic; (2) (of mycelia), being made up of heterokaryotic cells; see Parmeter *et al.* (*Ann. Rev. Phytopath.* **1**: 51, 1963; in plant pathogenic fungi), Caten & Jinks (*TBMS* **49**: 81, 1966; occurrence in nature).

heterokaryotic, (1) having two or more slightly (< 5%) genetically different nuclei in common cytoplasm (fungi); (2) showing nuclear dimorphism (protists).

heterokaryotise (of rusts and pyrenomycetes), fusion of haploid structures of opposite sex which does not give

a conjugate arrangement of the nuclei, cf. diploidization.

heterokont, (1) condition whereby a flagellum possesses two rows of tripartite tubular hairs (heterokont flagellum, syn. flimmergeissel, tinsel flagellum); (2) of an organism possessing flagella in pairs, the members of which differ in length; type of motion; external appendages. Usually with kinetosomes mutually attached at a wide angle. Cf. isokont.

Heterokonta, see *Chromista*.

heteromerous, (1) (of a lichen thallus), having the mycobiont and photobiont in well-marked layers, usually between the medulla and the upper cortex (Fig. 40A); (2) (of trama in *Russulaceae*), having sphaerocyst nests among filamentous hyphae; cf. homoiomerous.

heteromixis, see heterothallism.

heteromorphic (**heteromorphous**), (1) having variation from normal structure; (2) having organs of different length; (3) (of agaric lamella edge), sterile due to the pressure of cystidia; cf. homomorphous.

Heteromyces Müll. Arg. (1889), Cladoniaceae (L). 1, Brazil.

Heteromyces L.S. Olive (1957) ≡ Oliveonia (Ceratobasid.).

Heteronectria Penz. & Sacc. (1898) = Cercophora (Lasiosphaer.) fide Lundqvist (1972).

Heteropatella Fuckel (1873), Anamorphic Leotiaceae, 8.C1.10. Teleomorph *Heterosphaeria*. 5, temp. *Heteropatella antirrhini* (*Antirrhinum* shot hole). See Nag Raj (1993; key).

Heteropera Theiss. (1917) = Phomatospora (Ascomycota, inc. sed.). See Petrak (*Sydowia* **14**: 347, 1960).

Heterophracta (Sacc. & D. Sacc.) Theiss. & Syd. (1918) = Merismatium (Verrucar.).

heterophytic, the equivalent in the sporophyte generation of dioecious in the gametophyte generation (Blakeslee, *Bot. Gaz.* **42**: 161, 1906); cf. homophytic.

heteroplastic, see heterokaryotic.

Heteroplegma Clem. (1903) = Peziza (Peziz.) fide Eckblad (1968).

Heteroporimyces M.K. Elias (1966), Fossil fungi. 1, NZ.

Heteroporus Lázaro Ibiza (1916), Coriolaceae. 2, cosmop. See Fidalgo (*Rickia* **4**: 99, 1969). = Abortiporus (Polypor.) fide Donk (1974).

Heteroradulum Lloyd (1917) nom. dub. (Fungi, inc. sed.).

Heteroscypha Oberw. & Agerer (1979), Exidiaceae. 1, S. Afr.

Heteroseptata E.F. Morris (1972), Mitosporic fungi, 2.C2.15. 1, Costa Rica.

Heterospermales. Le Gal's (1953) term for inoperculate discomycetes; cf. *Homospermales*.

Heterosphaeria Grev. (1824), Leotiaceae. 6, Eur. See Dennis (1956), Leuchtmann (*Mycotaxon* **28**: 261, 1987; key 8 spp.). Anamorph *Heteropatella*.

Heterosphaeriopsis Hafellner (1979), Dothideales (inc. sed.). 1, Ecuador.

Heterosporiopsis Petr. (1950), Mitosporic fungi, 2.C2.?. 1, S. Am.

Heterosporium Klotzsch ex Cooke (1877) = Cladosporium (Mitosp. fungi) fide Hughes (1958).

Heterosporula (Singer) Kühner (1980), Tricholomataceae. 1, Eur.

heterospory, (1) having asexually produced spores of more than one kind (de Bary, 1887: 496) (obsol.); (2) having spores which differ in the mating type (+ or -) in heterothallic fungi (e.g. in *Mucorales*) (Blakeslee, *Bot. Gaz.* **42**: 161, 1906); (3) polymorphism of basidiospores in *Agaricales* associated with extreme conditions (Heim, *Rev. Mycol.* **8**: 32, 1943).

Heterostomum Fr. [not traced] nom. dub. (? 'Sphaeriales') fide Saccardo (*Syll. Fung.* **16**: 12, 1902).

Heterotextus Lloyd (1922), Dacrymycetaceae. 4, widespr. See McNabb (*N.Z. Jl Bot.* **3**: 215, 1965; key). = Guepiniopsis (Dacrymycet.) fide Martin (*Am. J. Bot.* **23**: 627, 1936).

heterothallism, condition of sexual reproduction in which 'conjugation is possible only through the interaction of different thalli' (Blakeslee, 1904). Heterothallism and homothallism (q.v.) were first applied by Blakeslee to the methods of zygospore formation in the Mucorales, and he considered the terms to correspond to dioecism and monoecism for higher plants. Heterothallism has, however, been used as the equivalent of haplodioecism or dioecism (as in *Dictyuchus monosporus*, where male and female organs are produced on different individuals), and self-sterility or self-incompatibility (as in *Ascobolus magnificus*, where male and female organs are developed on one individual). Whitehouse (*Biol. Rev.* **24**: 411, 1949) has distinguished the first type (haplodioecism) as '**morphological -**', and the second (haploid incompatibility) as '**physiological -**'. Physiological heterothallism may be determined either by two allelomorphs at one locus or by multiple allelomorphs at one or two loci (Whitehouse, *New Phytol.* **48**: 212, 1949). Multiple-allelomorph physiological heterothallism is characteristic of the hymenomycetes and gasteromycetes, among which Whitehouse has estimated 35% are heterothallic and bipolar (with one locus), 55% are heterothallic and tetrapolar (with two loci). Pontecorvo (*Adv. Genetics* **5**: 194, 1953) designated as **relative -** the formation of crossed asci in excess of 50% by the combination of certain homothallic strains of *Aspergillus nidulans*. Drayton & Groves (*Mycol.* **44**: 132, 1952) would restrict heterothallism to morphological heterothallism, Korf (*Nature* **170**: 534, 1952) would use heterothallism in a wide physiological sense and would distinguish a morphologically heterothallic organism as being both haplo-dioecious (dioecious) and heterothallic.

Burnett (*New Phytol.* **55**: 50, 1956) proposed the following terminology for mating systems of fungi: **heteromixis** (adj. heteromictic) for fusion of genetically different nuclei which includes **dimixis** when 2 types of nuclei (= heterothallism sensu Blakeslee, morphological, and 2-allelomorph physiological heterothallism), **diaphoromixis** when more than 2 types of nuclei (= multiple allelomorph physiological heterothallism), and **homoheteromixis** = secondary homothallism, amphithallism; **homomixis** (adj. homomictic) = homothallism; **amixis** (adj. amictic), apomixis in haploid organisms.

Heterothecium Flot. (1850) ≡ Megalospora (Megalospor.). See Santesson (*Symb. bot. upsal.* **12**(1), 1952).

Heterothecium Mont. (1852) nom. illegit. ≡ Lopadium (Ectolech.).

Heterotrichia Massee (1892) = Arcyria (Arcyr.).

heterotroph, Pfeiffer's name for a true saprophyte; cf. autotroph.

heterotrophic (of living organisms), using organic compounds as primary sources of energy, cf. autotrophic.

heteroxenous, having more than one host.

heterozygous, having heterokaryosis resulting from the fusion of gametes.

Heuflera Bail (1860), Ascomycota (inc. sed.). 1, Eur.

Heufleria Trevis. (1853) ≡ Astrothelium (Trypethel.).

Heufleria Trevis. (1861) [non *Heufleria* Trevisan (1853) fide Trevisan (1861)] = Cryptothelium (Trypethel.).

Heufleria Auersw. (1869) [non Trevis. (1853)], Rhytismatales (inc. sed.). 1 (on *Elyna*), Eur.

Heufleridium Müll. Arg. (1883), Pyrenulales (inc. sed., L). 2, trop., NZ.

Hexacladium D.L. Olivier (1983), Mitosporic fungi, 1.G1.1/10. 1 (on pollen grains), S. Afr.

Hexagona, see Hexagonia.

Hexagonella F. Stevens & Guba ex F. Stevens (1925), Schizothyriaceae. 1, Hawaii.

Hexagonia Pollini (1816) = Polyporus (Polypor.) fide Ryvarden (1991).

Hexagonia Fr. (1835), Coriolaceae. *c.* 20, esp. trop. See Pegler (1971), Fidalgo (*Taxon* **17**: 37, 1968), Ryvarden & Johansen (*Preliminary polypore flora of East Africa*: 366, 1980; key 11 Afr. spp.), Jülich (*Persoonia* **12**: 107; nomencl.). Sensu Bondartzev & Singer (1953) = Apoxona (Coriol.); sensu Fidalgo (*Taxon* **17**: 37, 1968; *Mem. N. Y. bot. Gdn* **17**: 35, 1968) = Polyporus (Polypor.) fide Donk (1974).

Hexajuga Fayod (1889) ≡ Clitopilus (Entolomat.).

Heydenia Fresen. (1852), Mitosporic fungi, ?.?.?. 4, N. temp. See Carmichael *et al.* (*Genera of Hyphomycetes*, 1980).

Heydeniopsis Naumov (1915) = Chaenotheca (Coniocyb.) fide Seifert & Brodo (*Sydowia* **45**: 101, 1993).

Heyderia Link (1833), Leotiaceae. 6, Eur., Chile. See Maas Geesteranus (*Persoonia* **3**: 87, 1964), Knudsen (*Bot. Tidsskr.* **69**: 248, 1978), Benkert (*Gleditschia* **10**: 141, 1983; key 5 spp.), Siepe (*Beitr. Kenntis Pilze Mitteleur.* **2**: 193, 1986; key 3 spp. Eur.), Malaval (*Docums mycol.* **19**: 9, 1989; ascus ultrastr.).

hiascent, becoming wide open.

Hiatula (Fr.) Mont. (1854) ? = Lepiota (Agaric.) fide Dennis (1952).

Hiatulopsis Singer & Grinling (1967), Agaricaceae. 1, Congo.

Hidakaea I. Hino & Katum. (1955), ? Microthyriaceae. 1, Japan.

Hiemsia Svrček (1969), Otideaceae. 1, Eur.

Higginsia Nannf. (1932) [non Pers. (1805), *Rubiaceae*] ≡ Blumeriella (Dermat.).

higher fungi, *Ascomycota*, *Basidiomycota*, Mitosporic fungi.

hilar appendage (of basidiospores), the small wart- or cone-like projection which connects the spore with the sterigma; sterigmatal appendage (Smith); apicule (Josserand). Cf. apophysis.

Hildebrandiella Naumov (1917) ? = Syzygites (Thamnid.) fide Hesseltine (1955).

Hilitzeria Dyr (1941) nom. nud. ? = Dichotomocladium (Thamnid.) fide Benny (1973).

Hillia, see Sillia.

hilum, a mark or scar, esp. that on a spore at the point of attachment to a conidiogenous cell or sterigma.

Himanthites, see Himantites.

Himantia Pers. (1801), Mitosporic fungi, ?.?.?. *c.* 4 (sterile mycelium), Eur. See Donk (*Taxon* **11**: 85, 1962; refs).

Himantia (Fr.) Zoll. (1844) = Cylindrobasidium (Hyphodermat.). See Donk (*Taxon* **12**: 161, 1963; nomencl.).

himantioid (of mycelium), in spreading fan-like cords, as in *Himantia*.

Himantites Debey & Ettingsh. (1859), Fossil fungi. 1, Eur.

Himantormia I.M. Lamb (1964), Parmeliaceae (L). 1, Antarctica.

Hindersonia Lév. (1846) ≡ Hendersonia (Mitosp. fungi) fide Holm (*Taxon* **24**: 275, 1975).

Hindersonia Moug. & Nestl. ex J. Schröt. (1897) ≡ Ceriospora (Hyponectr.).

Hinoa Hara & I. Hino (1961), Ascomycota (inc. sed.). 2, Japan.

Hiospira R.T. Moore (1962), Anamorphic Dothideales, 1.F2.1. Teleomorph *Brooksia*. 1, trop.

Hippocrepidium Sacc. (1874) = Hirudinaria (Mitosp. fungi) fide Hughes (1951).

Hippoperdon Mont. (1842) nom. rej. = Calvatia (Lycoperd.).

Hiratsukaia Hara (1948), Uredinales (inc. sed.). 1, Japan. Excluded by Cummins & Hiratsuka (1983).

Hiratsukamyces Thirum., F. Kern & B.V. Patil (1975), Uredinales (inc. sed.). 2, India.

Hirneola Fr. (1825) nom. rej. ≡ Mycobonia (Polypor.).

Hirneola Fr. (1848) nom. cons. = Auricularia (Auricular.) fide Lowy (1952).

Hirneola Velen. (1939) ≡ Clitopilopsis (Entolomat.).

Hirneolina (Pat.) Bres. (1905) = Heterochaete (Exid.) fide Donk (*Persoonia* **8**: 33, 1974). ≡ Eichleriella (Exid.) fide Lowy (1971),

Hirschioporus Donk (1933), Coriolaceae. 10, widespr. See Macrae & Aoshima (*Mycol.* **58**: 912, 1967), Pegler (1973). ≡ Trichaptum (Polypor.) fide Ryvarden (1978).

hirsute, having long hairs.

Hirsutella Pat. (1892), Mitosporic fungi, 1/2.A1.15. 20 (on insects), widespr. See Macleod (*CJB* **37**: 695, 819, 1959; nutrition), Minter & Brady (*TBMS* **74**: 271, 1980), Minter *et al.* (*TBMS* **81**: 455, 1983; on eriophyid mites), Samson *et al.* (*Persoonia* **12**: 123, 1984; Troglobiotic hyphomycetes), Evans & Samson (*CJB* **64**: 2098, 1986; with *Tetracrium* synanamorph), Strongman *et al.* (*Jl Invert. Pathol.* **55**: 11, 1990; spp. on spruce budworm), Fernández-Garcia *et al.* (*MR* **94**: 1111, 1990; on mealybug), Samson & Evans (*MR* **95**: 887, 1991; *Didymobotryopsis*, a synonym), Strongman & MacKay (*Mycol.* **85**: 65, 1993; RAPDs used for varietal discrimination), Tedford *et al.* (*MR* **98**: 1127, 1994; RAPDs analysis of variation in *H. rhossiliensis*).

hirtose (**hirtous**), having hairs; hirsute.

Hirudinaria Ces. (1856), Mitosporic fungi, 1.G2.1. 1, Eur., N. Am. See Hughes (*Mycol. Pap.* **39**, 1951).

hispid, having hairs or bristles.

Hispidicarpomyces Nakagiri (1993), Hispidicarpomycetaceae. 1 (on red alga, *Galaxaura*), Japan.

Hispidicarpomycetaceae Nakagiri (1993), Spathulosporales. 1 gen., 1 sp. Thallus composed of hyphal cells; ascomata without sterile setose hairs; peridium very thick, 3-layered; interascal tissue of simple paraphysis-like hyphae, the ostiole not periphysate; hymenium extending over the whole of the inner surface of the ascoma; ascospores without mucous appendages. Anamorph spermatial, with verticillately branched conidiophores. On red algae, Pacific Ocean (Japan).

Hispidocalyptella E. Horak & Desjardin (1994), Tricholomataceae. 1, Australia.

hispidulous, somewhat, or delicately; cf. hispid.

Histeridomyces Thaxt. (1931), Laboulbeniaceae. 5, Am., Eur. See Rossi (*Mycol.* **72**: 430, 1980).

histogenous, (1) produced from tissue; (2) (of spores), produced from hyphae or cells, without conidiogenous cells.

histolysis, the disappearance or solution of a wall or tissue.

Histoplasma Darling (1906), Anamorphic Onygenaceae, 1.A1.1. Teleomorphs *Ajellomyces*, *Emmonsiella*. 2, widespr. *H. capsulatum* on humans and animals (**histoplasmosis**). See Sweany (Ed.) (*Histoplasmosis*, 1960), Cooke (*Mycopathol.* **39**: 1, 1969; 2,300 ref.), Ajello *et al.* (Eds) (*Histoplasmosis symposium*, 1971), Vincent *et al.* (*J. Bact.* **165**: 813, 1986; RFLP and classification), Maresca & Kobayashi (*Microbiol. Rev.* **53**: 186, 1989; dimorphism and cell differentiation in *H. capsulatum*),

Kwong-Chung (*Science, N.Y.* **177**: 368, 1972; teleomorphs), Fukushima *et al.* (*Mycopathol.* **116**: 151, 1991; re-evaluation of teleomorph by ubiquinone).

histoplasmin, an antigen prepared from *Histoplasma capsulatum*, esp. for skin testing.

histoplasmosis, see *Histoplasma*.

History. Macrofungi and macrolichens have been used by man as food and in other ways since early times. There are references to lichenized and other fungi (see Buller, *TBMS* **5**: 21, 1915) in the Greek and Roman classics and illustrated accounts are offered in the European printed herbals of the sixteenth and seventeenth centuries (e.g. those of Dodoens, Bauhin, Parkinson) but their systematic investigation stems mainly from the studies of Tournefort, Dillenius, and, particularly, Micheli in the eighteenth century. Many of the earliest studies include a wide range of groups, but as the nature of lichens was not accepted until the end of the nineteenth century lichenology tended to be a rather independent study; integration of the two specialities has accelerated only since the 1970s, and the first well-integrated classifications are even later (see *Ascomycota*).

The first phase of mycology was mainly systematic. The chief contribution of Linnaeus was binomial nomenclature, but before 1800 advances were made by Tode and others. Outstanding among later workers before 1900 were Persoon, Fries, Berkeley, Corda, the Tulasne brothers, de Bary, Brefeld, and Saccardo — for all of whom (and many others) there are short biographical notices (see Authors). The twentieth century has seen major developments in the knowledge of cytology, sex, genetics, physiologic specialization, etc. in fungi and a great expansion in medical and industrial mycology and mycological aspects of plant pathology.

At the beginning of the nineteenth century Acharius put lichenology on a sound scientific basis and many terms and currently accepted genera were introduced by him. With improvements in the microscope by the mid-century spore characters were first used in lichen taxonomy, particularly by de Notaris, Massalongo, Trevisan, and Körber; and the dual nature of lichens was recognized by de Bary and Schwendener. During the latter part of the nineteenth century important regional and monographic studies were made by Nylander, Müller-Argoviensis, Hue, and Crombie. From the 1950s particularly metabolic products (q.v.) came to play an increasing role in lichen systematics, and physiological (see Physiology) and ultrastructural (see Ultrastructure) studies have increased the understanding of their symbiotic nature.

Lit.: **General**: Ainsworth (*Introduction to the history of mycology*, 1976). **Mycology**: Lutjeharms (*Meded. nederl. mycol. Vereen.* **23**, 1936; 18th century), Ramsbottom (*Proc. Linn. Soc. Lond.* **151**: 280, 1941; mycology since Linnaeus), Lazzari (*Storia della micologia italiana*, 1973), Krieger (*Mycol.* **14**: 311, 1922; mycological illustrations), Rea (*TBMS* **5**: 211, 1916), Rogers (*A brief history of mycology in North America*, edn 2, 1981), Singer (*Mycologists and other taxa*, 1984), Machol (*Mycologist* **4**: 129, 184, 1990; **5**: 28, 1991; early mushroom books). **Lichenology**: Krempelhuber (*Geschichte und Literatur der Lichenologie* **1**, 1867; **2**, 1869-72), Smith (*Lichens*, 1921 [reprint 1975]), Grummann (*Biographisch-bibliographisches handbuch der Lichenologie*, 1974), Hawksworth & Seaward (*Lichenology in the British Isles 1568-1975*, 1977).

See also Authors, Ethnomycology, Literature, Medical and Veterinary mycology, Plant pathogenic fungi.

HMAS. Mycology Herbarium, Systematic Mycology and Lichenology Laboratory, Institute of Microbiology (Beijing, China); founded 1953; a specialist Institute of Academia Sinica; the Laboratory was established in 1985; genetic resource collection **CGMCC** (Centre for General Microbiological Culture Collections).

hoary (esp. of a pileus or stipe), covered thickly with silk-like hairs; canescent.

Hobsonia Berk. ex Massee (1891), Mitosporic fungi, 3.F1.1. 3 (2 on lichens), widespr. See Lowen *et al.* (*Mycol.* **78**: 842, 1986; lichenicolous spp.).

Hochkultur, see Normkultur.

Hodophilus R. Heim (1965) = Hygrotrama (Hygrophor.) fide Singer (1965).

Hoehneliella Bres. & Sacc. (1902), Mitosporic fungi, 7.B2.15. 1, Austria. See Vasant Rao & Sutton (*Kavaka* **3**: 21, 1976).

Hoehnelogaster Lohwag (1926) nom. dub. (? Melanogasterales).

Hoehnelomyces Weese (1920) = Atractiella (Hoehnelomycet.) fide Oberwinkler & Bauer (*Sydowia* **41**: 224, 1989).

Hoehnelomycetaceae Jülich (1982), Atractiellales. 1 gen. (+ 2 syn.), 4 spp. Basidioma stilboid; hyphae with simple septal pores; metabasidium elongate; spores hyaline, thin-walled.

Hohenbuehelia Schulzer (1866), Tricholomataceae. 25, cosmop. See Barron & Dierkes (*CJB* **55**: 3054, 1977; anamorph). Anamorph *Nematoctonus*.

Höhnel (Franz Xavier Rudolf von; 1852-1920). Professor of Botany, etc., at the Vienna Technical College, 1884-1920. Proposed 250 new gen. and 500 spp. Chief writings: Fragmente zur Mykologie (*Sber. Akad. Wiss. Wien, Mat.-Nat. Kl.* I, **111-132**, 1902-23; [Index Mitt. 1-1000 printed separately 1916; to 1001-1225, see *Mitt. bot. Inst. Tech. Hochsch. Wien* **9**: 66, 1932]); Mykologische Fragmente (*Ann. Myc.* **1-18**, 1903-1920 [reprint 1967]), (by Weese) v. Höhnel's mykologischen Nachlass-Schriften 1-150 (*Mitt. bot. Inst. tech. Hochsch. Wien* **1-12** (index **12**: 33), 1924-35). See Weese (*Ber. dtsch. bot. Ges.* **38**: (103), 1921), Stafleu & Cowan (*TL-2* **2**: 229, 1979).

Holcomyces Lindau (1904) = Diplodia (Mitosp. fungi) fide Höhnel (*Ann. Myc.* **3**: 187, 1905).

holdfast, a process from the thallus for attachment, e.g. appressorium, hyphopodium, stigmatopodium, and stomatopodium; cf. rhizoid, hapteron, haustorium.

Holleya Y. Yamada (1986), Metschnikowiaceae. 1, Canada. See Yamada & Nagahama (*J. gen. appl. Microbiol.* **37**: 199, 1991).

Hollosia Gyeln. (1939) = Scutula (Micar.) fide Santesson (*Svensk bot. Tidskr.* **43**: 141, 1949).

Holmiella Petrini, Samuels & E. Müll. (1979), Patellariaceae. 1 (on *Juniperus*), Eur., N. Am., Pakistan. Anamorph *Corniculariella*.

Holmiodiscus Svrček (1992), Leotiaceae. 1, Sweden.

holo- (prefix), all; whole; entire.

Holobasidiomycetes, see *Basidiomycota*.

holobasidium, see basidium.

holoblastic, see blastic.

holocarpic, having all the thallus used for the fruit-body.

holocarpous (of lichen thalli), ones formed by colonies of a free-living photobiont being invaded by a mycobiont and developing directly into fruiting bodies (Henssen, *Lichenologist* **18**: 51, 1986).

Holocoenis Clem. (1909) = Coenogonium (Gyalect.).

holocoenocytic, see astatocoenocytic.

Holocoryne (Fr.) Bonord. (1851) = Clavaria (Clavar.) fide Corner (1950).

Holocotylon Lloyd (1906), Secotiaceae. 4, Mexico, Texas. See Zeller (*Mycol.* **39**: 282, 1947).

Holocyphis Clem. (1909) = Thelomma (Calic.).

hologamy, the condition in which all the thallus becomes a gametangium, i.e. there is fusion between two mature individuals as in *Polyphagus*.

Hologloea Pat. (1900) = Favolaschia (Tricholomat.).

Hologymnia Nyl. (1900) = Evernia (Parmel.).

holomorph, see States of fungi.

holomorphum, two or more accepted species, each comprising teleomorph and anamorph, but scarcely distinguishable in one of the morphs (Tribe, *Bull. BMS* **17**: 94, 1983).

holophyte, a physiologically self-supporting green plant.

holosaprophyte, a true saprophyte (Johow).

Holospora Tomas. & Cif. (1952) = Polyblastia (Verrucar.).

Holosporomyces Cif. & Tomas. (1953) ≡ Holospora (Verrucar.).

holosporous (of conidial maturation), the conidium approaches its final size and shape before delimiting cells and maturing as a whole (Luttrell, 1963).

Holothelis Clem. (1909) = Thelopsis (Stictid.).

holotype, see type.

Holstiella Henn. (1895) = Massarina (Lophiostom.) fide v. Arx & Müller (1975).

Holtermannia Sacc. & Traverso (1910), Tremellaceae. 6, Java to Japan, Brazil. See Kobayasi (*Sci. Rep. Tokyo Bunrika Daig.* B **50**: 75, 1937).

Holttumia Lloyd (1924), Xylariaceae. 1, China, S.E. Asia. See Læssøe (*SA* **13**: 43, 1994).

Holubovaea Mercado (1983), Mitosporic fungi, 1.C2.19. 1, Cuba.

Holubovaniella R.F. Castañeda (1985), Mitosporic fungi, 1.C2.1. 2, Cuba, Kenya.

Holwaya Sacc. (1889), Leotiaceae. 1, N. temp. See Korf & Abawi (*CJB* **49**: 1879, 1971), Krieglsteiner & Häffner (*Z. Mykol.* **51**: 131, 1985). Anamorph *Crinula*.

Holwayella H.S. Jacks. (1926), Uredinales (inc. sed.). 1, S. Am. = Chrysocyclus (Puccin.) fide Cummins & Hiratsuka (1983).

Homalopeltis Bat. & Valle (1961), Mitosporic fungi, 5.B2.?. 1, Brazil.

Homaromyces R.K. Benj. (1955), Laboulbeniaceae. 1, USA, Eur.

homeostasis, the maintenance of constant chemical and physical conditions within a living organism. See Whittenbury *et al.* (Eds) (*Homeostatic mechanisms in micro-organisms*, 1987).

homo- (prefix), one and the same.

Homobasidiomycetes = *Holobasidiomycetidae* (p.p.); see *Basidiomycota*.

homobasidium, see basidium.

homobium, a self-supporting association of a fungus and an alga, as in lichens.

Homodium Nyl. ex H. Olivier (1903) = Leptogium (Collemat.).

homohetromixis, see heterothallism.

homohylic vesicle, sporangial vesicle, the wall of which is continuous with, and of the same material as the wall layer, or one of the wall layers of the sporangium. See also plasmamembranic vesicle, precipitative vesicle.

homoiomerous, (1) (of a lichen thallus), having the mycobiont and phycobiont evenly intermixed throughout the thallus, as in *Collema* (Fig. 40B); (2) (of trama in agarics), composed of hyphal tissue only; cf. heteromerous.

homokaryotic, having genetically identical nuclei, as in a line of isolates without variation (Brierley, *Ann. appl. Biol.* **18**: 429, 1931).

homokaryotic diplospory, incorporation of one or more homogenetic diploid nuclei in an oospore formed in response to abortive meioses in the oogonium (Dick, 1972).

homologous, showing a resemblance in form or structure, but not necessarily function, which is considered to be evidence of evolutionary relatedness; cf. analogous. See Poelt (*in* Hawksworth (Ed.), *Frontiers in mycology*: 85, 1991; homologous characters in lichens). See also alternation of generations, Cladistics.

homomixis, haplo-monoecism or monoecism. Whitehouse (*Biol. Rev.* **24**: 428, 1949) recognized two types: **primary homothallism** such as occurs in a homokaryotic individual and **secondary homothallism** (= pseudo- or facultative heterothallism (Dodge), **amphithallism**, Lange, 1952) as in a thallus derived from one heterokaryotic spore containing nuclei of compatible mating types; cf. heterothallism.

homomorphous (of an agaric lamella edge), hymenium on edge not differentiated from that on the faces; cf. heteromorphous.

Homonemeae, the *Algae* and *Fungi* (obsol.).

homonym, a name which (under the *Code*) must not be used because of an earlier name in a different sense, i.e. the names are the same but the types different; Donk (*Bull. bot. Gard. Buitenzorg* ser. 3 **18**: 282, 1949) in addition to heterotypic homonyms recognizes homotypic, non-synonymous, synonymous, and monadelphous homonyms, the last being derived from a common source, particularly from one devalidated name.

Homophron (Britzelm.) W.B. Cooke (1953) = Psathyrella (Coprin.).

homophytic, the equivalent in the sporophyte generation of monoecious in the gametophyte generation (Blakeslee, *Bot. Gaz.* **42**: 161, 1906); cf heterophytic.

homoplasmic, see karyotic.

Homopsella Nyl. (1887) = Porocyphus (Lichin.) fide Henssen (1963).

Homopsellomyces Cif. & Tomas. (1953) ≡ Homopsella (Lichin.).

Homospermales. Le Gal's (1953) term for operculate discomycetes; cf. *Heterospermales*.

homospory, (1) having asexually produced spores of only one kind (= isospory); (2) having spores which are not differentiated according to mating type (c.f. heterospory).

Homostegia Fuckel (1870), Dothideales (inc. sed.). 1 or 2 (on lichens), Eur.

homothallism (adj. **homothallic**), the condition in which sexual reproduction can occur without the interaction of two differing thalli.

Homothecium A. Massal. (1853), Collemataceae (L). 3, temp. S. Am. See Henssen (*Lichenologist* **3**: 29, 1965).

homozygous, with identical alleles at the same locus on homologous chromosomes.

honey agaric (**honey fungus**), *Armillaria mellea*; shoestring or boot-lace fungus.

honey-dew, (1) a secretion, attractive to insects, associated with the *Sphacelia* phase of *Claviceps* (Mower & Hancock, *CJB* **53**: 2826, 1975); (2) a secretion by aphids.

Honoratia Cif., Vegni & Montemart. (1963) = Preussia (Spororm.) fide v. Arx & Storm (*Persoonia* **4**: 407, 1967).

Horakia Oberw. (1976) ≡ Verrucospora (Agaric.).

Horakiella Castellano & Trappe (1992), Sclerodermataceae. 1, Australia.

Horakomyces Raithelh. (1983), Crepidotaceae. 1, S. Am.

Hormiactella Sacc. (1886), Mitosporic fungi, 1.B2.3. 2, Eur., W. Indies. See Holubová-Jechová (*Folia geobot. phytotax.* **13**: 433, 1978).

Hormiactina Bubák (1916) ? = Hormiactis (Mitosp. fungi) fide Carmichael *et al.* (1980).

Hormiactis Preuss (1851), Mitosporic fungi, 1.B1.3. 5, Eur., Sri Lanka.

Hormiokrypsis Bat. & Nascim. (1957), Anamorphic Metacapnodiaceae, 1.G2.28. Teleomorph *Ophiocapnocoma*. 1, USA. See Hughes (*N.Z. Jl Bot.* **5**: 117, 1967; **10**: 225, 1972).

Hormisciella Bat. (1956) = Antennatula (Mitosp. fungi) fide Hughes (*N.Z. Jl Bot.* **8**: 153, 1970).

Hormiscioideus M. Blackw. & Kimbr. (1979), Mitosporic fungi, 1.?.?. 1 (on *Insecta*), USA.

Hormisciomyces Bat. & Nascim. (1957), Anamorphic Euantennariaceae, 1.A2.15. Teleomorph *Euantennaria*. 1, Cuba. See Hughes (*N.Z. Jl Bot.* **12**: 299, 1974).

Hormisciopsis Sumst. (1914), Mitosporic fungi, 1.A2.3. 1, N. Am.

Hormiscium Kunze (1817) = Torula (Mitosp. fungi) fide Hughes (1958).

Hormoascus Arx (1972), ? Endomycetaceae. 3 (fide Barnett *et al.*, 1990), Eur. = Ambrosiozyma (Endomycet.) fide Goto & Takami (*J. gen. appl. Microbiol.* **32**: 271, 1986). See van der Walt & v. Arx (*Syst. Appl. Microbiol.* **6**: 90, 1985).

Hormocephalum Syd. (1939), Mitosporic fungi, 1.C2.4. 1, Ecuador.

Hormocladium Höhn. (1923), Mitosporic fungi, 1.A/B2.?. 1, Japan.

Hormococcus Preuss (1852) nom. dub. fide Sutton (*Mycol. Pap.* **141**, 1977).

Hormococcus Robak (1956), Mitosporic fungi, 4.A1.15. 1, Norway.

Hormoconis Arx & G.A. de Vries (1973), Anamorphic Amorphothecaceae, 1.A2.3. Teleomorph *Amorphotheca*. 1, widespr. *H. resinae*, the creosote or kerosene fungus (q.v.).

hormocyst, a propagule or diaspore produced in a special **hormocystangium** comprising a few cyanobacterial cells and fungal hyphae (Fig. 22I); produced by a few gelatinous lichens, e.g. *Lempholemma cladodes*, *L. vesiculiferum*. See Degelius (*Svensk bot. Tidsk.* **39**: 419, 1945), Henssen (*Lichenologist* **4**: 99, 1969).

hormocystangium, see hormocyst.

Hormodendroides M. Moore & F.P. Almeida (1937) = Rhinocladiella (Mitosp. fungi).

Hormodendron, see *Hormodendrum*.

Hormodendrum Bonord. (1853) = Cladosporium (Mitosp. fungi) fide de Vries (1952).

Hormodochis Clem. (1909) = Trullulla (Mitosp. fungi) fide Sutton (1977).

Hormodochium (Sacc.) Höhn. (1911) = Trullula (Mitosp. fungi) fide Sutton (1977).

Hormodochium (Sacc.) Sacc. (1931), Mitosporic fungi, 3.A2.?. 1, Belgium.

Hormographiella Guarro & Gené (1992), Mitosporic fungi, 1.A1.38/39. 3 (from humans, coprophilous), Eur.

Hormographis Guarro, Punsola & Arx (1986), Mitosporic fungi, 2.A1.40. 1 (from soil), Spain.

Hormomitaria Corner (1950), Physalacriaceae. 2, trop. See Berthier (*Bibl. Mycol.* **98**: 81, 1985; key).

Hormomyces Bonord. (1851), Mitosporic fungi, 3.A1.3. 1, temp. See Tubaki (*Trans. mycol. Soc. Japan* **17**: 243, 1976; culture).

Hormonema Lagerb. & Melin (1927), Anamorphic Botryosphaeriaceae, Dothioraceae, 1.A1.1/10. Teleomorphs various. 5, widespr. See Hermanides-Nijhof (*Stud. Mycol.* **15**: 166, 1977), Funk *et al.* (*CJB* **63**: 1579, 1985).

hormones (sex or sexual), of fungi, see antheridiol, progamones, sirenin, trisporic acids; also erogen, etc.

Machlis (*Mycol.* **64**: 235, 1972), Gooday (*Ann. Rev. Biochem.* **43**: 35, 1974), reviews. Cf. pheromone.

Hormopeltis Speg. (1912) = Micropeltis (Micropeltid.) fide Clements & Shear (1931).

Hormosperma Penz. & Sacc. (1897) = Lasiosphaeria (Lasiosphaer.).

Hormosphaeria Lév. (1863), ? Arthoniales (inc. sed.). 1, S. Am. Nom. dub. fide Santesson (*Symb. bot. upsal.* **12**(1), 1952).

Hormospora De Not. (1844) [non Bréb. (1839), *Algae*] = Bombardioidea (Lasiosphaer.) p.p. and Sporormia (Spororm.) p.p.

Hormosporites Grüss (1928), ? Fossil fungi. 1 (Devonian), Eur.

Hormotheca Bonord. (1864) = Coleroa (Ventur.) fide Müller (*SA* **5**: 310, 1986). See Corlett & Barr (*Mycotaxon* **25**: 255, 1986).

Hormyllium Clem. (1909) ≡ Hormococcus (Mitosp. fungi) fide Sutton (1977).

horn of plenty, basidioma of the edible *Craterellus cornucopioides*.

Hornodermoporus Teixeira (1993), Polyporaceae. 1, Brazil.

horse mushroom, basidiomata of the edible *Agaricus arvensis*.

horse-hair blight fungi, the rhizomorphic mycelia of tropical species of *Marasmius*, e.g. *M. equicrinis* (E. trop.) and *M. sarmentosus* (W. Indies). See Petch (*Ann. R. bot. Gdns Peradeniya* **6**: 43, 1915). Cf. thread blight.

horse-tail lichen, see rock hair.

Hortaea Nishim. & Miyaji (1984), Mitosporic fungi, 1.B2.21. 1, Brazil. See Mittag (*Mycoses* **36**: 242, 1993; fine structure), Uijthof *et al.* (*Mycoses* **37**: 307, 1994; genotypes of *H. werneckii*).

Hosseusia Gyeln. (1940), ? Pannariaceae (L). 1, Argentina.

host, (1) a living organism harbouring a parasite; frequently in the sense of 'suscept' (q.v.); sometimes, as in 'host index', in a general sense covering certain substrata; (2) a commercial computer operator keeping databases other than their own on its machines and making them available to others for on-line searching for a fee (see Literature).

house fungus (or **dry rot fungus**), *Serpula lacrimans*.

Huangshania O.E. Erikss. (1992), Triblidiaceae. 1, China.

Hubbsia W.A. Weber (1965), Roccellaceae (L). 1, Guadaloupe Island (Mexico). = Reinkella (Roccell.) fide Follmann (*Nova Hedw.* **31**: 285, 1979).

Huea C.W. Dodge & G.E. Baker (1938) = Caloplaca (Teloschist.).

Hueella Zahlbr. (1926), Pannariaceae (L). 1, Asia.

Hughesiella Bat. & A.F. Vital (1956) = Chalara (Mitosp. fungi) fide Nag Raj & Kendrick (1975).

Hughesinia J.C. Lindq. & Gamundí (1970), Mitosporic fungi, 1.G2.28. 1, Chile.

Hugueninia J.L. Bezerra & T.T. Barros (1970) [non Rchb. (1832), *Cruciferae*], ? Microthyriaceae. 1, New Caledonia.

Huilia Zahlbr. (1930) = Porpidia (Porpid.).

hülle cells, terminal or intercalary thick-walled cells which occur in large numbers in association with the ascomata of, e.g. *Aspergillus nidulans*.

Humaria Fuckel (1870), Otideaceae. *c.* 15, N. temp. See Denison (*Mycol.* **51**: 612, 1961), Eriksson & Hawksworth (*SA* **11**: 179, 1993; nomencl.), Kanouse (*Mycol.* **39**: 635, 1947).

Humaria (Fr.) Boud. (1885) ≡ Octospora (Otid.).

Humariaceae, see Otideaceae.

Humariella J. Schröt. (1893) ≡ Scutellinia (Otid.).

Humarina Seaver (1927) ≡ Octospora (Otid.).

Humboldtina Chardón & Toro (1934) = Leptosphaeria (Leptosphaer.) fide Barr (*SA* **5**: 134, 1986).

Humicola Traaen (1914), Mitosporic fungi, 1.A2.1. 4, temp. See Fassatiová (*Česká Myk.* **18**: 102, 1964; key), de Bertoldi (*CJB* **54**: 2755, 1976; 13 spp., physiol. analysis), Nicoli & Russo (*Nova Hedw.* **25**: 737, 1974; 8 spp.).

Humicolopsis Cabral & S. Marchand (1976), Mitosporic fungi, 1.A1.?. 1, Argentina.

Humicolopsis Verona (1977) [non Cabral & S. Marchand (1976)], Mitosporic fungi, 1.A1.10. 1, Italy.

humicolous, living in or on decaying organic matter, soil.

Humidicutis (Singer) Singer (1959), Hygrophoraceae. 3, N. temp.

Humphreya Steyaert (1972), Ganodermataceae. 4, trop. See Ryvarden & Johansen (*Preliminary polypore flora of East Africa*: 94, 1980; key 4 Afr. spp.).

Husseia Berk. (1847) = Calostoma (Calostomat.).

Hyalacrotes (Korf & Kohn) Raitv. (1991), Hyaloscyphaceae. 2, Eur.

Hyalasterina Speg. (1919) nom. dub. (Fungi, inc. sed.).

Hyalina, see *Hyalinia*.

hyaline, transparent or nearly so; translucent; frequently used in the sense of colourless.

Hyalinia Boud. (1885) [non Stackh. (1809), *Algae*] = Orbilia (Orbil.) fide Baral (*SA* **13**: 113, 1994).

hyalo- (prefix) (of spores), hyaline or brightly coloured, esp. for groups of Mitosporic fungi (q.v.) (**hyalodidymae**, etc.).

Hyalobelemnospora Matsush. (1993), Mitosporic fungi, 1.A1.19. 1, Peru.

Hyalobotrys Pidopl. (1948) = Stachybotrys (Mitosp. fungi) fide Kendrick & Carmichael (1973).

Hyalocapnias Bat. & Cif. (1963) = Scorias (Capnod.).

Hyaloceras Durieu & Mont. (1849) = Seiridium (Mitosp. fungi) fide Höhnel (*Sber. Akad. Wiss. Wien* **125**: 27, 1916). See Sutton (1977).

Hyalochlorella Poyton (1970) nom. dub (? Algae). See Alderman (*Veröff. Inst. Meersforsch. Bremerh., Suppl.* **5**: 251, 1974).

Hyalocladium Mustafa (1977), Mitosporic fungi, 1.D1.10. 1, Kuwait.

Hyalococcus J. Schröt. (1889) ? = Trichosporon (Mitosp. fungi).

Hyalocrea Syd. & P. Syd. (1917), Dothideaceae. 4, trop. See Pirozynski (*Kew Bull.* **31**: 595, 1977), Rossman (*Mycol. Pap.* **157**, 1987; key).

Hyalocurreya Theiss. & Syd. (1915) = Uleomyces (Cookell.) fide v. Arx (1963).

Hyalocylindrophora J.L. Crane & Dumont (1978) = Dischloridium (Mitosp. fungi) fide Bhat & Sutton (*TBMS* **84**: 723, 1985).

Hyalodema Magnus (1910) = Coniodictyum (Cryptobasid.) fide Donk (*Reinwardtia* **4**: 115, 1956).

Hyalodendron Diddens (1934), Mitosporic fungi, 1.A1.3. 3, Eur., N. Am., Afr.

Hyaloderma Speg. (1884), Ascomycota (inc. sed.). 1, S. Am. See Pirozynski (*Kew Bull.* **31**: 595, 1977).

Hyalodermella Speg. (1918), Ascomycota (inc. sed.). 4, S. Am.

Hyalodictys Subram. (1962) = Miuraea (Mitosp. fungi) fide Deighton (*Mycol. Pap.* **133**, 1973).

Hyalodictyum Woron. (1916), Mitosporic fungi, 6.D1.1. 1, former USSR.

Hyalodothis Pat. & Har. (1893) nom. dub.; ? = Melanodothis (Mycosphaerell.). See Arnold (*Mycol.* **59**: 246, 1967).

Hyaloflorea Bat. & H. Maia (1955) = Cylindrocarpon (Mitosp. fungi) fide Carmichael *et al.* (1980).

Hyalohelicomina T. Yokoy. (1974), Mitosporic fungi, 1.F1.1. 1, Japan.

hyalohyphomycosis, a mycotic infection of humans or animals caused by a non-dematiaceous fungus. Cf. phaeohyphomycosis.

Hyalomelanconis Naumov (1954) = Melanconis (Melanconid.) p.p.

Hyalomeliolina F. Stevens (1924), Pseudoperisporiaceae. 2, C. & S. Am. See Hughes (*Mycol. Pap.* **166**, 1993).

Hyalopesotum H.P. Upadhyay & W.B. Kendr. (1975), Anamorphic Ophiostomataceae, 2.A1.10. Teleomorph *Ceratocystis*. 1, Canada.

Hyalopeziza Fuckel (1870), Hyaloscyphaceae. 8, Eur., N. Am. See Raschle (*Sydowia* **29**: 170, 1977; key), Korf & Kohn (*Mycotaxon* **10**: 503, 1980).

Hyalopsora Magnus (1902), Pucciniastraceae. 7 (+ 5), N. temp.; 0, I on *Abies*; II, III on *Pteridophyta*.

Hyalopus Corda (1838) ? = Acremonium (Mitosp. fungi) fide Kendrick & Carmichael (1973).

Hyalopycnis Höhn. (1918), Mitosporic Heterogastridiaceae, 4.A1.1. Teleomorph *Heterogastridium*. 1, Eur., N. Am. See Seeler (*Farlowia* **1**: 119, 1943).

Hyalorhinocladiella H.P. Upadhyay & W.B. Kendr. (1975), Anamorphic Ophiostomataceae, 1.A1.10. Teleomorph *Ophiostoma*. 1, USA.

Hyaloria A. Møller (1895), Hyaloriaceae. 2, Eur., Brazil. See Wells (*Mycol.* **61**: 77, 1969).

Hyaloriaceae Lindau (1897), Tremellales. 1 gen., 1 spp. Basidioma stilboid, gelatinous; monomitic; spores statismosporic; on leaves of dried palms.

Hyaloscolecostroma Bat. & J. Oliveira (1967), ? Capnodiaceae. 1, Brazil.

Hyaloscypha Boud. (1885) typ. cons., Hyaloscyphaceae. 27, widespr. See Huhtinen (*Karstenia* **29**: 45, 1990; key), Svrček (*Česká Myk.* **39**: 205, 1985; spp. descr. by Velenovský).

Hyaloscyphaceae Nannf. (1932), Leotiales. 59 gen. (+ 53 syn.), *c.* 370 spp. Stromata absent; ascomata usually small, flat or concave, the excipulum soft, fleshy, usually composed of prismatic or isodiametric cells, almost always with conspicuous, sometimes ornamented hairs surrounding the disc; interascal tissue of simple paraphyses, sometimes lanceolate; asci small, with an I+ or I-ring; ascospores small, usually elongated, sometimes septate. Anamorphs rarely noted, when so hyphomycetous. Saprobic on woody and herbaceous material, cosmop.
    Lit.: Dennis (*Mycol. Pap.* **32**, 1949; UK, *Kew Bull.* 1954: 289, 1954; trop. Am., *Kew Bull.* **14**: 418, 1960; Venezuela, *Kew Bull.* **13**: 32, 1958; Australia, *Kew Bull.* **17**: 319, 1963; redispos., *Persoonia* **2**: 171, 1972; *Lachneae*), Hein (*Nova Hedw.* **32**: 31, 1980; SEM hairs), Huhtinen (*Mycotaxon* **29**: 267, 1987; hairs, gen. limits), Korf (*Lloydia* **14**: 129, 1951; *Arachnopezizae*), Korf & Kohn (*Mycotaxon* **10**: 503, 1980; spp. glassy hairs), Raitviir (*Synopsis of the Hyaloscyphaceae*, 1970; *Mikol. Fitopat.* **21**: 200, 1987; subfams.), Sharma (*Nova Hedw.* **43**: 381, 1986; 76 spp. India), Spooner (*Bibl. Mycol.* **116**: 1987; Australasia), Svrček (*Česká Myk.* **41**: 193, 1987; key 50 gen. Eur.).

Hyalosphaera F. Stevens (1917), Dothideales (inc. sed.). 3, C. & S. Am. See Rossman (*Mycol. Pap.* **157**, 1987; key).

Hyalospora Nieuwl. (1916) ≡ Catharinia (Thelenell.).

Hyalostachybotrys Sriniv. (1958) = Stachybotrys (Mitosp. fungi) fide Barron (*Mycol.* **56**: 313, 1964).

Hyalostilbum Oudem. (1885) = Dictyostelium (Dictyost.) fide Saccardo (1888).

Hyalosynnema Matsush. (1975), Mitosporic fungi, 2.C1.10. 1, Japan.

Hyalotexis Syd. (1925) = Hyalosphaera (Dothideales) fide Petrak (*Ann. Myc.* **26**: 385, 1928).

Hyalotheles Speg. (1908), Elsinoaceae. 1, S. Am.

Hyalothyridium Tassi (1900), Mitosporic fungi, 4.D1.1. 3, widespr. See Latterell & Rossi (*Mycol.* **76**: 506, 1984; reinstatement of genus).

Hyalothyris Clem. (1909) ≡ Hyalothyridium (Mitosp. fungi).

Hyalotia Guba (1961) = Bartalinia (Mitosp. fungi) fide Nag Raj (*CJB* **53**: 1615, 1975).

Hyalotiella Papendorf (1967), Mitosporic fungi, 8.C2.10. 2, trop. to subtrop., S. Afr.

Hyalotiopsis Punith. (1970), Anamorphic Amphisphaeriaceae, 8.C2.19. Teleomorph *Ellurema*. 1, India.

Hyalotricha Dennis (1949) = Hyalopeziza (Hyaloscyph.) fide Raschle (1977).

Hyalotrochophora Finley & E.F. Morris (1967) ? = Delortia (Mitosp. fungi) fide Kendrick & Carmichael (1973).

Hybogaster Singer (1964), Hybogasteraceae. 1, Chile.

Hybogasteraceae Jülich (1982), Bondarzewiales. 1 gen., 1 sp. Gasterocarp angiocarpic, loculate; spores statismosporic, ornamentation amyloid.

hybrid, the result of a cross between organisms belong to different taxa which yield viable progeny. It is unclear how common the process is in nature with fungi; as with plants, hybridization can occur at levels from genus down; e.g. *Saccharomycopsis fibuligera* × *Yarrowia lipolytica* (Nga *et al.*, *J. gen. Microbiol.* **138**: 223, 1992), *Cladonia grayi* × *C. merochlorophaea* (Culberson *et al.*, *Am. J. Bot.* **75**: 1135, 1988); a **sexual -**. A **mechanical -** (in lichens) is formed by the growth together of propagules from different genera, species or genotypes to form a single thallus not involving sexual crossing (Hawksworth, *in* Street (Ed.), *Essays in plant taxonomy*: 211, 1978; see Lichens). The existing rules of nomenclature for hybrids and grafts are available to name both kinds of progeny (Hawksworth, *Internat. Lich. Newsl.* **21**: 59, 1988).

Hydatinophagus Valkanov (1931), Saprolegniaceae. 2, Eur. See Dick (*in press*; key).

Hydnaceae Chevall. (1926), Cantharellales. 9 gen. (+ 16 syn.), 137 spp. Basidioma pileate and stipitate; hymenophore spinose.
    Lit.: Donk (1964: 272), Grand & van Dyke (*J. Elisha Mitchell Sci. Soc.* **92**: 114, 1976; SEM spores), Harrison (*Mycol.* **65**: 277, 1973), Maas Geesteranus (*Verh. K. Ned. Akad. Wet., Natur.* **60**: 1975; Eur. spp.), Coker & Beers (*Stipitate hydnums of the Eastern United States*, 1951), Maas Geesteranus (*Hydnaceous fungi of the eastern old world*, 1971, *Die terrestrichen Stachelpilze Europas: the terrestrial hydnums of Europe*, 1975, *The stipitate hydnums of the Netherlands*, I-III [*Fungus* **26, 27**], IV [*Persoonia* **1**: 115, 1956-59]), Nikolaeva (*Fl. Pl. Crypt. URSS* **6**; *Fungi* **2**, 1961), Harrison (*The stipitate hydnums of Nova Scotia* [*Can. Dep. Agric. Res. Branch Publ.* **1099**] 1961).

Hydnangiaceae Gäum. & C.W. Dodge (1928), Agaricales. 4 gen., 6 spp. Gasterocarp hygogeous or epigeous; spores hyaline, subglobose, spinose. See Beaton *et al.* (*Kew Bull.* **39**: 499, 1984; key).

Hydnangium Wallr. (1839), Hydnangiaceae. 3, widespr. (possibly by dispersal with *Eucalyptus*). See Pegler & Young (*TBMS* **72**: 384, 1979), Mueller *et al.* (*Mycol.* **85**: 850, 1993; spores). S. str., gasteroid relatives of *Laccaria*.

Hydnellum P. Karst. (1879), Thelephoraceae. 34, widespr. See Coker & Beers (1951: 56), Maas Geesteranus (*Persoonia* **2**: 388, 1962), Stalpers (*Stud. Mycol.* **35**, 1993; key).

Hydnellum P. Karst. (1896) = Hyphoderma (Hyphodermat.).

Hydnites Mesch. (1892), Fossil fungi. 2 (Tertiary), Eur.

Hydnobolites Tul. & C. Tul. (1843), Pezizaceae. 2 (hypogeous), N. Am., Eur. See Kimbrough *et al.* (*Bot. Gaz.* **152**: 408, 1992; posn).

Hydnocaryon Wallr. (1833) = Genea (Otid.) fide Trappe (1975).

Hydnochaete Bres. (1896), Hymenochaetaceae. 8, mainly trop. See Ryvarden (*Mycotaxon* **15**: 425, 1982).

Hydnochaete Peck (1897) ≡ Hydnochaetella (Hymenochaet.).

Hydnochaetella Sacc. (1898) = Asterodon (Asterostromat.).

Hydnocristella Petersen (1971), Corticiaceae. 1, temp.

Hydnocystis Tul. (1844), Otideaceae. 2 (hypogeous), Eur., Japan. See Burdsall (*Mycol.* **60**: 503, 1968).

Hydnodon Banker (1913), Thelephoraceae. 1, Eur.

Hydnofomes Henn. (1900) = Echinodontium (Echinodont.) fide Banker (1913).

Hydnogloea Curr. (1871) = Pseudohydnum (Exid.) fide Donk (*Persoonia* **4**: 173, 1966).

Hydnophlebia Parmasto (1967), Meruliaceae. 1, N. Am.

Hydnophysa Clem. (1909) ≡ Hydnofomes (Echinodont.).

Hydnoplicata Gilkey (1955) = Peziza (Peziz.) fide Trappe (*Mycotaxon* **2**: 109, 1975).

Hydnopolyporus D.A. Reid (1962), Coriolaceae. 2, widespr. See Fildago (*Mycol.* **55**: 713, 1963).

Hydnoporia Murrill (1907) = Hydnochaete (Hymenochaet.) fide Banker (1914).

Hydnopsis Tul. & C. Tul. (1865) nom. dub. (Mitosp. fungi) fide Sutton (1977).

Hydnopsis (J. Schröt.) Rea (1909) = Tomentella (Thelephor.). See Donk (1956).

Hydnospongos Wallr. (1839) = Gautieria (Gautier.).

Hydnotrema Link (1833) ≡ Sistotrema (Sistotremat.).

Hydnotrya Berk. & Broome (1846), Helvellaceae. *c.* 12 (hypogeous), N. Am., Eur. See Gilkey (*N. Am. Fl.* **2** (1), 1954; key), Zhang (*Mycotaxon* **42**: 155, 1991), Pegler *et al.* (1993; key 4 Br. spp.).

Hydnotryopsis Gilkey (1916), Pezizaceae. 2 (hypogeous), N. Am. See Trappe (*Mycotaxon* **2**: 115, 1975), Gilkey (*N. Am. Fl.* **2** (1), 1954; key, as *Choiromyces*).

Hydnum L. (1753) nom. cons., Hydnaceae. 120, widespr. Some spp. are a cause of heartwood-rot in living trees. See Petersen (*Taxon* **26**: 144, 1977; typification).

Hydrabasidium Park.-Rhodes ex J. Erikss. & Ryvarden (1978), Ceratobasidiaceae. 1, widespr.

Hydraeomyces Thaxt. (1896), Laboulbeniaceae. 1, Am., Asia, Eur. See Siemaszko & Siemaszko (*Polsk. Pismo Entom.* **12**: 125, 1933).

Hydrocina Scheuer (1991), Hyaloscyphaceae. 1 (aquatic), UK.

Hydrocybe (Fr. ex Rabenh.) Wünsche (1877) = Cortinarius (Cortinar.) fide Kauffman (1932).

Hydrocybium Earle (1909) = Cortinarius (Cortinar.) fide Kauffman (1932).

hydrofungi, aquatic fungi (Cooke, 1963).

Hydrogen-ion concentration (pH). Most fungi grow best at approximately pH 7 but tolerate a wide range from pH 3-10 (or even 11). This property is made use of for freeing fungal cultures from bacteria or inhibiting bacterial growth when making isolates from soil by using an acid medium (pH 4). See Webb (*Ann. Mo. bot. Gdn* **6**: 201, 1919), MacInnes (*Phytopath.* **12**: 290, 1922), acidiphilous, alkaphilic.

Hydrogera F.H. Wigg. ex Kuntze (1891) = Pilobolus (Pilobol.).

Hydrometrospora J. Gönczöl & Révay (1985), Mitosporic fungi, 1.G1.1. 1 (aquatic), Hungary.

Hydromycus Raf. (1808) = Dacrymyces (Dacrymcet.). See Merrill (*Index Rafin.*: 69, 1949).

Hydromyxales. An order of *Myxomycetes* fide Jahn (*Nat. PflFam.* Aufl. 2, **2**, 1928) but excluded from this *Dictionary* as protozoans not now studied by 'mycologists'. Cf. monads.

Hydronectria Kirschst. (1925), ? Hypocreaceae (L). 1 (aquatic), Eur. See Kohlmeyer & Volkmann-Kohlmeyer (*MR* **97**: 753, 1993).

Hydronectriaceae Riedl (1987), Ascomycota (inc. sed., L). 1 gen., 1 sp. Ascomata perithecioid, clypeate, hamathecium of apical paraphyses, ascospores colourless, 1-septate. Thallus crustose. See *Hydronectria*.

Hydronema Carus ex Rchb. (1828) ≡ Achlya (Saproleg.).

Hydrophilomyces Thaxt. (1908), Laboulbeniaceae. 10, Am., Afr., Eur.

Hydrophora Tode (1791) nom. rej. = Mucor (Mucor.) fide Hesseltine (1955).

Hydrophorus Battarra ex Earle (1909) = Hygrocybe (Hygrophor.).

Hydropisphaera Dumort. (1822), Hypocreaceae. *c.* 3, mainly trop. See Rossman *et al.* (*Mycol.* **85**: 685, 1993). Anamorph *Acremonium*.

Hydropus Kühner ex Singer (1948), Tricholomataceae. 89, cosmop. See Singer & Grinling (*Persoonia* **4**: 363, 1967; key), Machol & Singer (*Mycol.* **69**: 1162, 1977; numerical taxonomy), Singer (*Fl. Neotrop. Monogr.* **32**, 1982).

Hydrothyria J.L. Russell (1856), Peltigeraceae (L). 1, N. Am. See Feige *et al.* (*Herzogia* **8**: 69, 1989).

hydrotropism, see tropism.

Hygramaricium Locq. (1979), Cortinariaceae. 1, Eur.

Hygroaster Singer (1955), Hygrophoraceae. 1, trop. Am.

Hygrochroma DC. [not traced] ? nom. dub. = Phloeospora (Mitosp. fungi) fide Mussat (*Syll. Fung.* **15**: 170, 1901).

Hygrocrocis C. Agardh (1824) ≡ Typhoderma (? Fungi).

Hygrocybe (Fr.) P. Kumm. (1871), Hygrophoraceae. 60, cosmop. See Singer (*Agaricales in modern taxonomy*, edn 3, 1975), Bon (*Docums mycol.* **7** (25), 1976; keys), Orton (*TBMS* **43**: 248, 1960; key), Heinemann (*Fl. Icon. Champ. Congo* **15**: 280, 1966; keys 13 spp. Congo).

Hygromitra Nees (1816) = Leotia (Leot.) fide Fries (*Syst. mycol.* **1**, 1821).

Hygromyxacium Locq. (1979), Cortinariaceae. 1, Eur.

hygrophanous, having a water-soaked appearance when wet.

hygrophilous, preferring a moist habitat.

Hygrophoraceae Lotsy (1907), Agaricales. 10 gen. (+ 11 syn.), 207 spp. Hymenophoral trama bilateral or regular; basidia long; spores white, thin-walled.

Hygrophoropsidaceae Kühner (1980), Boletales. 1 gen., 4 spp. Basidioma agaricoid, decurrent lamellate; spores hyaline, ellipsoid, dextrinoid.

Hygrophoropsis (J. Schröt.) Maire ex Martin-Sans (1929), Hygrophoropsidaceae. 4, widespr., esp. temp. See Singer *et al.* (*Beih. Nova Hedw.* **98**: 6, 1990; key C. Am. spp.).

Hygrophorus Fr. (1836), Hygrophoraceae. 80 (usually mycorrhizal), widespr., esp. N. temp. See Hesler & Smith (*North American species of Hygrophorus*, 1963; keys), Bon (*Docums mycol.* **7** (27-28), 1977; keys), Bird & Grund (*Proc. Nov. Scotia Inst. Sci.* **29**: 1, 1979; keys 53 Can. spp.), Heinemann (*Bull. Jard. bot. Etat. Belg.* **33**: 421, 1963; keys Centr. Afr. spp.).

Hygrophyllum, see *Hypophyllum*.

hygroscopic, (1) becoming soft in wet air, hard in dry; (2) (of a sporocarp), opening and discharging spores in dry air.

Hygrotrama Singer (1959), Hygrophoraceae. 15, widespr.

Hylophila Quél. (1886) [non Lindl. (1833), *Orchidaceae*] ≡ Hebeloma (Cortinar.).

Hylostoma Pers. (1822), Pezizales (inc. sed.). 1, Eur.

Hymenagaricus Heinem. (1981), Agaricaceae. 19, widespr. See Heinemann & Sister Little Flower (*Bull. Jard. bot. nat. Belg.* **54**: 151, 1984).

Hymenangiaceae Corda (1842), Cortinariales. 1 gen., 1 sp. Gasterocarp multiloculate; peridium epithelial; spores brown, mucronate, verruculose.

Hymenangium Klotzsch (1839), Hymenangiaceae. 1, temp. = Rhizopogon (Rhizopogon.) fide Bougher & Castellano (*Mycol.* **85**: 285, 1993).

Hymenelia Kremp. (1852), Hymeneliaceae (L). *c.* 3, N. temp.

Hymeneliaceae Körb. (1855), Lecanorales (L). 9 gen. (+ 12 syn.), 148 spp. Thallus varied but usually crustose, without rhizoids; ascomata apothecial, deeply immersed, with a well-developed margin, the disc pale or dark; interascal tissue of paraphyses, branched and moniliform at the apex; asci with a well-developed usually I- apical cap, usually without an ocular chamber, with an outer layer of I+ gelatinous material; ascospores large, aseptate, hyaline, thin-walled. Anamorphs pycnidial. Lichenized with green algae, usually on rocks, widespr.
    Lit.: Janex-Favre (*Cryptog., bryol. lich.* **6**: 25, 1985; ontogeny).

Hymenella Fr. (1822), Mitosporic fungi, 4.A1.?. 20, widespr. See Tulloch (*Mycol. Pap.* **130**, 1972; typification).

hymenial algae (or gonidia), algal cells in the hymenium of a lichenized ascomycete, e.g. *Endocarpon, Staurothele, Thelendia*; - **cystidium**, see cystidium; - **veil**, see annulus.

hymeniderm (of basidiomata), an outer layer composed of an unstratified layer of single cells or hyphal tips; see derm.

Hymeniopeltis Bat. (1959), Mitosporic fungi, 5.A1.?. 1, Brazil.

hymenium, the spore-bearing layer of a fruit-body; (of basidiomycetes) **euhymenium** (Donk, *Persoonia* **3**: 210, 1964), a hymenium in which the basidia and their sterile homologues are the first elements to be formed, as a palisade; in a **static** (non-thickening) **euhymenium** the exhausted basidia are replaced at the same level by intercalation; in a **thickening euhymenium** the tramal hyphae grow between the exhausted basidia to form a new hymenium above the old, as in the Cantharellaceae; **catahymenium** (Lemke, *CJB* **42**: 218, 1964), a hymenium in which hyphidia are the first-formed elements and the basidia embedded at various levels elongate to reach the surface and do not form a palisade. Cf. thecium, apothecium.

Hymenoascomycetes. A class proposed by Kimbrough (*in* Wheeler & Blackwell (Eds), *Fungus-insect relationships*: 184, 1984) for operculate and inoperculate discomycetes and pyrenomycetes; see *Ascohymeniales.*

Hymenobactron (Sacc.) Höhn. (1916), Mitosporic fungi, 3.A1.?. 1 or 2, Eur.

Hymenobia Nyl. (1854), Ascomycota (inc. sed.). 1 (on lichens), Eur. See Triebel (*Bibl. Lich.* **35**, 1989; sub *Hymenobiella*).

Hymenobiella Triebel (1989) ≡ Hymenobia (Ascomycota, inc. sed.) fide Eriksson & Hawksworth (*SA* **9**: 13, 1990).

Hymenobolina Zukal (1893) = Licea (Lic.).

Hymenobolus Durieu & Mont. (1845), Leotiales (inc. sed.). 1 (on *Agave*), N. Afr. See Rieuf (*Al Awamia* **4**: 127, 1962).

Hymenobolus Zukal (1893) ≡ Hymenobolina (Lic.).

Hymenochaetaceae Imazeki & Toki (1954), Hymenochaetales. 23 gen. (+ 37 syn.), 443 spp. Basidioma resupinate to pileate, clavarioid, annual or perennial; context mono- or dimitic, xanthochroic; hymenophore smooth to poroid, dichohyphidia or asterosetae present; clamp-connexions absent; spores hyaline to brown, usually smooth, rarely amyloid; typically lignicolous, causing white rots; widespr.
    Lit.: Donk (1964: 274), Fiasson & Niemelä (*Karstenia* **24**: 14, 1984; emend.).

Hymenochaetales, Basidiomycetes. 2 fam., 25 gen. (+ 40 syn.), 471 sp. Fams:
    (1) **Asterostromataceae**.
    (2) **Hymenochaetaceae**.

Hymenochaete Lév. (1846) nom. cons., Hymenochaetaceae. 100, widespr., esp. trop. *H. agglutinans* (canker of young hardwoods). See Cunningham (*Trans. R. Soc. N.Z.* **85**: 1, 1957; NZ spp.), Job (*Mycol. Helvet.* **5**: 1, 1990; temp. S. Hemisph. spp.).

Hymenochaetella P. Karst. (1889) = Hymenochaete (Hymenochaet.) fide Donk (1964).

Hymenoconidium Zukal (1888) ? = Marasmius (Tricholomat.) fide Donk (1962).

Hymenodecton Leight. (1854) = Phaeographis (Graphid.).

Hymenogaster Vittad. (1831), Hymenogasteraceae. 64, temp. See Soehner (*Beih. Nova Hedw.* **2**, 1962; key), Smith (*Mycol.* **58**: 100, 1966; keys N. Am.), Malençon (*BSMF* **75**: 99, 1959; sporogenesis), Pegler & Young (*Notes R. Bot. Gdn Edinb.* **44**: 437, 1987; key Br. spp.).

Hymenogasteraceae Vittad. (1831), Hymenogastrales. 8 gen. (+ 6 syn.), 87 spp. Gasterocarp multiloculate; peridium and epicutis present; spores smooth to verruculose, brown. Relationship to *Cortinariaceae* possible but still speculative.

Hymenogastrales, Basidiomycetes ('gasteromycetes'). 5 fam., 15 gen. (+ 7 syn.), 118 spp. Gasterocarp hypogeous; gleba typically fleshy, formed from tramal plates lined with hymenium differentiating in a primordial cavity, sometimes with columella; widespr., probably mycorrhizal.
    Whilst most included genera are obviously hypogeous relatives of other agaricoid *Basidiomycetes* their exact affiliation to a family (and even order) is not undisputable making it necessary to retain, at least temporarily, the present order. Fams:
    (1) **Gasterellaceae**.
    (2) **Gastrosporiaceae**.
    (3) **Hymenogasteraceae**.
    (4) **Octavianiaceae** (syn. *Octavianinaceae*).
    (5) **Protogastraceae**.
    Lit.: Smith (*in* Ainsworth *et al.* (Ed.), *The Fungi* **4B**: 421, 1973), Pegler & Young (*TBMS* **72**: 353, 1979), Singer & Smith (*Mem. Torrey bot. Cl.* **21**, 1960), Smith (*Mycol.* **54**: 626, 1962). See also *Lit.* under *Gasteromycetes.*

Hymenogloea Pat. (1900), Tricholomataceae. 1, trop. Am.

Hymenogramme Mont. & Berk. (1844), Grammotheleaceae. 2, Java, Philipp.

Hymenomarasmius Overeem (1927) ? = Marasmius (Tricholomat.).

Hymenomycetes, see *Basidiomycetes.*

Hymenophallus Nees (1816) = Phallus (Phall.).

hymenophore, a spore-bearing structure, esp. a basidioma, or that part of it bearing the hymenium. Cf. sporophore., incl.

Hymenopleella Munk (1953) = Lepteutypa (Amphisphaer.) fide Müller & v. Arx (1973). See also Shoemaker & Müller (*CJB* **43**: 1457, 1965).

hymenopodium (hymenopode), tissue under the hymenium; subhymenium or hypothecium.

Hymenopodium Corda (1837) = Clasterosporium (Mitosp. fungi) fide Hughes (*TBMS* **34**: 577, 1951).

Hymenopsis Sacc. (1886), Mitosporic fungi, 7.A2.15/19. 5, widespr. See Sutton (1980), Nag Raj (1993).

Hymenoscypha (Fr.) W. Phillips (1887) ≡ Hymenoscyphus (Leot.).

Hymenoscyphaceae, see *Leotiaceae*.

Hymenoscyphus Gray (1821), Leotiaceae. *c.* 100, widespr. See Baral (*SA* **13**: 113, 1994; gen. concept), Dennis (*Persoonia* **3**: 29, 1964), Dumont (*Mycotaxon* **12**: 313, 1981; neotrop. spp., **13**, 1981; temp. spp.), Thind & Sharma (*Nova Hedw.* **32**: 121, 1980; Indian spp.), Descals *et al.* (*TBMS* **83**: 541, 1984), Kimbrough & Atkinson (*Am. J. Bot.* **59**: 165, 1972), Lizon (*Mycotaxon* **45**: 1, 1992), Verkley (*Persoonia* **15**: 303, 1993; asci). Anamorphs *Geniculospora* or *Idriella*.

Hymenostilbe Petch (1931), Anamorphic Clavicipitaceae, 2.A1.?. Teleomorph *Cordyceps*. 9, widespr. See Samson & Evans (*Proc. K. ned. Akad. Wet.* C **78**: 73, 1975; key).

Hymenula Fr. (1825) ≡ Hymenella (Mitosp. fungi).

Hypasteridium Speg. (1923) nom. dub. (? Meliolaceae). See Eriksson & Hawksworth (*SA* **7**: 74, 1988).

hyper- (prefix), above.

Hyperomyxa Corda (1839) = Cheirospora (Mitosp. fungi) fide Hughes (1958).

hyperparasite, a parasite growing on another parasite (obsol.); Fungicolous fungi (q.v.) is a preferable general term where parasitism has not been established.

Hyperphyscia Müll. Arg. (1894), Physciaceae (L). 1 or 2, esp. trop. See Hafellner *et al.* (*Herzogia* **5**: 39, 1979), Kashiwadani (*Bull. natn. Sci. Mus. Tokyo*, B **11**: 91, 1985), Moberg (*Nord. Jl Bot.* **7**: 719, 1987; key 7 spp., E. Afr.).

hyperplasia, over-development of some sort (e.g. swellings, galls, witches' brooms) as a reaction to a disease-producing agent; cf. hypoplasia.

Hyperrhiza Spreng. (1827) ≡ Uperhiza (Melanogastr.).

hypersaprophyte, a saprophyte only found on substrates invaded by other saprophytes, e.g. *Herpotrichiellaceae, Lasiosphaeria, Nectria sanguinea*, etc. (Munk, *Sydowia, Beih.* **1**, 1957).

hypersensitivity, increased sensitivity, as in the condition in which there is death of the host tissue at the point of attack by a pathogen, so that the infection does not spread; esp. of reaction to rusts for which the word was first used by Stakman; intolerance; (and see Allergy, Medical fungi).

hypertonic (of culture media), having an osmotic pressure higher than that of the organism cultured; cf. hypotonic.

hypertrophy (of organs, etc.), the state of having growth greater than normal.

hypertrophyte, a parasitic fungus causing hyperplasia in a plant (Wakker).

Hyperus F. Stevens (1927) nom. dub. (Fungi, inc. sed.) fide v. Arx & Müller (1954).

hypha (pl. **hyphae**), one of the filaments of a mycelium; Vuillemin restricted the term to septate filaments; cf. siphon; **ampoule -**, a swollen hypha as in certain basidiomycetes; **arboriform -**, much branched skeletal - of *Ganoderma*; **ascogenous -**, a dikaryotic hypha from which an ascus develops; **binding -**, see Hyphal analysis; **costiferous -**, transverse ribs (costae) on the inner surface of the hyphal wall of the gill trama in *Paxillus involutus* and *P. filamentosus*, non-amyloid, stained by Congo red and calcofluor but not toluidine blue; **endo-** (intrahyphal -), vegetative or fertile element initiated by the differentiation within a - from the innermost wall layer (Cole & Samson, 1979); **inflated -**, one in which cells behind the growing apex enlarge and cause the apparent rapid rate of growth characteristic of most agaric and gasteromycete basidiocarps; in an **uninflated -** no change of cell size occurs as in most polypores; **mediate -** and **mycelial -**, see Corner (*TBMS* **17**: 54, 1932); **oleiferous -**, do not carry latex (cf. lactifer) but frequently resinous substances (Singer, 1960: 34); **oiliferous -**, a submerged hypha of an endolithic lichen having torulose, guttulate cells; see des Abbayes (*Traité de Lichenologie*, 1951); **racquet -**, one of racket cells (q.v.); **skeletal -**, see Hyphal analysis; **stuffing -**, see Corner (*TBMS* **17**: 54, 1932). See flexuous hypha, hyphal analysis, hyphal peg, textura, Woronin's hypha.

Hypha Pers. (1822) nom. dub. fide Donk (*Taxon* **11**: 86, 1962).

Hyphal analysis. A procedure by which the development and structure of basidiomata can be investigated, providing important taxonomic criteria. Three main types of hyphal systems of increasing complexity were recognized by Corner (*TBMS* **17**: 51, 1932) (Figs 16, 17): **monomitic**, having hyphae of one kind (generative hyphae which are branched, septate, with or without clamp-connexions, thin- to thick-walled, and of unlimited length; they give rise both to other hyphal types and to the hymenium) (Teixeira, *Mycol.* **52**: 30, 1961; gen. hyphae of polypores); **dimitic**, having hyphae of two kinds (generative and skeletal hyphae which are thick-walled, aseptate, and of limited length, with thin-walled apices, generally unbranched but when terminal they can develop arboriform branching or taper) or generative and binding (see below); **trimitic**, having hyphae of three kinds (generative, skeletal and binding (or ligative) hyphae which are aseptate, thick-walled, much branched, either *Bovista*-type with tapering branches or coralloid; they bind the skeletal and generative hyphae together). In *Polyporaceae* and *Lentinaceae*, intercalary skeletal hyphae can give rise to ligative branching, the entire element being termed a skeleto-binding cell (Corner, 1981) or skeleto-ligative hypha (Pegler, 1983). Corner also recognizes **sarco-dimitic** (in which the skeletal hyphae are replaced by thick-walled, long, inflating fusiform elements) and **sarcotrimitic** (in which the generative hyphae give both thick-walled inflated elements similar to binding hyphae but septate) types.

Most soft and fleshy basidiomata are monomitic, with hyphae which are generally inflated (most agaricoid and clavarioid fungi). Hard and tough basidiomata may be monomitic with the generative hyphae developing thickened walls, dimitic, with skeletal hyphae (e.g. *Phellinus* spp.) or (esp. when perennial) trimitic (e.g. *Fomes, Ganoderma, Microporus xanthopus*). Every species has a well-defined and constant construction, which is maintained regardless of changes in the external morphology of the basidioma due to environmental conditions, hence the importance of hyphal analysis in taxonomy.

Lit.: Corner (*Ann. Bot.* **46**: 71, 1932, *Phytomorphology* **3**: 152, 1953, *Beih. Nova Hedw.* **75**: 13, 1983, **78**: 13, 1984), Cunningham (*N. Z. Jl Sci. Tech.* **28**(A): 238, 1946, *TBMS* **37**: 44, 1954), Lentz (*Bot. Rev.* **20**: 135, 1954), Talbot (*Bothalia* **6**: 1, 1951).

hyphal fusions, see anastomosis. Vegetative hyphae of a mycelium may fuse, forming an interconnected network. Fusion between hyphae of different mycelia is controlled by genetic systems which determine sexual (see sex) or vegetative compatibility (q.v.).

hyphal net ('Hyphenfilz'), organ of attachment in some squamulose (placodioid) lichens (e.g. *Psora decipiens*) where a delicately branched reticulate net

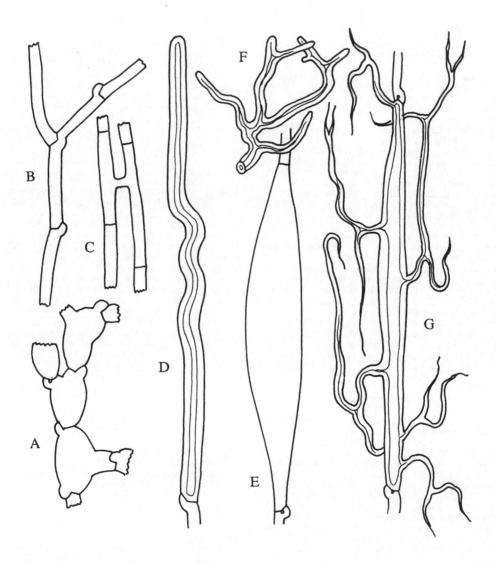

Fig. 16. Hyphal types. A, inflated generative hyphae; B, non-inflated generative hyphae with clamp connexions; C, generative hyphae without clamp connexions; D, unbranched skeletal hypha; E, sarco-hypha; F, highly branched ligative (binding) hypha; G, skeleto-ligative hypha. Not to scale.

penetrates the substrate (see Poelt & Baumgärtner, *Öst. bot. Z.* **111**: 1, 1964; rhizinose strand).

hyphal peg (of basidiomata), a bunch of somewhat interwoven hyphae extending from the trama (where it originates) to the hymenium from which it may project (Singer, 1960: 47); (of hyphae), projection from a hypha for fusion (Buller, **5**), peg-hypha.

hyphal rhizoid, a hypha acting as a rhizoid.

Hyphasma Rebent. ex Nocca & Balb. (1821) nom. dub. (Mitosp. fungi) fide Donk (*Taxon* **11**: 86, 1962).

Hyphaster Henn. (1903) = Asterostomella (Mitosp. fungi) fide Höhnel (*Sber. Akad. Wiss. Wien, mat.-nat. Abt.* I **119**: 21, 1910).

Hyphelia Fr. (1825) = Trichothecium (Mitosp. fungi) fide Hughes (1958). See also Donk (*Taxon* **11**: 86, 1962), Hennebert (*Persoonia* **7**: 195, 1973).

hyphidium (pl. **-ia**), (paraphysis, pseudoparaphysis, paraphysoid, dikaryoparaphysis, and pseudophysis sensu Singer (1962) are syn. or near syn.), a little, or strongly, modified terminal hypha in the hymenium of hymenomycetes (Fig. 18). Donk (*Persoonia* **3**: 229, 1964) distinguished; **haplo-** (simple -), unmodified, unbranched or little branched; **dendro-** (dendrophysis), irregularly strongly branched; **dicho-** (dichophysis), repeatedly dichotomously branched; **acantho-** (acanthophysis; bottle-brush paraphysis (Burt, 1918)), having pin-like outgrowths near the apex; in *Corticiaceae* may be botryose, clavate, coralloid, or cylindrical. Cf. cystidium.

Hyphochlaena Cif. (1962), Mitosporic fungi, 1.?.?. 1 (sterile mycelium), Dominican Republic.

Hyphochytriaceae A. Fisch. (1892), Hyphochytriales. 3 gen. (+ 1 syn.), 8 spp.

Hyphochytriales, Hyphochytriomycota. 2 fam., 7 gen. (+ 3 syn.), 24 spp. Fams:
(1) **Hyphochytriaceae**.
(2) **Rhizidiomycetaceae**.

Hyphochytriomycetes, see *Hyphochytriomycota*.

Hyphochytriomycetidae, see *Hyphochytriomycota*.

Hyphochytriomycota (Hyphochytriomycetes, Hyphochytridiomycetes). Chromista; the hyphochytrids. 1 ord., 2 fam., 7 gen. (+ 3 syn.), 24 spp. The whole (holocarpic) or part (eucarpic) of the thallus converting into a reproductive structure; holocarpic spp. often with branched rhizoids; zoospores with one

anterior flagellum with mastigonemes; lacing protoplasmic and nucleus-associated microtubules. On algae and fungi in freshwater and soil as parasites or saprobes, also saprobic on plant and insect debris. Ord.:

Hyphochytriales.
Lit.: Paterson (*in* Parker, 1982, **1**: 179), Sparrow (*in* Ainsworth *et al.* (Eds), *The Fungi* **4B**: 61, 1973).

Hyphochytrium Zopf (1884), Hyphochytriaceae. 6, N. temp. See Fuller (*in* Margulis *et al.* (Eds), 1990: 380), Dick (*in press*; key).

hyphocystidium, see cystidium.

Hyphoderma Wallr. (1833), Hyphodermataceae. *c.* 85, widespr. See Donk (*Fungus* **27**: 13, 1957; *Persoonia* **2**: 220, 1962); Eriksson & Ryvarden (*Corticiaceae of North Europe* **3**: 448, 1975; key 27 Eur. spp.), Wu (*Acta Bot. Fenn.* **142**: 64, 1990; key Taiwan spp.).

Hyphoderma Fr. (1849) ≡ Hyphelia (Mitosp. fungi) fide Donk (*Taxon* **11**: 88, 1962).

Hyphodermataceae Jülich (1982), Stereales. 33 gen (+ 18 syn.), 145 spp. Basidioma resupinate; hymenophore smooth to spinose; monomitic; cystidia present; spores inamyloid.

Hyphodermella J. Erikss. & Ryvarden (1976), Hyphodermataceae. 1, cosmop.

Hyphodermopsis Jülich (1982), Hyphodermataceae. 1, widespr.

Hyphodictyon Millardet (1866) = Atichia (Mitosp. fungi) fide v. Arx & Müller (1975).

Hyphodiscocioides Matsush. (1993), Mitosporic fungi, 1.A1.15/19. 1, Peru.

Hyphodiscosia Lodha & K.R.C. Reddy (1974), Mitosporic fungi, 1.B1.1/10. 3, India, Eur., Japan.

Hyphodiscus Kirschst. (1906), Hyaloscyphaceae. 3, Eur., N. Am. See Baral (*Z. Mykol.* **59**: 3, 1993), Raitviir & Galan (*SA* **13**: 159, 1995), Zhuang (*Mycotaxon* **31**: 411, 1988).

Hyphodontia J. Erikss. (1958), Hyphodermataceae. 54, widespr. See Weresub (*CJB* **39**: 1475, 1961), Eriksson & Ryvarden (**4**: 583, 1976; key 23 Eur. spp.), Wu (*Acta Bot. Fenn.* **142**: 85, 1990; key Taiwan spp.), Langer (*Bibl. Mycol.* **154**, 1994; world key).

Hyphodontiella Strid (1975), Hyphodermataceae. 1, Nordic.

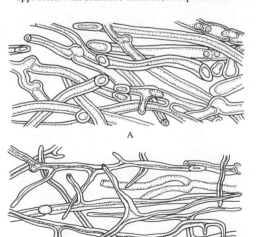

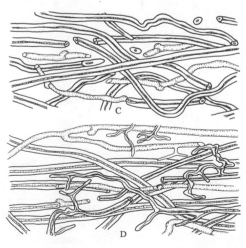

Fig. 17. Hyphal systems. See Pegler (*Bull. BMS* **7**(suppl.), 1973). A, monomitic hyphal system, with thick-walled generative hyphae; B, dimitic hyphal system, with generative and ligative (binding) hyphae; C, dimitic hyphal system, with generative and skeletal hyphae; D, trimitic hyphal system, with generative, skeletal and ligative hyphae.

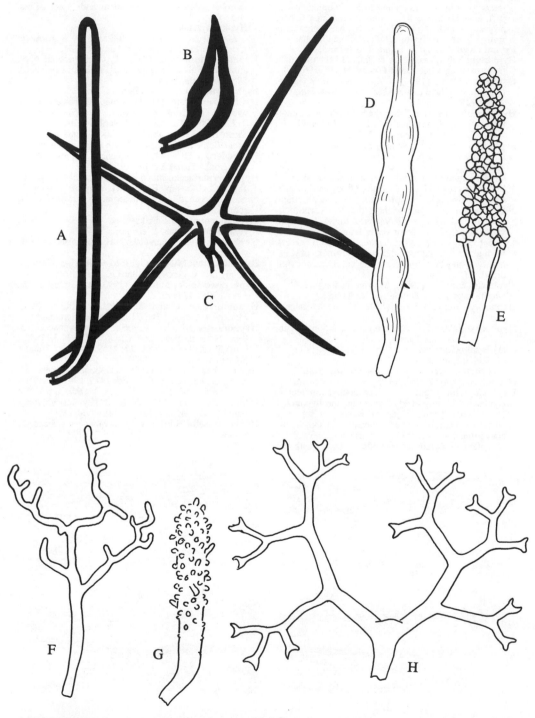

Fig. 18. Hyphidia. A, setal hypha (*Phellinus*); B, seta (*Inonotus*); C, asteroseta (*Asterostroma*); D, gloeo-hypha (*Gloeocystidiellum*); E, encrusted (*Peniophora*); F, dendrohyphidium (*Cytidia*); G, acanthohyphidium (*Aleurodiscus*); H, dendrohyphidim (*Vararia*). Not to scale.

hyphoid, (1) like hyphae; cobwebby; (2) (of aecia of *Dasyspora*), having aeciospores on hyphal projections from stomata (Arthur).

Hypholoma (Fr.) P. Kumm. (1871), Strophariaceae. 30 (esp. on wood), cosmop. = Nematoloma (Strophar.) fide Singer & Smith (1946). See Kühner (*BSMF* 52: 3, 1936).

Hypholomopsis Earle (1909) = Psathyrella (Coprin.) fide Singer (1951).

Hyphomucor Schipper & Lunn (1986), Mucoraceae. 1, India, Malaysia, Sri Lanka. See Schipper (*Mycotaxon* 27: 83, 1986).

Hyphomyces C.H. Bridges & C.W. Emmons (1961) nom. inval. (Fungi, inc. sed.).

Hyphomycetes (syn. **Hyphales**). Mitosporic fungi. 1700 gen. (+ 775 syn.), 11000 spp.; widespr. in most ecological niches. This artificial class is traditionally separated into three (or four, if agonomycetes are included) orders, based on the presence or absence of conidia and the degree of aggregation of the conidiophores into more complex structures (conidiomata). An alternative system of classification (now obsol.) for the Mitosporic fungi as a whole (including the coelomycetes, q.v.) was advanced by Sutton (1980). The traditional separation is:
(1) **Agonomycetales** (Agonomycetaceae) (Mycelia sterilia, q.v.). No conidia generally produced, but in some genera propagules which are liberated by multicellular secession are formed.
(2) **Hyphomycetales** (= Moniliaceae + Dematiaceae). This group comprises the main body of the hyphomycetes; conidiophores are separate, not organized on synnematal or sporodochial conidiomata. The full range of conidiogenous events occurs.
(3) **Stilbellales** (Coremiales, Synnematomycetes, Stilbellaceae). Conidiophores are aggregated as synnemata, and conidiogenous events are various but the 'basauxic' type is absent. See Morris (*Western Ill. Univ. Ser. biol. Sci.* 3, 1963; key 54 synnem. gen.), Benjamin (*Taxon* 17: 521, 1968; typification), Seifert (*Stud. Mycol.* 27, 1985; *Stilbella* and related genera).
(4) **Tuberculariales** (Tuberculariaceae). Conidiophores are aggregated on sporodochial conidiomata and a wide range of conidiogenous events occurs.
Lit.: Costantin (*Les mucedinées simples*, 1888; *Rabenh. Krypt.-Fl.* 1 (8-9), 1907-10), Wakefield & Bisby (*TBMS* 25: 49, 427, 1941; Brit. list), Hughes (*CJB* 36: 727, 1958; annotated list 400 gen. names), Delitsch (*Systematik der Schimmelpilze*, 1961), Kendrick & Carmichael (*in* Ainsowrth *et al.*, *The Fungi* 4: 323, 1973; synoptic keys, illustr., gen., gen. names), Litvinov ([Keys for the Identification of microscopic soil fungi. *Moniliales*], 1967), Barron (*The genera of hyphomycetes from soil*, 1968), Onions *et al.* (*Smith's Introduction to Industrial Mycology* edn 7, 1981), Ellis (*Dematiaceous Hyphomycetes*, 1971; *More Dematiaceous Hyphomyetes*, 1976), Subramanian (*Hyphomycetes. An account of Indian species, except Cercosporae*, 1971), Carmichael *et al.* (*Genera of Hyphomycetes*, 1980), Cole & Kendrick (Eds) (*Biology of conidial fungi*, 2 vols, 1981), Subramanian (*Hyphomycetes: taxonomy and biology*, 1983), Borowska (*Flora Polska, Grzyby (Mycota)*) 16 *Deuterom., Dematiaceae, Phialoconidiae*, 1986; keys to spp. in major genera), Domsch *et al.* (*Compendium of soil fungi*, 2 vols, 1980). See also under Mitosporic fungi.

Hyphonectria (Sacc.) Petch (1937) ≡ Nectriopsis (Hypocr.).

Hyphophagus Minden (1911) ≡ Hyphochytrium (Hyphochytr.).

hyphophore, erect stalked peltate asexual sporophores in the *Asterothyriaceae* (e.g. *Echinoplaca, Gyalideopsis, Tricharia*). See Sérusiaux & De Sloover (*Veröff. Geobot. Inst. ETH, Rübel* 91: 260, 1986; types), Vězda (*Čas. slez. Muz. Silesiae* A 22: 67, 1973; *Folia geobot. phytotax., Praha* 14: 43, 1979).

Hyphopichia Arx & Van der Walt (1976) = Pichia (Saccharomycet.) fide Kurtzman (*in* Hawksworth (Ed.), *Ascomycete systematics*: 361, 1994).

hyphopodium, a short branch of one or two cells on epiphytic mycelium of *Meliolales*, etc.; in a **capitate** - the end is rounded, = appressorium (fide Mibey & Hawksworth, *SA* 14: 25, 1995); **mucronate** - = conidiogenous cell (fide Hughes, *CJB* 59: 1514, 1981). A **stigmatopodium** (**stigmopodium**) is a hyphopodium in which the end cell or **stigmatocyst** has a haustorium (Arnaud). A stigmatocyst in a hypha is a **node cell**; cf. Doidge (*Bothalia* 4: 273, 1942). See also Walker (*Mycotaxon* 11: 1, 1980).

Hyphopolynema Nag Raj (1977), Mitosporic fungi, 3.A/B1.15. 1, Colombia.

Hyphoradulum Pouzar (1987), Hyphodermataceae. 1, Eur.

Hyphoscypha Bres. (1903) = Lachnum (Hyaloscyph.). See Huhtinen (*SA* 6: 131, 1987).

Hyphoscypha Velen. (1934), ? Leotiales (inc. sed.). 1, Eur.

Hyphosoma Syd. (1924) nom. conf. (Mitosp. fungi) fide Hughes (*N.Z. Jl Bot.* 8: 153, 1970).

Hyphostereum Pat. (1892), Mitosporic fungi, 3/7.A1.?. 1, S. Am.

Hyphothyrium B. Sutton & Pascoe (1989), Mitosporic fungi, 3.A/B2.19. 1, Australia.

Hyphozyma de Hoog & M.T. Sm. (1981), Mitosporic fungi, 1.A1.3/4. 5, Eur., N.Am. See de Hoog & Smith (*Ant. v. Leeuwenhoek* 52: 39, 1986; key), Hutchison, Sigler & Hiratsuka (*MR* 97: 1409, 1993; key).

hypnocyst, an *Alternaria*-like group of cells (Chippindale, *TBMS* 14: 203, 1929; Griffiths, *Nova Hedw.* 25: 511, 1974).

hypnospora, a resting spore.

Hypnotheca Tommerup (1970), Ascomycota (inc. sed.). 1, Australia. ? immature discomycete fide v. Arx & Müller (1975). Anamorph *Monochaetiellopsis*.

hypo- (prefix), under. See basidium.

hypobasidium, see basidium.

Hypoblema Lloyd (1902) = Calvatia (Lycoperd.) fide Zeller & Smith (*Lloydia* 27: 167, 1964).

Hypobryon Döbbeler (1983), Dothideales (inc. sed.). 5 (on hepatics), Eur.

Hypocapnodium Speg. (1918) = Aithaloderma (Capnod.) fide v. Arx & Müller (1975).

Hypocelis Petr. (1929) = Vizella (Vizell.) fide v. Arx & Müller (1975).

Hypocenia Berk. & M.A. Curtis (1874) = Topospora (Mitosp. fungi) fide Groves (*CJB* 43: 1195, 1965).

Hypocenomyce M. Choisy (1951), Lecideaceae (L). 10, N. & S. temp. See Timdal (*Nord. Jl Bot.* 4: 83, 1984; key).

Hypochanum Kalchbr. (1876) ≡ Macowanites (Elasmomycet.).

Hypochnella J. Schröt. (1888), Atheliaceae. 1, Eur. See Wakefield (*TBMS* 5: 127, 1915).

Hypochniciellum Hjortstam & Ryvarden (1980), Amylocorticiaceae. 3, S. Am.

Hypochnicium J. Erikss. (1958), Hyphodermataceae. 20, widespr.

Hypochnites Mesch. (1898), Fossil fungi. 1 (Oligocene), Baltic.

hypochnoid, having effused, resupinate, dry, rather loosely intertwined hyphae, as in *Tomentella* (formerly *Hypochnus*).

Hypochnopsis P. Karst. (1889), Atheliaceae. 1, Eur. See Wakefield (*TBMS* **35**: 43, 1952).

Hypochnus Fr. (1818) ? = Tomentella (Cortic.). See Donk (*Taxon* **6**: 75, 1957).

Hypochnus Fr. ex Ehrenb. (1820) = Cryptothecia (Arthon.) fide Thor (1991).

Hypocline Syd. (1939), Mitosporic fungi, 8.A1.10. 1, Afr.

Hypocopra (Fr.) J. Kickx f. (1867), Xylariaceae. 30 (esp. coprophilous), widespr. See Krug & Cain (*CJB* **52**: 809, 1974; key).

Hypocrea Fr. (1825), Hypocreaceae. 100, cosmop. See Boedijn (1964), Patil & Patil (*Indian Phytopath.* **36**: 635, 1983; India), Doi & Yamatoya (*Mem. N.Y. bot. Gdn* **49**: 233, 1989; *H. pallida* group. Anamorphs *Cylindrocladium, Trichoderma*.

Hypocreaceae De Not. (1844), Hypocreales. 81 gen. (+ 62 syn.), 614 spp. Ascomata usually brightly coloured, rarely non-ostiolate, often in or on a stroma or subiculum; asci clavate to cylindrical, apical apparatus often with an I- apical ring; ascospores very variable, O- multiseptate or muriform, colourless or pale brown, sometimes ornamented. Anamorphs prominent, hyphomycetous, e.g. *Acremonium, Cylindrocladium, Cylindrocarpon, Fusarium, Illosporium, Trichoderma, Tubercularia*. Saprobic or necrotrophic on plants.

Generic concepts formerly stressed spore septation; this is not now accepted as a valid criterion *per se* (Samuels & Rossman, 1979). See Helfer (*Pilze auf Pilzfruchtkörpern*, 1991; fungicolous spp., chemotaxonomy).

hypocreacous, fleshy and brightly coloured, like *Hypocrea*.

Hypocreales, Ascomycota. 3 fam., 115 gen. (+ 97 syn.), 862 spp. Ascomata perithecial, rarely cleistothecial, sometimes either in or on a stroma, ± globose, sometimes ornamented, rarely setose, the ostiole periphysate; peridium and stromatal tissues fleshy, usually brightly coloured; interascal tissue of apical paraphyses, often evanescent; asci ± cylindrical, thin-walled, sometimes with a small apical ring or a conspicuous apical cap, not blueing in iodine; ascospores varied, hyaline or pale brown, usually septate, sometimes muriform, sometimes elongate and fragmenting, without a sheath. Anamorphs prominent, hyphomycetous. Saprobes or parasites of plants, often fungicolous or lichenicolous, rarely coprophilous, cosmop. Fams:

(1) **Clavicipitaceae**.
(2) **Hypocreaceae**.
(3) **Niessliaceae**.

Malloch (*in* Kendrick (Ed.), *The whole fungus* **1**: 153, 1979) included the *Trichocomaceae* (*Eurotiales*) in this Order.

Lit.: Boedijn (*Persoonia* **3**: 1, 1964), Dingley (*Trans. Proc. R. Soc. N.Z.* **79**: 55, 177, 323, 403, **81**: 329, 489, **83**: 643, **84**: 467, 1951-57; NZ), Doi (*Bull. natn Sci. Mus. Tokyo* B **1**: 1, **2**: 119, 1975-76; S. Am.), Müller & v. Arx (1962, 1973), Petch (*TBMS* **21**: 243, 1938; UK), Rogerson (*Mycol.* **62**: 865, 1970; gen. names, key gen.), Rossmann (*Mycol. Pap.* **150**, 1983; key 52 phragmosporous spp.), Rossman *et al.* (*Mycol.* **85**: 685, 1993; key 8 gen. simple spores), Samuels & Rossman (*Whole fungus* **1**: 167, 1979; anamorphs, gen. concepts), Samuels & Seifert (*in* Sugiyama (Ed.), *Pleomorphic fungi*: 29, 1987; anamorphs), Samuels & Seifert (*Sydowia* **43**: 249, 1991; key 20 synnematous spp.), Samuels *et al.* (*Mem. N.Y. bot. Gdn* **59**: 6, 1990; 75 spp., Indonesia), Seaver (*Mycol.* **1**: 41, 177, **2**: 48, **3**: 207, 1909-1911; *N. Am. Fl.* **3** (1), 1910; N. Am.), under *Nectria* and anamorph gen. names.

Hypocrella Sacc. (1878), Clavicipitaceae. 30, widespr. Anamorph *Aschersonia*.

Hypocreodendron Henn. (1897), Anamorphic Xylariaceae, 2.A1.19. Teleomorph *Discoxylaria*. 1, S. Am.

Hypocreophis Speg. (1919) = Hypocrella (Clavicipit.) fide Clements & Shear (1931).

Hypocreopsis P. Karst. (1873), Hypocreaceae. 5, N. temp. See Candy & Webster (*Mycologist* **2**: 18, 1988), Niemelä & Nordin (*Karstenia* **25**: 75, 1985; Eur.).

Hypocreopsis G. Winter (1875) ≡ Selinia (Hypocr.).

hypocystidium, see cystidium.

Hypodendrum Paulet ex Earle (1909) = Pholiota (Strophar.) fide Singer (1951).

Hypoderma DC. (1805) nom. rej. = Lophodermium (Rhytismat.).

Hypoderma De Not. (1847) nom. cons., Rhytismataceae. 28, mostly temp. See Cannon & Minter (*Taxon* **30**: 572, 1983; gen. nomencl.), Johnston (*N.Z. Jl Bot.* **28**: 159, 1990; 12 spp. NZ, gen. concept.).

hypodermataceous (of asci), ones which are essentially unitunicate, lack any apical thickening, and which discharge the spores through a narrow pore; see ascus.

Hypodermella Tubeuf (1895), Rhytismataceae. 3, N. temp., Indian subcontinent. One causes needle-cast in conifers.

Hypodermellina Höhn. (1917), Rhytismatales (inc. sed.). 1, Eur.

Hypodermina Höhn. (1916), Mitosporic fungi, 8.A1.?. 1, Eur.

Hypodermium Link (1816) ≡ Caeoma (Anamorphic Uredinales). See Laundon (1965).

Hypodermium Link (1825) nom. dub. (Mitosp. fungi). See Sutton (1977).

Hypodermopsis Kuntze (1898) ≡ Hypoderma (Rhytismat.).

Hypodermopsis Earle (1902) = Hysterium (Hyster.) fide Nannfeldt (1932).

Hypodiscus Lloyd (1923) [non Nees (1836), Restionaceae] ≡ Neohypodiscus (Amphisphaer.).

Hypodrys Pers. (1825) ≡ Fistulina (Fistulin.).

Hypogaea E. Horak (1964), Secotiaceae. 1, Argentina.

Hypogaeum Pers. (1797) = Elaphomyces (Elaphomycet.) fide Fries (*Syst. mycol.* **3**: 57, 1829).

hypogean (**hypogeal**, **hypogeic**, **hypogeous**), in the earth; (of fungi) see Hypogeous fungi.

hypogenous, produced lower down.

Hypogeous fungi. Fungi having subterranean sporocarps include the truffles (q.v.; esp. *Elaphomycetales, Pezizales*), a few *Zygomycetes* (e.g. *Endogone*) and various gasteromycetes (e.g. *Hymenogaster, Rhizopogon*). Most are mycorrhizal. See Tulasne (*Fungi hypogaei*, 1851 [reprint 1970]), Hawker (*Biol. Rev.* **30**: 127, 1955). **British Isles**, Hawker (*Phil. Trans.* B**237**: 429, 1954; *TBMS* **63**: 67, 1974, list), Pegler *et al.* (*British truffles. A revision of British hypogeous fungi*, 1993); **Germany**, Hess (*Die Hypogaeen Deutschsland*, 1891-4 [reprint 1971]); **Hungary**, (Carpathian Basin), Szemere (*Die unteririschen Pilze des Karpatenbeckens*, 1965); **Malaysia**, Corner & Hawker (*TBMS* **36**: 125, 1953); **Mexico**, Trappe & Guzmán (*Mycol.* **63**: 317, 1971); **Spain**, Calonge *et al.* (*An. Inst. Bot. Cavanilles* **34**: 15, 1977). See Truffle.

Hypogloeum Petr. (1923), Mitosporic fungi, 6.A1.?. 1, Eur.

Hypogymnia (Nyl.) Nyl. (1896), Parmeliaceae (L). *c.* 40, cosmop. See Beltman (*Bibl. Lich.* **11**, 1978), Bitter (*Hedwigia* **40**: 171, 1901), Elix (*Brunonia* **2**: 175, 1980; Australasia), Elix & Jenkins (*Mycotaxon* **35**: 489, 1989; key 18 Australian spp.), Krog (*Skr. norsk. Polarinst.* **144**, 1968, *Lichenologist* **6**: 135, 1974), Luo (*Bull. Bot. Res.* **6**: 155, 1986; China), Wei (*Acta*

Mycol. Sin., Suppl. **1**: 379, 1987; isidiate spp. China), Awasthi (*Kavaka* **12**: 87, 1986; key 11 spp. India).

hypogyny (adj. **hypogynous**), the condition of having the antheridium under the oogonium and on the same hypha (Fig. 28A).

Hypohelion P.R. Johnst. (1990), Rhytismataceae. 2, Eur., N. Am.

Hypolepia Raf. (1808) ? nom. nud. fide Merrill (*Index Rafinesq.*, 1949).

hypolithic, see endolithic.

Hypolyssus Pers. (1825) nom. conf. Sensu Berk. = Caripia (Podoscyph.).

Hypomnema Britzelm. (1883) nom. dub. (Agaricales, inc. sed.).

Hypomyces (Fr.) Tul. (1860), Hypocreaceae. 30 (mostly on larger fungi), widespr. See Arnold (*Nova Hedw.* **21**: 529, 1972; classif., 1976; bibliogr.), Samuels (*Mem. N.Y. bot. Gdn* **26** (3), 1976), Tubaki (*Rept. Tottori mycol. Inst.* **12**: 161, 1975; Japanese spp.), Rogerson & Samuels (*Mycol.* **77**: 763, 1985; spp. on discomycetes, **81**: 413, 1989; key 10 spp. on *Boletales*, **86**: 839, 1994; key 13 spp. on agarics). Anamorphs *Cladobotryum*, *Papulaspora*, *Sibirina*, *Verticillium*.

Hypomycetales, see *Hypocreales*.

Hypomycopsis Henn. (1904) = Mycosphaerella (Mycosphaerell.) fide v. Arx & Müller (1975).

hyponecral, see necral layer.

Hyponectria Sacc. (1878), Hyponectriaceae. 9, widespr. See Barr (*Mycol.* **69**: 952, 1977; key).

Hyponectriaceae Petr. (1923), Ascomycota (inc. sed.). 17 gen. (+ 19 syn.), 135 spp. Stromatic tissues reduced, usually either clypeate or absent; ascomata immersed or erumpent, perithecial, usually thin-walled, the ostiole papillate, periphysate; interascal tissue of narrow or wide thin-walled paraphyses; asci cylindrical, persistent, thin-walled, with a small I+ or I- apical ring; ascospores variously shaped, hyaline to pale brown, simple or transversely septate, sometimes thick-walled, sometimes with a mucous sheath. Anamorphs hyphomycetous. Saprobic or necrotrophic in herbaceous and woody plant material, cosmop.
Placed in the *Xylariales* by Barr (1994), but perhaps has affinities with the *Phyllachorales*.

Hyponevris Paulet (1808) = Cantharellus (Cantharell.).

Hyponevris Earle (1909) ≡ Schizophyllum (Schizophyll.).

hyponym, a name only; one having no description or reference to a specimen.

hypoparasite, a hidden parasite; a pathogen dispersed along with another pathogen, such as a mycovirus in an *Ophiostoma ulmi* population.

Hypophloeda K.D. Hyde & E.B.G. Jones (1989), Melanconidaceae. 1, Brunei, Seychelles.

hypophloeodal, under the periderm or bark; endophloeodal; subcutical; within the bark.

hypophyllous, on the under surface of a leaf.

Hypophyllum Paulet (1808) ≡ Agaricus (Agaric.).

Hypophyllum Earle (1909) = Lactarius (Russul.) fide Singer (1951).

hypoplasia, the state of having growth less than normal, cf. hyperplasia.

Hypoplasta Preuss (1855) = Cytospora (Mitosp. fungi) fide Saccardo (*Syll. Fung.* **3**, 1884).

Hypoplegma Theiss. & Syd. (1917) = Perisporiopsis (Parodiopsid.) fide Müller & v. Arx (1962).

Hypopteris Berk. (1854) = Apiospora (Lasiosphaer.) fide Theissen & Sydow (*Ann. Myc.* **13**: 419, 1915).

Hyporhamma Corda (1854) nom. conf. = Hemitrichia (Trich.).

Hyporrhodius Staude (1857) = Rhodophyllus (Entolomat.) fide Singer (1951).

Hypospila Fr. (1825) Fungi (inc. sed.) fide v. Arx (*Ant. v. Leeuwenhoek* **17**: 257, 1951).

Hypospilina (Sacc.) Traverso (1913), Valsaceae. 2, Eur. See Barr (*Mycol. Mem.* **7**, 1978), Monod (1983).

Hypostegium Theiss. (1916) = Glomerella (Phyllachor.) fide v. Arx & Müller (1954).

Hypostigme Syd. (1925) = Parastigmatea (Polystomell.) fide v. Arx & Müller (1954).

Hypostomum Vuill. (1896) nom. dub (Fungi, inc. sed.).

hypostroma, see stroma.

Hypotarzetta Donadini (1985), Otideaceae. 1, Eur.

hypothallus, (1) (of lichens), the first hyphae of the thallus to grow, usually used of a crustaceous lichen which has no photobiont cells or cortex; = prothallus (protothallus), fide Maas Geesteranus (*Blumea* **6**: 47, 1947) who restricts hypothallus to the spongy tissue on the underside of the thallus in *Anzia*, *Pannaria* and *Pannoparmelia*, but see spongiostratum; (2) (of myxomycetes), the thin layer on the surface of the substratum not used up in sporangial development; Ross (*Mycol.* **65**: 477, 1973) distinguished epi- and subhypothallic development.

hypothecium, medullary excipulum; the hyphal layer under the subhymenium in an apothecium; sometimes used indiscriminately for all tissues below the hymenium (including the subhymenium).

Hypothele Paulet (1812) nom. inval. ≡ Hydnum (Hydn.).

hypotonic (of culture media), having an osmotic pressure lower than that of the organism cultured, cf. hypertonic.

Hypotrachyna (Vain.) Hale (1974) Parmeliaceae. *c.* 150, mainly sub- and montane trop. See Hale (*Smithson. Contr. bot.* **25**, 1975), Krog & Swinscow (*Norw. Jl Bot.* **26**: 11, 1979; E. Afr.). ? = Parmelia (Parmel.).

Hypoxylaceae, see *Xylariaceae*.

Hypoxylina Starbäck (1905) = Hypoxylon (Xylar.) fide Læssøe (*SA* **8**: 25, 1989).

Hypoxylites Kirschst. (1925), Fossil fungi. 2 (Neolithic), Germany.

Hypoxylon Adans. (1783) nom. rej. = Xylaria (Xylar.).

Hypoxylon Bull. (1791) nom. cons., Xylariaceae. 120, cosmop. *H. mammatum* (syn. *H. pruinatum*) (canker of *Populus*; see French *et al.*, *CJB* **47**: 223, 1969). See Miller (*A monograph of the world species of Hypoxylon*, 1961; keys), Abe (*Trans. mycol. Soc. Japan* **25**: 399, 1985; tissue types), Cherepanov (*Nov. Sist. niz. Rast.* **25**: 109, 1988; key 34 spp. former USSR), Dennis (*Bull. Jard. bot. Brux.* **33**: 317, 1963; Congo; Granmo *et al.* (*Opera Bot.* **100**: 59, 1989; Nordic spp.), Ju & Tzean (*Trans. Mycol. Soc. Rep. China* **1**: 13, 1985; key 13 spp. Taiwan), Kramer & Pady (*Mycol.* **62**: 1170, 1970; spore discharge), Jong & Rogers (*Tech. Bull. Wash. agric. Exp. Stn* **71**, 1972; anamorphs), Petrini & Müller (*Mycol. Helv.* **1**: 501, 1986; key 25 spp. Eur.), Læssøe (*SA* **13**: 43, 1994; gen. concept, nomencl.), van der Gucht & van der Veken (*Mycotaxon* **44**: 275, 1992; key 18 Papua New Guinea spp.), Whalley & Greenhalgh (*TBMS* **61**: 435, 1973, Br. spp., key, **64**: 369, 1975, numerical taxonomy). See also *Biscogniauxia*, *Nemania*.

Hypoxylonites Elsik (1990), Fossil fungi. *c.* 20 (Eocene-Oligocene), USA.

Hypoxylonopsis Henn. (1904) = Valsaria (Diaporthales) fide Saccardo (*Syll. Fung.* **24**: 538, 1926)

Hypoxylonsporites P. Kumar (1990), Fossil fungi. 2, India.

Hypoxylum Juss. (1789) nom. dub. = Cordyceps (Clavicipit.) p.p.

Hypsilophora Berk. (1879) nom. dub (? Fungi, inc. sed.).

Hypsizygus Singer (1947), Tricholomataceae. 3, N. temp.

Hypsolophora, see *Hysilophora*.

Hypsostroma Huhndorf (1992), Hypsostromataceae. 2, C. & S. Am. See Huhndorf (*Mycol.* **86**: 266, 1994; posn). Anamorph *Aposphaeria*- or *Pleurophomopsis*-like.

Hypsostromataceae Huhndorf (1994), Dothideales. 2 gen., 3 spp. Ascomata large, elongate, superficial; wall soft-textured, pseudoparenchymatic; interascal tissue of paraphysoids; asci stipitate and basally arranged, with an apical chamber and fluorescing ring; ascospores fusiform, septate.

Hypsotheca Ellis & Everh. (1885) = Caliciopsis (Corynel.) fide Fitzpatrick (*Mycol.* **34**: 464, 1942).

Hysterangiaceae E. Fisch. (1899), Phallales. 8 gen. (+ 4 syn.), 48 spp. Gasterocarp hypogeous, gelatinous layer poorly developed, often cartilaginous; gleba olivaceous, deliquescent.
   Lit.: Castellano & Beever (*N.Z. Jl Bot.* **32**: 305, 1994; NZ spp.).

Hysterangium Vittad. (1831), Hysterangiaceae. 40, esp. temp. See Soehner (*Sydowia* **6**: 246, 1952; Bayern), Schwarzel (*Schw. Z. Pilzkde* **10**: 154, 1979; key 18 spp.), Zeller & Dodge (*Ann. Mo. bot. Gdn* **16**: 83, 1929; key N. Am. spp.), Beaton *et al.* (*Kew Bull.* **40**: 435, 1985; key Austr. spp.).

Hysteriaceae Chevall. (1826), Dothideales. 14 gen. (+ 8 syn.), 44 spp. Ascomata erumpent or superficial, often aggregated, elongated, sometimes branched, wider than tall, opening by a longitudinal split; peridium black, very thick-walled, composed of small pseudoparenchymatous cells; interascal tissue of narrow cellular pseudoparaphyses; asci cylindrical, fissitunicate; ascospores hyaline to brown, variously septate, sometimes with a mucous sheath. Anamorphs varied. Mostly saprobic, cosmop.
   Lit.: Amano (*Trans. mycol. Soc. Japan* **24**: 283, 1983; Japan), Bellemère (*Ann. Sci. nat., Bot.* **12**: 429, 1971), van der Linde (*S. Afr. Jl Bot.* **58**: 491, 1992; key 5 gen., 12 spp., S. Afr.); Zogg (*Beitr. Krypt.-fl. schweiz* **11** (3), 1962; Eur.).

hysteriaceous (**hysterioid**, **hysteriiform**), long and cleft, like the **hysterothecium** (ascoma) of the *Hysteriaceae*; lirellate.

Hysteriales, see *Dothideales*; Formerly included a wide range of *Ascomycota* with lirelliform ascomata, but was later confined to bitunicate types, including lichen-forming taxa. Bitunicate types are here distributed between *Arthoniales*, *Dothideales*, and *Patellariales*; unitunicate types between *Ostropales* and *Rhytismatales*.
   Lit.: Bisby (*TBMS* **8**: 176, 1923, lit.; *Mycol.* **24**: 304, 1932, type specimens), Bisby & Hughes (*TBMS* **35**: 308, 1952; key Br. spp.), Munk (*Dansk bot. Arkiv.* **17**, 1957; Denmark), Zogg (*Beitr. Krypt.-Fl. Schweiz* **11** (3), 1962; Switzerland, detailed monogr.). See also *Ascomycota* and under Ords.

Hysteridium P. Karst. (1905), Mitosporic fungi, 8.A1.?. 1, Eur. See Sutton (1977).

Hysterina (Ach.) Gray (1821) = Opegrapha (Roccell.).

Hysteriopsis Geyl. (1887), Fossil fungi. 1 (Tertiary), Indonesia. = Hysterites (Fossil fungi) fide Meschinelli (1892).

Hysteriopsis Speg. (1906) ? = Hysterographium (Hyster.) fide Zogg (1962).

Hysterites Göpp. (1836), Fossil fungi. 15 (Tertiary), Eur.

Hysterites Unger (1841), Fossil fungi. 1 (Tertiary), former Yugoslavia.

Hysterium Tode (1791) = Colpoma (Rhytismat.).

Hysterium Pers. (1797), Hysteriaceae. 8, mainly temp. Anamorphs *Coniosporium*, *Hysteropycnis*.

Hysterocarina H. Zogg (1949), Hysteriaceae. 1, Brazil.

hysterochroic, see colour.

Hysterodiscula Petr. (1942), Mitosporic fungi, 8.A1.1. 2, N. temp.

Hysterodothis Höhn. (1909) = Sphaerodothis (Phyllachor.) fide v. Arx & Müller (1954).

Hysterogaster Zeller & C.W. Dodge (1928) = Hymenogaster (Hymenogastr.) fide Cunningham (*Gast. Austr. N.Z.*: 47, 1944).

Hysteroglonium Rehm ex Lindau (1896), ? Hysteriaceae. 2 or 3, N. Am., Eur.

Hysterographium Corda (1842), Hysteriaceae. 4, widespr. See Zogg (*Phytopath. Z.* **14**: 310, 1944). Anamorph *Hysteropycnis*.

Hysteromyces Vittad. (1844) = Rhizopogon (Rhizopogon.) fide Fischer (1938).

Hysteromyxa Sacc. & Ellis (1882) nom. dub. fide Petrak (*Sydowia* **5**: 196, 1951).

Hysteronaevia Nannf. (1984), Dermateaceae. 10, N. temp.

Hysteropatella Rehm (1890), ? Hysteriaceae. 3, Eur., N. Am.

Hysteropeltella Petr. (1923), Dothideales (inc. sed.). 1, Eur.

Hysteropeziza Rabenh. (1874) = Pyrenopeziza (Dermat.) fide Nannfeldt (1932).

Hysteropezizella Höhn. (1917), Dermateaceae. *c.* 18, widespr. See Défago (*Sydowia* **21**: 1, 1967), Hein (*Nova Hedw.* **34**: 449, 1981; paraphyses ornamentation).

hysterophyte, a saprophyte (obsol.).

Hysteropsis Rehm (1887), Dothideales (inc. sed.). 1, Eur.

Hysteropycnis Hilitzer (1929), Mitosporic fungi, 8.A1.?. 5, Eur., N. Am.

Hysterostegiella Höhn. (1917), Dermateaceae. 9, widespr. See Hein (*Nova Hedw.* **38**: 669, 1983; key).

Hysterostoma Theiss. (1913) = Dothidasteromella (Asterin.) fide Müller & v. Arx (1962).

Hysterostomella Speg. (1885), Parmulariaceae. 17, trop. See Hansford (*Mycol. Pap.* **15**, 1946).

Hysterostomina Theiss. & Syd. (1915) = Hysterostomella (Parmular.) fide Hansford (*Mycol. Pap.* **15**, 1946). See v. Arx & Müller (1975), Batista & Vital (*Atas do IMUR* **1**: 53, 1960).

hysterothecium, an elongated ascoma like that of the *Hysteriaceae* with a slit-like line of dehiscence.

Hystricapsa Preuss (1851) nom. dub. (? Myxomycetes) fide Saccardo (1906).

Hystrichosphaeridium Deflandre (1937), Fossil fungi (Ascomycota). 1, Eur.

Hystricula Cooke (1884) ? = Winterella (Vals.). See Saccardo (*Syll. Fung.* **1**: 471, 1882).

Hystrix Alstrup & Olech (1993) [non Moench (1794), Poaceae] = Acanthonitschkea (Nitschk.) fide Eriksson & Santesson (*SA* **14**: 54, 1995).

I, see iodine.

IAL, see Societies and organizations.

Ialomitzia Gruia (1964), ? Mitosporic fungi, 1.?1/2.?. 1, Romania. Alternatively a cyanobacterium.

IAP, see Index of Atmospheric Purity.

ibotenic acid, a metabolite of *Amanita muscaria*, etc., toxic to humans and flies (*Musca* spp.); amanita factor C.

Iceland moss. *Cetraria islandica*, habitually eaten by reindeer and caribou, and used as a substitute for flour during hard times in Scandinavia and Iceland, is rich in digestible carbohydrates; used in soups and pastilles. See Dahl (*Bot. Rev.* **20**: 463, 1954), Richardson

& Young (*in* Seaward (Ed.), *Lichen ecology*: 121, 1977), Medical uses of fungi.

Icerymyces Brain (1923), ? Mitosporic fungi, 8.A1.?. 1 (in insects). Nom. conf. fide Batra (1978).

Ichthyochytrium Plehn (1920), Chytridiales (inc. sed.). 1 (in *Cyprinus*), Germany.

Ichthyophonus Plehn & Mulsow (1911), Zygomycetes (inc. sed.). 2 or 3 (in fish). See Neish & Hughes (*Fungal diseases of fishes* [Diseases of fishes 6], 1980), Sprague (*Syst. zool.* **14**: 110, 1965), Rand & Whitney (*Mycol. Soc. America Newsletter*: 43 [abstract], 1987; ultrastr.), Grabda (*Marine fish parasitology. An outline*: 43, 1991).

Ichthyosporidium Caullery & Mesnil (1905) = Ichthyophonus (Zygomycetes, inc. sed.) fide Sprague (1965).

Icmadophila Trevis. (1853) nom. cons. prop., Icmadophilaceae (L). 3, N. Hemisph. See Frey (*Rabenh. Krypt.-Fl.* **9**, 4 (1): 819, 1933), Honegger (*Lichenologist* **15**: 57, 1983; asci), Rambold *et al.* (*Bibl. Lich.* **53**: 217, 1993).

Icmadophilaceae Triebel (1993), Leotiales (L). 5 gen. (+ 7 syn.), 7 spp. Thallus usually crustose or squamulose; ascomata sessile or shortly stipitate, sometimes clustered, formed on specialized often non-lichenized thalline branches, flat or convex; interascal tissue of simple or sparingly branched paraphyses, often swollen at the apices; asci with an I+ apical cap; ascospores hyaline, simple or transversely septate. Anamorph pycnidial. Lichenized with green algae.

Icmadophilomyces E.A. Thomas ex Cif. & Tomas. (1953) nom. illegit. ≡ Icmadophila (Icmadophil.)

icones (Latin), pictures; Figures; plates.

ICSU, see Societies and organizations.

ICTF, see Societies and organizations.

Idiocercus B. Sutton (1967), Mitosporic fungi, 4.A1.19. 3, Afr., India, Australia. See Nag Raj (1993).

Idiomyces Thaxt. (1893), Laboulbeniaceae. 1, Eur., Japan. See Benjamin (*Aliso* **10**: 345, 1983).

id-reaction, see dermatophytid.

Idriella P.E. Nelson & S. Wilh. (1956), Mitosporic fungi, 1.A2.10. 22, widespr. See v. Arx (*Sydowia* **34**: 30, 1981; key), Rodrigues & Samuels (*Mycotaxon* **43**: 271, 1992; endophytic spp. on palms), Castañeda Ruiz & Kendrick (*Univ. Waterloo Biol. Ser.* **35**: 1, 1991; 4 n.spp.).

IFO. Institute for Fermentation (Osaka, Japan); founded 1944 as Kōkū-Hakkō Kenyūsho and supported by the Japanese government and Takeda Chemical Industries Ltd, with a separate board of trustees; see Anon. (*Res. Comm. IFO 17*: 1, 1995; 50th anniv. issue).

Ijuhya Starbäck (1899) = Nectria (Hypocr.) fide Samuels (*N.Z. Jl Bot.* **14**: 231, 1976).

ikatake ('ika-take'), the basidiome of *Aseroë arachnoidea.*

IKI, see iodine.

Iledon Samuels & J.D. Rogers (1986), Anamorphic Dothideales, 8.C2.15. Teleomorph *Botryohypoxylon.* 1, Venezuela.

Ileodictyon Tul. ex M. Raoul (1844), Clathraceae. 2, mostly S. Hemisph.

illegitimate, see Nomenclature. Cf. legitimate.

Illosporium Mart. (1817), Mitosporic fungi, 4.A1.?. 1 (on lichens, *Peltigera*), N. temp. See Hawksworth (*Bull. Br. Mus. nat. Hist., Bot.* **6**: 181, 1979). See also *Marchandiomyces.*

Ilyomyces F. Picard (1917), Laboulbeniaceae. 2, Eur.

Ilytheomyces Thaxt. (1917), Laboulbeniaceae. 15, Am., Afr., Borneo.

IMA, see Societies and organizations.

Imazekia Tak. Kobay. & Kawabe (1992), Phyllachoraceae. 1, Japan.

Imbricaria (Schreb.) Michx. (1803) [non Juss. (1789), *Sapotaceae*] = Anaptychia (Physc.) p.p., Hypogymnia (Parmel.) p.p., Parmelia (Parmel.) p.p., etc.

imbricate (of pilei, scales, squamules, etc.), partly covering one another like the tiles on a roof.

IMC, see International Mycological Congresses.

IMI. International Mycological Institute (Imperial Bureau of Mycology, 1920-29; Imperial Mycological Institute, 1930-47; Commonwealth Mycological Institute, 1948-85; CAB International Mycological Institute, 1986-90); an institute of CABI (q.v.); relocated from Kew to Egham, Surrey, UK in 1992; see Aitchison & Hawksworth (*IMI: retrospect and prospect*, 1993).

immaculate, not spotted.

immarginate, having no well-defined edge.

Immersaria Rambold & Pietschm. (1989), Porpidiaceae (L). 1, widespr.

immersed, embedded in the substratum.

immobolisation, the controlled, intentional, attachment of fungal cells in fermentation technology (Webb, *Mycologist* **3**: 163, 1989). Cf. biomass support particles.

Immotthia M.E. Barr (1987), Dacampiaceae. 1, N. Am.

immune, the condition of having qualities which do not allow, or not having qualitites which allow, the development of a disease to take place; **natural immunity** is based on qualities natural to the organism, **acquired immunity**, on the development of such qualities in the course of its life-time, generally as a result of taking the disease naturally or experimentally (not certainly present in plants).

immune (from), exempt from infection (*TBMS* **33**: 155, 1950); having immunity.

immunization, the process of increasing the resistance of, or of giving resistance to, a living organism.

immunosuppressant, a substance such as the fungal metabolite cyclosporin (produced by *Tolypocladium inflatum*), which partly or completely suppresses the immune system; used to prevent rejection of transplanted organs.

imperfect state, see States of fungi; Mitosporic fungi.

imperforate, having no opening.

Imprimospora G. Norris (1986), Fossil fungi. 1 (Eocene), Canada.

impriorable, illegitimate (obsol.).

Imshaugia S.L.F. Mey. (1985), Parmeliaceae (L). 3, N. Hemisph.

in situ (Lat.), **in-situ** (Engl.) (of an organism), one living in its natural habitat; cf. *ex situ.*

inaequi-hymeniiferous (of hymenial development in agarics), having basidia which mature and shed their spores in zones; the coprinus type (Buller, *Researches* **2**: 19, 1922). cf. aequi-hymeniiferous.

Inapertisporites Hammen ex Rouse (1959), Fossil fungi (*f. cat.*). 28 (Jurassic, Cretaceous, Eocene, Tertiary), China, Colombia, N. Am.

incertae sedis (inc. sed.), of uncertain taxonomic position.

Incertisporites Hammen (1954), Fossil fungi (*f. cat.*). 1 (Cretaceous/Tertiary), Colombia.

Inciliaria Fr. (1825) nom. dub. (Fungi, inc. sed., L).

incised, as if cut into; esp. of a pileus margin or lobes of a foliose lichen thallus.

Incolaria Herzer (1893), ? Fossil fungi. 1 (Carboniferous), N. Am.

incompatible, (1) (of sex), unable to be cross-mated due to mating type or fertility barriers; (2) (of vegetative mycelia), unable to form a stable heterokaryon due to genetic differences at one or more vegetative compatibility (vc, het) loci. See vegetative compatibility.

incrassate, made thick.

Incrucipilum Baral (1985), Hyaloscyphaceae. *c.* 4, temp.

Incrupila Raitv. (1970), Hyaloscyphaceae. 5, Eur., N. Am.

Incrupilella Svrček (1986) = Hyphodiscus (Hyaloscyph.) fide Baral (*SA* **13**: 113, 1994).

incrusted (of hyphae), having matter excreted on the walls (Corner, 1950).

Incrustocalyptella Agerer (1983), Cyphellaceae. 2, Colombia, Papua New Guinea.

Incrustoporia Domański (1963), Coriolaceae. 1, Eur. = Poria (Coriol.) fide Lowe (1966).

incubation period, the time between inoculation and the development of visible symptoms.

indefinite, not sharply limited.

indehiscent (of sporocarps, sporangia, etc.), not opening, or with no special method of opening.

indeterminate, (1) having the edge not well-defined, esp. of fruit-bodies and leaf-spots; (2) (of conidiophores), continuing growth indefinitely.

Index of Atmospheric Purity (IAP). A numerical estimate of the purity of the air on the basis of the lichens present on trees (LeBlanc & DeSloover, *CJB* **48**: 1485, 1970). See Air pollution, Bioindicators.

Index of Ecological Continuity, see RIEC.

Indian bread, see tuckahoe.

Indian paint fungus, *Echinodontium tinctorium* (q.v.).

Indiella Brumpt (1906) = Madurella (Mitosp. fungi) fide Ciferri & Redaelli (1941).

indigenous, natural to a country or region; native.

indigenous property, see Bioprospecting, Patent protection.

indirect (of fruit-body development), see direct.

individualism in fungi, mechanisms may exist in nature to 'define' individuals involving co-operative (hyphal fusions, heterokaryosis) and individualistic methods. See Todd & Rayner (*Sci. Progr.* **66**: 331, 1980), incompatible.

indumentum, a covering, such as hairs, etc.

indurated, made hard.

Induratia Samuels, E. Müll. & Petrini (1987), Xylariaceae. 1, NZ. See Samuels & Rossman (*Mycol.* **84**: 26, 1992). Anamorph *Nodulisporium*.

indusium, (1) cover; (2) (of phalloids), a net-like structure hanging from the top of the stipe under the pileus.

Industrial mycology. Fungi are used in many industrial processes. Some of the more important substances produced, frequently from different forms of carbohydrates and starch, are:

By *yeasts*: alcohol (by *Saccharomyces cerevisiae* from sugar or, after hydrolysis, starch (e.g. cereals, potatoes) or cellulose (e.g. wood, waste sulphite liquor)); 'fat' (lipoids) (*Endomycopsis vernalis*; also *Geotrichum candidum*); glycerol (*S. cerevisiae* var. *ellipsoideus* by the sulphite process); riboflavin (various lactose-fermenting yeasts).

By *other fungi*: citric acid (*Aspergillus*, *Penicillium*, *Mucor*); enzyme mixtures such as takadiastase (*Aspergillus*); fats (*Penicillium*); fumaric (*Rhizopus*) and gluconic (*Aspergillus*) acids; itaconic acid (*A. terreus*, etc.) (used as a copolymer with acrylic resins); kojic acid (*A. flavus* group); lactic acid (*Rhizopus*); rennin (*Rhizomucor pusillus*, *Rhizopus oligosporus*, etc.; O'Leary & Fox, *J. Dairy Res.* **41**: 381, 1974); riboflavin (*Eremothecium*); steroid transformations are effected by hyphomycetes and other fungi (Peterson, *in* Rainbow & Rose (Eds), *Biochemistry of industrial micro-organisms*: 537, 1963). Ensilage is another fermentation process and the retting of flax, hemp, and other fibres is dependent on pectin-attacking bacteria and fungi; Thaysen & Bunker (1927). More recent developments include 'mycelial paper' (the addition of mycelium of *Phytophthora*

cinnamomi and other phycomycetes to wood pulp); see Johnson & Carlson (*Biotechnol. Bioeng.* **20**: 1063, 1978) and fuels from biomass (Jefferies *et al.*, *in* Smith *et al.*, 1980).

Lit.: Lafar (*Technische Mycologie*, 2 vols, 1896-1907 [English transl. 1989-1910], *Handbuch der technischen Mykologie*, 5 vols, 1904-14), Ramsbottom (*Rep. Br. Ass.* **1936**: 189; uses of fungi), Galloway (*Applied mycology and bacteriology*, edn 3, 1950), Prescott & Dunn (*Industrial microbiology*, 1940; edn 4, 1982), Gray (*The relation of fungi to human affairs*, 1959), Brian (*TBMS* **58**: 359, 1972; economic value of fungi), Smith & Berry (Eds) (*The filamentous fungi*, **1**, *Industrial mycology*, 1975), Beuchat (*Food and beverage mycology*, 1978), Smith *et al.* (Eds) (*Fungal biotechnology*, 1980), Thaysen & Bunker (*The microbiology of cellulose, hemicellulose, pectins and gums*, 1927; *Henrici's molds, yeasts and actinomycetes*, edn 2, 1947; *Jørgensen's Micro-organisms and fermentation* (rewritten by Hansen), 1948; *Smith's Introduction to industrial mycology* rewritten by Onions, Allsopp & Eggins), edn 7, 1981), Fulmer & Werkman (*An index to the chemical action of micro-organisms on the non-nitrogenous compounds*, 1930), Foster (*Chemical activities of fungi*, 1949), Bracken (*The chemistry of micro-organisms*, 1955),

See also antibiotic, Biodeterioration, Biodegradation, Biotechnology, Brewing, Cheese, Genetic engineering, Fermented food and drinks, Lichens (Economics), Edible fungi, Metabolic products, Mushroom cultivation, Nutrition, Pigments, Starters, Wine making.

Inermisia Rifai (1968), Otideaceae. 6, Eur., Australia. See Benkert (*Gleditschia* **15**: 173, 1987; status), Dennis & Itzerott (*Kew Bull.* **28**: 5, 1973), Caillet & Moyne (*Bull. Soc. Hist. nat. Doub* **84**: 9, 1991; key), Sivertsen (*SA* **9**: 23, 1991; nomencl.), *Byssonectria*.

inermous, having no spines or prickles.

infarctate, solid; turgid.

infect (of a pathogen), to enter and establish a pathogenic relationship with an organism; to enter and persist in a carrier (*TBMS* **33**: 155, 1950); to make an attack on an organism; (of an agent), to make infection of an organism take place; **-ed** (of an organism), attacked by a pathogen, cf. contaminated; **-ion**, the act of infecting; **-ious** (of diseases), resulting from infection; sometimes used in the sense of able to be handed on by touch (contagious) or by inoculum; **-ive** (of a pathogen), able to make an attack on a living organism; (of a vector, medium, etc.), having the power of effecting the transmission of a pathogen.

inferior (of an annulus), low down on the stipe.

infested, attacked by animals, esp. insects; sometimes used of fungi in soil or other substrata in the sense of 'contaminated'.

infissitunicate (Dughi, *C. r. hebd. Séanc. Acad. Sci., Paris* **243**: 750, 1956), see ascus.

inflated hypha, see hypha.

Inflatostereum D.A. Reid (1965), Podoscyphaceae. 2, S. Am., Asia.

inflexed (of pileus margin), turned down (Fig. 19F).

infra- (prefix); below; **-generic** (of ranks), all those below that of genus; **-specific** (of ranks), all those below that of species; used in a parallel way for other ranks.

Infrafungus Cif. (1951), Mitosporic fungi, 1.B/C1.?. 1 (on *Cladosporium*), Philipp.

infundibuliform, funnel-like in form.

Infundibulum Velen. (1934) = Peziza (Peziz.) fide Eckblad (1968). See Svrček (*Acta Mus. Nat. Prague* **32B** (2-4): 115, 1976).

Infundibura Nag Raj & W.B. Kendr. (1981), Mitosporic fungi, 3.A1.1. 1, NZ, UK.

Ingaderia Darb. (1897), Roccellaceae (L). 3, Eur., S. Am., N. Afr. See Feige & Lumbsch (*Mycotaxon* **48**: 381, 1993).

ingest, to obtain food by engulfing it; see phagotrophic. Cf. absorb.

Ingoldia R.H. Petersen (1962) = Gyoerffyella (Mitosp. fungi) fide Marvanová *et al.* (1967).

Ingoldiella D.E. Shaw (1972), Anamorphic Coriolaceae, Sistotremataceae, 1.G1.1. Teleomorph *Cerrena*. 1 (with clamp connexions), Australia. See Nawawi & Webster (*TBMS* **78**: 287, 1982).

inhabitant, see symbiosis.

inhibitory substances, see staling substances.

Inifatiella R.F. Castañeda (1985), Mitosporic fungi, 2.C2.1. 1, Cuba.

ink-caps, basidiomata of *Coprinus*.

innate, bedded in; immersed.

Innatospora J.F.H. Beyma (1929) = Arthrinium (Mitosp. fungi) fide Ellis (*Mycol. Pap.* **103**, 1965).

Inocephalus (Noordl.) P.D. Orton (1991), Entolomataceae. 1, Eur.

Inocibium Earle (1909) = Inocybe (Cortinar.) fide Kauffman (1924).

inoculate, to put a microorganism, or a substance containing one, into an organism or a substratum.

inoculation, the act of inoculating [of (an organism or substratum)] *with* (the inoculum); *of* (the inoculum) *into* (an organism or substratum); by (an agent or method).

inoculum, the substance, generally a pathogen, used for inoculating.

inoculum potential (of a fungus or other microorganism), the energy of growth available for colonization of a substratum at the surface of the substratum to be colonized (Garrett, 1956).

Inocutis Fiasson & Niemalä (1984), Hymenochaetaceae. 3, N. temp.

Inocybe (Fr.) Fr. (1863), Cortinariaceae. 150, widespr., esp. temp. See Heim (*Le genre Inocybe*, 1931), Alessio (*Bres. Icon. Mycol.* **29**: suppl. 3, 1980; col. illustr. & key Eur. spp.), Pegler & Young (*Kew Bull.* **26**: 499, 1972; basidiospores, key Br. spp.), Kauffman (*N. Am. Fl.* **10**: 227, 1924; key 105 N. Am. spp.), Roberts *et al.* (*Lloydia* **27**: 108, 1964; chemotaxonomic key), Pérez Silva (*An. Inst. Biol. Univ. Mexico*, ser. Bot. **38**: 1, 1967; Mexico), Horak (*N.Z. Jl Bot.* **15**: 713, 1977; key 24 NZ spp.), Stangl & Veselský (*Česká Myk.* **34**: 45, 1980; rough-spored spp.).

Inocybella Zerova (1974), Corticiaceae. 1, France.

Inocyclus Theiss. & Syd. (1915), Parmulariaceae. 6 (esp. on *Pteridophyta*), trop.

Inoderma (Ach.) Gray (1821) = Thrombium (Thromb.).

Inoderma P. Karst. (1879) ≡ Inodermus (Hymenochaet.).

Inoderma Berk. (1881) ≡ Mesophellia (Mesophell.).

Inodermus Quél. (1886) = Inonotus (Hymenochaet.) fide Donk (1960).

Inoloma (Fr.) Wünsche (1877) ≡ Cortinarius (Cortinar.).

Inonotopsis Parmasto (1973), Hymenochaetaceae. 1, USA.

Inonotus P. Karst. (1879), Hymenochaetaceae. 40, cosmop. See Pegler (*TBMS* **47**: 175, 1964; key), Gilbertson (*Mem. N.Y. bot. Gdn* **28**: 67, 1976).

inoperculate (of an ascus or sporangium), opening by an irregular apical split to discharge the spores, see ascus; cf. operculate.

Inopilus (Romagn.) Pegler (1983), Entolomataceae. 16, widespr., esp. trop.

inordinate, in no order.

inquinant, stained; blackened; dirty (obsol.).

Insecticola Mains (1950) = Akanthomyces (Mitosp. fungi) fide Samson & Evans (1974).

Insects (and invertebrates) and fungi. The relationships between insects and fungi are many and complex (Wilding *et al.* (Ed.), *Insect-fungus interactions*, 1989). Termites (q.v.) and ants (Uphof, *Bot. Rev.* **8**: 563, 1942; Lüscher, *Nature* **167**: 34, 1951; Weber, *Ecology* **38**: 480, 1957; *Lepiota* cultured by *Cyphomyrma*) and ambrosia beetles make 'cultures' of fungi (Bakshi, *TBMS* **33**: 111, 1950). 'Ambrosia galls' on plants caused by a mutualistic association between *Cecidomyiidae* flies and anamorphs of *Botryosphaeria* (see galls). Other insects use fungus spores as food (e.g. herbarium beetles, q.v.) as do slugs (Buller, *TBMS* **7**: 270, 1922). The transmission of fungus spores (esp. slime spores), in addition to pycniospores and other 'diploidizing agents', is frequently effected by insects. See Leach (*Insect transmission of plant diseases*, 1940), Carter (*Insects in relation to plant diseases*, 1962; edn 2, 1973); also stigmatomycosis.

Many invertebrates, particularly mites and molluscs, graze on lichens. Mites and ants may be important in the dispersal of soredia and ascospores ingested by rotifers may be viable after excretion. Some lichens have been reported from shells of land snails and some from marine limpets. Some foliose species have been found on *Coleoptera* (see Gressit, *Entom. News* **80**: 1, 1969). Many moths and other insects mimic lichens in larval or adult stages as well as using them for food. Lichen acids do little to deter grazing molluscs and may pass through the gut unchanged (Zopf, *Biol. Zbl.* **16**: 593, 1896). See Lawrey (*Biology of lichenized fungi*, 1984), Richardson (*The vanishing lichens*, 1975), Seaward & Gerson (*in* Seaward (Ed.), *Lichen ecology*: 69, 1977; detailed review), Smith (*Lichens*, 1921 [reprint 1975]); Lichens, soredia.

See Biological control, Coevolution, Entomogenous fungi, Symbiosis.

Insiticia Earle (1909) = Mycena (Tricholomat.) fide Singer (1951).

insititious, of inserted nature, introduced from without.

Insolibasidium Oberw. & Bandoni (1984), Platygloeaceae. 1, N. Am.

inspissate, made thick.

Institale Fr. (1825) = Hypoxylon (Xylar.) fide Læssøe (*SA* **13**: 43, 1994).

integrated (of conidiogenous cells), incorporated in the main axis or branches of the conidiophore; cf. discrete.

intellectual property, see Bioprospecting, Patent protection.

inter- (prefix), between; among.

interascal tissue, see hamathecium.

interascicular parenchyma, the paraphysis-like hyphae or paraphysoidal interthecial fibres (Stevens) (obsol.).

interbiotic, living as a parasite on or near one or more living organisms, as certain rhizoidal chytrids.

Intercalarispora J.L. Crane & Schokn. (1983), Mitosporic fungi, 1.D2.1. 1 (aquatic), USA.

intercalary, (1) (of growth), between the apex and the base; (2) (of cells, spores, etc.), between two cells.

intercellular, between cells.

International Congresses for Human and Animal Mycology (arranged by the International Society for Human and Animal Mycology). (1) Lisbon, Portugal (1958; *ISHAM Bull.* **4**: 1-30, 1959); (2) Montreal, Canada (1962; *Sabouraudia* **2**: 320, 1963); (3) Edinburgh, UK (1964; *Sabouraudia* **4**: 53-54, 1965); (4) New Orleans, USA (1967; *Sabouraudia* **6**: 83-86, 1967); (5) Paris, France (1971; *Sabouraudia* **10**: 103, 1972); (6) Tokyo, Japan (1975); (7) Israel (1979); (8) New Zealand (1982); (9) Atlanta, USA (1985); (10) Barcelona, Spain (1988).

International Mycological Congresses (arranged by the International Mycological Association, see Societies and organizations). (1) Exeter, UK (1971; *TBMS* **58**(2), Suppl.: 1-40, 1972); (2) Tampa, USA (1977; *Mycol.* **70**: 253-265, 1978); (3) Tokyo, Japan (1983); (4) Regensberg, Germany (1990); (5) Vancouver, Canada (1994); (6) Tel-Aviv, Israel (1998).

Internet. The Internet is a global communications network which connects together several million computers. It originated in the 1970's through the US Defense Department and grew rapidly during the 1980s within the educational sector. The use of the Internet by commercial organisations has recently stimulated another significant growth phase.

Connexion to the Internet provides remote access to a rapidly increasing number of scattered information resources which may consist of data files, computer programs, searchable on-line databases etc. There are a number of generic software tools for accessing these resources: TELNET allows users to directly access remote computers; FTP (File Transfer Protocol) allows files to be copied; E-MAIL (Electronic mail) lets you send and receive individual messages; USENET lets you access public news groups or bulletin boards. In addition, there are a number of ways of structuring data to facilitate access via the Internet; GOPHERs allow browsing and retrieval of remote files; WWW (World Wide Web) uses 'hypertext' technology to link together text and image components from geographically separate sources on the Internet. At the time of writing the WWW is the latest and most significant development on the Internet.

The biological community is well represented on the Internet and the WWW provides access to a number of genetic resource, molecular, floristic and bibliographic databases on most groups of organisms. The distribution of electronic books and journals is also becoming significant. Because of the nature of this rapidly changing area no details of mycological sources are provided here; user friendly search tools are available to locate any relevant data which are available.

Improved communication networks are being introduced (e.g. Information superhighway) which will allow video-conferencing and other facilities requiring large band-width to move very large amounts of data.

interspace (of a pileus), the space between the lamellae.

interthecial, between asci.

intervenose (of a pileus), veined in the interspaces.

Intextomyces J. Erikss. & Ryvarden (1976), Hyphodermataceae. 2, Eur., Am.

intra- (prefix), within; inside.

intracellular, within the cell.

intrahyphal, see hypha.

intramatrical, living in the matrix or substratum.

intramatrical spores, an alternative name for vesicles produced in host roots by most endomycorrhizal fungi.

intraparietal, within a wall or walls (e.g. of crystals amongst the tissues of an exciple).

Intrapes J.F. Hennen & Figueirado (1979), Uredinales (inc. sed.). 1, Brazil.

Intrasporangium Kalakoutskii, Kirillova & Krassiln. (1967), *Actinomycetes*, q.v.

intricate cortex (of lichen thalli), made up of hyphae twisted together; cf. textura intricata.

introrse, in the direction of the central axis; inwards.

intumescence, a swelling.

invaginated, covered by a sheath.

Inventories. Assessments of Biodiversity (q.v.) within specific areas, with the aim of cataloguing what is there and valueing its presence. The most prominent

inventorying activity to date is the proposed All-Taxa Biodiversity Inventory (ATBI) of the Guanacaste Conservation Area, Costa Rica (Janzen & Hallwachs, *Draft report of an NSF Workshop, Philadelphia*, 1993). See also Rossman (*in* Peng & Chou (Eds), *Biodiversity and terrestrial ecosystems*: 169, 1994) and Hawksworth *et al.* (*in* Janardhanan *et al.* (Eds), *Tropical mycology, in press*), Biodiversity, Collection.

involucrellum, tissue forming the upper part of a perithecioid ascoma surrounding the true exciple, not involving host or substrate materials (cfr. clypeus) and generally dimidiate (e.g. *Verrucaria*).

Involucrocarpon Servít (1953) = Catapyrenium (Verrucar.) fide Santesson (*Lichens and lichenicolous fungi of Sweden and Norway*, 1993).

Involucrothele Servít (1953) = Thelidium (Verrucar.).

involute, rolled in (Fig. 19I).

Involutisporonites R.T. Clarke (1965), Fossil fungi (*f. cat.*). 4 (Cretaceous, Paleocene, Tertiary), China, USA.

Inzengaea Borzí (1885) = Emericella (Trichocom.) fide v. Arx (1981).

Iocraterium E. Jahn (1904) = Craterium (Physar.).

iodine, (I, J), used as Lugol's solution (I 0.5g, KI 1.5 g, water 100 ml), in potassium iodide (**IKI**; I 1%, KI 3%), and formerly often Melzer's reagent (see Stains) giving blue, red, lavender or violet colours seen best after pre-treatment with potassium hydroxide in spores, asci, hymenial tissues, etc. Reactions can vary according to the kind of iodine solution used, its concentration, the age of the material, and the nature of any pretreatment; all need to be reported when referring to such tests. Extensively used in the systematics of lichenized and non-lichenized fungi since Nylander (*Flora, Jena* **48**: 465, 1865). See Baral (*Mycotaxon* **29**: 399, 1987; caution in use), Common (*Mycotaxon* **41**: 67, 1991; review I+ materials), Kohn & Korf (*Mycotaxon* **3**: 165, 1975; pre-treatment), Nannfeldt (*TBMS* **67**: 283, 1976; ascus plugs), amylomycan, amyloid, dextrinoid, lichenan.

Iodophanus Korf (1967), Pezizaceae. 14, widespr. See de Cachi & Ranalli (*Nova Hedw.* **49**: 59, 1989; ontogeny), Kimbrough *et al.* (*Am. J. Bot.* **56**: 1187, 1969, *in* Hawksworth (Ed.), *Ascomycete systematics*: 398, 1994), Kimbrough & Curry (*Mycol.* **77**: 219, 1985; posn.), Thind & Kaushal (*Ind. Phytopath.* **31**: 343, 1979), Valadon *et al.* (*TBMS* **74**: 187, 1980; carotenoids, taxonomy).

Iodosphaeria Samuels, E. Müll. & Petrini (1987), Amphisphaeriaceae. 2, temp. See Barr (*Mycotaxon* **39**: 43, 1990; posn.). Anamorphs *Ceratosporium*, *Selenosporella*.

Iola, see Jola.

Ionaspis Th. Fr. (1871), Hymeneliaceae (L). 25, cosmop. See Jørgensen (*Graphis Scripta* **2**: 118, 1989; key 11 spp. Scandinavia), Magnusson (*Acta Horti gothoburg.* **8**: 1, 1933).

ionomidotic reaction, release of a dark pigment into aqueous potassium hydroxide (KOH) mounts; an important taxonomic criterion in some dark coloured *Leotiales* apothecia (e.g. *Claussenomyces*; Oullette & Korf, *Mycotaxon* **10**: 255, 1979).

Ionomidotis E.J. Durand ex Thaxt. (1923) ? = Ameghiniella (Leot.) fide Gamundí (*SA* **11**: 62, 1992). See Zhuang (*Mycotaxon* **31**: 261, 1988; key 7 spp.).

Ionophragmium Peres (1961), Mitosporic fungi, 5.C2.?. 1, Brazil.

Ioplaca Poelt (1977), Teloschistaceae (L). 1, Nepal.

Iotidea Clem. (1909) = Peziza (Peziz.) fide Eckblad (1968).

ipomoearone, a phytoalexin (q.v.) from sweet potato (*Ipomoea batatas*).

Iraniella Petr. (1949), Ascomycota (inc. sed.). 1, Iran.

Irene Theiss. & P. Syd. (1917) = Asteridiella (Meliol.).

Irenina F. Stevens (1927) = Appendiculella (Meliol.) fide Hansford (1961).

Irenopsis F. Stevens (1927), Meliolaceae. 50, trop.

Iridinea Velen. (1934), ? Leotiales (inc. sed.). 2, former Czechoslovakia.

Iridionia Sacc. (1902) ≡ Irydyonia (Rhytismatales).

Irpex Fr. (1828), Steccherinaceae. 1, Eur., N. Am. See Maas Geesteranus (*Persoonia* 7: 443, 1974).

Irpiciochaete, see Irpicochaete.

Irpiciporus Murrill (1905) = Spongipellis (Coriol.) fide Pegler (1973).

Irpicium Bref. (1912) nom. dub. = Abortiporus (Coriol.) fide Donk (1974).

Irpicochaete Rick (1940) nom. dub. (Hydn.).

Irpicodon Pouzar (1966), Amylocorticiaceae. 1, Eur.

irpicoid, having teeth, or becoming toothed, as in *Irpex*.

Irpicomyces Deighton (1969), Mitosporic fungi, 1.C2.1. 1 (on *Schiffnerula*), Malaysia.

Irydyonia Racib. (1900), ? Rhytismatales (inc. sed.). 1, Java.

Isaria Fr. (1832), Mitosporic fungi, 2.A1.10. 2, widespr. See Petch (*TBMS* 19: 34, 1934; spp. on insects), de Hoog (*Stud. Mycol.* 1, 1972; descr. spp., nomencl.).

Isariella Henn. (1908), Mitosporic fungi, 3.A1.?. 2, C. & S. Am. See Seifert (*Mem. N.Y. bot. Gdn* 49: 202, 1989).

Isariopsella Höhn. (1929), Mitosporic fungi, 2.A/B1.?. 3, Eur. = Ramularia (Mitosp. fungi) fide Seifert (1985).

Isariopsis Fresen. (1863) ≡ Phacellium (Mitosp. fungi). See Braun (*Nova Hedw.* 47: 335, 1989; reassessment of spp.).

Ischnochaeta Sawada (1959) = Erysiphe (Erysiph.) fide Zheng (1985).

Ischnoderma P. Karst. (1879), Coriolaceae. 2, temp. See Jahn (*Westf. Pilzber.* 9: 99, 1973), Pouzar (*Česká Myk.* 25: 15, 1971).

Ischnostroma Syd. & P. Syd. (1914), Mitosporic fungi, 5.E1.?. 1, Philipp.

ISHAM, see Societies and organizations.

Isia D. Hawksw. & Manohar. (1978), Sordariales (inc. sed.). 2, India, Nepal, New Caledonia.

isidiate, having isidia.

isidiiferous, of a lichen thallus bearing isidia.

isidium (pl. **isidia**), a photobiont-containing protuberance of the cortex in lichens which may be warty, cylindrical, clavate, scale-like, coralloid, simple, or branched (Fig. 22D); occurring directly on the thallus (e.g. *Peltigera praetextata, Pseudevernia furfuracea*); may become sorediate (e.g. *Lobaria pulmonaria*). See Bailey (*in* Brown *et al. Lichenology: progress and problems*: 214, 1976; review), Du Rietz (*Svensk bot. Tidskr.* 18: 141, 1924; classification), Kershaw & Millbank (*Lichenologist* 4: 214, 1970; rôle), Puymaly (*Botaniste* 48: 237, 1965).

Isidium (Ach.) Ach. (1803) nom. rej. = Pertusaria (Pertusar.).

Isipinga Doidge (1921) = Symphaster (Asterin.) fide Müller & v. Arx (1962).

islandicin, toxins of *Penicillium islandicum*; a cause of hepatitis (yellow rice disease) in humans. See also luteoskyrin.

islanditoxin, see islandicin.

iso- (prefix), equal.

Isoachlya Kauffman (1921), Saprolegniaceae. 5, widespr. See Seymour (1970), Dick (1973).

isoenzyme, see isozyme.

isogamete, one of two sex cells like in form.

isogamy, the conjugation of isogametes.

isohaplont, a haplont of cells having genotypically like nuclei (Kniep, 1928), cf. miktohaplont.

isokont (**isokontous**) (of cilia and flagella), of equal length. Cf. heterokont.

isolate (v.), (1) to make an isolation; (n), (2) the culture itself, esp. the first 1-spore or pure isolation of a fungus from any place; Lotsy's 'species' (Brierley, *Ann. appl. Biol.* 18: 420, 1931).

isolation, (1) the process of getting a fungus or other organism into pure culture; (2) the culture itself.

Isolation methods. *Aquatic and marine fungi*: Couch (*Elisha Mitch. Sc. Soc.* 55: 208, 1939; chytrids), Fuller *et al.* (*Mycol.* 56: 745, 1964; marine 'phycomycetes'), Fuller & Jaworski (*Zoosporic fungi in teaching and research*, 1987), Jones (Ed.) (*Recent advances in aquatic mycology*, 1976), Sparrow (*Aquatic phycomycetes*, edn 2, 1960).

Single spore isolation: See Hildebrand (*Bot. Rev.* 4: 627, 1938; 16: 181, 1950), Badry (*Micromanipulators and micromanipulation*, 1963).

Soil fungi: Gams (*in* Winterhoff (Ed.), *Fungi in vegetation science*: 183, 1992; review); see also Soil fungi.

Spore trapping: See spora.

Use of antibiotics: Georg *et al.* (*Science* 114: 387, 1951) advocate penicillin 20 units/ml + streptomycin 40 units/ml to suppress unwanted bacteria when isolating fungi (see also cycloheximide); use of rose bengal in isolation media (Ottow, *Mycol.* 64: 304, 1972).

See also entries on particular habitats, and Culture and preservation.

isolichenin, see lichenin.

isomorphic, like in form but unlike in structure.

Isomunkia Theiss. & Syd. (1915), Fungi (inc. sed.). 1, Ecuador. See Petrak (*Sydowia* 5: 337, 1951).

Isomyces Clem. (1931) ≡ Debaryomyces (Saccharomycet.).

isonym, a homotypic synonym; see basionym.

isoplanogamete, one of two motile sex cells like in form.

Isosoma Svrček (1989) = Cudoniella (Leot.) fide Baral (*SA* 13: 113, 1994).

isospory, see homospory.

Isotexis Syd. (1931) = Elsinoë (Elsin.) fide v. Arx & Müller (1975).

Isothea Fr. (1849), Phyllachoraceae. 1, arctic-alpine. See v. Arx & Müller (1954).

isotomic dichotomous branching, branching in which both dichotomies are about the same thickness and length so that the dichotomic pattern is visible even in older parts of the thallus, as in *Cladonia evansii*, cf. anisotomic dichotomic branching.

isotype, see type.

isozyme (**isoenzyme**), one of a family of enzymes with different molecular weights and electron charges and so separating in electrophoresis (q.v.).

Issatchenkia Kudrjanzev (1960), Saccharomycetaceae. 4, widespr. See Kurtzman *et al.* (*Int. J. Syst. Bact.* 30: 503, 1980), Peterson & Kurtzman (*Ant. v. Leeuwenhoek* 58: 235, 1990; molec. phylogeny).

Isthmiella Darker (1967) [non Cleve (1873), *Bacillariaceae*], Rhytismataceae. 4 (on conifers), N. Am.

Isthmolongispora Matsush. (1971), Mitosporic fungi, 1.C1.10. 4, Papua New Guinea. See de Hoog & Hennebert (*Proc. K. Ned. Akad. Wet.* C 86: 343, 1983).

Isthmophragmospora Kuthub. & Nawawi (1992), Mitosporic fungi, 1.C2.10. 1 (aquatic), Malaysia.

Isthmospora F. Stevens (1918), Anamorphic Microthyriaceae, 1.G2.1. Teleomorph *Trichothyrium*. 1, trop.

isthmospore, a spore comprising two or more cells interconnected by a much narrower region, as in the ascospores of *Vialaea* (Cannon, *MR* 99: 367, 1995); a conidium composed of four more thick-walled cells separated by thin-walled cells as in *Isthmospora* (Hughes, *Mycol. Pap.* 50, 1953).

Isthmotricladia Matsush. (1971), Mitosporic fungi, 1.G1.10. 1, Papua New Guinea.

isthmus, (1) the narrower or thinner-walled portion of an isthmospore (q.v.); (2) the thickened medial perforated septum of a polarilocular ascospore.

isthmus disarticulation, protoplasmic retraction during spore formation, with the secretion of an endosporic wall membrane.

Isuasphaera H.D. Pflug (1978), Fossil fungi (? Mitosp. fungi). 1 (*c.* 3800 million yr old Isua quartzite), Greenland.

Itajahya A. Møller (1895), Phallaceae. 1, trop., subtrop. See Ahmad (*Lloydia* **8**: 238, 1945), Malençon (*BSMF* **100**: Atlas pl. 238, 1984).

italicization (of scientific names), see Classification.

Itersonia Rippel & Flehmig [not traced] nom. dub. (? Myxobacteria). 1 (cellulose decomposer).

Itersonilia Derx (1948), Mitosporic fungi, 1.A1.1. 3 (with clamp connexions), temp. See Sowell & Korf (*Mycol.* **52**: 934, 1960), Yamada & Konda (*J. Gen. Appl. Microbiol.* **30**: 313, 1984; significance of coenzyme Q system in classification), Channon (*Ann. appl. Biol.* **51**: 1, 1963; leaf spot and canker of parsnip by *I. pastinacae*), Webster *et al.* (*TBMS* **82**: 13, 1984; **91**: 193, 1988; ballistospore discharge), Ingold (*Mycologist* **5**: 35, 1991; development), Boekhout (*MR* **95**: 135, 1991; phenetic systematics), Boekhout *et al.* (*Can. J. Microbiol.* **37**: 188, 1991; genomic characterization), Boekhout & Jille (*Syst. Appl. Microbiol.* **14**: 117, 1991; mitosis and DNA content in *I. perplexans*).

Ithyphallus Gray (1821) nom. rej. ≡ Mutinus (Phall.).

ITS, see Molecular biology.

Ityorhoptrum P.M. Kirk (1986), Mitosporic fungi, 1.D2.19. 1, UK.

IUBS, see Societies and organizations.

IUCN. International Union for the Conservation of Nature and Natural Resources; the World Conservation Union. See Conservation.

IUMS, see Societies and organizations.

iwatake, see rock tripe.

Iwilsoniella E.B.G. Jones (1991), Halosphaeriaceae. 1 (in cooling tower), U.K.

Ixechinus R. Heim ex E. Horak (1968), Boletaceae. 1, Madagascar. See Wolfe (*Mycol.* **74**: 36, 1982). = Fistulinella (Bolet.) fide Guzmán (*Boln Soc. Mex. micol.* **8**: 53, 1974).

ixo- (prefix), sticky.

Ixocomus Quél. (1888) ≡ Suillus (Bolet.).

ixocutis (of a pileus), a slimy cuticle.

Ixodopsis P. Karst. (1870) ≡ Sordaria (Sordar.).

ixotrichoderm (ixotrichodermium) (of a pileus), a trichodermium composed of gelatinized hyphae (Snell, 1939; for a discussion, see Shaffer, *Mycol.* **58**: 486, 1966).

Iyengarina Subram. (1958), Mitosporic fungi, 1.G2.1. 1, India.

J, see iodine.

Jaapia Bres. (1911), Coniophoraceae. 2, Eur., N. Am. See Nannfeldt & Eriksson (*Svensk bot. Tidskr.* **47**: 177, 1953).

Jaapia Kirschst. (1938) ≡ Keisslerina (Dothior.).

Jack o'lantern, the basidioma of the luminous *Clitocybe illudens*.

jack-in-the-box discharge, see ascus.

Jacksonia J.C. Lindq. (1970) [non R. Br. ex Aiton (1811), *Papilionaceae*] ≡ Jacksoniella (Pucciniosir.). See Lindquist (*Revta Fac. Agron., Univ. Nac. La Plata* **47**: 303, 1972).

Jacksoniella Kamat & Sathe (1972) = Dietelia (Pucciniosir.) fide Cummins & Hiratsuka (1983).

Jacksoniella J.C. Lindq. (1972) = Dietelia (Pucciniosir.) fide Cummins & Hiratsuka (1983).

Jacksonomyces Jülich (1979), Meruliaceae. 8, widespr.

Jackya Bubák (1902) [non *Jackia* Wall. (1823), *Rubiaceae*] = Puccinia (Puccin.) fide Dietel (1928).

Jacobaschella Kuntze (1891) ≡ Diplosporium (Mitosp. fungi).

Jacobia Arnaud (1951), Mitosporic fungi, 1.A1.?. 1 (with clamp connexions), Eur.

Jacobsonia Boedijn (1935), Leotiaceae. 1, Sumatra.

Jaculispora H.J. Huds. & Ingold (1960), Mitosporic fungi, 1.A1.1. 1 (aquatic), Jamaica.

Jaczewski (Arthur Louis Arthurovič de; 1863-1932). First Director of the Bureau of Mycology and Phytopathology (later the Institut Jaczewski), St. Petersburg. Chief writings: [*Key to the Fungi*] (1913-17), [*Fundamentals of mycology*] (1933) [in Russian] and a number of other works on mycology and plant pathology. See Jones (*Phytopath.* **23**: 111, 1933; portr.), Grummann (1974: 548), Stafleu & Cowan (*TL-2* **2**: 413, 1979).

Jaczewskia Mattir. (1912) = Phallus (Phall.) fide Vasil'kov (*Bot. Zh.* **40**: 596, 1955).

Jaczewskiella Murashk. (1926) = Stigmina (Mitosp. fungi) fide Sutton (1977).

Jaffuela Speg. (1921), ? Dothioraceae. 1 (on *Puya*), Chile.

Jafnea Korf (1960), Otideaceae. 3, N. Am., Japan, India. See Berthet & Korf (*Nat. can.* **96**: 247, 1969).

Jafneadelphus Rifai (1968) = Aleurina (Otid.) fide Zhuang & Korf (1986).

Jahniella Petr. (1921), Mitosporic fungi, 4.E1.1. 1, former Czechoslovakia.

Jahnoporus Nuss (1980), Polyporaceae. 1, N. temp.

Jahnula Kirschst. (1936), Dothideales (inc. sed.). 2 (aquatic), Australia, Eur. See Hawksworth (*Sydowia* **37**: 43, 1984).

Jainesia Gonz. Frag. & Cif. (1925), Mitosporic fungi, 1.C2.?. 1, W. Indies.

Jamesdicksonia Thirum., Pavgi & Payak (1961), Tilletiaceae. 1 (on *Dichanthium*), India.

Janauaria Singer (1986), Agaricaceae. 1, Brazil.

Janetia M.B. Ellis (1976), Mitosporic fungi, 1.C2.10. 11, widespr. See Sutton & Pascoe (*Aust. Syst. Bot.* **1**: 127, 1988), Sivanesan (*MR* **94**: 566, 1990; synnematous sp.).

Jannanfeldtia Subram. & Sekar (1993), Coronophoraceae. 1, India.

Janospora (Starbäck) Höhn. (1923) = Stilbospora (Mitosp. fungi) fide Sutton (1977).

Janseella Henn. (1899) = Phaeotrema (Thelotremat.).

Jansia Penz. (1899) = Mutinus (Phall.) fide Dring (*in* Ainsworth *et al.*, *The Fungi* 4B: 458, 1973).

Japanese mushroom, see matsu-take.

Japewia Tønsberg (1990), Bacidiaceae (L). 3, N. temp.

Japonia Höhn. (1909), Mitosporic fungi, 4.B1.1. 1, Japan.

Japonochytrium Kobayasi & M. Ôkubo (1953), Thraustochytriaceae. 1 (on *Gracilaria*), Japan. See Dick (*in press*; key).

Japonogaster Kobayasi (1989), Lycoperdaceae. 1, Japan. ? a monstrosity of *Lycoperdon*.

Jaraia Němec (1912) nom. dub. (? Saprolegn.).

Jarxia D. Hawksw. (1989), Arthopyreniaceae. 1, C. Am.

Jattaea Berl. (1900), Calosphaeriaceae. 4, Eur., N. Am., N. Afr. See Barr (1985).

Jattaeolichen Tomas. & Cif. (1952) = Arthopyrenia (Arthopyren.).

Jattaeomyces Cif. & Tomas. (1953) ≡ Jattaeolichen (Arthopyren.).

Javaria Boise (1986), ? Melanommataceae. 2 (on *Palmae*), N. & S. Am.

Javonarxia Subram. (1992), Mitosporic fungi, 1.C2.19. 2, Malaysia.

Jeaneliomyces Lepesme (1945) = Dimeromyces (Laboulben.).

jelly fungi, a term sometimes applied to the *Tremellales* s.l.

Jenmania W. Wächt. (1897), Lichinaceae (L). 2, S. Am. See Henssen (*Lichenologist* 5: 444, 1973; key).

Jenmaniomyces Cif. & Tomas. (1953) ≡ Jenmania (Lichin.).

Jerainum Nawawi & Kuthub. (1992), Mitosporic fungi, 1.D2.1. 1 (aquatic), Malaysia.

Jew's ear, the basidioma of *Auricularia auricula-judae*.

Jobellisia M.E. Barr (1993), Clypeosphaeriaceae. 2, N. Am.

Joerstadia Gjaerum & Cummins (1982), Phragmidiaceae. 4 (on *Alchemilla*), Afr.

Johansonia Sacc. (1889), Saccardiaceae. c. 10, trop.

Johansoniella Bat., J.L. Bezerra & Cavalc. (1966) = Johansonia (Saccard.) fide v. Arx & Müller (1975).

Johncouchia S. Hughes & Cavalc. (1983), Anamorphic Septobasidiaceae,   1.G2.1.   Teleomorph *Septobasidium*. 1, widespr.

Johnkarlingia Pavgi & S.L. Singh (1979), Synchytriaceae. 1 (on *Brassica* roots), India.

Johnstonia M.B. Ellis (1971) [non Walkom (1925), fossil] ≡ Neojohnstonia (Mitosp. fungi).

Jola A. Møller (1895), Platygloeaceae. 1 (on *Musci*), trop. See Martin (*Mycol.* 31: 239, 1939).

Jonaspis see *Ionaspis*.

Jongiella M. Morelet (1971) = Camillea (Xylar.) fide Læssøe *et al.* (1989).

Jopex, see *Irpex*.

jordanon, Brierley's (*Ann. app. Biol.* 18: 420, 1931) application of Lotsy's term for a group of isolates constituting a race or strain, with similar cultural characters and behaviour (obsol.).

Jove's beard, the basidioma of *Odontia barba-jovis*.

Jrpex, see *Irpex*.

Jubispora B. Sutton & H.J. Swart (1986), Mitosporic fungi, 8.C2.19. 1, Afr.

Jugglerandia Lloyd (1923) ≡ Holwaya (Leot.).

Jugulospora N. Lundq. (1972), Lasiosphaeriaceae. 1, Eur., N. Am. See Barr (*Mycotaxon* 39: 43, 1990; posn).

Julella Fabre (1879), Thelenellaceae. 3 or 4, widespr. See Barr (*Sydowia* 38: 11, 1986).

Juliohirschhornia Hirschh. (1986), ? Tilletiaceae. 1 (on *Paspalum*), S. Am.

Junctospora Minter & Hol.-Jech. (1981), Mitosporic fungi, 1.A1.4. 1, former Czechoslovakia.

Jundzillia Racib. ex L.F. Čelak. (1893) nom. dub. (Myxom.)

Junghuhnia Corda (1842), Steccherinaceae. c. 8, widespr. See Ryvarden (*Persoonia* 7: 17, 1972).

Jungkultur, see Normkultur.

Junia Dumort. (1822) [non Adans. (1763), *Clethraceae*] = Phallus (Phall.).

juvenescence, the process of maturing at a stage of development normally immature.

K, (1) see Metabolic products; (2) Royal Botanic Gardens (Kew, Surrey, UK); founded 1841; managed by a Board of Trustees and mainly funded by the UK Ministry of Agriculture, Fisheries and Food (MAFF); most lichen material transferred to **BM** (q.v.); see Blunt (*In for a penny*, 1978), Hepper (Ed.) (*Royal Botanic Gardens, Kew: gardens for science and pleasure*, 1982); (3) see Sörensen coefficient.

Kabathia Nieuwl. (1916) = Sphaerellopsis (Mitosp. fungi) fide Sutton (1977).

Kabatia Bubák (1904), Mitosporic fungi, 8.A/B1.1. 9, widespr. See Sutton (1980).

Kabatiella Bubák (1907), Mitosporic fungi, 4.A1. 15, widespr., esp. N. temp. *K. caulivora* (clover (*Trifolium*) scorch). See v. Arx (*Revision of fungi classified as Gloeosporium*, 1970), Sampson (*TBMS* 13: 103, 1928).

Kabatina R. Schneid. & Arx (1966), Mitosporic fungi, 6.A1.15. 4, Eur., N. Am. See Butin & Pehl (*MR* 97: 1340, 1993).

Kacosphaeria Speg. (1887), ? Calosphaeriaceae. 1, Tierra del Fuego. See Petrak (*Sydowia* 5: 328, 1951).

Kafiaddinia Mekht. (1978), Mitosporic fungi, 1.C1.10. 1 (from nematodes), former USSR.

Kainomyces Thaxt. (1901), Laboulbeniaceae. 3, W. Afr., Asia. See Terada (*Trans. mycol. Soc. Japan* 19: 55, 1978).

Kakabekia Bargh. (1965), Fossil fungi. 2, Eur., N. Am.

Kalaallia Alstrup & D. Hawksw. (1990), Dacampiaceae. 1 (on lichens), Greenland.

Kalbiana Henssen (1988), ? Verrucariaceae (L). 1, Brazil.

Kalchbrennera Berk. (1876) = Lysurus (Clathr.) fide Dring (*Kew Bull.* 35: 68, 1980).

Kaleidosporium Van Warmelo & B. Sutton (1981), Mitosporic fungi, 6.D2.15. 1, N. Am.

Kallichroma Kohlm. & Volkm.-Kohlm. (1993), Hypocreaceae. 2 (marine), widespr.

Kalmusia Niessl (1872), Dothideales (inc. sed.). c. 6, widespr.

Kamatella Anahosur (1969), Mitosporic fungi, 4.B2.1. 1, Afr., India.

Kamatia V.G. Rao & Subhedar (1976) = Pseudodictyosporium (Mitosp. fungi) fide Kendrick in Carmichael *et al.* (1980).

Kamatomyces Sathe (1966), Uredinales (inc. sed.). 1, India. Kulkarni & Hardikar (*Ind. Phytopath.* 29: 28, 1977).

kamé, truffle (Arabic); **black -**, **brown -**, the edible *Terfezia boudieri* and *T. claveryi*, respectively, of the Middle East (Awameh & Alsheikh, *Mycol.* 72: 50, 494, 1980).

kames, see terfas.

Kameshwaromyces Kamal, R.V. Verma & Morgan-Jones (1986), Mitosporic fungi, 1.D2.1. 1, India.

Kananascus Nag Raj (1984), Trichosphaeriaceae. 2, India, W. India. Anamorph *Koorchaloma*.

Kanousea Bat. & Cif. (1962) = Microcallis (Chaetothyr.) fide v. Arx & Müller (1975).

Kapooria J. Reid & C. Booth (1989), ? Valsaceae. 1, India.

Karakulinia N.P. Golovina (1964) = Fusicladium (Mitosp. fungi) fide Deighton (1967).

Karlingia A.E. Johanson (1944), Spizellomycetaceae. 10, widespr.

Karlingiomyces Sparrow (1960), Chytridiaceae. 6, widespr. = Karlingia (Chytrid.) fide Karling (1977).

Karoowia Hale (1989) Parmeliaceae. 16, S. Afr., Australia. ? = Parmelia (Parmel.)

Karschia Körb. (1865), Dothideales (inc. sed.). 2 (on lichens), Eur. See Hafellner (*Beih. Nova Hedw.* 62, 1979; key; disp. excl. spp.).

Karsten (Petter Adolf; 1834-1917). Lecturer in botany, Agricultural and Dairy Institute, Mustiala, Finland, 1864-1908 (Professor from 1900). Chief writings: Books and papers on the fungi of Finland, e.g. *Mycologia fennica* (1871-78 [reprint 1967]), *Symbolae ad mycologiam fennicam* [33 parts] (1870-95 [reprint 1967]), *Rysslands, Finlands och den Skandinaviska Halföns Hattsvampar* (1879-82), *Kritisch öfversigt af Finlands Basidsvampar (Basidiomycetes; Gastero- et*

Hymenomycetes) (1889); collected mycological papers (1859-1909 [reprint 4 vols, 1973]). See Lowe (*Mycol.* **48**: 99, 1956; polypore types), Grummann (1974: 609), Stafleu & Cowan (*TL-2* **2**: 505, 1979).

Karstenella Harmaja (1969), Karstenellaceae. 1, Finland.

Karstenellaceae Harmaja (1974), Pezizales. 1 gen., 1 sp. Ascomata apothecial, flat, very thin, formed on a thick-walled papillate hyaline cyanophilic hyphal mat; peridium composed of intertwined hyphae; interascal tissue of simple paraphyses, reddish brown at the apices; asci cylindric, operculate, not blueing in iodine; ascospores hyaline, smooth, without oil droplets or appendages, binucleate. Anamorph not known. Saprobic on leaf litter, Finland.
    Lit.: Harmaja (*Karstenia* **9**: 20, 1969, **14**: 109, 1974).

Karstenia Fr. (1885) nom. cons., ? Rhytismatales (inc. sed.). 8, Eur., Am. See Sherwood (*Mycotaxon* **5**: 1, 1977; *Occ. Pap. Farlow Herb.* **15**, 1980).

Karstenia Britzelm. (1897) [non Göpp. (1836), fossil *Polypodiaceae*] ≡ Prillieuxia (Thelephor.).

Karsteniomyces D. Hawksw. (1980), Mitosporic fungi, 4.B1.1. 2 (on lichens, esp. *Peltigera*), Eur. See Hawksworth (*Bull. Br. Mus. nat. Hist., Bot.* **9**: 22, 1981).

Karstenula Speg. (1879), Melanommataceae. 3, temp. See Eriksson & Hawksworth (*SA* **10**: 140, 1991), Barr (*N. Am. Fl.* II **13**, 1990; N. Am.), Munk (1957), Constantinescu, *MR* **97**: 377, 1993; anamorph). Anamorph *Microdiplodia*.

karyochorisis, somatic nuclear division resulting from a constriction of the nuclear membrane (Moore, *Z. Zellforsch.* **63**: 921, 1964).

karyogamy, the fusion of two sex nuclei after cell fusion, i.e. after plasmogamy.

karyotype, the size and number of chromosomes in an organism. Generally determined by deduction from mating studies, microscopically or from specialized electrophoresis methods. See McCluskey & Mills (*Mol. Pl.-Microbe Interactions* **3**: 366, 1990), pulsed field gel-electrophoresis.

Kaskaskia Born & J.L. Crane (1972) = Gyrostroma (Mitosp. fungi) fide Bedker & Wingfield (*TBMS* **81**: 179, 1983).

Kathistaceae Malloch & M. Blackw. (1990), Ophiostomatales. 1 gen., 3 spp. Ascomata small, perithecial, long-necked, hyaline, with ostiolar setae; interascal tissue absent; asci formed in a basal fascicle, ± ellipsoidal, evanescent; ascospores elongated, septate, hyaline to pale brown, without a sheath. Anamorphs yeast-like, formed directly from ascospores; putative sporidiomata also formed. Coprophilous, temp.

Kathistes Malloch & M. Blackw. (1990), Kathistaceae. 3 (coprophilous), N. temp.

katothecium, see catathecium.

katsuobushi, a Japanese food obtained by fermenting cooked bonito fish (*Sarda sarda*) with *Aspergillus* spp.

Kauffman (Calvin Henry; 1869-1931). Michigan University, 1904-31 (Professor from 1912). Chief writings: books and papers on the *Agaricaceae* (q.v.). See Mains (*Phytopath.* **22**: 271, 1932; *Mycol.* **24**: 265, 1932; portr.), Stafleu & Cowan (*TL-2* **2**: 507, 1979).

Kaufmannwolfia Galgoczy & E.K. Novák (1962) = Trichophyton (Mitosp. fungi).

Kavinia Pilát (1938), Ramariaceae. 1, Eur.

Kawakamia Miyabe (1903) = Phytophthora (Pyth.) fide Fitzpatrick (1930).

Kazachstania Zubcova (1971) = Saccharomyces (Saccharomycet.) fide v. Arx *et al.* (1977).

Kazulia Nag Raj (1977), Anamorphic Chaetothyriaceae, 3.G1.10. Teleomorph ? *Chaetothyrium*. 1, Brazil.

kb (kilobases), an abbreviation for 1000 base pairs of DNA.

KC, see Metabolic products.

kefiran, a water soluble polysaccharide from kefir grains (Kooiman, *Carboh. Res.* **7**: 200, 1968).

Keinstirschia J. Reid & C. Booth (1989), Diaporthales (inc. sed.). 1, Germany.

Keisslerellum Werner (1944) nom. inval. = Mycoporellum (Dothideales).

Keissleria Höhn. (1918) = Broomella (Amphisphaer.) fide Höhnel (*Ann. Myc.* **18**: 71, 1920).

Keissleriella Höhn. (1919), Lophiostomataceae. *c.* 20, widespr. See Bose (*Phytopath. Z.* **41**: 179, 1961; key 11 spp.). Anamorphs *Ascochyta, Dendrophoma*.

Keisslerina Petr. (1920) = Dothiora (Dothior.) fide Froidevaux (1973).

Keissleriomyces D. Hawksw. (1981), Mitosporic fungi, 4.C1.15. 1 (on *Cladonia*), Eur.

Keithia Sacc. (1892) [non Spreng. (1822), ? *Capparaceae*] = Didymascella (Hemiphacid.).

Kelleria Tomin (1926) [non Endl. (1848), *Thymelaeceae*] ? = Thelocarpon (Acarospor.) fide Salisbury (1966).

Kellermania Ellis & Everh. (1885), Anamorphic Dothideales, 4.A/C1.1. Teleomorph *Planistromella*. 7, USA. See Nag Raj (1993), Ramaley (*Mycotaxon* **47**: 259, 1993; teleomorphs, **55**: 255, 1995; key to 7 spp.).

Kellermaniites Babajan & Tasl. (1977), Fossil fungi. 1 (Tertiary), former USSR.

Kellermanniopsis Edward, Kr.P. Singh, S.C. Tripathi, M.K. Sinha & Ranade (1974), Mitosporic fungi, 4.C1/2.?. 1, India. ? = Monochaetina (Mitosp. fungi) fide Sutton (1977).

Kelleromyxa Eliasson (1991), ? Physarales (inc. sed.). 1, Eur., N. Am.

Kemmleria Körb. (1861) ? = Buellia (Physc.).

Kendrickomyces B. Sutton, V.G. Rao & Mhaskar (1976), Mitosporic fungi, 8.A1.15. 1, India, Australia.

Kensinjia J. Reid & C. Booth (1989), Melanconidaceae. 1, USA.

Kentingia Sivan. & W.H. Hsieh (1989), Parmulariaceae. 1, Taiwan, India. Anamorph ?*Excipulariopsis*.

Kentrosporium Wallr. (1844) ≡ Claviceps (Clavicip.).

kephir, see Fermented food and drinks.

keratin, protein which is the main component of skin, hair, nails, feathers and horns.

Keratinomyces Vanbreus. (1952) = Trichophyton (Mitosp. fungi) fide Ajello (1968).

keratinophylic, (1) capable of decomposing keratin; (2) of many fungi causing superficial mycoses in humans; see dermatophyte, ringworm, tinea.

Keratinophyton Randhawa & R.S. Sandhu (1964) = Aphanoascus (Onygen.) fide Cano & Guarro (1990).

keratomycosis, a fungus infection of the cornea of the eye.

Keratosphaera H.B.P. Upadhyay (1964) ? = Koordesiella (Dothideales) fide Eriksson & Hawksworth (1987).

kerion, an inflammatory form of ringworm of the scalp; tinea kerion.

Kermincola Šulc (1907) nom. dub. (? Mitosp. fungi).

Kernella Thirum. (1949), Pucciniaceae. 1 (on *Litsea*), India.

Kernia Nieuwl. (1916), Microascaceae. 5, widespr. See Malloch & Cain (*CJB* **49**: 855, 1971; key), Locquin-Linard (*Revue mycol.* **41**: 509, 1977).

Kernia Thirum. (1947) ≡ Kernella (Puccin.).

Kerniomyces Toro (1939), Schizothyriaceae. 1, S. Am.

Kernkampella Rajendren (1970), Raveneliaceae. 8 (on *Euphorbiaceae*), widespr. See Laundon (*Mycotaxon* **3**: 142, 1975).

kerosene fungus (creosote fungus), *Amorphotheca resinae* (anamorph *Hormoconis resinae*); grows on creosoted wood and petrochemical fuels, also isolated from soil; utilizes n-alkanes with chain lengths between C_9-C_{19} grows esp. well at fuel/water interfaces forming hyphal mats that can break loose and cause blockages in aircraft wing and other fuel storage tanks, often associated also with accelerated corrosion. See Hendley (*TBMS* **47**: 467, 1964), David & Kelley (*Mycopath.* **129**: 159, 1995 [*IMI Descr.* **1230**]), Parbery (*Austr. J. Bot.* **17**: 331, 1969; natural occurrence), Ratledge (*in* Wilkinson (Ed.), *Developments in biodegradation of hydrocarbons*: 1, 1987; hydrocarbon uptake), Shennon (*in* Houghton *et al.* (Eds), *Biodeterioration* **7**, 248, 1988; biocides).

ketjap, shoyu sauce (q.v.) as made in Indonesia.

Ketubakia Kamat, Varghese & V.G. Rao (1987), Mitosporic fungi, 3.C2.1. 1, India.

Keys, construction of, see Hawksworth (*Mycologist's handbook*, 1974), Pankhurst (*Biological identification*, 1978), Parker-Rhodes (*Bull. BMS* **8**: 68, 1974), Rayner (*Bull. BMS* **10**: 31, 1976), Tilling (*J. Biol. Educ.* **18**: 293, 1984), Korf (*Mycol.* **64**: 937, 1972); also Numerical taxonomy.

Khekia Petr. (1921) = Herpotrichia (Lophiost.) fide v. Arx & Müller (*Sydowia* **37**: 6, 1984).

Khuskia H.J. Huds. (1963), Ascomycota (inc. sed.). 1, widespr. See Sivanesan & Holliday (*CMI Descr.* **311**, 1971). Anamorph *Nigrospora*.

Kickxella Coem. (1862), Kickxellaceae. 1, Eur., N. Am. See Linder (*Farlowia* **1**: 49, 1943), Benjamin (*Aliso* **4**: 149, 1958), Young (*Ann. Bot.* **38**: 873; spore ultrastr.).

Kickxellaceae Linder (1943), Kickxellales. 8 gen. (+ 2 syn.), 21 spp. Asexual reproduction by unisporous sporangiola, dry or wet spored; sexual reproduction by zygospores; cosmop., saprobes from soil and coprophilous (rarely mycoparasites on *Isaria* and *Vesiculomyces*).

Kickxellales, Zygomycetes. 1 fam., 8 gen. (+ 2 syn.), 21 spp. Fam.:
Kickxellaceae.
Lit.: Benjamin (*Aliso* **4**: 149, 321; **5**: 11, 273, 1958-1963; *The whole fungus* **2**: 573, 1979), Young (*New Phytol.* **67**: 823, 1968; spore ultrastr.), Moss & Young (*Mycol.* **70**: 944; phylogeny).

Kiehlia Viégas (1944), Parmulariaceae. 1 or 2, Brazi, Panama. See Farr (*Mycol.* **60**: 924, 1968).

kievitone, an isoflavonoid phytoalexin produced by the French bean (*Phaseolus vulgaris*).

Kiliasia Hafellner (1984) = Toninia (Catillar.) fide Timdal (*Opera Bot.* **110**, 1991).

Kilikiostroma Bat. & J.L. Bezerra (1961), Mitosporic fungi (L), 4.B2,?. 1, Brazil.

Kiliophora Kuthub. & Nawawi (1993), Mitosporic fungi, 1.16.A2. 2, Malaysia, Ivory Coast.

Kimbropezia Korf & W.Y. Zhuang (1991), Pezizaceae. 1, Canary Isl.

Kimuromyces Dianese, L.T.P. Santos, R.B. Medeiros & Furlan. (1995), Pucciniaceae. 1, Brazil.

Kineosporia Pagani & Parenti (1978), Actinomycetes, q.v.

kinetid, flagellar apparatus including kinetosomes and their associated tubules and fibres.

kinetosome, intracellular, non-membrane bound organelles; microtubular cylinders (0.2 μm diam.) organized in the 9 (A, B C- microtubules) + 0 pattern and with axoneme extensions in the 9 (A, B-microtubules) + 2 pattern.

KINGDOMS OF FUNGI. In the seventh edition of this *Dictionary*, the fungi were accepted as a single kingdom along with bacteria (*Monera*), plants (*Plantae*), animals (*Animalia*) and protists (*Protista*) according to the five-kingdom scheme of Whittaker (*Science*

163: 160, 1969). However, with advances in ultrastructural, biochemical, and especially molecular biology, the treatment of the fungi as one of the five kingdoms of life has become increasingly untenable (see Phylogeny). The organisms studied by mycologists are now established as polyphyletic (i.e. with different phylogenies) and have to be referred to at least three different kingdoms.

This situation has arisen because of similarities in biology and structure. Essentially, 'fungi', pragmatically defined as organisms studied by mycologists, are eukaryotic, heterotrophic, develop branching filaments (or are more rarely single-celled), and reproduce by spores. Where it is useful to speak of fungi in this polyphyletic sense, the name can be used non-italicized and not capitalized to differentiate this from the kingdom *Fungi*; it is also possible to use other informal names for 'fungi', e.g. eumycotan or protoctistan fungi if considered desirable (Kendrick, 1992), but their replacement by scientific names that could imply either a common ancestry (cf. *Panomycetes*) or more formal rank (e.g. union; Barr, 1992) is not advocated here. Retention of the fungal phylum termination -**mycota** regardless of the kingdom to which the 'fungal' phyla are refered fulfills the same purpose. See Table 5 for a comparison, by use of selected characters, of the main kingdoms of *Eukaryota*.

Selected systems proposed for fungi since 1950 are summarized in Table 4 (Fungi, q.v.) for comparison with the three kingdoms of fungi accepted in this edition of the *Dictionary*:
(1) **Chromista**.
(2) **Fungi**.
(3) **Protozoa**.
Their distinguishing features and names are discussed under the entries for each of the kingdoms. Of the three kingdoms, only *Fungi* consist exclusively of fungi (see above); the *Chromista* and *Protozoa* mainly comprise non-fungal phyla. Some authors unite the *Chromista* and *Protozoa* into a single highly polyphylectic kingdom *Protoctista* (syn. *Protista*), but that conclusion is not supported by molecular, biochemical, and other evidence (see under kingdom entries). Cavalier-Smith (1993) and Corliss (1994) both retain *Chromista* and *Protozoa* as separate kingdoms. Dick (*Mycol. Pap.*, *in press*) segregates the *Straminopila* as a kingdom separate from the *Chromista* and including the fungal phyla of that kingdom.

Lit.: Ainsworth *et al.* (Eds) (*The Fungi* **4A-4B**, 1973), v. Arx (*The genera of fungi sporulating in pure culture*, edn 3, 1981), Barr (*Mycol.* **84**: 1, 1992), Bessey (*Morphology and taxonomy of fungi*, 1950), Cavalier-Smith (*in* Osawa & Honjo (Eds), *Evolution of life*: 271, 1991; *Microbiol. Rev.* **37**: 953, 1993), Corliss (*Acta Protozool.* **33**: 1, 1994), Hawksworth (*MR* **95**: 641, 1991), Kendrick (*The fifth kingdom*, edn 2, 1992), Kriesel (*Grundzüge eines natürlichen Systems det Pilze*, 1969; *Biol. Rundsch.* **26**: 65, 1988), Leedale (*Taxon* **23**: 261, 1974; review early systems), Margulis (*Symbiosis in cell evolution*, edn 2, 1993), Moore (*Recent advances in microbiology*: 49, 1971; *Bot. Mar.* **23**: 361, 1980; *in* Hawksworth (Ed.), *Identification and characterization of pest organisms*, 1994). See also Classification, Phylogeny.

Kionocephala P.M. Kirk (1986), Mitosporic fungi, 1.C2.4. 1, UK.

Kionochaeta P.M. Kirk & B. Sutton (1986), Mitosporic fungi, 1.A1.15. 7, trop.

Kirchbaumia Schulzer (1866) = Phallus (Phall.).

Kirkia Benny (1995) [non *Kirkia* Oliv. (1868), *Simaroubaceae*] ≡ Kirkomyces (Thamnid.).

TABLE 5. Selected characters distinguishing the main kingdoms of *Eukaryota*.

Character	Animalia	Chromista	Fungi	Plantae	Protozoa
Nutrition	Heterotrophic (phagotrophic or osmotrophic)	Autotrophic (photosynthetic or absorptive)	Heterotrophic (absorptive/ osmotrophic)	Autotrophic (photosynthetic)	Heterotrophic (phagotrophic) or autotrophic (photosynthetic)
Cell walls	Absent; cellulosic material not produced	Often cellulose; chitin and ß-glucans absent	Chitin and ß-glucans	Cellulose and other polysaccharides	Absent when trophic; various when present
Mitochondrial cristae	Flattened (rarely tubular)	Tubular	Flattened	Flattened	Tubular
Flagellar mastigonemes	Absent	Tubular	Absent	Absent	Not tubular

Kirkomyces Benny (1995), Thamnidiaceae. 1, India. See Benny (*Mycol.* **87**: 922, 1995).

Kirramyces J. Walker, B. Sutton & Pascoe (1992), Mitosporic fungi, 4.C2.19. 4, widespr., esp. Australia.

Kirschsteinia Syd. (1906) = Nitschkia (Nitschk.). See Müller & v. Arx (1962).

Kirschsteiniella Petr. (1923) = Cyclothyrium (Mitosp. fungi) fide Hawksworth (*Bot. J. Linn. Soc.* **82**: 35, 1981).

Kirschsteiniothelia D. Hawksw. (1985), Pleomassariaceae. 7, Eur., N. & S. Am., NZ, ? Japan. See Barr (*SA* **12**: 27, 1993, *Mycotaxon* **49**: 129, 1993; posn), Shearer (*Mycol.* **85**: 963, 1994; asci). Anamorph *Dendryphiopsis*.

Kjeldsenia Colgan, Castellano & Bougher (1995), Cortinariaceae. 1, USA.

Klasterskya Petr. (1940), Ophiostomataceae. 1, Eur. See Minter (*TBMS* **80**: 162, 1983; posn). Anamorph *Hyalorhinocladiella*.

Klastopsora Dietel (1904) = Pucciniostele (Phakopsor.) fide Dietel (1928).

Klebahnia Arthur (1906) = Uromyces (Puccin.) fide Arthur (1934).

Klebahnopycnis Kirschst. (1939) = Hoehneliella (Mitosp. fungi) fide Petrak (*Ann. Myc.* **38**: 199, 1940).

Kleidiomyces Thaxt. (1908), Laboulbeniaceae. 3, Am.

Kleistobolus C. Lippert (1894) = Licea (Lic.).

Klöckeria, see *Kloeckera*.

Kloeckera Janke (1923), Anamorphic Saccharomycodaceae, 1.A1.?. Teleomorph *Hanseniaspora*. 6 (1 in Barnett *et al.*, 1990), widespr. See Meyer *et al.* (*Ant. v. Leeuwenhoek* **44**: 79, 1978).

Kloeckeraspora Niehaus (1932) = Hanseniaspora (Saccharomycod.) fide Lodder (1970).

Kluyveromyces Van der Walt (1956), Saccharomycetaceae. 14 (fide Barnett *et al.* , 1990), widespr. See Fuson *et al.* (*Int. J. Syst. Bact.* **37**: 371, 1987; DNA relatedness), Kreger van Rij (1984; key), v. Arx *et al.* (1977), Kämper (*Bibl. Mycol.* **136**, 1990; DNA killer plasmid), Kock *et al.* (*Syst. Appl. Microbiol.* **10**: 293, 1988; Martini & Martini (*Int. J. Syst. Bact.* **37**, 380, 1987; Martini & Rosini (*Mycol.* **81**: 317, 1989; killer relationships), Sidenberg & Lachance (*Int. J. Syst. Bact.* **36**: 94, 1986; isozymes, DNA homology).

Kmetia Bres. & Sacc. (1902), Mitosporic fungi, 3.E1.?. 1, Eur.

Kmetiopsis Bat. & Peres (1960), Mitosporic fungi, 3.E1.15. 1, Brazil.

Kneiffia Fr. (1836) [non Spach (1835), *Oenotheraceae*] ≡ Kneiffiella Underw. (1898) [non P. Karst. (1889)], ≡ Kneiffiella Henn. (1898), ≡ Neokneiffia Sacc. (Aug. 1898) (q.v.), ≡ Pycnodon Underw. (Dec. 1898), = Hyphoderma (Cortic.) fide Jülich (*Persoonia* **8**: 59, 1972).

Kneiffiella P. Karst. (1889), Thelephoraceae. 1, Eur. = Hyphoderma (Hyphodermat.) fide Donk (1964).

Kneiffiella Underw. (1897) ≡ Noekneiffia (Hyphodermat.).

Knemiothyrium Bat. & J.L. Bezerra (1960), Mitosporic fungi, 5.B/C2.?. 1, Jamaica.

Kniep (Karl Johannes Hans; 1881-1930). Director of the Institute of Plant Physiology, Berlin-Dahlem. Noted for his work on the genetics of fungi. Chief writings: *Die Sexualität der niederen Pflanzen* (1928), *Vererbungserscheinungen bie Pilzen* (*Bibliogr. genet.*,

1929). See Harder (*Ber. dtsch. bot. Ges.* **48**: (164), 1930), Grummann (1974: 97).

Knightiella Müll. Arg. (1886), Icmadophilaceae (L). 1, Australasia. See Galloway (*N.Z. Jl Bot.* **18**: 481, 1980).

Knoxdaviesia M.J. Wingf., P.S. van Wyk & Marasas (1988), Anamorphic Ophiostomataceae, 1.A1.15. Teleomorph *Ophiostoma.* 1, S. Afr.

Knyaria Kuntze (1891) ≡ Tubercularia (Mitosp. fungi).

Kobayasia S. Imai & A. Kawam. (1958), Protophallaceae. 1, Japan.

Kochiomyces D.J.S. Barr (1980), Spizellomycetaceae. 1. USA.

Koch's postulates. Criteria for proving the pathogenicity of an organism. Steps include: (1) the suspected causal organism must be constantly associated with the disease; (2) it must be isolated and grown in pure culture; (3) when a healthy plant is inoculated with it the original disease must be reproduced; (4) the same organism must be reiosolated from the experimentally infected plant. See Evans (*Yale Jl biol. Medic.* **49**: 175, 1976).

Kockovaella Nakase (1990) nom. inval. Mitosporic fungi, 1.A1.?. 1 (ballistosporic yeast).

Kodonospora K. Ando (1993), Mitosporic fungi, 1.G2.15. 1, Japan.

Koerberia A. Massal. (1854), Placynthiaceae (L). 2, Eur., N. Am., N. Afr. See Henssen (*CJB* **41**: 1347, 1963).

Koerberiella Stein (1879), Porpidiaceae (L). 2, Eur. See Rambold *et al.* (*Lichenologist* **22**: 225, 1990), Navarro-Rosinés & Hafellner (*Bibl. Lich.* **53**: 179, 1993).

Kohlmeyera S. Schatz (1980) = Mastodia (Mastod.) fide Eriksson (1981).

Kohlmeyeriella E.B.G. Jones, R.G. Johnson & S.T. Moss (1983), Halosphaeriaceae. 1 (marine), N. Hemisph.

koji mould, *Aspergillus oryzae* and allied spp. used as starter for various fermented Japanese foods (saké, miso, shoyu, mirin, amazaké) by inoculating rice. See Murakami (*J. gen. Microbiol.* **17**: 281, 1971), mould rice, miso, Industrial mycology.

kojic acid, characteristic metabolic product of *Aspergillus flavus, oryzae* and *tamarii* groups (Birkinshaw *et al., Phil. Trans.* **B220**: 127, 1931). It gives a blood-red colour with ferric chloride ($FeCl_3$).

Kokkalera Ponnappa (1970) = Sphaerotheca (Erysiph.) fide Zheng (1985).

Koleroga Donk (1958) = Ceratobasidium (Ceratobasid.) fide Talbot (1965) but see Muthappa (*TBMS* **73**: 159, 1979).

Kolman Adans. (1763) nom. rej. = Collema (Collemat.).

Komagataea Y. Yamada, M. Matsuda, K. Maeda, Sakak. & Mikata (1994), Saccharomycopsidaceae. 1, Asia, Eur.

Kommamyce Nieuwl. (1916) ≡ Biscogniauxia (Xylar.).

Kompsoscypha Pfister (1989), Sarcosomataceae. 4, trop.

Kondoa Y. Yamada, Nakagawa & I. Banno (1989), Filobasidiaceae. 1, Antarctic Ocean.

Konenia Hara (1913), Ascomycota (inc. sed.). 1, Japan, USA. See Petrak (*Sydowia* **6**: 358, 1952).

Konradia Racib. (1900), Clavicipitaceae. 1, Asia. See Boedijn (*Ann. Myc.* **33**: 229, 1935).

Kontospora A. Roldán, Honrubia & Marvanová (1990), Mitosporic fungi, 1.C1.19. 1 (aquatic), Spain.

Koorchaloma Subram. (1953), Anamorphic Trichosphaeriaceae, 3.A1.15. Teleomorph *Kananascus.* 5, W. Indies, Japan, India. See Nag Raj (*Mycotaxon* **19**: 167, 1984).

Koorchalomella Chona, Munjal & J.N. Kapoor (1958), Mitosporic fungi, 3.A1.15. 1, India, USA. See Nag Raj (*Mycotaxon* **19**: 167, 1984).

Koordersiella Höhn. (1909), Dothideales (inc. sed.). 3 (on lichens), trop. See Eriksson & Hawksworth (*SA* **6**: 133, 1987). Anamorph *Hansfordiellopsis.*

Koralionastes Kohlm. & Volkm.-Kohlm. (1987), Koralionastetaceae. 5 (marine, trop.), C. Am., Caribbean, Australia, Pacific. See Volkmann-Kohlmeyer & Kohlmeyer (*Mycotaxon* **44**: 417, 1992).

Koralionastetaceae Kohlm. & Volkm.-Kohlm. (1987), Ascomycota (inc. sed.). 1 gen., 5 spp. Stromata absent; ascomata formed from a dark hyphal layer, perithecial, superficial, black, thick-walled, the ostiole papillate, periphysate; peridium composed of pseudoparenchymatous tissue; interascal tissue of simple thin-walled paraphyses; asci cylindrical, thin-walled, without apical structures, evanescent; ascospores hyaline, transversely septate, thick-walled, without a mucous sheath. Anamorphs hyphomycetous, probably spermatial. On marine sponges, Caribbean & Australasia.

Körber (Gustav Wilhelm; 1817-85). Outstanding lichenologist, born in Schlesien, and who studied in Wrocław (Breslau; now Poland) and Berlin; lecturer and later Professor at the University of Wrocław. Made major contributions to the understanding of the nature of lichens (*De gonidiis lichenum*, 1840), and to the systematics of crustose genera using microscopical features. Although his work was often eclipsed by the more prolific Nylander (q.v.), many genera Körber recognized in the 1850s-60s have come into use once more only in the 1980s (e.g. *Lecidella, Porpidia, Pyrrhospora*). Main works: *Systema lichenum germaniae* (1854-55), *Parerga lichenologica* (1859-65). Main collections in Leiden (**L**; incl. 'Typenherbar'), other material in Wrocław (**WRSL**), etc. See Grummann (1974: 23), Stafleu & Cowan (*TL-2* **2**: 603, 1979).

Kordyana Racib. (1900), Exobasidiaceae. 5 (parasites esp. of *Commelinaceae*), trop. See Donk (*Reinwardtia* **4**: 117, 1956), Gäumann (*Ann. Myc.* **20**: 257, 1922).

Kordyanella Höhn. (1904) = Tremellidium (Mitosp. fungi). See Rogers (*Mycol.* **49**: 902, 1957).

Korfia J. Reid & Cain (1963), Hemiphacidiaceae. 1, USA.

Korfiella D.C. Pant & V.P. Tewari (1970), Sarcosomataceae. 1, India.

Korkir Adans. (1763) ? = Ochrolechia p.p. (Pertusar.).

Korunomyces Hodges & F.A. Ferreira (1981), Mitosporic fungi, 1.G1.1. 2, Brazil, Cuba.

Kostermansinda Rifai (1968), Mitosporic fungi, 2.D2.1. 1, Java.

Kostermansindiopsis R.F. Castañeda (1986), Mitosporic fungi, 2.D2.1. 1, Cuba.

Kotlabaea Svrček (1969), Otideaceae. 2, Eur.

Kramabeeja G.V. Rao & K.A. Reddy (1981), Mitosporic fungi, 1.C2.1. 1, India. = Brachysporiella (Mitosp. fungi) fide Sutton (*in litt.*).

Kramasamuha Subram. & Vittal (1973), Mitosporic fungi, 1.C2.11. 1, India, Brazil.

Kravtzevia Shvartsman (1961), Ascomycota (inc. sed.). 1, former USSR.

Kreiseliella Braun (1991), Mitosporic fungi, 6.A/B1.4/?3/10. 1, former USSR.

Krempelhuberia A. Massal. (1854) nom. rej. = Pseudographis (Triblid.).

Kretzschmaria Fr. (1849), Xylariaceae. 15, warmer areas. See Dennis (*Kew Bull.* 1957: 297, 1957, *Bull. Jard. bot. Brux.* **31**: 144, 1961), Narula & Rawla (*Nova Hedw.* **40**: 241, 1986; key 3 spp., Himalayas), Silveira & Rogers (*Acta Amazonia* Suppl. **15**: 7, 1985; key 4 spp. Brazil).

Kretzschmariella Viégas (1944), Xylariaceae. 1, Brazil. See Ju & Rogers (*Mycotaxon* 51: 241, 1994), Læssøe & Spooner (*Kew Bull.* 49: 1, 1994).

Kriegeria Bres. (1891), Platygloeaceae. 1, Eur., N. Am. See Kao (*Mycol.* 48: 288, 1956).

Kriegeria G. Winter ex Höhn. (1914) [non Bres. (1891)] = Rutstroemia (Sclerotin.) fide Dennis (*Mycol. Pap.* 62, 1956).

Kriegeriella Höhn. (1918), Pleosporaceae. 5, Eur., N. Am. See v. Arx & Müller (1975), Barr (*Mycotaxon* 2: 104, 1975; 29: 501, 1987, posn).

Krieglsteinera Pouzar (1987), Platygloeaceae. 1, Eur.

Krispiromyces T.N. Taylor, Hass & W. Remy (1993), Fossil fungi (Paleomastigomycetes). 1 (Devonian), UK.

Krombholzia P. Karst. (1881) [non Rupr. ex E. Fourn. (1876), *Gramineae*] ≡ Leccinum (Bolet.).

Krombholziella Maire (1937) = Leccinum (Bolet.) fide Singer (1945).

Kruphaiomyces Thaxt. (1931), Laboulbeniaceae. 1, Fiji.

Kryptastrina Oberw. (1990), Platygloeaceae. 1, Colombia.

K-selection, adaptation to the long-term colonization of habitats already occupied by other species; in fungi generally involving limited numbers of sexually produced and often long-lived propagules, e.g. large thick-walled ascospores; cf. r-selection (q.v. for *Lit.*).

Kubickia Svrček (1957) ? = Ombrophila (Leot.) fide Korf (*in* Ainsworth *et al.*, *The Fungi* 4A: 249, 1973).

Kubinyia Schulzer (1866) = Mamiania (Vals.) fide Saccardo (*Syll. Fung.* 16: 1263, 1902).

Kuehneola Magnus (1898), Phragmidiaceae. 8 (on *Rosaceae*), temp. *K. uredinis* (syn. *K. albida*) (blackberry (*Rubus*) stem (or yellow) rust).

Kuehneromyces Singer & A.H. Sm. (1946), Strophariaceae. 9, temp. See Mykoholz.

Kuehniella G.F. Orr (1976), Onygenaceae. 1, N. Am. See Currah (*SA* 7: 1, 1988).

Kuettlingeria Trevis. (1857) = Caloplaca (Teloschist.).

Kühn (Julius Gotthelf; 1825-1910). Became Professor of Agriculture, Halle Univ. in 1862. Noted for his important book on plant diseases, *Die Krankheiten der Kulturgewächse, ihre Ursachen und ihre Verhütung* (1858). See Whetzel (*An outline of the history of phytopathology*: 122, 1918, portr.), Wilhelm & Tietz (*Ann. Rev. Phytopath.* 16: 343, 1978), Grummann (1974: 101), Stafleu & Cowan (*TL-2* 2: 682, 1979).

Kuhneria P.A. Dang. (1933) nom. dub. (? 'phycomycetes').

Kulkarniella Gokhale & Patel (1952) = Monosporidium (Phakopsor.) fide Cummins & Hiratsuka (1983).

Kullhemia P. Karst. (1878), Dothideales (inc. sed.). 1, Eur.

Kumanasamuha P. Rag. Rao & D. Rao (1964), Mitosporic fungi, 1.A2.10. 1, India.

kumiss, see Fermented food and drinks.

Kunkelia Arthur (1917), Uredinales (inc. sed.). 1, N. Am. ≡ Gymnoconia (Uredinales) fide Laundon (*Mycotaxon* 3: 135, 1975). See Buriticá (*Rev. Acad. Colomb. Cienc.* 18(69): 131, 1991).

Kuntzeomyces Henn. ex Sacc. & P. Syd. (1899), Tilletiaceae. 1 (on *Rhynchospora*), S. Am. See Ling & Stevenson (*Mycol.* 41: 87, 1949), Molina-Valero (*Caldesia* 13: 49, 1980).

Kupsura Lloyd (1924) = Lysurus (Clathr.) fide Zeller (*Mycol.* 40: 646, 1948).

Kuraishia Y. Yamada, K. Maeda & Mikata (1994), Saccharomycetaceae. 1, Canada.

Kurosawaia Hara (1954) = Sphaceloma (Mitosp. fungi) fide Sutton (1977).

Kurssanovia Pidopl. (1948) = Fusariella (Mitosp. fungi) fide Sutton (*in litt.*).

Kurssanovia Kravtzev (1955) [non Pidopl. (1948)], Ascomycota (inc. sed.). 1, former USSR.

Kurtzmanomyces Y. Yamada, Itoh, H. Kawas., I. Banno & Nakase (1989), Mitosporic fungi, 1.A1.10. 1, Eur. See Yamada *et al.* (*Agric. Biol. Chem.* 53: 2993, 1989; phylogeny).

Kusano (Shunsuke; 1874-1962). At the University of Tokyo (1896-1934), Chief writings: papers on archimycetes, esp. *Olpidium* and *Synchytrium* (*Jap. J. Bot.* 5: 35, 1930, 8: 155, 1936), Grummann (1974: 593).

Kusanoa Henn. (1900) = Uleomyces (Cookell.) fide v. Arx (1963).

Kusanobotrys Henn. (1904), Dothideales (inc. sed.). 1, Japan. See Eriksson & Hawksworth (*SA* 13: 141, 1994; posn).

Kusanoopsis F. Stevens & Weedon (1923) = Uleomyces (Cookell.) fide Thirumalachar & Jenkins (*Mycol.* 45: 781, 1953).

Kusanotheca Bat. & Cif. (1963) = Dysrhynchis (Parodiopsid.) fide v. Arx & Müller (1975).

Kutchiathyrites R.K. Kar (1979), Fossil fungi. 1 (Oligocene), India.

Kutilakesa Subram. (1956) = Sarcopodium (Mitosp. fungi) fide Sutton (1981).

Kutilakesopsis Agnihothr. & G.C.S. Barua (1957) = Sarcopodium (Mitosp. fungi) fide Sutton (1981).

Kuzuhaea R.K. Benj. (1985), Piptocephalidaceae. 1, Japan. See Benjamin (*Bot. J. Linn. Soc.* 91: 117, 1985).

Kweilingia Teng (1940), Uredinales (inc. sed.) fide Cummins & Hiratsuka, 1983) or Ustilaginales (inc. sed.). 1 (on *Bambusa*), China. See Vánky (1987).

Kylindria DiCosmo, S.M. Berch & W.B. Kendr. (1983), Mitosporic fungi, 1.C1.15. 5, widespr. See Rambelli & Onofri (*TBMS* 88: 393, 1987; emend.).

Kymadiscus Kohlm. & E. Kohlm. (1971) = Dactylospora (Dactylospor.) fide Kohlmeyer (*in litt.*).

Kyphomyces I.I. Tav. (1985), Laboulbeniaceae. 14, C. & S. Am., Indonesia.

Kyphophora B. Sutton (1991), Mitosporic fungi, 8.A1.38. 1 (on mangroves), Australia.

L, Onderzoekinstituut Rijksherbarium/Hortus Botanicus (Leiden, Netherlands); founded 1575; part of the University of Leiden; see Anon. (*Meded. Rijks Herb.* 62-69, 1931), Kalkman & Smit (*Blumea* 25: 1, 1979).

Laaseomyces Ruhland (1900) ? = Scopinella (Ceratostomat.). See Benny & Kimbrough (*Mycotaxon* 12: 1, 1980).

labiate, having lips; lip-like.

Labidulomyces, see Labiduromyces.

Labiduromyces Ishik. (1941) = Filariomyces (Laboulben.) fide Tavares (1985).

labium, a lip.

Laboratory safety, see Safety, Laboratory.

Laboulbenia Mont. & C.P. Robin (1853), Laboulbeniaceae. *c.* 546 (on *Insecta, Arachnida*), widespr.

Laboulbeniaceae G. Winter (1886), Laboulbeniales. 116 gen. (+ 30 syn.), 1713 spp. Stroma (thallus) present; ascomata formed directly from successive cells of a lateral appendage of the primary thallus, the lateral appendage not extending beyond the base of the ascoma; outer wall cells of ascoma usually large and unequal; asci 4-spored; ascospores with a submedian septum; usually monoecious.

Laboulbeniales, Ascomycota. 5 fam., 140 gen. (+ 43 syn.), 1855 spp. Stromata usually present, composed of a basal black haustorium and a dark cellular thallus, formed under tight developmental control; ascomata perithecial, often surrounded by complex appendages, translucent, ovoid, thin-walled; interascal tissue absent; asci few, clavate, thin-walled, evanescent, usually 4-spored; ascospores hyaline, elongate, 1-septate. Anamorph hyphomycetous, spermatial, fleeting. Ectoparasites of *Insecta* (few on *Arachnida* and *Diplopoda*), a few coprophilous, cosmop.

The host is only rarely seriously damaged by the fungus, which generally does not extend below the chitin. Tavares (*Mycol. Mem.* 9, 1985; key gens., illustr.) is largely followed for generic and family concepts here, with the addition of *Pyxidiophoraceae*. Fams:

(1) **Ceratomycetaceae**.
(2) **Euceratomycetaceae**.
(3) **Herpomycetaceae**.
(4) **Laboulbeniaceae**.
(5) **Pyxidiophoraceae**.

Lit.: Thaxter (*Mem. Am. Acad. Arts Sci.* **12-16**, 1896-1931 [reprint 1971]; monogr.), Shanor (*Mycol.* **47**: 1, 1955; general review), Scheloske (*Parasitol SchrReihe* **19**, 1969; biology, ecology, systematics), Benjamin (*in* Thaxter reprint, 1971; *in* Ainsworth *et al.* (Eds), *The Fungi* **4A**: 223, 1973), Balazuc (*Bull. mens. Soc. linn. Lyon* **40**: 134, 1971, bibliogr.; **42** (9)-**43** (9), 1973-1974; France), Blackwell (*Mycol.* **86**: 1, 1994; molec. relationships), Lee (*Korean J. Pl. Taxon.* **16**: 89, 1986; fams.), Majewski (*Polish Bot. Stud.* **7**, 1994; 179 spp. Poland, 116 pl.), Rossi & Cesari Rossi (*Giorn. Bot. ital.* **112**: 63, 1978, Italy; *CJB* **57**: 993, 1979, W. Africa), Santamaria (*L'ordre Laboulbeniales ..... a la Península Ibèrica*, 1990; Spain), Santamaria *et al.* (*Trebalb d'Inst. Bot. Barcelona* **14**: 1-123, 1991; checklist, host list, and distribution European spp.), Sugiyama (*Ginkgoana* **2**, 1973; Japan), Tavares (*in* Batra (Ed.), *Insect-fungus symbiosis*, 1979; genera on host groups), Terada (*Trans. mycol. Soc. Japan* **19**: 55, 1978; Taiwan), Tavares (*Laboulbeniales*, 1985 [*Mycol. Mem.* **9**]), Blackwell & Rossi (*Mycotaxon* **25**: 581, 1986; on termites), Huldén (*Karstenia* **23**: 31, 1983, Finland; **25**: 1, 1985, palearctic).

Laboulbeniella Speg. (1912) = Laboulbenia (Laboulben.) fide Thaxter (1914).

Laboulbeniomycetes, see *Laboulbeniales*.

Laboulbeniomycetidae, see *Laboulbeniales*.

Laboulbeniopsis Thaxt. (1920) Ascomycota (inc. sed.). 1 (on termites), W. Indies. See Kimbrough & Gouger (*J. Invert. Path.* **16**: 205, 1970; *Mycol.* **68**: 541, 1976; ultrastructure).

Labrella Fr. (1828) nom. dub. fide Sutton (*Mycol. Pap.* **141**, 1977).

Labridella Brenckle (1929), Anamorphic Amphisphaeriaceae, 4.C2.1. Teleomorph *Griphosphaerioma*. 1, N. Am.

Labridium Vestergr. (1897) ? = Seimatosporium (Mitosp. fungi) fide Sutton (1977).

labriform, lip-shaped; frequently used for terminal soralia of lichens having this shape.

Labrintha Malcolm, Elix & Owe-Larss. (1995), Porpidiaceae (L). 1, NZ.

Labyrinthista, see *Labyrinthulomycota*.

Labyrinthodictyon Valkanov (1969) = Labyrinthula (Labyrinthul.) fide Dick (*in press*).

Labyrinthomorpha, see *Labyrinthomycota*.

Labyrinthomyces Boedijn (1939), Otideaceae. 1 (hypogeous), Australia. See Trappe *et al.* (*Austr. Syst. Bot.* **5**: 597, 1992).

Labyrinthomyxa Duboscq (1921) nom. conf. (Myxomycota) fide Olive (*The mycetozoans*: 215, 1975).

Labyrinthorhiza Chadef. (1956) nom. dub. (Protista, inc. sed.) fide Olive (*The mycetozoans*: 215, 1975).

Labyrinthula Cienk. (1867), Labyrinthulaceae. 10 (in salt and fresh water), N. temp. *L. macrocystis* (wasting disease of eel-grass, *Zostera* spp.). See Porter (*Protoplasma* **61**: 1, 1969; ultrastr., *in* Margulis *et al.* (Eds), 1990; 388), Muehlstein *et al.* (*Mycol.* **82**: 180, 1991), Dick (*in press*; key).

Labyrinthulaceae Cienk. (1867) (Labyrinthulidae), Labyrinthulales. 1 gen. (+ 4 syn.), 11 spp. Trophic phase spindle-shaped cells that glide through channels of an ectoplasmic net. See Dangeard (*Le Botaniste* **24**: 217, 1932).

Labyrinthulales, Labyrinthulomycota. 1 fam., 2 gen. (+ 5 syn.), 12 spp. Fam.:

Labyrinthulomycetes.

Labyrinthuloides F.O. Perkins (1973), Thraustochytriaceae. 4, N. Am.

Labyrinthulomycetes, see *Labyrinthulomycota*.

Labyrinthulomycota (Labyrinthomorpha, Labyrinthista, Labyrinthulia, Labyrinthulea, Labyrinthulomycetes), Chromista; labyrinthids and thraustochytrids. 2 ord., 2 fam., 13 gen. (+ 5 syn.), 42 spp. Trophic stage with an ectoplasmic network, and spindle-shaped or sphaerical cells that move by gliding within the network; uniquely contain bothrosomes; zoospores when present with two flagellae, one with mastigonemes; sexual reproduction known in some spp.

The spp. occur in salt and freshwater, often associated with plants and algal chromists, some as pathogens; culturable. Ords:

(1) **Labyrinthulales** (*Labyrinthulidae*).
(2) **Thraustochytriales**.

Lit.: Barr & Allen (*CJB* **63**: 138, 1985; flagella), Corliss (1994), Karling (*Predominantly holocarpic and eucarpic simple biflagellate phycomycetes*, 1981), Moss (*Bot. J. Linn. Soc.* **91**: 329, 1985), Olive (1975, 1982), Porter (*in* Margulis *et al.*, *Handbook of the Protoctista*: 388, 1990), Sparrow (*in* Ainsworth *et al.*, *The Fungi* **4B**: 61, 1973). See also Myxomycota.

Laccaria Berk. & Broome (1883), Tricholomataceae. 21, cosmop. See Singer (*BSMF* **83**: 104, 1967; *Pl. Syst. Evol.* **126**: 347, 1977), Singer & Moser (*Mycopath. Mycol. appl.* **26**: 191, 1965; key S. Am. spp.), McNabb (*N.Z. Jl Bot.* **10**: 461, 1972; key 12 NZ spp.), Aguirre-Acosta & Perez-Silva (*Bol. Soc. Mex. Mic.* **21**: 33, 1978; key Mexican spp.), Mueller (*Fieldiana* NS **30**, 1992; key N. Am spp., prov. key to world spp.).

laccate, polished; varnished; shining.

Laccocephalum McAlpine & Tepper (1895) = Polyporus (Polypor.) fide Cleland (1935).

Lacellina Sacc. (1913), Mitosporic fungi, 1.A2.13. 4, trop. See Ellis (*Mycol. Pap.* **67**, 1957).

Lacellinopsis Subram. (1953), Mitosporic fungi, 1.A2.3. 4, trop. See Ellis (*Mycol. Pap.* **67**, 1957).

lacerate, as if roughly cut or torn.

Lachnaster Höhn. (1917) = Lachnum (Hyaloscyph.). See Korf (1973).

Lachnea (Fr.) Gillet (1879) [non L. (1753), *Thymeleaceae*] ≡ Scutellinia (Otid.) fide Eckblad (1968). See Eriksson & Hawksworth (*SA* **10**: 40, 1991).

Lachnea Boud. (1885) ≡ Humaria (Otid.).

Lachnella Fr. (1836), Tricholomataceae. 5, cosmop. See Singer (*Flora neotrop.* **17**: 58, 1976).

Lachnella Boud. (1885) = Lachnum (Hyaloscyph.) fide Nannfeldt (1932).

Lachnellula P. Karst. (1884), Hyaloscyphaceae. c. 35 (on conifers), widespr. temp. *L. willkommii* (larch (*Larix*) canker). See Manners (*TBMS* **36**: 362, 1953;

40: 500, 1957; as *Trichoscyphella willkommii*), Dennis (*Persoonia* **2**: 171, 1962), Dharne (*Phytopath. Z.* **53**: 101, 1965; key), Raitviir (1970; key), Baral (*Beitr. Kenntis Pilze Mittelew.* **1**: 143, 1986; key 22 spp., Eur.).

Lachnidium Giard (1891) = Fusarium (Mitosp. fungi) fide Mussat (*Syll. Fung.* **15**: 184, 1901).

Lachnobelonium Höhn. (1926) = Lachnum (Hyaloscyph.) fide Nannfeldt (1932).

Lachnobolus Fr. (1825) nom. conf. (Arcyr.).

Lachnobolus (Fr.) Fr. (1849) = Arcyodes (Myxom.).

Lachnocaulon Clem. & Shear (1931) [non Kunth (1841), *Eriocaulaceae*], Cladoniaceae (L). 1, S. Hemisph.

Lachnocaulum, see *Lachnocaulon*.

Lachnocladiaceae D.A. Reid (1965), Lachnocladiales. 6 gen. (+ 3 syn.), 107 spp. Basidioma erect, ramarioid to pileate.

 Lit.: Hallenberg (*Lachnocladiaceae and Coniophoraceae of North Europe*, 1985).

Lachnocladiales, Basidiomycetes. 2 fam., 7 gen. (+ 3 syn.), 133 spp. Fams:
(1) **Dichostereaceae**.
(2) **Lachnocladiaceae**.
 Lit.: Pegler & Young (*Kew Bull.* **48**: 37, 1993; basidiome structure).

Lachnocladium Lév. (1846) nom. cons., Lachnocladiaceae. *c.* 100, widespr. See Corner (1950: 416), Hallenberg (*Lachnocladiaceae and Coniophoraceae of north Europe*, 1985).

Lachnodochium Marchal (1895), Mitosporic fungi, 3.A1.10. 1, Eur.

Lachnum Retz. (1779), Hyaloscyphaceae. 150, widespr. See Eriksson & Hawksworth (*SA* **10**: 40, 1991), Haines (*Nova Hedw.* **54**: 97, 1992; key 13 spp. Guyana); Haines & Dumont (*Mycotaxon* **19**: 1, 1984; key 9 long-spored spp.), Raitviir (*Eesti NSV Tead. Akad. Toim., Biol.* **36**: 313, 1987; concept), Spooner (*Bibl. Mycol.* **116**, 1987; 37 spp. Australasia). See also *Dasyscyphus*.

lacinia, a delicate branch of a foliose lichen thallus having an anatomical structure typical of foliose lichens.

laciniate (of an edge, etc.), as if cut into delicate bands (Fig. 37.47).

Laciniocladium Petri (1917), Mitosporic fungi, 1.A1.?15. 1, Eur.

Lacrimasporonites R.T. Clarke (1965), Fossil fungi (*f. cat.*). 9 (Cretaceous, Paleocene, Tertiary), China, Colombia, India, USA.

lacrimiform (lacrimoid), like a tear drop.

Lacrymaria Pat. (1887), Coprinaceae. 10, cosmop. = Psathyrella (Coprin.) fide Smith (1972).

Lacrymospora Aptroot (1991), ? Pyrenulaceae. 1 (on lichen), Madagascar.

Lactarelis Earle (1909) = Russula (Russul.) fide Singer (1951).

Lactaria Pers. (1797) [non Rafin. (1838), *Apocynaceae*] ≡ Lactarius (Russul.).

Lactariella J. Schröt. (1889) = Lactarius (Russul.) fide Singer (1951).

Lactariopsis Henn. (1901) = Lactarius (Russul.) fide Singer (1951).

Lactarius Pers. (1797), Russulaceae. 120, esp. temp. The milk-caps, *L. deliciosus* (saffron milk cap) and other spp. are edible. See Pearson (*Naturalist* **81**, 1950; key Br. spp.), Neuhoff (*Die Milchlinge*, 1956), Heinemann (*Naturalistes Belges* **41**: 133, 1960), Blum (*Les lactaires*, 1976), Bon (*Docums Mycol.* **10**(40); key 164 spp.), Smith & Hesler (*Brittonia* **12**: 119, 306; **14**: 369, 1960-62; *Lactarius of North America*, 1979), Heim (*Prodr. Fl. Mycol. Madag.* **1**: 31, 1937; keys 12 spp., Madagascar, *Bull. Jard. Bot. Brux.* **25**: 1, 1955, *Fl. Icon. Champ. Congo* **4**: 83,

1955; Congo and West Africa spp.), Pegler & Fiard (*Kew Bull.* **33**: 601, 1979; keys 11 L. Antilles spp.), McNabb (*N.Z. Jl Bot.* **9**: 46, 1971; key NZ spp.), Kytövuori (*Karstenia* **24**: 41, 1984; sect. *Scrobiculati* in Eur.).

Lactella Maessen (1955) = Enterobryus (Eccrin.) fide Manier & Lichtwardt (*Ann. Sci. nat., Bot. sér.* 12 **9**: 519, 1968).

lacteous, like milk.

lactescent, becoming like milk.

lactifer, a latex-carrying hypha (Singer, 1960: 33).

lactiferous, having a milk-like juice.

lactiferous hypha, see hypha.

Lactifluus (Pers.) Roussel (1806) ≡ Lactarius (Russul.).

Lactocollybia Singer (1939), Tricholomataceae. 7, trop. See Singer (*Flora neotrop.* **3**: 55, 1970).

Lactomyces Boul. (1899) ? = Mucor (Mucor.) or Rhizopus (Mucor.) fide Foster (*Chemical activities of fungi*, 1949).

Lactydina Subram. (1978), Mitosporic fungi, 1.C1.1. 1 (on amoebae), USA.

lacuna, a hole or hollow.

lacunose, having lacunae.

Lacunospora Cailleux (1969) ? = Apiosordaria (Lasiosphaer.).

Lacustromyces Longcore (1993), Cladochytriaceae. 1, USA.

Ladrococcus Locq. (1985), Fossil fungi. 1, Sahara.

Laestadia Auersw. (1869) [non Kunth ex Less. (1832), *Compositae*] = Plagiostoma (Vals.).

Laestadiella Höhn. (1918) = Guignardia (Mycosphaerell.) fide v. Arx & Müller (1954).

Laestadites Mesch. (1892), Fossil fungi. 1 (Pliocene, Pleistocene), Japan.

Laeticorticium Donk (1956) ≡ Corticium (Cortic.) fide Boidin & Lanquetin (*BMSF* **99**: 269, 1983).

Laetinaevia Nannf. (1932) nom. cons., Dermateaceae. 13, Eur. See Hein (*Willdenowia Beih.* **9**, 1976; key).

Laetiporus Murrill (1904), Coriolaceae. 4, widespr. *L. sulphureus* on oak. See Corner (*Beih. Nova Hedw.* **78**: 181, 1984).

Laetisaria Burds. (1979), Corticiaceae. 1, widespr.

laevigate, smooth (Fig. 29.1).

Laeviomeliola Bat. (1960), Meliolaceae. 1, Brazil.

Laeviomyces D. Hawksw. (1981), Mitosporic fungi, 4.A2.19. 2 (on lichens), Eur., N. Am.

Lagarobasidium Jülich (1974) = Hyphodontia (Hyphodermat.) fide Langer (*Bibl. Mycol.* **154**, 1994).

Lagena Vanterp. & Ledingham (1930), Lagenaceae. 1 (on wheat roots), N. Am. See Dick (*in press*; key). The correct name for this genus when the *Chromista* are included in the *Protozoa* is *Lagenocystis*.

Lagenaceae M.W. Dick (1994), Oomycota (inc. sed.). 3 gen. (+ 1 syn.), 4 spp.

Lagenidiaceae, see *Pythiaceae*.

Lagenidiales, see *Pythiales*.

Lagenidicopsis N.J. Artemczuk (1972) nom. dub. (? Oomycota).

Lagenidiopsis De Wild. (1896) = Cystosiphon (Pyth.).

Lagenidium Schenk (1857), Pythiaceae. 2, widespr. Dick (*in press*; key).

lageniform, swollen at the base, narrowed at the top; like a Florence flask (Fig. 37.28).

Lageniformia Plunkett (1925) ? = Eutypa (Diatryp.) fide Petrak (*Sydowia* **5**: 169, 1951).

Lagenisma Drebes (1968), Lagenismaceae. 1 (marine on *Coscinodiscus*), N. Hemisph. See Schnepf & Drebes (*Helgol. wiss Meersunters.* **29**: 291, 1977; development).

Lagenismaceae M.W. Dick (1995), Lagenismales. 1 gen., 1 sp.

Lagenismales, Oomycota. 1 fam., 1 gen., 1 sp. Fam.: **Lagenismaceae**.

Lagenocystis H.F. Copel. (1956) ≡ Lagena (Lagen.).

Lagenomyces Cavalc. & A.A. Silva (1972), Mitosporic fungi (L), 8.A1,?. 1, Brazil.

Lagenula G. Arnaud (1930) [non Lour. (1790), *Vitaceae*] = Caliciopsis (Corynel.) fide Fitzpatrick (*Mycol.* **34**: 464, 1942).

Lagenulopsis Fitzp. (1942), Coryneliaceae. 1 (on *Podocarpus*), C. Am., S. Afr., W. Indies, Fiji. See Benny *et al.* (*Bot. Gaz.* **146**: 431, 1985).

lager, see beer, brewing.

Lagerbergia J. Reid (1971) = Ascocalyx (Leot.) fide Müller & Dorworth (*Sydowia* **36**: 193, 1983).

Lagerheima Sacc. (1892), Leotiaceae. 3 or 4, widespr. See Gamundí (*Sydowia* **34**: 82, 1981).

Lagerheimina Kuntze (1891) = Diploschistes (Thelotremat.).

lagynocarpus ascomycetes, Pyrenomycetes (Moreau, 1953).

Lagynodella Petr. (1922) = Cryptosporiopsis (Mitosp. fungi) fide Wollenweber (*Arb. Biol. Reichsans. Berl.* **22**: 521, 1938).

Lahmia Körb. (1861), Lahmiaceae. 1, Eur., N. Am. See Eriksson (*Mycotaxon* **27**: 347, 1986).

Lahmiaceae O.E. Erikss. (1986), Lahmiales. 1 gen. (2 syn.), 1 sp. Ascomata cleistothecial, scattered, turbinate, shortly stipitate, black, uniloculate, opening by irregular splits in the upper part of the wall; peridium hyphal, gelatinized; interascal tissue of trabecular pseudoparaphyses, secondarily producing paraphysis-like elements in the upper part of the locule; asci clavate, long-stalked, probably with separate wall layers but not fissitunicate, without distinct apical structures, not blueing in iodine; ascospores hyaline, falcate, transversely septate, without a sheath.

Lahmiales, Ascomycota. 1 fam., 1 gen. (+ 2 syn.), 1 sp. Fam.:
Lahmiaceae.

Lahmiomyces Cif. & Tomas. (1953) Patellariaceae. 1, Italy. See Eriksson (*Mycotaxon* **27**: 347, 1986; posn).

Lajassiella Tuzet & Manier ex Manier (1968), Parataeniellaceae. 1 (in *Coleoptera*), France. See Lichtwardt (1986).

Lalaria R.T. Moore (1990), Anamorphic Taphrinaceae, 1.A1.10. Teleomorph *Taphrina*. 23, widespr. Yeast phase anamorphs.

Lambdasporium Matsush. (1971), Mitosporic fungi, 1.G1.10. 2, S.E. Asia.

Lambertella Höhn. (1918), Sclerotiniaceae. 47, widespr. See Dumont (*Mem. N.Y. bot. Gdn* **33**(1), 1971; monogr.), Korf & Zhuang (*Mycotaxon* **24**: 361, 1985; key, *Mycotaxon* **39**: 477, 1990; 7 spp. China).

Lambertellinia Korf & Lizoň (1993), Sclerotiniaceae. 1, Japan. Anamorph *Idriella*.

Lambiella Hertel (1984) = Rimularia (Rimular.) fide Hertel (*Mitt. bot. StSamml., München* **23**: 321, 1987).

Lambottiella (Sacc.) Sacc. (1913) = Lophiostoma (Lophiostomat.) fide v. Arx & Müller (1975).

Lambro Racib. (1900), Valsaceae. 1, Indonesia. See Müller & v. Arx (1962), Monod (1983).

lamella (pl. **lamellae**) (of an agaric), one of the characteristic hymenium-covered vertical plates on the underside of the pileus (Fig. 4B); gill.

lamellate, (1) having lamellae; (2) made up of thin plates.

Lamelloporus Ryvarden (1987), Coriolaceae. 1, trop. Am.

lamellula (pl. **lamellulae**), a small lamella which runs from the edge of the pileus towards the stipe, as in *Russula*.

Lamia Nowak. (1884) = Entomophthora (Entomophthor.) fide Remaudière & Keller (1980).

lamina (pl. **laminae**), (1) blade; (2) the main part of the thallus in foliose lichens; (3) epithecium + hymenium

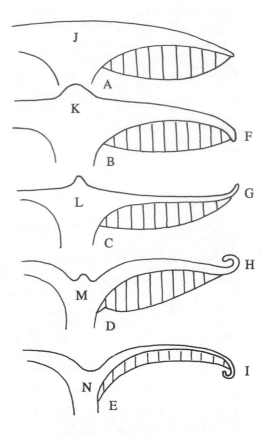

Fig. 19. A-E. Insertion of lamellae (gills). A, free; B, adnexed; C,adnate; D, sinuate; E, decurrent. F-I. Margin of pileus (cap). F, inflexed; G, reflexed; H, revolute; I, involute. J-N. Pileus. J, plano-convex; K, bullate; L, umbonate; M, umbilicate; N, depressed.

+ subhymenium in an apothecium (Hertel, *Beih. Nova Hedw.* **24**, 1967); (4) (of leaves) the flat surface; **-l**, on the lamina.

Lamproconium (Grove) Grove (1937), Mitosporic fungi, A2.19. 1, Eur.

lamprocystidium, see cystidium.

Lamproderma Rostaf. (1873), Stemonitidaceae. 28, cosmop. See Kowalski (*Mycol.* **62**: 623, 1970; key).

Lamprodermataceae, see *Stemonitidaceae*.

Lamprodermopsis Meyl. (1910) = Dianema (Dianemat.).

Lamprospora De Not. (1863), Otideaceae. *c.* 40, N. temp. See Schumacher (*Sydowia* **45**: 307, 1993; key 13 arctic-alpine spp.), Caillet & Moyne (*Bull. Soc. Hist. nat. Doubs* **84**: 9, 1991; keys), Benkert (*Z. Mykol.* **53**: 195, 1987; 29 spp.).

Lampteromyces Singer (1947), Tricholomataceae. 1 (poisonous), Japan.

Lamyella Fr. (1849) = Cytospora (Mitosp. fungi) fide Clements & Shear (1931).

Lamyella Berl. (1899) ≡ Neolamya (Ascomycota, inc. sed.).

Lamyxis Raf. (1820) nom. dub. (? Boletales). See Donk (*Persoonia* **1**: 173, 1960).

lanate, like wool; covered with short hair-like processes.

Lange (Jakob Emanuel; 1864-1941). Danish agaricologist, author of *Flora agaricina Danica*, 5 vols (1935-

40; preceded by 'Studies on the agarics of Denmark', I-XII, *Dansk Bot. Arkiv*, 1914-38; see M. Lange, *Friesia* 9: 121, 1969), was a teacher at the Agricultural Folk High School, Dalum (1888-1918), then (-1934), principal, Smallholders Agricultural School, Odense. *Obit.*: Buchwald (*Bot. Tidskr.* 46: 81, 1942; portr.), Pearson (*Mycol.* 39: 1, 1947; portr.). See also Stafleu & Cowan (*TL-2* 2: 745, 1979).

Langermannia Rostk. (1839), Lycoperdaceae. 3, widespr. The basidioma of *L. gigantea* (giant puffball) may be 120 cm across and have 10^{13} spores (see record fungi). Nom. rej. vs. *Calvatia* s.l. (q.v.) but used for segregate of it.

Langeron (Maurice Charles Pierre; 1874-1950). Doctor of medicine, botanist, parasitologist, and mycologist, who worked at the Institut de Parasitologie, Paris, from 1903 until his death, made many important contributions to medical and general mycology. His two major publications were *Précis de microscopie* (1913; edn 7, 1949) and *Précis de mycologie* (1945; edn 2 (by Vanbreuseghem), 1952). *Obit.*: Vanbreuseghem (*Mycopath.* 6: 58, 1951; portr., bibl.). See Stafleu & Cowan (*TL-2* 2: 750, 1979).

Langeronia Vanbreus. (1950) = Trichophyton (Mitosp. fungi) fide Ajello (1968).

Langloisula Ellis & Everh. (1889) nom. conf. fide Rogers & Jackson (*Farlowia* 1: 263, 1943) and Donk (*Fungus* 26: 3, 1956).

languid, feeble; hanging down.

Lanolea Nieuwl. (1916) ≡ Nolanea (Entolomat.).

Lanomyces Gäum. (1922) = Cystotheca (Erysiph.) fide Katumoto (1973).

Lanopila Fr. (1848) = Langermannia (Lycoperd.) fide Demoulin & Dring (*Bull. Jard. bot. nat. Belg.* 45: 361, 1975).

Lanosa Fr. (1825) nom. dub. (? Fungi, inc. sed.).

lanose, see lanate.

Lanspora K.D. Hyde & E.B.G. Jones (1986), Halosphaeriaceae. 1 (marine), Seychelles.

lanuginose, see lanate; see also nematogenous.

Lanzia Sacc. (1884) = Rutstroemia (Sclerotin.) fide Baral (*SA* 13: 113, 1994). See Sharma & Sharda (*Int. J. Mycol. Lichen.* 2: 95, 1985; key 15 spp., India), Spooner (*Bibl. Mycol.* 116, 1987; 7 spp. Australasia).

lao-chao, see Fermented food and drinks.

Lappodochium Matsush. (1975), Mitosporic fungi, 3.A1.10. 1, Brazil.

lapsus (Lat.), a slip; -calami, a slip of the pen.

Laquearia Fr. (1849), ? Rhytismatales (inc. sed.). 1, Eur.

Larger fungi, see Macromycetes.

largest fungi, see longevity, record fungi.

Laricifomes Kotl. & Pouzar (1957) ≡ Agaricum (Coriol.). See Donk (1971).

Laricina Velen. (1934), ? Leotiales (inc. sed.). 1, former Czechoslovakia.

Laridospora Nawawi (1976), Mitosporic fungi, 1.C1.1. 1 (aquatic), widespr.

Larseniella Munk (1942) = Neorehmia (Ascomycota inc. sed.) fide Barr (*Mycotaxon* 39: 43, 1990).

Lasallia Mérat (1821), Umbilicariaceae (L). 12, cosmop. (mainly temp.). See Blum (*Dokl. Akad. Nauk. Ukr. SSR*, B, 12: 58, 1986; DNA homology, status), Dombrovskaya (*Bot. Zh. SSSR* 63: 233, 1978), Krog & Swinscow (*Nordic Jl Bot.* 6: 75, 1986; E. Afr.), Llano (*A monograph of the lichen family Umbilicariaceae in the Western Hemisphere*, 1950; *Hvalrôd Skr.* 48: 112, 1965), Posner *et al.* (*Z. Naturf.* 46c: 19, 1991; chemotax.), Sancho & Balaquer (*An. Jard. Bot. Madrid* 46: 273, 1989; anatomy 4 spp.), Wei (*Proc. First Korea-China Jt Sem. Mycol.*: 5, 1994; molec. syst.), Wei & Jiang (1988, key 8 spp. China; 1993, key 10 spp. Asia).

Laschia Fr. (1830) nom. rej. = Campanella (Tricholomat.) fide Donk (1960

Laschia Jungh. (1838) ≡ Junghuhnia (Steccherin.).

Lascoderma, see Lasioderma.

Lasiella Quél. (1875) = Lasiosphaeria (Lasiosphaer.).

Lasiobelonis Clem. & Shear (1931) = Lasiobelonium (Hyaloscyph.) fide Spooner (1987).

Lasiobelonium Ellis & Everh. (1897), Hyaloscyphaceae. 3, Eur., Australasia. See Baral (*SA* 13: 113, 1994), Korf (*Mycotaxon* 7: 399, 1978), Raitviir (*Scripta Mycol.* 9, 1980; key), Spooner (*Bibl. Mycol.* 116, 1987; 3 spp. Australasia, concept).

Lasiobelonium (Sacc.) Sacc. & P. Syd. (1899) ≡ Belonidium (Hyaloscyph.) fide Korf (*Mycotaxon* 7: 399, 1978).

Lasiobertia Sivan. (1978), Hyponectriaceae. 1, Ghana. See Hyde (*Sydowia* 45: 204, 1993; posn).

Lasiobolidium Malloch & Cain (1971), Eoterfeziaceae. 5 (coprophilous), N. Afr., N. Am. See van Brummelen (*in* Hawksworth (Ed.), *Ascomycete systematics*: 400, 1994; posn), Loquin-Linard (*Cryptogamie, Mycol.* 4: 283, 1983).

Lasiobolus Sacc. (1884), Thelebolaceae. 11, temp. See Bezerra & Kimbrough (*CJB* 53: 1206, 1975; key), van Brummelen (*in* Hawksworth (Ed.), *Ascomycete systematics*: 400, 1994; posn).

Lasiobotrys Kunze (1823), Venturiaceae. 2 or 3 (on *Lonicera*), temp. See Müller (*Adv. Front. Myc. Pl. Path.* 11, 1981).

Lasiochlaena Pouzar (1990), Coriolaceae. 1, Eur. = Ischnoderma fide Ryvarden (1991).

Lasioderma Mont. (1845) ? = Phleogena (Phloegen.).

Lasiodiplodia Ellis & Everh. (1896), Mitosporic fungi, 8.B2.1. 1, widespr. *L. theobromae* (syn. *Botryodiplodia theobromae*), a common trop. saprobe and wound parasite. See Punithalingam (*CMI Descript.* 519, 1976; *Bibl. Mycol.* 71, 1980), Robell & Forester (*Sabouraudia* 14: 155, 1976; keratomycosis in humans), Yaguchi & Nakamura (*Jl Agr. Sci.* 35: 282, 1991; TEM of conidial wall).

Lasiodiplodiella Zambett. (1955), Mitosporic fungi, 1.B2.?. 3, trop.

Lasioloma R. Sant. (1952), Ectolechiaceae (L). 4, trop.

Lasionectria (Sacc.) Cooke (1884) = Nectria (Hypocr.) fide Rogerson (1970).

Lasiophoma Naumov (1916) = Pyrenochaeta (Mitosp. fungi) fide Sutton (1977).

Lasiophoma Speg. (1918) [non Naumov (1916)], Mitosporic fungi, 4.A1.?. 5, widespr. See Sutton (1977).

Lasiosordaria Chenant. (1919) = Cercophora (Lasiosphaer.) fide Lundqvist (1972).

Lasiosordariella Chenant. (1919) = Lasiosphaeria (Lasiosphaer.).

Lasiosordariopsis Chenant. (1919) = Cercophora (Lasiosphaer.) fide Lundqvist (1972).

Lasiosphaera Reichardt (1870) = Langermannia (Lycoperd.) fide Demoulin & Dring (*Bull. Jard. bot. nat. Belg.* 45: 362, 1975).

Lasiosphaeria Ces. & De Not. (1863), Lasiosphaeriaceae. 40, widespr. *L. pezizula* (greyish-olive stain of hardwoods). See Hubert (*Phytopathology* 11: 214, 1921), Lundqvist (1972; nomencl.), Hilber & Hilber (*Sydowia* 36: 105, 1983).

Lasiosphaeriaceae Nannf. (1932), Sordariales. 48 gen. (+ 43 syn.), 303 spp. Ascomata dark, ostiolate or not; interascal tissue often present (of paraphyses) but inconspicuous; asci cylindric-clavate to clavate, usually with an I- apical ring; ascospores variable, usually with at least one dark brown and one hyaline cell, occasionally ornamented, with gelatinous appendages or caudae (often long), normally lacking a sheath. Anamorphs varied.

Barr (*Mycotaxon* **39**: 43, 1990) separated the *Tripterosporaceae* from the fam., adding also some further genera from other fams.

The classification of this group is in some confusion, and genera which have been recently included (e.g. *Apiospora*, *Chaetosphaeria*) may not belong here.

Lit.: Cain (1934), Lundqvist (1972), Moreau (1953).

Lasiosphaeriella Sivan. (1975), Lasiosphaeriaceae. 2, Afr., Pacific.

Lasiosphaeriopsis D. Hawksw. & Sivan. (1980), Nitschkiaceae. 3 (on lichens), widespr. See Alstrup & Hawksworth (*Meddr. Grønl., Biosci.* **31**, 1990), Eriksson & Santesson (*Mycotaxon* **25**: 569, 1986).

Lasiosphaeris Clem. (1909) = Lasiosphaeria (Lasiosphaer.).

Lasiostemma Theiss., Syd. & P. Syd. (1917), Pseudoperisporiaceae. *c.* 15, trop. See Farr (*Mycol.* **71**: 243, 1979; sects.).

Lasiostemmella Petr. (1950) = Epipolaeum (Pseudoperispor.) fide Müller & v. Arx (1962).

Lasiostictella Sherwood (1986), Rhytismatales (inc. sed.). 1, France.

Lasiostictis (Sacc. & Berl.) Sacc. (1889) = Naemacyclus (Rhytismat.) fide Di Cosmo *et al.* (1983).

Lasiostroma Griffon & Maubl. (1911) = Phomopsis (Mitosp. fungi) fide Sutton (1977).

Lasiothelebolus Kimbr. & Luck-Allen (1974) = Thelebolus (Thelebol.) fide Lundqvist (*SA* **7**: 77, 1988).

Lasiothyrium Syd. & P. Syd. (1913), Mitosporic fungi, 5.E2.?. 1, Philipp.

Lasmenia Speg. (1886), Mitosporic fungi, 8.A2.1. 5, S. Am., Philipp.

Lasmeniella Petr. & Syd. (1927), Mitosporic fungi, 8.A2.19. 13, esp. trop.

Lasseria Dennis (1960), ? Leotiales (inc. sed.). 1, Venezuela.

lateral, at the side.

Latericonis G.V. Rao, K.A. Reddy & de Hoog (1984), Mitosporic fungi, 3.D2.10. 1, India.

Lateriramulosa Matsush. (1971), Mitosporic fungi, 1.G1.10. 1, New Britain.

Laterispora Uecker, W.A. Ayers & P.B. Adams (1982), Mitosporic fungi, 1.E2.1. 1 (on sclerotia), USA.

lateritiin, see enniatin A.

Laternea Turpin (1822), Clathraceae. 2, trop. Am.

Lateropeltis Shanor (1946) = Kiehlia (Parmular.) fide Farr (1968).

Laterotheca Bat. (1963) = Acrogenotheca (Dothideales) p.p. and Antennariella (Dothideales) p.p. fide Hughes (*Mycol.* **68**: 693, 1976).

Laterradea Raspail (1824) nom. dub. (Lycoperdales).

latex, a milk-like juice, as in *Lactarius*.

Lathagrium (Ach.) Gray (1821) = Collema (Collemat.) fide Degelius (1954).

Lathraeodiscus Dissing & Sivertsen (1989), Otideaceae. 1, arctic.

Latin. Acquaintance with Latin is an asset to any biologist and a working knowledge of the language or its rudiments are particularly useful to systematists because scientific names of all taxa are latinized. Latin descriptions or diagnoses are currently required under the International Code of Botanical Nomenclature (see Nomenclature) to validate the names of fungi at all formal ranks, and many early works on systematic mycology are in Latin, as are the descriptions in Saccardo's *Sylloge Fungorum*. Botanical Latin is derived from the classical language with liberal borrowings from Greek, but with simplified syntactical requirements. The pronunciation and stress is controversial, but guidance can usefully be found at the beginning of Latin dictionaries or in Allen (Ed.) (*Vox Latina*, 1978).

Stearn (*Botanical Latin*, 1966 [edn 4, 1992]) provides an excellent introduction to grammar, syntax, and vocabulary and gives much helpful guidance together with examples of mycological descriptions. This work may be supplemented by Cash (*A mycological English-Latin glossary*, 1965 [*Mycol. Mem.* **1**]), Baranov (*Botanical Latin for plant taxonomists*, 1968 [reprint 1971]), Borror (*Dictionary of word roots and combining forms*, 1960 [11th print, 1971]), and there is a useful Latin/English glossary in Clements & Shear (*The genera of fungi*, 1931). Nybakken (*Greek and Latin in scientific terminology*, 1960), gives much information on the derivation of scientific and medical terms.

Further articles which may be consulted for guidance on the correct formulation of Latin names are Zabinkova (*Taxon* **17**: 19, 1968; generic stems ending in -is), Nicholson (*Taxon* **43**: 97, 1994; gender of generic names, particularly those ending in -ma), Manara (*Taxon* **41**: 52, 1992; geographical epithets), Manara (*Taxon* **40**: 301, 1991; gender of generic names), Nicholson & Brooks (*Taxon* **23**: 163, 1974; orthography, stems and compounds).

See Classification, Colour, Nomenclature.

Latrostium Zopf (1894), Rhizidiomycetaceae. 1, Eur.

latticed, cross-barred; like a network.

lattice-work fungus, basidioma of *Clathrus* spp.

Latzeiia, see *Latzelia*.

Latzelia Zahlbr. (1926) = Epiphloea (Hepp.) fide Gyelnik (1940).

Latzinaea Kuntze (1891) ≡ Nolanea (Entolomat.).

Laudatea Johow (1884) = Dictyonema (Thelephor.) fide Parmasto (1978).

Lauderlindsaya J.C. David & D. Hawksw. (1989), Verrucariaceae (±L). 3 (on lichens or lichenized), widespr. See Diederich & Sérusiaux (*Lichenologist* **25**: 97, 1993). = Normandina (Ascom.) fide Aptroot (*Willdenowia* **21**: 263, 1991).

Laurera Rchb. (1841) nom. cons. prop., Trypetheliaceae (L). 35, trop. See Harris (*Acta Amazon.* Suppl. **14**: 55, 1986; gen. concept., key 4 spp. Brazil), Letrouit-Galinou (*Revue bryol. lichén.* **26**: 207, 1957, **27**: 66, 1958), Makhija & Patwardham (*Mycotaxon* **31**: 565, 1988; key 17 spp. India), Upreti & Singh (*Bull. Jard. bot. nat. Belg.* **57**: 367, 1987; key 10 spp. India).

Laureriella Hepp (1867) = Glypholecia (Acarospor.).

Laureromyces Cif. & Tomas. (1953) = Laurera (Trypethel.).

Laurilia Pouzar (1959), Stereaceae. 2, Eur. See Eriksson & Ryvarden (*Corticiaceae of north Europe* **4**: 787, 1976), Chamuris (*Mycol. Mem.* **14**, 1988; key). = Echinodontium (Echinodont.) fide Gross (1964).

Lauriomyces R.F. Castañeda (1990), Mitosporic fungi, 1.A1.3. 4, widespr.

Laurobasidium Jülich (1982), Exobasidiaceae. 1, Eur.

Lauterbachiella Henn. (1898) = Rhagadolobium (Parmular.) fide v. Arx & Müller (1975).

Lautisporopsis E.B.G. Jones, Yusoff & S.T. Moss (1994), Halosphaeriaceae. 1 (marine), Atlantic & Pacific Oceans.

Lautitia S. Schatz (1984), ? Phaeosphaeriaceae. 1 (on marine algae), USA.

Lautospora K.D. Hyde & E.B.G. Jones (1989), Lautosporaceae. 2, Brunei, N. Am. See Kohlmeyer *et al.* (*Bot. Mar.* **38**: 165, 1995).

Lautosporaceae Kohlm., Volkm.-Kohlm. & O.E. Erikss. (1995), Dothideales. 1 gen., 2 spp. Ascomata perithecioid, immersed, elongate, with an elongated neck; interascal tissue septate pseudoparaphyses; asci thick-walled with a large ocular chamber; ascospores muriform, with numerous A-trans-septa, but no euseptum, all septa develop simultaneously and form a mesh-like pattern. On plant material in marine and maritime situations.

Lawalreea Dieder. (1990), Mitosporic fungi, 4.A1.15. 1 (on lichens), Luxemburg.

Laxitextum Lentz (1956), Gloeocystidiellaceae. 3, widespr. See Lentz (*Sydowia* **14**: 123, 1960), Chamuris (*Mycol. Mem.* **14**, 1988; key), Ginns & Freeman (*Bibl. Mycol.* **157**, 1994). = Echinodontium (Echinodont.) fide Gross (1964).

Lazarenkoa Zerova (1938), Dothideales (inc. sed.). 1 (on *Selaginella*), Siberia.

Lazaroa Gonz. Frag. [not traced] = Phellinus (Hymenochaet.) fide Donk (1974).

Lazuardia Rifai (1988), Otideaceae. 1, W. Indies, Sri Lanka, Indonesia.

Lazulinospora Burds. & M.J. Larsen (1974), Corticiaceae. 2, N. Am., UK.

LCP, see PC.

LD, lethal dose; **LD$_{50}$** (of the concentration of a fungicide, etc.), that which kills 50% of the spores (cells or individuals) of the test organism. Cf. ED.

LE. Komarov Botanical Institute (St Petersburg, Russia); founded 1714; directed by the Academy of Sciences.

leaf curl, (1) of peach and almond (*Taphrina deformans*); (2) of cherry (*T. minor*).

Leaia Banker (1906) = Gloiodon (Gloeocystidiell.) fide Banker (1910).

Leandria Rangel (1915), Mitosporic fungi, 1.D2.1. 1, S. Am.

Leangium Link (1809) = Diderma (Didym.).

leather fungi, members of the *Thelephoraceae* s. lat.

Lecanactidaceae, see *Opegraphaceae*.

Lecanactiomyces Cif. & Tomas. (1953) = Lecanactis (Roccell.).

Lecanactis Eschw. (1824) nom. rej. = Phaeographis (Graphid.) fide Tehler (*Taxon* **35**: 382, 1986).

Lecanactis Körb. (1855) nom. cons., Roccellaceae (L). c. 20, cosmop. See Egea et al. (*Pl. Syst. Evol.* **187**: 103, 1993; *L. grymulosa* group), Tehler (*Taxon* **35**: 382, 1986; nomencl., *Willdenowia* **22**: 201, 1992), Torrente & Egea (1989).

Lecanephebe Frey (1929) = Zahlbrucknerella (Lichin.) fide Henssen (1963).

Lecania A. Massal. (1853), Bacidiaceae (L). c. 40, cosmop. (mainly temp.). See Mayrhofer (*Bibl. Lich.* **28**, 1988; 19 spp. on rock, Eur.), van den Boon (*Nova Hedw.* **54**: 229, 1992; key 11 spp. on rock, Netherlands).

Lecaniascus Moniez (1887) ? Fungi (inc. sed.).

Lecanidiaceae, see *Patellariaceae*.

Lecanidiales, see *Patellariales*.

Lecanidiella Sherwood (1986), Patellariaceae. 1, USA.

Lecanidion Endl. (1830) ≡ Patellaria (Patell.). See Hawksworth (1986).

Lecanidium A. Massal. (1856) = Pertusaria (Pertusar.).

Lecaniella Jatta (1889) = Lecania (Bacid.).

Lecaniella Vain. (1896) ≡ Gonolecania (Pilocarp.).

Lecaniocola Brain (1923), Mitosporic fungi, ?.A1.10. 15 (in scale insects), trop.

Lecaniomyces E.A. Thomas (1939) nom. inval. = Lecania (Bacid.).

Lecaniopsis (Vain.) Zahlbr. (1926) = Dimerella (Gyalect.).

Lecanocaulon Nyl. (1860) = Stereocaulon (Stereocaul.).

Lecanora Ach. (1810), Lecanoraceae (L). c. 300, cosmop. (mainly temp.). See Brodo (*Beih. Nova Hedw.* **79**: 63, 1984; key 38 spp., N. Am. *subfusca*-group), Eigler (*Diss. Bot., Lehre* **4**, 1969; gen. concept), Imshaug & Brodo (*Nova Hedw.* **12**: 1, 1966; *L. pallida*

group), Leuckert & Poelt (*Nova Hedw.* **49**: 121, 1989; *L. rupicola*-group), Lumbsch (*J. Hattori bot. Lab.* **77**: 1, 1994; key 46 spp. Australasia), Nimis & Bolognini (*Not. Soc. Lich. Ital.* **6**: 29, 1993; key spp. Italy), Poelt (*Mitt. bot. StSamml., Münch.* **2**: 411, 1958), Poelt & Grube (*Nova Hedw.* **57**: 305, 1993; key 22 spp. subgen. *Placodium* in Himalaya).

Lecanoraceae Körb. (1855), Lecanorales (L). 25 gen. (+ 29 syn.), 441 spp. Thallus crustose to foliose, rarely lacking, usually with rhizoids; ascomata apothecial, sessile, rarely immersed, often brightly pigmented, with a well-developed lecanorine margin; interascal tissue of simple or anastomosing paraphyses, often with a pigmented epithecium; asci with a well-developed I+ apical cap, a usually well-developed ocular chamber and apical cushion, and an outer layer of I+ gelatinized material; ascospores hyaline, aseptate or septate. Anamorphs pycnidial, varied. Lichenized with green algae, occasionally lichenicolous, cosmop.

lecanoralean, (1) (of asci), of asci which are essentially bitunicate in structure, generally thick-walled with an especially strongly thickened apex, and in which discharge is by a rostrate eversion of the endoascus; see ascus; (2) (of apothecioid ascomata), see lecanorine.

Lecanorales, Ascomycota (±L). 40 fam., 347 gen. (+ 407 syn.), 7,108 spp. Thallus very varied. Ascomata almost always apothecial, flat to strongly cup-shaped, with or without a thalline margin; interascal tissue of paraphyses, usually branched and swollen at the apices, often with a pigmented or I+ epithecium; hymenial gel often prominent; asci with a single wall layer visible in LM but thick-walled, almost always with a conspicuous thick cap-like apical part, often with an internal beak (ocular chamber) and sometimes also with complex apical structures, ascus walls and/or apical structures often I+; ascospores very varied. Anamorphs pycnidial, poorly known. Mainly lichen-forming (almost all with protococcoid green photobionts), some lichenicolous or saprobes (then esp. on wood in xeric situations).

The second largest order of the *Ascomycota*, including most lichen-forming fungi with many large foliose and fruticose as well as crustose spp. The saprobic and lichenicolous species may have been secondarily derived from lichen-forming taxa. The *Peltigerales*, *Pertusariales* and *Teloschistales* have often been included within the order (e.g. Henssen & Jahns, 1973; Poelt, 1974), but this has not been widely accepted in the 1980s. However, an extended subordinal classification has been proposed by Rambold & Triebel (*Bibl. Lich.* **48**, 1992) which recognizes 8 suborders (*Lichineae*, *Peltigerineae*, *Cladoniineae*, *Pertusariineae*, *Acarosporineae*, *Lecanorineae*, *Teloschistineae*, *Umbilicariineae*); this has stimulated fresh debate (Hawksworth (Ed.), *Ascomycete systematics*, 1994) but there is as yet no general agreement on an improved system.

The *Lecanorales* remain heterogeneous and the circumscription and placement of many families (and in some cases of genera to families) is currently uncertain. The following fams are accepted here; all are exclusively lichen-forming unless otherwise indicated. Fams:

(1) **Acarosporaceae**.
(2) **Agyriaceae** (-L).
(3) **Alectoriaceae**.
(4) **Arctomiaceae**.
(5) **Bacidiaceae**.
(6) **Biatorellaceae**.
(7) **Brigantiaeaceae**.
(8) **Candelariaceae**.
(9) **Catillariaceae**.
(10) **Cladoniaceae**.

(11) **Coccocarpiaceae.**
(12) **Collemataceae.**
(13) **Crocyniaceae.**
(14) **Ectolechiaceae.**
(15) **Eigleraceae.**
(16) **Gypsoplacaceae.**
(17) **Haematommataceae.**
(18) **Heppiaceae**
(19) **Heterodeaceae.**
(20) **Hymeneliaceae.**
(21) **Lecanoraceae.**
(22) **Lecideaceae** (±L).
(23) **Megalosporaceae.**
(24) **Micareaceae.**
(25) **Miltideaceae.**
(26) **Mycoblastaceae.**
(27) **Ophioparmaceae.**
(28) **Pannariaceae.**
(29) **Parmeliaceae** (syn. *Hypogymniaceae,* *Usneaceae*).
(30) **Physciaceae** (syn. *Buelliaceae, Pyxinaceae*).
(31) **Pilocarpaceae.**
(32) **Porpidiaceae.**
(33) **Psoraceae.**
(34) **Ramalinaceae.**
(35) **Rhizocarpaceae.**
(36) **Rimulariaceae.**
(37) **Stereocaulaceae.**
(38) **Trapeliaceae.**
(39) **Umbilicariaceae.**
(40) **Vezdaeaceae** (±L).

Diagnoses of most of the families listed above are provided by Henssen & Jahns (1973) or Poelt (1974). Further references to family treatments are included in the fam. entries and under the type genera.

Lit.: Brodo *et al.* (Eds) (*J. Hattori bot. Lab.* **52**: 303, 1982; papers from symp. 'Evolution in Lichenized Fungi'), Eriksson (1981; nomencl. fams), Forssell (*Nova Acta R. Soc. Scient. upsal.* III **13** (6), 1885; cyanophilous gen.), Hafellner (*Beih. Nova Hedw.* **79**: 241, 1984, fams, asci; *in* Galun (Ed.), *Handbook of lichenology* 3: 41, 1988; fams; *in* Hawksworth (Ed.), 1994: 315), Henssen & Jahns (1973), Hertel (*Ergebn. ForschUnternehems Nepal Himalaya* **6**: 154, 1977; lecideine gen.), Honegger (see under ascus), Ozenda & Clauzade (*Les lichens*, 1970; key 2188 spp. France), Poelt (*Bestimmungschlüssel europäischer Flechten*, 1969, & Vězda, *Erganzungsheft* I-II, 1977-81; keys Eur. spp.), Poelt (1974), Purvis *et al.* (*Lichen flora of Great Britain and Ireland*, 1992), Santesson (*Symb. bot. upsal.* **12** (1), 1952; foliicolous taxa), Timdal (*Opera Bot.* **110**, 1991; fam. concepts), Wirth (*Die Flechten Baden-Württembergs*, edn 2, 2 vols, 1995), Zahlbruckner (*Naturl. PflFam.*, edn 2, **8**: 61, 1926; fams and gen.).

See also under *Ascomycota*, ascus, Geographical distribution, Lichens, and Literature.

Lecanorella Frey (1926) = Koerberiella (Porpid.) fide Hafellner (*Beih. Nova Hedw.* **79**: 241, 1984).

lecanorine (of an apothecium), having an excipulum (q.v.) thallinum (e.g. *Lecanora, Parmelia*) (Fig. 20A); lecanoroid.

Lecanoromyces E.A. Thomas (1953) ≡ Lecanora (Lecanor.).

Lecanoropsis M. Choisy (1949) = Lecanora (Lecanor.).

Lecanosticta Syd. (1922), Anamorphic Mycosphaerellaceae, 6.C2.19. Teleomorph *Mycosphaerella.* 2 (on pines), N. & C. Am. See Wolf & Barbour (*Phytopath.* **31**: 61, 1941), Petrak (*Sydowia* **15**: 252, 1961), Han *et al.* (*Jl Nanjing For. Univ.* **15**: 1, 1991; pathotypes of *L. acicola*), Evans (*Mycol. Pap.* **153**, 1984).

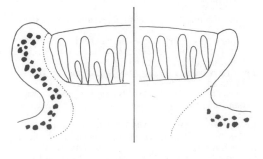

A B

Fig. 20. Ascoma structure in *Lecanorales.* A, section of lecanorine ascoma in *Lecanora chlarotera*; B, section of lecideine ascoma of *Buellia.* Not to scale.

Leccinum Gray (1821), Boletaceae. 30, N. temp. See Thiers (*Mycol.* **63**: 261, 1971; keys), Engel *et al.* (*Rauhstielröhrlinge*, 1978; 25 Eur. spp.).

Lecidea Ach. (1803), Lecideaceae (± L). *c.* 400, cosmop. See Hertel (*Beih. Nova Hedw.* **24**, 1967; *Herzogia* **1**: 25, 1968; 321, 1969; 405, 1970, parasitic spp.; **2**: 231, 1971; 479, 1973; **3**: 365, 1975; *Decheniana* **127**: 37, 1975, key saxic. holarctic spp.; *Ergebn. Forsch. Unternehems Nepal Himalaya* **6**: 145, 1977, Asia; *Mitt. Bot. StSamml. München* **30**: 297, 1991, arctic spp.), Inque (*J. Sci. Hiroshima Univ.* B(2), **18**: 1, 1982; Japan), Lowe (*Lloydia* **2**: 225, 1939), Magnusson (*Göteborgs K. Vetensk.-o. VitterhSamh. Handl.* ser. 4, **29** (4), 1925; *Acta Horti gothoburg.* **6**: 93, 1931; **16**: 125, 1945; *Svensk bot. Tidskr.* **46**: 178, 313, 1952), Poelt (*Ber. bayer. bot. Ges.* **34**: 82, 1961), Thomson *et al.* (*Bryologist* **72**: 137, 1969), Vainio (*Acta Soc. Fauna Flora fenn.* **53** (1), 1922), Schwab (*Mitt. bot. Münch.* **22**: 221, 1986; key 17 spp.).

Lecideaceae Chevall. (1826), Lecanorales (L). 10 gen. (+ 10 syn.), 418 spp. Thallus crustose to squamulose, rarely absent, with poorly developed rhizoids; ascomata apothecial, sessile, the margin absent or weakly developed, usually domed, light or dark; hymenium blueing in iodine; interascal tissue of branched or anastomosing paraphyses, usually swollen at the apices, often pigmented or with an epithecium; asci with a I+ apical cap, often with a more strongly staining shallow or tube-like apical ring, and an outer I+ gelatinized layer; ascospores hyaline, aseptate, thin-walled, without a sheath. Anamorphs pycnidial. Lichenized with green algae, rarely lichenicolous, cosmop.

Lit.: Hertel (*Bibl. Lich.* **25**: 219, 1987; gen. concepts).

lecideine (of an apothecium), one having no excipulum thallinum (q.v.) and the margin usually consisting only of the excipulum proprium (e.g. *Lecidea, Bacidia*) (Fig. 20B).

Lecidella Körb. (1855), Lecanoraceae (L). 30, mainly temp. See Hertel & Leuckert (*Willdenowia* **5**: 369, 1969), Knoph (*Bibl. Lich.* **36**, 1990; key 15 spp. rock), Leuckert & Knoph (*Lichenologist* **24** : 383, 1992; chloroxanthone spp., *Bibl. Lich.* **53**: 161, 1993; chemistry).

Lecidellomyces E.A. Thomas (1939) nom. inval. = Lecidella (Lecanor.).

Lecideola A. Massal. (1861) = Lecidella (Lecanor.) fide Knoph (1990).

Lecideomyces E.A. Thomas ex Cif. & Tomas. (1953) = Lecidea (Lecid.).

Lecideopsella Höhn. (1909), Schizothyriaceae. *c.* 10, mainly trop.

Lecideopsis (Almq.) Rehm (1891) = Arthonia (Arthon.).

Lecidocollema Vain. (1890) = Homothecium (Collemat.).

Lecidoma Gotth. Schneid. & Hertel (1981), ? Psoraceae (L). 1, Eur. See Pietschmann (*Nova Hedw.* **51**: 521, 1990; posn).

Lecidopyrenopsis Vain. (1907), Lichinaceae (L). 1, Thailand.

Leciographa A. Massal. (1854) = Opegrapha (Roccell.) fide Hafellner (*Beih. Nova Hedw.* **62**, 1979). See also *Dactylospora*.

Leciophysma Th. Fr. (1865), Collemataceae (L). 2, Eur., N. Am. See Henssen (*Lichenologist* **3**: 29, 1965).

Lecithium Sacc. (1895) ≡ Lecythium (Hypocr.).

Lecoglyphis Clem. (1909) = Opegrapha (Roccell.).

Lecophagus M.W. Dick (1990), Zoopagales (inc. sed.). 2, N. Am. See Dick (*MR* **94**: 347, 1990; key). = Cephaliophora (Mitosp. fungi) fide Morikawa *et al.* (*MR* **97**: 421, 1993).

Lecotheciaceae, see *Placynthiaceae*.

Lecothecium Trevis. (1851) = Placynthium (Placynth.).

Lecozonia Trevis. (1857) = Placynthium (Placynth.) fide Hafellner (*Beih. Nova Hedw.* **62**: 94, 1979).

lectotype, see type.

Lectularia Stirt. (1878) = Diploschistes (Thelotremat.).

Lecythea Lév. (1847) = Phragmidium (Phragmid.) etc. fide Dietel (1928).

lecythiform, like a stoppered bottle; ninepin-shaped (Fig. 37.13).

Lecythispora Chowdhry (1985) = Stylopage (Zoopag.) fide Sutton (*in litt.*).

Lecythium Zukal (1893), Hypocreaceae. 1, Austria.

Lecythophora Nannf. (1934) = Phialophora (Mitosp. fungi) fide Goidanich (*Ent. naz. Cell. e Cart., Roma* **112**, 1938).

Leeina Petr. (1923), Mitosporic fungi, 8.A1.?. 1, Philipp.

Legeriomyces Pouzar (1972), Legeriomycetaceae. 3 (in *Ephemeroptera*), USA. See Lichtwardt (1986; key), Williams & Lichtwardt (*CJB* **71**: 1109, 1993).

Legeriomycetaceae Pouzar (1972), Harpellales. 21 gen. (+ 4 syn.), 77 spp. Thalli branched, producing trichospores; attached to the hindgut cuticle of aquatic insect larvae.
  Lit.: Lichtwardt (1986), Lichtwardt *et al.* (*Trans. Mycol. Soc. Japa* **28**: 376, 1987), Longcore (*Mycol.* **81**: 482, 1989), Lichtwardt & Williams (*CJB* **68**: 1045, 1990, **70**: 1196, 1992), Williams & Lichtwardt (*CJB* **71**: 1109, 1993), Peterson *et al.* (*Mycol.* **83**: 389, 1991).

legitimate, (1) (of validly published names or epithets), in accordance with the *Code*, priorable; **illegitimate**, contrary to the *Code*, improriable. A recognized taxon of uncertain taxonomic position may have more than one legitimate name according to the position or rank given to it; (2) (of mating types), compatible. See Nomenclature.

Leifidium Wedin (1993), Sphaerophoraceae (L). 1, Australia, S. Am.

Leightonia Trevis. (1853) = Endocarpon (Verrucar.).

Leightonia Trevis. (1861) = Trypethelium (Trypethel.).

Leightoniella Henssen (1965), Collemataceae (L). 1, Sri Lanka.

Leightoniomyces D. Hawksw. & B. Sutton (1977), Mitosporic fungi, 1.A2.19. 1 (on lichens), Azores, UK.

Leioderma Nyl. (1888), Pannariaceae (L). 5, S. Hemisph. See Jørgensen & Galloway (*Lichenologist* **21**: 295, 1989; key), Galloway & Jørgensen (*Lichenologist* **19**: 345, 1987).

leiodisc (of an apothecium), having a smooth glazed disc.

Leiogramma Eschw. (1833) ≡ Leiorreuma (Graphid.).

Leiophloea (Ach.) Gray (1821) = Arthopyrenia (Arthopyren.). See Riedl (*Sydowia* **15**: 257, 1961; **16**: 263, 1962).

Leiophloea Trevis. (1860) ≡ Lejophloea (Arthopyren.).

Leiopoda Velen. (1947) ? = Mycena (Tricholomat.) fide Singer (1951).

Leiorreuma Eschw. (1824) = Phaeographis (Graphid.).

Leiosepium Sacc. (1900) = Sepedonium (Mitosp. fungi) fide Damon (*Mycol.* **44**: 95, 1952).

Leiosphaerella Höhn. (1919), Amphisphaeriaceae. 5, widespr. See Müller & v. Arx (1962), Samuels & Rossman (*Mycotaxon* **28**: 461, 1987).

leiosporous, having smooth spores.

Leiostigma Kirschst. (1944) = Sordaria (Sordar.) fide v. Arx & Müller (1954).

Leiothecium Samson & Mouch. (1975), Eurotiales (inc. sed.). 1, Greece.

Lejophallus (Fr.) Nees (1858) = Phallus (Phall.).

Lejophlea, see *Leiophloea*.

Lejosphaerella, see *Leiosphaerella*.

Lelum Racib. (1900) ? = Kordyana (Exobasid.). See Donk (*Reinwardtia* **4**: 117, 1956; affinities).

Lemalis Fr. (1825), ? Leotiales (inc. sed.). 3, widespr.

Lembidium Körb. (1855) nom. rej. = Anisomeridium (Monoblast.).

Lembopodia Bat. (1960) = Cirsosia (Asterin.) fide v. Arx & Müller (1975).

Lembosia Lév. (1845), Asterinaceae. c. 40, esp. warmer parts.

Lembosidium Speg. (1923) = Lembosia (Asterin.) fide v. Arx & Müller (1975).

Lembosiella Sacc. (1891), ? Microthyriaceae. 1, Afr.

Lembosiellina Bat. & H. Maia (1960) = Lembosia (Asterin.) fide v. Arx & Müller (1975).

Lembosina Theiss. (1913), ? Asterinaceae. 16, widespr.

Lembosiodothis Höhn. (1917) = Echidnodes (Asterin.) fide v. Arx & Müller (1975).

Lembosiopeltis Bat. & J.L. Bezerra (1967), Dothideales (inc. sed.). 1, Brazil. = Uleothyrium (Asterin.) fide v. Arx & Müller (1975).

Lembosiopsis Theiss. (1918), ? Asterinaceae. 1, N. Am. See Farr (*Sydowia* **38**: 65, 1986).

Lembuncula Cif. (1954), Mitosporic fungi, 5.A2.?. 1, Santo Domingo.

Lemkea Morgan-Jones & R.C. Sinclair (1983), Mitosporic fungi, 1.A2.1. 1, S. Afr., N. Am.

Lemmopsis (Vain.) Zahlbr. (1906), Lichinaceae (L). 3, Eur. See Ellis (*Lichenologist* **13**: 123, 1981; key).

Lemniscium Wallr. (1827) = Leptogium (Collemat.) fide Jørgensen (*Opera bot.* **45**, 1978).

Lemonniera De Wild. (1894), Mitosporic fungi, 1.G1.15. 6 (aquatic), widespr. See Descals *et al.* (*TBMS* **69**: 89, 1977; key).

Lempholemma Körb. (1855), Lichinaceae (L). c. 30, temp. See Schiman-Czeika (*Pl. Syst. Evol.* **158**: 283, 1988).

Lennisia Nieuwl. (1916) ≡ Monochaetia (Mitosp. fungi) fide Sutton (1977).

Lenormandia Delise (1841) nom. rej. = Normandina (Ascomycota, inc. sed.).

Lentaria Corner (1950), Lentariaceae. 17, widespr. See Sharda (*Biovigyanam* **10**: 131, 1984; Himalayan spp.).

Lentariaceae Jülich (1982), Gomphales. 1 gen., 17 spp. Basidioma ramarioid; spores hyaline, smooth.

Lentescospora Linder (1944) nom. dub. (Fungi, inc. sed.) fide Kohlmeyer & Kohlmeyer (1979).

lenthionine, an odorous metabolic product of *Lentinula edodes* (Nishikawa, *Chem. Pharm. Bull., Tokyo* **15**: 756, 1967).

lentic, habitat still water (lakes). Cf. lotic.

lenticular, like a double convex lens in form (Fig. 37.5c).

lentiginose (lentiginous), having very small spots as though freckled.

Lentinaceae Jülich (1982), Poriales. 9 gen. (+ 15 syn.), 145 spp. Basidioma mostly stipitate, lamellate; monomitic or dimitic; spores hyaline, cylindrical.

Lentinaria Pilát (1941) nom. nud. = Lentinellus (Lentinell.).

Lentinellaceae Locq. (1972), Hericiales. 1 gen. (+ 2 syn.), 8 spp. Basidioma agaricoid, lamellate with serrated edges.

Lentinellus P. Karst. (1879), Lentinellaceae. 8, widespr. See Miller & Stewart (*Mycol.* **63**: 333, 1971; key), Printz (*Svampe* **14**: 59, 1986; key).

Lentinopanus Pilát (1941), Lentinaceae. 2, widespr. = Panus (Lentin.) fide Singer (1951).

Lentinula Earle (1909), Tricholomataceae. 5, trop., *L. edodes* is the edible shii-take mushroom. See Pegler (1975; *Sydowia* **36**: 227, 1983).

Lentinus Fr. (1825), Lentinaceae. 70, widespr., esp. trop. Wood (e.g. railway sleepers) is attacked by *L. lepideus*. See Pegler (*Kew Bull. Addit. Ser.* **10**, 1983; world monogr.), Corner (*Beih. Nova Hedw.* **69**, 1981).

Lentispora Fayod (1889) = Coprinus (Coprin.) fide Singer (1951).

Lentodiellum Murrill (1915) = Pleurotus (Lentin.) fide Singer (1951).

Lentodiopsis Bubák (1895) = Pleurotus (Lentin.) fide Singer (1951).

Lentodium Morgan (1895) = Panus (Lentin.) fide Singer (1951).

Lentomita Niessl (1876) = Chaetosphaeria (Lasiosphaer.) fide Kohlmeyer & Kohlmeyer (*Marine mycology*, 1979).

Lentomitella Höhn. (1906) = Endoxyla (Bolin.).

Lentus Lloyd ex Torrend (1920) ≡ Polyporus (Polypor.) fide Donk (1974).

Lenzitella Ryvarden (1991), Thelephoraceae. 1, Morocco, Spain.

Lenzites Fr. (1836), Coriolaceae. 12, widespr. *L. sepiaria* causes timber rot. See Singer (1951).

Lenzitina P. Karst. (1889) ≡ Gloeophyllum (Coriol.) fide Donk (1974).

Lenzitites Mesch. (1892), Fossil fungi. 1 (Miocene), Italy.

Lenzitopsis Malençon & Bertault (1963) nom. inval. = Lenzitella (Thelephor.) fide Ryvarden (1991).

Leocarpus Link (1809), Physaraceae. 1, cosmop.

Leolophia Klotzsch [not traced] nom. dub. (? Fungi, inc. sed.).

Leotia Pers. (1794), Leotiaceae. 3, widespr., temp. See Honegger (*Lichenologist* **15**: 57, 1983; asci), Imai (*Bot. Mag. Tokyo* **50**: 9, 1936), Tai (*Lloydia* **7**: 146, 1944), Verkley (*Persoonia* **15**: 405, 1994; asci).

Leotiaceae Rehm (1886), Leotiales. 104 gen. (+ 69 syn.), 540 spp. Stromata usually absent; ascomata small to medium-sized, often brightly coloured; excipulum usually composed of parallel or intertwoven, or pseudoparenchymatous, sometimes gelatinized hyphae with ± remote septa, almost always glabrous or downy (with poorly differentiated hairs); hymenium sometimes gelatinous; interascal tissue of simple paraphyses; asci with an I+ or I- apical ring; ascospores usually hyaline, ellipsoidal or elongate, septate or not. Anamorphs very varied, only known for a few taxa. Usually saprobic on herbaceous or woody tissue, some spp. fungicolous.
    Lit.: Dennis (*Mycol. Pap.* **62**, 1956; UK), Verkley (*Persoonia* **15**: 405, 1994; asci), Zhuang (*Mycotaxon* **32**: 97, 1988; key 19 gen. *Encoelioideae*).

Leotiales, Ascomycota (±L). 13 fam., 392 gen. (+ 230 syn.), 2,036 spp. Stromata (thalli) usually absent, if present sclerotial (stroma) or with a morphologically varied lichenized thallus; ascomata apothecial, usually small, often brightly coloured, sessile or stipitate, sometimes surrounded by conspicuous hairs; interascal tissue of simple paraphyses, variously shaped, the apices sometimes swollen; asci usually small, thin-walled, without separable wall layers, with an apical pore surrounded by an I+ or I- ring, apical apparatus variable (14 types in Verkley, 1995); ascospores usually small, simple or transversely septate, mostly hyaline, usually not quite longitudinally symmetrical, usually smooth. Anamorphs varied, not known from many, hyphomycetous or coelomycetous. Saprobes and plant parasites, few lichenized or lichenicolous.
    Inoperculate discomycetes; cup fungi, formerly referred to as *Helotiales*. The taxonomy of the order is unsettled and 88 gen. are referred to order only. Korf (1973) accepted 8 Fams, and Eriksson & Hawksworth (1993) 9 Fams. Fams:
    (1) **Ascocorticiaceae**.
    (2) **Baeomycetaceae**.
    (3) **Dermateaceae**.
    (4) **Geoglossaceae**.
    (5) **Hemiphacidiaceae**.
    (6) **Hyaloscyphaceae**.
    (7) **Icmadophilaceae**.
    (8) **Leotiaceae** (syn. *Gelatinodiscaceae*).
    (9) **Loramycetaceae**.
    (10) **Orbiliaceae**.
    (11) **Phacidiaceae**.
    (12) **Sclerotiniaceae**.
    (13) **Vibrisseaceae**.
    Lit.: Arendholz (*Blattbewohnenden Ascomyceten aus der Ordnung der Helotiales*, 1979), Baral (*Z. Mykol.* **53**: 119, 1987; ascus ultrastr., *Mycotaxon* **44**: 333, 1992; microscopic preparation), Carpenter (*Mycol.* **80**: 127, 1988; ordinal nomencl.), Dennis (1968, 1978), Eriksson & Hawksworth (1993), Gamundí (*Fl. Cript. Tierra de Fuego* **10**(4), 1987), Hennebert & Bellemère (1979; anamorphs), Huhtinen (*in* Hawksworth (Ed.), 1994: 295), Korf (*in* Ainsworth *et al.* (Eds), *The Fungi* **4A**: 249, 1973; keys gen.), Nannfeldt (1932), Spooner (*Bibl. Mycol.* **116**, 1987), Verkley (*Persoonia* **15**: 303, 1993, *Hymenoscyphoideae* asci; *The ascus apical apparatus in Leotiales*, 1995, ultrastr.), Weber (*Bibl. Mycol.* **140**, 1992; reprod. system), see also under *Ascomycota*, *Discomycetes*, *Macromycetes*.

Leotiella Plöttn. (1900) ? = Cudonia (Geogloss.) fide Clements & Shear (1931). See Nannfeldt (1932).

Lepadolemma Trevis. (1853) = Haematomma (Haematommat.).

Lepidella J.-E. Gilbert (1925) [non Tiegh. (1911), *Loranthaceae*] = Gilbertia (Amanit.) fide Donk (1971). ≡ Aspidella (Amanit.), = Amanita (Amanit.) fide Bas (*Persoonia* **5**: 285, 1969).

Lepidocollema Vain. (1890) = Collema (Collemat.).

Lepidoderma de Bary (1873), Didymiaceae. 6, widespr. See Kowalski (*Mycol.* **63**: 492, 1971; key).

Lepidodermopsis Höhn. (1909) = Didymium (Didym.), but used by Lakhanpal (*Norw. Jl Bot.* **25**: 195, 1978).

Lepidodermopsis Wilczek & Meyl. (1934) Myxomycetes (inc. sed.). 1, Switzerland.

Lepidogium Clem. & Shear (1931) ≡ Lepidoleptogium (Pannar.).

Lepidoleptogium A.L. Sm. (1922) = Pannaria (Pannar.).

Lepidoma Link (1809) ≡ Rhizocarpon (Rhizocarp.).

Lepidoma (Ach.) Gray (1821) ≡ Lecidoma (Psor.).

Lepidomyces Jülich (1979), Xenasmataceae. 1, Eur.

Lepidonectria Speg. (1910) ? = Nectria (Hypocr.).

Lepidophyton Trib. (1899) nom. dub. (Mitosp. fungi). No spp. described.

Lepidopterella Shearer & J.L. Crane (1980), Argynnaceae. 1 (aquatic), USA.

Lepidosphaeria Parg.-Leduc (1970), Testudinaceae. 1, Sahara. See Hawksworth (*CJB* **57**: 91, 1979).

Lepidostroma Mägd. & S. Winkl. (1967), Stereales (inc. sed., L). 1, Eur. See Oberwinkler (*Dtsch. bot. Ges.* N.F. **4**: 139, 1970).

lepidote, covered with small scales.

Lepidotia Boud. (1885) = Peziza (Peziz.) fide Korf (*Persoonia* **7**: 205, 1973).

Lepidotus Clem. (1902) ≡ Lepiota (Agaric.).

lepiochlorin, an antibacterial antibiotic from *Lepiota* cultivated by the gardening ant (*Cyphomyrmex costatus*); see Hervey & Nair (*Mycol.* **71**: 1064, 1979).

Lepiota (Pers.) Gray (1821), Agaricaceae. 150, widespr. *L. scobinella* and several pink, red or orange spp. are very poisonous. See Kühner (*BSMF* **52**: 177, 1936; classific.), Locquin (*Bull. Soc. Linn. Lyon* **14**: 89, 1945; classific., *BSMF* **67**: 365, 1951; list spp., *BSMF* **68**: 267, 1952; interpretation of illustrations), Boiffard (*Docum. Mycol.* **8**: 39, 1973; microscopical characters), Bon (*Docum. Mycol.* **10**; key Eur. spp.), Beeli (*Bull. Soc. Bot. Belg.* **64**: 206, 1931; *Fl. Icon. Champ. Congo* **2**: 29, 1936, Congo spp.), Kauffman (*Pap. Mich. Acad. Sci.* **4**: 311, 1924; N. Am. spp.), Aberdeen (*Kew Bull.* **16**: 129, 1962; Australian type studies), Pegler (*Kew Bull.* **27**: 155, 1972; key 34 Sri Lanka spp., *Kew Bull., add. ser.* **6**, 1977; key 30 E. Afr. spp.), Dennis (*Kew Bull.* **7**: 459, 1952; W. Indian spp.).

Lepiotasporites T.C. Huang (1981), Fossil fungi. 1 (Miocene), Taiwan.

Lepiotella (J.-E. Gilbert) Konrad (1934) = Chamaemyces (Agaric.) fide Singer (1975).

Lepiotella Rick (1938) ≡ Volvolepiota (Agaric.).

Lepiotula (Maire) Locq. ex E. Horak (1968) = Lepiota (Agaric.) fide Singer (1975), but see Horak (*N.Z. Jl Bot.* **18**: 183, 1980; key 5 NZ spp.).

Lepista (Fr.) W.G. Sm. (1870), Tricholomataceae. 19, cosmop.; blewits. See Pegler & Young (*Kew Bull.* **29**: 659, 1974; basidiospore form), Harmaja (*Karstenia* **14**: 82, 1974; separation, **18**: 49, 1978; gen. division). = Clitocybe (Tricholomat.) fide Bigelow & Smith (*Brittonia* **21**: 144, 1969).

Lepocolla Eklund (1883) nom. dub. (? Actinomycetes or ? Mitosporic fungi).

Lepolichen Trevis. (1853), Coccotremataceae (L). 1, S. Am. See Henssen (*in* Brown *et al.* (Eds), *Lichenology: progress and problems*: 107, 1976).

Leporina Velen. (1947) = Coprotus (Thelebol.) fide van Brummelen (*in* Hawksworth (Ed.), *Ascomycete systematics*: 400, 1994).

Lepra Scop. (1777) nom. rej. = Pertusaria (Pertusar.).

Lepra Willd. (1787) = Illosporium (Mitosp. fungi).

Lepra Chevall. (1826) nom. dub. (Mitosp. fungi).

Leprantha Dufour ex Körb. (1855) = Arthonia (Arthon.).

Lepraria Ach. (1803) nom. cons., Ascomycota (inc. sed., L). *c.* 20, widespr. See Laundon (*Lichenologist* **24**: 315, 1992; key 9 Br. spp.), Lohtander (*Ann. Bot. fenn.* **31**: 223, 1994; key 11 spp. Finland).

Leprieuria Laessøe, J.D. Rogers & Whalley (1989), Xylariaceae. 1, trop. C. & S. Am. Anamorph *Geniculosporium*.

Leprieurina G. Arnaud (1918), Mitosporic fungi, 5.A2.1. 1, S. Am.

Leprieurinella Bat. & H. Maia (1961), Mitosporic fungi, 5.?.?. 1, Brazil.

Leprocaulon Nyl. (1878), Mitosporic fungi (L), -.-.-. 8, mainly S. Hemisph. See Lamb & Ward (*J. Hattori bot. Lab.* **38**: 499, 1974; key).

Leprocollema Vain. (1890), Lichinaceae (L). 3, Eur., Brazil, New Caledonia. See Henssen (*SA* **5**: 137, 1986), Magnusson (*Hedwigia* **78**: 219, 1938).

Leproloma Nyl. ex Cromb. (1883), ? Pannariaceae (L). 4, Eur., N. & S. Am., Africa, Australasia, Antarct. See Jørgensen (*in* Hawksworth (Ed.), *Ascomycete systematics*: 392, 1994; posn), Laundon (*Lichenologist* **21**: 1, 1989; key), Leuckert & Kümmerling (*Nova Hedw.* **52**: 17, 1991; chemotax.).

Leproncus Vent. (1799) = Pertusaria (Pertusar.).

Lepropinacia Vent. (1799) = Buellia (Physc.) p.p., Lecanora (Lecanor.) p.p., and Ochrolechia (Pertusar.) p.p.

Leproplaca (Nyl.) Nyl. ex Hue (1888) = Caloplaca (Telosch.) fide Hafellner & Vězda (*Nova Hedw.* **55**: 183, 1992), Jørgensen & Tønsberg (*Nordic Jl Bot.* **8**: 293, 1988), Kärnefelt (*Crypt. Bot.* **1**: 147, 1989).

leprose (of lichens), having the surface or the whole thallus entirely dissolved into soredia (e.g. *Lepraria*) (Fig. 21A).

Leprosis Neck. ex Kremp. (1869) Fungi (inc. sed., L).

Leptascospora Speg. (1918), Meliolaceae. 1, Java.

Leptasteromella Petr. (1968) ? = Leptodothiorella (Mitosp. fungi) fide Sutton (1977).

Lepteutypa Petr. (1923), ? Amphisphaeriaceae. 4, widespr. See Shoemaker & Müller (*CJB* **43**: 1459, 1965), v. Arx (1981), Nag Raj & Kendrick (*Sydowia* **38**: 178, 1986; key). Anamorph *Seiridium*.

Lepteutypella Petr. (1925) = Lepteutypa (Amphisphaer.) fide Müller & v. Arx (1973).

Leptina Bat. & Peres (1960) ? = Discosia (Mitosp. fungi) fide Sutton (1977).

Leptinia Juel (1897) = Puccinia (Puccin.) fide Dietel (1928).

Leptobelonium Höhn. (1924) = Strossmayeria (Leot.) fide Iturriaga (*Mycotaxon* **20**: 169, 1984).

Leptocapnodium (G. Arnaud) Cif. & Bat. (1963) = Scorias (Capnod.) fide Reynolds (*Bull. Torrey bot. Cl.* **98**: 157, 1971).

Leptochaete Lév. (1846) ≡ Hymenochaete (Hymenochaet.).

Leptochidium M. Choisy (1952), Placynthiaceae (L). 1, Eur., N. Am.

Leptochlamys Died. (1921), Mitosporic fungi, 8.E1.10. 1 (on *Musci*), Eur.

Leptocladia Marvanová & Descals (1985) [non J. Agardh (1892), *Rhodophyta*] ≡ Stenocladiella (Mitosp. fungi).

Leptocoryneum Petr. (1925) = Seimatosporium (Mitosp. fungi) fide Sutton (1977).

Leptocrea Syd. & P. Syd. (1916) = Polystigma (Phyllachor.) fide v. Arx & Müller (1954).

leptocystidium, see cystidium.

Leptodendriscum Vain. (1890) = Polychidium (Placynth.) fide Henssen (1963).

Leptoderma G. Lister (1913), Stemonitidaceae. 1, Eur., N. Am. See Martin *et al.* (1983).

leptodermatous (of hyphae), having the outer wall thinner than the lumen; cf. mesodermatous.

Leptodermella Höhn. (1915), Mitosporic fungi, 6.A1.1. 1, Eur.

Leptodermopsis Speg. ex Höhn. (1923) nom. dub. (Mitosp. fungi) fide Sutton (1977). Cited in Falk (*Mykol. Unters.* **1**: 331, 1923).

Leptodiscella Papendorf (1969), Mitosporic fungi, 3/6.B1.10. 1, S. Afr.

Leptodiscus Gerd. (1953) [non Hertwig (1877), *Algae*] ≡ Mycoleptodiscus (Mitosp. fungi).

Leptodon Quél. (1886) [non D. Mohr (1803), nom. cons., *Musci*] ≡ Mycoleptodon (Steccherin.).

Leptodontidium de Hoog (1979), Mitosporic fungi, 1.A1.10. 7, widespr.

Leptodontium de Hoog (1977) [non (Müll. Hal.) Hampe ex Lindb. (1864), *Musci*] ≡ Leptodontidium (Mitosp. fungi).

Leptodothiora Höhn. (1918) = Dothiora (Dothior.) fide Froidevaux (1973).

Leptodothiorella Höhn. (1918), Anamorphic Mycosphaerellaceae, 4.A1.15. Teleomorph *Guignardia*. 3, trop.

Leptodothiorella Aa (1973) = Leptodothiorella (Mitosp. fungi) fide Sutton (1977).

Leptodothis Theiss. & Syd. (1914) nom. dub. (Fungi, inc. sed.) fide Müller & v. Arx (1962).

lepto-form, see *Uredinales*.

Leptogidiomyces Cif. & Tomas. (1953) ≡ Leptogidium (Placynth.).

Leptogidium Nyl. (1873) = Polychidium (Placynth.) fide Henssen (1963).

Leptogiomyces E.A. Thomas ex Cif. & Tomas. (1953) = Leptogium (Collemat.).

Leptogiopsis Trevis. (1880) = Leptogium (Collemat.).

Leptogiopsis Müll. Arg. (1882) = Leptogium (Collemat.).

Leptogiopsis Nyl. (1884) = Mastodia (Mastod.).

Leptogium (Ach.) Gray (1821), Collemataceae (L). *c. c.* 160, cosmop. See Awasthi (*Geophytology* **8**: 189, 1979; India), Jørgensen (*Svensk bot. Tidskr.* **67**: 53, 1973; *Mallotium, in* Poelt & Vězda, 1977; Eur., *Lichenologist* **26**: 1, 1994; key small spp. Eur.), Poelt & James (*Lichenologist* **15**: 109, 1983; W. Eur.), Sierk (*Bryologist* **67**: 245, 1964; N. Am.).

Leptoglossum P. Karst. (1879) = Arrhenia (Tricholomat.) fide Redhead (1984).

Leptoglossum (Cooke) Sacc. (1884) = Geoglossum (Geogloss.). See Maas Geesteranus (1964).

leptogonidium, a photobiont consisting of small-sized cells (obsol.); cf. gonidium.

Leptographa Jatta (1892) = Toninia (Catillar.).

Leptographa (Th. Fr.) M. Choisy (1950) = Ptychographa (Agyr.).

Leptographium Lagerb. & Melin (1927), Anamorphic Ophiostomataceae, 1.A1.21. Teleomorph *Ophiostoma*. 28, temp. See Shaw & Hubert (*Mycol.* **44**: 693, 1952), Wingfield (*TBMS* **85**: 81, 1985; gen. relationships), Harrington & Cobb (*Leptographium root diseases of conifers* (APS Symposium Ser.), 1988), Zambino & Harrington (*Mycol.* **81**: 122, 1989; isozyme variation), Ayer *et al.* (*J. Nat. Prod.* **52**: 119, 1989; metabolites from *L. wageneri* var. *pseudotsugae*), Zambino & Harrington (*Mycol.* **84**: 12, 1992; isozyme characterization), Wingfield (*in* Wingfield *et al.* (Eds), *Ceratocystis and Ophiostoma*, 43, 1993; anamorphs).

Leptoguignardia E. Müll. (1955), Dothideales (inc. sed.). 1, France. See Barr (*Prodromus to class Loculoascomycetes*, 1987). Anamorph *Dothichiza*-like.

Leptogydiomyces, see *Leptogidiomyces*.

Leptojum Beltr. (1858) = Leptogium (Collemat.).

Leptokalpion Brumm. (1977), Thelebolaceae. 1, Thailand. See van Brummelen (*in* Hawksworth (Ed.), *Ascomycete systematics*: 398, 1994; posn).

Leptolegnia de Bary (1888), Saprolegniaceae. 5 or 6, N. temp. (Dick, 1973).

Leptolegniella Huneycutt (1952), Leptolegniellaceae. 5, USA. See Dick (*in press*; key).

Leptolegniellaceae M.W. Dick (1971), Leptomitales. 5 gen., 16 spp. Oospore wall thick, two-layered, separable endospore membrane.
Lit.: Dick (*TBMS* **57**: 417, 1971, *in press*; key).

Leptomassaria Petr. (1914) ? = Anthostomella (Xylar.) fide Læssøe (*SA* **13**: 43, 1994).

Leptomelanconium Petr. (1923), Mitosporic fungi, 6.A/B2.15. 3, Eur., N. Am., Australia. See Sutton (1980).

Leptomeliola Höhn. (1919), Parodiopsidaceae. 5, trop. See Hughes (*Mycol. Pap.* **166**, 1993).

Leptomitaceae Kütz. (1843), Leptomitales. 3 gen. (+ 1 syn.), 5 spp. Thallus lacking pedicellate segments; principal form zoospores not developed in zoosporangium; oosporogenesis not periplasmic.

Leptomitales, Oomycota. 4 fam., 13 gen. (+ 2 syn.), 30 spp. Freshwater saprobes and animal parasites, N. temp. Dick (*in* Margulis *et al.* (Eds), 1990: 661, *in* Ainsworth *et al.* (Eds), *The Fungi* **4B**: 145, 1973). Fams:
(1) **Apodachlyellaceae**.
(2) **Ducellieriaceae**.
(3) **Leptolegniellaceae**.
(4) **Leptomitaceae**.

Leptomitus C. Agardh (1824), Leptomitales (inc. sed.). 1, Eur., N. Am. See sewage fungus.

Leptomyces Mont. (1856) nom. dub. (Agaric.) fide Donk (1962).

Leptomycorhaphis Cif. & Tomas. (1953) nom. rej. = Leptorhaphis (Arthopyren.).

Leptonema J. Sm. (1896) [non Juss. (1824), *Euphorbiaceae*], Fossil fungi (mycel.). 1 (Carboniferous), UK.

Leptonia (Fr.) P. Kumm. (1871), Entolomataceae. 40, widespr. See Largent (*Mycol.* **65**: 987, 1974, *Bibl. Mycol.* **55**, 1977; monogr., keys). = Rhodophyllus (Entolomat.) fide Singer (1951).

Leptoniella Earle (1909) ≡ Leptonia (Entolomat.).

Leptopeltella Höhn. (1917) = Leptopeltis (Leptopeltid.) fide Holm & Holm (1977).

Leptopeltidaceae Höhn. ex Trotter (1928), Dothideales. 7 gen. (+ 7 syn.), 32 spp. Ascomata formed within a small subcuticular stroma, elongate, opening by a longitudinal split; covering wall composed of a single layer of radially arranged thick-walled isodiametric cells; interascal hyphae present; asci probably bitunicate but apparently not fissitunicate; ascospores hyaline, septate, without a sheath. Anamorphs unknown. Saprobic, esp. on ferns, mostly temperate.
Lit.: v. Arx (*Acta bot. neerl.* **13**: 182, 1964), Holm & Holm (*Bot. Notiser* **130**: 215, 1977).

Leptopeltina Speg. (1924) = Stomiopeltis (Micropeltid.) fide v. Arx & Müller (1975).

Leptopeltina Petr. (1947) ≡ Leptopeltinella (Leptopeltid.).

Leptopeltinella Petr. (1951) ≡ Leptopeltis (Leptopeltid.).

Leptopeltis Höhn. (1917), Leptopeltidaceae. 7 or 8, N. temp. See Holm & Holm (*Bot. Notiser* **130**: 215, 1977; key), Eriksson (1981; posn).

Leptopeltopsis Petr. (1947) = Leptopeltis (Leptopeltid.) fide Holm & Holm (1977).

Leptoperidia Rappaz (1987), Diatrypaceae. 4, Eur., Afr., C. Am.

Leptopeza G.H. Otth (1871) = Plicaria (Peziz.) fide Eckblad (1968).

Leptopeziza Rostr. (1888) ? = Pyrenopeziza (Dermat.) fide Nannfeldt (1932).

Leptophacidium Höhn. (1918) = Guignardia (Mycosphaerell.) fide v. Arx & Müller (1954).

Leptophoma Höhn. (1915) = Phoma (Mitosp. fungi) fide Sutton (1977).

Leptophrys Hertwig & Lesser (1885), *Hydromyxales*, q.v.

Leptophyllosticta I.E. Brezhnev (1939), Mitosporic fungi, 4.A1.?. 1, former USSR.

Leptophyma Sacc. (1889) = Helostroma (Mitosp. fungi) fide Moore (*Mycotaxon* **14**: 13, 1982).

Leptopodia Boud. (1885) = Helvella (Helvell.) fide Seaver (1928).

Leptopora Raf. (1808) = Poria (Coriol.) fide Merrill (*Index Rafinesq.*, 1949).

Leptoporus Quél. (1886) = Tyromyces (Coriol.) fide Kotlaba & Pouzar (1959).

Leptopterygium Zahlbr. (1930) = Zahlbrucknerella (Lichin.) fide Henssen (1963).

lepto-Puccinia, a *Puccinia* of lepto-form; see *Uredinales*.

Leptopuccinia (G. Winter) Rostr. (1902) = Puccinia (Puccin.) fide Dietel (1928).

Leptorhaphiomyces Cif. & Tomas. (1953) = Leptorhaphis (Arthopyren.).

Leptorhaphis Körb. (1855) nom. cons., Arthopyreniaceae (±L). 12, N. Hemisph. See Aguirre-Hudson (*Bull. Br. Mus. nat. Hist., Bot.* **21**: 85, 1991; key).

Leptosacca Syd. (1928), Ascomycota (inc. sed.). 1, Chile.

Leptosillia Höhn. (1928), Valsaceae. 1, Eur. See Hawksworth (*SA* **6**: 136, 1987). Anamorph *Harpostroma*.

Leptosphaerella Speg. (1909), Ascomycota (inc. sed.). 7, S. Am.

Leptosphaerella (Sacc.) Hara (1913) = Phaeosphaeria (Phaeosphaer.)

Leptosphaeria Ces. & De Not. (1863) nom. cons., Leptosphaeriaceae. *c.* 100, widespr. *L. avenaria* (speckled blotch of oats), *L. bondari* (areolate spot of citrus in S. Am.), *L. coniothyrium* (rose stem canker, raspberry cane blight), *L. herpotrichoides* and *L. tritici* (on wheat), *L. maculans* (anamorph *Phoma lingam*; dry rot of turnips, canker of crucifers), *L. sacchari* (sugar-cane ring spot), *L. salvinii* (see *Magnaporthe*). See Müller (*Sydowia* **4**: 185, 1950, **5**: 49, 1951; Swiss spp.), Wehmeyer (*Mycol.* **44**: 621, 1952), Meyers (*Mycol.* **49**: 475, 1957), Holm (1957; *Taxon* **24**: 475, 1975; nomencl.), Shoemaker (*CJB* **62**: 2688, 1984; key 64 spp.), Crane & Shearer (*Illinois nat. Hist. Bull.* **34** (3), 1991; disposition 1689 names), Hundorf (*Illinois nat. Hist. Survey Bull.* **34**: 479, 1992; spp. on *Rosaceae*), Morales *et al.* (*MR* **99**: 593, 1995). Anamorphs *Camarosporium, Hendersonia, Phoma, Rhabdospora, Stagonospora*.

Leptosphaeriaceae M.E. Barr (1987), Dothideales. 4 gen. (+ 15 syn.), 219 spp. Ascomata perithecial, often ± conical, papillate, immersed or erumpent, sometimes aggregated into small stromata, with a well-developed usually periphysate ostiole; peridium black, well-developed, sometimes thicker at the base, composed of thick-walled pseudoparenchymatous cells; interascal tissue of cellular pseudoparaphyses; asci cylindrical, relatively narrow and thin-walled but fissitunicate; ascospores hyaline to brown, transversely septate, sometimes elongated, sometimes with a sheath. Anamorphs coelomycetous. Saprobic or necrotrophic on stems or leaves.

Leptosphaeriopsis Berl. (1892) = Ophiobolus (Leptosphaer.) fide Walker (1980).

Leptosphaerites Richon (1885), Fossil fungi. 2 (Tertiary), France, former USSR.

Leptosphaerulina McAlpine (1902), Pleosporaceae. 22, temp. See Crivelli (*Ueber die heterogene Ascomycetengattung Pleospora*, 1983; key), Graham & Luttrell (*Phytopath.* **51**: 680, 1961; key), Barr (*Contr. Univ. Mich. Herb.* **9**: 523, 1972), Irwin & Davis (*Austr. Jl Bot.* **33**: 233, 1985; Australia), Müller (*Ber. schweiz. bot. Ges.* **76**: 185, 1966), Roux (*TBMS* **86**: 319, 1986; anamorph).

Leptospora Raf. (1808) nom. dub. fide Donk (*Persoonia* **1**: 173, 1960).

Leptospora Rabenh. (1857), Dothideales (inc. sed.). 1, Eur., N. Am. See Walker (*Mycotaxon* **11**: 1, 1980).

leptospore (of *Uredinales*), a teliospore (q.v.) adapted for immediate germination without a dormant period.

Leptosporella Penz. & Sacc. (1897), Ascomycota (inc. sed.). 2 or 3, Java, S. Am.

Leptosporina Chardón (1939), Ascomycota (inc. sed.). 1, S. Am.

Leptosporium Bonord. (1857) = Vibrissea (Vibriss.) fide Sanchez & Korf (1968).

Leptosporium (Sacc.) Höhn. (1923) = Fusicolla (Mitosp. fungi) fide Clements & Shear (1931).

Leptosporomyces Jülich (1972), Atheliaceae. 9, N. temp.

Leptosporopsis Höhn. (1920) = Leptosphaeria (Leptosphaer.) p.p., Ophiobolus (Leptosphaer.) p.p. and Ophiosphaerella (Phaeosphaer.) p.p. fide Walker (1980).

Leptostroma Fr. (1815), Anamorphic Rhytismataceae, 8.A1.10. Teleomorph *Lophodermium*. 200, widespr. See Minter (*CJB* **58**: 906, 1980; spp. on *Pinus*), Sieber-Canavesi *et al.* (*Mycol.* **83**: 89, 1991; ecology endophytic spp.).

Leptostromataceae (obsol.). See Mitosporic fungi.

Leptostromella (Sacc.) Sacc. (1884), Mitosporic fungi, 8.A1-E1.?. 20, temp.

Leptoteichion Kleb. (1933) = Phacidiopycnis (Mitosp. fungi) fide Rupprecht (*Sydowia* **13**: 10, 1959).

Leptothyrella Sacc. (1885), Mitosporic fungi, 5.B1.?. 10, widespr.

Leptothyrina Höhn. (1915), Mitosporic fungi, 8.A1.10. 1, Switzerland.

Leptothyriomyces Kräusel (1929), Fossil fungi (Asterin.). 2 (Miocene), Sumatra.

Leptothyrium Kunze (1823), Mitosporic fungi, 5.A1.15. 2, widespr. See Sutton (*Mycol. Pap.* **141**, 1977).

leptotichous (of tissue), thin-walled.

Leptotrema Mont. & Bosch (1855) = Ocellularia (Thelotremat.) fide Hale (1981).

Leptotrichum Corda (1842) nom. dub. (Mitosp. fungi) fide Holubová-Jechová (*Sydowia* **45**: 95, 1993).

Leptotrimitus Pouzar (1966) = Incrustoporia (Coriol.) fide Donk (1974).

Leptotrochila P. Karst. (1871), Dermateaceae. 14, widespr. See Schüepp (*Phytopath. Z.* **36**: 236, 1959; key).

Leptotus P. Karst. (1879) = Arrhenia (Tricholomat.) fide Redhead (1984).

Leptoxyphium Speg. (1918), Anamorphic Capnodiaceae, 2.A/B1/2.15. Teleomorph *Aithaloderma*. 1, trop. See Hughes (*Mycol.* **68**: 693, 1976), Roquebert & Bury (*CJB* **66**: 2265, 1988; ultrastr. conidiomata).

Leptuberia Raf. (1808), Fungi (inc. sed., L). 1, N. Am.

Le-Ratia Pat. (1907) [non Broth. & Paris (1909), *Musci*] nom. inval., Secotiaceae. 4, New Caledonia. See Heim (*Revue Mycol., Paris* **33**: 137, 1968).

Lesdainea Harm. (1910) = Trimmatothele (Verrucar.).

lesion, a wound; a well-marked but limited diseased-area.

Lespiaultinia Zobel (1854) = Tuber (Tuber.) fide Trappe (1979).

Letendraea Sacc. (1880), Tubeufiaceae. 2, Eur., Afr. See Rossman (1987; key), Samuels (*CJB* **51**: 1275, 1973).

Letendraeopsis K.F. Rodrigues & Samuels (1994), Tubeufiaceae. 1, Brazil.

Lethagrium A. Massal. (1853) ≡ Lathagrium (Collemat.).

Letharia (Th. Fr.) Zahlbr. (1892) nom. cons., Parmeliaceae (L). 10, Eur., Asia, N. Am. See Schade (*Ber. bayer. bot. Ges.* **30**: 108, 1954, *Feddes Repert.* **58**: 179, 1955).

Lethariella (Motyka) Krog (1976), Parmeliaceae (L). 6, Asia, S. Eur., Macaronesia.

Lethariicola Grummann (1969), Odontotremataceae. 2 (on lichens), Eur., N. Am. fide Lumbsch & Hawksworth (*Bibl. Lich.* **38**: 325, 1990; key).

Lethariopsis Zahlbr. (1926) = Caloplaca (Teloschist.) fide Eriksson & Hawksworth (*SA* **10**: 141, 1991).

Letrouitia Hafellner & Bellem. (1982), Letrouitiaceae (L). 15, widespr. See Hafellner (*Nova Hedw.* **35**: 645, 1983; key).

Letrouitiaceae Hafellner & Bellem. (1982), Teloschistales (L). 1 gen., 15 spp. Thallus crustose, green or brownish, lacking anthraquinone pigments; ascomata wuithout a well-developed thalline margin; asci with a diffuse I+ outer apical layer and a well-developed internal apical structure which blues diffusely in iodine; ascospores hyaline, multiseptate, sometimes muriform, with strongly thickened internal walls, often very large.
Lit.: Hafellner & Bellemère (*Nova Hedw.* **35**: 263, 1982).

Lettauia D. Hawksw. & R. Sant. (1990), ? Fuscideaceae. 1 (on lichens, esp. *Cladonia*), Eur.

Leucangium Quél. (1883) = Picoa (Balsam.) fide Fischer (1938).

Leucinocybe, see Asproinocybe.

Leucoagaricus (Locq.) Singer (1948), Agaricaceae. 25, widespr. See Heinemann (*Fl. Illustr. Champ. Afr. Centr.* **2**: 30, 1973; key 18 Afr. spp.).

Leucobolbitius (J.E. Lange) Locq. (1952) nom. inval. ? = Leucocoprinus (Agaric.).

Leucobolites Beck (1923) = Boletus (Bolet.) fide Singer (1945).

Leucocarpia Vězda (1969), Verrucariaceae (L). 1, Eur.

Leucocarpopsis G. Salisb. (1975), Verrucariaceae (L). 1, UK. ? = Verrucaria (Verruc.) fide Hawksworth (*in* Purvis *et al.* (Eds), *Lichen flora of Great Britain and Ireland*: 358, 1992).

Leucoconiella Bat., H. Maia & Peres (1960), Ascomycota (inc. sed.). 1, Paraguay.

Leucoconis Theiss. & Syd. (1918), Ascomycota (inc. sed.). 1, India.

Leucoconius Beck (1923) ≡ Gyroporus (Gyrodont.).

Leucocoprinus Pat. (1888), Agaricaceae. 12, trop. See Babos (*Beih. Sydowia* **8**: 33, 1979), Pegler (*Kew Bull., Addit. Ser.* **6**, 1977), Heinemann (*Fl. Illustr. Champ. Afr. Centr.* **5**: 87, 1977), Reid (*MR* **93**: 413; Br. spp.).

Leucocortinarius (J.E. Lange) Singer (1945), Cortinariaceae. 1, Eur.

Leucocrea Sacc. & P. Syd. ex Lindau (1900), Hypocreaceae. 1, S. Am.

Leucocytospora (Höhn.) Höhn. (1927) = Cytospora (Mitosp. fungi) fide Petrak (*Ann. Myc.* **19**: 17, 1921).

Leucodecton A. Massal. (1860) = Thelotrema (Thelotremat.) fide Thor (*Opera Bot.* **103**, 1990).

Leucodochium Syd. & P. Syd. (1917), Mitosporic fungi, 4.A1.?. 1, Philipp.

Leucofomes Kotl. & Pouzar (1957) = Rigidoporus (Coriol.) fide Pegler (1973).

Leucogaster R. Hesse (1882), Leucogastraceae. 10, N. Hemisph. See Fogel (*CJB* **57**: 1718, 1979; sep. order).

Leucogastraceae Moreau ex Fogel (1979), Melanogastrales. 2 gen. (+ 1 syn.), 14 spp. Gasterocarp loculate; gleba lacunose, tending to be compact and gelatinous at maturity; spores hyaline, thick-walled with reticulate ornamentation and gelatinous myxosporium.

Leucogastrales, see Melanogastrales.

Leucoglossum S. Imai (1942), ? Geoglossaceae. 1, Japan. ? = Trichoglossum (Geogloss.). See Rifai (*Lloydia* **28**: 113, 1965).

Leucogomphidius Kotl. & Pouzar (1972), Gomphidiaceae. 9, USA, Eur.

Leucogramma G. Mey. (1825) nom. rej. prop. = Graphina (Graphid.).

Leucogramma A. Massal. (1860) = Phaeographina (Graphid.).

Leucographa Nyl. (1857) nom. nud. (Ascomycota, L).

Leucogymnospora Fink (1930), ? Ostropales (inc. sed., L). 1, Puerto Rico.

Leucogyrophana Pouzar (1958), Coniophoraceae. 9, widespr. See Ginns (*CJB* **56**: 1953, 1978; key), Ginns & Weresub (*Mem. N.Y. bot. Gdn* **28**: 86, 1976; sclerotial spp.).

Leucogyroporus Snell (1942) = Tylopilus (Strobilomycet.) fide Singer (1945).

Leucoinocybe Singer (1943), ? Tricholomataceae. 1, Eur.

Leucolenzites Falck (1909) ≡ Lenzites (Coriol.).

Leucoloma Fuckel (1870) [non Brid. (1827) nom. cons., Musci] ≡ Octospora (Otid.).

Leucomyces Earle (1909) = Amanita (Amanit.) fide Singer (1951).

Leuconectria Rossman, Samuels & Lowen (1993), Hypocreaceae. 1, Afr., N. & S. Am., Asia. Anamorph Gliocephalotrichum.

Leuconeurospora Malloch & Cain (1970), Microascaceae. 2, Eur., Asia. See Udagawa (*in* Subramanian (Ed.), *Taxonomy of fungi* **1**: 225, 1978), v. Arx *et al.* (*Beih. Nova Hedw.* **94**, 1988; posn).

Leucopaxillus Boursier (1925), Tricholomataceae. 18, temp., subtrop. See Singer & Smith (*Pap. Mich. Acad. Sci.* **28**: 85, 1943; key 12 spp., *Mycol.* **39**: 725, 1947).

Leucopenicillifer G.R.W. Arnold (1971), Mitosporic fungi, 1.A1.15. 1 (on myxomycete), Russia. ? = Paecilomyces (Mitosp. fungi) fide Carmichael *et al.* (1980).

Leucopezis Clem. (1909) ? = Humaria (Otid.). fide Eckblad (1968).

Leucophellinus Bondartsev & Singer (1944), Coriolaceae. 1, Eur. See Pegler (1973).

Leucophlebs, see Leucophleps.

Leucophleps Harkn. (1899), Leucogastraceae. 4, Eur., N. Am. See Fogel (*CJB* **57**: 1718, 1979).

Leucophomopsis Höhn. (1917) = Phomopsis (Mitosp. fungi) fide Clements & Shear (1931).

Leucoporus Quél. (1886) ≡ Polyporellus (Polypor.).

Leucopus P. Kumm. (1871) = Cortinarius (Cortinar.) fide Donk (1949).

Leucorhizon Velen. (1925) = Gastrosporium (Gastrospor.) fide Pilát (*Fl. ČSR* B **1**: 227, 1958).

Leucoscypha Boud. (1885), Otideaceae. 8, temp. See Harmaja (*Karstenia* **17**: 73, 1977).

Leucosphaera Arx, Mukerji & N. Singh (1978) [non Gilg (1897), *Spermatophyta*] ≡ Leucosphaerina (Pseudeurot.).

Leucosphaerina Arx (1986), Pseudeurotiaceae. 2 (coprophilous), India, USA. See Malloch (*Stud. Mycol.* **31**: 107, 1989).

Leucosporidium Fell, Statzell, I.L. Hunter & Phaff (1970), Sporidiobolaceae. 2 (from sea water), Antarctica. See Yamada & Komagata (*J. Gen. Appl. Microbiol.* **33**: 456, 1987).

Leucosporium Corda (1833) nom. dub. (Mitosp. fungi) fide Holubová-Jechová (*Sydowia* **45**: 97, 1993).

leucosporous, having spores white in the mass.

Leucostoma (Nitschke) Höhn. (1917), Valsaceae. *c.* 10, temp. See Kern (*Phytopath. Z.* **40**: 303, 1961).

Leucotelium Tranzschel (1935), Uredinales (inc. sed.). 3, N. temp; I on Ranunculaceae, II, III on Prunus. See Savile (*CJB* **67**: 2983, 1989).

Leucothallia Trevis. (1853) ≡ Sphaerotheca (Erysiph.).

Leucothecium Arx & Samson (1973), Gymnoascaceae. 1, Netherlands. See Valldosera *et al.* (*MR* **95**: 243, 1991).

Leucothyridium Speg. (1909) = Cucurbitaria (Cucurbitar.) fide Petrak & Sydow (*Ann. Myc.* **32**: 1, 1934).

Leucovibrissea (A. Sánchez) Korf (1990), Vibrisseaceae. 1, USA.

Leuliisinea Matsush. (1985), Mitosporic fungi, 2.B1/2.19. 1, Taiwan.

Léveillé (Joseph-Henri; 1796-1870). A medical man of Paris. Chief writings: paper on the hymenium (*Ann. Sci. nat.* sér. 2, **8**: 321, 1837) in which the names basidium and cystidium are first used; on the taxonomy of *Erysiphe* (*Ann. Sci. nat.* sér. 3, **15**: 109, 1851); *Iconographie des champignons de Paulet* (1855). See Grummann (1974: 286), Stafleu & Cowan (*TL-2* **2**: 965, 1979).

Leveillea Fr. (1849) [non Decne. (1839), *Algae*] ? = Phylacia (Xylar.) fide Læssøe (1994).

Leveillella Theiss. & Syd. (1915), Asterinaceae. 1, Chile. See Eriksson (*SA* **7**: 78, 1988).

Leveillina Theiss. & Syd. (1915), ? Dothideales (inc. sed.). 2, Afr., S. Am.

Leveillinopsis F. Stevens (1924) = Coccodiella (Phyllachor.) fide Müller & v. Arx (1973).

Leveillula G. Arnaud (1921), Erysiphaceae. 8, widespr. See Durrieu & Rostam (*Cryptogamie, Mycol.* **5**: 279, 1985), Gelyuta & Simonian (*Biol. Zh. Armenii* **39**: 20, 1987; subgen. divis.), Palti (*Bot. Rev.* **54**: 423, 1988; monogr.). Anamorph *Oidiopsis*.

Levieuxia Fr. (1848) nom. dub. (Mitosp. fungi) fide Sutton (*Mycol. Pap.* **141**, 1977).

levigate, see laevigate.

Levispora Routien (1957) = Pseudeurotium (Pseudeurot.) fide Malloch (*in litt.*).

Lewia M.E. Barr & E.G. Simmons (1986), Pleosporaceae. 5, widespr. See *Lit.* under *Pleospora*. Anamorph *Alternaria*.

Liaoningnema S.L. Zheng & W. Zhang (1986), Fossil fungi. 1, China.

Libartania Nag Raj (1979), Mitosporic fungi, 8.C1.10. 3, widespr. See Nag Raj (1993).

Libellus Lloyd (1913) [non Cleve (1873), *Algae*] = Hymenogloea (Tricholomat.) fide Singer (1951).

Libertella Desm. (1830), Anamorphic Diatrypaceae, Xylariaceae, 8.E1.10. Teleomorphs *Diatrypella, Barrmaelia*. 20, temp.

Libertiella Speg. & Roum. (1880), Mitosporic fungi, 8.A1.15. 1 (on *Peltigera*), Eur. See Hawksworth (*Bull. Br. Mus. nat. Hist., Bot.* **9**: 30, 1981).

Libertina Höhn. (1920) = Phomopsis (Mitosp. fungi) fide Sutton (*TBMS* **50**: 355, 1967).

liberty cap, basidioma of the hallucinogenic *Psilocybe semilanceata*.

Licaethalium Rostaf. (1873) = Reticularia (Lycogal.).

Licea Schrad. (1797), Liceaceae. 41, N. Hemisph. See Keller & Brooks (*Mycol.* **69**: 669, 1977; key).

Liceaceae Chevall. (1826), Liceales. 2 gen. (+ 9 syn.), 42 spp.

Liceales. Myxomycota. 3 fam., 8 gen. (+ 27 syn.), 93 spp. Spore mass dingy or colourless, capillitium absent. Fams:
  (1) **Cribrariaceae**.
  (2) **Liceaceae**.
  (3) **Lycogalaceae**.
  Lit.: Martin (*Brittonia* **13**: 109, 1961; fam. key).

Licentia Pilát (1940) = Lopharia (Ster.) fide Donk (1956).

Liceopsis Torrend (1908) = Reticularia (Lycogal.).

lichen (pl. lichens; pronounced 'lie'ken') [Lat. *lichen*, from Gk λειχήν, tree moss] (cen (Welsh), Flechten (Germ.), gil-i-sang (Persian), huidmos and korsmos (Afrikaans), jäkälä (Finnish), kerpés (Lith.), korstmos (Dutch), lav (Danish, Norweg., Swed.), lichène (Ital.), ligen (Afrikaans), líquen and liquen (Portug., Span.,

S. Am.), lišaj (Croat.), lišejník (Bohem., Czech. Slovak), liszaj and porost (Pol.), mareru (Kikuyu), лишаj (Serb.), лишай (Russian), лишей (Bulgar.), ragu (Old English), ライケンtii (Jap.), 582.29 (Universal Decimal Classification)), one of the Lichens (q.v.); (in Medicine), any of various eruptive skin diseases. **- acids**, see Metabolic products; **- alga**, phycobiont (q.v.); photobiont (q.v.); **-biont interaction** see Ahmadjian (*in* Reisser, *Algae and symbiosis*: 675, 1992); **- desert**, the area in a town or around an air pollution source from which all lichens, or at least all foliose and fruticose lichens are absent (obsol.); see Air pollution; **-icolous**, inhabiting lichens; **-iform**, having the form of a lichen; **-iverous**, lichen-eating; **-oglyph**, a centuries-old picture made by Canadian Indians by scraping lichens off the surface of a large vertical rock face; **-oid**, resembling a lichen; **-ologist**, one engaged in the pursuit of lichenology; **-ology**, the scientific study of lichens; **- products, - substances**, see Metabolic products.

Lichen L. (1753) nom. rej. ≡ Parmelia (Parmel.).

Lichenagaricus Micheli (1729) nom. inval. = Xylaria (Xylar.), etc.

lichenen (**lichenin**), an I+ red linear polymer of β-d-glucose with 1,3 and 1,4 linkages in the ratio 3:2; **isolichenan** (isolichenin), an isomer of lichenan, is I+ lilac or lavender. These carbohydrates occur in the walls of the hyphae of many lichen-forming fungi; lichenan tends to occur in higher concentrations and its presence/absence is taxonomically significant (Common, *Mycotaxon* **41**: 67, 1991). See amylomycan, Iodine.

Lichenes (obsol.). Name of a class for all lichen-forming fungi used when these were regarded as quite separate from *Fungi*; *Mycophycophyta*. **- imperfecti** (obsol.), lichen-forming mitosporic fungi; *Deuterolichenes* (q.v.), often used inclusive of lichen-forming fungi in which the sporocarps are unknown and the position is uncertain.

Lichenicolous fungi. Fungi dwelling on or in lichens as parasites (pathogens or gall-forming), commensals (see also parasymbiont) or saprobes. There are about 300 gen. and 1000 spp. of obligately lichenicolous fungi now known; many of the genera are exclusively lichenicolous, and they are proving a rich source of novel genera. Common saprobic moulds are very scarce on lichens with depsides or depsidones, but a wide range of non-obligate species can be recovered by isolation (Petrini *et al.*, *Mycol.* **82**: 441, 1990).

Ascomycota, Basidiomycetes and Mitosporic fungi all include obligately lichenicolous spp. Some lichenicolous fungi can take over the algae from other lichens and form new thalli (Hawksworth, *Bot. J. Linn. Soc.* **96**: 3, 1988).

Many groups are undergoing a critical re-appraisal in what is proving to be one of the most unexplored of all ecological niches occupied by fungi. Some may affect the host chemistry or produce separate novel metabolites (Hawksworth *et al.*, *Bibl. Lich.* **53**: 101, 1993).

Lit.: **General**: Clauzade *et al.* (*Bull. Soc. linn. Provence, num.-spéc.* **1**, 1989; keys 682 spp.; amendments in *Bull. Ass. Fr. lichén.* **16**(2): 71, 1991), Diederich & Christiansen (*Lichenologist* **26**: 47, 1994; basidiomycetes), Grummann (*Bot. Jb.* **80**: 101, 1960; galls), Hafellner (*Herzogia* **6**: 289, 299, **7**: 145, 163, 343, 353, 1982-87, *Nova Hedw.* **48**: 357, 1989), Hawksworth (*Bull. Br. Mus. nat. Hist., Bot.* **6**: 183, 1979; hyphomycetes, *Bull. Br. Mus. nat. Hist., Bot.* **9**: 1, 1981; coelomycetes, *TBMS* **74**: 363, 1980; on *Peltigera, J. Hattori bot. Lab.* **52**: 357, 1982; review biol., *J. Hattori bot. Lab.* **52**: 323, 1982; coevolution), Keissler (*Beih. bot. Zbl.* **50**: 380, 1933), Triebel (*Bibl.*

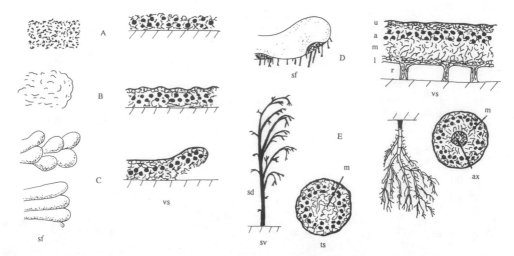

Fig. 21. Growth forms and thallus structure in lichens. A, leprose; B, crustose; C, squamulose (lower placodioid); D, foliose; E, fruticose. a - algal layer, ax - axis, l, - lower cortex, m - medulla, r - rhizinae, sd - soredium, sf - surface view, sv - side view, ts - transverse section, u - upper layer, vs - vertical section. Not to scale.

Lich. 35, 1989; on lecideoid spp.), Vouaux (*BSMF* 28: 177, 29: 33, 399, 30: 135, 281, 1912-14; keys, descr., monogr.), Zopf (*Hedwigia* 35: 312, 1896; host index).

Regional: Europe, Keissler (*Rabenh. Krypt.-Fl.* 8, 1930). **Austria,** Hafellner (*Herzogia* 10: 1, 1994). **Belgium,** Diederich *et al.* (*Dumortiera* 42: 17, 1988). **British Isles,** Hawksworth (*Kew Bull.* 30: 183, 1975, *Notes R. bot. Gdn Edinb.* 36: 181, 1978, 38: 165, 1980, 40: 375, 1982, 43: 497, 1986, 46: 391, 1990, *Lichenologist* 15: 1, 1983; key 218 spp., 26: 337, 1994). **Former Czechoslovakia,** Vězda (*Česká Myk.* 17: 149, 1960; 23: 104, 1969; 24: 220, 1970). **Denmark,** Alstrup (*Graphis Scripta* 5: 60, 1993). **France,** Olivier (*Bull. internat. géogr. Bot.* 15-17, 1905-07), Roux [*et al.*] (*Bull. Mus. Hist. nat. Marseille* 36: 19, 1976; 37: 83, 1977, *Bull. Soc. linn. Provence* 36: 195, 1984, 41: 117, 1990, 43: 81, 1992). **Greenland,** Alstrup & Hawksworth (*Meddr Grønl., Biosci.* 31, 1990; keys 124 spp.). **Luxembourg,** Diederich (*Lejennia* 119: 1, 1986, *Mycotaxon* 37: 297, 1990). **Norway and Sweden,** Santesson (*The lichens and lichenicolous fungi of Norway and Sweden*, 1993; checklist). **Poland,** Faltynowicz (*Polish Bot. Stud.* 6, 1993; checklist). **Sardinia,** Nimis & Poelt (*Studia Geobot.* 7 (Suppl.), 1987). **Spain,** Santesson (*Svensk bot. Tidskr.* 54: 499, 1960), Hafellner & Sancho (*Herzogia* 8: 363, 1990). **Sweden,** see Norway. **North America,** Egan (*Bryologist* 90: 77, 1987), Triebel *et al.* (*Mycotaxon* 42: 263, 1991; 58 spp.). Also often included in fungus and/or lichen catalogues or 'floras', see Geographical distribution, Lichenicolous lichens, Lichens.

Lichenicolous lichens, Lichens which grow on (or in) other lichens, either as commensals or parasites; 4-biont symbioses (see symbiosis). Each fungal partner in such associations has an independent algal or cyanobacterial partner, whereas no additional photosynthetic partner occurs in the obligately lichenicolous fungi (q.v.). The algae of the lichenicolous lichens occur either inside the host in e.g. *Buellia pulverulenta* on *Physconia distorta* (Hafellner & Poelt, *Phyton* 20: 129, 1980) or as discrete thalli on its surface as in *Caloplaca epithallina* on 13 host lichens

(Poelt, *Bot. Jahrb.* 167: 457, 1985). Parasitic spp. are known esp. in *Acarospora, Diploschistes, Rhizocarpon* (Poelt, *Mitt. bot. StSamml., Münch.* 29: 515, 1990), and *Verrucaria*. See also Poelt & Döbbeler (*Planta* 46: 467, 1956), Rambold & Triebel (*Bibl. Lich.* 48, 1992; inter-lecanoralean assocs.).

Lichenochora Hafellner (1989), Phyllachoraceae. 7 (on lichens), mainly temp.

Lichenoconium Petr. & Syd. (1927), Mitosporic fungi, 4.A2.19. 9 (on lichens), widespr. See Hawksworth (*Persoonia* 9: 159, 1977, *Bull. Br. Mus. nat. Hist., Bot.* 9: 33, 1981).

Lichenodiplis Dyko & D. Hawksw. (1979), Mitosporic fungi, 4.B2.19. 2 (on lichens), Eur., Asia. See Hawksworth (*Bull. Br. Mus. nat. Hist., Bot.* 9: 37, 1981).

Lichenoides Hoffm. (1789) ≡ Anaptychia (Physc.).

Lichenometry. Technique for study of exposure age of rock surfaces based on the size/diameter (proportional to age) of lichen thalli. Pioneered by Beschel in the 1950s (see Raasch (Ed.), *Geology of the arctic* 2: 1044, 1961) and now used extensively by glaciologists but also applicable to the minimum dating of many stone surfaces. See Webber & Andrews (*Arctic Alpine Res.* 5: 293, 1973; review, bibliogr.), Matthews (*Norsk geogr. Tidsskr.* 29: 97, 1975; Jotunheimen), Luckman (*Can. J. Earth Sci.* 14: 1804, 1977; Alberta), Innes (Lichenometry [*Prog. phys. Geogr.* 9: 187], 1985; *in* Galun, *CRC Handbook of lichenology* 3: 75, 1988).

A critical review of the technique is provided by Worsley (*in* Goudie *et al., Geomorphological techniques*: 302, 1981) and a size-frequency approach addressing some criticisms of the technique has been developed by Winchester & Harrison (*Earth surface processes & landforms* 19: 137, 1994). See Ecology, Growth rates.

Lichenomyces Trevis. (1853) = Plectocarpon (Roccell.) fide Hawksworth & Galloway (1984).

Lichenopeltella Höhn. (1919), Microthyriaceae. c. 20 (some on lichens), widespr. See Santesson (*SA* 9: 15, 1990; nomencl.), Spooner & Kirk (*MR* 94: 223, 1990; as *Micropeltopsis*).

Lichenopeziza Zukal (1884), Ascomycota (inc. sed., ?L). 1, Eur.

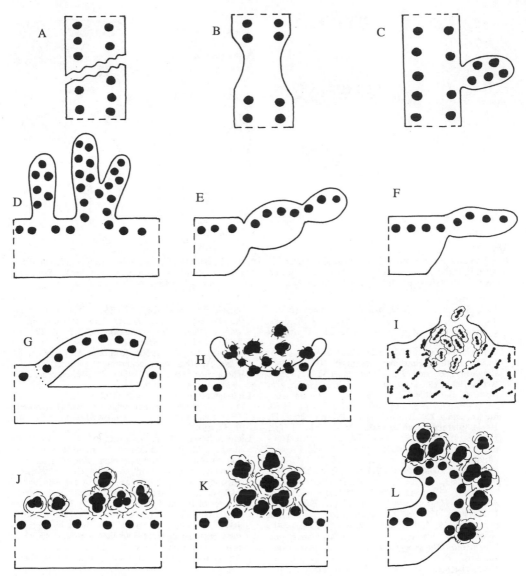

Fig. 22. Types of vegetative propagule in lichens. A, thallus fragmentation; B, fragmentation region; C, lateral spinule; D, isidia; E, blastidia; F, phyllidium; G, schizidium; H, goniocyst; I, hormocyst; J, soredia formed from eroded surface; K, soredia formed in soralium; L, soredia formed from recurved lower cortex (labriform).

Lichenophoma Keissl. (1911), Mitosporic fungi, 4.A1.?. 1 (on lichens), Eur. Nom. dub. fide Hawksworth (*Bull. Br. Mus. nat. Hist., Bot.* **9**: 77, 1981).

Lichenopsis Schwein. (1832) = Stictis (Stictid.).

Lichenopuccinia D. Hawksw. & Hafellner (1984), Mitosporic fungi, 3.C1.1. 1 (on lichens, esp. *Parmelia*), Austria, UK, Greenland.

Lichenosphaeria Bornet (1873) ? = Didymella (Dothideales) fide Henssen (*Symb. bot. upsal.* **18** (1), 1963).

Lichenosticta Zopf (1898), Mitosporic fungi, 4.A1.15. 1 (on lichens, esp. *Cladonia*), N. temp. See Hawksworth (*Bull. Br. Mus. nat. Hist., Bot.* **9**: 38, 1981).

Lichenostigma Hafellner (1983), Lichenotheliaceae. 2 (on lichens), widespr. See Thor (*Lichenologist* **17**: 269, 1985).

Lichenothelia D. Hawksw. (1981), Lichenotheliaceae. 18, widespr. See Henssen (*Bibl. Lich.* **25**: 257, 1987).

Lichenotheliaceae Henssen (1986), Dothideales. 2 gen. (+ 1 syn.), 20 spp. Ascomata formed in a dark crustose pseudoparenchymatous stroma, uniloculate, perithecial (without a clearly defined ostiole) or apothecial, globose to pulvinate; peridium of dark pseudoparenchymatous cells; interascal tissue present (? pseudoparaphyses), asci clavate, ? fissitunicate; ascospores dark brown, septate, thick-walled, sometimes ornamented, with a mucous sheath. Anamorphs hyphomycetous or coelomycetous. Saprobic or biotrophic, on rocks, lichens, or cyanobacterial mats.

Lichenothrix Henssen (1964) = Pyrenothrix (Pyrenothric.) fide Eriksson (1981).

Lichens. A lichen is essentially a stable self-supporting association of a fungus (mycobiont) and an alga or cyanobacterium (photobiont). More precisely, a lichen is an ecologically obligate, stable mutualism between an exhabitant fungal partner and an inhabitant population of extracellularly located unicellular or filamentous algal or cyanobacterial cells (Hawksworth & Honegger, *in* Williams (Ed.), *Plant galls*: 77, 1994). Many other definitions have been proposed (Hawksworth, *Bot. J. Linn. Soc.* **96**: 3, 1988; review), and Ahmadjian (*The lichen symbiosis*, edn 2, 1993) stresses the formation of a 'thallus or lichenized stroma that may contain unique secondary compounds' as the key feature. The partners involved may have coevolved (Ahmadjian, *Ann. N.Y. Acad. Sci.* **503**: 307, 1987; Hawksworth, *in* Pirozynski & Hawksworth (Eds), *Coevolution of fungi with plants and animals*: 125, 1988).

Lichens are a biological and not a systematic group, and are unique in that in many (but not all) cases the resulting life form and behaviour differ markedly from that of the isolated components. In most lichens the fungal partners, which do not occur free-living (except in some facultatively lichenized fungi), and appear to be responsible for the overall form of the thallus and its fruiting bodies (ascomata), but there is an interplay between both components to produce the final form (see below), an a wide range of biological interactions is involved - these may involve 4-, 5- or more partners (Hawksworth, 1988; review).

Fungal partners: Almost one fifth (19%) of all *Fungi* and two fifths (42%) of all *Ascomycota* are lichenized. These statements are based on addition of the totals of lichen-forming genera in this edition of the *Dictionary* which give a total of almost 13,500 spp.; that this figure is unchanged from the 7th edition (1983) suggests that rates of description and synonymy are roughly equal. Larger estimates of currently known species (e.g. 18,000 by Nash & Egan, *Bibl. Lich.* **30**: 11, 1988) should nevertheless not automatically be discounted as many exceptionally large genera lack modern revisions (see Numbers of fungi). Galloway (*Biodiv. Conserv.* **1**: 312, 1992) estimated that the world total would probably be in the range 17,000-20,000 when all had been described. Consequently, while lichenization is one of the principle biological strategies of the described fungi, it will be a much smaller proportion of the total biota if the estimate of 1.5 million is vindicated (see Numbers of fungi). Lichens are polyphyletic in origin, as established beyond doubt by molecular methods (Gargas *et al.*, *Science* **269**: 1492, 1995). The fungi of most species belong to the *Ascomycota* (q.v.), 13 of the 46 orders which include lichen-forming species. However, only 4 orders are exclusively lichenized. Lichenization also occurs in a few *Basidiomycota* (e.g. *Dictyonema, Multiclavula, Omphalina*), and also Mitosporic fungi (e.g. *Blarneya, Cystocoleus*; see Vobis & Hawksworth, *in* Cole & Kendrick (Eds), *The biology of conidial fungi* **1**: 245, 1981; review). However, while many families are exclusively lichen-forming, a mixture of lichenized and non-lichenized (and also sometimes lichenicolous) species can occur within the same family or even genus (e.g. *Arthonia, Arthothelium, Bacidia, Mycomicrothelia, Omphalina, Toninia*). What is or is not regarded as a 'lichen' is partly a matter of history; for example, *Omphalina, Orbilia*, and *Pezizella* are not usually studied by lichenologists, while some taxa generally regarded as the preserve of lichenologists are evidently not lichenized (e.g. *Arthopyrenia, Chaenothecopsis, Leptorhaphis, Stenocybe*). The existence of genera crossing biological boundaries, and of species that may be viewed as primitive lichens (Kohlmeyer & Kohlmeyer, *Marine mycology*, 1979) or are facultatively lichenized, makes it clear that the lichen method of nutrition (q.v.) is at various stages of evolution in different groups; evolving in some (e.g. lichenized *Leotiales, Agaricales*) and devolving in others (e.g. non-lichenized *Arthoniales* and *Lecanorales*). Axenic cultures of the fungal partner can be obtained from ascospores and thallus fragments; Crittenden *et al.* (*New Phytol.* **130**: 267, 1995) had success with 493 spp. (42%) of 1183 attempted. The cultures are generally slow-growing, have little organized structure, and do not produce ascomata (although some form conidiomata).

Algal and cyanobacterial partners: The number of photosynthetic partners involved in lichen formation is relatively small; only 40 genera are represented, 25 algae and 15 cyanobacteria (see Ahmadjian, *Phycologia* **6**: 128, 1967, identification; Ahmadjian, 1993; Hildreth & Ahmadjian, *Lichenologist* **13**: 65, 1981, keys *Trebouxia* spp.; Honegger, *Exper. Phycol.* **1**: 40, 1990; Tschermak-Woess, *in* Galun, *CRC Handbook of lichenology* **1**: 29, 1988). Most belong to genera which are also free-living both in the cyanobacteria (e.g. *Calothrix, Gloeocapsa, Nostoc, Scytonema, Stigonema*) and the green algae (*Cephaleuros, Coccomyxa, Myremecia, Trentepohlia*). Only *Trebouxia* (incl. *Pseudotrebouxia*) appears to be primarily lichen-forming, with doubt cast on its free occurrence in nature (Ahmadjian, 1993; Tschermak-Woess, 1988). The same species, but often as different strains, may occur in a very wide range of lichenized genera. In addition more than one strain of a single alga (or species of an algal genus) can be involved in the formation of a single lichen thallus. However, the systematics of the algal and cyanobacterial partners of lichens is currently inadequate to make more definite assertions as to specificity. However, some lichens (e.g. *Lobaria, Placopsis, Psoroma, Stereocaulon*) with an algal partner, cyanobacteria also occur in 'cephalodia' (q.v.) which may, according to the species, be warts on the upper or lower surface or circumscribed areas within the thallus. In some cases cephalodia can be separated and persist as independent lichens (e.g. *Dendriscocaulon*); occasionally, two species with different types of algae previously regarded as distinct have been found joined together as a composite ('chimeroid') thallus showing these 'photomorphs' (q.v.; phycosymbiodemes, photosymbiodeme, phycotypes) are the result of interactions between a single mycobiont and distinct photobionts (Armaleo & Clerc, *Exp. Mycol.* **15**: 1, 1991; James & Henssen, *in* Brown *et al.*, 1976: 27; Stöcker-Wörgötter & Türk, *Crypt. Bot.* **4**: 300, 1994; Tønsberg & Holtan-Hartwig, *Nordic Jl Bot.* **3**: 681, 1983), White & James, *Lichenologist* **20**: 103, 1988).

Nomenclature: Lichens do not have independent scientific names; the fungal and photosynthetic partners each have separate names, and names given to lichens are considered as referring to the fungal partner alone (see Nomenclature). The classification of 'lichens' therefore has to be completely integrated into the system of *Fungi*. The names of lichen-forming fungi have one other special provision in the Code, exemption from the provisions for the naming of pleomorphic fungi. These fungi are covered in the *Index of Fungi* (**5**(4), 1982 on. Current nomenclatural practice is consistent with the recognition of lichens as a nutritional rather than a taxonomic group.

Synthesis: Although the bionts can be separated and cultured, resynthesis under laboratory conditions is difficult. Some success has been achieved by lowering the nutrient supply and modifying the moisture requirements, features which suggest that lichens behave as two organisms united in adversity. Synthesis was first achieved experimentally by Stahl in 1877 (repeated by Ahmadjian & Heikkilä, *Lichenologist* **4**: 259, 1970), and has now also been carried out in *Dermatocarpon*, *Usnea* and *Xanthoria* (Ahmadjian, 1993). Novel syntheses have also been attempted; Ahmadjian *et al.* (*Mycol.* **72**: 73, 1980) found that *Cladonia cristatella* could be re-synthesized with 13 different *Trebouxia* isolates, but none of 10 '*Pseudotrebouxia*' isolates or free-living *Pleurastrum*; squamules, pycnidia and short podetia with immature hymenia were obtained and the metabolic products (q.v.) were identical in all cases (Ahmadjian, *The lichen symbiosis*, edn 2, 1993; *in* Cook (Ed.), *Cellular interactions with symbiosis and parasitism*: 3, 1980; *Progress Phycol. Res.* **1**: 179, 1982). Techniques for growing lichens as tissue cultures with intermixed fungi and algae have been developed for about 200 spp. since 1981 (Yamamoto *et al.*, *Bryologist* **96**: 383, 1993).

Structure: The algae or cyanobacteria are either distributed at random, often in a gelatinous matrix, throughout the thallus (**homoiomerous**; unlayered; unstratified), or in a compact layer below the upper or outer cortex (**heteromerous**; layered; stratified); see Fig. 40. There are a few **filamentous** lichens (*Coenogonium*, *Cystocoleus*, *Racodium*) in which the filamentous form of the alga predominates; in the majority, the fungus forms the outer structure and gives the lichen its shape. Six life-form categories are generally recognized (Fig. 21): shrub- or beard-like (**fruticose**), leaf-like (**foliose**), scale-like (**squamulose**; **placodioid** if rosette-like in surface view with only the tips free), crust-like (**crustose**, **crustaceous**), **filamentous** (see above), and **leprose** (powdery, loose and often powdery aggregations of algal cells entwined with hyphae and with no cortex). Fruticose thalli are the most complex and largest; these may be erect or pendulous, hair- or strap-like and are often richly branched. They have a radial structure arranged around a central cavity (e.g. *Bryoria*) or a tough axial strand (*Usnea*), and are attached to the substratum by small disc-like holdfasts or clusters of rhizoids. Some *Usnea* spp. on trees attain a length of 10 m. In foliose and crustose thalli the structure is essentially dorsiventral and growth most pronounced at the circumference; the former have a well-developed lower cortex and are attached to the substratum by puckering of the underside (*Hypogymnia*), a fine felted tomentum (*Lobaria*), or by coarse bundles of compacted hyphae or rhizines (*Parmelia*). Crustose lichens may be more or less immersed in the substratum and, like squamules, lack a lower cortex.

The surface of crustose species is often characteristically cracked or warted, and may be bounded by a pale or black marginal **prothallus** which contains no photobiont cells. The gelatinous homoiomerous lichens (*Collema*, *Placynthium*) also occur in similar life forms. Any particular life-form is not necessarily confined to a genus, family, or order (e.g. all except the filamentous life-form occur in *Teloschistales*). In a few genera, such as *Cladonia*, *Pilophorus* and *Stereocaulon*, there is a combination of different growth forms: in many species of *Cladonia*, the basal part of the thallus is squamulose, from which arise fruticose, hollow, simple, or branched structures (podetia; q.v.), often also bearing squamules, the apices of which are either pointed (subulate) or bear terminal cups

(scyphi) on which are borne the ascocarps. In *Stereocaulon* the basal, peltate, or coralloid phyllocladia give rise to simple or branched pseudopodetia (see podetia), also more or less also covered in phyllocladia. The pseudopodetia bear lateral or terminal ascocarps.

A few genera have additional thalline structures, such as aeration pores in the form of **cyphellae** (*Sticta*) and pseudocyphellae (*Punctelia*, *Pseudocyphellaria*); marginal eye-lash-like cilia occur in *Heterodermia* and in some species of *Physcia* and *Parmotrema*. See Beltmann (*Bibl. Lich.* **11**, 1978; Parmeliaceae), Hannemann (*Bibl. Lich.* **1**, 1973; attachment organs), Henssen & Jahns (1973), Jahns (*in* Galun, *CRC Handbook of lichenology* **1**: 95, 1988), Ozenda (*Handb. Pflanzenanatomie* **6** (9), 1963).

Reproduction: Asexual or vegetative reproduction of lichens involving both partners (dual propagules) may be achieved by a wide range of methods (Fig. 22), including: (1) simple fragmentation, (2) development of delimited or widespread areas of cortical breakdown (**soralia**) which contain minute powdery propagules (**soredia**), and (3) the development of numerous small, simple, or branched-coralloid, corticate papillae (**isidia**). These methods involve the dissemination of the united bionts. The position, shape, and sometimes the colour of the soralia and isidia is often diagnostic and is usually accompanied by a marked suppression of ascomatal development (see Phylogeny).

A number of species also develop innate, flask-shaped **pycnidial conidiomata** in which numerous, often characteristically shaped conidia develop. Although conidia of some species germinate in culture their functional role is not fully understood and may, in some species at least, be sexual. Only the mycobiont reproduces sexually, a process culminating in the development of ascomata. The nature of the ascomata depends on the order to which the fungal partner belongs. However, the ascomata are mainly perennial with serial development of the asci. Two main apothecial types are often distinguished: **lecanorine** with a thalline exciple (containing algae and generally the same colour as the main thallus); and **lecideine** only with a true exciple (lacking algae and usually differing in colour from the thallus). In some asexually reproducing species ascomata can be absent or very rare (and then often with few or no well-formed ascospores); see also species pairs.

Establishment: Where dual propagules are the means of dispersal, locating a compatible partner does not present difficulties. However, where several dual propagules of the same or sometimes different lichens start to grow close together they can become intermixed and form interspecific or even intergeneric **mechanical hybrids** (see hybrid); e.g. *Physcia adscendens* + *P. tenella* (Schuster *et al.*, *Lichenologist* **13**: 247, 1985), *P. tenella* + *Xanthoria parietina* (Ott, *Bibl. Lich.* **25**: 81, 1987). The extent to which 'individual' lichen thalli represent the product of a single propagule and belong to one genotype merits investigation.

Where ascospores are the dispersal agent, various strategies for establishment are documened: the ascospore may land on and re-shape an already growing algal or cyanobacterial colony (e.g. *Collema*), invade an established lichen killing the fungal partner and taking over the algae (e.g. *Diploschistes*), land near and out-compete the fungal partner in a dual propagule derived from another lichen, or persist in a loose unstructured association with other algae until a truely compatible algal species arrives (e.g.

Xanthoria; Ott, *Nordic Jl Bot.* **7**: 219, 1987). See Spore discharge and dispersal.

Interactions between the bionts: In the early stages of synthesis, the partners can be bound by a common sheath (Ahmadjian *et al.*, *Science* **200**: 1062, 1978); similar ultrastructural-level patterns on the surfaces of both partners can enable them to mesh together (Honegger *et al.*, *Lichenologist* **16**: 111, 1984). The actual nature of the cell-to-cell contact varies in different groups; these range from wall-to-wall apposition to a variety of kinds of intraparietal haustoria (see Honegger, *New Phytol.* **103**: 785, 1986; see haustorium, Fig. 15). Intracellular haustoria completely penetrating the walls of healthy algal cells are unusual, but the extent of penetration may vary even within a single thallus (Galun *et al.*, *J. Microscopie* **9**: 801, 1970). Dead algal cells occuring in thalli may be utilized, leading to suggestions that 'controlled parasitism' may more accurately explain the relationship than 'mutualism' (see above, Ahmadjian, 1993). The fungus obtains from the alga growth substances (thiamine, biotin), as well as products of its photosynthesis, including simple sugars and polyalcohols. Cyanobacterial partners, whether the main photobiont or in cephalodia, fix atmospheric nitrogen which is passed to the fungus. Some of the green algae involved may be distingushed from non-lichenized relatives by an accelerated leakage of sugars across their cell walls in the presence of a compatible fungus.

Whole-thallus physiology: See Kershaw (*Physiological ecology of lichens*, 1985), Vincente *et al.* (Eds) (*Surface physiology of lichens*, 1985), Brown (Ed.) (*Lichen physiology and cell biology*, 1985), Galun (Ed.), 1985), Nutrition, Physiology.

Chemistry: See Chemotaxonomy, chromatography, Metabolic products, Microcrystal tests, Pigments.

Economics: See Air pollution, Antibiotics, Biodeterioration, Bioindication, Dyeing, Edible fungi, Insects and fungi, lichenometry, Medical uses of fungi and lichens, RIEC. See also Llano (*Bot. Rev.* **10** (1), 1944), Richardson (*The vanishing lichens*, 1975; *in* Galun, *CRC Handbook of lichenology* **3**: 93, 1988), Seaward (Ed.) (*Lichen ecology*, 1977).

Lit.: **General**: Ahmadjian (*The lichen symbiosis*, edn 2, 1993), Ahmadjian & Hale (Eds) (*The lichens*, 1974), Bates & Farmer (Eds) (*Bryophytes and lichens in a changing environment*, 1992), Brown *et al.* (Eds) (*Lichenology: progress and problems*, 1978), Dalby *et al.* (Eds) (*Horizons in lichenology* [*Bot. J. Linn. Soc.* **96**], 1988), Galun (Ed.) (*Handbook of lichenology*, 3 vols, 1988), Hale (*The biology of lichens*, 1983), Hawksworth (*Crypt. Bot.* **4**: 117, 1994; advances 1972-92), Hawksworth & Hill (*The lichen-forming fungi*, 1984), Henssen & Jahns (*Lichenes*, 1973), Lawrey (*Biology of lichenized fungi*, 1984), Smith (*Lichens*, 1921 [reprint 1975); history], Zahlbruckner (*Naturl. Pflanzenfam.* **8**: 61, 1926; keys gen.).

Regional: Hawksworth & Ahti (*Lichenologist* **22**: 1, 1990; 1390 refs. by continent and country); the following list is restricted to major modern works, esp. of value in identification. **Africa, East**, Swinscow & Krug (*Macrolichens of East Africa*, 1988). **Antarctica**, Rédon (*Liquenes Antárticos*, 1985). **Austria**, Türk & Poelt (*Bibliographie de Flechten* .... *Österreich*, 1993). **Australia**: *Flora of Australia* 54 (1992 ongoing), McCarthy (*Checklist of Australian lichens*, edn 4, 1991), Rogers (*The genera of Australian lichens (lichenized fungi)*, 1980). **British Isles**, Purvis *et al.* (*The lichen flora of Great Britain and Ireland*, 1992; keys 1700 spp.; *Bull. Br. lichen. Soc.* **72** (Suppl.), 1993; checklist). **Canada**, Brodo

(*Lichens of the Ottawa region*, edn 2, 1988), Goward *et al.* (*The lichens of British Columbia, Illustrated keys*, 1994); see also N. America. **China**, Wei (*An enumeration of lichens in China*, 1991), Zhao *et al.* (*Prodromus Lichenum Sinicorum*, 1982). **Estonia**, Trass & Randlane (*Eesti suur-samblikud*, 1994). **Europe**, Clauzade & Roux (*Bull. Soc. Bot. Centre-Ouest, n.s.* **7**:1, 1985), Jahns (*Farne, Moose, Flechten Mittel-, Norde- und Westeuropas*, 1980), Poelt (*Bestimmungschfüssel europäischer Flechten*, 1969), Poelt & Vězda (*Bibl. Lich.* **9**: 1, 1977; **16**: 1, 1981). **France**, Ozenda & Clauzade (*Les Lichens. Étude biologique et flore illustrée*, 1970). **Germany**, Wirth (*Die Flechten Baden-Württembergs*, edn 2, 2 vols, 1995; keys 1500 spp., 555 col. pl.). **Hungary**, Verseghy (*Magyaroszág zuzmó flórájának Kézikönyve*, 1994). **India**, Awasthi (*J. Hattori Bot. Lab.* **65**: 207, 1988; macrolichens; *Bibl. Lich.* **40**, 1991; microlichens), Singh & Sinha (*Lichen flora of Nagaland*, 1994; keys, descr.). **Italy**, Nimis (*The lichens of Italy: an annotated catalogue*, 1993). **Japan**, Yoshimura (*Lichen flora of Japan in colour*, 1974). **Netherlands**, Aptroot & van Herk (*Veldgids Korstmossen*, 1994), Brand *et al.* (*Weten. Med. KNNV* **118**, 1988; checklist). **N. America**, Egan (*Bryologist* **90**: 77, 1987), Hale (*How to know the lichens*, edn 2, 1979), Thomson (*American Arctic Lichens. I. The Macrolichens*, 1984), Vitt *et al.* (*Mosses, lichens and ferns of northwest North America*, 1988). **New Zealand**, Galloway (*Flora of New Zealand lichens*, 1985). **Norway**, Krog *et al.* (*Lavflora. Norske buskog bladiav*, 1980); Tønsberg (*Sommerfeltia* **14**: 1, 1992; keys 128 corticolous spp.). **Poland**, Nowak & Tobolewski (*Porosty Polskie*, 1975). **Russia**, Abramov, (Ed.) (*Opredéltel' lishaĭnikov SSSR*, 5 vols, 1971-78). **Saudi Arabia**, Abu-zinada *et al.* (*Arab Gulf J. Sci. Res., s.p.* **2**: 1, 1986). **S.E. Asia**, Aguirre-Hudson & Wolseley (*J. Hattori Bot. Lab.* **76**: 313, 1994; bibliogr.). **Spain**: Llimona (*História natural des Països Catalans* **5**, *Fongsi líquens*, 1991). **Sweden**, Foucard (*Svensk skorplavs flora*, 1990), Moberg & Holmåsen (*Lavar. Enfälthandbok*, edn 2, 1984), Santesson (*The lichens and lichenicolous fungi of Sweden and Norway*, 1993). **Ukraine**, Oxner (*Flora lishaĭnik[i]v Ukraïni*, **1**, 1956; **2**(1), 1968; **2**(2), 1993). **U.S.A.**, Hale & Cole (*Lichens of California*, 1988); see also N. America. **Venezuela**, López Figueiras (*Censo de macroliquenes Venezolanos*, 1986).

See also *Ascomycota*, Lichenicolous fungi, Literature, and under individual subjects.

Lichina C. Agardh (1817) nom. cons., Lichinaceae (L). 7, temp. See Henssen (*Symb. bot. upsal.* **18**(1), 1963, *Lichenologist* **4**: 88, 1969, *Nova Hedw.* **15**: 543, 1969).

Lichinaceae Nyl. (1854), Lichinales (L). 39 gen. (+ 51 syn.), 253 spp. Thallus gelatinous; interascal tissue of simple or branched paraphyses, often swollen at the apices; asci thin-walled at all stages, evanescent or persistent. Lichenized with cyanobacteria, widespr.

Lit.: Moreno & Egea (*Acta Bot. Barcin.* **41**: 1, 1992; Iberia, N. Afr., keys 14 spp., gen. concepts).

Lichinales, Ascomycota (L). 3 fam., 44 gen. (+ 51 syn.), 281 spp. Thallus crustose, fruticose or foliose, sometimes peltate, often gelatinous; ascomata eventually apothecial but initially ± perithecial, often formed from pycnidia, opening by a pore, ± sessile or immersed; peridium often not well-defined; interascal tissue varied; hymenium often I+; asci thin-walled, without separable wall layers, without well-defined apical structures, usually with an I+ outer gelatinized layer; sometimes evanescent, sometimes polysporous; ascospores usually hyaline, aseptate. Anamorphs pycnidial. Lichenized with cyanobacteria, widespr. Fams:

(1) **Gloeoheppiaceae**.
(2) **Lichinaceae**.
(3) **Peltulaceae**.
   Lit.: Moreno & Egea (*Biología Taxonomía de la Familia Lichinaceae*, 1991), Henssen (*Symb. bot. upsal.* **18**(1), 1963, *Lichenologist* **27**: 261, 1995).
Lichinella Nyl. (1873), Lichinaceae (L). 11, Eur., N. Afr., S. Am. See Moreno & Egea (*Cryptogamie, bryol. lich.* **13**: 237, 1992; key 7 spp. Spain & N. Afr.).
Lichingoldia D. Hawksw. & Poelt (1986), Mitosporic fungi (L), 8.E1.15. 1 (aquatic), Norway.
Lichiniza Nyl. (1881) = Porocyphus (Lichin.) fide Henssen (1963).
Lichinodium Nyl. (1875), Lichinaceae (L). 2, Eur. See Henssen (*Symb. bot. upsal.* **18** (1), 1963).
Lichinomyces E.A. Thomas (1939) nom. inval. = Lichina (Lichin.).
Lichtheimia Vuill. (1903) = Absidia (Mucor.) fide Hesseltine (1955).
Licipenicillium, see *Lysipenicillium*.
Licopolia Sacc., Syd. & P. Syd. (1900), ? Dothideales (inc. sed.). 2, Brazil, Kenya.
Licrostroma P.A. Lemke (1964), Stereaceae. 1, N. Am., Japan.
Lidophia J. Walker & B. Sutton (1974), Dothideales (inc. sed.). 1, Germany. ? = Leptosphaeria (Leptosphaer.) fide v. Arx & Müller (1975), but see Walker (*Mycotaxon* **11**: 38, 1980).
Liesgang phenomenon, see zonation.
life history (life-cycle) (in fungi), the stage or series of stages (frequently characterized by different spore states, see States of fungi) between one spore form and the development of the same spore form again. Cf. sex.
ligative hyphae, Pouzar's term for binding hyphae; see hyphal analysis.
Light and fungi. Spore germination, growth, sporocarp development, sporulation, spore discharge, and movement in fungi are all influenced by light. See Marsh *et al.* (*Pl. Dis. Reptr Suppl.* **261**: 251-312, 1959; guide to lit.), Carlisle (*Ann. Rev. Pl. Physiol.* **16**: 175, 1965; photobiology of fungi). Cf. coolplate, photo-, zonation.
Ligiella J.A. Sáenz (1980), Clathraceae. 1, Costa Rica.
ligneous (lignose), wood-like.
lignicolous, living on or in wood.
Lignidium, see *Lignydium*.
Ligniella Naumov (1926) = Discula (Mitosp. fungi) fide Robak (*Sydowia* **6**: 378, 1952).
Ligniera Maire & A. Tison (1911), Plasmodiophoraceae. 7, N. temp. See Karling (1968; key).
Lignincola Höhnk (1955), Halosphaeriaceae. 1 (marine), widespr. See Kohlmeyer & Kohlmeyer (*Marine mycology*, 1979).
lignituber, see papillae.
Lignosus Lloyd ex Torrend (1920), Polyporaceae. 4, Afr., Asia, Australia. See Ryvarden & Johansen (*Preliminary polypore flora of East Africa*: 405, 1980).
Lignydium Link (1809) = Fuligo (Physar.) fide Martin (1966).
Lignyota Fr. (1849) nom. dub. (Didym.),
ligulate (liguliform), flat and narrow; strap-like in form; lorate.
Lillicoa Sherwood (1977), Stictidaceae. 1, N. Am.
Lilliputeana Sérus. (1993) nom. inval. = Scoliciosporum (Lecanor.).
Lilliputia Boud. & Pat. (1900) = Roumegueriella (Hypocr.) fide Malloch & Cain (1972).
Limacella Earle (1909), Amanitaceae. 2, widespr. See Smith (*Pap. Mich. Acad. Sci.* **30**: 125, 1944; key).
Limacinia Neger (1895) Fungi (inc. sed.) fide Hughes (*Mycol.* **68**: 693, 1976), but see Reynolds (*Mycotaxon* **23**: 153, 1985).

Limaciniella J.M. Mend. (1925) = Actinocymbe (Chaetothyr.) fide v. Arx & Müller (1975).
Limaciniopsis J.M. Mend. (1925), Dothideales (inc. sed.). 1, Hawaii. ? = Phragmocapnias (Capnod.) or Capnodium (Capnod.) fide v. Arx & Müller (1975).
Limacinula Höhn. (1907), Coccodiniaceae. 6, trop. See Reynolds (*Mycol.* **63**: 1173, 1973; key).
Limacinula (Sacc. & D. Sacc.) Höhn. (1909) ≡ Limacinula Höhn. (Coccodin.).
Limacinus (Quél.) Marchand (1896) ? = Hygrophorus (Hygrophor.).
Limacium (Fr.) P. Kumm. (1871) = Hygrophorus (Hygrophor.) fide Singer (1951).
Limbalba Nieuwl. (1916) ≡ Wainioa (Pilocarp.).
limbate, (1) edged with another colour; (2) (of a volva), adnate to base of stipe and having a narrow, free, membranous margin (Bas, 1969).
Limboria Ach. (1815) nom. conf. (Ascomycota, inc. sed.). = Catillaria (Catillar.), Diploschistes (Thelotremat.), Lithographa (Rimular.) and Schismatomma (Opegraph.) p.p.; sensu Stein (1873) = Mycoglaena (Dothideales) fide v. Arx & Müller (1975).
Limnaiomyces Thaxt. (1900), Laboulbeniaceae. 3, C. Am., W. Afr., E. Indies.
Limnomyces Lohammar (1953) nom. dub. ('phycomycetes').
Limnoperdon G.A. Escobar (1976), Cyphellaceae. 1, USA, Japan. See Escobar & McCabe (*Mycotaxon* **9**: 48, 1979; gast. rel. Cyphellaceae).
limoniform, lemon-like in form.
Limonomyces Stalpers & Loer. (1982), Sistotremataceae. 2, Eur., N. Am.
Lindau (Gustav; 1866-1923). Keeper, Royal Botanical Museum, and Privatdocent at Berlin Univ. Chief writings: *Fungi Imperfecti: Hyphomycetes* (Rabenh. *Krypt.-Fl.*, 1904-6), *Pflanzliche Parasiten* (*Sorauer Handb. PflKr.*, 1905-8), *Hemiasci* (*KryptFl. Mark Brandenb.*, 1906), and the much used *Hilfsbuch für das Sammeln parasitischer Pilze*, 1901; *.... der Ascomyceten*, 1903; (with P. Sydow) (*Thesaurus litteraturae mycologicae*, 5 vols, 1908-18). See Loesener (*Ber. dtsch. bot. Ges.* **41**: (93), 1923), Grummann (1974: 29), Stafleu & Cowan (*TL-2* **3**: 23, 1981).
Lindauella Rehm (1900), Phyllachoraceae. 1, S. Am. See Sherwood (*Mycotaxon* **5**: 1, 1977).
Lindauomyces Koord. (1907) = Arthrobotryum (Mitosp. fungi) fide Clements & Shear (1931).
Lindauopsis Zahlbr. (1907) = Caloplaca (Teloschist.) fide Reidl (*Sydowia* **28**: 166, 1976).
Lindavia Nieuwl. (1916) [non (F. Schütt) De Toni & Forti (1900), *Algae*] ≡ Scopularia (Mitosp. fungi).
Lindbladia Fr. (1849), Cribrariaceae. 2, N. temp., Sri Lanka.
Linderia G. Cunn. (1931) [non Lindera Thunb. (1783), *Lauraceae*] ≡ Linderiella (Clathr.).
Linderiella G. Cunn. (1942) ≡ Clathrus (Clathr.) fide Dring (*Kew Bull.* **35**: 12, 1980).
Linderina Raper & Fennell (1952), Kickxellaceae. 2, Afr., Hong Kong, Malaysia. See Chang (*TBMS* **50**: 312, 1967), Benjamin (1959), Young (*Ann. Bot.* **33**: 211, 1969, *TBMS* **54**: 15, 1970, **55**: 29, 1970; ultrastr.), Benny & Aldrich (*CJB* **53**: 2325, 1975; ultrastr. spore ontogeny).
Linderomyces Singer (1947) = Gloeocantharellus (Gomph.) fide Petersen (1972).
Lindquistia Subram. & Chandrash. (1977), Anamorphic Xylariaceae, 2.A1.6. Teleomorph *Podosordaria*. 1, India.
Lindquistomyces Aramb., E. Müll. & Gamundí (1982), Amphisphaeriaceae. 1, Argentina.
Lindra I.M. Wilson (1956), Halosphaeriaceae. 3 (marine), Eur. See Kohlmeyer & Kohlmeyer (1979),

Nakagiri (*Trans. mycol. Soc. Japan* **25**: 377, 1985). Anamorph *Anguillospora*.

Lindroth, see Liro.

Lindrothia Syd. (1922) = Puccinia (Puccin.) fide Arthur (1934).

Lindsay (William Lauder; 1829-1880) graduated in medicine from University of Edinburgh in 1852, resident physician at Murray's Royal Institution for the insane in Perth from 1854-79. A proponent of nonrestraint treatments, he studied the practices of psychiatric hospitals in other countries, and mental aspects of animals (*Mind in the lower animals*, 2 vols, 1879). He regarded natural history as therapeutic and popularized lichenology in particular (e.g. *A popular history of British lichens*, 1856). A polymath with a rigorous scientific approach he studied chemical variation within lichens, dyeing, and opened up little explored areas of science, including the diversity of lichenicolous fungi (e.g. Observations on new lichenicolous microfungi, *Trans. R. Soc. Edinb.* **25**: 513, 1869) and conidiomata in lichens (Memoirs on the spermagones and pycnides of lichens, *Trans. R. Soc. Edinb.* **22**: 101, 1859 [macrolichens]; *Trans. linn. Soc. Lond.* **28**: 189, 1872 [crustose lichens]). Visited Otago in 1861, thereafter publishing profusely not only on lichens and other fungi but also mineralogy and plants from that country (e.g. *Contributions to New Zealand botany*, 1868) and even colonialism. Collections now in **E**; many new taxa based on specimens loaned from **K** (now in **BM**). *Obit.*: *Proc. R. Soc. Edinb.* 1881-82: 736, 1882. See also Grummann (1974: 376), Hawksworth & Seaward (1977), Stafleu & Cowan (*TL-2* **3**: 63, 1981).

Lindtneria Pilát (1938), Lindtneriaceae. 11, widespr. See Hjortstam (*Mycotaxon* **28**: 19, 1987; key).

Lindtneriaceae Locq. (1984), Stereales. 3 gen. (+ 2 syn.), 4 spp. Basidioma resupinate; hymenophore smooth to poroid; spores yellowish, ellipsoid, spinose to ridged.

line (as a measure of length), 2.1167 mm ($^1/_{12}$ inch); **Paris -** (Parisier Linie, P.L.), 2.2558 mm; **Paris inch**, 27.9 mm (fide Mason, *Mycol. Pap.* **3**: 24, 1933); 'p.p.' and 'p.p.p.' (parts per Paris inch) are abbreviations used by Corda.

linear, long and narrow.

Linearistroma Höhn. (1910), Clavicipitaceae. 1, Brazil, India. See Phelps *et al.* (*Mycotaxon* **48**: 165, 1993). Anamorph *Ephelis*.

Lineolata Henssen, Döring & Kantvilas (1992), Dothideales (inc. sed.). 1 (marine), France.

lineolate, marked with lines.

Lineostroma H.J. Swart (1988), ? Venturiaceae. 1, Australia.

linguiform, see lingulate.

lingulate, tongue-like in form.

Linhartia Sacc. & P. Syd. (1902), Solorinellaceae (L). 4 (on leaves), trop. See Vězda (1979; key), Vězda & Poelt (*Phyton* **30**: 47, 1990).

Link (Johann Heinrich Friedrich; 1767-1851). Born at Hildesheim, educated there and at Göttingen; M.D., 1790. Professor at Rostock 1792-1811; of Botany at Breslau 1811-15, and at Berlin 1815-, where he was also Director of the Botanic Garden and of the Royal Herbarium. Chief writings: *Observationes in Ordines Plantarum naturales* (**1**, 1809; **2**, 1815), *Hyphomycetes* (1824), and *Gymnomycetes* (1825) in Linné (*Species Plantarum*, edn 4 [by Willdenow], **6**), *Handbuch zur Erkennung der .... Gewächse* (1831-3 [Fungi **3**: 274-486] [= Willdenow, *Grundiss der Kräuterkunde ....* , **3**]). See Grummann (1974: 29), Stafleu & Cowan (*TL-2* **3**: 65, 1981).

Linkiella Syd. (1921) = Puccinia (Puccin.) fide Arthur (1934).

Linkomyces Golovin (1958) = Erysiphe (Erysiph.) fide Zheng (1985). See Gelyuta (*Biol. Zh. Armen.* **41**: 351, 1988).

Linnaeus (Carl; 1707-1778; Carl von Linné; L.). Professor of Medicine and Botany at the University of Uppsala 1741-78, renowned for the introduction of the binomial system of nomenclature, and which he applied to all botanical groups in his *Species plantarum* (1753); this work is the starting point for the current nomenclature of all fungi (except for fossils). Linnaeus drew extensively on the works of Dillenius (*Historia muscorum*, 1742) for his treatments of lichen-forming, and of Micheli (1729: q.v.) for other fungi, but recognized only about 170 fungal spp. in *Species plantarum*. Amongst the generic names used were *Agaricus, Clathrus, Hydnum, Mucor, Phallus* and *Tremella*; all lichens were in the genus *Lichen* (80 spp.), except for some inadvertently placed in *Mucor* and *Tremella*. Acharius (q.v.) was Linnaeus' last pupil, defending his dissertation in 1776, Linnaeus' personal library and collections can be consulted at the Linnean Society of London (**LINN**). See Blunt (*The compleat naturalist*, 1971), Jørgensen *et al.* (*Bot. J. Linn. Soc.* **115**: 261, 1994; typif. 109 lichen names), Grummann (1974: 477), Ramsbottom (*Proc. Linn. Soc. Lond.* **151**: 280, 1941; contrib. to mycology), Stafleu (*Linnaeus and the Linnaeans*, 1971), Stafleu & Cowan (*TL-2* **3**: 71, 1981), Stearn (*in* Linnaeus, *Species Plantarum*, 2 vols, 1753 [reprint 1957]; biogr., herbaria, history of work, etc.), Uggla (*Linnaeus*, 1957).

Linobolus Syd. & P. Syd. (1917) ? = Tubeufia (Tubeuf.) fide v. Arx & Müller (1975).

Linocarpon Syd. & P. Syd. (1917), Hyponectriaceae. *c.* 10 (mainly on *Palmae*), trop. See Hyde (*Sydowia* **44**: 32, 1992; key). Anamorph *Phialophora*-like.

Linochora Höhn. (1910), Anamorphic Phyllachoraceae, 8.E1.?. Teleomorph *Phyllachora*. 5, esp. trop.

Linochorella Syd. & P. Syd. (1912), Mitosporic fungi, 8.C1.?. 1, S. Afr.

Linodochium Höhn. (1909), Mitosporic fungi, 3.E1.10. 1, Eur., N. Am. See Dyko & Sutton (*CJB* **57**: 370, 1979).

Linopeltis I. Hino & Katum. (1961), Schizothyriaceae. 1, Japan.

Linopodium Earle (1909) = Mycena (Tricholomat.) fide Singer (1951).

Linospora Fuckel (1870), Valsaceae. 4, Eur., N. Am. See Barr (*Mycol. Mem.* **7**, 1978), Walker (*Mycotaxon* **11**: 1, 1980), Monod (1983).

Linosporoidea R. Keller (1895), Fossil fungi (? perithecia). 1 (Miocene), Switzerland.

Linostoma Höhn. (1918) [non Wall. (1831), *Thymelaeaceae*] ≡ Ophiostoma (Ophiostomat.).

Linostomella Petr. (1925) ? = Endoxyla (Bolin.).

Linotexis Syd. & P. Syd. (1917) = Parenglerula (Englerul.) fide Müller & v. Arx (1962). See Deighton (*Mycol. Pap.* **78**, 1960).

lipids, esters of higher aliphatic alcohols (e.g. oils, fats, waxes); constituents of fungi (see Weete & Weber, *Lipid biochemistry of fungi and other organisms*, 1980).

lipochroic, see colour.

Lipocystis Cummins (1937), Raveneliaceae. 1 (on *Mimosa*), W. Indies.

Lipomyces Lodder & Kreger (1952), Lipomycetaceae. 5 (in soil), N. Am., Eur. See Smith *et al.* (*Ant. v. Leeuwenhoek* **67**: 177, 1995), Bab'eva & Gorin (*Pochvennye Drozhzhi*, 1987), Nieuwdorp *et al.* (*Ant. v. Leeuwenhoek* **40**: 241, 1974; key), Smith *et al.* (*Int. J. Syst. Bact.* **34**: 80, 1984; ultrastr.).

Lipomycetaceae E.K. Novák & Zsolt (1961), Saccharomycetales. 4 gen. (+ 2 syn.), 9 spp. Mycelium usually

poorly developed; vegetative cells usually proliferating by multilateral budding, in a few spp. by thallic fragmentation, covered with an I+ gelatinous matrix; asci formed either directly by budding from daughter cells or from conjugation, often multispored, sometimes elongate, usually evanescent; ascospores usually cylindrical or allantoid, smooth or ornamented. Fermentation negative; coenzyme system Q8-Q10; from soil, some insect-associated.
Lit.: Van der Walt et al. (Syst. Appl. Microbiol. 9: 115, 1987), Weijman & van der Walt (Stud. Mycol. 31: 193, 1989; cell wall carbohydrates, Ant. v. Leeuwenhoek 62: 247-250, 1992; review), Kock et al. (Ant. v. Leeuwenhoek 62: 251, 1992; phylogeny gen.).

Lipospora Arthur (1921) = Tranzschelia (Uropyxid.) fide Arthur (1934).

lipsanenchyma, primordial tissue of a basidioma, other than the universal veil, covering the hymenium (Reijnders, 1963; Singer, 1962: 29).

lipstick mould, Sporendonema purpurascens, an invader of mushroom beds.

liquid nitrogen, see Genetic resource collections.

Lirasporis Potonié & Sah (1960), ? Fossil fungi. 1 (Miocene), India.

lirella, a long, narrow apothecium as in Graphis and Hysterium.

Liro (Johan Ivar; 1872-1943) [also Lindroth, his stepfather's name under which he published up to 1906 when he took the Finnish name Liro]. Was the first Professor of Plant Biology and Plant Pathology at Helsinki Univ., Finland (1922-43), and one of the founders of the Vanamo Society (1896). Noted for more than 250 publications on taxonomic mycology and plant pathology, esp. of the mycoflora of Finland, e.g. Die Ustilagineen Finnlands (2 vols, 1924, 1938). See Jamalainen (Arch. Bot. Soc. zool.-bot. fenn. 'Vanamo' 3, 1949; biogr., bibl.), Grummann (1974: 620), Stafleu & Cowan (TL-2 3: 62, 1981).

Liroa Cif. (1933), Ustilaginaceae. 1 (on Polygonum), Asia. See Kamat & Viswanathan (Mycopath. 26: 289, 1965), Vánky (1987).

Lirula Darker (1967), Rhytismataceae. 7, N. Am., Eur., India, Japan, Pakistan.

Lisea Sacc. (1877) = Gibberella (Hypocr.) fide Müller & v. Arx (1962).

Lisiella (Cooke & Massee) Sacc. (1891), Hypocreaceae. 2, Australia, Eur.

Lister (Arthur, 1830-1908; and Gulielma, 1860-1949). Father and daughter, were eminent English amateur naturalists noted for their many publications on the Myxomycetes and in particular for A monograph of the Mycetozoa (1894; edn 3, 1925). See Lister (Proc. R. Soc. B 88: i, 1915), Hawker (TBMS 35: 177, 1952; portr. A. Lister), Wakefield (TBMS 33: 165, 1950; portr. G. Lister), Ainsworth (TBMS 35: 188, 1952; notebooks), Grummann (1974: 377), Stafleu & Cowan (TL-2 3: 118, 120, 1981).

Listerella E. Jahn (1906), Liceaceae. 1 (on Cladonia), Eur., former USSR, USA. See Kowalski (Mycol. 59: 1078, 1968), Nannenga-Bremekamp (1982).

Listeromyces Penz. & Sacc. (1901), Mitosporic fungi, 3.C2.1. 1, Java, Hawaii. See Goos (Mycol. 63: 213, 1971).

Literature. The mycological literature is vast (not less than 500 000 books and papers) and the current annual output is c. 5000 items distributed through not less than 3500 publications; approximately half this output is in English. Much of the literature up to 1930 is compiled by Lindau & Sydow (1908-18) and Ciferri (1957-60). Novak & Moore's 'One step mycology' (MR 98: 816, 1994-) lists selected papers pertinent to all aspects of mycology under broad subject headings.

Current systematic literature is covered in Bibliography of Systematic Mycology (1946-; twice-yearly, indexed to gen.; on CD-ROM 1985-1995). Most literature is ephemeral, but systematic literature is longer-lived and must be readily available; the valid publication of fungal names dates from 1753 (see Nomenclature).

Online and other computer retrieval systems make recent literature more accessible. The Review of plant pathology (1920-; monthly) and Review of medical and veterinary mycology (1943-; quarterly) have been included in the CAB ABSTRACTS database since 1973, and like BIOSIS (the database including Mycological Abstracts; 1967-), are available for searching on an increasing number of commercial database hosts (e.g. DIALOG, DIMDI, EURONET-DIANNE) and in CD-ROM format; see also Internet. Photocopies of original papers may sometimes be ordered over the computer systems, accessed by normal telephone lines.

Frequent use has been made here of the general works listed below. These are referred to by author and year or part, e.g. Buller (5), Clements & Shear (1931). In addition, literature is given under the names of genera and higher ranks, and special topics (e.g. Geographical distribution, Industrial mycology, Lichens, Macromycetes, Nomenclature, Physiology, Plant pathogenic fungi).

General Textbooks:

de Bary, A., Vergleichende Morphologie und Biologie der Pilze (1884) [Engl. transl. Comparative morphology and biology of the Fungi, Mycetozoa, and Bacteria, by Garnsey & Balfour, 1887].

Buller, A.H.R., Researches on fungi, 1 (1909), 2 (1911), 3 (1924), 4 (1931), 5 (1933), 6 (1934), 7 (1950). See also TBMS 57: 5 (1971).

Gäumann, E.A., Vergleichende Morphologie der Pilze, 1926. [Engl. transl., revision, Comparative morphology of fungi, by C.W. Dodge, 1928].

Bessey, E.A., A text-book of mycology, 1935; Morphology and taxonomy of fungi, 1951 [reprint 1961]. [Many refs. systematic lit.].

Lohwag, H., Anatomie der Asco- und Basidiomyceten, 1941 (Linsbauer, Tischler & Pascher, Handbuch der Pflanzenanatomie, II, Abt. 3, Teilband c: Eumyceten, Band VI) [reprint 1965].

Langeron, M., Précis de mycologie, 1945; edn 2, by Vanbreuseghem, 1952 [Engl. transl., 2 vol.; Outline of mycology, 1965].

Wolf, F.A. & F.T., The fungi, 2 vol., 1947 [reprint, 1970].

Gäumann, E.A., Die Pilze, 1949 [Engl. transl., The fungi by Wynd, 1952]; edn 2, 1965.

Alexopoulos, C.J., Introductory mycology, 1952; edn 2, 1962 [Germ. transl., Einführung in die Mykologie, by Farr, 1965]; edn 3 (& Mimms), 1979.

Moreau, F., Les champignons, 2 vols, 1953.

Chadefaud, M., Les végé non vasculaires cryptogamie (Traité de botanique systématique, 3), 1960 [Fungi pp. 429-902].

Ingold, C.T., The biology of Fungi, 1961 (edn 6, by Ingold, C.T. & Hudson, H.J., 1993) [a good introduction].

Ainsworth, G.C. & Sussman, A.S. (Eds), The fungi, an advanced treatise, 1 (The fungal cell), 1965; 2 (The fungal organism), 1966; 3 (The fungal population), 1968; (& Sparrow, F.K.) 4A and 4B (A taxonomic review), 1973.

v. Arx, J.A., Pilzkunde, 1967; edn 3, 1976.

Müller, E. & Loeffler, W., Mykologie, 1968; edn 2, 1971 [Engl. transl., Mycology. An outline for science *and medical students*, by B. Kendrick & F. Baerlocher, 1976].

Moore-Landecker, E., *Fundamentals of the Fungi*, 1972; edn 3, 1990.

Burnett, J.H., *Fundamentals of mycology*, 1968; edn 2, 1976.

Ubrizsy, G. & Vörös, J., *Mezögazdasági Mykologia*, 1968.

Kreisel, H., *Grundzuge eines natürlichen Systems der Pilze*, 1969.

Webster, J., *Introduction to fungi*, 1970; edn 2, 1980.

Talbot, P.H.B., *Principles of fungal taxonomy*, 1971.

Gams, W., *et al.*, *CBS course of mycology*, 1975; edn 2, 1980.

Esser, K., *Kryptogamen*, 1976. [Engl. transl. *Cryptogams*, by M.G. Hackston & J. Webster, 1981].

Deacon, J.W., *Introduction to modern mycology*, 1980; edn 2, 1984.

Parker, S.P. (Ed.), *Synopsis and classification of living organisms*, 2 vols, 1982.

Kendrick, B., *The fifth kingdom*, edn 2, 1992.

Hudson, H.J., *Fungal biology*, 1986.

Govi, G., *Introduzione alla Micologia*, 1986.

Herrera, T. & Ulloa, M., *El reino los fungos*, 1990.

Margulis, L., Corliss, J.O., Melkonian, M. & Chapman, D.J., *Handbook of Protoctista*, 1990.

Hawksworth, D.L. (Ed.), *Frontiers in mycology*, 1991.

Carlile, M.J. & Watkinson, S.C. *The fungi*, 1994.

Gravesen, S., Frisvad, J.C. & Samson, R.A., *Microfungi*, 1994.

Esser, K.A. & Lemke, P.A. (Eds), *The Mycota*, **1**, *Growth, differentiaion and sexuality*, 1994; **2**, *Genetics and biotechnology*, 1995; (2 vols, of 7 planned, including **3**, Biochemistry and molecular biology; **4**, *Environmental and microbial relationships*; **5**, *Plant relationships*; **6**, *Animal and human relationships*; **7**, *Systematics and cell structure*).

See also Genetics, Methods, Physiology, Sex.

General systematic works

v. Arx, J.A., *The genera of fungi sporulating in pure culture*, 1970; edn 3, 1981 [keys, *c.* 850 gen.].

Engler, A. & Prantl, K., *Die naturlichen Pflanzenfamilien* 1, 1*, 1**; edn 2, **2**, **5b**, **6**, **7a**, **8**, 1889-1933. [Dates publ. parts, Stafleu, *Taxon* 21: 501 (1972).] Descriptions fams. etc., illustr., incl. lichens; multi-authored.

Martin, G.W., *Outline of the fungi*, 1950. Family keys, this *Dictionary* edn 5: 597-519 (1961).

See also family keys in this volume.

See also Clements & Shear (1931) and Saccardo (1882-72) under Catalogues of names; and Ainsworth *et al.* (1973), under General Textbooks; and Floras.

Bibliographies

Culberson, W.L., Recent literature on lichens 1-100; Egan, R.S., 101-143; Esslinger, T.L. 144- [*Bryologist* **54**-, 1951-]. Annotated bibliogr., 2-4 times per year.

Grummann, V.J., *Biographisch-bibliographisches Handbuch der Lichenologie*, 1974.

Lindau, G. & Sydow, P., *Thesaurus litteraturae mycologicae et lichenologicae*, 5 vols, 1908-18; *Supplementum*, 1911-1930, 4 vols, by R. Ciferri, 1957-60. Lists books and papers published up to 1930.

Poelt, J., *Systematik der Flechten [Fortschr. Bot. Berl.* **17-36**, 1955-74]; by H. Hertel, **38-42**, 1976-80. Annotated bibliogr. by group.

Stafleu, F.A., *Taxonomic literature [Regnum veg.* **52**], 1967.

Stafleu, F.A. & Cowan, R.S., *Taxonomic literature*, edn 2 [Regnum veg. **94, 98, 105, 110, 112, 115, 116**),

1976-88; Stafleu & Mennega (*Supplementum [Regnum veg.* **125, 132, 130**], 1992-. Exhaustive treatment of major systematic lit. incl. dates publ., collections, types; good mycological and lichenological coverage.

Just's Bot. Jahresb. (63 vols., 1874-1944) included annotated lists of papers on fungi and lichens. Petrak contributed 2185 pp. citing lit. (**55-59, 63**, 1930-44), and Zahlbruckner prepared comprehensive lichen lists (**13-59**, 1884-1931).

Bibliography of systematic mycology, 1946-; see above)

Mycological abstracts, 1967-; see above

Biographies

See under Authors' names.

Catalogues of names

Clements, F.E. & Shear, C.L., *The genera of fungi*, 1931. Gen. names; keys.

David, J.C., Fungal family names in current use. *Regnum veg.* **126**: 71-91, 1993.

Farr, D.F., Bills, G.F., Chamuris, G.P. & Rossman, A.Y., *Fungi on plants and plant products in the United States*, 1989.

Farr, E.R., Leussink, J.A. & Stafleu, F.A. (Eds). *Index nominum genericorum (plantarum)*, 3 vols [*Regnum veg.* **100-102**], 1979. Catalogue validly published gen. names (incl. fungi and lichens).

Greuter, W., Brummitt, R.K., Farr, E., Kilian, N., Kirk, P.M. & Silva, P.C., *NCU-3, Names in current use for extant plant genera [Regnum veg.* **129**.], 1993; includes fungi.

Index of fungi, 1940-. Catalogue of new fam., gen., spp., vars., ff., comb. novs., nom. novs. publ.; lichens incl. since **4**- (1970-). The *Index* has also published four supplements: *A Supplement to Petrak's lists 1920-1939* (1969), *Lichens 1961-1969* (1972), *Saccardo's omissions* (1985), *Family names* (1989), and *Authors of fungal names* (1992).

Jong, S.C., *Stedman's ATCC fungus names*, 1993.

Lamb, I.M., *Index nominum lichenum*, 1963. Names publ. 1932-1960.

Petrak, F., Verzeichnis der neuen Arten, Varietäten, Formen, Namen und wichstigsten Synonyme. *Just's bot. Jber.* **48**(3): 184[1], **49**(2): 267 [2], **56**(2): 291 [3], **57**(2): 592 [4], **58**(1): 447 [5], **60**(1): 449 [6], **63**(2): 805 [7], 1930- 44; *Index of fungi 1936-1939*, 1950 [8]. Usually cited as 'Petrak's lists' nos. **1- 8**; fungal names 1920-40; all reprinted and available from IMI.

Pfeiffer, L.J.G., *Nomenclator botanicus*, 2 vols, 1871-74. Names gen.

Saccardo, P.A., *Sylloge Fungorum*, 1882-1972, 26 vols [reprints 1944, 1967; microfiche edn, 1972]. Names gen., spp. with Latin descriptions to 1950. Phycomycetes, vol. 7, 79, 11, 14, 16, 17, 21, 24. Pyrenomycetes, vol. 1, 2, 9, 11, 14, 16, 17, 24. Discomycetes, vol. 8, 10, 11, 14, 16, 18, 22, 24. Uredinales and Ustilaginales, vol. 7, 9, 11, 14, 16, 17, 21, 23. Hymenomycetes, vol. 5, 6, 9, 11, 14, 16. 17. 21. 23. Gasteromycetes, vol. 7, 9, 11, 14, 16, 17, 21. Coelomycetes, vol. 3, 10, 11, 14, 16, 18, 22, 25. Hyphomycetes, vol. 4, 10, 11, 14, 16, 18, 22, 25. Index, vol. 12; Host index, vol. 13; Synonyms, vol. 15; Index to illustrations, vol. 19; Generic index, vol. 18.

[Reed, C.F. & Farr, D.F., *Index*, 1993 (to scientific names).]

Steudel, E.G., *Nomenclator botanicus*, **2**, 1824. Esp. useful for pre-1821 names.

Streinz, W.M., *Nomenclator fungorum*, 1862. Alphabetical listing gen. spp. with syns.

Zahlbruckner, A., *Catalogus lichenum universalis*, 10 vols, 1921- 40. Lichen gen. and infraspecific taxa to 1932; incl. full listings of citations of names by later authors.

Floras
The following, mostly multi-authored, floras are valuable in including detailed descriptions of taxa and full synonymies. See also Geographical distribution.
Rabenhorst's Kryptogamen Flora von Deutschland, Deutsch-Österreich und der Schweiz, edn 2, 1881-1960.
Kryptogamenflora der Mark Brandenburg, 1905-.
Migula, W., *Kryptogamen-Flora von Deutschland, Österreich und der Schweiz*, edn 2, 1910-34.
Beiträge zur Kryptogamenflora der Schweiz, 1901-.
Flora italica Cryptogama, 1905-.
North American Flora, 1906-.

Methods.
See Methods.

Dictionaries and glossaries
Dörfelt, H., *Bi-Lexicon Mykologue Pilzkunde*, 1988.
Holliday, P., *A Dictionary of plant pathology*, 1989.
Jackson, B.D., *A glossary of botanic terms*, 1900; edn 4, 1928 [reprint 1960].
Tootill (Ed.) *The Penguin dictionary of botany*, 1988).
Martinez, A.T., *Terminos cientifica relacionadascen los micromicetos. Revista Ibérica Micol.* **2**: 36, 1985.
Snell, W.H., *Three thousand mycological terms*, 1936; & Dick, E.A., *A glossary of mycology*, 1957; edn 2, 1971.
Josserand, M., *La description des champignons supérieurs*, 1952.
Viégas, A.P., *Dicionario Almão-Português de micologia e fitopatologia*, 1958; *Dicionario de fitopatologia e micologia*, 1979.
Fidalgo, O. & Maria, E.P.K. *Dicionário micologico*, 1967 [*Rickia*, Suppl. 2].
Cowan, S.T., *A dictionary of microbial taxonomy*, 1978.
Singleton, P. & Sainsbury, D., *Dictionary of microbiology and molecular biology*, edn 2, 1987.
Berger, K. (Ed.), *Mykologisches Wörterbuch*, 1980 [Germ., Engl., Fr., Span., Latin, Czech., Polish, Russ. glossary].
   See also glossaries in Alexopoulos & Mimms (1979), Clements & Shear (1931), Kendrick (1992), Seaward (Ed.) (*Lichen ecology*, 1977), and Smith (1921), and *lit*. cited under Nomenclature.

Periodicals
African Journal of Mycology and Biotechnology (Cairo), 1993-.
Annales Mycologici (Berlin), 1903-44; 1947- as *Sydowia* (Vienna).
The Bryologist (Brooklyn), 1898-. [incl. lichenology]
Bulletin Trimestriel de la Société Mycologique de France (Paris), 1885- [Indexes **1- 40** (1884-1924), 1934; **41-70** (1925-54), 1972].
Cryptogamie (Paris), 1980- (in 3 ser., incorporating *Revue bryologique et lichénologique*, 1928-79; *Revue de mycologie*, 1936-79).
Experimental Mycology (New York), 1977-.
Friesia (Copenhagen), 1932-80 (1981-, incorporated in *Nordic Journal of Botany*, Copenhagen).
Herzogia (Vaduz), 1968-.
International Journal of Mycology and Lichenology (Braunschweig), 1982-1992.
The Lichenologist (London), 1958-.
Micologia Italiana, 1972-.
Micologia i Neotropica Applicada (Mexico), 1988-.

Mikologia i Fitopatologia (Leningrad), 1967-.
Mycologia (New York), 1909- [Index, 1-58 (1909-66), 1968].
Mycological Papers, 1925- (*Mycol. Pap.* **100**, Index to Papers 1-99, 1925-65].
Mycopathologia (den Haag), 1938-.
Mycorrhiza (Heidelberg), 1992-.
Mycoscience (Tokyo), 1994-; formerly *Transactions of the Mycological Society of Japan*, 1956-1993.
Mycosystema (Beijing), 1988-.
Mycotaxon (Ithaca, NY), 1974-.
Nova Hedwigia (Lehre, Germany), 1959-.
Novosti systematiki nizshikh rastenii (Leningrad), 1964-.
Persoonia (Leiden), 1959-.
Studies in Mycology (Baarn), 1973-.
Systema Ascomycetum (Egham), 1982-.
Transactions of the British Mycological Society (Cambridge, UK) 1898-1988 [Indexes, 1-30 (1898-1946), 1952; 31-40 (1947-57), 1961; 41- 67 (1958-76), 1979; 68-91 (1977-1988), 1991]; as *Mycological Research*, 1989-.

See also under Bibliographies and Catalogues of names above; and under Medical and veterinary mycology, Plant pathogenic fungi, and Societies and organizations.

litho- (prefix), pertaining to rocks.
Lithocia Gray (1821) = Verrucaria (Verrucar.) p.p. and Polyblastia (Verruc.) p.p.
Lithoecis Clem. (1909) = Verrucaria (Verrucar.).
Lithoglypha Brusse (1988), Acarosporaceae (L). 1, Natal.
Lithographa Nyl. (1857), Rimulariaceae (L). 3, Eur., N. Am., Asia, Antarctic Is. See Hertel & Rambold (*Bibl. Lich.* **38**: 145, 1990; key).
Lithographomyces Cif. & Tomas. (1953) ≡ Lithographa (Rimular.).
Lithomyces Viala & Marsais (1930) = Melanospora (Ceratostomat.) fide Udagawa & Cain (*CJB* **47**: 1915, 1969).
lithophyte, a plant living on rocks; see saxicolous.
lithophytic, the habit of a lithophyte (q.v.).
Lithopythium Bornet & Flahault (1891), ? Ascomycota (inc. sed., ?L). 1, Eur.
Lithosphaeria Beckh. ex Körb. (1863) = Psoroglaena (Verrucar.).
Lithothelidium M. Choisy (1954) nom. inval. = Polyblastia (Verrucar.).
Lithothelium Müll. Arg. (1885), Pyrenulaceae (L). 16, trop. See Aptroot (*Bibl. Lich.* **44**, 1991), Singh (*Curr. Sci.* **55**: 198, 1986).
lithotroph, utilizing rocks as nourishment.
litmus, an amphoteric dye obtained from depside-containing lichens, esp. *Roccella montagnei*; See dyeing.
Litschauerella Oberw. (1966), Tubulicrinaceae. 4, widespr. See Eriksson & Ryvarden (*Bot. Notiser* **130**: 461, 1977).
Litschaueria Petr. (1923), Lasiosphaeriaceae. 1 (on polypore), Eur., N. Am. See Barr (*Rhodora* **78**: 53, 1976, *Mycotaxon* **39**: 43, 1990; posn).
littoral, growing on sea or lake shores.
Lituaria Riess (1853) = Helicoma (Mitosp. fungi) fide Linder (*Ann. Mo. bot. Gdn* **16**: 298, 1929).
lituate, forked, and having the points turned out a little.
liverwort, sometimes used in common names for large foliose lichens, e.g. ash-coloured ground liverwort (*Peltigera canina*).
Livia Velen. (1947), Leotiales (inc. sed.). 1, Eur.
Lizonia (Ces. & De Not.) De Not. (1864), Pseudoperisporiaceae. 1, Eur. See Hansford (*Mycol. Pap.* **15**, 1946).

Lizoniella Henn. ex Sacc. & D. Sacc. (1905) = Mirocyclus (Mycosphaerell.) fide Barr (*Mycotaxon* **39**: 43, 1990).

Llanoa C.W. Dodge (1968) = Umbilicaria (Umbilicar.) fide Wei & Jiang (*Mycosystema* **2**: 135, 1989).

Llanolichen Tomas. & Cif. (1952) = Placynthium (Placynth.).

Llanomyces Cif. & Tomas. (1953) ≡ Llanolichen (Placynth.).

Llimonaea Egea & Torrente (1991), Arthoniales (inc. sed., L). 2, S. Eur., N. Afr., Cape Verde. See Egea & Torrente (*Mycotaxon* **53**: 63, 1995).

Llimoniella Hafellner & Nav.-Ros. (1993), Leotiaceae. 3 (on lichens), Spain. See Kümmerling *et al.* (*Bibl. Lich.* **53**: 147, 1993).

Lloyd (Curtis Gates; 1859-1926). A somewhat eccentric American businessman and amateur mycologist noted for his important studies on the *Gasteromycetes*, etc. Founded, with his brothers, the Lloyd Library and Museum, Cincinnati (see *Lloydia* **27**: 141, 1964), and issued *Mycological Notes*, 1898-1925 (index, 1932; Index to new names proposed by Lloyd (by Stevenson & Cash), 1936), in which he published results of his mycological investigations. Collections in **BPI** (*Mycol.* **24**: 247, 1932).

Lloyd poked fun at 'name changers' by making certain taxonomic and nomenclatural proposals under the name of a hypothetical 'McGinty'. Such proposals are generally attributed to Lloyd without comment, but perhaps wrongly (Donk, *Reinwardtia* **1**: 205, 1951).

Obit.: Fitzpatrick (*Mycol.* **19**: 153, 1927; portr.), Simons (*John Uri Lloyd, His life and works, 1849-1936, with a history of the Lloyd Library*, 1972); see Stafleu & Cowan (*TL-2* **3**: 123, 1981).

Lloydella Bres. (1901) = Lopharia (Ster.) fide Donk (1956).

Lloydellopsis Pouzar (1959) ≡ Amylostereum (Ster.).

Lloydia C.H. Chow (1935) [non Salisb. (1812), *Liliaceae*] ≡ Sinolloydia (Clathr.) fide Dring (*Kew Bull.* **35**: 68, 1980).

LN, liquid nitrogen; see Genetic resource collections.

LO-analysis. (L, lux [light]; O, obscuritas [darkness]). A method introduced by Erdtman (see *An introduction to pollen analysis*, 1943; edn 2, 1954) for pollen, and used by Payak (*Nature* **184**: 738, 1959, *Mycopath.* **16**: 70, 1962, *Recent advances in palynology*: 27, 1964) for fungus spores, by which different types of surface ornamentation may be distinguished microscopically by the differences in appearance of the spore surface at upper and lower focus. See ornamentation, Fig. 29.

Lobaca Vězda (1983) nom. inval. = Fellhanera (Pilocarp.) fide Vězda (*Folia geobot. phytotax.* **21**: 199, 1986).

Lobaria (Schreb.) Hoffm. (1796), Lobariaceae (L). *c.* 60, cosmop.; liverwort lichens. See Hakulinen (*Annls bot. fenn.* **1**: 202, 1964), Hale (*Bryologist* **60**: 35, 1957), Yoshimura (*J. Hattori bot. lab.* **34**: 231, 1971), Jordan (*Bryologist* **73**: 669, 1970; internal cephalodia, *Bryologist* **76**: 225, 1973; N. Am. spp.), Wei *et al.* (*Acta Mycol. Sin., Suppl.* **1**: 363, 1989; 17 spp. China).

Lobariaceae Chevall. (1826), Peltigerales (L). 4 gen. (+ 22 syn.), 382 spp. Thallus foliose, usually corticate only on the upper surface, often with areolate openings on the lower surface; ascomata marginal or laminar, initially deeply immersed, the covering layer becoming stretched and eventually fragmenting; hymenium strongly concave when immature, often with a conspicuous margin; asci with a thickened apex with an I+ apical ring and an outer I+ gelatinized layer; ascospores elongated, hyaline or brown, transversely septate. Anamorphs coelomycetous.

Lichenized with green algae or cyanobacteria; if the former then usually with cyanobacteria in cephalodia; widespr.

Lit.: Yoshimura & Hurutani (*Bull. Kochi Gakuen Coll.* **18**: 345, 1987; SEM).

Lobarina Nyl. ex Cromb. (1894) = Lobaria (Lobar.).

Lobariomyces E.A. Thomas (1939) nom. inval. ≡ Lobaria (Lobar.).

lobate, lobed.

Lobatopedis P.M. Kirk (1979), Mitosporic fungi, 1.C2.3. 3, UK, Malawi, Tanzania.

Lobiona H. Kilias & Gotth. Schneid. (1978) = Toninia (Catillar.) fide Timdal (1991).

Loboa Cif., P.C. Azevedo, Campos & Carneiro (1956), ? Entomophthorales (inc. sed.). 1, S. Am. *L. loboi* (Jorge Lobo's disease in man).

Lobodirina Follmann (1967) = Roccellina (Roccell.) fide Tehler (1983).

Lobomyces Borelli (1968), ? Mitosporic fungi, ?.?.?. 1 (lobomycosis of humans), C. & S. Am. See Grigoriu *et al.* (*Medical mycology*, 1987).

lobomycosis, keloidal blastomycosis caused by *Loboa loboi*.

Loborhiza A.M. Hanson (1944), Chytridiaceae. 1 (on *Volvox*), USA.

Lobothallia (Clauzade & Cl. Roux) Hafellner (1991), Hymeneliaceae (L). 4, N. Hemisph.

Lobularia Velen. (1934) [non Desv. (1815), *Cruciferae*], ? Leotiales (inc. sed.). 1, former Czechoslovakia.

lobulate, having small lobes.

Lobulicium K.H. Larss. & Hjortstam (1982), Atheliaceae. 1, Eur.

Locelliderma Tehon (1935) = Hypoderma (Rhytismat.).

Locellina Gillet (1876) nom. dub. (Plut.) fide Singer & Smith (1946).

Lockerbia K.D. Hyde (1994), Sordariales (inc. sed.). 1, Australia.

locule (**loculus**), a cavity, esp. one in a stroma.

Loculistroma F. Patt. & Charles (1910), Hypocreaceae. 1 (on *Phyllostachys*), China.

Loculoascomycetes (syn. Ascoloculares, Bitunicatae). This taxon was first proposed by Luttrell (*Mycol.* **47**: 511, 1955), as a subclass for *Ascomycota* with fissitunicately discharging asci, producing ascospores which are generally septate, and borne in unwalled locules (pseudothecia) in ascostromatic ascomata with an ascolocular ontogeny. Orders commonly recognized included the *Myriangiales* (asci globose), *Pleosporales* (asci clavate; pseudoparaphyses present), *Hysteriales* (ascocarp opening by a longitudinal split), *Dothideales* (asci clavate; pseudoparaphyses absent), *Capnodiales* (sooty-moulds; q.v.) and *Microthyriales* (ascomata shield-shaped). Most of these orders are not accepted by v. Arx & Müller (1975) and Eriksson (1982). For usages of the name see Tables 1-2 under *Ascomycota*. This class is not accepted in this edition of the *Dictionary*.

See Barr (*Prodromus to class Loculoascomycetes*, 1987).

Loculohypoxylon M.E. Barr (1976), Dothideales (inc. sed.). 1, N. Am.

Loculomycetes. Class introduced by Gorovii (*Dopov. Akad. Nauk. ukr. RSR* B 1977(8): 742) for *Zerovaemyces* (q.v.), an agaric similar to *Coprinus* but in which the gills are replaced by spherical multiloculate structures containing spores; each chamber contains one **loculospore**. Considered to be a new type of sexual sporulation in fungi.

loculospore, see *Loculomycetes*.

Loculotuber Trappe, Álv. García & Parladé (1993), ? Terfeziaceae. 1, S. Eur., N. Afr.

Lodderomyces Van der Walt (1966), Saccharomycetaceae. 1, USA, S. Afr. Utilizes higher alkanes. See James *et al.* (*Lett. Appl. Microbiol.* **19**: 308, 1994; anamorph). Anamorph *Candida.*

Loflammia Vězda (1986), Ectolechiaceae (L). 3 (on leaves), trop.

Logilvia Vězda (1986), Ectolechiaceae (L). 1 (on leaves), trop.

Lohwagia Petr. (1942), ? Phyllachoraceae. 3, C. & S. Am.

Lohwagiella Petr. (1970), Ascomycota (inc. sed.). 1, N. Am.

Lojkania Rehm (1905), Fenestellaceae. 10, temp. See Barr (*Mycotaxon* **20**: 1, 1984, *N. Am. Fl.* II **13**, 1990; N. Am.), Yuan & Barr (*Sydowia* **46**: 338, 1995; key 4 spp. China. = Herpotrichia (Lophiost.) fide v. Arx & Müller (1984).

Loliomyces Maire (1937), Mitosporic fungi, -.-.-. 1, Morocco.

Lolium endophyte. The seed-borne *Acremonium lolii* occuring in *Lolium perenne* producing lolitrem and peramine alkaloids and implicated in ryegrass staggers in sheep (Fletcher & Harvey, *N.Z. vet. J.* **29**: 185, 1981; Gallagher *et al.*, **29**: 189, 1981; Rowan *et al.*, *J. chem. Soc., Chem. Comm.* **142**: 935, 1986).

Lomaantha Subram. (1954), Mitosporic fungi, 1.C2.1. 1, India.

Lomachashaka Subram. (1956), Mitosporic fungi, 3.A1/2.15. 4, W. Afr., India. See Nag Raj (*Mycotaxon* **53**: 311, 1995).

lomasome, a vesicle derived from an intracytoplasmic membrane (Marchant *et al.*, *New Phytol.* **66**: 623, 1967). Cf. plasmalemmasome.

Lomatia (Fr.) P. Karst. (1889) [non R. Br. (1810), *Proteaceae*] ≡ Lomatina (Cortic.).

Lomatina P. Karst. (1892) = Cytidia (Cortic.) fide Cooke (1961).

Lomentospora Hennebert & B.G. Desai (1974), Mitosporic fungi, 1.A1.10. 1, Belgium.

Lonchospermella Speg. (1908) nom. dub. fide Sutton (*Mycol. Pap.* **141**, 1977).

Longevity. The life-span of an individual fungus in nature is often difficult to ascertain; the longest-lived fungi may be mycorrhizal or other soil-dwelling species whose mycelia radiate out for hugh distances; it has been claimed that an 'individual' *Armillaria bulbosa* thallus in Michigan weighs in excess of 10,000 kg, occupies 15 ha, and has been genetically stable for over 1,500 yrs (Smith *et al.*, *Nature* **356**: 428, 1992; see critique by Brasier, *Nature* **356**: 382, 1992). Some crustose lichens have been claimed to be even older, and 3,700 yrs has been presumed for a *Rhizocarpon geographicum* thallus in Alaska (Denton & Karlén, *Arctic Alp. Res.* **5**: 347, 1973). However, the tendency of individuals of some lichen species to coalesce into larger thalli makes such conclusions suspect. Studies on historical surfaces do, however, leave no doubt that some crustose lichens can live several centuries, e.g. a 620 yr *Aspicilia calcarea* thallus in Oxfordshire (Winchester, *Bot. J. Linn. Soc.* **96**: 57, 1988). In contrast, some lichens may exist for a few months or be seasonal (Poelt & Vězda, *Bibl. Lich.* **38**: 377, 1990). Fungal spores or sclerotia or other vegetative structures may retain their viability under natural conditions or in the laboratory under various conditions (e.g. after lyophilization) for periods of minutes or up to 50 years or more. See Sussman (*in* Ainsworth & Sussman (Eds), *The Fungi* **3**: 447, 1968; review, refs.), Ellis & Roberson (*Mycol.* **60**: 399, 1968; lyophilized cultures), Hwang *et al.* (*Mycol.* **68**: 377, 1976; at ultra low temp.). See also Genetic resource collections, Lichenometry, Record fungi.

Longia Syd. (1921) = Haploravenelia (Ravenel.) fide Dietel (1928).

Longia Zeller (1943) ≡ Longula (Secot.).

longicollous, having long beaks or necks.

longiseptum, see septum.

Longoa Curzi (1927) = Pleurostoma (Calosphaer.). See Eriksson (*SA* **5**: 139, 1986).

longuinose, see lanate.

Longula Zeller (1945), Secotiaceae. 1, N. Am. See Harding (*Mycol.* **49**: 273, 1957).

Lopacidia Kalb (1984), Lecideaceae (L). 2, S. Am.

Lopadiomyces Cif. & Tomas. (1953) nom. illegit. ≡ Lopadium (Ectolech.).

Lopadiopsidomyces Cif. & Tomas. (1953) ≡ Lopadiopsis (Gomphill.).

Lopadiopsis Vain. (1896) = Gyalectidium (Gomphill.).

Lopadium Körb. (1855) nom. cons., ? Ectolechiaceae (L). *c.* 50, cosmop. See Santesson (*Symb. bot. upsal.* **12**(1), 1952), Patwardhan & Makhija (*Indian J. Bot.* **4**: 20, 1981; 10 spp.).

Lopadostoma (Nitschke) Traverso (1906), Xylariaceae. 1, Eur. See Eriksson (*Svensk. bot. Tidskr.* **60**: 315, 1966).

Lopezaria Kalb & Hafellner (1990), Lecanorales (inc. sed., L). 1, trop.

Lopharia Kalchbr. & MacOwan (1881), Stereaceae. 2, cosmop. See Hjortstam & Ryvarden (*Synopsis Fung.* **4**: 19, 1990; key).

Lophiaceae, see *Mytilinidiaceae.*

Lophidiopsis Berl. (1890) = Lophiostoma (Lophiostomat.) fide v. Arx & Müller (1975).

Lophidium P. Karst. (1873) [non Rich. (1792), *Schizaeaceae*] = Lophium (Mytilinid.) fide Zogg (1962).

Lophidium Sacc. (1878) ≡ Platystomum (Lophiostomat.).

Lophiella Sacc. (1878), Lophiostomataceae. 1, Eur. See Parbery (*Austr. J. Bot.* **15**: 271, 1967).

Lophionema Sacc. (1883), Lophiostomataceae. 2 or 3, Eur., N. Am., Java. See Chesters & Bell (*Mycol. Pap.* **120**, 1970).

Lophiosphaera Trevis. (1877) = Lophiostoma (Lophiostomat.) fide v. Arx & Müller (1975).

Lophiosphaerella Hara (1948), ? Dothideales (inc. sed.). 1, Asia.

Lophiostoma Ces. & De Not. (1863) nom. cons., Lophiostomataceae. 30, widespr. See Chesters & Bell (1970; key 23 spp.), Leuchtmann (*Sydowia* **38**: 158, 1986; anamorph). Anamorph *Pleurophomopsis*-like.

Lophiostomataceae Sacc. (1883), Dothideales. 14 gen. (+ 14 syn.), 231 spp. Ascomata perithecial, immersed or erumpent, often strongly flattened, often aggregated, black, usually papillate, with a round or slit-like sometimes periphysate lysigenous pore; peridium of small pseudoparenchymatous cells, usually thickened above, sometimes clypeate; interascal tissue copious, of narrow cellular pseudoparaphyses; asci ± widely cylindrical, fissitunicate; ascospores transversely septate, ocasionally muriform, constricted at the primary and secondary septa, often with a sheath or appendages. Anamorphs coelomycetous. Saprobic or necrotrophic on herbaceous and woody stems, widespr.

Lit.: Chesters & Bell (*Mycol. Pap.* **120**, 1970), Holm & Holm (*Symb. bot. upsal.* **28**(2), 1988; keys 28 spp.), Yuan & Zhao (*Symbiosis* **46**: 112, 1994; China).

Lophiotrema Sacc. (1878), Lophiostomataceae. 5, widespr. See Holm & Holm (*Symb. bot. upsal.* **28**(2), 1988).

Lophiotricha Richon (1885) = Lophiostoma (Lophiostomat.) fide v. Arx & Müller (1975).

Lophium Fr. (1823), Mytilinidiaceae. 3 (on conifers), Eur., Am. See Zogg (1962).

Lophoderma Chevall. (1822) nom. rej. ≡ Hypoderma DC. (Rhytismat.).

Lophodermella Höhn. (1917), Rhytismataceae. 9, Eur., N. Am., Asia. See Millar *in* Peterson (*Recent Research in Conifer Needle Diseases*: 45, 1986; 7 spp.).

Lophodermellina Höhn. (1917) = Lophodermium (Rhytismat.).

Lophodermina Höhn. (1917) ? = Lophodermium (Rhytismat.).

Lophodermium Chevall. (1826) nom. cons., Rhytismataceae. 103, widespr. *L. seditiosum* (needle cast of *Pinus sylvestris*; Lanier, *BSMF* **83**: 959, 1968). See Darker (*Contr. Arnold Arb.* **1**: 65, 1932; *CJB* **45**: 1399, 1967), Johnston (*N.Z. Jl Bot.* **27**: 243, 1989; key 21 spp.), Minter (*Mycol. Pap.* **147**, 1981; monogr. 16 spp. on pines, key), Tehon (*Ill. biol. Monogr.* **13** (4), 1935), Morgan-Jones & Hulton (*Mycol.* **71**: 1043, 1979; ontogeny). Cannon & Minter (*Taxon* **30**: 572, 1983; gen. nomencl.).

Lophodermopsis Speg. (1910) nom. dub. fide Petrak & Sydow (*Ann. Myc.* **34**: 29, 1936).

Lophodiscella Tehon (1933) = Colletotrichum (Mitosp. fungi) fide Sutton (1977).

Lopholeptosphaeria Sousa da Câmara (1932), nom. dub (Dothideales, inc. sed.). No spp. included.

Lophomerum Ouell. & Magasi (1966), Rhytismataceae. 7, N. & S. temp. ? = Lophodermium (Rhytismat.).

Lophophacidium Lagerb. (1949), Phacidiaceae. 2, Eur., N. Am. See Corlett & Shoemaker (*CJB* **62**: 1836, 1984).

Lophophyton Matr. & Dassonv. (1899) = Trichophyton (Mitosp. fungi) fide Carmichael *et al.* (1980).

Lophothelium Stirt. (1887) = Polycoccum (Dacamp.) fide Hawksworth (*Notes R. bot. Gdn Edinb.* **36**: 181, 1978).

Lophotrichaceae, see Microascaceae.

lophotrichous (**lophotrichate**), having several flagella at one or both ends.

Lophotrichus R.K. Benj. (1949), Microascaceae. 4, widespr. See Seth (*Nova Hedw.* **19**: 591, 1971; key), v. Arx *et al.* (*Beih. Nova Hedw.* **94**, 1988; key).

Loramyces W. Weston (1929), Loramycetaceae. 2, USA, Eur. See Ingold & Chapman (*TBMS* **35**: 268, 1952), Digby & Goos (*Mycol.* **79**: 821, 1988; posn). Anamorph *Anguillospora*.

Loramycetaceae Dennis ex Digby & Goos (1988), Leotiales. 1 gen., 4 spp. Stromata absent; ascomata deeply cupulate, almost perithecial, formed within an exogenous gelatinous matrix; peridium hyaline, thin-walled; interascal tissue of thin-walled simple paraphyses; asci with a wide pore surrounded by a poorly developed I- ring; ascospores 2-septate, hyaline, with a long basal cellular appendage, with a gelatinous sheath. Anamorphs hyphomycetous, *Anguillospora*-like. Saprobic on submerged decaying plant tissue, N. temp.

Loranitschkia Lar. N. Vassiljeva (1990), Nitschkiaceae. 1, Sachalin.

Loranthomyces Höhn. (1909) = Actinopeltis (Microthyr.) fide Spooner & Kirk (1990).

lorate, like a narrow band; strap-like in form; ligulate.

Loratospora Kohlm. & Volkm.-Kohlm. (1993), Dothideales (inc. sed.). 1, USA.

lorchel (**false morel, lorel**), the ascoma of *Gyromitra esculenta*, which is poisonous; see gyromitrin.

lorel, see lorchel.

Loricella Velen. (1934), Leotiales (inc. sed.). 2, former Czechoslovakia.

Loten Adans. (1763) nom. dub. (? Ascomycota, inc. sed.).

lotic, habitat running water (streams). Cf. lentic.

Loweomyces (Kotl. & Pouzar) Jülich (1982), Coriolaceae. 2, USA, UK.

Loweporus J.E. Wright (1976), Coriolaceae. 5, pantrop. See Ryvarden & Eriksson (*Preliminary polypore flora of East Africa*: 410, 1980).

lower fungi, *Chytridiomycota, Hyphochytriomycota, Myxomycota, Oomycota, Plasmodiophoromycota* and *Zygomycota*. See also phycomycetes.

Loxophyllum Klotzsch (1831) [non Blume (1826), *Scrophulariaceae*] = Cyclomyces (Hymenochaet.) fide Saccardo (1887).

Loxospora A. Massal. (1852), Haematommataceae (L). c. 5, widespr.

LPS, Instituto de Botánica C. Spegazzini (q.v.; La Plata, Argentina); founded 1935; a part of the Museo de La Plata.

LSD, see lysergic acid.

Lucidium Lohde (1874) nom. dub. (Pyth.) fide Lohde (1874).

Luciotrichus R. Galán & Raitv. (1994), Otideaceae. 1, Spain.

Ludovicia Trevis. (1857) [non Coss. (1856), *Leguminosae*] = Baeomyces (Baeomycet.).

Ludwigiella Petr. (1922) = Selenophoma (Mitosp. fungi) fide Sutton (1977).

Ludwigomyces Kirschst. (1939), Ascomycota (inc. sed.). 1, Eur.

Luellia K.H. Larss. & Hjortstam (1974), Atheliaceae. 2, Eur., N. Am.

Lulesia Singer (1970), Tricholomataceae. 2, S. trop.

Lulworthia G.K. Sutherl. (1916), Halosphaeriaceae. 9 (marine), widespr. See Kohlmeyer & Kohlmeyer (1979), Nakagiri (*Trans. mycol. Soc. Japan* **25**: 377, 1985).

lumen, the central cavity of a cell or other structure.

Luminescent fungi. Some fungi (40 listed by Wassink, 1979), such as species of *Panus* and *Pleurotus*, *Clitocybe illudens* (also mycelium; see Carey, *Mycol.* **66**: 991, 1974), *Armillaria mellea* (mycelium in wood exposed to air), and *Xylaria hypoxylon* (mycelium) and certain bacteria, give out light, sometimes causing the attacked wood or leaves to become luminous. See Buller (*Researches* **3**, **6**), Corner (*Mycol.* **42**: 423, 1950; *TBMS* **37**: 256, 1954), Ramsbottom (*Mushrooms and toadstools*, 1953), Berliner (*Mycol.* **53**: 84, 1962), Calleja & Reynolds (*TBMS* **55**: 149, 1970), Wassink (*Meded. Landbouwhogeschool Wageningen* **79** (5), 1979; review, incl. physiol. basis), Glawe & Solberg (*Mycol.* **81**: 196, 1989; early records).

lunate, like a new moon; crescentic (Fig. 37.9).

lungwort, the lichen *Lobaria pulmonaria* from its external resemblance to a human lung; lung lichen; lungs of oak. Formerly used in folk-lore as a cure for pulmonary diseases.

Lunospora Frandsen (1943) = Pseudoseptoria (Mitosp. fungi) fide Sutton (1977).

Lunulospora Ingold (1942), Mitosporic fungi, 1.E1.11. 1 (aquatic), UK.

lupinosis, a mycotoxicosis of sheep, caused by *Phoma leptostromiformis*; see van Warmelo & Marasas (*Mycol.* **64**: 316, 1972); teleomorph *Diaporthe* (q.v.).

luteoskyrin, a carcinogenic toxin of *Penicillium islandicum*; a cause of hepatitis (yellow rice disease) in man. See also islandicin; cf. skyrin.

luteous, yellow.

Luttrell (Everett Stanley, 1916-88). American mycologist and plant pathologist, received PhD from Duke University 1940 under F.A. Wolf; mainly worked at Georgia Experiment Station (1942-44, 1949-66) and University of Georgia (1966-86) where he was finally D.W. Brooks Distinguished Professor of Plant Pathology. His *Taxonomy of pyrenomycetes* (1951) was a major step forward in ascomycete systematics, linking the bitunicate ascus to the ascolocular ontogeny whose importance had been stressed by Nannfeldt (q.v.), and his keys to all non-lichenized loculoascomycete genera (*in* Ainsworth *et al.* (Eds), *The Fungi* 4A: 133, 1973) were the first critical ones prepared. Published detailed accounts of many *Helminthosporium*-like fungi and plant pathogenic ascomycetes. *Obit.*: Hanlin & Garrett (*Phytopath.* 78: 1388, 1988; portr.); Hanlin & Mims (*Mycol.* 82: 9, 1990; bibliogr., portr.).

Luttrellia Khokhr. & Gornostaĭ (1978) = Exserohilum (Mitosp. fungi) fide Kirk (*in litt.*).

Luttrellia Shearer (1978), Halosphaeriaceae. 1 (aquatic), USA.

Lutypha Khurana, K.S. Thind & Berthier (1977), Typhulaceae. 1, India.

Lutziomyces O.M. Fonseca (1939) ≡ Paracoccidioides (Mitosp. fungi).

Luxuriomyces R.F. Castañeda (1988) = Diplococcium (Mitosp. fungi) fide Sutton (*in litt.*).

Luykenia Trevis. (1860) = Thelenella (Thelenell.).

Luzfridiella R.F. Castañeda & W.B. Kendr. (1991), Mitosporic fungi, 2.B2.24/27. 1, Cuba.

Lycogala Pers. (1794), Lycogalaceae. 4, cosmop. See Martin (*Mycol.* 59: 158, 1967; key).

Lycogalaceae Corda (1828), Liceales. 4 gen. (+ 16 syn.), 19 spp.

Lycogalopsis E. Fisch. (1886), Lycoperdaceae. 1, trop. See Martin (*Lilloa* 4: 69, 1939).

lycomarasmin, a dipeptide wilt toxin from *Fusarium bulbigenum* f.sp. *lycopersici* (Gäumann *et al.*, *Phytopath.* Z. 16: 257, 1950, 30: 87, 1957); **lycomarasmic acid** is a derivative (Gäumann & Naef-Roth, *Phytopath.* Z. 34: 426, 1959). Cf. fusaric acid.

Lycoperdaceae Chevall. (1826), Lycoperdales. 19 gen. (+ 27 syn.), 193 spp. Gasterocarp epigeous, with apical dehiscence; hyphae without clamps; gleba powdery; spores brown, verruculose.
Lit.: Kreisel (*Feddes Repert.* 64: 89, 1962, *Bibl. Mycol.* 36, 1973; Germany).

Lycoperdales, Basidiomycetes ('gasteromycetes'). 5 fam., 33 gen. (+ 46 syn.), 272 spp. The Puff Balls and Earth Stars. Gleba with cavities of independent origin lined by an hymenium, powdery at maturity with capillitium (seldom paracapillitium only) mixed with spores; peridium has, usually, at least two layers; cosmop. saprobes of soil and rotting wood, some perhaps mycorrhizal. Fams:
(1) **Broomeiaceae**.
(2) **Geastraceae**.
(3) **Lycoperdaceae**.
(4) **Mesophelliaceae**.
(5) **Mycenastraceae**.
Lit.: Boiffard (*Docums mycol.* 6 (24): 1, 1976; *Geastraceae*, France), Ponce de Leon (*Fieldiana, Bot.* 31: 303, 1968; *Geastraceae*, world monogr., broad species concept), Demoulin (*Bull. Jard. bot. nat. Belg.* 38: 1, 1968; Belgium), Dring (1973; key gen.), Kreisel (*Feddes Repert.* 64: 89, 1962, *Bibl. Mycol.* 36, 1973; *Lycoperdaceae*, Germany). See also *Lit.* under *Gasteromycetes*.

Lycoperdastrum P. Micheli (1729) nom. inval. = Scleroderma (Sclerodermat.).

Lycoperdastrum Haller ex Kuntze (1891) = Elaphomyces (Elaphomycet.) fide Dodge (1929).

Lycoperdellon Torrend (1913) = Ostracoderma (Mitosp. fungi) fide Hennebert (*Persoonia* 7: 200, 1973).

Lycoperdodes Kuntze (1891) = Pisolithus (Sclerodermat.).

Lycoperdoides P. Micheli ex Kuntze (1891) = Pisolithus (Sclerodermat.) fide Fischer (1938).

Lycoperdon Tourn. ex L. (1753) ≡ Lycogala (Myxom.).

Lycoperdon Pers. (1801), Lycoperdaceae. 50, cosmop. The common Puff Balls. See Demoulin (*Mycotaxon* 3: 275, 1976; key spp. peridial spherocysts, *Revista Biol.* 12: 65, 1983; key S. Eur.), Kreisel (*Khumbu Himal.* [*Ergebn. ForschUnternehmens Nepal Himalaya*] 6: 32, 1969; key spp. pedicellate spores).

Lycoperdopsis Henn. (1899), Lycoperdaceae. 1, trop. Asia.

Lylea Morgan-Jones (1975), Mitosporic fungi, 1.C2.3. 2, widespr, esp. N. temp. See Holubová-Jechová (*Folia geobot. phytotax.* 13: 437, 1978).

lymabiont, an organism only found in sewage.

lymaphile, an organism commonly found in sewage (Cooke, *Sydowia, Beih.* 1, 1957).

lymaphobe, an organism never found in sewage (Cooke, *Sydowia, Beih.* 1, 1957).

lymaxene, an organism rarely found in sewage (Cooke, *Sydowia, Beih.* 1, 1957).

Lymphocystidium Weiser (1943) ? Fungi or Protozoa. 1, Eur. See Weiser (*Zool. Anzeiger* 142: 200, 1943).

lyocystidium, see cystidium.

Lyomyces P. Karst. (1881) = Hyphoderma (Hyphodermat.) fide Donk (1956).

Lyomyces P. Karst. (1882) = Laeticorticium (Cortic.) fide Donk (1964).

Lyonella Syd. (1925) [non Raf. (1818), *Polygonaceae*], Ascomycota (inc. sed.). 1, Hawaii.

Lyonomyces T.N. Taylor, Hass & W. Remy (1993), Fossil fungi (Paleomastigomycetes). 1 (Devonian), UK.

lyophilization, freeze-drying, a technique used to preserve fungal cultures in a state of suspended animation. See Genetic resource collections.

Lyophyllopsis Sathe & J.T. Daniel (1981), Tricholomataceae. 1, India.

Lyophyllum P. Karst. (1881), Tricholomataceae. 25, N. temp. See Clemençon (*Mycotaxon* 15: 67, 1982; staining spp.).

Lyromma Bat. & H. Maia (1965), Mitosporic fungi (L), 8.E1.?. 1, Brazil.

lysergic acid, and derivatives, some of which are hallucinogenic (e.g. lysergic acid diethylamide; LSD) occur in *Claviceps* sclerotia and are the cause of paspalum staggers (q.v.); see also hallucinogenic fungi.

lysigenous, formed by the breaking down of cells. Cf. schizogenous.

Lysipenicillium Bref. (1908) = Gliocladium (Mitosp. fungi) fide Raper & Thom (*A manual of the Penicillia*, 1949).

lysis, (1) dissolution of a cell, e.g. by a lysin, see antigen; (2) (of conidial liberation), secession by the dissolution of the wall of the adjacent cell; cf. fission, fracture.

Lysospora Arthur (1906) = Puccinia (Puccin.) fide Dietel (1928).

Lysotheca Cif. (1962), Mitosporic fungi, 4.E1.?. 1 (on *Balladynella*), Dominican Republic.

Lysurus Fr. (1823), Clathraceae. 5, mostly trop.

M, Botanischen Staatssammlung München (Munich, Germany); founded 1813; funded by the State of Bavaria; see Hertel & Schreiber (*Mitt. Bot. StSamml., Münch.* 28: 81, 1988; history, collectors index).

MA, see Media.

Maasoglossum K.S. Thind & R. Sharma (1985), ? Geoglossaceae. 1, Himalaya.

Macabuna Buriticá & J.F. Hennen (1994), Phakopsoraceae. 7, widespr.

macaedium, see mazaedium.

Macalpinia Arthur (1906) = Uromycladium (Pileolar.) fide Dietel (1928).

Macalpinomyces Langdon & Full. (1977), Ustilaginaceae. 1 (on *Eriachne*), Australia.

Macbridella Seaver (1909) = Byssosphaeria (Melanommat.) fide Barr (1984).

Macbrideola H.C. Gilbert (1934), Stemonitidaceae. 13, Eur., C. & N. Am., Japan, India. See Alexopoulos (*Mycol.* **59**: 103, 1967; key).

Maccagnia Mattir. (1922), ? Hydnangiaceae. 1, Italy.

Macentina Vězda (1973), Verrucariaceae (L). 6, Afr., Eur. See Coppins & Vězda (*Lichenologist* **9**: 47, 1977), Orange (*Lichenologist* **23**: 15, 1991; key).

Macilvainea Nieuwl. (1916) ≡ Drudeola (Mitosp. fungi). See Sutton (1977).

Macmillanina Kuntze (1898) = Disculina (Mitosp. fungi) fide Sutton (1977).

Macowania Kalchbr. ex Berk. (1876) [non Oliv. (1870), *Compositae*] ≡ Macowanites (Elasmomycet.).

Macowaniella Doidge (1921), Asterinaceae. 2, S. Afr.

Macowanites Kalchbr. (1882), Elasmomycetaceae. 22, S. Afr., N. Am. See Smith (*Mycol.* **55**: 435, 1963; key).

Macraea Subram. (1952) [non Lindl. (1828), *Geraniaceae*] ≡ Prathigada (Mitosp. fungi).

macro- (prefix), long, but commonly used in the sense of mega- (q.v.).

Macroallantina Speer (1987), Anamorphic Ascodichaenaceae, 8.A1.1. Teleomorph *Delpinoina*. 1, France.

Macrobasis Starbäck (1893) = Leptosphaeria (Leptosphaer.) fide v. Arx & Müller (1975).

Macrobiotophthora Reukauf (1912), Ancylistaceae. 2, Germany, Australia. See Reukauf (*Centralbl. Bakt. Abt. Orig.* **63**: 390, 1912), McCulloch (*TBMS* **68**: 173, 1977), Tucker (1981; emend., key).

macrocephalic, see septum.

Macrochytrium Minden (1902), ? Chytridiaceae. 1, Eur., N. Am.

macroconidium, (1) the larger, and generally more diagnostic conidium of a fungus which has microconidia (and sometimes also mesoconidia) in addition; (2) (infrequent), a long large conidium.

macrocyclic, see Uredinales.

macrocyst, see cyst.

Macrocystidia Joss. (1934), Tricholomataceae. 3, Eur., S. Am. See Capellano (*BSMF* **92**: 221, 1971; basidiospore structure), Kühner (*Hyménomycètes agaricoïdes*: 428, 1980).

Macrocystis R. Heim (1931) [non C. Agardh (1820), *Algae*] ≡ Macrocystidia (Tricholom.).

Macrodendrophoma T. Johnson (1904) nom. inval. (Mitosp. fungi) fide Sutton (*Mycol. Pap.* **141**, 1977).

Macroderma Höhn. (1917), Cryptomycetaceae. 2, N. Am., Brazil.

Macrodiaporthe Petr. (1920), Melanconidaceae. 2, Eur., N. Am.

Macrodictya A. Massal. (1852) ≡ Lasallia (Umbilicar.).

Macrodiplis Clem. & Shear (1931) ≡ Macrodiplodiopsis (Mitosp. fungi).

Macrodiplodia Sacc. (1884), Mitosporic fungi, ?.B2.?. 2, Eur. See Sutton (1977).

Macrodiplodina Petr. (1962) = Ascochyta (Mitosp. fungi) fide Punithalingam (1979).

Macrodiplodiopsis Petr. (1922), Mitosporic fungi, 8.C/D2.19. 1, Eur., N. Am. See Glawe (*Mycol.* **77**: 880, 1985).

macrofungi, see Macromycetes.

macrogonidium, a large gonidium (obsol.); megalogonidium (obsol.).

Macrohilum H.J. Swart (1988), Mitosporic fungi, 4.B2.19. 1, Australia.

Macrohyporia I. Johans. & Ryvarden (1979), Coriolaceae. 2, trop.

Macrolepiota Singer (1948), Agaricaceae. 12, cosmop. *M. procera*, the edible parasol mushroom. See Heinemann (*Bull. Jard. bot. nat. Belg.* **39**: 201, 1969; key 11 Congo spp.), Pegler (*Kew Bull., Addit. Ser.* **6**, 1977; E. Afr. spp.).

macrolichen, one of the larger lichens of squamulose, foliose, or fruticose habit incl. spp. *Cladonia, Parmelia, Usnea*, etc.

Macrometrula Donk & Singer (1948), Coprinaceae. 1 (in glasshouse), UK.

Macromycetes. Fungi having large (macroscopic) sporocarps; larger fungi. Because of the interest shown by amateur mycologists, mycophagists, and others in macrofungi there are for many countries popular or semi-popular texts covering basidiomycetes (mushrooms and gasteroid fungi) and larger discomycetes (cup fungi and truffles). These works are frequently of a high mycological standard and are useful both to amateur and professional mycologists. They include:

Regional: **Africa**, Morris (*Common mushrooms of Malawi*, 1987), Zoberi (*Tropical macrofungi*, 1972). **N. America**, Phillips (*Mushrooms of North America*, 1991). **Australia**, Aberdeen (*Introduction to the mushrooms ... of Queensland*, 1979), Fuhrer (*Field companion to Australian fungi*, 1985), Shepherd & Totterdell (*Mushrooms and toadstools of Australia*, 1988), Young (*Common Australian fungi*, 1982). **Bulgaria**, Sechanov ([*Fungi of Bulgaria*] edn 2, 1957). **Canada**, Groves (*Edible and poisonous mushrooms of Canada*, 1979), Pomerleau (*Mushrooms of eastern Canada and the United States*, 1951; *Flore des champignons de Quebec*, 1980). **China**, Pfister (*Champignons du Tonkin*, 1985). **Europe**, Gams (*Kleine Kryptogamenflora* Bd IIa; Moser, *Phycomyceten und Ascomyceten*, 1963; Bd IIb1; Julich, *Nichtblatterpilze, Gallertpilze und Bauchpilze*, Bd IIb2; Moser, *Rohrlinge und Blatterpilze*, edn 5, 1983 [English edn, Phillips, 1978]), Peter (*Das grosse Pilzbuch. Eine Pilzkunde Mitteleuropas*, 1964), *Fungi Europeai*, **1** Cappelli, *Agaricus*, 1984, **2** Alessio, *Boletus*, 1985, suppl. 1991, **3** Riva, *Tricholoma*, 1988, **4** Candusso & Lanzoni, *Lepiota*, 1990, **5** Noordeloos, *Entoloma*, 1992), Pegler *et al.* (*Fungi of Europe*, 1993). **Former Czechoslovakia**, Pilat (*Nase Houby*, 1952 [English transl., *Mushrooms*, 1954]; Houby *Ceskoslovenska*, 1969; spp. grouped ecologically). **Denmark**, Lange (*Illustreret svampeflora*, 1961 [also English, French and German versions]; *Soppflora*, 1991). **France**, Costantin & Dufour (*Nouvelle flore des champignons*, 1891 [and later editions]), Rolland (*Atlas des champignons de France, Suisse et Belgique*, 1910), Heim (*Les champignons d'Europe*, 2 vols, 1957 [edn 2 (1 vol.), 1969]), Maublanc (*Les champignons comestibles et veneneux de France*, edn 6 [by Viennot-Bourgin], 2 vols, 1971), Romagnesi (*Petit atlas des champignons*, 3 vols, 1962-63; *Nouvelle atlas des champignons*, 4 vols, edn 2, 1970), Marchand (*Champignons de Nord et du Midi*, 9 vols, 1972-86), Moreau (*Larousse des champignons*, 1978; *Guide des champignons comestibles et vénéneux*, 1984). **Germany**, Ricken (*Die Blatterpilze (Agaricaceae) Deutschlands*, 1915), Gerhardt (*Pilze*, 2 vols, 1984-5), Hennig (*Taschenbuch fur Pilzfreunde*, 1964 [Aufl. 2, 1966]), Michael & Hennig (*Handbuch fur Pilzfreunde*, 6 vols, 1964-75 [and later edns by Kreisel]), Kreisel (Ed.) (*Pilzflora der DDR*, 1987). **Great Britain**, Buczacki (*Fungi of Britain and Europe*, 1989), Pegler (*Field guide to mushrooms and toadstools of Britain and Europe*, 1990), Wakefield & Dennis (*Common British fungi*, 1950; edn 2, 1981), Watling (*Identification of larger fungi*, 1973), see also Holden, *Guide to the* *literature for the identification of British fungi*, edn 4, 1982. **Greece**, Pantidou (*Mushrooms in the forests of Greece*, 1991). **Italy**, Cetto (*Enzyklopadie der Pilze*, 4 vols, 1987-8), Goidanich & Govi (*Funghi e ambiente*, 1982), Pacioni (*I funghi nostrani*, 1980). **Malawi**, Morris (*Checklist of the macrofungi of Malawi* [*Kirkia* **13**: 323, 1990]). **Mexico**, Blanco *et al.*, in Vega & Bousquets (Eds) (*Hongos Macroscopicos*, 1993). **Netherlands**, Arnolds *et al.* (*List of fungi in the Netherlands*, 1989), Bas *et al.* (*Flora Agaricina Neerlandica*, 1988-). **Scandinavia**, Hansen & Knudsen (*Nordic macromycetes*, 2 vols, 1992-3). **Spain**, Calonge (*Setas (Hongos), guia ilustrada*, 1990). **Sri Lanka**, Pegler (*Kew Bull. Addn. Ser.* **12**). **Sweden**, Cortin (*Svampplockarens handbok*, 1942), Nilsson & Persson (*Svampar i naturen*, 2 vols, 1977), Ryman & Holmåasen (*Svampar*, 1984). **Switzerland**, Breitenbach & Kranzlin (*Fungi of Switzerland* **1** Ascomycetes, 1981, **2** Non-gilled fungi, 1986, **3** Boletes and agarics, 1991 [English, French and German edns]), Clémençon (*Les quatre saisons des champignons*, 2 vols, 1980). **New Zealand**, Taylor (*Mushrooms and toadstools in New Zealand*, 1970). **USA**, Atkinson (*Studies of American fungi: mushrooms, edible, poisonous, etc.*, 1903 [reprint 1961], Krieger (*The mushroom handbook*, edn 2, 1936 [reprint 1967]), Lincoff (*Audubon field guide to North American mushrooms*, 1981), Christensen (*Common fleshy fungi*, 1946 [edn 3, 1965]), Metzler & Metzler (*Texas mushrooms*, 1992), Miller (*Mushrooms of North America*, 1974), Phillips (*Mushrooms of North America*, 1991), Smith (*The mushroom hunter's field guide*, 1958 [edn 3, 1980]; *Field guide to western mushrooms*, 1975), States (*Mushrooms and truffles of the south-west*, 1990).

Coloured plates. The following are important series of coloured reference plates (mostly of agarics but other groups sometimes included): Fries (*Icones selectae Hymenomycetum*, 1867- 84), Cooke (*Illustrations of British fungi (Hymenomycetes)*, 8 vols, 1881-1891; index, *TBMS* **20**: 33, 1935), Konrad & Maublanc (*Icones selectae fungorum*, 6 vols, 1924-36), *BSMF* **41-** (1925-) includes supplementary coloured plates constituting an Atlas), Bresadola (*Iconographia mycologica*, 28 vols, 1927-60; vol. 2 by Gilbert, + 3 suppl.), Cetto (*Der Grosse Pilzfuhrer*, 3 vols, 1979), Dahncke & Dahncke (*700 Pilze in Farbfotos*, 1980), Lange (*Flora agaricina danica*, 5 vols, 1935-41), Romagnesi (*Nouvelle atlas des champignons*, 4 vols, edn 2, 1970), Reid *et al.* (*Fungorum rariorum icones coloratae*, 1966-), Phillips (*Mushrooms and other fungi of Great Britain and Europe*, 1981), Moser & Julich (*Coloured atlas of Basidiomycetes*, 1986-). See further Bolton, Boudier, Bulliard, Greville and Sowerby. See also Watling & Watling (*A literature guide for identifying mushrooms*, 1980; by fam., gen. and geogr. region), Moser & Jülich (*Farbatlas der Basidiomyceten*, 1985).

macronematous (of conidiophores), morphologically different from vegetative hyphae (cf. micronematous).
Macronemeae, hyphomycetes having conidia unlike the hyphae and the conidiophores. Cf. Micronemeae.
Macronodus G.F. Orr (1977) = Gymnoascus (Gymnoasc.) fide v. Arx (*Persoonia* **9**: 393, 1977).
Macroon Corda (1833) = Helminthosporium (Mitosp. fungi) fide Saccardo (*Syll. Fung.* **4**: 403, 1886) but doubtful fide Sutton (*in litt.*).
Macrophoma (Sacc.) Berl. & Voglino (1886) = Sphaeropsis (Mitosp. fungi) fide Sutton (1980). See Petrak & Sydow (*Beih. rep. spec. nov. regni veg.* **42**, 1927; redispositions).
Macrophomella Died. (1916) = Botryodiplodia (Mitosp. fungi) fide Petrak & Sydow (1927).

Macrophomina Petr. (1923), Mitosporic fungi, 4.A1.15. 1, widespr. *M. phaseolina* (syn. *Rhizoctonia bataticola*), an important root-parasite in warmer regions. See Goidànich (*Ann. Sper. agr.* N.S. **1**: 449, 1947), Dhingra & Sinclair (Eds) (*An annotated bibliography of Macrophomina phaseolina, 1905-1976*, 1977; 904 refs.), Punithalingam (*Nova Hedw.* **38**: 339, 1983; cytology).
Macrophomopsis Petr. (1924), Mitosporic fungi, 8.A1.15. 1, France.
Macrophomopsis N.E. Stevens & Baechler (1926) = Botryodiplodia (Mitosp. fungi) fide Petrak & Sydow (1927).
macrophylline (of foliose lichens), having large lobes (obsol.).
Macrophyllosticta Sousa da Câmara (1929) = Phyllosticta (Mitosp. fungi) fide Sutton (1977).
macroplasmodium, used for the large plasmodium of *Physarum polycephalum*.
Macroplodia Westend. (1857) nom. rej. = Sphaeropsis (Mitosp. fungi).
Macroplodiella Speg. (1908) = Phoma (Mitosp. fungi) fide Clements & Shear (1931).
Macropodia Fuckel (1869) = Helvella (Helvell.) fide Dissing (1966).
Macroporia Johanson & Ryvarden (1979), Polyporaceae. 3, widespr.
Macropyrenium Hampe ex A. Massal. (1860) = Ocellularia (Thelotremat.) fide Hale (1981).
macroscopic, visible without a lens.
Macroscyphus Gray (1821) = Helvella (Helvell.) fide Dissing (1966). See Rifai (1968).
Macroseptoria Petr. (1923) = Trichoseptoria (Mitosp. fungi) fide Clements & Shear (1931).
Macrospora Fuckel (1870), Diademaceae. 3, Eur. See Barr (*SA* **12**: 27, 1993; posn). Anamorph *Nimbya*.
macrospore, a large spore when there are spores of two sizes.
Macrosporites Renault (1899), Fossil fungi. 3 (Carboniferous), Germany.
Macrosporium Fr. (1832) nom. rej. = Alternaria (Mitosp. fungi) fide Hughes (1958).
Macrostilbum Pat. (1898), Mitosporic fungi, 2.A1.?. 1, Java. Nom. dub. fide Seifert (*TBMS* **85**: 123, 1985).
Macrothelia M. Choisy (1949) = Amphisphaeria (Amphisphaer.) fide Santesson (*Bull. Soc. linn. Lyon* **23**: 103, 1954).
Macrotrichum Grev. (1825), Mitosporic fungi, 1.A-C1.?. 2, UK.
Macrotyphula R.H. Petersen (1972), Clavariaceae. 4, widespr. See Berthier (*Bull. Soc. linn. Lyon* num. spéc. **46**: 213, 1976).
Macrovalsaria Petr. (1962), Dothideales (inc. sed.). 1, trop. See Barr (*Mycotaxon* **49**: 129, 1993; posn), Sivanesan (*TBMS* **65**: 395, 1975).
Macroventuria Aa (1971), ? Pleosporaceae. 2, Afr., USA.
Macruropyxis Azbukina (1972), Uropyxidaceae. 1, former USSR.
maculate, spotted; blotched.
Maculatipalma J. Fröhl. & K.D. Hyde (1995), Valsaceae. 1 (on *Palmae*), Australia.
maculicole (adj. -icolous), an organism living on spots, e.g. leaf spots.
Madura foot, see mycetoma.
maduramycosis, see mycetoma.
Madurella Brumpt (1905), Mitosporic fungi, 2.A2.?. 10 (on humans causing mycetoma, q.v.), widespr., esp. trop. See Ciferri & Redaelli (*Mycopathol.* **3**: 182, 1941).
Magdalaenaea Arnaud (1952), Mitosporic fungi, 1.G1.?. 1, Eur.

magic mushrooms, typically hallucinogen-containing species of *Psilocybe*, and also *Gymnopilus*, *Panaeolus*, *Conocybe*, and *Amanita muscaria*; hallucinogenic fungi (q.v.).

Magmopsis Nyl. (1875) = Arthopyrenia (Arthopyren.).

Magnaporthaceae P.F. Cannon (1994), Ascomycota (inc. sed.). 5 gen. (+ 3 syn.), 13 spp. Stromata absent; ascomata perithecial, black, usually with long hairy necks; interascal tissue of thin-walled tapering paraphyses; asci ± cylindrical, persistent, fairly thickwalled, without separable layers, with a large apical pore surrounded by an often I+ apical ring; ascospores septate, often filiform, often with the middle cells pigmented, without germ pores, without a sheath. Anamorphs varied, hyphomycetous, usually with pigmented conidiogenous cells. Sclerotia sometimes formed. Usually necrotrophic on roots, esp. of grasses, widespr. Possibly close to *Phyllachorales*.

Magnaporthe R.A. Krause & R.K. Webster (1972), Magnaporthaceae. 4, mainly trop. *M. grisea* (rice blast; anamorph *Pyricularia oryzae*; Zeigler *et al.*, *Rice blast disease*, 1994). See Barr (*Mycol.* **69**: 952, 1977), Kato (*Ann. Phytopath. Soc. Japan* **60**: 266, 1994; *M. grisea* phylogeny), Kato *et al.* (*Ann. Phytopath. Soc. Japan* **60**: 175, 1994; *M. grisea* microconidia), Tsuda & Ueyama (*Trans. mycol. soc. Japan* **19**: 425, 1978; teleomorph formation). Anamorphs *Nakataea*, *Phialophora*, *Pyricularia*.

Magninia M. Choisy (1929) = Lecanora (Lecanor.).

Magnosporites Rouse (1962), ? Fossil fungi. 1 (Paleogene), Canada.

Magnusia Sacc. (1878) [non Klotzsch (1854), Begoniaceae] ≡ Kernia (Microasc.).

Magnusiella Sadeb. (1893) = Taphrina (Taphrin.) fide Mix (1949).

Magnusiomyces Zender (1925) = Dipodascus (Dipodasc.) fide de Hoog *et al.* (*Stud. Mycol.* **29**, 1986).

Magnusson (Adolf Hugo; 1885-1964). School teacher at Gothenburg, Sweden, from 1909-48. Devoted his spare time to lichenology making invaluable contributions to knowledge of the lichen floras of Hawaii, Scandinavia and China and the genera *Lecidea*, *Lecanora*, *Caloplaca* and *Acarospora*. He described about 900 new species (collection of 70,000 specimens in UPS) in *c.* 150 papers. See Degelius (*Svensk bot. Tidskr.* **59**: 393, 1965), Almborn (*Bot. Notiser* **117**: 428, 1964), Grummann (1974: 478), Stafleu & Cowan (*TL-2* **3**: 247, 1981).

Magnussoniolichen Tomas. & Cif. (1952) = Polyblastia (Verrucar.).

Magnussoniomyces Cif. & Tomas. (1953) ≡ Magnussoniolichen (Verrucar.).

Magoderna Steyaert (1972), Ganodermataceae. 1, trop.

Mahabalella B. Sutton & S.D. Patil (1966), Mitosporic fungi, 1.A1.15. 1, India.

Mahevia Lagarde (1917) = Hirsutella (Mitosp. fungi) fide Kendrick & Carmichael (1973).

Mainsia H.S. Jacks. (1931) = Gerwasia (Phragmid.) fide Cummins & Hiratsuka (1983).

Maire (René Charles Joseph Ernest; 1878-1949). Professor of Botany, University of Algiers, 1911-49; noted for his contributions to fungal cytology, the anatomy of *Russula*, and the fungi of N. Afr. See Kühner (*BSMF* **69**: 7, 1953; portr.), Feldmann & Guinier (*Bull. Soc. Hist. nat. Afr. Nord* **41**: 65, 1952; bibl.), Grummann (1974: 336), Stafleu & Cowan (*TL-2* **3**: 257, 1981).

Maireella Syd. ex Maire (1908) = Gibbera (Ventur.) fide Müller & v. Arx (1962).

Maireina W.B. Cooke (1961), Cyphellaceae. 27, widespr. = Cyphellopsis (Cyphell.) fide Donk (1951).

Maireomyces Feldmann (1941) ? Fungi (inc. sed.) fide Kohlmeyer & Kohlmeyer (1979).

Majewskia Y.B. Lee & K. Sugiy. (1986), Laboulbeniaceae. 1, Japan. See Eriksson & Hawksworth (*SA* **7**: 78, 1988; nomencl.).

Malacaria Syd. (1930), Tubeufiaceae. 2 (on *Meliola*), Philip., Uganda, Venezuela. See Rossman (*Mycol. Pap.* **157**, 1987; key).

malaceoid venation, see veins.

Malacharia Fée (1843) ? = Cerebella (Mitosp. fungi).

Malacodermis Bubák & Kabát (1912) = Glutinium (Mitosp. fungi) fide Höhnel (*Sber. Akad. wiss. Wien, mat.-nat. Kl.* **125**: 27, 1916).

Malacodermum Marchand (1896) nom. dub. fide Donk (*Taxon* **6**: 84, 1957).

Malacodon J. Bataille (1923) nom. dub. See Donk (*Taxon* **5**: 102, 1956).

malacoid, like mucilage.

Malacosphaeria Syd. (1924) = Eriosphaeria (Trichosphaer.) fide Müller & v. Arx (1962).

Malacostroma Höhn. (1920) [non Gürich (1906), fossil ? *Algae*] = Phomopsis (Mitosp. fungi) fide Petrak (*Ann. Myc.* **19**: 176, 1921).

Malajczukia Trappe & Castellano (1992), Mesophelliaceae. 8, Australia, NZ.

Malassezia Baill. (1889), Mitosporic fungi, 1.A1.19. 2 (on humans), widespr. See Salkin & Gordon (*Can. J. Microbiol.* **23**: 471, 1977; polymorphism), Bastide *et al.* (*Bull. Soc. Fr. Mycol. Med.* **17**: 233, 1988; morphology and physiology), Guého (*Bull. Soc. Fr. Mycol. Med.* **17**: 245, 1988; re-evaluation by TEM), Midgely (*Mycopathol.* **106**: 143, 1989; diversity), Guého & Meyer (*Ant. v. Leeuwenhoek* **55**: 245, 1989; genome comparison), Simmons & Guého (*MR* **94**: 1146, 1990), Marcon & Powell (*Clin. Microbiol. Rev.* **5**: 101, 1992; human infections), Ingham & Cunningham (*J. med. vet. Mycol.* **31**: 265, 1993), Hernández-Molina (*Rev. Iber. Mic.* **10**: 24, 1993; bibliogr.), Ingham & Cunningham (*Jl Med. Vet. Mycol.* **31**: 265, 1993; review of *M. furfur*), Howell, Quin & Midgley (*Mycoses* **36**: 263, 1993; karyotypes of *M. furfur*), Boekhout & Bosboom (*Syst. Appl. Microbiol.* **17**: 146, 1994; karyotyping), Guillot & Guého (*Ant. v. Leeuwenhoek* **67**: 297, 1995; rRNA sequence & nuclear DNA comparisons). See pityriasis versicolor.

Malasseziales. Used by Moore (1980) for *Basidioblastomycetes* lacking ballistospores (e.g. *Cryptococcus*, *Rhodotorula*, *Schizoblastosporon*, *Trichosporon*, *Vanrija*). Included in Mitosporic fungi in this *Dictionary*. See Yeasts for Lit.

Malbranchea Sacc. (1882), Anamorphic Myxotrichaceae and Onygenaceae, 1.A1.40. Teleomorphs *Myxotrichum*, *Auxarthron*, *Uncinocarpus*. 17, widespr. See Sigler & Carmichael (*Mycotaxon* **4**: 412, 1976; key).

Malenconia Bat. & H. Maia (1960) = Coccomyces (Rhytismat.) fide v. Arx (*in litt.*).

malformin, a plant-malforming cyclic pentapeptide from *Aspergillus niger* (Takahashi & Curtis, *Plant Physiol.* **36**: 30, 1961).

Malinvernia Rabenh. (1857) = Podospora (Lasiosphaer.) fide Lundqvist (1972).

Malleomyces Hallier (1870) nom. dub. (? Mitosporic fungi). See Rivolta (*Parass. Veg.* 2nd edn, 1884).

Mallochia Arx & Samson (1986), Gymnoascaceae. 1, Pakistan.

Mallotium (Ach.) Gray (1821) = Leptogium (Collemat.).

Malmella C.W. Dodge (1933) = Erioderma (Pannar.) fide Galloway & Jørgensen (*Lichenologist* **7**: 139, 1975).

Malmeomyces Starbäck (1899), Hypocreaceae. 1, Brazil. See Rossman (*Mycotaxon* **8**: 537, 1979).

Malmgrenia Vain. (1939) = Cryptothele (Lichin.) fide Jørgensen & Henssen (*Taxon* **39**: 343, 1990).

Malmia M. Choisy (1931) = Rinodina (Physc.).

Malotium Velen. (1934) [non *Mallotium* (Ach.) Gray (1921)], Leotiales (inc. sed.). 9, former Czechoslovakia.

malting, see brewing.

maltoryzine, a metabolite of *Aspergillus oryzae* var. *microsporus*; toxic for cattle (Iizuka & Iida, *Nature* **196**: 681, 1962).

Malupa Y. Ono, Buriticá & J.F. Hennen (1992), Uredinales (inc. sed.). 5, trop.

Malustela Bat. & J.A. Lima (1960) = Curvularia (Mitosp. fungi) fide Ellis (1966).

Mamiania Ces. & De Not. (1863), Valsaceae. 3, N. temp. See Barr (*Mycol. Mem.* **7**, 1978), Monod (1983).

Mamianiella Höhn. (1917), Valsaceae. 1 (on *Corylus*), Eur., N. Am. See Barr (*Mycol. Mem.* **7**, 1978).

Mammaria Ces. ex Rabenh. (1854), Mitosporic fungi, 1.A2.10. 1, Eur., N. Am. See Hughes (*Sydowia, Beih.* **1**: 359, 1957), Park (*TBMS* **60**: 351, 1973).

Mammariopsis L.J. Hutchison & J. Reid (1988), Mitosporic fungi, 1.A1.38, A/B1.1/10, A1.15. 1 (3 conidial types), NZ.

mammiform, breast-like in form.

Manaustrum Cavalc. & A.A. Silva (1972), Mitosporic fungi (L), 5.B1.?. 1, Brazil.

Manginella Bat. & H. Maia (1961), Mitosporic fungi, 5.A2.?. 2, Brazil.

Manginia Viala & Pacottet (1904) = Sphaceloma (Mitosp. fungi) fide Jenkins & Bitancourt (*Mycol.* **33**: 338, 1941).

Manginiella, see *Mauginiella*.

Manginula G. Arnaud (1918) nom. dub. (? Mitosp. fungi). 5 (recent & fossil), N. Am., Australia. See Lange (*Aust. J. Bot.* **17**: 565, 1969), Sutton (1977).

Manginulopsis Bat. & Peres (1963) = Leptothyrium (Mitosp. fungi) fide v. Arx (*K. Ned. Akad. Wet. Amst.* **66**: 172, 1963).

Manglicola Kohlm. & E. Kohlm. (1971), Hypsostromataceae. 2, Guatemala, Guyana. See Huhndorf (*Mycol.* **86**: 266, 1994).

Mangrovispora K.D. Hyde & Nakagiri (1991), Hyponectriaceae. 1, Australia.

Manikinipollis Krutzsch (1970), ? Fossil fungi (*f. cat.*). 1 (Miocene), former USSR.

Manilaea Syd. & P. Syd. (1914) = Arthonia (Arthon.) fide Santesson (1952).

Manina Adans. (1763) nom. dub. (Clavar.) fide Donk (*Reinwardtia* **2**: 441, 1954).

Manina Banker (1912) = Hericium (Heric.) fide Donk (1956).

manna, see *Sphaerothallia*.

Mannia Trevis. (1857) [non Opiz (1829), *Hepaticae*] = Buellia (Physc.).

mannitol, a polyhydric alcohol, often found as a storage compound in ectotrophic mycorrhizal mantles and lichens.

manocyst (of *Phytophthora*), a projection (receptive papilla) from the oogonium which undergoes fusion with the antheridium.

Manokwaria K.D. Hyde (1993), Amphisphaeriaceae. 1 (on *Palmae*), Australia, Irian Jaya.

mantle, a compact layer of hyphae enclosing short feeder roots of ectomycorrhizal plants, connected to the Hartig (q.v.) net on the inside, and to the extramatrical hyphae (q.v.) on the outside; acts as a nutrient sink.

Manuripia Singer (1960), Tricholomataceae. 1, Bolivia.

Manzonia Garov. (1866) = Aspicilia (Hymenel.).

map lichen, species of *Rhizocarpon* (e.g. *R. geographicum*) which have yellow or yellow-green areolae separated by dark black lines (the prothallus) and forming mosaics. See Lichenometry.

Mapea Pat. (1906) = Uredo (Anamorphic Uredinales).

Mapea Boedijn (1957) ≡ Telomapea (Uredinales, inc. sed.).

Mapletonia B. Sutton (1991), Mitosporic fungi, 4.B1.15. 1, Australia.

Mapping distribution, see Geographical distribution.

Marasmiellus Murrill (1915), Tricholomataceae. *c.* 200, widespr. *M. inoderma*, root rot of maize. See Sabet *et al.* (*TBMS* **54**: 123, 1970), *M. paspali* var. *americana*, 'borde blanco' of maize. Pegler (*Kew Bull., Addit. Ser.* **6**, 1977), Singer (*Beih. Nova Hedw.* **26**: 847, 1976; **44**: 1, 1973; 134 neotrop. spp., extralimital spp.).

Marasmiopsis Henn. (1898) = Phaeomarasmius (Cortinar.) fide Singer (1951).

Marasmius Fr. (1836) nom. cons., Tricholomataceae. *c.* 350, cosmop. *M. oreades* (fairy-ring champignon), *M. perniciosus*, (see *Crinipellis*), *M. plicatus* (Sugarcane root rot). See horse-hair blight fungi. See Singer (*Mycol.* **50**: 103, 1958; key sections, *Bull. Jard. bot. Brux.* **34**: 317, 1964; key 50 Congo spp., *Sydowia* **18**: 106, 1965; keys 154 S. Am. spp., *Fl. Neotrop.* **17**: 62, 1976; key 233 neotrop. spp.), Gilliam (*Mycotaxon* **4**: 1, 1976; 30 N. Am. spp.), Pegler (*Kew Bull., Addit. Ser.* **6**, 1977; 43 E. Afr. spp.), Antonín & Noordeloos (*Libri Botanici* **8**, 1993; Eur. spp.).

Maravalia Arthur (1922), Chaconiaceae. 31, trop. See Mains (*Bull. Torrey bot. Club* **66**: 173, 1939), Ono (*Mycol.* **76**: 892, 1984; key).

Marcelleina Brumm., Korf & Rifai (1967), Otideaceae. 7, Eur. See Dissing *et al.* (*Taxon* **39**: 130, 1990; nomencl.), Moravec (*Mycotaxon* **30**: 473, 1987; key), Pfister (*Sydowia* **38**: 235, 1986; N. Am.).

Marceloa Bat. & Peres (1962) = Treubiomyces (Chaetothyr.) fide v. Arx & Müller (1975).

marcescent (of basidiomata), withering, drying up *in situ*. Cf. putrescent.

Marchalia Sacc. (1889) nom. conf. See v. Arx (*Sydowia* **12**: 400, 1959).

Marchaliella G. Winter ex E. Bommer & M. Rousseau (1891) = Testudina (Testudin.) fide Hawksworth (1979).

Marchandiomyces Dieder. & D. Hawksw. (1990), Mitosporic fungi, 3/9.A1.1. 1 (on lichens), Eur., N. Am.

Marcosia Syd. & P. Syd. (1916) = Stigmina (Mitosp. fungi) fide Sutton (*TBMS* **58**: 164, 1972).

Margarinomyces Laxa (1930) = Phialophora (Mitosp. fungi) fide Schol-Schwarz (1970), but see Cole & Kendrick (*Mycol.* **65**: 682, 1973).

Margarita Lister (1894) [non Gaudin (1829), *Compositae*] ≡ Calomyxa (Dianemat.).

Margaritispora Ingold (1942), Mitosporic fungi, 1.C1.15. 5 (aquatic), Eur. See Marvanová & Descals (*Bot. J. Linn. soc.* **41**: 1, 1985).

marginal frill (of a conidium), the periclinal wall left attached to a spore after secession. Cf. basal frill.

marginal veil (of agarics), an incurving proliferation of the margin of the pileus which protects the developing hymenium. Cf. partial veil.

marginate, (1) having a well-marked edge; (2) (of basal bulb of agaric stipe), having a gutter-like rim as in *Leucocortinarius bulbiger*.

margo proprius, see excipulum proprium, proper margin.

margo thallinus, see excipulum thallinum, thalline margin.

Mariaella Šutara (1987), Strobilomycetaceae. 1, N. temp.

Mariannaea G. Arnaud ex Samson (1974), Mitosporic fungi, 1.A1.15. 2, widespr.

Marine fungi. Fungi are of widespread occurrence in the sea, 800-1000 spp. being known from this habitat. These parasitize marine algae or animals or are saprobes on timber, algae (Nakagiri, *Mycologia* **85**: 638, 1993), sea grasses (Cuomo *et al.*, *Prog. Oceanogr.* **21**: 189, 1988), protozoal cysts and corals (Kohlmeyer & Volkmann-Kohlmeyer, *CJB* **68**: 1554, 1990, Raghukumar & Ragukumar, *Mar. Ecol.* **12**: 251, 1991), sea foam (Nakagiri, *IFO Res. Comm.* **14**: 52, 1989) and other substrata; spores of many (esp. *Halosphaeriales*) have special appendages for attachment (Hyde & Jones, *Bot. Mar.* **32**: 205, 1989; 10 types).

Other fungi occur in brackish water (Tubaki & Ito, *Rep. Tottori mycol. Inst.* **10**: 523, 1973), salt marshes (Bayliss Elliott, *Ann. appl. biol.* **17**: 284, 1930; Apinis & Chesters, *TBMS* **47**: 419, 1964), mangrove swamps (Hyde & Jones, *Marine Ecology* **9**: 15, 1988, Tan Leong & Jones, *CJB* **67**: 2686, 1989), in the salt of salt pans (Quinta, *Food Technol.* **22**: 102, 1968) and on sand (Rees, *Bot. Mar.* **23**: 375, 1980, Nakagiri & Tokura, *Trans. mycol. Soc. Japan* **28**: 413, 1987); 27-36 spp. exclusively tropical (Kohlmeyer, *Marine Ecology* **5**: 329, 1984).

Lichens form distinctive band-like zones on siliceous rocky shores (Fletcher, *Lichenologist* **5**: 368, 401, 1973, terminology of zones; *in* Brown *et al.*, *Lichenology: progress and problems*: 359, 1976, nutrition; *in* Price *et al.*, *The shore environment* **2**: 789, 1980, compr. review).

Lit.: General: Fell *et al.* (*Molecular marine biol. biotech.* **1**: 175, 1992; molecular detection marine microeukaryotes), Johnson & Sparrow (*Fungi in oceans and estuaries*, 1961 [reprint, 1970]), Jones (Ed.) (*Recent advances in aquatic mycology*, 1976; compr. review), Moss (Ed.) (*The biology of marine fungi*, 1986), Kohlmeyer (*in* Moss, 1986: 199; checklist marine Ascomycota), Kohlmeyer & Volkmann-Kohlmeyer (*Bot. Mar.* **34**: 1, 1991).

Regional: **Aldabra, Galapagos,** Kohlmeyer & Volkmann-Kohlmeyer (*Can J. Bot.* **65**: 571, 1987). **Argentina,** Malacalza & Martinez (*Boln Soc. Argent. Bot.* **14**: 57, 1971). **Australia,** Kohlmeyer (*Aust. J. Mar. Freshwater Res.* **42**: 91, 1991), Hyde (*Aust. Syst. Bot.* **3**: 711, 1990). **British Isles,** Fletcher (*Lichenologist* **7**: 1 [siliceous rocky shore lichens], 73 [calcareous and terricolous lichens], 1975; keys Br. spp.). **Brunei,** Hyde (*Bot. J. Linn. Soc.* **98**: 135, 1988). **Denmark,** Koch & Jones (*Svampe* **8**: 49, 1983). **India,** Borse (*Trans. mycol. Soc. Japan* **28**: 55, 1987; *Indian Bot. Reptr* **7**: 18, 1988; key 55 spp.). **Malaysia,** Jones & Kuthubutheen (*Sydowia* **41**: 160, 1990). **Pacific,** Volkmann-Kohlmeyer & Kohlmeyer (*Mycol.* **85**: 337, 1993; biogeogr.). **Philippines,** Jones *et al.* (*Asian Marine Biol.* **5**: 103, 1988). **Seychelles,** Hyde & Jones (*Marine Ecol.* **9**: 15, 1988). **South Africa,** Steinke & Jones (*S. Afr. J. Bot.* **59**: 385, 1993). **USA,** Kohlmeyer, (*TMBS* **57**: 473, 1971; New England checklist), Jones (*Bot. J. Linn. Soc.* **91**: 219, 1985); and many other countries.

Marinosphaera K.D. Hyde (1989), Phyllachoraceae. 1, Indian and Pacific Oceans.

Marinospora A.R. Caval. (1966), Halosphaeriaceae. 3 (marine), widespr. See Johnson *et al.* (*Bot. Mar.* **27**: 557, 1984).

Marisolaris J. Koch & E.B.G. Jones (1989), Ascomycota (inc. sed.). 1 (marine), Denmark.

Mariusia D. Pons & Boureau (1977), Fossil fungi (Microthyr.). 1 (Cretaceous), France.

Maronea A. Massal. (1856), ? Fuscideaceae (L). 13. Eur., N. & S. Am. See Magnusson (*Acta Horti gothob.*

9: 41, 1934), Singh (*Geophytology* **10**: 34, 1980; India).

Maronella M. Steiner (1959) = Biatorella (Biatorell.).

Maroneomyces Cif. & Tomas. (1953) ≡ Maronea (Fuscid.).

Maronina Hafellner & R.W. Rogers (1990), Lecanoraceae (L). 2, Australia, S. Am.

Marssonia J.C. Fisch. (1874) [non H. Karst. (1860), *Gesneriaceae*] ≡ Marssonina (Mitosp. fungi).

Marssoniella Höhn. (1916) [non Lemmerm. (1900), Algae] = Neomarssoniella (Mitosp. fungi).

Marssonina Magnus (1906) nom. cons. prop., Anamorphic Dermateaceae, 6.B1.15/19. Teleomorph *Diplocarpon*. 70, esp. temp. *M. ochroleuca* (chestnut (*Castanea*) leaf spot), *M. rosae* (black spot of rose). See v. Arx (1957), Sutton *et al.* (*TBMS* **86**: 619, 1986; redisposition), Spiers (*Eur. J. For. Path.* **18**: 140, 1988; key 4 spp. on *Populus*), Spiers (*N.Z. Jl Bot.* **27**: 503, 1989; conidial morphology of poplar spp. on host and in culture), Spiers (*Eur. J. For. Path.* **20**: 154, 1990; conidial morphology variation spp. on *Populus*), Farr (*Mycol.* **85**: 814, 1993; rel. to *Septogloeum*).

Marssoninites Babajan & Tasl. (1970), Fossil fungi. 1 (Tertiary), former USSR.

Martella Adans. (1763) = Hericium (Heric.) fide Donk (*Reinwardtia* **2**: 466, 1954).

Martella Endl. (1836) = Hericium (Heric.).

Martellia Mattir. (1900), Elasmomycetaceae. 31, widespr. See Singer & Smith (*Mem. Torrey bot. Cl.* **21**: 26, 1960; key), Smith (*Mycol.* **55**: 421, 1963).

Martensella Coem. (1863), Kickxellaceae. 2, Eur., N. Am. See Benjamin (1959), Jackson & Dearden (*Mycol.* **40**: 168, 1948).

Martensiomyces J.A. Mey. (1957), Kickxellaceae. 1, Congo. See Benjamin (1959).

Martin (George Willard; 1886-1971). Professor of botany, Univ. Iowa, 1929-55. Esp. noted for studies on the *Tremellales* (q.v.) and *Myxomycetes* (q.v.). He contributed keys to the families of fungi to the first five editions of this *Dictionary*. See Wells & Lentz (*Mycol.* **65**: 985, 1973; portr., bibl.), Stafleu & Cowan (*TL-2* **3**: 320, 1981).

Martindalia Sacc. & Ellis (1885) = Phleogena (Phleogen.) fide Barr & Bigelow (*Mycol.* **60**: 456, 1968).

Martinella (Cooke) Sacc. (1892) [non Baill. (1888), *Bignoniaceae*], nom. dub. fide Sutton (*Mycol. Pap.* **141**, 1977).

Martinellisia V.G. Rao & Varghese (1977), Mitosporic fungi, 1.B2.10. 1, India.

Martinia Whetzel (1942) [non Vaniot (1903), *Compositae*] ≡ Martininia (Sclerotin.).

Martininia Dumont & Korf (1970), Sclerotiniaceae. 1, Eur., N. & S. Am. See Dumont (*Mycol.* **65**: 175, 1973).

Mason (Edmund William; 1890-1975). Mycologist, IMI, 1921-60. Noted for the developments he initiated in hyphomycete taxonomy and his interest in pyrenomycetes. See Ellis & Hughes (*TBMS* **66**: 371, 1976; portr.), Booth (*Bull. BMS* **9**: 114, 1975).

Masonhalea Kärnefelt (1977), Parmeliaceae (L). 1, arctic. See Kärnefelt *et al.* (*Pl. Syst. Evol.* **183**: 113, 1992).

Masonia Hansf. (1944) = Dictyonella (Saccard.) fide v. Arx (1963).

Masonia G. Sm. (1952) ≡ Masoniella (Mitosp. fungi).

Masoniella G. Sm. (1952) = Scopulariopsis (Mitosp. fungi) fide Morton & Smith (*Mycol. Pap.* **86**, 1963).

Masoniomyces J.L. Crane & Dumont (1975), Mitosporic fungi, 1.A1.10. 1, Jamaica.

Massalongia Körb. (1855), Ascomycota (inc. sed., L). 2, Eur., N. Am. See Henssen (*CJB* **41**: 1331, 1963), Jørgensen (*in* Hawksworth (Ed.), *Ascomycete systematics*: 390, 1994; posn).

Massalongiella Speg. (1880) = Enchnoa (Nitschk.) fide Petrak & Sydow (*Ann. Myc.* **34**: 11, 1936).

Massalongina Bubák (1916), Mitosporic fungi, 8.A1.?. 1, Eur.

Massalongo (Abramo Bartolommeo; 1824-60). Italian palaeobotanist and lichenologist, and Professor at the University of Verona, renowned for his contribution to an understanding of the systematics of fungi forming crustose lichens by the use of microscopic characters, introducing 114 gen. of which many have been consistently in use (e.g. *Arthopyrenia, Catillaria, Haematomma, Pyrenula, Solenopsora*) or recently reintroduced (e.g. *Bactrospora, Celothelium, Psilolechia*). Major works incl. *Ricerche sull'autonomia dei licheni crustosi* (1852), *Systema lichenum novorum* (1855); a selection of these reprinted (Lazzarini, *Selezione di lavori lichenologici di A.B. Massalongo*, 1991). Often involved in controversies with Trevisan (q.v.). Main collections in Verona (**VER**). See De Toni (*L'Opera lichenologica di Abramo Massalongo*, 1933), Grummann (1974: 518), Poelt (*in* Lazzarini, 1991: 13), Stafleu & Cowan (*TL-2* **3**: 349, 1981).

Massalongomyces Cif. & Tomas. (1953) ≡ Massalongia (Ascomycota, inc. sed.).

Massaria De Not. (1844), Massariaceae. 20, widespr. See Barr (*Mycotaxon* **9**: 17, 1979; key 4 N. Am. spp.), Wehmeyer (*Revision of Melanconis*, 1941), Shoemaker & Le Clair (*CJB* **53**: 1568, 1975; type studies).

Massariaceae Nitschke (1869), Pyrenulales. 5 gen. (+ 3 syn.), 38 spp. Ascomata immersed to erumpent, often aggregated within pseudostromatic tissues or beneath a clypeus, ± globose, black; peridium composed of small-celled pseudoparenchymatous tissue; interascal tissue of trabeculate pseudoparaphyses, true paraphyses not formed; asci cylindrical, persistent, with a conspicuous refractive apical ring; ascospores hyaline to dark brown, transversely septate or muriform, often with thickened septa, often with a sheath. Anamorphs unknown. Saprobic on wood and bark, cosmop.
  Lit.: Eriksson (*SA* **5**: 140, 1986).

Massariella Speg. (1880) = Amphisphaeria (Amphisphaer.) fide Eriksson & Hawksworth (*SA* **7**: 80, 1988). See Dzagania (*BSMF* **102**: 199, 1986).

Massariellops Curzi (1927) = Didymosphaeria (Didymosphaer.) fide Aptroot (1995).

Massarina Sacc. (1883), Lophiostomataceae. 125, widespr. See Bose (*Phytopath. Z.* **41**: 156, 1961; key), Srinivasulu & Sathe (*Sydowia* **26**: 83, 1974; Indian spp.), Webster (*TBMS* **48**: 449, 1965), Leuchtmann (*Sydowia* **37**: 75, 1984), Hyde (*Mycol.* **83**: 839, 1992, key marine spp.; **99**: 291, 1995, list 132 names). Anamorphs *Coniothyrium, Ceratophoma, Microsphaeropsis*, etc.

Massarinaceae, see *Lophiostomataceae*.

Massarinula Géneau (1894) = Massarina (Lophiostom.) fide Bose (1961).

Massariola Füisting (1868), Dothideales (inc. sed.). 3 (on bark), Eur.

Massariopsis Niessl (1876) = Amphisphaeria (Amphisphaer.) fide Müller & v. Arx (1973).

Massariosphaeria (E. Müll.) Crivelli (1983), Lophiostomataceae. 14, widespr. See Crivelli (*Ueber die heterogene Ascomycetengattung Pleospora Rbh.*, 1983; key 7 spp.), Holm & Holm (*Symb. bot. upsal.* **28**(2), 1988), Huhndorf *et al.* (*Mycotaxon* **37**: 203, 1990), Leuchtmann (*Sydowia* **37**: 75, 1985; key). = Chaetomastia (Dacamp.) fide Barr (*Mycotaxon* **34**: 507, 1989).

Massariothea Syd. (1939), Mitosporic fungi, 8.C2.15. 8, widespr. See Alcorn (*MR* **97**: 429, 1993; key).

Massariovalsa Sacc. (1882), Melanconidaceae. *c.* 5, widespr. See Wehmeyer (1941), Petrak (*Sydowia* **19**: 279, 1966), Speer (*BSMF* **102**: 363, 1986). Anamorph *Melanconiopsis*.

Massartia De Wild. (1897), ? Zoopagales (inc. sed.). 1 (in *Algae*), Java. See Hesseltine (1955).

Massee (George Edward; 1850-1917). In charge of the 'Lower Cryptogams' at the Kew Herbarium, 1893-1915; first President of the British Mycological Society, 1896. Chief writings: *British fungus flora* (1892-5), *Text book of fungi* (1906), *Diseases of cultivated plants and trees* (1910), *British fungi and lichens* (1911), (with C. Crossland) *Fungus flora of Yorkshire* (1902-5). See Ramsbottom (*TBMS* **5**: 469, 1917; *J. Bot. Lond.* **55**: 223, 1915), Anon. (*Kew Bull.* 1922: 335, 1922; bibliogr.), Grummann (1974: 377), Stafleu & Cowan (*TL-2* **3**: 359, 1981).

Masseea Sacc. (1899), Leotiales (inc. sed.). 1, N. Am. See Baral (*SA* **13**: 113, 1994).

Masseeella Dietel (1895), Uredinales (inc. sed.). 3, India, Philipp. See Sathe (*Sydowia* **19**: 187, 1966).

Masseeola Kuntze (1891) ≡ Sparassis (Sparassid.).

Masseerina Lloyd (1920) nom. nud. (Podoscyph.) fide Donk (*Taxon* **6**: 64, 1957).

Massospora Peck (1879), Entomophthoraceae. 11 (on cicadas), widespr. *M. cicadina* is a parasite of the 17-year cicada in N. Am. See Soper (*Mycotaxon* **1**: 13, 1974; key), Soper *et al.* (*Ann. Ent. Soc. Am.* **69**: 89, 1976; biology).

Mastigochytrium Lagerh. (1892) ? = Rhizophydium (Chytrid.) fide Sparrow (1960).

Mastigocladium Matr. (1911) = Acremonium (Mitosp. fungi) fide Gams (1971).

Mastigomyces Imshen. & Kriss (1933), Mitosporic fungi, 1.?.?. 1, former USSR. See Babjeva & Levin (*Mikrobiologiya* **48**: 541, 1979; neotypification).

Mastigomycètes, fungi having a motile phase; cf. **Amastigomycètes**.

Mastigomycotina (obsol.). Used for fungi with motile zoospores now dispersed in the phyla *Chytridiomycota, Hyphochytriomycota, Oomycota, Plasmodiophoromycota*.

Mastigonema Speg. (1926) = Chaetospermum (Mitosp. fungi) fide Petrak & Sydow (*Ann. Myc.* **34**: 11, 1936).

mastigoneme, see flagellum.

Mastigonetron Kleb. (1914) = Harknessia (Mitosp. fungi) fide Sutton (1977). See also Petrak (*Sydowia* **24**: 253, 1971).

mastigopod (of myxomycetes), a swarm cell (obsol.).

Mastigosporella Höhn. (1914), Anamorphic Melanconidaceae, 4.A1.19. Teleomorph *Wuestneiopsis*. 2, USA. See Nag Raj (1993).

Mastigosporium Riess (1852), Mitosporic fungi, 2.C1.19. 5 (causing leaf spots of grasses), temp. See Bollard (*TBMS* **33**: 250, 1950), Hughes (*Mycol. Pap.* **36**, 1951), Austwick (*TBMS* **37**: 161, 1954), Huss *et al.* (*Pflanzensch.* **49**: 97, 1988; on *Dactylis* spp.), Mayrhofer *et al.* (*Mitt. naturwiss. Ver. Steiermark* **121**: 73, 1991; key).

Mastocephalus Batt. ex Kuntze (1891) ≡ Leucocoprinus (Agaric.).

Mastodia Hook. f. & Harv. (1847), Mastodiaceae (L). 7, Antarctic. See Brodo (*Bryologist* **79**: 385, 1976), Gremmen *et al.* **27**: 387, 1995; ecology).

Mastodiaceae Zahlbr. (1907), Ascomycota (inc. sed., ± L). 2 gen. (+ 3 syn.), 9 spp. Stromata absent, or reduced to a small clypeus-like structure; ascomata perithecial, ± globose, black, thick-walled, weakly papillate, ostiole periphysate; interascal tissue of simple paraphyses, sometimes evanescent, or of apical paraphyses; asci clavate, thin-walled, without apical structures, evanescent or ? persistent; ascospores hyaline, simple or transversely septate, thin-walled, without a mucous sheath. Anamorphs unknown. Parasitic or lichenized with large marine algae, widespr.

Mastoleucomyces Batt. ex Kuntze (1891) nom. nud. = Tricholoma (Tricholomat.) fide Singer (1951).

Mastomyces Mont. (1848) ≡ Topospora (Mitosp. fungi) fide Groves (*CJB* **43**: 495, 1965).

matrix, (1) the substratum in or on which an organism is living; (2) mucilaginous material in which conidia and some ascospores are produced, influences dissemination, survival, germination etc. See Louis & Cooke (*TBMS* **84**: 661, 1985).

Matruchotia Boulanger (1893) nom. dub. (Mitosp. fungi). See Donk (*Reinwardtia* **2**: 466, 1954).

Matruchotia Skup. (1924) = Amaurochaete (Stemonitid.).

Matruchotiella Grigoraki (1924) ≡ Ateleothylax (Mitosp. fungi). ? Nom. dub. fide Dodge (*Medical mycology*: 431, 1935).

Matruchotiella Skup. ex G. Lister (1925) ≡ Amaurochaete (Stemonitid.).

Matsushimaea Subram. (1978), Mitosporic fungi, 1.G2.10. 1, Japan.

Matsushimomyces V.G. Rao & Varghese (1979), Mitosporic fungi, 1.B1.?10. 1, India.

matsu-take, *Tricholoma matsutake*, an important edible fungus in Japan. Nisikado *et al.* (*Ber. Ohara Inst.* **8**(4), 1941); pine mushroom. 'Matsutake' of N. Am., *T. ponderosum, T. murrillianum* (fide Singer, 1961).

Mattickiolichen Tomas. & Cif. (1952) = Buellia (Physc.).

Mattickiomyces Cif. & Tomas. (1953) ≡ Mattickiolichen (Physc.).

Mattirolella S. Colla (1929), Mitosporic fungi, 8.A1.15. See Khan & Kimbrough (*Am. J. Bot.* **61**: 395, 1974).

Mattirolia Berl. & Bres. (1889) = Thyronectria (Hypocr.) fide Rogerson (1970).

Mattirolomyces E. Fisch. (1938) = Terfezia (Terfez.) fide Trappe (1971).

Matula Massee (1888), Anamorphic Corticiaceae. ?.?.?. Teleomorph *Cytidia*. 1, trop. See Martin (*Lloydia* **5**: 158, 1942).

Maublancia G. Arnaud (1918), Microthyriaceae. 6, S. Am., S. Afr.

Maublancomyces Herter (1950) = Gyromitra (Helvell.). See Eckblad (1968).

Mauginiella Cavara (1925), Mitosporic fungi, 1.A1.38. 1 (on *Phoenix*), N. Afr., Middle East. See Sigler & Carmichael (*Mycotaxon* **4**: 349, 1976), v. Arx *et al.* (*Sydowia* **34**: 42, 1981; ultrastr.).

Maurinia Niessl (1876) = Anthostomella (Xylar.) fide Francis (1975).

Mauritzia Gyeln. (1935) = Pyrenopsis (Lichin.).

Maurodothella G. Arnaud (1918) = Echidnodes (Asterin.) fide v. Arx & Müller (1975).

Maurodothina G. Arnaud ex Piroz. & Shoemaker (1970) = Eupelte (Asterin.) fide v. Arx & Müller (1975).

Maurodothis Sacc., Syd. & P. Syd. ex Syd. & P. Syd. (1904) = Cycloschizon (Parmular.) fide Müller & v. Arx (1962).

Maurya Pat. (1898) [non Kunth (1824), *Anacardiaceae*] Fungi (inc. sed.) fide Petrak (*Sydowia* **5**: 345, 1951).

Mawsonia C.W. Dodge (1948), ? Lichinaceae (L). 1, Antarct. See Kärnefelt (*Crypt. Bot.* **1**: 147, 1989).

Maxillospora Höhn. (1914) = Tetracladium (Mitosp. fungi) fide Ingold (*TBMS* **25**: 371, 1942).

Mazaediothecium Aptroot (1991), Pyrenulaceae (L). 2, Costa Rica, Papua New Guinea. See Eriksson (*Lichenologist* **25**: 307, 1993).

mazaedium, a spore mass formed by an ascoma, as in *Caliciales* and *Onygenaceae*, in which the spores, generally with sterile elements, become free from the asci as a dry, loose powdery mass on the fruiting surface.

Mazosia A. Massal. (1854), ? Roccellaceae (L). 10, trop. See Eriksson (*SA* **14**: 58, 1995; posn), Kalb & Vězda (*Folia geobot. phytotax.* **23**: 199, 1988; key), Santesson (*Symb. bot. upsal.* **12** (1), 1952), Batista *et al.* (*Atas Inst. Mic. Recife* **5**: 429, 1967).

mazu, see Fermented food and drinks.

Mazzantia Mont. (1855), Valsaceae. 5, Eur., Am. See Monod (1983).

Mazzantiella Höhn. (1925) ? = Hypodermina (Mitosp. fungi) fide Clements & Shear (1931).

McAlpine (Daniel; 1849-1932). Vegetable Pathologist to the Department of Agriculture, Victoria, Australia from 1890. Chief writings: *Rusts* (1906) and *Smuts of Australia* (1910), *Fungous diseases of stone-fruit trees of Australia* (1902), *of the potato in Australia* (1911). See Fish (*Ann. Rev. Phytopath.* **8**: 14, 1970), Stafleu & Cowan (*TL-2* **3**: 207, 1981).

McGinty, see Lloyd.

MCZ, see Media.

MEA, see Media.

Mebarria J. Reid & C. Booth (1989), Melanconidaceae. 1, USA.

medallion clamp, a clamp connexion with a space between the main hypha and the hook.

Medastina Dodart [not traced] nom. dub. (? Fungi).

Medeolaria Thaxt. (1922), Medeolariaceae. 1, N. Am.

Medeolariaceae Korf (1982), Medeolariales. 1 gen., 1 sp. Stroma absent; ascomata ± absent, the hymenium exposed in an indefinite palisade on surface of the host, originating from epidermal tissues; interascal tissues well-developed, of simple or branched pigmented rather irregular paraphyses; asci clavate, thin-walled, without apical structures, persistent, 8-spored; ascospores aseptate, brown, fusiform, flattened on one side, striate. Anamorph unknown. Biotrophic and gall-forming on *Medeola*, N. Am.

Lit.: Thaxter (*Proc. Am. Acad. Arts Sci.* **57**: 425, 1922).

Medeolariales, Ascomycota. 1 fam., 1 gen., 1 sp. Fam.: **Medeolariaceae**.

Media (in microbiology), artificially prepared substrata on which fungi and other microorganisms are cultured. An enormous range of media recipies is available, and this entry provides an introduction to the selection of appropriate media and recipies for some of the most commonly employed. For further media see Levine & Schoenlein (*A compilation of culture media for the cultivation of micro-organisms*, 1930), Booth (*in* Booth (Ed.), *Methods in microbiology* **4**: 57, 1971; 185 formulae), Smith & Onions (*Preservation and maintenance of living fungi* edn 2, 1994).

The selection of a satisfactory medium for stimulating growth and sporulation of fungi may vary from strain to strain, although cultures of the same species and genera tend to grow best on similar media. The source of isolates can give an indication of suitable growth conditions.

A wide range of media is used by different workers. Raper & Thom (1949) use Czapek's Agar, Steep Agar and Malt Extract Agar for the growth of penicillia and aspergilli, while Pitt (1980) recommends Czapek Yeast Autolysate (CYA) and Malt Extract Agar (MEA).

Preferences for growth on particular media are normally developed over many years and knowledge of them is the result of experience. The standardization of media formulae is necessary for most work. Media will affect colony morphology and colour, whether particular structures are formed or not and may affect the retention of properties. Examples of particular preferences are given below. *Mucorales* do well on

Malt Agar (MA) and will not grow in Czapek Agar (CZ) as they lack enzymes to digest sucrose. Many fungi thrive on Potato Dextrose Agar (PDA), but this can be too rich, encouraging the growth of mycelium with ultimate loss of sporulation, so a period on Potato Carrot Agar (PCA), a starvation medium, may encourage sporulation. *Fusarium* species grow well on Potato Sucrose Agar (PSA). Wood-inhabiting fungi and dematiaceous fungi often spore better on Cornmeal Agar (CMA) and Oat Agar (OA), both of which have less easily digestible carbohydrate. Cellulose destroying fungi and spoilage fungi, such as *Trichoderma*, *Chaetomium* and *Stachybotrys* retain their ability to produce cellulase when grown on a weak medium such as TWA or PCA with a piece of sterile filter paper, wheat straw or lupin stem placed on the agar surface.

All sorts of vegetable decoctions are available and apart from the advantages of standardization it is reasonable to use what is readily accessible, e.g. yam media might be preferable to potato media in the tropics.

Entomophthora species can be grown in culture on several media but are reported to do best on an egg yolk medium. The introduction of pieces of tissue, such as rice, grains, leaves, wheatstraw or dung, often produces good sporulation dependent on the organism grown. The use of hair for some dermatophytes has proved very successful (Al-Doory, 1968). Animal hair or feathers should be de-fatted in organic solvents first to ensure good growth.

Corn Meal Agar (CMA)

Maize meal	30 g
Oxoid Agar N° 3	20 g
Tap Water	1 l

Place the maize and water in a saucepan. Heat over a double saucepan until boiling, continue heating for one hour stirring occasionally. Filter the decoction through muslin, add the agar, and boil until it is dissolved. Autoclave at 121°C for 20 min.

Czapek Dox Agar (CZ)
Made with stock Czapek solution
Solution A

Sodium nitrate (NaNO$_3$)	40.0 g
Potassium chloride (KCl)	10.0 g
Magnesium sulphate (MgSO$_4$.7H$_2$O)	10.0 g
Ferrous sulphate (FeSO$_4$.7H$_2$O)	0.2 g
Distilled water	1 l

(store in a refrigerator).
Solution B

Di-potassium hydrogen phosphate (K$_2$HPO$_4$)	20 g
Distilled water	1 l

(store in a refrigerator).
For 1 litre

Stock solution A	50 ml
Stock solution B	50 ml
Distilled water	900 ml
Sucrose (Analar)	30 g
Oxoid Agar N° 3	20 g

Dissolve agar in distilled water then add sucrose and stock solutions just before autoclaving.
To each litre add 1 ml of following stock solutions:
Zinc sulphate (ZnSO$_4$.7H$_2$O) Analar 1.0 g in 100 ml distilled water
Cupric sulphate (CuSO$_4$.5H$_2$O) Analar 0.5 g in 100 ml distilled water
Autoclave at 121°C for 20 min. See Smith (1954).

Czapek Yeast Autolysate Agar (CYA)

Di-Potassium hydrogen phosphate (K$_2$HPO$_4$)	1.0 g
Czapek concentrate	10.0 ml
Oxoid Yeast extract or autolysate	5.0 g
Sucrose (Analar)	30.0 g
Oxoid Agar N° 3	15.0 g
Distilled water	1.0 l

Autoclave at 121°C for 15 min. See Pitt (1973).

Egg Yolk Medium
Soak eggs in 90% alcohol with a little acetone added, for 2 h. Flame off, puncture a 5 mm hole in each end, pour white away. Puncture yolk membrane, pour into bottles or plates. Steam at 80°C for 30 to 45 min in a closed water bath.

Malt Extract Agar (MA)

White bread malt extract	20 g
Oxoid Agar N° 3	20 g
Tap water	1 l

Add agar to water and dissolve over a double saucepan, add malt and dissolve; pH should be 6.5. Autoclave at 121°C for 20 min.

Malt-Czapek Agar (MCZ)

Stock Czapek solution A	50 ml
Stock Czapek solution B	50 ml
Sucrose	30 g
Toffee malt extract	40 g
Oxoid Agar N° 3	20 g
Distilled water	900 ml

Dissolve malt extract and agar in water. Heat over a double saucepan until dissolved. Then add sucrose, when dissolved add stock solutions. Adjust pH to 5.0. Autoclave at 121°C for 20 min.

Malt Extract Agar plus Sucrose
For organisms requiring high osmotic pressure for sporulation.
M20

Malt extract	20 g
Sucrose	200 g
Oxoid Agar N° 3	20 g
Tap water	1 l

For M40 use 400g sucrose, for M60 use 600g sucrose. Prepare in the same way as Malt Extract Agar but add sugar last to reduce caramelization.

Malt Extract Agar (MEA)
Blakeslee's formulation.

Malt extract (powdered, Difco or Oxoid)	20 g
Peptone, bacteriological	1 g
Glucose (Analar)	20 g
Oxoid Agar N° 3	15 g
Distilled water	1 l

Sterilize by autoclaving at 121°C for 15 min. See Raper & Thom (1949), Pitt (1973).

Oat Agar (OA)

Oat Meal ground	30 g
Oxoid Agar N° 3	20 g
Tap water	1 l

Add oat meal to 500 ml of water in a saucepan. Heat for 1 h. To the other 500 ml water add agar and dissolve in a double saucepan. Pass cooked oat meal through a fine strainer and add to agar mixture. Stir thoroughly. Autoclave at 121°C for 20 min.

Potato Carrot Agar (PCA)

Avoid new crop potatoes, which do not make good media. Red Désirée potatoes have been found to be best. Wash, peel and grate vegetables.

Grated potato	20 g
Grated carrot	20 g
Oxoid Agar N° 3	20 g
Tap water	1 l

Boil vegatables for about 1 h in 500 ml tap water, then pass through a fine sieve keeping the liquid. The agar is added to 500 ml of water in a double saucepan. When the agar has dissolved add the strained liquid and stir. Pour through a funnel into bottles. Sterilize at 121°C for 20 min.

Potato Dextrose Agar (PDA)

Avoid new crop potatoes. Red Désirée have been found to be best.

Potatoes	200 g
Oxoid Agar N° 3	20 g
Dextrose	15 g
Tap water	1 l

Scrub potatoes clean and cut into 12 mm cubes (do not peel). Weigh out 200 g and rinse rapidly under a running tap, and drop into 1 l of tap water in a saucepan. Boil until potatoes are soft (about 1 h) then put through blender. Add 20 g of agar, and heat in a double saucepan until dissolved. Then add 15 g of dextrose and stir until dissolved. Make up to 1 l. Pour into bottles, stiring occasionally to ensure that each bottle has a percentage of solid matter. Autoclave at 121°C for 20 min.

Potato Sucrose Agar (PSA)

To make 1 litre of medium.

Potato water	500 ml
Sucrose	20 g
Oxoid Agar N° 3	20 g
Distilled water	500 ml

Heat in double saucepan until agar is dissolved. Autoclave at 121°C for 15 min. 2.3 kg potatoes makes 7 l. Adjust to pH 6.5 with calcium carbonate if necessary. To make 2 litres of potato water:

Tap water	1125 ml
Potato	450 g

Peel and dice potatoes, suspend in double cheesecloth and boil in the tap water until almost cooked.

PSA using powdered potato

Powdered potato	5 g
Sucrose	20 g
Oxoid Agar N° 3	20 g
Distilled water	1 l
Calcium carbonate	5 g

Autoclave at 121°C for 15 min.

Rabbit Dung Agar (RDA)

The rabbit dung must be from wild rabbits, and dried before use.

Oxoid Agar N° 3	15 g
Tap water	1 l

Heat agar to dissolve. Pour into bottles containing the rabbit dung. Autoclave at 126°C for 20 min.

Sabouraud's Agar

Glucose	20 g
Peptone	10 g
Oxoid Agar N° 3	15 g
Water	1 l

Autoclave 114°C for 15 min.

Soil Extract Agar (SEA)

(Flentje's formula for *Corticium praticola*, promoting formation of basidia). To make extract:

Soil	1 kg
Water	1 l

Agitate frequently for a day or two; pour extract through glass wool, and make up to 1 l again. Prepare SEA as follows:

Extract	1.0 l
Sucrose	1.0 g
Potassium dihydrogen phosphate (KH$_2$PO$_4$)	0.2 g
Dried yeast	0.1 g
Oxoid Agar N° 3	25.0 g

NB. It may be advisable to test pH before pouring into bottles.

Starch Agar (SA)

Soluble starch	40 g
Marmite	5 g
Oxoid Agar N° 3	20 g
Tap water	1 l

Place all the constituents in water and heat in a double saucepan until dissolved. Bottle and sterilize (pH is 6.5 - 7 and requires no adjustment). Autoclave at 121°C for 20 min.

Tap Water Agar (TWA)

Tap water	1 l
Oxoid Agar N° 3	15 g

Dissolve agar in water. Autoclave at 121°C for 20 min.

V8 Agar (V8A)

V8 Vegetable juice	200 ml
Oxoid Agar N° 3	20 g
Distilled water	800 ml

Dissolve agar in water and add vegetable juice. Adjust pH to 6.0 with 10% sodium hydroxide. Autoclave at 121°C for 20 min. (pH after autoclaving should be 5.8.) See Diener (1955).

V8 Agar (as recommended for *Actinomycetes*)

V8 Vegetable juice	200 ml
Calcium carbonate	4 g
Oxoid Agar N° 3	20 g
Water	800 ml

Adjust to pH 7.3 with KOH. Autoclave at 121°C for 20 min. See Galindo & Gallegly (1960).

Yeast Phosphate Soluble Starch (YPSS)

Difco yeast extract	4.0 g
Soluble starch	15.0 g
Di-Potassium hydrogen phosphate (K$_2$HPO$_4$)	1.0 g
Magnesium sulphate (MgSO$_4$.7H$_2$O)	0.5 g
Oxoid Agar N° 3	20.0 g
Distilled water	1.0 l

Mix together, dissolve and dispense. Autoclave at 121°C for 15 min.

Waksman's special medium for counting soil fungi

Glucose	10.0 g
Peptone	5.0 g
Potassium dihydrogen phosphate (KH$_2$PO$_4$)	1.0 g
Oxoid Agar N° 3	25.0 g
Water	1 l

Make to pH 4 by addition of normal H$_2$SO$_4$ or H$_3$PO$_4$.

Hansen's medium for yeasts

Peptone	1.0 g
Maltose	5.9 g
Potassium dihydrogen phosphate (KH$_2$PO$_4$)	0.3 g
Magnesium sulphate (MgSO$_4$.7H$_2$O)	0.2 g
Water	1 l

Sporulation media for yeasts

Among the many media and techniques described for the induction of ascospore development in yeasts are: (1) Culture on sterile slices of carrot or potato. (2) Gorodkwa's agar (after Kreger-van Rij). Glucose 1.0 g, Neopeptone (Difco) 10.0 g, Sodium chloride (NaCl) 5.0 g, Agar 30.0 g, Water to 1000.0 ml. (3) Inoculation of 0.5 per cent solution of potato starch and incubation at 37°C. (This is also used as a test for mycelium development.) (Almeida & da Silva Lacaz, *Folia Clin., Biol.* **12** (4), 1940). (4) Acetate agar, Sodium acetate 1.0 g, Raffinose 0.2 g (or Glucose 0.4 g), Agar 20.0 g, adjust pH to 5.8. Adams (*Can. J. Res.* **C27**: 179, **C28**: 413, 1949-50) recommends subculture twice on a 'presporulation medium' (e.g. Difco nutrient agar + glucose 5 per cent, and tartaric acid 0.5 per cent) and transfer to the sporulation medium from a 24-hr culture. See also Fowell (*Nature* **170**: 578, 1952).

Medical and veterinary mycology. Though not so important as bacteria, fungi cause a number of major diseases of humans and higher animals. The best known mycosis is ringworm (tinea), of the skin and hair (caused by gymnoascaceous fungi) but there are other widespread and important systemic mycoses (e.g. coccidioidomycosis, cryptococcosis, histoplasmosis) which are frequently fatal.

Fungi pathogenic for humans are often polymorphic (see dimorphic) or 'pleomorphic' and this, in addition to the number of species proposed on clinical grounds and without adequate descriptions, made the earlier literature very confused.

Mycoses (with the exception of dermatophytoses) are generally not contagious but originate from the microbial flora of both the external and internal environments where many potential fungal pathogens are present as saprobes (e.g. *Candida albicans* in the human mouth, alimentary and genital tracts; *Coccidioides immitis* and *Histoplasma capsulatum* in soil); see opportunistic fungi.

Therapy for dermatophytoses is by topical application of fungicides, oral administration of griseofulvin, or X-ray epilation. Systemic mycoses, which are frequently difficult to cure, may respond to surgery, X-rays, potassium iodide, or antibiotics (see amphotericin, nystatin).

For some important mycoses see: adiaspiromycosis, aspergillosis, blastomycosis, candidiasis, chromomycosis, *Coccidioides* (coccidioidomycosis), cryptococcosis, dermatophyte (ringworm), epizootic lymphangitis, *Histoplasma* (histoplasmosis), mycosis, mycetoma, *Paracoccidioides* (paracoccidiomycosis), phycomycosis, piedra, rhinosporidiosis, sporotrichosis, tinea; see also Allergy.

Lit.: General: *Henrici's Molds, yeasts and actinomycetes* (1947), Ainsworth (*Medical mycology. An introduction to its problems*, 1952).

Nomenclature: *Nomenclature of fungi pathogenic to man and animals* (*Med. Res. Council Mem.* **23**, edn 4, 1977), ISHAM (*Sabouraudia* **18**: 78, 1980; internat. nomencl., Engl., Fr.), Grigoriu *et al.* (*Medical mycology*, 1987), Odds *et al.* (Fungal disease nomenclature, *J. med. vet. mycol.* **30**: 1-10, 1992).

Mycology: Nannizzi (*Repertorio sistematico dei miceti dell'uomo e degli animale*, 1934), Dodge (*Medical mycology*, 1935), Brumpt (*Précis de parasitologie*, edn 6, **2**, 1949), Fragner (*Parasitische Pilze beim Menschen*, 1958), Ciferri (*Manuale di micologia medica*, 2 vols, 1958-60), Gedek (*Hefen als Krankheitserreger bei Tieren*, 1968), Koch (*Leitfadender medizinschen Mykologie*, 1973), Rippon (*Medical mycology*, 1974 (edn 3, 1988), Howard

(Ed.) (*Fungi pathogenic for humans and animals*, 3 vol., 1983-5), Larone (*Medically important fungi: a guide to identification*, 1987), Wentworth (Ed.) (*Diagnostic procedures for mycotic and parasitic infections*, edn 7, 1988), Evans & Richardson (Eds) (*Medical mycology: A practical approach*, 1989), Smith (*Opportunistic mycoses of man and other animals*, 1989), Hay (Ed.) (*Tropical fungal infections*, 1989), Warnock & Richardson (Eds) (*Fungal infection of the compromised patient*, edn 2, 1990), Jacobs & Nall (Eds) (*Antifungal drug therapy*, 1990).

Methods: Beneke & Rogers (*Medical mycology manual*, edn 3, 1971), Ajello *et al.* (*Laboratory manual for medical mycology*, edn 2, 1962), Vanbreuseghem (*Guide practique de mycologie médicale et vétérinaire*, 1966), Golvin *et al.* (*Techniques en parasitologie et en mycologie*, 1970), Segretain *et al.* (*Diagnostic de laboratoire en mycologie*, edn 3, 1974), Evans (Ed.) (*Serology of fungus infections. A laboratory manual*, 1976).

Humans: Conant *et al.* (*Manual of clinical mycology*, edn 3, 1971), Lewis *et al.* (*An introduction to medical mycology*, edn 4, 1958), Emmons *et al.* (*Medical mycology* edn 3, 1977), Hildick-Smith *et al.* (*Fungus diseases and their treatment*, 1964), da Silva Lacaz (*Compendio de micologia medica*, 1967), McGinnis (*Laboratory handbook of medical mycology*, 1980), Sarosi & Scott (*Fungal diseases of the lung*, 1986), Land & McCracken (*Handb. Appl. Mycol.*: Humans, animals & insects **2**, 1991; infection in the compromised host), de Hoog & Guarro (Eds) (*Atlas of clinical fungi*, 1995; 300 spp.).

Other vertebrates: Ainsworth & Austwick (*Fungal diseases of animals*, edn 2, 1973), Chute, O'Meara & Barden (*A bibliography of avian mycoses* [*Misc. Publ. Me agric. Exp. Stn* **655**], 1962 [edn 3, 1971]), Jungerman & Schwartzman (*Veterinary medical mycology*, 1972), Welsh & Hughes (*Fungal diseases of fishes*, 1980).

Bibliography & journals: Ciferri & Redaelli (*Bibliographia mycopathologica 1800-1940*, 2 vols, 1958); journals, *Review of medical and veterinary mycology*, 1943- (abstracts; also available in electronic formats 1973-); *Mykosen*, 1957-; *Sabouraudia* (then *Journal of Medical & Veterinary Mycology*), 1961-; *J. de mycologie medicale*, 1991-.

History: Ainsworth (*Introduction to the history of medical and veterinary mycology*, 1991), Ainsworth & Stockdale (*RMVM* **19**: 1, 1984; biographical notices).

See also Safety, Laboratory.

Medical uses of Fungi. Ergot (q.v.), the sclerotia of *Claviceps purpurea*, is the only fungus now in the British Pharmaceutical Codex. Yeast is a part of some patent medicines and takadiastase has medical uses. At one time a number of other fungi have been put to medical use. *Fomes officinalis* (Agaricum, the female, white, or purging agaric) was a noted 'universal remedy' (see Faull, *Mycol.* **11**: 267, 1919; 'pineapple fungus'). *Lycoperdon* spores and capillitium were used for stopping blood from wounds, while in the Far East *Cordyceps sinensis* (attached to the larva ('caterpillar') of which it is a parasite) and bukuryo (sclerotia of ? *Pachyma hoelen*) have general approval. Many spp. are used today in Chinese traditional medicine (Ying *et al.*, *Icones of medicinal fungi for China*, 1987; 272 spp.). The medieval Doctrine of Signatures, whereby an organism's appearance was considered to indicate the diseases organs it could treat, led to the use of a variety of lichens for medicinal purposes (see dog lichen, lungwort); some lichens are still supplied by pharmacies and health shops, either loose or in pastilles, e.g. Iceland moss (*Cetraria islandica*); see Richardson (*in* Galun, *CRC Handbook of lichenology* **3**: 93, 1988; review). For further applications of products from lichens see Fahselt (*Symbiosis* **16**: 117, 1994; review).

medium culture, a substance or solution for the culture of microorganisms (see Media).

medulla, (1) (of lichen thalli), the loose layer of hyphae below the cortex and algal layer; (2) (of sporocarps of macromycetes), the part composed mainly or entirely of longitudinal hyphae.

medullary excipulum (of ascomata), tissue below the generative layer in an apothecium; hypothecium. See excipulum.

Medusamyces G.L. Barron & Szijarto (1990), Mitosporic fungi, 1.A1.?. 1 (on rotifers), Canada.

Medusina Chevall. (1826) = Hericium (Heric.) fide Saccardo (1887).

Medusomyces Lindau (1913) nom. conf. fide Lindner (*Ber. dtsch. Bot. Ges.* **31**: 364, 1913).

Medusosphaera Golovin & Gamalizk. (1962), Erysiphaceae. 1, Asia. See Braun (*Beih. Nova Hedw.* **89**, 1987). Anamorph *Oidium*.

Medusula Tode (1790) nom. dub. (Mitosp. fungi) fide Holubová-Jechová (*Sydowia* **46**: 244, 1994).

Medusula Eschw. (1824) [non Pers. (1807), *Violaceae*] = Sarcographa (Graphid.).

Medusulina Müll. Arg. (1894), Graphidaceae (L). 6, trop.

mega- (prefix), of great size; large; cf. macro-.

Megacapitula J.L. Chen & Tzean (1993), Mitosporic fungi, 1.D2.1. 1, Taiwan.

Megachytriaceae, see *Cladochytriaceae*.

Megachytrium Sparrow (1931), Cladochytriaceae. 1 (on *Elodea*), N. Am.

Megacladosporium Vienn.-Bourg. (1949) = Fusicladium (Mitosp. fungi) fide Hughes (*Mycol. Pap.* **36**, 1951).

Megacollybia Kotl. & Pouzar (1972), Tricholomataceae. 1, Eur.

Megalaria Hafellner (1984), Lecanoraceae (L). 1, temp.

Megaloblastenia Sipman (1983), Megalosporaceae (L). 2, Australasia, S. Am.

Megalocitosporides Wernicke (1892) = Coccidioides (Mitosp. fungi) fide Dodge (1935).

Megalocystidium Jülich (1978), Gloeocystidiellaceae. 3, Eur., N. Am.

Megalodochium Deighton (1960), Mitosporic fungi, 3.A2.1. 2, trop. Afr.

megalogonidium (obsol.), a macrogonidium (q.v.).

Megalographa A. Massal. (1860) = Phaeographina (Graphid.).

Megalonectria Speg. (1881) = Nectria (Hypocr.) fide Seifert (*Stud. Mycol.* **27**, 1985).

Megalopsora Vain. (1921) = Physcidia (Bacid.) fide Kalb & Elix (1995).

Megaloseptoria Naumov (1925), Mitosporic fungi, 4.E1.15. 1, Russia. See Shoemaker (*CJB* **45**: 1297, 1967).

Megalospora Meyen (1843), Megalosporaceae (L). 27, mainly trop. See Kantvilas (1994; key 16 spp. Australia), Sipman (*Bibl. Lich.* **18**, 1983; monogr., key), Hafellner & Bellemère (*Nova Hedw.* **35**: 207, 1982; ultrastr.).

Megalospora A. Massal. (1852) = Mycoblastus (Mycoblast.).

Megalospora Naumov (1927) ≡ Gemmamyces (Cucurbitar.).

Megalosporaceae Vězda ex Hafellner & Bellem. (1982), Lecanorales (L). 3 gen. (+ 5 syn.), 31 spp. Thallus crustose; ascomata apothecial, sessile, with a wide margin, the disc brown or black; interascal tissue

of very narrow anastomosing pseudoparaphysis-like hyphae, often with a pigmented epithecium; asci with a well-developed I+ apical cap and an outer I+ gelatinized layer, 1- to 8-spored; ascospores large, hyaline, thick-walled, without a sheath. Anamorphs pycnidial. Lichenized with green algae, mainly trop.

Lit.: Kantvilas (*Lichenologist* **26**: 349, 1994; key 19 spp. Australia), Sipman (*Bibl. Lich.* **18**, 1983, monogr.; *Willdenowia* **15**: 557, 1986).

Megalosporon, see *Endothrix*.

Megalotremis Aptroot (1991), Trypetheliaceae (L). 2, Australia, India.

Megaloxyphium Cif., Bat. & Nascim. (1956) = Leptoxyphium (Mitosp. fungi) fide Hughes (*Mycol.* **68**: 693, 1976).

Megaspora (Clauzade & Cl. Roux) Hafellner & V. Wirth (1987), Megasporaceae (L). 2, N. temp. See Lumbsch *et al.* (1994; posn).

Megasporaceae Lumbsch, Feige & K. Schmitz (1994), Pertusariales (L). 1 gen., 2 spp. Thallus crustose; ascomata immersed within verrucose swellings of the thallus, apothecial, deeply immersed, cupulate, with a differentiated ascomatal wall and a separate thalline margin; interascal tissue of narrow anastomosing pseudoparaphyses; asci thick-walled, with a strongly thickened apical cap without an ocular chamber, the outer layer weakly I+, 8-spored; ascospores large, hyaline, aseptate, with a two-layered wall, lacking a gelatinous sheath. Anamorph pycnidial. Lichenized with green algae. N. temp.

Lit.: Lumbsch *et al.* (*J. Hattori bot. Lab.* **75**: 295, 1994).

megaspore, see macrospore.

Megasporoporia Ryvarden & J.E. Wright (1982), Coriolaceae. 4, pantropical.

Megaster Cif., Bat., Nascim. & P.C. Azevedo (1956), Mitosporic fungi, 1.G2.1. 2, Brazil.

Megathecium Link (1826) nom. rej. ≡ Ceratostoma (Ceratostomat.).

Megatricholoma G. Kost (1984), Tricholomataceae. 1, Eur.

Megatrichophyton Neveu-Lem. (1921) = Trichophyton (Mitosp. fungi).

Mehtamyces Mundk. & Thirum. (1945), Uredinales (inc. sed.). 1, India, Sri Lanka. See Ramachar & Rao (*Mycol.* **73**: 778, 1981). = Phragmidiella (Puccin.) fide Thirumalachar & Mundkur (1949).

meiocyte, a cell in which meiosis takes place. Cf. gonotokont.

Meionomyces Thaxt. (1931), Laboulbeniaceae. 5, trop.

meiophase, the part of a life cycle in which a diploid nucleus undergoes reduction.

Meiorganum R. Heim (1965), Coniophoraceae. 2, Malaysia, New Caledonia.

meiosporangium, a thick-walled diploid sporangium of certain *Blastocladiales* producing uninucleate, haploid zoospores (**meiospores**, q.v.) (Emerson, 1950); see Dick (*Mycologist* **1**: 166, 1987); cf. mitosporangium.

meiospore, (1) a spore from a meiosporangium (q.v.); (2) (of ascomycetes and basidiomycetes), a basidiospore or ascospore which is the product of meiosis (see Kendrick & Watling, *in* Kendrick (Ed.), *The whole fungus* **2**: 473, 1979).

meiotangium, for the sporangium or gametangium in which meiosis occurs. See Corner (*TBMS* **15**: 336, 1931).

Meissneria Fée (1837) [non *Meisneria* DC. (1828), *Melastomataceae*] ≡ Laurera (Trypethel.).

Meixner test, for amatoxins. Express juice from a fresh basidioma onto a piece of newspaper, allow to dry, add a drop of conc. hydrochloric acid (HCl) when a blue colour indicates presence of amatoxins. See

Meixner (*Z. Mykol.* **45**: 137, 1979), Beutler & Vergeer (*Mycol.* **72**: 1142, 1981).

Melachroia Boud. (1885) = Podophacidium (Dermat.) fide Seaver (*North American cup-fungi, inoperculates*, 1951).

Melaleuca Pat. (1887) [non L. (1767), *Myrtaceae*] ≡ Melanoleuca (Tricholomat.).

Melampsora Castagne (1843), Melampsoraceae. 80, mostly N. temp., autoecious or heteroecious. Incl. pathogens of flax (*Linum; M. lini*), poplar (*Populus*: a number; O, I on *Allium, Larix, Mercurialis, Tsuga,* etc.), willow (*Salix*: a number; 0, I on *Allium, Larix, Ribes*, etc.), and other plants.

Melampsoraceae Dietel (1897), Uredinales. 1 gen. (+ 5 syn.), 80 spp. Pycnia discoid, without bounding structures, subcuticular to subepidermal; aeciospores catenulate (*Caeoma*); teliospores sessile, unicellular, laterally adherent as crusts, pigmented.

Melampsorella J. Schröt. (1874), Pucciniastraceae. 3, N. temp.; O, I on *Abies*; II, III on *Caryophyllaceae* and *Boraginaceae*.

Melampsoridium Kleb. (1899), Pucciniastraceae. 4, esp. N. temp.; O, I on *Larix*; II, III on *Betulaceae*.

Melampsoropsis (J. Schröt.) Arthur (1906) = Chrysomyxa (Coleospor.) fide Arthur (1934).

Melampydiomyces Cif. & Tomas. (1953) ≡ Melampydium (Roccell.).

Melampydium Zahlbr. (1905) = Melampylidium (Roccell.) fide Eriksson & Hawksworth (1993).

Melampylidium Stirt. ex Müll. Arg. (1894) = Bactrospora (Roccell.) fide Egea & Torrente (*Mycotaxon* **53**: 57, 1995).

Melanamphora Lafl. (1976), Melanconidaceae. 2, Eur., Asia, N. Am. See Prasil *et al.* (*Česká Myk.* **28**: 1, 1974; anamorph). Anamorph *Cytosporina*.

Melanaria Erichsen (1936) = Pertusaria (Pertusar.) fide Hawksworth *et al.* (*Lichenologist* **12**: 1, 1980).

Melanaspicilia Vain. (1909) = Buellia (Physc.) fide Lamb (*Sci. Repts Br. Antarct. Surv.* **61**, 1968).

Melanchlenus Calandron (1953) = Rhinocladiella (Mitosp. fungi) fide de Hoog (1977).

Melanconiales (obsol.). Traditionally used for deuteromycetes (Mitosporic fungi) with acervular conidiomata. Not accepted by von Höhnel (1923) who treated *Melanconiales* in the Hyphomycetes, *Tuberculariales*, or by Sutton (1973, 1980) who discussed relationships of suprageneric taxa in Deuteromycotina. See Mitosporic fungi.

Melanconidaceae G. Winter (1886), Diaporthales. 35 gen. (+ 17 syn.), 100 spp. Ascomata often large, erect, oblique or horizontal, necks central, oblique or lateral, erumpent singly or converging and erumpent through a stromatic disc or immersed in a stroma; asci persistent, thick-walled, remaining attached; ascospores mainly broad and brown, simple or septate, sometimes distoseptate, end cells sometimes hyaline. Anamorphs coelomycetous. Saprobic or weakly parasitic, usually on woody tissue. Some recent treatments separate the *Pseudovalsaceae* and *Melogrammataceae* from this fam.
Lit.: Wehmeyer (1942), Barr (1978), v. Arx (*in* Kendrick (Ed.), *The whole fungus* 1: 201, 1979).

Melanconidium (Sacc.) Kuntze (1898) ≡ Melanconis (Melanconid.).

Melanconiella Sacc. (1882) = Melanconis (Melanconid.) fide Wehmeyer (1941).

Melanconiopsis Ellis & Everh. (1900), Mitosporic fungi, 6.A2.19. 2, N. Am., Eur.

Melanconis Tul. & C. Tul. (1863), Melanconidaceae. 26, widespr. *M. juglandis* (die-back of *Juglans*). See Wehmeyer (*Revision of Melanconis*, 1941). Anamorph *Melanconium*.

Melanconites Göpp. (1852), Fossil fungi. 1 (Tertiary), Eur.

Melanconium Link (1809), Anamorphic Melanconidaceae, 6.A2.19. Teleomorph *Melanconis*. 50, widespr. See Sutton (*Persoonia* **3**: 193, 1964; typification), Sieber *et al.* (*CJB* **69**: 2170, 1991; biochem. charact. *Alnus* isol.), Shamoun & Sieber (*Mycotaxon* **49**: 151, 1993; isozyme, protein patterns *Alnus* isol.).

Melanelia Essl. (1978) Parmeliaceae. *c.* 40, mainly temp. and boreal. See Esslinger (*Taxon* **29**: 692, 1980), Lumbsch *et al.* (*Mycotaxon* **33**: 447, 1988). ? = Parmelia (Parmel.)

melanin, a black pigment (tyrosine derivative) produced by fungi, animals etc. (Wheeler, *TBMS* **81**: 29, 1983; synthesis).

melanized, containing dark brown pigments.

Melanobasidium Maubl. (1906) = Sphaceloma (Mitosp. fungi) fide Jenkins & Bitancourt (*Mycol.* **33**: 338, 1941).

Melanobasis Clem. & Shear (1931) = Sphaceloma (Mitosp. fungi) fide Sutton (1977).

Melanobotrys Rodway (1926) = Xylobotryum (Ascomycota) fide Clements & Shear (1931).

Melanocarpus Arx (1975), Sordariales (inc. sed.). 1 (thermophilic), widespr. See Maheshwari & Kamalam (*J. gen. Microbiol.* **131**: 3017, 1985).

Melanocephala S. Hughes (1979), Mitosporic fungi, 1.A2.1. 3, widespr.

Melanochaeta E. Müll., Harr & Sulmont (1969), Lasiosphaeriaceae. 1, Eur., Sri Lanka. See Barr (*Mycotaxon* **39**: 43, 1990; posn). Anamorph *Sporoschisma*.

Melanochlamys Syd. & P. Syd. (1914) = Gilletiella (Dothideales) fide v. Arx & Müller (1975).

Melanocryptococcus Della Torre & Cif. (1964) = Cryptococcus (Mitosp. fungi) fide v. Arx *et al.* (*Stud. Mycol.* **14**, 1977).

Melanodecton A. Massal. (1860) = Chiodecton (Roccell.) fide Thor (1990).

Melanodiscus Höhn. (1918) [non Radlk. (1889), *Sapindaceae*], Mitosporic fungi, 3.A2.?. 1, Eur.

Melanodochium Syd. (1938) ? = Sphaceloma (Mitosp. fungi) fide Sutton (1977).

Melanodothis R.H. Arnold (1972), Mycosphaerellaceae. 1, Asia, N. Am.

Melanogaster Corda (1831) nom. cons., Melanogastraceae. 12, temp., N. Hemisph. (introduced in S.). See Zeller & Dodge (*Ann. Mo. bot. Gdn* **23**: 639, 1936). Relationship to *Agaricales* possible but still speculative.

Melanogastraceae E. Fisch. (1933), Melanogastrales. 3 gen. (+ 6 syn.), 28 spp. Gasterocarp hypogeous, real hymenium absent, mucilaginous at maturity; spores black-brown, fusoid, smooth.
Lit.: Montecchi & Lazzari (*Micol. Ital.* **18**: 33, 1989).

Melanogastrales, Basidiomycetes ('gasteromycetes'). 3 fam., 6 gen. (+ 7 syn.), 44 spp. Gleba with small locules without real hymenium, mucilaginous at maturity; often overlooked (hypogeous or marine) but probably cosmop. Fams:
(1) **Leucogastraceae**.
(2) **Melanogastraceae**.
(3) **Niaceae**.
Lit.: Dring (1973; key gen.), Hypogeous fungi, *Gasteromycetes*.

Melanogone Wollenw. & Ha. Richt. (1934) = Humicola (Mitosp. fungi) fide Mason (*Mycol. Pap.* **5**: 113, 1941).

Melanogramma Pers. [not traced] nom. dub. (Fungi, inc. sed.). See Keissler (*Nyt Mag. naturv.* **66**: 79, 1927).

Melanographa Müll. Arg. (1882) ? = Melaspilea (Melaspil.).

Melanographium Sacc. (1913), Mitosporic fungi, 1/2.A2.10. 5, Philipp. See Ellis (*DH*).

Melanolecia Hertel (1981), ? Hymeneliaceae (L). 8, widespr. See Poelt & Vězda (1981).

Melanoleuca Pat. (1897), Tricholomataceae. 38, cosmop. See Bresinsky & Stangl (*Z. Pilzk.* **43**: 145, 1977), Gillman & Miller (*Mycol.* **69**: 927, 1977), Kühner (*Bull. Soc. linn. Lyon* **47**: 12, 1978).

Melanomma Nitschke ex Fuckel (1870), Melanommataceae. *c.* 20, widespr. See Chesters (*TBMS* **22**: 116, 1938), Holm (*Symb. bot. upsal.* **14**(3), 1957), Barr (*N. Am. Fl.* II **13**, 1990; N. Am.). Anamorphs *Aposphaeria, Pseudospiropes*.

Melanommataceae G. Winter (1885), Dothideales. 17 gen. (+ 7 syn.), 70 spp. Ascomata perithecial, erumpent, often becoming superficial, often strongly aggregated, black, variable in form, with a well-developed lysigenous pore; peridium composed of small thick-walled pseudoparenchymatous cells, roughly equal in thickness; interascal tissue of trabeculate pseudoparaphyses (paraphysoids), usually immersed in gel; asci cylindrical, fissitunicate; ascospores brown, septate, sometimes muriform, often with a sheath. Anamorphs coelomycetous. Saprobic on woody substrata, widespr.

Melanommatales 1 , see *Dothideales*; Barr (*N. Am. Fl.* ser. 2 **13**, 1990), Eriksson & Hawksworth (*SA* **5**: 141, 1986; status).

Melanomphalia M.P. Christ. (1936), Crepidotaceae. 20, Eur., Am.

Melanomyces Syd. & P. Syd. (1917) Fungi (inc. sed.) fide Petrak (*Sydowia* **1**: 169, 1947).

Melanopelta Kirschst. (1939) = Gnomonia (Vals.) fide Bolay (1972).

Melanopeziza Velen. (1939), ? Leotiales (inc. sed.). 1, former Czechoslovakia.

Melanophloea P. James & Vězda (1971), Acarosporaceae (L). 1, Br. Solomon Isl.

Melanophoma Papendorf & J.W. du Toit (1967), Mitosporic fungi, 4.A1.15. 1, S. Afr.

Melanophora Arx (1957) = Sphaceloma (Mitosp. fungi) fide Jenkins (*Aq. Inst. Biol. S. Paulo* **38**: 83, 1971).

Melanophthalmum Fée (1825) = Strigula (Strigul.).

Melanophyllum Velen. (1921), Agaricaceae. 2, Eur., Am.

Melanoplaca Syd. & P. Syd. (1917) = Dothidasteroma (Parmular.) fide Müller & v. Arx (1962).

Melanoporella Murrill (1907) = Poria (Coriol.) fide Lowe (1966), but used by Pegler (1973).

Melanoporia Murrill (1907) = Poria (Coriol.) fide Lowe (1966), but used by Pegler (1973).

Melanoporthe Wehm. (1938) = Diaporthe (Vals.) fide Müller & v. Arx (1962).

Melanops Nitschke ex Fuckel (1870) = Botryosphaeria (Botryosphaer.). See Holm (*Taxon* **24**: 475, 1975).

Melanosamma Niessl (1876), Niessliaceae. 40, widespr. See Barr (*Mycotaxon* **39**: 43, 1990; status, posn).

Melanopsammella Höhn. (1920), ? Lasiosphaeriaceae. 1, Eur. See Barr (*Mycotaxon* **39**: 43, 1990).

Melanopsammina Höhn. (1919) = Lentomita (Lasiosphaer.) fide Holm (*Svensk bot. Tidskr.* **62**: 217, 1968).

Melanopsammopsis Stahel (1915) = Microcyclus (Mycosphaerell.) fide v. Arx & Müller (1975).

Melanopsichium Beck (1894), Ustilaginaceae. 4, Am., India. See Zundel (*Mycol.* **35**: 180, 654, 1943, **52**: 189, 1961), Vánky (1987).

Melanopus Pat. (1887) = Polyporus (Polypor.) fide Singer (1951).

Melanormia Körb. (1865), ? Leotiales (inc. sed., ?L). 1, Germany.

Melanosella Örösi-Pál (1936), Mitosporic fungi, 1.A1.?. 1 (yeast-like), widespr. *M. mors-apis* (melanosis of bees).

Melanosorus De Not. (1847) ≡ Rhytisma (Rhytismat.).

Melanosphaeria Sawada (1922) ? = Sirosphaera (Mitosp. fungi) fide Petch (*TBMS* **11**: 258, 1926).

Melanosphaerites Grüss (1928), Fossil fungi. 2 (Devonian), Eur.

Melanospora Corda (1837) nom. cons., Ceratostomataceae. 25, widespr. See Doguet (*Botaniste* **39**, 1955), Cannon & Hawksworth (*Bot. J. Linn. Soc.* **84**: 115, 1982; key 12 Br. spp.), Goh & Hanlin (*Mycol.* **86**: 357, 1994; ontogeny).

Melanospora Mudd (1861) ≡ Poeltinula (Rhizocarp.).

Melanosporaceae, see *Ceratostomataceae*.

Melanosporites Pampal. (1902), Fossil fungi. 1 (Miocene), Italy.

Melanosporopsis Naumov (1927) = Melanospora (Ceratostomat.) fide Clements & Shear (1931).

melanosporous, black-spored.

Melanostigma Kirschst. (1939) = Herpotrichiella (Herpotrichiell.) fide Barr (*Rhodora* **78**: 67, 1976).

Melanostroma Corda (1829) = Ceuthospora (Mitosp. fungi) fide Nag Raj (*in* Sherwood, *Mycotaxon* **5**: 1, 1977).

Melanostromella Petr. (1953) = Antennularia (Ventur.) fide Müller & v. Arx (1962).

Melanotaenium de Bary (1874), Tilletiaceae. 25, Eur., Am., Asia. *M. selaginellae* and *M. oreophilum* (parasites of *Selaginella*). See Zambettakis & Joly (*BSMF* **88**: 193, 1972; key, numerical taxonomy), Vánky (1987).

Melanotheca Fée (1837) = Pyrenula (Pyrenul.) fide Harris (1989).

Melanothecomyces Cif. & Tomas. (1953) = Laurera (Trypethel.) fide Letrouit-Galinou (1957).

Melanothecopsis C.W. Dodge (1967), Pyrenulales (inc. sed., L). 5, Asia, S. Am., Eur.

Melanotrichum Corda (1833) = Trichosporium (Mitosp. fungi) fide Saccardo (*Syll. Fung.* **4**: 292, 1886).

Melanotus Pat. (1900), Strophariaceae. 21, mainly trop. See Horak (*Persoonia* **9**: 305, 1977; key).

Melascypha Boud. (1885) = Pseudoplectania (Sarcosomat.) fide Seaver (1928).

Melasmia Lév. (1846), Anamorphic Rhytismataceae, 8.A1.15. Teleomorph *Rhytisma*. 20, widespr.

Melaspilea Nyl. (1857), Melaspileaceae (±L). *c.* 60 (s.l.), cosmop. (mainly trop.). See Eriksson (1981; applic. name), Redinger (*Rabenh. Krypt.-Fl.* **9**, 1(2): 434, 1937).

Melaspileaceae Walt. Watson (1929), Ascomycota (inc. sed., ±L). 1 gen., *c.* 60 spp. Thallus crustose or immersed, often evanescent; ascomata apothecial, immersed to superficial, elongate, sometimes branched, the hymenium exposed permanently or by a longitudinal slit; peridium black, well-developed; interascal tissue of narrow sparsely branched and anastomosing paraphyses, sometimes pigmented at the apex; asci cylindrical, persistent, with a poorly developed apical cap and an ocular chamber, usually not blueing in iodine; ascospores usually 1-septate, becoming brown, sometimes ornamented. Anamorphs unknown. Lichenized with green algae, occ. lichenicolous, widespr. esp. trop.

Melaspileella (P. Karst.) Vain. (1921) ? = Melaspilea (Melaspil.).

Melaspileomyces Cif. & Tomas. (1953) = Melaspilea (Melaspil.).

Melastiza Boud. (1885), Otideaceae. 12, N. temp. See Maas Geesteranus (*Persoonia* **4**: 418, 1967; key).

Lassuer (*Doc. Mycol.* **11** (42): 1, 1980; key), Häffner (*Beitr. Kenntis Pilze Mottelew.* **2**: 183, 1986; key 9 spp.), Dissing (*Svampe* 1980: 29, 1980; key Danish spp.), Arroyo & Calonge (*Boln Soc. micol. Madrid* **12**: 23, 1988). = Aleuria (Otid.) fide Moravec (*Czech Mycol.* **47**: 237, 1994).

Melastiziella Svrček (1948) = Scutellinia (Otid.) fide Eckblad (1968).

Melchioria Penz. & Sacc. (1897), Trichosphaeriaceae. 1, Java.

Meliderma Velen. (1920) = Cortinarius (Cortinar.) fide Singer (1951).

Melidium Eschw. (1822) = Thamnidium (Thamnid.) fide Benny (*Mycol.* **84**: 834, 1992).

Meliola Fr. (1825), Meliolaceae. 1,000, esp. trop. See Stevens (1927-28), Hansford (1961, 1963), Luttrell (*Mycol.* **81**: 192, 1989; ontogeny), Mibey & Hawksworth (*SA* **14**: 25, 1995; capitate hyphopodia), Mueller *et al.* (*CJB* **69**: 803, 1991; mucronate hyphopodia), Reynolds (*Mycotaxon* **42**: 99, 1991). Each species is commonly given a 'Beeli formula' (q.v.).

Meliolaceae G.W. Martin ex Hansf. (1946), Meliolales. 24 gen. (+ 15 syn.), 1586 spp. Mycelium superficial, dark, thick-walled, frequently branched, with appressoria ('capitate hyphopodia') and conidiogenous cells ('mucronate hyphopodia'); ascomata superficial on lateral branches, ± spherical, black, thin-walled, cleistothecial, breaking down irregularly or forming a large irregular apical hole; interascal tissue poorly developed, of wide thin-walled often evanescent paraphyses; asci very thin-walled, clavate or ± globose, without apical structures, evanescent, usually 2-spored; ascospores brown, usually 4-septate, constricted at the septa, often with a narrow sheath. Anamorphs inconspicuous, probably spermatial, with conidiogenous cells formed directly from vegetative hyphae. Biotrophic on leaves and stems, mostly trop. *Lit.*: Stevens (*Ann. Myc.* **25**: 405, **26**: 165, 1927-28), Hansford (*Sydowia, Beih.* **2**, 1961; 1814 taxa, *Sydowia* **16**: 302, 1963, *Sydowia, Beih.* **5**, 1963; 1814 figs), Doidge & Sydow (*Bothalia* **2**: 424, 1928; S. Afr.), Deighton (*Mycol. Pap.* **9**, 1944), Hansford (*Mycol. Pap.* **23**, 1948; W. Afr.), Hawksworth & Eriksson (*SA* **5**: 142, 1986; status), Mueller *et al.* (*CJB* **69**: 803, 1991; ultrastr. hyphopodia), Yamamoto (*Spec. Publ. Agric. nat. Taiwan Univ.* **10**: 197, 1961; Taiwan), Boedijn (*Persoonia* **1**: 393, 1961; Indonesia), Goos & Anderson (*Sydowia* **26**: 73, 1974; Hawaii), Stevenson (*Sydowia* **22**: 225, 1969; host index).

Meliolales, Ascomycota. 1 fam., 24 gen. (+ 15 syn.), 1586 spp. Fam.:
  Meliolaceae.
Lit.: Katumoto & Hosagoudar (*J. Econ. Tax. Bot.* **13**: 615, 1989; lists 169 taxa supplemental to Hansford's monograph), Schmiedeknecht (*Wiss. Zeitschr. Friedrich-Schiller Univ. Jena* **38**, 185, 1989; 109 spp. Cuba), and see under fam.

Meliolaster Höhn. (1918), Asterinaceae. 1, trop.

Meliolaster Doidge (1920) = Amazonia (Meliol.) fide Stevens (*Ann. Myc.* **25**: 405, 1927).

Meliolidium Speg. (1924) = Perisporiopsis (Parodiopsid.) fide Müller & v. Arx (1962).

Meliolina Syd. & P. Syd. (1914), Meliolinaceae. 39, Afr., Asia, Australasia, Pacific. See Hughes (*Mycol. Pap.* **166**, 1993; key), Reynolds (*Cryptogamie, Mycol.* **10**: 305, 1989; asci). Anamorph *Briania*.

Meliolinaceae S. Hughes (1994), Dothideales. 1 gen., 39 spp., trop. Colonies hypophyllous, velvety, black; hyphae superficial, stomatopodiate and brown, and also immersed and generally hyaline (sometimes re-emerging through stomata to form new colonies); ascomata perithecioid, subglobose, superficial on

hyphae, setose; interascal tissue of paraphyses; asci extenditunicate, not fissitunicate, 8-spored; ascospores ellipsoid to subcylindrical, brown, 3-septate, with a darker band over the septa and the terminal cells usually with 1-2 subhyaline bands. Anamorph *Briania*.

melioline, one of the *Meliolaceae*, esp. *Meliola* spp.

Meliolinella Hansf. (1946) = Scolionema (Parodiopsid.) fide v. Arx & Müller (1975).

Meliolinites Selkirk ex Janson. & Hills (1978), Fossil fungi. 3 (Eocene, Miocene), Australia, USA.

Meliolinopsis Beeli (1920) ≡ Patouillardina (Asterin.).

Meliolinopsis F. Stevens (1924) = Scolionema (Parodiopsid.).

Melioliphila Speg. (1924), Tubeufiaceae. 7, trop. See Rossman (*Mycol. Pap.* **157**, 1987; key).

Meliolopsis (Sacc.) Sacc. (1891) nom. dub. (Fungi, inc. sed.) fide Theissen & Sydow (*Ann. Myc.* **15**: 465, 1917).

Meliothecium Sacc. (1901) ≡ Myxothecium (Meliol.).

Melittiosporiopsis Höhn. (1913) ≡ Mellitosporiopsis (Ectolech.).

Melittosporiella Höhn. (1919), Rhytismatales (inc. sed.). 2, Eur., N. Am. See Sherwood (*Mycotaxon* **5**: 1, 1977; *Sydowia* **38**: 267, 1986).

Melittosporiopsis Sacc. (1902) ≡ Mellitosporiopsis (Ectolech.)

Melittosporium Sacc. (1884) ≡ Melittiosporium (Rhytismatales).

Mellitiosporium Corda (1838), Rhytismatales (inc. sed.). 4, Eur., N. Am. See Hawksworth & Kinsey (*SA* **14**: 59, 1995; nomencl.).

Mellitosporiopsis Rehm (1900) = Tapellaria (Ectolech.).

Meloderma Darker (1967), Rhytismataceae. 4, widespr. See Johnston (*Mycotaxon* **33**: 423, 1988; gen. concept).

Melogramma Fr. (1846), Melanconidaceae. 2, Eur. See Barr (*Mycol. Mem.* **7**, 1978), Cannon (*SA* **7**: 23, 1988; posn), Laflamme (*Sydowia* **28**: 237, 1977), Barr (1978).

Melogrammataceae, see Melanconidaceae.

Melomastia Nitschke ex Sacc. (1875), Clypeosphaeriaceae. 3 or 4, widespr.

melophase, the part of a life cycle in which a diploid nucleus undergoes reduction.

Melophia Sacc. (1884), Mitosporic fungi, 8.E1.?. 10, widespr. See Sutton (1977).

Melzericium Hauerslev (1975) nom. nud. (Amylocortic.) fide Burdsall (*Taxon* **26**: 327, 1977).

Melzerodontia Hjortstam & Ryvarden (1980), Corticiaceae. 1, Tanzania.

Membranatheca Matsush. (1995) = Amerosporium (Mitosp. fungi) fide Sutton (*in litt.*).

Membranicium J. Erikss. (1958) = Phanerochaete (Merul.) fide Donk (*Persoonia* **2**: 223, 1962), but see Hayashi (*Bull. Govt. For. Res. Stn* **88**: 260, 1974).

Membranomyces Jülich (1975), Clavulinaceae. 3, Eur., Canada.

Membranosorus Ostenf. & H.E. Petersen (1930), Plasmodiophoraceae. 1, USA.

membranous (membranaceous), like a thin skin or parchment.

Memnoniella Höhn. (1923), Mitosporic fungi, 1.A2.15. 3, widespr. See Verona & Mazzucchetti (*Publ. Ente naz. Cellul. Carta*, 1968).

Memnonium Corda (1833) = Trichosporium (Mitosp. fungi) fide Saccardo (*Syll. Fung.* **4**: 294, 1886).

memnospore, a spore remaining at its place of origin (Gregory, *in* Madelin (Ed.), *The fungus spore*, 1966). Cf. xenospore.

Mendogia Racib. (1900), Schizothyriaceae. 2, Java, Philipp., S. Am.

Mendoziopeltis Bat. (1959), Micropeltidaceae. 2, C. Am.

Menegazzia A. Massal. (1854), Parmeliaceae (L). *c.* 40, cosmop. (mainly S. temp.). See Santesson (*Ark. Bot.* **30A** (11), 1943).

Menezesia Torrend (1913), ? Saccharomycetales (inc. sed.). 1, Madeira.

Menidochium R.F. Castañeda & W.B. Kendr. (1990), Mitosporic fungi, 3.A2.15. 1, Cuba.

Menispora Pers. (1822), Anamorphic Lasiosphaeriaceae, 1.A1.15. Teleomorph *Chaetosphaeria*. 6, esp. temp. See Hughes (*CJB* **41**: 693, 1963; key), Holubová-Jechová (*Folia geobot. phytotax.* **8**: 317, 1973; key 4 spp.).

Menisporella Agnihothr. (1962) = Dictyochaeta (Mitosp. fungi) fide Sutton (*in litt.*).

Menisporopsis S. Hughes (1952), Mitosporic fungi, 2.A/B1.15. 2, Afr., NZ.

Menoidea L. Mangin & Har. (1907), Mitosporic fungi, 3.A1.?. 1, France, UK. See Wilson & Waldie (*TBMS* **13**: 151, 1928).

Mensularia Lázaro Ibiza (1916) = Inonotus (Hymenochaet.) fide Donk (1974).

Merarthonis Clem. (1909) = Arthonia (Arthon.) fide Santesson (1952).

Mercadomyces J. Mena (1988), Mitosporic fungi, 2.C2.25. 1, Cuba.

merenchyma, see plectenchyma.

Meria Vuill. (1896), Mitosporic fungi, 1.A1.?. 1, Eur. *M. laricis* on *Larix*. See Peace & Holmes (*Oxf. forest. Mem.* **15**, 1933), Drechsler (*Phytopath.* **31**: 773, 1941), Jansson *et al.* (*Ant. v. Leeuwenhoek* **50**: 321, 1984; ultrastr., life history).

Merimbla Pitt (1979), Anamorphic Trichocomaceae, 1.A1.15. Teleomorph *Talaromyces*. 1, widespr.

Meringosphaeria Peyronel (1918) = Acerbiella (Ascomycota) fide Clements & Shear (1931).

Meripilus P. Karst. (1882), Coriolaceae. 4, N. temp, S.E. Asia. See Corner (*Beih. Nova Hedw.* **78**: 193, 1984), Larson & Lombard (*Mycol.* **80**: 612, 1988; key).

Merisma Pers. (1797) = Thelephora (Thelephor.) fide Donk (1954).

Merisma (Fr.) Gillet (1878) ≡ Grifola (Polypor.).

Merismatium Zopf (1898), Verrucariaceae. *c.* 5 (on lichens), Eur. See Triebel (*Bibl. Lich.* **35**, 1989).

merismatoid (of a pileus), made up of smaller pilei.

Merismella Syd. (1927), Anamorphic Chaetothyriaceae, 5.A1.38. Teleomorph *Chaetothyrium*. 5, C. Am.

Merismodes Earle (1909), Crepidotaceae. 11, cosmop. See Singer (1975).

merispore, see sporidesm.

Meristacraceae Humber (1989), Entomophthorales. 3 gen. (+ 1 syn.), 6 spp. See Tucker (1981; key).

Meristacrum Drechsler (1940), Meristacraceae. 2, USA, India, Ukraine. See Drechsler (*J. Wash. Acad. Sci.* **30**: 250, 1940, *Sydowia* **14**: 98, 1960; spore discharge), Davidson & Barron (*CJB* **51**: 231, 1973), Couch *et al.* (*Proc. Natn Acad. Sci. USA* **76**: 2299, 1979), Prasad & Dayal (*Curr. Sci.* **55**: 321, 1986), Tucker (1981; key).

meristem arthrospore, one of the chain of conidia maturing in basipetal succession and originating by apical wall building at the tip of the condiogenous cell; **- blastospore**, a conidium arising either apically or laterally from a conidiogenous cell which elongates through ring wall building at the base (a basauxic conidiophore; Hughes, 1953). See Mitosporic fungi.

meristematic (of conidiophores), see wall-building.

meristogenous (of pycnidia, etc.), formed by growth and division of one hypha; **symphogenous**, formed

from a number of hyphae. See Sutton (*in* Ainsworth *et al.* (Eds), *The Fungi* **4A**: 1973).

Meristosporum A. Massal. (1860) = Laurera (Trypethel.).

Merodontis Clem. (1909), Leotiales (inc. sed.). 1, Java. See Carpenter (*Mem. N.Y. bot. Gdn* **33**, 1981; status).

merogamy, copulation between special sex cells or gametes.

meront (of *Myxomycota*), one of the daughter myxamoebae cut off in succession by a parent myxamoeba.

Merophora Clem. (1909) = Umbilicaria (Umbilicar.).

Meroplacis Clem. (1909) = Caloplaca (Teloschist.).

Merorinis Clem. (1909) = Rinodina (Physc.) fide Zahlbruckner (1931).

merosporangium (of *Zygomycetes*), a cylindrical outgrowth from the swollen end of a sporangiophore in which a chain-like series of sporangiospores is generally produced. See Benjamin (*Mycol.* **58**: 1, 1966; review, *in* Kendrick (Ed.), *The whole fungus* **2**: 573, 1979).

Merosporium Corda (1831) Fungi (inc. sed.) fide Hughes (*CJB* **36**: 784, 1958).

Merostictina Clem. (1909) = Pseudocyphellaria (Lobar.).

Merostictis Clem. (1909) = Diplonaevia (Dermat.) fide Hein (1983).

Merrilliopeltis Henn. (1908) = Oxydothis (Hyponectr.) fide Barr (1976), Hyde (1995).

Merugia Rogerson & Samuels (1990), Lasiosphaeriaceae. 1, Guyana.

Meruliaceae P. Karst. (1881), Stereales. 20 gen. (+ 23 syn.), 155 spp. Basidioma resupinate to reflexed; monomitic; leptocystidia present; spores hyaline, smooth, inamyloid.

Merulicium J. Erikss. & Ryvarden (1976), Corticiaceae. 1, Nordic.

Merulioporia Bondartsev & Singer (1943) = Poria (Coriol.) fide Lowe (1966), = Merulius (Cortic.) fide Donk (1956).

Meruliopsis Bondartsev (1959) ≡ Merulioporia (Coriol.).

Meruliporia Murrill (1942) = Poria (Coriol.) fide Lowe (1966), = Serpula (Coniophor.) fide Donk (1948).

Merulius Haller ex Boehm. (1760) = Cantharellus (Cantharell.).

Merulius Fr. (1821) = Merulius (Merul.) fide Nakasone & Burdsall (*Mycotaxon* **21**: 241, 1984).

Mesenterica Tode (1790) nom. dub.; used for plasmodia fide Fries (1829).

Mesniera Sacc. & P. Syd. (1902), Mesneriaceae. 1, Java. See Petrak (*Ann. Myc.* **39**: 344, 1941).

Mesnieraceae Arx & E. Müll. (1975), Dothideales. 4 gen., 5 spp. Ascomata perithecial, thin-walled, brightly coloured, ostiolate but the upper wall layers sometimes also degenerating; peridium consisting of thin-walled ± flattened cells; interascal tissue of cellular pseudoparaphyses; asci cylindrical, ? fissitunicate, often polysporous; ascospores aseptate or septate, often ornamented, without a sheath. Anamorphs unknown. Necrotrophic or saprobic on leaves, esp. trop.

Mesobotrys Sacc. (1880) = Gonytrichum (Mitosp. fungi) fide Hughes (1958).

mesochroic, see colour.

mesodermatous (of hyphae), having the outer wall and lumen of about the same thickness. Cf. leptodermatous.

Mesonella Petr. & Syd. (1924) ? = Guignardia (Mycosphaerell.).

Mesophellia Berk. (1857), Mesophelliaceae. 4, Australasia. See Beaton & Weste (1983; key).

Mesophelliaceae (G. Cunn.) Jülich (1982), Lycoperdales. 4 gen. (+ 2 syn.), 15 spp. Gasterocarp epigeal; hyphae clamped; spores large, ellipsoid.
Lit.: Beaton & Weste (*TBMS* **82**: 665, 1984).

Mesophelliopsis Bat. & A.F. Vital (1957), Lycoperdales (inc. sed.). 1, Brazil. ? = Geastrum (Geastr.).

mesophile, see thermophily.

Mesopsora Dietel (1922) = Melampsora (Melampsor.).

Mesopyrenia M. Choisy (1931) = Arthopyrenia (Arthopyren.).

mesospore, (1) a 1-celled teliospore among 2-celled ones; (2) an amphispore (obsol.); (3) the middle layer of a three-layered spore wall.

meta- (prefix), changed in form or position; between; with; after.

Metabasidiomycetidae. Subclass in *Basidiomycota* (Lowy, *Taxon* **17**: 125, 1968), for fungi considered intermediate between *Hetero-* and *Homobasidiomycetes*.

metabasidium, see basidium.

metabiosis, the association of two organisms acting or living one after the other; cf. synergism.

Metabolic products. Fungal metabolites are many and diverse. In addition to those associated with protein synthesis and respiration many special products ('secondary metabolites') have been isolated and, frequently, chemically defined. Some of these are waste products while others such as pigments, toxins, and antibiotics clearly have biological functions. Because of their synthetic abilities fungi are used in industry for the production of alcohol, citric acid and other organic acids, various enzymes, riboflavin, etc. (Fig. 23) (see Industrial mycology).

Many fungal products, even when chemically defined, have been given trivial names derived from the scientific names of the fungi involved, e.g. 'griseofulvin' from *Penicillium griseofulvum*. More than 400 such names were compiled in the 6th edition of this *Dictionary*, but most of these have been omitted since the 7th because of the monograph by Turner (1971) and Turner & Aldridge (1982) although representative antibiotics, hormones, mycotoxins, pigments, and other interesting metabolites are still included. See also Hegnauer (*Chemotaxonomie der Pflanzen* **7**: 277, 1986).

About 400 compounds have been reliably reported in lichens of which about 230 are only known in this biological group (lichen products, lichen substances; Culberson & Elix, *Meth. Pl. Biochem.* **1**: 509, 1989; Huneck, *Beih. Nova Hedw.* **79**: 793, 1984). The lichen substances are mainly derivatives of orcinol and β-orcinol and are weak phenolic acids. The most important groups of these are depsides (e.g. olivetoric acid), depsidones (e.g. physodic acid) and dibenzonfuran derivatives (e.g. usnic acid). Most are colourless but some are brightly coloured: red, yellow, orange or emerald green (e.g. pulvic acid derivatives such as vulpinic acid). These are deposited on the surfaces of hyphae in the medulla and cortex (different substances often occurring in different regions of the thalli; e.g. hymenium, thalline exciple, medulla, cortex) and are produced by the fungal partner; the hypothesis that production was dependent on the presence of an alga (Culberson & Ahmadjian, *Mycol.* **72**: 90, 1980), has not been upheld by later work (Leuckert *et al.*, *Mycol.* **82**: 370, 1990). However, their position and quantitative expression can be affected by the thallus environment (Fahselt, 1994).

Some depsides and depsidones give characteristic colours with 10% caustic potash (K), bleach (C), K followed by C (KC), C followed by KC (CK), iodine (I, q.v.), and *p*-phenylenediamine (P, PD; see Steiner's stable PD solution); diagnostic microcrystal tests (q.v.); and characteristic colours in UV-light. The metabolic products generally characterize particular lichens and are routinely used in identification of antibiotic properties. Their role in the lichen thallus is largely unknown, but may be partly antimicrobial and antifeedent (see Lawrey, *Biology of lichenized fungi*, 1984).

Lit.: Raistrick *et al.* (Studies in the biochemistry of microorganisms [mainly devoted to fungi, 116 parts], *Phil. Trans.* **B220**: 1, 1931; *Biochem. J.* **25-93**, 1931-64; Asahina & Shibata (*Chemistry of lichen substances*, 1954 [reprint 1971], WHO (*WHO Chron.* **17**: 389, 1963; nomenclature pharmaceutical preparations), Miller (*The Pfizer handbook of microbial metabolites*, 1961), Shibata *et al.* (*List of fungal products*, 1964), Culberson (*Chemical and botanical guide to lichen products*, 1969; *Bryologist* **73**: 177, 1970 [suppl. 1]; *et al.*, Second supplement, 1977), Turner (*Fungal metabolites*, 1971), Turner & Aldridge, (*Fungal metabolites* **II**, 1983), Pidoplichko (*Metabolity pochvennykh mikromitsetov*, 1971; of soil fungi), Weete (*Fungal lipid biochemistry*, 1974), Eugster (*Z. Pilzk.* **39**: 45, 1973; review), Sussman (*Taxon* **23**: 301, 1974; trends in metabolic specialization), Fahselt (*Symbiosis* **16**: 117, 1994).

See also antibiotics, Chemotaxonomy, chromatography, dyeing, ergot, Hallucinogenic fungi, hormones, litmus, Mycetisms, Mycotoxicoses, Phytotoxic mycotoxins, Pigments, Toxins.

Metabotryon Syd. (1926) = Sphaerellopsis (Mitosp. fungi) fide Sutton (1977).

Metabourdotia L.S. Olive (1957), Ceratobasidiaceae. 1, Tahiti.

Metacapnodiaceae S. Hughes & Corlett (1972), Dothideales. 1 gen. (+ 2 syn.), 6 spp. Mycelium superficial, dark, copious, hyphae strongly constricted at the septa with ± spherical cells, tapering towards the apices; ascomata superficial, small, ± globose, black, thin-walled, with a periphysate ostiole; peridium of pseudoparenchyma; interascal tissue of periphysoids; asci ± saccate, fissitunicate; ascospores brown, transversely septate, sometimes ornamented. Anamorph hyphomycetous, conidiogenous cells tretic. Epiphytic, ? on insect exudates or resin, widespr.
Lit.: Hughes (*N.Z. Jl Bot.* **10**: 225, 1972, *Mycol.* **68**: 693, 1976; gen. names, anamorphs).

Metacapnodium Speg. (1918), Metacapnodiaceae. *c.* 6, trop. See Eriksson & Hawksworth (*SA* **6**: 138, 1987; nomencl.), Hughes (*N.Z. Jl Bot.* **10**: 239, 1972). = Limacinia (Dothid.) fide Reynolds (1985). Anamorphs *Capnophialophora*, *Capnobotrys*, *Capnosporium*.

metacellulose, a cellulose in certain fungi.

Metachora Syd., P. Syd. & E.J. Butler (1911) = Phyllachora (Phyllachor.) fide Cannon (*Mycol. Pap.* **163**, 1991).

metachroic, see Colour.

metachromic, giving a red reaction to cresyl blue. See Singer (*The Agaricales (mushrooms) in modern taxonomy*: 77, 1951).

Metacoleroa Petr. (1927), Venturiaceae. 1, N. temp.

Metadiplodia Syd. (1937), Mitosporic fungi, 4.B2.19. 38, widespr. See Zambettakis (*BSMF* **70**: 219, 1955), Sutton (*Sydowia* **43**: 264, 1991; redescr. type sp.).

Metadothella Henn. (1904), ? Hypocreaceae. 1, Peru.

Metadothis (Sacc.) Sacc. (1892) = Dothiora (Dothior.) fide Saccardo (*Syll. Fung.* **12**: 430, 1897).

Metamelanea Henssen (1989), Lichinaceae (L). 1, Eur., N. Am.

Metameris Theiss. & Syd. (1915), Phaeosphaeriaceae. 3 (on ferns), N. Hemisph. See Barr (*Mycotaxon* **43**: 371, 1992). = Scirrhia (Dothid.) fide Holm & Holm (*Bot. Notiser* **131**: 97, 1978).

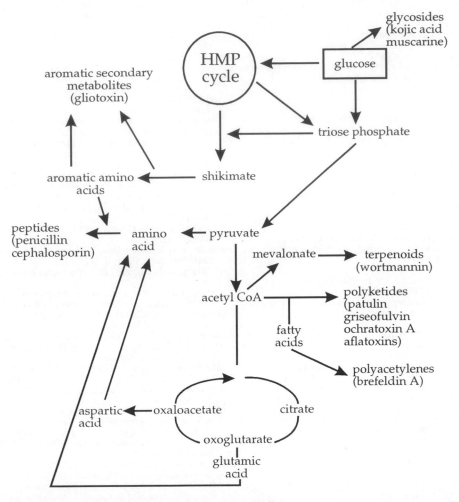

Fig. 23. Metabolic pathways for some fungal secondary metabolites.

Metanectria Sacc. (1878), Hypocreaceae. 1, S. Am.

Metapezizella Petr. (1968), Leotiaceae. 1, Mexico.

metaphysis (Petrak), see paraphysis.

metaplasm, see epiplasm.

Metarhizium Sorokīn (1879), Mitosporic fungi, 2.A1.15. 5 (on *Insecta*), widespr. See Tulloch (*TBMS* **66**: 407, 1976; key), Rombach *et al.* (*Mycotaxon* **27**: 87, 1986; *M. flavo-viride* var. *minus* on leaf- and plant-hoppers on rice), Guo *et al.* (*Acta Mycol. Sin.* **5**: 177, 1986), Rombach *et al.* (*TBMS* **88**: 451, 1987; *M. album* on leaf- and plant-hoppers on rice), Shimizu *et al.* (*J. Invert. Pathol.* **60**: 185, 1992; electrophoretic karyotype of *M. anisopliae*), Rakotonirainy *et al.* (*MR* **98**: 225, 1994; rRNA and separation of taxa), Bridge *et al.* (*J. Gen. Microbiol.* **139**: 1163, 1993; morph., biochem., molecular differentiation), Pipe *et al.* (*MR* **99**: 485, 1995; RFLPs in *M. anisopliae*).

Metasphaerella Speg. (1924) Fungi (inc. sed.) fide Petrak & Sydow (*Ann. Myc.* **34**: 42, 1936).

Metasphaeria Sacc. (1883) nom. ambig. = Saccothecium (Dothior.).

Metasteridium Speg. (1923) nom. dub. (? Meliolaceae); no spp. named. See Eriksson & Hawksworth (*SA* **7**: 74, 1988).

metathallus, assimilative (photobiont-containing) part of a lichen thallus, esp. where there is also a prothallus (q.v.).

Metathyriella Syd. (1927), Schizothyriaceae. 3 or 4, trop. Am.

Metatrichia Ing (1964), Trichiaceae. 4, Eur., trop. See Lakhanpal & Mukerji (*Proc. Indian natn Sci. Acad.* B, **42**: 125, 1977).

Metazythia Petr. (1950), Mitosporic fungi, 4.A1.15. 1, S. Am.

Metazythiopsis M. Morelet (1988), Mitosporic fungi, 4.A1.15. 1, France.

Methods. General literature: Johanssen (*Plant microtechnique*, 1940), McLean & Cooke (*Plant science formulae*, 1941), Spector (Ed.) (*Handbook of biological data*, 1956), Koch (*Fungi in the laboratory*, 1966), IMI (*Plant pathologists' pocketbook*, 1968; edn 2, 1983), Smith (*An introduction to industrial mycology*, edn 7, 1981), Booth (Ed.) (*Methods in microbiology* **4**, 1971), Constantinescu (*Metode și technici în micologie*, 1974), Hawksworth (*Mycologist's handbook*, 1974), Hawksworth & Kirsop (Eds) (*Filamentous fungi*, 1985 [source book]), Koneman & Roberts (*Practical laboratory mycology*, edn 3,

1985), Stevens (Ed.) (*Mycology guidebook*, 1974), Fuller (*Lower fungi in the laboratory*, 1978), Malloch (*Moulds: their isolation, cultivation and identification*, 1981). See also Lichens, Literature, Medical and veterinary mycology, Plant pathogenic fungi.

Special topics (q.v.):
Abbreviation of authors' names, see Authors' names
Arsenic detection, see *Scopulariopsis*
Auxanogram
Bauch test
Beeli formulae
Biomass determination, see biomass
Block culture, see Culture methods
Cellulolysis adequacy index
Chemotaxonomy
Chromatography
Collection and preservation
Colour nomenclature, see Colour
Continuous culture, see Culture methods
Coolplate
Culture media, see Media
Culture methods
Culture preservation, see Genetic resource collections
Electron microscopy (EM), see Ultrastructure
Electrophoresis
Fixatives
Fungicide testing, see Fungicides
Herbarium beetle control
Herbarium management, see Reference collections
Hydrogen-ion concentration
Iodine
Ionomidotic reaction
Isolation methods
Keys, construction of
Laboratory safety, see Safety, laboratory
Lyophilization, see Genetic resource collections
Media, culture
Meixner test
Melzer's solution, see Stains
Microchemical tests for lichen products, see Metabolic products, microcrystal tests
Microscopy
Mite infestation control, see mites
Molecular biology
Mounting media
Nomenclature
Normal saline solution
Numerical taxonomy
Phytosociology
Preservatives
Protoplasts
Reference collections
RIEC
Scanning electron microscopy (SEM)
Serological Techniques, see Serology
Single-spore isolation, see Isolation methods
Spore
Spore print
Spore trapping, see Air spora
Stains
Statistical methods and design of experiments
Steiner's stable PD solution
Sterilization
Ultrastructure
Zymogram.

Methysterostomella Speg. (1910) ? = Phragmopeltis (Mitosp. fungi).

metoecious, used by de Bary; see heteroecious.

Metraria Cooke & Massee (1891), Agaricaceae. 2, Australia, Nigeria.

Metrodia Raithelh. (1971), Agaricaceae. 1, Argentina.

Metschnikowia T. Kamieński (1899), Metschnikowiaceae. 8 (from sea water and arthropods), Eur.,

N. Am. See Barnett *et al.* (1990),van Uden & Castelo-Branco (*J. gen. Microbiol.* **26**: 141, 1961, *Revta Biol.* **3**: 95, 1962; synonymy), Kreger van Rij (1984; key), Pitt & Miller (*Mycol.* **60**: 682, 1968; anamorphs, *Mycol.* **62**: 462, 1970; parasexual cycle), Batra (*USDA Tech. Bull.* **1469**, 1973; key); Giménez-Jurado (*Syst. Appl. Microbiol.* **15**: 432, 1992), Mendonça-Hagler *et al.* (*Inst. Jl Syst. Bact.* **43**: 368, 1993; rRNA phylogeny). Anamorphs *Candida*, *Nectaromyces*.

Metschnikowiaceae T. Kamieński (1899), Saccharomycetales. 10 gen. (+ 8 syn.), 16 spp. Mycelium often well-developed, usually remotely septate; vegetative cells usually proliferating by multilateral budding; asci often formed an short lateral branches, elongate, sometimes curved or clavate, evanescent or persistent; ascospores narrowly fusiform to filiform, sometimes curved, sometimes with filiform appendages; fermentation usually negative; mostly either marine or necrotrophic on plant tissue, widespr. esp. trop.
Lit.: Batra (*USDA Tech. Bull.* **1469**, 1973; monogr.).

Metschnikowiella Genkel (1913) = Metschnikowia (Metschnikow.).

metula, a conidiophore branch having phialides, e.g. of *Penicillium* and *Aspergillus* (obsol.).

Metulocyphella Agerer (1983), Tricholomataceae. 2, S. Am.

Metulodontia Parmasto (1968) = Hyphoderma Wallr. (Hyphodermat.) fide Jülich (*Persoonia* **8**: 78, 1974).

metuloid, an encrusted cystidium thick-walled at maturity, as in *Peniophora*.

Metuloidea G. Cunn.(1965), Coriolaceae.1, Australasia.

Metus D.J. Galloway & P. James (1987), Cladoniaceae (L). 3, Chile, Australasia.

Miainomyces Corda (1833) = Sporotrichum (Mitosp. fungi) fide Saccardo (*Syll. Fung.* **4**: 106, 1886).

micaceous (of a pileus surface), covered with bright particles.

Micarea Fr. (1825) nom. cons., Micareaceae (L). *c.* 70, widespr. See Coppins (*Bull. Br. Mus. nat. Hist., Bot.* **11**: 17, 1983; monogr. 45 Eur. spp.), Vězda & Wirth (*Folia geobot. phytotax.* **11**: 93, 1976; key 31 spp.). See also *Hastifera*.

Micarea Fr. (1825) nom. rej. = Placynthiella (Trapel.).

Micareaceae Vězda ex Hafellner (1984), Lecanorales (L). 5 gen. (+ 2 syn.), 86 spp. Thallus crustose; ascomata apothecial, flat or domed, without a well-developed margin, pale or dark; interascal tissue of relatively wide branched and anastomosing paraphyses, sometimes with a pigmented epithecium; asci with a well-developed I+ apical cap, with a large tube-like more strongly staining ring surrounding a usually well-developed apical cushion, and an outer I+ gelatinized layer; ascospores small, hyaline, simple or septate, without a sheath. Anamorphs pycnidial. Lichenized with green algae, widespr.

Micheli (Pier Antonio; 1679-1737). Botanist to Cosmo III (Grand Duke of Tuscany) and keeper of the public gardens in Florence, made important discoveries in connexion with higher plants, Bryophyta, and lichens, while his additions to the knowledge of fungi were the greatest made by any one man before the time of Persoon and Fries. He gave an account of his discoveries in *Nova plantarum genera*, his most important printed work. This book was completed in 1719, but the first part was not published till 1729 [reprint 1976] and the second (the Figures for which are still in existence) was never printed.

Micheli gave the names to *Mucor*, *Tuber*, *Polyporus*, *Clathrus*, *Aspergillus* and other common genera of today. He made a new systematic arrangement with keys to genera and species. With the help of the microscope (then coming into use), he was the first to see cystidia on the lamella-edge and between the lamellae of agarics and he saw the arrangement of the spores (which he took to be seeds) in groups of four in the *Agaricaceae*, and of ascospores in aci (in *Pertusaria*). He was the first to make experiments on the culture of moulds by placing spores of *Botrytis*, *Aspergillus* and *Mucor* on freshly-cut bits of melon, quince and pear and noting their growth and development (see Buller, *Trans. r. Soc. Can.* sect. 4, ser. 3, **9**: 1915; Engl. transl. [reprinted by Ainsworth (1976)] of 'observations' on culturing moulds). See Ainsworth (1976), Hawksworth (introduction, 1976 *Nova pl. gen.* reprint), Targioni-Tozzetti (*Notizie della vita e delle opere di Pier Antonio Micheli*, 1858), Stafleu & Cowan (*TL-2* **3**: 446, 1981).

Michenera Berk. & M.A. Curtis (1868), Anamorphic Aleurodiscaceae, 7.A1.?. Teleomorph *Aleurodiscus*. 3, Cuba, Brazil, Sri Lanka. See Donk (*Taxon* **11**: 89, 1962).

Micraspis Darker (1963), Rhytismatales (inc. sed.). 3 (on *Rhytismatales*), N. Am., UK. See Darker, *CJB* **41**: 1389, 1963). Anamorph *Periperidium*.

micro- (prefix), small; one-thousandth (Système International d'Unités); see micron.

microaerophilic, making best growth under lowered oxygen pressure.

Microallomyces R. Emers. & J.A. Robertson (1974), Blastocladiaceae. 1, Costa Rica.

Microanthomyces Grüss (1926) = Candida (Mitosp. fungi) fide Lodder (*The yeasts*, 1970).

Microascaceae Luttr. ex Malloch (1970), Microascales. 11 gen. (+ 5 syn.), 61 spp. Ascomatal wall composed entirely of black small-celled pseudoparanchymatous cells, perithecial or cleistothecial; interascal hyphae absent; ascospores smooth-walled. Anamorphs prominent, hyphomycetous, sometimes synnematous, annellidic or forming resting spores.
Lit.: v. Arx (*Persoonia* **10**: 23, 1978), v. Arx *et al.* (*Beih. Nova Hedw.* **94**, 1988), Barr (*Mycotaxon* **39**: 43, 1990; concept), Benny & Kimbrough (1980), Malloch (*Mycol.* **62**: 729, 1970, *in* Reynolds, 1981).

Microascales, Ascomycota. 2 fam., 15 gen. (+ 11 syn.), 79 spp. Stromata absent; ascomata solitary, perithecial or cleistothecial, usually black, thin-walled, sometimes with well-developed smooth setae; interascal tissue absent or rarely of undifferentiated hyphae; centrum absent; asci formed in chains, ± globose, without a stalk, very thin-walled, evanescent, 8-spored; ascospores yellow or reddish brown, aseptate, often curved, often with very inconspicuous germ pores, without a sheath. Anamorphs hyphomycetous, prominent. Saprobic from soil or rotting vegetation, a few opportunistic human and animal pathogens, cosmop. Fams:
(1) **Chadefaudiellaceae**.
(2) **Microascaceae**.

Microascus Zukal (1885), Microascaceae. 13, widespr. See v. Arx (*Persoonia* **8**: 191, 1975; key), Corlett (*CJB* **41**: 253, 1963; ontogeny, sexuality, **44**: 79, 1966; perithecial development), Nishimura & Miyaji (*Mycopath.* **90**: 29, 1985; yeast-like state). Anamorphs *Scopulariopsis*, *Wardomyces*.

Microascus Sacc. (1916) ≡ Microdiscus (Leotiales).

Microasellaria Tuzet, Manier & Jolivet (1957) nom. dub. (Trichomycetes) fide Lichtwardt (1986).

Microbasidium Bubák & Ranoj. (1914) = Hadrotrichum (Mitosp. fungi) fide Höhnel (*Sber. Akad. Wiss. Wien* **125**: 111, 1916).

microbe, a microorganism (q.v.).

microbial (adj.), pertaining to microbes (q.v.).

microbiology, the study of microorganisms (q.v.).

microbiota, all the microorganisms present in the area or habitat specificied, including algae, bacteria and protozoa as well as fungi; see mycobiota.

Microbispora Nonom. & Y. Ohara (1957), *Actinomycetes*, q.v.

Microblastosporon Cif. (1930), Mitosporic fungi, 1.A1.?. 1, Eur.

microbodies, see peroxisome.

Microbotryodiplodia Sousa da Câmara (1951) = Microdiplodia (Mitosp. fungi) fide Petrak (*Sydowia* **16**: 353, 1963), but see Sutton (1977).

Microbotryum Lév. (1847) Ustilaginaceae. 7 (on *Caryophyllaceae*), temp. See Deml & Oberwinkler (*Phytopath. Z.* **104**: 345, 1982).

Microcaliciaceae Tibell (1984), Caliciales. 1 gen. (+ 2 syn.), 4 spp. Thallus absent; ascomata stalked; asci ellipsoidal, formed in chains, evanescent; ascospores 1- to 7-septate, dark brown, with an ornamentation of helical ridges, forming a dry mazaedial mass. Saprobic or fungicolous.

Microcalicium Vain. (1927), Microcaliciaceae. 4 (mainly on lichens), boreal & temp. See Tibell (*Bot. Notiser* **131**: 229, 1978; key), Hawksworth (*Bull. Br. Mus. Nat. Hist., Bot.* **9**: 1, 1981; anamorphs).

Microcalliomyces Bat. & Cif. (1962) = Microcallis (Chaetothyr.) fide v. Arx & Müller (1975).

Microcalliopsis Bat. & Cif. (1962) = Microcallis (Chaetothyr.) fide v. Arx & Müller (1975).

Microcallis Syd. (1926), Chaetothyriaceae. 10, trop. See Hansford (*Mycol. Pap.* **15**, 1946).

Microcarpon Schrad. ex J.F. Gmel. (1792) nom. dub. (? Myxomycetes).

microcephalic, see septum.

Microcephalis Bainier (1882) = Syncephalis (Piptocephalid.) fide Benjamin (1959).

Microcera Desm. (1848) = Fusarium (Mitosp. fungi) fide Wollenweber & Reinking (*Die Fusarien*: 6, 1935), but used by Petch (*TBMS* **7**: 89, 1921).

Microclava F. Stevens (1917), Mitosporic fungi, 1.B2.1. 3, W. Indies, Philipp., Brazil. See Deighton (*TBMS* **52**: 315, 1969; key).

Microcollybia Métrod (1952) nom. nud. = Collybia (Tricholomat.) fide Singer (1975).

Microcollybia Lennox (1979), Tricholomataceae. 1, N. temp.

microconidium, (1) the smaller conidium of a fungus which also has macroconidia; (2) a spermatium (q.v.).

microcrystal test, method of identification of phenolic metabolites in lichens developed by Asahina (q.v.) involving re-crystalizations on microscope slide from a range of solvents and the formation of salts with diagnostic shapes. The crystals are examined microscopically for identification. Many of these tests are extremely sensitive and can detect some compounds at concentrations close to the resolution possible by thin-layer chromatography (q.v.).
Lit.: Asahina (*J. Jap. Bot.* **12-16**, 1936-40), Hale (1974), Hawksworth (*Bull. Br. lichen Soc.* **28**: 5, 1971), Thomson (*The lichen genus Cladonia in North America*, 1968). See Metabolic products.

microculture, a culture of an organism under the microscope, as in a hanging drop.

Microcyclella Theiss. (1914), Dothideales (inc. sed.). 1, Afr.

Microcyclephaeria Bat. (1958), Ascomycota (inc. sed.). 1, Australia.

microcyclic (of conidiation), germination of spores by the direct function of the conidia without the intervention of mycelial growth (Hanlin, *Mycoscience* **35**: 113, 1994); (of rusts), see Uredinales.

Microcyclus Sacc., Syd. & P. Syd. (1904), Mycosphaerellaceae. c. 13, mainly trop. *M. ulei* (anamorphs *Aposphaeria ulei*, *Fusicladium macrosporum*) (syn.

Dothidella ulei, Melanopsammopsis ulei); S. Am. leaf blight of *Hevea*; Holliday (*Phytopath. Pap.* **12**, 1970), Dean (*Brazil and the struggle for rubber*, 1987), Chee & Holliday (*Malaysian Rubber Res. Board Monogr.* **13**, 1986), Hashim & de Almeida (*J. Nat. Rubber Res.* **2**: 111, 1987; physiologic races). See also Cannon *et al.* (*MR* **99**: 353, 1995; key 7 S. Am. spp.).

microcyst, see cyst.

Microcyta Petr. & Syd. (1927) ≡ Ceuthosporella (Mitosp. fungi) fide Sutton (1977).

microcytospore, (1) an encysted zoospore (planospore); (2) an encysted gametangium (*Blastocladiales*).

Microdiplodia Allesch. (1901), Mitosporic fungi, 4.B2.?. 27, widespr. See Zambettakis (*BSMF* **70**: 219, 1955), Sutton (1977).

Microdiplodia Tassi (1902), Mitosporic fungi, 4.B2.?. 79, widespr. See Sutton (1977).

Microdiplodiites Babajan & Tasl. (1973), Fossil fungi. 1 (Tertiary), former USSR.

Microdiscula Höhn. (1915), Mitosporic fungi, 4.A1.?. 1, Eur.

Microdiscus Sacc. (1916), Leotiales (inc. sed.). 1, N. Am.

Microdiscus Steinecke (1916) nom. dub (? Fungi).

Microdochium Syd. (1924), Anamorphic Hyponectriaceae, 3.A-C1.21. Teleomorph *Monographella*. 6, widespr. *M. nivale* (pink snow mould of turf grasses), *M. panattonianum* (lettuce ring spot). See Sutton *et al.* (*CJB* **50**: 1899, 1972), Sutton & Hodges (*Nova Hedw.* **27**: 215, 1976), v. Arx (*Sydowia* **34**: 30, 1981), Samuels & Hallett (*TBMS* **81**: 473, 1983; teleomorphs), Litschko & Burpee (*TBMS* **89**: 252, 1987; variation in *M. nivale*), Parman & Price (*Austral. Pl. Path.* **20**: 41, 1991; microsclerotia in *M. panattonianum*).

Microdothella Syd. & P. Syd. (1914), Dothideales (inc. sed.). 2, N. Am., Philip.

Microdothiorella C.A.A. Costa & Sousa da Câmara (1955), Mitosporic fungi, 8.A1.?. 1, Portugal.

Microeccrina Maessen (1955) = Arthromitaceae (Bacteria) fide Manier (*Ann. Parasit. hum. comp.* **38**: 1, 1961). See *Arthromitus*.

microendospore, minute cytoplasmic particles behaving like spores in *Ophiostoma ulmi* (Ouellette & Gagnon, *CJB* **38**: 235, 1960).

Microeurotium Ghatak (1936), ? Pezizales (inc. sed.). 1 (? coprophilous), UK.

microflora, sometimes used, inappropriately, for all the microorganisms present in a specified site or habitat; see microbiota, mycobiota.

microform, see *Uredinales*.

microfungi, see *Micromycetes*.

Microglaena Körb. (1855) [non *Microglena* Ehrenb. (1831, *Chlorophyta*] = Thelenella (Thelenell.) fide Eriksson (1981).

Microglaenomyces Cif. & Tomas. (1953) = Chromatochlamys (Thelenell.) fide Mayrhofer & Poelt (*Herzogia* **7**: 13, 1985).

Microglena Lönnr. (1858) ≡ Microglaena (Thelenell.).

Microgloeum Petr. (1922), Anamorphic Dermateaceae, 6.A1.10. Teleomorph *Blumeriella*. 1, widespr.

Microglomus L.S. Olive & Stoian. (1977), Protosteliaceae. 1, Hawaii. See Olive *et al.* (*TBMS* **81**: 449, 1983).

Microglossum Gillet (1879), Geoglossaceae. 8, temp. See Mains (*Mycol.* **47**: 846, 1955; N. Am. spp., key), Maas Geesteranus (*Persoonia* **3**: 82, 1964), Benkert (*Gleditschia* **10**: 141, 1983), Nitare & Ryman (*Svensk bot. Tidskr.* **78**: 63, 1984; Sweden).

Microglossum Sacc. (1884), Geoglossaceae. 3, Eur., Japan. See Maas Geesteranus (*Persoonia* **3**: 81, 1964), Spooner (*Bibl. Mycol.* **116**, 1987; key Australian spp.).

microgonidia, very small green bodies in lichen hyphae (Minks) (obsol.).

Micrographa Müll. Arg. (1890) ? = Lembosia (Asterin.) fide Santesson (*Svensk bot. Tidskr.* **43**: 547, 1949),? = Melaspilea (Arthoniales) fide Müller & v. Arx (1962).

Micrographina Fink (1930) = Mazosia (Roccell.) fide Santesson (*Svensk bot. Tidskr.* **43**: 547, 1949).

Micrographomyces Cif. & Tomas. (1953) nom. illegit. ≡ Micrographa (Asterin.).

Microhaplosporella Sousa da Câmara (1949) ? = Aplosporella (Mitosp. fungi) fide Sutton (*in litt.*).

Microhendersonula Dias & Sousa da Câmara (1952), Mitosporic fungi, 8.C2.?. 1, Portugal.

Microhilum H.Y. Yip & A.C. Rath (1989), Mitosporic fungi, 1.A1.10. 1 (on *Oncoptera* larvae), Australia.

Microides, see *Endothrix*.

Microlecia M. Choisy (1949) = Catillaria (Catillar.) fide Hafellner (1984).

microlichens, lichens in which the whole of their morphological characteristics can be seen only with a magnifier equal or larger than × 10. See Messuti (*Br. Lich. Soc. Bull.* **73**: 49, 1993).

Microlychnus A. Funk (1973), Anamorphic Asterothyriaceae, 2.E1.15 (L). 1, Canada. See Sérusiaux & De Sloover (*Veröff. Geobot. Inst. ETH, Rübel* **91**: 260, 1986).

Micromastia Speg. (1909) Ascomycota (inc. sed.).

micrometre, one-thousandth of a millimetre (0.001 mm; 1 μm) (Système International d'Unités); formerly often as 'μ'.

Micromium Pers. (1811) ≡ Phlyctis (Phlyctid.).

Micromma A. Massal. (1860), Pyrenulales (inc. sed., L). 1, Indonesia.

Micromonospora Orskov (1923), *Actinomycetes*, q.v.

Micromphale Gray (1821), Tricholomataceae. 11, widespr.

Micromucor Malchevsk. (1939) = Micromucor (Mortierell.) fide Kirk (*in litt.*).

Micromucor (W. Gams) Arx (1982), Mortierellaceae. 3, widespr. See v. Arx (*Sydowia* **35**: 10, 1982; key), Amano *et al.* (*Mycotaxon* **44**: 257, 1992; chemotaxonomy), Ruiter *et al.* (*MR* **97**: 690, 1993; chemotaxonomy). = Umbelopsis (Mortierell.) fide Yip (*TBMS* **86**: 334, 1986).

Micromyces P.A. Dang. (1889), Synchytriaceae. 2, widespr. See Sparrow (1960: 194; key).

Micromycetes, fungi having small (microscopic) sporocarps; microfungi. Most handbooks for the identification of microfungi are based on systematic groups (e.g. ascomycetes, hyphomycetes, coelomycetes), orders (*Laboulbeniales, Saprolegniales*), or genera (*Aspergillus, Penicillium*), q.v. Ellis & Ellis (*Microfungi on land plants: an identification handbook*, 1985; and *Microfungi on miscellaneous substrates*, 1988) cover most temperate groups. See also Literature.

Micromycopsidaceae, see *Synchytriaceae*.

Micromycopsis Scherff. (1926) ? = Micromyces (Synchytr.) fide Sparrow (1960) but see Subramanian (*Curr. Sci.* **43**: 723, 1974).

Micromyriangium Petr. (1929) = Uleomyces (Cookell.) fide v. Arx (1963).

micron, see micrometre.

Micronectria Speg. (1885), Hypocreaceae. 3, S. Am., Asia, Eur.

Micronectriella Höhn. (1906) = Sphaerulina (Mycosphaerell.) fide v. Arx & Müller (1975).

Micronectriopsis Höhn. (1918), Hypocreaceae. 1, Philipp.

Micronegeria, see *Mikronegeria*.

micronematous (micronemeous), (1) having hyphae of small diameter; (2) (of conidiophores) similar morphologically to vegetative hyphae (cf. Micronemeae).

Micronemeae, hyphomycetes with conidiophores or conidia like the hyphae (e.g. *Oospora*), or having no hyphae. Cf. Macronemeae.

microorganism, an organism which belongs to a phylum many members of which either cannot be seen with the unaided eye or require microscopic examination and/or growth in pure culture for their identification; a microbe; includes all unicellular prokaryotes and eukaryotes, and also some multicellular eukaryotes, i.e. microscopic algae, bacteria, fungi (including yeasts), protozoa and viruses; sometimes used (incorrectly) only for prokaryotes (bacteria and viruses). See Cowan (*A dictionary of microbial taxonomy*: 162, 1978; inappropriateness of term), Zavarzin (*in* Allsopp *et al.* (Eds), *Microbial diversity and ecosystem function*: 17, 1995; concept).

Micropeltella Syd. & P. Syd. (1913) = Micropeltis (Micropeltid.) fide v. Müller (1975).

Micropeltidaceae Clem. & Shear (1931), Dothideales. 25 gen. (+ 26 syn.), 168 spp. Mycelium at least partially superficial, often not strongly pigmented; ascomata strongly flattened, occasionally coalescing, opening by a wide irregular ostiole; peridium of one to several layers of pseudoparenchymatous cells, sometimes epidermoid, the lower wall indistinct; interascal tissue of cellular pseudoparaphyses, often sparse or evanescent; asci clavate to saccate, fissitunicate, sometimes faintly I+ blue at the apex; ascospores hyaline or pale brown, transversely septate or muriform, sometimes attenuated at the base.
Lit.: Batista (*Publs Inst. micol. Recife* **56**, 1959; 445 spp.)

Micropeltidium Speg. (1919) = Micropeltis (Micropeltid.) fide v. Arx & Müller (1975).

Micropeltis Mont. (1842), Micropeltidaceae. c. 100, trop. See Batista (*Publs Inst. micol. Recife* **56**, 1959).

Micropeltopsis Vain. (1921) = Lichenopeltella (Microthyr.) fide Santesson (1990).

Micropera Lév. (1846) [non Lindl. (1832), *Orchidaceae*] = Foveostroma (Mitosp. fungi) fide DiCosmo (*CJB* **56**: 1682, 1978).

Microperella Höhn. (1909), Mitosporic fungi, 8.B1.1. 1, Japan.

Micropeziza Fuckel (1870), Dermateaceae. 5, Eur. See Nannfeldt (*Bot. Notiser* **129**: 323, 1976; key).

Microphiale (Stizenb.) Zahlbr. (1902) = Dimerella (Gyalect.).

Microphiodothis Speg. (1919) = Ophiodothella (Phyllachor.) fide Petrak (*Sydowia* **5**: 352, 1951).

Microphlyctis J. Schröt. (1889) nom. dub. (Chytridiales).

Microphoma Buchw. (1958) = Dothichiza (Mitosp. fungi) fide Sutton (1977).

microphylline (of lichen thalli), composed of minute lobes or scales.

Microphyma Speg. (1889) = Phillipsiella (Phillipsiell.) fide Müller & v. Arx (1962).

Micropodia Boud. (1885), ? Leotiaceae. 1 or 2, Eur.

Microporellus Murrill (1905), Coriolaceae. 2, trop. See Reid (*Microscopy* **32**: 452, 1975).

Microporus P. Beauv. (1805), Polyporaceae. 15, widespr. See Pegler (1973), Ryvarden & Johansen (*Preliminary polypore flora of East Africa*: 429, 1980; key 7 Afr. spp.).

Micropsalliota Höhn. (1914), Agaricaceae. c. 73, widespr. See Heinemann (*BSMF* **106**: 1, 1990; key 73 spp.), Heinmann & Leelavathy (*MR* **95**: 341, 1991; key spp. of Kerala).

micro-Puccinia, a microform of *Puccinia*.

Micropuccinia Rostr. (1902) = Puccinia (Puccin.) fide Dietel (1928). See Laundon (*Mycol. Pap.* **99**, 1965).

Micropustulomyces R.W. Barreto (1995), Mitosporic fungi, 6.C1.1/4. 1, Brazil.

Micropyrenula Vain. (1921) = Microtheliopsis (Microtheliopsid.) fide Santesson (*Svensk bot. Tidskr.* **43**: 547, 1949).

Micropyxis Seeler (1943) [non Duby (1930), *Algae*] = Gelatinopsis (Leot.).

microsclerid, see cystidium.

microsclerotium, a very small sclerotium, as in *Verticillium dahliae*; pseudosclerotium.

Microscopy. Light microscopes. The compound microscope has two sets of lenses (objective and eyepiece) which magnify the object at each step. The image has a maximum final magnification of around x1000. Stereomicroscopes (also known as dissecting microscopes) are lower power and have a long free working distance for manipulation of the specimen. For use in the field, portable instruments are available. For viewing, a small sample of the subject is mounted on a glass slide with a stain or clear mountant, and overlaid with a coverslip. Common stains in mycology are cotton blue or fuchsin in lactic acid, or lactic acid alone if no stain is needed. Air bubbles can be eliminated by gentle heating. To keep the slide for examination at a later date, the coverslip must be sealed with a preparatory sealant or nail varnish.

Brightfield: The most common method of viewing subjects in the compound microscope. Light travels from beneath the subject, passing through a condenser and through the subject, illuminating translucent structures.

Differential Interference Contrast (DIC; Nomarski): Gives an almost three dimensional image of the subject. Especially useful for hyaline structures, it uses special polarizing prisms arranged according to a design by Georges Nomarski. High quality objectives are required in addition to a dedicated condenser and adapter.

Phase contrast: Special objectives and condenser are required for this technique. The image suffers from optical imperfections, such as a halo, but this effect can be of some value in drawing attention to small objects or details. Low contrast material benefits greatly from this technique, but it can be difficult to measure accurately in phase contrast.

Darkfield: Direct light is prevented from passing through the objective aperture and the image is formed from light scattered by features in the object, the detail appearing bright against a dark background.

Fluorescence: Microscopy in which the image is formed by fluorescence emitted from the specimen when illuminated with ultra-violet radiation. The sample may autofluoresce naturally, but fluorochromes such as mithramycin and calcafluor are typically used to enhance specific structures.

Electron microscopes. Scanning electron microscopy (SEM; q.v.) provides higher magnification images (to around x80,000) of the surface morphology of the specimen, and also gives large depth of field. Transmission electron microscopy (TEM) images provide a view similar to that of light microscopy at much higher magnifications (over x100,000) to show ultrastructural details. Specimens are prepared by fixing, dehydrating, staining, embedding in resin and cutting ultrathin sections on an ultramicrotome; see also Ultrastructure.

Lit.: Beckett & Read (*in* Todd (Ed.), *Ultrastructure techniques for microorganisms*, 1986), Bradbury (*An introduction to the optical microscope* (rev. edn), 1989), Ploem & Tanke (*Introduction to fluorescence microscopy*, 1987), Rawlins (*Light microscopy,*

1992), Sanderson (*Biological microtechnique*, 1994), Slayter & Slayter (*Light and electron microscopy*, 1982).

Microscypha Syd. & P. Syd. (1919), Hyaloscyphaceae. 3, Eur. See Huhtinen (*Karstenia* **29**: 45, 1990).

Microsebacina P. Roberts (1993), Exidiaceae. 2, Eur.

Microsomyces Thaxt. (1931), Laboulbeniaceae. 2, Afr., Sumatra, N. Am. See Benjamin (*Aliso* **11**: 127, 1986).

Microspatha P. Karst. (1889), Anamorphic Asterothyriaceae, 2.A1.? (L). 1, S. Am. See Sérusiaux & De Sloover (*Veröff. Geobot. Inst. ETH, Rübel* **91**: 260, 1986).

Microsphaera Lév. (1851), Erysiphaceae. 125, widespr. *M. grossulariae* (European mildew of gooseberry and currant), *M. alphitoides* (oak mildew). See Braun (*Nova Hedw.* **39**: 211, 1984; key 47 N. Am. spp., *Beih. Nova Hedw.* **89**, 1987; key). Anamorph *Oidium*.

Microsphaeropsis Höhn. (1917), Mitosporic fungi, 4.A2.15. 44, widespr. See Sutton (*Mycol. Pap.* **123**, 1971), Morgan-Jones (*CJB* **52**: 2575, 1974), Morgan-Jones & White (*Mycotaxon* **30**: 177, 1987; ultrastr. conidiogenesis *M. concentricum*).

Microsphaeropsis Sousa da Câmara, Oliveira & Luz (1936) nom. dub. (Mitosp. fungi).

Microspora Velen. (1934) [non Thur. (1850), *Algae*], ? Leotiales (inc. sed.). 1, former Czechoslovakia.

microsporangium, secondary sporangium formed from a zoospore cyst, either without any mitoses, or with very few mitoses.

microspore, (1) a small spore, where there are spores of two sizes; (2) a spore from a microsporangium (q.v.).

Microsporella Höhn. (1918) ? = Microsphaeropsis (Mitosp. fungi) fide Sutton (1977).

Microsporon, see *Microsporum*.

Microsporonites K.P. Jain (1968), ? Fossil fungi (*f. cat.*). 1 (Triassic), Argentina. See Pirozynski & Weresub (1979).

Microsporum Gruby (1843), Anamorphic Onygenales, 2.C1.2. Teleomorph *Nannizzia*. 14 (on humans and other mammals, causing **microsporoses**), widespr. *M. audouinii* (tinea capitis in humans, esp. children), *M. canis* (ringworm in cats, dogs and humans). See Ajello (1968), Conant (*Arch. Derm. Syph. Chicago* **36**: 781, 1937), Morace *et al.* (*Mycopath.* **94**: 53, 1986; serotyping by monoclonal antibodies), Vismer *et al.* (*Mycopath.* **98**: 149, 1987; ultrastr. conidia).

Microstelium Pat. (1899) nom. dub.; ? based on moss rhizoids fide Walker (*Mycotaxon* **11**: 1, 1980).

Microstella K. Ando & Tubaki (1984), Mitosporic fungi, 1.G1.10. 1 (aquatic), Japan.

Microsticta Desm. (1849) = Schizothyrium (Schizothyr.) fide Müller & v. Arx (1962).

Microstoma Bernstein (1852), Sarcoscyphaceae. 2, N. temp., Java. See Boedijn (*Sydowia* **5**: 211, 1951).

Microstoma Auersw. (1860) = Valsa (Vals.).

Microstroma Niessl (1861), Microstromataceae. 3, widespr. See v. Arx (*Genera of fungi sporulating in pure culture*: 104, 1981).

Microstromataceae Jülich (1982), Exobasidiales. 1 gen., 3 spp. Basidia erumpent through host stomata; 6-8 sterigmate; statismosporic.

Microtetraspora Thiemann, Pagani & Beretta (1968), *Actinomycetes*, q.v.

Microthallites Dilcher (1965), Fossil fungi (? Microthyr.). 3 (Eocene, Miocene), India, USA. But see Blanz (*Z. Mycol.* **44**: 91, 1978; annellides in), Hansen (*Grana* **19**: 67, 1980). = Phragmothyrites (Fossil fungi) fide Selkirk (1975).

Microthecium Corda (1842) = Melanospora (Ceratostomat.) fide v. Arx (*Genera of fungi sporulating in pure culture*, 1981). See also *Sphaerodes*.

Microthelia Körb. (1855) nom. rej. = Anisomeridium (Monoblast.) fide Hawksworth & Sherwood (*Taxon*

30: 339, 1981). See Hawksworth (*Bull. Br. Mus. nat. Hist., Bot.* **14**: 1, 1985; redisp. names).

Microtheliomyces Cif. & Tomas. (1953) = Leptorhaphis (Arthopyren.) fide Aguirre-Hudson (1991).

Microtheliopsidaceae O.E. Erikss. (1981), Dothideales (L). 1 gen. (+ 2 syn.), 1 sp. Thallus crustose, effuse, non-corticate; ascomata scattered, immersed, conical, black, ostiolate; interascal tissue absent; asci ovoid to saccate, fissitunicate, blue in KOH/IKI, the inner layer thickened towards the apex; ascospores brown, thin-walled, septate, without a sheath. Anamorph unknown. Algal partner *Phycopeltis*. Trop.
    Lit.: Eriksson (1981).

Microtheliopsidomyces Cif. & Tomas. (1953) ≡ Microtheliopsis (Microtheliopsid.).

Microtheliopsis Müll. Arg. (1890), Microtheliopsidaceae (L). 1, S. Am. See Santesson (*Symb. bot. upsal.* **12** (1), 1952).

Microthyriaceae Sacc. (1883), Dothideales. 50 gen. (+ 41 syn.), 229 spp. Superficial mycelium indistinct; ascomata small, strongly flattened, superficial, with a small central ostiole sometimes surrounded by thickened tissue; peridium composed of radiating rows of ± isodiametric cells, the basal layer hyaline, poorly developed or absent; interascal tissue trabeculate pseudoparaphyses, often deliquescing; asci saccate, fissitunicate; ascospores hyaline or brown, transversely septate, sometimes ciliate, without a sheath. Anamorphs hardly studied. Saprobic or epiphytic on leaves and stems, cosmop.
    Lit.: Doidge (*Bothalia* **4**: 273, 1942; S. Afr.), Stevens & Ryan (*Ill. biol. monogr.* **17** (2), 1939).

Microthyriacites Cookson (1947), Fossil fungi (Microthyr.). 8 (Cretaceous, Tertiary), widespr. = Phragmothyrites (Fossil fungi) fide Selkirk (1975).

Microthyriales, see *Dothideales*.

Microthyriella Höhn. (1909) = Schizothyrium (Schizothyr.) fide Müller & v. Arx (1962).

Microthyrina Bat. (1960) = Microthyrium (Microthyr.) fide v. Arx & Müller (1975).

Microthyriolum Speg. (1917) = Ferrarisia (Parmular.) fide Müller & v. Arx (1962).

Microthyris Clem. (1931) ≡ Lichenopeltella (Microthyr.).

Microthyrites Pampal. (1902), Fossil fungi. 1 (Miocene), Italy.

Microthyrium Desm. (1841), Microthyriaceae. 50, widespr. See Ellis (*TBMS* **67**: 382, 1977; key 13 Br. spp.), Doi & Uemura (*Bull. natn. Sci. Mus. Tokyo* B **11**: 127, 1985), Barr (*N. Am. Fl.* II **13**, 1990; N. Am.).

Microtrichella Maessen (1955) = Arthromitus (Bacteria) fide Manier & Lichtwardt (*Ann. Sci. Nat., Bot.* sér. 12 **9**: 519, 1968).

Microtrichophyton (Castell. & Chalm.) Neveu-Lem. (1921) = Trichophyton (Mitosp. fungi).

Microtyle Speg. (1919), Mitosporic fungi, 4.B1.?. 1, S. Am. See Müller & v. Arx (*Beitr. Krypt.-Fl. schweiz* **11** (2): 830, 1962).

Microtypha Speg. (1910) = Arthrinium (Mitosp. fungi) fide Subramanian (1971).

Microxiphium (Harv. ex Berk. & Desm.) Thüm. (1879), Anamorphic Coccodinaceae, 2.A1.?. Teleomorph *Dennisiella*. 1, widespr. See Hughes (*Mycol.* **68**: 693, 1976).

Microxyphiella Speg. (1918), Mitosporic fungi, 4.B1.?. 7, trop. See Hughes (*Mycol.* **68**: 788, 1976).

Microxyphiomyces Bat., Valle & Peres (1961), Mitosporic fungi (L), 8.A1.15. 5, Brazil.

Microxyphiopsis Bat. (1963), Mitosporic fungi, 4.B1.?. 2, Brazil.

Microxyphium Speg. (1918), Mitosporic fungi, ?.?.?. 10, widespr. = Microxiphium (Mitosp. fungi).

Micula Duby (1858) ? = Foveostroma (Mitosp. fungi) fide Sutton (1980).

Micularia Boedijn (1961), Elsinoaceae. 1, Java.

Midotiopsis Henn. (1902), Leotiales (inc. sed.). 2, trop. Am.

Midotis Fr. (1828) ? = Wynnella (Helvell.). See Nannfeldt (*TBMS* **23**: 239, 1939).

migration pseudoplasmodium (of *Acrasiales*), the migration stage after the massing of the myxamoebae (Raper, *J. agric. Res.* **50**: 135, 1935).

Mikronegeria Dietel (1899), Mikronegeriaceae. 2 (on *Nothofagus, Araucaria* and *Austrocedrus*), S. Am. See Peterson & Oehrens (*Mycol.* **70**: 321, 1978).

Mikronegeriaceae Cummins & Y. Hirats. (1983), Uredinales. 1 gen., 2 sp. Pycnia ampulliform (often lobed) with a long neck, deeply immersed in the host tissue; aeciospores catenulate (*Caeoma*); urediniospores pedicellate; teliospores sessile (or pedicellate), unicellular, often waxy, pale.

miktohaplont, a haplont made up of cells having genotypically different nuclei (Kniep, 1928); cf. isohaplont.

Miladina Svrček (1972), Otideaceae. 1, Eur., N. Am. See Pfister & Korf (*CJB* **52**: 1643, 1974), Descals & Webster (*TBMS* **70**: 466, 1978). Anamorph *Actinospora.*

mildew, (1) a plant disease in which the pathogen is seen as a growth on the surface of the host. A **powdery** ('**true**') - is caused by one of the *Erysiphaceae* (e.g. the American (*Sphaerotheca mors-uvae*) and European - (*Microsphaera grossulariae*) of *Ribes*); a **downy** ('**false**') - by one of the *Peronosporaceae* (the first may be controlled by sulphur, the second by copper fungicides); a **dark** -, or **black** -, by one of the *Meliolales* or *Capnodiaceae*; (2) the staining, and frequently the breaking up, of cloth and fibres, paint, etc., by fungi and bacteria (cf. mould); (3) a fungus causing (1) or (2).

Milesia F.B. White (1878), Anamorphic Uredinales. 25, widespr. A name for (II) only. Teleomorph *Milesina* and *Phakopsora.* See Ono *et al.* (*MR* **96**: 828, 1992).

Milesina Magnus (1909), Pucciniastraceae. 34 (+ 22 *Uredo/Milesia*), temp.; O, I on *Abies*, II, III on ferns. See Cummins & Hiratsuka (1983), Faull (*Contr. Arnold Arbor.* **2**, 1932; *J. Arnold Arbor.* **15**: 50, 1934; as '*Milesia*'). Ono *et al.* (*MR* **96**: 828, 1992) expanded the anamorph concept to include species on legumes.

milk-cap, basidioma of *Lactarius* spp.

Millerburtonia Cif. (1951), Anamorphic Valsaceae, 6.E1.?. Teleomorph *Plagiostoma*. 1, Venezuela.

Milleria Peck (1879) [non L. (1753), *Compositae*] = Testicularia (Ustilagin.) fide Dietel (1928).

Milleromyces T.N. Taylor, Hass & W. Remy (1993), Fossil fungi (Paleomastigomycetes). 1 (Devonian), UK.

Milospium D. Hawksw. (1975), Mitosporic fungi, 1.A2.1. 1 (on lichens), Eur. See Hawksworth (*Nova Hedw.* **79**: 373, 1984).

Milowia Massee (1884) = Thielaviopsis (Mitosp. fungi) fide Wakefield & Bisby (*TBMS* **25**: 63, 1941). Nom. dub. fide Nag Raj & Kendrick (*Monograph of Chalara and allied genera*: 43, 1975).

Miltidea Stirt. (1898), Miltideaceae (L). 1, NZ. See Hafellner (*Beih. Nova Hedw.* **79**: 241, 1984).

Miltideaceae Hafellner (1984), Lecanorales (L). 1 gen., 1 sp. Thallus crustose; ascomata apothecial, ± sessile, pale, without a well-developed margin; interascal tissue of branched and anastomosing paraphyses, swollen at the apices, with a reddish crystalline epithecium; asci with a weakly I+ apical cap and a well-developed ocular chamber, with an I+ gelatinized outer layer; ascospores hyaline, aseptate, with a gelatinous sheath. Lichenized with green algae, Australasia.

Mimema H.S. Jacks. (1931) Uredinales (inc. sed.). 2 (on *Leguminosae*), S. Am. See Dianese *et al.* (*MR* **98**: 786, 1994).

Mimeomyces Thaxt. (1912), Laboulbeniaceae. 16, S. Am., Eur., N. Afr., NZ.

Minakatella G. Lister (1921), Trichiaceae. 1, Japan, N. Am.

Mindeniella Kanouse (1927), Rhipidiaceae. 2, N. Am.

Mindoa Petr. (1949), Mitosporic fungi, 5.B1.?. 1, S. Am.

MINE. Microbial Information Network Europe; see Genetic resource collections.

Miniancora Marvanová & Bärl. (1989), Mitosporic fungi, 1.G1.15. 1 (aquatic), Canada.

minimal medium, the simplest chemically defined medium on which the wild type (prototroph) of a species will grow and which must be supplemented by one or more specific substances for the growth of auxotrophic mutants derived from the wild type.

Minimedusa Weresub & P.M. LeClair (1971), Mitosporic fungi, ?.?.?. 1 (sterile mycelium with basidiomycete affinities), Cuba.

Minimidochium B. Sutton (1970), Mitosporic fungi, 3.A1.15. 3, trop.

Minksia Müll. Arg. (1882), Roccellaceae (L). 3, trop.

Minutoexcipula V. Atienza & D. Hawksw. (1994), Mitosporic fungi, 3.B2.19. 2, Eur., N. Am., Australia. See Hafellner (*Herzogia* **10**: 1, 1994).

Minutophoma D. Hawksw. (1981), Mitosporic fungi, 4.A1.15. 1 (on *Chrysothrix*), UK.

Minutularia P.A. Dang. [not traced] nom. dub. (based on a protozoan) fide Dangeard [not traced].

Mirandia Toro (1934) nom. dub. (Fungi, inc. sed.) fide Müller & v. Arx (1962). See Petrak (*Sydowia* **5**: 328, 1951).

Mirandina G. Arnaud ex Matsush. (1975), Mitosporic fungi, 1.E1.10. 6, Eur., Papua New Guinea. = Dactylaria (Mitosp. fungi) fide de Hoog (*Stud. Mycol.* **26**: 1, 1985).

MIRCEN. Microbial Resource Centre. Since 1975 a network of centres for information on the resources available for diverse aspects of microbiology has been established throughout the world under UNESCO (United Nations Educational, Scientific, and Cultural Organization)/UNEP (United Nations Environment Programme). The mycology MIRCEN is based on IMI. See Kirsop & Da Silva (*in* Hawksworth & Kirsop (Eds), *Filamentous fungi*: 173, 1966), *MIRCEN NEWS*.

Mirimyces Nag Raj (1993), Mitosporic fungi, 8.A1.15. 1, Cuba.

Miriquidica Hertel & Rambold (1987), Lecanoraceae (L). 13, esp. arctic-alpine. See Rambold & Schwab (*Nordic Jl Bot.* **10**: 117, 1990; rust-coloured spp.).

Mischoblastia A. Massal. (1852) = Rinodina (Physc.). See Rambold *et al.* (*Pl. Syst. Evol.* **192**: 31, 1994).

mischoblastiomorph (of ascospores), ones similar to the polarilocular type but either without a septum or only with an incomplete septum.

Mischolecia M. Choisy (1931) = Rinodina (Physc.).

Misgomyces Thaxt. (1900), Laboulbeniaceae. 2, Afr., N. Am., Eur., Asia. See Tavares & Balazuc (*Mycol.* **69**: 1069, 1977).

miso, an oriental food product, used for soups and as a flavouring agent, composed of rice and cereals + soybeans fermented by *Aspergillus oryzae* and *Saccharomyces rouxii* (Hesseltine, *Mycol.* **57**: 168, 1965); see Fermented food and drinks.

Mison Adans. (1763) = Phellinus (Hymenochaet.) fide Donk (1960).

Mites. spp. of *Tyrophagus* and *Tarsonemus* sometimes infest fungus cultures. They are a serious problem as they transfer spores from culture to culture, causing cross-contamination that may be irredeemable and often introduce bacteria into the culture.

To prevent mite invasion, work surfaces must be kept clean and cultures protected from aerial infestation by storage in cabinets or incubators. To clean work surfaces, wash with a non-fungicidal acaricide (0.2% v/v Actelic is used at IMI). Mites can be detected by scrutiny of cultures at twice-weekly intervals; they appear as white objects, just detectable with the naked eye. Ragged colony margins or growth of contaminant fungi or bacteria forming trails mey denote their presence. If mites are detected, contaminated cultures should be destroyed by autoclaving (121°C for 15 min.).

Where contamination is not severe and the original fungus is sporulating it will normally be possible to recover an uncontaminated culture. If mite-infested cultures cannot be reisolated, they can be stored at −18°C for 1-3 days before being subcultured. This procedure will kill both eggs and adult mites. Fungi which would not survive short-term cold storage may be covered with a layer of mineral oil and subcultured after 24 hours. However, this latter procedure does not kill the eggs.

A cigarette paper fastened on the necks of universal bottles or test-tubes provides an effective barrier against mites. The cigarette papers are sterilized by dry heat at 180°C for 2 h. and aseptically stuck to the rim of the bottle using a copper sulphate/gelatin glue (20 g gelatin; 2 g copper sulphate; 100 mls distilled water) (Snyder & Hansen, 1946).

Standing the cultures on liquid paraffin and treating the wool plugs with kerosene (crude, such as 'tractor vapourizing oil') has proved an effective control. See Area Leão *et al.* (*Mem. Inst. Oswaldo Cruz* **42**: 559, 1946), Smith (*Mycol.* **59**: 600, 1967). Cf. insects, invertebrates and lichens.

mitochondrion, membrane bound intracellular organelle containing enzymes and electron transport chains for oxidative respiration of organic acids and the concomitant production of ATP. Possesses DNA, messenger RNA and small ribosomes and thus capable of protein synthesis.

Mitochytridium P.A. Dang. (1911), ? Endochytriaceae. 2, Eur., N. Am. See Couch (*J. Elisha Mitchell sci. Soc.* **51**: 293, 1935).

Mitopeltis Speg. (1921), ? Micropeltidaceae. 1, S. Am. See Petrak & Sydow (*Ann. Myc.* **33**: 169, 1935).

Mitosis in fungi. For reviews see Fuller (*Internat. Rev. Cytology* **45**: 113, 1976), Heath (*Mycol.* **72**: 229, 1980).

mitosporangium, a thin-walled diploid sporangium of certain *Blastocladiales* producing uninucleate diploid zoospores (**mitospores**) (Emerson, 1950); cf. meiosporangium.

mitospore, (1) a spore from a mitosporangium (q.v.); (2) (of ascomycetes and basidiomycetes), any non-basidiosporous or -ascosporous propagule (see Kendrick & Watling, *in* Kendrick (Ed.), *The whole fungus* **2**: 473, 1979). See Mitosporic fungi.

Mitosporic fungi (Deuteromycotina, Deuteromycetes, Fungi Imperfecti, asexual fungi, conidial fungi). Fungi. 2600 gen. (+ 1500 syn.), 15,000 spp. (a few L). This artificial assemblage comprises the residue (>95%) of known mitosporic fungi which have not been correlated with any meiotic states. Mitosporic fungi which have been correlated with teleomorphs in the ascomycetes and basidiomycetes can be termed anamorphs or anamorphic states of those groups. For the majority of ascomycetes and basidiomycetes anamorphs are unknown; for some they are still undescribed or unrecognized ('unconnected'); for others anamorphs have appeared to have lost sexuality and its functions sometimes replaced by such mechanisms as the parasexual cycle. Some mitosporic fungi have taken independent evolutionary paths. The *Code* (see Nomenclature) provides for the use of separate names for the different states of pleomorphic fungi, but rules that the name of the holomorph (the whole fungus in all its correlated states) is that of the teleomorph. See Kendrick (*Sydowia* **41**: 6, 1989), Sutton (*in* Reynolds & Taylor, *The fungal holomorph*: 27, 1993).

Characteristics are a combination of (1) absence, or presumed absence, of a state producing asci/ascospores, basidia/basidiospores, teliospores, or other basidium-bearing organs (teleomorph, perfect or sexual state), (2) absence, or presumed absence, of any meiotic or mitotic reproductive structures (agonomycetes, mycelia sterilia), (3) presence of conidia formed by mitosis (or presumed mitosis). They have been treated previously as subdivision *Deuteromycotina*, but other subdivisions of the *Eumycota* are separated by characteristics of the meiotic states (teleomorph), not states producing mitospores. The general term Mitosporic fungi is therefore used in this *Dictionary*.

It now seems probable that though many more teleomorph/anamorph state connexions will be established a permanent residue of mitosporic fungi will remain unconnected. Advancing molecular technology will eventually make it feasible to place these remaining taxa with the groups of teleomorphic fungi from which they are or once were derived. On morphological grounds this has already been done for it is traditional to exclude certain recognizable anamorphs from the Mitosporic fungi. Those of the *Zygomycetes*, *Erysiphales*, and *Uredinales*, for example, are treated in association with their teleomorphic states. Even with such exclusions the Mitosporic fungi is the second largest group of fungi and includes many common moulds and many organisms of considerable economic significance.

Three classes are traditionally recognized:

(1) **Hyphomycetes** - mycelial forms which bear conidia on separate hyphae or aggregations of hyphae (as synnematous or sporodochial conidiomata) but not inside discrete conidiomata.

(2) **Agonomycetes** - mycelial forms which are sterile, but may produce chlamydospores, sclerotia and/or related vegetative structures.

(3) **Coelomycetes** - forms producing conidia in pycnidial, pycnothyrial, acervular, cupulate or stromatic conidiomata.

The arrangement of the Mitosporic fungi in any phylogenetic way is at present impossible and previous hierarchical suprageneric taxa are not used in this *Dictionary*. Integrated systems for Mitosporic fungi as a whole were suggested by Höhnel (1923) and Sutton (1980); see also Luttrell *in* Kendrick (1977). Arrangement of correlated anamorphs with ascomycete systematics has been reviewed by Kendrick & DiCosmo (*in* Kendrick (Ed.), *The whole fungus*: 283, 1979) and Sutton & Hennebert (*in* Hawksworth (Ed.), *Ascomycete systematics*: 77, 1994).

Coding system in entries for mitosporic genera. Three categories of information are coded:

(i) **Conidiomatal types** 1-9, listed in Table 6, Fig. 10, e.g. 1, indicates hyphal, 2, synnematal etc.

(ii) **Saccardo's spore groups**. Saccardo arranged 'imperfect' fungi (and also many ascomycetes, particularly those of the Sphaeriales) according to the septation or form of the spores and their colour — whether

Table 6. Mitosporic fungi coding for conidiomata and conidia (for conidiogenous events see text).

Conidiomata	
1 hyphal	6 acervular
2 synnematal	7 cupulate
3 sporodochial	8 stromatic (complex conidiomata
4 pycnidial	excluded from other categories)
5 pycnothyrial	9 sclerotia

Conidial shape and septation			
		1 conidia hyaline or bright (hyalo-)	**2** conidia pigmented or dark (phaeo-)
A 1-celled (not E, F or G)	amerosporae	hyalosporae	phaeosporae
B 2-celled	didymosporae	hyalodidymae	phaeodidymae
C 2-many-celled (multiseptate)	phragmosporae	hyalophragmae	phaeophragmae
D muriform	dictyosporae	hyalodictyae	phaeodictyae
E filiform	scolecosporae		
F helical	helicosporae		
G branched	staurosporae		

dark or hyaline — and the coined Latin names for these different groupings are set out in Table 6, e.g. C1, indicates multiseptate hyaline conidia, F2, helical brown etc.

(iii) **Conidiogenous events**. The matrix system used is based on Minter *et al.* (1982, 1983a,b) who showed a continuum of developmental processes associated with conidial production, including ontogeny, delimitation and secession of conidia and proliferation and regeneration of the cells bearing them (see conidiogensis). For the 43 combinations of events so far recognized see Figs 24-26, e.g. 15, indicates a succession of holoblastic conidial ontogeny, delimitation by a transverse septum, schizolytic secession, percurrent enteroblastic conidiogenous cell proliferation followed by holoblastic conidial ontogeny, successive conidia seceding at the same level.

Use of '?', means that insufficient information is available for the feature to be coded, and '-', that the feature is absent, e.g. 9.-.-. indicates presence of sclerotia but no conidia, and 4.B2.?, that pycnidial conidiomata produce 1-septate brown conidia but their genesis is not known.

Lit.: General works on the mitosporic fungi include: Saccardo (*Syll. Fung.* **3, 4, 10, 11, 14, 16, 18, 22, 25, 26**, 1884-1972), Lindau (*Naturlichen Pflanzenfam.*, 1900), Jaczewski (*Key to Fungi* 2, *Fungi Imperfecti*, 1917), v. Höhnel (*Mykol. Unters.* **3**: 301-369, 1923), Clements & Shear (1931), Kendrick (Ed.) (*Taxonomy of Fungi Imperfecti*m 1971), Barnett & Hunter (*Illustrated genera of imperfect fungi*, 3 edn, 1972), Ainsworth *et al.* (Eds) (*The Fungi* 4, 1973), Cole & Kendrick (*Biology of conidial fungi*, 1981), Minter *et*

al. (*TBMS* **79**: 75, 1982; **80**: 39, 1983; **81**: 109, 1983), Stewart *et al.* (*Deuteromycotina and selected Ascomycotina from wood and wood products*, 1988; bibliogr. and guide to taxonomic lit.), Wilken-Jensen & Gravesen (*Atlas of moulds in Europe causing respiratory allergy*, 1984), Matsumoto & Ajello (*Handb. Appl. Mycol.: Humans, animals & insects* **2**: 117, 1991; dematiaceous fungi pathogenic to humans and lower animals), Campbell (*Handb. Appl. Mycol.: Humans, animals & insects* **2**: 395, 1991; conidiogenesis in fungi pathogenic to man and animals), McGinnis *et al.* (*Jl Med. Vet. Mycol.* **30**(Suppl. 1): 261, 1992), Howard (Ed.) (*Fungi pathogenic for humans and animals* A, 1993), Reynolds & Taylor (Eds), *The fungal holomorph*, 1993). See also under *Coelomycetes* and *Hyphomycetes*.

Mitosporium Clem. & Shear (1931) ≡ Aciculosporium (Clavicipit.).

Mitrasphaera Dumort. (1822) nom. rej. = Cordyceps (Clavicipit.).

mitrate (mitriform), mitre-like in form.

Mitremyces Nees (1816) = Calostoma (Calostomat.).

mitriform, see mitrate.

Mitrophora Lév. (1846) = Morchella (Morchell.) fide Seaver (*North American cup fungi*, 1928). See Eckblad (1968).

Mitrorhizopeltis Bat. & Cavalc. (1964) = Tracylla (Mitosp. fungi) fide Nag Raj (*CJB* **53**: 2435, 1975).

Mitrula Pers. (1794) = Heyderia (Leot.).

Mitrula Fr. (1821), ? Sclerotiniaceae. 7, temp. See Benkert (*Gleditschia* **10**: 141, 1983; key 4 spp.), Knudsen (*Bot. Tidsskr.* **69**: 248, 1978; Danish spp.),

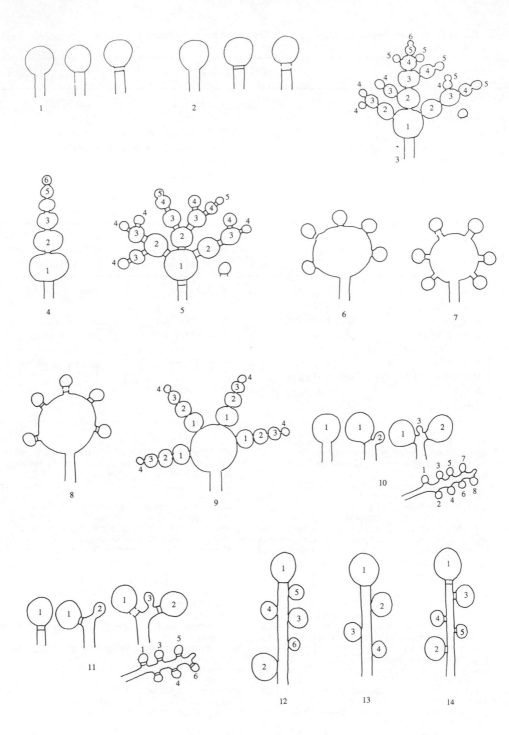

Fig. 24. Conidiogenous events (cc - conidiogenous cell). 1, conidial ontogeny holoblastic, 1 locus per cc, solitary conidia, delimited by 1 septum, maturation by diffuse wall-building, secession schizolytic, no proliferation of cc; 2, conidial ontogeny holoblastic, 1 locus per cc, solitary conidia, delimitation by 2 septa (or a separating cell), secession rhexolytic or by fracture of the cc, maturation by diffuse wall-building, no proliferation of cc; 3, conidial ontogeny holoblastic, apical wall-building random at more than one locus per cc and conidia becoming conidiogenous to form connected branched chains, each conidium delimited by 1 septum, maturation by diffuse wall-building, secession schizolytic, no cc proliferation; 4, conidial ontogeny holoblastic, apical wall-building at 1 locus per cc and each conidium with 1 locus to form a connnected unbranched chain, each conidium delimited by 1 septum, maturation by diffuse wall-building, secession schizolytic, no proliferation of cc; 5, conidial ontogeny holoblastic, apical wall-building randomly at more than 1 locus per cc and conidia becoming conidiogenous to form connected branched chains, each conidium delimited by 2 septa (or a separating cell), secession rhexolytic or by fracture of the cc, maturation by diffuse wall-building, no cc proliferation; 6, conidial ontogeny holoblastic, with localized apical wall-building simultaneously at different loci over the whole cc, each locus forming 1 conidium, delimited by 1 septum, maturation by diffuse wall-building, secession schizolytic, no cc proliferation; 7, conidial ontogeny holoblastic, with localized apical wall-building simultaneously at different loci on denticles over the whole cc, each locus forming 1 conidium, delimited by 1 septum, maturation by diffuse wall-building, secession by rupture of denticle, no cc proliferation; 8, conidial ontogeny holoblastic, with localized apical wall-building simultaneously at different loci over the whole cc, each conidium delimited by 2 septa (or a separating cell), secession rhexolytic or by fracture of the cc, each locus forming 1 conidium, maturation by diffuse wall-building, no cc proliferation; 9, conidial ontogeny holoblastic, apical wall-building simultaneously at several loci per cc and conidia becoming conidiogenous to form connected branched chains, each conidium delimited by 1 septum, maturation by diffuse wall-building, secession schizolytic, no cc proliferation; 10, conidial ontogeny holoblastic, regularly alternating with holoblastic sympodial cc proliferation, maturation by diffuse wall-building, each conidium delimitated by 1 septum, secession schizolytic; 11, conidial ontogeny holoblastic, regularly alternating with holoblastic sympodial cc proliferation, maturation by diffuse wall-building, each conidium delimited by 2 septa (or a separating cell), secession rhexolytic or by fracture of the cc; 12, conidial ontogeny holoblastic, each from apical or lateral loci, delimited by 1 septum, secession schizolytic, holoblastic cc proliferation sympodial or irregular, maturation by diffuse wall-building; 13, conidial ontogeny holoblastic, first from an apical locus, delimited by 1 septum, secession schizolytic, other conidia from lateral loci proceeding down the cc, maturation by diffuse wall-building; 14, conidial ontogeny holoblastic, first from an apical locus, each conidium delimited by 2 septa (or a separating cell), secession rhexolytic or by fracture of the cc, other conidia from lateral loci proceeding down the cc, maturation by diffuse wall-building.

Redhead (*CJB* **55**: 307, 1977; N. Am. spp.), Verkeley (*Persoonia* **15**: 405, 1994; posn).

Mitrulinia Spooner (1987), Sclerotiniaceae. 1, S. Hemisph.

Mitruliopsis Peck (1903) = Spathularia (Geogloss.) fide Durand (*Ann. Myc.* **6**: 387, 1908).

Mitteriella Syd. (1933), Anamorphic Englerulaceae, 1.C2.10. Teleomorph *Schiffnerula*. 2, India, Uganda. See Hughes (*CJB* **61**: 1727, 1984).

Miuraea Hara (1948), Anamorphic Mycosphaerellaceae, 1.C1.?. Teleomorph *Mycosphaerella*. 2, Asia. See Hara (*Manual of pests & diseases*: 779, 1948), Deighton (*Mycol. Pap.* **133**: 3, 1973; relationship to *Cercosporella*).

Mixia C.L. Kramer (1959), Mixiaceae. 1, Japan, USA. See Ando *et al.* (*Abstracts IMC5*: 5, 1994; posn), Kramer (1987).

Mixiaceae C.L. Kramer (1987), Basidiomycota (inc. sed.). 1 gen., 1 sp. Sporogenous cells ('spore sacs') formed directly from the germination of specialized cells of the internal mycelium, with a wall separating a peripheral layer of sporogenous multinucleate protoplasm.
Lit.: Kramer (*Stud. Mycol.* **30**: 151, 1987), Nishida *et al.* (*CJB* **73**(*Suppl.* 1): in press, 1995; posn).

Mixtoconidium Etayo (1995), Mitosporic fungi, 4.B1.19. 1 (on lichens), Spain.

Mixtura O.E. Erikss. & J.Z. Yue (1990), Phaeosphaeriaceae. 1, S. Am.

Miyabella S. Ito & Homma (1926) = Synchytrium (Synchytr.) fide Karling (1964).

Miyagia Miyabe ex Syd. & P. Syd. (1913), Pucciniaceae. 6, Eur., Asia.

Miyakeomyces Hara (1913), Hypocreaceae. 1, Japan. See Eriksson & Yue (*SA* **8**: 9, 1989), Rossman (*Mycotaxon* **8**: 492, 1979).

Miyoshia Kawam. (1907) [non Makino (1903), *Liliaceae*] ≡ Miyoshiella (Lasiosphaer.).

Miyoshiella Kawam. (1929) = Chaetosphaeria (Lasiosphaer.) fide Hino (*J. Jap. Bot.* **10**: 527, 1932).

mizaume, see Fermented food and drinks.

MLC, microbody-liquid globule complex; see *Chytridiales*.

Mniaecia Boud. (1885), Leotiaceae. 2 (on *Hepaticae*), Eur. See Purvis *et al.* (*Lichen flora of Great Britain and Ireland*, 1992; key).

Mniopetalum Donk & Singer (1962), Tricholomataceae. 2, Eur. ? = Leptoglossum (Tricholomat.) fide Donk (1964).

Moana Kohlm. & Volkm.-Kohlm. (1989), Halosphaeriaceae. 1 (marine), Pacific.

Mobergia H. Mayrhofer & Sheard (1992), Physciaceae (L). 2, C. & N. Am.

Models of larger fungi, were made by James Sowerby during 1796 and 1815 for his museum. These were later restored by Worthington G. Smith (q.v.) and are now at The Natural History Museum, London (see Tribe, *Bull. BMS* **15**: 161, 1984). The Ware Collection of Glass Models of Plants, made by B. Blaschka in 1929, at the Botanical Museum of Harvard University include a series showing fungal diseases of fruit; the Department of Applied Biology, Cambridge University has a collection of glass models of macro- and microfungi made by W.A.R. Dillon Western (1899-1953) which were exhibited at the 50th Anniversary Meeting of the British Mycological Society (*TBMS* **30**: 21, 1948). The longest more recent series of coloured pottery models (some 200) are those designed by the late Mr and Mrs Lovenzens of Lantz, Nova Scotia, of which there is a complete set in the Nova Scotia Museum, Halifax.

Modicella Kanouse (1936), Mortierellaceae. 2, N. & S. Am. See Gerdemann & Trappe (1974; key), Benny *et al.* (in Sylvia *et al.* (Eds), *Mycorrhizae in the next decade, practical applications and research priorities*: 311, 1987; syst. posn).

Moelleria Bres. (1896) [non Scop. (1777), *Algae*] ≡ Moelleriella (Clavicipit.).

Moelleriella Bres. (1897), Clavicipitaceae. 1, Brazil. See Rogerson (*Mycol.* **62**: 865, 1970).

Moelleroclavus Henn. (1902) = Xylaria (Xylar.) fide Clements & Shear (1931).

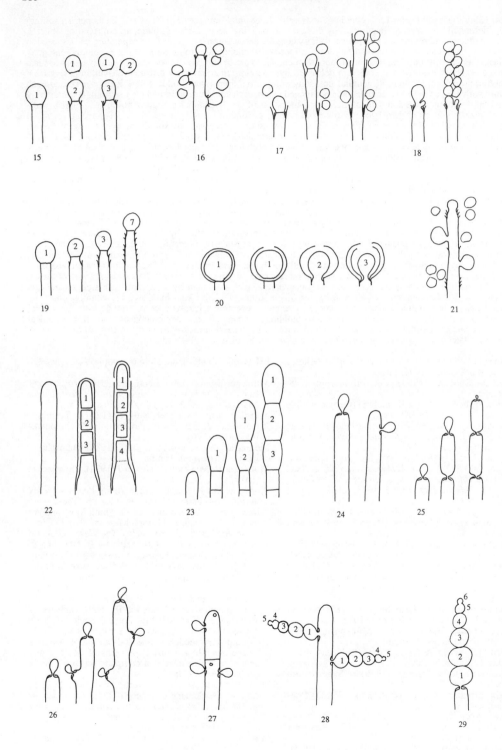

Fig. 25. Conidiogenous events (cc - conidiogenous cell). 15, conidial ontogeny holoblastic, delimitation by 1 septum, schizolytic secession, maturation by diffuse wall-building, percurrent enteroblastic cc proliferation followed by conidial ontogeny by replacement apical wall-building, successive conidia seceding at the same level, sometimes in unconnected chains, collarette variable; 16, same as 15 but with several random or irregular conidiogenous loci to each cc; 17, conidial ontogeny holoblastic, delimitation by 1 septum, schizolytic secession, maturation by diffuse wall-building, percurrent enteroblastic cc proliferation followed by conidial ontogeny by replacement apical wall-building, successive conidia seceding at the same level, collarette variable, conidiogenous activity interspersed periodically with percurrent vegetative proliferation; 18, conidial ontogeny holoblastic, delimitation by 1 septum, schizolytic secession, maturation by diffuse wall-building, percurrent and sympodial enteroblastic cc proliferation followed by conidial ontogeny by replacement apical wall-building, successive conidia seceding at the same level, collarette variable; 19, conidial ontogeny holoblastic, delimitation by 1 septum, schizolytic secession, maturation by diffuse wall-building, percurrent enteroblastic cc proliferation followed by conidial ontogeny by replacement apical wall-building, successive conidia seceding at progressively higher levels, sometimes in unconnected chains, collarette variable; 20, conidial ontogeny enteroblastic, delimitation by 1 septum, schizolytic secession, maturation by diffuse wall-building, outer wall of the cc remaining as a conspicuous collarette, percurrent enteroblastic cc proliferation followed by conidial enteroblastic ontogeny by replacement apical wall-building, successive conidia seceding at the same level, a succession of collarettes formed; 21, combination of 10, 12 and 19, where the sequences occur at random, irregularly or interchangeably; 22, conidial ontogeny holoblastic with new inner walls constituting the conidia laid down retrogressively by diffuse wall-building, delimitation retrogressive, loss of apical wall-building followed by replacment ring wall-building at the base of the cc adding more retrogressively delimited conidia, the outer (original) cc wall breaks as a connected chain of conidia is formed, collarette variable, 1 locus per cc, secession schizolytic; 23, conidial ontogeny holoblastic, 1 locus per cc, first conidium delimited by 1 septum, maturation by diffuse wall-building, loss of apical wall-building, replaced by ring wall-building below the delimiting septum which produces conidia in a connected unbranched chain, secession schizolytic, no proliferation of cc; 24, conidial ontogeny holoblastic, simultaneous with minimal enteroblastic percurrent proliferation at the preformed pore in the outer cc wall, conidia solitary, delimited by 1 septum, secession schizolytic, maturation by diffuse wall-buiilding, 1 locus per cc; 25, conidial ontogeny holoblastic, simultaneous with minimal enteroblastic percurrent proliferation at the preformed pore in the outer cc wall, conidia solitary, delimited by 1 septum, secession schizolytic, maturation by diffuse wall-buiilding, after one conidium formed extensive enteroblastic percurrent proliferation by apical wall-building occurs until the next apical locus is formed; 26, same as 24 but with holoblastic sympodial proliferation of the cc with conidiogenesis occurring between loci; 27, same as 24 but with several conidiogenous loci produced in the apical cc and laterally below septa in other ccs constituting the conidiophore; 28, same as 24 but several loci to each cc and first and subsequent conidia becoming conidiogenous by apical wall-building to form unbranched connected chains; more than one locus to a conidium will produce branched chains; 29, same as 24 but first conidium becoming conidiogenous by apical wall-building to form an unbranched connected chain;

Moellerodiscus Henn. (1902) = Ciboria (Sclerotin.) fide Spooner (*Bibl. Mycol.* **116**, 1, 1987).

Moelleropsis Gyeln. (1939), Pannariaceae (L). 1, N. temp. See Maas (*Proc. Nova Scotia Int. Sci.* **37**: 21, 1987).

Moeszia Bubák (1914) = Cylindrocarpon (Mitosp. fungi) fide v. Arx (*Genera of fungi sporulating in pure culture*, 1970).

Moesziella Petr. (1927) = Leptopeltis (Leptopeltid.) fide Holm & Holm (1977).

Moesziomyces Vánky (1977), Ustilaginaceae. 2, widespr. *M. penicillariae*, pearl millet seed smut. See Vánky (*Nordic Jl Bot.* **6**: 67, 1986), Rao & Thacker (*TBMS* **81**: 597, 1983), Mordue (*IMI Descript.* **1245**, 1995).

Moeszopeltis Petr. (1947) = Leptopeltis (Leptopeltid.) fide Holm & Holm (1977).

Mohgaonidium Singhai (1974), Fossil fungi (Mitosp. fungi). 1 (Eocene), India.

Mohortia Racib. (1909) = Septobasidium (Septobasid.) fide Couch (1938).

Molecular biology. Techniques in molecular biology are making a major contribution to our understanding of fungal biology and relationships (see Phylogeny; Bruns *et al.*, 1991), and also opening new horizons for their utilization. Varieties of different molecular techniques are available, and the one(s) most appropriate for a particular task must be selected. Most are DNA based, and the DNA has first to be extracted; of the methods available those of Raeder & Broda (*Lett. app. Microbiol.* **1**: 17, 1985) and Zolan & Pukkila (*Molec. Cell. Biol.* **6**: 195, 1986) are now well-tested. The DNA can then be studied by a variety of methods:

DNA-DNA hybridisation involves the separation of the DNA from two fungi to be compared into separate strands and then noting the extent to which they can reassociate; the extent of successful pairing is expressed as % DNA relatedness; formerly used extensively in yeasts and mould-fungi (see Kurtzman, 1985).

DNA fingerprinting relies on repetetive sequences dispersed through the genome and which have a high variability; valuable in the manufacture of genotype-specific probes for species or pathogenic races, e.g. in *Magnaporthe grisea* (Hamer *et al.*, *Proc. natn Acad. Sci., USA* **86**: 9981, 1989). Of particular use in characterization of individual strains and populations, especially in medical mycology (e.g. Girordin *et al., J. Infect. Dis.* **169**: 683, 1994).

DNA probes, labelled fragments of DNA that are used to identify particular regions of DNA in other organims, including fungi. Typically, a DNA probe will be used to generate RFLPs (q.v.) from DNA digestions. DNA probes may be constructed for regions specific to individual species and can provide valuable identification tools (e.g. Dobrowolski & O'Brien, *FEMS Microbiol. Letts* **113**: 43, 1993).

mol % G + C (guanine + cytosine, **GC**) contents in the DNA are determined by thermal denaturation profiles and expressed as a percentage; widely used in the 1970s (see Kurtzman, 1985) but now largely replaced by more sensitive methods; see base ratio.

PCR (polymerase chain reaction), a series of heating and cooling steps that allowd for the amplification of DNA. DNA is melted and primers attach to the separated single strands; new complementary strands are formed from the primers by the addition of dNTPs and heat stable DNA polymerase. Further heating melts this new dimer and the process is repeated through a number of cycles.

RFLP (restriction fragment length polymorphisms), used on both nuclear and mitochondrial DNA involves the use of restriction enzymes (e.g. *Hae*III,

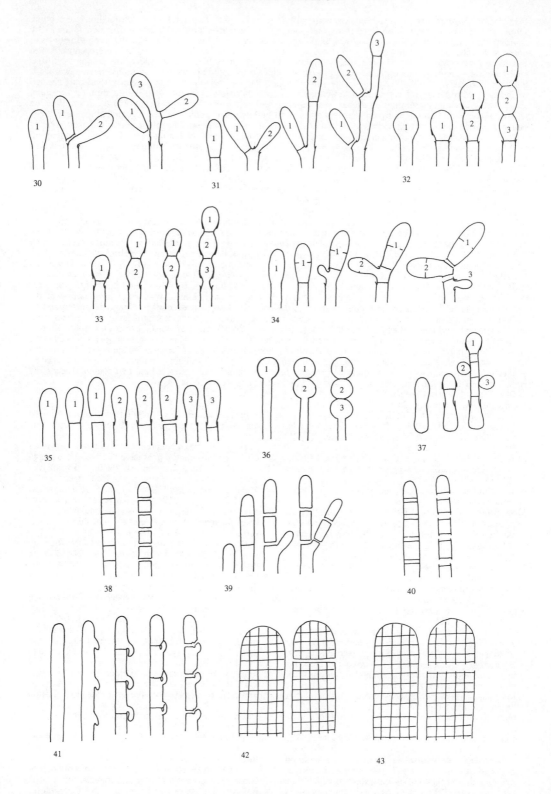

Fig. 26. Conidiogenous events (cc - conidiogenous cell). 30, conidal ontogeny holoblastic, delimitation by 1 septum, maturation by apical and diffuse wall-building, secession schizolytic and coincident with enteroblastic sympodial cc proliferation below the previous locus; subsequent conidia formed similarly but with holoblastic sympodial cc proliferation; 31, conidial ontogeny holoblastic, delimitation by 1 septum, maturation by apical and diffuse wall-building, secession schizolytic and coincident with enteroblastic sympodial cc proliferation below the previous conidiogenous locus, the sequence giving geniculate conidiophores; 32, conidial ontogeny holoblastic, with new inner walls continuous with all conidia laid down by diffuse wall-building, delimitation by 1 septum, loss of apical wall building followed by replacement continuous ring wall-building immediately below delimiting septum, the outer cc wall breaks between the first conidium and the cc to produce a variable collarette, followed by alternation of holoblastic conidial ontogeny by ring wall-building giving connected chains of conidia, maturation by diffuse wall-building, retrogressive delimitation, secession schizolytic; 33, conidial ontogeny holoblastic with new inner walls laid down by diffuse wall-building, delimitation by 1 septum, loss of apical wall-building followed by replacement ring wall-building immediately below delimiting septum, the outer cc wall breaks between the first conidium and the cc to produce a variable collarette, subsequent conidia formed by new inner walls for each conidium by ring wall-building giving connected chains of conidia, maturation by diffuse wall-building, retrogressive delimitation, secession schizolytic; 34, conidial ontogeny holoblastic, delimitation by 1 septum, secession schizolytic, enteroblastic sympodial cc proliferation below the previous locus and delimiting septum, the second and subsequent conidia formed from proliferations and delimited retrogressively, cc reduced in length with each conidium formed; 35, conidial ontogeny holoblastic, maturation by diffuse wall-building, delimitation by 1 septum, secession schizolytic, enteroblastic percurrent cc prolferation with retrogressive delimitation of next conidium, producing unconnected chains of conidia, the cc reduced in length with each conidium formed; 36, conidial ontogeny holoblastic, delimitation by 1 septum with loss of apical wall-building but replaced by diffuse wall-building below the previous conidium to form the next conidium which is retrogressively delimited giving an unconnected chain of conidia, secession schizolytic, cc reduced in length with each conidium formed; 37, conidial ontogeny holoblastic, delimitation by 1 septum with loss of apical wall-building, replaced by ring wall-building below the delimiting septum, outer wall of first conidium and cc breaks, followed by enteroblastic percurrent proliferation by ring wall-building, succeeding conidia holoblastic, delimited laterally and retrogressively, secession schizolytic, several loci per cc; 38, conidial ontogeny holothallic, ccs formed by apical wall-building coincident with conidial ontogeny, random delimitation by 1 septum at each end, no maturation during conidiogenesis, secession randomly schizolytic; 39, conidial ontogeny holothallic, ccs formed by apical wall-building coincident with conidial ontogeny, random delimitation by 1 septum at each end, no maturation during conidiogenesis, secession randomly schizolytic, cc proliferation holoblastic, irregular or sympodial, constituent cells conidiogenous; 40, same as 38 but conidial delimitaiton by 2 septa or separating cells at each end, secession rhexolytic; 41, conidial ontogeny holothallic, ccs formed in association with clamp connexions, random delimitation by septa in cc and the backwardly directed branch in the clamp connexion, maturation by diffuse and localized apical wall-building, secession randomly schizolytic, individual conidia comprised of part of the preceding and following clamp connexions; 42, conidial ontogeny holoblastic by simultaneous apical wall-building in adjacent cells, delimitation by septa in each of these cells, maturation by diffuse wall-building, secession simultaneous, multicellular, schizolytic, no cc proliferation; 43, conidial ontogeny holoblastic by simultaneous apical wall-building in adjacent cells, delimitation by septa in each of these cells, maturation by diffuse wall-building, followed by replacement apical wall-building in conidia to form additional conidia in connected chains, secession simultaneous, multicellular, rhexolytic, no cc proliferation.

*Msp*I) which cut the molecules into fragments at particular sites; the size of the fragments will vary, and these are then separated by electrophoresis (q.v.); electrophoresis of digested total cellular DNA will produce a 'smear' of fragments of different sizes and polymorphisms associated with individual genes or gene clusters will require hybridization to a labelled DNA probe (q.v.; e.g. Correll *et al.*, *Phytopath.* **83**1199, 1993). In some instances RFLPs can be detected without probes such as with PCR products or purified mitochondrial DNA (e.g. Varga *et al.*, *Can. J. Microbiol.* **4O**: 612, 1994).

RAPD (random amplified polymorphic DNA) analysis uses the polymerase chain reaction (PCR) with short primers to produce sufficient amounts of particular parts of a target DNA or RNA sequence that can then be compared in agarose gels; especially useful in studies of closely allied fungi where species-specific patterns can be found (Schaad *et al.*, *in* Hawksworth, 1994: 461), and when dealing with small samples (e.g. a single spore or old type collection).

DNA sequencing involves the comparision of the actual sequences of bases in particular parts of DNA (or RNA) molecules; the ribosomal gene clusters have been most used in phylogenetic studies by mycologists, particularly the genes coding for 5.8S, 18S and 28S rRNA (ribosomal RNA) genes and adjacent non-coding spacers (e.g. ITS; internally transcribed spacer); the method involves the use of the PCR method (see above), and the base sequences are determined either manually or by automated sequencing machines; the sequences obtained can be subjected to cladistic analyses, including already available sequences accessible from on-line databanks; a basic introduction to the method, which is now pivotal in phylogentic studies with fungi, is provided by Mitchell *et al.* (*Mycologist* **9**: 67, 1995). **Total sequencing** of DNA in fungus chromosomes have only been achieved to date in *Saccharomyces cerevisiae* (Dijon *et al.*, *Nature* **369**: 371, 1994).

In the case of all molecular studies it is advisable to deposit reference material of the fungi actually used in an appropriate Reference Collection.

DNA sequence libraries: EMBL (European Molecular Biology Laboratory, Heidelberg; datalib@embl), GenBank (National Institutes of Health, Bethesda, Md; info@ncbi.nlm.nih.gov); see also *J. Nucleic Acid Res.* **21**: 2963 (1993).

Lit.: Bennett & Lasure (Eds) (*Gene manipulations in fungi*, 1985; *More gene manipulations in fungi*, 1991), Bresinsky *et al.* (*Z. Mykol.* **53**: 303, 1987; DNA contents nuclei), Bruns *et al.* (*Ann. Rev. Ecol. Syst.* **22**: 525, 1991), Clutterbuck (*in* Gow & Gadd, *The growing fungus*: 255, 1995), Guthrie & Fink (Eds) (*Guide to yeast genetics and molecular biology*, [*Methods Enzym. Biol.* **194**], 1991), Hawksworth (Ed.) (*Identification and characterization of pest organisms*, 1994), Hibbett (*Trans. mycol. Soc. Japan* **33**: 533, 1992; *in* systematics), Kinghorn & Turner (Eds) (*Applied molecular genetics of filamentous fungi*, 1992), Kurtzman (*in* Samson & Pitt (Eds), *Advances in Penicillium and Aspergillus systematics*: 233, 1985, *Proc. Indian* *natn Acad. Sci., Pl. sci.* **97**: 185, 1987; DNA comparisons), Oliver (*in* Fox, *Principles of diagnostic techniques in plant pathology*: 153, 1993), Paterson & Bridge (*Biochemical techniques for fungi* [*IMI Techn. Handbk* 1], 1994), Reynolds & Taylor (Eds) (*The fungal holomorph*, 1993; examples use in systematics), Schots *et al.* (Eds) (*Modern assays for plant pathogenic fungi*, 1994), Rolfs *et al.* (Eds) (*Methods in DNA amplification*, 1994), Towner & Cockayne (*Molecular methods for microbial identification and typing*, 1993).

See Biotechnology, Genetics of fungi, Genetic engineering.

Molgosphaera Dumort. (1822) nom. dub. (Ascomycota, inc. sed.); used for diverse perithecioid fungi.

Molgosporidium Thor (1930) nom. dub. (? Fungi, inc. sed.).

Molinea Doub. & D. Pons (1975), Fossil fungi (Asterin.). 1 (Cretaceous), Colombia.

Molleriella G. Winter (1886), Elsinoaceae. 13 (on glandular hairs), trop.

Mölleropsis, see *Moelleropsis*.

Molliardia Maire & A. Tison (1911) = Tetramyxa (Plasmodiophor.).

Molliardomyces Paden (1984), Anamorphic Sarcoscyphaceae, 1.A1.10. 7, C. & N. Am.

Mollicarpus Ginns (1984), Coriolaceae. 1, S.E. Asia.

mollicute, see mycoplasma.

Mollisia (Fr.) P. Karst. (1871) nom. cons., Dermateaceae. *c.* 100, widespr. See Dennis (*Kew Bull.* 1950: 171, 1950), Le Gal & Mangenot (*Rev. Mycol.* **25**: 135, **26**: 263, **31**: 3, 1960-66), Thind & Sharma (*Bibl. Mycol.* **91**: 221, 1983; India), Wallwork & Spooner (*TBMS* **91**: 703, 1988; anamorph). Anamorphs *Phialophora*-like, *Pseudocercosporella*.

Mollisiaster Kirschst. (1939), Leotiales (inc. sed.). 1, Eur. = Microscypha (Hyaloscyph.) fide Arendholz (*in litt.*).

Mollisiella Boud. (1885) = Mollisia (Dermat.) fide Korf & Zhuang (*SA* **6**: 139, 1987).

Mollisiella (W. Phillips) Massee (1895) ≡ Unguiculariopsis (Leot.).

Mollisina Höhn. ex Weese (1926), Hyaloscyphaceae. 4, Eur. See Arendholz (1979).

Mollisinopsis Arendh. & R. Sharma (1984), Leotiaceae. 1, India.

Mollisiopsis Rehm (1908), Dermateaceae. 6, Eur. See Dennis (*Persoonia* **2**: 171, 1962), Graddon (*TBMS* **58**: 153, 1972), Svrček (*Česká Myk.* **41**: 88, 1987).

Molybdoplaca Nieuwl. (1916) ≡ Steinera (Coccocarp.).

Monacrosporiella Subram. (1978), Mitosporic fungi, 1.C1.10. 1, USA.

Monacrosporium Oudem. (1885), Mitosporic fungi, 2.C2.10. 11, widespr. See Subramanian (*J. Ind. bot. Soc.* **42**: 292, 1963; key), Cooke (*TBMS* **50**: 517, 1967; key 7 nematode-trapping spp.; **53**: 475, 1969), Liu & Zhang (*MR* **98**: 862, 1994; checklist of spp.).

Monadelphus, see *Monodelphus*.

monads (*Monadineae*), a group of flagellate *Protozoa* some of which (including *Hydromyxales*) were compiled by Saccardo (*Syll. Fung.* **7**: 453, 1888).

monandrous (of oospores), formed when only one functioning antheridium is present. Cf. polyandrous.

Monascaceae J. Schröt. (1894), Eurotiales. 2 gen. (+ 5 syn.), 7 spp. Ascomata small, cleistothecial, globose, thin-walled; peridium composed of flattened hyphae; interascal tissue absent; asci evanescent at a very early stage; ascospores hyaline, aseptate, ellipsoidal, thick-walled. Anamorphs prominent, hyphomycetous (*Basipetospora*). Saprobes, esp. on substrata with high water tension, e.g. dried foods.

Lit.: Landvik & Eriksson (*SA* **12**: 34, 1993; posn.).

Monascella Guarro & Arx (1986), Onygenaceae. 1, Spain.

Monascostroma Höhn. (1918) = Hendersonia (Mitosp. fungi) fide Holm & Holm (*in* Larsen *et al.* (Eds), *Arctic and alpine mycology* **2**: 109, 1987).

Monascus Tiegh. (1884), Monascaceae. 6, widespr. *M. ruber* turns silage light red; *M. purpureus* produces ang-kak and other Asian foods and drinks. See Cannon *et al.* (*MR* **99**: 659, 1995; key), Hawksworth & Pitt (*Austr. J. Bot.* **31**: 51, 1983; key), Bridge & Hawksworth (*Lett. Appl. Microbiol.* **1**: 25, 1985; biochem. tests), Wong & Chien (*Mycol.* **78**: 713, 1986; ultrastr.), Nishikawa *et al.* (*J. gen. app. Microbiol.* **34**: 467, 1988; proteinase, *J. Basic Microbiol.* **29**: 369, 1989; fatty acids, **33**: 331, 1993; isozymes). Anamorph *Basipetospora*.

monaxial, having one individual stem or axis.

Monera. Kingdom embracing *Bacteria* and *Cyanobacteria* in the five-Kingdom system of Whittaker (1969); see Kingdoms of fungi for current treatment.

Monerolechia Trevis. (1857) = Buellia (Physc.) fide Hafellner (*Beih. Nova Hedw.* **62**, 1979).

Monilia Link (1809) nom. rej. = Monilia (Mitosp. fungi).

Monilia Hill ex Pers. (1822) nom. rej. = Monilia (Mitosp. fungi).

Monilia Hill (1832) nom. rej. = Monilia (Mitosp. fungi).

Monilia Bonord. (1851) nom. cons., Mitosporic fungi, 1.A1.3. Teleomorph *Neurospora. c.* 10, widespr. See Turian & Bianchi (*Arch. Mikrobiol.* **77**: 262, 1971; conidiation), Donk (*Taxon* **12**: 266, 1963; nomencl.), Anon. (*Taxon* **23**: 419, 1974; nomencl.), Funk (*CJB* **65**: 23, 1987; pleoanamorphism *M. versiformis*). Formerly widely used by medical mycologists for *Candida*; see candidiasis.

Moniliaceae (obsol.). Mitosporic fungi (hyphomycetes) with hyaline-coloured or pale hyphae and/or conidia.

moniliaceous (of mycelium, spores, etc.), hyaline or brightly coloured; mucedinaceous.

Moniliales = blastomycetes + hyphomycetes.

moniliasis, see candidiasis.

Moniliella Stolk & Dakin (1966), Mitosporic fungi, 1.A2.3+38. 2, Eur. See de Hoog (*Stud. Mycol.* **19**, 1979; key).

moniliform (**monilioid**), having swellings at regular intervals like a string of beads.

Moniliger Letell. [not traced] nom. dub. ? = Penicillium (Mitosp. fungi).

Moniliites Babajan & Tasl. (1973), Fossil fungi. 1 (Tertiary), former USSR.

Monilinia Honey (1928), Sclerotiniaceae. 22, widespr. See v. Arx (*Sydowia* **34**: 13, 1981; anamorphs), Batra (*Mycotaxon* **8**: 476, 1979; N. Am.), Harada (*Bull. Fac. Agr. Hirosaki Univ.* **27**: 30, 1977; Japan), Willetts *et al.* (*J. gen. Microbiol.* **103**: 77, 1977; chemotaxonomy), Buchwald (*Friesia* **11**: 287, 1987; key 12 spp.), Batra (*World species of Monilinia*, 1991).

Moniliophthora H.C. Evans, Stalpers, Samson & Benny (1978), Mitosporic fungi, 1.A1.23. 1 (with dolipore septa), trop. *M. roreri* (synanamorph *Monilia roreri*) on *Cacao* (pod rot). See Evans (*Phytopath. Pap.* **24**, 1981).

Moniliopsis Ruhland (1908) ? = Rhizoctonia (Mitosp. fungi) fide v. Arx (*Genera of fungi sporulating in pure culture*, 1970). But see Moore (*Mycotaxon* **29**: 91, 1987; gen. *Rhizoctonia*-like fungi).

Monilites Pampal. (1902), Fossil fungi. 1 (Miocene), Italy.

Monilochaetes Halst. ex Harter (1916), Mitosporic fungi, 2.A2.1. 1, trop., esp USA. *M. infuscans* on *Ipomoea batatas* (sweet potato scurf).

Monka Adans. (1763) = Verpa (Morchell.).

mono- (prefix), one.

Monoblastia Riddle (1923), Monoblastiaceae (L). 3, trop. See Eriksson (*SA* **11**: 178, 1993), Harris (*Some Florida lichens*, 1990).

Monoblastiaceae Walt. Watson (1929), Dothideales (L). 2 gen. (+ 5 syn.), 48 spp. Thallus crustose, often immersed; ascomata sometimes aggregated, usually clypeate, ± globose, papillate, the ostiole sometimes lateral; interascal tissue of narrow trabeculate pseudoparaphyses, true paraphyses not developing; asci cylindrical, fissitunicate, with a large ocular chamber but without an apical ring; ascospores usually 1-septate, hyaline, smooth or ornamented, sometimes with a sheath. Anamorphs pycnidial where known. Lichenized with green algae, widespr.
   Lit.: Eriksson & Hawksworth, *SA* **11**: 178, 1993), Harris (*Some Florida lichens*, 1990).

monoblastic (of a conidiogenous cell), producing a blastic conidium at one locus.

Monoblepharella Sparrow (1940), Gonapodyaceae. 5, C. & N. Am. See Sparrow (1960: 721; key).

Monoblepharidaceae A. Fisch. (1892), Monoblepharidales. 1 gen. (+ 2 syn.), 11 spp. Thallus differentiated into a well developed hyphal, vegetative system bearing numerous reproductive structures, remaining in oogonium or encysting on the opening of the oogonium; male gamete completely engulfed at fertilization, oogamous.

Monoblepharidales, Chytridiomycetes. 3 fam., 4 gen. (+ 2 syn.), 19 spp. Thallus hyphal with oogamous reproduction (the only chytrids with non-motile female gametes and smaller, motile, male gametes), or thallus simple and no sexual stage known; zoospores possessing a non-flagellated centriole parallel to the kinetosome, a plate of electron opaque material in the core of the axonema at the of the transitional zone, a striated disk extending part way aroun the kinetosome, microtubules arising from the disk and extending into the cytoplasm, rumposome backed by microbody, and numerous lipid globules; aquatic saprobes. Fams:
(1) **Gonapodyaceae**.
(2) **Monoblepharidaceae**.
(3) **Oedogoniomycetaceae**.
   Lit.: Sparrow (1960: 713, *in* Ainsworth *et al.* (Eds), *The Fungi* **4B**: 85, 1973), Gauriloff *et al.* (*CJB* **58**: 2098, 1980; ultrastr.), Barr (1990), Mollicone & Longcore (*Mycol.* **86**: 615, 1994; zoospore ultrastr.).

Monoblephariopsis Laib. (1927) = Monoblepharis (Monobleph.) fide Sparrow (1960).

Monoblepharis Cornu (1871), Monoblepharidaceae. 11, Eur., N. Am. See Sparrow (1960: 727; key), Perrott (*TBMS* **38**: 247, 1955).

monocarpic (of *Exobasidium* infections), circumscribed and annual (Nannfeldt, 1981: 10), cf. polycarpic, surculicolous.

monocentric (of a chytrid thallus), having one centre of growth and development; see polycentric and cf. reproductocentric.

monocephalic (adj. **monocephalous**), 1-headed.

Monocephalis Bainier (1882) = Syncephalis (Piptocephalid.) fide Benjamin (1959).

Monoceras Guba (1961) [non Gothan (1909), fossil *Phanerogamae*] = Seimatosporium (Mitosp. fungi) fide Sutton (*Mycol. Pap.* **97**, 1964).

Monochaetia (Sacc.) Allesch. (1902), Mitosporic fungi, 6.C2.19. *c.* 100, widespr. See Guba (*Monograph of Monochaetia and Pestalotia*, 1961), Sutton (1980), Nag Raj (1993).

Monochaetiella E. Castell. (1943), Mitosporic fungi, 6.A1.15. 1, Afr., India, Australia.

Monochaetiellopsis B. Sutton & DiCosmo (1977), Anamorphic Dothioraceae, 6.B1.?. Teleomorph

Hypnotheca. 2, widespr. See Punithalingam (*Nova Hedw.* **54**: 255, 1992; nuclei in conidial appendages).

Monochaetina Subram. (1961) = Bleptosporium (Mitosp. fungi) fide Sutton (*Mycol. Pap.* **88**, 1963).

Monochaetinula Muthumary, Abbas & B. Sutton (1986), Mitosporic fungi, 4.C2.19. 5, trop. See Nag Raj (*Coelomycetous anamorphs*, 1993).

Monochaetites Babajan & Tasl. (1973), Fossil fungi. 1 (Tertiary), former USSR.

Monochaetopsis Pat. (1931), Mitosporic fungi, ?.?.?. 1, N. Afr.

Monochytrium Griggs (1910) ? = Olpidium (Olpid.) fide Fitzpatrick (1930).

Monocillium S.B. Saksena (1955), Anamorphic Trichosphaeriaceae, 1.A1/2.15. Teleomorph *Niesslia. c.* 15, widespr. See Gams (*Cephalosporium-artige Schimmelpilze*, 1971; key 12 spp.).

monoclinous, having the antheridium on the oogonial stalk (Fig. 28A-E). Cf. androgynous.

Monoconidia Roze (1897) ? = Acremonium (Mitosp. fungi).

Monodelphus Earle (1909) = Omphalotus (Paxill.) fide Singer (1951).

Monodia Breton & Faurel (1970), Mitosporic fungi, 4.C1.1. 1 (coprophilous), Chad.

Monodictyites Barlinge & Paradkar (1982), Fossil fungi (Mitosp. fungi). 1 (Cretaceous), India.

Monodictys S. Hughes (1958), Mitosporic fungi, 3.D2.1. 14, widespr. See Rao & de Hoog (*Stud. Mycol.* **28**: 25, 1986; key).

Monodidymaria U. Braun (1994), Mitosporic fungi, 1.C1.1. 3, Canada, India.

Monodisma Alcorn (1975), Mitosporic fungi, 1.C1.10. 1, widespr.

monoecism (adj. **monoecious**), having the male and female sex organs on one thallus; cf. dioecism, heterothallism, homothallism.

monogeocentric, see reproductocentric.

Monogrammia F. Stevens (1917) = Titaea (Mitosp. fungi) fide Damon (*J. Wash. Acad. Sci.* **42**: 365, 1952).

Monographella Petr. (1924), ? Hyponectriaceae. 4, widespr. See Müller (*Rev. Mycol.* **41**: 129, 1977; *M. nivalis*), Parkinson *et al.* (*TBMS* **76**: 59, 1981; *M. albescens*, rice leaf scald), Subramanian & Bhat (*Rev. Mycol.* **42**: 293, 1978; ontogeny). Anamorph *Microdochium*.

Monographos Fuckel (1875) = Scirrhia (Dothid.). See Eriksson & Hawksworth (*SA* **11**: 178, 1993), Holm & Holm (*Bot. Notiser* **131**: 97, 1978).

Monographus Clem. & Shear (1931) ≡ Monographos (Dothid.).

Monoicomyces Thaxt. (1900), Laboulbeniaceae. 41, widespr.

monokaryon, see haplont.

monokaryotic, having genetically identical haploid nuclei; cf. dikaryon.

Monoloculia Hara (1927) [ante, not traced] nom. dub. (Ascomycota, inc. sed.). Anamorph *Japonia*.

monomitic, see hyphal analysis.

monomorphic, having one structure or form; not pleomorphic.

monomycelial (of an isolate), from one spore or hyphal tip.

Monomyces Battarra ex Earle (1909) = Tricholoma (Tricholomat.) fide Singer (1951).

Mononema Balbiani (1889), ? Trichomycetes (inc. sed.). 3, Afr., Eur. See Manier *et al.* (*Biol. Gabon.* **8**: 323, 1972).

mononematous (of conidiophores), solitary or in tufts or loose fascicles; cf. synnematous.

monophagy (adj. **monophagous**) (of *Chytridiales*), the condition of having the thallus in one host cell; the opposite of polyphagy (q.v.), in which the thallus branches invade more than one host cell.

monophialidic (of a conidiogenous cell), having one locus through which conidia are produced. Cf. polyphialidic.

monophyllous (of foliose lichens), having a single leaf-like thallus.

monoplanetism (of zoospores in oomycetes), the condition of having one motile phase, with no resting period.

Monoplodia Westend. (1859) = Coniothyrium (Mitosp. fungi) fide Saccardo & Trotter (*Syll. Fung.* **22**, 1913).

monopodial, a type of branching in which a persistant main axis gives off branches, one at a time and frequently in alternate or spiral series.

Monopodium Delacr. (1890) = Acremoniella (Mitosp. fungi) fide Mason (*Mycol. Pap.* **3**: 29, 1933).

Monoporisporites Hammen (1954), Fossil fungi (*f. cat.*). 15 (Cretaceous, Tertiary), Colombia, India, USA.

Monopus Theiss. & Syd. (1915) = Rosenscheldiella (Ventur.) fide Hansford (1946).

Monopycnis Naumov (1916) = Cytospora (Mitosp. fungi) fide Petrak (*Ann. Myc.* **23**: 1, 1925).

monoreproductocentric, see reproductocentric.

Monorhiza Theiss. & Syd. (1915) = Rhagadolobium (Parmular.) fide Müller & v. Arx (1962).

Monorhizina Theiss. & Syd. (1915), Microthyriaceae. 1, Sri Lanka.

Monospermella Speg. (1924) = Psorotheciopsis (Gomphill.).

monospermous, 1-spored.

Monospora Metschn. (1884) [non Hochst. (1841), *Algae*] ≡ Metschnikowia (Metschnikow.).

Monosporascus Pollack & Uecker (1974), ? Sordariales (inc. sed.). 3 (thermophilic), Asia, N. Am., N. Afr. See Hawksworth & Ciccarone (*Mycopath.* **66**: 147, 1978).

monospore, used by Corner (*New Phytol.* **46**: 195, 1947) for the one spore maturing on a 2-spored basidium normally bearing 2 dispores (q.v.).

Monosporella Keilin (1920) ≡ Metschnikowia (Metschnikow.).

Monosporella S. Hughes (1953) ≡ Monotosporella (Mitosp. fungi).

monosporic (adj. **monosporous**), 1-spored.

Monosporidium Barclay (1888), Phakopsoraceae. 3 (on *Euphorbiaceae, Rubiaceae*), India. Telial aeciospores produce basidia on germination. See Buriticá (*Rev. Acad. Colomb. Cienc.* **18**(no. 69): 131, 1991).

Monosporiella Speg. (1918), Mitosporic fungi, 1.A1.?. 1, Argentina. See Deighton & Pirozynski (*Mycol. Pap.* **128**, 1972). ? = Acremonium (Mitosp. fungi) fide Sutton (*in litt.*).

Monosporium Bonord. (1851) nom. illegit. (Mitosp. fungi) fide Hughes (*CJB* **36**: 727, 1958).

Monosporonella Oberw. & Ryvarden (1991), Ceratobasidiaceae. 1 (in termite nests), Zambia.

Monostachys Arnaud (1954) = Chloridium (Mitosp. fungi) fide Gams & Holubová-Jechová (1976).

Monostichella Höhn. (1916), Mitosporic fungi, 6.A1.15. 9, widespr. See v. Arx (*Bibl. Mycol.* **24**, 1970).

monostichous, in one line or series.

Monothecium Lib. [not traced] [non Hochst. (1842), *Acanthaceae*] nom. nud. et dub. = Mastigosporium (Mitosp. fungi) fide Saccardo (*Syll. Fung.* **4**: 220, 1886).

Monotospora Corda (1837) nom. dub. (Mitosp. fungi) fide Carmichael (*CJB* **40**: 1137, 1962).

Monotospora Sacc. (1880) ≡ Acrogenospora (Mitosp. fungi).

Monotosporella S. Hughes (1958) = Brachysporiella (Mitosp. fungi) fide Ellis (1959). Used by Rao & de Hoog (*Stud. Mycol.* **28**: 6, 1986).

Monotoyella Castell. & Chalm. (1913) nom. dub. (Mitosp. fungi) fide de Hoog & Hermanides-Nijhof (*Stud. Mycol.* **15**: 186, 1977).

monotretic, see tretic.

Monotretomyces Morgan-Jones, R.C. Sinclair & Eicker (1987) = Corynesporopsis (Mitosp. fungi) fide Sutton (*in litt.*).

monotrichous (monotrichiate) (of bacteria), having one polar flagellum.

Monotrichum Gäum. (1922), Mitosporic fungi, 1.B1.?. 1, Celebes, India. See Carmichael *et al.* (1980; relationships).

Monotropomyces Costantin & Dufour (1921) nom. dub. (Mitosp. fungi) fide de Hoog & Hermanides-Nijhof (*Stud. Mycol.* **15**: 185, 1977).

monotypic, having only one representative, as a genus having only one species.

monoverticillate (of a penicillus), composed of phialides only.

Montagne (Jean Pierre François Camille; 1784-1866). A medical man in Napoleon's army; later, when living in Paris, gave his attention to fungi. Chief writings: papers on the fungi of France and other countries (incl. 'Centuries des plantes cellulaires' in *Ann. Sci. nat., Bot.* 1837-60 [reprint 1970]; see Stafleu, *Taxon* **19**: 633, 1970) and on plant diseases; *Sylloge generum specierumque Cryptogamarum* (1856), See also Stafleu & Cowan (*TL-2* **3**: 557, 1981).

Montagnea Fr. (1836), Podaxaceae. 2, warm dry regions. See Reid & Eicker (*S. Afr. J. Bot.* **57**: 161, 1991).

Montagneaceae, see *Podaxaceae*.

Montagnella Speg. (1881) Dothideales (inc. sed.). 9, widespr. Based on immature ascomata fide v. Arx & Müller (1975). See also *Gillotia*.

Montagnellina Höhn. (1912) ? = Guignardia (Mycosphaerell.) fide v. Arx & Müller (1954).

Montagnina Höhn. (1910) ≡ Gibbera (Ventur.) fide v. Arx & Müller (1975).

Montagnites Fr. (1838) ≡ Montagnea (Podax.).

Montagnula Berl. (1896), Phaeosphaeriaceae. 12, widespr. See Aptroot (*Nova Hedw.* **60**: 325, 1995; 1-septate spp.), Crivelli (*Ueber die heterogene Ascomycetengattung Pleospora*, 1983; key 6 spp.), Holm (*SA* **11**: 29, 1992; nomencl.), Leuchtmann (*Sydowia* **37**: 75, 1985; key 3 spp.).

Montemartinia Curzi (1927) = Chaetosphaeria (Lasiosphaer.) fide Kohlmeyer & Kohlmeyer (*Marine mycology*, 1979).

Montinia A. Massal. (1855) [non Thunb. (1776), *Saxifragaceae*] ≡ Pyrenocarpon (Lichin.).

Montoyella Castell. & Chalm. (1907) nom. dub. (Mitosp. fungi) fide de Hoog & Hermanides-Nijhof (*Stud. Mycol.* **15**: 186, 1977).

Moorella P. Rag. Rao & D. Rao (1964), Mitosporic fungi, 1.F2.10. 1, India.

Moralesia Urries (1956), Mitosporic fungi, 8.G1.?. 1, Canary Isles.

Moravecia Benkert, Caillet & Moyne (1987), Otideaceae. 1, Eur. See Caillet & Moyne (*Bull. Soc. Hist. nat. Doubs* **84**: 9, 1991).

Morchella Dill. ex Pers. (1794), Morchellaceae. *c.* 28, esp. temp.; the Morels, important edible fungi. See Bunyard *et al.* (*Mycol.* **86**: 762, 1994; molec. syst.), Gilbert (*Mycol.* **52**: 201, 1961; submerged culture), Waraitch (*Kavaka* **4**: 69, 1978; India), Hein (*BSMF* **82**: 422, 1966; trop. spp.), Jacquetant (*Les Morilles*, 1984; monogr., 28 col. pl.), Galli (*Il Genre Morchella in Lombardia*, 1984; 32 col. pl.), Royse & May (*Biochem. Syst. & Ecology* **18**: 475, 1990; allozymes),

Weber (*A morel hunters companion* .... , 1988), Volk & Leonard (*MR* **94**: 399, 1990; cytology).

Morchellaceae Rchb. (1834), Pezizales. 3 gen. (+ 8 syn.), 38 spp. Ascomata large, apothecial, usually distinctly stalked, the hymenium usually everted, often lobed or corrugated, usually darker than the stalk; interascal tissue well-developed; asci cylindrical, persistent, not blueing in iodine; ascospores ellipsoidal, hyaline, smooth, eguttulate but with a minutely guttulate capitate appendage at each apex, multinucleate. Anamorphs hyphomycetous. Saprobic on soil and leaf mould, esp. temp.

Moreaua Liou & H.C. Cheng (1949), = Tolyposporium (Ustilagin.) fide Vánky (1987).

morel, the edible ascoma of *Morchella*; **false -**, see lorchel.

Morella Pérez Reyes (1964) [non Lour. (1790), *Myricaceae*], ? Olpidiaceae. 1, USA. See Karling (*Bull. Torrey bot. Cl.* **99**: 223, 1972).

Morellus Eaton (1818) ≡ Phallus (Phall.).

Morenoella Speg. (1885) = Lembosia (Asterin.) fide v. Arx & Müller (1975).

Morenoina Theiss. (1913), ? Asterinaceae. 10, widespr. See Ellis (*TBMS* **74**: 297, 1980; key Br. spp.).

Morfea Roze (1867) = Polychaeton (Mitosp. fungi) fide Hughes (*Mycol.* **68**: 792, 1976).

Morfea (G. Arnaud) Cif. & Bat. (1963) = Polychaeton (Mitosp. fungi) fide Punithalingam (*Mycol. Pap.* **149**, 1981).

Morganella Zeller (1948), Lycoperdaceae. 9, widespr., mostly trop. See Ponce de Leon (*Fieldiana*, Bot. **34**: 27, 1971), Kreisel & Dring (*Feddes Repert.* **74**: 109, 1967; key 7 spp.).

moriform, like a mulberry (*Morus*) fruit in form.

Morilla Quél. (1886) ≡ Morchella (Morchell.).

Morinia Berl. & Bres. (1889), Mitosporic fungi, 6.D2.?. 1, Eur., Asia.

Moriola Norman (1872), Moriolaceae (±L). 4, Eur. See Bachman (*Nyt mag. naturvid.* **64**: 170, 1926), Eriksson (1981), Triebel (*Bibl. Lich.* **35**, 1989).

Moriolaceae Zahlbr. (1903), Ascomycota (inc. sed., ±L). 1 gen., 4 spp. Mycelium dark, superficial, ? sometimes associated with algal cells; stromata absent; ascomata perithecial, ± flattened to globose, superficial, black, ± thin-walled; interascal tissue of ? narrow anastomosing pseudoparaphyses; asci cylindrical, persistent, ± thick-walled, with separable wall layers, fissitunicate, with an indistinct ocular chamber surrounded by an inconspicuous I- ring; ascospores brown, transversely septate, without a sheath. Anamorphs hyphomycetous. Saprobic on bark, or lichenized with green algae, Eur.

Poorly known; the only species with a modern description probably does not belong to *Moriola*.

Moriolomyces Cif. & Tomas. (1953), Dothideales (inc. sed.). 1, Eur. See Eriksson (*Opera Bot.* **60**, 1981), Eriksson & Hawksworth (*SA* **9**: 27, 1991).

Moriolopis Norman ex Keissl. (1927) nom. inval. = Melanomma (Melanommat.) etc. fide Keissler (*Rabenh. Krypt.-Fl.* **9**, 1(1), 1933).

Moristroma A.I. Romero & Samuels (1991), ? Dacampiaceae. 1, Argentina.

Morobia E. Horak (1979), Agaricaceae. 1, Papua New Guinea.

Moronopsis Delacr. (1891) ? = Cheirospora (Mitosp. fungi) fide Sutton (1977).

Morosporium Renault & Roche (1898), Fossil fungi (Mitosp. fungi). 1 (Eocene), France.

morph, form.

morphotype, a group of morphologically differentiated individuals of a species of unknown or of no taxonomic significance.

Morqueria Bat. & H. Maia (1963) = Cirsosia (Asterin.) fide v. Arx & Müller (1975).

Morrisiella Saikia & A.K. Sarbhoy (1985) [non Aellen (1938), *Chenopodiaceae*] = Janetia (Mitosp. fungi) fide Sutton (*in litt.*).

Morrisographium M. Morelet (1968), Mitosporic fungi, 2.C2.1. 7, N. Am., former USSR. See Illman & White (*CJB* **63**: 423, 1985; key), Murvanishvili & Svanidze (*Soobshch. Akad. Nauk Gruz.* **137**: 573, 1990).

Morthiera Fuckel (1870) = Entomosporium (Mitosp. fungi) fide Saccardo (*Syll. Fung.* **3**: 657, 1884).

Mortierella Coem. (1863), Mortierellaceae. *c.* 90 (mainly in soil), widespr. See Gams (*Persoonia* **9**: 381, 1977; key spp. in cult.), Kuhlman (*Mycol.* **64**: 325, 1972, **67**: 678, 1975; zygospores), Ansell & Young (*TBMS* **91**: 221, 1988; zygospores), Benjamin (*Aliso* **9**: 157, 1978), Chen (*Mycosystema* **5**: 23, 1992; Chinese spp.).

Mortierellaceae A. Fisch. (1892), Mucorales. 7 gen. (+ 7 syn.), 106 spp. Sporangia and sporangiola typically with columella absent or rudimentary; zygospores smooth or angular, borne on opposed suspensors.

Morularia Nann. (1925) ≡ Heterobotrys (Seurat.).

mosaic fungus, a network resembling disorganized dermatophyte mycelium sometimes seen in skin scales cleared in potassium hydroxide. An artifact, fide Weidman, who first described the effect in 1927, or the extracellular deposit of a dermatophyte (fide Dowding, *Arch. Derm. Chicago* **66**: 470, 1952).

Moschomyces Thaxt. (1894) = Compsomyces (Laboulben.) fide Tavares (1985).

Moserella Pöder & Scheuer (1994), ? Sclerotiniaceae. 1, Austria.

Mosigia Fr. (1845) [non Spreng. (1826), *Compositae*] = Rimularia (Rimular.) fide Hertel & Rambold (*Bibl. Lich.* **38**: 145, 1990).

Mothesia Oddo & Tonolo (1967) = Claviceps (Clavicipit.).

mould, (1) a microfungus having a well-marked mycelium or spore mass, esp. an economically important saprobe; **anther -** of clover, *Botrytis anthophila*; **black -**, *Aspergillus niger*; **blue -**, *Penicillium*; of apple, *P. expansum*; of citrus, *P. italicum*; of tobacco, *Peronospora hyoscyami* (syn. *P. tabacina*; Shepherd, *TBMS* **55**: 253, 1970), McKeen (*Blue mold of tobacco*, 1989); **bread -**, *Chrysomilia sitophila*; also used of *Mucorales* on bread; **green -** of citrus, *Penicillium digitatum*; **grey -**, *Botrytis cinerea*; of snowdrop, *B. galanthina*; **pin -**, *Mucor* and other *Zygomycetes*; **plaster -**, brown, *Papulospora byssina*; white, '*Oospora*' (q.v.) *fimicola*; **slime -s**, *Myxomycetes*; **snow -s**, low temp. tolerant pathogens growing on unfrozen soil surface below snow cover, causing diseases of winter cereals, grasses and forage legumes, *Monographella nivalis* (pink snow mould), *Sclerotinia borealis*, *Typhula* spp.; **sooty -**, *Atichiaceae*, *Capnodiales*, etc.; **tomato leaf -**, *Fulvia fulva*; **water -s**, aquatic *Chytridiomycetes* and *Oomycetes*, esp. *Saprolegniales*; **white -** of sweet pea, *Hyalodendron album*; (2) see mildew (2). Illman (*Mycol.* **62**: 1214, 1970) advocated an American usage 'mould' for a fungus, 'mold' for a shape.

Mouliniea C.P. Robin (1853), Ascomycota (inc. sed.). 1, EUr.

Mounting media.

Lactophenol (Amann, 1896):

Phenol (pure crystals)	20.0 g
Lactic acid (S.G. 1.21)	20.0 g
Glycerol	40.0 g
Water	20.0 ml

And a little dye, such as cotton blue, if desired. Phenol is now regarded as a possible carcinogen; this can be omitted from the above to reduce the hazard.

'Necol' (cellulose acetate) technique for mounting microfungi:
See Ellis (*TBMS* **33**: 22, 1950).

Andre & Hoyer's fluid:
Arabic gum or gum guiac	15.0 g
Chloral hydrate	100.0 g
Glycerol	10.0 g
Water	25.0 ml

(Cunningham, *Mycol.* **64**: 1906, 1972)

Glycerine jelly:
Gelatin	1.0 g
Glycerol	7.0 g
Water	6.0 ml

With the addition of phenol, 1 per cent.
Water (distilled or tap): is also a valuabe mounting medium, particularly for observing gelatinous material and non-cellular appendages.
See also stains.

mouse favus, a skin disease in humans (*Trichophyton quinckeanum*).

Moutoniella Penz. & Sacc. (1901), Rhytismataceae. 1, Java. See Sherwood (*Mycotaxon* **5**: 1, 1977).

Mrakia Y. Yamada & Komag. (1987), Filobasidiaceae. 4, Antarctica.

MSDN. Microbial Strain Data Network (Cambridge, UK); started 1985; see Krichevsky *et al.* (*in* Hawksworth & Kirsop, 1988: 31).

Mucedinaceae (obsol.) = Moniliaceae (Mitosporic fungi).

mucedinaceous, see moniliaceous.

mucedinous, white or pale in colour and mould-like; mucedinoid.

Mucedites Bertrand & Renault (1896), Fossil fungi (mycel.). 1 (Carboniferous), France.

Mucedo Pers. (1794) ≡ Mucor (Mucor.).

Muchmoria Sacc. (1906) Fungi (inc. sed.) fide Hughes (*CJB* **36**: 787, 1958).

Mucidula Pat. (1887) = Oudemansiella (Tricholomat.) fide Singer (1951).

mucilaginous, sticky when wet; slimy.

Mucilago Battarra (1755), Didymiaceae. 1, cosmop.

Mucilopilus Wolfe (1979), Strobilomycetaceae. 6, widespr. See Wolf (*Mycol.* **74**: 36, 1982).

Muciporus Juel (1897) = Tulasnella (Tulasnell.) fide Juel (1914), Donk (1966).

Muciturbo P.H.B. Talbot (1989), Pezizaceae. 3 (hypogeous), Australia.

MUCL. Mycothèque de l'Université Catholique de Louvain (Louvain-la-Neuve, Belgium); founded 1894; an institute of the Catholic University of Louvain, with special funding for the genetic resource collection from the Belgian Science Policy Office.

Mucobasispora Mustafa & Abdul-Wahid (1990), Mitosporic fungi, 1.A2.35. 1 (from soil), Egypt.

Mucomassaria Petr. & Cif. (1932), Dothideales (inc. sed.). 1, C. Am. See Petrak (*Sydowia* **13**: 1, 1956).

Mucophilus Plehn (1920), ? Chytridiales (inc. sed.). 1 (in *Cyprinus*), Germany.

Mucor L. (1753) = Calicium (Calic.) p.p. and Coniocybe (Coniocyb.) p.p.

Mucor P. Micheli ex Fr. (1832) nom. rej. = Rhizopus (Mucor.) fide Kirk (*Taxon* **35**: 371, 1986).

Mucor Fresen. (1850) nom. cons., Mucoraceae. *c.* 50, cosmop. See Zycha *et al.* (1969), Schipper (*Stud. Mycol.* **17**: 48, 1978; key 39 spp.), Benjamin & Mehrotra (*Aliso* **5**: 235, 1963; azygosporic spp.), Mehrotra & Mehrotra (*Sydowia* **31**: 94, 1974;

azygosporic sp.), Schipper *et al.* (*Persoonia* **8**: 321, 1975; zygospore ornamentation), O'Donnell *et al.* (*CJB* **55**: 2712, 1977; azygospore ultrastr.), Stalpers & Schipper (*Persoonia* **11**: 39, 1980; zygospore ornamentation), James & Gauger (*Mycol.* **74**: 744, 1982; genetics), Hesseltine & Rogers (*Mycol.* **79**: 289, 1987; zygospore formation), Michailides & Spotts (*Mycologia* **80**: 837, 1988; zygospore germination), Ginman & Young (*MR* **93**: 314, 1989; azygospore ultrastr.), Schipper (*Stud. Mycol.* **31**: 151, 1989), Bärschi *et al.* (*MR* **94**: 373, 1991; selective medium), Orlowski (*Microb. Rev.* **55**: 234, 1991; dimorphism), Schipper & Samson (*Mycotaxon* **50**: 475, 1994), Watanabe (*Mycol.* **86**: 691, 1994; key homothallic spp.), Nagy *et al.* (*Curr. Genetics* **26**: 45, 1994; karyotype).

Mucoraceae Dumort. (1822), Mucorales. 20 gen. (+ 27 syn.), 122 spp. Sporangia columellate, specialized sporangiola absent; zygospores smooth to warty, borne on opposed, tongs-like or apposed, naked or appendaged suspensors; polyphyletic.

Mucorales, Zygomycetes. 13 fam., 56 gen. (+ 54 syn.), 299 spp. Asexual reproduction by multi-spored or few- (to one) spored sporangia (sporangiola), forcibly discharged in *Pilobolus*, sexual reproduction by zygospores; cosmop. saprobes (rarely mycoparasites), or few facultative parasites of plants or animals (incl. humans). Fams:
(1) **Chaetocladiaceae**.
(2) **Choanephoraceae**.
(3) **Cunninghamellaceae**.
(4) **Gilbertellaceae**.
(5) **Mortierellaceae**.
(6) **Mucoraceae**.
(7) **Mycotyphaceae**.
(8) **Phycomycetaceae**.
(9) **Pilobolaceae**.
(10) **Radiomycetaceae**.
(11) **Saksenaeaceae**.
(12) **Syncephalastraceae**.
(13) **Thamnidiaceae**.

Lit.: Benny (PhD thesis, Claremont-Graduate School, USA, 1973), Benny *et al.* (*Mycotaxon* **22**: 119, 1985), Hesseltine & Ellis (*in* Ainsworth *et al.* (Eds), *The Fungi* **4B**: 187, 1973), Zycha *et al.* (*Mucorales*, 1969), Benjamin (*in* Kendrick (Ed.), *The whole fungus*, 1979), Hesseltine (*Mycol.* **47**: 344, 1955, **57**: 149, 1965; *Mycologist* **5**: 162, 1991; food fermentations), Kirk (*Mycol. Pap.* **152**, 1984), Benny & Benjamin (*Mycol.* **83**: 713, 1992; **85**: 660, 1983), v. Arx (*Sydowia* **35**: 10, 1982), Scholer *et al.* (*Fungi pathogenic for humans and animals* **3A**: 9, 1983), Newsham & Gauger (*Exp. Mycol.* **8**: 314, 1984; heterokaryons), Mikawa (*in* Watanabe & Malla (Eds), *Cryptogams of the Himalaya* **1**: 77, 1988), Zhou *et al.* (*Mycosystema* **4**: 1, 1991; DNA base composition), Mikawa (*in* Nakaike & Malik (Eds), *Cryptogamic flora of Pakistan*: 119, 1992; **2**: 65, 1993), Orlowski (*in* Wessels & Meinhardt (Eds), *The Mycota* **1**: 143, 1994; dimorphism), Gooday (*in* Wessels & Meinhardt (Eds), *The Mycota* **1**: 401, 1994; sex hormones).

Mucoralites Patel (1979), Fossil fungi (Zygomycetes). 1 (Tertiary), India.

Mucoricola Nieuwl. (1916) ≡ Piptocephalis (Piptocephalid.) fide Benjamin (1959).

Mucorites Mesch. (1898), Fossil fungi. 1 (Carboniferous), France.

mucormycosis, strictly, a disease of humans or animals caused by one of the *Mucorales* (e.g. *Absidia corymbifera*) but sometimes also applied to infections caused by members of the *Entomophthorales*. Cf. phycomycosis, zygomycosis.

Mucorodium K.M. Zalessky (1915), Fossil fungi (? Mucorales). 1 (Permian), former USSR.

Mucosetospora M. Morelet (1972), Mitosporic fungi, 8.A1.1. 1, France.

mucronate, pointed; ending in a short, sharp point. (Fig. 37.39)

Mucronella Fr. (1874), Hericiaceae. 10, widespr. See Corner (1950).

Mucronia Fr. (1849) [non *Mucronea* Benth. (1836), *Polygonaceae*] ≡ Mucronella (Heric.).

Mucronoporus Ellis & Everh. (1889), Hymenochaetaceae. 10, widesp. See Pegler (1973). = Phellinus (Hymenochaet.) fide Donk (1974).

Mucrosporium Preuss (1851) = Cladobotryum (Mitosp. fungi) fide de Hoog (*Persoonia* **10**: 33, 1978).

Muda Adans. (1763) nom. dub. (? Algae (Rhodophy.) or Fungi, L).

Muellerella Hepp (1862), Verrucariaceae. *c.* 12 (on lichens and *Hepaticae*), esp. N. temp. See Hawksworth (*Bot. Notiser* **132**: 283, 1979), Döbbeler & Triebel (*Bot. Jahrb. Syst.* **107**: 503, 1985; on *Hepaticae*).

Muellerellomyces Cif. & Tomas. (1953) ≡ Muellerella (Verrucar.).

Muellerites L. Holm (1968), Dothideales (inc. sed.). 1 (on *Juniperus*), Eur. See Casagrande (*Phytopath. Z.* **66**: 97, 1969). Anamorph *Aureobasidium*-like.

Muelleromyces Kamat & Anahosur (1968), Amphisphaeriaceae. 1, India. See Barr (1978).

mu-erh, the edible cultivated *Auricularia* sp., esp. *A. polytricha* (China) and *A. auricula* (Japan) (Chang & Hayes, 1978).

Muhria P.M. Jørg. (1987), Stereocaulaceae (L). 1, Scandinavia.

Muiaria Thaxt. (1914), Mitosporic fungi, 1.D2.1. 4 (on *Insecta*), Afr., Borneo.

Muiogone Thaxt. (1914), Mitosporic fungi, 3.D2.1. 1 (on *Insecta*), Afr.

Muirella R. Sprague (1959), Mitosporic fungi, 3.C1.10. 1, USA.

Mukagomyces S. Imai (1940) = Tuber (Tuber.) fide Trappe (1979).

Mukhakesa Udaiyan & V.S. Hosag. (1992), Amphisphaeriaceae. 1, India.

Müller Argoviensis (Jean; 1828-96). Worked with de Candolle in Geneva becoming Professor of botany there in 1868; did important work on tropical lichens and especial association with the herbarium of Delessert. Published 104 papers on lichens including many new species [reprint *Gesammelte Lichenologische Schriften*, 2 vols, 1967]. Main collection in **G**, other material in **BM**, **M**. See Briquet (*Bull. Herb. Boissier* **4**: 111, 1896), Grummann (1974: 633), Stafleu & Cowan (*TL-2* **3**: 628, 1981).

multi- (prefix), a great number; many; much.

multiallelic (of an incompatibility system), having more than 2 alleles per locus; cf. biallelic.

Multicellaesporites Elsik (1968), Fossil fungi (*f. cat.*). 42 (Tertiary), widespr.

Multicladium K.B. Deshp. & K.S. Deshp. (1966), Mitosporic fungi, 1.G2.?. 1, India.

Multiclavula R.H. Petersen (1967), Clavariaceae. 13 (3 L; some assoc. with *Myxomycetes* and *Musci*), widespr. See Petersen (*Am. midl. Nat.* **77**: 205, 1967; key), Oberwinkler (*Votr. bot. Ges.* N.F. **4**: 139, 1970), Petersen & Kantvilas (*Austr. Jl Bot.* **34**: 217, 1986), Poelt & Obermayer (*Herzogia* **8**: 289, 1990; bulbil diaspores).

multifid, having division into a number of parts or lobes.

multiguttulate, containing many oil-like drops.

Multipatina Sawada (1928), Mitosporic fungi, 1.?.?. 1, Taiwan. *M. citricola* ('leaf felt' of citrus).

multiperforate, of a septum with many pores connecting compartments. See Cole & Samson (*Patterns of development in conidial fungi*, 1979). (Fig. 35D).

multiseptate, having a number of septa.

multisporous, having a number of spores.

multivesicular bodies, small vesicles limited by a membrane which in *Sclerotinia fructigena* originate from the endoplasmic reticulum and are possibly related to extracellular enzyme secretion (Calonge *et al.*, *J. gen. Microbiol.* **55**: 177, 1969).

Mundkur (Balchendra Bhavanishankar; 1896-1952). Indian plant pathologist and mycologist. Indian Agricultural Research Institute, 1931-47; subsequently Directorate of Plant Protection, Quarantine and Storage, then Poona Univ. (Professor of botany). Noted for work on rusts and smuts. Major publications include *Fungi and plant disease* (1949), *Ustilaginales of India* (with Thirumalachar; 1952). See Joshi *et al.* (*Rev. Trop. Pl. Path.* **7**: 91, 1993), Mehta (*Indian Phytopath.* **5**: 1, 1953; portr., bibl.), Stafleu & Cowan (*TL-2* **3**: 660, 1981).

Mundkurella Thirum. (1944), Ustilaginaceae. 3 (on *Aralia*), India, N. Am, Korea. See Savile (*Mycol.* **67**: 273, 1975), Vanky (*MR* **94**: 269, 1990; key).

Munk pores. Small (*c.* 1 µm) pores, each surrounded by a ring of thickening, between cells of the ascoma wall in the *Nitschkeaceae*.

Munkia Speg. (1886), Mitosporic fungi, 8.A1.?. 2, S. Am. See Marchionatto (*Rev. argent. Agron.* **7**: 172, 1940).

Munkiella Speg. (1885), Polystomellaceae. 3, S. Am.

Munkiellaceae, see *Polystomellaceae*.

Munkiodothis Theiss. & Syd. (1915) = Rehmiodothis (Phyllachor.) fide Müller & v. Arx (1962).

Munkovalsaria Aptroot (1995), ? Dacampiaceae. 2, mainly trop.

Murangium Seaver (1951), Patellariaceae. 1, N. Am. See Eriksson & Hawksworth (*SA* **6**: 140, 1987).

Murashkinskija Petr. (1928) = Mytilinidion (Mytilinid.) fide Barr (*SA* **6**: 144, 1986).

Muratella Bainier & Sartory (1913) = Cunninghamella (Cunninghamell.) fide Hesseltine (1955).

Murciasporidium Thor (1930) Fungi (inc. sed.). 1 (in mites), Spitzbergen.

Muribasidiospora Kamat & Rajendren (1968), Exobasidiaceae. 3, India. See Rajendren (*Mycol.* **61**: 1159, 1969, *Mycopath.* **41**: 287, 1970; culture). = Exobasidium (Exobasid.) fide Donk (*Persoonia* **8**: 33, 1974).

muricate, rough with short, hard outgrowths.

Muricopeltis Viégas (1944), Micropeltidaceae. 1, Brazil. See Petrak (*Sydowia* **5**: 341, 1951).

Muricularia Sacc. (1877) nom. dub. (? Ascomycota, inc. sed.). See Sutton (*Mycol. Pap.* **141**, 1977).

muriculate, delicately muricate

muriform (of spores), see dictyospore.

muriform cell, a thick-walled, dark, muriform cell (frequently referred to as a sclerotic cell or body), found in tissues affected by chromoblastomycosis (Matsumoto *et al.*, *Mycol.* **76**: 244, 1984).

Murogenella Goos & E.F. Morris (1965) = Coryneum (Mitosp. fungi) fide Sutton (*TBMS* **86**: 1, 1986).

Muroia I. Hino & Katum. (1958), Lophiostomataceae. 1, Japan.

Murrill (William Alphonso; 1869-1957). A leading American agaricologist; on the staff of the New York Botanical Garden, 1904-24 (assistant director, 1908-19) to which he contributed more than 70,000 specimens of hymenomycetes; editor of *Mycologia*, 1909-24; author of more than 500 papers and articles, incl. *Agaricaceae of tropical North America* (1911-18 [reprint 1971]). See Weber (*Mycol.* **53**: 543, 1961; portr., bibl.), also *Mycol. Index* (1968), Halling (.... *index to spp. and infraspecific taxa of Agaricales & Boletales described by W.A. Murrill*, 1980 [*Mem. N.Y. bot. Gdn* **40**]), Stafleu & Cowan (*TL-2* **3**: 672, 1981).

Murrilloporus Ryvarden (1985), Coriolaceae. 1, N. Am.

Musaespora Aptroot & Sipman (1993), Aspidotheliaceae (L). 4, Java, Papua New Guinea.

muscardine fungus (green), *Metarhizium anisopliae*; - - (yellow), *Paecilomyces farinosus*. Pathogens of silkworms and other insects.

muscaridin, and **muscarin**(e), toxic quaternary ammonium compounds from *Amanita muscaria*; muscarin also from *Inocybe patouillardii*.

muscazone, an insecticidal toxin from *Amanita muscaria*. Cf. tricholomic acid.

Muscia Gizhitsk. (1929), ? Leotiales (inc. sed.). 1, former USSR.

Muscicola Velen. (1934), Leotiales (inc. sed.). 1, Eur. See Svrček (*Česká Myk.* **43**: 65, 1989).

muscicolous, growing on *Musci*; see Bryophilous fungi.

muscimol, see pantherine.

mushroom, (1) an agaric (or a bolete), basidioma, esp. an edible one; (2) any agaric; a macrofungus with a distinctive fruiting body which can be either hypogeous or epigeous, large enough to be seen with the naked eye and to be picked by hand (Chang & Miles, *Mycologist* **6**: 65, 1992). Cf. toadstool.

mushroom bodies (entom.), corpora pedunculata, paired lobes of neurophile in dorsal brain of insects.

Mushroom cultivation. Apart from the button mushroom, *Agaricus bisporus*, several other species are cultivated now, but only very few at a commercial scale, e.g. oyster mushrooms (*Pleurotus*), shii-take (*Lentinula edodes*), padi straw mushroom (*Volvariella volvacea*).

Cultivation of the button mushroom starts from a master culture which is used to inoculate spawn. Well-composted horse manure is used as substrate (or sometimes 'artificial compost'). Gypsum and superphosphate may be added before or during decomposition and the compost has to be turned occasionally. Mushrooms do not need light but for the best results they have to be kept at an even temperature and the beds may be in any place meeting these conditions. Indoor beds, where space is limited, such as those in special houses ('sheds'), are frequently of the 'flat' type and are made up of a 15-23 cm layers of compost in boxes or trays; those outdoors are generally made as ridges about .75 m high. In place of the old method of using natural, 'virgin' spawn to make the 'brick' spawn for spawning the beds, 'pure culture' spawn (made by inoculating a sterile compost with cultures from spores or tissues of a good type of mushroom) is now used for inoculating bricks of horse manure, cow manure, and rich soil, or, better, the beds themselves. Spawning is done when the temperature has become 80°F or less and some 10 days later, when the spawn is 'running', the bed is covered ('cased') with an inch of soil and, if necessary, a layer of straw in addition. Mushrooms are first seen 6-8 weeks after spawning and basidioma production is stimulated by bacteria (Hayes *et al.*, *Ann. appl. Biol.* **64**: 177, 1969). A bed may go on producing for 4 months; the mean weight of mushrooms is about 6 kg for every m^2 of bed, though higher weights are not uncommon.

Lit.: Chang *et al.* (Eds) (*Mushroom biology and mushroom products*, 1993), Ware (*Bull. Min. Agric., Lond.* **34**, edn 4, 1938), Lambert (*Fmr's Bull. US Dep. Agric.* **1875**, 1941; *Bot. Rev.* **4**: 397, 1938), Maher (Ed.) (*Mushroom Science 13, Science and cultivation of edible fungi*, 1991), Stamets (*Growing gourmet and medicinal mushrooms*, 1993), Stoller (*Pl. Physiol.* **18**: 397, 1943; artificial composts), Atkins (*Mushroom growing today*, edn 5, 1966), Singer & Harris (*Mushrooms and truffles*, 1987), Chang & Hayes (Eds) (*The biology and cultivation of edible mushrooms*, 1978; compr. review).

Mushroom parasites: *Mycogone perniciosa* (white mould, wet bubble; *IMI Descr.* **499**), *Verticillium lamellicola* (dry bubble, brown spot, mole; flock), *V. fungicola*, *Cladobotryum dendroides* (mildew or cobweb diseases; *IMI Descr.* **498**, *Fusarium solani*, *Myceliophthora* (q.v.). See also Hawksworth (*in* Cole & Kendrick (Eds), *The biology of conidial fungi* **1**: 171, 1981), Jeffries & Young (*Interfungal parasitic relationships*, 1994).

Mushroom losses are sometimes caused by the presence of other fungi which inhibit the development of the mushrooms. Examples of such fungi ('invaders') are: *Xylaria vaporaria* (= *X. pedunculata*), *Scopulariopsis fimicola* (see *Oospora*) and *Papulospora byssina* (white and brown plaster moulds), *Diehliomyces microsporus* (q.v.), *Clitocybe dealbata*, *Clitopilus augeana*; see also cat's ear.

Mushroom pests: see Ware (*loc. cit.*; *J.S.-E. agric. Coll., Wye* **31**: 15, 1933 *et seq.*).

mushroom stones, mushroom-like effigies perhaps associated with Mayan religious cults, mainly S. Am. (see Lowy, *Mycol.* **63**: 983, 1971).

mushroom sugar, see trehalose.

musiform, banana-shaped (basidiospores in *Exobasidium*, fide Nannfeldt, *Symb. bot. upsal.* **23**: 27, 1981).

must, (1) unfermented or fermented grape juice; new wine; (2) = mould.

mutagen, a chemical or physical agent which promotes or increases the mutation rate.

mutant, a strain that differs by an induced or natural mutation of at least one genetic locus.

Mutatoderma (Parmasto) C.E. Gómez (1976), Corticiaceae. 4, widesp.

muticate (adj. **muticous**), having no point; not sharp at the ends.

Mutinus Fr. (1849) nom. cons., Phallaceae. 12, widespr.

mutualism, persistent and intimate association between organisms of different size in which the larger organism (the host) utilizes novel or enhanced properties possessed by the smaller partner(s) (symbionts), e.g. lichens, mycorrhizas. See Douglas & Smith (*in* Smith, *Pap. Proc. R. Soc. Tasmania* **123**: 1, 1989), symbiosis.

Muyocopron Speg. (1881), Microthyriaceae. 5, trop.

myc- (mycet-, myceto-, myco-) (prefix), pertaining to fungi.

Mycacolium Reinke (1895) nom. inval. = Acolium (Ascomycota).

mycangium (pl. **mycangia**), a sac or cup-shaped fungal repository of ectodermal origin located in or on an ambrosia beetle (Batra, *Trans. Kansas Acad. Sci.* **66**: 226, 1963).

Mycardothelium Vain. (1928) ≡ Arthothelium (Arthoniales).

Mycarthonia Reinke (1895) ? = Arthonia (Arthon.).

Mycarthopyrenia Keissl. (1921) = Arthopyrenia (Arthopyren.).

Mycarthothelium Vainio (1928) = Arthothelium (Arthon.).

Mycasterotrema Räsänen (1943) ≡ Asterotrema (Pyrenulales).

Mycastrum Raf. (1813) = Scleroderma (Sclerodermat.) fide de Toni (Saccardo, *Syll. Fung.* **7**: 134, 1888).

Mycaureola Maire & Chemin (1922) nom. dub. ('basidiomycetes', inc. sed.) fide Porter & Farnham (*TBMS* **87**: 575, 1986). Nom. dub. (Ascomycota, inc. sed.) fide v. Arx & Müller (1954), Kohlmeyer & Kohlmeyer (*Marine mycology*, 1979).

Mycelia sterilia, see Agonomycetales.

mycelial cord, a discrete filamentous aggregation of hyphae which, in contrast to a rhizomorph (q.v.), has no apical meristem; syrrotia. Thompson & Rayner (*TBMS* **78**: 193, 1982) prefer not to use 'mycelial strand' for such structures.

mycelial muff, a subterranean hyphal system surrounding a living root (Buscott & Roux, *TBMS* **89**: 249, 1987).

Myceliochytrium A.E. Johanson (1946), Algae (*Actinoplanateae*). See Sparrow (1960), Karling (1977).

Myceliophthora Costantin (1892), Anamorphic Arthrodermataceae and Ceratostomataceae, 1.A1/2.1/3/10/13. Teleomorphs *Ctenomyces, Corynascus, Arthroderma*. 9, temp. *M. lutea* (vert-degris disease ('mat disease') of mushrooms in culture). See van Oorschot (*Stud. Mycol.* **20**, 1980; key), Guarro & Figueras (*Int. Jl Myc. & Lichen.* **3**: 135, 1986; ultrastr. conidia).

Mycelites Roux (1887), ? Fossil fungi. 1 (Devonian to Recent), widespr.

Mycelithe Gasp. (1841) = Polyporus (Polypor.) fide Donk (1974).

Mycelium auct. (obsol.), nom. dub. *M. radicis* sometimes still used for certain mycorrhizal fungi.

mycelium, a mass of hyphae; the thallus of a fungus; 'spawn'; **mycelioid**, like mycelium. See Jennings & Rayner (Eds) (*The ecology and physiology of the fungal mycelium*, 1984), Gregory (*TBMS* **82**: 1, 1984; review).

Myceloblastanon M. Ota (1924) = Candida (Mitosp. fungi) fide Diddens & Lodder (*Die anaskosporogenen Hefen* **2**, 1942), v. Arx (1981; anamorph of *Pichia*).

myceloconidium, see stylospore.

Myceloderma Ducomet (1907), Mitosporic fungi, 1.A2/C2.1. 1 (with pycnothyrial state), Eur.

Mycelophagus L. Mangin (1903) = Phytophthora (Pyth.) fide Clements & Shear (1931).

Mycena (Pers.) Roussel (1806), Tricholomataceae. *c.* 200, cosmop. *M. citricolor* on coffee, etc. is luminous (see Buller, **6**). See Kühner (*Le genre Mycena*, 1938), Smith (*North American species of Mycena*, 1947 [reprint 1971]), Maas Geesteranus (*Persoonia* **11**: 93, 1980; subdiv., *Proc. K. Ned Akad. Wet.* C **86**: 401; sect. *Sacchariferae, Basipedes, Bulbosae, Clavulares, Exiguae, Longisetae*, **87**: 131; sect. *Viscipellies, Amictae, Supinae*, **87**: 413; sect. *Filipedes*, **88**: 339; sect. *Mycena*, **89**: 83; sect. *Lucentae, Carolineses, Monticola*, **89**: 159; sect. *Polyadelphia, Saetulipedes*, **89**: 279; sect. *Intermediae, Rubromarginatae*, **91**: 129; sect. *Fragilipedes*, **91**: 377; sect. *Lactipedes, Sanguinolentae, Galactopoda, Crocatae*, **92**: 89; sect. *Hygrocyboideae*: 93: 163; sect. *Adonidae, Aciculae, Oregonenses*, **94**: 81; sect. *Hiemales, Exornatae*), Métrod (*Les Mycènes de Madagascar*, 1949), Pearson (*Naturalist*, Hull 1955: 41; key Br. spp.), Charbonnel (*Docums mycol.* **7** 26, 1977; microsc. char.), Haluwyn (*Docums mycol.* **5**: 17, 1972; ecology), Pegler (*Kew Bull., Addit. Ser.* **6**, 1977; key 12 E. Afr. spp.), Emmett (*Mycologist* **6**: 72, 114, 164, 1992, **7**: 4, 1993; Br. spp., list & key), Desjardin (*Bibl. Mycol.* **159**:1, 1995; sect. *Sacchariferae*, world-wide), Maas Geesteranus & Horak (*Bibl. Mycol.* **159**: 143, 1995; Papua New Guinea, New Caledonia ssp.).

Mycenastraceae Zeller (1948), Lycoperdales. 1 gen. (+ 2 syn.), 1 spp. Gasterocarp epigeal; hyphae clamped; spores large, reticulate.

Mycenastrum Desv. (1842), Mycenastraceae. 1, widespr. See Heim (*Rev. mycol.* **36**: 81, 1971; Patouillard's spp.), Homrich & Wright (*Mycol.* **65**: 779, 1973).

Mycenella (J.E. Lange) Singer (1938), Tricholomataceae. 10, temp. See Boekhout (*Persoonia* **12**: 425, 1985; key Eur. spp.).

Mycenitis, see *Mycetinis*.

Mycenoporella Overeem (1926) = Filoboletus (Tricholom.) fide Singer (1951).

Mycenopsis Velen. (1947), Agaricaceae. 1, Eur.

Mycenula P. Karst. (1889) = Mycena (Tricholomat.) fide Singer (1951).

Mycepimyce Nieuwl. (1916) = Sphaerellopsis (Mitosp. fungi) fide Sutton (1977).

Mycerema Bat., J.L. Bezerra & Cavalc. (1963), Schizothyriaceae. *c.* 5, trop. See Farr (*Mycol.* **79**: 97, 1987. Anamorph *Plenotrichaius*-like.

Myces Paulet (1808) ≡ Fungus (Agaric.).

mycetal, a fungus or a lichen (obsol.).

Myceteae, see FUNGI.

Mycetes (obsol.), (1) *Fungi*. (2) **mycetes**, a general term (obsol.) for minute vegetable organisms or microbes. Hence **mycetology** = mycology; schizomycetes = bacteria; etc.

-mycetes (suffix), indicating the rank of a fungal class (see Classification).

Mycetinis Earle (1909) = Marasmius (Tricholomat.) fide Singer (1951).

Mycetism (mycetismus) (poisoning by larger fungi). Is of common occurrence and has a long history. Some people are allergic (see Allergy) or intolerant to mushrooms and illness may result from eating decayed or mouldy specimens but a number of larger fungi contain toxins and *c.* 6 species are deadly poisonous. Right diagnosis is the only certain way for determining poisonous fungi. Categories of mushroom poisoning include:

(1) *Cyclopeptide poisoning* (see amatoxins, phallotoxins) (by *Amanita phalloides, A. virosa, A. verna*). Symptoms first occur 4-6 h or more after ingestion. Early gasteroenteric symptoms may obscure hepatic and renal damage. This, the most dangerous type, is responsible for most deaths by mushroom poisoning in Eur. and N. Am. but the chance of recovery has been enhanced by the introduction of antiserum therapy and dialysis. Orollanine and Cortinarins (from *Cortinarius* spp.) have an incubation period of 2-20+ days and cause renal failure (see Muchelot & Tebbett, *MR* **94**: 289, 1990).

(2) *Haemolytic poisoning* (*A. rubescens, A. vaginata*). Characterized by anaemia resulting from eating raw or undercooked mushrooms containing thermolabile haemotoxins.

(3) *Muscarine poisoning* (*A. muscaria, A. pantherina*). Early symptoms, within 2 h, include increased perspiration, salivation, dehydration, and nausea.

(4) *Coprine poisoning* (*Coprinus atramentarius*). See antabuse.

(5) *Psychotropic poisoning*. Hallucinations and delirium 2-4 h after ingestion. (a) Ibotenic acid, muscimol group (*A. muscaria, A. pantherina*): sleep, torpidity, or coma in extreme cases; (b) Indole group (psilocin, psilocybin): stimulates psychic perception, see Hallucinogenic fungi.

(6) Gasteroenteric irritants (*Entoloma sinuatum, Paxillus involutus, Agaricus xanthoderma, Boletus satanus, Hebeloma crustuliniforme*, many *Tricholoma* spp., *Hypholoma fasciculare*, several *Lactarius* and *Russula* spp.).

The ascomycete *Gyromitra esculenta* contains gyromitrin (q.v.) which causes gasteroenteric discomfort, followed by hepatic and renal attack. This is the only fungus poisoning known to induce fever. The ascoma is edible if used without the cooking water or after drying.

Lit.: Dujarric de la Rivière & Heim (*Les champignons toxiques*, 1938; 600 refs), Heim (*Champignons toxiques et hallucinogènes*, 1963; edn 2, 1978), Arietti

& Tomasi (*Funghi velenosi*, 1969), Duffy & Vergeer (*California toxic fungi*, 1977), Lincoff & Mitchell (*Toxic and hallucinogenic mushroom poisoning. A handbook for physicians and mushroom hunters*, 1977), Rumack & Salzman (Eds) (*Mushroom poisoning: diagnosis and treatment*, 1978), Pegler & Watling (*Bull. BMS* **16**: 66, 1982; Br. toxic fungi), Ammirati *et al.* (*Poisonous fungi of the Northern United States and Canada*, 1985), Bresinsky & Besl (*Giftpilze mit einer Entführung in die Pilzbestimmung*, 1985 [Engl. transl. *A colour atlas of poisonois fungi*, 1990]), Oldridge *et al.* (*Wild mushroom and toadstool poisons*, 1989). See also ergot, Mycotoxicoses.

mycetismus, see Mycetism.

mycetobionts, fungus-dependent, used of obligate fungus-feeding arthropods.

mycetocyte, see mycetosome.

Mycetodium A. Massal. (1856) ≡ Gomphillus (Gomphill.).

mycetology, see *Mycetes*.

mycetoma (maduramycosis, madura foot), a disease, esp. tropical, of the foot or other part of man resulting in tumefactions and characterized by mycotic granules ('grains') in the infected tissues. Although a clinical entity, many different fungi (**eumycetoma**) and actinomycetes (**actinomycetoma**) are involved. Mycetomas can be roughly classified according to whether the grains are white or yellow (*Nocardia madurae*, *Pseudallescheria boydii*, *Aspergillus* spp., etc.), red (*Streptomyces pelletieri*, *S. somaliensis*) or black (*Madurella mycetomatis*, etc.). See Mahgoub & Murray (*Mycetoma*, 1973).

mycetophagy, see mycophagy.

mycetophiles, fungus-loving, facultative fungus-feeding arthropods.

mycetophilous, see mycophilic.

mycetosome, a sac-like structure in the gut of Anobiid beetles lined with cells (**mycetocytes**) containing yeast cells.

Mycetosporidium L. Léger & E. Hesse (1905) ? Fungi. 2 (on insects), Eur. See Tate (*Parasitology* **32**: 462, 1940).

Mycetozoa, see *Myxomycota*.

mycid, a secondary effect (manifested as eczema, urticaria, etc.) which is an allergic reaction to spores or toxin of a dermatophyte; dermatophytid. Cf. mycosis (2). A mycid may be a trichophytid (caused by *Trichophyton*); microsporid (*Microsporum*); epidermophytid (*Epidermophyton*).

-mycin (suffix), the recommended ending for names coined for antibiotics derived from actinomycetes.

Mycinema C. Agardh (1824) nom. dub., based on mycelium of *Corticium* (Cortic.) fide Saccardo [not traced].

myco- (prefix), pertaining to fungi.

Mycoacia Donk (1931), Meruliaceae. 4, widespr. See Bourdot & Galzin (*Hymenomycetes de France*: 414, 1928; as *Acia*), Ragab (*Mycol.* **43**: 459, 1951), Eriksson & Ryvarden (**4**: 873, 1976; key Eur. spp.).

Mycoaciella J. Erikss. & Ryvarden (1978), Steccherinaceae. 3, widespr. See Hjortstam *et al.* (*Kew Bull.* **45**: 303, 1990; key).

Mycoalvimia Singer (1981), Tricholomataceae. 1, Brazil.

Mycoamaranthus Castellano, Trappe & Malajczuk (1992), Hymenogastraceae. 1, Australasia.

Mycoarachis Malloch & Cain (1970), ? Pseudeurotiaceae. 1 (coprophilous), Afr. See Benny & Kimbrough (*Mycotaxon* **12**: 1, 1980; posn).

Mycoarctium K.P. Jain & Cain (1973), Thelebolaceae. 1 (coprophilous), Canary Isl., U.S.A. See Korf & Zhuang (*Mycotaxon* **40**: 79, 1991), van Brummelen

(*in* Hawksworth (Ed.), *Ascomycete systematics*: 400, 1994; posn).

Mycoarthopyrenia Cif. & Tomas. (1953) nom. illegit. ≡ Arthopyrenia (Arthopyren.).

Mycobacidia Rehm (1890) = Arthrorhaphis (Arthrorhaphid.).

Mycobacillaria Naumov (1915), Mitosporic fungi, 1.C2.1/10. 1, former USSR.

Mycobanche Pers. (1818) ≡ Mycogone (Mitosp. fungi).

Mycobilimbia Rehm (1890), Porpidiaceae (L). *c.* 12, Eur., N. Am., Asia. See Awasthi & Mathur (*Proc. Indian Acad. Sci.* (Pl. Sci.) **97**: 481, 1987; key 3 spp. India), Hafellner (*Herzogia* **8**: 53, 1989; key 8 spp. Eur.).

mycobiont, the fungal component of a lichen (Scott, *Nature* **179**: 486, 1957); cf. phycobiont, photobiont.

mycobiota, (1) the total fungal inventory of the area under consideration (e.g. all the species present); (2) the fungal mass present (e.g. in a soil sample).

Mycoblastaceae Hafellner (1984), Lecanorales (L). 1 gen. (+ 3 syn.), 9 spp. Thallus crustose; ascomata apothecial, black, flat or domed, without a well-developed margin; interascal tissue of branched and anastomosing paraphyses, with a pigmented epithecium; asci clavate, with an enormous I+ apical cap and a very large ocular chamber, and a very thick (esp. at the apex) outer I+ gelatinized layer, 1- or few-spored; ascospores usually very large, thick-walled, hyaline, aseptate, without a sheath. Anamorphs pycnidial. Lichenized with green algae, widespr., esp. temp.

Mycoblastomyces Cif. & Tomas. (1953) ≡ Mycoblastus (Mycoblast.).

Mycoblastus Norman (1853), orth. cons., Mycoblastaceae (L). 9, Eur., Asia, Australia. See Anders (*Hedwigia* **68**: 87, 1928), James (*Lichenologist* **5**: 114, 1971; key 4 Br. spp.).

Mycobonia Pat. (1894) nom. cons., Polyporaceae. 1 or 2, trop. See Martin (*Mycol.* **31**: 247, 1939), Corner (*Beih. Nova Hedw.* **78**: 102, 1984; key).

Mycobystrovia Goujet & Locq. (1979), Fossil fungi (mycel.). 1 (Devonian), Eur.

Mycocalia J.T. Palmer (1961), Nidulariaceae. 5, widespr. See Burnett & Boulter (*New Phytol.* **62**: 217, 1963; mating systems), Jeppson (*Agarica* **6**: 228, 1985; key).

Mycocaliciaceae Alf. Schmidt (1970), Caliciales. 4 gen. (+ 3 syn.), 55 spp. Thallus immersed, often absent; ascomata stalked, brown or black; asci cylindrical, thick-walled at least at the apex, not evanescent; ascospores brown, smooth-walled, not liberated in a mazaedial mass. Mainly saprobic on wood or bark, some lichenicolous, fungicolous or ? lichen-forming.

Mycocalicium Vain. (1890), Mycocaliciaceae. 10, mainly temp. See Schmidt (1970), Tibell (1984, *Nordic Jl Bot.* **10**: 221, 1990; anamorphs), Samuels & Buchanan (*N.Z. Jl Bot.* **21**: 163, 1983; anamorph). Anamorph *Phialophora*.

Mycocandida Langeron & Talice (1932) = Candida (Mitosp. fungi) fide Diddens & Lodder (*Die anaskosporogenen Hefen* **2**, 1942).

Mycocarpon S.A. Hutch. (1955), Fossil fungi (? Ascomycota). 4 (Carboniferous), Eur., USA. See Baxter (*Paleont. Contrib. Univ. Kansas* **77**, 1975), Pirozynski (1976).

mycocecidium, see cecidium.

Mycocentrospora Deighton (1972), Mitosporic fungi, 1.C/E1.10. 10, widespr. See Constantinescu (*Revue Mycol.* **42**: 105, 1978; conidial polymorphism), Braun (*Mycotaxon* **48**: 275, 1993; reassessment).

Mycochaetophora Hara & Ogawa (1931), Mitosporic fungi, 1.E1.?. 1, Japan.

Mycochlamys S. Marchand & Cabral (1976), Mitosporic fungi, 1.A2.10. 1 (from soil), Argentina.

Mycociferria Tomas. (1953) ≡ Ciferriolichen (Arthopyren.).

Mycocitrus A. Møller (1901), Hypocreaceae. 1 (on bamboo), Asia.

Mycocladus Beauverie (1900) ≡ Absidia (Mucor.). See Hesseltine & Ellis (*Mycol.* **56**: 569, 1964), Mirza *et al.* (1979).

Mycoclelandia Trappe & G.W. Beaton (1984), Pezizaceae. 2 (hypogeous), Australia.

mycoclena (orig. 'micoclena'), term coined by Peyronel (1922) for the 'fungus mantle' of an ectotrophic mycorrhiza having a loose structure; cf. mycoderm.

Mycocoelium Kütz. (1843) nom. dub. (? Fungi).

Mycocoenogonium Cif. & Tomas. (1954) = Coenogonium (Gyalect.)

mycocoenosis, the complete assemblage of fungi within a certain plant community and its environment, or in the absence of green plants another defined habitat; see mycosociology, phytosociology.

Mycoconiocybe Reinke (1895) nom. inval. ? = Coniocybe (Coniocyb.).

Mycocoscoma Bref. (1912), Ustilaginales (inc. sed.). 1, Eur.

Mycocryptospora J. Reid & C. Booth (1987), Dothideales (inc. sed.). 1, Germany.

Mycodendron Massee (1891) nom. dub. ('basidiomycetes'); ? based on an abnormal polypore (see Donk, *Fungus* **28**: 13, 1958).

mycoderm, coined by Ziegenspeck (1929) for a compact, tissue-like, ectotrophic mycorrhiza; cf. mycolena.

Mycoderma Pers. (1822), Mitosporic fungi, ?.?.?. 10, widespr. Some (e.g. *M. vini* and *M. cerevisiae*) are damaging to wine and beer. See *Mycokluyveria*.

mycodextran, an unbranched polysaccharide from *Aspergillus niger*, etc. (Barker *et al.*, *J. chem. Soc.* 1957: 2488, 1957); nigeran; see Bobbitt & Nordin (*Mycol.* **70**: 1201, 1979; as phylogenetic marker).

mycoecology, ecology of fungi; see Ecology.

Mycoenterolobium Goos (1970), Mitosporic fungi, 1.D2.1. 1 (conidia flattened and fan-shaped), Hawaii, N. Am.

Mycofalcella Marvanová, Om-Kalth. & J. Webster (1993), Mitosporic fungi, 1.?.?. 1, UK.

mycoflora, see mycobiota (a more appropriate term as fungi are not plants).

mycofungicides, see Mycopesticides.

Mycogala Rostaf. ex Sacc. (1884) = Orbicula (Otid.) fide Hughes (1951).

Mycogalopsis Gjurašin (1925), Otideaceae. 1, Eur.

Mycogemma K.M. Zalessky (1915), Fossil fungi. 1 (Carboniferous), former USSR.

mycogenous, coming from, or living on, fungi.

mycogeography, study of the geographical distribution of fungi (q.v.).

Mycoglaena Höhn. (1909), Dothideales (inc. sed.). *c.* 8, Eur., N. Am. See Harris (1973), Riedl (*Öst. bot. Z.* **119**: 41, 1971; key 6 spp.).

Mycogloea L.S. Olive (1950), Platygloeaceae. 3, widespr. See McNabb (*TBMS* **48**: 187, 1965).

Mycogone Link (1809), Mitosporic fungi, 1.B1.1. 10, widespr. See Mushroom diseases.

mycohaemia (**mycohemia**), a condition in which fungi are present in the blood stream.

mycoherbicide, (1) a herbicide derived from a fungus; (2) a preparation of fungal spores used in the biocontrol of weeds (see Mycopesticides).

Mycohypallage B. Sutton (1963), Anamorphic Phyllachoraceae, 8.C2.1. Teleomorph *Deshpandiella.* 1, Afr., Sri Lanka.

mycoin, see patulin.

mycoinsecticide, a fungus used to control insects; see Entomogenous fungi, Mycopesticide.

Mycokidstonia D. Pons & Locq. (1981), Fossil fungi. 1, France.

Mycokluyveria Cif. & Redaelli (1947) = Candida (Mitosp. fungi) fide v. Arx *et al.* (*Stud. Mycol.* **14**, 1977).

Mycolachnea Maire (1937) ≡ Humaria Fuckel (Otid.). See Eriksson & Hawksworth (*SA* **11**: 179, 1993), Wu & Kimbrough (*Int. J. Pl. Sci.* **153**: 128, 1992).

Mycolangloisia G. Arnaud (1918) = Actinopeltis (Microthyr.). fide v. Arx & Müller (1975).

mycolatry, the worship of fungi (introduced by Wasson, 1980).

Mycolecidia P. Karst. (1888) ? = Dactylospora (Dactylospor.).

Mycolecis Clem. (1909) ? = Dactylospora (Dactylospor.).

Mycoleptodiscus Ostaz. (1968), Anamorphic Magnaporthaceae, 3.A/B1.15. Teleomorph *Omnidemptus.* 9, widespr. *M. indicus* on vanilla. See Sutton & Alcorn (*MR* **94**: 564, 1990; key), Bezerra & Ram (*Fitopat. bras.* **12**: 717, 1986), Alcorn (*Austr. Syst. Bot* **7**: 591, 1994; incl. review of appressoria).

Mycoleptodon Pat. (1897) = Steccherinum (Steccherin.) fide Donk (1964).

Mycoleptodonoides M.I. Nikol. (1952), Steccherinaceae. 2, former USSR.

Mycoleptorhaphis Cif. & Tomas. (1953) = Leptorhaphis (Arthopyren.).

Mycolevis A.H. Sm. (1965), Cribbeaceae. 1, N. Am. See Fogel (*Mycol.* **68**: 1097, 1976).

Mycolindtneria Rauschert (1987), Lindtneriaceae. 5, widespr.

mycoliths, sand grains bound together with mycelium, forming structures 5-6 cm long, esp. by *Melanospora tulasnei.*

mycologist, one engaged in the pursuit of mycology.

mycology, the scientific study of fungi.

mycolysis, the lysis of a fungus.

Mycomalus A. Møller (1901), Clavicipitaceae. 1, Brazil.

Mycomater Fr. (1825) nom. dub.; based on mycelium fide Fries (*Summ. veg. Scand.*, 1849).

Mycomedusa R. Heim (1966) = Favolaschia (Tricholomat.) fide Pegler (1977).

Mycomedusiospora G.C. Carroll & Munk (1964), Lasiosphaeriaceae. 1, Brazil, Costa Rica.

Mycomelanea Velen. (1947), ? Leotiales (inc. sed.). 1, former Czechoslovakia.

Mycomelaspilea Reinke (1895) ? = Melaspilea (Melaspil.) fide Müller & v. Arx (1962).

Mycomicrothelia Keissl. (1936), Arthopyreniaceae (±L). 30, mainly trop. See Aptroot (*Bibl. Lich.* **44**, 1991), Hawksworth (*Bull. Br. Mus. nat. Hist., Bot.* **14**: 43, 1985; key 25 spp.).

Mycomyces Wyss-Chod. (1927), ? Mitosporic fungi, ?.?.?. 1 (on humans), Eur.

Mycomycophytes, see *Mycophytes.*

mycomyringitis, a fungal inflammation of the ear-drum.

mycomysticism, mystical state induced by eating hallucinogenic fungi.

Myconeesia Kirschst. (1936), Xylariaceae. 1, Taiwan. See Læssøe & Spooner (*Kew Bull.* **49**: 1, 1994).

myconematicides, see Mycopesticides.

Mycopandora Velen. (1947) = Unguicularia (Hyaloscyph.) fide Svrček (*Česká Myk.* **40**: 215, 1986).

Mycopappus Redhead & G.P. White (1985), Mitosporic fungi, 9.D2/E1.42. 2, N. Am.

Mycopara Bat. & J.L. Bezerra (1960), Mitosporic fungi, 4.A1.15. 1 (on *Dimerosporiopsis*), USA.

mycoparasitism, the parasitism of one fungus by another (the **mycoparasite**); preferable to hyperparasitism which has been used for the same phenomenon; see Fungicolous fungi.

mycopathology, the study of disease caused by fungi.

Mycopegrapha Vain. (1921) ? = Opegrapha (Roccell.).

Mycopepon Boise (1987), Melanommataceae. 1, French Guiana, Nicaragua. See Boise (*Mycotaxon* **52**: 303, 1994; nomencl.).

Mycopesticides. Are mass-produced, usually commercially formulated and marketed products based on fungi which are pathogens, parasites or antagonists, arthropod pests (mycoinsecticides), plant parasitic nematodes (myconematicides), weeds (mycoherbicides) or crop pathogens (mycofungicides), applied inundatively, like a chemical pesticide.

The potential of mycoinsecticides was first investigated more than 100 years ago when *Metarhizium anisopliae* was applied against weevil pests in the former USSR (Steinhaus, *Hilgardia* **26**: 107, 1965). Currently, *M. anisopliae* is reaching the final stages of commercialization for control of termites (Milner, *in* Lomer & Prior, *Biological control of locusts and grasshoppers* 200, 1992), scarab beetles (Rath, *in* Glare & Jackson, *Use of pathogens in scarab pest management*, 217, 1992) and black vine weevils (Wolfram, *in Proc. V Int. Colloq. Invert. Path. & Microb. Contr.* 2, 1990). *M. flavoviride* is also being field tested against locusts and grasshoppers in Africa (Lomer & Prior, 1992). *Beauveria bassiana* has been extensively employed against a range of crop pests but mainly at the semi-commercial or cottage-industry level, under the trade name Boverin™ in the former USSR (Ferron, *in* Burge, *Microbial control of pests and plant diseases 1970-1980*, 45, 1981), and in a commune-produced system in China (Hussey & Tinsley, *in* Burge, 1981: 785). *B. brongniartii* has been mass-produced and aerially applied against cockchafer pests in Switzerland and, despite problems with fungal stability and formulation (Zimmermann, *in* Glare & Jackson, 1992), is now a registered product (Engerlingspilz^R).

The first commercial mycoinsecticides became available in the 1980s and were based on strains of *Verticillium lecanii* for aphid (Vertalec™) and whitefly (Mycotal™) control in glasshouses (Quinlan, *in* Burge, *Fungi in biological control systems* 19, 1988). *Aschersonia aleyrodis* is also being assessed for control of whitefly pests (Samson & Rombach, *in* Hussey & Scopes, *Biological pest control* 34, 1985). Production and marketing problems are discussed by Samson *et al.* (*Atlas of entomopathogenic fungi*, 1988) and Prior & Moore (*Biocontrol News Inf.* **14**: 31, 1993).

Few myconematicides have been commercialized and, because of problems with quality control, inconsistent performance, potential health hazards and uneconomic application rates, none is currently available on the market (Stirling, *Biological control of plant parasitic nematodes*, 1991). Previous commercial or semi-commercial products have been based on: *Arthrobotrys robusta* (Royal 300^R), *A. superba* (Royal 350^R) and *Verticillium chlamydosporium* (Kerry *et al.*, *Ann. appl. Biol.* **105**: 509, 1984; Kerry, *in* Brown & Kerry, *Principles and practice of nematode control in crops*, 233, 1987).

The use of mycoherbicides is a relatively recent concept but a number of products are already on the market: COLLEGO (*Colletotrichum gloeosporioides* f.sp. *aeschynomenes* for control of northern jointvetch, DeVine (*Phytophthora palmivora*) for control of milkweed vine; CASST (*Alternaria cassiae*) for control of sicklepod; Bio Mal (*C. gloeosporioides* f.sp. *malvae*) against round-leaved mallow; whilst a number of others are in the final stages of commercialization (Charudattan, *in* Burge (Ed.), *Fungi in biological control systems* 86, 1988; TeBeest & Templeton, *Plant Disease* **69**: 6,

1985; Charudattan, *in* TeBeest (Ed.), *Microbial control of weeds*, 24, 1991). Although most are readily culturable, necrotrophic pathogens, marketing of the rust *Puccinia canaliculata*, as a formulated product ('Dr Biosedge') for control of yellow nutsedge is imminent (Phatak, *Proc. I. Int. Weed Control Conf.* 388, 1992). Considerable research is now being concentrated on fermentation technology and formulation to improve performances of potential mycoherbicides (*Pl. Prot. Quart.* **7** (4): 30 pp., 1992; Auld, *Crop Protection* **12**: 477, 1993).

The first mycofungicide, and indeed one of the first commercially available mycopesticides, arose from the pioneering work of Rishbeth (*Ann. appl. Biol.* **52**: 63, 1963) who investigated the potential of *Phlebia gigantea* for control of butt rot of conifers caused by *Heterobasidion annosum* (Deacon, *Microbial control of plant pests and diseases*, 1983; Campbell, *Biological control of microbial plant pathogens*, 1989). Most work has concentrated on *Trichoderma* and *Gliocladium* as antagonists; reviewed by Lynch & Ebben (*J. appl. Bact. Symp. Suppl.* 1986, 115), Papavizas (*Ann. Rev. Phytopath.* **23**: 23, 1985), Whipps & Lumsden (Eds) (*Biotechnology of fungi for improving plant growth* 1989) and Whipps (*Aspects Appl. Biol.* **24**: 211, 1990). Apart from Binab T, based on *T. viride*, for control of forest diseases, Glio-Gard™ based on *Gliocladium virens* has been marketed recently for control of *Rhizoctonia solani* and *Pythium ultimum* in horticultural crops (Mink & Walker, *in* Lumsden & Vaughn (Eds), *Pest management: biologically based technologies*, 398, 1993).

Mycophaga F. Stevens (1924) ? = Hyaloderma (Ascomycota, inc. sed.) fide Pirozynski (1977).

mycophage, (1) a mycophagist; (2) a phage-like antibacterial substance produced by certain actinomycetes.

mycophagist, an eater of fungi.

mycophagy (adj. **mycophagous**), (1) the use of fungi as food; mycetophagy; (2) the lysis of a fungus by a phage.

Mycopharus Petch (1926) = Lysurus (Clathr.) fide Dring (1980).

mycophilic, (1) fond of fungi (or mushrooms); mycetophilous; see Fungicolous fungi; (2) growing on fungi.

mycophobia, fear of mushrooms.

mycophthorous (of a fungus), parasitic on another fungus; see mycoparasitism.

mycophycobiosis, an obligate symbiosis between a systemic (inhabitant) marine fungus and a marine alga in which the alga is the exhabitant and dominates (Kohlmeyer & Kohlmeyer, *Bot. Mar.* **15**: 109, 1972), e.g. *Mycosphaerella ascophylli* on *Ascophyllum nodosum*; cf. Lichens.

Mycophycophila Cribb & J.W. Cribb (1960) = Chadefaudia (Halosphaer.) fide Müller & v. Arx (1973).

Mycophycophyta. Phylum name used by Margulis (*Symbiosis in cell evolution*, edn 2, 1993) for *all* lichens regardless of the systematic position of either the fungal, algal or cyanobacterial partners; i.e. embracing lichen-forming *Ascomycota* and *Basidiomycota* as well as their photosynthetic partners. = *Lichenes*. See Lichens.

Mycophycophytes, see *Mycophytes*.

Mycophyta, see *Eumycetes*.

Mycophytes (obsol.) = Mycomycophytes (i.e. non-lichenized fungi) + Mycophycophytes (i.e. lichenized fungi) (Marchand, *Énumération méthodique des Mycophytes*, 1896).

Mycoplacographa Reinke (1895) ? = Lithographa (Rimular.).

mycoplasm, a symbiotic phase of rust fungus and host protoplasm (Eriksson; now taken as an error); see also mycoplasma.

mycoplasma, (1) an intimate relationship between plant-invading fungi or other microorganisms and their host cells (Frank, *Ber. Deutsch. bot. Ges.* **7**: 332, 1889) (obsol.); (2) bacterium-like organisms without a cell wall living inside cells of a host, mollicutes (Krass & Gardner, *Internat. J. Syst. Bact.* **23**: 62, 1973).

Mycoporaceae Zahlbr. (1903), Dothideales. 1 gen. (+ 3 syn.), *c.* 20 spp. Ascomata erumpent or superficial, globose or aggregated into small pulvinate stromata, seated in a loose copiously branched pale mycelium; peridium of pseudoparenchymatous cells; interascal tissue absent or scarcely differentiated from stromatal tissue; hymenium often gelatinous; asci saccate, fissitunicate; ascospores becoming brown, variably septate (usually muriform), often with a sheath. Anamorphs not prominent. Saprobic on woody tissue, widespr.

Lit.: Harris (1973), Eriksson (1981).

Mycoporellum Müll. Arg. (1884), Dothideales (inc. sed.). 7 (+ ? 6), cosmop. See Riedl (*Sydowia* **15**: 257, 1962).

Mycoporis Clem. (1909), Dothideaceae. 1, Austria.

Mycoporopsis Müll. Arg. (1885), Dothideales (inc. sed., ?L). 6 (+ ? 3), cosmop. See Riedl (*Sydowia* **16**: 215, 1963).

Mycoporum G. Mey. (1825) nom. rej., Pyrenulales (inc. sed., L). 2, Eur.

Mycoporum Flot. ex Nyl. (1855) nom. cons., Mycoporaceae (± L). *c.* 20, widespr. See Coppins & James (*Lichenologist* **11**: 27, 1979), Harris (*Mich. Bot.* **12**: 3, 1973).

mycoprotein, fungal protein, e.g. 'Quorn' (q.v.); commercially processed mycelium of non-pathogenic *Fusarium graminearum* A35 for human consumption (Newark, *Nature* **287**: 6, 1980); see Gray & Staff (*Econ. Bot.* **21**: 341, 1966), Trinci (*MR* **96**: 1, 1992).

Mycopyrenium Hampe ex A. Massal. (1860) = Thelotrema (Thelotremat.) fide Zahlbruckner (1923).

Mycopyrenula Vain. (1921) = Pyrenula (Pyrenul.) fide Harris (1989).

Mycorhizonium F.E. Weiss (1904), Fossil fungi (mycorrhizal). 1 (Cretaceous), UK.

Mycorhynchella Höhn. (1918) = Ceratocystis (Microascales) fide Sutton (*Mycol. Pap.* **141**, 1977).

Mycorhynchidium Malloch & Cain (1971), Pyxidophoraceae. 1 (coprophilous), Kenya.

Mycorhynchus Sacc. & D. Sacc. (1906) = Pyxidiophora (Pyxidiophor.) fide Lundqvist (1980).

Mycorrhaphium Maas Geest. (1962), Steccherinaceae. 2, USA, Eur. See Ryvarden (*Mem. N.Y. bot. Gdn* **49**: 344, 1989; key).

Mycorrhiza (pl. **mycorrhizas**, **mycorrhizae**) (fungus root), a symbiotic, non-pathogenic or feebly or weakly pathogenic association of a fungus and the roots of a plant. Found in the majority, perhaps 85% of plant species. The resulting dual organism (cf. lichen) was first described in detail by Frank (1895) who observed it on the roots of the major tree species of temperate forests. Frank (1887) then recorded two types of mycorrhiza, which he called: (a) **ectotrophic** (the characteristic mycorrhiza of temperate and boreal forest trees with different basidiomycetes esp. spp. of *Amanita, Boletus, Cortinarius, Russula, Suillus,* and some ascomycetes, e.g. *Tuber*), in which the fungus forms a sheath on the surface of the root from which hyphae extend outward into the soil and inwards between the outer cortical cells with which they interface to form a 'Hartig net'; and: (b) **endotrophic** (e.g. of orchid-basidiomycete and ericoid-ascomycete associations) in which the fungal hyphae enter the cortical cells of the root, enveloped by the plasmalemma of host.

The mycorrhiza now known to be most widely distributed both through the plant kingdom (see Trappe, *in* Safir *et al.*, 1987), and geographically, is a type of endo-infection formed by zygomycetes of the order *Glomales* and called vesicular-arbuscular (**VA**, or **VAM**). In this, the penetrating hyphae produce finely branched haustorial branches (arbuscules) or coils (pelotons) and vesicles (Schlicht, 1889; Gallaud, 1905; Mosse, 1956). The taxonomic status of VA fungi is reviewed by Morton & Benny (*Mycotaxon* **37**: 471, 1990).

Current classifications of mycorrhizal types avoid use of the suffix 'trophic' and recognize the following categories: **Ectomycorrhiza, Vesicular-arbuscular** (occasionally simply '**arbuscular**'; **AM**), **Ericoid, Orchid, Arbutoid** and **Monotropoid** (Lewis, *Biol. Rev.* **48**: 261, 1973; Read, *CJB* **61**: 985, 1983).

Early development of some plants, especially those with very small seeds (e.g. orchids) are completely dependent upon fungal infection. This dependence may be retained where, as in *Monotropaceae*, and some members of *Orchidaceae* and *Gentianaceae*, plants lack chlorophyll throughout their lives, and hence continue to require all carbon and most mineral supplies from their fungal symbiont. In green plants the main function of infection differs according to mycorrhizal type. Enhancement of phosphorus (P) supply to the 'host' is characteristic of VA infections, that of nitrogen (N) of ericoid infections, while both N and P supplies can be supplemented by ecto, ectendo and probably orchid infections. The global distribution of mycorrhizal types largely reflects these functional attributes, selection favouring hosts of VA fungi in primarily P limited tropical and temperate grasslands, ectomycorrhizal plants in predominantly N and P limited temperate and boreal forests, and ericoid hosts in N-limited tundra (Read, *Experientia* **47**: 376, 1991). In each of these circumstances host plants show differing levels of responsiveness to infection depending upon the extent of nutrient limitation and the structure of their root systems. Those with evolutionarily advanced 'fibrous' root systems, e.g. grasses, are less responsive than those with primitive coarsely branched so called 'magnolioid' roots (Baylis, see Sanders *et al.*, 1975).

Lit.: Frank (*Ber. dtsch bot. Ges.* **3**: 128, 1885; **5**: 248, 1887), Schlicht (*Landwirtsch. Jahrb.* **18**: 478, 1889), Gallaud (*Rev. Gen. Bot.* **17**: 5, 1905), Melin (*Untersuchungen uber die Bedeutung der Baummycorrhiza*, 1925), Kelly (*Mycotrophy in plants*, 1950), Mosse (*Ann. Bot. Lond.* **20**: 349, 1956), Harley (*Biology of mycorrhiza*, 1959, edn 2, 1970), Trappe (*Bot. Rev.* **28**: 538, 1962; fungus & tree lists, 407 refs.), Marks & Kozlowski (Eds) (*Ectomycorrhizae their physiology and ecology*, 1973), Sanders *et al.* (Eds) (*Endomycorrhiza*, 1975), Harley & Smith (*Mycorrhizal Symbiosis*, 1983), Safir (Ed.) (*Ecophysiology of VA mycorrhizal plants*, 1987), Harley & Harley (*A check-list of mycorrhiza in the British flora*, 1987; *New Phytol.* Suppl. **105**), Morton (*Mycotaxon* **32**: 32: 267, 1988; checklist 126 VA spp.), Allen (*Ecology of ectomycorrhizae*, 1990), Mejstřík *et al.* (Eds) (*Agric., Ecosyst. Environ.* **28-29**, 1990; Eur. Congr. Mycorrh.), Agerer (*Colour atlas of ectomycorrhizae*, 1991), Sieverding (*VA mycorrhiza management in tropical agroecosystems*, 1991), Norris *et al.* (Eds) (*Techniques for the study of mycorrhiza. Methods in Microbiology* **23** & **24**, 1991), Read *et al.* (Eds) (*Mycorrhizas in ecosystems*, 1992), Currah & Zelmer (*Rep. Tottori mycol. Inst.* **30**: 43, 1992; key 15 gen.

with orchids), *Mycorrhiza* (1991-), Khalil *et al.* (*Soil Biol. Biochem.* **26**: 1587, 1994; recovery VAM spores from soil), Gianinnazzi & Schüepp (Eds) (*Impact of arbusclar mycorrhizas on sustainable agriculture and natural ecosystems*, 1994), Varma & Hoch (Eds) (*Mycorrhiza: structure, function, molecular biology and biochemistry*, 1995).

Mycosarcoma Bref. (1912) = Ustilago (Ustilagin.).

mycosclerid, see cystidium.

mycose, see trehalose.

mycosin, a nitrogenous substance like animal chitin in the cell wall of fungi.

mycosis (pl. **mycoses**), (1) a fungus disease of humans, animals, or, rarely, plants (e.g. tracheomycosis). Mycoses are frequently named after the part attacked (**broncho-**, respiratory tract; **dermato-**, skin; **onycho-**, nails; **oto-**, ear; **pneumo-**, lungs), or the pathogen **blasto-** (q.v.) (*Blastomyces*); **coccidioido-**, coccidioidal granuloma (*Coccidioides immitis*); (2) the first limited infection of a dermatophyte; cf. mycid. Nomenclature, see Odds (*J. med. vet. Mycol.* **30**: 21, 1992). See AIDS, Medical and veterinary mycology.

Mycosisymbrium Carris (1994) = Scolecobasidiella (Mitosp. fungi) fide Sutton (*in litt.*).

mycosocieties, (1) communities of fungi occurring in a special habitat; see Winterhoff (1992), mycosociology; (2) mycological societies (see Societies and organizations).

mycosociology, the study of fungal communities; these can be named according to the International Code of Phytosociological Nomenclature (see Phytosociology), e.g. the association *Clitocybo-Phellodonetum nigrae* (Smarda, *Acta Nat. Acad. Sci. Bohemoslov.* **7**, 1973). Communities including lichens have a long tradition of being named by the rules developed for those of plants.

Darimont (*Inst. roy. Sci. nat. Brussels, Mém.* **170**, 1975), introduced a separate system for fungi in which the basic unit was the **sociomycie** (the name for which terminates in '**-ecium**', e.g. *Amanitecium muscariae*) (± = 'association') which are grouped (in ascending order) as 'alliances' ('*-ecion*'; *Boletecion scabri*), 'orders' ('*-ecia*'; *Boleto-Amanitecia*), and 'classes' ('*-ecea*'; *Cortinario-Boletacea*).

See Benkert (*Boletus* **2**: 37, 1978; synopsis approach, bibliogr.), Winterhoff (Ed.) (*Fungi in vegetation science* [*Handbook of Vegetation Science* **19**(1)], 1992), mycocoenosis, mycosynusium, Phytosociology.

Mycosphaerella Johanson (1884), Mycosphaerellaceae. *c.* 500, cosmop. *M. ascophylli* (mycophycobiosis with *Ascophyllum* and *Pelvetia*; Fries, *J. Phycol.* **24**: 333, 1988; Kohlmeyer & Kohlmeyer, *Bot. Mar.* **15**: 109, 1972), *M. brassicicola* (ring spot of *Cruciferae*), *M. carinthiaca* (mid-vein spot of clover), *M. fragariae* (strawberry leafspot), *M. grossulariae* (*Ribes* leafspot), *M. laricina* (needle cast of larch; Patton & Spear, *Plant Disease* **67**: 149, 1983), *M. musicola* (banana leaf spot disease, Sigatoka (q.v.); Meredith (*Phytopath. Pap.* **11**, 1970), *M. pinodes* (pea foot rot), *M. rubi* (*Rubus* (blackberry and dewberry) leaf spot; Demaree & Wilcox, *Phytopath.* **33**: 986, 1943), *M. sentina* (pear leaf fleck), *M. ulmi* (elm leaf spot). See v. Arx (*Sydowia* **3**: 28, 1949), Corlett (*Mycol. Mem.* **18**, 1991; *Mycotaxon* **53**: 37, 1995; catalogue names), Barr (*Contr. Univ. Mich. Herb.* **9**: 523, 1972; N. Am.), Corlett (*Mycotaxon* **31**: 59, 1988; on *Brassicaceae*), Niyo *et al.* (*Mycol.* **78**: 202, 1986; ultrastructure), Evans (*Mycol. Pap.* **153**, 1984; on *Pinus*), Crous *et al.* (*IMI descriptions of fungi and bacteria*: 1231-1240, 1995; spp. on *Eucalyptus*). Anamorphs may be hyphomycetes (*Cercoseptoria*, *Cercosporella*, *Cladosporium*, *Passalora*, *Polythrincium*, *Ramularia*,

Ovularia, *Cercospora*, etc.) or coelomycetes (*Phoma*, *Ascochyta*, *Septoria*, *Phloeospora*, etc.); some 'genera' (e.g. *Ovosphaerella*, *Septorisphaerella*) have been segregated because of these states; see v. Arx (*Proc. K. ned. Akad. Wet.* C **86**: 15, 1983; anamorph gen.), Sutton & Hennebert (*in* Hawksworth (Ed.), *Ascomycete systematics*: 77, 1994; anamorphs).

Mycosphaerellaceae Lindau (1897) nom. cons., Dothideales. 13 gen. (+ 39 syn.), 634 spp. Ascomata small, immersed in plant tissue, often strongly aggregated or on a weakly developed basal stroma, black, papillate, with a well-developed lysigenous ostiole; peridium usually thin, composed of pseudoparenchymatous cells; interascal tissue lacking; asci ovoid to saccate, fissitunicate; ascospores usually hyaline and transversely septate, without a sheath. Anamorphs very varied. Biotrophic, necrotrophic or saprobic on various plant tissues, cosmop.

Mycosphaerellopsis Höhn. (1918) = Didymella (Dothideales) fide v. Arx & Müller (1975).

Mycospongia Velen. (1939) nom. dub. (? 'gasteromycetes').

Mycospraguea U. Braun (1993), Mitosporic fungi, 6.C1.1. 1, USA.

mycostasis (adj. **mycostatic**), inhibition of fungal growth; fungistasis (fungistatic); sporostasis; **mycostatic** in soil, see Dobbs *et al.* (*Nature* **172**: 197, 1953), Parkinson & Waid (Eds) (*The ecology of fungi*, 130, 1960).

mycostatin, trade name for nystatin.

Mycostevensonia Bat. & Cif. (1962) = Treubiomyces (Chaetothyr.) fide v. Arx & Müller (1975).

Mycosticta Höhn. (1918), Mitosporic fungi, ?.A1.?. 1, Eur.

Mycostigma Jülich (1976), Corticiaceae. 1, Eur.

mycostratum, see perisporium; Spore wall.

Mycosylva M.C. Tulloch (1973), Mitosporic fungi, 2.A1.3. 2, Eur. See Samson & Hintikka (*Karstenia* **14**: 133, 1974).

mycosymbiont, see mycobiont.

mycosymbiosis, symbiosis of two or more fungi (Vainio, 1921; cf. *Gonidiomyces*).

mycosynusium (pl. **-iae**), a part of the total number of fungal communities in a site or region studied; e.g. **macrofungussynusiae** (only macromycetes considered); see Winterhoff (1992), mycosociology.

Mycosyrinx Beck (1894), Ustilaginaceae. 4 (on *Vitaceae*, *Sterculiaceae*), trop. Afr., Am., Asia. See Mordue (*Mycopath.* **103**: 171, 1988).

Mycota, see FUNGI.

mycothallus (pl. **-lli**), a mutualistic symbiosis of a fungus with a hepatic (liverwort) or fern gametophyte (Boullard, *Syllogeus* **19**: 1, 1979; *in* Pirozynski & Hawksworth (Eds), *Coevolution of fungi with plants and animals*: 107, 1988).

Mycothamnion Kütz. (1843) nom. dub. (? Fungi, inc. sed.).

mycotheca, a distributed set of dried specimens of fungi.

Mycothele Jülich (1976), Epitheliaceae. 1, NZ.

Mycothelocarpon Cif. & Tomas. (1953) nom. illegit. ≡ Thelocarpon (Acarospor.).

Mycothyridium Petr. (1962), Dothideales (inc. sed.). 36, widespr. See Eriksson & Hawksworth (*SA* **13**: 191, 1994; nomencl.), Petrak, *Sydowia* **15**: 288, 1972).

Mycothyridium E. Müll. (1973) = Thyridium (Thyrid.).

mycotic (esp. of disease), caused by fungi.

Mycotodea Kirschst. (1936), Ascomycota (inc. sed.). 14 (on *Musci*), widespr. See Hawksworth (*SA* **8**: 71, 1989, posn).

mycotope, a major fungal association of a particular type of woodland (Darimont, 1975); see mycosociology.

Mycotorula H. Will (1916) = Syringospora (Mitosp. fungi) fide v. Arx *et al.* (*Stud. Mycol.* **14**, 1977).

Mycotoruloides Langeron & Talice (1932) = Syringospora (Mitosp. fungi) fide v. Arx *et al.* (*Stud. Mycol.* **14**, 1977).

Mycotoxicoses. Literally fungus poisonings but in current usage limited to poisoning of animals and esp. humans by feed and food products contaminated (and sometimes rendered carcinogenic) by toxin-producing microfungi. See aflatoxins, citreoviridin, citrinin, fumonisins, islanditoxin, luteoskyrin, lysergic acid, lupinosis, maltoryzine, ochratoxin, patulin, roridins, rubratoxin, satratoxins, slaframine, sporidesmin, sterigmatocystin, tremorgen, trichothecenes, zearalenone.
Lit.: Reviews: Forgacs & Carll (*Adv. vet. Sci.* **7**: 272, 1962), Wheeler & Luke (*Ann. Rev. Microbiol.* **17**: 223, 1963), Wright (*Ann. Rev. Microbiol.* **22**: 269, 1968). Books: Wogan (Ed.) (*Mycotoxins in foodstuffs*, 1965), Moreau (*Moisissures toxiques dans l'alimentation*, 1968; edn 2, 1974, 2938 refs), Kadis (Ed.) (*Microbial toxins*, **6**, *Fungal toxins*, 1971), Krogh (Ed.) (*Mycotoxins in food*, 1987), Wyllie & Morehouse (Eds) (*Mycotoxic fungi, mycotoxins, mycotoxicoses. An encyclopaedic handbook.* **1**, *Mycotoxic fungi and chemistry of mycotoxicoses*, 1977; **2**, *Mycotoxicoses of domestic and laboratory animals, poultry, and aquatic invertebrates and vertebrates*, 1978; **3**, *Mycotoxicoses of man and plants; mycotoxin control and regulatory aspects*, 1978), Steyn (Ed.) (*The biosynthesis of mycotoxins*, 1980), Cole & Cox (Eds) (*Handbook of toxic fungal metabolites*, 1981), Chelkowski (Ed.) (*Mycotoxin research. Fusarium* **7**(I & II), 1991), Smith & Moss (Eds) (*Mycotoxins*, 1985), Cole (Ed.) (*Modern methods in the analysis structural elucidation of mycotoxins*, 1986), Joffe (*Fusarium species; their biology and toxicology*, 1986), Krogh (*Mycotoxins in food*, 1987), Betina (*Mycetotoxins*, 1989), Chelkowski (Ed.) (*Fusarium mycotoxins*, 1989), Matossian (*Poisons of the past*, 1989), Smith & Henderson (*Mycotoxins and animal foods*, 1991), Chelkowski (Ed.) (*Cereal grain mycotoxins, fungi and quality in drying and storage*, 1991), Champ *et al.* (Eds) (*Fungi and mycotoxins in stored products*, 1992). O'Neill *et al.* (Eds) (*Relevance of human cancer of N-Nitroso compounds, tobacco and mycotoxins*, 1991), Castegnaro *et al.* (Eds) (*Mycotoxins, endemic nephropathy and urinary tract tumours*, 1991). See also phytoalexins.

mycotoxicosis, one of the Mycotoxicoses (q.v.).

mycotoxin, see toxin.

Mycotribulus Nag Raj & W.B. Kendr. (1970), Mitosporic fungi, 8.A1.4. 1, India, Brazil, W. Indies.

mycotroph, a fungus which obtains its nutrients from another fungus; cf. mycoparasitism and see Fungi on fungi.

mycotrophein, a 'growth factor' from fungi needed by a mycoparasite (Waley & Barnett, *Mycol.* **55**: 209, 1963).

mycotrophic (of plants), having mycorrhiza.

Mycotypha Fenner (1932), Mycotyphaceae. 3, widespr. See Benny *et al.* (*Mycotaxon* **22**: 119, 1985; key), Benny & Benjamin (*Aliso* **8**: 391, 1976), Brain & Young (*Microbios* **25**: 93, 1979; ultrastr.), Young (*J. gen. Microbiol.* **55**: 243, 1969; sporangiolum ultrastr.).

Mycotyphaceae Benny & R.K. Benj. (1985), Mucorales. 2 gen. (+ 1 syn.), 6 spp. Sporangiola borne on dehiscent pedicels.
Lit.: Benny *et al.* (*Mycotaxon* **22**: 119, 1985; key).

Mycousteria M.L. Farr (1986), Mitosporic fungi, 5.A2.?. 2, Brazil.

Mycovellosiella Rangel (1917), Mitosporic fungi, 1.C2.10. 30, mainly trop. See Deighton (*Mycol. Pap.*

137, 1974), Liu & Guo (*Mycosystema* **1**: 241, 1988; 21 spp. from China).

mycoviruses, see Viruses in fungi.

Mycowinteria Sherwood (1986), Protothelenellaceae. 1, temp. See David (*SA* **6**: 217, 1987).

Mydonosporium Corda (1833) = Cladosporium (Mitosp. fungi) fide Saccardo (*Syll. Fung.* **4**: 354, 1886).

Mydonotrichum Corda (1831) = Helminthosporium s.l. (Mitosp. fungi) fide Saccardo (*Syll. Fung.* **4**: 407, 1886).

Myelochroa (Asahina) Elix & Hale (1987) Parmeliaceae. 19, mainly N.E. Asia. See Hale (*Smithson. Contr. Bot.* **33**, 1976). ? = Parmelia (Parmel.).

Myelorrhiza Verdon & Elix (1986), Cladoniaceae (L). 2, Australia.

Myelosperma Syd. & P. Syd. (1915), ? Lasiosphaeriaceae. 1 (on *Palmae*), Brunei, Irian Jaya. See Hyde (*Sydowia* **45**: 241, 1993).

Myiocopraloa Cif. (1958) = Schizothyrium (Schizothyr.) fide Müller & v. Arx (1962).

Myiocoprella Sacc. (1916) = Rhagadolobium (Parmular.).

Myiocopron Speg. (1881) ≡ Muyocopron (Microthyr.).

Myiocoprula Petr. (1955), Mitosporic fungi, 5.A1.?. 1, USA.

Myiophagus Thaxt. ex Sparrow (1939), ? Chytridiales (inc. sed.). 1, N. temp. See Karling (*Am. J. Bot.* **35**: 246, 1948).

Myiophyton Lebert (1857) = Empusa (Entomophthor.) fide Saccardo (1888).

Mykoblastus Norman (1853) orth. rej. ≡ Mycoblastus (Mycoblast.).

mykoholz, basidioma of the edible *Kuehneromyces mutabilis* cultivated in Germany (Chang & Hughes, 1978).

mylitta, a large sclerotium, e.g. that of blackfellows' bread (q.v.).

Mylitta Fr. (1825) nom. dub.; based on bacterial root nodules on *Robinia* fide Mattirolo (*Bull. Soc. bot. ital.* 1924: 13, 1924).

Mylittopsis Pat. (1895), Auriculariaceae. 1, USA, Malaysia. See Rogers & Martin (*Mycol.* **47**: 891, 1955).

myosin, see actin.

Myriadoporus Peck (1884) = Bjerkandera (Coriol.) fide Donk (1960).

Myriangiaceae Nyl. (1854), Dothideales. 5 gen. (+ 9 syn.), 18 spp. Stromata crustose or pulvinate, composed of subhyaline or brown thin-walled pseudoparenchymatous tissue, with ± globose sessile or short-stalked fertile outgrowths of similar tissue containing scattered asci in individual locules, becoming gelatinous at maturity; asci ± globose, sessile, fissitunicate; ascospores pale brown, transversely septate or muriform. Anamorphs unknown. Associated with scale insects or resinous exudates, widespr.
Lit.: Boedijn (*Persoonia* **2**: 63, 1963; Indonesia), v. Arx (*Persoonia* **2**: 421, 1963; gen. names).

Myriangiales, see Dothideales.

Myriangiella Zimm. (1902), ? Schizothyriaceae. c. 5, trop. See v. Arx & Müller (1975).

Myriangina (Henn.) Höhn. (1909) = Uleomyces (Cookell.) fide v. Arx (1963).

Myrianginella F. Stevens & Weedon (1923) = Uleomyces (Cookell.) fide v. Arx & Müller (1975).

Myriangiomyces Bat. (1958) = Saccardia (Saccard.) fide v. Arx (1963).

Myriangiopsis Henn. (1902), Dothideales (inc. sed.). 1, Mexico.

Myriangium Mont. & Berk. (1845), Myriangiaceae. c. 7, widespr. See v. Arx (*Persoonia* **2**: 241, 1963),

Miller (*Mycol.* **30**: 158, **32**: 587, 1938-40), Petch (*TBMS* **10**: 45, 1924; on *Insecta*).

Myriapodophila Speg. (1918) ? = Pyxidiophora (Pyxidioph.) fide Blackwell (*Mycol.* **86**: 1, 1994).

Myridium Clem. (1909) nom. rej. = Laetinaevia (Dermat.).

Myriellina Höhn. (1915), Mitosporic fungi, 6.C1.15. 2, Eur., Papua New Guinea. See Sankaran & Sutton (*MR* **95**: 1021, 1991).

Myrillium Clem. (1931) = Gymnoascus (Gymnoasc.) fide Kuehn (*Mycol.* **51**: 665, 1959).

Myrioblastus Trevis. (1857) ≡ Biatorella (Biatorell.).

Myrioblepharis Thaxt. (1895) nom. dub.; based on a protozoan on *Pythium* or *Phytophthora* fide Waterhouse (*TBMS* **28**: 94, 1945), Koch (*Mycol.* **56**: 436, 1964).

Myriocarpa Fuckel (1870) [non Benth. (1846), *Urticaceae*] ? = Guignardia (Mycosphaerell.) fide v. Arx & Müller (1954).

Myriocarpium Bonord. (1864) = Leptosphaeria (Leptosphaer.) fide Saccardo (*Syll. Fung.* **2**: 13, 1883).

Myriocephalum De Not. ex Corda (1842) = Cheirospora (Mitosp. fungi) fide Hughes (1958).

Myriococcum Fr. (1823), ? Ascomycota (inc. sed.). 1 or 2, widespr. *M. albomyces* is included in *Melanocarpus*.

Myrioconium Syd. (1912), Anamorphic Sclerotiniaceae, 6.A1.15. Teleomorph *Myriosclerotinia*. 7, Eur., N. Am.

Myriodiscus Boedijn (1935), Leotiales (inc. sed.). 1, Sumatra.

Myriodontium Samson & Polon. (1978), Mitosporic fungi, 1.A1.12. 1, Italy.

Myriogenis Clem. & Shear (1931) ≡ Myriogenospora (Clavicipit.).

Myriogenospora G.F. Atk. (1894), Clavicipitaceae. 1 (on *Gramineae*, incl. *Saccharum*), Am. See Diehl (*USDA agric. Monogr.* **4**, 1950), Hanlin & Tortolero (*Mycotaxon* **39**: 237, 1990), Luttrell & Bacon (*CJB* **55**: 2090, 1977), Phelps & Morgan-Jones (*Mycotaxon* **47**: 41, 1993).

Myriogonium Cain (1948), Ascomycota (inc. sed.). 1 (on *Dacryobolus*), Eur., N. Am. See Kreger-van Rij (*in* Ainsworth *et al.* (Eds), *The Fungi* **4A**: 15, 1973), Carpenter & Krapp (*Mycotaxon* **21**: 487, 1984).

Myriolecis Clem. (1909) = Lecanora (Lecanor.).

Myrionora R.C. Harris (1988), Lecanoraceae (L). 1, USA.

Myriophacidium Sherwood (1974), Rhytismataceae. 4, N. & S. Am., Pakistan. See Sherwood (1980; key).

Myriophysa Fr. (1849) ? = Atichia (Mitosp. fungi).

Myriophysella Speg. (1910) = Seuratia (Seurat.).

Myrioclerotinia Buchw. (1947), Sclerotiniaceae. 9, widespr. See Schumacher & Kohn (*CJB* **63**: 1610, 1985; key), Palmer (*Friesia* **9**: 193, 1969), Schwegler (*Schweiz. Z. Pilzk.* **56**: 49, 1978), Swart-Velthuysen (*Coolia* **27**: 71, 1984). Anamorph *Myrioconium*.

Myriosperma Nägeli (1853) = Sarcogyne (Acarospor.).

Myriospora Nägeli (1853) = Acarospora (Acarospor.).

myriosporous, having many spores. Cf. oligosporous.

Myriostigma Kremp. (1874) = Cryptothecia (Arthon.).

Myriostigma G. Arnaud (1925) ≡ Myriostigmella (Dothideales).

Myriostigmella G. Arnaud (1952), Dothideales (inc. sed.). 1, Brazil. See Müller & v. Arx (1962).

Myriostoma Desv. (1809), Geastraceae. 1, widespr.

Myriotrema Fée (1824), Thelotremataceae (L). 150, mainly trop. See Hale (*Mycotaxon* **11**: 130, 1980; *Bull. Br. Mus. nat. Hist., Bot.* **8**: 227, 1981).

Myriotrichum, see *Myxotrichum*.

Myrmaeciella Lindau (1897), Hypocreaceae. 1, N. Am. See Samuels & Seifert (*in* Sugiyama (Ed.), *Pleomorphic fungi*: 29, 1987). Anamorph *Patellina*.

Myrmaecium Nitschke ex Fuckel (1870) Ascomycota (inc. sed.). 2, Eur. = Valsaria (Diaporthales) fide Clements & Shear (1931).

Myrmaecium Sacc. (1880) ≡ Myrmaeciella (Hypocr.).

Myrmecocystis Harkn. (1899) = Genabea (Otid.) fide Trappe (1975).

Myrmecomyces Jouvenaz & Kimbr. (1991), Mitosporic fungi, 1.A1.10. 1 (endoparasitic on fire ants), USA, Argentina.

myrmecophilous (of fungi), being a covering or food for ants.

Myrophagus Thaxt. ex Sparrow (1939) = Myiophagus (Chytridiales) fide Karling (1977).

Myropyxis Ces. ex Rabenh. (1851), ? Mitosporic fungi, ?.?.?. 1 or 2, Eur.

Myrosporium Corda (1831) ? = Lachnobolus (Myxom.).

Myrotheciastrum Abbas & B. Sutton (1988), Mitosporic fungi, 8.A2.19. 1, Pakistan, India.

Myrotheciella Speg. (1910) = Myrothecium (Mitosp. fungi) fide Tulloch (1972).

Myrothecium Tode (1790), Mitosporic fungi, 4.A2.15. 14, widespr. See Tulloch (*Mycol. Pap.* **130**, 1972; key), Nag Raj (*Mycotaxon* **53**: 295, 1995; *M. prestonii* heterogeneous).

Mystrosporiella Munjal & Kulshr. (1969), Mitosporic fungi, 1.D2.10. 1, India.

Mystrosporium Corda (1837), Mitosporic fungi, 1.D2.?. 5, temp. *M. adustum* (ink disease of iris). Nom. dub. fide Hughes (*CJB* **36**: 788, 1958).

Mythicomyces Redhead & A.H. Sm. (1986), Strophariaceae. 1, N. temp.

Mytilidion Sacc. (1875) ≡ Mytilinidion (Mytilinid.).

mytiliform, like a mussel shell in form.

Mytilinidiaceae Kirschst. (1924), Dothideales. 7 gen. (+ 7 syn.), 36 spp. Ascomata superficial or with the base immersed, elongate, sometimes branched, usually taller than wide, opening by a long slit in the upper surface, black, thick-walled; peridium carbonaceous, composed of intertwined hyphal tissue; interascal tissue of trabeculate pseudoparaphyses and/or periphysoids, sometimes evanescent; asci ± cylindrical, at least in most cases fissitunicate; ascospores becoming brown, septate, sometimes muriform. Anamorphs usually coelomycetous. Usually saprobic on woody tissue, esp. on gymnosperms, widespr.

Mytilinidion Duby (1861), Mytilinidiaceae. 12, Eur., N. Am. See Speer (*BSMF* **102**: 97, 1986), Zogg (*Beitr. Krypt.-fl. schweiz* **11**(3), 1962), Barr (*N. Am. Fl.* II **13**, 1990; N. Am.). Anamorph *Camaroglobulus*.

Mytilodiscus Kropp & S.E. Carp. (1984), Leotiaceae. 1, N. Am.

Mytilostoma P. Karst. (1880), Dothideales (inc. sed.). 2, Finland.

Myxacium (Fr.) P. Kumm. (1871) = Cortinarius (Cortinar.) fide Singer (1951), but used by Moser (*Schweiz. Z. Pilzk.* **40**: 181, 1962; key C. Eur. spp.).

myxamoeba, (1) (of myxomycetes), a zoospore after becoming amoeba-like; (2) a myxopod (obsol.).

myxarioid (of basidia), having a stalk-like portion separated by a wall from the globose metabasidial portion, as in *Myxarium* (Donk, *Persoonia* **4**: 232, 1966).

Myxarium Wallr. (1833), Exidiaceae. 14, widespr. See Raitviir (*Key to Heterobasidiomycetes found in the USSR*, 1967), Donk (*Persoonia* **4**: 232, 1966), Hauerslev (*Mycotaxon* **49**: 235, 1993; key Danish spp.).

Myxasterina Höhn. (1909) = Asterina (Asterin.) fide Stevens & Ryan (1939).

Myxobacterales, an order of Bacteria (gliding bacteria) bearing extracellular slime; lyse filamentous and yeast fungi and bacteria in soil, dung and decaying vegetation.

Myxocephala G. Weber, Spaaij & Oberw. (1989) = Phialocephala (Mitosp. fungi) fide Sutton (*in litt.*).

Myxochytridiales, see *Archimycetes*.

Myxocladium Corda (1837) = Cladosporium (Mitosp. fungi) fide Saccardo (*Syll. Fung.* **4**: 364, 1886).

Myxocollybia Singer (1936) = Flammulina (Tricholomat.) fide Singer (1951).

Myxocybe Fayod (1889) = Hebeloma (Cortinar.) fide Singer (1951).

Myxocyclus Riess (1852), Mitosporic fungi, 3/6.D2.1. 1, Eur., N. Am.

Myxoderma Fayod ex Kühner (1926) ≡ Limacella (Amanit.).

Myxodictyon A. Massal. (1860) = Brigantiaea (Brigant.) fide Hafellner & Bellemère (*Nova Hedw.* **35**: 237, 1982).

Myxodictyonomyces Cif. & Tomas. (1953) ≡ Myxodictyon (Brigant.).

Myxodiscus Höhn. (1906) = Leptothyrium (Mitosp. fungi) fide Höhnel (*Mykol. Unters.* **1**: 301, 1913).

Myxodochium Arnaud (1951), Mitosporic fungi, 3.A1.1. 1 (with clamp connexions), Eur.

Myxofusicoccum Died. (1912) = Pseudodiscula (Mitosp. fungi) fide Sutton (1977).

Myxogastres, see *Myxomycetes*.

Myxolibertella Höhn. (1903) nom. rej. prop. = Phomopsis (Mitosp. fungi) fide Sutton (1977).

myxolichens, synthetically produced loose associations of myxomycete plasmodia and a green alga (*Chlorella*).

Myxomonas Brzez. (1906) nom. dub.; based on dead or partly formed cells fide Trzebinski (*Z. PflanKrKh.* **17**: 321, 1908).

Myxomphalia Hora (1960) = Fayodia (Tricholomat.) fide Singer (1962).

Myxomphalos Wallr. (1833) = Agyrium (Agyr.) fide Saccardo (*Syll. Fung.* **8**: 635, 1889).

Myxomycetes (Myxogastrea, Myxogastres, Myxomycotina, Myxothallophyta, Myxomycophyta, Phytosarcodina), Myxomycota; true slime moulds. 6 ord., 12 fams., 60 gen. (+ 91 syn.), 690 spp. Trophic phase a free-living, multinucleate, coenocytic, saprobic plasmodium with a shuttle-movement of the protoplasm; plasmodium sometimes becoming a resting body or sclerodium under poor conditions, and the swarm spores of myxamoebae microcysts; sporangia sessile or stalked, often bright coloured; spores developed after meiosis, produced in masses with a persistent or evanescent peridium; swarm cells usually with two-anterior flagellae and no cell wall, forming myxamoebae directly or after loss of the flagella, sometimes undergoing division before capulation; reproduction usual, the resultant zygote becoming the plasmodium. Esp. on old wood or other plant material undergoing decomposition (see Madelin, *TBMS* **83**: 1, 1984; ecological significance).

The chief diagnostic characters are the type of sporocarp, the structure of the peridium and capillitium of the sporangium, calcium carbonate ('lime'), if present, and the size, colour and ornamentation of the spores. Ords:

(1) **Echinosteliales**.
(2) **Echinosteliopsidales**.
(3) **Liceales**.
(4) **Physarales**.
(5) **Stemonitales**.
(6) **Trichiales**.

Frederick (1990) and Newbert *et al.* (1993) include the *Ceratiomyxaceae* in this class, but see under *Protosteliomycetes*.

Lit.: *General*: Lister & Lister (*Mycetozoa*, edn 3, 1925), Jahn (*Nat. PflFam.* **2** (Aufl. 2), 1928), MacBride & Martin (*Myxomycetes*, 1934), Gray & Alexopoulos (*Biology of myxomycetes*, 1968), Martin (*Univ. Iowa Studies nat. Hist.* **20** (8), 1966; gen. list) Martin & Alexopoulos (*The myxomycetes*, 1969). Martin *et al.* (*The genera of Myxomycetes*, 1983), Olive (*Bot. Rev.* **36**: 59, 1970; keys, *The mycetozoans*, 1975), Teixeira (*Rickia* suppl. **4**, 1971; gen.), Alexopoulos (*in* Subramanian (Ed.), *Taxonomy of the fungi* **1**: 1, 1978; evolution), Martin *et al.* (*The genera of Myxomycetes*, 1983), Frederick (*in* Margulis *et al.* (Eds) 1990: 467), Corliss (1994), Stephenson & Stempen (*Myxomycetes, A handbook of slime molds*, 1994).

Regional: **Argentina**, Deschamps (*Physis* Sect. C **34**: 159, **35**: 319, 1975-77). **Australia**, Mitchell (*Nova Hedw.* **60**: 269, 1995: list). **British Isles**, Ing (*A census catalogue of British myxomycetes*, 1968; revised, *Bull. BMS* **14**: 97, **16**: 26, 1980-82; **19**: 109, 1985). **Central & S. America**, Farr (*Rickia* **3**: 45, 1968; key, *Flora neotropica* **16**, 1976). **Costa Rica**, Alexopoulos & Sáenz (*Mycotaxon* **2**: 223, 1975). **Denmark**, Bjornekaer & Klinge (*Friesia* **7**: 149, 1964). **Germany**, etc. Neubert *et al.* (*Die Myxomyceten* **1**, 1993; keys, col. pls), Schinz (*Rabenh. Krypt.-Fl.* **1** (10), 1920). **Hawaii**, Eliasson (*MR* **95**: 257, 1991). **India**, Thind (*The myxomycetes of India*, 1977), Lakhanpal & Mukerji (*Indian myxomycetes*, 1981). **Italy**, Pirola & Credaro (*Giorn. bot. Ital.* **105**: 157, 1971). **Japan**, Emoto (*The myxomycetes of Japan*, 1977). **Korea**, Nakagawa (*Jl Chosen Nat. Hist. Soc.* **17**: 17, 1934). **Mexico**, Braun & Keller (*Mycotaxon* **3**: 297, 1976). **Netherlands**, Nannenga-Bremekamp (*De Nederlandse Myxomyceten,* 1975; *A guide to temperate myxomycetes* [Engl. transl.], 1992). **Nigeria**, Ing (*TBMS* **47**: 45, 1964). **N. America**, Hagelstein (*The Mycetozoa of North America*, 1944), Martin (*N. Am. Fl.* **1**(1), 1949). **Sierra Leone**, Ing (*TBMS* **50**: 549, 1967). **Sweden**, Santesson (*Svensk bot. Tidskr.* **58**: 113, 1964; list). **Tierra del Fuego**, Arambarri (*Fl. cript. Tierra del Fuego* **2**, 1975). **Uruguay**, Garcia-Zorron (*Mixomycetos del Uruguay*, 1967). See also Myxomycota.

Myxomycetes Renault (1895), Fossil fungi (? Myxomycetes). 1 (Carboniferous), France.

myxomyceticolous, growing on myxomycetes. See Fungi on fungi.

Myxomycidium Massee (1901) = Mucronella (Heric.) fide Petersen (*Mycol.* **72**: 301, 1980).

Myxomycites Mesch. (1898) ≡ Myxomycetes (Fossil fungi).

Myxomycota. (Myzetozoa, Myxostelida, Myxobionta, Gymnomycota, Gymnomyxa), Protozoa; plasmodial slime moulds; acellular slime moulds. 2 classes, 7 orders, 15 fam., 74 gen. (+ 94 syn.), 719 spp. Free-living, unicellular or plasmodial, non-flagellate in single- or multi-celled phagotrophic stages; mitochondrial cristae tubular; with uni- or multicellular sporocarps with single to many spores; spore walls cellulosic or chitinous; spores germinating to produce 1- or 2-flagellate cells.

The circumscription and ranking of the higher taxa of plasmodial slime moulds varies according to different authors. No formal name at all was used in Margulis *et al.* (1990), while Corliss (1994) used *Mycetozoa* inclusive of the dictyostelids (see *Dictyosteliomycota*). Two classes are recognized in this edition of the *Dictionary*:

(1) **Myxomycetes.**
(2) **Protosteliomycetes.**
Corliss (1994) suggested that the phylum could be expanded to include the *Promycetozoida*, certain enigmatic marine protists with reticulopodia such as *Corallomyxa, Megamoebomyxa* and *Thalassomyxa* (see Grell, *Protistologica* **21**: 215, 1985; *Arch. Protistenkde* **140**: 303, 1991); these taxa have not been studied by mycologists and are not treated elsewhere in this *Dictionary.*
Lit.: Corliss (*Acta Protozool.* **33**: 1, 1994), Keller (*in* Parker (Ed.), 1982, **1**: 165), Margulis *et al.* (1990), Teixeira (*Rickia* suppl. **3**, 1970; gen. except Myxomycetes), Olive (*The mycetozoans*, 1975; *in* Parker, *Synopsis and classification of living organisms*: 521, 1982), Raper (*The dictyostelids*, 1994). See also *Acrasomycota, Dictyosteliomycota, Myxomycetes, Protozoa.*

Myxomyriangium Theiss. (1913) = Saccardinula (Elsin.) fide v. Arx (1963).

Myxonema Corda (1837) [non Fr. (1825), *Algae*] ≡ Myxonyphe (Inc. sed.).

Myxonyphe Rchb. (1841) nom. dub. (Fungi, inc. sed.).

Myxoparaphysella Caball. (1941), Mitosporic fungi, 6.A1.?. 1, Spain.

Myxophacidiella Höhn. (1917) = Pseudophacidium (Ascodichaen.) fide v. Arx & Müller (1954).

Myxophacidium Höhn. (1917) = Pseudophacidium (Ascodichaen.) fide v. Arx & Müller (1954), DiCosmo *et al.* (*Mycotaxon* **21**: 68, 1984).

Myxopholis Locq. (1979), Cortinariaceae. 1, Eur.

Myxophora Döbbeler & Poelt (1978), Pseudoperisporiaceae. 1, Austria.

Myxopuntia Mont. (1846) = Leptogium (Collemat.).

Myxormia Berk. & Broome (1850) = Myrothecium (Mitosp. fungi) fide Tulloch (1972).

myxospore, a myxomycete spore (obsol.).

Myxosporella Sacc. (1881), Mitosporic fungi, 6.A1.23. 1, Italy. See Sutton (1977).

Myxosporidiella Negru (1960), Mitosporic fungi, 7.A1.?. 1, Romania.

Myxosporina Höhn. (1927) = Rhodesia (Mitosp. fungi) fide Sutton (1977).

myxosporium, see perisporium; Spore wall.

Myxosporium Link (1825) nom. conf. fide Höhnel (*Z. Garungs.* **5**: 191, 1915), see also Weindlmayr (*Sydowia* **17**: 74, **18**: 26, **19**: 193, 1964-66).

Myxostomella Syd. (1931) = Campoa (Parmular.) fide Müller & v. Arx (1962).

Myxostomellina Syd. (1931), Mitosporic fungi, 8.A1.?. 1, Philipp.

Myxothallophyta, see *Myxomycetes.*

Myxotheca Ferd. & Winge (1910) = Cryptothecia (Arthon.).

Myxotheciella Petr. (1959) = Scolionema (Parodiopsid.) fide Müller & v. Arx (1962).

Myxothecium Kunze (1823) = Meliola (Meliol.) fide Hansford (1961).

Myxothyriopsis Bat. & A.F. Vital (1956), Mitosporic fungi, 4.A1.?. 1, Brazil.

Myxothyrium Bubák & Kabát (1915), Mitosporic fungi, 8.A1.15. 1, Eur.

Myxotrichaceae Locq. ex Currah (1985), Onygenales. 4 gen. (+ 6 syn.), 12 spp. Ascomata not stipitate; peridium often with complex thick-walled appendages; ascospores ± fusiform, smooth or striate. Anamorphs *Geomyces, Malbranchea* or *Oidiodendron.* Cellulolytic, on paper, straw etc., cosmop.

Myxotrichella Sacc. (1892) = Myxotrichum (Myxotrich.) fide Kendrick & Carmichael (*in* Ainsworth *et al.* (Eds), *The Fungi* **4A**: 390, 1973).

Myxotrichum Kunze (1823), Myxotrichaceae. 7, widespr. See Orr *et al.* (*CJB* **41**: 1463, 1963; key),

Apinis (1964), Currah (1985), Hughes (*CJB* **46**: 939, 1968; synonymy), v. Arx (1981), Dalpé (*MSA News* **38**: 22, 1987; anamorph). Anamorph *Oidiodendron.*

Myxozyma Van der Walt, Weijman & Arx (1981), Mitosporic fungi, 1.A1.3/4. 5, N. Am., S. Afr. See Yamada (*J. Gen. Appl. Microbiol.* **32**: 259, 1986; coenzyme Q system), v. d. Walt *et al.* (*Syst. Appl. Microbiol.* **9**: 121, 1987; key 4 spp.), Cottrell & Kock (*Syst. Appl. Microbiol.* **12**: 291, 1989; history, delimitation, phenotypic characters), Spaaij *et al.* (*Syst. Appl. Microbiol.* **13**:182, 1990).

Myxyphe, see *Myxonyphe.*

Myzeloblastanon, see *Myceloblastanon.*

Myzocytiopsidaceae M.W. Dick (1995), Myzocytiopsidales. 5 gen. (+ 2 syn.), 36 spp. See Dick (*in press*; key).

Myzocytiopsidales, Oomycota. 6 fam., 13 gen. (+ 4 syn.), 72 spp. Fams:
(1) **Crypticolaceae.**
(2) **Ectrogellaceae.**
(3) **Eurychasmataceae.**
(4) **Myzocytiopsidaceae.**
(5) **Pontismataceae.**
(6) **Sirolpidiaceae.**
The *Ectrogellaceae* may not belong here (the type is of uncertain application); other species may be close to *Haptoglossa.*
Lit.: Dick (*in press*; key).

Myzocytiopsis M.W. Dick (1995), Myzocytiopsidaceae. 19 (endobiotic in *Aschelminthes*), widespr.

Myzocytium Schenk (1858), Pythiaceae. 2, widespr. See Dick (*in press*; key)

nacreous, like mother-of-pearl.

Nadsonia Syd. (1912), Saccharomycodaceae. 2, N. temp. See Golubev *et al.* (*Ant. v. Leeuwenhoek* **55**: 369, 1989; key), Kreger van Rij (1984).

Nadsoniella Issatsch. (1914), Mitosporic fungi, 1.A1.6. 1 (from seawater), former USSR.

Nadsoniomyces Kudrjanzev (1932) = Trichosporon (Mitosp. fungi) fide Carmo-Sousa *in* Lodder (1970).

Nadvornikia Tibell (1984), Thelotremataceae (L). 2, N. & S. Am., Hawaii, Australia. See Harris (*Some Florida lichens*, 1990; posn).

Naegelia Rabenh. (1844) nom. dub. (Fungi, inc. sed.).

Naegelia Reinsch (1878) ≡ Sapromyces (Leptomit.). See Coker & Mathews (1937).

Naegeliella J. Schröt. (1893) [non Correns (1892), *Algae*] ≡ Sapromyces (Leptomit.). See Coker & Mathews (1937).

Naegiella (Rehm) Clem. (1909) = Eupropolella (Dermat.) fide Défago (1967).

Naemacyclus Fuckel (1873), Rhytismataceae. 4, Eur., India, N. Am.

Naemaspora Willd. (1787) ≡ Bombardia (Lasiosphaer.).

Naemaspora Pers. (1796) nom. conf. fide Sutton (*Mycol. Pap.* **141**, 1977).

Naemaspora Sacc. (1880) = Roscoepoundia (Mitosp. fungi) fide Sutton (1977).

Naematelia Fr. (1822) nom. dub.; based on parasitized *Stereum* fide Bandoni (*Am. midl. Nat.* **66**: 319, 1961).

Naematoloma P. Karst. (1880) = Hypholoma (Strophar.). See Smith (*Mycologia* **43**: 467, 1951; key 21 N. Am. spp.).

Naemosphaera P. Karst. (1888), Mitosporic fungi, 4.A2.?. 1, Finland.

Naemosphaera (Sacc.) Sacc. (1892) = Naemosphaera (Mitosp. fungi) fide Sutton (*in litt.*).

Naemosphaerella Höhn. (1923), Mitosporic fungi, ?.?.?. 2, Brazil, Japan. See Petrak (*Sydowia* 6: 302, 1952).

Naemospora Roth ex Kuntze (1898) nom. dub. (? Mitosp. fungi).

Naemostroma Höhn. (1919) = Septoriella (Mitosp. fungi) fide Sutton (1973).

Naetrocymbe Körb. (1865) = Arthopyrenia (Arthopyren.) fide Eriksson (*SA* 6: 141, 1987).

Naetrocymbe Bat. & Cif. (1963) = Limacinula (Coccodin.) fide Eriksson (*Opera Bot.* 60, 1981).

Naevala B. Hein (1976), Dermateaceae. 6, Eur. See Hein (*Willdenowia, Beih.* 9, 1976).

Naevia Fr. (1825) = Arthonia (Arthon.).

Naevia Fr. (1849) ≡ Naevala (Dermat.).

Naeviella (Rehm) Clem. (1909), Dermateaceae. 4, Eur. See Nannfeldt (*Sydowia* 35: 162, 1982).

Naeviopsis B. Hein (1976), Dermateaceae. 12, Eur.

Naganishia Goto (1963) = Cryptococcus (Mitosp. fungi) fide Phaff & Fell (*in* Lodder, 1970: 1088).

Nagrajia R.F. Castañeda & W.B. Kendr. (1991), Mitosporic fungi, 4.A1.15. 1, Cuba.

Nagrajomyces Melnik (1984), Mitosporic fungi, 4.D2.19. 1, Russia.

Naiadella Marvanová & Bandoni (1987), Mitosporic fungi, 1.A1.1. 1 (aquatic, basidiomycete anamorph), Canada, former Czechoslovakia.

Nailisporites T.C. Huang (1981), Fossil fungi. 1 (Miocene), Taiwan.

Naïs Kohlm. (1962), Halosphaeriaceae. 2 (marine), Eur., N. Am. See Crane & Shearer (*TBMS* 86: 509, 1986).

Nakaiomyces Kobayasi (1939) nom. dub.; based on a parasitized *Tremella* fide Olive (*Bull. Torrey bot. Cl.* 85: 99, 1958).

Nakataea Hara (1939) = Pyricularia (Mitosp. fungi) fide Sivanesan (*in litt.*).

Nakazawaea Y. Yamada, K. Maeda & Mikata (1994), Saccharomycetaceae. 1, USA.

Nalanthamala Subram. (1956), Mitosporic fungi, 3.A1.15. 2, India.

Namakwa Hale (1988) Parmeliaceae. 1, S. Afr. ? = Parmelia (Parmel.).

nameko, basidioma of the edible *Pholiota nameko* cultivated in Japan (Chang & Hayes, 1978).

names, (1) of fungi, see Common names, Nomenclature; (2) Escallon (*Precis de myconymie*, 1989; meaning of scientific names).

Nannfeldt (John Axel Frithiof; 1904-1985), Swedish botanist and mycologist, Professor and Director of the Botanical Garden and Herbarium at the University of Uppsala 1939-70, producing critical accounts of numerous groups of discomycete fungi, and later some pyrenomycetes (esp. *Sordariales* and *Xylariales*) and smuts, best known for his *Studien über die Morphologie und Systematik der nicht lichenisitten inoperculaten Discomyceten* [*Nova Acta reg. Soc sci. upsal.*, VI, 8(2)] (1932) that established that there were two main types of ascomycete ontogeny, the ascohymenial and ascolocular (see also Luttrell). Also made pioneering investigations into blue-stain fungi with E. Melin (*Researches into the blueing of ground wood-pulp*, 1934). *Obit.*: Holm (*Mycol.* 78: 692, 1986). See Grummann (1974: 495).

Nannfeldtia Petr. (1947), ? Leptopeltidaceae. 1, Austria.

Nannfeldtiella Eckblad (1968), Otideaceae. 2, Eur. See Nannfeldt (*Sydowia* 35: 166, 1982). = Pseudombrophila (Otid.) fide Harmaja (*Ann. bot. fenn.* 16: 159, 1979).

Nannfeldtiomyces Vánky (1981), Tilletiaceae. 2 (on *Sparganium*), Eur., N. Am.

Nannizzia Stockdale (1961) = Arthroderma (Arthrodermat.) fide Kawasaki *et al.* (*Mycopath.* 118:

95, 1992), Weitzman *et al.* (*Mycotaxon* 25: 505, 1986).

Nannizziopsis Currah (1985), Onygenaceae. 3, Eur., Afr., N. Am. See Guarro *et al.* (*Mycotaxon* 42: 193, 1991; key). Anamorph *Chrysosporium*.

Nanomyces Thaxt. (1931), Laboulbeniaceae. 3, Pacific.

Nanoschema B. Sutton (1980), Mitosporic fungi, 8.B1.15. 1, Solomon Isl.

Nanoscypha Denison (1972), Sarcoscyphaceae. 3, C. & N. Am., Afr.

Nanostictis M.S. Christ. (1954), Stictidaceae. 2 (on lichens, esp. *Peltigera*), Denmark, N. Am. See Sherwood (*Sydowia* 38: 267, 1986).

Nanstelocephala Oberw. & R.H. Petersen (1990), Crepidotaceae. 1, USA.

Naohidea Oberw. (1990), Platygloeaceae. 1, N. temp.

Naohidemyces S. Sato, Katsuya & Y. Hirats. (1993), Pucciniastraceae. 2, widespr.; 0, I on *Tsuga*, II, III on *Ericaceae*.

Naothyrsium Bat. (1960), Mitosporic fungi, 5.B2.1. 1, Brazil.

Napicladium Thüm. (1875) = Spilocaea (Mitosp. fungi) fide Hughes (1958).

napiform, turnip-like in form (Fig. 37.26).

Napomyces Setch. ex Clem. & Shear (1931) ≡ Daleomyces (Peziz.). See Korf (*Mycol.* 48: 711, 1956).

Naranus Ts. Watan. (1995), Mitosporic fungi, ?.?.?. 1, Japan.

Narasimhania Thirum. & Pavgi (1952), Tilletiaceae. 1 (on *Alisma*), India. See Vánky (*Sydowia* 34: 167, 1981).

Narasimhella Thirum. & P.N. Mathur (1966), Gymnoascaceae. 2, trop. See v. Arx (1981, *Persoonia* 13: 173, 1986). = Gymnascella (Gymn.) fide v. Arx & Samson (*Persoonia* 13: 185, 1988).

Nascimentoa Cif. & Bat. (1956) = Spiropes (Mitosp. fungi) fide Ellis (1968).

nassace (nasse), the finger-like protrusion of the inner part of a bitunicate ascus into the inner tunica; internal apical beak; see ascus.

nasse, see nassace.

Nassula Fr. (1849) = Arcyria (Arcyr.).

Natalia Fr. (1847) [non Hochst. (1841), *Melianthaceae*] ≡ Levieuxia (Mitosp. fungi).

native bread, see blackfellows' bread.

Nattrassia B. Sutton & Dyko (1989), Mitosporic fungi, 8.C2.15. 1, widespr. *N. mangiferae* (syn. *Hendersonula toruloidea*), a common plant and animal pathogen. See Sutton & Dyko (*MR* 93: 483, 1989).

Naucoria (Fr.) P. Kumm. (1871), Cortinariaceae. 20, cosmop. See Orton (*TBMS* 43: 308, 1960), Pegler & Young (*Kew Bull.* 30: 325, 1975; 15 Br. spp.), Reid (*TBMS* 82: 191, 1984; key 17 Br. spp.); not used by Singer (1962).

Naumovela Kravtzev (1955), Ascomycota (inc. sed.). 2 (on wood), former USSR.

Naumovia Dobrozr. (1928) = Rosenscheldia (Dothideales) fide v. Arx & Müller (1975).

Naumoviella Novot. (1950) = Mortierella (Mortierell.) fide Hesseltine (1955).

Nautosphaeria E.B.G. Jones (1964), Halosphaeriaceae. 1 (marine), UK.

navel, see umbilicus.

Navicella Fabre (1879), Massariaceae. 1, Eur. See Barr (*N. Am. Fl.* II, 13, 1990; posn), Eriksson (1981), Holm & Holm (*Symb. bot. upsal.* 28(2), 1988).

navicular (naviculate), boat-like in form; cymbiform (Fig. 37.32).

Navisporus Ryvarden (1980), Coriolaceae. 1, Afr.

Nawawia Marvanová (1980), Mitosporic fungi, 1.A1.15. 2 (aquatic), Malaysia. See Kuthubutheen *et al.* (*CJB* 70: 96, 1992).

Necator Massee (1898), Anamorphic Corticiaceae, 3.A1.?. Teleomorph *Corticium*. 1, S.E. Asia. See Brooks & Sharples (*Ann. appl. Biol.* **2**: 58, 1915).

Necium Arthur (1907), = Melampsora (Melampsor.) fide Davis (1915).

Necol technique, see Mounting media.

necral layer, a layer of horny dead fungal hyphae with indistinct lumina in or near the cortex of lichens; **epinecral layer** if above the algal layer, **hyponecral layer** if below.

Necraphidium Cif. (1951), Mitosporic fungi, 1.A/B1.?. 1 (on larvae of *Callipterus*), Italy.

necrophagous, saprobic.

necrophyte, an organism living on dead material (Münch); cf. perthophyte, saprophyte.

necrosis, death of plant cells, esp. when resulting in the tissue becoming dark in colour; commonly a symptom of fungus infection.

Necrosis Paulet (1793) = Ustilago (Ustilagin.) fide Dietel (1928).

necrotroph, a parasite that derives its energy from dead cells of the host (Thrower, *Phytopath. Z.* **56**: 258, 1966). Cf. biotroph.

nectar, the sticky, sometimes sweet, secretion (esp. of pycnia of rusts) in which spores or spermatia may be freed, and which has an attraction for insects.

Nectaromyces Syd. & P. Syd. (1919) = Metschnikowia (Metschnikow.) fide v. Arx *et al.* (1977).

Nectria (Fr.) Fr. (1849) nom. cons., Hypocreaceae. 200, cosmop. *N. cinnabarina* (coral spot of woody plants; Lohman & Watson, *Lloydia* **6**: 77, 1943; *Nectria* spp. on hardwoods); *N. galligena* (canker and eye rot of apple and pear; Flack & Swinburne, *TBMS* **68**: 186, 1977; host range). See Dingley (1951, 1957), Booth (*Mycol. Pap.* **73**, 1959; UK, keys), Brayford & Samuels (*Mycol.* **85**: 612, 1993; *Cylindrocarpon* anamorphs), Passeur (*Sydowia* **27**: 7, 1975; morphology, substratum connexion), Perrin (*BSMF* **92**: 335, 1976; key Eur. spp.), Rossman (*Mem. N.Y. bot. Gdn* **49**: 253, 1989; key 28 spp.), Samuels (*Mem. N.Y. bot. Gdn* **26** (3), 1976; *Hyphonectria*, key 33 spp., *N.Z. Jl Bot.* **14**: 231, 1976; *Acremonium* anamorphs, **16**: 73, 1978; *Cylindrocarpon* anamorphs, *Brittonia* **40**: 306, 1988; spp. striate ascospores, *Mem. N.Y. bot. Gdn* **49**: 266, 1989; *Sesquicillina* anamorphs), Doyle (*Mycol.* **70**: 355, 1978; metabolites), Rossman (*Mycotaxon* **8**: 485, 1979, former *Calonectria* spp.; *Mycol. Pap.* **150**, 1983, phragmosporous spp.), Brayford & Samuels (*Mycol.* **85**: 612, 1993), Samuels & Brayford (*Sydowia* **45**: 55, 1993; phragmosporous spp., **46**: 75, 1994; key 29 spp. red perith., striate spores), Samuels *et al.* (*Mycol. Pap.* **164**, 1991; keys 40 spp. subgen. *Dialonectria*). See also Fungicolous fungi. Anamorph *Tubercularia* (in *Nectria* s.str.); others (in *Nectria* s.l.) include *Acremonium, Cylindrocarpon, Fusarium*.

Nectriaceae, see *Hypocreaceae*.

Nectriella Nitschke ex Fuckel (1870), Hypocreaceae. *c.* 5, temp. See Lowen (*Mycotaxon* **39**: 461, 1990; concept).

Nectriella Sacc. (1877) ≡ Pseudonectria (Hypocr.).

Nectrioidaceae (*Zythiaceae*), nom. inval. and obsol. Mitosporic fungi. 58 gen. (38 monotypic) + 15 syn., 150 spp. Conidiomata brightly coloured, and more or less fleshy. See Petch (*TBMS* **26**: 53, 1943; genera), Sutton (*in* Ainsworth *et al.* (Eds), *The Fungi* **4A**, 1973; discussion family names).

Nectriopsis Maire (1911), Hypocreaceae. 42, widespr. See Samuels (*Mem. N.Y. bot. Gdn* **48**, 1988).

needle cast (of conifers), loss of leaves caused by spp. of *Hypoderma, Lophodermium, Rhabdocline*, or other *Rhytismatales*.

Neelakesa Udaiyan & V.S. Hosag. (1992), Pseudeurotiaceae. 1, India.

Nees (von Esenbeck, Christian Gottfried Daniel; 1776-1858) was author of *Das System der Pilze und Schwämme* (1817; 366 col. fig. on 44 pl.). He also issued papers with his brother (Theodore Friedrich Ludwig **Nees**; 1787-1837) and the latter also published (with Henry) *Das System der Pilze* (1837, with coloured plates [Abt. 2 by Bail (1858)]). See Grummann (1974: 34), Stafleu & Cowan (*TL-2* **3**: 712, 1981).

Neesiella Kirschst. (1935) [non Schiffn. (1893), *Hepaticae*] ≡ Myconeesia (Xylar.).

Negeriella Henn. (1897), Mitosporic fungi, 2.C2.?. 2, S. Am. See Morris (*Mycopathol.* **33**: 183, 1967).

Neilreichina Kuntze (1891) ≡ Polytrichia (Pleospor.).

Nellymyces Batko (1972), Rhipidiaceae. 1, Poland.

Nemaclada J. Sm. (1896), Fossil fungi (mycel.). 1 (Carboniferous), UK.

Nemacola A. Massal. (1855) ? Ascomycota (inc. sed.; (L) or *Algae*).

Nemadiplodia Zambett. (1955) = Botryodiplodia (Mitosp. fungi) fide Petrak (*Sydowia* **16**: 353, 1963).

Nemania Gray (1821), Xylariaceae. 11, mainly temp. See Læssøe (*SA* **13**: 113, 1994), Petrini & Rogers (*Mycotaxon* **26**: 40, 1986; key, as *Hypoxylon serpens* group), Pouzar (*Česká Myk.* **39**: 129, 1985).

Nemaplana J. Sm. (1896), ? Fossil fungi (mycel.). 1 (Carboniferous), UK.

Nemaria Navas (1909) = Roccella (Roccell.).

Nematidia Stirt. (1879) = Eremothecella (Arthon.) fide Sérusiaux (1992).

Nematococcus Kütz. (1833) nom. dub. (Fungi, inc. sed.).

Nematocolla Link (1833) = Myxosporium (Mitosp. fungi) fide Link (1833).

Nematoctonus Drechsler (1941), Anamorphic Tricholomataceae, 1.A2.10 + A1.1. Teleomorph *Hohenbuehelia*. 9 (with clamp connexions, on nematodes), widespr. See Thorn & Barron (*Mycotaxon* **25**: 321, 1986; key 14 spp.), Giuma *et al.* (*TBMS* **60**: 49, 1973; nematoxins).

nematodes (fungi pathogenic to), see Mycopesticides, Predacious fungi, *Zoopagaceae*.

nematogenous, of conidiogenous cells arising at all levels from single hyphae; lanose. See Gams (*Cephalosporium-artige Schimmelpilze*, 1971).

Nematogonum Desm. (1834), Mitosporic fungi, 1.A1.3. 2 (contact mycoparasites), temp. See Gams (*Rev. mycol.* **39**: 273, 1975), Walker & Minter (*TBMS* **77**: 299, 1981).

Nematographium Goid. (1935), Mitosporic fungi, 2.A1.?. 5, Eur.

Nematoloma P. Karst. (1880) = Hypholoma (Strophar.) fide Dennis *et al.* (*TBMS* **37**: 33, 1954), used by Singer (1962).

Nematomyces Faurel & Schotter (1965) = Gliomastix (Mitosp. fungi) fide v. Arx (1970).

nematophagous, nematode-feeding; see Predacious fungi. Nematophagous fungi are also reported as fossils (Jansson & Poinar, *TBMS* **87**: 471, 1986).

Nematophagus Mekht. (1975), Mitosporic fungi, 1.C1.8. 1, Azerbaidzhan.

Nematophthora Kerry & D.H. Crump (1980), Leptolegniellaceae. 1 (on nematodes), UK.

Nematora Fée (1825) = Strigula (Strigul.) fide Santesson (1952).

Nematospora Peglion (1897), Metschnikowiaceae. 1 (or 3), mainly trop. *N. coryli* (in cotton-bolls, beans (*Phaseolus*), etc.), inoculated by insects. See Batra (*USDA Tech. Bull.* **1469**, 1973), Kreger van Rij (1984), stigmatomycosis, *Ashbya*.

Nematospora Tassi (1904) ≡ Giulia (Mitosp. fungi).

Nematosporaceae, see *Metschnikowiaceae*.

Nematosporangium (A. Fisch.) J. Schröt. (1893) = Pythium (Pyth.) fide Fitzpatrick (1930).

Nematostelium L.S. Olive & Stoian. (1970), Protosteliaceae. 2, widespr.

Nematostigma Syd. & P. Syd. (1913) [non A. Dietr. (1833), *Iridaceae*], Pseudoperisporiaceae. 1, S. Afr. See Hansford (*Mycol. Pap.* **15**, 1946), Petrak (*Sydowia* **3**: 251, 1949).

Nematostoma Syd. & P. Syd. (1914), Pseudoperisporiaceae. *c.* 10, trop. See Hansford (*Mycol. Pap.* **15**, 1946), Petrak (*Sydowia* **3**: 251, 1949), Rossman (*Mycol. Pap.* **157**, 1987).

Nematothecium Syd. & P. Syd. (1912), ? Pseudoperisporiaceae. *c.* 2, trop. See Rossman (*Mycol. Pap.* **157**, 1987).

nematotoxin, a metabolic product toxic to nematodes, e.g. **nematoctonins** from *Nematoctonus* spp. (Kennedy & Taplin, *TBMS* **70**: 140, 1978); see also Mycopesticides.

Nemecomyces Pilát (1933) = Pholiota (Strophar.) fide Singer (1975).

nemin, a principle from a nematode-free culture filtrate of *Neoplectana glaseri*, which caused *Arthrobotrys conoides* to differentiate traps (Pramer & Stoll, *Science, N.Y.* **129**: 966, 1959); see Predacious fungi.

nemoral, living in woods or groves.

Nemostroma Höhn. (1925) , see *Naemostroma*.

Nemozythiella Höhn. (1925), Mitosporic fungi, 8.B1.?. 1, Eur.

neo- (prefix), new.

Neoalpakesa Punith. (1981), Mitosporic fungi, 4.A1.19. 1, UK.

Neoarbuscula B. Sutton (1983), Mitosporic fungi, 1.D2.42. 1, trop. See Subramanian & Vittal (*CJB* **51**: 977, 1973).

Neoarcangelia Berl. (1900) = Pleurostoma (Calosphaer.) fide Shear (*Mycol.* **29**: 355, 1937).

Neoballadyna Boedijn (1961) = Dysrhynchis (Parodiopsid.) fide Müller & v. Arx (1962).

Neobarclaya Sacc. (1899), Mitosporic fungi, 6.B2.19. 1, USA. See Nag Raj (1993).

Neobarya Lowen (1986), Clavicipitaceae. 5, Eur., N. Am., Java. See Seifert (*SA* **5**: 310, 1986; status). Anamorph *Acremonium*.

Neobroomella Petr. (1947), Amphisphaeriaceae. 1, Syria. See Eriksson (*SA* **5**: 145, 1986).

Neobulgaria Petr. (1921), Leotiaceae. 7, Eur. See Dennis (*Mycol. Pap.* **62**, 1956), Verkley (*Persoonia* **15**: 3, 1992).

Neocallimastigaceae I.B. Heath (1983), Neocallimastigales. 5 gen. (+ 1 syn.), 16 spp. Mono- or polycentric anaerobes in the digestive tracts of vertebrates.
Lit.: Heath *et al.* (1989), Dore & Stahl (*CJB* **69**: 1964, 1991; RNA, phylogeny), Coleman (*in* Levett (Ed.), *Anaerobic microbiology. A practical approach*, 1991), Li & Heath (1992).

Neocallimastigales, Chytridiomycetes. 1 fam., 5 gen. (+ 1 syn.), 16 spp. Thallus mono- or polycentric; zoospores mono- or polyflagellate, without mitochondria; kinetosomes with a skirt, spur, cylinder and microtubule root, lacking props; anaerobic saprobes occurring in the guts of herbivores. Fam.:
Neocallimastigaceae.
Lit.: Heath (*CJB* **67**: 2815, 1989; emend.), Heath *et al.* (*CJB* **71**: 393, 1993), Li & Heath (*CJB* **70**: 1738, 1992; RNA, cladistics).

Neocallimastix Vavra & Joyon ex I.B. Heath (1983), Neocallimastigaceae. 4 (in rumen of sheep and cattle), widespr. Wubah *et al.* (*CJB* **69**: 835, 1991; morphology); Wubah *et al.* (*Mycol.* **83**: 40, 1991; resistant body).

Neocapnodium W. Yamam. (1955) = Phragmocapnias (Capnod.) fide Reynolds (1979).

Neocatapyrenium H. Harada (1993), Verrucariaceae (L). 1, Japan.

Neochaetospora B. Sutton & Sankaran (1991), Mitosporic fungi, 4.C1.1. 1 (coprophilous), Sahara.

Neoclitocybe Singer (1962), Tricholomataceae. 14, widespr.

Neocoleroa Petr. (1934) = Wentiomyces (Pseudoperispor.) fide v. Arx & Müller (1975).

Neocordyceps Kobayasi (1984), Clavicipitaceae. 1, Japan.

Neocosmospora E.F. Sm. (1899), Hypocreaceae. 9, widespr. (mainly trop). See Cannon & Hawksworth (*TBMS* **82**: 673, 1984; key), Udagawa *et al.* (*Sydowia* **41**: 349, 1989; key 7 spp.).

Neocryptospora Petr. (1959), Ascomycota (inc. sed.). 1, Brazil.

Neocudoniella S. Imai (1941), Leotiaceae. 1, Japan, USA. Kohn *et al.* (*Mycol.* **78**: 934, 1986).

Neodasyscypha Spooner (1987) nom. inval. = Lachnum (Hyaloscyph.). See Eriksson & Hawksworth (*SA* **7**: 83, 1988; status).

Neodeightonia C. Booth (1970) = Botryosphaeria (Botryosphaer.) fide v. Arx & Müller (1975).

Neodimerium Petr. (1950) = Epipolaeum (Pseudoperispor.) fide Müller & v. Arx (1962).

Neodiplodina Petr. (1954), Mitosporic fungi, 8.B1.?. 1, Australia.

Neofabraea H.S. Jacks. (1913) = Pezicula (Dermat.) fide Nannfeldt (1932).

Neoflageoletia J. Reid & C. Booth (1966), Phyllachoraceae. 1, Philipp.

Neofracchiaea Teng (1938), ? Nitschkiaceae. 1, China.

Neofuckelia Zeller & Goodd. (1935), Mitosporic fungi, 8.A1.?. 1, N. Am.

Neofuscelia Essl. (1978) Parmeliaceae. *c.* 75, mainly S. temp. Close to *Xanthoparmelia* fide Elix (1993). ? = Parmelia (Parmel.).

Neogeotrichum O. Magalh. (1932) = Trichosporon (Mitosp. fungi) fide Lodder & van Rij (*The yeasts*, 1952).

Neogibbera Petr. (1947) = Acantharia (Ventur.) fide Müller & v. Arx (1975).

Neogodronia Schläpf.-Bernh. (1969) = Encoeliopsis (Leot.) fide Korf (*in* Ainsworth *et al.*, *The Fungi* **4A**: 249, 1973).

Neogymnomyces G.F. Orr (1970), Onygenaceae. 1, USA.

Neogyromitra S. Imai (1932) = Gyromitra (Helvell.). See Eckblad (1968).

Neohaplomyces R.K. Benj. (1955), Laboulbeniaceae. 3, USA, Cuba.

Neohendersonia Petr. (1921), Mitosporic fungi, 8.C2.1/19. 2, widespr. See Sutton (1980).

Neohenningsia Koord. (1907), Hypocreaceae. 4, Asia, Am., Eur. See Gamundí (*Darwiniana* **18**: 548, 1974), Samuels (*N.Z. Jl Bot.* **14**: 231, 1976).

Neoheppia Zahlbr. (1909), Peltulaceae (L). 2, Brazil, Zaïre. See Büdel (*Mycotaxon* **54**: 137, 1995).

Neoheteroceras Nag Raj (1993), Mitosporic fungi, 4/6.C2.19. 1, France.

Neohoehnelia Theiss. & Syd. (1918) = Dysrhynchis (Parodiopsid.) fide Müller & v. Arx (1962).

Neohygrocybe Herink (1959) = Hygrocybe (Hygrophor.) fide Singer (1962).

Neohygrophorus Singer (1962) = Hygrophorus (Hygrophor.) fide Hesler & Smith (1963).

Neohypodiscus (Lloyd) J.D. Rogers, Y.M. Ju & Læssøe (1994), ? Amphisphaeriaceae. 3, Afr., N., S. & C. Am., Asia.

Neojohnstonia B. Sutton (1983), Mitosporic fungi, 1.B2.1/10. 1, S.W. Asia, Pacific.

Neokeissleria Petr. (1920) = Melanconis (Melanconid.) fide Müller & v. Arx (1973), but see Barr (1978).

Neokellermania Punith. (1981) = Pseudorobillarda (Mitosp. fungi) fide Nag Raj (1993).

Neokneiffia Sacc. (1898) = Hyphoderma Wallr. (Hyphodermat.) fide Donk (1964).

Neolamya Theiss. & Syd. (1918), Ascomycota (inc. sed.). 1 (on lichens, esp. *Peltigera*), Eur.

Neolecta Speg. (1881), Neolectaceae. 3, N. temp. See Eriksson (1981), Redhead (*CJB* **55**: 301, 1977).

Neolectaceae Redhead (1977), Neolectales. 1 gen. (+ 2 syn.), 3 spp. Stromata absent; ascomata apothecial, pale, stalked, clavate or spathulate; peridium and stalk composed of interwoven hyphae, often with large I+ septal plugs; interascal tissue absent; asci not formed from croziers, cylindrical, thin-walled, the walls blueing in iodine, slightly thickened at the apex, with a wide apical pore, persistent, 8-spored; ascospores ellipsoidal, hyaline, aseptate. Anamorphs unknown. On soil and leaf mould, esp. N. temp.
For *Lit.* see *Neolectales*.

Neolectales, Ascomycota. 1 fam., 1 gen. (+ 2 syn.), 3 spp. Fam.:
Neolectaceae.
Lit.: Landvik *et al.* (*SA* **11**: 107, 1993).

Neolentinus Redhead & Ginns (1985), Lentinaceae. 9, widespr.

Neolichina Gyeln. (1940) = Lichina (Lichin.).

Neoligniella Naumov (1951), Mitosporic fungi, 8.A1.?. 1, former USSR.

Neolinocarpon K.D. Hyde (1992), Hyponectriaceae. 1 (on *Nypa*), Brunei.

Neolysurus O.K. Mill., Ovrebo & Burk (1991), Clathraceae. 1, Costa Rica.

Neomarssoniella U. Braun (1991), Mitosporic fungi, 6.B1.?. 2, Eur., N. Am.

Neomelanconium Petr. (1940), Mitosporic fungi, 8.A2.19. 2, Eur., Afr.

Neomichelia Penz. & Sacc. (1901) = Pithomyces (Mitosp. fungi) fide Ellis (1970).

Neomunkia Petr. (1947), Mitosporic fungi, 8.A1.?. 1, S. Am.

Neo-Naumovia, see *Neonaumovia*.

Neonaumovia Schwarzman (1959) = Lophophacidium (Phacid.) fide Korf (1973).

Neonectria Wollenw. (1917) = Nectria (Hypocr.) fide Clements & Shear (1931).

Neonorrlinia Syd. (1923) = Cercidospora (Dothideales, inc. sed.) fide Hafellner & Obermayer (*Crypt., bryol. lich.* **16**: 177, 1995).

Neoovularia U. Braun (1992), Mitosporic fungi, 3.A2.10. 3, Eur., Japan.

Neoparodia Petr. & Cif. (1932), Parodiopsidaceae. 1, W. Indies. Anamorphs *Chuppia*- or *Sarcinella*-like.

Neopatella Sacc. (1908) = Selenophoma (Mitosp. fungi) fide Petrak (*Sydowia* **5**: 354, 1951).

Neopaxillus Singer (1948), Paxillaceae. 2, C. & S. Am. See Singer *et al.* (*Beih. Nova Hedw.* **98**: 17, 1990).

Neopeckia Sacc. (1883), Dothideales (inc. sed.). 1, N. Am. See Barr (*Mycotaxon* **20**: 1, 1984); = Herpotrichia (Massarin.) fide Bose (1961) and v. Arx & Müller (1984).

Neopeltella Petr. (1950), ? Schizothyriaceae. 1, S. Am.

Neopeltis Syd. (1937), Mitosporic fungi, 8.A1.?. 1, C. Am.

Neopeltis Petr. (1949) = Microcallis (Chaetothyr.) fide v. Arx & Müller (1975).

Neopericonia Kamal, A.N. Rai & Morgan-Jones (1983), Mitosporic fungi, 1.D2.10. 1, India.

Neophacidium Petr. (1950), ? Rhytismatales (inc. sed.). 2, Pakistan, S. Am.

Neophoma Petr. & Syd. (1927), Mitosporic fungi, 8.A1.?. 2, Eur., N. Am.

Neophyllis F. Wilson (1891) Cladoniaceae (L). 2, Australasia. See Ahti (1993). = Gymnoderma (Cladon.)

fide Hawksworth & Yoshimura (*Taxon* **22**: 503, 1973).

Neophyscia M. Choisy (1959) nom. nud. = Physcia (Physc.).

Neopiptostoma Kuntze (1898) ≡ Piptostoma (Mitosp. fungi).

Neoplaconema B. Sutton (1977), Mitosporic fungi, 8.A1.15. 1, Germany.

Neoplacosphaeria Petr. (1921) = Sphaeriothyrium (Mitosp. fungi) fide Petrak (*Ann. Myc.* **22**: 1, 1924).

Neopsoromopsis Gyeln. (1940), ? Parmeliaceae (L). 1, Argentina.

Neopycnodothis Tak. Kobay. (1965) = Cytoplea (Mitosp. fungi) fide Sutton (1977).

Neoramularia U. Braun (1991), Mitosporic fungi, 8.A/B1.3. 3, widespr.

Neoravenelia Long (1903), = Ravenelia (Ravenel.) fide Arthur (1934).

Neorehmia Höhn. (1902), Ascomycota (inc. sed.). 1, Austria. See Barr (*Mycotaxon* **39**: 43, 1990).

Neosaccardia Mattir. (1921) ? = Scleroderma (Sclerodermat.) fide Guzmám (*Ciencia* **25**: 195, 1967).

Neosartorya Malloch & Cain (1973), Trichocomaceae. 12, widespr. See Guarro & Figueras (*Boln Soc. micol. Madrid* **11**: 217, 1987; ontogeny), Peterson (*MR* **96**: 547, 1992; 6 spp. DNA homology), Takada *et al.* (*Trans. mycol. Soc. Japan* **27**: 415, 1986; mating). Anamorph *Aspergillus fumigatus* group.

Neosecotium Singer & A.H. Sm. (1960), Secotiaceae. 2, N. Am., Afr.

neosexual, see protosexual.

Neoskofitzia Schulzer (1880), Hypocreaceae. 5, widespr. See Weese (*Mitt. bot. Lab. tech. Hochs.* **2**: 35, 1924).

Neosolorina (Gyeln.) Räsänen (1943) = Solorina (Peltiger.).

Neospegazzinia Petr. & Syd. (1936), Mitosporic fungi, 4.E1.?. 1, S. Am.

Neosphaeropsis Petr. (1921) ? = Sphaeropsis (Mitosp. fungi).

Neosporidesmium Mercado & J. Mena (1988), Mitosporic fungi, 2.C2.1. 1, Cuba.

Neostomella Syd. (1927), Asterinaceae. 4, C. Am. See Hansford (*Mycol. Pap.* **15**, 1946).

Neotapesia E. Müll. & Hütter (1963), Dermateaceae. 3, Eur. See Nannfeldt (*TBMS* **67**: 283, 1976). ? = Trichobelonium (Dermat.) fide Korf (1973).

Neotestudina Segretain & Destombes (1961), Testudinaceae. 2, Afr., Asia, Eur. *N. rosatii* (mycetoma of foot). See Hawksworth (*CJB* **57**: 91, 1979), Hawksworth & Booth (*Mycol. Pap.* **135**, 1974).

Neothyridaria Petr. (1934), Ascomycota (inc. sed.). 1, Eur.

neotony, a process in which normal development of cells, except those involved in reproduction, is arrested; results in a sexually mature organism with juvenile features; see Kreisel (*in* Hawksworth (Ed.), *Frontiers in mycology*: 69, 1991; in fungal phylogeny); cfr. pedogenesis.

Neotremella Lowy (1979), Tremellaceae. 1, Mexico.

Neotrichophyton Castell. & Chalm. (1919) = Trichophyton (Mitosp. fungi).

Neotrotteria Sacc. (1921) = Acanthonitschkea (Nitschk.) fide Nannfeldt (1975).

Neottiella (Cooke) Sacc. (1889) = Octospora (Otid.) fide Caillet & Moyne (*BSMF* **103**: 179, 1987, *Bull. Soc. Hist. nat. Doubs* **84**: 9, 1991; keys), but see Kristiansen (*SA* **14**: 61, 1995).

Neottiopezis Clem. (1903) ≡ Neottiella (Otid.).

Neottiospora Desm. (1843), Mitosporic fungi, 4.A1/2.15. 1, widespr.

Neottiosporella Höhn. ex Falck (1923) nom. dub (Mitosp. fungi). No spp. described.

Neottiosporina Subram. (1961), Mitosporic fungi, 4.C1.1. 6, S. Am., Afr., Australia. See Sutton & Alcorn (*Austr. Jl Bot.* **22**: 517, 1974; key).
Neottiosporis Clem. & Shear (1931) ≡ Neottiosporella (Mitosp. fungi).
neotype, see type.
Neotyphula Wakef. (1934), Auriculariaceae. 1, Guyana. See Martin (*Lloydia* **11**: 121, 1948).
Neournula Paden & Tylutki (1969), Sarcosomataceae. 1, USA.
Neoventuria Syd. & P. Syd. (1919), Dothideales (inc. sed.). 1, S. Am.
Neovossia Körn. (1879), Tilletiaceae. *c.* 10 (on *Gramineae*), widespr. See Vánky (1987); cf. *Tilletia.*
Neoxenophila Apinis & B.M. Clark (1974) = Aphanoascus (Onygen.) fide Cano & Guarro (1990).
Neozimmermannia Koord. (1907) = Glomerella (Phyllachor.) fide v. Arx & Müller (1954).
Neozygitaceae Ben Ze'ev, R.G. Kenneth & Uziel (1987), Entomophthorales. 2 gen., 10 spp. See Humber (1989), Ben-Ze'ev et al. (*Mycotaxon* **28**: 313, 1987).
Neozygites Witlaczil (1885), Neozygitaceae. 9 (on *Arthropoda*), widespr. See Keller & Wuest (*Entomophaga* **28**: 123, 1983), Butt & Heath (*Eur. Jl Cell. Biol.* **46**: 499, 1988; cell division), Butt & Humber (*Protoplasma* **151**: 115, 1989; mitosis), Keller (1991; key Swiss spp.), Dick et al. (*Mycol.* **84**: 729, 1992).
Neozythia Petr. (1958), Mitosporic fungi, 8.A1.38. 1, Iran.
Nephlyctis Arthur (1907) = Prospodium (Uropyxid.) fide Dietel (1928).
Nephrochytrium Karling (1938), Endochytriaceae. *c.* 8, N. Am.
Nephroma Ach. (1810), Nephromataceae (L). 35, cosmop. See Eriksson & Strans (*SA* **14**: 33, 1995; molec. phylogeny), Wetmore (*Publs Mich. St. Univ. Mus., ser. biol.* **1**: 369, 1960; N. Am.), White & James (*Lichenologist* **20**: 103, 1988; key 15 S. temp. spp.).
Nephromataceae Wetmore ex J.C. David & D. Hawksw. (1990), Peltigerales (L). 1 gen. (+ 6 syn.), 35 spp. Thallus foliose, corticate on both surfaces; ascomata initially immersed with a vegetative covering layer splitting open at a late stage of development, produced on the lower surface of the thallus which subsequently turns up to expose the hymenium; interascal tissue of simple paraphyses; asci with an I- apical cap lacking an apical ring, with a well-developed ocular chamber but no I+ gelatinized outer layer; ascospores brown, elongate, transversely septate. Conidiomata pycnidial. Lichenized with green algae or cyanobacteria, the latter sometimes in cephalodia; on soil, bark etc., esp. temp.
Nephromatomyces E.A. Thomas (1939) nom. inval. = Nephroma (Nephromat.).
Nephromiomyces E.A. Thomas ex Cif. & Tomas. (1953) = Nephroma (Nephromat.).
Nephromium Nyl. (1860) = Nephroma (Nephromat.).
Nephromopsis Müll. Arg. (1891), Parmeliaceae (L). 16, Eur., Asia, N. & S. Am. See Kärnefelt et al. (1992), Randlane & Saag (*Mycotaxon* **44**: 485, 1992).
Nephromyces Giard (1888), ? Cladochytriaceae. 3 (in ascidians), Eur. See Saffo & Nelson (*CJB* **61**: 3230, 1983).
Nephromyces Sideris (1927) ≡ Rhizidiocystis (? Chytridiales, inc. sed.).
Nephrospora Loubière (1923) = Microascus (Microasc.) fide v. Arx (1975).
Nepotatus Lloyd (1925) = Scleroderma (Sclerodermat.) fide Stevenson & Cash (*Bull. Lloyd Libr.* **35**: 130, 1936).

Nereiospora E.B.G. Jones, R.G. Johnson & S.T. Moss (1983), Halosphaeriaceae. 2 (marine), N. Hemisph. See Mouzouras & Jones (*CJB* **63**: 2444, 1985). Anamorph *Monodictys.*
nervicolous, living on veins of leaves or stems.
Nesolechia A. Massal. (1856) = Phacopsis (Lecanor.) fide Triebel et al. (1995).
Nesophloea Fr. (1849) nom. dub. (Fungi, inc. sed.).
Neta Shearer & J.L. Crane (1971), Mitosporic fungi, 1.B/C1.10. 6 (terrestrial and aquatic), N. temp. See de Hoog (*Stud. Mycol.* **26**: 44, 1985; key).
Neural networks (Artificial Neural Networks, ANNs), allow for the identification or classification of input patterns (characters). In practice this approach generally requires a considerable amount of data in order to set up the network. Final placement is arrived at after the passage of the input values through a series of nodes arranged in layers. The method has not yet been used widely with fungal data; examples include Boddy & Morris (*Binary* **5**: 17, 1991), Bridge et al. (*in* Hawksworth (Ed.), *Identification and characterization of pest organisms*: 153, 1994), Morris et al. (*MR* **96**: 967, 1992). For methodology see Boddy et al. (*Binary* **2**: 179, 1990), Specht (*Neural networks* **3**: 109, 1990), Zaruda (*Artificial neural systems*, 1992). See Numerical taxonomy.
Neuroecium Kunze (1823) nom. dub.; not a fungus fide Juel (*Dansk. bot. Ark.* **5**: 1, 1928).
Neuronectria Munk (1957) ≡ Hydropisphaera (Hypocr.).
Neuropogon Nees & Flot. (1835) = Usnea (Parmel.) fide Lamb (1964).
Neurospora Shear & B.O. Dodge (1927), Sordariaceae. 12, widespr. See Frederick et al. (*Mycol.* **61**: 1083, 1970; key), Bachman & Strickland (*Neurospora bibliography and index*, 1965; 2,310 refs to 1964), *Fungal Genetics Newsletter* (*Neurospora Newsletter*), Dutta et al. (*J. Elisha Mitchell Sci. Soc.* **97**: 126, 1981; SEM, isozymes), Perkins (*Genetics* **130**: 687, 1992; history as a model organism), Perkins & Turner (*Exp. Mycol.* **12**: 91, 1988; natural populations), Raju (*Mycol.* **79**: 696, 1987; mutant ascospores & 'salt-shaker asci', **80**: 825, 1988; multipored ascus apices, *MR* **96**: 241, 1992; genetic control of sexual cycle). See mould, bread, Genetics. Anamorph *Chrysonilia.*
neurotoxin, a toxin which affects the nervous system.
Nevrophyllum Pat. (1886) ≡ Gomphus (Gomph.).
Newinia Thaung (1973), Uropyxidaceae. 3 (on *Bignoniaceae*), Asia, Afr.
Nia R.T. Moore & Meyers (1961), Niaceae. 2 (marine), N. Hemisph. See Leightley & Eaton (*TBMS* **73**: 35, 1979), Doguet (*BSMF* **84**: 343, 1969; **85**: 93, 1969; cultures, development), Kohlmeyer & Kohlmeyer (*Marine mycology*, 1979), Jones & Jones (*MR* **97**: 1, 1993), Rosello et al. (*MR* **97**: 68, 1993; key).
Niaceae Jülich (1982), Melanogastrales. 1 gen., 2 spp. Gasterocarp small, marine; hyphae clamped; spores appendaged.
Nicholsoniella Kuntze (1891) ≡ Libertiella (Mitosp. fungi).
nidose (**nidorose**), having an unpleasant smell.
Nidula V.S. White (1902), Nidulariaceae. 4, widespr. (excl. Eur.).
nidulant, lying free in a cavity.
Nidularia With. (1787) nom. dub. (Myxomycetes).
Nidularia Bull. (1791) nom. rej. = Cyathus (Nidular.).
Nidularia Fr. (1817) nom. cons., Nidulariaceae. 3, Eur., N. Am., Australasia. See Cejp & Palmer (*Česká Myk.* **17**: 125, 1963; key).
Nidulariaceae Dumort. (1822), Nidulariales. 5 gen. (+ 6 syn.), 59 spp. Gasterocarp epigeous, obconical, with circumscissile epiphragm containing one to several peridioles; gleba consisting of individualized global

chambers with hardened wall (peridioles); spores smooth, hyaline. In the most typical genera, the basidiocarp is made up of a nest- or funnel-like peridium the top of which is first covered by a membrane, the epiphragm. Inside are the peridioles ('eggs'), sometimes fixed to the wall by thread-like funiculi. Distribution of the peridioles is effected by a splash-cup mechanism.
    Lit.: Brodie (*The birds' nest fungi*, 1975; monogr., *Lejeunia* NS **112**, 1984: suppl.).

Nidulariales, Basidiomycetes ('gasteromycetes'). 1 fam., 5 gen. (+ 6 syn.), 59 spp. Fam.: **Nidulariaceae**.
    Lit.: see under *Gasteromycetes*.

Nidulariopsis Greis (1935), Sphaerobolaceae. 1, Eur., N. Am. See Zeller (*Mycol.* **40**: 639, 1948).

Nidulispora Nawawi & Kuthub. (1990), Mitosporic fungi, 1.G1.1. 1 (aquatic), Malaysia.

Niebla Rundel & Bowler (1978) ? = Ramalina (Ramalin.). See Bowler *et al.* (*Phytologia* **77**: 23, 1994).

NIEC, see RIEC.

Nielsenia Syd. (1921) = Uromyces (Puccin.) fide Dietel (1928).

Niessl von Meyendorf (Gustav; 1839-1919). Born in Vienna. In 1859 appointed Professor of Geometry and Astronomy, Polytechnical Institute, Brunn, Moravia, where he worked until 1907 when he retired to Vienna. Published on myxomycetes, pyrenomycetes, smuts, etc. with particular reference to Moravian and Silesian fungi. See Nožička (*Česká Myk.* **18**: 185, 1964; portr.), Grummann (1974: 436).

Niesslella Speg. (1880) ? = Sporormiella (Spororm.).

Niesslella Höhn. (1919) = Micropeziza (Dermat.) fide Nannfeldt (1976).

Niesslia Auersw. (1869), Niessliaceae. 7, widespr. See Gams (*Cephalosporium-artige Schimmelpilze*, 1971). Anamorph *Monocillium*.

Niessliaceae Kirschst. (1939), Hypocreales. 6 gen. (+ 3 syn.), 10 spp. Ascomata perithecial, superficial on a subiculum or crustose stroma, small, ± globose, pale to dark brown, often setose, the ostiole periphysate; peridium thin, soft; interascal tissue of apical paraphyses; asci ± clavate, thin-walled, with an I-apical ring; ascospores ± hyaline, 0- or 1-septate, sometimes fragmenting. Anamorphs hyphomycetous, esp. *Acremonium*-like. Saprobic on herbaceous and woody tissue, widespr.
    Lit.: Barr (*Mycotaxon* **39**: 43, 1990).

nietsuki, a product of failure in the drying of *Lentinula edodes* basidiomata (Kawai & Kawai, *Rep. Tottori mycol. Inst.* **1**: 29, 1961).

nigeran, see mycodextran.

Nigredo (Pers.) Roussel (1806) nom. rej. ≡ Uredo (Anamorphic Uredinales).

Nigrococcus E.K. Novák & Zsolt (1961) [non Castell. & Chalm. (1919), *Bacteria*] = Phaeococcus (Mitosp. fungi).

Nigrocupula Sawada (1944) = Graphiola (Graphiol.) fide Sutton (*in litt.*).

Nigrodiplodia Kravtzev (1955) nom. dub. fide Sutton (*Mycol. Pap.* **141**, 1977).

Nigrofomes Murrill (1904), Coriolaceae. 1, pantrop.

Nigrohydnum Ryvarden (1987), Hydnaceae. 1, S. Am.

Nigropogon Coker & Couch (1928) = Richoniella (Richoniell.) fide Dring & Pegler (*Kew Bull.* **32**: 563, 1978).

Nigroporus Murrill (1905), Coriolaceae. 3, widespr. See Pegler (1973).

Nigropuncta D. Hawksw. (1981), Mitosporic fungi, 4.A2.23. 1 (on lichens), Austria, Norway.

Nigrosabulum Malloch & Cain (1970), Pseudeurotiaceae. 1 (coprophilous), USA.

Nigrosphaeria N.L. Gardner (1905) ? = Melanospora (Ceratostomat.) fide v. Arx & Müller (1954).

Nigrospora Zimm. (1902), Anamorphic Trichosphaeriales, 1.A2.1. Teleomorph *Khuskia. c.* 3, widespr. *N. oryzae* (syn. *Basisporium gallarum*) on maize (*Zea*) and other hosts. See Standen (*Ia St. Coll. J. Sci.* **17**: 263, 1943).

Nimbomollisia Nannf. (1983) = Niptera (Dermat.) fide Baral (*SA* **13**: 113, 1994).

Nimbospora J. Koch (1982), Halosphaeriaceae. 3 (marine), Hawaii, Sri Lanka, Seychelles. See Hyde & Jones (*CJB* **63**: 611, 1985).

nimbospore, a spore having a gelatinous, apparently many-layered wall, e.g. *Histoplasma capsulatum* (Nielsen & Evans, *J. Bact.* **68**: 261, 1954).

Nimbya E.G. Simmons (1989), Anamorphic Diademaceae, 1.C2.26. Teleomorph *Macrospora*. 5, widespr.

Nimisia Kärnefelt & Thell (1993), Parmeliaceae (L). 1, Tierra del Fuego.

Niopsora A. Massal. (1861) = Caloplaca (Teloschist.).

Niorma A. Massal. (1861) = Teloschistes (Teloschist.) fide Zahlbruckner (1931).

Niospora Kremp. (1869) ≡ Niopsora (Teloschist.).

Nipholepis Syd. (1935), Arthoniales (inc. sed.). 1, S. Am.

Nipicola K.D. Hyde (1992), ? Sordariales (inc. sed.). 1, Brunei. See Eriksson & Hawksworth (*SA* **13**: 198, 1994; posn), Hyde (*Sydowia* **46**: 257, 1995).

Niptera Fr. (1849), Dermateaceae. 2, Eur. See Dennis (*Kew Bull.* **26**: 439, 1972), Nannfeldt (*Sydowia* **38**: 194, 1986).

Nipterella Starbäck ex Dennis (1962), Leotiaceae. 3, Eur., N. Am.

nitid (adj. **nitidous**), smooth and clear; lustrous.

nitrophilous, having a preference for habitats rich in nitrogen; chionophilous; **nitrophobous**, having a preference for habitats poor in nitrogen.

Nitschkia G.H. Otth ex P. Karst. (1873), Nitschkiaceae. *c.* 10, Eur., N. Am., Java. *N. cupularis* (syn. *N. fuckelii*) on Nectria cinnabarina. See Fitzpatrick (*Mycol.* **15**: 23, 1923, **16**: 101, 1924), Nannfeldt (*Svensk bot. Tidskr.* **69**: 49, 1971).

Nitschkiaceae (Fitzp.) Nannf. (1932), Sordariales. 12 gen. (+ 25 syn.), 39 spp. Ascomata dark, thick-walled, opening with an irregular lysigenous pore, cells with 1 μm diam. perforations ('Munk pores'); interascal tissue absent; asci clavate, long-stalked, without apical apparatus; ascospores hyaline or brown, usually allantoid, sometimes septate, smooth, sheath absent. On wood, a few species lichenicolous. Retained as order *Coronophorales* by some authors, but separation from *Lasiosphaeriaceae* often difficult.
    Lit.: Fitzpatrick (*Mycol.* **15**: 23, **16**: 101, 1923-24), Hawksworth (*TBMS* **74**: 363, 1980), Nannfeldt (*Svensk bot. Tidskr.* **69**: 49, 289, 1975), Sivanesan (*TBMS* **62**: 40, 1974).

Nitschkiopsis Nannf. & R. Sant. (1975), ? Sordariales (inc. sed.). 1 (on lichens, esp. *Pseudocyphellaria*), Kenya.

Nivatogastrium Singer & A.H. Sm. (1959), Strophariaceae. 1, N. Am.

Niveostoma Svrček (1988) = Solenopezia (Hyaloscyph.) fide Raitvïr *et al.* (*Sydowia* **43**: 219, 1991).

nm, nanometer; one billionth of a metre.

noble rot, a condition in which the mould *Botrytis* grows on overripe grapes. A rich, sweet wine is made in small quantities from the affected grapes (Sauternes, Trockenbeerenlauslese, Botrytis-wine).

Nocardia Trevis. (1889), *Actinomycetes*, q.v.

Nocardioides Prauser (1976), *Actinomycetes*, q.v.

Nocardiopsis (Brocq-Rouss.) J. Mey. (1976), *Actinomycetes*, q.v.

Nochascypha Agerer (1983), Tricholomataceae. 4, S. Am.

node cell, see hyphopodium.

Nodobryoria Common & Brodo (1995), Parmeliaceae (L). 3, N.Am., Greenland.

Nodocrinella Scheer (1977), Parataeniellaceae. 1 (in *Isopoda*), Germany.

nodose-septum, see clamp-connexion.

Nodotia Hjortstam (1987) = Hypochnicium (Hyphodermat.) fide Hjortstam & Larsson (*Windahlia* 21, 1994).

nodular bodies (of dermatophytes), rounded bodies made up of massed hyphae.

Nodularia Peck (1872) [non Link ex Lyngb. (1819), *Algae*] = Aleurodiscus (Aleurodisc.) fide Rogers & Jackson (*Farlowia* 1: 263, 1943).

Nodulisporium Preuss (1849), Anamorphic Xylariales, 1.A2.10. Teleomorphs *Hypoxylon, Xylaria, Rosellinia*, etc. 19, widespr. See Deighton (*TBMS* 85: 391, 1985).

nodulose (of spores), having broad-based, blunt, wart-like excrescences.

Nodulosphaeria Rabenh. (1858) typ. cons., Phaeosphaeriaceae. *c.* 40, widespr. See Crivelli (*Ueber die heterogene Ascomycetengattung Pleospora*, 1983), Holm (*Svensk bot. Tidskr.* 55: 63, 1961), Shoemaker & Babcock (*CJB* 65: 1921, 1987; key 5-septate spp.), Barr & Holm (*Taxon* 33: 109, 1984; nomencl.), Shoemaker (*CJB* 62: 2730, 1985; key 23 spp.).

nodum (pl. **noda**) (in phytosociology), particular well-defined plant communities. See also Phytosociology.

Nohea Kohlm. & Volkm.-Kohlm. (1991), Halosphaeriaceae. 1 (marine), Hawaii.

Nolanea (Fr.) P. Kumm. (1871), Entolomataceae. 75, N. temp. See Largent (*Mycol.* 66: 987, 1974; subgen. class.), Pegler (*Kew Bull., Addit. Ser.* 6, 1977; key 5 E. Afr. spp.), Orton (*TBMS* 43: 328, 1960; key 15 Br. spp.), Noordeloos (*Persoonia* 10: 427, 1980; key 58 Eur. spp.).

nomen (Latin), name; - **ambiguum**, one having different senses; - **anamorphosis**, see States of fungi; - **confusum**, one of a taxonomic group based on two or more different elements; - **conservandum**, one authorized for used by a decision of an International Botanical Congress (see Nomenclature); - - **propositum**, one put up for conservation; - **dubium**, one of uncertain sense; - **holomorphosis**, see States of fungi; - **illegitimum** (**nom. illegit.**), a validly published name contravening particular Articles in the *Code* (see Nomenclature); - **invalidum** (**nom. inval.**), one not validly published (see Nomenclature); - **monstrositatis**, one based on an abnormality; - **novum**, a new name; a replacement [nom. nov. should replace the author's name only at the first publication]; - **nudum**, one for a taxon having no description; - **provisorium**, one proposed provisionally; - **rejiciendum**, one rejected by a Botanical Congress. A generic name may be a nomen ambiguum (etc.), but a binomial under such a name may be without ambiguity. - **species** (of bacteria), a type species.

Nomenclature. The allocation of scientific names to the units a systematist considers to merit formal recognition. The nomenclature of fungi (including slime-moulds, lichen-forming fungi and yeasts) is governed by the *International code of botanical nomenclature* (latest edn, Greuter *et al.* (Eds) [*Regnum veg.* 131], 1994), as adopted by each International Botanical Congress. Any proposals to change the Code are published in *Taxon*, debated, and voted on at the Nomenclature Section of such a Congress. As rules can change in different editions, the latest should always be consulted. The contents of this entry are based on

the 1994 (Tokyo) edition. The Congress appoints the Committee for Fungi (CF) which advises on action to be taken on proposals concerned with fungi. The Code aims at the provision of a stable method of naming taxonomic groups, avoiding and rejecting the use of names which may cause error or ambiguity or throw science into confusion.

The Code comprises six **Principles**, 62 **Articles** (which are mandatory), **Recommendations** (non-mandatory but good practice) and various Appendices. The Code is designed to allow any taxon to have as many correct names as there may be opinions as to its classification. When the taxonomic decisions have been taken, the Code provides the rules to determine the name that should be applied; each taxon in a given position and rank can have only one nomenclaturally correct name.

In determining the correct name for a taxon, five steps must be followed (Fig. 27) in what is in effect a nomenclatural filter.

Effective publication (A): In order to be effectively published names must now be in printed matter (journals, books) distributed to the public or at least botanical institutions through sale, exchange, or gift. See Nicolson (*Taxon* 29: 485, 1980) for categories and special provisions.

Valid publication (B): In order to be validly published, names of newly described taxa must now also simultaneously fulfil the following requirements: (1) have a correct form. (2) have a description or diagnosis (in Latin (q.v.) after 1 Jan. 1935); (3) be accepted by the author and comply with any relevant special provisions elsewhere in the Code; (4) have a clear indication of rank (after 1 Jan. 1953); and (5) indicate the type (after 1 Jan. 1958) and its place of conservation (after 1 Jan. 1990); and (6) be registered (after 1 Jan. 2000; details to be announced). Replacement names and combinations have to give full bibliographic details of the place of publication of and cite the replaced name or basionym (after 1 Jan. 1953). Names which are not validly published (**nom. inval.**) need not be considered further. Many are deliberately included in this *Dictionary* as they are best avoided by future workers and also as they can be encountered in the literature. The inclusion of a name in the *Dictionary* does not therefore automatically mean that the name is validly published and available for use.

Typification (C): The linking of each name to a nomenclatural type is the keystone of stability in the application of names. All ranks from fam. downwards are ultimately based on a single collection; e.g. *Erysiphaceae* on *Erysiphe* on *E. graminis* on a single collection. A holotype is required (see type) but where none exists, the order of priority is isotype - lectotype - neotype. Should the type not be critically identifiable, an interpretative type (epitype) can be designated. See 'type' for definitions of kinds of types. Type specimens must be dried specimens, specimens in a liquid preservative, dried cultures preserved in an herbarium, cultures preserved in a metabolically inactive state (e.g. freeze-dried, in liquid nitrogen) or microscopic preparations; actively growing cultures are not permitted. If a specimen cannot be preserved an illustration or description can suffice. If a type collection is mixed, one part must be selected as lectotype.

Legitimacy (D): Validly published names not in accordance with certain provisions of the Code are **illegitimate** (**nom. illegit.**) and to be rejected, i.e. (1) **superfluous**, i.e. included the type of a name that should have been used; or (2) **homonyms**, i.e. spelled like a previously validly published name.

Priority (E): Priority of publication determines the correct name for a taxon. The **correct name** (i.e. name in accordance with the Code) of a species is the combination of the earliest available legitimate epithet in the same rank with the correct generic name. This may change if the generic placement or rank is altered; i.e. a species can have more than one correct name according to different taxonomies. The principle of priority can only be set aside through **conservation** (q.v.), or where the rules on pleomorphic fungi apply (see below). Names for a taxon other than the correct one are **synonyms**; **heterotypic synonyms** (taxonomic synonyms) are names based on different nomenclatural types, and **homotypic synonyms** (nomenclatural synonyms) are names based on the same type.

The starting point for botanical nomenclature is Linnaeus' *Species plantorum* (1 May 1753), with later exceptions for some groups. Prior to 1981, different groups of fungi had dates later than 1753: 31 Dec. 1801 (Persoon, *Synopsis methodica fungorum*) for *Gasteromycetes* (s.l.), *Uredinales* and *Ustilaginales*; and 1 Jan. 1821 (Fries, *Systema mycologicum* 1 (1)) for the remaining fungi (other than slime-moulds and lichen-forming species). Reasons for the change in starting point dates are discussed in Demoulin *et al.* (*Taxon* **30**: 52, 1981). Names used in the previous starting point books of Fries and Persoon are **sanctioned** and not affected by, and take priority over, homonymous and synonymous names published earlier (listed by Gams, *Mycotaxon* **19**: 219, 1984). Names given to lichens are ruled as applying only to their fungal components, i.e. to the lichen-forming fungus. Algae in lichens have separate names, and the composite 'lichen' strictly has no name. Many names ending in -*myces* introduced for lichen mycobionts by Thomas (*Beitr. Kryptog.-Fl. Schweiz.* **9** (1), 1939) and Ciferri & Tomaselli (*Atti Ist. Bot. Univ. Lab. crittog. Pavia* ser. V **10**, 1953) are thus superfluous (see above).

Pleomorphic fungi:

The Code permits the different states of fungi (q.v.), other than lichen-forming fungi, with pleomorphic life-cycles to be given separate names; it does not apply to fungi in which pleomorphism is not confirmed (Reynolds & Taylor, *Taxon* **41**: 98, 1992). Whether a name can be considered as that of an anamorph or teleomorph depends on its original description and nomenclatural type, *not* the genus in which it was placed; if a teleomorph is present, the name automatically refers to that morph even if the anamorph is also evident (e.g. *Penicillium brefeldianum* B.O. Dodge (1933) included both states in its type and description and is thus a teleomorph name that has correctly been combined as *Eupenicillium brefeldianum* (B.O. Dodge) Stolk & D.B. Scott (1967); the name of the anamorph is *P. dodgei* Pitt (1980), the type and description of which are only of the anamorph). The correct name of a holomorph is that of its teleomorph.

Name changing:

In order to avoid name changes for nomenclatural reasons through a strict application of the Code, special provisions have been made. Names of families, genera and species can be **conserved** (**nomina conservanda, nom. cons.**) against names threatening their retention, and names in any rank can be **rejected** (**nomina rejicienda, nom. rej.**) if their use would cause disadvantageous nomenclatural change. If a name has been widley used for a taxon in a sense conflicting with its type, it is not to be used unless and until a proposal to deal with it has been submitted and the name rejected. If a particular work would cause

instability if it were used as a source of valid names, it can be proposed for inclusion in a list of **suppressed works**. Proposals for conservation, rejection, or suppression have to be published in *Taxon* and voted on by appropriate Committees. Appendices to the Code list names and works in these categories.

Lists of Names in Current Use (NCUs) are being developed through a Committee appointed by the 1994 Congress, and widespread protection of around 30,000 generic names is expected to be sought at the next Congress in 1999. Lists published include those for names of fungal families (David, *Regnum Veg.* **126**: 71, 1993; NCU-1), genera (Greuter *et al.*, *Regnum Veg.* **129**, 1993, NCU-3), and species in *Cladoniaceae* (Ahti, *Regnum Veg.* **128**: 58, 1983) and *Trichocomaceae* (Pitt & Samson, *Regnum Veg.* **128**: 13, 1993; subject of a special Nomenclatural Section Resolution). The 1993 Congress, noting improvements being made in the systems of nomenclature to promote stability, urged taxonomists 'to avoid displacing well established names for purely nomenclatural reasons' (see Greuter & Nicolson, *Taxon* **42**: 925, 1993; Hawksworth, *SA* **12**: 1, 1993).

Authorities:

The authority of a name, usually abbreviated (see Authors) is cited after a name for precision and is intended to be a much abbreviated bibliographic reference. If a species is move from one genus to another, or has its rank changed, the original author, i.e. of the basionym (q.v.), is given in brackets and the one making the change cited outside the brackets (e.g. *Fusarium poae* (Peck) Wollenw., based on the earlier *Sporotrichum poae* Peck). Where a name not validly published by one author is take up and validated by a second, '**ex**' is used to link the names of the two authors. In the case of names sanctioned by Persoon or Fries (see above), a colon (**:**) can be used where it is considered appropriate to indicate the special status of that name or epithet (see Korf, *Mycol.* **74**: 250, 1982; *Mycotaxon* **14**: 476, 1982); '**ex**', '**per**' or '**[ ]**' had previously been used for devalidated names taken up again after the former later starting dates; '**in**' was sometimes formerly used where one author contributed an account of a taxon to the work of another but this is ruled as part of the bibliographic reference by the 1994 Code and does not form a part of author citations.

Other Codes:

Separate Codes exist for zoology, bacteriology, cultivated plants, and viruses. These are independent of the Botanical Code, but discussions are underway as to how harmonization between the Codes can be improved (Hawksworth *et al.*, 1994) and a draft International Code of Bionomenclature planned to apply to names in all groups introduced after (provisionally) 1 Jan. 2000 is currently under discussion (see Hawksworth, *Taxon* **44**: 447, 1995). Note that taxa are treated under a particular Code by tradition and not because of changed opinions on phylogeny; e.g. the 'Botanical' Code covers slime-moulds although they belong in the Kingdom *Protozoa*; see also ambiregnal organisms.

Lit.: Davis & Heywood (*Principles of angiosperm taxonomy*, 1963), Greuter *et al.* (*Regnum Veg.* **131**, 1994; Tokyo Code), Hawksworth (*Mycologist's handbook*, 1974 [incl. relevant parts of 1972 Code with mycological examples, glossary]; (Ed.), *Improving the stability of names; needs and options*, [*Regnum Veg.* **123**], 1991; *Bot. J. Linn. Soc.* **109**: 543, 1992; *A draft glossary of terms used in bionomenclature*, [IUBS Monogr. **8**], 1994 [1,175 entries]), Hawksworth *et al.* (*Towards a harmonized bionomenclature for life on Earth*, [*Biol. Internat., Sp.*

Issue **30**, 1994), Jeffrey (*Biological nomenclature*, edn 3, 1989; general survey all Codes), Weresub (*Sydowia, Beih.* **8**: 416, 1979; history, problems).

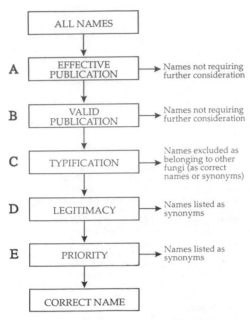

Fig. 27. The nomenclatural filter.

Nomuraea Maubl. (1903), Mitosporic fungi, 1.A1.15. 2, widespr. See Samson (*Stud. Mycol.* **6**, 1974).

nonfissitunicate (of asci), ones in which discharge does not involve a separation of the wall layers; see ascus.

non-target organisms, organisms found with or near those being treated with chemical or biological control agents.

Normal saline solution, sodium chloride (NaCl), 8.5 g, water, 1000.0 ml.

Normandina Nyl. (1855), Ascomycota (inc. sed., L). 3, widespr. See Aptroot (*Willdenowia* **21**: 263, 1991), Eriksson & Hawksworth (*SA* **11**: 67, 1992), Henssen (*in* Brown *et al.*, 1976: 107), Tschermak-Woess (*Nova Hedw.* **35**: 63, 1981; photobiont), *Lauderlindsaya*.

Normandinomyces Cif. & Tomas. (1953) ≡ Normandina (Ascomycota).

Normkultur (of cultures or states), one in which all the forms characteristic of a fungus are present and of good development; **Ankultur**, one of poor development; **Jungkultur**, one when young; **Hochkultur**, one when mature; **Altkultur**, one over-mature; **Abkultur**, a degenerate culture. These names were first used by Appel & Wollenweber (*Ark. biol. BundAnst. Land-u. Forstw.* **8**(1): 209, 1910) for *Fusarium*.

Norrlinia Theiss. & Syd. (1918), Verrucariaceae. 1 (on lichens), Eur. See Eriksson & Hawksworth (*SA* **8**: 109, 1990), Hawksworth (*TBMS* **74**: 363, 1980), Santesson (*Nordic Jl Bot.* **9**: 97, 1989).

Norrlinia Vain. (1921) ≡ Neonorrlinia (Dothideales, inc. sed.).

Nosophloea Fr. (1849), ? Mitosporic fungi, ?.?.?. 3, Eur.

Nostoclavus Paulet (1791) nom. dub. ('basidiomycetes', inc. sed.).

Nostocotheca Starbäck (1899) = Molleriella (Elsin.) fide v. Arx (1963).

Notarisiella (Sacc.) Clem. & Shear (1931) ≡ Pseudonectria (Hypocr.).

notate (of surfaces), marked by straight or curved lines.

Nothocastoreum G.W. Beaton (1984), Mesophelliaceae. 1, Australia.

Nothoclavulina Singer (1970), Anamorphic Tricholomataceae, 2.A1.38. Teleomorph *Arthrosporella*. 1, Argentina.

Nothodiscus Sacc. (1917) nom. dub. fide Petrak (*Ann. Myc.* **38**: 253, 1940).

Nothojafnea Rifai (1968), Otideaceae. 2, Australia, S. Am.

Notholepiota E. Horak (1971), Secotiaceae. 1, NZ.

Nothomitra Maas Geest. (1964), Geoglossaceae. 2, Eur., former USSR.

Nothopanus Singer (1944), Tricholomataceae. 2, trop.

Nothopatella Sacc. (1895) = Botryodiplodia (Mitosp. fungi) fide Petrak & Sydow (*Beih. Rep. spec. nov. regni veg.* **42**, 1927).

Nothophacidium J. Reid & Cain (1962), Dermateaceae. 1, N. Am. *N. phyllophilum* (snow blight of conifers). See Smerlis (*CJB* **44**: 563, 1966).

Nothopodospora Mirza (1963) nom. inval. = Arnium (Lasiosphaer.) fide Lundqvist (1972).

Nothoporpidia Hertel (1984) = Lecidea (Lecid.) fide Hertel (*Mitt. bot. StSamml. München* **23**: 321, 1987).

Nothoravenelia Dietel (1910), Phakopsoraceae. 2 (on *Euphorbiaceae*), Japan, Afr. See Thirumalachar (*Sydowia* **5**: 23, 1951).

Nothospora Peyronel (1913), ? Mitosporic fungi, 1.A1.?. 1, Italy.

Nothostrasseria Nag Raj (1983), Mitosporic fungi, 4.B2.15. 1, Australia.

Nothostroma Clem. (1909) = Tomasellia (Arthopyren.).

Notolecidea Hertel (1984), Lecanorales (inc. sed., L). 1, Antarctic.

Notothyrites Cookson (1947), Fossil fungi (Microthyr.). 12 (Cretaceous, Tertiary), widespr. See Eriksson (*Ann. bot. fenn.* **15**: 122, 1978).

Nowakowskia Borzí (1885), Chytridiaceae. 1, Eur.

Nowakowskiella J. Schröt. (1893), Cladochytriaceae. c. 15, widespr. See Karling (*TBMS* **44**: 453, 1961), Sparrow (1960: 581; key).

Nowellia F. Stevens (1924) [non Mitt. (1870), *Hepaticae*] Fungi (inc. sed.) fide Petrak (*Sydowia* **5**: 169, 1951).

Nozemia Pethybr. (1913) = Phytophthora (Pyth.) fide Lafferty & Pethybridge (1922).

NRRL. Agricultural Research Service Culture Collection (Peoria, Ill, USA); founded 1941; a part of the United States Department of Agriculture's (USDA) Agricultural Research Service; see Kurtzman (*Enzyme Microbiol. Tech.* **8**: 328, 1986).

nubilated, cloudy and semi-opaque as viewed by transmitted light.

nuclear cap (of *Blastocladiaceae*), a body at one side of the nucleus of a zoospore or gamete.

Nuclear division, see Heath (Ed.) (*Nuclear division in fungi*, 1978); **- status**, of fungal cells (fide Jinks & Simchen, *Nature* **210**: 778, 1966) may be considered (1) numerically: mono-, di-, multi-karyon (= uni-, bi-, multinucleate) and (2) by gene content: homokaryon or heterokaryon according as to whether the nuclei are genetically identical or not. See Chromosome numbers.

Nucleophaga P.A. Dang. (1895), ? Olpidiaceae. 1 (on *Amoeba*), France. See Karling (*Bull. Torrey bot. Cl.* **99**: 223, 1972).

Numbers of fungi. The development of reliable estimates of both the numbers of accepted species of fungi and the actual number on Earth is fraught with uncertainties. The difficulties involved in the enumeration of fungi as individuals, kinds (i.e. as genera and species) or even names are considerable (Ainsworth, 1968). For the 7th edition of this *Dictionary*, the numbers of genera and species accepted were derived by upwards summation of the totals from the generic entries; this gave a total of 64,200 species. New species of fungi are now being catalogued in the *Index of Fungi* at the rate of *c.* 800 each year. Assuming that mycologists still inadvertently redescribe already known species at the rate of about 2.5:1 (Hawksworth, 1991) this means that an 6,800 additional 'good' species would be expected to have been added in the period 1983-95; i.e. a total of 71,000 species. The close correspondence between this figure and the 72,065 in Table 7 obtained by upwards addition in entries in this edition of the *Dictionary* supports this assumption. However, it is also important to note that amongst the approximately 250,000 names at species rank proposed for fungi, many have never been reassessed since their first description nor have they been adopted by modern workers. Names in many accepted genera, especially some of the larger, have not been critically revised; in this edition of the *Dictionary* we have not added in all species names proposed and catalogued in the *Index of Fungi* in the absence of the opinions of specialists.

In the absence of a world checklist of accepted fungi (a topic starting to be addressed by IMI/BPI as a part of the IUBS/IUMS SPECIES 2000 project), the possibility that as many as 100,000 or even 150,000 (Rossman, 1994) fungi are already described cannot be excluded.

Estimating the total number of fungi on Earth is even more problematic. Bisby & Ainsworth (1943) estimated that there are at least 100,000 species of fungi. Martin (*Proc. Iowa Acad. Sci.* **58**: 175, 1951) considered this estimate 'excessively conservative' and suggested that the number of species of fungi is at least as great as the number of 'good' species of phanerogams, then believed to be not less than 250,000. Based on extrapolations from three independent data sets (ratios of the numbers of fungi in all habitats to plants in the British Isles, numbers restricted to particular hosts, and a community studied in depth) Hawksworth (1991) suggested that a 'conservative' working figure of *c.* 1.5 million be adopted; this has been widely accepted (Heywood, *Global biodiversity assessment*, 1995; Rossman, 1994) and is being supported by new data (Hawksworth, 1993; Cannon & Hawksworth, 1995), although Hammond (1992) recommended 1 million for general use.

The gap between described and estimated species of fungi is immense, and new species are regularly found in all countries of the world. Of the 16,013 new species recorded in the *Index of Fungi* in 1981-90, 51% were from countries outside the tropics; the individual countries providing most species were India and the USA (*c.* 10% each) (Hawksworth, 1993). In the tropics, around 15-25% of the fungi collected in short studies can be expected to be undescribed, the percentage rising to 60-85% in more prolonged intensive investigations (depending on the groups and habitats). Pirozynski (*Mycol. Pap.* **129**, 1972) reported a ratio of microfungi *alone* to plants of at least 3:1 (and possibly 5:1) in Tanzania, and detailed (but still incomplete) site inventories in temperate regions yield ratios of 3-4:1 when all groups are considered (Hawksworth, 1991).

Lit.: Ainsworth (*in* Ainsworth & Sussman, *The Fungi* **3**: 505, 1968), Bisby (*Am. J. Bot.* **20**: 246, 1933), Bisby & Ainsworth (*TBMS* **26**: 16, 1943), Cannon & Hawksworth (*Adv. Pl. Path.* **11**: 277, 1995), Hammond (*in* Groombridge, *Global biodiversity*: 17, 1992), Hawksworth (*MR* **95**: 641, 1991; *in* Isaac *et al.*,

Aspects of tropical mycology: 265, 1994), Pascoe (*in* Short, *History of systematic mycology in Australia*: 259, 1990), Rossman (*in* Hawksworth (Ed.), *Identification and characterization of pest organisms*: 35, 1994).

Numerical taxonomy (phenetics, taxometrics), the derivation of computer-based assessments of resemblance, for both classification and identification, has not been as widely used for fungi as for bacteria. Examples of this approach applied to fungal classification are the papers by Kendrick & Proctor (*CJB* **42**:65, 1964; mitosporic fungi), Kendrick & Weresub (*Syst. Zool.* **15**: 307, 1966; ordinal level in basidiomycetes), Ibrahim & Threlfall (*Proc. R. Soc.* **B165**: 362, 1966; graminicolous *Helminthosporium*), Joly (*BSMF* **85**: 213, 1969; *Alternaria*), Kiefer (*Mycol.* **71**: 343, 1979; methodology of Machol & Singer, *Nova Hedw.* **21**: 353, 1971), Hêls & Raitviïr (*Scripta mycol.* **6**, 1974; morphometrics), Parker-Rhodes & Jackson (*in* Cole (Ed.), 1969: 181; ecology of basidiomycetes), Bridge *et al.* (*J. Gen. Microbiol.* **135**: 2941, 1989; *Penicillium*). For examples of identification schemes see Pankhurst (*Nature* **227**: 1269, 1970; *Biological identification*, 1978; key generation), Barnett *et al.* (*Yeasts: characteristics and identification*, edn 2, 1990), Bridge *et al.* (*Mycol. Pap.* **165**, 1992).

For methods see Carmichael & Sneath (*Syst. Zool.* **18**: 402, 1969), Cole (Ed.) (*Numerical taxonomy*, 1969), Cutbill (Ed.) (*Data processing in biology and geology*, 1971), Sneath & Sokal (*Numerical taxonomy*, edn 2, 1973), Felsenstein (*Numerical taxonomy*, 1983), Bridge & Sackin (*Mycopath.* **115**: 105, 1991). See Cladistics, Neural networks.

Nummospora E. Müll. & Shoemaker (1964), Mitosporic fungi, 4.D2.?. 1, Switzerland.

Nummularia Tul. & C. Tul. (1863) [non Hill (1756), *Primulaceae*] ≡ Biscogniauxia (Xylar.). See Jong & Benjamin (*Mycol.* **63**: 862, 1971; N. Am. spp.).

Nummariella Eckblad & Granmo (1978) = Biscogniauxia (Xylar.) fide Pouzar (1979).

Nummularioidea (Cooke & Massee) Lloyd (1924) = Camillea (Xylar.) fide Læssøe *et al.* (1989).

Nummularoidea, see *Nummularioidea*.

Numulariola House (1925) ≡ Nummularia (Xylar.).

nurse cells (in *Scleroderma*), hyphae supplying food material to spores which have come away from the basidia.

Nusia Subram. (1993), Mitosporic fungi, 1.C2.24. 2, Singapore.

nutant, nodding.

nutrilite, any organic compound necessary in small amounts for the nutrition of an organism (Williams, *Biol. Rev.* **16**: 49, 1941).

nutriocyte (of *Ascosphaera*), the inflated part of the ascogonium which eventually develops into a spore cyst (Spiltoir & Olive, *Mycol.* **47**: 240, 1955).

Nutrition. Fungi are able to degrade and subsequently metabolize many widely different materials. Some parasites (obligate parasites, e.g. *Uredinales, Erysiphales, Peronosporaceae*) have such special needs that full development takes place only on the right host; but the growth of other parasites, like that of most fungi, will take place on a synthetic medium (see Methods). The growth of fungi is dependent on carbon (C), hydrogen (H), oxygen (O), nitrogen (N), potassium (K), phosphorus (P), magnesium (Mg), and sulphur (S), together with very small amounts of iron (Fe), zinc (Zn), copper (Cu), and/or possibly other minor (or trace) elements. Calcium (Ca) is probably a necessary element but it is not always possible to demonstrate that this is so. In addition, complex 'growth substances' are sometimes needed.

TABLE 7. The Numbers of Fungi.

	gen.	syn.	spp.	gen.	syn.	spp.
PROTOZOA						
Acrasiomycota				4	3	12
Dictyosteliomycota				4	1	46
Myxomycota				74	94	719
Myxomycetes	60	91	690			
Protosteliomycetes	14	3	29			
Plasmodiophoromycota				16	9	45
				115	60	760
CHROMISTA						
Hyphochytriomycota				7	3	24
Labyrinthulomycota				13	5	42
Oomycota				95	52	694
				115	60	760
FUNGI						
Ascomycota				3255	3173	32267
Basidiomycota				1428	1237	22244
Basidiomycetes	473	373	13857			
Teliomycetes	167	145	7134			
Ustomycetes	63	28	1064			
Chytridiomycota				112	41	793
Zygomycota				173	105	1056
Trichomycetes	40	22	189			
Zygomycetes	125	83	867			
Mitosporic fungi				2547	1462	14104
TOTAL						72065

Fungi are heterotrophic in needing their carbon in a complex (organic) form. In general, aliphatic carbon compounds (esp. carbohydrates) are more readily used by fungi than aromatic ones. Nevertheless, basidiomycete fungi are the key organisms in breaking down the ubiquitous aromatic polymer lignin. There are some yeasts which can grow on the one carbon methanol. Some fungi are dependent on organic nitrogen (esp. amino acids and proteins), others can use ammonium or nitrate nitrogen. Fungi are able to adapt and regulate their metabolism according to the nutrients available. The complex bios (q.v.) was the first **'growth substance'** to be noted for fungi, yeast making better growth with it. Thiamin [aneurin, vitamin B_1] is one of the substances necessary for the growth of some fungi; for certain species the complete molecule is needed but others are able to synthesize thiamin if given one, or both, of its components (thiazole and pyrimidine). Fungi are often able to tolerate relatively high concentrations of toxic metals and high concentrations of salt and sugar (which can be lethal to many other microorganisms) and the species which are able to do this cause spoilage of food.

Lichen-forming fungi obtain the carbohydrates they require in the form of sugars and sugar-alcohols (polyols) produced by the algal partner; the nature of the mobile carbohydrate depends on the kind of alga present (e.g. glucose with *Nostoc*, ribitol with *Myrmecia* and *Trebouxia*, erythritol with *Trentepohlia*); cyanobacteria fix atmospheric nitrogen. The carbohydrates are commonly stored as mannitol by the fungal component. The mineral requirements of lichen fungi are met by ions dissolved in rain and from the deposition of dust; in some cases diffusion from the substrate can occur but is usually very limited in extent.

See Jennings (*The physiology of fungal nutrition*, 1995), Lichens, Physiology.

NY. New York Botanical Garden (Bronx, New York, USA); founded 1891; a private institution.

Nyctalina Arnaud (1952), Mitosporic fungi, 1.A1.?. 1 (with clamp connexions), Eur.

Nyctalis Fr. (1825), Tricholomataceae. 4 (2 on other agarics, esp. *Russula*), temp. Chlamydospores on the pileus or even in place of basidiospores (Ingold, *TBMS* **24**: 29, 1940). Anamorph *Asterophora*.

Nyctalospora E.F. Morris (1972), Mitosporic fungi, 2.C2.1. 1, Costa Rica.

Nycteromyces Thaxt. (1917), Laboulbeniaceae. 1, S. Am.

Nyctomyces Hartig (1833), Fossil fungi (mycel.). 4 (Tertiary), Eur., N. Afr.

Nylander (William; 1822-1899). Professor of botany, Helsinki Univ. until 1863 when he moved to Paris. Visited England in 1857 to examine Hooker's lichen collection at Kew (now in **BM**). Published 330 works of which the first few were on ants. Introduced the use of chemical reagents (C, I and K) into lichenology (see Metabolic products). Described over 5,000 new species of lichens from all parts of the world. Main collections in **H**, but material in most of the major European institutions. See Hue (*Bull. Soc. bot. Fr.* **46**: 153, 1899; portr., bibl.), Ahti (Ed.) (*Collected papers of William Nylander*, 6 vols, 1967-90; biogr. in **1**: viii, 1990), Grummann (1974: 611), Stafleu & Cowan (*TL-2* **3**: 788, 1981).

Nylanderaria Kuntze (1891) = Letharia (Parmel.).

Nylanderiella Hue (1914) = Siphula (Lecanorales, inc. sed.).

Nylanderopsis Gyeln. (1935) = Heppia (Hepp.) fide Henssen (*Acta bot. fenn.* **150**: 57, 1994).

Nymanomyces Henn. (1899), Rhytismataceae. 1, Java.

Nypaella K.D. Hyde & B. Sutton (1992), Mitosporic fungi, 4.A1.15. 1, Brunei.

Nyssopsora Arthur (1906), Sphaerophragmiaceae. 9, widespr. See Lohsomboon *et al. MR* **94**: 907, 1990; key), Lutjeharms (*Blumea* suppl. **1**, 1937).

Nyssopsorella Syd. (1921) = Triphragmiopsis (Sphaerophragm.) fide Dietel (1928).

nystatin (mycostatin), an antibiotic from the actinomycete *Streptomyces noursei*; antifungal, widely used against *Candida albicans* infections of man. See Brown & Hazen (*Trans. N.Y. Acad. Sci.* ser. 2 **19**: 447, 1957), Baldwin (*The fungus fighters*, 1981).

O, Botanical Garden and Museum, University of Oslo (Norway); founded 1812.

OA, see Media.

oak-moss (**oakmoss, oak moss**), *Evernia prunastri* (mousse de chêne); extracts of which are used in perfumes to reduce the rate of evaporation of other ingredients. Cf. tree hair.

ob- (prefix), inversely or oppositely.

obclavate, inversely clavate (widest at the base) (Fig. 37.17).

Obconicum Velen. (1939), ? Leotiales (inc. sed.). 2, former Czechoslovakia.

Obelidium Nowak. (1877), Chytridiaceae. 2 or 3, N. temp.

Obeliospora Nawawi & Kuthub. (1990), Mitosporic fungi, 1.A/G1.15. 1 (on submerged decaying wood), Malaysia.

Oberwinkleria Vánky & R. Bauer (1995), Tilletiaceae. 1, Venezuela.

obligate, (1) necessary; essential; (2) (of a parasite), living as a parasite in nature, sometimes of one that has not been cultured on laboratory media, cf. facultative; see synonym.

oblique septum, see septum.

oblong (of spores), twice as long as wide and having somewhat truncate ends (Fig. 37.4*b*, *c*); **- ellipsoid** (of spores), rounded-oblong; having long sides parallel and ends almost hemispherical.

Obolarina Pouzar (1986), Xylariaceae. 1, Eur. See Candoussau & Rogers (*Mycotaxon* **39**: 345, 1990), Eriksson (*SA* **14**: 61, 1995; posn), Hawksworth (*SA* **13**: 198, 1985). Anamorph *Rhinocladiella*-like.

obovate, inversely ovate.

obovoid, inversely ovoid (Fig. 37.12).

obpyriform, the reverse of pear-shaped (Fig. 37.15).

Obryzaceae see *Valsaceae*.

Obryzum Wallr. (1825), Valsaceae. 1 (on lichen, *Leptogium*), Eur. See Eriksson (*Opera Bot.* **60**, 1981).

obsolete, (1) (of organs or parts), rudimentary or absent; (2) (of terms), no longer in use.

Obstipipilus B. Sutton (1968), Mitosporic fungi, 6.B2.19. 1, India.

Obstipispora R.C. Sinclair & Morgan-Jones (1979), Mitosporic fungi, 1.E1.1. 1 (aquatic), USA.

obsubulate, very narrow; pointed at the base and a little wider at the tip.

Obtectodiscus E. Müll., Petrini & Samuels (1979), Dermateaceae. 1, Switzerland.

obtuse, (1) rounded or blunt (Fig. 37.4*b*); (2) greater than a right angle.

occluded, closed; often used of the lumina of hyphae or pseudoparenchymatous cells.

Occultifur Oberw. (1990), Platygloeaceae. 1, N. temp.

Oceanites Kohlm. (1977), Ascomycota (inc. sed.). 1, Atlantic.

Ocellaria (Tul. & C. Tul.) P. Karst. (1871), Dermateaceae. c. 5, N. temp. Anamorph *Cryptosporiopsis*.

Ocellariella Petr. (1947) ? = Naevia (Dermat.) fide Korf (*in* Ainsworth *et al., The Fungi* **4A**: 249, 1973).

ocellate, having rounded marks, like eyes.

Ocellis Clem. (1909) = Ocellularia (Thelotremat.) fide Zahlbruckner (1923).

Ocellularia G. Mey. (1825) nom. cons., Thelotremataceae (L). 200, cosmop. (mainly trop.). See Hale (*Mycotaxon* **11**: 130, 1980; limits), Nagarker *et al.* (*Biovigynam* **14**: 24, 1988; key 32 spp. India), Redinger (*Ark. Bot.* **28A** (8), 1936; S. Am.).

ocellus, an eyespot functioning as a lens and concentrating light rays on a sensitive spot.

Ochotrichobolus Kimbr. & Korf (1983), Thelebolaceae. 1, U.S.A. See van Brummelen (*in* Hawksworth (Ed.), *Ascomycete systematics*: 400, 1994; posn).

Ochraceospora Fiore (1930) = Nectria (Hypocr.) fide Hawksworth (*SA* **5**: 146, 1986).

ochratoxin (**A, B**), toxins of *Aspergillus ochraceus*, *Penicillium viridicatum*, etc.; the cause of nephrotoxicosis in sheep, cattle, and pigs; also carcinogenic and has been found in coffee.

Ochroconis de Hoog & Arx (1974), Mitosporic fungi, 1.B2.11. 8, widespr. See de Hoog (*Stud. Mycol.* **26**: 51, 1985).

Ochroglossum S. Imai (1955) = Microglossum (Geogloss.) fide Maas Geesteranus (1964).

Ochrolechia A. Massal. (1852), Pertusariaceae (L). 40, cosmop. (mainly temp.). See Awasthi & Tewari (*Kavaka* **15**: 23, 1989; key 12 spp. Indian subcont.), Brodo (*CJB* **66**: 1264, 1988; N. Am.), Poelt (*Ergebn. Forsch.-Unternehmen Nepal Himal.* **1**: 251, 1966; Himalaya), Schmitz *et al.* (*Acta bot. fenn.* **150**: 153, 1994; gen. concept), Verseghy (*Beih. Nova Hedw.* **1**, 1962; monogr., keys).

Ochromitra Velen. (1934) = Pseudorhizina (Helvell.). See Petrak (*Sydowia* **1**: 61, 1947).

Ochroporus J. Schröt. (1888) = Phellinus (Hymenochaet.) fide Donk (1960).

Ochropsora Dietel (1895), Chaconiaceae. 3, N. temp.

Ochrosphaera Sawada (1959) [non Schussnig (1930), *Algae*], Ascomycota (inc. sed.). 1, Taiwan.

Ochrosporellus (Bondartseva & S. Herrera) Bondartseva & S. Herrera (1992), Hymenochaetaceae. 1, Eur.

ochrosporous, having yellow or yellow-brown spores.

Ocostaspora E.B.G. Jones, R.G. Johnson & S.T. Moss (1983), Halosphaeriaceae. 1 (marine), USA.

Ocotomyces H.C. Evans & Minter (1985), Rhytismatales (inc. sed.). 1 (on *Pinus*), Honduras. Anamorph *Uyucamyces*.

Octaviania Vittad. (1831) = Melanogaster (Melanogastr.).

Octavianina Kuntze (1893) nom. cons. prop., Octavianinaceae. 15, Eur., N. Am, Asia, Australasia.

Octavianinaceae Locq. ex Pegler & T.W.K. Young (1979), Hymenogastrales. 3 gen., 27 spp. Gasterocarp hypogeous; spores globose, coarsely tuberculate.

octo- (in combinations), 8.

Octojuga Fayod (1889) = Clitopilus (Entolomat.) fide Patouillard (1900).

Octomyces Mello & L.G. Fern. (1918) ? = Saccharomyces (Saccharomycet.) fide Batra (1978).

Octomyxa Couch, J. Leitn. & Whiffen (1939), Plasmodiophoraceae. 2 (on *Achlya* and other *Saprolegniales*), USA. See Sherwood (*J. Elisha Mitchell sci. Soc.* **84**: 52, 1968).

octophore, see haerangium.

octopolar (of incompatability systems), having 3 loci, as in *Psathyrella coprobia* (Jurand & Kemp, *Genetical Res., Cambr.* **22**: 125, 1973); cf. tetrapolar.

Octospora Hedw. (1789), Otideaceae. *c.* 50, widespr. See Dennis & Itzerott (*Kew Bull.* **28**: 5, 1973; key, **31**: 497, 1977, W. Eur.), Khare & Tewari (*Mycol.* **67**: 972, 1975), Caillet & Moyne (*BSMF* **103**: 277, 1987; key 32 spp., *Bull. Soc. Hist. nat. Doubs* **84**: 9, 1991; keys), Döbbeler (*Nova Hedw.* **31**: 817, 1980), Döbbeler & Itzerott (*Nova Hedw.* **34**: 127, 1981; biology).

octospore, one spore of an 8-spored ascus.

Octosporella Döbbeler (1980), Otideaceae. 4, Eur., Venezuela.

Octosporomyces Kudrjanzev (1960) = Schizosaccharomyces (Schizosaccharomycet.) fide Sipiczki (*in* Nasim *et al.* (Eds), *Molecular biology of the fission yeast*, 1989), Kurtzman & Robnett (*Yeast* **71**: 61, 1991).

Octosporonites Locq. & Sal.-Cheb. (1980), Fossil fungi. 1, Cameroon.

octosporous, producing spores in 8s.

ocular chamber, see ascus.

Odontia Gray (1821) = Caldesiella (Thelephor.) fide Donk (1956), = Tomentella (Thelephor.).

Odontia Fr. (1835) = Steccherinum (Steccherin.) fide Donk (1964), but used by Furukawa (*Bull. Govt. For. Exp. Stn* **261**, 1974).

Odonticium Parmasto (1968), Hyphodermataceae. 4, widespr.

Odontina Pat. (1887) = Steccherinum (Steccherin.) fide Donk (1956).

Odontiochaete Rick (1940) nom. dub. (Basidiomycetes, inc. sed.). See Donk (*Taxon* **5**: 107, 1956).

Odontiopsis Hjortstam & Ryvarden (1980), Hyphodermataceae. 1, trop.

Odontium Raf. (1817) = Odontia Fr. (Steccherin.) fide Merrill (*Index Rafinesq.*, 1949), = Steccherinum (Steccherin.).

Odontodictyospora Mercado (1984), Mitosporic fungi. 1.D1.11. 1, Cuba.

odontoid, tooth-like; dentate.

Odontoschizon Syd. & P. Syd. (1914) = Patinella (Dermat.) fide Clements & Shear (1931).

Odontotrema Nyl. (1858), Odontotremataceae. *c.* 6, temp.

Odontotremataceae D. Hawksw. & Sherwood (1982), Ostropales. 11 gen. (+ 6 syn.), 35 spp. Stroma sometimes present; ascomata apothecial, immersed or erumpent, with a well-developed hyphal margin, often black and carbonized; interascal tissue of simple paraphyses, sometimes branched at the base; asci with a thickened apex, with or without a well-developed I-pore; ascospores ellipsoidal to filiform, variously septate. Anamorphs not definitely known. Saprobic on wood, esp. xeric, or lichenicolous; a few spp. lichenized with green algae; esp. temp.
Lit.: Sherwood-Pike (*Mycotaxon* **28**: 137, 1987).

Odontotremella Rehm (1912) ≡ Odontura (Odontotremat.).

Odontura Clem. (1909), Odontotremataceae. 1, Eur. See Sherwood *et al.* (*TBMS* **75**: 479, 1980).

odour, see Smell.

Oedemium Link (1824), Mitosporic fungi, 1.B/C2.9. 2, widespr. See Hughes & Hennebert (*CJB* **41**: 773, 1963).

Oedemocarpus Trevis. (1857) = Mycoblastus (Mycoblast.) fide Hafellner (*Beih. Nova Hedw.* **79**: 241, 1984).

Oedipus J. Bataille (1908) = Boletus (Bolet.) fide Singer (1945).

oedocephaloid, having a swelling at the end or tip, as conidiophores of *Oedocephalum* and *Cunninghamella*.

Oedocephalum Preuss (1851), Anamorphic Ascobolaceae, Pezizaceae and Pyronemataceae, 1.A1.6. Teleomorphs *Ascophanus*, *Cleistoiodophanus*, *Iodophanus*, *Peziza*, *Pyronema*. 8, widespr. See Stalpers (*Proc. K. Ned. Akad. Wet.* C **77**: 383, 1974; key).

Oedogoniomyces Tak. Kobay. & M. Ôkubo (1954), Oedogoniomycetaceae. 1, N. trop. See Emerson & Whisler (*Arch. Mikrobiol.* **61**: 195, 1968).

Oedogoniomycetaceae D.J.S. Barr (1990), Monoblepharidales. 1 gen., 1 sp. Thallus an unbranched filament attahced to substratum by a holdfast; sporangia maturing in basipetal succession, forming H-shaped segments; saprophytic.

Oedomyces Sacc. ex Trab. (1894) = Physoderma (Physodermat.) fide Karling (1950), = Urophlyctis (Physodermat.) fide Ciferri (1963).

Oedothea Syd. (1930), Mitosporic fungi, 3.B2.19. 1, S. Am.

oenology, the study of wines; see wine making.

Oerskovia Prauser, M.P. Lechev. & H. Lechev. (1970), *Actinomycetes*, q.v.

Oesophagomyces Manier & Ormières (1980) Fungi (inc. sed.). 1, France.

Ogataea Y. Yamada, K. Maeda & Mikata (1994), Saccharomycetaceae. 5, widesp.

Ohleria Fuckel (1868), Melanommataceae. 3 (on wood), Eur., N. & S. Am., NZ. See Samuels (*N.Z. Jl Bot.* **18**: 515, 1980). Anamorph *Monodictys*.

Ohleriella Earle (1902), Fenestellaceae. 1, N. Am. See Barr (*SA* **6**: 142, 1987).

Oichitonium Durieu & Mont. [not traced] nom. dub. (? Fungi, inc. sed.).

-oid (suffix), like; having the form of. Most of the many mycological terms ending in this suffix (e.g. achlyoid, as in *Achlya*; daedaleoid, as *Daedalea*) have not been compiled in this *Dictionary*.

Oideum Ehrenb. (1818) ≡ Oidium (Mitosp. fungi).

Oidiodendron Robak (1932), Anamorphic Myxo-trichaceae, 2.A2.38. Teleomorphs *Byssoascus, Myxo-trichum*. 11, N. temp. See Morrall (*CJB* **46**: 204, 1968), Barron (*CJB* **40**: 589, 1962; key).

oidiomycin, an antigen prepared from *Candida albicans*, esp. for skin testing.

oidiophore, a structure producing oidia.

Oidiopsis Scalia (1902), Anamorphic Erysiphaceae, 1.A1.1. Teleomorph *Leveillula*. 2, widespr.

oidiospore, see oidium.

Oidites Mesch. (1892), Fossil fungi. 2 (Oligocene), Baltic.

oidium (pl. **oidia**), (1) spermatia formed on hyphal branches, esp. in heterothallic hymenomycetes; (2) flat-ended conidia formed by the breaking up (usually centripetally) of a hypha into cells, as in *Geotrichum candidum*; arthrospore; (3) a mildew.

Oidium Link (1809) nom. rej. = Oidium Link (Mitosp. fungi).

Oidium Link (1824) nom. cons., Anamorphic Erysiphaceae, 1.A1.23. Teleomorph *Erysiphe* etc. *c.* 120, widespr. See also Carmichael (1980), Ialongo (*Mycotaxon* **47**: 193, 1993; statistical characterization).

Oidium Sacc. (1880) nom. rej. = Oidium Link (Mitosp. fungi).

oidization, dikaryotization by the fusion of an oidium with a haploid hypha.

Oidospora Will. (1878), ? Fossil fungi. 1 (Carbonif.).

Ojibwaya B. Sutton (1973), Mitosporic fungi, 7.A2.23. 1, Canada, Malawi.

old man's beard, see beard moss.

Oleina Tiegh. (1887), Saccharomycetales (inc. sed.). 2, Eur.

Oleinis Clem. (1931) ≡ Oleina (Saccharomycetales) fide Batra (1978).

oleocystidium, see cystidium.

oleoso-locular (of spores), having cells like drops of oil.

Oligonema Rostaf. (1875), Trichiaceae. 4, N. Am., Eur., N. Afr.

Oligoporus Bref. (1888), Coriolaceae. 1, Eur. See Donk (*Persoonia* **6**: 210, 1971; anamorph).

oligosporous, having few spores. Cf. myriosporous.

Oligostroma Syd. & P. Syd. (1914) = Mycosphaerella (Mycosphaerell.) fide v. Arx & Müller (1975).

oligotropic, poor in nutrients; cf. eutrophic.

Olivea Arthur (1917) [non Sch.-Bip. ex Benth. (1872), *Compositae*], Chaconiaceae. 8, trop. See Ono & Hennen (*Trans. Mycol. Soc. Japan* **24**: 369, 1983).

Oliveonia Donk (1958), Ceratobasidiaceae. 2, widespr.

Olla Velen. (1934), Hyaloscyphaceae. 10, Eur. See Baral (*SA* **13**: 113, 1994; concept).

Ollula Lév. (1863) = Tubercularia (Mitosp. fungi) fide Sutton (*Mycol. Pap.* **141**, 1977).

Olpidiaceae J. Schröt. (1889), Spizellomycetales. 5 gen. (+ 6 syn.), 28 spp. Thallus monocentric, holocarpic, or eucarpic; sporangium and resting spore developed exogenous to the zoospore cyst; usually endobiotic. For Lit. see under *Spizellomycetales*.

Olpidiaster Pascher (1917) ≡ Asterocystis (Olpid.).

Olpidiella Lagerh. (1888) = Olpidium (Olpid.) fide Minden (1911).

Olpidiomorpha Scherff. (1926), Rozellopsidales (inc. sed.). 1 (on *Pseudospora*), Eur. See Dick (*in press*; key).

Olpidiopsidaceae Sparrow ex Cejp (1959), Olpidiopsidales. 2 gen. (+ 5 syn.), 15 spp.

Olpidiopsidales, Oomycota. 1 fam., 3 gen. (+ 5 syn.), 17 spp. Fam.:
    Olpidiopsidaceae.

Olpidiopsis Cornu (1872), Olpidiopsidaceae. 12 (on *Oomycetes*), widespr. See Dick (*in press*; key).

Olpidium (A. Braun) J. Schröt. (1886), Olpidiaceae. 25 (in algae, aquatic fungi, rotifers etc.), widespr. *O. brassicae* (lettuce big vein virus vector); *O. uredinis* on rust spores. See Litvinov (*Trudy Bot. Inst. Akad. nauk URSS* ser. 2 **12**: 188, 1959), Sparrow (1960: 128; key), Karling (1977), Sampson (*TBMS* **23**: 199, 1939), Sahtiyanci (*Arch. Mikrobiol.* **41**: 187, 1962), Lange & Insunza (*TBMS* **69**: 377, 1977).

Olpitrichum G.F. Atk. (1894), Mitosporic fungi, 1.A1.10. 3, widespr. See Holubová-Jechová (*Folia geobot. phytotax.* **9**: 425, 1974).

Omalycus Raf. (1814) ? = Calvatia (Lycoperd.).

Ombrophila Fr. (1849), Leotiaceae. *c.* 10, widespr. See Verkley (*Persoonia* **15**: 3, 1992).

Ombrophila Quél. (1892) nom. conf. fide Donk (*Persoonia* **4**: 219, 1966).

Omega B. Sutton & Minter (1988), Mitosporic fungi, 7.E1.19. 1, Greece.

Ommatomyces Kohlm., Volkm.-Kohlm. & O.E. Erikss. (1995), Amphisphaeriaceae. 1 (marine), N. Am.

Ommatospora Bat. & Cavalc. (1964) = Microclava (Mitosp. fungi) fide Deighton (1969).

Ommatosporella Bat., J.L. Bezerra & Poroca (1967), Mitosporic fungi, 1.B2.1. 1, Brazil.

Omnidemptus P.F. Cannon & Alcorn (1994), Magnaporthaceae. 1, Australia. Anamorph *Mycoleptodiscus*.

omnivorous (of parasites), attacking a number of different hosts.

Omoriza Paulet (1812) nom. conf. (Fungi), used for a range of *Ascomycota* and *Basidiomycetes*. See Donk (*Taxon* **6**: 85, 1957).

Omorrhiza, see Omoriza.

Omphalaria A. Massal. (1855) nom. rej. = Anema (Lichin.).

Omphalaria Girard & Dunal ex Nyl. (1855) = Thyrea (Lichin.).

Omphalia (Pers.) Gray (1821) [non *Omphalea* L. (1759) nom. cons., *Euphorbiaceae*] = Pseudoclitocybe (Tricholomat.). See Donk (*Beih. Nova Hedw.* **5**: 203, 1963), Redhead & Weresub (*Mycol.* **70**: 556, 1978).

Omphalia (Fr.) Staude (1857) ≡ Omphalina (Tricholomat.).

Omphaliaster Lamoure (1971), Tricholomataceae. 2, N. temp.

Omphalina Quél. (1886) nom. cons. prop., Tricholomataceae. 19 (±L), widespr. See Singer (1962), Bigelow (*Mycol.* **62**: 1, 1970, **70**: 556, 1978; N. Am.), Redhead & Weresub (*Mycol.* **70**: 556, 1978; nomencl.), Redhead & Kuyper (*Environm. Sci. Res.* **34**: 319, 1987; nomencl.), Watling (*Bull. Br. lichen Soc.* **49**: 28, 1982; key 7 Br. L spp.), Lange (*Nordic Jl Bot.* **1**: 691, 1981; typification), Clémençon (*Zeitschr. Mykol.* **48**: 203, 1982; classific.). Cf. *Coriscium, Botrydina, Phytoconis*.

Omphaliopsis (Noordel.) P.D. Orton (1991), Entolomataceae. 3, Eur.

Omphalius Roussel (1806) = Clitocybe (Tricholomat.) fide Murrill (1915).

Omphalocystis Balbiani (1889), Fungi (inc. sed.). 1 (on *Cryptopus*), France.

Omphalodiella Henssen (1991), Parmeliaceae (L). 1, Argentina.

Omphalodina M. Choisy (1929) = Rhizoplaca (Lecanor.).

omphalodisc, (1) an orbicular conical shaped disk; (2) (of *Umbilicaria* [*Omphalodiscus*]), an apothecium with a central knob of sterile hyphae.

Omphalodiscus Schol. (1934) = Umbilicaria (Umbilicar.).

Omphalodium Meyen & Flot. (1843), Parmeliaceae (L). *c.* 4, N. & S. Am.

Omphalodium Rabenh. (1845) = Umbilicaria (Umbilicar.).

Omphalomyces Battarra ex Earle (1909) = Russula (Russul.) fide Singer (1951).

Omphalophallus Kalchbr. (1883) = Phallus (Phall.).

Omphalopsis Earle (1909) [non Grev. (1863), *Algae*] ≡ Xeromphalina (Tricholom.).

Omphalora T.H. Nash & Hafellner (1990), Parmeliaceae (L). 1, N. Am.

Omphalosia Neck. ex Kremp. (1869) = Umbilicaria (Umbilicar.).

Omphalospora Theiss. & Syd. (1915), Dothideaceae. 4, Eur. See Obrist (*Phytopath. Z.* **35**: 383, 1959). Anamorph *Podoplaconema*.

Omphalotus Fayod (1889), Paxillaceae. 2, cosmop. *O. olearius* is luminescent when fresh. See Bigelow *et al.* (*Mycotaxon* **3**: 363, 1976; discussion).

Onakawananus Radforth (1958), Fossil fungi (mycel.). 1 (Cretaceous), Canada.

Onchopus P. Karst. (1879) = Coprinus (Coprin.) fide Singer (1951).

Oncidium Nees (1823) [non Sw. (1800), *Orchidaceae*] ≡ Myxotrichum (Myxotrich.).

Oncobasidium P.H.B. Talbot & Keane (1971), Ceratobasidiaceae. 1, Papua New Guinea. *O. theobromae*, vascular-streak dieback (Keane & Prior, *Phytopath. Pap.* **33**, 1991).

Oncobyrsa C. Agardh (1827) nom. dub. (? Fungi).

Oncocladium Wallr. (1833), Anamorphic Gymnoascaceae, 1.A1.38. Teleomorph *Gymnoascus*. 1 (from soil), Eur., Canada. See Sigler *et al.* (*Mycotaxon* **28**: 119, 1987; relationship to *Malbranchea flava*), Hughes (*CJB* **46**: 941, 1968).

oncom merah (**oncom hitah**), Javanese fermented soya bean products in which the principal fungi are *Neurospora sitophila* and *Rhizopus oligosporus*, respectively (Hedger, *Bull. BMS* **12**: 53, 1978); see Fermented food and drinks.

Oncomyces Klotzsch (1843) = Auricularia (Auricular.) fide Saccardo (*Syll. Fung.* **6**: 762, 1888).

Oncopodiella G. Arnaud ex Rifai (1965), Mitosporic fungi, 1.D2.10. 1, Eur.

Oncopodium Sacc. (1904), Mitosporic fungi, 2.D2.1. 4, Eur., N. Am. See Hudson (*TBMS* **44**: 406, 1961), Sutton (*Mycol.* **70**: 793, 1978).

Oncopus, see *Onchopus*.

Oncospora Kalchbr. (1880), Mitosporic fungi, 7.B2.1. 5, S. Afr., Eur., Java. See Chevassut (*BSMF* **106**: 107, 1990; genus rev.).

Oncosporella P. Karst. (1887), Mitosporic fungi, 7.C1.10. 1, Finland.

Oncosporomyces Bat. (1965), Mitosporic fungi (L), 8.E1.?. 1, Brazil.

Oncostroma Bat. & Marasas (1966), Mitosporic fungi, 5.A1.?. 1, S. Afr.

Ondiniella E.B.G. Jones, R.G. Johnson & S.T. Moss (1984), Halosphaeriaceae. 1 (marine), Eur., N. Am., Chile.

Onnia P. Karst. (1889) ≡ Mucronoporus (Hymenochaet.) but maintained by Jahn (*Westf. Pilzkde* **11**: 79, 1978; key 3 spp.).

ontjom, an Indonesian fermented food prepared from peanut press cake, surface inoculated with *Neurospora sitophila*.

ontomycosis see mycosis.

Ontostheca Bat. (1963) = Eudimeriolum (Pseudoperispor.) fide v. Arx (*in litt.*).

Ontotelium Syd. (1921) = Uromyces (Puccin.) fide Dietel (1928).

Onychocola Sigler (1990), Mitosporic fungi, 1.A/B1.38. 1 (from humans), Canada.

onychomycosis, see mycosis.

Onychophora W. Gams, P.J. Fisher & J. Webster (1984), Mitosporic fungi, 1.A1.15. 1 (coprophilous), UK.

Onygena Pers. (1799), Onygenaceae. 5 (on feathers, bones, etc.), Eur., N. Am. See Rammeloo (*Dumort.* **6**: 1, 1977), Currah (1985).

Onygenaceae Berk. (1857), Onygenales. 23 gen. (+ 7 syn.), 57 spp. Ascomata sometimes stipitate; peridium varied, usually hyphal but occ. pseudoparenchymatous, sometimes with complex appendages; ascospores oblate or allantoid, pitted or reticulate. Anamorphs mostly *Chrysosporium* or *Malbranchea*. Keratinophilic, on soil or coprophilous, occ. on hair or horn, cosmop.

Onygenales, Ascomycota. 4 fam., 36 gen. (+ 22 syn.), 90 spp. Stromata absent; ascomata formed from coiled initials, cleistothecial, sometimes aggregated, rarely stipitate, pale; peridium composed of loosely woven usually thick-walled hyphae, sometimes with complex appendages; interascal tissue absent; asci ? formed from croziers, ± globose, small, evanescent, 8-spored; ascospores small, often brightly coloured, usually oblate, often ornamented, esp. with equatorial ridges. Anamorphs prominent, hyphomycetous, arthric. Keratinophilic or cellulolytic, some parasitic on humans and other animals (see ringworm, dermatophytes), also in soil, cosmop.

Von Arx (*Persoonia* **9**: 393, 1977) accepted a much wider circumscription for the group. Included in *Pezizales* by Malloch (1981). Fams:

(1) **Arthrodermataceae**.
(2) **Gymnoascaceae**.
(3) **Myxotrichaceae**.
(4) **Onygenaceae**.

Lit.: Apinis (*Mycol. Pap.* **96**, 1964; Br. spp.), v. Arx (*Persoonia* **6**: 371, 1971; key, 1977; gen. syns., 1981; key gen.), Benjamin (*Aliso* **3**: 301, 1956; review), Benny & Kimbrough (1980), Currah (*Mycotaxon* **24**: 1, 1985; *SA* **7**: 1, 1988; key gen.; *in* Hawksworth, 1994: 281), Dalpé (*New Phytol.* **113**: 523, 1989; as ericoid mycorrhizas), Hughes (*CJB* **46**: 939, 1968), Kuehn (*Mycol.* **50**: 417, 51: 665, 1958-59; survey), Malloch (*in* Reynolds, 1981), Malloch & Cain (*CJB* **50**: 61, 1972), Orr (*Mycotaxon* **5**: 283, 1977; gen. septal swellings, *et al.*, 5: 466, 1977; gen. discoid-oblate spores), Pugh (*Sabouraudia* **5**: 49, 1966, & Evans, *TBMS* **54**: 233, 1970; birds' nests and birds), Sigler & Carmichael (*Mycotaxon* **4**: 349, 1976; anamorphs), Takizawa *et al.* (*Mycoscience* **35**: 327, 1994; ubiquinones), Vanbreuseghem (*Ann. Soc. belge Méd. trop.* **32**: 173, 1952; soil isolation by hair bait), Visset (*Mycopath. Mycol. appl.* **54**: 377, 1974; SEM).

Onygenopsis Henn. [not traced], Sordariales (inc. sed.). 1, trop. See Eriksson & Hawksworth (*SA* **5**: 147, 1986; posn), Petch (*Ann. R. bot. Gdn Peradeniya* **5**: 265, 1912).

Oochytrium Renault (1895), Fossil fungi (Chytridiomycetes). 1 (Carboniferous), France.

oocyst, the product of aposporous spore development in *Oomycetes* (Dick, *New Phytol.* **71**: 1151, 1972).

oogamy, heterogamy in which the gametes are a non-motile egg and a small, motile sperm.

Oogaster Corda (1854) = Tuber (Tuber.) fide Fischer (1938).

oogenesis, the development of the oogonium after being fertilized.

oogoniols, *Achlya* hormones which induce oogonial formation. Cf. antheridiol.

oogonium (**oogone**), uninucleate or coenocytic cell producing female gametes (oospheres) (Fig. 28A-F).

Oolithinia M. Choisy & Werner (1932) = Protoblastenia (Psor.).

Oomyces Berk. & Broome (1851), Acrospermaceae. 1, Eur. See Petch (*J. Bot., Lond.* **75**: 217, 1937), Eriksson (1981; posn).

Oomycetes, see *Oomycota*.

Oomycota (Oomycetes, Peronosporomycotina, Peronosporomycetes), Chromista. 9 ord., 25 fam., 95 gen. (+ 52 syn.), 694 spp. Aquatic or terrestrial, freshwater or marine, saprobic or parasitic (some economically important on higher plants); thallus unicellular to mycelial (hyphae coenocytic), mainly aseptate; assimilative phase diploid (as in plants); zoospores with unequal (anisokont, heterokont) flagella, a tinsel one diverted with 2 rows of mastigonemes formed, and a whiplash one smooth or with fine flexous hairs backwards; with protoplasmic and nucleus-associated microtubules; the cell walls are a glucan-cellulose, rarely with minor amounts of chitin. Cosmop. and widespr. Ords:
(1) **Leptomitales**.
(2) **Myzocytiopsidales**.
(3) **Olpidiopsidales**.
(4) **Peronosporales**.
(5) **Pythiales**.
(6) **Rhipidiales**.
(7) **Salilagenidales**.
(8) **Saprolegniales**.
(9) **Sclerosporales**.
      Lit.: Buczaki (Ed.) (*Zoosporic plant pathogens: a modern perspective*, 1983), Cavalier-Smith (1987, *in* Green *et al.* (Eds), *The chromophyte algae*: 379, 1989, 1993), Cejp (*Oomycetes* 1, 1959 [Flora ČSR]), Corliss (1994), Dick (*New Phytol.* **71**: 1151, 1972; morphology and taxonomy; *in* Subramanian (Ed.), *Taxonomy of fungi*: 82, 1978, *in* Parker, 1982, **1**: 179), Dick & Win-Tin (*Biol. Rev.* **48**: 133, 1973; cytology), Fitzpatrick (1930), Fuller (*in* Margulis *et al.* (Eds), 1990: 380), Fuller & Jaworski (Eds) (*Zoosporic fungi in teaching and research*, 1986), Sparrow (*Aquatic phycomycetes* edn 2, 1960; *in* Ainsworth *et al.* (Eds), *The Fungi* **4B**: 61, 1973; ordinal classification, *Bot. J. Linn. Soc.* **99**: 97, 1989, *in* Margulis *et al.* (Eds), 1990: 661). See also *Lit.* under Orders.

ooplasm (of *Peronosporales*), the protoplasm, at the centre of the oogonium, which becomes the oosphere; cf. periplasm and gonoplasm.

ooplast (of *Saprolegniaceae*), a large membrane bound inclusion of oospores formed by the fusion of dense body vesicles. See Howard & Moore (*Bot. Gaz.* **131**: 311, 1970).

oosphere (of *Oomycetes*), the female gamete; the 'egg' of the oogonium; **compound -**, one having many functional nuclei.

Oospora Wallr. (1833) ≡ Oidium (Mitosp. fungi) fide Donk (*Taxon* **12**: 270, 1963). *O. citri-aurantii* (in citrus fruits); *O. pustulans*, see *Polyscytalina*; *O. fimicola*, see *Scopulariopsis*; *O. lactis*, see *Geotrichum*. See also Sigler & Carmichael (*Mycotaxon* **4**: 349, 1976).

oospore (of oomycetes), the resting spore from a fertilized oosphere; a like structure produced by parthenogenesis (Fig. 28A-B).

Oosporidium Stautz (1931), Mitosporic fungi, 1.A1.10. 1, Eur., N. Am.

Oosporoidea Sumst. (1913) = Geotrichum (Mitosp. fungi) fide Carmichael (*Mycol.* **49**: 820, 1957).

Oostroma Bonord. (1864) = Pseudovalsa (Melanconid.) fide Læssøe (*SA* **13**: 43, 1994).

Oothecium Speg. (1919) = Asterostomella (Mitosp. fungi) fide Farr (*Bibl. Mycol.* **35**, 1973).

Oothyrium Syd. (1939), Mitosporic fungi, 5.A1.?. 1, Afr.

Oovorus Entz (1930) nom. dub. ('phycomycetes').

Opasterinella Speg. (1917) = Asterinella (Microthyr.) fide v. Arx & Müller (1975).

Opeasterina Speg. (1919) = Asterina (Asterin.) fide Müller & v. Arx (1962).

Opegrapha Humb. (1793) nom. rej. = Graphis (Graphid.).

Opegrapha Ach. (1809) nom. cons., Roccellaceae (±L). *c.* 300, widespr. See Hafellner (*Herzogia* **10**: 1, 1994; key 19 spp. on lichens), Hawksworth (*Bot. J. Linn. Soc.* **109**: 543, 1992; gen. nomencl.), Redinger (*Rabenh. Krypt.-Fl.* **9**, 2 (1): 246, 1938; Eur., *Ark. Bot.* **29A** (19), 1940; S. Am.), Santesson (1952; foliicolous spp.), Sérusiaux (*Lichenologist* **17**: 1, 1985; spp. with goniocysts), Torrente-Paños (*Cryptogamie, Mycol.* **8**: 159, 1987; asci), Torrente & Egea (1989).

Opegraphaceae, see *Roccellaceae*.

Opegraphales, see *Arthoniales*.

Opegraphella Müll. Arg. (1890) = Opegrapha (Roccell.) fide Santesson (1952).

Opegraphellomyces Cif. & Tomas. (1953) ≡ Fouragea (Roccell.).

Opegraphites Debey & Ettingsh. (1859), Fossil fungi (Ascomycota, L). 1 (Cretaceous), Belgium.

Opegraphoidea Fink (1933) = Opegrapha (Roccell.).

Opegraphomyces E.A. Thomas ex Cif. & Tomas. (1953) = Opegrapha (Roccell.).

Opercularia Stirt. (1878) [non Gaertn. (1788), *Rubiaceae*] = Phyllobathelium (Phyllobathel.) fide Santesson (1952).

operculate (of an ascus or sporangium), opening by an apical lid to discharge the spores, as in the ascus of the *Pezizales*; see ascus; cf. inoperculate.

Operculella Khesw. (1941) = Phacidiopycnis (Mitosp. fungi) fide Sutton (1977).

operculum, a cover or lid.

Opethyrium Speg. (1919) nom. dub. (Fungi, inc. sed.). No spp. listed.

ophiobolin (cochliobolin), an antibiotic from *Cochliobolus miyabeanus* and *C. heterostrophus*; antibacterial, antifungal, anti-*Trichomonas vaginalis* (Ishibashi, *J. agric. Chem. Soc. Japan* **35**: 257, 1961); phytotoxic to rice (Orsenigo, *Phytopath. Z.* **29**: 189, 1957). See Tsuda *et al.* (*Tetrahedron Lett.* **35**: 3369, 1967; nomencl.).

Ophiobolus Riess (1854), Leptosphaeriaceae. *c.* 100, widespr. *O. herpotrichus* (on cereals), *O. heterostrophus*, *O. miyabeanus*, see *Cochliobolus*; *O. graminis*, see *Gaeumannomyces*. See Holm (1957), Shoemaker (*CJB* **54**: 2365, 1976; key 31 Can. spp.), Walker (1980; type, disposition 60 names).

Ophiocapnocoma Bat. & Cif. (1963) = Metacapnodium (Metacapnod.). See Eriksson & Hawksworth (*SA* **6**: 124, 1987). = Limacinia (Dothid.) fide Reynolds (1985).

Ophiocapnodium Speg. (1918) ? = Euantennaria (Euantennar.) fide v. Arx & Müller (1975).

Ophiocarpella Theiss. & Syd. (1915) = Sphaerulina (Mycosphaerell.) fide Barr (1972).

Ophioceras Sacc. (1883), Lasiosphaeriaceae. 5, widespr. See Conway & Barr (*Mycotaxon* **5**: 376, 1977).

Ophiochaeta (Sacc.) Sacc. (1895) ≡ Acanthophiobolus (Dothid.).

Ophiociliomyces Bat. & I.H. Lima (1955), Meliolaceae. 4, Brazil.

Ophiocladium Cavara (1893) = Ramularia (Mitosp. fungi) fide Sutton & Waller (*TBMS* **90**: 55, 1988).

Ophiocordyceps Petch (1931) = Cordyceps (Clavicipit.) fide Rogerson (1970).

Ophiodeira Kohlm. & Volkm.-Kohlm. (1988), Halosphaeriaceae. 1 (marine), C. Am.

Ophiodendron Arnaud (1952) nom. dub. (Mitosp. fungi) fide Hennebert (*TBMS* **51**: 13, 1968).

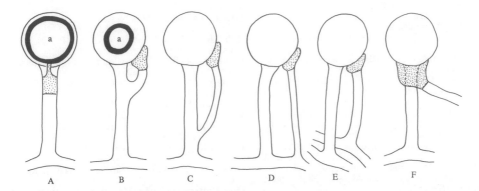

Fig. 28. Relationship of oogonium and antheridium (shaded): A, hypogynous (a, plerotic oospore); B, paragynous (a, aplerotic oospore); B,C,D, monoclinous; E, diclinous; F, amphigynous.

Ophiodictyon Sacc. & P. Syd. (1902) = Trichothelium (Trichothel.).

Ophiodothella (Henn.) Höhn. (1910), Phyllachoraceae. 25, widespr., trop. See Boyd (*Mycol.* **26**: 456, 1934), Hanlin (*Mycotaxon* **39**: 1, 1990), Hanlin *et al.* (*Mycotaxon* **44**: 103, 1992; key 26 spp.).

Ophiodothis Sacc. (1883) ? = Balansia (Clavicipit.) fide Diehl (1950).

Ophiogene Petr. (1931) = Nematothecium (Pseudoperispor.) fide v. Arx & Müller (1975).

Ophiogloea Clem. (1903) = Vibrissea (Vibriss.) fide Sánchez & Korf (1966).

Ophiognomonia (Sacc.) Sacc. (1899), Valsaceae. 1, Eur., N. Am. See Barr (*Mycol. Mem.* **7**, 1978), Monod (1983).

Ophioirenina Sawada & W. Yamam. (1959), Meliolaceae. 1, Taiwan.

Ophiomassaria Jacz. (1894), Ascomycota (inc. sed.). 1 (on *Alnus*), Eur. See v. Arx & Müller (*Stud. Mycol.* **9**, 1975).

Ophiomeliola Starbäck (1899), ? Meliolaceae. 1 or 2, trop.

Ophionectria Sacc. (1878), Hypocreaceae. 2, trop. See Rossman (*Mycol.* **69**: 355, 1977; key redisposed spp., *Mycol. Pap.* **150**, 1983), Subramanian & Bhat (*Kavaka* **6**: 55, 1979; anamorph, ontogeny).

Ophioparma Norman (1852), Ophioparmaceae (L). 3, N. Hemisph. See Rogers & Hafellner (*Lichenologist* **20**: 167, 1988).

Ophioparmaceae R.W. Rogers & Hafellner (1988), Lecanorales (L). 1 gen., 3 spp. Thallus crustose; ascomata apothecial, sessile, with a well-developed margin, disc reddish or brown; interascal tissue of thick sparingly branched paraphyses, with swollen apices; asci with a well-developed I+ apical cap, without an ocular chamber, with an outer I+ gelatinized layer; ascospores hyaline, elongated, transversely septate, without a sheath, coiled within the ascus. Anamorph pycnidial. Lichenized with green algae, boreal.
   Lit.: See also under *Haematomma*.

Ophioparodia Petr. & Cif. (1932), Parodiopsidaceae. 1, W. Indies. Anamorph *Septoideum*.

Ophiopeltis J.V. Almeida & Sousa da Câmara (1903) ? = Micropeltis (Micropeltid.). See Clements & Shear (1931).

Ophiopodium Arnaud (1954) nom. inval. = Grallomyces (Mitosp. fungi) fide Deighton & Pirozynski (1966).

Ophiosira Petr. (1955), Mitosporic fungi, 8.E1.?. 1, Austria.

Ophiosphaerella Speg. (1909), Phaeosphaeriaceae. *c*.6, widespr. See Walker (*Mycotaxon* **11**: 1, 1980).

Ophiosphaeria Kirschst. (1906) = Acanthophiobolus (Dothideales) fide v. Arx & Müller (1975).

Ophiosporella Petr. (1947) = Phloeosporina (Mitosp. fungi) fide Sutton (1977).

Ophiostoma Syd. & P. Syd. (1919), Ophiostomataceae. 102, widespr. *O. novo-ulmi* (Dutch elm disease; Brasier *Mycopath.* **115**: 151, 1991). See Seifert *et al.* (*in* Wingfield *et al.*, 1993: 269), Grylls & Seifert (*in* Wingfield *et al.*, 1993: 261; key); Hausner *et al.* (*Mycol.* **84**: 870, 1992; posn), Hausner *et al.* (*CJB* **71**: 52, 1993; gen. concept), van Wyk & Wingfield (*Mycol.* **83**: 698, 1991; ascospores), Berbee & Taylor (*Exp. Mycol.* **16**: 87, 1992; 18S rRNA), Spatafora & Blackwell (*MR* **91**: 1, 1994; posn), Wingfield (*et al.*, 1993; many papers), Wolfaardt *et al.* (*S. Afr. Jl Bot.* **58**: 277, 1992; synoptic key, database). Anamorphs *Graphium, Leptographium, Pesotum* and *Sporothrix*.

Ophiostomataceae Nannf. (1932), Ophiostomatales. 4 gen. (+ 4 syn.), 106 spp. Ascomata perithecial, rarely cleistothecial, dark, usually long-necked, with ostiolar setae; interascal tissue absent; asci ? usually formed in chains from a fertile layer lining the ascomatal cavity, ± saccate, very thin-walled, evanescent, 8-spored; ascospores hyaline, varied in shape, often with eccentric wall thickening or sheaths. Anamorphs hyphomycetous, *Graphium, Leptographium, Pesotum*, or *Sporothrix*.
   Close to *Xylariales* on both morphological (Samuels, *in* Wingfield *et al.*, 1993: 15) and molecular data (Andersson *et al.*, SA **14**: 1, 1995).
   Lit.: v. Arx (1981), Benny & Kimbrough (1980), de Hoog (*Stud. Mycol.* **7**, 1974), Redhead & Malloch (*CJB* **55**: 1701, 1977), Weijman (*Ant. v. Leeuwenhoek* **42**: 315, 1976; chemistry), Wingfield *et al.* (*Ceratocystis and Ophiostoma*, 1993).

Ophiostomatales, Ascomycota. 2 fam., 5 gen. (+ 4 syn.), 109 spp. Stromata absent; ascomata perithecial, rarely cleistothecial, hyaline or black, thin-walled, membranous, usually long-necked, with ostiolar

setae; interascal tissue absent; asci small, evanescent, formed in chains; ascospores usually small, hyaline, mostly aseptate, often with eccentric wall thickening or sheaths. Anamorphs hyphomycetous, very varied. Necrotrophs of a wide range of plants, many economically important; some arthropod-associated, a few coprophilous; cosmop. Fams:
(1) **Kathistaceae**.
(2) **Ophiostomataceae**.
Lit.: Wingfield *et al.* (1993; *in* Hawksworth (Ed.), 1994: 333).

Ophiostomella Petr. (1925) ? = Scopinella (Ceratostomat.) fide Cannon (*in litt.*).

Ophiotexis Theiss. (1916) = Schweinitziella (Trichosphaer.) fide v. Arx (*Acta bot. neerl.* **7**: 503, 1958).

Ophiotheca Curr. (1854) = Perichaena (Trich.).

Ophiotrichia Berl. (1893) = Acanthophiobolus (Dothideales) fide Walker (1980).

Ophiotrichum Kunze (1849), Mitosporic fungi, 1.C2.?. 2, Eur., W. Indies. Nom. dub. fide Nannfeldt (*in litt.*).

Ophiovalsa Petr. (1966) = Winterella (Vals.). See Glawe & Jensen (*Mycotaxon* **25**: 645, 1986), Reid & Booth (*CJB* **65**: 1320, 1987).

Ophisthomastigomycota, see *Chytridiomycota*.

Ophiuridium Hazsl. (1877) = Dictydiaethalium (Lycogal.).

Ophryomyces L. Léger & E. Hesse (1909) nom. dub. (? Fungi).

Ophthalmidium Eschw. (1824) = Porina (Trichothel.).

Opisteria (Ach.) Vain. (1909) ≡ Nephroma (Nephromat.).

opisthokont, having one or more flagella at the posterior end.

Oplophora Syd. (1921) = Nyssopsora (Sphaerophrag.) fide Dietel (1928).

Oplotheciopsis Bat. & Cif. (1963) = Trichosphaerella (Niessl.) fide Müller & v. Arx (1973).

Oplothecium Syd. (1923), Trichosphaeriaceae. 2, Guyana, Philipp. = Trichosphaerella (Niessliac.) fide Müller & v. Arx (1962).

opportunistic (of fungi), normally saprobic and frequently common but on occasion able to cause disease in or grow on a host (esp. of humans or other animals), rendered susceptible by some predisposing factor(s).

opsis-form, see *Uredinales*.

Opuntiella L. Léger & M. Gauthier (1932) [non Kylin (1925), *Algae*], nom. dub. (Harpellales).

Oramasia Urries (1956) = Vermiculariopsiella (Mitosp. fungi) fide Nawawi *et al.* (*Mycotaxon* **37**: 173, 1990).

Oraniella Speg. (1909) = Massarina (Lophiostom.) fide Bose (1960).

Orbicula Cooke (1871), ? Otideaceae. 1, Eur. See Hughes (*Mycol. Pap.* **42**, 1951).

Orbilia Fr. (1849), Orbiliaceae (±L). 30, widespr. See Baral (*SA* **13**: 113, 1994; concept), Benny *et al.* (*CJB* **56**: 2006, 1978; biol.), Spooner (*Bibl. Mycol.* **116**, 1987; 7 spp. Australasia), Pfister (*Mycol.* **86**: 451, 1994; anamorph). Anamorph *Arthrobotrys*.

Orbiliaceae Nannf. (1932), Leotiales (±L). 1 gen. (+ 5 syn.), 30 spp. Stromata absent; ascomata apothecial, small, usually convex, brightly coloured or translucent; excipulum usually of thin-walled isodiametric cells, usually without marginal hairs; hymenium waxy, rarely setose; interascal tissue of simple paraphyses, usually with knob-like apices; asci very small, the apex truncate, with I- apical rings, often forked at the base; ascospores small, hyaline, often aseptate. Anamorphs hyphomycetous where known. Saprobic, esp. on wet wood, a few spp. lichenized, widespr.

Perhaps better included in *Gyalectales* (Sherwood, *in litt.*).

Lit.: Spooner (*Bibl. Mycol.* **116**, 1987; key gen.).

Orbiliaster Dennis (1954) = Orbilia (Orbil.) fide Baral (1994).

Orbiliella Kirschst. (1938) = Orbilia (Orbil.) fide Korf (1973).

Orbiliopsis (Sacc.) Syd. (1924), ? Leotiales (inc. sed.). 1, NZ.

Orbiliopsis Höhn. (1926) ? = Parorbiliopsis (Leot.) fide Spooner & Dennis (*Sydowia* **38**: 294, 1986).

Orbiliopsis Velen. (1934) [non Höhn. (1926)], ? Leotiales (inc. sed.). 1, former Czechoslovakia.

orbilla (pl. **orbillae**), an apothecium (obsol.), (Sprengel, *Intr. study cryptog.*, 1807).

Orbimyces Linder (1944), Mitosporic fungi, 1.G2.10. 1 (marine), USA.

Orcadella Wingate (1889) = Licea (Lic.).

Orcadia G.K. Sutherl. (1915), Ascomycota (inc. sed.). 1 (on *Algae*), UK, Norway. See Kohlmeyer & Kohlmeyer (*Marine mycology*: 454, 1979).

Orcella Kuntze ex Earle (1909) ≡ Clitopilus (Entolomat.).

Orceolina Hertel (1970), Trapeliaceae (L). 1, Kerguelen.

Orcheomyces Burgeff (1909) = Thanatephorus (Ceratobasid.) fide Talbot (1965).

Orchesellaria Manier ex Manier & Lichtw. (1968), Asellariaceae. 4 (in *Collembola*), widespr. See Manier (*Rev. Mycol.* **43**: 341, 1979), Moss (*TBMS* **65**: 115, 1975; ultrastr.), Lichtwardt (1986; key).

orchil (**orchill**), see Dyeing.

orchinol (and **hircinol**), dihydrophenanthrenes produced by orchids as a response to infection by mycorrhizal fungi (Gäumann, *Phytopath. Z.* **49**: 212, 1964).

orculiform, see polarilocular.

Ordonia Racib. (1909) = Septobasidium (Septobasid.) fide Couch (1938).

Ordovicimyces M.K. Elias (1966), ? Fossil fungi. 2 (Ordovician, ? Recent), USA.

Ordus K. Ando & Tubaki (1983), Mitosporic fungi, 1.G1.1. 1, Japan.

orellanine, see Mycetism (1).

Oreophylla Cif. (1954), Mitosporic fungi, 1.C1.?. 1, Santo Domingo.

Organizations, see Societies and organizations.

Ormathodium Syd. (1928) nom. dub. fide Deighton (*Mycol. Pap.* **137**: 4, 1974).

Ormomyces I.I. Tav. (1985), Laboulbeniaceae. 1, Afr., Indonesia.

ornamented (of organs, esp. spores), having the surface marked or sculptured with lines, wrinkles, warts, striations, ridges, reticulations, fibrils, scales etc.; not smooth. See Fig. 29 for terminology; also basidiospore, LO-analysis.

Ornasporonites Ramanujam & Rao (1979), Fossil fungi (f. cat.). 1 (Miocene), India.

Ornatinephroma Gyeln. (1934) = Nephroma (Nephromat.).

Ornatopyrenis Aptroot (1991), Trypetheliaceae (?L). 1, Australia.

Ornithascus Velen. (1934) = Saccobolus (Ascobol.) fide v. Brummelen (1967).

ornithocoprophilous, preferring habitats rich in bird droppings.

Oropogon Th. Fr. (1861), Alectoriaceae (L). 30, C. & S. Am., West Indies, Asia. See Esslinger (*Syst. Bot. Monogr.* **28**, 1989; key).

Orphanocoela Nag Raj (1989), Mitosporic fungi, 4.D1/2.1. 3 (on *Gramineae*), widespr.

Orphanomyces Savile (1974), Ustilaginaceae. 3 (on *Cyperaceae*), Eur., N. Am., Asia. See Vánky (1994; key).

Orphella L. Léger & M. Gauthier (1932), Legeriomycetaceae. 3 (in *Plecoptera*), Eur. See Lichtwardt

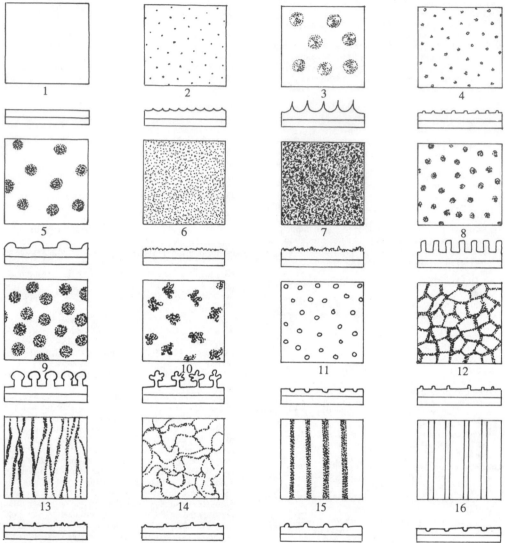

Fig. 29. Terminology for ornamentation (upper in surface view, lower in side view). 1, smooth (laevigate); 2, spinose (echinate); 3, aculeate; 4, granulate; 5, punctate (tuberculate); 6, verruculose; 7, verrucose; 8, baculate; 9, capitate; 10, irregularly projecting (see Sato & Sato, *Trans. Mycol. Soc. Jap.* **23**: 51, 1982, for additional terms, esp. for aeciospores); 11, foveate; 12, reticulate; 13, striate; 14, rugose; 15, cicatricose; 16, canaliculate.

(1986), Williams & Lichtwardt (*Mycol.* **79**: 473, 1987), Lichtwardt *et al.* (*Mycol.* **83**: 214, 1991).

Orphniospora Körb. (1874), ? Fuscideaceae (L). 1, Eur. See Hertel & Rambold (*Mitt. Bot. StSamml. München* **27**: 111, 1988; posn).

Orphniosporomyces Cif. & Tomas. (1953) ≡ Orphniospora (Fuscid.).

Orpinomyces D.J.S. Barr, H. Kudo, Jakober & K.J. Cheng (1989), Neocallimastigaceae. 2 (anaerobic rumen fungi), widespr.

Orromyces Sur & G.R. Ghosh (1987), Gymnoascaceae. 1, India. Anamorph *Chrysosporium*-like.

orthidium (obsol.), conidiomata of *Pyrenotrichum*, sometimes incorrectly listed as if a generic name. See Hawksworth (*Bull. Br. Mus. nat. Hist., Bot.* **9**: 59, 1981).

Orthobellus A.A. Silva & Cavalc. (1973), ? Schizothyriaceae. 2, S. Am.

Orthochaeta Sawada (1943) = Erysiphe (Erysiph.) fide Zheng (1985).

orthographic variant, a variant spelling; under the Code the original spelling must be retained but typographic or orthographic errors should be corrected; orthographic variants should not be listed as synonyms, but when two or more generic names are so similar as to be confused they are treated as orthographic variants or, when based on different types, as homonyms. See Nomenclature.

Orthomycotina. Subphylum proposed for all *Basidiomycota* other than *Urediniomycetes* (as treated in this Dictionary) by Cavalier-Smith (*in* Rayner *et al.* (Eds), *Evolutionary biology of the fungi*: 339, 1987).

Orthoscypha Syd. (1927) = Pocillum (Leot.) fide Petrak (*Sydowia* **5**: 328, 1951).

Orthotricha Wingate (1886) = Clastoderma (Echinostel.).

orthotrophy, Corda's (1842) term for the condition in which a basidiospore primordium develops at the apex of the apophysis in contrast to **heterotrophy** when the primordium develops laterally.

Oscarbrefeldia Holterm. (1898), Endomycetaceae. 1, Java.

oscule, a pore of a rust spore (Tulasne).

-osis (suffix), condition of; state caused by.

osmophily (adj. **-ilic, -ilous**), making growth under conditions of high osmotic pressure, as in *Xeromyces* (q.v.) and some yeasts on conc. sugar solutions. See Moustafa (*Can. J. Microbiol.* **21**: 1573, 1975; osmophilous spp. of Kuwait).

Osmoporus Singer (1944), Coriolaceae. 1, widespr. See Fildago (*Rickia* **1**: 95, 1962; key ssp. and vars., **6**: 27, 89, 1974). = Gloeophyllum (Coriol.) fide Donk (1974).

osmotolerant, capable of growing in high osmotic pressure, e.g. some yeasts and filamentous fungi on concentrated sugar solutions.

osmotrophic, exhibiting absorptive nutrition as do fungi.

Osoriomyces Terada (1981), Laboulbeniaceae. 1, Taiwan.

Ospriosporium Corda (1831) nom. dub. (? Fungi).

Ossicaulis Redhead & Ginns (1985), Tricholomataceae. 1, N. temp.

Osteina Donk (1966), Coriolaceae. 1, Eur., N. Am.

Ostenfeldiella Ferd. & Winge (1914) = Plasmodiophora (Plasmodiophor.).

Osteomorpha G. Arnaud ex Watling & W.B. Kendr. (1979), Mitosporic fungi, 1.A1.41. 1, Eur.

ostiole (sing. **ostiolum**), the schizogenous, paraphysis-lined cavity, ending in a pore, in the papilla or neck of a perithecium (Miller, *Mycol.* **20**: 196, 1928); any pore by which spores are freed from an ascigerous or pycnidial fruit-body. See v. Arx (*K. ned. Akad. Wet.*, C, **76**: 289, 1973; taxonomic importance in *Ascomycota*).

Ostracoblabe Bornet & Flahault (1891), ? Fungi (inc. sed.). 1 (on oyster shells), Eur. See Eckblad & Kirstiansen (*MR* **94**: 706, 1990).

Ostracococcum Wallr. (1833) nom. dub. (? Myxomycetes).

Ostracoderma Fr. (1825), Anamorphic Pezizaceae, 8.A1.6. Teleomorph *Peziza*. 3, widespr. See Hennebert (*Persoonia* **7**: 183, 1973).

Ostracodermidium Mukerji (1973), Mitosporic fungi, 1.A2.6. 1, India.

Ostracodermis Locq. (1982), Fossil fungi. 1, Libya.

Ostreichnion Duby (1861), Mytilinidiaceae. 3, N. Am. See Barr (*Mycotaxon* **3**: 81, 1975; key). = Mytilinidion (Mytilinid.) fide v. Arx & Müller (1975).

Ostreion Sacc. (1883) ≡ Ostreichnion (Mytilinid.).

Ostreionella Seaver (1926) = Actidium (Mytilinid.) p.p. and Ostropella (Melanomm.) p.p. fide v. Arx & Müller (1975).

Ostreola Darker (1963), Mytilinidiaceae. 4, Canada, India. See Tilak & Kale (*Indian Phytopath.* **21**: 289, 1968; key).

Ostropa Fr. (1825), Stictidaceae. 1, N. temp. See Sherwood (*Mycotaxon* **5**: 1, 1977).

Ostropaceae, see Stictidaceae.

ostropalean (of asci), ones essentially unitunicate in structure and with a thickened apex penetrated by a narrow pore; see ascus.

Ostropales, Ascomycota (±L). 6 fam., 76 gen. (+ 162 syn.), 1854 spp. Stromatic development usually weak, reduced to intramatrical hyphae, or absent, or with usually crustose thalli; ascomata ± apothecial, often deeply immersed and appearing perithecial, with rather varied peridial structure; interascal tissue of ± simple paraphyses, occ. gelatinized; asci usually narrow, cylindrical, with a well-developed capitate apical thickening often pierced by an I- pore; ascospores varied, often filiform. Anamorphs pycnidial, varied, sometimes ? absent. Usually saprobic on stems, bark or wood, some lichenized with green algae or lichenicolous, cosmop. Fams:

(1) **Gomphillaceae**.
(2) **Graphidaceae**.
(3) **Odontotremataceae**.
(4) **Solorinellaceae**.
(5) **Stictidaceae**.
(6) **Thelotremataceae**.
    Lit.: Korf (1973), Sherwood (*Mycotaxon* **5**: 1, 1977; monogr., keys).

Ostropella (Sacc.) Höhn. (1918), Melanommataceae. 3, widespr. See v. Arx & Müller (1975), Huhndorf (*Mycol.* **85**: 490, 1993).

Oswaldia Rangel (1921) = Apiosphaeria (Phyllachor.) fide Müller & v. Arx (1973).

Oswaldina Rangel (1921), Mitosporic Phyllachoraceae, 8.E1.15. Teleomorph *Apiosphaeria*. 1, Brazil. See Dianese *et al.* (*Sydowia* **46**: 233, 1994; reinstatement gen.).

Oswaldoa Bat. & I.H. Lima (1959) ? = Myriangiella (Schizothyr.).

Otagoa Lloyd (1922) nom. nud. (? Fungi, inc. sed.).

Otidea (Pers.) Bonord. (1851), Otideaceae. c. 15, N. temp. See Kanouse (*Mycol.* **41**: 660, 1949), Harmaja (*Karstenia* **14**: 138, 1974; gen. concept), Cao *et al.* (*Mycol.* **82**: 734, 1990; key Chinese spp.).

Otideaceae Eckblad (1968), Pezizales. 73 gen. (+ 58 syn.), 450 spp. Ascomata apothecial, rarely cleistothecial, sessile or shortly stipitate, small to large, flat to cupulate, occasionally asymmetrical, often brightly coloured due to carotenoid pigments, sometimes with hyaline or dark spinose hairs surrounding the hymenium; if cleistothecial usually large, hypogeous or emergent, thick-walled, hollow, solid or chambered; interascal tissue composed of simple paraphyses, sometimes branched, often with swollen pigmented apices; asci cylindrical, persistent, not blueing in iodine (sometimes ± globose in cleistothecial taxa); ascospores usually ellipsoidal, hyaline or brown, often ornamented, without a sheath, uninucleate, non-septate, guttulate or not. Anamorphs hyphomycetous where known. Saprobic on soil or rotten wood, some species hypogeous, cosmop.

The *Humariaceae* and most genera of the *Pyronemataceae* sensu edn 7 of this *Dictionary* are included here; see Kimbrough (*Mem. N.Y. bot. Gdn* **49**: 323, 1989). Many previous authors have separated the *Humariaceae* from this fam., based largely on the presence of carotenoid pigments. Trappe (1979) separated the *Geneaceae* from cleistothecial *Otideaceae* (as *Pyronemataceae*), based primarily on a hollow rather than solid or chambered ascoma.

Lit.: Hohmeyer (*Mitt. Arb. Pilzk. Niederrhein* **6**: 32, 1988; synoptic key *Aleuriae*), Wu & Kimbrough (*Bot. Gaz.* **152**: 421, 1991, *Int. Jl Pl. Sci.* **153**: 128, 1992; ascosporogenesis), Pegler *et al.* (*British truffles*, 1993; hypogeous taxa), and under *Pezizales*.

Otidella Sacc. (1889) ≡ Caloscypha (Otid.).

Otideopsis B. Liu & J.Z. Cao (1987), Otideaceae. 2, China, Eur. See Moravec (*Myc. Helv.* **3**: 135, 1988).

Otomyces Hallier (1869) nom. dub. (? Mitosp. fungi).

Otomyces Wreden (1874) ? = Aspergillus (Mitosp. fungi).

otomycosis, see mycosis.

Otthia Nitschke ex Fuckel (1870), Dothideales (inc. sed.). *c.* 10, widespr. See Scheinpflug (*Ber. Schweiz. bot. Ges.* **68**: 325, 1958). Anamorphs *Stigmina*.

Otthiella (Sacc.) Sacc. & D. Sacc. (1905) = Otthia (Dothideales). See Barr (*Rept. Kevo Subarct. Res. Stn* **11**: 12, 1974).

Otwaya G.W. Beaton (1978), Hyaloscyphaceae. 1 (on *Nothofagus*), Australia.

Oudemansia Speg. (1880) [non Miq. (1854), *Byttneriaceae*] ≡ Oudemansiella (Tricholomat.).

Oudemansiella Speg. (1881), Tricholomataceae. 9, trop., temp. See Pegler & Young (*TBMS* **87**: 583, 1987; key).

Outhovia Nieuwl. (1916) ≡ Scopularia (Mitosp. fungi).

Ovadendron Sigler & J.W. Carmich. (1976), Mitosporic fungi, 1.A1.38. 1 (from humans), Eur.

oval, widely elliptical (Fig. 37.1).

ovariicolous, living in ovaries.

ovate (of a surface [or sometimes a solid]), **ovoid** (of a solid) (Fig. 37.11), like a hen's egg with the narrower end at the top (Fig. 37.1).

Overeemia Arnaud (1954) = Brooksia (Mitosp. fungi) fide Deighton & Pirozynski (*Mycol. Pap.* **105**, 1966).

Ovinus (Lloyd) Torrend (1920) ≡ Albatrellus (Scutiger.).

Ovosphaerella Laib. (1922) = Mycosphaerella (Mycosphaerell.) fide v. Arx & Müller (1975).

Ovularia Sacc. (1880) = Ramularia (Mitosp. fungi) fide Hughes (*TBMS* **32**: 34, 1949).

Ovulariella Bubák & Kabát (1912) nom. nud. (Mitosp. fungi).

Ovulariopsis Pat. & Har. (1900), Anamorphic Erysiphaceae, 2.A1.1. Teleomorph *Phyllactinia*. 5, widespr.

Ovularites Whitford (1916), Fossil fungi. 1 (Cretaceous), USA.

Ovulinia F.A. Weiss (1940), Sclerotiniaceae. 2, Eur., N. Am.

Ovulitis N.F. Buchw. (1970), Anamorphic Sclerotiniaceae, 1.A1.hb-10. Teleomorph *Ovulinia*. 2, USA, UK.

oxydated (of crustose lichens), having thalli tinged rust-red by iron oxides; oxydized.

Oxydontia L.W. Mill. (1933) ≡ Sarcodontia (Hyphodermat.) fide Donk (1956).

Oxydothis Penz. & Sacc. (1898), Hyponectriaceae. 41, mainly trop. See Hyde (*Sydowia* **46**: 265, 1995), Barr (*Mycol.* **68**: 611, 1976; nomencl.), Hyde (*Sydowia* **45**: 204, 226, 1993; posn). Anamorph *Selenosporella*.

Oxyflavus Ryvarden (1973) = Oxyporus (Coriol.) fide Reid (*Microscopy* **32**: 448, 1975).

Oxyporus (Bourdot & Galzin) Donk (1933), Coriolaceae. 10, widespr.

Oxysporium Lév. (1863) = Helminthosporium (Mitosp. fungi) fide Saccardo (*Syll. Fung.* **10**: 613, 1892).

Oxystoma Eschw. (1824) = Graphis (Graphid.).

oyster mushroom (- **cap fungus**), basidiomata of the edible *Pleurotus ostreatus*.

Ozocladium Mont. (1851) = Polistroma (Thelotremat.).

Ozonium Link (1809) nom. dub. (Mitosp. fungi) fide Hughes (*CJB* **36**: 727, 1958); used for *Coprinus* anamorphs.

P, see Metabolic products, Steiner's Stable PD solution.

Paathramaya Subram. (1956), Mitosporic fungi, 3.A2.10. 3, India. See Bhat (*TBMS* **85**: 101, 1985; key).

Pachnocybaceae Oberw. & R. Bauer (1989), Atractiellales. 1 gen., 1 sp. Basidioma stilboid; hyphae with simple septal pores, clamp-connexions absent; spores brown, thick-walled, smooth, sessile.

Pachnocybe Berk. (1836), Pachnocybaceae. 1, Eur.

Pachnodium H.P. Upadhyay & W.B. Kendr. (1975), Anamorphic Ophiostomataceae, 3.A1.3. Teleomorph *Ceratocystis*. 1, Sweden.

Pachnolepia A. Massal. (1855) = Arthonia (Arthon.).

Pachyascaceae Poelt (1974) nom. inval. Ascomycota (inc. sed., L). 1 gen., 1 sp. Thallus granular, much reduced; ascomata apothecial, pulvinate, brownish, the peridium hardly developed; hymenium blueing in iodine; interascal tissue of irregular anastomosing paraphyses, not swollen at the apices; asci cylindrical, persistent, with distinct wall layers, not fissitunicate, with an enormously thickened apical cap which blues in iodine and a well-developed ocular chamber at least when immature, 12-spored; ascospores brown, 1-septate, blueing in iodine, without a sheath. Anamorph unknown. Lichenized with ? green algae, arctic Eur.

Pachyascus Poelt & Hertel (1968), Pachyascaceae (L). 1 (in leaf axils of *Andreaea*), Eur. See Poelt & Hertel (*Ber. dt. bot. Ges.* **81**: 210, 1968).

Pachybasidiella Bubák & Syd. (1915) = Aureobasidium (Mitosp. fungi) fide de Hoog & Hermanides-Nijhof (*Stud. Mycol.* **15**, 1977).

Pachybasium Sacc. (1885) = Trichoderma (Mitosp. fungi) fide Hughes (1958).

Pachycladina Marvanová (1987), Mitosporic fungi, 1.G1.10. 1 (from foam), former Czechoslovakia.

Pachycudonia S. Imai (1950), ? Geoglossaceae. 3, Japan.

Pachyderma Schulzer (1876) [non Blume (1826), *Oleaceae*] = Mycenastrum (Mycenastr.).

pachydermatous, (1) thick-skinned; (2) (of hyphae), having the outer wall thicker than the lumen.

Pachydisca Boud. (1885), ? Leotiaceae. 1, Eur. See Dumont (*Mycol.* **67**: 162, 1975).

Pachydiscula Höhn. (1915) = Cryptosporiopsis (Mitosp. fungi) fide Petrak (*Ann. Myc.* **21**: 182, 1923).

Pachyella Boud. (1907), Pezizaceae. 6, widespr. See Pfister (*CJB* **51**: 2009, 1973; key).

Pachykytospora Kotl. & Pouzar (1963), Coriolaceae. 3, Eur. See Kotlaba & Pouzar (*Česká Myk.* **33**: 129, 1979; key).

Pachylepyrium Singer (1958), Strophariaceae. 5, Am.

Pachyma Fr. (1822), Anamorphic Coriolaceae and Lentinaceae, 9.-.-. Teleomorph *Macrohyporia, Lentinus*. 1, N. Am. See Ginns & Lowe (*CJB* **61**: 1672, 1984), Weber (*Mycol.* **21**: 113, 1929; sclerotia).

Pachymetra B.J. Croft & M.W. Dick (1989), Verrucalvaceae. 1, Australia.

Pachyospora A. Massal. (1852), Hymeneliaceae (L). 1, arctic-alpine, widespr. See Hertel (*Willdenowia* **6**: 225, 1971).

Pachypatella Theiss. & Syd. (1915), Parmulariaceae. 2, Asia.

Pachyphiale Lönnr. (1858), Gyalectaceae (L). 6, Eur., Asia, N. Am. See Lettau (*Beih. Feddes Repert.* **69**(2), 1937), Vězda (*Sborn. Vysoké Školy Zeměd. Lesn. Brno C* **1**: 22, 1958, *Ergebn. Forsch. Unternehmens Nepal Himalaya* **6**: 127, 1974; key 5 spp.).

Pachyphlodes Zobel (1854) = Pachyphloeus (Terfez.) fide Fischer (1938).

Pachyphloeus Tul. & C. Tul. (1844) [non Göpp. (1836), fossil *Pteridophyta*], Terfeziaceae. 6 (hypogeous), Eur., N. Am., Japan. See Gilkey (*N. Am. Fl.* 2 **1**, 1954; key N. Am. spp.), Lange (*Dansk Bot. Ark.* **16**: 32, 1956; key Eur. spp.).

pachypleurous, thick-walled.

Pachyrhytisma Höhn. (1917) = Rhytisma (Rhytismat.) fide Nannfeldt (1932).

Pachysacca Syd. (1930), Dothideaceae. 1, Australia.

Pachysolen Boidin & Adzet (1958), Endomycetaceae. 1, Eur. See Kreger van Rij (1984).

Pachyspora Kirschst. (1906) = Delitschia (Spororm.) fide v. Arx & Müller (1975).

Pachysporaria (Malme) M. Choisy (1949) = Rinodina (Physc.).

Pachysterigma Johan-Olsen ex Bref. (1888) = Tulasnella (Tulasnell.) fide Rogers (*Ann. Myc.* **31**: 183, 1933).

Pachythyrium G. Arnaud ex Spooner & P.M. Kirk (1990), Microthyriaceae. 1, Eur.

Pachytichospora Van der Walt (1978), Saccharomycetaceae. 1, S. Afr.

Pachytrichum Syd. (1925) = Periconia (Mitosp. fungi) fide Linder (*Mycol.* **29**: 659, 1937).

Pachytrype Berl. ex M.E. Barr, J.D. Rogers & Y.M. Ju (1993), Calosphaeriaceae. 2, N. Am., Hawaii, Java, Philipp., Papua New Guinea.

Pactilia Fr. (1837), Mitosporic fungi, ?3.?.?. 2 or 3, Eur., S. Am.

PAD. Istituto di Botanica e Fisiologia Vegetale, Padova (Italy); founded 1837; a state institution; see Montemartini Corte (*in* Minelli (Ed.), *The botanical garden of Padua 1545-1995*: 271, 1995; mycol. collns esp. Saccardo).

paddy (paddi) straw mushroom, see straw mushrooms.

Padixonia Subram. (1972), Anamorphic Xylariaceae, 2.A1.9. Teleomorph *Xylaria*. 1, Ghana.

Paecilomyces Bainier (1907), Anamorphic Trichocomaceae, 1.A1.15. Teleomorph *Byssochlamys* etc. 41, widespr. See Samson (*Stud. Mycol.* **6**, 1974; key), Castro *et al.* (*Jl Med. Vet. Mycol.* **28**: 15, 1990; review of *Paecilomyces* in human infections), Mountfort & Rhodes (*Appl. Env. Microbiol.* **57**: 1963, 1991; anaerobic growth of marine *P. lilacinus*).

Paepalopsis J.G. Kühn (1882), Anamorphic Tilletiaceae. Teleomorph *Urocystis*. 3 (on *Primulaceae*), Eur., N. Am., Asia.

Pagidospora Drechsler (1960), Mitosporic fungi, 2.A1.1/41. 1 (with clamp connexions, on amoebae), USA.

Paidania Racib. (1909) = Erikssonia (Phyllachor.) fide v. Arx & Müller (1954).

Paipalopsis, see *Paepalopsis*.

Palaeachlya F.R.S. Duncan (1876), Fossil fungi (Oomycota). 5 (Silurian to Recent), widespr.

Palaeancistrus R.L. Dennis (1970), Fossil fungi (Basidiomycetes). 1 (Carboniferous), USA.

Palaeoamphisphaerella Ramanujam & Srisailam (1980), Fossil fungi (Ascomycota). 2 (Miocene), India.

Palaeocephala Singer (1962), Tricholomataceae. 1, Sierra Leone.

Palaeocirrenalia Ramanujam & Srisailam (1980), Fossil fungi (Mitosp. fungi). 2 (Miocene), India.

Palaeocytosphaera R.B. Singh & G.V. Patil (1980), Fossil fungi (Mitosp. fungi). 1 (Cretaceous), India.

Palaeofibula J.M. Osborn, T.N. Taylor & J.F. White (1989), Fossil fungi. 1 (middle Triassic), Antarctica.

Palaeoleptosphaeria Barlinge & Paradkar (1982), Fossil fungi (Ascomycota). 1 (Cretaceous), India.

Palaeomycelites Bystrov (1956), Fossil fungi (mycel.). 1 (Devonian), former USSR.

Palaeomyces Renault (1896), Fossil fungi (mycorrhizal endomycobiont). 3 (Carboniferous), France, UK. See Butler (*TBMS* **22**: 274, 1939).

Palaeomycites Mesch. (1902) ≡ Palaeomyces (Fossil fungi).

Palaeopede D.E. Ether. (1899), ? Fossil fungi. 1 (Devonian), Australia.

Palaeoperone D.E. Ether. (1891), ? Fossil fungi. 1 (Permo-Carboniferous), Australia.

Palaeophoma Singhai (1975), Fossil fungi (Mitosp. fungi). 1 (Eocene), India.

Palaeophthora Singhai (1975), Fossil fungi (Oomycota). 1 (Eocene), India. See Singhai (*Palaeobotanist* **25**: 481, 1978).

Palaeosclerotium G.W. Rothwell (1972), Fossil fungi (? Ascomycota). 1 (Carboniferous), USA. See Dennis (*Science, N.Y.* **192**: 66, 1976), Singer (*Mycol.* **69**: 850, 1977), Pirozynski & Weresub (1979: 687).

Palaeoslimacomyces Kalgutkar & Sigler (1995), Fossil fungi (Mitosp. fungi). 1 (Palaeocene-Eocene), Canada.

Palaeosordaria Sahni & H.S. Rao (1943), Fossil fungi (Xylariales). 1 (Tertiary), India.

Palambages Wetzel (1961), Fossil fungi (? sclerotia). 4 (Cretaceous, Tertiary), Canada, Eur., Malaysia.

Palavascia Tuzet & Manier ex Lichtw. (1964), Palavasciaceae. 3 (in *Isopoda*), widespr. See Lichtwardt (*Mycol.* **56**: 318, 1964, 1986; key).

Palavasciaceae Manier ex Manier & Lichtw. (1968), Eccrinales. 1 gen., 3 spp. Thalli unbranched, producing only multinucleate, thick-walled sporangiospores; attached to hindgut cuticole of *Isopoda*.

Palawania Syd. & P. Syd. (1914), Microthyriaceae. 1, Asia.

Palawaniella Doidge (1921), Parmulariaceae. 5, S. Am., S. Afr.

Palawaniopsis Bat., Cif. & Nascim. (1959), Mitosporic fungi, 5.A1.?. 1, S. Afr.

Paleoarcyria M. Jacq.-Fél., C.N. Mill. & Locq. (1983), Fossil fungi. 1, France.

Paleobasidiospora Locq. (1983), Fossil fungi. 1, France.

Paleoblastocladia W. Remy, T.N. Taylor & Hass (1994), Fossil fungi. 1 (Lower Devonian).

Paleocatenaria Locq. (1983), Fossil fungi. 1, UK.

Paleoguttulina Locq. & Mišík (1983), Fossil fungi. 1, former Czechoslovakia.

Paleomastigomycetes. Form class name proposed (Taylor *et al. Mycol.* **84**: 901, 1992) for fossil chytrid-like fungi.

paleomycology, the study of fossil fungi (q.v.).

Palifer Stalpers & P.K. Buchanan (1991), Corticiaceae. 1, NZ.

Paliphora Sivan. & B. Sutton (1985), Mitosporic fungi, 1.A/B1.27. 2, Australia, Malaysia.

palisade fungi, basidiomycetes.

palisade plectenchyma, plectenchyma in the cortex of a lichen thallus composed of hyphae arranged perpendicular to the surface.

palisade-cells (of lichens), the end cells of the hyphae of a fastigiate cortex.

palisoderm (of basidiomata), an outer layer composed of several strata of cells or hyphal tips; see derm.

pallid, light-coloured; pale.

pallisadoplectenchyma, see plectenchyma.

palmate, having lobes radiating from a common centre but not extending to the point of insertion. Cf. chiroid, digitate.

Palmellathyrites Locq., D. Pons & Sal.-Cheb. (1981), Fossil fungi. 1, France.

Palmicola K.D. Hyde (1993), Lasiosphaeriaceae. 1, Australia.

Palmomyces Maire (1926) ≡ Andreaeana (Mitosp. fungi).

Palomyces Höhnk (1955) = Halosphaeria (Halosphaer.) fide Kohlmeyer (*CJB* **50**: 1951, 1972).

paludal, living in wet places (marshes).

Palynomorphites L.R. Moore (1963), Fossil fungi (*f. cat.*). 1 (Carboniferous), UK.

Pampolysporium Magnus (1900) = Guignardia (Mycosphaerell.) fide Eriksson (*SA* **5**: 147, 1986).

Panaeolina Maire (1933) = Psathyrella (Coprin.) fide Smith (1972).

Panaeolopsis Singer (1969), Podaxaceae. 1, Argentina.

Panaeolus (Fr.) Quél. (1872) nom. cons., Strophariaceae. 10 (coprophilous), cosmop. See Hora (*Naturalist, Hull* 1957: 77, 1957), Ola'h (*Rev. Mycol. Mém. hors sér.* **10**, 1970), Pegler (*Kew Bull., Addit. Ser.* **6**, 1977; E. Afr. spp.).

Panama disease. A disease of banana caused by *Fusarium oxysporum* f.sp. *cubense*. See Stover (*Phytopath. Pap.* **4**, 1962), Stover & Buddenhagen (*Fruits* **41**: 175-191, 1986), Pegg & Langdon (*in* Persley & De Langhe (Eds), *Banana and plantain breeding strategies*: 119, 1987), Ploetz (Ed.) (*Fusarium wilt of banana*, 1990, *Int. Jl of Pest Manag.* **40**: 326, 1994).

Panchanania Subram & N.G. Nair (1966), Mitosporic fungi, 2.B2.10. 1, India. = Paathramaya (Mitosp. fungi) fide Bhat (1985).

Pandanicola K.D. Hyde (1994), Xylariales (inc. sed.). 1, Australia, Philipp.

Pandora Humber (1989), Entomophthoraceae. 16, widespr. See Humber (1989), Miller & Keil (*Mycotaxon* **38**: 227, 1990).

Panellus P. Karst. (1879), Tricholomataceae. 55, cosmop. See Miller (*Mich. Bot.* **9**: 17, 1970), Burdsall & Miller (*Beih. Nova Hedw.* **51**, 1975), Corner (*Gard. Bull., Singapore* **39**: 1986; Malaysian spp.), O'Kane *et al.* (*Mycol.* **82**: 595, 1990; bioluminescence).

Pannaria Delise ex Bory (1828), Pannariaceae (L). 12, mainly trop. (some temp.). See Jørgensen (*J. Hattori bot. Lab.* **76**: 197, 1994, *Op. bot.* **45**, 1978; Eur. spp.).

Pannariaceae Tuck. (1872), Lecanorales (L). 14 gen. (+ 8 syn.), 155 spp. Thallus crustose or foliose, usually dark grey; ascomata apothecial, sessile, usually brown or black, with a well-developed margin composed of isodiametric cells; interascal tissue of rigid thick-walled sparingly branched paraphyses; asci rather variable, usually with a well-developed apical cap and an I+ outer gelatinized layer, sometimes with a deeply staining tube-like apical ring; ascospores usually small, aseptate, hyaline, sometimes ornamented. Anamorphs pycnidial. Lichenized with cyanobacteria, widespr. esp. S. temp.

Pannariella (Vain.) Gyeln. (1935) = Heppia (Hepp.).

Pannoparmelia (Müll. Arg.) Darb. (1912), Parmeliaceae (L). 5, S. Am., Antarctic, China, Australia. See Sato (*Miscnea bryol. lichen., Nichinan* **4**: 45, 1966), Yoshimura & Elix (*J. Hattori bot. Lab.* **74**: 287, 1993).

pannose (**panniform**), having the appearance of felt or woollen cloth.

Pannucia P. Karst. (1879) = Psathyrella (Coprin.) fide Singer (1951).

Pannularia Nyl. (1886) = Placynthium (Placynth.) fide Jørgensen (*Op. bot.* **45**, 1986).

Panomycetes. Class introduced by Wei (*Chin. Biodiv.* **1**: 23, 1993) for 'fungi' in the broad sense, i.e. inclusive of those referred to the kingdoms *Chromista*, *Fungi*, and *Protozoa* in this edition of the *Dictionary*.

pantherine, a metabolite of *Amanita pantherina*, etc., toxic to humans and flies (*Musca* sp.); muscimol; amanita factor B.

pantonemic (of e.g. *Oomycota*, *Hyphochytriomycota*), presumed to have evolved from algae. Pantonemic 'fungi' exhibit the diaminopimelate pathway for lysine synthesis; the dominant sterols are cholesterols (not ergosterol); and the storage product is mycolaminarin (not glycogen).

Pantonemomycota. Phylum proposed by Olive (*The mycetozoans*, 1975) for the fungi placed in *Oomycota* and *Hyphochytriomycota* in this *Dictionary*.

Pantospora Cif. (1938), Mitosporic fungi, 3.D2.21. 1, W. Indies, S. Am. See Deighton (*Mycol. Pap.* **140**, 1976).

Panus Fr. (1838) nom. cons., Lentinaceae. 10 (on wood), widespr. See Luminescent fungi.

paper (fungi on), see Biodeterioration.

papilionaceous, variegated; mottled; marked with different colours as the lamellae of certain *Panaeolus* spp.

Papilionospora V.G. Rao & B. Sutton (1976), Mitosporic fungi, 2.A2.10. 1, Myanmar, Italy.

papilla, a small rounded process.

papillae (of host tissue), localized wall thickenings on the inner surface of plant cell walls at sites penetrated by fungi; callosity, lignituber, callus, 'sheath', are whole or part synonyms (Aist, *Ann. Rev. Phytopath.* **14**: 145, 1976).

Papillaria J. Kickx f. (1835) nom. rej. ≡ Pycnothelia (Cladon.).

papillate, having a papilla (Fig. 37.40).

Pappimyces B. Sutton & Hodges (1975), Mitosporic fungi, 1.E1.6. 1, Brazil.

Papulare Tomas. & Cif. (1952) ≡ Phragmothele (Verrucar.).

Papularia Fr. (1825) [non Forssk. (1775), *Aizoaceae*] = Arthrinium (Mitosp. fungi) fide Ellis (1965).

Papulariomyces Cif. & Tomas. (1953) ≡ Phragmothele (Verrucar.).

Papulaspora Preuss (1851), Mitosporic fungi, 1.D2.1. 11, widespr. *P. byssina* is the brown plaster mould of mushroom beds; wood pulp is damaged by other spp.; *P. stoveri* is a mycoparasite of *Rhizoctonia*. The hyphae bear 'papulospores' (q.v.), small rounded propagules comprising one or more enlarged central cells surrounded by a sheath of smaller cells. Forms having 'bulbils' (q.v.) are excluded as *Burgoa* and *Minimedusa* by Weresub & LeClair (*CJB* **49**: 2203, 1971; gen. key). See Tubaki *et al.* (*Trans. Mycol. Soc. Jap.* **32**: 31, 1991).

Papulasporites Shete & A.R. Kulk. (1978), Fossil fungi. 1 (Tertiary), India.

Papulosa Kohlm. & Volkm.-Kohlm. (1993), Ascomycota (inc. sed.). 1, USA.

papulose, covered with pimples or pustules.

Papulospora, see *Papulaspora*.

papulospore, asexual spore in e.g. *Papulaspora sepedonioides* (Weresub & LeClair, *CJB* **49**: 2203, 1971).

Papulosporonites Schmied. & A.J. Schwab (1964), Fossil fungi. 1 (Eocene), Germany.

Papyrodiscus D.A. Reid (1979), Corticiaceae. 1, Papua New Guinea.

Paraaoria R.K. Verma & Kamal (1987), Mitosporic fungi, 8.A2.19. 1, Nepal.

Paraarthrocladium Matsush. (1993), Mitosporic fungi, 1.C1.1/10. 1, Peru.

Parabotryon Syd. (1926) ? = Eudarluca (Phaeosphaer.) fide Müller & v. Arx (1962).

Paracainiella Lar.N. Vassiljeva (1983), Amphisphaeriaceae. 1 (on *Dryas*), Siberia.

paracapillitium (of *Lycoperdales*), a capillitium composed of thin-walled, hyaline, septate hyphae in contrast to a true capillitium of thick-walled, brown, aseptate hyphae (Kreisel, 1962).

Paracapnodium Speg. (1909) = Scorias (Capnod.) fide Hughes (1976).

Paracarpidium Müll. Arg. (1883) = Endocarpon (Verrucar.) fide Zahlbruckner (1921).

paracephalodium (pl. **-ia**), a hyphal mat covering cyanobacteria arising from a squamulose lichen thallus which has a green alga as the photobiont (Poelt & Mayrhofer, *Pl. Syst. Evol.* **158**: 265, 1988).

Paraceratocladium R.F. Castañeda (1987), Mitosporic fungi, 1.B1.15. 2, Cuba.

Paracercospora Deighton (1979), Anamorphic Mycosphaerellaceae, 1.E2.10. Teleomorph *Mycosphaerella.* 2, widespr.

Paracesatiella Petr. (1929) = Schweinitziella (Trichosphaer.) fide v. Arx (*Acta bot. neerl.* **7**: 503, 1958).

Parachinomyces Thaung (1979), Mitosporic fungi, 1.C1.10. 1 (on *Acroconidiellina*), Myanmar.

Parachnopeziza Korf (1978), Hyaloscyphaceae. 2, N. & S. Am.

Paracoccidioides F.P. Almeida (1930), Mitosporic fungi, 1.A1.3. 1 (on humans), S. Am. *P. brasiliensis* (paracoccidioidomycosis; q.v.). See Restrepo *et al.* (*RMVM* **8**: 97, 1973; 183 refs), Takeo *et al.* (*MR* **94**: 1118, 1990; cytoplasmic and plasma membrane ultrastructure), San-Blas & San-Blas (*Jap. J. Med. Mycol.* **32**: 75, 1991; morphogenesis). Formerly referred to 'Entomophthorales Imperfecti' (Ciferri, 1956), Franco *et al.* (Eds) (*Paracoccidioidomycosis*, 1994).

paracoccidiomycosis, see *Paracoccidioides.*

Paracoreomyces R. Poiss. (1929) = Coreomyces (Laboulben.) fide Thaxter (1931).

Paracostantinella Subram. & Sudha (1989), Mitosporic fungi, 1.A1.10. 1, India.

Paracryptophiale Kuthub. & Nawawi (1994), Mitosporic fungi, 1.D1/2.15. 1, Malaysia.

Paracudonia Petr. (1927) = Roesleria (Caliciales, inc. sed.) fide Nannfeldt (*Nova Acta R. Soc. Scient. Upsal.* 4, **8**(2), 1932).

Paracytospora Petr. (1925), Mitosporic fungi, 8.A1.?. 1, N. Am.

Paradactylaria Subram. & Sudha (1989), Mitosporic fungi, 2.B1.10. 1, India.

Paradactylella Matsush. (1993), Mitosporic fungi, 1.C1.2. 1, Peru.

Paradendryphiopsis M.B. Ellis (1976), Mitosporic fungi, 1.C2.25. 1, UK, Hungary.

Paradiachea Hertel (1956) = Comatricha (Stemonitid.).

Paradiacheopsis Hertel (1954) = Macbrideola (Stemonitid.) p.p., = Comatricha (Stemonitid.) p.p.

Paradidymella Petr. (1927) = Discostroma (Amphisphaer.). See Dennis (1968), Hawksworth & Sivanesan (*TBMS* **67**: 39, 1976).

Paradidymobotryon C.J.K. Wang & B. Sutton (1984), Mitosporic fungi, 2.B2.4. 1, USA.

Paradiplodia Speg. ex Trotter (1931), Mitosporic fungi, 8.B2.?. 7, widespr. See Zambettakis (*BSMF* **70**: 219, 1955; as *Paradiplodiella*). = Botryodiplodia (Mitosp. fungi) fide Petrak (1963).

Paradiplodiella Zambett. (1955) = Botryodiplodia (Mitosp. fungi) fide Sutton (1977).

Paradischloridium Bhat & B. Sutton (1985), Mitosporic fungi, 1.C2.15. 1, Ethiopia, N.Z.

Paradiscina Benedix (1969) = Gyromitra (Helvell.) fide Harmaja (*Karstenia* **15**: 33, 1976). See Benedix (*Kulturpflanz.* **19**: 163, 1972).

Paradiscula Petr. (1941), Mitosporic fungi, 4.A1.15. 1, Eur. See Morgan-Jones (*Mycotaxon* **2**: 167, 1975).

Paradoxa Mattir. (1935), Tuberaceae. 1 (hypogeous), Italy.

Paraepicoccum Matsush. (1993), Mitosporic fungi, 3.D2.2. 1, Peru.

Paraeutypa Subram. & Ananthap. (1988) = Leptoperidia (Diatryp.) fide Rappaz (*Mycol. Helv.* **3**: 281, 1989).

Parafulvia Kamal, A.N. Rai & Morgan-Jones (1983), Mitosporic fungi, 1.C2.10. 1, India.

paragynous (of *Pythiaceae*), having the antheridium at the side of the oogonium (Fig. 28B).

Paragyrodon (Singer) Singer (1942), Gyrodontaceae. 1, N. Am.

Parahelminthosporium Subram. & Bhat (1989) = Polytretophora (Mitosp. fungi). = Spadicoides (Mitosp. fungi) fide Sutton (*in litt.*).

Parahyalotiopsis Nag Raj (1976), Mitosporic fungi, 4.C2.19. 1, Myanmar.

Parahydraeomyces Speg. (1915) = Hydraeomyces (Laboulben.) fide Siemaszko & Siemaszko (*Polsk. Pismo Entom.* **12**: 125, 1933).

Paraisaria Samson & B.L. Brady (1983), Anamorphic Clavicipitaceae, 2.A1.15. Teleomorph *Cordyceps.* 1 (on larvae of *Hepialus*), UK, Germany.

Paralaestadia Sacc. ex Vain. (1921) ? = Guignardia (Mycosphaerell.).

Paraleptonia (Romagn. ex Noordel.) P.D. Orton (1991), Entolomataceae. 2, Eur.

Paraliomyces Kohlm. (1959), Dothideales (inc. sed.). 1 (marine), India. See Read *et al.* (*CJB* **70**: 2223, 1993; ultrastr.).

Parallobiopsis Collin (1913) ? Fungi. 1 (on invertebrate), Eur.

Paramacrinella Manier & Grizel (1971), Eccrinaceae. 1 (in *Amphipoda*), France. See Manier *et al.* (*Ann. Sci. Nat., Bot.* sér. 12, **12**: 1, 1971), See Lichtwardt (1986).

Paramassariothea Subram. & Muthumary (1979), Mitosporic fungi, 8.C1.19. 1, India.

Paramazzantia Petr. (1927) = Mazzantia (Vals.) fide Barr (1978).

Paramicrothallites K.P. Jain & R.C. Gupta (1970), Fossil fungi (Microthyr.). 2 (Miocene), India.

Paramitra Benedix (1962) = Peziza (Peziz.) fide Nannfeldt (*Ann. Bot. Fenn.* **3**: 309, 1966).

Paramoebidium L. Léger & Duboscq (1929), Amoebidiaceae. 5 (in *Diptera, Ephemeroptera, Plecoptera*), widespr. See Dang & Lichtwardt (*Am. J. Bot.* **66**: 1093, 1979; ultrastr.), Lichtwardt (1986; taxonomy), Williams & Lichtwardt (*CJB* **68**: 1045, 1990), Lichtwardt *et al.* (*Mycol.* **83**: 389, 1991), Lichtwardt & Williams (*Mycol.* **84**: 376, 1992).

Paramoeciella Zebrowski (1936), ? Fossil fungi (Chytrid.). 1 (Cambrian to ? Recent), Australia.

paramorph, a neutral term proposed by Huxley (*The new systematics*: 37, 1940) for any form differing from the mean of the group; advocated for use in fossil fungi lacking characters essential for their proper placement, but believed to have affinities with particular non-fossil groups by Reynolds (*Mycotaxon* **23**: 141, 1985).

paramorphogen, a compound (e.g. validomycin A, L-sorbose) able to induce a reversible morphological change in fungi (Tatum *et al., Science* **109**: 509, 1949, Jejewolo *et al., TBMS* **91**: 653, 661, 1988).

Paramyces Oehm (1937) = Ptychogaster (Mitosp. fungi).

Paranectria Sacc. (1878), Hypocreaceae. 3 (on lichens), Eur. See Hawksworth & Pirozynski (*CJB* **55**: 2555, 1977; nomencl.), Hawksworth (*Notes R. bot. Gdn Edinb.* **40**: 375, 1982).

Paranectriella (Henn. ex Sacc. & D. Sacc.) Höhn. (1910), Tubeufiaceae. 5, trop. See Rossman (*Mycol. Pap.* **157**, 1987; key), Hawksworth & Pirozynski (*CJB* **55**: 2555, 1977; nomencl.).

Paranthostomella Speg. (1910) ? = Anthostomella (Xylar.) fide Læssøe (*SA* **13**: 43, 1994).

Paraparmelia Elix & J. Johnst. (1986) Parmeliaceae. *c.* 53 spp., widespr. See Elix & Johnston (*Mycotaxon* **32**: 399, 1988; 15 spp. S. Hemisph.). Close to *Xanthoparmelia* fide Elix (1993). ? = Parmelia (Parmel.).

Parapaxillus Singer (1942) nom. inval. (nom. prov.). = Paxillus (Paxill.).

Parapeltella Speg. (1919) = Micropeltis (Micropeltid.) fide v. Arx & Müller (1975).

Parapericonia M.B. Ellis (1976), Mitosporic fungi, 3.A2.1/10. 2, Afr., Australia. See Alcorn & Kirk (*TBMS* **85**: 561, 1985; amendment gen.).

Paraperonospora Constant. (1989), Peronosporaceae. 8, widespr.

Paraphaeoisaria de Hoog & Morgan-Jones (1978), Mitosporic fungi, 1.A1.10. 1, USA.

Paraphaeosphaeria O.E. Erikss. (1967), Phaeosphaeriaceae. 6, Eur., N. Am. See Crivelli (*Ueber die heterogene Ascomycetengattung Pleospora*, 1983; key), Shoemaker & Babcock (*CJB* **63**: 1284, 1985; key). Anamorph *Coniothyrium*.

Paraphelaria Corner (1966), Auriculariaceae. 1, Java, S. Pacific.

Paraphoma Morgan-Jones & J.F. White (1983) = Phoma (Mitosp. fungi) fide van der Aa *et al.* (1990).

paraphysis, (pl. **paraphyses**), a sterile upward growing, basally attached hyphal element in a hymenium, esp. in ascomycetes where they are generally filiform, unbranched or branched, and the free ends frequently make an epithecium over the asci, see hamathecium; in basidiomycetes, see hyphidium; **apical paraphyses**, the downward growing hyphae with free tips in the centrum of hypocrealean fungi (Luttrell, *TBMS* **48**: 135, 1965); periphysoids; **asco-**, a multicellular diploid storage hypha originating from the base of the ascus in *Erysiphaceae* (Speer, *Sydowia* **27**: 1, 1976); **paraphysate**, having paraphyses.

paraphysoid, (1) (of ascomycetes), see hamathecium; (2) (of basidiomycetes), a sterile accessory hymenial structure; see basidiole, cystidiole, hyphidium; **- network**, branched and anastomosing paraphysoids surrounding asci in some ascolocular ascomycetes.

Paraphysorma A. Massal. (1852) nom. rej. = Staurothele (Verrucar.).

Paraphysotheca Bat. (1961) = Schizothyrium (Schizothyr.) fide v. Arx & Müller (1975).

Paraphysothele Zschacke (1934) = Thelidium (Verrucar.) p.p. and Arthopyrenia (Arthopyren.) p.p.

Parapithomyces Thaung (1976), Mitosporic fungi, 1.D2.10. 2, Myanmar, Australia. See Alcorn (*Aust. Syst. Bot.* **5**: 711, 1992).

Paraplacidiopsis Servít (1953) = Placidiopsis (Verrucar.).

paraplectenchyma, plectenchyma (q.v.) composed of cells which have isodiametric lumina and unthickened walls (see Yoshimura & Shimada, *Bull. Kochi Gakuen Jun. Coll.* **11**: 13, 1980).

Parapleurotheciopsis P.M. Kirk (1982), Mitosporic fungi, 1.C1/2.3. 2, UK, Japan.

Parapolyporites Tanai (1987), Fossil fungi. 1, Japan.

Paraporpidia Rambold & Pietschm. (1989), Porpidiaceae (L). 3, Australasia, China, Indonesia.

Parapterulicium Corner (1952), Pterulaceae. 2, Brazil.

Parapyrenis Aptroot (1991), Pyrenulaceae. 7, trop. See Aptroot (*Nova Hedw.* **60**: 625, 1995; key), Eriksson (*Lichenologist* **25**: 307, 1993).

Parapyricularia M.B. Ellis (1972), Mitosporic fungi, 1.C2.10. 1, Asia.

Parasaccharomyces Beurm. & Gougerot (1909) nom. rej. = Candida (Mitosp. fungi).

Parasclerophoma Petr. (1924) ? = Dothichiza (Mitosp. fungi) fide Clements & Shear (1931).

Parascorias J.M. Mend. (1930) = Rizalia (Trichosphaer.) fide Pirozynski (1977).

Parascutellinia Svrček (1975), Otideaceae. 3, Eur. See Benkert (*Gleditschia* **13**: 147, 1985), Donadini (*Boll. Gruppo micol. Bresad.* **29**: 273, 1986).

parasexual cycle, a mechanism discovered by Pontecorvo & Roper in 1952 in filamentous fungi by which re-combination of hereditary properties is based not on sexual reproduction (meiosis) but on the mitotic cycle. The essential features of the process are: (1) the production of diploid nuclei in a heterokaryotic haploid mycelium; (2) the multiplication of the diploid nuclei along with haploid nuclei in a heterokaryotic mycelium; (3) the sorting out of a diploid homokaryon; (4) segregation and recombination by crossing-over at mitosis; (5) haploidization of the diploid nuclei.

The results are similar to those achieved by meiosis but instead of a regular sequence of events in time as in the meiotic cycle, in the parasexual cycle the various processes may all be occurring at one time in one mycelium, at rates which have been estimated. The parasexual cycle may (as in *Aspergillus nidulans*) or may not (*A. niger*) be accompanied by a sexual cycle. Although the details of the mechanism of the parasexual cycle are unknown, Pontecorvo and his school obtained data from the parasexual cycle for mapping the 8 chromosomes of *Aspergillus nidulans*, results which showed good agreement with those for the same fungus obtained by the analysis of the sexual cycle. See Pontecorvo *et al.* (*Advances in genetics* **5**: 141, 1953), Pontecorvo & Käfer (*Advances in Genetics* **9**: 71, 1958). For other aspects see Pontecorvo (*Ann. Rev. Microbiol.* **10**: 393, 1956), Pontecorvo *et al.* (*J. gen. Microbiol.* **8**: 198, 1953; *Aspergillus niger*, **11**: 94, 1954; *Penicillium chrysogenum*, Buxton (*J. gen. Microbiol.* **15**: 133, 1956; *Fusarium oxysporum*). See protosexual.

parasite, an organism living on or in, and obtaining its nutrients from, its host, another living organism; a biotroph or necrotroph. Cf. parasymbiont, pathogen, symbiosis.

Parasitella Bainier (1903), Mucoraceae. 1 (facultative parasite of *Mucorales*), N. temp. See Schipper (*Stud. Mycol.* **17**: 65, 1978), Kellner *et al.* (*Mycologist* **5**: 120, 1990; parasit., genetic transfer), Burmester & Wöstemeyer (*Curr. Genetics* **26**: 456, 1994; genetics).

parasol mushroom, basidioma of the edible *Macrolepiota procera* (syn. *Lepiota procera*).

parasoredium, soredium-like structure, originally used for a structured plectenchyma in the upper thallus layer of the *Umbilicaria hirsuta* aggr. (see Codogno *et al.*, *Pl. Syst. Evol.* **165**: 55, 1989).

Parasphaeria Syd. (1924) = Massarina (Lophiostom.) fide v. Arx & Müller (1975).

Parasphaeropsis Petr. (1953), Mitosporic fungi, 4.E1.?. 1, Hawaii.

Paraspora Grove (1884), Mitosporic fungi, 3.C1.?. 2, Eur.

Parastenella J.C. David (1991), Mitosporic fungi, 1.C2.10. 1, USA.

Parastereopsis Corner (1976), Cantharellaceae. 1, Borneo.

Parasteridiella H. Maia (1960) = Asteridiella (Meliol.) fide Hughes (*in litt.*).

Parasteridium Speg. (1923) nom. dub. (? Meliolaceae); no spp. named. See Eriksson & Hawksworth (*SA 7*: 74, 1987).

Parasterina Theiss. & Syd. (1917) = Asterina (Asterin.) fide Doidge (1942).

Parasterinella Speg. (1924), Asteraceae. 2, S. Am. See Arambarri (*Sydowia* **38**: 1, 1986).

Parasterinopsis Bat. (1960), Asterinaceae. 3, trop.

Parastigmatea Doidge (1921), ? Polystomellaceae. 3, trop. fide Luttrell (1973).

Parastigmatellina Bat. & C.A.A. Costa (1959), Mitosporic fungi, 5.A1.?. 1, Philipp.

parasymbiont (obsol.), a fungus or lichen living on a lichen thallus but not causing any obvious damage (Zopf, *Ber. dtsch. bot. Ges.* **15**: 90, 1897); a commensalistic or gall-forming lichenicolous fungus (Hawksworth, *Bot. J. Linn. Soc.* **96**: 3, 1988). See Fungicolous fungi, Lichenicolous fungi.

Parasympodiella Ponnappa (1975), Mitosporic fungi, 1.C1.39. 2, cosmop. See Tokumasu (*Trans. Mycol. Soc. Jap.* **28**: 19, 1987).

Parataeniella R. Poiss. (1929), Parataeniellaceae. 5 (in *Isopoda*), widespr. See Lichtwardt & Chen (*Mycol.* **56**: 163, 1964), Scheer (*Arch. Protistenk.* **118**: 202, 1976), Lichtwardt (1986; key), Lichtwardt & Williams (*CJB* **68**: 1057, 1990).

Parataeniellaceae Manier & Lichtw. (1968), Eccrinales. 2 gen. 7 spp. Thalli unbranched, entire or part of thallus developing into a multispored sporangium or series of unispored sporangia; in hindgut of *Isopoda* or *Insecta*.

Parathalle Clem. (1909) = Arthrorhaphis (Arthrorhaphid.) fide Poelt & Hafellner (*Phyton* **17**: 213, 1976).

parathecium, (1) (of apothecia), the outside hyphal layer, esp. if darker in colour (obsol.); (2) ectal excipulum (q.v.).

Parathelium Nyl. (1862) = Pyrenula (Pyrenul.) fide Harris (1989). See Upreti & Singh (*J. Econ. Tax. Bot.* **10**: 236, 1987; India).

Paratomenticola M.B. Ellis (1976), Mitosporic fungi, 1.C2.10. 1, USA.

Paratorulopsis Novák & Zsolt (1961) = Torulopsis (Mitosp. fungi) fide Lodder (1970).

Paratrichaegum Faurel & Schotter (1966) = Epicoccum (Mitosp. fungi) fide Kendrick & Carmichael (1973).

Paratrichaptum Corner (1987), Coriolaceae. 1, Sumatra.

Paratrichella Manier (1947) = Enterobryus (Eccrin.) fide Manier & Lichtwardt (*Ann. Sci. Nat., Bot.*, sér. 12, **9**: 519, 1968).

Paratrichoconis Deighton & Piroz. (1972), Mitosporic fungi, 1.C1.11. 1, Sierra Leone.

Paratrichophaea Trigaux (1985), Otideaceae. 3, Europ., N. Am. See Pfister (*Mycol.* **80**: 515, 1988; key 3 spp.).

paratype, see type.

Paraulocladium R.F. Castañeda (1986), Mitosporic fungi, 1.D2.26. 1, Cuba.

Paravalsa Ananthap. (1990), Valsaceae. 1, India.

paraxonemal body (PAR), proteinaceous structure restricted to particular region of the flagellum in chromophytes and euglenoids; also manifest as flagellar spines in male gametes of oogamous phaeophytes.

Parencoelia Petr. (1950), Leotiaceae. 3 (on *Phyllachoraceae*), C. Am., W. Indies. See Zhuang (*Mycotaxon* **32**: 85, 1988).

Parendomyces Queyrat & Laroche (1909) nom. rej. = Candida (Mitosp. fungi).

Parenglerula Höhn. (1910), Englerulaceae. 5, trop.

parenthesome, a curved double membrane (which may be perforate, imperforate, or vesiculate; see Moore, *Bot. Marina* **23**: 362, 1980) on each side of a dolipore septum (Moore & McAlear, *Am. J. Bot.* **49**: 86, 1962); see Fig. 13 'dolipore'; septal pore cap.

parietal, fixed to the wall, e.g. of asci in a perithecium.

parietin (physcion), a bright yellow-orange to red anthraquinone pigment (in *Teloschistaceae*) giving a crimson-purple reaction with K (q.v.). Also known in some species of *Aspergillus*, *Penicillium*, *Polygonum* (Angiosp.), *Rheum* (Angiosp.) and *Ventilago* (Angiosp.). See Metabolic products, Pigments.

pariobasidium, see basidium.

Paris inch (Paris line), see line.

Parkerella A. Funk (1976) = Lahmia (Lahm.) fide Eriksson (1986).

Parksia E.K. Cash (1945) = Chloroscypha (Leot.) fide Korf (*in* Ainsworth *et al.*, *The Fungi* **4A**: 249, 1973).

Parmastomyces Kotl. & Pouzar (1964), Coriolaceae. 1, Siberia, E. Eur.

Parmathyrites K.P. Jain & R.C. Gupta (1970), Fossil fungi (Microthyr.). 2 (Tertiary), India.

Parmelaria D.D. Awasthi (1987) Parmeliaceae. 2, Himalaya. ? = Parmelia (Parmel.).

Parmelia Ach. (1803) nom. cons., Parmeliaceae (L). 38 (s. str.) [1,322 s. l.], cosmop., esp. Asia & Australasia. See Hale (*Smithson. Contr. bot.* **66**, 1987; monogr. s. str., key), Alder (1990), Elix (1993), Eriksson & Hawksworth (*SA* **8**: 72, 1989; typific., *SA* **13**: 146, 1994; concept), Galloway & Elix (*N.Z. Jl Bot.* **21**: 397, 1983), Hawksworth (*in* Hawksworth (Ed.), *Ascomycete systematics*: 383, 1994; gen. concepts), Krog (*J. Hattori bot. Lab.* **52**: 303, 1982; conidia, gen. concepts), Kurokawa (*J. Jap. Bot.* **69**: 61, 121, 204, 1994; Jap. spp.).

A broad generic concept is maintained in this edition of the *Dictionary* following the *Outline* (Eriksson & Hawksworth, 1993) but some segregates currently based on secondary metabolite, anatomical and morphological differences can be expected to warrant acceptance as more data on the fungi involved are obtained. See also *Lit.* under segregate names, esp. *Bulbothrix*, *Flavoparmelia*, *Hypotrachyna*, *Melanelia*, *Parmelina*, *Parmotrema*, *Pseudoparmelia*, *Punctelia*, *Relicina* and *Xanthoparmelia*.

Parmeliaceae Zenker (1827), Lecanorales (L). 85 gen. (+ 31 syn.), 2,319 spp. Thallus foliose or fruticose, corticate on both surfaces, usually with rhizoids, often brightly pigmented; ascomata apothecial, sessile, occasionally ± immersed or stalked, sometimes marginal, with a well-developed margin; interascal tissue of sparingly branched paraphyses, sometimes pigmented, the apices sometimes swollen; asci with a well-developed I+ apical cap with a more weakly I+ plug; ascospores varied, usually small, hyaline, and aseptate, without a sheath. Conidiomata pycnidial. Lichenized with green algae, cosmop.

Lit.: Beltman (*Bibl. Lich.* **11**, 1978; thallus str.), Elix (*Bryologist* **96**: 359, 1993; syn. key 64 gen.), Alder (*Mycotaxon* **38**: 331, 1990; key gen. *Parmelia* s.l.), Orchard (Ed.) (*Flora of Australia* **55**, 1994; keys 31 gen.), Smith (*Bryologist* **96**: 326, 1993; key 115 spp. Hawaii).

Parmeliella Müll. Arg. (1862), Pannariaceae (L). *c.* 40, cosmop. (mainly trop.). See Jørgensen (*Op. bot. Soc. bot. Lund* **44**, 1978; Eur.).

Parmelina Hale (1974) Parmeliaceae. 10, mainly Eur., Asia. See Hale (*Smithson. Contr. bot.* **33**, 1976), *Parmelinella*. ? = Parmelia (Parmel.).

Parmelinella Elix & Hale (1987) Parmeliaceae. 3, India, E. Asia, E. Afr., Australia. See Elix (*SA* **7**: 110, 1988; nomencl.). ? = Parmelia (Parmel.).

Parmelinopsis Elix & Hale (1987) Parmeliaceae. 18, temp. & trop. ? = Parmelia (Parmel.).

Parmeliomyces E.A. Thomas ex Cif. & Tomas. (1953) ? = Parmelia (Parmel.).

Parmeliopsis (Nyl. ex Stizenb.) Nyl. (1866) nom. cons., Parmeliaceae (L). 2, N. temp. See Meyer (*Mycol.* **74**: 592, 1982).

Parmentaria Fée (1824) = Pyrenula (Pyrenul.) fide Harris (1989). See Upreti & Singh (*Candollea* **43**: 109, 1988; key 11 spp. India).

Parmentariomyces Cif. & Tomas. (1953) ≡ Parmentaria (Pyrenul.).

Parmentieria Trevis. (1860) ≡ Parmentaria (Pyrenul.).

Parmocarpus Trevis. (1861) = Xanthoria (Teloschist.).

Parmophora M. Choisy (1950) = Umbilicaria (Umbilicar.).

Parmosticta Nyl. (1875) ≡ Crocodia (Lobar.).

Parmostictina Nyl. (1875) nom. rej. = Pseudocyphellaria (Lobar.).

Parmotrema A. Massal. (1860) Parmeliaceae. *c.* 250, mainly trop. See Hale (*Contr. U.S. natn. Herb.* **36**:

193, 1965), Krog & Swinscow (*Bull. Br. Mus. nat. Hist., Bot.* **9**: 143, 1981, *Lichenologist* **15**: 127, 1983). ? = Parmelia (Parmel.).

Parmotremopsis Elix & Hale (1987) Parmeliaceae. 2, C. Am., West Indies. ? = Parmelia (Parmel.).

Parmularia Lév. (1846), Parmulariaceae. 2, S. Am. See Batista & Vital (*Atas Ist. Micol. Recife* **1**: 159, 1960).

Parmularia Nilson (1907) = Lecanora (Lecanor.).

Parmulariaceae E. Müll. & Arx ex M.E. Barr (1979), Dothideales. 30 gen. (+ 24 syn.), 103 spp. Superficial mycelium absent. Ascomata superficial or subcuticular, developing from a more deeply immersed stromatic layer, sometimes as locules in a compound structure, strongly flattened, discoid or pulvinate, often elongate, opening by irregular disintegration or longitudinal splits; peridium composed of one or few layers of irregular or radiating cells; interascal tissue absent or poorly developed; asci ovoid to clavate, with fissitunicate or rostrate dehiscence, immersed in I-gel; ascospores hyaline to brown, septate, often with a sheath. Anamorphs poorly known. Biotrophic on leaves and stems, esp. trop.
 Lit.: Doidge (*Bothalia* **4**: 273, 1942; S. Afr.), Stevens & Ryan (*Ill. biol. monogr.* **17** (2), 1939).

Parmulariella Henn. (1904), Dothideales (inc. sed.). 1, S. Am.

Parmulariopsella Sivan. (1970), Parmulariaceae. 1, Sierra Leone.

Parmulariopsis Petr. (1954), Parmulariaceae. 1, N. Borneo.

parmuliform, shield-shaped with the margins slightly upturned.

Parmulina Theiss. & Syd. (1914), Parmulariaceae. 5, Asia, Am.

Parodiella Speg. (1880), Parodiellaceae. 1 (on *Leguminosae*), widespr. See Sydow & Petrak (*Ann. Myc.* **29**: 190, 1931), v. Arx & Müller (1975).

Parodiellaceae Theiss. & Syd. ex M.E. Barr (1987), Dothideales. 1 gen. (+ 2 syn.), 1 sp. Ascomata superficial, perithecial, ± globose with a thickened base, black, thick-walled, strongly aggregated, with a well-developed apical lysigenous pore; peridium soft, composed of small pseudoparenchymatous cells; interascal tissue of cellular pseudoparaphyses; asci cylindrical, fissitunicate; ascospores hyaline to brown, septate, striate, without a sheath. Anamorph coelomycetous. Biotrophic on leaves of *Leguminosae*, esp. trop.

Parodiellina Henn. ex G. Arnaud (1918), Parodiopsidaceae. 1, trop. Anamorph *Exosporinella*.

Parodiellina Viégas (1944) = Botryostroma (Ventur.) fide v. Arx & Müller (1975).

Parodiellinaceae, see *Parodiopsidaceae*.

Parodiellinopsis Hansf. (1946) ? = Chevalieropsis (Parodiopsid.) fide v. Arx & Müller (1975).

Parodiodia Bat. (1960) = Apiosporina (Ventur.) fide Müller & v. Arx (1962).

Parodiopsidaceae Toro (1952), Dothideales. 16 gen. (+ 29 syn.), 69 spp. Mycelium superficial, dark, sometimes setose or hyphopodiate; ascomata superficial, sometimes stalked, globose, small, thin-walled, opening by breakdown of apical cells to form an irregular ostiole; peridium composed of 1-2 layers of dark pseudoparenchymatous cells; interascal tissue absent; asci saccate, fissitunicate, sometimes I+ blue; ascospores brown, septate, sometimes with a sheath or gelatinous appendages. Anamorphs hyphomycetous. Biotrophic on leaves, esp. trop.

Parodiopsis Maubl. (1915) = Perisporiopsis (Parodiopsid.) fide Müller & v. Arx (1962).

Paropodia Cif. & Bat. (1956), Dothideales (inc. sed.). 1, Brazil. = Triposporiopsis (Capnod.) fide Hughes (1976).

Paropsis Speg. (1918) = Treubiomyces (Chaetothyr.).

Parorbiliopsis Spooner & Dennis (1986), Leotiaceae. 1, U.K.

parsimonious (**parsimony**), see Cladistics.

part spore, one of the 1-celled spores resulting from the breaking up of a 2 or more celled ascospore.

parthenogamy, the state of an oospore formed with a diploid nucleus resulting from restitution at telophase I or telophase II of meiosis (note that the parthenogametic state cannot be presumed from the absence of an antheridium).

parthenogenesis, the apomitic development of haploid cells (Gäumann).

parthenomixis, see parthenogamy.

Parthenope Velen. (1934), ? Leotiales (inc. sed.). 1, former Czechoslovakia.

parthenospore, an oospore (aboospore) or zygospore (azygospore) produced by parthenogenesis.

partial veil (or **inner veil**) (of agarics), a layer of tissue, developed from the stipe, which joins the stipe to the pileus edge during hymenium development, and which later may become an annulus or cortina; = velum (Persoon).

partridge wood, (1) wood attacked by a pocket rot, e.g. one caused by *Stereum frustulatum*; [(2) wood of *Caesalpinia*].

Paruephaedria Zukal (1891) = Dactylopsora (Dactylospor.) fide Döbbeler & Triebel (1985).

Parvacoccum R.S. Hunt & A. Funk (1988), Rhytismataceae. 1, Canada.

Parvobasidium Jülich (1975), Corticiaceae. 2, Eur.

Parvomyces Santam. (1995), Laboulbeniaceae. 1, Spain.

Paryphydria, see *Paruephaedria*.

Pasania fungus, the edible *Cortinellus shiitake*. See shiitake.

Paschelkiella Sherwood (1987), Odontotremataceae. 1, Eur., N. Am.

Pascherinema De Toni (1936) nom. dub. (? Mitosp. fungi).

Pasithoe M.J. Decne. (1840) ≡ Paulia (Lichin.).

paspalitrem, a tremorgenic mycotoxin from *Phomopsis* sp. (Bills *et al.*, *MR* **96**: 977, 1992).

Paspalomyces Linder (1933), Mitosporic fungi, 1.B2.10. 1, USA.

paspalum staggers, see ergot.

passage, the experimental infection of a host with a parasite which is subsequently re-isolated; [a method used to increase the virulence of the parasite].

Passalora Fr. (1849), Anamorphic Mycosphaerellaceae, 1.B/C2.10. Teleomorphs *Mycosphaerella*, *Microcyclus*. 3, Eur., N. Am. See Deighton (*Mycol. Pap.* **112**, 1967; key).

Passeriniella Berl. (1891), Dothideales (inc. sed.). 2, Eur. See Kohlmeyer & Volkmann-Kohlmeyer (*Bot. Mar.* **34**: 1991; nom. dub.).

Passerinula Sacc. (1875), Hypocreaceae. 1 or 2, Eur.

Pasteur (Louis; 1822-95). The work of this noted Frenchman, who was by training a chemist, has had a very great effect on biological (and much mycological) thought and work. Among Pasteur's early investigations was that on the isomeric forms of tartaric acid from which he made the discovery that only one of the two forms (the *d*-) in a racemic (*dl*-) mixture was used by '*Penicillium glaucum*'. This gave him an interest in fermenting processes and going to Lille in 1854 (after positions at Dijon and Strasbourg) as Professor of Chemistry and Dean of the new Faculté des Sciences, he made detailed studies (from 1857 at the École Normal, Paris) on wine and beer making (*Études sur le vin*, 1866, ... *le vinaigre*, 1868; ... *la bière*, 1876). One important outcome of this work was the proof by experiment of the general principle 'Omne vivum e vivo' (Every living thing from a living thing).

In 1865-69, as a result of an investigation at the suggestion of the chemist Dumas, he made clear the nature of the 'pébrine' and 'flacherie' diseases of silkworms and the methods for their control (*Études sur la maladie des vers à soie*, 1870).

Pasteur's most important work was a development from Jenner's vaccination against small-pox (1796). His studies on chicken cholera (*Pasteurella aviseptica*), anthrax (*Bacillus anthracis*), and rabies in dogs and man made it clear that active immunization of man and animals against pathogenic organisms and viruses was a general method. In 1888 the Institut Pasteur was opened in Paris for the development of his discoveries. See Valléry-Radot (*Vie de Pasteur*, 1900 [Eng. transl. Devonshire, 1911]), Dubos (*Louis Pasteur*, ..., 1951).

Pasteur effect. Increased respiration and decreased fermentation in the presence of oxygen and vice versa in the absence of oxygen; **- filter**, an unglazed porcelain tube for sterilization by filtration; **- pipette**, a short length of glass tubing with one end drawn out into a sealed capillary and the other plugged with cotton wool and sterilized, the tip of the capillary is broken before use; **pasteurization**, freeing a liquid or other material from pathogenic microorganisms by heat.

Patella F.H. Wigg. ex Seaver (1928) nom. rej. = Scutellinia (Otid.) fide Korf (*Taxon* **35**: 378, 1986).

Patellaria Hoffm. (1789) = Diploschistes (Thelotremat.) fide Santesson (*in* Eriksson, *Opera Bot.* **60**, 1981).

Patellaria Pers. (1794) = Caloplaca (Teloschist.) fide Riedl (*Taxon* **27**: 302, 1978).

Patellaria Fr. (1822), Patellariaceae. *c*. 10, mainly temp. See Bellemère *et al.* (*Cryptog., Mycol.* **7**: 47, 1986; asci), Butler (*Mycol.* **331**: 612, 1939), Hawksworth (*Taxon* **35**: 787, 1986; nomencl.), Tilak & Srinivasulu (*Beih. Nova Hedw.* **47**: 459, 1974; Indian spp.).

Patellariaceae Corda (1838), Patellariales. 18 gen. (+ 8 syn.), 34 spp. Ascomata sessile, round or elongated, usually discoid but with inrolled margins when dry; excipulum composed of pseudoparenchymatous tissue; ascospores hyaline or brown. Anamorphs coelomycetous where known. Mostly saprobic on bark or wood, widespr.
Lit.: Bellemère (*Ann. Sci. Nat., Bot.* sér. 12 **12**: 429, 1971; ultrastr.), Kamat *et al.* (*in* Subramanian, 1978; ascocarp structure), Nannfeldt (1932: 318), Müller & v. Arx (1962: 249; in *Dothiorales*), Luttrell (1963).

Patellariales, Ascomycota (±L). 2 fam., 20 gen. (+ 13 syn.), 46 spp. Stromata absent or weakly developed, composed of intramatrical hyphae; ascomata erumpent, eventually apothecial, sometimes strongly cupulate, sometimes elongated, the margin well-developed, frequently inrolled when dry; interascal tissue of narrow anastomosing pseudoparaphyses, initially attached at apex and base but eventually becoming free at the apex; asci cylindrical, fissitunicate, I-, often with a well-developed ocular chamber; ascospores hyaline or brown, transversely septate or muriform, usually without a sheath. Anamorphs coelomycetous where known. Saprobes on bark or wood or lichenized with green algae, a few lichenicolous, cosmop. Fams:
(1) **Arthrorhaphidaceae**.
(2) **Patellariaceae** (syn. *Lecanidiaceae*).
Lit.: Bellemère (*Ann. Sci. nat., Bot.* sér. 12 **12**: 429, 1971; asci), Eriksson (1981, 1982), Honegger (*Ber. dtsch. bot. Ges.* **91**: 579, 1978, *Lichenologist* **12**: 157, 1980; asci), Kamat (*in* Subramanian (Ed.), *Taxonomy of fungi* **1**: 292, 1978; ontogeny), Nannfeldt (1932), Pirozynski & Reid (*CJB* **44**: 655, 1966).

Patellariopsis Dennis (1964), Dermateaceae. 4, Eur., Australia.

Patellea (Fr.) Sacc. (1884) = Xylogramma (Leot.). See Nannfeldt (*Nova Acta R. Soc. Scient. upsal.*, 4, **8**(2), 1932).

patelliform, like a round plate having a well-marked edge.

Patellina Speg. (1880) = Catinula (Mitosp. fungi) fide Höhnel (*Sber. Akad. Wiss. Wien* **124**: 49, 1915).

Patellina Grove ex Petch (1943) nom. dub. (Mitosp. fungi) fide Sutton (1977).

Patellonectria Speg. (1919) = Aspidothelium (Aspidothel.) fide Santesson (1952).

patent, stretching out; spreading.

Patent protection. The principle under which an 'inventor' of a new product or process by publicly disclosing details of his/her 'invention' is granted for a limited period a legally enforceable right to exclude others from exploiting the invention. Microbiologists may patent discoveries (and under certain circumstances microorganisms) which are novel, show evidence of an inventive step, and are of utility (industrial applicability). The procedure for taking out a patent is complex and professional advice should be sought. It is recommended that the patent disclosure procedure should include the depositing in a genetic resource collection of a culture any microorganism which forms part of the 'invention'. The Budapest Treaty allows for a single deposit in a recognized International Depository Authoritiy (IDA), removing the need for deposits to be made in all countries where patent protection is sought.

The Convention on Biological Diversity (1992) also has implications for the patenting of genetic material from nature (Fritze, 1994).

Lit.: *Budapest treaty on the international recognition of the deposit of microorganisms for the purpose of patent procedures* (World International Property Organization, 1981), Bousfield (*in* Hawksworth & Kirsop (Eds), *Filamentous fungi* : 115, 1988), Crespi (*Patents: a basic guide to patenting in biotechnology* [*Cambridge studies in biotechnology* **6**], 1988), Fritze (*in* Kirsop & Hawksworth (Eds), *The biodiversity of micro-organisms and the role of microbial resource centres*: 37, 1994). See Bioprospecting.

pathogen, a parasite able to cause disease in a particular host or range of hosts (*TBMS* **33**: 155, 1950); **-ic**, disease-causing or able to be so; **-icity**, the condition of being pathogenic.

pathotoxin, see toxin.

pathotype, see pathovar.

pathovar (pv.) (of bacteria), an infrasubspecific subdivision characterized by a pathogenic reaction in one or more hosts; a pathotype (not recommended) See Young *et al.* (*ROPP* **70**: 213, 1991).

Patila Adans. (1763) = Oncomyces (Auricular.).

Patinella Sacc. (1875), Dermateaceae. 2, widespr. See Spooner (*Bibl. Mycol.* **116**, 1987).

Patinellaria H. Karst. (1885), Leotiaceae. 1, Eur., Mexico. See Dennis (*Mycol. Pap.* **62**, 1956), Medel & Chacón (*Rev. Mex. Micol.* **4**, 251, 1988).

Patouillard (Narcisse Théophile; 1854-1926). A French pharmacist; from 1922 Assistant de la Chaire de Cryptogamie du Muséum d'Histoire Naturelle, Paris. Chief writings: *Les Hyménomycètes d'Europe* (1887), *Essai taxonomique sur les familles et les genres des Hyménomycètes* (1900 [reprint, 1963]) and more than 200 papers [reprint, 3 vols, 1978]. Collections in **FH** but some types in **PC**. See Mangin (*BSMF* **43**: 8, 1927; portr., bibliogr.), Heim (*Ann. Crypt. exot.* **1**: 25, 1928), Pfister (Annotated index to fungi described by N. Patouillard [*Contrib. Reed Herb.* **25**], 1977), Heim

(*Rev. Mycol.* **36** (2): i, 1972; portr., hist.), Grummann (1974: 292), Stafleu & Cowan (*TL-2* **3**: 94, 1981).

Patouillardea Roum. (1885) = Dendrodochium (Mitosp. fungi) fide Höhnel *in* Falk (*Mykol. Unters.*: 360, 1923).

Patouillardiella Speg. (1889), Mitosporic fungi, 3.B1.?. 1, Brazil. See Petrak & Sydow (*Ann. Myc.* **34**: 11, 1936).

Patouillardina Bres. (1906), Exidiaceae. 1, trop. See Martin (*Mycol.* **31**: 507, 1939), Donk (*Taxon* **7**: 238, 1958).

Patouillardina G. Arnaud (1917) = Meliolaster (Asterin.).

Patoullardiella, see *Patouillardiella*.

patulin (clavicin, clavatin, claviformin, expansin, mycoin, penicidin), a mycotoxin from *Aspergillus clavatus*, *Penicillium patulum*, *P. claviforme*, *P. expansum*, etc.; also antibacterial and antifungal (Raistrick *et al.*, *Lancet* **2**: 625, 1943); toxic to plants and animals (carcinogenic to mice) and the cause of neurotoxicosis in cattle; can also occur in apple and pear juice (from *P. expansum*).

Pauahia F. Stevens (1925), Meliolaceae. 1, Hawaii. See Petrak (*Sydowia* **5**: 432, 1951).

Paucithecium Lloyd (1923), ? Xylariaceae. 1, Brazil.

Paulia Fée (1836), Lichinaceae (L). 8, trop. See Henssen (*Lichenologist* **18**: 201 1986; key).

Paulia Lloyd (1916) ≡ Xenosoma ('gasteromycetes' or Trichosphaeriales).

Paullicorticium J. Erikss. (1958), Sistotremataceae. 5, N. Am., Eur. See Liberta (*Brittonia* **13**: 219, 1962; key).

Paurocotylis Berk. ex Hook. f. (1855), Otideaceae. 1 (hypogeous), Eur., NZ. See Dennis (*Kew Bull.* **30**: 345, 1974), Pegler *et al.* (1993).

Paxillaceae Lotsy (1907), Boletales. 4 gen. (+ 8 syn.), 17 spp. Basidioma agaricoid; lamellae decurrent, gelatinized; spores olive-brown, smooth.

Paxillogaster E. Horak (1966), Xerocomaceae. 1, Argentina.

Paxillopsis J.-E. Gilbert (1931) nom. nud. = Paxillus (Paxill.) fide Singer (1951).

Paxillopsis J.E. Lange (1940) ≡ Clitopilus (Entolomat.) fide Donk (1962).

Paxillus Fr. (1836), Paxillaceae. 12, cosmop. See Kühner (*Bull. Soc. linn. Lyon* **46**: 81, 1977; classification), Singer (*Farlowia* **2**: 527, 1946; classification, *Nova Hedw.* **7**: 93, 1964; S. Am. spp.), McNabb (*N.Z. Jl Bot.* **7**: 349, 1969; NZ spp.), Corner (*Nova Hedw.* **20**: 793, 1970; Malaysia), Horak (*Sydowia* **32**: 154, 1980; Australasia), Watling (*British fungus flora* **1**, 1970; key Br. spp.), Singer *et al.* (*Beih. Nova Hedw.* **98**: 12, 1990; key C. Am. spp.).

Paxina Kuntze (1891) ≡ Acetabula (Helvell.).

Payosphaeria W.F. Leong (1990), Hypocreales (inc. sed.). 1, Malaysia.

Pazschkea Rehm (1898) = Tapellaria (Ectolech.).

Pazschkeella Syd. & P. Syd. (1901), Anamorphic Mycosphaerellaceae, 8.B1.?. Teleomorph *Microcyclus*. 2, Brazil, Philipp.

PC. Laboratoire de Cryptogamie, Museum National d'Histoire Naturelle (Paris, France); founded 1904; supported by the Ministry for Education; genetic resource collection **LCP**; see Anon. (*La Chaire de Cryptogamie du Museum National d'Histoire Naturelle*, 1954).

PCA, see Media.

PCP, see *Pneumocystis*.

PCR, see Molecular biology.

PD, see Metabolic products; Steiner's Stable PD solution.

PDA, see Media.

peach leaf curl, leaf hypertrophy caused by *Taphrina deformans*.

peat mould, *Chromelosporium fulvum*.

Peccania A. Massal. ex Arnold (1858) nom. cons., Lichinaceae (L). 4, Eur. N. Afr. See Moreno & Egea (*Acta bot. Barcin.* **41**: 1, 1992).

Peccaniomyces Cif. & Tomas. (1953) ≡ Peccania (Lichin.).

Peccaniopsis M. Choisy (1949) = Peccania (Lichin.).

Pecila Lepell. (1822) nom. dub. (Myxomycetes, inc. sed.).

Peck (Charles Horton; 1833-1917). Botanist to the State Cabinet of Natural History (later the New York State Museum), Albany, 1867-1915. He proposed more than 2,700 fungal names [for list 1867-1908, see Peck, *N.Y. St. Mus. Bull.* **131**: 59, 1909; 1905-15, see Gilbertson, *Mycol.* **54**: 460, 1962], and is noted for his long series of yearly Reports (1868-1913) [reprint, 6 vol., 1980-]. See Barr *et al.* (*An annotated catalogue of the pyrenomycetes described by Peck* [*N.Y. St. Mus. Bull.* **459**, 1986]), Burnham (*Mycol.* **11**: 33, 1919), Jenkins (*Mycotaxon* **7**: 23, 1978; *Amanita* types), Lloyd (*Mycol. Notes* **4**: 509, 1912; portr.), Petersen (*Mycol.* **68**: 304, 1976; cantharelloid spp.), Grummann (1974: 194), Stafleu & Cowan (*TL-2* **4**: 135, 1983).

Peckia Clinton (1878) [non Vell. (1825), *Myrsinaceae*] ≡ Drudeola (Mitosp. fungi).

Peckiella (Sacc.) Sacc. (1891), Hypocreaceae. 22 (fungicolous), N. Hemisph. = Hypomyces (Hypocr.) fide Rogerson & Samuels (1989).

Peckifungus Kuntze (1891) ≡ Appendiculina (Laboulben.).

Peckiomyces Sacc. & Trotter (1913) nom. nud. See Welch (*Mycol.* **18**: 82, 1926).

pecky cypress, decay of *Taxodium distichum* by *Stereum taxodii* (Davidson *et al.*, *Mycol.* **52**: 260, 1961).

pectic enzymes, components of the macerating enzymes of a number of fungal parasites; two types distinguished by Wood (*Physiological plant pathology*, 1967: 138; q.v. for details) are **pectinesterases** (syn. pectinmethylesterases), specific enzymes which saponify methyl ester groups of pectinic acid; and **polygalacturonases**, which break polygalacturonide chains by hydrolysis at the glycosidic linkages.

pectinate, like the teeth of a comb.

pectinate hyphae, comb-like hyphae of e.g. *Microsporum audouinii*.

Pectinotrichum Varsavsky & G.F. Orr (1971), Onygenaceae. 1, Venezuela.

Pediascus Chardón & Toro (1934) = Vestergrenia (Dothid.) fide v. Arx & Müller (1975).

pedicel, a small stalk.

pedicellate, having a pedicel.

Pedilospora Höhn. (1902) = Dicranidion (Mitosp. fungi) fide Hughes (*CJB* **31**: 607, 1953).

pedogamy, pseudomixis between mature and immature cells, e.g. copulation between a yeast mother cell and its bud.

pedogenesis, (1) reproduction in young or immature organisms; (2) soil formation (see Weathering).

Pedumispora K.D. Hyde & E.B.G. Jones (1992), Diaporthales (inc. sed.). 1 (on *Rhizophora*), Seychelles.

Peethambara Subram. & Bhat (1978), Hypocreaceae. 1, India. Anamorph *Putagraivam*.

Peethasthabeeja P. Rag. Rao (1981), Mitosporic fungi, 1.A2.10. 1, India.

peg, see hyphal peg.

Peglerochaete Sarwal & Locq. (1983), Tricholomataceae. 1, Sikkim.

Pegleromyces Singer (1981), Tricholomataceae. 1, Brazil.

Peglionia Goid. (1937) = Gyrothrix (Mitosp. fungi) fide Hughes & Pirozynski (*N.Z. Jl Bot.* **9**: 39, 1971).

peitschengeissel, unornamented flagellum.

Pekilo, protein from *Paecilomyces* sp. used in animal feeds.

Pelicothallos Dilcher (1965), Fossil fungi (Microthyr.). 1 (Eocene), USA. See Sherwood-Pike (*Lichenologist* **17**: 114, 1985), Pirozynski (1976).

pellet, a three dimensional colony in a liquid culture (particularly a shaken culture).

pellicle, (1) outermost living layer lying below any non-living secreted material, containing plasma membrane plus underlying epiplasm or other membranes and may show ridges, folds or distinct crests; (2) a growth on the surface of a liquid culture; (3) (of agaric basidiomata), a detachable, skin-like cuticle of the pileus.

pellicular veil, a very thin partial veil of a sporophore not having a stipe (Singer, 1962). Cf. cortina.

Pellicularia Cooke (1876) nom. conf. fide Donk (*Reinwardtia* **2**: 425, 1954).

pelliculose, like a thin crust, as is the hymenial layer in *Thelephoraceae*.

Pellidiscus Donk (1959), Crepidotaceae. 1, Eur.

Pellionella Sacc. (1902), Mitosporic fungi, 8.B2.1. 1, Java.

Pellionella (Sacc.) Petch (1924) nom. inval. Cited in *ING* (1979) but Petch (*Ann. R. bot. Gdn Peradeniya* **8**: 170, 1924) did not make the transfer.

pellis, the cellular cortical layers, not belonging to the veils, of a basidioma (Bas, *Persoonia* **5**: 327, 1969); cuticle. Bas distinguished different layers of the pellis as **supra-**, **medio-**, and **sub-**, and different topographies of the pellis as **pilei-**, etc.

Pelloporus Quél. (1886) ≡ Coltricia (Hymenochaet.).

pellucid-striate (of a pileus), having a somewhat transparent top so that the gills are seen through it as rays.

Pelodiscus Clem. (1901) ? Fungi (inc. sed.) fide Eckblad (1968). = Patella (Pyronemat.) fide Seaver (1928).

Peloronectria A. Møller (1901), Hypocreaceae. 1 or 2, S. Am.

Peloronectriella Yoshim. Doi (1968), Hypocreaceae. 1, Japan.

peloton, see mycorrhiza.

Peltaster Syd. & P. Syd. (1917), Mitosporic fungi, 5.A1.?. 5, Philipp., S. Am.

Peltasterales (obsol.). Mitosporic fungi with superficial shield-like conidiomata; anamorphs of *Microthyriaceae*. Batista & Ciferri (*Mycopath.* **11**: 6, 1959; 6 fam., keys, host index), Batista & Ciferri (*Atti Ist. bot. Univ. Pavia* ser. 5 **16**: 86, 1959).

Peltasterella Bat. & H. Maia (1959), Mitosporic fungi, 5.A1.?. 6, trop.

Peltasterinostroma Punith. (1975), Mitosporic fungi, 5.A1.15. 1, UK.

Peltasteropsis Bat. & H. Maia (1959), Mitosporic fungi, 5.A1.?. 7, trop.

peltate, in the form of a round plate with a stalk from the centre of the underside (Fig. 37.29).

Peltella Syd. & P. Syd. (1917) = Muyocopron (Microthyr.) fide v. Arx & Müller (1975).

Peltidea Ach. (1803) = Peltigera (Peltiger.).

Peltideomyces E.A. Thomas (1939) nom. inval. = Peltigera (Peltiger.).

Peltidium Kalchbr. (1862) [non Zollik. (1820), Compositae] ≡ Pachyella (Peziz.).

Peltigera Willd. (1787) nom. cons., Peltigeraceae (L). *c.* 45, cosmop. See Goward *et al.* (*CJB* **73**: 91-11, 1995; key 28 N. Am. spp.), Kurokawa *et al.* (*Bull. Natn. Sci. Mus. Tokyo* **9**: 101, 1966), Thomson (*Am. Midl. Nat.* **44**: 1, 1950), Vitikainen (*Acta Bot. Fenn.* **152**: 1, 1994; key 29 spp. Eur.), Chen (*Acta mycol. Sin.* **5**: 18, 1986; key 28 spp., China), Holtan-Hartwig (*Somerfeltia* **15**, 1993; key 17 spp. Norway).

Peltigeraceae Dumort. (1822), Peltigerales (L). 5 gen. (+ 14 syn.), 59 spp. Thallus foliose, corticate only on the upper surface; ascomata formed on the upper surface, with a thin covering layer which breaks apart at an early stage to expose the hymenium; asci with a thickened apex, an I+ apical ring and an I+ gelatinized outer layer; ascospores hyaline to brown, elongate, transversely septate. Anamorphs pycnidial. Lichenized with cyanobacteria or green algae (then often with cyanobacterial cephalodia), widespr.

Peltigerales, Ascomycota (L). 4 fam., 17 gen. (+ 66 syn.), 514 spp. Thallus usually foliose, occ. crustose, fruticose or squamulose, corticate on the upper or both surfaces, the lower surface sometimes characteristically areolate; ascomata apothecial, initially immersed, usually with a covering layer which splits to expose the hymenium at a late stage; interascal tissue of simple paraphyses; asci usually with a thickened apex (usually not capitate), an I+ apical ring, a well-developed ocular chamber and an I+ gelatinized outer coat; ascospores hyaline or brown, often elongated, often multiseptate. Anamorphs pycnidial. Lichenized with green algae or cyanobacteria, if the former then often with cyanobacteria in cephalodia, widespr. esp. temp. Fams:

(1) **Lobariaceae** (syn. *Stictaceae*).
(2) **Nephromataceae**.
(3) **Peltigeraceae** (syn. *Solorinaceae*).
(4) **Placynthiaceae** (syn. *Lecotheciaceae*).

The *Solorinaceae* have sometimes been separated (Hafellner, *in* Galun (Ed.), *Handbook of Lichenology* **3**: 41, 1988) but that is not strongly supported by molecular data which do support the retention of *Nephromataceae* (Eriksson & Strand, *SA* **14**: 33, 1995).

Lit.: Galloway (*Symbiosis* **11**: 327, 1991; chemical evol.), Hawksworth (*J. Hattori bot. Lab.* **52**: 323, 1982; evolutionary significance), Henssen & Jahns (1973), Honegger (*Lichenologist* **10**: 47, 1978; asci), Letrouit-Galinou & Lallement (*Lichenologist* **5**: 59, 1971; asci, ontogeny), Poelt (1974), and under *Ascomycota*.

Peltigeromyces A. Møller (1901), Leotiales (inc. sed.). 2, Afr. See Le Gal (*Bull. Jard. bot. de l'État* **29**: 73, 1959).

Peltigeromyces E.A. Thomas ex Cif. & Tomas. (1953) ≡ Peltigera (Peltiger.).

Peltiphylla M. Choisy (1950) = Psora (Psor.).

Peltistroma Henn. (1904), Mitosporic fungi, 5.A2.?. 1, Brazil.

Peltistromella Höhn. (1907), Mitosporic fungi, 5.C2.?. 2, Brazil, Sri Lanka.

Peltomyces L. Léger (1909), ? Plasmodiophoraceae. 3, Eur.

Peltophora Clem. (1909) ≡ Peltigera (Peltiger.).

Peltophoromyces Clem. (1909) ≡ Peltigeromyces (Leotiales).

Peltopsis Bat. (1960) = Muyocopron (Microthyr.) fide v. Arx & Müller (1975).

Peltosoma Syd. (1925), Mitosporic fungi, 5.C2.?1. 1, Philipp. See Sutton & Pascoe (*MR* **92**: 210, 1989; re-evaluation).

Peltosphaeria Berl. (1888) = Julella (Thelenell.) fide Mayrhofer (*Bibl. Lich.* **26**, 1987).

Peltostromellina Bat. & A.F. Vital (1959), Mitosporic fungi, 4.A1.?. 1, Sri Lanka.

Peltostromopsis Bat. & A.F. Vital (1959), Mitosporic fungi, 5.B2.?. 1, Sri Lanka.

Peltula Nyl. (1853), Peltulaceae (L). *c.* 18, cosmop. (mainly temp.). See Büdel & Henssen (*Int. Jl mycol. lich.* **2**: 235, 1986), Egea (*Bibl. Lich.* **31**, 1989; W. Eur. & N. Afr.), Filson (*Muelleria* **6**: 495, 1988; 10 Australian spp.), Swinscow & Krog (*Norw. J. Bot.* **26**:

213, 1979; key 10 E. Afr. spp.), Wetmore (*Ann. Mo. Bot. Gdn* **57**: 158, 1970; N. Am.).

Peltulaceae Büdel (1986), Lichinales (L). 2 gen., 20 spp. Thallus not gelatinized; asci with a thickened apex when young, thin-walled when mature, polysporous. Lichenized with cyanobacteria, widespr. *Lit.*: Büdel (*Bibl. Lich.* **25**: 209, 1987; ontogeny).

Peltularia R. Sant. (1944), Coccocarpiaceae (L). 2, S. Am., Australia. See Jørgensen & Galloway (*Lichenologist* **16**: 189, 1984).

Pemphidium Mont. (1840), Hyponectriaceae. 1, S. Am. See Hyde (*Sydowia* **45**: 204, 1993; posn), Petrak (*Sydowia* **7**: 354, 1953).

Penardia Cash (1904), *Hydromyxales*, q.v.

penatin (corylophillin, notatin, penicillin B), an antibiotic from *Penicillium notatum* and *P. chrysogenum*; antibacterial (Kocholary, *J. Bact.* **44**: 143, 1942).

pendant, (1) (of a fruticose lichen), one hanging downwards; (2) (in an ascus) see ascus.

Pendulispora M.B. Ellis (1961), Anamorphic Tubeufiaceae, 1.D2.10. Teleomorph *Tubeufia*. 1, Venezuela.

penicidin, see patulin.

Penicillaria Chevall. (1826) [non Willd. (1809), *Gramineae*] = Pterula (Pterul.) fide Corner (1950).

penicillate, like a little brush.

penicillic acid, a tetronic acid from *Penicillium puberulum*, *P. cyclopium*, etc. and antibacterial (Birkinshaw *et al.*, *Biochem. J.* **30**: 394, 1936).

Penicillifer Emden (1968), Anamorphic Hypocreaceae, 1.C2.32/33. Teleomorph *Nectria*. 4, widespr. See Samuels (*Mycol.* **81**: 347, 1989; key).

penicillin, one of a group of antibiotic substances (q.v.) produced by *Penicillium chrysogenum* (syn. *P. notatum*), active against Gram+ bacteria and of low toxicity to humans. Discovered by Fleming (*Br. J. exp. Path.* **10**: 226, 1929). See Hare (*The birth of penicillin*, 1970), *Penicillin 1929-43* [*Br. med. Bull.* **2**, 1944], Fleming (Ed.) (*Penicillin; its practical application*, 1946), Waksman (*Microbial antagonisms and anti-bacterial substances*, edn 2, 1947), Knox (*Nature* **192**: 492, 1961; survey of new penicillins).

penicillin B B, see penatin; **F**, see flavicin; **dihydro-F**, see gigantic acid; **N**, see cephalosporin N. See Knox (1961) under penicillin.

penicillinase, a bacterial enzyme which inhibits the action of penicillin.

Penicilliopsis Solms (1887), Trichocomaceae. 2, trop. See Samson & Seifert (*in* Samson & Pitt (Eds), *Advances in Penicillium and Aspergillus systematics*: 397, 1985), Udagawa & Takada (*Bull. Natn. Sci. Mus. Tokyo* **14**: 501, 1971). Anamorphs *Sarophorum*, *Stilbodendron*.

Penicillites Mesch. (1892), Fossil fungi. 1 (Oligocene), Eur.

Penicillium Link (1809), Anamorphic Trichocomaceae, 1.A2.15. Teleomorphs *Eupenicillium*, *Talaromyces*. 223, widespr. Common moulds, many first named on cultural, biochemical, or pathogenic characters. See Fig. 30. *P. expansum* (blue mould of apples), *P. digitatum* and *P. italicum* (green and blue moulds of citrus fruits), *P. gladioli* (storage rot of *Gladiolus*).
Lit.: **General**, Pitt & Samson, *Regnum veg.* **128**: 13, 1993; list of spp. names), Pitt (*The genus Penicillium and its teleomorphic states Eupenicillium and Talaromyces*, 1980 ['1979'], monogr., keys), Frisvad & Filtenborg (*Appl. Env. Microbiol.* **46**: 1301, 1983; secondary metabolites and classification), Pitt (*A laboratory guide to common Penicillium species*, 1985), Samson & Pitt (Eds) (*Advances in Penicillium and Aspergillus systematics*, 1985), Ström (*J. Appl. Bact.* **60**: 22, 1986; cross points of conidia and classification), Pitt (*CJB* **52**: 2231, 1974; sclerotigenic spp.),

Frisvad *et al.* (*CJB* **65**: 765, 1987; terverticillate taxa associated with rats), Newby & Gadd (*TBMS* **89**: 381, 1987; synnema induction), Fassatiová (*Česká Myk.* **42**: 12, 1988; key to toxigenic spp. in Bohemia), Peberdy (*Biology of industrial micro-organisms*, 1985; biology), Pitt *et al.* (*Food Microbiol.* **3**: 363, 1986; white cheese moulds), El-Banna *et al.* (*System. Appl. Microbiol.* **10**: 42, 1987; mycotoxin production), Peberdy (Ed.) (*Biotech. Handb.* **1**: 1987; taxonomy and morphology), Kozakiewicz (*Bot. J. Linn. Soc.* **99**: 273, 1989; conidial ornamentation), Ström *et al.* (*J. Appl. Bact.* **66**: 461, 1989; chemotaxonomy of symmetric spp.), Pitt (*J. Appl. Bact.* Suppl. Symp. Ser. **18**: 37S, 1989; recent developments in taxonomy), Bridge *et al.* (*J. Gen. Microb.* **135**: 2941, 2967, and 2979, 1989; numerical taxonomy, identification, isoenzymes of terverticillate spp.), Frisvad & Filtenborg (*Mycol.* **81**: 837, 1989; chemotaxonomy and mycotoxin production by terverticillate spp.), Frisvad *et al.* (*Persoonia* **14**: 193, 1990; typification of spp.), Frisvad *et al.* (*Persoonia* **14**: 209, 1990; redisposition of recent spp.), Samson & Pitt (Eds) (*Modern concepts in Penicillium and Aspergillus classification*, 1989), Qi *et al.* (*Mycosystema* **3**: 9, 1990; taxonomy of *P. marneffei*), Jimenez *et al.* (*Appl. Env. Microbiol.* **56**: 3718, 1990; differentiation of *P. griseofulvum*), Pitt (*J. Appl. Bact.* **71**: 86, 1991; PENNAME, a computer key to spp.), Pitt *et al.* (*Syst. appl. Microbiol.* **13**: 304, 1990; differentiation of *P. glabrum* from *P. spinulosum*), Chan & Chow (*Ultrastr. Path.* **14**: 439, 1990; ultrastr. *P. marneffei*), Samson & Frisvad (*Proc. Jap. Assoc. Mycotoxic.* **32**: 3, 1990; sp. concepts, mycotoxins), Pitt (*Jap. J. Med. Mycol.* **32**: 31, 1991; chemotaxonomy), Kozakiewicz (*Proc. 5th Int. Working Conf. on Stored Product Prot.* **1**: 347, 1990; classification mycotoxin-producing spp.), Schutte (*Bothalia* **22**: 77, 1992; spp. in S. Afr.), Bridge *et al.* (*Mycol. Pap.* **165**: 1, 1992; PENIMAT, computer-assisted identification), Kozakiewicz *et al.* (*Occurence and significance of mycotoxins*: 64, 1993; identification of mycotoxins), Drouhet (*J. Mycol. Med.* **3**: 195, 1993; penicilliosis by *P. marneffei* in SE Asia), Pitt (*J. Appl. Bact.* **75**: 559, 1993; creatine sucrose medium for spp. differentiation in subgen. *Penicillium*), Holmes *et al.* (*Phytopath.* **84**: 719, 1994; revised description of *P. ulaiense*), Tzean *et al.* (*Mycol. Monogr.* **1**, 1994; spp. in Taiwan).

Techniques, Ho & Smith (*J. Gen. Microbiol.* **132**: 3479, 1986; effect of CO_2 on morphology), Bridge *et al.* (*TBMS* **87**: 389, 1986; morphological and biochemical variation in single isolates), Scott *et al.* (*Can. J. Microbiol.* **32**: 687, 1986; ultrastr. changes and antibiotic synthesis), Polonelli *et al.* (*Appl. Env. Microbiol.* **53**: 2917, 1987; antigenic characterization of *P. camembertii*), Bridge *et al.* (*J. Gen. Microbiol.* **133**: 995, 1987; variation in phenotype and DNA content), Pitt (*Appl. Env. Microbiol.* **53**: 266, 1987; *P. viridicatum, P. verrucosum* and ochratoxin A), Cruickshank & Pitt (*Microbiol. Sci.* **4**: 14, 1987; isozyme patterns in classification), Bridge (*TBMS* **88**: 569, 1987; nitrate assimilation tests), Cruickshank & Pitt (*Mycol.* **79**: 614, 1987; identification by enzyme electrophoresis), Petruccioli *et al.* (*Mycol.* **80**: 726, 1988; extracellular enzymes), Bridge & Hudson (*Bot. J. Linn. Soc.* **99**: 11, 1989; continuous flow microfluorometry for DNA determination), Sekhon *et al.* (*J. Med. Vet. Mycol.* **27**: 105, 1989; antigenic relationships between spp.), Fuhrmann *et al.* (*Can. J. Microbiol.* **35**: 1043, 1989; immunological differentiation of spp.), Qi & Sun (*Mycosystema* **2**: 151, 1989; biochemistry and electrophoresis of *P. citrinum*), Paterson *et al.* (*MR* **94**: 152, 1990; high resolution thermal denaturation of DNA), Paterson & Kemmelmeier

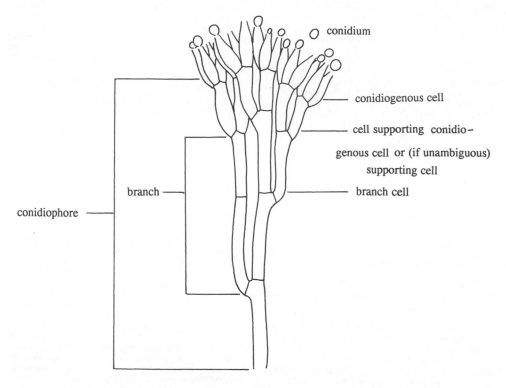

Fig. 30. Terms recommended to describe the different conidiogenous structures in *Penicillium*. See Minter *et al.* (*in* Samson & Pitt (Eds), *Advances in Penicillium and Aspergillus systematics*: 71, 1985).

(*Jl Chromat.* **511**: 195, 1990; u.v. spectra of secondary metabolites), Paterson & Kemmelmeier (*Jl Chromat.* **483**: 153, 1989; HPLC of insecticidal extracts), Jain *et al.* (*Enz. Microb. Tech.* **12**: 691, 1990; polysaccharidases in *P. occitanis*), Schubert & Kreisel (*Persoonia* **14**: 341, 1991; ubiquinones), Paterson & Buddie (*Lett. Appl. Microbiol.* **13**: 133, 1991; ubiquinone profiles in spp. by reverse phase HPLC), Nozawa *et al.* (*Phytochem.* **28**: 929, 1989; dioxopiperazine from *P. megasporum*), Meyer *et al.* (*Curr. Gen.* **19**: 239, 1991; DNA fingerprinting and strain differentiation), Kuraishi *et al.* (*MR* **95**: 705, 1991; ubiquinones in *Penicillium* and related genera), Frisvad & Lund (*Occurrence and significance of mycotoxins*: 146, 1993; secondary metabolite production), Lobuglio *et al.* (*MR* **98**: 250, 1994; origins of synnematous spp. by ribosomal DNA), Berbee *et al.* (*Mycol.* **87**: 210, 1995; monophylly explored, 18S, 5.8S, ITS rDNA).

Penicillium Fr. (1832) ? = Botrytis (Mitosp. fungi) fide Hawksworth *et al.* (*Taxon* **25**: 665, 1976).

penicillus, the brush-like conidiogenous apparatus of *Penicillium* and related genera composed of a stipe bearing a tuft of conidiogenous cells and other elements formerly on the rami and metulae (Fig. 30).

Peniophora Cooke (1879), Peniophoraceae. *c.* 40, widespr. See Slysh (*Tech. Publ. State Univ. Coll. Forestry, Syracuse*, **83**, 1960; N.Y. State), Weresub (*CJB* **39**: 1453, 1961; *Tubuliferae*), Hayashi (*Bull. Govt. Forest exp. Stn. Tokyo* **260**, 1974; Japan), Leger & Poncet (*BSMF* **92**: 229, 1976; numerical tax.), Leger (*BSMF* **92**: 377, 1976; electrophoresis). Many authors use *Peniophora* s.s. for *Peniophora* sect. *Coloratae*

(see Eriksson, *Symb. bot. upsal.* **10** (5), 1950; key), Boidin (*Bull. Soc. linn. Lyon* **34**: 161, 213, 1965; key), Eriksson *et al.* (*Corticiaceae of North Europe* **5**: 916, 1978; key 21 Eur. spp.). ? ≡ Corticium Pers. (Cortic.).

Peniophoraceae Lotsy (1907), Stereales. 4 gen. (+ 4 syn.), 48 spp. Basidioma resupinate to reflexed; monomitic; gloeocystidia or metuloids present; spore print pinkish; spores smooth, inamyloid.
  Lit.: Rattan (*Bibl. Mycol.* **60**, 1977; NW Himalayas), Liberta (*Mycol.* **60**: 827, 1968; *Peniophora*), Ginns (*Mycol.* **60**: 1211, 1969; *Merulius*), Donk (1951-63, VII; gen. names), Reid (*Beih. Nova Hedw.* **18**, 1965; stipitate steroid spp.), Jülich & Stalpers (*Verh. Kon. Ned. Akad. Wetensch., Afd. Natuurk.* sect. 2 **74**, 1980; temp. N. Hemisph.), Boidin *et al.* (*BSMF* **107**: 91, 1991; trop. spp.).

Peniophorella P. Karst. (1889) nom. dub. (Peniophor.) fide Donk (1962).

Peniophorina Höhn. (1917) Fungi (inc. sed.) fide Donk (*Reinwardtia* **1**: 216, 1951).

penitrem A, a mycotoxin produced by some *Penicillium* spp., including *P. cyclopium*, which affects the nervous system, causing tremors; see neurotoxin, tremorgen.

Pennella Manier (1968), Legeriomycetaceae. 6 (in *Diptera*), Eur., N. Am., Japan. See Lichtwardt (1986; key), Williams & Lichtwardt (*CJB* **68**: 1045, 1990).

Penomyces Giard ex Sacc. & Trotter (1913), Mitosporic fungi, 1.?.?. 2, Eur. See Petch (*TBMS* **19**: 190, 1935; nom. dub.).

Pentagenella Darb. (1897), Roccellaceae (L). 1, ? Chile.

Pentaposporium Bat. (1957) ? = Tripospermum (Mitosp. fungi) fide Hughes (*Mycol.* **68**: 810, 1976).

Pentasporium, see *Pentaposporium*.
Penzigia Sacc. (1888) = Xylaria (Xylar.) fide Læssøe (*SA* **8**: 25, 1989; **13**: 43, 1994).
Penzigina Kuntze (1891) ≡ Eriosphaeria (Trichosphaer.).
Penzigomyces Subram. (1992), Mitosporic fungi, 1.C2.19. 13, widespr.
Peplopus (Quél.) Quél. ex Moug. & Ferry (1887) ≡ Suillus (Bolet.).
per (in author citations) (obsol.), see Nomenclature.
percurrent, (1) extending throughout the entire length, as of the columella of a gasteromycete basidioma; (2) growing through in the direction of the long axis, as of a conidial germ tube emerging through the hilum or of a proliferation growing through the tip of the conidiogenous cell (Luttrell, 1963).
Perelegamyces R.F. Castañeda & W.B. Kendr. (1990), Mitosporic fungi, 1.A2.11. 1, Cuba.
perennial, living for a number of years.
Perenniporia Murrill (1942), Coriolaceae. c. 20, widespr. See Ryvarden (*Polyporaceae of North Europe* **2**: 305, 1978; key Eur. spp.), Ryvarden & Johansen (*Preliminary polypore flora of East Africa*: 463, 1980; key 10 Afr. spp.).
Peresia H. Maia (1960) = Colletotrichum (Mitosp. fungi) fide Sutton (1977).
Peresiopsis Bat. (1960) = Yamamotoa (Asterin.) fide v. Arx & Müller (1975).
perfect state, see States of fungi.
Perforaria Müll. Arg. (1891) = Coccotrema (Coccotremat.).
Perforariomyces Cif. & Tomas. (1953) ≡ Perforaria (Coccotremat.).
perforation lysis, the process by which degradation of resistant fungus propagules in soil is initiated (Old & Wong, *Soil Biol. Biochem.* **8**: 285, 1976; review).
perfume, see oak-moss, smell.
pergameneous, (**pergamenous**, **pergamentaceous**), like paper.
Periamphispora J.C. Krug (1989), Lasiosphaeriaceae. 1, Spain.
Periascomycetes. Class introduced for *Protascus*, *Protomyces*, *Ashbya*, and other 'primitive' ascomycetes (Moreau, 1953).
Periaster Theiss. & Syd. (1917) = Erikssonia (Phyllachor.) fide Petrak (*Ann. Myc.* **29**: 390, 1931).
Peribotryon Fr. (1832) nom. rej. = Chrysothrix (Chrysothric.).
Pericauda, see *Piricauda*.
Perichaena Fr. (1817), Trichiaceae. 17, cosmop. See Cavalcanti (*Rickia* **6**: 99, 1974; key), Keller & Eliasson (*MR* **96**: 1095, 1992; key).
Perichaenaceae, see *Trichiaceae*.
Perichlamys Clem. & Shear (1931) ≡ Kuntzeomyces (Tillet.).
Pericladium Pass. (1875), Ustilaginaceae. 2 (forming galls on *Grewia* and *Piper*), Afr., Asia, Austr. See Mordue (*Mycopath.* **103**: 173, 1988).
periclinal, curved in the direction of, or parallel to, the surface or the circumference; cf. anticlinal; - **thickening**, zone of increased material surrounding the protoplasmic channel at the apex of a 'phialide' (see Sutton, *The Coelomycetes*, 1980).
Pericoccis Clem. (1931) = Leptogium (Collemat.).
Pericoelium Bonord. (1851) = Ustilago (Ustilagin.) fide Saccardo (1888).
Periconia Tode (1791), Mitosporic fungi, 1.A2.10. 30, widespr. See Mason & Ellis (*Mycol. Pap.* **56**, 1953; Br. spp., key), Bunning & Griffiths (*TBMS* **78**: 147, 1984; ultrastr.).
Periconiella Sacc. (1885), Mitosporic fungi, 1.A-C2.10. 23 (some fungicolous), widespr. See Ellis (*Mycol.*

Pap. **111**, 1967; key), McKenzie (*Mycotaxon* **39**: 229, 1990), Priest (*MR* **95**: 924, 1991; spp. on *Proteaceae*).
Pericystales, see *Ascosphaerales*.
Pericystis Betts (1912) [non J. Agardh (1847), Algae] ≡ Ascosphaera (Ascosphaer.).
peridermium, an aecium as in the form genus *Peridermium*.
Peridermium (Link) J.C. Schmidt & Kunze (1817) nom. cons., Anamorphic Uredinales (inc. sed.). 20, widespr.; O, I on gymnosperms.
peridiole (**peridiolum**) (esp. of *Nidulariaceae*), a division of the gleba having a separate wall, frequently acting as a unit for distribution.
Peridiomyces H. Karst. (1843) nom. dub. (? Mitosp. fungi).
Peridiopsora Kamat & Sathe (1969), Anamorphic Uredinales. A name for (II) only. 1, India.
Peridiothelia D. Hawksw. (1985), Pleomassariaceae. 3, Eur. See Barr (*Mycotaxon* **43**: 371, 1992; posn).
Peridipes Buriticá & J.F. Hennen (1994), Phakopsoraceae. 1, S. Am.
peridium, the wall or limiting membrane of a sporangium or other fruit-body, an excipule; **peridial cells** (esp. of aecia), the cells of the peridium.
Peridoxylon Shear (1923) = Camarops (Bolin.) fide Nannfeldt (1972), but see Dargan & Thind (*J. Indian Bot. Soc.* **63**: 177, 1984; India).
Peridoxylum Clem. & Shear (1931) ≡ Peridoxylon (Bolin.).
perifulcrium, the wall of a pycnidium in a lichen thallus (obsol.).
Periline Syd. (1939) = Antennularia (Ventur.) fide Müller & v. Arx (1962).
Perinidium Cromb. (1870) ≡ Pyrenidium (Dacamp.).
Periola Fr. (1822), Mitosporic fungi, 3.?.?. 1 or 2, Eur., Am.
Periolopsis Maire (1913) = Sarcopodium (Mitosp. fungi) fide Sutton (1981).
Periperidium Darker (1963), Anamorphic Leotiaceae, 8.C1.15. Teleomorph *Micraspis*. 1, Canada.
Peripherostoma Gray (1821) nom. rej. ≡ Daldinia (Xylar.).
periphysis (pl. **periphyses**), an upward pointing hypha inside, or near, the ostiole of a perithecium, pycnidium, or pycnium; see hamathecium.
periphysoid, see hamathecium.
periphyton, the 'assemblage of organisms growing upon free surfaces of submerged objects in water and covering them with a slimy coat.' Young (1945) fide Cooke (*Bot. Rev.* **22**: 616, 1956).
periplasm (of *Peronosporales*), the outer, non-functional protoplasm of an oogonium or antheridium; cf. gonoplasm, ooplasm.
Perirhiza Karling (1946) = Catenophlyctis (Catenar.) fide Karling (1977).
Perischizon P. Syd. (1914), Parmulariaceae. 1, S. Afr.
Perisperma, see *Pyrisperma*.
Perisphaeria Roussel (1806) = Hypoxylon (Xylar.) fide Læssøe (*SA* **13**: 43, 1994).
perispore (**perisporium**), sheath outside the true spore wall. See Harmaja (*Karstenia* **14**: 123, 1974; cyanophilic in *Pezizales*); ascospore, spore wall.
Perisporiacites Félix (1894), Fossil fungi (Ascomycota). 2 (Cretaceous, Eocene), former USSR. See Sahni & Rao (*Proc. natn. Acad. Sci. India* B **13**: 45, 1943).
perisporial sac, a perispore forming a loose envelope around the spore as in *Coprinus* sp.
Perisporiales, see *Dothideales*.
Perisporiella Henn. (1902) = Hypocrella (Clavicipit.) fide Clements & Shear (1931).
Perisporina Henn. (1904) = Perisporiopsis (Parodiopsid.) fide Müller & v. Arx.

Perisporiopsella Bat., J.L. Bezerra, Castr. & Matta (1964) = Pilgeriella (Parodiopsid.) fide v. Arx & Müller (1975).

Perisporiopsidaceae, see *Parodiopsidaceae*.

Perisporiopsis Henn. (1904), Parodiopsidaceae. *c.* 15, trop. Anamorph *Septoidium*.

Perisporiopsis F. Stevens (1917) ≡ Stevensea (Myriang.).

Perisporites Pampal. (1902), Fossil fungi. 2 (Miocene), Italy.

perisporium see perispore.

Perisporium Fr. (1821) nom. dub. (Fungi, inc. sed.), sensu Corda (1842) = Preussia (Spororm.).

Peristemma Syd. (1921) = Miyagia (Puccin.) fide Cummins & Hiratsuka (1983).

peristome, an edging round an opening, esp. of basidiomata of certain gasteromycetes.

Peristomialis (W. Phillips) Boud. (1907), Hypocreaceae. 7, widespr. See Samuels (*Mem. N.Y. bot. Gdn* **48**, 1988).

Peristomium Lechmere (1912) = Microascus (Microasc.) fide v. Arx (1975).

perithecium (pl. **perithecia**), a subglobose or flask-like ostiolate ascoma; sometimes limited to ascohymenial types formed from the development of an ascogonium (not of stromatic origin), but now widely used as a general term regardless of the ontogenetic type. See Cherepanova (*Vestn. Leningrad Univ. Biol.* **3**: 39, 1986; types and evol. pathways).

Peritrichospora Linder (1944) = Corollospora (Halosphaer.) fide Kohlmeyer (*Ber. dtsch. bot. Ges.* **75**: 125, 1962).

peritrichous (**peritrichiate**), having hairs or flagella all over the surface.

Perizomatium Syd. (1927) = Phaeofabraea (Leot.) fide Pfister (*Occ. Pap. Farlow Herb.* **11**, 1977).

Perizomella Syd. (1927), Mitosporic fungi, 8.A2.1. 1, C. Am.

Perkinsus Levine. [not traced], Protozoa, see monads. See also Dick (*in press*).

Perona Pers. (1825) = Omphalina (Tricholomat.) fide Donk (1962).

peronate, sheathed; having a boot or covering, esp. of the lower part of a stipe covered by a volva or veil.

Peroneutypa Berl. (1902) = Eutypella (Diatryp.) fide Rappaz (1987).

Peroneutypella Berl. (1902) ≡ Scoptria (Diatryp.).

Peronium Cohn (1854) ? = Olpidiopsis (Olpidiopsid.).

Peronophythora C.C. Chen (1961), Pythiaceae. 1, New Guinea, Taiwan. See Hall (*Mycopath.* **106**: 189, 1989).

Peronophythoraceae, see *Pythiaceae*.

Peronoplasmopara (Berl.) G.P. Clinton (1905) ≡ Pseudoperonospora (Peronosp.).

Peronosclerospora (S. Ito) Hara (1927), Sclerosporaceae. 9, widespr.

Peronosclerospora C.G. Shaw (1978) = Peronosclerospora (Sclerospor.) fide Shaw & Waterhouse (*Mycol.* **72**: 425, 1980).

Peronospora Corda (1837), Peronosporaceae. 75, widespr. Downy mildews, e.g. of beet (*Beta*) (*P. farinosa* f.sp. *betae*), clover (*P. trifoliorum*), crucifers (*P. parasitica*), onion (*Allium*) (*P. destructor*), spinach (*Spinacia*) (*P. farinosa* f.sp. *spinaciae*), tobacco (*P. hyoscyami* f.sp. *tabacina*). See Gäumann (*Beitr. Kryptog.-fl. Schweiz* **5**, 1923; 260 spp. recognized), Yerkes & Shaw (*Phytopath.* **49**: 502, 1959; spp. on *Cruciferae*, *P. parasitica*, and *Chenopodiaceae*, *P. farinosa*), Rao (*Nova Hedw.* **16**: 269, 1968; India), Gustavsson (*Op. Bot. bot. Soc. Lund* **3**, 1959; Scandinavia), Constantinescu (*Thunbergia* **15**, 1991; annotated list names).

Peronosporaceae Warm. (1884), Peronosporales. 7 gen. (+ 8 syn.), 118 spp. The common Downy Mildews. *Lit.*: Ciferri (*Riv. Pat. veg., Pavia* ser. 3 **1**: 333, 1961; Italy), Gaponenko (*Semeĭstvo Peronosporacae Sredneĭ Azii i Yuzhnogo Kazakhstana*, 1972; C. Asia, etc., keys), Francis & Waterhouse (*TBMS* **91**: 1, 1988; Br. list).

Peronosporales, Oomycota. 2 fam., 8 gen. (+ 9 syn.), 148 spp. Mycelium intercellular with haustoria; asexual reproduction by zoospores, but not in *Peronospora* in which the spore is a 'conidium'; sexual reproduction by oospores from oospheres fertilized by nuclei in an antheridium, oospore wall thick and smooth or ornamented; cosmop., obligate parasites, almost totally confined to dicots (except. *P. destructor*), some economically important plant pathogens. Fams:
(1) **Albuginaceae**.
(2) **Peronosporaceae**.
	Lit.: Fitzpatrick (1930), Kochman & Majewski (*Grzyby (Mycota)*, **4** *Glonowce (Phycomycetes)*, *Wroslikowe (Peronosporales)* [*Flora Polska*], 1970), Spencer (Ed.) (*The downy mildews*, 1981).

Peronosporites W.G. Sm. (1877), Fossil fungi. 3 (Carboniferous, Miocene), Italy, UK.

Peronosporoides J. Sm. (1896), Fossil fungi (mycel.). 1 (Carboniferous), UK.

Peronosporoides E.W. Berry (1916), Fossil fungi (Peronospor.). 1 (Oligocene), USA.

Peronosporomycetes, see *Oomycota*.

Peronosporomycetidae, see *Oomycota*.

Peronosporomycotina, see *Oomycota*.

Peroschaeta Bat. & A.F. Vital (1957), Dothideales (inc. sed.). 1, Philipp.

peroxisome, one of the subcellular organelles which have indispensable functions in the metabolism of *n*-alkalenes, fatty acids, methanol, and several nitrogen-containing compounds in eukaryotic microorganisms. See Tanaka & Ueda (*MR* **97**: 1025, 1993).

Perrotia Boud. (1901), Hyaloscyphaceae. *c.* 8, widespr. See Dennis (*Persoonia* **2**: 82, 1962), Raitviir (1970; key), Spooner (*Bibl. Mycol.* **116**, 1987; 5 spp. Australasia).

Perrotiella Naumov (1916) = Nectria (Hypocr.) fide Nannfeldt (1932).

perrumpent, breaking through. Cf. erumpent.

Persiciospora P.F. Cannon & D. Hawksw. (1982), ? Ceratostomataceae. 4, widespr. See Horie *et al.* (*Mycotaxon* **25**: 229, 1986), Krug (*Mycol.* **80**: 414, 1988; key 3 spp.).

persistent, (1) (of interascal tissues) still evident at maturity; (2) (of spores), non-deciduous; (3) (of teliospore pedicels), remaining firmly attached to the spore after liberation.

Persoon (Christiaan Hendrik; 1761-1836) came from South Africa. In 1775 he was at school in Germany where he seems to have gone on with his studies and mycological work until about 1801, when he went to Paris, living there for the rest of his life.

There had been little development in the taxonomy of fungi from the low level at the time of Linnaeus till Persoon's first printed work in 1793. His *Observationes mycologicae* (1795-9) and other papers, and his great *Synopsis methodica Fungorum* (1801 [reprint 1952 + index by G.H. L(ünemann, 1808]) in which names of *Ustilaginales*, *Uredinales*, and 'gasteromycetes' have sanctioned status (see Nomenclature), formed the framework on which Fries and all later systematists based their classifications.

Persoon's working conditions in Paris were hard, but some papers and icones and the *Mycologia Europaea* (1822-8) were printed, the last getting less attention because Fries's *Systema mycologicum* came out at almost the same time.

Persoon's collection, a very important one because it has a great number of common species, was taken over by the Dutch government before his death and is in **L**.

Lit.: Lloyd (*Mycol. Notes* **1**: 158, 1903; **7**: 1301, 1924), Ainsworth (*Nature* **193**: 22, 1962), Petersen (*Kew Bull.* **31**: 695, 1977), Grummann (1974: 704), Stafleu & Cowan (*TL-2* **4**: 178, 1983). See also Bolton.

Persooniana Britzelm. (1897) = Tyromyces (Coriol.) fide Donk (1974).

Persooniella Syd. (1922) = Puccinia (Puccin.) fide Dietel (1928).

Perspicinora Riedl (1990) = Koerberiella (Porpid.) fide Coppins (*SA* **10**: 48, 1991).

perthophyte (**perthotroph**), a necrophyte on dead tissues of living hosts (Münch); cf. saprophyte.

Pertusaria DC. (1805) nom. cons., Pertusariaceae (L). *c*. 250, cosmop. See Archer (*Telopea* **4**: 165, 1991, 28 spp., Australia; *Mycotaxon* **41**: 223, 1991, key 63 Australian spp.; *Bibl. Lich.* **53**: 1, 1993, subgen.), Archer & Elix (*Myxotaxon* **50**: 203, 1994; key 26 NZ spp.), Dibben (*Milwaukee Public Mus., Publs Biol. Geol.* **5**, 1980; N. Am.), Erichsen (*Rabenh. Krypt.-Fl.* **9**, 5 (1), 1935; Eur.), Hanks *et al.* (*Nova Hedw.* **42**: 165, 1986), Kantvilas (*Lichenologist* **22**: 289, 1990; key 5 spp. Tasmania), Malme (*Ark. Bot.* **28A** (9), 1936), Oshio (*J. Sci. Hiroshima Univ., ser.* B, *div.* 2, **12**: 81, 1968; Japan), Niebel-Lohmann & Feurer (*Mitt. Inst. Allg. Bot. Hamburg* **24**: 199, 1992), Schmitz *et al.* (*Acta bot. fenn.* **150**: 153, 1994; gen. concept).

Pertusariaceae Körb. ex Körb. (1855), Pertusariales (L). 4 gen. (+ 13 syn.), 294 spp. Thallus crustose; ascomata apothecial, sessile or slightly sunken, the hymenium usually exposed from an early stage; interascal tissue of basal paraphyses only; hymenium amyloid; cephalodia absent. Anamorphs pycnidial. Lichenized with green algae, widespr.

Lit.: Dibben (*Milwaukee Public Mus., Publs Biol. Geol.* **5**, 1980, N. Am.; *J. Hattori bot. Lab.* **52**: 343, 1982, evol. trends), Erichsen (*Rabenh. Krypt.-Fl.* **9**, 5 (1), 1935; Eur.), Henssen (1976), Honegger (*Lichenologist* **14**: 205, 1982; asci), Letrouit-Galinou (*Revue bryol. lichén.* **34**: 413, 1966; ontogeny), Oshio (*J. Sci. Hiroshima Univ.*, B (2), **12**: 81, 1968; Japan).

Pertusariales, Ascomycota (L). 3 fam., 7 gen. (+ 16 syn.), 305 spp. Thallus crustose, rarely lobate or minutely foliose; ascomata apothecial, often deeply cupulate, usually initially immersed, opening widely or with a poroid aperture and appearing perithecial, usually with a well-developed thalline margin; interascal tissue of basal paraphyses, sometimes also with apical paraphyses in the poroid taxa; asci short, widely cylindrical, with a thick multilayered usually I+ wall, the apex often more strongly thickened, releasing spores through a ± vertical apical split, often less than 8-spored; ascospores very large, hyaline, aseptate, with a very thick multilayered wall. Anamorphs pycnidial. Lichenized with green algae, cosmop. Fams:

(1) **Coccotremataceae**.
(2) **Megasporaceae**.
(3) **Pertusariaceae**.

Many spp. reproduce mainly by soredia or isidia, forming 'species-pairs' with their meiotic counterparts. The *Trapeliaceae*, included here by Henssen & Jahns (1973), are referred to the *Lecanorales* in view of the differences in ascus structure (Honegger, 1982).

Pertusariomyces E.A. Thomas ex Cif. & Tomas. (1953) ≡ Pertusaria (Pertusar.).

pervious (of lichenized scyphi), open or perforate basally.

Pesavis Elsik & Janson. (1974), Fossil fungi (Mitosp. fungi). 2 (Paleogene), Australia, Canada, UK. See Smith & Crane (*Bot. J. linn. Soc.* **79**: 243, 1979).

Pesotum J.L. Crane & Schokn. (1973), Anamorphic Ophiostomataceae, 2.A1.10. Teleomorph *Ophiostoma*. 2, widespr. = Graphium (Mitosp. fungi) fide Wingfield *et al.* (*MR* **95**: 1328, 1991).

pest control, in reference collections of specimens, see Hall (*Taxon* **37**: 885, 1988).

Pestalopezia Seaver (1942), Leotiaceae. 3, N. Am.

Pestalosphaeria M.E. Barr (1975), ? Amphisphaeriaceae. 4, N. & S. Am., Hawaii. See Nag Raj (*Mycotaxon* **22**: 52, 1985; key), Zhu *et al.* (*Acta Agric. Univ. Zhejiang* **16**: 163, 1990; key 7 spp.). = Lepteutypa (Amphisphaer.) fide v. Arx (1981). Anamorph *Pestalotiopsis*.

Pestalotia De Not. (1841), Mitosporic fungi, 7.C2.19. 1, Italy. See Guba (*Monograph of Monochaetia and Pestalotia*, 1961; 222 spp. accepted, key), Steyaert (*Bull. Jard. bot. Brux.* **19**: 285, 1949; *s.str.* monotypic), Sutton (*CJB* **47**: 2083, 1969).

Pestalotiopsis Steyaert (1949), Anamorphic Amphisphaeriaceae, 6.C2.19. Teleomorph *Pestalosphaeria*. 50, widespr. *P. theae* (grey blight of tea; *CMI Descr.* **318**, 1971). See Steyaert (*TBMS* **36**: 181, 1953), Sutton (*Mycol. Pap.* **80**, 1961; conidium development), Zhu *et al.* (*Mycotaxon* **40**: 129, 1991; teleomorphs in China), Zhu *et al.* (*Acta Mycol. Sin.* **10**: 273, 1991; China), Zhu, Xu & Ge (*Acta Agr. Univ. Zhejiangensis* **16**: 173, 1990; soluble protein patterns).

Pestalozzia, see *Pestalotia.*

Pestalozziella Sacc. & Ellis ex Sacc. (1882), Mitosporic fungi, 4.A1.10. 4, temp. See Nag Raj & Kendrick (*CJB* **50**: 607, 1972).

Pestalozzina P. Karst. & Roum. (1890) ? = Bartalinia (Mitosp. fungi) fide Nag Raj (1993).

Pestalozzina (Sacc.) Sacc. (1895) ≡ Zetiasplozna (Mitosp. fungi) fide Nag Raj (1993).

Pestalozzites E.W. Berry (1917), Fossil fungi. 2 (Oligocene, Miocene), USA.

Petaloides Lloyd ex Torrend (1920) = Polyporus (Polypor.) fide Donk (1960).

Petalosporus G.R. Ghosh, G.F. Orr & Kuehn (1963) = Arachniotus (Gymnoasc.) fide v. Arx (*Persoonia* **9**: 393, 1977).

Petasodes Clem. (1909) = Phomopsis (Mitosp. fungi) fide Sutton (1977).

Petasospora Boidin & Abadie (1955) = Pichia (Endomycet.) fide v. Arx *et al.* (1977), Batra (1978).

Petch (Thomas; 1870-1948). Schoolmaster, then mycologist at the Royal Botanic Gardens, Peradeniya, Sri Lanka (1905-25), and first director of the Tea Research Institute, Sri Lanka (1925-28). A leading student of tropical mycology (esp. the fungi of Sri Lanka) and plant pathology (*Diseases and pests of the rubber tree*, 1921; .... *of the tea bush*, 1923). Married a daughter of Charles Plowright (q.v.) and on retirement lived in Plowright's old house at North Wootton, nr King's Lynn and published many papers on entomogenous fungi (q.v.) and on the British *Hypocreales* (*TBMS* **21**, **25**, **27**, 1938-45). Collections in **K** and Peradeniya, slides in **BM**. *Obit*.: Ainsworth (*TBMS* **67**: 179, 1976; portr.). See Stafleu & Cowan (*TL-2* **4**: 168, 1983).

Petchiomyces E. Fisch. & Mattir. (1938), Otideaceae. 2, N. Am., Sri Lanka. See Eckblad (1968).

Petelotia Pat. (1924) = Acanthonitschkea (Nitschk.) fide Nannfeldt (1975).

Petersenia Sparrow (1934), Pontismaceae. 3 (on marine *Rhodophyceae*), widespr. See Miller (*Mycol.* **54**: 422, 1963), Dick (*in press*; key).

Petractis Fr. (1845), ? Stictidaceae (L). 5, Eur., C. & N. Am. See Vězda (*Preslia* **37**: 127, 1965).

Petrak (Franz; 1886-1973). Austrian mycological taxonomist (esp. ascomycetes and mitosporic fungi) and bibliographer and indexer (see Literature) who from 1939 worked at the Naturhistorischen Museums, Vienna. Founded *Sydowia* (1947) to replace the *Annales Mycologici*. See Riedl (*Sydowia* **26**: xxix, 1974), v. Arx (*Persoonia* **9**: 95, 1976), Samuels (*An annotated index to the mycological writings of Franz Petrak* **1-5** A-O, 1981-86), Grummann (1974: 938), Stafleu & Cowan *TL-2* **4**: 204, 1983).

Petrakia Syd. & P. Syd. (1913), Mitosporic fungi, 3.D2.1. 2, Eur. See v. d. Aa (*Acta Bot. Neerl.* **17**l: 221, 1968). = Echinosporium (Mitosp. fungi) fide Carmichael *et al.* (1980).

Petrakiella Syd. (1924), ? Phyllachoraceae. 1, Brazil.

Petrakina Cif. (1932), ? Asterinaceae. 2, C. Am. See Farr & Palm (*Mycotaxon* **24**: 275, 1985).

Petrakiopeltis Bat., A.F. Vital & Cif. (1957), ? Microthyriaceae. 1, Guyana.

Petrakiopsis Subram. & K.R.C. Reddy (1968), Mitosporic fungi, 3.G1.1. 1, India.

Petrakomyces Subram. & K. Ramakr. (1953) = Ciliochora (Mitosp. fungi) fide Nag Raj (1993).

Petriella Curzi (1930), Microascaceae. 5, Eur., N. Am. See v. Arx *et al.* (*Beih. Nova Hedw.* **94**, 1988), Barron *et al.* (*CJB* **39**: 837, 1961; key), Corlett (*CJB* **44**: 79, 1966; ontogeny). = Microascus (Microasc.) fide Lodah (*in* Subramanian (Ed.), *Taxonomy of fungi* **1**: 241, 1978), Valmaseda *et al* (*CJB* **65**: 1802, 1987; concept). Anamorphs *Graphium, Sporotrichum.*

Petriellidium Malloch (1970) = Pseudallescheria (Microasc.) fide McGinnis *et al.* (*Mycotaxon* **14**: 94, 1982).

Petromyces Malloch & Cain (1973), Trichocomaceae. 2, Australia, Canada. Anamorph *Aspergillus.*

Petrona Adans. (1763) ? = Schizophyllum (Schizophyll.). See Donk (*Beih. Nova Hedw.* **5**: 220, 1962).

petrophilous, see saxicolous.

Petrosphaeria Stopes & Fujii (1910), Fossil fungi (mycel.). 1 (Cretaceous), Japan.

Peyerimhoffiella Maire (1916), Laboulbeniaceae. 1, N. Afr.

Peylia Opiz (1857) = Botryosporium (Mitosp. fungi) fide Saccardo & Trotter (*Syll. Fung.* **22**: 1252, 1913).

Peyritschiella Thaxt. (1890), Laboulbeniaceae. 46, widespr.

Peyronelia Cif. & Gonz. Frag. (1927), Anamorphic Lophiaceae, 1.D2.4. Teleomorph *Glyphium.* 4, Am., Eur.

Peyronelina G. Arnaud (1952) nom. inval. (Mitosp. fungi).

Peyronelina P.J. Fisher, J. Webster & D.F. Kane (1976), Mitosporic fungi, 1.D/G1.1. 1 (aquatic), UK.

Peyronellaea Goid. (1952) = Phoma (Mitosp. fungi) fide Boerema *et al.* (*Persoonia* **4**: 47, 1965).

Peyronellula Malan (1953) = Emericellopsis (Trichocom.) fide Stolk (*TBMS* **38**: 419, 1955).

Pezicula Paulet (1791) nom. rej. prop. = Craterellus (Craterell.) fide Cannon & Hawksworth (*Taxon* **32**: 477, 1983).

Pezicula Tul. & C. Tul. (1865) nom. cons., Dermateaceae. *c.* 30, esp. temp. See Groves (*Can. J. Res.* C, **17**: 125, 1939), Dennis (*Kew Bull.* **29**: 158, 1974; key Br. spp.). Anamorph *Cryptosporiopsis.*

Peziotrichum (Sacc.) J. Lindau (1900), Anamorphic Hypocreaceae, 2.A2.?. Teleomorph *Ophionectria.* 1 (entomogenous), Australia to Sri Lanka. See Petch

(*TBMS* **12**: 44, 1927). Nom. dub. fide Downing (*Mycol.* **45**: 938, 1953).

Peziza Fuckel (1870) = Aleuria (Otid.).

Peziza Fr. (1822), Pezizaceae. *c.* 80, widespr. See Romagnesi (*Bull. trim. Fed. Mycol. Dauphine-Savoie* **18**: 19, 1978; keys), Le Gal (*Rev. Mycol. Suppl.* **6**: 56, 1941; key), Donadini (*Doc. Mycol.* **9** (36): 1, 1979, key; **11** (41): 25, 1980; (53): 57, 1984, **14** (56): 39, 1984; *BSMF* **96**: 239, 247, 1980; key *nivalis* group; *Bull. Soc. linn. Prov.* **31**: 9, 1978), Donadini (*Bull. Soc. linn. Provence* **36**: 153, 1985; spores 74 spp.), Dyby & Kimbrough (*Bot. Gaz.* **148**: 283, 1987; ascospore ontogeny), Hohmeyer (*Z. Mykol.* **52**: 161, 1986; key Eur. spp.), Häffner (*Z. Mykol.* **52**: 189, 1986; subgen. *Phaeoporia*), Turnau (*TBMS* **91** 338, 1988; ultrastr.), Yao *et al.* (*SA* **14**: 17, 1995; gen. nomencl.). Anamorph *Ostracoderma.*

Peziza L. (1870) = Cyathus (Nidular.) fide Dennis (*Kew Bull.* **37**: 643, 1983).

Pezizaceae Dumort. (1829), Pezizales. 19 gen. (+ 28 syn.), 145 spp. Ascomata large, apothecial or cleistothecial, formed from a single ascogonium or ascogonial coil, some shade of brown, violaceous, yellow etc. (without carotenoid pigments), usually sessile, flat or cupulate, without setose hairs; if cleistothecial, ascomata solid or chambered, hypogeous or emergent; interascal tissue of simple or moniliform paraphyses, often swollen and pigmented at the apex; asci cylindrical, persistent, operculate, the wall blueing in iodine at least the the tip (sometimes globose in cleistothecial taxa), ascal pores simple; ascospores hyaline or pale brown, smooth or ornamented, often guttulate, uninucleate. Anamorphs hyphomycetous. Saprobic on soil or very rotten wood, cosmop.

Lit.: Curry & Kimborough (*Mycol.* **75**: 781, 1983; septal ultrastr.), Kimbrough (1989; delimitation).

Pezizales, Ascomycota. 17 fam., 177 gen. (+ 179 syn.), 1029 spp. Operculate discomycetes; cup-fungi. Stroma absent; ascomata apothecial or cleistothecial, rarely absent, often large, discoid, cupulate or ± globose, sometimes stalked, often brightly coloured; excipulum usually thick-walled, fleshy or membranous, composed of thin-walled pseudoparenchymatous cells; interascal tissue of simple or moniliform paraphyses, often pigmented and swollen at the apices, absent in esp. cleistothecial taxa; asci elongated, persistent, thin-walled, usually without obvious apical thickening, opening by a circular pore (operculum) or vertical split, the wall sometimes blueing in iodine, the asci ± globose and usually indehiscent in cleistothecial taxa; ascospores usually ellipsoidal, aseptate, hyaline to strongly pigmented, often ornamented, usually without a sheath. Anamorphs hyphomycetous where known, usually with sympodial proliferation. Saprobes on soil and very rotten wood, some coprophilous, some hypogeous (and then mycorrhizal); esp. temp.

Dennis (1978) and Korf (1973) accepted 7 fams.; Trappe (1979) added 5 fams. from the former *Tuberales*, and Malloch (1981) added two non-ostiolate fams. Kimbrough (*Mem. N.Y. bot. Gdn* **49**: 326, 1989) recognized 3 suborders: *Sarcoscyphineae* (asci suboperculate), *Pyronemineae* (asci operculate, dehiscence in exoascal layer, septal plugs of radiating tubules), and *Pezizineae* (asci operculate, dehiscence in endoascal layer, septal plugs not tubular). Eriksson (1982) adopted 17 fams. The *Monascaceae* is here referred to the *Eurotiales.* Fams:

(1) **Ascobolaceae**.
(2) **Ascodesmidaceae** (? syn. *Hemiascosporiaceae*).
(3) **Balsamiaceae**.
(4) **Carbomycetaceae**.
(5) **Eoterfeziaceae**.

(6) **Glaziellaceae**.
(7) **Helvellaceae**.
(8) **Karstenellaceae**.
(9) **Morchellaceae**.
(10) **Otideaceae** (syn. *Humariaceae*).
(11) **Pezizaceae** (syn. *Aleuriaceae* auct.).
(12) **Pyronemataceae**.
(13) **Sarcoscyphaceae**.
(14) **Sarcosomataceae**.
(15) **Terfeziaceae**.
(16) **Thelebolaceae**.
(17) **Tuberaceae**.
Lit.: *General*: van Brummelen (*Persoonia* 10: 113, 1978; *in* Reynolds, 1981; ascus ultrastr., *in* Hawksworth (Ed.), 1994: 303), Cabello (*Boln. Soc. Argent. Bot.* 25: 394, 1988; *Sarcoscyphineae*), Eckblad (*Nytt. Mag. Bot.* 15: 1, 1968; nomencl. gen., fam., etc.), Dennis (1968, 1978), Donadini (*Bull. Soc. Linn. Provence* 35: 53, 1984; tax. criteria, fams), Eriksson (1982), Harmaja (*Karstenia* 14: 109, 123, 1974; tetranucleate spored, cyanophilic perispored spp.), Hennebert & Bellemère (1979; anamorphs), Korf (*Mycol.* 64: 937, 1972; synoptic key, gen. names, *in* Ainsworth *et al.* (Eds), *The Fungi* 4A: 249, 1973; gen. keys, lit.), Kimbrough (*in* Reynolds (Ed.), 1981; centrum types), Malloch (*in* Kendrick (Ed.), *The whole fungus* 1: 153, 1979; *in* Reynolds, 1981, disposition cleistothecial gen.), Pfister (*Mycol.* 65: 326, 1973; psilopezioid gen.), Schumacher & Jenssen, *Arctic Alp. Fungi* 4, 1992; montane spp.); Svrček (*Česká Myk.* 30: 129, 135, 1976; *Sborník narod. Muz. Praza* 32B: 115, 1976; Velenovský spp.), Thind & Kaushal (*in* Subramanian (Ed.), *Taxonomy of fungi* 1: 283, 1978; tissue types), Trappe (*Mycotaxon* 9: 297, 1979; hypogeous taxa); Weber (*Bibl. Mycol.* 140, 1992; reprod. system).
Regional: **Australasia**: Rifai (*Verh. K. ned. Akad. wet.* ser. 2 57 (3), 1968). **Caribbean**: Pfister (*J. Agric. Univ. Puerto Rico* 58: 358, 1974). **Denmark**: Petersen (*Op. Bot.* 77, 1985; ecology). **N. America**: Larsen & Denison (*Mycotaxon* 7: 68, 1978; checklist), Seaver (1928). **Tierra del Fuego**: Gamundí (*Fl. cript. Tierra del Fuego* 10 (3), 1975). **Ukraine**: Smits'ka (*Petsitsovi gribi Ukraini*, 1975).
See also under *Ascomycota*, Discomycetes, Macrofungi, truffle.
Pezizasporites T.C. Huang (1981), Fossil fungi. 1 (Miocene), Taiwan.
Pezizella Fuckel (1870) = Calycina (Leot.) fide Baral (*SA* 13: 113, 1994). See Arendholz (*Mycotaxon* 36: 283, 1989; typific.).
Pezizella P. Karst. (1872) ≡ Pezizula (Thelebol.).
Pezizellaster Höhn. (1917) = Lachnum (Hyaloscyph.). See Dennis (*Mycol. Pap.* 32, 1949).
Pezizites Göpp. & Berendt (1845), Fossil fungi. 3 (Tertiary), Eur.
Pezizula P. Karst. (1871) = Thelebolus (Thelebol.) fide Kimbrough & Korf (1967).
Pezolepis Syd. (1925), Dermateaceae. 2, trop.
Pezoloma Clem. (1909), Leotiaceae. 5, N. temp. See Korf (*Phytol.* 21: 201, 1971).
Pezomela Syd. (1928), ? Leotiales (inc. sed.). 1, Chile.
PFGE, see pulsed field gel-electrophoresis.
Pfistera Korf & W.Y. Zhuang (1991), Otideaceae. 1, Canary Is.
pH, see Hydrogen-ion concentration.
Phacellium Bonord. (1860), Mitosporic fungi, 2.C2.10. 12, widespr. See Braun (*Nova Hedw.* 50: 499, 1990).
Phacellula Syd. (1927), Mitosporic fungi, 1.C1.?. 1, C. Am., Greece.
Phacidiaceae Fr. (1825), Leotiales. 4 gen. (+ 3 syn.), 9 spp. Stroma black, immersed (usually subcuticular), uniloculate, usually circular; ascomata apothecial,

without a separate wall, developing from central stromatal cells, opening by radial or longitudinal splits; interascal tissue of simple basal, and branched and anastomosing apical paraphyses; asci with an I+ apical ring; ascospores hyaline, aseptate, without a sheath. Anamorphs coelomycetous (*Phacidium*). Parasitic or saprobic on leaves, esp. of conifers, temp.
Sometimes confused with *Rhytismataceae* which do not have a distinct I+ blue ring.
Lit.: Di Cosmo *et al.* (*Mycotaxon* 21: 1, 1984).
Phacidiales, see *Rhytismatales*.
Phacidiella P. Karst. (1884), Mitosporic fungi, 8.A1.40. 3, widespr. See Sutton (1980).
Phacidiella Potebnia (1912) ≡ Potebniamyces (Cryptomycet.).
Phacidina Höhn. (1917), ? Leptopeltidaceae. 1, Eur. See Nannfeldt (*Nova Acta R. Soc. Scient. upsal.* 4, 8(2), 1932).
Phacidiopsis Hazsl. (1873) = Triblidium (Triblid.). See Saccardo (*Syll. Fung.* 8: 804, 1889).
Phacidiopsis Geyl. (1887), Fossil fungi. 1 (Tertiary), Indonesia. = Phacidites fide Meschinelli (1892).
Phacidiopycnis Potebnia (1912), Anamorphic Cryptomycetaceae, 8.A1.15. Teleomorph *Potebniamyces*. 3, widespr. *P. pseudotsugae* (syn. *Phomopsis pseudotsugae*), canker of *Pseudotsuga* and other conifers. See Sutton (1980), Rupprecht (*Sydowia* 13: 10, 1959).
Phacidiostroma Höhn. (1917) = Phacidium (Phacid.) fide Sherwood (*in litt.*).
Phacidiostromella Höhn. (1917) ? Ascomycota (inc. sed.).
Phacidites Mesch. (1892), Fossil fungi. 19 (Tertiary), Eur.
Phacidium Fr. (1815) typ. cons., Phacidiaceae. 5, temp. *P. infestans* (snow blight of conifers). See DiCosmo *et al.* (*Mycotaxon* 21: 1, 1984), Roll-Hansen (*Eur. J. Forest Path.* 17: 311, 1987, 19: 237, 1989; review).
Phacobolus Fr. (1849) = Stictis (Stictid.) fide Eriksson & Hawksworth (*SA* 5: 150, 1986).
Phacodothis, see *Placodothis*.
Phacopeltis Petch (1919) = Vizella (Vizell.) fide v. Arx & Müller (1975).
Phacopsis Tul. (1852), Lecanoraceae. 13 (on lichens), widespr. See Triebel *et al.* (*Bryologist* 98: 71, 1995; key), Hafellner (*Herzogia* 7: 343, 1987).
Phacorhiza Pers. (1822) = Typhula (Typhul.) fide Corner (1950).
Phacostroma Petr. (1955), Mitosporic fungi, 6.A1.15. 1, former Czechoslovakia.
Phacostromella Petr. (1955), Mitosporic fungi, 8.A1.15. 1, Germany.
Phacothecium Trevis. (1857) = Arthonia (Arthon.).
Phacotiella Vain. (1927) = Sphinctrina (Sphinctrin.) fide Tibell (1984).
Phacotium (Ach.) Trevis. (1821) ≡ Phacotrum (Calic.).
Phacotrum Gray (1821) = Calicium (Calic.) fide Tibell (1984).
Phaeangella (Sacc.) Massee (1895) ? = Phibalis (Leot.) fide Korf & Kohn (*Mem. N.Y. bot. Gdn* 28: 109, 1976).
Phaeangellina Dennis (1955), Leotiaceae. 1, Eur.
Phaeangium Pat. (1894), Otideaceae. 1, N. Afr., Arabian Peninsula. See Alsheikh & Trappe (*CJB* 61: 1923, 1983).
Phaeangium (Sacc.) Sacc. (1902) [non Pat. (1894)] = Pezicula (Dermat.) fide Nannfeldt (1932).
Phaeaspis Clem. & Shear (1931) = Vizella (Vizell.) fide v. Arx & Müller (1975).
Phaeaspis Kirschst. (1939) = Phomatospora (Ascomycota, inc. sed.) fide Francis (*Mycol. Pap.* 139, 1975).
Phaedropezia Le Gal (1953) = Acervus (Otid.) fide Pfister (*Occ. Pap. Farlow Herb.* 8, 1975).

Phaeidium Clem. (1931) ≡ Laaseomyces (Ceratostomat.).

phaeo- (prefix), dark-coloured or swarthy, esp. of spores; cf. Mitosporic fungi.

Phaeoannellomyces McGinnis & Schell (1985), Mitosporic fungi, 1.A1.19. 2 (from humans), widespr.

Phaeoantenariella Cavalc. (1969) nom. dub.; based on mycelium fide Hawksworth (*Bull. Br. Mus. nat. Hist., Bot.* **9**: 78, 1981).

Phaeoaphelaria Corner (1953), Aphelariaceae. 1, Australia.

Phaeoapiospora (Sacc. & P. Syd.) Theiss. & Syd. (1915) = Anisomyces (Vals.) fide Petrak (*Sydowia* **1**: 35, 1947).

Phaeobarlaea Henn. (1903) nom. nud. = Plicaria (Peziz.) fide Eckblad (1968).

Phaeobotryon Theiss. & Syd. (1915) = Botryosphaeria (Botryosphaer.) fide v. Arx & Müller (1975).

Phaeobotryosphaeria Speg. (1908) = Botryosphaeria (Botryosphaer.) fide v. Arx & Müller (1975).

Phaeobulgaria Seaver (1932) ≡ Bulgaria (Leot.) fide Korf (*Mycol.* **49**: 102, 1957).

Phaeocalicium Alb. Schmidt (1970), Mycocaliciaceae. c. 10, temp. See Tibell (1984), Titov (*Bot. Zhurnal* **71**: 384, 1986; key 4 spp.).

Phaeocapnias Cif. & Bat. (1963) = Euantennaria (Euantennar.) fide Hughes (1976).

Phaeocapnodinula Speg. (1924) = Phaeostigme (Pseudoperispor.) fide Hawksworth (*in litt.*).

Phaeocapnodinula Clem. & Shear (1931) ≡ Phaeocapnodinula (Pseudoperispor.).

Phaeocarpus Pat. (1887) [non Mart. & Zucc. (1824), *Sapindaceae*] ≡ Cymbella (Crepidot.).

Phaeochaetia Bat. & Cif. (1962) = Aithaloderma (Capnod.).

Phaeochora Höhn. (1909), ? Phyllachoraceae. 6 (on *Palmae*), trop. See Cannon (*SA* **11**: 181, 1993), Carrai-Giovanni d'Aghiano (*Sperimentozione Applicata* **4**: 4, 1992; col. pls), Müller (*Sydowia* **86**, 1965), *IMI Descr.* 1132.

Phaeochorella Theiss. & Syd. (1915), Phyllachoraceae. 2, S. Afr., Philipp. See Hyde (*Trans. mycol. Soc. Japan* **32**: 265, 1991).

Phaeociboria Höhn. (1918) = Lambertella (Sclerotin.) fide Dumont (1971).

Phaeociliospora Bat. & Peres (1967) = Ciliochorella (Mitosp. fungi) fide Sutton (1977).

Phaeoclavulina Brinkmann (1897) = Ramaria (Ramar.) fide Corner (1950).

Phaeococcomyces de Hoog (1979), Mitosporic fungi, 1.A1/2.3. 3, widespr.

Phaeococcus de Hoog (1977) [non Borzí (1892), *Algae*] ≡ Phaeococcomyces (Mitosp. fungi).

Phaeocollybia R. Heim (1931) nom. cons., Cortinariaceae. 45, temp, subtrop. See Horak (*Sydowia* **29**: 28, 1978), Singer (*Flora neotrop.* **4**: 3, 1970; monogr. neotrop. spp.), Smith & Trappe (*Mycol.* **64**: 1141, 1972; key N. Am. spp.), Singer (*Mycol. Helvet.* **2**: 247, 1987; key Costa Rica spp.).

Phaeoconis Clem. (1909) ≡ Nigrospora (Mitosp. fungi).

Phaeocoriolellus Kotl. & Pouzar (1957) = Gloeophyllum (Coriol.) fide David (1968).

Phaeocreopsis Sacc. & P. Syd. ex Lindau (1900), Dothideales (inc. sed.). 2, S. Am., Krakatau. See Kobayasi & Doi (*Mem. nat. Sci. Mus. Tokyo* **2**: 51, 1969).

Phaeocryptopus Naumov (1915), Venturiaceae. 5 or 6 (on conifers), Eur., N. Am. *P. gaeumannii* (Swiss needle cast of *Pseudotsuga taxifolia* in Eur.). See Butin (*Phytopath. Z.* **68**: 269, 1970), Stone & Carroll (*Sydowia* **38**: 317, 1986; ontogeny *P. gaeumannii*).

Phaeocyphella Pat. (1900) ≡ Chromocyphella (Crepidot.).

Phaeocyphella Speg. (1909) = Chromocyphella (Crepidot.) fide Singer (1975).

Phaeocyphellopsis W.B. Cooke (1961) = Merismodes (Crepidot.) fide Reid (1964).

Phaeocyrtidula Vain. (1921), ? Dothideales (inc. sed.). 2, Eur.

Phaeocyrtis Vain. (1921) = Merismatium (Verrucar.) fide Triebel (1989).

Phaeocytosporella G.L. Stout (1930) = Phaeocytostroma (Mitosp. fungi) fide Petrak (*Ann. Myc.* **39**: 252, 1941).

Phaeocytostroma Petr. (1921), Mitosporic fungi, 8.A2.15. 4, widespr. See Sutton (1980).

Phaeodactylella Udaiyan (1992), Mitosporic fungi, 1.C2.10. 1 (from water cooling tower), India.

Phaeodactylium Agnihothr. (1968), Mitosporic fungi, 1.A/C1.10. 1, India.

Phaeodaedalea M. Fidalgo (1962), Coriolaceae. 1, C. & S. Am.

Phaeodepas D.A. Reid (1961), Tricholomataceae. 1, Venezuela.

Phaeoderris (Sacc.) Höhn. (1907) = Leptosphaeria (Leptosphaer.) fide v. Arx & Müller (1975).

Phaeodiaporthe Petr. (1920) = Prosthecium (Melanconid.) fide Barr (*Mycol. Mem.* **7**, 1978).

Phaeodictyon M. Choisy (1929) = Anthracothecium (Pyrenul.).

Phaeodimeriella Speg. (1908), Pseudoperisporiaceae. c. 10 (on *Asterinaceae*, *Meliolaceae*), trop. See Hansford (1946).

Phaeodimeriella Theiss. (1912) = Phaeodimeriella Speg. (Pseudoperispor.).

Phaeodimeris Clem. & Shear (1931) ≡ Phaeodimeriella (Pseudoperispor.).

Phaeodiscula Cub. (1891), Mitosporic fungi, ?.?.?. 1, Eur. See Petch (*Ann. Bot., Lond.* **22**: 389, 1908).

Phaeodiscus L.R. Batra (1968) = Lambertella (Sclerotin.) fide Dumont (1971).

Phaeodochium M.L. Farr (1968) ? = Hymenopsis (Mitosp. fungi) fide Sutton (*in litt.*).

Phaeodomus Höhn. (1909), Mitosporic fungi, 8.A2.19. 1, Cuba, S. Am.

Phaeodon J. Schröt. (1888) = Hydnellum (Thelephor.) fide Donk (1956).

Phaeodothiopsis Theiss. & Syd. (1914), Ascomycota (inc. sed.). 3, trop. Nom. dub. fide Müller & v. Arx (1962).

Phaeodothiora Petr. (1948) = Saccothecium (Dothior.) fide Eriksson & Hawksworth (*in litt.*), v. Arx & Müller (1975).

Phaeodothis Syd. & P. Syd. (1904), Phaeosphaeriaceae. 2, widespr. See Aptroot (*Nova Hedw.* **60**: 325, 1995; key).

Phaeofabraea Rehm (1909), Leotiaceae. 2, S. Am. See Pfister (*Occ. Pap. Farlow Herb.* **11**, 1977; key).

Phaeogalera Kühner (1973), Strophariaceae. 3, Eur. See Pegler & Young (*Kew Bull.* **30**: 225, 1975; 3 Br. spp.), Reid (*TBMS* **82**: 228, 1984; 2 Br. spp.). = Galerina (Cortinar.) fide Singer (*Nova Hedw.* **26**: 436, 1975).

Phaeoglabrotricha W.B. Cooke (1961), Cyphellaceae. 8, widespr. = Pellidiscus (Cyphell.) fide Reid (1964).

Phaeoglaena Clem. (1909) nom. dub. (? Dothideales, inc. sed., ?L). No spp. included. See Mayrhofer & Poelt (*Herzogia* **7**: 13, 1985).

Phaeoglossum Petch (1922) [non Skottsb. (1907), *Algae*], Geoglossaceae. 1, Sri Lanka.

Phaeographidomyces Cif. & Tomas. (1953) = Phaeographis (Graphid.).

Phaeographina Müll. Arg. (1882), Graphidaceae (L). c. 90, cosmop. (mainly trop.). See Awasthi & Singh (*Kavaka* **1**: 87, 1974; India), Nakanishi (*J. Sci. Hiroshima Univ.* ser. B, div. 2 **11**: 51, 1966; Japan), Wirth

& Hale (*Contr. U.S. natn. Herb.* **36**: 63, 1963; C. Am.), Redinger (*Ark. Bot.* **26A** (1), 1933; S. Am.).

Phaeographinomyces Cif. & Tomas. (1953) = Phaeographina (Graphid.).

Phaeographis Müll. Arg. (1882), Graphidaceae (L). c. 200, cosmop. (mainly trop.). See Nakanishi (*J. Sci. Hiroshima Univ.* ser. B, div. 2 **11**: 51, 1966; Japan), Wirth & Hale (*Contr. U.S. natn. Herb.* **36**: 63, 1963; C. Am.), Redinger (*Ark. Bot.* **27A** (3), 1935; S. Am.), Singh & Awasthi (*Bull. Bot. Surv. India* **21**: 97, 1981; key 28 Indian spp.).

Phaeogyroporus Singer (1944), Gyrodontaceae. 4, trop. Afr., S. Am.

Phaeoharziella Loubière (1924) ? = Arthrinium (Mitosp. fungi) fide Sutton (*in litt.*).

Phaeohelotium Kanouse (1935), Leotiaceae. c. 12, N. temp. See Dennis (*Kew Bull.* **25**: 335, 1971; key Br. spp.). = Hyaloscypha (Leot.) fide Baral & Krieglsteiner (*Beih. Z. Mykol.* **6**, 1985).

Phaeohendersonia Höhn. (1918) = 'Hendersonia *sensu* Sacc.' (Mitosp. fungi) fide Sutton (*in litt.*).

Phaeohiratsukaea Udagawa & Iwatsu (1990), Mitosporic fungi, 1.A2.15. 1 (from a stained closet), Japan.

Phaeohydnochaete Lloyd (1916), Hymenochaetaceae. 1, N. Am.

Phaeohygrocybe Henn. (1901) = Russula (Russul.) fide Pegler (*in litt.*).

Phaeohymenula Petr. (1954), Mitosporic fungi, 3.A1.?. 1, Australia.

phaeohyphomycosis, a mycotic infection in humans or other animals caused by a dematiaceous fungus (Ajello *et al.*, *Mycol.* **66**: 490, 1974); cf. hyalohyphomycosis.

Phaeoisaria Höhn. (1909), Anamorphic Calosphaeriaceae, 2.A1.10. Teleomorph *Scoptria*. 6, widespr. See de Hoog & Papendorf (*Persoonia* **8**: 407, 1976; key).

Phaeoisariopsis Ferraris (1909), Mitosporic fungi, 2.E2.10. 10, widespr. See Ellis (*MDH*), Deighton (*MR* **94**: 1096, 1990; redisposition spp. from *Cercospora*), Walker & White (*MR* **95**: 1005, 1991; on *Wikstroemia*).

Phaeolabrella Speg. (1912), Mitosporic fungi, 6.A1.1. 1, S. Am. See Petrak & Sydow (*Ann. Myc.* **33**: 157, 1935).

Phaeolepiota Maire ex Konrad & Maubl. (1928), Cortinariaceae. 1, N. temp.

Phaeolimacium Henn. (1899) = Oudemansiella (Tricholomat.).

Phaeolopsis Murrill (1905), Coriolaceae. 1, trop. Am.

Phaeolus (Pat.) Pat. (1900), Coriolaceae. 2, widespr. See Pegler (1973).

Phaeomacropus Henn. (1899) = Helvella (Helvell.). See Eckblad (1968).

Phaeomarasmius Scherff. (1897), Cortinariaceae. 26, widespr. See Singer (*Schw. Z. f. Pilzk.* **34**: 44, 53, 1956; monogr.), Horak (*Sydowia* **32**: 167, 1979; key 8 Papua New Guinea spp.).

Phaeomarsonia, see *Phaeomarssonia*.

Phaeomarssonia Speg. (1908) nom. dub. (Mitosp. fungi). See Petrak & Sydow (*Ann. Myc.* **34**: 32, 1936).

Phaeomarssonia Bubák (1915) = Didymosporina (Mitosp. fungi) fide Sutton (1977).

Phaeomassaria Speg. (1880) ? = Massaria (Massar.).

Phaeomeris Clem. (1909) = Spheconisca (Verrucar.).

Phaeomonostichella Keissl. ex Petr. (1941), Mitosporic fungi, 4/6.A2.?. 1, China. See Sutton (1977).

Phaeomycena R. Heim ex Singer & Digilio (1952), Tricholomataceae. 1, Madagascar.

phaeomycosis, see phaeohyphomycoses.

Phaeonaevia L. Holm & K. Holm (1977), Dermateaceae. 1 (on *Rubus*), N. Eur.

Phaeonectria (Sacc.) Sacc. & Trotter (1913) = Nectria (Hypocr.) fide Samuels (*CJB* **51**: 1275, 1973).

Phaeonectriella R.A. Eaton & E.B.G. Jones (1971), Lasiosphaeriaceae. 1, UK. See Lowen (*SA* **5**: 150, 1986; posn).

Phaeonema Kütz. (1843) nom. dub. (? Fungi).

Phaeonematoloma (Singer) Bon (1994), Strophariaceae. 2, Eur.

Phaeopeltis Clem. (1909) ≡ Phaeosaccardinula (Chaetothyr.).

Phaeopeltis Petch (1919) ≡ Phaeaspis (Vizell.).

Phaeopeltium Clem. & Shear (1931) ≡ Phaeopeltosphaeria (Dothideales).

Phaeopeltosphaeria Berl. & Peglion (1892), Dothideales (inc. sed.). 2, Afr., Galapagos.

Phaeopeziza (Vido) Sacc. (1884) = Peziza (Peziz.) fide Eckblad (1968).

Phaeophacidium Henn. & Lindau (1897), Rhytismatales (inc. sed.). 1, Am.

Phaeophelaria Corner (1953), Clavariaceae. 1, Australia.

Phaeophlebia W.B. Cooke (1956) = Punctularia (Cortic.) fide Talbot (1958).

Phaeophleospora Rangel (1916) nom. dub. (Mitosp. fungi) fide Sutton (1977).

Phaeopholiota Locq. & Sarwal (1983), Agaricaceae. 1, Sikkim.

Phaeophomatospora Speg. (1909) = Anthostomella (Xylar.) fide Petrak & Sydow (*Ann. Myc.* **23**: 209, 1925).

Phaeophomopsis Höhn. (1917), Mitosporic fungi, 8.A2.?. 1, France. See Sutton (1977).

Phaeophragmeriella Hansf. (1944) = Leptomeliola (Pseudoperispor.) fide Hughes (1993).

Phaeophragmocauma F. Stevens (1931) = Dermatodothis (Dothideales) fide Müller (*Sydowia* **28**: 149, 1976).

Phaeophycopsis Bat. & Peres (1967) = Seuratia (Seurat.) fide Meeker (1975).

Phaeophyscia Moberg (1977), Physciaceae (L). 8, widespr. See Kashiwadani (*Bull. natn. Sci. Mag. Tokyo*, B, **10**: 127, 1984; Japan). = Physcia (Physc.) fide Hafellner *et al.* (*Herzogia* **5**: 39, 1979), but see Hale (*Lichenologist* **15**: 157, 1983).

Phaeopolynema Speg. (1912) = Hymenopsis (Mitosp. fungi) fide Sutton (1977).

Phaeopolystomella Bat. & H. Maia (1960) = Microdothella (Dothideales) fide v. Arx & Müller (1975).

Phaeoporotheleum (W.B. Cooke) W.B. Cooke (1961), Cyphellaceae. 2, Cuba, Argentina.

Phaeoporus J. Schröt. (1888) ? = Inonotus (Hymenochaet.) fide Donk (1974).

Phaeoporus J. Bataille (1908) ≡ Porphyrellus (Strobilomycet.).

Phaeopterula (Henn.) Sacc. & D. Sacc. (1905) = Pterula (Pterul.) fide Corner (1950).

Phaeopyxis Rambold & Triebel (1990), Leotiales (inc. sed.). 4 (mainly on lichens), N. temp.

Phaeoradulum Pat. (1900), Hydnaceae. 1, W. Indies.

Phaeoramularia Munt.-Cvetk. (1960), Mitosporic fungi, 1.C2.3. 25, widespr. See Ellis (*MDH*), Deighton (*Mycol. Pap.* **144**: 26, 1979).

Phaeorhytisma Henn. (1899) ? = Criella (Rhytismat.). Non-sporulating fide Müller (*Sydowia* **12**: 160, 1959).

Phaeorobillarda Bat. & J.L. Bezerra (1961) = Robillarda (Mitosp. fungi) fide Nag Raj (1993).

Phaeorrhiza H. Mayrhofer & Poelt (1979), Physciaceae (L). 2, Eur., N. Am., Asia, Antarctic.

Phaeosaccardinula Henn. (1905), Chaetothyriaceae. 14, trop. See Eriksson & Yue (*Mycotaxon* **22**: 269, 1985).

Phaeoschiffnerula Theiss. (1914) = Schiffnerula (Englerul.) fide Petrak [not traced].

Phaeoschizophyllum W.B. Cooke (1962) = Schizophyllum (Schizophyll.) fide Donk (1964).

Phaeosclera Sigler, Tsuneda & J.W. Carmich. (1981), Mitosporic fungi, 1.D2.1. 1, Canada.

Phaeosclerotinia Hori (1916), Sclerotiniaceae. 1, Japan. Anamorph *Monilia.*

Phaeoscopulariopsis Ota (1928) nom. provis. = Scopulariopsis (Mitosp. fungi) fide Morton & Smith (1963).

Phaeoscutella Henn. (1904) nom. dub. (Fungi, inc. sed.).

Phaeoscypha Spooner (1984), Hyaloscyphaceae. 1, UK. Anamorph *Chalara.*

Phaeoseptoria Speg. (1908), Mitosporic fungi, 4.C/E2.?. 17, widespr. See Punithalingam (*Nova Hedw.* **32**: 585, 1980; key), Petrak (*Ann. Myc.* **39**: 292, 1941), Knipscheer *et al.* (*S.Afr. For. Jl* **154**: 56, 1990; *P. eucalypti* in S. Afr.), Walker *et al.* (*MR* **96**: 911, 1992; redisp. of *P. eucalypti*, redescr. of *P. papayae*).

Phaeosiphonia Kütz. (1849) nom. dub. (? Fungi).

Phaeosolenia Speg. (1902), Crepidotaceae. 1, S. Am.

Phaeosperma Nitschke ex G.H. Otth (1869) [non *Phaiosperma* Raf. (1883)] = Camarops (Bolin.) fide Nannfeldt (1972).

Phaeosperma Nitschke ex Fuckel (1870), ? Dothideales (inc. sed.). 1 (on *Alnus*), Switzerland.

Phaeosperma (Sacc.) Traverso (1906) = Valsaria (Diaporthales) fide Shear (*Mycol.* **30**: 589, 1938).

Phaeosphaera Bat. & Cif. (1963) [non W. West & G. West (1903), *Algae*], Mitosporic fungi, 4.A2.?. 2, Philipp., USA.

Phaeosphaerella P. Karst. (1888) = Venturia (Ventur.) fide v. Arx & Müller (1975).

Phaeosphaeria I. Miyake (1909), Phaeosphaeriaceae. *c.* 45 (esp. on *Gramineae*), widespr. See Eriksson (*Ark. Bot.* II **6**: 339, 1967), Holm (1957), Koponen & Mäkelä (*Annls bot. fenn.* **12**: 141, 1975; Finland), Otani (*Bull. natn. Sci. Mus. Tokyo* B **2**: 87, 1976; Japan), Leuchtmann (*Sydowia* **37**: 75, 1984; key 45 spp., *in* Laursen *et al.* (Eds), *Arctic and alpine mycology* II: 153, 1987; key arctic alpine 16 spp.), Shoemaker & Babcock (*CJB* **67**: 1500, 1989; monogr., keys). Anamorphs *Scolecosporiella, Stagonospora.*

Phaeosphaeriaceae M.E. Barr (1979), Dothideales. 18 gen. (+ 8 syn.), 133 spp. Ascomata perithecial, immersed to erumpent, gregarious or occasionally as locules in small stromata, ± globose, with a well-developed lysigenous often periphysate ostiole; peridium soft, composed of relatively small thin-walled pseudoparenchymatous cells; interascal tissue sparse, of narrowly cellular pseudoparaphyses; asci cylindrical, fissitunicate; ascospores brown, transversely septate, constricted only at the primary septum, occasionally muriform, sometimes with a sheath. Anamorphs coelomycetous. Necrotrophic or saprobic on a wide range of plants, occasionally on other fungi, cosmop. See Barr (*Mycotaxon* **43**: 371, 1992; keys 14 N. Am. gen.).

Phaeospora Hepp ex Stein (1879), ? Verrucariaceae. 10 (on lichens), widespr.

Phaeosporella Keissl. (1922) ≡ Phaeosphaerella (Ventur.).

Phaeosporis Clem. (1909), ? Sordariales (inc. sed.). 1, France. See Hawksworth (*SA* **6**: 145, 1987), Krug *et al.* (*Mycol.* **86**: 581, 1994).

Phaeosporobolus D. Hawksw. & Hafellner (1986), Mitosporic fungi, 8.A2.3. 1 (on lichens), Eur., N. & S. Am., Australia, Antarctica.

Phaeostagonosporopsis Woron. (1925) = Stenocarpella (Mitosp. fungi) fide Sutton (1977).

Phaeostalagmus W. Gams (1976), Mitosporic fungi, 1.A1.15. 6, widespr. See Sutton & Melnik (*MR* **96**: 908, 1992).

Phaeosticta Trevis. (1869) nom. rej. = Pseudocyphellaria (Lobar.).

Phaeostigme Syd. & P. Syd. (1917), Pseudoperisporiaceae. *c.* 20, trop. See Hughes (*Mycol. Pap.* **166**, 1993).

Phaeostilbella Höhn. (1919) = Myrothecium (Mitosp. fungi) fide Tulloch (1972).

Phaeostoma Arx & E. Müll. (1954) [non Spach (1835), *Onagraceae*] ≡ Arxiomyces (Ceratostomat.).

Phaeostomiopeltis Bat. & Cavalc. (1963) = Haplopeltheca (Micropeltid.) fide v. Arx & Müller (1975).

Phaeotellus Kühner & Lamoure (1972), Coriolaceae. 3, Eur. = Arrhenia (Tricholomat.) fide Redhead (1984).

Phaeotheca Sigler, Tsuneda & J.W. Carmich. (1981), Mitosporic fungi, 3.A2.endoconidia. 2, Canada. See DesRochers & Ouellette (*CJB* **72**: 808, 1994; sp. inhibiting *Ophiostoma ulmi*).

Phaeothecium, see *Phacothecium.*

Phaeothrombis, see *Phaeotrombis.*

Phaeothyriolum Syd. (1938), Microthyriaceae. 1 (on *Eucalyptus*), Australia. See Swart (*TBMS* **87**: 81, 1986).

Phaeothyrium Petr. (1947), Mitosporic fungi, 5.A2.?. 1, China.

Phaeotomasellia Katum. (1981), Dothideales (inc. sed.). 1, Uganda.

Phaeotrabutia Orejuela (1941) = Phyllachora (Phyllachor.) fide v. Arx & Müller (1954).

Phaeotrabutiella Theiss. & Syd. (1915) = Phyllachora (Phyllachor.) fide v. Arx & Müller (1954).

Phaeotrametes Lloyd ex J.E. Wright (1966), Polyporaceae. 1, S. Hemisph.

Phaeotrema Müll. Arg. (1887) nom. cons., ? Thelotremataceae (L). *c.* 30, cosmop. (mainly trop.). See Redinger (*Ark. Bot.* **28A** (8), 1936; S. Am.). = Thelotrema (Thelotremat.) fide Salisbury (*Lichenologist* **7**: 59, 1975).

Phaeotremella Rea (1912) = Tremella (Tremell.) fide Donk (*Taxon* **7**: 238, 1958).

Phaeotrichaceae Cain (1956), Dothideales. 2 gen., 5 spp. Ascomata perithecial or cleistothecial, black, thin-walled, where ostiolate with a well-developed non-periphysate pore, setose; peridium thin, composed of small pseudoparenchymatous cells; interascal tissue absent or of ± evanescent cellular pseudoparaphyses; asci saccate and evanescent or cylindrical and fissitunicate; ascospores dark brown, septate, sometimes fragmenting, with terminal germ pores, sometimes with a sheath. Anamorphs unknown. Coprophilous, widespr.

Phaeotrichoconis Subram. (1956), Mitosporic fungi, 1.C2.26. 1, India.

Phaeotrichosphaeria Sivan. (1983), Lasiosphaeriaceae. 3, India, UK, NZ. See Barr (*Mycotaxon* **39**: 43, 1990; posn). Anamorph *Endophragmiella.*

Phaeotrichum Cain & M.E. Barr (1956), Phaeotrichaceae. 2, widespr.

Phaeotrombis Clem. (1909) = Thrombium (Thromb.).

Phaeotrype Sacc. (1920) = Diatrype (Diatryp.) fide Petrak (*Ann. Myc.* **23**: 46, 1925).

Phaeoxyphiella Bat. & Cif. (1963), Anamorphic Capnodiaceae, 4.C2.?. Teleomorph *Capnodium.* 7, widespr. See Hughes (*Mycol.* **68**: 693, 1976).

Phaeoxyphium Bat. & Cif. (1963) nom. dub. fide Sutton (*Mycol. Pap.* **141**, 1977).

Phaffia M.W. Mill., Yoney. & Soneda (1976), Mitosporic fungi, 1.A1.3/19. 1, Japan, USA.

Phagodinium Kristiansen (1993) nom. dub. (? Fungi, inc. sed.).

Phagomyxa Karling (1944), Plasmodiophorales (inc. sed.). 1, USA.

phagosome, a membrane surrounding an endosymbiont to form a distinctive structure, as *Nostoc*-containing vesicles of *Geosiphon*.

phagotrophic, feeding by ingestion, engulfing food.

Phakopsora Dietel (1895), Phakopsoraceae. 50, widespr. trop. *P. vitis* (rust of *Vitis*); *P. gossypii* (of cotton). See Cummins & Hiratsuka (1983), Thirumalachar & Kern (*Mycol.* **41**: 283, 1949), Ono *et al.* (*MR* B**96**: 825, 1992).

Phakopsoraceae (Arthur) Cummins & Hirats. f. (1983), Uredinales. 17 gen. (+ 8 syn.), 116 spp. Pycnia discoid, with bounding structures, subepidermal or subcuticular; aeciospores catenulate (*Aecidium* or *Caeoma*) or pedicellate (*Uraecium*); urediniospores borne singly, (shortly) pedicellate, uredinia paraphysate; teliospores sessile, usually catenulate and unicellular, adhering in crusts several cells deep, usually pigmented.

Phalacrichomyces R.K. Benj. (1992), Laboulbeniaceae. 2, Venezuela.

phalacrogenous, of conidiogenous cells arising at the same level from single hyphae to form a turf-like layer; velvety. See Gams (*Cephalosporium-artige Schimmelpilze*, 1971).

Phalangispora Nawawi & J. Webster (1982), Mitosporic fungi, 3.G1.10. 2, Malaysia. See Kuthubutheen (*TBMS* **89**: 414, 1987).

Phallaceae Corda (1842), Phallales. 7 gen. (+ 26 syn.), 35 spp. Gasterocarp epigeous, unipileate; receptacle cylindrical, unbranched; gleba external.
Lit.: Boedijn (*Bull. Jard. bot. Buitenz.* ser. 3 **12**: 71, 1932; Indonesia).

Phallales, Basidiomycetes ('gasteromycetes'). 6 fam., 32 gen. (+ 52 syn.), 137 spp. Basidioma at first subglobose, ovoid, or pyriform and limited by a peridium covering a more or less well developed gelatinous layer; subsequent development differs; widespr., esp. Australasia and trop.; saprobic in soil and rotting wood. Fams:
(1) **Clathraceae**.
(2) **Claustulaceae**.
(3) **Gelopellaceae**.
(4) **Hysterangiaceae**.
(5) **Phallaceae**.
(6) **Protophallaceae**.
The *Clathraceae* and *Phallaceae* comprise the Stink-horns, the gleba frequently having an offensive smell attractive to the insects by which the smooth, hyaline, bacilliform spores are dispersed.
Lit.: Fischer (*Nat. Pflanzenfam.* **7A**: 76, 1933) summarizes his own important contributions and those of his contemporaries. Burk (*Bibl. Mycol.* **73**, 1980; bibliogr. N. Am.), Dring (1973; key gen.). See also Lit. under *Gasteromycetes*.

Phallobata G. Cunn. (1926) = Hysterangium (Hysterang.) fide Cunningham (*Gast. Austr. N.Z.*: 65, 1944).

Phalloboletus Adans. (1763) ≡ Morchella (Morchell.).

Phallogaster Morgan (1893), Hysterangiaceae. 1, N. Am., Eur. (? introd.).

phalloid, one of the *Phallales*.

Phalloidastrum Batt. (1755) nom. inval. = Phallus (Phall.).

phallolysin, a protein of *Amanita phalloides*, has cytolytic effects *in vitro*, and is toxic to animals (Odenthal *et al.*, *Naunyn-Schmiederberg's Arch. Pharmacol.* **290l**: 133, 1975).

Phallomyces Bat. & Valle (1961), Mitosporic fungi (L), 8.E1.?. 1, Brazil.

phallotoxins, cyclic heptapeptides (phallcidin, phalloidin, phallicin, phallicidin, phallin B) toxic to humans from *Amanita phalloides* etc., see Wieland

(*Science* **159**: 950, 1968; *Peptides of Amanita mushrooms*, 1986). Cf. amatoxins.

Phallus Junius ex L. (1753), Phallaceae. 18, widespr. The Stinkhorns.

Phalodictyum Clem. (1909) = Rhizocarpon (Rhizocarp.).

Phalomia Nieuwl. (1916) ≡ Omphalia (Tricholomat.).

Phalostauris Clem. (1909) ≡ Willeya (Verrucar.).

Phalothrix Clem. (1909) = Unguicularia (Hyaloscyph.) fide Nannfeldt (1932).

Phaneroascus Baudyš (1919) = Cookella (Cookell.) fide v. Arx (1963).

Phanerochaete P. Karst. (1889), Meruliaceae. *c.* 50, widespr. See Donk (*Persoonia* **2**: 223, 1962), Eriksson *et al.* (*Taxon* **27**: 299, 1978; nomencl.), Eriksson *et al.* (*Corticiaceae of North Europe* **5**: 987, 1978; key 12 Eur. spp.), Burdsall (*Mycol. Mem.* **10**: 165, 1985; key 46 spp.), Wu (*Acta Bot. Fenn.* **142**: 37, 1990; key Taiwan spp.).

Phanerococculus Cif. (1954), Anamorphic Dothideales, 4.C1.?. Teleomorph *Koordersiella*. 1, Santa Domingo.

Phanerococcus Theiss. & Syd. (1918) ? = Koordersiella (Dothideales) fide Hansford (*Mycol. Pap.* **15**, 1946).

Phanerocorynella Höhn. (1923) ≡ Exosporiella (Mitosp. fungi).

Phanerocoryneum Höhn. (1923) = Clasterosporium (Mitosp. fungi) fide Sutton (*Mycol. Pap.* **141**, 1977).

Phaneromyces Speg. & Hara (1889), ? Patellariaceae. 1, S. Am.

phaneroplasmodium, see plasmodium.

Phanosticta Clem. (1909) = Pseudocyphellaria (Lobar.).

Phanotylium Clem. (1909) = Tremotylium (Thelotremat.).

Pharcidia Körb. (1865) = Stigmidium (Mycosphaerell.) fide Santesson (*Svensk bot. Tidskr.* **54**: 499, 1960).

Pharcidiella (Sacc.) Clem. & Shear (1931) = Phaeospora (Verrucar.) fide Hawksworth (*in litt.*).

Pharcidiopsis Sacc. (1905) = Stigmidium (Mycosphaerell.). See Keissler (*Rabenh. Krypt.-Fl.* **8**, 1930).

pharmaceutical prospecting, see Bioprospecting, Biotechnology, Screening.

Pharus Petch (1919) [non P. Browne (1756), *Gramineae*] ≡ Mycopharus (Clathr.).

Phascolomyces Boedijn (1959), Thamnidiaceae. 1, C. Am., Taiwan, Java. See Benny & Benjamin (*Aliso* **8**: 391, 1976), Jeffries & Young (*CJB* **56**: 747, 1978; sporangiolum ultrastr., **56**: 2449, 1978; response to mycoparasitism), Balasubramanian & Manocha (*CJB* **64**: 2441, 1986; biochemistry).

phaseolin, a phytoalexin (q.v.) from bean (*Phaseolus vulgaris*).

Phasya Syd. (1934) = Venturia (Ventur.) fide Müller & v. Arx (1962).

Phaulomyces Thaxt. (1931), Laboulbeniaceae. 8, N. Am., Pacific, Indonesia.

Phellinidium (Kotl.) Fiasson & Niemalä (1984), Hymenochaetaceae. 2, Eur.

Phellinites Singer & S. Archang. (1958), Fossil fungi (Basidiomycota). 1 (Jurassic), Argentina.

Phellinus Quél. (1886), Hymenochaetaceae. *c.* 220, widespr. See Larsen & Coldo-Poulle (*Synopsis Fung.* **3**, 1990; world key), Ryvarden & Johansen (*Preliminary polypore flora of East Africa*: 129, 1980; key 62 Afr. spp.), Wright & Blumenfeld (*Mycotaxon* **21**: 413, 1984; key 26 spp. Argentina).

Phellodon P. Karst. (1881), Bankeraceae. 13, Eur., N. Am. See Stalpers (*Stud. Mycol.* **35**, 1993; key).

Phellomyces A.B. Frank (1898) = Colletotrichum (Mitosp. fungi) fide Husz (*Z. Pflanzenkr. Pflanzensch.* **44**: 186, 1934).

Phellomycetes Renault (1896), Fossil fungi (mycel.). 1 (Carboniferous), France.

Phellomycites Mesch. (1896) ≡ Phellomycetes (Fossil fungi).

phellophagy, ability to attack cork cells (Speer, *Mycotaxon* **21**: 235, 1984).

Phellorinia Berk. (1843), Phelloriniaceae. 1, warm dry areas. See Malençon (*Ann. Crypt. Exot.* **8**: 5, 1935), Long (*Lloydia* **9**: 132, 1946), Kreisel (*Česká Myk.* **15**: 195, 1961).

Phelloriniaceae Ulbr. (1951), Tulostomatales. 2 gen. (+ 5 syn.), 2 spp. Gasterocarp stipitate; basidia persisting in bundles in the mature gleba.

Phellostroma Syd. & P. Syd. (1914), Ascomycota (inc. sed.). 1, Philipp.

Phelonites Fresen. (1861), Fossil fungi. 1 (Miocene), Germany.

Phelonitis Chevall. (1826) ? Ascomycota (inc. sed.).

Phenacopodium Debey (1849) = Melanospora (Ceratostomat.) fide Mussat (*Syll. Fung.* **15**: 279, 1901).

Pherima Raf. (1819) ≡ Phorima (Coriol.).

pheromone, a substance secreted to the outside by an individual and received by a second individual of the same species, in which it induces a specific reaction, e.g. a definite behaviour or developmental process (Karlson & Luscher, *Nature* **183**: 55, 1959).

Phiala Raf. (1815) nom. dub. (Fungi, inc. sed.). No spp. included.

Phialastrum Sunhede (1989), Geastraceae. 1, trop. Afr.

Phialea (Fr.) Gillet (1879) nom. dub. (Fungi, inc. sed.). See Dumont & Korf (*Taxon* **26**: 598, 1977).

Phialemonium W. Gams & McGinnis (1983), Mitosporic fungi, 1.A1.15. 3, USA. See King *et al.* (*J. Clin. Microbiol.* **31**: 1804, 1993; re-evaluation).

Phialetea Bat. & Nascim. (1960) = Grallomyces (Mitosp. fungi) fide Deighton & Pirozynski (*Mycol. Pap.* **105**, 1966).

Phialicorona Subram. (1993) = Kionochaeta (Mitosp. fungi) fide Sutton (*in litt.*).

phialide (after Vuillemin), a cell which develops one or more (the **polyphialide** of Hughes, *Mycol. Pap.* **45**, 1951) open ended conidiogenous loci from which a basipetal succession of conidia, **phialospores**, develops without an increase in length of the phialide itself (Hughes, *loc. cit.*); cf. annellophore; sterigma. In some fungi, e.g. *Acremonium*, the phialide may be the conidiophore; more frequently the phialide is either an end cell of a conidiophore or attached to a conidiophore (or **phialophore**). See Roquebert (*Rev. Myc.* **40**: 417, 1976; review); terminus phialospore, and Minter *et al.* (*TBMS* **81**: 109, 1983; history).

phialidic (of conidiogenesis, obsol.), the sort of conidiogenesis in which each conidium (**phialoconidium**, phialidic conidium, phialospore) originates by the laying down of new wall material not from existing walls or layers of the wall of the conidiogenous cell (**phialide**). A *basipetal* succession of conidia is formed from a *fixed* conidiogenous locus (cf. tretic). **mono-, poly-**, (of phialides), producing conidia through a single opening or a sympodial irregular or synchronous succession of openings, respectively, in the conidiogenous cell wall.

Phialina Höhn. (1926) = Calycellina (Hyaloscyph.) fide Baral (*Z. Mykol.* **59**: 3, 1993). See also Huhtinen (*Karstenia* **29**: 545, 1990; key 8 spp.).

Phialisphaera Dumort. (1822) nom. dub. (Ascomycota, inc. sed.).

Phialoarthrobotryum Matsush. (1975), Mitosporic fungi, 2.C2.15. 1, Japan.

Phialoascus Redhead & Malloch (1977), Cephaloascaceae. 1, Canada.

Phialocephala 1W.B. Kendr. (1961), Mitosporic fungi, 1.A2.15. 15, widespr. See Wingfield *et al.* (*TBMS* **89**:

509, 1987; classification), Currah & Tsuneda (*Trans. mycol. Soc. Jap.* **34**: 345, 1993; sporulation *P. fortinii* in culture), Onofri *et al.* (*MR* **98**: 745, 1994; key); Kowalski & Kehr (*CJB* **73**: 26, 1995; n.spp.).

Phialocladus Kreisel (1972) nom. inval. ≡ Escovopsis (Mitosp. fungi).

Phialoconidiophora M. Moore & F.P. Almeida (1937) = Rhinocladiella (Mitosp. fungi) fide de Hoog (1977).

Phialocybe P. Karst. (1879) ? = Crepidotus (Crepidot.) fide Singer (1951).

Phialogangliospora Udaiyan & V.S. Hosag. (1992), Mitosporic fungi, 1.A1.40 + A1.15. 1 (from cooling tower), India.

Phialogeniculata Matsush. (1971), Mitosporic fungi, 1.C1.15. 1, Guadaloupe. = Dictyochaeta (Mitosp. fungi) fide Kuthubutheen & Nawawi (1991).

Phialographium H.P. Upadhyay & W.B. Kendr. (1974), Anamorphic Ophiostomataceae, 2.A1.15. Teleomorph *Ophiostoma*. 1, Canada. = Graphium (Mitosp. fungi) fide Wingfield *et al.* (*MR* **95**: 1328, 1991).

Phialomyces P.C. Misra & P.H.B. Talbot (1964), Mitosporic fungi, 1.A1/2.32/33. 2, NZ, India, Cuba.

Phialophaeoisaria Matsush. (1995), Mitosporic fungi, 2.A1.16. 1, Japan.

Phialophora Medlar (1915), Anamorphic Magnaporthaceae, 1.A1/2.15. Teleomorph *Ophiobolus*. 12, widespr. *P. verrucosa* (a cause of chromoblastomycosis (q.v.) in humans), *P. cinerescens* (fan mould of carnation). See Schol-Schwarz (*Persoonia* **6**: 63, 1970; key), Gams & Holubová-Jechová (*Stud. Mycol.* **13**, 1976), Yamamoto *et al.* (*Ann. Phytopath. Soc. Jap.* **56**: 584, 1990; isozyme polymorphism in *P. gregata*), Kobayashi *et al.* (*Ann. Phytopath. Soc. Jap.* **57**: 225, 1991; f.spp. in *P. gregata*), Yamamoto *et al.* (*Jl mycol. Soc. Jap.* **34**: 465, 1993; RFLPs in *P. gregata*), Yan *et al.* (*Mycol.* **87**: 72, 1995; rDNA supports morphol. spp. separation).

phialophore, see phialide.

Phialophorophoma Linder (1944), Mitosporic fungi, 4.A1.15. 1 (marine), USA.

Phialophoropsis L.R. Batra (1968), Mitosporic fungi, 3.A1.15. 2, USA, UK. See Sutton & Brady (*TBMS* **72**: 337, 1979).

Phialopsis Körb. (1855) = Gyalecta (Gyalect.).

Phialoscypha Raitv. (1977) = Phialina (Hyaloscyph.) fide Huhtinen (1990).

Phialoselanospora Udaiyan (1992), Mitosporic fungi, 1.A1.15. 1 (from water cooling tower), India.

Phialospora Raf. (1832) ≡ Cucurbitaria (Cucurbitar.).

phialospore, see phialide.

Phialosporostilbe Mercado & J. Mena (1985), Mitosporic fungi, 2.A1.15. 1, Cuba.

Phialostele Deighton (1969), Mitosporic fungi, 2/3.A1.15. 1, Afr.

Phialotubus R.Y. Roy & Leelav. (1966), Mitosporic fungi, 1.A1.32/33. 1, India.

Phibalis Wallr. (1833) nom. rej. = Encoelia (Leot.). fide Korf & Kohn (*Mem. N.Y. bot. Gdn* **28**: 109, 1976).

Philately. More than 650 postage stamps, of 102 countries, illustrating 350 spp. fungi issued up to 1991 are catalogued by Greenewich (*Collect fungi on stamps*, 1991). Macromycetes predominate but some medically important fungi, lichens and mycorrhizas are covered. See also Ing (*Bull. BMS* **10**: 32, 1976), Moss & Dunkley (*Bull. BMS* **15**: 61, 1981), Moss (*Mycologist* **6**: 68, 1992, **7**: 28, 1993), Coetzee (*Mycologist* **7**: 29, 1993).

Phillippiregis Cif. & Tomas. (1953) ≡ Polyblastidea (Verrucar.).

Phillipsia Berk. (1881) nom. cons., Sarcoscyphaceae. *c.* 10, warmer areas. See Boedijn (*Bull. Jard. bot. Buitenz.* **13**: 58, 1933; **16**: 358, 1940), Denison

(*Mycol.* **61**: 289, 1969; key C. Am. spp.), Le Gal (1953), Paden (*Mycotaxon* **25**: 165, 1986; anamorph), Romero & Gamundí (*Darwiniana* **27**: 43, 1986; key 4 spp. Argentina, SEM). Anamorph *Molliardiomyces*.

Phillipsiella Cooke (1878), Phillipsiellaceae. 5, Am. See Rossman *et al.* (*Sydowia* **46**: 66, 1994).

Phillipsiellaceae Höhn. (1909), Ascomycota (inc. sed.). 1 gen. (+ 1 syn.), 5 spp. Stromata absent; ascomata small, pulvinate, ? apothecial, greenish black, sometimes surrounded by weakly developed mycelium; peridium scarcely developed, of a few rows of pseudoparenchymatous cells; interascal tissue of narrow trabeculate pseudoparaphyses which are enlarged at the apex to form an epithecium-like layer; asci saccate, sessile, thick-walled at least at the apex, with a wide ocular chamber, not fissitunicate; ascospores hyaline to brown, simple, transversely septate or muriform. Anamorphs unknown. Saprobic, epiphytic on leaves, sometimes associated with leaf hairs, Am.

Philliscidiopsis, see *Phylliscidiopsis*.

Philobryon Döbbeler (1988), Dothideales (inc. sed.). 1 (on *Bryophyta*), Papua New Guinea.

Philocopra Speg. (1880) = Podospora (Lasiosphaer.) fide Lundqvist (1972).

Philonectria Hara (1914), Dothideales (inc. sed.). 3, Japan, Uganda. See Eriksson & Hawksworth (*SA* **6**: 146, 1987).

Philophora Wallr. (1833) = Rhizopus (Mucor.) fide Hesseltine (1955).

Phlebia Fr. (1821), Meruliaceae. *c.* 50, widespr. See Cooke (*Mycol.* **48**: 386, 1956), Donk (*Fungus* **27**: 8, 1957), Wu (*Acta Bot. Fenn.* **142**: 25, 1990; key Taiwan spp.), Corner (*Gdn's Bull. Singapore* **25**: 355, 1971; Malaysian spp.), Ginns (*CJB* **54**: 100, 1976; disposition spp.), Nakasone & Burdsall (*Mycotaxon* **21**: 241, 1984; synonymy with *Merulius*).

Phlebiella P. Karst. (1890), Xenasmataceae. *c.* 12, widespr. See Hjortstam *et al.* (*Corticiaceae of northern Europe* **8**, 1988; key Eur. spp.).

Phlebiopsis Jülich (1978), Meruliaceae. 3, widespr.

Phlebogaster Fogel (1980), Hysterangiaceae. 1, Canary Isl.

Phlebomarasmius R. Heim (1967) ? = Xeromphalina (Tricholomat.) fide Singer (1975).

Phlebomorpha Pers. (1822) nom. dub.; based on myxomycete plasmodia fide Martin (1966).

Phlebomycena R. Heim (1966) = Mycena (Tricholomat.) fide Singer (1951).

Phlebonema R. Heim (1929), Agaricaceae. 1, Madagascar.

Phlebophora Lév. (1841) nom. dub.; based on a deformed *Tricholoma* (Tricholomat.) fide Boedijn (*Sydowia* **5**: 211, 1951).

Phlebophyllum R. Heim (1969) [non Nees (1832), *Acanthaceae*], ? Strophariaceae. 1, Gabon.

Phlebopus (R. Heim) Singer (1936), Gyrodontaceae. 12, pantrop. See Groves (*Mycol.* **54**: 319, 1962, *Fl. Illustr. Champ. Afr. centr.* **7**: 128, 1980, *Mycotaxon* **15**: 384, 1982).

Phleboscyphus Clem. (1903) = Helvella (Helvell.) fide Eckblad (1968).

Phlebriella, see *Phlebiella*.

Phlegmacium (Fr.) Wünsche (1877) = Cortinarius (Cortinar.) fide Kauffman (1924), but used by Moser (1978).

Phlegmatium Fr. (1819) nom. dub. ('basidiomycetes', inc. sed.). See Horníček (*Mykol. Sborn.* **4**: 121, 1984).

Phlegmographa A. Massal. (1860) = Sarcographa (Graphid.).

Phlegmophiale Zahlbr. (1926) = Arthonia (Arthon.) fide Santesson (1952).

Phleogena Link (1833), Phleogenaceae. 1, N. temp. See McNabb (*in* Ainsworth *et al.*, *The Fungi* **4B**, 1973), Donk (*Persoonia* **4**: 160, 1966).

Phleogenaceae Gäum. (1926), Atractiellales. 2 gen. (+ 4 syn.), 4 spp. Basidioma stilboid; hyphae with simple septal pores, clamp-connexions present; spores brown, thick-walled.
Lit.: Oberwinkler & Bauer (*Sydowia* **41**: 224, 1989).

Phloeochora Höhn. (1917) = Phloeospora (Mitosp. fungi) fide Sutton (1977).

Phloeoconis Fr. (1849), ? Mitosporic fungi. 3, temp.

Phloeopannaria Zahlbr. (1941) = Psoroma (Pannar.).

Phloeopeccania J. Steiner (1902), Lichinaceae (L). 1, Arabia.

Phloeopeccaniomyces Cif. & Tomas. (1953) ≡ Phloeopeccania (Lichin.).

Phloeophthora Kleb. (1905) = Phytophthora (Pyth.) fide Klebahn (1909).

Phloeoscoria Wallr. (1825) ≡ Polymorphum (Mitosp. fungi).

Phloeospora Wallr. (1833), Anamorphic Mycosphaerellaceae, 6.C1.19. Teleomorph *Mycosphaerella*. 160, widespr. See Sutton (1980).

Phloeosporella Höhn. (1924), Anamorphic Dermateaceae, 6.C1.10. Teleomorph *Blumeriella*. 5, widespr. See Sutton (1980).

Phloeosporina Höhn. (1924), Mitosporic fungi, 6.C1.?. 1, former USSR.

Phlogiotis Quél. (1886) = Tremiscus (Exid.) fide Donk (*Persoonia* **4**: 185, 1966).

Phlyctaena, see *Phlyctema*.

Phlyctaeniella Petr. (1922), Mitosporic fungi, 8.E1.15. 1 or 2, Eur., Australia.

Phlyctella Kremp. (1876) = Phlyctis (Phlyctid.) fide Galloway & Guzmán Grimaldi (1988).

Phlyctellomyces Cif. & Tomas. (1953) ≡ Phlyctella (Phlyctid.).

Phlyctema Desm. (1847), Anamorphic Dermateaceae, 8.A1.15. Teleomorph *Pezicula*. 30, widespr. See Sutton (1980), Spiers & Hopcroft (*N.Z. Jl Bot.* **28**: 67, 1990; TEM *P. vagabunda*).

Phlyctibasidium Jülich (1974), Thelephoraceae. 1, Eur.

Phlyctidaceae Poelt ex J.C. David & D. Hawksw. (1991), Ascomycota (inc. sed., L). 2 gen. (+ 9 syn.), 11 spp. Thallus crustose, sometimes areolate; ascomata apothecial, immersed, sometimes emergent, cupulate, thin-walled; interascal tissue of branched and anastomosed paraphyses; asci clavate, severallayered, with an I+ apical cap and an outer I+ gelatinized layer, 1- to 8-spored; ascospores hyaline, transversely septate or muriform. Anamorphs pycnidial. Lichenized with green algae, usually on bark, esp. trop.

Phlyctidia Müll. Arg. (1895) = Phlyctis (Phlyctid.) fide Galloway & Guzmán Grimaldi (1988).

Phlyctidiaceae, see *Chytridiaceae*.

Phlyctidium Wallr. (1833) = Spilocaea (Mitosp. fungi) fide Sutton (*Mycol. Pap.* **141**, 1977).

Phlyctidium (A. Braun) Rabenh. (1868) [non Wallr. (1833)] = Rhizophydium (Chytrid.) fide Karling (1977). See Sparrow (1960: 211; key 17 N. temp spp.).

Phlyctidium Müll. Arg. (1888) [non Wallr. (1833)] = Calenia (Gomphill.).

Phlyctidomyces E.A. Thomas ex Cif. & Tomas. (1953) = Phlyctis (Phlyctid.).

Phlyctis (Wallr.) Flot. (1850) [non Rafin. (1810), *Algae*], nom. cons., Phlyctidaceae (L). *c.* 10, Eur., Asia, N. Am. See Erichsen (*Hedwigia* **70**: 216, 1930), Galloway & Guzmán Grimaldi (*Lichenologist* **20**: 393, 1988).

Phlyctochytrium J. Schröt. (1892), Chytridiaceae. 31, temp. See Sparrow (1960: 324; key), Barr (*CJB* **62**: 1171, 1984).

Phlyctomia A. Massal. (1860) = Phlyctis (Phlyctid.).

Phlyctorhiza A.M. Hanson (1946), Chytridiaceae. 1, Am.

Phlyctospora Corda (1841) = Scleroderma (Sclerodermat.).

Phocys Niessl (1876) = Massariella (Amphisphaer.).

phoenicoid fungi, fungi growing amongst the ashes of former fires (Carpenter & Trappe, *Mycotaxon* **23**: 203, 1985); see Pyrophilous fungi.

Phoenicostroma Syd. (1925) = Coccodiella (Phyllachor.) fide Müller & v. Arx (1973).

Pholidotopsis Earle (1909) = Galerina (Cortinar.) fide Singer (1951).

Pholiota (Fr.) P. Kumm. (1871) nom. cons., Strophariaceae. 50 (lignicolous causing heartwood rot), esp. temp. See Singer (1975), Smith & Hesler (*The North American species of Pholiota*, 1968; 205 taxa, keys). See nameko.

Pholiotella Speg. (1889) nom. rej. = Conocybe (Bolbit.).

Pholiotina Fayod (1889) nom. rej. = Conocybe (Bolbit.). Used by Singer (1962), for 16 spp.

Phoma Fr. (1821) nom. rej. = Plagiostoma (Vals.).

Phoma Sacc. (1880) nom. cons., Anamorphic Pleosporaceae, 4.A/B1.15. Teleomorph *Pleospora*. *c.* 40, widespr. See Boerema (*Persoonia* **3**: 9, 1963; typification), van der Aa *et al.* (*Stud. Mycol.* **32**: 1, 1990; sect. classn), de Gruyter & Nordeloos (*Persoonia* **15**: 71, 1992; sect. *Phoma*), Boerema (*Persoonia* **15**: 197, 1993; sect. *Peyronellaea*), de Gruyter *et al.* (*Persoonia* **15**: 369, 1993; sect. *Phoma*, taxa with small conidia), Boerema *et al.* (*Persoonia* **15**: 431, 1994; sect. *Plenodomus*), Boerema (*TBMS* **67**: 289, 1976; spp. studied by Dennis), Boerema & Bollen (*Persoonia* **8**: 111, 1975; sep. from *Ascochyta*), Boerema & Dorenbosch (*Stud. Mycol.* **3**, 1973; fruit rotting spp.), Dorenbosch (*Persoonia* **6**: 1, 1970; soil spp.), Hawksworth (*Bull. Br. Mus. nat. Hist., Bot.* **9**: 49, 1981; lichenicolous spp.), Sutton (1980; key 27 spp.), Rajak & Rai (*J. Econ. Tax. Bot.* **7**: 588, 1986; key to spp. in culture), Monte & Garcia-Acha (*TBMS* **91**: 133, 1988; germination of *P. betae*), Monte & Garcia-Acha (*TBMS* **90**: 659, 1988; ultrastr. conidiogenesis *P. betae*), Morgan-Jones *et al.* (*Mycotaxon* **16**: 1983 et seq.; studies on *Phoma*), Upadhyay *et al.* (*CJB* **68**: 2059, 1990; *P. cyperi* phytotoxin production), Monte *et al.* (*Mycopath.* **115**: 89, 1991; integrated systematics), Pons (*Fitopat. Venez.* **3**: 34, 1990; spp. on *Saccharum*), Schäfer & Wöstemeyer (*J. Phytopathol.* **136**: 124, 1992; aggressive and non-aggressive strains of *P. lingam* separated by PCR), Nordeloos *et al.* (*MR* **97**: 1343, 1993; dendritic crystals and taxonomy). See lupinosis.

Phomachora Petr. & Syd. (1925), Mitosporic fungi, 8.A1.?. 2 or 3, Am., Australia.

Phomachorella Petr. (1947), Mitosporic fungi, 8.A1.?. 1, S. Afr. See Swart (*TBMS* **48**: 463, 1965).

Phomatosphaeropsis Ribaldi (1953) = Sphaeropsis (Mitosp. fungi) fide Sutton (1980).

Phomatospora Sacc. (1875), Ascomycota (inc. sed.). 20, widespr. See Rappaz (*Mycotaxon* **45**: 323, 1992). Anamorphs *Dinemasporium*, *Phomatosporella*, *Sporothrix*.

Phomatosporella Tak. Kobay. & K. Sasaki (1982), Anamorphic Ascomycota (inc. sed.), 8.A1.?. Teleomorph *Phomatospora*. 1, Japan.

Phomatosporopsis Petr. (1925) = Phomatospora (Ascomycota, inc. sed.) fide v. Arx & Müller (1954).

phomin, a cytostatic antibiotic from *Phoma* sp. (S 298) (Rothweiler & Tamm, *Experientia* **22**: 750, 1966); cytochalasin B (q.v.).

Phomites Fritel (1910), Fossil fungi. 2 (Cretaceous, Palaeocene), France, India.

Phomopsella Höhn. (1920) = Phomopsis (Mitosp. fungi) fide Petrak (*Ann. Myc.* **23**: 1, 1925).

phomopsin, see lupinosis.

Phomopsina Petr. (1922) = Phoma (Mitosp. fungi) fide Clements & Shear (1931).

Phomopsioides M.E.A. Costa & Sousa da Câmara (1954) = Phomopsis (Mitosp. fungi) fide Sutton (1977).

Phomopsis (Sacc.) Bubák (1905) nom. cons. prop., Anamorphic Valsaceae, 8.A1.15. Teleomorph *Diaporthe*. 100, widespr. *P. cinerescens* (fig canker). See Tomaz *et al.* (*Publ. Lab. Patol. Veg. 'Ver. de Alm.' Lisboa* : 54, 1989; *P. mali* on almonds), Uecker (*Mycol. Mem.* **13**, 1988; list sp. names), Wechtl (*Linzer biol. Beitr.* **22**: 161, 1990; spp. on *Compositae* and *Umbelliferae*), Brayford (*MR* **94**: 691 and 745, 1990; variation and vegetative incompatibility in spp. from *Ulmus*), Shivas *et al.* (*MR* **95**: 320, 1991; variation in *P. leptostromiformis* from lupin), Uecker & Johnson (*Mycol.* **83**: 192, 1991; spp. on *Asparagus*), Rehner & Uecker (*CJB* **72**: 1666, 1994; DNA and species concepts).

Phomyces Clem. (1931), Mitosporic fungi, 4.A1.?. 1 (on *Meliola*), Brazil.

Phorcys Niessl (1876) = Amphisphaeria (Amphisphaer.).

Phorima Raf. (1830) nom. dub. = Hexagonia (Polypor.) fide Cooke (1953).

phorophyte, the 'host' tree of an epiphyte.

photo- (prefix), pertaining to light.

photobiont, a photosynthetic symbiont in a lichen which may be a eukaryotic alga (phycobiont), see Gärtner (*in* Reisser (Ed.), *Algae and symbiosis*: 325, 1992; review systematics), or a cyanobacterium (bactobiont, cyanobiont) (Ahmadjian, *Internat. Lich. Newsl.* **15** (2): 19, 1982; Büdel (*in* Reisser (Ed.), *Algae and symbiosis*: 301, 1992; review systematics).

photomorph, an organism whose form is determined by the nature of its photosynthesis (Laundon, *Taxon* **44**: 387, 1995); see also Lichens, phycotype, phototype, photosymbiodeme, phycosymbiodeme.

photophilous, having a preference for well-illuminated habitats; cf. heliophilous, anheliophilous.

photophobous, having a preference for shaded habitats.

photosporogenic, requiring light for sporogenesis.

photosymbiodeme, a replacement term for phycosymbiodeme to allow for one biont being a cyanobacterium (Stocker-Wörgötter & Türk, *Crypt. Bot.* **4**: 300, 1994); a photomorph. See Lichens.

phototaxis, movement (e.g. of zoospores) influenced by light.

phototropism, see tropism.

Phragmaspidium Bat. (1960), ? Microthyriaceae. 3, widespr.

Phragmeriella Hansf. (1946), Pseudoperisporiaceae. 1 (on *Meliolaceae*), Afr.

Phragmidiaceae Corda (1837), Uredinales. 10 gen. (+ 12 syn.), 96 spp. Pycnia discoid or flat and of indeterminate growth, normally without bounding structures; aeciospores catenulate (*Caeoma*) or pedicellate (*Uraecium*); urediniospores pedicellate, uredinia paraphysate; teliospores pedicellate, usually several celled by horizontal septation, pigmented.

Phragmidiella Henn. (1905), Phakopsoraceae. 4 or 5 (on *Bignoniaceae*), Afr., India.

Phragmidiites Babajan & Tasl. (1970), Fossil fungi. 1 (Tertiary), former USSR.

Phragmidium Link (1816), Phragmidiaceae. 60, esp. temp.; autoecious on *Rosaceae*. The cause of rusts of *Rubus* (blackberry (*P. violaceum*), raspberry (*P. rubiidaei*), and rose (*P. mucronatum*).

Phragmiticola Sherwood (1987), ? Leotiales (inc. sed.). 1, Eur.

Phragmobasidiomycetes, see *Basidiomycetes*.

phragmobasidium, see basidium.

Phragmocalosphaeria Petr. (1923), Calosphaeriaceae. 1, Czech Republic. See Wehmeyer (*Revision of Melanconis*, 1941).

Phragmocapnias Theiss. & Syd. (1918), Capnodiaceae. c. 2, trop. See Reynolds (*Mycotaxon* **8**: 917, 1979).

Phragmocarpella Theiss. & Syd. (1915) = Phyllachora (Phyllachor.). See Petrak (*Ann. Myc.* **29**: 349, 1931).

Phragmocauma Theiss. & Syd. (1915) = Phyllachora (Phyllachor.) fide Cannon (*SA* **7**: 111, 1988).

Phragmocephala E.W. Mason & S. Hughes (1951), Mitosporic fungi, 1/2.C2.1. 5, widespr. See Hughes (*N.Z. Jl Bot.* **17**: 163, 1979).

Phragmodiaporthe Wehm. (1941), Melanconidaceae. 2 or 3, N. Am. See Barr (*Mycol. Mem.* **7**, 1978).

Phragmodimerium Petr. & Cif. (1932) = Philonectria (Dothideales) fide v. Arx & Müller (1975).

Phragmodiscus Hansf. (1947), Lasiosphaeriaceae. 1 (or 2), Afr. See Eriksson & Yue (*SA* **8**: 17, 1989).

Phragmodochium Höhn. (1924), Mitosporic fungi, 3.C1.?. 1, Java. = Fusarium (Mitosp. fungi) fide Clements & Shear (1931).

Phragmodothella Theiss. & Syd. (1915) = Discostroma (Amphisphaer.).

Phragmodothidea Dearn. & Barthol. (1926) = Scirrhia (Dothid.).

Phragmodothis Theiss. & Syd. (1914) = Dothidea (Dothid.) fide Barr (*Mycotaxon* **43**: 371, 1992).

Phragmogibbera Samuels & Rogerson (1990), Venturiaceae. 1, Venezuela.

Phragmogloeum Petr. (1954), Mitosporic fungi, 4.C2.?. 1, Australia.

Phragmographium E.F. Morris (1966) ≡ Morrisographium (Mitosp. fungi).

Phragmographum Henn. (1905) = Opegrapha (Roccell.).

Phragmonaevia Rehm (1896) nom. dub. (Fungi, inc. sed.) fide Sherwood (*Mycotaxon* **5**: 1, 1977).

Phragmopeltheca L. Xavier (1976) = Mazosia (Roccell.) fide Eriksson (*Opera bot.* **60**, 1981).

Phragmopelthecaceae, see *Roccellaceae* (Eriksson, *SA* **14**: 58, 1995).

Phragmopeltis Henn. (1904), Mitosporic fungi, 5.B2.?. 5, trop. Am.

Phragmoporthe Petr. (1934), Melanconidaceae. 1, Eur., N. Am. See Barr (*Mycol. Mem.* **7**, 1978), Monod (1983).

Phragmopsora Magnus (1875) = Pucciniastrum (Pucciniastr.) fide Sydow (1915).

Phragmopyxine Clem. (1909) = Pyxine (Physc.).

Phragmopyxis Dietel (1897), Uropyxidaceae. 4 (on *Leguminosae*), Am., Afr.

Phragmoscutella Woron. & Abramov (1926), Dothideales (inc. sed.). 1, former USSR.

Phragmospathula Subram. & N.G. Nair (1966), Mitosporic fungi, 3.C2.19. 1, India.

Phragmospathulella J. Mena & Mercado (1987), Mitosporic fungi, 1.C2.28. 1, Cuba.

Phragmosperma Theiss. & Syd. (1917) = Massarina (Lophiostom.) p.p. and Sphaerulina (Mycosphaerell.) p.p. fide v. Arx & Müller (1975).

phragmospore, differs from an amerospore (q.v.) and didymospore (q.v.) in having 2 to many transverse septa. See Mitosporic fungi.

Phragmosporonema Moesz (1924) ≡ Diplosporonema (Mitosp. fungi).

Phragmostachys Costantin (1888) = Sterigmatobotrys (Mitosp. fungi) fide Bisby (*TBMS* **26**: 138, 1943).

Phragmostele Clem. (1909) ≡ Pucciniostele (Phakopsor.).

Phragmostilbe Subram. (1959) = Arthrosporium (Mitosp. fungi) fide Wang (1972).

Phragmotelium Syd. (1921), Uredinales (inc. sed.). 10 (on *Rubus*), Asia, Australia. See Thirumalachar (*Proc. Indian Acad. Sci.* **15B**: 186, 1942). = Phragmidium (Phragmid.) fide Cummins & Hiratsuka (1983).

Phragmothele Clem. (1909) = Thelidium (Verrucar.).

Phragmothyriella Höhn. (1912) ≡ Myriangiella (Schizothyr.).

Phragmothyriella Speg. (1919) nom. dub. fide Petrak (*Sydowia* **5**: 169, 1951).

Phragmothyrites W.N. Edwards (1922), Fossil fungi. 13 (Tertiary), widespr. See Selkirk (1975).

Phragmothyrium Höhn. (1912) = Microthyrium (Microthyr.) fide v. Arx & Müller (1975).

Phragmotrichum Kunze (1823), Mitosporic fungi, 6/8.C/D2.23. 3, Eur. See Sutton & Pirozynski (*TBMS* **48**: 349, 1965).

Phragmoxenidiaceae Oberwinkler & R. Bauer (1990), Tremellales. 1 gen., 1 sp. Basidia auricularioid; hyphal septa with dolipores but lacking parenthesomes.

Phragmoxenidium Oberw. (1990), Phragmoxenidiaceae. 1, Eur.

Phragmoxyphium Bat. & Cif. (1963) = Ciferrioxyphium (Mitosp. fungi) fide Hughes (*Mycol.* **68**: 693, 1976).

Phrototecha, see *Prototheca*.

Phthora d'Hérelle (1909), Ascomycota (inc. sed.). 1 (on *Coffea*), Guatemala.

Phurmomyces Thaxt. (1931), Ceratomycetaceae. 1, Asia.

phyco- (prefix), pertaining to algae.

Phycoascus A. Møller (1901) ? = Pyronema (Pyronemat.). See Eckblad (1968).

phycobiont, the algal partner in a lichen (Scott, *Nature* **179**: 486, 1957), photobiont (q.v.); cf. mycobiont.

Phycodiscus Clem. (1909) = Knightiella (Icmadophil.).

phycolichenes (obsol.), lichens in which the vegetative thallus morphology is determined by the photobiont and which are of uncertain systematic position as the sporocarps are unknown (e.g. *Cystocoleus*, *Racodium*).

Phycomater Fr. (1825) nom. dub. (? Fungi).

Phycomelaina Kohlm. (1968), Phyllachoraceae. 1 (on *Laminaria*), Eur., N. Am. See Schatz (*Mycol.* **75**: 762, 1983).

Phycomyces Kunze (1823), Phycomycetaceae. 3, widespr. See Benjamin & Hesseltine (*Mycol.* **51**: 751, 1959; key); Carlile (*J. gen. Microbiol.* **28**: 161, 1962; sporangiophore phototropism); Cerdá-Olmedo & Lipson (*Phycomyces*, 1987; review, lit.).

Phycomycetaceae Arx (1982), Mucorales. 1 gen., 3 spp. Sporangiophores large, unbranched; zygospores with coiled tongs-like suspensors bearing branched appendages

Phycomycetes (obsol.). Class formerly used for fungi now treated in *Chromista* (q.v.) and some *Fungi* (*Chytridiomycota* and *Zygomycota*); best used only as a trivial term, 'phycomycetes'; an approx. syn. of 'lower fungi'.

Phycomycites D.E. Ellis (1915), Fossil fungi. 3 (Jurassic), Germany, UK.

phycomycosis, a general term for a disease of humans or animals caused by a phycomycete. Cf. mucormycosis, zygomycosis.

Phycomycotera. Superdivision used by Moore (1971) for all fungi with non-septate hyphae (i.e. *Oomycota* and *Zygomycota* in the *Dictionary*, this edition); cf. *Septomycotera*.

phycophilous, growing with or on algae.

Phycopsis L. Mangin & Pat. (1912) [non (Fisch.-Oost.) Rothpletz (1896), *Algae*] = Seuratia (Seurat.) fide Meeker (1975).

Phycorella Döbbeler (1980), Dothideales (inc. sed.). 1 (biotrophic parasite of *Scytonema*), Australia.

Phycosiphon A. Massal. (1859) ≡ Brachycarphium (Fossil fungi).

phycosymbiodeme, joined lichen thalli with a single mycobiont but different photobionts (Renner & Galloway, *Mycotaxon* **16**: 197, 1982); photomorph, phycotype.

phycosymbiont, phycobiont.

phycotrophic (of fungi), obtaining nutrients from algae (Dobbs, *Lichenologist* **4**: 323, 1970).

phycotype, see type.

Phylacia Lév. (1845), Xylariaceae. 6, trop. Am. See Dennis (*Kew Bull.* 1957: 320, 1957), Pérez-Silva (*Boln Soc. Mex. Micol.* **6**, **9**, 1972), Rodrigues & Samuels (*Mem. N.Y. bot. Gdn* **49**: 290, 1989), Silveira & Rogers (*Acta Amazonia* Suppl. **15**: 7, 1985; key 4 spp. Brazil). Anamorph *Geniculosporium*.

Phylaciaceae, see *Xylariaceae*.

Phylacteria (Pers.) Pat. (1887) ≡ Scyphopilus (Thelephor.).

Phyllachora Nitschke ex Fuckel (1867) nom. rej. = Scirrhia (Dothid.) fide Holm (*Taxon* **24**: 475, 1973).

Phyllachora Nitschke ex Fuckel (1870) nom. cons., Phyllachoraceae. *c.* 500, widespr. See Cannon (*Mycol. Pap.* **163**, 1991; spp. on *Leguminosae*), Parbery (*Austr. J. Bot.* **15**: 271, 1967; key, **19**: 207, 1971; spp. on *Gramineae*), Parbery & Langdon (*Austr. J. Bot.* **12**: 265, 1964; sp. concept), Kamat *et al.* (*Univ. Agric. Sci. Hebbal Monogr.* **4**, 1978; key 88 Indian spp.), Malençon (*BSMF* **74**: 423, 1959; spore dev.), Parbery (*in* Subramanian (Ed.), *Taxonomy of fungi* **1**: 263, 1978; fungi on).

Phyllachoraceae Theiss. & Syd. (1915), Phyllachorales. 42 gen (+ 59 syn.), 1150 spp. Stromata often well-developed, immersed in plant tissue, often clypeate, usually black; ascomata perithecial, usually thin-walled, the ostioles periphysate, the peridium usually composed of thin-walled compressed hyaline or brown tissue; interascal tissue of simple rather wide thin-walled paraphyses; asci ± cylindrical, thin-walled, persistent, usually with an inconspicuous I-apical ring; ascospores usually hyaline, aseptate, occ. ornamented. Anamorphs coelomycetous, spermatial or disseminative. Biotrophic, sometimes becoming necrotrophic, on leaves and stems, widespr. esp. trop. *Lit.*: see *Phyllachorales*.

Phyllachorales, Ascomycota. 1 fam., 42 gen. (+ 59 syn.), 1150 spp. Fam.:
Phyllachoraceae.
   Lit.: Cannon (*SA* **7**: 23, 1987; posn), Eriksson & Hawksworth (*SA* **11**: 181, 1993).

Phyllachorella Syd. (1914), ? Dothideaceae. 1, India. See Kar & Maity (*Mycol.* **63**: 1024, 1971). = Vestergrenia (Dothid.) fide v. Arx & Müller (1954).

Phyllactinia Lév. (1851), Erysiphaceae. 24, widespr. *P. corylea* (mildew of hazel (*Corylus*), birch (*Betula*), ash (*Fraxinus*), and other trees). See Braun (*Beih. Nova Hedw.* **89**, 1987; key). Anamorphs *Ovulariopsis, Streptopodium*.

Phyllerites Mesch. (1892), ? Fossil fungi. 16 (Tertiary), Eur.

Phyllerium Fr. (1832) nom. dub.; based on leaf outgrowths fide Fries (*Syst. mycol.* **3**: 523, 1832).

Phylleutypa Petr. (1934), Phyllachoraceae. 2, Afr., N. temp. fide v. Arx & Müller (1973).

phyllidium, lichen propagule formed by abstriction of a leaf- or scale-like portion of the thallus. See Poelt (*Flora, Jena* **169**: 23, 1980) (Fig. 22F).

Phyllis F. Wilson (1889) [non L. (1753), *Rubiaceae*] ≡ Neophyllis (Cladon.).

Phylliscidiopsis Sambo (1937), Lichinaceae (L). 1, Ethiopia.

Phylliscidium Forssell (1885), Lichinaceae (L). 1, Brazil.

Phyllisciella Henssen (1984), Lichinaceae (L). 3, S. Hemisph.

Phylliscum Nyl. (1855), Lichinaceae (L). 2, Eur., N. Am., Asia. See Forssell (*Nova Acta reg. Soc. sci. upsal.* ser. 3, **13** (6), 1885).

Phyllobaeis Kalb & Gierl (1993), Baeomycetaceae (L). 4, trop.

Phyllobatheliaceae Bitter & F. Schill. (1927), Ascomycota (inc. sed., L). 2 gen. (+ 1 syn.), 4 spp. Thallus crustose, poorly developed; ascomata perithecial, erumpent, the peridium thin-walled, ± hyaline but covered with dark melanized powdery material; interascal tissue of simple paraphyses, not gelatinized; asci saccate, ± thick-walled, fissitunicate, with a distinct ocular chamber, not blueing in iodine, 1- to 8-spored; ascospores hyaline, muriform, without a sheath. Anamorphs pycnidial. Lichenized with green algae, foliicolous, C. and S. Am.

Phyllobathelium (Müll. Arg.) Müll. Arg. (1890), Phyllobatheliaceae (L). 3, C. & S. Am. See Santesson (*Symb. bot. upsal.* **12**(1), 1952).

Phylloblastia Vain. (1921), ? Trichotheliaceae (L). 1, Afr., Asia. See Aptroot (*in* Galloway (Ed.), *Systematics, conservation and ecology of tropical lichens*: 253, 1991), Santesson (*Symb. bot. upsal.* **12**(1), 1952).

Phylloboletellus Singer (1952), Xerocomaceae. 1, Argentina.

Phyllobolites Singer (1942) nom. dub. (Agaric.) fide Singer (1945).

Phyllobrassia Vain. (1921) = Chroodiscus (Thelotremat.).

Phyllocarbon Lloyd (1921) = Polyozellus (Thelephor.).

Phyllocardium Korshikov (1927), Algae.

Phyllocarpos Poir. (1813) = Cladonia (Cladon.).

Phyllocaulon (Tuck.) Vain. (1909) = Stereocaulon (Stereocaul.) fide Zahlbruckner (1927).

Phyllocelis Syd. (1925), Ascomycota (inc. sed.). 1 or 2, C. Am.

Phyllocharis Fée (1825) = Strigula (Strigul.).

phyllocladia, the granular, verrucose, coralloid, squamuliform, digitate, peltate, or foliaceous parts of the thallus of *Stereocaulon* which contain the photobiont.

Phyllocrea Höhn. (1918), Phyllachoraceae. 1, S. Am. fide v. Arx & Müller (1973).

Phyllodontia P. Karst. (1883) = Cerrena (Coriol.) fide Donk (1974).

Phylloedia Fr. (1849) ≡ Phylloedium (Mitosp. fungi).

Phylloedium Fr. (1825), Mitosporic fungi, ?8.A2.?. 1, Eur. See Sutton (*Mycol. Pap.* **141**, 1977).

Phyllogaster Pegler (1969), Secotiaceae. 1, Ghana.

Phyllogloea Lowy (1961) [non P.C. Silva (1959), *Algae*], Platygloeaceae. 4, trop.

Phyllographa (Müll. Arg.) Räsänen (1943) nom. inval. = Opegrapha (Roccell.).

Phyllohendersonia Tassi (1902), Mitosporic fungi, 4.C2.?. 28, widespr. See Sutton (1977).

Phyllomyces Lloyd (1921) = Cordierites (Leot.) fide Zhuang (*Mycotaxon* **31**: 261, 1988).

Phyllonochaeta Gonz. Frag. & Cif. (1927) Fungi (inc. sed.) fide Petrak (*Ann. Myc.* **32**: 317, 1934).

Phyllopezis Petr. (1949), ? Leotiales (inc. sed.). 1, S. Am.

Phyllophiale R. Sant. (1952), Mitosporic fungi (L), ?.-.-. 1, widespr.

Phyllophthalmaria (Müll. Arg.) Zahlbr. (1905) ? = Leptosphaeria (Leptosphaer.). See Santesson (*Symb. bot. upsal.* **12**(1), 1952).

phylloplane, the leaf surface; Last & Deighton (*TBMS* **48**: 83, 1965), the non-parasitic biotas of the leaf surface.

Phylloporia Murrill (1904), Hymenochaetaceae. 5, pantrop. See Ryvarden & Johansen (*Preliminary polypore flora of East Africa*: 230, 1980; key).

Phylloporina (Müll. Arg.) Müll. Arg. (1890) = Porina (Trichothel.).

Phylloporina C.W. Dodge (1948), Ascomycota (inc. sed., L). 1, Kerguelan.

Phylloporis Clem. (1909) = Strigula (Strigul.) fide Hawksworth (*SA* **5**: 151, 1986). See Vězda (*Folia geobot. phytotax.* **19**: 177, 1984).

Phylloporthe Syd. (1925), Valsaceae. 1, trop. Am. See Barr (*Mycol. Mem.* **7**, 1978).

Phylloporus Quél. (1888), Xerocomaceae. 10, widespr. See Corner (*Nova Hedw.* **20**: 793, 1970; Malaysia), Singer & Gómez (*Brenesia* **22**: 163, 1984; Costa Rica).

Phyllops Raf. (1817) nom. dub. (Fungi, inc. sed.).

Phyllopsora Müll. Arg. (1894), Bacidiaceae (L). 65, esp. trop. See Brako (*Mycotaxon* **35**: 1, 1989), Swinscow & Krog (*Lichenologist* **13**: 203, 1981).

Phyllopta (Fr.) Fr. (1825), Tremellaceae. 1, Eur. See Donk (*Taxon* **7**: 239, 1958, *Persoonia* **4**: 306, 1966).

Phyllopyrenia C.W. Dodge (1948) nom. dub. (Ascomycota, inc. sed.).

Phyllopyreniaceae, see *Coccotremataceae*.

Phyllosphaera Dumort. (1822) nom. rej. ≡ Phyllosticta (Mitosp. fungi).

phyllosphere, the zone immediately surrounding a leaf; frequently used in the sense of phylloplane (q.v.). See Preece & Dickinson (Eds) (*Ecology of leaf surface micro-organisms*, 1970), Dickinson & Preece (Eds) (*Microbiology of aerial plant surfaces*, 1976), Blakeman (Ed.) (*Microbial ecology of the phylloplane*, 1981), Fokkema & van der Heuvel (Eds) (*Microbiology of the phyllosphere*, 1986). Cf. rhizoplane; spermoplane.

Phyllosticta Pers. (1818) nom. cons., Anamorphic Mycosphaerellaceae, 4.A1.1/19. Teleomorph *Guignardia.* 46, widespr. See van der Aa (*Stud. Mycol.* **5**, 1973), Punithalingam (*Mycol. Pap.* **136**, 1974), Punithalingam & Woodhams (*Nova Hedw.* **36**: 151, 1982; conidial appendages), Yip (*MR* **93**: 489, 1989; 5 n.spp. from Australia), Petrini *et al.* (*Sydowia* **43**: 148, 1991; key spp. on conifers), Leuchtmann *et al.* (*MR* **96**: 287, 1992; isozyme polymorphism in endophytic spp.).

Phyllostictella Tassi (1901) ? = Microsphaeropsis (Mitosp. fungi) fide Sutton (1977).

Phyllostictina Syd. & P. Syd. (1916) = Phyllosticta (Mitosp. fungi) fide van der Aa (*Stud. Mycol.* **5**, 1973).

Phyllostictites Babajan & Tasl. (1970), Fossil fungi. 1 (Tertiary), former USSR.

Phyllothelium Trevis. (1861) = Trypethelium (Trypethel.).

Phyllotopsis J.-E. Gilbert & Donk ex Singer (1936), Lentinaceae. 2, temp.

Phyllotremella Lloyd (1920) = Hohenbuehelia (Tricholomat.) fide Singer (1951).

Phyllotus P. Karst. (1879) = Pleurotus (Lentin.) fide Killermann (1908).

Phylogeny. Mycologists' views on phylogeny, the evolution of a group of organisms through time, have been based traditionally on comparative morphology, cytology, hyphal wall chemistry, ultrastructure, and to a lesser degree fossils (see Fossil fungi). The advent of cladistic and molecular approaches since the last edition of this *Dictionary* have led to major advances in understanding of the relationships of fungal phyla. These methods are now accepted as effective for this purpose (Hillis *et al., Science* **264**: 671, 1994).

The three kingdoms including fungi (see Kingdoms of fungi) are shown by molecular studies of 16S-like ribosome RNA sequences to be phylogenetically remote (Fig. 31). According to the molecular evidence, the *Fungi* may have originated from protozoan ancestors before the kingdoms *Animalia* and *Plantae* split (Gouy & Li, 1989); there are indications that *Fungi* are closer to *Animalia* than *Plantae*, but the confidence in the branching point in that part of the tree so far produced is low for rRNA sequence alone (Embley *et al.*, 1995). However, studies of 23 other protein sequences clearly support the hypothesis that *Fungi* are much closer to *Animalia* (Nikoh *et al.*, 1994).

Molecular and ultrastructural data support Atkinson's (*Ann. Myc.* **7**: 441, 1909) opinion that *Fungi* were derived from colourless organisms 'lower' than the *Chytridiomycota*. The hypothesis of a polyphyletic origin from algae has less support now than it had; the case for a red algal (*Rhodophyta*) ancestry of the *Ascomycota* and *Basidiomycota* is strongly argued (Demoulin, *Bot. Rev.* **40**: 315, 1974, *BioSystems* **18**: 347, 1985) but apparently not supported by the molecular evidence to date.

Within the kingdom *Fungi*, 18S-rRNA sequences prove to be of particular value in elucidating the relationship between the four phyla recognized in this *Dictionary* (Fig. 32). The basal positions of the *Chytridiomycota* and *Zygomycota* are in agreement with the cladistic study of Tehler (1988) based on non-molecular data.

Suggestions that the main ascus types might have originated separately from different algae (Cain, *Mycol.* **64**: 1, 1972) are not supported by the molecular data. The rate of substitutions in molecular sequences can be used to indicate the date of evolutionary radiations. Berbee & Taylor (1993) calculated that the three main fungal phyla diverged from the *Chytridiomycota* approximately 550 Myr ago, that the *Ascomycota/Basidiomycota* split occurred at about 400 Myr ago after plants invaded the land, and that many ascomycetous yeasts and moulds (e.g. *Eurotiales*) evolved after the origin of angiosperms in the last 200 Myr. These results are broadly supported by fossil evidence, the main difference being that remains of *Ascomycota* go back to the Silurian (*c.* 440 Myr ago) while the oldest definite *Basidiomycota* occur at about 380 Myr ago; see Fossil fungi, Taylor (1994).

Fungi are associated with some of the earliest remains of land plants. Church (*J. Bot., Lond*, **59**: 7, 1921) considered that 'lichens' might be transmigrants, the earliest colonizers of land (see also Corner, *The life of plants*, 1964). Land plants may, however, have their origins in a green alga-oomycete symbiosis (Pirozynski & Malloch, *BioSystems* **6**: 153, 1975), and it has even been argued that plants are of biphyletic origin with some parts of the fungal genome (Jørgensen, *BioSystems* **31**: 193, 1993) or even 'inside-out' lichens as they contain organelles questionably of fungal origin (Atstatt, *Ecology* **69**: 17, 1988).

Lit.: Berbee & Taylor (*CJB* **71**: 1114, 1993), Bowman *et al.* (*Molec. Biol. Evol.* **9**: 285, 1992), Bruns *et al.* (*Molec. Phyl. Evol.* **1**: 231, 1992), Cavalier-Smith (1993), Embley *et al.* (*in* Hawksworth (Ed.), *Biodiversity: measurement and estimation*: 21, 1995), Gouy &

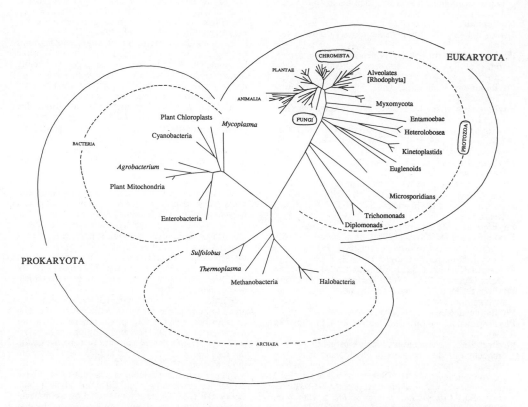

Fig. 31. An inferred unrooted phylogenetic tree of eukaryotes based on a distance analysis of all positions that can be unambiguously aligned among complete 16S-like rRNA molecular sequences from 75 taxa, modified from Patterson & Sogin, *in* Hartman & Matsumo (Eds) (*On the origin and evolution of prokayryotes and eukaryotes,* 1992). Shows relative positions of the Domains *Eukaryota* and *Prokaryota* and the kingdoms including fungi (Kingdom names ringed).

Li (*Molec. Biol. Evol.* **6**: 109, 1989), Hartman & Matsumo (1992), Kandler (*Mycoses* **37** (Suppl.): 13, 1994), Martin (*in* Ainsworth & Sussman (Eds), *The Fungi* **3**: 635, 1968; bibliogr.), Nikoh *et al.* (*Molec. Biol. Evol.* **11**: 762, 1994), Patterson & Sogin (*in* Hartman & Matsumo (Eds), *The origin and evolution of prokaryotic and eukaryotic cells*: 13, 1992), Taylor (*in* Hawksworth (Ed.), *Ascomycete systematics*: 175, 1994), Wilmotte *et al.* (*Syst. appl. Microbiol.* **16**: 436, 1993; molec. phylogeny), Tehler (*Cladistics* **4**: 227, 1988), Woess *et al.* (*Proc. natn. Acad. Sci., USA* **87**: 4576, 1990).

See Cladistics, Classification, Coevolution, Kingdoms of fungi.

phylum, a taxonomic rank between kingdom and class; a division; see Classification.

Phymatium Chevall. (1826) = Elaphomyces (Elaphomycet.) fide Fries (*Syst. mycol.* **3**: 57, 1829).

Phymatodiscus Speg. (1919) = Myriangium (Myriang.) fide v. Arx (1963).

Phymatomyces Kobayasi (1937) = Barssia (Balsam.) fide Trappe (1979).

Phymatopsis Tul. ex Trevis. (1857) = Abrothallus (Ascomycota, inc. sed.) fide Keissler (*Rabenh. Krypt.-Fl.* **8**, 1930).

Phymatosphaeria Pass. (1875) = Myriangium (Myriang.) fide v. Arx (1963).

Phymatostroma Corda (1837) = Leucosporium (nom. dub.) fide Holubová-Jechová (*Sydowia* **45**: 97, 1993).

Phymatotrichopsis Hennebert (1973), Anamorphic Sistotremataceae, 1.A1.6. Teleomorph *Trechispora*. 1, N. Am. *Phymatotrichum omnivorum* (teleomorph *Trechispora brinkmanii*), root rot of cotton and other plants. See Baniecki & Bloss (*Mycol.* **61**: 1054, 1969), Mouton (*Rev. Myc.* **18** (*suppl. colon.* 2): 69, 1953).

Phymatotrichum Bonord. (1851) = Botrytis (Mitosp. fungi) fide Hennebert (*Persoonia* **7**: 183, 1973).

Physalacria Peck (1882), Physalacriaceae. 28, trop. & S. Hemisph. See Berthier (*Bibl. Mycol.* **98**, 1985, world monogr.).

Physalacriaceae Corner (1970), Cantharellales. 4 gen. (+ 2 syn.), 35 spp. Basidioma small, with inflated head and solid stipe; oleocystidia present; monomitic; lignicolous or epiphyllous.

Physalidiella Rulamort (1990), Mitosporic fungi, 1.G2.1. 1, Italy.

Physalidiopsis R.F. Castañeda & W.B. Kendr. (1990), Mitosporic fungi, 1.G2.1. 1, Cuba.

Physalidium Mosca (1965) [non Fenzl (1866), *Cruciferae*] ≡ Physalidiella (Mitosp. fungi).

Physalospora Niessl (1876), Hyponectriaceae. 30, widespr. See Barr (*Mycol.* **68**: 611, 1976; nomencl.), v. Arx & Müller (1954), also *Botryosphaeria* (Botryosph.).

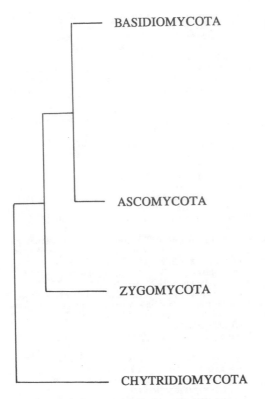

BASIDIOMYCOTA

ASCOMYCOTA

ZYGOMYCOTA

CHYTRIDIOMYCOTA

Fig. 32. Diagnostic unrooted phylogenetic tree showing relative positions of the phyla recognized in Kingdom *Fungi* by 18S-r RNA molecular sequences. Note that the confidence values are low for the *Chytridiomycota* + *Zygomycota* linkages. Simplified from Bruns *et al.*, 1992; Cavalier-Smith, 1993.

Physalosporella Speg. (1910) Fungi (inc. sed.) fide Petrak & Sydow (*Ann. Myc.* **32**: 28, 1934).

Physalosporina Woron. (1911) = Stigmatula (Phyllach.). See Cannon (*Mycol. Pap.* **163**, 1991).

Physalosporopsis Bat. & H. Maia (1955), ? Dothideales (inc. sed.). 1, Brazil. See Müller & v. Arx (1962).

Physaraceae Chevall. (1826), Physarales. 10 gen. (+ 19 syn.), 170 spp.

Physarales, Myxomycota. 2 fam., 17 gen. (+ 29 syn.), 295 spp. Spore mass dark-coloured, peridium or capillitium calcareous. Fams:
(1) **Didymiaceae**.
(2) **Physaraceae**.

Physarella Peck (1882), Physaraceae. 1, widespr.

Physarina Höhn. (1909), Didymiaceae. 3, Indonesia, India, Mexico, Japan.

Physarum Pers. (1794), Physaraceae. *c.* 120, cosmop. See Hutterman (Ed.) (*Physarum polycephalum*, 1973; use in cell-biology), Aldrich & Daniel (Eds) (*Cell biology of Physarum and Didymium*, 2 vols, 1982).

Physcia (Schreb.) Michx. (1803), Physciaceae (L.). *c.* 50, cosmop. See Awasthi (*J. Ind. bot. Soc.* **39**: 1, 1960; India), Maas Geesteranus (*Blumea* **7**: 206, 1952), Frey (*Ber. schweiz. bot. Ges.* **73**: 389, 1963), Moberg (*Symb. bot. upsal.* **22** (1), 1977; *Herzogia* **8**: 249, 1989; Eur.; *Nordic Jl Bot.* **10**: 319, 1990; key 34 spp. C. & S. Am.), Moberg & Hansen (1986; keys), Thomson (*Beih. Nova Hedw.* **7**, 1963; N. Am.),

Moberg (*Nordic Jl Bot.* **6**: 843, 1986; key 21 spp. E. Afr.).

Physciaceae Zahlbr. (1898) nom. cons., Lecanorales (L). 24 gen. (+ 45 syn.), 858 spp. Thallus crustose, foliose or fruticose; ascomata apothecial, sessile, occasionally immersed, usually with a distinct margin; interascal tissue composed of ± branched and anastomosing paraphyses, often with swollen and/or pigmented apices; asci with a I+ apical cap, with a conical or funnel-shaped more weakly staining apical cushion, usually with an I+ gelatinized outer layer; ascospores usually septate, usually dark brown with thickened walls and septa. Anamorphs pycnidial. Lichenized with green algae, often on nitrogen-rich substrata, widespr.
Lit.: Aptroot (*Fl. Guianas*, E, **1**, 1987; 48 spp. Guianas), Aptroot & Berendsen (*Proc. K. ned. Akad. Wet.* C **92**: 409, 1989; pruina), Hafellner *et al.* (*Herzogia* **5**: 39, 1979; key gen.), Moberg & Hansen (*Meddr. Grønl.* Biosci. **22**, 1986; keys 17 foliose spp. Greenland), Rambold *et al.* (*Pl. Syst. Evol.* **192**: 31, 1994; ascus types).

Physcidia Tuck. (1862), Bacidiaceae (L). 7, Afr., Asia, Australasia, C. & S. Am. See Kalb & Elix (*Bibl. Lich.* **57**: 265, 1995; key).

Physciella Essl. (1986) ? = Phaeophyscia (Physc.) fide Hawksworth (*SA* **5**: 151, 1986).

Physciomyces E.A. Thomas ex Cif. & Tomas. (1953) = Physcia (Physc.).

physcion, see parietin.

Physciopsis M. Choisy (1950) = Hyperphyscia (Physc.) fide Hafellner *et al.* (*Herzogia* **5**: 39, 1979).

Physconia Poelt (1965) typ. cons., Physciaceae (L). *c.* 15, Eur., N. Am., Asia. See Moberg (*Symb. bot. upsal.* **22** (1), 1977), Moberg (*Nord. Jl Bot.* **7**: 719, 1987; key 3 spp., E. Afr.).

physiologic form, (Stakman & Levine), see physiologic race.

physiologic race, one of a group of forms alike in morphology but unlike in certain cultural, physiological, biochemical, pathological, or other characters. The use of this term in place of biologic form, etc. was a recommendation of the International Botanical Congress, 1935. The term 'race' has been used in different senses by plant pathologists, the classical approach of using tests on differential hosts now being replaced by smaller flexible groups of host genotypes permitting characterization by virulence. See Caten (*in* Wolfe & Caten, *Populations of plant pathogens: their dynamics and genetics*: 21, 1987; review), Dennis (*Proc. Linn. Soc. Lond.* **163**: 47, 1952; taxonomic treatment), Johnson (*Biol. Rev.* **28**: 105, 1953; variation), Classification, special form, Species.

Physiology. Good introductory texts are: Berry (*Physiology of industrial fungi*, 1988), Garraway & Evans (*Fungal nutrition and physiology*, 1984), Griffin (*Fungal physiology*, 1981). At the more encyclopaedic level, Smith & Berry (Eds) (*The filamentous fungi*, 4 vols, 1975-83) is recommended, though now somewhat dated; the most up-to-date texts are the first volumes of Esser & Lemke (Eds) (*The Mycota*, 1994-). Other specialist treatments include: Ayers & Boddy (*Water, fungi and plants,* 1986), Bennett & Ciegler (*Secondary metabolism and differentiation in fungi*, 1983), Boddy *et al.* (*Nitrogen, phosphorus and sulphur utilisation by fungi*, 1989), Burnett & Trinci (*Fungal walls and hyphal growth*, 1989), Cooke & Whipps (*Ecophysiology of fungi*, 1983), Elliott (*Reproduction in fungi: genetical and physiological aspects*, 1994), Frankland *et al.* (*Decomposer basidiomycetes: their biology and ecology*, 1982), Jennings & Rayner (*The ecology and physiology of the fungal mycelium*, 1984), Jennings (*Stress tolerance of fungi*,

1993; *The physiology of fungal nutrition*, 1994), Kershaw (*Physiological ecology of lichens*, 1985), Lewis (*MR* **95**: 897, 1991; sugars), Moore *et al.* (*Developmental biology of higher fungi*, 1985), Smith (*Fungal differentiation*, 1983); Szaniszlo (*Fungal dimorphism*, 1985), Vicente *et al.* (*Surface physiology of lichens*, 1985), Weete (*Lipid biochemistry of fungi and other organisms*, 1980), Winkelmann & Winge (*Metal ions in fungi*, 1994).

Bakers' yeast is by far the most studied fungus, physiologically and biochemically (see *Saccharomyces*), and also because yeasts in general are relatively easy to handle experimentally, there is an extensive literature on their physiology and biochemistry, the best entry into which is Berry *et al.* (*Yeast biotechnology*, 1987) and Rose & Harrison (*The yeasts* **3-4**, 1989-91). *Advances in Microbial Physiology* contain significant review articles. Because of the power of molecular biology, many of the key physiological findings about fungi, albeit for a restricted number of genera (particularly *Aspergillus*, *Neurospora*, *Saccharomyces*), are in such journals as *J. biol. Chem.* and *Mol. cell. Biol.*

General reviews of lichen physiology are given by Smith (*Biol. Rev.* **37**: 537, 1962), Quispel (*Encycl. Pl. Physiol.* **11**: 577, 1959), Ahmadjian (*The lichen symbiosis*, edn 2, 1993), Ahmadjian & Hale (Eds) (*The lichens*, 1974), Brown *et al.* (Eds) (*Lichenology: progress and problems*, 1976), and Seaward (Ed.) (*Lichen ecology*, 1977; ecophysiology). All these publications have extensive bibliographies.

See also Air pollution, Lichens, Light and fungi, Metabolic products, Nutrition, Pigments, Antibiotic, Spore discharge, Symbiosis, diurnal cycles, anaerobic fungi, etc.

Physisporinus P. Karst. (1889), Coriolaceae. 1, Eur. See Donk (*Persoonia* **4**: 341, 1966).

Physisporus Chevall. (1826) ≡ Poria (Coriol.).

Physma A. Massal. (1854), Collemataceae (L). *c.* 10, cosmop. See Degelius (*Svensk bot. Tidskr.* **49**: 136, 1955), Verdon & Elix (*Acta bot. fenn.* **150**: 209, 1994; key 4 Australasian spp.).

Physmatomyces Rehm (1900), ? Leotiaceae. 1, Brazil.

Physmatomyces E.A. Thomas ex Cif. & Tomas. (1953) ≡ Physma (Collemat.).

Physocladia Sparrow (1932), Cladochytriaceae. 1, N. Am.

Physocystidium Singer (1962), Tricholomataceae. 1, Trinidad.

Physoderma Wallr. (1833), Physodermataceae. 50, widespr. *P. maydis* (maize brown spot), *P. alfalfae* (on *Medicago*). See Karling (*Lloydia* **13**: 29, 1950), Lange & Olson (1980), Sparrow (*TBMS* **60**: 340, 1973; type).

Physodermataceae Sparrow (1952), Blastocladiales. 1 gen. (+ 1 syn.), 50 spp. Thallus a polycentric endobiotic system bearing septate, turbinate cells which form thick-walled resting spores, or monocentric with a epibiotic zoosporangium; parasites of phanerogams.

Physodontia Ryvarden & H. Solheim (1977), Corticiaceae. 1, Sweden.

Physokermincola Brain (1923) nom. dub. (? Mitosp. fungi).

Physomitra Boud. (1885) ≡ Gyromitra (Helvell.).

Physomyces Harz (1890) = Monascus (Monasc.).

Physonema Lév. (1847), Anamorphic Uredinales. 3, temp. A name for (I) only. Teleomorph *Phragmidium*.

Physopella Arthur (1906), Uredinales (inc. sed.). 18, widespr. See Cummins & Ramachar (*Mycol.* **50**: 741, 1958). Used by Ono *et al.* (*MR* **96**: 825, 1992) for anamorphs of *Phakopsora* and *Cerotelium*.

Physorhizophidium Scherff. (1926), Chytridiaceae. 1 (on diatoms), Eur., N. Am.

Physospora Fr. (1835) = Rhinotrichum (Mitosp. fungi) fide Sumstine (*Mycol.* **3**: 45, 1911). Nom. dub. fide Donk (*Taxon* **11**: 95, 1962).

Physosporella Höhn. (1919), Hyponectriaceae. 1, Germany. See Barr (*Mycol.* **68**: 611, 1976).

Physotheca, see *Rhysotheca*.

phytoalexin, a metabolite produced by a plant in response to infection by a fungus or other pathogen (or by an abiotic factor) inhibitory to the invading pathogen. Reviews: Cruickshank, *Ann. Rev. Phytopath.* **1**: 351, 1963; Kuć, *Ann. Rev. Phytopath.* **10**: 207, 1972; Ingram, *Bot. Rev.* **38**: 343, 1972. Cf. antibiotics.

Phytoalexins include: capsidiol, glyceollin, ipomearone, kievitone, phaseollin, pisatin, rishitin, wyerone (q.v.).

phytoalternarin (**A**, **B**, **C**), host-specific toxins produced by *Alternaria kikuchiana* (black spot of Japanese pears; *Pyrus serotina*) (Park *et al.*, *Physiol. Pl. Pathol.* **9**: 167, 1976).

Phytoceratiomyxa Sawada (1929) nom. dub. (? Myxomycetes).

Phytoconis Bory (1797) nom. rej. prop., Tricholomataceae (L). 7, N. temp. See Jørgensen & Ryman (*Taxon* **38**: 305, 1989), Redhead & Kuyper (*Mycotaxon* **31**: 221, 1988), Omphalina.

Phytocordyceps C.H. Su & H.H. Wang (1986) = Cordyceps (Clavicipit.) fide Eriksson (*SA* **5**: 152, 1986).

phytolysine, a plant tissue macerating enzyme from *Plowrightia ribesia* (Neaf-Roth *et al.*, *Phytopath. Z.* **40**: 283, 1961).

Phytomyxa J. Schröt. (1886) nom. dub. (? Plasmodiophor.).

Phytomyxinae. An order for *Plasmodiophora*, *Phytomyxa*, *Sorosphaera* and *Tetramyxa*.

phytopathology, the branch of science for plant disease; see Plant pathogenic fungi.

Phytophthora de Bary (1876), Pythiaceae. 39, cosmop. Includes important pathogens of apple and pear (*P. cactorum* and *P. syringae*, fruit rot), cacao etc. (*P. palmivora*; taxonomy, Brasier & Griffin, *TBMS* **72**: 111, 1979; pod rot and canker; Gregory (Ed.), *Phytophthora disease of cocoa*, 1974), chestnut (*Castanea*) (*P. cambivora*, ink disease), citrus (*P. citrophthora*, gummosis, foot and brown fruit rots), lima bean (*P. phaseoli*, downy mildew), *P. erythroseptica* and *P. megasperma*, pink rot), tomato and potato (*P. infestans*, blight; Ingram & Williams (*Adv. Pl. Path.* **7**, 1991); Lucas *et al.* (Eds) (*Phytophthora*, 1991); *P. cryptogea*, foot rot; *P. nicotianae*, foot and buck-eye rots in tobacco, tomato, and many other plants; see Rangaswami (1962); hosts of *P. cinnamomi* (Newhook & Podger, *Ann. Rev. Phytopath.* **10**: 299, 1972; in Australia, NZ; Zentmyer, *P. cinnamomi and the diseases it causes* [*Am. phytopath. Soc. Monogr.* **10**], 1980), *P. palmivora* Chee (*RAM* **48**: 337, 1969).

See Waterhouse (*Mycol. Pap.* **92**, 1963; key), Tucker (*Res. Bull. Mo agric. Exp. Stn* **153**, 1931, taxonomy; **184**, 1933, distribution), Hickman (*TBMS* **41**: 1, 1958; gen. review), Waterhouse (*Mycol. Pap.* **122**, 1970), Stamps *et al.* (*Mycol. Pap.* **162**, 1990; tabular key), Waterhouse & Blackwell (*Mycol. Pap.* **57**, 1954; key Br. spp.), Blackwell (*Mycol. Pap.* **30**, 1949; terminology), Ribeiro (*A source book of the genus Phytophthora*, 1978).

Phytosarcodina, see *Myxomycetes*.

Phytosociology. The scientific study of plant communities; these can be given latinized names based on the dominant or characteristic species and classified by a hierarchial system; an internationally agreed Code of Phytosociological Nomenclature (Barkman *et al.*, *Vegetatio* **67**: 145, 1986) has 51 Articles governing the valid publication, legitimacy, and application of names to communities. The basal unit is the **association** ('union'), the first name of which ends in the suffix '*-etum*' (e.g. *Graphidetum scriptae*, from the species *Graphis scripta*). These can be referred to an **alliance** ('federation'), the first name of which ends in the suffix '*-ion*' (e.g. *Graphidion scriptae*). A subdivision of associations can be recognized as a **sociation** ('society') or **variant**. These systems are much used by some European lichenologists, but less so by other mycologists (some of whom have used different suffixes; see Mycosociology).

Lit. Barkman (*Phytosociology and ecology of cryptogamic epiphytes*, 1958), Delzenne-van Haluwyn (*Bibliogr. phytosociol.*, *Suppl.* **1**, 1976; catalogue of names), Hawksworth (*Mycologist's handbook*, 1974), James *et al.* (*in* Seaward (Ed.), *Lichen ecology*: 295, 1977; principles, UK communities), Roux (*Bibl. Lich.* **15**, 1981; calcareous rocks), Wirth (*Dissnes bot.* **17**, 1972; siliceous rocks).

Phytotoxic mycotoxins (toxins injurious to plants; see phytotoxin (2) under toxin). May be host specific (e.g. helminthosporoside, phytoalternarins, victorin, q.v.) or non-host specific (e.g. alternaric acid, baccatin, culmomarasmin, diaporthin, fusaric acid, fusicoccin, helminthosporal, ophiobolin (cochliobolin), skyrin, tentoxin, q.v.). Reviews: Pringle & Scheffer (*Ann. Rev. Phytopath.* **2**: 133, 1964), Wright (*Ann. Rev. Microbiol.* **22**: 269, 1968), Wood *et al.* (Eds) (*Phytotoxins in plant disease*, 1972), Strobel (*Ann. Rev. Pl. Physiol.* **25**: 541, 1974), Durbin (Ed.) (*Toxins in plant diseases*, 1981).

phytotoxin, see toxin.

Picardella I.I. Tav. (1985), Laboulbeniaceae. 2, Eur.

Piccolia A. Massal. (1856) = Biatorella (Biatorell.).

Piceomphale Svrček (1957) = Ciboria (Sclerotin.) fide Baral (*SA* **13**: 113, 1994).

Pichia E.C. Hansen (1904), Saccharomycetaceae. 94 (fide Barnett *et al.*, 1990), widespr. See v. Arx *et al.* (*Stud. Mycol.* **14**, 1977), Boidin *et al.* (*BSMF* **80**: 396, **81**: 5, 197, 566, 1964-1965; key), Kreger van Rij (1984; key), Kurtzman (*Mycol.* **84**: 72, 1992; DNA-relatedness), Miller *et al.* (*Syst. appl. Microbiol.* **12**: 191, 1989; OFAGE), Poncet (*Mycopath.* **57**: 99, 1975; numer. tax.), Yamada *et al.* (*Biosc. Biotech. Biochem.* **58**: 1245, 1994; gen. concept).

Picoa Vittad. (1831), Balsamiaceae. 5 (hypogeous), Eur., Asia, N. Am., N. Afr.

Picromyces Battarra ex Earle (1909) ≡ Hebeloma (Cortinar.).

Pidoplitchkoviella Kiril. (1975), Microascaceae. 1, former USSR.

Piedraia Fonseca & Leão (1928), Piedraiaceae. 2 (on hair), trop. *P. hortae* (**black piedra** of humans). See Boedijn (*Mycopath.* **11**: 354, 1959), Takashio & Vanbreuseghem (*Mycol.* **63**: 612, 1971), van Uden *et al.* (*Revta bras. Biol.* **3**: 271, 1963). Anamorph *Trichosporon.*

Piedraiaceae Viégas ex M.E. Barr (1979), Dothideales. 1 gen. (+ 2 syn.), 2 spp. Stromata small, black, multiloculate, erumpent; locules opening by irregular lysigenous breakdown of apical cells; interascal tissue absent; asci subglobose, one or few per locule, dehiscence unknown; ascospores hyaline, aseptate, with a gelatinous sheath which is strongly attenuated at each apex. Anamorph unknown. Keratinophilic, trop.

piedra, black, see *Piedraia*; **white -**, see *Trichosporon*.

Piemycus Raf. (1813) ≡ Piesmycus (Lycoperd.).

Piersonia Harkn. (1899) = Choiromyces (Helvell.) fide Trappe (1975).

Piesmycus Raf. (1808) ? = Bovista (Lycoperd.).

Pietria fungaia, see stone-fungus.

Piggotia Berk. & Broome (1851), Mitosporic fungi, 6.A2.19. 2, temp. See Sutton (1980).

Pigments. Characteristic pigments are produced by a wide variety of fungi. Species of *Drechslera* s.l. give hydroxyanthraquinones (e.g. helminthosporin ('maroon' in colour), catenarin (red), cynodontin ('bronze'), tritisporin (red brown)); the similar compound, erythroglaucin (red) is produced by forms of *Aspergillus glaucus* which gives in addition, auroglaucin (orange) and flavoglaucin (yellow). Among other pigments, investigations have been made on aurofusarin (orange yellow) and rubrofusarin from *Fusarium culmorum*; aurantin (yellow) and oosporin (purple-brown with ferric chloride ($FeCl_3$)) from *Chrysonilia sitophila* (teleomorph *Neurospora sitophila*); boletol (blue) from *Boletus luridus* and other species; citromycetin, chrysogenin, citrinin, fulvic acid, and other yellow pigments from *Penicillium*. Many mycotoxins are pigmented, e.g. naphthoquinones from *Penicillium* and *Aspergillus*. Also 'yellow rice' refers to the colour of rice infected by pigment and toxin producing penicillia. The important fungal sex hormones such as trisporic acid from *Blakeslea trispora* are carotenoid pigments (Turner & Aldridge, *Fungal metabolites* II, 1983). Pigments from fungi are involved in biodeterioration of commodities, such as paint. The Chinese red rice (angkak, q.v.) is used for colouring food. See Wolf (*J. Elisha Mitch. Sci. Soc.* **89**: 184, 1973; synthesis, 261 refs).

Many pigments occur in lichens and may be red (e.g. rhodophyscin), yellow (e.g. usnic acid (q.v.), stictaurin), orange (e.g. parietin, q.v.), bright emerald green (e.g. vulpinic acid), or brown (e.g. *Parmelia*-brown). The chemical structures of pigments in thalli are mainly known whilst those of apothecial tissues (e.g. epithecium colours) are less understood. Lichen pigments are mainly pulvic acid derivatives, usnic acids, and anthraquinones.

See see Carotene, Dyeing, Metabolic products, Wood attacking fungi.

Pila Speg. (1923) [non C.E. Bertrand & Renault (1892), fossil *Algae*] ≡ Gastropila (Lycoperd.).

Pilacre Fr. (1825) nom. dub. (Basidiomycetes) fide Donk (*Persoonia* **4**: 325, 1966).

Pilacrella J. Schröt. (1887) = Atractiella (Hoehnelomycet.) fide Oberwinkler & Bauer (*Sydowia* **41**: 224, 1989).

Pilaira Tiegh. (1875), Pilobolaceae. 5 (esp. coprophilous), N. temp. See Mil'ko (*Mikol. i Fitopatol.* **4**: 262, 1970), Fletcher (*J. Biol. Educ.* **5**: 229, 1971, *TBMS* **61**: 553, 1973; phototropism), Zycha *et al.* (1969; key), Wood & Cooke (*TBMS* **86**: 672, 1986; parasitism, **88**: 247, 1987; nutrition).

Pilát (Albert; 1903-74). From 1930 worked at the National Museum, Prague (head of Mycological Dept, 1965). Noted for *Atlas des champignons de l'Europe*, 1934-48 (including *Pleurotus*, 1935; *Polyporaceae*, 1936-42); *Agaricus*, 1951; *Houby Československa ve svém životním prostředí*, 1969. See Kotlaba & Pouzar (*TBMS* **65**: 163, 1975; portr.), Singer (*Mycol.* **67**: 445, 1975; portr.), Stafleu & Cowan (*TL-2* **4**: 265, 1983).

Pilatia Velen. (1934) = Urceolella (Hyaloscyph.) fide Raschle (1977).

Pilatoporus Kotl. & Pouzar (1990), Coriolaceae. 2, Eur.

pileate, having a pileus.

pileipellis, see pellis.

pileocystidium, see cystidium.

Pileolaria Castagne (1842), Pileolariaceae. 20 (on *Anacardiaceae*), widespr.

Pileolariaceae (Arthur) Cummins & Y. Hirats. (1983), Uredinales. 3 gen. (+ 2 syn.), 32 spp. Pycnia discoid, with bounding structures, subcuticular or subepidermal; Aeciospores pedicellate (Uraecium); urediniospores pedicellate; teliospores pedicellate, unicellular; the spore states often distinctively ornamented.

pileolus, a small pileus.

pileus, the hymenium-supporting part of the basidioma of non-resupinate *Basidiomycetes*; cap (Fig. 4A).

Pilgeriella Henn. (1900), Parodiopsidaceae. 2, S. Am. Anamorph *Septoidium*.

Pilidiella Petr. & Syd. (1927) = Coniella (Mitosp. fungi) fide Sutton (1980).

Pilidium Kunze (1823), Mitosporic fungi, 8.A1.15. 2, widespr. See Sutton (1980), Palm (*Mycol.* **83**: 787, 1992; synanamorph *Hainesia*).

Piligena Schumach. (1821) = Onygena (Onygen.) fide Saccardo (*Syll. Fung.* **20**: 230, 1911).

Pilimelia W.D. Kane (1966), *Actinomycetes*, q.v.

Piline Theiss. (1917) = Perisporiopsis (Parodiopsid.) fide Müller & v. Arx (1962).

Pillulasclerotes Stach & Pickh. (1957), Fossil fungi (*f. cat.*). 1 (Carboniferous), Germany.

Pilobolaceae Corda (1842), Mucorales. 3 gen. (+ 3 syn.), 13 spp. Sporangia columellate, specialized liberation mechanism present; zygospores smooth, borne on tongs-like or apposed suspensors.

Pilobolus Tode (1784), Pilobolaceae. 7 (coprophilous), widespr. See Grove (*in* Buller, **6**: 190, 1934), Nand & Mehrotra (*Sydowia* **30**: 283, 1978; spp.), Zycha *et al.* (1969), Hu *et al.* (*Mycosystema* **2**: 111, 1989; revision).

Pilocarpaceae Zahlbr. (1905), Lecanorales (L). 3 gen. (+ 8 syn.), 36 spp. Thallus crustose; ascomata apothecial, brightly coloured, with a poorly-developed margin composed of loosely intertwined hyphae, sometimes appearing woolly; interascal tissue of branched and anastomosing paraphyses, usually immersed in gel, the apices not swollen; asci with a well-developed I+ apical cap and a strongly I+ tube-like apical ring, usually without a well-developed ocular chamber, with an outer I+ gelatinized layer; ascospores hyaline, often elongated, with one or more septa, usually without a sheath. Anamorphs pycnidial. Lichenized with green algae, widespr.
  Doubtfully distinct from *Ectolechiaceae* (Lücking *et al., Bot. Acta* **107**: 393, 1994).

Pilocarpon Vain. (1890) = Byssoloma (Pilocarp.).

Pilocratera Henn. (1891) ≡ Cookeina (Sarcoscyph.).

Piloderma Jülich (1969), Atheliaceae. 7, widespr.

Pilonema Nyl. (1860) = Polychidium (Placyth.) fide Hawksworth (*in litt.*).

Pilopeza Fr. (1849) ≡ Psilopezia (Otid.).

Pilophora Wallr. (1833) [non Jacq. (1802), *Palmae*] = Rhizopus (Mucor.) fide Hesseltine (1955). See *Crinofera*.

Pilophoron (Tuck.) Tuck. (1858) = Pilophorus (Stereocaul.).

Pilophorum Nyl. (1857), ≡ Pilophorus (Stereocaul.).

Pilophorus Th. Fr. (1857), Stereocaulaceae (L). 11, cosmop. (mainly temp.). See Jahns (*Lichenologist* **4**: 199, 1970; Eur., *Mycotaxon* **13**: 289, 1981; monogr., key), Hawksworth *et al.* (*Taxon* **21**: 327, 1972; nomencl.), Hertel & Rambold (*Pl. Syst. Evol.* **158**: 289, 1989; gen. concept).

Piloporia Niemalä (1982), Coriolaceae. 1, former USSR.

Pilosace (Fr.) Quél. (1873) nom. dub. (Agaric.) fide Singer (1951).

pilose, covered with hairs.

Pilula Massee (1910) = Dimerina (Pseudoperispor.) fide Clements & Shear (1931).

Pilula Harker, Sarjeant & Caldwell (1990) [non Massee (1910)], Fossil fungi (? mycorrhizal). 1, Eur.

Pilulina Arnaud (1954), Mitosporic fungi, 1.A2.?. 1, France.

Pimina Grove (1888) = Zygosporium (Mitosp. fungi) fide Mason (*Mycol. Pap.* **5**, 1941).

Piminella Arnaud (1954) = Gonytrichum (Mitosp. fungi) fide Gams & Holubová-Jechová (*Stud. Mycol.* **13**, 1976).

Pinacisca A. Massal. (1854) = Hymenelia (Hymenel.).

Pindara Velen. (1934), Sarcoscyphaceae. 1, Eur. See Svrček (*Česká Myk.* **1**: 45, 1947), Kristiansen (*Agarica* **5**: 105, 1984).

pine moss, species of *Alectoria* and *Bryoria*.

pine mushroom, see matsu-take.

Pinoyella Castell. & Chalm. (1919) = Trichophyton (Mitosp. fungi).

Pinuzza Gray (1821) = Suillus (Bolet.) fide Singer (1945).

pionnotes (of *Fusarium*), a spore mass having a fat- or grease-like appearance; **pseudo-**, see sporodochium.

Pionnotes Fr. (1849) = Fusarium (Mitosp. fungi) fide Wollenweber & Reinking (*Die Fusarien*: 6, 1935).

Pionospora Th. Fr. (1859) = Pertusaria (Pertusar.).

Piptarthron Mont. ex Höhn. (1918), Anamorphic ? Dothideaceae, 4.A/C1.1. Teleomorph *Planistroma*. 6 (on *Agavaceae*), widespr. See Sutton (1980), Sutton (*TBMS* **81**: 407, 1983), Ramaley (*Mycotaxon* **42**: 63, 1991; teleomorph, **55**: 255, 1995; key 6 spp.).

Piptocephalidaceae J. Schröt. (1886), Zoopagales. 3 gen. (+ 9 syn.), 51 spp. Sporangiospores borne in merosporangia (q.v.) on branched or unbranched aerial sporangiophores; zygospores warty, borne on tongs-like suspensors which are often spirally wound and with globose outgrowths; obligate mycoparasites of *Mucorales* (rarely *Penicillium* and *Wynnea*).

Piptocephalis de Bary (1865), Piptocephalidaceae. *c.* 20 (mycoparasites esp. of *Mucorales*), widespr. See Dobbs & English (*TBMS* **37**: 375, 1954; *P. xenophila* on non-mucoralean hosts), Leadbetter & Mercer (*TBMS* **39**: 17, 1956, **40**: 109, 1957; zygospores), Berry (*Mycol.* **51**: 824, 1961; parasitism), Zycha *et al.* (1969; key), Benjamin (*Aliso* **4**: 321, 1959), Kuzuha (*J. Jap. Bot.* **51**: 123, 1976), Kirk (*TBMS* **70**: 335, 1978), Jeffries (*Mycol.* **71**: 209, 1979), Manocha (*CJB* **63**: 772, 1985; mycoparasitism), Manocha & McCullough (*CJB* **63**: 1985; mycoparasitism).

Piptoporus P. Karst. (1881), Coriolaceae. 3, widespr. *P. betulinus*, razor strop fungus, on birch. See Corner (*Beih. Nova Hedw.* **78**: 140, 1984).

Piptostoma Berk. & Broome (1875) nom. dub. fide Petch (*Ann. R. bot. Gdns Peradeniya* **6**: 341, 1917); ? lapsus for *Piptostomum*.

Piptostomum Lév. (1845) nom. dub. (Mitosp. fungi) fide Höhnel (*Sber. Akad. Wiss. Wien* **119**: 617, 1910).

Piptostroma Fr. (1849) ≡ Piptostomum (Mitosp. fungi).

Pirella Bainier (1882), Thamnidiaceae. 2, Eur., USA, former USSR, India. See Benny (1973), Benny & Schipper (*Mycol.* **84**: 52, 1992).

Pirex Hjortstam & Ryvarden (1985), Hyphodermataceae. 1, N. Am.

Piricauda Bubák (1914), Mitosporic fungi, 1.D2.24. 1, S. Am. See Hughes (*CJB* **38**: 921, 1960).

Piricaudilium Hol.-Jech. (1988), Mitosporic fungi, 1.D2.24. 1, Cuba.

Piricaudiopsis J. Mena & Mercado (1987), Mitosporic fungi, 1.D2.24. 1, Cuba.

Piricularia, see *Pyricularia*.

piricularin, a phytotoxin from *Pyricularia oryzae* (Togashi *et al., Ann. phytopath. Soc. Japan* **25**: 142, 1960).

piriform, see pyriform.

Piringa Speg. (1910) [non Juss. (1820), *Rubiaceae*] = Camarosporium (Mitosp. fungi) fide Petrak & Sydow (*Ann. Myc.* **33**: 157, 1935).

Pirispora Faurel & Schotter (1966), Mitosporic fungi, 4.A1.?. 1, Algeria.

Piriurella Cookson & Eisenack (1979), Fossil fungi. 2 (Cretaceous - Tertiary), Australia, Canada. ? = Alternaria (Hyphom.) fide Smith & Chaloner (*N. Jb. Geol. Paläont. Mh.* **11**: 701, 1979).

Pirobasidium Höhn. (1902) = Coryne (Mitosp. fungi) fide Groves & Wilson (1962).

Pirogaster Henn. (1901), Sclerodermataceae. 1, Java. ? = Scleroderma (Sclerodermat.).

Piromonas E. Liebet. (1910), Chytridiomycetes (inc. sed.). 3, Eur. See Orpin (*J. gen. Microbiol.* **99**: 107, 1977; life history).

Piromyces J.J. Gold, I.B. Heath & Bauchop (1988), Neocallimastigaceae. 6 (anaerobic rumen fungi), widespr. See Gaillard-Martinie *et al.* (*Curr. Microbiol.* **24**: 159, 1992).

Pirostoma (Fr.) Fuckel (1870) nom. dub. (Mitosp. fungi) fide Sutton (1977).

Pirostoma (Fr.) Sacc. (1896) nom. dub. (Mitosp. fungi) fide Sutton (1977).

Pirostomella Sacc. (1914), Mitosporic fungi, 2.A2.?. 2, Philipp.

Pirottaea Sacc. (1878), Dermateaceae. *c.* 16, esp. N. temp. See Nannfeldt (*Symb. bot. upsal.* **25**(1): 1, 1985; key 23 spp.).

Pirozynskia Subram. (1972), Anamorphic Asterinaceae, 1.B2.24. Teleomorph *Maurodothina*. 2, Canada, USA.

pisatin, a phytoalexin (q.v.) from pea (*Pisum sativum*).

Pisocarpium Link (1808) ≡ Pisolithus (Sclerodermat.).

Pisolithus Alb. & Schwein. (1805), Sclerodermataceae. 2, widespr.

Pisomyxa Corda (1837), ? Eurotiales (inc. sed.). 2, S. Am., Madeira.

Pistillaria Fr. (1821), Typhulaceae. 50, N. temp. = Typhula (Typhul.) fide Berthier (1976).

Pistillaria Jeekel, Tuzet, Manier & Jolivet (1959) = Enterobryus (Eccrin.) fide Manier & Lichtwardt (*Ann. Sci. Nat. Bot.*, sér. 12, **9**: 519, 1968).

Pistillina Quél. (1880), Typhulaceae. 4, N. temp. See Corner (1950).

Pithoascina Valmaseda, A.T. Martínez & Barrasa (1987) = Eremomyces (Eremomycet.) fide Malloch & Sigler (*CJB* **66**: 1929, 1988).

Pithoascus Arx (1973), Microascaceae. 6, widespr. See v. Arx (*Persoonia* **7**: 367, 1973; key), v. Arx *et al.* (*Beih. Nova Hedw.* **94**, 1988), Malloch (*SA* **6**: 147, 1987; status), Valmaseda *et al.* (*CJB* **65**: 1802, 1987; concept).

Pithomyces Berk. & Broome (1875), Anamorphic Pleosporaceae, 1.C2/D2.1. Teleomorph *Leptosphaerulina*. 11, widespr. See Ellis (*Mycol. Pap.* **76**, 1960; key), Rao & de Hoog (*Stud. Mycol.* **28**: 1, 1986; key), Morgan-Jones (*Mycotaxon* **30**: 29, 1987; synonymy), Brewer *et al.* (*Proc. N.S. Inst. Sci.* **38**: 73, 1989; production of sporidesmin and sporidesmolides), Roux (*TBMS* **86**: 319, 1986; teleomorph). See facial eczema.

Pithosira Petr. (1949), Mitosporic fungi, 3.A2.?. 1, S. Am. See v. Arx (*Persoonia* **11**, 1981; redescr.).

Pithospermum Mont. ex Berk. & Broome (1856) = Sporoschisma (Mitosp. fungi) fide Mussat (*Syll. Fung.* **15**: 294, 1901).

Pithya Fuckel (1870), Sarcoscyphaceae. 5, N. temp. See Nannfeldt (*Svensk bot. Tidskr.* **43**: 468, 1949), Denison (*Mycol.* **64**: 616, 1972), Meléndez-Howell *et al.* (*Anales Inst. Biol., Univ. Nac. México*, **60**: 9, 1990).

Pithyella Boud. (1885), Hyaloscyphaceae. 5, Eur., N. & S. Am. See Korf & Zhuang (*Mycotaxon* **29**: 1, 1987).

Pittierodothis Chardón (1939) = Protoscypha (Patellar.) fide v. Arx & Müller (1975).

Pittocarpium Link (1815) ? = Fuligo (Myxom.) fide Martin (1966).

Pittostroma Kowalski & T.N. Sieber (1992), Mitosporic fungi, ?.?.?. 1, Poland.

pityriasis versicolor ('tinea versicolor'), a superficial skin disease of humans (*Malassezia furfur*).

Pityrosporum Sabour. (1904) = Malassezia (Mitosp. fungi) fide Keddie (*XIII Int. Congr. Derm. (Munich)*: 867, 1968). See pityriasis versicolor.

Placella Syd. (1938), Mitosporic fungi, 5.A1.?. 1, Australia.

Placentaria Auersw. ex Rabenh. (1851) ? = Periola (Mitosp. fungi).

Placidiopsis Beltr. (1858), Verrucariaceae (L). 10, Eur., N. Afr., N. Am., Asia. See Breusse (*Nova Hedw.* **58**: 229, 1994; key 2 N. Afr. spp.), Thomson (*Bryologist* **90**: 27, 1987; key 2 N. Am. spp.), Zschacke (*Rabenh. Krypt.-Fl.* **9**, 1 (1): 638, 1933).

Placidium A. Massal. (1855) = Catapyrenium (Verrucar.) fide Brusse (*Stapfia* **23**, 1990).

Placoasterella Sacc. ex Theiss. & Syd. (1915), Asterinaceae. 3, Afr., Australia. See Tyson & Griffiths (*TBMS* **66**: 249, 1976; morphology, ultrastr.).

Placoasterina Toro (1930), ? Asterinaceae. 1, S. Am.

Placocarpus Trevis. (1860), Verrucariaceae (L). 1, Eur., See Breuss (*Pl. Syst. Evol.* **148**: 313, 1985).

Placocrea Syd. (1939), ? Mycosphaerellaceae. 1, Ecuador. See Petrak (*Sydowia* **6**: 336, 1952).

Placoderma (Ricken) Ulbr. (1928) ≡ Piptoporus (Coriol.) fide Donk (1974).

Placodes Quél. (1886) ≡ Fomes (Coriol.).

Placodiella (Zahlbr.) Szatala (1941) = Solenopsora (Bacid.) fide Hafellner (1984).

placodioid (of a crustose lichen thallus), disc-shaped with plicate lobes at the circumference (Fig. 21C).

placodiomorph, a 2-celled spore with a thickened septum which may or may not have a pore. Cf. polarilocular.

Placodiomyces E.A. Thomas ex Cif. & Tomas. (1953) = Lecanora (Lecanor.).

Placodion Adans. (1763) nom. rej. = Peltigera (Peltiger.).

Placodiplodia Bubák (1916), Mitosporic fungi, 8.B2.15. 1, Philipp.

placodium, see stroma, thyriothecium.

Placodium Hill (1773) nom. dub. (Fungi, inc. sed., L) fide Laundon (*in litt.*).

Placodium F.H. Wigg. (1780) = Xanthoria (Teloschist.).

Placodium (Ach.) DC. (1805) = Caloplaca (Teloschist.).

Placodothis Syd. (1928), ? Ascomycota (inc. sed.). 1, W. Indies.

Placographa Th. Fr. (1860) = Lithographa (Rimular.).

Placoheppia (Zahlbr.) Oxner (1956) nom. inval. = Heppia (Hepp.).

placoid, see placodioid.

Placolecania (J. Steiner) Zahlbr. (1906) = Solenopsora (Bacid.).

Placolecanora Räsänen (1940) ≡ Protoparmeliopsis (Lecanor.).

Placolecis Trevis. (1857), Catillariaceae (L). 1, Eur. See Schneider (*Bibl. Lich.* **13**, 1980).

Placomaronea Räsänen (1944), Candelariaceae (L). 2 (or 3), S. Am. See Osorio (*Bryologist* **77**: 463, 1974), Poelt (*Phyton* **16**: 189, 1974; key).

Placomelan Cif. (1962), Anamorphic Parmulariaceae, 5.A1.?. Teleomorph *Dothidasteroma*. 1, Dominican Republic.

placomycetoid, pileus with diam : stipe ratio <1. Cf. campestroid.

Placonema (Sacc.) Petr. (1921), Mitosporic fungi, 8.A/B1.19. 3, S. Am. See Nag Raj (1993).

Placonemina Petr. (1921), Mitosporic fungi, 8.B1.15. 1, former Czechoslovakia.

Placoparmelia Henssen (1992), Parmeliaceae (L). 1, Argentina.

Placopezizia Höhn. (1916) = Leptotrochila (Dermat.) fide Schüepp (1959).

Placophomopsis Grove (1921) = Phomopsis (Mitosp. fungi) fide Sutton (1977).

Placopsis (Nyl.) Linds. (1867), Trapeliaceae (L). 34, cosmop. See Brodo (*Bibl. Lich.* **57**: 59, 1995; key 3 N. Am. spp.), Lamb (*Lilloa* **13**: 151, 1947).

Placopyrenium Breuss (1987), Verrucariaceae (L). 2, Eur. See Ménard & Roux (*Mycotaxon* **53**: 129, 1995).

Placosoma Syd. (1924), Asterinaceae. 1, NZ. See Hughes (*N.Z. Jl Bot.* **16**: 311, 1978).

Placosphaerella Pat. (1897) nom. conf. fide Petrak (*Sydowia* **5**: 195, 1951).

Placosphaeria (De Not.) Sacc. (1880), Anamorphic Dothideales, 8.A1.?. Teleomorphs various. 50, widespr.

Placosphaerina Maire (1917), Mitosporic fungi, 8.C1.?. 1, Algeria.

Placostroma Theiss. & Syd. (1914) = Phyllachora (Phyllachor.) fide Cannon (*Mycol. Pap.* **163**, 1991).

Placostromella Petr. (1947), Dothideales (inc. sed.). 1, China.

Placothallia Trevis. (1853) = Rinodina (Physc.).

Placothea Syd. (1931), Mitosporic fungi, 8.A1.?. 1, Philip.

Placotheliomyces Cif. & Tomas. (1953) ≡ Placothelium (Verrucar.).

Placothelium Müll. Arg. (1893), Verrucariaceae (L). 1, USA.

Placothyrium Bubák (1916), Mitosporic fungi, 8.E1.10. 1, Eur.

Plactogene Theiss. (1917) = Dimerina (Pseudoperisp.) fide v. Arx & Müller (1975).

Placuntium Ehrenb. (1818) = Rhytisma (Rhytism.) fide Minter (*in litt.*).

Placynthiaceae A.E. Dahl (1950) nom. cons. prop., Peltigerales (L). 7 gen. (+ 24 syn.), 38 spp. Thallus crustose to squamulose or minutely fruticose; ascomata on the upper surface, sessile, dark, without a well-developed thalline margin; asci with a thickened apex, an I+ apical ring, and ocular chamber and an I+ outer gelatinized coat; ascospores simple or transversely septate. Anamorphs pycnidial. Lichenized with cyanobacteria, mainly on soil or rocks, esp. N. temp.

Placynthiella Elenkin (1909), Trapeliaceae (L). 4, Eur., N. Am. See Coppins & James (*Lichenologist* **16**: 241, 1984; key), Coppins *et al.* (*Lichenologist* **19**: 93, 1987; nomencl.).

Placynthiella Gyeln. (1939) = Placynthiella Elenkin (Trapel.).

Placynthiomyces Cif. & Tomas. (1953) ≡ Placynthium (Placynth.).

Placynthiopsis Zahlbr. (1932), Placynthiaceae (L). 1, C. Afr.

Placynthium (Ach.) Gray (1821), Placynthiaceae (L). 25, cosmop. (mainly temp.). See Gyelnik (*Rabenh. Krypt.-Fl.* **9**, 2 (2), 1940; Eur.), Henssen (*CJB* **41**: 1687, 1963; N. Am.), Weber (*Muelleria* **3**: 250, 1970; Australia).

plage, (1) a smooth, paler-coloured, or colourless spot on a surface; (2) (of basidiospores), esp. a smooth spot above the hilar appendage.

Plagiocarpa R.C. Harris (1973) ? = Lithothelium (Pyrenul.) fide Aptroot (*Bibl. Lich.* **44**, 1991).

Plagiographis C. Knight & Mitt. (1860) ? = Opegrapha (Roccell.) fide Redinger (1938).

Plagiolagynion Schrantz (1962) = Oxydothis (Hyponectr.) fide Barr (1976), Hyde (1995).

Plagionema Subram. & K. Ramakr. (1953) = Ciliochorella (Mitosp. fungi) fide Chona & Munjal (*Ind. Phytopath.* **8**: 74, 1955).

Plagiophiale Petr. (1955), Melanconidaceae. 1, Eur., N. Am. See Barr (*Mycol. Mem.* **7**, 1978).

Plagiorhabdus Shear (1907) = Strasseria (Mitosp. fungi) fide Höhnel (*Mitt. bot. Inst. Techn. Hochsch. Wien* **2**: 26, 1925).

Plagiosphaera Petr. (1941), Lasiosphaeriaceae. 5, Eur., N. & S. Am., Japan. See Barr (*Mycotaxon* **39**: 43, 1990; posn), Walker (*Mycotaxon* **11**: 94, 1980).

Plagiostigme Syd. (1925), Melanconidaceae. 1, C. Am. See Petrak (*Sydowia* **18**: 382, 1965).

Plagiostigmella Petr. (1949), Mitosporic fungi, 4.A1.?. 1, S. Am.

Plagiostoma Fuckel (1870), Valsaceae. 15, Eur., N. Am. See Barr (*Mycol. Mem.* **7**, 1978; key 13 spp.), Bolay (*Ber. schweiz. bot. Ges.* **81**: 398, 1972), Monod (1983).

Plagiostomella Höhn. (1917) = Plagiostoma (Vals.) fide Barr (1978).

Plagiostromella Höhn. (1917), Dothideales (inc. sed.). 4, India, Japan. See Bose & Müller (*Sydowia* **31**: 1, 1979; key).

Plagiothecium Schrantz (1961) [non Bruch & Schimp. (1851), *Musci*] ≡ Plagiolagynion (Hyponectr.).

Plagiothelium Stirt. (1881), Pyrenulales (inc. sed., L). 1, Australia. = Parmentaria (Pyrenul.) fide Filson (*Checklist of Australian lichens*, 1988).

Plagiotrema Müll. Arg. (1885) = Pseudopyrenula (Trypthel.) fide Harris (*in litt.*).

Plagiotus Kalchbr. ex Roum. (1879) nom. inval. = Anthracophyllum (Tricholomat.).

Planctomyces M. Gimesi (1924) nom. dub. (? Fungi). See Wawrik (*Sydowia* **6**: 443, 1952). Also considered a bacterium by Skerman *et al.* (*Approved list of bacterial names*, 1980), Schlesner & Stackebrandt (*Syst. Appl. Microbiol.* **8**: 174, 1986; assigned to new order *Planctomycetales*).

plane, flat.

Planetella Savile (1951), Ustilaginaceae. 1 (on *Cyperaceae*), Canada. Resembles *Anthracoidea* but germination unknown. See Savile (*CJB* **29**: 326, 1951), Vánky (1987).

planetism (of oomycetes), the condition of having motile stages.

Planistroma A.W. Ramaley (1991), Dothideales (inc. sed.). 1 (on *Yucca*), N. Am. See Eriksson & Hawksworth (*SA* **12**: 38, 1993; posn). Anamorph *Piptarthron*.

Planistromella A.W. Ramaley (1993), Dothideales (inc. sed.). 1, U.S.A. See Eriksson & Hawksworth (*SA* **12**: 38, 1993; posn). Anamorph *Kellermania*.

Planobispora Thiemann & Beretta (1968), *Actinomycetes*, q.v.

planoconvex, a flat zygote.

planocyte (**planont**), a motile cell.

planogamete, a motile gamete; zoogamete.

Planomonospora Thiemann, Pagani & Beretta (1967), *Actinomycetes*, q.v.

planont, see planocyte.

Planoprotostelium L.S. Olive & Stoian. (1971), Cavosteliaceae. 1, Brazil.

planospore, a zoospore.

planozygote, (1) a motile zygote; (2) a flat zygote (of a pileus or spore), convex but somewhat flat.

Plant pathogenic fungi. A large proportion of fungal species live in association with plants, as saprobes on plant surfaces or dead plant tissue, as mutualists in mycorrhizal associations or as endophytes or parasites of living plants. A few, e.g. *Uredinales*, *Erysiphaceae*, are obligate parasites of plants. Many parasitic fungi are highly phytopathogenic and induce severe and frequently economically important diseases. They are the most important agents of plant disease; more than 60 per cent of the literature on plant diseases is devoted to fungal infections. Plant pathogens are represented in all the major groups of fungi and some groups, e.g. *Ustilaginales* (smuts), *Uredinales* (rusts), *Peronosporaceae* (downy mildews), *Erysiphaceae* (powdery mildews), are known by the common names of the diseases for which they are responsible. Other groups important as plant pathogens are *Basidiomycetes* (frequently wood-attacking), certain ascomycetes, and particularly Mitosporic fungi. (The teleomorph of a pathogen frequently develops only on dead plant tissue in the terminal phase of the disease.)

The symptoms of fungus diseases in plants are diverse. They may be roughly classified as **necrosis** - death of tissues (cf. anthracnose, blight, canker, damping-off, scab, shot hole); **wilting** - loss of turgor due to the mechanical plugging of the vessels by hyphae or to the action of a toxin secreted by the pathogen; **hyperplasia** - overgrowth (cf. galls, witches' brooms); and **hypoplasia** - dwarfing (and chlorosis).

Transmission of plant pathogenic fungi may be by air (either dry or moist, as in droplet transmission), water, soil, seeds and other propagating material, plant debris, and insects and other animal agents. Humans have spread many important diseases within one country, from one country to another, and from one part of the world to another by transporting infected material in the course of trade.

Control of parasitic disease is (fide Whetzel) based on exclusion, eradication, protection, and immunization. Quarantine procedures, inspection at ports, and legislation governing the movement of plants and plant products, the growth of susceptible plants under conditions unfavourable for the development of the pathogen, and the use of disease-free seed or planting material are based on exclusion. Eliminating ('roguing') diseased plants, the cutting-out of diseased parts, heat treatment of seed material, and the use of eradicant fungicides on infected plants (i.e. against the pathogen) are examples of methods based on eradication. Among protective measures are the dusting or spraying of healthy plants with fungicides. The use of immune or resistant varieties is an example of 'immunization', but active immunization, which is so important in the control of many diseases of animals and man, is of little value against plant diseases. The three main strategies for plant disease control are: phytosanitation (quarantine, removal of inoculum sources, crop rotation, clean seed etc.); protection (use of fungicides, avoidance of predisposing cultural conditions) and resistance (use of disease resistance cultivars).

Lit.: The literature on plant disease is large, but well covered by review journals and databases (see below). The standard general phytopathological texts all deal with fungal diseases, as do most monographs on diseases of particular crop plants. Here a selection has been made of general works and review journals. For additional references see Moreau (*Rev. mycol., Paris* **27**: 41, 1962), *RAM* **46**: 1 and 113, 1967 (select bibliogr.), IMI (*Plant pathologist's pocketbook*, 1968; edn 2, 1983). See also Fungicides, gene-for-gene, Genetics, Geographical distribution, physiologic race, Seed-borne fungi, Special forms.

General: Heald (*Manual of plant diseases*, 1926, edn 2, 1933), Butler & Jones (*Plant pathology*, 1949), Gaumann (*Principles of plant infection*, 1950) [trans. of *Pflanzliche Infektionslehre*, 1946; edn 2, 1951], Walker (*Plant pathology*, 1950; edn 3, 1969),

Stakman & Harrar (*Principles of plant pathology*, 1957), Holton *et al.* (Eds) (*Plant pathology, problems and progress, 1908-1958*, 1959), Horsfall & Dimond (Eds) (*Plant pathology, an advanced treatise*, 3 vols, 1959- 60), Vanderplank (*Plant diseases: epidemics and control*, 1963), Wood (*Physiological plant pathology*, 1967), Tarr (*Principles of plant pathology*, 1971), Horsfall & Cowling (Eds) (*Plant disease, an advanced treatise* 5 vols, 1977-80), Brown & Brotzman (*Phytopathogenic fungi: a scanning electron stereoscopic survey*, 1979), Agrios (*Plant pathology*, 1969; edn 3, 1985), Bruell (*Soilborne plant pathogens*, 1987), Smith *et al.* (*European handbook of plant diseases*, 1988), APS (*Compendia of plant diseases*, series), Schumann (*Plant diseases: their biology and social impact*, 1991).

Tropical: Roger (*Phytopathologie des pays chauds*, 3 vol. 1951-54), Weber (*Bacterial and fungal diseases of plants in the tropics*, 1973) Holliday (*Fungus diseases of tropical crops*, 1980), Thurston (*Tropical plant diseases*, 1984).

Forest pathology: Boyce (*Forest pathology*, edn 3, 1961), Peace (*Pathology of trees and shrubs with special reference to Britain*, 1961), *Diseases of widely planted forest trees* (1964; lists diseases and their distribution), Browne (*Pests and diseases of forest plantation trees. An annotated list of the principle species occurring in the British Commonwealth*, 1968), Hepting (*Diseases of forest and shade trees in the United States* [*Agric. Handb. Forest Serv. US* **386**], 1971), *Forestry abstracts* 1939- [also as TREE-CD]. See also Wood-attacking fungi.

Legislation and quarantine: Ebbels & King (Eds) (*Plant health*, 1979), CABI/EPPO (*Quarantine pests from Europe*, 1992).

Methods: Riker & Riker (*Introduction to research on plant diseases*, 1936), IMI (*Plant pathologist's pocketbook*, 1968; edn 2 1983), Ingram & Helgeson (Eds) (*Tissue culture methods for plant pathologists*, 1980), Fox (*Principles of diagnostic techniques in plant pathology*, 1993), Schots *et al.* (Eds) (*Modern assays for plant pathogenic fungi: identification, detection and quantification*, 1994).

Identification: Stevens (*The fungi which cause plant disease*, 1913; *Plant disease fungi*, 1925 [reprint, 1966]), Viennot-Bourgin (*Les champignons parasites des plantes cultivées*, 1949), Brooks (*Plant diseases*, edn 2, 1953), Kreisel (*Die phytopathogen-Grosspilze Deutschlands*, 1961), *IMI Descriptions of fungi and bacteria* (1964-), Lindquist (*Revta Fac. Agron. Univ. nac. La Plata* **43**: 1, 1967; key), Rossman *et al.* (*A literature guide for the identification of plant pathogenic fungi*, 1987), Farr *et al.* (*Fungi on plants and plant products in the United States*, 1989), *Names of British plant diseases and their control* [*Phytopath. Pap.* **28**, 1984], v. Arx (*Plant pathogenic fungi*, 1987), Holliday (*A dictionary of plant pathology*, 1989), Hodge (*The fungal spore and disease initiation in plants and animals*, 1991), Brandenburger (*Parasitische Pilze au Gefassplanzen in Europa*, 1985).

History: Whetzel (*An outline of the history of phytopathology*, 1918), *Phytopathological classics* (1926-), Large (*The advance of the fungi*, 1940), Parris (*A chronology of plant pathology*, 1968), Ainsworth (*Introduction to the history of plant pathology*, 1981), Schumann (*Plant diseases: their biology and social impact*, 1991).

Review journals: *Review of applied mycology*, 1922- 69 (abstracts; *Index* **1-40** (1922-61), 1968),1970- as *Review of plant pathology* [also available in a variety of electronic formats, incl. PEST-CD;

see Literature]; *Annual review of phytopathology*, 1963-; *Advances in plant pathology*, 1982-95.

plaque, a clear area in a bacterial colony caused by localized viral lysis; also applied to similar areas in fungal cultures (e.g. Riegman & Wessels, *TBMS* **75**: 325, 1980).

Plasia Sherwood (1981), Anamorphic Leotiaceae, 8.C1.15/16. Teleomorph *Durella*. 1, Eur.

plasmalemma, outer membrane composed of phospholipids and proteins which surrounds a cell and regulates exchange of materials between the cell and its environment; cell, cytoplasmic, or plasma membrane.

plasmalemmasome, an intracytoplasmic vesicle (formed by invagination of the plasmalemma) filled with tubular diverticula. Cf. lomasome.

plasmamembranic vesicle, sporangial vesicle, the membrane of which is the same membrane bounding the protoplasm within the sporangium, c.f. homohylic vesicle, precipitative vesicle.

plasmatoogosis (of *Pythiaceae*), a bud-like outgrowth (like a prosporangium in form) in the host tissue (Sideris, *Mycol.* **23**: 255, 1931).

plasmodesma (pl. **-ta**), an isthmus-like strand of protoplasm connecting adjacent cells.

plasmodia, see plasmodium.

plasmodic granules, very small, dark-coloured particles on the surface of the peridium, and frequently of the spores, in the *Cribrariaceae*; dictydine granules.

plasmodiocarp (of myxomycetes), a sessile and vein-like sporangium, like part of the larger veins of a plasmodium.

Plasmodiophora Woronin (1877), Plasmodiophoraceae. 6 (obligate endoparasites of plants), widespr. *P. brassicae* (club root of crucifers). See Karling (1968; key), Colhoun (*Phytopath. Pap.* **3**, 1958), Ingram & Tommerup (*Proc. r. Soc.* B **180**: 103, 1972; life history), Buczaki *et al.* (*TBMS* **65**: 295, 1975; specialization), Jousson *et al.* (*Acta Agric. Scand.* **25**: 261, 1975; bibliogr., 612 refs).

Plasmodiophoraceae Zopf ex Berl. (1888), Plasmodiophorales. 13 gen. (+ 8 syn.), 42 spp. Obligate endoparasites of phanerogams, algae and fungi, frequently inducing hypertrophy of infected cells. Cosmop.

Plasmodiophorales (Plasmodiophorida), Plasmodiophoromycota. 2 fam., 15 gen. (+ 10 syn.), 46 spp. Obligate endoparasites of flowering plants (*Plasmodiophora*, *Spongospora*), algae (*Woronina* in *Vancheria*; *Sorodiscus* in *Chara*), and fungi (*Woronina* in *Saprolegniales*); frequently inducing hypertrophy of the infected cells. Fams:
(1) **Endemosarcaceae**.
(2) **Plasmodiophoraceae**.

Plasmodiophoromycetes, see *Plasmodiophoromycota*.

Plasmodiophoromycota (Plasmodiophoromycetes, Plasmodiophorida, Plasmodiophorina). Protozoa; plasmodiophorids. 1 ord., 2 fam., 16 gen. (+ 10 syn.), 46 spp. Trophic phase intracellular in algal, fungal or plant host cells; plasmodia multinucleate, unwalled, cell division mitotic and cruciform (division producing a cross-like structure at metaphase), developing into either sporangia (forms zoospores or cytosori; resting structures); zoospores with two anteriorly directed whiplash flagella; flagellae smooth, lacking mastigonemes, equal in length. Obligate symbionts in soil or freshwater habitats; require co-culturing with their hosts needed for laboratory studies. A single order.

Plasmodiophorales.
The placement of these fungi has varied. Dylewski (1990) kept them with the mastigomycete fungi (e.g. *Chromista*); they are retained in the *Protozoa* following Barr (1992) and Corliss (1994). Olive (1975)

treated them as a separate class, while Corliss (1994) placed them as one of several orders in the phylum *Opalozoa*.
Lit.: Corliss (1994), Dylewski (*in* Margulis *et al.* (Eds), *Handbook of the Protoctista*: 399, 1990), Karling (*The Plasmodiophorales*, 1943; edn 2, 1968), Olive (1975), Waterhouse (*in* Ainsworth *et al.*, *The Fungi* **4B**: 75, 1973; key gen.).

plasmodium (pl. **plasmodia**) (of myxomycetes), a multinucleate, motile mass of protoplasm bounded by a plasma membrane but lacking a wall, generally reticulate, characteristic of the growth phase; Alexopoulos (*Mycol.* **52**: 17, 1961) distinguishes: **protoplasmodium**, an undifferentiated microscopic plasmodium which gives rise to a single sporangium, as in *Echinosteliales*; **aphanoplasmodium**, a plasmodium composed of a network of undifferentiated strands of non-granular protoplasm, as in *Stemonitis*; **phaneroplasmodium**, a plasmodium composed of a well-differentiated advancing fan and thick strands of granular protoplasm exhibiting ecto- and endoplasmic regions, as in *Physarales*; **aggregate plasmodium**, **pseudoplasmodium** (of *Acrasales*), a structure formed by the massing of separate myxamoebae or cells before reproduction (a grex); **filoplasmodium**, the net-like pseudoplasmodium of the *Labyrinthulales*.

plasmogamospore, aeciospore (Laundon, *TBMS* **58**: 345, 1972).

plasmogamy, fusion of two cells or plasmodial cytoplasms without karyogamy (nuclear fusion) or a precursor to karyogamy.

Plasmopara J. Schröt. (1886), Peronosporaceae. 20, widespr. *P. viticola* (vine downy mildew).

Plasmoparopsis De Wild. (1896) = Pythiogeton (Pyth.) fide Dick (*in press*).

Plasmophagus De Wild. (1895), Rozellopsidales (inc. sed.). 3, Eur., N. Am.

Platisma P. Browne ex Adans. (1763) = Ramalina (Ramalin.) fide Santesson & Culberson (*Bryologist* **69**: 100, 1966).

Platisma Hoffm. (1790) = Cetraria (Parmel.) fide Santesson & Culberson (*Bryologist* **69**: 100, 1966).

Platismatia W.L. Culb. & C.F. Culb. (1968), Parmeliaceae (L). 10, cosmop. See Culberson & Culberson (*Contr. U.S. natn. Herb.* **34**: 449, 1968; monogr.).

Platisphaera Dumort. (1822) ≡ Lophiostoma (Lophiostomat.) fide Holm (1975).

Platistoma, see *Platystoma*.

Platistomum, see *Platystomum*.

Platycarpa Couch (1949), Platygloeaceae. 1 (on *Pteridophyta*), Am. See Couch (*Mycol.* **41**: 427, 1949).

Platycarpium P. Karst. (1905) = Fusamen (Mitosp. fungi) fide v. Arx (*Verh. K. ned. Akad. Wet.* **51**: 1, 1957).

Platychora Petr. (1925), Venturiaceae. 2, N. temp.

Platygloea J. Schröt. (1887), Platygloeaceae. 27, widespr. See Bandoni (*Mycol.* **48**: 821, 1956; key).

Platygloeaceae Racib. (1909), Platygloeales. 21 gen. (+ 8 syn.), 78 spp. Basidioma pezizoid, effused, putular or intrahymenial, waxy, gelatinous or pellicular; basidia auricularioid, with septate metabasidium; hyphal septal pores simple; saprobic or mycoparasitic.

Platygloeales, Ustomycetes. 1 fam., 24 gen. (+ 14 syn.), 80 spp. Fam.:
Platygloeaceae.

platygonidia, photobionts occurring in stellately or orbicular spreading colonies (e.g. *Cephaleuros*) (obsol.).

Platygramma G. Mey. (1825) = Phaeographis (Graphid.).

Platygramma Leight. (1854) = Enterographa (Roccell.).

Platygramme Fée (1874), Ostropales (inc. sed., L). 9, S. Am.

Platygrapha Nyl. (1855) = Schismatomma (Roccell.) fide Tehler (1993).

Platygrapha Berk. & Broome (1870) [non Nyl. (1855)] ? = Ocellularia (Thelotrem.). See Sherwood (*Mycotaxon* **3**: 233, 1976).

Platygraphis Hook. f. (1867) ≡ Platygrapha (Roccell.).

Platygraphomyces Cif. & Tomas. (1953) ≡ Platygrapha (Roccell.).

Platygraphopsis Müll. Arg. (1887), Ostropales (inc. sed., L). 1, trop. Am., Papua New Guinea.

Platypeltella Petr. (1929), Microthyriaceae. 2, C. Am. See Farr (*Mycotaxon* **15**: 448, 1982).

platyphyllous, broadly lobed.

Platyphyllum Vent. (1799) nom. rej. ≡ Cetraria (Parmel.).

Platysma Hill (1773) = Ramalina (Ramalin.) fide Santesson & Culberson (*Bryologist* **69**: 100, 1966).

platysmoid, (1) (of lichen thalli), loosely attached foliose thalli with ascending lobes, as in *Platismatia* (obsol.); (2) (of tissues), a scleroplectenchyma in which the hyphae are brown (Yoshimura & Shimada, *Bull. Kochi Gakuen Jun. Coll.* **11**: 13, 1980).

Platysphaera Trevis. (1877) ≡ Platisphaera (Lophiostomat.).

Platysphaeria, see *Platisphaera*.

Platyspora Wehm. (1961) = Graphyllium (Hyster.) fide Barr (*SA* **12**: 27, 1993).

Platysporoides (Wehm.) Shoemaker & C.E. Babc. (1992), Pleosporaceae. 11, widespr. = Clathrospora (Diadem.) and Graphyllium (Diadem.) fide Barr (*SA* **11**: 27, 1993).

Platysticta Cooke (1889) = Stictis (Stictid.) fide Sherwood (1977).

Platystoma (Fr.) Bonord. (1851) = Lophiostoma (Lophiostomat.) fide Mussat (*Syll. Fung.* **15**: 295, 1901).

Platystomum Trevis. (1877) = Lophiostoma (Lophiostomat.). See Chesters & Bell (1970), Eriksson (*SA* **10**:144, 1991).

Pleamphisphaeria Höhn. (1919) ≡ Titanella (Pyrenul.). See Harris (1989).

Plearthonis Clem. (1909) = Arthonia (Arthon.).

Plecostoma Desv. (1809) ≡ Geastrum (Geastr.).

Plecotrichum Corda (1831) nom. dub. (Mitosp. fungi). See Hughes (*CJB* **36**: 798, 1958).

Plectania Fuckel (1870), Sarcosomataceae. 4, N. temp. See Korf (*Mycol.* **49**: 107, 1957).

Plectascales. *Ascomycota* with ascoma in which the asci were irregularly distributed; sometimes extended and including *Erysiphales*, *Eurotiales* (q.v.) and *Myriangiaceae*. See also *Plectomycetes*.

plectenchyma, a thick tissue formed by hyphae becoming twisted and fixed together; synchyma (Vuillemin); it is **prosenchyma (proso-)** when the hyphal elements are seen to be hyphae; **pseudoparenchyma (para-)**, when they are not; Vuillemin (1912) distinguished as **merenchyma** tissue derived by cell division in several planes, and Degelius (1954) used **euthyplectenchyma** for hyphal tissue with no cellular structure. Yoshimura & Shimada (*Bull. Kochi Gakuen Jun. Coll.* **11**: 13, 1980) key nine categories of plectenchyma, **chalaroplectenchyma** (hyphal walls not united, lumina wide), **pallisadoplectenchyma** (hyphae parallel, not coherent, with intercellular spaces), **platysmoid** (a scleroplectenchyma with brown hyphal walls), **scleroplectenchyma** (cell-walls thickened, lumina narrow), **scleroprosoplectenchyma** (hyphae parallel, cohering, walls thick, lumina narrow), and **serioplectenchyma** (hyphae parallel, cohering, walls not thick, lumina wide).

Plectobasidiales, see *Sclerodermatales*, *Melanogastrales* and *Tulostomatales*.

Plectocarpon Fée (1825), Roccellaceae. 10 (on lichens, esp. *Peltigerales*), N. & S. temp. See Diederich & Etayo (*Nordic Jl Bot.* **14**: 589, 1994; key 7 spp.), Hawksworth & Galloway (*Lichenologist* **16**: 85, 1984).

Plectodiscella Woron. (1914) = Elsinoë (Elsin.) fide Jenkins (*J. agric. Res.* **44**: 689, 1932).

Plectolitus Kohlm. (1960) = Amylocarpus (Ascomycota, inc. sed.) fide Kohlmeyer & Kohlmeyer (*Syn. Pl. Marine Fungi*, 1971).

Plectomyces Thaxt. (1931), Ceratomycetaceae. 1, Philipp.

Plectomycetes. Class of *Ascomycota* with ± globose non-ostiolate ascomata; formerly frequently used for *Elaphomycetales*, *Erysiphales*, *Eurotiales*, *Meliolales*, *Microascales* and *Onygenales* but now recognized as heterogeneous, the non-ostiolate habit having repeatedly evolved. Reintroduced by Berbee & Taylor (1992) in a restricted sense (see *Ascomycota*).

Lit.: Benny & Kimbrough (*Mycotaxon* **12**: 1, 1980; keys 90 gen., 6 orders, 12 fams, excluded gen. names), Malloch (*in* Kendrick (Ed.), *The whole fungus* **1**: 153, 1979; *in* Reynolds (Ed.), 1981: 73, placement of *all* fams in orders also characterized by ostiolate or apothecioid taxa).

Plectomyriangium C. Moreau & M. Moreau (1950) = Lecideopsella (Schizothyr.) fide Eriksson & Hawksworth (*SA* **9**: 15, 1991).

Plectonaemella Höhn. (1915), Mitosporic fungi, 4.A1.?. 1, Eur.

plectonematogenous, of conidiogenous cells arising from rope-like strands of interwoven hyphae (not from single hyphae), the strands intertwined and not synnematous; funiculose. See Gams (*Cephalosporium-artige Schimmelpilze*, 1971).

Plectopeltis Syd. (1927), Mitosporic fungi, 5.A1.?. 1, C. Am.

Plectophoma Höhn. (1907) = Asteromella (Mitosp. fungi) fide Sutton (1977).

Plectophomella Moesz (1922), Mitosporic fungi, 8.A1.15. 1, Eur.

Plectophomopsis Petr. (1922), Mitosporic fungi, 8.A1.?. 1, Eur.

Plectopsora A. Massal. (1860) = Lempholemma (Lichin.).

Plectopycnis Bat. & A.F. Vital (1959) = Plectopeltis (Mitosp. fungi) fide Arx (*in litt.*).

Plectosclerotes Stach & Pickh. (1957), Fossil fungi (*f. cat.*). 2 (Carboniferous), Germany.

Plectosira Petr. (1929), Mitosporic fungi, 4/8.A1.?. 1, Eur.

Plectosphaera Theiss. (1917) = Phyllachora (Phyllachor.) fide Cannon (*Mycol. Pap.* **166**, 1991).

Plectosphaerella Kleb. (1930), Hypocreaceae. 2, widespr. See Gams & Gerlagh (*Persoonia* **5**: 177, 1968), Uecker (*Mycol.* **85**: 470, 1993; ontogeny). Anamorph *Plectosporium*.

Plectosphaerella Kirschst. (1938) ≡ Plectosphaerina (Dothid.).

Plectosphaerina Kirschst. (1938) = Omphalospora (Dothid.) fide v. Arx & Müller (1975).

Plectospira Drechsler (1927), Saprolegniaceae. 4, N. Am., Japan.

Plectosporium M.E. Palm, W. Gams & Nirenberg (1995), Anamorphic Hypocreaceae, 1.B1.15/16/19. Teleomorph *Plectosphaerella*. 1, widespr.

Plectronidiopsis Nag Raj (1979), Mitosporic fungi, 6.B1.19. 1, Chile.

Plectronidium Nag Raj (1977), Mitosporic fungi, 6.A1.19. 3, India, USA, Australia. See Sutton & Pascoe (*TBMS* **87**: 249, 1986).

Plectrothrix Shear (1902), Mitosporic fungi, 1.A1.?. 1, N. Am.

Pleiobolus, see *Plejobolus*.

Pleiochaeta (Sacc.) S. Hughes (1951), Mitosporic fungi, 1.C1/2.10. 2, widespr. *P. setosa* (lupin leaf spot).

Pleiomorpha (Nann.-Bremek.) Dhillon (1978) = Licea (Lic.) fide Pando (*in litt.*).

Pleiopatella Rehm (1908), Leotiales (inc. sed.). 1, N. Am.

Pleiopyrenis, see *Pyrenopsis*.

pleiosporous, having many spores.

Pleiostictis Rehm (1882) = Melittiosporium (Rhytismatales, inc. sed.) fide Sherwood (*Mycotaxon* **5**: 1, 1977).

Pleiostomella Syd. & P. Syd. (1917) = Mendogia (Schizothyr.) fide v. Arx & Müller (1975).

Pleiostomellina Bat., J.L. Bezerra & H. Maia (1964), Dothideales (inc. sed.). 1, Brazil.

pleioxeny, the condition of plurivorous parasitism.

Plejobolus (E. Bommer, M. Rousseau & Sacc.) O.E. Erikss. (1967), Dothideales (inc. sed.). 1 (on *Ammophila*), Eur.

Plenocatenulis Bat. & Cif. (1959), Mitosporic fungi, 5.A1.?. 2, USA.

Plenodomus Preuss (1851) = Phoma (Mitosp. fungi) fide Boerema & Kesteren (*Persoonia* **3**: 17, 1963), Boerema *et al.* (*TBMS* **77**: 61, 1981).

Plenophysa Syd. & P. Syd. (1920), Mitosporic fungi, 5.A1.?. 1 or 2, Philipp., Sri Lanka.

Plenotrichaius Bat. & Valle (1961), Mitosporic fungi, 5.A1.?. 1, Brazil.

Plenotrichella Bat. & A.F. Vital (1959) = Myxothyriopsis (Mitosp. fungi) fide v. Arx (*in litt.*).

Plenotrichopsis Bat. (1961), Mitosporic fungi, 5.A1.38. 1, Brazil.

Plenotrichum Syd. (1927), Mitosporic fungi, 5.A1.39. 2, C. Am., China.

Plenozythia Syd. & P. Syd. (1916), Mitosporic fungi, 4.A1.?. 1 or 2, Asia.

pleoanamorphy, having more than one anamorph. See Hennebert (*in* Sugiyama (Ed.), *Pleomorphic fungi: the diversity and its taxonomic implications*, 1987), States of fungi.

Pleochaeta Sacc. & Speg. (1881), Erysiphaceae. 5, N. & S. Am., Asia, S. Afr. See Braun (*Beih. Nova Hedw.* **89**, 1987; key), Kimbrough & Korf (*Mycol.* **55**: 619, 1963), Kimbrough (*Mycol.* **55**: 608, 1963; development). Anamorph *Ovulariopsis*, *Streptopodium.*

Pleochroma Clem. (1909) ≡ Candelariella (Candelar.).

Pleococcum Desm. & Mont. (1849) nom. dub. (Mitosp. fungi) fide Höhnel (*Sber. Akad. Wiss. Wien* **119**: 617, 1910).

Pleoconis Clem. (1909) = Peccania (Lichin.).

Pleocouturea G. Arnaud (1911), Mitosporic fungi, ?.?.?. 1, France. See Sutton (1977).

Pleocryptospora J. Reid & C. Booth (1969), Ascomycota (inc. sed.). 1, Brazil.

Pleocystidium C. Fisch (1884), Olpidiopsidaceae. 3 (parasitic in freshwater algae), widespr. See Dick (*in press*; key).

Pleocyta Petr. & Syd. (1927) = Phaeocytostroma (Mitosp. fungi) fide Sutton (*Mycol. Pap.* **97**, 1964).

Pleodothis Clem. (1909) = Sydowia (Dothior.) fide v. Arx & Müller (1975).

Pleogibberella Sacc. ex Berl. & Voglino (1886), Hypocreaceae. 2, India, Brazil.

Pleoglonis Clem. (1909) ≡ Hariota (Dothior.).

Pleolecis Clem. (1909) ≡ Steinia (Aphanopsid.).

Pleolpidium A. Fisch. (1892) = Rozella (Rozellopsid.) fide Sparrow (1960).

Pleomassaria Speg. (1880) = Splanchnonema (Pleomassar.) fide Barr (1993). See Barr (1982), Crivelli (*Ueber die heterogene Ascomycetengattung Pleospora*, 1983).

Pleomassariaceae M.E. Barr (1979), Dothideales. 5 gen. (+ 2 syn.), 57 spp. Ascomata immersed to erumpent, often large, spherical or conical, black, with a well-developed lysigenous ostiole; peridium composed of large-celled pseudoparenchymatous tissue, narrower at apex and base; interascal tissue of cellular pseudoparaphyses, immersed in gel; asci cylindrical to clavate, fissitunicate; ascospores large, brown, transversely septate or muriform, sometimes distoseptate, with a large gelatinous sheath. Anamorphs coelomycetous, with complex conidia. Saprobic or necrotrophic on wood and bark, temp.
      Lit.: Barr (*Mycotaxon* **15**: 349, 1982; keys 24 spp., **49**: 129, 1993; cladistic analysis, key gen.).

Pleomeliola (Sacc.) Sacc. (1899), ? Meliolaceae. 1 or 2, Eur., N. Am.

Pleomelogramma Speg. (1909), ? Herpotrichiellaceae. 1, S. Am., fide Barr (1978). See Petrak & Sydow (*Ann. Myc.* **32**: 21, 1934).

Pleomeris Syd. (1921) = Puccinia (Puccin.) fide Dietel (1928).

Pleomerium Speg. (1918), Meliolaceae. 1, Brazil.

pleomorphic, (1) of fungi having more than one independent form or spore-stage in the life cycle, especially of holomorphs comprising a teleomorph and one or more anamorphs; polymorphic; see Savile (*Mycol.* **61**: 1161, 1970), Sugiyama (Ed.) (*Pleomorphic fungi: their diversity and its taxonomic implications*, 1987), States of fungi; (2) (of dermatophytes), changes due to 'degeneration' in culture.

pleomorphism, the condition of being pleomorphic.

Pleonectria Sacc. (1876) = Thyronectria (Hypocr.) fide Seeler (1940).

pleont, any one of the two or more states of a pleomorphic fungus (Delphino, 1887).

Pleophalis Clem. (1909) = Spheconisca (Verrucar.).

Pleophragmia Fuckel (1870) = Sporormia (Spororm.) fide v. Arx & Müller (1975).

Pleopsidium Körb. (1855), Lecanoraceae (L). 3, widespr. See Hafellner (*Nova Hedw.* **56**: 281, 1993), *Acarospora*.

Pleopus Paulet (1808) ? = Gyromitra (Helvell.).

Pleopyrenis Clem. (1909) = Pyrenopsis (Lichin.).

Pleoravenelia Long (1903) = Ravenelia (Ravenel.) fide Dietel (1928).

Pleorinis Clem. (1909) = Rinodina (Physc.).

Pleoscutula Vouaux (1913), ? Dermateaceae. 1 (on *Heterodermia*), Mexico. See Hafellner (*Herzogia* **6**: 289, 1983).

Pleoseptum A.W. Ramaley & M.E. Barr (1995), ? Leptosphaeriaceae. 1 (on *Yucca*), N. Am.

Pleosphaerellula Naumov & Czerepan. (1952), ? Dothideales (inc. sed.). 1 (on *Cornus*), former USSR.

Pleosphaeria Speg. (1881), Ascomycota (inc. sed.). 15, widespr. See Petrak (*Ann. Myc.* **38**: 197, 1940).

Pleosphaeria Henssen (1964), Ascomycota (inc. sed.). 1, USA.

Pleosphaeropsis Died. (1916) = Aplosporella (Mitosp. fungi) fide Petrak & Sydow (*Beih. Rep. spec. nov. regni veg.* **42**, 1927).

Pleosphaeropsis Vain. (1921) ≡ Norrlinia (Dothideales).

Pleosphaerulina Pass. (1891) = Saccothecium (Dothior.).

Pleospilis Clem. (1909) = Spirographa (Odontotrem.) fide Santesson (*The lichens and lichenicolous fungi of Sweden and Norway*, 1993).

Pleospora Rabenh. ex Ces. & De Not. (1863) nom. cons., Pleosporaceae. *c.* 50 cosmop. *P. herbarum* (anamorph *Stemphylium botryosum*), plurivorous, causes leaf spots of legumes (Smith, *J. agric. Res.* **61**:

831, 1940), lettuce and storage fruit spot of apple. See Crivelli (*Ueber die heterogene Ascomycetengattung Pleospora*, 1983; gen. concept, keys), Henssen (*SA* **13**: 202, 1995; ontogeny), Holm & Holm (*Sydowia* **45**: 167, 1993; 9 spp. Svalbard), Müller (*Sydowia* **5**: 248, 1951), Wehmeyer (1961; keys), Simmons (*Mycotaxon* **25**: 287, 1986), Simmons (*Sydowia* **38**: 284, 1986). Anamorph *Stemphylium*.

Pleosporaceae Nitschke (1869), Dothideales. 15 gen. (+ 11 syn.), 149 spp. Ascomata perithecial, ± globose, thick-walled, immersed or erumpent, black, opening by a well-developed lysigenous ostiole, sometimes hairy or setose; peridium usually thick, with several layers of large thick-walled pseudoparenchymatous cells; interascal tissue of cellular pseudoparaphyses; asci ± cylindrical, fissitunicate, the inner wall often thickened in the apical region; ascospores brown, septate, sometimes muriform, often with a gelatinous sheath. Anamorphs hyphomycetous. Saprobic or necrotrophic on leaves and stems, cosmop.
Lit.: Corlett (*Nova Hedw.* **24**: 347, 1975; ontogeny), Holm (*Symb. bot. upsal.* **14** (3), 1957), Wehmeyer (*A world monograph of the genus Pleospora and its segregates*, 1961; keys).

Pleosporales, see *Dothideales*.

Pleosporites Y. Suzuki (1910), Fossil fungi. 2 (Cretaceous), India, Japan.

Pleosporonites R.T. Lange & P.H. Sm. (1971), Fossil fungi. 1 (Eocene), Australia.

Pleosporopsis Oerst. (1866) ? = Coniochaeta (Coniochaet.) fide Læssøe (*SA* **13**: 43, 1994), Petrini (*Sydowia* **44**: 169, 1993).

Pleostigma Kirschst. (1939), Dothideales (inc. sed.). 9, Eur., N. & S. Am., N. Afr. = Teichosporella (Patellar.) fide Petrak (*Ann. Myc.* **38**: 197, 1940).

Pleostomella, see *Pleiostomella*.

Pleothelis M. Choisy (1949) = Muellerella (Verrucar.).

Pleotrachelus Zopf (1884), Rozellopsidales (inc. sed.). 3, Eur. See Dick (*in press*).

Pleotrichiella Sivan. (1984), Dothideales (inc. sed.). 1, Australia.

Pleovalsa Kirschst. (1936) = Fenestella (Fenestell.) fide Petrak [not traced].

Plerogone M.W. Dick (1986), Leptomitaceae. 1, USA, UK.

plerotic, of an oospore which completely fills the oogonial cavity; but also of an oospore which occupies >65% of the oogonium volume (Dick *et al.*, 1992) (Fig. 28A).

Plesiospora Drechsler (1971), Mitosporic fungi, 1.A1.1. 1, USA.

pleuracrogenous, formed at the end and on the sides.

Pleurage Fr. (1849) ≡ Schizothecium (Lasiosphaer.).

Pleurella E. Horak (1971), Tricholomataceae. 1, NZ. = Hydropus (Tricholomat.) fide Singer (1975).

pleuro- (in combination), side, at the side.

Pleuroascus Massee & E.S. Salmon (1901), Pseudeurotiaceae. 10, widespr. See Lodha (*in* Subramanian (Ed.), *Taxonomy of fungi* **1**: 241, 1978), Barrasa & Moreno (*An. Jard. bot. Madrid* **41**: 31, 1984).

pleurobasidium, see basidium.

Pleurobasidium Arnaud (1951), Thelephoraceae. 1, Eur.

Pleurobotrya Berk. [not traced] ? = Botrytis (Mitosp. fungi).

Pleurocatena G. Arnaud ex Aramb. (1981), Mitosporic fungi, 1.A2.32/33. 1, Argentina, France.

Pleuroceras Riess (1854), Valsaceae. 10, N. temp. See Barr (*Mycol. Mem.* **7**, 1978), Monod (1983).

Pleurocolla Petr. (1924), Mitosporic fungi, 3.A1.?. 2, Eur., N. Am.

Pleurocollybia Singer (1947), Tricholomataceae. 2, USA.

Pleurocybe Müll. Arg. (1884) = Bunodophoron (Sphaerophor.) fide Wedin (1993).

Pleurocybella Singer (1947), Tricholomataceae. 1, N. temp.

Pleurocybomyces Cif. & Tomas. (1953) ≡ Pleurocybe (Sphaerophor.).

pleurocystidium, see cystidium.

Pleurocystis Bonord. (1851) ≡ Helicostylum (Thamnid.).

Pleurocytospora Petr. (1923), Mitosporic fungi, 8.A1.15. 2, Eur.

Pleurodesmospora Samson, W. Gams & H.C. Evans (1979), Mitosporic fungi, 1.A1.16. 1 (on *Insecta*), Sri Lanka, China. See Li & Han (*Acta Mycol. Sin.* **10**: 166, 1991; China).

Pleurodiscula Höhn. (1926), Mitosporic fungi, 8.A1.?. 1, Eur.

Pleurodiscus Petr. (1931) [non Lagerh. (1895), *Algae*], Mitosporic fungi, 4.?.?. 1, Austria.

Pleurodomus Petr. (1934), Mitosporic fungi, 8.A1.?. 1, Siberia.

Pleurodon Quél. ex P. Karst. (1881) ≡ Auriscalpium (Auriscalp.).

Pleuroflammula Singer (1946), Strophariaceae. 9, Am., Asia. See Horak (*Persoonia* **9**: 439, 1978; key 10 spp.).

Pleurogala Redhead & Norvell (1993), Russulaceae. 4, Brazil, Japan, W. Indies.

pleurogenous, formed on the side.

Pleurographium Goid. (1935) = Dactylaria (Mitosp. fungi) fide de Hoog & v. Arx (1973). ? = Nodulisporium (Mitosp. fungi) fide Sutton (*in litt.*).

Pleuromycenula Singer (1974) = Rimbachia (Tricholomat.) fide Redhead (1984).

Pleuronaema Höhn. (1917) = Cytospora (Mitosp. fungi) fide Gvritishvili (*Mikol. i Fitopat.* **3**: 207, 1969).

Pleuropedium Marvanová & S.H. Iqbal (1973), Mitosporic fungi, 1.G1.?. 1, UK. ? = Gyoerffiella (Mitosp. fungi) fide Ingold (*Guide to aquatic hyphomycetes*: 40, 1975).

Pleurophoma Höhn. (1914), Mitosporic fungi, 8.A1.15. 8, widespr.

Pleurophomella Höhn. (1914) = Sirodothis (Mitosp. fungi) fide Sutton & Funk (*CJB* **53**: 521, 1975).

Pleurophomopsis Petr. (1924), Mitosporic fungi, 4.A1.15. 6, widespr.

Pleurophragmium Costantin (1888), Mitosporic fungi, 1.A/C1.10. 15, widespr. See Hughes (*CJB* **36**: 796, 1958), Ellis (*Mycol. Pap.* **114**: 42, 1968). = Dactylaria (Mitosp. fungi) fide de Hoog (*Stud. Mycol.* **26**: 1, 1985).

Pleuroplaconema Petr. (1923), Mitosporic fungi, 8.A1.15. 2, Eur., India.

Pleuroplacosphaeria Syd. (1928), Mitosporic fungi, 8.A1.?. 1, Chile.

Pleuropus (Pers.) Gray (1821) nom. rej. = Panus (Polypor.).

Pleuropyxis Corda (1837) nom. conf. fide Hughes (*CJB* **36**: 798, 1958).

Pleurosordaria Fernier (1954) = Arnium (Lasiosphaer.) fide Lundqvist (1972).

pleurosporous, having spores on the sides, e.g. a basidium of the *Uredinales*.

Pleurosticta Petr. (1931) ? = Parmelia (Parmel.) fide Lumbsch *et al.* (*Mycotaxon* **33**: 447, 1988). = Melanelia (Parmel.) fide Esslinger (*Taxon* **29**: 692, 1980).

Pleurostoma Tul. & C. Tul. (1863), Calosphaeriaceae. 5, Eur., N. Am. See Barr (1985), Barr *et al.*. (*Mycotaxon* **48**: 529, 1993).

Pleurostromella Petr. (1922), Mitosporic fungi, 8.A1.?. 15, Eur.

Pleurotellus Fayod (1889), Crepidotaceae. 2, temp. Eur., Am.

Pleurotheciopsis B. Sutton (1973), Mitosporic fungi, 1.C1.3. 2, Hungary, UK.

Pleurothecium Höhn. (1919), Mitosporic fungi, 1.C1/2.10. 1, Eur., N. Am. See Goos (*Mycol.* **61**: 1048, 1970).

Pleurotheliopsis Zahlbr. (1922) = Pyrenula (Pyrenul.) fide Harris (1989). See Singh & Upreti (*Geophytology* **16**: 261, 1986; India).

Pleurothelium Müll. Arg. (1877) = Parapyrenis (Pyrenul.).

Pleurothelium Müll. Arg. (1885) = Pyrenula (Pyrenul.).

Pleurothyriella Petr. & Syd. (1925), Mitosporic fungi, 6.A1.?. 1, Eur.

Pleurothyrium Bubák (1916) [non Nees (1836), *Lauraceae*], Mitosporic fungi, 8.E1.10. 1, Eur.

Pleurotopsis (Henn.) Earle (1909) = Resupinatus (Tricholomat.) fide Singer (1975).

Pleurotrema Müll. Arg. (1885), Pleurotremataceae. 1, mainly trop. See Barr (*Mycotaxon* **51**: 191, 1994), Harris (*Acta Amazonica Suppl.* **14**: 55, 1986; status). = Lithothelium (Pyrenul.) fide Aptroot (*Bibl. Lich.* **44**, 1991).

Pleurotremataceae Walt. Watson (1929), Pyrenulales (? L). 1 gen. (+ 2 syn.), 1 spp. Ascomata immersed, clypeate, flattened, ± papillate, the ostiole periphysate, lateral; interascal tissue of narrow anastomosing paraphyses in a gelatinous matrix; asci cylindrical, thick-walled but not with separable wall layers, with a narrow I- apical ring; ascospores hyaline, with thickened septa and ± cuboid lumina, without germ pores or slits, with a thin mucous sheath. Anamorphs unknown. ? Lichenized or saprobic, neotropics.
Barr (*Mycotaxon* **51**: 191, 1994), who gave the family a different circumscription, used it for 5 genera with non-fissitunicate asci (*Daruvedia, Melomastia, Phomatospora, Pleurotrema,* and *Saccardoella*.

Pleurotrematomyces Cif. & Tomas. (1957) ≡ Pleurotrema (Pleurotrem.).

Pleurotus (Fr.) P. Kumm. (1871) nom. cons., Lentinaceae. 50 (esp. on wood), cosmop. *P. ostreatus*, the Oyster Cap, which damages wood, is edible and commercially cultivated; *P. olearius* is poisonous (Maretic, *Toxicon* **13**: 379, 1975). See Hilber (*Z. Mykol.* **44**: 31, 1976, *Bibl. Mycol.* **87**, 1982), Pegler (*Kew Bull.* **31**: 501, 1977; key Indian spp.).

Pliariona A. Massal. (1860) = Phaeographina (Graphid.).

Plicaria Fuckel (1870), Pezizaceae. *c.* 10, temp. See Egger (*Mycotaxon* **29**: 183, 1986), Hirsch (*Agarica* **6**: 241, 1985; key 4 Eur. spp.).

Plicariella (Sacc.) Rehm (1894) = Plicaria (Peziz.) fide Korf (1960).

plicate, folded into pleats; **plica**, a pleat.

Plicatura Peck (1872), Amylocorticiaceae. 1, N. temp.

Plicaturella Murrill (1910) = Paxillus (Paxill.) fide Singer (1951).

Plicaturopsis D.A. Reid (1964), Schizophyllaceae. 1, Eur.

Plocaria Nees (1820) nom. rej. (Algae).

Plochmopeltidella J.M. Mend. (1925) = Chaetothyrina (Micropeltid.) fide v. Arx & Müller (1975).

Plochmopeltinites Cookson (1947), Fossil fungi. 2 (Oligocene, Miocene), India.

Plochmopeltis Theiss. (1914), Schizothyriaceae. 4, N. & C. Am.

Plochmothea Syd. (1939) = Xenostomella (Microthyr.) fide v. Arx & Müller (1975).

Ploettnera Henn. (1899), Dermateaceae. 4, Eur. See Hein (*Willdenowia, Beih.* **9**, 1976; key).

Ploettnerula Kirschst. (1924) = Pirottaea (Dermat.) fide Nannfeldt (1932).

Ploioderma Darker (1967), Rhytismataceae. 6, India, N. Am. Anamorph *Cryocaligula*.

Plokamidomyces Bat., C.A.A. Costa & Cif. (1958), Anamorphic Euantennariaceae, 1.A1.15. Teleomorph *Trichopeltheca*. 1, NZ.

Plowrightia Sacc. (1883), Dothioraceae. *c.* 6, widespr. *P. ribesia* (black pustule of gooseberry and currants (*Ribes*)). See Wakefield (*TBMS* **24**: 286, 1940).

Plowrightiella (Sacc.) Trotter (1926) = Sydowia (Dothior.) fide v. Arx & Müller (1975).

Pluesia Nieuwl. (1916) ≡ Maurya (nom. conf.).

plug (in an ascus), see ascus.

Plunkettomyces G.F. Orr (1977) = Arachniotus (Gymnasc.) fide v. Arx (*Persoonia* **9**: 393, 1977).

Pluricellaesporites Hammen (1954), Fossil fungi (*f. cat.*). 30 (Cretaceous, Tertiary), widespr.

Pluricellulites Hammen (1954), ? fossil fungi

plurilocular, (1) (of ascospores), many-celled; (2) (of stromata), having several locules.

Pluriporus F. Stevens & R.W. Ryan (1925) = Dothidella (Polystomell.) fide Müller & v. Arx (1962).

Plurisperma Sivan. (1970), ? Verrucariales (inc. sed.). 1, Pakistan. See Matzer (*Nova Hedw.* **56**: 203, 1993).

plurivorous, attacking a number of hosts or substrates; not specialized.

Pluteaceae Kotl. & Pouzar (1972), Agaricales. 5 gen. (+ 9 syn.), 80 spp. Hymenophoral trama bilateral, convergent; spores dull pink.

Pluteolus (Fr.) Gillet (1876) = Bolbitius (Bolbit.) fide Singer (1951).

Pluteopsis Fayod (1889) = Psathyrella (Coprin.) fide Singer (1951).

Pluteus Fr. (1836), Pluteaceae. 50 (esp. on wood), cosmop. See Singer (*TBMS* **39**: 222, 1956; key), Vellinga & Schreurs (*Persoonia* **12**: 337, 1985; key W. Eur. spp.), Orton (*British fungus flora* **4**: 4, 1986; key Br. spp.), Smith & Stuntz (*Lloydia* **21**: 115, 1959; key S. Am. spp.), Singer (*Sydowia* **15**: 114, 1962; suppl. S. Am. spp.), Homola (*Mycol.* **64**: 1211, 1972; sect. *Celluloderma* N. Am., keys), Pegler (*Kew Bull., Addit. Ser.* **6**, 1977; key 6 E. Afr. spp.), Horak (*Fl. Illustr. Champ. Afr. Centr.* **6**: 107, 1976; key 10 C. Afr. spp.).

Pneumatospora B. Sutton, Kuthub. & Muid (1984), Mitosporic fungi, 1.G2.1. 1 (rudimentary clamp connexions, dolipore septa), Malaysia, Japan. See Fisher & Wakley (*TBMS* **86**: 507, 1986; TEM of septa).

Pneumocystidaceae O.E. Erikss. (1994), Pneumocystidales. 1 gen., 1 sp. Mycelium absent; vegetative cells thin-walled, irregularly shaped, uninucleate, dividing by fission, sometimes becoming thick-walled and cyst-like (? transformed into asci), with 4-8 endogenously produced daughter cells (? ascospores), at first globose but becoming falcate. Parasitic in lungs of mammals, widespr.

Pneumocystidales, ? Ascomycota (inc. sed.). 1 fam., 1 gen., 1 sp. Fam: **Pneumocystidaceae**.

Pneumocystis P. Delanoë & Delanoë (1912), Pneumocystidaceae. 1, widespr. *P. carinii* causes pneumonia (PCP) in immunocompromized (esp. AIDS) patients. See Eriksson (*SA* **13**: 165, 1995), Mackenzie (*Rev. Ib. Micol.* **7**: 3, 1990; aetiology), Kovacs *et al.* (*Exp. Parasitol.* **71**: 60, 1990; characterization of dihydrofolate reductase), Hong *et al.* (*Jl Clin. Microbiol.* **28**: 1785, 1990; karyotypes), Wakefield *et al.* (*Mol. Microbiol.* **8**: 426, 1993; basidiomycete affinity), Tamburrini *et al.* (*J. Clin. Microbiol.* **31**: 2788, 1993; rapid diagnosis by PCR).

pneumomycosis, see mycosis.

Pocheina A.R. Loebl. & Tappan (1961) = Guttulina (Guttulin.) fide Pando (*in litt.*).

Pochonia Bat. & O.M. Fonseca (1965) = Diheterospora (Mitosp. fungi) fide Barron & Onions (1966).

Pocillaria P. Browne (1756) ≡ Lentinus (Lentin.).

Pocillopycnis Dyko & B. Sutton (1979), Mitosporic fungi, 8.E1.10. 1, Sweden. See DiCosmo (*Mycotaxon* **10**: 288, 1980).

Pocillum De Not. (1863), Leotiaceae. 2 or 3, Eur., Am. See Petrak (*Sydowia* **5**: 345, 1951).

pocket plums, plums swollen then 'mummified' by *Taphrina pruni*.

pocket rot, localized rotting of trunks of trees or roots by wood-destroying fungi.

Pocosphaeria (Sacc.) Berl. (1892) = Nodulosphaeria (Phaeosphaer.).

Pocsia Vězda (1975), Ascomycota (inc. sed., L). 1, Afr. See Hafellner & Kalb (*Bibl. Lich.* **57**: 161, 1995; posn).

Poculinia Spooner (1987), Sclerotiniaceae. 1 (on *Nothofagus*), Tasmania.

Poculopsis Kirschst. (1935), Leotiaceae. 1, Eur.

Poculum Velen. (1934) = Rutstroemia (Sclerotin.) fide Baral (*SA* **13**: 113, 1994).

Podabrella Singer (1945), Amanitaceae. 1 (associated with termites; edible), trop.

Podaleuris Clem. (1909) = Peziza (Peziz.) fide Eckblad (1968).

Podaxaceae Corda (1842), Agaricales. 6 gen. (+ 8 syn.), 13 spp. Gasterocarp subagaricoid, stipitate; gleba powdery; spores brown-black, ellipsoid, smooth.
 Lit.: Smith (*in* Ainsworth *et al.* (Eds), *The Fungi* **4B**: 421, 1973), Heim (*in* Petersen (Ed.), *Evolution in the higher Basidiomycetes*: 505, 1973; keys gen.).

Podaxales, see *Agaricales*.

Podaxis Desv. (1809), Podaxaceae. 5, widespr., warm dry areas. See Morse (*Mycol.* **25**: 1, 1933), Martinez (*Boln Soc. Argent. Bot.* **14**: 73, 1971; Argentina), McKnight & Stransky (*Mycol.* **72**: 195, 1980), McKnight (*Mycol.* **77**: 24, 1985; key small-spored spp.).

Podaxon Fr. (1829) ≡ Podaxis (Podax.).

podetium, lichenized stem-like portion (stipe, or disco-podium) bearing the hymenial discs and sometimes conidiomata in a fruticose apothecium (Ahti, *Lichen-ologist* **14**: 105, 1982), esp. as in *Cladonia*. Cf. pseudopodetium.

Podisoma Link (1809) = Gymnosporangium (Puccin.) fide Saccardo (*Syll. Fung.* **7**: 737, 1887).

Podobactridium Penz. & Sacc. ex Petch (1916) = Bactridium (Mitosp. fungi) fide Hughes (1966).

Podobelonium (Sacc.) Sacc. & D. Sacc. (1906) = Crocicreas (Leot.). See Nannfeldt (*Nova Acta R. Soc. Scient. upsal.* ser. 4 **8**(2), 1932).

Podocapsa Tiegh. (1887), Endomycetaceae. 1, Eur.

Podocapsium Clem. (1909) ≡ Podocapsa (Endomycet.).

Podochytrium Pfitzer (1870), Chytridiaceae. 7 (on diatoms or saprobes), N. temp. See Canter (*Bot. J. Linn. Soc.* **63**: 47, 1970).

Podoconis Boedijn (1933) = Sporidesmium (Mitosp. fungi) fide Ellis (1958).

Podocratera Norman (1861) ≡ Tholurna (Calic.).

Podocrea (Sacc.) Lindau (1897) = Podostroma (Hypocr.) fide Atkinson (*Bot. Gaz.* **40**: 401, 1905).

Podocrella Seaver (1928), Clavicipitaceae. 1, Trinidad.

Podocystis Fr. (1849) nom. rej. = Melampsora (Melampsor.) fide Saccardo (*Syll. Fung.* **18**: 812, 1906); based on uredinia of *M. caprearum*.

Pododimeria E. Müll. (1959), ? Pseudoperisporiaceae. 4, Eur., N. Am. See Luttrell & Barr (*Am. J. Bot.* **65**: 251, 1978; key).

Podofomes Pouzar (1966) = Ischnoderma (Coriol.) fide Donk (1974).

Podohydnangium G.W. Beaton, Pegler & T.W.K. Young (1984), Hydnangiaceae. 1, Australia.

Podonectria Petch (1921), Tubeufiaceae. 8 (on scale insects), mainly trop. See Pirozynski (*Kew Bull.* **31**: 595, 1977), Rossman (*Mycotaxon* **7**: 163, 1978; key, *Mycol. Pap.* **157**, 1987).

Podophacidium Niessl (1868), Dermateaceae. 1, Eur., N. Am. See Seaver (*Mycol.* **31**: 350, 1939).

Podoplaconema Petr. (1921), Mitosporic fungi, 8.A1.15. 1, Eur.

Podoporia P. Karst. (1892), Coriolaceae. 6, widespr. See Pegler (1973).

Podoscypha Pat. (1900), Podoscyphaceae. 26, widespr., esp. trop. See Reid (1965: 150; key).

Podoscyphaceae D.A. Reid (1965), Stereales. 10 gen. (+ 8 syn.), 60 spp. Basidioma spathulate to infundibuliform; dimitic; skeletocystidia present; spores inamyloid.

Podoserpula D.A. Reid (1963), Coniophoraceae. 1, Australia, NZ, Venezuela.

Podosordaria Ellis & Holw. (1897), Xylariaceae. 17 (esp. coprophilous), widespr. See Krug & Cain (*CJB* **52**: 589, 1974; key), Rogers & Læssøe (*Mycotaxon* **44**: 435, 1992), Krug & Jeng (*CJB* **73**: 65, 1995; key 17 spp.). Anamorph *Lindquistia*.

Podosphaera Kunze (1823), Erysiphaceae. 12, widespr. *P. leucotricha* (mildew of apple and other pome fruits) and *P. oxyacanthae* (hawthorn (*Crataegus*) mildew). See Braun (*Beih. Nova Hedw.* **89**, 1987; key). Anamorph *Oidium*.

Podospora Ces. (1856) nom. cons., Lasiosphaeriaceae. 78 (coprophilous), cosmop. See Mirza & Cain (*CJB* **47**: 1999, 1969; key), Lundqvist (1972), Krug & Khan (*CJB* **67**: 1174, 1989; culture, records), Guarro *et al.* (*Syst. Ascom.* **10**: 79, 1991), Schmidt (*Bibl. Mycol.* **127**, 1989; *P. anserina*); also *Schizothecium*.

Podosporiella Ellis & Everh. (1894), Mitosporic fungi, 2.C2.?. 1, N. Am., Australia. See Shoemaker (*CJB* **44**: 145, 1966).

Podosporium Schwein. (1832), Mitosporic fungi, 2.C2.24. 10, esp. trop. See Ellis (*DH*: 291; *MDH*: 303).

Podosporium Lév. (1847) = Melampsora (Melampsor.) fide Dietel (1928).

Podosporium Bonord. (1851) = Aplosporella (Mitosp. fungi) fide Sutton (1977).

Podosporium Sacc. & Schulzer (1884) = Aplosporella (Mitosp. fungi) fide Saccardo (*Syll. Fung.* **3**, 1884).

Podostictina Clem. (1909) = Pseudocyphellaria (Lobar.) fide Galloway (1988).

Podostroma P. Karst. (1892), Hypocreaceae. 9, widespr. See Boedijn (*Ann. Myc.* **36**: 314, 1938).

Podostrombium Kunze ex Rchb. (1828) nom. inval. (Clavar.). See Donk (*Taxon* **6**: 110, 1957).

Podoxyphiomyces Bat., Valle & Peres (1961), Mitosporic fungi (L), 8.A1,?. 1, Brazil.

Podoxyphium Speg. (1918) = Conidiocarpus (Mitosp. fungi) fide Hughes (*Mycol.* **68**: 693, 1976).

Poecylomyces, see *Paecilomyces*.

Poelt (Josef; 1924-95). Bavarian mycologist and bryologist, obtained doctorate in 1950 from University of Munich, where he remained to 1965; Professor of Systematic Botany, Free University of Berlin, 1965-72; Professor and Director of the Botanical Garden, University of Graz, from 1972-91, where he remained active after retirement. His wide-ranging interests included lichenized fungi, rusts and smuts, and oomycetes, as evident in over 320 publications, incl. his pioneering keys to European lichens (*Bestimmungsschlussel europaischer Flechten*, 1969; *Erganzungsheft I-II* with A. Vězda, 1977, 1981). He travelled extensively, making especially important contributions to Himalayan lichens (1966-77) and the Mediterranean (esp. Greek) mycobiotas. Renowned for his determination to tackle 'difficult' groups, for

insights into lichen biology, phylogeny and structure
(see species pairs), and also work on lichenicolous
and bryophilous lichens. An inspiring teacher, his
PhD students included P. Döbbeler, J. Hafellner, B.
Hein, H. Hertel, W. Jülich, H. Mayrhofer and F.
Oberwinkler; also editor/co-editor of jnls. incl.
Herzogia, Nova Hedwigia. See De Priest (*Inoculum*
46(3): 3, 1995), Hertel (*Internat. Lichen. Newsl.* **26**:
25, 1993), Grummann (1974: 36).
Poeltia Petr. (1972) [non Grolle (1966), *Hepaticae*] ≡
Poeltiella (Dothid.).
Poeltiaria Hertel (1984), Porpidiaceae (L). 3, S. Am.,
Australia.
Poeltidea Hertel & Hafellner (1984), Porpidiaceae (L).
1, Kerguelen.
Poeltiella Petr. (1974) = Hyalocrea (Dothid.) fide Ross-
man (1987).
Poeltinula Hafellner (1984), Rhizocarpaceae (L). 1,
Eur. See Hawksworth (*SA* **10**: 36, 1991; nomencl.).
Poetschia Körb. (1861), Patellariaceae. 3, Eur., S. Am.
See Hafellner (*Beih. Nova Hedw.* **62**, 1979).
Pogonomyces Murrill (1904) = Hexagonia (Coriol.) fide
Pegler (1973).
Pogonospora Petr. (1957) ? = Endoxylina (Ascomycota,
inc. sed.) fide Müller & v. Arx (1962).
Poikiloderma Füisting (1868) = Amphisphaeria
(Amphisphaer.) fide Eriksson & Hawskworth (*SA* **7**:
80, 1988).
Poikilosperma Bat. & J.L. Bezerra (1961), Mitosporic
fungi, 5.A1.?. 1, Brazil.
Poikilosporium Dietel (1897) = Thecaphora (Ustilagin.) fide Dietel (1928).
Poisonous fungi. Diverse fungi produce toxins (q.v.)
which affect humans, animals and plants. For larger
fungi poisonous for humans, see **Mycetisms**;
microfungi which produce toxins affecting humans
and higher animals, see Mycotoxicoses, plants, Phytotoxic mycotoxins. See also antibiotic, Hallucinogenic
fungi.
Poisonous lichens, see Edible fungi.
Poitrasia P.M. Kirk (1984), Choanephoraceae. 1,
widespr. in tropics. See Kirk (*TBMS* **68**: 429, 1977;
zygospore ultrastr.).
polar (of bacteria, spores, etc.), at the ends or poles.
Polar and alpine mycology, see Laursen *et al.* (Eds)
(*Arctic and alpine mycology* **1**, 1985; **2**, 1987), Gulden *et al.* (*Arctic and alpine fungi* **1**, 1985; **2**, 1988),
Longton (*Biology of polar bryophytes and lichens*,
1988), Trappe (*Mycol.* **80**: 1, 1988; alpine fungi).
polar-diblastic, see polarilocular.
polaribilocular, polarilocular (q.v.) with two cells.
polarilocular (of ascospores), bicellular and the two
cells separated by a central perforated septum; orculiform, polaribilocular, polar-diblastic. See Sheard
(*Lichenologist* **3**: 328, 1967). (Fig. 35B).
Polaroscyphus Huhtinen (1987), Hyaloscyphaceae. 1,
Svalbard.
poleophilous, town loving; sometimes used of lichens
which thrive in urban areas (e.g. *Lecanora
conizaeoides* in Eur.).
Polhysterium Speg. (1912) ? = Hysterographium
(Hyster.). See Clements & Shear (1931).
Polioma Arthur (1907), Pucciniaceae. 4, N. & S. Am.
See Baxter & Cummins (*Bull. Torrey bot. Cl.* **78**: 51,
1951).
Poliomella Syd. (1922) = Eriosporangium (Puccin.) fide
Dietel (1928).
Poliomopsis A.W. Ramaley (1987), Uredinales (inc.
sed.). 1, USA.
Poliotelium Syd. (1922), Uredinales (inc. sed.). 3, Am.,
Asia. See Mains (*Bull. Torrey bot. Cl.* **66**: 173, 1939).
Polistophthora Lebert (1858) = Cordyceps (Clavicipit.)
fide Tulasne & Tulasne (*Sel. Fung. Carp.* **3**: 4, 1865).

Polistroma Clemente (1807), Thelotremataceae (L). 1,
trop. Am.
Pollaccia E. Bald. & Cif. (1937), Anamorphic Ven-
turiaceae, 3.B/C2.19. Teleomorph *Venturia.* 6,
widespr. *P. radiosa,* leaf and shoot blight of poplars.
See Funk (*Can. J. Pl. Path.* **11**: 353, 1989;
P. borealis), Wu & Sutton (*MR* **99**: 983, 1995; rela-
tionships of *P. mandshurica,* key).
Poloniodiscus Svrček & Kubička (1967) ? =
Ameghiniella (Leot.). See Zhuang (*Mycotaxon* **31**:
261, 1988).
poly- (prefix), a great number; many.
Polyactis Link (1809) = Botrytis (Mitosp. fungi) fide
Hennebert (*Persoonia* **7**: 183, 1973).
Polyadosporites Hammen (1954), Fossil fungi (*f. cat.*).
6 (Cretaceous, Paleogene), Colombia.
Polyandromyces Thaxt. (1920), Laboulbeniaceae. 1,
Afr., Pacific Isl., Japan.
polyandrous (of oospores), formed when more than one
functioning antheridium is present; cf. monandrous.
Polyangium Link (1809) nom. dub. (Myxobacteriales).
Polyascomyces Thaxt. (1900), Laboulbeniaceae. 1, UK.
See Tavares (*Mycol.* **65**: 929, 1973).
polyascous, having many asci; esp. having the asci in
one hymenium, not separated by sterile bands.
Polyblastia A. Massal. (1852) nom. cons., Verru-
cariaceae (L). *c.* 120, cosmop. See Servít (*Českos-
lovenské lišejníky čeledi Verrucariaceae*, 1954),
Swinscow (*Lichenologist* **5**: 92, 1971; Br. spp.),
Zschacke (*Rabenh. Krypt.-Fl.* **9**, 1 (1), 1933; Eur.).
polyblastic (of conidiogenous cell), producing blastic
conidia at several conidiogenous loci, either synchro-
nously or irregularly.
Polyblastidea (Zschacke) Tomas. & Cif. (1952) =
Polyblastia (Verrucar.).
Polyblastiomyces E.A. Thomas (1939) nom. inval. =
Staurothele (Verrucar.).
Polyblastiomyces Cif. & Tomas. (1953) = Polyblastia
(Verrucar.).
Polyblastiopsis Zahlbr. (1903) = Julella (Thelenell.)
fide Mayrhofer (*Bibl. Lich.* **26**, 1987).
Polycarpella Theiss. & Syd. (1918) = Muellerella (Ver-
rucar.) fide Hawksworth (*in litt.*).
polycarpic (of *Exobasidium* infections), systemic (or
circumscribed) and perennial (Nannfeldt, 1981: 15);
cf. monocarpic.
Polycarpum Stempel [not traced] nom. dub. (protozoa
or fungi).
Polycauliona Hue (1908) = Caloplaca (Teloschist.).
Polycellaesporonites A. Chandra, R.K. Saxena & Setty
(1984), Fossil fungi. 1, Arabian Sea.
Polycellaria H.D. Pflug (1965), Fossil fungi (*f. cat.*). 1
(Tertiary), USA.
polycentric, having a number of centres of growth and
development and more than one reproductive organ,
as in the *Cladochytriaceae* (Karling, *Mycol.* **36**: 528,
1934); see monocentric; cf. reproductocentric.
Polycephalomyces Kobayasi (1941), Anamorphic
Clavicipitaceae, 2.A1.15. Teleomorph *Byssostilbe.* 4,
widespr. See Seifert (*Stud. Mycol.* **27**: 168, 1985;
key).
polycephalous, many-headed.
Polycephalum Kalchbr. & Cooke (1880) nom. dub.,
Myxobacteria fide Seifert (*TBMS* **85**: 123, 1985).
Polychaetella Speg. (1918), Mitosporic fungi. 3, Eur.,
N. Am. See Sutton (1977).
Polychaeton (Pers.) Lév. (1846) nom. dub. (?
Capnodiaceae). See Hughes (*Mycol.* **68**: 693, 1976),
Sutton (*Mycol. Pap.* **141**, 1977; nomencl.).
Polychidium (Ach.) Gray (1821), ? Placynthiaceae (L).
5, cosmop. See Henssen (*Symb. bot. upsal.* **18** (1),
1963).

polychotomous, having an apex dividing simultaneously into more than two branches (Corner, 1950).

Polychytrium Ajello (1942), Cladochytriaceae. 1, USA, Brazil.

Polycladium Ingold (1959), Mitosporic fungi, 1.G1.1. 1 (aquatic), UK.

Polyclypeolina Bat. & I.H. Lima (1959), Aulographaceae. 1, Uganda.

Polyclypeolum Theiss. (1914) = Schizothyrium (Schizothyr.) fide Müller & v. Arx (1962).

Polycoccum Saut. ex Körb. (1865), Dacampiaceae. 27 (on lichens), widespr. See Hawksworth & Diederich (TBMS 90: 293, 1988; key 23 spp.). Anamorph Cyclothyrium (van der Aa, Stud. Mycol. 31: 15, 1989).

Polycyclina Theiss. & Syd. (1915), Parmulariaceae. 1, S. Am.

Polycyclinopsis Bat., A.F. Vital & I.H. Lima (1958), Microthyriaceae. 1, Brazil.

Polycyclus Höhn. (1909), Parmulariaceae. 2, S. Am. See Petrak (Sydowia 4: 533, 1950).

Polycystis Lév. (1846) nom. rej. = Urocystis (Tillet.).

Polycytella C.K. Campb. (1987), Mitosporic fungi, 1.C1.1. 1 (causing human mycetoma), India.

polydactyloid venation, see veins.

Polydesmia Boud. (1885), Hyaloscyphaceae. 4, widespr. See Raitviir & Galan (Mycotaxon 53: 447, 1995; key).

Polydesmus Mont. (1845) ? = Sporidesmium (Mitosp. fungi) fide Hughes (CJB 36: 799, 1958). See Ellis (Mycol. Pap. 70: 61, 1958).

Polydiscidium Wakef. (1934), Leotiaceae. 1, Guyana, Afr. See Dennis (Bull. Jard. bot. Brux. 31: 154, 1961).

Polydiscina Syd. (1930), ? Leotiales (inc. sed.). 1, Bolivia.

polyenergid, coenocytic.

Polyetron Bat. & Peres (1963), Mitosporic fungi, 5.A/B1/2.?. 1, Brazil.

Polygaster Fr. (1823) nom. dub. (? 'gasteromycetes').

Polylagenochromatia Sousa da Câmara (1929) ? = Polystigmina (Mitosp. fungi).

Polymarasmius Murrill (1915) = Marasmius (Tricholomat.) fide Singer (1951).

Polymeridium (Müll. Arg.) R.C. Harris (1980), Trypetheliaceae (L). 19, mainly trop. See Harris (Boln Mus. Parense Emílio Goeldi, Bot. 7: 619, 1993; key).

polymorphic, having different forms; pleomorphic.

polymorphic species, species with a series of intergrading morphological features; e.g. resulting from inbreeding or automictic sexual reproduction.

Polymorphomyces Coupin (1914) = Geotrichum (Mitosp. fungi) fide Kendrick & Carmichael (1973).

Polymorphum Chevall. (1822), Anamorphic Ascodichaenaceae, 8.A1/2.1. Teleomorph Ascodichaena. 1, temp. See Hawksworth & Punithalingam (TBMS 60: 501, 1973), Hawksworth (Taxon 32: 212, 1983; nomencl.).

Polymyces Battarra ex Earle (1909) ≡ Armillaria (Tricholomat.).

Polymyxa Ledingham (1933), Plasmodiophoraceae. 2, Eur. N. Am. See Karling (1968).

Polynema Lév. (1846), Mitosporic fungi, 7.B1/2.15. 3, Italy, Brazil, USA. See Nag Raj (1993).

Polyopeus A.S. Horne (1920) = Phoma (Mitosp. fungi) fide Boerema & Dorenbosch (Persoonia 6: 49, 1970).

Polyorus, see Polyozus.

polyoxin antibiotics, inhibit chitin synthesis, e.g. polyoxin D inhibits synthetase formation (Endo et al., J. Bact. 104: 189, 1970).

Polyozellus Murrill (1910), Thelephoraceae. 1, USA. See Imazeki (Mycol. 45: 555, 1953).

Polyozosia A. Massal. (1855) = Lecania (Bacid.) fide Hafellner (Beih. Nova Hedw. 79: 214, 1984).

Polyozus P. Karst. (1881) = Tremellodendropsis (Tremellodendropsid.) fide Donk (Persoonia 4: 184, 1966).

Polypaecilum G. .Sm. (1961), Mitosporic fungi, 1.A1.19. 2, Eur., S. Am. See Piontelli et al. (Bol. Micol. 4: 155, 1989; hyalohyphomycosis caused by P. insolitum).

Polypedia Bat. & Peres (1959), Micropeltidaceae. 1, Brazil.

Polypera Pers. (1818) ≡ Pisolithus (Sclerodermat.).

polyphagous, see polyphagy.

Polyphagus Nowak. (1877), Chytridiaceae. c. 9, widespr. See Bartch (Mycol. 37: 553, 1945).

polyphagy (adj. polyphagous), see monophagy.

polyphialidic (of a conidiogenous cell), having more than one conidiogenous locus at which conidia are produced. Cf. monophialidic.

Polyphlyctis Karling (1968), Chytridiaceae. 1 or 2, widespr.

polyphyllous (of foliose lichen thalli), having many connected leaf-like lobes.

Polypilus P. Karst. (1881) ≡ Grifola (Coriol.).

polyplanetism, sequence of two or more motile flagellate phases with interspersed mobile aplanosporic phases in the zoosporic part of the life history; the aplanosporic phase may be naked or as a walled cyst; motile phases may be monomorphic or dimorphic.

Polyplocium Berk. (1843) = Gyrophragmium (Podax.) fide Zeller (Mycol. 35: 409, 1943).

polyploidy (in fungi), see Chromosome numbers.

Polyporaceae Fr. ex Corda (1839), Poriales. 18 gen. (+ 22 syn.), 98 spp. Basidioma stipitate, annual or perennial; dimitic with skeleto-ligative hyphae; hymenophore tubulate; no setae (cf. Hymenochaetaceae); spores cylindrical; lignicolous (and causing decay of standing and structural timber) or terrestrial (humicolous); cosmop.

Lit.: General: Corner (Phytomorphology 3: 152, 1953; construction of polypores), Nobles (CJB 40: 987, 1958; cultural characters), Cooke (Lloydia 22: 163, 1959), Donk (1951-63; gen. lists), Teixeira (Biol. Rev. 37: 51, 1962; taxon. rev.), Lowe (Mycol. 55: 1, 1963; rev.), Pegler (1973; key 123 gen., Brit. spp.), Nuss (Bibl. Mycol. 45, 1975; ecology, sporulation; Hoppea 39: 127, 1980, structure, taxonomy), Corner (Ad Polyporaceae I, II & III [Beih. Nova Hedw. 75, 1983; 78, 1984).

Regional: N. America, Overholts (Polyporaceae of the United States, Alaska, & Canada, 1953), Lowe & Gilbertson (J. Elisha Mitchell sci. Soc. 77: 43, 1961; keys gen., 293 spp. S.E. USA), Martin & Gilbertson (Mycotaxon 7: 337, 1978; key 99 wood-rotting spp.), Gilbertson & Ryvarden (N. Am. Polypores, 2 vol. 1986-7). E. Africa, Ryvarden (Norw. Jl Bot. 19: 229, 1972; checklist), Ryvarden & Johansen (A preliminary polypore flora of East Africa, 1980). Europe, Pilát (Atlas des champignons de l'Europe. III. Polyporaceae, 2 vols, 1936-42), Domanski (Fungi, Polyporaceae, 2 vols, 1972-73 [Engl. transl.]), Jahn (Westfälische Pilzbrief 4, 1963; C. Eur.), Donk (Checklist of European polypores [Verh. K. ned. Akad. Wet. 62], 1974), Ryvarden (The Polyporaceae of North Europe 1, 1976, 2, 1978). New Zealand, Cunningham (Polyporaceae of New Zealand [N.Z. Dep. sci. industr. Res. Bull. 164], 1965). Trinidad & Tobago, Fidalgo & Fidalgo (Mycol. 58: 862, 1967). former USSR, Bondartsev ([Polyporaceae of European Russia and the Caucasus], 1953).

Polyporales, see Poriales.

Polyporasclerotes Stach & Pickh. (1957), Fossil fungi (f. cat.). 1 (Carboniferous), Germany.

polypore, one of the Polyporaceae.

Polyporellus P. Karst. (1879) = Polyporus (Polypor.) fide Donk (1974).

Polyporisporites Hammen (1954), Fossil fungi (*f. cat.*). 1 (Cretaceous to Paleogene), Colombia.

Polyporites Lindl. & Hutton (1833), Fossil fungi. 12 (Carboniferous to Quaternary), Eur., USA.

Polyporites Daugherty (1941), Fossil fungi. 1 (Triassic), USA.

Polyporoletus Snell (1936) = Scutiger (Scutiger.) fide Singer *et al.* (*Mycol.* **37**: 124, 1945).

Polyporus Fr. (1815), Polyporaceae. *c.* 50 (mostly on wood), cosmop. The sclerotium (*Mylitta australis*) of *P. mylittae* is 'blackfellows' bread' of Australia. See Donk (*Persoonia* **1**: 262, 1960; typific.), Stahl (*Bibl. mycol.* **50**, 1970; genetics sporocarp formation *P. ciliatus*), Jahn (*Schw. Z. Pilzk.* **47**: 218, 1969), Nuss (1980), Pouzar (*Česká Myk.* **26**: 82, 1972), Ryvarden (1978: 378; key 8 Eur. spp.), Ryvarden & Johansen (*Preliminary polypore flora of East Africa*: 481, 1980; key 16 Afr. spp.), Corner (*Beih. Nova Hedw.* **78**: 12, 1984; key subgen. classif.).

Polyporus Pers. (1821) ≡ Laetiporus (Coriol.).

Polyporus (Pers.) Gray (1821) ≡ Polyporus (Polypor.).

Polypyrenula D. Hawksw. (1985), Pyrenulaceae (?L). 1, W. Indies.

Polyrhina Sorokīn (1876) ≡ Harposporium (Mitosp. fungi). See Karling (*Mycol.* **30**: 512, 1938).

Polyrhizium Giard (1889) ? = Cladosporium (Mitosp. fungi) fide Petch (*TBMS* **19**: 190, 1935).

Polyrhizon Theiss. & Syd. (1914), ? Venturiaceae. 2, India, S. Afr.

Polysaccopsis Henn. (1898), Tilletiaceae. 1 (inducing galls on *Solanum*), S. Am. Resembles *Urocystis*. See Vánky (1987).

Polysaccum F. Desp. & DC. (1807) = Pisolithus (Sclerodermat.).

Polyschema H.P. Upadhyay (1966), Mitosporic fungi, 1.C2.24/27. 6, widespr. See Ellis (*MDH*).

Polyschismium Corda (1842) = Diderma (Didym.).

Polyschistes J. Steiner (1898) = Diploschistes (Thelotremat.).

Polyscytalina Arnaud (1954), Mitosporic fungi, 1.B2.3. 1, France. *P. pustulans* (syn. *Oospora pustulans*), skin spot of potato tubers. See Ciferri & Caretta (*Mycopathol.* **16**: 304, 1962). = Hormiactella (Mitosp. fungi) fide Holubová-Jechová (*Fol. Geobot. Phyt. Praha* **13**: 433, 1976).

Polyscytalum Riess (1853), Mitosporic fungi, 1.A/B1.3. 5, widespr. See Sutton & Hodges (*Nova Hedw.* **28**: 487, 1976).

polysidia, specialized dual vegetative propagules in certain *Pyxine* spp. formed in depressions at the tips of coral-like structures (**polysidiangia**) which recall isidia but are not themselves propagules (Kalb, *Bibl. Lich.* **24**, 1987).

Polysphondylium Bref. (1884), Dictyosteliaceae. 6, widespr. See Harper (*Bull. Torrey bot. Cl.* **59**: 49, 1932), Traub & Hohl (*Am. J. Bot.* **63**: 664, 1976), Raper (1984).

Polyspora Laff. (1921) [non Sweet ex Don (1831), *Theaceae*] = Aureobasidium (Mitosp. fungi) fide Hermanides-Nijhof (*Stud. Mycol.* **15**: 141, 1977).

Polysporidiella Petr. (1960), Dothideales (inc. sed.). 1, Iran.

Polysporidium Syd. & P. Syd. (1908) = Guignardia fide Eriksson (*SA* **5**: 197, 1986).

Polysporina Vězda (1978) nom. cons. prop., Acarosporaceae (L). 3, Eur.

polysporous, many-spored.

Polystema Raf. (1815) nom. dub. (Fungi, inc. sed.). No spp. included.

Polystictoides Lázaro Ibiza (1916) ≡ Inonotus (Hymenochaet.).

Polystictus Fr. (1851) ≡ Coltricia (Hymenochaet.).

Polystigma DC. (1815), Phyllachoraceae. 5 (on *Rosaceae*), temp. See Hyde & Cannon (*Aust. Syst. Bot.* **5**: 415, 1992).

Polystigmatales, see *Phyllachorales*.

Polystigmella Jacz. & Namalina (1931) ? = Polystigma (Phyllachor.) fide Cannon (*in litt.*).

Polystigmina Sacc. (1884), Anamorphic Phyllachoraceae, 8.E1.10. Teleomorph *Polystigma.* 2, Eur.

Polystigmites A. Massal. (1857), Fossil fungi. 1 (Miocene), Italy.

Polystoma Gray (1821) ≡ Myriostoma (Geastr.).

Polystomella Speg. (1888) = Dothidella (Polystomell.) fide Müller & v. Arx (1962), v. Arx & Müller (1975).

Polystomellaceae Theiss. & P. Syd. (1915), Dothideales. 3 gen. (+ 6 syn.), 9 spp. Ascomata superficial or subcuticular, strongly flattened, black, multiloculate; peridium composed of several layers of pseudoparenchymatous cells; interascal tissue of cellular pseudoparaphyses; asci cylindrical, ? fissitunicate; ascospores hyaline to brown, septate, without a sheath. Anamorphs unknown. Biotrophic on leaves, trop.

Polystomellina Bat. & A.F. Vital (1958), ? Microthyriaceae. 1, Brazil.

Polystomellomyces Bat. (1959), Mitosporic fungi, 4.A1.?. 1, Greece.

Polystomellopsis F. Stevens (1924), Dothideales (inc. sed.). 1, S. Am. See Ciferri & Batista (*An. Soc. Biol. Pernambuco* **14**: 79, 1956).

Polystratorictus Matsush. (1993), Mitosporic fungi, 1.C2.15. 2, Peru.

Polystroma Fée (1824) ≡ Polistroma (Thelotremat.).

Polysynnema Constant. & Seifert (1988), Mitosporic fungi, 2.A1.10. 1, Hawaii, Grenada.

Polythecium Bonord. (1861) = Fusicoccum (Mitosp. fungi) fide Saccardo (*Syll. Fung.* **3**, 1884).

Polythelis Arthur (1906) = Tranzschelia (Uropyxid.) fide Arthur (1934).

Polythelis Clem. (1909) ≡ Polypyrenula (Pyrenul.).

Polythrinciella Bat. & H. Maia (1960), Mitosporic fungi, 3.B2.?. 1, Brazil.

Polythrinciopsis J. Walker (1966), Mitosporic fungi, 1.B1.10. 1, Australia. = Polythrincium (Mitosp. fungi) fide Kendrick & Carmichael (1973).

Polythrincium Kunze (1817), Anamorphic Mycosphaerellaceae, 1.B2.10. Teleomorph *Cymadothea.* 2, temp.

Polythyrium Syd. (1929) = Neostomella (Asterin.) fide Müller & v. Arx (1962).

Polytolypa J.A. Scott & Malloch (1993), Onygenaceae. 1 (coprophilous), Canada.

polytomous, dividing into many branches, usually at one node or point.

polytretic, see tretic.

Polytretophora Mercado (1983), Mitosporic fungi, 1.B2.27. 2, Cuba, Malaysia. See Kuthubutheen & Nawawi (*MR* **95**: 623, 1991), ? = Spadicoides (Mitosp. fungi) fide Sutton (*in litt.*).

Polytrichia Sacc. (1882) = Pyrenophora (Pleospor.) fide v. Arx & Müller (1975).

Polytrichiella M.E. Barr (1972) = Capronia (Herpotrichiell.) fide Müller *et al.* (1987).

polyxeny, see pleioxeny.

Pomatomyces Oerst. (1864) = Thekopsora (Pucciniastr.) fide Sydow (1915).

Pompholyx Corda (1834) = Scleroderma (Sclerodermat.) fide Guzmám (*Darwiniana* **16**: 270, 1970).

Pontisma H.E. Petersen (1905), Pontismaceae. 7, Eur., USA, Japan.

Pontismaceae H.E. Petersen (1909), ? Myzocytiopsidales. 2 gen., 10 spp. Dick (*in press*; key).

Pontogeneia Kohlm. (1975), Ascomycota (inc. sed.). 6 (marine), widespr. See Kohlmeyer & Kohlmeyer (1979).

Pontomyxa Topsent (1892), *Hydromyxales*, q.v.

Pontoporeia Kohlm. (1963) = Zopfia (Zopf.) fide Malloch & Cain (*CJB* **50**: 61, 1972).

Population biology. The study of patterns of distribution and variation in space and time within fungal species and the interpretation of these patterns in terms of genetic, developmental and environmental influences on phenotype and modes of proliferation. Some means of distinguishing between individual population components is necessary, and this can cause difficulties in view of the indeterminacy (indefinite growth potential) and capacity of anastomosis of fungal mycelia. However, it is usually possible to determine whether fungal samples have the same or different genetic origins (i.e. arise from the same or different 'genets') on the basis of somatic and/or sexual incompatibility, phenotypic differences (including isoenzyme polymorphism) or DNA polymorphisms. The latter are sometimes regarded as most definitive in that they do not depend on gene expression. However, they can be relatively expensive to detect, and identify differences and similarities at varied levels of resolution which may or may not be correlated with 'biological' entities or groups.

Often a variety of approaches is needed before deciding on which, or which combination, of the above approaches resolves population components at the level most appropriate to the biological question being addressed. Once population components have been located, their distribution can be quantified by mapping their extent and varied content within regional boundaries. The resulting information can help to provide meaningful answers to fundamentally important questions concerning where the organisms (and their offspring) are and what they do there, how they arrived, whether they will persist and how they are likely to change in character and/or distribution. It is also vital for assessment of the levels at which population variation should be recognized for taxonomic purposes.

Distributional patterns in fungal populations can be interpreted in terms of resource relationships, modes of development and modes of reproduction. Resource relationships, the ways in which fungi interact with those non-living or living materials (substrata, hosts) which provide them with a source of organic nutrients, are determined both by selective influences and by the mode of fungal arrival (propagules or migratory mycelium) at colonization sites. Genets capable of arriving only as propagules have 'resource-unit-restricted' regional boundaries; those with migratory mycelium are 'non-unit-restricted' and can become very large. At the individual level, the heterogeneity of microenvironmental conditions encountered by a fungal genet may be reflected in the range of developmental modes or alternative phenotypes that it produces, allowing it to vary functional role with circumstances. At the population level, whether reproduction is clonal (non-recombinatorial) or diversifying (recombinatorial) may partly reflect habitat heterogeneity. The abilities of fungi to vary both their developmental pattern and degree of commitment to clonal and recombinatorial modes of reproduction has important implications for understanding gene flow and genetic and epigenetic diversity within and betwen populations. These abilities are also a crucial consideration when attempting to predict or interpret the evolutionary responses of fungi to environmental change.

Lit.: Brasier (*Adv. Pl. Pathol.* **5**: 53, 1986, *Nature* **332**: 438, 1988), Carroll & Wicklow (Eds) (*The fungal community*, edn 2, 1992), Jacobson *et al.* (*PNAS* **90**: 9159, 1993), McDonald *et al.* (*Ann. Rev. Phytopathol.* **27**: 77, 1989), Rayner (*Mycol.* **83**: 48, 1991, *Ann. Rev. Phytopathol.* **29**: 305, 1993), Rayner & Todd (*Adv. Bot. Res.* **7**: 333, 1979), Smith *et al.* (*Nature* **356**: 428, 1992).

pore, (1) a small opening, as in tretic (q.v.) conidiogenesis; (2) in - **fungi** (*Polyporaceae* and *Boletaceae*), the mouth of a tube.

Poria Pers. (1794), Coriolaceae. 250 (on wood; s. str., 2), cosmop. *P. obliqua* (decay of standing birch and other hardwoods), *P. incrassata* and *P. vaillantii* (decay of structural timbers). See Lowe (*Tech. Pub. Sta. Univ. Coll. Forestry, Syracuse*, **90**, 1966; syns, key 133 N. Am. spp., *Mycol.* **55**: 453, 1963; key 162 trop. spp.), Donk (1960: 266; nomencl.), Wright (*Mycol.* **56**: 694, 1964; dextrinoid reaction).

Poriales, Basidiomycetes. 4 fam., 140 gen. (+ 129 syn.), 1070 spp.
(1) **Coriolaceae.**
(2) **Grammotheleaceae.**
(3) **Lentinaceae.**
(4) **Polyporaceae.**
Lit.: Ryvarden (*Genera of polypores* [*Synopsis Fungorum* 5], 1991), Teixeira (*Genera of Polyporaceae, an objective approach* [*Bol. Chácara Bot. Itu* 1], 1994).

poricidal (of asci), see ascus.

poricin, an antitumour antibiotic from *Poria corticola* (Ruelius *et al.*, *Arch. Biochem. Biophys.* **125**: 126, 1968).

poriform, pore-like.

Porina Ach. (1809) nom. rej. ≡ Pertusaria (Pertusar.).

Porina Müll. Arg. (1883) nom. cons., Trichotheliaceae (L). *c.* 200, cosmop. (mainly trop.). See Hafellner & Kalb (*Bibl. Lich.* **57**: 161, 1995; nomencl., posn), Janex-Favre (*Cryptogamie, bryol. lichén.* **2**: 253, 1981; ontogeny), Malme (*Ark. Bot.* **23A** (1), 1929; S. Am.), McCarthy (*Bibl. Lich.* **52**, 1993; S. Hemisph. rock spp.), McCarthy & Kantvilas (*Lichenologist* **25**: 137, 1993; key 6 spp., Tasmania), Santesson (*Symb. bot. upsal.* **12** (1), 1952; foliicolous spp.), Singh (*Revue bryol. lichén.* **37**: 973, 1971; India), Swinscow (*Lichenologist* **2**: 6, 1962; Br. spp.), Upreti (*Bryologist* **97**: 73, 1994; key 8 spp. India).

Porinopsis Malme (1928) = Aspidothelium (Aspidothel.).

Porinula Vězda (1975) = Strigula (Strigul.) fide Harris (*in* Hafellner & Kalb, *Bibl. Lich.* **57**: 161, 1995).

Poriodontia Parmasto (1982), Hyphodermataceae. 1, former USSR.

Poroauricula McGinty (1917) = Favolaschia (Tricholomat.) fide Donk (1960).

Porocladium Descals (1976), Mitosporic fungi, 1.G1.24. 1 (aquatic), UK.

poroconidium, see tretic.

Poroconiochaeta Udagawa & Furuya (1979), Coniochaetaceae. 2, Japan. See Læssøe (*SA* **13**: 43, 1994; posn).

Porocyphus Körb. (1855), Lichinaceae (L). 7, cosmop. See Henssen (*Symb. bot. upsal.* **18** (1), 1963).

Porodaedalea Murrill (1905), Hymenochaetaceae. 4, N. temp. See Fiasson & Niemalä (*Karstenia* **24**: 14, 1984).

Porodiscella Viégas (1944) ? = Xylaria (Xylar.) fide Læssøe (*SA* **13**: 43, 1994).

Porodisculus Murrill (1907), Coriolaceae. 1, USA.

Porodiscus Murrill (1903) [non Grev. (1863), *Algae*] ≡ Porodisculus (Coriol.).

Porodiscus Lloyd (1919) [non Grev. (1863), *Algae*], Ascomycota (inc. sed.). 1, Brazil.

Porodothion Fr. (1825) ≡ Porothelium (Pyrenulales).
Porogramme (Pat.) Pat. (1900), Grammotheleaceae. 8, trop. See Lowe (*Pap. Mich. Acad. Sci.* **49**: 33, 1964; key), Ryvarden (*TBMS* **73**: 9, 1979).
Poroidea Göttinger ex G. Winter (1884) = Craterocolla (Exid.) fide Donk (*Persoonia* **4**: 165, 1966).
Poroisariopsis M. Morelet (1971), Mitosporic fungi, 1.C2.24. 2, N. Am.
Porolaschia Pat. (1900) = Favolaschia (Tricholomat.) fide Singer (1945).
Poromycena Overeem (1926) = Mycena (Tricholomat.) fide Singer (1951).
Poronea Raf. (1815) nom. dub. (Fungi, inc. sed.). No spp. included.
Poronia Willd. (1787), Xylariaceae. 2 (coprophilous), widespr. See Jong & Rogers (*Mycol.* **61**: 853, 1969), Lohmeyer & Benkert (*Z. Mykol.* **54**: 93, 1988), Paden (*CJB* **56**: 1774, 1976; culture, anamorph).
Poronidulus Murrill (1904), Coriolaceae. 1, USA.
Poroniopsis Speg. (1922) = Hypocreodendron (Mitosp. fungi) fide Lindquist & Wright (*Darwiniana* **11**: 598, 1959).
Poropeltis Henn. (1904), Mitosporic fungi, 5.A2.?. 1, S. Am.
Porophora G. Mey. (1825) ≡ Ascidium (Thelotremat.). See Hafellner & Kalb (*Bibl. Lich.* **57**: 161, 1995).
Porophora Zenker ex Göbelez (1827) = Trypethelium (Trypethel.).
Porophoromyces Thaxt. (1926), Laboulbeniaceae. 1, Afr.
Poroptyche Beck (1888) ? = Poria (Coriol.) fide Donk (1974).
Porosphaera Dumort. (1822) nom. dub. (Ascomycota, inc. sed.).
Porosphaerella E. Müll. & Samuels (1982), Trichosphaeriaceae. 1, Switzerland.
Porosphaerellopsis Samuels & E. Müll. (1982), Lasiosphaeriaceae. 1, Brazil. See Barr (*Mycotaxon* **39**: 43, 1990; posn). Anamorph *Sporoschisma*-like.
Porosphaeria Samuels & E. Müll. (1979) ≡ Porosphaerellopsis (Lasiosphaer.).
porospore, see tretic.
Porostereum Pilát (1936), Stereaceae. 15, widespr. See Hjortstam & Ryvarden (*Synopsis fung.* **4**: 25, 1990; key).
Porostigme Syd. & P. Syd. (1917) = Phaeostigme (Pseudoperispor.). See Müller & v. Arx (1962).
Porosubramaniania Hol.-Jech. (1985), Mitosporic fungi, 1.B2.27. 2, former Czechoslovakia, India.
Porotenus Viégas (1960), Uropyxidaceae. 4, C. Am. and Brazil.
Porotheleum Fr. (1818), Stromatoscyphaceae. 1, Eur.
Porothelium Eschw. (1824), ? Pyrenulales (inc. sed., L). 4, trop.
Porphyrellus J.-E. Gilbert (1931) = Tylopilus (Strobilomycet.) fide Smith & Thiers (1971). Used by Singer (1962), Wolfe & Petersen (*Mycotaxon* **7**: 152, 1978; nomencl. suprasp. taxa).
Porphyriospora A. Massal. (1852) = Polyblastia (Verrucar.).
Porphyrosoma Pat. (1927), ? Hypocreaceae. 1 (on *Amphisphaeria*), Madagascar.
Porpidia Körb. (1855), Porpidiaceae (L). 16, temp. See Gowan (*Bryol.* **92**: 25, 1989; key 21 spp. N. Am.), Gowan & Ahti (*Ann. Bot. Fenn.* **30**: 53, 1993; key 15 spp. Scand.), Hertel & Knoph (*Mitt. Bot. StSamml. Münch.* **20**: 467, 1984), Hertel (*Beih. Nova Hedw.* **79**: 399, 1984; Antarct.).
Porpidiaceae Hertel & Hafellner (1984), Lecanorales (L). 16 gen. (+ 7 syn.), 59 spp. Thallus crustose; ascomata apothecial, sessile or immersed, pale to black, sometimes irregularly shaped, with variably defined marginal tissue; interascal tissue of branched and anastomosing paraphyses, often swollen at the apices and with a pigmented epithecium; asci with a well-developed I+ apical cap and a conspicuous strongly blueing tube-like apical ring, and an outer I+ gelatinized layer; ascospores hyaline, aseptate, often with a sheath. Anamorphs pycnidial. Lichenized with green algae, widespr. esp. S. temp.
Lit.: Gowan (*Syst. Bot.* **14**: 77, 1989; chemistry).
Porpoloma Singer (1952), Tricholomataceae. 7 (esp. under *Nothofagus*), S. Am.
Porpomyces Jülich (1982), Coriolaceae. 1, Eur.
Portalites Hemer & Nygreen (1967), Fossil fungi. 1 (Carboniferous), Libya.
porter, ale (q.v.) produced by fermenting malt that has been charred by heat.
Porterula Speg. (1920) = Asteromella (Mitosp. fungi) fide Clements & Shear (1931).
Posadasia Cantón (1898) = Coccidioides (Mitosp. fungi) fide Dodge (1935).
posterior, (1) at or in the direction of the back; (2) (of a lamella), the end at or near the stipe.
Postia Fr. (1874) = Tyromyces (Coriol.) fide Donk (*Persoonia* **1**: 274, 1960).
Potebniamyces Smerlis (1962), ? Cryptomycetaceae. 1, Eur, N. Am. *P. discolor* (apple bark canker). See Hahn (*Mycol.* **49**: 226, 1957).
Potoromyces Müll. bis ex Hollós (1902) = Mesophellia (Mesophell.).
Pouzarella Mazzer (1976), Entolomataceae. 31, mainly N. Am. See Mazzer (*Bibl. Mycol.* **46**, 1976; monogr.). ≡ Pouzaromyces (Entolomat.).
Pouzaromyces Pilát (1953) = Rhodophyllus (Entolomat.) fide Singer (1962). Used by Noordeloos (*Persoonia* **10**: 207, 1979; key 7 Eur. spp.).
Pouzaroporia Vampola (1992), Polyporaceae. 1, USA.
Powellia Bat. & Peres (1964) [non Mitt. (1868), *Musci*], Mitosporic fungi. 1, Jamaica.
ppb, parts per billion; a measure of concentration.
ppm, parts per million; a measure of concentration.
Prachtflorella Matr. (1903) ≡ Gonatobotrys (Mitosp. fungi).
praemorse (of the stipe base), as if broken off; truncate.
Pragmoparopsis Höhn. (1917) = Colpoma (Rhytismat.) fide Holm & Holm (*Symb. bot. upsal.* **21**(3), 1977).
Pragmopora A. Massal. (1855), Leotiaceae. 6, Eur., N. Am. See Groves (*CJB* **45**: 169, 1967; key).
Pragmopycnis B. Sutton & A. Funk (1975), Anamorphic Leotiaceae. 8.A1.16. Teleomorph *Pragmopora*. 1, Canada.
Prataprajella Hosag. (1992), Meliolaceae. 2, India, Taiwan.
Pratella (Pers.) Gray (1821) ≡ Agaricus (Agaric.).
Prathigada Subram. (1956), Mitosporic fungi, 1.C2.10. 1, Asia.
Prathoda Subram. (1956) = Alternaria (Mitosp. fungi) fide Cejp & Deighton (*Mycol. Pap.* **117**, 1969).
precipitative vesicle, sporangial vesicle, the membrane of which is formed after extrusion of naked sporangial protoplasm, whether partially cleaved into planonts or not, possibly as a precipitation reaction between periprotoplasmic colloids and the environment; amorphous or fibrillar, c.f. homohylic vesicle, plasmamembranic vesicle.
precipitin, see antigen.
Predacious fungi. Parasites of amoebae, nematodes, rotifers and other small terrestrial or aquatic animals. 150 + spp. fungi are involved. The most important are *Zoopagales* (q.v.); others are hyphomycetes (*Harposporium, Monacrosporium, Rotiferophthora, Verticillium*), and in the *Saprolegniales* (*Sommerstorffia*). Comandon & de Fonbrune (see Lloyd &

Madison, *Discovery* 7: 303, 1946) studied the trapping mechanism of *Arthrobotrys*, etc. by cinemicrography.

Lit.: Dolfus (*Parasites* .... *des helminthes*, 1946), Duddington (*Bot. Rev.* 21: 377, 1955, (& Wyborn) *Bot. Rev.* 38: 545, 1972, *Biol. Rev.* 31: 152, 1956; *TBMS* 38: 97, 1955; technique, *Friendly fungi*, 1956), Cooke & Godfrey (*TBMS* 47: 61, 1964; key to nematode-destroying fungi), Soprunov (*Predacious hyphomycetes and their application to the control of pathogenic nematodes*, 1966 [transl. 1958, Russ. edn]), Castner (*CJB* 46: 764, 1966; key to 14 nematode-destroying spp. with constricting rings), Verona & Lepidi (*Agric. ital.*, A, 71: 204, 1971; identif., review), Newell *et al.* (*Bull. mar. Sci.* 17: 177, 1977; keys to marine nematode-destroying spp.), Barron (*The nematode-destroying fungi*, 1977), Poinar & Jansson (Ed.) (*Diseases of nematodes* 2, 1988; 150+ fungi known), Ilyaletdinov *et al.* (*Griby Gifomitsety Regulyatory Chislennosti Paraziticheskikh Nematod*, 1990, in Russian), Dürschner (*Mitt. biol. Bundes. Land-Forstw.* 217, 1983; nematode endoparasites), Gray (*Biol. Rev.* 62: 245, 1987; review), Carris & Glawe (*Fungi colonizing cysts of Heterodera glycines*, 1990), Barron (*CJB* 63: 211, 1985, *CJB* 69: 494, 1991; rotifer endoparasites), Gams (*Neth. J. Plant Pathol.* 94: 123, 1988; nematophagous species of *Verticillium*), Dix & Webster (*Fungal ecology*, 1995).

Predaldinia P. Briot, Lar.-Coll. & Locq. (1983), Fossil fungi. 1, Australia.

PREM. National Collection of Fungi of the Republic of South Africa (Pretoria, South Africa); founded 1908; a government institute; genetic resource collection **PPRI**; see Baxter (*S. Afr. J. Sci.* 82: 348, 1986).

Premyxomyces Locq. (1979), Fossil fungi. 1, France.

Preservation. To keep alive, retaining quality and condition, preventing deterioration (*ex situ* conservation). See Genetic resource collections, Reference collections.

Preservatives. *General preservative for reference specimens*: 5% formaldehyde (40%) in water; or 25 ml formaldehyde (40%) and 150 ml alcohol (95%) in 1000 ml water.

Preuss (Carl Gottlieb Traugott; 1795-1855). Apotheker and Sanitätsrath of Hoyerswerda, Germany. Between 1843 and 1853 described 344 new spp. (incl. approx. 40 new gen.) from C. Germany. See Meyer (*Willdenowia* 1: 573, 1956), Jülich (*Willdenowia* 7: 261, 1974; list collections), Stafleu & Cowan (*TL-2* 4: 396, 1983).

Preussia Fuckel (1867), Sporormiaceae. 10, widespr. See Cain (*CJB* 39: 1633, 1961; key), Valldosera & Guarro (*Boln Soc. Micol. Madrid* 14: 81, 1990; key 20 spp.).

Preussiaster Kuntze (1891) ≡ Cordana (Mitosp. fungi). See Ellis (*DH*).

Preussiella Lodha (1978) [non Gilg (1897), *Spermatophyta*] = Pycnidiophora (Spororm.) fide Hawksworth (*in litt.*).

Prévost (Isaac-Bénédict; 1755-1819). Amateur scientist who was from 1810 Professor of philosophy at the Faculté de Théologie protestante, Montauban, Depart. du Lot, France. Noted for his *Mémoire sur la cause immèdiate de la carie ou charbon des blés*, 1807 [Engl. transl. *Phytopath. Classics* 6, 1939; biogr.] in which he described his experiments on wheat bunt (*Tilletia caries*) by which he was the first to demonstrate the pathogenicity of a microorganism. See Keitt (*Phytopath.* 46: 1, 1956), Stafleu & Cowan (*TL-2* 4: 398, 1983).

Priapus Raf. (1808) ? = Lycoperdon (Lycoperd.) fide Donk (*Taxon* 5: 109, 1956).

Prillieuxia Sacc. & P. Syd. (1899) = Tomentella (Thelephor.) fide Donk (1964).

Prillieuxina G. Arnaud (1918), Asterinaceae. 40, esp. trop. See Doidge (*Bothalia* 4: 273, 1942). Anamorph *Leprieurina*.

primary, first; first-formed; **- homothallism**, see homothallism; **- mycelium** (of basidiomycetes), the haploid mycelium from a basidiospore; **- septum**, see septum; **- squamules**, the first formed squamules of *Cladonia* from which the podetia arise; **- universal veil**, see protoblem; **- uredo** (uredium, uredinium), see `Uredinales.`

primordial, first in order of appearance; pertaining to the earliest stages of development; **- covering** or **cuticle** = blematogen; **- hypha**, intensely coloured hyphae of the epicutis in *Russula* (Melzer, 1934); **- shaft**, the monaxial basidioma initial, as in *Clavariaceae* (Corner, 1950); **- tissue**, undifferentiated tissue of a basidioma initial; cf. lipsanenchyma; **- veil**, protoblem; **primordium**, the earliest stage of development of an organ.

primospore, a spore very like a cell of the thallus (MacMillan).

Pringsheimia Schulzer (1866) = Saccothecium (Dothior.) fide Holm (*in litt.*).

Pringsheimiella Couch (1939) [non Höhn. (1920), *Algae*] ≡ Dictyomorpha (Rozellopsid.).

Pringsheimina Kuntze (1891) ≡ Achlya (Saproleg.).

priorable, see legitimate.

Prismaria Preuss (1851), Mitosporic fungi, 1.G1.?. 2, Eur.

Pritzeliella Henn. (1903) = Penicillium (Mitosp. fungi) fide Seifert & Samson (*in* Samson & Pitt (Eds), *Advances in Penicillium and Aspergillus systematics*: 143, 1985).

Proabsidia Vuill. (1903) = Absidia (Mucor.) fide Hesseltine (1955).

Proactinomyces (K. Lehm. & Haag) H.L. Jensen (1934), = Nocardia (Actinomycetes).

probasidium, see basidium.

Probilimbia Vain. (1899) ≡ Mycobilimbia (Porpid.).

Proboscispora Punith. (1984), Mitosporic fungi, 4.B1.10. 1, Malaysia.

Procandida Novák & Zsolt (1961) = Syringospora (Mitosp. fungi) fide v. Arx *et al.* (*Stud. Mycol.* 14, 1977).

pro-diploidization hypha, a hypha which may be diploidized, cf. flexuous hypha.

progametangium (of *Mucorales*), a hyphal branch forming a gametangium and suspensor cell (Fig. 43C).

progamones, a group of sex hormones of zygomycetes (Reschke & Plempel, *Z. Pflanzenphysiol.* 67: 343, 1972).

prohybrid, a mycelium having additional nuclei from hyphal fusions (after Dodge, *Mycol.* 28: 407, 1936).

Prokaryota, see *Eukaryota*.

prokaryote (adj. **-otic**), an organism lacking membrane-limited nuclei and not exhibiting mitosis, e.g. bacteria; cf. eukaryote.

prolate (of a spore, sporocarp, etc.), elongated in the direction of the poles. See subglobose.

proliferation, successive development of new parts, esp. of new sporangia within the old wall in *Oomycota*, or of new wall material in conidiogenous cells.

proliferin, an anti-tubercle bacillus antibiotic from *Aspergillus proliferans* (Gupta & Viswanathan, *Antibiot. & Chemotherapy* 5: 496, 1955).

Proliferobasidium J.L. Cunn. (1976), Brachybasidiaceae. 1, W. Indies.

Proliferodiscus J.H. Haines & Dumont (1983), Hyaloscyphaceae. 6, neotrop., S.E. Asia, Australasia. See Spooner (*Bibl. Mycol.* 116, 1987; 4 spp. Australasia).

Prolisea Clem. (1931) = Cercidospora (Dothideales) fide Müller & v. Arx (1962).

Prolixandromyces R.K. Benj. (1970), Laboulbeniaceae. 6, C. Am., Eur.

Promicromonospora Krassiln., Kalak. & Kirillova ˙ (1961), *Actinomycetes*, q.v.

promitosis, the special type of nuclear division during the growth stage in *Plasmodiophoraceae*; cruciform division.

promycelium, Tulasne's term for the germ tube of the teliospore (*Uredinales*) or ustilospore (*Ustilaginales*) from which **promycelial spores** (Plowright) (sporidia) are produced. The teliospore has been interpreted as a probasidium, the - as a metabasidium (after septation a phragmobasidium) and the ballistisporous promycelial spores basidiospores, as have the nonballistosporic smut sporidia (see Donk, *K. ned. Akad. Wet.* C 76: 109, 1973).

Promycetes = *Uredinales* and *Ustilaginales* (Clements & Shear, 1931).

Promycetozoida, see *Myxomycota*.

Pronectria Clem. (1931), Hypocreaceae. 11 (on lichens and algae), widespr. See Lowen (*Mem. N.Y. bot. Gdn* 49: 243, 1989; key, *Mycotaxon* 39: 461, 1990; nomencl.). Anamorph *Acremonium*.

proper exciple (**margin**), see excipulum (proprium).

prophialide, see metula; primary sterigma.

Prophytroma Sorokīn (1877), Mitosporic fungi, 2.A2.?. 1, former USSR.

Propolidium Sacc. (1884), Rhytismatales (inc. sed.). 2, Afr., N. Am. See Sherwood (*Mycotaxon* 5: 1, 1977).

Propolina Sacc. (1884), ? Dothideales (inc. sed.). 1, Eur.

Propoliopsis Rehm (1914), ? Stictidaceae. 1, Asia, N. Am. See Sherwood (*Mycotaxon* 5: 1, 1977).

Propolis Fr. (1849), Rhytismataceae. c. 10, widespr. See Sherwood (*Mycotaxon* 5: 320, 1977; key), Johnston (*N.Z. Jl Bot.* 24: 84, 1986; key 4 NZ spp.).

Propolomyces Sherwood (1977) ≡ Propolis (Rhytismat.).

Proprioscypha Spooner (1987), Hyaloscyphaceae. 1 (on *Eucalyptus*), Australia.

Propythium M.K. Elias ex Janson. & Hills (1979), Fossil fungi (Oomycota). 1 (Carboniferous), USA.

Prosaccharomyces E.K. Novák & Zsolt (1961) = Saccharomycopsis (Saccharomycopsid.) fide v. Arx *et al.* (1977).

prosenchyma, see plectenchyma.

prosoplectenchyma, see plectenchyma.

prosorus (of *Chytridiales*), a cell giving a group of sporangia (the sorus).

Prospodium Arthur (1907), Uropyxidaceae. 40 (on *Bignoniaceae*, *Verbenaceae*), warmer Am. See Cummins (*Lloydia* 3: 1, 1940; key).

prosporangium (in *Oomycota*), a sporangium-like body which puts out a vesicle (sporangium) in which zoospores may undergo development and from which they are freed; presporangium.

Prosporobolomyces Novák & Zsolt (1961) = Sporobolomyces (Mitosp. fungi) fide Lodder (1970).

Prosthecium Fresen. (1852), Melanconidaceae. 7, Eur., N. Am. See Merezhko (*Ukr. Bot. Zh.* 45: 57, 1988).

Prosthemiella Sacc. (1881), Mitosporic fungi, 6.G1.?. 3, Eur., N. Am., Afr.

Prosthemium Kunze (1817), Mitosporic fungi, 6.G2.1. 5, temp.

Prostratus Sivan., W.H. Hsieh & C.Y. Chen (1993), Melanconidaceae. 1, Taiwan.

Protascus P.A. Dang. (1903) nom. conf. (Myzocytiopsid.).

Protascus Wolk (1913) ≡ Wolkia (Ascomycota, inc. sed.).

Protasia Racib. (1900) nom. nud. (? Fungi, inc. sed.).

Protendomycopsis Windisch (1965) = Trichosporon (Mitosp. fungi) fide Carmo Sousa (*in* Lodder, 1970).

proteoglycan, an antitumour metabolite from *Coriolus pubescens* active against sarcoma-180.

Proteomyces Moses & Vianna (1913) = Trichosporon (Mitosp. fungi) fide Diddens & Lodder (*Die anaskosporogenen Hefen* 2, 1942).

Proteophiala Cif. (1958), Anamorphic Ceratostomataceae, 1.A1.15. Teleomorph *Melanospora*. 3, widespr.

proteophilous fungi, fungi associated with ammoniarich (e.g. urea affected) soils (Sagara, *Trans. mycol. Soc. Japan* 14: 41, 1973).

proterospore, a spore formed at the start of the sporulation period in *Ganoderma* and able to germinate easily without passing through the gut of a fly larva (Nuss, *Pl. Syst. Evol.* 141: 53, 1982).

prothallus, see hypothallus.

prothecium, a primitive or rudimentary perithecium, as in the *Gymnoascaceae*.

Protista, see *Protoctista*.

proto- (prefix), primitive; primordial.

Protoabsidia Naumov (1935) = Absidia (Mucor.) fide Hesseltine (1955).

Protoachlya Coker (1923), Saprolegniaceae. 3, N. Am., Eur.

protoaecium, a haploid structure which, after diploidization, becomes a fruiting structure.

Protoascon L.R. Batra, Segal & Baxter (1964), Fossil fungi (? Ascomycota). 1 (Carboniferous), USA. See Baxter (*Paleont. Contrib. Univ. Kansas* 77, 1975), Pirozynski (1976).

Protobagliettoa Servít (1955) = Verrucaria (Verrucar.).

protobasidium, see basidium.

Protoblastenia (Zahlbr.) J. Steiner (1911), ? Psoraceae (L.). 11, cosmop.

protoblem, a loose flocculent mycelial layer covering the universal veil, as in *Amanita*; primordial veil.

Protocalicium Woron. (1927), Ascomycota (inc. sed., L). 1, former USSR.

Protochytrium Borzí (1884) nom. dub. (? Protozoa).

protoconidium, see hemispore.

Protocoronis Clem. & Shear (1931) = Aureobasidium (Mitosp. fungi) fide Sutton (*in litt.*).

Protocoronospora G.F. Atk. & Edgerton (1907) = Aureobasidium (Mitosp. fungi) fide Hermanides-Nijhof (*Stud. Mycol.* 15: 141, 1977).

Protocrea Petch (1937) = Hypocrea (Hypocr.) fide Rogerson (1970).

Protocreopsis Yoshim. Doi (1977), Hypocreaceae. 7, trop. See Doi (*Bull. natn Sci. Mus. Tokyo* 4: 113, 1978).

Protoctista (Protista), kingdom of eukaryotic microorganisms exclusive of the Kingdoms *Animalia*, *Fungi*, and *Plantae*, but including *Myxomycota* and *Oomycota* and related groups (Margulis *et al.* (Eds), *Handbook of Protoctista*, 1991). Molecular data has shown the 'kingdom' to be exceptionally heterogenous and recent authors accept *Chromista* and *Protozoa* instead. See *Chromista*, Kingdoms of fungi, *Protozoa*.

Protocucurbitaria Naumov (1951), Ascomycota (inc. sed.). 1, former USSR.

Protodaedalea Imazeki (1955), Exidiaceae. 1, Japan.

Protoderma Rostaf. (1874) [non Kütz. (1854), *Algae*] ≡ Licea (Lic.).

Protodermium Rostaf. ex Berl. (1888) ≡ Licea (Lic.).

Protodermodium Kuntze (1891) ≡ Licea (Lic.).

Protodontia Höhn. (1907), Exidiaceae. 5, Eur., N. Am. See Martin (*Mycol.* **24**: 508, 1932), Donk (*Persoonia* **4**: 172, 1966).

Protogaster Thaxt. (1934), Protogastraceae. 1 (on roots of *Viola*), USA.

Protogastraceae Zeller (1934), Hymenogastrales. 1 gen., 1 sp. Gasterocarp minute, a hollow sphere lined with hymenium; spores ellipsoid, smooth.

Protogautieria A.H. Sm. (1965), Gautieriaceae. 1, USA.

Protogenea Kobayasi (1963) = Hydnocystis (Otid.) fide Trappe (1975).

Protoglossum Massee (1891) = Hymenogaster (Hymenogastr.) fide Cunningham (*Gast. Austr. N.Z.*: 50, 1944).

protogonidium, the first of a series of gonidia (obsol.).

Protograndinia Rick (1933) = Patouillardina (Exid.) fide Rogers (*Mycol.* **31**: 513, 1939).

Protohydnum A. Møller (1895), Exidiaceae. 1, trop. Am. See Martin (*Lloydia* **4**: 262, 1941).

protohymenial, having a primitive hymenium (Maire).

protologue, everything associated with a name on its first publication, i.e. diagnosis, description, references, synonymy, geographical data, citation of specimens, discussion, comments, illustrations (see Stearn, in Linnaeus, *Species plantarum* [Reprint] **1**: 126, 1957). See also Nomenclature.

Protomarasmius Overeem (1927) ? = Marasmius (Tricholomat.).

Protomerulius A. Møller (1895), Exidiaceae. 1, trop. Am., Java.

Protomyces Unger (1832), Protomycetaceae. *c.* 10, widespr. See Reddy & Kramer (1975; key).

Protomycetaceae Gray (1821), Protomycetales. 5 gen., 20 spp. Mycelium intercellular, producing thick-walled smooth or ornamented resting-spores either throughout all gall tissue or in a continuous subepidermal layer, the resting spores either converted directly into a spore sac, or rupturing to form the spore sac from an emergent internal membrane. *Lit.*: Reddy & Kramer (*Mycotaxon* **3**: 1, 1975).

Protomycetales, Ascomycota. 1 fam., 5 gen., 201 spp. Mycelium apparently diploid, growing intercellularly within plant tissue, sometimes forming thick-walled resting spores; ascogenous cells ('spore sacs') formed either directly or via rupture and the subsequent emergence of an internal membrane, with a multinucleate protoplast which becomes oriented in a peripheral layer, the nuclei undergoing meiosis to form four small hyaline aseptate ascospores ('endospores'), the endospores discharged in a single mass. Anamorphs yeast-like. Biotrophic on leaves and stems, usually gall-forming. Fam.: **Protomycetaceae.** *Lit.*: Blanz & Unseld (*Stud. Mycol.* **30**: 247, 1987; 5s rRNA suggests close to *Taphrinales*).

Protomycites Mesch. (1892), Fossil fungi. 1 (Carboniferous), UK.

Protomycocladus Schipper & Samson (1994), Mucoraceae. 1, Pakistan. See Schipper & Samson (*Mycotaxon* **50**: 487, 1994).

Protomycopsis Magnus (1905), Protomycetaceae. 5, widespr. See Reddy & Kramer (1975; key), Haware & Pavgi (*Caryologia* **30**: 313, 1977; development).

protonym (in nomenclature), a name effectively but not validly published after the starting point for the group (Donk, *Persoonia* **1**: 175, 1960).

Protoparmelia M. Choisy (1929), Parmeliaceae (L). 10, N. temp. See Hafellner *et al.* (in Hawksworth (Ed.), *Ascomycete systematics*: 379, 1994; posn), Miyawaki (*Hikobia* **11**: 29, 1991), Poelt & Leuckert (*Nova*

Hedw. **52**: 39, 1991; key 6 spp., mostly on lichens), Sancho & Crespo (*Actas Simp. Nac. bot. Crypt.* **6**: 441, 1987; 3 spp. Spain).

Protoparmeliopsis M. Choisy (1929) = Lecanora (Lecanor.).

Protopeltis Syd. (1927) = Myriangiella (Schizothyr.) fide v. Arx & Müller (1975).

protoperithecium, a young but walled perithecium before ascus formation (Ellis, *Mycol.* **51**: 416, 1960).

Protophallaceae Zeller (1939), Phallales. 4 gen. (+ 1 syn.), 11 spp. Gasterocarp hypogeous; peridium gelatinized; columella branched.

Protophallus Murrill (1910) = Protubera (Protophall.) fide Furtado & Dring (*TBMS* **50**: 500, 1967).

Protophysarum M. Blackw. & Alexop. (1975), Physaraceae. 1, USA.

protophyte, see antithetic.

Protopistillaria Rick (1933) = Eocronartium (Platygl.) fide Martin (*Lloydia* **5**: 158, 1942).

protoplasmodium, see plasmodium.

protoplast, traditionally the totality of the living cell constituents, whether walled or not, but now frequently used for the cell protoplasm after experimental removal of the cell wall, a usage to which Brenner *et al.* (*Nature* **181**: 1713, 1958) proposed to restrict the term. Fungal protoplasts are proving to have value in the study of cell organelles, biochemistry, genetics; they also have potential applications in biotechnology, esp. since protoplasts from different strains or even species can be fused (first achieved in 1975) into an aggregate protoplast and produce a heterokaryon with changed properties. See Villanueva (in Ainsworth & Sussman (Eds), *The Fungi* **2**: 3, 1966), Perberdy (*Sci. Progr.* **60**: 73, 1972; review methods), Villanueva *et al.* (Eds) (*Yeast, mould and protoplasts*, 1973), Perberdy *et al.* (Eds) (*Microbial and plant protoplasts*, 1976), Peberdy & Ferenczy (Eds) (*Fungal protoplasts. Applications in biochemistry and genetics*, 1985), Peberdy (*Microbiol. Sci.* **4**: 108, 1987; review). **sphaeroplast** a - enclosed by a modified or fragmentary cell wall. [**gymnoplast** was used by Küster (1935) for plant cells without a cell wall and Frey-Wyssling (*Nature* **216**: 516, 1967) prefered **semi-gymnoplast** to sphaeroplast.]

Protoradulum Rick (1933), Tremellaceae. 1, Brazil.

Protoschistes M. Choisy (1928) = Diploschistes (Thelotremat.).

Protoscypha Syd. (1925), ? Patellariaceae. 1, trop. See Petrak (*Ann. Myc.* **32**: 317, 1934).

protosexual (of yeasts or other organisms), having diploid or dikaryotic cells which produce haploid or unisexual cells in the absence of fructifications or sexual spores; in contrast to parasexual (q.v.), which is redefined to cover organisms having both protosexual and sexual cycles, and **neo-sexual**, for organisms having a sexual but not a protosexual cycle (Wickerham, *Mycol.* **56**: 254, 1964, cf. **57**: 134, 1965, **58**: 943, 1967).

Protosporangium L.S. Olive & Stoian. (1972), Cavosteliaceae. 3, USA. See Raper (1984).

protospore, (1) the multinucleated mass of cytoplasm cut out by primary cleavage planes, to be followed by further cleavage to form the uninucleate spores of *Phycomyces* and other *Mucorales* (Harper), and the sporangiospores of *Coccidioides*; (2) (of *Synchytriaceae*), a 1-nucleate portion of protoplasm which becomes the sporangium.

Protostegia Cooke (1880), Mitosporic fungi, 4.E1.1. 1, S. Afr. See Dyko *et al.* (*Mycol.* **71**: 918, 1979).

Protostegiomyces Bat. & A.F. Vital (1955), Mitosporic fungi, 7.E1.?. 1 (on *Lembosia*), Brazil.

Protosteliaceae L.S. Olive (1962) (Protosteliidae, Protostelidaceae), Protosteliales. 9 gen., 19 spp. Flagellate motile cells absent, sporocarps simple. See Spiegel (1990).

Protosteliales, (Protostelida; incl. Ceratiomyxales), Protosteliomyces. 3 fam., 14 gen. (+ 3 syn.), 29 spp. Fams:
(1) **Cavosteliaceae**.
(2) **Ceratiomyxaceae**.
(3) **Protosteliaceae**.
Lit.: see *Protosteliomycetes*.

Protosteliomycetes (Protostelia, Protostelida, Protostelea), Myxomycota; protostelid slime moulds, protostelids. 1 ord., 3 fam., 14 gen. (+ 3 syn.), 29 spp. Trophic phase of simple amoeboid cells with filose pseudopodia, plasmodia not with a shuttle movement of the protoplasm; ± flagellate cells (present in 'amoeboflagellates'); sporulation not proceeding by the long grex action of myxamoebae; sporocarps stalked, stalks simple and often very delicate; spores formed singly or several together in some spp. germinating to produce 8 haploid flagellate spores). Isolated mainly by moist-chamber culture of aerial dead or decaying plant parts, especially bark; also on dung. Ord.:
Protosteliales.

The *Ceratiomyxaceae* were recognized as the basis of a separate class (*Ceratiomyxomycetes*) in the seventh edition of this *Dictionary*, but Spiegel (1990) did not distinguish any rank above family, and Corliss (1994) included *Ceratiomyxa* in the same class as *Protostelium*; consequently, no separate class, subclass or ordinal name is used for this group of slime moulds in this edition.

Spiegel (1990) noted that the protostelid genera could be placed in five more natural groups, based on the number of kinetoids per cell, the presence/absence and nature of the nuclear attachments, and the type of cell coats in addition to colonial morphology and other macroscopic features; these were not given formal names.

Lit.: Corliss (1994), Olive (1975, 1982), Raper (1973), Spiegel (*in* Margulis *et al.* (Eds), *Handbook of the Protoctista*: 484, 1990).

Protosteliopsis L.S. Olive & Stoian. (1966), Protosteliaceae. 2, widespr. (warmer regions).

Protostelium L.S. Olive & Stoian. (1960), Protosteliaceae. 5, widespr.

protosterigma, see basidium.

Protostroma Bat. (1957), Mitosporic fungi, 8.A1.?. 1, USA.

Prototheca Krüger (1894) nom. dub. (? achloric algae). 5, widespr. See Arnold & Ahearn (*Mycol.* **64**: 265, 1972; key), Smith (*Mycopath.* **71**: 95, 1980), Capriotti (*Arch. Microbiol.* **42**: 409, 1962; in fish). Associated with disease in humans (**protothecosis**), bovine mastitis, etc.; see Sudman (*Am. J. clin. Path.* **61**: 10, 1974).

protothecium, an incompletely differentiated ascoma containing neither asci nor ascospores (Shoemaker, *CJB* **34**: 641, 1955).

Protothelenella Räsänen (1943), Protothelenellaceae (±L). 9 (on lichens and *Musci*), Eur., N. Am. See Mayrhofer (*Herzogia* **7**: 313, 1987; key).

Protothelenellaceae Vězda, H. Mayhofer & Poelt (1985), Ascomycota (inc. sed., ±L). 2 gen. (+ 2 syn.), 10 spp. Thallus crustose but poorly developed or absent; stromata absent; ascomata intermediate between apothecial and perithecial, immersed, sometimes becoming erumpent, dark green to black, opening by a broad pore; peridium composed of interwoven hyphae in a gelatinous matrix; interascal

tissue of sparingly branched and anastomosing trabeculate pseudoparaphyses, becoming free at the apices, the apices not swollen; hymenium blueing in iodine; asci cylindrical, persistent, at first thick-walled but usually without separable layers, the apex somewhat thickened and blueing diffusely in iodine, with an I+ apical ring, splitting at the apex to release the spores; ascospores hyaline, usually muriform, without a sheath. Anamorphs pycnidial where known. Saprobic on bark, or lichenized with green algae, rarely lichenicolous or bryophilous, temp.

Protothyrium G. Arnaud (1917), Parmulariaceae. 4, trop.

Prototrema M. Choisy (1928) ? = Thelotrema (Thelotremat.).

Prototremella Pat. (1888) = Tulasnella (Tulasnell.) fide Rogers (*Ann. Myc.* **31**: 183, 1933).

Prototrichia Rostaf. (1876), Trichiaceae. 1, temp.

prototroph, see wild type.

prototunicate (of asci), basically unitunicate, but with the wall lysing at or before maturity and lacking an differentiated apical structures (e.g. *Saccharomycetales*); such asci may develop in an hymenium or be distributed randomly in the interior of the ascoma. See ascus.

Prototylium M. Choisy (1929) ? = Pseudopyrenula (Trypethel.).

Protounguicularia Raitv. & R. Galán (1986), Hyaloscyphaceae. 3, Eur. See Huhtinen (*Mycotaxon* **30**: 9, 1987). ? = Olla (Hyaloscyph.) fide Baral (*SA* **13**: 113, 1994).

protouredinium, see protoaecium.

Protousnea (Motyka) Krog (1976), Parmeliaceae (L). 7, S. Am.

Protoventuria Berl. & Sacc. (1887) ? = Antennularia (Ventur.) fide Müller & v. Arx (1962). See Barr (*Prodr. Class Ascom.*, 1989, *Sydowia* **41**: 25, 1989; key 23 N. Am. spp.), Richiteana & Bontea (*Rev. Roum. Biol.*, biol. vég. **32**: 15, 1987; nomencl., on *Ericaceae*).

PROTOZOA. Kingdom of *Eukaryota*. Predominantly unicellular, plasmodial, or colonial phagotropic, wall-less in the trophic state; ciliary hairs never rigid or tubular; chloroplasts, where present lacking starch and phycobilisomes, with stalked thylakoid and 3 membranes; multicellular species with minimal cell differentiation and lacking collagenous connective tissue sandwiched between dissimilar epithelia.

The *Protozoa* have been included with the *Chromista* in a broader kingdom *Protoctista* (syn. *Protista*) by some (e.g. Margulis *et al.*, 1990) but the classification has been rejected by recent authors on both molecular and non-molecular grounds (Cavalier-Smith, 1993; Corliss, 1994).

Various arrangements and ranks have been proposed for the main groupings of 'fungi' within the *Protozoa*; selected ones are compared in Table 4 (see FUNGI).

Cavalier-Smith (1993) acceps 18 phyla in the *Protozoa* and Corliss (1994) 14; however, only one has been traditionally studied by mycologists: *Mycetozoa* (incl. 3 classes *Protostelea*, *Myxogastrea* and *Dictyostelea*). Other protozoan phyla are not treated further in this account or elsewhere in this edition of the *Dictionary*. Olive (1975) arranged the slime moulds into five classes; that system was adopted (using fungal terminations) in the seventh edition of the *Dictionary* which accepted *Acrasomycetes*, *Ceratiomyxomycetes*, *Dictyosteliomycetes*, *Myxomycetes* and *Protosteliomycetes*. Olive regarded the *Acrasomycetes* as isolated from the other classes. Molecular data support that view (see Phylogeny) but further the remoteness of the *Dictyosteliomycetes* from the other classes. In this

edition of the *Dictionary* four phyla are distinguished to reflect that new information:
(1) **Acrasomycota** (syn. *Acrasomycetes*).
(2) **Dictyosteliomycota** (syn. *Dictyosteliomycetes*).
(3) **Myxomycota** (incl. *Ceratiomyxomycetes, Myxomycetes, Protosteliomycetes*).
(4) **Plasmodiophoromycota** (syn. *Plasmodiophoromycetes*).
Lit.: Bonner (*The cellular slime molds*, 1959; edn 2, 1967), Cavalier-Smith (*Microbiol. Rev.* **57**: 953, 1993), Corliss (*Acta Protozool.* **33**: 1, 1994), Heywood & Rothschild (*Biol. J. Linn. Soc.* **30**: 91, 1987; nomencl. higher taxa), Olive (*The mycetozoans*, 1975), Teixeira (*Rickia, suppl.* **3**, 1970; gen. except *Myxomycetes*).
See Classification, Kingdoms of fungi, Phylogeny.
Protubera A. Møller (1895), Protophallaceae. 8, trop., subtrop. See Furtado & Dring (*TBMS* **50**: 500, 1967), Castellano & Beever (*N.Z. Jl Bot.* **32**: 305, 1994; NZ spp.).
protuberate (of conidia), having short projections.
Protuberella S. Imai & A. Kawam. (1958), Protophallaceae. 1, Japan. See Malloch (*Mycotaxon* **34**: 133, 1989).
pruinose, having a frost-like or flour-like surface covering of **pruina**. Often caused by calcium oxalate hydrates on lichen thalli; see Wadsten & Moberg (*Lichenologist* **17**: 239, 1985).
Prunulus Gray (1821) = Mycena (Tricholomat.) fide Donk (1962).
PSA, see Media.
Psalidosperma Syd. & P. Syd. (1914) = Ypsilonia (Mitosp. fungi) fide Clements & Shear (1931).
Psaliota, see *Psalliota*.
Psalliota (Fr.) P. Kumm. (1871) ≡ Agaricus (Agaric.).
Psalliotina Velen. (1939) = Psathyrella (Coprin.) fide Pilát (1951).
Psammina Sacc. & M. Rousseau ex E. Bommer & M. Rousseau (1891), Mitosporic fungi, 6.G2.1. 2, Eur.
Psammocoparius Delile ex De Seynes (1863) nom. inval. = Psathyrella (Coprin.).
Psammomyces Lebedeva (1932) = Galeropsis (Galeropsid.) fide Zeller (*Mycol.* **35**: 409, 1943).
Psammospora Fayod (1889) nom. inval. ≡ Melaleuca (Tricholomat.).
Psathyra (Fr.) P. Kumm. (1871) [non Spreng. (1818), *Rubiaceae*] = Psathyrella (Coprin.) fide Singer (1951).
Psathyrella (Fr.) Quél. (1872), Coprinaceae. *c.* 600, cosmop. See Kits van Waveren (*Persoonia* **6**: 249, **7**: 23, **8**: 345, 1971, **9**: 199, 1977, *Suppl.* **2**, 1985; monogr. Dutch, French, Br. spp.), Pegler (*Kew Bull., Addit. Ser.* **6**, 1977; E. Afr.), Smith (*Mem. N.Y. bot. Gdn* **24**, 1972; 414 N. Am. spp.).
Psathyrodon Maas Geest. (1977), Sistotremataceae. 1, Zambia.
Psathyromyces Bat. & Peres (1964), Mitosporic fungi (L), 2.B1,?. 1, Amazon.
Psathyrophlyctis Brusse (1987), Phlyctidaceae (L). 1, S. Africa.
Psecadia Fr. (1849) ≡ Cytospora (Mitosp. fungi). See Sutton (1977).
Pselaphidomyces Speg. (1917), Laboulbeniaceae. 1, N. Am.
Pselliophora P. Karst. (1879) = Coprinus (Coprin.) fide Earle (1909).
Pseudacoliomyces Cif. & Tomas. (1953) ≡ Pseudacolium (Calic.).
Pseudacolium Stizenb. ex Clem. (1909) = Cyphelium (Calic.).
Pseudaegerita J.L. Crane & Schokn. (1981), Mitosporic fungi, 3.A2.3. 1, N. Am.
Pseudaleuria Lusk (1987), Otideaceae. 1, USA.

Pseudallescheria Negr. & I. Fisch. (1944), Microascaceae. 7, widespr. *P. boydii* on humans (white grain mycetoma). See v. Arx *et al.* (*Beih. Nova Hedw.* **94**, 1988), Campbell & Smith (*Mycopath.* **78**: 145, 1982; conidiogenesis), McGinnis *et al.* (*Mycotaxon* **14**: 94, 1982). Anamorph *Scedosporium*.
Pseudapiospora Petr. (1928) = Pseudomassaria (Hyponectr.) fide Müller & v. Arx (1962).
Pseudarctomia Gyeln. (1939), ? Lichinaceae (L). 1, France.
Pseudarthopyrenia Keissl. (1935) = Pyrenocollema (Pyrenul.) fide Harris (*in litt.*).
Pseudascozonus Brumm. (1985), Thelebolaceae. 1 (coprophilous), France. See van Brummelen (*Persoonia* **13**: 369, 1987; ultrastr., *in* Hawksworth (Ed.), *Ascomycete systematics*: 398, 1994; posn).
Pseudasterodon Rick (1959) = Asterodon sensu Rick (Hymenochaet.); nom. dub. fide Donk (1964).
Pseudephebe M. Choisy (1930), Parmeliaceae (L). 2, montane-arctic, bipolar. See Brodo & Hawksworth (*Op. bot. Soc. bot. Lund* **42**, 1977).
Pseuderiospora Keissl. (1924) = Eriosporella (Mitosp. fungi) fide Sutton (1977).
Pseuderiospora Petr. (1959) ≡ Suttoniella (Mitosp. fungi).
Pseudeurotiaceae Malloch & Cain (1970), Ascomycota (inc. sed.). 14 gen. (+ 2 syn.), 28 spp. Stromata absent; ascomata cleistothecial, formed from coiled initials, usually brown to black, the peridium either pseudoparenchymatous or cephalothecoid, fragmenting into predetermined polygonal plates; interascal tissue absent; asci irregularly disposed, ± globose, thin-walled, without apical structures, evanescent; ascospores small, hyaline or brown, usually aseptate, smooth-walled, without germ pores. Anamorphs hyphomycetous, mostly *Acremonium*-like. Saprobic on woody tissue and rotting vegetation, rarely fungicolous, widespr.
Lit.: Booth (*Mycol. Pap.* **83**, 1961), Lodha (*in* Subramanian (Ed.), *Taxonomy of fungi* **1**: 241, 1978), Malloch & Cain (*CJB* **48**: 1815, 1970; gens).
Pseudeurotium J.F.H. Beyma (1937), Pseudeurotiaceae. 4, widespr. See Booth (*Mycol. Pap.* **83**, 1961; key). = Pleuroascus (Pseudeurot.) fide Lodha (1978).
Pseudevernia Zopf (1903), Parmeliaceae (L). 5 (or 6), Eur., C. & N. Am. See Hale (*Bryologist* **71**: 1, 1968).
Pseudhaplosporella Speg. (1920) = Botryodiplodia (Mitosp. fungi) fide Petrak & Sydow (1927).
Pseudhydnotrya E. Fisch. (1897) = Geopora (Otid.) fide Fischer (1938).
pseudo- (prefix), false; spurious.
Pseudoabsidia Bainier (1903) = Absidia (Mucor.) fide Hesseltine (1955).
pseudoaethium (of myxomycetes), a group of separate sporangia looking like an aethalium.
pseudoamyloid, see dextrinoid.
pseudoangiocarpous (of a basidioma), hymenial surface at first exposed but later covered by an incurving pileus margin and/or excrescences from the stipe (Singer, 1975: 26).
Pseudoanguillospora S.H. Iqbal (1974), Mitosporic fungi, 1.E1.1/10. 2, UK.
Pseudoarachniotus Kuehn (1957) = Arachniotus (Gymnoasc.) fide v. Arx (1981).
Pseudoaristastoma Suj. Singh (1979), Mitosporic fungi, 4.C1.?. 1, India.
Pseudoarmillariella (Singer) Singer (1956), Tricholomataceae. 1, N. Am.
Pseudoarthonia Marchand (1896) = Arthonia (Arthon.).
Pseudoauricularia Kobayasi (1982), Agaricaceae. 1, Papua New Guinea.

Pseudobaeomyces M. Satô (1940), Icmadophilaceae (L). 1, Asia.

Pseudobaeospora Singer (1942), Agaricaceae. 4, C. Asia, S. Am. See Singer (*Mycologia* **60**: 13, 1963, Horak (*Rev. Myc.* **29**: 72, 1964; key).

Pseudobalsamia E. Fisch. (1907) = Balsamia (Balsam.) fide Trappe (1975).

Pseudobasidiospora Dyko & B. Sutton (1978), Mitosporic fungi, 4.A1.1. 1 (aquatic), Australia, USA.

Pseudobasidium Tengwall (1924) = Arthrinium (Mitosp. fungi) fide Ellis (*Mycol. Pap.* **103**, 1965).

Pseudobeltrania Henn. (1902), Mitosporic fungi, 1.A2.10. 2, S. Am., Nepal. See Sutton (*TBMS* **55**: 506, 1970).

Pseudoboletus Šutara (1991), Boletaceae. 1, N. temp.

Pseudobotrytis Krzemien. & Badura (1954), Mitosporic fungi, 1.B2.10. 1, Poland.

Pseudobuellia de Lesd. (1907) = Rinodina (Physc.).

Pseudocalicium Marchand (1896) = Chaenothecopsis (Mycocalic.) p.p. and Mycocalicium (Mycocalic.) p.p.

Pseudocamptoum Gonz. Frag. & Cif. (1925) = Melanographium (Mitosp. fungi) fide Ellis (*Mycol. Pap.* **93**, 1963).

Pseudocanalisporium R.F. Castañeda & W.B. Kendr. (1991), Mitosporic fungi, 3.D2.1. 1, Cuba.

pseudocapillitium (of myxomycetes), a sterile structure in the fruit-body which has had no connexion with the sporogenous protoplasm.

Pseudocenangium P. Karst. (1886), Mitosporic fungi, 7.E1.19. 2, Eur., N. Am. See Dyko & Sutton (*CJB* **57**: 370, 1979).

Pseudocenangium A. Knapp (1924) = Ascocoryne (Leot.) fide Dennis (*Mycol. Pap.* **62**, 1956).

Pseudocercophora Subram. & Sekar (1986), Lasiosphaeriaceae. 1, India. Anamorph *Mammaria*.

Pseudocercospora Speg. (1910), Mitosporic fungi, 1.E2.10. *c.* 250, widespr, esp. trop. See Deighton (*Mycol. Pap.* **140**, 1976, *TBMS* **88**: 365, 1987), Guo & Liu (*Mycosystema* **2**: 225, 1989; China), Anon. (*Progr. Rep. Asian Veg. Res. Develop. Centre, 1990*: 150, 1991; in vitro data on *P. fuligena*), Guo & Hsieh (*The genus Pseudocercospora in China*, 1995).

Pseudocercosporella Deighton (1973), Anamorphic Mycosphaerellaceae and Dermateaceae, 1.E1.10. Teleomorphs *Mycosphaerella*, *Tapesia*. 10, widespr. *P. herpotrichoides* causing eyespot of cereals. See Wallwork & Spooner (*TBMS* **91**: 703, 1988; *Tapesia* teleomorph of *P. herpotrichoides*), Fitt (*in* Singh *et al.* (Eds), *Plant diseases of international importance: diseases of cereals and pulses*, 1992), Creighton (*Pl. Path.* **38**: 484, 1989; identification of W-type and R-type isolates of *P. herpotrichoides*), Inman *et al.* (*MR* **95**: 1334, 1991; *Mycosphaerella* teleomorph of *P. capsellae*), Nicholson *et al.* (*Pl. Path.* **40**: 584, 1991; DNA markers for classification of *P. herpotrichoides*), Priestley *et al.* (*Pl. Path.* **41**: 591, 1992; comparison of isoenzyme and DNA markers for *P. herpotrichoides*), Thomas *et al.* (*J. Gen. Microbiol.* **138**: 2305, 1992; identification of R- and W-types of *P. herpotrichoides*), Hocart, Lucas & Peberdy (*MR* **97**: 967, 1993; parasexuality), Frei & Wenzel (*J. Phytopath.* **139**: 229, 1993; genomic DNA gene probes), Nicholson & Rezanoor (*MR* **98**: 13, 1994; identification of pathotypes of *P. herpotrichoides* by random amplified polymorphic DNA). = Ramulispora (Mitosp. fungi) fide v. Arx (*Proc. K. Akad. Wet. Amst.* **86**: 36, 1983).

Pseudocercosporidium Deighton (1973), Mitosporic fungi, 1.E2.10. 1, Venezuela.

Pseudochaetosphaeronema Punith. (1979), Mitosporic fungi, 4.A1.15. 1 (from humans), Venezuela.

Pseudochuppia Kamal, A.N. Rai & Morgan-Jones (1984), Mitosporic fungi, 1.D2.1/10. 1, India.

Pseudociboria Kanouse (1944), Sclerotiniaceae. 1, widespr. See Dumont (*Mycol.* **66**: 706, 1974), Schumacher (*Mycotaxon* **38**: 233, 1990).

Pseudoclathrus B. Liu & Y.S. Bau (1980), Clathraceae. 1, China.

pseudocleistothecium, see discrete body.

Pseudoclitocybe (Singer) Singer (1956), Tricholomataceae. 6, N. temp., S. Am.

Pseudococcidioides O.M. Fonseca (1928) = Coccidioides (Mitosp. fungi) fide Dodge (1935).

Pseudocochliobolus Tsuda, Ueyama & Nishih. (1978) = Cochliobolus (Pleospor.) fide Alcorn (*Mycotaxon* **16**: 353, 1983).

Pseudocoelomomyces E.A. Nam & Dubitskiĭ (1977) ≡ Tabanomyces (Entomophthor.).

Pseudocollema Kanouse & A.H. Sm. (1940) = Byssonectria (Otid.) fide Sivertsen (*SA* **9**: 23, 1990).

pseudocolumella (of *Physaraceae*), lime-knots massed like a columella at the centre of the sporangium.

Pseudocolus Lloyd (1907), Clathraceae. 2, trop. (locally introd. warm temp. areas).

Pseudoconiocybe Marchand (1896) = Coniocybe (Coniocyb.).

Pseudoconium Petr. (1969), Mitosporic fungi, 4.A2.?. 1, Eur.

Pseudoconocybe Hongo (1967) = Conocybe (Bolbit.) fide Watling (*Kew Bull.* **31**: 593, 1977).

Pseudocoprinus Kühner (1928) = Coprinus (Coprin.) fide Singer (1975).

Pseudocordyceps Hauman (1936), Mitosporic fungi, 2.A1.?. 1, Congo.

Pseudocornicularia Gyeln. (1933) = Coelocaulon (Parmel.) fide Kärnefelt (*Opera Bot.* **86**, 1986).

pseudocortex (of lichen thalli), a false cortex, used for the outer layer of the pseudopodetia in Pycnothelia papillaria.

Pseudocraterellus Corner (1957) = Craterellus (Craterell.) fide Peterson (*Česká Myk.* **29**: 199, 1975).

Pseudocryptosporella J. Reid & C. Booth (1969), Diaporthales (inc. sed.). 1, Venezuela.

pseudocyphella (pl. -e), an opening in the cortex of lichens where the medulla is exposed to the open air but lacking specialized cells surrounding the cavity; they provide valuable taxonomic characters in e.g. *Alectoria, Bryoria, Pseudocyphellaria*.

Pseudocyphellaria Vain. (1890) nom. cons., Lobariaceae (L). 112, mainly S. temp. See Coppins & James (*Lichenologist* **11**: 139, 1979; key Eur. spp.), Galloway (*Lichenologist* **18**: 105, 1986; S. Am. nonglabrous spp., *Bull. Br. Mus. nat. hist., Bot.* **17**: 1-267, 1988; key 48 spp.), Galloway & James (*Lichenologist* **12**: 291, 1980; *NZ*, *Nova Hedw.* **42**: 423, 1986; Delise's spp., *Bibl. Lich.* **46**, 1992; 53 S. Am. spp.), Galloway *et al.* (*Lichenologist* **15** : 135, 1983; NZ), Galloway & Laundon (*Taxon* **37**: 48, 1988; gen. nomencl.), Kondratyuk & Galloway (*Bibl. Lich.* **57**: 327, 1985; lichenicolous spp. on).

pseudocystidium, (1) (in agarics), see cystidium; (2) (of *Entomophthora*), the organ penetrating the insect cuticle allowing conidiophores to emerge.

Pseudocytospora Petr. (1923), Mitosporic fungi, 8.A1.?. 1, Eur.

Pseudodasyscypha Velen. (1939) ? = Cyphellopsis (Crepidot.) fide Donk (1962).

Pseudodeconica Overeem (1927) nom. nud. = Agrocybe (Bolbit.) fide Singer (1951).

Pseudodelitschia J.N. Kapoor, Bahl & S.P. Lal (1976) = Neotestudina (Testudin.) fide Hawksworth (1979).

Pseudodescolea Raithelh. (1980) = Descolea (Bolbit.) fide Bougher (*MR* **94**: 287, 1990).

Pseudodiaporthe Speg. (1909) = Massarina (Lophiostom.) fide Bose (1961).

pseudodiblastic (of ascospores), having oil- drops at the poles so that they superficially resemble polarilocular spores (q.v.).

Pseudodichomera Höhn. (1918), Mitosporic fungi, 8.D2.?. 3, Eur. See Arnold & Russell (*Mycol.* 52: 509, 1960).

Pseudodictya Tehon & G.L. Stout (1929) nom. dub. fide Sutton (*Mycol. Pap.* 141, 1977).

Pseudodictyosporium Matsush. (1971), Mitosporic fungi, 1.D2.24. 2, Papua New Guinea.

Pseudodidymaria U. Braun (1993), Mitosporic fungi, 3.B2.10/12. 2, USA.

Pseudodimerium Petr. (1924) = Phaeostigme (Pseudoperispor.). See Hansford (*Mycol. Pap.* 15, 1946).

Pseudodiplodia (P. Karst.) Sacc. (1884), Mitosporic fungi, 4.B2.15. 45, widespr. See Buchanan (*Mycol. Pap.* 156: 1, 1987).

Pseudodiplodia Speg. (1920) [non (P. Karst.) Sacc. (1884)] = Botryodiplodia (Mitosp. fungi) fide Petrak & Sydow (1927).

Pseudodiplodiella Bender (1932) = Botryodiplodia (Mitosp. fungi) fide Sutton (1977).

Pseudodiscinella Dennis (1956) ≡ Ciliatula (Leot.).

Pseudodiscosia Hösterm. & Laubert (1921) = Heteropatella (Mitosp. fungi) fide Sutton (1977).

Pseudodiscula Laubert (1911), Mitosporic fungi, 8.A1.?. 2, Eur. See Sutton (1977).

Pseudodiscus Arx & E. Müll. (1959), Saccardiaceae. 1, Eur.

Pseudodoassansia (Setch.) Vánky (1981), Tilletiaceae. 1 (on *Sagittaria*), N. Am.

Pseudoecteinomyces W. Rossi (1977), Euceratomycetaceae. 1, Afr., Asia.

Pseudoendococcus Marchand (1896) = Endococcus (Dothideales).

Pseudoepicoccum M.B. Ellis (1971), Mitosporic fungi, 3.A2.10. 1, trop.

pseudoepithecium, an amorphous or granular layer overlying paraphyses in an apothecium and in which their tips are immersed, but not forming a separate tissue.

Pseudofarinaceus Battarra ex Kuntze (1891) ≡ Amanitopsis (Amanit.).

Pseudofarinaceus Earle (1909) = Volvariella (Plut.).

Pseudofavolus Pat. (1900), Polyporaceae. 3, pantrop.

pseudofissitunicate (of asci), see ascus.

Pseudofistulina O. Fidalgo & M. Fidalgo (1963), Fistulinaceae. 1, Brazil. = Fistulina (Fistulin.) fide Gibertson & Ryvarden (1987: 261).

Pseudofomes Lázaro Ibiza (1916) = Phellinus (Hymenochaet.) fide Donk (1974).

Pseudofumago Briosi & Farneti (1906) nom. dub. fide de Hoog & Hermanides-Nijhof (*Stud. Mycol.* 15: 186, 1977).

Pseudofungi. A subdivision in the *Chromista*; including the fungi treated as belonging to the oomycetes in a broad sense, and as *Oomycota, Hyphochytriomycota, Labyrinthulomycota*, and *Thraustochytriales* in this edition of the *Dictionary*; see Cavalier-Smith (*Progr. Phycol. Res.* 4: 309, 1986), Pseudomycotina.

Pseudofusarium Matsush. (1971) = Fusarium (Mitosp. fungi) fide Booth & Sutton (*in litt.*).

Pseudofuscophialis Sivan. & H.S. Chang (1995), Mitosporic fungi, 1.C2.15. 1, Taiwan.

Pseudofusidium Deighton (1969) = Acremonium (Mitosp. fungi) fide Gams (1971).

Pseudogaster Höhn. (1907), Mitosporic fungi, 2.A2.?. 1 (on lichens), S. Am.

Pseudogenea Buchholz (1901) = Genabea (Otid.) fide Trappe (1975).

Pseudogibellula Samson & H.C. Evans (1973), Mitosporic fungi, 2.A1.32/33. 1, Ghana.

Pseudogliomastix W. Gams (1985), Anamorphic ? Niessliaceae, 1.A2.15. Teleomorph *Wallrothiella*. 1, Italy.

Pseudogliophragma Phadke & V.G. Rao (1980), Mitosporic fungi, 2.A1.10. 1, India.

Pseudogloeosporium Jacz. (1917) = Kabatia (Mitosp. fungi) fide Sutton (1977).

Pseudogomphus R. Heim (1970), Gomphaceae. 1, Gabon.

Pseudographiella E.F. Morris (1966), Mitosporic fungi, 2.B1.?. 1, UK.

Pseudographis Nyl. (1855) nom. cons., Triblidiaceae. 3, temp.

Pseudographium Jacz. (1898) ≡ Sphaerographium (Mitosp. fungi) fide Sutton (1977).

Pseudographium Höhn. (1915), Mitosporic fungi, 2.?.?. 1, USA. See Sutton (1977).

Pseudoguignardia Gutner (1927) = Physalospora (Hyponectr.) fide Müller & v. Arx (1962).

Pseudogyalecta Vězda (1975) nom. rej. prop. = Badimia (Ectolech.) fide Lücking & Vězda (1995).

Pseudogymnoascus Raillo (1929), Myxotrichaceae. 3, Eur., N. Am., former USSR. See Ito & Yokoyama (*IFO Res. Comm.* 13: 83, 1987), Samson (*Acta Bot. neerl.* 21: 517, 1972), Orr (*Mycotaxon* 8: 165, 1979). Anamorph *Geomyces*.

Pseudogymnopilus Raithelh. (1974), Cortinariaceae. 1, S. Am.

Pseudogyrodon Heinem. & Rammeloo (1983), Gyrodontaceae. 1, Afr.

Pseudohalonectria Minoura & T. Muroi (1978), Ascomycota (inc. sed.). 6 (on submerged wood), Japan, N., C. & S. Am. See Shearer (*CJB* 67: 194, 1989; key, posn).

Pseudohansenula Mogi (1939) nom. dub. (Fungi, inc. sed.).

Pseudohansenula E.K. Novák & Zsolt (1961) Fungi (inc. sed.) fide Lodder (1970: 236).

Pseudohansfordia G.R.W. Arnold (1970), Mitosporic fungi, 1.B/C1.10. 9 (on fungi), widespr. See de Hoog (*Persoonia* 10: 57, 1978), Eicker et al. (*Bot. Bull. Acad. Sin.* 31: 205, 1990; disease of *Auricularia mesenterica*).

Pseudohansfordia S.M. Reddy & Bilgrami (1975) [non G.R.W. Arnold (1970)], Mitosporic fungi, 1.A1.?. 1, India. See de Hoog (*Persoonia* 10: 58, 1978).

Pseudohaplis Clem. & Shear (1931) = Botryodiplodia (Mitosp. fungi) fide Sutton (1977).

Pseudohaplosporella, see *Pseudohaplosporella*.

Pseudohelotium Fuckel (1870), Leotiaceae. 4, Eur. See Arendholz (*Mycotaxon* 36: 283, 1989; nomencl.).

Pseudohepatica P.M. Jørg. (1993), Ascomycota (inc. sed., L). 1 (ascomata unknown), Venezuela.

Pseudoheppia Zahlbr. (1903), Heppiaceae (L). 1, Eur.

pseudoheterothallism, see heterothallism.

Pseudohiatula (Singer) Singer (1938), Tricholomataceae. 1, trop. See Singer (*Persoonia* 2: 407, 1962).

Pseudohydnotrya E. Fisch. (1897) = Geopora (Otid.) fide Fischer (1938).

Pseudohydnum P. Karst. (1868), Exidiaceae. 1, widespr. See Donk (*Persoonia* 3: 199, 1964).

Pseudohydnum Rick (1904) = Hydnodon (Thelephor.) fide Donk (1956).

Pseudohygrocybe (Bon) Kovalenko (1988), Hygrophoraceae. 18, widespr.

Pseudohygrophorus Velen. (1939) nom. dub. (Agaric., inc. sed.).

Pseudohypocrea Yoshim. Doi (1972), Hypocreaceae. 1, N. Am.

pseudoidia, separated hyphal cells able to be germinated (Bensaude).

Pseudoidium Y.S. Paul & J.N. Kapoor (1986) = Oidium (Mitosp. fungi) fide Sutton (*in litt.*).

pseudoisidium (pl. -ia), (1) an outgrowth from the surface of a lichen thallus resembling an isidium (e.g. *Gyalideopsis*; see Vězda, *Folia geobot. phytotax.* **14**: 48, 1979); (2) isidium without photosynthetic cells in *Pseudocyphellaria* (see Galloway, *Bull. Br. Mus. nat. Hist., Bot.* **17**: 1, 1988).

Pseudokarschia Velen. (1934) = Dactylospora (Dactylospor.) fide Hafellner (*Beih. Nova Hedw.* **62**, 1979).

Pseudolachnea Ranoj. (1910), Mitosporic fungi, 6/7.B/C1.15. 5, widespr. See Sutton (1980).

Pseudolachnea Velen. (1934) [non Ranoj. (1910)] ? = Hyalopeziza (Hyaloscyph.). See Eckblad (*Nytt Mag. Bot.* **15**: 174, 1978).

Pseudolachnella Teng (1936) = Pseudolachnea (Mitosp. fungi) fide Sutton (1977).

Pseudolachnum Velen. (1934), ? Leotiales (inc. sed.). 1, former Czechoslovakia.

Pseudolagarobasidium J.C. Jang & T. Chen (1985), Hyphodermataceae. 2, S.E. Asia, Australia. See Wu (*Acta Bot. Fenn.* **142**, 1990; key).

Pseudolasiobolus Agerer (1983), Podoscyphaceae. 1, Eur.

Pseudolecanactis Zahlbr. (1907), ? Roccellaceae (L). 1, Samoa.

Pseudolecidea Marchand (1896) = Abrothallus (Ascomycota, inc. sed.).

Pseudolecidea Clauzade & Cl. Roux (1984) ≡ Claurouxia (Candelar.).

Pseudolembosia Theiss. (1913), Parmulariaceae. 2, Australia, C. Am. See Petrak (*Sydowia* **8**: 297, 1954).

Pseudoleptogium Müll. Arg. (1885) = Leptogium (Collemat.). See Degelius (*Svensk bot. Tiskr.* **37**: 65, 1943).

Pseudoleptogium Jatta (1900) = Polychidium (Placynth.).

Pseudolizonia Pirotta (1889) = Lizonia (Pseudoperispor.) fide v. Arx & Müller (1975).

Pseudolpidiella Cejp (1959) = Sirolpidium (Sirolpid.) fide Dick (*in press*).

Pseudolpidiopsis Minden (1911) = Pleocystidium (Olpidiopsid.) fide Dick (*in press*).

Pseudolpidium A. Fisch. (1892) = Olpidiopsis (Olpidiopsid.) p.p. fide Shanor (*J. Elisha Mitch. sci. Soc.* **55**: 179, 1939).

Pseudolycoperdon Velen. (1947) = Bovista (Lycoperd.) fide Kreisel (*Beih. Nova Hedw.* **25**: 63, 1967).

Pseudolyophyllum (Singer) Raithelh. (1980), Tricholomataceae. 14, S. Am.

Pseudomassaria Jacz. (1894), Hyponectriaceae. 15, widespr. See Barr (*Mycol.* **56**: 841, 1964; keys, **68**: 611, 1976; nomencl.).

Pseudomassariella Petr. (1955) = Leiosphaerella (Amphisphaer.) fide Müller & v. Arx (1962).

Pseudobrophila Boud. (1885), Otideaceae. 28 (esp. coprophilous), Eur., N. & S. Am. See van Brummelen (*Libri Bot.* **14**, 1995; monogr., key).

Pseudomelasmia Henn. (1902) = Phyllachora (Phyllachor.) fide Höhnel (*Sber. Akad. wiss. Wien, mat.-nat. Kl.* **119**: 54, 1910).

Pseudomeliola Speg. (1889), ? Hypocreaceae. 4, trop. Am. See Petrak & Sydow (*Ann. Myc.* **34**: 17, 1936).

Pseudomeria G.L. Barron (1980), Mitosporic fungi, 1.A1.10. 1 (on nematodes), Canada.

Pseudomerulius Jülich (1979), Coniophoraceae. 2, widespr.

Pseudomicrocera Petch (1921) = Fusarium (Mitosp. fungi) fide Wollenweber & Reinking (*Die Fusarien*: 7, 1935).

Pseudomicrodochium B. Sutton (1975), Mitosporic fungi, 1.B/C1.15. 6, widespr. See Sutton *et al.* (*Mycopath.* **114**: 159, 1991; from humans).

Pseudomitrula Gamundí (1979), Geoglossaceae. 1, S. Am.

pseudomixis (-gamy), the type of fertilization in which the copulating elements are not special sexual cells.

Pseudomonilia A. Geiger (1910) nom. rej. = Candida (Mitosp. fungi).

Pseudomorfea Punith. (1981), Dothideales (inc. sed.). 1, India. See Venedikian (*Boln Soc. Argent. Bot.* **25**: 495, 1988). Anamorph *Chaetasbolisia*.

pseudomorph, a stroma made up of plant parts kept together by plectenchyma.

pseudomycelium (of *Candida*, etc.), loosely united, catenulate groups of cells (see Zobl, *RAM* **23**: 177, 1944).

Pseudomycena Cejp (1929) = Mycena (Tricholomat.) fide Smith (1947).

Pseudomycoderma H. Will (1916), Mitosporic fungi, 1.A1.?. 1, Eur.

Pseudomycoporon Marchand (1896) ? = Mycoporum (Mycoporaceae).

pseudomycorrhiza, see mycorrhiza.

Pseudomycotina. Subphylum including *Hyphochytriomycota* and *Oomycota* as used in this *Dictionary*; proposed by Barr (*Mycol.* **84**: 1, 1992) to replace *Pseudofungi* (q.v.) but excluding the *Labyrinthulomycota* which he treated as subphylum *Labyrinthista*.

Pseudonaevia Dennis & Spooner (1993), Dermateaceae. 1, UK.

Pseudonectria Seaver (1909), Hypocreaceae. 3, Eur., Am. See Rossman *et al.* (*Mycol.* **85**: 685, 1993; key). Anamorph *Volutella*.

Pseudonectriella Petr. (1959) = Catabotrys (Catabotryd.) fide v. Arx (*in litt.*).

Pseudoneottiospora Faurel & Schotter (1965), Mitosporic fungi, 4.G1.19. 2 (coprophilous), Algeria, Italy.

Pseudoniptera Velen. (1947), Dermateaceae. 1, former Czechoslovakia.

Pseudonitschkia Coppins & S.Y. Kondr. (1995), ? Dacampiaceae. 1 (on *Parmotrema*), Afr., S. Am., Nepal.

Pseudonocardia Henssen (1957), *Actinomycetes*, q.v.

Pseudoolla Velen. (1934) = Olla (Hyaloscyph.) fide Baral (*SA* **13**: 113, 1994).

Pseudoomphalina (Singer) Singer (1956), Tricholomataceae. 5, N. temp.

pseudooperculate (of asci), ones which are essentially unitunicate in structure and with a thickened apical cap which splits completely away at discharge (e.g. *Odontotrema*); see ascus.

Pseudopannaria (de Lesd.) Zahlbr. (1924), ? Lecideaceae (L). 1, France.

Pseudopapulaspora N.D. Sharma (1977), Mitosporic fungi, 1.D2.1. 1, India.

pseudoparaphyses (of ascomycetes), see hamathecium; (of basidiomycetes), see hyphidium.

pseudoparenchyma, see plectenchyma.

Pseudoparmelia Lynge (1914) Parmeliaceae. 5, trop. Afr. & Am. See Hale (*Smithson. Contr. bot.* **31**, 1976; *Mycotaxon* **25**: 603, 1986). ? = Parmelia (Parmel.).

Pseudoparodia Theiss. & Syd. (1917), Venturiaceae. 1 (on *Vaccinium*), S. Am. See Petrak (*Sydowia* **1**: 169, 1947).

Pseudoparodiella F. Stevens (1927), Venturiaceae. 1, Costa Rica. ? = Gibbera (Ventur.) fide Müller & v. Arx (1962).

Pseudopatella Sacc. (1884) = Cystotricha (Mitosp. fungi) fide Höhnel (*Sber. Akad. Wiss. Wien* **119**: 617, 1910).

Pseudopatella Speg. (1891) = Botryodiplodia (Mitosp. fungi) fide Petrak & Sydow (1927).

Pseudopatellina Höhn. (1908), Mitosporic fungi, 8.A1.?. 1, Eur.

Pseudopeltis L. Holm & K. Holm (1978), ? Leotiales (inc. sed.). 1 (on *Dryopteris*), Sweden.

Pseudopeltistroma Katum. (1975), Mitosporic fungi, 5.A2.?. 1, Japan.

Pseudopeltula Henssen (1995), Gloeoheppiaceae (L). 3, Mexico, Isle of Pines.

pseudoperidium, a false peridium; covering membrane of the aecium in the *Uredinales*.

Pseudoperis Clem. & Shear (1931) ≡ Pseudoperisporium (Pseudoperispor.).

Pseudoperisporiaceae Toro (1926), Dothideales. 23 gen. (+ 40 syn.), 176 spp. Mycelium superficial or within surface layers of substratum; ascomata superficial, small, ± globose, with a clearly defined ostiole, sometimes setose; peridium thin-walled, composed of pseudoparenchymatous cells; interascal tissue of narrowly cellular pseudoparaphyses, often deliquescing at maturity; asci saccate, fissitunicate; ascospores hyaline to brown, variously shaped, transversely septate. Anamorphs coelomycetous where known; saprobic or biotrophic on plant tissue, or on other fungi. *Lit.*: Farr (*Mycol.* **55**: 226, 1963; on *Pinaceae*, **58**: 221, 1966; on *Gramineae*, **71**: 243, 1979; on *Asteraceae*, **76**: 793, 1984; on *Rubiaceae*).

Pseudoperisporium Toro (1926) = Lasiostemma (Pseudoperispor.) fide Farr (*Mycol.* **71**: 250, 1979).

Pseudoperitheca Elenkin (1922), Ascomycota (inc. sed., L). 1, former USSR.

pseudoperithecium (of *Laboulbeniales*), a perithecium-like structure in which the asci and spores become free.

Pseudoperonospora Rostovzev (1903), Peronosporaceae. 7, widespr. *P. cubensis* (cucurbit downy mildew); *P. humuli* (hop (*Humulus*) downy mildew). See Waterhouse (*Mycol. Pap.* **148**, 1981; monogr., keys), Bedlan (*Pflanzenschutzberichte* **50**: 119, 1989; oospores).

Pseudopestalotia Elenkin & Ohl (1912) = Truncatella (Mitosp. fungi) fide Steyaert (1949). See Sutton (1977).

Pseudopetrakia M.B. Ellis (1971), Mitosporic fungi, 2.D2.10. 1, India.

Pseudopezicula Korf (1986), Leotiaceae. 2, Eur., N. Am. Anamorph *Phialophora*.

Pseudopeziza Fuckel (1870), Dermateaceae. 3, widespr. *P. trifolii* (clover leaf spot). See Schüepp (*Phytopath. Z.* **36**: 224, 1959; key).

Pseudopezizites Fiore (1932), Fossil fungi. 1 (Eocene), Italy.

Pseudophacidium P. Karst. (1885), ? Ascodichaenaceae. 4, Eur., N. Am., Asia.

Pseudophaeolus Ryvarden (1975), Coriolaceae. 1, Ghana. See Westhuizen (*Bothalia* **11**: 143, 1973), Ofosu-Asiedu (*TBMS* **65**: 285, 1975).

Pseudophaeotrichum Aue, E. Müll. & C. Stoll (1969) = Neotestudina (Testudin.) fide Hawksworth (1979).

pseudophialide, a cell bearing a sporangiolum in the *Kickxellaceae*.

Pseudophloeosporella U. Braun (1993), Mitosporic fungi, 6.C1.10. 1, Japan.

Pseudophoma Höhn. (1916) = Chaetosphaeronema (Mitosp. fungi) fide Petrak (1944).

Pseudophomopsis Höhn. (1926) = Phomopsis (Mitosp. fungi) fide Clements & Shear (1931).

Pseudophyllachora Speg. (1919) Fungi (inc. sed.) fide Petrak (*Sydowia* **5**: 350, 1951).

Pseudophysalospora Höhn. (1918) = Physalospora (Hyponectr.) fide Barr (1976).

Pseudophyscia Müll. Arg. (1894) ≡ Heterodermia (Physc.).

pseudophyse, see cystidium.

pseudophysis, see hyphidium.

Pseudopileum Canter (1963), Chytridiaceae. 1 (on *Mallomonas*), UK.

pseudopionnotes, see sporodochium.

Pseudopiptoporus Ryvarden (1980), Coriolaceae. 1, Mozambique.

Pseudopithyella Seaver (1928), Sarcoscyphaceae. 1, Eur., Bermuda. See Donadini *et al.* (*Cryptogamie, Mycol.* **10**: 283, 1989; asci).

pseudoplasmodium, see plasmodium.

Pseudoplasmodium Molisch (1925) = Labyrinthula (Labyrinth.) fide Olive (1975).

Pseudoplasmopara Sawada (1922) = Plasmopara (Peronospor.) fide Fitzpatrick (1930).

Pseudoplea Höhn. (1918) = Leptosphaerulina (Pleospor.) fide Barr (1972).

Pseudoplectania Fuckel (1870), Sarcosomataceae. 2, Eur., N. Am. See Donadini (*Mycol. Helv.* **2**: 217, 1987; 4 spp.).

Pseudopleospora Petr. (1920), Dothideales (inc. sed.). 3, Eur. See Crivelli (*Ueber die heterogene Ascomycetengattung Pleospora*, 1983; key).

pseudopodetium, a lichenized podetium-like structure of vegetative origin, ascogonia arising on this not on the pre-formed granular or squamulose thallus initials (e.g. *Cladia, Stereocaulon*).

pseudopodium (of myxomycetes), a protoplasmic process from a myxamoeba or plasmodium.

Pseudopolyporus Hollick (1910), Fossil fungi (? Basidiomycota). 1 (Carboniferous), USA.

Pseudopolystigmina Murashk. (1928), Mitosporic fungi, 8.A1.?. 1, Russia.

Pseudoprotomyces Gibelli (1874) = Phloeoconis (Mitosp. fungi) fide Saccardo (*Syll. Fung.* **14**: 1197, 1899).

Pseudopuccinia Höhn. (1925) = Stigmina (Mitosp. fungi) fide Laundon (*Mycol. Pap.* **99**: 14, 1965).

pseudopycnidium (obsol.), a pycnidium-like structure of hyphal tissue, as in certain Mitosporic fungi.

Pseudopyrenula Müll. Arg. (1883), ? Trypetheliaceae (±L). *c.* 45, trop. See Riedl (*Sydowia* **16**: 215, 1963).

Pseudopythium Sideris (1930) nom. nud. See Merlich (*Mycol.* **24**: 453, 1932).

Pseudoramularia Matsush. (1983), Mitosporic fungi, 1.C1.3. 2, Uganda, Pacific Isl.

pseudorhiza, a rooting base, as in *Collybia radicata* (Buller, **4**).

Pseudorhizina Jacz. (1913), Helvellaceae. 2, N. Am. See Harmaja (*Karstenia* **13**: 48, 1973).

Pseudorhizopogon Kobayasi (1983), Mitosporic fungi, 4.A1.?23. 1, Japan.

Pseudorhynchia Höhn. (1909) = Trichosphaeria (Trichosphaer.) fide Clements & Shear (1931).

Pseudorhytisma Juel (1895), Cryptomycetaceae. 1, Eur., Greenland, India, ? Brazil. See Schüepp (*Phytopath. Z.* **36**: 262, 1959).

Pseudorobillarda M. Morelet (1968), Mitosporic fungi, 4.C1.15. 8, widespr. See Nag Raj (1993).

Pseudorobillarda Nag Raj, Morgan-Jones & W.B. Kendr. (1972) = Pseudorobillarda (Mitosp. fungi) fide Nag Raj *et al.* (*CJB* **51**: 688, 1973).

Pseudosaccharomyces Briosi & Farneti (1906), Mitosporic fungi, 1.A1.?. 1, Italy.

Pseudosaccharomyces Klöcker (1912) [non Briosi & Farneti (1906)] ≡ Kloeckera (Mitosp. fungi).

Pseudosagedia (Müll. Arg.) M. Choisy (1949), Trichotheliaceae (L). 22, widespr. See Hafellner & Kalb (*Bibl. Lich.* **57**: 161, 1995; status).

Pseudosarcophoma Urries (1952) ? = Selenophoma (Mitosp. fungi) fide Sutton (1977).

Pseudoschizothyra Punith. (1980), Mitosporic fungi, 5.B1.15. 1, Myanmar.

Pseudosclerophoma Petr. (1923) = Phoma (Mitosp. fungi) fide Sutton (1977).

pseudosclerotium, a compacted mass of intermixed substratum (soil, stones, etc.) held together by mycelium, as in *Polyporus tuberaster* (see stone-fungus; also zone lines).

Pseudoscypha J. Reid & Piroz. (1966), ? Hysteriaceae. 1 (on *Abies*), Canada.

Pseudoseptoria Speg. (1910), Mitosporic fungi, 4.A1.19. 5 (on *Gramineae*), temp. widespr. See Sutton (1980).

pseudoseptum, (1) (obsol.) a protoplasmic or vacuolar membrane looking like a septum (= distoseptum, distoseptate) as in *Corynespora*; (2) (of *Blastocladiales*), a septum having pores.

pseudosetae (false setae), the upturned free-ends of context hyphae in the hymenium of *Duportella*.

Pseudosolidum Lloyd (1923) = Hypocreopsis (Hypocr.).

Pseudosphaerella Höhn. (1911) = Microcyclus (Mycosphaerell.) fide v. Arx & Müller (1975).

Pseudosphaeria Höhn. (1907) = Wettsteinina (Pleospor.) fide Müller & v. Arx (1962). See also Shoemaker & Babcock (*CJB* **65**: 373, 1987).

Pseudosphaeriales, see *Dothideales*.

Pseudosphaerialites Venkatach. & R.K. Kar (1968), Fossil fungi (? Ascomycota). 1 (Tertiary), India.

Pseudosphaerita P.A. Dang. (1895), Pseudosphaeritaceae. 4 (on *Algae*), France, Mexico. See Karling (*Bull. Torry bot. Club* **99**: 223, 1973).

Pseudosphaeritaceae M.W. Dick (1995), Rozellopsidales. 2 gen., 14 spp.

Pseudosphaerophorus M. Satô (1968) = Bundophoron (Sphaerophor.) fide Wedin (1993).

Pseudospiropes M.B. Ellis (1971), Mitosporic fungi, 1.C2.10. 10, widespr. See Ellis (*DH, MDH*).

Pseudospora Schiffn. (1931) Algae.

pseudospore, (1) (of *Acrasales*), an encysted myxamoeba; (2) (of *Ustilaginales*), a basidiospore (obsol.); (3) a chlamydospore, as in *Rhizoctonia rubi*.

Pseudosporopsis Scherff. (1925) ? Protozoa, see monads. See also Dick (*in press*).

Pseudostegia Bubák (1906), Mitosporic fungi, 6.A1.15. 1, N. Am.

pseudostem (of gasteromycete basidiomata), spongy tissue in which hyphae are not orientated parallel to the stipe axis (Dring, 1973).

Pseudostemphylium (Wiltshire) Subram. (1961) = Ulocladium (Mitosp. fungi) fide Simmons (*Mycol.* **59**: 80, 1967).

Pseudostictis Lambotte (1887) = Cryptodiscus (Stictid.) fide Sherwood (1977).

pseudostroma, a stroma formed of thalline tissue and remnants of host tissue (see Eriksson, *Opera bot.* **60**: 14, 1981), an aggregation of perithecial ascomata into a pustule some partly of bark cells altered by the fungus (see Johnson, *Ann. Mo bot. Gdn* **27**: 31, 1940), a coelomycetous conidioma of fungal and host tissue (see Sutton, *The Coelomycetes*, 1980), Fig. 10S). cf. substroma.

Pseudostypella McNabb (1969), Exidiaceae. 1, NZ.

Pseudotaeniolina J.L. Crane & Schokn. (1986), Mitosporic fungi, 1.A/B2.3. 1, Iran.

Pseudotapesia Velen. (1939), ? Leotiales (inc. sed.). 1, former Czechoslovakia.

pseudothecium, (1) an ascostromatic ascoma having asci in numerous unwalled locules, as in loculoascomycetes; cf. euthecium; (2) a protoperithecium.

Pseudothelephora Lloyd (1919) = Thelephora (Thelephor.) fide Donk (1957).

Pseudothiella Petr. (1928), Phyllachoraceae. 1, Brazil. See Hyde (*Trans. mycol. Soc. Japan* **32**: 265, 1991).

Pseudothiopsella Petr. (1928), Anamorphic Phyllachoraceae, 8.A1.?. Teleomorph *Pseudothiella*. 1, Brazil.

Pseudothis Theiss. & Syd. (1914), Diaporthales (inc. sed.). 1, trop. Am.

Pseudothyridaria Petr. (1925) = Valsaria (Diaporth.) fide Clements & Shear (1931).

Pseudothyrium Höhn. (1927), Mitosporic fungi, 8.A1.15. 1, Eur.

Pseudotis (Boud.) Boud. (1907) = Otidea (Otid.) fide Eckblad (1968).

Pseudotomentella Svrček (1958), Thelephoraceae. 13, widespr. See Larsen (*Mycol.* **66**: 167, 1974; key). = Tomentella (Thelephor.) fide Donk (1963).

Pseudotorula Subram. (1958), Mitosporic fungi, 1.C/E2.3. 1, India.

Pseudotracylla B. Sutton & Hodges (1976), Mitosporic fungi, 5.A1.15. 2, Brazil, USA. See Carris (*Mycol.* **84**: 534, 1992, *Mycotaxon* **50**: 93, 1994).

Pseudotrametes Bondartsev & Singer (1944) = Trametes (Coriol.) fide Donk (1974).

Pseudotremellodendron D.A. Reid (1957) = Tremellodendropsis (Tremellodendropsid.) fide Corner (*TBMS* **49**: 205, 1966).

Pseudotrichia Kirschst. (1939), Melanommataceae. 3, N. temp. See Barr (*Mycotaxon* **20**: 1, 1984), Huhndorf (*Mycol.* **86**: 134, 1994; key 4 spp. neotropics).

Pseudotrochila Höhn. (1917), ? Rhytismatales (inc. sed.). 1, Java.

Pseudotryblidium Rehm (1890), Leotiales (inc. sed.). 1, Eur.

Pseudotrype Henn. (1900) = Eutypella (Diatryp.) fide Höhnel (*Sber. Akad. Wiss. Wien* **119**: 926, 1910).

Pseudotthia Henn. (1900) = Gibbera (Ventur.) fide v. Arx & Müller (1975).

Pseudotulasnella Lowy (1964), Tulasnellaceae. 1, Guatemala.

Pseudotyphula Corner (1953), Physalacriaceae. 1, N. Am. See Berthier (*Bibl. Mycol.* **98**: 89, 1985).

Pseudovalsa Ces. & De Not. (1863), Melanconidaceae. 4, N. temp. See Barr (*Mycol. Mem.* **7**, 1978), Wehmeyer (1941).

Pseudovalsaceae, see *Melanconidaceae*.

Pseudovalsaria Spooner (1986), Clypeosphaeriaceae. 1, Europ., China. See Barr (*Mycotaxon* **46**: 45, 1993; **51**: 191, 1994; posn).

Pseudovalsella Höhn. (1918), Melanconidaceae. 1, Eur., N. Am., Japan. See Barr (*Mycol. Mem.* **7**, 1978).

Pseudovularia Speg. (1910) = Ramularia (Mitosp. fungi) fide Deighton (*TBMS* **59**: 419, 1972).

Pseudoxenasma K.H. Larss. & Hjortstam (1976), Gloeocystidiellaceae. 1, Sweden.

Pseudoxylaria Boedijn (1959) = Xylaria (Xylar.) fide Læssøe (*SA* **13**: 43, 1994).

Pseudoyuconia Lar.N. Vassiljeva (1983), Pleosporaceae. 1, Germany.

Pseudozyma Bandoni (1985), Mitosporic fungi, 1.G1.1/10. 1, Canada.

Pseudozythia Höhn. (1902), Mitosporic fungi, 4.A1.?. 1, Eur.

Psiammopomopiospora Locq. & Sal.-Cheb. (1980), Fossil fungi. 3, Cameroon.

Psiamspora Locq. & Sal.-Cheb. (1980), Fossil fungi. 1, Cameroon.

Psidimobipiospora Locq. & Sal.-Cheb. (1980), Fossil fungi. 4, Cameroon.

Psilachnum Höhn. (1926), Hyaloscyphaceae. *c.* 14, Eur., N. Am., former USSR. See Dennis (*Persoonia* **2**: 171, 1962), Raitviir (1970; key), Sharma (*Nova Hedw.* **46**: 369, 1988; 4 spp. on *Pteridophyt*).

Psiloboletinus Singer (1945), Gyrodontaceae. 1, temp. See Smith (*Mycol.* **58**: 332, 1966).

Psilobotrys Sacc. (1879) = Chloridium (Mitosp. fungi) fide Hughes (*CJB* **36**: 727, 1958).

psilocin, a hallucinatory indole derivative from *Psilocybe mexicana* (Hoffman *et al.*, *Experientia* **14**: 11, 397, 1958). See also psilocybin, hallucinogenic fungi.

Psilocistella Svrček (1977), Hyaloscyphaceae. 5, Eur.

Psilocybe (Fr.) P. Kumm. (1871), Strophariaceae. 60, widespr. See Guzmán & Vergier (*Mycotaxon* **6**: 464, 1978; epithet list), Guzmán (*Mycotaxon* **7**: 225, 1978; C. & S. Am., *Nova Hedw.* **29**: 625, 1978; Mexico), Høiland (*Norw. Jl Bot.* **25**: 111, 1978; Norway), Guzmán & Horak (*Sydowia* **31**: 44, 1978; Australasia), Guzmán (*The genus Psilocybe*, 1982; 144 spp. accepted, 90 syn., 119 doubtful), Guzmán (*Bibl. Mycol.* **159**: 91, 1995; suppl. to monogr.). See also hallucinogenic fungi.

Psilocybe Fayod (1889) = Psathyrella (Coprin.) fide Smith (1972).

psilocybin, a hallucinatory indole derivative from *Psilocybe mexicana*. See Bazanté (*Rev. Mycol.* **36**: 25, 1971; action). See also psilocin, hallucinogenic fungi.

Psilodiporites C.P. Varma & Rawat (1963), Fossil fungi. 7, Cameroon, India.

Psiloglonium Höhn. (1918) = Glonium (Hyster.) fide v. Arx & Müller (1975).

Psilolechia A. Massal. (1860), Micareaceae (L). 4, widespr. See Coppins & Purvis (*Lichenologist* **19**: 29, 1987; key).

Psilonia Fr. (1825) ? = Volutella (Mitosp. fungi).

Psiloniella Costantin (1888) = Catenularia (Mitosp. fungi) fide Mason (*Mycol. Pap.* **5**: 120, 1941).

Psiloparmelia Hale (1989) Parmeliaceae. 12, S. Am., S. Afr. See Elix & Nash (*Bryologist* **95**: 377, 1992; key). ? = Parmelia (Parmel.).

Psilopezia Berk. (1847), Otideaceae. 3, Eur., Am. See Pfister (*Am. J. Bot.* **60**: 355, 1973; key).

Psilophana Syd. (1939), ? Leotiales (inc. sed.). 1, Ecuador.

Psilosphaeria Cooke (1879), Ascomycota (inc. sed.). 1, UK.

Psilospora Rabenh. (1856) ≡ Polymorphum (Mitosp. fungi) fide Hawksworth & Punithalingam (1973).

Psilosporina Died. (1913) = Polymorphum (Mitosp. fungi) fide Hawksworth & Punithalingam (1973).

Psilothecium Fuckel (1866) = Cercospora (Mitosp. fungi) fide Sutton (1977).

Psilothecium Clem. (1903), ? Leotiales (inc. sed.). 1, N. Am.

Psora Hoffm. (1789) nom. rej. = Physcia (Physc.) fide Hawksworth & Sherwood (*Taxon* **30**: 338, 1981).

Psora Hoffm. (1796) nom. cons., Psoraceae (L). 25, widespr. See Schneider (*Bibl. Lich.* **13**, 1980; key), Timdal (*Nord. Jl Bot.* **4**: 525, 1984; gen. concept, *Bryol.* **89**: 253, 1987; key 18 spp. N. Am.).

Psora Link (1833) = Toninia (Catillar.).

Psoraceae Zahlbr. (1898), Lecanorales (L). 5 gen. (+ 6 syn.), 41 spp. Thallus crustose or squamulose; ascomata apothecial, sessile or immersed, flat or domed, sometimes on the thallus margin, without well-defined marginal tissue, pale or dark; paraphyses usually wide, occasionally anastomosing, with swollen apices and a crystalline epithecium, often agglutinated; asci with an I+ apical cap, a usually well-developed strongly staining tubular ring and a sometimes very well-developed outer I+ gelatinized layer; ascospores hyaline, aseptate, without a sheath. Anamorphs pycnidial. Lichenized with green algae, widespr.

Psorella Müll. Arg. (1894) ? = Bacidia (Bacid.) fide Brako (*Mycotaxon* **35**: 1, 1989).

Psorinia Gotth. Schneid. (1980), Lecanoraceae (L). 2, Eur., former USSR.

Psoroglaena Müll. Arg. (1891), Verrucariaceae (L). 3, N., C. and S. Am., Afr., Madeira, China. See Eriksson (*SA* **11**: 11, 1992, & Hawksworth, *SA* **14**: 65, 1995), Henssen (*Bibl. Lich.* **57**: 161, 1995; key).

Psoroglaenomyces Cif. & Tomas. (1953) ≡ Psoroglaena (Verrucar.).

Psorographis Clem. (1909) = Acanthothecis (Graphid.).

Psoroma Ach. ex Michx. (1803), Pannariaceae (L). *c.* 50, cosmop. (mainly S. temp.). See Jørgensen (*Op. bot. Soc. bot. Lund* **45**, 1978), Malme (*Ark. Bot.* **20A** (3), 1925), Quilhot *et al.* (*J. Nat. Products* **52**: 191, 1989; chemistry).

Psoromaria Nyl. ex Hue (1891) = Psoromidium (Pannar.) fide Galloway & James (1985).

Psoromatomyces Cif. & Tomas. (1953) ≡ Psoroma (Pannar.).

Psoromella Gyeln. (1940), Parmeliaceae (L). 1, Argentina.

Psoromidium Stirt. (1877), Pannariaceae (L). 2, S. Am., Australasia. See Galloway & James (*Lichenologist* **17**: 173, 1985; key).

Psoromopsis Nyl. (1863) ? = Phyllopsora (Bacid.).

Psoropsis Nyl. ex Zwackh (1883) nom. inval. = Porocyphus (Lichin.).

Psorotheciella Sacc. & P. Syd. (1902) = Asterothyrium (Thelotremat.).

Psorotheciopsis Rehm (1900), ? Gomphillaceae (L). 3, trop. Am., Afr., Asia. See Santesson (*Symb. bot. upsal.* **12** (1), 1952).

Psorothecium A. Massal. (1860), nom. dub. (Fungi, inc. sed.) fide Sipman (*Willdenowia* **15**: 557, 1986).

Psorotichia A. Massal. (1855), Lichinaceae (L). *c.* 50, cosmop. See Forssell (*Nova Acta Soc. sci. upsal.* ser. 3, **13** (6), 1885).

Psorotichiella Werner (1955), ? Lecanorales (inc. sed., L). 1, Lebanon.

Psorotichiomyces Cif. & Tomas. (1953) ≡ Psorotichia (Lichin.).

Psorula Gotth. Schneid. (1980), ? Psoraceae (L). 3, widespr. See Pietschmann (*Nova Hedw.* **51**: 521, 1990; posn).

psychrophile, see thermophily.

psychrotolerant, growing at temperatures below 10°C (opt. below 20°C).

Psyllidomyces Buchner (1912), ? Saccharomycetaceae. 1 (in insects), Eur.

psylocybin, see psilocybin.

Ptechetelium Oberw. & Bandoni (1984), Platygloeaceae. 1, Ecuador.

pterate, having wings; alate.

Pteridiosperma J.C. Krug & Jeng (1979), Ceratostomataceae. 1, Japan.

Pteridiospora Penz. & Sacc. (1897), Dothideales (inc. sed.). 3, widespr. See Filer (*Mycol.* **61**: 167, 1969).

Pteridomyces Jülich (1979) = Athelopsis (Athel.) fide Hjortstam (*Mycotaxon* **42**: 149, 1991).

Pteroconium Sacc. ex Grove (1914), Anamorphic Lasiosphaeriaceae, 3.A2.37. Teleomorph *Apiospora*. 2, widespr.

Pterodinia Chev. (1837) = Botrytis (Mitosp. fungi).

Pteromaktron Whisler (1963), Legeriomycetaceae. 1 (in *Ephemeroptera*), France. See Lichtwardt (1986).

Pteromyces E. Bommer, M. Rousseau & Sacc. (1906), ? Leotiales (inc. sed.). 1, Eur.

Pterophyllus Lév. (1844) nom. rej. = Pleurotus (Lentin.).

Pterospora Métrod (1949) [non Nutt. (1818), *Pyrolaceae*] = Marasmiellus (Tricholomat.) fide Singer (1962). = Tetrapyrgos (Tricholomat.) fide Horak (1987).

Pterula Fr. (1832), Pterulaceae. 50, esp. trop.

Pterulaceae Corner (1970), Cantharellales. 7 gen. (+ 3 syn.), 65 spp. Basidioma filiform, branched; hymenium amphigenous; dimitic.

Pterulicium Corner (1950), Pterulaceae. 1, S.E. Asia.

Pterulopsis Wakef. & Hansf. (1943), Mitosporic fungi, 2.E1.?. 1, Uganda.

Pterygellus Corner (1966), Cantharellaceae. 4, trop. Asia.

Pterygiopsidomyces Cif. & Tomas. (1953) ≡ Pterygiopsis (Lichin.).

Pterygiopsis Vain. (1890), Lichinaceae (L). 2, Eur., S. Am. See Henssen (*Symb. bot. upsal.* **18** (1), 1963), Jørgensen (*Lichenologist* **22**: 213, 1990).

Pterygium Nyl. (1854) = Placynthium (Placynth.).

Pterygosporopsis P.M. Kirk (1983), Mitosporic fungi, 1.A2.4. 1, UK.

Ptychella Roze & Boud. (1879), Bolbitiaceae. 1, Eur.

Ptychogaster Corda (1838), Anamorphic Coriolaceae, 1.A1.40. Teleomorph *Tyromyces. c.* 10, widespr. See Sigler & Carmichael (*Mycotaxon* **4**: 394, 1976), Donk (1974).

Ptychographa Nyl. (1874), ? Agyriaceae (L). 2, Eur. See Redinger (*Rabenh. Krypt.-Fl.* **9**, 2(1): 217, 1938).

Ptychographomyces Cif. & Tomas. (1953) ≡ Ptychographa (Agyr.).

Ptychopeltis Syd. (1927) = Calothyriopsis (Microthyr.) fide Müller & v. Arx (1962), v. Arx & Müller (1975).

Ptychoverpa Boud. (1907) = Verpa (Peziz.) fide Eckblad (1968).

ptyophagous (of endotrophic mycorrhiza), the young hyphae rupturing and extruding plasmal masses (**ptyosomes**) which are digested by the host cells (Burgeff, 1924); **tolypophagous**, the penetrating hyphae killed and digested by the host (Burgeff, 1924); **thamnisophagous**, forming haustorial arbuscles which are finally digested by the host (Burgeff, 1938). Cf. halmophagous.

pubescent, having soft hairs.

Puccinella Fuckel (1860) [non Parl. (1848) nom. cons., *Gramineae*] = Uromyces (Puccin.) fide Dietel (1928).

Puccinia Pers. (1801), Pucciniaceae. *c.* 4,000, cosmop., on many families of Angiosperms. Heteroecious or autoecious, macro- or micro-cyclic; teliospores 2-celled, though in some species (e.g. *P. heterospora* on *Malvaceae*) most of the spores are 1-celled (mesospores). There are important pathogens of cereals (*P. hordei*) (barley, brown or leaf rust; I on *Ornithogalum*), *P. coronata* (oat, crown rust; I on *Rhamnus*), *P. striiformis* (syn. *P. glumarum*) (yellow or stripe rust; Johnson *et al.*, *TBMS* **58**: 475, 1972; physiologic races), *P. graminis* (black or stem rust; I on *Berberis*; see Lehmann *et al.*, *Der Schwarzrost*, 1937), *P. recondita* (syn. *P. dispersa*) (rye, brown, or leaf rust; I on *Anchusa*) and (syn. *P. triticina*) (wheat, brown, or leaf rust; I on *Isopyrum*; see Chester, *Cereal rusts....*, 1946); *P. polysora* (maize rust; Cammack, *TBMS* **41**: 89, 1958), *P. kuehnii*, *P. melanocephata* (sugarcane, Mordue, *TBMS* **84**: 758, 1985), grasses, groundnut (*P. arachidis*), cotton (*P. cacabata*), and a great number of other crop plants. See Bushnell & Roelfs (1984, 1985), Cummins (1971, 1978).

Pucciniaceae Chevall. (1826), Uredinales. 17 gen. (+ 52 syn.), *c.* 4,700 spp. Pycnia ampulliform, with bounding structures, subepidermal; aeciospores catenulate (*Aecidium*, rarely *Caeoma*) or pedicellate (*Uraecium*); teliospores pedicellate, 1- or more (commonly 2) celled by horizontal septation, usually pigmented.

Pucciniastraceae Gäum. ex Leppik (1972), Uredinales. 8 gen. (+ 3 syn.), 127 spp. Pycnia either discoid or ampulliform, without bounding structures, subcuticular or subepidermal; aeciospores catenulate (*Peridermium*) or, rarely, pedicellate (*Uraecium*); urediniospores pedicellate, uredinia with peridium;

teliospores sessile, unicellular or multicellular by vertical septation, laterally adherent, formed either in epidermal cells or subepidermally, pale or pigmented.

Pucciniastrum G.H. Otth (1861), Pucciniastraceae. 37 (+ 16), mostly N. temp.; O, I on *Abies*, *Picea* and *Tsuga*, II, III on Dicots, and *Orchidaceae*. See Cummins & Hiratsuka (1983).

Puccinidia Mayr (1890) = Rostrupia (Puccin.) fide Saccardo (*Syll. Fung.* **9**: 316, 1891).

Pucciniola L. Marchand (1829) nom. rej. = Uromyces (Puccin.).

Pucciniopsis Speg. (1888) nom. ambig. (Mitosp. fungi) fide Sutton (*TBMS* **60**: 515, 1973).

Pucciniosira Lagerh. (1892), Pucciniosiraceae. 14, Am., Afr., Asia; O, III. See Buriticá (*Rev. Acad. Colomb. Cienc.* **19**(69): 131, 1991).

Pucciniosiraceae (Dietel) Cummins & Y. Hirats. (1983), Uredinales. 9 gen. (+ 8 syn.), 53 spp. Pycnia ampulliform, with bounding structures; aeciospores and urediniospores never developed; teliospores sessile, often catenulate with intercalary cells, 1- or 2-celled by horizontal septation; telia often resemble aecial states of other genera, especially *Aecidium* and *Caeoma* states.

Pucciniospora Speg. (1886), Mitosporic fungi, 4.B1.?. 1, S. Am.

Pucciniostele Tranzschel & Kom. (1899), Phakopsoraceae. 5, Asia. See Cummins & Thirumalachar (*Mycol.* **45**: 572, 1953).

Puccinites Ettingsh. (1853), Fossil fungi. 3 (Cretaceous, Tertiary), Eur., USA.

Puciola De Bert. (1976) = Dicyma (Mitosp. fungi) fide Arx (1981).

puff-ball, basidioma of the *Lycoperdales*.

puffing, a phenomenon in which thousands of asci in an apothecial ascoma discharge their ascospores simultaneously, producing a visible cloud.

Puiggariella Speg. (1881) = Strigula (Strigul.) fide Santesson (1952).

Puiggarina Speg. (1919) = Phyllachora (Phyllachor.) fide Cannon (*Mycol. Pap.* **163**, 1991).

Pulcherricium Parmasto (1968), Corticiaceae. 1, Eur.

Pulchromyces Hennebert (1973), Mitosporic fungi, 1.A1.7. 1 (coprophilous), Ghana. See Pfister *et al.* (*Mycotaxon* **1**: 137, 1974).

Pulina Adans. (1763) nom. rej. ≡ Lepraria (Mitosp. fungi).

Pullospora Faurel & Schotter (1965), Mitosporic fungi, 4.A1.15. 2 (coprophilous), Algeria, USA. See Nag Raj (1993).

Pullularia Berkhout (1923) = Aureobasidium (Mitosp. fungi).

pullulation, budding, as in yeasts.

Pulmonaria Hoffm. (1789) [non L. (1753), *Boraginaceae*] = Lobaria (Lobar.).

Pulparia P. Karst. (1871) nom. rej. = Pulvinula (Otid.). See also *Marcelleina*.

pulque, a Mexican alcoholic drink made by yeast fermentation of the juice of *Agave* spp.; *Lactobacillus* and *Leuconostoc* spp. add acidity and viscosity; when distilled yields the spirit **tequila**.

pulsed field gel-electrophoresis (PFGE), a term used to describe a number of different techniques for separating large pieces of DNA, e.g. chromosomes. Different proprietary systems available, e.g. CHEF, FIGE, PFGE, but all involve applying electric field to an electrophoresis gel as a series of pulses or variations rather than a single continuous constant field. See Boekhout *et al.* (*Ant. v. Leeuwenhoek* **63**: 157, 1993); Cytology.

pulveraceo-delitescent, covered with a layer of powdery granules.

Pulveraria Ach. (1803) nom. rej. = Chrysothrix (Chrysothric.).

Pulveria Malloch & Rogerson (1977), Xylariaceae. 1, N. Am. See Læssøe (*SA* **13**: 43, 1994; nomencl.).

Pulveroboletus Murrill (1909), Boletaceae. 14, widespr.

Pulverolepiota Bon (1993), Agaricaceae. 3, Eur.

pulverulent, powdered; as if powdered over.

Pulvinaria Bonord. (1851), Ascomycota (inc. sed.). 1, Eur.

Pulvinaria Rodway (1918) ≡ Waydora (Mitosp. fungi).

Pulvinaria Velen. (1934) ? = Pachyella (Peziz.) fide Pfister (*CJB* **51**: 2009, 1973).

pulvinate, cushion-like in form.

Pulvinotrichum Gamundí, Aramb. & Giaiotti (1981), Mitosporic fungi, 1.A/B1.15. 2, Argentina, Australia. See Sutton (*Sydowia* **41**: 330, 1989), Crous *et al.* (*Mycotaxon* **50**: 441, 1994). = Cylindrodendrum (Mitosp. fungi) fide Summerbell *et al.* (*CJB* **67**: 577, 1989).

Pulvinula Boud. (1885) nom. cons., Otideaceae 17, widespr. See Dissing (*Mycotaxon* **32**: 365, 1988; nomencl.), Pfister (*Occ. Pap. Farlow Herb.* **9**, 1976; key), O'Donnell & Hooper (*CJB* **56**: 101, 1978; ontogeny), Korf & Zhuang (*Mycotaxon* **20**: 607, 1984).

Pumilus Viala & Marsais (1934), Ascomycota (inc. sed.). 1 (on *Vitis*), Eur.

punctate, marked with very small spots (Fig. 29.5); **puncta**, small spots.

Punctelia Krog (1982) Parmeliaceae. 25, widespr. See Galloway & Elix (*N.Z. J. Bot.* **22**: 441, 1984; Australasia, key), Modensi (*Nova Hedw.* **45**: 423, 1987; histochemistry), Wilhelm & Lodd (*Mycotaxon* **44**: 495, 1992; key 13 N. Am. spp.). ? = Parmelia (Parmel.).

punctiform (of rust sori, bacterial colonies, etc.), very small, but seen without a lens.

Punctillina Toro (1934), Mitosporic fungi, 5.A1.?. 1, S. Am.

Punctillum Petr. & Syd. (1924), Dothideales (inc. sed.). 2 (on *Musci*), Afr., Eur., N.Z. See Döbbeler (1978).

Punctularia Pat. (1895), Corticiaceae. 3, widespr.

Punctulariaceae, see *Corticiaceae*.

punk, see touchwood or amadou; **punky**, soft and tough.

Pureke P.R. Johnst. (1991), Rhytismataceae. 1, NZ.

Pustularia Bonord. (1851) nom. dub. (Ascomycota, inc. sed.); used for a wide range of taxa.

Pustularia Fuckel (1870) ≡ Pustulina (Otid.).

pustule, a blister-like, frequently erumpent, spot or spore-mass.

Pustulina Eckblad (1968) = Tarzetta (Otid.) fide Rogers *et al.* (1971).

Pustulipora P.F. Cannon (1982), Ceratostomataceae. 1, UK.

Putagraivam Subram. & Bhat (1978), Anamorphic Hypocreales, 3.B1.15. Teleomorph *Peethambara*. 1, India.

putrescent (of basidiomata), decaying, rotting. Cf. marcescent.

Puttemansia Henn. (1902), Tubeufiaceae. 6 (on fungi), trop. See Rossman (*Mycol. Pap.* **157**, 1987; key).

Puttemansiella Henn. (1908) Fungi (inc. sed.).

pv., see pathovar.

Pycnidiales (obsol.), see *Sphaeropsidales* (Mitosp. fungi).

Pycnidiella Höhn. (1915), Anamorphic Agyriaceae, 8.A1.15. Teleomorph *Sarea*. 1 (on resin), widespr. See Hawksworth & Sherwood (*CJB* **59**: 357, 1981).

Pycnidioarxiella Punith. & N.D. Sharma (1980), Mitosporic fungi, 4.2B.3. 1, India.

Pycnidiochaeta Sousa da Câmara (1950) = Dinemasporium (Mitosp. fungi) fide Sutton (1977).

Pycnidiopeltis Bat. & C.A.A. Costa (1959), Mitosporic fungi, 5.A1.?. 1, USA, Brazil.

Pycnidiophora Clum (1956), Sporormiaceae. 5, widespr. See Cain (*CJB* **39**: 1633, 1961; as *Preussia*).

pycnidiospore, a conidium in or from a pycnidium (obsol.).

Pycnidiostroma F. Stevens (1927) = Phomachora (Mitosp. fungi) fide Petrak (*Ann. Myc.* **27**: 324, 1929).

pycnidium (pl. -ia), a frequently ± flask-shaped conidioma of fungal tissue with a circular or longitudinal ostiole, the inner surface of which is lined entirely or partially by conidiogenous cells; pycnidial conidioma (Fig. 10A-F).

pycniospore (of *Uredinales*), a spore from a pycnium; spermatium; sometimes used in error for pycnidiospore.

Pycnis Bref. (1881), Mitosporic fungi, 4.A1.?. 1, Eur.

pycnium (in *Uredinales*), the pycnidium-like haploid fruit-body, or spermogonium. See Hiratsuka & Cummins (*Mycol.* **55**: 487, 1963), Savile (*Mycol.* **63**: 1089, 1971).

pycnoascocarp, an ascoma arising from a pycnidial conidioma.

Pycnocalyx Naumov (1916) = Bothrodiscus (Mitosp. fungi) fide Sutton (1977).

Pycnocarpon Theiss. (1913), Dothideales (inc. sed.). 4, India, Philip.

Pycnochytrium (de Bary) J. Schröt. (1892) = Synchytrium (Synchytr.) fide Fitzpatrick (1930).

Pycnociliospora Bat. (1962), Mitosporic fungi (L), 8.B1.?. 4, Brazil.

pycnoconidium, see pycnidiospore (obsol.).

Pycnodactylus Bat., A.A. Silva & Cavalc. (1967), Mitosporic fungi, 4.D2.?. 1, Brazil.

Pycnoderma Syd. (1914), Cookellaceae. 2, trop. See Petrak (*Sydowia* **1**: 108, 1947).

Pycnodermella Petr. (1947) = Saccardinula (Elsin.) fide v. Arx (1963).

Pycnodermellina Bat. & H. Maia (1957) ? = Echinoplaca (Gomphill.) fide v. Arx & Müller (1975).

Pycnodermina Petr. (1954) ? = Pycnoderma (Cookell.) fide v. Arx & Müller (1975).

Pycnodon Underw. (1898) ≡ Neokneiffia (Hyphodermat.).

Pycnodothis F. Stevens (1924) Fungi (inc. sed.) fide Petrak (*Sydowia* **5**: 169, 1951).

Pycnofusarium Punith. (1973) = Fusarium (Mitosp. fungi) fide Sutton (*TBMS* **86**: 1, 1986).

pycnogonidium, see pycnidiospore, pycniospore, or stylospore (obsol.).

Pycnographa Müll. Arg. (1890) = Parmularia (Parmular.). See Santesson (*Svensk bot. Tidskr.* **43**: 547, 1949).

Pycnomma Syd. (1924), Mitosporic fungi, 8.A1.?. 1, Canary Isl.

Pycnomoreletia Rulamort (1990), Mitosporic fungi, 8.C2.1. 2, Afr., Pakistan.

Pycnopeltis Syd. & P. Syd. (1916) = Saccardinula (Elsin.) fide v. Arx & Müller (1975).

Pycnopeziza W.L. White & Whetzel (1938), Sclerotiniaceae. 3, Eur., N. Am. See Whetzel & White (*Mycol.* **32**: 616, 1940). Anamorph *Acarosporium.*

Pycnopodium Corda (1842) ≡ Pilobolus (Pilobol.). See Hesseltine (1955).

Pycnoporellus Murrill (1905), Coriolaceae. 2, widespr. See Ryvarden (1978).

Pycnoporus P. Karst. (1881), Coriolaceae. 3, widespr. See Nobles & Frew (*CJB* **40**: 987, 1962).

Pycnorostrum Golovin (1950) nom. dub. fide Sutton (*Mycol. Pap.* **141**, 1977).

pycnosclerotium, a more or less hard-walled structure resembling a pycnidial conidioma but having no spores.

Pycnoseynesia Kuntze (1898), Mitosporic fungi, 5.A1.?. 5, trop.

pycnosis (of *Microthyriaceae*), the process by which a part of the stroma is arched up and becomes thick while an ascigerous hymenium is formed under it.

pycnospore, formerly occasionally used for pycniospore or pycnidiospore (obsol.).

Pycnosporium Siegel (1909) nom. dub. (? Mitosp. fungi).

Pycnostemma Syd. (1927) = Poropeltis (Mitosp. fungi) fide Clements & Shear (1931).

pycnostroma, see stroma.

Pycnostroma Clem. (1909) = Munkia (Mitosp. fungi) fide Höhnel (*Sber. Akad. Wiss. Wien* **120**: 379, 1911).

Pycnostysanus Lindau (1904), Mitosporic fungi, 2.A2.3. 2, Eur., N. Am. *P. azaleae* (*Rhododendron* leaf and bud scorch).

pycnothecium (of *Microthyriaceae*), an ascoma formed by pycnosis.

Pycnothele Sommerf. (1826) ≡ Dufourea (Teloschist.).

Pycnothelia (Ach.) Dufour (1821), Cladoniaceae (L). 2, Eur., N. Am., NZ. See Ahti (1993), Laundon (*Lichenologist* **18**: 169, 1986; nomencl.).

Pycnotheliomyces Cif. & Tomas. (1953) ≡ Pycnothelia (Cladon.).

Pycnothera N.D. Sharma & G.P. Agarwal (1974), Mitosporic fungi, 5.A1.38. 2, India. See Punithalingam (*Nova Hedw.* **31**: 95, 1979).

Pycnothyriella Bat. (1952), Mitosporic fungi, 5.A1.?. 1, Brazil.

pycnothyrium, a superficial flattened shield-shaped conidioma with radiate upper and sometimes lower walls; pycnothyrial conidioma. See Hughes (*Mycol. Pap.* **50**: 7, 1953). Characteristic of *Microthyriaceae* (q.v.) etc. (Fig. 10G-M).

Pycnothyrium Died. (1913), Mitosporic fungi, 5.A1.?. 6, Eur., Philipp. See Sutton (1977). Sensu v. Arx (1964) = Leptopeltis (? Phacid.) fide Holm & Holm (*Bot. Notiser* **130**: 115, 1977).

Pycnovellomyces R.F. Castañeda (1987), Mitosporic fungi, 4.A1.41. 1 (with clamp connexions), Cuba. See Nag Raj *et al.* (*CJB* **67**: 3386, 1989; redescription).

Pygmaea Stackh. (1809) nom. rej. = Lichina (Lichin.).

Pygmomyces Arnaud (1949) nom. inval. (Myxomycota).

Pyonema Nieuwl. (1916) ≡ Myxonyphe (Fungi, inc. sed.).

Pyramidospora Sv. Nilsson (1962), Mitosporic fungi, 1.G1.1. 2 (aquatic), S. Am., Cuba, Sierra Leone.

Pyrenastromyces Cif. & Tomas. (1953) ? = Pyrenula (Pyrenul.).

Pyrenastrum Eschw. (1824) = Pyrenula (Pyrenul.) fide Harris (1989).

Pyrenidiaceae, see Dacampiaceae.

Pyrenidiomyces Cif. & Tomas. (1953) ≡ Pyrenidium (Dacamp.).

Pyrenidium Nyl. (1865), Dacampiaceae. 3 (on lichens), Afr., Australia, Eur. See Hawksworth (*TBMS* **80**: 547, 1983).

Pyreniella Theiss. (1916) = Botryosphaeria (Botryosphaer.) fide v. Arx & Müller (1954).

Pyrenillium, see *Pyrenyllium*.

Pyreniococcus Wheldon & A. Wilson (1915) ? = Phaeospora (Verrucar.).

Pyreniopsis Kuntze (1898) ≡ Trichoderma (Mitosp. fungi).

Pyrenisperma, see *Pyrisperma*.

pyrenium, a pyrenomycete ascoma (obsol.).

Pyrenium Tode (1790) = Trichoderma (Mitosp. fungi).

Pyrenobotrys Theiss. & Syd. (1914), Venturiaceae. 3, Am., Scandinavia. See Barr (*Sydowia* **41**: 25, 1989; key 3 N. Am. spp.), Eriksson (*Svensk. bot. Tidskr.* **68**: 223, 1974).

pyrenocarp, a perithecium (s. l., q.v.); used colloquially as a term for any fungus with a perithecium-like ascoma.

Pyrenocarpon Trevis. (1855), Lichinaceae (L). 1, Eur. See Jørgensen & Henssen (*Taxon* **39**: 343, 1990), Jørgensen (*SA* **10**: 51, 1991).

pyrenocarpous, see pyrenocarp.

Pyrenochaeta De Not. (1849), Anamorphic Lophiostomataceae, 4.A1.15. Teleomorph *Herpotrichia*. 10, widespr. See Schneider (*Mitt. biol. Bundes. Land.-Forstw. Beih.* **189**, 1979), Schneider (*in* Subramanian (Ed.), *Taxonomy of fungi*: 513, 1984; disposition of spp.), Datnoff *et al.* (*TBMS* **87**: 297, 1986; sclerotial state), Last *et al.* (*Ann. appl. Biol.* **57**: 95, **62**: 55, **64**: 449, 1966-69; *P. lycopersici*, tomato brown rot), Ferreira *et al.* (*J. Phytopath.* **133**: 289, 1991; variability in *P. terrestris* using isozyme polymorphism etc.), Sieber (*MR* **99**: 274, 1995; *P. ligni-putridi* in *Abies* butt rot).

Pyrenochaetella P. Karst. ex Höhn. (1917) nom. dub. (Mitosp. fungi) fide Sutton (*Mycol. Pap.* **141**, 1977).

Pyrenochaetina Syd. & P. Syd. (1916) = Parodiella (Parodiell.) fide Müller & v. Arx (1962).

Pyrenochium Link (1833), ? Dothideales (inc. sed.). 1, Eur.

Pyrenocollema Reinke (1895), ? Pyrenulaceae (L). *c.* 10, Eur., N. Am. See Puymaly (*Botaniste* **36**: 331, 1952), Tucker & Harris (*Bryologist* **83**: 1, 1980).

Pyrenocyclus Petr. (1955), Dothideales (inc. sed.). 1, Hawaii.

Pyrenodermium Bonord. (1851) = Hypoxylon (Xylar.) fide Læssøe (*SA* **13**: 43, 1994).

Pyrenodesmia A. Massal. (1853) nom. rej. = Caloplaca (Teloschist.).

Pyrenodiscus Petr. (1927) = Diplonaevia (Dermat.) fide Hein (1983).

Pyrenodium Fée (1837) = Asterothelium (Trypethel.) fide Zahlbruckner (1921).

Pyrenodochium Bonord. (1851) = Diatrype (Diatryp.) p.p.

Pyrenogaster Malençon & Riousset (1977), Geastraceae. 1, France.

Pyrenographa Aptroot (1991), Pyrenulaceae. 1, Australia.

pyrenomycete, one of the *Pyrenomycetes*.

Pyrenomycetes. Class of *Ascomycota*; used in various senses, but mostly for fungi with perithecioid ascomata which are also ascohymenial in ontogeny and have unitunicate asci, often with an apical annulus (e.g. *Diaporthales*, *Hypocreales*, *Sordariales*, *Xylariales*). The *Dothideales*, *Erysiphales*, *Meliolales* and *Laboulbeniales* also included by some authors. For selected usages of the name see Table 1 under *Ascomycota*.

The class is not generally accepted but was reintroduced by Berbee & Taylor (1992) in a restricted sense (see *Ascomycota*).

However, the term 'pyrenomycetes' still has value as a colloquial term for all ascomycetes with flask-shaped ascomata. See Hanlin (*in* Parker, 1982, **1**: 225), Rogers (*in* Hawksworth, 1994: 321).

Pyrenomyxa Morgan (1895) = Pulveria (Xylar.) fide Læssøe (*SA* **13**: 43, 1994).

Pyrenopeziza Fuckel (1870), Dermateaceae. *c.* 50, widespr. Gremmen (*Fungus* **28**: 37, 1958; key sects.), Hütter (*Phytopath. Z.* **33**: 1, 1958; key), Nannfeldt (1932; 32 spp.). Anamorph *Phialophora*-like.

Pyrenopezizopsis Höhn. (1917) = Cenangiopsis (Leot.) fide Dennis (*Persoonia* **2**: 171, 1962).

Pyrenophora Fr. (1849), Pleosporaceae. 7, widespr. Plant pathogens include: *P. chaetomioides*, anamorph *Drechslera avenacea* (oat leaf spot and seedling blight); *P. graminea*, anamorph *D. graminea* (barley leaf stripe); *P. teres*, anamorph *D. teres* (net blotch of barley). See Crivelli (*Ueber die heterogene Ascomycetengattung Pleospora*, 1983; key), Wehmeyer (1961), Amici & Rollo (*Nucleic Acids Res.* **19**: 5073, 1991; SS rDNA), Ammon (*Phytopath. Z.* **47**: 269, 1963; keys), Barr (*Contr. Univ. Mich. Herb.* **9**: 523, 1972), Medd (*Rev. Pl. Path.* **71**: 891, 1992; review *P. semeniperda*), Shoemaker (*CJB* **39**: 901, 1963; status), Sutton (*Taxon* **21**: 319, 1972; nomencl.). Anamorph *Drechslera*.

Pyrenophoraceae, see *Pleosporaceae*.

Pyrenophoromyces Cif. & Tomas. (1953) ≡ Pyrenophorum (Verrucar.).

Pyrenophoropsis C. Ramesh (1988), Fungi (inc. sed.). 1, India. See Eriksson & Hawksworth (*SA* **8**: 75, 1989).

Pyrenophorum Tomas. & Cif. (1952) = Thelidium (Verrucar.).

Pyrenopolyporus Lloyd (1917) = Hypoxylon (Xylar.) fide Miller (1961).

Pyrenopsidium (Nyl.) Forssell (1885) = Cryptothele (Lichin.) fide Henssen & Büdel (1984), but see Moreno & Egea (*Biología y taxonomía de la familia Lichinaceae*, 1991).

Pyrenopsis Nyl. (1858) typ. cons., Lichinaceae (L). *c.* 40, cosmop.

Pyrenostigme Syd. (1926), Dothideales (inc. sed.). 2, C. Am. See Hansford (*Mycol. Pap.* **15**, 1946).

Pyrenotea Fr. (1821) nom. rej. = Lecanactis (Roccell.) fide Hawksworth & David (*Taxon* **38**: 493, 1989).

Pyrenothamnia Tuck. (1883) = Endocarpon (Verrucar.) fide Santesson (*Svensk bot. Tidskr.* **43**: 547, 1949).

Pyrenothamniomyces Cif. & Tomas. (1953) ≡ Pyrenothamnia (Verrucar.).

Pyrenothea, see *Pyrenotea*.

Pyrenotheca Pat. (1886) = Myriangium (Myriang.) fide v. Arx & Müller (1975).

Pyrenothricaceae Zahlbr. (1926), Dothideales (L). 2 gen. (+ 1 syn.), 2 spp. Thallus dark brown, indeterminate, gelatinous; ascomata perithecial, pyriform, the ostiole small, ? lysigenous; peridium thin, dark, composed of pseudoparenchymatous cells; interascal tissue lacking; hymenium I+ blue; asci clavate, ? fissitunicate; ascospores dark grey, muriform. Anamorphs unknown. Associated with cyanobacteria, trop. and subtrop.

Pyrenothrix Riddle (1917), Pyrenothricaceae (L). 1, N. Am., NZ. See Eriksson (1981), Tschermak-Woess *et al.* (*Pl. Syst. Evol.* **143**: 293, 1983; ultrastr.).

Pyrenotrichum Mont. (1843), Mitosporic fungi (L), 8.E/G1.15. 9, esp. trop. See Hawksworth (*Bull. Br. Mus. nat. Hist., Bot.* **9**: 59, 1981), Kalb & Vězda (*Folia geobot. phytotax.* **22**: 286, 1987; nomencl.), Sérusiaux (*Lichenologist* **18**: 1, 1986; conidioma campylidia). Teleomorphs *Badimia, Calopadia, Lasioloma, Sporopodium, Tapellaria*.

Pyrenotrochila Höhn. (1917) = Trochila (Dermat.). See Dennis (*British ascomycetes*, 1968; sub *Trochila*).

Pyrenowilmsia R.C. Harris & Aptroot (1991), Pyrenulaceae (L). 1, S. Afr.

Pyrenula Ach. (1809) nom. rej. = Thelidium (Verrucar.).

Pyrenula A. Massal. (1814) nom. cons., Pyrenulaceae (L). *c.* 300, cosmop. (mainly trop.). See Harris (*Mem. N.Y. bot. Gdn* **49**: 74, 1989; gen. concept, key 47 N. Am. spp.), Singh & Upreti (*Geophytology* **17**: 75, 1987; key 21 spp. Andaman Isl.), Upreti (*Bull. Soc. bot. Fr., lett. bot.* **3**: 241, 1991; key 12 spp. *P. brunnea* group, *Feddes Repert.* **103**: 279, 1992; 24 spp. India), Upreti & Singh (*Geophytology* **18**, 67, 1988; key 14 spp. Sri Lanka).

Pyrenulaceae Rabenh. (1870), Pyrenulales (±L). 16 gen. (+ 29 syn.), 429 spp. Thallus (where present) inconspicuous, immersed; ascomata sometimes aggregated, papillate, the ostioles sometimes lateral; interascal tissue initially of narrow anastomosing trabeculate pseudoparaphyses in a faintly I+ gel, with true paraphyses subsequently formed from the basal layers; asci cylindrical, with a small or large ocular chamber, sometimes staining differentially, not blueing in iodine, without an apical ring, sometimes evanescent; ascospores hyaline, or brown, transversely septate or muriform, the septa often thickened. Anamorphs coelomycetous where known. Usually lichenized with green algae (esp. *Trentepohliaceae*), on bark, cosmop.
Lit.: Aptroot (*Bibl. Lich.* **44**, 1991), Eriksson & Hawksworth (*SA* **11**: 71, 1992; concept), Harris (*Mem. N.Y. bot. Gdn* **49**: 74, 1989).

Pyrenulales, Ascomycota (±L). 5 fam., 47 gen. (+ 57 syn.), 666 spp. Thallus (where present) crustose, often immersed, often inconspicuous; ascomata perithecial, immersed or erumpent, ± globose or flattened, often clypeate, sometimes aggregated into stromata, the peridium usually thick-walled, usually composed of small-celled pseudoparenchymatous tissue, sometimes crystalline; interascal tissue initially of thin-walled anastomosing pseudoparaphyses, subsequently ± unbranched paraphyses often developing from the base of the ascoma, occasionally only of true paraphyses; asci cylindrical, persistent, usually multi-layered and fissitunicate, sometimes with a conspicuous apical ring, not blueing in iodine; ascospores hyaline or brown, septate, sometimes muriform, often with strongly thickened septa and angular lumina, sometimes with a sheath. Anamorphs pycnidial where known. Saprobes on bark, or lichenized with green algae (esp. *Trentepohliaceae*), mainly trop.
Sometimes united with *Dothideales* but with significant ontogenetic differences reflected in the hamathecial tissues (Janex-Favre, 1971). Eriksson (1982) accepted two fams and tentatively referred others here. Fams.:
(1) **Massariaceae**.
(2) **Monoblastiaceae** (syn. *Acrocordiaceae*).
(3) **Pleurotremataceae**.
(4) **Pyrenulaceae**.
(5) **Trypetheliaceae** (syn. *Astrotheliaceae*).
Barr (*Mycotaxon* **9**: 17, 1979; key 6 gen.) used the *Massariaceae* inclusive of the *Zopfiaceae*, here placed in *Dothideales*. The *Trichotheliaceae* has been recently transferred to the *Trichotheliales*.
Lit.: Eriksson (1981, 1982), Harris (*Michigan Bot.* **12**: 3, 1973; N. Am.), Janex-Favre (1971; ontogeny), Johnston (*Ann. Mo. bot. Gdn* **27**: 1, 1940; *Trypetheliaceae*), Keissler (*Rabenh. Krypt.-Fl.* **9**, 1(2), 1936-38; Eur.), Santesson (1952; foliicolous spp.), and under *Ascomycota*.

Pyrenulella Fink (1935) ≡ Pharcidiella (Verrucar.).

Pyrenulomyces E.A. Thomas ex Cif. & Tomas. (1953) = Pyrenula (Pyrenul.).

Pyrenyllium Clem. (1909) Ascomycota (inc. sed., ?L). See Aguirre-Hudson (*Bull. Br. Mus. nat. Hist., Bot.* **21**: 85, 1991).

Pyrgidiomyces Cif. & Tomas. (1953) ≡ Pyrgidium (Sphinctrin.).

Pyrgidium Nyl. (1867), Sphinctrinaceae (?L). 1, Australia, Italy, India, C. & S. Am. See Tibell (*Lichenologist* **14**: 219, 1982, 1987).

Pyrgillocarpon Nádv. (1942) = Pyrgillus (Pyrenul.) fide Aptroot (*Bibl. Lich.* **44**, 1991).

Pyrgillomyces Cif. & Tomas. (1953) ≡ Pyrgillus (Pyrenul.).

Pyrgillus Nyl. (1858), Pyrenulaceae (L). 12, Java, Australia, Am. See Tibell (*Beih. Nova Hedw.* **79**: 597, 1984).

Pyrgostroma Petr. (1951), Mitosporic fungi, 8.C2.?. 1, USA.

Pyricularia Sacc. (1880), Anamorphic Magnaporthaceae, 1.C2.10. Teleomorph *Magnaporthe*. 5, esp. temp. *P. oryzae* (sometimes as *P. grisea*; conservation to be sought), rice blast. See Ou (Ed.) (*The rice blast disease*, 1965), Leung & Williams (*CJB* **65**: 112, 1987; nuclear division and chromosomes), Vales *et al.* (*L'Agron. Tropic.* **41**: 242, 1986; electrophoresis in *P. oryzae* identification), Rossman *et al.* (*Mycol.* **82**: 509, 1990; nomencl. rice blast pathogen), Zhu *et al.* (*Mycosystema* **5**: 89, 1992; DNA fingerprinting and pathotype identification), Zeigler *et al.* (*Rice blast disease*, 1994).

Pyriculariopsis M.B. Ellis (1971), Mitosporic fungi, 1.C2.10. 2, trop. See Lai & Gao (*Acta Mycol. Sin.* **10**: 79, 1991; synonymy of *Gonatopyricularia*).

pyriform, pear-like in form (Fig. 37.14).

Pyriomyces Bat. & H. Maia (1965), Mitosporic fungi (L), 4.A1.?. 1, Brazil.

Pyripnomyces Cavalc. (1972), Mitosporic fungi (L), 5.?G1.?. 1, Brazil.

Pyrisperma Raf. (1808) nom. dub. ('gasteromycetes'). = Hymenogaster (Hymenogastr.) fide Soehner, 1962; or 'discomycetes', hypogeous).

Pyrispora, see *Pyrisperma*.

Pyrobolus Kuntze (1891) ≡ Eurotium (Trichocom.).

Pyrochroa Eschw. (1824) = Phaeographis (Graphid.).

Pyroctonum Prunet (1897) = Cladochytrium (Cladochytr.) p.p. fide Karling (1977).

Pyrofomes Kotl. & Pouzar (1964), Coriolaceae. 3, widespr. See Ryvarden & Johansen (*Preliminary polypore flora of East Africa*: 528, 1980; key 3 Afr. spp.). = Truncospora (Polypor.) fide Donk (1974).

Pyrographa Fée ex A. Massal. (1860) = Phaeographis (Graphid.).

Pyronema Carus (1835), Pyronemataceae. 2, widespr. See Moore & Korf (*Bull. Torrey bot. Cl.* **90**: 33, 1963; key).

Pyronemataceae Corda (1842), Pezizales. 2 gen. (+ 2 syn.), 3 spp. Ascomata on a well-developed hyphal mat, formed from clustered antherida and ascogonia, small, discoid or pulvinate, not setose, pale to orange, sometimes coalescing; excipulum sometimes poorly-developed, the hymenium exposed from a very early stage; interascal tissue of simple or branched paraphyses, sometimes swollen and pigmented at the apices; asci ± cylindrical, persistent, operculate, not blueing in iodine, with complex ascal pores; ascospores ellipsoidal, usually hyaline, smooth or ornamented. Anamorphs not known. Saprobic, esp. on burnt ground and coprophilous, widespr.
    Lit.: Kimbrough & Curry (*Mycol.* **78**: 735, 1986; septal ultrastr.), Kimbrough (*Mem. N.Y. bot. Gdn* **49**: 326, 1989; limits).

Pyronemella (Vido) Sacc. (1889) = Lasiobolus (Thelebol.) fide Pfister (*Mycol.* **76**: 843, 1984).

pyrophilous, growing on burnt ground, steam sterilized soil etc.; carbonicolous; - **fungi**, fireplace fungi, phoenicoid fungi (see Ramsbottom, *Mushrooms and toadstools*: 231, 1953; list of some larger fungi, Webster *et al.*, *TBMS* **47**: 445, 1964; discomycetes); see also *Pyronema*.

Pyropolyporus Murrill (1903) ≡ Phellinus (Hymenochaet.).

Pyropyxis Egger (1984), Otideaceae. 1, N. Am. Anamorph *Dichobotrys*.

pyroxylophilous, living on burnt wood.

Pyrrhoderma Imazeki (1966), Hymenochaetaceae. 2, Japan.

Pyrrhoglossum Singer (1944), Cortinariaceae. 9, trop.

Pyrrhosorus Juel (1901), ? Labyrinthulales (inc. sed.). 1, Eur.

Pyrrhospora Körb. (1855), Lecanoraceae (L). 7, Eur., N. Am., Macaronesia. See Hafellner (*Herzogia* **9**: 725, 1993; key).

Pythiaceae J. Schröt. (1893), Pythiales. 10 gen. (+ 14 syn.), 183 spp.
    Lit.: See Middleton (*Tijdschr. PlZiekt.* **58**: 226, 1952; generic concepts), Rangaswami (*Pythiaceous fungi*, 1962; host list), Barr *et al.* (*CJB* **70**: 2163, 1992; flagellum ultrastr.), Dick (*in* Margulis *et al.* (Eds), 1990: 661).

Pythiacystis R.E. Sm. & E.H. Sm. (1906) = Phytophthora (Pyth.) fide Leonian (1925).

Pythiales, Oomycota. 1 fam., 11 gen. (+ 15 syn.), 184 spp. Fam.:
    Pythiaceae.

Pythiella Couch (1935), Lagenaceae. 2 (in *Pythium*), N. Am.

Pythiogeton Minden (1916), Pythiaceae. 5, N. temp. See Batko (*Acta mycol.* **7**: 241, 1971; key), Drechsler (*J. Wash. Acad. Sci.* **22**: 421, 1932).

Pythiomorpha H.E. Petersen (1909) = Phytophthora (Pyth.) fide Blackwell *et al.* (*TBMS* **25**: 148, 1941), Waterhouse (*TBMS* **41**: 196, 1958).

Pythiopsis de Bary (1888), Saprolegniaceae. 3, N. temp.

pythiosis, a cosmopolitan disease of horses, cattle, and dogs caused by *Pythium insidiosum* (De Cock *et al.*, *J. clin. Microbiol.* **25**: 344, 1987). See also Mendoza *et al.* (*J. med. vet. Mycol.* **26**: 5, 1988).

Pythites Pampal. (1902), Fossil fungi. 1 (Eocene), Italy.

Pythium Nees (1823) nom. rej. (Saprolegniales, inc. sed.).

Pythium Pringsh. (1858) nom. cons., Pythiaceae. 120, cosmop. Causing damping-off, root diseases, and mycoparasitic; a few marine (for host list see Rangaswami, 1962); *P. thalassium* (Atkins, *TBMS* **38**: 31, 1955; on the marine pea-crab, *Pinnotheres*). See van der Plaats-Niterink (*Stud. Mycol.* **21**, 1981; key), Middleton (*Mem. Torrey bot. Cl.* **20**(1), 1943), Waterhouse (*Mycol. Pap.* **110**, 1968; diagnoses, descriptions, figs. from original papers), Hendrix & Campbell (*Mycol.* **66**: 681, 1974), Dick (1990; key).

pyxidate, provided with a lid, pertaining to, of having the character of a box; box-like

Pyxidiophora Bref. & Tavel (1891), Pyxidiophoraceae. 15 (mainly coprophilous), widespr. See Lundqvist (*Bot. Notiser* **133**: 121, 1980), Barrassa & Moreno (*Cryptogamie, Mycol.* **4**: 251, 1983; sections), Blackwell (*Mycol.* **86**: 1, 1994; life-cycle, molec. relationships), Blackwell & Malloch (*CJB* **67**: 2552, 1989; life-histories), Blackwell *et al.* (*MR* **92**: 397, 1989; spp. on mites), Blackwell *et al.* (*Science* **232**: 993, 1986; anamorphs). Anamorphs *Chalara*, *Thaxteriola*.

Pyxidiophoraceae Arnold (1971), Laboulbeniales. 2 gen. (+ 13 syn.), 16 spp. Stroma (thallus) absent. Ascomata perithecial, rarely cleistothecial, ± hyaline, often long-necked; asci clavate, without an apical ring, evanescent; ascospores broadly fusiform to clavate, 1- to 3-septate, often caudate, with ± pigmented patches; anamorph if present *Chalara*-like; mainly coprophilous.
    The family is confirmed as a member of *Laboulbeniales* by molecular data (Blackwell, *Mycol.* **86**: 1, 1994); it was included in the *Hypomycetales* by Eriksson (1982).
    Lit.: Lundqvist (*Bot. Notiser* **133**: 121, 1980).

Pyxidium Hill (1771) = Cladonia (Cladon.).

Pyxidium Gray (1821) = Perichaena (Trich.).

Pyxinaceae, see *Physciaceae*.

Pyxine Fr. (1825), Physciaceae (L). 45, trop. See Awasthi (*Phytomorphology* **30**: 359, 1982; key 21 spp. India), Imshaug (*Trans. Am. microsc. Soc.* **76**: 246, 1957), Kalb (*Bibl. Lich.* **24**, 1987; key 24 spp. Brazil,

Herzogia **10**: 61, 1994; key 21 spp. Australia), Kashiwadani (*J. Jap. Bot.* **52**: 137, 1977, Japan; *Bull. natn. Sci. Mus. Tokyo* B **3**: 63, 1977, Papua New Guinea), Malme (*Bih. K. Svenska Vet. Akad. Handl.* **23**, 3(13), 1897; S. Am.), Rogers (*Austr. J. Bot.* **34**: 131, 1986; key 15 spp. Australia; *Brunonia* **9**: 229, 1986; Australia), Sammy (*Nuytsia* **6**: 279, 1988; key 4 spp. W. Australia), Swinscow & Krog (*Norw. Jl Bot.* **22**: 43, 1975; E. Afr.).

Q, the ratio of length to breadth of elongate spores of agarics; spores ellipsoidal or ovoid when Q = < 2, ellipsoidal-oblong, fusoid, cylindrical, etc. when Q = > 2 (Singer, 1962: 68). Cf. E, sporograph.

quadrangular, see rhomboidal.

Quadricladium Nawawi & Kuthub. (1989), Mitosporic fungi, 1.G1.10. 1 (aquatic), Malaysia.

Quadrispora Bougher & Castellano (1993), Hymenogastraceae. 2, Australia.

Quadrisporomyces Sekunova (1960) = Schizosaccharomyces (Schizosaccharomycet.) fide Batra (1978).

Quasiconcha M.E. Barr & M. Blackw. (1981), Mytilinidiaceae. 1, USA. See Blackwell & Gilbertson (*Mycol.* **77**: 50, 1985; anamorph). Anamorph *Chalara*-like.

Quasidiscus B. Sutton (1991), Mitosporic fungi, 7.A1.15. 1, Australia.

Quaternaria Tul. & C. Tul. (1863) nom. rej. = Eutypella (Diatryp.) fide Rappaz (1987).

Queenslandia Bat. & H. Maia (1959), Mitosporic fungi, 5.A1.?. 4, widespr.

Queirozia Viégas & Cardoso (1944) = Pleochaeta (Erysiph.)

Quélet (Lucien; 1832-99). A medical man; first President of Société Mycologique de France, 1885. Chief writings: *Les champignons du Jura et des Voges* (1872-5) a number of Fries' sub-genera of *Agaricaceae* first used as genera [reprint 1964 + 1972 Suppl., 1872-1902; index to reprint, *BSMF* **88**: cxi, 1973], (with M.C. Cooke) *Clavis synoptica Hymenomycetum Europaeorum* (1878), *Flore mycologique de la France...* (1888 [reprint 1962]), Mangin & Chomette (*Essai d'une table de concordance des principales espèces mycologiques avec la Flore de France* (1906) [reprint 1963]). See Boudier (*BSMF* **15**: 321, 1899), Gilbert (*BSMF* **65**: 5, 1949; bibl.), Grummann (1974: 294), Stafleu & Cowan (*TL-2* **4**: 453, 1983).

Queletia Fr. (1872), Tulostomataceae. 2, Eur., USA, S. Am. See Wright (*Cryptog. Bot.* **1**: 26, 1989).

quellkörper, a mucilaginous mass of thick-walled cells within the ascoma of *Nitschkeaceae*; believed to induce rupture of the ascoma.

Quercella Velen. (1921), Cortinariaceae. 1, Eur.

Questieria G. Arnaud (1918) = Schiffnerula (Englerul.) fide v. Arx & Müller (1975).

Questieriella G. Arnaud ex S. Hughes (1983), Anamorphic Englerulaceae, 1.C2.1. Teleomorph *Questieria*. 3, widespr. See Hughes (*CJB* **61**: 1729, 1983).

Quezelia Faurel & Schotter (1965), Mitosporic fungi, 5.B1.?. 1 (coprophilous), Sahara.

Quilonia K.P. Jain & R.C. Gupta (1970), Fossil fungi (Mitosp. fungi). 1 (Miocene), India.

quinine fungus, *Fomes officinalis*.

Quintaria Kohlm. & Volkm.-Kohlm. (1991), Lophiostomataceae. 1 (marine), Belize.

Quorn. The trade name under which mycoprotein (q.v.) manufactured from *Fusarium graminearum* by Marlow Foods (a subsidiary of Rank Hovis McDougall and ICI) is marketed.

Rabdosporium Chev. (1826) = Cheirospora (Mitosp. fungi) fide Hughes (*CJB* **36**: 800, 1958).

Rabenhorst (Gottlob Ludwig; 1806-81). A pharmacist in Lukau (1830-40), then took up botany at Dresden; first editor of *Hedwigia* (1852-78). Noted for his *Deutschlands Kryptogamen-Flora* (1844-53; 2 Aufl. *Kryptogamen-Flora von Deutschland, Oesterreich und der Schweiz*, 1884 on) and the *Fungi Europaei Exsiccati* (1859 on; see Kohlmeyer, *Beih. Nova Hedw.* **4**, 1963, Stevenson, *Taxon* **16**: 112, 1967; numbering of specimens). See Richter (*Hedwigia* **20**: 113, 1881), Grummann (1974: 38), Stafleu & Cowan (*TL-2* **4**: 460, 1983).

Rabenhorstia Fr. (1849) [non Rchb. (1841), *Brunoniaceae*], Anamorphic Melanconidaceae, 8.A1.1. Teleomorph *Hercospora*. 10, esp. Eur.

Rabenhorstiites Teterevn. & Tasl. (1977), Fossil fungi. 1 (Tertiary), former USSR.

Rabenhorstinidium R.B. Singh & G.V. Patil (1980), Fossil fungi (Mitosp. fungi). 1 (Cretaceous), India.

race, see physiologic race and Classification.

Racemella Ces. (1861) = Cordyceps (Clavicipit.) fide Tulasne & Tulasne (*Sel. Fung. Carp.* **3**: 4, 1865).

Racemosporium Moreau & Moreau (1941) = Arthrinium (Mitosp. fungi) fide Ellis (*Mycol. Pap.* **103**, 1965).

rachis, a geniculate or zig-zag holoblastic extension of a conidiogenous cell (as in *Tritirachium*) resulting from sympodial conidiogenous cell development; **rachiform** (of conidiogenous cells), having a rachis; cf. raduliform.

Rachisia Lindner (1913) = Fusarium (Mitosp. fungi).

Raciborski (Marjan; 1863-1917). After working in Poland, Germany, Indonesia (1896-1900), and at the Agricultural Academy, Dublany, became Professor of Botany at Lwow, later at Cracow Univ. Chief writings: papers on the fungi of Poland, Switzerland, and esp. Java. See Goebel (*Ber. dtsch. bot. Ges.* **35**: (97), 1918; bibl.), Kornasia (*Marian Raciborski*, 1986), Grummann (1974: 437), Stafleu & Cowan (*TL-2* **4**: 529, 1983).

Raciborskia Berl. (1888) = Comatricha (Stemonitid.).

Raciborskiella Höhn. (1909) = Strigula (Strigul.) fide Harris (*in litt.*).

Raciborskiella Speg. (1919) ≡ Trichopeltella (Microthyr.).

Raciborskiomyces Siemaszko (1925) = Epipolaeum (Pseudoperispor.) fide Müller & v. Arx (1962).

racket (racquette) cell (of dermatophytes), a hyphal cell having a swelling at one end; cf. hypha.

Racoblenna A. Massal. (1852) = Placynthium (Placynth.).

Racodium Pers. (1794) nom. rej. = Rhinocladiella (Mitosp. fungi) fide Hawksworth & Riedl (1977); see cellar fungus.

Racodium Fr. (1829) nom. cons., Mitosporic fungi (L), 1.A2.?. 1, Eur., N. Am. See Hawksworth & Riedl (*Taxon* **26**: 208, 1977; nomencl.).

Racoplaca Fée (1825) = Strigula (Strigul.).

Racovitziella Döbbeler & Poelt (1978) [non De Wild. (1900), *Algae*], Dothideales (inc. sed.). 1 (on *Musci*), Austria.

racquette cell, see racket cell.

Radaisiella Bainier (1910) = Botryosporium (Mitosp. fungi). See Mason (*Mycol. Pap.* **2**: 27, 1928).

Raddetes P. Karst. (1887) nom. rej. = Conocybe (Bolbit.).

radial (of lichen thalli), radially symmetrical in transverse section (e.g. *Alectoria*, *Bryoria*, *Coelocaulon*, *Usnea*).

radiate, spreading from a centre.

Radiatispora Matsush. (1995), Mitosporic fungi, 1.G1.1. 1, China.

radicating (of stipes), like a root; rooting (Fig. 4H).
Radiciseta Sawada & Katsuki (1959), Mitosporic fungi, 1.C1.?. 1, Taiwan.
Radiigera Zeller (1944), Geastraceae. 5, N. & S. Am., Eur. See Kers (*Bot. Notiser* **129**: 173, 1976).
Radiogaster Lloyd (1924) nom. inval. ? = Hymenogaster (Hymenogastr.).
Radiomyces Embree (1959), Radiomycetaceae. 3, USA. See Benjamin (*Aliso* **4**: 523, 1960), Ellis & Hesseltine (*Mycol.* **66**: 85, 1974), Kitz *et al.* (*Mycol.* **74**: 110, 1982), Benny & Benjamn (1991; key).
Radiomycetaceae Hesselt. & J.J. Ellis (1974), Mucorales. 2 gen. (+ 1 syn.), 4 spp. Sporangiola borne on complex ampullae, simple or branched, often stoloniferous sporangiophores, sporangia absent; zygospores smooth, borne on apposed, appendaged suspensors.
Lit.: Benny & Khan (*Scanning Microscopy* **2**: 1199, 1988; ultrastr., sporangiolar appendage chemistry), Benny & Samson (*Mem. N.Y. bot. Gdn* **49**: 11, 1989; ultrastr.), Benny & Benjamin (*Mycol.* **83**: 713, 1991; key).
Radiomycopsis Pidopl. & Milko (1971) = Radiomyces (Radiomycet.) fide Benny & Benjamin (1991).
Radotinea Velen. (1934), ? Leotiales (inc. sed.). 1, former Czechoslovakia.
radula spore (**radulaspore, radulospore**), one of the slimy spores borne over the surface of ascospores as in *Nectria coryli* while still in the ascus; **dry - - =** sympoduospore (Mason, 1933, 1937).
raduliform (of conidiogenous cells), the elongating conidiogenous axis resulting from holoblastic sympodial development, clavate or somewhat inflated rather than zig-zag; cf. rachiform.
Radulochaete Rick (1940) nom. dub. ('basidiomycetes') fide Donk (*Persoonia* **3**: 199, 1964).
Radulodon Ryvarden (1972), Hyphodermataceae. 6, widespr. See Ryvarden (*Česká Myk.* **30**: 38, 1976; key), Hjortstam *et al.* (*Kew Bull.* **45**: 303, 1990; key).
Radulomyces M.P. Christ. (1960), Hyphodermataceae. *c.* 6, widespr.
Radulum Fr. (1825) = Xenotypa (Vals.) fide Donk (*Taxon* **5**: 109, 1956).
Raesaeneniolichen Tomas. & Cif. (1952) = Polyblastia (Verrucar.).
Raesaeneniomyces Cif. & Tomas. (1953) ≡ Raesaeneniolichen (Verrucar.).
Raffaelea Arx & Hennebert (1965), Mitosporic fungi, 1.A1.10. 7, Eur., N. Am., S. Afr. See Scott & du Toit (*TBMS* **55**: 181, 1970).
ragi, a starter for arrack, etc. composed of small balls of rice flour containing *Mucor, Rhizopus*, yeasts and bacteria (Hesseltine, *Mycol.* **57**: 163, 1965).
Ragnhildiana Solheim (1931) = Mycovellosiella (Mitosp. fungi) fide Deighton (*Mycol. Pap.* **137**, 1974).
Raizadenia S.L. Srivast. (1981), Mitosporic fungi, 2.C2.1. 1, India.
Rajapa Singer (1945) = Termitomyces (Amanit.) fide Singer (1951).
Ramacrinella Manier & Ormières (1962), Eccrinaceae. 1 (in *Amphipoda*), France. See Manier *et al.* (*Ann. Sci. nat., Bot.* sér. 12, **2**: 625, 1961), Lichtwardt (1986).
Ramakrishnanella Kamat & Ullasa ex Ullasa (1970), Mitosporic fungi, 4.A2.?. 1, India.
Ramakrishnania Ramachar & Bhagyan. (1979), Uredinales (inc. sed.). 1 (on *Ixora*), India. = Puccinia (Puccin.) fide Cummins & Hiratsuka (1983).
Ramalea Nyl. (1866), Cladoniaceae (L). 4, C. & S. Am., NZ.
Ramalia Bat. (1957) = Fusicladium (Mitosp. fungi) fide Sutton & Pascoe (*Austr. Syst. Bot.* **1**: 79, 1988).
Ramalina Ach. (1810) nom. cons., Ramalinaceae (L). *c.* 200, cosmop. See Howe (*Bryologist* **16**: 65, 81, 1913,

17: 1, 17, 33, 49, 81, 1914), Kashiwadani & Kalb (*Lichenologist* **25**: 1, 1993; key 22 spp. Brazil), Krog & James (*Norw. Jl Bot.* **24**: 15, 1977; Fennoscandia, UK), Krog & Østhagen (*Norw. Jl Bot.* **27**: 255, 1980; Canary Is.), Krog & Swinscow (*Norw. Jl Bot.* **23**: 153, 1976; E. Afr.), Stevens (*Bull. Br. Mus. nat. Hist., Bot.* **16**: 107, 1987; key 20 spp. Australia).
Ramalinaceae C. Agardh (1821), Lecanorales (L). 2 gen. (+ 11 syn.), 201 spp. Thallus fruticose, much-branched, erect or pendulous, usually distinctly flattened, greyish to yellow, not gelatinous; ascomata apothecial, usually shortly stipitate, often on short lateral branches, usually pale; interascal tissue of sparingly branched paraphyses; asci with an I+ apical cap, and usually a well-developed non-staining apical cushion, without an outer I+ gelatinized layer; ascospores hyaline, 1-septate, often curved. Anamorphs pycnidial. Lichenized with green algae, widespr.
Ramalinomyces E.A. Thomas ex Cif. & Tomas. (1953) ≡ Ramalina (Ramalin.).
Ramalinopsis (Zahlbr.) Follmann & Huneck (1969), Ramalinaceae (L). 1, Hawaii. See Riedl (*Sydowia* **28**: 134, 1977).
Ramalodium Nyl. (1879), Collemataceae (L). 3, Japan, New Caledonia, Tasmania. See Henssen (*Lichenologist* **3**: 29, 1965).
Ramaraomyces N.K. Rao, Manohar. & Goos (1989), Mitosporic fungi, 2.C1.1. 1, India.
Ramaria Holmsk. (1790) nom. rej. = Clavulinopsis (Clavar.) fide Donk (1959).
Ramaria Fr. ex Bonord. (1851) nom. cons., Ramariaceae. 100, cosmop. See Corner (1950: 124, 542), Mann & Stuntz (*Ramaria of western Washington*, 1973; keys), Petersen (*Bibl. Mycol.* **43**, 1975; subgen. *Lentoramaria*, *Bibl. Mycol.* **79**, 1981; subgen. *Echinoramaria*).
Ramariaceae Corner (1970), Gomphales. 4 gen. (+ 7 syn.), 105 spp. Basidioma ramarioid; spores yellowish-brown, ornamented.
Ramaricium J. Erikss. (1954), Ramariaceae. 3, Eur., N. Am., Colombia. See Ginns (*Bot. Notiser* **132**: 93, 1979).
ramarioid, with a form similar to that of the basidioma of *Ramaria*.
Ramariopsis (Donk) Corner (1950), Clavariaceae. 20, widespr. See Petersen (*Mycol.* **61**: 557, 1967; key N. Am. spp.), Pegler & Young (*TBMS* **84**: 207, 1985; basidiospore structure).
Ramboldia Kantvilas & Elix (1994), Lecanoraceae (L). 5, Australia.
Ramgea Brumm. (1992), Thelebolaceae. 1, Netherlands.
Ramichloridium Stahel ex de Hoog (1977), Mitosporic fungi, 1.A2.10. 15, widespr. See de Hoog *et al.* (*TBMS* **81**: 485, 1983; relationships with *Stenella* and *Veronaea*), Campbell & Al-Hedaithy (*Jl Med. Vet. Mycol.* **31**: 325, 1993; *R. mackenziei*, phaeohyphomycosis of the brain). = Periconiella (Mitosp. fungi) fide Ellis (*Mycol. Pap.* **111**, 1967).
Ramicola Velen. (1929) = Agrocybe (Bolbit.) fide Romagnesi (1963).
ramicolous, living on branches.
ramoconidium, an apical branch of a conidiophore which secedes and functions as a conidium, as in *Cladosporium* and *Subramaniomyces*.
Ramonia Stizenb. (1862), Gyalectaceae (L). 10, Eur., N. & S. Am., Afr. See Coppins (*Lichenologist* **19**: 409, 1987; key 4 Br. spp.), Vězda (*Folia geobot. phytotax.* **2**: 311, 1967).
Ramosiella Syd. & P. Syd. (1917) = Anhellia (Myriang.) fide v. Arx (1963).

Ramosphaerella Laib. (1921) = Mycosphaerella (Mycosphaerell.).

Ramsaysphaera H.D. Pflug (1978), Fossil fungi (? Mitosp. fungi). S. Afr. (c. 3400 million yr old Swartkoppie chert).

Ramsbottom (John; 1885-1974). Worked at the British Museum (Natural History) 1910-50 (keeper of botany from 1930). Wrote extensively on many aspects of mycology including taxonomy, nomenclature, and history (q.v.) and did much to popularize the study of fungi esp. by his *Mushrooms and toadstools*, 1953. See Gregory (*TBMS* **65**: 1, 1975; portr.), Grummann (1974: 411), Stafleu & Cowan (*TL-2* **4**: 574, 1983).

Ramsbottomia W.D. Buckley (1923), Otideaceae. 3, boreal and temp. See Benkert & Schumacher (*Agarica* **6**: 28, 1985), Kullman & van Brummelen (*Persoonia* **15**: 93, 1992), Caillet & Moyne (*Bull. Soc. Hist. nat. Doubs* **84**: 9, 1991; keys).

Ramularia Unger (1833) [non Roussel (1806), *Algae*] nom. cons. Anamorphic Dothideaceae, 1.A/C1.3/10. Teleomorph *Mycosphaerella*. 300 (causing leaf spots), esp. temp. *R. vallisumbrosae* (*Narcissus* white mould). See Braun & Sutton (*Taxon* **112**: 656, 1991; conservation), Hughes (*TBMS* **32**: 50, 1949), Hughes (*TBMS* **32**: 50, 1949), Vimba ([The flora of the genus *Ramularia* Sacc. in the Latvian SSR], 1970), Deighton (*TBMS* **90**: 330, 1988; spp. on *Salix*), Sutton & Waller (*TBMS* **90**: 55, 1988; spp. on *Gramineae*), Braun (*Int. Jl Myc. Lich.* **3**: 271, 1988, *Nova Hedw.* **47**: 335, 1989, *Nova Hedw.* **53**: 291, 1991; related genera, *A monograph of Cercosporella, Ramularia and allied genera* **1**, 1995). See also *IMI Descr.* **851-869**, 1986.

Ramularia Sacc. (1880) = Ramularia Unger (Mitosp. fungi).

Ramulariites Babajan & Tasl. (1973), Fossil fungi. 1 (Tertiary), former USSR.

Ramulariopsis Speg. (1910), Mitosporic fungi, 1.B/C1.3. 1, S. Am. See Deighton (*TBMS* **59**: 185, 1972).

Ramulariospora Bubák (1914) = Phacidiella (Mitosp. fungi) fide Sutton (1977).

Ramularisphaerella Kleb. (1918) = Mycosphaerella (Mycosphaerell.) fide v. Arx & Müller (1975).

Ramularites Pia (1927), Fossil fungi. 1 (Oligocene), Eur.

Ramulaspera Lindr. (1902), Mitosporic fungi, 1.A1.?. 2, Eur. See Griffiths (*TBMS* **40**: 232, 1957).

Ramulina Velen. (1947) [non Thurm. (1863), fossil ? *Algae*] = Pseudombrophila (Otid.) fide Svrček (*Acta Mus. Nat. Prague* **32**(2-4): 115, 1976).

Ramulispora Miura (1920), Mitosporic fungi, 4.C1.10). 4, temp. See Rawla (*TBMS* **60**: 285, 1973), Robbertse, Campbell & Crous (*S.Afr. Jl Bot.* **61**: 43, 1995; spp. causing eyespot of wheat). See *Pseudocercosporella*.

ramus, (1) (pl. **-i**), a branch (Lat.); (2) (of a penicillus), a cell bearing a verticil of 'metulae' and phialides.

ramycin, see fusidic acid.

rangiferoid, branched like a reindeer's horn.

Ranojevicia Bubák (1910), Mitosporic fungi, 3.A1.?. 1, Eur.

RAPD, see Molecular biology.

Raperia Subram. & Rajendran (1976) = Aspergillus (Mitosp. fungi) fide Samson (*Stud. Mycol.* **18**: 1, 1979).

Raphanozon P. Kumm. (1871) nom. rej. ≡ Telamonia (Cortinar.).

raphe (of *Chaetomella*), the longitudinal dehiscence mechanism. See Di Cosmo & Cole (*CJB* **58**: 1129, 1980; ultrastr. development).

raphides, needle-shaped crystals; as hyphae inside some lichen thalli.

Raphiospora A. Massal. (1853) nom. rej. prop. = Arthrorhaphis (Arthrorhaphid.) fide Jørgensen & Santesson (*Taxon* **42**: 881, 1993).

Rasutoria M.E. Barr (1987), Euantennariaceae. 4, trop.

Ravenelia Berk. (1853), Raveneliaceae. *c.* 200 (on *Leguminosae*), warmer parts. Arthur (1934) and subsequent workers take the genus in the older, wider sense in which it includes *Cystingophora, Haploravenelia, Pleoravenelia* and others (Dietel, 1906). *R. esculenta* is edible (Narasimhan & Thirumalachar, *Phytopath. Z.* **41**: 97, 1961). See Cummins (1978: key 61 spp.), Hennen & Cummins (*Rept. Tottori Mycol. Inst.* **28**: 1, 1990), Mordue (*TBMS* **90**: 473, 1988).

Raveneliaceae Leppik (1972), Uredinales. 15 gen. (+ 9 syn.), *c.* 3800 spp. Pycnia discoid, with bounding structures, subcuticular or subepidermal; aeciospores either catenulate (*Aecidium*) or pedicellate (*Uraecium*); urediniospores pedicellate, uredinia often paraphysate; teliospores pedicellate (the pedicels often fused in groups, with apical cells or cysts, or otherwise specialized), vertically septate or vertically or radially arranged on the pedicels, often forming complex spore heads.

Ravenelula Speg. (1881), ? Lecanorales (inc. sed.). 1 or 2, Am.

ray fungi, members of the *Actinomycetes* (Bacteria), q.v..

razor-strop fungus, basidioma of *Piptoporus betulinus*.

RDA, see Media.

Readeriella Syd. (1908), Mitosporic fungi, 4.A2.19. 1, Australia, UK. See Macauley & Thrower (*TBMS* **48**: 105, 1965).

Rebentischia P. Karst. (1869), Tubeufiaceae. 2, Eur., N. Am. See Barr (*Mycotaxon* **12**: 137, 1980; key).

receptacle, an axis having one or more organs, as the stem in *Phallales*; any hymenium-supporting structure.

receptive body, a small branched or unbranched process from the stroma (as in *Stromatinia gladioli*) able to be 'spermatized' by microconidia (see Drayton, *Mycol.* **26**: 46, 1934); **receptive hypha** = flexuous hypha, trichogyne, and possibly other like structures.

Rechingeria Servít (1931) ? = Lichinella (Lichin.).

Rechingeriella Petr. (1940), ? Zopfiaceae. 2, Eur., Asia. See Hawksworth (*CJB* **57**: 91, 1979).

Recifea Bat. & Cif. (1962) = Microcallis (Chaetothyr.) fide v. Arx (*in litt.*).

recognition (of mutualistic symbionts), the process by which two compatible potential symbionts initiate the relationship.

Reconditella Matzer & Hafellner (1990), Sordariales (inc. sed.). 1 (on lichens, esp. *Physconia*), Eur.

record fungi, *highest spore count*, 161,037 fungus spores per m³, Cardiff 1971; *largest area occupied*, *Armillaria ostoyae* occupying 600 ha in Washington State, USA; *largest fruit body*, *Rigidoporus ulmarius*, 163 cm long × 140 cm wide × 50 cm tall, circumference 4.8 m, in IMI's former grounds at Kew, Surrey, UK (measurements in Feb. 1995; still growing; unweighed); *heaviest fruit body*, *Laetiporus sulphureus*, 45.4 kg, New Forest, Hants, UK (found 1990); *heaviest mycelium*, *Armillaria bulbosus*, 100+ tonnes, Michigan, USA, reported April 1992 (see Longevity); *largest edible*, *Langermannia gigantea*, circumference 2.64 m, 22 kg, from Canada, 1987; *oldest thallus*, *Rhizocarpon geographicum* (see Longevity); *most poisonous*, *Amanita phalloides*, 5-7 mg amanitoxins lethal. See Matthews (*The new Guinness book of records*, 1995), also acidophily, alkilophily, Longevity, spore discharge, thermophily, water activity.

Recticharella Scheer (1944) = Asellaria (Asellar.) fide Scheer (*Arch. Protistenk.* **114**: 343, 1972).

Recticoma Scheer (1935) = Enterobryus (Eccrin.) fide Manier & Lichtwardt (*Ann. Sci. nat., Bot.,* sér. 12, **9**: 519, 1968).

Rectipilus Agerer (1973), Schizophyllaceae. 6, cosmop.

recurved (of a pileus), convexo-expanded (q.v.).

Red List, see Conservation.

red rice, see ang-kak; **- bread mould,** see *Chrysonilia*; **- rust,** (1) urediniospore state of rusts, especially of cereals; (2) (of tea), the alga *Cephaleuros*; **- truffle,** *Melanogaster variegatus*.

Redaellia Cif. (1930) = Aspergillus (Mitosp. fungi) fide Alecrim (*Ann. Fac. Med. Univ. Recife* **18**: 81, 1958).

Redbia Deighton & Piroz. (1972), Mitosporic fungi, 1.C1.10. 1 (on *Puccinia*), trop.

Reddellomyces Trappe, Castellano & Malajczuk (1992), Otideaceae. 4, Australasia, Mediterranean.

Redonia C.W. Dodge (1973), ? Physciaceae (L). 1, S. Am.

Reduviasporonites L.R. Wilson (1962), Fossil fungi (*f. cat.*). 1 (Permian), USA.

Reessia C. Fisch (1883), Rhizidiomycetaceae. 2, Eur. See Dick (*in press*; key).

Reference collections, of fungi are essential to secure the application of scientific names (see Nomenclature), to serve as vouchers supporting the occurrence of species or material used in research, and as the raw materials for new systematic and other investigations. Reference collections of fungi are of two main types: (1) Genetic resource collections (q.v.; culture collections) hold material in a living or metabolically inactive state from which they can be resuscitated; and (2) Reference collections of non-living material, including dried specimens attached to plant, rock or other substrata, dried cultures, microscope slides, colour transparencies, drawings and paintings, and ultrastructural mounts. A collection of dried material is often referred to as an herbarium (q.v.), but as fungi are not plants, and not always kept with plants, the more general term is now used at IMI; the practice is especially appropriate where living and dried material are both being stored.

An immense amount of information related to knowledge of the distribution and ecology of fungi is stored in the world's mycological collections; this is increasingly being captured in computerized databases so that it can be made more accessible.

Most institutional collections have unique acronyms (generally retained unchanged despite name changes of the institutions) assigned by the International Association for Plant Taxonomy, a selection of which are included in this edition of the *Dictionary* (e.g. **BPI, DAOM, IMI, K, L, LE, UPS**); collections of living cultures are given acronyms by the World Data Center for Micro-organisms (see Genetic resource collections) - where possible the same acronym is used under both systems. Acronyms are commonly associated with collection accession numbers in mycological publications to make it clear what material was employed.

Lit.: **General**: Bartz *et al.* (*Museums of the world,* edn 4, 1992), Bridson & Forman (*The herbarium handbook,* edn 2, 1992), Fosberg & Sachet (*Manual for tropical herbaria* [*Regnum. veg.* **39**], 1965), Hall & Minter (*International mycological directory,* edn 3, 1994), Hawksworth (*Mycologist's handbook,* 1974), Holmgren *et al.* (*Index Herbariorum. Part 1. Herbaria of the world,* edn 8 [*Regnum veg.* **120**], 1990), Laundon (*Lichenologist* **11**: 1, 1974), Pinniger (*Biodet. Abstr.* **5**: 125, 1991; insect control in), Stafleu & Cowan (*TL-2,* 7 vols. [+ 3 suppl.], 1976-).

Regional: **Australia,** Walker (*Mycological herbaria and culture collections in Australia,* 1980); **British Isles,** Hawksworth & Seaward (*Lichenology in the British Isles 1568-1975,* 1977), Kent & Allen (*British and Irish herbaria,* 1984); **Germany,** Scholz (*Boletus* **12**(2): 33, 1989; lichens); **Israel,** Ferber (*Israel national collections of natural history,* 1985); **Italy,** Ciferri (*Taxon* **1**: 126, 1952), Tretiach & Passadore (*Notiz. Soc. Lich. Ital.* **3** (*Suppl.*), 1990); **New Zealand,** Wright (*N.Z. Jl Bot.* **22**: 323, 1984); **former USSR,** Parmasto (*Taxon* **34**: 359, 1985).

See also Collection and preservation, Genetic resource collections, herbarium beetle.

reflexed (of an edge), turned up or back (Fig. 19G).

Refractohilum D. Hawksw. (1977), Mitosporic fungi, 1.A/C1.19. 3 (on lichens), temp.

Rehmia Kremp. (1861) = Rhizocarpon (Rhizocarp.).

Rehmiella G. Winter (1883) = Gnomonia (Vals.) fide Barr (1978).

Rehmiellopsis Bubák & Kabát (1910) = Delphinella (Dothior.) fide Barr (1972).

Rehmiodothis Theiss. & Syd. (1914), Phyllachoraceae. 1 (on leaves of *Melastomataceae*), Asia. See Katumoto (*Trans. mycol. Soc. Japan* **32**: 37, 1991).

Rehmiomycella E. Müll. (1962), Ascomycota (inc. sed.). 1, Brazil.

Rehmiomyces Henn. (1904) = Dictyonia (Leot.) fide Clements & Shear (1931).

Rehmiomyces (Sacc. & P. Syd.) Syd. (1904) ≡ Rehmiomycella (Ascomycota, inc. sed.).

reindeer lichen (-moss), mainly *Cladonia stellaris* and *C. rangiferina,* species grazed by reindeer and caribou.

Reinkella Darb. (1897) ? = Roccella (Roccell.) fide Tehler (*CJB* **68**: 2458, 1990).

Reinkellomyces Cif. & Tomas. (1953) ≡ Reinkella (Roccell.).

Reisneria Velen. (1922) = Gloeophyllum (Coriol.) fide Donk (1974).

Relhanum Gray (1821) ≡ Verpa (Morchell.).

Relicina (Hale & Kurok.) Hale (1974) Parmeliaceae. *c.* 40, temp. & trop., esp. S.E. Asia, Australia. See Hale (*Smithson. Contr. bot.* **26**, 1975; monogr.). ? = Parmelia (Parmel.).

Relicinopsis Elix & Verdon (1986) Parmeliaceae. 5, S. Afr., Australasia. ? = Parmelia (Parmel.).

religion (use of mushrooms in), see entheogen, ethnomycology, hallucinogenic fungi, soma.

Remispora Linder (1944), Halosphaeriaceae. 5 (marine), widespr. See Johnson *et al.* (*Bot. Mar.* **27**: 557, 1984), Manimohan *et al.* (*MR* **97**: 1190, 1993).

remote (of lamellae), proximal end free and at some distance from the stipe.

Renatobasidium Hauerslev (1993), Exidiaceae. 1, Eur.

reniform, kidney-like in form; fabiform (Fig. 37.7).

Reniforma Pore & Sorenson (1990), Mitosporic fungi, 1.A1.19. 1 (basidiomycetous yeast), USA.

Renispora Sigler & J.W. Carmich. (1979), Onygenaceae. 1, USA.

repand (of a pileus), having a waved edge which is turned back.

repeating spore, a spore which gives rise to the same type of mycelium as that on which it developed.

repetition (spore germination by), producing a new spore like the first.

Repetobasidiellum J. Erikss. & Hjortstam (1981), Sistotremataceae. 1, N. Eur.

Repetobasidium J. Erikss. (1958), Sistotremataceae. 10, widespr.

Repetophragma Subram. (1992), Mitosporic fungi, 1.C2.19. 9, widespr.

reproductocentric (of *Chytridiales*), having development of one (**mono-, monogenocentric**) or more reproductive structures at the centre of gravity of the thallus; genocentric. See Karling (*Am. J. Bot.* **19**: 54, 1932). Cf. mono- and polycentric.

Requienella Fabre (1883), Pyrenulaceae. 2, N. temp. See Boise (*Mycol.* **78**: 37, 1986).

Resendea Bat. (1961), ? Microthyriaceae. 2, S. Am.

Resinicium Parmasto (1968), Meruliaceae. 4, widespr.

Resinomycena Redhead & Singer (1981), Tricholomataceae. 5, N. Am.

resistance, the power of an organism to overcome, completely or in some degree, the effect of a pathogen or other damaging factor. **acquired -**, a non-inherited resistance response in a normally susceptible host following a predisposng treatment. See immunity, axeny.

resistant sporangium, see meiosporangium.

Resticularia P.A. Dang. (1890) = Syzygangia (Myzocytiopsid.) fide Dick (*in press*).

resting sporangium, see meiosporangium.

resting spore, (1) a spore germinating after a resting period (frequently after overwintering), as does an oospore or a teliospore; a 'winter spore'; (2) an encysted zygote formed after the fusion of gametes, meiosis occuring on germination to produce planospores which encyst to produce gametes (*Blastocladiales*).

resupinate (of basidiomata), flat on the substrate with the hymenium on the outer side .

Resupinatus (Nees) Gray (1821), Tricholomataceae. 13, cosmop.

retention (of surfaces), the ability of a surface to hold a fungicide or other crop protectant. Cf. adherance.

Retiarius D.L. Olivier (1978), Mitosporic fungi, 1.G1.10. 2 (on pollen), S. Afr.

Reticellites Glass, D.D. Br. & Elsik (1987), Fossil fungi. 1 (upper Eocene), USA.

Reticularia Baumg. (1790) = Lobaria (Lobar.).

Reticularia Bull. (1791) = Enteridium (Lycogal.) fide Farr (1976).

Reticulariaceae see *Lycogalaceae*.

reticulate, like a net; netted (Fig. 29.12).

Reticulatisporonites Elsik (1968), Fossil fungi (*f. cat.*). 1 (Paleocene), USA.

Reticulocephalis Benny, R.K. Benj. & P.M. Kirk (1992), Sigmoideomycetaceae. 2, USA, UK. See Benny *et al.* (*Mycol.* **84**: 635, 1992; key).

Reticulomyxa Nauss (1949), ? Myxomycetes (inc. sed.). 1 (aquatic), USA.

Reticulosphaeria Sivan. & Bahekar (1982), ? Amphisphaeriaceae. 1, India.

Reticulosporidium Thor (1930) nom. dub. (? Fungi, inc. sed.).

Retidiporites C.P. Varma & Rawat (1963), ? Fossil fungi (*f. cat.*). 1 (Eocene), India.

Retigerus Raddi (1829) = Phallus (Phall.).

Retihelicosporonites Ramanujam & Rao (1979), Fossil fungi (*f. cat.*; Hiospira). 1 (Miocene), India. ? = Hiospira (Mitosp. fungi).

Retinocyclus Fuckel (1871) ≡ Tromera (Agyr.).

Retocybe Velen. (1947) ? = Mycena (Tricholomat.) fide Singer (1951).

retraction septum, see septum (adventitious).

Retroa P.F. Cannon (1991), Phyllachoraceae. 2, S. Am.

Retroconis de Hoog & Bat. Vegte (1989), Mitosporic fungi, 1.A2.3. 1, India, Pakistan, Mozambique.

retroculture, a reisolate of a pathogen from a host into which it had been experimentally introduced.

retroneme, a bipartite tubular hair.

retrorse, backward.

revalidated, see devalidated.

revolute, having the edge rolled back or up (Fig. 19H).

Reyesiella Sacc. (1917) ≡ Anthomycetella (Ravenel.).

Reymanella Marcink. (1979), Fossil fungi. 1 (Jurassic), Poland.

RFLP, see Molecular biology.

Rhabdium P.A. Dang. (1903) [non Wallr. (1833), *Algae*] = Harpochytrium (Harpochytr.) fide Atkinson (1903).

Rhabdoclema Syd. (1939), Mitosporic fungi, 4.A1.?. 2, S. Am.

Rhabdocline Syd. (1922), Rhytismatales (inc. sed.). 3, N. temp. *R. pseudotsugae* (needle cast of *Pseudotsuga menziesii*). See Parker & Reid (*CJB* **47**: 1533, 1969; infraspecific taxa), Sherwood-Pike *et al.* (*CJB* **64**: 1849, 1986), v. Vloten (*Proefschr. Landbouwhoogesch., Wageningen*, 1932).

Rhabdocystis Arnaud (1949) nom. inval. (? Myxomycota).

Rhabdodiscus Vain. (1921) = Ocellularia (Thelotremat.) fide Hale (1981).

Rhabdogloeopsis Petr. (1925), Mitosporic fungi, 6.A1.10. 1, USA.

Rhabdogloeum Syd. (1922), Mitosporic fungi, 6.A1.15. 3, N. Am.

Rhabdomyces Balbiani (1889), Mitosporic fungi, 1.?.?. 1, France.

Rhabdopsora Müll. Arg. (1888), Verrucariaceae (L). 1, Brazil. See Hawksworth (*SA* **9**: 24, 1991).

Rhabdopsoromyces Cif. & Tomas. (1953) ≡ Rhabdopsora (Verrucar.)

Rhabdospora (Durieu & Mont. ex Sacc.) Sacc. (1884) nom. cons., Mitosporic fungi, 4.C/E1.?. 150, widespr.

Rhabdosporium, see *Rabdosporium*.

Rhabdostroma Syd. & P. Syd. (1916) = Apiospora (Lasiosphaer.) fide Barr (1976).

Rhabdostromella Höhn. (1915), Mitosporic fungi, 8.A1.?. 1, Eur.

Rhabdostromellina Höhn. (1917) = Coleophoma (Mitosp. fungi) fide Sutton (1977).

Rhabdostromina Died. (1921) = Coleophoma (Mitosp. fungi) fide Petrak (*Ann. Myc.* **27**: 324, 1929).

Rhabdothyrella Höhn. (1917) = Leptothyrium (Mitosp. fungi) fide Clements & Shear (1931).

Rhabdothyrium Höhn. (1915) = Leptothyrium (Mitosp. fungi) fide Clements & Shear (1931).

Rhachomyces Thaxt. (1895), Laboulbeniaceae. *c.* 62, widespr. See Balazuc (*Ann. Soc. ent. Fr.* n.s. **6**: 677, 1970).

Rhacodiella Peyronel (1919) = Myrioconium (Mitosp. fungi) fide Kendrick & Carmichael (1973).

Rhacodiopsis Donk (1975) ≡ Racodium Fr. (Mitosp. fungi).

Rhacodium Pers. ex Wallr. (1833) , ≡ Racodium (Mitosp. fungi).

Rhacophyllus Berk. & Broome (1871), Coprinaceae. 1, Asia, Afr. See Petch (*TBMS* **11**: 238, 1926).

Rhadinomyces Thaxt. (1893) = Rhachomyces (Laboulben.).

rhagadiose, having deep cracks.

Rhagadolobium Henn. & Lindau (1896), Parmulariaceae. 10 (on *Pteridophyta*), esp. trop. See Dingley (*N.Z. Jl Bot.* **10**: 74, 1972; NZ), Swart (*TBMS* **91**: 581, 1988; *R. dicksoniifolium* on *Dicksonia*).

Rhagadostoma Körb. (1865), Nitschkiaceae. 1 (on lichens), Eur. See Hawksworth (*TBMS* **74**: 363, 1980).

Rhagidiasporidium Thor (1930) nom. dub. (? Fungi, inc. sed.).

Rhamphoria Niessl (1876), ? Diaporthales (inc. sed.). 9, mainly temp. See Sivanesan (*TBMS* **67**: 469, 1976).

Rhamphosphaeria Kirchst. (1936), Ascomycota (inc. sed.). 1, Eur.

Rhamphospora D.D. Cunn. (1888), Tilletiaceae. 1 (on *Nymphaceae*), widespr. See Reid (*TBMS* **52**: 25, 1969), Vánky (1987).

Rhaphidicyrtis Vain. (1921), Pyrenulales (inc. sed., L). 1, temp. See Aguirre-Hudson (*Bull. Br. Mus. nat. Hist., Bot.* **21**: 85, 1991).

Rhaphidophora Ces. & De Not. (1863) [non Hassk. (1842), *Araceae*] = Gaeumannomyces (Magnaporth.) p.p., Ophiosphaerella (Phaeosphaer.) p.p. etc. See Walker (1980).

Rhaphidospora Fr. (1849) [non Nees (1832), *Acanthaceae*] ≡ Rhaphidophora (Magnaporth.).

Rheophila Cépède & F. Picard (1907) = Peyritschiella (Laboulben.) fide Cépède & Picard (*Bull. Sci. Fr. Belg.* **42**: 251, 1908).

Rheosporangium Edson (1915) = Pythium (Pyth.) fide Fitzpatrick (1923).

Rheumatopeltis F. Stevens (1927) = Trabutia (Phyllachor.) fide v. Arx & Müller (1975).

Rhexoampullifera P.M. Kirk (1982), Mitosporic fungi, 1.C2.40. 1, UK.

Rhexocercosporidium U. Braun (1994), Mitosporic fungi, 1.C2.11. 1, Norway.

rhexolytic, secession of conidia involving the circumscissile splitting of the periclinal wall of the cell below the basal conidial septum (Cole & Samson, *Patterns of development in conidial fungi*, 1979) rather than the septum itself; fracture. Cf. schizolytic.

Rhexophiale Th. Fr. (1860) = Sagiolechia (Gomphill.).

Rhexosporium Udagawa & Furuya (1977), Sordariales (inc. sed.). 1, Japan.

Rhexothecium Samson & Mouch. (1975), Eremomycetaceae. 1, Egypt, Kenya. See Malloch & Sigler (*CJB* **66**: 1929, 1988).

Rhinocephalum Kamÿschko (1961) = Arthrinium (Mitosp. fungi) fide Kendrick & Carmichael (1973).

Rhinocladiella Nannf. (1934), Anamorphic Herpotrichiellaceae, 1.A1.10. Teleomorph *Capronia*. 4, widespr. See de Hoog (*Stud. Mycol.* **15**, 1977), Onofri & Castagnola (*Mycotaxon* **18**: 337, 1983; generic synonymy), Iwatsu *et al.* (*Mycotaxon* **28**: 199, 1987; conidiogenesis). See cellar fungus.

Rhinocladiella Kamÿschko (1960) [non Nannf. (1934)] = Chrysosporium (Mitosp. fungi) fide Kendrick & Carmichael (1973).

Rhinocladiopsis Kamÿschko (1961) = Chrysosporium (Mitosp. fungi) fide Carmichael (*in litt.*).

Rhinocladium Sacc. & Marchal (1885), Mitosporic fungi, 1.A2.1. 5, widespr.

Rhinodinomyces E.A. Thomas ex Cif. & Tomas. (1953) = Rinodina (Physc.).

rhinosporidiosis, polypoid growths in the nose and other organs of humans, horse, etc., caused by *Rhinosporidium seeberi*.

Rhinosporidium Minchin & Fantham (1905) ? Hyphochytriales (inc. sed.). 1 or 2 (on humans and other animals causing rhinosporidiosis, q.v.), widespr. See Ashworth (*Trans. R. Soc. Edinb.* **53**: 301, 1923), Stoddart *et al.* (*J. Med. Microbiol.* **20**: 'x', 1985).

Rhinotrichella G. Arnaud ex de Hoog (1977), Mitosporic fungi, 1.A2.10. 1, Japan. = Nodulisporium (Mitosp. fungi) fide Carmichael (*in litt.*).

Rhinotrichum Corda (1837) nom. dub. See Carmichael *et al.* (1980).

Rhipidiaceae Sparrow ex Cejp (1959), Rhipidiales. 6 gen. (+ 2 syn.), 15 spp. Thallus with a basal cell, pedicellate segments, and thicker-walled constrictions; principal zoospores developed in zoosporangium; oosporogenesis periplasm.

Rhipidiales, Oomycota. 1 fam., 6 gen. (+ 2 syn.), 15 spp. Fam.:
Rhipidiaceae.

Rhipidiomyces Thaxt. (1926), Laboulbeniaceae. 1, S. Am.

Rhipidium Wallr. (1833) nom. rej. ≡ Schizophyllum (Schizophyll.).

Rhipidium Cornu (1871) nom. cons., Rhipidiaceae. 4, widespr.

Rhipidocarpon (Theiss.) Theiss. & Syd. (1915), Parmulariaceae. 1, Java, Philipp.

Rhipidocephalum Trail (1888), Mitosporic fungi, 1.A1.?. 1, UK.

Rhipidonema Mattir. (1881) = Dictyonema (Merul.) fide Parmasto (1978).

Rhipidonematomyces Cif. & Tomas. (1954) ≡ Rhipidonema (Merul.).

Rhizellobiopsis Hovasse (1926), Algae.

Rhizidiaceae, see Chytridiaceae.

Rhizidiocystis Sideris (1929), ? Chytridiales (inc. sed.). 1, Hawaii.

Rhizidiomyces Zopf (1884), Rhizidiomycetaceae. 12, widespr. See Sparrow (1960: 751; key), Fuller (*Am. J. Bot.* **49**: 64, 1962; culture), Dick (*in press*; key).

Rhizidiomycetaceae Karling (1943), Hyphochytriales. 3 gen. (+ 1 syn.), 15 spp. See Dick (*in press*; key).

Rhizidiomycopsis Sparrow (1960) = Rhyzidiomyces (Rhyzidiomycet.) fide Dick (*in press*).

Rhizidiopsis Sparrow (1933) = Podochytrium (Chytrid.) fide Sparrow (1960).

Rhizidium A. Braun (1856), Chytridiaceae. *c.* 17, widespr. See Sparrow (1960; key), Dick (*in press*; key).

rhizina (pl. -ae), a root-like hair or thread; the attachment organs of many foliose lichens (e.g. *Parmelia*); rhizine; they may be divided into several types, for details see Gyelnik (*Bot. Közl.* **24**: 122, 1927), Hale & Kurokawa (*Contr. U.S. natn Herb.* **36**: 122, 1964), Hannemann (*Bibl. Lich.* **1**, 1973). Cf. hyphal net.

Rhizina Fr. (1815), Helvellaceae. 1 or 2, N. temp.

rhizinose-strand ('Rhizinenstränge'), a rhizine-like organ of attachment in squamulose (placodioid) lichens (e.g. *Toninia*), which is tough and much branched. At least 3 types occur. See Poelt & Baumgartner (*Öst. bot. Z.* **111**: 1, 1964), Hannemann (*Bibl. Lich.* **1**, 1973). Cf. hyphal net.

Rhizoblepharia Rifai (1968), Otideaceae. 2, Australia, Jamaica. See Erb (*Phytologia* **24**: 11, 1972).

Rhizoblepharis P.A. Dang. (1900) nom. dub. ('phycomycetes').

Rhizocalyx Petr. (1928), Leotiaceae. 1 (on *Abies*), Siberia.

Rhizocarpaceae M. Choisy ex Hafellner (1984), Lecanorales (±L). 4 gen. (+ 15 syn.), 203 spp. Thallus crustose or squamulose, rarely absent; ascomata apothecial, sometimes angular or elongated, sessile or immersed, flat or domed, black, marginal tissue variably developed; interascal tissue of branched and anastomosed agglutinated paraphyses, usually not swollen at the apices, usually with a pigmented epithecium; asci clavate, with a weakly I+ apical cap, with a discoid more strongly staining apical region and an outer I+ gelatinized layer; ascospores hyaline or brown, 1-septate to muriform, sometimes with a sheath. Anamorphs pycnidial. Lichenized with green algae, occ. lichenicolous, widespr.

Rhizocarpon Ramond ex DC. (1805), Rhizocarpaceae (±L). *c.* 200, cosmop., mainly temp., Arctic and Antarctic. See Feurer (*Ber. bayer. bot. Ges.* **49**: 59, 1978; C. Eur.), Geyer *et al.* (*Pl. Syst. Evol.* **145**: 41, 1984; key *R. superficale* group), Hertel & Leuckert (*Herzogia* **5**: 25, 1979; lichenicolous spp.), Honegger (*Lichenologist* **12**: 157, 1980, ultrastr., posn), Innes (*Boreas* **14**: 83, 1985; lichenometry nomencl.), Poelt (*Mitt. bot. Staatssamml. München* **29** : 515, 1990; key 28 parasitic spp.), Runemark (*Op. bot . Soc. Lund.* **2** (1-2), 1956; yellow spp. Eur.), Timdal & Holtan-Hartwig (*Graphis Scripta* **2** : 41, 1988; key Scandinavian spp.), Thomson (*Nova Hedw.* **14**: 421, 1967; arctic); map lichen.

Rhizocarponomyces Cif. & Tomas. (1953) = Rhizocarpon (Rhizocarp.).

Rhizoclosmatium H.E. Petersen (1903), Chytridiaceae. 4, widespr.

Rhizoctonia DC. (1816), Anamorphic Corticiaceae, Ceratobasidiaceae, Otideaceae, 1/9.-.-. Teleomorphs *Ceratobasidium, Thanatephorus, Tricharina*, etc. 7, widespr. See Parmeter (Ed.) (*Rhizoctonia solani. Biology and pathogenicity*, 1970), Moore (*Mycotaxon* **29**: 91, 1987; segregated genera), Carling *et al.* (*Phytopath.* **77**: 1609, 1987; new anastomosis group AG-9), Mordue *et al.* (*MR* **92**: 78, 1989; integrated approach to taxonomy), Ogoshi (*Ann. Rev. Phytopath.* **25**: 125, 1987; ecology and pathology), Liu *et al.* (*Phytopath.* **79**: 1205, 1989; isozyme phylogeny), Yang *et al.* (*MR* **93**: 429, 1989; sclerotial morphogenesis), Andersen (*Mycotaxon* **37**: 25, 1990; hyphal morphology), Cruickshank (*MR* **94**: 938, 1990; pectic zymograms), Liu *et al.* (*Can. J. Pl. Path.* **12**: 376, 1990; *R. solani* group 2 relationships by isozyme analysis), Jabaji-Hare *et al.* (*Can. J. Pl. Path.* **12**: 393, 1990; anastomosis group relationships using cloned DNA probes), Cubeta *et al.* (*Phytopath.* **81**: 1395, 1991; anastomosis groups characterized by RA analysis of amplified rRNA gene), Wako *et al.* (*J. Gen. Microbiol.* **137**: 2817, 1991; unique DNA plasmid pRS64 associated with chromosomal DNA), Sneh *et al.* (*Identification of Rhizoctonia species*, 1991), Burpee & Martin (*Pl. Dis.* **76**: 112, 1991; spp. associated with turfgrasses), Kellens & Peumans (*MR* **95**: 1235, 1991; biochemistry and serology of lectins), Johnk & Jones (*Phytopathol.* **83**: 278, 1993; analysis of fatty acids), Vilgalys & Cubeta (*Ann. Rev. Phytopathol.* **32**: 135, 1994; molecular systematics).

Rhizodiscina Hafellner (1979), Patellariaceae. 1 (on wood), temp.

Rhizogaster Reinsch (1875) nom. dub. (? Fungi).

Rhizogene Syd. & P. Syd. (1921), Venturiaceae. 1 (on *Symphoricarpos*), N. Am. See Hafellner (*Nova Hedw.* **27**: 903, 1976).

Rhizohypha Chodat & Sigr. (1911), Mitosporic fungi, 9.-.-. 1 (mycorrhizal), Eur.

rhizoid, a root-like structure consisting of anucleate filaments; branched, extension of a chytrid thallus acting as a feeding organ (Karling, *Am. J. Bot.* **19**: 44, 1932), cf. haustorium, rhizina, and holdfast; **-al**, having, or made up of, rhizoids.

Rhizolecia Hertel (1984), ? Lecideaceae (L). 1, NZ.

rhizomorph, a root-like aggregation of hyphae having a well-defined apical meristem (cf. mycelial cord) and frequently differentiated into a rind of small dark-coloured cells surrounding a central core of elongated colourless cells. See Snider (*Mycol.* **51**: 693, 1961; *Armillaria mellea*), Jacques-Felix (*BSMF* **83**: 1, 1967; *Marasmius*, **84**: 161, 1969; agarics).

Rhizomorpha Roth (1791) nom. dub.; rhizomorphs of *Armillaria mellea* fide Donk (*Taxon* **11**: 97, 1962).

Rhizomorpha Ach. (1809) nom. dub. (Fungi, inc. sed., ?L).

Rhizomorphites Göpp. (1848), Fossil fungi. 3 (Carboniferous, Jurassic), Eur., USA.

Rhizomucor Lucet & Costantin (1900), Mucoraceae. 5, widespr. See Schipper (*Stud. Mycol.* **17**: 53, 1978), Zheng & Chen (*Mycosystema* **4**: 45, 1991, **6**: 1, 1993; key), Deploey (*Mycol.* **84**: 77, 1992; spore germination).

rhizomycelium, a rhizoidal system which resembles mycelium, e.g. thallus of the *Cladochytriaceae* (Karling, *Am. J. Bot.* **19**: 53, 1932).

Rhizomyces Thaxt. (1896), Laboulbeniaceae. 9, Afr.

Rhizomyxa Borzí (1884) ? = Ligniera (Plasmodiophor.) etc. fide Karling (1968).

Rhizonema Thwaites (1849) = Dictyonema (Merul.).

Rhizophagites Rosend. (1943), Fossil fungi. 3 (Cretaceous, Pleistocene), N. Am. See Butler (*TBMS* **22**: 274, 1939). = Glomus (Glom.) fide Yao *et al.* (1995).

Rhizophagus P.A. Dang. (1896) ? = Glomus (Glom.) fide Gerdemann & Trappe (1974).

Rhizophidites Daugherty (1941), Fossil fungi (Chytridiomycetes). 1 (Triassic), USA.

Rhizophila K.D. Hyde & E.B.G. Jones (1989), Ascomycota (inc. sed.). 1 (marine), Seychelles.

Rhizophlyctis A. Fisch. (1891), ? Spizellomycetaceae. 9, cosmop. See Sparrow (1960: 438; key), Barr & Désaulniers (*CJB* **64**: 561, 1986).

Rhizophoma Petr. & Syd. (1927) = Rhizosphaera (Mitosp. fungi) fide Kobayasi (*Bull. Govt For. Exp. Stn* **204**: 91, 1969).

Rhizophydium Schenk (1858), Chytridiaceae. 100, cosmop. See Sparrow (1960: 231; key).

Rhizophyton Zopf (1887) = Rhizophydium (Chytrid.) fide Minden (1911).

Rhizoplaca Zopf (1905), Lecanoraceae (L). 7, Eur., Asia, N. Am. See Leuckert *et al.* (*Nova Hedw.* **28**: 71, 1977), McCune (*Bryol.* **90**: 6, 1987; N. Am.), Roux *et al.* (*CJB* **71**: 1660, 1993; posn), Weber (*Mycotaxon* **8**: 559, 1979), Wei (*Acta Mycol. Sin.* **3**: 207, 1984; key 4 spp. China).

rhizoplane, the surface of a root. See Lynch (Ed.) (*The rhizosphere*, 1990).

rhizoplast, see blepharoplast.

Rhizopodella (Cooke) Boud. (1885) ≡ Plectania (Sarcosomat.).

rhizopodium, a branched process (pseudopodium) from a plasmodium. Cf. filopodium.

Rhizopodomyces Thaxt. (1931), Laboulbeniaceae. 7, Am. See Benjamin (*Aliso* **9**: 379, 1979; key).

Rhizopodopsis Boedijn (1959), Mucoraceae. 1, Java.

Rhizopogon Fr. & Nordholm (1817), Rhizopogonaceae. *c.* 150, widespr. See Smith & Zeller (*Mem. N.Y. bot. Gdn* **14**(2), 1966; key N. Am. spp.), Harrison & Smith (*CJB* **46**: 881, 1968), Talbot (*Persoonia* **7**: 339, 1973; gen. concept), Hosford & Trappe (*Trans. mycol. Soc. Japan* **29**: 63, 1988; Japan).

Rhizopogonaceae Gäum. & C.W. Dodge (1928), Boletales. 3 gen. (+ 4 syn.), 154 spp. Gasterocarp hypogeous; no columella; spores statismosporic, stramineous, thin-walled, smooth.
Lit.: Bruns *et al.* (*Nature* **339**: 140, 1989; relat. with *Suillus*).

Rhizopogoniella Soehner (1953) = Hymenogaster (Hymenogastr.) fide Gross *et al.* (*Beih. Z. Mykol.* **2**: 210, 1980).

Rhizopus Ehrenb. (1821) nom. cons., Mucoraceae. *c.* 10, widespr. *R. stolonifer* (syn. *R. nigricans*) is a common saprobe and facultative parasite of mature fruits and vegetables. See Dabinett & Wellman (*CJB* **51**: 2053, 1973; numerical taxonomy), Gauger (*J. Gen. Microbiol.* **101**: 211, 1977; genetics), Ellis (*Mycol.* **77**: 243, 1985, **78**: 508, 1986; DNA), Liou *et al.* (*Trans. Mycol. Soc. Japan* **32**: 535, 1991; SEM of sporangiospores), Seviour *et al.* (*CJB* **61**: 2374, 1983), Schipper (*Stud. Mycol.* **25**: 1, 1984), Schipper & Stalpers (*Stud. Mycol.* **25**: 20, 1984), Schipper *et al.* (*J. Gen. Microbiol.* **131**: 2359, 1985; hybridization), Sevior *et al.* (*TBMS* **84**: 701, 1985), Yuan & Jong (*Mycotaxon* **20**: 397, 1984), Polonelli *et al.* (*Ant. v. Leeuwenhoek* **54**: 5, 1988; antigens), Huang & Yu (*Mycosystema* **1**: 61, 1988; electrophoresis), Jong & McManus (*Mycotaxon* **47**: 161, 1993; computer coding).

Rhizosiphon Scherff. (1926), Chytridiales (inc. sed.). 2, Eur.

Rhizosphaera L. Mangin & Har. (1907), Anamorphic Venturiaceae, 4.A1.15. Teleomorph *Phaeocryptopus*.

4 (on conifers), widespr. *R. kalkhoffii* (on pine needles). See Sutton (1980), Petrak (*Ann. Myc.* **36**: 9,
1938).
Rhizosphaerella Höhn. (1917) ? = Phoma (Mitosp.
fungi) fide Sutton (1977).
Rhizosphaerina B. Sutton (1986), Mitosporic fungi,
8.A2.15. 1, Australia, S.E. Asia.
rhizosphere, the region immediately surrounding a root
and influenced by its presences; the microbiota is frequently richer than that of the soil away from a root.
See Katznelson *et al.* (*Bot. Rev.* **14**: 543, 1948).
Rhizosporium Rabenh. (1844) ? = Phloeoconis
(Mitosp. fungi).
Rhizostilbella Wolk (1914), Anamorphic Ascobo-
laceae, 2.?A1.?. Teleomorph *Ascobolus*. 1, Java. See
also Seifert (*Stud. Mycol.* **27**: 1, 1985; used for
anamorph, *Sphaerostilbe repens*, of *Nectria mauriti-
icola*).
Rhizostroma Fr. (1819) nom. dub. (? Mitosp. fungi).
See Donk (*Taxon* **11**: 98, 1962).
Rhizotexis Theiss. & Syd. (1917), Englerulaceae. 1,
Natal.
Rhizothyrium Naumov (1915), Mitosporic fungi,
5.C1.?. 2, Russia, Chile.
Rhodesia Grove (1937), Mitosporic fungi, 6.A1.15. 1,
Germany.
Rhodesiopsis B. Sutton & R. Campb. (1979),
Mitosporic fungi, 6.A1.19. 1, UK, Australia.
Rhodoarrhenia Singer (1964), Cyphellaceae. 6, trop.
See Singer (*Sydowia* **17**: 142, 1964; key).
Rhodobolites Beck (1923) ? = Tylopilus (Strobilo-
mycet.) fide Singer (1945).
Rhodocarpon Lönnr. (1858) ≡ Endocarpon (Verrucar.).
Rhodocephalus Corda (1837) ? = Aspergillus (Mitosp.
fungi) fide Thom (1965).
Rhodococcus Zopf (1891), *Actinomycetes*, q.v.
Rhodocollybia Singer (1939) = Collybia (Tricholomat.)
fide Singer (1951).
Rhodocybe Maire (1926), Entolomataceae. 25, widespr.
See Pegler & Young (*Kew Bull.* **30**: 19, 1975; key 9
Br. spp.), Horak (*Sydowia* **31**: 58, 1978, *N.Z. Jl Bot.*
17: 275, 1979; key 9 NZ spp.), Baroni (*Beih. Nova
Hedw.* **67**, 1982; revision), Baroni & Halling (*Mycol.*
84: 419, 1992; key N. Am. spp.).
Rhodocybella T.J. Baroni & R.H. Petersen (1987),
Entolomataceae. 1, N. Am.
Rhodocyphella W.B. Cooke (1961), Cyphellaceae. 2,
USA, Bermuda, Sri Lanka. = Stigmatolemma
(Tricholomat.) fide Donk (1962).
Rhodofomes Kotl. & Pouzar (1990), Coriolaceae. 3,
widespr.
Rhodogaster E. Horak (1964), Richoniellaceae. 1,
Chile.
Rhodomyces Wettst. (1885) ? = Candida (Mitosp.
fungi).
Rhodopaxillus Maire (1913) = Lepista (Tricholomat.)
fide Singer (1951).
Rhodopeziza Hohmeyer & J. Moravec (1994),
Pezizaceae. 1, Tierra del Fuego.
Rhodophana Kühner (1971), Entolomataceae. 1, Eur.
Rhodophyllus Quél. (1886), Entolomataceae. *c.* 600,
cosmop. See Singer (1975), Hesler (*Entoloma in
southeastern North America*, 1967; key 200 spp.),
Largent & Benedict (*Madrono* **21**: 32, 1971; gen.
defin.), Kühner (*BSMF* **93**: 445, 1978; alpine spp.),
Romagnesi (*Bull. Soc. linn. Lyon* **43**: 325, 1974),
Romagnesi & Gilles (*Beih. Nova Hedw.* **59**, 1979; key
200 spp. Ivory Coast etc.). ≡ Entoloma (Entolomat.).
Rhodoporus Quél. ex J. Bataille (1908), Boletaceae. 1,
Eur.
Rhodoscypha Dissing & Sivertsen (1983), Otideaceae.
1, Eur., N. Am., India. See Lohmeyer (*Z. Mykol.* **50**:
147, 1984).

Rhodoseptoria Naumov (1913) = Polystigmina
(Mitosp. fungi) fide Clements & Shear (1931).
Rhodosporidium Banno (1967), Sporidiobolaceae. 6
(often from sea water), widespr. See Fell *et al.* (*Can.
J. Microbiol.* **19**: 643, 1973; key). Anamorphs
Cryptococcus (?), *Rhodotorula*.
rhodosporous, having light-red spores.
Rhodosporus J. Schröt. (1889) ≡ Pluteus (Plut.).
Rhodosticta Woron. (1911), Anamorphic Phyl-
lachoraceae, 8.A1.15. Teleomorph *Stigmatula*. 3,
Asia, USA.
Rhodotarzetta Dissing & Sivertsen (1983), Otideaceae.
1 (on burnt ground), Eur.
Rhodothallus Bat. & Cif. (1959), Mitosporic fungi, 5.-.-
. 1, Brazil.
Rhodothrix Vain. (1921) = Nectria (Hypocr.) fide
Santesson (*Svensk bot. Tidskr.* **43**: 547, 1949).
Rhodotorula F.C. Harrison (1927), Mitosporic fungi,
1.A1.?. 34 (fide Barnett *et al.*, 1990), widespr. See
Lodder (1970), Hamamoto *et al.* (*J. Gen. Appl.
Microbiol.* **32**: 315, 1986; DNA base composition),
Hamamoto *et al.* (*J. Gen. Appl. Microbiol.* **33**: 57,
1987; DNA-DNA reassociation), Westhuizen *et al.*
(*Syst. Appl. Microbiol.* **10**: 35, 1987; long chain fatty
acid composition), Weijman *et al.* (*Ant. v.
Leeuwenhoek* **54**: 55, 1988; emend.), Golubev
(*Mikrobiol.* **58**: 99, 1989; killer toxins), Almendros *et
al.* (*MR* **94**: 211, 1990; pyrolysis-gas chromatogra-
phy-mass spectrometry), van der Westhuizen (*Syst.
Appl. Microbiol.* **14**: 282, 1991; cellular long-chain
fatty acid composition, taxonomy).
Rhodotus Maire (1926), Tricholomataceae. 1, Eur., N.
Am. See Kühner (1980), Pegler & Young (*Kew Bull.*
30: 19, 1975; spore ornament.).
Rhombiella Liro (1939), Anamorphic Ustilaginaceae. 1,
Eur. Teleomorph *Ustilago*.
rhomboidal, resembling an equilateral not right-angled
parallelogram (a rhomboid); quadrangular (Fig.
37.22).
Rhombostilbella Zimm. (1902), Mitosporic fungi,
2.A1.10. 2, Java, Brazil. See Pirozynski (*Mycol. Pap.*
90, 1963; key).
Rhopalidium Mont. (1856) = Alternaria (Mitosp. fungi)
fide Höhnel (*Sber. Akad. Wiss. Wien* **119**: 670, 1910).
Rhopaloconidium Petr. (1952), Mitosporic fungi,
3.C1.?. 1, USA.
Rhopalocystis Grove (1911) = Aspergillus (Mitosp.
fungi) fide Raper & Fennell (1965).
Rhopalogaster J.R. Johnst. (1902), Hysterangiaceae. 1,
USA.
Rhopalomyces Corda (1839), Helicocephalidaceae. 7,
widespr. See Costantin (*Bull. Soc. Bot. Fr.* **33**: 489,
1886), Thaxter (*Bot. Gaz.* **16**: 14, 1891), Berlese
(*BSMF* **10**: 94, 1892), Marchal (*Rev. Mycol.* **15**: 7,
1893), Drechsler (*Bull. Torrey Bot. Cl.* **82**: 473, 1955),
Ellis (*Mycol.* **55**: 183, 1963), Barron (*CJB* **51**: 2505,
1973, *Mycol.* **72**: 427, 1980; biology), Stalpers (*Proc.
Ned. Akad. Wet.* C **77**: 383, 1974), Benjamin (1979),
Cano *et al.* (*Nova Hedw.* **49**: 427, 1989).
Rhopalomyces Harder & Sörgel (1938) [non Corda
(1839)] = Blastocladiella (Blastoclad.) fide Couch &
Whiffen (1942).
Rhopalophlyctis Karling (1945), Chytridiaceae. 1, Am.
Rhopalopsis Cooke (1883) ≡ Kretzschmaria (Xylar.).
Rhopalospora, see *Ropalospora*.
Rhopalostroma D. Hawksw. (1977), Xylariaceae. 9,
Afr., Asia. See Hawksworth & Whalley (*TBMS* **84**:
560, 1985). Anamorph *Nodulisporium*-like.
Rhopographella (Henn.) Sacc. & Trotter (1913), Asco-
mycota (inc. sed.). 1, Brazil.
Rhopographina Theiss. & Syd. (1915) = Ophiodothella
(Phyllachor.) fide v. Arx & Müller (*Stud. Mycol.* **9**,
1975).

Rhopographus Nitschke ex Fuckel (1870), ? Phaeosphaeriaceae. 6 (on *Pteridophyta*), widespr. See Barr (*Mycotaxon* **43**: 371, 1992; posn), Obrist (*Phytopath. Z.* **35**: 367, 1959).

Rhydochytrium Lagerh. (1893) Algae (? Chytridiales). 1, N. temp.

Rhymbocarpus Zopf (1896), Ascomycota (inc. sed.). 1 (on lichens, esp. *Rhizocarpon*), Eur. See Coppins *et al.* (*SA* **10**: 51, 1991).

Rhymovis Pers. ex Rabenh. (1844) ≡ Paxillus (Paxill.).

Rhynchodiplodia Briosi & Farneti (1906), Mitosporic fungi, 4.B2.?. 1, Italy.

Rhynchogastrema B. Metzler & Oberw. (1989), Rhynchogastremataceae. 1, Eur.

Rhynchogastremataceae Oberw. & B. Metzler (1989), Tremellales. 1 gen., 1 spp. Basidioma absent or synnematous; metabasidium with quadripartite apex; spores statismosporic, brown, encrusted, fused into tetrad.

Rhynchomelas Clem. (1909) = Melanospora (Ceratostomat.) fide v. Arx & Müller (1954).

Rhynchomeliola Speg. (1884), Trichosphaeriaceae. 1, S. Am.

Rhynchomyces Willk. (1866), Mitosporic fungi, 1.C2.?. 1, Eur.

Rhynchomyces Sacc. & Marchal ex Marchal (1885) ≡ Mycorhynchus (Pyxidophor.).

Rhynchonectria Höhn. (1902) ? = Pyxidiophora (Pyxidophor.). See Hawksworth & Webster (*TBMS* **68**: 329, 1977).

Rhynchophoma P. Karst. (1884) = Ceratocystis (Microascales). See Petrak (*Sydowia* **7**: 298, 1953).

Rhynchophoromyces Thaxt. (1908), Ceratomycetaceae. 8, Am., Afr., Pacific, Eur.

Rhynchophorus Hollós (1926) = Ceratopycnis (Mitosp. fungi) fide Clements & Shear (1931).

Rhynchoseptoria Unamuno (1940), Mitosporic fungi, 4.C1.?. 1, Morocco.

Rhynchosphaeria (Sacc.) Berl. (1891), Ascomycota (inc. sed.). 7, widespr.

Rhynchosporina Arx (1957), Mitosporic fungi, 1.A1.15. 1, N. Am.

Rhynchosporium Heinsen ex A.B. Frank (1897), Mitosporic fungi, 1.B1.19. 2, temp. *R. secalis* (barley and rye leaf blotch). See Owen (*TBMS* **41**: 99, 1958, **46**: 604, 1963), Williams & Owen (*TBMS* **60**: 223, 1973; physiol. specialization on barley), Ryan *et al.* (*Rhynchosporium secalis. A keyword index to the literature*, 1987), Goodwin *et al.* (*Phytopath.* **80**: 1330, 1990; nomencl. *R. secalis* pathotypes), Beer (*Zentralbl. Mikrobiol.* **146**: 339, 1991; review *R. secalis*), Goodwin *et al.* (*MR* **97**: 49, 1993; isozyme variation).

rhynchosporous, having beaked spores.

Rhynchostoma P. Karst. (1870), ? Trichosphaeriaceae. 10, widespr. Anamorph *Arthropycnis*. See Constantinescu & Tibell (*Nova Hedw.* **55**: 169, 1992).

Rhynchostomopsis Petr. & Syd. (1923) = Amphisphaeria (Amphisphaer.) fide Aptroot (*Stud. Mycol.* **37**, 1995).

Rhynchotheca Kleb. (1933) [non Ruiz & Pav. (1794), *Geraniaceae*] nom. dub. fide Sutton (*Mycol. Pap.* **141**, 1977).

Rhyncomeliola, see *Rhynchomeliola*.

Rhyparobius, see *Ryparobius*.

Rhysotheca G.W. Wilson (1907) = Plasmopara (Peronospor.) fide Fitzpatrick (1930).

Rhytidenglerula Höhn. (1918), Englerulaceae. *c.* 10, widespr. Anamorph *Capnodiastrum*.

Rhytidhysterium Sacc. (1883) ≡ Rhytidhysteron (Patellar).

Rhytidhysteron Speg. (1881), Patellariaceae. 2, mainly trop. See Samuels & Müller (*Sydowia* **32**: 277, 1979).

Rhytidiella Zalasky (1968), ? Cucurbitariaceae. 3, Canada, Sweden. See Aguirre-Hudson (*Bull. Br. Mus. nat. Hist., Bot.* **21**: 85, 1991). Anamorph *Phaeoseptoria*.

Rhytidocaulon Nyl. ex Elenkin (1916) nom. rej. = Letharia (Parmel.).

Rhytidopeziza Speg. (1885) = Rhytidhysteron (Patellar.) fide v. Arx & Müller (1975).

Rhytidospora Jeng & Cain (1977), Ceratostomataceae. 2, Venezuela, Spain. See Guarro (*Mycol.* **75**: 927, 1983).

Rhytisma Fr. (1818), Rhytismataceae. 15, widespr. *R. acerinum* and *R. punctatum* (tar spot of *Acer*), *R. salicinum* (of *Salix*). See Duravetz & Morgan-Jones (*CJB* **49**: 1267, 1971; ontogeny), Rath (*Rivista Micol.* **35**: 43, 1992; key 6 Italian spp.), air pollution.

Rhytismataceae Chevall. (1826) nom. cons., Rhytismatales. 43 gen. (+ 26 syn.), 344 spp. Ascomata apothecial, immersed in plant tissue or an often well-developed clypeate stroma, often elongated, opening by radial or longitudinal splits; upper wall usually black, thick-walled, composed of fungal and degraded plant cells; lower wall variously developed, sometimes black; interascal tissue of simple paraphyses, sometimes anastomosing near the base, often swollen at the apex, often immersed in a gelatinous matrix, occasionally absent; asci cylindrical, usually thin-walled, not fissitunicate, usually not thickened at the apex, occasionally with a minute refractive pore, not blueing in iodine, often splitting irregularly; ascospores almost always hyaline, usually aseptate, often elongate, often with mucous sheaths. Anamorphs coelomycetous, usually spermatial. Biotrophic, necrotrophic or saprobic on leaves and bark, esp. temp.

Lit: Cannon & Minter (*Mycol. Pap.* **155**, 1986; Indian subcont., key 14 gen.), Johnston (*Mycotaxon* **52**: 221, 1994, ascospore sheaths; *N.Z. Jl Bot.* **29**: 395, 405, 1991, NZ spp.), Minter (*in* Peterson (Ed.), *Recent research on conifer needle diseases*: 71, 1986; illustr. 14 gen.).

Rhytismatales, Ascomycota. 3 fam., 71 gen. (+ 36 syn.), 411 spp. Ascomata apothecial, immersed, sometimes erumpent, opening by longitudinal or radial splits, often within black clypeate pseudostromatic tissue; interascal tissue of simple paraphyses, often anastomosing near the base, often with mucous coating, often swollen at the apices; asci cylindrical, thin-walled, usually not differentiated at the apex, not blueing in iodine, often releasing spores through an irregular split; ascospores usually hyaline and aseptate, often elongated, often with a mucous sheath. Anamorphs coelomycetous. Saprobes and necrotrophic parasites, sometimes endophytic, on leaves, also on bark and wood, mainly temp. Fams:
(1) **Ascodichaenaceae**.
(2) **Cryptomycetaceae**.
(3) **Rhytismataceae** (syn. *Phacidiaceae* auct.).

Lit.: Cannon & Minter (*Mycol. Pap.* **155**: 1986; Indian subcont.), Darker (*Contr. Arnold Arb.* **1**, 1932; on conifers, *CJB* **45**: 1399, 1967; gen. revis.), Hunt & Ziller (*Mycotaxon* **6**: 481, 1978; host-gen. keys on conifers), Korf (1973; gen. key), Nannfeldt (1932), Livsey & Minter (*CJB* **72**: 549, 1994; circumscription, fams), Minter (*in* Capretti *et al.* (Eds), *Shoot and foliage diseases in forest trees*: 65, 1995; Eur. spp. on conifers, illustr., key, 20 gen.), Pirozynski & Weresub (*in* Kendrick (Ed.), *The whole fungus* **1**: 93, 1979; biogeogr.), Sherwood (*Occ. Pap. Farlow Herb.* **15**, 1980; key 13 gen.), Tehon (*Ill. Biol. Monogr.* **13** (4), 1935; *Lophodermium* s.l.), Minter & Cannon (*TBMS* **83**: 65, 1984; ascospore discharge).

Rhytismella P. Karst. (1884) ≡ Cliostomum (Bacid.).

Rhytismites Mesch. (1892), Fossil fungi. 27 (Cretaceous, Tertiary), Eur.

Rhytismopsis Geyl. (1887), Fossil fungi. 1 (Eocene), Indonesia. = Rhytismites (Fossil fungi) fide Meschinelli (1892).

Ribaldia Cif. (1954) = Asteroma (Mitosp. fungi) fide Sutton (1977).

Ricasolia De Not. (1846) = Lobaria (Lobar.).

Ricasolia A. Massal. (1855) = Placolecania (Bacid.).

Ricasoliomyces E.A. Thomas ex Cif. & Tomas. (1953) ≡ Ricasolia (Lobar.).

Riccoa Cavara (1903) = Heydenia (Mitosp. fungi) fide Höhnel (*Sber. Akad. Wiss. Wien* **124**: 56, 1915).

Richonia Boud. (1885), Zopfiaceae. 1, France. See Hawksworth (*CJB* **57**: 91, 1979).

Richoniella Costantin & L.M. Dufour (1916), Richoniellaceae. 5, cosmop. See Dring & Pegler (*Kew Bull.* **32**: 563, 1978; key).

Richoniellaceae Jülich (1982), Agaricales. 2 gen. (+ 1 syn.), 6 spp. Gasterocarp loculate; spores faceted, pink.

Rickella Locq. (1952) ≡ Volvolepiota (Agaric.).

Rickenella Raithelh. (1973), Tricholomataceae. 4, cosmop.

Rickia Cavara (1899), Laboulbeniaceae. 122, mainly trop.

Rickiella Syd. (1904) = Phillipsia (Sarcoscyph.) fide Korf (*Austr. Jl Bot., Suppl.* **10**: 77, 1983). See Pfister (*Mycotaxon* **29**: 329, 1987).

Riclaretia Peyronel (1915), Mitosporic fungi, 3.A1.15. 1, Eur.

Ricnophora Pers. (1825) = Phlebia (Merul.) fide Saccardo (1957).

Riddlea C.W. Dodge (1953) = Laurera (Trypethel.) fide Harris (*Acta Amazon.* Suppl. **14**: 55, 1986).

RIEC (Revised Index of Ecological Continuity). The percentage occurrence of up to a maximum of 20 out of a total list of 30 selected old-forest indicator lichens that require a continuity of mature trees to persist in a site. This value has proved useful in identifying woodland of particular antiquity and long environmental continuity in the UK. See Rose (*in* Brown *et al.* (Eds), *Lichenology: progress and problems*: 279, 1976). An *NIEC* (New Index of Ecological Continuity) for use in wider areas of W. Europe uses 'main' (counted to 70) plus 'bonus' species to give a *T* value (Rose, *in* Bates & Farmer (Eds), *Bryophytes and lichens in a changing environment*: 211, 1992).

Riedera Fr. (1849), ? Leotiales (inc. sed.). 1, Russia. Nom. dub. fide Saccardo (*Syll. Fung.* **8**: 842, 1889).

Riessia Fresen. (1852), Mitosporic fungi, 2.G1.1. 4, Eur., S. Am., S.E. Asia, former USSR. See Goos (*Mycol.* **59**: 718, 1967; dikaryotic basidiomycete).

Riessiella Jülich (1985), Mitosporic fungi, 1.C2.1. 2 (with clamp connexions), S.E. Asia.

Rigidoporopsis I. Johans. & Ryvarden (1979), Coriolaceae. 1, Afr.

Rigidoporus Murrill (1905), Coriolaceae. *c.* 24, widespr. See Donk (*Persoonia* **4**: 341, 1966), Ryvarden (1978: 401; key 4 Eur. spp.), Ryvarden & Johansen (*Preliminary polypore flora of East Africa*: 537, 1980; key 6 Afr. spp.), Corner (*Beih. Nova Hedw.* **86**: 152, 1987; key Malaysia spp., *Mycologist* **9**: 127, 1995; gen. concept). See also record fungi.

Rikatlia P.F. Cannon (1993), ? Phyllachoraceae. 1, E. Afr.

Rileya A. Funk (1979), Mitosporic fungi, 4/8.C1.19. 1, Canada.

Rimbachia Pat. (1891), Tricholomataceae. 6, trop. See Singer (*Boln Soc. argent. Bot.* **10**: 210, 1963; key), Redhead (*CJB* **62**: 865, 1984; key).

Rimelia Hale & A. Fletcher (1990) Parmeliaceae. 12, mainly trop. S. Am. ? = Parmelia (Parmel.).

Rimeliella Kurok. (1991) Parmeliaceae. 8, mainly trop. Close to *Parmotrema* fide Elix (1993). ? = Parmelia (Parmel.).

Rimella Raf. (1819) nom. dub. (? 'gasteromycetes').

rimose (1) cracked; (2) (of a pileus surface), cracked; originally, cracked in all directions (the recommended usage); frequently, cracked by radial fissures as in *Inocybe*. Cf. rimulose.

rimose, having small cracks. Cf. rimose.

Rimula Velen. (1934), ? Leotiales (inc. sed.). 1, former Czechoslovakia.

Rimularia Nyl. (1868), Rimulariaceae (L). 13, temp. See Hertel & Rambold (*Bibl. Lich.* **38**: 145, 1990; key), Muhr & Tønsberg (*Nordic. Jl Bot.* **8**: 649, 1989).

Rimulariaceae Hafellner (1984), Lecanorales (L). 3 gen. (+ 5 syn.), 17 spp. Thallus crustose; ascomata apothecial, often contorted, black, sessile, without well-developed marginal tissue; interascal tissue of anastomosing paraphyses, the apices hardly swollen, sometimes evanescent, usually with a pigmented epithecium; asci with an I+ apical cap, a very well-developed less strongly-staining apical cushion and a strongly I+ funnel-shaped apical ring, with an outer layer of I+ gelatinized tissue; ascospores ellipsoidal, hyaline, aseptate, without a sheath. Anamorphs pycnidial. Lichenized with green algae, temp.

Lit.: Hertel & Rambold (*Bibl. Lich.* **38**: 145, 1989).

rind, sometimes used for the firm outer layer of a rhizomorph or other organ; cortex (q.v.).

ring, (1) see annulus (Fig. 4C); (2) (of liquid cultures, esp. of bacteria), growth at the surface, sticking to the glass; **- wall building**, see wall building.

ringworm, see tinea.

Rinia Penz. & Sacc. (1901) = Erikssonia (Phyllachor.) fide Petrak (*Ann. Myc.* **29**: 390, 1931).

Rinodina (Ach.) Gray (1821), Physciaceae (L). *c.* 200, cosmop. See Giralt *et al.* (*Mycotaxon* **50**: 47, 1994; pannarin-cont. spp.), Giralt & Matzer (*Lichenologist* **26**: 319, 1994; S. Eur.), Hecklau *et al.* (*Herzogia* **5**: 489, 1981; chemistry), Lamb (*Br. Antarct. Surv. Sci. Rep.* **61**, 1968; Antarctica), Mayrhofer (*J. Hattori bot. Lab.* **55**: 327, 1984; Eur., key 95 spp., *Beih. Nova Hedw.* **79**: 571, 1984; key 20 spp. Austral.), Mayrhofer & Poelt (*Bibl. Lich.* **12**, 1979; Eur. saxic. spp.), Poelt & Mayrhofer (*Sydowia Beih.* **8**: 312, 1979; spore types), Sheard (*Lichenologist* **3**: 328, 1967; Br. spp.).

Rinodinella H. Mayrhofer & Poelt (1978), Physciaceae (L). 2, Eur.

Rinomia Nieuwl. (1916) ≡ Morinia (Mitosp. fungi).

Riopa D.A. Reid (1969) = Poria (Coriol.) fide Donk (1974).

Ripartitella Singer (1947), Tricholomataceae. 1, trop. Am.

Ripartites P. Karst. (1879), Tricholomataceae. 4, N. Hemisph., Argentina. See Huijsman (*Persoonia* **1**: 335, 1960), Pegler & Young (*Kew Bull.* **29**: 659, 1974; spores).

Ripexicium Hjortstam (1995), Corticiaceae. 1, Solomon Isl.

rishitin (**rishitinol**), terpenoid phytoalexins (q.v.) from potato (*Solanum tuberosum*).

rivulose, marked with lines like little rivers.

Rizalia Syd. & P. Syd. (1914), Trichosphaeriaceae. 2 (on *Meliolaceae*), trop. See Pirozynski (*Kew Bull.* **31**: 595, 1977).

Rizaliopsis Bat., Castr., J.L. Bezerra & Matta (1964) = Rizalia (Trichosphaer.) fide Pirozynski (1977).

RNA, see Molecular biology.

Roannaisia T.N. Taylor (1994), Fossil fungi. 1 (Carboniferous), France.

Robakia Petr. (1952), Mitosporic fungi, 8.A1.?. 1, Norway.

Robergea Desm. (1847), Stictidaceae. 8, Eur., N. Am., Afr. See Sherwood (*Mycotaxon* **5**: 1, 1977).

Robertomyces Starbäck (1905) = Bagnisiella (Dothid.) fide Nannfeldt (1932).

Robillarda Castagne (1845) nom. rej. prop. = Pestalotiopsis (Mitosp. fungi) fide Nag Raj & Morgan-Jones (*Taxon* **21**: 535, 1972), Nag Raj (1993).

Robillarda Sacc. (1882), Mitosporic fungi, 4.B2.10. 4, widespr. See Nag Raj (1993).

Robillardiella S. Takim. (1943) [ante] nom. dub. (? Dothideales, inc. sed.).

Robincola Velen. (1947), ? Leotiales (inc. sed.). 1, former Czechoslovakia.

Robledia Chardón (1929) = Botryostroma (Ventur.) fide Müller & v. Arx (1962).

Roburnia Velen. (1947), ? Leotiales (inc. sed.). 1, former Czechoslovakia.

Roccella DC. (1805) nom. cons., Roccellaceae (L). *c.* 35, cosmop. See Ahti (*Taxon* **33**: 330, 1984; nomencl.), Darbishire (1898), Tavares (*Revta Fac. Cienc. Lisb.* sect. C **6**: 125, 1958); dyeing.

Roccellaceae Chevall. (1826), Arthoniales (L). 39 gen. (+ 63 syn.), 554 spp. Thallus crustose, often thin, or fruticose, usually pruinose, sometimes with a thick cortex; ascomata with well-developed walls, round or elongated, often in groups on a receptacle. Many coastal and trop.
    Lit.: Darbishire (*Bibl. bot.* **45**, 1898), Follmann (*Nova Hedw.* **31**: 285, 1979), Follman et al. (*Herzogia* **9**: 653, 1993; chemistry), Torrente & Egea (*Bibl. Lich.* **32**. 1989; Mediterranean, N. Afr.).

Roccellaria Darb. (1897), Roccellaceae (L). 1, S. Am.

Roccellina Darb. (1898), Roccellaceae (L). 23, Afr., N. & S. Am., Australia, Japan. See Tehler (*Op. Bot.* **70**, 1983; key), Tehler (*Acta bot. fenn.* **150**: 185, 1994; phylogeny).

Roccellinastrum Follmann (1968), ? Micareaceae (L). 1, Chile, Tasmania. See Kantvilas (*Lichenologist* **22**: 79, 1990).

Roccellodea Darb. (1932), ? Roccellaceae (L). 1, Galapagos Is. See Weber (*Mycotaxon* **27**: 451, 1986).

Roccellographa J. Steiner (1902), Roccellaceae (L). 1, Socotra.

Roccellographomyces Cif. & Tomas. (1953) ≡ Roccellographa (Roccell.).

Roccellomyces E.A. Thomas ex Cif. & Tomas. (1953) = Roccella (Roccell.).

Roccellopsis Elenkin (1929) = Roccella (Roccell.) fide Eriksson & Hawksworth (*SA* **6**: 150, 1987).

rock hair, pendent brown, grey to black species of *Bryoria* which resemble human hair.

rock tripe, edible lichens of the genus *Umbilicaria*; *U. esculenta*, 'Iwatake', is still eaten in Japan; for details of its use see Sato (*Nova Hedw.* **16**: 505, 1969), Mattick (*Nova Hedw.* **16**: 511, 1969).

rodlet, structural unit of conidial and some hyphal walls composed of particles *c.* 50 Å diam. arranged in linear series (Hess et al., *Mycol.* **60**: 290, 1968).

Rodwaya Syd. & P. Syd. (1901) = Gyrodon (Gyrodont.) fide Singer (1945).

Rodwayella Spooner (1986), Hyaloscyphaceae. *c.* 5, temp.

Roesleria Thüm. & Pass. (1877), ? Caliciales (inc. sed.). 1 (on *Vitis* roots), Eur., N. Am. See Arnaud (*Ann. Epiphyt.* **16**: 235, 1930), Nieder (*Pflanzenschitz* **3**: 24, 1987), Redhead (*CJB* **62**: 2514, 1985).

Roeslerina Redhead (1985), Caliciales (inc. sed.). 2, Eur., N. Am.

Roestelia Rebent. (1804), Anamorphic Uredinales. 14, widespr. Aecial states of *Gymnosporangium*. See Kern (*Gymnosporangium*, 1973).

roestelioid (of an aecium), long and tube-like, as in *Gymnosporangium*.

Rogersella Liberta & A.J. Navas (1978) = Hyphodontia (Hyphodermat.) fide Langer (*Bibl. Mycol.* **154**, 1994).

Rogersia Shearer & J.L. Crane (1976) = Filosporella (Mitosp. fungi) fide Crane & Shearer (*Mycotaxon* **6**: 27, 1977).

Rogersiomyces J.L. Crane & Schokn. (1978), Sporidiobolaceae. 1, USA.

rohr, extra-cellular infection apparatus of plasmodiophorids. See also schlauch and stachel.

Roigiella R.F. Castañeda (1984), Mitosporic fungi, 2.B1.10. 1, Cuba.

Rolfidium Moberg (1986), Bacidiaceae (L). 1, Sri Lanka.

Rollandina Pat. (1905) = Arachniotus (Gymnoasc.). See Roy et al. (*in* Subramanian (Ed.), *The taxonomy of fungi* **1**: 215, 1978), Ghosh et al. (*Mycotaxon* **10**: 21, 1979; SEM).

Romanoa Thirum. (1954) [non Trevis. (1848), *Euphorbiaceae*], Clavicipitaceae. 1, Italy.

Romellia Berl. (1900), Calosphaeriaceae. 3, Eur., N. Am. = Pleurostoma (Calosphaer.) fide Barr et al. (*Mycotaxon* **48**: 529, 1993).

Romellia Murrill (1904) ≡ Phaeolus (Coriol.).

Romellina Petr. (1955), Ascomycota (inc. sed.). 1 (on insects), Java.

Ronnigeria Petr. (1947), Leptopeltidaceae. 1, Eur. See Holm & Holm (*Bot. Notiser* **130**: 115, 1977).

root nodules, of legumes are caused by nitrogen-fixing bacteria of the genus *Rhizobium*; those of spp. of *Alnus*, *Elaeagnus*, *Hippophaë* and *Myrica* by members of the *Plasmodiophorales* fide Hawker & Fraymouth (*J. gen. Microbiol.* **5**: 369, 1951).

root rots of cereals, Simmonds (*Bot. Rev.* **7**: 308, 1941; **19**: 131, 1953); of certain non-cereal crops: Berkeley (*Bot. Rev.* **10**: 67, 1944). And see Garrett (*Root disease fungi*, 1944, *Biology of root-infecting fungi*, 1956, *Pathogenic root-infecting fungi*, 1970), take-all.

Ropalospora A. Massal. (1860), ? Fuscideaceae (L). 1, arctic-alpine. See Hertel (*Mitt. bot. StSamml., München* **17**: 537, 1981). = Fuscidea (Fuscid.) fide Purvis et al. (*Lichen flora of Great Britain and Ireland*, 1992).

Ropalosporia, see *Rhopalospora*.

roridins, terpinoid toxins of *Myrothecium roridum* and *M. verrucaria*; the cause of dendrochiotoxicosis (illthrift) in sheep, pigs, and humans.

roridous, covered with drops of liquid like dew.

Rosasporina Beneš (1956), Fossil fungi (*f. cat.*). 1 (Carboniferous), former Czechoslovakia.

Roscoepoundia Kuntze (1898), Mitosporic fungi, ?8.A1.?. 1, Eur.

Rosellinia De Not. (1844), Xylariaceae. 100, widespr. Root rots are caused by *R. aquila* (mulberry, *Morus*), *R. arcuata* (tea), *R. bunodes* (tropical crops), *R. necatrix* (with *Dematophora* state) (apple, vine (*Vitis*), etc.), *R. pepo* (cacao), *R. quercina* (oak, *Quercus*). See Dargan & Thind (*Mycol.* **71**: 1010, 1979; key Indian spp.), Francis (*Sydowia* **38**: 75, 1985; *R. necatrix* group, *TBMS* **87**: 397, 1986; conifer needle blights), Gonzalez & Rogers (*Mycotaxon* **53**: 115, 1995; key 10 spp. Mexico), Læssøe & Spooner (*Kew Bull.* **49**: 1, 1994), Petrini (*Sydowia* **44**: 169, 1993; key 20 temp. spp.), Petrini et al. (*Sydowia* **41**: 257, 1989; *R. mammaeformis*-group). Anamorph *Dematophora*.

Roselliniella Vain. (1921), Sordariales (inc. sed.). 12 (lichenicolous), Eur., Asia, C. & S. Am., NZ. See Matzer & Hafellner (*Bibl. Lich.* **37**, 1990).

Roselliniomyces Matzer & Hafellner (1990), Sordariales (inc. sed.). 1 (on lichen, *Trichothelium*), Costa Rica. See Matzer (*Crypt. Mycol.* **14**: 11, 1993).

Roselliniopsis Matzer & Hafellner (1990), Sordariales (inc. sed.). 4 (on lichens), Afr., N. temp. See Matzer (*Crypt., Mycol.* **14**: 11, 1993).

Rosellinites Mesch. (1892), Fossil fungi. 2 (Tertiary), Eur.

Rosellinites Potonié (1893), Fossil fungi. 1 (Oligocene), Germany.

Rosellinula R. Sant. (1986), Dothideales (inc. sed.). 4 (on lichens), widespr.

Rosenscheldia Speg. (1885), Dothideales (inc. sed.). 3 or 4, widespr. See Holm (*Svensk bot. Tidskr.* **62**: 217, 1968).

Rosenscheldiella Theiss. & Syd. (1915), Venturiaceae. *c.* 10, trop. See Hansford (*Mycol. Pap.* **15**, 1946), Swart (*TBMS* **58**: 417, 1972).

Rosenschoeldia L. Holm (1968) ≡ Rosenscheldia (Dothid.).

Rostafiński (Jœf Thomaz; 1850-1928). Born in Warsaw and after studying there, at Jena (under Strasburger), and at Halle (under de Bary) became lecturer (1876) and in 1881 Professor of botany at Cracow Univ., a position he held until 1924. Noted for his work on myxomycetes (*Śluzowce (Mycetozoa) Monografia*, 1874 [reissued with supplement, 1875]) and algae. See Kulczynski (*Acta Bot. Soc. Polon.* **6**: 391, 1929), Stafleu & Cowan (*TL-2* **4**: 908, 1983).

Rostafinskia Speg. (1880) ? Ascomycota (inc. sed.).

Rostafinskia Racib. (1884) ≡ Raciborskia (Stemonitid.).

Rostania Trevis. (1880) = Collema (Collemat.).

Rostkovites P. Karst. (1881) = Suillus (Bolet.) fide Singer (1945).

rostrate, (1) beaked; (2) (of asci), see ascus; bent tip of macroconidia of *Microsporum canis* and other Mitosporic fungi.

Rostrella Fabre (1879) = Ceratocystis (Microascales) fide de Hoog (1974).

Rostrella Zimm. (1900) = Ceratocystis (Microascales) fide Bakshi (1951).

Rostrocoronophora Munk (1953) = Gnomonia (Vals.) fide Bolay (1972).

Rostronitschkia Fitzp. (1919), Diatrypaceae. 1, Puerto Rico. See Rappaz (1987). = Eutypa (Diatryp.) fide Petrak (*Sydowia* **5**: 169, 1951).

Rostrosphaeria Tehon & E.Y. Daniels (1927) = Botryosphaeria (Botryosphaer.) fide v. Arx & Müller (1954).

Rostrospora Subram. & K. Ramakr. (1952) = Colletotrichum (Mitosp. fungi) fide Nag Raj (*CJB* **51**: 2463, 1973).

rostrum, any beak-like process.

Rostrup (Emil; 1831-1907). Danish mycologist and plant pathologist. School teacher at Skaarup (1858-83) then at Royal Veterinary and Agricultural College, Copenhagen, as lecturer (Professor, 1902) and Chief Consulting Pathologist. Chief work: *Plantepatologi* (1902). His collections (now in Botanical Museum, Copenhagen; **C**) was catalogued by Lind (*Danish fungi*, 1913). See Rosenvinge (*Bot. Tidsskr.* **28**: 185, 1908; portr.), Ravn (*Ber. dtsch. bot. Ges.* **26A**: (47), 1909), Lind (1913), Grummann (1974: 679), Stafleu & Cowan (*TL-2* **4**: 912, 1983).

Rostrupia Lagerh. (1889) = Puccinia (Puccin.).

rostrupioid (of *Uredinales*), having teliospores as in *Rostrupia*.

rosulate, in a rosette.

Rosulomyces S. Marchand & Cabral (1976), Mitosporic fungi, 1.A1.39. 1, Argentina. See de Hoog (*Stud. Mycol.* **15**, 1977).

Rota Bat., Cif. & Nascim. (1959), Mitosporic fungi, 5.B1.?. 1, Brazil.

Rotaea Ces. ex Schltdl. (1851), Mitosporic fungi, 2.C1.?. 1, Eur.

Rotiferophthora G.L. Barron (1991), Mitosporic fungi, 1.A1.15/16. 22 (on rotifers), widespr. See Glockling & Dick (*MR* **98**: 833, 1994).

rots (types of), see Wood-attacking fungi.

Rotula Raf. (1815) nom. dub. (Agaric.) fide Merrill (*Index Rafinesq.*, 1949).

Rotula (Müll. Arg.) Müll. Arg. (1890) [non Lour. (1790), *Boraginaceae*] ≡ Mazosia (Roccell.).

Rotularia (Vain.) Zahlbr. (1923) [non Sternb. (1825), fossil *Phanerogamae*] = Mazosia (Roccell.).

Roumegueria (Sacc.) Henn. (1908), Dothideales (inc. sed.). *c.* 2, trop.

Roumegueriella Speg. (1880), ? Hypocreaceae. 2 (1 esp. in mushroom beds), temp. See Malloch & Cain (*CJB* **50**: 61, 1972), Udagawa *et al.* (*Mycoscience* **35**: 409, 1994). Anamorph *Gliocladium*.

Roumeguerites P. Karst. (1879) = Hebeloma (Cortinar.) fide Singer (1951).

Roussoella Sacc. (1888), Amphisphaeriaceae. 4 (on bamboos), trop. See Aptroot (*Nova Hedw.* **60**: 325, 1995; key).

Roussoellopsis I. Hino & Katum. (1965), Dothideales (inc. sed.). 3 (on bamboo), Japan.

Royella R.S. Dwivedi (1960) nom. nud. = Dichotomomyces (Trichocom.) fide Scott (*TBMS* **55**: 313, 1970).

Royoungia Castellano, Trappe & Malajczuk (1992), Boletaceae. 1, Australia.

Rozella Cornu (1872), Rozellopsidaceae. *c.* 24, widespr. See Sparrow (1960: 167; key), Dick (*in press*; key).

Rozellopsidaceae M.W. Dick (1995), Rozellopsidales. 3 gen., 27 spp.

Rozellopsidales, Oomycota. 2 fam., 8 gen. (+ 4 syn.), 45 spp. Fams:
(1) **Pseudosphaeritaceae**.
(2) **Rozellopsidaceae**.

Rozellopsis Karling ex Cejp (1959), Rozellopsidaceae. 4, Eur., Japan. See Dick (*in press*; key).

Rozia Cornu (1872) [non *Rozea* Besch. (1872), *Musci*] ≡ Rozella (Rozellopsid.).

Rozites P. Karst. (1879), Cortinariaceae. 3, temp. See Moser & Horak (*Beih. Nova Hedw.* **52**: 513, 1975).

r-selection, adaptation to the rapid colonization and exploitation of newly opened or uncolonized habitats; in fungi generally involving large numbers of usually small and asexually produced and short-lived propagules, e.g. conidia, soredia; cf. K-selection. See Andrews, *in* Carroll & Wicklow (1992: 119), Armstrong (*Ecology* **57**: 953, 1976), Boyce (*Ann. Rev. Ecol.Syst.* **15**: 427, 1984), population biology, secondary species.

Rubelia Nieuwl. (1916) ≡ Sphaerosporula (Otid.).

Rubetella Tuzet, Rioux & Manier ex Manier (1964) = Smittium (Legeriomycet.) fide Lichtwardt (*Am. J. Bot.* **51**: 836, 1964).

Rubigo (Pers.) Roussel (1806) = Uredo (Anamorphic Uredinales).

Rubikia H.C. Evans & Minter (1985), Mitosporic fungi, 8.D2.1. 1, Honduras.

Rubinoboletus Pilát & Dermek (1969), Strobilomycetaceae. 1, Eur.

rubratoxin B, a toxic metabolite of *Penicillium rubrum* P-13 (Hayes & Wilson, *Appl. Microbiol.* **16**: 1163, 1968) causing hepatitis in cattle and pigs.

Rubromadurella Talice (1935) = Madurella (Mitosp. fungi) fide Ciferri & Redaelli (1941).

ruderal, (1) living in waste places; (2) (of fungi) having a high growth rate, rapidly germinating spores, and a short life expectancy due to exhaustion of the available nutrients; cf. zymogenous (see autochthonous), sugar fungus.

Rudetum Lloyd (1919) = Septobasidium (Septobasid.). See Stevenson & Cash (*Bull. Lloyd Libr. Mus.* **35**: 47, 1936).

Ruggieria Cif. & Montemart. (1958), Mitosporic fungi, 8.E1.?. 1, Italy. *R. glaucescens* on *Citrus* (melanose).

Rugosaria Raf. (1833) ≡ Gemmularia (Pezizales).

rugose, wrinkled (Fig. 29.14). Cf. rugulose.

Rugosomyces Raithelh. (1979), Tricholomataceae. 10, N. temp.

Rugosospora Heinem. (1973), Agaricaceae. 2, trop.

rugulose, delicately wrinkled. Cf. rugose.

Ruhlandiella Henn. (1903), Pezizaceae. 1, USA, Macaronesia, Australia. See Dissing & Korf (*Mycotaxon* **12**: 290, 1980).

rum, see spirits.

rumen fungi, see anaerobic fungi.

Ruminomyces Y.W. Ho (1990) = Anaeromyces (Neocallimastig.) fide Ho (*Mycotaxon* **47**: 283, 1983).

Rumpomycetes. Class within the *Chytridiomycota* distinguished by the presence of a rumposome (q.v.) including *Chytridiales* and *Monoblepharidales* (Cavalier-Smith, *in* Rayner *et al.* (Eds), *Evolutionary biology of the fungi*: 339, 1987).

rumposome, an organelle in zoospores of certain *Chytridiomycota* located close to the cell wall; tooth-like in section and honey-comb like in surface view; see *Rumpomycetes*.

rupestral (rupestrine), living on walls or rocks; cf. saxicolous.

Rupinia Speg. & Roum. (1879) [non L. f. (1782), *Hepaticae*] = Heydenia (Mitosp. fungi) fide Saccardo (*Syll. Fung.* **4**: 625, 1886).

Ruspoliella Sambo (1937) = Solorina (Peltiger.).

Russula Pers. (1796), Russulaceae. 280, cosmop. See Watson (*TBMS* **49**: 11, 1966; pigments), Burge (*Mycol.* **71**: 977, 1979; spore structure), Crawshay (*The spore ornamentation of the Russulas*, 1930), Pearson (*Naturalist*, Hull: 85, 1948; key Br. spp.), Schaeffer (*Russula Monographie*, 1951), Hesler (*Mem. Torrey bot. Cl.* **21**, 1960, *Mycol.* **53**: 605, 1962; N. Am. types), Blum (*Les Russules*, 1962; France), Romagnesi (*Les Russules d'Europe et Afrique du Nord*, 1967), Rayner (*Bull. BMS* **2**: 76, **3**: 59, 89, 1968-69; keys, **4**: 19, 1970; descr. Br. spp., *Keys to British species of Russula*, edn 3, 1985), Pegler & Young (*TBMS* **72**: 353, 1979).

Russulaceae Lotsy (1907), Russulales. 4 gen. (+ 12 syn.), 398 spp. Basidioma agaricoid or gasteroid; heteromerous with sphaerocytes; spores hyaline, amyloid, ornamented.

Russulales, Basidiomycetes. 2 fam., 10 gen. (+ 15 syn.), 484 spp. Basidioma epigeal or hypogeal, agaricoid or gasteroid; saprobic, mycorrhizal, cosmop.

A relationship between the agaricoid *Russulaceae* and certain gasteroid genera was first proposed by Bucholtz (1902), and Malençon (1931) postulated a natural series, the *Asterospores*. The relationship was based on the heteromerous nature of the flesh, with sphaerocytes, and the amyloid, ornamented spores. The concept was expanded by Heim (1948), whilst Singer & Smith (1960) monographed the gasteroid series calling them 'Astrogastraceous Series'. The order was formally recognized by Kreisel (1969). Fams:

(1) **Elasmomycetaceae**.

(2) **Russulaceae**.

Lit.: Heim (*TBMS* **30**: 161, 1948; phylogeny and classification), Malençon (*Recueil trav. crypt. L. Mangin*: 377, 1931), Pegler & Young (*TBMS* **72**: 353, 1979; classification, spore structure), Singer *et al.* (*Beih. Nova Hedw.* **77**, 1983), Singer & Smith (*Mem. Torrey bot. Cl.* **21**: 1, 1960; gasteroid genera), Reijnders (*Persoonia* **9**: 65, 1976; development), Beaton *et al.* (*Kew Bull.* **39**: 669, 1984; Austral. gasteroid spp.).

Russulina J. Schröt. (1889) = Russula (Russul.) fide Singer & Smith (1946).

Russuliopsis J. Schröt. (1889) ≡ Laccaria (Tricholomat.).

rust, (1) a disease caused by one of the *Uredinales*; (2) one of the *Uredinales*; (3) a disease with 'rusty' symptoms; **black (stem)** - of cereals, *Puccinia graminis*; **blister** - of *Pinus* and *Ribes*, *Cronartium ribicola*; **brown (leaf)** - of barley, *P. hordei*; of rye and wheat, *P. recondita*; **crown** - of oats, *P. coronata*; **red** -, (1) urediniospore state of cereal rusts, esp. *Puccinia graminis*; (2) (of tea) the alga *Cephaleuros*; **white** - (1) (esp *Cruciferae*) = white blister; (2) (of *Chrysanthemum*) = *P. horiana*; **yellow (stripe)** - of cereals, *P. striiformis*.

Ruthea Opat. (1836) ≡ Paxillus (Paxill.).

Rutola J.L. Crane & Schokn. (1978), Mitosporic fungi, 1.C2.1. 1, widespr.

Rutstroemia P. Karst. (1871) typ. cons., Sclerotiniaceae. c. 70, widespr. s.l. fide White (*Lloydia* **4**: 153, 1941). See Baral (*SA* **13**: 113, 1994; concept), Holm (*TBMS* **67**: 333, 1976, *Mycotaxon* **7**: 139, 1978; status), Korf & Dumont (*Mycotaxon* **5**: 517, 1977; status), Kohn & Schumacher (*Mycotaxon* **18**: 531, 1983, *Taxon* **33**: 507, 1984; nomencl.).

Ryparobius Boud. (1869) = Thelebolus (Thelebol.) fide Kimbrough & Korf (*Am. J. Bot.* **54**: 9, 1967).

Ryssospora Fayod (1889) ? = Gymnopilus (Cortinar.) fide Singer & Smith (1946).

SA, see Media.

Sabouraud (Raymond; 1864-1938). A dermatologist of Paris noted for his monumental researches on the dermatophytes, summarized in *Les teignes* (1910) and other works. See Pautrier (*Annls Derm.*, sér. 7, **9**: 275, 1938; portr.), Grigoraki (*Mycopath.* **2**: 171, 1939; portr., bibl.), Pignot (*Mycopath.* **7**: 348, 1956; portr.).

Sabouraudiella Boedijn (1953) = Trichophyton (Mitosp. fungi).

Sabouraudites M. Ota & Langeron (1923) ≡ Microsporum (Mitosp. fungi).

Saccardaea Cavara (1894) nom. dub. fide Tulloch (*Mycol. Pap.* **130**: 36, 1972).

Saccardia Cooke (1878), Saccardiaceae. 2, warmer parts. See v. Arx (*Persoonia* **2**: 421, 1963).

Saccardiaceae Höhn. (1909), Ascomycota (inc. sed.). 10 gen. (+ 7 syn.), 27 spp. Mycelium superficial, sometimes setose, often inconspicuous; stromata absent or small and basal; ascomata superficial, discoid, thin-walled, the upper part of the wall gelatinous, deliquescing to release the spores; interascal tissue of usually narrow pseudoparaphyses, becoming free at the apex, the apices swollen and sometimes pigmented; asci ± saccate, sessile, thick-walled, fissitunicate, without an apical ring; ascospores hyaline or pale brown, muriform, without a sheath.

Saccardinula Speg. (1885), Elsinoaceae. 5, trop.

Saccardo (Pier Andrea; 1845-1920). Lived for much of his time in Padua, where he was for many years Professor of Botany. Among his printed works are the series of papers *Fungi veneti novi vel critici* (1873-78), the *Fungi italici autographice delineati* (1877-86; 1,500 small coloured figs), and *Michelia* (1877-82). In addition, he put out Exsiccati and made a special study of pyrenomycetes and Mitosporic fungi, but the work by which his name will be kept living is his '*Sylloge*', the *Sylloge Fungorum omnium hucusque cognitorum* (1882-1925 [-72]). His collection is at Padua, **PAD** (see Gola, *L'erbario di P.A. Saccardo, Catalogo*, 1930).

By Saccardo's time the amount of printed work on systematic mycology was great. Much was hard to get, and it was in no order. For these reasons, Saccardo took up the work of listing with Latin descriptions, all the genera and species of fungi of which there was then knowledge. To do this it was frequently necessary to make new arrangements of old groups, and 'emendations' of numbers of genera and species. He made special use of his system of 'spore groups' (see Mitosporic fungi) and his classification has had to be taken into account by all later systematic mycologists. The first volume of the *Sylloge* was printed in 1882 and, with the help of P. Sydow, Berlese, de Toni, Mussat, D. Saccardo (son), Traverso and others, the work went on till his death; vols 24-26 (1926-1972) were prepared and printed later. Most of the genera and species to 1920 are listed. See Stafleu & Cowan (*TL-2* **4**: 1023, 1983), Zalin & Lazzari (*Carteggio Bresadola-Saccardo*, 1987; corresp.), Paganelli, *in* Minelli (Ed.) (*The Botanical Garden of Padua 1545-1995*: 118, 1995; Pref. of Garden, 1879-1915), Grummann (1974: 521). See Colour, Literature.

Saccardoa Trevis. (1869) nom. rej. = Pseudocyphellaria (Lobar.).

Saccardoella Speg. (1879), Ascomycota (inc. sed.). 2 or 3, Eur., N. Am. See Barr (*Mycotaxon* **51**: 191, 1994; posn), Hawksworth & Eriksson (*SA* **10**: 52, 1991), Hyde (*Mycol.* **84**: 803, 1992).

Saccardomyces Henn. (1904) = Schweinitziella (Trichosphaer.) fide Pirozynski (1977).

saccate, like a sac or bag.

Saccharomyces Meyen ex E.C. Hansen (1838), Saccharomycetaceae. 10, cosmop. *S. cerevisiae* (with 25 races), 'brewer's yeast', is used in beer- and bread-making, and other fermentations. See Vaughan-Martini (*Syst. Appl. Microbiol.* **16**: 113, 1993; key), Barnett (*Yeasts* **8**: 1, 1992; review taxonomy), Kreger-van Rij (1984; key), v. Arx *et al.* (*Stud. mycol.* **14**, 1977), Dujon *et al.* (*Nature* **369**: 371, 1994; complete DNA sequence chromosome XI, and refs others), Kock *et al.* (*Appl. Microbiol. Biotech.* **23**: 499, 1986; fatty acids), Martini & Martini (*Ant. v. Leeuwenhoek* **53**: 77-84, 1987; DNA reassociation), Neumann (*Beih. Nova Hedw.* **40**, 1972), Strathern *et al.* (Eds) (*The molecular biology of the yeast Saccharomyces, Cold Spring Harb. Monogr.* **11A** (*Life cycle and inheritance*), 1981, **11B** (*Metabolism and gene expression*), 1982), Martini & Kurtzman (*Int. J. Syst. Bact.* **35**: 508, 1985; DNA homology), Martini & Martini (*in* Cantarelli & Lanzarini (Eds), *Biotechnology applications in beverage production*: 1, 1989; nomencl. domesticated spp.), Stewart & Russell (*in* Demain & Solomon (Eds), *Biology of industrial micro-organisms*: 511, 1985; review biology), Herskowitz (*Microbiol. Rev.* **52**: 536, 1988; *S. cerevisiae* life-cycle, 166 refs.), Tuite & Oliver (Eds) (*Saccharomyces*, 1991 [*Biotech. Handb.* **4**]), Industrial mycology.

Saccharomycetaceae G. Winter (1881), Saccharomycetales. 28 gen. (+ 31 syn.), 159 spp. Mycelium ± absent; vegetative cells reproducing by multilateral budding, ± ellipsoidal, without mucus; asci morphologically similar to vegetative cells, not in well-defined chains, ± globose, thin-walled, 1- to 4-spored, evanescent or persistent; ascospores usually ± spherical, often ornamented with equatorial ridges etc.; fermentation positive; coenzyme system usually Q-6. Cosmop., in a very wide range of habitats.
Lit.: v. Arx (*Ant. v. Leeuwenhoek* **38**: 289, 1972), v. Arx *et al.* (*Stud. Mycol.* **14**, 1977; gen. names, 1981; keys gen.), Batra (*in* Subramanian (Ed.), *Taxonomy of fungi* **1**: 187, 1978), Gams *et al.* (*CBS course of mycology*, edn 2, 1980), Kreger-van Rij (*in* Ainsworth *et al.* (Eds), *The Fungi* **4A**: 1973; Ed., *The yeasts*, edn 3, 1984), Redhead & Malloch (*CJB* **55**: 1701, 1977), Viljoen *et al.* (*Ant. v. Leeuwenhoek* **52**: 45, 1986; fatty acid composition), and also under Yeasts (q.v.).

Saccharomycetales, Ascomycota. 8 fam., 75 gen. (+ 57 syn.), 273 spp. Mycelium absent or poorly developed, where present usually with septa which have a series of minute pores rather than a single simple pore; vegetative cells proliferating by budding (blastically) or by fission (thallically), the walls usually lacking chitin except around bud scars, sometimes with I+ gel; ascomata absent; asci formed singly or in chains, sometimes not strongly differentiated morphologically from vegetative cells, usually at least eventually evanescent; ascospores varied in shape, sometimes with equatorial or asymmetric thickenings. The teleomorphic ascomycetous yeasts, cosmop. in a very wide range of habitats. Fams:
(1) **Cephaloascaceae**.
(2) **Dipodascaceae**.
(3) **Endomycetaceae**.
(4) **Lipomycetaceae**.
(5) **Metschnikowiaceae**.
(6) **Saccharomycetaceae**.
(7) **Saccharomycodaceae**.
(8) **Saccharomycopsidaceae**.
Lit.: Eriksson *et al.* (*SA* **11**: 119, 1993), Viljoen & Kock (*Syst. Appl. Microbiol.* **14**: 178, 1991; pyrolysis g.l.c. identification of yeasts).

Saccharomycetes, formerly used for most yeast-like fungi, both anamorphic and teleomorphic (obsol.).

Saccharomycetes Grüss (1928), ? Fossil fungi. 1 (Devonian), Spitsbergen.

Saccharomycodaceae Kudrjanzev (1960), Saccharomycetales. 4 gen. (+ 8 syn.), 10 spp. Mycelium poorly developed; vegetative cells usually lemon-shaped, proliferating by bipolar budding; asci either formed directly from an apparently undifferentiated vegetative cell, or by conjugation between a mother cell and its bud; ascospores varied in form, sometimes brown, sometimes ornamented or with a median or eccentric flange; fermentation usually positive; from a wide range of sources, cosmop.

Saccharomycodes E.C. Hansen (1904), Saccharomycodaceae. 2, N. temp. See Barnett *et al.* (1990), Kreger-van Rij (1984).

Saccharomycopsidaceae Arx & Van der Walt (1987), Saccharomycetales. 5 gen. (+ 1 syn.), 18 spp. Mycelium poorly to well developed; vegetative cells ± ellipsoidal, proliferating by multilateral budding, mycelial fragmentation also occurring; asci ellipsoidal to ± globose, formed directly from vegetative cells, usually evanescent, usually 4-spored; ascospores hyaline or pale brown, with ± equatorial flanges; coenzyme system usually Q-8. From a wide range of sources, widespr.

Saccharomycopsis Schiønning (1903), Saccharomycopsidaceae. 7 (or 2 fide Barnett *et al.*, 1990), temp. See v. Arx *et al.* (1977, 1981), Kreger van Rij (1984; key).

Saccharomycopsis Guillierm. (1912) ≡ Cyniclomyces (Saccharomycet.).

Saccharopolyspora J. Lacey & Goodfellow (1975), *Actinomycetes*, q.v.

Saccoblastia A. Møller (1895) = Helicogloea (Platygl.) fide Baker (*Ann. Mo. bot. Gdn* **23**: 69, 1936) but see Donk (*Persoonia* **4**: 217, 1966).

Saccobolus Boud. (1869), Ascobolaceae. 36, widespr. See v. Brummelen (1967; key, *Persoonia* **8**: 421, 1976).

Saccomorpha Elenkin (1912) = Placynthiella (Trapel.).

Saccomyces Serbinow (1907), ? Chytridiaceae. 1, Eur.

Saccopodium Sorokīn (1877), ? Cladochytriaceae. 1, former USSR.

Saccothecium Fr. (1836), Dothioraceae. *c.* 10, widespr. See Froidevaux (*Nova Hedw.* **23**: 679, 1973; key),

Holm (*Taxon* **24**: 475, 1975; status). Anamorph *Hormonema*-like.

Sachsia Bay (1894) [non Griseb. (1866), *Compositae*], ? Saccharomycetaceae. 1, Eur.

Sachsia Lindner (1895) [non Griseb. (1866), *Compositae*], nom. dub. (Ascomycota, inc. sed.).

Sacidium Nees (1823) = Pilobolus (Mucor.) fide v. Höhnel (*Sber. Akad. Wiss. Wein* **119**: 617, 1910).

Sacidium Sacc. (1880) = Pilobolus (Pilobol.) fide Höhnel (*Sber. Akad. Wiss. Wien* **119**: 617, 1910). See Sutton (*Mycol. Pap.* **141**, 1977).

Sackea Rostk. (1844) = Bovista (Lycoperd.).

Sadasivanella Agnihothr. (1964) = Neottiosporina (Mitosp. fungi) fide Sutton & Alcorn (*Austr. Jl Bot.* **22**: 517, 1974).

Sadasivania Subram. (1957), Mitosporic fungi, 1.A2.3. 3, India, USA. See Sundberg & Wicklow (*Mycol.* **65**: 925, 1973).

saddle fungus, ascomata of *Helvella* spp.

saddle-back fungus (**dryad's saddle**), basidioma of *Polyporus squamosus*.

Saeenkia Kudrjanzev (1960) = Saccharomycodes (Saccharomycod.) fide Lodder (1970).

Safety, Laboratory. Precautions should be taken to avoid mycological laboratory hazards to humans. It must be remembered that many fungi pathogenic for humans are 'opportunistic', widespread saprobes, and their full potential to effect human health is rarely known (see Smith, *Opportunistic mycoses of man and other animals*, 1989). Fungi and other microorganisms can enter the body through the mouth, lungs, broken or unbroken skin, and the conjunctiva. However, in the laboratory the route of infection may not be the same as when the disease is acquired naturally. The potential hazards are greater in the laboratory than in nature as the organisms are grown in vast numbers, transferred from one container to another, and manipulated, increasing the risk of infection. The main routes of infection are accidental inoculation, accidental ingestion, splashing into the face and eyes, spillage and direct contact. Good laboratory practice to keep cultures pure normally prevents their escape to cause infection, but note that the conditions provided for the growth of a particular fungus may also be suitable for the growth of a potentially hazardous contaminant. Eating and smoking in laboratories is best prohibited to reduce risk of ingestion and inhalation.

Special precautions should be taken when harvesting bulk cultures or large quantities of spores. Fungi may also cause allergic reactions or mycotoxicoses (poisoning) and therefore contact with them and the materials or equipment they have been in contact with must be avoided.

Great care must be taken in handling the pathogens of coccidioidomycosis, histoplasmosis, and other systemic mycoses for there are many instances of laboratory infections by these fungi resulting in illness or death (see Hanel & Kruse, *Misc. Publ. Fort Detrich, Dept. Army* **27**, 1967; laboratory acquired mycoses).

Microorganisms are divided into four hazard groups by the Advisory Committee on Dangerous Pathogens (*Categorization of pathogens according to hazard and categories of containment*, edn 2, 1990):

Group 1: A biological agent that is most unlikely to cause human disease.

Group 2: A biological agent that may cause human disease and which might be a hazard to laboratory workers but is unlikely to spread in the community. Laboratory exposure rarely produces infection and effective prophylaxis or treatment is available.

Group 3: A biological agent that may cause severe human disease and present a serious hazard to laboratory workers. It may present a risk of spread in the community but there is usually effective prophylaxis or treatment.

Group 4: A biological agent that causes severe human disease and is a serious hazard to laboratory workers. It may present a high risk of spread in the community and there is usually no effective prophylaxis or treatment.

Most fungi belong to category 1, and the second EEC directive on *Community classification of biological agents* (93/88/EEC), lists the following species in category 2: *Aspergillus fumigatus, Candida* spp., *Filobasidiella neoformans, Emmonsia parva, Epidermophyton floccosum, Fonsecaea* spp., *Madurella* spp., *Microsporum* spp., *Neotestudina rosatii, Sporothrix schenkii, Trichophyton* spp., and *Xylohypha bantiana*. This directive includes only 5 fungi in category 3: *Ajellomyces capsulatus, A. dermatidis, Coccidioides immitis, Paracoccidioides brasiliensis*, and *Penicillium marneffei* [*Note*: teleomorph names only listed where the fungi are pleomorphic; see generic entries for those of the anamorphs]. No fungi are placed in category 4. Other classifications of hazardous organisms are available (e.g. US Public Health Service, *Classification of etiologic agents on the basis of hazard*, 1974; World Health Organization, *Laboratory biosafety manual*, 1983).

Increasing containment levels are prescribed by Health and Safety regulations for these hazard categories in most countries, and information on those which apply should be sought from appropriate national agencies before starting work with listed organisms.

Lit.: Collins (*Biologist* **23**: 83, 1976; biological hazards), Fuscaldo *et al.* (*Laboratory safety. Theory and practice*, 1980; general, biological, medical), Collins (*Laboratory-acquired infections*, 1983), Miller (*Laboratory safety: principles and practices*, 1986), Simpson & Simpson (*The COSHH regulations: a practical guide*, 1991), Stricoff & Walters (*Handbook of laboratory health and safety*, edn 2, 1995).

See also Allergy, Human and veterinary mycology, Mycetism, Mycotoxins.

saffron milk-cap, basidioma of the edible *Lactarius deliciosus*.

Sagedia Ach. (1809) nom. rej. = Aspicilia (Hymenel.) fide Laundon & Hawksworth (1988).

Sagedia A. Massal. (1852) nom. rej. = Porina (Trichothel.).

Sagediomyces Cif. & Tomas. (1953) = Strigula (Strigul.) fide Harris (*in litt.*).

Sagediopsis (Sacc.) Vain. (1921), Adelococcaceae. 7 (on lichens), mainly temp. See Alstrup & Hawksworth (*Meddr. Grønl., Biosci.* **31**, 1990; nomencl.), Hafellner (*Herzogia* **9**: 749, 1993).

Sagema Poelt & Grube (1993), Lecanoraceae (L). 1, Nepal.

sagenetosome, see sagenogen.

Sagenidiopsis R.W. Rogers & Hafellner (1987), Arthoniales (inc. sed.), L). 2, Australia, S. Am. See Egea *et al.* (*Lichenologist* **27**: 351, 1995; posn).

Sagenidium Stirt. (1877), Roccellaceae (L). 3, S. Am., Australia. See Henssen *et al.* (*Lichenologist* **11**: 263, 1979).

sagenogen (**sagenetosome**), see bothrosome.

sagenogenetosome, see bothrosome.

sagenogens, Olive's (1975) modification of Porter's (1972) 'sagenogenetosomes', organelles from which the ectoplasmic nets of *Labyrinthulales* are produced.

Sagenoma Stolk & G.F. Orr (1974), Trichocomaceae. 2, Australia, Japan. See Ueda & Udagawa (*Mycotaxon* **20**: 499, 1984). Anamorph *Sagenomella*.

Sagenomella W. Gams (1978), Anamorphic Trichocomaceae, 1.A1/2.3. Teleomorph *Sagenoma*. 9, widespr.

Sageria A. Funk (1975), Leotiaceae. 1 (on *Tsuga*), Canada. Anamorph *Ascoconidium*.

Sagiolechia A. Massal. (1854), ? Gomphillaceae (L). 2 (or 4), Eur. See Vězda (*Folia geobot. phytotax.* **2**: 383, 1968).

Sagittospora Lubinsky (1955), Chytridiales (inc. sed.). 1 (on *Eudiplodinium* (*Protozoa*) in goat rumen), Pakistan.

Sagrahamala Subram. (1972), Mitosporic fungi, 1.A1/2.15. 10, widespr.

Saitoa Rajendran & Muthappa (1980) = Neosartorya (Trichocom.) fide Samson (*in* Samson & Pitt (Eds), *Advances in Penicillium and Aspergillus systematics*: 365, 1985; coenzyme-Q system).

Saitoella Goto, Sugiy., Hamam. & Komag. (1987), Mitosporic fungi, 1.A1.3/10. 1 (from soil), Himalayas.

Saitomyces Ricker (1906) ≡ Actinocephalum (Cunninghamell.).

Sakaguchia Y. Yamada, K. Maeda & Mikata (1994), Sporidiobolaceae. 1 (from sea water), Antarctica.

Sakireeta Subram. & K. Ramakr. (1957) = Tiarosporella (Mitosp. fungi) fide Sutton (1977).

Saksenaea S.B. Saksena (1953), Saksenaeaceae. 1, N. & C. Am., Asia, Australia, Ethiopia. See Ajello *et al.* (*Mycol.* **68**: 53, 1976; on humans), Ellis & Hesseltine (1974), Ellis & Ajello (*Mycol.* **74**: 144, 1982), Chien *et al.* (*Trans. Mycol. Soc. Japan* **33**: 443, 1992), Mathews *et al.* (*J. Myc. Méd.* **3**: 95, 1993; causing zygomycosis, lit. review).

Saksenaeaceae Hesselt. & J.J. Ellis (1974), Mucorales. 1 gen., 1 sp. Sporangia lageniform and columellate; zygospores unknown.

Saliastrum Kujala (1946), Mitosporic fungi, 1.G2.1. 1, Eor., Canada. See Sutton (1977), Redhead & Perrin (*CJB* **50**: 2083, 1972).

Salilagenidiaceae M.W. Dick (1995), Salilagenidiales. 1 gen., 5 spp.

Salilagenidiales, Oomycota. 2 fam., 3 gen., 10 spp. Fams:
(1) **Haliphthoraceae**.
(2) **Salilagenidiaceae**.

Salilagenidium M.W. Dick (1995), Salilagenidiaceae. 5 (parasites in marine *Crustacea*), widespr.

Salmonia S. Blumer & E. Müll. (1964) = Brasiliomyces (Erysiph.) fide Zheng (*Mycotaxon* **22**: 209, 1985).

Salmonomyces Chidd. (1959) = Erysiphe (Erysiph.) fide Zheng (1985).

Salsuginea K.D. Hyde (1991), ? Dothideales (inc. sed.). 1 (marine), Thailand. See Eriksson & Hawksworth (*SA* **11**: 73, 1992; posn).

saltation (of fungi), mutation; dissociation; see Variation in fungi.

Samarospora Rostr. (1892), ? Eurotiales (inc. sed.). 1, Eur.

Samarosporella Linder (1944) nom. dub. (Fungi, inc. sed.) fide Kohlmeyer (*CJB* **50**: 1951, 1972).

Samboa Tomas. & Cif. (1952) = Buellia (Physc.).

Samboamyces Cif. & Tomas. (1953) ≡ Samboa (Physc.).

Sambucina Velen. (1947), ? Leotiales (inc. sed.). 1, former Czechoslovakia.

sambucinin, see enniatin.

Sampaioa Gonz. Frag. (1923), Lophiostomataceae. 1, Eur.

Samukuta Subram & K. Ramakr. (1957) = Neottiospora (Mitosp. fungi) fide Nag Raj (*CJB* **51**: 2463, 1973).

Sand dune fungi. Sand dunes have a characteristic mycobiota of both microfungi (Moreau & Moreau, *Rev. Mycol.* **6**: 49, 1941, Dickinson & Kent, *TBMS* **58**:

269, 1972) and macrofungi (Andersson, *Op. bot. Soc. bot. Lund* **2**, 1950, Bon, *BSMF* **86**: 79, 1970, Picardy; **88**: 15, 1972; *Lepiota*, Courtecuisse, *Docums Mycol.* fasc. 57-58, 66, 1984-1986, Høiland, *Blyttia* **33**: 127, 1975, **35**: 139, 1977, **36**: 69, 1978, Watling & Rotheroe, *Proc. R. Soc. Edinb.* **96**: 111, 1989) which, like that of flowering plants, exhibits a zonation from the shore sand to fixed dune (Wallace, *Trans. Devon. Ass. Advmt Sci.* **86**: 201, 1954). Macrofungi play a role in facilitating colonization by higher plants and in sustaining coastal dune systems, both by the action of fungal hyphae in accretion of sand particles and as mycorrhizal symbionts (see Rotheroe, *in* Pegler *et al.* (Eds), *Fungi of Europe*, 1993). Characteristic lichen communities frequently develop on stable sand dunes and are usually dominated by species of *Cladonia*; succession in the lichen communities is correlated with the stability of the substrate (see James *et al.*, *in* Seaward (Ed.), *Lichen ecology*: 400, 1977). See also Ecology, Marine fungi.

Sandersoniomyces R.K. Benj. (1968), Laboulbeniaceae. 1, USA. See Benjamin (*Aliso* **10**: 345, 1983).

sandimmum, see cyclosporin A.

Sanjuanomyces R.F. Castañeda & W.B. Kendr. (1991), Mitosporic fungi, 1.C2.24. 1, Cuba.

Santapauella Mundk. & Thirum. (1945) = Phragmidiella (Phakopsor.) fide Thirumalachar & Mundkur (*Indian Phytopath.* **2**: 193, 1949).

Santapauinda Subram. (1994), Mitosporic fungi, 1.D2.1. 1, India.

Santessonia Hale & Vobis (1978), Physciaceae (L). 3, Namibia. See Sérusiaux & Wessels (*Mycotaxon* **19**: 479, 1984; key).

Santessoniolichen Tomas. & Cif. (1952) = Arthopyrenia (Arthopyren.).

Santessoniomyces Cif. & Tomas. (1953) ≡ Santessoniolichen (Arthopyren.).

Santiella Tassi (1900) = Passeriniella (Dothideales) fide Sutton & Sellar (*CJB* **44**: 1505, 1966).

Sappinia P.A. Dang. (1896) nom. dub.; excluded from *Myxomycota* by Olive (1975: 6).

saprobe (**saprogen**, **saprotroph**), an organism using dead organic material as food, and commonly causing its decay (saprobe is the preferred term for fungi); **saprobic**, **saprogenic** (**-ous**), **saprophilous**, **saprotrophic** (adj.). See Hudson (*Fungal saprophytism*, edn 2, 1980), saprophyte.

Saprochaete Coker & Shanor ex D.T.S. Wagner & Dawes (1970), Mitosporic fungi, 1.E1.10. 1 (aquatic), widespr.

saprogen, a saprobe (q.v.).

Saprolegnia Nees (1823), Saprolegniaceae. 22 (in freshwater), esp. N. temp. *S. parasitica* and other spp. pathogenic to fish and other aquatic vertebrates See Mil'ko (*Mikol. i Fitopat.* **13**: 288, 1979), Scott *et al.* (*Virg. J. Sci.* **14**: 42, 1963), Seymour (*Nova Hedw.* **19**: 1, 1970; gen. account, key 19 spp.), Tiffney (*Mycol.* **31**: 310, 1939; of fish), Neish & Hughes (*Fungal diseases of fishes* [*Diseases of fishes* **6**], 1980), Willoughby (*in* Roberts (Ed.), *Microbial diseases of fish*, 1982; of salmon).

Saprolegniaceae Kütz. ex Warm. (1884), Saprolegniales. 19 gen. (+ 7 syn.), 138 spp.

Saprolegniales, Oomycota. 1 fam., 20 gen. (+ 9 syn.), 140 spp. 'water moulds'. Zoospores frequently diplanetic; freshwater, rarely marine, saprobic or parasitic; widespr. Fam.:
Saprolegniaceae.
Lit.: Coker (*The Saprolegniaceae with notes on other water molds*, 1923 [reprint 1968]), Fitzpatrick (1930), Cejp (1969), Seymour (1970), Dick (1973; *in* Margulis *et al.* (Eds), 1990: 661).

Sapromyces Fritsch (1893), Rhipidiaceae. 3, widespr.

Saprophragma K.B. Deshp. & K.S. Deshp. (1966), Mitosporic fungi, 1.C1.?. 1, India.

saprophyte, a plant feeding by external digestion of dead organic matter; commonly misapplied to fungi where saprobe (q.v.) is the preferred term.

Saprotaphrina Verona & Rambelli (1962) nom. nud. (Mitosp. fungi). Yeast phase of Taphrina (Taphrin.).

saprotroph, (1) a saprobe (q.v.); (2) a necrophyte on dead material which is not part of a living host (Münch), cf. perthophyte.

Sapucchaka K. Ramakr. (1956), ? Microthyriaceae. 1, India. See v. Arx & Müller (1975).

Sarawakus Lloyd (1924), Hypocreaceae. 10, Br. Isl., Asia, N. & S. Am. See Samuels & Rossman (*Mycol.* 84: 26, 1992; key). = Hypocrea (Hypocr.) fide Rossman (*in* Hawksworth (Ed.), *Ascomycete systematics*: 375, 1994). Anamorph *Gliocladium* or *Trichoderma*-like.

Sarbhoyomyces Saikia (1981), Mitosporic fungi, 2.A1.?. 1, India.

Sarcanthia Raf. (1817) nom. dub. (Fungi, inc. sed.). No spp. included.

Sarcinella Sacc. (1880), Anamorphic Englerulaceae, 1.D2.1. Teleomorph *Schiffnerula*. 8, widespr. See Hughes (*CJB* 61: 1727, 1984).

sarciniform, bundle-like, as the dictyospore of *Stemphylium botryosum*.

Sarcinodochium Höhn. (1905), Mitosporic fungi, 3.D1.?. 1, Eur.

Sarcinomyces Lindner (1898), Mitosporic fungi, 1.D1.21. 1, widespr. See Jong & King (*Mycotaxon* 3: 397, 1976), Hermanides-Nijhof (*Stud. Mycol.* 15: 173, 1977).

Sarcinomyces Oho (1926) ≡ Sarcinosporon (Mitosp. fungi) fide Jong & King (*Mycotaxon* 3: 397, 1976).

Sarcinosporon D.S. King & S.C. Jong (1975), Mitosporic fungi, 1.A1.10. 1 (from humans), widespr.

Sarcinulella B. Sutton & Alcorn (1983), Anamorphic Monoblastiaceae, 4.A1.15. Teleomorph *Anisomeridium*. 1, Eur., S. Hemisph.

sarcodimitic, see Hyphal analysis.

Sarcodon Quél. ex P. Karst. (1881), Thelephoraceae. 31, widespr. See Maas Geesteranus & Nannfeldt (*Svensk bot. Tidskr.* 63: 401, 1969), Singer (*Beih. Nova Hedw.* 77, 1983), Stalpers (*Stud. Mycol.* 35, 1993; key).

Sarcodontia Schulzer (1866), Hyphodermataceae. 1, boreal.

Sarcographa Fée (1824), Graphidaceae (L). *c.* 70, trop. See Pant (*Geophytology* 20: 48, 1991; 3 spp., India).

Sarcographina Müll. Arg. (1887), Graphidaceae (L). 8, cosmop. See Pant (*Geophytology* 20: 48, 1991; India).

Sarcographomyces Cif. & Tomas. (1953) = Sarcographa (Graphid.).

Sarcogyne Flot. (1850) nom. rej. prop. = Polysporina (Acarospor.).

Sarcogyne Flot. (1851) nom. cons. prop., Acarosporaceae (L). 28, cosmop. See Magnusson (*Rabenh. Krypt.-Fl.* 9, 5(1): 49, 1935, *Annls Crypt. Exot.* 7: 115, 1935), Jørgensen & Santesson (*Taxon* 42: 881, 1993; nomencl.).

Sarcoleotia S. Ito & S. Imai (1934), Leotiaceae. 1, Eur., N. Am., Tierra del Fuego. See Schumacher & Sivertsen (*in* Larsen *et al.* (Eds), *Arctic and alpine mycology* 2: 163, 1987).

Sarcoloma Locq. (1979), Cortinariaceae. 1, Eur.

Sarcomelas Raf. (1817) nom. dub. (Fungi, inc. sed.).

Sarcomyces Massee (1891), Leotiales (inc. sed.). 1, Am.

Sarcomyxa P. Karst. (1891) = Hohenbuehelia (Tricholomat.) fide Singer (1951).

Sarconemus Raf. (1815) nom. dub. (? Fungi).

Sarcophoma Höhn. (1916), Anamorphic Mycosphaerellaceae, 8.A1.15. Teleomorph *Guignardia*. 1, widespr. See van der Aa (*Persoonia* 8: 283, 1975).

Sarcopodium Ehrenb. (1818), Mitosporic fungi, 3.A/B1.15. 5, widespr. See Ellis (*MDH*), Sutton (*TBMS* 76: 97, 1981).

Sarcoporia P. Karst. (1894) = Podoporia (Coriol.) fide Pegler (1973), = Poria (Coriol.) fide Donk (1974).

Sarcopyrenia Nyl. (1858), Ascomycota (inc. sed., L). 3 (on lichens), N. Hemisph. See Aguirre-Hudson (*Bull. Br. Mus. nat. Hist., Bot.* 21: 85, 1991), Navarno-Rosinés & Hladún (*Candollea* 45: 469, 1990).

Sarcopyreniomyces Cif. & Tomas. (1953) = Sarcopyrenia (Ascomycota).

Sarcopyreniopsis Cif. & Tomas. (1953) ≡ Sarcopyrenia (Ascomycota).

Sarcorhopalum Rabenh. (1851) = Taphrina (Taphrin.) fide Mix (1949).

Sarcosagium A. Massal. (1856), Acarosporaceae (L). 1, temp. See Poelt & Vězda (1977).

Sarcoscypha (Fr.) Boud. (1885) typ. cons., Sarcoscyphaceae. *c.* 9, N. temp. See Denison (*Mycol.* 64: 609, 1972), Baral (*Z. Mykol.* 50: 117, 1984; key 5 spp.), Harrington (*Mycotaxon* 38: 417, 1990; 3 spp. N. Am.). Anamorph *Molliardiomyces*.

Sarcoscyphaceae Le Gal ex Eckblad (1968), Pezizales. 10 gen. (+ 7 syn.), 21 spp. Ascomata apothecial, ± sessile or stalked, leathery or somewhat gelatinous, usually brightly coloured due to carotenoid pigments; excipulum composed of hyphal cells often embedded in a gelatinous matrix, sometimes with hyaline or rarely melanized hairs surrounding the hymenium; interascal tissue composed of paraphyses, often anastomosing near the base and pigmented at the usually swollen apices; asci cylindrical, persistent, with an often slightly subapical operculum, not blueing in iodine; ascospores hyaline, smooth or ornamented, multinucleate. Anamorphs hyphomycetous. Saprobic on soil and rotten wood, cosmop.
Lit.: Benkert (*Gleditschia* 19: 173, 1991; keys E. German spp.).

Sarcosoma Casp. (1891), Sarcosomataceae. 1, N. temp. See Eckblad (1968), Korf (*Mycol.* 49: 102, 1957).

Sarcosomataceae Kobayasi (1937), Pezizales. 14 gen. (+ 3 syn.), 30 spp. Ascomata large, apothecial, ± sessile, leathery or somewhat gelatinous, the disc pale or dark (rarely with carotenoid pigments); peridium composed of intertwined hyphae often in a gelatinous matrix, dark brown, sometimes with dark brown setose hairs surrounding or interspersed with the hymenium; interascal tissue of paraphyses, often anastomosing near the base; asci cylindrical, persistent, with an often slightly subapical operculum; ascospores hyaline, rarely ornamented, multinucleate, sometimes with a sheath. Anamorphs hyphomycetous where known. Saprobic on rotten wood, esp. temp.
Lit.: Benkert (*Gleditschia* 19: 173, 1991; keys E. German spp.), Otani (*Trans. mycol. Soc. Japan* 21: 149, 1980; key Jap. spp.), Paden (*Fl. neotropica* 37, 1983; key).

Sarcosphaera Auersw. (1869), Pezizaceae. 1 (hypogeous), N. Am., Eur. See Brandrud *et al.* (*Blyttia* 44: 113, 1986), Trappe (*Mycotaxon* 2: 109, 1975).

Sarcostroma Cooke (1871) = Seimatosporium (Mitosp. fungi) fide Sutton (*Mycol. Pap.* 138, 1975), but see Nag Raj (1993).

Sarcostromella Boedijn (1959) = Camarops (Bolin.) fide Nannfeldt (1972).

sarcotrimitic, see Hyphal analysis.

Sarcotrochila Höhn. (1917), Hemiphacidiaceae. 4 (on conifers), N. Am., Eur. See Korf (*Mycol.* 54: 12, 1962), Dennis (*British Ascomycetes, addenda and corrigenda*: 22, 1981).

Sarcoxylon Cooke (1883), Xylariaceae. 2, Asia, Afr. See Rogers (*Mycol.* **73**: 28, 1981).

Sarcoxylum Clem. & Shear (1931) nom. nud. ≡ Sarcoxylon (Xylar.).

Sarea Fr. (1825), Agyriaceae. 2 (on resin), temp. See Hawksworth & Sherwood (*CJB* **59**: 357, 1981). Anamorphs *Epithyrium, Pycnidiella*.

Sarocladium W. Gams & D. Hawksw. (1976), Mitosporic fungi, 1.A1.15. 2 (on *Oryza*), Asia, USA. See Boa & Brady (*TBMS* **89**: 161, 1987), Bridge *et al.* (*Pl. Pathol.* **38**: 239, 1989; *Bambusa* blight), Chen *et al.* (*Acta Mycol. Sin. suppl.* **1**: 318, 1987; purple sheath disease of rice).

Sarophorum Syd. & P. Syd. (1916) = Penicilliopsis (Trichocom.) fide Malloch & Cain (*CJB* **50**: 2613, 1972). Accepted in Pitt & Samson (*Regnum veg.* **128**: 13, 1993).

Sarrameana Vězda & P. James (1973), ? Fuscideaceae (L). 2, New Caledonia, Tasmania. See Vězda & Kantvilas (*Lichenologist* **20**: 179, 1988).

Sartorya Vuill. (1927) nom. dub. (Ascomycota, inc. sed.). See Malloch & Cain (1972), *Neosartorya*.

Sartvellia Berk. (1857) ≡ Dasyspora (Uropyxid.).

Satchmopsis B. Sutton & Hodges (1975), Mitosporic fungi, 7.A1.15. 2, widespr. See Sutton & Pascoe (*TBMS* **88**: 169, 1987).

Sathropeltis Bat. & Matta (1959) = Myriangiella (Schizothyr.) fide v. Arx & Müller (1975).

satratoxins, toxins of *Stachybotrys atra* (syn. *S. alternans*); the cause of stachybotryotoxicosis in farm animals and humans.

saturnine (of ascospores), having a flat edge round the middle (as in some *Hansenula* spp.).

Saturnomyces Cain (1956) = Emericellopsis (Trichocom.) fide Gilman (*Manual of soil fungi*, 1956).

Saturnospora Z.W. Liu & Kurtzman (1991), ? Saccharomycetaceae. 4, widespr.

Satwalekera D. Rao, V.G. Rao & P.Rag. Rao (1970) = Torula (Mitosp. fungi) fide Kendrick & Carmichael (1973).

Satyrus Bosc (1811) = Phallus (Phall.).

Sauvageautia Har. (1892) = Urosporella (Amphisphaer.).

Savoryella E.B.G. Jones & R.A. Eaton (1969), ? Sordariales. 5, Eur., Asia. See Jones & Hyde (*Bot. Mar.* **35**: 83, 1992; posn), Read *et al.* (*CJB* **71**: 273, 1993; ultrastr.).

Săvulescu (Traian; 1889-1963). Romanian mycologist and plant pathologist esp. noted for his studies on rusts and smuts: *Monographia Uredinalelor din R.P.R.* (2 vols, 1953), *Ustilaginalele din R.P.R.* (2 vols, 1957). See Bontea (*Rev. Trop. Pl. Path.* **7**: 35, 1993), Sandu-Ville (*Sydowia* **18**: 1, 1965; portr., bibl.), Stafleu & Cowan (*TL-2* **5**: 91, 1985).

Savulescua Petr. (1959), Diaporthales (inc. sed.). 1, Puerto Rico.

Savulescuella Cif. (1959), Anamorphic Tilletiaceae, 3, widespr. Teleomorphs *Doassansia, Tracya*.

Sawadaea, see *Sawadaia*.

Sawadaia Miyabe (1914), Erysiphaceae. 5, Eur., Asia, NZ. See Braun (*Beih. Nova Hedw.* **89**, 1987), Zheng & Chen (*Acta microbiol. Sin.* **20**: 35, 1980).

saxicolous, growing on rocks. A few fungi are able to live on saxicolous substrata (Kobluk & Kahle, *Bull. Can. Pet. Geol.* **25**: 208, 1977; bibliogr.); *Lichenothelia* (q.v.) is primarily saxicolous. Rocks are one of the main substrata for lichens and the lit. on the latter is vast, see refs. cited under Ecology. Lichens can also change the mineral composition of rocks through the action of oxalic acid, etching being visible by SEM (see Syers & Iskandar, *in* Ahmadjian & Hale (Eds),

The Lichens: 225, 1974, Jones *et al.*, *Lichenologist* **12**: 277, 1980). See also Biodeterioration.

scab, as delimited by Jenkins (*Phytopath.* **23**: 389, 1933), a plant disease having hyperplastic scab-like lesions; cf. anthracnose. - of apple (*Venturia inaequalis*), cherry (*V. cerasi*), pear (*V. pirina*); cereals (*Gibberella zeae*, frequently as *G. saubinettii*, and *Fusarium* spp.), citrus (*Elsinoë fawcettii*), peach (*Fusicladium carpophilum*); **powdery** - of potato (*Spongospora subterranea*).

Scabradiporites Y.K. Mathur (1966), Fossil fungi (*f. cat.*). 1 (Paleocene), India.

scabrid, rough with delicate and irregular projections.

Scabropeziza Dissing & Pfister (1981), Pezizaceae. 2, Eur., N. Am. See Hirsch (*Agarica* **6**: 241, 1985; key Eur. spp.).

scabrous, rough.

Scalaria Lázaro Ibiza (1916) nom. dub. (Hymenochaet.).

Scalarispora Buriticá & J.F. Hennen (1994), Phakopsoraceae. 1, Taiwan.

Scalenomyces I.I. Tav. (1985), Laboulbeniaceae. 1, Sardinia.

Scalidium Hellb. (1867) = Scoliciosporum (Lecanor.) fide Poelt & Vězda (*Bibl. Lich.* **16**, 1981), Santesson (*Lichens of Norway and Sweden*, 1984).

Scanning electron microscopy (SEM). Technique for the examination of surface features at high magnifications by coating with a thin layer of gold, palladium or aluminium and studying the images (photographs) produced by a scanning electron beam. More recent technological developments allow specimens to be examined frozen or even uncoated. When linked to a computer, a digitization of the analogue signal allows specimens to be measured, counted and mapped.

Lit.: Heywood (Ed.) (*Scanning electron microscopy: systematic and evolutionary applications*, 1971), Claugher (Ed.) (*Scanning electron microscopy in taxonomy and functional morphology*, 1990). See Microscopy, Ultrastructure.

Scaphidiomyces Thaxt. (1912), Laboulbeniaceae. 4, S. Am., W. Afr.

Scaphidium Clem. (1901), Mitosporic fungi, 8.B1.19. 1, USA.

Scaphis Eschw. (1824) = Opegrapha (Roccell.).

Scaphophoeum Ehrenb. ex Wallr. (1833) ≡ Schizophyllum (Schizophyll.).

scar, (1) (of yeasts), **bud** -, on parent cell; **birth** -, on daughter cell; (2) (of Mitosporic fungi), at conidiogenous locus and conidial base/apex, left after secession of conidium.

scariose, thin; paper-like.

scarlet (elf) cup, the ascoma of *Sarcoscypha coccinea*.

Scedosporium Sacc. ex Castell. & Chalm. (1919), Anamorphic Microascales, 1.A1.19. Teleomorph *Pseudallescheria*. 2 (on humans, other animals and from hay), warmer areas. See Kendrick & Carmichael (1973), Salkin *et al.* (*J. Clin. Microb.* **26**: 498, 1988; *S. inflatum*, an emerging pathogen), Dykstra *et al.* (*Mycol.* **81**: 896, 1989; TEM of conidiogenesis). ? = Monosporium (Mitosp. fungi).

Scelobelonium (Sacc.) Höhn. (1905) = Crocicreas (Leot.) fide Carpenter (*Mem. N.Y. bot. Gdn* **33**: 1, 1981).

Scelophoromyces Thaxt. (1912), Laboulbeniaceae. 1, Am.

Scenidium (Klotzsch) Kuntze (1898) = Hexagonia (Coriol.). See Jülich (*Persoonia* **12**: 107, 1984).

Scenomyces F. Stevens (1927), Mitosporic fungi, ?5.-.-. 1, Panama.

Sceptrifera Deighton (1965), Mitosporic fungi, 2.B2.10/12. 1, Philipp.

Sceptromyces Corda (1831) = Aspergillus (Mitosp. fungi) fide Engelke (*Hedwigia* **41**: 219, 1902).

Schachtia Schulzer (1866) [non H. Karst. (1859), *Rubiaceae*] = Fenestella (Fenestell.).

Schadonia Körb. (1859), Bacidiaceae (L). 2, N. temp.

Schaereria Körb. (1855) typ. cons., Schaereriaceae (L). 4, mainly temp. See Eriksson & Hawksworth (*SA* **12**: 42, 1993; posn), Schneider (*Bibl. Lich.* **13**, 1980).

Schaereriaceae M. Choisy ex Hafellner (1984), Ascomycota (inc. sed., L). 2 gen., 5 spp. Thallus crustose or squamulose; ascomata apothecial, black, sessile or immersed, with a well-developed margin composed of globose cells; interascal tissue of paraphyses, branched near the apex, with a bright green or dark gelatinized epithecial layer; asci cylindrical, persistent, without separable wall layers, not thickened at the apex, with a thin external I+ gelatinized layer; releasing spores through a vertical split; ascospores usually globose to ellipsoidal, hyaline, aseptate, sometimes with a sheath. Anamorphs pycnidial. Lichenized with green algae, esp. temp.
Lit.: Eriksson & Hawksworth (*SA* **12**: 42, 1993; posn).

Scharifia Petr. (1955), Ascomycota (inc. sed.). 1, Iran.

Schasmaria (Ach.) Gray (1821) = Cladonia (Cladon.).

scheda (pl. **-ae**), **schedula** (pl. **-ae**), (Latin), label(s), esp. printed labels relating to sets of dried specimens (exsiccati; q.v.).

Scheleobrachea S. Hughes (1958) = Pithomyces (Mitosp. fungi) fide Ellis (*DH*).

Schenckiella Henn. (1893), Saccardiaceae. 1, S. Am.

Schenella T. Macbr. (1911), Stemonitidaceae. 2, USA.

Scherffelia Sparrow (1933) [non Pascher (1911), *Algae*] ≡ Scherffeliomyces (Chytrid.).

Scherffeliomyces Sparrow (1934), Chytridiaceae. 3, Eur., N. Am.

Scherffeliomycopsis Geitler (1962), Chytridiaceae. 1 (on *Coleochaete*), Austria.

Schiffnerula Höhn. (1909), Englerulaceae. *c.* 30, esp. warmer areas. See Hansford (*Mycol. Pap.* **15**, 1946), Hughes (*CJB* **62**: 2213, 1984). Anamorphs *Mitteriella, Questieriella, Sarcinella*.

Schinzia Nägeli (1842) = Entorrhiza (Tillet.) fide Vánky (1987).

Schinzinia Fayod (1889), Agaricaceae. 1, E. Afr.

Schismatomma Flot. & Körb. ex A. Massal. (1852), Roccellaceae (L). 8, Eur., N. Am., Asia, NZ. See Tehler (*Opera Bot.* **118**: 1, 1993; key), Torrente & Egea (1989).

Schismatommatomyces Cif. & Tomas. (1953) ≡ Schismatomma (Roccell.).

Schistodes Theiss. (1918) = Perisporiopsis (Parodiopsid.) fide Müller & v. Arx (1962).

Schistophoron Stirt. (1876), Caliciales (inc. sed., L). 2, E. Afr., C. & S. Am. See Tibell (1982, 1984).

Schistoplaca Brusse (1987) = Lecanora (Lecanor.) fide Lumbsch & Feige (*Mycotaxon* **52**: 429, 1994).

Schistostoma Stirt. (1878) = Graphis (Graphid.) fide Salisbury (*Lichenologist* **6**: 126, 1974).

Schizacrospermum Henn. (1899) nom. dub. (Fungi, inc. sed.) fide Walker (1980).

schizidium, a propagule formed by upper layers of a lichen thallus splitting off as scale-like segments from the main lobes (e.g. the lobule-like structures in *Fulgensia bracteata* subsp. *deformis*). See Poelt (*Flora, Jena* **159**: 23, 1980). Fig. 22G.

Schizmaturus (Corda) Kalchbr. (1880) = Lysurus (Clathr.) fide Dring (1980).

schizobiont, bacteria once considered to be additional symbionts of lichens.

Schizoblastosporion Cif. (1930), Mitosporic fungi, 1.A1.10. 2, Chile, N. Am. See Barnett (*et al.* (1990), Di Menna (*Mycopathol.* **25**: 205, 1965).

Schizoblastosporon Hara (1936) ≡ Schizoblastosporion (Mitosp. fungi).

Schizocapnodium Fairm. (1921) ? = Nitschkia (Nitschk.) fide Nannfeldt (1975).

Schizocephalum Preuss (1852) = Haplographium (Mitosp.fungi) fide Saccardo (*Syll. Fung.* **4**: 306, 1886).

Schizochora Syd. & P. Syd. (1913), Phyllachoraceae. 4, Philipp., Hawaii, NZ. See v. Arx & Müller (1973).

Schizochorella Höhn. (1918), Rhytismataceae. 1, Indonesia, Philipp.

Schizochytrium S. Goldst. & Belsky (1964), Thraustochytriaceae. 4 (marine), widespr. See Raghu-Kumar (*TBMS* **90**: 627, 1988; key), Dick (*in press*; key).

Schizoderma Ehrenb. (1818) = Leptostroma (Mitosp. fungi) fide Sutton (1977).

Schizoderma Kunze (1825) [non Ehrenb. (1818)], Mitosporic fungi, 6.?.?. 1, Germany.

Schizodiplodia Zambett. (1955) = Didymosporiella (Mitosp. fungi) fide Sutton (1977).

Schizodiscus Brusse (1988), Porpidiaceae (L). 1, Natal.

schizogenous, formed by cracking or splitting, cf. lysigenous.

schizogony, the process of division of a schizont.

Schizographa Nyl. (1857) nom. dub. (Fungi, inc. sed.).

Schizolaboulbenia Middelh. (1957) = Laboulbenia (Laboulben.) fide Benjamin (1971).

schizolytic, secession of conidia involving a splitting of the delimiting septum so that one half of the crosswall becomes the base of the seceding conidium and the other half remains at the apex of the conidiogenous cell (Cole & Samson, *Patterns of development in conidial fungi*, 1979). Cf. rhexolytic.

Schizoma Nyl. ex Cromb. (1894) = Thyrea (Lichin.) fide Henssen (1963).

Schizomeromyces Thaxt. (1931) = Clematomyces (Laboulben.) fide Tavares (1985).

Schizonella J. Schröt. (1877), Ustilaginaceae. 3 (on *Cyperaceae*), widespr., esp. Eur., N. Am.

Schizonia Pers. (1828) ≡ Schizophyllum (Schizophyll.).

schizont, a vegetative thallus, having no wall, which undergoes simple or multiple division.

Schizontopeltis Bat. & H. Maia (1962) = Schizothyrium (Schizothyr.) fide v. Arx & Müller (1975).

Schizoparme Shear (1923), Melanconidaceae. 7, Eur., N., C. & S. Am. See Samuels *et al.* (*Mycotaxon* **46**: 459, 1993). Anamorph *Coniella*.

Schizopelte Th. Fr. (1875), Roccellaceae (L). 1, N. Am.

Schizopeltis Bat. & I.H. Lima (1959) = Schizothyrium (Schizothyr.) fide Müller & v. Arx (1962).

Schizopeltomyces Cif. & Tomas. (1953) ≡ Schizopelte (Roccell.).

Schizophoma Kleb. (1933) ? = Sclerophoma (Mitosp. fungi) fide Sutton (1977).

Schizophyllaceae Quél. (1888), Schizophyllales. 4 gen. (+ 11 syn.), 42 spp. Basidioma pleurotoid; hymenophore of compressed radiate cupules appearing split-lamellate; lignicolous, rarely parasitic. In *Schizophyllum* marginal proliferations form 'split gills' composed of two adjacent pileal margins which in dry weather roll in to cover the hymenium; in *Plicaturopsis, Stromatoscypha* and *Henningsomyces* there is no proliferation of the cups.
Lit.: Donk (1964: 289), Cooke (*Mycol.* **53**: 575, 1961; *Schizophyllum*), Nuss (*Hoppea* **39**: 127, 1980).

Schizophyllales, Basidiomycetes. 2 fam. 6 gen. (+ 12 syn.), 46 spp. Fams:
(1) **Schizophyllaceae**.
(2) **Stromatoscyphaceae**.

Schizophyllum Fr. (1821), Schizophyllaceae. 5 (wood rot; pathogenic for humans), widespr. See Watling & Sweeney (*Sabouraudia* **12**: 214, 1974; cultural chars.).

Schizophyllus Fr. (1815) ≡ Schizophyllum (Schizophyll.).

Schizoplasmodiopsis L.S. Olive (1967), Protosteliaceae. 4, widespr. See Olive & Stoianovitch (*Mycol.* **67**: 1088, 1975; key, *Am. J. Bot.* **63**: 1385, 1976).

Schizoplasmodium L.S. Olive & Stoian. (1966), Protosteliaceae. 3, widespr. See Olive & Stoianovich (*Am. J. Bot.* **63**: 1385, 1976; ballistosporic).

Schizopora Velen. (1922), Hyphodermataceae. 4, widespr. See Ryvarden & Johansen (1980). = Hyphodontia (Hyphodermat.) fide Langer (*Bibl. Mycol.* **154**, 1994).

Schizosaccharis Clem. & Shear (1931) ≡ Schizosaccharomyces (Schizosaccharomycet.).

Schizosaccharomyces Lindner (1893), Schizosaccharomycetaceae. 1 or 4, widespr. See Barnett *et al.* (1990), Bridge & May (*J. gen. Microbiol.* **130**: 1921, 1984; numerical taxonomy), Kreger van Rij (1984; key), Mikata & Banno (*IFO Res. Comm.* **13**: 45, 1987; ascospores).

Schizosaccharomycetaceae Beij. ex Klöcker (1905), Schizosaccharomycetales. 2 gen. (+ 3 syn.), 4 spp. Mycelium absent or poorly developed; vegetative cells cylindrical with rounded ends, proliferating thallically by fission into 2 ± equal daughter cells; asci formed by somatic conjugation of vegetative cells, often irregularly shaped, 4- to 8-spored; ascospores globose to shortly cylindrical, blueing in iodine, smooth, without sheaths. Fermentation positive. *Lit.*: see *Schizosaccharomycetales*.

Schizosaccharomycetales, Ascomycota. 1 fam., 2 gen. (+ 3 syn.), 4 spp. Fam.: **Schizosaccharomycetaceae**. The name has been used by Prillinger *et al.* (*Z. Mykol.* **56**: 219, 1990) to include *Protomycetales* and *Taphrinales* which they considered intermediate between *Ascomycota* and *Basidiomycota*. *Lit.*: Eriksson *et al.* (*SA* **11**: 119, 1993).

Schizospora Dietel (1895) [non Reinsch (1875), *Algae*] = Pucciniosira (Pucciniosir.) fide Dietel (1928).

Schizostege Theiss. (1917) [non W.F. Hillebr. (1888), *Pteridophyta*] = Saccothecium (Dothior.).

Schizostoma Ehrenb. ex Lév. (1846), Tulostomataceae. 1, warm dry areas. See Long & Stouffer (*Mycol.* **35**: 21, 1943).

Schizostoma Ces. & De Not. ex Sacc. (1878) [non Ehrenb. ex Lév. (1846)] = Xenolophium (Melanomm.) fide Huhndorf (*Mycol.* **85**: 490, 1993). See Holm & Yue (*Acta mycol. Sin.* Suppl. **1**: 82, 1986).

Schizothecium Corda (1838), Lasiosphaeriaceae. *c.* 25 (coprophilous), cosmop. See Lundqvist (*Symb. bot. upsal.* **20** (1), 1972). = Podospora (Lasiosphaer.) fide Bell & Mahoney (*Mycol.* **87**: 375, 1995; key 8 spp. NZ).

Schizothyra Bat. & C.A.A. Costa (1957), Mitosporic fungi, 5.A1.?. 1, Brazil.

Schizothyrella Thüm. (1880), Mitosporic fungi, ?.?.?. 5, Eur., USA.

Schizothyrellina Petr. (1977) nom. nud. = Phacidiella (Mitosp. fungi) fide Sutton (1977).

Schizothyriaceae Höhn. ex Trotter *et al.* (1928), Dothideales. 15 gen. (+ 28 syn.), 51 spp. Ascomata strongly flattened or crustose, rounded or elongate, opening by irregular splits; upper wall composed of a single layer of ± epidermoid cells; interascal tissue absent, or composed only of remnants of stromatal cells; asci ± globose to saccate, fissitunicate; ascospores hyaline to pale brown, transversely septate, without a sheath. Anamorphs unknown. Saprobic, epiphytic on leaves or stems, widespr.

Schizothyrina Bat. & I.H. Lima (1959) = Schizothyrium (Schizothyr.) fide Müller & v. Arx (1962).

Schizothyrioma Höhn. (1917), Dermateaceae. 2, Eur. See Holm (*Svensk bot. Tidskr.* **65**: 208, 1971).

Schizothyrium Desm. (1849), Schizothyriaceae. *c.* 12, widespr., esp. trop. *S. pomi* (fly speck of apple and pear). See v. Arx & Müller (1975).

Schizothyropsis Bat. & A.F. Vital (1960), Mitosporic fungi, 5.A1.?. 1, Paraguay.

Schizotorula Krassiln. (1954) nom. inval. A lapsus for *Schizotorulopsis*.

Schizotorulopsis Cif. (1930) nom. dub.; ? based on bacteria.

Schizotrichella E.F. Morris (1956) = Colletotrichum (Mitosp. fungi) fide Sutton (1977).

Schizotrichum McAlpine (1903), Mitosporic fungi, 1.E1.10. 2, Australia, Bermuda.

schizotype, a syntype taken to be the type of name by the exclusion of other syntypes; an implicit lectotype, but not acceptable as a formal typification; see type.

Schizoxylon Pers. (1810), Stictidaceae. 32, esp. temp. See Sherwood (*Mycotaxon* **6**: 215, 1977; key).

Schizoxylum Pers. (1822) ≡ Schizoxylon (Stictid.).

schlauch, open-ended extension of the rohr, orientated toward the cytoplasm of encysted zoospore of plasmodiophorids. See also rohr and stachel.

Schmitzomia Fr. (1849) = Stictis (Stictid.) fide Sherwood (1977).

Schnablia Sacc. & P. Syd. (1899), ? Leotiales (inc. sed.). 1, Eur.

Schneepia Speg. (1885) = Parmularia (Parmular.) fide Müller & v. Arx (1962).

Schoenbornia Bubák (1906) = Hymenopsis (Mitosp. fungi) fide Sutton (1977).

Schoenleinium Johan-Olsen (1897) = Trichophyton (Mitosp. fungi) fide Kendrick & Carmichael (1973).

Schrakia Hafellner (1979), ? Patellariaceae. 1, N. & S. Am.

Schroeterella Syd. (1922) [non Herzog (1916), *Musci*] = Puccinia (Puccin.) fide Dietel (1928).

Schroeteria G. Winter (1881), Mitosporic fungi, 1.A1.15. 2 (on *Veronica*), Eur. See Nagler *et al.* (*Mycol.* **81**: 884, 1989; ascomycete affinities).

Schroeteriaster Magnus (1896), Uredinales (inc. sed.). 1, Eur.; II, III on *Rumex*. See Mains (*Ann. Myc.* **32**: 256, 1934). = Uromyces (Puccin.) fide Cummins & Hiratsuka (1983).

Schulzeria Bres. & Schulzer (1886) nom. dub. (Agaric.) fide Singer (1951).

Schwanniomyces Klöcker (1909) = Debaryomyces (Saccharomycet.) fide Kurtzman & Robnett (*Yeast* **7**: 61, 1991). See also Yamada *et al.* (*J. gen. appl. Microbiol.* **37**: 523, 1991).

Schwarzmannia Pisareva (1968), Mitosporic fungi, 6.B2.19. 1, Kazakhstan.

Schweinitz (Lewis David von; 1780-1834). The first important American mycologist. A Pennsylvanian by birth, Schweinitz went to Saxony in 1798 for a training in theology and mycology under Albertini, there producing with him the important *Conspectus Fungorum in Lusatiae* (1805). On going back to the United States in 1812 as an official of the church he was able to go on with his mycological work. In 1822 his *Synopsis Fungorum Carolinae superioris* was printed and in the *Synopsis Fungorum in America Boreali* (1832) he gave an account of all the fungi (about 4,000) then known in North America.

Schweinitz, who worked with care, had one of the best compound microscopes of his day. He sent parts of his specimens to numbers of co-workers in Europe and America. See Shear & Stevens (*Mycol.* **9**: 191, 1917), Rogers (*Mycol.* **69**: 223, 1977), Shear (*Pl. World* **5**: 45, 1902; portr.), Shear & Stevens (*Mem. Torrey bot. Cl.* **16**: 119, 1921; Schweinitz-Torrey correspondence), Stafleu & Cowan (*TL-2* **5**: 437, 1985).

Schweinitzia Grev. (1823) [non Elliott ex Nuttall (1818), *Pyrolaceae*] = Podaxis (Podax.).

Schweinitzia Massee (1895) ? = Velutaria (Hyaloscyph.) fide Nannfeldt (1932).

Schweinitziella Speg. (1888), Trichosphaeriaceae. 3 (on fungi), trop. See Pirozynski (*Kew Bull.* **31**: 359, 1977).

Schwendener (Simon; 1829-1919). Studied at Geneva under de Candolle. Professor of botany, Berlin. Elucidated true nature of the components of lichen symbiosis (*Verh. Schweiz. naturf. Ges. Aaran* **51**: 88, 1867, *Die Algentypen der Flechtengonidien*, 1869). See Ainsworth (*Introduction to the history of mycology*, 1976), Vines (*Proc. Linn. Soc.* 1919-20: 47, 1921), Stafleu & Cowan (*TL-2* **5**: 443, 1985).

Scindalma (Hill) Kuntze (1898) = Phellinus (Hymenochaet.) fide Pegler (1973).

Sciniatosporium Kalchbr. ex Morgan-Jones (1971) ≡ Seimatosporium (Mitosp. fungi) fide Sutton (*TBMS* **58**: 164, 1972).

Sciodothis Clem. (1909) = Tomasellia (Arthopyren.) fide Harris (*in litt.*).

Scirrhia Nitschke ex Fuckel (1870), Dothideaceae. 8, widespr. *S. acicola* (brown needle blight of pine; Siggers, *Tech. Bull. U.S. Dept. Agric.* **870**, 1944). See Obrist (*Phytopath. Z.* **35**: 370, 1959), Barr (*Contr. Univ. Mich. Herb.* **9**: 523, 1972). ? = Metameris (Dothid.) p.p. fide v. Arx & Müller (1975).

Scirrhiachora Theiss. & Syd. (1915) = Mycosphaerella (Mycosphaerall.) fide v. Arx & Müller (1975).

Scirrhiella Speg. (1885) = Apiospora (Lasiosphaer.) fide Barr (1976).

Scirrhiopsis Henn. (1905) nom. dub. (Fungi, inc. sed.).

Scirrhodothis Theiss. & Syd. (1915) = Metameris (Phaeosphaer.) fide Barr (*Mycotaxon* **43**: 371, 1992).

Scirrhophoma Petr. (1941), Mitosporic fungi, 8.A1.?. 1, Eur.

Scirrhophragma Theiss. & Syd. (1915) = Metameris (Phaeosphaer.) fide Barr (*Mycotaxon* **43**: 371, 1992).

scissile (of the flesh of a pileus), separating into horizontal layers.

Scitovszkya Schulzer (1866) = Mucor (Mucor.) fide Hesseltine (1955).

Sclerangium Lév. (1848) = Scleroderma (Sclerodermat.).

sclerobasidium, see basidium.

sclerocarps, sclerotium-like modified ascomata permanently lacking a sexual capacity and acting as sclerotia, as in *Varicosporina ramulosa* (Kohlmeyer & Charles, *CJB* **59**: 1787, 1981).

Sclerochaeta Höhn. (1917) = Chaetopyrena (Mitosp. fungi) fide Höhnel (*Hedwigia* **60**: 129, 1918).

Sclerochaetella Höhn. (1917) = Diploplenodomus (Mitosp. fungi) fide Petrak (*Ann. Myc.* **42**: 58, 1944).

Sclerocleista Subram. (1972) = Hemicarpenteles (Trichocom.) fide v. Arx (1981). Accepted in Pitt & Samson (*Regnum veg.* **128**: 13, 1993).

Sclerococcum Fr. (1819), Mitosporic fungi, 3.D2.1/10. 4 (on lichens), Eur. See Hawksworth (*TBMS* **65**: 219, 1975, *Bull. Br. Mus. nat. Hist., Bot.* **6**: 181, 1979), Hawksworth & Jones (*TBMS* **77**: 485, 1981; culture).

Scleroconium Syd. (1935), Mitosporic fungi, 3.A1.6. 1, S. Am.

Sclerocrana Samuels & L.M. Kohn (1987), Sclerotiniaceae. 1, NZ.

Sclerocystis Berk. & Broome (1873), Glomaceae. 1, trop. See Almeida & Schenk (*Mycol.* **82**: 703, 1990; revision), Wu (*Mycotaxon* **47**: 25, 1993, **49**: 327, 1993; revision), Yao *et al.* (1995).

Sclerodepsis Cooke (1890), Coriolaceae. 4, widespr. See Pegler (1973).

Scleroderma Pers. (1801), Sclerodermataceae. 25, widespr. See Guzmán (*Darwiniana* **16**: 233, 1970;

key), Richter (*Mycotaxon* **45**: 461, 1992; cultures), Rifai (*Trans. mycol. Soc. Japan* **28**: 97, 1987; Malesia, key), Sarasini (*Boll. Ass. Micol. Bresad.* **34**: 119, 1991; Italy).

Sclerodermataceae Corda (1842), Sclerodermatales. 6 gen. (+ 23 syn.), 31 spp. Gasterocarp epigeous, tuberous, fracturing irregularly; gleba finally powdery; spores brown, verrucose.

Sclerodermatales, Basidiomycetes ('gasteromycetes'). 4 fam., 11 gen. (+ 26 syn.), 38 spp. Gleba with small locules without real hymenium, powdery at maturity, no true stipe and no true capillitium; widespr. subcosmop., saprobic in soil or rotting wood or mycorrhizal. Fams:
(1) **Astraeaceae**.
(2) **Diplocystaceae**.
(3) **Sclerodermataceae**.
(4) **Sphaerobolaceae**.
Lit.: Dring (1973; key gen.). See also *Lit.* under *Gasteromycetes*.

Sclerodermatopsis Torrend (1923) ? = Xylaria (Xylar.) fide Læssøe (*SA* **13**: 43, 1994).

Scleroderris (Fr.) Rehm (1888) = Godronia (Leot.) fide Groves (*CJB* **43**: 1195, 1965).

Sclerodiscus Pat. (1890), Mitosporic fungi, 8.A2.?. 1, Asia.

Sclerodon P. Karst. (1889) ≡ Gloiodon (Gloeocystidiell.).

Sclerodothiorella Died. (1912), Mitosporic fungi, 8.A1.?. 1, Eur.

Sclerodothis Höhn. (1918) = Leptosphaeria (Leptosphaer.) fide v. Arx & Müller (1975).

Sclerogaster R. Hesse (1891), Octavianinaceae. 10, Eur., Am. See Dodge & Zeller (*Ann. Mo. bot. Gdn* **23**: 565, 1937), Fogel (*Mycol.* **82**: 655, 1990; N. Am. key).

Scleroglossum Pers. (1820) = Acrospermum (Acrosperm.).

Scleroglossum Hara (1948) = Scleromitrula (Sclerotin.) fide Korf (*in* Ainsworth *et al.*, *The Fungi* **4A**: 249, 1973).

Sclerogone Warcup (1990), Endogonaceae. 1, Australia. See Warcup (*in* Sanders *et al.*, *Endomycorrhizas*: 53, 1975; culture), Warcup (*MR* **94**: 173, 1990), Yao *et al.* (1995).

Sclerographa (Vain.) Zahlbr. (1923) = Opegrapha (Roccell.) fide Upreti & Singh (*Beitr. Bibl. Pflanzen* **62**: 233, 1987).

Sclerographiopsis Deighton (1973), Mitosporic fungi, 2.C2.3. 1, Sierra Leone. See Sutton & Pascoe (*Austr. Jl Bot.* **35**: 183, 1987; cf. *Tandonella*).

Sclerographium Berk. (1854), Mitosporic fungi, 2.D2.10. 3, Asia, Afr. See Deighton (*Mycol. Pap.* **78**, 1960).

Scleroma Fr. (1838) = Panus (Lentin.) fide Singer (1951).

Scleromeris Syd. (1926), Mitosporic fungi, 8.A2.?. 1, C. Am.

Scleromitra Corda (1829) = Pistillaria (Typhul.) fide Corner (1950).

Scleromitrula S. Imai (1941), Sclerotiniaceae. 1 (on *Morus*), Japan. See Kohn & Nagasawa (*Trans. mycol. Soc. Japan* **25**: 127, 1984), Spooner (*Bibl. Mycol.* **116**, 1987).

Scleromium Linds. (1869) nom. inval. ≡ Pseudographis (Tryblid.).

Scleroparodia Petr. (1934) = Ascochytopsis (Mitosp. fungi) fide Sutton (1977).

Sclerophoma Höhn. (1909), Anamorphic Dothioraceae, 8.A1.15. Teleomorph *Sydowia*. 30, widespr. See Hermanides-Nijhof (*Stud. Mycol.* **15**: 166, 1977; as *Dothichiza*).

Sclerophomella Höhn. (1917) = Phoma (Mitosp. fungi) fide Sutton (1977).

Sclerophomina Höhn. (1917) = Phoma (Mitosp. fungi) fide Sutton (1977).

Sclerophora Chevall. (1826), Coniocybaceae (L). 5, temp. See Tibell (1984).

Sclerophthora Thirum., C.G. Shaw & Naras. (1953), Verrucalvaceae. 4, widespr.

Sclerophyton Eschw. (1824), Roccellaceae (L). 8, cosmop. See Egea & Torrente (*Bryologist* **98**: 207, 1995; N. Am.), Redinger (*Feddes Repert.* **43**: 49, 1938).

Sclerophytonomyces Cif. & Tomas. (1953) ≡ Sclerophyton (Roccell.).

Scleroplea (Sacc.) Oudem. (1900) = Pyrenophora (Pleospor.) fide (*Sydowia* **5**: 248, 1951).

scleroplectenchyma, plectenchyma (q.v.) composed of very thick-walled conglutinate cells. See Yoshimura & Shimada (*Bull. Kochi Gakuen Jun. Coll.* **11**: 13, 1990); stereome.

Scleropleella Höhn. (1918) = Leptosphaerulina (Pleospor.) fide Barr (1972).

scleroprosoplectenchyma, see plectenchyma.

Scleropycnis Syd. & P. Syd. (1911), Mitosporic fungi, 8.A1.15. 2, Eur. See Sutton & Livsey (*TBMS* **88**: 271, 1987; gen. redescription).

Scleropycnium Heald & C.E. Lewis (1912) ? = Phomopsis (Mitosp. fungi) fide Sutton (1977).

Scleropyrenium H. Harada (1993), Verrucariaceae (L). 2, Japan.

Sclerosphaeropsis Bubák (1914) nom. dub. fide Sutton (*Mycol. Pap.* **141**, 1977).

Sclerospora J. Schröt. (1879), Sclerosporaceae. 16 (pathogens of *Gramineae*), widespr. See Waterhouse (*Misc. Publ. CMI* **17**, 1964; key), Narayanan (*Mycopath.* **20**: 315, 1963; India).

Sclerosporaceae M.W. Dick (1984), Sclerosporales. 2 gen. (+ 1 syn.), 25 spp.

Sclerosporales, Oomycota. 2 fam., 5 gen. (+ 1 syn.), 27 spp. Fams:
(1) **Sclerosporaceae**.
(2) **Verrucalvaceae**.

Sclerosporis Stach & A. Chandra (1956), Fossil fungi (*f. cat.*). 1 (Carboniferous, Tertiary), Eur. See Beneš (1969).

Sclerostagonospora Höhn. (1917) = Stagonospora (Mitosp. fungi) fide Petrak (*Ann. Myc.* **23**: 4, 1925), but see Sutton (1980).

Sclerostilbum Povah (1932) = Tilachlidiopsis (Mitosp. fungi) fide Stalpers (*CJB* **69**: 6, 1990).

Sclerotelium Syd. (1921) = Puccinia (Puccin.) fide Dietel (1928).

Sclerotheca Bubák & Vleugel (1908) [non DC. (1839), *Campanulaceae*] = Camarosporium (Mitosp. fungi) fide Sutton (1980).

sclerothionine, a plant growth promoting metabolic product of *Sclerotinia libertiana* (Matsu & Satomura, *Agr. biol. Chem.* **32**: 611, 1968).

Sclerothyrium Höhn. (1918) = Microsphaeropsis (Mitosp. fungi) fide Sutton (1977).

sclerotic cell, see muriform cell.

Sclerotiella A.K. Sarbhoy & A. Sarbhoy (1974), Mitosporic fungi, 9.-.-. 1, India.

Sclerotinia Fuckel (1870) nom. cons., Sclerotiniaceae. 3, widespr. *S. minor* and *S. sclerotiorum* (plurivorous), *S. trifoliorum* (clover rot). See Kohn (*Mycotaxon* **9**: 365, 1979; monogr., key, 259 epithets), Spooner (1987; concept), Tariq *et al.* (*TBMS* **84**: 381, 1985; biochem. systematics), Willetts & Wong (*Bot. Rev.* **46**: 101, 1980; compr. review).

Sclerotiniaceae Whetzel (1945), Leotiales. 33 gen. (+ 13 syn.), 225 spp. Stromata present, as sclerotia or mummified host tissue; ascomata apothecial, often long-stalked, usually brown, cupulate, without hairs,

the stalk often darker; interascal tissue of simple paraphyses; asci usually with an I+ apical ring; ascospores large or small, ellipsoidal, usually aseptate, hyaline or pale brown, often ± longitudinally symmetrical. Anamorphs often prominent, hyphomycetous, spermatial or disseminative. Pathogenic or saprobic on various plant parts, esp. seeds and fruits, cosmop.

Lit.: Buchwald (*Kgl. Veter. Landboh. Aarsskr.*, 1949), Dumont & Korf (*Mycol.* **63**: 157, 1971; gen. nomencl.), Kohn (*TBMS* **91**: 639, 1988; protein electroph.), Kohn & Grenulle (*CJB* **67**: 371, 1989; anatomy, histochem., **67**: 394, 1989; ultrastr.), Novak & Kohn (*Appl. Envir. Microbiol.* **57**: 525, 1991; protein electrophor.), Schwegler (*Schweiz. Z. Pilzkde* **56**: 49, 1978), Spooner (*Bibl. Mycol.* **116**, 1987; gen. concepts), Verkley (*MR* **97**: 179, 1993; ultrastr.), Whetzel (*Mycol.* **37**: 648, 1945; fams.), Zhuang (*Acta Mycol. Sin.* **13**: 13, 1994; China).

Sclerotiomyces Woron. (1926) ? = Sclerotium (Mitosp. fungi).

Sclerotiopsis Speg. (1882) = Pilidium (Mitosp. fungi) fide Petrak (*Sydowia* **5**: 328, 1951).

Sclerotites A. Massal. (1857), Fossil fungi. 16 (Tertiary), Eur., USA.

Sclerotites Jeffrey & Chrysler (1906), Fossil fungi (*f. cat.*). 6 (Tertiary), Eur., USA. See Beneš (1969).

sclerotium, (1) a firm, frequently rounded, mass of hyphae with or without the addition of host tissue or soil, normally having no spores in or on it (cf. bulbil, stroma); see Willetts (*Biol. Rev.* **46**: 387, 1971; survival, **47**: 515, 1972; morphogenesis), Willetts & Bullock (*MR* **96**: 801, 1992; developm. biology). A sclerotium may give rise to a fruit body, a stroma (as in ergot), or mycelium. See blackfellows' bread, stone fungi, tuckahoe, Mitosporic fungi. (2) (of *Myxomycetes*), the firm, resting condition of a plasmodium.

Sclerotium Tode (1790), Mitosporic fungi, 9.-.-. 100, widespr. *S. cepivorum* (onion (*Allium*) white rot), *S. oryzae* (on rice), *S. rolfsii* (plurivorous; Aycock, *Tech. Bull. N. Carol. agric. Exp. Stn* **174**, 1966), *S. tuliparum* (grey bulb rot of tulip). A number of the earlier spp. are anamorphs of *Basidiomycota*, *Ascomycota*, etc. See Backhouse & Stewart (*TBMS* **89**: 561, 1987; histochemistry), Hanlin & Tortolero (*CJB* **67**: 1852, 1989; morphology *S. coffeicola*).

Sclerozythia Petch (1937), Mitosporic fungi, 4.A1.?. 1, UK.

Sclerozythia Petr. (1955) [non Petch (1937)] ≡ Neozythia (Mitosp. fungi).

scobiculate, in fine grains, like sawdust.

scocho, see Fermented food and drinks.

Scodellina Gray (1821) ≡ Peziza (Peziz.).

Scolecactis Clem. (1909) = Bactrospora (Roccell.) fide Egea & Torrente (1993).

Scoleciasis Roum. & Fautrey (1889) nom. dub.; based on ascospores of *Leptosphaeria* (Phaeosphaer.) fide Sutton (*in litt.*).

Scoleciocarpus Berk. (1843) = Arachnion (Lycoperd.) fide Demoulin (*Nova Hedw.* **21**: 646, 1972).

scolecite, see Woronin's hypha.

Scolecites Stizenb. (1862) ≡ Skolekites (Catillar.).

Scolecobasidiella M.B. Ellis (1971), Mitosporic fungi, 1.B2.10. 1, Somalia.

Scolecobasidium E.V. Abbott (1927), Mitosporic fungi, 1.B2.11. 5, N. Am., Eur. See Barron & Busch (*CJB* **40**: 77, 1962; key), Graniti (*G. bot. Ital.* **69**: 364, 1963; gen. concept), de Hoog & v. Arx (*Kavaka* **1**: 55, 1973; monotypic), de Hoog (*Stud. Mycol.* **26**: 51, 1985; key).

Scolecobasis Clem. & Shear (1931) ≡ Scolecobasidium (Mitosp. fungi).

Scolecobonaria Bat. (1962), Dothideales (inc. sed.). 2, Taiwan, USA.

Scolecoccoidea F. Stevens (1927) = Coccodiella (Phyllachor.) fide Müller & v. Arx (1973). See also Pirozynski (*Mycol.* **65**: 164, 1973).

Scolecodothis Theiss. & Syd. (1914) = Ophiodothella (Phyllachor.).

Scolecodothopsis F. Stevens (1924) = Diatractium (Phyllach.) fide Cannon (*MR* **92**: 327, 1989).

Scoleconectria Seaver (1909), Hypocreaceae. 8, Eur., C. & N. Am., Australasia. = Ophionectria (Hypocr.) fide Clements & Shear (1931).

Scolecopeltella Speg. (1923) = Micropeltis (Micropeltid.) fide v. Arx & Müller (1975).

Scolecopeltidella J.M. Mend. (1925) nom. dub. fide Petrak (*Sydowia* **5**: 190, 1951), v. Arx & Müller (1975).

Scolecopeltidium F. Stevens & Manter (1925), ? Microthyriaceae. *c.* 80, trop.

Scolecopeltis Speg. (1889) = Micropeltis (Micropeltid.) fide v. Arx & Müller (1975).

Scolecopeltium Clem. & Shear (1931) ≡ Scolecopeltidium (Microthyr.).

Scolecopeltopsis Höhn. (1909) = Micropeltis (Micropeltid.) fide v. Arx & Müller (1975).

scolecospore, a spore resembling an amerospore (q.v.) but with or without septa and a length/width ratio > 15:1. See Mitosporic fungi.

Scolecosporiella Petr. (1921), Anamorphic Phaeosphaeriaceae, 4.C/D2.1. Teleomorph *Phaeosphaeria*. 5, Eur., N. Am., Afr. See Sutton (1980, *CJB* **46**: 189, 1968; key), Nag Raj (1993).

Scolecosporiella Höhn. (1923), Mitosporic fungi, 4/6.?C1. 1, Eur.

Scolecosporites R.T. Lange & P.H. Sm. (1971), Fossil fungi (*f. cat.*). 1 (Paleocene), Australia.

Scolecosporium, see *Scolicosporium*.

Scolecostroma Bat. & Peres (1960) = Lophium (Mytilinid.) fide v. Arx & Müller (1975).

Scolecotrichum, see *Scolicotrichum*.

Scolecoxyphium Cif. & Bat. (1956), Mitosporic fungi, 3.A1.?. 4, trop. See Hughes (*Mycol.* **68**: 693, 1976).

Scolecozythia Curzi (1927), Mitosporic fungi, 8.E1.?. 1, Italy.

Scoliciosporomyces Cif. & Tomas. (1953) = Scoliciosporum (Lecanor.).

Scoliciosporum A. Massal. (1852), Lecanoraceae (L). 7, temp. See Vězda (*Folia geobot. phytotax.* **13**: 397, 1978; key).

Scolicosporium Lib. ex Roum. (1880), Mitosporic fungi, 6.C2.1. 2, Eur. See Sutton (*Mycol. Pap.* **138**, 1975), Spooner & Kirk (*TBMS* **78**: 247, 1982; relationship with *Excipularia*), Constantinescu (*Mycotaxon* **42**: 467, 1991).

Scolicotrichum Kunze (1817), Mitosporic fungi, ?.?.?. 20, widespr. *S. graminis* (leaf spot of cereals and grasses). See Höhnel (*Zbl. Bakt.* Abt. 2 **60**: 1, 1932; nom. conf.).

Scoliocarpon Nyl. (1858), Ascomycota (inc. sed.). 1, N. Am.

Scoliolegnia M.W. Dick (1969), Saprolegniaceae. 3, Eur.

Scolionema Theiss. & Syd. (1918), Parodiopsidaceae. 1, S.E. Asia.

Scoliotidium Bat. & Cavalc. (1963), Mitosporic fungi, 5.C1.?. 1, Brazil.

Scopaphoma Dearn. & House (1925), Mitosporic fungi, 4.B1.1. 1, USA.

Scopella Mains (1939) = Maravalia (Chacon.) fide Ono (*Mycol.* **76**: 892, 1984).

Scopellopsis T.S. Ramakr. & K. Ramakr. (1947) = Maravalia (Puccin.) fide Thirumalachar & Mundkur (*Indian Phytopath.* **2**: 193, 1949).

Scopinella Lév. (1849), ? Ceratostomataceae. 5, temp. See Cannon & Hawksworth (*Bot. J. Linn. Soc.* **84**: 115, 1982; key), Tsuneda (*Rept. Tottori mycol. Inst.* **22**: 82, 1984; ontogeny), Tsuneda *et al.* (*Trans. mycol. Soc. Japan* **26**: 221, 1985; SEM, posn).

Scoptria Nitschke (1867) nom. rej. = Eutypella (Diatryp.) fide Rappaz (1987).

Scopula Arnaud (1952) = Gloiosphaera (Mitosp. fungi) fide Wang (*Mycol.* **63**: 890, 1971).

Scopularia Preuss (1851) [non Lindl. (1834), *Orchidaceae*] nom. dub. fide Kendrick (*CJB* **42**: 1119, 1964).

Scopulariella Gjaerum (1971), Mitosporic fungi, 2.B1.4. 1, Denmark.

Scopulariopsis Bainier (1907), Anamorphic Microascaceae, 1.A1.19. Teleomorph *Microascus*. 13, widespr. *S. fimicola*, white plaster mould of mushroom beds; *S. brevicaulis* used for the detection of arsenic (Lerrigo, *Analyst* **57**: 155, 1932) and proposed as an agent of cot death (q.v.). See Raper & Thom (*Manual of the penicillia*, 1949), Morton & Smith (*Mycol. Pap.* **86**, 1963; descr. & illustr. 18 spp.).

Scopulina Lév. (1846) [non Dumort. (1882), *Hepaticae*] ≡ Scopinella (Ceratostomat.).

Scopuloides (Massee) Höhn. & Litsch. (1908), Meruliaceae. 4, widespr. See Boidin *et al.* (*Cryptogamie Mycol.* **14**: 195, 1993; key Eur. spp.).

Scoriadopsis J.M. Mend. (1930), ? Capnodiaceae. 1, S. Am.

Scorias Fr. (1825), Capnodiaceae. 3, trop. See Reynolds (*Mycotaxon* **8**: 417, 1979).

Scoriomyces Ellis & Sacc. (1885) nom. dub. (Myxomycota) fide Carris (*Sydowia* **45**: 92, 1993).

scorpioid, a branching system in which the laterals are curved so that they all appear to arise from one side of the main stem, as in *Cladonia arbuscula*.

Scorpiosporium S.H. Iqbal (1974), Mitosporic fungi, 1.G2.1. 4, UK.

Scortechinia Sacc. (1885) = Nitschkia (Nitschk.) fide Nannfeldt (1975).

Scortechinia Sacc. (1891) = Nitschkia (Nitschk.) fide Nannfeldt (1975).

Scortechiniella Arx & E. Müll. (1954) = Nitschkia (Nitschk.) fide Nannfeldt (1975).

Scortechiniellopsis Sivan. (1974) = Nitschkia (Nitschk.) fide Nannfeldt (1975).

Scorteus Earle (1909) = Marasmius (Tricholomat.) fide Pennington (1915).

Scothelius Bat., J.L. Bezerra & Cavalc. (1965), Mitosporic fungi, 5.A1.?. 1, Brazil.

Scotiosphaeria Sivan. (1977), Ascomycota (inc. sed.). 1, UK.

Scotoderma Jülich (1974), Stereaceae. 1, UK.

Scotomyces Jülich (1978) = Hydrabasidium (Ceratobasid.) fide Roberts (*in litt.*).

screening, routine testing of organisms or chemical substances for a particular property (e.g. for antibiotic production, fungicidal effect, etc.). See Dreyfuss (*Sydowia* **39**: 22, 1987; pharmaceutical prospecting), Bioprospecting.

scrobiculate, (1) roughened; resembling sawdust; (2) (of lichens), coarsely pitted; foveolate.

scrupose, rough with very small hard points.

scutate, like a round plate or shield.

Scutellaria Baumg. (1790) [non L. (1753), *Labiatae*], nom. dub. (Lecanorales).

Scutellinia (Cooke) Lambotte (1887) nom. cons., Otideaceae. 45, widespr., esp. N. temp. See Schumacher (*Opera Bot.* **101**, 1990; key), van Brummelen (*Persoonia* **15**: 129, 1993; ultrastr.), Kaushal *et al.* (*Bibl. Mycol.* **91**: 583, 1983; key 14 spp. India), Schumacher (*Mycotaxon* **33**: 149, 1988; 144 excl. names),

Wu & Kimbrough (*Bot. Gaz.* **152**: 421, 1992; ascospores).

Scutelloidea Tim (1971) = Rosencheldiella (Ventur.) fide v. Arx & Müller (1975).

Scutellospora C. Walker & F.E. Sanders (1986), Gigasporaceae. 29, widespr. See Walker & Saunders (*Mycotaxon* **27**: 169, 1986), Koske & Walker (*Mycol.* **77**: 702, 1985, *Mycotaxon* **27**: 219, 1986), Morton & Koske (*Mycol.* **80**: 520, 1988), Koske & Halvorson (*Mycol.* **81**: 927, 1989), Spain *et al.* (*Mycotaxon* **35**: 219, 1989), Błaszkowski (*Mycol.* **83**: 537, 1991), Walker *et al.* (*Cryptogamie, Mycol.* **14**: 279, 1993), Franke & Morton (*CJB* **72**: 122, 1994; revision, phylogeny), Morton (*Mycol.* **87**: 127, 1995; development, phylogeny), Yao *et al.* (1995).

scutellum, see thyriothecium.

Scutellum Speg. (1881) nom. dub. (Fungi, inc. sed.) fide Müller & v. Arx (1962).

Scutiger Paulet (1793), Scutigeraceae. 1, Eur. ? = Albatrellus (Scutiger.). See Donk (1974).

Scutigeraceae Bondartsev & Singer (1941), Cantharellales. 3 gen. (+ 2 syn.), 14 spp. Basidioma pileate, tubulate; hymenium not thickening.

Scutisporium Preuss (1851) = Stemphylium (Mitosp. fungi) fide Saccardo (*Syll. Fung.* **4**: 519, 1886).

Scutisporus K. Ando & Tubaki (1985), Mitosporic fungi, 1.D1.1/10. 1, Japan.

Scutobelonium Graddon (1984), Dermateaceae. 1, UK.

Scutomollisia Nannf. (1976), Dermateaceae. 4, N. Eur.

Scutomyces J.L. Bezerra & Cavalc. (1972), Ascomycota (inc. sed., L). 1, Brazil.

Scutopeltis Bat. & H. Maia (1957), Mitosporic fungi, 5.A1.?. 1, Ghana.

Scutopycnis Bat. (1957), Mitosporic fungi, 5.A1.?. 1, USA.

Scutoscypha Graddon (1980) = Calycellina (Hyaloscyph.) fide Baral & Krieglsteiner (*Beih. Z. Mykol.* **6**, (1985).

Scutula Tul. (1852) nom. cons. prop. [non Lour. (1790), Melastomataceae], Micareaceae. 10 (in lichens), widespr. See Hawksworth (*TBMS* **74**: 363, 1980), Jørgensen & Santesson (*Taxon* **42**: 881, 1993; nomencl.), Rambold & Triebel (*Bibl. Lich.* **48**, 1992).

Scutularia P. Karst. (1885) = Xylogramma (Leot.). See Nannfeldt (*Nova Acta R. Soc. Scient. upsal.* 4 **8**(2), 1932).

Scutulopsis Velen. (1934), ? Leotiales (inc. sed.). 1, former Czechoslovakia.

scutulum (pl. -a), cup-like crust or mat of hyphae produced in the follicles of the scalp or the body in infections by *Trichophyton schoenleinii*.

Scypharia (Quél.) Quél. (1886) [non Miers (1860), *Rhamnaceae*] = Sarcoscypha (Sarcoscyph.) fide Denison (1972).

Scyphiphorus Vent. (1799) ≡ Pyxidium (Cladon.).

Scyphium Rostaf. (1874) = Badhamia (Physar.).

scyphoid, cup-like.

Scyphophora Gray (1821) ≡ Pyxidium (Cladon.).

Scyphophorus Ach. ex Michx. (1803) ≡ Pyxidium (Cladon.).

Scyphopilus P. Karst. (1881) = Thelephora (Thelephor.) fide Killermann (1916).

Scyphorus Raf. (1814) ≡ Pyxidium (Cladon.).

Scyphorus Raf. (1815) ≡ Scyphophora (Cladon.).

Scyphospora L.A. Kantsch. (1928), Mitosporic fungi, 6.A2.1. 1, former USSR.

Scyphostroma Starbäck (1899), = Subicicularium (Mitosp. fungi) fide Eriksson (*SA* **8**: 77, 1989).

scyphus, a cup-like apex of a lichenized podetium, as in *Cladonia fimbriata*.

Scytalidium Pesante (1957), Mitosporic fungi, 1.A/B2.38. 9, widespr. See Sigler & Carmichael (*Mycotaxon* **4**: 395, 1976; key 6 spp.), Sigler & Wang

(*Mycol.* **82**: 399, 1990), Straatsma & Samson (*MR* **97**: 321, 1993; taxonomy of *S. thermophilum*).

Scytenium Gray (1821) = Leptogium (Collemat.).

Scytinopogon Singer (1945), Clavariaceae. 1, Eur.

Scytinostroma Donk (1956), Lachnocladiaceae. *c.* 30, widespr. See Rattan (*TBMS* **63**: 1, 1974; key), Boidin & Lanquetin (*Bibl. Mycol.* **114**, 1987; key).

Scytinostromella Parmasto (1968), Gloeocystidiellaceae. 5, widespr. See Freeman & Petersen (*Mycol.* **71**: 85, 1979; key), Ginns & Freeman (*Bibl. Mycol.* **157**, 1994; N. Am. spp.).

Scytinotopsis Singer (1936) nom. nud. = Resupinatus (Tricholomat.) fide Singer (1951).

Scytinotus P. Karst. (1879) = Panellus (Tricholomat.) fide Singer (1951).

Scytopezis Clem. (1903) Fungi (inc. sed.). See Eckblad (*Nytt. Mag. Bot.* **15**: 1, 1968).

SEA, see Media.

Searchomyces B.S. Mehrotra & M.D. Mehrotra (1963) = Amblyosporium (Mitosp. fungi) fide Pirozynski (1969).

Seaver (Fred Jay; 1877-1970). N. Am. mycologist, New York Botanic Garden (1908-48); editor of *Mycologia* (1909-47); author of 125 papers mainly on discomycete systematics. See Rogerson (*Mycol.* **65**: 721, 1973; portr.), Stafleu & Cowan (*TL-2* **5**: 463, 1985).

Seaverinia Whetzel (1945), Sclerotiniaceae. 1, N. Am. Anamorph *Verrucobotrys*.

Sebacina Tul. & C. Tul. (1871), Exidiaceae. 6, cosmop. See McGuire (*Lloydia* **4**: 1, 1941), Oberwinkler (*Nova Hedw.* **7**: 489, 1965), Wells (*Mycol.* **53**: 317, 1962), Keller (*Z. Pilzk.* **51**: 74, 1973; ultrastr.).

Sebacinella Hauerslev (1977), Ceratobasidiaceae. 2, Eur.

seceding, (1) (of lamellae), at first joined to the stipe (adnate), then free; separating from the stipe; (2) (of conidia), at first attached to the conidiogenous cell, then separating by schizolysis or rhexolysis; secession.

Secoliga Norman (1853) = Gyalecta (Gyalect.) fide Hafellner (*Beih. Nova Hedw.* **79**: 241, 1984).

Secoligella (Müll. Arg.) Vain. (1896) = Aspidothelium (Aspidothel.).

secondary metabolite, a compound, generally produced in a phase subsequent to growth, which is not an essential intermediary of the central metabolism, often with an unusual chemical structure and found as a mixture of closely related chemicals (see *Secondary metabolites: their structure and function* [*CIBA Foundation Symp.* **171**], 1992); e.g. antibiotics, Metabolic products, mycotoxins. See also metabolic products.

secondary mycelium (of basidiomycetes), the dikaryotic mycelium resulting from plasmogamy in the primary mycelium (q.v.); mycelium developed from the base of a basidioma (Corner).

secondary spores (of basidiomycetes), spores other than basidiospores.

Secotiaceae Tul. (1845), Agaricales. 14 gen. (+ 2 syn.), 24 spp. Gasterocarp subagaricoid; lamellae distorted; spores yellowish-brown to black-brown, thick-walled, smooth. Probably related to Agaricaceae but some genera are likely to remain of uncertain application.
    Lit.: Smith (*in* Ainsworth *et al.* (Eds), *The Fungi* **4B**: 421, 1973), Heim (*in* Petersen (Ed.), *Evolution in the higher Basidiomycetes*: 505, 1973).

secotioid (of basidiomycetes), the margin of the pileus does not break free from the stipe (or if it does the pileus never fully expands), lamellae convoluted and anastomosed, basidiospores not ballistosporic; phylogeny of a - **syndrome**, see Thiers (*Mycol.* **76**: 1, 1984).

Secotium Kunze (1840), Secotiaceae. 1, S. Afr. See Singer & Smith (*Madroño* **15**: 152, 1960).

sectoring, 'mutation' or selection in plate cultures resulting in one or more sectors of the culture having a changed form of growth.

secund, having parts directed to one side only; cfr. scorpioid.

Sedecula Zeller (1941), ? Lycoperdales (inc. sed.). 1, USA.

Seed-borne fungi. Are frequently important in connexion with the transmission of plant diseases, esp. into new areas. Examples of such fungi are: *Ascochyta pisi*, *Colletotrichum lindemuthianum*, *Microdochium panattonianum*, *Gloeotinea temulenta*, *Phoma betae*, *P. lingam*, *Septoria apiicola*, *Sphaerella linorum*, *Uromyces betae* and *Ustilago nuda* in the seed (controlled by the use of hot water sterilization or sometimes by seed disinfectants); *Polyspora lini*, *Puccinia antirrhini*, *Tilletia tritici*, *Urocystis agropyri*, *Ustilago avenae*, *U. hordei*, and other smuts as spores on the surface of the seed (controlled by the use of seed disinfectants); *Claviceps purpurea*, *Sclerotinia trifoliorum* and *Sclerotium rolfsii* as sclerotia with the seed.
Lit.: Orton (*Bull. Virg. agric. Exp. Stn* **245**, 1931; bibliogr.), Doyer (*Manual of seed-borne diseases*, 1938), Dykstra (*Bot. Rev.* **27**: 445, 1961; production resistant seed), Malone & Muskett (*Proc. Internat. Seed Testing Assn* **29** (2), 1964; descr. 77 spp.), Neergaard (*Seed pathology*, 2 vols, 1977; monogr.), Agarwal & Sinclair (Eds) (*Principles of seed pathology*, 2 vols, 1987-9), Agarwal *et al.* (*Phytopath. Pap.* **30**, 1989; testing of rice), Singh *et al.* (*An illustrated manual on identification of some seed-borne aspergilli, fusaria, penicillia and their mycotoxins*, 1991), Martin (Ed.) (*Seed treatment: progress and prospects. British Crop Protection Council Monograph* **57**, 1994), Jeffs (Ed.) (*Seed treatment*, 1986), McGee (*Maize diseases: a reference source for seed technologists*, 2nd edn, 1990), McGee (*Soybean diseases: a reference source for seed technologists*, 1992), IRRI (*Rice seed health*, 1988).

Segestrella Fr. (1831) ≡ Segestria (Trichothel.).

Segestria Fr. (1825) = Porina (Trichothel.).

segment (of a dictyospore), a part of the spore cut off by an A-trans-septum (Eriksson, *Opera bot.* **60**, 1981); see septum.

segregate (in taxonomy), a taxon based on part of an earlier taxon.

Seimatosporiopsis B. Sutton, Ghaffer & Abbas (1972), Mitosporic fungi, 4.C2.19. 1, Pakistan.

Seimatosporium Corda (1833), Anamorphic Amphisphaeriaceae, 6/7.C2.19. Teleomorph *Discostroma*. 38, widespr. See Sutton (1980).

seiospore, a dry dispersal spore.

Seiridiella P. Karst. (1890) ? = Phragmotrichum (Mitosp. fungi) fide Sutton & Pirozynski (*TBMS* **48**: 363, 1965).

Seiridina Höhn. (1930) = Seimatosporium (Mitosp. fungi) fide Sutton (*Mycol. Pap.* **97**, 1964).

Seiridium Nees (1816), Anamorphic Amphisphaeriaceae, 6.C2.19. Teleomorphs *Lepteutypa*, *Blogiascospora*. 19, widespr. See Nag Raj (1993), Sutton (1980), Nag Raj & Kendrick (*Sydowia* **38**: 179, 1986), Chou (*Eur. J. For. Path.* **19**: 435, 1989; spp. on *Cupressus*), Viljoen *et al.* (*Exp. Mycol.* **17**: 323, 1993; spp. on cypress compared by sequence data).

Seirophora Poelt (1983), Teloschistaceae (L). 1, Syria.

Seismosarca Cooke (1889) = Auricularia (Auricular.) fide Donk (*Persoonia* **4**: 158, 1966).

Selenaspora R. Heim & Le Gal (1936), Otideaceae. 1, Eur.

Selenodriella R.F. Castañeda & W.B. Kendr. (1990), Mitosporic fungi, 1.A1.?12. 5, Cuba, N. Am. Cf. *Selenosporella* (Mitosp. fungi).

Selenophoma Maire (1907), Mitosporic fungi, 4.A1/2.16. 5, temp. See Sutton (1980).

Selenophomites Babajan & Tasl. (1973), Fossil fungi. 1 (Tertiary), former USSR.

Selenophomopsis Petr. (1924) = Selenophoma (Mitosp. fungi) fide v. Arx (*Verh. K. ned. Akad. Wet. C* **51**, 1957).

Selenosira Petr. (1957), Mitosporic fungi, 4.A1.?. 1, Pakistan.

Selenosporella G. Arnaud ex MacGarvie (1968), Mitosporic fungi, 1.A1.?16. 4, widespr. Often synanamorphs of dematiaceous mitosporic fungi. See Sutton & Hodges (*Nova Hedw.* **28**: 487, 1977, **29**: 602, 1978), Hughes (*N.Z. Jl Bot.* **17**: 139, 1979).

Selenosporium Corda (1837) = Fusarium (Mitosp. fungi) fide Holubová-Jechová *et al.* (*Sydowia* **46**: 247, 1994; revision).

Selenosporopsis R.F. Castañeda & W.B. Kendr. (1991), Mitosporic fungi, 1.E1.16. 1, Cuba.

Selenotila Lagerh. (1892), Mitosporic fungi, 1.A1?. 2, Am., Eur. See Yarrow (*Ant. v. Leeuwenhoek* **35**: 418, 1969; *S. nivalis* in snow), Yarrow (*Stud. Mycol.* **14**: 29, 1977; nom. dub.).

Selenozyma Yarrow (1977), Mitosporic fungi, 1.A1.4. 2 (from humans and bovines), Netherlands.

Selinia P. Karst. (1876), Hypocreaceae. 2, Eur., S. Am., Japan. See Ranalli & Mercuri (*Mycotaxon* **53**: 109, 1995; anamorph), Udagawa (*Trans. mycol. Soc. Japan* **21**: 283, 1980). Anamorph *Tubercularia*-like.

Seliniana Kuntze (1891) ≡ Selinia (Hypocr.).

Seliniella Arx & E. Müll. (1955) = Ascobolus (Ascobol.) fide Müller & v. Arx (1962).

SEM, scanning electron microscopy (q.v.).

semi- (prefix), half. Cf. hemi-.

Semidelitschia Cain & Luck-Allen (1969), Sporormiaceae. 1 (coprophilous), USA.

Semifissispora H.J. Swart (1982), Dothideales (inc. sed.). 3 (on *Eucalyptus*), Australia.

semifissitunicate, see ascus.

Semigyalecta Vain. (1921), Gyalectaceae (L). 1, Philipp., Java, Borneo. See Santesson (*Symb. bot. upsal.* **12** (1), 1952).

semigymnoplast, see protoplast.

semimacronematous, only slightly different from vegetative hyphae; frequently ascending, rarely erect. See macronematous.

Semimorula E.F. Haskins, McGuinn. & C.S. Berry (1983), Myxomycetes (inc. sed.). 1, USA.

Semiomphalina Redhead (1984), Tricholomataceae. 1, Papua New Guinea.

Semipseudocercospora J.M. Yen (1983) = Pseudocercospora (Mitosp. fungi) fide Sutton (*in litt.*).

Semisphaeria K. Holm & L. Holm (1991), ? Arthopyreniaceae. 1, Norway, Sweden.

Senegalosporites S. Jardiné & Magloire (1965), Fossil fungi. 2 (Cretaceous, Miocene), Senegal, Taiwan.

senescence (of fungi), the degeneration which makes indefinite propagation of certain fungi in culture impossible. See Holliday (*Nature* **221**: 1224, 1969; *Neurospora*, *Podospora*).

sensitive, reacting with severe symptoms to the attack of a given pathogen (*TBMS* **33**: 155, 1950); sensitivity, the tendency of an organism attacked by a disease to give more or less strong symptoms; sensibility (Wilbrink).

sensu lato (s.l.), in a broad sense; - stricto (s.str.), in a narrow sense.

separating, see seceding; - cell, a cell between a conidium and a conidiogenous cell, involved in rhexolytic secession, q.v.

Sepedonium Link (1809), Anamorphic Hypocreaceae, 1.A1.1. Teleomorph *Apiocrea*. 10, temp. See Rudakov (*Mik. Grib. Biol. Prakt. Znach.*, 1981; key).

septal pore apparatus, see dolipore septum (- - **swelling**) (Fig. 13), parenthesome (- - **cap**); - **plug**, an occlusion of a septal pore.

Septaria Fr. (1821) nom. rej. ≡ Septoria (Mitosp. fungi).

septate, having septa.

Septatium Velen. (1934) = Hymenoscyphus (Leot.) fide Dennis (*British Ascomycetes*, 2nd edn, 1978).

Septobasidiaceae Racib. (1909), Septobasidiales. 3 gen. (+ 6 syn.), 173 spp. Parasitic on scale-insects. Basidioma corticioid; basidia auricularioid.

Septobasidiales, Teliomycetes. 1 fam., 3 gen. (+ 6 syn.), 173 spp. Basidioma not gelatinous; complex relationship with insect hosts. Fam.:
> **Septobasidiaceae**.
> *Lit.*: Donk (1951-63, VIII, *Taxon* 7: 164, 193, 236, 1958, **12**: 165, 1963).

Septobasidium Pat. (1892) nom. cons., Septobasidiaceae. 170 (symbiotic with scale insects), esp. warmer areas. See Couch (*The genus Septobasidium*, 1938), Dykstra (*CJB* **52**: 971, 1974; ultrastr.).

Septocarpus Zopf (1888) = Podochytrium (Chytrid.) fide Fitzpatrick (1930).

Septochora Höhn. (1917) = Helhonia (Mitosp. fungi) fide Sutton (*SA* **6**: 151, 1987).

Septochytrium Berdan (1939), Cladochytriaceae. 4, N. Am.

Septocladia Coker & F.A. Grant (1922) = Allomyces (Blastoclad.) fide Coker (1923).

Septocolla Bonord. (1851) = Dacrymyces (Dacrymycet.) fide Saccardo (*Syll. Fung.* **6**: 798, 1888).

Septocylindrium Bonord. ex Sacc. (1880), Mitosporic fungi, 1.C1.3. 20, widespr.

Septocyta Petr. (1927), Mitosporic fungi, 8.E1.10. 1, Eur.

Septocytella Syd. (1929), Mitosporic fungi, 8.E1.?. 1, China.

Septodochium Matsush. (1971), Mitosporic fungi, 3.C1.15. 1, Solomon Is.

Septodothideopsis Henn. (1904) nom. dub. fide Höhnel (*Sber. Akad. Wiss. Wien* **119**: 877, 1910).

Septofusidium W. Gams (1971), Mitosporic fungi, 1.B1.?33. 1, Afr.

Septogloeum Sacc. (1880), Mitosporic fungi, 6.C1.15. 2, widespr. See Sutton & Pollack (*Mycopathol.* **52**: 331, 1974), Sutton & Webster (*TBMS* **83**: 59, 1984).

Septoidium G. Arnaud (1921), Anamorphic Parodiopsidaceae, 1.C2.19. Teleomorph *Alina*. 7, trop.

Septolpidium Sparrow (1933), ? Myzocytiopsidaceae. 1, Eur.

Septomazzantia Theiss. & Syd. (1915) = Diaporthe (Vals.) fide Petrak (*Sydowia* **8**: 294, 1954).

Septomycetes. Class within the *Orthomycotina* (q.v.) including *Brachybasidiales*, *Exobasidiales*, *Septobasidiales* and *Tilletiales* (Cavalier-Smith, *in* Rayner *et al.* (Eds), *Evolutionary biology of the fungi*, 1987).

Septomycotera. Superdivision used by Moore (1971) for all fungi with septate hyphae; cfr. *Phycomycotera*.

Septomyrothecium Matsush. (1971), Mitosporic fungi, 3.A2.15. 1, Papua New Guinea.

Septomyxa Sacc. (1884) = Diplodina (Mitosp. fungi) fide Sutton (1977).

Septomyxella (Höhn.) Höhn. (1923), Mitosporic fungi, 6.B1.?. 1, Italy.

Septonema Corda (1837), Mitosporic fungi, 1.C2.3. 10, widespr. See Hughes (*Naturalist*, Hull, 1951: 173, 1952: 7), Holubová-Jechová (*Folia geobot. phytotax.* **13**: 422, 1978).

Septopatella Petr. (1925), Mitosporic fungi, 7.E1.10. 1, N. Am. See Dyko & Sutton (*CJB* **57**: 370, 1979).

Septopezizella Svrček (1987), Leotiaceae. 1, Eur.

Septoplaca Petr. (1964) = Piptarthron (Mitosp. fungi) fide Sutton (1980).

Septorella Berk. [not traced] ? nom. nud. (Mitosp. fungi) fide Sutton (1977).

Septorella Allesch. (1897) = Fusarium (Mitosp. fungi) fide Höhnel (*Sber. Akad. Wiss. Wien* **121**: 339, 1912).

Septoria Fr. (1819) nom. rej. (Mitosp. fungi).

Septoria Sacc. (1884) nom. cons., Anamorphic Dothideaceae, 4.E1.10. Teleomorph *Mycosphaerella*. 1,000, widespr. *S. apiicola* (celery (*Apium*) leaf spot), *S. azaleae* (azalea (*Rhododendron*) leaf scorch), *S. gladioli* (hard rot and leaf spot of *Gladiolus*), *S. lycopersici* (tomato leaf spot), *S. nodorum* (wheat glume blotch), *S. tritici* (wheat leaf spot). See Sutton (1980; refs), MacMillan & Plunkett (*J. agric. Res.* **64**: 547, 1942; spore structure and germination), Jörstad (*Skr. Utg. Norske Vidensk.-Akad. Oslo* N.S. **22**: 1, 1965, and **24**: 1, 1967; Norwegian spp. on dicots and *Gramineae*), Scharen & Sanderson (*Septoria of Cereals*: 37, 1983), Andrianova (*Mikol. i Fitopatol.* **20**: 259, 1986; taxonomic criteria), Andrianova (*Mikol. i Fitopatol.* **21**: 393, 1987; subgen. classif.), Teterevnikova-Babayan (*Griby roda Septoria*, 1987; spp. former USSR), Constantinescu (*TBMS* **83**: 383, 1984; spp. on *Betulaceae*), Sutton & Pascoe (*TBMS* **89**: 521, 1987; spp. on *Acacia*), Sutton & Pascoe (*Stud. Mycol.* **31**: 177, 1989; spp. from Australia), McDonald & Martinez (*Phytopath.* **79**: 1186, 1989, **80**: 1368, 1990; RLFP's and *S. tritici*), Cooley *et al.* (*MR* **94**: 145, 1990; cotransformation in *S. nodorum*), Farr (*Mycol.* **83**: 611, 1991; spp. on *Cornus*), Newton & Caten (*Pl. Path.* **40**: 546, 1991; strains of *S. nodorum* adapted to wheat or barley), Farr (*Sydowia* **44**: 13, 1992; spp. on *Fabaceae* tribe *Genisteae*), Andrianova (*Mikol. i Fitopatol.* **26**: 425, 1992; spp. described by Hollós).

Septoriella Oudem. (1889), Mitosporic fungi, 8.C2.1. 6, widespr. See Nag Raj (*Coelomycetous anamorphs*, 1993).

Septoriomyces Cavalc. & A.A. Silva (1972), Mitosporic fungi (L), 4.C1.?. 1, Brazil.

Septoriopsis Gonz. Frag. & Paúl bis (1915) ? = Septoria (Mitosp. fungi).

Septoriopsis F. Stevens & Dalbey (1918) ≡ Cercoseptoria (Mitosp. fungi).

Septoriopsis Höhn. (1920) ≡ Groveolopsis (Mitosp. fungi).

Septorisphaerella Kleb. (1918) = Mycosphaerella (Mycosphaerell.) fide v. Arx & Müller (1975).

Septosperma Whiffen ex R.L. Seym. (1971), Chytridiaceae. 5 (on chytrids), temp.

Septosphaerella Laib. (1922) = Mycosphaerella (Mycosphaerell.).

Septosporiella, see *Septoriella*.

Septosporium Corda (1831), Mitosporic fungi, 1.D2.1. 5, Eur., N. Am.

Septothyrella Höhn. (1911), Mitosporic fungi, 5.C1.?. 5, Afr.

Septotinia Whetzel ex J.W. Groves & M.E. Elliott (1961), Sclerotiniaceae. 2, temp. Anamorph *Septotis*.

Septotis N.F. Buchw. ex Arx (1970), Anamorphic Sclerotiniaceae, 6.C1.39. Teleomorph *Septotinia*. 2, N. Am., Eur. See Sutton (*Mycol.* **72**: 208, 1980).

Septotrichum Corda (1840) nom. dub.; based on a leaf outgrowth fide Bonorden (1851).

Septotrullula Höhn. (1902), Mitosporic fungi, 3.C1.38. 1, Eur. See Sutton & Pirozynski (*TBMS* **48**: 355, 1965).

septum (pl. **septa**), a cell wall or partition; Talbot (*Taxon* **17**: 622, 1968) distinguishes **primary -**, a septum formed in direct association with nuclear division (by constriction, mitosis, or meiosis) separating the daughter cells and having a pore (see Markham, *MR* **98**: 1089, 1994; review) which may be modified as a dolipore (Fig. 13) (q.v.; in basidiomycetes) or be associated with Woronin bodies (in ascomycetes; see Kimbrough, *in* Hawksworth, 1994: 127, ultrastr.), and **adventitious -** (retraction or retaining septum, 'cloison de retrait'), a septum formed in the absence of, or independently of, nuclear division, esp. in association with the movement of cytoplasm from one part of the fungus to another; primary septa are characteristic of higher fungi, adventitious septa of lower fungi where nuclear division is by constriction; **angular -** (Eriksson, *Ark. Bot.* II **6**: 339, 1967), oblique septum, see Fig. 33; **longiseptum**, a longitudinal septum within a spore (Reynolds, *Mycol.* **63**: 1173, 1971); **oblique -**, a septum within a segment of a spore arising at an oblique angle to that delimiting the segment (Eriksson, *Opera bot.* **60**, 1981); **trans-**, a transverse septum within a spore (Reynolds, 1971), which may be an **A trans-septum** (forming a segment) or a **B trans-septum** (laid down in a segment after division by a longiseptum; never in macrocephalic spores) (Eriksson, 1981; Fig. 33). Ascospores in which the septation proceeds from the primary septum towards the poles, so that immature spores have longer end cells are termed **macrocephalic** (Fig. 34A); those in which it proceeds by each trans-septum dividing a segment into two of ± equal lengths are **microcephalic** (Fig. 34B) (Eriksson, 1967, 1981; Curry & Kimbrough (*Mycol.* **75**: 781, 1983; *Pezizaceae*). See also Mitosporic fungi, euseptum, distoseptum, multiperforate septum, and Fig. 35.

Sepultaria (Cooke) Boud. (1885) = Geopora (Otid.) fide Burdsall (*Mycol.* **60**: 504, 1968).

sequestrate, fungal fruit-bodies which have evolved from having exposed hymenia and forcibly discharged spores to a closed or even hypogeous habit in which spores are retained in the fruit-body until it decays or is eaten by an animal vector. Many sequestrate taxa can be clearly recognized as being derived from specific spore-discharging ancestors, e.g. *Rhizopogon* from *Suillus*.

Serda Adans. (1763) = Gloeophyllum (Coriol.) fide Donk (1974).

Serendipita P. Roberts (1993), Exidiaceae. 7, widespr.

Serenomyces Petr. (1952), ? Phyllachoraceae. 4 (on *Palmae*), trop. See Barr *et al.* (*Mycol.* **81**: 47, 1989; key).

Sericeomyces Heinem. (1978), Agaricaceae. 9, widespr.

sericeous, like silk.

Seriella Fr. (1849) nom. dub. (Ascomycota, inc. sed.).

serioplectenchyma, see plectenchyma.

Serology. Serological methods, which depend on the ability of a fungus to act as an antigen (q.v.), have two main mycological applications: (1) for the identification of fungi or investigations into the relationships between different fungi; (2) for the diagnosis of infections of humans and other vertebrates by fungal pathogens.

For identification, a preparation of the unknown fungus is tested with antiserum prepared against a named fungus when a reaction indicates that the two fungi have antigens in common and are therefore conspecific or closely related. An antiserum prepared from one fungus may, however, not infrequently give a reaction ('cross-reaction') with other fungi indicating the possession of some antigens in common. This may be taken to be evidence of phylogenetic relatedness. *Phymatotrichum omnivorum* antiserum, for example, gives a reaction with certain gasteromycetes. Because of such cross-reactions and the difficulty of obtaining purified fungal antigens serological identification of fungi is in a more primitive state than that for bacteria and viruses.

Serological diagnosis of infection (or past infection) of humans or animals by fungal pathogens is a medical mycological technique of proven utility for a number of important mycoses. In N. Am., for example, millions of humans and domestic and farm animals are marked for life by antibodies induced by subclinical or mild attacks of coccidioidomycosis and histoplasmosis as revealed by testing the reaction of the skin or blood serum to coccidioidin and histoplasmin (q.v.), antigens prepared from the pathogens. (See

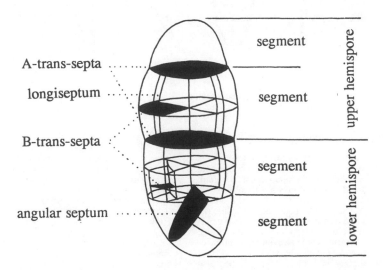

Fig. 33. Terminology of spore septation in dictyosporous ascospores, used by Eriksson (1981).

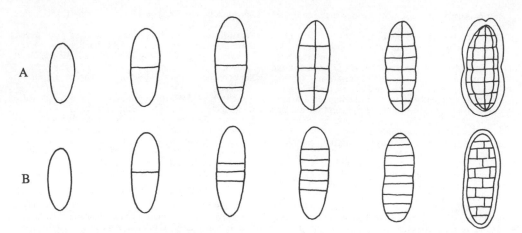

Fig. 34. Sequence of spore septation (left to right) in dictyosporous ascospores, after Eriksson (1981). A, microcephalic; B, macrocephalic.

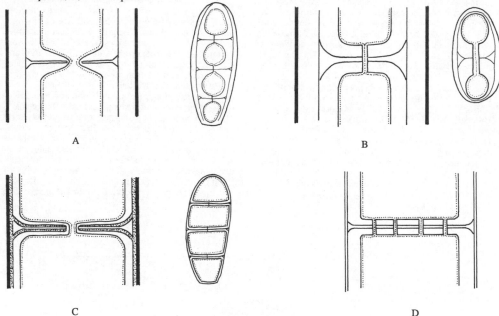

Fig. 35. Types of septum. A, distoseptum (pseudoseptum); B, distoseptum (polarilocular); C, euseptum; D, multiperforate septum. See also dolipore septum.

also blastomycin, oidiomycin, spherulin, sporotrichin.)

The antigen-antibody (antiserum) reaction may be tested in several ways. The most useful techniques are based on agglutination or precipitation (see antigen). In the first the antigenic particles are usually whole fungal cells but an alternative is to adsorb the antigen onto particles of latex or charcoal. In the second, the reaction is demonstrated by the formation of a precipitate when the antiserum (antibody) is added to a tube containing the corresponding antigen in solution or when antibody and antigen are allowed to diffuse towards one another in a clear agar gel. See also anaphylaxis, complement-fixation, ELISA.

Lit.: Chester (*Quart. Rev. Biol.* **12**: 19, 165, 294, 1937; plants), Seeliger (*Mykologische Serodiagnostik*, 1958), Mathews (*Plant virus serology*, 1956; methods), Proctor (*Progress in microbiological techniques*: 213, 1967), Kaufman (*Manual of clinical mycology*, 1975), Evans (Ed.) (*Serology of fungal infections and farmer's lung*, 1976).

serous (of latex), like serum; watery; opalescent.

Serpentisclerotes Beneš (1959), Fossil fungi (*f. cat.*). 1 (Carboniferous), former Czechoslovakia.

Serpula (Pers.) Gray (1821), Coniophoraceae. 2, cosmop. *S. lacrymans* (syn. *Merulius lacrymans*), the dry rot or 'house' fungus. See Cooke (1957: 201; key), Anon. (*Leafl. For. Prod. Res. Lab.* **6**, rev. 1964), Harmsen (*Friesia* **6**: 233, 1960; taxonomy, culture), Wood-attacking fungi.

serrate, edged with teeth, like a saw (Fig. 37.46).

serrulate, delicately toothed.

Servazziella J. Reid & C. Booth (1987), Ascomycota (inc. sed.). 1 (on *Araucaria*), Australasia. See Eriksson & Hawksworth (*SA* **6**: 250, 1987; posn).

Sesia Adans. (1763) ≡ Gloeophyllum (Coriol.) fide Murrill (1904). ≡ Serpula (Conioph.) fide Donk (1964).

Sesquicillium W. Gams (1968), Mitosporic fungi, 1.A1.15. 3, widespr. See Veenbas-Rijks (*Acta Bot. Neerl.* **19**: 323, 1970).

sessile, having no stem.

Sessiliospora D. Hawksw. (1979), Mitosporic fungi, 1.C2.24. 1 (on lichens), Malaysia.

seta (pl. -ae) (Lat., a bristle), (1) a stiff hair, generally thick-walled and dark in colour; in hyphomycetes, see Dev Rao (*in* Subramanian (Ed.), *Taxonomy of fungi*: 397, 1984); (2) (in hymenomycetes), a sterile hyphal end, thick-walled, darkening in KOH sol., found frequently projecting from the hymenium in xanthochroic basidiomata (Fig. 18A). Lentz (1954) distinguished 'seta', 'embedded seta', and 'stellate seta' (asteroseta) (Fig. 18C). Smith (1966) treated the last as a cystidium; see cystidium (3).

setaceous (Lat., setaceus), bristle-like; cf. setose.

Setaria Ach. ex Michx. (1803) nom. rej. ≡ Bryoria (Parmel.).

Setchellia Magnus (1896) = Doassansia (Tillet.) fide Dietel (1928).

Setchelliogaster Pouzar (1958), Cortinariaceae. 2, Am. See Singer & Smith (*Madroño* **15**: 73, 1959).

Seticyphella Agerer (1983), Cyphellaceae. 3, Eur.

Setiferotheca Matsush. (1995), Ceratostomataceae. 1, Japan.

setiform, see setaceous.

Setigeroclavula R.H. Petersen (1988), Clavariaceae. 1, NZ.

Setocampanula Sivan. & W.H. Hsieh (1989), Trichosphaeriaceae. 1, Taiwan.

Setodochium Bat. & Cif. (1958) ? = Wiesneriomyces (Mitosp. fungi) fide Maniotis & Strain (*Mycol.* **60**: 203, 1968).

Setoerysiphe Y. Nomura (1984), Erysiphaceae. 1, Japan. See Braun (*Beih. Nova Hedw.* **89**, 1987).

Setogyroporus Heinem. & Rammeloo (1982), Gyrodontaceae. 1, Burundi.

Setomelanoma, see *Setomelanomma*.

Setomelanomma M. Morelet (1980), Dothideales (inc. sed.). 1, France.

Setomyces Bat. & Peres (1961) nom. inval. Mitosporic fungi (L), ?.-.-. 4, Brazil.

Setopeltis Bat. & A.F. Vital (1959) = Chaetothyrina (Micropeltid.) fide v. Arx & Müller (1975).

Setophiale Matsush. (1995), Mitosporic fungi, 1.A1.15. 1, Peru.

Setoscypha Velen. (1934) = Phialina (Hyaloscyph.) fide Huhtinen (1990).

setose (Lat. setosus; bristly), covered with bristles; cf. setaceous, setulose.

Setosphaeria K.J. Leonard & Suggs (1974), Pleosporaceae. 4 (on *Gramineae*), widespr. Anamorph *Exserohilum*.

Setosporella Mustafa & Abdul-Wahid (1989), Mitosporic fungi, 1.A2.?36. 1 (from soil), Egypt.

Setosynnema D.E. Shaw & B. Sutton (1985), Mitosporic fungi, 2.E1.10. 1 (aero-aquatic), Australia, Papua New Guinea.

setula (Lat., a little bristle), (1) a delicate hair-like appendage arising from a conidium, as in *Dinemasporium*; (2) (in hymenomycetes), a thick-walled, pigmented, terminal element of a tramal cystidium.

setule (in hymenomycetes), a thin-walled, rarely pigmented, usually lageniform cystidium on the pileus or stipe.

Setulipes Antonín (1987), Tricholomataceae. 2, N. temp.

setulose (Lat., setulosus), covered with fine bristles or hairs; cf. setaceous.

Seuratia Pat. (1904), Seuratiaceae (?L). 4, Eur., Indonesia, C. & S. Am., Pacific. See Meeker (*CJB* **53**: 2485, 1975; key), Meeker (*CJB* **53**: 2462, 1975; morphology). Anamorph *Atichia*.

Seuratiaceae Vuill. ex M.E. Barr (1987), Ascomycota (inc. sed.). 3 gen. (+ 5 syn.), 6 spp. Stromata superficial or absent; ascomata crustose, dark, irregular in outline, the upper part deliquescing to expose the asci; peridium composed of pale chains of cells in a gelatinous matrix; interascal tissue poorly developed, consisting of irregular interthecial strands; asci scattered or in small groups, saccate, thick-walled, ? fissitunicate; ascospores hyaline to brown, transversely septate, without a sheath. Anamorph usually hyphomycetous with elaborate lobed or torulose conidia (see *Atichia*). Saprobic and epiphytic on leaves, mostly trop.

This family was placed in the *Myriangiales* near the *Saccardiaceae* by Meeker, and in the *Arthoniales* by Eriksson (1982).

Lit.: Meeker (*CJB* **53**: 2464, 2485, 1975).

Seuratiopsis Woron. (1934), ? Seuratiaceae. 1, Crimea.

sewage fungus, *Leptomitus lacteus*, found in polluted water, sometimes blocking sewage filters.

Sex. In some fungi reproduction is by asexual methods only, in others it is parthenogenetic, but in most there is a sexual process. Probably one third of all fungi have more than one type of reproduction, frequently in two well-marked phases (the 'teleomorph' and 'anamorph') which may have separate names (see Nomenclature).

As in the *Algae* and *Bryophyta*, the sexual or assimilative phase is generally haploid (i.e. a gametophytic generation), though in basidiomycetes (see under) the cells are commonly dikaryotic. In most eumycetes (but not myxomycetes) nuclear fusion is not till shortly before meiosis and the development of 'sexual' spores. For a life-cycle summary of an hypothetical typical filamentous fungus see Fig. 36, and for *Uredinales* see Fig. 41.

Fungi having sex organs are commonly monoecious, infrequently dioecious. If sex organs are not present or (as in basidiomycetes and *Mucorales*) are of like morphology the fungus is (1) homothallic (q.v.), when one haploid mycelium is able to give a sporocarp or (2) heterothallic (q.v.), when there are two or more haploid mycelial types and diploidization must be effected for sporocarp development to take place. Heterothallic mycelia may be different (sexually dimorphic) or, more commonly, alike in form so that, as the 'sexes' may be determined only by chemical methods, they are generally named + and -.

In bipolar species sporocarp development is dependent on two factors, and in tetrapolar species on four. When segregation in a tetrapolar species takes place at the second division of meiosis one basidium may give all four types of spore, but when segregation is at the first division only two types (A_1B_1 and A_2B_2 or A_1B_2 and A_2B_1) are produced by one basidium. Normal basidiocarps are generally produced only when the secondary mycelium resulting from the fusion of primary mycelia is of the genotype $A_1B_1A_2B_2$ (Buller, 1941). These various genetical devices favour or ensure outbreeding or inbreeding. As Esser has pointed out incompatibility is of 2 main types: (1)

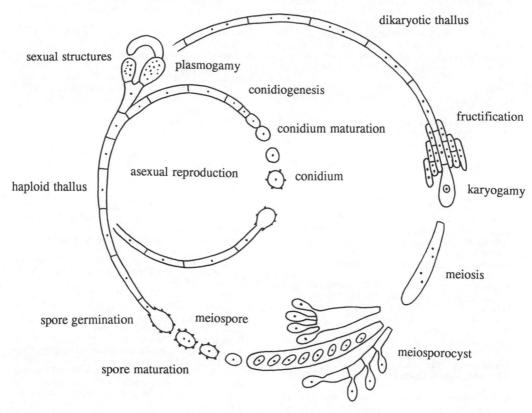

Fig. 36. Life-cycle of an hypothetical 'typical' filamentous fungus. Adapted from Kreisel (*in* Hawksworth (Ed.), *Frontiers in mycology*: 71, 1991).

homogenic incompatibility (e.g. bipolar and tetrapolar systems) in which mating is prevented between strains having the same factor(s) so that inbreeding is prevented and outbreeding encouraged, and (2) **heterogenic incompatibility** in which mating is prevented between strains having different factors so that outbreeding is prevented and inbreeding encouraged.

Lit.: Raper (*in* Wenrich (Ed.), *Sex in micro-organisms*, 1954, *Mycol.* **51**: 107, 1959, **55**: 79, 1963, *Genetics of sexuality in higher fungi*, 1966), Esser & Raper (Eds) (*Incompatibility in fungi*, 1965), Stabens (*in* Gow & Gadd (Eds), *The growing fungus*: 383, 1995). See also Genetics of fungi, hormones, parasexual cycle.

Seychellomyces Matsush. (1981), Mitosporic fungi, 1.C2.23. 1, Seychelles.

Seynesia Sacc. (1883), ? Xylariaceae. 1, trop. See Barr (*Mycotaxon* **39**: 43, 1990; posn), Læssøe (*SA* **13**: 43, 1994).

Seynesiella G. Arnaud (1918), Microthyriaceae. 2, Eur., N. Am.

Seynesiola Speg. (1919) = Arnaudiella (Microthyr.) fide Stevens & Ryan (1925). See also Müller & v. Arx (1962).

Seynesiopeltis F. Stevens & R.W. Ryan (1925), Microthyriaceae. 1, Hawaii.

Seynesiopsis Henn. (1904), Mitosporic fungi, 5.B2.?. 1, S. Am.

Seynesiospora Bat. (1960) = Cyclothea (Microthyr.) fide v. Arx & Müller (1975).

shaggy ink cap (or **mane**), basidioma of the edible *Coprinus comatus*.

Shanorella R.K. Benj. (1956), Onygenaceae. 1, USA, Eur.

Shanoria Subram. & K. Ramakr. (1956) = Strigula (Strigul.) fide Eriksson & Hawksworth (*SA* **11**: 56, 1992). See also Sutton (*Kew Bull.* **31**: 461, 1977).

Shanoriella Bat. & Cif. (1962) = Limacinula (Coccodin.) fide Reynolds (1971).

shape, configuration or form, total effect produced by the outline of the structure (Fig. 37).

Shawiella Hansf. (1957), Mitosporic fungi, 7.C2.1. 1, Papua New Guinea.

shaybah, Saudi Arabian name for *Parmelia austrosinensis* used in cooking.

Shear (Cornelius Lott; 1865-1956). Plant pathologist, mycologist; USDA Bureau of Plant Industry, 1901-35 (Head of the Division of Mycology and Disease Survey from 1923); second president of the Mycological Society of America. Noted for his studies in cranberry and other fruit diseases, pyrenomycetes and many publications on general mycology and nomenclature; *The genera of fungi* (1931; with F.E. Clements). See Stevenson (*Mycol.* **49**: 283, 1957; portr., bibliogr., *Taxon* **6**: 7, 1957; portr., *Phytopath.* **47**: 321, 1957), Petrak (*Sydowia* **11**: 1, 1957; portr., bibliogr.), Stafleu & Cowan (*TL-2* **5**: 585, 1985).

Shearia Petr. (1924), Mitosporic fungi, 8.D2.19. 2, N. Am. See Tubaki *et al.* (*Trans. mycol. Soc. Jap.* **24**: 121, 1983; cultural characters).

Sheariella Petr. (1952), Mitosporic fungi, 8.A2.1. 1, Hawaii.

Shecutia Nieuwl. (1916) ≡ Libertiella (Mitosp. fungi).

shield-lichens, formerly applied to lichens having large apothecia (obsol.).

shii-take. *Lentinula edodes*. For cultivation the wood is inoculated with spawn. Basidiomata are produced after 2 years, then there are two crops a year for a number of years. Alternatively, basidiomata are obtained 12 weeks after inoculation of a mixture of hardwood sawdust, nutrients and water (Jones & Liu, *Mycologist* **4**: 121, 1990). See Singer & Harris (*Mushrooms and truffles*, 1987), Maher (Ed.) (*Mushroom Science* **13**, *Science and cultivation of edible fungi*, 1991), Stamets (*Growing gourmet and medicinal mushrooms*, 1993), Chang & Hayes (*Biology and cultivation of edible mushrooms*, 1978), Przybylowicz & Donoghue (*Shiitake growers handbook*, 1988), Tan & Moore (*MR* **96**: 1077, 1992; *in vitro* cultivation).

Shimizuomyces Kobayasi (1981), Clavicipitaceae. 1, Japan.

Shiraia Henn. (1900), Dothideales (inc. sed.). 1, China, Japan. See Amano (*Bull. natn Sci. Mus. Tokyo*, B **6**: 55, 1980, *Trans. mycol. Soc. Japan* **24**: 35, 1983).

Shiraiella Hara (1914) = Mycocitrus (Hypocr.) fide Rogerson (1970).

Shitaker Lloyd (1924) nom. nud. ≡ Lentinula (Tricholomat.).

shoe-string fungus, the honey agaric, *Armillaria mellea*.

Shomea Bat. & J.L. Bezerra (1964) ? = Geastrumia (Mitosp. fungi) fide Sutton (*in litt.*).

Shortensis Dilcher (1965), Fossil fungi (Micropeltid.). 1 (Eocene), USA. = Manginula fide Lange (*Austr. J. Bot.* **17**: 565, 1969), = Vizella (Dothid.) fide Selkirk (1972).

shot-hole, a leaf spot disease characterized by holes made by the dead parts dropping out; of *Antirrhinum* (*Heteropatella antirrhini*); of peach (*Stigmina carpophila*); of cherry (*Blumeriella jaapii*).

shoyu (soy sauce), an oriental sauce of soybeans and wheat fermented by *Aspergillus*, yeasts, and bacteria (Hesseltine, *Mycol.* **57**: 174, 1965). See also ketjap.

Shropshiria F. Stevens (1927) = Munkia (Mitosp. fungi) fide Marchionatto (*Rev. Argent.-agron.* **7**: 172, 1940).

Shrungabeeja V.G. Rao & K.A. Reddy (1981), Mitosporic fungi, 1.D2.19. 1, India.

Shuklania J.N. Dwivedi (1959), Fossil fungi (Uredin.). (Tertiary), India.

Sibirina G.R.W. Arnold (1970), Anamorphic Hypocreaceae, 1.B1.15. Teleomorph *Hypomyces*. 1 (on *Lentinus*), Irkutsk.

sibling species, accepted synonym for cryptic species; species which are genetically distinct and not interbreeding, but are not morphologically distinct.

sicyospore, a thick-walled storage cell (obsol.).

siderophore, a metabolic product of a fungus (or other organism) which binds iron and facilitates its transport from the environment into the microbial cell. See Winkelmann (*in* Hawksworth (Ed.), *Frontiers in mycology*: 49, 1991; review roles).

SIDS, see cot death.

Siegertia Körb. (1860) = Rhizocarpon (Rhizocarp.).

Siemaszkoa I.I. Tav. & T. Majewski (1976), Laboulbeniaceae. 6, Eur., S. Am. See Majewski (*Polish Bot. Stud.* **2**: 219, 1991; key).

sigatoka. A major disease of banana (*Musa*) caused by *Mycosphaerella musicola* (anamorph *Cercospora musae*). See *IMI Descriptions* **414**, 1974; *IMI Map* **7**; Meredith (*Phytopath. Pap.* **11**, 1970; monogr.).

Sigmatomyces Sacc. & P. Syd. (1913), Mitosporic fungi, 3.A1.?. 1, Philipp.

sigmoid, curved like the letter 'S' (Fig. 37.6).

Sigmoidea J.L. Crane (1968), Mitosporic fungi, 1.C1.15. 1, N. Am.

Sigmoideomyces Thaxt. (1891), Sigmoideomycetaceae. 2, N. Am., Eur. See Benny *et al.* (1992; key).

Sigmoideomycetaceae Benny, R.K. Benj. & P.M. Kirk (1992), Zoopagales. 3 gen., 7 spp. Fertile hyphae dichotomously branched, septate; spores borne on vesicles, ± hyaline, small; parasitic on fungi.
Lit.: Benny *et al.* (*Mycol.* **84**: 615, 1992; key).

Sigridea Tehler (1993), Roccellaceae (L). 5, Afr., N. & S. Am., Asia. See Tehler (*Nova Hedw.* **57**: 417, 1993; key).

sikyotic cell (the 'Schröpfkopf-Zelle' of Burgeff, 1924), the terminal cell of a *Parasitella simplex* hypha by which the parasite anchors itself to its host (*Absidia glauca*). After several days this organ differentiates into a **sikospore** (sikyotic spore). See Kellner *et al.* (*Mycologist* **5**: 120, 1991).

Sillia P. Karst. (1873), Valsaceae. 1 or 2, Eur., Am. See Barr (*Mycol. Mem.* **7**, 1978).

silver ear, the basidioma of the edible *Tremella fuciformis*. See *Tremella*.

silver leaf, of plum and other fruit trees (*Chondrostereum purpureum*.

simblospore, Langeron's (1945) term for zoospore, q.v.

Simblum Klotzsch ex Hook. (1831) = Lysurus (Clathr.) fide Dring (1980).

Simocybe P. Karst. (1879), Cortinariaceae. 25, widespr. See Pegler & Young (*Kew Bull.* **30**: 225, 1975; 4 Brit. spp.), Singer (*Sydowia* **15**: 71, 1962, *Beih. Nova Hedw.* **44**: 485, 1973; key 18 neotrop. spp.), Horak (*Sydowia* **32**: 123, 1979; key 4 Papua New Guinea spp., *N.Z. Jl Bot.* **18**: 187, 1980; key 6 NZ spp.), Reid (*TBMS* **82**: 221, 1984; key Br. spp.).

Simoninus Roum. (1879) ≡ Chaenocarpus (Xylar.).

Simonyella J. Steiner (1902), Roccellaceae (L). 1, Oman, Socotra, S. Yemen. See Feige *et al.* (*Flora* **187**: 159, 1992; anatomy), Tehler (*CJB* **68**: 2458, 1990).

simple, unbranched; having no divisions.

Simuliomyces Lichtw. (1972), Legeriomycetaceae. 2 (in *Diptera*, *Plecoptera*), Eur., USA. See Lichtwardt (1986; key).

Singer (Rolf; 1906-1994). Born at Schliersee in S. Germany, Singer studied chemistry at the University of Munich, and in 1928 moved to Vienna where he was taught by R. Wettstein, receiving a PhD in 1931. He held positions in Barcelona (1933-35), St Petersburg (1935-40; Dr. Biol. Sci.), Harvard (1941-47), Tucumán (1948-60), Buenos Aires (1960-69), and Santiago (1967-68); from 1968 he lived in Chicago and was associated with the Field Museum and University of Illinois. Made major contributions to the overall understanding of *Agaricales* (esp. *The Agaricales in modern taxonomy*, 1951; last edn 4, 1986, treats 230 gen.) and to neotropical agaricology, describing some 2450 new fungi in about 400 publications in 9 languages. Also made important contributions on ethnomycology, mycorrhizas, nomenclature, and truffles. His collections are distributed through some 42 institutions. *Obit.*: Mueller (*Mycol.* **87**: 144, 1995). See also Singer (M., *Mycologists and other taxa*, 1984).

Singera Bat. & J.L. Bezerra (1960) = Vermiculariopsiella (Mitosp. fungi) fide Sutton (1977).

Singerella Harmaja (1974), Tricholomataceae. 1, Eur.

Singeriella Petr. (1959) = Blasdalea (Vizell.) fide v. Arx (*in litt.*).

Singerina Sathe & S.D. Deshp. (1981), Agaricaceae. 1, India.

Singerocybe Harmaja (1988), Tricholomataceae. 3, N. Eur.

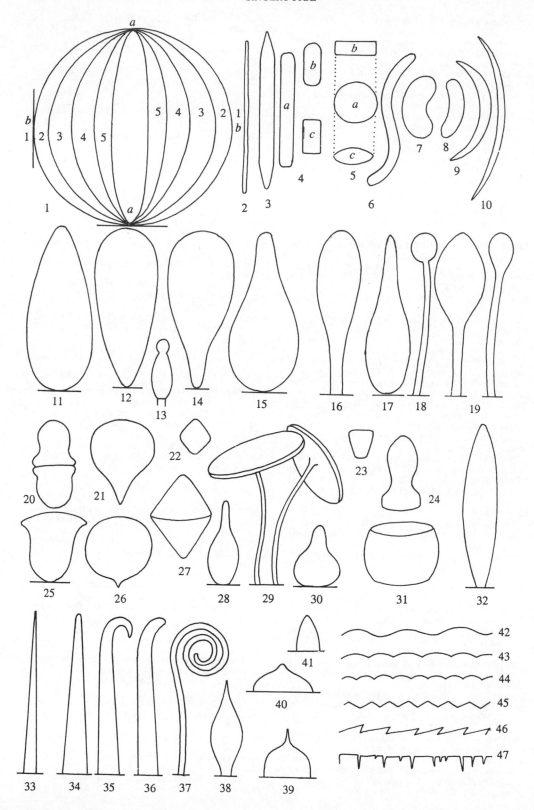

Fig. 37. Shapes. 1, shapes based on the sphere and ellipsoids, adapted from Payak (*Mycopath.* **16**: 72, 1962). Terms in parentheses are used by Erdtman (*An introduction to pollen analysis*, 1954). Ratios are those of Bas (*Persoonia* **5**: 321, 1969).

	axis aa	axis bb	length:breadth
1 by 1	globose (spherical)	(spherical)	1.0-1.05
1 to 2	subglobose (prolate spheroidal)	(oblate spheroidal)	1.05-1.15
2 to 3	broadly ellipsoidal (subprolate)	(suboblate)	1.15-1.3
3 to 4	ellipsoidal (prolate)	(oblate)	1.3-1.6; elongate
4 by 4	oval (perprolate)	(peroblate)	
5 by 5	fusiform		

2, filiform; 3, acerose; 4, cylindrical, restricted by Bas (1969) to cylinders with a length:breadth ratio of 2.0-3.0; a, bacilliform (1:b 3.0), b, c, oblong, b, apices rounded (obtuse), c, apices truncate; 5, a, discoid or lenticular in surface view, b, discoid in side view, c, lenticular in side view; 6, sigmoid; 7, reniform (fabiform); 8, allantoid; 9, lunate (crescentic); 10, falcate; 11, ovoid; 12, obovoid; 13, lecythiform; 14, pyriform; 15, obpyriform; 16, clavate; 17, obclavate; 18, capitate; 19, spathulate; 20, bicampanulate; 21, turbinate; 22, quadrangular (rhomboidal); 23, cuneiform; 24, dolabriform; 25, campanulate; 26, napiform; 27, biconic; 28, lageniform; 29, peltate; 30, ampulliform; 31, doliiform; 32, cymbiform (navicular); 33, acicular; 34, subulate; 35, hamate (uncinate); 36, corniform; 37, circinate; 38, ventricose. Apices. 39, mucronate; 40, papillate; 41, acute. Edges. 42, sinuate; 43, crenate; 44, crenulate; 45, dentate; 46, serrate; 47, laciniate. See also Systematics Association Committee for Descriptive Terminology (*Taxon* **9**: 245, 1960; list of works, **11**: 145, 1962; terminology of simple symmetrical plane shapes, chart).

Singeromyces M.M. Moser (1966), Xerocomaceae. 1, Argentina.

Sinoboletus M. Zang (1992), Boletaceae. 2, China.

Sinodidymella J.Z. Yue & O.E. Erikss. (1985), Dacampiaceae. 2, China, N. Am. See Barr (*Mycotaxon* **46**: 387, 1993).

Sinolloydia C.H. Chow (1936) = Lysurus (Clathr.) fide Dring (1980).

Sinosphaeria J.Z. Yue & O.E. Erikss. (1987) = Thyridium (Thyrid.) fide Eriksson & Yue (1989).

Sinotermitomyces M. Zang (1981), Pluteaceae. 2 (on termite nests), China.

sinuate, (1) (of lamellae), notched at the proximal end at junction with stipe (Fig. 19D) (cf. adnate); emarginate; (2) (of an edge), undulating (Fig. 37.42).

Siphomycetes, see *Phycomycetes*.

siphon, an aseptate hypha (Vuillemin, 1912).

Siphonaria H.E. Petersen (1903), Chytridiaceae. 3, Eur., Am. See Karling (*Am. J. Bot.* **32**: 580, 1945).

Siphonia Fr. (1820) [non Rich. ex Schreb. (1791), *Euphorbiaceae*] ≡ Dufourea (Teloschist.).

Siphopodium Reinsch (1875) nom. dub. ('phycomycetes').

Siphoptychium Rostaf. (1876) = Tubifera (Lycogal.).

Siphula Fr. (1825) nom. rej. ≡ Dufourea (Teloschist.).

Siphula Fr. (1831) nom. cons., Lecanorales (inc. sed., L). c. 25, cosmop. See Santesson (*Svensk Natur.* 1964: 176, 1964), Bendz et al. (*Acta Chem. scand.* **19**: 1250, 1965).

Siphulaceae, see *Lecanorales*.

Siphulastrum Müll. Arg. (1889), Peltigeraceae (L). 2, cosmop. (temp.).

Siphulella Kantvilas, Elix & P. James (1992), Icmadophilaceae (L). 1, Tasmania.

Siphulina (Hue) C.W. Dodge (1965), Peltigeraceae (L). 1, Antarctic.

sirenin, a hormone secreted by the female gamete of *Allomyces* which attracts male gametes.

Sirentyloma Henn. (1895) ? = Phyllachora (Phyllachor.).

Sirexcipula Bubák (1907), Mitosporic fungi, 4.A1/2.15. 2, Eur., N. Am.

Sirexcipulina Petr. (1923) = Topospora (Mitosp. fungi) p.p. and Sirococcus (Mitosp. fungi) p.p. fide Sutton (1977).

Sirobasidiaceae Lindau (1897), Tremellales. 2 gen., 9 spp. Basidioma gelatinous, pustulate; monomitic; basidia often catenulate, sterigmata deciduous; spores ballistosporic.

Sirobasidium Lagerh. & Pat. (1892), Sirobasidiaceae.

8, esp. trop. See Kobayashi (*Trans. mycol. Soc. Japan* **4**: 29, 1962), Boedijn (*Bull. Jard. bot. Buitenz.* **13**: 266, 1934), Lowy (*Mycol.* **48**: 324, 1956; key), Moore (*Ant. v. Leeuwenhoek* **45**: 113, 1979; septal ultrastr.).

Sirococcus Preuss (1855), Mitosporic fungi, 8.B1.15. 2, temp. See Sutton (1980).

Sirocrocis Kütz. (1843) nom. dub. (? Fungi).

Sirocyphis Clem. (1909), Mitosporic fungi, 4.A1.?. 1, USA.

Sirodesmites Pia (1927), Fossil fungi. 1 (Oligocene), Eur.

Sirodesmium De Not. (1849) = Coniosporium (Mitosp. fungi) fide Hughes (*CJB* **36**: 805, 1958).

Sirodiplospora Naumov (1915) = Topospora (Mitosp. fungi) fide Groves (*CJB* **43**: 1195, 1965).

Sirodochiella Höhn. (1925), Mitosporic fungi, 4.A1.?. 1, Eur.

Sirodomus Petr. (1947) = Sirexcipula (Mitosp. fungi) fide Sutton (1977).

Sirodothis Clem. (1909), Anamorphic Leotiaceae, 8.A1.15. Teleomorph *Tympanis*. 4, N. temp. Am. See Sutton & Funk (*CJB* **53**: 521, 1975).

Sirogloea Petr. (1923), Mitosporic fungi, 8.A1.?. 1, Eur.

Siroligniella Naumov (1926), Mitosporic fungi, 8.A1.3. 1, former USSR.

Sirolpidiaceae Sparrow ex Cejp (1959), ? Myzocytiopsidales. 1 gen. (+ 1 syn.), 6 spp.

Sirolpidium H.E. Petersen (1905), Sirolpidiaceae. 6 (marine), widespr. *S. zoophthorum* on clam and oyster larvae.

Siropatella Höhn. (1903) nom. dub. fide Sutton (*Mycol. Pap.* **141**, 1977).

Siropeltis Arx & R. Garnier (1960) = Sphaerobolus (Sphaerobol.) fide v. Arx (*in litt.*).

Sirophoma Höhn. (1917), Mitosporic fungi, 4.A1.?. 2 or 3, Eur.

Siroplacodium Petr. (1939), Mitosporic fungi, 8.A1.15. 3, Eur.

Siroplaconema Petr. (1922) ? = Ceuthospora (Mitosp. fungi).

Siropleura Petr. (1934), Mitosporic fungi, 8.A1.?. 1, Eur.

Siroscyphella Höhn. (1910) = Pseudocenangium (Mitosp. fungi) fide Dyko & Sutton (*CJB* **57**: 370, 1979).

Siroscyphellina Petr. (1923), Mitosporic fungi, 8.A1.?. 2, Austria, N. Am. See Wehmeyer (*Sydowia* **6**: 433, 1952).

Sirosiphon Kütz. (1843), Algae.

Sirosperma Syd. & P. Syd. (1916), Mitosporic fungi, 4.A1.23. 2, Papua New Guinea, Florida.

Sirosphaera Syd. & P. Syd. (1913), Mitosporic fungi, 4/8.A1.23. 1, Philipp.

Sirosporium Bubák & Serebrian. (1912), Mitosporic fungi, 1.C2.10. 11, widespr. See Ellis (*DH, MDH*).

Sirosporonaemella Naumov (1951), Mitosporic fungi, 4.A1.15. 1, Russia.

Sirostromella Höhn. (1916) = Pseudodiscula (Mitosp. fungi) fide Sutton (1977).

Sirothecium P. Karst. (1887), Mitosporic fungi, 8.G2.1. 2, Finland, Azores. See Sutton (*Proc. Ind. Acad. Sci.* (Pl. Sci.) **94**: 229, 1985).

Sirothyriella Höhn. (1910), Mitosporic fungi, 5.A1.?. 1, Eur.

Sirothyrium Syd. & P. Syd. (1916), Mitosporic fungi, 5.A1.?. 1, India.

Sirotrema Bandoni (1986), Tremellaceae. 3, N. temp. See Reid & Minter (*TBMS* **72**: 345, 1979; mycoparasitic on *Lophodermium*).

Sirozythia Höhn. (1904), Mitosporic fungi, 4.A1.?. 2, Eur.

Sirozythiella Höhn. (1909), Mitosporic fungi, 8.B/C1.?. 1, Eur.

Sisostrema, see *Sistotrema*.

Sistotrema Pers. (1794) ≡ Cerrena (Coriol.).

Sistotrema Raf. (1820) = Sistotrema (Sistotremat.).

Sistotrema Fr. (1821), Sistotremataceae. 34, widespr. See Donk (*Fungus* **26**: 3, 1956), Hallenberg (*Mycotaxon* **21**: 389, 1984), Boidin & Gilles (*BMSF* **110**: 185, 1994; key). Anamorph *Ingoldiella*.

Sistotremastrum J. Erikss. (1958), Sistotremataceae. 3, widespr.

Sistotremataceae Jülich (1982), Stereales. 13 gen. (+ 11 syn.), 89 spp. Basidioma resupinate to pileate; monomitic, hyphae inflated at septa; basidia urniform, 2-12 sterigmate; spores inamyloid.

Sistotremella Hjortstam (1984), Sistotremataceae. 4, widespr.

Sitochora H.B.P. Upadhyay (1964), Mitosporic fungi, 1.B1.?. 1 (on *Phyllachora*), Brazil.

Sivanesania W.H. Hsieh & C.Y. Chen (1994), Botryosphaeriaceae. 1, Taiwan.

skeletal hyphae, see hyphal analysis.

Skeletocutis Kotl. & Pouzar (1958), Coriolaceae. 1, Eur., Siberia.

skeletocystidium, see cystidium (1).

Skeletohydnum Jülich (1979), Epitheliaceae. 1, N.Z.

Skelophoromyces Thaxt. (1931) ≡ Scelophoromyces (Laboulben.).

Skepperia Berk. (1857), Thelephoraceae. 5, esp. trop.

Skepperiella Pilát (1927), Tricholomataceae. 4, widespr. See Pegler (*Kew Bull.* **28**: 257, 1973).

Skierka Racib. (1900), Uredinales (inc. sed.). 10 (on *Geraniales, Sapindales, Rhamnales*), trop. See Mains (*Mycol.* **31**: 175, 1939).

skiophilous, shade loving. Cf. photophobous.

Skirgiellia Batko (1978) = Rozella (Rozellopsid.) fide Dick (*in press*).

Skirgiellopsis Batko (1978) = Rozellopsis (Rozellopsid.) fide Dick (*in press*).

Skolekites Norman (1853) nom. rej. = Toninia (Catillar.) fide Hafellner (1984).

Skoteinospora Bat. (1962) = Euantennaria (Euantennar.) fide v. Arx & Müller (1975).

Skottsbergiella Petr. (1927), ? Valsaceae. 1, Juan Fernandez. See Petrak (*Sydowia* **24**: 264, 1971).

Skvortzovia Bononi & Hjortstam (1987), Corticiaceae. 1, S. Am.

Skyathea Spooner & Dennis (1986), Leotiaceae. 1 (on *Hedera*), U.K.

skyrin, an orange-yellow wilt toxin of *Cryphonectria parasitica* (Gäumann *et al.*, *Phytopath. Z.* **36**: 116, 1959). Cf. luteoskyrin.

Skyttea Sherwood, D. Hawksw. & Coppins (1981), Odontotremataceae. 14 (on lichens), widespr. See Coppins *et al.* (*SA* **10**: 51, 1991; gen. concept).

Skyttella D. Hawksw. & R. Sant. (1988), Leotiales (inc. sed.). 1 (on lichen, *Peltigera*), Eur., N. Am.

slaframine, a toxin of *Rhizoctonia leguminicola*; the cause of the slobber syndrome in livestock.

Slimacomyces Minter (1986), Mitosporic fungi, 1.F2.1. 1, UK.

slime, a wet, generally sticky, substance; mucus; - **flux**, a thick liquid from the stems or branches of trees made up of, or having a connexion with, fungi and bacteria (Ogilvie, *TBMS* **9**: 167, 1924; Stautz, *Phytopath. Z.* **3**: 163, 1931); - **moulds**, the *Acrasiomycota, Dictyosteliomycota*, and *Myxomycota*; - **spore**, a spore that becomes separated with slime from the cell producing it (Mason, 1937), cf. dry spore. See Gloiosporae.

slug (in slime moulds), the aggregated pseudoplasmodium of *Dictyosteliomycota*.

Smardaea Svrček (1969), Otideaceae. 2, Eur. See Dissing (*Sydowia* **38**: 35, 1986).

Smell. Many fungi have characteristic smells which are sometimes of diagnostic value. Gilbert (*Méthode de mycologie descriptive*, 1934; *BSMF* **48**: 241, 1932; **50**: 25, 1935) made a classification of the smells of natural basidiomata (see also Imler, *BSMF* **68**: 400, 1952; Josserand, 1952: 63; Locquin, *Petite flore des champignons de France* **1**: 97, 1956); Badcock (*TBMS* **23**: 188, 1939) describes those of wood-attacking fungi in culture. The smell of 'stinking' smut (bunt) of wheat (*Tilletia laevis*) is that of trimethylamine, $(CH_3)_3N$ (Hanna *et al.*, *J. biol. Chem.* **97**: 351, 1932; *Phytopath.* **22**: 978, 1932). Components of some perfumes are obtained from actinomycetes, lichens, and truffles.

Smeringomyces Thaxt. (1908), Laboulbeniaceae. 4, Am., Asia, Eur.

Smith (Annie Lorrain; 1854-1937). Originally a governess, worked at the British Museum (Natural History) as an 'unofficial worker'. Her first publications were on seaweeds, then she turned her attention to the fungi, becoming a President of the British Mycological Society. Later her attention focused on lichens, and she completed J.M. Crombie's *A monograph of lichens found in Great Britain* (2 vols, 1894, 1911), produced a second edition of this work (*A monograph of the British lichens*, vol. 1, 1918; vol. 2, 1926), and *Lichens* (1921 [reprint 1975]) which became the standard English introduction to the subject. See Gepp & Rendle (*J. Bot., Lond.* **75**: 328, 1937), Grummann (1974: 240), Stafleu & Cowan (*TL-2* **5**: 663, 1985).

Smith (Alexander Hanchett; 1904-1986). American agaricologist; University of Michigan, 1934-1972. Monographed *Mycena* (q.v.); see also Macromycetes, USA. See Rogers (*Mushroom* **5**: 17, 1987), Thiers (*Mycol.* **79**: 811, 1987; portr., bibl.), Stafleu & Cowan (*TL-2* **5**: 662, 1985).

Smith (Erwin Frink; 1854-1927). American plant pathologist and bacteriologist employed by the USDA from 1886-1927. First interested in peach yellows, he later specialized in bacterial pathogens of plants and was the author of *Bacteria in relation to plant disease*, 3 vols, 1911-14, and *Introduction to bacterial diseases of plants*, 1920. See Rogers (*Erwin Frink Smith. A story of North American plant pathology*, 1952), Stafleu & Cowan (*TL-2* **5**: 669, 1985), Verma (*Rev. Trop. Pl. Path.* **7**: 25, 1993).

Smith (Worthington George; 1835-1917). British mycologist, botanical artist and antiquarian. Artist for the *Gardeners' Chronicle*, 1869-1910, and of the *Floral Magazine*, 1870-1876. He regularly attended meetings of the Woolhope Club, making humorous reports in *GC* and a series of cartoons in other periodicals. A founder member of the British Mycological Society. See Ainsworth (*Mycologist* **4**(1): 32, 1990; portr.), Stafleu & Cowan (*TL-2* **5**: 706, 1985).

Smithiogaster J.E. Wright (1975), Secotiaceae. 1, Argentina.

Smithiomyces Singer (1944), Agaricaceae. 1, trop. Am.

Smittium R. Poiss. (1937), Legeriomycetaceae. 38 (in *Diptera*), widespr. See Lichtwardt (1986; biology, key), Horn (*Mycol.* **81**: 742, 1989; ultrastr., *MR* **93**: 303, 1989; spore extrusion, *Exp. Mycol.* **14**: 113, 1990; physiology & spore extrusion), Lichtwardt & Williams (*CJB* **68**: 1057, 1990, **70**: 1193, 1992, *Mycol.* **84**: 384, 392, 1992), Williams & Lichtwardt (*Mycol.* **79**: 832, 1987, *CJB* **68**: 1045, 1990), Williams (*Mycol.* **75**: 242, 1983; trichospore germination and zygospores), Sato *et al.* (*Trans. mycol. Soc. Japan* **30**: 51, 1989; ultrastr.).

smut, (1) a disease caused by one of the *Ustilaginales*, esp. a member of the *Ustilaginaceae* (cf. bunt); (2) a smut fungus. **- spore**, a chlamydospore of a - fungus; ustilospore; ustospore; **covered -**, a smut in which the mature spore mass keeps for a time within a covering of host (or fungal) tissue, frequently till after the sorus becomes free from the host: of barley, oats *Ustilago segetium* (syn. *U. hordei*, *U. kolleri*); sorghum, *Sporisorium sorghi*; wheat (**bunt, stinking -**), *Tilletia caries* and *T. laevis* (syn. *T. foetida*); (dwarf bunt), *T. controversa*. **'fig -'**, *Aspergillus niger*; **flag -** of wheat, see stripe smut. **loose -**, a smut in which the spores are as an uncovered mass of powder becoming free from the host plant by wind and rain; of barley and wheat, *Ustilago segetum* var. *tritici* (syn. *U. nuda*); oats, *U. segetum* var. *avenae*; sorghum, *S. cruenta*. **stripe -**, of grasses, *U. striiformis*; rye, *Urocystis occulta*; wheat, *U. agropyri*.

sociation, society, see Phytosociology.

Societies and organizations. The *International Mycological Directory* (Hall & Minter, edn 3, 1994) gives details of 43 societies, 10 associations, 2 unions and 6 groups of mycologists. Of these, the following list represents a selection of the largest and most active societies. Most are concentrated in Europe and North America, although the number of societies in Asia and Latin America is growing rapidly. There are few organizations for mycology in Africa.

Lichenological societies and their journals. **Argentina**: Sociedad Liquenologica Argentina, 1977- . **Australia**: Society of Australasian Lichenologists, established 1974 (*Australian Lichenological Newsletter*). **Czech Republic**: Czech Botanical Society, Section for Bryology & Lichenology (*Bryonora*). **France**: Association Française de Lichenologie, 1976- (*Bulletin d'Informations de l'Association Française de Lichenologie*). **Germany**: Bryologisch-lichenologische, Arbeitsgemeinschaft für Mitteleuropa, 1968- (*Herzogia*, 1968-). **International Association for Lichenology**, proposed 1964 and formally established at the XI International Botanical Congress, 1969 (*International Lichenological Newsletter*). **Italy**: Società Lichenologica Italiana, founded 1987 (*Notizario*, 1990-). **Japan**: Lichenological Society of Japan, 1972- (*Lichen*, 1972-). **Norway**: Nordisk Lichenologisk Forening, 1975- (*Graphis Scripta*). **U.K.**: British Lichen Society, 1958- (*The Lichenologist*, 1958-); The Lichen Exchange Club of the British Isles (1907-14) (*The Lichen Exchange Club Reports*, 1909-13). **USA**: The American Bryological and Lichenological Society, renamed 1969 (formerly the

Sullivan Moss Society (1899-1949); *The Bryologist*) and The American Bryological Society, 1898-1969.

Mycological societies and their journals. **Argentina**: Asociación Argentina de Micología, founded 1960 (*Revista Argentina de Micología*, 1978-). **Australia**: Australian Mycological Society, founded 1994 (*Quarterly Newsletter*). **Austria**: Österreichische Mykologische Gesellschaft, founded 1919 (*Österreichische Zeitschrift für Pilzkunde*). **Belgium**: Antwerpse Mycologische Kring (*Sterbeckia*). **former Czechoslovakia**: Česká Mykologické Společnosti, 1920- (Czech Mycological Society), (*Casopis československych Houbaru*); Česká Vedeka Společnosti, pro Mykologii, (Czech Scientific Society for Mycology) 1947- (successor to the Mykologický Klub, 1922-); (*Česká Mykologie*, Prague, 1947- ; continued as *Czech Mycology*, 1993-). **Denmark**: Foreningen til Svampekundskabens Fremme (*Friesia*, 1931-80; *Svampe*, 1980-). **Finland**: Societas Mycologica Fennica, establ. 1948 (*Sienilehti*, 1948-). **France**: Societè Mycologique de France, 1885 (*Bulletin Trimestriel de la Société Mycologique de France*, 1885-). **Germany**: Deutsche Gesellschaft für Mykologie, founded 1977 (*Zeitschrift für Mykologie*, 1939-). **Hungary**: Hungarian Mycological Society, founded in 1962 as a section of the Hungarian Forestry Association; became independent in 1992 (*Mikológiai Közlémenyk Clusiana*, 1962-). **India**: Mycological Society of India, 1971 (*Kavaka*, 1973-). **Italy**: Unione Mycologia Italiana (*Mycologia Italiana*); Associazone Micologica Bresadola, 1957- (*Revista de Micologia*). **Japan**: Mycological Society of Japan, (*Transactions of the Mycological Society of Japan*, 1956-1993; *Mycoscience*, 1994-). **Korea**: Korean Society for Mycology (*Korean Journal of Mycology*, 1973-). **Latin America**: Asociación Latinoamerica de Micología, founded 1990 (*Revista Iberoamericana de Micología*). **Mexico**: Sociedad Mexicana de Micologia (*Revista Mexicana de Micologa*, 1985-). **Netherlands**: Nederlandse Mycologische Vereniging (*Fungus*, 1929-58; *Coolia*, 1954-). **Norway**: Norsk Soppforening (1954-) (*Blekksoppen*). **Poland**: Polskie Towarzystwo Botaniczne (*Acta Mycologica*, 1965-). **Republic of China**: Mycological Society of the Republic of China, founded 1983 (*Trans. Myc. Soc., Rep. China*, 1985-). **Romania**: Mycological Society of Romania, founded 1990 (*Mycologica Romanica*, 1994-). **Spain**: Sociedad Micologica de Madrid (*Boletin de la S.M. de M.*). **Sweden**: Sveriges Mykologiska Forening, founded 1979 (*Jordstjärnen*). **Switzerland**: Schweizerische Verein für Pilzkunde (*Schweizerische Zeitschrift für Pilzkunde*). **U.K.**: British Mycological Society, 1896- (*Transactions of the BMS*, 1897-1988; continued as *Mycological Research*, 1989- , *Bulletin of the British Mycological Society*, 1967-1986 *Mycologist*, 1987-). **USA**: Mycological Society of America, 1931 (*Mycologia*, 1909- ; *Inoculum*, 1992-).

Medical mycology societies and their journals: There are some twenty national medical mycological Societies (for addresses see *ISHAM Newsletter*) including the following publishing societies: International Society for Human and Animal Mycology (*J. of Medical and Veterinary Mycology*, 1985- [as *Sabouraudia* 1961-84], *ISHAM Newsletter*); Société Française de Mycologie Medicale (*Bulletin de la Soc. franç de Mycol. med.*); Deutschsprachige Mykologische Gesellschaft (*Mykosen*); Japanese Society for Medical Mycology (*Japanese Journal of Medical Mycology*), Asociación Espãnola de Especialistas en Micología (*Revista Ibérica de Micología*).

International organizations. Most societies are affiliated to the International Mycological Association

(IMA), founded in 1971 at the First International Mycological Congress (Exeter, UK; *TBMS* **58** (Suppl.): 14, 1972) with the responsibility for arranging additional International Mycological Congresses (q.v.) at regular intervals. The IMA has Regional Committees for Africa (Committee for the Development of Mycology in Africa, CODMA; African Mycological Association, 1985-; Conferences 1990, 1992, 1995), Asia (1977-; congresses 1987, 1991, 1995), Europe (Congress of European Mycologists, CEM; congresses 3 yearly, XIIth in Wageningen 1995), Latin America (Associacion Latinoamericana de Micologia, 1993-; congresses 1993, 1996).

The IMA constitutes the Section for General Mycology of the International Union of Biological Societies (IUBS), the primary biological component of the International Council of Scientific Unions' (ICSU). Other mycological organizations associated with the IUBS include the International Society for Plant Pathology (ISPP; founded 1968), the International Commission on Mushoom Science, International Association for Lichenology (IAL, see above; participating in International Mycological Congresses). Regional medical mycological societies are integrated in the International Society for Human and Animal Mycology (ISHAM) (founded 1954) which arranges independent international congresses. Regional microbiological societies represented by the International Union of Microbiological Societies (IUMS, embraced in IUBS until 1980 when it became a separate ICSU union). Mycological interests within the IUMS are represented by the Mycology Division, and International Commissions on Mycotoxicology, Food Mycology, Yeasts, and Fungal Serology. In addition there are two IUBS/IUMS inter-Union commissions: the International Commission on the Taxonomy of Fungi (ICTF; founded 1982) and the World Federation for Culture Collections; and also an IUBS/IUMS Committee on Microbial Diversity (founded 1994).

Issues related to the formal system of naming fungi are considered through the Permanent Committee on Fungi (CF; Committee for Fungi and Lichens to 1993) which reports through the General Committee on Botanical Nomenclature to International Botanical Congresses; the International Association for Plant Taxonomy (*Taxon*, 1957-) also covers fungi.

See also BioNET-INTERNATIONAL, MIRCEN.

sociomycie, see mycosociology.

soft rot, a decomposition of plant parts (fruits, roots, stems, etc.) by fungi or bacteria resulting in the tissues becoming soft.

Soil fungi. The soil normally has large numbers of bacteria (and actinomycetes) and smaller but still large numbers of fungi (mitosporic fungi, *Mucorales*, *Ascomycota*, *Chytridiomycetes* and *Oomycetes*). These organisms are important for keeping the soil fertile, though some may be pathogens. Some components of the soil mycobiota are cosmopolitan, but environment, geographical location, climate and the soil physico-chemical properties have an effect on its composition; for example *Penicillium* is more frequent than *Aspergillus* in temperate areas while in warmer areas the opposite is true; and the mycelium of basidiomycetes is most frequent under trees. The fungal biomass almost everywhere exceeds that of bacteria, with the exception of rhizosphere soil. Usually more than half of the fungal biomass is of basidiomycetes which are overlooked by most isolation techniques. Peat bogs have large numbers of fungi (Dooley & Dickinson, *Irish J. agric. Res.* **10**: 195, 1971). See also Sand dune fungi.

The chief work of soil fungi relates to decomposition and recycling organic residues. Fungi are important in the early stages of the decomposition of plant material, immobilizing large amounts of nitrogen (N) as protein. Cellulose and lignin are the most important plant cell wall components that are mainly decomposed in litter and soil by fungi. For the latter mainly basidiomycetes but also some ascomycetes are responsible. Lichenized fungi with cyanobacterial partners take a part in the nitrogen cycle, as these fix nitrogen from the air.

There are complex biological relations between the different soil organisms themselves (see Antagonism) and between them and the green plants (see Mycorrhiza, rhizosphere) on which they may be parasites. Reinking and Mann (fide Garrett, 1939) have grouped soil fungi as, (1) having a wide distribution (the 'soil inhabitants') and (2) the 'soil invaders', having a limited distribution, e.g. most parasitic root-infecting fungi.

A common method for the investigation of the soil biota is that of plating a series of dilutions of the soil (for problems in quantification see James and Sutherland, *Can. J. Res.* C **17**: 73, 97, **18**: 347, 435, 1939-40). Other techniques used are 'direct observation' (Conn, *Soil Sci.* **26**: 257, 1928), the Cholodny slide (Conn, *Tech. Bull. N.Y. Sta. agric. Exp. Stn* **204**, 1932), and Warcup's soil plate (*Nature* **166**: 117, 1950). Washing roots (Harley & Waid, *TBMS* **38**: 105, 1955) or soil particles (Parkinson & Williams; Gams & Domsch, *Arch. Microbiol.* **58**: 134, 1967; Bååth, *CJB* **66**: 1566, 1988), Bills & Polishook (*Mycologia* **86**: 187, 1994) shifts the balance in favour of slow-growing and poorly sporulating fungi that may be active as mycelium in the soil.

Lit.: Waksman (*Principles of soil microbiology*, edn 2, 1931; *Soil microbiology*, 1952), Gilman (*A manual of soil fungi*, edn 2, 1957; systematics), Burges (*Micro-organisms in the soil*, 1958; gen. ecol. survey), Johnson & Curl (*Methods for studying soil microflora-plant disease relationships*, edn 2, 1972), Durbin (*Bot. Rev.* **27**: 522, 1961; isolation techniques), Parkinson & Waid (Eds) (*The ecology of soil fungi*, 1960), Garrett (*Soil fungi and soil fertility*, edn 2, 1981), Baker & Snyder (Eds) (*Ecology of soilborne plant pathogens*, 1965), Barron (*The genera of hyphomycetes from soil*, 1968), Carroll & Wicklow (Eds) (*The fungal community*, edn 2, 1992), Dix & Webster (*The ecology of fungi*, 1995), Domsch & Gams (*Pilze aus Agrarböden*, 1970 [Engl. transl; Hudson, *Fungi in agricultural soils*, 1972]), Gray & Williams (*Soil micro-organisms*, 1971), Litvinov (*Key to microscopic soil fungi*, 1967 [Russ.]), Griffin (*Ecology of soil fungi*, 1972), Lockwood (*Biol. Rev.* **52**: 1, 1977; fungistasis), Schippers & Gams (Eds) (*Soil-borne plant pathogens*, 1979), Domsch et al. (*Compendium of soil fungi*, 2 vols, 1980, [supplemented reprint, 1993]; keys, descr., monogr. 400 spp., 6593 refs), Bruell (*Soil-borne plant pathogens*, 1987), Arora et al. (*Handbook of Applied Mycology* **1**, 1991), Cooke & Rayner (*Ecology of saprotrophic fungi*, 1984), Frankland (*Trans. mycol. Soc. Japan* **31**: 89, 1990), Hornby (Ed.) (*Biological control of soil-borne plant pathogens*, 1990), Jensen et al. (Eds) (*Microbial communities in soil*, 1986), Rayner & Boddy (*Fungal decomposition of wood, its biology and ecology*, 1988), Wainwright (*TBMS* **90**: 159, 1988), Singleton et al. (*Methods for research on soilborne phytopathogenic fungi*, 1992), Watanabe (*Pictorial atlas of soil and seed fungi*, 1994). See dermatophytes, Isolation methods, Medical and Veterinary Mycology, Root rots.

soiling, the disfiguring of paint, fruit, or other materials by pigmented moulds.

Solanella Vaňha (1910), ? Leotiales (inc. sed.). 1, Eur.

Soleella Darker (1967), Rhytismataceae. 1, N. Am.

soleiform, shaped like the sole of a shoe, i.e. elongate-ellipsoid, with one cell much larger and broader than the other.

Solenarium Spreng. (1827) = Glonium (Hyster.) fide Fries (*Syst. mycol.* **2**: 594, 1823).

Solenia Pers. (1794) [non *Solena* Lour. (1790), *Cucurbitaceae*] ≡ Henningsomyces (Schizophyll.).

Solenia Hill ex Kuntze (1898) ≡ Pinuzza (Bolet.).

Solenodonta Castagne (1845) = Puccinia (Puccin.) fide Saccardo (*Syll. Fung.* **18**: 824, 1906).

Solenographa A. Massal. (1860) = Phaeographis (Graphid.).

Solenopezia Sacc. (1889), Hyaloscyphaceae. 2, Eur., N. Am., former USSR. See Raitviir *et al.* (*Sydowia* **43**: 219, 1991), Seaver (*Mycol.* **22**: 122, 1930).

Solenoplea Starbäck (1901) = Camarops (Bolin.) fide Nannfeldt (1972).

Solenopsora A. Massal. (1855), ? Bacidiaceae (L). 13, Eur., Asia, N. Am.

Solheimia E.F. Morris (1967), Mitosporic fungi, 2.A2.?. 1, Panama.

Solicorynespora R.F. Castañeda & W.B. Kendr. (1990), Mitosporic fungi, 1.C2.24. 6, widespr.

Soloacrospora W.B. Kendr. & R.F. Castañeda (1991), Mitosporic fungi, 1.B1.12. 1, Cuba.

solopathogenic (of a smut such as *Ustilago zeae*), a pathogenic monosporidial line (Christensen).

Solorina Ach. (1808), Peltigeraceae (L). 10, cosmop. See Hue (*Mém. Soc. natn Sci. nat. math. Cherbourg* **38**: 1, 1911), Jahns *et al.* (*Bibl. Lich.* **57**: 161, 1995; ontogeny), Thomson & Thomson (*Bryologist* **87**: 151, 1984; SEM).

Solorinaceae, see *Peltigeraceae.*

Solorinaria (Vain.) Gyeln. (1935), Heppiaceae (L). 2, Eur.

Solorinella Anzi (1860), Solorinellaceae (L). 2, Eur., Peru. See Farkas & Lökös (*Acta bot. fenn.* **150**: 21, 1994; distrib.), Thor (*Nordic Jl Bot.* **4**: 823, 1985), Vězda & Poelt (*Phyton* **30**: 47, 1990).

Solorinellaceae Vězda & Poelt (1990), Ostropales (L). 3 gen. (+ 4 syn.), 30 spp. Thallus crustose; ascomata apothecial, often deeply immersed, the wall composed of thick-walled gelatinized hyphae with very narrow lumina; interascal tissue of simple gelatinized paraphyses; asci ± cylindrical, with a slightly thickened apical cap; ascospores transversely septate or muriform, sometimes with a sheath. Anamorph pycnidial. Lichenized with green algae, widespr.

Solorinellomyces Cif. & Tomas. (1953) ≡ Solorinella (Solorinell.).

Solorinina Nyl. (1884) = Solorina (Peltiger.).

Solorinomyces E.A. Thomas ex Cif. & Tomas. (1953) ≡ Solorina (Peltiger.).

Solosympodiella Matsush. (1971), Mitosporic fungi, 1.A/B1.10. 6, Papua New Guinea. See de Hoog (*Stud. Mycol.* **26**: 54, 1985; key).

Solutopariës Whiffen (1942), Chytridiaceae. 1, USA.

soma, (1) body, excluding reproductive parts or phase; (2) in Aryan religion, *Amanita muscaria* (Wasson, *Soma: the divine mushroom of immortality*, 1968, *The wondrous mushroom*, 1980); **-tic,** pertaining to the soma. See also ethnomycology, hallucinogenic fungi.

Somatexis Toro (1934) = Aphanostigme (Pseudoperispor.). See v. Arx & Müller (1975).

somatogamy, fusion of somatic (vegetative) cells or hyphae involving plasmogamy but not karyogamy.

Somion Adans. (1763) = Spongipellis (Coriol.) fide Donk (1974).

Sommerfeltia Flörke ex Sommerf. (1827) nom. rej. = Solorina (Peltiger.) fide Zahlbruckner (1925).

Sommerstorffia Arnautov (1923), Saprolegniaceae. 1 (on rotifers), Eur., N. Am. See Dick (*in press*; key).

Sonderhenia H.J. Swart & J. Walker (1988), Mitosporic fungi, 4.C2.19. 2, widespr.

sooty moulds, used for fungi forming dense dark mats of hyphae on living leaves in the tropics, esp. *Dothideales* (fams: *Antennulariaceae, Capnodiaceae, Chaetothyriaceae, Euantennariaceae, Metacapnodiaceae*), and their anamorphs. See Hughes (*Mycol.* **68**: 693, 1976; genera and interrelationships). Also *Lit.* under *Dothideales.*

Sopagraha Subram. & Sudha (1979) = Arachnophora (Mitosp. fungi) fide Sutton (*in litt.*).

Sophronia Pers. (1827) [non Licht. ex Roem. & Schult. (1817), *Iridaceae*] = Phallus (Phall.).

Soppittiella Massee (1892) = Cristella (Exid.) fide Donk (*Persoonia* **4**: 329, 1966).

soralium (pl. **-ia**), decorticate portions of a lichen thallus where **soredia** are located. Usually formed from medullary tissues thrusting upwards through the cortical layers and so sometimes with the chemistry of the medulla rather than of the cortex. Soralia may be **diffuse** (the upper surface of the lichen becoming a continuous soredial mass), or **delimited** (i.e. confined to well-defined areas), and can be classified according to where they originate. They can arise on tubercles (**tuberculate -**) or as fissures (**fissural -**) in some genera (Hawksworth, *Lichenologist* **5**: 181, 1972). Soralia can arise at the tips of isidia, and in some taxa the soralia can contain a mixture of soredia and isidia-like structures.

Lit.: Du Rietz (*Svensk bot. Tidskr.* **18**: 371, 1924; soralium types).

See Spore discharge and dispersal, Spore germination.

Sorataea Syd. (1930), Uropyxidaceae. 6 (on *Leguminosae*), S. Am., Asia, W. Afr. See Eboh & Cummins (*Mycol.* **72**: 203, 1980), Cummins & Hiratsuka (1983).

Sorauer (Paul Carl Moritz; 1839-1916). Head of the Experimental Station for Plant Physiology, Imperial Cider Institute, Proskau; at the Royal Agricultural College, Berlin from 1881. First editor of the *Zeitschrift für Pflanzenkrankheiten* (1891-1916). Chief writings: *Handbuch der Pflanzenkrankheiten* (1874 [2 Aufl., 1886; 3 Aufl. (with Lindau & Reh), 1905-8], etc.) and a number of papers on fungus diseases of plants. See Wittmack (*Ber. dtsch bot. Ges.* **34**: (50), 1917), Stafleu & Cowan (*TL-2* **5**: 749, 1985).

Sordaria Ces. & De Not. (1863) nom. cons., Sordariaceae. 10 (esp. coprophilous), cosmop. See Moreau (*Le Botaniste* **39**: 1, 1953), Lundqvist (1972), Guarro & v. Arx (*Persoonia* **13**: 369, 1987; key 14 spp., list spp. names), Read & Beckett (*CJB* **63**: 281, 1985; ascoma ontogeny), Barrasa *et al.* (*Persoonia* **13**: 83, 1986; narrow-spored spp.).

Sordariaceae G. Winter (1885), Sordariales. 12 gen. (+ 6 syn.), 52 spp. Ascomata dark, usually thick-walled and ostiolate; interascal tissue composed of thin-walled undifferentiated cells, usually inconspicuous, often evanescent; asci cylindrical with a thickened I-apical ring; ascospores brown, simple or very rarely septate, sometimes ornamented, often with a gelatinous sheath but lacking caudae.

Lit.: Lundqvist (1972), Moreau (1953).

Sordariales, Ascomycota. 8 fam., 121 gen. (+ 124 syn.), 676 spp. Ascomata very rarely stromatic, perithecial or cleistothecial, thin- or thick-walled; interascal tissue very inconspicuous or lacking; asci cylindrical or clavate, persistent or evanescent, not fissitunicate; ascospores ± always with at least one dark cell, with germ pore, often with a gelatinous sheath or appendages. Anamorphs usually absent or spermatial. Saprobes on rotting wood and soil, coprophilous, some fungicolous, many cellulolytic. Fams:

(1) **Batistiaceae**.
(2) **Catabotrydaceae**.
(3) **Ceratostomataceae** (syn. *Melanosporaceae*).
(4) **Chaetomiaceae** (syn. *Achaetomiaceae*).
(5) **Coniochaetaceae**.
(6) **Lasiosphaeriaceae** (syn. *Tripterosporaceae*).
(7) **Nitschkiaceae** (syn. *Coronophoraceae*).
(8) **Sordariaceae** (syn. *Neurosporaceae*).

Subramanian (*Kavaka* **14**: 1, 1988) argues that *Coronophorales* should be retained and two families accepted (*Bertiaceae* and *Coronophoraceae*). The *Coniochaetaceae* have spores with germ slits and smooth setae and have previously been excluded (Hawksworth & Wells, *Mycol. Pap.* **134**, 1973). Malloch (1979) placed several of the above families in the *Diaporthales*.

Some taxa (8 gen. (+ 3 syn.), 17 spp.) referred to *Sordariales* here are not placed in any of the above families, including *Corynascus*, *Helminthosphaeria* and *Monosporascus*.

Lit.: v. Arx & Müller (1954), v. Arx (1981, *Sydowia* **34**: 13, 1982; fams.), v. Arx *et al.* (*Beih. Nova Hedw.* **94**, 1988), Cain (1934), Lundqvist (*Symb. bot. upsal.* **20**(1), 1972), Moreau (*Les genres Sordaria et Pleurage*, 1953), Müller & v. Arx (1973).

Sordariella J.N. Kapoor, S.P. Lal & Bahl (1975) = Herpotrichia (Lophiostom.) fide v. Arx & Müller (1984).

soredium (pl. **-ia**), a non-corticate combination of phycobiont cells and fungal hyphae having the appearance of a powdery granule, and capable of reproducing a lichen vegetatively (Fig. 22J-L). For their liberation and dispersal see Bailey (*J. Linn. Soc. (Bot.)* **59**: 479, 1966, *Revue bryol. lichén.* **36**: 314, 1969, *in* Brown *et al.* (Eds), *Lichenology: progress & problems*: 215, 1976). Cf. soralium.

Soredospora Corda (1837) = Fumago (Mitosp. fungi) fide Hughes (1958).

Sörensen coefficient (K). A numerical estimate of the affinity between two biotas; $K = {}^{200 \times} c/_{a+b}$, where a = species in one region, b = species in second region, and c = number of species in common. See Poore (*J. Ecol.* **43**: 606, 1955).

Soreymatosporium Sousa da Câmara (1930) = Stemphylium (Mitosp. fungi) fide Wiltshire (1938).

Sorica Giesenh. (1904) = Caliciopsis (Corynel.) fide Fitzpatrick (*Mycol.* **42**: 464, 1942).

sorocarp (of *Acrasiales*), a stalked fruiting structure.

Sorochytriaceae Dewel (1985), Blastocladiales. 1 gen., 1 spp. Colonial phase endobiotic; thallus polysporangiate through sequential septation, resulting segments forming rhizoids bearing sporangia giving rise to zoospores; a new colonial thallus developing intramatrically from encysted zoospores on host; if development extramatrical (or on agar) a branching, polycentric, rhizomycelial thallus results; parasites of tardigrades.

Sorochytrium Dewel (1985), Sorochytriaceae. 1, USA.

Sorocybe Fr. (1849) = Pycnostysanus (Mitosp. fungi) fide Hughes (1958).

sorocyst (of *Acrasiales*), a sorocarp lacking a stalk.

Sorodiplophrys L.S. Olive & Dykstra (1975), ? Labyrinthulomycota (inc. sed.). 1, Russia, USA. See Dick (*in press*).

Sorodiscus Lagerh. & Winge (1913) [non G.J. Allman (1847), *Algae*], Plasmodiophoraceae. 4 (in *Chara*), widespr. See Karling (1968; key).

Sorokina Sacc. (1892), Dermateaceae. *c.* 3, S. Am., Java, Australia. See Dennis (*Kew Bull.* **13**: 321, 1958).

Sorolpidium Němec (1911), ? Plasmodiophoraceae. 1 (on *Beta*), Eur., N. Am.

Sorophorae, see *Acrasiomycota*.

sorophore (of *Acrasiales*), a sorocarp stalk.

Sorosphaera J. Schröt. (1886), Plasmodiophoraceae. 2, Eur., N. Am.

Sorosporella Sorokīn (1888), Mitosporic fungi, 1.A1.?. 1 (on insects), N. temp. See Speare (*J. agric. Res.* **18**: 399, 1920), Pendland & Boucias (*Mycopathol.* **99**: 25, 1987; SEM and TEM).

Sorosporium F. Rudolphi (1829), Ustilaginaceae. 1 (on *Caryophyllaceae*), widespr. (esp. N. Hemisph.). See *Sporisorium*.

Sorosporonites X. Mu (1977), Fossil fungi. 1, China.

Sorothelia Körb. (1865) ? = Endococcus (Dothideales).

sorus, a fruiting structure in certain fungi, esp. the spore mass in Uredinales and Ustilaginales; a group of fruitbodies, as in *Synchytriaceae*.

Sowerby (James; 1757-1822). Famous English botanical artist (whose children and their descendants included a dozen naturalists and illustrators) best known mycologically for his *Coloured figures of English fungi or mushrooms* (3 vols, 1796-1803; suppl. 1809-15; for dates of publication see Ramsbottom, *TBMS* **18**: 167, 1933). See Ainsworth (*Mycologist* **2**: 125, 1988; portr.), *Dict. nat. Biogr.* **53**: 305, 1898. See Grummann (1974: 381), Stafleu & Cowan (*TL-2* **5**: 759, 1985).

Sowerbyella Nannf. (1938), Otideaceae. 12, Eur. See Moravec (*Mycol. Helv.* **1**: 427, 1985; key 9 spp., *Česká Myk.* **42**: 193, 1988; key 12 spp.), Jeppson (*Göteborgs Svampklubbs Årsskr* 1980: 9; key 3 spp.).

soy sauce, see shoyu.

Spadicesporium V.N. Boriss. & Dvoĭnos (1982) ? = Cladosporium (Mitosp. fungi) fide Sutton (*in litt.*).

Spadicoides S. Hughes (1958), Mitosporic fungi, 1.C2.27. 10, widespr. See Wang (*Mem. N.Y. bot. Gdn* **28**: 218, 1976), Sinclair *et al.* (*TBMS* **85**: 736, 1985), Kuthubutheen & Nawawi (*MR* **95**: 163, 1991).

Spadonia Fr. (1829) nom. dub. (? Fungi).

Spalovia Nieuwl. (1916) ≡ Spermotrichum (Mitosp. fungi).

spalted (of wood), with dark zone lines (q.v.) due to the interactions of different fungal colonies of the same species; as caused, e.g. by *Diatrype disciformis*; generally brittle, short-grained, and easily breaking through decay.

Sparassidaceae Herter (1910), Cantharellales. 2 gen., (+ 1 syn.), 6 spp. Basidioma coralloid, large, with flattened branches; monomitic; spores colourless (spore print cream), smooth, non-amyloid; lignicolous.

Lit.: Donk (1964: 291).

Sparassiella Schwarzman (1964), Sparassidaceae. 1, former USSR.

Sparassis Fr. (1819), Sparassidaceae. 5, N. temp. *Sparassis crispa* is edible.

sparassoid, composed of interlaced flabelliform branches forming ball-like structures recalling *Sparassis* basidiomata. Known also in *Pezizales* (Korf, *Rept Tottori mycol. Inst.* **10**: 389, 1973).

sparassol, orsellinic acid monomethyl ether from *Sparassis ramosa*; antifungal.

Sparrow (Frederick Kroeber; 1903-77). Professor of botany (1949) and director of the Biological Station (1968), Univ. Michigan, Ann Arbor. Noted for *Aquatic phycomycetes*, 1943 (edn 2, 1960). See Petersen (*Mycol.* **70**: 213, 1978; portr., bibliogr.). Grummann (1974: 297), Stafleu & Cowan (*TL-2* **5**: 775, 1985).

Sparrowia Willoughby (1963), Chytridiaceae. 2, UK.

Sparsitubus L.W. Hsu & J.D. Zhao (1980), Coriolaceae. 1, China.

Spartiella Tuzet & Manier ex Manier (1962), Legeriomycetaceae. 1 (in *Ephemeroptera*), Eur. See Lichtwardt (1986).

Spathularia Pers. (1797), ? Geoglossaceae. 2, N. temp. See Maas Geesteranus (*Proc. K. ned. Akad. Wet. C* **75**: 243, 1972), Benkert (*Gleditschia* **10**: 141, 1983), Ohenoja (*Opera Bot.* **100**: 193, 1989; nomencl.), Stellmacher (*Abh. Naturhist. Ges. Nüremberg* **40**: 78, 1985).

Spathulariopsis Maas Geest. (1972), ? Geoglossaceae. 1, N. Am.

spathulate, like a spoon in form (Fig. 37.19).

Spathulea Fr. (1849) ≡ Spathularia (Geogloss.).

Spathulina Pat. (1900) ? = Irpex (Steccherin.).

Spathulospora A.R. Caval. & T.W. Johnson (1965), Spathulosporaceae. 5 (on *Ballia* spp.), widespr. See Kohlmeyer & Kohlmeyer (1979; key), Walker *et al.* (*TBMS* **73**: 193, 1979; host-parasite interface).

Spathulosporaceae Kohlm. (1973), Spathulosporales. 1 gen., 5 spp. Thallus without well-developed hyphae; ascomata usually with dark setose hairs; peridium relatively thin, 2-layered; interascal tissue absent but ostiole periphysate; hymenium extending over the lower part only of the inner wall of the ascoma; ascospores with mucous appendages. Anamorph spermatial, without verticillate conidiophores. On *Ballia*, southern oceans.

Spathulosporales, Ascomycota. 2 fam., 2 gen., 6 spp. Vegetative tissue reduced to a small dark crustose thallus-like structure with a hyaline haustorial foot cell, the thallus cells irregular or hyphal, thick-walled, not formed in a strict developmental sequence; ascomata single or paired, perithecial, thick-walled, often surrounded by dark thick-walled hairs; interascal tissue absent or composed of simple paraphysis-like filaments, the ostiole periphysate or not; asci numerous, clavate to fusiform, without apical structures, 8-spored, evanescent; ascospores aseptate, sometimes with apical mucous appendages. Anamorph spermatial, conidiogenous cells formed from the thallus cells or from hair tips. Biotrophic on marine *Rhodophyta*, S. Hemisph. and Pacific. Fams:

(1) **Hispidocarpomycetaceae**.
(2) **Spathulosporaceae**.

An isolated order included in the *Laboulbeniomycetidae* by Eriksson (1982). Perhaps an ancient group with characters close to hypothetical ancestors of the *Ascomycota* (Kohlmeyer, *BioScience* **25**: 86, 1975).

spawn, (1) (n.), mycelium, esp. that used for starting mushroom cultures (q.v.); (2) (v.), to put inoculum (spawn) into a mushroom bed or other substrate.

special form (**forma specialis, f. sp.; formae speciales, ff. spp.**, pl.), an informal rank in a Classification (q.v.) not regulated by the Code (see Nomenclature) and used for parasitic fungi characterized from a physiological standpoint (e.g. by the ability to cause disease in particular hosts) but scarcely or not at all from a morphological standpoint; e.g. *Fusarium oxysporum* f.sp. *cubense* (Panama disease of banana) and f.sp. *elaeidis* (wilt of oil palm); first used by Eriksson (*Ber. dtsch bot. Ges.* **12**: 292, 1894) in rusts, and now extensively used in some genera (over 120 ff. spp. in *F. oxysporum*; Armstrong & Armstrong *in* Nelson *et al.* (Eds), *Fusarium: diseases, biology, taxonomy*: 391, 1981), although not always in a parallel manner; best viewed as a temporary category while further research as to taxonomic status is in progress (Hawksworth, *in* Hawksworth (Ed.), *The identification and characterization of pest organisms*: 93, 1994); comparable to 'pathovar' as used for plant pathogenic bacteria (Young *et al.*, *ROPP* **70**: 211, 1991). See Plant pathology.

species, (1) (colloquially), one sort of organism; (2) (scientifically), the lowest principal rank in the nomenclatural hierarchy (see Classification), consisting of two elements (a binomial): a generic name and a species epithet.

There is much debate as to how species should be defined, and a variety of concepts have been proposed: (1) **morpho - ** (**morphological -, phenetic -**), the traditional approach recognizing units that could be delimited on the basis of morphological characters, and ideally by discontinuities in several such; (2) **biological - ** (**cryptic-, sibling -**), actually or potentially interbreeding populations reproductively isolated from other such groups, whether or not they are distinguishable morphologically; (3) **phylogenetic - ** (**evolutionary -**), based on measureable differences in biochemical, molecular or any other characters assessed by cladistic analysis (see Cladistics) and especially well-suited for groups in which no sexual reproduction is known (e.g. **clonal -**, Mitosporic fungi); (4) **ecological -**, based on adaptation to particular niches rather than reproductive isolation (e.g. to particular hosts); (5) **polythetic -**, based on a combination of characters, not all of which are necessarily present in each individual. The **aggregate - ** (**aggr.**), rarely used in mycology, has been used for groups of closely related morphospecies only distinguishable with difficulty.

The morphospecies concept has predominated in mycology, although in the past some mycologists have frequently taken the host of a parasitic fungus into account (in some cases giving different specific names to like forms on unlike hosts). Population studies (see Population biology) and molecular data are, however, increasingly showing that many morphospecies comprise several biological and(or) phylogenetic species; it is likely that species characterized other than by morphology will increasingly be recognized, especially where these cause different plant diseases (see also special forms).

In lichen-forming fungi, **- pairs** (**Artenpaare**) are also recognized. Otherwise identical species one of which is fertile (the **primary -**), and the other reproduces vegetatively (the **secondary -**). Primary and secondary species generally have identical secondary metabolites, but the secondary species tend to have much wider geographical ranges. See Culberson & Culberson (*Science* **180**: 196, 1973), Poelt (*Vortr. bot. Ges.*, n.f. **4**: 187, 1970, *Bot. Notiser* **125**: 77, 1972), Mattsson & Lumbsch (*Taxon* **38**: 238, 1989), Tehler (*Taxon* **31**: 708, 1982).

Lit.: Andersson (*Taxon* **39**: 375, 1990; ecological spp.), Brasier (*in* Rayner *et al.* (Eds), *Evolutionary biology of the fungi*: 231, 1987; dynamics fungal speciation), Burges (*in* Lousley (Ed.), *Species studies in the British flora*: 65, 1955), Ciferri (*Ann. Myc.* **30**: 122, 1932), Claridge & Boddy (*in* Hawksworth (Ed.), *Identification and characterization of pest organisms*: 261, 1994; recognition systems), Burnett (*TBMS* **81**: 1, 1983; speciation in fungi), Ereshefsky (Ed.) (*The units of evolution: essays on the nature of species*, 1992; collected key papers), Hawksworth (*Mycologist's handbook*, 1974), Heywood (*Taxon* **27**: 26, 1963; aggregate spp.), Mason (*TBMS* **24**: 115, 1940), Perkins (*in* Bennett & Lasure (Eds), *More gene manipulations in fungi*: 3, 1991), Poelt (*in* Hawksworth (Ed.), *Ascomycete systematics*: 273, 1994; concepts in lichens), Regenmortel (*Intervirol.* **31**: 241, 1990; polythetic spp.), Rieppel (*in* Scotland *et al.* (Eds), *Models in phylogeny reconstruction*: 31, 1994; phylogenetic spp.).

Speerschneidera Trevis. (1861), Bacidiaceae (L). 1, N. Am. See Hafellner & Egan (*Lichenologist* **13**: 11, 1981).

Spegazzini (Carlo Luigi; 1858-1926). An Italian; became General Director of Studies and Professor, Univ. La Plata. Chief writings: Books and papers on the plants of S. Am., esp. the fungi of Argentina, Brazil, Chile, and Uruguay (see Farr, *An annotated list of Spegazzini's fungus taxa* [*Bibl. Mycol.* **35**], 2 vols, 1978). Collections at Instituto de Botanica C. Spegazzini, La Plata (LPS). See Molfino (*An. Soc. cient. argent.* **108**: 7, 1929), Grummann (1974: 522), Stafleu & Cowan (*TL-2* **5**: 776, 1985).

Spegazzinia Sacc. (1879), Mitosporic fungi, 1/3.G2.27. 6, widespr. See Hughes (*Mycol. Pap.* **50**, 1953), Roquebert (*Rev. Mycol.* **42**: 309, 1978; conidiogenesis).

Spegazziniella Bat. & I.H. Lima (1959) ? = Myriangiella (Schizothyr.) fide Luttrell (*in* Ainsworth *et al.*, *The Fungi* **4A**: 135, 1973).

Spegazzinites Félix (1894), Fossil fungi. 1 (Neogene, Pleistocene), Africa, Eur., India.

Spegazzinula Sacc. (1883) ≡ Dubitatio (Hypocr.) fide Wakefield (*TBMS* **24**: 282, 1940).

Speira Corda (1837) = Dictyosporium (Mitosp. fungi) fide Guéguen (*BSMF* **21**: 98, 1905).

Speiropsis Tubaki (1958), Mitosporic fungi, 1.G1.10. 6, Japan, USA. See Kuthubutheen & Nawawi (*TBMS* **89**: 584, 1987, key), Petersen (*Mycol.* **55**: 26, 1963).

Spelaeomyces Fresen. (1863) ? = Xylostroma (Mitosp. fungi).

sperm, a male sex cell, typically motile; **-atiophore**, a spermatia-producing or -supporting structure; **-atium** (pl. **-a**), a 'sex' (+ or -) cell, e.g. a pycniospore; a microconidium in discomycetes and pyrenomycetes; a non-motile gamete, as in *Laboulbeniales*; **-atization**, the placing of spermatia on structures (receptive hyphae, etc.) for diploidization; **-odochidium**, a fruitbody having spermodochia in a lysigenous cavity in the suscept tissue (Whetzel, *Mycol.* **35**: 337, 1943); **-odochium**, a spermogonium having no wall (Whetzel, *Mycol.* **29**: 135, 1937); **-ogonium** (**-agone**, **-agonium**), a walled structure in which spermatia are produced, as in ascomycetes; a pycnium of a rust; a lichen pycnidium (obsol.).

Spermatodermia Wallr. (1833) ≡ Spermodermia (Xylar.).

Spermatodium Fée ex Trevis. (1860) nom. dub. (Ascomycota, inc. sed.). See Aguirre-Hudson (*Bull. Br. Mus. nat. Hist., Bot.* **21**: 85, 1991). ≡ Endophis (Arthopyren.) fide Hafellner & Kalb (*Bibl. Lich.* **57**: 161, 1995).

Spermatoloncha Speg. (1908), Mitosporic fungi, 1.G1.10. 1, S. Am., Afr.

Spermochaetella Cif. (1954), Mitosporic fungi, 4.B2.?. 1, S. Am.

Spermodermia Tode (1790) ? = Hypoxylon (Xylar.) fide Höhnel (*Mykol. Unters.* **1**: 362, 1923).

Spermoedia Fr. (1822) = Claviceps (Clavicipit.).

Spermophthora S.F. Ashby & W. Nowell (1926), Metschnikowiaceae. 1, W. Indies. *S. gossypii*, taken by Nannfeldt (1932), in the light of Guilliermond's study, to be the most primitive ascomycete; Nannfeldt made the sub-class *Diplobionticae* for it; only a contaminated culture of *Crebrothecium ashbyi* fide v. Arx *et al.* (*Stud. Mycol.* **14**, 1977).

Spermophthoraceae, see *Metschnikowiaceae*.

Spermophthorales, see *Saccharomycetales*.

Spermophyllosticta Kamilov (1972) nom. nud. (Anamorphic Dothideaceae). Teleomorph *Mycosphaerella*.

spermoplane, the surface of a seed; **spermosphere**, the microhabitat around a seed in soil (Verona, *Ann. Inst. Pasteur* **105**: 75, 1963). Cf. phylloplane; rhizoplane.

Spermospora R. Sprague (1948), Mitosporic fungi, 1.C1.1/10. 6, widespr. See Deighton (*TBMS* **51**: 41, 1968), MacGarvie & O'Rourke (*Ir. Jl agric. Res.* **8**: 151, 1969).

Spermosporella Deighton (1969), Mitosporic fungi, 1.B/C1.10. 1 (on *Meliola*), Sierra Leone.

Spermosporina U. Braun (1993), Mitosporic fungi, 1.A/B/C1.1. 7, Eur., S. Am.

Spermotrichum Kuntze (1898), Mitosporic fungi, 7.E1.?. 1, Paraguay.

Sphacelia Lév. (1827), Anamorphic Clavicipitaceae, 4.A1.?. Teleomorph *Claviceps*. 5, widespr. See Loveless (*TBMS* **47**: 205, 1964; identification).

Sphaceliopsis Speg. (1910) = Ephelis (Mitosp. fungi) fide Diehl (*USDA Monogr.* **4**, 1950).

Sphaceloma de Bary (1874), Anamorphic Elsinoaceae, 6.A1.16. Teleomorph *Elsinoë*. 50, widespr. Causing anthracnose and scab diseases. *S. rosarum* (rose leaf spot). See Jenkins & Bitancourt (*Mycol.* **33**: 338, 1941), Gonzalez & Pons (*Ernstia* **40**: 1, 1986; *S. manihoticola* and disease).

Sphacelotheca de Bary (1884), Ustilaginaceae. *c.* 6 (on *Polygonaceae*), cosmop. See Langdon & Fullerton (*Mycotaxon* **6**: 421, 1978; gen. delimitation), Vánky & Oberwinkler (*Nova Hedw. Beih.* **107**, 1994); cf. *Sporisorium*.

Sphaerangium Seaver (1951) [non Schimp. (1860), *Musci*] = Pezicula (Dermat.).

Sphaerella (Fr.) Rabenh. (1856) [non Sommerf. (1824), *Algae*] ≡ Mycosphaerella (Mycosphaerell.). See Corlett (*Mycol. Mem.* **18**, 1991; list of names).

Sphaerella Ces. & De Not. (1863) = Mycosphaerella (Mycosphaerell.).

Sphaerellopsis Cooke (1883), Anamorphic ? Phaeosphaeriaceae, 8.B/C1.15/19. Teleomorph *Eudarluca*. 2 (on *Uredinales*), widespr. See Sutton (1977, 1980), Nag Raj (1993), Melnik (*Nov. Sist. niz. Rast.* **13**: 91, 1976).

Sphaerellopsis Kleb. (1918) = Venturia (Ventur.) fide v. Arx & Müller (1975) but see Holm (*Taxon* **24**: 275, 1975).

Sphaerellothecium Zopf (1897), Mycosphaerellaceae. 5 (on lichens), Eur., N., C. & S. Am. See Roux & Triebel (*Bull. Soc. linn. Provence* **45**: 451, 1994; descr. 4 spp.).

Sphaeria Haller (1768) nom. rej. = Hypoxylon (Xylar.). Formerly used for most fungi with either perithecia or pycnidia. See Donk (*Rgenum veg.* **34**: 16, 1964), Wakefield (*TBMS* **24**: 286, 1940).

Sphaeriaceae, s. str., see *Xylariaceae*.

Sphaerialea Sousa da Câmara (1926) ? = Sphaerulina (Mycosphaerell.).

Sphaeriales, s. str., see *Xylariales*; s. lat., see *Halosphaeriales*, *Trichosphaeriales*, and other perithecoid ascomycetes.

Sphaerialites Venkatach. & R.K. Kar (1969), ? Fossil fungi. 1 (Tertiary), India.

Sphaericeps Welw. & Curr. (1868) = Battarrea (Battarr.) fide Cunningham (*Gast. Austr. N.Z.*: 131, 1944).

Sphaeridiobolus (Boud.) Boud. (1885) = Ascobolus (Ascobol.) fide v. Brummelen (1967).

Sphaeridium Fresen. (1852), Mitosporic fungi, 2.A1.?. 5, widespr.

Sphaerioidaceae (syn. *Phomaceae*, *Sphaeropsidaceae*). See *Sphaeropsidales*.

Sphaeriopsis Geyl. (1887), Fossil fungi (Pyrenomat.). 1 (Eocene), Indonesia. = Sphaerites (Fossil fungi) fide Meschinelli (1892).

Sphaeriostromella Bubák (1916), Mitosporic fungi, 8.E1.?. 1, Eur.

Sphaeriothyrium Bubák (1916), Mitosporic fungi, 8.A1.15. 2, Eur.

Sphaerita P.A. Dang. (1886), Pseudosphaeritaceae. c. 10, widespr. See Sparrow (*Bull. Torrey bot. Cl.* **99**: 223, 1972).

Sphaerites Unger (1850), Fossil fungi. 110 (Cretaceous, Tertiary), Eur., N. Am.

Sphaerobasidioscypha Agerer (1983), Cyphellaceae. 2, Papua New Guinea, Venezuela.

Sphaerobasidium Oberw. (1966), Sistotremataceae. 3, widespr.

Sphaerobolaceae J. Schröt. (1849), Sclerodermatales. 2 gen. (+ 2 syn.), 3 spp. Gasterocarp minute, with stellate peridium; inner peridium everting to eject single, globose peridiole.

Sphaerobolales, see *Sclerodermatales*.

Sphaerobolus Tode (1790), Sphaerobolaceae. 2 (on wood or coprophilous), cosmop. See Ingold (*TBMS* **58**: 179, 1972; review), Flegler (*Mycol.* **76**: 944, 1984; cultural char.).

Sphaerocarpa Schumach. (1803) = Craterium (Physar.) fide Fries (1829).

Sphaerocarpus Bull. (1791) [non Adans. (1763), *Hepaticae*] = Physarum (Physar.) fide Martin (1966).

Sphaerocarpus Ehrh. (1793) [non Adans. (1763)], Caliciales (inc. sed.). 1, Eur.

Sphaerocephalum F.H. Wigg. (1780) ? = Calicium (Calic.).

Sphaerocephalus Battarra ex Earle (1909) [non Lag. ex DC. (1812), *Compositae*] = Tricholoma (Tricholomat.) fide Singer (1951).

Sphaerochaetia Bat. & Cif. (1962) = Chaetothyrium (Chaetothyr.) fide v. Arx & Müller (1975).

Sphaerocista Preuss (1852) = Topospora (Mitosp. fungi) fide Groves (*CJB* **43**: 1195, 1965).

Sphaerocladia Stüben (1939) = Blastocladiella (Blastoclad.) fide Couch & Whiffen (1942).

Sphaerocolla P. Karst. (1892), Mitosporic fungi, 3.A1.3. 2, Eur., S. Am. See Donk (*Taxon* **11**: 99, 1962). ? = Hormomyces (Mitosp. fungi).

Sphaerocordyceps Kobayasi (1981), Clavicipitaceae. 2, Eur., N.E. Asia, N. Am.

Sphaerocreas Sacc. & Ellis (1882) = Glomus (Glom.) fide Gerdemann & Trappe (1974).

Sphaerocybe Magrou & Marneffe (1946), Mitosporic fungi, 2.?.?. 1, Eur.

sphaerocysts, globose cells in tissues of fungi, e.g. *Russula* and *Lactarius*.

Sphaeroderma Fuckel (1875) = Melanospora (Ceratostomat.) fide Cannon & Hawksworth (1982).

Sphaerodermatella Seaver (1909) ? = Coniochaeta (Coniochaet.).

Sphaerodermella Höhn. (1907) = Coniochaeta (Coniochaet.) fide v. Arx & Müller (1954).

Sphaerodes Clem. (1909), Ceratostomataceae. 6, widespr. See Cannon & Hawksworth (*Bot. J. Linn. Soc.* **84**: 115, 1982; key).

Sphaerodothis (Sacc. & P. Syd.) Shear (1909), Phyllachoraceae. 8, esp. warmer parts. See Cannon (*Stud. Mycol.* **31**: 49, 1989; on *Gramineae*), Joly (*Bull. Res. Counc. Israel* D **10**: 187, 1961; key).

Sphaerognomonia Potebnia ex Höhn. (1917) = Apiosporopsis (Melanconid.) fide Reid & Dowsett (1990).

Sphaerognomoniella Naumov & Kusnezowa (1952), ? Diaporthales (inc. sed.). 1, former USSR.

Sphaerographium Sacc. (1884), Mitosporic fungi, 8.E1.15. 10, N. temp.

Sphaerolina Fuckel (1860) = Schizoxylon (Stictid.) fide Nannfeldt (1932) but see Sherwood (1977).

Sphaeromma H.B.P. Upadhyay (1964), Anamorphic Phragmopelthecaceae, 4.C1.?. Teleomorph ? *Mazosia*. 1, Brazil.

Sphaeromonas E. Liebet. (1910), Chytridiomycetes (inc. sed.). 1, Eur.

Sphaeromphale Rchb. (1828) ≡ Segestria (Trichothel.).

Sphaeromphale A. Massal. (1854) = Staurothele (Verrucar.).

Sphaeromyces Mont. (1845), Mitosporic fungi, 1.A2.?. 3, Afr., Am. ? = Aspergillus (Mitosp. fungi).

Sphaeromyces G. Arnaud (1954) nom. inval., nom. illegit. (Mitosp. fungi).

Sphaeromycetella G. Arnaud (1954) = Chloridium (Mitosp. fungi) fide Gams & Holubová-Jechová (1976).

Sphaeromyxa Spreng. (1827) ≡ Sphaeronaema (Mitosp. fungi).

Sphaeronaema Fr. (1815), Mitosporic fungi, 8.A1.?. 50, widespr. See Jaczewski (*Nouv. Mem. Soc. nat. Moscou* **15**: 277, 1898), Petch (*TBMS* **25**: 167, 1941; monotypic), Sutton (1977).

Sphaeronaemella P. Karst. (1884), ? Microascales. 1 (on *Gyromitra*) or 5, N. temp. See Cannon & Hawksworth (*Bot. J. Linn. Soc.* **84**: 115, 1982), Hausner *et al.* (*CJB* **71**: 52, 1993; posn), Seifert *et al.* (*in* Wingfield *et al.* (Eds), *Ceratocystis and Ophiostoma*, 1993: 269). Anamorph (?) *Gabarnaudia*. See also *Viennotidia*.

Sphaeronaemina Höhn. (1917) ≡ Sphaeronaema (Mitosp. fungi).

Sphaeronemopsis Speg. (1910) = Melanospora (Ceratostomat.) fide v. Arx & Müller (1954).

Sphaeropezia Sacc. (1884) = Odontotrema (Odontrem.) fide Sherwood-Pike (1987).

Sphaeropeziella P. Karst. (1885) ? = Xylogramma (Leot.).

Sphaerophoma Petr. (1924), Mitosporic fungi, 8.A1.?. 1, N. Am.

Sphaerophoraceae Fr. (1831), Caliciales (L). 3 gen. (+ 9 syn.), 24 spp. Thallus foliose or fruticose, well-developed; apothecia sessile, marginal or terminal, often ± enclosed by a thalline cup; asci cylindrical, thin-walled, evanescent; ascospores brown, simple, ornamented with carbonized material, liberated in a black mazaedial mass.
   Lit.: Wedin (*Pl. Syst. Evol.* **187**: 213, 1993; cladistics, *Acta univ. upsal.* **77**, 1994; keys 23 S. Hemisph. spp.).

Sphaerophoron Ach. (1803) ≡ Sphaerophorus (Sphaerophor.).

Sphaerophoronomyces Cif. & Tomas. (1953) ≡ Sphaerophorus (Sphaerophor.).

Sphaerophoropsis Vain. (1890), Cladoniaceae (L). 2, S. Am., NZ.

Sphaerophorum Schrad. (1794) nom. rej. = Sphaerophorus (Sphaerophor.).

Sphaerophorus Pers. (1794) nom. cons., Sphaerophoraceae (L). 3, N. Hemisph., Antarctica, NZ. mainly S. temp. See Wedin (1993, *Nordic Jl Bot.* **10**: 539, 1990; ontogeny, *Lichenologist* **24**: 119, 1992; S. hemisph.).

Sphaerophragmiaceae Cummins & Y. Hirats. (1983), Uredinales. 6 gen. (+ 4 syn.), 47 spp. Pycnia discoid, with bounding structures subcuticular or subepidermal(or without bounding structures and extensive, or unknown); aeciospores pedicellate (*Uraecium*) or catenulate (*Aecidium*); urediniospores pedicellate, uredinia paraphysate; teliospores pedicillate, 4 to several cells by vertical and horizontal septation.

Sphaerophragmium Magnus (1891), Sphaerophragmiaceae. 16 (most on *Leguminosae*), widespr., tropics. See Lohsomboon *et al.* (*MR* **98**: 907, 1994; key).

sphaeroplast, see protoplast.

Sphaeropleum Link (1833) nom. dub. (? Fungi).

Sphaeroporalites Hemer & Nygreen (1967), ? Fossil fungi. 1, Libya.

Sphaeropsidales (obsol.). Traditionally used for Mitosporic fungi with pycnidial conidiomata (*Phomales, Phyllostictales*, etc.). Not accepted by Sutton (1973, 1980) or by Nag Raj (1981).

Sphaeropsis Lév. (1842) nom. dub., nom. rej. See Hawksworth & Sherwood (*Taxon* **30**: 338, 1981).

Sphaeropsis Flot. (1847) ≡ Thelocarpon (Acarospr.).

Sphaeropsis Sacc. (1880) nom. cons., Mitosporic fungi, 4.B2.1. 30 (assoc. with canker or die-back), widespr. *S. sapinea* on conifers. See Wang *et al.* (*Mycol.* **78**: 960, 1986; TEM conidial wall *S. sapinea*), Swart *et al.* (*Phytopath.* **81** : 489, 1991; variation in *S. sapinea*).

Sphaeropus Paulet (1793) = Onygena (Onygen.) fide Saccardo (*Syll. Fung.* **20**: 230, 1911).

Sphaeropyxis Bonord. (1864) = Coniochaeta (Conioch.) fide Petrini (*Sydowia* **44**: 169, 1993).

Sphaerosoma Klotzsch (1839), Otideaceae. 3 (hypogeous), N. Am., Eur. See Dissing (1980), Gamundí (*Sydowia* **28**: 339, 1976; gen. concept).

Sphaerosperma Preuss (1853) = Diatrypella (Diatryp.) fide Saccardo (*Syll. Fung.* **1**: 203, 1882).

Sphaerospora (Sacc.) Sacc. (1889) [non Klatt (1864), *Iridaceae*] ≡ Sphaerosporula (Otid.).

Sphaerosporangium Sparrow (1931) = Phytophthora (Pyth.) p.p., Pythium (Pyth.) p.p. fide Dick (*in press*).

Sphaerosporella (Svrček) Svrček & Kubička (1961), Otideaceae. 2, widespr. See Häffner (*Beitr. Kennt. Pilze Mitteleur.* **3**: 413, 1987).

Sphaerosporiites Babajan & Tasl. (1977), Fossil fungi. 1 (Tertiary), former USSR.

Sphaerosporium Schwein. (1832), Mitosporic fungi, 1.A1.3. 1, widespr. See Damon & Downing (*Mycol.* **46**: 214, 1954).

Sphaerosporula Kuntze (1898) = Scutellinia (Otid.) fide Denison (*Mycol.* **51**: 605, 1959).

Sphaerostilbe Tul. & C. Tul. (1861) = Nectria (Hypocr.) fide Rogerson (1970).

Sphaerostilbella (Henn.) Sacc. & D. Sacc. (1905), Hypocreaceae. 3, Afr., Asia, Eur., N. & S. Am. See Seifert (*Stud. Mycol.* **27**, 1985). Anamorph *Gliocladium*.

Sphaerostylidium (A. Braun) Sorokīn (1882) = Rhizophydium (Chytrid.) fide Fischer (1892).

Sphaerothallia Nees ex Eversm. (1831) = Aspicilia (Hymenel.) fide Hafellner (*Acta Bot. Malac.* **16**: 133, 1991). Includes the 'manna lichens'; *S. esculenta* may be one of the types of manna in The Bible, and can be used in bread production (see Rogers, *in* Seaward (Ed.), *Lichen ecology*: 211, 1977); thalli unattached and blown in the wind. See Follmann & Crespo (*An. Inst. Bot. Cavanilles* **31**: 325, 1974), Crespo & Barreno (*Acta Bot. Malac.* **4**: 55, 1978).

Sphaerotheca Desv. (1817) nom. rej. (Puccin.).

Sphaerotheca Lév. (1851) nom. cons., Erysiphaceae. 50, widespr. *S. humuli* (hop mildew), *S. mors-uvae* (American mildew of gooseberry and currant (*Ribes*)), *S. pannosa* (rose mildew) and its var. *persicae* (peach mildew) are important pathogens. See Braun (*Beih. Nova Hedw.* **89**, 1987; key). Anamorph *Oidium*.

Sphaerothyrium Wallr. (1833) ≡ Eustegia (Fungi, inc. sed.).

Sphaerotrachys Fayod (1889) = Cortinarius (Cortinar.) fide Kauffman (1924).

Sphaerozone Zobel (1854), Pezizaceae. 4 (hypogeous), Eur., N. Am., Australia. See Beaton & Weste (*TBMS* **71**: 164, 1978), Dissing (*Mycotaxon* **30**: 2, 1980).

Sphaerozosma Corda (1842) = Sphaerosoma (Otid.).

Sphaerula Pat. (1883) = Pistillaria (Typhul.) fide Corner (1950).

Sphaerulina Sacc. (1878), Mycosphaerellaceae. 40, widespr. *S. rehmiana* (rose leaf scorch), *S. taxi* (yew

(*Taxus*) leaf spot), *S. rubi* (raspberry (*Rubus*) leaf spot); Boerema (*Neth. J. Pl. Path.* **69**: 76, 1963). Anamorphs *Cercospora, Septoria*.

Sphaerulomyces Marvanová (1977), Mitosporic fungi, 1.A1.3. 1 (aquatic), Slovakia.

Sphagnicola Velen. (1934) ≡ Ciliatula (Leot.).

Sphaleromyces Thaxt. (1894), Laboulbeniaceae. 3, Eur., Am., Asia.

Spheconisca (Norman) Norman (1876), ? Verrucariaceae (L). 20, Eur. See Bachmann (*Nyt. mag. naturv.* **64**: 170, 1926).

Spheconiscomyces Cif. & Tomas. (1953) = Spheconisca (Verrucar.).

Sphenospora Dietel (1892), Raveneliaceae. 10, trop. Am., Afr. See Linder (*Mycol.* **36**: 464, 1944).

spheridium, see capitulum.

Spherites Dijkstra (1949), Fossil fungi. 1 (Devonian), Netherlands.

spheroplast, see protoplast.

Spheropsis Raf. (1815) nom. dub. (? Ascomycota, inc. sed.). Nom. nov. for 'Sphaeria Pers.'; no spp. included.

spherule, (1) a sporangium-like structure in *Coccidioides* (Baker & Mrak, *Am. J. trop. Med.* **21**: 589, 1941); (2) a multinucleate cell of a resting myxomycete plasmodium.

spherulin, spherule phase coccidioidin (q.v.) (Scalarone *et al., Sabouraudia* **11**: 222, 1973).

Sphinctrina Fr. (1825), Sphinctrinaceae (±L). 5 (on lichen, *Pertusaria*), widespr. See Löfgren & Tibell (*Lichenologist* **11**: 109, 1979; key 5 Eur. spp.), Tibell (1984).

Sphinctrinaceae M. Choisy (1950), Caliciales. 2 gen. (+ 2 syn.), 6 spp. Thallus absent; ascomata stalked; asci cylindrical, thin-walled, evanescent; ascospores 0- to 1-septate, dark brown, ellipsoidal, with an ornamentation formed early in development beneath the plasmalemma. Lichenicolous or saprobic, rarely if ever lichenized.

Sphinctrinella Nádv. (1942) = Mycocalicium (Mycocalic.) fide Tibell (1984).

Sphinctrinopsis Woron. (1927), ? Caliciaceae. 1 (on *Pertusaria*), former USSR. See Tibell (*Beih. Nova Hedw.* **79**: 597, 1984).

Sphinctrosporium Kunze (1828) = Cladotrichum (Mitosp. fungi) fide Saccardo (*Syll. Fung.* **4**: 374, 1886).

Sphondylocephalum Stalpers (1974), Mitosporic fungi, 1.A1.6. 1, N. Am.

Sphyridiomyces E.A. Thomas ex Cif. & Tomas. (1953) = Baeomyces (Baeomycet.).

Sphyridium Flot. (1843) = Baeomyces (Baeomycet.).

Spicaria Harting (1846) nom. conf. fide Brown & Smith (*TBMS* **40**: 22, 1957).

Spicariopsis R. Heim (1939) = Paecilomyces (Mitosp. fungi) fide Samson (1974).

Spicellum Nicot & Roquebert (1976), Mitosporic fungi, 1.A1.10. 1, France.

Spicularia Pers. (1822), Mitosporic fungi, 1.A2.?. 1 or 2, Eur., Am.

Spicularia Chevall. (1826) ≡ Exidia (Exid.).

spicule (spiculum) (Tulasne; obol.), see sterigma.

spiculospore, a spore formed at the tip of a pointed structure often elongate and so resembling a spike, as in *Hirsutella* and *Akanthomyces* (Subramanian, *Curr. Sci.* **31**: 410, 1962).

Spiculostilbella E.F. Morris (1963) = Phaeoisaria (Mitosp. fungi) fide de Hoog & Papendorf (*Persoonia* **8**: 408, 1976).

spiculum (pl. **spicula**), see sterigma.

Spilobolus Link (1833) = Amphisphaerella (Amphisphaer.) fide Gams (*Taxon* **43**: 265, 1994).

Spilocaea Fr. (1819), Anamorphic Venturiaceae, 1.A/B2.19. Teleomorph *Venturia*. 6, widespr. *S. pomi* (syn. *S. dendriticum*), cause of apple scab. See Ellis (*DH, MDH*).

spilodium, introduced by Lettau (*Beih. Feddes Rep.* **69**: 62, 1932) for the minute round blackish structures on the thallus of *Dirina massiliensis* caused by *Milospium graphideorum*.

Spilodium A. Massal. (1856) = Arthonia (Arthon.) fide Hawksworth (*TBMS* **74**: 363, 1980).

Spilodochium Syd. (1927), Mitosporic fungi, 3.A/B2.?. 2, C. Am., Australia.

Spiloma Ach. (1803) = Xylographa p.p. (Agyr.).

Spilomela (Sacc.) Keissl. (1920) ≡ Pleospilis (Odontotremat.).

Spilomium Nyl. (1858) ≡ Sclerococcum (Mitosp. fungi).

Spilomyces Petr. & Syd. (1927) = Neottiospora (Mitosp. fungi) fide Sutton (1977).

Spilonema Bornet (1856), Coccocarpiaceae (L). 4, cosmop. See Henssen (*Symb. bot. upsal.* **18**(1), 1963).

Spilonematopsis Vain. (1909) nom. nud. = Spiloma (Agyr.) fide Henssen (*Symb. bot. upsal.* **18**(1), 1963).

Spilonematopsis Å.E. Dahl (1950) = Ephebe (Lichin.) fide Henssen (1963).

Spilonemopsis Vain. (1909) nom. nud. = Spilonema (Coccocarp.) fide Henssen (1963).

Spilopezis Clem. (1909) = Pseudopeziza (Dermat.) fide Nannfeldt (1932).

Spilopodia Boud. (1885), Dermateaceae. 3, Eur.

Spilopodiella E. Müll. (1989), Dermateaceae. 1, Switzerland.

Spilosphaeria Rabenh. (1857) = Septoria (Mitosp. fungi) fide Saccardo (*Syll. Fung.* **4**: 474, 1884).

Spilosphaerites A. Massal. (1857), Fossil fungi. 2 (Miocene), Italy.

Spilosticta Syd. (1923) = Venturia (Ventur.) fide v. Arx & Müller (1975).

Spinalia Vuill. (1904), ? Dimargaritales (inc. sed.). 1, Eur. See Benjamin (1959, 1979).

spindle, see fuseau.

spindle-organ, see turbinate cell.

spine, a narrow sharply pointed process; **spinule**, a small spine (in lichens see Fig. 22C); **spiny**, having spines; **spinose** (dim. **spinulose**), delicately spiny. Fig. 29.2.

Spinellus Tiegh. (1875), Mucoraceae. 5 (parasitic on agarics, esp. *Mycena*), N. temp. See Zycha *et al.* (1969; key).

Spiniger Stalpers (1974), Anamorphic Basidiomycetes, 1.A1.6. 2, widespr. See Carmichael *et al.* (1980; refs).

Spinomyces Bat. & Peres (1961) nom. inval. Mitosporic fungi (L), ?.-.-. 1, Brazil.

Spinulosphaeria Sivan. (1974), Lasiosphaeriaceae. 1, Liberia, Sierra Leone, USA. See Barr (*Mycotaxon* **39**: 43, 1990; posn), Nannfeldt (*Svensk bot. Tidskr.* **69**: 289, 1975).

Spinulospora Deighton (1973), Mitosporic fungi, 3.A1.1. 1 (on *Puccinia*), W. Afr.

spiral hypha, a hypha ending in a spiral or helical coil, as in *Trichophyton* (Davidson & Gregory, *TBMS* **21**: 98, 1937).

Spiralia Grigoraki (1925) [non Toula (1900), fossil ? Algae] = Trichophyton (Mitosp. fungi).

Spiralotrichum H.S. Yates (1918) nom. dub. fide Mulder (*TBMS* **65**: 518, 1975).

Spiralum J.L. Mulder (1975), Mitosporic fungi, 1.F2.10. 2, Asia.

Spirechina Arthur (1907) = Kuehneola (Phragmid.) fide Jackson (*Mycol.* **23**: 96, 1931).

spirits, high alcohol-content drinks obtained by distilling grape wine (**brandy**), malted barley (**whisky**), fermented rye or maize worts (**gin**), or fermented molasses (**rum**). See also pulque.

Spirodactylon R.K. Benj. (1959), Kickxellaceae. 1, USA.

Spirogramma Ferd. & Winge (1909) = Xylaria (Xylar.) fide Cannon (*SA* **6**: 171, 1987).

Spirographa Zahlbr. (1903), Odontotremataceae. 1 (on lichens), Eur. See Santesson (*The lichens and lichenicolous fungi of Sweden and Norway*, 1993).

Spirographomyces Cif. & Tomas. (1953) ≡ Spirographa (Odontotremat.).

Spirogyromyces Tzean & G.L. Barron (1981), ? Trichomycetes (inc. sed.). 1 (in gut of *Rhabditis*), Canada. See Tzean & Barron (*CJB* **59**: 1861, 1981).

Spiroidium Saito (1949), *Actinomycetes*, q.v.

Spiromastix Kuehn & G.F. Orr (1962) [non B.V. Perfil. (1929), *Algae*], Onygenaceae. 1, Afr., Australia. See Currah & Locquin-Linard (*CJB* **66**: 1135, 1988; key 2 spp.).

Spiromyces R.K. Benj. (1963), Kickxellaceae. 1, USA. See Benjamin (*Aliso* **5**: 273, 1963).

Spiropes Cif. (1955), Mitosporic fungi, 1.C2.10. 26 (on *Meliolaceae*), widespr. esp. trop. See Ellis (*Mycol. Pap.* **114**, 1968; key).

Spirosphaera Beverw. (1953), Mitosporic fungi, 1.F1.1. 5 (aquatic), Eur., Japan. See Hennebert (*TBMS* **51**: 13, 1968), Abdullah *et al.* (*Nova Hedw.* **43**: 507, 1986; 2 spp. from Japan).

Spirospora L. Mangin & Vincens (1920) = Acrospeira (Mitosp. fungi) fide Wiltshire (*TBMS* **21**: 233, 1938).

Spirospora Scherff. (1926), Protozoa, see monads. See also Dick (*in press*).

Spirotremesporites Dueñas (1979), Fossil fungi (f. cat.). 3 (Eocene-Pleistocene), Colombia, USA.

Spirotrichum Saito ex J.F.H. Beyma (1940) = Tritirachium (Mitosp. fungi) fide Langeron (*Rev. Mycol.* **14**: 133, 1949); see also Paulitz & Menge (*Mycol.* **78**: 99, 1984).

spitzenkörper, a small, densely staining body found close to the cytoplasmic membrane in the apices of actively growing septate hyphae (Grove & Bracker, *J. Bact.* **104**: 989, 1970, Girbardt, *Planta (Berl.)* **50**: 47, 1957).

Spizellomyces D.J.S. Barr (1980), Spizellomycetaceae. 8, Eur., S. Afr., Canada.

Spizellomycetaceae D.J.S. Barr (1980), Spizellomycetales. 6 gen., 31 spp. Thallus eucarpic, monocentric; sporangium, or prosporangium, and resting spore endogenous within the zoospore cyst.

Spizellomycetales, Chytridiomycetes. 4 fam., 13 gen. (+ 6 syn.), 66 spp. Monocentric; development endogenous or exogenous; saprobic or parasitic, predominantly from soil on organic substrata including other fungi, cellulose, keratin, pollen, plant debris, chitinous exoskeletons; cosmop.

A segregate from *Chytridiales* (q.v.) which they resemble except: zoospores of most species have more than one lipid globule, they may move in an amoeboid fashion, rhizoid tips are blunt.

The order was defined using ultrastructural characters of the zoospore (Barr, *CJB* **58**: 2380, 1980). Fams:

(1) **Caulochytriaceae**.
(2) **Olpidiaceae**.
(3) **Spizellomycetaceae**.
(4) **Urophlyctidaceae**.

Lit.: Barr (*CJB* **62**: 1171, 1984; key spp. in culture, in Margulis *et al.* (Eds), *Handbook of Protoctista*: 454, 1990).

Spizomycetes. Class within the *Chytridiomycota* (Cavalier-Smith, in Rayner *et al.* (Eds), *Evolutionary biology of the fungi*: 339, 1987) including only *Spizellomycetales*.

Splanchnomyces Corda (1831) = Rhizopogon (Rhizopogon.).

Splanchnonema Corda (1829), Pleomassariaceae. 35, Eur., N. Am. See Barr (1982; *Mycotaxon* **42**: 129, 1993, key 27 N. Am. spp.). Anamorph *Prosthemium*.

splash (splashing) cup, an open cup-like structure (as in *Cladonia, Cyathus*, and the liverwort *Marchantia*), from which the reproductive bodies are discharged by falling drops of water. See Brodie (*CJB* **29**: 593, 1951); bird's nest fungi.

Spogotteria Dyko & B. Sutton (1979) = Plectronidium (Mitosp. fungi) fide Sutton & Pascoe (1986).

Spolverinia A. Massal. (1856) = Phyllactinia (Erysiph.) fide Junell (*Svensk bot. Tidskr.* **58**: 55, 1964).

Spondylocladiella Linder (1934), Mitosporic fungi, 1.C2.24. 1 (on *Corticium*), Canada.

Spondylocladiopsis M.B. Ellis (1963), Mitosporic fungi, 1.C1.10. 1, UK.

Spondylocladium Mart. ex Sacc. (1880) = Stachylidium (Mitosp. fungi) fide Hughes (*CJB* **36**: 747, 1958).

Spongiasclerotes Stach & Pickh. (1957), Fossil fungi (*f. cat.*). 1 (Carboniferous), Germany. See Beneš (1969).

Spongioides Lázaro Ibiza (1916) nom. dub. (Coriol.).

spongiostratum, used for the hypothallus of *Anzia* and *Pannoparmelia* (Hannemann, *Bibl. Lich.* **1**, 1973). See Henssen & Dobelmann (*Bibl. Lich.* **25**: 103, 1987).

Spongiosus Lloyd ex Torrend (1920) ≡ Phaeolus (Hymenochaet.).

Spongipellis Pat. (1887), Coriolaceae. 3, N. temp. See Ryvarden (1978).

Spongiporus Murrill (1905), Coriolaceae. 14, widespr. David (*Bull. mens. Soc. linn. Lyon* **49**: 6, 1980). = Tyromyces (Coriol.) fide Pegler (1973).

Spongospora Brunch. (1887), Plasmodiophoraceae. 3 (obligate endoparasites of plants), widespr. *S. subterranea* (potato powdery scab; see Karling, 1968); *S. subterranea* var. *nasturtii* (crook root of watercress; Tomlinson, *TBMS* **41**: 491, 1958).

Sponheimeria Kirschst. (1941) = Lasiobelonium (Hyaloscyph.) fide Baral (*SA* **13**: 113, 1994).

Spooneromyces T. Schumach. & J. Moravec (1989), Otideaceae. 2, Eur.

spora, the spore content of a particular place or ecological niche; of air, see Air spora.

sporabola, the curve made by a basidiospore after discharge from its sterigma (Buller, *Researches* **1**; Ingold, 1971: 111).

Sporacestra A. Massal. (1860) ≡ Bacidia (Bacid.) fide Coppins (*Bull. Br. Mus. nat. Hist., Bot.* **11**: 203, 1983).

Sporadospora Reinsch (1875) nom. dub.; based on appressoria fide Lowen (*SA* **9**: 26, 1991). = Pseudonectria (Hypocr.) fide Racovitza (*Mém. Mus. natn Hist. nat. Paris, Bot.* **10**: 223, 1959).

sporangial vesicle, vesicle produced at the mouth of the sporangium during planont maturation and discharge. See homohylic vesicle, plasmamembranic vesicle, precipitative vesicle.

sporangiocyst (of *Chytridiales*), a resting sporangium (A. Fischer); cf. cystosorus.

sporangiolum (sporangiole), (1) (of *Mucorales*), a small sporangium with or without a columella, generally having a small number of spores; (2) a degenerating arbuscule (Janse, 1897) (obsol.).

sporangiophore, thallus element (usually morphologically differentiated) subtending one or more sporangia.

sporangiosorus, group of spherical sporangia fused together and formed from a single plasmodium; also one or more lobed sporangia formed from a single plasmodium.

sporangiospore, walled spore produced in a sporangium; **primary infestation** - (of *Eccrinales*), 1-4-nucleate, thich-walled spore which serves to transmit

an infestation from one host individual to another fater passage through the gut; **secondary infestation** - (of *Eccrinales*), multinucleate, thin-walled spore which germinates in the same gut as where they were produced.

sporangium (sporange), an organ enclosing endogenously generated spore(s), the walls of the spore(s) not being derived from the supporting or containing structure.

Sporastatia A. Massal. (1854), Acarosporaceae (L). 4, Asia, Eur., N. Am. See Grube & Poelt (*Fragm. Flor. geobot.* Suppl. **2**: 113, 1993).

spore, a general term for a reproductive structure in fungi, bacteria, and cryptogamic plants. In fungi, a differentiated morphological form which may be: (a) specialized for dissemination; (b) produced in response to, and resistant to, adverse conditions; and/or (c) produced during or as a result of a sexual or asexual reproductive process. Commonly 1-celled, but in fungi frequently a multicelled structure (e.g. phragmospore, spore ball, etc.) which is in effect a group of 1-celled spores because every cell may produce one or more germ tubes. Thick- or thin-walled, pigmented or not, motile or non-motile.

More attention has been given to the spore than to any other fungal structure. Spore morphology (e.g. flagellation of zoospores) and development (e.g. of ascospores and basidiospores; sexual and asexual spores) provide basic taxonomic criteria and biologically spores may be differentiated into groups disseminated by wind, water, insects and other animals, etc., and those which allow a fungus to survive conditions unfavourable for growth (e.g. resting spores), although one type of spore may serve several functions (Sutton, *TBMS* **86**: 1, 1986). Vuillemin (1912) defined spores morphologically (e.g. motile and non-motile) and biologically and borrowed or coined terms for his different spore types. Most of these terms have never been in current use but the naming of spores continues (see Spore terminology). See Chapela (*Sydowia* **43**: 1, 1991; measurement with Coulter counter), De Toledo (*Mycotaxon* **52**: 259, 1994; descriptions of), Madelin (Ed.) (*The fungus spore*, 1966), Weber & Hess (Eds) (*The fungal spore. Form and function*, 1976), Turian & Hohl (Eds) (*The fungal spore: morphogenetic controls*, 1981), Cole & Hoch (*The fungal spore and disease initiation in plants and animals*, 1991), Beakes (*in* Gow & Gadd (Eds), *The growing fungus* 339, 1995; sporulation), and the following entries.

spore ball, a unit of dispersal comprised of a more or less firmly aggregated group of spores (e.g. *Sorosporium, Tolyposporium*) or spores and sterile cells (e.g. *Urocystis*).

spore charge. Airborne basidiospores carry either a positive or negative electrostatic charge, fide Buller (**1**: 192, 1909). See Webster *et al.* (*TBMS* **91**: 193, 1988), Gregory (*Nature* **180**: 330, 1957; *Ganoderma applanatum*, etc.), Swinbank *et al.* (*Ann. Bot., Lond.* **28**: 239, 1964; *Serpula*).

Spore discharge and dispersal. Aquatic phycomycetes may be dispersed in water as self-motile zoospores but terrestrial fungi are mostly spread by animals or through the air. *Pilobolus* and other coprophilous fungi may be spread by animals and even require ingestion for their spores to germinate. However, they may also have an airborne dispersal phase. For spores to be dispersed through the air, they have to overcome the adhesive forces attaching them to the surface on which they are growing, cross the laminar boundary layer that surrounds all surfaces and enter the turbulent boundary layer above. Many mechanisms have developed in fungi to enable their spores readily to become airborne. Some (*Agaricales*, *Myxomycetes*) have tall sporophores which lift spores well into or through the laminar boundary layer, others forcibly project their spores into the air while still others rely on passive mechanisms such as mechanical disturbance and rainsplash.

Dry passive methods are independent of water availability and so can occur in the absence of rain and at low humidities. Shedding under gravity and in convection currents are of uncertain importance. Deflation without mechanical disturbance can occur at low wind speeds in myxomycetes with raised fruiting bodies, e.g., *Dictydium*, and with lichens with cup-shaped podetia in which double eddy systems may be set up by wind blowing across the top. However, removal of conidia from conidiophores by wind requires speeds of at least 0.4-2.0 m/s, which are seldom achieved in crops except in gusts, so that mechanical disturbance may often be of greater imortance. Insects may carry the anamorph of *Claviceps* spp. to new host plants. Although water is required to activate them, rain tap and puff and bellows mechanisms disperse dry spores. In the first, rain striking a stem causes vibration which loosens spores while the cushion of air which precedes the splash disturbs the laminar boundary layer and blows spores into the air. Bellows mechanisms are seen in *Lycoperdales* that have a thin, flexible, unwettable peridium covering the spore mass which is depressed momentarily by raindrops or drips from foliage, expelling a puff of air carrying millions of spores from the aperture.

Mist pickup has been described for *Cladosporium*. Rain and drip splash are important for fungi with spores embedded in slimy masses, e.g., *Colletotrichum lindemuthianum*, and for other non-slimy species whose spores are easily released by rain, e.g. *Venturia inaequalis*, or which also have active dispersal mechanisms, e.g. *Phytophthora infestans*.

Open cup-shaped structures ('splash cups', q.v.) lined with round periodioles produce by bird's nest fungi (*Nidulariales*), e.g., *Cyathus* spp., *Crucibulum vulgare*, and podetia lined with soredia produced by *Cladonia* utilize the energy of raindrops to project their contents up to 1 m.

Bubble scavenging concentrates particles, e.g. spores of aquatic fungi, from water suspension and so they are projected into the air when the bubbles burst at the surface. Droplets from breaking waves and waterfalls fulfil a similar function.

Active mechanisms activated by changing humidity and drying are seen in the violent twisting movements of some *Oomycota* and in water rupture mechanisms. Release through hygroscopic movement is characteristic of sporangiophores of *Phytophthora infestans* and *Peronospora hyoscyami* (syn. *P. tabacina*) and may also serve to loosen spores of some hyphomycetes, e.g. *Botrytis cinerea*, *Cladosporium*, *Drechslera*, before subsequent deflation. Water rupture occurs when tension is sufficient for the cohesion between water molecules or their adhesion to cell walls to fail resulting in the sudden formation of a gas bubble. A spore-bearing cell which had been collapsing as water was withdrawn by evaporation then suddenly returns to its former size as the gas bubble forms, flicking off the spores. The catapult effect may be enhanced by differential thickening of the conidiophore, as seen in *Deightoniella torulosa* and *Zygosporium oscheoides*.

Squirt gun mechanisms are characteristic of many *Ascomycota*, in which the spores are ejected explosively from turgid cells to a distance of 1-2 cm for small ascospores and up to 60 cm for larger ascospores, e.g. *Ascobolus immersus*. Asci may discharge singly from perithecia or many simultaneously from apothecia, e.g., *Sclerotinia sclerotiorum* producing a visible puff. About 0.2 mm rainfall is necessary for discharge of ascospores from perithecia of *Venturia inaequalis* in apple leaves or of *Mycospherella pinodes* in pea straw but at least 1.2 mm for ascospores from stromata of *Eutypa armeniacae*. Ascospores are usually dispersed as single spores but in *Pyrenopeziza brassicae* groups, usually of four spores, may be dispersed together.

Squirting mechanisms are also found in *Pilobolus* spp., *Basidiobolus ranarum*, *Entomophthora muscae*, *Pyricularia oryzae* and *Nigrospora sphaerica*. The sudden rounding of turgid cells can cause spores of *Conidiobolus coronata* to travel up to 4 cm and also occurs in *Epicoccum nigrum*, *Arthrinium cuspidatum*, *Sclerospora philippinensis* and the anamorph of *Xylaria furcata*.

Discharge of ballistospores and basidiospores is undoubtedly activated by water but the mechanism remains an enigma and a matter of controversy. The formation of a drop of liquid at the hilum end of the spore is characteristic; this is termed Buller's drop (q.v.) and has a crucial role in spore discharge. The spore is ejected 0.01-0.02 cm, sufficient for it to be able to fall freely between the gills or through pores into the free air below.

Dispersal of spores depends on their size, roughness and other characteristics and on wind speed and turbulence. Spore dispersal mechanisms determine the periodicity of spore release (see Air spora). Spore size and surface roughness affect the rate at which spores fall through the air and their ability to impact on stems and other obstacles. Fall rate or terminal velocity is a function of the square of the spore radius while the efficiency with which spores impact onto obstacles increases with their size. Many spores in crops and similarly dense vegetation either fall to the ground or impact onto stems and fail to escape into the air. Most that do escape are deposited within 100 m and few travel long distances. Spore clouds are spread both horizontally and vertically by eddies as they travel downwind from their source and concentrations decrease following theories of diffusion. Nevertheless, clouds of spores released in successive days and nights have been identified up to 600 km downwind of the British coast (Hirst & Hurst, *Symp. Soc. gen. Microbiol.* **17**: 307, 1967). Dispersal is ended by impaction of particles onto obstacles and by washout in rain.

Lit.: Buller (**1-6**), Ingold (*Fungal spores, their liberation and dispersal*), 1971), Gregory (*TBMS* **21**: 26, 1945, *Microbiology of the atmosphere*, edn 2, 1973), Gregory & Monteith (*Airborne microbes*, 1967), Edmonds (*Aerobiology, the ecological systems approach*, 1979), Malloch & Blackwell (*in* Carroll & Wicklow, *The fungal community*, edn 2 : 147, 1992).

spore dormancy, see Spore germination.

Spore germination, and dormancy in fungi have been reviewed by Sussman (*in* Ainsworth & Sussman (Eds), *The Fungi* **2**: 733, 1966), Sussman & Douthit (*Ann. Rev. Pl. Physiol.* **24**: 311, 1973), and Gottlieb (*The germination of fungus spores*, 1978).

In lichenized fungi the ascospores germinate readily in water films at moderate temperatures. Percentage germination is lowest in areas of high air pollution (Kofler *et al.*, *Mém. Soc. bot. Fr.* 1968: 219, 1969). In *Xanthoria parietina* ascospores remain dormant until contact with the photobiont is made (Werner, *Mém. Soc. Sci. nat. phys. Maroc* **27**: 7, 1931). See Lichens. See *Lit.* under Spore discharge and dispersal.

spore groups, for Saccardo's spore groups see Mitosporic fungi. Also see Conidial nomenclature, Sphaeriales.

spore horn, see cirrus.

spore longevity, See Longevity.

spore print, the deposit of basidioma obtained by allowing spores from a basidiocarp to fall onto a sheet of paper (white or coloured) placed below the lamellae or pores. See Singer (1975: 5).

spore specific gravity, Buller (**1**: 153, 1909) determined the Specific Gravity (SG) of various agaric spores as 1.02-1.43.

spore terminology. see aboo**spore**, acro-, adia-, aecio-, aleurio-, alpha-, amero-, amphi-, annello- (under annellidic), aplano-, arthro-, asco-, azygo-, ballisto-, basidio-, beta- -, blasto-, botryo-aleurio-, botryoblasto-, chlamydo-, clostero-, conidiole, conidium, cyst, deuteroconidium, dia-, dictyochlamydo-, dictyoporo-, dictyo-, didymo-, diploconidium, dispersal -, di-, dry -, ecto-, endoconidium, endo-, fragmentation -, fuseau, ganglio-, gasteroconidium, gastero-, gemma, gonosphere, haploconidium, helico-, hemi-asco-, hemi-, isthmo-, loculo- (see Loculomycetes), macroconidium, macro-, meio-, memno-, meri-, meristem arthro-, meristem blasto-, meso-, microconidium, micro-endo-, micro-, mischoblastiomorph, mito-, mono-, myceloconidium, myxo-, nimbo-, oidium, oo-, papulo-, part -, partheno-, peri-, phialoconidium (under phialidic), phragmo-, placodiomorph, plasmagamo-, polarilocular-, poroconidium, poro-, primo-, promycelial -, protero-, protoconidium, proto-, pseudoidium, pseudo-, pycnidio-, pycnio-, pycno-, radula -, ramoconidium, repeating -, resting -, scoleco-, secondary -, seio-, sicyo-, simblo-, slime -, smut -, spiculo-, sporangio-, sporidesm, sporidiole, sporidium, stalagmo-, statismo-, stauro-, stylo-, summer -, sympodioconidium, synchrono-, teleutosporodesm, telio-, terminus -, tetra-, texo-, thallo-, theca-, tretoconidium (under tretic), tricho-, ustilo-, usto-, winter -, xeno-, zoo-, zygo-.

Spore wall. Conventional or electron microscopy shows the spore wall to be layered. The terminology of these layers by different authors is somewhat confused (see the comparison by Payak, 1964: 33). Five layers (the spore wall proper; **eusporium**) which have been distinguished are, from within outwards: (1) **endosporium** (endospore, corium), which is usually thin and is the last to develop during sporogenesis; (2) **episporium**, the thick, fundamental layer which determines the shape of the spore; (3) **exosporium** (exospore, epitunica, trachytectum, tunica), a layer derived from (2) but chemically distinct and frequently responsible for the ornamentation; (4) **perisporium** (= mucostratum, myxosporium), a layer, frequently fugacious, enveloping the whole spore and limited by (5), the hardly visible **ectosporium** (sporothecium). On this disappearance of (4) and (5) (3) is the outer spore layer. (1)-(3) are thus the spore wall proper; (4) and (5) of extraspsoral origin. See Heim (*Rev. Mycol.* **27**: 199, 1962), Payak (*in* Nair (Ed.), *Recent advances in palynology*, 1964), Pyatt (*TBMS* **52**: 167, 1969; lichens), Pegler & Young (*TBMS* **72**: 356, 1979; basidiospores). See Fig. 5; cf. Cell wall chemistry.

Sporendocladia G. Arnaud ex Nag Raj & W.B. Kendr. (1975), Mitosporic fungi, 1.A1.22. 5, widespr. See Wingfield *et al.* (*TBMS* **89**: 609, 1987; separation from *Phialocephala*), Onofri *et al.* (*MR* **98**: 745, 1994; key).

Sporendonema Desm. (1827), Mitosporic fungi, 1.A2.40. 2, widespr. See Sigler & Carmichael (*Mycotaxon* **4**: 376, 1976).

Sporhaplus H.B.P. Upadhyay (1964), Anamorphic Roccellaceae p.p. 4.A1.?. Teleomorph *Mazosia*. 1, Brazil. See Hawksworth (*Bull. Br. Mus. nat. Hist., Bot.* **9**: 1, 1981).

Sporhelminthium Speg. (1918) = Clasterosporium (Mitosp. fungi) fide Hughes (1958).

Sporichthya M.P. Lechev., H. Lechev. & Holbert (1968), *Actinomycetes*, q.v.

sporidesm (**sporodesm**), a compound spore or sporeball, the components of which are merispores. See teleutosporodesm.

Sporidesmiella P.M. Kirk (1982), Mitosporic fungi, 1.B/C2.19. 17, widespr. See Kuthubutheen & Nawawi (*MR* **97**: 1305, 1993; review), Zhang *et al.* (*Mycotaxon* **18**: 243, 1983; conidiogenesis).

sporidesmin (and **sporodesmolides**), toxins (oligopeptides) of *Pithomyces chartarum* (teleomorph *Leptosphaerulina chartarum*); the cause of facial eczema in sheep and cattle, esp. in NZ.

Sporidesmina Subram. & Bhat (1989) = Janetia (Mitosp. fungi) fide Sutton (*in litt.*).

Sporidesmiopsis Subram. & Bhat (1989) = Brachysporiella (Mitosp. fungi) fide Sutton (*in litt.*).

Sporidesmium Link (1809), Mitosporic fungi, 1.C/D2.19. 50, widespr. See Ellis (*Mycol. Pap.* **70**, 1958; key), Moore (*Mycol.* **50**: 681, 1958), Hughes (1958: 807), Bullock *et al.* (*CJB* **67**: 313, 1989; morphol., histochem. and germination of *S. sclerotivorum*), Subramanian (*Proc. Ind. Nat. Sci. Acad.* **58**: 179, 1992; segregation).

Sporidiaceae, see *Sporidiobolaceae*.

Sporidiales, Ustomycetes. 1 fam., 8 gen., 26 spp. Yeast-like saprobes (teleomorphic). Fam: **Sporidiobolaceae**.

    Lit.: Moore (1980), Boekhout *et al.* (*Can. J. Microb.* **39**: 276, 1993). See also *Lit.* for Yeast.

Sporidiobolaceae Kobayasi (1961), Sporidiales. 8 gen., 26 spp. Metabasidium elongate with or without ballistoconidia; haploid yeast phase.

Sporidiobolus Nyland (1949), Sporidiobolaceae. 5 (fide Barnett *et al.,* 1990), N. & S. Am., Asia. See Bandoni *et al.* (*CJB* **49**: 683, 1971). Anamorph *Sporobolomyces*.

sporidiolae, spore-like bodies produced inside **sporidiomata**, perithecium-like structures, in *Kathistes* (Malloch & Blackwell, *CJB* **68**: 1712, 1990).

sporidiole, a small spore.

Sporidiomycetes. Class used by Moore (1980) for *Sporidiales*, q.v.

sporidium (pl. **sporidia**), (1) a basidiospore of the *Uredinales* and *Ustilaginales* or, in the latter, any spore other than an ustilospore; (2) ascospore (obsol.).

Sporigastrum Link [not traced] nom. dub. (? Myxomycetes).

Sporisorium Ehrenb. ex Link (1825), Ustilaginaceae. *c.* 85 (on *Gramineae*), cosmop. *S. sorghi* and *S. cruenta* (covered and loose smuts of sorghum); *S. reiliana* (head smut of maize (*Zea*) and sorghum). See Langdon & Fullerton (*Mycotaxon* **6**: 421, 1978). Many graminicolous smuts originally described as *Sorosporium*, *Sphacelotheca* or *Ustilago* belong here (Vánky, 1994).

Sporoacania A. Massal. (1855) ? = Lecidea (Lecid.).

Sporoblastia Trevis. (1851) = Cliostomum (Bacid.).

Sporobolomyces Kluyver & C.B. Niel (1924), Mitosporic fungi, 1.A1.3. 21, Eur., N. Am. See Lodder & van Rij (1952), Last (*TBMS* **38**: 453, 1955; on cereal leaves), Boekhout (*Stud. Mycol.* **33**: 1, 1991; spp. revision).

Sporobolomycetales (mirror or shadow yeasts), anamorphic yeasts with basidiomycetous affinities characteristically producing ballistospores following multiplication by budding (e.g. *Bullera, Itersonilia,*

Sporobolomyces, Tilletiopsis). Included as Mitosporic fungi in this *Dictionary*; the order accepted in *Basidioblastomycetes* by Moore (1980). See Yeast for *Lit.*, also Martin (*Univ. Iowa Stud. nat. Hist.* **19**(3), 1952).

Sporocadus Corda (1839) = Seimatosporium (Mitosp. fungi) fide Shoemaker & Müller (*CJB* **42**: 403, 1964), but retained for non-appendaged spp. by Brockmann (*Sydowia* **28**: 275, 1976) and Morelet (*Ann. Soc. Sci. nat. Arch. Toulon & Var* **37**: 233, 1985).

sporocarp, a general term for spore-bearing organs; fruit-body (q.v.). Used esp. of *Acrasiomycota, Myxomycota* and *Endogonaceae*.

Sporocarpon Will. (1878), Fossil fungi (? Ascomycota). 5 (Carboniferous), Eur., USA. See Hutchinson (*Ann. Bot.* n.s. **19**: 428, 1955), Baxter (*Paleont. Contrib. Univ. Kansas* **77**, 1975), Pirozynski (1976).

Sporocephalium Chevall. (1826) = Acladium (Mitosp. fungi) fide Hughes (1958).

Sporocephalum Arnaud (1952) ? = Oedocephalum (Mitosp. fungi) fide Kendrick & Carmichael (1973).

sporocladium, a special sporogenous branch in the *Kickxellaceae*.

Sporocladium Chevall. (1826) ≡ Cladosporium (Mitosp. fungi) fide Hughes (1958).

Sporoclema Tiesenh. (1912) nom. dub. ('phycomycete'). See Kendrick & Carmichael (*in* Ainsworth *et al.* (Eds), *The Fungi* **4A**: 323, 1973).

Sporoctomorpha J.V. Almeida & Sousa da Câmara (1903), Ascomycota (inc. sed.). 1, Eur.

Sporocybe Fr. (1825) = Periconia (Mitosp. fungi) fide Mason & Ellis (1953).

Sporocybomyces H. Maia (1967), Mitosporic fungi (L), 2.A2.?. 1, Brazil.

sporocyst, a cyst producing asexual spores.

Sporocystis Morgan (1902) [non Lesq. (1880), fossil], Mitosporic fungi, 3.D1.?. 2, N. Am., Sri Lanka.

Sporoderma Mont. (1856) = Trichoderma (Mitosp. fungi) fide Höhnel (*Sber. Akad. Wiss. Wien* **119**: 671, 1910).

Sporodesmium, see *Sporidesmium*.

Sporodictyon A. Massal. (1852) nom. rej. = Polyblastia (Verrucar.).

Sporodinia Link (1824) ≡ Syzygites (Mucor.).

Sporodiniella Boedijn (1959), Mucoraceae. 1 (on eggs, larvae & imago of *Membracidae* (*Diptera, Homoptera, Hymenoptera & Lepidoptera*), apparently parasitic), Ecuador, Malaysia, Sumatra. See Evans & Samson (*CJB* **55**: 2981, 1977), Gbaja & Young (*Microbios* **42**: 263, 1985; ultrastr.).

Sporodiniopsis Höhn. (1903) nom. dub. (Mitosp. fungi) fide Hawksworth (*Mycol. Pap.* **126**, 1971).

sporodochium, conidioma, typical of the *Tuberculariaceae* (q.v.) in which the spore mass is supported by a superficial cushion-like (pulvinate) mass of short conidiophores and pseudoparenchyma; **pionnote - (pseudopionnotes**, Sherbakoff, 1915) (of *Fusarium*), minute sporodochia near the surface of the substrate having no stroma, the spores forming a continuous slimy layer. Cf. acervulus, pionnotes.

Sporodum Corda (1837) = Periconia (Mitosp. fungi) fide Mason & Ellis (1953).

sporogenesis, spore development.

sporogenous, producing, having or supporting spores; cf. conidiogenous.

Sporoglena Sacc. (1894), Mitosporic fungi, 2.A2.?. 1, Papua New Guinea.

sporograph, the straight-line graph obtained by plotting the ratio (E) of the length (D) to width (d) against the length of the basidiospores of a species of agaric (Corner, *New Phytol.* **46**: 196, 1947). Cf. Q.

sporoma (pl. **-omata**), a multicellular structure specially developed to produce spores.

Sporomega Corda (1842), Rhytismataceae. 1, Eur., N. Am. See Eriksson (*Symb. bot. upsal.* **19**(4), 1970).

Sporomyxa L. Léger (1908), ? Plasmodiophoraceae. 2, Eur., Afr. See Karling (1968: 101).

Sporonema Desm. (1847), Mitosporic fungi, 8.A1.15. 15, N. temp. See Limber (*Mycol.* **47**: 389, 1955), Sutton (1980).

Sporonites R. Potonié (1931), ? Fossil fungi.

sporont, a thallus on which spores will be produced.

Sporopachydermia Rodr. Mir. (1978), Saccharomycetaceae. 3, widespr. See Kreger van Rij (1984; key), Yamada *et al.* (*J. gen. appl. Microbiol.* **38**: 179, 1992; molec.). Anamorph *Cryptococcus*.

Sporophaga Harkn. (1899) nom. dub. (? Fungi, inc. sed.).

sporophagous, feeding on spores, as in certain thrips species on fungus spores (Ananthakrishnan *et al., Proc. Ind. Acad. Sci.* Anim. Ser. **92**: 95, 1983), where the dimensions of the mouth parts are related to the sizes of the spores eaten (Ananthakrishnan & Dhileepan, *Proc. Ind. Acad. Sci.* Anim. Ser. **93**: 243, 1984); species lists (Ananthakrishnan *et al.*, 1984).

Sporophiala P. Rag. Rao (1970), Mitosporic fungi, 1.C2.3. 1, USA.

Sporophleum Nees ex Link (1824) = Arthrinium (Mitosp. fungi) fide Hughes (1958).

Sporophlyctidium Sparrow (1933), Chytridiaceae. 2, USA, N. Afr.

Sporophlyctis Serbinow (1907), ? Chytridiaceae. 2, widespr.

Sporophora Luteraan (1952), Mitosporic fungi, 1.A1.3. 1, France.

sporophore, (1) a spore-producing or -supporting structure, esp. a conidiophore; (2) (of macrofungi) ascoma, basidioma (see Fig. 4); cf. hymenophore; a basidium (sensu Berkeley).

Sporophormis Malloch & Cain (1973) ≡ Warcupiella (Trichocom.).

Sporophysa (Sacc.) Vain. (1921) = Physalospora (Hyponectr.) fide Keissler (*Rabenh. Krypt.-Fl.* **8**: 446, 1930).

Sporophyta, see *Cryptogamia*.

sporophyte, the diploid or asexual phase in the life-cycle of a plant; diplont; diplophase. Cf. gametophyte.

sporoplasm, the spore-producing protoplasm within the epiplasm in a sporangium or ascus (Guilliermond).

Sporopodium Mont. (1851), Ectolechiaceae (L). c. 9, trop. See Santesson (*Symb. bot. upsal.* **12**(1), 1952), Hafellner (*Beih. Nova Hedw.* **79**: 241, 1984).

Sporormia De Not. (1845), Sporormiaceae. 3, widespr. See Ahmed & Cain (*CJB* **50**: 419, 1972).

Sporormiaceae Munk (1957), Dothideales. 9 gen. (+ 9 syn.), 138 spp. Ascomata perithecial, black, thick-walled, sometimes hairy, with a well-developed lysigenous ostiole; interascal tissue copious, of cellular pseudoparaphyses; asci cylindrical, fissitunicate; ascospores usually dark brown, strongly constricted and fragmenting at the septa, with germ slits in each cell, occasionally ornamented, sometimes with a sheath.

    Lit.: v. Arx & v. d. Aa (*TBMS* **89**: 116, 1987; key 6 gen.).

Sporormiella Ellis & Everh. (1892), Sporormiaceae. c. 70 (mainly coprophilous), widespr. See Ahmed & Cain (*CJB* **50**: 419, 1972), Barr (*N. Am. Fl.* II **13**, 1990), Eriksson & Hawksworth (*SA* **10**: 144, 1991). Used as indicators of megafauna in Quaternary lake deposits (Davis, *Quatern. Res.* **28**: 290, 1987). = Preussia (Spororm.) fide v. Arx (1973).

Spororminula Arx & Aa (1987), Sporormiaceae. 1, Canary Isl.

Sporormiopsis Breton & Faurel (1964) = Sporormiella (Spororm.). See v. Arx (1973).

Sporoschisma Berk. & Broome (1847), Mitosporic fungi, 2.C2.22. 4, widespr. See Nag Raj & Kendrick (*A monograph of Chalara and allied genera*, 1975).

Sporoschismatites Babajan & Tasl. (1977), Fossil fungi. 1 (Tertiary), former USSR.

Sporoschismopsis Hol.-Jech. & Hennebert (1972), Mitosporic fungi, 1.C2.22. 2, Eur., N. Am.

Sporoschizon Riedl (1960), ? Arthopyreniaceae (L). 1, Austria.

Sporostachys Sacc. (1919) = Melanographium (Mitosp. fungi) fide Saccardo (*Syll. Fung.* **25**: 936, 1931).

sporostasis (adj. **sporostatic**), inhibition of spore germination; cf. mycostasis; sporostatic products of fungi, see Robinson *et al.* (*TBMS* **51**: 113, 1968).

sporothallus, a thallus producing spores; cf. gametothallus.

Sporotheca Corda (1829) ? = Melanconis (Melanconid.) fide Mussat (*Syll. Fung.* **15**: 398, 1901).

sporothecium, (1) the tip of a basidium bearing basidiospores when the basidiospores are sessile (Clémençon, *Z. Pilzk.* **36**: 113, 1970); (2) see spore wall.

Sporothrichum, see *Sporotrichum*.

Sporothrix Hektoen & C.F. Perkins (1901), Anamorphic Ophiostomataceae, and Saccharomycetales, 2.A1.10. Teleomorphs *Ophiostoma, Dolichoascus*. 23, widespr. See De Hoog (*Stud. Mycol.* **7**, 1974), Thibaut (*Ann. Parasit. hum. Comp.* **46**: 93, 1971; *S. schenckii*, see sporotrichosis), Findlay & Vismar (*Mycopath.* **96**: 115, 1986; SEM of morphology), Kreisel & Schauer (*Jl Basic Mycol.* **25**: 653, 1985; use of C sources), Smith & Batenburg-van der Vegte (*J. Gen. appl. Microb.* **32**: 549, 1986; ultrastructure of septa), Takeda *et al.* (*Mycopathol.* **116**: 9, 1991; phylogeny, molecular epidemiology), de Hoog (*in* Wingfield *et al.* (Eds), *Ceratocystis and Ophiostoma*: 53, 1993; anamorphs of *Ophiostoma*).

Sporotrichella P. Karst. (1887) = Fusarium (Mitosp. fungi) fide Hughes (1958).

Sporotrichites Göpp. & Berendt (1945), Fossil fungi. 4 (Oligocene, Miocene), Baltic.

Sporotrichopsis Guég. (1911) = Sporothrix (Mitosp. fungi) fide de Hoog (1974).

sporotrichosis, a lymphatic disease in humans and animals (*Sporothrix schenckii*; see de Beurmann & Gougerot, *Les Sporotrichoses*, 1912), Norden (*Acta Path. Microbiol. Scand., Suppl.* **84**, 1951).

Sporotrichum Link (1809), Mitosporic fungi, 1.A1.10. 40, widespr. See v. Arx (*Persoonia* **6**: 179, 1971), Stalpers (*Stud. Mycol.* **24**: 15, 1984; list 339 epithets). = Poria (Polyp.) fide Donk (1974).

spot anthracnose, see anthracnose.

Spragueola Massee (1896) = Neolecta (Neolect.) fide Korf (*Phytol.* **21**: 201, 1971).

Spumaria Pers. (1792) = Mucilago (Didym.).

Spumatoria Massee & E.S. Salmon (1901), ? Ophiostomataceae. 1, Eur. See Malloch & Blackwell (*CJB* **68**: 1712, 1990).

Spumula Mains (1935), Raveneliaceae. 5 (on *Leguminosae*), Mexico, Afr., Philipp.

spunk, see touchwood, amadou.

Squamacidia Brako (1989), Bacidiaceae (L). 1, trop.

Squamanita Imbach (1946), Agaricaceae. 7, widespr. See Bas (*Persoonia* **3**: 331, 1965; key).

Squamaphlegma Locq. (1979), Cortinariaceae. 1, Algeria.

Squamaria Hoffm. (1789) [non Ludw. (1757), *Orobanchaceae*] = Cetraria (Parmel.).

Squamarina Poelt (1958), ? Bacidiaceae (L). 20, Eur., Asia, N. Am. See Poelt & Krüger (*Feddes Repert.* **81**: 187, 1970).

Squamariomyces E.A. Thomas (1939) nom. inval. = Lecanora (Lecanor.).

Squammaria DC. (1805) = Squamarina (Bacid.).

squamose, having scales.

Squamotubera Henn. (1903) ? = Sarcoxylon (Xylar.) fide Rogers (1981).

squamule, a small scale.

Squamuloderma Kowalski (1973) = Didymium (Didym.) fide Martin *et al.* (1983).

squamulose, (1) having small scales; (2) growth form of a lichen thallus (Fig. 21C).

squarrose, rough with scales.

St Anthony's Fire, see ergot.

St George's mushroom. basidioma of the edible *Calocybe gambosum*.

stachel, bullet-like structure in the rohr with its pointed end orientated towards the appressorium and the host cell wall in plasmodiophorids. See also rohr and schlauch.

Stachybotryella Ellis & Barthol. (1902), Mitosporic fungi, 1.A2.?. 2, N. Am.

Stachybotryna Tubaki & T. Yokoy. (1971), Mitosporic fungi, 1.A1.15. 1, Japan.

stachybotryotoxin, a mycotoxin produced by *Stachybotrys* growing on hay; implicated in serious poisoning of horses (stachyobotrytoxicosis).

Stachybotrys Corda (1837), Mitosporic fungi, 2.A2.15. 10, widespr. See Barron (*CJB* **39**: 1566, 1961), Verona & Mazzucchetti (*Microfunghi della cellulosa e della carta .... I generi Stachybotrys e Memnoniella*, 1968), Jong & Davis (*Mycotaxon* **3**: 409, 1976; spp. in cult.), McKenzie (*Mycotaxon* **41**: 179, 1991; spp. from *Freycinetia*).

Stachycoremium Seifert (1986), Mitosporic fungi, 2.B1.16. 1, USA, Japan.

Stachylidium Link (1809), Mitosporic fungi, 1.A2.15. 1, widespr. See Hughes (*TBMS* **34**: 551, 1951).

Stachylina L. Léger & M. Gauthier (1932), Harpellaceae. 18 (in *Diptera*), widespr. See Lichtwardt (*Mycol.* **64**: 167, 1972), Moss (*Mycol.* **66**: 173, 1974; cytology, *in* Fuller & Lovelock (Eds), *Microbial ultrastructure*: 279, 1976; ultrastr.), Lichtwardt (1986; key), Lichtwardt *et al.* (1987), Lichtwardt & Williams (*Mycol.* **80**: 400, 1988; zygospores, *CJB* **68**: 1057, 1990), Williams & Lichtwardt (*CJB* **68**: 1045, 1990).

Stachyomphalina H.E. Bigelow (1979), Tricholomataceae. 1, widespr.

stage, a phase of the life cycle (q.v.). See States of fungi.

Staginospora Trivedi & C.L. Verma (1971), Fossil fungi (Mitosp. fungi). 1 (Tertiary), India.

Stagnicola Redhead & A.H. Sm. (1986), Cortinariaceae. 1, N. temp.

Stagonopatella Petr. (1927), Mitosporic fungi, 8.E1.?. 1, Eur.

Stagonopsis Sacc. (1884), Mitosporic fungi, 4.C1.?. 4, temp. See Petrak (*Sydowia* **3**: 139, 1949).

Stagonospora (Sacc.) Sacc. (1884) nom. cons., Mitosporic fungi, 4.C1.1. 200, widespr. *S. curtisii* (*Narcissus* leaf scorch). See Castellani & Germano (*Ann. Fac. Sci. Agric. Univ. Torino* **10**: 1, 1977; 77 graminicolous spp.), Philipson (*New Phytol.* **113**: 127, 1989; ultrastr.).

Stagonosporella Tassi (1902) = Stagonospora (Mitosp. fungi) fide Saccardo (*Syll. Fung.* **18**, 1906).

Stagonosporina Tassi (1902), Mitosporic fungi, 4.C1.?. 5, widespr.

Stagonosporites Babajan & Tasl. (1970), Fossil fungi. 1 (Tertiary), former USSR.

Stagonosporopsis Died. (1912) = Ascochyta (Mitosp. fungi) fide Petrak (*Ann. Myc.* **23**: 5, 1925).

Stagonostroma Died. (1914) = Fusarium (Mitosp. fungi) fide Sutton (1977).

Stagonostromella Petr. & Syd. (1927), Mitosporic fungi, 8.C1.?. 1, Brazil.

Staheliella Emden (1974), Mitosporic fungi, 1.A2.39. 1, Surinam.

Staheliomyces E. Fisch. (1921), Phallaceae. 1, C. &. S. Am.

Staibia Bat. & Peres (1966), Leptopeltidaceae. 1, Brazil.

Stains. Cotton blue (frequently added to the mounting medium; see Mounting media) is the most popular stain for the microscopical examination of fungi. Acid fuchsin (in lactic acid), orange G, picro-nigrosin, trypan blue, congo red, chlorazol black, brilliant chresyl blue are also used. See also Lillie (*Conn's biological stains*, edn 9, 1977).

Fungi in plant tissue: May be differentiated by lactophenol cotton blue (see above). Another useful method is thionin and orange G (Stoughton, *Ann. appl. Biol.* **17**: 162, 1930):

Stain in	
Thionin	0.1 g
Phenol (5 per cent in distilled water)	100.0 ml

Diazonium blue B, see Summerbell (*Mycol.* **77**: 587, 1985).

Periodic acid-Schiff technique (see below) has been adapted for fungi in plant tissues by Dring (*New Phytol.* **54**: 277, 1955).

Sulphovanillin reaction (for *Basidiomycota*)	
Pure vanillin	2.0 g
Distilled water	6.0 ml
Conc. sulphuric acid	16.0 ml

Fungi in animal tissue: May be differentiated by the usual haematoxylin and eosin stain. Good results are also obtained with the periodic acid-Schiff stain (Kligman & Mescon, *J. Bact.* **60**: 415, 1950), by Gram staining (see below), and with Gridley's technique [a combination of periodic acid-Schiff and Gomori-aldehyde-fuchsin stains (*Am. J. clin. Path.* **23**: 303, 1953)]. For details of the last 3 stains see Conant *et al.* (*Manual of clinical mycology*, edn 2, 1954), Emmons *et al.* (*Medical mycology*, 1970).

Melzer's Reagent (Langeron's modification):	
Chloral hydrate	100.0 g
Potassium iodide	5.0 g
Iodine	1.5 g
Distilled water	100.0 ml

For distinguishing amyloid spores of agarics (see Melzer, *BSMF* **40**: 78, 1924) and testing I reaction of asci, etc. (*see* iodine where other now preferred solutions are cited).

Gram's staining method:	
Solution A	
Crystal violet	4.0 g
Ethyl alcohol (95 per cent)	20.0 ml
Solution B	
Ammonium oxalate	0.8 g
Water	80.0 ml

Stain for 1 min in equal parts of A and B (diluting A as necessary), wash, flood with **Lugol's iodine** (*see* iodine), decolourize 1/2 min in 95 per cent alcohol, counter-stain with safranin 2.5, alcohol (95 per cent) 100.0, diluted 1 in 10 with water before use. See Kohn & Korf (*Mycotaxon* **3**: 165, 1975) and Nannfeldt (*TBMS* **67**: 283, 1976) for discussion.

Stakman (Elvin Charles; 1885-1979). After graduating at Minnesota Univ. (1906) he was appointed instructor (1909) in the Department of Plant Pathology where he worked until his retirement (1953), as head from 1913, and was for long director of the Federal Rust Laboratory at St Paul. A cultured and versatile man and the author of more than 300 papers and the coauthor of two books (e.g. *Principles of plant pathology*, 1957, with J.G. Harrar), Stakman is most famous for his work on the cereal rusts and their physiologic races which stemmed from his doctoral thesis (*Bull. Minn. Exp. Stn* **138**, 1913). See Christensen (*Phytopath.* **69**: 195, 1979; portr.); *Aurora sporalis. E.C. Stakman Day issue 17 May 1979* (Dept. Pl. Path., Univ. Minn.), Christensen (*E.C. Stakman, Statesman of science*, 1984; biogr., bibliogr., portr.), Stafleu & Cowan (*TL-2* **5**: 830, 1985), Wilcoxson & Kommedahl (*Rev. Trop. Pl. Path.* **7**: 223, 1993).

Stakmania Kamat & Sathe (1968), Uredinales (inc. sed.). 1, India. = Phakopsora (Phakopsor.) fide Cummins & Hiratsuka (1983).

Stalactocolumella S. Imai (1950) nom. nud. ≡ Circulocolumella (Hysterang.).

Stalagmites Theiss. & Syd. (1914) = Nectria (Hypocr.) fide Rogerson (1970).

Stalagmochaetia Cif. & Bat. (1963), Mitosporic fungi, 4.A1.?. 1, USA.

stalagmoid (of spores, **stalagmospores**), like a long tear or drop.

staling substances, substances produced by an organism which slow up or stop its growth (iso-antagonism) (see Brown, *Ann. Bot., Lond.* **37**: 106, 1923); **inhibitory substances** are similar substances which retard or inhibit the growth of *other* organisms (e.g. penicillin) (hetero-antagonism; cf. antibiotic substances). See Porter & Carter (*Bot. Rev.* **4**: 165, 1938).

Stamnaria Fuckel (1870), Leotiaceae. 3, Eur., N. Am. See Seaver (*Mycol.* **28**: 186, 1936).

stane crottle, a crottle (q.v.) growing on stone.

Stanglomyces Raithelh. (1986), Tricholomataceae. 1, S. Am.

Stanhughesia Constant. (1989), Anamorphic Chaetothyriaceae. 1.G1.1. Teleomorph *Ceramothyrium*. 3, Sweden.

Staninwardia B. Sutton (1971), Mitosporic fungi, 6.B2.38. 1, S.E. Asia.

Stanjehughesia Subram. (1992), Mitosporic fungi, 1.C2.1. 5, widespr.

Staphlosporonites Sheffy & Dilcher (1971), Fossil fungi (*f. cat.*). 5 (Eocene, Tertiary), China, USA.

Staphylocystis Coutière (1911) = Thalassomyces (Algae) fide Kane (1964).

Staphylotrichum J. Mey. & Nicot (1957), Mitosporic fungi, 1.A1.1. 1, Afr. See Maciejowska & Williams (*Mycol.* **55**: 221, 1963).

Starbaeckia Rehm ex Starbäck (1890), Leotiales (inc. sed.). 1, Eur. See Nannfeldt (*Nova Acta R. Soc. Scient. upsal.* 4, **8**(2), 1932).

Starbaeckiella (Sacc. & P. Syd.) Syd. & P. Syd. (1919) = Pyrenula (Pyrenul.) fide Harris (1989).

Starkeyomyces Agnihothr. (1956) = Myrothecium (Mitosp. fungi) fide Tulloch (1972).

starters, the pure cultures or mixtures of microorganisms used for starting fermenting processes. Pure culture starters are used in beer making, etc., and frequently in butter and cheese making. Examples of the more complex type are ragi (q.v.) used for Javanese arak (an alcoholic drink from rice starch), Chinese rice (Mingen or Men) (*Rhizopus oryzae*) the starter for rice wine, the Japanese Koji (preparations of *Aspergillus* used in soy and other fermenting processes), and kephir grains, a mixture of yeasts and bacteria which is the starter for kephir. See Fermented foods and beverages.

States of fungi. Since the publication of the Tulasnes' *Selecta carpologia fungorum*, 1861-5, it has been accepted that many fungi are pleomorphic, that is, one fungus may produce several sorts of spores which may or may not be coincident in time and may or may not be produced after a nuclear fusion followed by meiosis, a sequence that may be interpreted as sexual. The state characterized by sexual spores (ascospores, basidiospores, etc.) has traditionally been designated the **'perfect'** state (or stage), that characterized by asexual spores (conidia) or the absence of spores the **'imperfect'** state (or stage). Under the *Code* it is permissible to treat both 'perfect' and 'imperfect' states as species designated by latinized binomials, but when a 'perfect' and 'imperfect' species have been established to be states of one fungus the binomial applied to the 'perfect' state also covers that of the 'imperfect' and takes precedence (see Nomenclature). In order to increase precision in the terminology for states of pleomorphic fungi, Hennebert & Weresub (*Mycotaxon* **6**: 207, 1977) introduced new nouns and adjectives: **holomorph**, for the whole fungus in all its morphs and phases; **teleomorph**, for the sexual ('perfect') form or morph (e.g. that characterized by ascomata or basidiomata); and **anamorph**, for the asexual ('imperfect') form or morph (e.g. that characterized only by presence or absence of conidia). **Synanamorph** is applied to any one of two or more anamorphs which have the same teleomorph (Gams, *Mycotaxon* **15**: 459, 1982). A holomorph includes a teleomorph and frequently one, or rarely more anamorphs. The term **ana-holomorph** has been used for an 'imperfect' fungus (mitosporic fungus) which appears to lack a 'perfect' state (teleomorph). For such presumptively mitotic states not correlated with teleomorphs the general term 'Mitosporic fungi' is used in this *Dictionary*. A name applied under the Code to both perfect and imperfect states of a fungus is a **nomen holomorphosis**; that to the imperfect state a **nomen anamorphosis**. It has also been suggested that **anamorph-genus** and **anamorph-species** should replace the terms 'form genus' and 'form species'.

The morph terminology (which has been adopted for pleomorphic fungi in this *Dictionary*) is discussed in detail in Kendrick (Ed.) (*The whole fungus*, 2 vols, 1979) which also includes a reprint of Hennebert & Weresub's paper. See Nomenclature.

statismospore, a spore not forcibly discharged. Cf. ballistospore.

Statistical methods and design of experiments, see Johnston & Booth (Eds) (*Plant pathologist's pocketbook*: 353 edn 2, 1983).

statolon, an antiviral substance (which induces interferon formation) from *Penicillium stoloniferum*; the active principle of which is considered to be RNA of viral origin (Banks *et al.*, *Nature* **218**: 542, 1968).

Staurochaeta Sacc. (1875) ? = Staurophoma (Mitosp. fungi) fide Sutton (1977).

Staurolemma Körb. (1867), Lichinaceae (L). 1, S. Eur., Medit., Norway. See Jørgensen & Henssen (*Graphis Scripta* **5**: 12, 1993).

Stauronema (Sacc.) Syd., P. Syd. & E.J. Butler (1916), Mitosporic fungi, 7.A1.19. 6, widespr. See Nag Raj (1993).

Staurophallus Mont. (1845) nom. dub. (Phallaceae).

Staurophoma Höhn. (1907), Mitosporic fungi, 4.A1.15. 1, S. Am. See Morgan-Jones *et al.* (*Univ. Waterloo Biol. Ser.* **4**, 1972).

Staurosphaeria Rabenh. (1858) = Karstenula (Melanommat.) fide Saccardo (*Syll. Fung.* **3**: 1883).

staurospore (stauroconidium), a non-septate or septate spore with more than one axis; axes not curved through more than 180° (cf. helicospore); protuberances present and $>^1/_4$ spore body length (cf. amerospore). See Mitosporic fungi.

Staurothele Norman (1853) nom. cons., Verrucariaceae (L). *c.* 40, cosmop. (mainly trop.). See Harada (*Nat. Hist. Res.* **2**: 39, 1992; Japan), Swinscow (*Lichenologist* **2**: 152, 1963; Br. spp.), Thomson (*Bryologist* **94**: 351, 1991; key 17 spp., N. Am.).

Stearophora L. Mangin & Viala (1905), Ascomycota (inc. sed.). 1, N. Afr.

Stecchericium D.A. Reid (1963), Auriscalpiaceae. 2, S. Hemisph.

Steccherinaceae Parmasto (1968), Stereales. 7 gen. (+ 11 syn.), 37 spp. Basidioma resupinate to reflexed; dimitic; skeletocystidia present; spores inamyloid.

Steccherinum Gray (1821), Steccherinaceae. *c.* 45, cosmop. See Maas Geesteranus (*Persoonia* **7**: 443, 1974), Saliba & David (*Crypt. Myc.* **9**: 93, 1988; key Eur. spp.).

Steganopycnis Syd. & P. Syd. (1916) = Seynesia (Xylar.) fide Læssøe (*SA* **13**: 43, 1994).

Steganosporium, see *Stegonsporium*.

Stegasma Corda (1843) = Perichaena (Trich.).

Stegasphaeria Syd. & P. Syd. (1916), Mesneriaceae. 2, Philipp., Afr.

Stegastroma Syd. & P. Syd. (1916) = Anisomyces (Vals.) fide v. Arx & Müller 1973),

Stegia Fr. (1818) [non DC. (1805), *Malvaceae*] ≡ Stegilla (Fungi, inc. sed.).

Stegiacantha Maas Geest. (1966), Hydnaceae. 1, Madagascar.

Stegilla Rchb. (1828) nom. dub. (Fungi, inc. sed.) fide Sutton & Pirozynski (*TBMS* **46**: 517, 1964).

Stegites Mesch. (1892), Fossil fungi. 2 (Tertiary), Eur.

Stegobolus Mont. (1845) = Ocellularia (Thelotremat.) fide Hale (1981).

Stegonsporiopsis Van Warmelo & B. Sutton (1981), Mitosporic fungi, 8.D2.15. 1, N. Am.

Stegonsporium Corda (1827), Mitosporic fungi, 6.D2.19. 2, widespr. See van Warmelo & Sutton (*Mycol. Pap.* **145**, 1981).

Stegopeziza Höhn. (1917) = Hysterostegiella (Dermat.) fide Hein (1983).

Stegopezizella Syd. (1924) = Sarcotrochila (Hemiphacid.) fide Korf (1962).

Stegophora Syd. & P. Syd. (1916), Valsaceae. 1 (on *Ulmus*), N. Am. See McGranahan & Smalley (*Phytopath.* **74**: 1300, 1984; culture), Petrak (*Ann. Myc.* **38**: 267, 1940). Anamorph *Cylindrosporella*.

Stegophorella Petr. (1947), Ascomycota (inc. sed.). 1, Ecuador.

Stegothyrium Höhn. (1918), ? Microthyriaceae. 1, Eur.

Steinera Zahlbr. (1906), Coccocarpiaceae (L). 4, Australasia. See Henssen & James (*Bull. Br. Mus. nat. Hist., Bot.* **10**: 227, 1982).

Steiner's stable PD solution. *p*-phenylenediamine, 1 g; sodium sulphite (Na$_2$SO$_3$), 10 g liquid detergent, *c.* 10 drops; water, 100 ml. Stable for at least 6 months; see Scott (*Lichenologist* **1**: 88, 1958). See also Metabolic products.

Steinia Körb. (1873), Aphanopsidaceae (L). 1, Eur. See Faurel & Schotter (*Bull. Soc. Hist. nat. Afr. Noire* **25**: 126, 1954), Printzen & Rambold (*Lichenologist* **27**: 91, 1995).

Steirochaete A. Braun & Casp. (1853) = Colletotrichum (Mitosp. fungi) fide Southworth (*J. Mycol.* **6**: 115, 1890).

Stelechotrichum Ritgen (1831) nom. inval. ≡ Cephalotrichum (Mitosp. fungi).

steliogen, the structure which gives rise to the sporocarp stalk in protostelids.

Stella Massee (1889) [non Medik. (1787), *Leguminosae*] = Scleroderma (Sclerodermat.) fide Demoulin (*in litt.*).

Stellasclerotes Beneš (1959), Fossil fungi (f. cat.). 1 (Carboniferous), former Czechoslovakia.

stellate, like a star in form; - **seta**, a compound seta having several radiating arms; asterophysis.

Stellatospora Tad. Ito & Nakagiri (1994), Sordariaceae. 1, Japan.

Stellatostroma, see *Asterostroma*.

Stellifera Léman (1824) nom. dub. (? Anamorphic Agaric.).

Stellifraga Alstrup & Olech (1993), Ascomycota (inc. sed.). 1 (on lichens, esp. *Cladonia*), Spitsbergen.

Stelligera R. Heim ex Doty (1948) = Lachnocladium (Lachnoclad.) fide Corner (1950).

Stellomyces Morgan-Jones, R.C. Sinclair & Eicker (1987), Mitosporic fungi, 1.A1.6. 1, S. Afr.

Stellopeltis Bat. & A.F. Vital (1959), Mitosporic fungi, 5.A1.?. 2, Brazil.

Stellospora Alcorn & B. Sutton (1984), Mitosporic fungi, 1.A1.15. 1 (on *Appendiculella*), Australia, Philipp.

Stellothyriella Bat. & Cif. (1959), Mitosporic fungi, 5.B1.?. 1, USA.

Stemastrum Raf. (1808) nom. dub. ('gasteromycetes').

Stemmaria Preuss (1851) nom. dub. fide Seifert (*Sydowia* **45**: 103, 1993).

Stemmatomyces Thaxt. (1931), Laboulbeniaceae. 3, C. & S. Am., Philipp.

Stemonaria Nann.-Bremek., R. Sharma & Y. Yamam. (1984), Stemonitidaceae. 9, Eur., Japan, India.

Stemonitales, Myxomycota. 1 fam., 16 gen. (+ 13 syn.), 159 spp. Spore mass dark-coloured, peridium and capillitium non-calcareous. Fam.: **Stemonitidiaceae**.

Stemonitidaceae Fr. (1832), Stemonitales. 16 gen. (+ 13 syn.), 159 spp.

Stemonitis Gled. (1753), Stemonitidaceae. 23, cosmop.

Stemonitis Roth (1788) nom. dub. (Myxomycetes, inc. sed.).

Stemonitopsis (Nann.-Bremek.) Nann.-Bremek. (1975), Stemonitidaceae. 7, cosmop.

Stemphyliites Babajan & Tasl. (1973), Fossil fungi. 1 (Neogene), former USSR.

Stemphyliomma Sacc. & Traverso (1911) = Pithomyces (Mitosp. fungi) fide Kirk (*TBMS* **80**: 449, 1983).

Stemphyliopsis A.L. Sm. (1901), Mitosporic fungi, 1.D1.24. 1, UK. An albino mutant of Stemphylium (Mitosp. fungi). See also Petch (*TBMS* **23**: 146, 1939), Barron & Onions (*CJB* **44**: 861, 1966).

Stemphyliopsis Speg. (1910) [non A.L. Sm. (1910)] ≡ Stemphyliomma (Mitosp. fungi).

Stemphylium Wallr. (1833), Anamorphic Pleosporaceae, 1.D2.19. Teleomorph *Pleospora*. 20, cosmop. See Simmons (*Mycol.* **61**: 1, 1969), Wiltshire (*TBMS* **21**: 211, 1938).

Stenella Syd. (1930), Mitosporic fungi, 1.C2.10. 13, widespr. See Mulder (*TBMS* **65**: 514, 1975), de Hoog *et al.* (*TBMS* **81**: 485, 1983; relationships with *Ramichloridium*).

Stenellopsis B. Huguenin (1966), Mitosporic fungi, 3.C2.10. 1, New Caledonia.

Stenellopsis Morgan-Jones (1980) [non B. Huguenin (1966)] ≡ Parastenella (Mitosp. fungi).

Stenhammara Flot. ex Körb. (1855) ≡ Stenhammarella (Lecanorales).

Stenhammara Á. Massal. ex Zahlbr. (1924) = Lecidea (Lecid.) fide Hertel (*Beih. Nova Hedw.* **24**, 1967).

Stenhammarella Hertel (1967), ? Lecanorales (inc. sed., L). 1, Eur., C. Asia.

Stenocarpella Syd. & P. Syd. (1917), Mitosporic fungi, 4.B/C2.15. 2, trop. See Sutton (1980).

Stenocephalum Chamuris & C.J.K. Wang (1990), Mitosporic fungi, 1.A1/2.10. 1, widespr.

Stenocladiella Marvanová & Descals (1987), Mitosporic fungi, 1.G1.10. 2 (aquatic), Eur.

Stenocybe Nyl. ex Körb. (1855), Mycocaliciaceae. 10, widespr. See Schmidt (1970), Tschermak-Woess (*Lichenologist* **10**: 69, 1978; biology), Tibell (1984, 1987).

Stenocybella Vain. (1927) = Calicium (Calic.) fide Tibell (1984).

Stenographa Mudd (1861) = Graphina (Graphid.).

Stenospora Deighton (1969), Mitosporic fungi, 1.E1.10. 1 (on *Puccinia*), Sierra Leone.

Stephanoascus M.T. Sm., Van der Walt & Johannsen (1976), ? Endomycetaceae. 2, Brazil, S. Afr. See Giménez-Jurado *et al.* (*Syst. appl. microbiol.* **17**: 237, 1994), Traqueir *et al.* (*CJB* **66**: 926, 1988). Anamorphs *Candida, Sporothrix*.

Stephanocyclos Hertel (1983), Porpidiaceae (L). 1, Antarctic.

stephanocyst, a structure, typically bicellular (basal cell cup-like, terminal cell globose), found in certain basidiomycetes. See Burdsall (*Mycol.* **61**: 915, 1969).

Stephanoma Wallr. (1833), Mitosporic fungi, 1.G2.1. 3, widespr. See Butler & McCain (*Mycol.* **60**: 955, 1968).

Stephanomyces Speg. (1917) = Cucujomyces (Laboulben.) fide Thaxter (1931).

Stephanophallus MacOwan [not traced] = Anthurus (Clathr.) fide Saccardo (*Syll. Fung.* **7**: 23, 1888).

Stephanophoron Nádv. (1942) = Nadvornikia (Thelotremat.) fide Tibell (*Beih. Nova Hedw.* **79**: 597, 1984).

Stephanophorus Flot. (1843) = Leptogium (Collemat.).

Stephanopus M.M. Moser & E. Horak (1975), Cortinariaceae. 3, S. Am. See Moser & Horak (*Beih. Nova Hedw.* **52**: 608, 1975; key).

Stephanospora Pat. (1914), Stephanosporaceae. 4, Eurasia, NZ. See Oberwinkler & Horak (*Pl. Syst. Evol.* **131**: 157, 1979).

Stephanosporaceae Oberw. & E. Horak (1979), Stereales. 1 gen., 4 spp. Gasterocarp hypogeous; spores statismosporic, yellowish, verrucose with peri-appendicular corona.

Stephanosporium Dal Vesco (1961), Mitosporic fungi, 1.A2.39. 1, Italy.

Stephanotheca Syd. & P. Syd. (1914), Elsinoaceae. 2, Philipp.

Stephembruneria R.F. Castañeda (1988), Mitosporic fungi, 1.C2.15. 1, Cuba.

Stephensia Tul. (1851), Otideaceae. 6 (hypogeous), widespr. See Fischer (1938), de Vries (*Coolia* **28**: 96, 1985), Fontana & Giovannetti (*Mycotaxon* **29**: 37, 1987; anamorph).

Stephosia Bat. & H. Maia (1967), Mitosporic fungi (L), 7.A1.?. 1, Brazil.

Sterbeeckia Dumort. (1822) ≡ Craterellus (Cantharell.).

Stercum, see *Stereum*.

Stereaceae Pilát (1930), Stereales. 18 gen. (+ 12 syn.), 138 spp. Basidioma appressed, effused-reflexed or stalked; pileus often zoned; dimitic (rarely trimitic), typically differentiated into a trichoderm, a closely woven cortex, and loose hyphae curving up to the hymenium; hymenophore smooth to tuberculate; spores hyaline, smooth, amyloid or nonamyloid; lignicolous, terrestrial.

    Lit.: Donk (1964: 292), Reid (1965), Rattan (*The resupinate Aphyllophorales of the North West Himalayas* [*Bibl. Mycol.* **60**], 1977), Jahn (*Westfäl. Pilzbr.* **8**: 69, 1971), Reid (*Beih. Nova Hedw.* **18**, 1965; stipitate steroid spp.), Jülich & Stalpers (*The resupinate non-poroid Aphyllophorales of the temperate Northern Hemisphere*, [*Verh. Kon. Ned. Akad. Wetensch., Afd. Natuurk.* sect. 2 **74**], 1980).

Stereales Ferro (1907) , Basidiomycetes. 19 fam., 215 gen. (+ 115 syn.), 1136 spp. Fams:
(1) **Aleurodiscaceae**.
(2) **Amylocorticaceae**.

(3) **Atheliaceae.**
(4) **Botryobasidiaceae.**
(5) **Corticiaceae.**
(6) **Cyphellaceae.**
(7) **Echinodontiaceae.**
(8) **Epitheliaceae.**
(9) **Hyphodermataceae.**
(10) **Lindtneriaceae.**
(11) **Meruliaceae.**
(12) **Peniophoraceae.**
(13) **Podoscyphaceae.**
(14) **Sistotremataceae.**
(15) **Steccherinaceae.**
(16) **Stephanosporaceae.**
(17) **Stereaceae.**
(18) **Tubulicrinaceae.**
(19) **Xenasmataceae.**
Lit.: Bourdot & Galzin (1927), Donk (1964: 255), Hjortstam & Ryvarden (*Corticiaceae of North Europe*, 8 vols., 1973-88), Jülich (*Willdenowia Beih.* **7**, 1972; *Athelieae*), Parmasto (*Conspectus Systematis Corticiacearum*, 1968; 11 subfam.), Hjortstam (*Windahlia* **17**: 55, 1987; gen. & spp. checklist), Maekawa (*Rep. Tottori Mycol. Inst.* **31**: 1, 1993; Japanese spp.).

Sterellum P. Karst. (1889) = Peniophora (Peniophor.) fide Weresub & Gibson (1960).

Stereocaulaceae Chevall. (1826), Lecanorales (L). 4 gen. (+ 9 syn.), 136 spp. Thallus composed of a crustose to squamulose primary thallus, often degenerating at an early stage, and a solid erect shrubby secondary thallus; ascomata apothecial, flat or domed, usually brown, without well-developed marginal tissues; interascal tissue of simple sparingly branched paraphyses with hardly swollen apices; asci with a strongly I+ apical cap with a central dark tubular structure, without an outer I+ gelatinized layer; ascospores hyaline, often elongated, transversely septate, sometimes muriform. Cephalodia often present. Conidiomata pycnidial. Lichenized with green algae, the cephalodia with cyanobacteria, widespr., esp. temp.

Stereocauliscum Nyl. (1865) ? = Micarea (Micar.).

Stereocaulomyces E.A. Thomas ex Cif. & Tomas. (1953) ≡ Stereocaulon (Stereocaul.).

Stereocaulon (Schreb.) Schrad. (1794) nom. rej. = Pertusaria (Pertusar.).

Stereocaulon Hoffm. (1796) nom. cons., Stereocaulaceae (L). 123, cosmop. See Lamb (*J. Hattori bot. Lab.* **43**: 191, 1978, **44**: 209, 1978; keys), Smith & Øvstedal (*Polar Biology* **11**: 91, 1991; 6 Antarctic spp.).

Stereocaulum Clem. (1909) = Pertusaria (Pertusar.).

Stereochlamydomyces Cif. & Tomas. (1953) ≡ Stereochlamys (Trichothel.).

Stereochlamys Müll. Arg. (1885) = Trichothelium (Trichothel.).

Stereocladium Nyl. (1875) = Stereocaulon (Stereocaul.).

Stereoclamydomyces, see *Stereochlamydomyces*.

Stereocrea Syd. & P. Syd. (1917), Hypocreaceae. 2, Philipp., Sri Lanka.

Stereofomes Rick (1928), Lachnocladiaceae. 5, S. Am., Japan. Nom. dub. fide Donk (*Taxon* **6**: 114, 1957).

Stereogloeocystidium Rick (1940) = Podoscypha (Podoscyph.) fide Boidin (1959).

Stereolachnea Höhn. (1917) = Scutellinia (Otid.) fide Denison (*Mycol.* **51**: 605, 1959).

stereome (of lichens), a scleroplectenchyma which forms the main supporting tissue of the thallus, as in *Alectoria*, *Bryoria*, and *Cladonia*.

Stereonema Kütz. (1836) = Lecidea (Lecid.).

Stereopeltis Franzoni & De Not. (1861) = Sarcogyne (Acarospor.).

Stereophyllum P. Karst. (1889) [non Mitt. (1859), *Musci*] ≡ Cyphellostereum (Podoscyph.).

Stereopodium Earle (1909) ≡ Mycena (Tricholomat.).

Stereopsis D.A. Reid (1965), Podoscyphaceae. 10, widespr., esp. trop.

Stereosorus Sawada (1943) = Burrillia (Tillet.) fide Ling (*Mycologia* **41**: 252, 1949).

Stereosphaeria Kirschst. (1939) = Clypeosphaeria (Clypeosphaer.) fide Barr (1989).

Stereostratum Magnus (1899), Pucciniaceae. 1 (on *Bambusoideae*), China, Japan. See Thirumalachar (*Mycol.* **52**: 690, 1961).

Stereum Pers. (1794), Stereaceae. c. 25, cosmop. *S. purpureum* is included in *Chondrostereum*. See Welden (*Mycol.* **63**: 796, 1971; key genus s.s.), Boidin (*Rev. Mycol.* **23**: 318, **24**: 197, *Bull. Soc. linn. Lyon* **28**: 205, 1958-59, *Bull. Jard. bot. Brux.* **30**: 51, 283, 1960; Congo), Talbot (*Bothalia* **6**: 303, 1954; S. Afr.), Chamuris (*Mycotaxon* **22**: 105, 1985; key N. Am. spp.).

Stericium Raf. (1819) nom. dub. (Fungi, inc. sed.). No spp. included.

sterigma (pl. **-ata**), (1) (of a basidium, q.v.), an extension of the metabasidium composed of a basal filamentous or inflated part (the **proto-**; epibasidium) and an apical spore-bearing projection (the **spiculum**) (Fig. 6F); (2) (of *Aspergillus*, etc.) [a usage not recommended], phialide (**secondary -**); prophialide (**primary -**); metula; (3) (of lichens), a spermatiophore (Nylander).

Sterigmatobotrys Oudem. (1886), Mitosporic fungi, 1.C1.10. 1, N. temp. See Jong & Davies (*Norw. Jl Bot.* **18**: 177, 1971), Ellis (*DH*), Sutton (*Mycol. Pap.* **132**, 1973).

sterigmatocystin, a carcinogenic hepatotoxin (xanthone derivative) from *Aspergillus versicolor* (Van der Walt & Purchase, *Brit. J. exp. Path.* **51**: 183, 1970); precursor of aflatoxin B1 (Singh & Hsich, *Appl. environ. Microbiol.* **31**: 743, 1976).

Sterigmatocystis Cramer (1859) = Aspergillus (Mitosp. fungi) fide Raper & Fennell (*The genus Aspergillus*, 1965).

Sterigmatomyces Fell (1966), Mitosporic fungi, 1.A1.10. 3, Atlantic and Indian Oceans. See Lodder (1970: 1229), Yamada *et al.* (*J. Gen. appl. microbiol.* **32**: 157, 1986; enzyme systems), van der Walt *et al.* (*Ant. v. Leeuwenhoek* **53**: 137, 1987; key to 3 spp.), Kurtzman (*Int. Jl Syst. Bact.* **40**: 56, 1990; DNA relatedness), Guého *et al.* (*Int. Jl Syst. Bact.* **40**: 60, 1990; partial rRNA sequences), Yamada *et al.* (*Agric. Biol. Chem.* **53**: 2993, 1989; phylogeny).

Sterigmatosporidium G. Kraep. & U. Schulze (1983), Sporidiobolaceae. 1, Germany. Anamorph *Sterigmatomyces*.

sterile, (1) not producing spores or a sporocarp; (2) free from living microorganisms; sterilized.

sterilization, making free from living microorganisms, may be done by chemical (see Fungicides) or physical methods. Heat is the most widely used physical agent. Dry heat may be used for glass and other materials; death of even the most resistant bacterial spores takes place within an hour in a hot air oven at 160°C. A flame may frequently be used for the surface sterilization of instruments. Moist heat, especially if the pressure is increased, is even better. Most media may be made sterile by autoclaving for 15 min at 121°C, but when such severe heating is not possible 'discontinuous steaming' (steaming at atmospheric pressure for 20 min every day for 3 days) is frequently used. In addition, steam is of use for the 'sterilization' (partial (or incomplete) sterilization) of soil for controlling soil-living pathogens, and hot water (about 50°C) has an important application for the control of diseases such as loose smut of wheat (*Ustilago tritici*) in which the pathogen is inside the seed. Though cold will not let growth take place, even great cold may not be lethal to micro-organisms.

Among other physical agents are ultra-violet light (to which the sterilizing effect to sunlight is due) but certain fluorescent dyes, such as eosin, are able to make bacteria sensitive to light of longer wave lengths (Blum, *Physiol. Rev.* **12**: 23, 1932), electricity (the effect of low frequency currents being possibly that of the heat or the nascent oxygen (O) or chlorine (Cl) produced), X-rays, radium emanation, and supersonic waves. Desiccation (drying) is frequently lethal, specially to the vegetative phase, but some fungal and bacterial spores are very resistant. It is sometimes possible to make a liquid sterile by filtering out the microorganisms with a filter of unglazed porcelain. See Rahn (*Bact. Rev.* **9**: 1, 1945; physical methods), Sykes (*Disinfection and sterilization. Theory and practice*, 1958, *in* Norris & Ribbons (Eds), *Methods in microbiology* **1**: 77, 1969).

Sterrebekia Link (1816) [non *Sterbeckia* Schreb. (1789), *Papilionaceae*] = Scleroderma (Sclerodermat.).

Stevensea Trotter (1926) = Diplotheca (Myriang.) fide v. Arx & Müller (1975).

Stevensiella Trotter (1928) ≡ Diatractium (Phyllachor.).

Stevensomyces E.F. Morris & Finley (1965), Mitosporic fungi, 2.A1.?. 1, Panama.

Stevensonula Petr. (1952), Mitosporic fungi, 7.B2.1. 1, USA, New Caledonia. See Sutton (*Nova Hedw.* **26**: 1, 1975).

Stevensula Speg. (1924) = Leptomeliola (Parodiopsid.) fide Hughes (*Mycol. Pap.* **166**, 1993).

Steyaertia Bat. & H. Maia (1960) = Asterolibertia (Asterin.) fide Müller & v. Arx (1962).

Sthughesia M.E. Barr (1987) ? = Metacapnodium (Metacapnod.) fide Eriksson & Hawksworth (*SA* **7**: 91, 1988).

stichobasidium, see basidium.

Stichoclavaria Ulbr. (1928) ≡ Holocoryne (Clavar.).

Stichodothis Petr. (1927) = Auerswaldiella (Dothid.) fide v. Arx & Müller (1954).

Stichomyces Thaxt. (1901), Laboulbeniaceae. 7, Am., Afr., Sumatra, Eur., Japan.

Stichophoma Kleb. (1933) ? = Sclerophoma (Mitosp. fungi) fide Sutton (1977).

Stichopsora Dietel (1899) = Coleosporium (Coleospor.) fide Sydow (1915).

Stichoramaria Ulbr. (1928) ≡ Clavulina (Clavulin.).

Stichospora Petr. (1927) [non Heydr. (1900), *Algae*], Mitosporic fungi, 8.A1/2.?. 1, Eur.

Stichus D.E. Ether. (1904), ? Fossil fungi (mycel.). 1 (Cretaceous), Australia.

Sticta (Schreb.) Ach. (1803), Lobariaceae (L). *c*. 200, cosmop. (mainly trop.). See Delise (*Histoire des lichens. Genre Sticta*, 1822), Galloway (*Lichenologist* **26**: 223, 1994; key 12 spp. S. Am.), Joshi & Awasthi (*Biol. Mem.* **7**: 165, 1982; key 13 spp., India), Malme (*B.K. svenska Vet. Akad. Handl.* **25**, 3(6), 1899; S. Am.).

Stictaceae, see Lobariaceae.

Stictidaceae Fr. (1849), Ostropales. 20 gen. (+ 27 syn.), 149 spp. Stroma poorly developed, restricted to intramatrical hyphae; ascomata apothecial, often deeply immersed and almost perithecioid, usually with a well-developed margin formed largely of crystalline inclusions; interascal tissue usually of simple paraphyses, sometimes branched and pigmented, sometimes swollen at the apices; asci usually narrowly cylindrical with a strongly thickened apex;

ascospores often filiform, fragmenting. Anamorphs pycnidial, inconspicuous. Saprobic on wood and stems, a few lichenicolous.

Stictina Nyl. (1860) nom. rej. = Pseudocyphellaria (Lobar.) fide Galloway (1988).

Stictis Pers. (1799), Stictidaceae. 57, cosmop. See Sherwood, *Mycotaxon* **5**: 1, 1977; key, **6**: 215, 1977).

Stictochorella Höhn. (1917) = Asteromella (Mitosp. fungi) fide Clements & Shear (1931).

Stictochorellina Petr. (1922) = Asteromella (Mitosp. fungi) fide Sutton (1977).

Stictoclypeolum Rehm (1904) = Asterothyrium (Thelotremat.).

Stictographa Mudd (1861) = Melaspilea (Melaspil.).

Stictomyces E.A. Thomas ex Cif. & Tomas. (1953) ≡ Lobaria (Lobar.).

Stictopatella Höhn. (1918), Mitosporic fungi, 7.A1.?. 1, Eur.

Stictophacidium Rehm (1888), ? Stictidaceae. 1, Eur. See Sherwood (*Mycotaxon* **5**: 1, 1977).

Stictosepta Petr. (1964), Mitosporic fungi, 8.E1.10. 1, former Czechoslovakia.

Stictosphaeria Tul. & C. Tul. (1863) = Diatrype (Diatryp.).

Stictospora Cif. (1957), Anamorphic Stictidaceae, 8.A1.?. Teleomorph *Stictis*. 1, Eur.

Stictostroma Höhn. (1917) ? = Placuntium (Rhytismat.) fide Sherwood (*Mycotaxon* **5**: 1, 1977).

Stigeosporium C. West (1916) ? = Glomus (Glom.) fide Gerdemann & Trappe (1974).

Stigmagora Trevis. (1853) = Ocellularia (Thelotremat.) p.p. and Thelotrema (Thelotremat.) p.p.

Stigmastoma Bat. & H. Maia (1960) = Pycnothyrium (Mitosp. fungi) fide v. Arx & Müller (1975).

Stigmatea Fr. (1849), Ascomycota (inc. sed.). 1, Eur. ≡ Stigmea (Dothior.).

Stigmateacites S.L. Zheng & W. Zhang (1986), Fossil fungi. 2, China.

Stigmatella Berk. & M.A. Curtis (1857) ? = Chondromyces (Myxobacteria).

Stigmatella Mudd (1861) = Sclerophyton (Roccell.).

Stigmatellina Bat. & H. Maia (1960), Mitosporic fungi, 5.A1.?. 1, Belgium.

Stigmateopsis Bat. (1960) = Placonema (Mitosp. fungi). See Sutton (*Kew Bull.* **31**: 461, 1977), Nag Raj (1993).

Stigmatidium G. Mey. (1825) = Enterographa (Roccell.).

Stigmatisphaera Dumort. (1822) nom. dub. (Ascomycota, inc. sed.), used for diverse perithecioid fungi.

stigmatocyst, see hyphopodium.

Stigmatodothis Syd. & P. Syd. (1914), Micropeltidaceae. 1, Philipp.

Stigmatolemma Kalchbr. (1882), Tricholomataceae. 2, cosmop.

Stigmatomassaria Munk (1953) = Splanchnonema (Pleomassar.) fide Barr (1982, 1993).

Stigmatomma Körb. (1855) = Staurothele (Verrucar.).

Stigmatomyces H. Karst. (1869), Laboulbeniaceae. 110 (on *Diptera*), widespr. See Weir & Rossi (*MR* **99**: 841, 1995; key Br. spp.).

stigmatomycosis (of cotton (*Gossypium*) bolls, *Phaseolus*, and other plants), damage caused by insect-inoculated fungi such as *Nematospora gossypii* and *N. coryli*. See Leach (*Insect transmission of plant diseases*, 1940), Frazer (*Ann. appl. Biol.* **31**: 271, 1944).

Stigmatopeltis Doidge (1927) = Vizella (Vizell.) fide v. Arx & Müller (1975).

Stigmatophragmia Tehon & G.L. Stout (1929), Micropeltidaceae. 1, N. Am.

stigmatopodium (**stigmopodium**), see hyphopodium.

Stigmatopsis Traverso (1906) = Cryptosphaeria (Diatryp.) fide Barr (*The Diaporthales in North America*, 1978).

Stigmatoscolia Bat. & Peres (1960) = Lophodermium (Rhytismat.) fide v. Arx (*in litt.*).

Stigmatula (Sacc.) Syd. & P. Syd. (1901) Phyllachoraceae. 5 (on *Leguminosae*), widespr. = Polystigma (Phyllachor.) fide v. Arx & Müller (1954), but see Cannon (*Mycol. Pap.* **163**, 1991).

Stigme Syd. & P. Syd. (1917) = Dimerina (Pseudoperispor.) fide Müller & v. Arx (1962).

Stigmea Fr. (1836), Ascomycota (inc. sed.). 20, widespr.

Stigmea Bonord. (1864) = Dothiora (Dothior.).

Stigmella Lév. (1842), Mitosporic fungi, 4.D2.1. 26, widespr. See Hughes (*Mycol. Pap.* **49**, 1952).

Stigmidium Trevis. (1860), Mycosphaerellaceae. *c.* 20 (on lichens), cosmop. See David & Hawksworth (*SA* **5**: 158, 1986; posn), Roux & Triebel (*Bull. Soc. linn. Provence* **45**: 451, 1994; key 17 spp.).

Stigmina Sacc. (1880), Anamorphic Dothideales, 3.C/D2.19. Teleomorph Otthia. *c.* 50, widespr. *S. carpophila* (peach shot hole). See Ellis (*Mycol. Pap.* **72**, 1959; key), Sutton & Pascoe (*MR* **92**: 210, 1989; taxonomic re-evaluation).

Stigmochora Theiss. & Syd. (1914), Phyllachoraceae. 2 or 3, S. Am.

Stigmopeltella Syd. (1927) = Stigmopeltis (Mitosp. fungi) fide Clements & Shear (1931).

Stigmopeltis Syd. (1927), Mitosporic fungi, 5.E1.?. 2, C. Am.

Stigmopeltopsis Peres (1961) = Myxothyriopsis (Mitosp. fungi) fide v. Arx (*in litt.*).

Stigmopsis Bubák (1914) = Cheiromyces (Mitosp. fungi) fide Moore (*Mycol.* **50**: 681, 1959).

stilbaceous (obsol.), having synnemata; synnematous (q.v.).

Stilbechrysomyxa M.M. Chen (1984), Coleosporiaceae. 3 (on *Rhododendron*), Asia.

Stilbella Lindau (1900) nom. cons., Mitosporic fungi, 2.A1.15. 40 (from soil, coprophilous, on *Insecta*), widespr. See Morris (1963), Benjamin (*Taxon* **17**: 521, 1968), Sutton (*Mycol. Pap.* **132**: 18, 1973), Seifert (*Stud. Mycol.* **27**: 1, 1985; key).

Stilbellula Boedijn (1951), Mitosporic fungi, 2.A1.?. 1, Java.

Stilbites Pia (1927), Fossil fungi. 1 (Eocene), Baltic.

Stilbochalara Ferd. & Winge (1910) = Chalara (Mitosp. fungi) fide Nag Raj & Kendrick (1975).

Stilbocrea Pat. (1900) = Nectria (Hypocr.).

Stilbodendron Syd. & P. Syd. (1916), Anamorphic Trichocomaceae, 2.A2.38. Teleomorph *Penicilliopsis*. 1, Afr. = Corallodendron (Mitosp. fungi) fide Morris (*W. Ill. Univ. Ser. Biol. Sci.* **3**: 36, 1963).

Stilbodendrum Bonord. (1851) = Syzygites (Mucor.) fide Kirk (*in litt.*).

Stilbohypoxylon Henn. (1902) ? = Xylaria (Xylar.) fide Læssøe (*SA* **13**: 43, 1994).

stilboid, a sterile, basidioma-like structure (as in *Mycena citricolor* and other agarics) which functions as a propagule (Singer, 1962: 25); gemma (Buller). Cf. carpophoroid.

Stilbomyces Ellis & Everh. (1895) nom. dub. (AScomycota, inc. sed.); based on hyphophores of a lichen fide Seifert (*Sydowia* **45**: 103, 1993).

Stilbonectria P. Karst. (1889) = Nectria (Hypocr.) fide Rossman (*Myc. Pap.* **150**, 1983).

Stilbopeziza Speg. (1908), ? Leotiales (inc. sed., ?L). 1, S. Am.

Stilbophoma Petr. (1942), Mitosporic fungi, 8.A1.15. 2, India, Afr. See Sutton (1980).

Stilbospora Pers. (1797), Mitosporic fungi, 4/6.C2.19. 15, widespr.

Stilbothamnium Henn. (1896) = Aspergillus (Mitosp. fungi) fide Thom & Blochwitz (1965).

Stilbotulasnella Oberw. & Bandoni (1982), Tulasnellales (inc. sed.). 1, USA.

stilbum, the erect synnema (q.v.) of *Stilbella* with its head of slime spores.

Stilbum Tode (1790), Chionosphaeraceae. 1, temp. See Donk (*Taxon* **7**: 236, 1958; typification).

stink horns, basidiomata of certain *Phallales*.

Stioclettia Dennis (1975), Diaporthales (inc. sed.). 1 (on *Luzula*), UK.

stipe, a stalk (Fig. 4D, F-H).

Stipella L. Léger & M. Gauthier (1932), Legeriomycetaceae. 1 (in *Diptera*), Eur. See Moss (*TBMS* **54**: 1, 1970), Lichtwardt (1986).

Stipinella, see *Stypinella*.

stipitate, stalked.

Stipitochaete Ryvarden (1985), Hymenochaetaceae. 2, S. Am.

Stipiza Raf. (1815) nom. nud. See Merrill (*Index Rafinesq.*, 1949).

Stiptophyllum Ryvarden (1973), Polyporaceae. 1, Brazil.

Stirtonia A.L. Sm. (1926) [non R. Br. (1900), *Musci*], Arthoniaceae (L). *c.* 6, trop. See Santesson (1952).

Stirtoniopsis Groenh. (1938) nom. inval. Arthoniales (L). 2, Java, Morocco.

stock (of basidiomycetes), a dikaryotic mycelium (fide Raper, 1966). Cf. strain.

stolon, a 'runner', as in *Rhizopus*.

Stomatisora J.M. Yen (1971), Uredinales (inc. sed.). 1, Gabon. = Chrysocelis (Melampsor.) fide Cummins & Hiratsuka (1983).

Stomatogene Theiss. (1917), Parodiopsidaceae. 2, N. Am.

Stomatogenella Petr. (1955), Ascomycota (inc. sed.). 1, Australia.

stomatopodium (stomopodium), a hyphal branch (an appressorium; cf. hyphopodium) or 'plug' above or in a stoma.

Stomiopeltella Theiss. (1914) = Stomiopeltis (Micropeltid.) fide v. Arx & Müller (1975).

Stomiopeltina Bat. (1963) = Metathyriella (Schizothyr.) fide v. Arx & Müller (1975).

Stomiopeltis Theiss. (1914), Micropeltidaceae. 14, widespr. See Batista (1959), Ellis (*TBMS* **68**: 157, 1977; key Br. spp.).

Stomiopeltites Alvin & M.D. Muir (1970), Fossil fungi. 1 (Cretaceous), UK.

Stomiopeltopsis Bat. & Cavalc. (1963), Micropeltidaceae. 1, Brazil.

Stomiotheca Bat. (1959), Micropeltidaceae. 1, Brazil.

stone rag (stone raw), *Parmelia saxatilis*.

stone-fungus, the hard pseudosclerotium of *Polyporus tuberaster*; Pietraia fungaia. On being watered, an edible basidioma is produced. The Canadian tuckahoe is the same species (Vanterpool & Macrae, *CJB* **29**: 147, 1951).

stopper, the *Neurospora* phenotype characterized by irregular cycles of cessation and renewal of growth.

Straggaria Reinsch (1888) nom. dub. (? Fungi).

strain, (1) a group of clonally related individuals or cells. See Yoder *et al.* (*Phytopath.* **76**: 383, 1986); (2) a homokaryotic mycelium (fide Raper, 1966), cf. stock.

Straminella M. Choisy (1929) = Lecanora (Lecanor.) fide Brodo & Elix (*Bibl. Lich.* **53**: 19, 1993).

Straminipila. Kingdom proposed by Dick (*Mycol. Pap.*, *in press*) to accommodate most organisms previously generally referred to the Kingdom *Chromista* (q.v.) and primarily characterized by the presence of tripartite tubular hairs on flagella or cysts. If this classification becomes generally accepted, the

Hyphochytriomycota and *Oomycota* will need to be referred to it.

straminipile (straminopile), colloquial noun for an organism bearing tripartite tubular hairs (Patterson, 1989).

straminipilous (straminopilous), bearing tripartite tubular hairs; applicable to flagella and/or cells, whether uniflagellate, multiflagellate or non-flagellate (the auxiliary cyst of *Saprolegnia* which bears a tuft of tripartite tubular hairs would be a straminipilous cyst), cf. flimmergeissel; - fungi, see *Chromista*, *Straminipila*.

strand plectenchyma, plectenchyma in strands forming supporting tissues in a lichen thallus.

strangle fungus, *Epichloë typhina*; see choke.

Strangospora Körb. (1860), Acarosporaceae (L). *c.* 10, N. temp. See Harris *et al.* (*Evansia* **5**: 26, 1988; posn).

Strangulidium Pouzar (1967) = Poria (Coriol.) fide Donk (1974).

Strasseria Bres. & Sacc. (1902), Mitosporic fungi, 8.A1.15. 1, widespr. *S. geniculata* (black rot of apple). See Sutton (1980), Parmelee & Cauchon (*CJB* **57**: 1660, 1979), Dennis (*Gdnrs' Chron.* **114**: 221, 1943; black rot of apple).

Strasseriopsis B. Sutton & Tak. Kobay. (1970), Mitosporic fungi, 8.B1.15. 1, Japan.

Stratisporella Hafellner (1979), ? Patellariaceae. 1 (on lichen, *Tremotylium*), Angola.

stratose thallus, a lichen thallus having the tissue in horizontal layers.

Strattonia Cif. (1954), Lasiosphaeriaceae. 5 (coprophilous), widespr. See Barr (*Mycotaxon* **39**: 43, 1990; posn), Lundqvist (*Symb. bot. upsal.* **20**(1), 1972; key).

straw mushrooms (paddy straw or Chinese mushroom), the edible *Volvariella volvacea* and *V. diplasia*. These agarics are widely used in the tropics. In Myanmar, where *V. diplasia* is cultured, wet rice ('paddy') straw is made into a bed about $1 \times 1 \times 5$ m which is inoculated with 'pure culture' spawn (cf. mushroom culture), and kept wet. Mushrooms are first seen after 2-3 weeks; 4 kg or so being obtained from one bed. An air temperature of at least 21°C is necessary. See Thet & Seth (*Indian Fmg* **1**: 332, 1940), Chang (*Econ. Bot.* **31**: 374, 1977), Sukara *et al.* (*Bull. BMS* **19**: 129, 1985), Edible fungi.

Streblema Chevall. (1826) nom. dub.; stroma zone lines of *Xylariaceae*.

Streblocaulium Chev. (1837) = Conoplea (Mitosp. fungi) fide Kendrick & Carmichael (1973).

Streblomyces Thaxt. (1920) = Nycteromyces (Laboulben.) fide Tavares (1985).

Streimannia G. Thor (1991), Roccellaceae (L). 1, Australia.

Streptobotrys Hennebert (1973), Anamorphic Sclerotiniaceae, 1.A1.7. Teleomorph *Streptotinia*. 3, N. Am.

Streptomyces Waksman & Henrici (1943), *Actinomycetes*, q.v.

streptomycin, a broad spectrum aminoglycoside antibiotic produced by *Streptomyces griseus*. Active against *Mycobacterium tuberculosis*, staphylococci, some gram-negative bacteria, and inhibiting vegetative growth of some fungi.

Streptopodium R.Y. Zheng & G.Q. Chen (1978), Anamorphic Erysiphaceae, 1.A1.1. Teleomorph *Pleochaeta*. 1, S. Am.

Streptosporangium Couch (1955), *Actinomycetes*, q.v.

Streptotheca Vuill. (1887), Ascobolaceae. 4, N. temp. See Bergman & Shanor (*Mycol.* **49**: 879, 1957). = Ascozonus (Ascobol.) fide Eckblad (1968).

Streptothrix Corda (1839) [non Cohn (1875), *Actinomycetes*] = Conoplea (Mitosp. fungi) fide Hughes (1958).

Streptotinia Whetzel (1945), Sclerotiniaceae. 2, USA. Anamorph *Streptobotrys*.

Streptotrichites Mesch. (1892), Fossil fungi. 1 (Oligocene), Baltic.

Streptoverticillium E. Bald. (1958), *Actinomycetes*, q.v.

Striadiporites C.P. Varma & Rawat (1963), Fossil fungi (*f. cat.*). 3 (Tertiary), India, USA.

Striadyadosporites Dueñas (1979), Fossil fungi (*f. cat.*). 2 (Pleistocene), Colombia.

Striatasclerotes Stach & Pickh. (1957), Fossil fungi (*f. cat.*). 1 (Carboniferous), Germany. See Beneš (1969).

striate, marked with delicate lines, grooves or ridges (Fig. 29.13).

Striatosphaeria Samuels & E. Müll. (1979), Lasiosphaeriaceae. 1, Brazil. See Barr (*Mycotaxon* **39**: 43, 1990; posn). Anamorph *Dictyochaeta*.

Strickeria Körb. (1865), Ascomycota (inc. sed.). *c.* 7, widespr. See Eriksson (*SA* **10**: 144, 1991), Wakefield (*TBMS* **24**: 282, 1940).

Stridiporosporites Ke & Shi (1978), Fossil fungi (*f. cat.*). 6 (Oligocene, Tertiary), China.

Striglia Adans. (1763) = Daedalea (Coriol.).

Strigopodia Bat. (1957), Euantennariaceae. 3, temp.-boreal. See Hughes (*CJB* **46**: 1009, 1968), Reynolds (1986). Anamorph *Capnophialophora*.

strigose, rough with sharp-pointed hairs; hispid.

Strigula Fr. (1823), Strigulaceae (L). *c.* 50 (many foliicolous), widespr., esp. trop. See Hawksworth (*Taxon* **35**: 787, 1986; nomencl.), Santesson (*Symb. bot. upsal.* **12**(1), 1952; foliicolous spp.), Margot (*Lichenologist* **9**: 51, 1977; host relationship), Nag Raj (*CJB* **59**: 2519, 1981; asci, pycnidia).

Strigulaceae Zahlbr. (1898), Ascomycota (inc. sed., L). 1 gen. (+ 16 syn.), 50 spp. Thallus crustose, subcuticular, ± circular or emarginate, sometimes lobed; ascomata perithecial, usually strongly flattened, immersed or erumpent, uniloculate, sometimes clypeate, ± hyaline to black; interascal tissue of simple or sparingly branched paraphyses; hymenium blueing in iodine; asci ± clavate, persistent, ± thin-walled, fissitunicate, with a distinct ocular chamber, not blueing in iodine; ascospores hyaline, transversely septate, without a sheath. Anamorph pycnidial. Lichenized with green algae (esp. *Cephaleuros*; q.v.), on leaves, esp. trop.

Strigulomyces Cif. & Tomas. (1953) = Strigula (Strigul.).

Strilia Gray (1821) = Coltricia (Hymenochaet.) fide Parmasto (*in litt.*). See Donk (1974: 49).

Striodiplodia Zambett. (1955) = Lasiodiplodia (Mitosp. fungi) fide Sutton (1977).

Strionemadiplodia Zambett. (1955), Mitosporic fungi, 4.B2.?. 3, widespr.

strobiliform, like a fir-cone in form.

Strobilofungus Lloyd (1915) = Boletellus (Xerocom.) fide Singer (1945).

Strobilomyces Berk. (1851), Strobilomycetaceae. 9, widespr.

Strobilomycetaceae J.-E. Gilbert (1931), Boletales. 11 gen. (+ 5 syn.), 64 spp. Basidioma boletoid, stipitate; spores either brown-black and ornamented or pink-vinaceous and smooth.

Strobiloscypha N.S. Weber & Denison (1995), Sarcosomataceae. 1, USA.

Strobilurus Singer (1962), Tricholomataceae. 8, temp. See Redhead (*CJB* **58**: 68, 1980).

stroma, (pl. **stromata**), a mass or matrix of vegetative hyphae, with or without tissue of the host or substrate, sometimes sclerotium-like in form, in or on which spores or fruit bodies bearing spores are produced. Many ascomycetes (esp. *Xylariales*) and Mitosporic fungi have stromata; a few *Uredinales* and other fungi. **ecto-** (**epi-**, Fuisting), a -, normally conidial, formed in the periderm and frequently breaking through the bark; **endo-** (**ento-**) (**hypho-**, Fuisting), a perithecial - formed under the **ecto-**; **eu-**, one of fungal tissue only (Fig. 10P-R); **pseudo-** see under pseudostroma. Ruhland's name for the ostiolar disc is **placodium**, forms (as in *Diatrype*) from the **endo**-being **ento-placodial**, those (at least in part) from the **ecto-** being **ecto-placodial**. A species having ecto- and endo- is **diplostromatic**; one with only one **haplostromatic**. See Miller (*Mycol.* **20**: 188, 1928). Cf. sclerotium.

Stromaster Höhn. (1930), Phyllachoraceae. 1, S.Am. See Petrak (*Sydowia* **5**: 354, 1951).

Stromatella Henssen (1989), Lichinaceae (L). 1, Bermuda.

Stromateria Corda (1837) = Tubercularia (Mitosp. fungi.) fide Saccardo (*Syll. Fung.* **4**: 646, 1886).

Stromatinia (Boud.) Boud. (1907), Sclerotiniaceae. 6, Eur., N. Am. See Schumacher (*Agarica* **5**: 111, 1984).

Stromatocrea W.B. Cooke (1952), Anamorphic Hypocreaceae, 3.A1.15. Teleomorph *Selinia*. 1, USA. See Petrak (*Sydowia* **6**: 336, 1952), Cauchon & Quellette (*Mycol.* **56**: 453, 1964).

Stromatocyphaceae Jülich (1982), Schizophyllales. 2 gen. (1 syn.), 4 spp. Basidioma resupinate; hymenophore forming aggregated cupules.

Stromatocyphella W.B. Cooke (1961), Stromatocyphaceae. 3, N. Am.

Stromatographium Höhn. (1907), Anamorphic Trichosphaeriaceae, 2.A2.15. Teleomorph *Fluviostroma*. 1, trop. See Seifert (*CJB* **65**: 2196, 1987).

stromatolites (**lichen**), laminar calcretes formed abiotically in rock and sometimes wrongly interpreted as fossils (Klappa, *Sediment. Petrol.* **49**: 387, 1979).

Stromatoneurospora S.C. Jong & E.E. Davis (1973), Xylariaceae. 1, trop. See Rogers *et al.* (*Mycol.* **84**: 166, 1992).

Stromatopogon Zahlbr. (1897), Mitosporic fungi, 4.D1.?15. 1 (on lichen, *Usnea*), Australia. See Diederich (*Lichenologist* **24**: 371, 1992; redescr.).

Stromatopycnis A.F. Vital (1956), Mitosporic fungi, 4.A1.?. 1, Brazil.

Stromatoscypha Donk (1951) ≡ Porotheleum (Stromatoscyph.).

Stromatosphaeria Grev. (1824) nom. rej. ≡ Daldinia (Xylar.).

Stromatostilbella Samuels & E. Müll. (1980) ≡ Stromatographium (Mitosp. fungi) fide Seifert (1987).

Stromatostysanus Höhn. (1919), Mitosporic fungi, 2.A1.?. 1, Eur.

Stromatothecia D.E. Shaw & D. Hawksw. (1971), Odontotremataceae. 1 (on *Nothofagus*), Papua New Guinea.

Stromatothelium Trevis. (1861), ? Pyrenulales (inc. sed., L). 4, trop.

Stromne Clem. (1909) ≡ Engleromyces (Xylar.).

Strongwellsea Batko & Weiser (1965), Entomophthoraceae. 2, USA. See Batko & Weiser (*J. Invert. Path.* **7**: 455, 1965), Weiser & Batko (*Folia Parasit.* **13**: 144, 1966), Humber (*Mycol.* **68**: 1042, 1976; emend., key, *Mycotaxon* **15**: 167, 1982). = Erynia (Entomophthor.) fide Remaudière & Keller (1980), but see Humber (*Mycotaxon* **15**: 167, 1982).

Strongyleuma Vain. (1927) = Chaenothecopsis (Mycocalic.) fide Tibell (1984).

Strongylium Ditmar (1809) = Reticularia (Lycogal.).

Strongylium (Ach.) Gray (1821) = Calicium (Calic.) fide Tibell (1984).

Strongylopsis Vain. (1927) = Microcalicium (Mirocalic.) fide Tibell (1978).

Strongylothallus Bat. & Cif. (1959), Mitosporic fungi, 5.-.-. 1, Brazil.

Stropharia (Fr.) Quél. (1872), Strophariaceae. 15, widespr. The edible *S. rugoso-annulata* ('strophaire') cultivated in Eur. See Orton (*Notes R. bot. Gdn Edinb.* **35**: 148, 1976; key Br. spp.).

Strophariaceae Singer & A.H. Sm. (1946), Agaricales. 17 gen. (+ 16 syn.), 224 spp. Pileipellis never epithelial; spores cinnamon-brown, fuscous to lilaceous-black, with a germ-pore.

Lit: Watling & Gregory (*British fungus flora* **5**, 1987).

Stropholoma (Singer) Balletto (1989), Strophariaceae. 3, Eur.

Strossmayeria Schulzer (1881), Leotiaceae. 16, widespr. See Iturriaga & Korf (*Mycotaxon* **36**: 119, 1990; monogr.). Anamorph *Pseudospiropes*.

Strumella Fr. (1849), Anamorphic Sarcosomataceae, 3.A2.?. Teleomorph *Urnula*. 8, widespr. *S. coryneoidea*, canker of oak (*Quercus*) and sometimes of other trees. See Wolf (*Mycol.* **50**: 837, 1959).

Strumellopsis Höhn. (1909), Mitosporic fungi, 8.A2.?. 2, Java.

Stuartella Fabre (1879), Dothideales (inc. sed.). 3, Eur., N. Am. See LaFlamme & Müller (*Sydowia* **29**: 278, 1977), Funk & Shoemaker (*CJB* **61**: 2277, 1983). Anamorph *Bactrodesmium*.

stuffed (of a stipe), having the inside of a different structure from that of the outer layer (Fig. 4F).

stupose, of tissue formed from hyphae which are not gelatinized.

Stygiomyces Coppins & S.Y. Kondr. (1995), Mitosporic fungi, 4.E1.15. 1 (on *Pseudocyphellaria*), Tasmania.

Stylaspergillus B. Sutton, Alcorn & P.J. Fisher (1982), Mitosporic fungi, 1.E1.15. 1, widespr. Synanamorph Parasympodiella.

Stylina Syd. (1921), Graphiolaceae. 1 (on *Livistona*), China.

Stylobates Fr. (1837), Agaricaceae. 2, trop.

Stylodothis Arx & E. Müll. (1975), Dothideaceae. 2, widespr.

Styloletendraea Weese (1924) nom. nud. = Nectria (Hypocr.) anamorph.

Stylonectria Höhn. (1915) = Nectria (Hypocr.) fide Booth (*Mycol. Pap.* **73**, 1959).

Stylonectriella Höhn. (1915) ? = Nectria (Hypocr.) fide Sutton (*Mycol. Pap.* **141**, 1977).

Stylonites Fr. (1848) = Physarum (Physar.) fide Martin (1966).

Stylopage Drechsler (1935), Zoopagaceae. 15, N. Am., Kenya, Eur. See Drechsler (*Mycol.* **27**: 197, 206, 1935, **28**: 241, 1936, **30**: 137, 1938, **31**: 388, 1939, **37**: 1, 1945, **40**: 85, 1948), Dayal (1973; key), Duddington (*Ann. Bot.* **17**: 127, 1953, *Mycol.* **47**: 245, 1955), Peach & Juniper (*TBMS* **38**: 431, 1955), Wood (*TBMS* **80**: 368, 1983), Saikawa (*Mycol.* **78**: 309, 1986; ultrastr.), Blackwell & Malloch (*Mycol.* **83**: 360, 1991; life history).

stylospore, (1) a spore on a pedicel or hypha, esp. a urediniospore (obsol.); (2) an elongated pycnidiospore (obsol.); (3) the sporangiolum (the 'Stielgemmen' of Linnemann, 1941) of *Mortierella*.

Stypella A. Møller (1895), Exidiaceae. 2, Am., Eur. See Martin (*Stud. nat. Hist. Univ. Iowa* **16**: 143, 1934), Donk (*Persoonia* **4**: 241, 1966).

Stypinella J. Schröt. (1888) ≡ Helicobasidium (Platygl.).

Stysanopsis Ferraris (1909) = Cephalotrichum (Mitosp. fungi) fide Hughes (1958).

Stysanus Corda (1837) = Cephalotrichum (Mitosp. fungi) fide Hughes (1958).

suaveolent, having a sweet smell.

sub- (prefix), under; below; frequently in the sense of approximating to the condition qualified, slightly, somewhat.

Subbaromyces Hesselt. (1953), Ophiostomataceae. 2 (on filter beds), N. Am., India. See Cole *et al.* (*CJB* **52**: 2453, 1974; dev., ultrastr.), Malloch & Blackwell (*CJB* **68**: 1712, 1990).

subcentric, see centric.

subcutis, see cutis.

subglobose, not quite spherical (Fig. 37.1).

subhymenium, generative tissue below the hymenium; sometimes used as equivalent to medullary exciple or hypothecium.

Subicularium M.L. Farr & Goos (1989), Mitosporic fungi, 3.D2.1. 1, Venezuela. See also Eriksson (*SA* **8**: 77, 1989).

Subiculicola Speg. (1924) = Melioliphila (Tubeuf.) fide Rossmann (1987).

subiculum (**subicule**), a net-, wool-, or crust-like growth of mycelium under fruit-bodies.

suboperculate, see ascus.

Subramanella H.C. Srivast. (1962) = Phomopsis (Mitosp. fungi) fide Sutton (1977).

Subramania D. Rao & P. Rag. Rao (1964), Mitosporic fungi, 1.A2.3. 1, India.

Subramaniomyces Varghese & V.G. Rao (1980), Mitosporic fungi, 1.A2.3. 2, India.

Subramaniula Arx (1985), Chaetomiaceae. 2, S. Afr., India. See Cannon (*TBMS* **86**: 56, 1986; key).

substrate (**substratum**), although these two terms are frequently treated as synonyms by mycologists both have useful special senses: (1) **substrate**: (in enzymology) is applied to the substance on which an enzyme acts and in microbiology to the substances (e.g. culture medium constituents) utilized by a microorganism for growth in distinction to the material; (2) **substratum** (in ecology), the material on which an organism is growing or to which it is attached.

substroma, pseudostroma (q.v.) in which the vegetative hyphae of the host predominate (Johnston, *Ann. Mo. bot. Gdn* **27**: 31, 1940).

Subulariella Höhn. (1915) nom. dub. fide Sutton (*Mycol. Pap.* **141**, 1977).

subulate, slender and tapering to a point; awl-shaped (Fig. 37.34).

Subulicium Hjortstam & Ryvarden (1979), Hyphodermataceae. 3, N. temp.

Subulicystidium Parmasto (1968), Hyphodermataceae. 2, widespr.

Subulispora Tubaki (1971), Mitosporic fungi, 1.E1.10. 9, N. temp. See Sutton (*TBMS* **71**: 167, 1978), de Hoog (*Stud. Mycol.* **26**: 54, 1985; key).

subuniversal veil, see protoblem.

Sucinaria Syd. (1925) = Coccodiella (Phyllachor.) fide Müller & v. Arx (1973).

Sufa Adans. (1763) = Lycoperdon (Lycoperd.) fide Mussat (*Syll. Fung.* **15**: 408, 1901).

sufu (Chinese cheese), an oriental food composed of *Actinomucor* or *Mucor* fermented soybeans (Hesseltine, *Mycol.* **57**: 164, 1965, *Mycologist* **5**: 162, 1991); see Fermented food and drinks.

sugar fungus, a fungus attacking decaying substances and only able to utilize simple sugars, amino acids, and other relatively simple organic compounds (Thom & Morrow, *J. Bact.* **33**: 77, 1937).

Sugiyamaemyces I.I. Tav. & Balazuc (1989), Laboulbeniaceae. 1, Borneo.

Suillellus Murrill (1909) = Boletus (Bolet.) fide Singer (1945).

Suillosporium Pouzar (1958), Botryobasidiaceae. 1, Eur., N. Am.

Suillus P. Micheli ex Adans. (1763) = Boletus (Bolet.) fide Donk (*Persoonia* **8**: 279, 1975).

Suillus Gray (1821), Boletaceae. 50 (associated with conifers), N. temp. See Smith & Thiers (*A contribution towards a monograph of North American species*

of Suillus, 1964; key), Thiers (*Beih. Nova Hedw.* **51**, 1975; key N. Am. spp., *Mycotaxon* **9**: 285, 1979; key 40 W. USA spp.), Watling (*Trans. bot. Soc. Edinb.* **40**: 100, 1965; key Br. spp., *British fungus flora* **2**, 1979), Pantidou & Groves (*CJB* **44**: 1371, 1966; culture).

Suillus P. Karst. (1882) ≡ Gyroporus (Gyrodont.).

Suillus (Haller) Kuntze (1898) = Boletus (Bolet.).

Sulcaria Bystrek (1971), Alectoriaceae (L). 3, Asia, USA. See Brodo & Hawksworth (*Op. bot. Soc. bot. Lund* **42**, 1977).

sulcate, grooved.

Sulcatisclerotes Beneš (1959), Fossil fungi (*f. cat.*). 2 (Carboniferous), former Czechoslovakia.

Sulcispora, see *Sulcospora*.

Sulcispora Shoemaker & C.E. Babc. (1989), Ascomycota (inc. sed.). 1, Eur.

Sulcospora Kohlm. & Volkm.-Kohlm. (1993) ≡ Aropsiclus (Ascomycota).

sulcus, a furrow or groove.

sulphur polypore (**sulphur shelf-mushroom**), basidioma of the edible *Laetiporus sulphureus*.

Sulphurina Pilát (1953) = Cristella (Cortic.) fide Donk (1974).

summer spore, a spore germinating without resting, frequently living only a short time; cf. resting spore.

super- (prefix), above; used in combination with the names of taxonomic categories to give additional ranks in zoology (e.g. superfamily), but not permitted under the botanical Code.

superficial, on the surface of the substratum.

superior (of an annulus), near the top of the stipe.

supine (of fructifications), closely applied to the substratum.

supra- (prefix), above; **-generic**, all taxonomic ranks above that of genus; **-specific**, all taxonomic ranks above that of species; also used at other ranks of the hierarchy; cf. infra-.

suprahilar plage (of basidiospores, esp. of *Lactarius* and *Russula*), the area above the hilar appendage on which the eusporial ornamentation is lacking or reduced (Kühner, 1926).

surculicolous (of *Exobasidium* infections), monocarpic (q.v.) and systemic in annual shoots (Nannfeldt, 1981).

surfactant, an agent which reduces the surface tension of a liquid, e.g. detergent.

suscept, a living organism which is **susceptible** to (able to be attacked by; non-immune to) a given disease, pathogen, or toxin.

suspensor, a hypha supporting a gamete, gametangium, or esp. a zygospore (Fig. 43C).

Sutravarana Subram. & Chandrash. (1977), Mitosporic fungi, 2.A1.10. 1, India.

Suttonia S. Ahmad (1961) [non A. Rich. (1832), *Myrsinaceae*] ≡ Suttoniella (Mitosp. fungi).

Suttoniella S. Ahmad (1961), Mitosporic fungi, 8.G1.16. 3, Asia, Australia. See Sutton (*TBMS* **59**: 285, 1972).

Suttonina H.C. Evans (1984), Mitosporic fungi, 6.B/C2.38. 1, Guatemala.

Svrcekia Kubička (1960) = Boudiera (Peziz.) fide Dissing (*SA* **6**: 153, 1987).

Svrcekomyces J. Moravec (1976) = Pseudombrophila (Otid.) fide van Brummelen (1995).

Swampomyces Kohlm. & Volkm.-Kohlm. (1987), Ascomycota (inc. sed.). 2, Belize, Aldabra, Australia, Papua New Guinea. See Hyde & Nakagiri (*Sydowia* **44**: 122, 1992).

swarm-cell (of *Myxomycetes* and some *Chytridiales*), a motile cell acting, before or after division, as an isogamete.

swarm-spore (**swarmer**), see zoospore.

Syamithabeeja Subram. & Natarajan (1976), Mitosporic fungi, 1.A1.18. 1, India.
Sychnoblastia Vain. (1921) = Thelopsis (Stictid.).
Sychnogonia Körb. (1855) = Muellerella (Verruc.).
Sychnogonia Trevis. (1860) [non Körb. (1855)] = Thelopsis (Stictid.).
sycosis, a fungus disease of the hair follicles; esp. of the face; ringworm of the beard.
Sydow (Paul; 1851-1925). Chief writings and exsiccati: *Die Flechten Deutschlands* (1887), Index vols (**12-13**) to Saccardo's *Sylloge*, (with Lindau) *Thesaurus literaturae mycologicae* (1908-1918), exsiccati of rusts (1888-), smuts (1894-), (with **Hans Sydow** (employed at the Dresden Bank, Berlin), his son, 1879-1946) the exsiccati *Mycotheca germanica* (1903-), *Monographia Uredinearum* (1902-24), and the *Annales Mycologici* (1903-1944). See Grummann (1974: 49), Stafleu & Cowan (*TL-2* **6**: 132, 129, 1986).
Sydowia Bres. (1895), Dothioraceae. 5, Eur., N. Am. See Barr (*Contr. Univ. Mich. Herb.* **9**: 523, 1972), Froidevaux (*Nova Hedw.* **23**: 679, 1973), Funk & Finck (*CJB* **66**: 212, 1988), v. Arx & Müller (1975). Anamorph *Sclerophoma*.
Sydowiella Petr. (1923), Melanconidaceae. 2, Eur., N. Am.
Sydowiellina Bat. & I.H. Lima (1959) = Myriangiella (Schizothyr.) fide v. Arx & Müller (1975).
Sydowina Petr. (1923) = Lojkania (Fenestell.) fide Barr (1984).
Sydowinula Petr. (1923) = Acanthonitschkea (Nitschk.) fide Nannfeldt (1975).
Sylviacollaea Cif. (1963), Mitosporic fungi, 8.A1.15. 1 (on termites), Dominican Republic.
sym-, see **syn-**.
symbiosis, associations between unlike organisms, generally ones persisting for long periods (relative to the generation time of the interacting organisms); apparently first used by the lichenologist A.B. Frank in 1877 (often credited to de Bary, 1879) who later coined the word mycorrhiza (q.v.). At times equated with mutualism (q.v.), but correctly also covering parasitic (harmful) and commensalistic (unharmful) associations (Ahmadjian & Paracer, 1987).

There has been debate as to which partner might be regarded as host in mutualistic symbioses (e.g. de Bary regarded the algae in lichens as the host, and Douglas, 1994, the fungus as host to the algae); in order to circumvent this controversy, Law & Lewis (1983) used the neutral terms **exhabitant** (the organism forming the outer tissues) and **inhabitant** (the enclosed organism).

In lichens, Poelt (*Abstr. IMC2*, 1977) recognized: **two-membered -**, one alga + one fungus, **three-membered -**, one alga + two fungi (lichenicolous fungi; q.v.), or two algae + one fungus (cephalodium; q.v.); and **four-membered -**, two algae + two fungi (lichenicolous lichens); Hawksworth (*Bot. J. Linn. Soc.* **96**: 3, 1988) used '-**biont**' (q.v.) instead and discussed further fungal/algal interactions (see also Rambold & Triebel, *Bibl. Lich.* **48**, 1992).

Lit.: Ahmadjian & Paracer (*Symbiosis, an introduction to biological associations*, 1986), Carroll (in Carroll & Wicklow, *The fungal community*, edn 2 : 327, 1992; fungal mutualisms), Cook *et al.* (Eds) (*Symp. Soc. exp. Biol.* **29**, 1975), Cooke (*The biology of symbiotic fungi*, 1977), Cook *et al.* (Eds) (*Cellular interactions in symbiosis and parasitism*, 1980), Douglas (*Symbiotic interactions*, 1994), Law & Lewis (*Biol. J. Linn. Soc.* **20**: 249, 1983), Lewis (*Biol. Rev.* **48**: 261, 1973; *in* Rayner *et al.* (Eds), *Evolutionary biology of the fungi*: 161, 1987), Margulis & Fester (Eds) (*Symbiosis as a source of evolutionary innovation*, 1991), Sapp (*Evolution by association: a history of symbiosis*, 1984). Cf. Antagonism, Coevolution, Endophyte, Lichens, Mycorrhiza.

Symbiotaphrina Kühlw. & Jurzitza ex W. Gams & Arx (1980), Mitosporic fungi, 1.A1.?. 2 (beetle mycetomas), Germany.
sympatric, occurring in the same geographical region. Cf. allopatric.
Symperidium Klotzsch (1843) = Aecidium (Anamorphic Uredinales) fide Saccardo (*Syll. Fung.* **18**: 829, 1906).
Symphaeophyma Speg. (1912), Parmulariaceae. 1, S. Am.
Symphaster Theiss. & Syd. (1915), Asterinaceae. 2, S. Am., S. Afr.
symphogenous, see meristogenous.
Symphyosira Preuss (1853), Anamorphic Leotiaceae, 2.C1.39. Teleomorph *Symphyosirinia*. 3, Eur.
Symphyosirinia E.A. Ellis (1956), Leotiaceae. 3, Eur., N. Am. See Barr (*SA* **13**: 113, 1994; posn, *Z. f. Mykol.* **60**: 211, 1994; key). Anamorph *Symphyosira*.
Symphysos Bat. & Cavalc. (1967), Mitosporic fungi, 5.A1.?. 1, Brazil.
Symphytocarpus Ing & Nann.-Bremek. (1967), Stemonitidaceae. 7, N. temp.
symplastic, entering living cells; cf. apoplastic.
Symplectromyces Thaxt. (1908), Laboulbeniaceae. 1, N. Am., Eur., Asia. See Benjamin (*Aliso* **10**: 345, 1983).
symplesiomorphy (**-ies**), see Cladistics.
Symplocia A. Massal. (1854) nom. rej. ≡ Crocynia (Crocyn.).
Sympodia (Heim) W.B. Cooke (1952) nom. inval. = Marasmius (Tricholomat.) fide Kirk (*in litt.*).
sympodial (of conidiogenous cells), characterized by continued growth, after the main axis has produced a terminal spore, by the development of a succession of apices each of which originates below and to one side of the previous apex.
Sympodiella W.B. Kendr. (1958), Mitosporic fungi, 1.A1.39. 3, UK, Cuba.
Sympodina Subram. & Lodha (1964) = Veronaea (Mitosp. fungi) fide v. Arx (*Genera of fungi sporulating in pure culture*, 1970).
Sympodiocladium Descals (1982), Mitosporic fungi, 1.A1.1. 1 (aquatic), UK.
sympodioconidium (**sympodulospore**), a spore produced on a sympodula.
Sympodiomyces Fell & Statzell (1971), ? Saccharomycetaceae. 1 (marine), S. Hemisph.
Sympodiomycopsis Sugiy., Tokuoka & Komag. (1991), Mitosporic fungi, 1.A1.21. 1 (from nectar), Japan.
Sympodiophora G.R.W. Arnold (1970) = Pseudohansfordia (Mitosp. fungi) fide de Hoog (1978).
Sympodomyces R.K. Benj. (1973), Laboulbeniaceae. 1, Papua New Guinea.
sympodula, a sympodial conidiogenous cell.
syn- (**sym-**) (in compounds), growing together; adhesion; aggregation.
Synalissa Fr. (1825), Lichinaceae (L). 5, Eur., N. Am. See Forssell (*Nova Acta reg. Soc. sci. Upsal.*, ser. 3, **13**(6), 1885).
Synalissina Nyl. (1885) = Lempholemma (Lichin.) fide Zahlbruckner (*Nat. Pflanzenfam.* **8**: 167, 1926).
Synallisopsis Nyl. ex Stizenb. (1882) nom. inval. = Pyrenopsis (Lichin.).
synanamorph, see States of fungi.
Synandromyces Thaxt. (1912), Laboulbeniaceae. 9, Am., Afr., Sumatra to Fiji.
Synaphia Nees & T. Nees ex Rchb. (1841) nom. dub. (Fungi, inc. sed.).
synapomorphy (**-ies**), see Cladistics.

Synaptomyces Thaxt. (1912), Ceratomycetaceae. 1, S. Am.

synaptonemal complex, proteinaceous, longitudinally aligned structure which usually unites homologous chromosomes during the prophase of meiosis.

Synaptospora Cain (1957), ? Coniochaetaceae. 1, Eur. See Barr (*Mycotaxon* **39**: 43, 1990; posn), Matzer (*Crypt. Mycol.* **14**:11, 1993).

Synarthonia Müll. Arg. (1891), Arthoniales (inc. sed., L). 1, C. Am.

Synarthoniomyces Cif. & Tomas. (1953) ≡ Synarthonia (Arthoniales).

Synascomycetes. Class (Gäumann, *Die Pilze*, 1949) for *Ascomycota* with spores formed in a spore sac ('synascus', q.v.), interpreted as a group of fused asci; included fungi now in *Eurotiales* (*Ascosphaera*) and *Protomycetales*.

synascus, the gametangium of *Ascosphaera* (Varitchak, 1933).

Syncarpella Theiss. & Syd. (1915), Cucurbitariaceae. 6, widespr. See Barr & Boise (*Mem. N.Y. bot. Gdn* **49**: 298, 1989).

Syncephalastraceae Naumov ex R.K. Benj. (1959), Mucorales. 1 gen., 2 spp. Merosporangia present; zygospores warty, borne on opposed suspensors.

Syncephalastrum J. Schröt. (1886), Syncephalastraceae. 2, widespr. See Benjamin (*Aliso* **4**: 321, 1959), Misra (*Mycotaxon* **3**: 51, 1975), Hobot & Gull (*Protoplasma* **107**: 339, 1981; germination), Schipper & Stalpers (*Persoonia* **12**: 81, 1983), Zheng *et al.* (*Mycosystema* **1**: 39, 1988; key), Chen & Huang (*Mycosystema* **1**: 53, 1988; electrophoresis).

Syncephalidium Badura (1963) = Syncephalis (Piptocephalid.) fide Benjamin (1966).

Syncephalis Tiegh. & G. Le Monn. (1873), Piptocephalidaceae. *c.* 30 (mycoparasites of *Mucorales*), widespr. See Benjamin (*Aliso* **4**: 321, 1959), Zycha *et al.* (1969; key), Kuzuha (*Trans. Mycol. Soc. Japan* **14**: 237, 1973, *J. Jap. Bot.* **55**: 343, 1980; zygospore formation), Hunter & Butler (*Mycol.* **67**: 863, 1975), Baker *et al.* (*Mycol.* **69**: 1008, 1977; host range & culture), Bawcutt (*TBMS* **80**: 219, 1983), Benjamin (*Aliso* **11**: 1, 1985), Can *et al.* (*Nova Hedw.* **49**: 427, 1989), Gruhn & Petzold (*Can. J. Microbiol.* **37**: 355, 1991).

Syncephalopsis Boedijn (1959) = Syncephalis (Piptocephalid.) fide Benjamin (1966).

Syncesia Taylor (1836), Roccellaceae (L). 20, Eur., Afr., N., C. & S. Am., Indonesia, Australia. See Tehler (*Crypt. Bot.* **3**: 139, 1993).

Synchaetophagus Apstein (1911) Fungi (inc. sed.) fide Vishniac (*Mycol.* **50**: 74, 1958).

synchronized culture, a culture manipulated so that division of all the component cells is simultaneous; see Williamson & Scopes (*Symp. Soc. gen. Microbiol.* **11**: 217, 1961; *Saccharomyces cerevisiae*).

Synchronoblastia Uecker & F.L. Caruso (1988), Mitosporic fungi, 4.A1.6. 1, USA.

synchronospore, a spore produced simultaneously with other neighbouring spores.

Synchytriaceae J. Schröt. (1892), Chytridiales. 5 gen. (+ 5 syn.), 127 spp. Thallus holocarpic, exogenous to zoospore cyst, at maturity developing into a sorus, prosorus or resting spore.

Synchytrium de Bary & Woronin (1863), Synchytriaceae. 121 (plant parasites), cosmop. *S. aureum* reported from 198 host spp. (123 gen., 34 fam.); Karling experimentally infected 1465 spp. in more than 918 gen. of 176 fam. with *S. macrosporum*, *S. endobioticum* (potato wart disease). See Karling (*Synchytrium*, 1964; keys, *Adv. Frontier Pl. Sci.* **29**: 1, 1972).

Syncladium Rabenh. (1859), ? Mitosporic fungi, ?4.?.?. 1, Sri Lanka, Cambodia, France. See Hughes (*Mycol.* **68**: 693, 1976).

Syncleistostroma Subram. (1972) nom. conf. fide Malloch & Cain (*CJB* **51**: 1647, 1973); = Petromyces (Trichocom.) fide v. Arx (1981).

Syncoelium Wallr. (1833) ? Algae fide Hughes (*CJB* **36**: 747, 1958).

Syncollesia Nees (1823) ? = Fumago (Mitosp. fungi).

Syncomista Nieuwl. (1916) ≡ Toninia (Catillar.).

syncytium, see coenocyte.

Syndiplodia Peyronel (1915) = Microdiplodia (Mitosp. fungi) fide Sutton (1977).

syndrome, a complex of symptoms, especially that which constitutes the picture of a disease.

Synechoblastus Trevis. (1853) = Collema (Collemat.).

synecology, ecology of communities.

synergism, two organisms or environmental factors acting simultaneously to effect a change greater than either could alone; e.g. the increase of fungicidal value in certain mixtures of fungicides, fungicides and non-toxic materials, or of air pollutants. Cf. metabiosis.

syngamy, fertilization; the fusion of male and female cells to form a zygote.

Syngenosorus Trevis. (1860) = Tomasellia (Arthopyren.).

Syngliocladium Petch (1932), Mitosporic fungi, 2.A1.?. 3 or 4, N. temp. See Petch (*TBMS* **25**: 262, 1942).

Synglonium Penz. & Sacc. (1897) ? = Nymanomyces (Rhytismat.) fide Höhnel (*Ann. Myc.* **16**: 154, 1918).

synisonym, one of two or more names having the same basionym (Donk, *Persoonia* **1**: 175, 1960).

synkaryon, a diploid zygote nucleus.

synkaryotic (of a nucleus), having $2n$ chromosomes.

synnema, (pl. **synnemata**), a conidioma composed of a more or less compacted group of erect and sometimes fused conidiophores bearing conidia at the apex only or on both apex and sides. Seifert (*Stud. Mycol.* **27**: 1, 1985) distinguished 3 types: **determinate -**, with a terminal, non-elongated conidiogenous zone, growth ceasing after sporulation has begun, e.g. *Stilbella*; **indeterminate -**, with an elongated fertile zone, sometimes covering the whole conidioma, growth continuing after sporulation, e.g. *Doratomyces*; **compound -**, branched in which determinate or indeterminate branches are formed on a branched or unbranched axis, e.g. *Tilachlidiopsis*. Anatomical stipe types found in the three groups include **parallel -**, of primarily parallel hyphae; **intricate -**, of primarily or entirely textura intricata; **basistromatic -**, with well-defined basal stromata; **amphistromatic -**, well-defined basal stroma, stipe of parallel hyphae and an apical dome of textura angularis to globulosa with conidiogenous cells; **cupulate -**, conidiogenous zone concave. See Fig. 38.

synnema coremium, sometimes used for synnemata with looser fascicles as in 'coremioid' spp. of *Penicillium* and *Aspergillus* (obsol.).

Synnemadiplodia Zambett. (1955) = Botryodiplodia (Mitosp. fungi) fide Petrak (*Sydowia* **16**: 353, 1963).

Synnematium Speare (1920) = Hirsutella (Mitosp. fungi) fide Evans & Samson (*TBMS* **79**: 431, 1982).

Synnematomyces Kobayasi (1981), Mitosporic fungi, 2.A1.?. 1, Japan.

Synnematomycetes, mitosporic fungi having synnemata (Höhnel, 1923); *Coremiales* (Potebnia, 1909), (obsol.).

synnematous, (**synnematogenous**), having synnemata. See Seifert (*Stud. Mycol.* **27**: 1, 1985, *Stilbella* and allies, *Mem. N.Y. bot. Gdn* **59**: 109, 1990, keys).

Synnemellisia N.K. Rao, Manohar. & Goos (1989), Mitosporic fungi, 2.A1.15. 1, India.

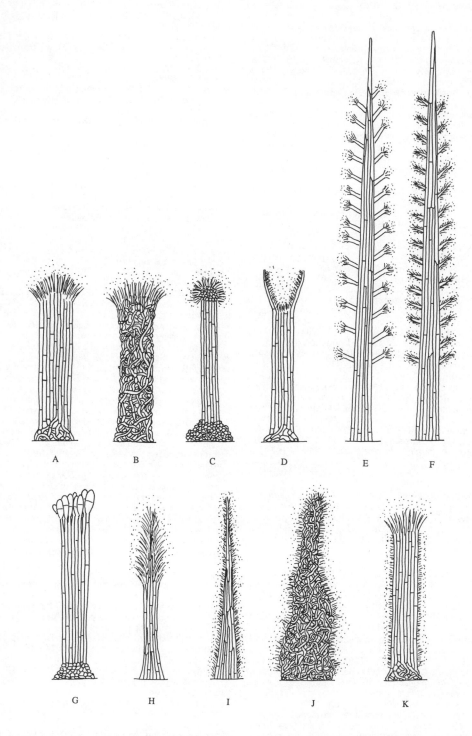

Fig. 38. Types of synnema. A-D, G, determinate synnemata. A, parallel; B, intricate; C, amphistromatic; D, cupulate; G, basistromatic. E-F, compound. E, indeterminate axis with determinate branches; F, indeterminate axis with indeterminate branches. H-J, indeterminate. H,I, parallel; J, intricate. L, determinate with A-conidia or indeterminate with B-conidia. After Seifert (*Stud. Mycol.* **27**, 1985).

Synomyces Arthur (1924) = Coleosporium (Coleospor.) fide Dietel (1928).

synonym, another name for a species or group, esp. a later or illegitimate name not currently employed for the taxon. If two or more names are based on the same type they are **homotypic** (**nomenclatural**, 'obligate', ≡) synonyms, if on different types they may be **heterotypic** (**taxonomic**, 'facultative', =) synonyms. See Nomenclature, and cf. basionym, homonym, orthographic synisonym, variant.

Synostomella Syd. (1927) = Cyclotheca (Microthyr.) fide Müller & v. Arx (1962).

Synostomina Petr. (1949), Mitosporic fungi, 8.A1.?. 1, S. Am.

Synpeltis Syd. & P. Syd. (1917) = Cyclotheca (Microthyr.) fide Müller & v. Arx (1962).

Synpenicillium Costantin (1888) = Doratomyces (Mitosp. fungi) fide Morton & Smith (1963).

Synphragmidium F. Strauss (1853) = Speira (Mitosp. fungi) fide Saccardo (*Syll. Fung.* **4**: 514, 1886).

Synsphaeria Bonord. (1851), Ascomycota (inc. sed.). 1, Eur.

Synsporium Preuss (1849) = Stachybotrys (Mitosp. fungi) fide Hughes (1958).

Synsterigmatocystis Costantin (1888) = Gibellula (Mitosp. fungi).

Syntexis Theiss. (1916) ? = Mollisia (Dermat.). See Petrak (*Ann. Myc.* **26**: 399, 1928).

Synthetospora Morgan (1892) = Stephanoma (Mitosp. fungi) fide Mattirolo.

syntype, see type.

synzoospore, multinucleate zoospore with many sets of flagella.

Syphosphaera Dumort. (1822), Ascomycota (inc. sed.). 1, Eur.

Syrigosis Neck. ex Kremp. (1869) = Sphaerophorus (Sphaerophor.).

Syringospora Quinq. (1868) nom. rej. = Candida (Mitosp. fungi). Retained in *Tulasnellales* by v. d. Walt (*Mycopathol.* **40**: 231, 1970), v. Arx *et al.* (*Stud. Mycol.* **14**, 1977).

Syrropeltis Bat., J.L. Bezerra & Matta (1964), ? Dothideales (inc. sed.). 1 (on *Xylopia*), Brazil.

syrrotium (pl. **-ia**), term coined by Falk (1912) for the mycelial cord of *Merulius*. See Thompson (*in* Jennings & Rayner (Eds), *The ecology and physiology of the fungal mycelium*: 185, 1984).

Syspastospora P.F. Cannon & D. Hawksw. (1982), Ceratostomataceae. 2 (on fungi), widespr. See Horie *et al.* (*Mycotaxon* **25**: 229, 1986).

systematics, (1) in biology, **biosystematics** (q.v.) the study of the relationships and classification of organisms and the processes by which it has evolved and by which it is maintained (includes the subdisciplines of nomenclature and taxonomy), see Hawksworth (Ed.) (*Prospects in systematics*, 1988), Minelli (*Biological systematics: the state of the art*, 1993), Ross (*Biological systematics*, 1974), Stevens (*The development of biological systematics*, 1995); (2) (in computing), a language for the systems analyst (Grindley, *Systematics: a new approach to systems analysis*, 1975); (3) **phylogenetic -**, see Cladistics. See also Classification, Nomenclature.

systemic, (1) (of a parasite), spreading throughout the host; (2) (of a fungicide), absorbed, esp. by the roots, and translocated to other parts of the plant.

Systremma Theiss. & Syd. (1915) = Dothidea (Dothid.) fide v. Arx & Müller (1975).

Systremmopsis Petr. (1923), Mitosporic fungi, 8.A1.?. 1, Eur.

Syzygangia M.W. Dick (1995), Myzocytiopsidaceae. 8 (intracellular parasites of freshwater algae), widespr.

Syzygites Ehrenb. (1818), Mucoraceae. 1 (on decaying *Boletaceae* (esp. *Boletus*), N. temp. See Hesseltine (*Lloydia* **20**: 228, 1957), Kaplan & Goos (*Mycol.* **74**: 684, 1982; zygospore formation), Ekpo & Young (*Microbios Lett.* **10**: 63, 1979; ultrastr.).

Syzygospora G.W. Martin (1937) Syzygosporaceae. 11, N. temp. See Donk (*Taxon* **11**: 101, 1962; ? nom. anam.), Ginns (*Mycol.* **78**: 619, 1986; key).

Syzygosporaceae Jülich (1982), Tremellales. 1 gen. (2 syn.), 11 spp. Basidioma pustulate or inducing host galls; basidiospores germinating by budding; mycoparasitic on agarics, corticioids and *Leotiales*.

Szczawinskia A. Funk (1984), Mitosporic fungi, 8.E1.15. 1 (L), Canada.

T, see RIEC.

T-2 toxin, a mycotoxin of the trichothecene group (q.v.) produced by some *Fusarium* spp., causing alimentary toxic aleukia.

Tabanomyces Couch, R.V. Andrejeva, Laird & Nolan (1979) = Meristacrum (Entomophthor.) fide Tucker (1981).

Tachaphantium Bref. (1888) = Platygloea (Platygl.) fide Bandoni (1956).

Taeniella L. Léger & Duboscq (1911), Eccrinaceae. 1 (in *Decapoda*), widespr. See Hibbets (*Syesis* **11**: 213, 1978), Lichtwardt (1986).

Taeniellopsis R. Poiss. (1927), Eccrinaceae. 3 (in *Amphipoda*), Eur. See Manier (*Ann. Sci. Nat., Bot. sér.* 12, **10**: 565, 1969), Lichtwardt (1986; key).

Taeniola Bonord. (1851) = Hormiscium (Mitosp. fungi) fide Saccardo (*Syll. Fung.* **4**: 263, 1886); nom. illegit. fide Hughes (1958).

Taeniolella S. Hughes (1958), Anamorphic Mytilinidiaceae, 1.C2.3/4. Teleomorph *Glyphium. c.* 25, widespr. See Ellis (*MDH*), Hawksworth (*Bull. Br. Mus. nat. Hist., Bot.* **6**: 183, 1979; on lichens), Sutton (*TBMS* **54**: 255, 1970; teleomorphs).

Taeniolina M.B. Ellis (1976), Mitosporic fungi, 1.C2.3. 1, Eur., Cuba.

Taeniophora P. Karst. (1886) = Phragmotrichum (Mitosp. fungi) fide Sutton & Pirozynski (*TBMS* **48**: 349, 1965).

Taeniospora Marvanová (1977), Anamorphic Atheliaceae, 1.G1.1. Teleomorph *Fibulomyces*. 2 (with clamp connexions), Czech Republic. See Marvanová & Stalpers (*TBMS* **89**: 489, 1987; key, teleomorphs).

take-all, a cereal disease (*Gaeumannomyces graminis*).

Talaromyces C.R. Benj. (1955), Trichocomaceae. 24, widespr. See Pitt (*The genus Penicillium*, 1980; keys), Frisvad *et al.* (*Ant. v. Leeuwenhoek* **57**: 179, 1990; chemotaxonomy). Anamorphs *Penicillium, Paecilomyces*.

Talekpea Lunghini & Rambelli (1979), Mitosporic fungi, 1.B2.19. 1, Ivory Coast.

Tamnidium, see *Thamnidium*.

Tandonea, see *Tandonia*.

Tandonella S.S. Prasad & R.A.B. Verma (1970), Mitosporic fungi, 2.C2.3. 2, India, Australia. See Sutton & Pascoe (*Austr. Jl Bot.* **35**: 183, 1987; on *Olearia*).

Tandonia M.D. Mehrotra (1991) [non *Tandonia* Moq. (1849), *Basellaceae*], Mitosporic fungi, 8.A1.15. 1, India.

Tanglella Höhn. (1918), ? Leotiales (inc. sed.). 1, Eur.

Tania Egea, Torrente & Sipman (1995) Arthoniales (inc. sed., L). 1, Malaysia.

tao-cho (tao-si), see hamanatto.

tapé, an Indonesian fermented food prepared by the action of *Rhizopus oryzae* and *Endomyces chodatii* on rice (Swan Djien Ko, *Appl. Microbiol.* **23**: 976, 1972);

- **ketala** (peuyeum), a Javanese food obtained by the fermentation of cassava tubers with *Mucor javanicus* (Hedger, *Bull. BMS* **12**: 54, 1978).

tape, see Fermented food and drinks.

Tapeinosporium Bonord. (1853) = Septocylindrium (Mitosp. fungi) fide Saccardo (*Syll. Fung.* **4**: 226, 1886).

Tapellaria Müll. Arg. (1890), ? Ectolechiaceae (L). 8, trop. See Santesson (*Symb. bot. upsal.* **12** (1), 1952).

Tapellariopsis Lücking (1992) nom. inval. Ectolechiaceae (L). 1, Costa Rica.

Tapesia (Pers.) Fuckel (1870) nom. rej. = Mollisia (Dermat.) fide Baral & Krieglsteiner (*Beih. Z. Mykol.* **6**, 1985). See also Baral (*SA* **13**: 113, 1994).

Tapesina Lambotte (1887), Hyaloscyphaceae. 1, Eur. See Höhnel (*Sber. Akad. wiss. Wien* Abt. 1 **130**: 110, 1923), Svrček (*Česká Myk.* **41**: 193, 1987).

Taphria Fr. (1821) = Taphrina (Taphrin.) fide Fries (*Syst. mycol.* **3**: 520, 1832).

Taphridium Lagerh. & Juel (1902), Protomycetaceae. 2, widespr. See Reddy (*Mycotaxon* **3**: 1, 1975; key).

Taphrina Fr. (1815), Taphrinaceae. 95, widespr. *T. populina* (syn. *T. aurea*) (poplar leaf blister), *T. betulina* (witches' broom of birch (*Betula*)), *T. bullata* (pear leaf blister), *T. cerasi* (witches' broom of cherry), *T. caerulascens* (oak leaf curl), *T. deformans* (peach leaf curl), *T. insititiae* (witches' broom of plum), *T. minor* (cherry leaf curl), *T. pruni* (pocket plums). See Mix (*Kansas Univ. Sci. Bull.* **33**: 1, 1949), Sałata (*Flora Polska, Grzyby (Mycota)* **6**, 1974), Kramer (*Mycol.* **52**: 295, 1960; dev. and nuclear behaviour), Snider & Kramer (*Mycol.* **66**: 743, 754, 1974; numerical taxonomy). Anamorph *Lalaria*.

Taphrinaceae Gäum. (1928) nom. cons., Taphrinales. 1 gen. (+ 7 syn.), 95 spp. Mycelium subcuticular or subepidermal, composed of dikaryotic ascogenous cells, vegetative tissue ± lacking; ascomata absent; interascal tissue absent; asci formed either directly from ascogenous cells or with a separating stalk cell, forming a palisade on the surface of the host tissue, ± cylindrical, the end often truncate, ± persistent, usually 8-spored, the ascospores discharged simultaneously; ascospores hyaline, aseptate, globose or ellipsoidal. Anamorph yeast-like, monokaryotic, formed from budding ascospores. Biotrophic on plants, usually causing hyperplasia (galls, witches' brooms) or lesions.
Lit.: see under *Taphrina*.

Taphrinales, Ascomycota. 1 fam., 1 gen. (+ 7 syn.), 95 spp. Fam.:
Taphrinaceae.
Cell walls are two-layered and conidiogenesis (ascospore budding) is basidiomycete-like. Has similarities with the red yeasts, *Exobasidiales* and *Ustilaginales*; included in the *Ustomycetes* by v. Arx (1981), but in *Ascomycota* class *Endomycetes* by v. Arx *et al.* (*Mycol.* **74**: 285, 1982; TEM cell walls etc.). Perhaps related to the *Protomycetales* (q.v.).
Lit.: Kramer (*in* Ainsworth *et al.* (Eds), *The Fungi* **4A**: 33, 1973; keys gen., *Stud. Mycol.* **30**: 151, 1987), v. Arx *et al.* (1982; relationships, ultrastr., wall chemistry).

Taphrophila Scheuer (1988), Tubeufiaceae. 2, Eur. See Scheuer (*MR* **95**: 811, 1991; key). Anamorph *Mirandina*.

Tapinella J.-E. Gilbert (1931) ≡ Paxillus (Paxill.).

Tapinia (Fr.) P. Karst. (1879) [non Steud. (1841), *Anacardiaceae*] ≡ Paxillus (Paxill.).

tapuy, a rice wine indigenous to the Philippines; similar to saki.

Tarbertia Dennis (1974), Arthoniales (inc. sed.). 1, UK.

Tarichium Cohn (1875), Entomophthoraceae. 29, widespr. See MacLeod & Müller-Kögler (*Mycol.* **62**: 33, 1970; key), Humber (1989).

Tarsodisporus Bat. & A.A. Silva (1965), Mitosporic fungi, 5.A2.?. 1, Brazil.

tartareous, having a thick rough crumbling surface.

Tartufa (Gray) Kuntze (1891) ≡ Choiromyces (Helvell.).

Tarzetta (Cooke) Lambotte (1888), Otideaceae. 8, N. temp. See Dumont & Korf (*Mycol.* **63**: 165, 1971; nomencl.), Harmaja (*Karstenia* **14**: 138, 1974; gen. concept), Lazzari (*Micol. ital.* **2**: 20, 1984; nomencl.), Rogers *et al.* (*Mycol.* **63**: 1084, 1971), Senn-Irlet (*Beitr. Kenntnis Pilze Mittel.* **5**: 191, 1989; key 6 spp.).

Tassia Syd. & P. Syd. (1919), Mitosporic fungi, 5.A1.?. 2, Brazil, Eur.

Taste. In addition to the use of fungi as flavouring for soups, etc., a number of fungi, esp. the fresh basidiomata of agarics, have characteristic tastes which are sometimes used to aid identification; see Gilbert (*Méthode de mycologie descriptive*, 1934), Josserand (1952: 68), Locquin (*Petite flore des champignons de France*, 1956: 100). Tasting unknown fungi should be practised with caution.

Tatraea Svrček (1993), Leotiaceae. 1, Eur.

Tauromyces Cavalc. & A.A. Silva (1972), Mitosporic fungi (L), 7.A1.3. 1, Brazil.

Tavaresiella T. Majewski (1980), ? Laboulbeniales (inc. sed.). 4, Eur., Madagascar, Asia. See Benjamin (*Aliso* **13**: 559, 1993).

Tawdiella K.B. Deshp. & K.S. Deshp. (1966), Mitosporic fungi, 1.D2.?. 1, India.

tawny grisette, basidioma of the edible *Amanita fulva*.

taxis (frequently a suffix), a movement of a plasmodium or zoospore as a reaction to a one-sided stimulus; + (positive) when the movement is in the direction of the stimulus, − (negative) when away from the stimulus. The following tactic movements of zoospores have been described; (1) **chemotaxis**, in response to root exudates (*Pythium* spp.) and amino acids (*Allomyces* spp.); (2) **gravitaxis** (*Phytophthora palmivora*); (3) **electrotaxis** (*Phytophthora palmivora*; Morris *et al.*, *Plant Cell & Environment* **15**: 645, 1992).

taxol, an antitumor diterpenoid used in treatment of some cancers originally obtained from bark of Pacific yew (*Taxus brevifolia*), but also produced by the endophytic fungus *Taxomyces andreanae* (Stierle *et al.*, *Science* **260**: 214, 1993).

Taxomyces Strobel, A. Stierle, D. Stierle & W.M. Hess (1993), Mitosporic fungi, 1.G1.1. 1 (endophytic), USA.

taxon (pl. **taxa**, **taxons**), a taxonomic group of any rank (Code, Art. 1). See Lam (*Taxon* **6**: 213, 1957; history and usage). See Classification.

taxonomy, the science of classification, in biology the arrangment of organisms into a classification; **idio-**, of organisms; **syn-**, of communities. See Classification, Nomenclature, systematics.

Tazzetta, see *Tarzetta*.

TDP, thermal death point; generally used for a 10 min. application of heat.

tea fungus, a symbiotic association of yeasts (*Saccharomycodes ludwigii*) and bacteria (esp. *Acetobacter xylinum*). See Stadelmann (*Sydowia* **11**: 380, 1958; bibliogr., *Zbl. Bakt.* Abt. I **180**: 401, 1961), Kappel & Anken (*The Mycologist* **7**: 12, 1993; analysis); cf. teekwass, tibi, ginger beer plant.

Tectacervulus A.W. Ramaley (1991), Mitosporic fungi, 6.A1.15. 1, USA.

Tectella Earle (1909), Tricholomataceae. 1, N. temp.

teekwass, a Russian drink obtained by fermenting tea with a symbiotic mixture of *Acetobacter xylinum* and *Schizosaccharomyces pombe*. Cf. ginger beer plant, tea fungus, tibi.

Tegillum Mains (1940), = Olivea (Chacon.) fide Ono & Hennen (*Trans. Mycol. Soc. Japan* **24**: 369, 1978).

Tegoa Bat. & Fonseca (1961) nom. inval., nom. dub. (Mitosp. fungi).

Teichospora Fuckel (1870), Dothideales (inc. sed.). *c.* 10 (on wood), widespr. See Barr (*Mem. N.Y. bot. Gdns* **62**:1, 1990; key 20 N. Am. spp.).

Teichosporella (Sacc.) Sacc. (1895), Patellariaceae. 1 or 2, temp. See also Routien (*Mycol.* **50**: 117, 1958).

Teichosporina (G. Arnaud) Cif. & Bat. (1962) = Limacinula (Coccodin.) fide v. Arx & Müller (1975).

Telamonia (Fr.) Wünsche (1877) nom. cons., = Cortinarius (Cortinar.) fide Kauffman (1924).

teleblem (**teleoblema**), see universal veil.

Telebolus Lindau (1905) = Thelebolus (Thelebol.).

Telemeniella Bat. (1955) = Phyllachora (Phyllachor.) fide Cannon (*SA* **11**: 75, 1992).

teleomorph, see States of fungi.

teleutosorus, see telium.

Teleutospora Arthur & Bisby (1921) = Uromyces (Puccin.) fide Arthur (1934).

teleutospore, see teliospore.

Teleutosporites Mesch. (1902), Fossil fungi (? Uredin.). 4 (Carboniferous, Tertiary), France, Malaysia. But see Pirozynski (*Biol. Mem.* **1**: 104, 1976).

teleutosporodesm, Donk's (*Proc. Kon. Nederl. Akad. Wetensch.* C **75**: 385, 1972) term for teliospore.

Telimena Racib. (1900), Phyllachoraceae. 2, Eur., Asia, S. Am. See Müller (*Sydowia* **27**: 74, 1975), Sivanesan (*TBMS* **88**: 473, 1987).

Telimenella Petr. (1940), Phyllachoraceae. 2, Eur., N. Am. See Barr (*Mycol.* **69**: 952, 1977), Sivanesan (*TBMS* **88**: 473, 1987).

Telimeniella Bat. (1955), Phyllachoraceae. 1, Brazil.

Telimenochora Sivan. (1987), Phyllachoraceae. 1, Mexico, Puerto Rico.

Telimenopsis Petr. (1950) = Telimena (Phyllachor.) fide Sivanesan (1987).

Telioclipeum Viégas (1962), Ascomycota (inc. sed.). 1 (on *Aspidiosperma*), Brazil.

Teliomycetes, Basidiomycota. 2 ord., 15 fam., 167 gen. (+ 145 syn.), 7,134 spp. Ord.:
(1) **Septobasidiales**.
(2) **Uredinales**.
For *Lit.* see under *Uredinales*.

teliospore (teleutospore; teleutosporidesm, Donk (1973)), the spore (commonly a winter or resting spore) of the *Uredinales* (or *Ustilaginales*) from which the basidium is produced (Fig. 8H).

Teliosporeae. A subclass (Bessey, 1935). = *Teliomycetes* and *Ustomycetes*.

telium, a sorus producing teliospores.

Telligia Hendr. (1948), Mitosporic fungi, 1.D2.?. 1, Congo.

Teloconia Syd. (1921) = Phragmidium (Phragmid.) fide Dietel (1928).

Telomapea G.F. Laundon (1967), Uredinales (inc. sed.). 1, Java, Polynesia. = Maravalia (Chacon.) fide Cummins & Hiratsuka (1983).

Teloschistaceae Zahlbr. (1898), Teloschistales (L). 11 gen. (+ 40 syn.), 525 spp. Thallus varied; ascomata brightly coloured, with anthraquinones, rarely other pigments, usually with a well-developed thalline margin; asci with a well-developed outer I+ layer, internal structures rudimentary or absent; ascospores 0- to 3-septate, the septa strongly thickened, polarilocular.
Lit.: Bellemère & Letrouit-Galinou (*Cryptogamie, bryol. lichén.* **3**: 95, 1982; asci), Bellemère *et al.* (*Cryptogamie, bryol. licheń.* **7**: 189, 1986), Filson

(*Muelleria* **2**: 65, 1969; Australia), Hillmann (*Rabenh. Krypt.-Fl.* **9**, 6(1), 1935; Eur.), Honegger (*Lichenologist* **10**: 47, 1978; asci), Kärnefelt (*in* Galloway (Ed.), *Tropical lichens*: 105, 1991), Poelt & Hafellner (*Mitt. bot. StSamml., München* **16**: 503, 1980; key gen.), Santesson (*Vortr. bot. Ges.*, n.f. **4**: 5, 1970; chemotaxonomy).

Teloschistales, Ascomycota (L). 3 fam., 18 gen. (+ 43 syn.), 577 spp. Thallus varied, usually foliose or fruticose, often brightly coloured; ascomata apothecial, usually strongly concave, usually orange or reddish, usually with a well-developed thalline margin, often with anthraquinone pigments (K+ crimson); interascal tissue of usually unbranched paraphyses, the apices often capitate; asci cylindrical, persistent, the apex strongly thickened, without separable wall layers, with a conspicuous outer I+ layer, sometimes also with an internal I+ structure, releasing ascospores through an apical split; ascospores mostly hyaline, often with very thick septa, reduced lumina, and polarilocular. Anamorphs pycnidial. Lichenized with green algae, cosmop., esp. on nutrient-rich or basic substrata.
Pycnidia with numerous small locules lined by ± doliiform conidiogenous cells support the separation from *Lecanorales* as well as characters of the ascomata and chemistry. Fams:
(1) **Fuscideaceae**.
(2) **Letrouitiaceae**.
(3) **Teloschistaceae** (syn. *Caloplacaceae*).
Lit.: Kärnefelt (*Crypt. Bot.* **1**: 147, 1989; gen. names, phylogeny).

Teloschistes Norman (1853), Teloschistaceae (L). *c.* 30, cosmop. See Almborn (*Nord. Jl Bot.* **8**: 521, 1989; key 9 spp., C. & S. Afr.), Filson (*Muelleria* **2**: 65, 1969; Australia), Hillmann (*Rabenh. Krypt.-Fl.* **9**, 6: 31, 1935; Eur.), Thomson (*Am. Midl. Nat.* **41**: 706, 1949; N. Am.).

Teloschistomyces Cif. & E.A. Thomas (1953) ≡ Teloschistes (Teloschist.).

Telospora Arthur (1906) = Uromyces (Puccin.) fide Arthur (1934).

TEM, transmission electron microscope. See ultrastructure.

Temerariomyces B. Sutton (1993), Mitosporic fungi, 1.A2.?3. 1, Malawi.

Temnospora A. Massal. (1860) = Chrysothrix (Chrysotric.) fide Laundon (1981).

tempeh (**tempé**), an oriental food composed of *Rhizopus oligosporus* fermented soybeans (Hesseltine, *Mycol.* **57**: 154, 1965).

tempeh-bongrek, an oriental food composed of *Rhizopus* and manioc prepared in Malaysia.

tenacle, see haerangium.

Tenorea Tornab. (1848) [non Rafin. (1814), *Rutaceae*, etc.] = Teloschistes (Teloschist.) p.p. and Pseudevernia (Parmel.) p.p.

tentoxin, a chlorosis-inducing cyclic tetrapeptide toxin from *Alternaria alternata* (Saad *et al.*, *Phytopath.* **60**: 415, 1970).

Tenuicutites Bertrand (1898), Fossil fungi (? Chytridiomycetes). 1 (Carboniferous), France.

teonanácatl, see hallucinogenic fungi.

Tephrocybe Donk (1962), Tricholomataceae. 20, widespr., mostly temp. See Orton (*Bull. BMS* **18**: 114, 1984; key Br. spp.).

Tephromela M. Choisy (1929), Bacidiaceae (L). *c.* 20, widespr. See Hertel & Rambold (*Bot. Jahrb. Syst.* **107**: 469, 1985), Poelt & Grube (*Nova Hedw.* **57**: 1, 1993; key 5 spp. Himalaya).

Tephrophana Earle (1909) = Marasmius (Tricholom.) fide Singer (1951) but used by Métrod (*BSMF* **75**: 184, 1959).

Tephrosticta (Sacc., Syd. & P. Syd.) Syd. & P. Syd. (1913) = Phaeosaccardinula (Chaetothyr.) fide v. Arx & Müller (1975).

tequila, see pulque.

Terana Adans. (1763) = Corticium (Cortic.).

teratogenic, causing abnormalities in fetus growth.

Teratomyces Thaxt. (1893), Laboulbeniaceae. 9, widespr. See Benjamin (*Aliso* **10**: 345, 1983).

Teratonema Syd. & P. Syd. (1917) = Nitschkia (Nitschk.) fide Nannfeldt (1975).

Teratoschaeta Bat. & O.M. Fonseca (1967), ? Dothideales (inc. sed.). 1 (on lichens), Brazil. See Hawksworth (*Bull. Br. Mus. nat. Hist., Bot.* **6**: 183, 1979; anamorph). Anamorph ? *Ampullifera*.

Teratosperma Syd. & P. Syd. (1909), Mitosporic fungi, 1.C2.1. 6, widespr. See Ellis (*Mycol. Pap.* **69**, 1957; key), Hughes (*N.Z. Jl Bot.* **17**: 139, 1979).

Teratosphaeria Syd. & P. Syd. (1912), Phaeosphaeriaceae. 3, S. Hemisph. See Müller & Oehrens (*Sydowia* **35**: 138, 1982).

teratum (pl. -ta), an abnormal modification; for terata in lichens see Grummann (*Feddes Repert., Beih.* **122**, 1941). See also galls.

terebrate, having scattered perforations.

terebrator (of lichens), a trichogyne (Lindau).

Terenodon Maas Geest. (1971), Gomphaceae. 1, Japan.

terete, cylindrical; frequently, circular in section but narrowing to one end.

terfas, see *Terfezia*.

Terfezia (Tul. & C. Tul.) Tul. & C. Tul. (1851), Terfeziaceae. 12 (hypogeous, esp. under *Helianthemum, Cistus*, etc.), widespr. The ascocarps (terfas, kamés) are edible. See Fischer (1938; key), Gilkey (*N. Am. Fl.* 2, **1**, 1954; key N. Am. spp.), Janex-Favre & Parguey-Leduc (*Cryptogamie, Mycol.* **6**: 87, 1985; ascus structure), Janex-Favre *et al.* (*BSMF* **104**: 145, 1988; ultrastr.), Trappe (*TBMS* **57**: 85, 1971; nomencl.).

Terfeziaceae E. Fisch. (1897), Pezizales. 4 gen. (+ 5 syn.), 20 spp. Ascomata large, cleistothecial, ± globose, thick-walled, solid, the asci formed in marbled veins interspersed with sterile issue; interascal tissue absent; asci cylindrical to globose, indehiscent, not blueing in iodine; ascospores hyaline to pale brown, globose, ornamented, uninucleate. Anamorphs unknown. Hypogeous, sometimes emergent, widespr. esp. in dry conditions.
    Lit.: Trappe (1979), Zhang (*SA* **11**: 31, 1992).

Terfeziopsis Harkn. (1899) = Tuber (Tuber.) fide Fischer (1938).

terminus (phialo) spore, a phialospore of a 1-spored phialide, i.e. one terminating the growth of the phialide (Mason, 1933).

Termitaria Thaxt. (1920), Mitosporic fungi, 8.A1.15. 2 or 3, Am., Sardinia. See Khan & Kimbrough (*Am. J. Bot.* **61**: 395, 1974).

Termitariopsis M. Blackw., Samson & Kimbr. (1980), Mitosporic fungi, 3.C1.?. 1 (on termites), USA.

Termite fungi. Termites cultivate fungi in their nests as food, see Heim (*Termites et champignons*, 1977). Termites are also attacked by fungi, see Blackwell & Kimbrough (*Mycol.* **70**: 1279, 1979; key gen. termite-infesting fungi), Zobel & Grace (*Mycol.* **82**: 289, 1990; on *Reticulitermes*), Thomas (*TBMS* **84**: 519, 1985; isolation medium); Van der Westhuizen & Eicker (*MR* **94**: 923, 1991; key S. Afr. spp.).

Termiticola E. Horak (1979), Agaricaceae. 1, Papua New Guinea.

Termitomyces R. Heim (1942), Amanitaceae. 30 (in termite nests), trop. Afr., Asia. See Heim (*Termites et champignons*, 1977), Pegler (*Kew Bull., Addit. Ser.* **6**, 1977; key), Jing & Ma (*Acta Mycol. Sin.* **4**: 103, 1985; key Chinese spp.). Anamorph *Termitosphaera*.

Termitosphaera Cif. (1935), Anamorphic Amanitaceae, 3.A1.3. Teleomorph *Termitomyces*. 1, trop.

terrestrial, growing on land as opposed to in water. Cf. terricolous.

terricolous, growing on the ground. Cf. terrestrial.

Terriera B. Erikss. (1970), Rhytismataceae. 1, Eur.

Terrostella Long (1945) ≡ Geasteroides (Geastr.).

terverticillate (of a penicillus), having branching at three levels, i.e. having rami bearing metulae and phialides.

tessellate, marked with a mosaic design; chequered.

Testicularia Klotzsch (1832), Ustilaginaceae. 2 (on *Cyperaceae*), N. & S. Am., W. Indies.

Testudina Bizz. (1885), Testudinaceae. 1, Eur. See Hawksworth (*CJB* **57**: 91, 1979).

Testudinaceae Arx (1971), Ascomycota (inc. sed.). 5 gen. (+ 3 syn.), 6 spp. Stromata absent; ascomata cleistothecial, black, thick-walled, fragmenting into predefined polygonal plates; interascal tissue of branched pseudoparaphyses; asci ± clavate, thick-walled when young, without apical structures, evanescent; ascospores relatively small, brown, usually septate, mostly ornamented. Anamorphs unknown. Saprobic, usually isol. ex soil, widespr.
    Lit.: Hawksworth (*CJB* **57**: 91, 1979; *SA* **6**: 153, 1987).

Tetena Raf. (1806) nom. dub (Fungi, inc. sed.).

tetra (prefix), four; **-cytes**, the spores resulting from meiosis; **-d**, a group of four; **-polar** (of incompatibility systems), having 2 loci; bifactorial; cf. bipolar; -**spore**, see dispore; **-tomic**, 4-times furcate at one node.

Tetrabrachium Nawawi & Kuthub. (1987), Mitosporic fungi, 1.G1.1. 1 (aquatic), Malaysia.

Tetrabrunneospora Dyko (1978), Mitosporic fungi, 1.G2.1. 1 (aquatic), USA.

Tetrachaetum Ingold (1942), Mitosporic fungi, 1.G1.1. 1 (aquatic), UK.

Tetrachia Berk. & M.A. Curtis (1882) nom. nud. = Spegazzinia (Mitosp. fungi) fide Saccardo (*Syll. Fung.* **12**: 775, 1897).

Tetrachia Sacc. (1921) = Spegazzinia (Mitosp. fungi) fide Boedijn (*in litt.*).

tetrachotomous, with four branches arising from the same point.

Tetrachytrium Sorokīn (1874), ? Chytridiomycetes (inc. sed.). 1, Eur.

Tetracium, see *Tetracrium*.

Tetracladium De Wild. (1893), Mitosporic fungi, 1.G1.1. 7 (aquatic), widespr. Roldán *et al.* (*MR* **93**: 452, 1989; culture characters, key), Petersen (*Mycol.* **54**: 140, 1962; key).

Tetracoccosporium Szabó (1905), Mitosporic fungi, 1.D2.1. 1, widespr.

Tetracolium Kunze ex Link (1824) = Torula (Mitosp. fungi) fide Petch (*TBMS* **24**: 56, 1940). Nom. dub. fide Hughes (1958).

Tetracrium Henn. (1902), Anamorphic Tubeufiaceae, 3.G1.?. Teleomorph *Puttemansia*. 3 (on insects), trop. See Martin (*Pacific Sci.* **2**: 71, 1948).

Tetracytum Vanderw. (1945) = Cylindrocladium (Mitosp. fungi) fide Kendrick & Carmichael (1973).

Tetradia T. Johnson (1904) [non Benn. & R. Br. (1849), *Sterculiaceae*] nom. dub. (Mitosp. fungi). See Sutton (1977).

Tetradium Schltdl. (1852) [non Lour. (1790), *Rutaceae*] = Bremia (Peronospor.) fide Fischer (1892).

Tetragoniomyces Oberw. & Bandoni (1981), Tetragoniomycetaceae. 1, N. Am.

Tetragoniomycetaceae Oberw. & Bandoni (1981), Tremellales. 1 gen., 1 sp. Mycoparasitic; probasidium producing tetrad, with thick walls, functioning as a propagule; basidiospores absent.

Tetramelaena (Trevis.) C.W. Dodge (1971) = Hyperphyscia (Physc.) fide Hafellner *et al.* (*Herzogia* **5**: 39, 1979).

Tetramelas Norman (1853) = Buellia (Physc.) fide Hawksworth *et al.* (*Lichenologist* **12**: 85, 1980).

Tetrameronycha Speg. ex W. Rossi & M. Blackw. (1990), Mitosporic fungi, 1.A1.1. 1 (on *Forficulidae*), W. Afr.

Tetramyxa K.I. Goebel (1884), Plasmodiophoraceae. 4, N. temp. See Karling (1968; key).

Tetranacrium H.J. Huds. & B. Sutton (1964), Mitosporic fungi, 8.G1.1. 1, W. Indies, India.

Tetrandromyces Thaxt. (1912), Laboulbeniaceae. 5, S. Am., W. Afr., Philipp.

Tetraploa Berk. & Broome (1850), Anamorphic Massarinaceae, 1.D2.1. Teleomorph *Massarina*. 6, widespr. See Ellis (*TBMS* **32**: 246, 1949), Rifai *et al.* (*Reinwardtia* **10**: 419, 1988; Javan spp.), Scheuer (*MR* **95**: 126, 1991; teleomorph).

Tetraposporium S. Hughes (1951), Mitosporic fungi, 1.G2.1. 2, Afr., USA.

Tetrapyrgos E. Horak (1987), Tricholomataceae. 15, trop.

Tettigomyces Thaxt. (1915), Ceratomycetaceae. 16, Asia, Afr., S. Am. See Ye & Shen (*Acta Mycol. Sin.* **11**: 285, 1992; key 10 spp. China).

Tettigorhyza G. Bertol. (1875) ? = Cordyceps (Clavicipit.).

texospore, ascospore coated with a layer of cells of paraphysal origin, as in *Texosporium* (Tibell & Hofsten, *Mycol.* **60**: 557, 1968).

Texosporium Nádv. ex Tibell & Hofsten (1968), Caliciaceae (L). 1, USA.

textura, see tissue types.

Thaelaephora, see *Thelephora*.

Thailandia Vardhan. (1959) = Candida (Mitosp. fungi) fide Orr & Kuehn (*Mycol.* **63**: 191, 1971).

thalamium, asci + hamathecium (obsol.).

thalassiomycetes, fungi living in marine environments; see Marine fungi.

Thalassoascus Ollivier (1926), Dothideales (inc. sed.). 3 (on *Algae*), Eur., N. Afr., Chile. See Kohlmeyer (*Mycol.* **73**: 833, 1981).

Thalassogena Kohlm. & Volkm.-Kohlm. (1987), Halosphaeriaceae. 1 (marine), Belize.

Thalassomyces Niezab. (1913) Algae. See Dick (*in press*), Kane (*N.Z. Jl Sci.* **7**: 289, 1964), Whisler (*in* Margulis *et al.* (Eds), *Handbook of Protoctista*: 715, 1990).

thallic (of conidiogenesis), one of the two basic sorts (cf. blastic) in which any enlargement of the recognizable conidial initial occurs *after* the initial has been delimited by one or more septa. The conidium is differentiated from a *whole* cell. **entero-**, thallic conidiogenesis in which the outer wall of the sporogenous cell is not involved in the formation of the spore wall (as for sporangiospores).

thalline exciple (margin), see excipulum thallinum.

Thallinocarpon Å.E. Dahl (1950) ? = Lichinella (Lichin.). See Henssen (*Ber. dtsch. bot. Ges.* **92**: 483, 1980).

Thallisphaera Dumort. (1822) nom. dub. (Ascomycota, inc. sed.), used for diverse perithecioid fungi.

Thallochaete Theiss. (1913) = Aphanopeltis (Asterin.).

thalloconidium, a mitosporic propagule produced and seceded directly from the lower cortex and(or) rhizines of certain *Umbilicaria* spp.; similar structures arise from the prothallus in *Protoparmelia*, *Rhizoplaca* and *Sporastatia* (Poelt & Obermayer, *Herzogia* **8**: 273, 1990); thalloconidia are dark brown, smooth to rugged, with 2-3 wall layers, and consist of

one to 2500 cells. See Hestmark (*Nord. Jl Bot.* **9**: 547, 1990; ultrastr., occurrence); see also thallyles.

thallodic, of, pertaining to, or belonging to a thallus (Weresub & LeClair, *CJB* **49**: 2203, 1971).

Thalloedematomyces E.A. Thomas ex Cif. & Tomas. (1953) ≡ Thalloidima (Catillar.).

Thalloidima A. Massal. (1852) nom. rej. prop. = Toninia (Catillar.).

Thalloidimatomyces, see *Thalloedematomyces*.

Thalloloma Trevis. (1853) = Graphina (Graphid.).

Thallomicrosporon Benedek (1964) = Microsporum (Mitosp. fungi) fide Ajello (1968).

Thallomyces H.J. Swart (1975), Parmulariaceae. 1, Australia.

Thallophyta, see *Fungi*; **thallophyte**, one of the *Thallophyta*

Thallospora L.S. Olive (1948), Mitosporic fungi, 1.G1.1. 1, USA.

thallospore, (1) an asexual spore having neither conidiophore nor conidiogenous cell, or one which is not separated from the hypha or conidiogenous cell producing it; i.e. an arthrospore, blastospore, or chlamydospore (and aleuriospore) (after Vuillemin; see Mason, 1933); (2) a thalloconidium (q.v.).

thallus, the vegetative body of a thallophyte; for thallus types in lichens see Lichens; **heteromerous-**, a layered thallus (Fig. 40A); **homoiomerous-**, an unlayered thallus (Fig. 40B).

thallyles, minute thallus-like propagules produced on the underside of certain *Umbilicaria* thalli (Krog & Swinscow, *Nordic Jl Bot.* **6**: 75, 1986); see also thalloconidia.

Thamnidiaceae Fitzp. (1930), Mucorales. 12 gen. (+ 9 syn.), 22 spp. Sporangia (diffluent and columellate) and sporangiola (few to one spored, persistent walled and columellate) borne on the same or separate, morphologically identical sporangiophores, or sporangiola only present; zygospores warty, borne on opposed suspensors; polyphyletic.

Thamnidium Link (1809), Thamnidiaceae. 1, widespr. See Benny (*Mycol.* **84**: 834, 1992).

Thamnidium Tuck. ex Schwend. (1860) = Lichina (Lichin.).

thamniscophagous, see ptyophagous.

Thamnium Vent. (1799) nom. rej. = Roccella (Roccell.) fide Ahti (*Taxon* **33**: 330, 1984).

Thamnocephalis Blakeslee (1905), Sigmoideomycetaceae. 3, N. Am., India. See Mehrotra (*Zbl. Bakt.* Abt. II **117**: 425, 1964), Benny *et al.* (1992; key).

Thamnochrolechia Aptroot & Sipman (1991), Pertusariaceae (L). 1, Papua New Guinea. See Lumbsch *et al.* (*Bibl. Lich.* **57**: 355, 1995).

Thamnogalla D. Hawksw. (1980), Phyllachoraceae. 1 (on *Thamnolia*), N. Hemisph. See Hafellner & Sancho (*Herzogia* **8**: 363, 1990; posn).

Thamnolecania (Vain.) Gyeln. (1933), Bacidiaceae (L). 5, Subantarctic.

Thamnolia Ach. ex Schaer. (1850) nom. cons., Lecanorales (inc. sed., L). 1 (or 2), sterile, cosmop. (montane). See Sheard (*Bryologist* **80**: 100, 1977).

Thamnoma, see *Thamnonoma*.

Thamnomyces Ehrenb. (1820), Xylariaceae. 5, mainly trop. See Dennis (*Kew Bull.* 1957: 297, 1957, *Bull. Jard. bot. Brux.* **31**: 150, 1961), Samuels & Müller (*Sydowia* **33**: 274, 1980; relationships). Anamorph *Geniculosporium*.

Thamnonoma (Tuck.) Gyeln. (1933) = Caloplaca (Teloschist.).

Thamnostylum Arx & H.P. Upadhyay (1970), Thamnidiaceae. 4, widespr. (esp warmer areas). See Benny & Benjamin (*Aliso* **8**: 301, 1975; key).

Thanatephorus Donk (1956), Ceratobasidiaceae. 9, widespr. *T. cucumeris* (stat. mycel. *Rhizoctonia solani*) is now used to replace *Corticium solani*. See Currah & Zelmer (*Rep. Tottori mycol. Inst.* **40**: 43, 1992; 4 spp. with orchids), Talbot (*Persoonia* **3**: 371, 1965).

Thanatophytum Nees (1816) ≡ Rhizoctonia (Mitosp. fungi).

Thanatostrea Franc & Arvy (1969) nom. inval. (? Labyrinthista).

Thaptospora B. Sutton & Pascoe (1987), Mitosporic fungi, 7.A1.15. 1, Australia.

Tharoopama Subram. (1956), Mitosporic fungi, 2.A2.10. 2, India, Panama.

Thaumasiomyces Thaxt. (1931), Ceratomycetaceae. 3, W. Afr., Borneo.

Thaxter (Roland; 1858-1932). Professor of Cryptogamic Botany, Harvard Univ., 1901-19; then Honorary Curator of the Farlow Herbarium. Chief writings: See *Laboulbeniales*. See Weston (*Mycol.* **25**: 69, 1933; portr., bibliogr.; *Phytopath.* **23**: 565, 1933), Pfister (*Mycotaxon* **20**: 225, 1984; index to non-*Laboulbeniales* names), Grummann (1974: 197), Stafleu & Cowan (*TL-2* **6**: 231, 1986).

Thaxteria Sacc. (1891), Lasiosphaeriaceae. 2, widespr. See Booth & Müller (*TBMS* **58**: 73, 1972), Nannfeldt (*Svensk bot. Tidskr.* **69**: 204, 1975; posn).

Thaxteria Giard (1892) = Laboulbenia (Laboulben.) fide Thaxter (*Proc. Am. Acad. Arts Sci.* **30**: 471, 1895).

Thaxteriella Petr. (1924) = Tubeufia (Tubeuf.) fide Barr (*Mycotaxon* **12**: 137, 1980).

Thaxteriellopsis Sivan., Panwar & S.J. Kaur (1977), ? Tubeufiaceae. 3, India. See Pande & Rao (*Indian Phytopath.* **40**: 36, 1987; key), Sivanesan (*TBMS* **81**: 325, 1983; key).

Thaxterina Sivan., R.C. Rajak & R.C. Gupta (1988), Tubeufiaceae. 1, India.

Thaxteriola Speg. (1918) = Pyxidiophora (Pyxidiophor.).

Thaxterogaster Singer (1951), Cortinariaceae. 33, Am., Australasia. See Horak & Moser (*Nova Hedw.* **10**: 211, 1965; key), Beaton *et al.* (*Kew Bull.* **40**: 171, 1985; key Australia).

Thaxterosporium Ben Ze'ev & R.G. Kenneth (1987), Neozygitaceae. 1, widespr. See Ben-Ze'ev *et al.* (*Mycotaxon* **28**: 323, 1987). = Neozygites (Entomophthor.) fide Keller (*Sydowia* **43**: 39, 1991).

theca, see ascus (obsol.).

Thecamycetes, see *Ascomycota* (Marchand, 1896).

Thecaphora Fingerh. (1836), Ustilaginaceae. *c*. 30 (on *Leguminosae, Compositae, Convolvulaceae,* etc.), widespr. *T. solani*, potato (*Solanum*) smut. Frequently has a mitosporic state, cf. *Thecaphorella*. See Zambettakis & Joly (*BSMF* **91**: 71, 1975; numerical taxonomy), Durán (*CJB* **60**: 1512, 1982), Mordue (*Mycopath.* **103**: 177, 1988).

Thecaphorella H. Scholz & I. Scholz (1988), Anamorphic Ustilaginales. Teleomorph *Thecaphora*. 1, temp.

Thecaria Fée (1824) = Phaeographina (Graphid.).

thecaspore, see ascospore (obsol.).

Theciopeltis F. Stevens & Manter (1925) = Micropeltis (Micropeltid.) fide Clements & Shear (1931).

thecium, the part of an apothecium containing the asci between the epithecium and hypothecium; sometimes used for the whole sporocarp or as equivalent to hymenium.

Theclospora Harkn. (1884) = Emericella (Trichocom.) fide Peek & Solheim (*Mycol.* **50**: 844, 1958).

Thecographa A. Massal. (1860) = Phaeographina (Graphid.).

Thecopsora, see *Theclospora* and *Thekopsora*.

Thecostroma Clem. (1909) = Bloxamia (Mitosp. fungi) fide Clements & Shear (1931).

Thecotheus Boud. (1869), Ascobolaceae. 10, Eur., Afr., N. Am. See Kimbrough (*Mycol.* **61**: 107, 1969; key, *in* Hawksworth (Ed.), *Ascomycete systematics*: 398, 1994; posn), Krug & Khan (*Mycol.* **79**: 200, 1987; key 10 spp.).

Thedgonia B. Sutton (1973), Mitosporic fungi, 3.B/C1.39. 3, N. temp. See Yoshikawa & Yokoyama (*Trans. Mycol. Soc. Jap.* **33**: 177, 1992), Braun (*Monograph of Cercosporella etc.* **1**: 211, 1995; key).

Theissen (Ferdinand; 1877-1919). German clergyman, school-master, and amateur mycologist died in the Voralberg Alps while collecting lichens. See Sydow (*Ann. Myc.* **17**: 134, 1919; bibliogr.), Stafleu & Cowan (*TL-2* **6**: 239, 1986).

Theissenia Maubl. (1914), Xylariaceae. 1, Brazil.

Theissenula Syd. & P. Syd. (1914) Fungi (inc. sed.) fide Hansford (*Mycol. Pap.* **15**, 1946).

Thekopsora Magnus (1875), Pucciniastraceae. 15, N. temp. See Hiratsuka *et al.* (*Rust flora of Japan*, 1992; key), Cummins & Hiratsuka (1983).

Thelactis Mart. (1821) = Mucor (Mucor.) fide Fries (1832).

Thelebolaceae (Brumm.) Eckblad (1968), Pezizales. 14 gen. (+ 6 syn.), 45 spp. Ascomata minute, ± globose or pulvinate, at least initially cleistothecial, the excipulum hyaline, poorly developed, ± glabrous; interascal tissue poorly-developed, composed of simple paraphyses; asci ± ellipsoidal, often multispored, ± persistent, opening with a rather irregular vertical split; ascospores usually small, hyaline, smooth or with ornamentation formed as an elaboration of an initially homogeneous secondary wall layer. Anamorphs unknown. Saprobic, usually coprophilous, widespr.
    Lit.: Kimbrough (*CJB* **44**: 685, 1966), Kimbrough & Korf (*Am. J. Bot.* **54**: 9, 1967), Moravec (*Ceská Myk.* **25**: 150, 1971), and under *Thelebolus*.

Thelebolus Tode (1790), Thelebolaceae. 7 (psychrophilic, coprophilous), N. temp. See Czymmek & Klomparens (*CJB* **70**: 1669, 1992; ascosp.), Kimbrough (*Mycol.* **73**: 1, 1981; ascus structure), Momol & Kimbrough (*SA* **13**: 1, 1994; posn), Wicklow & Malloch (*Mycol.* **63**: 118, 1971).

Thelenella Nyl. (1855), Thelenellaceae (L). 15, widespr. See Mayrhofer (1987; key), Mayrhofer & McCarthy (*Muelleria* **7**: 333, 1991; key 12 saxicolous spp.).

Thelenellaceae H. Mayrhofer (1986), Ascomycota (inc. sed., ±L). 3 gen. (+ 10 syn.), 35 spp. Thallus crustose, sometimes areolate; ascomata perithecial, ± immersed, thick-walled, pale to dark brown, the ostiole sometimes periphysate; interascal tissue of narrow branched and anastomosed pseudoparaphyses; asci thick-walled, with separable wall layers, the apex sometimes thickened and with a small ocular chamber, not blueing in iodine; ascospores hyaline to brown, thin-walled, muriform, the septa forming in a median position in each cell of the developing spore, without a sheath.
    Lit.: Mayrhofer (*Bibl. Lich.* **26**, 1987).

Thelenidia Nyl. (1886), ? Dothideales (inc. sed., L). 1, Greenland, Switzerland. See Topham & Swinscow (*Lichenologist* **4**: 294, 1970).

Thelenidiomyces Cif. & Tomas. (1953) ≡ Thelenidia (Dothideales).

Thelephora Ehrh. ex Willd. (1787), Thelephoraceae. 45, widespr. See Corner (*Beih. Nova Hedw.* **27**, 1968; monogr.), Stalpers (1993; key).

Thelephoraceae Chevall. (1926), Thelephorales. 17 gen. (+ 24 syn.), 220 spp. Basidioma resupinate to flabelliform; hymenophore smooth, papillate, or not strictly resupinate with hymenophore smooth, or stalked with hymenophore toothed; spores globose to ellipsoidal, often uneven in outline, ornamented, brownish to colourless, non-amyloid, non-cyanophilous; the usually dark context becomes greenish on treatment with KOH (due to thelephoric acid); terrestrial, humicolous, a few lignicolous.

Lit.: Bourdot & Galzin (1927), Donk (1951-63, VII, gen. list; 1964: 295), Burt (*Ann. Mo. bot. Gdn* 1-7, 11-13, 1914-26; N. Am. [reprinted as 1 vol. 1966; see Lentz, *Mycol.* **60**: 214, 1968]), Rogers & Jackson (*Farlowia* **1**: 263, 1943; re-examination of Burt's types), Rattan (*Bibl. Mycol.* **60**, 1977; Himalayas), Ginns (*Mycol.* **60**: 1211, 1969; *Merulius*), Donk (1951-63, VII; gen. names), Jahn (*Westfäl. Pilzbr.* **8**: 69, 1971), Svrček (*Sydowia* **14**: 170, 1960; subfam. *Tomentelloideae*, keys), Cunningham (*The Thelephoraceae of Australia and New Zealand* [*Bull. DSIR N.Z.* **145**], 1963), Reid (*Beih. Nova Hedw.* **18**, 1965; stipitate steroid spp.), Jülich & Stalpers *Verh. Kon. Ned. Akad. Wetensch., Afd. Natuurk.* sect. 2 **74**, 1980; N. Hemisph.), Stalpers (*Stud. Mycol.* **35**, 1993). See also *Lit.* under *Hydnaceae*.

Thelephorales, Basidiomycetes. 2 fam. 19 gen. (+ 24 syn.), 237 spp. Fams:
(1) **Bankeraceae**.
(2) **Thelephoraceae**.

Thelephorella P. Karst. (1889) nom. dub. (Thelephor.) fide Donk (1957).

Theleporus Fr. (1847), Grammotheleaceae. 2, trop. See Ryvarden (*TBMS* **73**: 9, 1979).

Thelidea Hue (1902) = Knightiella (Icmadophil.) fide Rambold *et al.* (*Bibl. Lich.* **53**: 217, 1993).

Thelidiella Fink (1933), Ascomycota (inc. sed.). 1 (on lichens), N. Am.

Thelidiola C.W. Dodge (1968) = Muellerella (Verrucar.) fide Castello & Nimis (*Bibl. Lich.* **57**: 71, 1995).

Thelidiomyces Cif. & Tomas. (1953) ≡ Thelidium (Verrucar.).

Thelidiopsis Vain. (1921), Verrucariaceae (L). 1, Eur., Asia.

Thelidium A. Massal. (1855), Verrucariaceae (L). *c.* 100, cosmop. See Servít (*Československé lišejníky Čeledi Verrucariaceae*, 1954), Zschacke (*Rabenh. Krypt.-Fl.* **9**, 1(2): 217, 1933).

Thelignya A. Massal. (1855), Lichinaceae (L). 1, Eur. See Henssen (1963), Jørgensen & Henssen (*Taxon* **39**: 343, 1990).

Thelis Clem. (1931) ≡ Hanseniaspora (Saccharomycod.).

Thelocarpon Nyl. (1853), Acarosporaceae (±L). 18, Eur., N. Am. See Poelt & Hafellner (*Phyton* **17**: 67, 1975), Salisbury (*Lichenologist* **3**: 175, 1966).

Thelocarponomyces Cif. & Tomas. (1953) ≡ Thelocarpon (Acarospor.).

Thelocarpum Clem. (1909) ≡ Thelocarpon (Acarospor.).

Thelochroa A. Massal. (1855) = Pyrenocarpon (Lichin.) fide Jorgensen & Henssen (*Taxon* **39**: 343, 1990).

Thelococcum Nyl. ex Hue (1888) = Thelocarpon (Acarospor.).

Thelographis Nyl. (1857) nom. nud. = Phaeographina (Graphid.).

Thelomma A. Massal. (1860), Caliciaceae (L). 7, Canary Isl., Eur., N. Am. See Tibell (*Bot. Notiser* **129**: 221, 1976, 1984).

Thelomphale Flot. (1863) = Thelocarpon (Acarospor.).

Thelophora Clem. (1902) ≡ Thelephora (Thelephor.).

Thelopsidomyces Cif. & Tomas. (1953) ≡ Thelopsis (Stictid.).

Thelopsis Nyl. (1855) nom. cons., ? Stictidaceae (L). 6, Eur., N. Am. See Sherwood (1977), Vězda (*Folia geobot. phytotax.* **4**: 363, 1968).

Theloschisma Trevis. (1860) = Phaeographis (Graphid.).

Theloschistes Th. Fr. (1861) ≡ Teloschistes (Teloschist.).

Theloschistomyces Cif. & Tomas. (1953) ≡ Theloschistes (Teloschist.).

Thelospora, see *Theclospora*.

Thelotrema Ach. (1803), Thelotremataceae (L). *c.* 100, cosmop. (mainly trop.). See Hale (*Smithson. Contr. bot.* **16**, 1974; Dominican Republic, **38**, 1978; Panama, **40**, 1980; limits, *Bull. Br. Mus. nat. Hist., Bot.* **8**: 227, 1981; Sri Lanka), Redinger (*Ark. Bot.* **28A**(8), 1936), Salisbury (*Lichenologist* **5**: 262, 1972; *lepadinum*-group, **7**: 59, 1975; gen. concept).

Thelotremataceae (Nyl.) Stizenb. (1862), Ostropales (L). 11 gen. (+ 43 syn.), 536 spp. Thallus ± crustose; ascomata apothecial, often deeply immersed and appearing perithecial, usually with a well-developed thalline margin, usually not carbonized; interascal tissue of simple paraphyses; asci usually with a well-developed apical cap and a distinct pore; ascospores hyaline or brown, transversely septate, sometimes muriform, the septa often strongly thickened. Conidiomata pycnidial. Lichenized with green algae, esp. *Trentopohlia*, mainly on bark, esp. trop.

Lit.: Hale (*Mycotaxon* **11**: 130, 1980, *Bull. Br. Mus. nat. Hist., Bot.* **8**: 227, 1981; gen. concepts), Lettau (*Beih. Repert. spec. nov. Regni veg.* **69**, 1932-37), Redinger (1936-38).

Thelotrematomyces E.A. Thomas ex Cif. & Tomas. (1953) ≡ Thelotrema (Thelotremat.).

Themisia Velen. (1939), ? Leotiales (inc. sed.). 1, former Czechoslovakia.

Thermoactinomyces Tsikl. (1899), Actinomycetes, q.v.

Thermoascaceae, see *Trichocomaceae*.

Thermoascus Miehe (1907), Trichocomaceae. 4, Eur., N. Am., Egypt. See Apinis (*TBMS* **50**: 573, 1967), Ellis (*TBMS* **76**: 457, 467, 1981; ultrastr.), Malloch (*in* Samson & Pitt (Eds), *Advances in Penicillium and Aspergillus systematics*: 365, 1985). Anamorphs *Paecilomyces*, *Polypaecilum*.

thermodury, withstanding high temperature, esp. when in a dormant state, e.g. as spores. Cf. thermophily.

Thermoidium Miehe (1907) = Malbranchea (Mitosp. fungi) fide Sigler & Carmichael (*Mycotaxon* **4**: 441, 1976).

Thermomonospora Henssen (1957), Actinomycetes, q.v.

Thermomucor Subraham., B.S. Mehrotra & Thirum. (1977), Mucoraceae. 1, India. See Subrahamanyam *et al.* (*Georgia J. Sci.* **35**: 1, 1977), Schipper (*Ant. v. Leeuwenhoek* **45**: 275, 1979).

Thermomyces Tsikl. (1899), Mitosporic fungi, 1.A2.1. 3 (soil), widespr. See Pugh *et al.* (*TBMS* **47**: 115, 1964).

thermophily, making active growth at high temperature. Cf. thermodury. Fungi may be classified as **thermophiles** (adj. **-ilic**), growth at 20-50+°C (opt. 40-50+°C). See Cooney & Emerson (*Thermophilic fungi*, 1964; descriptions), Emerson (*in* Ainsworth & Sussman (Eds), *The Fungi* **3**: 105, 1968), Crisan (*Mycol.* **65**: 1170, 1973; concepts), Bilaĭ & Zakharchenko (*Opredelitel' Termofill'nȳkh Gribov*, 1987; keys, illustr. 38 spp.); **thermotolerant fungi**, e.g. *Aspergillus fumigatus*, *Absidia ramosa*, max. *c.* 50°C, min. well below 20°C; **mesophiles** (adj. **-ilic**), growth 10-40°C (opt. 20-35°C); **psychrophiles** (adj. **-ilic**), growth below 10°C (opt. below 20°C).

Thermophymatospora Udagawa, Awao & Abdullah (1986), Mitosporic fungi, 1.A2.1. 1 (with clamp connexions, from soil), Iraq.

Thermutis Fr. (1825), Lichinaceae (L). 1, Eur. See Henssen (*Symb. bot. upsal.* **18**(1), 1963).

Thermutomyces Cif. & Tomas. (1953) ≡ Thermutis (Lichin.).

Thermutopsis Henssen (1990), Lichinaceae (L). 1, Antigua.

Therrya Sacc. (1882), Rhytismataceae. 5, Eur., N. Am. See Reid & Cain (*CJB* **39**: 1117, 1961).

Thielavia Zopf (1876), Chaetomiaceae. 25, widespr. See Malloch & Cain (*Mycol.* **65**: 1055, 1973), v. Arx (*Stud. Mycol.* **8**, 1975), v. Arx *et al.* (*Bibl. Mycol.* **94**, 1988).

Thielaviella Arx & T. Mahmood (1968) = Boothiella (Chaetom.).

Thielaviopsis Went (1893), Mitosporic fungi, 1.A2.38. 2, widespr. *T. basicola* (root rot of tobacco and other plants). See Johnson (*J. agric. Res.* **7**: 289, 1916; hosts), Rawlings (*Ann. Mo. bot. Gdn* **27**: 561, 1940; culture).

Thindia Korf & Waraitch (1971), Sarcoscyphaceae. 1 (on *Cupressus*), India.

Thindiomyces Arendh. & R. Sharma (1983), Leotiaceae. 1, Bhutan.

Thirumalacharia Rathaiah (1981), Mitosporic fungi, 1.B2.10. 1, India.

Thirumalachariella Sathe (1975), Uredinales (inc. sed.). 1, Brazil. = Phragmidiella (Phakopsor.) fide Buriticá & Hennen (1991).

Tholurna Norman (1861) nom. cons., Caliciaceae (L). 1, Eur., N. Am.

Tholurnaceae, see *Caliciaceae*.

Tholurnomyces Cif. & Tomas. (1953) ≡ Tholurna (Calic.).

tholus, see ascus.

Thom (Charles; 1872-1956). Mycologist at the Storrs (Connecticut) Experiment Station, 1904-13; then until 1942 in the US Department of Agriculture, Washington. Noted for his studies on *Penicillium* and *Aspergillus* (q.v.). See Raper (*Mycol.* **49**: 134, 1957; portr., bibl.), Stafleu & Cowan (*TL-2* **6**: 268, 1986).

Thomiella C.W. Dodge (1935) = Gonatobotryum (Mitosp. fungi) fide Kendrick & Carmichael (1973).

Thoracella Oudem. (1900), Mitosporic fungi, 8.A1.10. 1, Eur.

Thozetella Kuntze (1891), Mitosporic fungi, 3.A1.15. 4, Asia, S. Am., USA. See Pirozynski & Hodges (*CJB* **51**: 157, 1973), Sutton & Cole (*TBMS* **81**: 97, 1983).

Thozetellopsis Agnihothr. (1958) = Thozetella (Mitosp. fungi) fide Pirozynski & Hodges (1973).

Thozetia Berk. & F. Muell. (1881) [non F. Muell. ex Benth. (1868), *Asclepiadaceae*] ≡ Thozetella (Mitosp. fungi).

Thrauste Theiss. (1916), Englerulaceae. 1 or 2, Java, Philipp.

Thraustochytriaceae Sparrow ex Cejp (1959), Thraustochytriales. 10 gen., 29 spp. Cells of the rhizoidal endobiotic net not spindle-shaped and gliding, sporangia epibiotic. Some bactivorous (q.v.).
  Lit.: Sparrow (*in* Ainsworth *et al.*, The Fungi 4A: 69, 1973).

Thraustochytriales, Labyrinthulomycota. 1 fam., 10 gen., 29 spp. Fam.:
  Thraustochytriaceae.

Thraustochytrium Sparrow (1936), Thraustochytriaceae. 15 (on marine algae), N. Am. See Gaertner (*Encycl. Cinematogr.* E 1664/1970, 1971; life cycle), Ulken (*Bibl. Mycol.* **137**, 1990), Porter & Jennings (*MR* **92**: 470, 1989), Dick (*in press*; key).

Thraustotheca Humphrey (1893), Saprolegniaceae. 2 or 3, N. temp. See Salvin (*Mycol.* **34**: 48, 1942).

thread blight, (1) a disease caused by species of *Corticium* and *Marasmius* having mycelium running as well-marked threads over the leaves and stems of tropical plants; (2) a fungus causing - -; see Petch

(*Ann. R. bot. Gdns Peradeniya* **9**: 1-43, 1924). Cf. horse-hair blight fungi.

Thrinacospora Petr. (1948), Mitosporic fungi, 8.C1.1. 1, Ecuador.

Thripomyces Speg. (1915), Ceratomycetaceae. 1, Italy.

Thrombiaceae Poelt & Vĕzda ex J.C. David & D. Hawksw. (1991), Ascomycota (inc. sed., L). 1 gen. (+ 3 syn.), 1 sp. Thallus crustose, granular or film-like, ± gelatinized, often evanescent; ascomata perithecial, immersed, the peridium dark, blueing in iodine towards the apex, thickened towards the ? periphysate ostiole, not clypeate; interascal tissue of sparsely branched paraphyses; asci cylindrical, persistent, thinwalled, with a well-developed I+ apical cap and a narrow cylindrical apical ring; ascospores hyaline, aseptate, thin-walled, without a sheath. Anamorph not known. Lichenized with green algae (*Leptosira*, N. temp.

Thrombium Wallr. (1831), Thrombiaceae (L). 1, Eur., N. Am. See Swinscow (*Lichenologist* **2**: 276, 1964).

Thrombocytozoons Tchacarof (1963) = Candida (Mitosp. fungi) fide Desser & Barta (*Can. J. Microbiol.* **34**: 1096, 1988).

thrush, a throat and genital disease of humans caused by *Candida albicans*.

thryptogen (**thryptophyte**), an organism increasing the sensitivity of a suscept to outside factors, e.g. to cold (Langer, 1936).

Thryptospora Petr. (1947), Dothideales (inc. sed.). 1, Syria.

Thuchomyces Hallbauer & Jahns (1977), ? Fossil fungi (L). 1 (Precambrian), S. Afr. Formed abiotically fide Cloud (*Palaeobiology* **2**: 351, 1976). See also Klappa (*Sediment. Petrol.* **49**: 387, 1979).

Thuemenella Penz. & Sacc. (1898), Xylariaceae. 1, Afr., N. & C. Am., Asia, Eur. See Samuels & Rossman (*Mycol.* **84**: 26, 1992). Anamorph *Nodulisporium*.

Thuemenia Rehm (1878) = Botryosphaeria (Botryosphaer.) fide v. Arx & Müller (1975).

Thuemenidium Kuntze (1891) = Geoglossum (Geogloss.) fide Nitare (*Windahlia* **14**: 37, 1984). See Benkert (*Delitschia* **10**: 141, 1983; key), Spooner (*Bibl. Mycol.* **116**, 1987; nomencl.).

Thujacorticium Ginns (1988), Hyphodermataceae. 1, Canada.

Thwaitesiella Massee (1892) = Lopharia (Ster.) fide Donk (1957).

Thyrea A. Massal. (1856), Lichinaceae (L). *c.* 20, cosmop. See Zahlbruckner (1924), Yoshimura (*J. Jap. Bot.* **43**: 354, 500, 1968), Asahina (*J. Jap. Bot.* **45**: 65, 1970), Henssen (*Ber. dt. bot. Ges.* **81**: 176, 1968), Moreno & Egea (*Lichenologist* **24**: 215, 1992, *Acta Bot. Barcin.* **41**: 1, 1992; Eur. N. Afr.).

Thyriascus Schulzer (1877), Fungi (inc. sed.).

Thyridaria Sacc. (1875), ? Dothideales (inc. sed.). 16, widespr. See Wehmeyer (*Lloydia* **4**: 241, 1941), Barr (*N. Am. Fl.* II **13**, 1990; N. Am.).

Thyridella (Sacc.) Sacc. (1895), Ascomycota (inc. sed.). 10, widespr.

Thyridiaceae J.Z. Yue & O.E. Eriks. (1987), Ascomycota (inc. sed.). 2 gen. (+ 6 syn.), 17 spp. Stromata immersed to erumpent, usually yellow to brownish, soft-textured; ascomata perithecial, ± globose to pyriform, often long-necked, thin-walled, the ostioles sometimes convergent, periphysate; interascal tissue of thin-walled paraphyses and apical paraphyses; asci cylindrical, persistent, without separable wall layers, thickened at the apex, with an I- apical ring; ascospores at least partially brown, muriform, without a sheath. Anamorphs coelomycetous. Saprobic on wood and bark, widespr.
  Lit.: Eriksson & Yue (*SA* **8**: 9, 1989, **10**: 57, 1990).

Thyridium Nitschke (1867), Thyridiaceae. 15, widespr. See Eriksson & Yue (1989), Esfandiari & Petrak (*Sydowia* **4**: 11, 1950), Leuchtmann & Müller (*Bot. Helvetica* **96**: 283, 1986). Anamorph *Pleurocytospora*.

Thyridium Fuckel (1870) [non Mitt. (1868), *Musci*] ≡ Mycothyridium E. Müll. (Thyrid.).

Thyrinula Petr. & Syd. (1924), Anamorphic Asterinaceae, 5.E1.15. Teleomorph *Aulographina*. 1, S. Afr., Australia. See Swart (*TBMS* **90**: 286, 1988).

Thyriochaetum Frolov (1968) ? = Amerosporium (Mitosp. fungi) fide Sutton & Sarbhoy (*TBMS* **66**: 297, 1976).

Thyriodictyella Cif. (1962), Micropeltidaceae. 1, Dominican Republic.

Thyriopsis Theiss. & Syd. (1915), Asterinaceae. 1, Eur.

Thyriostroma Died. (1913) = Leptostroma (Mitosp. fungi) fide Höhnel (*Sber. Akad. Wiss. Wien* **124**: 49, 1915).

Thyriostromella Bat. & C.A.A. Costa (1959), Mitosporic fungi, 5.B1.?. 1, Malaysia.

thyriothecium, an inverted flattened ascoma, having the wall ('scutellum', 'placodium') more or less radial in structure, and lacking a basal plate, e.g. *Microthyrium*; cf. catathecium.

Thyrococcum (Sacc.) Sacc. (1913) nom. dub. fide v. Höhnel (*Sber. Akad. Wiss. Wien* **124**: 49, 1915).

Thyrodochium Werderm. (1924) = Stemphylium (Mitosp. fungi) fide Wiltshire (1938).

Thyronectria Sacc. (1875), Hypocreaceae. 20, esp. S. Am., Antarctic. See Seeler (*J. Arnold Arbor.* **21**: 429, 1940), Subramanian & Bhat (*Cryptogamie, Mycol.* **5**: 307, 1985; ontogeny).

Thyronectroidea Seaver (1909) = Thyronectria (Hypocr.) fide Seeler (1940).

Thyrosoma Syd. (1921) = Cyclotheca (Microthyr.) fide Müller & v. Arx (1962).

Thyrospora Tehon & E.Y. Daniels (1925) = Stemphylium (Mitosp. fungi) fide Wiltshire (1938). See Smith (*J. agric. Res.* **61**: 831, 1940).

Thyrospora Kirschst. (1938) [non Tehon & E.Y. Daniels (1925)], Dothideales (inc. sed.). 1, Germany.

Thyrostroma Höhn. (1911), Mitosporic fungi, 3.C/D2.19. 15, widespr. See Sutton & Pascoe (*MR* **92**: 210, 1989; relationship with *Stigmina*), Yuan & Old (*MR* **94**: 573, 1990; *T. eucalypti* on *Eucalyptus*).

Thyrostromella Höhn. (1919), Mitosporic fungi, 3.D2.10. 1, Australia, Eur.

Thyrostromella Syd. (1924) [non Höhn. (1919)] = Stigmina (Mitosp. fungi) fide Ellis (*DH*). = Thyrostroma (Mitosp. fungi) fide Sutton & Pascoe (*MR* **92**: 212, 1989).

Thyrotheca Kirschst. (1944), Ascomycota (inc. sed.). 1, N. Am.

Thyrsidiella Höhn. ex Höhn. (1909), Mitosporic fungi, ?.?.?. 2, Eur.

Thyrsidina Höhn. (1905), Mitosporic fungi, 6.D1.1. 1, Eur.

Thyrsidium Mont. (1849) ≡ Myriocephalum (Mitosp. fungi).

thyrsus (pl. **thyrsi**), (1) a type of inflorescence (Bot.); (2) the densely branched apices of some lichens, e.g. *Cladonia stellaris*.

thysanoblastic (of conidiogenesis), when 'the whole of the upper surface of the conidiogenous cell takes part in the process of conidium formation and secession, and both schizolysis and rhexolysis occur alternately in successively seceding conidia' (Roux & van Warmelo, *MR* **92**: 225, 1989).

Thysanophora W.B. Kendr. (1961), Mitosporic fungi, 1.A2.15. 5, Eur., N. Am. See Stolk & Hennebert (*Persoonia* **5**: 189, 1968).

Thysanophoron Stirt. (1883) = Sphaerophorus (Sphaerophor.) fide Wedin (1993).

Thysanopyxis Rabenh. ex Bonord. (1864) = Volutella (Mitosp. fungi) fide Saccardo (*Syll. Fung.* **4**: 684, 1886).

Thysanothecium Mont. & Berk. (1846), Cladoniaceae (L). 2, Australia, NZ, Asia. See Ahti (1993), Galloway (*Nova Hedw.* **28**: 499, 1977).

Tiarospora Sacc. & Marchal (1885), Mitosporic fungi, 4.B1.1. 1, Eur.

Tiarosporella Höhn. (1924), Anamorphic Rhytismatales, 4.A1.1. Teleomorph *Darkera*. 7, Eur., N. Am. See Sutton (1980), Roux *et al.* (*MR* **94**: 254, 1990; conidiomatal and conidial ontogeny), Nag Raj (1993).

Tiarosporellivora Punith. (1981), Mitosporic fungi, 4.A1.15. 1 (in *Tiarosporella*), Germany.

Tibellia Vězda & Hafellner (1992), Bacidiaceae (L). 1, Australia.

tibi, Swiss drink derived from a 15% sucrose sol. (+ other ingredients) by fermentation with 'Tibi grains', a symbiotic association of *Betabacterium vermiforme* and *Saccharomyces intermedius*. Cf. gingerbeer plant, tea fungus, teekwass.

Tichodea Körb. (1848) = Collema (Collemat.) fide Hawksworth (*in litt.*).

Tichospora A. Massal. ex Horw. (1912) nom. inval. ? = Psorotichia (Lichin.).

Tichothecium Flot. (1850) = Verrucaria (Verrucar.) fide Hawksworth (*Bot. Notiser* **132**: 283, 1979).

tichus, peripheral layer of cells of perithecial walls forming a dark protective layer as in *Pleospora herbarum* (Groenhart, *Persoonia* **4**: 11, 1965).

Ticogloea G. Weber, Spaaij & W. Gams (1994), Mitosporic fungi, 1.A2.10. 1, Costa Rica.

Ticomyces Toro (1952), Meliolaceae. 1, C. Am.

Tieghemella Berl. & De Toni (1888) = Absidia (Mucor.) fide Hesseltine (1955).

Tieghemiomyces R.K. Benj. (1959), Dimargaritaceae. 2 (mycoparasites of *Mucorales*), USA. See Benjamin (*Aliso* **5**: 11, 1961; key).

tiger's milk, *Polyporus sacer*; used as a medicine in Malaysia.

Tigria Trevis. (1853) = Erysiphe (Erysiph.) fide Braun (*SA* **7**: 57, 1989).

Tilachlidiopsis Keissl. (1924), Anamorphic Tricholomataceae, 2.A1.41. Teleomorph *Collybia*. 1 (with clamp connexions), N. temp. See Stalpers *et al.* (*CJB* **69**: 6, 1991; generic revision), Mains (*Bull. Torrey bot. Cl.* **78**: 122, 1951).

Tilachlidium Preuss (1851), Mitosporic fungi, 3.A1.15. 1, widespr. See Gams (*Cephalosporium-artige Schimmelpilze*, 1971).

Tilakidium Vaidya, C.D. Naik & Rathod (1986), ? Hypocreaceae. 1, India.

Tilakiella Srinivas. (1973), Dothideales (inc. sed.). 1, India.

Tilakomyces Sathe & Vaidya (1979) = Eutypella (Diatryp.) fide Rappaz (*Mycol. Helv.* **3**: 281, 1989).

Tilletia Tul. & C. Tul. (1847), Tilletiaceae. *c.* 100 (on *Gramineae*), cosmop. *T. caries* (syn. *T. tritici*) and *T. laevis* (syn. *T. foetida*) (wheat bunt); *T. controversa* (dwarf bunt); possible conspecificity of these, see Russell & Mills (*Phytopath.* **84**: 576, 1994); *T. barclayana* (syn. *T. horrida*; *Neovossia horrida*) (rice smut); *T. indica* (syn. *N. indica*) (kernal bunt of wheat); See Holton & Heald (*Bunt or stinking smut of wheat*, 1941), Durán & Fischer (*The genus Tilletia*, 1961; keys).

Tilletiaceae J. Schröt. (1887), Ustilaginales. 28 gen. (+ 9 syn.), 432 spp. Basidiospores produced at the apex of an aseptate or 1-septate promycelium; 4 or more (commonly 8) 1-nucleate 'primary conidia' produced by the promycelium often, after conjugation with one another, give 'secondary conidia' (Brefeld), but in some spp. primary conidia are multinucleate and do not conjugate (Duran, 1987). The 'primary conidia' have been interpreted either as basidiospore or as highly specialized sterigmata because the 'secondary conidia' are ballistosporic like typical basidiospores (see Buller & Vanterpool, *in* Buller, **5**; Kollmagen *et al.*, *TBMS* **75**: 461, 1980; Ingold, *TBMS* **88**: 75, 1987).

Tilletiales, see *Ustilaginales*.

Tilletiaria Bandoni & B.N. Johri (1972), Sporidiobolaceae. 1, Canada.

Tilletiella Zambett. (1970) nom. inval. (Anamorphic Tilletiaceae). Teleomorph *Tilletia*.

Tilletiopsis Derx (1948), Anamorphic Tilletiaceae. Teleomorphs *Entyloma*, *Melanotaenium*. 12, widespr. See Boekhout (*Stud. Mycol.* **33**: 1, 1991; species revision).

Tilmadoche Fr. (1849) = Physarum (Physar.).

Tilotus Kalchbr. (1881) = Phyllotopsis (Lentin.) fide Singer (1975).

Timgrovea Bougher & Castellano (1993), Hymenogastraceae. 5, Australasia, China.

Tinctoporellus Ryvarden (1979), Coriolaceae. 1, pan-trop.

Tinctoporia Murrill (1907) = Porogramme (Grammothel.) fide Ryvarden (1979).

tinder fungus, *Fomes fomentarius*; **false - -**, *Phellinus igniarius*. Cf. amadou.

tinea, ringworm or other skin diseases in humans or animals caused by a parasitic fungus (esp. a dermatophyte). **- barbae** (- sycosis), beard ringworm; **- capitis** (- tonsurans), head ringworm; **- corporus** (- circinata), body ringworm; **- cruris**, groin ringworm; **- favosa** = favus (q.v.); **- imbricata** (Tokelau) (*Trichophyton concentricum*); **- nigra**, pigmented cutaceous infection caused by a dematiaceous fungus; **- nodosa** = piedra (q.v.); **- pedis**, 'athlete's foot', foot ringworm; **- unguium**, ringworm of the nails; **- versicolor** = pityriasis versicolor (q.v.).

Tingiopsidium Werner (1939) = Koerberia (Placynth.) fide Henssen (1963).

tinophyses (Groenhart, *Persoonia* **4**: 11, 1965), paraphysoids; see hamathecium.

Tipularia Chevall. (1822) [non Nutt. (1818), *Orchidaceae*] ≡ Halterophora (? Myxom.).

Tirispora E.B.G. Jones & Vrijmoed (1994), Halosphaeriaceae. 1 (marine), Hong Kong.

Tirmania Chatin (1892), Pezizaceae. 2 (hypogeous), Asia, N. Afr. See Trappe (*TBMS* **57**: 185, 1971), Malençon (*Persoonia* **7**: 261, 1973), Alsheikh & Trappe (*TBMS* **81**: 83, 1983).

tissue types, Korf distinguished the types of hyphal tissues in discomycetes as different **textura**'s and this is now applied to all ascomycetes and coelomycetes. Tissue (textura) types (from Korf, *Sci. Rep. Yokohama nat. Univ.* II **7**: 13, 1958; which is derived from Starbäck, 1895). See also Dargan (*Nova Hedw.* **44**: 489, 1987; *Xylariaceae*), plectenchyma. See Fig. 39.

Titaea Sacc. (1876), Mitosporic fungi, 1.G1.10. 4 (on fungi), widespr. See Sutton (*TBMS* **83**: 399, 1984; key).

Titaeella G. Arnaud ex K. Ando & Tubaki (1985), Mitosporic fungi, 1.G1.1. 1 (with clamp connexions, aquatic), Eur., Japan. See Ando & Tubaki (*Trans. Mycol. Soc. Jap.* **26**: 151, 1985).

Titaeopsis B. Sutton & Deighton (1984), Mitosporic fungi, 1.G1.10. 1 (on leaf ascomycetes), Uganda.

Titaeospora Bubák (1916), Mitosporic fungi, 6.B1.15. 2, temp.

Titaeosporina Luijk (1920) = Asteroma (Mitosp. fungi) fide Sutton (1977).

Titanella Syd. & P. Syd. (1919) = Pyrenula (Pyrenul.) fide Harris (1989). See Barr (*Mycotaxon* **9**: 17, 1979).

Titania Berl. (1900) [non Endl. (1833), *Orchidaceae*] ≡ Freminaevia (Melanconid.).

Tjibodasia Holterm. (1898) = Platygloea (Platygl.) fide Martin (*Lloydia* **11**: 119, 1948).

toadstool, a basidioma, esp. an inedible one, of an agaric or a bolete; there is no unequivocal definition and the term is best avoided; in 1959 there was a controversial court case as to whether soup made from *Boletus edulis* should be referred to as 'toadstool soup' (Small, *Pl. Sci. Bull.* **21**(3): 34, 1975). Cf. mushroom.

Tode (Heinrich Julius; 1733-97). President of the Synod of Wittenburg. Noted for his *Fungi Mecklenburgenses selecti* (1790-91) in which he gave names to such common genera as *Acrospermum*, *Hysterium*, *Myrothecium*, *Stilbum*, *Sphaerobolus*, and *Volutella* and gave accounts of 54 spp. of *Sphaeria*. See Grummann (1974: 50), Stafleu & Cowan (*TL-2* **6**: 382, 1986).

Tofispora G. Langer (1994), Ceratobasidiaceae. 3, trop.

tofu, see sufu.

Togaria W.G. Sm. (1908) = Agrocybe (Bolbit.) fide Singer (1975).

Togninia Berl. (1900) = Pleurostoma (Calosph.) fide Barr *et al.* (*Mycotaxon* **48**: 529, 1993).

Tolediella Viégas (1943) = Phyllachora (Phyllachor.) fide Petrak (*Sydowia* **5**: 340, 1951).

tolerant (of an organism), giving little reaction to infection by a pathogen or to the effect of other factors (e.g. tolerant of heat, of a virus).

Tolypocladium W. Gams (1971), Mitosporic fungi, 1.A1.15. 10, widespr. See Holubová-Jechová (*Mykol. Listy* **31**: 8, 1988; key), Riba *et al.* (*J. Invert. Path.* **48**: 362, 1986; isozyme analysis), Aarnio & Agathos (*Appl. Microbiol. Biotech.* **33**: 435, 1990; pigmented variants in *T. inflatum*), Rakotonirainy (*Jl Invert. Path.* **57**: 17, 1991; rRNA sequence comparison with *Beauveria*), Stimberg *et al.* (*Appl. Microbiol. Biotechnol.* **37**: 485, 1992; electrophoretic karyotyping); see also cyclosporin. = Beauveria (Mitosp. fungi) fide v. Arx (*Mycotaxon* **25**: 153, 1986, *Persoonia* **13**: 467, 1988).

Tolypomyria Preuss (1852) = Trichoderma (Mitosp. fungi) fide Hughes (*Friesia* **9**: 64, 1969).

tolypophagous, see ptyophagous.

Tolyposporella G.F. Atk. (1897), Ustilaginaceae. 1 (on *Sorghastrum*), N. Am. See Thirumalachar *et al.* (*Mycol.* **59**: 389, 1967).

Tolyposporidium Thirum. & Neerg. (1978), = Moesziomyces (Ustilagin.) fide Vánky (1985).

Tolyposporium Woronin ex J. Schröt. (1887), Ustilaginaceae. *c.* 10, widespr. See Vánky (*Bot. Not.* **130**: 131, 1977); cf. *Meosziomyces*.

Tomasellia A. Massal. (1856), ? Arthopyreniaceae (±L). *c.* 50, cosmop. (mainly trop.). See Keissler (*Rabenh. Krypt.-Fl.* **9**, 1(2), 1937).

Tomaselliella Cif. (1952) = Arthonia (Arthon.).

Tomaselliellomyces Cif. (1953) ≡ Tomaselliella (Arthon.).

Tomaselliomyces Cif. & Tomas. (1953) ≡ Tomasellia (Arthopyren.).

Tomentella Pat. (1887) nom. cons., Thelephoraceae. 70, cosmop. See Larsen (*Mycol. Mem.* **4**, 1974; key 72 spp.), Wakefield (*TBMS* **53**: 161, 1969; key 40 Br. spp.), Stalpers (1993; key).

Tomentella Johan-Olsen (1888) [non Pat. (1887), nom. cons.] = Tomentella Pat. (Thelephor.).

Tomentella P. Karst. (1889) [non Pat. (1887), nom. cons.] = Trechispora (Sistotremat.).

Tomentellago Hjortstam & Ryvarden (1988), Thelephoraceae. 1, Colombia.

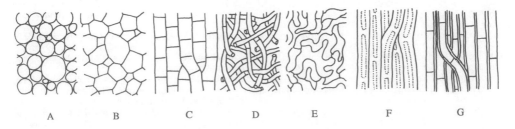

Fig. 39. Hyphal tissue (textura) types (Korf, 1958). A, textura globulosa; B, textura angularis; C, textura prismatica; D, textura intricata; E, textura epidermoidea; F, textura oblita; G, textura porrecta.

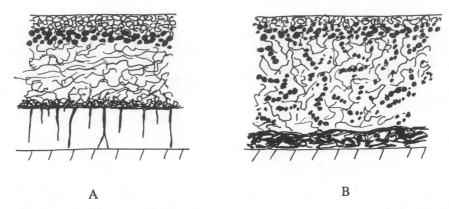

A   B

Fig. 40. Tissue organization in lichen thalli (diagrammatic). A, heteromerous (stratified) e.g. *Parmelia*; B, homoiomerous (unstratified) e.g. *Collema*.

Tomentellastrum Svrček (1958) = Tomentella (Thelephor.) fide Donk (1963).

Tomentelleopsis Orlova (1959) ? = Chromelosporium (Mitosp. fungi) fide Sutton (*in litt.*).

Tomentellina Höhn. & Litsch. (1906) = Kneiffiella (Thelephor.) fide Donk (1957).

Tomentellopsis Hjortstam (1970), Thelephoraceae. 5, widespr. See Hjortstam (*Svensk bot. Tidskr.* **68**: 51, 1974).

Tomenticola Deighton (1969), Mitosporic fungi, 1.C2.10. 1, trop. Afr.

Tomentifolium Murrill (1903) ≡ Phyllotopsis (Lentin.).

Tomentoporus Ryvarden (1973) = Microporus (Polypor.) fide Reid (*Microscopy* **32**: 448, 1975).

tomentose, having a covering of soft, matted hairs (a **tomentum**; downy.

Tomeoa I. Hino (1954), Dothideales (inc. sed.). 1, Japan.

Tomophagus Murrill (1905) = Ganoderma (Ganodermat.) fide Furtado (1965).

Tompetchia Subram. (1985), Mitosporic fungi, 1.D2.1. 1 (on scale insects), widespr.

Tonduzia F. Stevens (1927) [non Pittier (1908), *Apocynaceae*] ≡ Dontuzia (Ascomycota, inc. sed.).

Toninia A. Massal. (1852) nom. cons. prop., Catillariaceae (±L). 48 (a few on lichens), widespr. See Timdal (*Opera Bot.* **110**, 1991; key).

Toniniopsis Frey (1926), Bacidiaceae (L). 1, Eur.

tonophily (adj. **-ilic, -ilous**), the ability to grow under conditions of high osmotic pressure.

Topelia P.M. Jørg. & Vězda (1984), ? Stictidaceae (L). 4, Eur., N. Am.

Tophora Fr. (1825) ≡ Byssus (Algae).

Topospora Fr. (1836), Anamorphic Leotiaceae, 8.C1.15. Teleomorph *Godronia*. Widespr.

Tormentella H.D. Pflug (1966), Fossil fungi. 1, USA.

Tornabea Østh. (1980), Physciaceae (L). 2, Eur., S. Am., Afr., China, Canary Isl. See Kurokawa (*J. Jap. Bot.* **37**: 289, 1962).

Tornabenia A. Massal. (1853) [non *Tornabenea* Parl. (1850), *Umbelliferae*] = Anaptychia (Physc.).

Tornabenia Trevis. (1853) [non *Tornabenea* Parl. (1850), *Umbelliferae*] ≡ Tornabea (Physc.).

Tornabeniopsis Follmann (1980) nom. inval. ≡ Tornabea (Physc.).

Toroa Syd. (1926), Pseudoperisporiaceae. 2, trop. See Hansford (*Mycol. Pap.* **15**, 1946).

Torpedospora Meyers (1957), Ascomycota (inc. sed.). 2 (marine), widespr.

Torrendia Bres. (1902), Torrendiaceae. 2, mediterr. See Malençon (*Rev. Mycol.* **20**: 81, 1955; development), Horak (*Mycol.* **84**: 64, 1992), Bas (*Beih. Nova Hedw.* **51**: 53, 1975; relationship to *Amanita*).

Torrendiaceae Jülich (1982), Agaricales. 1 gen., 2 sp. Gasterocarp loculate, often with pseudostipe and volva; spores hyaline (spore print white), ellipsoid to tetraradiate.

Torrendiella Boud. & Torrend (1911), Sclerotiniaceae. 1, Portugal. See Galán *et al.* (*Mycotaxon* **48**: 229, 1993; posn), Spooner (*Bibl. Mycol.* **116**, 1987).

Torrubia Lév. (1863) ≡ Cordyceps (Clavicipit.).

Torrubiella Boud. (1885), Clavicipitaceae. 10 (on *Arachnida* and *Coccida*), widespr. See Petch (*TBMS* **27**: 81, 1944).

Torsellia Fr. (1849) = Cytospora (Mitosp. fungi) fide Défago (*Phytopath. Z.* **14**: 103, 1944).

torsive, spirally twisted.

Torula Pers. (1794), Mitosporic fungi, 1.A2.3. 6, widespr. *T. ligniperda* causes stain in hardwoods. See Rao & de Hoog (*Persoonia* **8**: 201, 1975; key), Crane & Schoknecht (*Mycol.* **78**: 86, 1986). Sensu Turpin (1838) = Torulopsis (Mitosp. fungi).

Torulaspora Lindner (1904), Saccharomycetaceae. 3, widespr. See Kreger-van Rij (1984), v.d. Walt & Johansen (*CSIR Res. Rept.* **325**, 1975), Yamada *et al.* (*J. gen. appl. Microbiol.* **37**: 503, 1991; molec. relations).

Torulella Gyeln. (1939) nom. dub. (Mitosp. fungi) fide Hawksworth (*Bull. Br. Mus. nat. Hist., Bot.* **6**: 181, 1979).

Torulina Sacc. & D. Sacc. (1906) = Gliomastix (Mitosp. fungi) fide Dickinson (1968).

Torulites Pia (1927), Fossil fungi. 1 (Oligocene), Eur.

Torulites Grüss (1927), Fossil fungi. 1 (Tertiary), Eur.

Toruloidea Sumst. (1913) = Oospora (Mitosp. fungi) fide Clements & Shear (1931).

Torulomyces Delitsch (1943), Mitosporic fungi, 1.A1.15. 2, Eur., N. Am. See Barron (*Mycol.* **59**: 716, 1967).

Torulopsidosira Geitler (1955), ? Algae (achlorotic) fide Batra (*in* Subramanian (Ed.), *The taxonomy of fungi* **1**: 187, 1978).

Torulopsiella Bender (1932), Mitosporic fungi, 1.A2.38. 2, S. Am. See Hughes (*N.Z. Jl Bot.* **10**: 232, 1972).

Torulopsis Berl. (1894) = Candida (Mitosp. fungi) fide Yarrow & Meyer (*Int. J. Syst. Bact.* **28**: 611, 1978), Odds *et al.* (*J. Gen. Microbiol.* **136**: 761, 1990; synonymy with *Candida*). = Saccharomyces (Endomycet.) fide v. d. Walt & Johannsen (*Ant. v. Leeuwenhoek* **40**: 281, 1974).

Torulopsis Oudem. (1903) = Gliomastix (Mitosp. fungi) fide Sutton (*in litt.*).

Torulopsis Speg. (1918) ≡ Torulopsiella (Mitosp. fungi).

torulose (**torulous**), cylindrical but having swellings at intervals; moniliform.

torulosis, see cryptococcosis.

Tothia Bat. (1960), ? Microthyriaceae. 1, Hungary.

totipotent, bisexual.

touchwood, (1) wood rotted by fungi (esp. *Polyporus squamosus*); (2) *Fomes fomentarius* or *Phellinus igniarius* basidiomata or the tinder ('amadou', q.v.) made from them.

Tovariella Syd. (1930), Leotiales (inc. sed.). 1, S. Am.

toxic, of, caused by, or acting as, poison; **-ity**, the power of acting as a poison.

toxigenic, toxin producing.

toxin, a non-enzymic metabolite of one organism which is injurious to another (cf. antibiotic); **myco-**, a toxin produced by a fungus, esp. one affecting humans or animals (see Mycetisms, Mycotoxicoses, Phytotoxic mycotoxins); **patho-** (Wheeler & Luke, *Ann. Rev. Phytopath.* **17**: 223, 1963), see vivotoxin; **phyto-**, (1) a toxin produced by a plant (cf. phytoalexins); (2) (frequently, but better avoided), a toxin injurious to plants (see Phytotoxic mycotoxins); **vivo-**, a toxin 'produced in the infected host by the pathogen and/or its host, which functions in the production of disease, but is not itself the initial inciting agent' (Dimond & Waggoner, *Phytopath.* **43**: 229, 1953); pathotoxin.

toxiphilous, favouring a polluted habitat (e.g. *Lecanora conizaeoides* in area of high sulphur dioxide pollution), cf. poleophilous; **toxiphobous**, not tolerating such a habitat, e.g. *Usnea* spp.); **toxitolerant**, tolerant of toxins. See Air pollution.

Toxosporiella B. Sutton (1986), Mitosporic fungi, 4.C2.1. 1, Australia.

Toxosporiopsis B. Sutton & Sellar (1966), Mitosporic fungi, 8.C2.1. 3, Sierra Leone, Italy, China, USA. See Sutton & Dyko (*MR* **93**: 476, 1989), Wu (*Acta Mycol. Sin.* **12**: 205, 1993; n.sp.).

Toxosporium Vuill. (1896), Mitosporic fungi, 3.C2.1. 2, widespr. See Sutton (*Mycol. Pap.* **138**, 1975).

Toxotrichum G.F. Orr & Kuehn (1964) = Myxotrichum (Myxotrich.) fide Apinis (1964).

trabecula (pl. **-ae**; adj. **-ate**), (1) a lamella primordium; (2) (of *Gymnoglossum* and other gasteromycetes), plates of undifferentiated primordial tissue in the developing gleba forming a branch of a dendroid columella; (3) (of pseudoparaphyses), paraphysoids, tinophyses, see hamathecium.

Trabecularia Bonord. (1857) = Merulius (Merul.) fide Donk (1951).

Trabrooksia H.W. Keller (1980), Didymiaceae. 1, USA.

Trabutia Sacc. & Roum. (1881), Phyllachoraceae. 5 (on *Quercus* and *Nothofagus*), temp. See Petrak (*Ann. Myc.* **27**: 385, 1929); Barr (*Mycol.* **79**: 188, 1987; key 3 N. Am. spp.), v. Arx (*SA* **6**: 213, 1987).

Trabutiella Theiss. & Syd. (1914) = Phyllachora (Phyllachor.) fide v. Arx & Müller (1954).

Trabutiella F. Stevens (1920) ≡ Diatractium (Phyllachor.).

trace elements, see nutrition of fungi.

tracheomycosis, see hadromycosis.

Trachyderma Norman (1853) = Pannaria (Pannar.) fide Jørgensen (1978).

Trachyderma (Imazeki) Imazeki (1952) = Ganoderma (Ganodermat.) fide Donk (1960).

Trachylia Fr. (1817) = Arthonia (Arthon.).

Trachylia Tuck. (1848) = Cyphelium (Calic.).

Trachypus J. Bataille (1908) [non Reinw. & Hornsch. (1826), *Musci*] ≡ Krombholzia (Bolet.).

Trachysphaera Tabor & Bunting (1923), Pythiaceae. 1 (*T. fructigena*, mealy pod of cacao), Afr.

Trachyspora Fuckel (1861), Phragmidiaceae. 4 (on *Alchemilla*), N. temp., Java. See Gäumann (*Boissiera* **7**: 105, 1943), Gjaerum & Cummins (*Mycotaxon* **15**: 420, 1982).

Trachysporella Syd. (1921) = Trachyspora (Phragmid.) fide Dietel (1928).

trachytectum, see exosporium; Spore wall.

Trachythyriolum Speg. (1919) nom. conf. fide Petrak & Sydow (*Ann. Myc.* **33**: 157, 1935).

Trachyxylaria A. Møller (1901) = Xylobotryum (Ascomycota) fide Clements & Shear (1931).

Tracya Syd. & P. Syd. (1901), Tilletiaceae. 2 (on aquatic plants), N. Hemisph.

Tracyella Zambett. (1970) nom. nud. (Mitosp. fungi). Teleomorph *Tracya*.

Tracylla (Sacc.) Tassi (1904), Mitosporic fungi, 5.A1.15/16. 2, widespr. See Petrak (*Sydowia* **1**: 202, 1947), Sutton (*Sydowia* **43**: 264, 1991), Nag Raj (1993).

Trailia G.K. Sutherl. (1915), Halosphaeriaceae. 1 (on *Ascophyllum*), UK.

Trailia Syd. (1922) = Puccinia (Puccin.) fide Dietel (1928).

trama, the layer of hyphae in the central part of a lamella of an agaric, a spine of *Hydnaceae*, or the dissepiment between pores in a polypore. Cf. context.

Trametella Pinto-Lopes (1952) = Funalia (Coriol.) fide Pegler (1970).

Trametes Fr. (1836), Coriolaceae. 60 s.str. *c.* 12; (on wood), cosmop. Living trees attacked by *T. pini* (Haddow, *TBMS* **22**: 182, 1938) and others. See Ryvarden (1978: 421; key 8 Eur. spp.), Ryvarden & Johansen (*Preliminary polypore flora of East Africa*: 555, 1980; key 20 Afr. spp. s.lat.).

Trametites Mesch. (1892), Fossil fungi. 3 (Cretaceous, Pliocene), Eur.

Transeptaesporites V.S. Ediger (1981), Fossil fungi. 5 (Upper Oligocene), Turkey, USA.

transformations, in fungi; see Biotechnology, Genetic engineering.

trans-septum, see septum.

Tranzschelia Arthur (1906), Uropyxidaceae. 7, widespr. (esp. N. temp.). *T. discolor*, *T. pruni-spinosae* (I on Ranunculaceae, II, III on *Prunus*). See Bennell & Henderson (*TBMS* **71**: 271, 1978; spore devel.), López-Franco & Hennen (*System. Bot.* **15**: 560, 1990).

Tranzscheliella Lavrov (1936), Ustilaginaceae. 1 (on *Stipa*), widespr. ? = Ustilago (Ustilag.) fide Gutner (*Golovnevye griby*, 1941), Yuan *et al.* (*Acta mycol. sin.* **6**: 250, 1987), Vánky (1987).

Tranzschel's Law, in essence, states that the telia of microcyclic rust species adopt the habit of the parent macrocyclic species and occur on the aecial host plants of the latter. There are no proven exceptions, the only one firmly suspected (*Chrysomyxa arctostaphyli*) having been disproved (Peterson, *Science*, *N.Y.* **134**: 468, 1961).

Trapelia M. Choisy (1929) nom. cons. prop., Trapeliaceae (L). 8, temp. See Coppins & James (*Lichenologist* **16**: 241, 1984), Hertel (*Herzogia* **1**: 111, 1968, *Vortr. bot. Ges.* n.f. **4**: 171, 1970), Honegger (*Lichenologist* **14**: 205, 1982; asci).

Trapeliaceae M. Choisy ex Hertel (1970) nom. cons., Lecanorales (L). 7 gen. (+ 5 syn.), 61 spp. Thallus crustose or lobed; ascomata apothecial, brightly coloured or brown, usually domed, with poorly-developed marginal tissue; interascal tissue of branched and anastomosing paraphyses, sometimes swollen and pigmented at the apices; asci with a weakly I+ or I- apical cap, without an ocular chamber or apical cushion, rarely with an internal more strongly staining apical region, sometimes with an outer I+ gelatinized layer; ascospores hyaline, aseptate, without a sheath. Cephalodia sometimes present. Anamorphs pycnidial. Lichenized with green algae (cyanobacteria in the cephalodia), widespr. esp. cold temp.

Included in *Agyriaceae* by Rambold & Triebel (*Notes R. bot. Gdn, Edin.* **46**: 375, 1990).

Trapeliopsis Hertel & Gotth. Schneid. (1980), Trapeliaceae (L). 12, Eur., Afr., NZ, N. Am. See Coppins & James (*Lichenologist* **16**: 241, 1984), Coppins (*SA* **7**: 93, 1988).

Trappea Castellano (1990), Hysterangiaceae. 1, Eur., N. Am.

Traquairia Carruth. (1873), Fossil fungi (Ascomycota). 2, USA. See Smith (*Am. J. Bot.* **70**: 387, 1983).

Traversoa Sacc. & Syd. (1913) = Lasiodiplodia (Mitosp. fungi) fide Sutton (1977).

Trechispora P. Karst. (1890), Sistotremataceae. 42, widespr. See Liberta (*CJB* **51**: 1871, 1973), Larsson (1992).

tree hair, (1) lichens, esp. fruticose spp. (*Bryoria*, *Usnea*), etc. growing on tree trunks (obsol.); (2) the lichen *Pseudevernia furfuracea* (mousse d'arbre); a source of perfume. Cf. oak-moss.

trehalose, a reserve disaccharide (α-D-glucopyanosyl-α-D-glucopyanoside) of fungi (esp. yeasts) and lichens which is hydrolyzed by the enzyme **trehalase**; mycose.

Treleasia Speg. (1896) = Pyxidiophora (Pyxidiophor.). See Petch (*Ann. Myc.* **34**: 74, 1936), Sutton (*Mycol. Pap.* **141**, 1977).

Treleasiella Speg. (1896) nom. dub. (Mitosp. fungi) fide Sutton (1977).

trellis rust, on pear (*Gymnosporangium fuscum*); on juniper (*G. sabinae*).

Tremateia Kohlm., Volkm.-Kohlm. & O.E. Erikss. (1995), ? Pleosporaceae. 1 (on *Juncus*), USA. Anamorph *Phoma*-like.

Trematomyces Schrantz (1961) = Requienella (Pyrenul.) fide Boise (1986).

Trematophlyctis Pat. (1918), ? Chytridiales (inc. sed.). 1, Madagascar.

Trematophoma Petr. (1924), Mitosporic fungi, 4.A1.19. 1, Austria.

Trematophora Eisenack (1965), ? Fossil fungi.

Trematosphaerella Kirschst. (1906) ? = Phaeosphaeria (Phaeosphaer.) fide Hara (1959) but see Holm (1957).

Trematosphaeria Fuckel (1870), ? Melanommataceae. 5, widespr. See Boise (*Mycol.* **77**: 230, 1985; key), Fisk & Webster (*Nova Hedw.* **54**: 77, 1992). Anamorph *Aposphaeria*, *Zalerion*.

Trematosphaeriopsis Elenkin (1901) ? = Homostegia (Dothideales) fide Theissen & Sydow (*Ann. Myc.* **13**: 603, 1915).

Trematosphaeris Clem. & Shear (1931) ≡ Trematosphaeriopsis (Dothideales).

Trematosphaerites Mesch. (1892), Fossil fungi. 1 (Tertiary), Eur. See Bužek & Holy (*Sbor. geol. ved.* **4**: 108, 1964).

Trematosphaerites Grüss (1924), ? Fossil fungi. 1 (Devonian), Spitzbergen.

Trematostoma (Sacc.) Shear (1942) = Exarmidium (Hyponectr.) fide Barr & Boise (1985).

Trematovalsa Jacobesco (1906), ? Diaporthales (inc. sed.). 1, France.

trembling fungi, the *Tremellales*.

Tremella L. (1753) nom. inval. (Cyanophyta).

Tremella Pers. (1794) nom. cons., Tremellaceae. *c.* 80, cosmop. The edible *T. fuciformis* ('Silver ear') is cultivated in China. See Donk (*Persoonia* **4**: 179, 1966; regional lists), Bandoni & Bisalputra (*CJB* **49**: 27, 1971; ultrastr. of haplonts), Barnett *et al.* (1990; yeast states 10 spp.), Diederich & Christiansen (*Lichenologist* **26**: 47, 1994; on lichens).

Tremellacantha Jülich (1980), Exidiaceae. 1, Indonesia.

Tremellaceae Fr. (1821), Tremellales. 13 gen. (+ 12 syn.), 95 spp. Basidioma intrahymenial to flabellate, gelatinous; spores mostly ballistosporic; haplophase blastosporic; mycoparasitic. See Barnett *et al.* (*Yeasts: characteristics and identification*, edn 2, 1990; yeast states).

Tremellales, Basidiomycetes. 10 fam., 60 gen. (+ 34 syn.), 256 spp. Saprobes on wood [used s.l. by Martin (1952) and other authors to include also *Auriculariales*, *Septobasidiales*, *Dacrymycetales* and *Tulasnellales*]. Fams:

(1) **Aporpiaceae.**
(2) **Exidiaceae.**
(3) **Hyaloriaceae.**
(4) **Phragmoxenidiaceae.**
(5) **Rhynchogastremataceae.**
(6) **Sirobasidiaceae.**
(7) **Syzygosporaceae.**
(8) **Tetragoniomycetaceae.**
(9) **Tremellaceae.**
(10) **Tremellodendropsidaceae.**

Lit.: Martin (1952), Donk (1951-1963; VIII, 1966: 218), Lowy (*Fl. Neotrop.* **6**, 1971; *Nova Hedw.* **19**: 407, 1971; key neotrop. spp.), Raitviir (*in* Parmasto, *Zhivaya priroda Dal'nego Vostoka*: 84, 1971; E. former USSR), McNabb (*in* Ainsworth *et al.* (Eds), *The Fungi* **4B**, 1973), Bandoni (*Trans. Mycol. Soc. Japan* **25**: 489, 1984; review, classific.; *Stud. Mycol.* **30**: 87, 1987; review), Wells (*Mycol.* **86**: 18, 1994; review, classif.).

Tremellastrum Clem. (1909) ≡ Crepidotus (Crepidot.).

Tremellidium Petr. (1927), Mitosporic fungi, ?8.A1.?. 1, Eur.

Tremellina Bandoni (1986), Mitosporic fungi, 1.?.?. 1, USA.

Tremellochaete Raitv. (1964) = Exidia (Exid.) fide Donk (*Persoonia* **4**: 166, 1966).

Tremellodendron G.F. Atk. (1902), Exidiaceae. 8, N. Am. See Bodman (*Am. Midl. Nat.* **27**: 203, 1942).

Tremellodendropsidaceae Jülich (1982), Tremellales. 1 gen. (+ 2 syn.), 7 spp. Basidioma ramarioid; basidia partially quadripartite; terrestrial.

Tremellodendropsis (Corner) D.A. Crawford (1954), Tremellodendropsidaceae. 7, widespr. See Corner (*TBMS* **49**: 205, 1966).

Tremellodiscus Lloyd (1925) nom. nud. ≡ Ruhlandiella (Peziz.) fide Dissing & Korf (1980).

Tremellodon (Pers.) Fr. (1874) ≡ Pseudohydnum (Exid.).

Tremellogaster E. Fisch. (1924), Sclerodermataceae. 1, Surinam and Guyana. See Linder (*Mycol.* **22**: 265, 1930).

tremelloid, (1) like jelly or wet gelatin; gelatinous; (2) *Tremella*-like.

Tremellopsis Pat. (1903) = Crepidotus (Crepidot.) fide Singer (1947).

Tremelloscypha D.A. Reid (1979), Exidiaceae. 1, Australia.

Tremellostereum Ryvarden (1986), Tremellaceae. 1, Bahamas.

Tremiscus (Pers.) Lév. (1846), Exidiaceae. 1, cosmop. See Donk (*Persoonia* **4**: 185, 1966).

Tremolecia M. Choisy (1953), ? Hymeneliaceae (L). 5, widespr. See Hertel (*Ergebn. Forsch Unternehmens Nepal Himalaya* **6**: 150, 1977).

tremorgen, a mycotoxin inducing a neurotoxicosis (tremor) in humans and higher animals, e.g. fumitremorgin, verruculotoxin.

Tremotyliomyces Cif. & Tomas. (1953) ≡ Tremotylium (Thelotremat.).

Tremotylium Nyl. (1865), Thelotremataceae (L). 6, Australasia, Asia, S. Am. See Redinger (*Ark. Bot.* **28A** (8), 1936).

Trenomyces Chatton & F. Picard (1908), Laboulbeniaceae. 11, widespr.

Tretendophragmia Subram. (1994) = Diplococcium (Mitosp. fungi) fide Sutton (*in litt.*).

tretic (of conidiogenesis), the sort of conidiogenesis in which each conidium (**tretoconidium**, tretic conidium, poroconidium, porospore) is delimited by an extension of the inner wall of the conidiogenous cell. Tretoconidia are solitary or in acropetal chains (cf. phialidic). **mono-, poly-**, (of conidiogenous cells), producing tretoconidia by the extrusion of the inner wall through one or several channels, respectively.

Tretocephala Subram. (1992), Mitosporic fungi, 1.A2.24. 1, Singapore.

Tretophragmia Subram. & Natarajan (1974), Mitosporic fungi, 2.C2.24. 2, India.

Tretopileus B.O. Dodge (1946), Mitosporic fungi, 2.D2.42. 3, USA, Afr. See Deighton (*Mycol. Pap.* **78**, 1960).

Tretospeira Piroz. (1972), Mitosporic fungi, 1.D2.24. 1, Uganda.

Tretospora M.B. Ellis (1976), Anamorphic Parodiopsidaceae, 1.C2.24. Teleomorph *Balladynopsis*. 4, widespr. See Khan et al. (*Mycotaxon* **49**: 477, 1993; key).

Tretovularia Deighton (1984), Mitosporic fungi, 1.A1.15. 1, N. Eur.

Treubiomyces Höhn. (1909), Chaetothyriaceae. 2, Java, S. Afr., USA. See Hughes (1976), Pohlad & Reynolds (*Mycol.* **66**: 521, 1974).

Trevisan (Vittore, Earl of San Leon; 1818-97). Born of a wealthy family in Padova, Trevisan devoted much of his time to advancing the study of a wide range of cryptogams, including plant pathogens and lichens; he was especially concerned in using new microscopic features, introducing 75 gen. names for lichenized taxa and making numerous critical and nomenclatural remarks on the works of Körber (q.v.) and esp. Massalongo (q.v.). He had various administrative and military charges, was Professor of Natural History and Popular Physics in Padova in 1851-53, but from 1853 worked from his estate in Mason and then moved to Monza and Milan (where he died). His collections were transferred to the University of Genoa after his death, but these were all destroyed in World War II. See Lazzarin (Ed.) (*L' opera lichenologica di Vittore Trevisan*, 1994; reprint collected works, incl. biogr.), Grummann (1974: 524), Stafleu & Cowan (*TL-2* **6**: 480, 1986).

tri- (in combination), three, triple.

Triactella Syd. (1921) = Hapalophragmium (Sphaerophragm.) fide Cummins & Hiratsuka (1983). See Lohsomboon et al. (*MR* **96**: 461, 1992).

Triacutus G.L. Barron & Tzean (1981), Mitosporic fungi, 1.G1.1. 1 (on bdelloid rotifer), Canada.

Triadelphia Shearer & J.L. Crane (1971), Mitosporic fungi, 1.C2.1. 7 (aquatic and terrestrial), widespr. See Constantinescu & Samson (*Mycotaxon* **15**: 472, 1982; key), Arx (*TBMS* **85**: 566, 1985; related genera), Révay (*Acta Bot. hung.* **33**: 67, 1987), Tzean & Chen (*Mycol.* **81**: 626, 1989; synoptic key).

Triandromyces Thaxt. (1931) = Tetrandromyces (Laboulben.) fide Tavares (1985).

Triangularia Boedijn (1934), Lasiosphaeriaceae. 5, widespr. See Cain & Farrow (*CJB* **34**: 689, 1956; key), v. Arx & Hennebert (*BSMF* **84**: 423, 1969), Guarro & Cano (*TBMS* **91**: 587, 1988; key 5 spp.).

Triblidiaceae Rehm (1888), Triblidiales. 3 gen. (+ 4 syn.), 5 spp. Stromata absent, the ascomata often strongly aggregated; ascomata erumpent, ultimately ± apothecial, opening by irregular cracks in the upper wall; ascomatal walls black, composed of strongly melanized isodiametric cells; interascal tissue of narrow trabeculate pseudoparaphyses, becoming free at the apex; asci thin-walled, ± cylindrical, persistent, without separable wall layers, without distinct apical structures, sometimes becoming mature before the covering layer ruptures; ascospores hyaline, multiseptate. Saprobic on bark, widespr.

For *Lit.* see *Triblidiales*.

Triblidiales, Ascomycota. 1 fam., 3 gen. (+ 4 syn.), 5 spp. Fam.:

Triblidiaceae.

Lit.: Eriksson (*SA* **11**: 1, 1992).

Triblidium Rebent. (1804), Triblidiaceae. 1 or 2, Eur. See Eriksson (*Opera Bot.* **60**, 1981).

Tribolites W.H. Bradley ex Janson. & Hills (1976), Fossil fungi (Mitosp. fungi). 1 (Eocene), USA.

Tribolospora D.A. Reid (1966), Mitosporic fungi, 8.A1.15. 1, Australia.

Tricella Long (1912) = Phragmopyxis (Uropyxid.) fide Dietel (1928).

Tricellaesporonites Sheffy & Dilcher (1971), Fossil fungi (f. cat.). 2 (Eocene), USA.

Tricellula Beverw. (1954), Mitosporic fungi, 2.C1.10. 2, Eur., USA. See Haskins (*Can. J. Microbiol.* **4**: 279, 1958), Petersen (*Bull. Torrey bot. Cl.* **89**: 287, 1962).

Tricellulortus Matsush. (1995) = Pneumatospora (Mitosp. fungi) fide Sutton (*in litt.*).

Triceromyces T. Majewski (1980), Laboulbeniales (inc. sed.). 1, Eur, N. Am., Asia. See Benjamin (*Aliso* **11**: 245, 1986; key 6 spp.).

Trichaegum Corda (1837), Mitosporic fungi, 1.D2.?. 3, Eur., N. Am.

Trichaleurina Rehm (1903) = Scutellinia (Otid.) fide Eckblad (1968).

Trichaleuris Clem. (1909) = Scutellinia (Otid.) fide Eckblad (1968).

Trichamelia Bat. (1960), Asterinaceae. 1, Brazil.

Trichamphora Jungh. (1838) = Physarum (Physar.).

Trichangium Kirschst. (1935), Leotiales (inc. sed.). 1, Eur.

Trichaptum Murrill (1904), Coriolaceae. 20, pantrop. See Macrae (*CJB* **45**: 1371, 1966), Corner (*Beih. Nova Hedw.* **86**: 197, 1987; key).

Tricharia Fée (1825), Gomphillaceae (L). 21, trop. See Sérusiaux (*Mycol.* **76**: 108, 1984; key), Vězda & Poelt (*Folia geobot. phytotax.* **22**: 179, 1987).

Tricharia Boud. (1885) ≡ Tricharina (Otid.).

Tricharina Eckblad (1968), Otideaceae. 12, temp. See Yang & Korf (*Mycotaxon* **24**: 467, 1985; key). Anamorph Ascorhizoctonia (Yang & Kristiansen, *Mycotaxon* **35**: 313, 1989).

Trichaster Czern. (1845) = Geastrum (Geastr.) fide Staněk (Pilát, *Flora ČSR* B **1**: 480, 1958).

Trichasterina G. Arnaud (1918), Asterinaceae. 6, trop.

Trichella Léger & Duboscq (1929) = Enterobryus (Eccrin.) fide Manier & Lichtwardt (*Ann. Sci. Nat., Bot.*, sér. 12, **9**: 519, 1968).

Trichellopsis Maessen (1955) = Enterobryus (Eccrin.) fide Manier & Lichtwardt (*Ann. Sci. Nat., Bot.* sér. 12, **9**: 519, 1968).

Trichia Haller (1768), Trichiaceae. 25, cosmop.

Trichiaceae Chevall. (1826), Trichiales. 10 gen. (+ 8 syn.), 75 spp.

Trichiales, Myxomycota. 3 fam. 14 gen. (+ 17 syn.), 155 spp. Spore mass bright coloured, columella absent, peridium persistent. Fams:
(1) **Arcyriaceae**.
(2) **Dianemataceae**.
(3) **Trichiaceae**.
Lit.: Blackwell & Busard (*Mycotaxon* **7**: 61, 1978; pigments in spp. separation).

trichidium, see sterigma.

Trichidium Raf. (1815) ≡ Trichia (Trich.).

Trichobacidia Vain. (1921) nom. dub. (Fungi, inc. sed.) fide Santesson (*Symb. bot. upsal.* **12** (1), 1952).

Trichobasis Lév. (1849) = Uredo (Anamorphic Uredinales) fide Saccardo (*Syll. Fung.* **7**: 838, 1887). See Laundon (*Mycotaxon* **3**: 133, 1975).

Trichobelonium (Sacc.) Rehm (1896) = Belonopsis (Dermat.). fide Aebi (1972).

Trichobolbus Bat. (1964), Mitosporic fungi, 4.A2.?. 1, Brazil.

Trichobolus (Sacc.) Kimbr. & Cain (1967), Thelebolaceae. 4, Eur., N. Am. See van Brummelen (*in* Hawksworth (Ed.), *Ascomycete systematics*: 400, 1994; posn), Krug (*CJB* **51**: 1497, 1973; key), Olsen (*Blyttia* **45**: 117, 1987; key 4 spp.).

Trichobotrys Penz. & Sacc. (1901), Mitosporic fungi, 1.A2.3. 2, trop. Asia.

Trichocarpus P. Karst. (1889) [non Schreb. (1789), *Tiliaceae*] = Amylostereum (Ster.) fide Donk (1964).

Trichocephalum Costantin (1888) = Periconia (Mitosp. fungi) fide Mason & Ellis (1953).

Trichoceridium R. Poiss. (1932) ? = Smittium (Legeriomycet.) fide Lichtwardt (1986).

Trichochora Theiss. & Syd. (1915) nom. dub. (Fungi, inc. sed.) fide Petrak (*Ann. Myc.* **27**: 324, 1929).

Trichocicinnus (Sacc.) Höhn. (1926) = Chaetosticta (Mitosp. fungi) fide Sutton (1977).

Trichocladia Stirt. (1882) [non Harv. (1836), *Algae*] = Heterodea (Heterod.).

Trichocladia (de Bary) Neger (1901) = Microsphaera (Erysiph.) fide Blumer (1933).

Trichocladium Harz (1871), Mitosporic fungi, 1.C2.1. 6, widespr. See Hughes (*TBMS* **35**: 152, 1952), Hughes (*N.Z. Jl Bot.* **7**: 153, 1969), Pidoplichko & Kirilenko (*Mikol. i Fitopatol.* **6**: 510, 1972; key), Kohlmeyer & Volkmann-Kohlmeyer (*Mycotaxon* **53**: 349, 1995; key 7 aquatic spp.).

Trichocollonema Höhn. (1902) = Zignoella (Lasiosphaer.) fide v. Höhnel (1917).

Trichocoma Jungh. (1838), Trichocomaceae. 1, warmer areas. See Martin (*Mycol.* **29**: 620, 1937).

Trichocomaceae E. Fisch. (1897), Eurotiales. 26 gen. (+ 20 syn.), 170 spp. Ascomatal walls varied, pseudoparenchymatous or hyphal, sometimes thick and sclerotioid, usually brightly coloured; ascogonia often coiled; interascal tissue absent; asci small, ± globose, often formed in chains; ascospores ± hyaline, usually bivalvate and often ornamented. Anamorphs *Aspergillus, Paecilomyces, Penicillium, Polypaecilum*, etc. Saprobes, many economically important, cosmop.
Lit.: Malloch (*in* Arai (Ed.), *Filamentous microorganisms*: 37, 1985; key gen.), Malloch & Cain (*CJB* **50**: 2613, 1972, **51**: 1647, 1973; gen. names), Malloch (*in* Samson & Pitt (Eds), *Advances in Penicillium and Aspergillus Systematics*: 365, 1985), Pitt (*The genus Penicillium*, 1979), Pitt & Samson (*Regnum veg.* **128**: 13, 1993; names in use), Raper & Fennell (*The genus Aspergillus*, 1965).

Trichoconiella B.L. Jain (1976) = Alternaria (Mitosp. fungi) fide Sutton (*in litt.*).

Trichoconis Clem. (1909), Mitosporic fungi, 1.C1.1. 12 (on fungi), widespr. See Deighton & Pirozynski (*Mycol. Pap.* **128**, 1972; keys).

Trichoconium Corda (1837) = Melanconium (Mitosp. fungi) fide Saccardo (*Syll. Fung.* **3**: 1884).

Trichocrea Marchal (1891) nom. dub. fide Minter & Caine (*TBMS* **74**: 434, 1980).

trichocyst, a subpellicular organelle of many ciliates and dinoflagellates; sometimes an offensive weapon able to disable prey, sometimes an anchoring device.

Trichodelitschia Munk (1953), Phaeotrichaceae. 3, widespr. See Barr (*Prodromus to class Loculoascomycetes*, 1987; posn), Eriksson (*SA* **7**: 93, 1988), Lundqvist (*Svensk Bot. Tidskr.* **58**: 267, 1964).

trichoderm (of basidiomata), an outer layer composed of hair-like elements projecting from the surface (Furtado, *Mycol.* **57**: 599, 1965). See derm.

Trichoderma Pers. (1794), Anamorphic Hypocreaceae, 1.A1.15. Teleomorph Hypocrea. 9 (esp. in soil), widespr. Sometimes of use for controlling pathogenic fungi (see antagonism). See Rifai (*Mycol. Pap.* **116**, 1969; key), Bissett (*CJB* **62**: 924, 1984; key to sect. *Longibrachiatum*), Hayes *et al.* (*Anal. Biochem.* **209**: 176, 1993; methods for karyotyping), Park *et al.* (*Korean Jl. Pl. Prot.* **23**: 102, 1984; electrophoresis and spp. differentiation), Barak *et al.* (*Can. J. Microbiol.* **31**: 810, 1985; serology), Doi *et al.* (*Bull. Natn Sci. mus. Tokyo* ser. B **13**: 1, 1987; sect. *Saturnisporum*), Doi & Doi (*Bull. Natn Sci. Mus. Tokyo* ser. B **5**: 117, 1979; list of teleomorphs), Eveleigh (*Biology of industrial micro-organisms*: 487, 1985; biology), Doi *et al.* (*Bull. Natn Sci. Mus. Tokyo* **15**: 27, 1989; SEM conidia), Stasz *et al.* (*Mycol.* **81**: 391, 1989; cladistic analysis of isoenzyme polymorphism), Meyer & Plaskowitz (*Mycol.* **81**: 312, 1989; SEM conidia and matrix), Meyer (*Phytopath.* **79**: 1212, 1989; mitochondrial and plasmid DNA), Meyer *et al.* (*Curr. Gen.* **19**: 239, 1991; DNA fingerprinting and differentiation of strains), Gilly & Sands (*Biotech. Lett.* **13**: 477, 1991; electrophoretic karyotype of *T. reesei*), Gullino (Ed.) (*Petria* **1**: 120, 1991, *IV International Trichoderma and Gliocladium Workshop*), Meyer (*Appl. Environ. Microbiol.* **57**: 2269, 1991; mDNA and plasmids in taxonomy of *T. viride*), Meyer *et al.* (*Curr. Genetics* **21**: 27, 1992; DNA fingerprint analysis and classification), Bissett (*CJB* **69**: 2357, 1992, infragen. classif., *CJB* **69**: 2373, 1992, sect. *Pachybasium*), Nevalainen *et al.* (*in* Leong & Berka (Eds), *Molecular industrial mycology*, 1991; molecular biology), Mills & Muthumeenakshi (*in* Blakeman & Williamson (Eds), *Ecology of plant pathogens*: 135, 1994; spp. from mushroom compost). ? = Reticularia (Myxom.) fide Martin (1966). See *Hypocrea*.

Trichoderma Pers. (1801) [non Pers. (1794)], nom. dub. (Myxomycetes).

Trichodermia Hoffm. (1795) = Trichothecium (Mitosp. fungi) fide Saccardo (*Syll. Fung.* **4**: 178, 1886).

Trichodesmium Chevall. (1826) nom. rej. ≡ Graphiola (Graphiol.).

Trichodiscula Vouaux (1910), Mitosporic fungi, ?.?.?. 1 (on cardboard), France.

Trichodiscus Kirschst. (1924) [non Welsford (1912), *Algae*] ≡ Dennisiodiscus (Dermat.).

Trichodochium Syd. (1927), Mitosporic fungi, 3.B2.19. 2, C. Am., India. See Ellis (*Mycol. Pap.* **111**, 1967).

Trichodothella Petr. (1946), Venturiaceae. 1, Eur.

Trichodothis Theiss. & Syd. (1914), Venturiaceae. 3, N. Am., Afr. See Barr (*Sydowia* **41**: 25, 1989).

Trichodytes Kleb. (1898) nom. dub. (? Mitosp. fungi). See Sutton (*Mycol. Pap.* **141**, 1977).

Trichofusarium Bubák (1906) = Fusarium (Mitosp. fungi). See Booth (1971).

Trichoglossum Boud. (1885), Geoglossaceae. *c.* 18, esp. temp. See Mains (*Mycol.* **46**: 61, 1954; key N. Am. spp.), Benkert (*Mycol. Mittl.* **20**: 47, 1977; Germany), Rifai (*Lloydia* **28**: 113, 1965; key Javanese spp.).

trichogyne, the receptive hypha of the female organ, esp. in certain ascomycetes.

Trichohelotium Killerm. (1935), Leotiales (inc. sed.). 2, Eur.

Trichohleria Sacc. (1908) ? = Trichosphaerella (Niessl.) fide v. Arx & Müller (*Stud. Mycol.* **9**, 1975).

Tricholechia A. Massal. (1853) ≡ Byssoloma (Pilocarp.).

Tricholeconium Corda (1837) = Sarcopodium (Mitosp. fungi) fide Saccardo (*Syll. Fung.* **4**: 312, 1886).

Tricholoma (Fr.) Staude (1857) nom. cons., Tricholomataceae. 90, widespr. See Bon (*Docums Mycol.* **3**: 1, **4**: 55, **5**: 111, **6**: 165, 1974-76, *Encycl. Mycol.* **36**, 1984), Haluwyn (*Docums Mycol.* **4**: 43, 1972; ecology), Horak (*Sydowia* **17**: 153, 1964; key 11 S. Am. spp.), Gulden (*Musseronflora Slekton Tricholoma sensu lato*, 1972). See blewitt, matsutake.

Tricholomataceae R. Heim ex Pouzar (1983), Agaricales. 147 gen. (+ 101 syn.), 2347 spp. Hymenophoral trama regular; spores hyaline to pink, never dark, lacking a germ-pore.

Tricholomella Zerova ex Kalaméés (1992), Tricholomataceae. 2, Europe.

tricholomic acid, an insecticidal amino-acid derivative produced by *Tricholoma muscarium* (Takemoto, *Jap. J. Pharm. Chem.* **33**: 252, 1961). Cf. muscazone.

Tricholomopsis Singer (1936), Tricholomataceae. 19, cosmop. See Smith (*Brittonia* **12**: 41, 1960).

Tricholosporum Guzmán (1975), Tricholomataceae. 3, Mexico.

Trichomaris Hibbits, G.C. Hughes & Sparks (1981), Halosphaeriaceae. 1 (on tanner crab), Alaska.

Trichomerium Speg. (1918), Capnodiaceae. 23, widespr. See Reynolds (*Mycotaxon* **14**: 189, 1982). = Phragmocapnias (Capnod.) fide Hughes (1976).

Trichometasphaeria Munk (1953), Lophiostomataceae. 4, N. Am. See Barr (*Mycotaxon* **45**: 191, 1992).

Trichomonascus H.S. Jacks. (1948), ? Eurotiales (inc. sed.). 1 (on *Corticium*), Canada. See Benny & Kimbrough (1980).

Trichomyces Malmsten (1848) = Trichophyton (Mitosp. fungi) fide Dodge (1935).

Trichomycetes, Zygomycota. 4 ord., 7 fam., 48 gen. (+ 22 syn.), 189 spp. Thallus simple or branched, attached to the host cuticle by a holdfast. Asexual reproduction by sporangiospores or arthrospores. Sexual reproduction known in the *Harpellales*, where the zygospore is characteristically biconical, and one genus of *Eccrinales*. Except for the ectozoic *Amoebidium*, parasites or commensals within digestive tracts of living arthropods. Manier & Lichtwardt (1968) recognize 4 Ords, (2) and (4) are considered to be closely related to the *Kickxellales* (Zygomycota), whereas the phyletic affinities of (1) and (3) are uncertain. Ords:
(1) **Amoebidiales**.
(2) **Asellariales**.
(3) **Eccrinales**.
(4) **Harpellales**.
Lit.: Lichtwardt (*The Trichomycetes. Fungal associates of arthropods*, 1986, *in* Ainsworth *et al.* (Eds), *The Fungi* **4B**: 651, 1973, *in* Jones (Ed.), *Recent advances in aquatic mycology*, 1976; keys, *Mycol.* **65**: 1, 1973; phylogeny, *in* Parker (Ed.), 1982, **1**: 195), Moss (*TBMS* **65**: 115, 1975; phylogeny), Moss & Young (*Mycol.* **70**: 944, 1978; phylogeny), Benjamin (*in* Kendrick (Ed.), *The whole fungus* **2**: 573, 1979; phylogeny), Porter & Smiley (*Exp. Mycol.* **3**: 188, 1979; phylogeny), Lichtwardt *et al.* (*Trans. mycol. Soc. Japan* **28**: 359, 1987; Japan).
Reviews: Duboscq *et al.* (*Arch. Zool. Exp. Gén.* **86**: 29, 1948), Manier (*Ann. Sci. nat., Bot.* sér. 12, **10**: 565, 1969), Moss (*in* Batra (Ed.), *Insect-fungus symbiosis*: 175, 1979), Lichtwardt & Williams (*CJB* **66**: 1259, 1988; spp. diversity & distrib., **68**: 1057, 1990; Australia), Williams & Lichtwardt (*CJB* **68**: 1045, 1990; NZ).

trichomycin, an antiobiotic from *Streptomyces hachijoensis* (Hosoya *et al.*, 1955); antifungal (esp. against *Candida albicans*) and anti-*Trichomonas*.

Trichonectria Kirschst. (1907), Hypocreaceae. 3 (on lichens and *Musci*), Eur. See Döbbeler (*Mitt. bot. StSamml., München* **14**: 1, 1978; on *Musci*), Hawksworth (*Notes R. bot. Gdn Edinb.* **36**: 181, 1978, *Lichenologist* **12**: 100, 1980; nomencl.).

Trichopeltella Höhn. (1910), ? Microthyriaceae. 1, Java.

Trichopeltheca Bat., C.A.A. Costa & Cif. (1958), Euantennariaceae. 2, Asia, Pacific. See Hughes (*N.Z. Jl Bot.* **10**: 230, 1972). Anamorphs *Plokamidomyces*, *Trichothallus*.

Trichopeltina Theiss. (1914), ? Microthyriaceae. 2, S. Am.

Trichopeltinites Cookson (1947), Fossil fungi. 4 (Cretaceous, Tertiary), widespr.

Trichopeltis Speg. (1889) = Trichothyrium (Microthyr.). See Spooner & Kirk (*MR* **94**: 1990).

Trichopeltium Clem. (1909) ≡ Trichopeltulum (Mitosp. fungi) fide Sutton (1977).

Trichopeltopsis Höhn. (1909) = Trichothyrium (Microthyr.). See Spooner & Kirk (1990).

Trichopeltospora Bat. & Cif. (1958), ? Microthyriaceae. 1, Brazil.

Trichopeltula Theiss. (1914) = Trichothyrium (Microthyr.). See Clements & Shear (1931).

Trichopeltulum Speg. (1889), Mitosporic fungi, 5.A1.?. 1, Brazil. See Petrak & Sydow (*Ann. Myc.* **33**: 173, 1935).

Trichopeltum Bat., Cif. & C.A.A. Costa (1958), ? Microthyriaceae. 1, Hawaii.

Trichopeziza Fuckel (1870), Hyaloscyphaceae. 20, mainly temp. See Raitviir (*Eesti NSV Tead. Akad. Toim., Biol.* **363**: 313, 1987).

Trichopezizella Dennis ex Raitv. (1969) = Lachnum (Hyaloscyph.). See Haines (*Mycol.* **66**: 216, 1974; key), Spooner (*Bibl. Mycol.* **116**, 1987; status).

Trichophaea Boud. (1885), Otideaceae. 10, N. temp. See Kanouse (*Mycol.* **50**: 128, 1958; key).

Trichophaeopsis Korf & Erb (1972), Otideaceae. 1, N. & S. Am., Eur., Asia.

Trichophila Oudem. (1889) ? = Piedraia (Piedr.) fide Sutton (*Mycol. Pap.* **141**, 1977).

Trichophyma Rehm (1904) = Arthothelium (Arthoniales) fide Santesson (*Svensk Bot. Tidskr.* **43**: 547, 1949).

Trichophysalospora Lebedeva (1933) = Physalospora (Hyponectr.) fide v. Arx & Müller (1954).

trichophytin, an antigen prepared from dermatophytes, esp. for use in skin testing. Commercial trichophytin is usually a mixture of antigens of several spp. of *Trichophyton* and *Microsporum*.

Trichophyton Malmsten (1848), Anamorphic Arthrodermataceae, 1.C2.2. Teleomorph *Arthroderma*. 20 (on humans and animals, causing trichophytoses, and in soil and river sediments), cosmop. *T. mentagrophytes* (syns. *T. asteroides*, *T. gypseum*); *T. schoenleinii* (human favus); *T. concentricum* (tinea imbricata); *T. verrucosum* (cattle ringworm). See Ajello (1968), Dodge (1935), Ajello & Padhye (*Mykosen* **30**: 258, 1987; stimulation of macroconidia in culture), Padhye *et al.* (*J. Med. & Vet. Mycol.* **25**: 195, 1987; mating behaviour), Symoens *et al.* (*Mycoses* **32**: 652, 1990; isoelectric focusing to differentiate *T. mentagrophytes* and *T. interdigitale*), Mochizuki *et al.* (*J. Med. Vet. Mycol.* **28**: 191, 1990; taxonomy *T. interdigitale*), Nishio *et al.* (*Mycopathol.* **117**: 127, 1992; phylogeny by mitochondrial DNA), Devliotou-Panagiotidou *et al.* (*Mycoses* **35**: 375, 1992; *T. rubrum* in Greece). See also tinea.

Trichopilus (Romagn.) P.D. Orton (1991), Entolomataceae. 6, Eur.

Trichoplacia A. Massal. (1853) ? Ascomycota (inc. sed.). See Zahlbruckner (*Nat. Pflanzenfam.* **8**: 61, 1926).

Trichopsora Lagerh. (1892), Pucciniosiraceae. 1 (on *Tournefortia*), Ecuador; O, III.

Trichoramalina Rundel & Bowler (1974) ? = Ramalina (Ramalin.).

Trichoscypha Boud. (1885) [non Hook. f. (1862), *Anacardiaceae*] ≡ Trichoscyphella (Hyaloscyph.).

Trichoscypha (Cooke) Sacc. (1889) ≡ Cookeina (Sarcoscyph.).

Trichoscyphella Nannf. (1932) = Lachnellula (Hyaloscyph.) fide Dennis (1962).

Trichoseptoria Cavara (1892), Mitosporic fungi, 4.E1.?. 2, Eur., N. Am. *T. fructigena* (soft rot of apples).

Trichosia Bat. & R. Garnier (1960) = Trichasterina (Asterin.) fide v. Arx & Müller (1975).

Trichoskytale Corda (1842) ≡ Trichocoma (Trichocom.).

Trichosperma Speg. (1888) [non *Trichospermum* Blume (1825), *Tiliaceae*] ≡ Spermotrichum (Mitosp. fungi).

Trichospermella Speg. (1912), Ascomycota (inc. sed.). 1, S. Am. See Petrak & Sydow (*Ann. Myc.* **33**: 175, 1935).

Trichosphaera Dumort. (1822) nom. dub. (Ascomycota, inc. sed.); used for diverse fungi with flasklike sporocarps in stromata.

Trichosphaerella E. Bommer, M. Rousseau & Sacc. (1891), Niessliaceae. 2, Eur., Afr. See Barr (*Mycotaxon* **39**: 43, 1990; posn).

Trichosphaeria Fuckel (1870), Trichosphaeriaceae. 25, widespr.

Trichosphaeriaceae G. Winter (1885), Trichosphaeriales. 26 gen. (+ 18 syn.), 371 spp. Stromata absent, or reduced to a hyphal subiculum; ascomata superficial, perithecial, ± globose, black, often thick-walled, sometimes setose, the ostiole periphysate; interascal tissue of narrow thin-walled paraphyses; asci cylindrical, persistent, thin-walled, without separable layers, usually with a small I- apical ring; ascospores variously shaped, hyaline or brown, usually septate, sometimes fragmenting at the septa, without germ pores, sometimes with a sheath. Anamorphs varied, hyphomycetous. Saprobic esp. on wood and bark, occ. on other fungi, cosmop.

Trichosphaeriales, Ascomycota. 1 fam., 26 gen. (+ 18 syn.), 371 spp. Fam.:
 Trichosphaeriaceae.

Trichosphaeropsis Bat. & Nascim. (1960), Ascomycota (inc. sed.). 1, Brazil.

trichospore, a caducous, dehiscent, monosporous sporangium with basal appendages characteristic of the *Harpellales* (Moss & Lichtwardt, *CJB* **54**: 2346, 1976).

Trichosporiella Kamyschko (1960), Mitosporic fungi, 1.A1.1/10. 4, widespr. See v. Oorschot (*Stud. Mycol.* **20**, 1980), de Hoog *et al.* (*Ant. v. Leeuwenhoek* **51**: 79, 1985).

Trichosporites Félix (1894), Fossil fungi. 1 (Cretaceous), Sweden.

Trichosporon Behrend (1890), Mitosporic fungi, 1.A1.10/38. 10, widespr. *T. beigelii* (**white piedra** of man). See Barnett *et al.*, 1990), Guého *et al.* (*Ant. v. Leeuwenhoek* **61**: 285, 1992; typification), King & Jong (*Mycotaxon* **6**: 391, 1977; key), Weijman (*Ant. v. Leeuwenhoek* **45**: 119, 1979; chemotaxonomy), Gienow *et al.* (*Zentralbl. Mikrobiol.* **145**: 3, 1990; molecular characterization), Lu & Li (*Acta Mycol. Sin.* **10**: 43, 1991; n.spp. from China), Guého *et al.* (*Ant. v. Leeuwenhoek* **61**: 289, 1992; systematics).

Trichosporonoides Haskins & J.F.T. Spencer (1967), Mitosporic fungi, 1.A1.38/39+3+6. 5, India, Thailand, N. Am. See de Hoog (*Stud. Mycol.* **15**: 20, 1979; gen. revision), Ramirez (*Mycopathol.* **108**: 25, 1989; key), Inglis *et al.* (*Mycol.* **84**: 555, 1992).

Trichosporum Fr. (1825) [non D. Don (1822), *Gesneriaceae*] nom. conf., nom. illeg. See Hughes (1958; nomencl.).

Trichosporum Vuill. (1901) [non D. Don (1822), *Gesneriaceae*] ? = Piedraia (Piedr.).

Trichosterigma Petch (1923) [non Klotzsch & Garcke (1859), *Euphorbiaceae*] = Hirsutella (Mitosp. fungi) fide Petch (*TBMS* **9**: 93, 1923).

Trichostroma Link (1826) nom. dub. (Mitosp. fungi). No spp. included.

Trichostroma Corda (1829) nom. dub. fide Hughes (*CJB* **36**: 747, 1958).

Trichothallus F. Stevens (1925), Anamorphic Euantennariaceae, 5.-.-. Teleomorph *Trichopeltheca*. 1, Hawaii.

Trichotheca P. Karst. (1887) ? = Bloxamia (Mitosp. fungi) fide Sutton (1977).

trichothecenes, toxins (scirpenes) of *Fusarium tricinctum*, *F. sporotrichioides*, *F. poae*, *Trichothecium*, etc.; the cause of alimentary toxic aleukia in farm animals and humans. See Beasley (*Trichothecene mycotoxicosis*, 2 vols, 1991).

trichothecin, an antifungal metabolic product of *Trichothecium roseum* (Freeman & Morrison, *Biochem. J.* **44**: 1, 1949).

Trichotheciopsis J.M. Yen (1979) = Trichothecium (Mitosp. fungi) fide Sutton (*in litt.*).

Trichothecium Link (1809), Mitosporic fungi, 1.B1.34. 5, widespr. *T. roseum* (pink rot of apples). See Park (*TBMS* **39**: 239, 1956; conidia), Rifai & Cooke (*TBMS* **49**: 147, 1966; status), Sesan (*Probl. Prot. Plant.* **13**: 381, 1985; bibliogr.).

Trichotheliaceae (Müll. Arg.) Bitter & F. Schill. (1927), Trichotheliales (L). 5 gen. (+ 14 syn.), 240 spp. Thallus crustose, sometimes poorly developed; ascomata perithecial, sometimes pale- and thin-walled, the ostioles ± periphysate, sometimes surrounded by agglutinated radiating hairs; interascal tissue of simple paraphyses; asci cylindrical, thin-walled, not fissitunicate, sometimes with an inconspicuous refractive ring, not blueing in iodine; ascospores hyaline, thin- or thick-walled, transversely septate. Anamorphs coelomycetous where known. Lichenized with green algae, on living leaves and bark, etc., esp. trop.

This fam. has recently been segregated from *Pyrenulales* (Hafellner & Kalb, 1995).

Trichotheliales, Ascomycota (L). 1 fam., 5 gen. (+ 14 syn.), 240 spp. Thallus crustose, non-areolate, or immersed; ascomata perithecia, partly or completely surrounded by an involucrellum; interascal filaments unbranched paraphyses, also with periphyses during development; asci functionally unitunicate, with a truncated apical tip, an external chitinoid ring structure, and poroid dehiscence; ascospores colourless, thinly septate, 3-septate to muriform; producing distinctive hyphal pigments not soluble in acetone; lichenized with *Phycopeltis* or other *Trentepohliales*; widespr., esp. trop. Fam.: **Trichotheliaceae**.

Lit. Hafellner & Kalb (*Bibl. Lich.* **57**: 161, 1995).

Trichotheliomyces Cif. & Tomas. (1953) ≡ Trichothelium (Trichothel.).

Trichothelium Müll. Arg. (1885), Trichotheliaceae (L). 10, trop. See Santesson (*Symb. bot. upsal.* **12**(1), 1952), Vězda (*Nova Hedw.* **58**: 123, 1995; key 4 spp.).

Trichothrauma Germ. (1850) nom. conf. (? Saprolegniales, inc. sed.).

Trichothyriaceae, see *Microthyriaceae*.

Trichothyriella Theiss. (1914), Microthyriaceae. 1, S.E. Asia. See Spooner & Kirk (*MR* **94**: 223, 1990).

Trichothyrina (Petr.) Petr. (1950) = Lichenopeltella (Microthyr.) fide Santesson (*SA* **9**: 15, 1991).

Trichothyrinula Petr. (1950), Microthyriaceae. 1, Ecuador.

Trichothyriomyces Bat. & H. Maia (1955), Microthyriaceae. 1, Brazil. See Spooner & Kirk (*MR* **94**: 223, 1990).

Trichothyriopsis Theiss. (1914), Microthyriaceae. 3 or 4, Indonesia, Brazil. See Spooner & Kirk (*MR* **94**: 223, 1990).

Trichothyrites Rosend. (1943), Fossil fungi. 2 (Eocene, Pleistocene), UK, USA. See Smith (1980).

Trichothyrium Speg. (1889), Microthyriaceae. 12, warmer areas. See Hughes (*Mycol. Pap.* **50**, 1953), Spooner & Kirk (*MR* **94**: 223, 1990). Anamorph *Isthmospora*.

trichotomous, with three branches arising from the same point.

Trichotrema Clem. (1909) ? = Pleurotrema (Pleurotrem.).

Trichozygospora Lichtw. (1972), Legeriomycetaceae. 1 (in *Diptera*), USA, Eur. See Moss & Lichtwardt (*CJB* **54**: 2346, 1976, **55**: 3099, 1977; ultrastr.), Lichtwardt (1986).

Trichulius Schmidel ex Corda (1842) ? = Trichia (Myxom.) fide Mussat (1901).

Trichurus Clem. (1896), Mitosporic fungi, 2.A2.19. 3, temp. ? = Doratomyces (Mitosp. fungi) fide Swart (*Ant. v. Leeuwenhoek* **30**: 257, 1964).

Tricladiella K. Ando & Tubaki (1984), Mitosporic fungi, 1.G1.1. 1 (aquatic), Japan.

Tricladiomyces Nawawi (1985), Mitosporic fungi, 1.G1.1. 1 (aquatic, dolipore septa), Malaysia.

Tricladiopsis Descals (1982), Mitosporic fungi, 1.G1.10. 2 (aquatic), UK.

Tricladiospora Nawawi & Kuthub. (1988), Mitosporic fungi, 1.G1.19. 4, Malaysia.

Tricladium Ingold (1942), Mitosporic fungi, 1.G1.1/10. 7 (aquatic), widespr. See Petersen (*Mycol.* **54**: 135, 1962; key), Marvanová & Bandoni (*TBMS* **85**: 747, 1985), Ando & Kawamoto (*Trans. mycol. Soc. Jap.* **26**: 471, 1985).

Triclinum Fée (1825), Pannariaceae (L). 1, trop. See Jørgensen (1978).

Tricornispora Bonar (1967) = Tridentaria (Mitosp. fungi) fide Kendrick (*in litt.*).

Tridens Massee (1901) [non Roem. & Schult. (1817), *Gramineae*], ? Rhytismatales (inc. sed.). 1, N. Am.

Tridentaria Preuss (1852), Mitosporic fungi, 1.G1.1. 2, temp. See Drechsler (*J. Wash. Acad. Sci.* **27**: 391, 1937), v. d. Aa & Oorschot (*Persoonia* **12**: 415, 1985; redisposition spp.).

Trifurcospora K. Ando & Tubaki (1988), Mitosporic fungi, 1.G1.10. 1 (aquatic), Japan, USA.

Triglyphium Fresen. (1852), Mitosporic fungi, 1.G1.?. 2, Eur., Asia.

Trigonia J.F.H. Beyma (1933) [non Aubl. (1775), *Trigoniaceae*] ≡ Triangularia (Lasiosphaer.).

Trigonipes Velen. (1939) = Clitocybe (Tricholomat.) fide Singer (1975).

Trigonopsis Schachner (1929), Mitosporic fungi, 1.A1.?. 1, Eur. See Sentheshanmuganathan & Nickerson (*J. gen. Microbiol.* **27**: 437, 1962; nutrition, form).

Trigonosporium Tassi (1900), Mitosporic fungi, 4.A2.?. 2, Australasia. See Sutton (*Mycol. Pap.* **123**, 1971).

Trihyphaecites Peppers (1970), Fossil fungi (*f. cat.*). 1 (Carboniferous), USA.

trimerous, in threes.

trimitic, see hyphal analysis.

Trimmatostroma Corda (1837), Mitosporic fungi, 3.D2.36. 9, widespr. See Ellis (*MDH*).

Trimmatothele Norman ex Zahlbr. (1903), Verrucariaceae (L). 5, Eur., N. Am.

Trimmatothelopsis Zschacke (1934), ? Verrucariaceae (L). 1, Eur.

Trimorphomyces Bandoni & Oberw. (1983), Tremellaceae. 1, N. Am. See Kang *et al.* (*Nucleic Acids Res.* **20**: 5229, 1992; 5S rRNA sequence).

Trinacrium Riess (1852), Mitosporic fungi, 1.G1.1. 6, Eur., Am. See Tzean & Chen (*MR* **93**: 391, 1989; comparison of spp.).

Triophthalmidium (Müll. Arg.) Gyeln. (1933) = Caloplaca (Teloschist.).

Triparticalcar D.J.S. Barr (1980), Spizellomycetaceae. 1, Arctic.

tripartite tubular hair, filamentous appendage on the flagellum composed of a tapered solid base, a hollow cylindrical shaft and one or more terminal filaments. See straminipile.

Tripedotrichum G.F. Orr & Kuehn (1964) = Gymnoascus (Gymnoasc.) fide Benny & Kimbrough (*Mycotaxon* **12**: 1, 1980).

Triphragmiopsis Naumov (1914), Sphaerophragmiaceae. 3 (on *Berberidaceae, Ranunculaceae, Larix*), Eur., Asia. See Monoson (*Mycopath.* **52**: 115, 1974), Lohsomboon *et al.* (*Trans. Mycol. Soc. Japan* **31**: 335, 1990).

Triphragmium Link (1825), Sphaerophragmiaceae. 4 (2 on *Rosaceae*, 2 (doubtful) on *Leguminosae*), N. temp. See Lohsomboon *et al.* (*Trans. Mycol. Soc. Japan* **31**: 215, 1990).

Triplicaria P. Karst. (1889), Anamorphic Xylariaceae, 3.A1.10. Teleomorph *Hypoxylon*. 1, Finland. See Petrak (*Sydowia* **7**: 299, 1953).

Triplosporium (Thaxt.) Batko (1964) = Entomophthora (Entomophthor.) fide Krejzova (*Česká Myk.* **30**: 207, 1976), = Neozygites (Entomophthor.) fide Remaudière & Keller (1980).

Tripoconidium Subram. (1978), Mitosporic fungi, 1.G1.1. 1 (on nematodes), USA.

Tripocorynelia Kuntze (1898) ≡ Tripospora (Corynel.).

Triporicellaesporites Ke & Shi (1978), Fossil fungi (*f. cat.*). 3 (Paleocene, Tertiary), China.

Triporisporites Hammen (1954), Fossil fungi (*f. cat.*). 1 (Cretaceous), Colombia.

Triporisporonites Sheffy & Dilcher (1971), Fossil fungi. 1 (Eocene), USA.

Tripospermum Speg. (1918), Anamorphic Capnodiaceae, 1.G2.1. Teleomorph *Trichomerium*. 6, widespr. See Hughes (*Mycol. Pap.* **46**, 1951), Ando (*MR* **98**: 879, 1994; behaviour in culture).

Tripospora Sacc. (1886), Coryneliaceae. 3 (on *Podocarpus*), E. & S. Afr., S. Am. See Benny *et al.* (*Bot. Gaz.* **146**: 431, 1985; key).

Triposporina Höhn. (1912), Mitosporic fungi, 1.G1.19. 2, Java, N. Am. See Deighton & Pirozynski (*Mycol. Pap.* **128**: 96, 1972).

Triposporiopsis W. Yamam. (1955) = Trichomerium (Capnod.) fide Reynolds (*Mycotaxon* **14**: 189, 1982).

Triposporium Corda (1837), Anamorphic Asterinaceae, 1.G2.19. Teleomorph *Batistinula*. 10, widespr. See Hughes (*Mycol. Pap.* **46**, 1951).

Triposporonites Sheffy & Dilcher (1971), Fossil fungi (*f. cat.*). 1 (Eocene), USA.

Tripotrichia Corda (1837) = Leocarpus (Physar.).

Tripterospora Cain (1956) = Zopfiella (Lasiosphaer.) fide Guarro *et al.* (*SA* **10**: 79, 1991).

Tripterosporaceae, see Lasiosphaeriaceae.

Tripterosporella Subram. & Lodha (1968), Lasiosphaeriaceae. 1 (coprophilous), India.

triquetrous, three-edged, three-cornered.

Triramulispora Matsush. (1975), Mitosporic fungi, 1.G1.1. 2, Japan.

Triscelophorus Ingold (1943), Mitosporic fungi, 1.G1.1. 2 (aquatic), widespr. See Petersen (*Mycol.* **54**: 162, 1962; key).

Triscelosporium Nawawi & Kuthub. (1987), Mitosporic fungi, 1.G2.2. 1 (aero-aquatic), Malaysia.

trisporic acid C, a hydroxy-keto acid from mated cultures of *Blakeslea trispora* able to induce carotenogenesis in separate strains (Caglioti *et al.*, *Chimica Industria* **46**: 961, 1964).

tristichous, in three rows.

Trisulcosporium H.J. Huds. & B. Sutton (1964), Mitosporic fungi, 1.G1.1. 1 (aquatic), UK.

Tritirachium Limber (1940), Mitosporic fungi, 1.A1.10. 10 (some on humans), widespr. See de Hoog (*Stud. Mycol.* **1**, 1972; 2 spp. accepted), MacLeod (*CJB* **32**: 818, 1954).

trivial name, (1) the zoological term for specific epithet; (2) a common name for a chemical, see Fungal metabolites.

Trochila Fr. (1849), Dermateaceae. *c.* 15, temp. See Greenhalgh & Jones (*TBMS* **47**: 311, 1964).

Trochodium Syd. & P. Syd. (1920) = Uromyces (Puccin.) fide Cummins & Hiratsuka (1983).

Trochoideomyces Thaxt. (1931), Laboulbeniaceae. 1, E. Indonesia.

Trochophora R.T. Moore (1955), Mitosporic fungi, 1.F2.10. 1, Sri Lanka. See Goos (*Mycol.* **78**: 744, 1986).

Trogia Fr. (1836), Tricholomataceae. 200, trop. See Corner (*Gdns Bull., Singapore* Suppl. 2, 1991).

Troglobiomyces Pacioni (1980) = Hirsutella (Mitosp. fungi) fide Samson *et al.* (1984).

troglobiotic, living in caves.

Troglomyces S. Colla (1932), Laboulbeniaceae. 1, Eur. See Rossi & Balazuc (*Revue mycol.* **41**: 525, 1977).

troll, a goblin or dwarf in Scandinavian mythology, said to carry off naughty children; dolls were traditionally made of *Alectoria*, *Bryoria* and *Usnea* spp. and given to children to remind them to behave.

Trolliomyces Ulbr. (1938) ≡ Teloconia (Phragmid.).

Trombetta Adans. (1763) ≡ Craterellus (Craterell.).

Tromera A. Massal. ex Körb. (1865) = Sarea (Agyr.) fide Hawksworth & Sherwood (1981).

Tromera Sacc. (1889) cited in *ING*, in error.

Tromeropsis Sherwood (1981), Ascomycota (inc. sed.). 1, Finland.

troop, a group of sporocarps (esp. basidiomata), generally from one mycelium.

trophocyst (of *Pilobolus*), a hyphal swelling from which a sporangiophore is produced.

trophogonium (**trophogone**) (of ascomycetes), an antheridium of which the only use is supplying food (Dangeard).

tropism (frequently as a suffix), a turning or growth as a reaction to a one-sided stimulus; + (positive) when in the direction of the stimulus, − (negative) when away from the stimulus. Among the more important tropisms seen in fungi are: (1) **autotropism**, an avoidance (−) response between neighbouring hyphae which in part is responsible for the spacing of hyphae at the colony margin (Trinci, *in* Jennings & Rayner (Eds), *The ecology and physiology of the fungal mycelium*: 23, 1984); (2) **chemotropism**, a reaction to a chemical, e.g. oxygen (Robinson, *New Phytologist* **72**: 1349, 1973), a hormone (Banbury, *J. Exper. Bot.* **6**: 235, 1975) or (in the case of water moulds but not other fungi) nutrients (Musgrove *et al.*, *J. Gen. Microbiol.* **101**: 65, 1977); (3) **galvanotropism**, a reaction to an electrical field (McGillivray & Gow, *J. Gen. Microbiol.* **132**: 2515, 1986); (4) **gravitropism**, a reaction to gravity, e.g. sporangiophores of *Mucorales* and stipes, gills and tubes of fruit bodies of *Basidiomycota* (Moore, *New Phytol.* **117**: 3, 1991); (5) **phototropism**, a reaction of conidiophores, sporangiophores, asci, stipes etc. to light (Bergman *et al.*, *Bact. Rev.* **33**: 99, 1969); (6) **thigmotropism**, a reaction of germ tubes and hyphae to plant and other sufaces (Dickson, *Phytopath. Z.* **66**: 38, 1969, Kwon & Hoch, *Exp. Mycol.* **15**: 116, 1991).

Troposporella P. Karst. (1892), Mitosporic fungi, 3.F2.1. 1, N. temp. See Sutton (*Mycol. Pap.* **132**, 1973).

Troposporium Harkn. (1884), Mitosporic fungi, 3.F1.1. 1, W. N. Am. See Peek & Solheim (*Mycol.* **50**: 847, 1959).

Trotteria Sacc. (1919) = Actinopeltis (Microthyr.) fide Sutton (*Mycol. Pap.* **141**, 1977).

Trotterula Speg. (1921) = Chaetothyrium (Chaetothyr.) fide Petrak & Sydow (*Ann. Myc.* **32**: 6, 1934), Hughes (1976).

TRTC. Cryptogamic Herbarium, Royal Ontario Museum (Toronto, Canada); founded 1887; transferred from the University of Toronto to the Museum in 1992.

truffle, an ascoma, generally subterranean, of *Tuber* or other *Pezizales* or *Elaphomycetales* (see below), or a basidioma of *Hymenogastrales*; about 180 truffle-forming spp. are known. **Burgundy -**, *Tuber* *uncinatum*; **false -**, *Hymenogaster*; **hart's -**, *Elaphomyces*; **Périgord (French) -**, *T. melanosporum*; **red -**, *Melanogaster variegatus*; **summer -**, *T. aestivum* (the best British sp.); **white -**, *Choiromyces meandriformis*; **white Piedmont -**, *T. magnatum*; **white winter -**, *T. hiemalbum*; **winter -**, *T. brumale*; **yellow -**, *Terfezia* and *Tirmania*.

Growth of truffles is best in well-drained, calcareous soils, and there is frequently an association between truffles and the roots of certain trees, esp. oaks (*Quercus petraea, Q. robur*, and the evergreen *Q. ilex* and *Q. coccifera*, among others). The position of the truffles is sometimes given by cracks in the soil, by the 'scorched' look of the plants over the truffles, by the look of the 'truffle trees', or by noting the habitat of the truffle fly, *Anistoma cinnamomea*; but they are generally looked for with the help of trained dogs or pigs.

For the last hundred years truffle culture has been undertaken in France where 'truffières' have been started by planting oak trees in good places, by inoculating the soil with soil from under truffle trees or with truffle tissue, and (the best method) by planting young trees taken from soil in which the truffle fungus is present. Truffles are first produced under such trees after 7 to 15 years; then generally for 20 to 30 or more years. In the south of France and Italy the crop is taken from December to March. In addition to their use as food, truffles have been used in liqueur making, for scenting tobacco, and in certain perfumes.

See Malençon (*Rev. Myc.* n.s. **3**(Mém. hors sér.), **1**, 1938, *Persoonia* **7**: 261, 1973), Kaltenbach (*Int. Rev. Agric.* **26**: 267T, 1935), Singer (*Mushrooms and truffles*, 1961), Chang & Hayes (Eds) (*Biology and cultivation of edible mushrooms*, 1978), Torini (*Jl Tartufo e la sua coltivazione*, 1984; cultivation), Bokhary (*Arab Gulf J. scient. Res., agric. biol. sci.* **B5**: 245, 1987; desert truffles).

The ascomycete truffles were formerly placed in a special order, the *Tuberales*, characterized by fleshy to leathery, ± globose ascomata with the hymenium lining a single or complex series of locules. They are now believed to be the result of the convergent evolution of various discomycete lines and are mainly referred to the *Pezizales* by Trappe (*Mycotaxon* **9**: 247, 1979) except for the isolated *Elaphomycetales* (for *Elaphomyces* only).

Lit.: Fischer (*Nat. PflFam.* **5b**, viii, 1938), Gilkey (*Ore. St. Monogr. Stud. Bot.* **1**, 1939, *Mycol.* **46**: 783, 1954; keys, *N. Am. Flora, ser.* 2, **1**: 1, 1954; N. Am.), Lange (*Dansk Bot. Arkiv.* **16**, 1956; Denmark), Hawker (*Phil. Trans.* **B237**: 453, 1954), Dennis (1968: 71; Br. Isl.), Trappe (*TBMS* **57**: 87, 1971; *Terfeziaceae, Carbomycetaceae, Mycotaxon* **2**: 109, 1975; gen. names, **9**: 297, 1979; re-classification), Ławrynowicz (*Fl. Polska, Grzyby* **18**, 1988; keys 75 spp. Poland, SEM). Trappe *et al.* (*Austr. Syst. Bot.* **5**: 597, 613, 617, 631, 693, 1992; Australian spp.), Montecchi & Lazzari (*Funghi ipogei*, 1993; col. pls, keys, bibliogr.), Pegler *et al.* (*British truffles*, 1993; keys, descriptions, illustrations, UK spp.).

Truittella Karling (1949), ? Endochytriaceae. 1, USA.

Trullula Ces. (1852), Mitosporic fungi, ?.?.?. 15, temp. See Sutton (*Mycol. Pap.* **141**, 1977; nomencl.).

truncate, ending abruptly, as though with the end cut off horizontally (Fig. 37.4c).

Truncatella Steyaert (1949), Mitosporic fungi, 6.C2.19. 3, widespr.

Truncicola Velen. (1934) = Hyaloscypha (Hyaloscyph.) fide Huhtinen (1990).

Truncocolumella Zeller (1939), Rhizopogonaceae. 3, N. Am., trop. Afr. See Smith & Singer (*Brittonia* **11**: 205, 1959).

Truncospora Pilát ex Pilát (1953), Coriolaceae. 2, Eur. See Pegler (1973).

Tryblidaria (Sacc.) Rehm (1904), ? Patellariaceae. 4 or 5, warmer areas.

Tryblidiaceae, see *Triblidiaceae*.

Tryblidiella Sacc. (1883) = Rhytidhysteron (Patellar.) fide v. Arx & Müller (1975).

Tryblidiopsis P. Karst. (1871), Rhytismataceae. 1, N. temp. See Gams (*Taxon* **41**: 99, 1992; nomencl.), Livsey & Minter (*CJB* **72**: 549, 1994). Anamorph *Tryblidiopycnis*.

Tryblidiopycnis Höhn. (1918), Anamorphic Rhytismataceae, 8.E1.10. Teleomorph *Tryblidiopsis*. 1, Eur.

Tryblidis Clem. (1909) ≡ Tryblidiopsis (Rhytismat.).

Tryblidium Wallr. (1833) , ≡ Triblidium (Triblid.).

Tryblis Clem. (1931), Leotiales (inc. sed.). 1, Eur.

trypacidin, an antitrypanosome antibiotic from *Aspergillus fumigatus* (Balan *et al., J. Antibiotics* A **16**: 157, 1963).

Trypetheliaceae Zenker (1827), Pyrenulales (L). 13 gen. (+ 18 syn.), 166 spp. Thallus crustose, often poorly developed or immersed; ascomata globose, sometimes aggregated within pseudostromatic tissues, papillate, the ostiole sometimes lateral; interascal tissue of narrow branched and anastomosed trabeculate pseudoparaphyses; asci cylindrical, fissitunicate, with a wide ocular chamber, without a refractive ring, not blueing in iodine; ascospores usually hyaline, transversely septate or muriform, the septa often thickened, with a mucous sheath. Anamorphs coelomycetous where known. Lichenized with green algae, cosmop.

Lit.: Harris (*Acta Amazonica* Suppl. **14**: 55, 1986; fam. concept, key 9 gen.), Makhija & Patwardhan (*Biovigynam* **15**: 61, 1989; India).

Trypetheliomyces Cif. & Tomas. (1953) = Trypethelium (Trypethel.).

Trypetheliopsis Asahina (1937), ? Trypetheliaceae (L). 1, Japan.

Trypethelium Spreng. (1804) nom. cons., Trypetheliaceae (L). *c.* 100, mainly trop. See Harris (*Acta Amazonica* Suppl. **14**: 55, 1986; key 11 spp. Amazonia), Lambright & Tucker (*Bryologist* **83**: 170, 1980; ultrastr., biology), Makhija & Patwardhan (*J. Hattori Bot. Lab.* **73**: 183, 1993; key 30 spp. India).

Tryssglobulus B. Sutton & Pascoe (1987), Mitosporic fungi, 1.A2.16. 1, Australia.

Tsuchiyaea Y. Yamada, H. Kawas., Itoh, I. Banno & Nakase (1988), Mitosporic fungi, 1.A1.1. 1, S. Afr. See Yamada *et al.* (*Agric. Biol. Chem.* **53**: 2993, 1989; phylogeny).

Tubakia B. Sutton (1973), Mitosporic fungi, 5.A1/B1.15. 5, N. temp. See Limber & Cash (*Mycol.* **37**: 129, 1945), Yokoyama & Tubaki (*Res. Comm. Inst. Ferm. Osaka* **5**: 43, 1971), Glawe & Crane (*Mycotaxon* **29**: 101, 1987; *T. dryina* revised), Belisario (*Inform. Fitopat.* **40**: 54, 1990; disease caused by *T. dryina*), Munkvold & Neely (*CJB* **69**: 1865, 1991; development of *T. dryina* in host tissue).

Tubaria (W.G. Sm.) Gillet (1876), Crepidotaceae. 15, temp. See Romagnesi (*Rev. Mycol.* **5**: 29, 1940; **8**: 26, 1943).

Tubariopsis R. Heim (1931), Bolbitiaceae. 1, Madagascar.

Tuber F.H. Wigg. (1780), Tuberaceae. *c.* 60 (hypogeous), cosmop. See Gross (*Z. Pilzkde* **41**: 143, 1975; key 13 spp., *Docums. Mycol.* **21**: 1, 1991; key 30 spp., Europe), Parguey-Leduc & Janex-Favre (*Rev. Mycol.* **41**: 1, 1977; ontogeny), Janex-Favre & Parguey-Leduc (*Crypt., Mycol.* **4**: 353, 1984; spore ultrastr.), Mostecchi & Lazzari (*Boln Grup. Micol. Bresadola* **27**: 196, 1984; col. pl.), Parguey-Leduc *et al.* (*C.R. Acad. Sci. Paris* III, **301**: 143, 1985; ontogeny, *Crypt., Mycol.* **8**: 173, 1987; ascoma, *CJB* **65**: 1491, 1987;

ascospores, *BSMF* **105**: 227, 1989; ascoma ontogeny, *Crypt., Mycol.* **12**: 165, 1991; ontogeny), Pacioni & Pomponi (*Micol. Veget. Medit.* **4**: 63, 1989; enzyme electrophoresis), Pegler *et al.* (1993; key 11 Br. spp.). See truffles.

Tuberaceae Dumort. (1822), Pezizales. 2 gen. (+ 9 syn.), 61 spp. Ascomata large, cleistothecial, ± globose, thick-walled, solid, often ornamented, the asci formed in irregular marbled veins separated by sterile strands; interascal tissue absent, at least at maturity; asci formed apparrently randomly within the fertile zones, ± globose, thick-walled, indehiscent, less than 8-spored; ascospores brown, usually strongly ornamented. Anamorphs unknown. Hypogeous, mycorrhizal with angiospermous trees, widespr.
Lit.: Trappe (1979).

Tuberales, see *Elaphomycetales, Pezizales* and truffles.

Tuberaster Boccone (1697) ≡ Polyporus (Polypor.).

Tubercolarites Barsanti (1903), Fossil fungi (Mitosp. fungi). 1 (Carboniferous), Italy.

Tubercularia F.H. Wigg. (1780) nom. rej. = Dibaeis (Icmadophil.) fide Gierl & Kalb (1993).

Tubercularia Tode (1790) nom. cons., Anamorphic Hypocreaceae, 3.A1.15. Teleomorph *Nectria*. 25, widespr. See Seifert (*Stud. Mycol.* **27**: 95, 1985).

Tuberculariaceae (obsol.). Mitosporic fungi (hyphomycetes) with conidiomata.

Tuberculariella Höhn. (1915) = Cryptosporiopsis (Mitosp. fungi) fide Petrak (*Sydowia* **19**: 227, 1966).

Tuberculariopsis Höhn. (1909), Mitosporic fungi, 3.A1.10. 1, Java.

Tubercularis Clem. & Shear (1931) ≡ Tuberculariopsis (Mitosp. fungi).

Tubercularites Arcang. (1903), Fossil fungi. 1 (Carboniferous), Italy.

tuberculate solarium, see solarium.

tubercule (tubercle), a small wart-like process; **tuberculate**, having tubercles, syn. of punctate (Fig. 29.5);

Tuberculina Tode ex Sacc. (1880), Mitosporic fungi, 3.A1.15. 10 (on rusts), widespr.

Tuberculinia Velen. (1922) nom. nud. = Grandinia (Hydn.).

Tuberculis Clem. & Shear (1931) ≡ Tuberculariella (Mitosp. fungi).

Tuberculispora Deighton & Piroz. (1972), Mitosporic fungi, 1.G1.15. 1 (on *Irenopsis*), Jamaica.

Tuberculopsis Sacc. (1914) nom. nud. See Saccardo (*Ann. Myc.* **12**: 303, 1914).

Tuberculostoma Sollm. (1864) = Robergea (Stictid.) fide Sherwood (1977).

Tuberium Raf. (1815) ≡ Tuber (Tuber.).

Tuberosurculus Paulet (1791) nom. dub. (Ascomycota, inc. sed.); used for diverse taxa, incl. *Clavicipitaceae* and *Xylariaceae*.

Tubeufia Penz. & Sacc. (1898), Tubeufiaceae. 14, mainly trop. (fide Rossman, *Mycol. Pap.* **157**, 1987; key). Anamorphs *Helicoma, Helicosporium, Monodictys*.

Tubeufiaceae M.E. Barr (1979), Dothideales. 16 gen. (+ 8 syn.), 59 spp. Ascomata perithecial, sometimes aggregated into stromata, ± globose, thick-walled, sometimes setose, opening by a well-developed lysigenous pore; peridium pale, fleshy, composed of small pseudoparenchymatous cells; interascal tissue of cellular pseudoparaphyses; asci cylindrical, fissitunicate; ascospores hyaline or pale brown, transversely septate, without a sheath. Anamorphs hyphomycetous, often helicosporous; biotrophic or saprobic on scale insects or other fungi, widespr.
Lit.: Barr (*Mycotaxon* **12**: 137, 1980), Rossman (*Mycol. Pap.* **157**, 1987; key gen.).

Tubifera J.F. Gmel. (1792), Lycogalaceae. 6, cosmop. See Nelson *et al.* (*Mycol.* **74**: 541, 1982; key).

Tubiferaceae see Lycogalaceae.

Tubipeda Falck (1923) ≡ Leptopodia (Helvell.).

Tubiporus P. Karst. (1881) ≡ Boletus (Bolet.).

Tubolachnum Velen. (1934), ? Leotiales (inc. sed.). 2, former Czechoslovakia.

Tubosaeta E. Horak (1967), Xerocomaceae. 3, trop.

Tubularia auct. [non Roussel (1806), *Algae*] = Tubifera (Lycogal.), etc.

Tubulicium Oberw. (1965), Tubulicrinaceae. 3, widespr.

Tubulicrinaceae Jülich (1982), Stereales. 3 gen. (+ 1 syn.), 14 spp. Basidioma resupinate; monomitic, lyocystidia present; spores inamyloid.

Tubulicrinis Donk (1956), Tubulicrinaceae. 14, widespr. See Weresub (*CJB* **39**: 1456, 1961), Hayashi (*Bull. Govt Forest Exp. Stn* **260**, 1974).

Tubulifera O.F. Müll. ex Jacq. (1779) = Tubifera (Lycogal.).

Tubulina Pers. (1794) = Tubifera (Lycogal.).

Tubulixenasma Parmasto (1965) ≡ Tubulicium (Tubulicrin.).

Tuburcinia Fr. (1832) [non Woronin (1882)], nom. rej. ≡ Urocystis (Tillet.).

Tuburcinia Woronin (1882) nom. dub. (Ustilagin.) fide Mordue (*in litt.*).

Tuburciniella Zambett. (1970), Anamorphic Ustilaginaceae. Teleomorph *Urocystis*.

Tucahus Raf. (1830) ≡ Gemmularia (Pezizales).

tuckahoe (or Indian bread), the sclerotia (*Pachyma cocos*) of *Poria cocos* (Weber, *Mycol.* **21**: 113, 1929), USA. **Canadian -**, see stone-fungus.

Tuckerman (Edward; 1817-1886). Father of American lichenology, visited Europe in 1841, greatly influenced by E.M. Fries (q.v.), appointed Professor of oriental history in 1855 and also of botany in 1858 at Harvard Univ., holding the latter position until his death. His main works included *Genera lichenum* (1872) and *Synopsis of the North American lichens* (2 vols, 1882, 1888). See Culberson (*Collected lichenological papers of Edward Tuckerman*, 2 vols, 1964), Reid (*Mycotaxon* **26**: 3, 1986), Grummann (1974: 197), Stafleu & Cowan (*TL-2* **6**: 523, 1986).

Tuckermannopsis Gyeln. (1933), Parmeliaceae (L). *c.* 20, widespr. See Kärnefelt *et al.* (*Pl. Syst. Evol.* **183**: 113, 1992, *Bryol.* **96**: 394, 1993).

Tuckneraria Randlane (1994), Parmeliaceae (L). 4, E. & S.E. Asia.

Tulasne (Louis René, 1815-85; and Charles, 1816-84). L.R. Tulasne, the 'reconstructor of mycology' was from 1842 in a position at the Paris Natural History Museum and in more than 50 papers he made additions to the knowledge of smuts, rusts, ergot, subterranean fungi (*Fungi hypogaei*, 1851), pyrenomycetes, lichens (*Ann. Sci. nat., Bot.* sér. 3, **27**: 225 pp., 17 pl., 1852), and higher plants while Charles, who gave his brother much help, made the beautiful and detailed Icones for the *Selecta fungorum carpologia* (3 vols, 1861-5 [Engl. transl. Grove, 1931; biogr. notes in vol. 1]) and other writings. An important new idea in the *Carpologia*, a work which has had a great effect on mycology, was that of pleomorphism in fungi. See Stafleu & Cowan (*TL-2* **6**: 530, 529, 1986).

Tulasneinia Zobel ex Corda (1854) ≡ Terfezia (Terfez.).

Tulasnella J. Schröt. (1888), Tulasnellaceae. 44, widespr. See Currah & Zelmer (*Rep. Tottori mycol. Inst.* **30**: 43, 1992; 6 spp. with orchids), Olive (*Mycol.* **49**: 663, 1957; key), Jülich & Jülich (*Persoonia* **9**: 49, 1976), Roberts (*MR* **98**: 1431, 1994; key Eur. spp.).

Tulasnellaceae Juel (1897), Tulasnellales. 2 gen. (+ 4 syn.), 28 spp. Basidioma resupinate, typically gelatinized or ephemeral; probasidium developing tetrad of lacrymoid epibasidia; anamorph *Epulorhiza*.

Tulasnellales, Basidiomycetes. 1 fam., 3 gen. (+ 4 syn.), 29 spp. Fam.:
Tulasnellaceae.
Lit.: Donk (1954-62, II), Martin (*Brittonia* **9**: 25, 1957), Talbot (*Persoonia* **3**: 371, 1965), Donk (1966: 255).

Tulasnia Lesp. [not traced] = Terfezia (Terfez.) fide Saccardo (*Syll. Fung.* **8**: 902, 1889).

Tulasnodea Fr. (1849) ≡ Tulostoma (Tulostomat.).

Tulostoma Pers. (1794), Tulostomataceae. 79, mainly dry regions. See Wright (*The genus Tulostoma (Gasteromycetes)*, 1987 [*Bibl. Mycol.* **113**], *Pap. Mich. Acad. Sci.* **40**: 79, 1955; sp. characters), Maas Geesteranus (*Gorteria* **5**: 189, 1971; Netherlands), Calonge & Wright (*Bol. Soc. Micol. Madrid* **13**: 119, 1989; Spain).

Tulostomataceae E. Fisch. (1900), Tulostomatales. 4 gen. (+ 1 syn.), 83 spp. Gasterocarp stipitate; peridium globose with apical dehiscence; gleba compact, finally powdery; basidia pleurosporus.

Tulostomatales, Basidiomycetes ('gasteromycetes'). 4 fam., 9 gen. (+ 14 syn.), 105 spp. Basidioma stipitate, basidia usually with lateral sterigmata dispersed in a compact gleba; cosmop., mostly warm dry regions; saprobic in soil. Fams:
(1) **Battarreaceae**.
(2) **Calostomataceae**.
(3) **Phelloriniaceae**.
(4) **Tulostomataceae**.
Lit.: Dring (1973; key gen.). See also Lit. under Gasteromycetes.

tumid, swollen; inflated.

Tumidapexus D.A. Crawford (1954), Aphelariaceae. 1, NZ.

Tumularia Marvanová & Descals (1987), Mitosporic fungi, 1.B/C1.21. 2 (aquatic), Europe.

Tunbridge ware. Ornaments made from coloured wood, usually using marquetry techniques; the green wood used is that of deciduous trees attacked by *Chlorociboria aeruginascens*.

tunic, see exospore.

tunica, a coat, esp. a thin white membrane round the peridiole in most species of the *Nidulariaceae*. See also spore wall (2) and basidiospore.

Tunicago B. Sutton & Pollack (1977), Mitosporic fungi, 4.B2.15. 1, USA.

Tunicatispora K.D. Hyde (1990), Halosphaeriaceae. 1 (marine), Australia.

Tunicopsora Suj. Singh & P.C. Pandey (1971), Uredinales (inc. sed.). 1, India. = Kweilingia (Uredinales) fide Cummins & Hiratsuka (1983).

Tunstallia Agnihothr. (1961), Ascomycota (inc. sed.). 1, India. *T. aculeata* (thorny stem blight of tea, *Camellia*).

Tupia L. Marchand (1830) nom. rej. prop. ≡ Icmadophila (Icmadophil.). See Gierl & Kalb (*Herzogia* **9**: 593, 1993).

turbid, not clear; cloudy.

turbinate, like a top in form (Fig. 37.21). - **organ** or **cell** (of *Cladochytriaceae*), a swelling on the vegetative thallus (see Karling, *Am. J. Bot.* **18**: 528, 1931); spindle organ.

Turbinellus Earle (1909) = Gomphus (Gomph.). See Donk (*Beih. Nova Hedw.* **6**: 291, 1962).

Tureenia J.G. Hall (1915) = Arthrinium (Mitosp. fungi) fide Höhnel (*Mitt. bot. Lab. Techn. Hochsch. Wien* **2**: 9, 1925).

turgid, tightly swollen.

Turgidosculum Kohlm. & E. Kohlm. (1972), Mastodiaceae. 2 (on algae), widespr. See Kohlmeyer & Kohlmeyer (*Marine mycology*, 1979), Schatz (*Mycol.* **72**: 110, 1980).

turgor pressure, of mycelium, see Adebayo *et al.* (*TBMS* **57**: 145, 1971).

TWA, see Media.

twist, a disease of cereals and grasses (*Dilophospora alopecuri*).

Tychosporium Spiegel (1995), Protosteliaceae. 1, USA.

Tylochytrium Karling (1939) ≡ Phlyctidium (Chytrid.).

Tylodon Banker (1902) ≡ Radulum (Vals.). See Donk (*Taxon* **12**: 155, 1963).

Tylomyces Cortini (1921), Mitosporic fungi, ?.?.?. 1, Italy.

Tylophoma Kleb. (1933) = Dothichiza (Mitosp. fungi) fide Petrak (*Sydowia* **10**: 201, 1957).

Tylophorella Vain. (1890), Ascomycota (inc. sed., L). 1, S. Am. See Egea & Tibell (*Nordic Jl Bot.* **13**: 207, 1993).

Tylophorellomyces Cif. & Tomas. (1953) ≡ Tylophorella (Caliciales).

Tylophoron Nyl. ex Stizenb. (1862), Caliciales (inc. sed., L). 3, trop. See Tibell (1982, 1984, 1987, *MR* **95**: 290, 1991; anamorph).

Tylophoropsis Sambo (1938) [non N.E. Br. (1894), *Asclepiadaceae*], ? Caliciaceae. 1, Afr. Nom. dub. fide Tibell (1984).

Tylopilus P. Karst. (1881), Strobilomycetaceae. 18, N. & S. temp. See Smith (*Mycol.* **60**: 954, 1968), Wolfe (*Bibl. Mycol.* **69**, 1979; N. Am.).

Tylosperma Donk (1957) [non Botsch. (1952), *Rosaceae*] ≡ Tylospora (Athel.).

Tylospora Donk (1960), Atheliaceae. 2, widespr.

Tylostoma, see *Tulostoma*; Long (*Mycol.* **36**: 320, 1944).

Tylothallia P. James & R. Kilias (1981), Lecanoraceae (L). 1, Eur.

Tympanella E. Horak (1971), Galeropsidaceae. 1, NZ.

Tympanicysta Malme (1980), Fossil fungi. 2 (Late Permian-Early Triassic), cosmop.

Tympanis Tode (1790), Leotiaceae. 29, temp. See Groves (*CJB* **30**: 571, 1952; monogr.), Ouellette & Pirozynski (*CJB* **52**: 1889, 1974; key).

Tympanopsis Starbäck (1894) = Nitschkia (Nitschk.) fide Nannfeldt (1975).

Tympanosporium W. Gams (1974), Mitosporic fungi, 1.A1.19. 1 (on *Tubercularia*), Netherlands.

type (in nomenclature), the element on which the descriptive matter fulfilling the conditions of valid publication of a scientific name is based, or is considered to have been based, and which fixes the application of the name; e.g. a family name on a genus, a generic name on a species (a - **species**; see also nomen species), a specific name generally a - **specimen**, which may be a slide, sometimes on a - **culture** (incorrectly for fungi if still metabolically active), a Figure, or a description. Numerous terms with the suffix '-type' have been used in nomenclature, both formally and informally (*see* Hawksworth, *A draft glossary of terms used in bionomenclature*, [IUBS Monogr. **9**], 1994), and only a selection of those most used by mycologists are included here: **epi-** a specimen or illustration used to serve as an interpretive type where the existing type material is inadequate for the precise application of the name; **ex-type**, out of the type, used especially for living cultures where the holotype is a dried culture or one preserved in a metabolically inactive state; **holo-**, the single element on which the describing author based a name; **iso-**, a duplicate or part of the type collection (other than the holotype) [(in immunology), part of the imunoglobin molecule from the mouse used in the characterization of sera]; [**histo-**, a reaction between different types of cells]; **lecto-**, an element selected in a later work from the original material where no holotype was designated; **mono-**, the only species included in a genus when first described; **neo-**, specimen or other material designated as nomenclatural type when all the original material is missing; **para-**, any specimen other than the holotype on which the first account of a species or other group is based; **patho-**, see pathovar; **phyco-**, each of the morphologically distinct structures derived by symbiosis between a single mycobiont and different photobionts (Swinscow, *Lichenologist* **9**: 89, 1977; see Lichens); **syn-** one of several elements cited by an author when originally proposing a name but where no holotype was selected ; **topo-**, a later collection from the original locality; **typo-**, the specimen used to prepare an illustration where the latter is the type. See Nomenclature, wild type.

Type culture collections, see Genetic resource collections.

Typhella L. Léger & M. Gauthier (1935) nom. nud. ? = Smittium (Legeriomycet.). See Manier & Lichtwardt (*Ann. Sci. nat., Bot. sér.* 12, **9**: 519, 1968).

Typhoderma Gray (1821) nom. dub. (? Fungi).

Typhodium Link [not traced] = Epichloë (Clavicipit.) fide Tulasne & Tulasne (*Sel. carp. fung.* **3**: 23, 1865).

Typhula (Pers.) Fr. (1818), Typhulaceae. 63, temp. *T. incarnata* injures cereals under snow. See Berthier (*Bull. Soc. linn. Lyon* num. spéc. **45**, 1976), Woodbridge & Coley-Smith (*MR* **95**: 995, 1991).

Typhulaceae Jülich (1982), Cantharellales. 4 gen. (+ 6 syn.), 119 spp. Basidioma small, clavarioid, with a differentiated head and stipe, often arising from a sclerotium; monomitic; lignicolous or epiphyllous.

Typhulochaeta S. Ito & Hara (1915), Erysiphaceae. 4, Asia, N. Am. See Braun (*Beih. Nova Hedw.* **98**, 1987; key), Solheim *et al.* (*J. Elisha Mitchell sci. Soc.* **84**: 236, 1968).

typonym, a name having the same type as another name which is neither its basionym nor synisonym (Donk, *Persoonia* **1**: 175, 1960).

Tyrannosorus Untereiner, Straus & Malloch (1995), Dothideales (inc. sed.). 1 (on wood), Canada, Pakistan. Anamorph *Helicodendron*.

Tyridiomyces W.A. Wheeler (1907), ? Saccharomycetales. 1 (from ants), *sine loc.*

Tyrodon P. Karst. (1881) ≡ Hydnum (Hydn.).

Tyromyces P. Karst. (1881), Coriolaceae. *c.* 50, widespr. See Lowe (*Mycotaxon* **2**: 1, 1975; key N. Am. spp.), David (*Bull. Soc. linn. Lyon* **49**: 596, 1980; key 16 Afr. spp.).

UAMH. University of Alberta Microfungus Collection and Herbarium (Edmonton, Alberta, Canada); founded 1960; see Sigler (*J. Ind. Microbiol.* **13**: 191, 1994).

Uberispora Piroz. & Hodges (1973), Mitosporic fungi, 1.G2.1. 1, Japan, N. Am.

ubiquinones, a class of terpenoid lipids involved in electron transport and of potential use in fungal systematics due to variation in the isoprenoid side chain in some taxa. See Kuraishi *et al.* (*MR* **95**: 705, 1991).

Ubrizsya Negru (1965), Mitosporic fungi, 8.A1.?. 1, Roumania.

Ucographa A. Massal. (1860) = Pragmopora (Leot.) fide Nannfeldt (*Nova Acta R. Soc. Scient. upsal.* 4, **8**(2), 1932).

Udeniomyces Nakase & Takem. (1992), Saccharomycetaceae. 3, Japan, Eur.

Ugola Adans. (1763) nom. dub. (Mitosp. fungi). See Stalpers (*Stud. Mycol.* **24**: 80, 1984; anamorph name).

ulcerose, ulcer-like.

Ulea J. Schröt. (1892), ? Uredinales (inc. sed.). 1 (on *Araucaria*), Brazil. Excluded from rusts by

Thirumalachar (*Bull. Torrey bot. Club* **76**: 339, 1949, *Indian Phytopath.* **3**: 4, 1950).

Uleiella J. Schröt. (1894) ≡ Ulea (? Uredinales, inc. sed.).

Uleodothella Syd. & P. Syd. (1921) = Tomasellia (Arthopyren.) fide Müller & v. Arx (1962).

Uleodothis Theiss. & Syd. (1915), Venturiaceae. 4, S. Am.

Uleomyces Henn. (1895), Cookellaceae. 10 (on rusts and ascomycetes), widespr.

Uleomycina Petr. (1954) = Elsinoë (Elsin.) fide v. Arx (1963).

Uleopeltis Henn. (1904) = Mendogia (Schizothyr.) fide Clements & Shear (1931).

Uleoporthe Petr. (1941), Valsaceae. 1, trop. Am. See Barr (*Mycol. Mem.* **7**, 1978).

Uleothyrium Petr. (1929), Asterinaceae. 1, Brazil. Anamorph *Septothyrella*.

Ulkenia A. Gaertn. (1977), Thraustochytriaceae. 5, Eur.

Ulocladium Preuss (1851), Mitosporic fungi, 1.D2.26. 9, widespr. See Simmons (*Mycol.* **59**: 77, 1967; key).

Ulocodium A. Massal. (1855) = Catillaria (Catillar.) fide Zahlbruckner (1926).

Ulocolla Bref. (1888) = Exidia (Exid.) fide Donk (*Persoonia* **4**: 166, 1966).

Ulocoryphus Michaelides, L. Hunter & W.B. Kendr. (1982), Mitosporic fungi, 2.E1.19. 1, NZ.

Uloploca Kleb. (1933) nom. dub. fide Sutton (*Mycol. Pap.* **141**, 1977).

Uloporus Quél. (1886) = Gyrodon (Gyrodont.) fide Patouillard (1900).

Uloseia Bat. (1963) ? = Chaetothyrium (Chaetothyr.). See also Hughes (1976).

Ulospora D. Hawksw., Malloch & Sivan. (1979), Testudinaceae. 1, India.

Ultrastructure. Studies with the electron microscope have shown fungi to have a typical eukaryotic structure but the nuclei of some show unusual features. See Hawker (*Rev. Biol.* **40**: 52, 1965), Bracker (*Ann. Rev. Phytopath.* **5**: 343, 1967; reviews), Beckett *et al.* (*An atlas of fungal ultrastructure*, 1975), Cole & Samson (*Patterns of development in conidial fungi*, 1979), Hess (*Shokubutsu Byogai Kenkyi* **8**: 71, 1973; of germination), Littlefield & Heath (*Ultrastructure of rust fungi*, 1979), Mimms (*Mycol.* **83**: 1, 1991; plant pathogens). For some organelles of which the structure has been elucidated by electron microscopy see: apical granule, concentric bodies, dictyosome, dolipore septum, flagellum, parenthesome. Lichen ultrastructure, see Jacobs & Ahmadjian (*J. Phycol.* **5**: 227, 1969; review), Hale (*in* Brown *et al.*, *Lichenology: progress and problems*: 1, 1976; scanning electron microscope).

See also ascus, Microscopy, Scanning electron microscopy.

Ulva L. (1753), Algae sensu Agardh p.p. = Phycomyces (Mucor.) fide Fries (1832).

Ulvella (Nyl.) Trevis. (1880) [non H. Crouan & P. Crouan (1859), *Algae*], nom. dub. (Ascomycota, inc. sed.), but see Santesson (*Symb. bot. upsal.* **12** (1), 1952).

Umbellidion B. Sutton & Hodges (1975), Mitosporic fungi, 1.A1.10. 1, Brazil.

Umbellula E.F. Morris (1955) = Pseudobotrytis (Mitosp. fungi) fide Subramanian (*Proc. Ind. Acad. Sci. B* **43**: 277, 1956).

Umbelopsis Amos & H.L. Barnett (1966), Mortierellaceae. 7 (in soil), widespr. See v. Arx (*Sydowia* **35**: 10, 1982), Yip (*TBMS* **86**: 334, 1986, **87**: 243, 1986), Kendrick *et al.* (*Mycotaxon* **51**: 15, 1994).

Umbilicaria Hoffm. (1789) [non Heist. ex Fabr. (1759), *Boraginaceae*], nom. cons. prop., Umbilicariaceae (L). *c.* 45, cosmop. (mainly temp. and arctic). See

Crespo & Sancho (*An. Inst. bot. Cavanilles* **35**: 79, 1978; Spain, adaptation, convergence), Filson (*Muelleria* **6**: 335, 1987; key 5 spp. Antarctica), Frey (*Ber. schweiz. bot. Ges.* **59**: 427, 1949), Hakulinen (*Annls bot. Soc. zool.-bot. fenn. Vanamo* **32**(6), 1962), Henssen (*Vortr. Gesgeb. Bot. n.f.* **4**: 103, 1970; ontogeny), Hestmark (*Lichenologist* **23**: 343, 361, 1991; anamorph-teleomorph relationships, *MR* **96**: 1033, 1992, conidiogenesis), Krog & Swinscow (*Nordic Jl Bot.* **6**: 75, 1986; E. Africa), Llano (*A monograph of the lichen family Umbilicariaceae in the Western Hemisphere*, 1950, *Hvalråd Skr.* **48**: 112, 1965), Wei (*Mycosystema* **5**: 1, 1993), Wei & Jiang (1989; key 29 spp. China, 1993; key 50 spp. Asia), Sipman & Topham (*Nova Hedw.* **54**: 63, 1992; key 6 spp., Colombia).

Umbilicariaceae Chevall. (1826), Lecanorales (L.). 2 gen. (+ 14 syn.), 57 spp. Thallus foliose, usually with a central attachment; ascomata apothecial, sessile or slightly stipitate, often convoluted, black, flat or domed, without well-developed marginal tissue; interascal tissue of sparingly branched paraphyses, sometimes swollen at the apices; asci with a relatively small I+ apical cap and an outer layer of I+ gelatinized material, 1- to 8-spored; ascospores hyaline to brown, variously septate. Thalloconidia often produced. Lichenized with green algae, widespr. esp. temp.
    Lit.: Hestmark (*Somerfeltia, Suppl.* 3, 1991; sexual strategies), Wei & Jiang (*Mycosystema* **2**: 135, 1989; cluster analysis, *The Asian Umbilicariaceae (Ascomycota) [Mycosystema, Monogr.* 1], 1993; keys, descr. 60 spp.).

Umbilicariomyces Cif. & Tomas. (1953) ≡ Umbilicaria (Umbilicar.).

umbilicate, having a small hollow; esp. of a pileus having a hollow on the top above the stipe (Fig. 19M).

umbilicus, (1) the central hold-fast occurring in some foliose lichens (e.g. *Umbilicaria*), navel, umbo; (2) the pore in the perispore of an ascospore (Eriksson, *Opera bot.* **60**, 1981).

umbo, a central swelling like the boss at the centre of a shield; esp. one on top of a pileus above the stipe; see also umbilicus; **-nate**, having an umbo (Fig. 19L).

Unamunoa Urries (1942) = Rechingeriella (Zopf.) fide Petrak (*Sydowia* **23**: 265, 1970).

Uncigera Sacc. (1885), Mitosporic fungi, 1.A1.15. 1, Eur.

uncinate (**uncate**), (1) hooked (Fig. 37.35); (2) of gill insertion, near sinuate (q.v.).

Uncinia Velen. (1934) [non Pers. (1807), *Cyperaceae*] ≡ Hamatocanthoscypha (Hyaloscyph.).

Unciniella K. Holm & L. Holm (1977) ≡ Hamatocanthoscypha (Hyaloscyph.).

Uncinocarpus Sigler & G.F. Orr (1976), Onygenaceae. 1, USA. See Currah (1985). The human pathogen *Coccidioides immitis* may be closely related (Pan *et al.*, *Microbiol.* **140**: 1481, 1994). Anamorph *Malbranchea*.

Uncinula Lév. (1851), Erysiphaceae. 81, widespr. *U. necator* (vine (*Vitis*) mildew); *U. bicornis* (syn. *U. aceris*; powdery mildew of sycamore, *Acer*). See Braun (*Beih. Nova Hedw.* **89**, 1987; key), Pirozynski (*Mycol. Pap.* **101**, 1965; Afr., key). Anamorph *Oidium*.

Uncinulella Hara (1936) = Uncinula (Erysiph.) fide Zheng (*Mycotaxon* **22**: 209, 1985).

Uncinuliella R.Y. Zheng & G.Q. Chen (1979), Erysiphaceae. 4, Asia, N. Am. See Braun (*Beih. Nova Hedw.* **89**, 1987; key).

Uncinulites Pampal. (1902), ? Fossil Fungi. 1 (Miocene), Italy. See Salmon (*J. Bot., Lond.* **41**: 127, 1903).

Uncinulopsis Sawada (1916) = Pleochaeta (Erysiph.) fide Kimbrough & Korf (1963).

Uncispora R.C. Sinclair & Morgan-Jones (1979), Mitosporic fungi, 1.C2.1. 1, USA.

Uncobasidium Hjortstam & Ryvarden (1978), Hyphodermataceae. 1, Norway.

under cortex, lower cortex in foliose lichens.

Underwoodia Peck (1890) = Helvella (Helvell.) fide Dissing (1966), Harmaja (*Karstenia* **14**: 102, 1974).

Underwoodina Kuntze (1891) ≡ Aschersonia (Mitosp. fungi).

undulate, wavy.

Unger (Franz; 1800-70). After a medical training became Professor of Botany at Graz; then Professor of Physiology at Vienna. Chief writings: *Die Exantheme der Pflanzen*, 1833 [an important book, but fungi are still taken to be the result of disease]; and a number of papers on general botany. See Anon. (*J. Bot., Lond.* **8**: 192, 1870), Stafleu & Cowan (*TL-2* **6**: 594, 1986).

Unguicularia Höhn. (1905), Hyaloscyphaceae. 7, Eur. See Baral (*SA* **13**: 113, 1994; posn), Korf & Kohn (*Mycotaxon* **10**: 503, 1980), Raschle (*Sydowia* **29**: 170, 1977; key).

Unguiculariella K.S. Thind & R. Sharma (1990), Hyaloscyphaceae. 1, Bhutan.

Unguiculariopsis Rehm (1909), Leotiaceae. 15 (on lichens and other fungi), widespr. See Zhuang (*Mycotaxon* **32**: 1, 1988; key). Anamorph *Deltosperma*.

Unguiculella Höhn. (1906), Hyaloscyphaceae. c. 10, Eur. See Dennis (*Mycol. Pap.* **32**, 1949), Raitviir (*Mikol. Fitopat.* **21**: 200, 1987).

Ungularia Lázaro Ibiza (1916) ≡ Piptoporus (Coriol.).

ungulate, shaped like a horse's hoof.

Ungulina Pat. (1900) ≡ Fomes (Coriol.).

unialgal (of cultures of lichen photobionts), ones in which a single algal species is present but which may also contain bacteria, fungi, or other organisms.

Unicellomycetales, yeasts (Kudriavtsev, 1954).

union, see Phytosociology.

unipolar, at one end only (esp. of a bacterial cell).

uniseriate, in one row.

Uniseta Ciccar. (1948), Mitosporic fungi, 8.B2.19. 1, USA.

universal veil (of agarics and gasteromycetes), a layer of tissue covering the basidioma while development takes place; teleblem; blematogen; cf. volva; **primary - -** = protoblem.

unorientated, not arranged in any particular direction.

unstratified (of lichen thalli), not layered; homoiomerous (q.v.).

Uperhiza Bosc (1811) ? = Melanogaster (Melanogastr.).

UPS. Botanical Museum, University of Uppsala (Uppsala, Sweden); founded 1785; genetic resource collection **UPSC**.

Uraecium Arthur (1933), Uredinales (inc. sed.). 11, cosmop., esp. trop. Established for aecial uredinia (II).

Urceola Quél. (1886) [non Roxb. (1799) nom. cons., *Apocynaceae*], Leotiales (inc. sed.). 1, Eur.

Urceolaria Ach. (1803) [non Molina ex Brandis (1786), *Gesneriaceae*], nom. rej. = Diploschistes (Thelotremat.).

Urceolaria Hook. (1821) [non Molina ex Brandis (1786), *Gesneriaceae*] = Aspicilia (Hymenel.).

Urceolaria Bonord. (1851) [non Molina ex Brandis (1786), *Gesneriaceae*], Pezizales (inc. sed.). 1, Eur.

Urceolariomyces Cif. & Tomas. (1953) = Diploschistes (Thelotremat.).

urceolate, pitcher-like in form.

Urceolella Boud. (1885), Hyaloscyphaceae. 9, Eur., N. Am. See Raschle (*Sydowia* **29**: 170, 1977; key), Huhtinen (*Karstenia* **27**: 8, 1988), Korf & Kohn (*Mycotaxon* **10**: 503, 1980).

Urceolina Tuck. (1875) [non Rchb. (1828), nom. cons., *Amaryllidaceae*] ≡ Orceolina (Trapel.).

Urceolus Velen. (1939) [non Mereschk. (1879), *Algae*] = Resupinatus (Tricholomat.). Used by Singer (1975).

Uredendo Buriticá & J.F. Hennen (1994), Phakopsoraceae. 1 (on *Gramineae*), trop. Am.

Uredinales, Teliomycetes. the rust fungi; rusts. 14 fam., 164 gen. (+ 139 syn. and uncertain), *c.* 7,000 spp. The rust fungi or rusts. Mycelium (without clamp connexions) generally intercellular (frequently with haustoria), limited to parts of leaves or other aerial organs of the host ('local' infection), sometimes perennial, if systemic overwintering in roots or other parts; cosmop. on seed plants and ferns, frequently causing major disease; obligate parasites but axenic culture reported for *Gymnosporangium juniperi-virginianae* (Cutter, *Mycol.* **51**: 248, 1959), *Puccinia graminis* f.sp. *tritici* (Williams *et al.*, *Phytopath.* **56**: 1418, 1966; **57**: 326, 1967; *TBMS* **57**: 129, 137, 1971), *Melampsora lini* (Turel, *CJB* **47**: 821, 1969), *Uromyces dianthi* (Jones, *TBMS* **58**: 29, 1972) and others, are now known to be possible from all states of the life cycle (Narisawa *et al.*, *Trans. Mycol. Soc. Japan* **33**: 35, 1992).

Rusts have up to five spore states (frequently numbered **O-IV**; these roman numerals can be ambiguous unless restricted to a morphological system. See Holm, *Notes R. Bot. Gard. Edin.* **44**: 433, 1987) (Fig. 41A). Traditionally the spore terminology was based on morphology. Arthur coined contractions for the original terms (e.g. telium for teleutosorus) and later linked them to the nuclear events in the life cycle, as did Cummins (1959). Other authors have used either the long or short spellings with interchanged definitions and much confusion resulted. In attempts to unite the merits of both schemes Laundon (*TBMS* **50**: 189, 1964; **58**: 344, 1972) proposed a basically morphological terminology which incorporated nuclear events by adding qualifiers where desirable and Holm (1987) devised a compromise in which 'short' terms were linked to nuclear events and 'long' terms to morphology. Essentially, Hiratsuka (*Mycol.* **65**: 432, 1973, *Rep. Tottori mycol. Inst.* **12**: 99, 1975), who, following Cummins, relates the spore states to the nuclear cycle, is followed here.

O. Spermatia (sing. -ium; pycniospores), monokaryotic gametes produced in **pycnia** (sing. -ium; spermogonia; morphological types of pycnia, see Hiratsuka & Hiratsuka, *Rept. Tottori Mycol. Inst.* **18**: 257, 1980) which are variable in form and position, contain a palisade of sporogenous cells which produce spores in nectar exuded through the ostiole and may have periphyses and flexuous hyphae (q.v.). Savile (*Mycotaxon* **33**: 387, 1988) emphasized that pycnia are 'hermaphroditic' structures.

I. Aeciospores (aecidiospores, plasmogamospores), produced in **aecia** (sing. -ium; aecidiosori; morphological types of aecia see Sato & Sato, *TBMS* **85**: 223, 1985), are unicellular, non-repeating vegetative spores, usually resulting from dikaryotization (and thus usually associated with pycnia), which germinate to give dikaryotic mycelium. Aeciospores (aecial aeciospore (I^I), Laundon) are typically catenulate, thin-walled, and verrucose but sometimes they resemble typical urediniospores when they are designated **uredinioid aeciospores** by Cummins (= aecial

TABLE 8. Nomenclature of rust life-cycles.

Example	Schröter	Arthur	Laundon	Durrieu
Puccinia graminis	0 I II III IV eu-form	0 I II III IV macrocyclic	0 I II III IV macrocyclic	macrocyclic
P. punctiformis	0 II III IV brachy-form	0 I II III IV macrocyclic	0 II^I II^{II} III IV brachycyclic	brachycyclic
Gymnosporangium sp. (most spp.)	0 I III IV opsis-form	0 I III IV macrocyclic	0 I III IV demicyclic	opsicyclic
Coleosporium sp.	0 I II III IV eu-form	0 I II III IV macrocyclic	0 I^I II^{II} III IV demicyclic	opsicyclic (pseudopsicyclic)

(The uredinial aecia of *Coleosporium* spp. have been traditionally known as uredosori because of their association with telia in a heteroecious life-cycle.)

P. lagenophorae	(0) I III IV opsis-form	(0 I) II III IV macrocyclic	(0 I^I) II^{II} III IV demicyclic	opsicyclic (pseudopsicyclic)

(The uredinial aecia of such rusts as this have not generally been recognized on account of their being morphologically indistinguishable from ordinary aecia. The pycnia and true aecia are rarely found.)

P. heterospora	(0) III IV micro-form	(0) III IV macrocyclic	(0) III IV demicyclic	microcyclic

(Rusts like *P. malvacearum* with spores which germinate immediatley were referred to as leptoforms by Schröter.)

Endophyllum sp.	0 I IV endo-form	0 III IV microcyclic	0 I^{III} IV microcyclic	endocyclic
P. chryanthemi	II III IV hemi-form			

urediniospores (II^I), Laundon; primary uredospores, Winter).

II. Urediniospores (uredospores, urediospores (for orthography see Savile, *Mycol.* **60**: 459, 1968), summer spores, red rust spores), repeating vegetative spores (which give urediniospores again or teliospores), usually on dikaryotic mycelium, in **uredinia** (uredosori, uredia; morphological types of uredinia, see Sathe, *Kavaka* **5**: 59, 1977; Hiratsuka & Sato, *in* Scott & Chakravorty (Eds), 1982). Typical urediniospores are unicellular, pedicellate, deciduous, with the pigmented echinulate wall showing two or more germ pores. Rarely they resemble typical aeciospores when they are designated **aecidioid urediniospores** by Cummins (= uredinial aeciospores (I^{II}), Laundon). **Amphispores** (II^{II};X) or resting urediniospores are produced by some rusts. These spores generally have thicker and darker walls than normal urediniospores.

III. Teliospores (teleutospores, teleutosporodesma, winter spores, black rust spores), produced in **telia** (sing. -ium; teleutosori; ontogeny and morphology, see Hiratsuka, *Mycotaxon* **31**: 517, 1988), are basidia-producing spores. Telia and teliospores, which characterize the teleomorph of rust fungi, show wide morphological variation but typically teliospores are resting spores, 2- or more celled, sessile or pedicellate but not deciduous, and the thick wall is variously ornamented. Rarely they resemble typical aeciospores when they are designated **aecidioid teliospores** by Cummins (= telial aeciospores (I^{III}), Laundon). Teliospores that germinate immediately, especially in species of genera that usually show dormancy, may be termed leptospores.

IV. Basidiospores (sporidia) are haploid, unicellular, thin-walled, short-lived spores produced on 2-4-celled **basidia** (sing. -ium; promycelium, metabasidium) after meiosis and liberated from sterigmata by abjection (Buller, **3**).

Hughes (*CJB* **48**: 2147, 1970) studied the development (ontogeny) of rust spores and concluded that in his hyphomycete spore terminology O spores are phialospores; I, meristem arthrospores; II, sympodioconidia (or, less common, meristem arthrospores); III, terminal chlamydospore-like cells, sympodioconidia, or meristem arthrospores. Savile (*in* Kendrick (Ed.), *The whole fungus* **2**: 547, 1979) has considered the evolution of anamorphs in rusts.

Rust life cycles (Petersen, *Bot. Rev.* **40**: 453, 1974) vary according to which stages are present or absent. Special terms designate the different life cycles as shown in Table 8 but have been applied in slightly different ways (see Durrieu, *BSMF* **95**: 379, 1979). Hiratsuka *et al.* (*Rust flora of Japan*, 1992) revert to terminology similar to Schröter.

A rust fungus may be **autoecious** (Fig. 41B) with its life cycle on one host (or group of closely related collateral hosts) or **heteroecious** (Fig. 41A) with O and I on one sort of host and II and III (or I^{II}, III or III only) on another sort (i.e. it has alternate hosts generally living in the same plant association).

The life-cycle of a rust is generally constant, though there may be no development of O, II, or sometimes I, because of weather or other conditions. A species with I, II, III, but not O, is sometimes given the name **cataspecies.** If there is no knowledge of III, the form-genus (e.g. *Aecidium, Uredo*) is, however, put in the Uredinales and not in the Mitosporic fungi. Physiologic specialization of *Uredinales* has had much attention. *Puccinia graminis* s.l. has 7 races (formae speciales) (*tritici, avenae*, etc.), and there are about 250 physiologic races of *P. graminis* f.sp. *tritici*,

and so on. Races are determined by the use of differential hosts.

Nuclear cycle: A rust may be heterothallic or homothallic. In a heterothallic macrocyclic species a basidium has two + and two − basidiospores (see Sex). A + (or −) spore, after infection of the right host, gives a haploid mycelium, pycnia with + (−) spermatia, and protoaecia. If taken (frequently by an insect) to a flexuous hypha of a − (+) pycnium, a + (−) spermatium may put out a 'peg' to make a connexion, its nucleus goes into (spermatizes) the hypha, and by division gives nuclei for the diploidization of cells down to the protoaecia. The cells of a protoaecium undergo conjugate division and an aecium with aeciospores is produced. An aeciospore and its mycelium, and urediniospores and their mycelia, are dikaryotic. There is nuclear fusion in the teliospore, meiosis in the basidium.

In homothallic species, where pycnia are not necessary and are frequently not present, the dikaryophase has its start from two cell nuclei at some point or points in the life-cycle. Nuclear fusion and reduction are as in heterothallic species. Following Dietel (1928) two families, *Melampsoraceae* and *Pucciniaceae*, have frequently been recognized. Concepts have been refined by subsequent workers (see Hennen & Buriticá, *Rept. Tottori Mycol. Inst.* **18**: 43, 1980). In this edition of the *Dictionary* fourteen familes are recognized as they are in current use by Cummins & Hiratsuka (1983) and in the major compilation Hiratsuka *et al.* (*Rust Flora of Japan*, 1992). Fams:

(1) **Chaconiaceae**.
(2) **Coleosporiaceae**.
(3) **Cronartiaceae**.
(4) **Melampsoraceae**.
(5) **Mikronegeriaceae**.
(6) **Phakopsoraceae**.
(7) **Phragmidiaceae**.
(8) **Pileolariaceae**.
(9) **Pucciniaceae**.
(10) **Pucciniastraceae**.
(11) **Pucciniosiraceae**.
(12) **Raveneliaceae**.
(13) **Sphaerophragmiaceae**.
(14) **Uropyxidaceae**.

Pucciniosiraceae is inherently heterogeneous, *Sphaerophragmiaceae* is probably polyphyletic and *Pileolariaceae* heterogeneous (Savile, *CJB* **67**: 2983, 1989). Placement of some genera is controversial and still others cannot be accommodated in any of the accepted families due to absence of key structures (especially pycnia, regarded as conservative and therefore valuable at the higher levels of classification) or otherwise incomplete information on their characters. It is acknowledged that family circumscription requires further investigation.

Control: Sulphur is used against a number of rusts. Destruction of an alternate host (e.g. *Berberis* for *Puccinia graminis*) is sometimes of value. The development of resistant varieties of plants is ever in view.

Lit. (*General*): Sydow (*Monographia Uredinearum*, 4 vols, 1902-24), Dietel (*Naturl. PflFam.* **6**, 1928), Arthur (*Plant rusts,* 1929), Guyot (*Les Uredinées*, 1939-57), Thirumalachar & Mundkur (*Indian Phytopath.* **2**: 65, **3**: 4, 203, 1949-50; gen.), Hiratsuka (*Revision of taxonomy of the Pucciniastreae*, 1958), Cummins (*Illustrated genera of rust fungi*, 1959; keys, bibliogr. [edn 2, Cummins & Hiratsuka, 1983]; *The rust fungi of cereals, grasses and bamboos*, 1971, keys; *Rust fungi on legumes and composites in North America*, 1978), Staples & Wynn (*Bot. Rev.* **31**: 537, 1965; urediniospore physiology), Preece & Hick (*Introductory scanning electron*

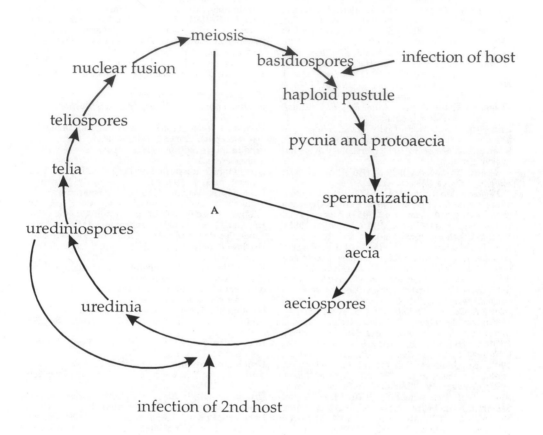

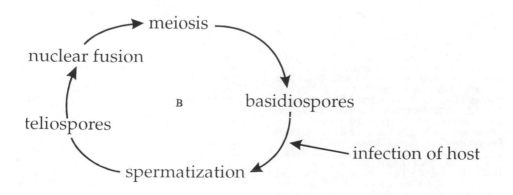

Fig. 41. Life cycles of A, a macrocyclic heteroecious rust; B, a microcyclic autoecious rust.

microscope atlas of rust fungi, 1990), Ziller (*Tree Rusts of Western Canada*, 1974), Dupias (*BSMF* **87**: 129, 1971; biogeogr.), Hart (*Cladistics* **4**: 339, 1988; coevolution), Laundon (*Mycol. Pap.* **89, 91, 102**, 1963-65; *Acanthaceae-Amaryllidaceae*; 99 gen. names), Leppik (*Mycol.* **45**: 46, phylogeny conifer rusts; **48**: 637, *Gymnosporangium*; **51**: 512, grass rusts; **53**: 378, stem rusts; 1953-61; *Annls bot. fenn.* **9**: 135, 1972; evolutionary specialization), Littlefield (*Biology of rust fungi*, 1981), Littlefield & Heath (*Ultrastructure of rust fungi*, 1979), Petersen (*Bot. Rev.* **40**: 453, 1975; life cycle), Savile (*Evol. Biol.* **9**: 137, 1976; *Rept. Tottori Mycol. Inst.* **28**: 15, 1990; evolution with hosts, *Nova Hedw.* **57**: 269, 1993; cladistics), Scott & Chakravorty (Eds) (*The rust fungi*, 1982), Bushnell & Roelfs (Eds) (*The cereal rusts.* **1** *Origin, specificity, structure and physiology*, 1984; **2** *Diseases, distribution, epidemiology, control*, 1985), Zhang, Dickinson & Pryor (*Ann. Rev. Phytopathol.* **32**: 115, 1994; double-stranded RNAs), Vogler & Bruns (*in* Reynolds & Taylor (Eds), *The fungal holomorph*: 273, 1993; molecular approaches), Swertz (*Stud. Mycol.* **36**, 1994; morphology of urediniospore germlings).

Serials: *Uredineana*, 1939 on (Paris); *Cereal rusts bulletin*, 1973 on (Wageningen).

Regional: **Australia,** McAlpine (*Rusts of Australia*, 1906). **Canada,** see USA. **Central Europe,** Gäumann (*Beitr. Kryptog.-fl. Schweiz.* **12**, 1959). **Austria,** Poelt (*Catalogus Florae Austriae* **3**(1) *Uredinales*, 1985). **former Czechoslovakia,** Markova & Urban (*Novit. Bot. Univ. Carol.* **3**: 25, 1987). **Finland,** Liro (*Uredineae Fennicae*, 1908), Makinen (*Annls bot. fenn.* **1**: 214, 1964). **Germany,** Braun (*Feddes Repert.* **93**: 213, 1982, Brandenburger (*Regensb. Mykol. Schriften* **3**, 1994). **British Isles,** Wilson & Henderson (*British rust fungi*, 1966), Henderson & Bennell (*Notes R. bot. Gdn Edinb.* **37**: 475, 1979; **38**: 184, 1980); see also Grove, Plowright. **Iceland,** Jørstad (*Skr. norske Vidensk Akad. I Mat. Nat. n.s.* **2**, 1952). **Indonesia,** Boedijn (*Nova Hedw.* **1**: 463, 1960). **Italy,** Ciferri & Camara (*Ist. Bot. Univ. Pavia Quaderno* **23**, 1962; list). **Jamaica,** Dale (*Mycol. Pap.* **60**, 1955). **Japan,** Hiratsuka *et al.* (*The rust flora of Japan*, 1992). **Madagascar,** Bouriquet & Bassino (*Prodr. Flor. mycol. Madagascar* **5**, 1966). **Majorca and Minorca,** Jørstad (*Skr. norske VidenskAkad., I Mat. Nat. n.s.* **2**, 1962). **Morocco,** Guyot & Malençon (*Uredinées du Maroc*, I, 1957). **Mexico,** León-Gallegos & Cummins (*Uredinales (Royas) de Mexico* **1+2**, 1981). **Nepal,** Durrieu (*Mycol.* **79**: 90, 1987). **New Caledonia,** Huguenin (*BSMF* **82**: 248, **83**: 941, 1966-68). **New Zealand,** Cunningham (*Rust fungi of New Zealand*, 1931), Dingley (*N.Z. Jl Bot.* **15**: 29, 1977). **Nigeria,** Eboh (*Mycol.* **70**: 1077; **73**: 445; **76**: 179, 1978-84). **Norway,** Jørstad (*Nytt. Mag. Bot.* **8**: 103, list; **9**: 61, distribution; **11**: 27, life cycles, etc., 109, distribution, 1960-64), Gjaerum (*Nordeus Rustsopper*, 1974). **Malawi,** Bisby & Wiehe (*Mycol. Pap.* **51**, 1953). **Pakistan,** Ahmad (*Biologia* **2**: 26, 1956). **Poland,** Majewski (*Flora Polska* **9,11** *Uredinales* **1**, 1977, **2**, 1979). **Romania,** Savulescu (*Monografia Uredinalelor.*, 2 vols, 1953). **Scandinavia,** Hylander, Jørstad & Nannfeldt (*Op. bot. Soc. bot. Lund* **1**, 1953). **Spain,** Gonzales-Fragoso (*Flora Iberica. Uredinales*, 2 vols, 1924-25). **South Africa,** Doidge (*Bothalia* **2, 3, 4**, 1927-48). **USA,** Arthur (*Manual of rusts in the United States and Canada*, 1934 [suppl. by Cummins, 1962]), Cummins & Stevenson (*Pl. Dis. Reptr, Suppl.* **240**, 1956; list), Gardner & Hodges (*Pacific Science* **43**: 41, 1989; Hawaii). **Cuba,** Urban (*Rept. Tottori Mycol. Inst.* **28**: 37, 1990). **Argentina,** Lindqvist (*Royas de la Republica Argentina*, 1981). **Brazil,**

Hennen, Hennen & Figueiredo (*Arq. Inst. Biol., Saõ Paulo* **49**(Suppl. 1), 1982). **former USSR,** Tranzschel (*Conspectus Uredinalium U.R.S.S.*, 1939), Kuprevich & Tranzschel ([*Crypt. Pl. USSR* **4**], 1957 [Engl. transl. 1970]. **Ukraine,** Gutsevich ([*Survey of the rust fungi of the Crimea*], 1952 [Russ.]). **Kazakhstan,** Nevodovsky (*Flora sporovykh rastenii Kazakhstana* [I. Uredinales], 1956), Kuprevich & Ulyanishchev ([*Keys to the rust fungi of USSR*] **1**, 1975, **2**, 1978). **Lithuania,** Minkevičius & Ignatavičiute (*Lietuvos Grybai* **5**, 1991). **Tadzhikistan,** Korbonskaya ([*Rust fungi of Tadzhikistan*], 1954). *Melampsoraceae*).

Uredinaria Chevall. (1826) nom. conf. See Sutton (*Mycol. Pap.* **141**, 1977).

Uredinella Couch (1937), Septobasidiaceae. 2 (on *Insecta*), USA, Sri Lanka, W. Afr. See Couch (*Mycol.* **33**: 405, 1941).

Urediniomycetes, see *Teliomycetes.*

urediniospore (uredinospore, urediospore), see *Urediniomycetes.*

Uredinites Velen. (1889), Fossil fungi (? Uredin.). 1 (Cretaceous), former Czechoslovakia.

uredinium (uredium, uredosorus), see *Urediniomycetes.*

Uredinophila Rossman (1987), Tubeufiaceae. 2 (on rusts), Asia, C. & S. Am.

Uredinopsis Magnus (1893), Pucciniastraceae. 25, chiefly N. temp.; heteroecious on *Abies* and *Pteridophyta*. See Faull (*J. Arnold Arbor.* **19**: 402, 1938).

Uredinula Speg. (1880) = Tuberculina (Mitosp. fungi) fide Spegazzini (*An. Soc. Cient. Arg.* **10**: 64, 1880).

Uredites Babajan & Tasl. (1970), Fossil fungi. 1 (Tertiary), former USSR.

Uredo Pers. (1801), Anamorphic Uredinales. 500, esp. trop. A name for uredinia only.

uredoconidium (of *Cumminsiella*), see Kuhnholtz-Lordat (*RAM* **26**: 469, 1947).

Uredopeltis Henn. (1908), Phakopsoraceae. 1 (on *Markhamia*), Afr. See Laundon (*TBMS* **46**: 503, 1964).

Uredostilbe Buriticá & J.F. Hennen (1994), Phakopsoraceae. 1, Honduras.

Urnobasidium Parmasto (1968) = Sistotrema (Sistotremat.) fide Eriksson *et al.* (*Corticiaceae of Northern Europe* **7**, 1984).

Urnula Fr. (1849), Sarcosomataceae. 4, Eur., N. Am. See Dissing (*Mycol.* **73**: 272, 1981; key).

Urnularia P. Karst. (1866) nom. rej. = Chaetosphaeria (Lasiosphaer.) fide Kohlmeyer & Kohlmeyer (*Marine mycology*, 1979).

Urobasidium Giesenh. (1893) = Zygosporium (Mitosp. fungi) fide Mason (*Mycol. Pap.* **5**: 134, 1941).

Uroconis Clem. (1909) ≡ Urohendersonia (Mitosp. fungi).

Urocystis Rabenh. ex Fuckel (1870) nom. cons., Tilletiaceae. *c.* 130, cosmop. Spore balls comprising spores completely or incompletely covered by sterile cells. Pathogens of anemone (*U. anemones*), onion (*Allium*) (*U. cepulae*), rye (*U. occulta*), violet (*U. violae*), wheat (*U. agropyri* (syn. *U. tritici*), flag smut). See Liro (*Ann. Univ. Ábo.* A **1** (1), 1922; monograph, as *Tuburcinia*), Vánky (1994).

Urocystites Babajan & Tasl. (1970), Fossil fungi. 1 (Tertiary), former USSR.

Urohendersonia Speg. (1902), Mitosporic fungi, 4.C2.19. 5, S. Am., India. See Nag Raj & Kendrick (*CJB* **49**: 1853, 1971; key).

Urohendersoniella Petr. (1955), Mitosporic fungi, 4.D2.19. 1, Australia. See Nag Raj (*CJB* **67**: 3169, 1989; revision).

Uromyces (Link) Unger (1832) nom. cons., Pucciniaceae. 600, cosmop. On *Leguminosae*, e.g. *U. appendiculatus* (*Phaseolus*), *U. fabae* (*Vicia*),

U. trifolii (*Trifolium*), *U. striatus* (*Medicago*). See Guyot (*Les Uredinées. Genre Uromyces*, 3 vols, 1938, 1951, 1957), Hiratsuka (*Rept Tottori Inst.* **10**: 1, 1973; *Rust flora of Japan*, 1992, Japan).

Uromycetites Braun (1840), Fossil fungi (? Uredin.). 1 (Triassic), Germany.

Uromycladium McAlpine (1905), Pileolariaceae. 7 (on *Acacia, Albizia*), Australasia.

Uromycodes Clem. (1909) ≡ Schroeteriaster (Uredinales, inc. sed.).

Uromycopsis Arthur (1906) = Uromyces (Puccin.) fide Arthur (*N. Am. Fl.* **7**: 440, 1921).

Urophiala Vuill. (1910) = Zygosporium (Mitosp. fungi) fide Mason (*Mycol. Pap.* **5**: 134, 1941).

Urophlyctidaceae Hadar (1982), ? Spizellomycetales. 1 gen., 5 spp. Thallus polycentric; parasitic on vascular plants.

Urophlyctis J. Schröt. (1886), Urophlyctidaceae. 5, widespr. See Sparrow (*Trans. mycol. Soc. Japan* **3**: 16, 1962, *TBMS* **60**: 341, 1973; type), Hadar (*C.r. Acad. Sci. Paris* ser. 3 **294**: 329, 1982). = Physoderma (Physodermat.) fide Karling (1977)

Urophlyctites Magnus (1903), Fossil fungi. 1 (Carboniferous), Eur. See Weiss (*New Phytol.* **3**: 68, 1904).

Urophora Sommerf. ex Arnold (1899) = Bacidia (Bacid.).

Uropolystigma Maubl. (1920), ? Hypocreaceae. 1, Brazil.

Uropyxidaceae (Arthur) Cummins & Y. Hirats. (1983), Uredinales. 11 gen. (+ 9 syn.), 83 spp. Pycnia discoid, without bounding structures, subcuticular or subepidermal; aeciospores pedicellate (*Uraecium*) or catenulate (*Aecidium, Caeoma*); urediniospores pedicellate, uredinia often paraphysate; teliospores pedicellate, 2 or more cells by horizontal septation.

Uropyxis J. Schröt. (1875), Uropyxidaceae. 13, Am.; all autoecious. See Baxter (*Mycol.* **51**: 210, 1959; key).

Urospora Fabre (1879) [non Aresch. (1866), nom. cons., *Algae*] ≡ Urosporella (Amphisphaer.).

Urospora Fayod (1889) = Panellus (Tricholomat.) fide Singer (1975).

Urosporella G.F. Atk. (1897), Amphisphaeriaceae. 1 or 2, Eur., N. Am. See Barr (*Mycol.* **58**: 690, 1966).

Urosporellina E. Horak (1968) = Panellus (Tricholomat.) fide Pegler (*in litt.*).

Urosporellopsis W.H. Hsieh, C.Y. Chen & Sivan. (1994), Amphisphaeriaceae. 1, Taiwan.

Urosporium Fingerh. (1836) nom. dub. (Mitosp. fungi).

Urupe Viégas (1944), Meliolaceae. 1, Brazil.

US. US National Herbarium (Washington, DC, USA); founded 1868; a part of the Smithsonian Institution.

Uslaria Nieuwl. (1916) ≡ Gautieria (Gautier.).

Usnea Dill. ex Adans. (1763), Parmeliaceae (L). *c*. 600, cosmop. See Asahina (*Lichens of Japan* **3**, 1956), Keissler (*Rabenh. Krypt.-Fl.* **9**, 5 (4): 403, 1959-60), Motyka (*Lichenum generis Usnea studium monographicum*, 1936, 1938, *Annls Univ. Mariae Curie-Skłodowska*, C, **1**: 277, 1947), Swinscow & Krog (*Norw. Jl Bot.* **25**: 221, 1978; *Lichenologist* **11**: 207, 1979; E. Afr., keys), Tavares (*Mycotaxon* **30**: 39, 1987; *U. strigosa*, N. Am., key), Walker (*Bull. Br. Mus. nat. Hist., Bot.* **13**: 1, 1985; *Neuropogon*, key 15 spp.), Awasthi (*Jl Hattori Bot. Lab.* **61**: 333, 1986; key 54 spp. India). See Old man's beard.

Usneomyces E.A. Thomas ex Cif. & Tomas. (1953) = Usnea (Parmel.).

usnic acid, yellow dibenzofuran derivative occurring in lichens (e.g. some spp. of *Cladonia, Usnea*) which has some antibiotic properties (anti-Gram + bacterial; antifungal). Usually occurs in lichen cortices. See antibiotics, Lichen products, Pigments.

Ussurithyrites Krassilov (1967), Fossil fungi (Ascomycota). 1 (Cretaceous), former USSR.

Ustacystis Zundel (1945), Ustilaginaceae. 1 (on *Rosaceae*), N. Am., Eur.

Ustalia Fr. (1825) ≡ Pyrochroa (Graphid.).

Usteria Bat. & H. Maia (1962) ≡ Mycousteria (Mitosp. fungi).

ustic acid, (1) a hydroxyquinol from *Aspergillus ustus* (*Biochem. J.* **48**: 53, 1951); (2) an antimycobacterial product from *Ustilago maydis* (Haskins *et al.*, *Can. J. Microbiol.* **1**: 749, 1955).

ustilagic acids, metabolic products from *Ustilago maydis*; antifungal and antibacterial (Haskins & Thorn, *CJB* **29**: 585, 1951).

Ustilagidium Herzberg (1895) = Ustilago (Ustilagin.) fide Dietel (1928).

Ustilaginaceae Tul. & C. Tul. (1847), Ustilaginales. 33 gen. (+ 19 syn.), 629 spp. Basidiospores usually produced on the sides of a 4-celled promycelium; at meiosis the promycelial nucleus forms four nuclei which may undergo division so that a number of basidiospores (which are not ballistosporic) are produced by each promycelial cell.

Ustilaginales, Ustomycetes; smut fungi; smuts. 2 fam., 50 gen. (+ 38 syn.), 950 spp., cosmop. Thick-walled probasidia (ustilospores) formed in sori on living plants, metabasidia (promycelia) non-septate or transversely septate, producing ballisto- or statismospores. Fams:

(1) **Tilletiaceae**
(2) **Ustilaginaceae**.

The *Ustilaginales*, which have long been regarded as a distinct group, have a mycelial parasitic phase and are yeast-like when cultured *in vitro*. Their taxonomic status has always been uncertain. Brefeld (*Unters. Gesammtgeb. Mykol.* **12**, 1895) introduced *Hemibasidiomycetes* (*Hemibasidii*) for them (the rusts being assigned to the *Protobasidiomycetes*) and although *Hemibasidiomycetes* was subsequently used in wider senses to include the rusts, Donk (see *Basidiomycota*) again restricted it to the smuts. Von Arx (1967) classified the smuts and the associated saprobic yeasts together with other yeasts in a class *Endomycetes* and in this he was followed by Kreisel (1969) who restricted the *Ustilaginales* to the *Ustilaginaceae*. Moore (*Ant. v. Leeuwenh.* **38**: 567, 1972) on the basis of ultrastructural studies and other considerations treated the group as two classes, *Ustomycetes* and *Sporidiomycetes*, of a division (*Ustomycota*) intermediate between ascomycetes and the basidiomycetes with which smuts have been traditionally associated. The current concept of *Ustomycetes* (q.v.) is broad and includes 7 orders, with the smuts divided into 2 families. Characters of the promycelium in these families are sufficiently distinct to indicate possible ordinal rank, but elevation requires additional support from molecular and ultrastructural data at present available for few species. In many species ustilospore germination (metabasidium development) has never been observed (or is mycelial) and some investigators (e.g. Vánky, 1994) therefore prefer not to emphasize the family division.

The smut fungi occur typically as host-specific endophytes which, but for 2 spp. of *Melanotaenium* on *Selaginella*, are parasitic on flowering plants, esp. *Gramineae* and *Cyperaceae*. Sori are commonly limited to the ovary, anthers, inflorescence, or leaves (*Entyloma*) and stem of the host, though the root is attacked by *Entorrhiza*. The mycelium of delicate hyaline hyphae made up of 1-, 2-, or multinucleate cells which may be throughout the plant or only at the points of infection; it is commonly annual but sometimes, as in *Ustilago segetum* var. *avenae* on *Arrhenatherum*, perennial. The hyphae are generally intercellular (in *U. maydis*) and frequently with haustoria and sometimes clamp connexions; dolipore septa lacking (Moore, *Ant. v. Leeuwenh.* **38**: 567, 1972); conidia may be formed on the surface of the host (esp. in *Entyloma*) although in most genera only the smut spores (chlamydospores, brand spores, resting spores, pseudospores, teliospores, ustospores, ustilospores) and basidiospores. Ustilospores (the preferred term, introduced by Donk, 1972), when mature, are generally exposed as a dark powder (sometimes as in *Anthracoidea* and *Cintractia* they are compacted); less frequently they are enclosed within host tissue or are unpigmented. The spores may be in ones (*Ustilago, Tilletia*) or twos (*Mycosyrinx, Schizonella*), or in balls made up of fertile spores only (*Sorosporium, Tolyposporium, Thecaphora*), a sterile layer covering fertile spores (*Urocystis, Doassansia*), or fertile spores covering sterile tissue (*Testicularia, Doassansiopsis*). Every mature spore has one diploid nucleus (Sampson, *TBMS* **23**: 1, 1939), and limited by a thin endospore and a thicker smooth or ornamented exospore (Zogg & Schwinn, *TBMS* **57**: 403, 1971; Mordue, *TBMS* **87**: 407, 1986). At germination meiosis takes place and a promycelium (basidium, hemi-, or metabasidium, germ tube) having 4 or more basidiospores (sporidia (q.v.), sporidiola, 'conidia', promycelial spores) is produced.

Smuts are facultative saprobes after ustilospore development when their growth on culture media is mycelial or yeast-like (and composed of budding cells variously known as sporidia or sprout cells). Completion of the life cycle in culture is infrequent. There is segregation for sex in the promycelium and conjugation between two basidiospores, a basidiospore and a promycelial cell, two cells of one promycelium, or infrequently, two promycelia (Durán, *Ustilaginales of Mexico*, 1987; Ingold, *Nova Hedw.* **55**: 153, 1992). Many physiologic races are on record. Hybridization of smuts takes place and methods for its experimental study have been perfected. See Whitehouse (*TBMS* **34**: 340, 1951; heterothallism); Carris & Gray (*Mycol.* **86**: 157, 1994), Bakkuren & Kronstad (*Plant Cell* **5**: 123, 1993).

Species have frequently been based on physiologic races and a number of these become conspecific if morphological characters are used for the differentiation of species.

The three chief types of infection are: (1) seedling infection from ustilospores on the seed; (2) seedling infection by mycelium in the seed as a result of ustilospore germination on the stigma at flowering time; (3) infection by wind-borne sporidia from promycelia among decaying plant material (as for *Ustilago maydis*, *Entyloma*). Dusting seed with a fungicide is of use against (1), hot water seed treatment against (2), and spraying and dusting susceptible plants is a possible control for (3). Resistant varieties are used whenever possible.

Lit.: **General**: Dietel (*Nat. Pflanzenfam.* 2 Aufl. **6**, 1928), Fischer (*The smut fungi: a guide to the literature with bibliography*, 1951; 3,300 ref.), Zundel (*The Ustilaginales of the world*, 1953; descript., syn.), Fischer & Holton (*Biology and control of smut fungi*, 1957; general account), Fischer & Shaw (*Phytopath.* **43**: 181, 1953; speciation), Savulescu (*Sydowia Beih.* **1**: 64, 1957; gen. key), Holm (*Svensk. bot. Tidskr.* **55**: 585, 1961; phylogeny), Savile (*Rep. Tottori mycol. Inst.* **28**: 15, 1990; coevolution), Thirumalachar (*Indian Phytopath.* **19**: 3, 1966; gen. rev.), Vánky (*Illustrated genera of smut fungi*, 1987), Zambettakis (*Rev. Mycol. Paris* **34**: 399, 1970, gen. key; *Bot. Rev.* **6**: 389, **19**: 187, **31**: 114, 1940-65; *Ann. Rev.*

Phytopath. **6**: 213, 1968, specialization, genetics, variation),

Regional: **Africa**, Zambettakis (*BSMF* **86**: 305-692, 1971; *c.* 300 spp., key gen., host index [also as reprint, *Les Ustilaginales des plantes d'Afrique*]; suppl., *BSMF* **95**: 393, 1979). **Argentina**, Hirschhorn (*Las Ustilaginales de la flora Argentina*, 1985). **Australia**, McAlpine (*The smuts of Australia*, 1910). **Baltic Region**, Ignataviciute ([*Smuts of the Baltic Region*], 1975). **Brazil**, Viégas (*Bragantia* **4**: 739, 1944). **Carpathians**, Vánky (*Carpathian Ustilaginales*, 1985). **China**, Lee Ling (*Mycol. Pap.* **11**, 1945; *Mycol.* **41**: 252, 1949). **Former Czechoslovakia**, Bubák (*Die Pilze Böhmens* **2**, 1916). **Finland**, Liro (*Die Ustilagineen Finnlands*, 2 vols, 1924-38). **Germany**, Scholz & Scholz (*Englera* **8**, 1988). **Great Britain**, Ainsworth & Sampson (*The British smut fungi*, 1950), Mordue & Ainsworth (*Mycol. Pap.* **154**, 1984); see also Plowright. **Hungary**, Moesz (*A Kárpát-Medence Üszöggombai. Les Ustilaginales du bassin du Carpathes*, 1950; Hung., Fr. summ.). **India**, Mundkur & Thirumalachar (*Ustilaginales of India*, 1952). **Italy**, Ciferri (*Flora Italica* **7** (17), 1938). **Japan**, Ito (*Mycological flora of Japan. 2. Basidiomycetes. 1. Ustilaginales*, 1936), Kakushima (*Mem. Inst. Agr. For. Univ. Tsukuba* **1**, 1982). **Kazakhstan**, Shvartsman ([*Flora of the sporing plants of Kazakhstan. 2. Smut fungi*, 1960; Russ.). **Mexico**, Durán (*Ustilaginales of Mexico*, 1987). **New Zealand**, Cunningham (*Trans. N.Z. Inst.* **55**: 397, 1924; suppl. *Trans. N.Z. Inst.* **57**, **59**, **61**). **N. America**, Fischer (*Manual of North American smut fungi*, 1953). **Norway**, Jørstad (*Nytt Mag. Bot.* **10**: 85, 1963; excl. *Anthracoidea*). **Pakistan**, Ahmad (*Mycol. Pap.* **64**, 1956). **Poland**, Kochman (*Acta Soc. Bot. Polon.* **29**: 413, 1960; list). **Romania**, Săvulescu (*Ustilaginalele*, 2 vols, 1957). **S. Africa**, Zundel (*Bothalia* **3**: 283, 1938). **Sweden**, Lindeberg (*Symb. bot. upsal.* **16** (2), 1959; excl. *Anthracoidea*). **Switzerland**, Zogg (*Cryptog. Helv.* **16**: 277, 1985 ['1986']). **former USSR**, Gutner (*Golozneoye griby*, 1941), Ul'yanischev ([*Key to the smut fungi of the USSR*, 1968; Russ.]). Ramazanova *et al.* (*Flora gribov Uzbekistana.4. Golonevge griby*, 1987), **former Yugoslavia**, Lindtner (*Bull. Mus. Hist. nat. Pays Serbe*, ser. B, **304**: 1, 1950).

Ustilaginoidea Bref. (1895), Mitosporic fungi, 1.A2.1. 5 (on *Gramineae*), warmer regions.

Ustilaginoidella Essed (1911) = *Fusarium* (Mitosp. fungi) fide Brandes (*Phytopath.* **9**: 373, 1919).

Ustilaginomycetes, see *Ustomycetes*.

Ustilaginula Clem. (1909) ≡ *Ustilagopsis* (Mitosp. fungi).

Ustilagites Babajan & Tasl. (1970), Fossil fungi. 1 (Tertiary), former USSR.

Ustilago (Pers.) Roussel (1806), Ustilaginaceae. *c.*350 (esp. on *Gramineae*), cosmop. Important diseases of cereals, the causal organisms biologically distinct but morphologically close and variously treated as species, varieties or even physiologic races: *U. segetum* (covered smut of barley (*Hordeum*; *U. hordei*) and oats (*Avena*; *U. kolleri, U. levis*)); *U. segetum* var. *avenae* (loose smut of oats; *U. avenae*); *U. segetum* var. *tritici* (loose smut of wheat; *U. nuda*, and barley). Pathogens of grasses: *U. bullata* (esp. on *Bromus*), *U. hypadytes*, *U. striiformis* (stripe smut, see Osner, *Bull. Cornell agric. exp. Stn* **381**, 1916; Mordue & Waller, *IMI Descript.* **717**, 1981); sugarcane: *U. scitaminea* (Trione, *MR* **94**: 489, 1990); maize (*Zea*): *U. maydis* (syn. *U. zeae*); miscellaneous hosts, *Tulipa*: *U. heufleri*; *Scilla* and related bulbs: *U. vaillantii*; *Scarzonera*: *U. scarzonerae*. *U. esculenta* on *Zizania* (Mordue, *Mycopath.* **16**: 227,

1991) and *U. maydis* are edible. Ustilospore germination, see Ingold (*MR* **93**: 405, 1989).

Ustilagopsis Speg. (1880) = Sphacelia (Mitosp. fungi) fide Langdon (*TBMS* **36**: 74, 1953).

Ustilentyloma Savile (1964), Ustilaginaceae. 1, Canada.

ustilospore, Donk's (*K. ned. Akad. Wet.* C **76**: 111, 1973) name for a smut spore; **ustospore** (Moore, *Ant. v. Leeuwenhoek* **38**: 579, 1972).

Ustomycetes (Ustilaginomycetes), Basidiomycota; smut fungi. 7 ord., 10 fam., 63 gen. (+ 28 syn.), 1064 spp. Basidiospores in sori, frequently with a yeast-like phase; Ords:
(1) **Cryptobasidiales**.
(2) **Cryptomycocolacales**.
(3) **Exobasidiales**.
(4) **Graphiolales**.
(5) **Platygloeales**.
(6) **Sporidiales**.
(7) **Ustilaginales**.
The smuts, which have long been regarded as a distinct group, have a mycelial parasitic phase and are yeast-like when cultured *in vitro*. Banno (*J. gen. appl. Microbiol.* **13**: 167, 19XX) by mating two haploid strains of *Rhodotorula glutinis* obtained a teleomorph which he described, as *Rhodosporidium toruloides*, in a new genus characterized by the development (after conjugation and plasmogamy) of a dikaryotic mycelium (with clamp connexions) bearing resting spores. Each of these on germination following meiosis gives a promycelium bearing sporidia by which the haploid yeast phase is restarted. This life-cycle resembles that of a smut and subsequently several more similar genera were described.
The taxonomic status of the smuts has always been uncertain. Brefeld (*Unters. Gesammtgeb. Mykol.* **12**, 1895) introduced *Hemibasidiomycetes* (*Hemibasidii*) for the smuts (the rusts being assigned to the *Protobasidiomycetes*) and although *Hemibasidiomycetes* was subsequently used in wider senses to include the rusts Donk (see *Basidiomycota*) recently again restricted this taxon to the smuts. Von Arx (1967) classified the smuts and the associated saprobic yeasts together with other yeasts in a class *Endomycetes* and in this he was followed by Kreisel (1969) who restricted the *Ustilaginales* to the *Ustilaginaceae*. Moore (*Ant. v. Leeuwenhoek* **38**: 567, 1972) on the basis of ultrastructural studies and other considerations treated the group as two classes, *Ustomycetes* and *Sporidiomycetes*, of a new division (*Ustomycota*) intermediate between ascomycetes and the basidiomycetes with which smuts have been traditionally associated.

Ustomycota. Phylum for *Septomycetes* rather like *Saccharomycetales* and *Uredinales* but not satisfactorily either (Moore, *Ant. v. Leeuwenhoek* **38**: 567, 1972); assimilative phase often yeast-like. See *Ustomycetes*.

ustospore, see ustilospore.

Ustulina Tul. & C. Tul. (1863), Xylariaceae. 5 or 6, widespr. *U. deusta* (butt rot) has a wide host range. See Wilkins (*TBMS* **18**: 320, 1934), Hawksworth (*CMI Descr.* **360**, 1972). = Kretzschmaria (Xylar.) fide Læssøe (*SA* **13**: 43, 1994).

Ustulinites Kirschst. (1925), Fossil fungi. 1, Germany.

Utharomyces Boedijn (1959), Pilobolaceae. 1 (esp. coprophilous), trop. & sub-trop. See Kirk & Benny (*TBMS* **75**: 123, 1980).

Uthatobasidium Donk (1956), Ceratobasidiaceae. 2, widespr.

Utraria Quél. (1873) ≡ Lycoperdon (Lycoperd.) fide Demoulin (*Persoonia* **7**: 152, 1973).

utricle, the bladder-like covering of certain fungi, e.g. *Dendrogaster*.

utriform, bag-like.

Uvarispora Goos & Piroz. (1975), Mitosporic fungi, 1.G1.1. 1, Panama.

Uvasporina Beneš (1956), Fossil fungi (*f. cat.*). 1 (Carboniferous), former Czechoslovakia.

Uyucamyces H.C. Evans & Minter (1985), Anamorphic Rhytismataceae, 7.A1.1. Teleomorph *Ocotomyces*. 1, Honduras.

V8A, see Media.

VA, VAM (of mycorrhizas), vesicular-arbuscular; see Mycorrhiza.

vagant (of lichens), see vagrant.

Vaginaria (Forq.) Sacc. (1887) [non Rich. ex Pers. (1805), *Cyperaceae*] = Amanitopsis (Amanit.).

Vaginarius Roussel (1806) ≡ Amanitopsis (Amanit.).

Vaginata Nees (1816) nom. rej. = Amanita (Amanit.).

vagrant (of lichens), unattached; erratic; vagant. See Rosentreter (*Bryologist* **96**: 333, 1993; N. Am.).

Vainio ('**Wainio**') (Edvard August; 1853-1929). Finnish lichenologist received PhD from Helsinki Univ. 1880, where he worked until 1922 when he moved to Turku. His most important publications are the *Monographia Cladoniarum universalis*, I-III (*Acta Soc. Fauna Fl. fenn.* **4**, 1887; **10**, 1894; **14**, 1897, [reprint 1978]), *Lichenes Insularum Philippinarum*, I-IV (*Philipp. J. Sci.*, C, **4**: 641, 1909, **8**: 99, 1913, *Anns Acad. Sci. fenn.* A, **15**(6), 1921, **19**(6), 1923), *Lichens du Brésil*, I-II (*Acta Soc. Fauna Fl. fenn.* **7**, 1890), *Lichenographia Fennica*, I-IV (*Lichenographia Fennica* **49**(2), 1921; **53** (1), 1922; **57**(1), 1927; **57**(2), 1934), together with important papers on Siberia, C. Afr., E. Asia, etc. Acquired an uncontested position as the 'Grand Old Man of lichenology' (Lynge). Collections of over 33,000 specimens now in Turku (**TUR**). *Obit.*: Magnusson (*Annls Crypt. Exot.* **3**: 5, 1930; portr., bibliogr.), Linkola (*Acta Soc. Fauna Fl. fenn.* **57** (3), 1934; portr., bibliogr.). See also Alava (*Publs Herb. Univ. Turku* **1**, 1986, journey to Brazil; **2**, 1988, types in **TUR**, etc.), Grummann (1974: 612), Stafleu & Cowan (*TL-2* **6**: 636, 1986).

Vainiocora, see *Wainiocora*.

Vainiona Werner (1943) ≡ Neonorrlinia (Dothideales, inc. sed.).

Vainionia Räsänen (1943) nom. inval. = Calicium (Calic.).

Vainionora Kalb (1991), Lecanoraceae. 5, Eur.

Vakrabeeja Subram. (1957) ≡ Nakataea (Mitosp. fungi) fide Ellis (*DH*).

Valdensia Peyronel (1923), Anamorphic Sclerotiniaceae, 1.E2.1. Teleomorph *Valdensinia*. 1, Italy. See Redhead & Perrin (*CJB* **50**: 409, 2083, 1972).

Valdensinia Peyronel (1953), Sclerotiniaceae. 1, Eur., N. Am. See Norvell & Redhead (*Can. J. For. Res.* **24**: 1981, 1994). Anamorph *Valdensia*.

Valentinia Velen. (1939), Agaricaceae. 1, Eur.

Valetoniella Höhn. (1909), Niessliaceae. 1, Java, NZ. See Barr (*Mycotaxon* **39**: 43, 1990; posn), Samuels (*N.Z. Jl Bot.* **21**: 157, 1983).

valid (of names), published in accord with the Code Arts 29-45; such names may be illegitimate or legitimate; **pre-** (of names or authors), published before 1753, the starting point for the nomenclature of fungi under the Code; devalidated; cf. Nomenclature.

Valsa Adans. (1763) nom. rej. = Diatrype (Diatryp.) fide Cannon & Hawksworth (*Taxon* **32**: 478, 1983), Læssøe (*SA* **13**: 43, 1994).

Valsa Fr. (1849) nom. cons., Valsaceae. 60, cosmop. See Barr (*Mycol. Mem.* **7**, 1978), Défago (*Phytopath. Z.* **14**: 103, 1944), Spielman (*CJB* **63**: 1355, 1985; N. Am.). Anamorph *Cytospora*.

Valsaceae Tul. & C. Tul. (1861), Diaporthales. 50 gen. (+ 34 syn.), 365 spp. Ascomata erect, oblique or horizontal, beaks central, oblique or lateral, erumpent singly or converging through a stromatic disc; asci numerous, evanescent at the base, becoming free within the ascomatal cavity; ascospores simple or septate, hyaline to yellowish. Anamorphs coelomycetous. Saprobes or weak parasites esp. of petioles, stems and bark.

Many accounts have separated the *Gnomoniaceae* from this fam., with upright rather than oblique ostioles. *Lit.*: Barr (1978), Monod (*Beih. Sydowia* 9, 1983), Vasil'eva (*Pyrenomycetes of the Russian Far East* 1, 1993, 2, 1994; keys).

Valsales, see *Diaporthales*.

Valsaria Ces. & De Not. (1863), Diaporthales (inc. sed.). 40, widespr. See Petrak (*Sydowia* 15: 299, 1962), Glawe (*Mycol.* 77: 62, 1985; anamorphs).

Valsarioxylon Höhn. (1929) nom. nud. (Ascomycota).

Valsella Fuckel (1870), Valsaceae. 2 (on *Salix*), Eur. See Barr (*Mycol. Mem.* 7, 1978).

Valseutypella Höhn. (1919), Valsaceae. 2, Eur., N. Am. See Hubbes (*Phytopath. Z.* 39: 389, 1960), Checa *et al.* (*Mycotaxon* 25: 323, 1988), Checa & Martínez (*Mycotaxon* 36: 43, 1989; anamorph). Anamorph *Cystospra*.

valsoid, having groups of perithecia with their beaks pointing inward (convergent), or even parallel to the surface, as in *Valsa*. Cf. eutypoid.

Valsonectria Speg. (1881), Thyridiaceae. 2, S. Am. See Huhndorf (*Mycol.* 84: 642, 1992). Petrak & Sydow (*Ann. Myc.* 34: 40, 1936).

VAM fungi, vesicular-arbuscular mycorrhizal fungi; VA fungi; see Mycorrhiza.

Vampyrella Cienk. (1865), *Hydromyxales*, q.v.

Vampyrellidium Zopf (1885), *Hydromyxales*, q.v.

Vampyrelloides Schepotieff [not traced] = Biomyxa (Hydromyxales) fide Jahn.

van Tieghem cell, a ring of glass or other material, fixed to a glass slide, over which is placed a cover-glass with a 'hanging drop' of the microorganism under investigation. See Duggar (*Fungous diseases of plants*, 1909).

Vanakripa Bhat, W.B. Kendr. & Nag Raj (1993), Mitosporic fungi, 3.A/B2.1. 2, India.

Vanbeverwijkia Agnihothr. (1961), Mitosporic fungi, 3.F1.15. 1, Assam, USA. See Shearer & Crane (*Mycol.* 63: 249, 1971).

Vanbreuseghemia Balab. (1965) ≡ Keratinomyces (Mitosp. fungi).

Vandasia Velen. (1922), ? Hymenogastrales (inc. sed.). 1, Eur.

Vanderbylia D.A. Reid (1973), Coriolaceae. 5, widespr. See Corner (*Beih. Nova Hedw.* 86: 241, 1987; key), Sikombwa & Piearce (*Bull. BMS* 19: 124, 1985; medicinal use *V. ungulata*).

Vanderwaltia E.K. Novák & Zsolt (1961) = Kloeckeraspora (Saccharomycod.) fide Batra (1978), = Hanseniaspora (Saccharomycod.) fide v. Arx (1981).

Vanderystiella Henn. (1908), Mitosporic fungi, 6.A2.15. 1, Afr.

Vanhallia L. Marchand (1828) = Chaetomium (Chaetom.) fide Mussat (*Syll. Fung.* 15: 449, 1901).

Vanrija R.T. Moore (1980) = Apiotrichum (Mitosp. fungi) fide Moore (1991).

Vanromburghia Holterm. (1898) ? = Marasmius (Tricholomat.) fide Singer (1962: 808).

Vanterpoolia A. Funk (1982), Mitosporic fungi, 3.G1.3. 1, Canada.

Vantieghemia Kuntze (1891) ≡ Syncephalis (Piptocephalid.).

Vanudenia Bat. & H. Maia (1963) = Schizothyrium (Schizothyr.) fide v. Arx & Müller (1975).

Vararia P. Karst. (1898), Lachnocladiaceae. *c.* 55, widespr. See Boidin & Lanquetin (*BSMF* 91: 457, 1975, 92: 247, 1976, *Mycotaxon* 6: 277, 1977), Pascoe *et al.* (*TBMS* 82: 723, 1984; white rot of *Rubus*).

Vargamyces Tóth (1980), Mitosporic fungi, 1.C2.1/10. 1, aquatic. See Gönczöl *et al.* (*Mycotaxon* 39: 301, 1990; growth and development in culture).

Variation in fungi. May be the result of environment and not heritable (modifications, Bauer) or may be heritable and (1) determined by the coming together or separating of heritable factors (as in combinations) or (2) not so determined (mutations). Modifications ('temporary' or 'reversible' variations, such as changes in mycelial growth and pathogenicity) frequently take place in culture, sometimes in association with non-reversible (discontinuous) variations of a mutation (saltation, dissociation) type. The composition of the medium may have an effect on frequency of saltation; *Fusarium*, for example, saltating less readily on a poor medium, *Aspergillus* more readily in a mannitol-nitrite solution.

The polymorphic variation (Snyder & Hansen) in a species in nature may be so great that the forms are put in three, four or more genera. The existence of the 'dual phenomenon' (Hansen) in Mitosporic fungi, has the effect of giving two kinds of culture.

Lit.: Brierley (*Proc. Int. Congr. Pl. Sci.* 2, 1929), Snyder & Hansen (*Proc. 6th Pacific Sci. Congr.* 4, 1940), Hansen (*Mycol.* 30: 242, 1938), Day (*Ann. Rev. Microbiol.* 14: 1, 1960; plant pathogenic fungi), Sugiyama (Ed.) (*Pleomorphic fungi*, 1987).

Varicellaria Nyl. (1858), Pertusariaceae (L). 3, Eur., Japan, N. & S. Am. See Erichsen (*Rabenh. Krypt.-Fl.* 9, 5(1): 687, 1935; Eur.), Oshio (*J. Sci. Hiroshima Univ.* B(2), 12: 81, 1968; Japan).

Varicellariomyces Cif. & Tomas. (1953) = Varicellaria (Pertusar.).

Varicosporina Meyers & Kohlm. (1965), Anamorphic Leotiaceae, 1.G1.10. Teleomorph *Hymenoscyphus*. 1 (marine), USA. See Nakagiri (*TBMS* 90: 265, 1988; conidia form and function).

Varicosporium W. Kegel (1906), Mitosporic fungi, 1.G1.1. 4, widespr. See Nawawi (*TBMS* 63: 27, 1974).

variecolin, an anti-tubercle bacillus antibiotic from *Aspergillus variecolor* (teleomorph, *Emericella variecolor*) (Gupta & Viswanathan, *Antibiot. & Chemother.* 5: 496, 1955).

variety, see Classification.

Variolaria Bull. [not traced] nom. dub. (? Fungi, inc. sed.).

Variolaria Pers. (1794) nom. dub. (? Fungi, inc. sed.).

Variolaria Gray (1821) nom. dub. (Fungi, inc. sed.).

variolarioid, having powdery or granular tubercules.

Vascellum F. Smarda (1958), Lycoperdaceae. 10, widespr. See Ponce de Leon (*Fieldiana, Bot.* 32: 109, 1970; key), Smith (*Bull. Soc. linn. Lyon* num. spéc. 43: 407, 1974; N. Am.), Homrich & Wright (*CJB* 66: 1285, 1988; S. Am.), Kreisel (*Blyttia* 51: 125, 1993; key).

Vasculomyces S.F. Ashby (1913), Mitosporic fungi, 1.-.-. 1, W. Indies. See Ciferri (*Sydowia* 8: 258, 1954).

Vasudevella Chona, Munjal & Bajaj (1957), Mitosporic fungi, 4.B1.19. 1, India. See Nag Raj (*CJB* 51: 1337, 1973).

vector, an organism that carries and transmits a pathogen to the host which it attacks, such as an insect carrying fungal mycelium or spores (e.g. *Scolytus* spp. beetles transmitting *Ophiostoma novo-ulmi* to *Ulmus* trees). A **fungus -** carries a virus infection with it when moving from its existing host to a new host; some plant viruses are transmitted to new host plants by association with root-infecting fungi; the vector phase is the motile zoospores in *Olpidium* sp. transmitting *Tombusvirus* and *Necrovirus* spp. and the unclassified lettuce big vein and tobacco stunt viruses; *Polymyxa graminis*, *P. betae* and *Spongospora subterranea* transmit viruses in the genus *Furovirus*, and *P. graminis* transmit viruses in the genus *Bymovirus* (see Viruses); **plasmid -**, a plasmid constructed to include a particular gene sequence, and inserted into another fungus where its properties are expressed (see Genetic engineering).

vegetable caterpillar, a mummified lepidopteran larva from which arise teleomorph stromata of *Cordyceps* spp.

vegetative, see assimilative.

vegetative compatibility, the ability of vegetative hyphae to anastomose and form a stable heterokaryon. This ability is restricted genetically by the vegetative incompatibility system such that hyphae differing at one or more loci, termed 'vegetative compatibility' (vc) or 'heterokaryon compatibility (het) loci, are unable to form a stable heterokaryon. Mycelia sharing the same vc loci belong to the same vegetative compatibility group (vcg). Vcg's have been used to determine genetic structures of fungal populations. In most fungi, the vegetative compatibility system is independant of the mating system, which controls sexual compatibility (see sex).
    Lit.: Anagnostakis (*Exp. Mycol.* **1**: 306, 1977), Beadle & Coonradlt (*Genetics* **29**: 291, 1944), Brayford (*MR* **94**: 745, 1990), Carlisle (*in* Rayner *et al.* (Eds), *Evolutionary biology of the fungi*, 1987), Caten (*J. Gen. Microbiol.* **72**: 221, 1972), Coates *et al.* (*TBMS* **76**: 41, 1981), Leslie (*Ann. Rev. Phytopath.* **31**: 127, 1993), Mylyk (*Genetics* **83**: 275, 1976), Puhalla (*CJB* **63**: 179, 1985), Todd & Rayner (*Sci. Progress* Oxford **66**: 331, 1980).

veil, see annulus (**apical -**, **hymenial -**), cortina, ma:ginal -, **partial -** (**inner -**), **pellicular -**, protoblem (**primordial -**), **universal -**.

veins (of lichens), strands of tissue on the lower surface of foliose lichens, esp. *Peltigera* where they may replace a lower cortex. Gyelnik (*Bot. Közlemén.* **24**: 122, 1927) distinguished 2 types: **caninoid -** where the strands are separated to the tips of the lobes; **polydactyloid -** where the strands are confluent towards the tips of the lobes. Maas Geesteranus also recognized **malaceoid -**, where the undersurface has a few whitish interstices faintly indicating venation.

velar, pertaining to a veil.

Veligaster Guzmán (1970) = Scleroderma (Sclerodermat.) fide Demoulin & Dring (*Bull. Jard. bot. nat. Belg.* **45**: 343, 1975).

Vellosiella Rangel (1915) [non Baill. (1887), *Scrophulariaceae*] ≡ Mycovellosiella (Mitosp. fungi).

Velolentinus Overeem (1927) nom. nud. (Polypor.). See Donk (1962).

Velomycena Pilát (1953) = Galerina (Cortinar.) fide Singer (1975).

Veloporphyrellus L.D. Gómez & Singer (1984), Strobilomycetaceae. 1, Costa Rica.

Veloporus Quél. (1888) = Boletus (Bolet.) fide Killermann (1928).

velum, see veil.

Velutaria Fuckel (1870), Hyaloscyphaceae. 1, Eur. See Korf (*Mycol.* **45**: 298, 1953).

Velutarina Korf ex Korf (1971), Leotiaceae. 2, widespr.

Veluticeps (Cooke) Pat. (1894), Peniophoraceae. 12, widespr. *V. berkeleyi* (wood rot of *Pinus*). See Nakasone (*Mycol.* **82**: 622, 1990; key), Gilbertson *et al.* (*Mycol.* **60**: 29, 1968), Hjortstam & Tellería (*Mycotaxon* **37**: 53, 1990).

velutinate (velutinous), thickly covered with delicate hairs; like velvet; see phalacrogenous.

Velutipila D. Hawksw. (1987), Mitosporic fungi, 1.E1.10. 1 (on *Algae*), Austria.

Venenarius Earle (1909) ≡ Amanita (Amanit.).

venose, having veins.

ventral, front, or lower surface; the surface facing the axis, cf. dorsal; frequently used for the lower surface of foliose lichens.

ventricose, swelling out in the middle or at one side; inflated (Fig. 37.38).

Ventrographium H.P. Upadhyay, Cavalc. & A.A. Silva (1986), Mitosporic fungi, 2.E1.15. 1, Brazil.

Venturia De Not. (1844) nom. rej. ≡ Protoventuria (Ventur.).

Venturia Sacc. (1882) nom. cons., Venturiaceae. 52, widespr. *V. cerasi* (cherry scab), *V. chlorospora* (willow (*Salix*) scab and canker), *V. inaequalis* (apple scab) (see *Spilocaea*), *V. pirina* (pear scab) (see *Fusicladium*). See Sivanesan (*Bibl. Mycol.* **59**, 1977; key), Barr (*Sydowia* **41**: 25, 1989; key 39 N. Am. spp.), Morelet (*Crypt., Mycol.* **6**: 101, 1985; key 10 spp. on *Populus*). Anamorph *Pollaccia*.

Venturiaceae E. Müll. & Arx ex M.E. Barr (1979), Dothideales. 28 gen. (+ 36 syn.), 178 spp. Ascomata perithecial, becoming superficial, usually small, ± globose, sometimes aggregated into stromata, often setose or hairy, opening by a well-defined lysigenous pore; peridium composed of ± small pseudoparenchymatous cells; interascal tissue of narrowly cellular pseudoparaphyses, sometimes evanescent; asci ± cylindrical, fissitunicate; ascospores variously pigmented, usually asymmetrical, 1-septate, sometimes with a sheath. Anamorphs prominent, hyphomycetous. Usually biotrophic or necrotrophic on leaves or stems, widespr.
    Lit.: Barr (*CJB* **46**: 799, 1968; N. Am., *Prodromus to Class Loculoascomycetes*, 1987, *Sydowia* **41**: 25, 1989; keys 12 gen. N. Am.).

Venturiella Speg. (1909) [non Müll. Hal. (1875), *Musci*] ≡ Neoventuria (Dothideales).

Venturiocistella Raitv. (1978), Hyaloscyphaceae. 7, Eur., N. Am. See Baral (*Z. Mykol.* **59**: 3, 1993; key).

Venularia Pers. (1822) nom. nud. (? Fungi).

Venustocephala Matsush. (1995), Mitosporic fungi, 1.A1.9. 1, Ecuador.

Venustusynnema R.F. Castañeda & W.B. Kendr. (1990), Mitosporic fungi, 2.A1.15. 1, Cuba.

Veralucia D.R. Reynolds & P.H. Dunn (1982), Mitosporic fungi, 1.A1.10. 1, Brazil. An alga fide Reynolds & Dunn (*Mycol.* **76**: 719, 1984).

Veramyces Subram. (1993) nom. inval. Anamorphic Lasiosphaeriaceae, 1.A1.15. Teleomorph *Thaxteria*. 1, Himalayas.

Veramyces Matsush. (1993), Mitosporic fungi, 1.C2.3/10. 1, Peru.

Verlandea Bat. & Cif. (1957) = Stomiopeltis (Micropeltid.) fide v. Arx & Müller (1975).

Verlotia Fabre (1879) = Heptameria (Dothideales) fide Lucas & Sutton (1971).

Vermicularia Tode (1790) [non Moench (1802), *Verbenaceae*] = Colletotrichum (Mitosp. fungi) fide Duke (*TBMS* **13**: 156, 1928). See Sutton (*in* Bailey & Jeger (Eds), *Colletotrichum: biology, pathology and control*, 1992).

Vermiculariella Oudem. (1898) nom. conf. fide Petrak (*Ann. Myc.* **42**: 58, 1944).

Vermiculariopsiella Bender (1932), Mitosporic fungi, 3.A1.15. 6, widespr. See Nawawi *et al.* (*Mycotaxon* **37**: 173, 1990; key).

Vermiculariopsis Torrend (1912) nom. conf. See Sutton (1977).

Vermiculariopsis Höhn. (1918) ≡ Vermiculariopsiella (Mitosp. fungi).

vermiform, worm-like.

Vermispora Deighton & Piroz. (1972), Mitosporic fungi, 1.C1.10. 1 (on *Irenopsis*), Sierra Leone.

Vermisporium H.J. Swart & M.A. Will. (1983), Mitosporic fungi, 6.E1.19. 14, Australia, S. Afr., India, Italy. See Nag Raj (1993).

Veronaea Cif. & Montemart. (1957), Mitosporic fungi, 1.B2.10. 8, widespr. See Ellis (*MDH*; key), de Hoog *et al.* (*TBMS* **81**: 485, 1983; relationship to *Ramichloridium*), Wang *et al.* (*Acta Mycol. Sin.* **10**: 159, 1991; *V. botryosa* from humans).

Veronaella Subram. & K.R.C. Reddy (1975), Mitosporic fungi, 5.G1.10. 1, India.

Veronaia Benedek (1961), Eurotiales (inc. sed.). 3, widespr.

Veronidia Negru (1964), Mitosporic fungi, 4.A1.?. 1, Roumania.

Verpa Sw. (1815), Morchellaceae. *c.* 5, N. temp.

Verpatinia Whetzel & Drayton (1945) ? = Cudoniopsis (Sclerotin.) fide Dumont & Korf (*Mycol.* **63**: 157, 1971).

verruca (pl. **verrucae**), a wart-like swelling.

Verrucalvaceae M.W. Dick (1984), Sclerosporales. 3 gen. 2 spp.

Verrucalvus P. Wong & M.W. Dick (1985), Verrucalvaceae. 1, Australia.

Verrucaria Scop. (1777) = Dibaeis (Icmadophil.) fide Griel & Kalb (*Herzogia* **9**: 593, 1993).

Verrucaria F.H. Wigg. (1780) = Lecanora (Lecanor.).

Verrucaria Schrad. (1794) nom. cons., Verrucariaceae (L). *c.* 300 (some on lichens), cosmop., esp. N. temp. See Fletcher (*Lichenologist* **7**: 1, 1975; coastal spp.), Hawksworth (*Lichenologist* **21**: 23, 1989; Brit., aquatic), McCarthy (*Lichenologist* **27**: 105, 1995; aquatic, Australia), Navarro-Rosinés & Roux (*Bull. Soc. linn. Provence* **39**: 129, 1988; key 8 spp. on lichens), Santesson (*Ark. Bot.* **29A** (10), 1939), Servít (*Československé lišejníky čeledi Verrucariaceae*, 1954), Swinscow (*Lichenologist* **4**: 34, 1968; freshwater spp.), Zschacke (*Rabenh. Krypt.-Fl.* **9**(1,1): 50, 1933), Zehetleitner (*Nova Hedw.* **29**: 683, 1978; spp. on lichen spp.).

Verrucariaceae Zenker (1827), Verrucariales. 37 gen. (+ 83 syn.), 711 spp. Thallus usually well-developed; ascomata usually clypeate, sometimes strongly aggregated, usually thin-walled; interascal tissue of narrow gelatinized paraphysis-like hyphae, sometimes evanescent, without apical paraphyses. Mostly lichenized with green algae, cosmop.

Verrucariales, Ascomycota (±L). 2 fam., 40 gen. (+ 83 syn.), *c.* 720 spp. Thallus usually crustose, rarely squamulose or umbilicate; ascomata perithecial, ± globose, sometimes strongly aggregated, ± superficial or immersed, usually black, thin-walled, sometimes clypeate; interascal tissue of gelatinized pseudoparaphysis-like hyphae, often evanescent and sometimes altogether absent, the ostiole periphysate; asci clavate, thin-walled, sometimes with two layers visible near the apex but these not separable, sometimes blueing in iodine, persistent or evanescent; ascospores usually ellipsoidal, aseptate to muriform, hyaline or brown. Anamorphs pycnidial. Lichenized with green algae, a few lichenicolous or saprobic, cosmop. Fams:
(1) **Adelococcaceae**.
(2) **Verrucariaceae**.
The *Dermatocarpaceae* are sometimes treated as distinct (e.g. Wanger, *CJB* **65**: 2441, 1987). Gen. concepts still largely based on ascospore colour and septation and in need of a modern revision taking into account ascomatal structure etc.
Lit.: Doppelbauer (*Planta* **53**: 246, 1959; ontogeny), Eriksson (1981), Janex-Favre (1970), McCarthy

(*Muelleria* **7**: 317, 1991; Australia), Servít (*Československé lišejníky čelidi Verrucariaceae*, 1954), Swinscow (*Lichenologist* **1-5**, 1960-71; many papers, UK), Zschacke (*Rabenh. Krypt.-Fl.* **9**, 1(1), 1933; Eur.).

Verrucariella S. Ahmad (1967), Mitosporic fungi, 4.B2.?. 1, Pakistan.

Verrucarina (Hue) Zahlbr. (1931) = Coccotrema (Coccotremat.).

verrucarioid (of asci), see ascus.

Verrucariomyces E.A. Thomas ex Cif. & Tomas. (1953) = Verrucaria (Verrucar.).

Verrucarites Göpp. (1845), Fossil fungi (Ascomycota, ?L). 2 (Tertiary), Eur. See Watelet (*Descr. Pl. Foss.*, 1866).

Verrucaster Tobler (1912) nom. dub. fide Hawksworth (*Bull. Br. Mus. nat. Hist., Bot.* **9**: 88, 1981).

Verrucispora D.E. Shaw & Alcorn (1967) [non *Verrucospora* E. Horak, 3 Aug. 1967] ≡ Verrucisporota (Mitosp. fungi).

Verrucisporota D.E. Shaw & Alcorn (1993), Mitosporic fungi, 1.C2.10. 2, Australia, Papua New Guinea.

Verrucobotrys Hennebert (1973), Anamorphic Sclerotiniaceae, 1.A2.7. Teleomorph *Seaverinia*. 1, N. Am.

Verrucophragmia Crous, M.J. Wingf. & W.B. Kendr. (1994), Mitosporic fungi, 1.C2.10. 1, S. Afr.

verrucose, having small rounded processes or 'warts' (Fig. 29.7).

Verrucosia Teng (1932) = Lycogala (Myxom.).

Verrucospora E. Horak (1967), Agaricaceae. 1, paleotrop.

Verrucula J. Steiner (1896) = Verrucaria (Verrucar.) p.p. and Dermatocarpon (Verrucar.) p.p.

Verruculina Kohlm. & Volkm.-Kohlm. (1990), Didymosphaeriaceae. 1 (marine), Liberia.

verruculose, delicately verrucose (Fig. 29.6).

verruculotoxin, a tremorgenic toxin from *Penicillium verruculosum* (Cole *et al.*, *Toxicol. appl. Pharmacol.* **31**: 465, 1975).

versiform, of different forms; changing form with age.

Versiomyces Coppins (1989), Xylariaceae. 1, Australia. = Hypoxylon (Xylar.) fide Læssøe (*SA* **13**: 43, 1994).

Versipellis Quél. (1886) nom. rej. ≡ Xerocomus (Xerocom.).

vertex, (1) the top of an organ; (2) pileus (obsol.).

Verticicladiella S. Hughes (1953), Mitosporic fungi, 1.A1.10. 7, N. Am., Eur. See Kendrick (*CJB* **40**: 771, 1962). = Leptographium (Mitosp. fungi) fide Wingfield (*TBMS* **85**: 81, 1985).

Verticicladium Preuss (1851), Anamorphic Sarcosomataceae, 1.A2.10. Teleomorph *Desmazierella*. 1, Eur. See Hughes (*Mycol. Pap.* **43**, 1951).

Verticicladus Matsush. (1993), Mitosporic fungi, 1.C2.1. 1, Peru.

verticillate, having parts in rings (**verticils**); whorled.

Verticilliastrum Dasz. (1912) ? = Trichoderma (Mitosp. fungi) fide Sutton (*in litt.*).

Verticilliodochium Bubák (1914) = Dendrodochium (Mitosp. fungi) fide Tulloch (*Mycol. Pap.* **130**, 1972).

Verticilliopsis Costantin (1892) nom. dub. (Mitosp. fungi) fide Gams (*Cephalosporium-artige Schimmelpilze*: 18, 1971).

Verticillis Clem. & Shear (1931) ≡ Verticilliodochium (Mitosp. fungi).

Verticillium Nees (1816), Mitosporic fungi, 1.A1.15. 40, widespr. *V. alboatrum* and *V. dahliae* cause wilt disease (hadromycosis) in many plants, *V. theobromae* causing cigar-end of *Musa*. See Isaac (*TBMS* **32**: 137, 1949, *CMI Descr.* **255-256**, 1970), Rudolph (*Hilgardia* **5**: 197, 1931; host list), Rudolph (*Plant Dis. Reptr, Suppl.* **244**, 1957, **255**, 1959; deciduous fruit trees). *V. fungicola* on mushrooms. See (*CMI Descr.* **498**, 1976), Chen & Fu (*Acta Mycol. Sin.*

8: 123, 1989; similar fungi on mushrooms). See also Isaac (*Ann. Rev. Phytopath.* **5**: 201, 1967; speciation), Gams & van Zaayen (*Neth. Jl Pl. Path.* **88**: 57, 1982; fungicolous spp., keys sects., 9 spp.), Evans & Samson (*in* Samson *et al.* (Eds), *Fundamental and applied aspects of invertebrate pathology*: 186, 1986; spp. taxonomy on invertebrate hosts), Gams (*Neth. Jl Pl. Path.* **94**: 123, 1988; key to 9 nematophagous spp.), Carder (*MR* **92**: 297, 1989; cellulase isoenzyme patterns of 5 spp.), Heale (*in* Ingrams & Williams (Eds), *Advances in Plant Pathology* **6**: 291, 1988; vascular wilt spp.), Jun *et al.* (*J. Gen. Microbiol.* **137**: 1437, 1991; integrated taxonomic approaches), Carder & Barbara (*MR* **95**: 935, 1991; RFLPs in 6 spp.), Williams *et al.* (*Mycopathol.* **119**: 101, 1992; biochemical and physiological aids ident. sect. *Nigrescentia*), Typas *et al.* (*FEMS Microbiol. Lett.* **95**: 157, 1992; RFLP analysis), Okoli *et al.* (*Pl. Pathol.* **43**: 33, 1994; RFLPs and host adaptation), Rowe (*Phytoparasitica* **23**: 31, 1995; relationships between spp. and subspp. groups), Morton *et al.* (*MR* **99**: 257, 1995; rRNA intergenic regions in *V. alboatrum* and *V. dahliae*).

Verticimonosporium Matsush. (1971), Mitosporic fungi, 1.A1.15. 1, Papua New Guinea.

Vertixore V.A.M. Mill. & Bonar (1941) = Aithaloderma (Capnod.) fide v. Arx & Müller (1975).

Vesicladiella Crous & M.J. Wingf. (1994), Mitosporic fungi, 3.A1.15. 1, Australia.

vesicle, (1) a bladder-like sac; (2) (of *Aspergillus*), the swollen apex of the conidiophore; (3) (of *Pythium*), the evanescent extra-sporangial structure in which zoospores are differentiated; **homohylic -**, sporangial vesicle, the wall of which is continuous with, and of the same material as the wall layer, or one of the wall layers of the sporangium; **vesiculose**, made from or full of vesicles.

vesicular bodies, thin-walled vesicles in the sub-hymenium of certain hymenomycetes (mostly *Thelephoraceae*) (Overholts); vesicular-arbuscular type of mycorrhiza, see Mycorrhiza.

Vesicularia I. Schmidt (1974) [non P. Micheli ex Targ. Tozz. (1826), *Algae*] = Basramyces (Mitosp. fungi).

Vesiculomyces E. Hagstr. (1977), Gloeocystidiellaceae. 3, widespr. = Gloiothele fide Ginns & Freeman (*Bibl. Mycol.* **157**, 1994).

Vestergrenia Rehm (1901), ? Dothideaceae. 10, widespr. See v. Arx & Müller (1954).

Vestergrenia (Sacc. & Syd.) Died. (1913) = Phomopsis (Mitosp. fungi) fide Sutton (1977).

Vestergrenopsis Gyeln. (1940), Placynthiaceae (L). 2, Eur., N. Am. See Henssen (*CJB* **41**: 1359, 1963).

Vestigium Piroz. & Shoemaker (1972), Mitosporic fungi, 6.G1.1. 1, USA.

Veterinary mycology, see Medical mycology.

Vezdaea Tscherm.-Woess & Poelt (1976), Vezdaeaceae (±L). 10, Eur. See Poelt & Döbbeler (*Bot. Jb.* **96**: 328, 1975), Giralt *et al.* (*Herzogia* **9**: 715, 1993; key 10 spp.).

Vezdaeaceae Poelt & Vĕzda (1981), Lecanorales (±L). 1 gen., 10 spp. Thallus crustose, initially subcuticular; ascomata apothecial, rounded or irregular, sessile or short-stalked, marginal tissue poorly developed, formed from paraphysis-like filaments; interascal tissue of branched and anastomosing hyphae, sometimes entwining asci; asci with an I+ apical cap and a well-developed more darkly staining tubular ring; ascospores hyaline, sometimes septate, sometimes ornamented. Conidia sometimes formed from thalline hyphae. Lichenized with *Leptosira*, Eur.

viable, living; able to make growth.

Vialaea Sacc. (1896), Vialaeaceae. 2, Eur., C. & N. Am., Thailand. See Cannon (*MR* **99**: 367, 1995).

Vialaeaceae P.F. Cannon (1995), Ascomycota (inc. sed.). 1 gen. (+ 1 syn.), 2 spp. Stromata absent or poorly developed; ascomata perithecial, black, ± immersed, long-necked, the necks central, surrounded by a clypeus; peridium with occasional pores in the cell walls; interascal tissue of short wide very thin-walled paraphyses, evanescent at an early stage; asci cylindrical, persistent, thin-walled, with a complex partially I+ apical structure; ascospores elongated, hyaline, strongly isthmoid, septate, without a sheath.

Vialina Curzi (1935) = Phoma (Mitosp. fungi) fide Sutton (1977).

Vibrissea Fr. (1822), Vibrisseaceae. 7 (aquatic or semi-aquatic), widespr. See Sanchéz & Korf (*Mycol.* **58**: 733, 1966; sects.), Sanchéz (*J. Agric. Univ. Puerto Rico* **51**: 79, 1967), Beaton & Weste (*TBMS* **69**: 323, 1977; Australia), Korf (*Mycosystema* **3**: 19, 1991; concept), Sherwood (*Mycotaxon* **5**: 1, 1977; posn).

Vibrisseaceae Locq. (1972), Leotiales. 3 gen. (+ 4 syn.), 27 spp. Stromata absent; ascomata apothecial, flat or pulvinate, the disc pale or brightly coloured, sessile or long-stalked; excipulum composed of thin-walled angular cells, narrow, usually dark, without hairs; interascal tissue of simple paraphyses, sometimes swollen at the apices; asci elongated, with a thick I-apical cap penetrated by a narrow pore; ascospores filiform, hyaline, multiseptate, fragmenting. Anamorphs unknown. Saprobic, usually on wood and bark, aquatic, cosmop.

victorin, a toxin from *Drechslera victoriae* which induces symptoms of leaf blight of oats.

victotoxinine, a constituent of victorin (Pringle & Brown, *Nature* **181**: 1205, 1958, *Phytopath.* **50**: 324, 1960).

Viégas (Ahmés Pinto; 1906-1986). Brazilian mycologist and linguist, born in Piracicaba, São Paulo; graduated in agronomy at ESALQ-Piracicaba (1932) and obtained a PhD from Cornell University (1938). Made important contributions to the study of fungi, particularly microfungi, from the Atlantic forests of Brazil, incl. 12 gen. nov. (mostly published in Bragantia 3-19, 1943-60). Major works incl. the classic compilation *Indice dos fungos da América do Sul* (1961) and *Dicionario de fitopatologia e micologia* (1979). *Obit.*: Anon. (*Summa Phytopath.* **13**: iv, 1987). See Grummann (1974: 772).

Viegasia Bat. (1951), Asterinaceae. 4, trop. Am., Afr.

Viennotidea, see *Viennotidia*.

Viennotidia Negru & Verona ex Rogerson (1970), Hypocreaceae. 4, widespr. See Hawksworth & Cannon (1982; key). = Sphaeronaemella (Hypocr.) fide Hutchinson & Reid (*N.Z. Jl Bot.* **26**: 63, 1988). See Dissing (*in* Hawksworth (Ed.), *Ascomycete systematics*: 375, 1994; status). Anamorph *Gabarnaudia*.

Viennotiella Negru (1964) ? = Heteropatella (Mitosp. fungi) fide Sutton (1977).

villi (sing. **villus**), long soft hairs.

villose (**villous**), covered with villi, which are not matted; cf. tomentose.

Vinculum R.Y. Roy, R.S. Dwivedi & P.K. Khanna (1965) = Barnettella (Mitosp. fungi) fide Rao & Rao (*Ind. Phytopath.* **26**: 233, 1973).

vinescent, turning wine-red.

violet root rot, of a number of plants (*Helicobasidium purpureum*).

virescent, turning green.

Virgaria Nees (1816), Mitosporic fungi, 1.A2.10. 1, widespr.

Virgariella S. Hughes (1953), Mitosporic fungi, 1.A2.10. 6, widespr. See Sutton (*Sydowia* **44**: 321, 1993; related genera).

Virgasporium Cooke (1875) = Cercospora (Mitosp. fungi) fide Saccardo (*Syll. Fung.* **4**: 435, 1886).

virgate, banded; streaked.

Virgatospora Finley (1967), Mitosporic fungi, 2.C2.?. 1, Panama.

Virgella Darker (1967), Rhytismataceae. 1, N. Am.

viridin, an antibiotic from *Gliocladium virens*; antifungal (Brian *et al.*, *Ann. appl. Biol.* **33**: 190, 1946).

virose, (1) poisonous; (2) having a strong and unpleasant smell.

virulence, the degree or measure of pathogenicity; virulent, strongly pathogenic.

Viruses in fungi (mycoviruses). Many fungi are infected by viruses which typically have double-stranded RNA genomes but 'mycoviruses' with DNA or single-stranded RNA genomes are known. Those viruses so far classified are in the families *Barnaviridae* (gen. *Barnavirus*) and *Totiviridae* (gen. *Totivirus*) or the genus *Rhizidiovirus*. Unclassified viruses are known which infect fungi in the genera *Agaricus*, *Allomyces*, *Aspergillus*, *Colletotrichum*, *Gaeumannomyces*, *Helminthosporium*, *Lentinus* and *Periconia*.

Most viruses of fungi induce only inapparent effects. However some notable effects are hypovirulence of chestnut blight fungus (see *Hypovirus*) and die-back of cultivated mushrooms.

Some fungi act as vectors in the transmission of plant viruses from plant to plant (see vectors).

The naming of viruses has undergone extensive revision in recent years, rendering much of the early literature difficult to interpret. The main viruses infecting fungi (see Murphy *et al.*, 1995) are:

Barnavirus (fam. *Barnaviridae*): bacilliform virions containing a single molecule of positive-sense single-stranded RNA, *c.* 4.4 kb in size; infecting *Agaricus* spp.

Chrysovirus (fam. *Partitiviridae*): isometric virions containing 1 of 3 molecules of linear 2-stranded RNA about 3 kbp in size; infecting *Penicillium* and probably *Helminthosporium*.

Hypovirus (fam. *Hypoviridae*): genomes of linear 2-stranded RNA of 10 to 13 kbp in size, no virions are formed by infected cells contain lipid vesicles which contain the genome RNA; infecting *Cryphonectria parasitica* and causing hypovirulence.

Partitivirus (fam. *Partitiviridae*): isometric particles and genomes which comprise 2 molecules of linear 2-stranded RNA, 1.4 to 2.2 kpb in size; infecting *Agaricus*, *Aspergillus*, *Gaeumannomyces*, *Penicillium*, *Rhizoctonia* and probably *Diplocarpon* and *Phialophora*; the infections are latent.

Rhizidiovirus (fam. uncertain): isometric virions which contain a single molecule of *c.* 25 kbp linear 2-stranded DNA; infecting *Rhizidiomyces*.

Totivirus (fam. *Totiviridae*): isometric virions which contain a single molecule of linear 2-stranded RNA, 4.6 to 6.7 kbp in size; infecting *Helminthosporium*, *Saccharomyces*, *Ustilago* and probably *Aspergillus*, *Gaeumannomyces* and *Mycogone*; the infections are often latent.

Lit.: Adams (*Ann. appl. biol.* **118**: 479, 1991; transmission plant viruses by fungi), Buck (Ed.) (*Fungal virology*, 1986), Cooper & Asher (Eds) (*Viruses with fungal vectors*, 1985), Hillings & Stone (*Ann. Rev. Phytopath.* **9**: 93, 1971), Koltin & Leibowitz (Eds) (*Viruses of fungi and simple procaryotes*, 1988), Lemke (*in* Molitoris *et al.* (Eds), *Fungal viruses*: 2, 1979; coevolutin), Murphy *et al.* (*Virus taxonomy - The classification and nomenclature of viruses* [*Archives of Virology, Suppl.* **10**], 1995; taxonomy).

Vischia C.W. Dodge (1971) = Coccocarpia (Coccocarp.) fide Arvidsson (1982).

viscid, slimy, sticky, glutinous, lubricous, mucilaginous, viscous. Cf. gelatinous, ixo-.

Viscipellis (Fr.) Quél. (1886) ≡ Suillus (Bolet.).

Viscomacula R. Sprague (1950), Mitosporic fungi, 1.A1.?. 1, USA.

Visculus Earle (1909) = Pholiota (Strophar.) fide Singer (1951).

Vitalia Cif. & Bat. (1962) = Dennisiella (Coccodin.) fide Hughes (1976).

viteline, yellow like egg yolk.

Vitreostroma P.F. Cannon (1991), Phyllachoraceae. 1, montane trop., Afr., Asia.

Vittadinion Zobel (1854) = Tuber (Tuber.) fide Fischer (1938).

Vittadinula (Sacc.) Clem. & Shear (1931) ≡ Sphaerodes (Ceratostomat.).

vittate, having longitudinal lines, bands, or ridges.

Vivianella (Sacc.) Sacc. (1898) = Lophiostoma (Lophiostomat.).

vivotoxin, see toxin.

Vizella Sacc. (1883), Vizellaceae. 7, trop. See Petrak (*Sydowia* **8**: 294, 1954), Batista & Ciferri (*Sydowia Beih.* **1**: 325, 1957), Swart (1971), Selkirk (*Proc. Linn. Soc. NSW* **97**: 141, 1972; fossil taxa).

Vizellaceae H.J. Swart (1971), Dothideales. 2 gen. (+ 10 syn.), 8 spp. Mycelium superficial, the hyphae hyaline, cylindrical, with thickened brown septa; ascomata strongly flattened, intra- or subcuticular, the upper wall composed of a single layer of pseudoparenchymatous cells, without a clearly defined ostiole; interascal tissue absent or poorly developed, sometimes with a central columella; asci ovoid to saccate, probably evanescent; ascospores brown, usually aseptate or with a small appendage-like cell, with a conspicuous hyaline band, without a sheath. Biotrophic on leaves, trop.

Lit.: Swart (*TBMS* **57**: 455, 1971).

Vizellopsis Bat., J.L. Bezerra & T.T. Barros (1969), Dothideales (inc. sed.). 1, New Caledonia.

Vladracula P.F. Cannon, Minter & Kamal (1986), Rhytismataceae. 1, Asia.

Vleugelia J. Reid & C. Booth (1969), Ascomycota (inc. sed.). 1, Sweden.

Voeltzkowiella Henn. (1908) ? = Bulgaria (Leot.) fide Korf (*Mycol.* **49**: 102, 1957).

Voglinoana Kuntze (1891) nom. dub. fide Lindau (*Rabenh. Krypt. Fl.* **1**(8): 673, 1907; sub *Cystophora*), but used by Hughes (*CJB* **36**: 824, 1958).

Volkartia Maire (1909), Protomycetaceae. 1, Eur. See Reddy & Kramer (1975).

Volucrispora Haskins (1958), Mitosporic fungi, 1.G1.10. 3, Canada, Eur. See Hawksworth (*TBMS* **67**: 51, 1976), Petersen (*Bull. Torrey bot. Cl.* **89**: 287, 1962).

Volutella Fr. (1832) nom. cons., Anamorphic Hypocreaceae, 3.A1.15. Teleomorph *Pseudonectria*. 20, widespr.

Volutellaria Sacc. (1886), Mitosporic fungi, 3.A1.?. 1, N. Am.

Volutellis Clem. & Shear (1931), Mitosporic fungi, 1.A1.?. 1, Congo.

Volutellopsis Speg. (1910), Mitosporic fungi, 7.C1.?. 1, S. Am.

Volutellopsis Torrend (1913) [non Speg. (1910)] ≡ Volutellis (Mitosp. fungi).

Volutellospora Thirum. & P.N. Mathur (1965) = Chaetomella (Mitosp. fungi) fide Petrak (*Sydowia* **18**: 373, 1965).

volutin, a reserve material of fungi, esp. yeasts, seen as electron-dense granules; metachromatic polymetaphosphate material. See Nagel (*Bot. Rev.* **14**: 174, 1948).

Volutina Penz. & Sacc. (1901), Mitosporic fungi, 3.A1.15. 1, Java. ? = Volutella (Mitosp. fungi) fide Sutton (*in litt.*).

volva (of agarics and gasteromycetes), the cup-like lower part of the universal veil round the base of the mature stipe or receptacle (Fig. 4E); sometimes = universal veil, which is the preferred usage, fide Bas (*Persoonia* **5**: 304, 1969), who should be consulted for details of volva types and terminology.

Volva Adans. (1763) = Volvariella (Plut.) fide Donk (*Beih. Nova Hedw.* **5**: 297, 1962).

Volvaria DC. (1805) = Gyalecta (Gyalect.) p.p., Petractis (Stictid.) p.p. and Trapelia (Trapel.) p.p.

Volvaria (Fr.) P. Kumm. (1871) ≡ Volvariopsis (Plut.).

Volvariella Speg. (1899), Pluteaceae. *c.* 25, widespr. *V. volvacea* and *V. diplasia*, the straw (or 'paddy straw') mushrooms (q.v.) are edible. See Orton (*Bull. Soc. linn. Lyon*, num. spéc. **43**: 313, 1974; Eur., *British Fungus Flora* **4**: 61, 1986), Heinemann (*Fl. Illustr. Champ. Afr. Centr.* **4**: 75, 1975; C. Afr., **6**: 119, 1978), Pegler (*Kew Bull., Add. Ser.* **6**, 1977; E. Afr.), Shaffer (*Mycol.* **49**: 545, 1957; N. Am.).

Volvariopsis Murrill (1911) = Volvariella (Plut.) fide Singer (1951).

Volvarius Roussel (1806) = Volvariella (Plut.) fide Donk (1962).

Volvella J.-E. Gilbert & Beeli (1941) = Amanita (Amanit.) fide Pegler (*in litt.*).

Volvigerum (E. Horak & M.M. Moser) R. Heim (1966), Secotiaceae. 1, NZ.

Volvoamanita (Beck) E. Horak (1968), Amanitaceae. 2, Eur., A. Am.

Volvoboletus Henn. (1898) nom. dub., ? = Amanita (Amanit.) parasitized. See Ulbrich (*Ber. dtsch. bot. Ges.* **57**: 389, 1939).

Volvolepiota Singer (1959), Agaricaceae. 2, S. Am.

Volvopolyporus Lloyd ex Sacc. & Trotter (1912) ≡ Coltricia (Hymenochaet.).

Volvycium Raf. (1808) nom. dub. (? 'gasteromycetes').

Vonarxella Bat., J.L. Bezerra & Peres (1965), Saccardiaceae. 1 (or 2), Brazil.

Vonarxia Bat. (1960), Mitosporic fungi, 5.E1.?. 1, Brazil.

Vorarlbergia Grummann (1969) = Epigloea (Epigl.) fide Döbbeler (1984).

Vossia Thüm. (1879) [non Wall. & Griff. (1836), *Gramineae*] ≡ Neovossia (Tillet.).

Vouauxiella Petr. & Syd. (1927), Mitosporic fungi, 4.A/B2.38. 4 (on lichens), Eur., S. Am. See Sutton (1980), Hawksworth (*Bull. Br. Mus. nat. Hist., Bot.* **9**: 64, 1981).

Vouauxiomyces Dyko & D. Hawksw. (1979), Anamorphic Dothideales, 4.A2.19. Teleomorph *Abrothallus*. 3 (on *Lecanorales*), widespr. See Hawksworth (*Bull. Br. Mus. nat. Hist., Bot.* **9**: 67, 1981).

Vrikshopama D. Rao & P. Rag. Rao (1964) ? = Dematophora (Mitosp. fungi) fide Sutton (*in litt.*).

Vuillemin (Paul; 1861-1932). Professor in the Faculty of Medicine, Nancy. Chief writings: papers on the classification of the hyphomycetes (he coined the names phialide and aleuriospore) [see Mason, 1933] and other writings on taxonomy such as *Les bases actuelles de la systématique en mycologie* (1907), *Les champignons; Essai de classification* (1912), papers on diseases of man and plants. *Obit.*: Joyeux (*Mycopathol.* **3**: 64, 1941; bibl.), Potron (*BSMF* **67**: 42, 1951; portr.), Grummann (1974: 356), Stafleu & Cowan (*TL-2* **6**: 801, 1986).

Vuilleminia Maire (1902), Corticiaceae. 8, widespr. See Boidin *et al.* (*BSMF* **110**: 91, 1994; key).

Vulpicida Mattsson & M.J. Lai (1993), Parmeliaceae (L). 6, arctic to N. temp. See Mattsson (*Opera Bot.* **119**, 1993; key).

W, Naturhistorisches Museum, Wien (Vienna, Austria); founded 1748; formerly known as the Hofnaturalienkabinett und Naturhistorisches Hofmuseum; a state institution.

Wadeana Coppins & P. James (1978), Catillariaceae (L). 2, Eur. See Hertel & Rambold (*Bibl. Lich.* **38**: 145, 1990; posn).

Wageria F. Stevens & Dalbey (1918) = Ballyadyna (Parodiopsid.). = Balladynopsis (Parodiopsid.) fide Sivanesan (1981).

Wainio, see Vainio.

Wainioa Nieuwl. (1916) ≡ Pilocarpon (Pilocarp.).

Wainiocora Tomas. (1950) = Dictyonema (Merul.) fide Parmasto (1978).

Waitea Warcup & P.H.B. Talbot (1962), Botryobasidiaceae. 1, Australia, Switzerland.

Wakefield (Elsie Maud; 1886-1972). Was born in Birmingham (daughter of H.R. Wakefield, schoolmaster and naturalist) and graduated with first class honours from Oxford (Somerville College). In 1910 she was appointed to the Royal Botanic Gardens, Kew, as assistant to G. Massee whom, in 1911 she succeeded, and held the post until 1941. She was the first graduate mycologist at Kew and was a pioneer woman in the Scientific Civil Service. She was influential. With A.D. Cotton she monographed *Clavaria* but her main interest was in the resupinate basidiomycetes. She was a skilful watercolourist and her *Observer's book of common fungi*, 1954, later editions of the Ministry of Agriculture *Bulletin* on edible and poisonous fungi, and *Common British fungi*, 1950 (with R.W.G. Dennis) included coloured illustrations of her drawings. She was secretary of the BMS for 17 years and president in 1928. *TBMS* **49**: 355, portr., 1966 (80th birthday); **60**: 167-174, portr., bibl., 1973; Stafleu & Cowan (*TL-2* **7**: 24).

Wakefieldia Corner & Hawker (1953), Octavianinaceae. 2, Malaysia, UK.

Wakefieldia G. Arnaud (1954) = Anaphysmene (Mitosp. fungi) fide Sutton (*TBMS* **59**: 285, 1972).

Wakefieldiomyces Kobayasi (1981), Hypocreaceae. 1, W. Indies.

Waksmania H. Lechev. & M.P. Lechev. (1957), *Actinomycetes*, q.v.

Waldemaria Bat., H. Maia & Cavalc. (1960) [non Klotzsch (1862), *Ericaceae*] = Arachniotus (Gymnoasc.) fide Orr & Kuehn (*Mycol.* **63**: 191, 1971).

Walkeromyces Thaung (1976) = Mycovellosiella (Mitosp. fungi) fide Deighton (*Mycol. Pap.* **144**, 1979).

wall building, descriptive of hyphal growth in which cell wall material is produced by certain ultrastructural secretory bodies in the cytoplasm. Three types of wall building may be distinguished (Fig. 42): **apical -** in which the bodies are concentrated at the hyphal tip, producing new wall by distal growth, forming a cylindrical hypha in which the youngest wall material is at the tip; **ring -**, in which the bodies are concentrated adjacent to the cell wall at some point below the tip, in the shape of an imaginary ring, producing new wall by proximal growth, forming a cylindrical hypha in which the youngest wall material is always at the base; **diffuse -**, in which the bodies occur throughout the cytoplasm at a low concentration, producing lateral growth (i.e. swelling of the cylindrical hypha) by alteration of pre-existing wall.

The terms assume special significance in conidial development, where they have been used to clarify the concepts of thallic and blastic. Apical wall building occurs in *Geniculosporium*, *Cladosporium* and *Scopulariopsis*, and 'phialides' where conidia are produced in gummy masses (e.g. *Trichoderma*) or false chains (e.g. *Mariannaea*). Ring wall building occurs in 'phialides' with conidia in true chains (e.g. *Penicillium*, *Chalara*), in so-called meristem arthrospores (e.g. *Wallemia*) and in conidiogenous cells of basauxic fungi (e.g. *Arthrinium*). Diffuse wall building occurs simultaneously with or shortly after apical or ring wall building in most of the preceding examples, but its occurrence is much delayed or even absent in thallic development (e.g. *Geotrichum*). Wall building is a preferable term to meristem which implies growth by cell division rather than within a single cell. See Minter *et al.* (*TBMS* **79**: 75, 1982, **80**: 39, 1983, **81**: 109, 1983). Also see Mitosporic fungi.

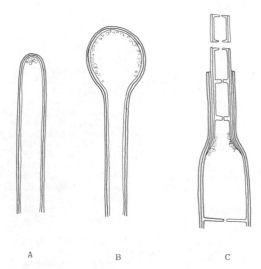

A   B   C

Fig. 42. Wall building in relation to conidiogenesis. A, apical; B, diffuse; C, ring. After Minter *et al.* (*TBMS* **80**: 39, 1983).

Wallemia Johan-Olsen (1887), Mitosporic fungi, 1.A1.?38. 1, widespr. *W. sebi* is osmophilic. See Moore (*Ant. v. Leeuwenhoek* **52**: 183, 1986; interpreted as a basidiomycete), Cole & Samson (*Patterns of development in conidial fungi*: 103, 1979; conidiogenesis).

Wallrothiella Sacc. (1882), ? Niessliaceae. 1, Indonesia. See Barr (*Mycotaxon* **39**: 43, 1990), Eriksson & Hawksworth (*SA* **10**: 59, 1991). Anamorph *Gliomastix*.

Waltiozyma H.B. Muller & J.L.F. Koch (1986) = Lipomyces (Lipomycet.) fide Kurtzman & Liu (*Curr. Microbiol.* **21**: 287, 1991).

Waltomyces Y. Yamada & Nakase (1985), Lipomycetaceae. 1, Eur. See Kock *et al.* (*Ant. v. Leeuwenhoek* **62**: 251, 1991). = Lipomyces (Lipomycet.) fide van der Walt (1992).

Waltonia Saho (1970), Dermateaceae. 1, Japan.

Walzia Sorokīn (1871) ? = Penicillium (Mitosp. fungi) fide Costantin (*Les mucedinées simples*: 201, 1888).

wandering lichens, lichens with an epigeic habit (e.g. *Parmelia afrorevoluta*; see Paulson & Hastings, *Knowledge* **37**: 319, 1914). Cf. manna.

Wangiella McGinnis (1977) = Exophiala (Mitosp. fungi).

Warcupia Paden & J.V. Cameron (1972), Otideaceae. 1, Canada.

Warcupiella Subram. (1972), Trichocomaceae. 1, Brunei. = Hamigera (Trichocom.) fide v. Arx (*Mycotaxon* **26**: 119, 1986). Anamorph *Aspergillus*.

Wardina G. Arnaud (1918) = Asterolibertia (Asterin.) fide Müller & v. Arx (1962).

Wardinella Bat. & Peres (1960), Mitosporic fungi, 5.B2.?. 1, Brazil.

Wardomyces F.T. Brooks & Hansf. (1923), Mitosporic fungi, 1.A2.13. 2, Eur., India. See Dickinson (*TBMS* **47**: 321, 1964).

Wardomycopsis Udagawa & Furuya (1978), Anamorphic Leotiaceae, 1.A2.19. Teleomorph *Microdiscus*. 1, Thailand.

Warkallisporonites Ramanujam & Rao (1979), Fossil fungi (*f. cat.*). 1 (Miocene), India.

wart disease, of potatoes (*Synchytrium endobioticum*) See Curtis (*Phil. Trans. roy. Soc.* **B210**: 409, 1921; life history).

Waste utilization, see Biodegradation, Bioremediation, Biotechnology, Industrial mycology.

water activity, some fungi can grow at extremely reduced levels of water activity, e.g. *Xeromyces bisporus* at 0.75 a_w; see Hocking (*Microbiol. Sci.* **5**: 280, 1988; review).

water moulds, see Aquatic fungi.

Wawea Henssen & Kantvilas (1985), Arctomiaceae (L). 1, NZ, Tasmania.

Wawelia Namysl. (1908), Xylariaceae. 2 (on coprophilous), Eur. See Minter & Webster (*TBMS* **80**: 370, 1983).

Waydora B. Sutton (1976), Mitosporic fungi, 8.A1.15. 1, widespr.

Waynea Moberg (1990), Bacidiaceae (L). 1, USA.

WDCM. World Data Center on Microorganisms (Saitama, Japan; formerly University of Queensland); see Genetic resource collections.

Weathering, of rocks can be brought about by physical, chemical and biological processes. All these factors ultimately result in fragmentation of the rock into smaller particles which can then contribute to the mineral fraction of soils. The biological weathering of rocks and their constituent minerals involves biogeochemical and biogeophysical weathering processes, the former bringing about a change in the chemical stability and composition of rocks and their minerals through the action of living organisms and their metabolic activities and the latter resulting in a mechanical disruption of the rocks and their constituent minerals (Silverman, in Trudinger & Swaine, *Studies in environmental science* **3**: 445, 1979).

One of the few comprehensive studies on the microbiota of rock surfaces was by Webley *et al.* (*J. Soil Sci.* **14**: 102, 1963) who recorded the numbers of microorganisms including bacteria, actinomycetes and fungi in the interior of porous weathered stones. No microorganisms were found in unweathered stones. Fungi, notably *Trichoderma* sp. were found to be growing directly on sandstone (Williams & Rudolph, *Mycol.* **66**: 648, 1974). *Alternaria* sp., *Acremonium* sp. and various species of *Penicillium* were also noted by these authors. *Penicillium simplicissimum* has been isolated from the surface of weathered basalt (Silverman & Munoz, *Science* **169**: 985, 1970). The biodegradation of basalt rock by *P. simplicissimum* has also been studied by Mehta *et al.* (*Biotechnol. & Bioengin.* **21**: 875, 1979) who showed that organic acids were responsible for the decomposition processes although the organism was not isolated from the rock substrate.

A high proportion of the organisms isolated by Webley *et al.* (1963) rendered silicates soluble when tested in pure culture and the authors (Henderson & Duff, *J. Soil Sci.* **14**: 236, 1963) concluded that fungi which produced citric and/or oxalic acid proved effective in decomposing certain natural silicates while an oxalic acid-producing strain also released metallic ions and silica from rocks and soils. Amongst the fungi studied were *Penicillium* spp. and species of *Trichoderma*, *Mucor* and *Spicaria*. Weathering of the edges of biotite flakes by *Aspergillus niger* was noted by Boyle & Voigt (*Science* **169**: 193, 1967) in pure culture studies and the activity was attributed to oxalic and citric acids. Cromack *et al.* (*Soil Biol. Biochem.* **11**: 463, 1979) have presented evidence for the accumulation of calcium oxalate in the fungal mats of the ectomycorrhizal fungus *Hysterangium crassum* associated with Douglas fir (*Pseudotsuga menziesii*) roots. This precipitation was brought about by oxalic acid secreted by the fungus and this was considered to be the reason for weathering of the igneous rock, andesite, in the root vicinity.

The weathering effects of crustose lichens on rock surfaces can manifest themselves as etch patterns on the rock minerals which can involve the transformation of certain minerals to siliceous relics and the formation of crystalline organic salts such as oxalates in the lichen thallus; the chemical composition of these oxalates is directly related to the substrate (Jones, *in* Galun (Ed.), *Handbook of lichenology* **3**: 109, 1988; Wilson & Jones, *in* Wilson (Ed.), *Residual deposits*, Sp. Publ. **11**: 5, 1983). See also Biodeterioration.

Weddellomyces D. Hawksw. (1986), Dacampiaceae. 8 (on lichens), Eur., Greenland. See Alstrup & Hawksworth (*Meddr. Grønl., Biosci.* **31**, 1990), Navarro-Rosinés & Roux (*Mycotaxon* **53**: 161, 1995).

Weesea Höhn. (1920), Hypocreaceae. 1, Brazil.

Wegelina Berl. (1900) = Calosphaeria (Calosphaer.) fide Barr *et al.* (*Mycotaxon* **48**: 529, 1993).

Wehmeyera J. Reid & C. Booth (1989), Melanconidaceae. 1, U.S.A.

Weinmannioscyphus Svrček (1977), Leotiaceae. 1, Eur.

Weinmannodora Fr. (1849) nom. dub. fide Petrak & Sydow (*Beih. Rep. spec. nov. regni veg.* **42**, 1927).

Weinzettlia Velen. (1921), Cortinariaceae. 1, former Czechoslovakia.

Weissia Bat. & M.M.P. Herrera (1964) [non Hedw. (1801), *Musci*], Mitosporic fungi, 5.B1.?. 1, Brazil.

Weitenwebera Opiz (1857) [non Opiz (1839), *Campanulaceae*] = Mycobilimbia (Porpid.) fide Hafellner (*Beih. Nova Hedw.* **79**: 241, 1984).

Weitenwebera Körb. (1863) = Chromatochlamys (Thelenell.).

Wentiomyces Koord. (1907), Pseudoperisporiaceae. *c.* 50, trop. See Hansford (1946, sub *Dimeriella*), Bose & Müller (*Indian Phytopath.* **17**: 3, 1964). Nom. dub. fide Farr (*Taxon* **14**: 18, 1965).

Weraroa Singer (1958), Galeropsidaceae. 4, NZ, S. Am. See Singer & Smith (*Bull. Torrey bot. Cl.* **85**: 324, 1958; key).

Wernera Zschacke ex Werner (1934) = Thrombium (Thromb.).

Westea H.J. Swart (1988), Dothideales (inc. sed.). 1, Australia.

Westerdijk (Johanna; 1883-1961). Plant pathologist and mycologist, was for many years concurrently director of the Phytopathological Laboratory 'Willie Commelin Scholten' (from 1906), Professor of plant pathology in the Universities of Utrecht (1917 on) and Amsterdam (1930 on), and the founder director of CBS (q.v.). See Ten Houten (*J. gen. Microbiol.* **32**: 1, 1963; portr., bibl.), Lohnis (*Johanna Westerdijk een markante persoonlijkheid*, 1963), Stafleu & Cowan (*TL-2* **7**: 210, 1988).

Westerdykella Stolk (1955), Sporormiaceae. 1, widespr. See Cejp & Milko (*Česká Myk.* **18**: 82, 1964; key), v. Arx & Storm (*Persoonia* **4**: 407, 1967).

Weston (William Henry; 1890-1978). Professor of cryptogamic botany, Harvard Univ. 1928-60, was an influential teacher. He was the first president of the Mycological Society of America. See Wilson (*Mycol.* **71**: 1103, 1980; portr.), Stafleu & Cowan (*TL-2* **7**: 217, 1988).

wet bubble, a mushroom disease (*Mycogone perniciosa*); white mould. See under Mushroom cultivation.

Wettsteiniella Kuntze (1891) ≡ Arthrobotryum (Mitosp. fungi).

Wettsteinina Höhn. (1907), Pleosporaceae. 23 Eur., N. Am. See Shoemaker & Babcock (*CJB* **65**: 373, 1987; key), Barr (*Contr. Univ. Mich. Herb.* **9**: 523, 1972).

Weufia Bhat & B. Sutton (1985), Mitosporic fungi, 1.G2.27. 1, Ethiopia.

WFCC, see Societies and organizations.

Whetstonia Lloyd (1906) = Dictyocephalos (Phellorin.) fide Long & Plunkett (*Mycol.* **32**: 637, 1940).

Whetzel (Herbert Hice; 1877-1944). Plant pathologist and mycologist (esp. *Sclerotiniaceae*) at Cornell Univ., Ithaca, N.Y., from 1902, where in 1907 he founded the first department of plant pathology in the USA. He was the author of more than 200 papers and books. See (*Mycol.* **37**: 393, 1945; portr.), Barrus & Stakman (*Phytopathol.* **35**: 659, 1945; bibl.), Stafleu & Cowan (*TL-2* **7**: 245, 1988).

Whetzelia Chardón & Toro (1934) = Vestergrenia (Dothid.) fide v. Arx & Müller (1954).

Whetzelia Zundel (1945) ≡ Ustacystis (Ustilagin.).

Whetzelinia Korf & Dumont (1972) nom. rej. = Sclerotinia (Sclerotin.) fide Kohn (1979).

Whetzeliomyces Viégas (1945) = Anhellia (Myriang.) fide v. Arx (1963).

whisky (whiskey, Ir, US), see spirits.

white blister (white 'rust'), a disease of plants caused by *Albugo*, esp. *A. candida* on crucifers.

white jelly fungus, basidioma of the edible *Tremella fuciformis*.

white piedraia, infection of hair shaft by *Trichophyton*.

whiteheads, a cereal disease (*Gaeumannomyces graminis*).

Whitfordia Murrill (1908) [non *Whitfordia* Elmer (1910), *Leguminosae*] = Amauroderma (Ganodermat.) fide Pegler (1970).

Wickerhamia Soneda (1960), Saccharomycodaceae. 1, Japan.

Wickerhamiella Van der Walt (1973), Saccharomycetaceae. 1, S. Afr.

Wielandomyces Raithelh. (1988), Bolbitiaceae. 1, Eur.

Wiesnerina Höhn. (1907), Cyphellaceae. 1, Brazil.

Wiesneriomyces Koord. (1907), Mitosporic fungi, 3.C1.10. 1, widespr. See Maniotis & Strain (*Mycol.* **60**: 203, 1968), Kirk (*TBMS* **82**: 748, 1984; nomencl.).

Wilcoxina Chin S. Yang & Korf (1985), Otideaceae. 3, Eur., N. Am. See Egger *et al.* (*MR* **95**: 866, 1991; as E-strain mycorrhizas).

Wilczekia Meyl. (1925) = Diderma (Didym.) for Kowalski (*Mycol.* **67**: 448, 1975).

wild type, (1) the naturally occuring species or taxon as opposed to morphological variants resulting from culture *in vitro* or biochemical mutants obtained therefrom; prototroph; (2) arbitrary designation for one or more strains chosen as genetic standards.

Willeya Müll. Arg. (1883) = Staurothele (Verrucar.) fide Zahlbruckner (1921).

Willia E.C. Hansen (1904) [non Müll. Hal. (1890), *Musci*] ≡ Hansenula (Endomycet.).

Williopsis Zender (1925), Saccharomycopsidaceae. 5, widespr. See Liu & Kurtzman (*Ant. v. Leeuwenhoek* **60**: 21, 1991; rRNA, key 5 spp.); Naumov *et al.* (*Mikrobiol.* **54**: 239, 1985; key 6 spp.), Naumov (*Molek. Genet., Mikrobiol. Virusol.* **2**: 3, 1987).

Willkommlangea Kuntze (1891), Physaraceae. 1, N. temp. See McHugh & Reed (*MR* **94**: 710, 1990; plasmocarp formation).

Wilmsia Körb. (1865) = Placynthium (Placynth.) fide Henssen (1963).

Wilsonia Cheel & Dughi (1944) [non R. Br. (1810), *Convolvulaceae*], nom. inval. = Polychidium (Placynth.).

Wilsonomyces Adask., J.M. Ogawa & E.E. Butler (1990) = Thyrostroma (Mitosp. fungi) fide Sutton (*in litt.*).

wilt, a plant disease (esp. *Verticillium*, *Fusarium*) characterized by loss of turgidity and collapse of leaves. See Mace *et al.* (Eds) (*Fungal wilt diseases of plants*, 1981). Cf. hadromycosis.

Wiltshirea Bat. & Peres (1962) = Phaeosaccardinula (Chaetothyr.) fide v. Arx & Müller (1975).

Wine making. E.C. Hansen (q.v.), later a director of the Carlsberg Physiological Laboratory, first showed that yeasts were invariably present in the soils of vineyards. Hansen's use of pure culture techniques in brewing were introduced into the wine industry between 1880 and 1890. The first distribution of pure yeast cultures for wine making was made from Geisenheimam Rhein in 1890. The term 'pure yeast culture' as used in wine making is not precise, since grape **must** (the juice produced after the grapes are crushed) is not sterile before the addition of the yeast starter. Sulphur dioxide and large amounts of yeast starter overwhelm the natural microbiota and result in a fermentation by *Saccharomyces cerevisiae*. Suitable strains must have: a high ethanol-producing ability (18-20% vol/vol), be cold resistant (fermenting at 4°C), be resistant to sulphite, ethanol (have an ability to start a new fermentation at 8-12% ethanol), tannin, and high concentrations of sugar (able to start fermentation above 30% wt/vol of sugar), be heat resistant (ability to ferment at 30-32°C), and be able to produce low volatile acidity (mostly as acetic acid). In wine making sulphur dioxide is added to the grape must at the rate of 50-200 ppm for 2 h prior to adding the yeast. The temperature of the fermentation (in vats) is controlled and usually below 30°C; high temperatures can destroy the wine. In general, red wines ferment for 3-6 days at 20-30°C, and white for 1-2 weeks at 10-21°C. The fermented juice is then drawn off from the residues and stored in a second vat for 7-11 days at 20-30°C. The wine is then racked, stored and aged to allow for clearing and flavour and colour development. Another form of wine depends on the 'noble rot' (*Botrytis cinerea*); grapes left on the vine until they are overripe become mouldy with *B. cinerea*, the grapes crack open, and the fungus lives on the juice and the sugar content increases; the mouldy and shrivelled grapes are hand-picked, and made into Sauterne (France), Tokai (Hungary), and Trockenbeerenauslesen (Germany).

Lit.: Ainsworth (1976), Amerine & Kunkee (*Ann. Rev. Microbiol.* **22**: 323, 1968), Benda (*in* Read, *Prescott & Dunn's industrial microbiology*, edn 4: 293, 1982), Beuchart (Ed.) (*Food and beverage mycology*, 1987), Kendrick (*The fifth kingdom*, edn 2, 1992), Rose (Ed.) (*Alcoholic beverages*, 1977), Varnam (*in* Jones, *Exploitation of micro-organisms*: 297, 1993) See also Brewing, Fermented foods and beverages, Food and beverage mycology, Food spoilage, Mycotoxicoses, Yeast.

Wingea Van der Walt (1967) = Debaryomyces (Saccharomycet.) fide Kurtzman & Robnett (*Ant. v. Leeuwenhoek* **66**: 337, 1994).

Wingina Kuntze (1891) ≡ Orthotricha (Echinostel.).

Winter (Heinrich Georg; 1848-87). Privatdocent, University of Leipzig after being at Halle and Zürich;

editor of *Hedwigia* (1878-87). Chief writings: *Pilze* in *Rabenh. Krypt.-Fl.* 1881-87. See Pazschke (*Hedwigia* **26**: 185, 1887; portr.), Grummann (1974: 55), Stafleu & Cowan (*TL-2* **7**: 380, 1988).

winter mushroom, see enokitake.

winter spore, a resting spore for overwintering, e.g. a teliospore of *Puccinia graminis*. See teliospore.

Winterella (Sacc.) Kuntze (1891), Valsaceae. 11, temp. See Reid & Booth (*CJB* **65**: 1320, 1987). = Montagnula (Didymosphaer.) fide Holm (*SA* **11**: 29, 1992).

Winterella Berl. (1893) [non (Sacc.) Kuntze (1891)] = Nitschkia (Nitschk.).

Winteria Sacc. (1878) ≡ Selinia (Hypocr.).

Winteria (Rehm) Sacc. (1883) ≡ Mycowinteria (Protothelenell.).

Winterina Sacc. (1891) nom. ambig. (Fungi, inc. sed.).

Winterina Sacc. (1899) ≡ Calyculosphaeria (Nitschk.).

Winteromyces Speg. (1912) = Gibbera (Ventur.) fide Müller & v. Arx (1962).

witches' brooms, massed outgrowths (proliferations) of the branches of woody plants caused by mites, viruses, etc., and fungi, esp. rusts (e.g. *Pucciniastrum goeppertianum* (*Vaccinium*), *Gymnosporangium ellisii* (*Chamaecyparis*), *Melampsorella cerastii* (*Abies* and *Picea*)) and *Taphrina* (e.g. *T. betulina* (birch), *T. cerasi* (cherry), *T. insititiae* (plum)); also the 'krulloten' of Cacao caused by *Crinipellis perniciosa*.

witches' butter, basidioma of *Exidia glandulosa*.

Witwateromyces Hallbauer, Jahns & Van Warmelo (1977), Fossil fungi (Ascomycota, ?L). 1 (Precambrian), S. Afr.

Woessia D. Hawksw. & Poelt (1986), Mitosporic fungi (L), 1.E1.15. 1, Austria.

Wojnowicia Sacc. (1899), Mitosporic fungi, 4.C2.15. 3, widespr. *W. hirta* (cereal foot-rot). See Sutton (*Česká Myk.* **29**: 97, 1975), Farr & Bills (*Mycol.* **87**: 518, 1995; key).

Woldmaria W.B. Cooke (1961), Cyphellaceae. 1, Eur., N. Am.

Wolfina Seaver ex Eckblad (1968), Sarcosomataceae. 1, USA.

Wolfiporia Ryvarden & Gilb. (1984), Coriolaceae. 2, N. & trop. Am.

wolf's-moss, *Letharia vulpina*, from Scandinavian belief that it was poisonous to wolves.

Wolkia Ramsb. (1915), Ascomycota (inc. sed.). 1, Java.

Wood-attacking fungi. Are (but for a few ascomycetes, e.g. *Ustulina*, *Xylaria*, *Chaetomium*) basidiomycetes, these genera being specially important: *Agaricales* (*Armillaria*, *Collybia*, *Pholiota*); *Boletales* (*Coniophora*, *Serpula*); *Cantharellales* (*Hydnum*); *Fistulinales* (*Fistulina*); *Ganodermatales* (*Ganoderma*); *Hymenochaetales* (*Hymenochaeta*); *Poriales* (*Coriolus*, *Daedalea*, *Fomes*, *Lentinus*, *Lenzites*, *Polyporus*, *Poria*, *Trametes*); *Schizophyllales* (*Schizophyllum*); *Stereales* (*Echinodontium*, *Merulius*, *Peniophora*, *Stereum*).

There are several generally-recognized types of timber decay, based on symptoms: (1) **Brown rots** - cellulose is utilized, brick-shaped cracks develop, the wood becomes brown and crumbles when handled; (2) **White rots** - caused by some ascomycetes and mitosporic fungi. Hardwoods are more susceptible than softwoods. The wood darkens and may cross crack; the surface erodes and elongated cavities form in secondary xylem areas. The mechanism of soft rot is the degradation of cellulose in less heavily lignified areas and therefore affects superficial layers of wood. High water content aids soft rot which is often found in structures such as jetty posts and cooling tower slats. Rots of standing timber, due to colonization by fungi via wounds, broken branches and dead roots are sometimes classified by rot position, e.g. **top rot** (branches), **core** or **heart rot** (trunk), **butt rot** (base).

Two other rot classification schemes should be noted: (1) True **dry rot** - a brown rot caused by *Serpula lacrimans*. This is a major decay fungus of buildings in temperate ares. The name is misleading as stable damp conditions are needed for initiation of growth. As the fungus develops, it conducts water to drier wood through hyphal strands which penetrate masonry and plaster though not feeding on these latter materials Eradication can be difficult and costly. See Jennings & Bravery (*Serpula lacrymans: fundamental biology and control strategies*, 1991). (2) **Wet rots** - caused by many fungal species, including both brown and white rot types. A common example is *Coniophora puteana*, the 'cellar fungus'.

Certain fungi are a cause of staining in wood. Stored wood may be stained by the growth of surface moulds (*Alternaria, Aspergillus, Mucor, Penicillium, Rhizopus*, etc.) or by fungi in the wood (e.g. *Ophiostoma*, a cause of 'blue stain'), sometimes of living trees. Brown oak (*Quercus*) wood stained by *Fistulina hepatica*; Cartwright, *TBMS* **21**: 68, 1937) is valued, and wood coloured green by *Chlorociboria aeruginascens* is used in making 'Tunbridge Ware'. Wood having 'zone lines' (q.v.) is sometimes used for ornament (see spalted wood). Numbers of the fungi staining wood in addition damage wood pulp; see Melin & Nannfeldt (*Svenska Skogvs Fören. Tidskr.* **314**: 397, 1934).

Forest hygiene and good forestry practice are the chief methods by which decay in living trees is controlled. In cut trees, in stored wood, and wood in buildings, steps may be taken to keep conditions such that growth of fungi does not take place. The water content of the wood and the air are kept low, and fungicides used. A range of preservatives is available for preservation of wood, which are subject to various mational regulations in their use. For example, in the UK only products cleared for use under the Control of Pesticides Regulations are used. Active ingredients include coal tar creosote a coal tar oil, copper and zinc naphthenates, organotin compounds, phenolic compounds, and copper, chromium and arsenic salts.

Lit.: Boyce (*Forest pathology*, edn 3, 1961), Carroll & Wicklow (Eds) (*The fungal community*, edn 2, 1992), Dix & Webster (*The ecology of fungi*, 1995), Easton & Hale (*Wood. Decay, pests and protection*, 1993), Hunt & Garratt (*Wood preservation*, 1938), Bravery *et al.* (*Recognizing wood rot and insect damage in buildings*, 1987), Cartwright & Findlay (*Decay of timber and its prevention*, edn 2, 1958), Findlay (*Dry rot and other timber troubles*, 1952), Fergus (*Illustrated genera of wood decay fungi*, 1960), Gilbertson (*Mycol.* **72**: 1, 1980; wood-rotting fungi of N. Am., 119 refs), Nobles (*CJB* **43**: 1097, 1965; identification of cultures; keys 149 spp.), Stalpers (*Stud. Mycol.* **16**, 1978; keys 550 spp. polypores in culture), Zainal (*Micro-morphological studies on soft rot fungi in wood* [*Bibl. Mycol.* **70**], 1981), Cockroft (Ed.) (*Some wood-destroying basidiomycetes*, 1981), Findlay (Ed.) (*Preservation of timber in the tropics*, 1985), Rayner & Boddy (*Fungal decomposition of wood*, 1988), Wang & Zabel (Eds) (*Identification manual for fungi from utility poles in Eastern United States*, 1990).

Woodiella Sacc. & P. Syd. (1899), Leotiales (inc. sed.). 1, S. Afr.

Wood's light. Ultra-violet light filtered through nickel oxide-containing soda glass. *Microsporum-* but not *Trichophyton*-infected hairs show a bright greenish fluorescence in Wood's light, an effect made use of in diagnosis (see Davison & Gregory, *Can. J. Res.* **7**: 378, 1932). For examination of agarics by Wood's light see Deysson (*BSMF* **74**: 207, 1958).

Wormald (Hugh; 1879-1955). Plant pathologist at Wye College, Kent (1911-22); Head of the Plant Pathology Section, East Malling Research Station (1923-39). Noted for his work on diseases of fruit trees, esp. bacterial diseases and brown rot. Chief writings: Brown rot diseases of fruit trees (*Bull. Min. Agric.* **88**, 1935; *Tech. Bull. Min. Agric.* **3**, 1954; *Diseases of fruits and hops*, 1939 [edn 3, 1955]). *Obit.*: Barnes (*TBMS* **39**: 289, 1956; portr.), Harris (*Rept E. Malling Res. Stn* **1955**: 15; portr.).

Woronichina Naumov (1951) [non Elenkin (1933), *Algae*], ? Hypocreaceae. 1, former USSR.

Woronin (Michael Stepanovitch; 1838-1903). A Russian mycologist who worked with de Bary. Chief writings: papers on *Puccinia helianthi* (1872), *Plasmodiophora brassicae* (1877 [in *English, Phytopath. Classics* **4**, 1934]), *Monoblepharis* (1901), *Sclerotinia* (1888, 1895, 1900); (with de Bary) the series *Beiträge zur Morphologie und Physiologie der Pilze* (1864 on). See Smith (*Phytopathol.* **2**: 1, 1912), Izbrannÿe proizvedeniya ([collected works], 1961), Grummann (1974: 568), Stafleu & Cowan (*TL-2* **7**: 455, 1988).

Woronin bodies. Rounded or elongated-oval highly refractive bodies in the cells of certain discomycetes, particularly in association with septa (Buller, **5**: 127, 1933); in *Erysiphe graminis*, see McKeen (*Can. J. Microbiol.* **17**: 1557, 1971); **Woronin's hypha** (of ascomycetes), a coiled hypha probably homologous with an archicarp (de Bary, 1887); a loosely coiled hypha of large diam., at the centre of a young perithecium, which later develops ascogenous hyphae (Miller, 1928); scolecite.

Woronina Cornu (1872), Plasmodiophoraceae. 4 (in *Vaucheria*), N. temp. See Karling (1968; key).

Woroninella Racib. (1898) = Synchytrium (Synchytr.) fide Gäumann (1927).

Woroninula Mekht. (1979), Mitosporic fungi, 1.C1.10. 3, widespr.

wort, see Brewing.

wortmannin, an antibiotic from *Talaromyces wortmannii* (anamorph); antifungal esp. against *Botrytis, Cladosporium* and *Rhizopus* (Brian *et al.*, *TBMS* **40**: 365, 1957).

Wrightiella Speg. (1923) [non F. Schmitz (1893), *Algae*] ≡ Leptodothis (Ascomycotina).

Wrightoporia Pouzar (1966), Hericiaceae. 14, widespr. See David & Rajchenberg (*CJB* **65**: 202, 1987; key).

Wuestneia Auersw. ex Fuckel (1863), Melanconidaceae. 10, temp. See Ananthapadmanaban (*TBMS* **91**: 517, 1988), Reid & Booth (*CJB* **67**: 909, 1989; key), Holm (*Taxon* **24**: 475, 1975).

Wuestneiopsis J. Reid & Dowsett (1990), Melanconidaceae. 1, N. Am.

WWW, see Internet.

wyerone, a phytoalexin (q.v.) from broad bean (*Vicia faba*).

Wynnea Berk. & M.A. Curtis (1867), Sarcosomataceae. 4, widespr. See Pfister (*Mycol.* **71**: 144, 1979), Kaushal (*Res. Bull. Punjab Univ.* **34**: 29, 1983), Liu *et al.* (*Mycotaxon* **30**: 465, 1987; key 6 spp.).

Wynnella Boud. (1885), Helvellaceae. 1, Eur. = Helvella (Helvell.) fide Harmaja (*Karstenia* **14**: 102, 1974).

Xanthocarpia A. Massal. & De Not. (1853) nom. rej. = Caloplaca (Teloschist.).

Xanthochroales. An order proposed for poroid basidiomycetes having uninflated hyphae lacking clamp connexions (Corner, *Clavaria*: 22, 1950).

xanthochroic (of a hymenomycete basidioma), having a reddish- or yellowish-brown context which darkens on treatment with KOH.

Xanthochrous Pat. (1897) ≡ Polystictus (Hymenochaet.) or Coltrichia (Hymenochaet.) fide Donk (1974).

Xanthoconium Singer (1944), Boletaceae. 7, N. Am. See Wolfe (*CJB* **65**: 2142, 1987).

Xanthodactylon P.A. Duvign. (1941), Teloschistaceae (L). 1, S.W. Afr. See Kärnfelt (*Crypt. Bot.* **1**: 147, 1989).

Xanthoglossum (Sacc.) Kuntze (1891) = Geoglossum (Geogloss.).

Xanthomaculina Hale (1985) Parmeliaceae. 3, S. Afr. See Büdel & Wessels (*Dinteria* **18**: 3, 1986; vagant *X. huena*). ? = Parmelia (Parmel.).

Xanthoparmelia (Vain.) Hale (1974) Parmeliaceae. 440, widespr., esp. S. Hemisph. See Elix *et al.* (*Bull. Br. Mus. nat. Hist., Bot.* **15**: 163, 1986; key 117 spp. Australasia), Hale (*Smithson. Contr. Bot.* **74**, 1990; monogr.), Nash *et al.* (*Bibl. Lich.* **56**, 1995; key 77 spp. S.Am.), Thomson (*Bryologist* **96**: 342, 1993; key 55 spp. N. Am.). ? = Parmelia (Parmel.).

Xanthopeltis R. Sant. (1949), Teloschistaceae (L). 1, Chile. See Follmann (*Revta Universitaria (Chile)* **47**: 33, 1962).

Xanthoporia Murrill (1916) = Inonotus (Hymenochaet.) fide Pegler (1964).

Xanthoporina C.W. Dodge (1948) = Pyrenocollema (Pyrenul.).

Xanthopsis Acloque (1893) [non (DC.) Koch (1851), *Compositae*] = Catolechia (Rhizocarp.).

Xanthopsora Speg. (1922) = Linochora (Mitosp. fungi) fide Petrak & Sydow (*Ann. Myc.* **33**: 176, 1935).

Xanthopsora Gotth. Schneid. & W.A. Weber (1980) [non Speg. (1922)] ≡ Xanthopsorella (Catillar.).

Xanthopsorella Kalb & Hafellner (1984), Catillariaceae (L). 1, Mexico, USA.

Xanthopyrenia Bachm. (1919) = Pyrenocollema (Pyrenul.) fide Harris (*in litt.*).

Xanthopyreniaceae, see *Pyrenulaceae*.

Xanthoria (Fr.) Th. Fr. (1860) nom. cons., Teloschistaceae (L). *c.* 30, cosmop. See Almborn (*Bot. Notiser* **16**: 161, 1963; S. Afr.), Hillmann (*Rabenh. Krypt.-Fl.* **9**: 6, 1936; Eur.), Janex-Favre & Ghaleb (*Cryptogamie, bryol. lich.* **7**: 457, 1986; ontogeny), Poelt & Pelztschnig (*Nova Hedw.* **54**: 1, 1992; key *X. candelaria* group), Steiner & Poelt (*Pl. Syst. Evol.* **140**: 151, 1982; sect. *Xanthoriella*), Thomson (*Am. Midl. Nat.* **41**: 706, 1949; N. Am.), Awasthi (*Proc. Ind. Acad. Sci. Pl. Sci.* **96**: 227, 1986; Indian spp.).

Xanthoriicola D. Hawksw. (1973), Mitosporic fungi, 3.A2.15. 1 (on lichens, *Xanthoria*), Eur. See Hawksworth (*Bull. Br. Mus. nat. Hist., Bot.* **6**: 183, 1979; SEM].

Xanthoriomyces E.A. Thomas ex Cif. & Tomas. (1953) ≡ Xanthoria (Teloschist.).

Xanthothecium Arx & Samson (1973), Onygenaceae. 1, widespr. See Currah (1985; posn).

Xeilaria Lib. (1830) ? = Placosphaeria (Mitosp. fungi) p.p. fide Saccardo (*Syll. Fung.* **10**: 236, 1892).

Xenasma Donk (1957), Xenasmataceae. 9, Eur., N. Am. See Liberta (*Mycol.* **52**: 884, 1962; key), Hjortstam *et al.* (*Corticiaceae of Northern Europe* **8**, 1988; key Eur. spp.).

Xenasmataceae Oberw. (1966), Stereales. 6 gen. (+ 2 syn.), 39 spp. Basidioma resupinate, ceraceous; hymenophore smooth; monomitic; basidia pleural; spores often amyloid.

Xenasmatella Oberw. (1966) = Phlebiella (Xenasmat.) fide Hjortstam *et al.* (*Corticiaceae of Northern Europe* **8**, 1988).

Xenidiocercus Nag Raj (1993), Mitosporic fungi, 4.A1.19. 3, India, W. Afr.

xenobiotic, (1) a chemical not normally synthesized or metabolized by living organisms, e.g. a manufactured drug; (2) chemical waste or other pollutant toxic to a living organism.

Xenobotrytis R.F. Castañeda & W.B. Kendr. (1990) = Paracostantinella (Mitosp. fungi) fide Kendrick (*in litt.*).

Xenochora Petr. (1948), Mitosporic fungi, 4.A1/2.?. 1, Ecuador.

Xenodiella Syd. (1935), Mitosporic fungi, 3.A1.?. 1, S. Am.

Xenodimerium Petr. (1947) ? = Eudarluca (Phaeosphaer.) fide Müller & v. Arx (1962).

Xenodiscella Petr. (1954) = Rhagadolobium (Parmular.) fide Müller & v. Arx (1962).

Xenodium Syd. (1935), Elsinoaceae. 1, S. Am.

Xenodochus Schltdl. (1826), Phragmidiaceae. 2 (on *Sanguisorba*), N. temp.

Xenodomus Petr. (1922), Mitosporic fungi, 4.A1.?. 1, N. Am.

Xenogliocladiopsis Crous & W.B. Kendr. (1993), Mitosporic fungi, 1.A1.15. 1, S. Afr.

Xenogloea Syd. & P. Syd. (1919) ≡ Kriegeria (Platygl.).

Xenoheteroconium Bhat, W.B. Kendr. & Nag Raj (1993), Mitosporic fungi, 1.C2.1. 1, India.

Xenokylindria DiCosmo, S.M. Berch & W.B. Kendr. (1983), Mitosporic fungi, 1.B1.19. 1, Japan.

Xenolachne D.P. Rogers (1947), Sirobasidiaceae. 1, USA.

Xenolecia Hertel (1984), Porpidiaceae (L). 1, Chile.

Xenolophium Syd. (1925), Melanommataceae. 4, trop. See Huhndorf (*Mycol.* **85**: 490, 1993).

Xenomeris Syd. (1924), Venturiaceae. *c.* 10, widespr. *X. abietis* (dieback of conifers). See Barr (*Sydowia* **41**: 25, 1989), Funk & Shoemaker (*Mycol.* **63**: 567, 1971). Anamorph *Aureobasidium*.

Xenomyces Ces. (1879) = Sclerocystis (Glom.) fide Zycha *et al.* (1969).

Xenomyxa Syd. (1939), Ascomycota (inc. sed.). 1, Ecuador.

Xenonectria Höhn. (1920) = Byssosphaeria (Melanommat.) fide Barr (1984).

Xenonectriella Weese (1919), Hypocreaceae. 1 (on *Solorina*), Eur.

Xenopeltis Syd. & P. Syd. (1919), Mitosporic fungi, 5.A1.?. 1, Philipp.

Xenoplaca Petr. (1949), Mitosporic fungi, 1.A2.?. 1, S. Am. See v. Arx (*Persoonia* **11**: 388, 1981; redescr.).

Xenopus Penz. & Sacc. (1901) nom. dub. ('basidiomycetes') fide Hughes (*CJB* **36**: 747, 1958).

Xenosoma Syd. & P. Syd. (1921) nom. dub. ('gasteromycetes' or ? Trichosphaeriales).

Xenosperma Oberw. (1966), Xenasmataceae. 4, widespr.

Xenosphaeria Trevis. (1860) = Dacampia (Dacamp.) fide Alstrup & Hawksworth (*Meddr. Grønl., Biosci.* **31**, 1990).

xenospore, a spore dispersed from its place of origin (Gregory, *in* Madelin, 1966). Cf. memnospore.

Xenosporella Höhn. (1923) = Xenosporium (Mitosp. fungi) fide Pirozynski (1966).

Xenosporium Penz. & Sacc. (1901), Mitosporic fungi, 1.D2.1. 12, mainly tropical. See Pirozynski (*Mycol. Pap.* **105**, 1966; key), Goos (*Mycol.* **82**: 742, 1990; key).

Xenostele Syd. & P. Syd. (1921) = Puccinia (Puccin.) fide Cummins (1959).

Xenostigme Syd. (1930), ? Meliolaceae. 1, S. Am.

Xenostigmella Petr. (1950) ? = Balladynopsis (Parodiopsid.) fide Sivanesan (1981).

Xenostilbum Petr. (1959) = Calostilbella (Mitosp. fungi) fide v. Arx (*Persoonia* **11**: 391, 1981).

Xenostomella Syd. (1930), Microthyriaceae. 2, S. Am.

Xenostroma Höhn. (1915), Mitosporic fungi, 8.A1/2.?. 1, Eur.

Xenothecium Höhn. (1919), ? Hyponectriaceae. 1, S. Am.

Xenotypa Petr. (1955), Valsaceae. 1, Eur., N. Am. See Barr (*Mycol. Mem.* **7**, 1978), Pirozynski (*CJB* **52**: 2129, 1974).

Xenus Kohlm. & Volkm.-Kohlm. (1992), Pyrenulales (inc. sed.). 1 (on coralline alga), Belize. See Eriksson & Hawksworth (*SA* **11**: 189, 1993).

Xepicula Nag Raj (1993), Mitosporic fungi, 7.A1/2.15. 2, widespr.

Xepiculopsis Nag Raj (1993), Mitosporic fungi, 7.A1/2.15. 2, widespr.

xero- (prefix), dry; drought.

Xerocarpus P. Karst. (1881) [non Guill., Perr. & A. Rich. (1832), *Papilionaceae*] = Phanerochaete (Merul.) fide Donk (1957).

Xerocomaceae (Singer) Pegler & T.W.K. Young (1981), Boletales. 9 gen. (+ 6 syn.), 105 spp. Basidioma boletoid, stipe slender, not reticulate; hymenophore adnate to decurrent, tubulate or lamellate, spores brown, fusoid, smooth or ridged.

Xerocomopsis Reichert (1940) ≡ Xerocomus (Xerocom.).

Xerocomus Quél. (1887) nom. cons., Xerocomaceae. 40, cosmop. See Thiers (*Madroño* **17**: 237, 1974; key N. Am. spp.).

Xeroconium D. Hawksw. (1981), Mitosporic fungi, 4.A2.19. 1, Finland.

Xerocoprinus Maire (1907), Coprinaceae. 1, Afr.

Xerodiscus Petr. (1943) = Arthonia (Arthon.) fide Müller & v. Arx (1962).

Xeromedulla Korf & W.Y. Zhuang (1987), Leotiaceae. 3, China, Philipp.

Xeromphalia, see *Xeromphalina*.

Xeromphalina Kühner & Maire (1934), Tricholomataceae. *c.* 20, widespr. See Singer (*Boln Soc. argent. Bot.* **10**: 302, 1965), Smith (*Pap. Mich. Acad. Sci. Arts* **38**: 53, 1953; N. Am.), Miller (*Mycol.* **60**: 156, 1968; key 12 N. Am. spp.), Horak (*Sydowia* **32**: 131, 1979; Indomalaya & Austral. spp.), Klán (*Česká Myk.* **38**: 205, 1984; Eur. spp.).

Xeromyces L.R. Fraser (1954), ? Monascaceae. 1 (osmophilic), widespr. See Pitt & Hocking (*CSIRO Food Res. Q.* **42**: 1, 1983), Hawksworth & Pitt (*Austr. Jl Bot.* **31**: 51, 1983; posn). Anamorph *Fraseriella*.

xerophilic, favouring habitats in which water is not available; either living in desert conditions or where water is not generally available because of the physiological status of cells (c.f. water activity).

xerophyte, a plant of dry habitats; sometimes incorrectly applied to fungi.

Xerosporae. Mitosporic fungi (hyphomycetes and coelomycetes) with dry spores. Cf. *Gloiosporae* See Wakefield & Bisby (*TBMS* **25**: 49, 1942).

Xerotinus Rchb. (1828), Polyporaceae. 2, Afr. See Pegler (1973).

xerotolerant, able to grow under dry conditions (Pitt *in* Duckworth (Ed.), *Water relations of foods*, 1975).

Xerotrema Sherwood & Coppins (1980), Odontotremataceae. 1, UK.

Xerotus Fr. (1828) [non *Xerotes* R. Br. (1810), *Xanthorrhoeaceae*] = Xerotinus (Polypor.) fide Donk (1960).

Xerula Maire (1933), Tricholomataceae. 6, Eur. See Dörfelt (*Mycotaxon* **15**: 62, 1982; nomencl., bibliogr.). ≡ Oudemansiella (Tricholomat.).

Xerulina Singer (1962), Tricholomataceae. 5, Am., Afr.

Xiambola Minter & Hol.-Jech. (1981), Mitosporic fungi, 7.B1.38. 1, Czech Republic.

Xiphomyces Syd. & P. Syd. (1916), Mitosporic fungi, 3.A2.?. 2, Philipp., Tristan da Cuhna.

Xylaria Hill ex Schrank (1789) nom. cons., Xylariaceae. 100, cosmop. *X. digitata* (root rot of hardwoods); *X. hypoxylon*, the Candle-snuff fungus (black root rot of apple); *X. vaporaria* (an invader of mushroom beds). See Dennis (*Kew Bull.* **1956**: 401, 1957; trop. Am., *Revta Biol. Lisbon* **1**: 175, 1958; trop. Afr., *Bull. Jard. bot. Brux.* **31**: 111, 1961; Congo, *in* Parmasto, *Zhivaya priroda Dal'nego Vostoka*: 42, 1972; E. former USSR), González & Rogers (*Mycotaxon* **34**: 283, 1989; key 63 spp. Mexico), Joly (*Revue Mycol.* **33**: 155, 1966; Vietnam), Rogers (*Mycotaxon* **26**: 85, 1986; key 30 spp. USA), Bertault (*Bull. Soc. mycol. Fr.* **100**: 139, 1984; key 45 spp., Eur., N. Afr.), Rogers (*Mycol.* **75**: 457, 1983; **76**: 23, 1984), Rogers & Callan (*Mycol.* **78**: 391, 1986; *X. polymorpha*-group, USA), Rogers (*Sydowia* **38**: 255, 1986; sectional classif., anamorphs), Rogers & Samuels (*N.Z. Jl Bot.* **24**: 615, 1986; key 19 NZ spp.), Rogers *et al.* (*Mycotaxon* **31**: 103, 1988; key 41 spp. Venezuela), Silveira & Rogers (*Acta Amazonia* Suppl. **15**: 7, 1985; key 11 spp. Brazil), Whalley (*Agarica* **8**: 68, 1987; on fallen fruits), Brunner & Petrini (*MR* **96**: 723, 1992; isozyme electrophoresis), Callan & Rogers (*Mycotaxon* **46**: 141, 1993; cultural features, 23 US spp.). Anamorph *Xylocoremium*.

Xylaria Hill ex Grev. (1823) nom. rej. ≡ Cordyceps (Clavicipit.).

Xylariaceae Tul. & C. Tul. (1861), Xylariales. 46 gen. (+ 70 syn.), 595 spp. Stromata usually well-developed, varied in morphology, sometimes stipitate and branched, usually black, the internal tissue white or concolorous with the surface; ascomata perithecial, black, globose, the ostiole periphysate; interascal tissue well-developed, of narrow ± thick-walled paraphyses; asci cylindrical, persistent, ± thick-walled but without separable layers, almost always with a large ± complex I+ apical structure; ascospores usually dark brown, aseptate (sometimes with a hyaline daughter cell), with a germ slit, sometimes with a mucous sheath. Saprobes or weak pathogens, many endophytic, mainly in wood and bark, cosmop.

Lit.: Dargan (*Beih. Nova Hedw.* **44**: 489, 1987), Læssøe, (*SA* **13**: 43, 1994), Petrini & Müller (*Mycol. Helv.* **1**: 501, 1986; keys 33 spp., culture), Petrini & Petrini (*Sydowia* **38**: 216, 1986; keys 22 spp.; culture, *Bibl. Mycol.* **150**: 193, 1993; arctic-alp. spp.), Rappaz (*Mycol. Helv.* **7**: 99, 1995; on hardwood, Eur., N. Am.), Rogers & Callan (*Mycotaxon* **29**: 113, 1987; keys 60 spp. Indonesia), Romero & Minter (*TBMS* **90**: 457, 1988; asci), Silveira & Rodrigues (*Acta Amazonia* Suppl. **15**: 7, 1985; Brazil), Whalley (*Sydowia* **38**: 369, 1986; ecology, distrib.), Whalley & Edwards (*in* Rayner *et al.* (Eds), *Evolutionary biology of the fungi*: 423, 1987; secondary metabolites).

Xylariales, Ascomycota. 3 fam., 92 gen. (+ 94 syn.), 795 spp. Stromata usually well-developed, mostly consisting only of fungal tissue; ascomata perithecial, rarely cleistothecial, ± globose, superficial or immersed in the stroma, usually black- and thick-walled, the ostiole usually papillate, periphysate. Interascal tissue well-developed, of narrow paraphyses; asci cylindrical, persistent, relatively thick-walled but without separable layers, with an often complex I+ apical ring, usually 8-spored; ascospores usually pigmented, sometimes transversely septate, with germ pores or slits, sometimes with a mucous sheath. Anamorphs varied, usually hyphomycetous. Saprobes and plant parasites, mainly on bark or wood, cosmop.

Barr (*Mycotaxon* **39**: 43, 1990) adopted a much broader view of the order and accepted 11 fams. Fams:
(1) **Amphisphaeriaceae** (syn. ? *Anthostomataceae*, *Cainiaceae*).
(2) **Clypeosphaeriaceae**.
(3) **Xylariaceae** (syn. *Hypoxylaceae*, *Phylaciaceae*, *Sphaeriaceae* s. str.).
Lit.: v. Arx (1981; *Sydowia* **34**: 13, 1982; fams.), v. Arx & Müller (1954), Dennis (1978), Müller & v. Arx (1962, 1973; keys gen.), Munk (1957), Wehmeyer (1975).

Xylariodiscus Henn. (1899) = Xylaria (Xylar.) fide Höhnel (*Sber. Akad.-wiss. Wien, math.-nat.* I, **119**: 928, 1910).

Xylariopsis F.L. Tai (1934) = Konradia (Clavicipit.) fide Boedijn (*Ann. Myc.* **33**: 229, 1935).

Xylasclerotes Stach & Pickh. (1957), Fossil fungi (*f. cat.*). 1 (Carboniferous), Germany. See Beneš (1969).

Xylastra A. Massal. (1855) = Opegrapha (Roccell.) fide Zahlbruckner (1923).

Xylissus Raf. (1808) nom. nud. See Merrill (*Index Rafinesq.*, 1949).

Xylobolus P. Karst. (1881), Stereaceae. 7, widespr. See Chamuris (*Mycol. Mem.* **14**, 1988; key N. Am. spp.).

Xylobotryum Pat. (1895), Ascomycota (inc. sed.). 2 or 3, trop. See Barr (*Prodromus to Class Loculascomycetes*, 1987), Eriksson & Hawksworth (*SA* **7**: 95, 1988), Ju & Rogers (*Crypt. Bot.* **4**: 346, 1994; ontogeny, posn), Rossman (*Mycotaxon* **4**: 179, 1976).

Xyloceras A.L. Sm. (1901) = Xylobotryum (Ascomycota) fide Müller & v. Arx (1973).

Xylochia B. Sutton (1983), Mitosporic fungi, 1.D1/2.1/10. 1, India.

Xylochoeras Fr. (1849) = Sclerotium (Mitosp. fungi) fide Saccardo (1899).

Xylochora Arx & E. Müll. (1954), Amphisphaeriaceae. 2, Eur.

Xylocladium P. Syd. ex Lindau (1900), Anamorphic Xylariaceae, 2.A1.8. Teleomorph *Hypoxylon*. 2, Indonesia, Jamaica.

Xylocoremium J.D. Rogers (1984), Anamorphic Xylariaceae, 2.A1.14. Teleomorph *Xylaria*. 1, USA.

Xylocoryneum Sacc. [not traced] (? Mitosp. fungi).

Xylocrea A. Møller (1901) ? = Sarcoxylon (Xylar.) fide Læssøe (*SA* **13**: 43, 1994).

Xylodon (Pers.) Gray (1821) nom. dub. (Hyphodermat.) fide Donk (1963).

xylogenous, living on wood.

Xyloglossum Pers. (1818) = Acrospermum (Acrosperm.) fide Bonorden (*Handb. Mycol.*, 1851).

Xyloglyphis Clem. (1909), Mitosporic fungi, 8.A1.15. 1, Eur. See Sherwood (*Mycotaxon* **5**: 87, 1977).

Xylogone Arx & T. Nilsson (1969), Ascomycota (inc. sed.). 1, Sweden. See Currah (*Mycotaxon* **21**: 1, 1985; posn).

Xylogramma Wallr. (1833), Leotiaceae. *c.* 15, N. temp. See Sherwood (*Mycotaxon* **5**: 1, 1977).

Xylographa (Fr.) Fr. (1836), Agyriaceae (±L). 6, Eur., N. Am.

Xylographomyces Cif. & Tomas. (1953) ≡ Xylographa (Agyr.).

Xylohypha (Fr.) E.W. Mason ex Deighton (1960), Mitosporic fungi, 1.A2.3. 5, widespr. See Hughes & Sugiyama (*N.Z. Jl Bot.* **10**: 447, 1972), Sekhon *et al.* (*Europ. Jl Epidemiol.* **8**: 387, 1992; cerebral phaeohyphomycosis by *X. bantiana*).

Xylohyphites Kalgutkar & Sigler (1995), Fossil fungi (Mitosp. fungi). 1 (Tertiary), India.

Xyloidium Czern. (1845) nom. dub. (? Myxomycetes) fide Streinz (1862).

xyloma (of *Dothideales*), a sclerotium-like body producing sporogenous structures inside (de Bary) (obsol.)

Xyloma Pers. (1794) = Rhytisma (Rhytismat.) fide Nannfeldt (1932).

Xyloma Raf. (1837) Fungi (inc. sed., L), No spp. included.

Xylometron Paulet (1812) = Pycnoporus (Coriol.) fide Donk (1974).

Xylomides Schimp. (1869) ≡ Xylomites (Fossil fungi).

Xylomites Unger (1841), Fossil fungi. 57 (Tertiary), Argentina, Eur.

Xylomyces Goos, R.D. Brooks & Lamore (1977), Mitosporic fungi, 1.C2.3. 1, USA.

Xylomyzon Pers. (1825) = Serpula (Coniophor.) fide Cooke (1953).

Xyloon, see *Xyloidium*.

Xylopezia Höhn. (1917), ? Dothideales (inc. sed.). 1, Eur., N. Am. See Sherwood (*Mycotaxon* **5**: 1, 1977), Sherwood-Pike & Boise (*Brittonia* **38**: 35, 1986).

Xylophagus Link (1809) ≡ Serpula (Coniophor.).

Xylophallus (Schltdl.) E. Fisch. (1933) = Mutinus (Phall.) fide Dring (*in* Ainsworth *et al.*, *The Fungi* **4B**: 458, 1973).

Xylopilus P. Karst. (1882) = Fomes (Coriol.) fide Donk (1960).

Xylopodium Mont. (1845) = Phellorinia (Phellorin.).

Xyloschistes Vain. ex Zahlbr. (1903), ? Ostropales (inc. sed., L). 1, Lapland. See Redinger (1938), Etayo (*Monogr. Inst. Pirenaico Ecol.* **5**: 43, 1990).

Xyloschistomyces Cif. & Tomas. (1953) ≡ Xyloschistes (Ostropales).

Xyloschizon Syd. (1922), Rhytismataceae. 2, N. Am. = Hysteroglonium (Hyster.) fide Clements & Shear (1931).

Xylosorium Zundel (1939) = Pericladium (Ustilagin.) fide Mundkur (*Mycol.* **36**: 287, 1944).

Xylosphaera Dumort. (1822) nom. rej. = Xylaria (Xylar.).

Xylosphaeria G.H. Otth (1869) ≡ Mycothyridium (Dothideales).

Xylosphaeria Cooke (1879) Ascomycota (inc. sed.). 1, UK. = Endoxyla (Bolin.), Kalmusia (Pleosp.) p.p., Ophiobolus (Phaeosphaer.) p.p. and Thyridium (Thyrid.) fide Hawksworth (*in litt.*).

Xylostroma Tode (1790) = Daedalea (Coriol.) fide Donk (1974; based on mycelium).

xylostromata, sheets of mycelium as in *Xylostroma*.

Xynophila Malloch & Cain (1971) = Aphanoascus (Onygen.) fide Cano & Guarro (1990).

Xyphasma Rebent. (1844) ≡ Hyphasma (nom. dub.). A misprint fide Carmichael *et al.* (1980).

Xystozukalia Theiss. (1916) ? = Scorias (Capnod.). See Luttrell (1973).

Yalodendron Capr. (1962) nom. nud. (? Fungi, inc. sed.); used for a yeast from fish (*Arch. Microbiol.* **42**: 407, 1962).

Yalomyces Nag Raj (1993), Mitosporic fungi, 4.D1.15/19. 1, India.

Yamadazyma Billon-Grand (1989), Endomycetaceae. 16, widespr. See Fiol & Claisse (*J. Gen. appl. Microbiol.* **37**: 309, 1991; cytochromes).

Yamamotoa Bat. (1960), Asterinaceae. 2 or 3, Brazil. Anamorphs *Clasterosporium, Mitteriella, Sarcinella*.

Yarrowia Van der Walt & Arx (1981), ? Saccharomycetaceae. 1, USA.

Yatesula Syd. & P. Syd. (1917), Chaetothyriaceae. 3 or 4, trop.

Ybotryomyces Rulamort (1986), Mitosporic fungi, ?1.A2/-.38. 1 (from humans), Spain. See Benoldi *et al.* (*Jl. Med. Vet. Mycol.* **29**: 9, 1991; cutaneous phaeohyphomycosis, as *Botryomyces*).

Yeasts. Unicellular, budding fungi. The yeasts are not a formal taxonomic unit but a growth form exhibited by a range of unrelated fungi, and one exhibited in some cases by primarily filamentous forms as a part of the life-cycle or under particular environmental conditions. Different genera are included depending of the authority and definition used; Kreger-van Rij (Ed.) (1984) accepts 54 gen. and 486 spp., and Barnett *et al.* (1990) 80 gen. and 597 spp.

Sporogenous yeasts have teleomorphs either in the *Ascomycota* (esp. *Saccharomycetales* and *Schizosaccharomycetales*) or in the *Basidiomycota* (*Sporidiales, Tremellales*); **asporogenous yeasts** are classified in the Mitosporic fungi; **apiculate -**, yeasts (e.g. *Saccharomycodes, Nadsonia, Hanseniaspora, Kloeckera*) having minute polar projections which are multiple scars (annellides); Streiblova *et al., J. Bact.* **88**: 1104, 1964); **baker's, brewer's** or **beer -**, *Saccharomyces cerevisiae*; **black -**, yeast-like states of *Aureobasidium, Cladosporium, Moniliella*, etc., and esp. anamorphs of *Herpotrichielleae* (q.v.) incl. *Exophiala, Ramichloridium* and *Rhinocladiella* (de Hoog & Hermanides-Nijhof, *Stud. Mycol.* **15**, 1977; de Hoog (Ed.), *Stud. Mycol.* **19**, 1979; *Ann. Rep. Res. Center Path. Fungi, Chiba* **7**: 50, 1993, surveys); **bottom -**, one settling out at the bottom of a fermented liquid (the wort), e.g. *S. uvarum*; **top -**, one accumulating at the surface of the fermented wort, e.g. *S. cerevisiae*; **Chinese -**, *Amylomyces rouxii* and other fungi (Ellis *et al., Mycol.* **68**: 131, 1976); **'flor' -**, one to which special qualities of wines (e.g. bouquet, taste) are due; **food -**, dry *Candida utilis* (q.v.) and other yeasts; **petite -**, a respiratory deficient mutant (Bulder, *Ant. v. Leeuwenhoek* **30**: 1, 1964); **scum -**, one (e.g. *Trichosporon cutaneum*) forming a surface scum or slime layer; **shadow (mirror) -**, *Bullera, Sporobolomyces* etc., producing ballistospores; **springer -**, the Institut Pasteur, Paris, strain of *S. cerevisiae*; **toddy -**, a mixture of yeasts which ferment the juice of the palmyra palm (*Borassus flabellifer*) (Ahmad *et al., Mycol.* **46**: 708, 1954); **wine -**, races of *S. cerevisiae*.

Lit. (covering anamorphs and teleomorphs): **Identification**: Lodder & Kreger-van Rij (*The yeasts*, 1952), Lodder (Ed.) (*The Yeasts*, edn 3, 1984; edn 2, 1970; edn 1, 1952), Kudrjanzev (*Sistematika drozheĭ*, 1954 [Russ., Germ., transl. *Die Systematik der Hefen*, 1960]), Barnett & Pankhurst (*A new key to the yeasts*, 1974), Barnett (*J. gen. Microbiol.* **99**: 183, 1977; nutritional tests), Barnett *et al.* (*A guide to identifying and classifying yeasts*, 1979; *Yeasts: characteristics and identification*, 1983; edn 2, 1990), v. Arx *et al.* (*Stud. Mycol.* **14**, 1977; genera), Payne *et al.* (*J. gen. Microbiol.* **128**: 1265, 1982; computer generated keys gen.), Moore (*Bot. mar.* **23**: 361, 1980; basidiomycetous yeasts), Belin (*Can. J. Microbiol.* **27**: 1235, 1981; spp. described since 1973), Yarrow & Nakase (*Ant. v. Leeuwenhoek* **41**: 81, 1975; DNA base composition), Nakase *et al.* (*Jap. J. Med. Myc.* **32**: 21, 1991; systematics of basidiomycetous yeasts).

General: Rose & Harrison (Eds) (*The yeasts*, **1**, *Biology of yeasts*, 1969, edn 2, 1987; **2**, *Physiology and biochemistry of yeasts*, edn 2, 1986; **3**, *Yeast technology*, 1970; **4**, *Yeast organelles*, edn 2, 1991), Rose *et al.* (*The Yeasts*, **6**, *Yeast genetics*, edn 2, 1995), Kurtzman (*Int. J. Syst. Bact.* **42**: 1, 1992; review), Prescott (Ed.) (*Methods in cell biology*, **11**, **12**, *Yeast cells*, 1975), Skinner *et al.* (Eds) (*Biology and activities of yeasts*, 1980), Fukazawa, Shimoda & Kagaya (*Handb. Appl. Mycol.: Humans, Animals & Insects* **2**: 425, 1991; serology and immunology of medically important yeasts), Fragner (*Česká Myk.* **39**: 234, 1985; spp. on humans), Herskowitz (*Nature* **357**:

190, 1992; regulation hyphal growth), Jong *et al.* (*Mycotaxon* **31**: 207, 1988; coding strain features), Seehaus et al. (*Current Genetics* **10**: 103, 1985; gene probes), Sherman *et al.* (Eds) (*Methods in yeast genetics*, 1987), Kirsop & Kertzman (Eds) (*Yeasts*, 1988; guide to sources), Odds (*J. med. vet. mycol.* **29**: 413, 1991; preservation in distilled water), Guthrir & Fink (Ed.) (*Guide to yeast genetics and molecular biology*, 1991).

See also *Blastomycota*, Mitosporic fungi, *Saccharomycetales, Schizosaccharomycetales, Ustomycetes.*

yellow rice, rice discoloured by *Penicillium islandicum* and rendered carcinogenic for rodents and possibly humans; see Mycotoxicoses.

yellows, of cabbage (*Fusarium oxysporum* f.sp. *conglutinans*).

Yenia Liou (1949) = Ustilago (Ustilagin.). See Yang & Lea (1978), Mordue (*Mycopathologia* **116**: 227, 1991).

Yoshinagaia Henn. (1904), ? Seuratiaceae. 1, Japan. See Eriksson & Hawksworth (*SA* **5**: 161, 1986; nomencl.).

Yoshinagamyces Hara (1912) = Japonia (Mitosp. fungi) fide Clements & Shear (1931).

Yoshinagella Höhn. (1913), Dothideales (inc. sed.). 4, Japan, Hawaii. ? = Gibberidea (Dothideales) fide v. Arx & Müller (1954).

Youngiomyces Y.J. Yao (1995), Endogonaceae. 4, N. Am., Australasia. See Yao *et al.* (1995).

Ypsilonia Lév. (1846), Anamorphic Sordariales, 7.G2.1. Teleomorph *Acanthotheciella*. 8, widespr. See Nag Raj (*CJB* **55**: 1599, 1977).

Ypsilonidium Donk (1972) = Thanatephorus (Ceratobasid.) fide Langer (1994).

Ypsilospora Cummins (1940), Raveneliaceae. 2 (on *Leguminosae*), W. Afr. See Ono & Hennen (*TBMS* **73**: 229, 1979). = Chaconia (Chacon.) fide Eboh (1985).

YPSS see Media.

Yuccamyces Gour, Dyko & B. Sutton (1979), Mitosporic fungi, 3.E1.3. 4, India, Cuba.

Yukonia R. Sprague (1962) = Buergenerula (Hyponectr.) fide Barr (*Mycol.* **68**: 611, 1976).

Zaghouania Pat. (1901), Pucciniaceae. 2 (on *Oleaceae*), Mediterr., India.

Zahlbruckner (Alexander; 1860-1938). Czech lichenologist, received PhD Vienna Univ. 1883, worked at the Naturhistorischen Museum, Vienna, becoming its director. Published important regional works on China, Easter Island, Juan Fernandez, C. Afr., Formosa, Hawaii (partly with Magnusson), Japan, Java, S. Am., Dalmatia, Samoa, etc.; was the first author to provide a comprehensive account of the lichenized fam. and gen. (*in* Engler & Prantl, *Naturlichen Pflanzenfamilien* **1**(1*): 49-249, 1903-07); and is owed a debt by all lichenologists for his *Catalogus Lichenum universalis* (10 vols, 1921-40) which compiled all lichen taxa, places they had been used, and details of synonymy, and remains the basic nomenclatural reference work on lichens. Planned the lichen sections of *Rabenhorst's Kryptogamenflora* (**9**, 1930-60) and published important exsiccata (e.g. *Lichenes rariores exsiccati*). Collections in **PAD** and **W**. See Lackovičova (*Dr. Alexander Zahlbruckner (1860-1938) osonosť a dielo*, 1988; biogr., bibliogr.), Redinger (*Annls Crypt. Exot.* **6**: 85, 1933; portr., bibliogr.), Grummann (1974: 444), Stafleu & Cowan (*TL-2* **7**: 500, 1988).

Zahlbrucknera Herre (1910) [non Rchb. (1832), *Saxifragaceae*] ≡ Zahlbrucknerella (Lichin.).

Zahlbrucknerella Herre (1912), Lichinaceae (L). 7, widespr. See Henssen (*Lichenologist* **9**: 17, 1977; key).

Zakatoshia B. Sutton (1973), Mitosporic fungi, 1.A1.15. 2 (on fungi), Canada, Austria. See Gams (*Windahlia* **16**: 59, 1986).

Zalerion R.T. Moore & Meyers (1962), Mitosporic fungi, 2.F2.1. 7 (mainly marine), N. Am., Eur. See Buczacki (*TBMS* **59**: 159, 1972).

Zamenhofia Clauzade & Cl. Roux (1985), Trichotheliaceae (L). 3, widespr. See Roux (*SA* **6**: 156, 1987). = Porina (Trichothel.) fide McCarthy (*Bibl. Lich.* **52**, 1993).

Zanchia Rick (1958), Tremellaceae. 3, Brazil. See Donk (*Taxon* **12**: 167, 1963).

Zanclospora S. Hughes & W.B. Kendr. (1965), Mitosporic fungi, 1.A1.15. 4, S. Hemisph.

Zasmidium Fr. (1849) nom. dub. (Mitosp. fungi). Used by de Hoog (*Stud. Mycol.* **15**, 1977), but see Hawksworth & Riedl (1977), cellar fungus.

zearalenone, a toxin of *Fusarium graminearum* (teleomorph *Gibberella zeae*); the cause of vulvovaginitis and infertility in cattle and pigs.

Zebrospora McKenzie (1991), Mitosporic fungi, 1.C2.10. 1, Australasia, Pacific Is.

Zelandiocoela Nag Raj (1993), Mitosporic fungi, 8.A1.15. 1, NZ.

Zelleromyces Singer & A.H. Sm. (1960), Elasmomycetaceae. 10, widespr. (excl. Afr., S. Am.).

Zelopelta B. Sutton & R.D. Gaur (1984), Mitosporic fungi, 5.G1.10. 1, India.

Zelosatchmopsis Nag Raj (1991), Mitosporic fungi, 7.A1.15. 1, Cuba.

Zendera Redhead & Malloch (1977) = Dipodascus (Dipodasc.) fide v. Arx (1977).

Zeora Fr. (1825) = Lecanora (Lecanor.).

zeorine (of apothecia), like those of *Zeora*.

Zephirea Velen. (1947) = Mycena (Tricholomat.) fide Horak (1968).

Zercosporidium Thor (1930) nom. dub. (? Fungi, inc. sed.).

Zernya Petr. (1947), Mitosporic fungi, 4.B1.?. 1, Brazil.

Zerovaemyces Gorovij (1977), Coprinaceae. 1, former USSR. See Loculomycetes.

Zeta Bat. & R. Garnier (1961) = Pseudomeliola (Hypocr.) fide v. Arx (*in litt.*).

Zetesimomyces Nag Raj (1988), Mitosporic fungi, 7.B1.15. 1, Cuba.

Zetiasplozna Nag Raj (1993), Mitosporic fungi, 4.C2.19. 4, widespr.

Zeuctomorpha Sivan., P.M. Kirk & Govindu (1984), Pleosporaceae. 1, India.

Zeugandromyces Thaxt. (1912), Laboulbeniaceae. 4, N. & S. Am., Asia.

zeugite, the organ in which fertilization is completed and the dikaryophase ends; e.g. an ascus or a basidium.

Zeus Minter & Diam. (1987), Rhytismataceae. 1, Greece.

Zignoëlla Sacc. (1878) = Chaetosphaeria (Lasiosphaer.) fide Müller (*SA* **6**: 156, 1987).

Zignoina Cooke (1885) Ascomycota (inc. sed.). 1, UK. See Eriksson & Hawksworth (*SA* **6**: 253, 1987).

Zilingia Petr. (1934), Mitosporic fungi, 8.A1.?. 1, Siberia.

Zimmermanniella Henn. (1902), Phyllachoraceae. 1, SE Asia. See *IMI Descr.* **1140**, Petrak (*Hedwigia* **68**, 1928).

Zinzipegasa Nag Raj (1993), Mitosporic fungi, 6.C1.15/19. 1, Argentina.

Zobelia Opiz (1855) = Choiromyces (Helvell.) fide Trappe (1975).

Zodiomyces Thaxt. (1891), Laboulbeniaceae. 2, widespr.

Zoellneria Velen. (1934), Hyaloscyphaceae. 4, Eur., Australia. See Beaton & Weste (*TBMS* **68**: 79, 1977), Dennis (*Kew Bull.* **1958**: 398, 1959).

Zografia Bogoyavl. (1922) = Coelomomyces (Coelomomycet.) fide Keilin (1927).

zonate, having concentric lines often forming alternating pale and darker zones near the margins; used of crustose lichen thalli, polypore surfaces, etc.

zonation (of cultures), regular concentric variation of texture, pigmentation or sporulation frequently associated with fluctuations (esp. diurnal) in light, temperature, or other factors; 'Liesegang' phenomenon. See Bisby (*Mycol.* **17**: 89, 1925), Hein (*Am. J. Bot.* **17**: 143, 1930), Kafi & Tarr (*TBMS* **46**: 549, 1964). **ecological -**, see Ecology.

zone lines, narrow, dark brown, or black, lines (pseudosclerotia) or plates (pseudosclerotial plates) in decayed wood (esp. hardwoods) generally caused by fungi (Lopez *et al.*, *TBMS* **64**: 465, 1975); see also spalted wood.

Zonilia Raf. (1815) nom. dub. (Fungi, inc. sed.). No spp. included.

Zonosporis Clem. (1931) ≡ Schwanniomyces (Saccharomycet.).

zoogametes, a motile gamete; planogametes.

zoogloea (of bacteria), a colony embedded in a slimy substance.

zoogonidium, (1) ? = zoospore (q.v.), (2) an aplanospore of a photobiont within the thallus of a lichen (obsol.).

Zoopagaceae Drechsler (1938), Zoopagales. 6 gen., 63 spp. Mycelium non-septate, produced outside the host; zygospores warty, borne on spirally twisted suspensors; predacious parasites of nematodes, amoebae and other small terrestrial animals; cosmop.
    Lit.: Drechsler (*Mycol.* **26**: 135, 1934, *et seq.* who proposed the family in **30**: 152, 1938), Duddington (1973; emend. & keys), Dayal (*Sydowia* **27**: 293, 1976; keys), Predacious fungi.

Zoopagales, Zygomycetes. 5 fam., 22 gen. (+ 9 syn.), 161 spp. Asexual reproduction by conidia or merosporangia, sexual reproduction by zygospores; cosmop. parasites of fungi (mycoparasites), nematodes, amoebae, and other small terrestrial animals. Fams:
    (1) **Cochlonemataceae**.
    (2) **Helicocephalidaceae**.
    (3) **Piptocephalidaceae**.
    (4) **Sigmoideomycetaceae**.
    (5) **Zoopagaceae**.
    Lit.: Duddington (in Ainsworth *et al.*, *The Fungi* **4B**: 231, 1973, *Biol. Rev.* **31**: 152, 1956), Drechsler (*Biol. Rev.* **16**: 265, 1941; review), Dyal (*Sydowia* **27**: 293, 1973; key spp. on amoebae & nematodes).

Zoopage Drechsler (1935), Zoopagaceae. 10, N. Am. See Drechsler (*Mycol.* **27**: 30, 1935, **28**: 363, 1936, **29**: 229, 1937, 1938, **39**: 379, 1947), Jones (*TBMS* **45**: 348, 1962), Dayal (1973; key).

Zoophagus Sommerst. (1911), Zoopagaceae. 3 (on *Algae*), Eur., N. Am. See Dick (1973, *MR* **94**: 347, 1990; key, emend. of genus, *et al.* (*Mycol.* **82**: 316, 460, 1990; status), Prowse (*TBMS* **37**: 134, 1954), Morikawa *et al.* (*MR* **97**: 421, 1993; status).

zoophilic (of dermatophytes, etc.), preferentially pathogenic for animals; cf. anthrophilic.

Zoophthora Batko (1964), Entomophthoraceae. 20, widespr. See Remaudière & Hennebert (*Mycotaxon* **11**: 269, 1980), Ben-Ze'ev & Kenneth (*Mycotaxon* **14**: 456, 1982), Glare *et al.* (*Aust. Jl Bot.* **35**: 49, 1987), Humber (1989), Keller (1991; key).

zoosporangium (**zoosporange**), a sporangium producing zoospores.

zoospore, a motile sporangiospore, i.e. one having flagella; swarm spore; swarmer; simblospore; planospore; planont; cf. swarm cell; see Waterhouse (*TBMS* **45**: 1, 1962), Fuller (*Mycol.* **69**: 1, 1977), Lange & Olson (*Dansk bot. Arkiv.* **33**, 1979; uniflagellate zoospores).

Zopf (Wilhelm; 1846-1909). Professor at Halle Univ. Made major contributions to knowledge of an exceptionally wide range of fungi and lichens, including accounts of *Chaetomium*, lichenicolous fungi, and lichen chemistry. Main works: *Die pilze in morphologischer, physiologischer, biologischer und systematischer Beziehung* (1890), *Untersuchungen über die durch parasitische Pilze hervorgerufenen Krankheiten der Flechten* (1897-98), *Die Flechtenstoffe* (1907). See Huneck *et al.* (*Willdenowia* **7**: 31, 1973; application of Zopf's chemical names), Tobler (*Ber. dtsch. bot. Ges.* **27**: (58), 1910; portr.), Grummann (1974: 55), Stafleu & Cowan (*TL-2* **7**: 553, 1988).

Zopfia Rabenh. (1874), Zopfiaceae. 3, Eur., N. Am., N. Afr. See Hawksworth (*CJB* **57**: 91, 1979).

Zopfiaceae G. Arnaud ex D. Hawksw. (1992), Dothideales. 8 gen. (+ 2 syn.), 10 spp. Ascomata usually superficial, ± globose, black, usually non-ostiolate; peridium usually thick-walled, cephalothecoid; interascal tissue of trabecular pseudoparaphyses; asci globose to saccate, evanescent; ascospores ellipsoidal, 1-septate, dark brown, often ornamented, without a sheath. Anamorphs unknown. Usually saprobic, especially on rhizomes and roots.
Lit.: Hawksworth (*CJB* **57**: 91, 1979, *SA* **6**: 153, 1987).

Zopfiella G. Winter (1884), Lasiosphaeriaceae. 20, widespr. Guarro *et al.* (*SA* **10**: 79, 1991; key), See Malloch & Cain (*CJB* **49**: 869, 1971).

Zopfinula Kirschst. (1939) = Keissleriella (Lophiostom.) fide Bose (1961).

Zopfiofoveola D. Hawksw. (1979), Zopfiaceae. 1, Sweden.

Zopheromyces B. Sutton & Hodges (1977), Mitosporic fungi, 1.B2.6. 1, Brazil.

Zosterodiscus Hertel (1984) = Lecidea (Lecid.) fide Hertel (*Mitt. bot. StSamml. München* **23**: 321, 1987).

Zschackea M. Choisy & Werner (1932) = Verrucaria (Verrucar.).

Zukalia Sacc. (1891) = Chaetothyrium (Chaetothyr.) fide v. Arx & Müller (1975).

Zukalina Kuntze (1891) ? = Thecotheus (Ascobol.) fide Korf (1973).

Zukaliopsis Henn. (1904) = Molleriella (Elsin.) fide v. Arx (1963).

Zundeliomyces Vánky (1987), Ustilaginaceae. 1, Kazakhstan.

Zundelula Thirum. & Naras. (1952) = Dermatosorus (Ustilagin.) fide Langdon (1977).

Zunura Nag Raj (1993), Mitosporic fungi, 6.A1.19. 1, India.

Zwackhia Körb. (1855) = Opegrapha (Roccell.).

Zwackhiomyces Grube & Hafellner (1990), Dothideales (inc. sed.). 9 (on lichens), widespr. See Grube & Hafellner (*Nova Hedw.* **51**: 283, 1990; monogr.).

Zychaea Benny & R.K. Benj. (1975), Thamnidiaceae. 1, Mexico.

Zygaenobia Weiser (1951), Entomophthorales (inc. sed.). 1, former Czechoslovakia.

zygangium, gametangium of a zygomycete.

Zygnemomyces K. Miura (1973), Meristacraceae. 1, Japan, Australia. See Miura (*Rep. Tottori Mycol. Inst.* **10**: 520, 1973), McCulloch (*TBMS* **68**: 173, 1977), Tucker (1981; key).

Zygoascus M.T. Sm. (1986), ? Endomycetaceae. 1, Eur., Asia. S. Afr. See v. Arx & v. de Walt (*Stud. Mycol.* **30**: 167, 1987; posn). Anamorph *Candida*.

Zygochytrium Sorokīn (1874), ? Chytridiomycetes (inc. sed.). 1, Eur.

Zygodesmella Gonz. Frag. (1917) ? nom. dub. fide Donk (*Taxon* **11**: 103, 1962).

Zygodesmus Corda (1837) nom. dub. fide Rogers (*Mycol.* **40**: 633, 1948).

Zygofabospora Kudrjanzev (1960), Saccharomycetaceae. 11, widespr. See Naumov (*Mikrobiologiya* **57**: 114, 1988; key).

Zygogloea P. Roberts (1994), Platygloeales (inc. sed.). 1, UK.

Zygohansenula Lodder (1932) = Pichia (Saccharomycet.).

Zygolipomyces Krassiln., Babeva & Meavahd (1967) = Lipomyces (Lipomycet.) fide Lodder (1970).

Zygomycetes, Zygomycota. 7 ord., 30 fam., 125 gen. (+ 83 syn.), 867 spp.; saprobes or parasites (esp. of arthropods). Ords:
(1) **Dimargaritales**.
(2) **Endogonales**.
(3) **Entomophthorales**.
(4) **Glomales**.
(5) **Kickxellales**.
(6) **Mucorales**.
(7) **Zoopagales**.
The *Asellariales* and *Harpellales*, retained in the *Trichomycetes*, may also belong here.
Lit.: Benjamin (*in* Kendrick (Ed.), *The whole fungus*: 579, 1979), Benny (*in* Parker, 1982, **1**: 184), O'Donnell (*Zygomycetes in culture*, 1979), Jeffries (*Bot. J. Linn. Soc.* **91**: 135, 1985; mycoparasitism), Schipper (*in* Rayner *et al.* (Ed.), *Evolutionary biology of the fungi*: 261, 1987), Morton & Benny (*Mycotaxon* **37**: 471, 1990; *Endogonales*, *Glomales*), and see under Orders.

zygomycosis, a mycosis caused by a member of the Zygomycetes. Cf. Mucormycosis, phycomycosis.

Zygomycota (Zygomycotina). Fungi. 2 class., 11 ord., 37 fam., 173 gen. (+ 105 syn.), 1056 spp. Classes:
(1) **Zygomycetes**.
(2) **Trichomycetes**.
Lit.: Walker (*Syst. Appl. Microbiol.* **5**: 448, 1984; 5S ribosomal RNA), and see under Classes and Orders.

Zygomycotina, see *Zygomycota*.

Zygophiala E.W. Mason (1945), Mitosporic fungi, 1.B2.10. 1, Jamaica.

zygophore (of *Mucorales*), a special hyphal branch producing copulation branches.

Zygopichia (Klöcker) Kudrjanzev (1960) = Pichia (Saccharomycet.) fide Batra (1978).

Zygopichia E.K. Novák & Zsolt (1961) = Pichia (Saccharomycet.) fide v. Arx (1981).

Zygopleurage Boedijn (1962), Lasiosphaeriaceae. 1 (coprophilous), Java.

Zygopolaris S.T. Moss, Lichtw. & Manier (1975), Legeriomycetaceae. 2 (in *Ephemeroptera*), USA. See Moss & Lichtwardt (*CJB* **55**: 3099, 1977; ultrastr.), Lichtwardt (1986; key).

Zygorenospora Krassiln. (1954) = Zygofabospora (Saccharomycet.) fide Naumov (*Mikol. i Fitopat.* **21**: 131, 1987).

Zygorhizidium Löwenthal (1904), Chytridiaceae. *c*. 11, Eur, N. Am. See Sparrow (1960: 548; key).

Zygorhynchus Vuill. (1903), Mucoraceae. 6 (esp. in soil), widespr. See Hesseltine *et al.* (*Mycol.* **51**: 173, 1959; key), Schipper *et al.* (*Persoonia* **8**: 321, 1975; zygospore ornamentation), O'Donnell *et al.* (*CJB* **56**: 1061, 1978; ontogeny), Heath & Rethoret (*Eur. J. Cell*

Biol. **28**: 180, 1982; mitosis, ultrastr.), Schipper (*Persoonia* **13**: 97, 1986; key), Taiwo *et al.* (*Microbios.* **51**: 23, 1987, **52**: 183, 1987; physiol.), Edelmann & Klomparens (*Mycol.* **87**: 304, 1995; ultrastr. zygospore dev.)

Zygorrhynchus, see *Zygorhynchus*.

Zygosaccharis Clem. & Shear (1931) ≡ Zygosaccharomyces (Saccharomycet.).

Zygosaccharomyces B.T.P. Barker (1901), Saccharomycetaceae. 8 (osmotolerant), widespr. See v. Arx *et al.* (*Stud. Mycol.* **14**, 1977), James *et al.* (*Yeast* **10**: 871, 1994; molec. syst.), Kreger-van Rij (1984; key), Kurtzman (*Yeasts* **6**: 213, 1990; 9 spp. by DNA complementarity).

Zygosaccharomycodes Nishiw. (1929) = Saccharomyces (Saccharomycet.) fide Batra (1978).

Zygospermella Cain (1935), Lasiosphaeriaceae. 2, N. Am., Eur.

Zygospermum Cain (1934) [non Thwaites ex Baill. (1858), *Euphorbiaceae*] ≡ Zygospermella (Lasiosphaer.).

zygospore, the resting spore resulting from the conjugation of isogametes or (in *Zygomycetes*), from the fusion of like gametangia (Fig. 43).

Zygosporites Will. (1880), Fossil fungi (Zygomycetes). 1 (Carboniferous), UK. See Pia (*in* Hirnier, *Handb. Paläobot.* **1**, 1927).

Zygosporium Mont. (1842), Mitosporic fungi, 1.A2.1. 8, widespr. See Hughes (*Mycol. Pap.* **44**, 1951), Ichinoe (*Bull. natn Inst. Hyg. Sci.* **89**: 135, 1971).

zygote, the result of fusion of two gametes; a cell in which two nuclei of opposite sex have undergone fusion (Buller, 1941).

Zygothrix Reinsch ex Rabenh. (1866) nom. dub. (based on sterile hyphae).

Zygowillia (Klöcker) Kudrjanzev (1960) = Pichia (Endomycet.) fide Lodder (1970).

Zygowilliopsis Kudrjanzev (1960) = Williopsis (Saccharomycopsid.) fide v. Arx *et al.* (1977), but see Yamada et al. (*Biosc. Biotech. Biochem.* **58**: 1236, 1994; status).

Zygozyma Van der Walt & Arx (1987), Lipomycetaceae. 1, Natal. See van der Walt *et al.* (*Syst. Appl. Microbiol.* **12**: 288, 1989; key 3 spp.; *Ant. v. Leeuwenhoek* **59**: 77, 1991; ultrastr.).

Zymodebaryomyces Novák & Zsolt (1961) = Torulaspora (Saccharomycet.) fide v.d. Walt & Johannsen (1975). = Debaryomyces (Saccharomycet.) fide Batra (1978).

zymogenous, ferment producing. See autochthonous.

zymogram, (1) the pattern of bands obtained by electrophoretic enzyme analysis (see electrophoresis); (2) a tabulation of carbohydrate fermentations test results.

Zymonema Beurm. & Gougerot (1909) nom. dub. (? Mitosp. fungi). See Batra (*in* Subramanian (Ed.), *Taxonomy of fungi* **1**: 187, 1978), v. Oorschot (*Stud. Mycol.* **20**, 1980).

Zymopichia E.K. Novák & Zsolt (1961) = Pichia (Endomycet.) fide Lodder (1970).

Zymoxenogloea D.J. McLaughlin & Doublés (1992), Mitosporic Auriculariaceae. ?.?.?. Teleomorph *Kriegeria*. 1, USA.

zymurgy, the practice of fermentation as in brewing and wine-making.

Zythia Fr. (1849), Mitosporic fungi, 4.A/B1.?. 25, widespr.

Zythiaceae, see *Nectrioidaceae*.

Zythiostroma Höhn. ex Falck (1923), Anamorphic Hypocreaceae, 8.A1.15. Teleomorphs *Nectria*, *Ophionectria*. 2 or 3, Eur., Java.

Zyxiphora B. Sutton (1981), Mitosporic fungi, 2.A2.1. 1, India.

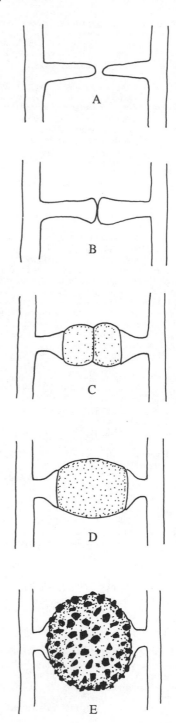

Fig. 43. Stages in zygospore formation in *Mucor mucedo*. A,B, young progametangia; C, gametangia and suspensors; D, young zygospore; E, mature zygospore. Not to scale.

Keys to the families of fungal phyla

The following series of keys cover all families in the eleven fungal phyla in the three kingdoms accepted in this edition of the *Dictionary*. A key to the phyla is followed by keys to the families within each phylum (in some cases split into classes for convenience). The phyla are placed in alphabetical sequence for ease of reference.

This is the first time such a comprehensive series of keys to families, including also the lichenized groups, has been published since Clements & Shear (1931); the keys included in editions 1-5 of this *Dictionary*, of Bessey (1950), or of Ainsworth *et al.* (1973), did not integrate the numerous lichen-forming families. Users should be aware that these keys have not been as extensively tested as we would wish, and that all genera are not assigned to families. Families not referred to an order are indiacted by FIS (familiae incertae sedis).

The key to *Ascomycota* has been mainly prepared by P.F. Cannon, with input from several specialists (see Acknowledgements), that for *Basidiomycota* mainly by D.N. Pegler, and that for chromistan phyla by M.W. Dick. The other keys have been prepared by DLH and PMK based on a variety of sources and with modifications suggested by the group specialists referred to in the Introduction. Mitosporic fungi cannot be treated in such keys as they cannot be assigned to families in the accepted phyla unless they are anamorphs.

The pages on which the keys for particular phyla and classes start are as follows:

The genera placed in each order and family, as well as those not referred to families, are indicated in the final section of this *Dictionary*.

KEY TO FUNGAL PHYLA

1	Phagotrophic phase present	**(Protozoa) 2**
	Phagotrophic phase absent	**5**
2 (1)	Phagotrophic phase extracellular	**3**
	Phagotrophic phase intracellular	**Plasmodiophoromycota**
3 (2)	Assimilative phase a saprobic plasmodium	**Myxomycota**
	Assimilative phase free-living myxamoebae which unite as a pseudoplasmodium before reproduction	**4**
4 (3)	Myxamoebae with lobose pseudopodia and nuclei with a centrally positioned nucleolus	**Acrasiomycota**
	Myxamoebae with mainly filose pseudopodia and nuclei with two or more peripheral nucleoli; aggregating by forming convergent streams in developing pseudoplasmodia	**Dictyosteliomycota**
5 (1)	Motile phase with flagellae which have mastigonemes; cell walls generally cellulosic	**(Chromista) 6**
	Motile phase when present with flagellae lacking mastigonemes; cell walls chitinose	**(Fungi) 8**
6 (5)	Trophic phase an ectoplasmic network with cell gliding on or within it	**Labyrinthulomycota**
	Trophic phase not an ectoplasmic network and gliding cells absent	**7**

7 (6)   Zoospores with one anterior flagellum   **Hyphochytriomycota**
              Zoospores with two flagellae   **Oomycota**

8 (5)   Motile zoospores present   **Chytridiomycota**
              Motile zoospores absent   **9**

9 (8)   Mitospores endogenous, formed in sporangia (or asexual propagules derived from
              sporangia), zygospores formed by hyphal conjugation   **Zygomycota**
              Mitospores (conidia) exogenous   **10**

10 (9)   Meiospores endogenous, formed in asci   **Ascomycota**
              Meiospores exogenous, borne on basidia (or equivalent structures)   **Basidiomycota**

ACRASIOMYCOTA

1   Sporocarp comprising a sporogenous zone and a stalk   **2**
              Sporocarp lacking any evidence of a stalks, consisting entirely of simple or branched
              columns forming encysted myxamoebae   **Copromyxaceae**

2 (1)   Sporogenous zone comprising a more or less well defined sorus   **Guttulinopsidaceae**
              Sporogenous zone comprising simple or branched chains   **Acrasiaceae**

ASCOMYCOTA

1   Mycelium usually absent, vegetative cells proliferating by budding or fission;
              ascomata absent; asci formed singly or in chains from morphologically
              undifferentiated cells   **2**
              Mycelium well-developed; ascomata usually present; asci usually formed from
              differentiated cells   **10**

2 (1)   Vegetative cells dividing exclusively by fission (thallic proliferation) into ± equal
              daughter cells   **3**
              Vegetative cells dividing at least primarily by budding (blastic proliferation), from any
              point on the cell surface   **(Saccharomycetales) 4**

3 (2)   Hyphal growth minimal, the vegetative cells separating almost immediately after
              formation; asci variable in shape, the walls breaking down irregularly; ascospores ±
              spherical,   blueing   in   iodine,   without
              sheaths   **(Schizosaccharomycetales) Schizosaccharomycetaceae**
              Hyphal growth sometimes extensive; asci usually persistent, one- or multispored,
              usually elongated, releasing ascospores from the apex; ascospores usually
              ornamented or with gelatinous sheaths, not blueing in iodine [If in lung tissue of
              mammals, see **(Pneumocystidales) Pneumocystidaceae**]   **(Saccharomycetales) Dipodascaceae**

4 (2)   Asci formed in chains from a cluster of cells at the tip of an erect seta-like diploid
              hypha   **Cephaloascaceae**
              Asci not in well-developed chains, formed from diploid daughter cells which are
              morphologically similar to vegetative cells   **5**

5 (4)   Asci elongate, usually persistent; ascospores narrowly fusiform to filiform, often
              curved   **Metschnikowiaceae**
              Asci and ascospores not significantly elongated   **6**

6 (5)   Colonies producing gel which stains blue in iodine; asci often multispored, evanescent;
              ascospores usually cylindrical or allantoid, smooth   **Lipomycetaceae**
              If producing gel, not staining in iodine; asci usually 1- to 4-spored, sometimes
              multispored; ascospores spherical or flattened, often ornamented   **7**

7 (6)   Vegetative cells lemon-shaped, proliferating at both ends; coenzyme system
              Q-6   **Saccharomycodaceae**
              Vegetative cells usually ellipsoidal, proliferating from any point on the cell surface;
              coenzyme system Q-6 or Q-8   **8**

8 (7) Asci evanescent or persistent, 1- to 4-spored; coenzyme system usually Q-6; hyphae
   rarely produced   **Saccharomycetaceae**
   Asci usually evanescent, sometimes multispored; coenzyme system usually Q-8;
   sometimes with significant hyphal development   **9**

9 (8) Ascospores with asymmetrical flanges or sheaths, appearing hat-shaped, usually
   hyaline   **Endomycetaceae**
   Ascospores with more or less equatorial flanges, often pale brown **Saccharomycopsidaceae**

10 (1) Vegetative tissue restricted to a small thallus with a haustorial foot cell and often with
   complex appendages; ascomata perithecia; asci evanescent; ascospores with mucous
   sheaths or appendages   **11**
   Vegetative tissue hyphal in construction, usually extensive and indeterminate **16**

11 (10) Thallus reduced but the cells not formed in a well-defined sequence; haustorial cells
   hyaline; ascomata thick-walled; asci numerous, 8-spored; ascospores aseptate;
   ectoparasites of *Rhodophyta*   **(Spathulosporales) 12**
   Thallus cells formed under tight developmental control; haustorial cells black;
   ascomata thin-walled; asci few, basally attached, usually 4-spored; ascospores
   hyaline, elongate, 1-septate; almost all ectoparasites of arthropods **(Laboulbeniales) 13**

12 (11) Thallus composed of hyphal cells; peridium three-layered; ascospores without mucous
   appendages; conidiophores verticillately branched   **Hispidicarpomycetaceae**
   Thallus mostly non-hyphal; peridium two-layered; ascospores with mucous
   appendages; conidiophores simple   **Spathulosporaceae**

13 (11) Dioecious; ascospore germinating to form a usually four-celled primary thallus;
   ascomata formed from secondary thalli; asci eight-spored; ascospores with median
   septa; on cockroaches   **Herpomycetaceae**
   Usually monoecious; ascomata formed directly from the primary thallus; asci almost
   always four-spored; ascospores with a submedian septum; not on cockroaches **14**

14 (13) Ascomata formed directly from successive intercalary cells of the primary thallus;
   outer wall layer of ascoma composed of many short usually ± equal cells **Ceratomycetaceae**
   Ascomata formed from successive cells of a lateral appendage of the primary thallus **15**

15 (14) Lateral appendage extending beyond base of the ascoma; outer wall cells of ascoma
   usually numerous, short, ± equal   **Euceratomycetaceae**
   Lateral appendage not extending beyond ascoma; outer wall cells of ascomata usually
   large and unequal   **Laboulbeniaceae**

16 (10) Ascomata absent or not clearly developed; parasitic on vascular plants, usually causing
   distinct hypertrophy and colour changes in the host tissue;   **17**
   Ascomata clearly developed, or if not, saprobic   **19**

17 (16) Ascogenous cells usually formed directly from germinating resting spores, usually
   containing a peripheral layer of multinucleate protoplasm, the nuclei undergoing
   meiosis to form four 'endospores' which are dispersed simultaneously in a single
   mass   **(Protomycetales) Protomycetaceae**
   Asci forming a definite superficial palisade; endospores not dispersed in a single mass **18**

18 (17) Asci small, usually cylindrical with truncate apices, paraphyses absent; ascospores
   small, hyaline   **(Taphrinales) Taphrinaceae**
   Asci large, clavate, developing within a well-developed layer of paraphyses;
   ascospores large, dark brown, striate   **(Mediolariales) Mediolariaceae**

19 (16) Ascomata absent; asci ± spherical, ? produced from croziers which are formed after
   hyphal anastomosis; ascospores small, hyaline, smooth, globose to ellipsoidal;
   anamorph of thick-walled resting spores, produced singly **Eurotiales: Eremascaceae**
   Ascomata present   **20**

20 (19) Ascomata cleistothecial , without a definite centrum; asci evanescent at an early stage;
   ascospores almost always aseptate   **21**
   Ascomata usually perithecial or apothecial (rarely poorly defined; see 130); if
   cleistothecial then with a definite centrum at least at early stages of development; asci
   persistent or evanescent   **37**

21 (20)   Ascomata large, thick-walled, subterranean or epigean; at least usually mycorrhizal   **22**
               Ascomata usually small and thin-walled; not mycorrhizal   **23**

22 (21)   Ascomata brightly coloured, hollow, epigean, with gelatinous walls; paraphyses
                   absent; asci single-spored; ascospores large, ellipsoidal, smooth, orange   **(Pezizales) Glaziellaceae**
               Ascomata variously coloured, subterranean, thick-walled; paraphyses present as
                   copious, apparently unordered sterile hyphae; asci two- to eight-spored; ascospores
                   globose,   strongly   pigmented   (usually   black)   and
                   ornamented   **(Elaphomycetales) Elaphomycetaceae**

23 (21)   Ascomata usually brightly coloured, composed of loosely intertwined hyphae, often
                   with thick-walled branched and/or ornamented hyphal appendages; asci never in
                   chains; ascospores very small, brightly coloured, usually spherical to lenticular, often
                   ornamented; anamorph usuallly with thallic proliferation; often keratinophilic   **(Onygenales) 24**
               Ascomata brightly coloured or dark (very rarely absent; see 26), the peridium usually
                   well-developed, usually composed of pseudoparenchymatous tissue, rarely
                   ornamented; asci sometimes formed in chains; ascospores varied; anamorph usually
                   with blastic proliferation; never keratinophilic   **27**

24 (23)   Ascospores fusiform to ellipsoidal, sometimes striate; cellulolytic   **Myxotrichaceae**
               Ascospores spherical or flattened (oblate or lenticular); smooth or variously
                   ornamented   **25**

25 (24)   Ascospores usually flattened, pitted or reticulate; always keratinophilic   **Onygenaceae**
               Ascospores smooth-walled; nutrition variable   **26**

26 (25)   Ascospores   oblate   to   discoid,   without   equatorial   thickenings;   always
                   keratinophilic   **Arthrodermataceae**
               Ascospores usually flattened, with equatorial thickenings; keratinophilic or cellulolytic **Gymnoascaceae**

27 (23)   Ascomata dark brown   **28**
               Ascomata hyaline or brightly coloured, rarely pale brown   **35**

28 (27)   Ascomatal walls not cellular   **29**
               Ascomatal walls clearly composed of cells, often fragmenting into predefined plates   **30**

29 (28)   Ascomata non-cellular, consisting merely of amorphous melanin-containing deposits
                   surrounding the developing asci; asci developing from croziers, small, evanescent;
                   anamorph hyphomycetous; associated with hydrocarbons   **Amorphothecaceae [FIS]**
               Ascomata cyst-like, the walls non-cellular; ascus walls deliquesce at a very early stage
                   (if they form at all); ascospores becoming clumped into variably sized 'spore balls'
                   acting as separate dispersal units; associated with bee larvae   **Eurotiales: Ascosphaeraceae**

30 (28)   Ascoma initials clusters of pseudoparenchymatous cells, an internal locule developed
                   in which the asci are formed   **(Dothideales) 31**
               Ascoma initials hyphal, the cavity formed from an early stage   **33**

31 (30)   Ascospores small, hyaline to pale brown, aseptate, smooth; anamorphs hyphomycetous
                   with thallic development   **Eremomycetaceae**
               Ascospores septate, often ornamented; anamorphs unknown   **32**

32 (31)   Ascospores usually very large; usually associated with roots   **Zopfiaceae**
               Ascospores usually small; isolated from soil or mycetomas   **Testudinaceae**

33 (30)   Ascospores strongly curved   **Argynnaceae [FIS]**
               Ascospores not curved   **34**

34 (33)   Ascomata with walls breaking apart into cephalothecoid plates, surrounded by a web of
                   pale yellow hyphae   **Cephalothecaceae [FIS]**
               Ascomata with cephalothecoid or pseudoparenchymatous walls, without surrounding
                   hyphae   **Pseudeurotiaceae [FIS]**

35 (27)   Ascomata brightly coloured, composed of thick-walled pseudoparenchymatous tissue, rarely of interwoven hyphae or absent; asci globose to ellipsoidal, often formed in chains; ascospores oblate, usually yellowish, usually with well-developed equatorial flanges, sometimes ornamented; anamorph producing conidia in blastic chains, usually on complex branched conidiophores; saprobic   **Trichocomaceae**

Ascomata composed of intertwined hyphae; ascospores ellipsoid, smooth, hyaline   **36**

36 (35)   Ascomata hyphae not flattened; ascomata cavity filled with radiating clusters of sterile hyphae, between which the asci form; fungicolous; anamorph unknown **(Pezizales) Eoterfeziaceae**

Ascomata composed of flattened hyaline to pale brown hyphae; asci evanescent at a very early stage; anamorph composed of simple basipetal thallic chains of thick-walled conidia; often osmophilic   **(Eurotiales) Monascaceae**

37 (20)   Ascomata vertically elongated, usually flattened or cup-shaped at least in the apical part, often stalked; asci deliquescing, releasing ascospores in a powdery mass (mazaedium) which is usually wind-dispersed; ascospores often ornamented   **38**

Ascomata variously shaped; if asci deliquescent ascospores not wind-dispersed   **47**

38 (37)   Upper part of ascomatal wall composed of anastomosing setae; the mass of ascospores interspersed with sterile hyphae; asci ± globose, formed in chains; ascospores striate; on dung   **Microascales: Chadefaudiellaceae**

Ascomatal wall composed of pseudoparenchymatous tissue; the mass of released ascospores not interspersed with hyphae; sometimes lichenized   **39**

39 (38)   Ascomata elongated, often opening by splits, often with a perithecial lower chamber containing asci and an upper one filled with released ascospores   **40**

Ascomata basically apothecial in form   **41**

40 (39)   Asci thin-walled, long-stalked, bitunicate but the outer layer peeling away at a very early stage; ascomata superficial on a basal stroma; biotrophic, especially on *Podocarpaceae*   **(Coryneliales) Coryneliaceae**

Asci thick-walled, short-stalked, clearly bitunicate, the outer layer separating at discharge; ascomata immersed, not on a basal stroma; either saprobic on bark or wood or lichenized   **(Pyrenulales) Pyrenulaceae**

41 (39)   Ascomata within a lichenized thallus, with a wide opening surrounded by separating corona-like thalline flaps   **Ostropales: Thelotremataceae**

Ascomata flattened or cup-shaped but often stalked; asci usually cylindrical very thin-walled, not bitunicate; fungicolous or lichenized   **(Caliciales) 42**

42 (41)   Lichenized, with a foliose thallus; ascomata sessile, marginal; mazaedium brown; ascospores spherical, aseptate, smooth-walled   **Calycidiaceae**

Lichenized or not, if so then thallus crustose or fruticose (rarely foliose; mazaedium black); ascospores spherical or ellipsoidal, often septate, usually ornamented   **43**

43 (42)   Lichenized, thallus foliose to fruticose; ascomata sessile or immersed, becoming exposed by rupture of the covering layer; ascospores spherical, aseptate, smooth and hyaline, but becoming covered with carbonized material originating either from the ascus epiplasm or formed within the mazaedium   **Sphaerophoraceae**

Lichenized or not, if so then thallus crustose; ascomata usually stalked; ascospores usually ornamented, the ornamentation originating from the epi- or perispore   **44**

44 (43)   Ascospores almost always spherical and aseptate, hyaline to mid brown, smooth, minutely verrucose or ornamented with irregular fissures   **Coniocybaceae**

Ascospores almost always ellipsoidal and septate, dark brown and conspicuously ornamented   **45**

45 (44)   Asci ellipsoidal, formed in chains without croziers; ascospores 1- to 7-septate, with an ornamentation of helical ridges; lichenicolous or saprobic   **Microcaliciaceae**

Asci cylindrical, formed from croziers; ascospores 0- to 1-septate; lichenized or lichenicolous   **46**

46 (45)   Ascospore ornamentation formed early in development beneath the plasmalemma; lichenicolous   **Sphinctrinaceae**

Ascospore ornamentation formed late in development, by rupturing of the outer wall layers; usually lichenized   **Caliciaceae**

47 (37) Asci developed within a true hymenium, with paraphyses (if present) apically free from a basal centrum; with only a single wall layer clearly visible by LM, though sometimes thick-walled, especially at the apex; sometimes with complex apical structures; ascus walls, apical structures and/or hymenial gel sometimes staining blue in iodine   **48**

Asci mostly developing within locules in an ascostroma; interascal tissue (if present) at least originally attached at the apex as well as the base; asci with two distinct wall layers visible by LM, the inner layer often extending as part of an ascospore release mechanism; apical structures usually restricted to an internal beak (ocular chamber); iodine reactions usually negative   **198**

48 (47) Asci with a conspicuously thickened apical cap, sometimes with an internal beak (ocular chamber), breaking at the apex, the apical cap partly pushed out to form a larger beak-like structure; ascus wall and/or hymenial gel usually staining blue in iodine; almost always apothecial; almost all lichenized   **49**

Asci variable in form, sometimes somewhat thickened at the apex, if so usually with complex apical structures; ocular chamber usually absent; ascomata varied in form; rarely lichenized   **105**

49 (48) Ascomata initially formed within the gelatinous thallus as a filamentous weft of hyphae, bursting through the covering layers, sometimes converted from conidiomata; ascospores hyaline, aseptate; photobiont usually cyanobacteria   **Lichinales: Lichinaceae**

Ascomata formed from surface layers of the thallus, sometimes bursting through the covering layer but never converted from conidiomata; ascospores hyaline or dark brown, often septate; photobiont an alga or cyanobacterium   **50**

50 (49) Ascomata with an initial covering layer, which usually splits open just before maturity; asci usually with an iodine positive pendant ring within the apical cap, discharge generally involving separation of wall layers; ascospores often elongated, multiseptate; usually foliose lichens with at least some cyanobacteria as photobionts **(Peltigerales) 51**

Ascomata without a covering layer, the developing asci exposed from an early stage; ascus ring if iodine positive not pendant; discharge generally involving an extension or eversion but no separating of wall layers   **54**

51 (50) Ascomata formed on the lower surface of the thallus, which is then reflexed; cortex well-developed on both sides; asci without a gelatinous coat or iodine positive apical ring   **Nephromataceae**

Ascomata formed on upper surface of thallus; cortex usually on upper surface surface only; asci with a gelatinous coat and at least some part iodine positive   **52**

52 (51) Thalli crustose to squamulose or minutely fruticose; ascomata lacking a thalline margin   **Placynthiaceae**
Thalli foliose; ascomata with or without a thalline margin   **53**

53 (52) Covering layer of ascomata very thick, becoming stretched and eventually fragmenting; hymenial layer strongly concave when immature, with a conspicuous margin; asci with an iodine positive cap   **Lobariaceae**

Covering layer thin, pushed apart at an early stage of development; hymenial layer usually ± flat at all stages, without a well-developed margin; asci with an iodine positive ring   **Peltigeraceae**

54 (50) Ascus apex strongly thickened, extending and releasing ascospores through an apical split; with an outer layer at the apex which blues strongly in iodine; ascospores often with extremely thick septa and reduced lumina (polarilocular)   **(Teloschistales) 55**

Ascus apex with an internal apical cap which everts at dehiscence to form a beak-like structure, the outer layer at the ascus apex not blueing strongly in iodine; ascospores usually with thin septa (if polarilocular then brown)   **57**

55 (54) Asci with an iodine-positive internal apical structure surrounding the ocular chamber; thallus crustose; ascomata black, lecideine; ascospores usually aseptate; anthraquinone pigments absent, not K+ crimson   **Fuscideaceae**

Asci iodine-negative around the ocular chamber; anthraquinone pigments usually present, K+ crimson; thallus crustose or not; ascomata orange-red or black, sometimes with a thalline margin; ascospores usually polarilocular   **56**

56 (55)   Internal apical structure well-developed, outer apical cap diffuse, not strongly blueing
in iodine; ascospores multiseptate to muriform, not polarilocular   **Letrouitiaceae**
Internal apical structure rudimentary, outer cap well-developed, strongly blueing in
iodine; ascospores 0- to 3-septate, polarilocular   **Teloschistaceae**

57 (54)   Thallus crustose, often poorly developed; ascomata usually elongated, sometimes
branched, the hymenium exposed or only visible as a slit; asci usually not staining
blue in iodine; ascospores brown, 1-septate   **Melaspileaceae [FIS]**
Thallus varied in form, usually well-developed; ascomata usually circular, the
hymenium exposed but lacking a distinct margin; asci usually blueing in iodine   **58**

58 (57)   Thallus very reduced; ascomata small, rounded; asci with an enormously thickened
apical cap which blues in iodine and a well-developed ocular chamber, 12-spored;
ascospores 1-septate, staining in iodine; in leaf axils of *Andreaea*
(*Bryophyta*)   **Pachyascaceae [FIS]**
Thallus usually well-developed; asci usually 8-spored (sometimes multi-spored);
ascospores variously septate, not blueing in iodine; not associated with mosses   **59**

59 (58)   Ascomata perithecia   **60**
Ascomata apothecia   **61**

60 (59)   Thallus inconspicuous, often disappearing; ascomata immersed, the upper part staining
blue in iodine; asci with an apical cushion which blues in iodine; ascospores aseptate;
on soil   **Thrombiaceae [FIS]**
Thallus crustose; ascomata erumpent, not blueing in iodine; asci saccate, not blueing in
iodine; ascospores muriform; foliicolous or corticolous   **Phyllobatheliaceae [FIS]**

61 (59)   Asci not blueing in iodine, though sometimes with a mucous coat that does blue   **(Lecanorales) 62**
At least some part of the ascus apex blueing in iodine   **65**

62 (61)   Asci polysporous   **Acarosporaceae**
Asci usually 8-spored, rarely to 32-spored   **63**

63 (62)   Paraphyses usually not swollen at the apex; asci not or weakly blueing in iodine   **Trapeliaceae**
Paraphyses moniliform towards the apex   **64**

64 (63)   Thallus rarely areolate, the upper cortex poorly developed; ascospores many-septate,
sometimes muriform; usually on wood and bark   **Phlyctidaceae [FIS]**
Thallus areolate, the upper cortex usually well-developed; ascospores 0- to 3-septate;
mainly on rocks by streams   **Hymeneliaceae**

65 (61)   Asci polysporous; ascospores aseptate, hyaline   **66**
Asci usually 8-spored; ascospores of various types   **(Lecanorales) 68**

66 (65)   Thallus very weakly developed, usually immersed within bark; most species not
lichenized   **Lecanorales: Agyriaceae**
Thallus well-developed, superficial; lichenized   **67**

67 (66)   Thallus foliose or elongated, sometimes with a short central stalk; asci with a well-
developed ocular chamber; lichenized with cyanobacteria   **Lichinales: Peltulaceae**
Thallus crustose; asci very strongly thickened at the apex while immature and with
distinct wall layers which stain differentially in iodine, but without a clear ocular
chamber; lichenized with green algae   **Lecanorales: Biatorellaceae**

68 (65)   Photobiont cyanobacteria   **69**
Photobiont usually exclusively green algae   **72**

69 (68)   Thallus gelatinized; paraphyses often branched   **70**
Thallus not gelatinized; paraphyses usually unbranched, rather rigid and thick-walled   **71**

70 (69)   Thallus crustose, with rhizoids; paraphyses anastomosing, swollen at the tips;
ascospores elongate, with attenuated apices; associated with bryophytes, arctic and
subarctic   **Arctomiaceae**
Thallus usually foliose to fruticose, dark grey or green to black, unstratified and
gelatinous; paraphyses simple or branched, immersed in a gelatinous matrix;
ascospores varied; on bark and especially calcareous rocks, widespread   **Collemataceae**

71 (69)   Ascomatal wall hardly developed, formed from aggregations of short cells derived
from adjacent thalline hyphae   **Coccocarpiaceae**
Ascomatal wall well-developed, composed of isodiametric cells   **Pannariaceae**

72 (68)   Thallus crustose, usually forming cortical granules (goniocysts) which are often
ornamented; paraphyses branched and anastomosing, merging into a rudimentary
ascomatal wall composed of similar cells; asci initially with an ocular chamber which
later disappears as an apical pore is formed; often associated with bryophytes,
sometimes with other lichens   **Vezdaeaceae**
Thallus various, cortical granules usually not formed; ascomatal wall not paraphysis-
like in structure; asci without a clear apical pore; not associated with bryophytes   73

73 (72)   Thallus foliose, attached to the substratum by a short central stalk (umbilicus);
ascomata lecideine, often strongly convoluted; often with thalloconidia; rhizoids
absent; mainly on rocks   **Umbilicariaceae**
Thallus various in form, but if umbilicate not strongly convoluted; if foliose usually
attached to substratum at many points, usually by rhizoids; on rocks, trees or soil   74

74 (73)   Thallus shrubby, much-branched, the branches usually strongly flattened with algae
incorporated in both surface layers; not gelatinous, usually grey or greenish grey   **Ramalinaceae**
Thallus not shrubby, or if so branches usually terete; if flattened, algae only
incorporated into one surface layer   75

75 (74)   Thallus foliose, erect or spreading; lobes sometimes strongly recurved, the upper
surface smooth, the lower one felted, sometimes veined, usually with rhizoids;
ascomata irregular in shape, on the margins of lobes   **Heterodeaceae**
Thallus various, if foliose, lobes usually not recurved, corticate on both surfaces and
ascomata not marginal   76

76 (75)   Thallus usually composed of a horizontal foliose to squamulose primary thallus, and a
± vertical simple or branched, hollow or solid secondary thallus or lichenized stripe
which bears the hymenium; ascomata often lacking a properly differentiated wall,
dark or brightly coloured   77
Thallus not differentiated into horizontal and vertical elements; ascomata with well-
developed walls   78

77 (76)   Primary thallus crustose to squamulose, degenerating at an early stage; secondary thalli
solid; cephalodia usually present; ascospores usually several-septate, sometimes
muriform   **Stereocaulaceae**
Primary thallus well-developed, squamulose, persistent or granular and degenerating at
an early stage; secondary thalli solid or hollow; cephalodia absent; ascospores usually
one-celled, never muriform   **Cladoniaceae**

78 (76)   Thallus crustose; ascomata black, without a clear rim, elongate, umbonate or contorted;
paraphyses anastomosing, moniliform; asci with a well-developed, weakly staining,
apical cushion and a strongly staining apical ring; ascospores aseptate, thin-walled,
sometimes becoming brown   **Rimulariaceae**
Thallus crustose, foliose or fruticose; ascomata ± circular; asci with one, or several thin
layers of iodine-positive material; ascospores of various types   79

79 (78)   Thallus and ascomata usually yellow to orange, K-, the wall composed of closely
septate twisted hyphae; asci with a thick, non-amyloid cap above the blue-staining
region, occasionally up to 32-spored; often in nitrogen-rich habitats   **Candelariaceae**
Thallus and ascomata usually grey, green or brown, K+ or K-; asci without a thick non-
amyloid cap; habits varied   80

80 (79)   Ascospores becoming dark brown, thick-walled, usually with thickened septa
(polarilocular); paraphyses often pigmented at the apex; often on nitrogen-rich
substrata   **Physciaceae**
Ascospores usually thin-walled, aseptate or with simple septa, not polarilocular   81

81 (80)   Asci with an apical cap blueing in iodine, usually eventually with a small ocular
chamber below a non-staining apical cushion, which is surrounded by a more strongly
stained ring or tube-like structure within the apical cap   82
Asci without a tube-like structure within the apical cap, often with an ocular chamber   87

82 (81)   Ascomata formed directly from the upper cortex, poorly defined, effuse without proper walls; hymenial gel not blueing in iodine   **Gypsoplacaceae**

Ascomata well-defined, usually circular, usually with clearly delimited walls; hymenial gel usually blueing in iodine   **83**

83 (82)   Paraphyses simple or anastomosing, not swollen at the tip, sometimes immersed in gel   **84**

Paraphyses simple or branched, at least slightly swollen at the tip, sometimes with a distinct epithecial gel or crust   **87**

84 (83)   Asci with a well-developed ocular chamber; thallus somewhat spongy; ascomata pale, convex, without a clear wall; tropical   **Crocyniaceae**

Asci without a clear ocular chamber; thallus not spongy; temperate or tropical   **85**

85 (84)   Ascomatal wall composed of loosely intertwined hyphae, sometimes woolly in appearance   **Pilocarpaceae**

Ascomatal wall composed of branched radiating hyphae, sometimes immersed in a pigmented gel; thallus mostly granular, at least in part   **86**

86 (85)   Paraphyses usually simple; apical cap almost completely filled by an iodine positive blue plug, the central canal hardly visible   **Aphanopsidaceae**

Paraphyses usually branched and anastomosing; apical cap iodine positive with a distinct darker-staining tube, the central canal clearly visible   **Micareaceae**

87 (81)   Ascomata usually convex; ascospores without gelatinous sheaths   **Psoraceae**

Ascomata usually ± flat; ascospores with gelatinous sheaths, at least when young   **Porpidiaceae**

88 (83)   Asci with an apical cap weakly blueing in iodine, rarely with a strongly staining inner layer; hymenium pale   **89**

Asci with apical caps blueing strongly in iodine   **90**

89 (88)   Thallus poorly developed, often immersed within the substratum; ascomata sometimes elongated; asci occasionally many-spored; paraphyses usually unbranched; not lichenized   **Agyriaceae**

Thallus crustose to squamulose; paraphyses simple or branched, sometimes anastomosing; lichenized   **Trapeliaceae**

90 (88)   Thallus fruticose, much-branched; branches usually terete, smooth; pseudocyphellae present; ascomata lateral; asci usually containing fewer than 8 ascospores; ascospores large, thick-walled, brown, aseptate to muriform   **Alectoriaceae**

Thallus various, fruticose, pseudocyphellae absent and asci 8-spored; ascospores of various types   **91**

91 (90)   Asci mostly 1- to 2-spored; ascospores hyaline, variously septate   **92**

Asci usually 8-spored; ascospores hyaline or brown, rarely muriform   **95**

92 (91)   Ascospores thick-walled, aseptate or transversely septate   **93**

Ascospores muriform, not conspicuously thick-walled   **94**

93 (92)   Ascomata black, convex; asci usually 1- to 2-spored, clavate, with an enormous apical cap and a very well-developed ocular chamber; ascospores very large and thick-walled, hyaline, aseptate, without a gelatinous sheath   **Mycoblastaceae**

Ascomata brown to black, flat to concave; asci with a poorly-developed ocular chamber; ascospores septate, ellipsoidal to elongate, hyaline or not, sometimes with a gelatinous sheath   **Megalosporaceae**

94 (92)   Thallus usually crustose or lobed (placodioid); ascomata with anthraquinone pigments; asci without an ocular chamber   **Brigantiaeaceae**

Thallus usually foliose; ascomata without anthraquinone pigments; asci with a very well-developed ocular chamber   **Ectolechiaceae**

95 (91)   Ascomata sessile, without a clear thalline margin, dark brown to black; paraphyses with swollen and strongly pigmented apices; asci with an amyloid apical cap and outer layer, usually without an ocular chamber or apical cushion   **Catillariaceae**

Paraphyses not strongly pigmented at the apex, though often with pigmented epithecial gel; asci not as above   **96**

96 (95) Ascomata ± immersed within the thallus, upper part of wall and hymenium blue-green;
asci broadly clavate to saccate   **Eigleraceae**
Ascomata ± superficial on the thallus, not blue-green; asci clavate to cylindric-clavate   **97**

97 (96) Ascospores elongate and multiseptate, often helically coiled within the ascus   **98**
Ascospores not elongate, variously septate   **99**

98 (97) Asci usually with a well-developed ocular chamber and apical cushion; paraphyses
narrow, much branched   **Haematommataceae**
Asci without an ocular chamber or apical cushion; paraphyses relatively wide, simple
or little-branched   **Ophioparmaceae**

99 (97) Ascomata reddish; asci with an apical cap only weakly blueing in iodine, though the
outer layer blues strongly; ocular chamber well-developed; ascospores with
gelatinous sheaths   **Miltideaceae**
Ascomata reddish to balck; asci with apical caps at least part of which blue strongly and
uniformly in iodine   **100**

100 (99) Ascomata black, round, angular or elongated, with well-developed walls; asci with
only the tip of the apical cap blueing strongly in iodine; ascospores septate, often
muriform, usually with a gelatinous sheath   **Rhizocarpaceae**
Ascomata usually round; asci clavate to cylindric-clavate; the whole of the apical cap
usually blueing in iodine, often with more strongly staining tubular rings; ascospores
of various types   **101**

101 (100) Ascospores often very large, very thick-walled   **Megalosporaceae**
Ascospores small, not conspicuously thick-walled   **102**

102 (101) Ascomata with well-developed thalline margins; asci with a plug in the apical cap
which weakly stains blue in iodine   **103**
Ascomata without thalline margins; asci without an apical plug   **104**

103 (102) Thallus corticate above and below, sometimes with rhizoids; squamulose, foliose or
fruticose   **Parmeliaceae**
Thallus corticate above, crustose to foliose, usually with rhizoids, sometimes granular
or minutely fruticose   **Lecanoraceae**

104 (102) Ascomata black; asci with an apical cap staining pale blue in iodine, with a strongly
staining apical part, and a wide, rather poorly defined ocular chamber; ascospores
aseptate   **Lecideaceae**
Ascomata pale to black; asci with an apical cap which stains strongly in iodine apart
from an apical cushion; ocular chamber usually narrow; ascospores usually septate   **Bacidiaceae**

105 (48) Ascomata apothecial   **106**
Ascomata perithecial or cleistothecial   **149**

106 (105) Ascus walls blueing in iodine   **107**
Ascus walls not blueing in iodine   **115**

107 (106) Ascomata brightly coloured, club-shaped, the hymenium covering the outer surface;
interascal tissue absent; asci not formed from croziers; ascospores aseptate, hyaline;
non-lichenized   **(Neolectales) Neolectaceae**
Ascomata not brightly coloured, disc- or cup-shaped, not stalked; paraphyses present;
asci at least usually formed from croziers; lichenized or not   **108**

108 (107) Lichenized with cyanobacteria; thallus gelatinous; ascomata initially formed as a
filamentous weft of hyphae, bursting through the covering layers of the thallus,
sometimes converted from conidiomata; asci often multi-spored; ascospores hyaline,
aseptate   **Lichinales: Lichinaceae**
Ascomata formed in the surface layers of the thallus; asci 8-spored; ascospores of
various types   **109**

109 (108) Lichenized with green algae; thallus crustose to squamulose; paraphyses branched near
the apex, with a dark epithecial layer; asci cylindrical, thin-walled even at the apex;
ascospores mostly spherical, aseptate, thin-walled, sometimes with a gelatinous
sheath **Schaereriaceae [FIS]**
Ascospores usually ellipsoidal **110**

110 (109) Asci thick-walled at the apex **111**
Asci thin-walled at the apex even when immature **112**

111 (110) Lichenized with green algae; thallus crustose; ascomata deeply embedded; paraphyses
anastomosing; asci cylindric-clavate, thick-walled, without an ocular chamber, the
outer layer faintly iodine positive; ascospores large, aseptate, hyaline, ellipsoidal, the
wall two-layered **(Pertusariales) Megasporaceae**
Ascomata brown or black; asci thick-walled at the apex when immature, the wall
becoming thinner at maturity through reabsorption of the inner layers; with an outer
gelatinized cap; breaking to release the ascospores, forming a beak-like structure;
ascospores septate, usually pigmented, sometimes ornamented; saprobic or
lichenicolous **Dactylosporaceae [FIS]**

112 (110) Lichenized, usually with a well-developed thallus **113**
Not lichenized; saprobic, usually on soil, dung or very rotten wood **(Pezizales) 115**

113 (112) Lichenized with cyanobacteria; ascomata superficial from inception, flat to domed,
with thick-walled straight anastomosing paraphyses which are swollen at the apex;
ascospores aseptate **Lecanorales: Heppiaceae**
Ascomata usually cupulate, initially completely covered by the thallus **114**

114 (113) Thallus filamentous or crustose; ascospores with severla transverse and sometimes
longitudinal septa; lichenized with green algae **Gyalectales: Gyalectaceae**
Thallus small, squamulose to peltate, often gelatinous, with a network of hyphae;
ascospores 0- 1-septate; lichenized with cyanobacteria **Lichinales: Gloeoheppiaceae**

115 (112) Ascomata usually small, cushion-shaped, rarely almost absent, never hairy; asci
usually clavate or ellipsoidal, rarely ± globose, usually significantly longer than the
paraphyses when mature; ascospores usually dark brown and strongly ornamented;
often on dung **Ascobolaceae**
Ascomata usually large and flat or cup-shaped, sometimes hairy; asci ±cylindrical, not
extending beyond the paraphyses; ascospores hyaline to pale brown, ornamented or
not; usually on soil or rotten wood **Pezizaceae**

115 (106) Ascomata multiple, placed within the surface tissues of a usually massive brightly
coloured erumpent stroma; causing galls on *Nothofagus* spp. **(Cyttariales) Cyttariaceae**
Stromata usually absent or inconspicuous; where well-developed, black and not
strongly erumpent **116**

116 (115) Ascomata immersed within plant tissue, breaking through the surface layers to expose
the hymenium **117**
Ascomata superficial, at least from an early stage of development **127**

117 (116) Ascomatal wall very poorly developed; asci small, hardly thickened at the apex, with
an apical pore which blues in iodine in some species; usually subepidermal; on
Coniferae, the covering layers pushed back, appearing as a lid **Leotiales: Hemiphacidiaceae**
Ascomatal wall well-developed **118**

118 (117) Developing ascomata large, pushing off the covering layers, leaving the hymenium
exposed at maturity; on bark of *Salix* **Rhytismatales: Cryptomycetaceae [FIS]**
Covering layers incorporated into the upper wall of the ascomata, sometimes pushed
aside to expose the hymenium but retracting in dry conditions **119**

119 (118) Asci distinctly thickened at the apex **(Ostropales) 120**
Asci not thickened at the apex, either splitting or releasing ascospores through a small
pore **124**

120 (119) Lichenized with green algae; ascomata elongate, often curved or branched, usually
black; ascospores usually large, variously septate, hyaline or brown **Graphidaceae**
Ascomata ± radially symmetrical **121**

121 (120) Thallus crustose; ascomatal walls composed of thick-walled gelatinized hyphae with
very narrow lumina; ascospores transversely septate to muriform; lichenized with
green algae   **Solorinellaceae**
Ascomatal walls not gelatinized, the hyphae not conspicuously thick-walled   **122**

122 (121) Lichenized with green algae; ascomata usually with a distinct thalline margin;
ascospores hyaline or brown, transversely and sometimes longitudinally septate, the
septa usually strongly thickened forming lenticular lumina   **Thelotremataceae**
Not lichenized, though sometimes lichenicolous; ascospores without strongly
thickened septa   **123**

123 (122) Ascomatal walls with conspicuous crystalline inclusions; asci narrow, cylindrical, with
a strongly thickened apex; ascospores filiform, often fragmenting   **Stictidaceae**
Ascomatal walls without crystalline inclusions; asci cylindrical to clavate; ascospores
usually ellipsoidal, variously septate   **Odontotremataceae**

124 (119) Asci small, with an iodine-positive apical ring   **Leotiales: Phacidiaceae**
Asci usually large, either without a ring, or with one which does not stain blue in iodine   **125**

125 (124) Ascomata erumpent, breaking open by irregular splits; paraphyses branched and
anastomosing, attached at the apex at least when immature; ascospores septate,
sometimes muriform   **(Triblidiales) Triblidiaceae**
Ascomata immersed or erumpent; paraphyses free at the apex; ascospores usually
aseptate, often with gelatinous sheaths   **(Rhytismatales) 126**

126 (125) Stromata erumpent; asci with a broad apical ring; ascospores without gelatinous
sheaths; anamorph disseminative; on *Fagaceae*   **Ascodichaenaceae**
Stromata usually remaining within the surface layers of the host; asci usually without
an apical ring; ascospores usually with gelatinous sheaths or appendages; anamorph
usually spermatial   **Rhytismataceae**

127 (116) Ascomata mostly sessile, usually disc-shaped; asci cylindrical to clavate, strongly
thickened at the apex, with separable wall layers according to some interpretations, 1-
to 8-spored; ascospores ranging from filiform to ellipsoidal, usually with many
transverse septa and often muriform; lichens with crustose thalli and distinctive
spine- or brush-like, or peltate conidiomata (hyphophores)   **Ostropales: Gomphillaceae**
Ascomata often stalked; not lichenized   **128**

128 (127) Asci conspicuously thickened at the apex   **129**
Asci not obviously thickened at the apex   **130**

129 (128) Ascomata pulvinate, sessile to long-stalked; usually brightly coloured; asci cylindrical,
elongated; ascospores filiform; aquatic; not lichenized   **Vibrisseaceae**
Ascomat often stalked, brown or black; asci usually small, with a thickened apex and
rarely an apical canal; ascospores dark brown, variously septate, smooth or weakly
ornamented; lichenicolous or saprobic   **Caliciales: Mycocaliciaceae**

130 (128) Ascomata usually small, sessile or shortly stipitate; asci usually small, not
conspicuously thickened, almost always with an apical pore, which sometimes blues
in iodine; ascospores small, sometimes septate, very rarely ornamented, usually with
the widest point towards the apex; saprobic or parasitic on plant tissue, rarely
lichenized   **(Leotiales) 131**
Ascomata usually large, sometimes stalked; asci usually large, ± truncate, with a wide
opening (operculum); ascospores aseptate, sometimes pigmented, often ornamented,
longitudinally symmetrical; on soil, dung or very rotten vegetation, sometimes
associated with bryophytes   **(Pezizales) 139**

131 (130) Lichenized; ascomata convex when mature, single or several on short stalks; hymenial
gel sometimes blueing in iodine   **Baeomycetaceae**
Very rarely lichenized; if so ascomata sessile   **132**

132 (131) Ascomata effuse, with an indefinite layer of asci and undifferentiated paraphyses;
usually on conifer bark and leaves   **Ascocorticiaceae**
Ascomata well-defined   **133**

133 (132) Ascomata strongly cupulate, immersed in gel; asci with a relatively wide pore; ascospores clavate, with a long cellular basal appendage, the apical part covered in a mucous sheath; aquatic, on Juncaceae **Loramycetaceae**

Ascomata variously shaped, not immersed in gel, although gelatinized tissues may be present; ascospores without appendages **134**

134 (133) Ascomata usually dark, clavate, spathulate or with an irregular stalk, the hymenium not clearly separated from the stalk; on soil, often in grassland, or on rotten wood or leaves **Geoglossaceae**

Ascomata usually disc- or cup-shaped, if stipitate then the stalk clearly differentiated from the hymenium **135**

135 (134) Ascomata usually stalked, arising from a sclerotium or from stromatized host tissue, often brown; asci with an apical ring which often blues in iodine; ascospores often longitudinally symmetrical **Sclerotiniaceae**

Ascomata stalked or sessile, not arising from well-developed stromatic tissue **136**

136 (135) Ascomata sessile, translucent, often strongly convex; hymenium usually immersed in a waxy matrix; asci small, usually forked at the base; apical pores inconspicuous or absent; paraphyses often with swollen apices; usually saprobic on wood **Orbiliaceae**

Ascomata sessile or stalked; asci usually with a conspicuous apical pore; hymenium not waxy; paraphyses rarely strongly swollen at the tip **137**

137 (136) Ascomata usually brown or black, the outer wall layers composed of darkly pigmented isodiametric cells **Dermateaceae**

Ascomata usually pale or brightly coloured, the outer wall layers composed of hyphal tissue **138**

138 (137) Ascomata usually small, with distinct hairs surrounding the hymenium and the outer wall layers; hyphae of ascomatal wall often closely and regularly septate, rarely gelatinized **Hyaloscyphaceae**

Ascomata small to medium-sized, almost always glabrous, or downy (with poorly-differentiated hairs); outer wall layers usually composed of hyphae with rather remote septa, sometimes gelatinized **Leotiaceae**

139 (130) Ascomata usually distinctly stalked, lacking carotenoid pigments, the hymenium often strongly convex and convoluted; asci not blueing in iodine **140**

Ascomata usually ± sessile and flat or cup-shaped; occasionally elongate but the hymenium not clearly differentiated from the stalk; asci blueing or not in iodine **141**

140 (139) Ascospores ellipsoidal, smooth, eguttulate but with a minutely guttulate cap-like gelatinous appendage at each apex, with many nuclei **Morchellaceae**

Ascospores ellipsoidal to fusiform, smooth or ornamented, usually guttulate, without appendages, with four nuclei **Helvellaceae**

141 (139) Ascomata usually small, cushion-shaped, rarely almost absent, very rarely hairy; asci usually clavate or ellipsoidal, rarely ± globose, usually significantly longer than the paraphyses when mature; often on dung **142**

Ascomata usually large and flat or cup-shaped, sometimes hairy; asci ±cylindrical, not extending beyond the paraphyses; usually on soil or rotten wood **144**

142 (141) Asci often blueing in iodine, typically with an operculum which is sharply delimited by an internal indentation within a thickened ring; ascospores thick-walled when immature, often dark brown or purplish, and ornamented **Ascobolaceae**

Asci not staining blue in iodine; ascomata usually reduced to a small clump of asci and paraphyses; **143**

143 (142) Asci with a very large operculum, sharply delimited by a weak indentation; ascospores brown, with ornamentation formed directly from the secondary wall layer **Ascodesmidaceae**

Ascus operculum poorly delimited or absent, usually opening by an irregular split, often multi-spored; ascospores smooth, or with ornamentation formed as a result of elaboration of an initially homogeneous secondary wall layer **Thelebolaceae**

144 (141) Ascomata usually large, some shade of brown (lacking in carotenoids); asci blueing in iodine; ascospores hyaline or pale brown **Pezizaceae**

Asci not blueing in iodine **145**

145 (144) Ascomata leathery or somewhat gelatinous; paraphyses often anastomosing; ascus
opening often subapical; ascospores often blunt-ended, sometimes flattened on one
side   **146**
Ascomata fleshy; paraphyses not anastomosing; asci with an apical opening;
ascospores radially symmetrical   **147**

146 (145) Ascomatal wall dark brown, sometimes with dark brown setose hairs; hymenium pale
or dark, very rarely with carotenoid pigments   **Sarcosomataceae**
Ascomatal wall usually brightly coloured, rarely with brown setose hairs; hymenium
with paraphyses containing carotenoid pigments   **Sarcoscyphaceae**

147 (145) Ascomata not formed on a mat of hyphae, the hymenium sometimes surrounded by
well-developed spiny hairs   **Otideaceae**
Ascomata formed on a well-developed mat of hyphae   **148**

148 (147) Ascomata small, discoid, sometimes coalescing; asci sometimes multispored;
especially on dung or burnt ground   **Pyronemataceae**
Ascomata relatively large, flat, very thin, the hymenium brownish red, greenish yellow
at the margin; asci 8-spored   **Karstenellaceae**

149 (105) Ascomata large, cleistothecial, subterranean   **(Pezizales) 150**
Ascomata not subterranean   **158**

150 (149) Asci blueing in iodine, either in the apical region or diffusely over the whole wall   **151**
Asci not blueing in iodine   **152**

151 (150) Ascomata small (<10 mm diam.), solid, with the hymenium on the outer surface   **Ascobolaceae**
Ascomata large, usually infolded, with the hymenium lining the inner walls   **Pezizaceae**

152 (150) Ascomata hollow, often infolded, the inner wall lined with hymenial tissue   **153**
Ascomata solid, often with a contorted gleba   **154**

153 (152) Ascospores tetranucleate   **Helvellaceae**
Ascospores uninucleate   **Otideaceae**

154 (152) Ascospores smooth or almost so   **Balsamiaceae**
Ascospores distinctly ornamented   **155**

155 (154) Asci 1- to 6-spored; ascospores spinose or reticulate, dark brown   **Tuberaceae**
Asci 4- to 8-spored   **156**

156 (155) Ascus walls becoming brown in KOH; asci 8-spored, formed between groups of large
inflated cells; ascospores hyaline to pale brown or pink   **Carbomycetaceae**
Ascus walls not becoming brown in KOH; asci not formed in pockets of inflated cells   **157**

157 (156) Ascospores medium to dark brown, ellipsoidal   **Helvellaceae**
Ascospores hyaline to pale brown, globose   **Terfeziaceae**

158 (149) Asci clavate, saccate or globose, very thin-walled, deliquescing to release the
ascospores   **159**
Asci usually cylindrical to clavate, relatively thick-walled, discharging ascospores
actively through a specialized apical structure   **170**

159 (158) Mycelium conspicuous, superficial, dark-brown, usually hyphopodiate; ascospores
usually with an even number of septa; ? all biotrophs of plants   **160**
Mycelium at least partly immersed not conspicuously thick-walled; ascospores
aseptate or with an odd number of septa; saprobes or necrotrophs of plants   **161**

160 (159) Asci usually 2-spored; ascospores dark brown, usually 4-septate, the septa strongly
constricted; biotrophic on plant leaves   **(Meliolales) Meliolaceae**
Asci 8-spored; ascospores with two subapical septa and a loose brown perispore;
associated with roots   **Diporothecaceae [FIS]**

161 (159) Marine; paraphyses absent; ascospores usually hyaline, with complex cellular or
gelatinous appendages   **(Halosphaeriales) Halosphaeriaceae**
Terrestrial or if marine ascospores lacking complex appendages   **162**

162 (161) Marine; either associated with algae or corals   **163**
Terrestrial; a few species associated with intertidal wood or mangroves   **164**

163 (162) Associated with corals, though possibly dependent on encrusting sponges; ascomata
ostiolate, developing on a subiculum; ascospores thick-walled, several-septate near
the apices   **Koralionastetaceae [FIS]**
Parasitic or in a lichen association with large marine algae; ascomata dark brown, at
least at the base and apex, at least sometimes with a gelatinous cushion sealing the
ostiole; ascospores ± ellipsoidal, 0- 1-septate, without a gelatinous
sheath   **Mastodiaceae [FIS]**

164 (162) Ascomata with an annulate stalk, globose, cleistothecial, thick-walled, fragmenting
into well-defined plates; ascospores small, pale brown, without germ pores or
slits   **Sordariales: Batistiaceae**
Ascomata not stalked   **165**

165 (164) Ascomata perithecia (often with long necks) or cleistothecia, thin-walled; paraphyses
absent; asci evanescent, formed in chains from fertile cells arranged throughout the
ascomatal cavity, ± globose, without a stalk   **166**
Asci persistent or evanescent, formed individually from croziers in a basal fascicle   **167**

166 (165) Ascospores hyaline, varied in shape, usually with an excentric wall thickening or
gelatinous appendage, not staining dark brown in iodine   **Ophiostomatales: Ophiostomataceae**
Ascospores golden yellow or orange, staining dark brown in iodine when young,
without appendages or sheaths   **Microascales: Microascaceae**

167 (165) Ascospores brown, with germ pores   **(Sordariales) 168**
Ascospores hyaline, without germ pores   **169**

168 (167) Ascospores usually small, with minute germ pores; ascomata often hairy   **Chaetomiaceae**
Ascospores usually large, with conspicuous germ pores, often ornamented; ascomata
not hairy, or with setae surrounding the ostiole; usually associated with other
fungi   **Ceratostomataceae**

169 (167) Ascospores large, fusiform, with a dark patch towards one end; anamorph not yeast-
like   **Laboulbeniales: Pyxidiophoraceae**
Ascospores completely colourless; a yeast-like anamorph present, and also conidia
(probably spermatia) formed from characteristic minute long-necked
conidiomata   **Ophiostomatales: Kathistaceae**

170 (158) Asci thin-walled, sometimes with an inconspicuous refractive apical ring; ascomata
hyaline to black, sometimes with radiating clusters of agglutinated hairs extending
horizontally from the ostiolar region; ascospores fusiform, hyaline, many-septate;
lichens, often foliicolous   **Trichotheliales: Trichotheliaceae**
Not definitely lichenized, though rarely associated with algae; if so not foliicolous   **171**

171 (170) Stromata (if present) and ascomata brightly coloured, usually yellow, orange or red;
not lichenized   **172**
Stromata (if present) and ascomata black (at least partially) at maturity   **176**

172 (171) Stromata superficial, elongate, membranous; ascomata solitary or few within the
stroma, often elongate; true paraphyses present; asci narrowly cylindrical, thick-
walled, with a thickened apex; ascospores filiform   **Acrospermaceae [FIS]**
If stromata superficial, containing many ascomata   **173**

173 (172) Stromata superficial, formed on a hyphal mat; ascospores aseptate; on
Palmae   **Sordariales: Catabotrydaceae**
Stromata immersed at least at the base; ascospores almost always septate   **174**

174 (173) Ascomata with a distinct neck, often protruding from the substratum; true paraphyses
and apical paraphyses present; ascospores usually muriform, at least partially
pigmented   **Thyridiaceae [FIS]**
Ascomata either papillate or with short necks which hardly protrude from the stroma;
apical paraphyses present but true paraphyses lacking; ascospores variable in
septation and pigmentation but never dark brown   **(Hypocreales) 175**

175 (174) Ascomata immersed in conspicuous brightly coloured stromata, sometimes formed
from dark resting bodies (sclerotia); asci narrow, cylindrical, with a strongly
thickened apex penetrated by a narrow pore; ascospores filiform; usually associated
with *Gramineae*, insects or other fungi   **Clavicipitaceae**
Stromata present or not; asci cylindric to cylindric-clavate, with a small iodine-
negative apical ring; ascospores mostly not filiform   **Hypocreaceae**

176 (171) Asci ± cylindrical, the apex truncate, thickened, with an apical ring, usually developing
in a spicate arrangement from proliferating ascogenous hyphae; paraphyses broad,
tapering; ascospores hyaline or pale brown, usually allantoid   **(Calosphaeriales) 177**
Asci formed in a basal fascicle   **178**

177 (176) Ascomata immersed in a stroma with the necks emerging separately; ascus apical ring
blueing in iodine; anamorph *Nodulisporium*   **Graphostromataceae**
Ascomata solitary, aggregated or in a stroma with convergent ostioles; ascus apical
ring blueing or not; anamorphs unknown   **Calosphaeriaceae**

178 (176) Ascus wall blueing in iodine; ascomata immersed in a gelatinous algal
mat   **Epigloeaceae [FIS]**
Asci not blueing in iodine   **179**

179 (178) Biotrophic parasites; mycelium primarily superficial; ascomata basically cleistothecial,
thick-walled, dark brown, sometimes with complex appendages, opening by
unordered and sometimes explosive disintegration of apical cells; asci one or few per
ascoma; anamorph *Oidium*-like   **(Erysiphales) Erysiphaceae**
If biotrophic, mycelium mostly immersed   **180**

180 (179) Clypeus (involucrellum) often present; periphysoids usually present; asci sometimes
deliquescent; ascospores usually aseptate; mostly lichenized   **(Verrucariales) Verrucariaceae**
Clypeus usually not present; rarely lichenized   **181**

181 (180) Ascomata superficial, sometimes aggregated, stromatic material if present poorly
developed and restricted to hyphae joining the bases of the ascomata   **182**
Ascomata usually immersed in the substratum and/or in a well-developed stroma
which is composed either entirely of fungal tissue or of mixed fungal and host
elements; the stroma sometimes erumpent   **188**

182 (181) Ascomata originally cleistothecial, the ostiole forming by the breakdown of a
gelatinous plug of cells; cells of the ascomatal wall with well-developed pores; asci
often polysporous; ascospores usually allantoid   **Sordariales: Nitschkiaceae**
Ostiole forming early in development; ascomatal cells almost always without pores **183**

183 (182) Asci with an apical ring which does not blue in iodine; ascospores pigmented, at least
in part, with germ pores or slits   **(Sordariales) 184**
Asci with or without an apical ring; ascospores usually hyaline, if pigmented then
without germ pores or slits   **186**

184 (183) Ascospores with germ slits   **Coniochaetaceae**
Ascospores with germ pores   **185**

185 (184) Ascospores aseptate, brown, with mucous sheaths absent or uniformly covering the
spore   **Sordariaceae**
Ascospores usually septate, with both brown and colourless cells and with complex
mucous appendages   **Lasiosphaeriaceae**

186 (183) Ascomata long-necked; asci with an apical ring which sometimes blues in iodine;
ascospores usually elongate, septate, the middle cells often pigmented; mycelium
with appresoria; necrotrophic, especially on roots of *Gramineae*   **Magnaporthaceae [FIS]**
Ascomata papillate; ascus apical ring small, iodine negative or absent; appresoria
absent; mostly saprobic   **187**

187 (186) Basal but not apical paraphyses present; ascomata usually carbonaceous, not
collapsing **(Trichosphaeriales) Trichosphaeriaceae**
Apical but not basal paraphyses present; ascomata membranous, often collapsing, often
setose **Hypocreales: Niessliaceae**

188 (181) Asci with an apical ring which does not blue in iodine **189**
Asci with an apical ring which blues in iodine **194**

189 (188) Paraphyses well-developed at maturity **190**
Paraphyses either absent or poorly developed at maturity; saprobes or necrotrophic
parasites **(Diaporthales) 192**

190 (189) Stroma poorly developed, usually reduced to a clypeus; ascomata with a wide thick-
walled papillate ostiole **191**
Stromata usually well-developed **193**

191 (190) Ascomata sometimes with lateral ostioles; ascospores hyaline, radially symmetrical,
sometimes with thickened septa, without germ pores **Pyrenulales: Pleurotremataceae**
Ascomata with central ostioles; ascospores brown, radially asymmetrical, sometimes
with germ pores **Xylariales: Clypeosphaeriaceae**

192 (189) Stroma mostly composed of fungal tissue, usually erumpent; ascospores brown,
sometimes with a germ pore; appressoria absent; saprobic **Boliniaceae [FIS]**
Stroma composed of a mixture of fungal and plant tissue, rarely absent; ascospores
mostly hyaline and aseptate; appressoria often present; biotrophic or necrotrophic
parasites of plants **(Phyllachorales) Phyllachoraceae**

193 (190) Asci usually evanescent at the base, becoming free within the ascomatal cavity;
ascospores usually thin-walled and hyaline **Valsaceae**
Asci not or hardly evanescent at the base; ascospores usually thick-walled and
pigmented **Melanconidaceae**

194 (188) Ascomata usually with distinctly furrowed ostioles; ascospores pale brown, allantoid;
asci long-stalked, the apex truncate, with a minute apical ring (often blueing
weakly) **(Diatrypales) Diatrypaceae**
Ostioles not furrowed; ascospores not allantoid **195**

195 (194) Ascospores aseptate (sometimes with a small, usually evanescent daughter cell),
usually dark brown with a conspicuous germ slit **Xylariales: Xylariaceae**
Ascospores without germ slits **196**

196 (195) Stromatal tissues usually well-developed; asci attached peripherally; ascospores
usually thick-walled, often with germ pores; where known anamorphs
coelomycetous, usually with appendaged conidia **Xylariales: Amphisphaeriaceae**
Asci attached basally; ascospores thin-walled, hyaline or pale brown; stromatal tissues
absent or reduced to a small clypeus; where known anamorphs hyphomycetous **197**

197 (196) Ascomata immersed in bark, long-necked, clypeate; asci with a complex partially
iodine positive ring; ascospores strongly isthmoid; anamorph absent **Vialaeaceae [FIS]**
Ascomata usually in herbaceous material, short-necked, clypeate or not; asci usually
with an inconspicuous iodine positive or iodine negative ring; ascospores not
isthmoid; anamorph hyphomycetous where known **Hyponectriaceae [FIS]**

198 (47) Ascus wall (at least at the apex) and/or apical structures blueing in iodine; associated
with algae, many lichenized **199**
Ascus wall not blueing in iodine; lichenized or not **209**

199 (198) Ascomata perithecial, sometimes strongly flattened **200**
Ascomata apothecial, though sometimes strongly concave with asci becoming exposed
only at a late stage of development; sometimes elongate, branched or curved **205**

200 (199) Ascomata not flattened; usually lichenized or lichenicolous; ascospores septate, often
muriform **201**
Ascomata usually strongly flattened, opening by disintegration of apical cells or
irregular splits; interascal tissue usually absent; ascospores usually 1-septate; on
living leaves, often with appressoria **(Dothideales) 203**

201 (200) Often lichenized, the thallus crustose; ascomatal walls greenish; paraphyses narrow, anastomosing; asci with a thickened apex, blueing diffusely in iodine   **Protothelenellaceae [FIS]**
          Non-lichenized; ascomata pulvinate, sometimes opening very widely   **(Dothideales) 202**

202 (201) Ascomata formed on a thick dark hyphal mat; asci saccate, with an outer layer of gel which blues in iodine; ascospores muriform   **Coccodiniaceae**
          Ascomata composed of dark thick-walled pseudoparenchymatous tissue; asci clavate; ascospores transversely septate   **Lichenotheliaceae**

203 (200) Mycelium formed within the substratum; ascomata usually developing as locules within flattened or crustose stromata   **Parmulariaceae**
          Mycelium primarily superficial; stromata usually absent   **204**

204 (203) Ascomata round or elongate, very strongly flattened, the upper wall composed of radiating cells, persistent   **Asterinaceae**
          Ascomata globose to pulvinate; ascomatal wall not composed of radiating cells, becoming gelatinized   **Englerulaceae**

205 (199) Ascomata strongly concave, sometimes appearing perithecial; hymenium blueing or not in iodine; asci ± cylindrical, releasing ascospores through an apical split; ascospores hyaline, aseptate, often very large and thick-walled   **(Pertusariales) 206**
          Ascomata very variable in form; hymenium usually blueing in iodine; asci clavate to saccate, the apex usually very strongly thickened, usually fissitunicate; ascospores usually septate   **(Arthoniales) 207**

206 (205) Ascomata almost closed, the poroid opening surrounded by periphysoids   **Coccotremataceae**
          Ascomata strongly cupulate but clearly apothecial; periphysoids absent   **Pertusariaceae**

207 (205) Thallus absent or poorly developed; ascomatal walls rudimentary   **208**
          Thallus often well-developed, crustose or fruticose; ascomatal rounded or elongate, the walls usually well-developed   **Rocellaceae**

208 (207) Thallus granular, without a cortex, yellow to yellow-green; ascomata ± round; disc usually yellowish   **Chrysothricaceae**
          Thallus usually almost absent, immersed in bark, rarely greyish and crustose; ascomata often elongated, sometimes branched, sometimes poorly defined; disc reddish, brownish or black   **Arthoniaceae**

209 (198) Clypeus (involucrellum) often present; periphysoids usually present; asci sometimes deliquescent; ascospores usually aseptate; mostly lichenized   **(Verrucariales) 210**
          Clypeus not present; rarely lichenized   **211**

210 (209) Clypeus absent; apical paraphyses present; lichenicolous   **Adelococcaceae**
          Clypeus usually present; interascal tissue only of simple narrow paraphysis-like hyphae; mostly lichenized   **Verrucariaceae**

211 (209) Ascomata stalked, membranous, initially cleistothecial but opening with irregular radial splits; asci bitunicate though apparently not fissitunicate, long-stalked; ascospores hyaline, curved, elongate, 3-septate   **(Lahmiales) Lahmiaceae**
          Ascomata either apothecial or perithecial   **212**

212 (211) Ascomata apothecial, not strongly flattened   **213**
          Ascomata perithecial, rarely cleistothecial, opening either with an ostiole or by deliquescing of the upper wall layers, rarely by radial or irregular splits   **218**

213 (211) Ascomata sessile, usually disc-shaped; asci cylindrical to clavate, strongly thickened at the apex, 1- to 8-spored; ascospores ranging from filiform to ellipsoidal, with many transverse septa and often muriform; tropical foliicolous lichens, often with crustose thalli and distinctive spine- or brush-like, or peltate conidiomata (hyphophores)   **Ostropales: Gomphillaceae**
          Hyphophores absent   **214**

214 (213) Asci ellipsoidal or ± globose, without an obvious stalk   **215**
          Asci elongate, usually distinctly stalked   **(Patellariales) 217**

215 (214) Stromata superficial, often radially lobed, or absent; ascomatal wall absent; asci ±
    globose, surrounded by the gelatinous stromatal tissue; ascospores 1-septate; on
    living leaves, often associated with honeydew   **Seuratiaceae [FIS]**
    Stromata absent   **216**

216 (215) Ascomata pulvinate, with a greenish black epithecial layer; subhymenium blueing in
    iodine; ascospores 1-septate or muriform   **Phillippsiellaceae [FIS]**
    Ascomata discoid, in a hyphal mat; upper part of ascomatal wall becoming gelatinous;
    ascospores muriform   **Saccardiaceae [FIS]**

217 (214) Ascomata sometimes strongly cupulate; ascomatal wall composed of hyphae with
    swollen walls; ascospores hyaline; mostly lichenized or lichenicolous **Arthrorhaphidaceae**
    Ascomata round or elongated; ascomatal wall composed of pseudoparenchymatous
    cells; ascospores usually pigmented; mostly saprobic   **Patellariaceae**

218 (212) Thallus crustose or effuse; ascomata usually slightly flattened, with a disc-like collar
    around the ostiole or with verrucose or ridged ornamentation, sometimes with flaps
    (campylidia) producing conidia; paraphyses usually branched; ascospores hyaline,
    transversely septate or muriform, with a gelatinous sheath; foliicolous
    lichens   **Aspidotheliaceae [FIS]**
    Ascomata without collars or ornamentation   **219**

219 (218) Ascomata strongly flattened; foliicolous lichens, mainly in the tropics   **220**
    Ascomata not flattened; if lichenized, then thallus not subcuticular   **221**

220 (219) Thallus subcuticular; ascomata opening by a pore; paraphyses branched or not; asci ±
    globose when young, becoming clavate;   **Strigulaceae [FIS]**
    Thallus superficial; ascomata opening by radiating splits which eventually form a wide
    opening; interascal hyphae anastomosing   **Arthoniales: Rocellaceae**

221 (219) Ascomata globose, often with a clypeus (involucrellum); interascal tissue connected at
    the apex at least at first, often anastomosing, very narrow with narrow lumina; asci
    cylindrical, very thick-walled; ascospores usually septate, the septa usually strongly
    thickened and the lumina lenticular; when aseptate, ornamented; mostly
    lichenized   **(Pyrenulales) 222**
    Ascomata without a clypeus; interascal tissue varied, sometimes absent; ascospores
    usually septate, the septa not conspicuously thickened; usually not lichenized **(Dothideales) 226**

222 (219) Ascomata with lateral ostioles; ascospores hyaline, with thickened septa but the lumina
    not lenticular   **Pleurotremataceae**
    Ascomata with vertical ostioles; ascospores hyaline or brown, when septate usually
    with lenticular leumina   **223**

223 (222) Interascal tissue at first composed of simple or sparingly branched trabecular
    pseudoparaphyses, true paraphyses then developing   **Pyrenulaceae**
    Interascal tissue composed of branched and anastomosing pseudoparaphyses **224**

224 (223) Asci with a large refractive apical ring; ascospores hyaline or pigmented; saprobic **Massariaceae**
    Asci without a refractive ring; ascospores usually hyaline; lichenized   **225**

225 (224) Ascomata with a clypeus; asci with a large ocular chamber; ascospores sometimes
    aseptate and ornamented   **Monoblastiaceae**
    Ascomata without a clypeus; ocular chamber small; ascospores always septate **Trypetheliaceae**

226 (221) Ascomata usually pulvinate, without a clearly defined ostiole; asci spherical or broadly
    clavate, usually one or few per locule   **227**
    Ascomata variously shaped; asci usually elongated, borne in a hymenium-like layer,
    often separated by interascal tissue   **230**

227 (226) Keratinophilic; ascospores aseptate, with both ends attenuated   **Piedraiaceae**
    Not keratinophilic; ascospores septate, not attenuated   **228**

228 (227) Ascostromata composed of a basal crustose or pulvinate stroma and short-stalked
    globose outgrowths containing asci singly or in small groups; usually in association
    with scale insects   **Myriangiaceae**
    Ascomata pulvinate or crust-like, without fertile outgrowths; parasitic   **229**

229 (228) Ascomata crust-like; parasitic on leaves   **Elsinoaceae**
          Ascomata crustose or pulvinate; parasitic on other fungi   **Cookellaceae**

230 (226) Ascomata without a clearly defined ostiole, opening by rupture or degradation of apical
          cells into a gelatinous mass; usually pulvinate or strongly flattened; ascospores
          mostly aseptate or with few septa   **231**
          Ascomata with a clearly defined ostiole, often lined with periphyses; opening
          sometimes elongated; ascospores often multiseptate   **238**

231 (230) Ascomata spherical, cleistothecial, thin-walled, often setose; ascospores dark brown;
          often on dung   **232**
          Ascomata not cleistothecial   **233**

232 (231) Ascomata usually setose; ascospores 1-septate, with prominent apical germ pores   **Phaeotrichaceae**
          Ascomata hairy but not setose; ascospores usually many-septate, usually fragmenting,
          with prominent straight or oblique germ slits   **Sporormiaceae**

233 (231) Ascomata formed within a very thin thallus-like stroma   **234**
          Stromatic material either absent or composed of hyphal mats; ascomata usually
          strongly flattened, sometimes elongated   **235**

234 (233) Ascomata usually crustose, sometimes appearing apothecial; stroma composed of dark
          thick-walled pseudoparenchymatous tissue; on rocks or lichens   **Lichenotheliaceae**
          Ascomata strongly flattened; stroma composed of weakly defined radiating tissue;
          parasitic on vascular plants   **Brefeldiellaceae [FIS]**

235 (233) Ascomata walls composed of epidermoid cells   **236**
          Ascomata composed of radially arranged cuboid cells   **237**

236 (235) Asci numerous, clavate to broadly cylindrical, without interascal tissue; ascomata
          elongated, opening by a split   **Aulographaceae**
          Asci globose to saccate, interspersed with ± gelatinized interascal cells; ascomata
          sometimes elongate   **Schizothyriaceae**

237 (235) Ascomata subcuticular; inner layer of ascus wall indistinct; saprobic, mainly on
          ferns   **Leptopeltidaceae**
          Ascomata superficial or subcuticular; asci immersed in gel; inner layer of ascus wall
          well-defined; usually parasitic on angiosperm leaves   **Parmulariaceae**

238 (230) Lichenized with cyanobacteria; interascal tissue absent   **239**
          Mostly non-lichenized; if lichenized, then with green algae; interascal tissue usually
          present   **240**

239 (238) Ascomata minute, pyriform; ascospores dark grey, muriform; thallus filamentous,
          gelatinous; on leaves and bark   **Pyrenothricaceae**
          Ascomata small, ± globose, thick-walled; ascospores brown, with transverse and
          sometimes longitudinal septa; thallus poorly developed, with goniocysts   **Moriolaceae [FIS]**

240 (238) Mycelium superficial, copious, strongly pigmented, rarely penetrating the plant tissue;
          simple hyphomycetous anamorphs usually prominent; ascospores usually
          multiseptate; mostly deriving nutrients from plant exudates   **241**
          Mycelium at least mostly immersed in plant tissue; saprobes or parasites, rarely
          lichenized   **244**

241 (240) Anamorphs coelomycetous, sometimes with hyphomycetous synanamorphs   **242**
          Anamorphs exclusively hyphomycetous   **243**

242 (241) Hyphae deeply pigmented, very irregular in form; conidiomata globose;
          hyphomycetous anamorphs with elongate multiseptate conidia often present   **Antennulariellaceae**
          Hyphae pale to dark brown, usually roughly cylindrical, but the cells often slightly
          inflated; conidiomata elongate, often stalked; hyphomycetous anamorph absent   **Capnodiaceae**

243 (241) Hyphae ± cylindrical; conidiogenous cells not tretic; interascal tissue absent   **Euantennariaceae**
          Hyphae strongly constricted at the septa with almost spherical cells, tapering towards
          the apices; conidiogenous cells tretic; short periphysoids present   **Metacapnodiaceae**

244 (240) Ascomata superficial, strongly flattened   **245**
Ascomata mostly immersed, either within plant material or stromatic tissue; if superficial then ± globose, sometimes stalked   **248**

245 (244) Asci arranged radially around a central column; superficial hyphae strongly flattened, often with thickened septa; ascospores aseptate or with a small appendage-like cell, with a pale transverse band   **Vizellaceae**
Central column absent; superficial hyphae not strongly flattened; ascospores septate, uniformly pigmented   **246**

246 (245) Ascomata weakly to strongly flattened, covered with an often setose layer of dark brown hyphae   **Chaetothyriaceae**
Ascomata strongly flattened; hyphal layer above ascomata absent   **247**

247 (246) Ascomatal wall composed of radiating hyphae, at least in the central region; ascospores usually 1-septate, sometimes with lateral cilia   **Microthyriaceae**
Ascomatal wall composed of epidermoid or irregular hyphal cells; ascospores usually multiseptate, without cilia   **Micropeltidaceae**

248 (244) Ascomata often strongly aggregated and erumpent, globose; asci widely clavate; ascospores mostly aseptate, hyaline or pale yellow   **Botryosphaeriaceae**
Ascospores almost always septate; if not then strongly pigmented   **249**

249 (248) Ascomata perithecia or cleistothecia; ascospores dark brown, with germ pores or slits; isolated from soil or dung   **250**
Ascomata usually stromatic, never cleistothecial; ascospores lacking germ pores and slits   **251**

250 (249) Ascomata usually setose; ascospores 1-septate, with prominent apical germ pores   **Phaeotrichaceae**
Ascomata hairy but not setose; ascospores usually many-septate, usually fragmenting, with prominent straight or oblique germ slits   **Sporormiaceae**

251 (249) Ascomata elongated, the opening slit-like, extending almost the whole length of the ascoma   **252**
Ascomata not elongated   **253**

252 (251) Ascomata rather thin-walled, higher than broad, the tissue around the aperture usually protruding slightly in a small crest; interascal tissue trabeculate, sometimes only periphysoids present   **Mytilinidiaceae**
Ascomata thick-walled, wider than tall; the aperture usually sunken; interascal tissue cellular   **Hysteriaceae**

253 (251) Ascomata depressed globose to pulvinate, apex plane, crumbling open to expose a single layer of asci separated by remnants of interthecial tissue, sometimes with a central sterile region   **Dothioraceae**
Ascomata globose, pyriform or conical, sometimes stalked   **254**

254 (253) Ascomatal cavity containing only asci, though periphysoids sometimes present   **255**
Ascomata containing asci and interascal tissue   **258**

255 (254) Inner wall of ascus remaining thick after ascospore release; ascospores greyish brown or greyish green, usually multiseptate   **256**
Inner wall of ascus thin after ascospore discharge; ascospores hyaline to yellowish, usually 1-septate   **257**

256 (255) Ascomata conical; interascal tissue absent; hymenium staining blue in iodine; lichenized   **Microtheliopsidaceae**
Ascomata usually ± spherical, papillate, often setose and superficial; periphysoids present; hymenium not blueing in iodine; not lichenized   **Herpotrichiellaceae**

257 (255) Ascomata immersed in a well-developed stroma   **Dothideaceae**
Ascomata mostly not stromatic; stroma if present basal   **Mycosphaerellaceae**

258 (254) Ascomata usually spherical, very small, borne on a superficial mycelium; ascomatal wall thin, pliable; saprobic on leaves or parasitic on other fungi   **Pseudoperisporiaceae**
Ascomata usually larger, immersed or superficial; ascomatal wall usually thick   **259**

259 (258) Ascomata immersed in leaf tissue, fleshy, pale; ascospores dark brown, aseptate or 1-
septate, warted or striate   **Mesnieraceae**
Ascomata immersed or superficial, usually with a thick rigid wall; ascospores rarely
ornamented   **260**

260 (259) Parasitic on vascular plants; subcuticular hyphae or stromata often present; ascomata
usually superficial, often hairy; ascospores small, 1-septate   **261**
Parasitic or saprobic, sometimes lichenized; ascomata usually without hairs;
ascospores often large, multiseptate   **265**

261 (260) Ascomata superficial, sometimes with a foot-like stroma   **262**
Ascomata originally immersed but becoming superficial   **264**

262 (261) Ascomata strongly flattened   **Polystomellaceae**
Ascomata not significantly flattened   **263**

263 (262) Ascomata ovoid to globose with a thickened base; ascospores hyaline to brown;
anamorphs coelomycetous   **Parodiellaceae**
Ascomata globose, sometimes stalked; ascospores olivaceous; anamorphs
hyphomycetous   **Parodiopsidaceae**

264 (261) Ascomata immersed in an erumpent pulvinate stroma with spongy walls; hairs absent   **Coccoideaceae**
Ascomata immersed, becoming superficial, not stromatic, often hairy, the wall not
spongy   **Venturiaceae**

265 (260) Commensals or saprobic on bark, some lichenized, loosely associated with algae or
lichenicolous   **266**
Saprobic or parasitic on vascular plants, not associated with algae or lichenicolous   **269**

266 (265) Ascomata clypeate; asci clavate to saccate; lichenized or loosely associated with algae   **267**
Ascomata not clypeate; asci cylindrical to clavate   **268**

267 (266) Ascomata with a single locule; interascal tissue pseudoparaphyses; ascospores
transversely septate   **Arthopyreniaceae**
Ascomata with many locules; interascal tissue paraphysoids; ascospores usually
muriform   **Mycoporaceae**

268 (266) Interascal tissue composed of paraphyses and periphysoids; ascospores muriform with
a large number of cells; often lichenized   **Thelenellaceae [FIS]**
Interascal tissue composed only of anastomosing pseudoparaphyses; ascospores
transversely septate or sometimes muriform   **Dacampiaceae**

269 (265) Ascomatal wall composed of pseudoparenchyma; pseudoparaphyses cellular, narrow
or broad, sometimes deliquescing at maturity; ascospores often radially asymmetrical
and mucous-coated   **270**
Ascomatal wall composed of hyphal cells; pseudoparaphyses trabeculate, narrow,
simple or branched and anastomosing, often becoming mucous; ascospores always
radially symmetrical   **279**

270 (269) Ascomata turbinate, globose or obovoid, the apex sometimes minutely papillate; the
wall often thickened at the base   **Cucurbitariaceae**
Ascomata variously shaped, usually obviously papillate; wall not thickened at the base **271**

271 (270) Ascomata usually strongly flattened, sometimes sphaerical or conical, the wall
thickened at the sides and/or the top   **272**
Ascomata globose to conical or vertically elongated; wall ± equal in thickness   **273**

272 (271) Ascomata wall thickened only laterally; ascospores mostly dark brown, usually with
thickened septa and lenticular lumina, with a large gelatinous sheath   **Pleomassariaceae**
Ascomatal wall thickened both at the sides and top; ascospores hyaline or light brown,
with thin septa and a mucous sheath or appendages   **Lophiostomataceae**

273 (271) Ascomata soft and fleshy, often brightly coloured; parasitic or saprobic on other fungi,
or associated with scale insects   **Tubeufiaceae**
Ascomata dark, the wall usually rather brittle; associated with plants   **274**

274 (273) Ascomatal wall composed of large, often thick-walled pseudoparenchymatous cells   **275**
Ascomatal wall composed of small, sometimes flattened pseudoparenchymatous cells   **278**

275 (274) Ascomata usually opening by removal of a disc-like operculum; ascospores usually
thick-walled, radially asymmetrical   **Diademaceae**
Ascomata opening by a typical lysigenous ostiole   **276**

276 (275) Ascospores muriform, the septa forming simultaneously in a net-like pattern; marine **Lautosporaceae**
Ascospores septate or muriform, with a primary septum followed by secondary septa;
usually terrestrial   **277**

277 (276) Asci rather wide, the inner wall often thickened near the apex; anamorphs
hyphomycetous   **Pleosporaceae**
Asci narrow, the inner wall not obviously thickened; anamorphs coelomycetous   **Leptosphaeriaceae**

278 (274) Ascomata often with a flattened ostiole; ascospores strongly constricted, especially at
the primary and secondary septum; pseudoparaphyses numerous, often rather
narrow   **Lophiostomataceae**
Ascomata with a circular ostiole; if ascospores constricted then only at the primary
septum; pseudoparaphyses sparse   **Phaeosphaeriaceae**

279 (269) Ascomata turbinate or stalked, globose to vertically elongates, the wall often thickened
at the base and/or the apex   **280**
Ascomata globose or slightly flattened, usually papillate; wall not thickened at the base   **281**

280 (279) Ascomata single or gregarious, superficial, vertically elongated or stalked, the wall
soft-textured   **Hypsostromataceae**
Ascomata in valsoid groups, often turbinate, immersed, not stalked, the wall not soft-
textured   **Fenestellaceae**

281 (279) Ascomata flattened, usually immersed beneath a clypeus; the wall thicker in the apical
region   **Didymosphaeriaceae**
Ascomata globose or subglobose, often erumpent; the wall equal in thickness   **Melanommataceae**

BASIDIOMYCOTA

1 Karyogamy somatogamous, often forming a visible basidiome; saprophytes, symbionts
(ectomycorrhizal) or parasites; basidiospores (meiospores) ballistosporic or
statismosporic; clamp-connexions present or absent   **2**
Karyogamy by spermatia, not forming a basidiome; obligate parasites of plants or
insects; basidiospores ballistosporic; clamp-connexions absent; hyphae with dolipore
septum, often with a plug, but lacking parenthesomes   **Teliomycetes**

2 (1) Hyphae with a dolipore septum and parenthesomes   **Basidiomycetes**
Hyphal septum generally lacking a dolipore and always lacking parenthesomes   **Ustomycetes**

Key to Subclasses of Basidiomycetes

1 Basidiospores generally germinating by replication, or by budding, or by formation of
conidia; basidium or sterigmata modified, often either transversely or vertically
septate; lignicolous or mycoparasitic; not ectomycorrhizal   **Phragmobasidiomycetidae**
Basidiospores forming a mycelium directly or rarely replicating in the gelatinous taxa;
basidium clavate, cylindrical, urniform or furcate, lacking internal septation;
epigeous, hypogeous or lignicolous, rarely mycoparasitic; sometimes
ectomycorrhizal   **Holobasidiomycetidae**

Key to orders and families of Phragmobasidiomycetidae

1 Basidium either inflated with the metabasidium cruciately divided by vertical septa, or
partially quadripartite at the apex or retaining an apical tetrad of sessile basidiospores   **(Tremellales) 2**
Basidium cylindrical, transversely septate   **11**

2 (1)   Basidium either partially quadripartite at the apex or statismosporic and retaining an
            apical tetrad of sessile basidiospores; tremelloid haustoria present or absent   **3**
            Basidium cruciately divided   **6**

3 (2)   Basidiospores ballistosporic   **4**
            Basidiospores statismosporic   **5**

4 (3)   Basidiome gelatinous and pustulate or absent and inducing galls on the host,
            mycoparasitic on agarics and corticioid basidiomycetes; basidiospores germinate by
            budding; blastoconidia present   **Syzygosporaceae**
            Basidiome erect, ramarioid, terrestrial; spore repetition absent   **Tremellodondropsidaceae**

5 (3)   Basidiospores hyaline, often catenulate; basidium very elongate, cyindrical, not
            quadripartite; mycoparasitic; basidium absent or rarely stilboid; yeast phase present **Filobasidiaceae**
            Basidiospores brown, encrusted, fused into a tetrad and remaining attached; basidium
            comprising a subglobose probasidium and tapering metabasidium with quadripartite
            apex; basidiome absent or synnematous   **Rhynchogastremaceae**

6 (2)   Basidiospores not produced; probasidium produces a tetrad of cells, with thick walls
            which functions as a deciduous propagule; mycoparasitic   **Tetragoniomycetaceae**
            Basidiospores or ballistoconidia produced   **7**

7 (6)   Basidiome resupinate to pileate and dimidiate; hymenophore poroid; hyphal system
            dimitic with skeletal hyphae; context coriaceous; hymenophore poroid; lignicolous **Aporpiaceae**
            Basidiome gelatinous to horny; hyphal system monomitic; hymenophore never poroid   **8**

8 (7)   Basidiospores statismosporic, symmetric; basidiome stalked with a fertile, capitate
            head; on leaves and dead palms   **Hyaloriaceae**
            Basidiospores ballistosporic   **9**

9 (8)   Basidia basipetally produced, catenulate (4-8); basidiome pustulate to tuberculate **Sirobasidiaceae**
            Basidia produced singly; basidiome resupinate to flabellate, lobate to clavarioid and
            substipitate; hymenophore smooth, papillate to spinose; with haustorial branches;
            lignicolous or mycoparasitic   **10**

10 (9) Haplophase (monokaryons) mycelial, lacking a conidial state   **Exidiaceae**
            Haplophase blastosporic, usually as a *Cryptococcus* anamorph   **Tremellaceae**

11 (1) Basidiospores tetraradiate, sessile, statismosporic; basidiome pycnidial; with
            simultaneous conidiogenesis; spores aggregate in a slime drop at basidiome
            apex   **Heterogastridiales: Heterogastridiaceae**
            Basidiospores unbranched, usually cylindrical, in some cases absent   **12**

12 (11) Basidiome resupinate, pulvinate, foliose, orbicular or auriform, gelatinous to waxy;
            basidiospores cylindrical, often curved, ballistosporic; repetition and budding, with
            curved microconidia   **Auriculariales: Auriculariaceae**
            Basidiome stilboid, rarely resupinate   **13**

13 (12) Basidiospores ballistosporic or statismosporic, thin- or thick-walled, often brown;
            yeast phase often present; single budding   **(Atractiellales) 14**
            Basidiospores statismosporic; multiple budding   **Agaricostilbales: Agaricostilbaceae**

14 (13) Basidiome stilboid, with a fertile head and cylindrical stipe   **15**
            Basidiome pulvinate, gelatinous; clamp-connexions present; basidiospores
            statismosporic, sessile, hyaline, smooth; germination by repetition   **Atractogloeaceae**

15 (14) Clamp-connexions present; basidiospores brown, thick-walled; branched hyphidia;
            conidial state frequent but no yeast phase   **Phleogenaceae**
            Clamp-connexions absent   **16**

16 (15) Basidiospores brown, sessile, thick-walled, smooth; basidia non-septate; conidia
            allantoid   **Pachnocybaceae**
            Basidiospores hyaline   **17**

17 (16)   Basidiospores thick-walled, subellipsoid, with rugose surface; metabasidia septate or
not, with short, straight sterigmata; yeast phase present   **Chionosphaeraceae**
Basidiospores thin-walled, narrowly ellipsoid, smooth; metabasidia elongate, 1-3
septate; conidia present   **Hoehnelomycetaceae**

Key to orders and families of Holobasidiomycetidae

1   Basidium undifferentiated, clavate, urniform to cylindrical, with 2 - 8 apical
sterigmata; basidiome not gelatinized; ballistosporic or statismosporic   **2**
Basidium either pyriform with very stout sterigmata or differentiated into a lower
probasidium and upper epibasidium; basidiome typically but not always gelatinized;
ballistosporic   **113**

2 (1)   Basidiospores constantly statismosporic; basidia with straight, often uneven
sterigmata; development angiocarpic   **3**
Basidiospores mostly ballistosporic but including related statismosporic families;
development gymnocarpic, hemiangiocarpic, rarely angiocarpic   **20**

3 (2)   Spores hyaline or nearly so, smooth or ornamented   **4**
Spores yellowish brown, ferruginous, fuscous brown   **11**

4 (3)   Spores thick-walled, with a reticulato-alveolate ornamentation overlaid by a thick,
gelatinous, inamyloid myxosporium   (**Melanogastrales** *p.p.*) **Leucogastraceae**
Spores with a slightly thickened wall, smooth, sometimes with a loosened
myxosporium   **5**

5 (4)   Gasterocarp small, 1 - 10 mm diam., globose to cupulate, sessile, often gregarious;
hymenium absent, basidia borne singly or in groups; gleba forming one to many
separate peridioles; spores often large, smooth, hyaline   (**Nidulariales**) **Nidulariaceae**
Gasterocarp large, pileate or multipileate, with mucilaginous or soft fleshy,
deliquescent gleba borne on a rapidly expanding pseudostipe or receptacle arising
within the peridium; hymenium present; peridioles absent; spores small, ellipsoid or
cylindrical, hyaline to pale green   (**Phallales**) **6**

6 (5)   Gasterocarp remaining hypogeous at maturity, globose to tuberiform, with indehiscent
peridium and lacking a receptacle   **7**
Gasterocarp epigeous a maturity, with a dehiscent peridium releasing a mucilaginous
gleba supported by a rapidly expanding receptacle   **9**

7 (6)   Peridium not gelatinized; gleba olivaceous grey with radiating tramal plates, finally
deliquescent and foetid   **Hysterangiaceae**
Peridium with a gelatinized layer   **8**

8 (7)   Columella white to translucent, gelatinized which branches radially in the gleba;
lacking a foetid odour   **Protophallaceae**
Columella white, fleshy, unbranched; gleba solid; odour foetid at maturity   **Gelopellaceae**

9 (6)   Gasterocarp multipileate, with the gleba borne on inner surfaces of a reticulate to
stellate receptacle   **Clathraceae**
Gasterocarp unipileate   **10**

10 (9)   Gleba borne externally on the upper part of unbranched, cylindrical, hollow receptacle   **Phallaceae**
Gleba borne internally within an unopened peridium   **Claustulaceae**

11 (3)   Gasterocarp truly stipitate, subtending the fertile body; hymenium absent, basidia in
scattered groups throughout the gleba; capillitium present   (**Tulostomatales**) **Tulostomataceae**
Gasterocarp sessile or forming a pseudostipe, lacking a differentiated stipe   **12**

12 (11)   Hymenium present, basidia maturing simultaneously; capillitium present; gleba at first
whitish; spore ornament obtusely verrucose   (**Lycoperdales**) **13**
Hymenium not developed, basidia borne singly or in groups, maturing at different
times; capillitium absent; gleba at first violaceous to purple; spore ornament spinose
to reticulate   (**Sclerodermatales**) **17**

13 (11)   Compound basidiome bearing numerous spore sacs on a more or less massive stroma;
              outer peridium forming a thin universal veil breaking away at maturity   **Broomeiaceae**
              Basidiome simple   **14**

14 (13)   Peridium usually indehiscent, rather thick, 3-4 layered; gleba usually radially arranged **Mesophelliaceae**
              Peridium dehiscent, thick or thin; gleba usually with pseudocolumella   **15**

15 (14)   Peridium dehiscing by irregular lobes from the apex; gleba homogenous   **Mycenastraceae**
              Peridium apically dehiscent; gleba with a prominent pseudocolumella or not   **16**

16 (15)   Outer peridium splitting radially to form stellate rays, revealing an endoperidium
              which is apically dehiscent; capillitium non-septate, typically unbranched and
              forming a pseudocolumella   **Geastraceae**
              Peridium not splitting radially, dehiscence either by an apical pore or by apical
              fragmentation and flaking away; capillitium septate or not, typicaly branched, not
              forming a columella   **Lycoperdaceae**

17 (12)   Compound gasterocarp bearing numerous spore sacs on a more or less massive stroma;
              outer peridium forming a thin universal veil breaking away at maturity   **Diplocystaceae**
              Gasterocarp simple   **18**

18 (17)   Gasterocarp up to 3 mm diam., on wood or dung; peridium with stellate, apical
              dehiscence to reveal a single, globose peridiole; inner peridiole wall everts to allow
              violent release of peridiole   **Sphaerobolaceae**
              Gasterocarp much larger, dehiscence variable, not stellate   **19**

19 (18)   Peridium complex and stratified; splitting stellately from the apex; gasterocarp
              geastroid, sessile   **Astraeaceae**
              Peridium not stratified, fracturing irregularly to expose the gleba; gasterocarp globose
              to tuberous, sometimes with a pseudostipe   **Sclerodermataceae**

20 (2)   Setoid structures or dichohyphidia present; context hyphae usually dextrinoid and
              mostly xanthochroic   **21**
              Neither setoid structures nor dichohyphidia present; context hyphae never
              xanthochroic   **24**

21 (20)   Setoid structures present; dichohyphae and dichohyphidia absent; generative hyphae
              always lacking clamp-connexions; euhymenial   **(Hymenochaetales) 22**
              Setoid structures absent; dichohyphae and/or dichohyphidia present; generative hyphae
              without or, more rarely, with clamp-connexions; catahymenial; mostly tropical or
              warm localities   **(Lachnocladiales) 23**

22 (21)   Asterosetae present in context and often in hymenium; haplosetae sometimes present;
              basidiome resupinate, creamy buff or tawny brown, thin and membranous; hyphal
              system monomitic, with xanthochroic generative hyhae; spores hyaline, globose,
              smooth or tuberculate, mostly amyloid, sometimes inamyloid   **Asterostromataceae**
              Asterosetae or dicho-hyphidia present; haplosetae and/or macrosetae often present;
              basidiome resupinate, clavarioid, hydnoid or poroid; context tawny brown; hyphal
              system monomitic or dimitic with euskeletal hyphae; spores hyaline or brown   **Hymenochaetaceae**

23 (21)   Basidiome erect, ramarioid to pileate; clamp-connexions present   **Lachnocladiaceae**
              Basidiome resupinate, pale buffy brown; clamp-connexions absent or more rarely
              present   **Dichostereaceae**

24 (20)   Generative hyphae inflated; hyphal system monomitic, rarely dimitic; basidiomes
              putrescent, mostly clavarioid, cantharelloid, agaricoid or boletoid   **25**
              Generative hyphae not inflated; hyphal system mono- or dimitic; basidiomes mostly
              persistent, variable, ranging from corticioid or clavarioid to pileate and stereoid,
              hydnoid or poroid, rarely pseudolamellate   **76**

25 (24)   Hymenium thickening to greater or lesser extent; hymenophore smooth, wrinkled or
              folded, with principal folds radially arranged, or hydnoid; hypogeous taxa not known **26**
              Hymenium not thickening unless the hymenophore is unspecialised (and then with
              smooth, brown basidiospores); hymenophore ranging from lamellate, tubulate; with
              many hypogeous relatives   **40**

26 (25)   Basidiospores hyaline, thin-walled, smooth; basidia 2 - 8 sterigmate; context negative to ferric chloride **(Cantharellales) 27**

Basidiospores yellowish brown to subhyaline, ornamented, rarely smooth, ellipsoid to elongate, cyanophilic; context green positive to ferric chloride **(Gomphales) 38**

27 (26)   Hymenophore amphigenous, at least on upper part of basidiome **28**

Hymenophore confined to lower or outer surface; basidiomes differentiated, not simple-clavarioid **33**

28 (27)   Context dimitic, with skeletal hyphae; clamp-connexions present or absent; basidiomes erect or decumbent, filiform, branched **Pterulaceae**

Context monomitic, with inflated generative hyphae **29**

29 (28)   Basidiome small, lignicolous or epiphyllous, with a differentiated head and stipe **30**

Basidiome larger, clavarioid, without a distinct head **31**

30 (29)   Basidiome with an inflated and hollow head and solid stipe, small, simple; oleocystidia present; erect or pendent **Physalacriaceae**

Basidiome with a solid head; cystidia absent; generally erect, often but not always arising from a sclerotium **Typhulaceae**

31 (29)   Basidia with two, strongly curved sterigmata; basidiome clavarioid, simple or branched **Clavulinaceae**

Basidia mostly with four straight or slightly curved sterigmata **32**

32 (31)   Basidiomes small to medium, simple or branched, usually gregarious; basidiospores smooth or echinulate, typically uniguttulate **Clavariaceae**

Basidiomes robust, clavarioid or truncate, solitary; basidiospores smooth, multiguttulate **Clavariadelphaceae**

33 (27)   Basidiome compound, with numerous branches **34**

Basidiome simple, with a single, differentiated pileus **35**

34 (33)   Basidiome ramarioid, small, slender, with flattened, erect branches; context with thickened, uninflated hyphae; terrestrial **Aphelariaceae**

Basidiome coralloid, large, with numerous, flattened, horizontal branches; context with inflated hyphae; lignicolous **Sparassidaceae**

35 (33)   Basidiome pseudoagaricoid, with smooth to wrinkled hymenophore and thickening hymenium **36**

Basidiome with spinose or tubulate hymenophore; hymenium not thickening **37**

36 (35)   Basidiome pileate with a solid stipe; fleshy; hymenophore developing radial folds and cross-veining **Cantharellaceae**

Basidiome infundibuliform, with a hollow stipe; subcoriaceous; hymenophore smooth to wrinkled **Craterellaceae**

37 (35)   Hymenophore tubulate **Scutigeraceae**

Hymenophore spinose **Hydnaceae**

38 (26)   Spore-print ochraceous; basidiospores ornamented, verrucose to ridged **39**

Spore-print white; basidiospores smooth; basidiomes ramarioid, erect or pendent, simple or branched **Lentariaceae**

39 (38)   Basidiome cantharelloid to agaricoid, turbinate to infundibuliform; lepto- or gloeocystidia present **Gomphaceae**

Basidiome ramarioid, erect, branched; cystidia absent but sometimes gloeoplerous hyphae **Ramariaceae**

40 (25)   Context heteromerous, with sphaerocytes and filamentous hyphae, lacking clamp connexions; basidiospores hyaline, ornamented, with an amyloid myxosporium; basidiome agaricoid or gasteroid **(Russulales) 41**

Context homoiomerous, not forming sphaerocytes **42**

41 (40)   Basidiospores ballistosporic, asymmetric, bearing a suprahilar plage and tapering hilar
      appendix; basidiome agaricoid to gasteroid, always stipitate   **Russulaceae**
      Basidiospores statismosporic, symmetric, lacking a suprahilar plage, with a
      subcylindrical hilar appendix, and often a sterigmal appendage; basidiome gasteroid,
      stipitate or sessile   **Elasmomycetaceae**

42 (40)   Hymenophoral trama bilateral when present, often mucilaginous; spore-print brown,
      pinkish brown or fuliginous black, ellipsoid to typically fusoid, smooth or
      ornamented; hymenophore tubulate or pseudolamellate; basidiome boletoid,
      agaricoid or gasteroid, but with hydnoid and merulioid relatives   **(Boletales) 43**
      Hymenophoral trama regular to irregular, never mucilaginous or if bilateral then
      basidiospores hyaline or pink; spore-print white, pink, brown or black; basidiome
      agaricoid, with true lamellae, cupulate or gasteroid; hymenophore truly lamellate on
      reduced; basidiome agaricoid or gasteroid   **53**

43 (42)   Basidiomes ranging from resupinate to pileate, boletoid to agaricoid; basidiospores
      ballistosporic, sometimes statismosporic   **44**
      Basidiomes always gasteroid, hypogeous to subepigeous; basidiospores statismosporic   **51**

44 (43)   Basidiome boletoid or agaricoid; hymenophore lamellate to tubulate   **45**
      Basidiome not boletoid or agaricoid, ranging from resupinate to pileate; hymenophore
      smooth to merulioid or spinose   **52**

45 (44)   Clamp-connexions present; basidiospores short, subglobose to ellipsoid, lacking a
      suprahilar depression   **46**
      Clamp-connexions absent; hymenophore tubulate or sometimes secondarily lamellate;
      basidiospores often with a suprahilar depression   **48**

46 (45)   Hymenophore truly lamellate, decurrent, at times furcate; pileal margin involute   **47**
      Hymenophore tubulate, sometimes radial or labyrinthoid, arcuato-decurrent to
      adnexed; basidiospores olivaceous brown, smooth   **Gyrodontaceae**

47 (46)   Basidiospores ferruginous or hyaline, never olivaceous brown, smooth or rarely
      echinate to verruculose; also includes gasteroid genera   **Paxillaceae**
      Basidiospores hyaline to yellowish, dextrinoid   **Hygrophoropsidaceae**

48 (45)   Spore-print olivaceous brown, never pinkish, vinaceous or fuscous black   **49**
      Spore-print black, fuscous brown, vinaceous or pinkish, never with an olivaceous tint **50**

49 (48)   Basidiome robust; stipe stout, with a scabrous or reticulate surface; hymenophore
      sinuato-adnexed, always tubulate; basidiospores either ovoid to fusoid-cylindric and
      smooth or subglobose with a reticulato-ridged ornamentation   **Boletaceae**
      Basidiome small to medium; stipe slender, not reticulate; hymenophore adnate to
      decurrent, tubulate or secondarily lamellate; basidiospores fusoid-cylindric, smooth
      or longitudinally striate to ridged   **50**

50 (48/49) Hymenophore tubulate; basidiospores either fuscous brown to black and ornamented or
      vinaceous to pinkish and smooth   **Strobilomycetaceae**
      Hymenophore uniformly secondarily lamellate, never tubulate; basidiospores fuscous
      to black, large, elongate cylindrical   **Gomphidiaceae**

51 (43)   Basidiospores allantoid to subglobose, stramineous, smooth; gasterocarp lacking a
      columella   **Rhizopogonaceae**
      Basidiospores fusoid with longitudinal ridges, golden brown; gasterocarp with reduced
      columella   **Chamonixiaceae**

52 (44)   Basidiome resupinate to pileate; hymenophore smooth, grandinioid or spinose;
      basidiospores hyaline to yellowish, ellipsoid, verrucose   **Beenakiaceae**
      Basidiome resupinate or effuso-reflexed; hymenophore smooth to merulioid or
      pseudoporoid; basidiospores yellowish to brownish, ellipsoid to fusoid, smooth   **Coniophoraceae**

53 (42)   Basidiome agaricoid, epigeous; basidiospores smooth, with a complex wall, with or
      without a germ-pore   **54**
      Basidiome gasterocarpic, hypogeous to subepigeous; tuberiform, mucilaginous,
      lacking a stipe-columella; basidiospores dark olive-brown to black, thick-walled,
      fusoid, smooth   **(Melanogastrales** *p.p.***) 75**

54 (53) Spore-print white to cream, pink, cinnamon brown, fuliginous, purplish brown or black; basidiospores thin- to thick-walled, with or without a germ-pore or ornamentation   **(Agaricales) 55**
Spore-print rusty brown to ferruginous brown; basidiospores ochraceous brown, thick-walled, typically with echinate to verrucose ornamentation, typicaly lacking a germ-pore   **(Cortinariales) 70**

55 (54) Hymenophoral trama bilateral, with lateral strata divergent or convergent   **56**
Hymenophoral trama regular to irregular, not bilateral   **59**

56 (55) Lamellae decurrent, thick, waxy, often distant; basidia elongate; spore-print white   **Hygrophoraceae**
Lamellae free, thin; spore-print whitish to pink   **57**

57 (56) Spore-print white to cream; hymenophoral trama with divergent lateral strata   **58**
Spore-print dull pink; hymenophoral trama bilateral, with convergent lateral strata   **Pluteaceae**

58 (57) Ballistosporic; basidiome agaricoid, with lamellae; often with velar layers; spores globose to elongate ellipsoid; terrestrial   **Amanitaceae**
Statismosporic; basidiome gasterocarpic, loculate, with or without a pseudostipe and volva; lignicolous, terrestrial or marine; spores ellipsoid or tetraradiate   **Torrendiaceae**

59 (55) Spores angular, not terete but either longitudinally ridged or with 5-13 facets; spore-print pink   **60**
Spores different, never both angular and pink   **61**

60 (59) Ballistosporic; basidiome agaricoid, lamellate   **Entolomataceae**
Statismosporic; basidiome gasterocarp, with or without stipe-columella, loculate   **Richoniellaceae**

61 (59) Lamellae thick and waxy; basidia elongate; spores hyaline   **Hygrophoraceae**
Lamellae not thickened; basidia not exceptionaly elongated   **62**

62 (61) Spore-print white, cream or pink, never dark; spores always lacking a germ-pore; when a partial veil is present the spores are neither amyloid and nor dextrinoid   **63**
Spore-print pale or dark, if pale then either the basidiome is either lepiotoid, has velar structures or the spores have a complex wall, a germ-pore or ornamentation   **64**

63 (62) Ballistosporic; basidiome agaricoid, lamellate; spores ornamentation various   **Tricholomataceae**
Statismosporic; basidiome gasterocarpic, sublamellate to loculate, hypogeous or epigeous; spores subglobose, spinose   **Hydnangiaceae**

64 (62) Basidiome with velar structures, frequently forming an annulus on the stipe or squamules on the pileus; spore-print variously coloured but never rusty- or cinnamon-brown   **65**
Not this combination of characters   **66**

65 (64) Ballistosporic; basidiome agaricoid, lamellate   **Agaricaceae**
Statismosporic; basidiome gasterocarpic, subagaricoid with distorted lamellae   **Secotiaceae**

66 (64) Ballistosporic; basidiome agaricoid   **67**
Statismosporic; basidiome gasterocarpic or sub agaricoid   **69**

67 (66) Pileipellis forming a true epithelium; spores with a germ-pore; basidia short, often pyriform   **68**
Pileipellis never a continuous epithelium; spore-print cinnamon-brown, fuscous or lilaceous-black; chrysocystidia often present   **Strophariaceae**

68 (67) Spore-print black, purplish brown, fuliginous to black   **Coprinaceae**
Spore-print argillaceous brown, ochraceous brown or rusty brown   **Bolbitiaceae**

69 (65) Peridium expanding; glebal lamellae powdery or autolysing; basidiospores reddish brown to black   **Podaxaceae**
Peridium narrowly conical, not expanding; gleba of anastomising tramal plates; basidiospores argillaceous- to rusty-brown   **Galeropsidaceae**

70 (54)   Basidiome agaricoid; gymnocarpic or hemiangiocarpic; hymenophore lamellate;
               ballistosporic   **71**
               Basidiome gasteroid; angiocarpic; hymenophore loculate; statismosporic   **72**

71 (70)   Spore-print pale cinnamon-brown; spores with a simple wall, although at times with an
               exosporial ornamentation   **Crepidotaceae**
               Spore-print rusty- to ferruginous brown; spore-wall thickened, either smooth or with an
               ornamentation which may be exosporial or myxosporial   **Cortinariaceae**

72 (70)   Gasterocarp secotioid, with a stipe and lamellate gleba; basidiospores pale yellowish
               brown, with a complex wall of an echinulate exosporium overlaid by a myxosporium   **Cribbeaceae**
               Gasterocarp hypogeous to subepigeous, lacking a stipe and gleba loculate   **73**

73 (72)   Basidiospores with conspicuous, meridional, solid costae; peridium evansescent; gleba
               labyrinthoid   **Gautieriaceae**
               Basidiospores smooth or veruculose, often with a membranous myxosporium but never
               costae   **74**

74 (73)   Peridiopellis an epithelium; basidiospores mucronate, verruculose, with a loose
               myxosporium   **Hymeniangiaceae**
               Peridiopellis an epicutis of repent hyphae; basidiospores coarsely ornamented, with
               tuberculate verrucae   **Octavianinaceae**

75 (53)   Gasterocarps large, multiloculate; basidiospores yellowish-brown   **Hymenogasteraceae**
               Gasterocarp minute, uniloculate, basidiospores dark-brown, thick-walled   **Gasterellaceae**

76 (24)   Basidiospores brown (very rarely hyaline) and ornamented   **77**
               Basidiospores hyaline and smooth (rarely with amyloid ornamentation)   **80**

77 (76)   Basidiospores golden brown (rarely hyaline), angular, with verrucose or nodulose
               ornamentation; context monomitic, rarly dimitic, releasing a green pigment with
               alkali; basidiome effused to pileate, lacking a surface crust; hymenophore smooth,
               papillate, hydnoid, pseudolamellate or tubulate   **(Thelephorales) 78**
               Basidiospores yellowish brown, subglobose to ellipsoid, with a complex wall of a
               verrucose or ridged eusporium overlaid by a persistent, membranous myxosporium;
               context dimitic with skeletal and/or skeleto-ligative hyphae; basidiome pileate, with a
               conspicuous, often waxy, crust; hymenophore tubulate   **(Ganodermatales) 79**

78 (77)   Spore-print white or pale; hymenophore spinose; basidome pileate and stipitate   **Bankeraceae**
               Spore-print deep brown; hymenophore smooth to papillate, very rarely tubulate or
               pseudolamellate   **Thelephoraceae**

79 (77)   Basidiospores subglobose to ellipsoid or truncate, with a verrucose to reticulate
               exosporium, overlaid by a persistent, smooth myxosporium; basidiome sessile to
               stipitate; context pale to dark brown   **Ganodermataceae**
               Basidiospores ellipsoid, with an exosporium of longitudinal crests; basidiome stipitate;
               context cream to ochraceous   **Haddowiaceae**

80 (76)   Hymenophore comprising crowded, individual tubules or cupules to appear
               pseudotubulate or pseudolamellate; basidiospores small, hyaline, thin-walled, smooth   **81**
               Hymenophore smooth, hydnoid, lamellate or tubulate   **83**

81 (80)   Basidiome pileate; context watery-fleshy context; hymenophore of densely crowded,
               free tubes, the individual tubes constricted at their bases   **(Fistulinales) Fistulinaceae**
               Basidiome resupinate to agaricoid with a lateral stipe; context membranous to firm-
               coriaceous; hymenophore at first smooth then cupulate, with the cupules sometimes
               proliferating and compressed to appear pseudolamellate   **(Schizophyllales) 82**

82 (81)   Basidiome resupinate; hymenophore at first smooth developing aggregated
               cupules   **Stromatoscyphaceae**
               Basidiome pleurotoid; hymenophore with aggregated cupules compressed radially to
               produce a 'split-lamellate' condition   **Schizophyllaceae**

83 (80)   Basidiospores amyloid, mostly ornamented   **84**
               Basidiospores inamyloid or amyloid (but never forming both amyloid spores and
               gloeoplerous hyphae)   **91**

84 (83)   Basidiospores with characteristic ridged ornament; cystidia absent; basidiome either
flabelliform, ramarioid or gasteroid   **(Bondarzewiales) 85**
Basidiospores finely echinate or smooth; gloeoplerous hyphae abundant; basidiome
resupinate, effuso-reflexed, clavarioid or pileate   **(Hericiales) 87**

85 (84)   Basidiospores statismosporic; gasterocarp loculate, with branched columella;
angiocarpic   **Hybogasteraceae**
Basidiospores ballistosporic; gymnocarpic   **86**

86 (85)   Basidiome clavarioid, with flabellate branching; hymenophore smooth   **Amylariaceae**
Basidiome compound, with flabellate pilei; hymenophore tubulate   **Bondarzewiaceae**

87 (84)   Hymenophore spinose   **88**
Hymenophore ranging from smooth to lamellate,not spinose   **89**

88 (87)   Context tough to coriaceous; hyphal system monomitic or dimitic with skeletal hyphae;
basidiome effuso-reflexed to pileate, sessile or stipitate   **Auriscalpiaceae**
Context soft-fleshy to membranous; hyphal system monomitic; basidiome pileate or of
isolated spines; sessile or with short stalk   **Hericiaceae**

89 (87)   Basidiome resupinate to effuso-reflexed; hymenophore smooth to tuberculate; context
hypochnoid to membranous   **Gloeocystidiellaceae**
Basidiome erect, stipitate   **90**

90 (89)   Basidiome clavarioid, with pyxidate branching; hymenophore smooth; hyphal system
monomitic or dimitic   **Clavicoronaceae**
Basidiome agaricoid ; hymenophore lamellate, serrulate; hyphal system monomitic   **Lentinellaceae**

91 (83)   Hymenophore tubulate, rarely pseudolamellate; basidiome resupinate to pileate and
stipitate; basidiospores inamyloid   **(Poriales) 92**
Hymenophore smooth to tuberculate, rarely tubulate; basidiome mosty resupinate to
effuso-reflexed, rarely stipitate   **(Stereales) 95**

92 (91)   Hymenophore lamellate; basidiome centrally to laterally stipitate, sometimes sessile;
hyphal system monomitic or dimitic with skeletal or skeleto-ligative hyphae;
basidiopores cylindrical   **Lentinaceae**
Hymenophore tubulate, very rarely secondarily lamellate; hyphal system dimitic   **93**

93 (92)   Basidiome resupinate; hymenophore irpicoid to poroid, pale to dark bluish grey;
skeleto-ligative hyphae present; coralloid dendrohyphidia often present   **Grammotheleaceae**
Basidiome resupinate to dimidiate or stipitate; hymenophore truly tubulate   **94**

94 (93)   Basidiome stipitate; stipe frequently with a black crust; skeleto-ligative hyphae present   **Polyporaceae**
Basidiome resupinate to pileate, rarely stipitate; skeletal and/or skeleto-ligative hyphae
present   **Coriolaceae**

95 (91)   Basidiospores hyaline, smooth or verrucose   **97**
Basidiospores yellowish, ellipsoid, with spinose to ridged ornamentation, inamyloid   **96**

96 (95)   Basidiome epigeous, resupinate on plant debris; hymenophore smooth to poroid;
basidiospores ballistosporic, lacking a peri-appendicular corona   **Lindtneriaceae**
Basidiome hypogeous, angiocarpic, with labyrinthoid chamber; basidiospores
statismosporic, with a peri-appendicular corona   **Stephanosporaceae**

97 (95)   Basidiospores amyloid   **98**
Basidiospores inamyloid   **102**

98 (97)   Hyphal system dimitic; skeletohyphidia usually present   **99**
Hyphal system monomitic; skeletohyphidia absent   **100**

99 (98) Hyphal system with skeletal or skeleto-ligative hyphae; skeletohyphidia abundant and
conspicuous; basidiome resupinate to pileate and ungulate, often perennial;
hymenophore smooth to spinose; basidiospores smooth or usually verrucose **Echinodontiaceae**
Hyphal system with skeletal hyphae, rarely monomitic; skeletohyphidia scattered,
often inconspicuous; basidiome resupinate to flabelliform, annual; hymenophore
smooth to tuberculate; basidiospores smooth **Stereaceae**

100 (98) Dendrohyphidia present; basidia terminal **101**
Dendrohyphidia absent, leptocystidia present and conspicuous; basidia pleural;
hymenophore smooth; basidiome resupinate, ceraceous; spores mostly amyloid **Xenasmataceae**

101 (100) Basidiospores very large, smooth or verrucose; hymenophore smooth to tuberculate;
basidiome resupinate, often discoid, to cupulate **Aleurodiscaceae**
Basidiospores small to medium, smooth; hymenophore smooth to poroid; basidiome
membranous, resupinate **Amylocorticaceae**

102 (97) Hyphal system dimitic **103**
Hyphal system monomitic **105**

103 (102) Basidiome spathulate to infundibuliform, often stipitate; hymenophore smooth,
tuberculate or ribbed; skeletal or skeleto- ligative hyphae present; skeletocystidia
often present **Podoscyphaceae**
Basidiome resupinate to effuso-reflexed; hymenophore odontoid, spinose to poroid;
skeletal hyphae present **104**

104 (103) Skeletocystidia abundant, often encrusted; hyphal pegs absent **Steccherinaceae**
Skeletocystidia absent; hymenophore odontoid, with hyphal pegs **Epitheleaceae**

105 (102) Hymenial cystidia present **106**
Hymenial cystidia absent **110**

106 (105) Hymenium thickening; hymenophore smooth, merulioid to pseudoporoid; basidiome
resupinate, effuso-reflexed to pileate, firm-gelatinous; cystidia leptocystidioid, at
times absent **Meruliaceae**
Hymenium not thickenng **107**

107 (106) Lyocystidia present, cylindrical, thick-walled, partly soluble in KOH, with subcapitate
apex, often encrusted; hymenophore smooth **Tubulicrinaceae**
Lyocystidia not developed **108**

108 (107) Cystidia gloeocystidioid, often becoming thick-walled and encrusted; spore deposit
pinkish; basidiospores smooth; basidiome resupinate to effuso-reflexed, ceraceous to
crustose; hymenophore smooth to papillate **Peniophoraceae**
Gloeocystidia absent; spore deposit white; basidiome resupinate **109**

109 (108) Lampro-, lepto- or septocystidia present and usually prominent; basidiospores smooth
or verrucose; hymenophore smooth to spinose **Hyphodermataceae**
Dendrohyphidia present, rarely absent; basidiospores smooth; hymenophore smooth to
tuberculate **Corticiaceae**

110 (105) Basidiome cupulate, sessile to slightly stipitate; membranous; basidiospores often large **Cyphellaceae**
Basidiome resupinate, rarely pileate, never cupulate **111**

111 (110) Context formed by broad, loosely woven hyphae, with right-angled branching; basidia
short, clustered, with 4-6 (-8) sterigmata; hymenophore smooth **Botryobasidiaceae**
Context hyphae not showing distinctive right-angled branching **112**

112 (111) Basidia urniform, 4-6-8 sterigmate; basidiome resupinate to pileate; hymenophore
smooth to spinose; context hyphae loosely arranged and inflated at the septa **Sistotremataceae**
Basidia narrowly clavate, 4 sterigmate, clustered; basidiome resupinate, loose,
pellicular; hymenophore smooth; hyphae normal **Atheliaceae**

113 (1) Probasidium absent; large, stout sterigmata present; spore replication sometimes
occurring; basidiome effused, pellicular, sometimes
gelatinous **(Ceratobasidiales) Ceratobasidiaceae**
Basidia forming probasidium and epibasidium **114**

114 (113) epibasidium of two cylindrical branches, each bearing a terminal sterigma;
basidiospores cylindrical, often transversely septate, germinating to produce
microconidia but no other secondary spores   **(Dacrymycetales) 115**
Probasidium subglobose, apically developing a tetrad of swollen, lacrymoid epibasidia,
each with a terminal sterigma; basidiospores subglobose, ellipsoid, allantoid to
helicoid; secondary spores common but no yeast phase; basidiome ephemeral,
membranous or gelatinized, resupinate   **Tulasnellales (Tulasnellaceae)**

115 (114) Basidiome pustulate, clavarioid, spathulate, simple or branched, with a narrow
attachment to substratum; gelatinous   **Dacrymycetaceae**
Basidiome resupinate, broadly effused over substratum with broad attachment;
membranous-ceraceous   **Cerinomycetaceae**

Key to orders and families of Teliomycetes

1   Parasitic with scale insects; corticioid basidiomes; basidia
auricularioid   **(Septobasidiales) Septobasidiaceae**
Parasitic with angiosperms, gymnosperms and pteridophytes; basidiome not formed;
probasidium (teliospore) persistent or not; complex life histories that include 1-4
morphologically distinct sori   **(Uredinales) 2**

2 (1)   Telium erumpent   **3**
Telium remaining embedded in host tissue   **4**

3 (2)   Teliospores sessile   **6**
Teliospores pedicellate   **10**

4 (2)   Telium composed of a crust of teliospores one cell deep   **5**
Telium composed of a crust of teliospores two or more cells deep   **Phakopsoraceae**

5 (4)   Aecium peridermioid   **Pucciniastraceae**
Aecium caemoid   **Melampsoraceae**

6 (3)   Pycnia discoid, without bounding structures, subepidermal; urediniospores catenulate **Coleosporiaceae**
Pycnia either discoid with bounding structures, or flat, intraperidermal, or ampulliform,
subepidermal; urediniospores, if formed, pedicellate   **7**

7 (6)   Teliospores catenulate   **8**
Teliospores not catenulate   **9**

8 (7)   Teliospores 1-celled; pycnia intra- or subperidermal   **Cronartiaceae**
Teliospores 2-celled by horizontal septation; pycnia simple, subepidermal,
ampulliform, with paraphyses   **Pucciniosiraceae**

9 (7)   Teliospores laterally free but grouped on sporogenous basal cells; pycnia ampulliform
and lobed, deep seated in host   **Mikronegeriaceae**
Teliospores laterally free, not grouped on sporogenous basal cells; pycnia discoid,
subcuticular to subepidermal, with bounding structures   **Chaconiaceae**

10 (3)   Pycnia flat or discoid   **11**
Pycnia ampulliform   **Pucciniaceae**

11 (10)   Pycnia with bounding structures   **12**
Pycnia without bounding structures   **Phragmidiaceae**

12 (11)   Teliospores of two or more cells, variously arranged   **13**
Teliospores unicellular   **Pileolariaceae**

13 (12)   Teliospores divided into two or more cells by vertical or muriform septation **14**
Teliospores two (occasionally more) celled by horizontal septation **Uropyxidaceae**

14 (13)   Teliospores one per pedicel, three or more celled by horizontal and vertical septation; pedicels unicellular   **Sphaerophragmiaceae**
              Teliospores either vertically septate or two or more radially arranged on each pedicel, sometimes forming complex discoid spore heads; pedicels commonly with specialized apical cells or cysts   **Raveneliaceae**

Key to orders and families of Ustomycetes

1   Hymenial, with palisadic basidia ('ustidia'); basidiome often present   **2**
              Not hymenial; spores solitary or clustered, or produced in sori   **6**

2 (1)   Phragmobasidioid, basidia auricularioid, with transversely septate metabasidia; basidiome resupinate, pezizoid to erect and pileate, fleshy, gelatinous or waxy, saprobic on wood, litter or dung or mycoparasitic; germination of sporidia by repetition   **(Platygloeales) Platygloeaceae**
              Holobasidioid; basidia clavate, with 2-8 apical sterigmata; hymenium formed on surface of hypertrophied or modified, angiosperm host tissue or erumpent through stomata   **3**

3 (2)   Basidiospores thick-walled, septate, statismosporic; basidia thin- or thick-walled; sterigmata very short   **(Cryptobasidiales) Cryptobasidiaceae**
              Basidiospores thin-walled, septate or not, ballistosporic or statismosporic; basidia thin-walled; sterigmata stout   **(Exobasidiales) 4**

4 (3)   Stromatic sheath formed below host epidermis; basidia narrowly clavate, with 2-8 sterigmata; basidiospores ballistosporic, aseptate or with 1-5 cross-walls; blastoconidia common   **Exobasidiaceae**
              Basidia erumpent through host stomata   **5**

5 (4)   Basidia clavate, with 6-8 sterigmata; basidiospores statismosporic, with yeast-like budding   **Microstromataceae**
              Probasidia persistent, pyriform to clavate; metabasidium cylindrical, with two short sterigmata; basidiospores ballistosporic, germinating with either a germ-tube or conidia   **Brachybasidiaceae**

6 (1)   Probasidia hyaline, thin-walled, smooth   **7**
              Probasidia colourless or pigmented, thick-walled, often ornamented   **8**

7 (6)   Obligate parasites of ascomycetes, causing botryose outgrowths of host; metabasidium subulate, non-septate; basidiospores statismosporic, fusoid; hyphal septa with associated Woronin-bodies   **Cryptomycocolacales (Cryptomycocolacaceae)**
              Obligate parasites of *Palmae*, forming black stromata with cupulate locules surrounding basipetally produced chains of hyaline fertile cells often interspersed with columns of sterile hyphae (elaters)   **(Graphiolales) Graphiolaceae**

8 (6)   Saprobic, on dead leaves, not forming sori; probasidium smooth, usually brown, terminal or intercalary, producing ballisto- or statismospores; mitotic yeast phase with multiple budding   **(Sporidiales) Sporidiobolaceae**
              Parasitic on plants; probasidium formed in sori   **(Ustilaginales) 9**

9 (8)   Probasidia almost invariably brown (or purplish) and ornamented; metabasidia transversely septate, producing lateral and terminal basidiospores; basidiospores globose to cylindrical   **Ustilaginaceae**
              Probasidia either hyaline or brown, smooth or ornamented; metabasidia non-septate, producing basidiospore apically, basidiospores mostly filiform   **Tilletiaceae**

CHYTRIDIOMYCOTA

1   Obligately anaerobic fungi from the guts of herbivores; zoospore mono- or polyflagellate   **(Neocallimastigales) Neocallimastigaceae**
              Saprobic or parasitic; predominantly aerobic; not from guts of herbivores   **2**

2 (1) Zoospores usually lacking lipid globules; with a definite nuclear cap surrounding the anterior of the nucleus; thallus usually bearing a thick-walled, punctate or ornamented, asexually formed resting spore at some stage in the life history; mono- or polycentric, hyphal or rhizomycelial; sexuality by iso- or anisoplanogametes, heterothallism common; aquatic or from soil, many parasites of plants or invertebrates **(Blastocladiales) 12**

Zoospores usually with lipid globules; thallus not forming thick-walled, ornamented resting spores **3**

3 (2) Zoospores somewhat apically pointed, lipid globules evident in anterior; thallus unbranched, eucarpic with basal or subbasal, disc-like holdfast, or differentiated into a well-developed hypha-like vegetative system with oogamous sexuality; saprobic **(Monoblepharidales) 16**

Not with above combination of characters **4**

4 (3) Zoospore usually with several (sometimes one) lipid globules, sometimes amoeboid when in motion; rhizoids when present with blunt tips >0.5 μm diam; predominantly from soil **(Spizellomycetales) 5**

Zoospore usually with a single, spherical lipid globule or with several lipid globules, not amoeboid when in motion; rhizoids, if present, tapering to a fine tip <0.5 μm diam. **(Chytridiales) 8**

5 (4) Thallus polycentric; parasitic on higher plants **Urophlyctidaceae**

Thallus monocentric **6**

6 (5) Aerial sporangiocarp absent, slender-stalked; thallus monocentric, with haustorium penetrating host; parasites of fungi **Caulochytriaceae**

Aerial sporangiocarp absent **7**

7 (6) Zoosporangium developing from expansion of zoospore cyst (endogenous development) **Spitzellomycetaceae**

Zoosporangium developing exterior to the zoospore cyst (exogenous development) **Olpidiaceae**

8 (4) Thallus uni-axial, eucarpic with basal or sub-basal holdfast, comprising upper region which forms zoospores and lower nucleated, vegetative region; capable of repeated sporulation **Harpochytriaceae**

Thallus eucarpic or holocarpic, monocentric or polycentric, but not as above **9**

9 (8) Thallus holocarpic, formed exogenous to zoospore cyst, at maturity developing a sorus, prosorus or resting spore **Synchytriaceae**

Thallus eucarpic or holocarpic, not forming a sorus **10**

10 (9) Thallus eucarpic, polycentric, comprising rhizomycelium and reproductive bodies **Cladochytriaceae**

Thallus eucarpic or holocarpic, monocentric **11**

11 (10) Thallus eucarpic, monocentric; zoospore cyst expanding to form sporangium (endogenous development) **Chytridiaceae**

Thallus eucarpic or holocarpic, monocentric; nucleus leaving zoospore cyst and swelling in germ tube to become sporangium or prosporangium (exogenous development) **Endochytriaceae**

12 (2) Specialized endoparasites of invertebrates; sporophyte thallus in dipterans, unwalled, branched or lobed, converting into a mass of thick-walled, variously ornamented resting spores which crack open and function as zoosporangium upon germination; gametophyte phase in microcrustaceans **Coelomomycetaceae**

Not specialized as above; thallus walled, mono- or polycentric, or colonial; **13**

13 (12) Endobiotic stage colonial, parasitic in tardigrades; epibiotic stage polycentric, branching, bearing rhizoids, extramatrical on previously infected tardigrade **Sorochytriaceae**

Lacking a colonial stage **14**

14 (13) Life history involving a monocentric, epibiotic zoosporangium and a polycentric, endobiotic system bearing septate, turbinate cells, from which many thick-walled resting spores are formed; parasites of phanerogams **Physodermataceae**

Thallus polycentric, hyphal, rhizomycelial, or with an expanded basal cell; or monocentric; not parasites of phanerogams **15**

15 (14)   Thallus polycentric, rhizomycelial with catenulate swellings separated by sterile
        isthmuses; saprophytic or parasitic   **Catenariaceae**
        Thallus mono- or polycentric, typically with a ± prominent basal part on which are
        borne one or more reproductive structures; or hyphal; saprophytic   **Blastocladiaceae**

16 (3)   Thallus an unbranched filament, attached to substratum by holdfast; sporangia
        maturing in basipetal succession, forming H-shaped segments; saprophytic   **Oedogoniomycetaceae**
        Thallus differentiated into a well-developed, hyphal, vegetative system bearing
        numerous reproductive structures; oogamous sexuality   **17**

17 (16)   Zygote undergoing a period of motility before encystment, propelled by flagellum of
        male gamete; mycelium with or without pseudosepta; oogonium with one or more
        oospheres   **Gonapodyaceae**
        Reproductive structures remaining in oogonium or encysting on the opening of the
        oogonium; male gamete completely engulfed at fertilization; mycelium not
        pseudoseptate   **Monoblepharidaceae**

DICTYOSTELIOMYCOTA

1   Sorophores composed of strongly vacuolate cells enveloped by a cellulosic sheath or
        tube   **(Dictyosteliales) Dictyosteliaceae**
        Sorophores very thin and consisting of acellular cellulosic tubes   **(Dictyosteliales) Acytosteliaceae**

HYPHOCHYTRIOMYCOTA

1   Thallus intramatrical, monocentric or polycentric, with proliferating branches,
        rhizoidal system absent; sporangia with thin walls   **(Hyphochytriales) Hyphochytriaceae**
        Thallus epibiotic, normally monocentric, rarely apophysate; sporangia with thin or
        thick walls, the latter sometimes bullately ornamented   **(Hyphochytriales) Rhizidiomycetaceae**

LABYRINTHULOMYCOTA

1   Trophic cells normally uninucleate, bounded by one or two layers of scales; trophic
        cells embedded within the ectoplasmic net   **(Labyrinthulales) Labyrinthulaceae**
        Trophic cells becoming multinucleate, bounded by several layers of scales; trophic
        cells superficial to the ectoplasmic net   **(Thraustochytriales) Thraustochytriaceae**

MYXOMYCOTA

1   Trophic phase simple amoebae or plasmodia having no shuttle movement of the
        protoplasm   **(Protosteliomycetes, Protosteliales) 2**
        Trophic phase a free-living, multinucleate, coenocytic, saprobic plasmodium having
        schuttle movement of the protoplasm   **(Myxomycetes) 4**

2 (1)   Flagellate cells absent   **Protosteliaceae**
        Flagellate cells present   **3**

3 (2)   Spores borne singly at the tips of hair-like stalks, on columnar, dendroid or morchelloid
        sporophores   **Ceratiomyxaceae**
        Spores (1-4) borne borne in sporangiate sporophores with short stalks not on hair-like
        stalks or morchelloid sporophores   **Cavosteliaceae**

4 (1)   Sporophore development subhypothallic, the plasmodium protoplast rising internally
        through the developing stalk in stipitate forms   **5**
        Sporophore development epihypothallic, stalk, when present, secreted internally,
        hollow, or partially filled with strands   **(Stemonitales) Stemonitidaceae**

5 (4) Spores in mass black, violet-brown, dark purple-brown, ferruginous, deep red or
purple; by transmitted light usually purple-brown or deeply tinted, rarely pallid **(Physarales) 6**
Spores in mass pallid, white or bright coloured, sometimes brown, rarely black; by
transmitted light pallid to bright coloured, yellow-brown, smoky-brown or dingy, but
never purple-brown 7

6 (5) Peridial lime granular; capillitium calcareous, rarely limeless, usually consisting either
of calcareous tubules or of limeless, slender tubules connecting calcareous nodes **Physaraceae**
Peridium lime granular or crystalline; capillitium consisting of typically limeless
thread-like, dark to pallid tubules, these occasionally exhibiting inclusions of
crystalline lime, sometimes scanty, rarely lacking **Didymiaceae**

7 (5) Peridium persistent in whole or in part; true capillitium lacking; pseudocapillitium
when present of tubular, irregular filaments or of perforated plates which may fray
out into threads **(Liceales) 8**
True capillitium typically present; if lacking then peridium early fugacious and
sporocarps minute 10

8 (7) Sporophores sporangiate or plasmodiocarpous, generally small, often minute;
pseudocapillitium and dictydine granules lacking **Liceaceae**
Sporophores sporangiate, pseudoaethalioid or aethalioid; pseudocapillitium or
dictydine granules may be present 9

9 (8) Dictydine granules lacking **Lycogalaceae**
Dictydine granules present; peridium usually netted, interstices usually fugacious **Cribrariaceae**

10 (7) Stalked, minute, rarely exceeding 0.5 mm in total height; peridium sometimes
persistent in part or in whole; stalk filled with granular material; columella typically
present, rarely lacking; capillitium, when present, scanty or forming an open globose
net or branching and anastomosing; spores in mass white, yellowish, pinkish, gray or
brown 11
Sessile or stalked, larger; columella absent; capillitium usually abundant, of individual
threads or a network, the threads sometimes delicate and solid, but mostly course,
hollow and conspicuously sculptured; spores in mass white, ochraceous, yellow,
orange , or red **(Trichiales) 12**

11 (10) Spores in mass white, cream-coloured, yellow, pink, gray, or rarely pink-brown;
peridium delicate, evanescent at an early stage 13
Spores in mass brown; peridium persistent, either as a whole or in fragment which cling
to the tips of the capillitium **(Echinosteliales) Clastodermataceae**

12 (10) Capillitium threads typically slender, rarely exceeding 2μm diam., appearing solid,
smooth or ornamented **Dianemataceae**
Capillitium threads typically coarser, rarely below 2μm diam., tubular, generally
conspicuously ornamented 14

13 (11) Flagellate trophic phase present; nuclei with a single, central
nucleolus **(Echinosteliales) Echinosteliaceae**
Flagellate trophic phase absent; nuclei with many peripheral
nucleoli **(Echinosteliopsidales) Echinosteliopsidaceae**

14 (12) Capillitium threads with thickenings in the form of spirals **Trichiaceae**
Capillitium threads smooth or thickened with cogs, rings, half rings, spines or warts,
never with spirales **Arcyriaceae**

OOMYCOTA

1 Motile cells with one straminipilous flagellum or two isokont or anisokont flagella, one
of which is straminipilous; thalli normally walled during the assimilative
phase **(Oomycetes *s.l.*) 2**
Motile cells with one or two anisokont flagella, but neither flagellum known to be
straminipilous; thalli naked during the assimilative phase [excluded from the
Oomycetes s.l.] **(Rozellopsidales) 27**

2 (1)   Thallus walled, at least towards the end of the assimilative phase; thallus, if
             endoparasitic, within the host plasmamembrane; gametes, if discrete,
             non-motile   **(Oomycetes** *s.str.***) 3**
             Thallus plasmodial throughout the assimilative phase; thallus between the host
             plasmamembrane and frustule or cell wall of the host; gametes motile
             [marine]   **Lagenismales: Lagenismaceae**

3 (2)   Thallus filamentous, mycelial with branched hyphae, occasionally septate (rarely of
             very narrow diameter (*Pythiogeton, Verrucalvus*), or allantoid with or without septa
             (*Lagenidium, Myzocytium*) but then with extrasporangial homohylic vesicular
             discharge); zoosporogenesis intrasporangial or extrasporangial in a homohylic
             vesicle; oogonia and antheridia usually differentiated   **4**
             Thallus polar, blastic, coralloid, allantoid or olpidioid (rarely pseudomycelial);
             zoosporogenesis usually intrasporangial but if extrasporangial then without a
             homohylic vesicle; gametangia frequently not morphologically differentiated   **9**

4 (3)   Oosporogenesis centripetal with persistent periplasm; sporangio-conidiophores well-
             differentiated [mostly stem and leaf parasites of dicotyledons]   **(Peronosporales) 5**
             Oosporogenesis centrifugal or centripetal with insignificant periplasm; sporangio-
             conidiophores rarely differentiated   **6**

5 (4)   Mycelium intercellular with large, lobate haustoria; sporangio-conidiophore frequently
             dichotomously branched, sporangia or sporangio-conidia borne on pedicels (downy
             mildews)   **Peronosporaceae**
             Mycelium intercellular with small spherical haustoria; sporangio-conidiophore
             percurrent with chains of zoosporangia (white blister rusts)   **Albuginaceae**

6 (4)   Hyphae slender (< 20 μm diam.) and of uniform diameter   **7**
             Hyphae often stout (*c.* 20 μm diam.) with diameter increasing with age (up to 150 μm
             diam.) [zoosporogenesis intrasporangial; oosporogenesis centrifugal; oogonia often
             pluriovulate, oospores usually aplerotic; oospore with a fluid and granular ooplast,
             lipid coalescence variable (limited or to a single eccentric
             globule)]   **(Saprolegniales) Saprolegniaceae**

7 (6)   Hyphae of extremely narrow diameter (<5 μm diam.); zoosporogenesis intrasporangial
             or absent; oogonial wall thick, often verrucate; oospores plerotic or nearly so; oospore
             with a homogeneous ooplast, lipid coalescence limited [parasites of *Poaceae*]   **(Sclerosporales) 8**
             Hyphae usually 6-10 μm diam.; zoosporogenesis intrasporangial with a
             plasmamembranic vesicle or extrasporangial in a homohylic vesicle, rarely absent;
             oogonial wall usually thin; oospores never strictly plerotic; oospore with a hyaline
             ooplast, lipid coalescence minimal   **(Pythiales) Pythiaceae**

8 (7)   Haustoria present; zoosporangia or conidia borne on inflated sporangio-conidiophores,
             pseudo-dichotomously branched (graminicolous downy mildews)   **Sclerosporaceae**
             Haustoria not known; zoosporangia without sporangiophores or asexual phase not
             known (graminicolous root parasites)   **Verrucalvaceae**

9 (3)   Oosporogenesis with nucleated periplasm; zoosporogenesis intrasporangial or with a
             plasmamembranic vesicle   **10**
             Oosporogenesis without nucleated periplasm; zoosporogenesis intrasporangial or with
             a precipitative vesicle   **11**

10 (9)   Thallus polar with an inflated basal cell and rhizoids; thalloid segments separated by
             short narrow thick-walled isthmuses; gametangia differentiated into oogonia and
             antheridia; saprophytic   **(Rhipidiales) Rhipidiaceae**
             Thallus allantoid, non-polar and without rhizoids; gametangia not differentiated,
             presumed to be automictic; parasitic, endobiotic   **Lagenaceae [FIS]**

11 (9)   Thallus blastic, allantoid or coralloid to pseudomycelial; freshwater or terrestrial
             families (rarely with marine species)   **12**
             Thallus tubular, saccate or olpidioid, rarely lobed but not blastic or coralloid; marine
             families   **19**

12 (11) Straminipilous flagella with tripartite tubular hairs arranged in two rows   **13**
             Straminipilous flagella with tripartite tubular hairs arranged in a single
             row   **Crypticolaceae [FIS]**

13 (12) Zoospores (or zoospore cysts or aplanospores) small or very small (<60 μm³ volume), with sub-apical flagellar insertion; obligate endobiotic parasites; thallus olpidioid or allantoid, never septate **14**

Zoospores (or zoospore cysts) medium-small (>60 μm³ volume) or large (>300 μm³ volume), with lateral flagellar insertion at least in the second phase in polyplanetic species; saprophytic or parasitic; thallus usually elongate, often partitioned by constrictions or septa, rarely olpidioid (*Eurychasmidium*) **15**

14 (13) Zoosporangia with a single exit tube; teleomorphic, sexual reproduction presumed to be heterothallic with donor gametangial thalli of very different volume from the receptor thalli (*c.* 10%); fertilization via an insipient fertilization tube, oospores plerotic; often inducing extramural deposits on oogonial, companion cell and sporangial thalli; parasites of *Peronosporomycetes* and *Chlorophyceae* **(Olpidiopsidales) Olpidiopsidaceae**

Zoosporangia frequently with more than one exit tube; anamorphic; extramural fibrillar deposits not known; parasites of *Bacillariophyceae* and *Aschelminthes* **Ectrogellaceae [FIS]**

15 (13) Thallus blastic, coralloid, lobed, rarely olpidioid, seldom septate, segments not disarticulating; zoospores medium-large or large (>175μm³ volume); sexual reproduction homothallic, automictic or heterothallic; donor gametangia (when present) smaller than the oogonium; fertilization by means of a fertilization tube or fertilization hypha; oospores with a nearly hyaline ooplast and diverse patterns of lipid coalescence **16**

Thallus initially tubular, sometimes branched, becoming septate and often disarticulating; zoospores medium-small (<150 μm³ volume); sexual reproduction homothallic or automictic with gametangia of more or less equal volume; fertilization by copulation through a pore; oospore ooplast and lipid reserve distribution not established; parasitic in nematodes and rotifers, or algae **(Myzocytiopsidales) Myzocytiopsidaceae**

16 (15) Thallus blastic (rarely olpidioid), composed of allantoid segments connected by simple constrictions, occasionally with a few rhizoids; donor and receptor gametangia differentiated, gametangia terminal or intercalary, apparently delimited by plugs at the constrictions, olpidioid taxa heterothallic **(Leptomitaceae and Apodachlyellaceae) 17**

Thallus tubular with swellings or coralloid (rarely of unconstricted hyphae of very narrow (<5 μm diam.); gametangia little differentiated **(Ducellieriaceae and Leptolegniellaceae) 18**

17 (16) Separable gametic membranes not distinguished; fertilization by means of a short fertilization tube; oospores single and plerotic [oogonia rarely pluriovulate (*Plerogone*), rarely anamorphic (*Leptomitus*)]; zoospore cysts sometimes with short hollow protrusions; often found on corticated twigs at lake margins **Leptomitaceae**

Separable gametic membranes present; fertilization by means of a fertilization hypha; oospores aplerotic; sometimes parasitic in aquatic protists **Apodachlyellaceae**

18 (16) Homothallic, often automictic; gametangial zones usually little or not differentiated, often without delimiting septa, intercalary, as lateral diverticula, or rarely differentiated and terminal (*Aphanodictyon*); pluriovulate; oospore wall layers separated by fluid zones; zoosporogenesis variable; oogonia often pluriovulate; oospores aplerotic; keratinophilic saprophytes or parasites of algae (*Aphanomycopsis*) or nematodes (*Nematophagus*) **Leptolegniellaceae**

Heterothallic (?); zoospores emerging to form a fenestrated hollow sphere of cysts, connected by hollow protrusions from each cyst; oogonia uniovulate; oospore plerotic (?); saprophytic in pollen floating on lakes **Ducellieriaceae**

19 (11) Parasitic in marine animals or saprophytic; zoospores medium small to large (mostly >60 μm³ volume) **20**

Parasitic in marine algae or fungi parasitic in marine algae; zoospores very small (<25 μm³ volume) [many diagnostic details not recorded] **23**

20 (19) Parasitic in *Crustacea* or saprophytic; thallus pseudomycelial or coralloid, cytoplasm often with prominent granulation; oospore with a multilayered wall and granular ooplast, lipid droplets condensed to varing degrees **(Salilagenidales) 21**

Parasitic in *Aschelminthes* [see key points 14, 15] **22**

21 (20) Thallus more or less tubular, rarely septate; zoospores medium-large (>275 μm³ volume); teleomorphic; oospores with granular ooplasts and lipid coalescence **Salilagenidiaceae**

Thallus tubular or saccate, sometimes septate; zoospores medium-small (<275 μm³ volume); anamorphic; facultatively parasitic or saprophytic **Haliphthoraceae**

22 (20) Thallus becoming septate; thalloid segments behaving as sporangia usually with a
single exit tube, or gametangia   **(Myzocytiopsidales) Myzocytiopsidaceae**
Thallus never septate; thallus behaving as a sporangium with one to several exit tubes
[sporangiospores transforming into glossoid (gun cell) spores]   **Ectrogellaceae**

23 (19) Zoospores biflagellate; flagellar insertion not well-defined   **24**
Zoospores reputedly uniflagellate; flagellar insertion not well-defined [resting spores
reported with or without companion cells; parasitic in
Phaeophyceae]   **(Anisolpidiales) Anisolpidiaceae**

24 (23) Rarely causing hypertrophy of the host cell; zoosporogenesis intrasporangial but
zoospores not known to encyst prematurely to form a 'net' sporangium   **25**
Causing marked hypertrophy of host cell; zoosporogenesis often with a 'net
sporangium' (intrasporangial encystment of the first planetic stage) [anamorphic;
parasitic in *Phaeophyceae* or *Rhodophyceae*]   **Eurychasmataceae [FIS]**

25 (24) Parasitic in *Phaeophyceae*, *Chlorophyceae*, or marine fungal parasites associated with
these algae   **26**
Parasitic in *Rhodophyceae* [thallus lobed and non-septate (*Petersenia*) or tubular and
more or less septate (*Pontisma*)]   **Pontismaceae [FIS]**

26 (25) Thallus often septate, segments not disarticulating; primarily parasitic in
Chlorophyceae; teleomorphic   **Sirolpidiaceae [FIS]**
Thallus never septate; parasitic in *Bacillariophyceae* [sexuality reported for *Ectrogella
perforans*, a species which may not be congeneric with the type
species]   **Ectrogellaceae [FIS]**

27 (1) Parasitic in protozoa; plasmodium not filling host cell, sometimes schizogenous;
zoospores uniflagellate or biflagellate [many diagnostic details imprecisely
recorded]   **Pseudophaeritaceae**
Parasitic in fungi (especially Peronosporomycetes and Chytridiomycetes) and algae;
plasmodium (and asexual sporangium) filling or almost filling the host compartment;
zoospores uniflagellate or biflagellate; resting spores formed without any known
union; often with fibrillar or spiny ornamentation; aplerotic in the host
compartment]   **Rozellopsidaceae**

PLASMODIOPHOROMYCOTA

1   Phagotrophic from the start; nuclear division cruciform; endoparasites of plant
cells   **(Plasmodiophorales) Plasmodiophoraceae**
Initially osmotrophic; nuclear division not cruciform; endoparasites of ciliate
protozoa   **(Plasmodiophorales) Endemosarcaceae**

ZYGOMYCOTA

1   Saprobic or, if parasitic or predacious, having mycelium immersed in the host tissue **(Zygomycetes) 2**
Associated with arthropods and attached to the cuticle or digestive tract by a holdfast
aand not immersed in the host tissue   **(Trichomycetes) 39**

2 (1)   Mycelium ± regularly septate, septa with plugs; zygospores, where known, formed on
undifferentiated hyphae   **3**
Mycelium, septa and zygospores not as above   **4**

3 (2)   Merosporangia bispored, septal plugs with polar protuberance, dissolving in 2-3%
KOH   **(Dimargaritales) Dimargaritaceae**
Merosporangia unispored, septal plugs without polar protuberance, not dissolving in
KOH   **(Kickxellales) Kickxellaceae**

4 (2)   Sporangia or 'conidia' forcibly released, or if not forcibly released then non-haustorial
parasites of cicadas or nematodes with the latter producing two or more pedicellate,
globose, spinose 'conidia' terminally and laterally from a coenocytic, erect pedicel **5**
Sporangia not forcibly released; if non-haustorial parasites of nematodes spores formed
in chains or if solitary, not both globose ans spinose; fertile hyphae often septate at
maturity   **6**

5 (4) Sporangiophore arising from trophocyst; sporangia multispored, thallus
mycelial   **(Mucorales,** *p.p.*) **11**
Sporangiophore not arising from trophocyst; 'conidia' few- or unispored; thallus
mycelial or consisting of hyphal bodies or protoplast-like
cells   **(Entomophthorales,** *p.p.*) **28**

6 (4) Spores formed in sporocarps produced in soil or on organic material above ground,
saprobes or ecto- or endomycorrhizal; or spores formed singly in soil,
endomycorrhizal and forming arbuscules in roots of phycobiont   **7**
Spores not formed as above; saprobes or parasites   **10**

7 (6) Sporocarps formed   **8**
Sporocarps not formed, spores formed singly in soil   **(Glomales,** *p.p.*) **33**

8 (7) Sporocarps containing only multispored sporangia   **(Mucorales,** *p.p.*) **11**
Sporocarps containing zygospores, azygospores or chlamydospores   **9**

9 (8) Only zygospores present in sporocarp; saprobes or ectomycorrhizal   **(Endogonales) Endogonaceae**
Sporocarps containing azygospores or chlamydospores;
endomycorrhizal   **Glomales,** *p.p.*) **33**

10 (6) Saprobes or facultative parasites; sporangiospores formed in sporangia, sporangiola, or
less commonly merosporangia; arthrospores, chlamydospores and/or yeast-like cells
sometimes present   **(Mucorales,** *p.p.*) **11**
Obligate haustorial or non-haustorial parasites of fungi or small animals or their eggs;
sporangiospores formed in multispored or unispored merosporangia; chlamydospores
sometimes formed   **(Zoopagales) 35**

11 (5) Sporangia ± lageniform, multispored; sporangiospores hyaline, smooth-walled,
without appendages   **12**
Sporangia usually globose to obpyriform, never lageniform   **13**

12 (11) Sporangia formed singly or in pairs; rhizoides present   **Saksenaeaceae**
Sporangia borne verticellately; rhizoids absent   **Mucoraceae** (*p.p.*)

13 (11) Sporangia formed in a sporocarp   **Mortierellaceae** (*p.p.*)
Sporangia not formed in a sporocarp   **14**

14 (13) Uni- or multispored merosporangia produced on a fertile vesicle   **Syncephalastraceae**
Merosporangia not produced; sporangia and/or sporangiola present   **15**

15 (14) Sporangiospores fusiform to broadly fusiform, often with hyaline, hair-like, polar
appendages; sporangia non-apophysate, wall persistent, fracturing regularly into 2 or
irregularly into 3-4 segments by a preformed suture   **16**
Sporangiospores not with the above combination of characters, usually smooth-walled,
rarely ornamented or appendaged; sporangium wall deliquescent' if persistent
fracturing irregularly; sporangia sometimes absent and only sporangia produced   **17**

16 (15) Sporangiospore wall pale brown to reddish-brown; sporangia only or sporangia and
sporangiola (on separate sporangiophores) produced; orange pigment (carotene)
usually produced in mycelium on rich media; zygosporangium smooth, thin, hyaline;
zygospore reddish-brown, wall striate; suspensors apposed   **Choanephoraceae**
Sporangiospore wall hyaline; sporangia only produced; mycelium hyaline;
zygosporangium verruculose, blackish-brown; suspensors opposed   **Gilbertellaceae**

17 (15) Sporangia uni- to multi-spored, acolumellate (columella-like structure septum-like or
dome-shaped, never protruding into the sporangium); mycelium usually extremely
fine   **Mortierellaceae**
Not with the above combination of characters; sporangia columellate; sporangiola
columellate or acolumellate, sometimes arising from a vesicle; mycelium more robust   **18**

18 (17) Sporangia never formed; sporangiola uni- or multispored, pedicellate, arising from a
fertile vesicle   **19**
Sporangia formed; sometimes sporangiola also present   **24**

19 (18) Fertile vesicles produced from primary vesicles present on uniseptate sporangiophore
branch; uni- or multi-spored pedicellate or denticulate , acolumellate sporangiola
present; zygospores where known smooth-walled, suspensors with appendages **Radiomycetaceae**
Not with the above combination of characters; fertile vesicles produced terminally on
the sporophore or its branches or on a lateral vesicle; sporangiola columellate or
acolumellate; sterile spines sometimes present   **20**

20 (19) Sporangiola columellate, uni- or multi-spored; saprophytic or gall-forming
mycoparasites; sporophore sometimes bearing sterile spines   **21**
Sporangiola acolumellate, uni- or multi-spored; sterile spines on sporophore absent   **22**

21 (20) Sporangiola uni-spored; sporophore bearing sterile spines   **Chaetocladiaceae**
Sporangiola multi-spored; sterile spines on sporophore absent   **Thamnidiaceae** (*p.p.*)

22 (20) Sporangiospore wall pale brown to reddish-brown, striate; zygospores with apposed
suspensors   **Choanephoraceae** (*p.p.*)
Sporangiospore wall hyaline, smooth or spinose; zygospore with opposed suspensors   **23**

23 (22) Sporophore mostly branched; fertile vesicles terminal and lateral; sporangiola unispored, pedicel monomorphic, sporangiospores released by random fracture of the
pedicel and/or dissolution of the sporangiolar wall; yeast-like cells absent   **Cunninghamellaceae**
Sporophore simple or rarely once or twice branched; fertile vesicles terminal;
sporangiola uni- or multispored, pedicel mono- or dimorphic, sporangiospores
released by fracture of the pedicel at preformed cicumscissile zone, sporangiolum
wall persistent, separable from the sporangiospore wall; yeast-like cells produced
after spore germination on rich media   **Mycotyphaceae**

24 (18) Deliquescent sporangia and persistent-walled sporangiola present   **Thamnidiaceae** (*p.p.*)
Only sporangia produced, either with persistent or deliquescent wall   **25**

25 (24) Sporangia ± apophysate   **Mucoraceae** (*p.p.*)
Sporangia non-apophysate; sporophore sometimes vesiculate or constricted below the
sporangium   **26**

26 (25) Sporangium wall blackish-brown to black, persistent; trophocyst sometimes produced;
subsporangial vesicle sometimes produced, globose or broadly ellipsoid; sporangium
liberated whole by collapse of the subsporangial vesicle or by cicumscissile rupture
of the junction between the sporangium wall and the sporophore, or sometimes
forcibly discharged; zygospore usually yellowish, suspensors apposed,
zygosporangium thin, hyaline   **Pilobolaceae**
Sporangium liberation not as described above; subsporangial vesicles absent or small
and hemispherical and then with a deliquescent sporangium; zygospore suspensors
opposed or tongs-like   **27**

27 (26) Sporophore unbranched, usually bluish-green; zygospore suspensors tongs-like, with
branched appendages   **Phycomycetaceae**
Sporophore branched or unbranched, never bluish-green; zygospore suspensors
opposed, without appendages   **Mucoraceae** (*p.p.*)

28 (5)   Nuclei large, >10 μm long, not staining with aceto-orcein, with a prominent nucleolus;
sporophores with a vesicle subtending the spore ('conidium'); gametangia, if present,
with small, beak-like accessory cells; contents of secondary spores and somatic
hyphae in some species may divide ad produce small protoplast-like cells (appearing
multispored); all cells uninucleate   **Basidiobolaceae**
Nuclei small, < 10 μm long, aceto-orcein nuclear staining and nucleolus visibility
variable; sporophores without a subtending vesicle; gametangia, if present, without
accessory cells; spores and mycelium not producing internal protoplast-like cells;
cells not all uninucleate   **29**

29 (28) Nuclei variable, 2-15 μm long during interphase, with much heterochromatin (staining
readily in aceto-orcein and bismark brown); nucleolus not prominent; nucleus
remaining visible during mitosis   **Entomophthoraceae**
Not as above   **30**

30 (29) Nuclei relatively large (>5 μm long), nucleoplasm appearing granular (chromatin condensed at interphase); sporophores short and unbranched, arising from sporogenous cells; spores forcibly discharged by papillar eversion; obligate intracellular parasites of fern gametophytes **Completoriaceae**

Nuclei 2-5 μm long, with little heterochromatin (not staining in aceto-orcein or bismark brown), usually inconspicuous during mitosis; nucleolus prominent; spores discharged by papillar eversion, fluid discharge, or passively released; saprobes or parasites of insects, freshwater algae or soil invertebrates **31**

31 (30) Sporophores upright, simple, each bearing several spores; parasitic on soil invertebrates (nematodes and tardigrades) **Meristacraceae**

Sporophores usually unbranched, one spore borne terminally; saprobes or parasites of insects, soil invertebrates, freshwater algae or causing mycotic infections of animals **32**

32 (31) Spores (asexual spores and zygospores) tending to be melanized (pale grey to black); zygospores budding from conjugation point of two hyphal bodies; obligate parasites of mites and insects (esp. *Homoptera*) **Neozygitaceae**

Spores hyaline; zygospores formed in axis of parential hyphae as a result of conjugation between adjacent cells or scalariform conjugations between two hyphae; saprobes or obligate parasites of desmid algae, soil invertebrates, or facultative mycotic disease agents of animals **Ancylistaceae**

33 (7) Only arbuscules formed in mycorrhizal roots; azygospore-like bodies formed at the apex of sporogenous cells of fertile hyphae; auxiliary cells present **Gigasporaceae**

Arbuscules and vesicles formed in mycorrhzal roots; chlamydospores formed terminally and laterally on or within fertile hyphae; auxiliary cells absent **34**

34 (33) Chlamydospores formed apically **Glomaceae**

Chlamydospores formed from within the neck of a sporiferous saccule **Acaulosporaceae**

35 (10) Spores solitary, short pedicellate, formed on the surface of a vesicle or sessile and formed in pairs or chains from the coiled apex of a sporophore; spores of some taxa large, brown **36**

Spores solitary (unispored) or in chains (multispored) and borne in merosporangia, long or short pedicellate, or sessile, or arising from a head-cell or a secondary branchlet, if arising from a vesicle usually multispored; unispored merosporangia not large and brown; sporophore apex not coiled **37**

36 (35) Fertile hyphae several times dichotomously branched, septate, terminating in sterile spines; fertile vesicles pedicellate, arising in pairs from the cell at the branching point of the fertile hyphae; spores ± hyaline, relatively small; parasites of fungi **Sigmoideomycetaceae**

Fertile hyphae simple, coenocytic, without sterile spines; sporophore apex forming a fertile vesicle or not swollen and coiled; spores brown, large; parasites of microfauna or their eggs **Helicocephalidaceae**

37 (35) Haustorial mycoparasites, especially of Mucorales **Piptocephalidaceae**

Haustorial ectoparasites or non-haustorial endoparasites of soil microfauna **38**

38 (37) Predacious, with adhesive material on hyphae to trap prey; vegetative mycelium formed outside host **Zoopagaceae**

Ecto- or endoparasitic; vegetative mycelium within host, only sporophores formed outside host **Cochlonemataceae**

39 (1) Spores (trichospores) produced exogenously, usually bearing one or more basal appendages; zygospores produced in most genera **(Harpellales) 40**

Spores (sporangiospores) or amoeboid cells produced endogenously, or thallus breaking up into arthrospores; zygospores not present **41**

40 (39) Thallus unbranched, usually attached to the peritrophic membrane of the midgut (less common to the hindgut cuticle) **Harpellaceae**

Thallus branched, attached to the hindgut cuticle **Legeriomycetaceae**

41 (39) Thallus branched and septate, producing arthrospores **(Asellariales) Asellariaceae**

Thallus unbranched, or branched only at the base; non-septate; producing sporangiospores or amoeboid cells **42**

42 (41)   No amoeboid cells; sporangiospores usually produced singly in basipetal series of
terminal sporangia **(Eccrinales) 43**
Amoeboid cells produced at some stage; entire thallus functioning as sporangium,
releasing spores or amoeboid cells more or less simultaneously   **(Amoebidiales) Amoebidiaceae**

43 (42)   Thalli producing directly only one type of spore; sporangia sometimes germinating in
situ **Palavasciaceae**
Thalli usually producing at least two types of spores; sporangia not germinating in situ **44**

44 (43)   Primary infestation spores produced in thalli that become converted entirely or partly
into multispored sporangia **Parataeniellaceae**
Primary infestation spores produced singly in series of terminal sporangia **Eccrinaceae**

Systematic arrangement

This section lists all generic names included in the *Dictionary* according to their systematic position. To facilitate reference, the phyla are arranged alphabetically, as are the orders within the phyla (or class), and the families within the orders. Families (incertae sedis) not referred to any order are listed at the end of the appropriate phylum, and genera (incertae sedis) not placed in families are compiled at the end of the appropriate phylum or order.

The only exception to this arrangement are for Mitosporic fungi and Fossil fungi. In these non-systematic categories, the genera are listed alphabetically without further subdivision.

In all cases, the names of genera treated as accepted in this edition of the *Dictionary* are placed in roman type. Synonymous generic names appear in *italic* type in the same alphabetical sequence together with an indication of the genus of which they have been considered a synonym.

The pages on which the listings for the phyla, classes, and other categories of names start are as follows:

Acrasiomycota	541
Ascomycota	541
Basidiomycota	574
Basidiomycetes	574
Teliomycetes	586
Ustomycetes	588
Chytridiomycota	589
Dictyosteliomycota	590
Hyphochytriomycota	590
Labyrinthulomycota	590
Myxomycota	590
Oomycota	591
Plasmodiophoromycota	592
Zygomycota	592
Trichomycetes	592
Zygomycetes	593
Mitosporic fungi	594
Fossil fungi	613
Excluded genera	615

By using these listings in conjunction with the keys, a compilation of the candidate genera for a particular fungus can be reached. The production of comprehensive keys to the generic level falls beyond the scope of the *Dictionary*; we anticipate that that task is for a future work.

ACRASIOMYCOTA

ACRASIALES
Acrasiaceae
Acrasis

Copromyxaceae
Copromyxa

Fonticulaceae
Fonticula

Guttulinaceae
Guttulina (= Pocheina)
Guttulinopsis
Pocheina (= Guttulina)

Incertae sedis
Copromyxella

ASCOMYCOTA

ARTHONIALES
Arthoniaceae
Allarthonia (= Arthonia)

Allarthoniomyces (= Arthonia)
Allarthotheliomyces (= Allarthothelium)
Allarthothelium (= Arthonia)
Amazonomyces (= Eremothecella)
Arthonia
Arthoniomyces (= Arthonia)
Arthoniopsis (= Arthonia)
Arthotheliomyces (= Arthothelium)
Arthothelium
Asterotrema (= Arthonia)
Bryostigma (= Arthonia)
Caldesia (= Arthonia)
Celidiopsis (= Arthonia)
Celidium (= Arthonia)
Charcotia (= Arthonia)
Coccopeziza (= Arthonia)
Coniangium (= Arthonia)
Conida (= Arthonia)
Conidella (= Arthonia)
Coniocarpon (= Arthonia)
Conioloma (= Arthonia)
Craterolechia (= Arthonia)
Cresponea
Cryptothecia

Dermatina (= Arthonia)
Diarthonis (= Arthonia)
Eremothecella
Glyphidium (= Arthonia)
Gymnographoidea
Herpothallon (= Cryptothecia)
Herpothallonomyces (= Herpothallon)
Hypochnus (= Herpothallon)
Lecideopsis (= Arthonia)
Leprantha (= Arthonia)
Manilaea (= Arthonia)
Merarthonis (= Arthonia)
Mycardothelium (= Arthothelium)
Mycarthonia (= Arthonia)
Myriostigma (= Cryptothecia)
Myxotheca (= Cryptothecia)
Naevia (= Arthonia)
Nematidia (= Arthonia)
Pachnolepia (= Arthonia)
Phacothecium (= Arthonia)
Phlegmophiale (= Arthonia)
Plearthonis (= Arthonia)
Pseudoarthonia (= Arthonia)
Pyrenotea (= Arthonia)

Spilodium (= Arthonia)
Stirtonia
Tomaselliella (= Arthonia)
Tomaselliellomyces (=
    Tomaselliella)
Trachylia (= Arthonia)
Trichophyma (= Arthothelium)
Xerodiscus (= Arthonia)

Chrysothricaceae
Amphilomopsis (= Chrysothrix)
Byssocaulon
Chrysothrix
Peribotryon (= Chrysothrix)
Pulveraria (= Chrysothrix)
Temnospora (= Chrysothrix)

Roccellaceae
Alyxoria (= Opegrapha)
Ancistroporella
Ancistrospora (= Ancistroporella)
Arthonaria (= Enterographa)
Arthoniactis (= Lecanactis)
Aulaxinomyces (= Opegrapha)
Bacidiactis (= Lecanactis)
Bactrospora
Bactrosporomyces (= Bactrospora)
Chiodecton
Chiodectonomyces (=
    Enterographa)
Combea
Cyrtographa (= Minksia)
Darbishirella (= Ingaderia)
Delisea (= Plectocarpon)
Dendrographa
Dichosporidium
Dichosporium (= Dichosporidium)
Dictyographa (= Ingaderia)
Dictyographa (= Darbishirella)
Dirina
Dirinastromyces (= Dirinastrum)
Dirinastrum
Dirinopsis (= Dirina)
Dolichocarpus
Enterodictyon
Enterodictyonomyces (=
    Enterodictyon)
Enterodictyum (= Enterodictyon)
Enterographa
Epiphora (= Plectocarpon)
Erythrodecton
Feigeana
Fouragea (= Opegrapha)
Gomphospora (= Schismatomma)
Gorgadesia
Graphidastra
Halographis
Haplodina
Hubbsia
Hysterina (= Opegrapha)
Ingaderia
Lecanactiomyces (= Lecanactis)
Lecanactis
Leciographa (= Opegrapha)
Lecoglyphis (= Opegrapha)
Lichenomyces (= Plectocarpon)
Lobodirina (= Roccellina)
Mazosia
Melampydiomyces (=
    Melampydium)
Melampydium (= Melampylidium)
Melampylidium (= Bactrospora)
Melanodecton (= Chiodecton)

Micrographina (= Mazosia)
Minksia
Mycopegrapha (= Opegrapha)
Nemaria (= Roccella)
Opegrapha
Opegraphella (= Opegrapha)
Opegraphellomyces (= Opegrapha)
Opegraphoidea (= Opegrapha)
Opegraphomyces (= Opegrapha)
Pentagenella
Phragmographum (= Opegrapha)
Phragmopeltheca (= Mazosia)
Phyllographa (= Opegrapha)
Plagiographis (= Opegrapha)
Platygramma (= Enterographa)
Platygrapha (= Schismatomma)
Platygraphis (= Schismatomma)
Platygraphomyces (=
    Schismatomma)
Plectocarpon
Pseudolecanactis
Reinkella (= Roccella)
Reinkellomyces (= Reinkella)
Roccella
Roccellaria
Roccellina
Roccellodea
Roccellographa
Roccellographomyces (=
    Roccellographa)
Roccellomyces (= Roccella)
Roccellopsis (= Roccella)
Rotula (= Mazosia)
Rotularia (= Mazosia)
Sagenidium
Scaphis (= Opegrapha)
Schismatomma
Schismatommatomyces (=
    Schismatomma)
Schizopelte
Schizopeltomyces (= Schizopelte)
Sclerophyton
Sclerophytonomyces (=
    Sclerophyton)
Scolecactis (= Lecanactis)
Sigridea
Simonyella
Stigmatella (= Sclerophyton)
Stigmatidium (= Enterographa)
Streimannia
Syncesia
Thamnium (= Roccella)
Xylastra (= Opegrapha)
Zwackhia (= Opegrapha)

Seuratiaceae
Atichiopsis (= Seuratia)
Euthrypton (= Seuratia)
Myriophysella (= Seuratia)
Phaeophycopsis (= Seuratia)
Phycopsis (= Seuratia)
Seuratia
Seuratiopsis
Yoshinagaia

Incertae sedis
Byssophoropsis (= Sagenidiopsis)
Catarraphia
Cyclographa (= Catarraphia)
Hormosphaeria
Llimonaea
Nipholepis
Sagenidiopsis

Stirtoniopsis
Synarthonia
Synarthoniomyces (= Synarthonia)
Tania
Tarbertia

CALICIALES
Caliciaceae
Acoliomyces (= Thelomma)
Acolium (= Cyphelium)
Acolium (= Cyphelium)
Acroscyphus
Caliciella (= Calicium)
Caliciomyces (= Calicium)
Calicium
Calycium (= Calicium)
Carlosia (= Thelomma)
Crateridium (= Cyphelium)
Cypheliomyces (= Cyphelium)
Cypheliopsis (= Thelomma)
Cyphelium
Descematia (= Sphaerocephalum)
Embolidium (= Calicium)
Holocyphis (= Thelomma)
Mucor (= Calicium)
Mycacolium (= Cyphelium)
Phacotium (= Calicium)
Phacotrum (= Calicium)
Podocratera (= Tholurna)
Pseudacoliomyces (= Cyphelium)
Pseudacolium (= Cyphelium)
Sphaerocephalum (= Calicium)
Sphinctrinopsis
Stenocybella (= Calicium)
Strongylium (= Calicium)
Texosporium
Thelomma
Tholurna
Tholurnomyces (= Tholurna)
Trachylia (= Cyphelium)
Tylophoropsis (nom. dub.)
Vainionia (= Calicium)

Calycidiaceae
Calycidiomyces (= Calycidium)
Calycidium
Coniophyllum (= Calycidium)

Coniocybaceae
Allodium (= Chaenotheca)
Chaenotheca
Chaenotheciella (= Chaenotheca)
Chaenothecomyces (=
    Chaenotheca)
Chroocybe (= Coniocybe)
Coniocybe
Coniocybomyces (= Coniocybe)
Cybebe (= Chaenotheca)
Cyphelium
Cyphelium (= Chaenotheca)
Embolus (= Chaenotheca)
Eucyphelis (= Chaenotheca)
Fulgia (= Coniocybe)
Heydeniopsis (= Chaenotheca)
Mycoconiocybe (= Coniocybe)
Pseudoconiocybe (= Coniocybe)
Sclerophora

Microcaliciaceae
Coniocybopsis (= Microcalicium)
Microcalicium
Strongylopsis (= Microcalicium)

Mycocaliciaceae
Chaenothecopsis
Mycocalicium
Phaeocalicium
Pseudocalicium (=
    Chaenothecopsis)
Sphinctrinella (= Mycocalicium)
Stenocybe
Strongyleuma (= Chaenothecopsis)

Sphaerophoraceae
Baeoderma (= Sphaerophorus)
Bunodophoron
Leifidium
Pleurocybe (= Bunodophoron)
Pleurocybomyces (=
    Sphaerophorus)
Pseudosphaerophorus (=
    Sphaerophorus)
Sphaerophoron (= Sphaerophorus)
Sphaerophoronomyces (=
    Sphaerophorus)
Sphaerophorum (= Sphaerophorus)
Sphaerophorus
Syrigosis (= Sphaerophorus)
Thysanophoron (= Sphaerophorus)

Sphinctrinaceae
Phacotiella (= Sphinctrina)
Pyrgidiomyces (= Pyrgidium)
Pyrgidium
Sphinctrina

Incertae sedis
Allophoron
Ditylis (= Tylophoron)
Heterocyphelium
Paracudonia (= Roesleria)
Roesleria
Roeslerina
Schistophoron
Sphaerocarpus
Tylophoron

CALOSPHAERIALES
Calosphaeriaceae
Cacosphaeria (= Kacosphaeria)
Calosphaeria
Erostella (= Pleurostoma)
Jattaea
Kacosphaeria
Longoa (= Calosphaeria)
Massalongiella (= Enchnoa)
Neoarcangelia (= Pleurostoma)
Pachytrype
Phragmocalosphaeria
Pleurostoma
Romellia
Togninia (= Pleurostoma)
Wegelina (= Calosphaeria)

Graphostromataceae
Graphostroma

CORYNELIALES
Coryneliaceae
Alboffia (= Corynelia)
Caliciopsis
Capnodiella (= Sorica)
Corynelia
Coryneliopsis
Coryneliospora
Endohormidium (= Corynelia)

Fitzpatrickella
Hypsotheca (= Caliciopsis)
Lagenula (= Caliciopsis)
Lagenulopsis
Sorica (= Caliciopsis)
Tripocorynelia (= Tripospora)
Tripospora

CYTTARIALES
Cyttariaceae
Cyttaria

DIAPORTHALES
Melanconidaceae
Allantoporthe
Apiosporopsis
Bioporthe (= Plagiostigme)
Calospora (= Macrodiaporthe)
Calospora (= Prosthecium)
Calosporella (= Prosthecium)
Ceratoporthe
Chapeckia
Cytomelanconis
Diatractium
Dicarpella
Dictyoporthe
Discodiaporthe (= Melanconis)
Disperma (= Dicarpella)
Ditopella
Fremineavia
Gibellia
Hapalocystis
Herbampulla
Hercospora
Hyalomelanconis (= Melanconis)
Hypophloeda
Kensinjia
Macrodiaporthe
Massariovalsa
Mebarria
Melanamphora
Melanconidium (= Melanconis)
Melanconiella (= Melanconis)
Melanconis
Melogramma
Neokeissleria (= Melanconis)
Phaeodiaporthe (= Prosthecium)
Phragmodiaporthe
Phragmoporthe
Plagiophiale
Plagiostigme
Prosthecium
Prostratus
Pseudovalsa
Pseudovalsella
Schizoparme
Sphaerognomonia (=
    Apiosporopsis)
Sporotheca (= Melanconis)
Stevensiella (= Diatractium)
Sydowiella
Titania (= Freminaevia)
Trabutiella (= Diatractium)
Wehmeyera
Wuestneia
Wuestneiopsis

Valsaceae
Amphiporthe
Anisogramma
Anisomyces
Apiognomonia
Apioplagiostoma

Apioporthe (= Anisogramma)
Apioporthella
Apiothecium (= Cryptodiaporthe)
Aporhytisma (= Diaporthopsis)
Bagcheea
Batschiella (= Clypeoporthella)
Ceuthocarpon (= Linospora)
Chadefaudiomyces
Chailletia
Chalcosphaeria (= Plagiostoma)
Chorostate (= Diaporthe)
Chorostella (= Cryptodiaporthe)
Circinaria (= Valsa)
Circinostoma (= Valsa)
Clypeocarpus (= Mazzantia)
Clypeoporthe
Clypeoporthella
Clypeorhynchus (= Diaporthe)
Cryphonectria
Cryptascoma
Cryptoderis (= Pleuroceras)
Cryptodiaporthe
Cryptospora (= Ophiovalsa)
Cryptosporella (= Winterella)
Dialytes (= Diaporthe)
Diaporthe
Diaporthella
Diaporthopsis
Diplacella
Ditopellina
Ditopellopsis
Durispora
Endothia
Engizostoma (= Valsa)
Gnomonia
Gnomoniella
Gnomoniopsis (= Gnomonia)
Hypospilina
Hystricula (= Cryptosporella)
Kapooria
Kubinyia (= Mamiania)
Lambro
Leptosillia
Leucostoma
Linospora
Maculatipalma
Mamiania
Mamianiella
Mazzantia
Melanopelta (= Gnomonia)
Melanoporthe (= Diaporthe)
Microstoma (= Valsa)
Obryzum
Ophiognomonia
Ophiovalsa
Paramazzantia (= Mazzantia)
Paravalsa
Phaeoapiospora (= Anisomyces)
Phoma (= Plagiostoma)
Phylloporthe
Plagiostoma
Plagiostomella (= Plagiostoma)
Pleuroceras
Radulum (= Xenotypa)
Rehmiella (= Gnomonia)
Rostrocoronophora (= Gnomonia)
Septomazzantia (= Diaporthe)
Sillia
Skottsbergiella
Stegastroma (= Anisomyces)
Stegophora
Tylodon (= Radulum)
Uleoporthe

Valsa
Valsella
Valseutypella
Winterella (= Ophiovalsa)
Xenotypa

Incertae sedis
Anisomycopsis
Caudospora
Clavatisporella
Exormatostoma (nom. dub.)
Hypoxylonopsis (= Valsaria)
Keinstirschia
Pedumispora
Phaeosperma (= Valsaria)
Pseudocryptosporella
Pseudothis
Pseudothyridaria (= Valsaria)
Rhamphoria
Savulescua
Sphaerognomoniella
Stioclettia
Trematovalsa
Valsaria

DIATRYPALES
Diatrypaceae
Anthostoma (= Cryptosphaeria)
Cladosphaeria (= Cryptosphaeria)
Cryptosphaeria
Cryptosphaerina (=
    Cryptosphaeria)
Diatrype
Diatrypella
Dothideovalsa (= Eutypa)
Echinomyces
Ectosphaeria (= Diatrype)
Epheliopsis (= Eutypa)
Eutypa
Eutypella
Fassia
Lageniformia (= Eutypa)
Leptoperidia
Paraeutypa (= Leptoperidia)
Peroneutypa (= Eutypella)
Peroneutypella (= Scoptria)
Phaeotrype (= Diatrype)
Pseudotrype (= Eutypella)
Pyrenodochium (= Diatrype)
Quaternaria (= Eutypella)
Rostronitschkia (= Eutypa)
Scoptria (= Eutypella)
Sphaerosperma (= Diatrypella)
Stictosphaeria (= Diatrype)
Stigmatopsis (= Cryptosphaeria)
Tilakomyces (= Eutypella)

DOTHIDEALES
Antennulariellaceae
Achaetobotrys
Antennulariella
Capnociferria (= Antennulariella)
Capnocrinum (= Antennulariella)

Argynnaceae
Argynna
Lepidopterella

Arthopyreniaceae
Arthopyrenia
Athrismidium (= Tomasellia)
Beckhausia (= Tomasellia)
Campylacia (= Leptorhaphis)

Chlorodothis (= Tomasellia)
Ciferriolichen (= Arthopyrenia)
Endophis (= Leptorhaphis)
Giacominia (= Arthopyrenia)
Jarxia
Jattaeolichen (= Arthopyrenia)
Jattaeomyces (= Arthopyrenia)
Leiophloea (= Arthopyrenia)
Leiophloea (= Arthopyrenia)
Leptomycorhaphis (=
    Leptorhaphis)
Leptorhaphiomyces (=
    Leptorhaphis)
Leptorhaphis
Magmopsis (= Arthopyrenia)
Mesopyrenia (= Arthopyrenia)
Microtheliomyces (=
    Mycomicrothelia)
Mycarthopyrenia (= Arthopyrenia)
Mycoarthopyrenia (=
    Arthopyrenia)
Mycociferria (= Arthopyrenia)
Mycoleptorhaphis (= Leptorhaphis)
Mycomicrothelia
Naetrocymbe (= Arthopyrenia)
Nothostroma (= Tomasellia)
Santessoniolichen (= Arthopyrenia)
Santessoniomyces (= Arthopyrenia)
Sciodothis (= Tomasellia)
Semisphaeria
Sporoschizon
Syngenosorus (= Tomasellia)
Tomasellia
Tomaselliomyces (= Tomasellia)
Uleodothella (= Tomasellia)

Asterinaceae
Allothyrium
Anariste
Aphanopeltis
Asterella (= Asterina)
Asterina
Asterinotheca (= Asterina)
Asterodothis
Asterolibertia
Asterotexis
Aulographella (= Morenoina)
Aulographina
Balansina (= Dothidasteromella)
Batistinula
Calothyriolum (= Asterina)
Cirsosia
Cirsosiella (= Cirsosia)
Dimerosporium (= Asterina)
Doguetia (= Trichasterina)
Dothiclypeolum (= Thyriopsis)
Dothidasteris (=
    Dothidasteromella)
Dothidasteromella
Echidnodella
Echidnodes
Englera (= Asterina)
Englerulaster (= Asterina)
Eupelte
Halbania
Halbanina (= Cirsosia)
Heraldoa (= Lembosia)
Hysterostoma (=
    Dothidasteromella)
Isipinga (= Symphaster)
Lembopodia (= Cirsosia)
Lembosia

Lembosidium (= Lembosia)
Lembosiellina (= Lembosia)
Lembosina
Lembosiodothis (= Echidnodes)
Lembosiopsis
Leveillella
Macowaniella
Maurodothella (= Echidnodes)
Maurodothina (= Eupelte)
Meliolaster
Meliolinopsis (= Patouillardina)
Micrographa (= Lembosia)
Micrographomyces (= Lembosia)
Morenoella (= Lembosia)
Morenoina
Morqueria (= Cirsosia)
Myxasterina (= Asterina)
Neostomella
Opeasterina (= Asterina)
Parasterina (= Asterina)
Parasterinella
Parasterinopsis
Patouillardina (= Meliolaster)
Peresiopsis (= Yamamotoa)
Petrakina
Placoasterella
Placoasterina
Placosoma
Polythyrium (= Neostomella)
Prillieuxina
Steyaertia (= Asterolibertia)
Symphaster
Thallochaete (= Aphanopeltis)
Thyriopsis
Trichamelia
Trichasterina
Trichosia (= Trichasterina)
Uleothyrium
Viegasia
Wardina (= Asterolibertia)
Yamamotoa

Aulographaceae
Aulographum
Polyclypeolina

Botryosphaeriaceae
Amarenomyces
Amerodothis (= Botryosphaeria)
Apomella (= Botryosphaeria)
Auerswaldiella
Botryosphaeria
Caumadothis (= Botryosphaeria)
Coutinia (= Botryosphaeria)
Creomelanops (= Botryosphaeria)
Cryptosporina (= Botryosphaeria)
Desmotascus (= Botryosphaeria)
Discochora (= Botryosphaeria)
Dothidotthia
Epiphyma (= Botryosphaeria)
Melanops (= Botryosphaeria)
Neodeightonia (= Botryosphaeria)
Phaeobotryon (= Botryosphaeria)
Phaeobotryosphaeria (=
    Botryosphaeria)
Pyreniella (= Botryosphaeria)
Rostrosphaeria (= Botryosphaeria)
Sivanesania
Thuemenia (= Botryosphaeria)

Brefeldiellaceae
Brefeldiella

Capnodiaceae
Aithaloderma
Algorichtera (= Scorias)
Anopeltis
Antennella (= Scorias)
Antennellina (= Scorias)
Antennellopsis (= Phragmocapnias)
Blastocapnias (= Aithaloderma)
Callebaea
Capnobatista (= Trichomerium)
Capnodaria
Capnodenia (= Capnodium)
Capnodina (= Aithaloderma)
Capnodium
Capnophaeum
Ceramoclasteropsis
Chaetopotius (= Aithaloderma)
Chaetoscorias (= Phragmocapnias)
Echinothecium
Hyalocapnias (= Scorias)
Hyaloscolecostroma
Hypocapnodium (= Aithaloderma)
Leptocapnodium (= Scorius)
Neocapnodium (=
   Phragmocapnias)
Paracapnodium (= Scorias)
Phaeochaetia (= Aithaloderma)
Phragmocapnias
Polychaeton (nom. dub.)
Scoriadopsis
Scorias
Trichomerium
Triposporiopsis (= Trichomerium)
Vertixore (= Aithaloderma)

Chaetothyriaceae
Actinocymbe
Ainsworthia (= Phaeosaccardinula)
Almeidaea (= Chaetothyrium)
Barnettia (= Microcallis)
Batistaella (= Phaeosaccardinula)
Capnites (= Phaeosaccardinula)
Ceramothyrium
Chaetasterina (= Chaetothyrium)
Chaetomeris (= Treubiomyces)
Chaetothyrium
Ciferriusia (= Yatesula)
Euceramia
Fraseria (= Microcallis)
Gilmania (= Chaetothyrium)
Hansfordina (= Microcallis)
Kanousea (= Microcallis)
Limaciniella (= Actinocymbe)
Marceloa (= Treubiomyces)
Microcalliomyces (= Microcallis)
Microcalliopsis (= Microcallis)
Microcallis
Mycostevensonia (=
   Treubiomyces)
Neopeltis (= Microcallis)
Paropsis (= Treubiomyces)
Phaeopeltis (= Phaeosaccardinula)
Phaeosaccardinula
Recifea (= Microcallis)
Sphaerochaetia (= Chaetothyrium)
Tephrosticta (= Phaeosaccardinula)
Treubiomyces
Trotterula (= Chaetothyrium)
Uloseia (= Ceramothyrium)
Wiltshirea (= Phaeosaccardinula)
Yatesula
Zukalia (= Chaetothyrium)

Coccodiniaceae
Coccodinium
Dennisiella
Deslandesia (= Limacinula)
Limacinula
Limacinula (= Limacinula)
Naetrocymbe (= Limacinula)
Shanoriella (= Limacinula)
Teichosporina (= Limacinula)
Vitalia (= Dennisiella)

Coccoideaceae
Apiodothina (= Coccoidea)
Coccodiscus (= Coccoidea)
Coccodothella (= Coccoidella)
Coccoidea
Coccoidella
Eumycrocyclus (= Coccoidella)

Cookellaceae
Ascomycetella (= Cookella)
Calolepis (= Pycnoderma)
Cookella
Dictyomollisia (= Uleomyces)
Hyalocurreya (= Uleomyces)
Kusanoa (= Uleomyces)
Kusanoopsis (= Uleomyces)
Micromyriangium (= Uleomyces)
Myriangina (= Uleomyces)
Myrianginella (= Uleomyces)
Phaneroascus (= Cookella)
Pycnoderma
Pycnodermina (= Pycnoderma)
Uleomyces

Cucurbitariaceae
Byssolophis
Crotonocarpia (= Cucurbitaria)
Cucurbidothis (= Curreya)
Cucurbitaria
Cucurbitariopsis (= Gemmamyces)
Curreya
Cyathisphaera (= Cucurbitaria)
Gemmamyces (= Cucurbitaria)
Gibberidea (= Cucurbitaria)
Leucothyridium (= Cucurbitaria)
Megalospora (= Gemmamyces)
Phialospora (= Cucurbitaria)
Rhytidiella
Syncarpella

Dacampiaceae
Byssothecium
Clypeococcum
Cocciscia
Dacampia
Dacampiosphaeria (= Pyrenidium)
Decampia (= Dacampia)
Immotthia
Kalaallia
Lophothelium (= Polycoccum)
Moristroma
Munkovalsaria
Perinidium (= Pyrenidium)
Polycoccum
Pseudonitschkia
Pyrenidiomyces (= Pyrenidium)
Pyrenidium
Sinodidymella
Weddellomyces

Diademaceae
Clathrospora

Comoclathris
Diadema
Diademosa
Macrospora

Didymosphaeriaceae
Cryptodidymosphaeria (=
   Didymosphaeria)
Didymascina (= Didymosphaeria)
Didymosphaerella (=
   Didymosphaeria)
Didymosphaeria
Haplovalsaria (= Didymosphaeria)
Massariellops (= Didymosphaeria)
Verruculina

Dothideaceae
Auerswaldia
Bagnisiella
Catacaumella (= Vestergrenia)
Coccostromella
Coscinopeltella (= Vestergrenia)
Dimeriellina (= Auerswaldiella)
Dothidea
Guignardiella (= Vestergrenia)
Haplodothella (= Vestergrenia)
Hyalocrea
Monographos (= Schirrhia)
Monographus (= Monographos)
Mycoporis
Omphalospora
Pachysacca
Pediascus (= Vestergrenia)
Phragmodothidea (= Scirrhia)
Phragmodothis (= Dothidea)
Phyllachora (= Scirrhia)
Phyllachorella
Planistroma
Plectosphaerella (=
   Omphalospora)
Plectosphaerina (= Omphalospora)
Robertomyces (= Bagnisiella)
Scirrhia
Scirrhodothis (= Scirrhia)
Stichodothis (= Auerswaldiella)
Stylodothis
Systremma (= Dothidea)
Vestergrenia
Whetzelia (= Vestergrenia)

Dothioraceae
Botryochora
Delphinella
Diplosphaerella (= Delphinella)
Dothiora
Elmerococcum (= Plowrightia)
Endodothiora
Hariotia (= Delphinella)
Jaapia (= Dothiora)
Jaffuela
Keisslerina (= Dothiora)
Leptodothiora (= Dothiora)
Metasphaeria (= Saccothecium)
Phaeodothiora (= Saccothecium)
Pleodothis (= Sydowia)
Pleoglonis (= Hariota)
Pleosphaerulina (= Pringsheimia)
Plowrightia
Plowrightiella (= Sydowia)
Pringsheimia (= Saccothecium)
Rehmiellopsis (= Delphinella)
Saccothecium
Schizostege (= Pringsheimia)

Stigmea (= Dothiora)
Sydowia

Elsinoaceae
Agyrona (= Molleriella)
Beelia
Bitancourtia (= Elsinoë)
Butleria
Capnodiopsis (= Molleriella)
Elachophyma (= Molleriella)
Elenkinella (= Molleriella)
Elsinoë
Hemimyriangium
Hyalotheles
Isotexis (= Elsinoë)
Micularia
Molleriella
Myxomyriangium (= Saccardinula)
Nostocotheca (= Molleriella)
Plectodiscella (= Elsinoë)
Pycnodermella (= Saccardinula)
Pycnopeltis (= Saccardinula)
Saccardinula
Stephanotheca
Uleomycina (= Elsinoë)
Xenodium
Zukaliopsis (= Molleriella)

Englerulaceae
Anatexis (= Englerula)
Clypeolella (= Schiffnerula)
Coniosporiella (= Schiffnerula)
Dialacenium (= Rhytidenglerula)
Diathrypton (= Schiffnerula)
Englerula
Englerulella (= Rhytidenglerula)
Linotexis (= Parenglerula)
Parenglerula
Phaeoschiffnerula (= Schiffnerula)
Questieria (= Schiffnerula)
Rhizotexis
Rhytidenglerula
Schiffnerula
Thrauste

Eremomycetaceae
Eremomyces
Pithoascina (= Eremomyces)
Rhexothecium

Euantennariaceae
Aithalomyces (= Euantennaria)
Chaetosaccardinula (=
   Strigopodia)
Euantennaria
Ophiocapnodium (= Euantennaria)
Phaeocapnias (= Euantennaria)
Rasutoria
Skoteinospora (= Euantennaria)
Strigopodia
Trichopeltheca

Fenestellaceae
Fenestella
Lojkania
Ohleriella
Pleovalsa (= Fenestella)
Schachtia (= Fenestella)

Herpotrichiellaceae
Acanthostigmella
Berkelella
Berlesiella (= Capronia)

Capronia
Caproniella (= Capronia)
Dictyotrichiella (= Capronia)
Herpotrichiella (= Capronia)
Melanostigma (= Herpotrichiella)
Pleomelogramma
Polytrichiella (= Capronia)

Hypsostromataceae
Hypsostroma
Manglicola

Hysteriaceae
Encephalographa
Farlowia (= Farlowiella)
Farlowiella
Fragosoa (= Hysterographium)
Gloniella
Gloniopsis
Glonium
Graphyllium
Hemigrapha
Hypodermopsis (= Hysterium)
Hysteriopsis (= Hysterographium)
Hysterium
Hysterocarina
Hysteroglonium
Hysterographium
Hysteropatella
Platyspora (= Graphyllium)
Polhysterium (= Hysterographium)
Pseudoscypha
Psiloglonium (= Glonium)
Solenarium (= Glonium)
Xenothecium

Lautosporaceae
Lautospora

Leptopeltidaceae
Dothiopeltis
Dothithyrella (= Leptopeltis)
Leptopeltella (= Leptopeltis)
Leptopeltina (= Leptopeltis)
Leptopeltinella (= Leptopeltis)
Leptopeltis
Leptopeltopsis (= Leptopeltis)
Moeszia (= Leptopeltis)
Moeszopeltis (= Leptopeltis)
Nannfeldtia
Phacidina
Ronnigeria
Staibia

Leptosphaeriaceae
Ampullina (= Leptosphaeria)
Baumiella (= Leptosphaeria)
Bilimbiospora (= Leptosphaeria)
Chaetoplea (= Leptosphaeria)
Dendroleptosphaeria (=
   Leptosphaeria)
Didymolepta
Dothideopsella (= Leptosphaeria)
Exilispora (= Leptosphaeria)
Humboldtina (= Ophiobolus)
Leptosphaeria
Leptosphaeriopsis (= Ophiobolus)
Leptosporopsis (= Leptosphaeria)
Macrobasis (= Leptosphaeria)
Myriocarpium (= Leptosphaeria)
Ophiobolus
Phaeoderris (= Leptosphaeria)

Phyllophthalmaria (=
   Leptosphaeria)
Pleoseptum
Sclerodothis (= Leptosphaeria)

Lichenotheliaceae
Anzia (= Lichenothelia)
Lichenostigma
Lichenothelia

Lophiostomataceae
Abaphospora (= Massarina)
Bertiella (= Massarina)
Brigantiella (= Lophiostoma)
Ceriosporella (= Lophiostoma)
Chaetopyrena (= Chaetopyrenis)
Chaetopyrenis (= Keissleriella)
Chaetosphaerulina (=
   Herpotrichia)
Cilioplea
Coenosphaeria (= Keissleriella)
Delacourea (= Lophiostoma)
Didymotrichia (= Herpotrichia)
Didymotrichiella (= Herpotrichia)
Enchnosphaeria (= Herpotrichia)
Entodesmium
Epiphegia (= Massarina)
Herpotrichia
Holstiella (= Massarina)
Keissleriella
Khekia (= Herpotrichia)
Lambottiella (= Lophiostoma)
Lophidiopsis (= Lophiostoma)
Lophidium (= Platystomum)
Lophiella
Lophionema
Lophiosphaera (= Lophiostoma)
Lophiostoma
Lophiotrema
Lophiotricha (= Lophiostoma)
Massarina
Massarinula (= Massarina)
Massariosphaeria
Muroia
Oraniella (= Massarina)
Parasphaeria (= Massarina)
Phragmosperma (= Massarina)
Platisphaera (= Lophiostoma)
Platysphaera (= Platisphaera)
Platystoma (= Lophiostoma)
Platystomum (= Lophiostoma)
Pseudodiaporthe (= Massarina)
Quintaria
Sampaioa
Schizostoma (= Lophiostoma)
Sordariella (= Herpotrichia)
Sydowina (= Herpotrichia)
Trematostoma (= Massarina)
Trichometasphaeria
Vivianella (= Lophiotrema)
Xenonectria (= Herpotrichia)
Zopfinula (= Keissleriella)

Melanommataceae
Acrocordiopsis
Anomalemma
Asterella
Asterella (= Astrosphaeriella)
Asterosphaeria (=
   Astrosphaeriella)
Asterotheca (= Astrosphaeriella)
Astrosphaeriella
Astrotheca (= Astrosphaeriella)

Bicrouania
Bimuria
Byssosphaeria
Caryosporella
Javaria
Karstenula
Macbridella (= Byssosphaeria)
Melanomma
Moriolopis (= Melanomma)
Mycopepon
Ohleria
Ostropella
Pseudotrichia
Staurosphaeria (= Karstenula)
Trematosphaeria
Xenolophium

Meliolinaceae
Meliolina

Mesnieraceae
Bondiella
Helochora
Mesniera
Stegasphaeria

Metacapnodiaceae
Metacapnodium
Ophiocapnocoma (=
    Metacapnodium)
Sthughesia (= Metacapnodium)

Micropeltidaceae
Akaropeltella (= Stomiopeltis)
Akaropeltis (= Stomiopeltis)
Akaropeltopsis (= Stomiopeltis)
Armata
Bonaria
Ceratochaetopsis (=
    Chaetothyrina)
Chaetopeltopsis (= Chaetothyrina)
Chaetothyrina
Clypeolina (= Stomiopeltis)
Clypeolina
Clypeolinopsis (= Stomiopeltis)
Clypeolopsis (= Stomiopeltis)
Cyclopeltis
Dictyopeltella
Dictyopeltis
Dictyostomiopelta
Dictyothyriella
Dictyothyriella (= Micropeltis)
Dictyothyrina
Dictyothyrium
Diplocarponella (= Stomiopeltis)
Hansfordiopsis
Haplopeltheca
Hormopeltis (= Micropeltis)
Leptopeltina (= Stomiopeltis)
Mendoziopeltis
Micropeltella (= Micropeltis)
Micropeltidium (= Micropeltis)
Micropeltis
Mitopeltis
Muricopeltis
Ophiopeltis (= Micropeltis)
Parapeltella (= Micropeltis)
Phaeostomiopeltis (=
    Haplopeltheca)
Plochmopeltidella (=
    Chaetothyrina)
Polypedia
Scolecopeltella (= Micropeltis)

Scolecopeltis (= Micropeltis)
Scolecopeltopsis (= Micropeltis)
Setopeltis (= Chaetothyrina)
Stigmatodothis
Stigmatophragmia
Stomiopeltella (= Stomiopeltis)
Stomiopeltis
Stomiopeltopsis
Stomiotheca
Theciopeltis (= Micropeltis)
Thyriodictyella
Verlandea (= Stomiopeltis)

Microtheliopsidaceae
Micropyrenula (= Microtheliopsis)
Microtheliopsidomyces (=
    Microtheliopsis)
Microtheliopsis

Microthyriaceae
Actinomyxa
Actinopeltella (= Actinopeltis)
Actinopeltis
Actinosoma (= Chaetothyriopsis)
Arnaudiella
Asteridiellina (= Actinopeltis)
Asterinella
Asterinema
Asterinopeltis (= Platypeltella)
Asteritea
Asteronia
Byssopeltis
Caenothyrium (= Actinopeltis)
Calopeltis (= Cyclotheca)
Calothyriella (= Microthyrium)
Calothyriopsis
Calothyrium (= Asterinella)
Caribaeomyces
Caudella
Caudellopeltis (= Maublancia)
Chaetothyriopsis (= Actinopeltis)
Cirsosina
Cirsosiopsis
Corynocladus (nom. dub.)
Cyclotheca
Dasypyrena (= Actinopeltis)
Dictyoasterina
Ellisiodothis (= Muyocopron)
Govindua
Halbaniella (= Actinopeltis)
Haplopeltis (= Myiocopron)
Hariotula (= Cyclotheca)
Helminthopeltis
Hidakaea
Hugueninia
Lembosiella
Lichenopeltella
Loranthomyces (= Trichothyrina)
Maublancia
Micropeltopsis (= Trichothyrina)
Microthyrina (= Microthyrium)
Microthyris (= Lichenopeltella)
Microthyrium
Monorhizina
Muyocopron
Mycolangloisia (= Actinopeltis)
Myiocopron (= Muyocopron)
Opasterinella (= Asterinella)
Pachythyrium
Palawania
Peltella (= Myiocopron)
Peltopsis (= Myiocopron)
Petrakiopeltis

Phaeothyriolum
Phragmaspidium
Phragmothyrium (= Microthyrium)
Platypeltella
Plochmothea (= Xenostomella)
Polycyclinopsis
Polystomellina
Ptychopeltis (= Asterinema)
Raciborskiella (= Trichopeltella)
Resendea
Sapucchaka
Scolecopeltidium
Scolecopeltium (=
    Scolecopeltidium)
Seynesiella
Seynesiola (= Arnaudiella)
Seynesiopeltis
Seynesiospora (= Cyclothea)
Stegothyrium
Synostomella (= Cyclotheca)
Synpeltis (= Cyclotheca)
Thyrosoma (= Cyclotheca)
Tothia
Trichopeltella
Trichopeltina
Trichopeltis (= Trichothyrium)
Trichopeltopsis (= Trichothyrium)
Trichopeltospora
Trichopeltula (= Trichothyrium)
Trichopeltum
Trichothyriella
Trichothyrina (= Lichenopeltella)
Trichothyrinula
Trichothyriomyces
Trichothyriopsis
Trichothyrium
Trotteria (= Actinopeltis)
Xenostomella

Monoblastiaceae
Acrocordia
Acrocordiomyces (= Acrocordia)
Amphididymella (= Acrocordia)
Anisomeridium
Ascocratera (= Anisomeridium)
Ditremis (= Anisomeridium)
Lembidium (= Anisomeridium)
Microthelia (= Anisomeridium)
Monoblastia

Moriolaceae
Moriola

Mycoporaceae
Bottaria (= Mycoporum)
Dictyothyrium (= Mycoporum)
Mycoporum
Pseudomycoporon (= Mycoporon)

Mycosphaerellaceae
Achorodothis
Ascospora (= Mycosphaerella)
Cercosphaerella (=
    Mycosphaerella)
Columnosphaeria (= Guignardia)
Cyclodothis (= Mycosphaerella)
Cymadothea
Didymellina (= Mycosphaerella)
Diplochora (= Diplochorella)
Diplochorella (= Microcyclus)
Discomycopsis (= Euryachora)
Discosphaerina
Epicymatia (= Stigmidium)

Euryachora
Gillotia
Gnomonina (= Guignardia)
Guignardia
Haplodothis (= Mycosphaerella)
Hyalodothis (= Melanodothis)
Hypomycopsis (= Mycosphaerella)
Laestadia (= Guignardia)
Laestadiella (= Guignardia)
Leptophacidium (= Guignardia)
Lizoniella (= Mirocyclus)
Melanodothis
Melanopsammopsis (=
  Microcyclus)
Mesonella (= Guignardia)
Microcyclus
Micronectriella (= Sphaerulina)
Montagnellina (= Guignardia)
Mycosphaerella
Myriocarpa (= Guignardia)
Oligostroma (= Mycosphaerella)
Ophiocarpella (= Sphaerulina)
Ovosphaerella (= Mycosphaerella)
Pampolysporium (= Guignardia)
Paralaestadia (= Guignardia)
Pharcidia (= Stigmidium)
Pharcidiopsis (= Stigmidium)
Placocrea
Polysporidium (=
  Pampolysporium)
Pseudosphaerella (= Microcyclus)
Ramosphaerella (=
  Mycosphaerella)
Ramularisphaerella (=
  Mycosphaerella)
Scirrhiachora (= Mycosphaerella)
Septorisphaerella (=
  Mycosphaerella)
Septosphaerella (=
  Mycosphaerella)
Sphaerella (= Mycosphaerella)
Sphaerella (= Mycosphaerella)
Sphaerellothecium
Sphaerialea (= Sphaerulina)
Sphaerulina
Stigmidium

Myriangiaceae
Agostaea (= Anhellia)
Anhellia
Diplotheca
Eurytheca
Myriangium
Perisporiopsis (= Diplotheca)
Phymatodiscus (= Myriangium)
Phymatosphaeria (= Myriangium)
Pyrenotheca (= Myriangium)
Ramosiella (= Anhellia)
Stevensea (= Diplotheca)
Whetzeliomyces (= Anhellia)

Mytilinidiaceae
Actidium
Bulliardella (= Actidium)
Glyphium
Lophidium (= Lophium)
Lophium
Murashkinskija (= Mytilinidion)
Mytilidion (= Mytilinidion)
Mytilinidion
Ostreichnion
Ostreion (= Ostreichnion)
Ostreionella (= Actidium)

Ostreola
Quasiconcha
Scolecostroma (= Lophium)

Parmulariaceae
Aldona
Aldonata
Apoa (= Pachypatella)
Aspidothea (= Inocyclus)
Aulacostroma
Byliana (= Palawaniella)
Campoa
Chaetaspis (= Rhagadolobium)
Clypeum (= Parmularia)
Coccodothis
Cocconia
Cocconiopsis (= Cyclostomella)
Cycloschizella (= Cycloschizon)
Cycloschizon
Cyclostomella
Dielsiella (= Cycloschizon)
Discodothis (= Rhagadolobium)
Dothidasteroma
Dothophaeis (= Englerodothis)
Ellimonia (= Inocyclus)
Englerodothis
Ferrarisia
Fraserula (= Inocyclus)
Hysterostomella
Hysterostomina (=
  Hysterostomella)
Inocyclus
Kentingia
Kiehlia
Lateropeltis (= Kiehlia)
Lauterbachiella (=
  Rhagadolobium)
Maurodothis (= Cycloschizon)
Melanoplaca (= Dothidasteroma)
Microthyriolum (= Ferrarisia)
Monorhiza (= Rhagadolobium)
Myiocoprella (= Rhagadolobium)
Myxostomella (= Campoa)
Pachypatella
Palawaniella
Parmularia
Parmulariopsella
Parmulariopsis
Parmulina
Perischizon
Polycyclina
Polycyclus
Protothyrium
Pseudolembosia
Pycnographa (= Schneepia)
Rhagadolobium
Rhipidocarpon
Schneepia (= Parmularia)
Symphaeophyma
Thallomyces
Xenodiscella (= Rhagadolobium)

Parodiellaceae
Diplodiopsis (= Parodiella)
Parodiella
Pyrenochaetina (= Parodiella)

Parodiopsidaceae
Alina
Asteromyxa (= Dimeriella)
Balladyna
Balladynastrum (= Balladynopsis)
Balladynella (= Dysrhynchis)

Balladynocallia
Balladynopsis
Ceratochaete (= Dysrhynchis)
Chevalieria (= Chevalieropsis)
Chevalieropsis
Chrysomyces (= Perisporiopsis)
Cicinnobella (= Perisporina)
Cleistosphaera
Dichothrix (= Perisporiopsis)
Dimeriella
Dimerosporiella (= Dysrhynchis)
Dimerosporina (= Dysrhynchis)
Dysrhynchis
Henningsomyces (= Dysrhynchis)
Hypoplegma (= Perisporiopsis)
Kusanotheca (= Dysrhynchis)
Leptomeliola
Meliolidium (= Perisporiopsis)
Meliolinella (= Scolionema)
Meliolinopsis (= Meliolinella)
Myxotheciella (= Scolionema)
Neoballadyna (= Dysrhynchis)
Neohoehnelia (= Dysrhynchis)
Neoparodia
Ophioparodia
Parodiellina
Parodiellinopsis (= Scolionema)
Parodiopsis (= Perisporiopsis)
Perisporina (= Perisporiopsis)
Perisporiopsella (= Pilgeriella)
Perisporiopsis
Pilgeriella
Piline (= Perisporiopsis)
Schistodes (= Perisporiopsis)
Scolionema
Setella (= Dysrhynchis)
Stevensula (= Leptomeliola)
Stomatogene
Wageria (= Balladynopsis)
Xenostigmella (= Balladynopsis)

Phaeosphaeriaceae
Barria
Botanamphora (=
  Trematosphaeria)
Bricookea
Carinispora
Cryptocrea (= Eudarluca)
Eudarluca
Hadrospora
Lautitia
Leptosphaerella (= Phaeosphaeria)
Lopholeptosphaeria
Metameris
Mixtura
Montagnula
Nodulosphaeria
Ophiosphaerella
Parabotryon (= Eudarluca)
Paraphaeosphaeria
Phaeodothis
Phaeosphaeria
Pocosphaeria (= Nodulosphaeria)
Rhopographus
Scirrhophragma (= Metameris)
Teratosphaeria
Trematosphaerella (=
  Phaeosphaeria)
Xenodimerium (= Eudarluca)

Phaeotrichaceae
Phaeotrichum
Trichodelitschia

Piedraiaceae
Piedraia
Trichophila (= Piedraia)
Trichosporum (= Piedraia)

Pleomassariaceae
Asteromassaria
Eopyrenula
Kirschsteiniothelia
Peridiothelia
Pleomassaria (= Splanchnonema)
Splanchnonema
Stigmatomassaria (=
  Splanchnonema)

Pleosporaceae
Allewia (= Lewia)
Ceuthospora (= Pyrenophora)
Cleistotheca (= Pleospora)
Cleistothecopsis (= Pleospora)
Cochliobolus
Extrawettsteinina
Falciformispora
Kriegeriella
Leptosphaerulina
Lewia
Macroventuria
Neilreichina (= Pyrenophora)
Platysporoides
Pleospora
Polytrichia (= Pyrenophora)
Pseudocochliobolus (=
  Cochliobolus)
Pseudoplea (= Leptosphaerulina)
Pseudosphaeria (= Wettsteinina)
Pseudoyuconia
Pyrenophora
Scleroplea (= Pyrenophora)
Scleropleella (= Leptosphaerulina)
Setosphaeria
Tremateia
Wettsteinina
Zeuctomorpha

Polystomellaceae
Apiotrabutia (= Munkiella)
Coscinopeltis (= Munkiella)
Dothidella
Hypostigme (= Parastigmatea)
Marchalia (nom. conf.)
Munkiella
Parastigmatea
Pluriporus (= Dothidella)
Polystomella (= Dothidella)

Pseudoperisporiaceae
Acanthostigma (= Nematostoma)
Acanthostigmella (= Nematostoma)
Acanthostoma (= Phaeodimeriella)
Acarothallium (= Wentiomyces)
Aphanostigme
Astiothyrium (= Eudimeriolum)
Bolosphaera (= Dimerium)
Bryochiton
Bryomyces
Calochaetis (= Wentiomyces)
Capnodinula (= Wentiomyces)
Ceratosperma (= Nematostoma)
Chaetostigme (= Wentiomyces)
Chaetostigmella (=
  Phaeodimeriella)
Chaetyllis (= Lasiostemma)
Chilemyces (= Dimerina)

Dichaetis (= Wentiomyces)
Dimeriellopsis (= Nematostoma)
Dimerina
Dimerinopsis (= Dimerina)
Dimeriopsis (= Dimerina)
Dimerium
Epibryon
Epiploca (= Lasiostemma)
Epipolaeum
Episoma (= Dimerium)
Episphaerella
Eudimeriolum
Eumela
Gomezina (= Aphanostigme)
Hyalomeliolina
Lasiostemma
Lasiostemmella (= Lasiostemma)
Lizonia
Myxophora
Nematostigma
Nematostoma
Nematothecium
Neocoleroa (= Wentiomyces)
Neodimerium (= Lasiostemma)
Ontostheca (= Eudimeriolum)
Ophiogene (= Nematothecium)
Phaeocapdinula (= Dimerium)
Phaeocapnodinula (=
  Phaeocapdinula)
Phaeodimeriella (=
  Phaeodimeriella)
Phaeodimeriella
Phaeodimeris (= Phaeodimeriella)
Phaeophragmeriella (=
  Leptomeliola)
Phaeostigme
Phragmeriella
Pilula (= Dimerina)
Plactogene (= Dimerina)
Pododimeria
Porostigme (= Dimerium)
Pseudodimerium (= Dimerium)
Pseudolizonia (= Lizonia)
Pseudoperis (=
  Pseudoperisporium)
Pseudoperisporium (=
  Lasiostemma)
Raciborskiomyces (=
  Lasiostemma)
Somatexis (= Nematostoma)
Stigme (= Dimerina)
Toroa
Wentiomyces

Pyrenothricaceae
Cyanoporina
Lichenothrix (= Pyrenothrix)
Pyrenothrix

Saccardiaceae
Angatia
Angiotheca (= Dictyonella)
Ascolectus
Calopeziza (= Dictyonella)
Creangium (= Saccardia)
Cyanodiscus
Dictyonella
Ekmanomyces (= Dictyonella)
Epibelonium
Johansonia
Johansoniella (= Johansonia)
Masonia (= Dictyonella)
Myriangiomyces (= Saccardia)

Pseudodiscus
Saccardia
Schenckiella
Vonarxella

Schaereriaceae
Hafellnera
Schaereria

Schizothyriaceae
Agyronella (= Schizothyrium)
Amazonotheca
Chaetoplaca
Ciferriotheca (= Metathyriella)
Didymopeltis (= Schizothyrium)
Didymothyriella (= Plochmopeltis)
Endocycla (= Schizothyrium)
Epipeltis (= Schizothyrium)
Eremotheca (= Schizothyrium)
Gymnopeltis (= Lecideopsella)
Gyrothyrium (= Schizothyrium)
Henningsiella
Hexagonella
Kerniomyces
Lecideopsella
Linopeltis
Mendogia
Metathyriella
Microsticta (= Schizothyrium)
Microthyriella (= Schizothyrium)
Mycerema
Myiocopraloa (= Schizothyrium)
Myriangiella
Neopeltella
Orthobellus
Oswaldoa (= Myriangiella)
Paraphysotheca (= Schizothyrium)
Phragmothyriella (= Myriangiella)
Plectomyriangium (=
  Lecideopsella)
Pleiostomella (= Mendogia)
Plochmopeltis
Polyclypeolum (= Schizothyrium)
Protopeltis (= Myriangiella)
Sathropeltis (= Myriangiella)
Schizontopeltis (= Schizothyrium)
Schizopeltis (= Schizothyrium)
Schizothyrina (= Schizothyrium)
Schizothyrium
Spegazziniella (= Myriangiella)
Stomiopeltina (= Metathyriella)
Sydowiellina (= Myriangiella)
Uleopeltis (= Mendogia)
Vanudenia (= Schizothyrium)

Sporormiaceae
Brochospora (= Sporormia)
Chaetopreussia
Delitschia
Delitschiella (= Delitschia)
Fleischhakia (= Preussia)
Honoratia (= Preussia)
Niesslella (= Sporormiella)
Pachyspora (= Delitschia)
Pleophragmia (= Sporormia)
Preussia
Preussiella (= Westerdykella)
Pycnidiophora
Semidelitschia
Sporormia
Sporormiella
Spororminula

Sporormiopsis (= Preussia)
Westerdykella

Testudinaceae
Eremodothis
Lepidosphaeria
Marchaliella (= Testudina)
Neotestudina
Pseudodelitschia (= Neotestudina)
Pseudophaeotrichum (=
   Neotestudina)
Testudina
Ulospora

Tubeufiaceae
Acanthostigmina (= Tubeufia)
Allonecte
Annajenkinsia (= Puttemansia)
Boerlagella (= Thaxteriellopsis)
Boerlagiomyces
Byssocallis
Letendraea
Letendraeopsis
Linobolus (= Tubeufia)
Malacaria
Melioliphila
Paranectriella
Podonectria
Poeltia (= Paranectriella)
Poeltiella (= Paranectriella)
Puttemansia
Rebentischia
Subiculicola (= Melioliphila)
Taphrophila
Thaxteriella (= Tubeufia)
Thaxteriellopsis
Thaxterina
Tubeufia
Uredinophila

Venturiaceae
Acantharia
Actinodothidopsis (= Venturia)
Adelopus (= Phaeocryptopus)
Aloysiella (= Antennularia)
Antennataria (= Antennularia)
Antennina (= Antennularia)
Antennularia
Aphysa (= Coleroa)
Apiosporina
Arkoola
Arnaudia (= Acantharia)
Asterula (= Venturia)
Atopospora
Botryostroma
Botryothecium (=
   Rosenscheldiella)
Coleroa
Crotone
Cryptoparodia (= Antennularia)
Cryptopus (= Phaeocryptopus)
Cyphospilea (= Coleroa)
Dibotryon
Didothis (= Uleodothis)
Dimerosporiopsis (= Antennularia)
Endocoleroa (= Venturia)
Endostigme (= Venturia)
Gibbera
Hormotheca (= Coleroa)
Lasiobotrys
Lineostroma
Maireella (= Gibbera)
Melanostromella (= Antennularia)

Metacoleroa
Monopus (= Rosenscheldiella)
Montagnina (= Gibbera)
Neogibbera (= Acantharia)
Parodiellina (= Botryostroma)
Parodiodia (= Apiosporina)
Periline (= Antennularia)
Phaeocryptopus
Phaeosphaerella (= Venturia)
Phaeosporella (= Venturia)
Phasya (= Venturia)
Phragmogibbera
Platychora
Polyrhizon
Protoventuria (= Antennularia)
Pseudoparodia
Pseudoparodiella
Pseudotthia (= Gibbera)
Pyrenobotrys
Rhizogene
Robledia (= Botryostroma)
Rosenscheldiella
Scutelloidea (= Rosenscheldiella)
Sphaerellopsis (= Venturia)
Spilosticta (= Venturia)
Trichodothella
Trichodothis
Uleodothis
Venturia
Venturia (= Protoventuria)
Winteromyces (= Gibbera)
Xenomeris

Vizellaceae
Blasdalea
Entopeltis (= Vizella)
Haplopyrenula (= Vizella)
Haplopyrenulomyces (= Vizella)
Haplospora (= Haplopyrenula)
Hypocelis (= Vizella)
Phacopeltis (= Vizella)
Phaeaspis (= Vizella)
Phaeopeltis (= Phaeaspis)
Singeriella (= Blasdalea)
Stigmatopeltis (= Vizella)
Vizella

Zopfiaceae
Caryospora
Celtidia
Coronopapilla
Halotthia
Pontoporeia (= Zopfia)
Rechingeriella
Richonia
Unamunoa (= Rechingeriella)
Zopfia
Zopfiofoveola

Incertae sedis
Aaosphaeria
Acanthophiobolus
Acerbia (= Rosenscheldia)
Achorella
Acrocorelia (nom. nud.)
Acrogenotheca
Allosoma
Amylirosa (nom. dub.)
Anthracostroma
Appendispora
Arcangelia (= Didymella)
Arthopyreniella (= Mycoglaena)
Ascagilis

Ascohansfordiellopsis (=
   Koordersiella)
Ascomycetella (= Myriangiopsis)
Ascostratum
Austropeltum
Belizeana
Bertossia (= Mycoglaena)
Biatriospora
Bifrontia
Botryohypoxylon
Brooksia
Bryopelta
Bryorella
Bryosphaeria
Bryostroma
Buelliella
Byssogene
Calyptra
Capillataspora
Capnodinula
Capnogonium (= Brooksia)
Caprettia
Catulus
Ceratocarpia
Cercidospora
Cerodothis
Chaetobotrys (= Kusanobotrys)
Chaetomelanops (= Pyrenostigme)
Chaetoscutula
Ciferriomyces (= Pyrenostigme)
Coccochora
Coccochorella (= Coccochora)
Coccochorina (nom. dub.)
Colensoniella
Collemopsidiomyces (=
   Collemopsidium)
Collemopsidium
Comesella
Crauatamyces
Cyrtidium
Cyrtidula
Cyrtopsis
Dangeardiella
Daruvedia
Dawsomyces
Dawsophila
Dermatodothella
Dermatodothis
Dexteria (= Hyalosphaera)
Diapleella (= Kalmusia)
Dictyodothis
Didymella
Didymellopsis
Didymocyrtidium
Didymocyrtis
Didymopleella
Dilophia (= Lidophia)
Dimerium (nom. dub.)
Diplochorina
Discothecium (= Endococcus)
Dolabra
Elmerinula
Endococcus
Epibotrys (= Gilletiella)
Gibberidea
Gibberinula (= Gibberidea)
Gilletiella
Globoa
Globulina
Gloeodiscus
Grandigallia
Griggsia
Haplotheciella (= Didymella)

Harknessiella
Hassea
Helicascus
Heptameria
Heterochlamys (= Gilletiella)
Heterophracta (= Merismatium)
Heterosphaeriopsis
Homostegia
Hyalosphaera
Hyalotexis (= Hyalosphaera)
Hypobryon
Hysteropeltella
Hysteropsis
Jahnula
Kalmusia
Karschia
Keisslerellum (= Mycoporellum)
Keratosphaera (= Koordesiella)
Koordersiella
Kullhemia
Kusanobotrys
Laterotheca (= Acrogenotheca)
Lazarenkoa
Lembosiopeltis
Leptoguignardia
Leptospora
Leveillina
Lichenosphaeria (= Didymella)
Licopolia
Lidophia
Limaciniopsis
Lineolata
Loculohypoxylon
Lophiosphaerella
Loratospora
Macrovalsaria
Massariola
Melanochlamys (= Gilletiella)
Melanopsammina (= Otthia)
Merismatium
Microdothella
Montagnella (nom. dub.)
Moriolomyces
Mucomassaria
Muellerites
Mycocryptospora
Mycoglaena
Mycoporellum
Mycoporopsis
Mycosphaerellopsis (= Didymella)
Mycothyridium
Myriangiopsis
Myriostigma (= Myriostigmella)
Myriostigmella
Mytilostoma
Naumovia (= Rosenscheldia)
Neonorrlinia (= Cercidospora)
Neopeckia
Neoventuria
Norrlinia (= Neonorrlinia)
Ophiochaeta (= Acanthophiobolus)
Ophiosphaeria (=
    Acanthophiobolus)
Ophiotrichia (= Acanthophiobolus)
Otthia
Otthiella (= Otthia)
Paraliomyces
Parmulariella
Paropodia
Passeriniella
Peroschaeta
Phaeocreopsis
Phaeocyrtidula

Phaeocyrtis (= Merismatium)
Phaeoglaena
Phaeopeltium (=
    Phaeopeltosphaeria)
Phaeopeltosphaeria
Phaeophragmocauma (=
    Dermatodothis)
Phaeopolystomella (=
    Microdothella)
Phaeosperma
Phaeotomasellia
Phanerococcus (= Koordersiella)
Philobryon
Philonectria
Phragmodimerium (= Philonectria)
Phragmoscutella
Phycorella
Physalosporopsis
Placostromella
Plagiostromella
Planistromella
Pleiostomellina
Plejobolus
Pleosphaerellula
Pleosphaeropsis (= Norrlinia)
Pleostigma
Pleotrichiella
Polysporidiella
Polystomellopsis
Prolisea (= Cercidospora)
Propolina
Pseudoendococcus (= Endococcus)
Pseudomorfea
Pseudopleospora
Pteridiospora
Punctillum
Pycnocarpon
Pyrenochium
Pyrenocyclus
Pyrenostigme
Racovitziella
Robillardiella (nom. dub.)
Rosellinula
Rosenscheldia
Rosenschoeldia
Roumegueria
Roussoellopsis
Salsuginea
Santiella (= Passeriniella)
Scolecobonaria
Semifissispora
Setomelanomma
Shiraia
Sorothelia (= Endococcus)
Stuartella
Syrropeltis
Teichospora
Teratoschaeta
Thalassoascus
Thelenidia
Thelenidiomyces (= Thelenidia)
Thryptospora
Thyridaria
Thyrospora
Tilakiella
Tomeoa
Trematosphaeriopsis (=
    Homostegia)
Trematosphaeris (= Homostegia)
Tyrannosorus
Vainiona (= Neonorrlinia)
Venturiella (= Neoventuria)
Verlotia (= Heptameria)

Vizellopsis
Westea
Xylopezia
Xystozukalia (nom. dub.)
Yoshinagella
Zwackhiomyces

ELAPHOMYCETALES
Elaphomycetaceae
Ascoscleroderma (= Elaphomyces)
Ceratogaster (= Elaphomyces)
Ceraunium (= Elaphomyces)
Elaphomyces
Hypogaeum (= Elaphomyces)
Lycoperdastrum (= Elaphomyces)
Phymatium (= Elaphomyces)

ERYSIPHALES
Erysiphaceae
Albigo (= Sphaerotheca)
Alphitomorpha (= Erysiphe)
Arthrocladia (= Arthrocladiella)
Arthrocladiella
Blumeria
Brasiliomyces
Bulbomicrosphaera
Bulbouncinula
Californiomyces (= Brasiliomyces)
Calocladia (= Microsphaera)
Cystotheca
Desetangsia (= Sphaerotheca)
Erysiphe
Erysiphella (= Erysiphe)
Erysiphopsis (= Erysiphe)
Furcouncinula (= Uncinula)
Golovinomyces (= Erysiphe)
Ischnochaeta (= Erysiphe)
Kokkalera (= Sphaerotheca)
Lanomyces (= Cystotheca)
Leucothallia (= Sphaerotheca)
Leveillula
Linkomyces (= Erysiphe)
Medusosphaera
Microsphaera
Orthochaeta (= Erysiphe)
Phyllactinia
Pleochaeta
Podosphaera
Queirozia (= Pleochaeta)
Salmonia (= Brasiliomyces)
Salmonomyces (= Uncinula)
Sawadaia
Setoerysiphe
Sphaerotheca
Spolverinia (= Phyllactinia)
Tigria (= Erysiphe)
Trichocladia (= Microsphaera)
Typhulochaeta
Uncinula
Uncinulella (= Uncinula)
Uncinuliella
Uncinulopsis (= Pleochaeta)

EUROTIALES
Ascosphaeraceae
Arrhenosphaera
Ascosphaera
Bettsia
Pericystis (= Ascosphaera)

Cephalothecaceae
Aposphaeriopsis (= Cephalotheca)
Ascocybe (= Cephaloascus)

Aureomyces (= Cephaloascus)
Carothecis (= Cephalotheca)
Cephalotheca
Crepinula (= Cephalotheca)
Erythrosphaera (= Cephalotheca)

Monascaceae
Allescheria (= Monascus)
Backusia (= Monascus)
Eurotiella (= Monascus)
Eurotiopsis (= Monascus)
Monascus
Physomyces (= Monascus)
Xeromyces

Pseudeurotiaceae
Albertiniella
Aporothielavia
Azureothecium
Bulbithecium
Connersia
Cryptendoxyla
Fragosphaeria
Hapsidospora
Leucosphaera (= Leucosphaerina)
Leucosphaerina
Levispora (= Pseudeurotium)
Mycoarachis
Neelakesa
Nigrosabulum
Pleuroascus
Pseudeurotium

Trichocomaceae
Byssochlamys
Capsulotheca
Carpenteles (= Eupenicillium)
Chaetosartorya
Chaetotheca
Chromocleista
Cleistosoma (= Emericella)
Clistosoma (= Cleistosoma)
Cristaspora
Dactylomyces
Dendrosphaera
Dichlaena
Dichotomomyces
Diplostephanus (= Emericella)
Edyuillia
Emericella
Emericellopsis
Erythrogymnotheca
Eupenicillium
Eurotium
Fennellia
Gymnoeurotium (= Eurotium)
Hamigera (= Talaromyces)
Harpezomyces (= Chaetosartorya)
Hemicarpenteles
Hemisartorya (= Neosartorya)
Inzengaea (= Emericella)
Neosartorya
Penicilliopsis
Petromyces
Peyronellula (= Emericellopsis)
Pyrobolus (= Eurotium)
Royella (= Dichotomomyces)
Sagenoma
Saitoa (= Neosartorya)
Sarophorum (= Penicilliopsis)
Saturnomyces (= Emericellopsis)
Sclerocleista (= Hemicarpenteles)
Sporophormis (= Warcupiella)

Syncleistostroma (= Petromyces)
Talaromyces
Theclospora (= Emericella)
Thermoascus
Trichocoma
Trichoskytale (= Trichocoma)
Warcupiella

Incertae sedis
Bryocladium (= Pisomyxa)
Leiothecium
Pisomyxa
Samarospora
Trichomonascus
Veronaia

GYALECTALES
Gyalectaceae
Bacidiopsis (= Pachyphiale)
Biatorinopsis (= Dimerella)
Bryophagus
Coenogoniomycella (= Coenogonium)
Coenogonium (= Coenogonium)
Coenogonium
Coenomycogonium (= Coenogonium)
Cryptolechia
Dimerella
Finerania (= Ramonia)
Flabellomyces (= Coenogonium)
Gloeolecta (= Bryophagus)
Gyalecta
Gyalectella (= Dimerella)
Gyalectina (= Cryptolechia)
Gyalectomyces (= Gyalecta)
Holocoenis (= Coenogonium)
Lecaniopsis (= Dimerella)
Microphiale (= Dimerella)
Mycocoenogonium (= Coenogonium)
Pachyphiale
Phialopsis (= Gyalecta)
Ramonia
Secoliga (= Gyalecta)
Semigyalecta
Volvaria (= Gyalecta)

Incertae sedis
Belonia
Beloniella (= Belonia)
Beloniomyces (= Belonia)

HALOSPHAERIALES
Halosphaeriaceae
Aniptodera
Antennospora
Appendichordella
Arenariomyces
Bathyascus
Bovicornua
Carbosphaerella
Ceriosporella (= Marinospora)
Ceriosporopsis
Chadefaudia
Corallicola
Corollospora
Cucullospora (= Cucullosporella)
Cucullosporella
Etheirophora
Fluviatispora
Haligena
Halophiobolus (= Lulworthia)

Halosarpheia
Halosphaeria
Halosphaeriopsis
Iwilsoniella
Kohlmeyeriella
Lanspora
Lautisporopsis
Lignincola
Lindra
Lulworthia
Luttrellia
Marinospora
Moana
Mycophycophila (= Chadefaudia)
Naïs
Nautosphaeria
Nereiospora
Nimbospora
Nohea
Ocostaspora
Ondiniella
Ophiodeira
Palomyces (= Halosphaeria)
Peritrichospora (= Corollospora)
Remispora
Thalassogena
Tirispora
Trailia
Trichomaris
Tunicatispora

HYPOCREALES
Clavicipitaceae
Aciculosporium
Acrophytum (= Cordyceps)
Akrophyton (= Cordyceps)
Albomyces (= Aciculosporium)
Ascopolyporus
Atkinsonella
Atricordyceps
Balansia
Balansiella (= Claviceps)
Balansiopsis
Barya (= Neobarya)
Baryella (= Neobarya)
Byssostilbe
Campylothecium (= Cordyceps)
Cavimalum
Claviceps
Cordyceps
Cordylia (= Cordyceps)
Cordyliceps (= Cordyceps)
Corynesphaera (= Cordyceps)
Dothichloë (= Balansia)
Dussiella
Echinodothis (= Dussiella)
Epichloë
Fleischeria (= Hypocrella)
Helminthascus
Hypocrella
Hypocreophis (= Hypocrella)
Hypoxylum (= Cordyceps)
Kentrosporium (= Claviceps)
Konradia
Linearistroma
Mitosporium (= Aciculosporium)
Mitrasphaera (= Cordyceps)
Moelleria (= Moelleriella)
Moelleriella
Mothesia (= Claviceps)
Mycomalus
Myriogenis (= Myriogenospora)
Myriogenospora

Neobarya
Neocordyceps
Ophiocordyceps (= Cordyceps)
Ophiodothis (= Balansia)
Perisporiella (= Hypocrella)
Phytocordyceps (= Cordyceps)
Podocrella
Polistophthora (= Cordyceps)
Racemella (= Cordyceps)
Romanoa
Shimizuomyces
Spermoedia
Sphaerocordyceps
Tettigorhyza (= Cordyceps)
Torrubia (= Cordyceps)
Torrubiella
Typhodium (= Epichloë)
Xylaria (= Cordyceps)
Xylariopsis (= Konradia)

Hypocreaceae
Allantonectria (= Nectria)
Aphysiostroma
Apiocrea
Aponectria (= Nectria)
Arachnocrea
Balzania
Battarrina
Bionectria
Biotyle (= Pseudomeliola)
Bonordenia (= Hypomyces)
Calonectria
Calostilbe (= Nectria)
Calyptronectria
Cesatiella
Charonectria (= Nectriella)
Chiajaea
Chilonectria (= Nectria)
Chitinonectria (= Nectria)
Chromocrea (= Creopus)
Chrysogluten (= Nectria)
Ciliomyces (= Paranectria)
Clintoniella (= Hypomyces)
Corallomyces (= Nectria)
Corallomycetella (= Nectria)
Cordycepioideus
Cosmospora (= Nectria)
Creonectria (= Nectria)
Creopus
Cryptoleptosphaeria
Cryptonectriella (= Nectriella)
Cryptothecium (= Calonectria)
Dasyphthora (= Nectria)
Debarya (= Hypocrea)
Dialhypocrea
Dialonectria (= Nectria)
Didymocrea
Dozya (= Hypocreopsis)
Dubitatio (= Passerinula)
Endocreas
Eolichen (= Nectria)
Eolichenomyces (= Eolichen)
Ephedrosphaera (= Nectria)
Epicrea
Epinectria
Erispora
Feracia
Gibberella
Gibberellulina
Halonectria
Heleococcum
Hydronectria
Hydropisphaera

Hyphonectria (= Nectria)
Hypocrea
Hypocreopsis (= Selinia)
Hypocreopsis
Hypomyces
Ijuhya (= Nectria)
Kallichroma
Lasionectria (= Nectria)
Lecithium (= Lecythium)
Lecythium
Lepidonectria (= Nectria)
Leucocrea
Leuconectria
Lilliputia (= Roumegueriella)
Lisea (= Gibberella)
Lisiella
Loculistroma
Malmeomyces
Mattirolia (= Thyronectria)
Megalonectria (= Thyronectria)
Metadothella
Metanectria
Micronectria
Micronectriopsis
Miyakeomyces
Mycocitrus
Myrmaeciella
Myrmaecium (= Hypocreopsis)
Nectria
Nectriella
Nectriella (= Pseudonectria)
Nectriopsis
Neocosmospora
Neohenningsia
Neonectria (= Nectria)
Neoskofitzia
Neuronectria (= Nectria)
Notarisiella (= Pseudonectria)
Ochraceospora (= Nectria)
Ophionectria
Paranectria
Passerinula
Peckiella
Peethambara
Peloronectria
Peloronectriella
Peristomialis
Perrotiella (= Nectria)
Phaeonectria (= Nectria)
Pleogibberella
Pleonectria (= Thyronectria)
Podocrea (= Podostroma)
Podostroma
Porphyrosoma
Pronectria
Protocrea (= Hypocrea)
Protocreopsis
Pseudohypocrea
Pseudomeliola
Pseudonectria
Pseudosolidum (= Hypocreopsis)
Rhodothrix (= Nectria)
Roumegueriella
Sarawakus
Scoleconectria
Selinia
Seliniana (= Selinia)
Shiraiella (= Mycocitrus)
Spegazzinula (= Dubitatio)
Sphaerostilbe (= Nectria)
Sphaerostilbella
Sporadospora (= Pseudonectria)
Stalagmites (= Nectria)

Stereocrea
Stilbocrea (= Hypocreopsis)
Stilbonectria (= Nectria)
Stylonectria (= Nectria)
Stylonectriella (= Nectria)
Thyronectria
Thyronectroidea (= Thyronectria)
Tilakidium
Trichonectria
Uropolystigma
Viennotidia
Wakefieldiomyces
Weesea
Winteria (= Selinia)
Woronichina
Xenonectriella
Zeta (= Pseudomeliola)

Niessliaceae
Cryptoniesslia
Larseniella (= Trichosphaerella)
Melanopsamma
Niesslia
Oplotheciopsis (=
   Trichosphaerella)
Trichohleria (= Trichosphaerella)
Trichosphaerella
Valetoniella
Wallrothiella

Incertae sedis
Payosphaeria

LABOULBENIALES
Ceratomycetaceae
Autoicomyces
Ceratomyces
Drepanomyces
Eusynaptomyces
Helodiomyces
Phurmomyces
Plectomyces
Rhynchophoromyces
Synaptomyces
Tettigomyces
Thaumasiomyces
Thripomyces

Euceratomycetaceae
Cochliomyces
Colonomyces
Euceratomyces
Euzodiomyces
Pseudoecteinomyces

Herpomycetaceae
Herpomyces

Laboulbeniaceae
Acallomyces
Acanthomyces (= Rhachomyces)
Acompsomyces
Acrogynomyces
Adelomyces (= Phaulomyces)
Amorphomyces
Amphimyces
Apatelomyces
Apatomyces
Aphanandromyces
Aporomyces
Appendicularia (= Stigmatomyces)
Appendiculina (= Stigmatomyces)
Arthrorhynchus

Asaphomyces
Autophagomyces
Balazucia
Barbariella (= Asaphomyces)
Blasticomyces
Bordea (= Autophagomyces)
Botryandromyces
Camptomyces
Cantharomyces
Carpophoromyces
Ceraiomyces (= Laboulbenia)
Chaetarthriomyces
Chaetomyces
Chitonomyces
Clematomyces
Clonophoromyces
Columnomyces
Compsomyces
Coreomyces
Corethromyces
Cryptandromyces
Cucujomyces
Cupulomyces
Dermapteromyces
Diandromyces
Diaphoromyces
Dichomyces (= Peyritschiella)
Diclonomyces
Dicrandromyces (= Tetrandromyces)
Dimeromyces
Dimorphomyces
Dioicomyces
Diphymyces
Diplomyces
Diplopodomyces
Dipodomyces
Distichomyces (= Rickia)
Distolomyces
Dixomyces
Ecteinomyces
Enarthromyces
Eucantharomyces
Eucorethromyces (= Corethromyces)
Eudimeromyces (= Dimeromyces)
Euhaplomyces
Eumisgomyces
Eumonoicomyces
Euphoriomyces
Fanniomyces
Filariomyces
Gloeandromyces
Haplomyces
Heimatomyces (= Chitonomyces)
Helminthophana (= Arthrorhynchus)
Hesperomyces
Histeridomyces
Homaromyces
Hydraeomyces
Hydrophilomyces
Idiomyces
Ilyomyces
Ilytheomyces
Jeaneliomyces (= Dimeromyces)
Kainomyces
Kleidiomyces
Kruphaiomyces
Kyphomyces
Labiduromyces (= Filariomyces)
Laboulbenia
Laboulbeniella (= Laboulbenia)

Limnaiomyces
Majewskia
Meionomyces
Microsomyces
Mimeomyces
Misgomyces
Monoicomyces
Moschomyces (= Compsomyces)
Nanomyces
Neohaplomyces
Nycteromyces
Ormomyces
Osoriomyces
Paracoreomyces (= Coreomyces)
Parahydraeomyces (= Hydraeomyces)
Parvomyces
Peckifungus (= Stigmatomyces)
Peyerimhoffiella
Peyritschiella
Phalacrichomyces
Phaulomyces
Picardella
Polyandromyces
Polyascomyces
Porophoromyces
Prolixandromyces
Pselaphidomyces
Rhachomyces
Rhadinomyces (= Rhachomyces)
Rheophila (= Peyritschiella)
Rhipidiomyces
Rhizomyces
Rhizopodomyces
Rickia
Sandersoniomyces
Scalenomyces
Scaphidiomyces
Scelophoromyces
Schizolaboulbenia (= Laboulbenia)
Schizomeromyces (= Clematomyces)
Siemaszkoa
Skelophoromyces (= Scelophoromyces)
Smeringomyces
Sphaleromyces
Stemmatomyces
Stephanomyces (= Cucujomyces)
Stichomyces
Stigmatomyces
Streblomyces (= Nycteromyces)
Sugiyamaemyces
Symplectromyces
Sympodomyces
Synandromyces
Teratomyces
Tetrandromyces
Thaxteria (= Laboulbenia)
Trenomyces
Triandromyces (= Tetrandromyces)
Trochoideomyces
Troglomyces
Zeugandromyces
Zodiomyces

Pyxidiophoraceae
Acariniola (= Pyxidiophora)
Amphoropsis (= Pyxidiophora)
Ascolanthanus (= Pyxidiophora)
Copranophilus (= Pyxidiophora)
Eleutherosphaera (= Pyxidiophora)
Endosporella (= Pyxidiophora)

Entomocosma (= Pyxidiophora)
Mycorhynchidium
Mycorhynchus (= Pyxidiophora)
Myriapodophila (= Pyxidiophora)
Pyxidiophora
Rhynchomyces (= Pyxidiophora)
Rhynchonectria (= Pyxidiophora)
Thaxteriola (= Pyxidiophora)
Treleasia (= Pyxidiophora)

Incertae sedis
Benjaminella
Coreomycetopsis
Tavaresiella
Triceromyces

LAHMIALES
Lahmiaceae
Lahmia
Lahmiomyces (= Lahmia)
Parkerella (= Lahmia)

LECANORALES
Acarosporaceae
Acarospora
Acarosporomyces (= Acarospora)
Ahlesia (= Thelocarpon)
Alinocarpon (= Thelocarpon)
Athelium (= Thelocarpon)
Cathisinia (= Sarcogyne)
Cyanocephalium (= Thelocarpon)
Endocena
Eschatogonia
Glypholecia
Glypholeciomyces (= Glypholecia)
Gussonea (= Acarospora)
Gyrothecium (= Sporostatia)
Kelleria (= Thelocarpon)
Laureriella (= Glypholecia)
Lithoglypha
Melanophloea
Mycothelocarpon (= Thelocarpon)
Myriosperma (= Sarcogyne)
Myriospora (= Acarospora)
Polysporina
Sarcogyne (= Polysporina)
Sarcogyne
Sarcosagium
Sphaeropsis (= Thelocarpon)
Sporastatia
Stereopeltis (= Sarcogyne)
Strangospora
Thelocarpon
Thelocarponomyces (= Thelocarpon)
Thelocarpum (= Thelocarpon)
Thelococcum (= Thelocarpon)
Thelomphale (= Thelocarpon)

Agyriaceae
Agyrium
Exogone (= Agyrium)
Leptographa (= Ptychographa)
Myxomphalos (= Agyrium)
Ptychographa
Ptychographomyces (= Ptychographa)
Retinocyclus (= Tromera)
Sarea
Spiloma (= Xylographa)
Spilonematopsis (= Spiloma)
Tromera (= Sarea)
Tromera

Xylographa
Xylographomyces (= Xylographa)

Alectoriaceae
Alectoria
Alectoriomyces (= Alectoria)
Atestia (= Oropogon)
Bryopogon (= Alectoria)
Ceratocladia (= Alectoria)
Eualectoria (= Alectoria)
Oropogon
Sulcaria

Arctomiaceae
Arctomia
Wawea

Arthrodermataceae
Arthroderma
Ctenomyces
Nannizzia (= Arthroderma)

Bacidiaceae
Adelolecia
Adermatis (= Lecania)
Aipospila (= Lecania)
Bacidia
Bacidiomyces (= Bacidia)
Bayrhofferia (= Lecania)
Biatorina (= Catinaria)
Byssopsora (= Bacidia)
Callopis (= Physcidia)
Catinaria
Cliostomum
Compsocladium
Dimerospora (= Lecania)
Diphratora (= Solenopsora)
Dyslecanis (= Lecania)
Echidnocymbium
Heppsora
Herteliana
Japewia
Lecania
Lecaniella (= Lecania)
Lecaniomyces (= Lecania)
Megalopsora (= Physcidia)
Phyllopsora
Physcidia
Placodiella (= Solenopsora)
Placolecania (= Solenopsora)
Polyozosia (= Lecania)
Psorella (= Bacidia)
Psoromopsis (= Physcidia)
Rhytismella (= Cliostomum)
Ricasolia (= Placolecania)
Rolfidium
Schadonia
Solenopsora
Speerschneidera
Sporacestra (= Bacidia)
Sporoblastia (= Cliostomum)
Squamacidia
Squamarina
Squammaria (= Squamarina)
Tephromela
Thamnolecania
Tibellia
Toniniopsis
Urophora (= Bacidia)
Waynea

Baeomycetaceae
Baeomyces

Baeomycomyces (= Baeomyces)
Cladoniopsis (= Baeomyces)
Cyanobaeis (= Baeomyces)
Ludovicia (= Baeomyces)
Phyllobaeis
Sphyridiomyces (= Baeomyces)
Sphyridium (= Baeomyces)
Verrucaria (= Baeomyces)

Biatorellaceae
Biatorella
Maronella (= Biatorella)
Myrioblastus (= Biatorella)
Piccolia (= Biatorella)

Brigantiaeaceae
Argopsis
Brigantiaea
Myxodictyon (= Brigantiaea)
Myxodictyonomyces (= Brigantiaea)

Candelariaceae
Caloplacopsis (= Candelariella)
Candelaria
Candelariella
Candelariellomyces (= Candelariella)
Candelariellopsis (= Candelariella)
Candelina
Claurouxia
Eklundia (= Candelariella)
Gyalolechia (= Candelariella)
Placomaronea
Pleochroma (= Candelariella)
Pseudolecidea (= Claurouxia)

Catillariaceae
Arthrospora (= Arthrosporum)
Arthrosporum
Astroplaca (= Placolecis)
Austrolecia
Bacillina (= Toninia)
Bibbya (= Toninia)
Catillaria
Catillariomyces (= Catillaria)
Diphloeis (= Toninia)
Diplosis (= Toninia)
Halecania
Kiliasia (= Toninia)
Leptographa (= Toninia)
Lobiona (= Toninia)
Microlecia (= Catillaria)
Placolecis
Psora (= Toninia)
Scolecites (= Skolekites)
Skolekites (= Toninia)
Syncomista (= Toninia)
Thalloedematomyces (= Thalloidima)
Thalloidima (= Toninia)
Toninia
Ulocodium (= Catillaria)
Wadeana
Xanthopsora (= Xanthopsorella)
Xanthopsorella

Cladoniaceae
Biscladinomyces (= Cladonia)
Biseucladinomyces (= Cladonia)
Calathaspis
Capitularia (= Cladonia)
Cenomyce (= Cladonia)

Cenomyces (= Cladonia)
Cladia
Cladina (= Cladonia)
Cladinomyces (= Cladonia)
Cladona (= Cladonia)
Cladonia
Cladoniomyces (= Cladonia)
Clathrina (= Cladia)
Eucladoniomyces (= Cladonia)
Gymnoderma
Gymnodermatomyces (= Gymnoderma)
Helopodium (= Cladonia)
Heteromyces
Lachnocaulon
Metus
Myelorrhiza
Neophyllis
Papillaria (= Pycnothelia)
Phyllis (= Gymnoderma)
Phyllocarpos (= Cladonia)
Pycnothelia
Pycnotheliomyces (= Pycnothelia)
Pyxidium (= Cladonia)
Ramalea
Schasmaria (= Cladonia)
Scyphiphorus (= Cladonia)
Scyphophora (= Cladonia)
Scyphophorus (= Cladonia)
Scyphorus (= Cladonia)
Scyphorus (= Cladonia)
Sphaerophoropsis
Thysanothecium

Coccocarpiaceae
Asirosiphon (= Spilonema)
Circinaria (= Coccocarpia)
Coccocarpia
Molybdoplaca (= Steinera)
Peltularia
Spilonema
Spilonemopsis (= Spilonema)
Steinera
Vischia (= Coccocarpia)

Collemataceae
Blennothallia (= Collema)
Chiastosporum (= Collema)
Collema
Collemis (= Collema)
Collemodes (= Collema)
Collemodiopsis (= Collema)
Collemodium (= Leptogium)
Colleptogium (= Leptogium)
Dichodium (= Physma)
Dicollema (= Collema)
Enchylium (= Leptogium)
Epidrolithus (= Leptogium)
Eucollema (= Collema)
Gabura (= Collema)
Homodium (= Leptogium)
Homothecium
Kolman (= Collema)
Lathagrium (= Collema)
Lecidocollema (= Homothecium)
Leciophysma
Leightoniella
Lemniscium (= Leptogium)
Lepidocollema (= Collema)
Leptogiomyces (= Leptogium)
Leptogiopsis (= Leptogium)
Leptogiopsis (= Leptogium)
Leptogium

Leptojum (= Leptogium)
Lethagrium (= Lathagrium)
Mallotium (= Leptogium)
Myxopuntia (= Leptogium)
Pericoccis (= Leptogium)
Physma
Physmatomyces (= Physma)
Pseudoleptogium (= Leptogium)
Ramalodium
Rostania (= Collema)
Scytenium (= Leptogium)
Stephanophorus (= Leptogium)
Synechoblastus (= Collema)
Tichodea (= Collema)

Crocyniaceae
Crocynia
Symplocia (= Crocynia)

Ectolechiaceae
Badimia
Badimiella
Barubria
Calopadia
Ectolechia (= Sporopodium)
Heterothecium (= Lopadium)
Lasioloma
Loflammia
Logilvia
Lopadiomyces (= Lopadium)
Lopadium
Melittiosporiopsis (= Tapellaria)
Melittosporiopsis (=
   Mellitosporiopsis)
Mellitosporiopsis (= Tapellaria)
Pazschkea (= Tapellaria)
Pseudogyalecta (= Badimia)
Sporopodium
Tapellaria
Tapellariopsis

Eigleraceae
Eiglera

Gypsoplacaceae
Gypsoplaca

Haematommataceae
Haematomma
Haematommatomyces (=
   Haematomma)
Lepadolemma (= Haematomma)
Loxospora

Heppiaceae
Amphidium (= Epiphloea)
Corynecystis
Endocarpiscum (= Heppia)
Epiphloea
Guepinella (= Heppia)
Guepinia (= Heppia)
Heppia
Heppiomyces (= Heppia)
Heterina (= Heppia)
Latzelia (= Epiphloea)
Nylanderopsis (= Heppia)
Pannariella (= Heppia)
Placoheppia (= Heppia)
Pseudoheppia
Solorinaria

Heterodeaceae
Heterodea
Trichocladia (= Heterodea)

Hymeneliaceae
Agrestia (= Aspicilia)
Aphragmia (= Ionaspis)
Aspicilia
Aspiciliella
Aspiciliomyces (= Pachyospora)
Athecaria (= Aspicilia)
Bouvetiella
Chlorangium (= Sphaerothallia)
Circinaria (= Aspicilia)
Durietzia (= Ionaspis)
Hymenelia
Ionaspis
Lobothallia
Manzonia (= Aspicilia)
Melanolecia
Mosigia (= Aspicilia)
Pachyospora
Sagedia (= Aspicilia)
Sphaerothallia (= Aspicilia)
Tremolecia
Urceolaria (= Aspicilia)

Lecanoraceae
Arctopeltis
Bacidina
Biatora
Biatorellopsis (= Pleopsidium)
Boreoplaca
Bryonora
Byssiplaca (= Lecanora)
Carbonea
Circinaria (= Lecanora)
Cladidium
Cladodium (= Cladidium)
Courtoisia (= Lecanora)
Diomedella
Diplophragmia (= Lecidella)
Dirinella (= Lecanora)
Edrudia
Glaucomaria (= Lecanora)
Lecanora
Lecanoromyces (= Lecanora)
Lecanoropsis (= Lecanora)
Lecidella
Lecidellomyces (= Lecidella)
Lecideola (= Lecanora)
Lilliputeana (= Scoliciosporum)
Magninia (= Lecanora)
Maronina
Megalaria
Miriquidica
Myriolecis (= Lecanora)
Myrionora
Nesolechia (= Phacopsis)
Omphalodina (= Rhizoplaca)
Parmularia (= Lecanora)
Phacopsis
Pinacisca (= Lecanora)
Placodiomyces (= Lecanora)
Placolecanora (= Lecanora)
Pleopsidium
Protoparmeliopsis (= Lecanora)
Psorinia
Pyrrhospora
Ramboldia
Rhizoplaca
Sagema
Scalidium (= Scoliciosporum)

Schistoplaca (= Lecanora)
Scoliciosporomyces (=
   Scoliciosporum)
Scoliciosporum
Squamariomyces (= Lecanora)
Straminella (= Lecanora)
Tylothallia
Vainionora
Verrucaria (= Lecanora)
Zeora (= Lecanora)

Lecideaceae
Amphischizonia (= Cryptodictyon)
Bacidiopsora
Bahianora
Cecidonia
Cladopycnidium (= Lecidea)
Cryptodictyon
Diplotheca (= Lecidea)
Glyphopeltis
Hypocenomyce
Lecidea
Lecideomyces (= Lecidea)
Lopacidia
Nothoporpidia (= Lecidea)
Pseudopannaria
Rhizolecia
Saccomorpha (= Lecidea)
Sporoacania (= Lecidea)
Stenhammara (= Lecidea)
Stereonema (= Lecidea)
Zosterodiscus (= Lecidea)

Megalosporaceae
Austroblastenia
Bombyliospora (= Megalospora)
Bombyliosporomyces (=
   Megalospora)
Byssophragmia (= Megalospora)
Dumoulinia (= Megalospora)
Heterothecium (= Megalospora)
Megaloblastenia
Megalospora

Micareaceae
Helocarpon
Hollosia (= Scutula)
Micarea
Psilolechia
Roccellinastrum
Scutula
Stereocauliscum (= Micarea)

Miltideaceae
Miltidea

Mycoblastaceae
Megalospora (= Mycoblastus)
Mycoblastomyces (= Mycoblastus)
Mycoblastus
Oedemocarpus (= Mycoblastus)

Ophioparmaceae
Ophioparma

Pannariaceae
Amphinomium (= Pannaria)
Degelia
Erioderma
Fuscoderma
Fuscopannaria
Hosseusia
Hueella

Leioderma
Lepidogium (= Pannaria)
Lepidoleptogium (= Pannaria)
Leproloma
Malmella (= Erioderma)
Moelleropsis
Pannaria
Parmeliella
Phloeopannaria (= Psoroma)
Psoroma
Psoromaria (= Psoromidium)
Psoromatomyces (= Psoroma)
Psoromidium
Trachyderma (= Pannaria)
Triclinum

Parmeliaceae
Ahtia (= Cetrariopsis)
Ahtiana
Alectoria (= Usnea)
Allantoparmelia
Allocetraria
Almbornia
Anzia
Arctocetraria
Arctoparmelia
Asahinea
Aspidelia (= Parmelia)
Brodoa
Bryocaulon
Bryopogon (= Bryoria)
Bryoria
Bulborrhizina
Bulbothrix
Canomaculina
Canoparmelia
Cavernularia
Ceratophyllum (= Hypogymnia)
Ceteraria (= Cetraria)
Cetraria
Cetrariastrum
Cetrariella
Cetrariomyces (= Cetraria)
Cetrariopsis
Cetrelia
Cetreliopsis (= Cetrelia)
Chlorea (= Letharia)
Chondropsis
Chondrospora (= Anzia)
Coelocaulon
Coelopogon
Concamerella
Coralloides (= Coelocaulon)
Cornicularia (= Usnea)
Cornicularia
Coronoplectrum
Dactylina
Dufoureomyces (= Dactylina)
Esslingeriana
Eumitria (= Usnea)
Evernia
Everniastrum
Everniomyces (= Evernia)
Everniopsis
Flavocetraria
Flavoparmelia
Flavopunctelia
Foraminella (= Parmeliopsis)
Geissodea (= Parmelia)
Hendrickxia (= Everniopsis)
Himantormia
Hologymnia (= Evernia)
Hypogymnia

Hypotrachyna
Imshaugia
Karoowia
Letharia
Lethariella
Lichen (= Parmelia)
Masonhalea
Melanelia
Menegazzia
Myelochroa
Namakwa
Neofuscelia
Neopsoromopsis
Nephromopsis
Neuropogon (= Usnea)
Nimisia
Nodobryoria
Nylanderaria (= Letharia)
Omphalodiella
Omphalodium
Omphalora
Pannoparmelia
Paraparmelia
Parmelaria
Parmelia
Parmelina
Parmelinella
Parmelinopsis
Parmeliomyces (= Parmelia)
Parmeliopsis
Parmotrema
Parmotremopsis
Placoparmelia
Platisma (= Cetraria)
Platismatia
Platyphyllum (= Cetraria)
Pleurosticta (= Melanelia)
Protoparmelia
Protousnea
Pseudephebe
Pseudevernia
Pseudocornicularia (= Cornicularia)
Pseudoparmelia
Psiloparmelia
Psoromella
Punctelia
Relicina
Relicinopsis
Rhytidocaulon (= Letharia)
Rimelia
Rimeliella
Setaria (= Bryoria)
Squamaria (= Cetaria)
Tuckermannopsis
Tuckneraria
Usnea
Usneomyces (= Usnea)
Vulpicida
Xanthomaculina
Xanthoparmelia

Physciaceae
Abacina (= Diplotomma)
Amandinea
Anaptychia
Anaptychiomyces (= Anaptychia)
Anapyrenium (= Buellia)
Aplotomma (= Buellia)
Awasthia
Beltraminia (= Dimelaena)
Berengeria (= Rinodina)
Buellia

Buelliomyces (= Diploicia)
Buelliopsis (= Buellia)
Chaudhuria (= Heterodermia)
Dermatiscum
Dermiscellum
Dictyorinis (= Rinodina)
Dimelaena
Diploicia
Diplotomma (= Diploicia)
Diplotomma
Dirinaria
Gassicurtia (= Buellia)
Hafellia
Hagenia (= Anaptychia)
Heterodermia
Hyperphyscia
Imbricaria (= Anaptychia)
Kemmleria (= Buellia)
Lepropinacia (= Buellia)
Lichenoides (= Anaptychia)
Malmia (= Rinodina)
Mannia (= Buellia)
Mattickiolichen (= Buellia)
Mattickiomyces (= Buellia)
Melanaspicilia (= Buellia)
Merorinis (= Rinodina)
Mischoblastia (= Rinodina)
Mischolecia (= Rinodina)
Mobergia
Monerolechia (= Buellia)
Neophyscia (= Physcia)
Pachysporaria (= Rinodina)
Phaeophyscia
Phaeorrhiza
Phragmopyxine (= Pyxine)
Physcia
Physciella (= Phaeophyscia)
Physciomyces (= Physcia)
Physciopsis (= Hyperphyscia)
Physconia
Placothallia (= Rinodina)
Pleorinis (= Rinodina)
Pseudobuellia (= Rinodina)
Pseudophyscia (= Heterodermia)
Psora (nom. rej.)
Pyxine
Redonia
Rhinodinomyces (= Rinodina)
Rinodina
Rinodinella
Samboa (= Buellia)
Sambomyces (= Buellia)
Santessonia
Tetramelaena (= Hyperphyscia)
Tetramelas (= Buellia)
Tornabea
Tornabenia (= Tornabea)
Tornabenia (= Anaptychia)
Tornabeniopsis (= Tornabea)

Pilocarpaceae
Byssolecania
Byssoloma
Calidia (= Byssoloma)
Fellhanera
Gonolecania (= Byssolecania)
Lecaniella (= Byssolecania)
Limbalba (= Byssoloma)
Lobaca (= Fellhanera)
Pilocarpon (= Byssoloma)
Tricholechia (= Byssoloma)
Wainioa (= Byssoloma)

Porpidiaceae
Amygdalaria
Bellemerea
Bilimbia (= Mycobilimbia)
Catarrhospora
Clauzadea
Farnoldia
Haplocarpon (= Huilia)
Huilia (= Porpidia)
Immersaria
Koerberiella
Labyrintha
Lecanorella (= Koerberiella)
Mycobilimbia
Paraporpidia
Perspicinora (= Koerberiella)
Poeltiaria
Poeltidea
Porpidia
Probilimbia (= Mycobilimbia)
Schizodiscus
Stephanocyclos
Weitenwebera (= Mycobilimbia)
Xenolecia

Psoraceae
Blasteniomyces (= Protoblastenia)
Chrysopsora (= Psora)
Eremastrella
Fritzea (= Psora)
Lecidoma
Lepidoma (= Lecidoma)
Oolithinia (= Protoblastenia)
Peltiphylla (= Psora)
Protoblastenia
Psora
Psorula

Ramalinaceae
Alectoriopsis (= Ramalina)
Cenozosia (= Ramalina)
Chlorodictyon (= Ramalina)
Desmazieria (= Ramalina)
Dievernia (= Ramalina)
Fistulariella (= Ramalina)
Niebla (= Ramalina)
Platisma (= Ramalina)
Platysma (= Ramalina)
Ramalina
Ramalinomyces (= Ramalina)
Ramalinopsis
Trichoramalina (= Ramalina)

Rhizocarpaceae
Catillariopsis (= Rhizocarpon)
Catocarpon (= Rhizocarpon)
Catocarpus (= Rhizocarpon)
Catolechia
Cormothecium (= Rhizocarpon)
Dimaura (= Catolechia)
Diphaeis (= Rhizocarpon)
Diphanis (= Rhizocarpon)
Encephalographomyces (= Poeltinula)
Epilichen
Lepidoma (= Rhizocarpon)
Melanospora (= Poeltinula)
Phalodictyum (= Rhizocarpon)
Poeltinula
Rehmia (= Rhizocarpon)
Rhizocarpon
Rhizocarponomyces (= Rhizocarpon)

Siegertia (= Rhizocarpon)
Xanthopsis (= Catolechia)

Rimulariaceae
Amylora
Haplographa (= Lithographa)
Lambiella (= Rimularia)
Lithographa
Lithographomyces (= Lithographa)
Mycoplacographa (= Lithographa)
Placographa (= Lithographa)
Rimularia

Stereocaulaceae
Austropeltum
Cereolus (= Stereocaulon)
Chlorocaulum (= Stereocaulon)
Corynophoron (= Stereocaulon)
Gymnocaulon (= Stereocaulon)
Lecanocaulon (= Stereocaulon)
Muhria
Phyllocaulon (= Stereocaulon)
Pilophoron (= Pilophorus)
Pilophorus
Stereocaulomyces (= Stereocaulon)
Stereocaulon
Stereocladium (= Stereocaulon)

Trapeliaceae
Anamylopsora
Anzina
Aspiciliopsis (= Placopsis)
Discocera (= Trapelia)
Micarea (= Placynthiella)
Orceolina
Placopsis
Placynthiella (= Placynthiella)
Placynthiella
Trapelia
Trapeliopsis
Urceolina (= Orceolina)

Umbilicariaceae
Actinogyra (= Umbilicaria)
Agyrophora (= Umbilicaria)
Gyromium (= Umbilicaria)
Gyrophora (= Umbilicaria)
Gyrophoromyces (= Umbilicaria)
Gyrophoropsis (= Umbilicaria)
Lasallia
Llanoa (= Umbilicaria)
Macrodictya (= Lasallia)
Merophora (= Umbilicaria)
Omphalodiscus (= Umbilicaria)
Omphalodium (= Umbilicaria)
Omphalosia (= Umbilicaria)
Parmophora (= Umbilicaria)
Umbilicaria
Umbilicariomyces (= Umbilicaria)

Vezdaeaceae
Vezdaea

Incertae sedis
Auriculora
Bapalmuia
Bartlettiella
Biatora (= Stenhammarella)
Biatoridium
Buelliastrum
Cerania (= Thamnolia)
Chiliospora (= Biatoridium)
Chlorangium (nom. dub.)

Clauzadeana
Corticifraga
Haploloma
Lopezaria
Notolecidea
Nylanderiella (= Siphula)
Psorotichiella
Ravenelula
Scutellaria
Siphula
Stenhammara (= Stenhammarella)
Stenhammarella
Thamnolia

LEOTIALES
Ascocorticiaceae
Ascocorticium

Dermateaceae
Actinoscypha (= Micropeziza)
Aivenia
Aleuriella (= Mollisia)
Angelina
Ascluella
Asteronaevia (= Diplonaevia)
Atropellis
Beloniella (= Calloria)
Belonium (= Pyrenopeziza)
Belonopeziza (= Calloria)
Belonopsis
Belonopsis
Blumeriella
Briardia (= Duebenia)
Bulbomollisia
Bulgariastrum (= Dermea)
Calloria
Calloriella
Callorina (= Calloria)
Calycellinopsis
Cashiella
Catinella
Cejpia
Cenangella (= Dermea)
Chaetonaevia
Chlorosplenium
Chrysosplenium (= Chlorosplenium)
Coleosperma
Coronellaria
Crustula (= Mollisia)
Cryptohymenium
Dennisiodiscus
Dermatea (= Dermea)
Dermatella (= Dermea)
Dermateopsis
Dermatina (= Pezicula)
Dermea
Dibeloniella
Dibelonis (= Fabraea)
Dibelonis (= Fabraea)
Diplocarpon
Diplonaevia
Discocurtisia
Discohainesia
Discorehmia
Drepanopeziza
Duebenia
Durandia (= Durandiella)
Durandiella
Echinella (= Pirottaea)
Entomopeziza (= Diplocarpon)
Ephelina (= Leptotrochila)
Eupropolella

Excipula (= Pyrenopeziza)
Fabraea (= Leptotrochila)
Felisbertia
Graddonia
Haglundia
Higginsia (= Blumeriella)
Hysteronaevia
Hysteropeziza (= Pyrenopeziza)
Hysteropezizella
Hysterostegiella
Laetinaevia
Leptopeziza (= Pyrenopeziza)
Leptotrochila
Melachroia (= Podophacidium)
Merostictis (= Diplonaevia)
Micropeziza
Mollisia
Mollisiopsis
Naegiella (= Eupropolella)
Naevala
Naevia (= Naevala)
Naeviella
Naeviopsis
Neofabraea (= Pezicula)
Neotapesia
Niesslella (= Micropeziza)
Nimbomollisia (= Niptera)
Niptera
Nothophacidium
Obtectodiscus
Ocellaria
Ocellariella (= Naevia)
Odontoschizon (= Patinella)
Patellariopsis
Patinella
Pezicula
Pezolepis
Phaeangium (= Pezicula)
Phaeonaevia
Pirottaea
Placopeziza (= Leptotrochila)
Pleoscutula
Ploettnera
Ploettnerula (= Pirottaea)
Podophacidium
Pseudonaevia
Pseudoniptera
Pseudopeziza
Pyrenodiscus (= Diplonaevia)
Pyrenopeziza
Pyrenotrochila (= Trochila)
Schizothyrioma
Scutobelonium
Scutomollisia
Sorokina
Sphaerangium (= Pezicula)
Spilopezis (= Pseudopeziza)
Spilopodia
Spilopodiella
Stegopeziza (= Hysterostegiella)
Syntexis (= Mollisia)
Tapesia (= Mollisia)
Trichobelonium (= Belonopsis)
Trichodiscus (= Dennisiodiscus)
Trochila
Waltonia

Geoglossaceae
Bagnisimitrula
Bryoglossum
Cibalocoryne (= Geoglossum)
Corynetes (= Thuemenidium)
Cudonia

Geoglossum
Gloeoglossum (= Geoglossum)
Helote (= Microglossum)
Hemiglossum
Leotiella (= Cudonia)
Leptoglossum (= Thuemenidium)
Leucoglossum
Maasoglossum
Microglossum
Microglossum
Mitruliopsis (= Spathularia)
Nothomitra
Ochroglossum (= Microglossum)
Pachycudonia
Phaeoglossum
Pseudomitrula
Spathularia
Spathulariopsis
Spathulea (= Spathularia)
Thuemenidium (= Geoglossum)
Trichoglossum
Xanthoglossum (= Thuemenidium)

Hemiphacidiaceae
Didymascella
Fabrella
Hemiphacidium
Keithia (= Didymascella)
Korfia
Sarcotrochila
Stegopezizella (= Sarcotrochila)

Hyaloscyphaceae
Albotricha (= Lachnum)
Amicodisca
Arachnopeziza
Arachnopezizella (= Arachnopeziza)
Arachnoscypha (= Arachnopeziza)
Arenaea (= Dasyscyphus)
Asperopilum
Austropezia
Belonidium (= Lachnum)
Betulina
Brunnipila (= Lachnum)
Calycellina
Calyptellopsis
Capitotricha (= Lachnum)
Catinella (= Unguicularia)
Chaetoscypha (= Lachnum)
Chrysothallus
Chytrella (= Unguicularia)
Ciliolarina
Ciliosculum
Cistella
Cistellina
Clavidisculum
Dasypezis (= Lachnum)
Dasyscypha (= Dasyscyphus)
Dasyscyphella
Dasyscyphus (= Lachnum)
Debaryoscyphus (= Hamatocanthoscypha)
Dematioscypha
Dendrotrichoscypha (= Mollisina)
Dimorphotricha
Diplocarpa
Discocistella (= Cistella)
Dyslachnum (= Belonidium)
Echinula
Encoeliella (= Unguiculariopsis)
Erinella (= Lachnum)
Erinella (= Lachnum)

Erinellina (= Lachnum)
Eriopezia
Erioscypha (= Dasyscyphus)
Erioscyphella (= Lachnum)
Eupezizella (= Hyaloscypha)
Farinodiscus (= Proliferodiscus)
Fuscolachnum
Fuscoscypha
Globulina (= Unguiculella)
Graddonidiscus
Hamatocanthoscypha
Helolachnum (= Lachnum)
Helotiopsis (= Pithyella)
Hyalacrotes
Hyalopeziza
Hyaloscypha
Hyalotricha (= Hyalopeziza)
Hydrocina
Hyphodiscus
Hyphoscypha (= Lachnum)
Incrucipulum
Incrupila
Incrupilella (= Incrupila)
Lachnaster (= Dasyscyphus)
Lachnella (= Lachnum)
Lachnellula
Lachnobelonium (= Lachnum)
Lachnum
Lasiobelonis (= Lasiobelonium)
Lasiobelonium
Lasiobelonium (= Belonidium)
Microscypha
Mollisiella (= Cistella)
Mollisiella (= Unguiculariopsis)
Mollisina
Mycopandora (= Unguicularia)
Neodasyscypha (= Lachnum)
Niveostoma (= Solenopezia)
Olla
Otwaya
Parachnopeziza
Perrotia
Pezizellaster (= Lachnum)
Phaeoscypha
Phalothrix (= Unguicularia)
Phialina (= Calycellina)
Phialoscypha (= Phialina)
Pilatia (= Urceolella)
Pithyella
Polaroscyphus
Polydesmia
Proliferodiscus
Proprioscypha
Protounguicularia
Pseudolachnea (= Hyalopeziza)
Pseudoolla (= Unguicularia)
Psilachnum
Psilocistella
Rodwayella
Schweinitzia (= Velutaria)
Scutoscypha (= Calycellina)
Setoscypha (= Phialina)
Solenopezia
Sponheimeria (= Lasiobelonium)
Tapesina
Trichopeziza
Trichopezizella (= Lachnum)
Trichoscypha (= Lachnellula)
Trichoscyphella (= Lachnellula)
Truncicola (= Hyaloscypha)
Uncinia (= Hamatocanthoscypha)
Unciniella (= Hamatocanthoscypha)

Unguicularia
Unguiculariella
Unguiculella
Urceolella
Velutaria
Venturiocistella
Zoellneria

Icmadophilaceae
Cystolobis (= Knightiella)
Dibaeis
Glossodium (= Icmadophila)
Icmadophila
Icmadophilomyces (= Icmadophila)
Knightiella
Phycodiscus (= Thelidea)
Pseudobaeomyces
Siphulella
Thelidea (= Knightiella)
Tubercularia (= Dibaeis)
Tupia (= Icmadophila)

Leotiaceae
Allophylaria
Ameghiniella
Antinoa (= Pezizella)
Aquadiscula
Ascocalyx
Ascoclavulina
Ascocoryne
Ascotremella
Ascoverticillata (= Crocicreas)
Banksiamyces
Belonioscypha (= Cyathicula)
Belonioscyphella
Belospora (= Hymenoscyphus)
Bioscypha
Bispora (= Bisporella)
Bisporella
Bryoscyphus
Bulgaria
Bulgariella
Bulgariopsis
Burcardia (= Bulgaria)
Calloriopsis
Calycella
Calycella (= Bisporella)
Calycella (= Hymenoscyphus)
Calycina
Capillipes
Carneopezizella
Cenangiella (= Godronia)
Cenangina (= Cenangium)
Cenangiopsis
Cenangium
Chlorencoelia
Chlorociboria
Chloroscypha
Ciboriella (= Hymenoscyphus)
Ciliatula (= Pezoloma)
Claussenomyces
Clithris (= Cenangium)
Conchatium (= Cyathicula)
Cordierites
Corynella (= Claussenomyces)
Crinula (= Holwaya)
Crocicreas
Crumenella
Crumenula (= Godronia)
Crumenula (= Crumenulopsis)
Crumenulopsis
Crustomollisia
Cucullaria (= Leotia)

Cudoniella
Cyathicula
Cystopezizella (= Calycina)
Davincia (= Crocicreas)
Dencoeliopsis
Dictyonia
Diehlia (= Phaeangellina)
Diplothrix (= Calloriopsis)
Discinella
Ditiola (= Holwaya)
Durella
Encoelia
Encoeliopsis
Epiglia (= Mniaecia)
Episclerotium
Erikssonopsis
Eubelonis
Eubelonis (= Bisporella)
Evulla (= Neobulgaria)
Exotrichum (= Crocicreas)
Fungodaster (= Leotia)
Gelatinodiscus
Gelatinopsis
Geocoryne
Gloeopeziza
Godronia
Godroniopsis
Gorgoniceps
Grahamiella
Gremmeniella
Grimmicola
Grovesia
Grovesiella (= Erikssonopsis)
Grovesiella
Gymnomitrula (= Heyderia)
Haplocybe (= Ombrophila)
Helotidium (= Allophylaria)
Helotium (= Cudoniella)
Heterosphaeria
Heyderia
Holmiodiscus
Holwaya
Hygromitra (= Leotia)
Hymenoscypha (=
    Hymenoscyphus)
Hymenoscyphus
Ionomidotis (= Ameghiniella)
Isosoma (= Cudoniella)
Jacobsonia
Jugglerandia (= Holwaya)
Kubickia (= Ombrophila)
Lagerbergia (= Ascocalyx)
Leotia
Leptobelonium (= Strossmayeria)
Llimoniella
Metapezizella
Micraspis
Micropodia
Micropyxis (= Gelatinopsis)
Mitrula (= Heyderia)
Mniaecia
Mollisinopsis
Mytilodiscus
Neobulgaria
Neocudoniella
Neogodronia (= Encoeliopsis)
Nipterella
Ombrophila
Orbiliopsis (= Parorbiliopsis)
Orthoscypha (= Pocillum)
Pachydisca
Parencoelia

Parksia (= Chloroscypha)
Parorbiliopsis
Patellea (= Durella)
Patinellaria
Perizomatium (= Phaeofabraea)
Pestalopezia
Pezizella (= Calycina)
Pezoloma
Phaeangella (= Phibalis)
Phaeangellina
Phaeobulgaria (= Bulgaria)
Phaeofabraea
Phaeohelotium
Phibalis (nom. rej. prop.)
Phyllomyces (= Cordierites)
Physmatomyces
Pocillum
Poculopsis
Podobelonium (= Belonioscypha)
Poloniodiscus (= Ameghiniella)
Polydiscidium
Pragmopora
Pseudocenangium (= Ascocoryne)
Pseudodiscinella (= Pezoloma)
Pseudohelotium
Pseudopezicula
Pyrenopezizopsis (= Cenangiopsis)
Rehmiomyces (= Dictyonia)
Rhizocalyx
Sageria
Sarcoleotia
Scelobelonium (= Crocicreas)
Scleroderris (= Godronia)
Scutularia (= Durella)
Septatium (= Hymenoscyphus)
Septopezizella
Skyathea
Sphaeropeziella (= Durella)
Sphagnicola (= Pezoloma)
Stamnaria
Strossmayeria
Symphyosirinia
Tatraea
Thindiomyces
Tympanis
Ucographa (= Pragmopora)
Unguiculariopsis
Velutarina
Voeltzkowiella (= Bulgaria)
Weinmannioscyphus
Xeromedulla
Xylogramma

Loramycetaceae
Loramyces

Orbiliaceae
Cheilodonta (= Habrostictis)
Habrostictis (= Orbilia)
Hyalinia (= Orbilia)
Myridium (= Orbilia)
Orbilia
Orbiliaster (= Orbilia)
Orbiliella (= Orbilia)

Phacidiaceae
Ascocoma
Gremmenia (= Phacidium)
Lophophacidium
Neonaumovia (= Lophophacidium)
Phacidiostroma (= Phacidium)
Phacidium

Sclerotiniaceae
Asterocalyx
Botryotinia
Ciboria
Ciborinia
Ciboriopsis (= Moellerodiscus)
Coprotinia (= Lambertella)
Cudoniopsis
Dicephalospora
Dumontinia
Elliottinia
Gloeotinia
Grovesinia
Kriegeria (= Rutstroemia)
Lambertella
Lambertellinia
Lanzia (= Rutstroemia)
Martinia (= Martininia)
Martininia
Mitrula
Mitrulinia
Moellerodiscus (= Ciboria)
Monilinia
Moserella
Myriosclerotinia
Ovulinia
Phaeociboria (= Lambertella)
Phaeodiscus (= Lambertella)
Phaeosclerotinia
Piceomphale (= Ciboria)
Poculinia
Poculum (= Rutstroemia)
Pseudociboria
Pycnopeziza
Rutstroemia
Sclerocrana
Scleroglossum (= Scleromitrula)
Scleromitrula
Sclerotinia
Seaverinia
Septotinia
Streptotinia
Stromatinia
Torrendiella
Valdensinia
Verpatinia (= Cudoniopsis)
Whetzelinia (= Sclerotinia)

Vibrisseaceae
Apostemidium (= Vibrissea)
Apostemium (= Vibrissea)
Chlorovibrissea
Leptosporium (= Vibrissea)
Leucovibrissea
Ophiogloea (= Vibrissea)
Vibrissea

Incertae sedis
Adelodiscus
Algincola
Ambrodiscus
Ascographa
Ascotremellopsis (= Myriodiscus)
Atrocybe
Benguetia
Brachyascus (= Microdiscus)
Capricola
Cenangiopsis
Chlorospleniella
Chondroderris
Ciliella
Comesia
Cornuntum

Criserosphaeria
Cryptopezia
Davinciella (= Merodontis)
Dawsicola
Didonia
Didymocoryne
Discomycella
Endoscypha
Endotryblidium
Favraea
Helotiella
Hymenobolus
Hyphoscypha
Iridinea
Lagerheima
Laricina
Lasseria
Lemalis
Livia
Lobularia
Loricella
Malotium
Masseea
Melanopeziza
Melanormia
Merodontis
Microascus (= Microdiscus)
Microdiscus
Microspora
Midotiopsis
Mollisiaster
Muscia
Muscicola
Mycomelanea
Myriodiscus
Obconicum
Orbiliopsis
Orbiliopsis
Parthenope
Peltigeromyces
Peltophoromyces (= Peltigeromyces)
Pezomela
Phaeopyxis
Phragmiticola
Phyllopezis
Pleiopatella
Polydiscina
Pseudolachnum
Pseudopeltis
Pseudotapesia
Pseudotryblidium
Psilophana
Psilothecium
Pteromyces
Radotinea
Riedera
Rimula
Robincola
Roburnia
Sambucina
Sarcomyces
Schnablia
Scutulopsis
Skyttella
Solanella
Starbaeckia
Stilbopeziza
Tanglella
Themisia
Tovariella
Trichangium
Trichohelotium

Tryblis
Tubolachnum
Urceola
Woodiella

LICHINALES
Gloeoheppiaceae
Gloeoheppia
Gudelia
Pseudopeltula

Lichinaceae
Amphopsis (= Pyrenopsis)
Anema
Arctoheppia (= Theligyna)
Arnoldia (= Lempholemma)
Calotrichopsis
Cladopsis (= Pyrenopsis)
Collemopsis (= Psorotichia)
Collemopsis
Corinophoros (= Peccania)
Cryptothele
Cryptotheliomyces (= Cryptothele)
Digitothyrea
Edwardiella
Enchylium (= Forssellia)
Ephebe
Ephebeia (= Ephebe)
Ephebomyces (= Ephebe)
Euopsis
Fernaldia (= Arctoheppia)
Finkia
Forssellia (= Pterygiopsis)
Gonionema (= Thermutis)
Gonohymenia (= Lichinella)
Gonohymeniomyces (= Gonohymenia)
Gyrocollema
Harpidium
Homopsella (= Porocyphus)
Homopsellomyces (= Porocyphus)
Jenmania
Jenmaniomyces (= Jenmania)
Lecanephebe (= Zahlbrucknerella)
Lecidopyrenopsis
Lemmopsis
Lempholemma
Leprocollema
Leptopterygium (= Zahlbrucknerella)
Lichina
Lichinella
Lichiniza (= Porocyphus)
Lichinodium
Lichinomyces (= Lichina)
Malmgrenia (= Pyrenopsis)
Mauritzia (= Pyrenopsis)
Metamelanea
Montinia (= Pyrenocarpon)
Neolichina (= Lichina)
Omphalaria (= Thyrea)
Omphalaria (= Anema)
Pasithoe (= Paulia)
Paulia
Peccania
Peccaniomyces (= Peccania)
Peccaniopsis (= Peccania)
Phloeopeccania
Phloeopeccaniomyces (= Phloeopeccania)
Phylliscidiopsis
Phylliscidium
Phyllisciella

Phylliscum
Plectopsora (= Lempholemma)
Pleoconis (= Peccania)
Pleopyrenis (= Pyrenopsis)
Porocyphus
Pseudarctomia
Psoropsis (= Porocyphus)
Psorotichia
Psorotichiomyces (= Psorotichia)
Pterygiopsidomyces (=
   Pterygiopsis)
Pterygiopsis
Pygmaea (= Lichina)
Pyrenocarpon
Pyrenopsidium (= Cryptothele)
Pyrenopsis
Rechingeria (= Thyrea)
Schizoma (= Thyrea)
Spilonematopsis (= Ephebe)
Staurolemma
Stromatella
Synalissa
Synalissina (= Lempholemma)
Synallisopsis (= Pyrenopsis)
Thallinocarpon (= Gonohymenia)
Thamnidium (= Lichina)
Thelignya
Thelochroa (= Psorotichia)
Thermutis
Thermutomyces (= Thermutis)
Thermutopsis
Thyrea
Tichospora (= Psorotichia)
Zahlbrucknera (=
   Zahlbrucknerella)
Zahlbrucknerella

Peltulaceae
Neoheppia
Peltula

MEDEOLARIALES
Medeolariaceae
Medeolaria

MELIOLALES
Meliolaceae
Actinodothis (= Amazonia)
Amazonia
Amazoniella (= Amazonia)
Appendiculella
Arberia (= Meliola)
Ariefia
Armatella
Artallendea (= Armatella)
Asteridiella
Asteridium (= Meliola)
Ceratospermopsis
Chaetomeliola (= Meliola)
Endomeliola
Haraea
Hypasteridium (nom. dub.)
Irene (= Asteridiella)
Irenina (= Appendiculella)
Irenopsis
Laeviomeliola
Leptascospora
Meliola
Meliolaster (= Amazonia)
Meliothecium (= Meliola)
Metasteridium (nom. dub.)
Myxothecium (= Meliola)
Ophiociliomyces

Ophioirenina
Ophiomeliola
Parasteridiella (= Asteridiella)
Parasteridium (nom. dub.)
Pauahia
Pleomeliola
Pleomerium
Prataprajella
Ticomyces
Urupe
Xenostigme

MICROASCALES
Chadefaudiellaceae
Chadefaudiella
Faurelina

Microascaceae
Anekabeeja
Canariomyces
Enterocarpus
Fairmania (= Microascus)
Kernia
Leuconeurospora
Lophotrichus
Magnusia (= Kernia)
Microascus
Nephrospora (= Microascus)
Peristomium (= Microascus)
Petriella
Petriellidium (= Pseudallescheria)
Pidoplitchkoviella
Pithoascus
Pseudallescheria

Incertae sedis
Ceratocystis
Chaetonaemosphaera (=
   Ceratocystis)
Endoconidiophora (= Ceratocystis)
Mycorhynchella (= Ceratocystis)
Rhynchophoma (= Ceratocystis)
Rostrella (= Ceratocystis)
Rostrella (= Ceratocystis)
Sphaeronaemella

NEOLACTALES
Neolectaceae
Ascocorynium (= Neolecta)
Neolecta
Spragueola (= Neolecta)

ONYGENALES
Gymnoascaceae
Acitheca
Arachniotus
Disarticulatus (= Arachniotus)
Gymnascella
Gymnoascoideus
Gymnoascus
Leucothecium
Macronodus (= Gymnoascus)
Mallochia
Myrillium (= Gymnoascus)
Narasimhella
Orromyces
Petalosporus (= Arachniotus)
Plunkettomyces (= Arachniotus)
Pseudoarachniotus (=
   Arachniotus)
Rollandina (= Arachniotus)
Tripedotrichum (= Gymnoascus)
Waldemaria (= Arachniotus)

Myxotrichaceae
Actinospira (= Myxotrichum)
Byssoascus
Campsotrichum (= Myxotrichum)
Eidamella (= Myxotrichum)
Gymnostellatospora
Myxotrichella (= Myxotrichum)
Myxotrichum
Oncidium (= Myxotrichum)
Pseudogymnoascus
Toxotrichum (= Myxotrichum)

Onygenaceae
Ajellomyces
Amauroascus
Anixiopsis (= Aphanoascus)
Aphanoascus
Apinisia
Arachnotheca
Ascocalvatia
Auxarthron
Bifidocarpus
Brunneospora
Byssoonygena
Emmonsiella (= Ajellomyces)
Keratinophyton (= Aphanoascus)
Kuehniella
Monascella
Nannizziopsis
Neogymnomyces
Neoxenophila (= Aphanoascus)
Onygena
Pectinotrichum
Piligena (= Onygena)
Polytolypa
Renispora
Shanorella
Sphaeropus (= Onygena)
Spiromastix
Uncinocarpus
Xanthothecium
Xynophila (= Aphanoascus)

OPHIOSTOMATALES
Kathistaceae
Kathistes

Ophiostomataceae
Ceratocystiopsis (= Ophiostoma)
Europhium (= Ophiostoma)
Grosmannia (= Ophiostoma)
Klasterskya
Linostoma (= Ophiostoma)
Ophiostoma
Spumatoria
Subbaromyces

OSTROPALES
Gomphillaceae
Actinoplaca
Actinoplacomyces (= Echinoplaca)
Artheliopsis (= Echinoplaca)
Arthotheliopsidomyces (=
   Echinoplaca)
Arthotheliopsis (= Echinoplaca)
Aulaxina
Baeopodium (= Gomphillus)
Bullatina
Calenia
Caleniomyces (= Calenia)
Caleniopsis
Conicosolen (= Psorotheciopsis)
Diploschistella (= Gyalideopsis)

Echinoplaca
Gomphillus
Gonothecis (= Gyalectidium)
Gonothecium (= Gyalectidium)
Gyalectidium
Gyalideopsis
Lopadiopsidomyces (= Gyalectidium)
Lopadiopsis (= Gyalectidium)
Monospermella (= Psorotheciopsis)
Mycetodium (= Gomphillus)
Phlyctidium (= Calenia)
Psorotheciopsis
Pycnodermellina (= Echinoplaca)
Rhexophiale (= Sagiolechia)
Sagiolechia
Tricharia

Graphidaceae
Acanthographina (= Acanthothecis)
Acanthographis (= Acanthothecis)
Acanthotheciopsis (= Acanthothecis)
Acanthothecis
Acanthothecium (= Acanthothecis)
Acanthothecomyces (= Acanthothecis)
Actinoglyphis (= Sarcographa)
Allographa (= Graphina)
Anomalographis
Anomomorpha (= Graphis)
Asterisca (= Sarcographa)
Aulacographa (= Graphis)
Chiographa (= Phaeographis)
Creographa (= Phaeographina)
Ctesium (= Graphina)
Cyclographina
Digraphis (= Graphis)
Diorygma (= Graphina)
Diplogramma
Diplogrammatomyces (= Diplogramma)
Diplographis (= Graphis)
Diplolabia (= Graphis)
Dyplolabia (= Graphis)
Ectographis (= Phaeographina)
Emblemia (= Phaeographina)
Fissurina (= Graphis)
Flegographa (= Sarcographa)
Glaucinaria (= Graphina)
Glyphis
Glyphomyces (= Glyphis)
Graphidomyces (= Graphis)
Graphidula (= Phaeographis)
Graphina
Graphinella (= Graphis)
Graphinomyces (= Graphina)
Graphis
Gymnographa (= Sarcographa)
Gymnographomyces (= Sarcographa)
Gymnographopsis
Gymnotrema (= Gyrostomum)
Gyrostomomyces (= Gyrostomum)
Gyrostomum
Helminthocarpon
Helminthocarponomyces (= Helminthocarpon)
Hemithecium (= Graphina)
Hymenodecton (= Phaeographis)
Lecanactis (= Phaeographis)
Leiogramma (= Phaeographis)

Leiorreuma (= Phaeographis)
Leucogramma (= Phaeographina)
Leucogramma (= Graphina)
Medusula (= Sarcographa)
Medusulina
Megalographa (= Phaeographina)
Opegrapha (= Graphis)
Oxystoma (= Graphis)
Phaeographidomyces (= Phaeographis)
Phaeographina
Phaeographinomyces (= Phaeographina)
Phaeographis
Phlegmographa (= Sarcographa)
Platygramma (= Phaeographis)
Pliariona (= Phaeographina)
Psorographis (= Acanthothecis)
Pyrochroa (= Phaeographis)
Pyrographa (= Phaeographis)
Sarcographa
Sarcographina
Sarcographomyces (= Sarcographa)
Schistostoma (= Graphis)
Sclerographa (= Opegrapha)
Solenographa (= Phaeographis)
Stenographa (= Graphina)
Thalloloma (= Graphina)
Thecaria (= Phaeographina)
Thecographa (= Phaeographina)
Thelographis (= Phaeographina)
Theloschisma (= Phaeographis)
Ustalia (= Phaeographis)

Odontotremataceae
Beloniella (= Odontura)
Bryodiscus
Coccomycetella
Geltingia
Lethariicola
Odontotrema
Odontotremella (= Odontura)
Odontura
Paschelkiella
Pleospilis (= Spirographa)
Skyttea
Sphaeropezia (= Odontotrema)
Spilomela (= Pleospilis)
Spirographa
Spirographomyces (= Spirographa)
Stromatothecia
Xerotrema

Solorinellaceae
Actinopelte (= Solorinella)
Aglaothecium (= Gyalidea)
Cappellettia (= Gyalidea)
Gyalidea
Linhartia
Solorinella
Solorinellomyces (= Solorinella)

Stictidaceae
Absconditella
Acarosporina
Agyriella (= Schizoxylon)
Agyriopsis (= Schizoxylon)
Bergorea (= Robergea)
Biostictis
Bisbyella (= Schizoxylon)
Carestiella
Clathroporinopsis (= Topelia)

Conotrema
Conotrematomyces (= Conotrema)
Conotremopsis
Cryptella (= Robergea)
Cryptodiscus
Cyanoderma (= Cyanodermella)
Cyanodermella
Cyanospora (= Robergea)
Cycledium (= Schizoxylon)
Cyclostoma (= Stictis)
Cylindrina (= Stictis)
Delpontia
Diplocryptis (= Cryptodiscus)
Dithelopsis (= Thelopsis)
Haplothelopsis (= Thelopsis)
Holothelis (= Thelopsis)
Lichenopsis (= Stictis)
Lillicoa
Nanostictis
Ostropa
Petractis
Phacobolus (= Stictis)
Platysticta (= Stictis)
Propoliopsis
Pseudostictis (= Cryptodiscus)
Robergea
Schizoxylon
Schizoxylum (= Schizoxylon)
Schmitzomia (= Stictis)
Sphaerolina (= Schizoxylon)
Stictis
Stictophacidium
Sychnoblastia (= Thelopsis)
Sychnogonia (= Thelopsis)
Thelopsidomyces (= Thelopsis)
Thelopsis
Topelia
Tuberculostoma (= Robergea)

Thelotremataceae
Antrocarpon (= Thelotrema)
Antrocarpum (= Thelotrema)
Ascidium (= Ocellularia)
Asteristion (= Thelotrema)
Asteristium (= Thelotrema)
Asterothyriomyces (= Asterothyrium)
Asterothyrium
Brassia (= Thelotrema)
Chapsa (= Ocellularia)
Chroodiscus
Coniochila (= Ocellularia)
Coscinedia (= Myriotrema)
Diplopeltopsis (= Asterothyrium)
Diploschistes
Ectolechia (= Ocellularia)
Enterostigma (= Thelotrema)
Enterostigmatomyces (= Thelotrema)
Eupropolis (= Phaeotrema)
Gyalecta (= Diploschistes)
Janseella (= Phaeotrema)
Lagerheimina (= Diploschistes)
Lectularia (= Diploschistes)
Leptotrema (= Ocellularia)
Leucodecton (= Thelotrema)
Macropyrenium (= Ocellularia)
Mycopyrenium (= Thelotrema)
Myriotrema
Nadvornikia
Ocellis (= Ocellularia)
Ocellularia
Ozocladium (= Polistroma)

Patellaria (= Diploschistes)
Phaeotrema
Phanotylium (= Tremotylium)
Phyllobrassia (= Thelotrema)
Platygrapha
Polistroma
Polyschistes (= Diploschistes)
Polystroma (= Polistroma)
Porophora (= Ascidium)
Protoschistes (= Diploschistes)
Prototrema (= Thelotrema)
Psorotheciella (= Asterothyrium)
Rhabdodiscus (= Ocellularia)
Stegobolus (= Ocellularia)
Stephanophoron (= Nadvornikia)
Stictoclypeolum (= Asterothyrium)
Stigmagora (= Ocellularia)
Thelotrema
Thelotrematomyces (= Thelotrema)
Tremotyliomyces (= Tremotylium)
Tremotylium
Urceolaria (= Diploschistes)
Urceolariomyces (= Diploschistes)

Incertae sedis
Leucogymnospora
Platygramme
Platygraphopsis
Xyloschistes
Xyloschistomyces (= Xyloschistes)

PATELLARIALES
Arthrorhaphidaceae
Arthrorhaphis
Gongylia (= Arthrorhaphis)
Mycobacidia (= Arthrorhaphis)
Parathalle (= Arthrorhaphis)
Raphiospora (= Arthrorhaphis)

Patellariaceae
Ascoporia
Baggea
Bruneaudia (= Rhytidhysteron)
Caldesia (= Holmiella)
Eutryblidiella (= Rhytidhysteron)
Globuligera
Haematomyxa
Holmiella
Lecanidiella
Lecanidion (= Patellaria)
Murangium
Patellaria
Phaneromyces
Pittierodothis (= Protoscypha)
Poetschia
Protoscypha
Rhizodiscina
Rhytidhysterium (=
    Rhytidhysteron)
Rhytidhysteron
Rhytidopeziza (= Rhytidhysteron)
Schrakia
Stratisporella
Teichosporella
Tryblidaria
Tryblidiella (= Rhytidhysteron)

Phillipsiellaceae
Microphyma (= Phillipsiella)
Phillipsiella

PELTIGERALES
Lobariaceae
Crocodia (= Pseudocyphellaria)
Cyanisticta (= Pseudocyphellaria)
Dendriscocaulon
Diclasmia (= Sticta)
Diphaeosticta (=
    Pseudocyphellaria)
Diphanosticta (= Sticta)
Dysticta (= Sticta)
Dystictina (= Sticta)
Lobaria
Lobarina (= Lobaria)
Lobariomyces (= Lobaria)
Merostictica (= Pseudocyphellaria)
Parmosticta (= Pseudocyphellaria)
Parmostictina (=
    Pseudocyphellaria)
Phaeosticta (= Pseudocyphellaria)
Phanosticta (= Pseudocyphellaria)
Podostictina (= Sticta)
Pseudocyphellaria
Pulmonaria (= Lobaria)
Reticularia (= Lobaria)
Ricasolia (= Lobaria)
Ricasoliomyces (= Ricasolia)
Saccardoa (= Pseudocyphellaria)
Sticta
Stictina (= Sticta)
Stictomyces (= Lobaria)

Nephromataceae
Dermatodea (= Nephroma)
Nephroma
Nephromatomyces (= Nephroma)
Nephromiomyces (= Nephroma)
Nephromium (= Nephroma)
Opisteria (= Nephroma)
Ornatinephroma (= Nephroma)

Peltigeraceae
Antilyssa (= Peltigera)
Byrsalis (= Peltigera)
Chloropeltigera (= Peltigera)
Chloropeltis (= Peltigera)
Hydrothyria
Neosolorina (= Solorina)
Peltidea (= Peltigera)
Peltideomyces (= Peltigera)
Peltigera
Peltigeromyces (= Peltigera)
Peltophora (= Peltigera)
Placodion (= Peltigera)
Ruspoliella (= Solorina)
Siphulastrum
Siphulina
Solorina
Solorinina (= Solorina)
Solorinomyces (= Solorina)
Sommerfeltia (= Solorina)

Placynthiaceae
Antarctomia (= Placynthium)
Anziella (= Placynthium)
Calkinsia (= Placynthium)
Callolechia (= Collolechia)
Collolechia (= Placynthium)
Garovaglia (= Polychidium)
Garovaglina (= Garovaglia)
Ginzbergerella (= Polychidium)
Hertella
Koerberia
Lecothecium (= Placynthium)

Lecozonia (= Placynthium)
Leptochidium
Leptodendriscum (= Polychidium)
Leptogidiomyces (= Polychidium)
Leptogidium (= Polychidium)
Llanolichen (= Placynthium)
Llanomyces (= Placynthium)
Pannularia (= Placynthium)
Pilonema (= Polychidium)
Placynthiomyces (= Placynthium)
Placynthiopsis
Placynthium
Polychidium
Pseudoleptogium (= Polychidium)
Pterygium (= Placynthium)
Racoblenna (= Placynthium)
Tingiopsidium (= Koerberia)
Vestergrenopsis
Wilmsia (= Placynthium)
Wilsonia (= Polychidium)

PERTUSARIALES
Coccotremataceae
Coccotrema
Lepolichen
Perforaria (= Coccotrema)
Perforariomyces (= Coccotrema)
Verrucarina (= Coccotrema)

Megasporaceae
Megaspora

Pertusariaceae
Clausaria (= Pertusaria)
Isidium (= Pertusaria)
Korkir (= Ochrolechia)
Lecanidium (= Pertusaria)
Lepra (= Pertusaria)
Leproncus (= Pertusaria)
Melanaria (= Pertusaria)
Ochrolechia
Pertusaria
Pertusariomyces (= Pertusaria)
Pionospora (= Pertusaria)
Porina (= Pertusaria)
Stereocaulon (= Pertusaria)
Stereocaulum (= Pertusaria)
Thamnochrolechia
Varicellaria
Varicellariomyces (= Varicellaria)

Thelenellaceae
Acrorixis (= Thelenella)
Catharinia (= Julella)
Chromatochlamys
Hyalospora (= Julella)
Julella
Luykenia (= Thelenella)
Microglaena (= Thelenella)
Microglaenomyces (= Thelenella)
Microglena (= Microglaena)
Peltosphaeria (= Julella)
Polyblastiopsis (= Julella)
Thelenella
Weitenwebera (= Thelenella)

PEZIZALES
Ascobolaceae
Anserina (= Ascobolus)
Ascobolus
Ascophanella
Ascophanopsis (= Thecotheus)
Ascophanus

Ascozonus
Cleistoiodophanus
Crouaniella (= Ascobolus)
Cubonia
Dasybolus (= Ascobolus)
Dasyobolus (= Ascobolus)
Gymnodiscus (= Zukalina)
Ornithascus (= Saccobolus)
Saccobolus
Seliniella (= Ascobolus)
Sphaeridiobolus (= Ascobolus)
Streptotheca
Thecotheus
Zukalina (= Thecotheus)

Ascodesmidaceae
Amaurascopsis
Ascodesmis
Eleutherascus
Hemiascosporium (= Eleutherascus)

Balsamiaceae
Balsamia
Barssia
Leucangium (= Picoa)
Phymatomyces (= Barssia)
Picoa
Pseudobalsamia (= Balsamia)

Carbomycetaceae
Carbomyces

Eoterfeziaceae
Cleistothelebolus
Eoterfezia
Lasiobolidium

Glaziellaceae
Endogonella (= Glaziella)
Glaziella

Helvellaceae
Acetabula (= Helvella)
Biverpa (= Helvella)
Boletolichen (= Helvella)
Cazia
Choeromyces (= Choiromyces)
Choiromyces
Cidaris
Coelomorum (= Helvella)
Costapeda (= Helvella)
Cowlesia (= Helvella)
Cyathipodia (= Helvella)
Discina (= Gyromitra)
Fastigiella (= Discina)
Fischerula
Fuckelina (= Helvella)
Geomorium (= Helvella)
Geoporella (= Hydnotrya)
Globipilea (= Helvella)
Globopilea (= Helvella)
Gymnohydnotrya
Gyrocephalus (= Gyromitra)
Gyrocratera (= Hydnotrya)
Gyromitra
Gyromitrodes (= Pseudorhizina)
Helvella
Helvellella (= Pseudorhizina)
Hydnotrya
Leptopodia (= Helvella)
Macropodia (= Helvella)
Macroscyphus (= Helvella)

Maublancomyces (= Discina)
Midotis (= Helvella)
Neogyromitra (= Discina)
Ochromitra (= Pseudorhizina)
Paradiscina (= Gyromitra)
Paxina (= Helvella)
Phaeomacropus (= Helvella)
Phleboscyphus (= Helvella)
Physomitra (= Gyromitra)
Piersonia (= Choiromyces)
Pleopus (= Gyromitra)
Pseudorhizina
Rhizina
Tartufa (= Choiromyces)
Tubipeda (= Helvella)
Underwoodia (= Helvella)
Wynnella
Zobelia (= Choiromyces)

Karstenellaceae
Karstenella

Morchellaceae
Boletus (= Morchella)
Disciotis
Eromitra (= Mitrophora)
Mitrophora (= Morchella)
Monka (= Verpa)
Morchella
Morilla (= Morchella)
Phalloboletus (= Morchella)
Ptychoverpa (= Verpa)
Relhanum (= Verpa)
Verpa

Otideaceae
Acervus
Aleuria
Aleurina
Angiophaeum (= Phaeangium)
Anixia (= Orbicula)
Anthracobia
Aparaphysaria
Arpinia
Ascocalathium
Ascosparassis
Barlaea (= Lamprospora)
Barlaeina (= Lamprospora)
Boubovia
Boudierella
Byssonectria (= Inermisia)
Caloscypha
Cheilymenia
Ciliaria (= Scutellinia)
Cochlearia (= Otidea)
Coprobia (= Cheilymenia)
Crouania (= Lamprospora)
Densocarpa (= Stephensia)
Dingleya
Discomycetella (= Inermisia)
Elderia (= Stephensia)
Eoaleurina
Fimaria (= Pseudombrophila)
Flavoscypha (= Otidea)
Fleischhakia (= Psilopezia)
Galeoscypha
Genabea
Genea
Geneosperma (= Scutellinia)
Geopora
Geopyxis
Greletia (= Smardaea)
Hiemsia

Humaria
Humaria (= Octospora)
Humariella (= Scutellinia)
Humarina (= Octospora)
Hydnocaryon (= Genea)
Hydnocystis
Hypotarzetta
Inermisia
Jafnea
Jafneadelphus (= Aleurina)
Kotlabaea
Labyrinthomyces
Lachnea (= Humaria)
Lachnea (= Scutellinia)
Lamprospora
Lathraeodiscus
Lazuardia
Leucoloma (= Octospora)
Leucopezis (= Octospora)
Leucoscypha
Luciotrichus
Marcelleina
Melastiza
Melastiziella (= Scutellinia)
Miladina
Moravecia
Mycogala (= Orbicula)
Mycogalopsis
Mycolachnea (= Humaria)
Myrmecocystis (= Genabea)
Nannfeldtiella
Neottiella (= Octospora)
Neottiopezis (= Leucoscypha)
Nothojafnea
Octospora
Octosporella
Orbicula
Otidea
Otidella (= Caloscypha)
Otideopsis
Parascutellinia
Paratrichophaea
Patella (= Scutellinia)
Paurocotylis
Petchiomyces (nom. nud.)
Peziza (= Aleuria)
Pfistera
Phaeangium
Phaedropezia (= Acervus)
Pilopeza (= Psilopezia)
Protogenea (= Hydnocystis)
Pseudaleuria
Pseudhydnotrya (= Geopora)
Pseudocollema (= Byssonectria)
Pseudogenea (= Genabea)
Pseudohydnotrya (= Geopora)
Pseudombrophila
Pseudotis (= Otidea)
Psilopezia
Pulparia (= Pulvinula)
Pulvinula
Pustularia (= Tarzetta)
Pustulina (= Tarzetta)
Pyropyxis
Ramsbottomia
Ramulina (= Pseudombrophila)
Reddellomyces
Rhizoblepharia
Rhodoscypha
Rhodotarzetta
Rubelia (= Scutellinia)
Scutellinia
Selenaspora

Sepultaria (= Geopora)
Smardaea
Sowerbyella
Sphaerosoma
Sphaerospora (= Scutellinia)
Sphaerosporella
Sphaerosporula (= Scutellinia)
Sphaerozosma (= Sphaerosoma)
Spooneromyces
Stephensia
Stereolachnea (= Scutellinia)
Svrcekomyces (=
    Pseudombrophila)
Tarzetta
Trichaleurina (= Scutellinia)
Trichaleuris (= Scutellinia)
Tricharia (= Tricharina)
Tricharina
Trichophaea
Trichophaeopsis
Warcupia
Wilcoxina

Pezizaceae
Aleuria (= Peziza)
Aleurina (= Peziza)
Amylascus
Boudiera
Carnia
Caulocarpa (= Sarcosphaera)
Clelandia (= Mycoclelandia)
Curreyella (= Plicaria)
Daleomyces (= Peziza)
Detonia (= Plicaria)
Discaria (= Plicaria)
Durandiomyces (= Peziza)
Galactinia (= Peziza)
Geoscypha (= Peziza)
Gonzala (= Peziza)
Gorodkoviella (= Pachyella)
Hapsidomyces
Heteroplegma (= Peziza)
Hydnobolites
Hydnoplicata (= Peziza)
Hydnotryopsis
Infundibulum (= Peziza)
Iodophanus
Iotidea (= Peziza)
Kimbropezia
Lepidotia (= Peziza)
Leptopeza (= Plicaria)
Muciturbo
Mycoclelandia
Napomyces (= Peziza)
Pachyella
Paramitra (= Peziza)
Peltidium (= Pachyella)
Peziza
Phaeobarlaea (= Plicaria)
Phaeopezia (= Peziza)
Plicaria
Plicariella (= Plicaria)
Podaleuris (= Peziza)
Pulvinaria (= Pachyella)
Rhodopeziza
Ruhlandiella
Sarcosphaera
Scabropezia
Scodellina (= Peziza)
Sphaerozone
Svrcekia (= Boudiera)
Tirmania
Tremellodiscus (= Ruhlandiella)

Pyronemataceae
Dictyocoprotus
Leporina (nom. dub.)
Phycoascus (= Pyronema)
Pyronema

Sarcoscyphaceae
Anthopeziza (= Microstoma)
Articulariella (= Microstoma)
Aurophora
Boedijnopeziza (= Cookeina)
Cookeina
Geodina
Microstoma
Phillipsia
Pilocratera (= Cookeina)
Pindara
Pithya
Pseudopithyella
Rickiella (= Phillipsia)
Sarcoscypha
Scypharia (= Sarcoscypha)
Thindia
Trichoscypha (= Cookeina)

Sarcosomataceae
Chorioactis
Desmazierella
Galiella
Gloeocalyx (= Plectania)
Kompsoscypha
Korfiella
Melascypha (= Pseudoplectania)
Nanoscypha
Neournula
Plectania
Pseudoplectania
Rhizopodella (= Plectania)
Sarcosoma
Strobiloscypha
Urnula
Wolfina
Wynnea

Terfeziaceae
Cryptica (= Pachyphloeus)
Delastria
Loculotuber
Mattirolomyces (= Terfezia)
Pachyphlodes (= Pachyphloeus)
Pachyphloeus
Terfezia
Tulasneinia (= Terfezia)
Tulasnia (= Terfezia)

Thelebolaceae
Caccobius
Chalazion
Coprobolus
Coprotiella
Coprotus
Dennisiopsis
Lasiobolus
Lasiothelebolus (= Thelebolus)
Leptokalpion
Mycoarctium
Ochotrichobolus
Pezizella (= Thelebolus)
Pezizula (= Thelebolus)
Pseudascozonus
Pyronemella (= Lasiobolus)
Ramgea
Ryparobius (= Thelebolus)

Telebolus (= Thelebolus)
Thelebolus
Trichobolus

Tuberaceae
Aschion (= Tuber)
Delastreopsis (= Tuber)
Ensaluta (= Tuber)
Lespiaultinia (= Tuber)
Mukagomyces (= Tuber)
Oogaster (= Tuber)
Paradoxa
Terfeziopsis (= Tuber)
Tuber
Tuberium (= Tuber)
Vittadinion (= Tuber)

Incertae sedis
Discinella
Gemmularia
Hylostoma
Microeurotium
Rugosaria (= Gemmularia)
Tucahus (= Gemmularia)
Urceolaria

PHYLLACHORALES
Phyllachoraceae
Anisochora (= Placostroma)
Apiosphaeria
Bagnisiopsis (= Coccodiella)
Camarotella (= Coccodiella)
Catacauma (= Phyllachora)
Caulochora (= Glomerella)
Causalis (= Coccodiella)
Chiloella (= Glomerella)
Clypeostigma (= Phyllachora)
Clypeotrabutia (= Phyllachora)
Coccodiella
Coccoidiopsis (= Coccodiella)
Coccostroma (= Coccodiella)
Coccostromopsis (= Coccodiella)
Deshpandiella
Diachora
Diplosporis (= Phyllachora)
Discomycopsella (= Phyllachora)
Dothidina (= Coccodiella)
Endodothella (= Phyllachora)
Endophyllachora (= Phyllachora)
Endotrabutia (= Phyllachora)
Erikssonia
Geminispora
Gibellia (= Gibellina)
Gibellina
Glomerella
Gnomoniopsis (= Glomerella)
Haloguignardia
Halstedia (= Phyllachora)
Haplostroma (= Coccodiella)
Haplothecium (= Glomerella)
Hypostegium (= Glomerella)
Hysterodothis (= Sphaerodothis)
Imazekia
Isothea
Leptocrea (= Polystigma)
Leveillinopsis (= Coccodiella)
Lichenochora
Lindauella
Lohwagia
Marinosphaera
Metachora (= Phyllachora)
Microphiodothis (= Ophiodothella)
Munkiodothis (= Rehmiodothis)

Neoflageoletia
Neozimmermannia (= Glomerella)
Ophiodothella
Oswaldia (= Apiosphaeria)
Paidania (= Erikssonia)
Periaster (= Erikssonia)
Petrakiella
Phaeochora
Phaeochorella
Phaeotrabutia (= Phyllachora)
Phaeotrabutiella (= Phyllachora)
Phoenicostroma (= Coccodiella)
Phragmocarpella (= Phyllachora)
Phragmocauma (= Phyllachora)
Phycomelaina
Phyllachora
Phylleutypa
Phyllocrea
Physalosporina (= Stigmatula)
Placostroma (= Phyllachora)
Plectosphaera (= Phyllachora)
Polystigma
Polystigmella (= Polystigma)
Pseudomelasmia (= Phyllachora)
Pseudothiella
Puiggarina (= Phyllachora)
Rehmiodothis
Retroa
Rheumatopeltis (= Trabutia)
Rhopographina (= Ophiodothella)
Rikatlia
Rinia (= Erikssonia)
Schizochora
Schizochorella (= Phyllachora)
Scolecoccoidea (= Coccodiella)
Scolecodothis (= Ophiodothella)
Scolecodothopsis (=
    Ophiodothella)
Serenomyces
Sirentyloma (= Phyllachora)
Sphaerodothis
Stigmatula (= Polystigma)
Stigmochora
Stromaster
Sucinaria (= Coccodiella)
Telemeniella (= Phyllachora)
Telimena
Telimenella
Telimeniella
Telimenochora
Telimenopsis (= Telimena)
Thamnogalla
Tolediella (= Phyllachora)
Trabutia
Trabutiella (= Phyllachora)
Vitreostroma
Zimmermanniella

PNEUMOCYSTIDALES
Pneumocystidaceae
Pneumocystis

PROTOMYCETALES
Protomycetaceae
Burenia
Protomyces
Protomycopsis
Taphridium
Volkartia

PYRENULALES
Massariaceae
Aglaospora (= Massaria)

Aigialus
Bathystomum (= Massaria)
Decaisnella
Dothivalsaria
Massaria
Navicella
Phaeomassaria (= Massaria)

Phyllobatheliaceae
Opercularia (= Phyllobathelium)
Phyllobathelium

Pleurotremataceae
Pleurotrema
Pleurotrematomyces (=
    Pleurotrema)
Trichotrema (= Pleurotrema)

Pyrenulaceae
Acrocordiella
Anthracothecium
Anthracothecomyces (= Pyrenula)
Bottariomyces (= Pyrenula)
Bunodea (= Pyrenula)
Chrooicia (= Pyrenula)
Clypeopyrenis
Coenicia (= Pyrenula)
Coenoicia (= Melanotheca)
Dipyrenis (nom. dub.)
Distopyrenis
Granulopyrenis
Lacrymospora
Lithothelium
Mazaediothecium
Melanotheca (= Pyrenula)
Mycopyrenula (= Pyrenula)
Parapyrenis
Parmentaria (= Pyrenula)
Parmentariomyces (= Parmentaria)
Parmentieria (= Parmentaria)
Phaeodictyon (= Anthracothecium)
Plagiocarpa (= Lithothelium)
Pleamphisphaeria (= Pyrenula)
Pleurotheliopsis (= Pyrenula)
Pleurothelium (= Pyrenula)
Pleurothelium (= Parapyrenis)
Polypyrenula
Polythelis (= Polypyrenula)
Pseudarthopyrenia (=
    Pyrenocollema)
Pyrenastromyces (= Pyrenula)
Pyrenastrum (= Pyrenula)
Pyrenocollema
Pyrenographa
Pyrenowilmsia
Pyrenula
Pyrenulomyces (= Pyrenula)
Pyrgillocarpon (= Pyrgillus)
Pyrgillomyces (= Pyrgillus)
Pyrgillus
Requienella
Starbaeckiella (= Pyrenula)
Titanella (= Pyrenula)
Trematomyces (= Requienella)
Xanthoporina (= Pyrenocollema)
Xanthopyrenia (= Pyrenocollema)

Trypetheliaceae
Architrypethelium
Astrothelium
Bathelium (= Laurera)
Bathelium (= Trypethelium)
Buscalionia

Campylothelium
Cryptothelium (= Laurera)
Exiliseptum
Heufleria (= Laurera)
Heufleria (= Astrothelium)
Laurera
Laureromyces (= Laurera)
Leightonia (= Trypethelium)
Megalotremis
Meissneria (= Laurera)
Melanothecomyces (= Laurera)
Melanothecopsis
Meristosporum (= Laurera)
Ornatopyrenis
Parathelium (= Campylothelium)
Phyllothelium (= Trypethelium)
Plagiotrema (= Pseudopyrenula)
Polymeridium
Porophora (= Trypethelium)
Prototylium (= Pseudopyrenula)
Pseudopyrenula
Pyrenodium (= Asterothelium)
Riddlea (= Laurera)
Trypetheliomyces (= Trypethelium)
Trypetheliopsis
Trypethelium

Incertae sedis
Asteroporomyces (= Asteroporum)
Asteroporum
Blastodesmia
Celothelium
Heufleridium
Micromma
Mycasterotrema (= Asterotrema)
Mycoporum
Plagiothelium
Porodothion (= Porothelium)
Porothelium
Rhaphidicyrtis
Stromatothelium
Xenus

RHYTISMATALES
Ascodichaenaceae
Ascodichaena
Delpinoina
Diplochora (= Pseudophacidium)
Henriquesia (= Delpinoina)
Myxophacidiella (=
    Pseudophacidium)
Myxophacidium (=
    Pseudophacidium)
Pseudophacidium

Cryptomycetaceae
Cryptomyces
Macroderma
Phacidiella (= Potebniamyces)
Potebniamyces
Pseudorhytisma

Rhytismataceae
Aporia (= Lophodermium)
Biatorellina (= Tryblidiopsis)
Bifusella
Bifusepta
Bivallum
Ceratophacidium
Cerion
Coccomycella (= Coccomyces)
Coccomyces
Coccophacidium (= Therrya)

Colpoma
Criella
Cyclaneusma
Davisomycella
Dermascia (= Lophodermium)
Discocainia
Duplicaria
Duplicariella
Elytroderma
Epidermella (= Hypoderma)
Hadotia (= Lophodermium)
Hypoderma
Hypoderma (= Hypoderma)
Hypodermella
Hypodermopsis (= Hypoderma)
Hypohelion
Hysterium (= Colpoma)
Isthmiella
Lasiostictis (= Naemacyclus)
Lirula
Locelliderma (= Hypoderma)
Lophoderma (= Hypoderma)
Lophodermella
Lophodermellina (=
   Lophodermium)
Lophodermina (= Lophodermium)
Lophodermium
Lophomerum (= Lophodermium)
Malenconia (= Coccomyces)
Melanosorus (= Rhytisma)
Meloderma
Moutoniella
Myriophacidium
Naemacyclus
Nymanomyces
Pachyrhytisma (= Rhytisma)
Parvacoccum
Phaeorhytisma
Placuntium
Ploioderma
Pragmoparopsis (= Colpoma)
Propolis
Propolomyces (= Propolis)
Pureke
Rhabdocline
Rhytisma
Soleella
Sporomega
Stictostroma (= Placuntium)
Stigmatoscolia (= Lophodermina)
Synglonium (= Nymanomyces)
Terriera
Therrya
Tryblidiopsis
Tryblidis (= Tryblidiopsis)
Virgella
Vladracula
Xyloma (= Rhytisma)
Xyloschizon
Zeus

Incertae sedis
Apiodiscus
Bonanseja
Cavaraella
Chailletia (= Karstenia)
Darkera
Didymascus
Gelineostroma
Haplophyse
Heufleria
Hypodermellina
Iridionia (= Irydyonia)

Irydyonia
Karstenia
Laquearia
Lasiostictella
Melittiosporium
Melittosporiella
Melittosporium (=
   Melittiosporium)
Metadothis
Neophacidium
Ocotomyces
Phaeophacidium
Pleiostictis (= Melittiosporium)
Propolidium
Pseudotrochila
Tridens

SACCHAROMYCETALES
Cephaloascaceae
Cephaloascus
Phialoascus

Dipodascaceae
Dipodascus
Endyllium (= Dipodascus)
Galactomyces
Magnusiomyces (= Dipodascus)
Zendera (= Dipodascus)

Endomycetaceae
Ambrosiozyma
Ascocephalophora
Ascoidea
Botryoascus
Conidiascus
Endomyces
Hormoascus
Oscarbrefeldia
Pachysolen
Podocapsa
Podocapsium (= Podocapsa)
Stephanoascus
Yamadazyma
Zygoascus

Eremascaceae
Eremascus

Lipomycetaceae
Dipodascopsis
Lipomyces
Waltiozyma (= Lipomyces)
Waltomyces
Zygolipomyces (= Lipomyces)
Zygozyma

Metschnikowiaceae
Anthomyces (= Metschnikowia)
Ashbia (= Ashbya)
Ashbya
Chlamydozyma (= Metschnikowia)
Clavispora
Coccidiascus
Crebrothecium
Endomycopsis (= Guillermondella)
Eremothecium
Guilliermondella
Holleya
Metschnikowia
Metschnikowiella (=
   Metschnikowia)
Monospora (= Metschnikowia)
Monosporella (= Metschnikowia)

Nectaromyces (= Metschnikowia)
Nematospora
Spermophthora

Saccharomycetaceae
Aphidomyces
Arxiozyma
Azymohansenula (= Hansenula)
Azymomyces (= Debaryomyces)
Babjevia
Byrrha (= Pichia)
Cicadomyces
Citeromyces
Coccidomyces
Cyniclomyces
Debaryolipomyces (nom. nud.)
Debaryomyces (= Torulaspora)
Debaryomyces
Debaryozyma (= Debaryomyces)
Dekkera
Dekkeromyces (= Kluyveromyces)
Dekkeromyces (= Kluyveromyces)
Fabospora (= Kluyveromyces)
Hansenula (= Pichia)
Hyphopichia (= Pichia)
Isomyces (= Debaryomyces)
Issatchenkia
Kazachstania (= Saccharomyces)
Kluyveromyces
Kuraishia
Lodderomyces
Nakazawaea
Octomyces (= Saccharomyces)
Ogataea
Pachytichospora
Petasospora (= Hansenula)
Pichia
Prosaccharomyces (=
   Saccharomycopsis)
Psyllidomyces
Saccharomyces
Saccharomycopsis (=
   Cyniclomyces)
Sachsia
Saturnospora
Schwanniomyces (=
   Debaryomyces)
Sporopachydermia
Sympodiomyces
Torulaspora
Udeniomyces
Wickerhamiella
Willia (= Hansenula)
Wingea (= Debaryomyces)
Yarrowia
Zonosporis (= Schwanniomyces)
Zygofabospora
Zygohansenula (= Pichia)
Zygopichia (= Pichia)
Zygopichia (= Pichia)
Zygorenospora (= Saccharomyces)
Zygosaccharis (=
   Zygosaccharomyces)
Zygosaccharomyces
Zygosaccharomycodes (=
   Saccharomyces)
Zygowillia (= Pichia)
Zymodebaryomyces (=
   Torulaspora)
Zymopichia (= Pichia)

Saccharomycodaceae
Carpozyma (= Hanseniaspora)

Guilliermondia (= Nadsonia)
Hansenia (= Hanseniaspora)
Hansenia (= Kloeckeraspora)
Hanseniaspora
Kloeckeraspora (= Hanseniaspora)
Nadsonia
Saccharomycodes
Saeenkia (= Saccharomycodes)
Thelis (= Hanseniaspora)
Vanderwaltia (= Kloeckeraspora)
Wickerhamia

Saccharomycopsidaceae
Arthroascus
Endomycopsella
Komagataea
Saccharomycopsis
Williopsis
Zygowilliopsis (= Williopsis)

Incertae sedis
Bacillopsis (nom. dub.)
Dabaryomyces (nom. alt.)
Dolichoascus
Endomycodes
Entelexis
Fragosia
Helicogonium
Menezesia
Oleina
Oleinis (= Oleina)

SCHIZOSACCHAROMYCETALES
Schizosaccharomycetaceae
Hasegawaea
Octosporomyces (=
    Schizosaccharomyces)
Quadrisporomyces (=
    Schizosaccharomyces)
Schizosaccharis (=
    Schizosaccharomyces)
Schizosaccharomyces

SORDARIALES
Batistiaceae
Batistia

Boliniaceae
Apiocamarops
Bolinia (= Camarops)
Camarops
Cerastomis (= Endoxyla)
Endoxyla
Eutypopsis (= Endoxyla)
Lentomitella (= Endoxyla)
Linostomella (= Endoxyla)
Peridoxylon (= Camarops)
Peridoxylum (= Peridoxylon)
Phaeosperma (= Camarops)
Sarcostromella (= Camarops)
Solenoplea (= Camarops)

Catabotrydaceae
Catabotrys
Pseudonectriella (= Catabotrys)

Ceratostomataceae
Ampullaria (= Melanospora)
Arxiomyces
Auerswaldia (= Melanospora)
Ceratostoma (= Melanospora)
Ceratostoma (= Arxiomyces)
Chaetoceratostoma (= Scopinella)

Chaetoceris (= Scopinella)
Corynascus
Erostrotheca (= Melanospora)
Erythrocarpon
Erythrocarpum (= Erythrocarpon)
Gibsonia (= Melanospora)
Guttularia (= Melanospora)
Laaseomyces (= Scopinella)
Lithomyces (= Melanospora)
Megathecium (= Melanospora)
Melanospora
Melanosporopsis (= Melanospora)
Microthecium (= Melanospora)
Nigrosphaeria (= Melanospora)
Ophiostomella (= Scopinella)
Persiciospora
Phaeidium (= Scopinella)
Phaeostoma (= Arxiomyces)
Phenacopodium (= Melanospora)
Pteridiosperma
Pustulipora
Rhynchomelas (= Melanospora)
Rhytidospora
Scopinella
Scopulina (= Scopinella)
Setiferotheca
Sphaeroderma (= Melanospora)
Sphaerodes
Sphaeronemopsis (= Melanospora)
Syspastospora
Vittadinula (= Sphaerodes)

Chaetomiaceae
Achaetomiella (= Chaetomium)
Achaetomium
Bolacotricha (= Chaetomium)
Bommerella
Boothiella
Chaetomidium
Chaetomiopsis
Chaetomiotricha (= Chaetomium)
Chaetomium
Cladochaete (= Chaetomium)
Corynascella
Emilmuelleria
Farrowia
Subramaniula
Thielavia
Thielaviella (= Boothiella)
Vanhallia (= Chaetomium)

Coniochaetaceae
Ascotrichella
Coniochaeta
Coniochaetidium
Coniomela (= Coniochaeta)
Cucurbitariella (= Coniochaeta)
Ephemeroascus
Germslitospora (=
    Coniochaetidium)
Pleosporopsis (= Coniochaeta)
Poroconiochaeta
Sphaerodermatella (=
    Coniochaeta)
Sphaerodermella (= Coniochaeta)
Synaptospora

Lasiosphaeriaceae
Acrospermoides
Adomia
Andreanszkya (= Podospora)
Annulatascus
Anopodium

Apiosordaria
Apiospora
Apodospora
Aposphaeriella (= Zignoella)
Arniella
Arnium
Biconiosporella
Bizzozeria
Bizzozeria (= Thaxteria)
Bombardia
Bombardioidea
Bovilla (= Cercophora)
Cainea (= Apiosordaria)
Camptosphaeria
Ceratosphaeria
Cercophora
Chaetolentomita (=
    Chaetosphaeria)
Chaetosphaerella
Chaetosphaeria
Cleistobombardia (=
    Tripterosporella)
Coscinospora (= Jugulospora)
Detonina (= Apiospora)
Didymopsamma (=
    Chaetosphaeria)
Diffractella
Echinopodospora (= Apiosordaria)
Echinospora (= Apiosordaria)
Emblemospora
Eosphaeria
Fimetariella
Garethjonesia
Hansenia (= Strattonia)
Herminia (= Eosphaeria)
Heteronectria (= Cercophora)
Hormosperma (= Lasiosphaeria)
Hormospora (= Bombardioidea)
Hypopteris (= Apiospora)
Jugulospora
Lacunospora (= Apiosordaria)
Lasiella (= Lasiosphaeria)
Lasiosordaria (= Cercophora)
Lasiosordariella (= Lasiosphaeria)
Lasiosordariopsis (= Cercophora)
Lasiosphaeria
Lasiosphaeriella
Lasiosphaeris (= Lasiosphaeria)
Lentomita (= Chaetosphaeria)
Malinvernia (= Podospora)
Melanopsammella
Merugia
Miyoshia (= Chaetosphaeria)
Miyoshiella (= Chaetosphaeria)
Montemartinia (= Chaetosphaeria)
Mycomedusiospora
Myelosperma
Naemaspora (= Bombardia)
Nothopodospora (= Arnium)
Ophioceras
Palmicola
Phaeonectriella
Philocopra (= Podospora)
Phragmodiscus
Plagiosphaera
Pleurage (= Schizothecium)
Pleurosordaria (= Arnium)
Podospora
Pseudocercophora
Rhabdostroma (= Apiospora)
Savoryella
Schizothecium
Scirrhiella (= Apiospora)

Spinulosphaeria
Strattonia
Thaxteria
Triangularia
Trichocollonema (= Zignoella)
Trigonia (= Triangularia)
Tripterospora (= Zopfiella)
Tripterosporella
Urnularia (= Chaetosphaeria)
Zignoella (= Chaetosphaeria)
Zopfiella
Zygopleurage
Zygospermella
Zygospermum (= Zygospermella)

Nitschkiaceae
Acanthonitschkea
Bertia
Bertiella (= Fracchiaea)
Biciliospora (= Nitschkia)
Biciliosporina
Botryola
Calyculosphaeria (= Nitschkia)
Castagnella (= Rhagadostoma)
Coelosphaeria (= Nitschkia)
Coronophora
Coronophorella (= Tympanopsis)
Cryptosphaerella (= Coronophora)
Echusias (= Fracchiaea)
Enchnoa
Euacanthe (= Acanthonitschkea)
Fitzpatrickia (= Acanthonitschkea)
Fracchiaea (= Nitschkia)
Gaillardiella
Groenhiella
Hystrix (= Acanthonitschkea)
Jannanfeldtia
Kirschsteinia (= Fracchiaea)
Lasiosphaeriopsis
Loranitschkia
Neofracchiaea
Neotrotteria (= Acanthonitschkea)
Nitschkia
Petelotia (= Acanthonitschkea)
Rhagadostoma
Schizocapnodium (= Gaillardiella)
Scortechinia (= Nitschkia)
Scortechinia (= Nitschkia)
Scortechiniella (= Nitschkia)
Scortechiniellopsis (= Nitschkia)
Sydowinula (= Acanthonitschkea)
Teratonema (= Nitschkia)
Tympanopsis (= Nitschkia)
Winterella (= Nitschkia)
Winterina (= Calyculosphaeria)

Sordariaceae
Anixiella (= Gelasinospora)
Apodus
Asordaria (= Sordaria)
Cainiella
Copromyces
Diplogelasinospora
Effetia
Eusordaria (= Sordaria)
Fimetaria (= Sordaria)
Gelasinospora
Guilliermondia
Ixodopsis (= Sordaria)
Leiostigma (= Sordaria)
Lockerbia
Neurospora
Periamphispora

Sordaria
Stellatospora

Incertae sedis
Acanthotheca (= Acanthotheciella)
Acanthotheciella
Bitrimonospora (= Monosporascus)
Bombardiella
Globosphaeria
Helminthosphaeria
Isia
Melanocarpus
Monosporascus
Nipicola
Nitschkiopsis
Onygenopsis (nom. dub.)
Phaeosporis
Reconditella
Rhexosporium
Roselliniella
Roselliniomyces
Roselliniopsis

SPATHULOSPORALES
Hispidicarpomycetaceae
Hispidicarpomyces

Spathulosporaceae
Spathulospora

TAPHRINALES
Taphrinaceae
Ascomyces (= Taphrina)
Ascosporium (= Taphrina)
Entomospora (= Taphrina)
Exoascus (= Taphrina)
Magnusiella (= Taphrina)
Sarcorhopalum (= Taphrina)
Taphria (= Taphrina)
Taphrina

TELOSCHISTALES
Fuscideaceae
Biatorinella (= Fuscidea)
Fuscidea
Lettauia
Maronea
Maroneomyces (= Maronea)
Orphniospora
Orphniosporomyces (= Orphniospora)
Ropalospora
Sarrameana

Letrouitiaceae
Letrouitia

Teloschistaceae
Aglaopisma (= Caloplaca)
Amphiloma (= Caloplaca)
Apatoplaca
Blastenia (= Caloplaca)
Blasteniospora (= Xanthoria)
Borrera (= Teloschistes)
Callopisma (= Caloplaca)
Caloplaca
Caloplacomyces (= Caloplaca)
Candelariopsis (= Caloplaca)
Cephalophysis
Chrysomma (= Caloplaca)
Diblastia (= Xanthoria)
Dufourea (= Xanthoria)

Dufouria (= Xanthoria)
Follmannia (= Caloplaca)
Fulgensia
Gasparrinia (= Caloplaca)
Huea (= Blastenia)
Ioplaca
Kuettlingeria (= Caloplaca)
Leproplaca (= Caloplaca)
Lethariopsis (= Caloplaca)
Lindauopsis (= Caloplaca)
Mawsonia
Meroplacis (= Caloplaca)
Niopsora (= Caloplaca)
Niorma (= Teloschistes)
Niospora (= Niopsora)
Parmocarpus (= Xanthoria)
Patellaria (= Caloplaca)
Placodium (= Caloplaca)
Placodium (= Xanthoria)
Polycauliona (= Caloplaca)
Pycnothele (= Dufourea)
Pyrenodesmia (= Caloplaca)
Seirophora
Siphonia (= Dufourea)
Siphula (= Dufourea)
Teloschistes
Teloschistomyces (= Teloschistes)
Tenorea (= Teloschistes)
Thamnonoma (= Caloplaca)
Theloschistomyces (= Theloschistes)
Triophthalmidium (= Caloplaca)
Xanthocarpia (= Blastenia)
Xanthodactylon
Xanthopeltis
Xanthoria
Xanthoriomyces (= Xanthoria)

TRIBLIDIALES
Triblidiaceae
Blitridium (= Triblidium)
Huangshania
Krempelhuberia (= Pseudographis)
Phacidiopsis (= Triblidium)
Pseudographis
Scleromium (= Pseudographis)
Triblidium

TRICHOSPHAERIALES
Trichosphaeriaceae
Acanthosphaeria
Bakeromyces (= Trichosphaeria)
Bresadolella (= Trichosphaerella)
Cantharosphaeria (= Eriosphaeria)
Ceratostomina (= Rhynchomeliola)
Chaetodimerina (= Rizalia)
Collematospora
Cresporhaphis
Debaryella
Doratospora (= Rizalia)
Eriosphaerella (= Eriosphaeria)
Eriosphaeria
Exomassarinula (= Melchioria)
Fluviostroma
Gaeumannia (= Melchioria)
Kananascus
Litschaueria
Malacosphaeria (= Eriosphaeria)
Melanochaeta
Melchioria
Ophiotexis (= Schweinitziella)
Oplothecium
Paracesatiella (= Schweinitziella)

Parascorias (= Rizalia)
Penzigina (= Eriosphaeria)
Phaeotrichosphaeria
Plectosphaerella
Porosphaerella
Porosphaerellopsis
Porosphaeria (=
  Porosphaerellopsis)
Pseudorhynchia (= Trichosphaeria)
Pyrgillocarpon
Rhynchomeliola
Rhynchostoma
Rizalia
Rizaliopsis (= Rizalia)
Saccardomyces (= Schweinitziella)
Schweinitziella
Setocampanula
Striatosphaeria
Trichosphaeria

TRICHOTHELIALES
Trichotheliaceae
Actiniopsis (= Trichothelium)
Asteropeltis (= Trichothelium)
Clathroporina
Cryptopeltis (= Porina)
Ophiodictyon (= Trichothelium)
Ophthalmidium (= Porina)
Phylloblastia
Phylloporina (= Porina)
Porina
Pseudosagedia (= Porina)
Sagedia (= Porina)
Segestrella (= Porina)
Segestria (= Porina)
Sphaeromphale (= Porina)
Stereochlamydomyces (=
  Trichothelium)
Stereochlamys (= Trichothelium)
Trichotheliomyces (=
  Trichothelium)
Trichothelium
Zamenhofia

VERRUCARIALES
Adelococcaceae
Adelococcus
Sagediopsis

Verrucariaceae
Acrotellomyces (= Thelidium)
Acrotellum (= Thelidium)
Actinothecium (= Verrucaria)
Agonimia
Agonimiella
Amphoridium (= Verrucaria)
Amphoridium (nom. alt.)
Amphoroblastia (= Polyblastia)
Awasthiella
Bachmannia (= Verrucaria)
Bagliettoa (= Verrucaria)
Bogoriella
Bogoriellomyces (= Bogoriella)
Bohleria (= Placidiopsis)
Capnia (= Dermatocarpon)
Catapyrenium
Catopyrenium (= Catapyrenium)
Coniothele (= Trimmatothele)
Dermatocarpella
Dermatocarpon (= Endocarpon)
Dermatocarpon
Dimerisma (= Spheconisca)
Encliopyrenia (= Verrucaria)

Endocarpidium (= Placidiopsis)
Endocarpomyces (=
  Dermatocarpon)
Endocarpon
Endopyreniomyces (= Endocarpon)
Endopyrenium (= Catapyrenium)
Entosthelia (= Dermatocarpon)
Epistictum (= Dermatocarpon)
Glomerilla
Goidanichia (= Staurothele)
Goidanichiomyces (= Goidanichia)
Haleomyces
Henrica
Heterocarpon
Holospora (= Polyblastia)
Holosporomyces (= Polyblastia)
Involucrocarpon (=
  Dermatocarpon)
Involucrothele (= Thelidium)
Kalbiana
Lauderlindsaya
Leightonia (= Endocarpon)
Lesdainea (= Trimmatothele)
Leucocarpia
Leucocarpopsis
Lithocia (= Verrucaria)
Lithoecis (= Verrucaria)
Lithosphaeria (= Psoroglaena)
Lithothelidium (= Polyblastia)
Macentina
Magnussoniolichen (= Polyblastia)
Magnussoniomyces (= Polyblastia)
Microcyclella
Muellerella
Muellerellomyces (= Muellerella)
Neocatapyrenium
Norrlinia
Papulare (= Thelidium)
Papulariomyces (= Thelidium)
Paracarpidium (= Endocarpon)
Paraphysorma (= Staurothele)
Paraphysothele (= Thelidium)
Paraplacidiopsis (= Placidiopsis)
Phaeomeris (= Spheconisca)
Phaeospora
Phalostauris (= Willeya)
Pharcidiella (= Phaeospora)
Philippiregis (= Polyblastidea)
Phragmothele (= Thelidium)
Placidiopsis
Placidium (= Dermatocarpon)
Placocarpus
Placopyrenium
Placotheliomyces (= Placothelium)
Placothelium
Pleophalis (= Spheconisca)
Pleothelis (= Muellerella)
Polyblastia
Polyblastidea (= Polyblastia)
Polyblastiomyces (= Staurothele)
Polyblastiomyces (= Polyblastia)
Polycarpella (= Muellerella)
Porphyriospora (= Polyblastia)
Protobagliettoa (= Verrucaria)
Psoroglaena
Psoroglaenomyces (= Psoroglaena)
Pyreniococcus (= Phaeospora)
Pyrenophoromyces (= Thelidium)
Pyrenophorum (= Thelidium)
Pyrenothamnia (= Endocarpon)
Pyrenothamniomyces (=
  Endocarpon)
Pyrenula (= Thelidium)

Pyrenulella (= Phaeospora)
Raesaeneniolichen (= Polyblastia)
Raesaeneniomyces (= Polyblastia)
Rhabdopsora
Rhabdopsoromyces (=
  Rhabdopsora)
Rhodocarpon (= Endocarpon)
Scleropyrenium
Sphaeromphale (= Staurothele)
Spheconisca
Spheconiscomyces (= Spheconisca)
Sporodictyon (= Polyblastia)
Staurothele
Stigmatomma (= Staurothele)
Sychnogonia (= Muellerella)
Thelidiola (= Muellerella)
Thelidiomyces (= Thelidium)
Thelidiopsis
Thelidium
Tichothecium (= Verrucaria)
Trimmatothele
Trimmatothelopsis
Verrucaria
Verrucariomyces (= Verrucaria)
Verrucula (= Verrucaria)
Willeya (= Staurothele)
Xenosphaeria (= Phaeospora)
Zschackea (= Verrucaria)

Incertae sedis
Plurisperma

XYLARIALES
Amphisphaeriaceae
Acanthodochium
Amphisphaerella
Amphisphaeria
Ascotaiwania
Atrotorquata
Blogiascospora
Broomella
Cainia
Chitonospora
Clathridium (= Discostroma)
Clethridium
Clypeophysalospora
Curreyella (= Discostroma)
Discostroma
Discostromopsis (= Discostroma)
Dyrithium
Ellurema
Fabreola (= Urosporella)
Fasciatispora
Flagellosphaeria
Frondispora
Griphosphaeria (= Discostroma)
Griphosphaerioma
Hymenopleella (= Lepteutypa)
Hypodiscus (= Neohypodiscus)
Iodosphaeria
Keissleria (= Broomella)
Leiosphaerella
Lepteutypa
Lepteutypella (= Lepteutypa)
Lindquistomyces
Macrothelia (= Amphisphaeria)
Manokwaria
Massariella (= Poikiloderma)
Massariopsis (= Amphisphaeria)
Muelleromyces
Mukhakesa
Neobroomella
Neohypodiscus

Ommatomyces
Paracainiella
Paradidymella (= Discostroma)
Pestalosphaeria
Phocys (= Massariella)
Phorcys (= Amphisphaeria)
Phragmodothella (= Clethridium)
Poikiloderma (= Amphisphaeria)
Pseudomassariella (=
  Leiosphaerella)
Reticulosphaeria
Rhynchostomopsis (=
  Amphisphaeria)
Roussoella
Sauvageautia (= Urosporella)
Spilobolus (= Amphisphaerella)
Steganopycnis (= Seynesia)
Urospora (= Urosporella)
Urosporella
Urosporellopsis
Xylochora

Clypeosphaeriaceae
Apiorhynchostoma
Ceratostomella
Clypeosphaeria
Crassoascus
Duradens
Entosordaria (= Clypeosphaeria)
Frondicola
Jobellisia
Melomastia
Pseudovalsaria
Stereosphaeria (= Clypeosphaeria)

Xylariaceae
Acrosphaeria (= Xylaria)
Albocrustum (= Hypoxylon)
Alocospora
Anthostomella
Areolospora
Ascostroma (= Ustilina)
Ascotricha
Astrocystis
Bacillaria (= Camillea)
Barrmaelia
Biscogniauxia
Byssitheca (= Rosellinia)
Calceomyces
Camillea
Carnostroma (= Xylaria)
Cerillum (= Engleromyces)
Chaenocarpus (= Thamnomyces)
Chaenocarpus (nom. dub.)
Chaenocarpus
Chromocreopsis (= Thuemenella)
Coelorhopalon (= Xylaria)
Coenocarpus (= Chaenocarpus)
Colletomanginia (= Engleromyces)
Collodiscula
Coprolepa (= Hypocopra)
Creosphaeria
Cryptosordaria (= Anthostomella)
Cryptothamnium (nom. dub.)
Cucurbitula (= Rosellinia)
Cytispora
Daldinia
Diatrypeopsis (= Camillea)
Discosphaera (= Hypoxylon)
Discoxylaria
Engleromyces
Entoleuca (= Hypoxylon)
Entonaema

Entosordaria (= Anthostomella)
Epixylon (= Hypoxylon)
Euepixylon
Euhypoxylon (= Hypoxylon)
Fuckelia (= Lopadostoma)
Gamosphaera (= Hypoxylon)
Helicogermslita
Hemisphaeria (= Daldinia)
Henningsinia (= Phylacia)
Holttumia
Hypocopra
Hypoxylina (= Hypoxylon)
Hypoxylon
Hypoxylon (= Xylaria)
Induratia
Institale (nom. dub.)
Jongiella (= Camillea)
Kommamyce (= Biscogniauxia)
Kretzschmaria
Kretzschmariella
Leprieuria
Leptomassaria (= Anthostomella)
Leveillea (= Hypoxylon)
Lichenagaricus (= Xylaria)
Lopadostoma
Maurinia (= Anthostomella)
Moelleroclavus (= Xylaria)
Myconeesia
Neesiella (= Anthostomella)
Nemania
Nummularia (= Biscogniauxia)
Nummulariella (= Biscogniauxia)
Nummularioidea (= Camillea)
Numulariola (= Nummularia)
Obolarina
Oostroma (= Hypoxylon)
Paranthostomella (=
  Anthostomella)
Paucithecium
Penzigia (= Xylaria)
Peripherostoma (= Daldinia)
Perisphaeria (= Daldinia)
Phaeophomatospora (=
  Anthostomella)
Phylacia
Podosordaria
Porodiscella (= Hypoxylon)
Poronia
Pseudoxylaria (= Xylaria)
Pulveria
Pyrenodermium (= Hypoxylon)
Pyrenomyxa (= Pulveria)
Pyrenopolyporus (= Hypoxylon)
Rhopalopsis (= Kretzschmaria)
Rhopalostroma
Rosellinia
Sarcoxylon
Sarcoxylum (= Sarcoxylon)
Sclerodermatopsis (= Xylaria)
Seynesia
Simoninus (= Chaenocarpus)
Spermatodermia (= Hypoxylon)
Spermodermia (= Hypoxylon)
Sphaeria (= Hypoxylon)
Sphaeropyxis (= Rosellinia)
Spirogramma (= Hypoxylon)
Squamotubera (= Sarcoxylon)
Stilbohypoxylon (= Xylaria)
Stromatoneurospora
Stromatosphaeria (= Daldinia)
Stromne (= Engleromyces)
Thamnomyces
Theissenia

Thuemenella
Ustulina
Valsa (= Hypoxylon)
Versiomyces
Wawelia
Xylaria
Xylariodiscus (= Xylaria)
Xylocrea (= Sarcoxylon)
Xylosphaera (= Xylaria)

Incertae sedis
Pandanicola

FAMILIA INCERTAE SEDIS
Acrospermaceae
Acrospermum
Coscinaria (= Oomyces)
Oomyces
Scleroglossum (= Acrospermum)
Xyloglossum (= Acrospermum)

Aphanopsidaceae
Agyrina (= Steinia)
Aphanopsis
Pleolecis (= Steinia)
Steinia

Aspidotheliaceae
Aspidopyrenis (= Aspidothelium)
Aspidopyrenium (= Aspidothelium)
Aspidotheliomyces (=
  Aspidothelium)
Aspidothelium
Musaespora
Patellonectria (= Aspidothelium)
Porinopsis (= Aspidothelium)
Secoligella (= Aspidothelium)

Dactylosporaceae
Abrothallomyces (= Dactylospora)
Dactylospora
Kymadiscus (= Dactylospora)
Mycolecidia (= Dactylospora)
Mycolecis (= Dactylospora)
Paruephaedria
Pseudokarschia (= Dactylospora)

Diporothecaceae
Diporotheca

Epigloeaceae
Epigloea
Epigloeomyces (= Epigloea)
Vorarlbergia (= Epigloea)

Hyponectriaceae
Acanthorhynchus (= Physalospora)
Anisostomula (= Hyponectria)
Apioclypea
Apiosporella (= Pseudomassaria)
Apiosporina (= Pseudapiospora)
Apiothyrium
Aplacodina (= Pseudomassaria)
Arwidssonia
Benedekiella (= Physalospora)
Ceriophora (= Ceriospora)
Ceriospora
Chaetapiospora (=
  Pseudomassaria)
Chamaeascus
Clypeothecium (= Exarmidium)
Exarmidium

Griphosphaerella (=
  Monographella)
Hindersonia (= Ceriospora)
Hyponectria
Lasiobertia
Linocarpon
Mangrovispora
Merrilliopeltis (= Oxydothis)
Monographella
Neolinocarpon
Oxydothis
Pemphidium
Physalospora
Physosporella
Plagiolagynion (= Oxydothis)
Plagiothecium (= Plagiolagynion)
Pseudapiospora (=
  Pseudomassaria)
Pseudoguignardia (=
  Physalospora)
Pseudomassaria
Pseudophysalospora (=
  Physalospora)
Sporophysa (= Physalospora)
Trichophysalospora (=
  Physalospora)

Koralionastaceae
Koralionastes

Magnaporthaceae
Buergenerula
Clasterosphaeria
Gaeumannomyces
Magnaporthe
Omnidemptus
Rhaphidophora (=
  Gaeumannomyces)
Rhaphidospora (=
  Gaeumannomyces)
Yukonia (= Buergenerula)

Mastodiaceae
Dermatomeris (= Mastodia)
Kohlmeyera (= Mastodia)
Leptogiopsis (= Mastodia)
Mastodia
Turgidosculum

Melaspileaceae
Hazslinszkya (= Melaspilea)
Melanographa (= Melaspilea)
Melaspilea
Melaspileella (= Melaspilea)
Melaspileomyces (= Melaspilea)
Mycomelaspilea (= Melaspilea)
Stictographa (= Melaspilea)

Pachyascaceae
Pachyascus

Phlyctidaceae
Comocheila (= Phlyctis)
Dactyloblastus (= Phlyctella)
Dyctyoblastus (= Phlyctis)
Micromium (= Phlyctis)
Phlyctella (= Phlyctis)
Phlyctellomyces (= Phlyctella)
Phlyctidia (= Phlyctis)
Phlyctidomyces (= Phlyctis)
Phlyctis
Phlyctomia (= Phlyctis)
Psathyrophlyctis

Protothelenellaceae
Gloeopyrenia (= Protothelenella)
Mycowinteria
Protothelenella
Winteria (= Mycowinteria)

Strigulaceae
Craspedon (= Strigula)
Dichoporis (= Strigula)
Diporina (= Strigula)
Discosiella (= Strigula)
Geisleria (= Strigula)
Geisleriomyces (= Strigula)
Haploblastia (= Strigula)
Heterodothis (= Strigula)
Melanophthalmum (= Strigula)
Nematora (= Strigula)
Phyllocharis (= Strigula)
Phylloporis (= Strigula)
Porinula (= Strigula)
Puiggariella (= Strigula)
Raciborskiella (= Strigula)
Racoplaca (= Strigula)
Sagediomyces (= Strigula)
Shanoria (= Strigula)
Strigula
Strigulomyces (= Strigula)

Thrombiaceae
Inoderma (= Thrombium)
Phaeotrombis (= Thrombium)
Thrombium
Wernera (= Thrombium)

Thyridiaceae
Bivonella (= Thyridium)
Mycothyridium (= Thyridium)
Sinosphaeria (= Thyridium)
Thyridium (= Mycothyridium)
Thyridium
Valsonectria
Xylosphaeria (= Mycothyridium)

Vialaeaceae
Boydia (= Vialaea)
Vialaea

Genera Incertae sedis
Abrothallus
Abyssomyces
Acerbiella
Allescherina (= Cryptovalsa)
Allonectella
Ameromassaria
Amorphotheca
Amphisphaerellula
Amphisphaerina
Amphitrichum (nom. dub.)
Amphorulopsis
Amylis
Amylocarpus
Antennaria (nom. ambig.)
Anthostomaria
Anthostomellina
Antimanoa
Apharia
Apiotypa
Apodothina
Arachnomyces
Aropsiclus
Arthopyreniomyces (=
  Pyrenyllium)
Ascocorticiellum

Ascorhiza
Ascosorus
Ascoxyta
Assoa
Astomella
Atractobolus
Aulospora
Azbukinia
Bactrosphaeria
Baculospora
Banhegyia
Batistospora
Berggrenia
Biflua
Bombardiastrum
Brenesiella
Bresadolina
Byrsomyces
Byssophytum
Byssotheciella
Caleutypa
Calosphaeriopsis
Caproniella
Cerastoma
Ceratospermum (nom. dub.)
Chaetoamphisphaeria
Chaetomastia
Ciliofusospora
Cladosphaera
Clibanites
Clypeoceriospora
Clypeolum
Clypeosphaerulina
Collema
Collematomyces (= Collema)
Collonema (nom. dub.)
Coniothyriella (nom. conf.)
Coryneliella
Coscinocladium
Crinigera
Cryptoascus
Cryptomycina
Cryptonectriopsis (=
  Phomatospora)
Cryptovalsa
Cucurbitopsis
Cyanopyrenia
Cylindrotheca
Dasysphaeria
Delpinoella
Diaboliumbilicus
Diacrochordon
Diehliomyces
Dimerosporiella
Dipyrgis
Dontuzia
Dryinosphaera (nom. dub.)
Dryosphaera
Eiona
Endocolium
Endoxylina
Esfandiaria (= Esfandiariomyces)
Esfandiariomyces
Farriolla
Farriollomyces (= Farriolla)
Flageoletia (= Phomatospora)
Flakea
Fuckelia (nom. conf.)
Gaeumanniella
Glabrotheca
Gonidiomyces
Gyrophthorus
Haplotrichum (nom. dub.)

Hapsidascus
Harmandiana (nom. dub.)
Heliastrum
Herpothrix (= Chaetomastia)
Heteropera (= Phomatospora)
Heterostomum (nom. dub.)
Heuflera
Hyaloderma
Hyalodermella
Hymenobia
Hymenobiella (= Hymenobia)
Hypnotheca
Iraniella
Khuskia
Konenia
Kravtzevia
Kurssanovia
Laboulbeniopsis (nom. dub.)
Lamyella (= Neolamya)
Leptosacca
Leptosphaerella
Leptosporella
Leptosporina
Leucoconiella
Leucoconis
Leucographa (nom. nud.)
Lichenopeziza
Limboria
Lohwagiella
Lophodermopsis (nom. dub.)
Loten (nom. dub.)
Ludwigomyces
Lyonella
Marisolaris
Massalongia
Massalongomyces (= Massalongia)
Melanobotrys (= Xylobotryum)
Meringosphaeria (= Acerbiella)
Microcyclephaeria
Molgosphaera (nom. dub.)
Monoloculia (nom. dub.)
Mouliniea
Mycophaga (= Hyaloderma)
Mycotodea
Myriococcum
Myriogonium
Myrmaecium
Naumovela
Neocryptospora
Neolamya
Neorehmia
Neothyridaria
Normandina
Normandinomyces (= Normandina)
Oceanitis
Ochrosphaera
Ophiomassaria
Orcadia
Papulosa
Phaeaspis (= Phomatospora)
Phaeodothiopsis
Phellostroma
Phialisphaera (nom. dub.)
Phomatospora
Phomatosporopsis (=
  Phomatospora)
Phthora
Phyllocelis
Phylloporina
Phyllopyrenia (nom. dub.)
Phymatopsis (= Abrothallus)
Pilophorum
Placodothis

Plectolitus (= Amylocarpus)
Pleocryptospora
Pleosphaeria
Pleosphaeria
Pocsia
Pogonospora (= Endoxylina)
Pontogeneia
Porodiscus
Porosphaera (nom. dub.)
Protascus (= Wolkia)
Protocalicium
Protocucurbitaria
Pseudohalonectria
Pseudohepatica
Pseudolecidea (= Abrothallus)
Pseudoperitheca
Psilosphaeria
Pulvinaria
Pumilus
Pustularia (nom. dub)
Pyrenyllium
Rehmiomycella
Rehmiomyces (= Rehmiomycella)
Rhamphosphaeria
Rhizophila
Rhopographella
Rhymbocarpus
Rhynchosphaeria
Romellina
Saccardoella
Sachsia (nom. dub.)
Sarcopyrenia
Sarcopyreniomyces (=
  Sarcopyrenia)
Sarcopyreniopsis (= Sarcopyrenia)
Sartorya (nom. dub.)
Scharifia
Scolecopeltidella (nom. dub.)
Scoliocarpon
Scotiosphaeria
Scutomyces
Seriella (nom. dub.)
Servazziella
Spermatodium (nom. dub.)
Spheropsis (nom. dub.)
Sporoctomorpha
Stearophora
Stegophorella
Stellifraga
Stigmatea
Stigmatisphaera (nom. dub.)
Stigmea
Stomatogenella
Strickeria
Sulcispora
Sulcispora (= Aropsiclus)
Swampomyces
Synsphaeria
Syphosphaera
Telioclipeum
Thallisphaera (nom. dub.)
Thelidiella
Theloschistes (nom. dub.)
Thyridella
Thyrotheca
Tonduzia (= Dontuzia)
Torpedospora
Trachyxylaria (= Xylobotryum)
Trichospermella
Trichosphaera (nom. dub.)
Trichosphaeropsis
Tromeropsis
Tuberosurculus (nom. dub.)

Tunstallia
Tylophorella
Tylophorellomyces (=
  Tylophorella)
Ulvella (nom. dub.)
Valsarioxylon (nom. nud.)
Vleugelia
Wolkia
Wrightiella (= Leptodothis)
Xenomyxa
Xylobotryum
Xyloceras (= Xylobotryum)
Xylogone
Xylosphaeria
Zignoina (= Zignoella)
Cilicia
Amerostege
Batistamnus
Clypeomyces
Coccidophthora
Darwiniella
Dictyochorina
Halonia
Haplosporium
Hypospila
Limacinia
Maurya
Melanomyces
Metasphaerella
Nowellia
Physalosporella
Pseudohansenula
Pseudophyllachora
Scytopezis
Theissenula
Trichochora (nom. dub.)

BASIDIOMYCOTA

BASIDIOMYCETES

AGARICALES
Agaricaceae
Agaricus
Amanita (= Agaricus)
Arenicola
Chamaemyces
Cheilophlebium (nom. dub.)
Chitonia (= Clarkeinda)
Chitoniella (= Clarkeinda)
Chitonis (= Chitonia)
Chlorolepiota
Chlorophyllum
Chlorosperma (= Melanophyllum)
Chlorospora (= Melanophyllum)
Clarkeinda
Constricta
Coolia (= Squamanita)
Crucispora
Cystoagaricus
Cystoderma
Cystolepiota
Dissoderma
Drosella (= Chamaemyces)
Fungus (= Agaricus)
Fungus (= Agaricus)
Fusispora (= Lepiota)
Glaucospora (= Melanophyllum)
Graminicola
Hiatula (= Lepiota)
Hiatulopsis
Horakia (= Verrucospora)

Hymenagaricus
Hypophyllum (= Agaricus)
Janauaria
Lepidotus (= Lepiota)
Lepiota
Lepiotella (= Chamaemyces)
Lepiotella (= Volvolepiota)
Lepiotula (= Lepiota)
Leptomyces (nom. dub.)
Leucoagaricus
Leucobolbitius (= Leucocoprinus)
Leucocoprinus
Macrolepiota
Mastocephalus (= Leucocoprinus)
Melanophyllum
Metraria
Metrodia
Micropsalliota
Morobia
Mycenopsis
Myces (= Fungus)
Phaeopholiota
Phlebonema
Phyllobolites (nom. dub.)
Pilosace (nom. dub.)
Pratella (= Agaricus)
Psalliota (= Agaricus)
Pseudoauricularia
Pseudobaeospora
Pulverolepiota
Rickella (= Volvolepiota)
Rotula (nom. dub.)
Rugosospora
Schinzinia
Schulzeria (nom. dub.)
Sericeomyces
Singerina
Smithiomyces
Squamanita
Stylobates
Termiticola
Valentinia
Verrucospora
Volvolepiota

Amanitaceae
Agaricus (= Amanita)
Amanita
Amanitaria (= Amanita)
Amanitella (= Limacella)
Amanitella (= Amanita)
Amanitina (= Amanita)
Amanitopsis (= Amanita)
Amidella (= Amanita)
Amplariella (= Amanita)
Ariella (= Amanita)
Aspidella (= Amanita)
Boletium (= Volvoboletus)
Gilbertia (nom. inval.)
Lepidella (= Amanita)
Leucomyces (= Amanita)
Limacella
Myxoderma (= Limacella)
Podabrella
Pseudofarinaceus (= Amanita)
Rajapa (= Termitomyces)
Termitomyces
Vaginaria (= Amanitopsis)
Vaginarius (= Amanita)
Vaginata (= Amanita)
Venenarius (= Amanita)
Volvella (= Amanita)

Volvoamanita
Volvoboletus (= Amanita)

Bolbitiaceae
Agrocybe
Agrogaster
Bolbitius
Bulla (= Agrocybe)
Conocybe
Cyclopus (= Agrocybe)
Descolea
Galerella
Gymnoglossum
Pholiotella (= Conocybe)
Pholiotina (nom. rej.)
Pluteolus (= Bolbitius)
Pseudoconocybe (= Conocybe)
Pseudodeconica (= Agrocybe)
Pseudodescolea (= Descolea)
Ptychella
Raddetes (= Conocybe)
Ramicola (= Agrocybe)
Togaria (= Agrocybe)
Tubariopsis
Wielandomyces

Coprinaceae
Annularius (= Coprinus)
Astylospora (= Psathyrella)
Copelandia
Coprinellus (= Coprinus)
Coprinopsis (= Coprinus)
Coprinus
Coprinusella
Cortiniopsis (= Lacrymaria)
Drosophila (= Psathyrella)
Ephemerocybe (= Coprinus)
Gasteroagaricoides
Glyptospora (= Psathyrella)
Gymnochilus (= Psathyrella)
Hemigaster
Homophron (= Psathyrella)
Hypholomopsis (= Psathyrella)
Lacrymaria
Lentispora (= Coprinus)
Macrometrula
Onchopus (= Coprinus)
Panaeolina (= Psathyrella)
Pannucia (= Psathyrella)
Pluteopsis (= Psathyrella)
Psalliotina (= Psathyrella)
Psammocoparius (= Psathyrella)
Psathyra (= Psathyrella)
Psathyrella
Pselliophora (= Coprinus)
Pseudocoprinus (= Coprinus)
Psilocybe (= Psathyrella)
Rhacophyllus
Xerocoprinus
Zerovaemyces

Entolomataceae
Acurtis (= Entoloma)
Alboleptonia
Claudopus
Clitopilopsis (= Rhodocybe)
Clitopilus
Eccilia
Entoloma
Hexajuga (= Clitopilus)
Hirneola (= Rhodocybe)
Hyporrhodius (= Rhodophyllus)
Inocephalus

Inopilus
Lanolea (= Nolanea)
Latzinaea (= Nolanea)
Leptonia
Leptoniella (= Leptonia)
Nolanea
Octojuga (= Clitopilus)
Omphaliopsis
Orcella (= Clitopilus)
Paraleptonia
Paxillopsis (= Clitopilus)
Pouzarella
Pouzaromyces (= Rhodophyllus)
Rhodocybe
Rhodocybella
Rhodophana
Rhodophyllus
Trichopilus

Galeropsidaceae
Cytophyllopsis (= Weraroa)
Cyttarophyllopsis
Cyttarophyllum (= Galeropsis)
Galeropsis
Gastrocybe
Psammomyces (= Galeropsis)
Tympanella
Weraroa

Hydnangiaceae
Gigasperma
Hydnangium
Maccagnia
Podohydnangium

Hygrophoraceae
Aeruginospora (= Hygrophorus)
Austroomphaliaster
Bertrandia
Camarophyllopsis
Camarophyllus (= Hygrophorus)
Cuphophyllus
Gliophorus (= Hygrocybe)
Godfrinia (= Hygrocybe)
Hodophilus (= Hygrotrama)
Humidicutis
Hydrophorus (= Hygrocybe)
Hygroaster
Hygrocybe
Hygrophorus
Hygrotrama
Limacinus (= Hygrophorus)
Limacium (= Hygrophorus)
Neohygrocybe (= Hygrocybe)
Neohygrophorus (= Hygrophorus)
Pseudohygrocybe
Pseudohygrophorus (nom. dub.)

Pluteaceae
Annularia (= Chamaeota)
Braunia (= Brauniella)
Brauniella
Chamaeota
Locellina (nom. dub.)
Pluteus
Pseudofarinaceus (= Volvariella)
Rhodosporus (= Pluteus)
Sinotermitomyces
Volva (= Volvariella)
Volvaria (= Volvariella)
Volvariella
Volvariopsis (= Volvariella)
Volvarius (= Volvariella)

Podaxaceae
Catachyon (= Podaxis)
Cauloglossum (= Podaxis)
Chainoderma (= Podaxis)
Gasterellopsis
Gymnogaster
Gyrophragmium
Herculea (= Podaxis)
Montagnea
Montagnites (= Montagnea)
Panaeolopsis
Podaxis
Podaxon (= Podaxis)
Polyplocium (= Gyrophragmium)
Schweinitzia (= Podaxis)

Richoniellaceae
Nigropogon (= Richoniella)
Rhodogaster
Richoniella

Secotiaceae
Araneosa
Artymenium (= Secotium)
Clavogaster
Endolepiotula
Endoptychum
Holocotylon
Hypogaea
Le-ratia
Longia (= Longula)
Longula
Neosecotium
Notholepiota
Phyllogaster
Secotium
Smithiogaster
Volvigerum

Strophariaceae
Anellaria
Campanularius (= Panaeolus)
Chalymmota (= Panaeolus)
Coprinarius (= Panaeolus)
Deconica (= Psilocybe)
Delitescor (= Psilocybe)
Derminus (= Pholiota)
Dryophila (= Pholiota)
Flammopsis (= Pholiota)
Flammula (= Pholiota)
Galeropsina (= Psilocybe)
Geophila (= Stropharia)
Hemipholiota
Hypholoma
Hypodendrum (= Pholiota)
Kuehneromyces
Melanotus
Mythicomyces
Naematoloma (= Hypholoma)
Nematoloma (= Hypholoma)
Nemecomyces (= Pholiota)
Nivatogastrium
Pachylepyrium
Panaeolus
Phaeogalera
Phaeonematoloma
Phlebophyllum
Pholiota
Pleuroflammula
Psilocybe
Stropharia
Stropholoma
Visculus (= Pholiota)

Torrendiaceae
Torrendia

Tricholomataceae
Acanthocystis (= Hohenbuehelia)
Agaricochaete
Amparoina
Amyloflagellula
Anastrophella
Androsaceus (= Marasmius)
Anthracophyllum
Aphyllotus
Armillaria
Armillariella (= Armillaria)
Arrhenia
Arthrosporella
Asproinocybe
Aspropaxillus (= Leucopaxillus)
Asterotus (= Resupinatus)
Austroclitocybe
Bactroboletus (= Filoboletus)
Baeospora
Basidopus (= Mycena)
Bertrandiella (= Lactocollybia)
Biannularia (= Catathelasma)
Boehmia (= Arrhenia)
Botrydina (= Omphalina)
Caesposus
Calathella
Callistodermatium
Callistosporium
Calocybe
Calyptella
Campanella
Cantharellopsis
Cantharellula
Cantharocybe
Catathelasma
Catatrama
Caulorhiza
Cellypha
Chaetocalathus
Chamaeceras (= Marasmius)
Cheimonophyllum
Chrysobostrychodes
Chrysomphalina
Clavomphalia
Clitocybe
Clitocybula
Collopus (= Mycena)
Collybia
Collybidium (= Flammulina)
Collybiopsis (nom. dub.)
Collyria (nom. dub.)
Conchomyces (= Hohenbuehelia)
Coprinopsis (= Oudemansiella)
Coriscium (= Omphalina)
Corniola (= Arrhenia)
Corrugaria (= Mycena)
Cortinellus (= Tricholoma)
Crinipellis
Cryptomphalina
Cymatella
Cymatellopsis
Cynema
Cyphellocalathus
Cyptotrama
Dactylosporina
Delicatula
Dennisiomyces
Dermoloma
Dictyolus (= Arrhenia)
Dictyopanus (= Panellus)

Dictyoploca
Discocyphella (= Gloiocephala)
Echinoderma
Echinosporella
Eomycenella (= Mycena)
Epicnaphus
Favolaschia
Fayodia
Filoboletus
Flabellimycena
Flagelloscypha
Flammulina
Floccularia
Galactopus (= Mycena)
Galeromycena (= Macrocystidia)
Gamundia
Geotus (= Arrhenia)
Gerhardtia
Gerronema
Gloiocephala
Glutinaster (= Tricholoma)
Gymnopus (= Mycena)
Gyrophila (= Tricholoma)
Haasiella
Heimiomyces
Heliomyces (= Marasmius)
Helotium (= Omphalina)
Hemimycena
Heterosporula
Hispidocalyptella
Hohenbuehelia
Hologloea (= Favolaschia)
Hydropus
Hymenoconidium (= Marasmius)
Hymenogloea
Hymenomarasmius (= Marasmius)
Hypsizygus
Insiticia (= Mycena)
Laccaria
Lachnella
Lactocollybia
Lampteromyces
Laschia (= Campanella)
Leiopoda (= Mycena)
Lentinula
Lepista
Leptoglossum (= Arrhenia)
Leptotus (= Arrhenia)
Leucoinocybe
Leucopaxillus
Libellus (= Hymenogloea)
Linopodium (= Mycena)
Lulesia
Lyophyllopsis
Lyophyllum
Macrocystidia
Macrocystis (= Macrocystidia)
Manuripia
Marasmiellus
Marasmius
Mastoleucomyces (= Tricholoma)
Megacollybia
Megatricholoma
Melaleuca (= Melanoleuca)
Melanoleuca
Metulocyphella
Microcollybia (= Collybia)
Microcollybia
Micromphale
Mniopetalum
Monomyces (= Tricholoma)
Mucidula (= Oudemansiella)
Mycena

Mycenella
Mycenoporella (= Filoboletus)
Mycenula (= Mycena)
Mycetinis (= Marasmius)
Mycoalvimia
Mycomedusa (= Favolaschia)
Myxocollybia (= Flammulina)
Myxomphalia (= Fayodia)
Neoclitocybe
Nochascypha
Nothopanus
Nyctalis
Omphalia (= Pseudoclitocybe)
Omphalia (= Omphalina)
Omphaliaster
Omphalina
Omphalius (= Clitocybe)
Omphalopsis (= Xeromphalina)
Ossicaulis
Oudemansia (= Oudemansiella)
Oudemansiella
Palaeocephala
Panellus
Peglerochaete
Pegleromyces
Perona (= Omphalina)
Phaeodepas
Phaeolimacium (= Oudemansiella)
Phaeomycena
Phalomia (= Omphalia)
Phlebomarsmius (= Xeromphalina)
Phlebomycena (= Mycena)
Phlebophora (nom. dub.)
Phyllotremella (= Hohenbuehelia)
Physocystidium
Phytoconis
Plagiotus (= Anthracophyllum)
Pleurella
Pleurocollybia
Pleurocybella
Pleuromycenula (= Rimbachia)
Pleurotopsis (= Resupinatus)
Polymarasmius (= Marasmius)
Polymyces (= Armillaria)
Poroauricula (= Favolaschia)
Porolaschia (= Favolaschia)
Poromycena (= Mycena)
Porpoloma
Protomarasmius (= Marasmius)
Prunulus (= Mycena)
Psammospora (= Melaleuca)
Pseudoarmillariella
Pseudoclitocybe
Pseudohiatula
Pseudolyophyllum
Pseudomycena (= Mycena)
Pseudoomphalina
Pterospora (= Marasmiellus)
Resinomycena
Resupinatus
Retocybe (= Mycena)
Rhizomorpha (nom. dub.)
Rhodocollybia (= Collybia)
Rhodopaxillus (= Lepista)
Rhodotus
Rickenella
Rimbachia
Ripartitella
Ripartites
Rugosomyces
Russuliopsis (= Laccaria)
Sarcomyxa (= Hohenbuehelia)

Scorteus (= Marasmius)
Scytinotopsis (= Resupinatus)
Scytinotus (= Panellus)
Semiomphalina
Setulipes
Shitaker (= Lentinula)
Singerella
Singerocybe
Skepperiella
Sphaerocephalus (= Tricholoma)
Stachyomphalina
Stanglomyces
Stereopodium (= Mycena)
Stigmatolemma
Strobilurus
Sympodia (= Marasmius)
Tectella
Tephrocybe
Tephrophana (= Marasmius)
Tetrapyrgos
Tricholoma
Tricholomella
Tricholomopsis
Tricholosporum
Trigonipes (= Clitocybe)
Trogia
Urceolus (= Resupinatus)
Urospora (= Panellus)
Urosporellina (= Panellus)
Vanromburghia (= Marasmius)
Xeromphalina
Xerula
Xerulina
Zephirea (= Mycena)

Incertae sedis
Acetabularia (= Cyphellopus)
Amanita (nom. dub.)
Cyphellopus
Hypomnema (nom. dub.)

AGARICOSTILBALES
Agaricostilbaceae
Agaricostilbum
Amerobotryum (= Agaricostilbum)

ATRACTIELLALES
Atractogloeaceae
Atractogloea

Chionosphaeraceae
Chionosphaera
Fibulostilbum
Stilbum

Hoehnelomycetaceae
Atractiella
Hoehnelomyces (= Atractiella)
Pilacrella (= Atractiella)

Pachnocybaceae
Pachnocybe

Phleogenaceae
Botryochaete
Botryochaete (= Phleogena)
Ecchyna (= Phleogena)
Lasioderma (= Phleogena)
Martindalia (= Phleogena)
Phleogena

AURICULARIALES
Auriculariaceae
Anthoseptobasidium
Auricula (= Auricularia)
Auricularia
Auriculariella (= Auricularia)
Conchites (= Auricularia)
Hirneola (= Auricularia)
Mylittopsis
Neotyphula
Oncomyces (= Auricularia)
Paraphelaria
Patila (= Oncomyces)
Seismosarca (= Auricularia)

BOLETALES
Beenakiaceae
Beenakia

Boletaceae
Aureoboletus (= Pulveroboletus)
Boletochaete
Boletopsis (= Suillus)
Boletus
Buchwaldoboletus (= Pulveroboletus)
Ceriomyces (= Boletus)
Cricunopus (= Suillus)
Dictyopus (= Boletus)
Gastroleccinum
Gastrosuillus
Gymnopus (= Rostkovites)
Ixechinus
Ixocomus (= Suillus)
Krombholzia (= Leccinum)
Krombholziella (= Leccinum)
Leccinum
Leucobolites (= Boletus)
Oedipus (= Boletus)
Peplopus (= Suillus)
Pinuzza (= Suillus)
Pseudoboletus
Pulveroboletus
Rhodoporus
Rostkovites (= Suillus)
Royoungia
Sinoboletus
Solenia (= Pinuzza)
Suillellus (= Boletus)
Suillus (= Boletus)
Suillus (= Boletus)
Suillus
Trachypus (= Leccinum)
Tubiporus (= Boletus)
Veloporus (= Boletus)
Viscipellis (= Suillus)
Xanthoconium

Chamonixiaceae
Chamonixia

Coniophoraceae
Boninohydnum (= Gyrodontium)
Chrysoconia
Coniobotrys (= Jaapia)
Coniophora
Coniophorafomes (nom. dub.)
Coniophorella (= Coniophora)
Coniophoropsis
Corneromyces
Gyrodontium
Gyrophana (= Serpula)
Gyrophora (= Serpula)

Jaapia
Leucogyrophana
Meiorganum
Podoserpula
Pseudomerulius
Serpula
Xylomyzon (= Serpula)
Xylophagus (= Serpula)

Gomphidiaceae
Brauniellula
Chroogomphus
Cystogomphus
Gomphidius
Gomphogaster
Gomphus (= Gomphidius)
Gymnogomphus (nom. dub.)
Leucogomphidius

Gyrodontaceae
Boletinellus (= Gyrodon)
Boletinus
Campbellia (= Gyrodon)
Coelopus (= Gyroporus)
Eryporus (= Boletinus)
Euryporus (= Boletinus)
Gilbertiella (= Gyrodon)
Gilbertina (= Gyrodon)
Gyrodon
Gyroporus
Leucoconius (= Gyroporus)
Paragyrodon
Phaeogyroporus
Phlebopus
Pseudogyrodon
Psiloboletinus
Rodwaya (= Gyrodon)
Setogyroporus
Suillus (= Gyroporus)
Uloporus (= Gyrodon)

Hygrophoropsidaceae
Hygrophoropsis

Paxillaceae
Austrogaster
Monodelphus (= Omphalotus)
Neopaxillus
Omphalotus
Parapaxillus (nom. inval.)
Paxillopsis (= Paxillus)
Paxillus
Plicaturella (= Paxillus)
Rhymovis (= Paxillus)
Ruthea (= Paxillus)
Tapinella (= Paxillus)
Tapinia (= Paxillus)

Rhizopogonaceae
Amogaster
Anthracophlous (= Rhizopogon)
Dodgea (= Truncocolumella)
Hysteromyces (= Rhizopogon)
Rhizopogon
Splanchnomyces (= Rhizopogon)
Truncocolumella

Strobilomycetaceae
Afroboletus
Austroboletus
Chalciporus
Eriocorys (= Strobilomyces)
Fistulinella

Fuscoboletinus
Leucogyroporus (= Tylopilus)
Mariaella
Mucilopilus
Phaeoporus (= Tylopilus)
Porphyrellus (= Tylopilus)
Rhodobolites (= Tylopilus)
Rubinoboletus
Strobilomyces
Tylopilus
Veloporphyrellus

Xerocomaceae
Boletellus
Boletogaster (= Boletellus)
Frostiella (= Boletellus)
Gastroboletus
Gymnopaxillus
Heimiella (= Boletellus)
Paxillogaster
Phylloboletellus
Phylloporus
Singeromyces
Strobilofungus (= Boletellus)
Tubosaeta
Versipellis (= Xerocomus)
Xerocomopsis (= Xerocomus)
Xerocomus

Incertae sedis
Anastomaria
Lamyxis (nom. dub.)

BONDARZEWIALES
Amylariaceae
Amylaria

Bondarzewiaceae
Bondarzewia

Hybogasteraceae
Hybogaster

CANTHARELLALES
Aphelariaceae
Aphelaria
Phaeoaphelaria
Tumidapexus

Cantharellaceae
Alectorolophoides (= Cantharellus)
Cantharellus
Chanterel (= Cantharellus)
Goossensia
Hyponevris (= Cantharellus)
Merulius (= Cantharellus)
Parastereopsis
Pterygellus

Clavariaceae
Capitoclavaria (= Clavaria)
Clavaria
Clavulinopsis
Cornicularia (= Clavulinopsis)
Donkella (= Clavulinopsis)
Holocoryne (= Clavaria)
Macrotyphula
Manina (nom. dub.)
Multiclavula
Phaeophelaria
Podostrombium (nom. inval.)
Ramaria (= Clavulinopsis)
Ramariopsis

Scytinopogon
Setigeroclavula
Stichoclavaria (= Holocoryne)

Clavariadelphaceae
Araeocoryne
Ceratella (= Ceratellopsis)
Ceratellopsis
Chaetotyphula
Clavaria (= Clavariadelphus)
Clavariadelphus

Clavulinaceae
Clavulicium
Clavulina
Membranomyces
Stichoramaria (= Clavulina)

Craterellaceae
Craterellus
Fungoidaster (= Craterellus)
Pezicula (= Craterellus)
Pseudocraterellus (= Craterellus)
Sterbeeckia (= Craterellus)
Trombetta (= Craterellus)

Hydnaceae
Amylodontia
Bidonia (= Hydnum)
Climacodon
Corallofungus
Cystidiodendron (nom. dub.)
Dacrina (nom. dub.)
Dentinum
Dicarphus (= Hydnum)
Donkia (= Climacodon)
Echinus (= Hydnum)
Erinaceus (= Hydnum)
Gloeomucro
Grandinia (nom. dub.)
Grandiniochaete (nom. dub.)
Hydnum
Hypothele (= Hydnum)
Irpicochaete (nom. dub.)
Malacodon (nom. dub.)
Nigrohydnum
Odontiochaete (nom. dub.)
Phaeoradulum
Radulochaete (nom. dub.)
Stegiacantha
Tuberculinia (= Grandinia)
Tyrodon (= Hydnum)

Physalacriaceae
Baumanniella (= Physalacria)
Deigloria
Eoagaricus (= Physalacria)
Hormomitaria
Physalacria
Pseudotyphula

Pterulaceae
Actiniceps
Allantula
Ceratella
Deflexula
Dimorphocystis (= Actiniceps)
Parapterulicium
Penicillaria (= Pterula)
Phaeopterula (= Pterula)
Pterula
Pterulicium

Scutigeraceae
Albatrellopsis
Albatrellus
Ovinus (= Albatrellus)
Polyporoletus (= Scutiger)
Scutiger

Sparassidaceae
Masseeola (= Sparassis)
Sparassiella
Sparassis

Typhulaceae
Cnazonaria (= Pistillaria)
Dacryopsella (= Pistillina)
Gliocoryne (= Pistillaria)
Lutypha
Phacorhiza (= Typhula)
Pistillaria
Pistillina
Scleromitra (= Pistillaria)
Sphaerula (= Pistillaria)
Typhula

CERATOBASIDIALES
Ceratobasidiaceae
Aquathanatephorus
Cejpomyces (= Thanatephorus)
Ceratobasidium
Heteroacanthella
Heteromyces (= Oliveonia)
Hydrabasidium
Koleroga (= Ceratobasidium)
Metabourdotia
Monosporonella
Oliveonia
Oncobasidium
Orcheomyces (= Thanatephorus)
Scotomyces (= Hydrabasidium)
Sebacinella
Thanatephorus
Tofispora
Uthatobasidium
Ypsilonidium (= Thanatephorus)

CORTINARIALES
Cortinariaceae
Agmocybe (= Inocybe)
Alnicola (= Naucoria)
Astrosporina (= Inocybe)
Bulbopodium (= Cortinarius)
Cereicium
Clypeus (= Inocybe)
Cortinaria (= Cortinarius)
Cortinarius
Cuphocybe
Cyanicium
Cyclocybe (= Inocybe)
Cystocybe (= Cortinarius)
Dermocybe
Epicorticium (= Phaeomarasmius)
Fissolimbus
Flammulaster (= Phaeomarasmius)
Flocculina (= Phaeomarasmius)
Fulvidula (= Gymnopilus)
Galera (= Galerina)
Galerina
Galerula (= Galerina)
Gomphos (= Cortinarius)
Gymnocybe (nom. dub.)
Gymnopilus
Hebeloma
Hebelomatis (= Hebeloma)

Hebelomina
Hydrocybe (= Cortinarius)
Hydrocybium (= Cortinarius)
Hygramaricium
Hygromyxacium
Hylophila (= Hebeloma)
Inocibium (= Inocybe)
Inocybe
Inoloma (= Cortinarius)
Kjeldsenia
Leucocortinarius
Leucopus (= Cortinarius)
Marasmiopsis (= Phaeomarasmius)
Meliderma (= Cortinarius)
Myxacium (= Cortinarius)
Myxocybe (= Hebeloma)
Myxopholis
Naucoria
Phaeocollybia
Phaeolepiota
Phaeomarasmius
Phlegmacium (= Cortinarius)
Pholidotopsis (= Galerina)
Picromyces (= Hebeloma)
Pseudogymnopilus
Pyrrhoglossum
Quercella
Raphanozon (= Cortinarius)
Roumeguerites (= Hebeloma)
Rozites
Ryssospora (= Gymnopilus)
Sarcoloma
Setchelliogaster
Simocybe
Sphaerotrachys (= Cortinarius)
Squamaphlegma
Stagnicola
Stephanopus
Telamonia (= Cortinarius)
Thaxterogaster
Velomycena (= Galerina)
Weinzettlia

Crepidotaceae
Calathinus (= Pleurotellus)
Chromocyphella
Crepidotus
Cymbella (= Chromocyphella)
Cyphellathelia (= Pellidiscus)
Cyphellopsis
Dochmiopus (= Crepidotus)
Episphaeria
Horakomyces
Melanomphalia
Merismodes
Nanstelocephala
Pellidiscus
Phaeocarpus (= Chromocyphella)
Phaeocyphella (= Chromocyphella)
Phaeocyphella (= Chromocyphella)
Phaeocyphellopsis (= Merismodes)
Phaeosolenia
Phialocybe (= Crepidotus)
Pleurotellus
Pseudodasyscypha (= Cyphellopsis)
Tremellastrum (= Crepidotus)
Tremellopsis (= Crepidotus)
Tubaria

Cribbeaceae
Cribbea
Mycolevis

Gautieriaceae
Austrogautieria
Ciliciocarpus (= Gautieria)
Gautieria
Hydnospongos (= Gautieria)
Protogautieria
Uslaria (= Gautieria)

Hymenangiaceae
Hymenangium

DACRYMYCETALES
Cerinomycetaceae
Cerinomyces

Dacrymycetaceae
Arrhytidia (= Dacrymyces)
Calocera
Caloceras (= Calocera)
Calopposis (= Calocera)
Corynoides (= Calocera)
Dacrymyces
Dacryomitra (= Calocera)
Dacryonaema
Dacryopinax
Dacryopsis (= Ditiola)
Ditiola
Femsjonia
Guepiniopsis
Heterotextus
Hydromycus (= Dacrymyces)
Septocolla (= Dacrymyces)

Incertae sedis
Dacryomycetopsis
Dicellomyces

FISTULINALES
Fistulinaceae
Agarico-carnis (= Fistulina)
Buglossus (= Fistulina)
Fistulina
Hypodrys (= Fistulina)
Pseudofistulina

GANODERMATALES
Ganodermataceae
Amauroderma (= Amauroderma)
Amauroderma
Dendrophagus (= Ganoderma)
Elfvingia
Friesia (= Ganoderma)
Ganoderma
Humphreya
Magoderna
Tomophagus (= Ganoderma)
Trachyderma (= Ganoderma)
Whitfordia (= Amauroderma)

Haddowiaceae
Haddowia

GOMPHALES
Gomphaceae
Chloroneuron (= Gomphus)
Chlorophyllum (= Gomphus)
Gloeocantharellus
Gomphora (= Gomphus)
Gomphus

Linderomyces (=
    Gloeocantharellus)
Nevrophyllum (= Gomphus)
Pseudogomphus
Terenodon
Turbinellus (= Gomphus)

Lentariaceae
Lentaria

Ramariaceae
Cladaria (= Ramaria)
Clavariella (= Ramaria)
Corallium (= Ramaria)
Coralloidea (= Ramaria)
Coralloides (= Ramaria)
Delentaria
Dendrocladium (= Ramaria)
Kavinia
Phaeoclavulina (= Ramaria)
Ramaria
Ramaricium

HERICIALES
Auriscalpiaceae
Amylonotus
Auriscalpium
Dentipellis
Pleurodon (= Auriscalpium)
Stecchericium

Clavicoronaceae
Clavicorona

Gloeocystidiellaceae
Amylosporomyces
Boidinia
Conferticium
Confertobasidium (=
    Scytinostromella)
Gloeocystidiellum
Gloeocystidiopsis (=
    Gloeocystidiellum)
Gloeodontia
Gloiodon
Gloiothele
Laxitextum
Leaia (= Gloiodon)
Megalocystidium
Pseudoxenasma
Sclerodon (= Gloiodon)
Scytinostromella
Vesiculomyces

Hericiaceae
Amylosporus
Artomyces
Creolophus
Dentipratulum
Dryodon (= Hericium)
Friesites (= Hericium)
Hericium (= Hericium)
Hericium
Hericius (= Hericium)
Manina (= Hericium)
Martella (= Hericium)
Martella (= Hericium)
Medusina (= Hericium)
Mucronella
Mucronia (= Mucronella)
Myxomycidium (= Mucronella)
Wrightoporia

Lentinellaceae
Hemicybe (= Lentinellus)
Lentinaria (= Lentinellus)
Lentinellus

HETEROGASTRIDIALES
Heterogastridiaceae
Heterogastridium

HYMENOCHAETALES
Asterostromataceae
Aciella (= Asterodon)
Asterodon
Asterostroma
Hydnochaete (= Asterodon)
Hydnochaetella (= Asterodon)

Hymenochaetaceae
Aurificaria
Boletus (= Phellinus)
Boudiera (= Phellinus)
Cerrenella (= Inonotus)
Clavariachaeta (= Clavariachaete)
Clavariachaete
Coltricia
Coltriciella
Coltriciopsis
Cryptoderma
Cyclomyces
Cyclomycetella (= Cyclomyces)
Cycloporellus (= Cyclomyces)
Cycloporus
Daedaloides (= Cryptoderma)
Erythromyces
Fomitiporella (= Phellinus)
Fomitiporia (= Phellinus)
Fulvifomes (= Phellinus)
Fuscoporella (= Phellinus)
Fuscoporia (= Phellinus)
Hydnochaete
Hydnoporia (= Hydnochaete)
Hymenochaete
Hymenochaetella (=
    Hymenochaete)
Inocutis
Inoderma (= Inonotus)
Inodermus (= Inonotus)
Inonotopsis
Inonotus
Lazaroa (= Phellinus)
Leptochaete (= Hymenochaete)
Loxophyllum (= Cyclomyces)
Mensularia (= Inonotus)
Mison (= Phellinus)
Mucronoporus
Ochroporus (= Phellinus)
Ochrosporellus
Onnia (= Mucronoporus)
Pelloporus (= Coltricia)
Phaeohydnochaete
Phaeoporus (= Inonotus)
Phellinidium
Phellinus
Phylloporia
Polystictoides (= Inonotus)
Polystictus (= Coltricia)
Porodaedalea
Pseudasterodon (nom. dub.)
Pseudofomes (= Phellinus)
Pyropolyporus (= Phellinus)
Pyrrhoderma
Scalaria (nom. dub.)
Scindalma (= Phellinus)

Spongiosus (= Phaeolus)
Stipitochaete
Strilia (= Coltricia)
Volvopolyporus (= Coltricia)
Xanthochrous (= Coltricia)
Xanthoporia (= Inonotus)

HYMENOGASTRALES
Gasterellaceae
Gasterella

Gastrosporiaceae
Gastrosporium
Leucorhizon (= Gastrosporium)

Hymenogastraceae
Chondrogaster
Cortinomyces
Dendrogaster (= Hymenogaster)
Descomyces
Destuntzia
Fechtneria (= Hymenogaster)
Hymenogaster
Hysterogaster (= Hymenogaster)
Mycoamaranthus
Protoglossum (= Hymenogaster)
Quadrispora
Radiogaster (= Hymenogaster)
Rhizopogoniella (= Hymenogaster)
Timgrovea

Octavianinaceae
Octavianina
Sclerogaster
Wakefieldia

Protogastraceae
Protogaster

Incertae sedis
Vandasia

LACHNOCLADIALES
Dichostereaceae
Dichostereum

Lachnocladiaceae
Asterostromella (= Vararia)
Dichantharellus
Dichopleuropus
Eriocladus (= Lachnocladium)
Lachnocladium
Scytinostroma
Stelligera (= Lachnocladium)
Stereofomes
Vararia

LYCOPERDALES
Broomeiaceae
Broomeia

Geastraceae
Anthropomorphus (= Geastrum)
Astrocitum (= Astrycum)
Astrycum (= Geastrum)
Coilomyces (= Geastrum)
Geaster (= Geastrum)
Geasteroides
Geasteropsis (= Geastrum)
Geastrum
Glycydiderma (= Geastrum)
Myriostoma
Phialastrum

Plecostoma (= Geastrum)
Polystoma (= Myriostoma)
Pyrenogaster
Radiigera
Terrostella (= Geasteroides)
Trichaster (= Geastrum)

Lycoperdaceae
Abstoma
Acutocapillitium
Anixia (nom. dub.)
Arachnion
Arachniopsis
Bovista
Bovistaria (= Langermannia)
Bovistella
Bovistina (= Disciseda)
Bovistoides (= Disciseda)
Calbovista
Calvatia
Calvatiella (= Bovistella)
Calvatiopsis
Capillaria (= Lycoperdon)
Catastoma (= Disciseda)
Cerophora (= Lycoperdon)
Disciseda
Enteromyxa (= Lycogalopsis)
Eriosphaera (= Lasiosphaera)
Gastropila
Globaria (= Bovista)
Glyptoderma
Handkea (= Calvatia)
Hippoperdon (= Calvatia)
Hypoblema (= Calvatia)
Japonogaster
Langermannia
Lanopila (= Langermannia)
Lasiosphaera (= Langermannia)
Laterradea (nom. dub.)
Lycogalopsis
Lycoperdon
Lycoperdopsis
Morganella
Omalycus (= Calvatia)
Piemycus (= Piesmycus)
Piesmycus (= Bovista)
Pila (= Gastropila)
Priapus (= Lycoperdon)
Pseudolycoperdon (= Bovista)
Sackea (= Bovista)
Scoleciocarpus (= Arachnion)
Sufa (= Lycoperdon)
Utraria (= Lycoperdon)
Vascellum

Mesophelliaceae
Castoreum
Inoderma (= Mesophellia)
Malajczukia
Mesophellia
Nothocastoreum
Potoromyces (= Mesophellia)

Mycenastraceae
Endonevrum (= Mycenastrum)
Mycenastrum
Pachyderma (= Mycenastrum)

Incertae sedis
Cycloderma (nom. dub.)
Eriosperma (nom. dub.)
Mesophelliopsis
Mycospongia (nom. dub.)

Sedecula
Stemastrum (nom. dub.)

MELANOGASTRALES
Leucogastraceae
Cremeogaster (= Leucophleps)
Leucogaster
Leucophleps

Melanogastraceae
Alpova
Argylium (= Melanogaster)
Bulliardia (= Melanogaster)
Corditubera
Hoehnelogaster (nom. dub.)
Hyperrhiza (= Melanogaster)
Melanogaster
Octaviania (= Melanogaster)
Uperhiza (= Melanogaster)

Niaceae
Nia

NIDULARIALES
Nidulariaceae
Crucibulum
Cyathia (= Cyathus)
Cyathodes (= Cyathus)
Cyathus
Granularia (= Nidularia)
Mycocalia
Nidula
Nidularia (= Cyathus)
Nidularia
Peziza (= Cyathus)
Polygaster (nom. dub.)

PHALLALES
Clathraceae
Anthurus (= Clathrus)
Aseroë
Aserophallus (= Clathrus)
Blumenavia
Calathiscus (= Lysurus)
Clathrella (= Clathrus)
Clathrus
Cletria (= Clathrus)
Colonnaria (= Clathrus)
Colus
Desmaturus (= Lysurus)
Dictyobole (= Lysurus)
Dycticia (= Clathrus)
Foetidaria (= Lysurus)
Ileodictyon
Kalchbrennera (= Lysurus)
Kupsura (= Lysurus)
Laternea
Ligiella
Linderia (= Clathrus)
Linderiella (= Clathrus)
Lloydia (= Sinolloydia)
Lysurus
Mycopharus (= Lysurus)
Neolysurus
Pharus (= Lysurus)
Pseudoclathrus
Pseudocolus
Schizmaturus (= Lysurus)
Simblum (= Lysurus)
Sinolloydia (= Lysurus)
Stephanophallus (= Anthurus)

Claustulaceae
Claustula

Gelopellaceae
Gelopellis

Hysterangiaceae
Boninogaster
Circulocolumella
Clathrogaster
Gallacea (= Hysterangium)
Hysterangium
Phallobata (= Hysterangium)
Phallogaster
Phlebogaster
Rhopalogaster
Rimella (nom. dub.)
Stalactocolumella (=
  Circulocolumella)
Trappea

Phallaceae
Aedycia (= Mutinus)
Alboffiella (= Itajahya)
Aporophallus
Caromyxa (= Mutinus)
Clautriavia (= Phallus)
Corynites (= Mutinus)
Cryptophallus (= Phallus)
Cynophallus (= Mutinus)
Dictyopeplos (= Phallus)
Dictyophallus (= Phallus)
Dictyophora (= Phallus)
Echinophallus
Endophallus
Floccomutinus (= Mutinus)
Hymenophallus (= Phallus)
Itajahya
Ithyphallus (= Mutinus)
Jaczewskia (= Phallus)
Jansia (= Mutinus)
Junia (= Phallus)
Kirchbaumia (= Phallus)
Lejophallus (= Phallus)
Morellus (= Phallus)
Mutinus
Omphalophallus (= Phallus)
Phalloidastrum (= Phallus)
Phallus
Retigerus (= Phallus)
Satyrus (= Phallus)
Sophronia (= Phallus)
Staheliomyces
Staurophallus (nom. dub.)
Xylophallus (= Mutinus)

Protophallaceae
Calvarula
Kobayasia
Protophallus (= Protubera)
Protubera
Protuberella

PORIALES
Coriolaceae
Abortiporus
Agaricon (= Agaricum)
Agarico-igniarium (= Fomes)
Agarico-pulpa (= Agaricum)
Agarico-suber (= Daedalea)
Agaricum
Agaricum (= Laricifomes)
Agaricus (= Daedalea)

Amylocystis
Amyloporia
Amyloporiella
Anisomyces (= Osmoporus)
Anomoporia (= Poria)
Antrodia
Antrodiella
Apoxona (= Hexagonia)
Artolenzites (= Trametes)
Aurantiporellus (= Pycnoporellus)
Aurantiporus (= Tyromyces)
Auriporia
Australoporus
Baeostratoporus (= Flaviporus)
Bjerkandera
Bornetina (= Diacanthodes)
Buglossoporus
Bulliardia (= Cerrena)
Cartilosoma (= Antrodia)
Cautinia
Cellularia (= Coriolus)
Ceraporia (= Ceriporia)
Ceraporus (= Ceriporia)
Ceratophora (= Gloeophyllum)
Ceriporia
Ceriporiopsis (= Poria)
Cerrena
Chaetoporus
Choriphyllum (= Phaeolus)
Cinereomyces
Cladodendron (= Grifola)
Cladomeris (= Grifola)
Cladoporus (= Laetiporus)
Climacocystis
Coriolellus (= Antrodia)
Coriolopsis
Coriolus
Cryptoporus
Cubamyces (= Trametes)
Cystidiophorus (nom. inval.)
Cystostiptoporus
Daedalea
Daedaleopsis
Datronia
Dedalea (= Daedalea)
Dextrinosporium
Diacanthodes
Diplomitoporus
Donkioporia
Earliella (= Trametes)
Echinoporia
Elfvingiella (= Fomes)
Enslinia (= Porodisculus)
Fibroporia (= Poria)
Flabellophora
Flabellopilus (= Meripilus)
Flaviporellus (= Flaviporus)
Flaviporus
Fomes
Fomitella (= Coriolopsis)
Fomitopsis
Funalia
Fuscocerrena
Gelatoporia
Globifomes
Gloeophyllum
Grammothelopsis
Grifola
Griseoporia
Hansenia (= Coriolus)
Hapalopilus
Haploporus (= Haploporus)
Haploporus

Hemidiscia (= Tyromyces)
Henningsia
Heterobasidion
Heteroporus
Hexagonia
Hirschioporus
Hydnopolyporus
Incrustoporia
Irpiciporus (= Spongipellis)
Irpicium (nom. dub.)
Ischnoderma
Laetiporus
Lamelloporus
Laricifomes (= Agaricum)
Lasiochlaena
Lenzites
Lenzitina (= Gloeophyllum)
Leptopora (= Poria)
Leptoporus (= Tyromyces)
Leptotrimitus (= Incrustoporia)
Leucofomes (= Rigidoporus)
Leucolenzites (= Lenzites)
Leucophellinus
Loweomyces
Loweporus
Macrohyporia
Megasporoporia
Melanoporella (= Poria)
Melanoporia (= Poria)
Meripilus
Merisma (= Grifola)
Merulioporia (= Poria)
Meruliopsis (= Poria)
Meruliporia (= Poria)
Metuloidea
Microporellus
Mollicarpus
Murrilloporus
Mycodendron (nom. dub.)
Myriadoporus (= Bjerkandera)
Navisporus
Nigrofomes
Nigroporus
Oligoporus
Osmoporus
Osteina
Oxyflavus (= Oxyporus)
Oxyporus
Pachykytospora
Paratrichaptum
Parmastomyces
Perenniporia
Persooniana (= Tyromyces)
Phaeocoriolellus (=
    Gloeophyllum)
Phaeodaedalea
Phaeolopsis
Phaeolus
Phaeotellus
Pherima (= Phorima)
Phorima (= Hexagonia)
Phyllodontia (= Cerrena)
Physisporinus
Physisporus (= Poria)
Pilatoporus
Piloporia
Piptoporus
Placoderma (= Piptoporus)
Placodes (= Fomes)
Podofomes (= Ischnoderma)
Podoporia
Pogonomyces (= Hexagonia)
Polypilus (= Grifola)

Polyporus (= Laetiporus)
Poria
Porodisculus
Porodiscus (= Porodisculus)
Poronidulus
Poroptyche (= Poria)
Porpomyces
Postia (= Tyromyces)
Pseudophaeolus
Pseudopiptoporus
Pseudotrametes (= Trametes)
Pycnoporellus
Pycnoporus
Pyrofomes
Reisneria (= Gloeophyllum)
Rhodofomes
Rigidoporopsis
Rigidoporus
Riopa (= Poria)
Romellia (= Phaeolus)
Sarcoporia (= Podoporia)
Scenidium (= Hexagonia)
Sclerodepsis
Serda (= Gloeophyllum)
Sesia (= Gloeophyllum)
Sistotrema (= Cerrena)
Skeletocutis
Somion (= Spongipellis)
Sparsitubus
Spongioides (nom. dub.)
Spongipellis
Spongiporus
Strangulidium (= Poria)
Striglia (= Daedalea)
Tinctoporellus
Trametella (= Funalia)
Trametes
Trichaptum
Truncospora
Tyromyces
Ungularia (= Piptoporus)
Ungulina (= Fomes)
Vanderbylia
Wolfiporia
Xylometron (= Pycnoporus)
Xylopilus (= Fomes)
Xylostroma (= Daedalea)

Grammotheleaceae
Grammothele
Hymenogramme
Porogramme
Theleporus
Tinctoporia (= Porogramme)

Lentinaceae
Austrolentinus
Crepidopus (= Pleurotus)
Cyclopleurotus (= Pleurotus)
Digitellus (= Lentinus)
Faerberia
Gelona (= Pleurotus)
Geopetalum
Geopetalum (= Geopetalum)
Lentinopanus
Lentinus
Lentodiellum (= Pleurotus)
Lentodiopsis (= Pleurotus)
Lentodium (= Panus)
Neolentinus
Panus
Phyllotopsis
Phyllotus (= Pleurotus)

Pleuropus (= Panus)
Pleurotus
Pocillaria (= Lentinus)
Pterophyllus (= Pleurotus)
Scleroma (= Panus)
Tilotus (= Phyllotopsis)
Tomentifolium (= Phyllotopsis)

Polyporaceae
Asterochaete (= Echinochaete)
Atroporus
Bresadolia (= Polyporus)
Caloporus (= Polyporus)
Cerioporus (= Polyporus)
Cyanosporus (= Polyporus)
Dendrochaete (= Echinochaete)
Dendropolyporus
Dichomitus
Echinochaete
Favolus (= Polyporus)
Favolus
Grandinioides (= Mycobonia)
Heliocybe
Hexagonia (= Polyporus)
Hirneola (= Mycobonia)
Hornodermoporus
Jahnoporus
Laccocephalum (= Polyporus)
Lentus (= Polyporus)
Leucoporus (= Polyporus)
Lignosus
Macroporia
Melanopus (= Polyporus)
Microporus
Mycelithe (= Polyporus)
Mycobonia
Petaloides (= Polyporus)
Phaeotrametes
Polyporellus (= Polyporus)
Polyporus
Polyporus (= Polyporus)
Pouzaroporia
Pseudofavolus
Stiptophyllum
Tomentoporus (= Microporus)
Tuberaster (= Polyporus)
Velolentinus (nom. nud.)
Xerotinus
Xerotus (= Xerotinus)

RUSSULALES
Elasmomycetaceae
Arcangeliella
Buchholtzia (= Macowanites)
Elasmomyces
Gymnomyces
Hypochanum (= Macowanites)
Macowania (= Macowanites)
Macowanites
Martellia
Zelleromyces

Russulaceae
Cystangium
Dixophyllum (= Russula)
Galorrheus (= Lactarius)
Gloeocybe (= Lactarius)
Hypophyllum (= Lactarius)
Lactarelis (= Russula)
Lactaria (= Lactarius)
Lactariella (= Lactarius)
Lactariopsis (= Lactarius)
Lactarius

Lactifluus (= Lactarius)
Omphalomyces (= Russula)
Phaeohygrocybe (= Russula)
Pleurogala
Russula
Russulina (= Russula)

SCHIZOPHYLLALES
Schizophyllaceae
Apus (= Schizophyllum)
Ditiola (= Schizophyllum)
Flabellaria (= Schizophyllum)
Henningsomyces
Hyponevris (= Schizophyllum)
Petrona (= Schizophyllum)
Phaeoschizophyllum (=
  Schizophyllum)
Plicaturopsis
Rectipilus
Rhipidium (= Schizophyllum)
Scaphophorum (= Schizophyllum)
Schizonia (= Schizophyllum)
Schizophyllum
Schizophyllus (= Schizophyllum)
Solenia (= Henningsomyces)

Stromatoscyphaceae
Porotheleum
Stromatocyphella
Stromatoscypha (= Porotheleum)

SCLERODERMATALES
Astraeaceae
Astraeus
Diploderma (= Astraeus)

Diplocystaceae
Diplocystis

Sclerodermataceae
Actigea (= Scleroderma)
Actigena (= Actigea)
Actinodermium (= Sterbeeckia)
Caloderma (= Scleroderma)
Durosaccum (= Pisolithus)
Endacinus (= Pisolithus)
Eudacnus (= Endacinus)
Favillea
Goupilia (= Scleroderma)
Horakiella
Lycoperdastrum (= Scleroderma)
Lycoperdodes (= Pisolithus)
Lycoperdoides (= Pisolithus)
Mycastrum (= Scleroderma)
Neosaccardia (= Scleroderma)
Nepotatus (= Scleroderma)
Phlyctospora (= Scleroderma)
Pirogaster
Pisocarpium (= Pisolithus)
Pisolithus
Polypera (= Pisolithus)
Polysaccum (= Pisolithus)
Pompholyx (= Scleroderma)
Sclerangium (= Scleroderma)
Scleroderma
Stella (= Scleroderma)
Sterrebekia (= Scleroderma)
Tremellogaster
Veligaster (= Scleroderma)

Sphaerobolaceae
Carpobolus (= Sphaerobolus)
Nidulariopsis

Siropeltis (= Sphaerobolus)
Sphaerobolus

Incertae sedis
Endogonopsis

STEREALES
Aleurodiscaceae
Acanthobasidium
Acanthophysellum
Acanthophysium
Aleurobotrys
Aleurocystidiellum
Aleurocystis
Aleurodiscus
Cyphella (= Aleurodiscus)
Dendrocyphella (= Cyphella)
Dendrophysellum
Gloeosoma (= Aleurodiscus)
Nodularia (= Aleurodiscus)

Amylocorticaceae
Amyloathelia
Amylocorticium
Digitatispora
Hypochniciellum
Irpicodon
Melzericium (nom. nud.)
Plicatura

Atheliaceae
Aldrigiella (nom. dub.)
Amphinema
Athelia
Athelicium
Athelidium
Athelopsis
Butlerelfia
Byssocorticium
Byssoporia
Caerulicium
Diplonema (= Amphinema)
Fibulomyces
Hypochnella
Hypochnopsis
Leptosporomyces
Lobulicium
Luellia
Piloderma
Pteridomyces (= Athelopsis)
Tylosperma (= Tylospora)
Tylospora

Botryobasidiaceae
Botryobasidium
Botryodontia
Botryohypochnus (=
  Botryobasidium)
Candelabrochaete
Cyanohypha (= Botryobasidium)
Suillosporium
Waitea

Corticiaceae
Aleurocorticium (= Dendrothele)
Amaurodon
Ambivina
Amethicium
Auricula (= Punctularia)
Byssocristella
Cericium
Chrysoderma
Corticirama

Corticium
Crustomyces
Cystidiodontia
Cystostereum
Cytidia
Dendrocorticium
Dendrodontia
Dendrothele
Dentocorticium
Dextrinocystis
Dextrinodontia
Entomocorticium
Gloeocorticium
Gyrophanopsis
Hydnocristella
Inocybella
Laeticorticium (= Corticium)
Laetisaria
Lazulinospora
Lomatia (= Lomatina)
Lomatina (= Cytidia)
Lyomyces (= Laeticorticium)
Melzerodontia
Merulicium
Mutatoderma
Mycinema (nom. dub.)
Mycostigma
Palifer
Papyrodiscus
Parvobasidium
Pellicularia (nom. conf.)
Phaeophlebia (= Punctularia)
Physodontia
Pulcherricium
Punctularia
Ripexicium
Skvortzovia
Terana (= Corticium)
Vuilleminia

Cyphellaceae
Asterocyphella
Catilla
Cephaloscypha
Glabrocyphella
Halocyphina
Incrustocalyptella
Limnoperdon
Maireina
Phaeoglabrotricha
Phaeoporotheleum
Rhodoarrhenia
Rhodocyphella
Seticyphella
Sphaerobasidioscypha
Wiesnerina
Woldmaria

Echinodontiaceae
Echinodontium
Hydnofomes (= Echinodontium)
Hydnophysa (= Hydnofomes)

Epitheliaceae
Epithele
Epithelopsis
Mycothele
Skeletohydnum

Hyphodermataceae
Adustomyces
Amylobasidium
Atheloderma

Basidioradulum
Bulbillomyces
Cerocorticium
Chaetoporellus (= Hyphodontia)
Conohypha
Crustoderma
Cyanodontia
Cylindrobasidium
Erythricium
Fibrodontia (= Hyphodontia)
Flavophlebia (= Radulomyces)
Globulicium
Gloeohypochnicium (=
   Hypochnicium)
Granulobasidium (=
   Hypochnicium)
Himantia (= Cylindrobasidium)
Hydnellum (= Hyphoderma)
Hyphoderma
Hyphodermella
Hyphodermopsis
Hyphodontia
Hyphodontiella
Hyphoradulum
Hypochnicium
Intextomyces
Kneiffia (= Hyphoderma)
Kneiffiella (= Hyphoderma)
Lagarobasidium (= Hyphodontia)
Lyomyces (= Hyphoderma)
Metulodontia (= Hyphoderma)
Neokneiffia (= Hyphoderma)
Nodotia (= Hypochnicium)
Odonticium
Odontiopsis
Oxydontia (= Sarcodontia)
Pirex
Poriodontia
Pseudolagarobasidium
Pycnodon (= Hyphoderma)
Radulodon
Radulomyces
Rogersella (= Hyphodontia)
Sarcodontia
Schizopora
Subulicium
Subulicystidium
Thujacorticium
Uncobasidium
Xylodon (nom. dub.)

Lindtneriaceae
Cristinia
Cyanobasidium (= Lindtneria)
Dacryobasidium (= Cristinia)
Lindtneria
Mycolindtneria

Meruliaceae
Acia (= Mycoacia)
Auriculariopsis
Byssomerulius
Caloporia (= Caloporus)
Caloporus (= Merulius)
Castanoporus
Ceraceomerulius (=
   Byssomerulius)
Ceraceomyces
Chondrostereum
Columnodontia
Cora (= Dictyonema)
Coraemyces (= Dictyonema)
Corella (= Dictyonema)

Corticium (= Phanerochaete)
Cytidiella (= Auriculariopsis)
Dacryobolus
Dichonema (= Dictyonema)
Dictyonema
Dictyonematomyces (=
   Dictyonema)
Efibula
Gloeocystidium (= Dacryobolus)
Gloeoporus
Grandiniella
Granulocystis
Gyrolophium (= Dictyonema)
Hydnophlebia
Jacksonomyces
Laudatea (= Dictyonema)
Membranicium (= Phanerochaete)
Merulius (= Phlebia)
Mycoacia
Phanerochaete
Phlebia
Phlebiopsis
Resinicium
Rhipidonema (= Dictyonema)
Rhipidonematomyces (=
   Dictyonema)
Rhizonema (= Dictyonema)
Ricnophora (= Phlebia)
Scopuloides
Trabecularia (= Merulius)
Wainiocora (= Dictyonema)
Xerocarpus (= Phanerochaete)

Peniophoraceae
Cryptochaete (= Peniophora)
Dendrophora
Duportella
Gloeopeniophora (= Peniophora)
Peniophora
Peniophorella (nom. dub.)
Sterellum (= Peniophora)
Veluticeps

Podoscyphaceae
Actinostroma (= Cymatoderma)
Aquascypha
Beccaria (= Cymatoderma)
Beccariella (= Cymatoderma)
Caripia
Cladoderris (= Cymatoderma)
Coralloderma
Cotylidia
Cymatoderma
Cyphellostereum
Hypolyssus (= Caripia)
Inflatostereum
Masseerina (nom. nud.)
Podoscypha
Pseudolasiobolus
Stereogloeocystidium (=
   Podoscypha)
Stereophyllum (=
   Cyphellostereum)
Stereopsis

Sistotremataceae
Brevicellicium
Cristelloporia (= Trechispora)
Echinotrema (= Trechispora)
Elaphocephala
Fibriciellum (= Trechispora)
Fibuloporia (= Trechispora)
Galzinia

Galziniella (= Sistotrema)
Heptasporium (= Sistotrema)
Hydnotrema (= Sistotrema)
Limonomyces
Paullicorticium
Psathyrodon
Repetobasidiellum
Repetobasidium
Sistotrema (= Sistotrema)
Sistotrema
Sistotremastrum
Sistotremella
Sphaerobasidium
Sulphurina (= Cristella)
Tomentella (= Trechispora)
Trechispora
Urnobasidium (= Sistotrema)

Steccherinaceae
Aschersonia (= Junghuhnia)
Ceraceohydnum (= Mycoaciella)
Etheirodon (= Steccherinum)
Fibricium
Flavodon (= Irpex)
Irpex
Junghuhnia
Laschia (= Junghuhnia)
Leptodon (= Steccherinum)
Mycoaciella
Mycoleptodon (= Steccherinum)
Mycoleptodonoides
Mycorrhaphium
Odontia (= Steccherinum)
Odontina (= Steccherinum)
Odontium (= Steccherinum)
Spathulina (= Irpex)
Steccherinum

Stephanosporaceae
Stephanospora

Stereaceae
Amaurohydnum
Amauromyces
Amylohyphus
Amylostereum
Australohydnum
Boreostereum
Bresadolina
Chaetocarpus (= Columnocystis)
Chaetoderma
Chaetodermella
Columnocystis
Craterella (= Bresadolina)
Crystallocystidium (nom. dub.)
Friesula (nom. dub.)
Gloeoasterostroma (nom. conf.)
Gloeopeniophorella (nom. dub.)
Gloeostereum
Haematostereum (= Stereum)
Laurilia
Licentia (= Lopharia)
Licrostroma
Lloydella (= Lopharia)
Lloydellopsis (= Amylostereum)
Lopharia
Porostereum
Scotoderma
Stereum
Thwaitesiella (= Lopharia)
Trichocarpus (= Amylostereum)
Xylobolus

Tubulicrinaceae
Litschauerella
Tubulicium
Tubulicrinis
Tubulixenasma (= Tubulicium)

Xenasmataceae
Aphanobasidium
Coronicium
Cunninghammyces (= Xenasma)
Lepidomyces
Phlebiella
Xenasma
Xenasmatella (= Phlebiella)
Xenosperma

Incertae sedis
Globosomyces
Lepidostroma

THELEPHORALES
Bankeraceae
Bankera
Phellodon

Thelephoraceae
Boletopsis
Botryobasidium
Bubacia (nom. nud.)
Caldesiella (= Tomentella)
Calodon (= Hydnellum)
Clitopilina (nom. nud.)
Cyphellina (nom. dub.)
Entolomina
Gymnoderma (nom. rej.)
Hydnellum
Hydnodon
Hydnopsis (= Tomentella)
Hypochnus (= Tomentella)
Karstenia (= Prillieuxia)
Kneiffiella
Lenzitella
Lenzitopsis (= Lenzitella)
Merisma (= Thelephora)
Odontia (= Tomentella)
Phaeodon (= Hydnellum)
Phlyctibasidium
Phylacteria (= Thelephora)
Phyllocarbon (= Polyozellus)
Pleurobasidium
Polyozellus
Prillieuxia (= Tomentella)
Pseudohydnum (= Hydnodon)
Pseudothelephora (= Thelephora)
Pseudotomentella
Sarcodon
Scyphopilus (= Thelephora)
Skepperia
Thelephora
Thelephorella (nom. dub.)
Thelophora (= Thelephora)
Tomentella (= Tomentella)
Tomentella
Tomentellago
Tomentellastrum (= Tomentella)
Tomentellina (= Kneiffiella)
Tomentellopsis

TREMELLALES
Aporpiaceae
Aporpium
Elmeria (= Elmerina)
Elmerina

Exidiaceae
Atkinsonia (= Sebacina)
Atractobasidium (= Patouillardina)
Basidiodendron
Bourdotia
Ceratosebacina
Collodendrum (=
   Tremellodendron)
Craterocolla
Cristella (= Sebacina)
Ductifera
Efibulobasidium
Eichleriella
Endoperplexa
Exidia
Exidiopsis
Fibulosebacea
Gloeosebacina (= Stypella)
Gloeotromera (= Ductifera)
Guepinia (= Tremiscus)
Gyrocephalus (= Tremiscus)
Heterochaete
Heterochaetella
Heteroscypha
Hirneolina (= Heterochaete)
Hydnogloea (= Pseudohydnum)
Microsebacina
Myxarium
Patouillardina
Phlogiotis (= Tremiscus)
Poroidea (= Craterocolla)
Protodaedalea
Protodontia
Protograndinia (= Patouillardina)
Protohydnum
Protomerulius
Pseudohydnum
Pseudostypella
Renatobasidium
Sebacina
Serendipita
Soppittiella (= Cristella)
Spicularia (= Exidia)
Stypella
Tremellacantha
Tremellochaete (= Exidia)
Tremellodendron
Tremellodon (= Pseudohydnum)
Tremelloscypha
Tremellostereum
Tremiscus
Ulocolla (= Exidia)

Filobasidiaceae
Cystofilobasidium
Filobasidiella
Filobasidium
Kondoa
Mrakia

Hyaloriaceae
Hyaloria

Rhynchogastremataceae
Rhynchogastrema

Sirobasidiaceae
Fibulobasidium
Sirobasidium

Syzygosporaceae
Carcinomyces (= Syzygospora)

Nigredo (= Uredo)
Peridermium
Phragmotelium
Physonema
Physopella
Poliomopsis
Poliotelium
Ramakrishnania
Roestelia
Rubigo (= Uredo)
Schroeteriaster
Skierka
Stakmania
Stomatisora
Symperidium (= Aecidium)
Telomapea
Thirumalachariella
Trichobasis (= Uredo)
Tunicopsora
Ulea
Uleiella (= Ulea)
Uraecium
Uredo
Uromycodes (= Schroeteriaster)

USTOMYCETES

CRYPTOBASIDIALES
Cryptobasidiaceae
Botryoconis
Clinoconidium
Coniodictyum
Cryptobasidium (= Botryoconis)
Drepanoconis
Hyalodema (= Coniodictyum)

CRYPTOMYCOCOLACALES
Cryptomycocolacaceae
Cryptomycocolax

EXOBASIDIALES
Brachybasidiaceae
Brachybasidium
Ceraceosorus
Proliferobasidium

Exobasidiaceae
Arcticomyces (= Exobasidium)
Endobasidium
Exobasidiellum
Exobasidium
Kordyana
Laurobasidium
Lelum (= Kordyana)
Muribasidiospora

Microstromataceae
Microstroma

GRAPHIOLALES
Graphiolaceae
Elpidophora (= Graphiola)
Graphiola
Nigrocupula (= Graphiola)
Stylina
Trichodesmium (= Graphiola)

PLATYGLOEALES
Platygloeaceae
Achroomyces (= Platygloea)

Aphelariopsis
Auriculoscypha
Camptobasidium
Colacogloea
Collopezis (= Tjibodasia)
Cystobasidium
Eocronartium
Helicobasidium
Helicobasis (= Helicobasidium)
Helicogloea
Herpobasidium
Insolibasidium
Jola
Kriegeria
Krieglsteinera
Kryptastrina
Mycogloea
Naohidea
Occultifur
Phyllogloea
Platycarpa
Platygloea
Protopistillaria (= Eocronartium)
Ptechetelium
Saccoblastia (= Helicogloea)
Stypinella (= Helicobasidium)
Tachaphantium (= Platygloea)
Tjibodasia (= Platygloea)
Xenogloea (= Kriegeria)

Incertae sedis
Biatoropsis
Zygogloea

SPORIDIALES
Sporidiobolaceae
Aessosporon
Leucosporidium
Rhodosporidium
Rogersiomyces
Sakaguchia
Sporidiobolus
Sterigmatosporidium
Tilletiaria

USTILAGINALES
Tilletiaceae
Burrillia
Clintamra
Conidiosporomyces
Cornuella (= Tracya)
Didymochlamys (= Kuntzeomyces)
Doassansia
Doassansiella
Doassansiopsis
Entorrhiza
Entyloma
Entylomella
Georgefischeria
Ginanniella
Glomosporium
Heterodoassansia
Jamesdicksonia
Juliohirschhornia
Kuntzeomyces
Melanotaenium
Nannfeldtiomyces
Narasimhania
Neovossia
Oberwinkleria

Perichlamys (= Kuntzeomyces)
Polycystis (= Urocystis)
Polysaccopsis
Pseudodoassansia
Rhamphospora
Savulescuella
Schinzia (= Entorrhiza)
Setchellia (= Doassansia)
Stereosorus (= Burrillia)
Tilletia
Tilletiella
Tracya
Tuburcinia (= Urocystis)
Urocystis
Vossia (= Neovossia)

Ustilaginaceae
Angiosorus (= Thecaphora)
Anthracocystis (= Ustilago)
Anthracoidea
Bauhinus
Cintractia
Cintractiella
Cintractiomyxa (= Anthracoidea)
Crotalia
Crozalsiella
Dermatosorus
Elateromyces (= Farysia)
Farinaria (= Ustilago)
Farysia
Franzpetrakia
Granularia (= Urocystis)
Liroa
Macalpinomyces
Melanopsichium
Microbotryum (= Ustilago)
Milleria (= Testicularia)
Moesziomyces
Moreaua
Mundkurella
Mycosarcoma (= Ustilago)
Mycosyrinx
Necrosis (= Ustilago)
Orphanomyces
Pericladium
Pericoelium (= Ustilago)
Planetella
Poikilosporium (= Thecaphora)
Rhombiella
Schizonella
Sorosporium
Sphacelotheca
Sporisorium
Testicularia
Thecaphora
Tolyposporella (nom. inval.)
Tolyposporidium
Tolyposporium
Tranzscheliella
Tuburcinia (nom. dub.)
Ustacystis
Ustilagidium (= Ustilago)
Ustilago
Ustilentyloma
Whetzelia (= Ustacystis)
Xylosorium (= Pericladium)
Yenia (= Ustilago)
Zundeliomyces
Zundelula (= Dermatosorus)

Incertae sedis
Endothlaspis
Mycocoscoma

CHYTRIDIOMYCOTA

BLASTOCLADIALES
Blastocladiaceae
Allomyces
Blastocladia
Blastocladiella
Blastocladiopsis
Clavochytridium (=
  Blastocladiella)
Microallomyces
Rhopalomyces (= Blastocladiella)
Septocladia (= Allomyces)
Sphaerocladia (= Blastocladiella)

Catenariaceae
Catenaria
Catenomyces
Catenophlyctis
Perirhiza (= Catenophlyctis)

Coelomomycetaceae
Coelomomyces
Zografia (= Coelomomyces)

Physodermataceae
Oedomyces (= Physoderma)
Physoderma

Sorochytriaceae
Sorochytrium

Incertae sedis
Coelomycidium
Endoblastidium

CHYTRIDIALES
Chytridiaceae
Amphicypellus
Arnaudovia (= Polyphagus)
Blyttiomyces
Chytridium
Chytriomyces
Coralliochytrium (=
  Scherffeliomyces)
Cylindrochytrium
Dangeardia
Dangeardiana
Diplochytridium
Diplochytrium (= Diplochytridium)
Ectochytridium (= Zygorhizidium)
Hapalopera (= Phlyctidium)
Karlingiomyces
Loborhiza
Macrochytrium
Mastigochytrium (=
  Rhizophydium)
Nowakowskia
Obelidium
Phlyctidium (= Rhizophydium)
Phlyctochytrium
Phlyctorhiza
Physorhizophidium
Podochytrium
Polyphagus
Polyphlyctis
Pseudopileum
Rhizidiopsis (= Podochytrium)

Rhizidium
Rhizoclosmatium
Rhizophydium
Rhizophyton (= Rhizophydium)
Rhopalophlyctis
Saccomyces
Scherffelia (= Scherffeliomyces)
Scherffeliomyces
Scherffeliomycopsis
Septocarpus (= Podochytrium)
Septosperma
Siphonaria
Solutoparies
Sparrowia
Sphaerostylidium (=
  Rhizophydium)
Sporophlyctidium
Sporophlyctis
Tylochytrium (= Phlyctidium)
Zygorhizidium

Cladochytriaceae
Amoebochytrium
Cladochytrium
Coenomyces
Deckenbachia (= Coenomyces)
Lacustromyces
Megachytrium
Nephromyces
Nowakowskiella
Physocladia
Polychytrium
Pyroctonum (= Cladochytrium)
Saccopodium
Septochytrium

Endochytriaceae
Allochytridium
Canteria
Catenochytridium
Diplophlyctis
Endochytrium
Endocoenobium
Entophlyctis
Mitochytridium
Nephrochytrium
Truittella

Synchytriaceae
Carpenterella
Chrysophlyctis (= Synchytrium)
Endodesmidium
Johnkarlingia
Micromyces
Micromycopsis (= Micromyces)
Miyabella (= Synchytrium)
Pycnochytrium (= Synchytrium)
Synchytrium
Woroninella (= Synchytrium)

Incertae sedis
Achlyella
Achlyogeton
Amoebosporus
Aphanistis
Asterophlyctis
Bicricium (nom. conf.)
Chytridiopsis
Dermomycoides
Gamolpidium
Haplocystis
Ichthyochytrium
Microphlyctis (nom. dub.)

Mucophilus
Myiophagus
Myrophagus (= Myiophagus)
Nephromyces (= Rhizidiocystis)
Rhizidiocystis
Rhizosiphon
Sagittospora
Trematophlyctis

HARPOCHYTRIALES
Harpochytriaceae
Fulminaria (= Harpochytrium)
Harpochytrium
Rhabdium (= Harpochytrium)

MONOBLEPHARIDALES
Gonapodyaceae
Gonapodya
Monoblepharella

Monoblepharidaceae
Diblepharis (= Monoblepharis)
Monoblephariopsis (=
  Monoblepharis)
Monoblepharis

Oedogoniomycetaceae
Oedogoniomyces

NEOCALLIMASTIGALES
Neocallimastigaceae
Anaeromyces
Caecomyces
Neocallimastix
Orpinomyces
Piromyces
Ruminomyces (= Anaeromyces)

SPIZELLOMYCETALES
Caulochytriaceae
Caulochytrium

Olpidiaceae
Asterocystis (= Olpidium)
Chytridhaema
Cibdelia
Cyphidium (= Olpidium)
Endolpidium (= Olpidium)
Monochytrium (= Olpidium)
Morella
Nucleophaga
Olpidiaster (= Asterocystis)
Olpidiella (= Olpidium)
Olpidium

Spizellomycetaceae
Gaertneriomyces
Karlingia
Kochiomyces
Rhizophlyctis
Spizellomyces
Triparticalcar

Urophlyctidaceae
Urophlyctis

GENERA INCERTAE SEDIS
Bertramia
Chytridioides
Endospora (nom. dub.)
Piromonas
Sphaeromonas

Tetrachytrium
Zygochytrium

DICTYOSTELIOMYCOTA

DICTYOSTELIALES
Acytosteliaceae
Acytostelium

Dictyosteliaceae
Coenonia
Dictyostelium
Hyalostilbum (= Dictyostelium)
Polysphondylium

HYPHOCHYTRIOMYCOTA

HYPHOCHYTRIALES
Hyphochytriaceae
Canteriomyces
Cystochytrium
Hyphochytrium
Hyphophagus (= Hyphochytrium)

Rhizidiomycetaceae
Latrostium
Reessia
Rhizidiomyces
Rhizidiomycopsis (=
  Rhizidiomyces)

LABYRINTHULOMYCOTA

LABYRINTHULALES
Labyrinthulaceae
Chlamydomyxa (= Labyrinthula)
Labyrinthodictyon (= Labyrinthula)
Labyrinthomyxa (nom. conf.)
Labyrinthula
Pseudoplasmodium (=
  Labyrinthula)

Incertae sedis
Pyrrhosorus
Thanatostrea (nom. inval.)

THRAUSTOCHYTRIALES
Thraustochytriaceae
Althornia
Aplanochytrium
Corallochytrium
Diplophrys
Elina
Japonochytrium
Labyrinthuloides
Schizochytrium
Thraustochytrium
Ulkenia

GENERA INCERTAE SEDIS
Sorodiplophrys
Rostafinskia

MYXOMYCOTA

ECHINOSTELIALES
Clastodermataceae
Barbeyella
Clastoderma

Echinosteliaceae
Echinostelium

Endodromia (= Echinostelium)
Heimerlia (nom. dub.)
Orthotricha (= Clastoderma)
Wingina (= Orthotricha)

ECHINOSTELIOPSIDALES
Echinosteliopsidaceae
Bursulla
Echinosteliopsis

LICEALES
Cribrariaceae
Cribraria
Dictydium (= Cribraria)
Heterodictyon (= Dictydium)
Lindbladia

Liceaceae
Cylichnium (= Licea)
Hymenobolina (= Licea)
Hymenobolus (= Hymenobolina)
Kleistobolus (= Licea)
Licea
Listerella
Orcadella (= Licea)
Pleiomorpha (= Licea)
Protoderma (= Licea)
Protodermium (= Licea)
Protodermodium (= Licea)

Lycogalaceae
Alwisia (= Tubifera)
Antonigeppia (= Lycogala)
Clathroptychium (=
  Dictydiaethalium)
Dermodium (= Lycogala)
Dictydiaethalium
Diphtherium (= Lycogala)
Enteridium
Galoperdon (= Lycogala)
Licaethalium (= Reticularia)
Liceopsis (= Reticularia)
Lycogala
Lycoperdon (= Lycogala)
Ophiuridium (= Dictydiaethalium)
Reticularia (= Enteridium)
Siphoptychium (= Tubifera)
Strongylium (= Reticularia)
Tubifera
Tubularia (= Tubifera)
Tubulifera (= Tubifera)
Tubulina (= Tubifera)
Verrucosia (= Lycogala)

PHYSARALES
Didymiaceae
Amphisporium (= Didymium)
Carcerina (= Diderma)
Chondrioderma (= Diderma)
Cionium (= Didymium)
Diderma
Didymium
Leangium (= Diderma)
Lepidoderma
Lepidodermopsis (= Didymium)
Lignyota (nom. dub.)
Mucilago
Physarina
Polyschismium (= Diderma)
Spumaria (= Mucilago)
Trabrooksia
Wilczekia (= Diderma)

Physaraceae
Aethaliopsis (= Fuligo)
Aethalium (= Fuligo)
Angioridium (= Physarum)
Badhamia
Badhamiopsis (= Badhamia)
Cienkowskia
Claustria (= Physarum)
Crateriachea (= Physarum)
Craterium
Cupularia (= Craterium)
Cytidium (= Physarum)
Dichosporium (nom. dub.)
Erionema
Fuligo
Iocraterium (= Craterium)
Leocarpus
Lignydium (= Fuligo)
Physarella
Physarum
Pittocarpium (= Fuligo)
Protophysarum
Scyphium (= Badhamia)
Sphaerocarpa (= Craterium)
Sphaerocarpus (= Physarum)
Stylonites (= Physarum)
Tilmadoche (= Physarum)
Trichamphora (= Physarum)
Tripotrichia (= Leocarpus)
Willkommlangea

Incertae sedis
Kelleromyxa

STEMONITALES
Stemonitidaceae
Amaurochaete
Ancyrophorus (= Enerthenema)
Brefeldia
Clathroidastrum (= Stemonitis)
Clathroidastrum (= Stemonitis)
Collaria (= Comatricha)
Colloderma
Comatricha
Comatrichoides (= Comatricha)
Coscinium (= Lamproderma)
Diachaeella (= Diachea)
Diachea
Diacheopsis
Elaeomyxa
Enerthenema
Lamproderma
Leptoderma
Macbrideola
Matruchotia (= Amaurochaete)
Matruchotiella (= Amaurochaete)
Paradiachea (= Comatricha)
Paradiacheopsis (= Macbrideola)
Raciborskia (= Comatricha)
Rostafinskia (= Raciborskia)
Schenella
Stemonaria
Stemonitis
Stemonitopsis
Symphytocarpus

TRICHIALES
Arcyriaceae
Arcyrella (= Arcyria)
Arcyria
Clathroides (= Arcyria)
Cornuvia
Heterotrichia (= Arcyria)

Lachnobolus (nom. conf.)
Myrosporium (= Lachnobolus)
Nassula (= Arcyria)

Dianemataceae
Calomyxa
Dianema
Dianemina (= Dianema)
Lamprodermopsis (= Dianema)
Margarita (= Calomyxa)

Trichiaceae
Arcyodes
Arcyriatella
Calonema
Hemiarcyria (= Hemitrichia)
Hemitrichia
Hyporhamma (= Hemitrichia)
Lachnobolus (= Arcyodes)
Metatrichia
Minakatella
Oligonema
Ophiotheca (= Perichaena)
Perichaena
Prototrichia
Pyxidium (= Perichaena)
Stegasma (= Perichaena)
Trichia
Trichidium (= Trichia)
Trichulius (= Trichia)

GENERA INCERTAE SEDIS
Amylotrogus (nom. dub.)
Calyssosporium (nom. dub.)
Coccospora (nom. dub.)
Coniocephalum (nom. dub.)
Cribraria (nom. dub.)
Demordium
Dermodium (nom. conf.)
Disporium
Halterophora (nom. dub.)
Hystricapsa (nom. dub.)
Jundzillia (nom. dub.)
Lepidodermopsis
Mesenterica (nom. dub.)
Microcarpon (nom. dub.)
Nidularia (nom. dub.)
Ostracococcum (nom. dub.)
Pecila (nom. dub.)
Phlebomorpha (nom. dub.)
Phytoceratiomyxa (nom. dub.)
Polyangium (nom. dub.)
Reticulomyxa
Scoriomyces (nom. dub.)
Semimorula
Sporigastrum (nom. dub.)
Squamuloderma (= Didymium)
Stemonitis (nom. dub.)
Tipularia (= Halterophora)
Trichoderma (nom. dub.)
Xyloidium (nom. dub.)
Calospeira (nom. inval.)
Embolus (nom. dub.)
Enigma (nom. inval.)
Heliomycopsis (nom. inval.)
Pygmomyces (nom. inval.)
Rhabdocystis (nom. inval.)

PROTOSTELIALES
Cavosteliaceae
Cavostelium
Ceratiomyxella

Planoprotostelium
Protosporangium

Ceratiomyxaceae
Ceratiomyxa
Ceratiopsis (= Ceratiomyxa)
Ceratium (= Ceratiomyxa)
Famintzinia (= Ceratiomyxa)

Protosteliaceae
Clastostelium
Endostelium
Microglomus
Nematostelium
Protosteliopsis
Protostelium
Schizoplasmodiopsis
Schizoplasmodium
Tychosporium

OOMYCOTA

LAGENISMALES
Lagenismaceae
Lagenisma

LEPTOMITALES
Apodachlyellaceae
Apodachlyella
Eurychasmopsis

Ducellieriaceae
Ducellieria

Leptolegniellaceae
Aphanodictyon
Aphanomycopsis
Brevilegniella
Leptolegniella
Nematophthora

Leptomitaceae
Apodachlya
Apodya (= Leptomitus)
Leptomitus
Plerogone

Incertae sedis
Blastulidiopsis (= Blastulidium)
Blastulidium
Cornumyces

MYZOCYTIOPSIDALES
Crypticolaceae
Crypticola

Ectrogellaceae
Cymbanche (= Ectrogella)
Ectrogella
Haptoglossa

Eurychasmaceae
Eurychasma
Eurychasmidium

Myzocytiopsidaceae
Chlamydomyzium
Gonimochaete
Myzocytiopsis
Protascus (nom. conf.)
Resticularia (= Syzygangia)

Septolpidium
Syzygangia

Pontismaceae
Petersenia
Pontisma

Sirolpidiaceae
Pseudolpidiella (= Sirolpidium)
Sirolpidium

OLPIDIOPSIDALES
Olpidiopsidaceae
Bicilium (= Olpidiopsis)
Diplophysa (= Olpidiopsis)
Olpidiopsis
Peronium (= Olpidiopsis)
Pleocystidium
Pseudolpidiopsis (= Pleocystidium)
Pseudolpidium (= Olpidiopsis)

Incertae sedis
Gracea

PERONOSPORALES
Albuginaceae
Albugo
Cystopus (= Albugo)

Peronosporaceae
Actinobotrys (= Bremia)
Basidiophora
Bremia
Bremiella
Chlorospora (= Peronospora)
Dicksonomyces (nom. conf.)
Gilletia (= Basidiophora)
Paraperonospora
Peronoplasmopara (=
  Pseudoperonospora)
Peronospora
Plasmopara
Pseudoperonospora
Pseudoplasmopara (= Plasmopara)
Rhysotheca (= Plasmopara)
Tetradium (= Bremia)

PYTHIALES
Pythiaceae
Artotrogus (= Pythium)
Blepharospora (= Phytophthora)
Cystosiphon
Diasporangium
Eupythium (= Pythium)
Halophytophthora
Kawakamia (= Phytophthora)
Lagenidiopsis (= Cystosiphon)
Lagenidium
Lucidium (nom. dub.)
Mycelophagus (= Phytophthora)
Myzocytium
Nematosporangium (= Pythium)
Nozemia (= Phytophthora)
Peronophythora
Phloeophthora (= Phytophthora)
Phytophthora
Pythiacystis (= Phytophthora)
Pythiogeton
Pythiomorpha (= Phytophthora)
Pythium
Rheosporangium (= Pythium)

Sphaerosporangium (=
Phytophthora)
Trachysphaera

Incertae sedis
Achlyopsis (nom. dub.)
Endosphaerium

RHIPIDIALES
Rhipidiaceae
Aqualinderella
Araiospora
Mindeniella
Naegelia (= Sapromyces)
Naegeliella (= Sapromyces)
Nellymyces
Rhipidium
Sapromyces

SALILAGENIDIALES
Haliphthoraceae
Atkinsiella
Haliphthoros

Salilagenidiaceae
Salilagenidium

SAPROLEGNIALES
Saprolegniaceae
Achlya
Aphanomyces
Aplanes
Aplanopsis
Archilegnia (nom. conf.)
Brevilegnia
Calyptralegnia
Cladolegnia (= Saprolegnia)
Couchia
Dictyuchus
Diplanes (= Saprolegnia)
Geolegnia
Hamidia (nom. dub.)
Hydatinophagus
Hydronema (= Achlya)
Isoachlya
Jaraia (nom. dub.)
Leptolegnia
Plectospira
Pringsheimina (= Achlya)
Protoachlya
Pythiopsis
Saprolegnia
Scoliolegnia
Sommerstorffia
Thraustotheca

Incertae sedis
Branchiomyces
Pythium (nom. rej.)
Trichothrauma (nom. conf.)

SCLEROSPORALES
Sclerosporaceae
Peronosclerospora (=
Peronosclerospora)
Peronosclerospora
Sclerospora

Verrucalvaceae
Pachymetra
Sclerophthora
Verrucalvus

FAMILIA INCERTAE SEDIS
Anisolpidiaceae
Anisolpidium

Lagenaceae
Ciliomyces
Lagena
Lagenocystis (= Lagena)
Pythiella

Genera Incertae sedis
Lagenidicopsis (nom. dub.)

ORDO INCERTAE SEDIS

ROZELLOPSIDALES
Pseudosphaeritaceae
Pseudosphaerita
Sphaerita

Rozellopsidaceae
Dictyomorpha
Pleolpidium (= Rozella)
Pringsheimiella (= Dictyomorpha)
Rozella
Rozellopsis
Rozia (= Rozella)
Skirgiellia (= Rozella)
Skirgiellopsis (= Rozellopsis)

Incertae sedis
Olpidiomorpha
Plasmophagus
Pleotrachelus

PLASMODIOPHOROMYCOTA

PLASMODIOPHORALES
Endemosarcaceae
Endemosarca

Plasmodiophoraceae
Anisomyxa (= Ligniera)
Clathrosorus (= Spongospora)
Frankia (= Frankiella)
Frankiella (= Plasmodiophora)
Ligniera
Membranosorus
Molliardia (= Tetramyxa)
Octomyxa
Ostenfeldiella (= Plasmodiophora)
Peltomyces
Phytomyxa (nom. dub.)
Plasmodiophora
Polymyxa
Rhizomyxa (= Ligniera)
Sorodiscus
Sorolpidium
Sorosphaera
Spongospora
Sporomyxa
Tetramyxa
Woronina

Incertae sedis
Acrocystis (nom. dub.)

Cystospora (nom. dub.)
Phagomyxa

ZYGOMYCOTA

TRICHOMYCETES

AMOEBIDIALES
Amoebidiaceae
Amoebidium
Paramoebidium

ASELLARIALES
Asellariaceae
Asellaria
Orchesellaria
Recticharella (= Asellaria)

ECCRINALES
Eccrinaceae
Alacrinella
Andohaheloa (= Enterobryus)
Arundinella (= Arundinula)
Arundinula
Astreptonema
Capillus (= Enterobryus)
Cestodella (= Enterobryus)
Daloala (= Enterobryus)
Eccrina (= Enterobryus)
Eccrinella (= Astreptonema)
Eccrinidus
Eccrinoides
Eccrinopsis (= Enterobryus)
Enterobryus
Enteromyces
Enteropogon
Lactella (= Enterobryus)
Paramacrinella
Paratrichella (= Enterobryus)
Pistillaria (= Enterobryus)
Ramacrinella
Recticoma (= Enterobryus)
Taeniella
Taeniellopsis
Trichella (= Enterobryus)
Trichellopsis (= Enterobryus)

Palavasciaceae
Palavascia

Parataeniellaceae
Lajassiella
Nodocrinella
Parataeniella

HARPELLALES
Harpellaceae
Carouxella
Furculomyces
Harpella
Harpellomyces
Stachylina

Legeriomycetaceae
Allantomyces
Austrosmittium
Bojamyces
Capniomyces
Caudomyces
Dixidium (= Smittium)
Ejectosporus
Gauthieromyces

Genistella (= Legeriomyces)
Genistelloides
Genistellospora
Glotzia
Graminella
Legeriomyces
Orphella
Pennella
Pteromaktron
Rubetella (= Smittium)
Simuliomyces
Smittium
Spartiella
Stipella
Trichoceridium (= Smittium)
Trichozygospora
Zygopolaris

Incertae sedis
Opuntiella (nom. dub.)
Typhella (nom. dub.)

GENERA INCERTAE SEDIS
Microasellaria (nom. dub.)
Mononema
Spirogyromyces

ZYGOMYCETES

DIMARGARITALES
Dimargaritaceae
Dimargaris
Dispira
Tieghemiomyces

Incertae sedis
Spinalia

ENDOGONALES
Endogonaceae
Endogone
Sclerogone
Youngiomyces

ENTOMOPHTHORALES
Ancylistaceae
Ancylistes
Boudierella (= Delacroixia)
Conidiobolus
Delacroixia (= Conidiobolus)
Macrobiotophthora

Basidiobolaceae
Amphoromorpha (= Basidiobolus)
Basidiobolus

Completoriaceae
Completoria

Entomophthoraceae
Batkoa
Culicicola (= Lamia)
Empusa (= Entomophthora)
Entomophaga
Entomophthora (= Entomophthora)
Entomophthora
Erynia
Eryniopsis
Furia
Lamia (= Entomophthora)
Massospora
Myiophyton (= Empusa)
Pandora

Strongwellsea
Tarichium
Triplosporium (= Entomophthora)
Zoophthora

Meristacraceae
Ballocephala
Botryobolus (= Ballocephala)
Meristacrum
Pseudocoelomomyces (= Tabanomyces)
Tabanomyces (= Meristacrum)
Zygnemomyces

Neozygitaceae
Neozygites
Thaxterosporium

Incertae sedis
Loboa
Zygaenobia

GLOMALES
Acaulosporaceae
Acaulospora
Entrophospora

Gigasporaceae
Gigaspora
Scutellospora

Glomaceae
Ackermannia (= Sclerocystis)
Glomus
Rhizophagus (= Glomus)
Sclerocystis
Sphaerocreas (= Glomus)
Stigeosporium (= Glomus)
Xenomyces (= Sclerocystis)

KICKXELLALES
Kickxellaceae
Coemansia
Coemansiella (= Kickxella)
Coronella (= Kickxella)
Dipsacomyces
Kickxella
Linderina
Martensella
Martensiomyces
Spirodactylon
Spiromyces

MUCORALES
Chaetocladiaceae
Chaetocladium
Dichotomocladium
Hilitzeria (= Dichotomocladium)

Choanephoraceae
Blakeslea
Choanephora
Choanephorella (= Choanephora)
Choanephoroidea (= Choanephora)
Cunninghamia (= Choanephora)
Poitrasia

Cunninghamellaceae
Actinocephalum (= Cunninghamella)
Cunninghamella
Muratella (= Cunninghamella)
Saitomyces (= Actinocephalum)

Gilbertellaceae
Gilbertella

Mortierellaceae
Actinomortierella (= Mortierella)
Aquamortierella
Azygozygum (= Mortierella)
Carnoya (= Mortierella)
Dissophora
Echinosporangium
Haplosporangium (= Mortierella)
Herpocladium (nom. dub.)
Micromucor
Modicella
Mortierella
Naumoviella (= Mortierella)
Umbelopsis

Mucoraceae
Absidia
Actinomucor
Amylomyces
Apophysomyces
Ascidiophora (= Mucor)
Ascophora (= Mucor)
Azygites (= Syzygites)
Calyptromyces (= Mucor)
Chionyphe (= Mucor)
Chlamydoabsidia
Chlamydomucor (= Mucor)
Circinella
Circinomucor (= Mucor)
Circinumbella (= Circinella)
Crinofera (= Pilophora)
Glomerula (= Actinomucor)
Gongronella
Halteromyces
Herpocladiella (nom. dub.)
Hydrophora (= Mucor)
Hyphomucor
Lactomyces (= Mucor)
Lichtheimia (= Absidia)
Micromucor (= Micromucor)
Mucedo (= Mucor)
Mucor
Mucor (= Rhizopus)
Mycocladus (= Absidia)
Parasitella
Philophora (= Rhizopus)
Pilophora (= Rhizopus)
Proabsidia (= Absidia)
Protoabsidia (= Absidia)
Protomycocladus
Pseudoabsidia (= Absidia)
Rhizomucor
Rhizopodopsis
Rhizopus
Scitovszkya (= Mucor)
Spinellus
Sporodinia (= Syzygites)
Sporodiniella
Stilbodendrum (= Syzygites)
Syzygites
Thelactis (= Mucor)
Thermomucor
Tieghemella (= Absidia)
Zygorhynchus

Mycotyphaceae
Benjaminia (= Benjaminiella)
Benjaminiella
Mycotypha

Phycomycetaceae
Phycomyces

Pilobolaceae
Hydrogera (= Pilobolus)
Pilaira
Pilobolus
Pycnopodium (= Pilobolus)
Sacidium (= Pilobolus)
Utharomyces

Radiomycetaceae
Hesseltinella
Radiomyces
Radiomycopsis (= Radiomyces)

Saksenaeaceae
Saksenaea

Syncephalastraceae
Syncephalastrum

Thamnidiaceae
Backusella
Bulbothamnidium (= Helicostylum)
Chaetostylum (= Helicostylum)
Chordostylum (nom. dub.)
Cokeromyces
Dicranophora
Ellisomyces
Fennellomyces
Haynaldia (= Helicostylum)
Helicostylum
Hildebrandiella (= Syzygites)
Kirkia (= Kirkomyces)
Kirkomyces
Melidium (= Thamnidium)
Phascolomyces
Pirella
Pleurocystis (= Helicostylum)
Thamnidium
Thamnostylum
Zychaea

ZOOPAGALES
Cochlonemataceae
Amoebophilus
Aplectosoma
Cochlonema
Endocochlus
Euryancale

Helicocephalidaceae
Brachymyces
Helicocephalum
Rhopalomyces

Piptocephalidaceae
Acephalis (= Syncephalis)
Calvocephalis (= Syncephalis)
Clavocephalis (= Syncephalis)
Kuzuhaea
Microcephalis (= Syncephalis)
Monocephalis (= Syncephalis)
Mucoricola (= Piptocephalis)
Piptocephalis
Syncephalidium (= Syncephalis)
Syncephalis
Syncephalopsis (= Syncephalis)
Vantieghemia (= Syncephalis)

Sigmoideomycetaceae
Reticulocephalis

Sigmoideomyces
Thamnocephalis

Zoopagaceae
Acaulopage
Bdellospora
Cystopage
Lecythispora (= Stylopage)
Stylopage
Zoopage
Zoophagus

Incertae sedis
Lecophagus
Massartia

GENERA INCERTAE SEDIS
Adlerocystis
Blastocystis
Geosiphon
Ichthyophonus
Ichthyosporidium (=
   Ichthyophonus)

MITOSPORIC FUNGI
Abgliophragma
Abropelta
Acanthoderma
Acanthorus
Acanthothecium (= Ypsilonia)
Acarella
Acarellina
Acarocybe
Acarocybella
Acarocybellina
Acaropeltis
Acarosporium
Acaulium (= Scopulariopsis)
Acerviclypeatus
Achitonium (= Pactilia)
Achorion (= Trichophyton)
Achoropeltis
Aciculariella
Aciculoconidium
Aciesia
Acinula
Acladium
Acleistia
Acleistomyces
Acmosporium (= Aspergillus)
Acontiopsis (= Cylindrocladium)
Acontium
Acremoniella
Acremoniula (= Acremoniula)
Acremoniula
Acremonium
Acrocalymma
Acrocladium (= Periconiella)
Acroconidiella
Acroconidiellina
Acrocylindrium
Acrodesmis (= Periconiella)
Acrodictyopsis
Acrodictys
Acrodontium
Acrogenospora
Acrophialophora
Acrophragmis
Acrospeira
Acrospira
Acrosporella (= Cladosporium)
Acrosporium (= Oidium)

Acrosporium
Acrostalagmus (= Verticillium)
Acrostaphylus
Acrostaurus
Acrostroma
Acrotheca (= Pleurophragmium)
Acrotheciella
Acrothecium
Actinochaete (nom. conf.)
Actinochaete
Actinocladium
Actinodendron (= Oncocladium)
Actinodochium
Actinomma (= Atichia)
Actinonema (nom. dub.)
Actinonema (= Spilocaea)
Actinonemella (= Asteroma)
Actinopelte (= Tubakia)
Actinospora
Actinostilbe (= Sarcopodium)
Actinoteichus
Actinotexis
Actinothecium
Actinothyrella (nom. dub.)
Actinothyrium
Actinotrichum (nom. nud.)
Acumispora
Adea (= Seiridium)
Adella (= Wojnowicia)
Aderkomyces
Adhogamina (= Gilmaniella)
Aegerita
Aegeritella
Aegeritina
Aegeritopsis (= Aegerita)
Aerophyton (nom. dub.)
Agaricodochium
Agarwalia
Agarwalomyces
Aglaocephalum (= Pulchromyces)
Agyriella
Agyriellopsis
Ahmadia
Ahmadinula (= Truncatella)
Ajrekarella
Akanthomyces
Akenomyces (= Akenomyces)
Akenomyces
Alatosessilispora
Alatospora
Albophoma
Albosynnema
Aleurisma (= Trichoderma)
Aleurodomyces
Aleurophora (= Aleurisma)
Aleurosporia (= Trichophyton)
Algonquinia
Allantophoma (nom. inval.)
Allantophomopsis
Allantospora (= Cylindrocarpon)
Allantozythia (= Phlyctema)
Allantozythiella (= Endothiella)
Allelochaeta (= Seimatosporium)
Allescheria (= Hartigiella)
Allescheriella
Alliospora (= Aspergillus)
Allonema
Alloneottiosporina
Allosphaerium (= Rhizoctonia)
Allothyriella
Allothyrina
Allothyriopsis
Alpakesa

Alphitomyces (= Paecilomyces)
Alternaria
Alveophoma
Alysia (= Vouauxiella)
Alysidiopsis
Alysidium
Alysisporium (= Phragmotrichum)
Amallospora
Amarenographium
Amastigis (= Amastigosporium)
Amastigosporium (= Mastigosporium)
Amblyosporiopsis (= Oedocephalum)
Amblyosporium
Ambrosiella
Amerodiscosiella
Amerodiscosiellina
Ameropeltomyces
Amerosporiella
Amerosporina (= Amerosporium)
Amerosporiopsis
Amerosporis (= Amerosporiella)
Amerosporium
Amoebomyces
Ampelomyces
Amphiblistrum (= Oidium)
Amphichaeta (= Seimatosporium)
Amphichaete (= Amphichaetella)
Amphichaetella
Amphichorda (= Isaria)
Amphiciliella (nom. dub.)
Amphicytostroma
Amphiernia (= Sporobolomyces)
Amphiloma (= Lepraria)
Amphitiarospora (= Dinemasporium)
Amphobotrys
Amphoropycnium
Amphorula (= Chaetoconis)
Ampullifera
Ampulliferella (= Ampullifera)
Ampulliferina
Ampulliferopsis (= Ampullifera)
Anaphysmene
Anarhyma
Anavirga
Anconomyces
Ancoraspora
Ancylospora (= Pseudocercospora)
Andreaea (= Andreaeana)
Andreaeana
Anematidium
Angiopoma (= Drechslera)
Angiopomopsis
Anguillospora
Angulimaya
Angulospora
Ankistrocladium (= Casaresia)
Annellodentimyces
Annellodochium
Annellolacinia
Annellophora
Annellophorella
Annellophragmia
Anodotrichum (= Blastotrichum)
Anomomyces (nom. dub.)
Ansatospora (= Mycocentrospora)
Antenaglium
Antennariella
Antennatula
Antennopsis
Anthasthoopa (= Coniella)

Anthina
Anthopsis
Anthracoderma
Antimanopsis (= Monostichella)
Antipodium
Antromyces
Antromycopsis
Anulohypha
Anulosporium (= Dactylaria)
Anungitea
Anungitopsis
Aorate (= Titaea)
Aoria
Aphanocladium
Aphanofalx
Aphotistus (= Rhizomorpha)
Apiocarpella
Apiosporella (= Aplosporidium)
Apiosporella (= Apiocarpella)
Apiosporium (= Sclerotium)
Apiotrichum (= Candida)
Aplosporella
Aplosporidium (= Asteromella)
Apocoryneum (= Massariothea)
Apocytospora (= Plectophomella)
Apogloeum
Apomelasmia
Aporella (= Aporellula)
Aporellula
Aposphaeria
Aposphaeria (nom. rej.)
Aposporella
Apostrasseria
Apotemnoum (= Clasterosporium)
Appelia (= Trichoconis)
Apyrenium (nom. dub.)
Arachnophora
Araneomyces
Arborispora
Arbuscula (= Neoarbuscula)
Arbusculidium (= Neoarbuscula)
Arbusculina
Arcuadendron
Ardhachandra (= Rhinocladiella)
Argopericonia
Aristadiplodia (nom. dub.)
Aristastoma
Arnaudina
Arnoldia (= Arnoldiomyces)
Arnoldiella
Arnoldiomyces
Arthrinium
Arthrobotryella
Arthrobotryomyces
Arthrobotrys
Arthrobotryum (= Gonyella)
Arthrobotryum
Arthrocladium
Arthrocristula
Arthrodochium
Arthrographis
Arthrographium (= Arthrobotryum)
Arthropsis
Arthropycnis
Arthrosporia (= Trichophyton)
Arthrosporium
Articularia
Articulis (= Articulariella)
Articulophora
Articulospora
Arualis
Arxiella

Arxula
Asbolisia (nom. dub.)
Asbolisia
Asbolisiomyces
Aschersonia
Aschersoniopsis (= Munkia)
Aschizotrichum (= Wiesneriomyces)
Ascochyta
Ascochytella (= Ascochyta)
Ascochytopsis
Ascochytula (= Ascochyta)
Ascochytulina
Ascoconidium
Ascorhizoctonia
Ascospora (nom. conf.)
Ashtaangam
Aspergilloides (= Penicillium)
Aspergillopsis (= Aspergillus)
Aspergillopsis (nom. dub.)
Aspergillus
Asperisporium
Aspilaima
Asporomyces (= Torulopsis)
Asporothrichum (nom. dub.)
Astelechia
Asterinothyriella
Asterinothyrium
Asterinula (= Leptothyrella)
Asterobolus (= Valdensia)
Asteroconium
Asteroma
Asteromella
Asteromellopsis
Asteromidium
Asteromyces
Asteronectrioidea
Asterophoma
Asterophora
Asteropsis
Asteroscutula
Asterosporium
Asterostomella
Asterostomidium (= Asteromidium)
Asterostomopora
Asterostomopsis
Asterostomula
Asterostomulina
Asterostromina
Asterothecium (= Stephanoma)
Asterothyrium (= Septothyrella)
Asterotrichum (= Asterophora)
Astoma (= Sclerotium)
Astrabomyces
Astragoxyphium (= Leptoxyphium)
Astrodochium
Astronatelia
Ateleothylax
Atelosaccharomyces (= Cryptococcus)
Atichia
Atractilina
Atractina (= Sterigmatobotrys)
Atractium
Atrichophyton (= Chrysosporium)
Atroseptaphiale
Attamyces
Auerswaldiopsis (= Patouillardiella)
Aulographopsis (nom. nud.)
Aurantiosacculus
Aureobasidium
Aureobasis (= Aureobasidium)

Avettaea
Azosma (= Cladosporium)
Azymocandida (= Candida)
Azymoprocandida (= Candida)
Bachmanniomyces
Bacillispora
Bacillopeltis
Bactrexcipula (= Rhizothyrium)
Bactridiopsis (= Coccospora)
Bactridiopsis (= Phillipsiella)
Bactridium
Bactrodesmiastrum
Bactrodesmiella
Bactrodesmium
Bactropycnis (= Coleophoma)
Badarisama
Baeumleria (= Coniella)
Bahuchashaka
Bahugada
Bahukalasa
Bahupaathra (= Cladorrhinum)
Bahusaganda
Bahusakala
Bahusandhika
Bahusutrabeeja
Bainieria
Bakerophoma (nom. dub.)
Balaniopsis
Balanium
Ballistosporomyces
Barbarosporina
Bargellinia (= Wallemia)
Barklayella (= Polynema)
Barnettella
Bartalinia
Bartaliniopsis (= Doliomyces)
Bartheletia
Baryeidamia (= Papulaspora)
Basauxia
Basiascella (= Piggottia)
Basiascum (= Spilocaea)
Basididyma
Basidiella (= Aspergillus)
Basidiobotrys (= Xylocladium)
Basidiolum (nom. dub.)
Basifimbria
Basilocula (= Ceuthospora)
Basipetospora
Basipilus (= Seimatosporium)
Basisporium (= Nigrospora)
Basitorula (= Gliomastix)
Basramyces
Batcheloromyces (= Stigmina)
Batistina
Beauveria
Beauveriphora
Beccopycnidium
Beejadwaya
Beejasamuha
Belaina (= Polynema)
Belainopsis (nom. dub.)
Belemnospora
Bellulicauda
Beltrania
Beltraniella
Beltraniopsis
Benekea (= Geastrumia)
Beniowskia
Benjaminia
Benjpalia
Bensingtonia
Berkeleyna (= Cephalotrichum)
Berkleasmium

Berteromyces (= Cercosporidium)
Beverwykella
Bhargavaella
Biatoridina (= Epithyrium)
Bibanasiella
Bidenticula (= Fusarium)
Biflagellospora
Biflagellosporella
Biharia (= Stenella)
Bilboque
Bilgramia
Bimeris
Bioconiosporium
Biophomopsis
Bipolaris
Bisbyopeltis
Bispora
Bisporomyces (= Chloridium)
Bisporostilbella
Bisseomyces
Bitunicostilbe
Bizozzeriella
Blarneya
Blastacervulus
Blastobotrys
Blastocatena
Blastoconium
Blastodendrion (= Candida)
Blastoderma (nom. ambig.)
Blastodictys
Blastomyces (= Histoplasma)
Blastomyces (= Zymonema)
Blastomycoides (= Zymonema)
Blastophoma (= Sclerophoma)
Blastophorella
Blastophorum
Blastophragma
Blastoschizomyces
Blastosporidium (= Coccidioides)
Blastotrichum
Blennorella (= Colletotrichum)
Blennoria
Blennoriopsis
Blepharia (= Dematium)
Bleptosporium
Blistum (= Polycephalomyces)
Blodgettia (= Blodgettia)
Blodgettia
Blodgettiomyces (= Blodgettia)
Bloxamia
Bodinia (= Trichophyton)
Bomplandiella
Bonordeniella (= Coniosporium)
Bostrichonema
Bostrychia (= Cytospora)
Bothrodiscus
Botrydiella (= Staphylotrichum)
Botrydiplis (= Botryodiplodia)
Botryella (= Sphaerellopsis)
Botryocladium (= Nematogonum)
Botryocrea (= Fusarium)
Botryoderma
Botryodiplis (= Botryodiplodia)
Botryodiplodia
Botryodiplodina
Botryogene (nom. dub.)
Botryomonilia
Botryomyces (= Ybotryomyces)
Botryomyces
Botryonipha (nom. rej. prop.)
Botryophialophora (= Myrioconium)
Botryophoma (= Sclerodothiorella)

Botryosphaerostroma (= Sphaeropsis)
Botryosphaerostroma (= Botryodiplodia)
Botryosporium
Botryosporium (= Dictyosporium)
Botryotrichum
Botryoxylon (= Conoplea)
Botryozyma
Botrypes (= Ciliciopodium)
Botrysphaeris (= Sphaeropsis)
Botrytis
Botrytoides (= Rhinocladiella)
Bovetia
Brachiosphaera
Brachycladium (= Dendryphion)
Brachyconidiella
Brachydesmiella
Brachydesmium (= Clasterosporium)
Brachyhelicoon
Brachysporiella
Brachysporiellina
Brachysporium
Brefeldiopycnis
Brencklea (= Scolecosporiella)
Brettanomyces
Briania
Briarea (= Aspergillus)
Briosia
Bromicolla (= Sclerotium)
Broomeola
Brunchorstia
Brycekendrickia
Bryochysium (= Rhizoctonia)
Bryophytomyces
Bullaserpens
Bullera
Burgoa
Byssocladiella (= Byssocladium)
Byssocladium (nom. dub.)
Byssocystis (= Ampelomyces)
Cacahualia (= Arachnophora)
Cacumisporium
Cadophora (= Phialophora)
Calcarispora
Calcarisporiella
Calcarisporium
Calceispora
Caldariomyces (= Leptoxyphium)
Callistospora
Callosisperma (nom. dub.)
Calocline
Calogloeum (= Fusamen)
Calopactis (= Endothiella)
Calostilbella
Camaroglobulus
Camarographium
Camaropycnis
Camarosporellum
Camarosporium
Camarosporula
Camarosporulum
Camposporidium
Camposporium
Camptomeris
Camptosporium (nom. dub.)
Camptoum (= Arthrinium)
Campylospora
Canalisporium
Cancellidium
Candelabrella
Candelabrum

Candelospora (= Cylindrocladium)
Candelosynnema
Candida
Capillaria
Capitorostrum
Capnobotryella
Capnobotrys
Capnocheirides
Capnocybe
Capnodendron
Capnodiastrum
Capnogoniella (nom. conf.)
Capnokyma
Capnophialophora
Capnosporium
Capnostysanus (= Stysanus)
Capsicumyces
Carlosia (= Isthmospora)
Carmichaelia
Carrionia (= Rhinocladiella)
Casaresia
Castellania (= Candida)
Catenella
Catenomycopsis
Catenophora
Catenophoropsis
Catenospegazzinia
Catenosubulispora
Catenularia
Catenulaster
Catenuloxyphium (nom. dub.)
Catinopeltis
Catinula
Catosphaeropsis (= Sphaeropsis)
Cattanea (= Dictyosporium)
Caudophoma (= Phyllosticta)
Caudosporella (= Harknessia)
Cellulosporium (nom. dub.)
Cenangiomyces
Cenococcum
Centrospora (= Mycocentrospora)
Cephaliophora
Cephalocladium (= Botrytis)
Cephalodiplosporium
Cephalodochium
Cephaloedium (= Exosporium)
Cephalomyces (= Cephaliophora)
Cephalophorum (nom. dub.)
Cephalosporiopsis
Cephalosporium (= Acremonium)
Cephalothecium (= Trichothecium)
Cephalotrichum
Cephalotrichum (=
   Trichocephalum)
Ceracea
Ceratocladium (= Xylocladium)
Ceratocladium
Ceratonema (nom. dub.)
Ceratophoma
Ceratophorum
Ceratopodium (= Graphium)
Ceratopycnidium
Ceratopycnis
Ceratopycnium (=
   Ceratopycnidium)
Ceratorhiza
Ceratosporella
Ceratosporium
Cercocladospora (=
   Pseudocercospora)
Cercodeuterospora (=
   Mycovellosiella)

Cercoseptoria (=
   Pseudocercospora)
Cercosperma
Cercospora (= Pseudocercospora)
Cercospora
Cercosporella
Cercosporidium
Cercosporina (= Cercospora)
Cercosporiopsis (=
   Pseudocercospora)
Cercosporula
Cercostigmina
Cerebella (= Epicoccum)
Cerinosterus
Ceriomyces (= Ptychogaster)
Cesatia (= Trullula)
Ceuthodiplospora
Ceuthosira
Ceuthospora
Ceuthosporella
Ceuthosporella (= Helhonia)
Chaetalysis (= Acarosporium)
Chaetantromycopsis
Chaetasbolisia
Chaetendophragmia
Chaetendophragmiopsis
Chaetobasidiella (nom. dub.)
Chaetobasis (= Chaetobasidiella)
Chaetoblastophorum
Chaetochalara
Chaetoconidium
Chaetoconis
Chaetocytostroma
Chaetodiplis
Chaetodiplodia
Chaetodiplodina
Chaetodiscula (= Hymenopsis)
Chaetodochis (= Chaetostroma)
Chaetodochium (= Volutella)
Chaetomelasmia (= Diachorella)
Chaetomella
Chaetomonodorus
Chaetopatella (= Pseudolachnea)
Chaetopeltaster
Chaetopeltiopsis (nom. dub.)
Chaetopeltis (= Tassia)
Chaetophiophoma
Chaetophoma
Chaetophomella (nom. conf.)
Chaetopsella (= Chaetopsis)
Chaetopsina
Chaetopsis
Chaetopyrena
Chaetosclerophoma
Chaetoseptoria
Chaetosira (= Wiesneriomyces)
Chaetospermella (=
   Chaetospermum)
Chaetospermella (=
   Spermochaetella)
Chaetospermopsis
Chaetospermum
Chaetosphaeronema
Chaetosphaeropsis (=
   Coniothyriopsis)
Chaetospora (= Neochaetospora)
Chaetosticta
Chaetostroma
Chaetostromella (nom. conf.)
Chaetothyriolum (nom. dub.)
Chaetotrichum (= Annellophora)
Chaetozythia (nom. non fung.)
Chailletia (= Truncatella)

Chalara
Chalarodendron
Chalarodes
Chalaropsis (= Chalara)
Chantransiopsis
Characonidia
Chardonia
Charomyces
Chaunopycnis
Cheilaria
Cheilariopsis (= Apomelasmia)
Cheiroconium (= Sirothecium)
Cheiromoniliophora
Cheiromycella
Cheiromyceopsis
Cheiromyces
Cheiromycina
Cheiropodium (= Clasterosporium)
Cheiropolyschema
Cheirospora
Chelisporium (= Sirothecium)
Chiastospora
Chikaneea
Chionomyces
Chithramia
Chlamydoaleurospora (=
   Trichophyton)
Chlamydomyces
Chlamydopsis
Chlamydorubra
Chlamydosporium (= Phoma)
Chlamydotomus (nom. non fung.)
Chloridiella (= Idriella)
Chloridium
Chlorocyphella (= Pyrenotrichum)
Chmelia
Choanatiara
Chondroplea (= Discosporium)
Chondropodiella
Chondropodiola (nom. dub.)
Chondropodium (= Cornulariella)
Chondrostroma (= Phacidiopycnis)
Choreospora
Christiaster (= Gonatobotryum)
Chromatium (= Dematium)
Chromelosporium (=
   Ostracoderma)
Chromocytospora (= Phomopsis)
Chromosporium
Chromostylium (= Metarhizium)
Chromotorula (= Rhodotorula)
Chroolepus (= Cystocoleus)
Chroostroma (= Pactilia)
Chrysachne
Chrysalidopsis
Chryseidea
Chrysogloeum
Chrysonilia
Chrysosporium
Chuppia
Cicadocola
Cicinobolus (= Ampelomyces)
Ciferria
Ciferriella
Ciferrina
Ciferriopeltis
Ciferrioxyphium
Ciliciopodium
Ciliciopus (= Ciliciopodium)
Ciliochora
Ciliochorella
Ciliofusa (= Ciliofusarium)
Ciliofusarium (= Menispora)

Ciliophora
Ciliophorella
Ciliosira (= Acarosporium)
Ciliospora (= Chaetospermum)
Ciliosporella
Circinastrum (= Weinmannodora)
Circinoconis
Circinotrichum
Cirrenalia
Cirrhomyces (= Chloridium)
Cirrosporium
Cissococcomyces
Citromyces (= Penicillium)
Civisubramaniania
Cladaspergillus (= Aspergillus)
Cladobotryum
Cladobyssus (= Hypha)
Cladoconidium
Cladographium
Cladophialophora
Cladorrhinum
Cladosarum (= Aspergillus)
Cladosporiella
Cladosporium
Cladosporothyrium (= Zelopelta)
Cladosterigma
Cladotrichum (= Oedemium)
Clasteropycnis
Clasterosporium
Clathrococcum (= Epicoccum)
Clathroconium
Clathrosphaera (nom. conf.)
Clathrosphaerina
Clathrosporium
Clathrotrichum (= Beniowskia)
Clavariana
Clavariopsis
Clavatospora
Clavularia (= Ciliciopodium)
Cleistocystis
Cleistonium
Cleistophoma
Clinotrichum (= Acladium)
Clinterium (= Topospora)
Clisosporium (= Coniothyrium)
Clithramia
Clohesyomyces
Clonostachyopsis (= Clonostachys)
Clonostachys
Closteroaleurosporia (=
  Microsporum)
Closterosporia (= Microsporum)
Closterosporium
Clypeispora
Clypeochorella
Clypeodiplodina (= Ascochytulina)
Clypeopatella
Clypeophialophora
Clypeopycnis
Clypeoseptoria
Clypeostagonospora
Coccidioides
Coccobolus (= Ceuthospora)
Coccobotrys
Coccogloeum
Coccopleum (= Sclerotium)
Coccosporella (= Mycogone)
Coccosporium (nom. dub.)
Coccotrichum (= Botrytis)
Coccularia (nom. dub.)
Codinaea (= Dictyochaeta)
Codinaeopsis

Coeloanguillospora (=
  Filosporella)
Coelographium
Coelosporium
Colemaniella
Coleodictyospora
Coleodictys (= Coleodictyospora)
Coleomyces (= Cylindrocarpon)
Coleonaema (= Coleophoma)
Coleophoma
Coleoseptoria
Colispora
Collacystis (nom. dub.)
Collarium (nom. dub.)
Collecephalus
Colletoconis
Colletogloeum
Colletosporium
Colletostroma (= Colletotrichum)
Colletotrichella (= Kabatia)
Colletotrichopsis (=
  Colletotrichum)
Colletotrichum
Collodochium
Collonaemella (= Cornularia)
Collostroma
Colpomella (nom. dub.)
Columnodomus
Columnophora
Columnothyrium
Coma
Comatospora
Combodia (= Lasiodiplodia)
Cometella (= Clasterosporium)
Comocephalum
Complexipes
Compsosporiella
Condylospora
Confertopeltis (nom. dub.)
Confistulina
Conia (= Lepraria)
Conicomyces
Conidiocarpus
Conidioxyphium (= Conidiocarpus)
Coniella
Conioscypha
Coniosporiopsis (nom. conf.)
Coniosporium
Coniotheciella (= Coniothecium)
Coniothecium (nom. dub.)
Coniothyriella (= Coniothyrina)
Coniothyrina
Coniothyrinula (= Coniothyrium)
Coniothyriopsiella (=
  Cyclothyrium)
Coniothyriopsis (= Cyclothyrium)
Coniothyriopsis (=
  Microsphaeropsis)
Coniothyris (= Coniothyriella)
Coniothyrium
Conoplea
Conostoma
Conostroma
Consetiella
Cooksonomyces
Coprotrichum (= Sporendonema)
Corallinopsis
Corallocytostroma
Corallodendron (= Corallomyces)
Corallomorpha (nom. dub.)
Corallomyces
Cordalia (= Tuberculina)
Cordana

Cordella
Coremiella
Coremiopsis (= Paecilomyces)
Coremium (= Penicillium)
Corethropsis
Corethrostroma
Corniculariella
Cornucopiella
Cornularia (= Corniculariella)
Cornutispora
Cornutostilbe
Corollium (= Paecilomyces)
Coronium (nom. dub.)
Coronospora
Corticomyces
Corymbomyces (= Gliocladium)
Coryne
Coryneopsis (= Seimatosporium)
Corynespora
Corynesporella
Corynesporopsis
Coryneum
Corynodesmium (nom. nud.)
Cosmariospora
Costanetoa
Costantinella
Coutourea (nom. dub.)
Crandallia
Craneomyces
Craspedodidymum
Cremasteria
Creodiplodina
Creonecte
Creoseptoria
Creothyriella
Creothyrium (= Cylindrocolla)
Cribropeltis
Crinitospora
Crinium (= Crinula)
Crinula
Cristidium (nom. nud.)
Cristula
Cristularia (= Botrytis)
Cristulariella
Crocicreomyces
Crocysporium (= Aegerita)
Crucella
Crucellisporiopsis
Crucellisporium
Crustodiplodina
Cryocaligula
Cryptoceuthospora
Cryptocline
Cryptococcus
Cryptococcus
Cryptocoryneopsis
Cryptocoryneum
Cryptogene (= Ascochytopsis)
Cryptogenella (= Ascochytopsis)
Cryptomela (= Cryptosporium)
Cryptomycella
Cryptophaeella (=
  Microsphaeropsis)
Cryptophiale
Cryptophialoidea
Cryptorhynchella (=
  Sphaerographium)
Cryptosphaeria
Cryptosporiopsis
Cryptosporium
Cryptostictella (= Discosia)
Cryptostictis (= Seimatosporium)
Cryptostroma

Dinemasporiella (nom. dub.)
Dinemasporiella (=
    Pseudolachnea)
Dinemasporiopsis (=
    Pseudolachnea)
Dinemasporis (= Dinemasporiella)
Dinemasporium
Dionysia (= Candelabrum)
Dioszegia (= Cryptococcus)
Diploceras (= Seimatosporium)
Diplocladiella
Diplocladium (= Cladobotryum)
Diplococcium
Diplodia
Diplodiella (nom. dub.)
Diplodina
Diplodinis
Diplodinula
Diplodothiorella (= Sphaerellopsis)
Diploidium (= Septoidium)
Diplolaeviopsis
Diploöspora
Diplopeltis (= Pycnoseynesia)
Diploplacis (= Sphaerellopsis)
Diploplacosphaeria (=
    Sphaerellopsis)
Diploplenodomopsis (= Diplodina)
Diploplenodomus
Diplorhinotrichum (= Dactylaria)
Diplorhynchus
Diplosclerophoma (= Diplodina)
Diplosporium (= Oedemium)
Diplosporonema
Diplozythia (nom. conf.)
Diplozythiella
Dirimosperma (nom. dub.)
Disaeta (= Seimatosporium)
Discella (= Rhabdospora)
Dischloridium
Discocolla
Discofusarium (= Fusarium)
Discogloeum
Discomycetoidea
Discosia
Discosiellina
Discosiopsis (= Pestalotiopsis)
Discosiospora (= Discosia)
Discosporella (= Conostroma)
Discosporiella (=
    Cryptosporiopsis)
Discosporina
Discosporiopsis (= Phacidiopycnis)
Discosporium
Discosporium
Discostromella (= Leptostromella)
Discotheciella
Discothecium (= Discotheciella)
Discozythia
Discula
Disculina
Disporotrichum
Dissitimurus
Dissoacremoniella
Dissoconium
Distocercospora
Ditangium
Dithozetia (= Didymothozetia)
Divinia
Dochmolopha (= Seimatosporium)
Doliomyces
Domingoella
Doratomyces
Dothichiza

Dothideodiplodia
Dothiomyces
Dothiopsis (nom. dub.)
Dothiorella
Dothiorellina (nom. dub.)
Dothiorina
Dothioropsis
Dothisphaeropsis (=
    Microsphaeropsis)
Dothistroma
Drechmeria
Drechslera
Drechslerella
Drechsleromyces
Drepanispora
Drudeola
Drummondia (= Usteria)
Drumopama
Dualomyces
Dubiomyces (= Ustilaginoidea)
Duddingtonia (= Arthrobotrys)
Dumortieria
Duosporium
Dwayaangam
Dwayabeeja
Dwayaloma
Dwayalomella
Dwayamala (= Dendryphiella)
Dwibahubeeja
Dwibeeja
Dwiroopa
Dwiroopella
Ebollia
Echinobotryum
Echinocatena
Echinochondrium
Echinodia
Echinosporium (= Petrakia)
Ectomyces (= Termitaria)
Ectosticta (= Rhizosphaera)
Ectostroma (nom. dub.)
Ectotrichophyton (= Trichophyton)
Edmundmasonia (=
    Brachysporiella)
Eeniella
Eidamia (= Harzia)
Elachopeltella
Elachopeltis
Eladia
Elaeodema
Elattopycnis
Eleutheris (= Eleutheromycella)
Eleutheromycella
Eleutheromyces
Elletevera
Ellisembia
Ellisia
Ellisiella (= Colletotrichum)
Ellisiella
Ellisiellina (= Ellisiella)
Ellisiopsis (= Beltraniella)
Ellula
Elosia (= Alternaria)
Embellisia
Embolidium (= Coniella)
Emmonsia (= Chrysosporium)
Enantioptera
Enantiothamnus
Endoblastoderma (= Candida)
Endoblastomyces (= Trichosporon)
Endobotrya
Endobotryella
Endocalyx

Endocladis (= Endoramularia)
Endoconidium
Endoconospora
Endocoryneum
Endodermophyton (=
    Trichophyton)
Endodesmia (= Broomeola)
Endogloea (= Phomopsis)
Endomelanconium
Endonema (= Pascherinema)
Endophragmia
Endophragmiella
Endophragmiopsis
Endoplacodium
Endoramularia
Endosporostilbe (= Bloxamia)
Endostilbum
Endothiella
Endotrichum (nom. dub.)
Endozythia
Enerthidium
Engelhardtiella
Engyodontium
Enthallopycnidium
Entoderma
Entomopatella (= Chaetospermum)
Entomosporium
Entomyclium (= Dendryphion)
Eoetvoesia (nom. dub.)
Epaphroconidia
Ephelidium
Ephelidium (nom. conf.)
Ephelis
Epicladonia
Epiclinium
Epicoccum
Epicyta (= Aplosporella)
Epidermomyces (=
    Epidermophyton)
Epidermophyton
Epidochiopsis
Epilithia (nom. dub.)
Epinyctis (= Lepraria)
Episporogoniella
Epistigme
Epithyrium
Epochniella (= Stemphylium)
Epochnium (= Monilia)
Epulorhiza
Ergotaetia (= Sphacelia)
Ericianella (= Bactridium)
Eriocercospora
Eriocercosporella
Eriomene (= Menispora)
Eriomenella (= Menispora)
Eriomycopsis
Erionema (= Menispora)
Eriospora
Eriosporella
Eriosporina (= Sirothecium)
Eriosporopsis
Eriothyrium
Erysiphopsis
Erythrobasidium
Erythrogloeum
Escovopsis
Esdipatilia
Euaspergillus (= Aspergillus)
Euoidium (= Oidium)
Eurasina
Euricoa (= Cylindrocarpon)
Eurotiopsis (nom. dub.)
Eustilbum (= Dendrostilbella)

Eutorula (= Torulopsis)
Eutorulopsis (= Torulopsis)
Evanidomus
Everhartia
Everniicola
Eversia
Excioconidium (= Chalara)
Excioconis (= Excioconidium)
Excipularia
Excipulariopsis
Excipulella (= Heteropatella)
Excipulina (= Heteropatella)
Exobasidiopsis (= Kabatiella)
Exochalara
Exophiala
Exophoma
Exosporella
Exosporiella
Exosporina
Exosporina (= Arnaudina)
Exosporinella (= Arnaudina)
Exosporium
Exserohilum
Exserticlava
Fairmaniella
Falcipatella (= Heteropatella)
Falcipatellina (= Heteropatella)
Falcispora (= Selenophoma)
Falcocladium
Favostroma
Favotrichophyton (=
  Trichophyton)
Fellneria (= Colletotrichum)
Fellomyces
Feltgeniomyces
Fermentotrichon (= Trichosporon)
Fibrillaria (nom. dub.)
Fibulochlamys
Fibulocoela
Fibulotaeniella
Filaspora (= Rhabdospora)
Filosporella
Fioriella (= Diplodina)
Fissuricella
Flabellocladia
Flabellospora
Flagellospora
Flahaultia
Floccaria (= Penicillium)
Flosculomyces
Fluminispora (= Dimorphospora)
Fominia (= Colletotrichum)
Fonsecaea (= Rhinocladiella)
Fontanospora
Foveostroma
Foxia (= Exophiala)
Fragosoella (= Scleropycnium)
Frankiella (= Greeneria)
Fraseriella
Fresenia (= Graphiothecium)
Freynella (= Coccosporium)
Fuckelia
Fuckelina (= Stachybotrys)
Fujimyces
Fuligomyces
Fulvia
Fumago (nom. conf.)
Fumagopsis (= Tridentaria)
Fumagospora
Funicularius
Furcaspora
Fusamen
Fusariella

Fusariopsis
Fusarium
Fuscophialis
Fusella
Fusichalara
Fusicladiella
Fusicladina
Fusicladiopsis (= Stemphylium)
Fusicladiopsis (= Karakulinia)
Fusicladium
Fusicoccum
Fusicolla
Fusicytospora (= Phomopsis)
Fusidium (= Cylindrocarpon)
Fusidomus (= Fusarium)
Fusisporella
Fusisporium (= Fusarium)
Fusoma (nom. dub.)
Fusticeps
Gabarnaudia
Galeraicta (= Rabenhorstia)
Gamonaemella (= Gamospora)
Gamospora (= Wiesneriomyces)
Gamosporella (nom. dub.)
Gampsonema
Gamsia
Gangliophora
Gangliophragma
Gangliostilbe
Garnaudia
Gaubaea
Geastrumia
Gelatinocrinis
Gelatinopycnis
Gelatinosporis (=
  Gelatinosporium)
Gelatinosporium
Gelatosphaera (= Rhizosphaera)
Geminella (= Schroeteria)
Geminoarcus
Gemmophora (nom. dub.)
Genicularia (= Geniculifera)
Geniculifera
Geniculodendron
Geniculospora
Geniculosporium
Geniopila
Geomyces
Geosmithia
Geotrichella
Geotrichoides (= Trichosporon)
Geotrichopsis
Geotrichum
Gerlachia (= Microdochium)
Gerulajacta (nom. dub.)
Gibbago
Gibellula
Gibellulopsis (= Verticillium)
Gilchristia (= Zymonema)
Gilletia (= Telligia)
Gilmaniella
Giulia
Glaphyriopsis
Glenosporella (= Chrysosporium)
Glenosporopsis
Glioannellodochium
Glioblastocladium
Gliobotrys (= Stachybotrys)
Gliocephalis
Gliocephalotrichum
Gliocladiopsis (=
  Cylindrocladium)
Gliocladium

Gliocladochium (= Periola)
Gliodendron (= Sterigmatobotrys)
Gliomastix
Gliophragma
Gliostroma (nom. dub.)
Glischroderma
Globuliroseum
Gloeocercospora
Gloeocoryneum
Gloeodes
Gloeosporidiella
Gloeosporidina
Gloeosporidium (= Discula)
Gloeosporiella
Gloeosporina (= Asteroma)
Gloeosporiopsis (=
  Colletotrichum)
Gloeosporium (= Marssonina)
Gloeosynnema
Gloeotrochila (= Cryptocline)
Gloiosphaera
Glomerularia (= Glomopsis)
Glomerularia (= Gonatobotrys)
Glomerulomyces
Glomopsis
Glomospora
Glutinium
Glutinoagger
Glycyphila
Godroniella (= Myxormia)
Goidanichia (nom. inval.)
Goidanichiella
Goidanichiella (= Haplographium)
Golovinia
Gomphinaria (= Acrotheca)
Gonatobotrys
Gonatobotryum
Gonatophragmiella
Gonatophragmium
Gonatopyricularia (=
  Pyriculariopsis)
Gonatorhodis (= Gonatorrhodiella)
Gonatorrhodiella (=
  Gonatobotryum)
Gonatorrhodum
Gonatosporium (= Arthrinium)
Gonatotrichum (= Gonytrichum)
Gongromeriza (= Chloridium)
Gongylocladium (= Oedemium)
Goniopila
Goniosporium (= Arthrinium)
Gonyella
Gonytrichella
Gonytrichum
Goosiella
Goosiomyces
Gorgomyces
Grallomyces
Granmamyces (= Weufia)
Granularia
Granulodiplodia (= Sphaeropsis)
Granulomanus
Graphidium (= Paecilomyces)
Graphilbum
Graphiocladiella
Graphiopsis (= Phaeoisaria)
Graphiopsis (= Cladosporium)
Graphiothecium
Graphium
Greeneria
Groveolopsis
Grubyella (= Trichophyton)
Guceviczia (= Wojnowicia)

Guedea
Gueguenia (= Amblyosporium)
Guelichia
Gutturomyces (= Aspergillus)
Gymnodochium
Gymnosphaera (= Stagonospora)
Gymnosporium (nom. dub.)
Gymnoxyphium
Gyoerffyella
Gyratylium (= Sphaeropsis)
Gyrocerus (= Circinotrichum)
Gyrostroma
Gyrothrix
Gyrotrichum (= Circinotrichum)
Habrostictis
Hadronema
Hadrosporium
Hadrotrichum
Hainesia
Halobyssus (= Monilia)
Halysiomyces
Halysium (nom. conf.)
Hansfordia
Hansfordiella
Hansfordiellopsis
Hansfordiopeltis
Hansfordiopeltopsis
Hansfordiula (= Phaeoisaria)
Hantzschia (= Phialocephala)
Hapalosphaeria
Haplaria (= Botrytis)
Haplariella
Haplariella (= Haplariella)
Haplariopsis (= Haplariella)
Haplariopsis
Haplobasidion
Haplochalara (= Catenularia)
Haplographium
Haplolepis
Haplomela (= Melanconium)
Haplophoma (= Phomopsis)
Haplosporidium
Haplostromella (= Strasseria)
Haplotrichella (= Gliomastix)
Haplotrichum (= Acladium)
Haptocara
Haptospora
Haraella (= Hinoa)
Harikrishnaella (= Chaetomella)
Harknessia
Harmoniella
Harpagomyces
Harpocephalum (= Periconia)
Harpographium
Harposporella (nom. dub.)
Harposporium
Harpostroma
Hartiella (= Calostilbella)
Hartigiella (= Meria)
Harzia
Harziella (= Trichocladium)
Harziella
Hasskarlinda (= Corallodendron)
Hastifera
Hawksworthiana
Heimiella
Heimiodiplodia (= Botryodiplodia)
Heimiodora
Helhonia
Helicia (= Cylindrocolla)
Helicobolomyces
Helicobolus (= Phloeospora)
Helicoceras

Helicocoryne (= Helicoma)
Helicodendron
Helicodesmus (= Helicodendron)
Helicofilia
Helicogoosia
Helicoma
Helicomina (= Pseudocercospora)
Helicominopsis
Helicomyces
Helicoon
Helicopsis (= Helicoma)
Helicorhoidion
Helicosingula
Helicosporangium (= Papulospora)
Helicosporella
Helicosporina (nom. inval.)
Helicosporina
Helicosporium
Helicostilbe
Helicostilbe (= Trochophora)
Helicothyrium
Helicotrichum (= Helicosporium)
Helicoubisia
Heliocephala
Heliscella
Heliscina
Heliscus
Helminthophora
Helminthosporiopsis (= Podosporium)
Helminthosporium
Helmisporium (= Helminthosporium)
Helostroma (= Microstroma)
Hemialysidium
Hemibeltrania
Hemicorynespora
Hemicorynesporella
Hemidothis
Hemisphaeropsis
Hemispora (= Wallemia)
Hemisynnema
Hendersonia (= Stagonospora)
Hendersonia (= Hendersonia)
Hendersoniella (= Hendersoniella)
Hendersoniella
Hendersonina
Hendersoniopsis
Hendersoniopsis (= Stenocarpella)
Hendersonula
Hendersonulina
Henicospora
Hennebertia
Heptaster
Hermatomyces
Herposira
Herpotrichiopsis (= Pyrenochaeta)
Herreromyces
Heterobotrys (= Atichia)
Heterocephalum
Heteroceras (= Neoheteroceras)
Heteroconidium
Heteroconium
Heterographa (= Polymorphum)
Heteropatella
Heteroseptata
Heterosporiopsis
Heterosporium (= Cladosporium)
Hexacladium
Heydenia
Himantia
Hindersonia (= Hendersonia)
Hinoa

Hiospira
Hippocrepidium (= Hirudinaria)
Hirsutella
Hirudinaria
Histoplasma
Histoplasma
Hobsonia
Hoehneliella
Holcomyces (= Diplodia)
Holubovaea
Holubovaniella
Homalopeltis
Hormiactella
Hormiactina (= Hormiactis)
Hormiactis
Hormiokrypsis
Hormisciella (= Antennatula)
Hormiscioideus
Hormisciomyces
Hormisciopsis
Hormiscium (= Torula)
Hormocephalum
Hormocladium
Hormococcus
Hormococcus (nom. dub.)
Hormoconis
Hormodendroides (= Rhinocladiella)
Hormodendrum (= Cladosporium)
Hormodochis (= Trullula)
Hormodochium (= Trullula)
Hormodochium
Hormographiella
Hormographis
Hormomyces
Hormonema
Hormyllium (= Hormococcus)
Hortaea
Hughesiella (= Chalara)
Hughesinia
Humicola
Humicolopsis
Humicolopsis
Hyalobelemnospora
Hyalobotrys (= Stachybotrys)
Hyaloceras (= Seiridium)
Hyalocladium
Hyalococcus (= Trichosporon)
Hyalocylindrophora (= Dischloridium)
Hyalodendron
Hyalodictys (= Miuraea)
Hyalodictyum
Hyaloflorea (= Cylindrocarpon)
Hyalohelicomina
Hyalopesotum
Hyalopus (= Acremonium)
Hyalopycnis
Hyalorhinocladiella
Hyalostachybotrys (= Stachybotrys)
Hyalosynnema
Hyalothyridium
Hyalothyris (= Hyalothyridium)
Hyalotia (= Bartalinia)
Hyalotiella
Hyalotiopsis
Hyalotrochophora (= Delortia)
Hydnopsis (nom. dub.)
Hydrometrospora
Hygrochroma (= Phloeospora)
Hymenella
Hymeniopeltis

Hymenobactron
Hymenopodium (=
  Clasterosporium)
Hymenopsis
Hymenostilbe
Hymenula (= Hymenella)
Hyperomyxa (= Cheirospora)
Hypha (nom. dub.)
Hyphasma (nom. dub.)
Hyphaster (= Asterostomella)
Hyphelia (= Trichothecium)
Hyphochlaena
Hyphoderma (= Hyphelia)
Hyphodictyon (= Atichia)
Hyphodiscocioides
Hyphodiscosia
Hyphopolynema
Hyphosoma (nom. conf.)
Hyphostereum
Hyphothyrium
Hyphozyma
Hypocenia (= Topospora)
Hypocline
Hypocreodendron
Hypodermina
Hypodermium (nom. dub.)
Hypogloeum
Hypoplasta (= Cytospora)
Hysteridium
Hysterodiscula
Hysteropycnis
Ialomitzia
Icerymyces
Idiocercus
Idriella
Iledon
Illosporium
Indiella (= Madurella)
Infrafungus
Infundibura
Ingoldia (= Gyoerffyella)
Ingoldiella
Inifatiella
Innatospora (= Arthrinium)
Insecticola (= Akanthomyces)
Intercalarispora
Ionophragmium
Irpicomyces
Isaria
Isariella
Isariopsella
Isariopsis (= Phacellium)
Ischnostroma
Isthmolongispora
Isthmophragmospora
Isthmospora
Isthmotricladia
Itersonilia
Ityorhoptrum
Iyengarina
Jacobaschella (= Diplosporium)
Jacobia
Jaculispora
Jaczewskiella (= Stigmina)
Jahniella
Jainesia
Janetia
Janospora (= Stilbospora)
Japonia
Javonarxia
Jerainum
Johncouchia
Johnstonia (= Neojohnstonia)

Jubispora
Junctospora
Kabathia (= Sphaerellopsis)
Kabatia
Kabatiella
Kabatina
Kafiaddinia
Kaleidosporium
Kamatella
Kamatia (= Pseudodictyosporium)
Kameshwaromyces
Karakulinia (= Fusicladium)
Karsteniomyces
Kaskaskia (= Gyrostroma)
Kaufmannwolfia (= Trichophyton)
Kazulia
Keissleriomyces
Kellermania
Kellermanniopsis
Kendrickomyces
Keratinomyces (= Trichophyton)
Kermincola (nom. dub.)
Ketubakia
Kilikiostroma
Kiliophora
Kionocephala
Kionochaeta
Kirramyces
Kirschsteiniella (= Cyclothyrium)
Klebahnopycnis (= Hoehneliella)
Kloeckera
Kmetia
Kmetiopsis
Knemiothyrium
Knoxdaviesia
Knyaria (= Tubercularia)
Kockovaella
Kodonospora
Kontospora
Koorchaloma
Koorchalomella
Kordyanella (= Tremellidium)
Korunomyces
Kostermansinda
Kostermansindiopsis
Kramabeeja
Kramasamuha
Kreiseliella
Kumanasamuha
Kurosawaia (= Sphaceloma)
Kurssanovia (= Fusariella)
Kurtzmanomyces
Kutilakesa (= Sarcopodium)
Kutilakesopsis (= Sarcopodium)
Kylindria
Kyphophora
Labrella (nom. dub.)
Labridella
Labridium (= Seimatosporium)
Lacellina
Lacellinopsis
Lachnidium (= Fusarium)
Lachnodochium
Laciniocladium
Lactydina
Laeviomyces
Lagenomyces
Lagynodella (= Cryptosporiopsis)
Lalaria
Lambdasporium
Lamproconium
Lamyella (= Cytospora)
Langeronia (= Trichophyton)

Lappodochium
Laridospora
Lasiodiplodia
Lasiodiplodiella
Lasiophoma
Lasiophoma (= Pyrenochaeta)
Lasiostroma (= Phomopsis)
Lasiothyrium
Lasmenia
Lasmeniella
Latericonis
Lateriramulosa
Laterispora
Lauriomyces
Lawalreea
Leandria
Lecaniocola
Lecanosticta
Lecythophora (= Phialophora)
Leeina
Leightoniomyces
Leiosepium (= Sepedonium)
Lembuncula
Lemkea
Lemonniera
Lennisia (= Monochaetia)
Lepidophyton (nom. dub.)
Lepra (nom. dub.)
Lepra (= Illosporium)
Lepraria
Leprieurina
Leprieurinella
Leprocaulon
Leptasteromella (=
  Leptodothiorella)
Leptina (= Discosia)
Leptochlamys
Leptocladia (= Stenocladiella)
Leptocoryneum (=
  Seimatosporium)
Leptodermella
Leptodermopsis (nom. dub.)
Leptodiscella
Leptodiscus (= Mycoleptodiscus)
Leptodontidium
Leptodontium (= Leptodontidium)
Leptodothiorella
Leptodothiorella (=
  Leptodothiorella)
Leptographium
Leptomelanconium
Leptophoma (= Phoma)
Leptophyllosticta
Leptophyma (= Helostroma)
Leptosporium (= Fusicolla)
Leptostroma
Leptostromella
Leptoteichion (= Phacidiopycnis)
Leptothyrella
Leptothyrina
Leptothyrium
Leptotrichum (nom. dub.)
Leptoxyphium
Leucocytospora (= Cytospora)
Leucodochium
Leucopenicillifer
Leucophomopsis (= Phomopsis)
Leucosporium (= Pactilia)
Leuliisinea
Levieuxia (nom. dub.)
Libartania
Libertella
Libertiella

Libertina (= Phomopsis)
Lichenoconium
Lichenodiplis
Lichenophoma
Lichenopuccinia
Lichenosticta
Lichingoldia
Ligniella (= Discula)
Lindauomyces (= Arthrobotryum)
Lindavia (= Scopularia)
Lindquistia
Linochora
Linochorella
Linodochium
Listeromyces
Lituaria (= Helicoma)
Lobatopedis
Lobomyces
Loliomyces
Lomaantha
Lomachashaka
Lomentospora
Lonchospermella (nom. dub.)
Lophodiscella (= Colletotrichum)
Lophophyton (= Trichophyton)
Ludwigiella (= Selenophoma)
Lunospora (= Pseudoseptoria)
Lunulospora
Luttrellia (= Exserohilum)
Lutziomyces (= Paracoccidioides)
Luxuriomyces (= Diplococcium)
Luzfridiella
Lycoperdellon (= Ostracoderma)
Lylea
Lyromma
Lysipenicillium (= Gliocladium)
Lysotheca
Macilvainea (= Drudeola)
Macmillanina (= Disculina)
Macraea (= Prathigada)
Macroallantina
Macrodendrophoma (nom. inval.)
Macrodiplis (= Macrodiplodiopsis)
Macrodiplodia
Macrodiplodina (= Ascochyta)
Macrodiplodiopsis
Macrohilum
Macroon (= Helminthosporium)
Macrophoma (= Sphaeropsis)
Macrophomella (= Botryodiplodia)
Macrophomina
Macrophomopsis
Macrophomopsis (=
    Botryodiplodia)
Macrophyllosticta (= Phyllosticta)
Macroplodia (= Sphaeropsis)
Macroplodiella (= Phoma)
Macroseptoria (= Trichoseptoria)
Macrosporium (= Alternaria)
Macrostilbum
Macrotrichum
Madurella
Magdalaenaea
Mahabalella
Mahevia (= Hirsutella)
Malacharia (= Cerebella)
Malacodermis (= Glutinium)
Malacostroma (= Phomopsis)
Malassezia
Malbranchea
Malleomyces (nom. dub.)
Malustela (= Curvularia)
Mammaria

Mammariopsis
Manaustrum
Manginella
Manginia (= Sphaceloma)
Manginula (nom. dub.)
Manginulopsis (= Leptothyrium)
Mapletonia
Marchandiomyces
Marcosia (= Stigmina)
Margarinomyces (= Phialophora)
Margaritispora
Mariannaea
Marssonia (= Marssonina)
Marssoniella (= Neomarssoniella)
Marssonina
Martinella (nom. dub.)
Martinellisia
Masonia (= Masoniella)
Masoniella (= Scopulariopsis)
Masoniomyces
Massalongina
Massariothea
Mastigocladium (= Acremonium)
Mastigomyces
Mastigonema (= Chaetospermum)
Mastigonetron (= Harknessia)
Mastigosporella
Mastigosporium
Mastomyces (= Topospora)
Matruchotia (nom. dub.)
Matruchotiella (= Ateleothylax)
Matsushimaea
Matsushimomyces
Mattirolella
Matula
Mauginiella
Maxillospora (= Tetracladium)
Mazzantiella (= Hypodermina)
Medusamyces
Medusula (nom. dub.)
Megacapitula
Megacladosporium (=
    Fusicladium)
Megalocitosporides (=
    Coccidioides)
Megalodochium
Megaloseptoria
Megaloxyphium (= Leptoxyphium)
Megaster
Megatrichophyton (=
    Trichophyton)
Melanchlenus (= Rhinocladiella)
Melanconiopsis
Melanconium
Melanobasidium (= Sphaceloma)
Melanobasis (= Sphaceloma)
Melanocephala
Melanocryptococcus (=
    Cryptococcus)
Melanodiscus
Melanodochium (= Sphaceloma)
Melanogone (= Humicola)
Melanographium
Melanophoma
Melanophora (= Sphaceloma)
Melanosella
Melanosphaeria (= Sirosphaera)
Melanostroma (= Ceuthospora)
Melanotrichum (= Trichosporium)
Melasmia
Melophia
Membranatheca (=
    Amerosporium)

Memnoniella
Memnonium (= Trichosporium)
Menidochium
Menispora
Menisporella (= Codinaea)
Menisporopsis
Menoidea
Mercadomyces
Meria
Merimbla
Merismella
Mesobotrys (= Gonytrichum)
Metabotryon (= Sphaerellopsis)
Metadiplodia
Metarhizium
Metazythia
Metazythiopsis
Methysterostomella (=
    Phragmopeltis)
Miainomyces (= Sporotrichum)
Michenera
Microanthomyces (= Candida)
Microbasidium (= Hadrotrichum)
Microblastosporon
Microbotryodiplodia (=
    Microdiplodia)
Microcera (= Fusarium)
Microclava
Microcyta (= Ceuthosporella)
Microdiplodia
Microdiplodia
Microdiscula
Microdochium
Microdothiorella
Microgloeum
Microhaplosporella (=
    Aplosporella)
Microhendersonula
Microhilum
Microlychnus
Micropera (= Foveostroma)
Microperella
Microphoma (= Dothichiza)
Micropustulomyces
Microspatha
Microsphaeropsis
Microsphaeropsis (nom. dub.)
Microsporella (=
    Microsphaeropsis)
Microsporum
Microstella
Microtrichophyton (=
    Trichophyton)
Microtyle
Microtypha (= Arthrinium)
Microxiphium
Microxyphiella
Microxyphiomyces
Microxyphiopsis
Microxyphium
Micula (= Foveostroma)
Millerburtonia
Milospium
Milowia (= Thielaviopsis)
Mindoa
Miniancora
Minimedusa
Minimidochium
Minutoexcipula
Minutophoma
Mirandina
Mirimyces
Mitrorhizopeltis (= Tracylla)

Mitteriella
Miuraea
Mixtoconidium
Moeszia (= Cylindrocarpon)
Molliardiomyces
Monacrosporiella
Monacrosporium
Monascostroma (= Hendersonia)
Monilia (= Monilia)
Monilia (= Monilia)
Monilia
Monilia (= Monilia)
Moniliella
Moniliger (= Penicillium)
Moniliophthora
Moniliopsis (= Rhizoctonia)
Monilochaetes
Monoceras (= Seimatosporium)
Monochaetia
Monochaetiella
Monochaetiellopsis
Monochaetina (= Bleptosporium)
Monochaetinula
Monochaetopsis
Monocillium
Monoconidia (= Acremonium)
Monodia
Monodictys
Monodidymaria
Monodisma
Monogrammia (= Titaea)
Monoplodia (= Coniothyrium)
Monopodium (= Acremoniella)
Monopycnis (= Cytospora)
Monosporella (= Monotosporella)
Monosporiella
Monosporium (nom. illegit.)
Monostachys (= Chloridium)
Monostichella
Monothecium (= Mastigosporium)
Monotospora (nom. dub.)
Monotospora (= Acrogenospora)
Monotosporella (=
  Brachysporiella)
Monotoyella (nom. dub.)
Monotretomyces (=
  Corynesporopsis)
Monotrichum
Monotropomyces (nom. dub.)
Montoyella (nom. dub.)
Moorella
Moralesia
Morfea (= Polychaeton)
Morfea (= Polychaeton)
Morinia
Moronopsis (= Cheirospora)
Morrisiella (= Janetia)
Morrisographium
Morthiera (= Entomosporium)
Morularia (= Heterobotrys)
Mucobasispora
Mucosetospora
Mucrosporium (= Cladobotryum)
Muiaria
Muiogone
Muirella
Multicladium
Multipatina
Munkia
Muricularia (nom. dub.)
Murogenella (= Coryneum)
Myceliophthora
Myceloblastanon (= Candida)

Myceloderma
Mycepimyce (= Sphaerellopsis)
Mycobacillaria
Mycobanche (= Mycogone)
Mycocandida (= Candida)
Mycocentrospora
Mycochaetophora
Mycochlamys
Mycoderma
Mycoenterolobium
Mycofalcella
Mycogone
Mycohypallage
Mycokluyveria (= Candida)
Mycoleptodiscus
Mycomater (nom. dub.)
Mycomyces
Mycopappus
Mycopara
Mycosisymbrium (=
  Scolecobasidiella)
Mycospraguea
Mycosticta
Mycosylva
Mycotorula (= Syringospora)
Mycotoruloides (= Syringospora)
Mycotribulus
Mycousteria
Mycovellosiella
Mydonosporium (= Cladosporium)
Mydonotrichum (=
  Helminthosporium)
Myiocoprula
Myriellina
Myriocephalum (= Cheirospora)
Myrioconium
Myriodontium
Myriophysa (= Atichia)
Myrmecomyces
Myropyxis
Myrotheciastrum
Myrotheciella (= Myrothecium)
Myrothecium
Mystrosporiella
Mystrosporium
Myxocephala (= Phialocephala)
Myxocladium (= Cladosporium)
Myxocyclus
Myxodiscus (= Leptothyrium)
Myxodochium
Myxofusicoccum (= Pseudodiscula)
Myxolibertella (= Phomopsis)
Myxoparaphysella
Myxormia (= Myrothecium)
Myxosporella
Myxosporidiella
Myxosporina (= Rhodesia)
Myxosporium (nom. conf.)
Myxostomellina
Myxothyriopsis
Myxothyrium
Myxozyma
Nadsoniella
Nadsoniomyces (= Trichosporon)
Naemaspora (= Roscoepoundia)
Naemaspora (nom. conf.)
Naemosphaera (= Rabenhorstia)
Naemosphaera
Naemosphaerella
Naemospora (nom. dub.)
Naemostroma (= Septoriella)
Naganishia (= Cryptococcus)
Nagrajia

Nagrajomyces
Naiadella
Nakataea (= Pyricularia)
Nalanthamala
Nanoschema
Naothyrsium
Napicladium (= Spilocaea)
Naranus
Nascimentoa (= Spiropes)
Natalia (= Levieuxia)
Nattrassia
Nawawia
Necator
Necraphidium
Negeriella
Nemadiplodia (= Botryodiplodia)
Nematocolla (= Myxosporium)
Nematoctonus
Nematogonum
Nematographium
Nematomyces (= Gliomastix)
Nematophagus
Nematospora (= Giulia)
Nemostroma
Nemozythiella
Neoalpakesa
Neoarbuscula
Neobarclaya
Neochaetospora
Neodiplodina
Neofuckelia
Neogeotrichum (= Trichosporon)
Neohendersonia
Neoheteroceras
Neojohnstonia
Neokellermania (=
  Pseudorobillarda)
Neoligniella
Neomarssoniella
Neomelanconium
Neomichelia (= Pithomyces)
Neomunkia
Neoovularia
Neopatella (= Selenophoma)
Neopeltis
Neopericonia
Neophoma
Neoplaconema
Neoplacosphaeria (=
  Sphaeriothyrium)
Neopycnodothis (= Cytoplea)
Neoramularia
Neospegazzinia
Neosphaeropsis (= Sphaeropsis)
Neosporidesmium
Neotrichophyton (= Trichophyton)
Neottiospora
Neottiosporella (nom. dub.)
Neottiosporina
Neottiosporis (= Neottiosporella)
Neozythia
Neta
Nicholsoniella (= Libertiella)
Nidulispora
Nigrococcus (= Phaeococcus)
Nigrodiplodia (nom. dub.)
Nigropuncta
Nigrospora
Nimbya
Nodulisporium
Nomuraea
Nosophloea
Nothoclavulina

Nothopatella (= Botryodiplodia)
Nothospora
Nothostrasseria
Nummospora
Nusia
Nyctalina
Nyctalospora
Nypaella
Obeliospora
Obstipipilus
Obstipispora
Ochroconis
Odontodictyospora
Oedemium
Oedocephalum
Oedothea
Oideum (= Oidium)
Oidiodendron
Oidiopsis
Oidium (= Oidium)
Oidium (= Oidium)
Oidium
Ojibwaya
Ollula (= Tubercularia)
Olpitrichum
Omega
Ommatospora (= Microclava)
Ommatosporella
Oncocladium
Oncopodiella
Oncopodium
Oncospora
Oncosporella
Oncosporomyces
Oncostroma
Onychocola
Onychophora
Oospora (= Oidium)
Oosporidium
Oosporoidea (= Geotrichum)
Oothecium (= Asterostomella)
Oothyrium
Operculella (= Phacidiopycnis)
Ophiocladium (= Ramularia)
Ophiodendron (nom. dub.)
Ophiopodium (= Grallomyces)
Ophiosira
Ophiosporella (= Phloeosporina)
Ophiotrichum
Oramasia (= Vermiculariopsiella)
Orbimyces
Ordus
Oreophylla
Ormathodium (nom. dub.)
Orphanocoela
Osteomorpha
Ostracoderma
Ostracodermidium
Oswaldina (= Hemidothis)
Otomyces (nom. dub.)
Otomyces (= Aspergillus)
Outhovia (= Scopularia)
Ovadendron
Overeemia (= Brooksia)
Ovularia (= Ramularia)
Ovulariella (nom. nud.)
Ovulariopsis
Ovulitis
Oxysporium (= Helminthosporium)
Ozonium (nom. dub.)
Paathramaya
Pachnodium

Pachybasidiella (= Aureobasidium)
Pachybasium (= Trichoderma)
Pachycladina
Pachydiscula (= Cryptosporiopsis)
Pachyma
Pachytrichum (= Periconia)
Pactilia
Padixonia
Paecilomyces
Paepalopsis
Pagidospora
Palawaniopsis
Paliphora
Palmomyces (= Andreaeana)
Panchanania
Pantospora
Papilionospora
Pappimyces
Papularia (= Arthrinium)
Papulaspora
Paraaoria
Paraarthrocladium
Paraceratocladium
Paracercospora
Parachinomyces
Paracoccidioides
Paracostantinella
Paracryptophiale
Paracytospora
Paradactylaria
Paradactylella
Paradendryphiopsis
Paradidymobotryum
Paradiplodia
Paradiplodiella (= Botryodiplodia)
Paradischloridium
Paradiscula
Paraepicoccum
Parafulvia
Parahelminthosporium (= Polytretophora)
Parahyalotiopsis
Paraisaria
Paramassariothea
Paramyces (= Ptychogaster)
Parapericonia
Paraphaeoisaria
Paraphoma (= Phoma)
Parapithomyces
Parapleurotheciopsis
Parapyricularia
Parasaccharomyces (= Candida)
Parasclerophoma (= Dothichiza)
Parasphaeropsis
Paraspora
Parastenella
Parastigmatellina
Parasympodiella
Paratomenticola
Paratorulopsis (= Torulopsis)
Paratrichaegum (= Epicoccum)
Paratrichoconis
Paraulocladium
Parendomyces (= Candida)
Pascherinema (nom. dub.)
Paspalomyces
Passalora
Patellina (= Catinula)
Patellina (nom. dub.)
Patouillardea (= Dendrodochium)
Patouillardiella
Pazschkeella

Peckia (= Drudeola)
Pedilospora (= Dicranidion)
Peethasthabeeja
Peglionia (= Gyrothrix)
Pellionella (nom. inval.)
Pellionella
Peltaster
Peltasterella
Peltasterinostroma
Peltasteropsis
Peltistroma
Peltistromella
Peltosoma
Peltostromellina
Peltostromopsis
Pendulispora
Penicillifer
Penicillium (= Botrytis)
Penicillium
Penomyces
Pentaposporium (= Trispermum)
Penzigomyces
Perelegamyces
Peresia (= Colletotrichum)
Periconia
Periconiella
Peridiomyces (nom. dub.)
Periola
Periolopsis (= Sarcopodium)
Periperidium
Perizomella
Pesotum
Pestalotia
Pestalotiopsis
Pestalozziella
Pestalozzina (= Bartalinia)
Pestalozzina (= Mastigosporium)
Petasodes (= Phomopsis)
Petrakia
Petrakiopsis
Petrakomyces (= Ciliochora)
Peylia (= Botryosporium)
Peyronelia
Peyronelina (nom. inval.)
Peyronelina
Peyronellaea (= Phoma)
Peziotrichum
Phacellium
Phacellula
Phacidiella
Phacidiopycnis
Phacostroma
Phacostromella
Phaeoannellomyces
Phaeoantenariella (nom. dub.)
Phaeociliospora (= Ciliochorella)
Phaeococcomyces
Phaeococcus (= Phaeococcomyces)
Phaeoconis (= Nigrospora)
Phaeocytosporella (= Phaeocytostroma)
Phaeocytostroma
Phaeodactylella
Phaeodactylium
Phaeodiscula
Phaeodochium (= Hymenopsis)
Phaeodomus
Phaeoharziella (= Arthrinium)
Phaeohendersonia (= Hendersonia)
Phaeohiratsukaea
Phaeohymenula
Phaeoisaria

Phaeoisariopsis
Phaeolabrella
Phaeomarssonia (=
  Didymosporina)
Phaeomarssonia (nom. dub.)
Phaeomonostichella
Phaeophleospora (nom. dub.)
Phaeophomopsis
Phaeopolynema (= Hymenopsis)
Phaeoramularia
Phaeorobillarda (= Robillarda)
Phaeosclera
Phaeoscopulariopsis (=
  Scopulariopsis)
Phaeoseptoria
Phaeosphaera
Phaeosporobolus
Phaeostagonosporopsis (=
  Stenocarpella)
Phaeostalagmus
Phaeostilbella (= Myrothecium)
Phaeotheca
Phaeothyrium
Phaeotrichoconis
Phaeoxyphiella
Phaeoxyphium (nom. dub.)
Phaffia
Phalangispora
Phallomyces
Phanerococculus
Phanerocorynella (= Exosporiella)
Phanerocoryneum (=
  Clasterosporium)
Phellomyces (= Colletotrichum)
Phialemonium
Phialetea (= Grallomyces)
Phialicorona (= Kionochaeta)
Phialoarthrobotryum
Phialocephala
Phialocladus (= Escovopsis)
Phialoconidiophora (=
  Rhinocladiella)
Phialogangliospora
Phialogeniculata
Phialographium
Phialomyces
Phialophaeoisaria
Phialophora
Phialophorophoma
Phialophoropsis
Phialoselanospora
Phialosporostilbe
Phialostele
Phialotubus
Phloeochora (= Phloeospora)
Phloeoconis
Phloeoscoria (= Polymorphum)
Phloeospora
Phloeosporella
Phloeosporina
Phlyctaeniella
Phlyctema
Phlyctidium (= Spilocaea)
Phoma
Phomachora
Phomachorella
Phomatosphaeropsis (=
  Sphaeropsis)
Phomatosporella
Phomopsella (= Phomopsis)
Phomopsina (= Phoma)
Phomopsioides (= Phomopsis)
Phomopsis

Phomyces
Phragmocephala
Phragmodochium
Phragmogloeum
Phragmographium (=
  Morrisographium)
Phragmopeltis
Phragmospathula
Phragmospathulella
Phragmosporonema (=
  Diplosporonema)
Phragmostachys (=
  Sterigmatobotrys)
Phragmostilbe (= Arthrosporium)
Phragmotrichum
Phragmoxyphium (=
  Ciferrioxyphium)
Phylloedia (= Phylloedium)
Phylloedium
Phyllohendersonia
Phyllophiale
Phyllosphaera (= Phyllosticta)
Phyllosticta
Phyllostictella (=
  Microsphaeropsis)
Phyllostictina (= Phyllosticta)
Phymatostroma (= Pactilia)
Phymatotrichopsis
Phymatotrichum (= Botrytis)
Physalidiella
Physalidiopsis
Physalidium (= Physalidiella)
Physokermincola (nom. dub.)
Physospora (= Rhinotrichum)
Piggotia
Pilidiella (= Coniella)
Pilidium
Pilulina
Pimina (= Zygosporium)
Piminella (= Gonytrichum)
Pinoyella (= Trichophyton)
Pionnotes (= Fusarium)
Piptarthron
Piptostomum (nom. dub.)
Piptostroma (= Piptostomum)
Piricauda
Piricaudilium
Piricaudiopsis
Piringa (= Camarosporium)
Pirispora
Pirobasidium (= Coryne)
Pirostoma (nom. dub.)
Pirostoma (nom. dub.)
Pirostomella
Pirozynskia
Pithomyces
Pithosira
Pithospermum (= Sporoschisma)
Pittostroma
Pityrosporum (= Malassezia)
Placella
Placentaria (= Periola)
Placodiplodia
Placomelan
Placonema
Placonemina
Placophomopsis (= Phomopsis)
Placosphaerella (nom. conf.)
Placosphaeria
Placosphaerina
Placothea
Placothyrium
Plagionema (= Ciliochorella)

Plagiorhabdus (= Strasseria)
Plagiostigmella
Plasia
Platycarpium (= Fusamen)
Plecotrichum (nom. dub.)
Plectonaemella
Plectopeltis
Plectophoma (= Asteromella)
Plectophomella
Plectophomopsis
Plectopycnis (= Plectopeltis)
Plectosira
Plectosporium
Plectronidiopsis
Plectronidium
Plectrothrix
Pleiochaeta
Plenocatenulis
Plenodomus (= Phoma)
Plenophysa
Plenotrichaius
Plenotrichella (= Myxothyriopsis)
Plenotrichopsis
Plenotrichum
Plenozythia
Pleococcum (nom. dub.)
Pleocouturea
Pleocyta (= Phaeocytostroma)
Pleosphaeropsis (= Aplosporella)
Plesiospora
Pleurobotrya (= Botrytis)
Pleurocatena
Pleurocolla
Pleurocytospora
Pleurodesmospora
Pleurodiscula
Pleurodiscus
Pleurodomus
Pleurographium (= Dactylaria)
Pleuronaema (= Cytospora)
Pleuropedium
Pleurophoma
Pleurophomella (= Sirodothis)
Pleurophomopsis
Pleurophragmium
Pleuroplaconema
Pleuroplacosphaeria
Pleuropyxis (nom. conf.)
Pleurostromella
Pleurotheciopsis
Pleurothecium
Pleurothyriella
Pleurothyrium
Plokamidomyces
Pneumatospora
Pochonia (= Diheterospora)
Pocillopycnis
Podobactridium (= Bactridium)
Podoconis (= Sporidesmium)
Podoplaconema
Podosporiella
Podosporium
Podosporium (= Aplosporella)
Podosporium (= Aplosporella)
Podoxyphiomyces
Podoxyphium (= Conidiocarpus)
Poikilosperma
Pollaccia
Polyactis (= Botrytis)
Polycephalomyces
Polychaetella
Polycladium
Polycytella

Polydesmus (= Sporidesmium)
Polyetron
Polylagenochromatia (=
　Polystigmina)
Polymorphomyces (= Geotrichum)
Polymorphum
Polynema
Polyopeus (= Phoma)
Polypaecilum
Polyrhina (= Harposporium)
Polyrhizium (= Cladosporium)
Polyschema
Polyscytalina
Polyscytalum
Polyspora (= Aureobasidium)
Polystigmina
Polystomellomyces
Polystratorictus
Polysynnema
Polythecium (= Fusicoccum)
Polythrinciella
Polythrinciopsis
Polythrincium
Polytretophora
Porocladium
Poroisariopsis
Poroniopsis (= Hypocreodendron)
Poropeltis
Porosubramaniania
Porterula (= Asteromella)
Posadasia (= Coccidioides)
Powellia
Prachtflorella (= Gonatobotrys)
Pragmopycnis
Prathigada
Prathoda (= Alternaria)
Preussiaster (= Cordana)
Prismaria
Pritzeliella (= Penicillium)
Proboscispora
Procandida (= Syringospora)
Prophytroma
Prosporobolomyces (=
　Sporobolomyces)
Prosthemiella
Prosthemium
Protendomycopsis (=
　Trichosporon)
Proteomyces (= Trichosporon)
Proteophiala
Protocoronis (= Aureobasidium)
Protocoronospora (=
　Aureobasidium)
Protostegia
Protostegiomyces
Protostroma
Psalidosperma (= Ypsilonia)
Psammina
Psathyromyces
Psecadia (= Cytospora)
Pseudaegerita
Pseuderiospora (= Eriosporella)
Pseuderiospora (= Suttoniella)
Pseudhaplosporella (=
　Botryodiplodia)
Pseudoanguillospora
Pseudoaristastoma
Pseudobasidiospora
Pseudobasidium (= Arthrinium)
Pseudobeltrania
Pseudobotrytis
Pseudocamptoum (=
　Melanographium)

Pseudocanalisporium
Pseudocenangium
Pseudocercospora
Pseudocercosporella
Pseudocercosporidium
Pseudochaetosphaeronema
Pseudochuppia
Pseudococcidioides (=
　Coccidioides)
Pseudoconium
Pseudocordyceps
Pseudocytospora
Pseudodichomera
Pseudodictya (nom. dub.)
Pseudodictyosporium
Pseudodidymaria
Pseudodiplodia (= Botryodiplodia)
Pseudodiplodia
Pseudodiplodiella (=
　Botryodiplodia)
Pseudodiscosia (= Heteropatella)
Pseudodiscula
Pseudoepicoccum
Pseudofumago (nom. dub.)
Pseudofusarium (= Fusarium)
Pseudofuscophialis
Pseudofusidium (= Acremonium)
Pseudogaster
Pseudogibellula
Pseudogliomastix
Pseudogliophragma
Pseudogloeosporium (= Kabatia)
Pseudographiella
Pseudographium (=
　Sphaerographium)
Pseudographium
Pseudohansfordia
Pseudohansfordia
Pseudohaplis (= Botryodiplodia)
Pseudoidium (= Oidium)
Pseudolachnea
Pseudolachnella (=
　Pseudolachnea)
Pseudomeria
Pseudomicrocera (= Fusarium)
Pseudomicrodochium
Pseudomonilia (= Candida)
Pseudomycoderma
Pseudoneottiospora
Pseudopapulaspora
Pseudopatella (= Botryodiplodia)
Pseudopatella (= Cystotricium)
Pseudopatellina
Pseudopeltistroma
Pseudopestalotia (= Truncatella)
Pseudopetrakia
Pseudophloeosporella
Pseudophoma (=
　Chaetosphaeronema)
Pseudophomopsis (= Phomopsis)
Pseudopolystigmina
Pseudoprotomyces (= Phloeoconis)
Pseudopuccinia (= Stigmina)
Pseudoramularia
Pseudorhizopogon
Pseudorobillarda (=
　Pseudorobillarda)
Pseudorobillarda
Pseudosaccharomyces
Pseudosaccharomyces (=
　Kloeckeria)
Pseudosarcophoma (=
　Selenophoma)

Pseudoschizothyra
Pseudosclerophoma (= Phoma)
Pseudoseptoria
Pseudospiropes
Pseudostegia
Pseudostemphylium (=
　Ulocladium)
Pseudotaeniolina
Pseudothiopsella
Pseudothyrium
Pseudotorula
Pseudotracylla
Pseudovularia (= Ramularia)
Pseudozyma
Pseudozythia
Psilobotrys (= Chloridium)
Psilonia (= Volutella)
Psiloniella (= Catenularia)
Psilospora (= Polymorphum)
Psilosporina (= Polymorphum)
Psilothecium (= Cercospora)
Pteroconium
Pterodinia (= Botrytis)
Pterulopsis
Pterygosporopsis
Ptychogaster
Pucciniopsis (nom. ambig.)
Pucciniospora
Puciola (= Dicyma)
Pulchromyces
Pulina (= Lepraria)
Pullospora
Pullularia (= Aureobasidium)
Pulvinaria (= Waydora)
Pulvinotrichum
Punctillina
Putagraivam
Pycnidiella
Pycnidioarxiella
Pycnidiochaeta (=
　Dinemasporium)
Pycnidiopeltis
Pycnidiostroma (= Phomachora)
Pycnis
Pycnocalyx (= Bothrodiscus)
Pycnociliospora
Pycnodactylus
Pycnofusarium (= Fusarium)
Pycnomma
Pycnomoreletia
Pycnorostrum (nom. dub.)
Pycnoseynesia
Pycnosporium (nom. dub.)
Pycnostemma (= Poropeltis)
Pycnostroma (= Munkia)
Pycnostysanus
Pycnothera
Pycnothyriella
Pycnothyrium
Pycnovellomyces
Pyramidospora
Pyreniopsis (= Trichoderma)
Pyrenium (= Trichoderma)
Pyrenochaeta
Pyrenochaetella (nom. dub.)
Pyrenotrichum
Pyrgostroma
Pyricularia
Pyriculariopsis
Pyriomyces
Pyripnomyces
Quadricladium
Quasidiscus

Queenslandia
Questieriella
Quezelia
Rabdosporium
Rabenhorstia
Racemosporium (= Arthrinium)
Rachisia (= Fusarium)
Racodium
Radaisiella (= Botryosporium)
Radiatispora
Radiciseta
Raffaelea
Ragnhildiana (= Mycovellosiella)
Raizadenia
Ramakrishnanella
Ramalia (= Fusicladium)
Ramaraomyces
Ramichloridium
Ramularia (= Ramularia)
Ramularia
Ramulariopsis
Ramulariospora (= Phacidiella)
Ramulaspera
Ramulispora
Ranojevicia
Raperia (= Aspergillus)
Readeriella
Redaellia (= Aspergillus)
Redbia
Refractohilum
Reniforma
Repetophrgamia
Retiarius
Retroconis
Rhabdoclema
Rhabdogloeopsis
Rhabdogloeum
Rhabdomyces
Rhabdospora
Rhabdostromella
Rhabdostromellina (=
  Coleophoma)
Rhabdostromina (= Coleophoma)
Rhabdothyrella (= Leptothyrium)
Rhabdothyrium (= Leptothyrium)
Rhacodiella (= Myrioconium)
Rhacodiopsis (= Racodium)
Rhacodium (= Racodium)
Rhexoampullifera
Rhexocercosporidium
Rhinocephalum (= Arthrinium)
Rhinocladiella (= Chrysosporium)
Rhinocladiella
Rhinocladiopsis (=
  Chrysosporium)
Rhinocladium
Rhinotrichella
Rhipidocephalum
Rhizoctonia
Rhizohypha
Rhizophoma (= Rhizosphaera)
Rhizosphaera
Rhizosphaerella (= Phoma)
Rhizosphaerina
Rhizosporium (= Phloeoconis)
Rhizostilbella
Rhizostroma (nom. dub.)
Rhizothyrium
Rhodesia
Rhodesiopsis
Rhodocephalus (= Aspergillus)
Rhodomyces (= Candida)
Rhodoseptoria (= Polystigmina)

Rhodosticta
Rhodothallus
Rhodotorula
Rhombostilbella
Rhopalidium (= Alternaria)
Rhopaloconidium
Rhopalocystis (= Aspergillus)
Rhynchodiplodia
Rhynchomyces
Rhynchophorus (= Ceratopycnis)
Rhynchoseptoria
Rhynchosporina
Rhynchosporium
Rhynchotheca (nom. dub.)
Ribaldia (= Asteroma)
Riccoa (= Heydenia)
Riclaretia
Riessia
Riessiella
Rileya
Rinomia (= Morinia)
Robakia
Robillarda (= Pestalotiopsis)
Robillarda
Rogersia (= Filosporella)
Roigiella
Roscoepoundia
Rostrospora (= Colletotrichum)
Rosulomyces
Rota
Rotaea
Rotiferophthora
Rubikia
Rubromadurella (= Madurella)
Ruggieria
Rupinia (= Heydenia)
Rutola
Sabouraudiella (= Trichophyton)
Sabouraudites (= Microsporum)
Saccardaea (nom. dub.)
Sacidium (= Pilobolus)
Sadasivanella (= Neottiosporina)
Sadasivania
Sagenomella
Sagrahamala
Saitoella
Sakireeta (= Tiarosporella)
Saliastrum
Samukuta (= Neottiospora)
Sanjuanomyces
Santapauinda
Saprochaete
Saprophragma
Saprotaphrina (nom. nud.)
Sarbhoyomyces
Sarcinella
Sarcinodochium
Sarcinomyces (= Sarcinosporon)
Sarcinomyces
Sarcinosporon
Sarcinulella
Sarcophoma
Sarcopodium
Sarcostroma (= Seimatosporium)
Sarocladium
Satchmopsis
Satwalekera (= Torula)
Scaphidium
Scedosporium
Scenomyces
Sceptrifera
Sceptromyces (= Aspergillus)
Scheleobrachea (= Pithomyces)

Schizoblastosporion
Schizocephalum (=
  Haplographium)
Schizoderma (= Leptostroma)
Schizoderma
Schizodiplodia (=
  Didymosporiella)
Schizophoma (= Sclerophoma)
Schizothyra
Schizothyrella
Schizothyrellina (= Phacidiella)
Schizothyropsis
Schizotrichella (= Colletotrichum)
Schizotrichum
Schoenbornia (= Hymenopsis)
Schoenleinium (= Trichophyton)
Schroeteria
Schwarzmannia
Sciniatosporium (=
  Seimatosporium)
Scirrhophoma
Sclerochaeta (= Chaetopyrena)
Sclerochaetella (=
  Diploplenodomus)
Sclerococcum
Scleroconium
Sclerodiscus
Sclerodothiorella
Sclerographiopsis
Sclerographium
Scleromeris
Scleroparodia (= Ascochytopsis)
Sclerophoma
Sclerophomella (= Phoma)
Sclerophomina (= Phoma)
Scleropycnis
Scleropycnium (= Phomopsis)
Sclerosphaeropsis (nom. dub.)
Sclerostagonospora (=
  Stagonospora)
Sclerostilbum (= Tilachlidiopsis)
Sclerotheca (= Camarosporium)
Sclerothyrium (=
  Microsphaeropsis)
Sclerotiella
Sclerotiomyces (= Sclerotium)
Sclerotiopsis (= Pilidium)
Sclerotium
Sclerozythia (= Neozythia)
Sclerozythia
Scolecobasidiella
Scolecobasidium
Scolecobasis (= Scolecobasidium)
Scolecosporiella
Scolecosporiella
Scolecoxyphium
Scolecozythia
Scolicosporium
Scolicotrichum
Scoliotidium
Scopaphoma
Scopula (= Gloiosphaera)
Scopularia (nom. dub.)
Scopulariella
Scopulariopsis
Scorpiosporium
Scothelius
Scutisporium (= Stemphylium)
Scutisporus
Scutopeltis
Scutopycnis
Scyphospora
Scyphostroma

Scytalidium
Searchomyces (= Amblyosporium)
Seimatosporiopsis
Seimatosporium
Seiridiella (= Phragmotrichum)
Seiridina (= Seimatosporium)
Seiridium
Selenodriella
Selenophoma
Selenophomopsis (= Selenophoma)
Selenosira
Selenosporella
Selenosporium (= Fusarium)
Selenosporopsis
Selenotila
Selenozyma
Semipseudocercospora (=
    Pseudocercospora)
Sepedonium
Septaria (= Septoria)
Septochora (= Helhonia)
Septocylindrium
Septocyta
Septocytella
Septodochium
Septodothideopsis (nom. dub.)
Septofusidium
Septogloeum
Septoidium
Septomyrothecium
Septomyxa (= Diplodina)
Septomyxella
Septonema
Septopatella
Septoplaca (= Piptarthron)
Septorella (nom. nud.)
Septorella (= Fusarium)
Septoria
Septoria (nom. rej.)
Septoriella
Septoriomyces
Septoriopsis (= Groveolopsis)
Septoriopsis (= Cercoseptoria)
Septoriopsis (= Septoria)
Septosporium
Septothyrella
Septotis
Septotrullula
Sesquicillium
Sessiliospora
Setodochium (= Wiesneriomyces)
Setomyces
Setophiale
Setosporella
Setosynnema
Seychellomyces
Seynesiopsis
Shawiella
Shearia
Sheariella
Shecutia (= Libertiella)
Shomea (= Geastrumia)
Shropshiria (= Munkia)
Shrungabeeja
Sibirina
Sigmatomyces
Sigmoidea
Singera (= Oramasia)
Sirexcipula
Sirexcipulina (= Topospora)
Sirococcus
Sirocyphis
Sirodesmium (= Coniosporium)

Sirodiplospora (= Topospora)
Sirodochiella
Sirodomus (= Sirexcipula)
Sirodothis
Sirogloea
Siroligniella
Siropatella (nom. dub.)
Sirophoma
Siroplacodium
Siroplaconema (= Ceuthospora)
Siropleura
Siroscyphella (=
    Pseudocenangium)
Siroscyphellina
Sirosperma
Sirosphaera
Sirosporium
Sirosporonaemella
Sirostromella (= Pseudodiscula)
Sirothecium
Sirothyriella
Sirothyrium
Sirozythia
Sirozythiella
Sitochora
Slimacomyces
Solheimia
Solicorynespora
Soloacrospora
Solosympodiella
Sonderhenia
Sopagraha (= Arachnophora)
Soredospora (= Fumago)
Soreymatosporium (=
    Stemphylium)
Sorocybe (= Pycnostysanus)
Sorosporella
Spadicesporium (= Cladosporium)
Spadicoides
Spalovia (= Spermotrichum)
Spegazzinia
Speira (= Dictyosporium)
Speiropsis
Spelaeomyces (= Xylostroma)
Spermatoloncha
Spermochaetella
Spermophyllosticta (nom. dub.)
Spermospora
Spermosporella
Spermosporina
Spermotrichum
Sphacelia
Sphaceliopsis (= Ephelis)
Sphaceloma
Sphaerellopsis
Sphaeridium
Sphaeriostromella
Sphaeriothyrium
Sphaerocista (= Topospora)
Sphaerocolla
Sphaerocybe
Sphaerographium
Sphaeromma
Sphaeromyces
Sphaeromyces (nom. inval.)
Sphaeromycetella (= Chloridium)
Sphaeromyxa (= Sphaeronaema)
Sphaeronaema
Sphaeronaemina (=
    Sphaeronaema)
Sphaerophoma
Sphaeropsis
Sphaeropsis (nom. dub.)

Sphaerosporium
Sphaerulomyces
Sphinctrosporium (=
    Cladotrichum)
Sphondylocephalum
Spicariopsis (= Paecilomyces)
Spicellum
Spicularia
Spiculostilbella (= Phaeoisaria)
Spilocaea
Spilodochium
Spilomium (= Sclerococcum)
Spilomyces (= Neottiospora)
Spilosphaeria (= Septoria)
Spiniger
Spinomyces
Spinulospora
Spiralia (= Trichophyton)
Spiralotrichum (nom. dub.)
Spiralum
Spiropes
Spirosphaera
Spirospora (= Acrospeira)
Spirotrichum (= Tritirachium)
Spogotteria (= Plectronidium)
Spondylocladiella
Spondylocladiopsis
Spondylocladium (= Stachylidium)
Sporendocladia
Sporendonema
Sporhaplus
Sporhelminthium (=
    Clasterosporium)
Sporidesmiella
Sporidesmina (= Janetia)
Sporidesmiopsis (=
    Brachysporiella)
Sporidesmium
Sporobolomyces
Sporocadus (= Seimatosporium)
Sporocephalium (= Acladium)
Sporocephalum (= Oedocephalum)
Sporocladium (= Cladosporium)
Sporocybe (= Periconia)
Sporocybomyces
Sporocystis
Sporoderma (= Trichoderma)
Sporodiniopsis (nom. dub.)
Sporodum (= Periconia)
Sporoglena
Sporonema
Sporophiala
Sporophleum (= Arthrinium)
Sporophora
Sporoschisma
Sporoschismopsis
Sporostachys (= Melanographium)
Sporothrix
Sporotrichella (= Fusarium)
Sporotrichopsis (= Sporothrix)
Sporotrichum
Stachybotryella
Stachybotryna
Stachybotrys
Stachycoremium
Stachylidium
Stagonopatella
Stagonopsis
Stagonospora
Stagonosporella (= Stagonospora)
Stagonosporina
Stagonosporopsis (= Ascochyta)
Stagonostroma (= Fusarium)

Stagonostromella
Staheliella
Stalagmochaetia
Stanhughesia
Staninwardia
Stanjehughesia
Stanjehughesia
Staphylotrichum
Starkeyomyces (= Myrothecium)
Staurochaeta (= Staurophoma)
Stauronema
Staurophoma
Stegonsporiopsis
Stegonsporium
Steirochaete (= Colletotrichum)
Stelechotrichum (= Cephalotrichum)
Stellifera (nom. dub.)
Stellomyces
Stellopeltis
Stellospora
Stellothyriella
Stemmaria (nom. dub.)
Stemphyliomma (= Pithomyces)
Stemphyliopsis (= Stemphyliomma)
Stemphyliopsis
Stemphylium
Stenella
Stenellopsis
Stenellopsis (= Parastenella)
Stenocarpella
Stenocephalum
Stenocladiella
Stenospora
Stephanoma
Stephanosporium
Stephembruneria
Stephosia
Sterigmatobotrys
Sterigmatocystis (= Aspergillus)
Sterigmatomyces
Stevensomyces
Stevensonula
Stichophoma (= Sclerophoma)
Stichospora
Stictochorella (= Asteromella)
Stictochorellina (= Asteromella)
Stictopatella
Stictosepta
Stictospora
Stigmastoma (= Pycnothyrium)
Stigmatellina
Stigmateopsis (= Placonema)
Stigmella
Stigmina
Stigmopeltella (= Stigmopeltis)
Stigmopeltis
Stigmopeltopsis (= Myxothyriopsis)
Stigmopsis (= Cheiromyces)
Stilbella
Stilbellula
Stilbochalara (= Chalara)
Stilbodendron
Stilbomyces (nom. dub.)
Stilbophoma
Stilbospora
Stilbothamnium (= Aspergillus)
Strasseria
Strasseriopsis
Streblema (nom. dub.)
Streblocaulium (= Botrytis)

Streptobotrys
Streptopodium
Streptothrix (= Conoplea)
Striodiplodia (= Lasiodiplodia)
Strionemadiplodia
Stromateria (= Tubercularia)
Stromatocrea
Stromatographium
Stromatopogon
Stromatopycnis
Stromatostilbella (= Stromatographium)
Stromatostysanus
Strongylothallus
Strumella
Strumellopsis
Stygiomyces
Stylaspergillus
Styloletendraea (= Nectria)
Stysanopsis (= Cephalotrichum)
Stysanus (= Cephalotrichum)
Subicularium
Subramanella (= Phomopsis)
Subramania
Subramaniomyces
Subulariella (nom. dub.)
Subulispora
Sutravarana
Suttonia (= Suttoniella)
Suttoniella
Suttonina
Syamithabeeja
Sylviacollaea
Symbiotaphrina
Symphyosira
Symphysos
Sympodiella
Sympodina (= Veronaea)
Sympodiocladium
Sympodiomycopsis
Sympodiophora (= Pseudohansfordia)
Synchronoblastia
Syncladium
Syncollesia (= Fumago)
Syndiplodia (= Microdiplodia)
Syngliocladium
Synnemadiplodia (= Botryodiplodia)
Synnematium (= Hirsutella)
Synnematomyces
Synnemellisia
Synostomina
Synpenicillium (= Doratomyces)
Synphragmidium (= Speira)
Synsporium (= Stachybotrys)
Synsterigmatocystis (= Gibellula)
Synthetospora (= Stephanoma)
Syringospora (= Candida)
Systremmopsis
Szczawinskia
Taeniola (= Hormiscium)
Taeniolella
Taeniolina
Taeniophora (= Phragmotrichum)
Taeniospora
Talekpea
Tandonella
Tandonia
Tapeinosporium (= Septocylindrium)
Tarsodisporus
Tassia

Tauromyces
Tawdiella
Taxomyces
Tectacervulus
Tegoa (nom. inval.)
Telligia
Temerariomyces
Teratosperma
Termitaria
Termitariopsis
Termitosphaera
Tetrabrachium
Tetrabrunneospora
Tetrachaetum
Tetrachia (= Spegazzinia)
Tetrachia (= Spegazzinia)
Tetracladium
Tetracoccosporium
Tetracolium (= Torula)
Tetracrium
Tetracytum (= Cylindrocladium)
Tetradia (nom. dub.)
Tetrameronycha
Tetranacrium
Tetraploa
Tetraposporium
Thailandia (= Candida)
Thallomicrosporon (= Microsporum)
Thallospora
Thanatophytum (= Rhizoctonia)
Thaptospora
Tharoopama
Thecaphorella
Thecostroma (= Bloxamia)
Thedgonia
Thermoidium (= Malbranchea)
Thermomyces
Thermophymatospora
Thielaviopsis
Thirumalacharia
Thomiella (= Gonatobotryum)
Thoracella
Thozetella
Thozetellopsis (= Thozetella)
Thozetia (= Thozetella)
Thrinacospora
Thrombocytozoons (= Candida)
Thyrinula
Thyriochaetum (= Amerosporium)
Thyriostroma (= Leptostroma)
Thyriostromella
Thyrococcum (nom. dub.)
Thyrodochium (= Stemphylium)
Thyrospora (= Stemphylium)
Thyrostroma
Thyrostromella (= Stigmina)
Thyrostromella
Thyrsidiella
Thyrsidina
Thyrsidium (= Myriocephalum)
Thysanophora
Thysanopyxis (= Volutella)
Tiarospora
Tiarosporella
Tiarosporellivora
Ticogloea
Tilachlidiopsis
Tilachlidium
Tilletiopsis
Titaea
Titaeella
Titaeopsis

Titaeospora
Titaeosporina (= Asteroma)
Tolypocladium
Tolypomyria (= Trichoderma)
Tomentelleopsis (=
   Chromelosporium)
Tomenticola
Tompetchia
Topospora
Torsellia (= Cytospora)
Torula
Torulella (nom. dub.)
Torulina (= Gliomastix)
Toruloidea (= Oospora)
Torulomyces
Torulopsiella
Torulopsis (= Torulopsiella)
Torulopsis (= Candida)
Torulopsis (= Gliomastix)
Toxosporiella
Toxosporiopsis
Toxosporium
Trachythyriolum (nom. conf.)
Tracyella (nom. nud.)
Tracylla
Traversoa (= Lasiodiplodia)
Treleasiella (nom. dub.)
Trematophoma
Tremellidium
Tremellina
Tretendophragmia (=
   Diplococcium)
Tretocephala
Tretophragmia
Tretopileus
Tretospeira
Tretospora
Tretovularia
Triacutus
Triadelphia
Tribolospora
Tricellula
Tricellulortus (= Pneumatospora)
Trichaegum
Trichobolbus
Trichobotrys
Trichocephalum (= Periconia)
Trichocicinnus (= Chaetosticta)
Trichocladium
Trichoconiella (= Alternaria)
Trichoconis
Trichoconium (= Melanconium)
Trichocrea (nom. dub.)
Trichoderma
Trichodermia (= Trichothecium)
Trichodiscula
Trichodochium
Trichodytes (nom. dub.)
Trichofusarium (= Fusarium)
Tricholeconium (= Sarcopodium)
Trichomyces (= Trichophyton)
Trichopeltium (= Trichopeltulum)
Trichopeltulum
Trichophyton
Trichoseptoria
Trichosperma (= Spermotrichum)
Trichosporiella
Trichosporon
Trichosporonoides
Trichosporum (nom. conf.)
Trichosterigma (= Hirsutella)
Trichostroma (nom. dub.)
Trichothallus

Trichotheca (= Bloxamia)
Trichotheciopsis (= Trichothecium)
Trichothecium
Trichurus
Tricladiella
Tricladiomyces
Tricladiopsis
Tricladiospora
Tricladium
Tricornispora (= Tridentaria)
Tridentaria
Trifurcospora
Triglyphium
Trigonopsis
Trigonosporium
Trimmatostroma
Trinacrium
Triplicaria
Tripoconidium
Tripospermum
Triposporina
Triposporium
Triramulispora
Triscelophorus
Triscelosporium
Trisulcosporium
Tritirachium
Trochophora
Troglobiomyces (= Hirsutella)
Troposporella
Troposporium
Trullula
Truncatella
Tryblidiopycnis
Tryssglobulus
Tsuchiyaea
Tubakia
Tubercularia
Tuberculariella (=
   Cryptosporiopsis)
Tuberculariopsis
Tubercularis (= Tuberculariopsis)
Tuberculina
Tuberculis (= Tuberculariella)
Tuberculispora
Tuberciniella
Tumularia
Tunicago
Tureenia (= Arthrinium)
Tylomyces
Tylophoma (= Dothichiza)
Tympanosporium
Uberispora
Ubrizsya
Ugola
Ulocladium
Ulocoryphus
Uloploca (nom. dub.)
Umbellidion
Umbellula (= Pseudobotrytis)
Uncigera
Uncispora
Underwoodina (= Aschersonia)
Uniseta
Uredinaria (nom. conf.)
Uredinula (= Tuberculina)
Urobasidium (= Zygosporium)
Uroconis (= Urohendersonia)
Urohendersonia
Urohendersoniella
Urophiala (= Zygosporium)
Urosporium (nom. dub.)
Usteria (= Mycousteria)

Ustilaginoidea
Ustilaginoidella (= Fusarium)
Ustilaginula (= Ustilagopsis)
Ustilagopsis (= Sphacelia)
Uvarispora
Uyucamyces
Vakrabeeja (= Nakataea)
Valdensia
Vanakripa
Vanbeverwijkia
Vanbreuseghemia (=
   Keratinomyces)
Vanderystiella
Vanrija (= Apiotrichum)
Vanterpoolia
Vargamyces
Varicosporina
Varicosporium
Vasculomyces
Vasudevella
Vellosiella (= Mycovellosiella)
Velutipila
Ventrographium
Venustocephala
Venustusynnema
Veralucia
Veramyces
Veramyces
Vermicularia (= Colletotrichum)
Vermiculariopsiella
Vermiculariopsis (=
   Vermiculariopsiella)
Vermiculariopsis (nom. conf.)
Vermispora
Vermisporium
Veronaea
Veronaella
Veronidia
Verrucariella
Verrucaster (nom. dub.)
Verrucispora (= Verrucisporota)
Verrucisporota
Verrucobotrys
Verrucophragmia
Verticicladiella
Verticicladium
Verticicladus
Verticilliastrum (= Trichoderma)
Verticilliodochium (=
   Dendrodochium)
Verticilliopsis (nom. dub.)
Verticillis (= Verticilliodochium)
Verticillium
Verticimonosporium
Vesicaldiella
Vesicularia (= Basramyces)
Vestergrenia (= Phomopsis)
Vestigium
Vialina (= Phoma)
Viennotiella (= Heteropatella)
Vinculum (= Barnettella)
Virgaria
Virgariella
Virgasporium (= Cercospora)
Virgatospora
Viscomacula
Voglinoana (nom. dub.)
Volucrispora
Volutella
Volutellaria
Volutellis
Volutellopsis
Volutellopsis (= Volutellis)

Volutellospora (= Chaetomella)
Volutina
Vonarxia
Vouauxiella
Vouauxiomyces
Vrikshopama (= Dematophora)
Wakefieldia (= Anaphysmene)
Walkeromyces (= Mycovellosiella)
Wallemia
Walzia (= Penicillium)
Wangiella (= Exophiala)
Wardinella
Wardomyces
Wardomycopsis
Waydora
Weissia
Wettsteiniella (= Arthrobotryum)
Weufia
Wiesneriomyces
Wilsonomyces (= Thyrostroma)
Woessia
Wojnowicia
Woroninula
Xanthopsora (= Linochora)
Xanthoriicola
Xeilaria (= Placosphaeria)
Xenidiocercus
Xenobotrytis (= Paracostantinella)
Xenochora
Xenodiella
Xenodomus
Xenogliocladiopsis
Xenoheteroconium
Xenokylindria
Xenopeltis
Xenoplaca
Xenosporella (= Xenosporium)
Xenosporium
Xenostilbum (= Calostilbella)
Xenostroma
Xepicula
Xepiculopsis
Xeroconium
Xiambola
Xiphomyces
Xylochia
Xylochoeras (= Sclerotium)
Xylocladium
Xylocoremium
Xylocoryneum (nom. nud.)
Xyloglyphis
Xylohypha
Xylomyces
Xyphasma (nom. nud.)
Yalodendron (nom. nud.)
Yalomyces
Ybotromyces
Yoshinagamyces (= Japonia)
Ypsilonia
Yuccamyces
Zakatoshia
Zalerion
Zanclospora
Zasmidium (nom. dub.)
Zebrospora
Zelandiocoela
Zelopelta
Zelosatchmopsis
Zernya
Zetesimomyces
Zetiasplozna
Zilingia
Zinzipegasa

Zopheromyces
Zunura
Zygophiala
Zygosporium
Zymonema (nom. dub.)
Zymoxenogloea
Zythia
Zythiostroma
Zyxiphora

FOSSIL FUNGI
Abeliella
Achlyites
Acremonites
Actinomycites
Actinomycodium
Aecidites
Agaricites
Alleppeysporonites
Amepiospora
Ampulliferinites
Anatolinites
Annella
Anthracomyces
Arborella
Arbusculites
Archagaricon
Archecribraria
Archeomycelites
Archeplax
Archeterobasidium
Arthriniites
Arthroon
Arthurella
Ascochytites
Ascochytites
Ascodesmisites
Aspergillites
Asterinites
Asterinites
Asterothyrites
Asyregraamspora
Bactrodesmiites
Basidiosporites
Biporisporites
Bireticulasporis
Birsiomyces
Boletellites
Botrytites
Brachycarphium
Brachycladites (=
   Brachycarphium)
Brachycladium (=
   Brachycarphium)
Brachysporisporites
Brefeldiellites
Bretonia
Cadyexinis
Caenomyces
Callimothallus
Callorites
Cannanorosporonites
Capsulasclerotes
Cellulasclerotes
Celyphus
Cenangites
Centonites
Cercosporites
Cercosporites
Chaetomites
Chaetophorites
Chaetosphaerites

Chlamydosporites
Chordecystia
Cladosporites
Clasterosporites
Clavascina
Coleocarpon
Colligerites
Conchyliastrum
Coniothyriites
Coprinites
Coronasclerotes
Corynelites
Coulterella
Crenasclerotes
Cribrites
Cryptocolax
Ctenosporites
Cucurbitariaceites
Cytosporites
Dactyloporus
Dactylosporites
Daedaleites
Deccanodia
Dendromyceliates
Depazites
Dicellaesporites
Dictyosporites
Dictyotopileos
Didymoporisporonites
Didymosphaerites
Didymosporonites
Dihyphis
Diplodites
Diploneurospora
Diplosporites
Diporicellaesporites
Diporisporites
Diporopollis
Discoascina
Dodgella
Dothideites
Dothidites
Dremuspora
Dubiocarpon
Dyadosporites
Dyadosporonites
Elsikisporonites
Endochaetophora
Entopeltacites
Eomycetopsis
Ephedracetes
Erysiphites
Erysiphites
Eurotites
Euthythyrites
Excipulites
Exesisporites
Falciascina
Faveoletisporonites
Felixites
Foliopollenites
Fomesporites
Fomites
Foveodiporites
Foveoletisporonites
Fractisporonites
Fungites
Fungites
Fusariellites
Fusellites
Fusicladiites
Fusidites
Fusiformisporites

Gastromyces
Geasterites
Gelasinosporites
Geleenites
Geotrichites
Giraffachitina
Globoasclerotes
Globosasclerotes
Glomites
Glossifungites
Gonatobotrytites
Graamspora
Granatisporites
Graphiolites
Grilletia
Guizhounema
Gyromyces
Haplographites
Helicominites
Helicoonites
Helicosporiates
Helminthosporites
Helminthosporites
Heteroporimyces
Himantites
Hormosporites
Hydnites
Hypochnites
Hypoxylites
Hypoxylonites
Hypoxylonsporites
Hysteriopsis
Hysterites
Hysterites
Hystrichosphaeridium
Imprimospora
Inapertisporites
Incertisporites
Incolaria
Involutisporonites
Isuasphaera
Kakabekia
Kellermaniites
Krispiromyces
Kutchiathyrites
Lacrimasporonites
Ladrococcus
Laestadites
Lenzitites
Lepiotasporites
Leptonema
Leptosphaerites
Leptothyriomyces
Liaoningnema
Linosporoidea
Lirasporis
Lyonomyces
Macrosporites
Magnosporites (nom. alt.)
Manikinipollis
Mariusia
Marssoninites
Melanconites
Melanosphaerites
Melanosporites
Meliolinites
Microdiplodiites
Microsporonites
Microthallites
Microthyriacites
Microthyrites
Milleromyces
Mohgaonidium

Molinea
Moniliites
Monilites
Monochaetites
Monodictyites
Monoporisporites
Morosporium
Mucedites
Mucoralites
Mucorites
Mucorodium
Multicellaesporites
Mycelites
Mycobystrovia
Mycocarpon
Mycogemma
Mycokidstonia
Mycorhizonium
Myxomycetes
Myxomycites (= Myxomycetes)
Nailisporites
Nemaclada
Nemaplana
Notothyrites
Nyctomyces
Octosporonites
Oidites
Oidospora
Onakawananus
Oochytrium
Opegraphites
Ordovicimyces
Ornasporonites
Ostracodermis
Ovularites
Palaeachlya
Palaeancistrus
Palaeoamphisphaerella
Palaeocirrenalia
Palaeocytosphaera
Palaeofibula
Palaeoleptosphaeria
Palaeomycelites
Palaeomyces
Palaeomycites (= Palaeomyces)
Palaeopede
Palaeoperone
Palaeophoma
Palaeophthora
Palaeosclerotium
Palaeoslimacomyces
Palaeosordaria
Palambages
Paleoarcyria
Paleobasidiospora
Paleoblastocladia
Paleocatenaria
Paleoguttulina
Palmellathyrites
Palynomorphites
Papulasporites
Papulosporonites
Paramicrothallites
Paramoeciella
Parapolyporites
Parmathyrites
Pelicothallos
Penicillites
Perisporiacites
Perisporites
Peronosporites
Peronosporoides
Peronosporoides

Pesavis
Pestalozzites
Petrosphaeria
Pezizasporites
Pezizites
Phacidiopsis
Phacidites
Phellinites
Phellomycetes
Phellomycites (= Phellomycetes)
Phelonites
Phomites
Phragmidiites
Phragmothyrites
Phycomycites
Phycosiphon (= Brachycarphium)
Phyllerites
Phyllostictites
Pillulasclerotes
Pilula
Piriurella
Plectosclerotes
Pleosporites
Pleosporonites
Plochmopeltinites
Pluricellaesporites
Pluricellulites
Polyadosporites
Polycellaesporonites
Polycellaria
Polyporasclerotes
Polyporisporites
Polyporites
Polyporites
Polystigmites
Portalites
Predaldinia
Premyxomyces
Propythium
Protoascon
Protomycites
Pseudopezizites
Pseudopolyporus
Pseudosphaerialites
Psiammopomopiospora
Psiamspora
Psidimobipiospora
Psilodiporites
Puccinites
Pythites
Quilonia
Rabenhorstiites
Rabenhorstinidium
Ramsaysphaera
Ramulariites
Ramularites
Reduviasporonites
Reticellites
Reticulatisporonites
Retidiporites
Retihelicosporonites
Reymanella
Rhizomorphites
Rhizophagites
Rhizophidites
Rhytismites
Rhytismopsis
Roannaisia
Rosasporina
Rosellinites
Rosellinites
Saccharomycetes
Scabradiporites

Sclerosporis
Sclerotites
Sclerotites
Scolecosporites
Selenophomites
Senegalosporites
Serpentisclerotes
Shortensis
Shuklania
Sirodesmites
Sorosporonites
Spegazzinites
Sphaerialites
Sphaeriopsis
Sphaerites
Sphaeroporalites
Sphaerosporiites
Spherites
Spilosphaerites
Spirotremesporites
Spongiasclerotes
Sporocarpon
Sporonites
Sporoschismatites
Sporotrichites
Staginospora
Stagonosporites
Staphlosporonites
Stegites
Stellasclerotes
Stemphyliites
Stichus
Stigmateacites
Stilbites
Stomiopeltites
Streptotrichites
Striadiporites
Striadyadosporites
Striatasclerotes
Stridiporosporites
Sulcatisclerotes
Teleutosporites
Tenuicutites
Thuchomyces
Tormentella
Torulites
Torulites
Trametites
Transeptaesporites
Traquairia
Trematophora
Trematosphaerites
Trematosphaerites
Tribolites
Tricellaesporonites
Trichopeltinites
Trichosporites
Trichothyrites
Trihyphaecites
Triporicellaesporites
Triporisporites
Triporisporonites
Triposporonites
Tubercolarites
Tubercularites
Tympanicysta
Uncinulites
Uredinites
Uredites
Urocystites
Uromycetites
Urophlyctites
Ussurithyrites

Ustilagites
Ustulinites
Uvasporina
Verrucarites
Warkallisporonites
Witwateromyces
Xylasclerotes
Xylohyphites
Xylomides (= Xylomites)
Xylomites
Zygosporites

EXCLUDED GENERA
Actinomyce (nom. dub.)
Actonia (nom. dub.)
Actycus (nom. dub.)
Aetnensis (nom. nud.)
Agonium (nom. conf.)
Alveolinus (nom. dub.)
Alysphaeria
Alytosporium (nom. dub.)
Ambrosiaemyces (nom. nud.)
Amphinectria
Angiococcus (nom. dub.)
Annularia (nom. dub.)
Anomothallus (nom. dub.)
Antlea (nom. dub.)
Anulomyces (nom. dub.)
Arctosporidium (nom. dub.)
Artocreas (nom. ambig.)
Ascospermum (nom. dub.)
Aseimotrichum (nom. dub.)
Asteroides (nom. dub.)
Asteronema (nom. dub.)
Asterothrix (nom. dub.)
Atractodorus (nom. dub.)
Boerlagellopsis (nom. dub.)
Borinquenia (nom. dub.)
Calothyriopeltis
Calothyris (= Calothyriopeltis)
Campylostylus (nom. conf.)
Canceromyces (nom. dub.)
Carlia (nom. dub.)
Chaetocrea (nom. dub.)
Chaetocypha
Chaetosporium (nom. dub.)
Chaetotrichum (= Chaetosporium)
Chamaenema (nom. conf.)
Charrinia (nom. dub.)
Chermomyces
Cirrholus (nom. dub.)
Cladophytum (nom. dub.)
Clypeolaria (nom. dub.)
Clypeostroma (nom. dub.)
Coelosporidium (nom. dub.)
Coenogoniomyces (= Coenogonium)
Conisphaeria (nom. dub.)
Coralloides (nom. dub.)
Corenohydnum (nom. dub.)
Corneohydnum (nom. nud.)
Cryptodesma (nom. dub.)
Cyanotheca (nom. dub.)
Cyathela (nom. dub.)
Cyathella (nom. dub.)
Cylindrotaenium (nom. conf.)
Cynicus (nom. dub.)
Cystodium
Detonisia (nom. dub.)
Diamphora (nom. conf.)
Dicranotropis (nom. dub.)
Dicteridium (nom. dub.)

Dictyochora
Dictyochorella (nom. dub.)
Dictyonema (nom. dub.)
Didymocrater (nom. conf.)
Didymosamarospora (nom. dub.)
Didymostilbe (nom. dub.)
Diplodiella (nom. illegit.)
Discina (nom. dub.)
Ditmaria (nom. conf.)
Dothiora (= Melanogramma)
Elaeomyces
Ellobiocystis (nom. dub.)
Ellobiopsis (nom. dub.)
Elytrosporangium (nom. dub.)
Endaematus (nom. dub.)
Endematus (= Endaematus)
Endonius (nom. dub.)
Enduria
Enterobotryum (nom. dub.)
Ephebella (nom. dub.)
Epichloea (nom. dub.)
Epixyla (nom. dub.)
Erebonema (nom. dub.)
Euactinomyces (nom. dub.)
Eustegia (= Stegella)
Farringia (nom. dub.)
Favaria (nom. dub.)
Favomicrosporon
Fungoides (nom. inval.)
Fusitheca
Gelatinaria (nom. dub.)
Geosiphonomyces (= Geosiphon)
Gliotrichum (nom. dub.)
Globosopyreno (nom. conf.)
Gloeoradulum (nom. dub.)
Glukomyces (nom. conf.)
Glutinisporidium (nom. dub.)
Gnaphalomyces (= Lanosa)
Godal (nom. dub.)
Gremlia (= Polycephalum)
Gymnoascopsis (= Ascosorus)
Haematomyces
Halisaria (nom. conf.)
Haplosporidium (nom. dub.)
Hectocerus (nom. nud.)
Helicosporidium
Hemicyphe (nom. conf.)
Heringia (nom. nud.)
Hermanniasporidium (nom. dub.)
Heterobasidium (nom. conf.)
Heteroradulum (nom. dub.)
Hyalasterina (nom. dub.)
Hygrocrocis (= Typhoderma)
Hyperus (nom. dub.)
Hyphomyces (nom. inval.)
Hypolepia (nom. dub.)
Hypostomum (nom. dub.)
Hypsilophora (nom. conf.)
Hysteromyxa (nom. dub.)
Inciliaria (nom. dub.)
Isomunkia
Kuhneria (nom. dub.)
Langloisula (nom. conf.)
Lanosa (nom. conf.)
Lecaniascus
Lenormandia (= Normandina)
Lentescospora (nom. dub.)
Leolophia (nom. dub.)
Leprosis
Leptodothis (nom. dub.)
Leptospora (nom. dub.)
Leptuberia (nom. dub.)
Limnomyces (nom. dub.)

Lithopythium
Lymphocystidium (nom. conf.)
Maireomyces
Malacodermum (nom. dub.)
Medastina (nom. dub.)
Medusomyces (nom. conf.)
Melanogramma (nom. dub.)
Meliolopsis (nom. dub.)
Merosporium
Microdiscus (nom. dub.)
Micromastia
Mirandia (nom. dub.)
Molgosporidium (nom. dub.)
Muchmoria
Murciasporidium (nom. dub.)
Mycelium (nom. dub.)
Mycetosporidium (nom. dub.)
Mycocoelium (nom. dub.)
Mycothamnion (nom. conf.)
Myrioblepharis (nom. dub.)
Myxonema (nom. dub.)
Myxonyphe (nom. dub.)
Naegelia (nom. dub.)
Nemacola
Nematococcus (nom. dub.)
Neopiptostoma (= Piptostoma)
Nesophloea (nom. dub.)
Nothodiscus (nom. dub.)
Oesophagomyces (nom. dub.)
Oichitonium (nom. dub.)
Ombrophila (nom. conf.)
Omoriza (nom. conf.)
Omphalocystis (nom. dub.)
Oncobyrsa (nom. conf.)
Oovorus (nom. dub.)
Opethyrium (nom. dub.)
Ophryomyces (nom. dub.)
Ospriosporium (nom. dub.)
Ostracoblabe
Otagoa (nom. nud.)
Parallobiopsis (nom. conf.)
Peckiomyces (nom. nud.)
Pelodiscus

Peniophorina (nom. dub.)
Perisporium (nom. dub.)
Phacidiostromella
Phaeonema (nom. dub.)
Phaeoscutella (nom. dub.)
Phaeosiphonia (nom. dub.)
Phelonitis
Phiala (nom. dub.)
Phialea (nom. dub.)
Phragmonaevia (nom. dub.)
Phragmothyriella (nom. dub.)
Phycomater (nom. dub.)
Phyllonochaeta
Phyllops (nom. dub.)
Piptostoma (nom. dub.)
Placodium (nom. dub.)
Planctomyces (nom. dub.)
Plasmoparopsis (= Pythiogeton)
Pluesia (= Maurya)
Polycarpum (nom. conf.)
Polycephalum (nom. dub.)
Polystema (nom. dub.)
Poronea (nom. dub.)
Protasia (nom. nud.)
Pseudohansenula (nom. dub.)
Pseudopythium (nom. nud.)
Pseudosporopsis (nom. non fung.)
Psorothecium (nom. dub.)
Puttemansiella
Pycnodothis
Pyonema (= Myxonyphe)
Pyrenophoropsis
Reticulosporidium (nom. dub.)
Rhagidiasporidium (nom. dub.)
Rhinotrichum (nom. dub.)
Rhizoblepharis (nom. dub.)
Rhizogaster (nom. dub.)
Rhizomorpha (nom. dub.)
Samarosporella (nom. dub.)
Sappinia (nom. dub.)
Sarcanthia (nom. dub.)
Sarcomelas (nom. dub.)
Sarconemus (nom. dub.)

Schizacrospermum (nom. dub.)
Schizographa (nom. dub.)
Schizotorula (nom. inval.)
Scirrhiopsis (nom. dub.)
Scoleciasis (nom. dub.)
Scutellum (nom. dub.)
Siphopodium (nom. dub.)
Sirocrocis (nom. conf.)
Spadonia (nom. conf.)
Sphaeropleum (nom. conf.)
Sphaerothyrium (= Stegilla)
Spicaria (nom. conf.)
Sporoclema (nom. dub.)
Sporophaga (nom. conf.)
Stegia (= Stegilla)
Stegilla (nom. dub.)
Stericium (nom. dub.)
Stipiza (nom. nud.)
Straggaria (nom. conf.)
Synaphia (nom. dub.)
Synchaetophagus (nom. dub.)
Tetena (nom. dub.)
Thyriascus
Tremella (nom. inval.)
Trichobacidia (nom. dub.)
Trichoplacia (nom. dub.)
Tuberculopsis (nom. nud.)
Typhoderma (nom. dub.)
Tyridiomyces (nom. dub.)
Variolaria
Variolaria (nom. dub.)
Variolaria (nom. dub.)
Venularia (nom. nud.)
Vermiculariella (nom. conf.)
Weinmannodora (nom. dub.)
Winterina (nom. ambig.)
Xylissus (nom. nud.)
Xyloma (nom. dub.)
Zercosporidium (nom. dub.)
Zonilia (nom. dub.)
Zygodesmella (nom. dub.)
Zygodesmus (nom. dub.)
Zygothrix (nom. dub.)